Phosphorus-Carbon Heterocyclic Chemistry: The Rise of a New Domain

Phosphorus-Carbon Heterocyclic Chemistry: The Rise of a New Domain

Edited by

Professor François Mathey

Ecole Polytechnique,

Palaiseau, France

2001

PERGAMON

An Imprint of Elsevier Science

Amsterdam – London – New York – Oxford – Paris – Shannon – Tokyo

ELSEVIER SCIENCE Ltd
The Boulevard, Langford Lane
Kidlington, Oxford OX5 1GB, UK

First edition 2001

Library of Congress Cataloging in Publication Data
A catalog record of the Library of Congress has been applied for.

British Library Cataloguing in Publication Data
Phosphorus-carbon heterocyclic chemistry: the rise of a new domain
1. Heterocyclic chemistry 2. Phosphorus compounds 3. Carbon compounds
I. Mathey, François
547.5′9
ISBN 0080439527

ISBN: 0 08 043952 7

⊗ The paper used in this publication meets the requirements of ANSI/NISO Z39.48-1992 (Permanence of Paper).
Printed in The Netherlands.

Contents

Preface

As soon as Elsevier published *Comprehensive Heterocyclic Chemistry II* in 1996, it became obvious that phosphorus heterocycles had reached a state of development where they had joined Oxygen, Nitrogen, and Sulfur heterocycles as the fourth major strand of heterocyclic chemistry. It seemed natural to collect all the phosphorus chapters of CHEC II and to offer them, after updating, as a single volume to specialists of coordination chemistry, catalysis, and other areas who are involved in this field. In practice however, this idea proved to be more difficult to achieve than expected. For instance, CHEC II treated both saturated and insaturated rings in the same chapters and often did not cover the literature prior to 1982. Thus, this book finally became much more than the initially planned compilation. In fact, it includes seven new chapters, which cover the history of the domain, catalytic applications, phospholanes and phospholenes, phosphinanes and phosphinines, early work on phosphetanes, together with recent work on phospholes. Nine other chapters are updated versions of their CHEC II counterparts. The chapter on heterophospholes comprises an updated combined version of three chapters of CHEC II. Finally, phosphole chemistry is treated in two separate sections. The first one reproduces the initial chapter of CHEC II, with new schemes and corrections, whilst the second one is an updating which can be read independently. The final result is a huge book which comprehensively covers the chemistry of phosphorus–carbon heterocycles up to 1999 and also treats the most significant P-heterocycles including other heteroatoms such as heterophospholes, heterophosphinines, and spirocyclic compounds. We hope that it will serve as the reference work for all those interested in the domain for the years to come.

During this work, I have been fortunate enough to have the enthusiastic collaboration of most of the authors who had already written the phosphorus chapters of CHEC II, including several of the founding fathers of the field. Six new authors have joined the team. I also owe many thanks to Dr. Duncan Carmichael who revised my English and to Mme. Françoise Girard who, in addition to drawing the schemes in Chapters 4.2.1 and 4.3, helped me to manage the flow of manuscripts. At Elsevier, Dr. Guido Zosimo-Landolfo promoted the project , whilst Drs. Adrian Shell and Dinali de Silva took care of the publishing. Without the help of all these dedicated people, the book would have not existed.

F. Mathey
Palaiseau, January 2001

Journal Codes for References

ABC	Agric. Biol. Chem.
ACH	Acta Chim. Acad. Sci. Hung.
ACR	Acc. Chem. Res.
AC(R)	Ann. Chim. (Rome)
ACS	Acta Chem. Scand.
ACS(A)	Acta Chem. Scand., Ser.A
ACS(B)	Acta Chem. Scand., Ser. B
AF	Arzneim.-Forsch.
AG	Angew. Chem.
AG(E)	Angew. Chem., Int. Ed. Engl.
AHC	Adv. Heterocycl. Chem.
AJC	Aust. J. Chem.
AK	Ark. Kemi
ANY	Ann. N.Y. Acad. Sci.
AP	Arch. Pharm. (Weinheim, Ger.)
APO	Adv. Phys. Org. Chem.
AX	Acta Crystallogr.
AX(B)	Acta Crystallogr., Part B
AX(C)	Acta Crystallogr., Part C
AQ(C)	An. Quim.
B	Biochemistry
BAP	Bull. Acad. Pol. Sci., Ser.
BAU	Bull. Acad. Sci. USSR, Div. Chem. Sci.
BBA	Biochim. Biophys. Acta
BBR	Biochem. Biophys. Res. Commun.
BCJ	Bull. Chem. Soc. Jpn.
BEP	Belg. Pat.
BJ	Biochem. J. CRV Chem. Rev.
BJP	Br. J. Pharmacol.
BKS	Bull. Korean Chem.Soc.
BMS	Bioorg. Med. Chem. Lett.
BRP	Br. Pat.
BSB	Bull. Soc. Chim. Belg.
BSF	Bull. Soc. Chim. Fr.
BSF(2)	Bull. Soc. Chim. Fr., Part 2
C	Chimia
CA	Chem. Abstr.
CAR	Carbohydr. Res.
CB	Chem. Ber.
CBR	Chem. Br.
CC	J. Chem. Soc., Chem. Commun.
CCA	Croat. Chem. Acta
CCC	Collect. Czech. Chem. Commun.
CCR	Coord. Chem. Rev.
CEJ	Chem. Eur. J.
CHE	Chem. Heterocycl. Compd. (Engl. Transl.)
CHEC	Comp. Heterocycl. Chem., 1st edition
CHEC-II	Comp. Heterocycl. Chem., 2nd edition
CI(L)	Chem. Ind. (London)
CJC	Can. J. Chem
CL	Chem. Lett.
COS	Comp. Org. Synth.
CPB	Chem. Pharm. Bull.
CPH	Chem. Phys.
CPL	Chem. Phys. Lett.
CR	C.R. Hebd. Seances Acad. Sci.
CR(B)	C.R. Hebd. Seances Acad. Sci., Ser. B.
CR(C)	C.R. Hebd. Seances Acad. Sci., Ser. C.
CS	Chem. Scr.
CSC	Cryst. Struct. Commun.
CSR	Chem. Soc. Rev.
CZ	Chem.-Ztg.
DIS	Diss. Abstr.
DIS(B)	Diss. Abstr. Int. B
DOK	Dokl. Akad. Nauk SSSR
E	Experientia
EGP	Ger. (East) Pat.
EJIC	Eur. J. Inorg. Chem.
EJOC	Eur. J. Org. Chem.
EUP	Eur. Pat.
FES	Farmaco Ed. Sci.
FOR	Fortschr. Chem. Org. Naturst.
FRP	Fr. Pat.
G	Gazz. Chim. Ital.
GEP	Ger. Pat.
GEP(O)	Ger. Pat. Offen.
H	Heterocycles
HC	Chem. Heterocycl. Compd. [Weissberger-Taylor series]
HCA	Helv. Chim. Acta
HOU	Methoden Org. Chem. (Houben-Weyl)
IC	Inorg. Chem.
ICA	Inorg. Chem. Acta
IJC	Indian J. Chem.
IJC(B)	Indian J Chem., Sect. B
IJS	Int. J. Sulfur Chem.
IJS(B)	Int. J. Sulfur Chem., Part B
IZV	Izv. Akad. Nauk SSSR, Ser. Khim.
JA	J. Am. Chem. Soc.
JAP	Jpn. Pat.

JAP(K)	Jpn. Kokai
JBC	J. Biol. Chem.
JC	J. Chromatogr.
JCC	J. Coord. Chem.
JCI	I. Chem. Inf. Comput. Sci.
JCO	J. Comput. Chem.
JCP	J. Chem. Phys.
JCR(M)	J. Chem. Res. (M)
JCR(S)	J. Chem. Res. (S)
JCS	J. Chem. Soc.
JCS(C)	J. Chem. Soc. (C)
JCS(D)	J. Chem. Soc., Dalton Trans.
JCS(F1)	J. Chem. Soc., Faraday Trans. 1
JCS(P1)	J. Chem. Soc., Perkin Trans. 1
JCS(P2)	J. Chem. Soc., Perkin Trans.2
JGU	J. Gen. Chem. USSR (Engl. Transl.)
JHC	J. Heterocycl. Chem.
JIC	J. Indian Chem. Soc.
JMC	J. Med. Chem.
JMR	J. Magn. Reson.
JOC	J. Org. Chem.
JOM	J. Organomet. Chem.
JOU	J. Org. Chem. USSR (Engl. Transl.)
JPC	J. Phys. Chem.
JPO	J. Phys. Org. Chem.
JPR	J. Prakt. Chem.
JPS	J. Pharm. Sci.
JSP	J. Mol. Spectrosc.
JST	J. Mol. Struct.
K	Kristallografiya
KFZ	Khim.-Farm. Zh.
KGS	Khim. Geterotsikl. Soedin.
LA	Liebigs Ann. Chem.
M	Monatsh. Chem.
MC	Mendeleev Communications
MI	Miscellaneous [book or journal]
MIP	Miscellaneous Pat.
MRC	Magn. Reson. Chem.
MS	Q.N. Porter and J. Baldas, 'Mass Spectrometry of Heterocyclic Compounds', Wiley, New York, 1971
N	Naturwissenschaften
NAT	Nature
NEP	Neth. Pat.
NJC	Nouv. J. Chim.
NKK	Nippon Kagaku Kaishi
NMR	T.J. Batterham, 'NMR Spectra of Simple Heterocycles', Wiley, New York, 1973
OM	Organometallics
OMR	Org. Magn. Reson.
OMS	Org. Mass Spectrom.
OPP	Org. Prep. Proced. Int.
OR	Org. React.
OS	Org. Synth.
OSC	Org. Synth., Coll. Vol.
P	Phytochemistry
PAC	Pure Appl. Chem.
PC	Personal Communication
PH	'Photochemistry of Heterocyclic Compounds', ed. O. Buchardt, Wiley, New York, 1976
PHA	Pharmazie
PHC	Prog. Heterocycl. Chem.
PIA	Proc. Indian Acad. Sci.
PIA(A)	Proc. Indian Acad. Sci., Sect. A
PJC	Pol. J. Chem.
PMH	Phys. Methods Heterocycl. Chem.
PNA	Proc. Natl. Acad. Sci. USA
PS	Phosphorus Sulfur
QR	Q. Rev., Chem. Soc.
RCR	Russ. Chem. Rev. (Engl. Transl.)
RRC	Rev. Roum. Chim.
RTC	Recl. Trav. Chim. Pays-Bas
S	Synthesis
SA	Spectrochim. Acta
SA(A)	Spectrochim. Acta, Part A
SAP	S. Afr. Pat.
SC	Synth. Commun.
SH	W.L.F. Armarego, 'Stereochemistry of Heterocyclic Compounds', Wiley, New York, 1977, parts 1 and 2
SL	Synlett
SR	Sulfur Reports
SST	Org. Compd. Sulphur, Selenium, Tellurium [R. Soc. Chem. series]
SUL	Sulfur Letters
T	Tetrahedron
TA	Tetrahedron: Asymmetry
TAL	Talanta
TCA	Theor. Chim. Acta
TH	Thesis IJC(B)
TL	Tetrahedron Lett.
UKZ	Ukr. Khim. Zh. (Russ. Ed.)
UP	Unpublished Results
URP	USSR Pat.
USP	U.S. Pat.
YZ	Yakugaku Zasshi
ZAAC	Z. Anorg. Allg. Chem.
ZC	Z. Chem.
ZN	Z. Naturforsch.
ZN(A)	Z. Naturforsch., Teil A
ZN(B)	Z. Naturforsch., Teil B
ZOB	Zh. Obshch. Khim.
ZOR	Zh. Org. Khim.
ZPC	Hoppe-Seyler's Z. Physiol. Chem.
ZPH	Zh. Phys. Khim.
ZPK	Zh. Prikl. Khim.

1
The Birth of a New Domain

F. MATHEY
Ecole Polytechnique, Palaiseau Cedex, France

1.1 INTRODUCTION

Phosphorus is certainly one of the most important non-metals of the Periodic Table. It plays a central role in biochemistry, organic synthesis, coordination chemistry, homogeneous catalysis, solid state chemistry and material sciences. With such a scope, one might wonder why its heterocyclic chemistry has been underdeveloped for such a long time. It is startling that, whilst the first known P–C heterocycle, i.e. the 1-phenylphosphinane, was prepared relatively early (in 1915 ⟨15CB1473⟩) (Equation (1)), the domain really only began in 1953 with the discovery of the McCormack reaction between conjugated dienes and halophosphines ⟨53USP2663736-9⟩ (Equation (2)).

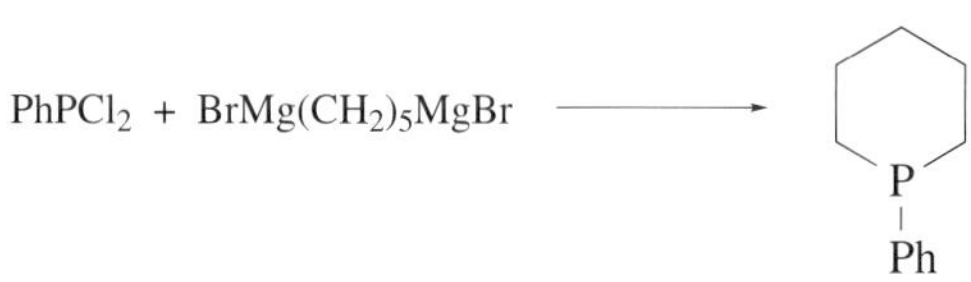

(1)

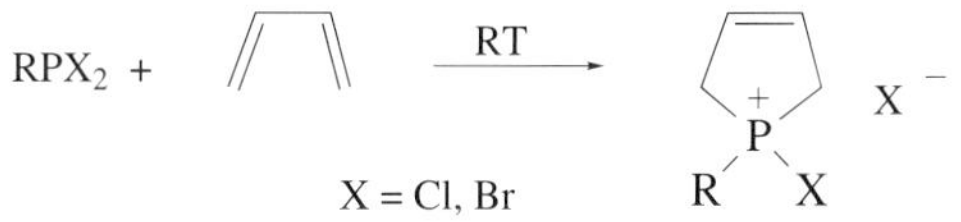

(2)

Why this should be is not obvious. It is clear that organophosphorus chemistry as a whole did not progress significantly between the two World Wars, probably because the inherently complex nature of this chemistry was not matched by the appropriate analytical tools. Without doubt, the discovery of the ^{31}P NMR spectroscopy in the early fifties was a major turning point for phosphorus chemistry. More subtly, organophosphorus chemistry in general, and its heterocyclic version in particular, does not mimic organonitrogen chemistry. Whilst stimulating, the absence of a reliable model is obviously a major handicap when trying to develop a new field. As an illustration, the McCormack reaction has no counterpart in nitrogen chemistry. Thus, the development of phosphorus heterocyclic chemistry had to rely on the imagination of a few dedicated chemists, and early developments were, inevitably, slow. The 1950 edition of Mann's book describing the heterocyclic derivatives of P, As, Sb and Bi comprehensively surveyed all phosphorus heterocycles in 15 pages and 18 references ⟨B-50MI(1)9⟩. Twenty years later, the second edition needed 354 pages and 503 references to cover the same topic ⟨B-70MI(3)1⟩. By way of comparison, the 1984 edition of *Comprehensive Heterocyclic Chemistry* ⟨B-84MI5⟩ which deals almost exclusively with nitrogen, oxygen and sulphur heterocycles, needed ca. 8 000 pages and 30 000 references to cover the fundamental aspects of its subject. The discrepancy reflects the fact that these more classical heterocycles were essentially discovered during the 19th century, and are present in a number of natural substances and in drugs, dyes and numerous other materials. Phosphorus heterocyclic chemistry cannot yet compete at such a level of complexity but, nevertheless, is undergoing very rapid development. In 1981, Quin's book dealing with P–C heterocycles ⟨B-81MI4⟩ already needed 430 pages and ca. 1100 references to treat the topic. The size of the present work confirms the continuing expansion in the field. Furthermore, the first applications in homogeneous catalysis do not lie far ahead (see Chapter 7). The future is thus quite exciting. Few specialized fields of molecular chemistry present such a potential for growth and applications.

1.2 THE FOUNDATIONS

After the initial discovery of 1-phenylphosphinane by Grüttner and Wiernik ⟨15CB1473⟩ (Equation (1)), Grüttner and Krause prepared 1-phenylphospholane using the same procedure ⟨16CB437⟩. In 1920, Conant and Cook showed that phosphorus trichloride readily reacts with α,β-unsaturated ketones to give 1,2-oxaphospholenes ⟨20JA830⟩ (Equation (3)).

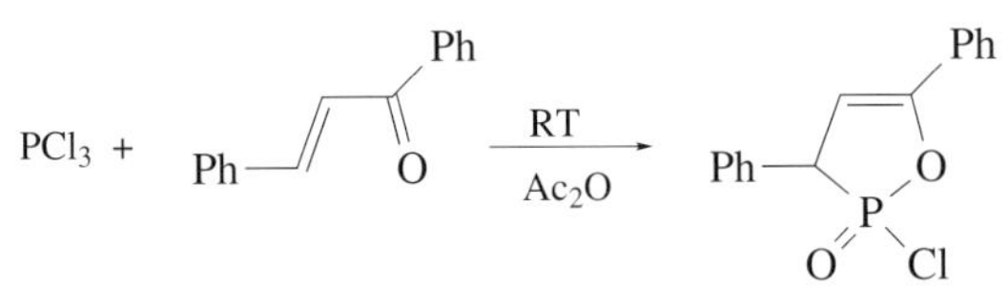

(3)

Later, phosphorus trichloride was replaced by phenyldichlorophosphine ⟨21JA1665⟩. With hindsight, this reaction is the forerunner of the McCormack reaction discussed above (Equation (2)). Several years elapsed before the next important step was made by Wittig and Geissler in 1953 ⟨53MI2(580)44⟩, when they isolated 5-phenyldibenzophosphole as a side-product of their classic work on pentaphenylphosphorane (Equation (4)).

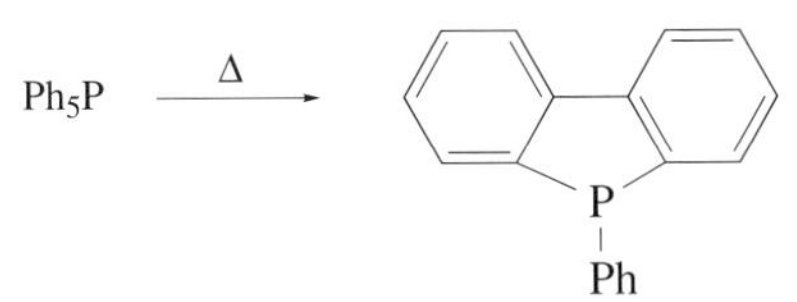

(4)

The same product was also obtained through the more conventional reaction of 2,2′-dilithiobiphenyl with phenyldichlorophosphine. The phosphorus atom in 5-phenyl-dibenzophosphole is pyramidal and has a fairly normal inversion barrier of 26 kcal mol^{-1} ⟨71JA6205⟩. Thus no electronic delocalization of the lone pair into the 5-membered ring occurs. Since it is readily available, 5-phenyldibenzophosphole is used by coordination chemists as a non-flexible equivalent of triphenylphosphine. More significantly, the dibenzophosphole unit served as a building block for Hellwinkel's synthesis of the first well-characterized hexacoordinate phosphates with six P–C bonds ⟨65CB576⟩ (Equation (5)).

But, of course, the landmark of 1953 was the Dupont work describing the McCormack reaction for the first time ⟨53USP2663736-9⟩ (Equation (2)). Suffice it to say that these patents formed the basis for almost all progress in the field of 5-membered P–C rings for the next two decades. It is also interesting to note that the phospholene oxides resulting from the hydrolysis of the McCormack cycloadducts immediately found application in the catalytic decarboxylation of isocyanates into carbodiimides ⟨62JA3673⟩. This reaction has also been applied to generate heat-resistant polycarbodiimides from diisocyanates ⟨62JA1493⟩.

The next event was the synthesis of pentaphenylphosphole by two groups working independently in 1959 ⟨59JA3163, 59CI(L)1250⟩. The easiest preparation, from diphenylacetylene, is depicted in Equation (6). Of course, the interest engendered by phospholes, concerned the degree of aromaticity in the ring. With its heavy substitution, pentaphenylphosphole was obviously not the ideal species for such a study. Some time later, Quin and his group prepared 1-methyl and 1-benzyl-phospholes and obtained an X-ray crystal structure of the 1-benzyl derivative ⟨67JA5984, 70JA5779⟩ (Equation (7)).

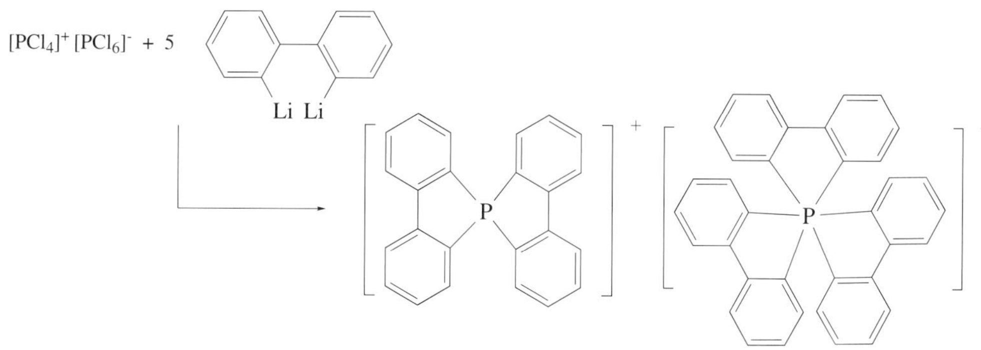

(5)

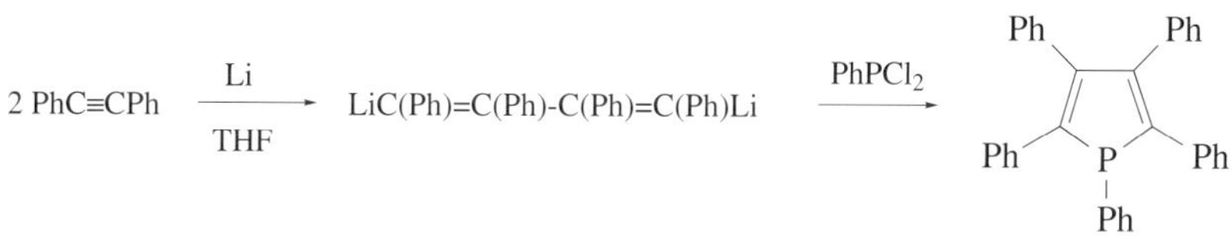

(6)

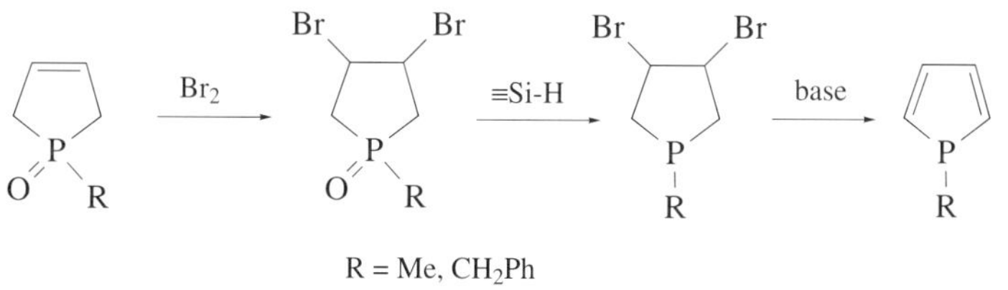

(7)

The non-planarity of phosphorus – whose P-benzyl bond makes an angle of 67° with the mean plane of the ring – casts serious doubt upon the aromaticity of the system. Almost at the same time, Mislow and his group demonstrated that phospholes display a very low pyramidal inversion barrier, of ca. 16 kcal mol^{-1} ⟨71JA6205⟩. The most logical way to reconcile these data was to admit that the pyramidal ground state of phospholes was weakly aromatic, whereas full delocalization took place in the planar transition state as a result of the extensive overlap between the lone pair orbital and the π orbitals of the diene. Recent high-level calculations fully confirm this analysis ⟨95AG(E)337⟩ and show that the aromaticity of the pyramidal parent phosphole is ca. 7 kcal mol^{-1}. Phospholes are not planar because the aromaticity of the planar state is insufficient to overcome the intrinsically high inversion barrier of phosphorus. Here again, we see a sharp contrast with the corresponding nitrogen heterocycles. In the meantime, both the synthesis and the chemistry of phospholes had been thoroughly investigated. Mathey's discovery that it is possible to dehydrohalogenate the McCormack cycloadducts directly to phospholes at room temperature using tertiary amines provided a useful practical advance ⟨69CR(C)1066⟩ (Equation (8)).

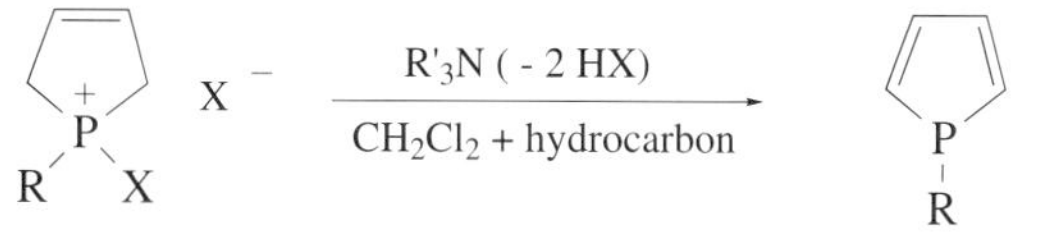

(8)

This method, which has been refined several times ⟨81S983⟩, remains the simplest and most general access to phospholes. Its mechanism, which was fully understood only much later, involves a reversible [1,5] shift of hydrogen from carbon to phosphorus. These shifts are the key feature of phosphole chemistry, but only in 1981 was the first unambiguous demonstration of a thermal equilibrium between 1*H*- and 2*H*-phospholes brought to light ⟨81JA4595⟩ (Equation (9)). Trapping of the reactive 2*H*-phospholes by diphenylacetylene gave the first known 1-phosphanorbornadienes. Very soon, the same type of transformation was shown to occur with P-unsubstituted 1*H*-phospholes at low temperature ⟨83JA6871⟩, thus explaining why it had been impossible to characterize such species previously. In the absence of trapping reagents, these 2*H*-species give P–P bonded [4 + 2] Diels–Alder dimers ⟨82CC1272⟩ (Equation (10)).

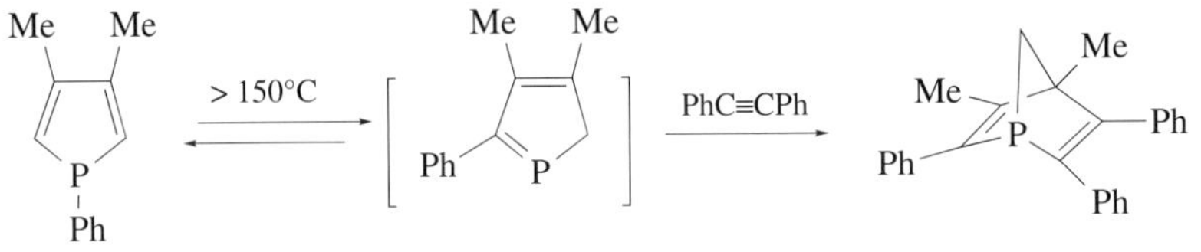

(9)

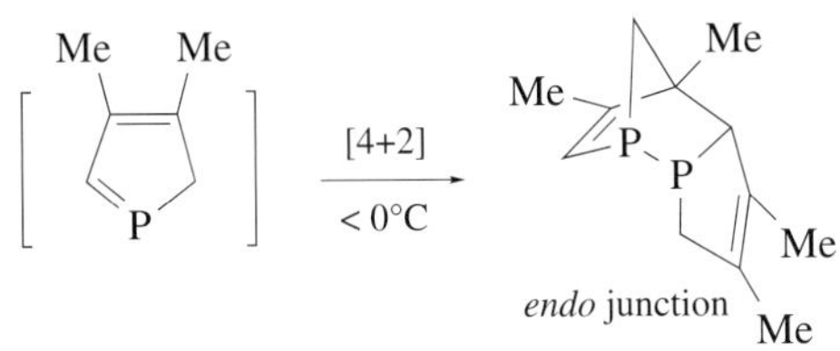

(10)

Eventually, the unstable parent phosphole was fully characterized by ^1H, ^{13}C and ^{31}P NMR spectroscopy in 1987 ⟨87TL5025⟩. It must be stressed here that these [1,5] sigmatropic shifts are similar to those observed for cyclopentadienes. From a theoretical standpoint, they reflect σ,π* overlap between the dienic system and the exocyclic P–R bond existing as a result of the pyramidal geometry of phospholes. Surprisingly, high-level computations later indicated that the parent 2*H*-phosphole appears to be slightly *more stable* than its 1*H*-isomer ⟨93JOC5414⟩.

The sixties saw an acceleration in the development of phosphorus heterocyclic chemistry. 1962 saw the first authenticated synthesis of phosphetanes from methyl-substituted olefins and halophosphines ⟨62JOC606⟩ (Equation (11)).

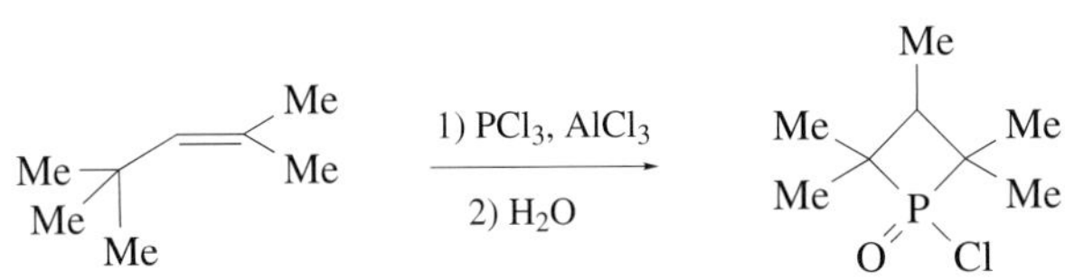

(11)

The ionic mechanism implies a [1,2] methyl shift (Equation (12)).

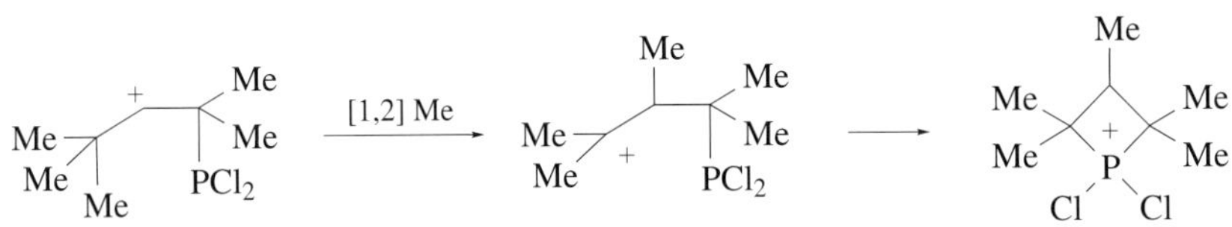

(12)

Having an internal angle in the range of 80–90°, which favours the formation of pentacoordinate TBP intermediates, phosphetanes have been ideal models for mechanistic studies of reactions involving P(III) and P(IV) centres. Among these are nucleophilic substitutions at their chloro derivatives in the P(III) and P(IV) states ⟨72JCS(P1)713, 74JCS(D)633⟩.

Wagner's work describing a very simple synthesis of phosphiranes appeared just one year later, in 1963 ⟨63USP3086053, 67JA1102⟩ (Equation (13)).

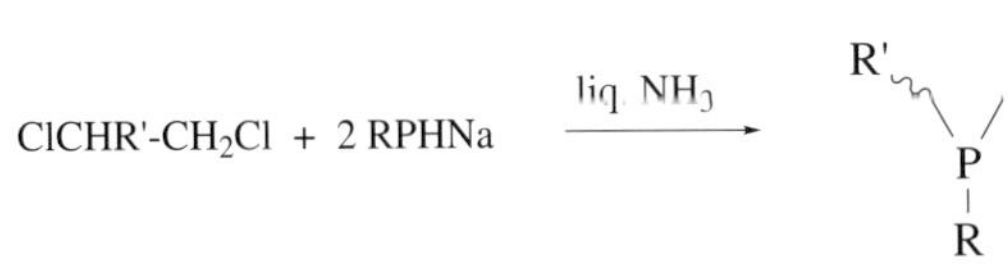

(13)

The ^{31}P resonance of the parent phosphirane appears at −341 ppm ⟨83OMR494⟩ and, of diamagnetic phosphorus derivatives, only P_4 itself resonates at higher field. The ring, having an internal angle of 47°, is highly strained ⟨69JA17⟩, but less than aziridine (C_2H_4X: X = PH, ΔE −22 kcal mol^{-1}; X = NH, ΔE −29.5 kcal mol^{-1}) according to recent theoretical studies ⟨89JPC3025⟩. Another feature brought to light recently concerns the inversion of phosphorus in halophosphiranes, which apparently occurs via a turnstile rotation ⟨97CC1033⟩. Clearly, these rings will provide rich pickings for theoreticians! The cyclic strain induces a number of unusual ring-opening and ring-expansion reactions. The phosphorus lone pair displays a very low nucleophilicity, but it has been possible to prepare the first well-characterized phosphiranium salts by reaction with methyl triflate. These salts appear to be convenient precursors for non-stabilized phosphenium ions which can be trapped by alkynes ⟨95CC257⟩.

A year later, in 1964, Mahler described the first 1,2-dihydro-1,2-diphosphete ⟨64JA2306⟩ (Equation (14)).

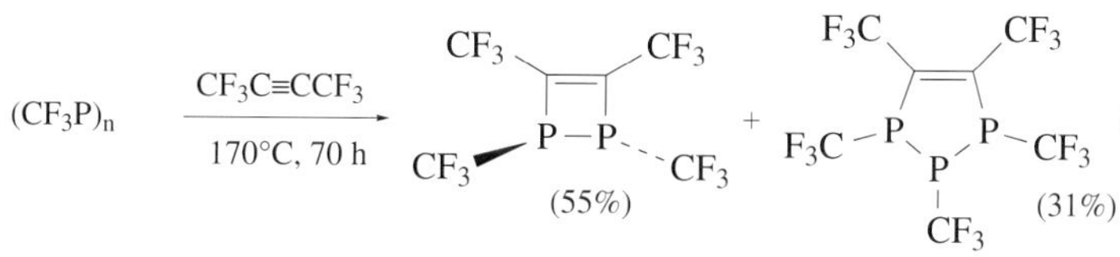

(14)

Later, this very simple synthesis was shown to be more general, but extensive studies of these rather interesting molecules had to await the discovery of a more convenient route involving the ring-expansion of phosphirenes ⟨87AG(E)548⟩ (Equation (15)).

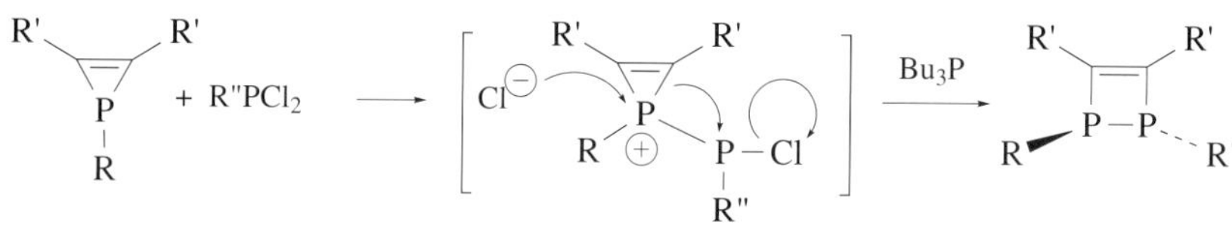

(15)

However, the major event of 1964 unquestionably took place during a IUPAC meeting on organophosphorus chemistry in Heidelberg. On that occasion, Braye described the synthesis of phospholide ions by cleavage of the exocyclic P–R bond of phospholes. His results were published much later ⟨71T5523⟩ (Equation (16)).

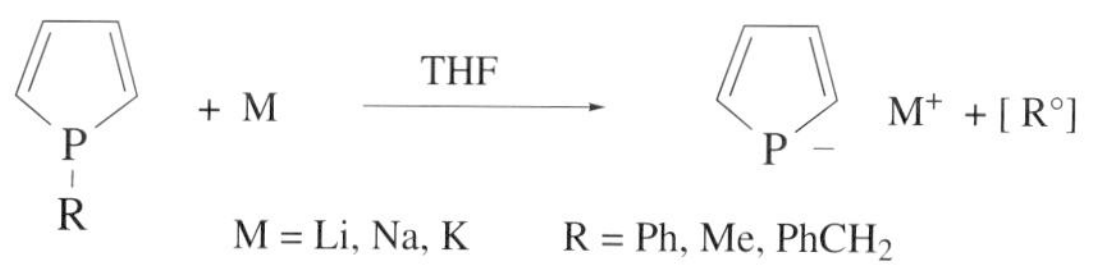

(16)

It was subsequently demonstrated ⟨71T5705⟩ that the reaction proceeds through an unstable ion-radical resulting from the addition of one electron to the ring. The cleavage takes place exclusively at the exocyclic P–R bond. For some time, the phospholide ions were used merely as synthetic tools for the preparation of new phospholes by reaction with various electrophiles (RX, H^+), and not characterized as such. Then in 1979, Quin and Orton reported the first ^{31}P data on these ions ⟨79CC401⟩. The ^{31}P resonances appear at low field, i.e. +77 ppm for the parent (more classical phosphides resonate around 0 ppm). This deshielding is associated with the presence of an in-plane lone pair at phosphorus which, through being weakly coupled to the ring, induces a large paramagnetic shift ⟨94JA9638⟩. Full ^{1}H, ^{13}C and ^{31}P NMR characterization came in 1987 ⟨87TL5025⟩. Subsequently, the aromatic nature of these ions was demonstrated by the X-ray crystal structure analysis of lithium tetramethylphospholide ⟨89AG(E)1367⟩. The ring is planar, the P–C bonds are short at 1.75 Å, and the C–C bonds are equal at ca. 1.40–1.42 Å. It is interesting to note that the lithium is η^5-bonded. Recent theoretical studies calculate an aromatic stabilization energy for the free parent ion at 29 kcal mol^{-1} ⟨96OM1755⟩, which is close to that of cyclopentadienide. Thus, phospholide ions appear to be genuine aromatic species, even though all of their chemistry to date takes place at phosphorus. Nonetheless, development of chemistry at the ring α-carbons is possible through a very simple functionalization method that has been discovered recently ⟨97AG(E)98, 98OM2996⟩ (Equation (17)). The key step is the [1,5] sigmatropic shift of the functional group from phosphorus to the α-carbon. The resulting 2*H*-phosphole is then deprotonated by the base as soon as it is formed. Obviously, the functional group must be compatible with the base at temperatures where the migration takes place.

Perhaps the most important discovery of all appeared in 1966 when Märkl described the first phosphinine and paved the way for the development of the chemistry of dicoordinate phosphorus. 2,4,6-Triphenylphosphinine was obtained through a very simple $O^+ \rightarrow$ P exchange process in suitable pyrylium salts ⟨66AG(E)846⟩ (Equation (18)). Several sources of PH_3 such as $P(CH_2OH)_3$, $P(SiMe_3)_3$ and PH_4I have been used. This initial discovery was later complemented by the preparation of the parent compound by Ashe ⟨71JA3293⟩ (Equation (19)).

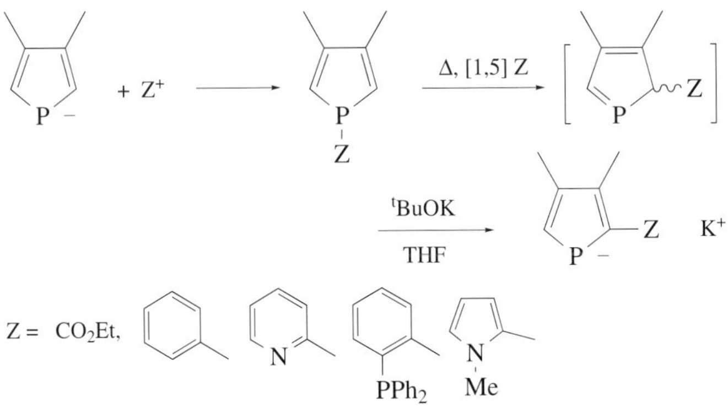

(17)

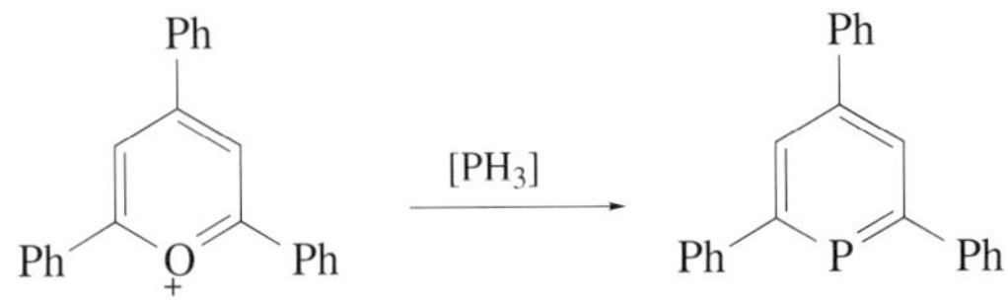

(18)

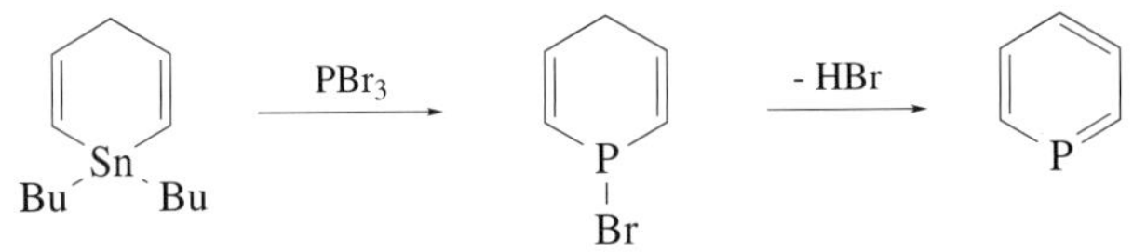

(19)

The phosphinine ring is planar with equalized C–C bonds between 1.38 and 1.41 Å and short P–C bonds at 1.73 Å. The internal CPC angle is ca. 101° ⟨74JCP2840⟩. The aromaticity has been estimated as 88% of that of benzene ⟨88JA4204⟩. The in-plane lone pair is only the third highest occupied level and the LUMO is heavily localized at P. Unlike pyridine, which is basic and nucleophilic, phosphinine shows no basicity — phosphininium salts are unknown — and is electrophilic. This result is very easy to understand: P is less, whereas N is more electronegative than carbon. Since the easily available 2,4,6-derivatives are substituted on the most active positions of the ring, their chemistry initially developed almost exclusively at the phosphorus atom. Strong nucleophiles attack at P to give a delocalized pentadienyl carbanion. Then, electrophiles either react at P to generate λ^5-phosphinines which behave as delocalized ylides ⟨82ACR58⟩, or at the electron-rich α or γ-carbons to give dihydrophosphinines ⟨67AG(E)87, 68TL3611, 77TL407⟩ (Equation (20)).

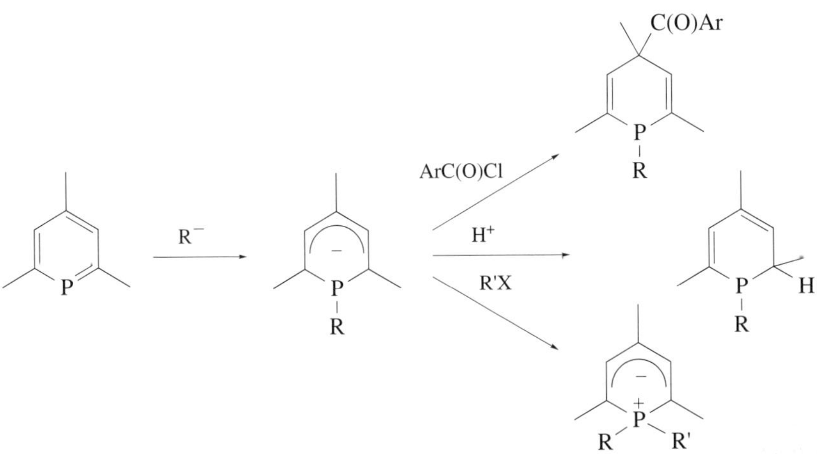

(20)

In practice, the development of any useful chemistry at the carbons of the ring only appeared with the discovery of a straightforward access to 2-halogenophosphinines ⟨89TL817⟩ (Equation (21)).

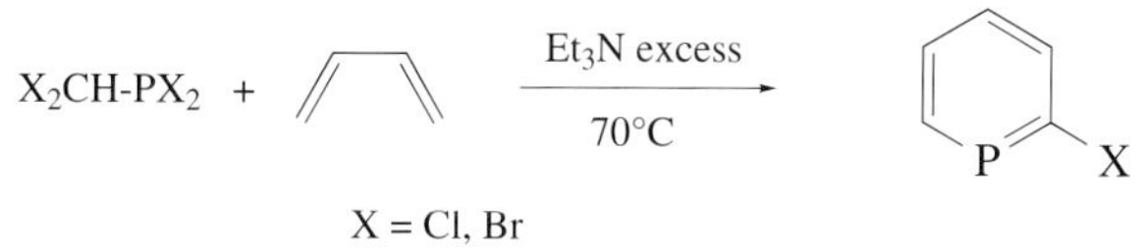

(21)

The mechanism involves the initial formation of a trihalogenophosphaalkene which is trapped by the diene as a [4 + 2] cycloadduct. These 2-halogenophosphinines were shown later to insert [PdL_2] ⟨93JA10665⟩ and [$ZrCp_2$] ⟨97JA9417⟩, thus providing a direct access to 2-functional or 2,6-difunctional phosphinines (Equations (22) and (23)).

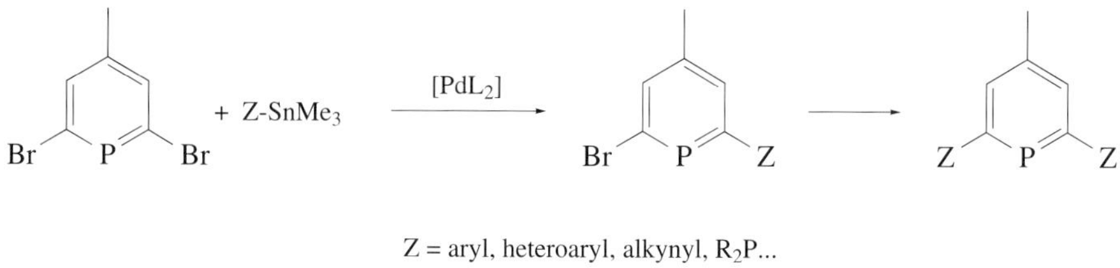

(22)

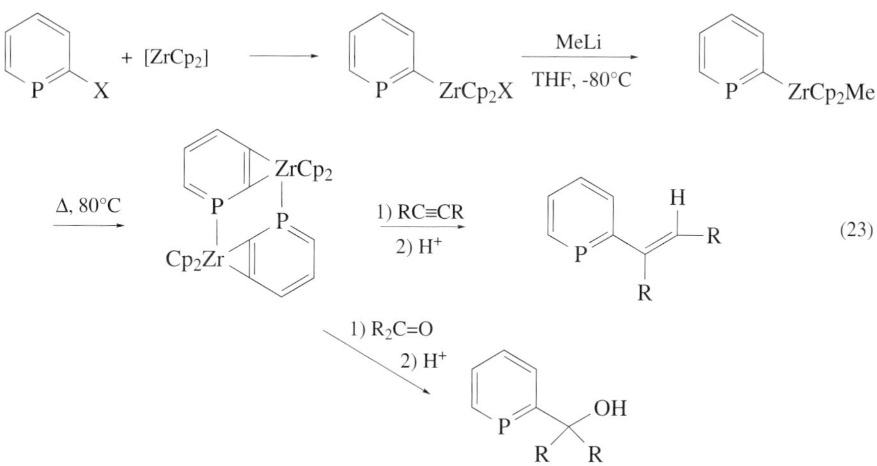

(23)

These η^2-phosphabenzyne–zirconocene complexes have no equivalent in pyridine chemistry. The ready synthesis of 2,2′-biphosphinines through the zirconium reagent must also be underlined ⟨98JOC4826⟩. These new chelating ligands, which were discovered in 1991 ⟨91JA667⟩, display very promising coordination chemistry ⟨98CCR(179)771⟩.

The discovery of 1,2,3-diazaphospholes by Russian chemists in 1967 ⟨67KGS753⟩ further demonstrated that dicoordinate phosphorus could be stabilized efficiently within a cyclic aromatic system. The synthesis is remarkably simple (Equation (24)) and paved the way to a huge variety of heterophospholes, see Section 3 in Chapter 4.

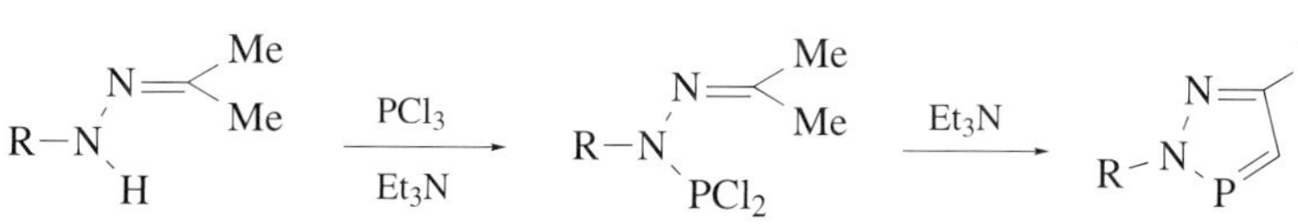

(24)

The seventies were essentially a period of consolidation. The most significant events took place in 1976 and 1977 with the discovery of the first phosphorus analogues of metallocenes, i.e. the phosphacymantrenes ⟨76TL4155⟩ and phosphaferrocenes ⟨77JA3537⟩ (Equation (25)). The reaction probably involves a radical mechanism more or less analogous to the formation of phospholide ions (Equation (16)). This discovery was further substantiated by the preparation of 1,1′-diphosphaferrocenes through a more conventional ionic route ⟨80JA994⟩ (Equation (26)).

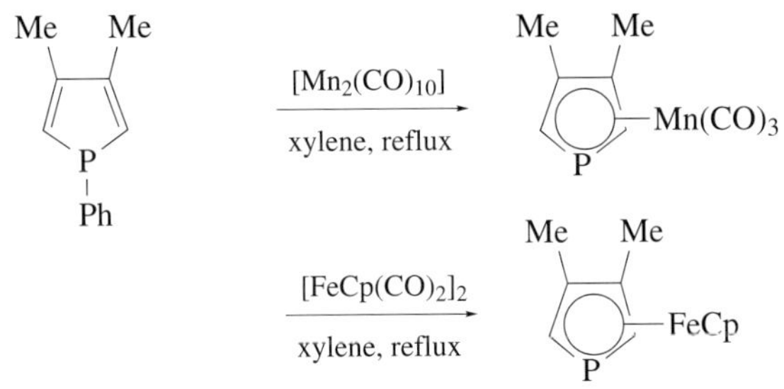

(25)

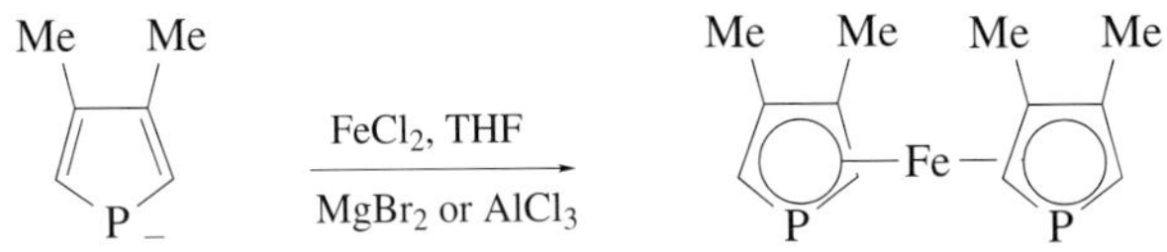

(26)

Subsequently, a whole range of phosphorus analogues of η^5-cyclopentadienyl complexes were prepared ⟨94CCR(137)1⟩. But for organophosphorus chemists, the most fascinating property of these new systems lies in an array of electrophilic substitution reactions such as Friedel–Crafts acylations and Vilsmeier formylation which have no equivalent for any other phosphorus heterocycle ⟨80JA994⟩. Thus functional derivatives are easily accessible (Equation (27)).

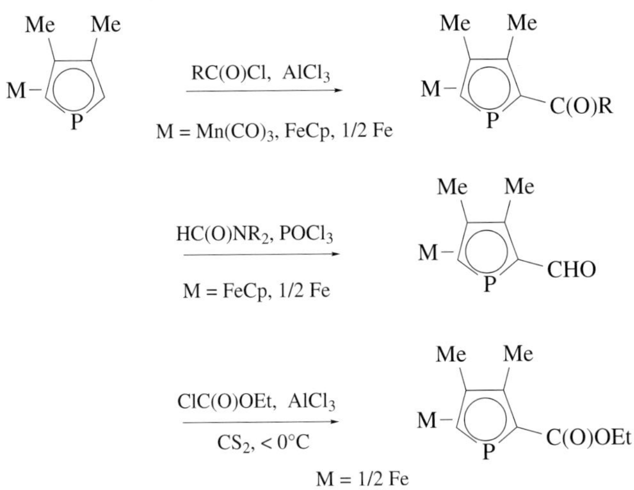

(27)

In these systems, the in-plane lone pair is so low in energy that it can only interact with the most reactive electrophiles ⟨83OM1008⟩. Nevertheless, the LUMO is low in energy and significantly localized at phosphorus, so these phosphametallocenes can coordinate to transition metal centres through the heteroatom. In this respect, their behaviour — weak σ-donors, strong π-acceptors — is very similar to other sp^2-phosphorus ligands. Some applications in homogeneous catalysis have already been found, see Chapter 7.

The last foundation of phosphorus heterocyclic chemistry was laid in 1982 with the structural characterization of the first phosphirene derivatives ⟨82JA4484⟩. Stable phosphirene complexes were obtained by trapping transient terminal phosphinidene complexes with alkynes (Equation (28)). Free trivalent phosphirenes were later obtained by mild oxidative decomplexation of the tungsten complexes ⟨84CC45⟩. In the meantime, a very simple synthesis of phosphirenium cations had appeared ⟨83TL643⟩ (Equation (29)). Reduction of these phosphirenium salts provides a straightforward access to phosphirenes ⟨86TL2635⟩. Unlike 1*H*-azirines, the high pyramidality of phosphorus prevents any antiaromatic destabilization within the 1*H*-phosphirene ring. An intriguing question concerned a possible 2π-aromatic stabilization of phosphirenylium cations. After much effort, these species were finally obtained by abstraction of a triflate ion from the appropriate phosphirene derivatives by using a very powerful Lewis acid ⟨94JA9407⟩ (Equation (30)).

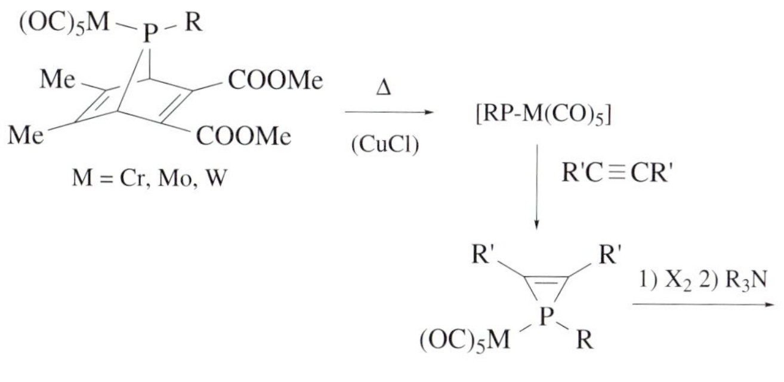

(28)

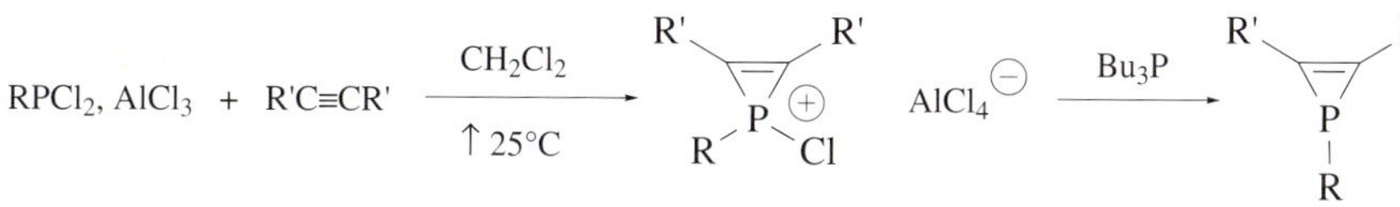

(29)

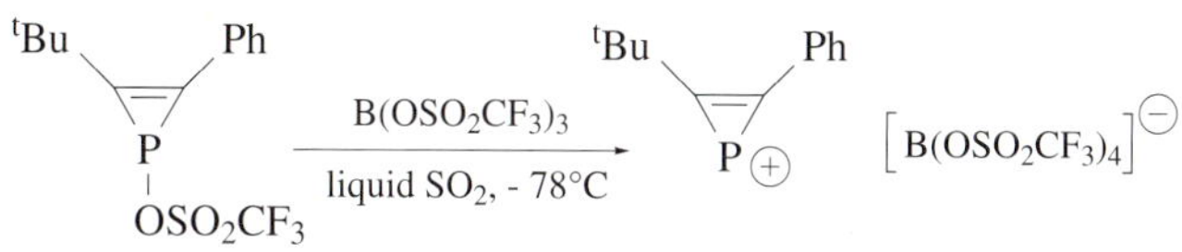

(30)

These very reactive species have been stabilized as η^3-nickel complexes ⟨94AG(E)2330⟩. According to recent theoretical calculations ⟨98JOC2814⟩, these ions are highly aromatic, but quite unstable and reactive. More work will be needed to understand their chemistry.

1.3 RECENT TRENDS

The discovery of phosphirenes brought the first era in phosphorus heterocyclic chemistry to a close. Since then, investigation has continued at an unabated pace. Two main trends emerge from the flood of data that have been collected during the eighties and the nineties. A first trend concerns molecules having unusual electronic structures. Amongst 3-membered rings, the most surprising species is certainly a dimer of the diphosphirenyl radical ⟨98SCI(279)2080⟩ (Equation (31)).

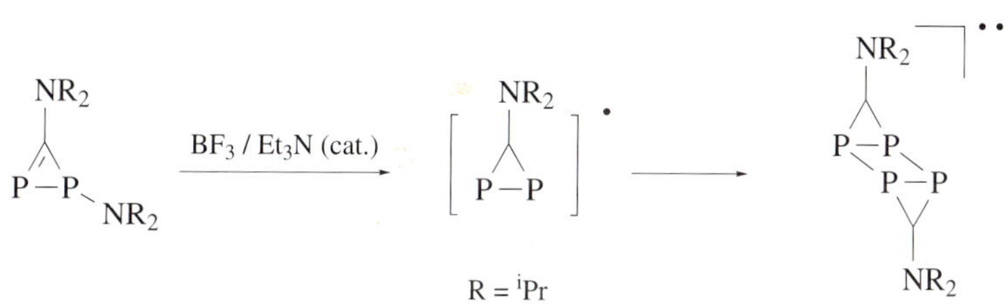

(31)

The stabilization of this dimer results from sharing of two electrons. Hence, each P–P bridge is a formal one-electron bond. Accordingly, these bridges are exceptionally long at 2.634(3) Å vs 2.2 to 2.3 Å for a normal two-electron P–P bond.

In 1986, the groups of Nixon ⟨86CC737⟩ and Binger ⟨86AG(E)644⟩ independently described the synthesis of stable η^4-1,3-diphosphete complexes (Equation (32)).

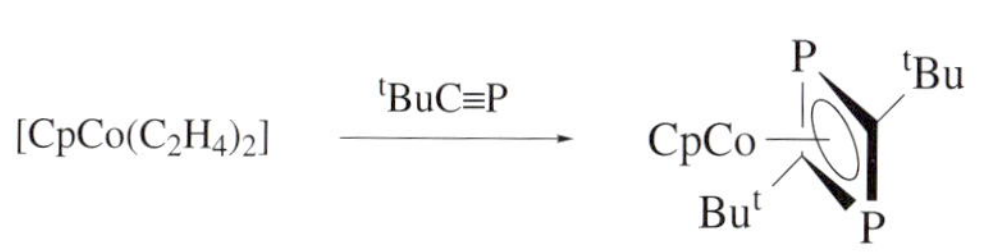

(32)

Stable η^4-1,2-diphosphete complexes were later reported by Binger ⟨95CB1261⟩. However, both 1,2- and 1,3-diphosphetes have remained elusive molecules in the free state, as might be anticipated from the Hückel rules. This is not the case when a pentavalent phosphorus is introduced within the ring. The λ^5P-unit disrupts the electronic delocalization and the antiaromaticity is switched off. This point is discussed in a recent review ⟨98AG(E)271⟩. An example of stable λ^3,λ^5-1,2-diphosphete is presented in Equation (33) ⟨95JA10785⟩.

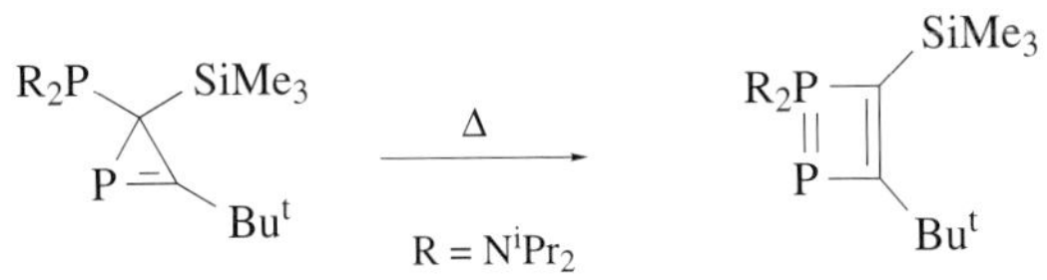

(33)

The [P(II)–W(CO)$_5$] complex displays a long P=P bond at 2.176(3) Å which indicates an ylidic structure $R_2P^{\delta+}$–$P^{\delta-}$.

Another unusual species among 4-membered rings is a stable singlet biradical, which was synthesized by Niecke's group ⟨95AG(E)555⟩ (Equation (34)).

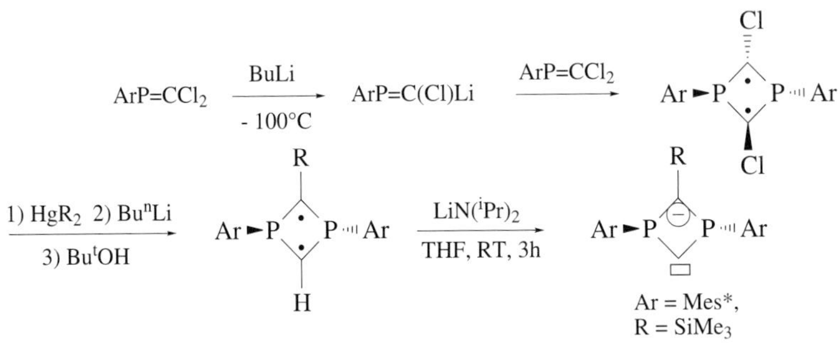

(34)

This deep red product can be considered to be a 6π-aromatic system (2 lone pairs + 2 lone electrons). The P–C bonds are short at 1.75 Å, the ring is planar, but the two phosphorus and carbon atoms are pyramidal. When the aryl groups at phosphorus are replaced by amino substituents, the product rearranges to the more stable 1,2-dihydro-1,2-diphosphete ⟨98AG(E)949⟩. Alternatively, it is possible to transform an aryl derivative into a stable cyclic anionic bis(phosphino)carbene ⟨99AG(E)3031⟩.

Polyphospholides form another intriguing class of heterocyclic structures. 1,2-P_2C_3 ⟨95AG(E)590⟩, 1,3-P_2C_3 ⟨87CC1146⟩, 1,2,3-P_3C_2 ⟨94BSF397⟩, 1,2,4-P_3C_2 ⟨85MI(6)55⟩, 1,2,3,4-P_4C ⟨87ZAAC(544)87⟩ and 1,2,3,4,5-P_5 ⟨87ZAAC(544)87⟩ are all known. The most fascinating species is undoubtedly the pentaphospholide ion which was discovered by Baudler in 1987. This cyclic anion is best made according to Equation (35) ⟨91CB1159⟩.

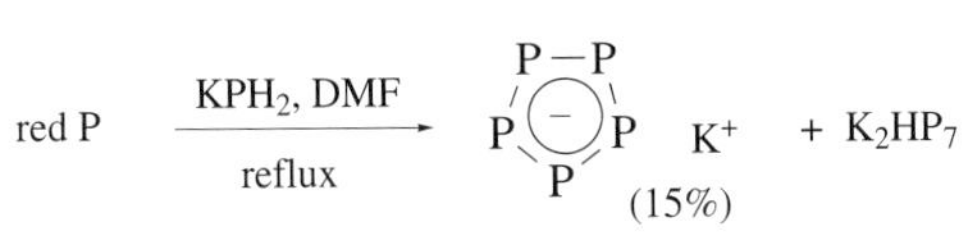

(35)

It must be stressed here that phosphorus is the only heteroelement known to give a stable homoatomic aromatic ring. According to a recent theoretical study, the aromaticities of $[P_5]^-$ and Cp^- are equal ⟨92JOC3694⟩. $[P_5]^-$ can form stable η^5-complexes such as the 1,2,3,4,5-pentaphosphaferrocene independently discovered by the Scherer's group ⟨87AG(E)59⟩ (Equation (36)).

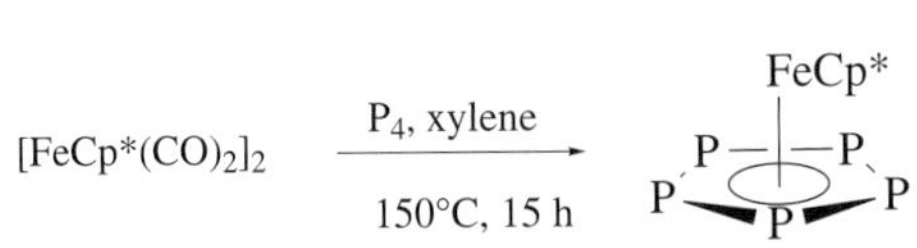

(36)

The phosphinine ring has also served as a basis for interesting structural variations. With several adjacent bulky substituents such as *tert*-butyl, the ring becomes distorted and its aromaticity is reduced. Such a substitution scheme has been used to prepare a stable Dewar-phosphinine whose rearrangement into the normal monocyclic species is prevented by its steric destabilization ⟨86AG(E)280⟩ (Equation (37)).

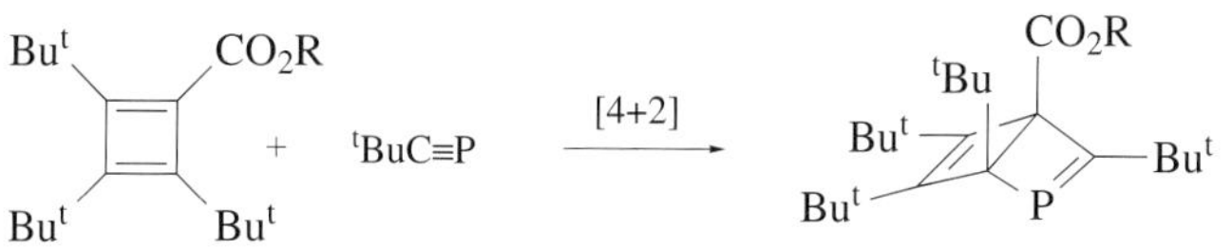

(37)

At 160°C, the 2-P-Dewar-phosphinine rearranges to the 1-P-isomer via the unstable monocyclic phosphinine ⟨87AG(E)85⟩.

Another line of research deals with polyphosphinines. 1,2-Diphosphinines are unknown and 1,4-diphosphinines appear to be unstable ⟨76TL3715⟩, but stable 1,3-diphosphinines have been prepared by Zenneck ⟨95AG(E)198⟩. In the same vein, the long sought-after 1,3,5-triphosphinines have been recently obtained by cyclotrimerization of phosphaalkynes in the coordination sphere of hafnium ⟨95AG(E)2227⟩ (Equation (38)).

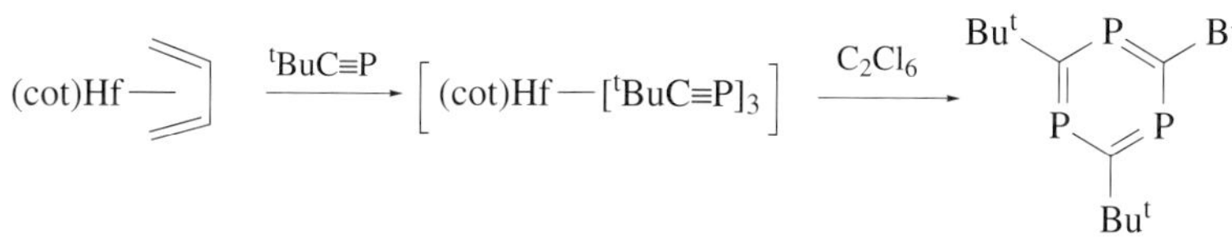

(38)

The exact polycyclic structure of the intermediate hafnium complex was established later ⟨99MI(8)763⟩. A more recent and simpler synthesis of triphosphinines involves a cyclotrimerization of phosphaalkynes in the presence of a vanadium catalyst ⟨98AG(E)1233⟩ (Equation (39a)).

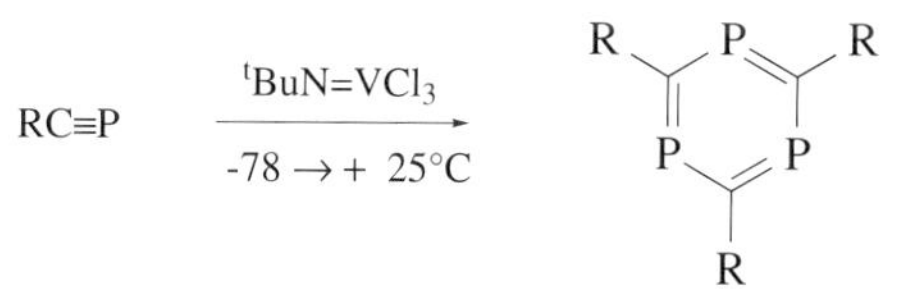

(39a)

In the same area, the discovery of a stable 1,3-diphosphinine moiety incorporating an allene unit within the 6-membered ring deserves to be noted ⟨2000AG(E)1261⟩ (Equation (39b)).

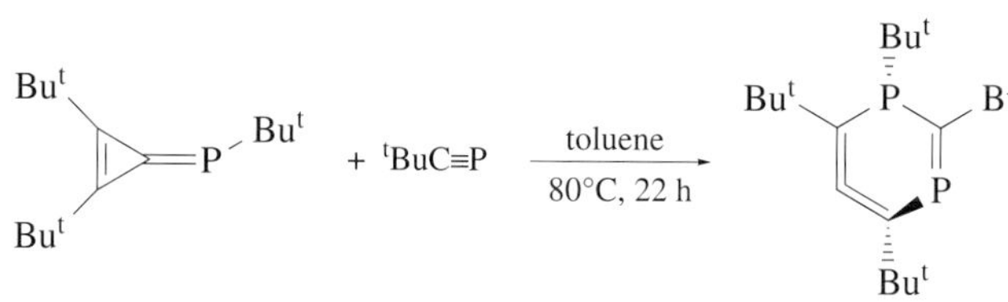

(39b)

A [3 + 2] cycloadduct involving the P=C double bond shows a twisted allenic unit with a C=C=C angle of 155.8(3)°. The chemistry of these new species remains to be explored.

The second major trend that can be observed in recent work is directed toward complexity. Most of the results concern macro- and poly-cycles. Since phosphines are excellent ligands for transition metals, it is not surprising that much effort has been devoted to the synthesis of polyphosphine macrocycles; the inspiration coming from nitrogen chemistry through porphyrins and other widely used polyaza-macrocycles. A major problem arises from the high pyramidal inversion barrier of sp^3-phosphorus (>30 kcal mol^{-1}). Each polyphospha-macrocycle exists as a mixture of isomers reflecting the relative orientations of the phosphorus lone pairs. Thus, the use of such products in coordination chemistry and homogeneous catalysis is somewhat difficult. Three approaches have been used to tackle this major problem. The first relies on a template synthesis giving rise to a single isomer whose geometry is dictated by the central metal. The main difficulty lies in the liberation of the polyphosphine macrocycle from the tightly bound metal. This problem has been solved in some cases ⟨82JA4700, 96CC293⟩ (Equation (40)).

Sometimes, racemization occurs during the demetallation as shown by the study of the oxidative decomplexation of macrocyclic tetraphosphines built in the coordination sphere of Ni(II) ⟨2000T157⟩. The second approach takes advantage of the low pyramidal inversion barrier of phospholes, ca. 16 kcal mol^{-1}

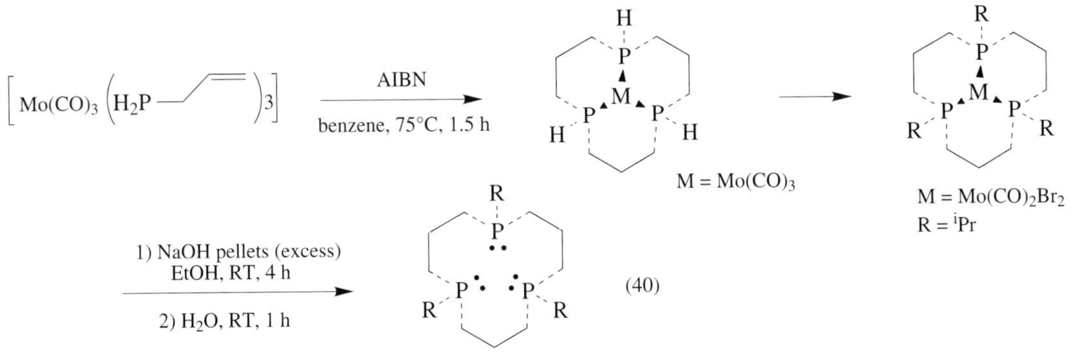

(40)

⟨71JA6205⟩. Each phosphorus of the macrocycle is incorporated into an adjacent phosphole ring. Thus, an equilibrium between the various isomers of the macrocycle is reached around room temperature. When allowed to react with a transition metal derivative, the macrocycle selects its best geometry to coordinate the metal. An example is depicted in Equation (41) ⟨94JA3306⟩.

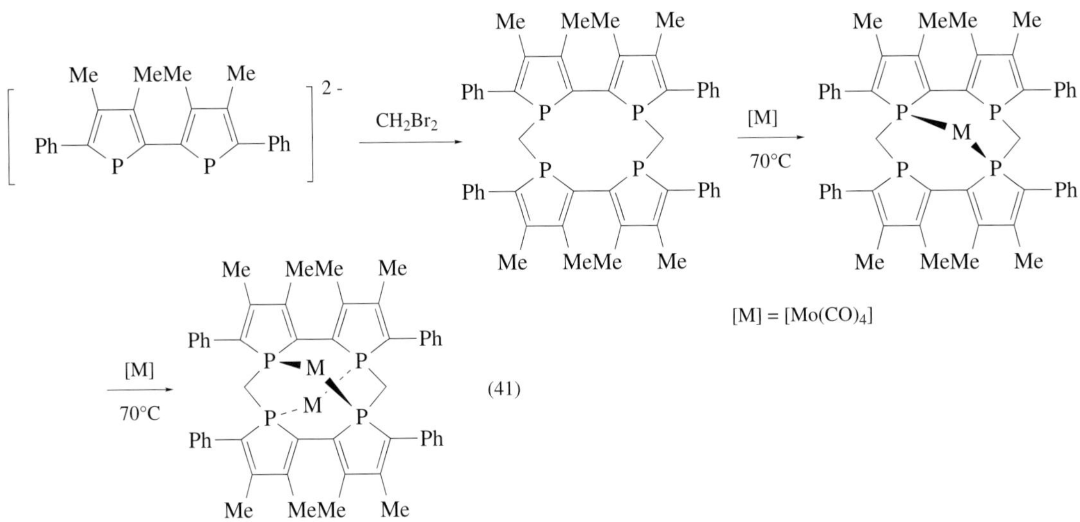

(41)

In the last approach, planar sp^2 rather than pyramidal sp^3-hybridized phosphorus centres are used. Thus, tri- and tetra-phosphinine macrocycles have been prepared by a novel [4 + 2] cycloaddition–cycloreversion route from a 1,3,2-diazaphosphinine ⟨98SCI(280)1587, 99MI(7)2109⟩ (Equation (42)). These macrocycles display an outstanding capability to stabilize low oxidation states. Thus a $[AuP_4]$ gold(0) complex which is stable up to 253 K has been characterized ⟨99AG(E)3194⟩. It must be pointed out here that gold carbonyls are kept in matrices and decompose above 77 K.

In a different vein, numerous phosphorus polycycles have been synthesized in recent years, essentially by cyclooligomerization of phosphaalkynes. The most significant one is undoubtedly the tetraphosphacubane discovered by the group of Regitz ⟨89AG(E)1013⟩. Its best synthesis is depicted in Equation (43) ⟨92AG(E)758⟩. The product shows a ^{31}P resonance at exceptionally low field for a sp^3-P derivative (δ + 257 ppm). The lone pairs participate in the σ-framework of the cube ⟨90CB757⟩. As a result, the phosphorus atoms have a partial positive charge and have diminished reactivity towards electrophiles. Thus, only the mono-methylphosphonium salt and the mono-P–$Fe(CO)_4$ complex can be obtained, even under forcing conditions ⟨92AG(E)879⟩.

Inevitably, the selection of data in this brief survey of phosphorus heterocyclic chemistry reflects a personal bias. Nevertheless, we hope to have convincingly demonstrated that this domain has a unique character, far removed from that of its nitrogen counterpart. Let us recall that 1*H*-phosphirenes are stable whereas 1*H*-azirines are not, that 1*H*-phospholes are pyramidal and essentially non-aromatic whereas 1*H*-pyrroles are planar and aromatic, and that phosphinines are electrophilic whereas pyridines are nucleophilic. One clear similarity between the two domains, however, is their exceptional versatility for

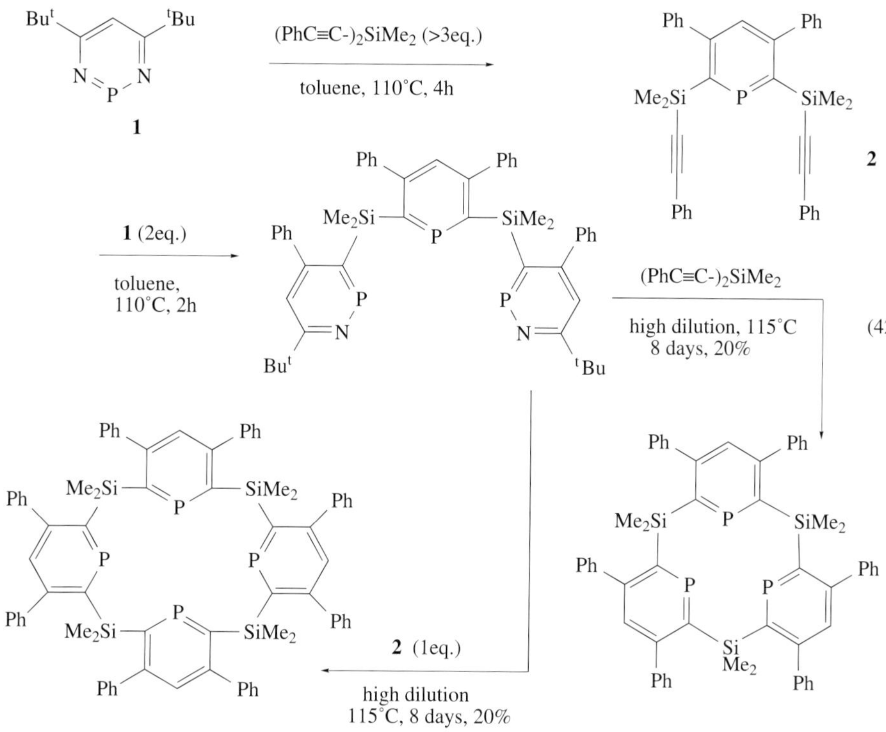

(42)

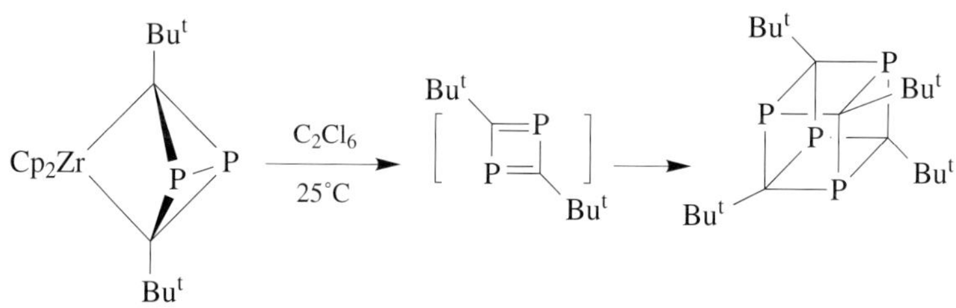

(43)

the formation of varied and complex structures. Bearing in mind the novelty of phosphorus heterocyclic chemistry, much remains to be discovered in this field.

1.4 REFERENCES

15CB1473	G. Grüttner and M. Wiernik; *Chem. Ber.*, 1915, **48**, 1473.
16CB437	G. Grüttner and E. Krause; *Chem. Ber.*, 1916, **49**, 437.
20JA830	J.B. Conant and A.A. Cook; *J. Am. Chem. Soc.*, 1920, **42**, 830.
21JA1665	J.B. Conant and S.M. Pollack; *J. Am. Chem. Soc.*, 1921, **43**, 1665.
B-50MI(1)9	F.G. Mann; "The Heterocyclic Derivatives of Phosphorus, Arsenic, Antimony, Bismuth and Silicon", Interscience, New York, 1950, p. 9.
53MI2(580)44	G. Wittig and G. Geissler; *Annalen,* 1953, **580**, 44.
53USP2663736-9	W.B. McCormack; *US Pat.* 2 663 736-9 (1953) (*Chem. Abstr.*, 1955, **49**, 7602).
59JA3163	F.C. Leavitt, T.A Manuel and F. Johnson; *J. Am. Chem. Soc.*, 1959, **81**, 3163.
59CI(L)1250	E.H. Braye and W. Hübel; *Chem. Ind. (London),* 1959, 1250.
62JA1493	T.W. Campbell and J.J. Monagle; *J. Am. Chem. Soc.*, 1962, **84**, 1493.
62JA3673	T.W. Campbell, J.J. Monagle and V.S. Foldi; *J. Am. Chem. Soc.*, 1962, **84**, 3673.
62JOC606	E. Jungermann, J.J. McBride Jr., R. Clutter and A. Mais; *J. Org. Chem.*, 1962, **27**, 606.
63USP3086053	R.I. Wagner; *US Pat.* 3 086 053 (1963) (*Chem. Abstr.*, 1963, **59**, 10, 124).
64JA2306	W. Mahler; *J. Am. Chem. Soc.*, 1964, **86**, 2306.
65CB576	D. Hellwinkel; *Chem. Ber.*, 1965, **98**, 576.
66AG(E)846	G. Märkl; *Angew. Chem., Int. Ed. Engl.*, 1966, **5**, 846.

67JA1102 R.I. Wagner, L.V.D. Freeman, H. Goldwhite and D.G. Rowsell; *J. Am. Chem. Soc.*, 1967, **89**, 1102.
67JA5984 L.D. Quin and J.G. Bryson; *J. Am. Chem. Soc.*, 1967, **89**, 5984.
67AG(E)87 G. Märkl, F. Lieb and A. Merz; *Angew. Chem., Int. Ed. Engl.*, 1967, **6**, 87.
67KGS753 N.P. Ignatova, N.N. Melnikov and N.I. Shvetsov-Shilovskii; *Khim. Geterotsikl. Soedin,* 1967, 753.
68TL3611 G. Märkl and A. Merz; *Tetrahedron Lett.*, 1968, 3611.
69CR(C)1066 F. Mathey; *C.R. Acad. Sci., Ser. C,* 1969, **269**, 1066.
69JA17 M.T. Bowers, R.A. Beaudet, H. Goldwhite and R. Tang; *J. Am. Chem. Soc.*, 1969, **91**, 17.
70JA5779 P. Coggon, J.F. Engel, A.T. McPhail and L.D. Quin; *J. Am. Chem. Soc.*, 1970, **92**, 5779.
B-70MI(3)1 F.G. Mann; "The Heterocyclic Derivatives of Phosphorus, Arsenic, Antimony, Bismuth," 2nd ed., Interscience, New York, 1970, p. 1.
71JA3293 A.J. Ashe III; *J. Am. Chem. Soc.*, 1971, **93**, 3293.
71JA6205 W. Egan, R. Tang, G. Zon and K. Mislow; *J. Am. Chem. Soc.*, 1971, **93**, 6205.
71T5523 E.H. Braye, I. Caplier and R. Saussez; *Tetrahedron,* 1971, **27**, 5523.
71T5705 D. Kilcast and C. Thomson; *Tetrahedron,* 1971, **27**, 5705.
72JCS(P1)713 J.R. Corfield, R.K. Oram, D.J.H. Smith and S. Trippett; *J. Chem. Soc., Perkin Trans. I,* 1972, 713.
74JCP2840 T.C. Wong and L.S. Bartell; *J. Chem. Phys.*, 1974, **61**, 2840.
74JCS(D)633 J. Emsley, T.B. Middleton and J.K. Williams; *J. Chem. Soc., Dalton Trans.*, 1974, 633.
76TL3715 Y.Kobayashi, I. Kumadaki, A. Ohsawa and H. Hamana; *Tetrahedron Lett.*, 1976, 3715.
76TL4155 F. Mathey; *Tetrahedron Lett.*, 1976, 4155.
77JA3537 F. Mathey, A. Mitschler and R. Weiss; *J. Am. Chem. Soc.*, 1977, **99**, 3537.
77TL407 A.J. Ashe III and T.W. Smith; *Tetrahedron Lett.*, 1977, 407.
79CC401 L.D. Quin and W.L. Orton; *J. Chem. Soc., Chem. Commun.*, 1979, 401.
80JA994 G. de Lauzon, B. Deschamps, J. Fischer, F. Mathey and A. Mitschler; *J. Am. Chem. Soc.*, 1980, **102**, 994.
B-81MI4 L.D. Quin; "The Heterocyclic Chemistry of Phosphorus", Wiley-Interscience, New York, 1981.
81JA4595 F. Mathey, F. Mercier, C. Charrier, J. Fischer and A. Mitschler; *J. Am. Chem. Soc.*, 1981, **103**, 4595.
81S983 A. Brèque, F. Mathey and P. Savignac; *Synthesis,* 1981, 983.
82ACR58 K. Dimroth; *Acc. Chem. Res.*, 1982, **15**, 58.
82CC1272 G. de Lauzon, C. Charrier, H. Bonnard, F. Mathey, J. Fischer and A. Mitschler; *J. Chem. Soc., Chem. Commun.*, 1982, 1272.
82JA4484 A. Marinetti, F. Mathey, J. Fischer and A. Mitschler; *J. Am. Chem. Soc.*, 1982, **104**, 4484.
82JA4700 B.N. Diel, R.C. Haltiwanger and A.D. Norman; *J. Am. Chem. Soc.*, 1982, **104**, 4700.
83JA6871 C. Charrier, H. Bonnard, G. de Lauzon and F. Mathey; *J. Am. Chem. Soc.*, 1983, **105**, 6871.
83OM1008 N.M. Kostic and R.F. Fenske; *Organometallics,* 1983, **2**, 1008.
83OMR494 H. Goldwhite, D. Rowsell, L.E. Vertal, M.T. Bowers, M.A. Cooper and S.L. Manatt; *Org. Magn. Reson.*, 1983, **21**, 494.
83TL643 K.S. Fongers, H. Hogeveen and R.F Kingma; *Tetrahedron Lett.*, 1983, **24**, 643.
B-84MI5 A.R. Katritzky and C.W. Rees (eds.); "Comprehensive Heterocyclic Chemistry", Pergamon, Oxford, 1984, Vol. 1–8.
84CC45 A. Marinetti, F. Mathey, J. Fischer and A. Mitschler; *J. Chem. Soc., Chem. Commun.*, 1984, 45.
85MI(6)55 G. Becker, W. Becker, R. Knebl, H. Schmidt, U. Weeber and M. Westerhausen; *Nova Acta Leopold.*, 1985, **59**, 55.
86AG(E)280 J. Fink, W. Rösch, U.-J. Vogelbacher and M. Regitz; *Angew. Chem., Int. Ed. Engl.*, 1986, **25**, 280.
86AG(E)644 P. Binger, R. Milczarek, R. Mynott, M. Regitz and W. Rosch; *Angew. Chem., Int. Ed. Engl.*, 1986, **25**, 644.
86CC737 P.B. Hitchcock, M.J. Maah and J.F. Nixon; *J. Chem. Soc., Chem. Commun.*, 1986, 737.
86TL2635 S. Lochschmidt, F. Mathey and A. Schmidpeter; *Tetrahedron Lett.*, 1986, **27**, 2635.
87AG(E)59 O.J. Scherer and T. Brück; *Angew. Chem., Int. Ed. Engl.*, 1987, **26**, 59.
87AG(E)85 K. Blotter, W. Rösch, U.-J. Vogelbacher, J. Fink and M. Regitz; *Angew. Chem., Int. Ed. Engl.*, 1987, **26**, 85.
87AG(E)548 L. Ricard, N. Maigrot, C. Charrier and F. Mathey; *Angew. Chem., Int. Ed. Engl.*, 1987, **26**, 548.
87CC1146 R. Bartsch, P.B. Hitchcock and J.F. Nixon; *J. Chem. Soc., Chem. Commun.*, 1987, 1146.
87TL5025 C. Charrier and F. Mathey; *Tetrahedron Lett.*, 1987, **28**, 5025.
87ZAAC(544)87 M. Baudler, D. Düster and D. Ouzounis; *Z. Anorg. Allg. Chem.*, 1987, **544**, 87.
88JA4204 K.K. Baldridge and M.S. Gordon; *J. Am. Chem. Soc.*, 1988, **110**, 4204.
89AG(E)1013 T. Wettling, J. Schneider, O. Wagner, C.G. Kreiter and M. Regitz; *Angew. Chem. Int. Ed. Engl.*, 1989, **28**, 1013.
89AG(E)1367 T. Douglas and K.H. Theopold; *Angew. Chem. Int. Ed. Engl.*, 1989, **28**, 1367.
89JPC3025 J.A. Boatz and M.S. Gordon; *J. Phys. Chem.*, 1989, **93**, 3025.
89TL817 P. Le Floch and F. Mathey; *Tetrahedron Lett.*, 1989, **30**, 817.
90CB757 R. Gleiter, K.-H. Pfeifer, M. Baudler, G. Scholz, T. Wettling and M. Regitz; *Chem. Ber.*, 1990, **123**, 757.
91CB1159 M. Baudler and T. Etzbach; *Chem. Ber.*, 1991, **124**, 1159.
91JA667 P. Le Floch, D. Carmichael, L. Ricard and F. Mathey; *J. Am. Chem. Soc.*, 1991, **113**, 667.
92AG(E)758 T. Wettling, B. Geissler, R. Schneider, S. Barth, P. Binger and M. Regitz; *Angew. Chem., Int. Ed. Engl.*, 1992, **31**, 758.
92AG(E)879 M. Birkel, J. Scholz, U. Bergsträsser and M. Regitz; *Angew. Chem., Int. Ed. Engl.*, 1992, **31**, 879.
92JOC3694 E.J. Padma Malar; *J. Org. Chem.*, 1992, **57**, 3694.
93JA10665 P. Le Floch, D. Carmichael, L. Ricard and F. Mathey; *J. Am. Chem. Soc.*, 1993, **115**, 10665.
93JOC5414 S.M. Bachrach; *J. Org. Chem.*, 1993, **58**, 5414.

94AG(E)2330	A.G. Avent, F.G.N. Cloke, K.R. Flower, P.B. Hitchcock, J.F. Nixon and D.M. Vickers; *Angew. Chem. Int. Ed. Engl.*, 1994, **33**, 2330.
94BSF397	N. Maigrot, M.L. Sierra, C. Charrier and F. Mathey; *Bull. Soc. Chim. Fr.*, 1994, **131**, 397.
94CCR(137)1	F. Mathey; *Coord. Chem. Rev.*, 1994, **137**, 1.
94JA3306	F. Laporte, F. Mercier, L. Ricard and F. Mathey; *J. Am. Chem. Soc.*, 1994, **116**, 3306.
94JA9407	K.K. Laali, B. Geissler, O. Wagner, J. Hoffmann, R. Armbrust, W. Eisfeld and M. Regitz; *J. Am. Chem. Soc.*, 1994, **116**, 9407.
94JA9638	D.B. Chesnut and L.D. Quin; *J. Am. Chem. Soc.*, 1994, **116**, 9638.
95AG(E)198	D. Böhm, F. Knoch, S. Kummer, U. Schmidt and U. Zenneck; *Angew. Chem. Int. Ed. Engl.*, 1995, **34**, 198.
95AG(E)337	P.v.R. Schleyer, P.K. Freeman, H. Jiao and B. Goldfuss; *Angew. Chem. Int. Ed. Engl.*, 1995, **34**, 337.
95AG(E)555	E. Niecke, A. Fuchs, F. Baumeister, M. Nieger and W.W. Schoeller; *Angew. Chem. Int. Ed. Engl.*, 1995, **34**, 555.
95AG(E)590	N. Maigrot, N. Avarvari, C. Charrier and F. Mathey; *Angew. Chem., Int. Ed. Engl.*, 1995, **34**, 590.
95AG(E)2227	P. Binger, S. Leininger, J. Stannek, B. Gabor, R. Mynott, J. Bruckmann and C. Krüger; *Angew. Chem. Int. Ed. Engl.*, 1995, **34**, 2227.
95CB1261	P. Binger, G. Glaser, S. Albus and C. Kruger; *Chem. Ber.*, 1995, **128**, 1261.
95CC257	D.C.R. Hockless, M.A. McDonald, M. Pabel and S.B. Wild; *J. Chem. Soc., Chem. Commun.*, 1995, 257.
95JA10785	R. Armbrust, M. Sanchez, R. Réau, U. Bergsträsser, M. Regitz and G. Bertrand; *J. Am. Chem. Soc.*, 1995, **117**, 10785.
96CC293	S.J. Coles, P.G. Edwards, J.S. Fleming, M.B. Hursthouse and S.S. Liyanage; *J. Chem. Soc., Chem. Commun.*, 1996, 293.
96OM1755	B. Goldfuss, P.v.R. Schleyer and F. Hampel; *Organometallics,* 1996, **15**, 1755.
97AG(E)98	S. Holand, M. Jeanjean and F. Mathey; *Angew. Chem., Int. Ed. Engl.*, 1997, **36**, 98.
97CC1033	A. Göller and T. Clark; *Chem. Commun.*, 1997, 1033.
97JA9417	P. Rosa, P. Le Floch, L. Ricard and F. Mathey; *J. Am. Chem. Soc.*, 1997, **119**, 9417.
98AG(E)271	G. Bertrand; *Angew. Chem., Int. Ed. Engl.*, 1998, **37**, 271.
98AG(E)949	O. Schmidt, A. Fuchs, D. Gudat, M. Nieger, W. Hoffbauer, E. Niecke, and W.W. Schoeller; *Angew. Chem. Int. Ed. Engl.*, 1998, **37**, 949.
98AG(E)1233	F. Tabellion, A. Nachbauer, S. Leininger, C. Peters, F. Preuss and M. Regitz; *Angew. Chem. Int. Ed. Engl.*, 1998, **37**, 1233.
98CCR(179)771	P. Le Floch and F. Mathey; *Coord. Chem. Rev.*, 1998, **179–180**, 771.
98JOC2814	W. Eisfeld and M. Regitz; *J. Org. Chem.*, 1998, **63**, 2814.
98JOC4826	P. Rosa, N. Mézailles, F. Mathey and P. Le Floch; *J. Org. Chem.*, 1998, **63**, 4826.
98OM2996	S. Holand, N. Maigrot, C. Charrier and F. Mathey; *Organometallics,* 1998, **17**, 2996.
98SCI(279)2080	Y. Canac, D. Bourissou, A. Baceiredo, H. Gornitzka, W.W. Schoeller and G. Bertrand; *Science,* 1998, **279**, 2080.
98SCI(280)1587	N. Avarvari, N. Mézailles, L. Ricard, P. Le Floch and F. Mathey; *Science,* 1998, **280**, 1587.
99AG(E)3031	E. Niecke, A. Fuchs, M. Nieger, O. Schmidt and W.W. Schoeller; *Angew. Chem. Int. Ed.*, 1999, **38**, 3031.
99AG(E)3194	N. Mézailles, N. Avarvari, N. Maigrot, L. Ricard, F. Mathey, P. Le Floch, L. Cataldo, T. Berclaz and M. Geoffroy; *Angew. Chem. Int. Ed.*, 1999, **38**, 3194.
99MI(7)2109	N. Avarvari, N. Maigrot, L. Ricard, F. Mathey and P. Le Floch; *Chem. Eur. J.*, 1999, **5**, 2109.
99MI(8)763	P. Binger, S. Stutzmann, J. Stannek, K. Günther, P. Phillips, R. Mynott, J. Bruckmann and C. Krüger; *Eur. J. Inorg. Chem.*, 1999, 763.
2000T157	T. Lebbe, P. Machnitzki, O. Stelzer and W.S. Sheldrick; *Tetrahedron*, 2000, **56**, 157.
2000AG(E)126	M.A. Hofmann, U. Bergsträsser, G.J. Reiss, L. Nyulaszi and M. Regitz; *Angew. Chem. Int. Ed.*, 2000, **39**, 1261.

2.1 Three-membered Rings. 1. Phosphiranes and Phosphirenes

F. MATHEY
Ecole Polytechnique, Palaiseau Cedex, France

and

M. REGITZ
Universität Kaiserslautern, Kaiserslautern, Germany

2.1.1 INTRODUCTION

The first phosphiranes, which included the parent compound (**1**), were described by Wagner in two patents issued in 1963 ⟨63USP3086053, 63USP3086056⟩. Although initially viewed with some scepticism, their discovery was fully confirmed in 1967 ⟨67JA1102⟩. Almost two decades later, the first stable phosphirene derivative (**2**) was prepared by Marinetti and Mathey and unambiguously characterised on the basis of a X-ray crystal structure analysis ⟨82JA4484⟩. The same authors obtained the first, potentially antiaromatic, tervalent 1*H*-phosphirene (**3**) two years later ⟨84CC45⟩. The story was completed in 1987 with the discovery of the first stable 2*H*-phosphirene (**4**) by Wagner, Maas and Regitz ⟨87AG(E)1257⟩. At present, the parent 1*H*- and 2*H*-phosphirenes remain unknown.

(1)

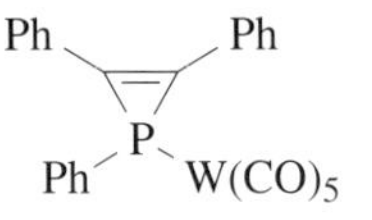

(2)

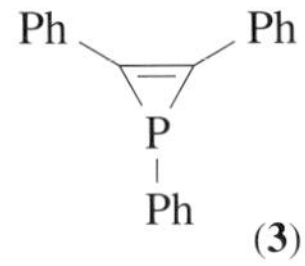

(3)

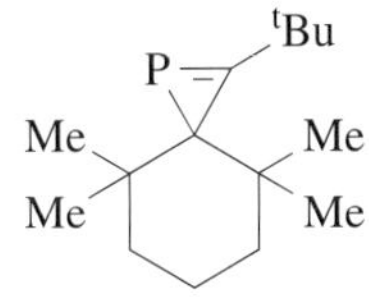

(4)

It is striking to compare the relative novelty of these phosphorus–carbon three-membered rings with the well-established history of aziridines, which dates back to 1888. However, progress has been extremely rapid since 1963, in part due to the powerful tool of ^{31}P NMR, and the overall shape of the field is now definitively established. In the next sections, we shall often refer to the most accessible and relatively recent comprehensive review on these rings which was published in 1990 ⟨90CRV997⟩.

2.1.2 STRUCTURE OF PHOSPHIRANES AND PHOSPHIRENES

2.1.2.1 Theoretical Data

Only the most recent theoretical data will be treated here. For a presentation of older results, the reader is referred to ⟨90CRV997⟩. *Ab initio* calculations using a 4-31G(*) basis set have been performed on the parent phosphirane by Gonbeau and Pfister-Guillouzo ⟨87IC1799⟩. The computed geometry is satisfactory: P–C 1.844 Å (1.867), C–C 1.504 Å (1.502), P–H 1.407 Å (1.428); <C–P–C 48.1° (47.4), <H–C–P 96.4° (95.2) by comparison with the experimental data in parentheses which come from a microwave study ⟨69JA17⟩. The calculations indicate that phosphorus bears a significant positive charge (+0.207), and the carbons of the ring are negative (−0.478). The HOMO at −9.85 eV corresponds to the lone pair, whilst the NHOMO and the LUMO, which result from the bonding and antibonding combinations of $p_{(P)}$ and $\pi^*_{(C=C)}$, are associated with the σ(P–C) bent bonds. The P–C bonds are weak according to a Mulliken population analysis. A similar picture was obtained some time later by Dobbs *et al.* ⟨88JPC4886⟩, whose use of a 6-31G* basis set gave a better geometry: P–C 1.853 Å, <C–P–C 47.4°. The replacement of H by Cl at phosphorus induces substantial shortening of the P–C bonds (1.840 Å). This was explained in terms of a large increase of the positive charge at P (from +0.24 for PH to +0.50 for PCl) which leads to a stronger ionic interaction between P and C. Some time later, a calculation of the inversion barrier at P was reported by Bachrach ⟨89JPC7780⟩.The extremely high value of 68.3 kcal mol^{-1} at the MP2/6-31G*//HF/6-31G*+ZPE(3-21G(*) level was seen as a consequence of the large increase of the H–P–C angle, combined with a stretch of the C–C bond by 0.088 Å during the inversion process. Even later, Göller and Clark ⟨97CC1033⟩ computed a similar value of 66.5 kcal mol^{-1} [CCSD(T)/6-31+G*//(MP2/6-31+G*)] but, more importantly, found that the 1-fluoro, 1-chloro, and 1-bromo-phosphiranes invert by turnstile rotation. Schematically, the P–X unit rotates around the symmetry axis of the three-membered ring. The inversion barriers thus calculated, 45.5 (X=F), 55.4 (X=Cl), and 50.9 kcal mol^{-1} (X=Br), are 10–20 kcal mol^{-1} lower than those for a pyramidal inversion mechanism. In a different vein, Boatz and Gordon ⟨89JPC3025⟩ calculated some thermodynamic parameters for phosphirane at the MP2/6-31G(d) level. Their data include the heat of formation from the heteroatomic subunit (here PH) and ethylene. The computed ΔH of −72.3 kcal mol^{-1} may be compared with −109.7 for aziridine, −128.7 for oxirane and −84.3 kcal mol^{-1} for thiirane; the corresponding strain energies are ΔE −22.0 kcal mol^{-1} for phosphirane vs −29.5 for aziridine, −30.0 for oxirane and −17.7 kcal mol^{-1} for thiirane. A remarkable agreement for the fragmentation energy of phosphirane into phosphinidene and ethylene exists between the results of Boatz and Gordon and the more recent calculations of Nguyen, Landuyt and Vanquickenborne ⟨94JCS(F)1771⟩: ΔE +76.6 vs +74.2 kcal mol^{-1}. This latter paper deals more precisely with the various ring-chain rearrangements of phosphirane. The transformation of phosphirane into vinylphosphine is predicted to be a concerted process with an energy barrier of only 56.2 kcal mol^{-1} at the QCISD(T)/6-31G(d,p)+ZPE level (see later, Equation (28)). Quite interestingly, more recent calculations at the GVB/6-31G** level ⟨95JOC3723⟩ indicate that the transition state for the opening of phosphirane into bis-(methylene)phosphorane lies on a disrotatory reaction trajectory in contradiction to the Woodward–Hoffmann rules. The transition state is highly pyramidal and has diradical character. If phosphorus is constrained to be planar in the transition state, a concerted conrotatory opening is favoured. Hence, the experimental observations pointing to a conrotatory process ⟨83CC1171, 92CB771⟩ could be explained by the influence of bulky substituents at P which impose a planarisation of the transition state. Phosphiranium salts have also been thoroughly investigated from a theoretical standpoint. It has been experimentally proven that they easily undergo alkene exchange with alkynes to give phosphirenium salts

⟨97JOM(529)189⟩. This exchange reaction has been shown to proceed through a spirobicyclic transition state having C_{2v} symmetry whose barrier was calculated to be 6.7 kcal mol^{-1} at the G2 level ⟨98JA7063⟩. Three related works deal with the phosphiranium cation $[C_2H_4PH_2]^+$ and the phosphiranylium cation $[C_2H_4P]^+$. The first one predicts a substantial increase in the strain energy of phosphirane upon protonation: ΔE -47.3 kcal mol^{-1} at the 6-31G*//3-21G level ⟨87JST(152)69⟩. The second predicts an easy disrotatory ring opening of the phosphiranylium ion into the bis(methylene)-phosphonium cation. The open cation is more stable than its cyclic isomer by 17.5 kcal mol^{-1} and the activation energy for the opening of the ring is only 15.2 kcal mol^{-1} ⟨93JOC1379⟩. Conversely, the third study predicts that the electrocyclic opening of the phosphiranylium ion into the 2-phosphaallyl cation is difficult. The cyclic species appears to be more stable than the open one by 9.9 kcal mol^{-1} and the barrier to the cyclisation of the open cation was found to be only 1.2 kcal mol^{-1} ⟨93BSF636⟩. Finally, various studies have dealt with the valence-bond description ⟨95JST(341)13⟩ and the molecular polarizability of phosphirane ⟨97MI(2)255⟩, the electronic structure of the anions formed through electron capture by phosphiranes ⟨96JST(371)69, 98JST(424)67⟩, and some properties of triphospha[1.1.1]-propellane ⟨94IC1153⟩, including calculations of its molecular structure and its relatively low ring strain by comparison with [1.1.1]propellane.

The first *ab initio* (Hondo) analysis of the electronic structure of phosphirene was made by Gonbeau and Pfister-Guillouzo ⟨85NJC71⟩. The ring was treated as a combination of phosphinidene and acetylene: the HOMO at -9.13 eV is a combination of $n_{p(P)}$ and $\pi_{1(C\equiv C)}$ (perpendicular to the ring plane). The NHOMO and LUMO result from the bonding and antibonding combinations of $p^*_{(P)}$ and $\pi^*_{2(C\equiv C)}$ (in the ring plane). The stabilisation of the phosphirene ring by complexation with $Cr(CO)_5$ was ascribed to a weakening of the $n_{p(P)}\pi_{1(C\equiv C)}$ antiaromatic interaction and an increased $p^*_{(P)}\pi^*_{2(C\equiv C)}$ interaction which together serve to stabilise the P–C ring bonds ⟨85IC4133⟩. Subsequent studies have mainly addressed the question of the relative stability of 1*H*- and 2*H*-phosphirenes. For example, Nguyen *et al.* ⟨92CB923⟩ calculated that 1*H*-phosphirene is higher in energy than 2*H*-phosphirene by 11.2 kcal mol^{-1} at the MP4/6-31G** level. Similar values have been obtained by Bachrach ⟨91JOC2205⟩ (12.3 kcal mol^{-1}) and Berger *et al.* ⟨94OM640⟩ (10 kcal mol^{-1}). Furthermore, all of these authors confirm experimental observations that 1-halophosphirenes are more stable than their 2*H*-isomers. Nguyen *et al.* found that 1-fluorophosphirene is more stable by 21.3 and 1-chlorophosphirene by 10.3 kcal mol^{-1} than the corresponding P-dicoordinated species. The same authors calculated that the presence of a P-halogen substituent induces a substantial shortening of the P–C ring bonds, from 1.821 (PH) to 1.773 (PF) and 1.778 Å (PCl). This stabilising effect of halogens on 1*H*-phosphirenes was ascribed to a greater contribution of the aromatic phosphirenylium cation $[C_2H_2P]^+$ to the electronic structure of the ring because of the increased electronegativity of the P-substituent. Similar results and interpretations were provided by Bachrach ⟨92JST(255)207⟩. Another interesting point concerns the pyramidal inversion barrier of phosphorus in 1*H*-phosphirene. According to the calculations of Bachrach at the HF/6-31G* level ⟨91JOC2205⟩, this barrier is very high at 92.6 kcal mol^{-1}. The antiaromatic character of the planar 1*H*-phosphirene structure explains why pyramidal inversion at phosphorus is even more difficult than in phosphirane and why the preferred inversion mechanism involves a turnstile rotation having a barrier of only 52.2 kcal mol^{-1} at the [CCSD(T)/6-31+G*//(MP2/6-31+G*)] level ⟨97CC1033⟩. For 1-halophosphirenes, the barrier to turnstile rotation appears to be even lower at 38.4 (X=F), 43.4 (X=Cl), and 40.2 kcal mol^{-1} (X=Br). In a different vein, it has been shown that a BH_2 substituent at P considerably lowers the inversion barrier of phosphirene ⟨94JPC6481⟩. Other studies have dealt with the vertical electron affinities of various 1-R-phosphirenes (R=H, F, CN) ⟨97BSF1033⟩ and the electronic structure of the lowest excited states of phosphirene ⟨96T4709⟩. More significant investigations on the electronic structures of phosphirenium and phosphirenylium cations have also appeared. The stability of phosphirenium cations $[C_2H_2PY_2]^+$ has been ascribed to the bonding interaction between the $[PY_2]$ σ^*orbital of p-type character at phosphorus and the π orbital of the olefinic fragment. The strain energy difference between the saturated and unsaturated rings gives an indication of the σ^* aromaticity of the unsaturated species. In the reference couple cyclopropane/cyclopropene which shows no σ^* aromaticity, the difference is 22.1 kcal mol^{-1}, and between 5.3 and 8.0 kcal mol^{-1} for phosphiranium/phosphirenium couples as a function of the nature of Y ⟨96JOC5840⟩. The first *ab initio* treatment of the phosphirenylium cation was released by Maclagan ⟨89CPL(163)349⟩ some time ago. Of the possible $[C_2H_2P]^+$ isomers, the cyclic species was predicted to be the most stable. The optimised geometry at the MP3/6-31G* level suggested substantial cyclic delocalisation within the ring: P–C 1.729, C=C 1.349 Å and the formation of this ion was found to be possible in interstellar clouds when C_2H_2 is present together with PH_n^+. These data have been corroborated by a more recent study at the HF/6-31+G* level ⟨94JA9407⟩ indicating P–C and C=C bond lengths of 1.7032 and 1.3406 Å, respectively. The missing electron density at the C=C double bond is predicted to be localised in the phosphorus lone pair orbital and, consequently, the C=C bond order is reduced by comparison with 1-fluorophosphirene (bond orders of 1.776 and 1.943 respectively). A third

Table 1. Structural parameters of monocyclic phosphiranes

1-R-phosphirane	*Bond angles (°)*			*Bond lengths (Å)*		
	C–P–C	*C–P–R*	*P–C–C*	*P–C*	*C–C*	*P–R*
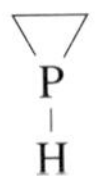 (**1**)	47.4	95.2	66.3	1.867	1.502	1.428 (P–H)
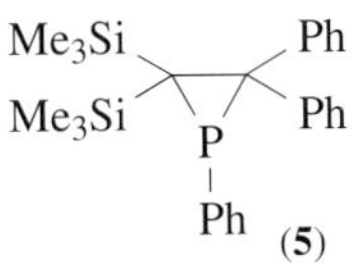 (**5**)	49	110.6 115.1		1.87 1.89	1.56	
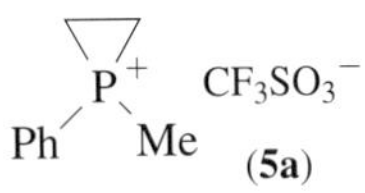 (**5a**)	51.7	117.6–118.8 (Me) 119.1 (Ph)	64.0–64.3	1.756 1.761	1.533	1.779 (Me) 1.763 (Ph)

(**1**): microwave study ⟨69JA17⟩; (**5**): X-ray crystal structure analysis ⟨86CB1977⟩; (**5a**): X-ray crystal structure analysis ⟨95CC257⟩.

study at the MP4SDTQ(fc)/6-31+g*//RHF/6-31g* level ⟨95JST(343)49⟩ gives a precise description of the electronic structure of this ion. The HOMO at −18.7 eV corresponds to the in-plane lone pair at P, whereas the LUMO at −5.5 eV is an antibonding π orbital delocalised all over the ring. The net positive charge at phosphorus of only 0.678 e is smaller than in the more localised phosphiranylium cation (0.865 e). The fourth study indicates that the P–X bond of 1-halophosphirenes has a strong ionic character, especially when X=F. The aromaticity of the phosphirenylium ion, evaluated by comparison of the heterolysis energies of 1-halo-phosphiranes and phosphirenes, has been estimated at 32.5 kcal mol^{-1}. The introduction of two amino substituents at the carbons of the ring stabilises the cation by 40–42 kcal mol^{-1}. Interaction with SO_2 molecules (liquid SO_2 is used as the reaction medium for the synthesis of these ions) reduces the delocalisation but also lowers the heterolysis energy by ca. 18 kcal mol^{-1} ⟨98JOC2814⟩. The structure computed in the last study differs slightly from the two preceding ones: P–C 1.737, C=C 1.353 Å (MP2/6-31+G*).

2.1.2.2 Structural Data

A listing of all the structural data concerning phosphiranes and phosphirenes that were available in 1989 has been published ⟨90CRV997⟩. Since then, numerous phosphirane complexes have been characterised by X-ray analysis, but no new features have been brought to light. In Table 1, we list the structural parameters of two representative monocyclic phosphiranes and one phosphiranium salt. Typically, the <C–P–C> intracyclic angles of phosphirane derivatives fall in a range from 47 ⟨89CC973⟩ to 51° ⟨89CB453⟩. In fact, the ring is quite supple, according to the range of intracyclic C–C bond lengths which vary between 1.460 ⟨89CC973⟩ and 1.588 Å ⟨89AG(E)900⟩. The same effect is found for the intracyclic P–C bonds whose lengths range from 1.78 ⟨89POL2671⟩ to 1.89 Å ⟨86CB1977⟩. These data nicely explain the observation that the ease of dissociation of the phosphirane ring into alkene and phosphinidene is highly dependent on the nature of the ring substituents.

Some representative X-ray crystal structure analyses of phosphirene derivatives are collected in Table 2. As in the phosphirane case, it is clear that significant variations in structural parameters take place with changes in the substitution pattern. A comparison of the structures of (**3**) and (**6**) supports the theoretical assertion that the halogen substituent stabilises the phosphirene ring: a substantial shortening of the P–C bonds is observed in (**6**). In the same vein, the structure of the phosphirenium salt (**8**) suggests some kind of electronic delocalisation within the ring. The very short P–C ring bonds of the pentacoordinate phosphirene (**9**) are also noteworthy, but no obvious explanation of this phenomenon can be offered at present. Finally, a 2*H*-phosphirene (**4**) has been structurally characterised as its P–W(CO)$_5$ complex. The structure shows an intracyclic <C–P–C angle of 49.2° and a P=C double bond of 1.634 Å ⟨87AG(E)1257⟩.

Table 2. Structural parameters of some monocyclic phosphirene derivatives

1-R-phosphirene	*Bond angles (°)*			*Bond lengths (Å)*		
	C–P–C	*C–P–R*	*P–C–C*	*P–C*	*C=C*	*P–R*
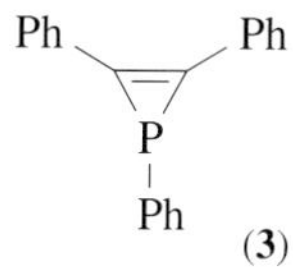 (**3**)	41.8	104.0 104.4	69.1	1.820 1.821	1.299	1.842
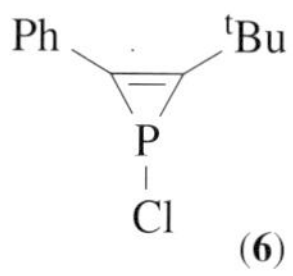 (**6**)	42.9	103.7 104.2	68.4 68.7	1.781 1.784	1.303	2.166
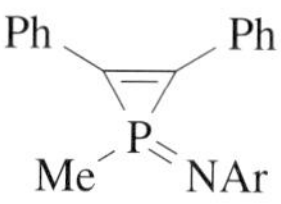 (**7**)	43.5		68.2	1.75	1.30	1.80 (P–Me)
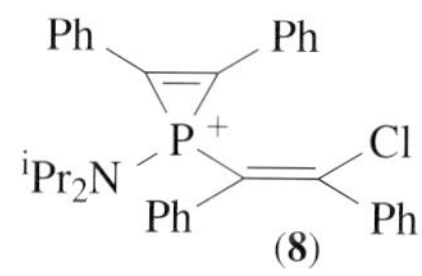 (**8**)	46.1		66.9 67.0	1.730 1.732	1.36	1.640 (P–N) 1.799 (P–C)
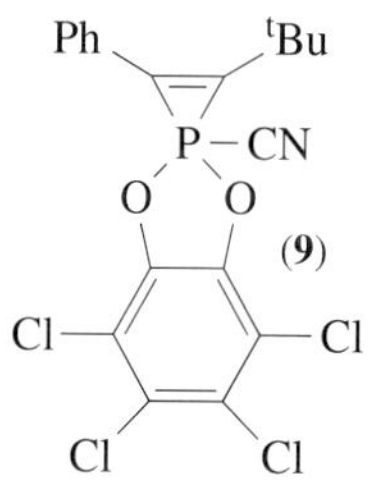 (**9**)	46.8		66.3 66.9	1.709 1.715	1.360	1.784 (P–C) 1.668 (P–O) 1.700 (P–O)

(**3**): ⟨84CC45⟩; (**6**): ⟨91CB1207⟩; (**7**): ⟨93CB649⟩; (**8**) ⟨88CC462⟩; (**9**): ⟨90TL3429⟩.

2.1.2.3 Spectroscopic Data

^{1}H, ^{13}C and ^{31}P NMR data for two representative phosphiranes are listed in Table 3. The main spectroscopic features associated with phosphiranes include: (a) an extremely high field ^{31}P NMR shift; (b) correspondingly high field shifts of the ^{1}H and ^{13}C resonances, as expected for any 3-membered ring; (c) a large negative ^{1}J (C–P) coupling within the ring; and (d) a ^{1}J (H–P) coupling which is small in magnitude, the values for acyclic secondary phosphines being in the range between 169 and 217 Hz. On the basis of a theoretical analysis, the high field shift of the ^{31}P resonance has been correlated with the large HOMO/LUMO gap and the high p character of the phosphorus lone pair ⟨97MI451⟩. Studies of the correlation between δ^{31}P and the nature of R for a series of 1-R-phosphiranes ⟨89CB453⟩, a similar work on the correlation between the ^{31}P shifts and the Hammett σ constants of the *para*-substituents in phosphirane *p*-X–C_6H_4–P–W$(CO)_5$ complexes ⟨95JOM(489)1⟩ and the measurement of the ^{31}P magnetic shielding tensor components of phosphirane (**5**) ⟨88CPL(149)363⟩ have appeared.

The NMR characteristics of phosphirenes once again include a large upfield shift of the ^{31}P resonance and very large ^{1}J (C–P) coupling constants. The parameters for a representative phosphirene are listed in Table 4. Measurements of the ^{31}P magnetic shielding tensor components of phosphirene (**3**) have also been made ⟨87CPL(136)224⟩.

A description of the IR and mass spectra of phosphiranes has appeared ⟨69T1097⟩. A combined

Table 3. NMR data for some representative phosphiranes

Phosphirane	*^1H*			*^{13}C*	*^{31}P*
	PH	*Ha*	*Hb*	*CH_2*	
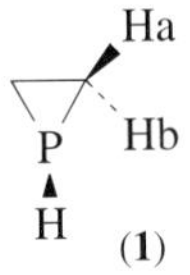 (**1**)	−0.59 (158.34)	1.21 (−2.64)	0.66 (16.14)	0.7 (−33)	−341
 (**10**)	–	0.3–1.0		11.04 (−39.7)	−234

δ, ppm from TMS or external 85% H_3PO_4 (couplings to ^{31}P in Hz). (**1**): ⟨83OMR494⟩; (**10**): ⟨69T1097, 76JA2109⟩.

Table 4. NMR data of a representative phosphirene

Phosphirene	*^1H*	*^{13}C*			*^{31}P*
	=CH	*=CH*	*=CMe*	*Cipso (Mes)*	
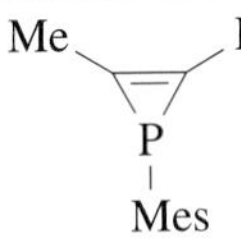 (**11**)	7.43 (22)	113.38 (45)	129.33 (47.5)	140.43 (72.5)	−192.0

δ, ppm from TMS or external 85% H_3PO_4 (couplings to ^{31}P in Hz). (**11**): ⟨91T6673⟩.

IR–Raman study of a deuterated phosphirane (**1**) has shown that the P–C stretching vibrations are shifted to low frequencies relative to those of Me_2PH. The relative weakness of the P–C ring bonds of (**1**) is thus established ⟨69SA(A)819⟩. For further IR data, see also ⟨70SA(A)249⟩ and ⟨73JST(18)75⟩. Other spectroscopic studies of phosphiranes include microwave ⟨70JCP2831⟩ and ion cyclotron resonance spectrometric measurements ⟨75ZN(A)1470⟩.

2.1.3 REACTIVITY OF PHOSPHIRANES AND PHOSPHIRENES

2.1.3.1 Reactions at Phosphorus with Retention of the Cyclic Structure

The nucleophilic reactivity of the lone pair of phosphiranes is very low. Thus, acids or quaternising agents (RX) do not generally give the corresponding $[PH]^+$ or $[PR]^+$ phosphiranium salts. The first well-characterised phosphiranium salts were prepared from phospha-[3]radialenes ⟨91CB933⟩ (Equation (1)).

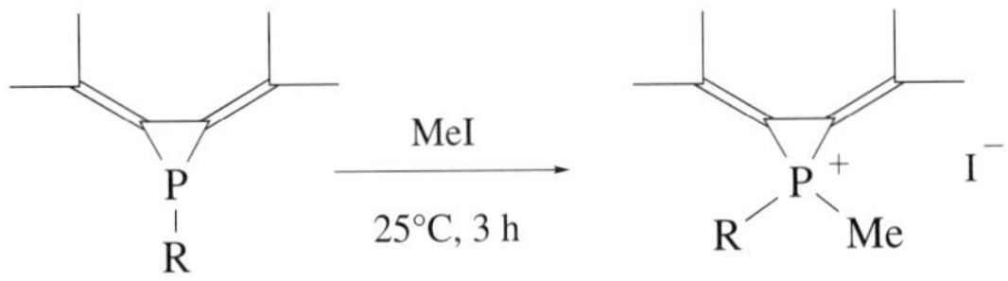

(1a)

R = Et: $\delta^{31}P$ - 89.2
R = ^iPr: $\delta^{31}P$ - 82.7
R = ^tBu: $\delta^{31}P$ - 78.1

These salts display a characteristic upfield shifted ^{31}P resonance and are sensitive to moisture. In some instances, ring-opening is observed during the quaternisation reaction (Equation (1a)).

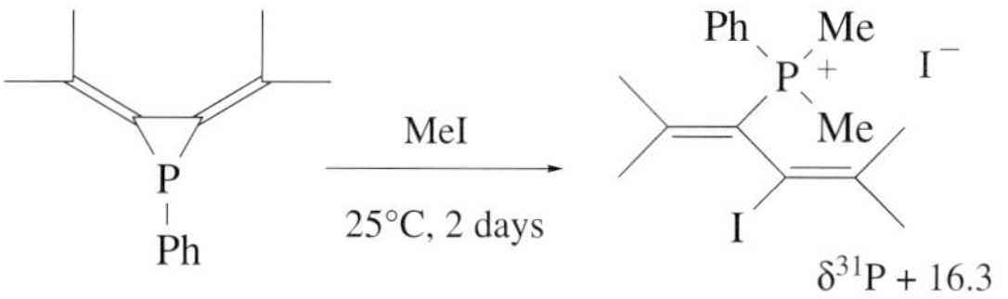

$\delta^{31}P$ + 16.3 (1a)

Of course, the substitution pattern found here induces peculiar reactivity which need not be representative of phosphiranes as a class. However, 1-phenylphosphirane was later shown to react with methyl triflate to give a corresponding phosphiranium salt which was fully characterised by X-ray crystal structure analysis, see Table 1 ⟨95CC257, 97JOM(529)189⟩ (Equation (1b)).

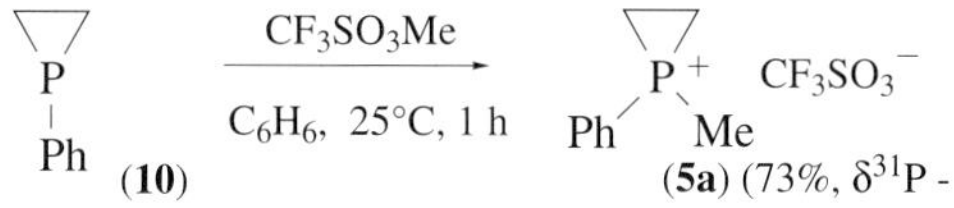

(1b)

(10) **(5a)** (73%, $\delta^{31}P$ - 96.8)

In a further demonstration of the influence of the substitution pattern, it has been found that the bridgehead phosphorus of the polycyclic phosphirane dubbed *BABAR*-Phos is not quaternised by methyl triflate ⟨99AG(E)1623⟩. In other studies, the first phosphirane oxide was prepared by Quin from a sterically protected phosphirane ⟨87TL1077⟩ (Equation (1c),

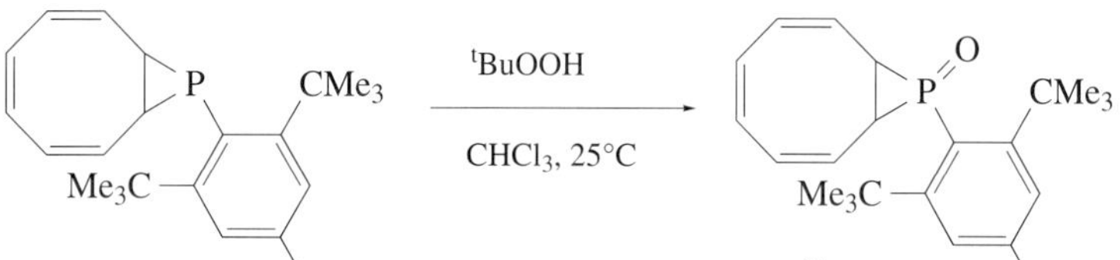

(1c)

(12) **(13)** ($\delta^{31}P$ -38)

and Denney ⟨78CC854⟩ was able to monitor the formation of an unstable phosphorane (**14**) upon low temperature reaction of a dioxetane with phosphirane (**10**) (Equation (2)).

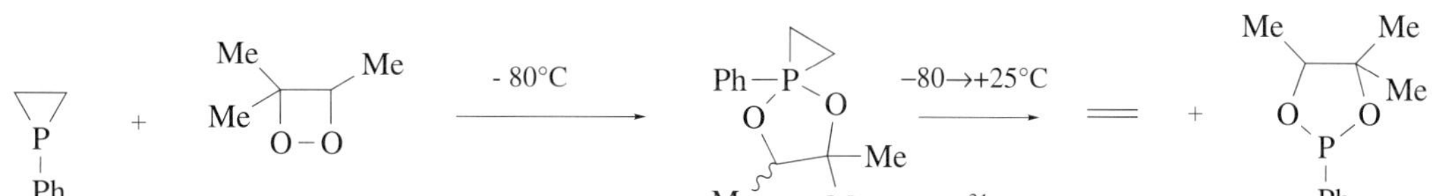

(2)

(10) **(14)**($\delta^{31}P$ -87)

Numerous stable complexes of phosphiranes, especially with molybdenum and tungsten carbonyls, have been described. Most of them have been obtained through building the phosphirane within the metal coordination sphere, thus taking advantage of the stabilising effect of the metal (see later). Nevertheless, some of these complexes have been prepared from preformed phosphiranes. For example, the first phosphirane–BH_3 complex was prepared from 1-*tert*-butylphosphirane ⟨97OM1526⟩. Similarly, *fac*-[Mo(CO)$_3$L$_3$] ⟨96OM1301⟩ and [Ir(Cp*)(Cl)L] complexes ⟨97OM1526⟩ were obtained from 1-phenyl- and 1-*tert*-butyl-phosphirane respectively and characterised by X-ray crystal structure analysis. Some particularly interesting cationic rhodium (I) complexes of optically active 1-*l*-menthyl-2-phenylphosphirane induce the asymmetric hydrogenation of dehydro-amino acids with enantiomeric excesses up to 76% ⟨93OM1207⟩.

As a general rule, tetra- and pentacoordinated derivatives of phosphirenes display much higher stability than the corresponding phosphirane derivatives. Some kind of electronic stabilisation probably operates in phosphirenium salts (see the structural data of one such salt: ⟨88CC462⟩), which can be synthesised directly (see later) or from tervalent phosphirenes ⟨85JA4700, 94RHA61, 98SYN175⟩ (Equations (3), (3a)).

Bromine and $AlCl_3$ react in the same way with (**3**) to give the corresponding bromophosphirenium salt ⟨85JA4700⟩.

At the moment, the only known phosphirene oxide was transiently observed during the oxidation of a tervalent phosphirene at low temperature ⟨94RHA61⟩ (Equation (4)).

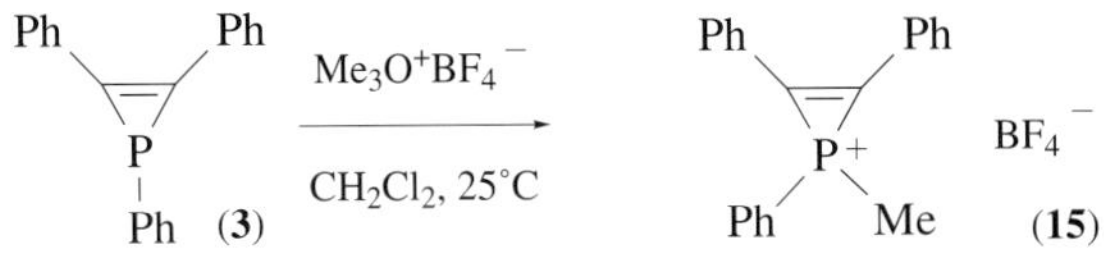

(15) (δ^{31}P -110) (3)

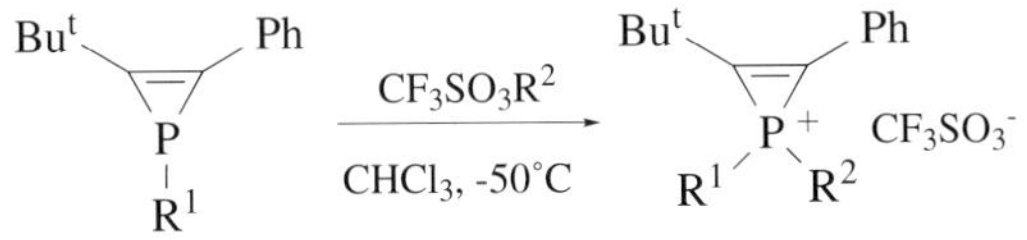

(3a)

R^1= Me, Ph, Bu^t, CH_2Tms, $CHTms_2$, OMe, NPr^i_2
R^2= Me, CH_2Tms

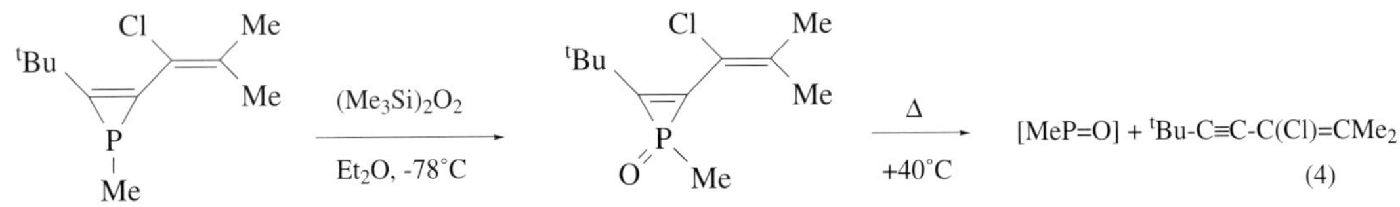

(4)

However, several sulphides and selenides have been prepared by the direct reaction of sulphur or selenium with phosphirenes in the presence of a basic tertiary amine as a catalyst 〈85JA4700, 94RHA61〉, for example, (Equation (5)).

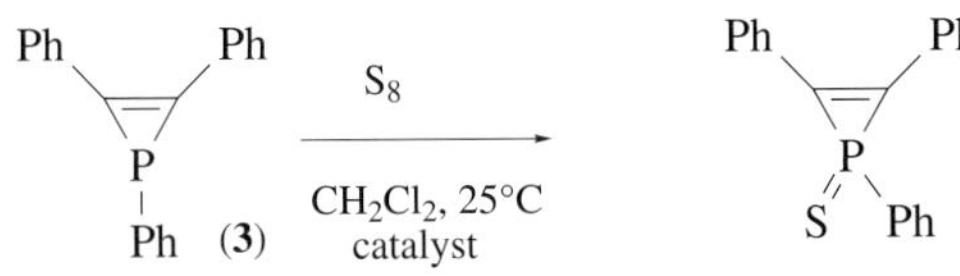

(16) (δ^{31}P -79) (5)

Similarly, tosyl azide readily reacts with phosphirenes at room temperature in ether to give the expected phosphirene imines 〈94RHA61〉: these display good thermal stability and can be also synthesised directly 〈93CB649, 95ZOR404, 95ZOR495, 96ZOR433〉. Their protonation by HCl or HBr takes place at nitrogen to give 1-aminophosphirenium halides 〈95ZOR400〉. One of the most surprising reactions in this area is undoubtedly the synthesis of very stable pentacoordinate derivatives by oxidative addition of an orthoquinone to tervalent phosphirenes 〈90TL3429〉 (Equation (6)).

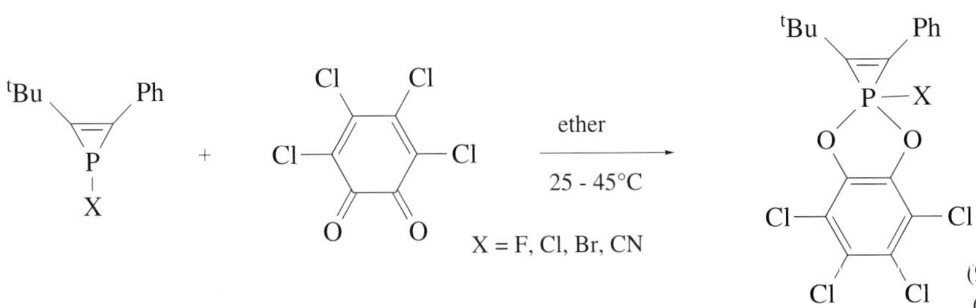

(6)

(9: X = CN)
(δ^{31}P -128)

Finally, as in the case of phosphiranes, numerous stable complexes of phosphirenes are known both with boranes 〈97CB711〉 and transition metals. Some of them will be discussed later.

The availability of 1-chlorophosphiranes 〈87TL2693, 88ZN(B)1285, 89CB453, 91SL743〉 and 1-chlorophosphirenes either in the free state 〈89AG(E)225, 89TL3951, 95SYN941, 98OM2677〉 or as P–W(CO)$_5$ complexes 〈89POL2671, 88NJC755〉 has led to a series of studies on the reactions of nucleophiles with such species. Some significant results are given in Equations (7)–(9) 〈89CB453, 89POL2671, 89AG(E)225, 91OM3949, 94RHA61, 97CB711, 99MI1581〉.

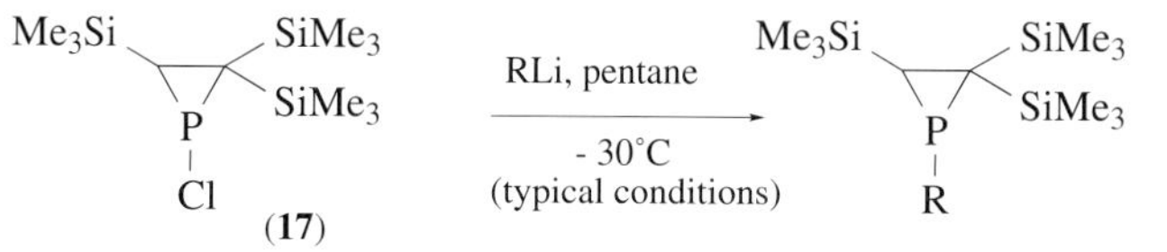

(7)

R = Me, ^{n}Bu, ^{s}Bu, ^{t}Bu, Me_3SiCH_2, $(Me_3Si)_2CCl$,
Et_2N, MeO, ^{t}BuS, ${}^{n}Bu_2P$, H, etc...see <89CB453>
R = Cp*, $CpFe(CO)_2$, $Cp^*Mo(CO)_3$ <91OM3949>

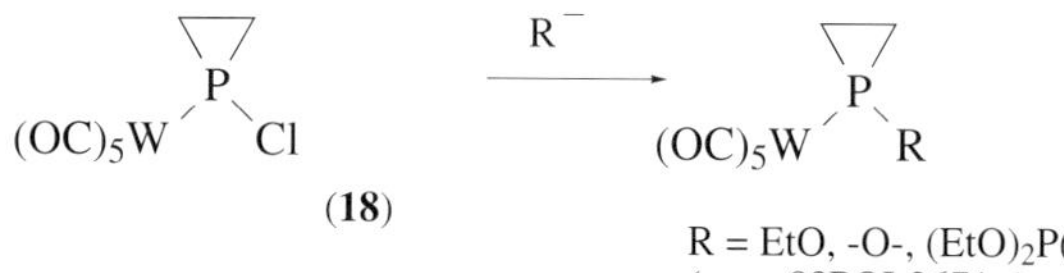

(8)

R = EtO, -O-, $(EtO)_2P(O)$, H
(see <89POL2671>)

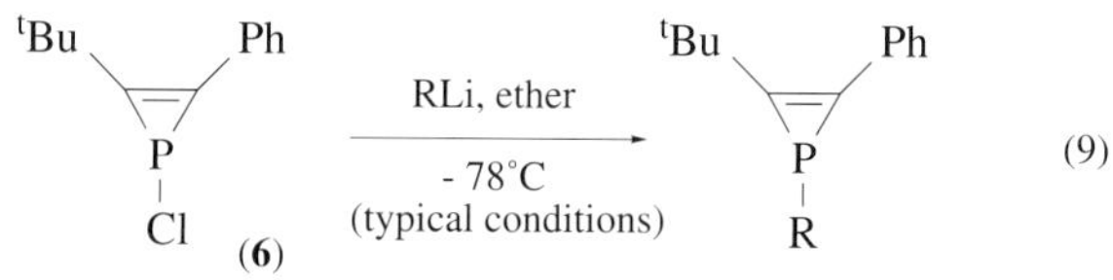

(9)

R = Me, ^{t}Bu, ^{t}BuC≡C, $CHTms_2$, $CTms_3$, MeO, Et_2N, ${}^{i}Pr_2N$,
$(Me_3Si)_2P$, $R'_2C{=}P$, N_3, $Me_3SiC(N_2)$, $SiTms_3$, $GeTms_3$, etc...
(see <89AG(E)225><94RHA61><97CB711><99MI1581>)

The isolation of the first *P*-unsubstituted phosphirene (Equation (9), R = H) by reaction of $LiAlH_4$ or $LiBEt_3H$ with (**6**) ⟨97CB711⟩ must be mentioned. The product displays a ^{31}P resonance at high field and a small ^{1}J(P–H) coupling: δ^{31}P −247, ^{1}J(P–H) = 107 Hz. It decomposes easily upon heating to give the corresponding alkyne. A systematic study of the exchange of chlorine for other halogen (F, Br, I) or pseudohalogen substituents, as part of an investigation into the stability of the phosphirenylium cation, was carried out on (**6**) by Regitz's group ⟨91CB1207⟩. Some representative experiments are depicted in Equation (10).

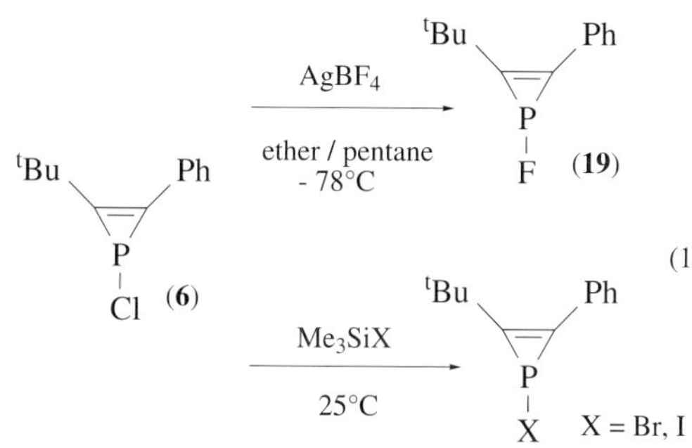

(10)

The formation of the PF compound (**19**) from (**6**) and $AgBF_4$ probably involves the abstraction of F^- from BF_4^- by a transient phosphirenylium cation. A similar reaction was observed between (**6**) and BPh_4^- (Equation (11)). The NMR data for the triflate derivative (**21**) also suggest the reversible formation of a phosphirenylium cation (Equation (12)). The ^{31}P chemical shift of (**21**) is at relatively low field: δ^{31}P −9.5 vs −73.4 ppm for (**6**) in [D_8] toluene; no ^{1}J (^{17}O–^{31}P) is observed; both ^{31}P and ^{13}C ring resonances are shifted to low fields when the polarity of the solvent increases, i.e. δ^{31}P −9.5 → +10.5 ppm when toluene is replaced by nitromethane. The highly electrophilic nature of the system (**21**)/(**22**) is shown by the easy formation of a salt with pyridine and its ability to undergo electrophilic aromatic *para*-substitution into the

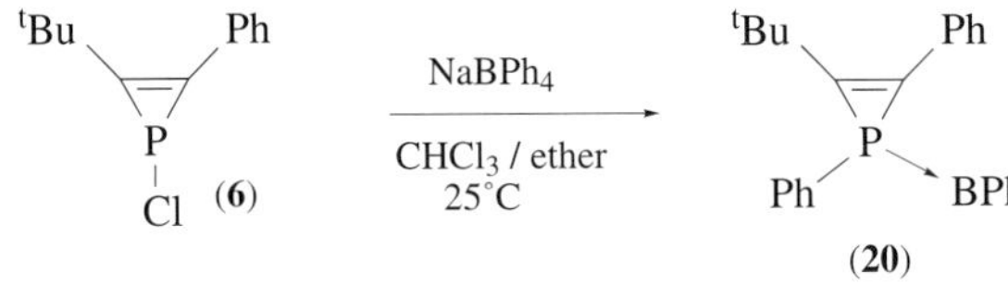

(11)

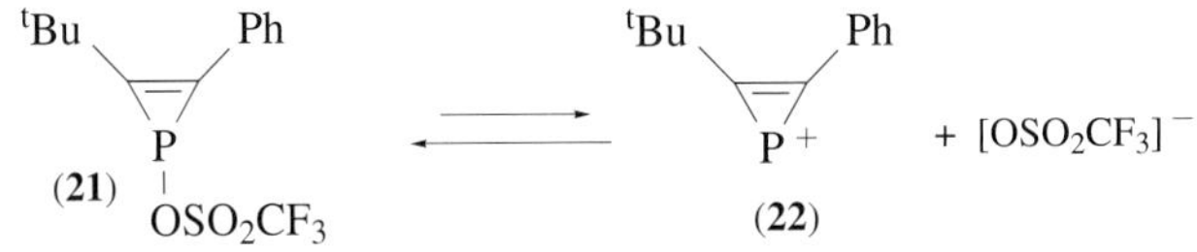

(12)

phenyl ring of *N,N*-dimethylaniline ⟨94RHA61⟩. Definitive evidence for the formation of a phosphirenylium cation was produced subsequently ⟨94JA9407⟩ (Equation (12a)).

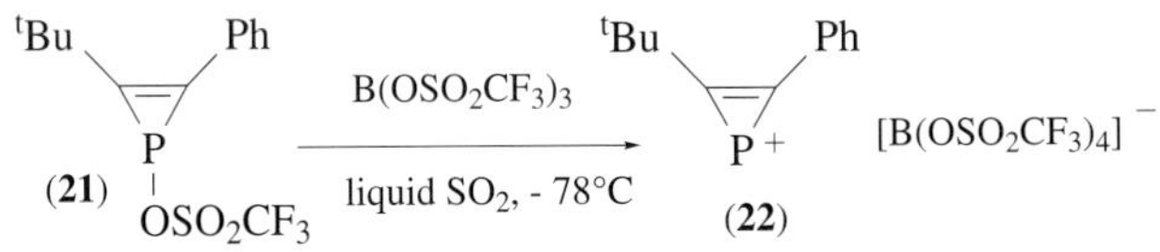

(12a)

The phosphirenylium cation (**22**) displays a very strong and characteristic downfield shift of the ^{31}P resonance: δ^{31}P +309.7 ppm. The ^1J (P–C) couplings increase from 77.0 and 66.0 Hz in (**21**) to 94.9 and 83.3 Hz in (**22**). Due to its high reactivity, (**22**) must be kept below −40°C. The phosphirenylium cation has also been produced in the coordination spheres of transition metals. Early experiments suggested the intermediacy of phosphirenylium–$W(CO)_5$ complexes in dealkylation reactions ⟨85TL4595, 88NJC755⟩ (Equation (12b)).

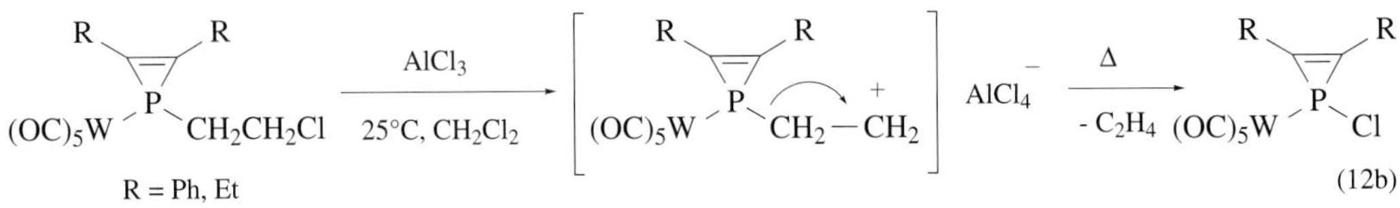

(12b)

The driving force permitting the elimination of C_2H_4 was thought to be the stabilisation associated with the transient phosphirenylium complex. Such a phosphirenylium complex was identified some time later ⟨99OM817⟩ (Equation (12c)).

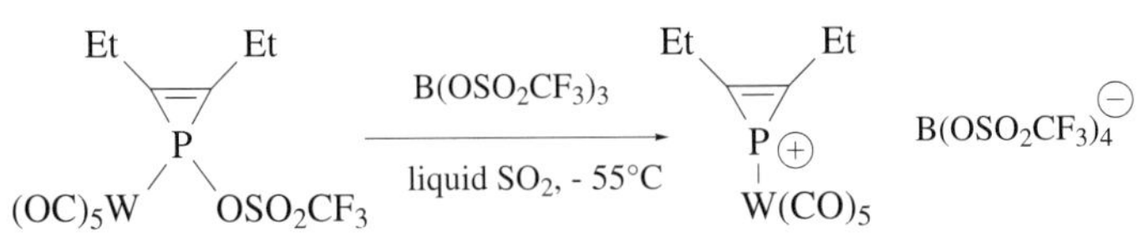

(12c)

As expected, this complex displays a ^{31}P resonance at low field (+232) and a huge $^1J(^{31}P–^{183}W)$ coupling constant (392 Hz) implying η^1-coordination. A more spectacular demonstration was provided by the structural characterisation of the first known η^3-phosphirenyl complex ⟨94AG(E)2330⟩ (Equation (12d)).

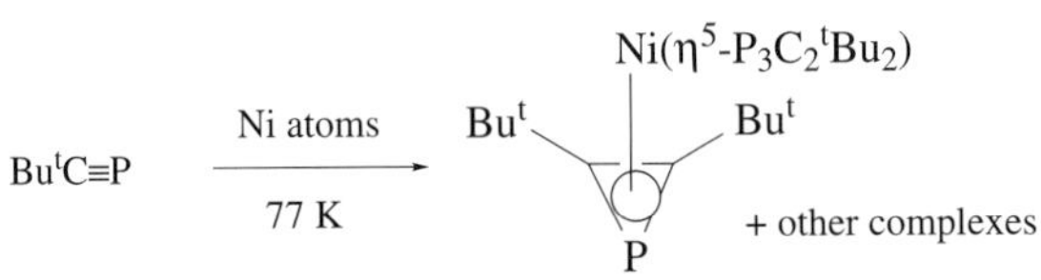

(12d)

The ^{31}P resonance of the phosphirenyl ring appears at −161 ppm. The structure of its η^1-P–$W(CO)_5$ complex, showing rather long P–C phosphirenyl bonds at 1.78–1.82(2) Å, and a short C=C bond at 1.33(3)

Å, confirms the η^3-coordination at nickel. Finally, two curious reactions involving a chlorophosphirane ⟨89CB453⟩ and a chlorophosphirene ⟨98CC867⟩ are depicted in Equations (13) and (13a).

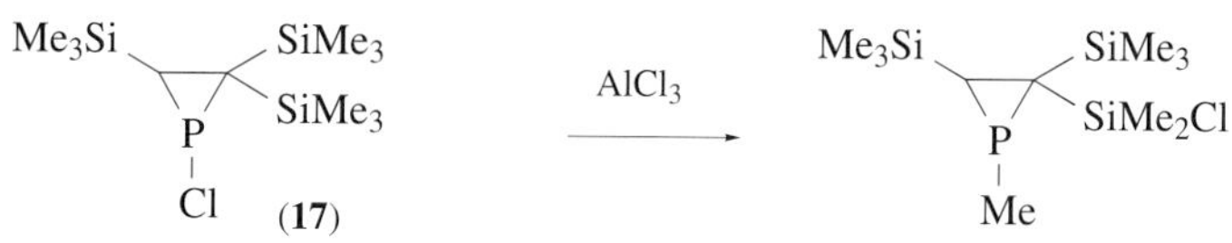

(13)

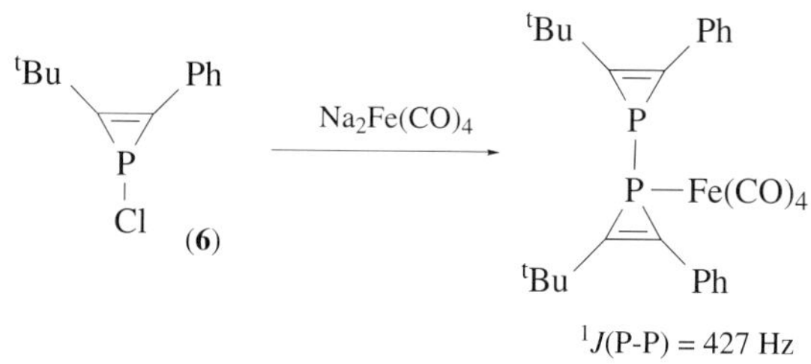

(13a)

As it can be seen, the reactivity of the P–Cl bond of 1-chloro-phosphiranes and phosphirenes has been developed fairly thoroughly. By comparison, the chemistry of the less stable *P*-unsubstituted phosphiranes and phosphirenes has been almost untouched. The metallation of the PH bond of *P*-unsubstituted phosphiranes has only been achieved in the coordination sphere of a transition metal ⟨71AG(E)125⟩ (Equation (14)).

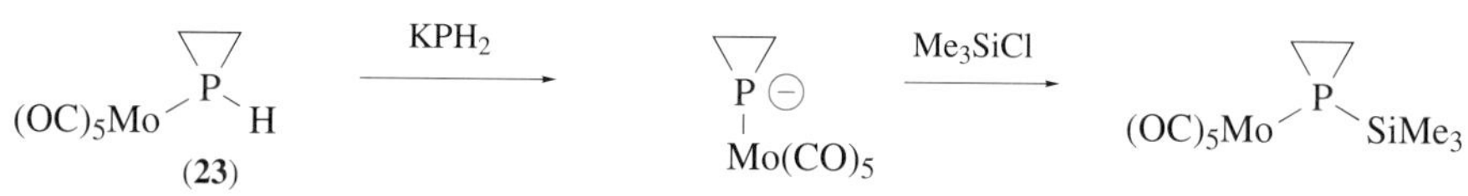

(14)

Analogous chemistry has been carried out on a *P*-unsubstituted phosphirene complex ⟨88NJC755⟩ (Equation (15)).

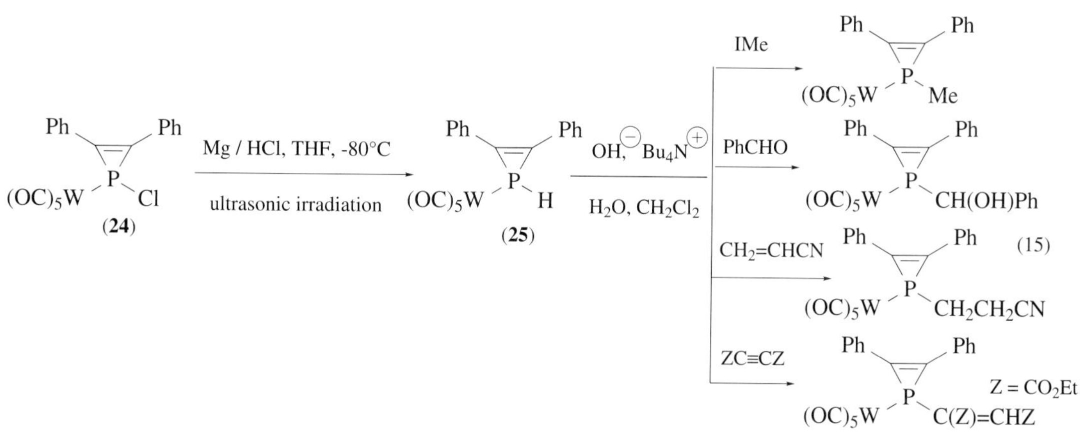

(15)

Without the stabilising effect of the transition metal, the free phosphiranide and phosphirenide anions appear to be very unstable and cannot be used for synthetic purposes. Even the anion of (**25**) displays some instability, perhaps because of some antiaromaticity, and must be used under phase-transfer conditions.

2.1.3.2 Reactions at the Carbons of the Ring with Retention of the Cyclic Structure

Only a few such reactions are known for phosphirenes. In one instance, it proved possible to carry out a [4 + 2] cycloaddition involving the phosphirene double bond by employing an activated phosphirene complex ⟨85JA4700⟩ (Equation (16)).

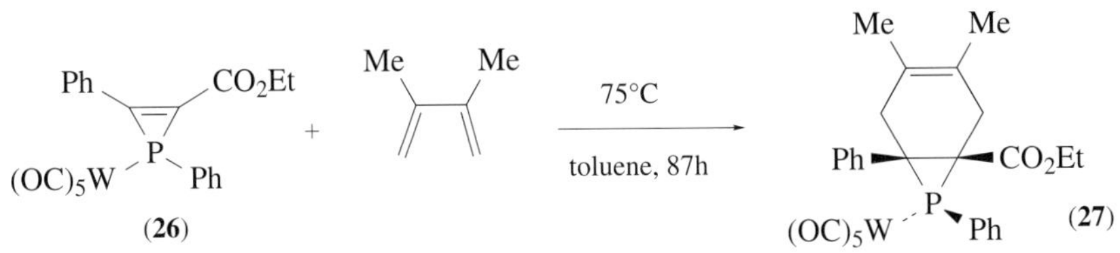

(16)

The cycloaddition selectively takes place on the tungsten side of the ring.

In another instance, a phosphirene carbanion has been made and trapped ⟨91T6673⟩ (Equation (17)).

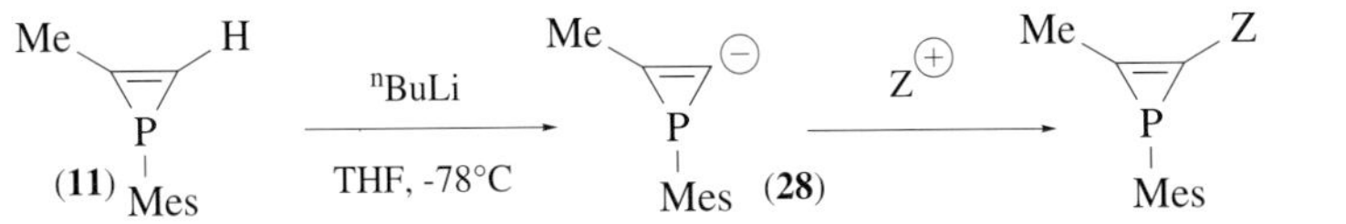

(17)

Z = D, CH_2OH, $C(Me)_2OH$

At room temperature, carbanion (**28**) rearranges to give an open-chain alkynylphosphide (see later, Equation (37)). Finally, the thermal conversion of a phosphirane into a phosphirene has been reported ⟨91TL6537⟩ (Equation (18)).

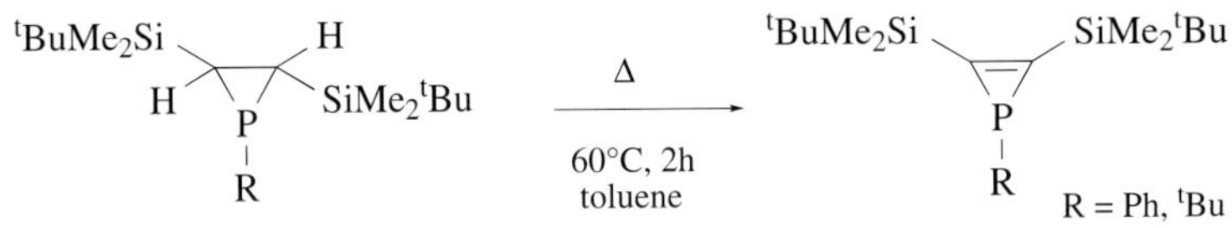

(18)

2.1.3.3 Reactions with Cleavage of one P–C Bond of the Ring

It should be recalled here that theoretical calculations have shown that the P–C bonds of the phosphirane ring are high in energy and polarised in the sense $P^{+\delta}–C^{-\delta}$. In fact, their behaviour strongly resembles classical π-bonds. Three main mechanisms leading to their cleavage have been recognised; these involve electrophilic or nucleophilic (basic) attack at one carbon of the ring or nucleophilic attack at phosphorus (Equations (19)–(21)).

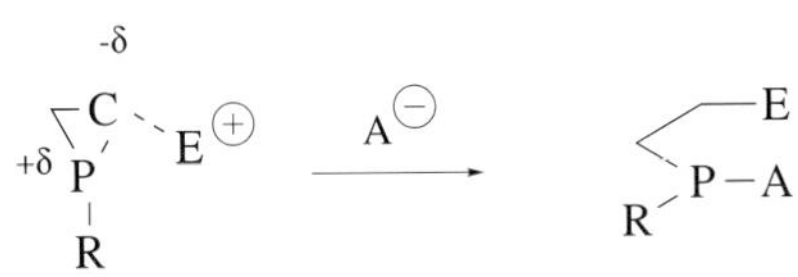

(19)

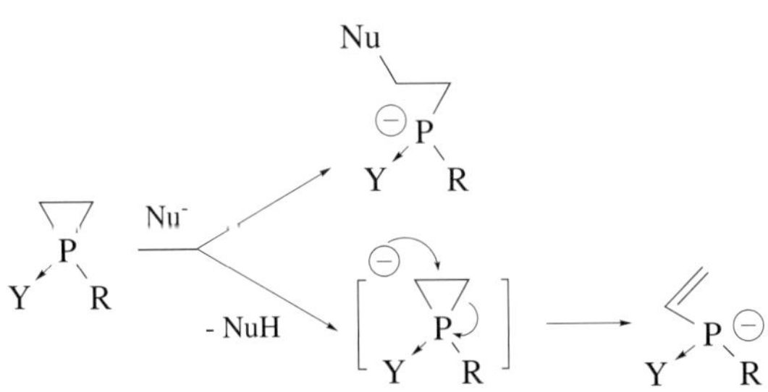

(20)

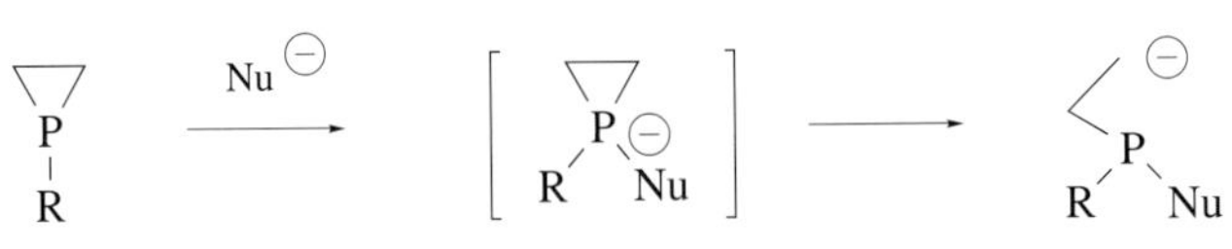

(21)

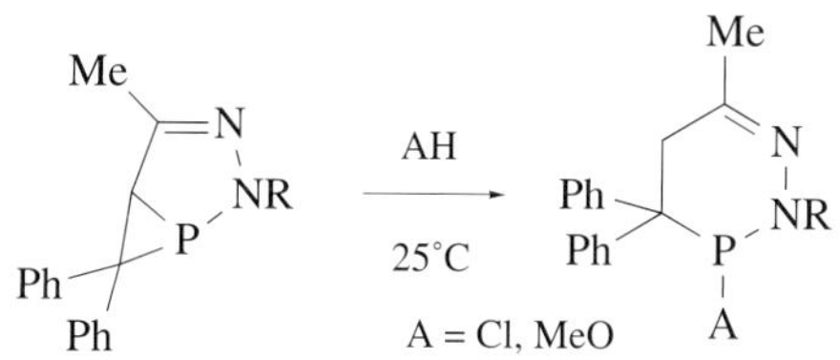

(22)

The first mechanism is probably operative in the cleavage of a bicyclic phosphirane whose P–C bond polarisation is enhanced by the nitrogen substituent at phosphorus ⟨79DOK(244)117⟩ (Equation (22)). The cleavage of phospha[3]radialenes by dichlorophosphines ⟨91CB933⟩ probably belongs in the same category (Equation (22a)).

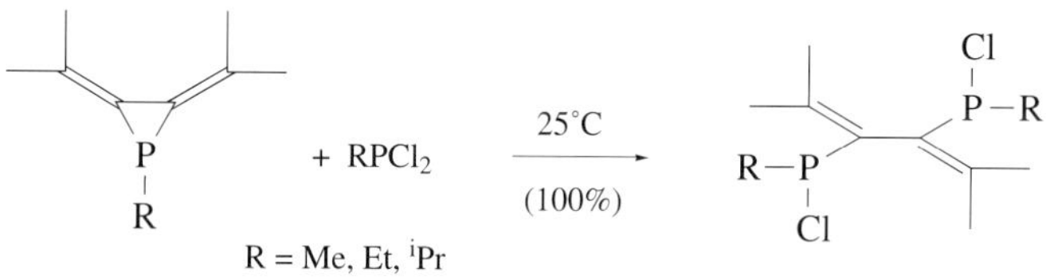

(22a)

The polarity is reversed when a substituent bearing a lone pair is located on the carbon of the ring and, upon hydrolysis, the proton is found at phosphorus ⟨84OM456, 84OM1492⟩ (Equation (23)).

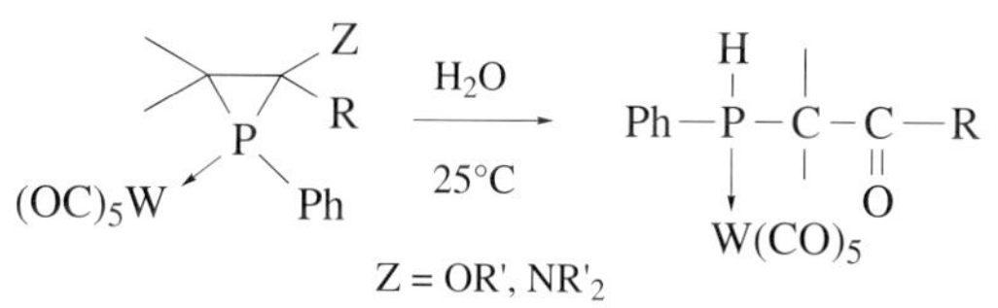

(23)

The second mechanism has been illustrated by Quast ⟨81LA977⟩ and Marinetti ⟨89T3061⟩ with lithium amides used as bases (Equations (24), (25)).

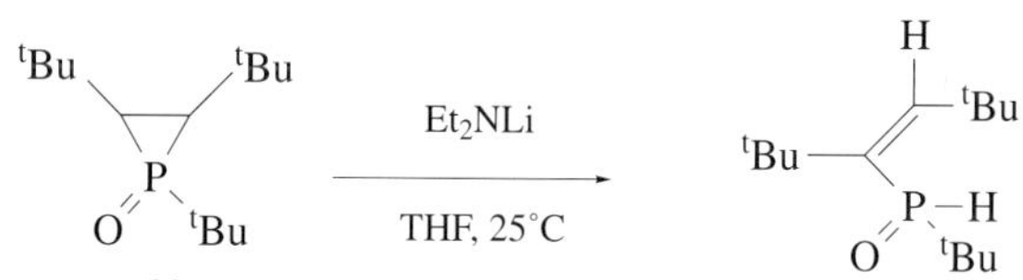

(24)

(**29**)

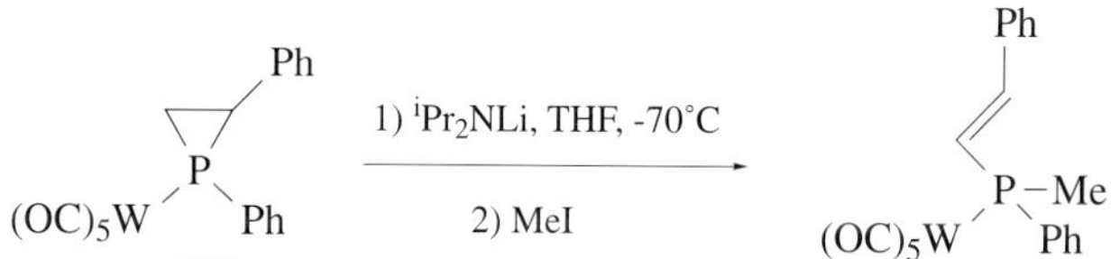

(25)

(**30**)

The chemistry depicted in Equation (25) has been transposed to prepare a 1-phospha-pentadienyl complex from a 2-vinylphosphirane via a metallation at C_3 ⟨95OM4048⟩. When the nucleophile is the tervalent phosphirane itself, polymerisation occurs ⟨94MI567⟩ (Equation (25a)). Methyl triflate is thus able to catalyse the cationic polymerisation of 1-(tris-*tert*-butylphenyl)phosphirane to give a polyphosphine with a molecular weight of 7800. The third mechanism operates with (**30**) ⟨89T3061⟩ and some free 2-vinyl-phosphiranes ⟨84T3273⟩ (Equations (26), (27)). It also governs the chemistry of phosphiranium salts ⟨95CC257, 97JOM(529)189⟩ (Equation (27a)). In fact, the reaction with water is so easy that the phosphiranium salt must be kept under dry argon to prevent its hydrolysis.

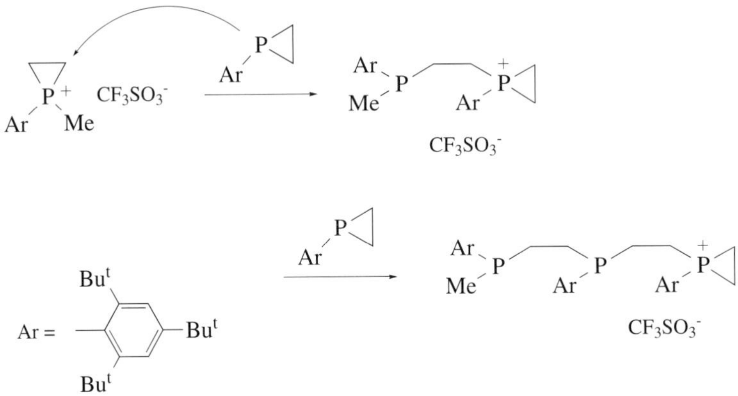

(25a)

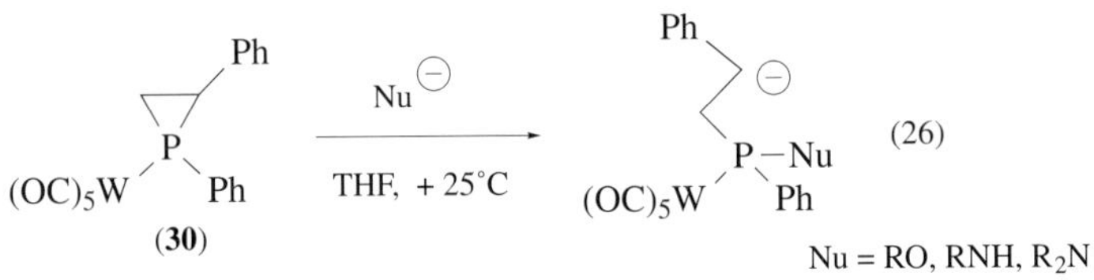

(26)

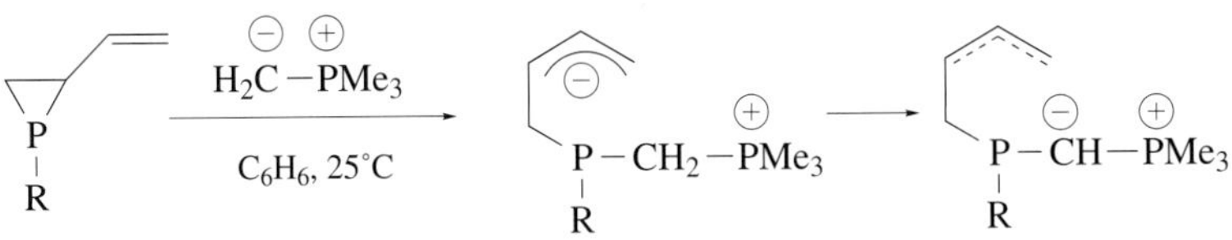

(27)

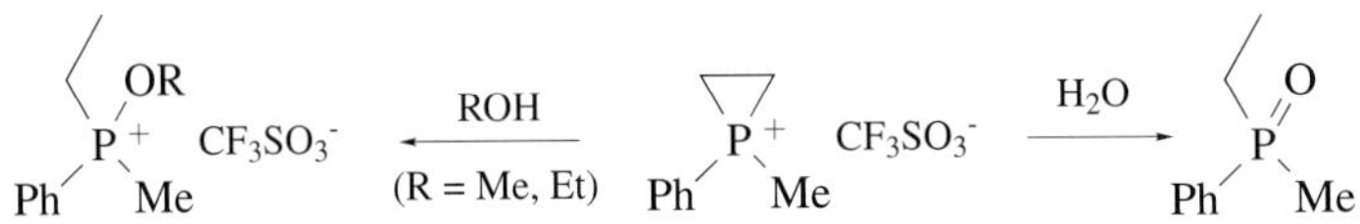

(27a)

At higher temperature, the behaviour of phosphiranes is entirely different. Around 500°C, an equilibrium takes place between phosphiranes and secondary vinylphosphines ⟨92CC1799, 93PS(75)225⟩ (Equation (28)).

(28)

Nguyen has investigated this equilibrium from a theoretical standpoint ⟨94JCS(F)1771⟩ and found a concerted process with an activation energy of 56.2 kcal mol^{-1} (R=H). A related [1,3] halogen shift from C_2 to P was found to occur in 2-halogenophosphirane complexes at ca. 100°C ⟨95SL353, 2000JOC652⟩ (Equation (28a)).

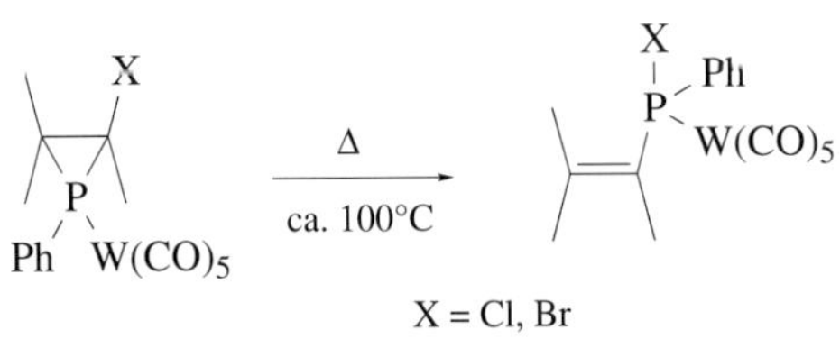

(28a)

The stereochemistry of the C–C unit is preserved during the shift, thus suggesting a concerted mechanism. The reaction is facilitated by a Pd(0) catalyst.

Ring opening is also frequently observed in phosphirene chemistry. For example, phosphirenium salts are readily opened upon hydrolysis ⟨84TL1345, 85JA4700, 94RHA61, 98SYN175⟩ (Equation (29)).

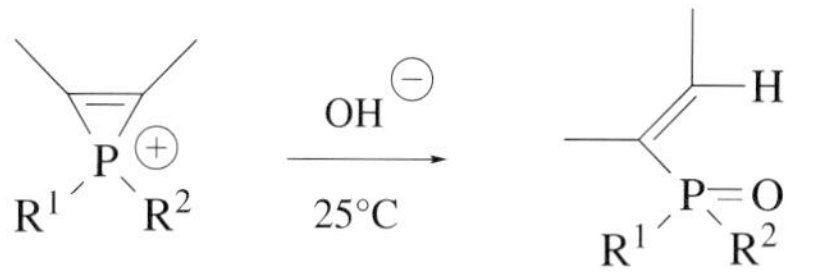

(29)

The same type of reaction has been observed for a transient phosphirene oxide 〈72CC1085, 79CC390〉 (Equation (30)), phosphirene imines 〈95ZOR400, 95ZOR404〉 (Equation (30a)), an activated phosphirene complex 〈85JA4700〉 (Equation (31)) and 1-chloro-phosphirenes 〈91CB1207, 91SL433〉 (Equation (32)), where, in all cases, the phosphorus atom displays an enhanced electrophilicity over more conventionally substituted phosphirenes.

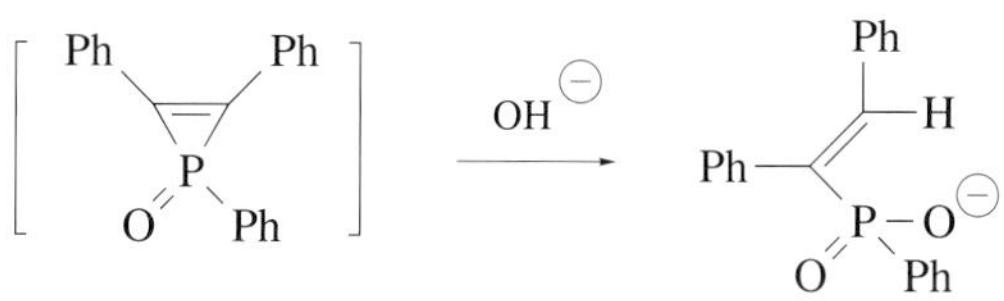

(30)

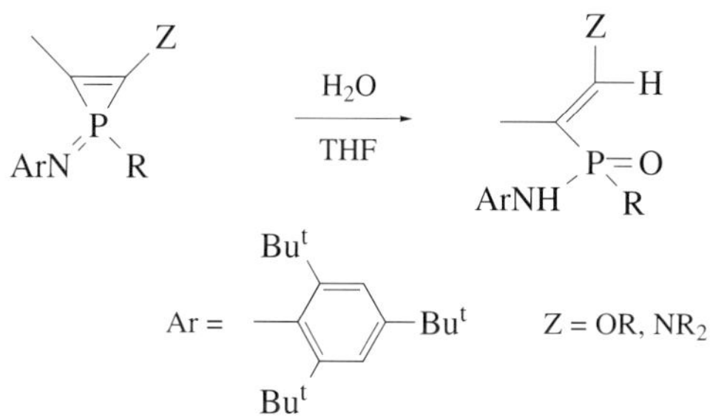

(30a)

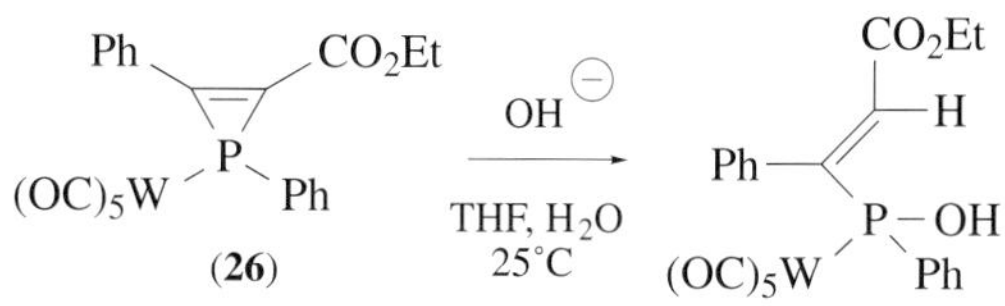

(**26**)

(31)

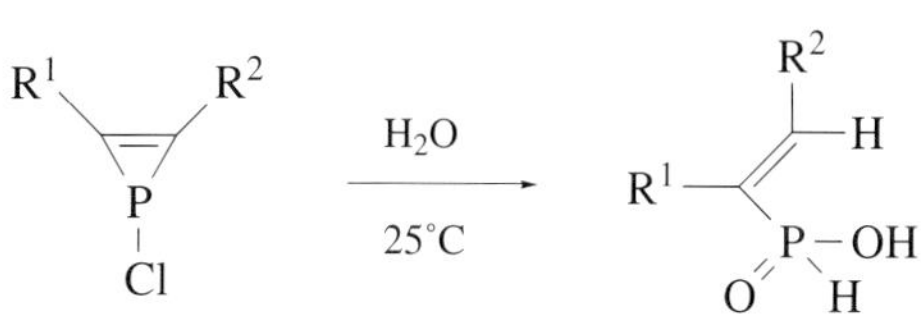

(32)

When the electrophilicity of phosphorus is low, it is still possible to promote the opening of the ring using UV irradiation 〈85JA4700〉 (Equation (33)).

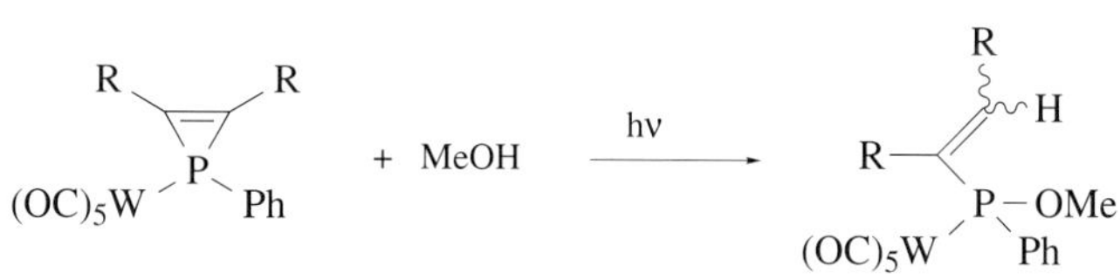

(33)

Another type of reaction showing same fundamental mechanism involves organolithium compounds 〈85JA4700〉. The general scheme is depicted in Equation (34). The vinyl carbanion isomerises rapidly even at −70°C, so a mixture of *E* and *Z* isomers is obtained upon reaction with electrophiles. This kind of chemistry has been used to prepare some unusual phosphino-1,2,3-butatrienes 〈92SL635〉 (Equation (35)).

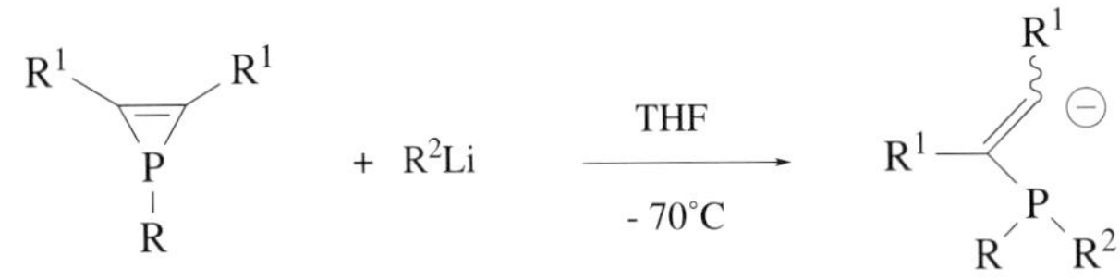

(34)

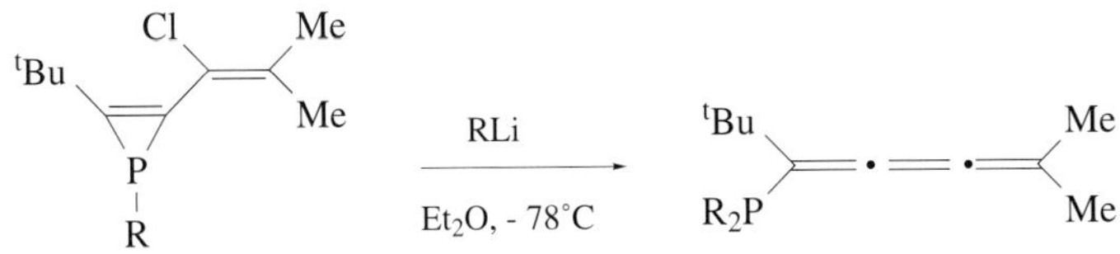

(35)

The reactions of some phosphirene imines with butyllithium or borane belong to the same general scheme ⟨95ZOR400, 94CB313⟩ (Equations (36), (36a)).

(36)

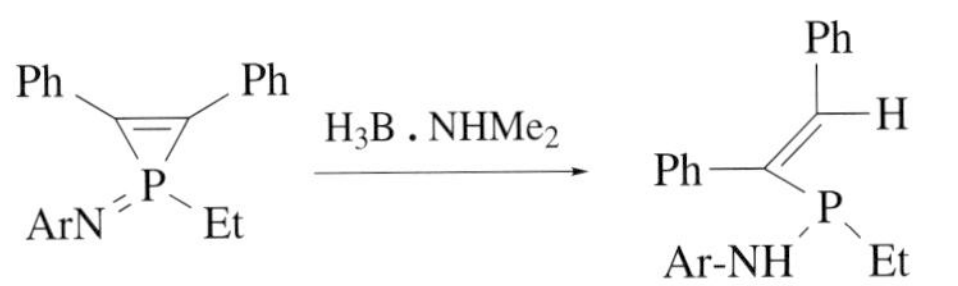

(36a)

Ar = 2,4,6-^{t}Bu$_3$C$_6$H$_2$

Apart from nucleophilic attack at phosphorus, other mechanisms can lead to the opening of the phosphirene ring. For instance, the phosphirene carbanion (**28**) depicted in Equation (17) rearranges at around room temperature to give an open-chain phosphide ⟨91T6673⟩ (Equation (37)).

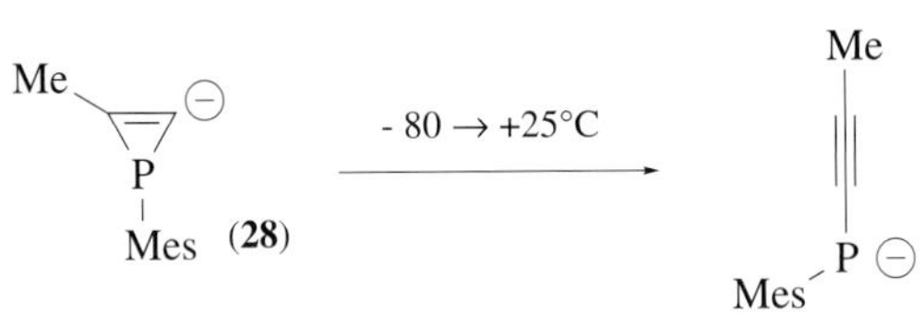

(37)

The mechanism is similar to that described in Equation (20) for the opening of the phosphirane ring. In the same work ⟨91T6673⟩, it was also shown that phosphirenylmethyl carbanions rearrange to give allenylphosphides (Equation (38)).

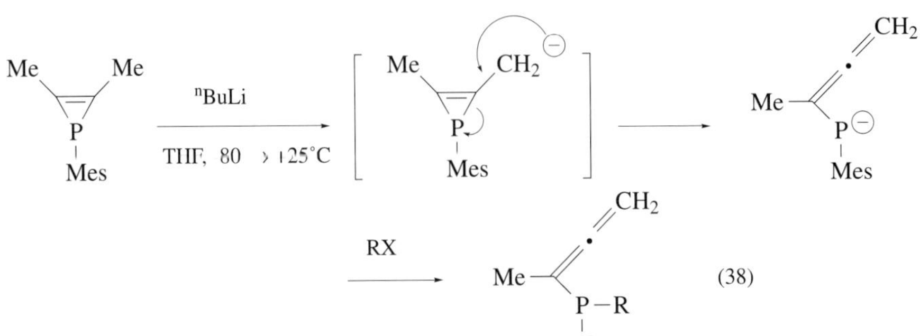

(38)

R = H, Me, SiMe$_3$

It is clear that the mesityl group serves to block the nucleophilic attack of butyllithium at phosphorus in the chemistry depicted in Equations (17), (37) and (38).

Another work describes the reductive cleavage of the phosphirene ring by alkali metals ⟨85JA4700⟩ (Equation (39)).

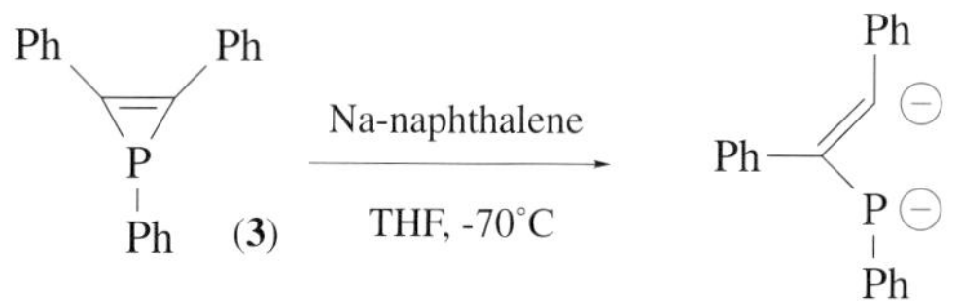

(39)

We shall close this section by a mention of the ring-chain tautomerism observed in some phosphirenium salts ⟨90TL819, 93TL1331⟩ (Equations (40) and (41)).

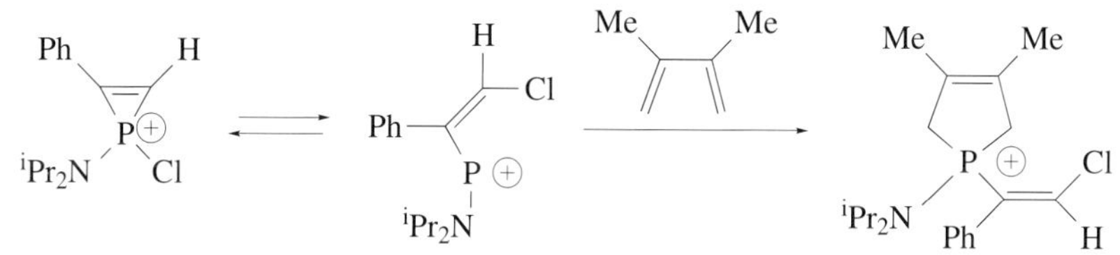

(40)

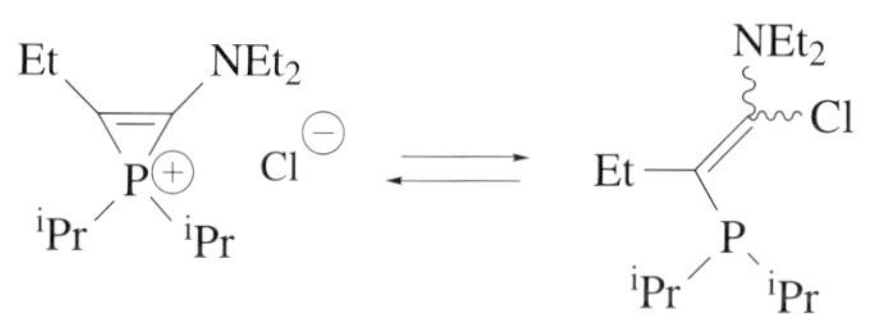

(41)

In the first case, the existence of the phosphenium salt was demonstrated through its trapping by a conjugated diene.

2.1.3.4 Reactions with Cleavage of two P–C Bonds of the Ring

According to theoretical calculations, the fragmentation of phosphirane to yield ethylene plus singlet phosphinidene is a viable proposition (ΔH ca. 74 kcal mol^{-1} ⟨94JCS(F)1771⟩). In the same vein, the competition between fragmentation and insertion reactions involving phosphiranium salts or phosphirane–borane adducts has been studied at the G2 level of theory and shown to favour the fragmentation processes ⟨98JA7063, 99JOM(580)320⟩. Thus, any phosphirane derivative can be viewed as a potential phosphinidene precursor but experimental data clearly demonstrate that clean and easy fragmentation is profoundly dependent upon the substitution pattern of the ring, both at phosphorus and carbon. In the case of tervalent phosphiranes, the first serious evidence of such a splitting came from the work of Niecke ⟨89CB453⟩ (Equation (42)).

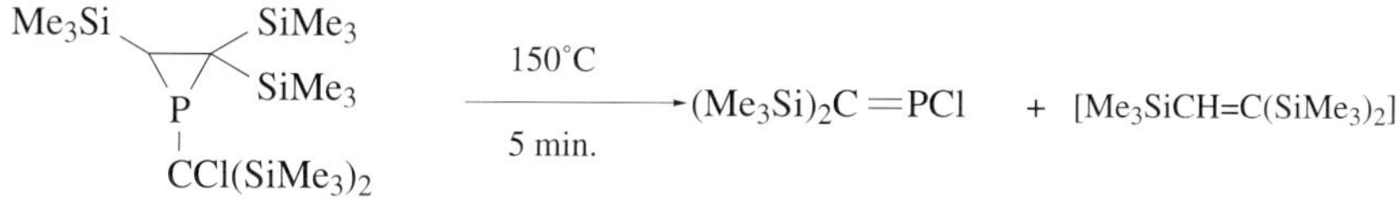

(42)

More recently, a thorough study of the thermal and photochemical decomposition of 1-mesitylphosphirane (**31**) by Gaspar ⟨92JA8526, 94JA7899⟩ yielded convincing evidence for the formation of mesitylphosphinidene. Some of the most significant results of that study are summarised in Equation (43). Additional data on the photolysis of bulky 1-arylphosphiranes are available ⟨98MI607⟩. Almost simultaneously with Gaspar's work, the flash thermolysis of 1-vinylphosphirane (**32**) was shown to give phosphapropyne, probably via the rearrangement of vinylphosphinidene ⟨92CC1799⟩ (Equation (44)). The mechanism of this reaction has been studied in some depth from a theoretical standpoint ⟨94OM640, 96OM4904⟩. A similar flash thermolysis of 1-trimethylsilyl-diazomethylphosphirane at 350ºC led to 1-trimethylsilylphosphaacetylene in 59% yield ⟨96OM4904⟩.

The first splitting of a phosphirane derivative into an alkene and a phosphinidene fragment involved an oxide (**29**) and was described by Quast ⟨82CB901⟩ (Equation (45)). The retention of the stereochemistry

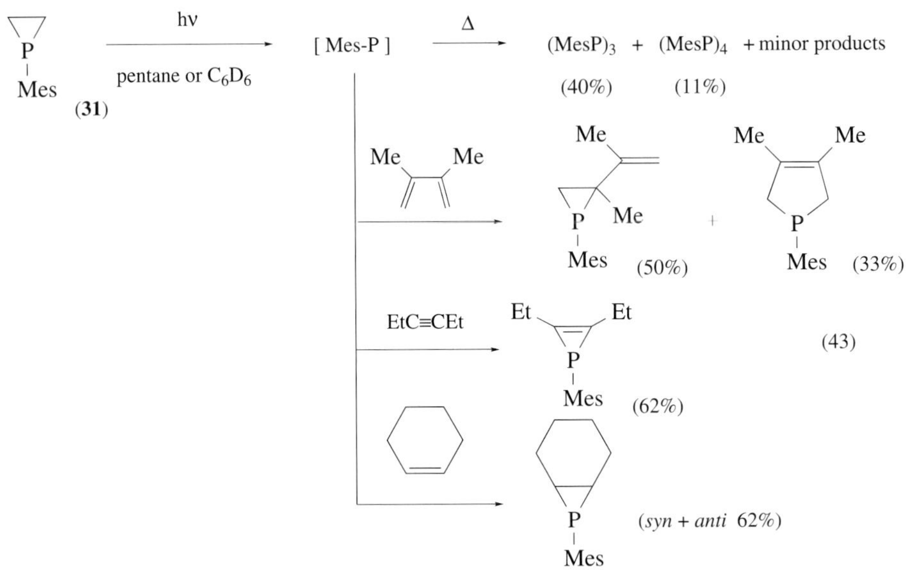

(43)

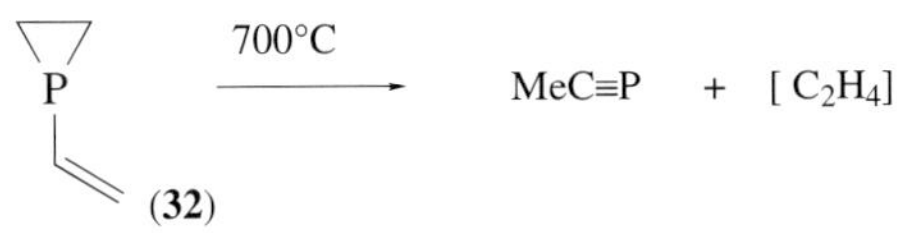

(44)

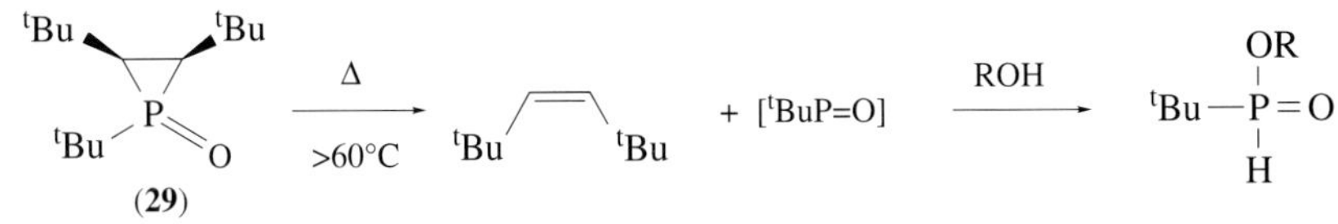

(45)

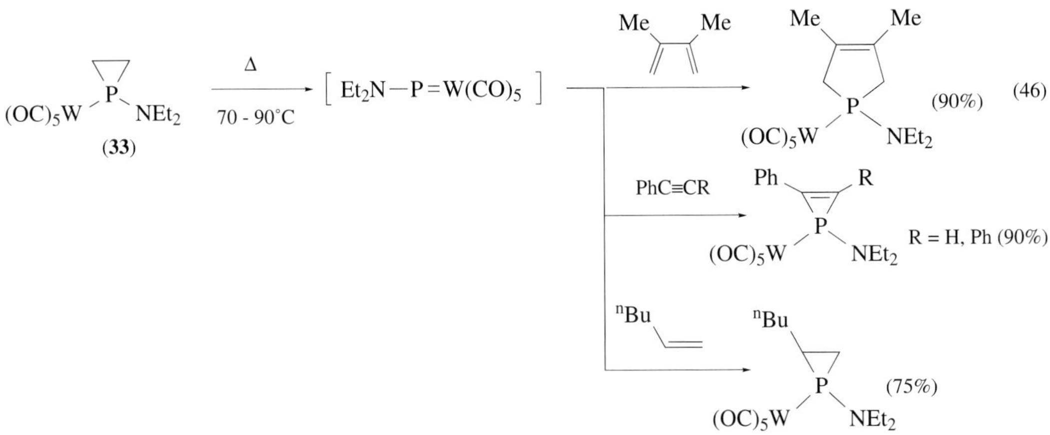

(46)

of the olefinic fragment suggests a concerted mechanism. Clean thermolysis has been also observed for several phosphirane complexes ⟨89JA9098, 92SYN157⟩ (Equations (46), (47)). Finally, an exchange process has been observed between phosphiranium salts and alkynes ⟨95CC257, 97JOM(529)189⟩ (Equation (47a)).

Another type of splitting has been found in the reaction of some nucleophiles with phosphirane derivatives. For example, the reaction of aryllithium compounds with a heavily substituted 1-chlorophosphirane (**35**) leads to the collapse of the ring ⟨87TL2693⟩ (Equation (48)). The reaction probably involves a transient destabilised 1-arylphosphirane. The finding that phosphirane complex (**30**) also splits upon reaction with soft nucleophiles is more significant ⟨89T3061⟩ (Equation (49)). It is interesting to compare this observation with the data of Equation (26) where hard nucleophiles were used. The proposed explanation involves a common transient phosphorane (**36**) whose behaviour depends on the nature of the nucleophile.

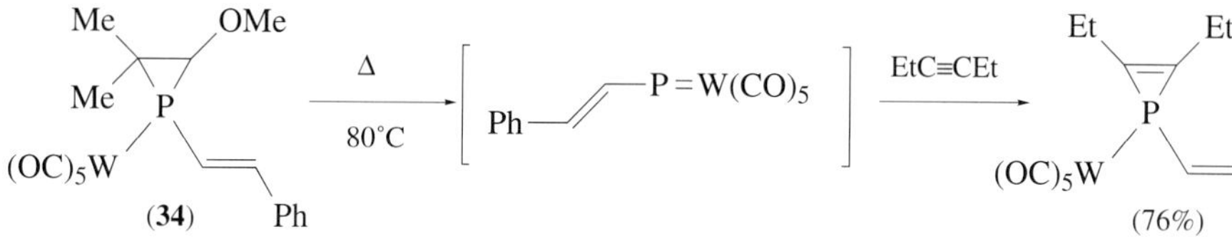
(47)

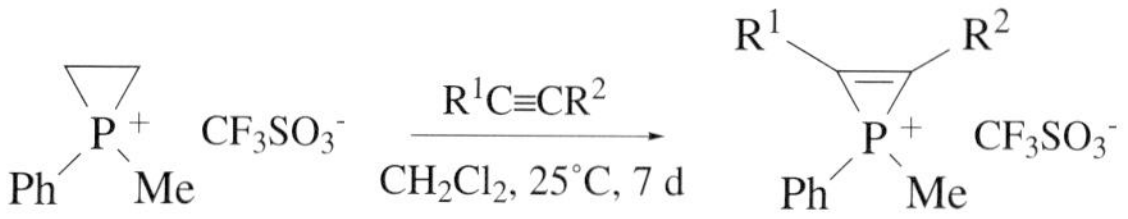
(47a)

(48)

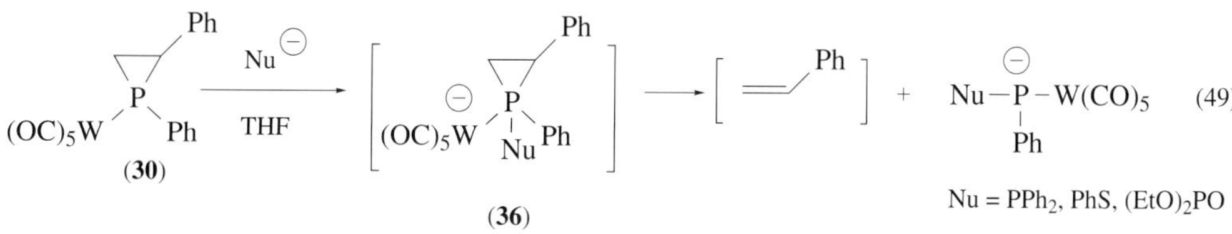
(49)

In contrast to phosphiranes, phosphirenes do not tend to dissociate into phosphorus and C_2 units. Only one well documented case of such a splitting has been described ⟨89JA9098⟩ (Equation (50)). However,

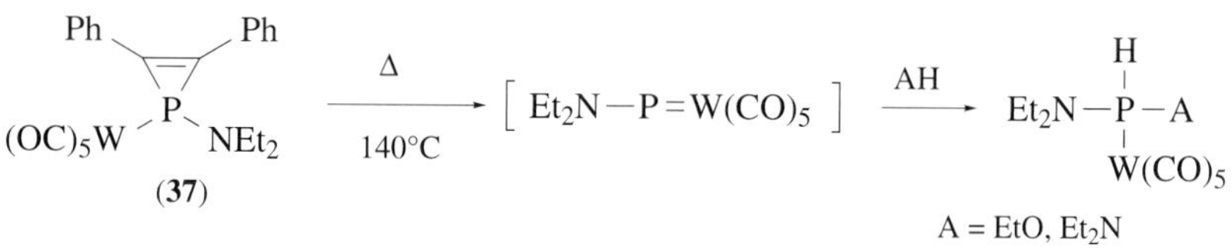
(50)

as a phosphinidene precursor, complex (**37**) is far less efficient than its phosphirane counterpart (**33**), see Equation (46).

2.1.3.5 Ring Expansions

The ring expansion of phosphiranes and phosphirenes can occur either via internal reorganisation or insertion reactions. The former was first illustrated for 2-vinylphosphiranes by Richter ⟨82AG(E)292, 83CB3293⟩ (Equation (51)).

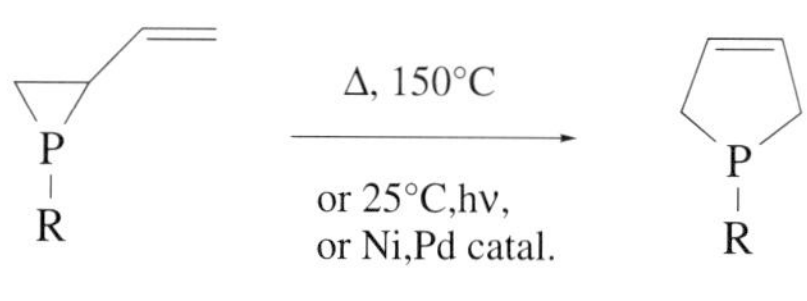
(51)

A two-step biradical mechanism was proposed on the basis of a kinetic study ⟨85CB1575⟩. A similar rearrangement was found to occur for 2-vinylphosphirane complexes ⟨84OM456⟩. In this latter case, it was shown on the basis of structural studies that both concerted and biradical mechanisms were available to the rearrangement ⟨92JOC6557, 96JA1690⟩. As expected, the concerted pathway proceeds with complete inversion at phosphorus ⟨92JOC6557⟩. A parallel with carbon chemistry is even more striking in the case of 1,2-divinylphosphirane complexes. They undergo a concerted phospha-Cope rearrangement to give a 7-membered ring which has been identified through its [4 + 2] cycloadducts with conjugated dienes ⟨92OM1438⟩ (Equation (52)).

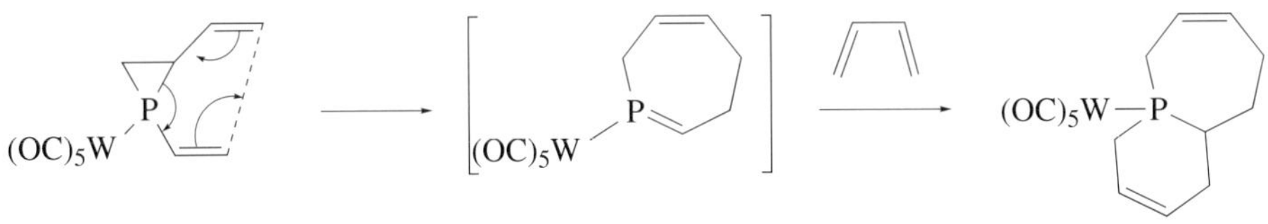

(52)

Several ring reorganisations have been also shown to occur with 9-phosphabicyclo[6.1.0]nonatrienes. The first one is a thermal [1,5] sigmatropic shift with initial retention of stereochemistry at phosphorus: discovered by Katz ⟨66JA3832, 73JA4292⟩, it has also been studied by Richter ⟨85CB97⟩ and Quin ⟨87TL1077⟩ (Equation (53)).

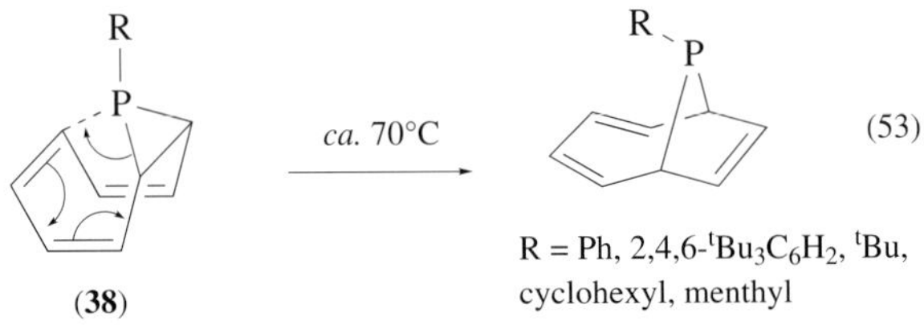

(**38**)

(53)

The second is a 8π-conrotatory opening of the 8-membered ring to a phosphonin which recyclises through 6π-disrotatory ring closure to give a dihydrophosphindole. This second pathway was first demonstrated by Quin using P-oxides such as (**39**) ⟨83JA5960, 86JA4519, 86PS(27)109⟩ (Equation (54)) and has subsequently been shown to occur with tervalent derivatives ⟨85CB97⟩.

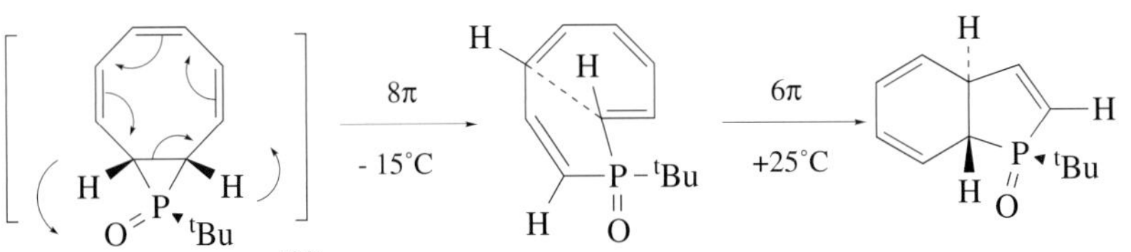

(**39**)

(54)

Finally, quaternary salts of (**38**) have been shown to rearrange into 9-phosphabarbaralanes, probably via phosphonin intermediates ⟨83TL3981⟩ (Equation (55)).

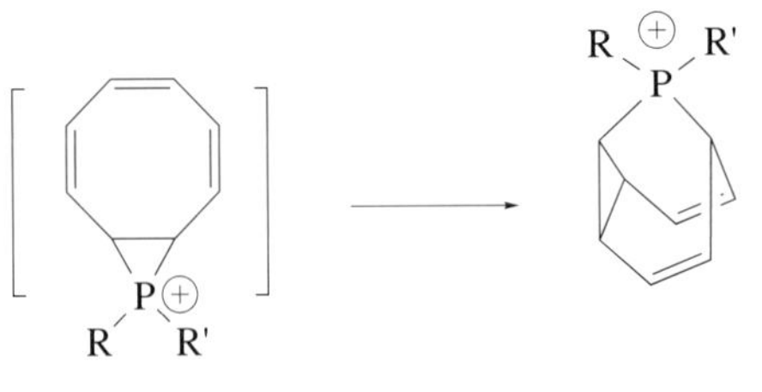

(55)

These products behave as degenerate Cope systems. A further type of internal reorganisation has been described by Märkl who has shown that 1-phosphanorcaradienes are in equilibrium with 2*H*-phosphepines ⟨87AG(E)1134⟩ (Equation (56)).

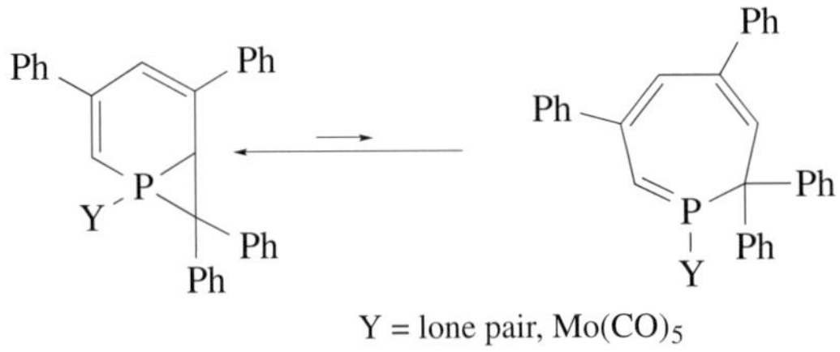

(56)

In contrast to the various phosphirane rearrangements above, only a few insertion reactions have been described with phosphiranes. An activated phosphirane complex (**40**) has been shown to insert a molecule of enol ether ⟨84OM456⟩ (Equation (57)).

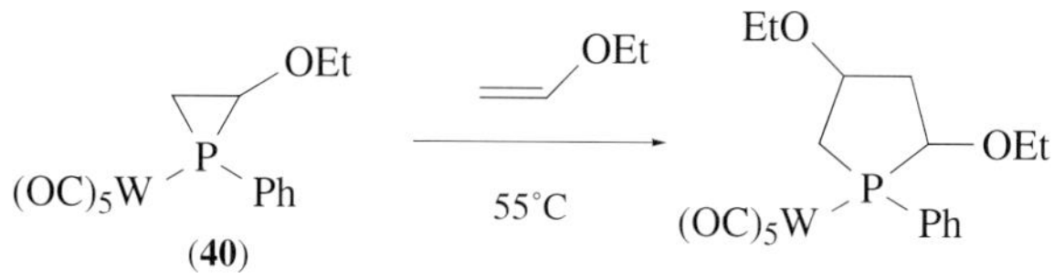

(57)

Similarly, several 14-electron metallic fragments can be inserted into phosphirane complexes ⟨89CC1389, 93JCS(D)1811⟩ (Equation (58)).

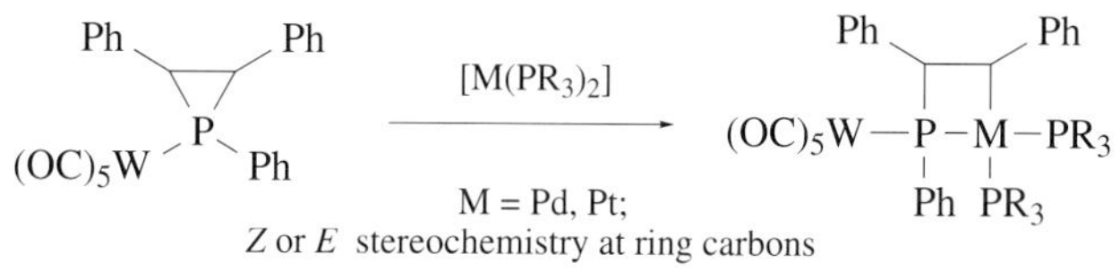

(58)

The retention of stereochemistry at the C–C bond of the ring suggests a concerted mechanism which probably involves η^2-interaction with the P–C ring bond.

Four intramolecular reorganisations provoking an expansion of the phosphirene ring have been reported. In the first, a 1-allylphosphirene complex is transformed into a bicyclic phosphirane ⟨86OM1161⟩ (Equation (59)).

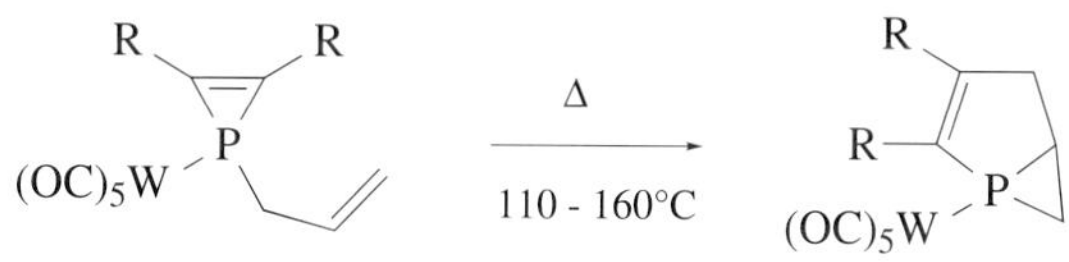

(59)

In the second, an intermediate 1-diazoalkylphosphirene, generated from the corresponding 1-chloro-1*H*-phosphirene and ^{t}BuC(=N$_2$)Li, rearranges to a 1,2,4-diazaphosphinine ⟨94RHA61⟩ (Equation (60)).

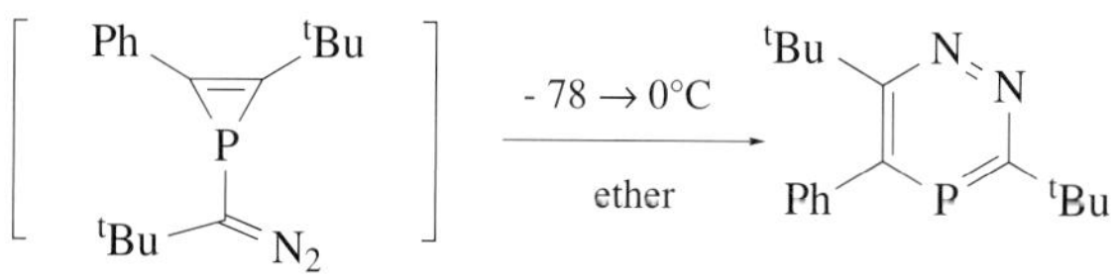

(60)

In the third, 1-silyl and 1-germylphosphirenes give 4-membered rings upon UV-irradiation ⟨90AG(E)1436, 99MI1581⟩ (Equation (61)).

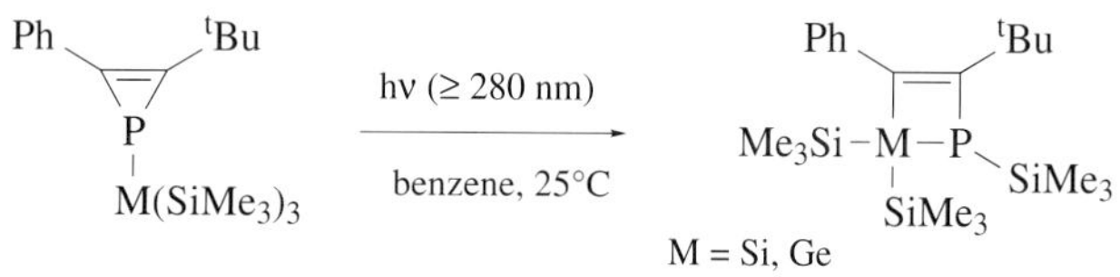

(61)

Finally, a 2-phosphino-2*H*-phosphirene, produced by [1 + 2] cyclocondensation of a phosphinocarbene with a phosphaalkyne, gives a novel λ^3,λ^5-1,2-diphosphete ⟨95JA10785, 99MI274⟩ (Equation (61a)).

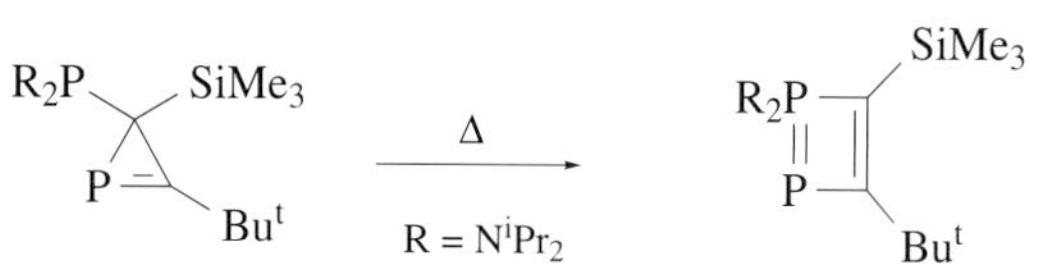

(61a)

The corresponding 1-phosphino-1*H*-phosphirene does not rearrange similarly.

Several other insertion reactions have also been described. At high temperature, phosphirene pentacarbonylmetal complexes insert CO ⟨85JA5001⟩ (Equation (62)).

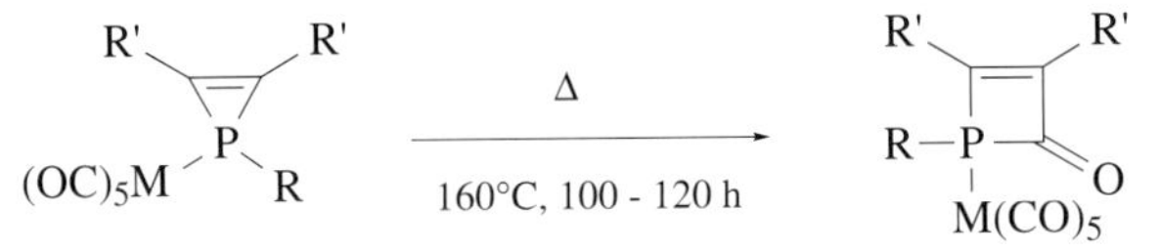
(62)

M = Cr, Mo, W; R, R' = Ph, Et

The carbon monoxide originates from the $M(CO)_5$ complexing group, and the mechanism probably involves an intermediate 4-membered metallacycle. This hypothesis receives some support from the fact that phosphirenes and their complexes readily insert a variety of 14-electron metallic fragments ⟨86CC762, 88CC1156, 91JCS(D)905, 92CC750⟩ (Equation (63)).

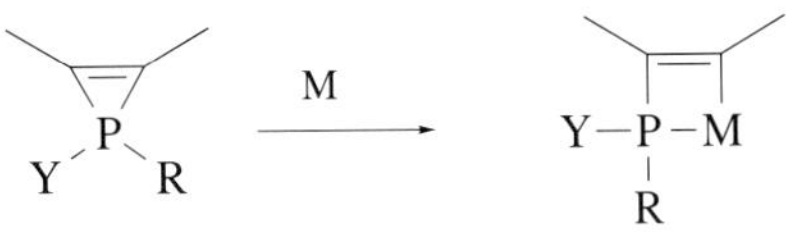
(63)

Y = lone pair, $W(CO)_5$
M = NiL_2, PdL_2, PtL_2, $RhClL_2$
(L = phosphine)

The nickelacycle has indeed been shown to insert CO at room temperature. It is noteworthy that Pt(II) species may react with tervalent phosphirenes to give either normal η^1-P-complexes or dimeric insertion products through halogen transfer from Pt to P ⟨93CC142⟩ (Equation (64)).
Transition metals are also able to promote the insertion of monosubstituted alkynes into the phosphirene ring ⟨87TL5021, 93BSF695⟩ (Equation (65)).

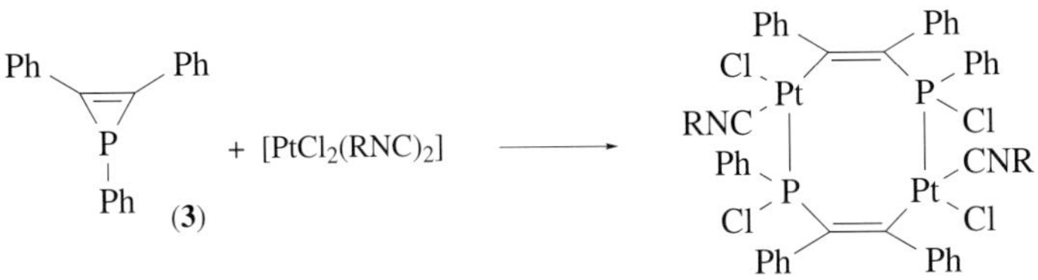
(3)
(64)

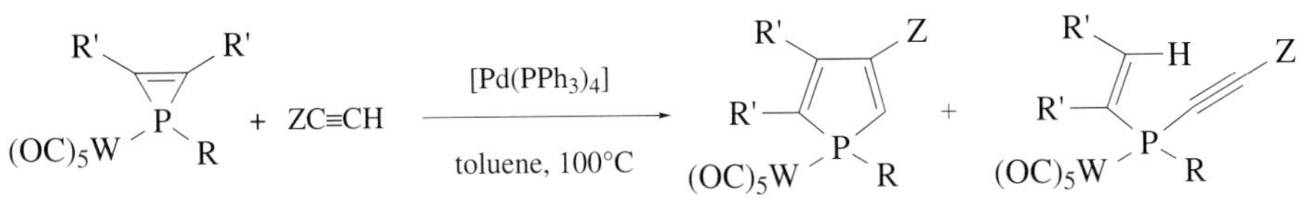
(65)

Once again, a transient 4-membered metallacycle is probably involved. This also rationalises the formation of the alkynylphosphine complex as a side-product of the reaction.

Without assistance from transition metals, it is also possible to insert a phosphinidene or arsinidene unit into the phosphirene ring ⟨87AG(E)548, 93BSF521⟩ (Equation (66)).

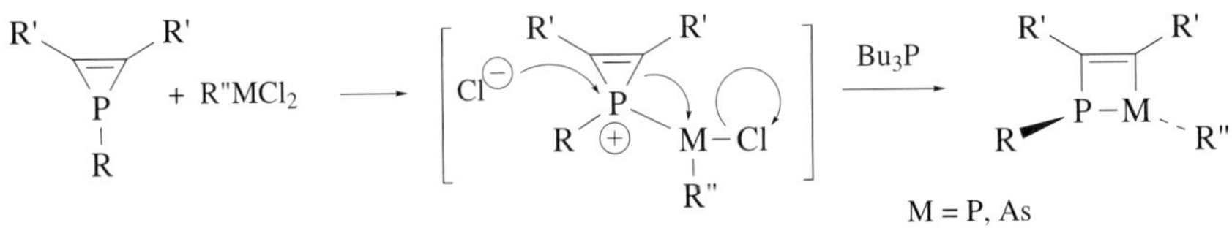
(66)

M = P, As

The yields are moderate, but the reaction is rather general and represents the best (M = P), or the only (M = As) access to the resulting 4-membered heterocycles. The intermediacy of a phosphirenium salt has been demonstrated for M = P by using $AlCl_3$ as a trap for the chloride ion. It is interesting to compare the reactivity of phosphirenes and phosphiranes toward dichlorophosphines. For phosphiranes, a ring-opening has been observed only when the phosphirane has a very specific substitution pattern, see Equation (22a). This different behaviour may reflect the lower nucleophilicity of phosphorus and the lower stability of the phosphonium salts in the phosphirane series. In a similar fashion, two groups ⟨93MC68, 94CB313⟩ have shown that it is sometimes possible to insert the phosphorus of an iminophosphane into the 3-membered ring of a phosphirene imine (Equation (66a)). Similarly, terminal phosphinidene complexes can insert into the conjugatively destabilised P–C_2 ring bond of 2-alkynylphosphirenes ⟨97OM4501, 99OM796⟩ (Equation (66b)).

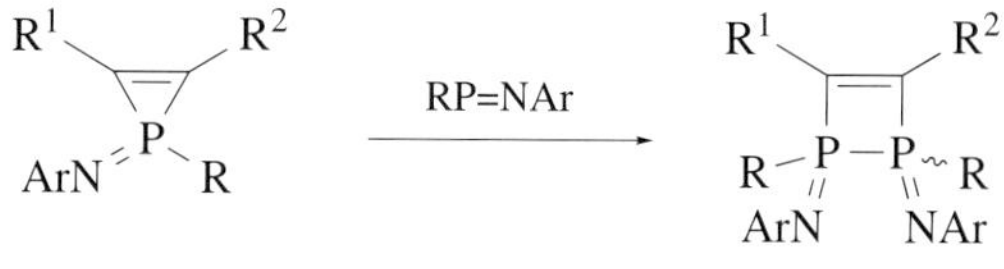

(66a)

Ar = 2,4,6-$^tBu_3C_6H_2$

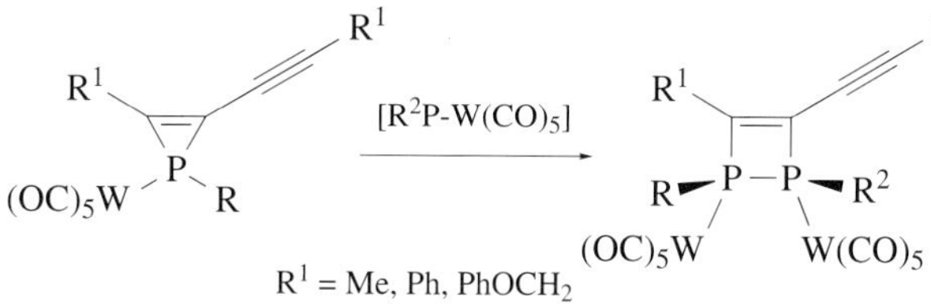

(66b)

R^1 = Me, Ph, $PhOCH_2$

The insertion of phosphinidene favours the *cis*-stereochemistry. An intramolecular version of the same reaction which does not require the destabilising alkynyl substituents has been described ⟨99JOM(582)53⟩.

Finally, four miscellaneous results must be mentioned. A spontaneous dimerisation of a series of 2-alkoxy-phosphirenium salts ⟨87ZOB2482⟩ has been reported to give 5-membered cyclic ylides (Equation (67)).

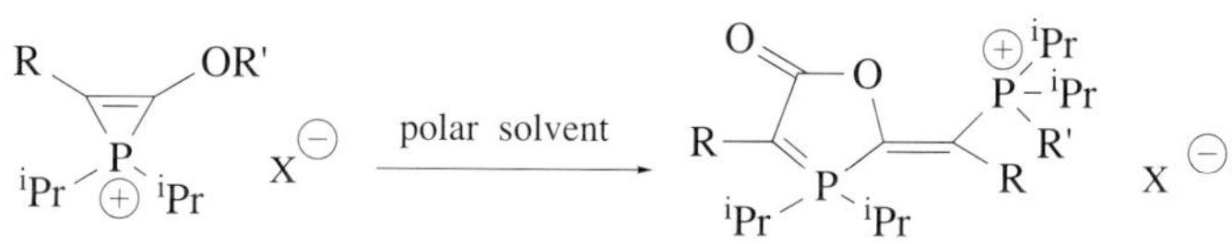

(67)

The reaction of triphenylphosphirene with a mesoionic dithioliumolate gives a thiaphosphacyclohexadiene ⟨95H(40)311⟩ (Equation (67a)).

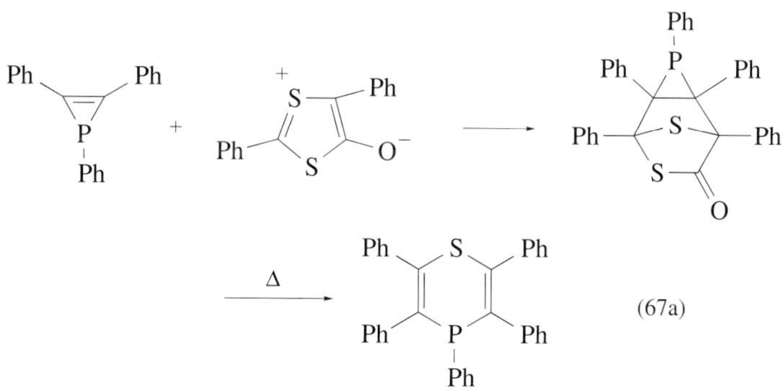

(67a)

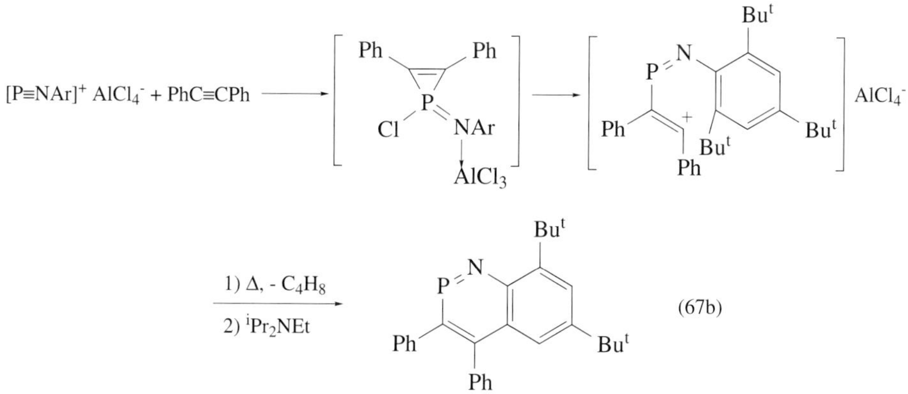

(67b)

Aluminum trichloride induces the rearrangement of an intermediate phosphirene imine into a novel phosphaquinoline ⟨92TL2335⟩ (Equation (67b)). The $[Ru_3(CO)_{12}]$ cluster inserts into the P–C ring bond of triphenylphosphirene ⟨95OM3592⟩ (Equation (67c)).

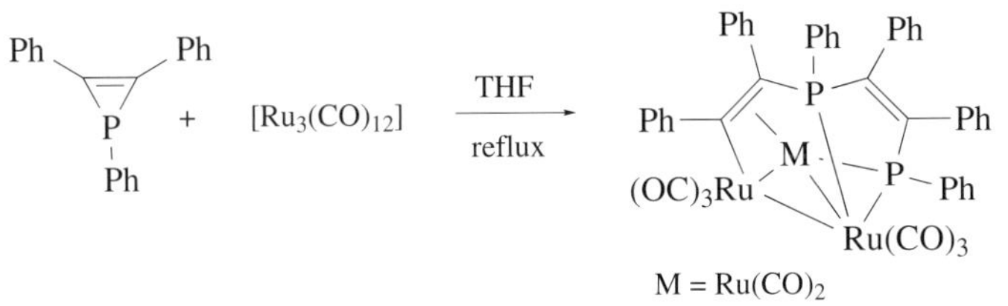

(67c)

2.1.4 SYNTHESES OF PHOSPHIRANES AND PHOSPHIRENES

2.1.4.1 Synthesis from P and C_2 Units

Basically, syntheses of phosphiranes from P and C_2 units fall into three formal categories: (a) RP^{2-} + XC–CX; (b) RPX_2 + $^-$C–C$^-$; (c) RP + C=C (X = halogen or pseudohalogen). The original synthesis, due to Wagner, ⟨63USP3086053, 63USP3086056, 67JA1102, 69T1097⟩ belongs in the first category (Equation (68)).

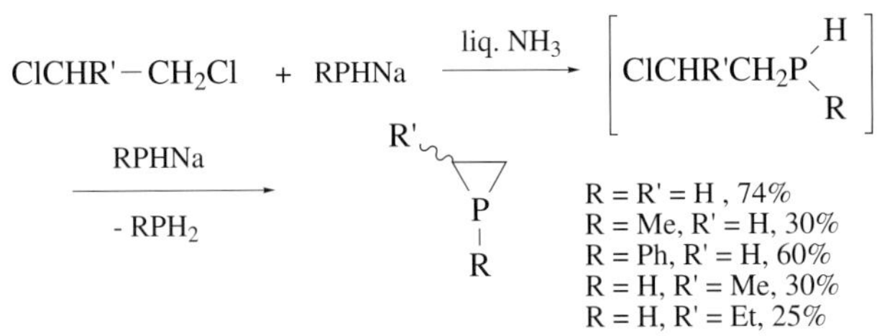

(68)

Variations include the use of HMPA as the solvent to obtain the P–D and C–D derivatives ⟨69JA17⟩, the replacement of chlorine by the mesyloxy leaving group in the C_2 unit ⟨74JA317⟩ and the replacement of the sodium derivatives by dilithiophosphines ⟨85CB4285, 94CC475, 96OM1301, 97OM1526⟩. A related chemistry involves the reaction of a nucleophilic zirconocene–phosphinidene complex with 1,2-dichloroethane ⟨95JA11914⟩ (Equation (68a)).

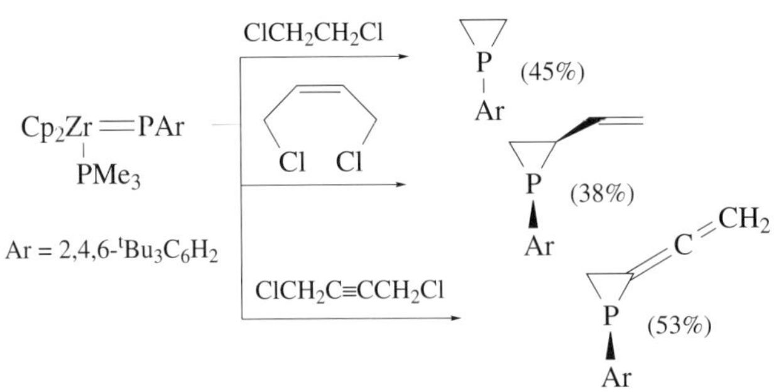

(68a)

A *cis*-2-vinylphosphirane and a 2-vinylidenephosphirane have been also obtained using the same reagent. A similar approach can be carried out in the coordination sphere of tungsten ⟨89JA9098⟩ (Equation (69)).

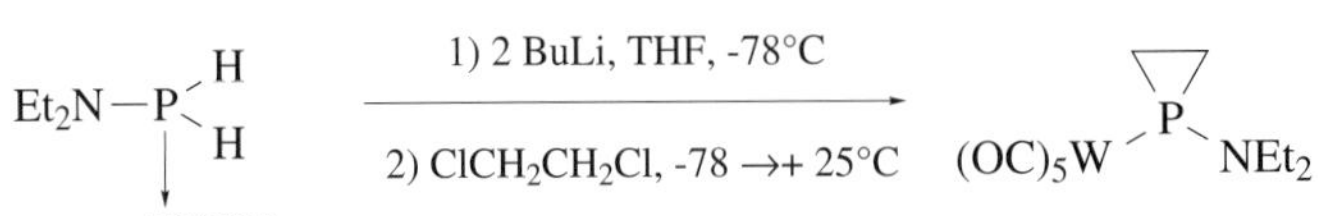

(33) (40%)

(69)

More recently, it has been shown that oxiranes can replace 1,2-dihaloalkanes in the synthesis of phosphiranes, provided that appropriate anionic RP reagents are used. The first example was described by Yoshifuji ⟨85CL441⟩ (Equation (70)).

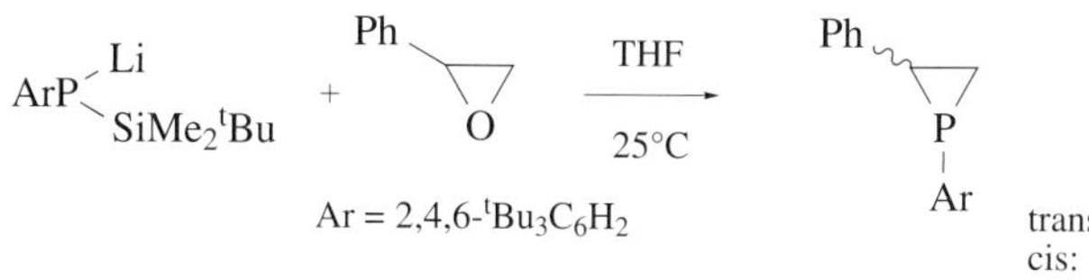

(70)

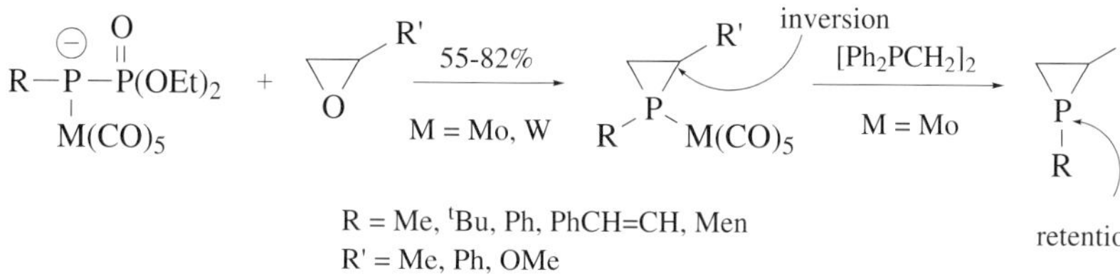

(71)

The same transformation can be performed with the zirconocene–phosphinidene complex in Equation (68a). The epoxide stereochemistry is retained in the product phosphirane ⟨95JA11914⟩. Alternatively, it has been shown by Marinetti *et al.* that so-called "phospha-Wittig" reagents transform oxiranes into phosphiranes with an inversion of configuration at carbon ⟨92SYN157, 93OM1207⟩ (Equation (71)). Decomplexation by $Ph_2PCH_2CH_2PPh_2$ is easily achieved in the molybdenum series. Optically active phosphiranes are readily accessible by this route.

The use of the second approach is restricted by the poor availability of 1,2-dimetallic derivatives of alkanes. The first example of a synthesis of 2-vinyl-substituted phosphiranes was described by Richter, using the polymeric magnesium–butadiene adduct ⟨82AG(E)292, 83CB3293⟩ (Equation (72)).

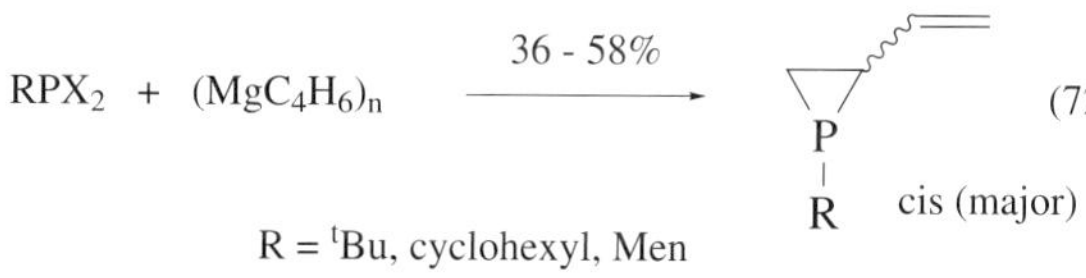

(72)

A second example concerning the synthesis of phospha[3]radialenes was reported more recently ⟨91CB933⟩ (Equation (73)).

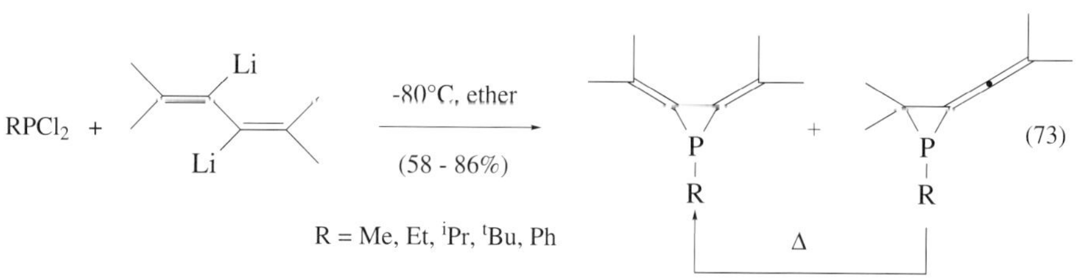

(73)

The allenyl-substituted phosphiranes isomerise into phospha[3]radialenes upon distillation at standard pressure. A third example makes use of P=O-stabilised zirconacyclopropanes as shown in Equation (73a) ⟨98CC1177⟩.

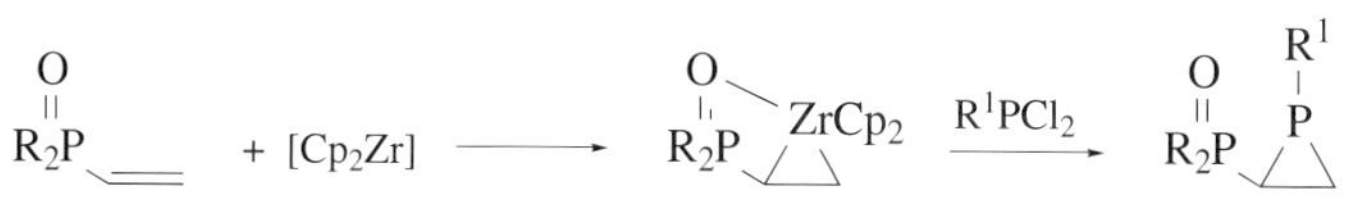

(73a)

Bicyclic phosphiranes are thus obtained from 2-phospholene oxides.

A third approach under neutral conditions involves either free or complexed phosphinidenes. With free phosphinidenes, the problem is to find an efficient precursor. The first example was disclosed by Fritz ⟨89AG(E)315, 89ZAAC(570)54⟩ (Equation (74)). More recently, Gaspar ⟨92JA8526⟩ has proposed a method employing the photolysis of 1-arylphosphiranes (Equation (43) which has the potential to be more general. Using this technique, triplet mesitylphosphinidene is generated in a methylcyclohexane glass at 77 K and may be detected at 4 K by ESR ⟨94JA7899⟩. None of these methods, however, can compete for generality and efficiency with the condensation of electrophilic terminal phosphinidene complexes with alkenes.

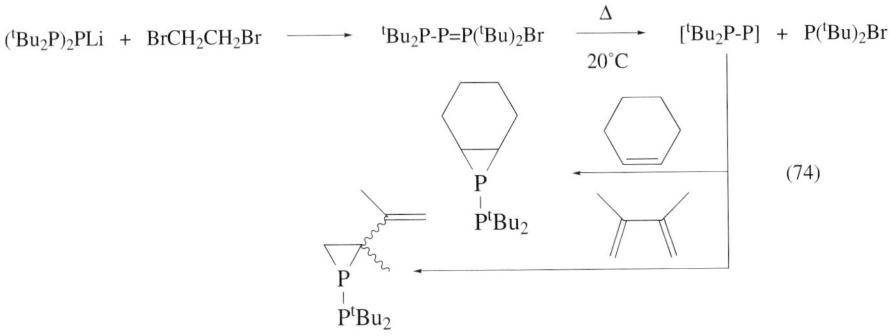

(74)

These transient species are usually generated by thermal decomposition of 7-phosphanorbornadiene complexes, themselves prepared from phosphole complexes by [4 + 2] cycloaddition ⟨82CC667⟩ (Equation (75)).

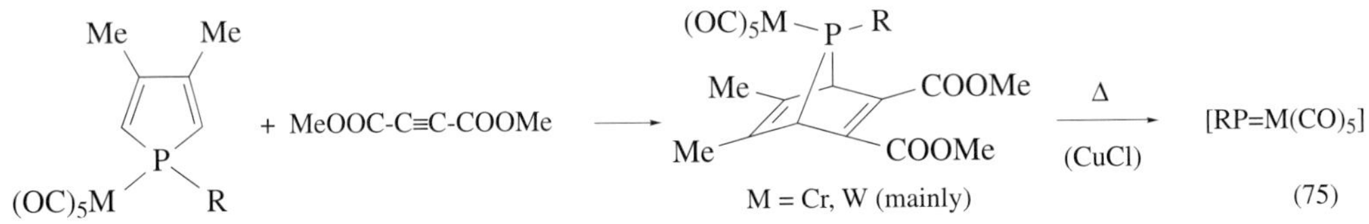

(75)

Alternatively, some phosphirane complexes can also be used as precursors (Equations (46), (47)). The condensation of these terminal phosphinidene complexes with alkenes takes place with full retention of stereochemistry at the C_2 unit ⟨84OM456⟩ (Equation (76)).

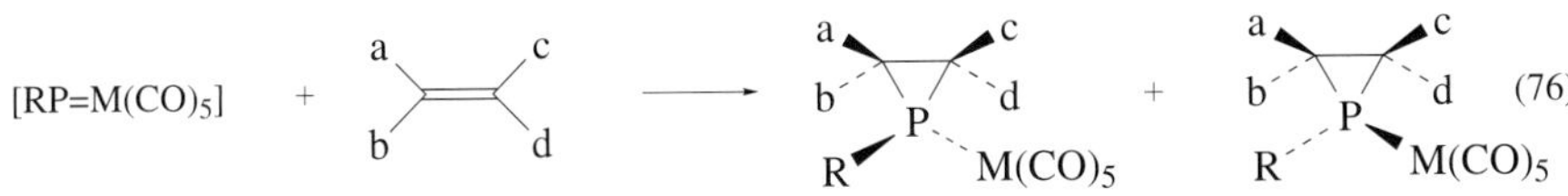

(76)

Many alkenes have been successfully used in this condensation, including ethylene itself, methylenecycloalkanes ⟨93JOC6786⟩, conjugated dienes ⟨84OM456, 92JOC6557⟩, cyclic dienes ⟨93JOC1800, 96JA1690⟩, norbornadiene ⟨93OM1401⟩, allenes ⟨97JA8432⟩, bicyclopropylidene ⟨99JA11650⟩, enol ethers, acrylic esters ⟨84OM456⟩ and enamines ⟨84OM1492⟩. The Hammett reaction constants ρ^+, obtained through a study of the addition of various styrenes to $[PhP{=}W(CO)_5]$ and $[MeP{=}W(CO)_5]$ (-0.76 and -0.60, respectively), indicate a mild electrophilicity ⟨88OM1875, 92OM4365⟩. The relative reactivity of various substituted olefins toward $[PhP{=}W(CO)_5]$ has also been quantified ⟨94JA10966⟩. The reaction with conjugated dienes gives 2-vinylphosphirane complexes which rearrange upon mild heating to the corresponding 5-membered phospholene complexes via a concerted [1,3] shift ⟨92JOC6557⟩ or a biradical mechanism ⟨96JA1690⟩. Many substituents at phosphorus including alkoxy ⟨86NJC321⟩ and amino groups ⟨89JA9098⟩ have also been used. Double condensation of alkenes with $[(PCH_2CH_2P)W(CO)_4]$ ⟨87OM586⟩ and self-condensation with bifunctional phosphinidenes bearing olefinic substituents have been observed ⟨86OM1161⟩ (Equation (77)).

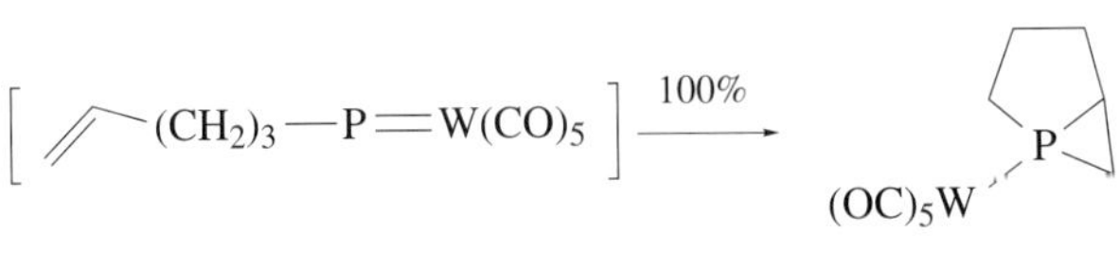

(77)

Decomplexation of phosphorus has been achieved in a limited number of cases. Initially, halogenation at low temperature with I_2 or Br_3^- transforms the $W(CO)_5$ complexing group into $W(CO)_4X_2$. Displacement of the phosphorus ligand from the coordination sphere of tungsten may then be performed with *N*-methylimidazole or 2,2′-bipyridyl, for examples, see ⟨84CC45, 97JA8432⟩. Finally, phosphirane complexes have been obtained as major by-products in the reaction of terminal phosphinidene complexes with oxiranes ⟨87OM2189⟩ and ethoxyacetylene ⟨90PS(47)477⟩. See also ⟨86JOM(306)29⟩.

The first synthesis of phosphirenes also involved electrophilic terminal phosphinidene complexes. Besides the precursors already mentioned for the synthesis of phosphiranes, azaphosphirene complexes have been proposed by Streubel's group as a convenient alternative when bulky substituents at phosphorus are needed ⟨94AG(E)2427⟩ (Equation (77a)).

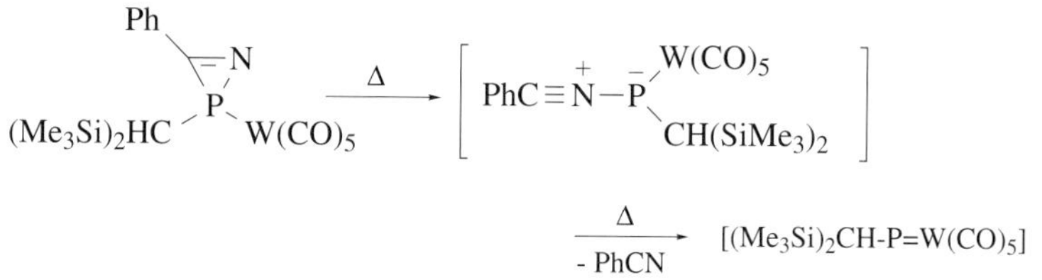

(77a)

Subsequent work by the same group provided some evidence in favour of a 1,3-dipolar intermediate in the decomposition of these species ⟨97AG(E)1492, 98CC1529⟩. Whatever their precursors, electrophilic terminal phosphinidene complexes react efficiently with a variety of alkynes giving phosphirene complexes whose decomplexation is easily achieved via the halogen–tertiary amine procedure ⟨82JA4484, 84CC45, 85JA4700⟩ or by displacement with $Ph_2PCH_2CH_2PPh_2$ ⟨95SYN941⟩ (Equation (78)).

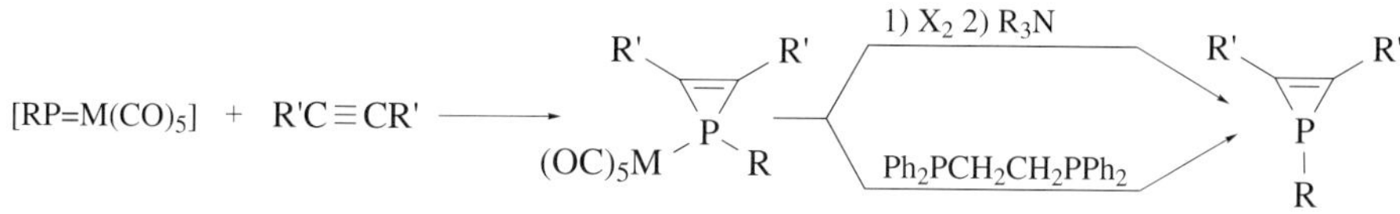

(78)

M = Cr, Mo, W

Besides classical alkyl and aryl groups, many functional substituents at phosphorus have been used, including allyl ⟨86OM1161⟩, alkoxy ⟨86NJC321, 88OM1796⟩, amino ⟨89JA9098⟩, cyano ⟨88OM1796⟩, chloro ⟨89JA9098, 95SYN941⟩, chloroalkyl ⟨85TL4595, 88JOM(354)83⟩, ethoxycarbonyl ⟨88OM1796⟩ and (ethoxycarbonyl)alkyl groups ⟨93BSF695⟩. The reaction works with acetylene ⟨95SYN941⟩, mono and disubstituted alkynes and conjugated diynes ⟨97OM4501, 99OM796, 99JCS(D)2409⟩. The bulky phosphinidene in Equation (77a) shows a high tolerance of functional groups on the alkyne. These include alkoxy, silyl, stannyl and alkoxycarbonyl substituents ⟨95CC2507, 97ZAAC(623)1897, 98CC1761, 99ZAAC(625)102⟩. Sometimes the reaction follows a different pathway. For example, amino and alkoxy-substituted alkynes and non-bulky phosphinidenes give phosphole complexes directly, via a formal [1 + 2 + 2] cycloaddition ⟨90PS(47)477⟩. Ethoxy(stannyl) and ethoxy(silyl)-acetylenes and the bulky phosphinidene in Equation (77a) give the expected phosphirene plus an isomeric phosphaallene complex ⟨98CC1761, 99ZAAC(625)102⟩. Other phosphinidene generating systems have been proposed which do not rely on a cycloreversion reaction. For example, a transient aminophosphinidene complex is probably involved in the chemistry depicted in Equation (79) ⟨86TL1323⟩.

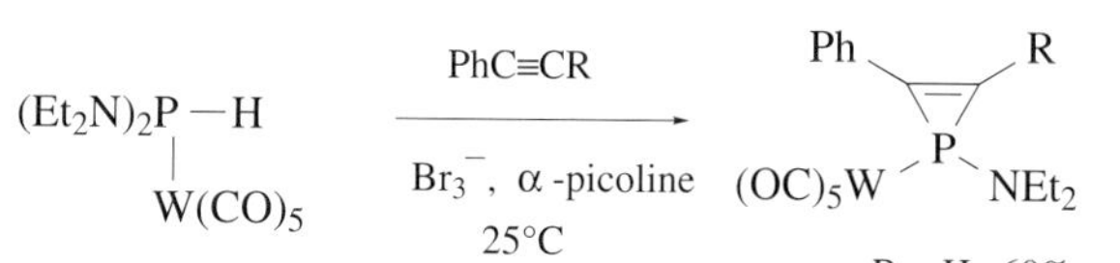

(79)

R = H, 60%
R = Ph, 20%

This basic scheme has been subsequently extended to free arylphosphinidenes by Gaspar ⟨92JA8526⟩ (Equation (43)). A related chemistry has also been developed by Niecke ⟨88TL605, 93CB649⟩, using the phosphinidene-like behaviour of iminophosphanes (Equation (80)).

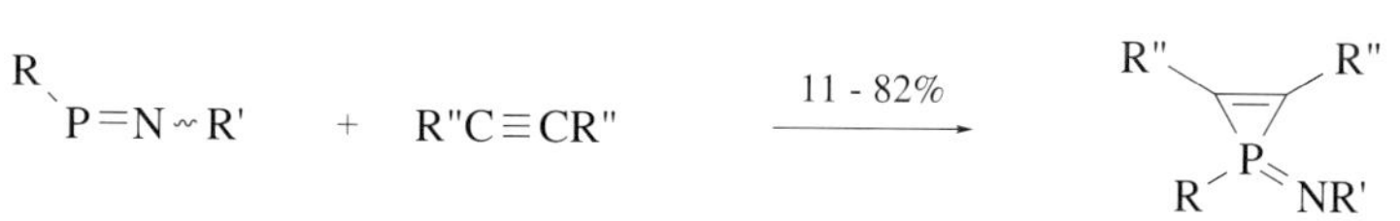

(80)

R = Me, Et, PhCH=CH, Cl; R' = 2,4,6-$^tBu_3C_6H_2$

R = Et_3C; R' = ^tBu

R'' = Ph, COOMe (also PhC≡CH)

The reaction has been extended to 1,4-diphenylbutadiyne for which mono and double-condensation together with insertion (see Equations (66a,b)) have been observed ⟨94CB313⟩ and to 1-amino and

1-alkoxyalkynes ⟨93MC68, 93TL1331, 95ZOR404, 95ZOR495, 96ZOR433, 96PS(109)609⟩. In these two cases, both phosphirenes and 1,2-dihydro-1,2-diphosphetes are obtained.

Another P + C_2 approach for the synthesis of phosphirenes involves the electrophilic addition of dihalophosphine–$AlCl_3$ complexes to alkynes as discovered by Hogeveen ⟨83TL643⟩ (Equation (81)).

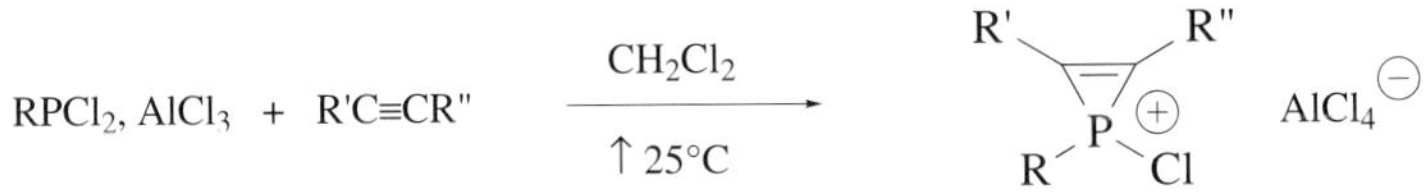

(81)

R = Me, Ph <83TL643>; ^{t}Bu, C(Et)=C(Et)Cl, Me_2N <86TL2635>; Mes <91T6673>; 4,5-dimethylphosphinin-2-yl <96OM1597>; R', R" = H, Me, Et, Ph

The method has been extended to monohalophosphines by Breslow ⟨84TL1345⟩ and to stable and transient phosphenium ions by Cowley ⟨88CC462⟩ and Wild ⟨97JOM(529)189⟩, see Equation (47a). The chlorophosphirenium salts thus obtained are easily reduced by tributylphosphine to afford the corresponding tervalent phosphirenes ⟨86TL2635, 91T6673, 96OM1597⟩ (Equation (82)).

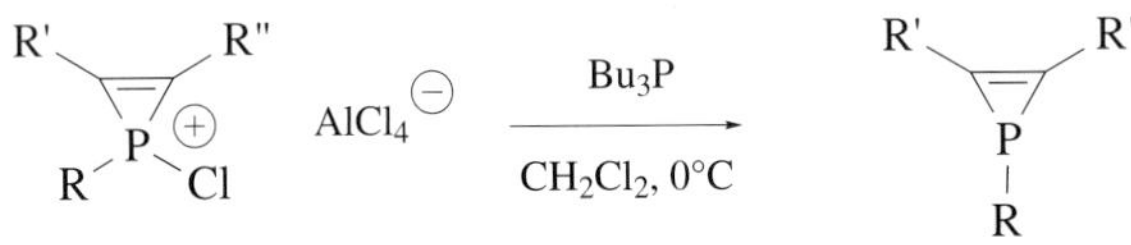

(82)

In the cases where it is applicable, this method is the most convenient route to substantial quantities of phosphirenes. The availability of titanacyclopropenes and P=O-stabilised zirconacyclopropenes has allowed a third P + C_2 route to phosphirenes ⟨98OM2677, 98CC1177⟩ (Equations (82a,b)).

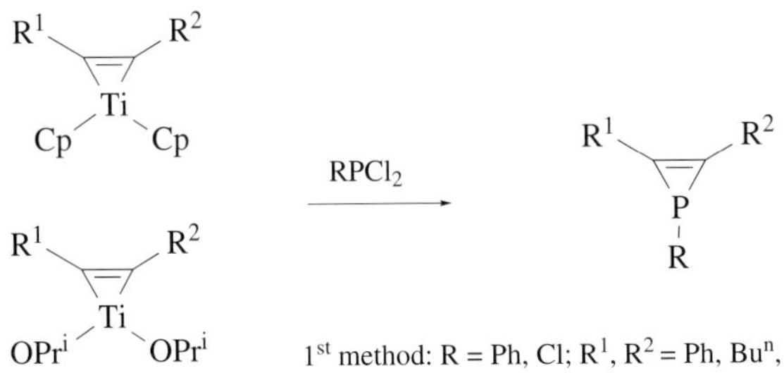

(82a)

1st method: R = Ph, Cl; R^1, R^2 = Ph, Bu^n, $SiMe_3$, Fc (65-95%)

2nd method: R = Ph, Cl→OPr^i; R^1, R^2 = Me, Et, Bu^t (60-72%)

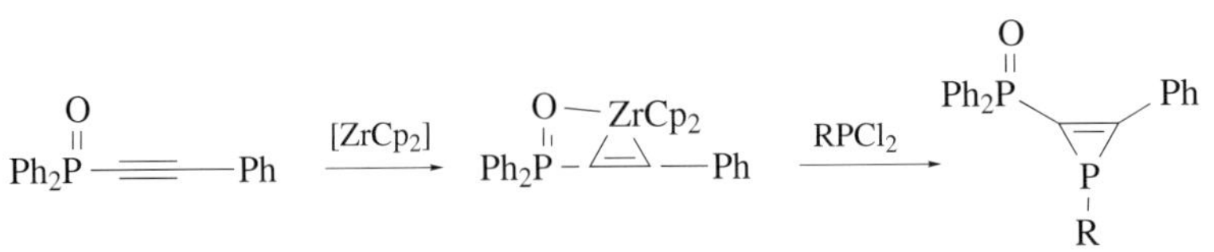

(82b)

R = Me, Ph

The first titanium method is especially useful for the synthesis of 1-chlorophosphirenes. The second method does not allow the isolation of the 1-chloro species which are transformed *in situ* into the corresponding isopropoxy derivatives. However, this synthesis is far more efficient than the first one for electron-rich alkynes ⟨98OM2677⟩.

2.1.4.2 Synthesis from PC and C Units

The availability of phosphaalkenes RP=CR'_2 and phosphaalkynes RC≡P has led to the development of another route to phosphiranes and phosphirenes involving the formal addition of carbenes to phosphorus–carbon multiple bonds. When using diazoalkanes as the carbene source, the reaction mechanism involves the series of steps demonstrated by Niecke in Equation (83) ⟨81AG(E)131, 83CC1171⟩; see also ⟨75AG(E)363⟩.

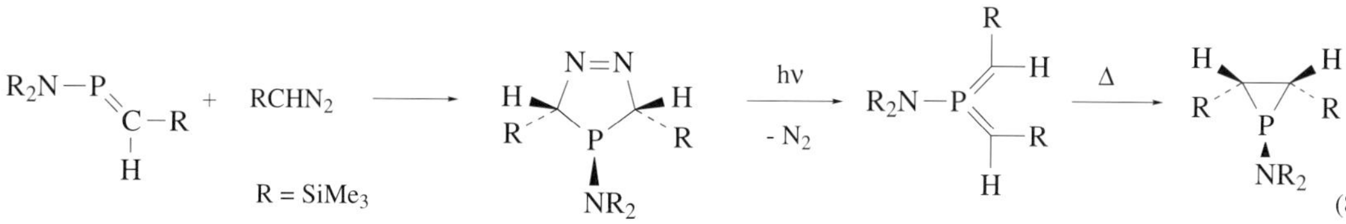

(83)

The 4π-electrocyclic closure of the intermediate bis(methylene)phosphorane is conrotatory according to the Woodward–Hoffmann rules ⟨83CC1171⟩; see, however, the results of a recent theoretical study ⟨95JOC3723⟩. Several other authors have developed similar approaches ⟨85CB3419, 86CB1977, 87TL2693, 88ZN(B)1285⟩. The first stable 1-chlorophosphiranes were prepared by a similar route ⟨87TL2693, 88ZN(B)1285, 89TL3951⟩ (Equation (84)).

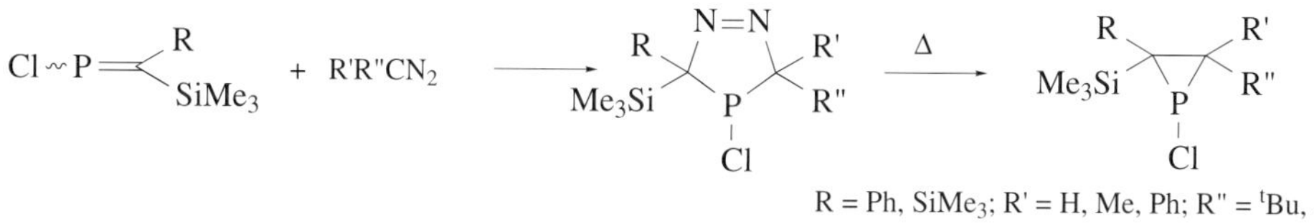

(84)

Similarly, alkylidenediazaphospholes have served as sources of alkylidenephosphiranes ⟨95CC25, 97CB779⟩ (Equation (84a)).

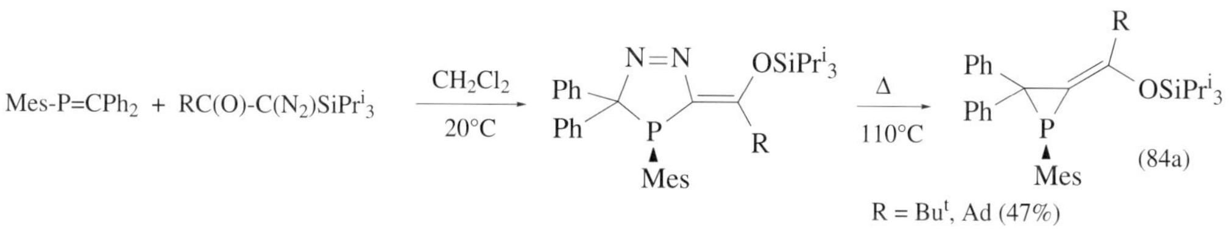

(84a)

Other carbene sources which have also been used with some success include several organometallic carbenoids ⟨89CB453⟩ (Equation (85)), ⟨90CL827⟩ (Equation (86)), ⟨91CC124, 94MI195⟩ (Equation (87)) and ⟨91TL6879⟩ (Equation (88)).

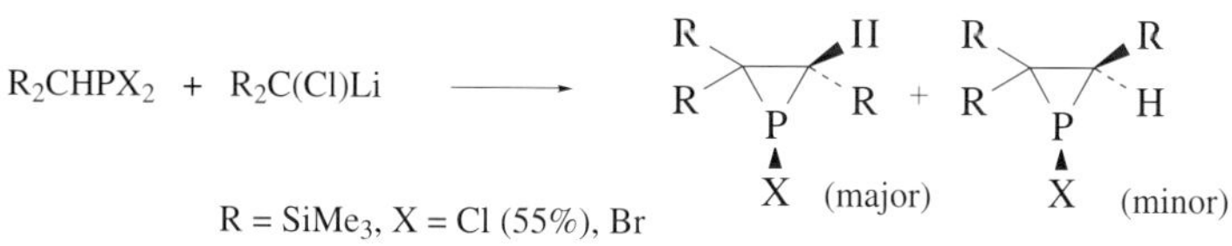

(85)

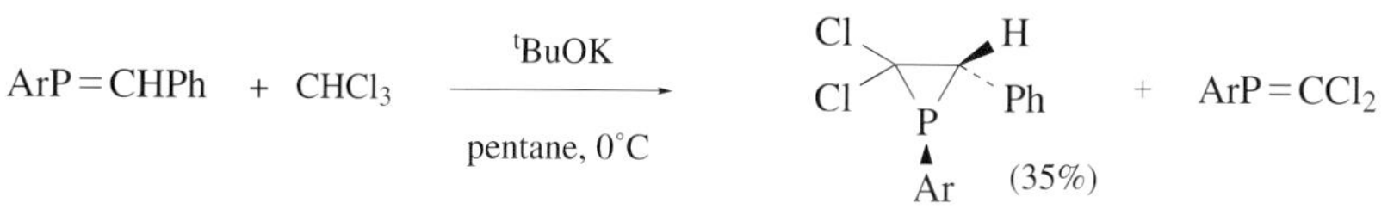

(86)

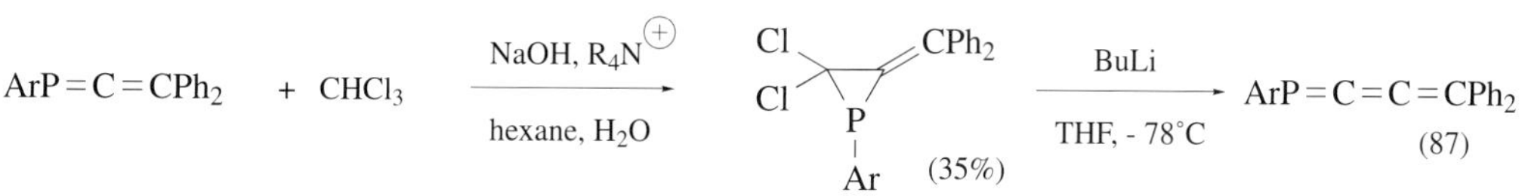

(87)

The formation of phospharadialene (**41**) results from the isomerisation of an intermediate 2-(ethenylidene)phosphirane. Its structure was confirmed by X-ray analysis ⟨92POL385⟩. A carbenoid was also used in the synthesis of an unusual phosphirene from a siloxy-substituted phosphaalkene ⟨97JOM(529)127⟩ (Equation (88a)).

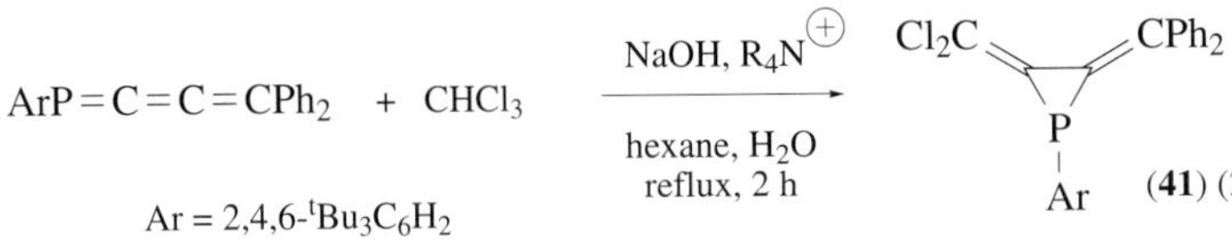

(41) (28%) (88)

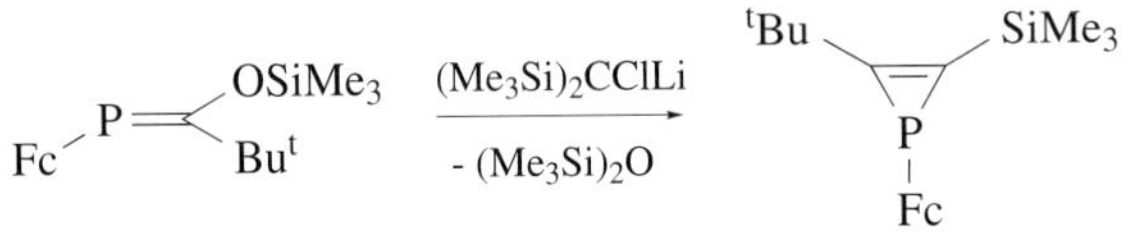

(88a)

In the same vein, Weber has shown that isonitriles can replace carbenes and readily react with transition metal-substituted phosphaalkenes ⟨93OM4653⟩ (Equation (89)).

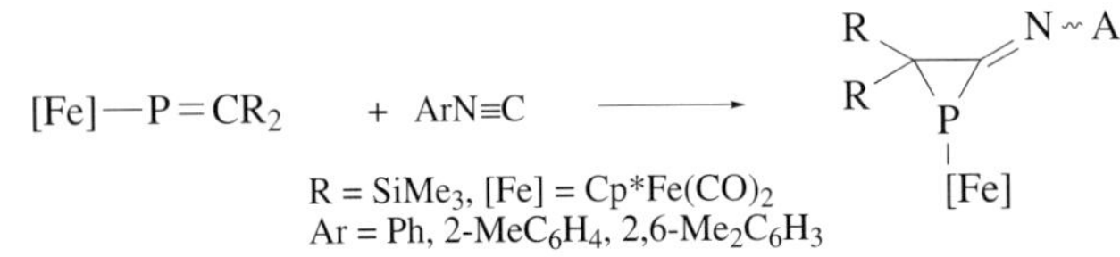

(89)

The condensation of a carbene with a P≡C triple bond theoretically affords a 2*H*-phosphirene. This was shown to be the case in practice by Regitz in the synthesis of the first characterised 2*H*-phosphirene (**4**) ⟨87AG(E)1257⟩ (Equation (90)).

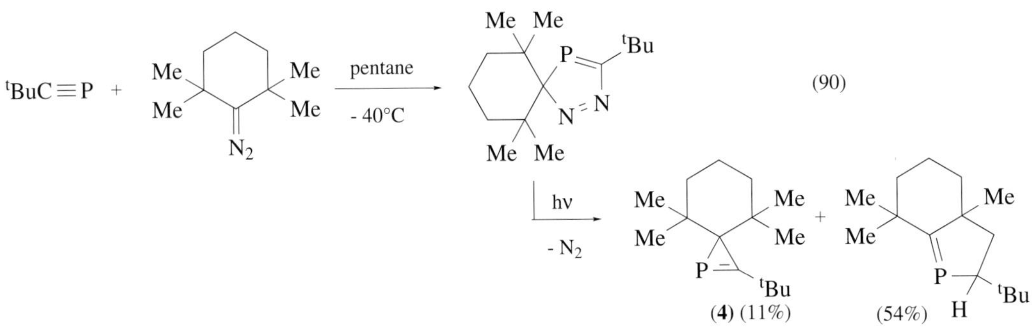

(90) (**4**) (11%) (54%)

A similar chemistry was performed later with a stable phosphinocarbene ⟨99MI274⟩ (Equation (90a)).

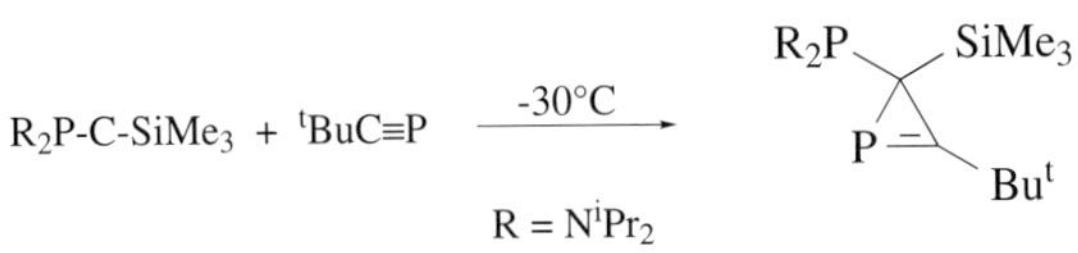

(90a)

The same type of cyclocondensation was shown to occur also with a coordinated carbene ⟨98OM893⟩ (Equation (90b)).

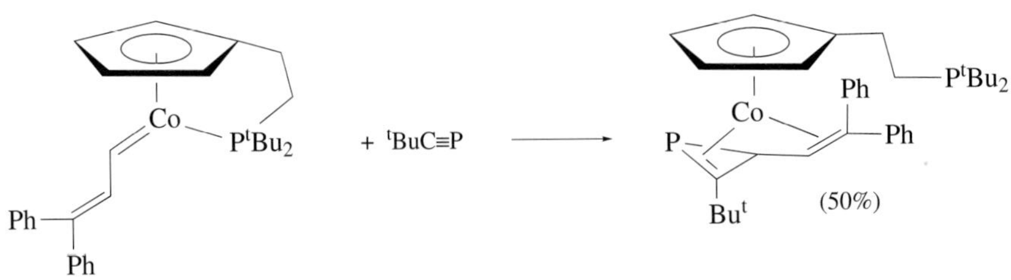

(90b)

In a similar fashion, Regitz showed that the use of a chlorocarbene precursor leads to 1-chlorophosphirenes through a formal [1,3] chlorine shift in the transient 2*H*-phosphirenes ⟨89AG(E)225, 91CB1207⟩ (Equation (91)).

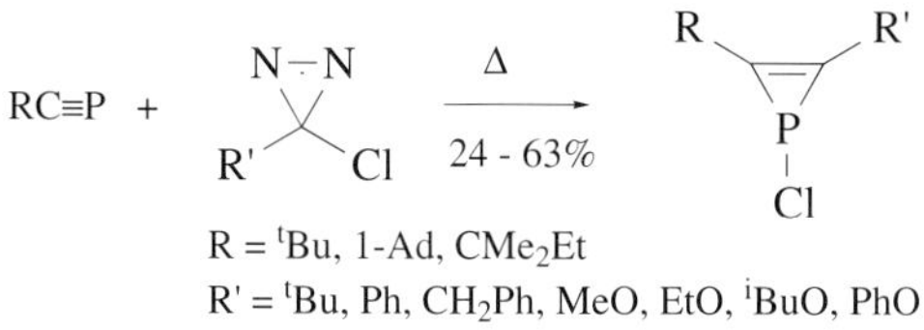

(91)

Chlorodiazirines can be replaced by 1,2-dichlorocyclopropenes ⟨91SL433⟩ in this kind of chemistry (Equation (92)).

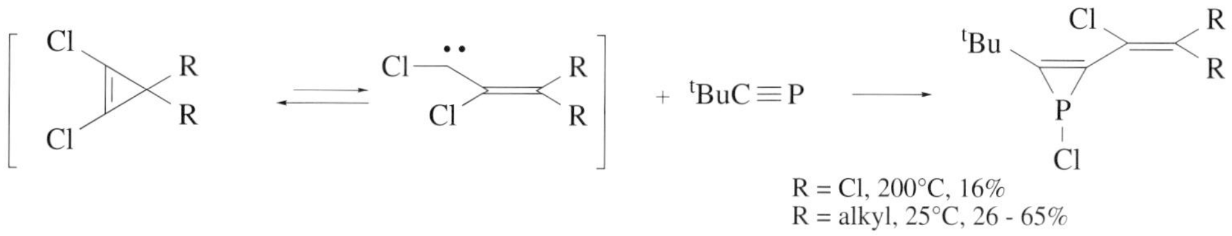

(92)

Finally, phosphaalkynes can also be replaced by silylated chlorophosphaalkenes in these reactions ⟨89TL3951⟩ (Equation (93)).

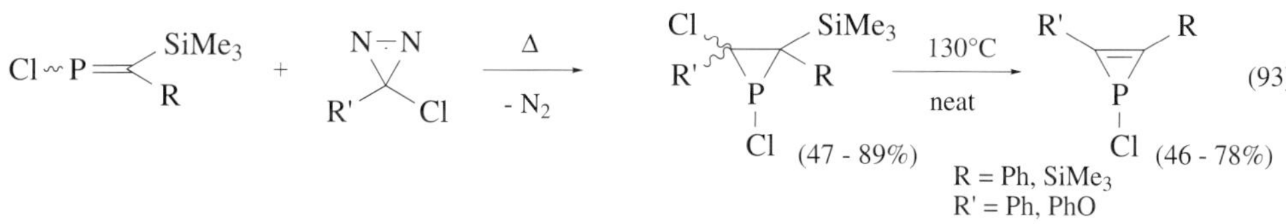

(93)

2.1.4.3 Synthesis by Cyclisation of a P–C–C Unit

As shown in Equation (68), Wagner's synthesis of phosphiranes involves an intermediate secondary *β*-chloroalkylphosphine. A preformed XCCPH unit is thus a potential phosphirane precursor. This approach was first developed by Oshikawa ⟨85SYN290⟩, see also ⟨98MI607⟩ (Equation (94)).

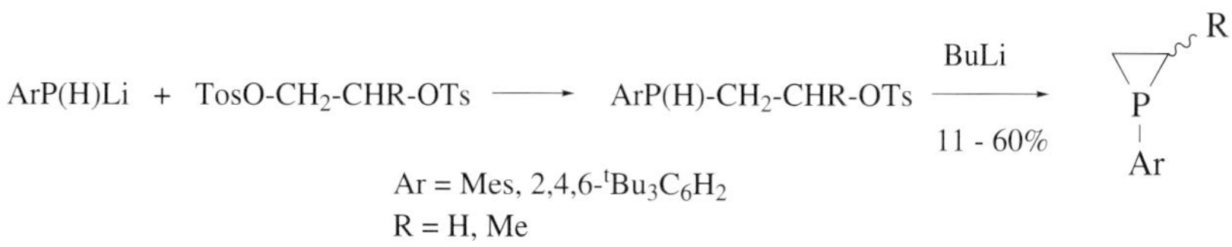

(94)

This technique was later optimised by Gaspar ⟨92JA8526, 96JOC7702⟩ and used for the synthesis of chiral phosphiranes from optically pure diols. A related chemistry was also used for the synthesis of parent phosphirane ⟨94ZN(B)1511⟩ (Equation (94a)).

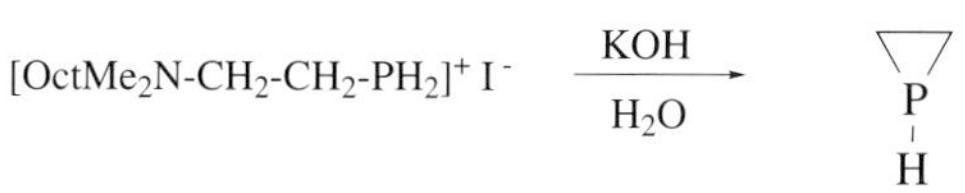

(94a)

The availability of a 7-phosphanorbornadiene precursor of the transient terminal phosphinidene complex [$ClCH_2CH_2P{=}W(CO)_5$] (see Equation (75)), also allowed the development of similar routes in the coordination sphere of tungsten. The first one ⟨86OM1159⟩ involves an initial insertion of the phosphinidene into one of the C–H bonds of ferrocene (Equation (95)).

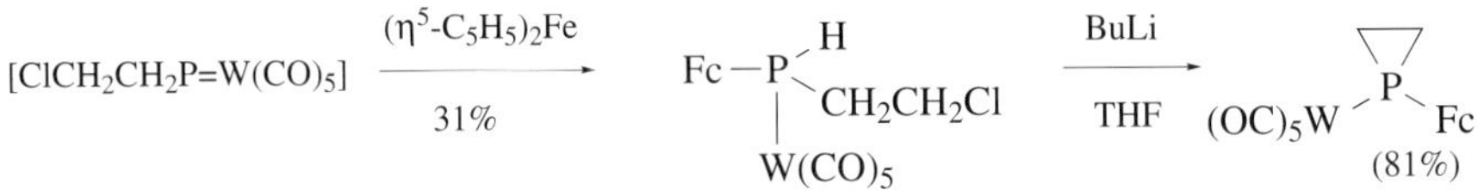
(95)

The second includes an initial reaction with aniline ⟨89POL2671⟩ (Equation (96)).

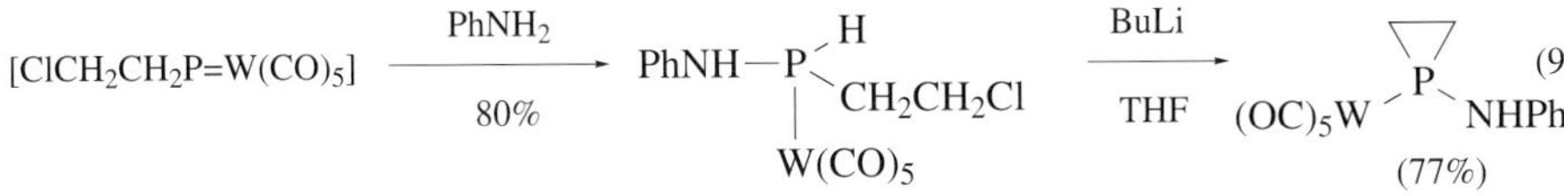
(96)

The third one involves the initial formation of a "phospha-Wittig" reagent (to transform the reactivity of the phosphinidene from electrophilic to nucleophilic) which undergoes a self-alkylation ⟨91SL743⟩ (Equation (97)).

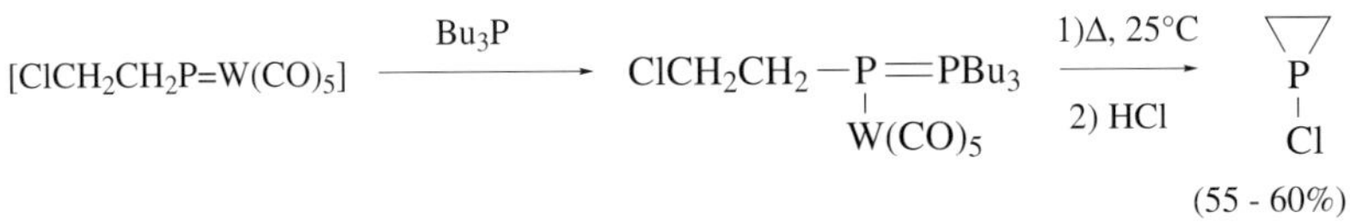
(97)

At the moment, this scheme represents the only synthesis of the parent 1-chlorophosphirane.

2.1.4.4 Synthesis by Cyclisation of a C–P–C Unit

This approach was initially developed by Quast ⟨78AG(E)867, 81LA977⟩ for the synthesis of the first stable, sterically hindered, phosphirane oxide (**29**) (Equation (98)).

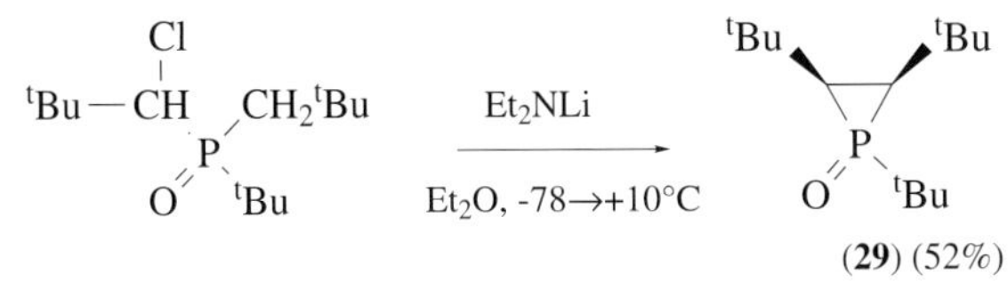
(98)

Subsequently, Oshikawa ⟨86BCJ3293⟩ reported a similar chemistry (Equation (99)).

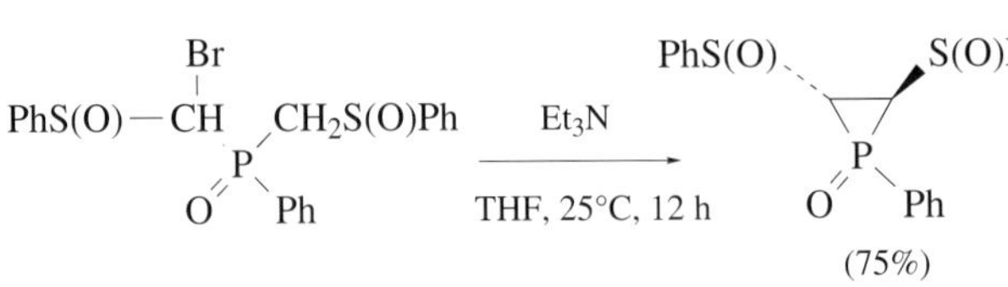
(99)

Curiously, the stereochemistries of these two phosphirane oxides are different.

As already obvious from Equation (83), another potential route to phosphiranes involves the cyclisation of bis-ylids. The thermal cyclisation of such species has been studied in some depth by Niecke ⟨92CB771⟩ (Equation (100)).

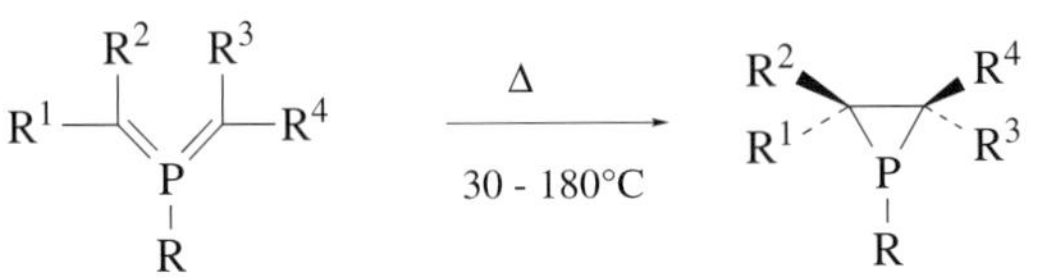
(100)

R = ^{t}Bu, Cp*, Ph, Et_2N, $^iPr_2N$
R^1- R^4= H, Ph, $SiMe_3$...

This conrotatory cyclisation does not work for highly hindered bis-ylids, e.g. R = Et_2N, $R^1 = R^2 =$ Ph, $R^3 = R^4 = SiMe_3$. Some time earlier, Fluck ⟨86CZ(110)379, 89PS(44)291⟩ had also described some applications of this scheme, e.g. (Equation (101)).

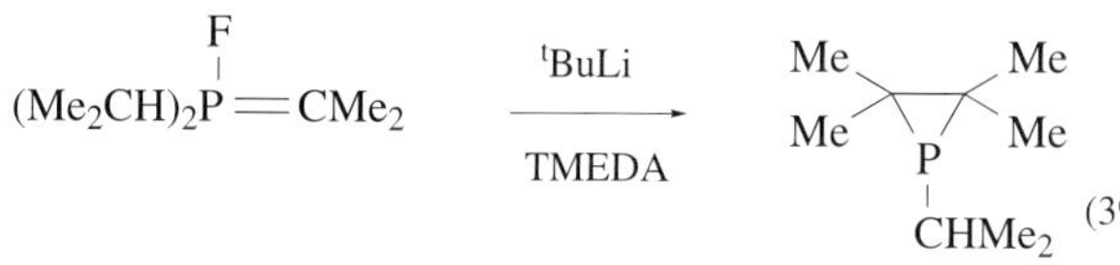

(101)

A similar chemistry was also described by Kolodyazhnyi ⟨92ZOB464, 92ZOB1418⟩ (Equations (102), (103)).

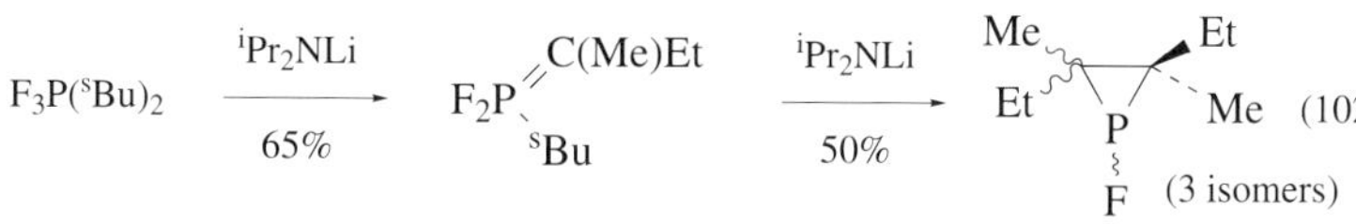

(102)

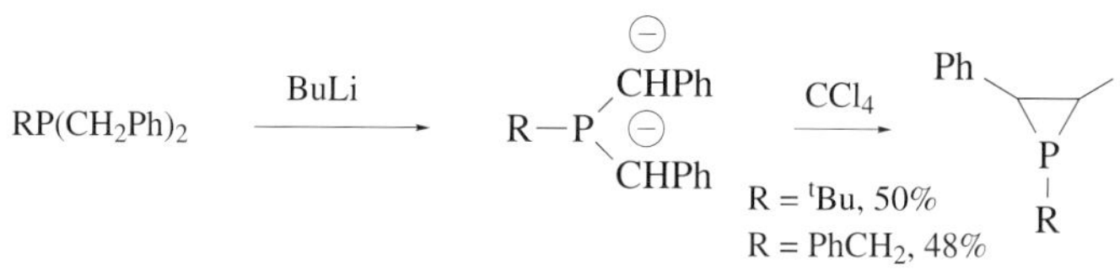

(103)

Finally, Niecke has shown that the controlled metallation of a gem-dichloro-bis-ylid provides another access to phosphirenes ⟨95AG(E)1849⟩ (Equation (103a)).

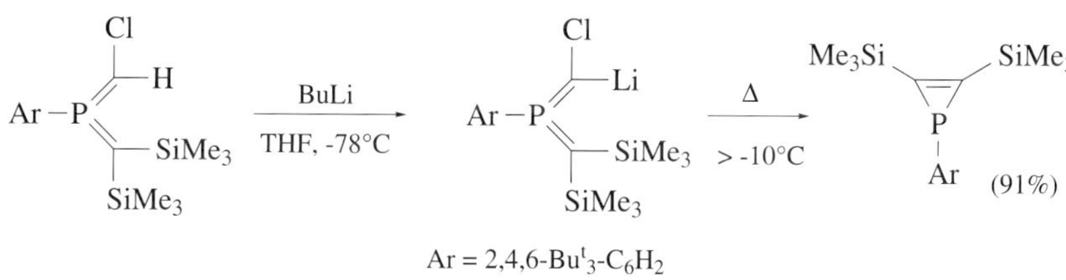

(103a)

2.1.4.5 Specific Syntheses of Polycyclic Phosphiranes

The first specific synthesis of a polycyclic phosphirane was depicted by Katz in 1966 ⟨66JA3832, 73JA4292⟩, soon after the discovery of phosphirane itself by Wagner. It involves the reaction of cyclooctatetraene dianion with dihalophosphines ⟨82TL4915, 85CB97, 87TL1077, 89CC973⟩ (Equation (104)).

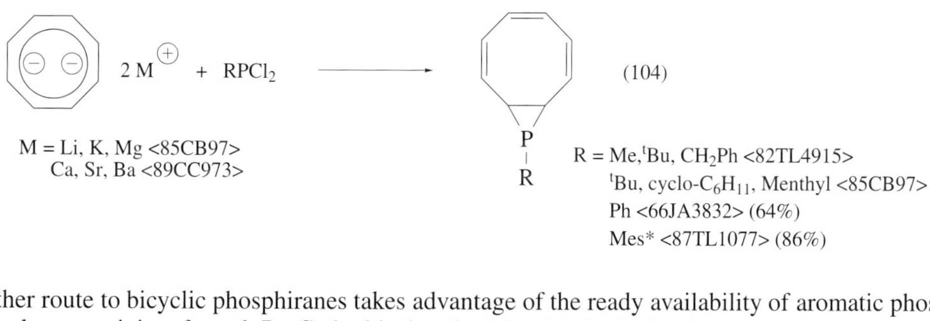

(104)

Another route to bicyclic phosphiranes takes advantage of the ready availability of aromatic phosphorus heterocycles containing formal P=C double bonds. This kind of chemistry was first exemplified by Arbuzov and Dianova ⟨77IZV1453, 79DOK(244)117, 82IZV2730, 83DOK(273)613, 87CB597⟩ (Equation (105)). The reaction proceeds via an intermediate [2 + 3] cycloadduct ⟨81IZV1113, 83IZV2395⟩. A similar chemistry was performed by Märkl with the phosphinine ring, either in the free or the complexed state ⟨87AG(E)236, 87AG(E)1134⟩ (Equation (106)).

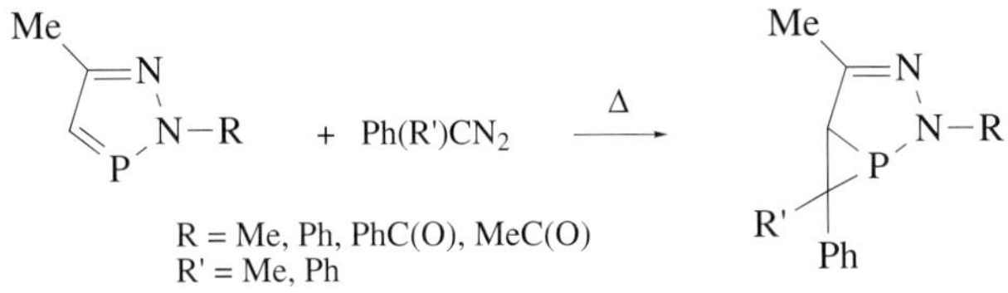

(105)

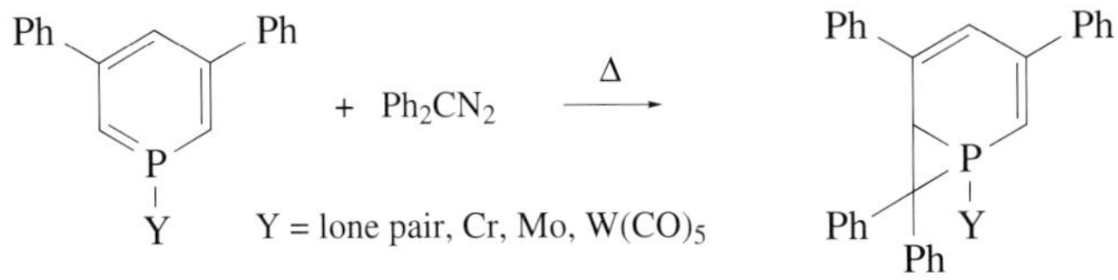

(106)

The reaction does not give these 1-phosphanorcaradienes when CH_2N_2, $MeCHN_2$ or Me_2CN_2 are used as the carbene precursors. An excess of Ph_2CN_2 leads to a complicated rearrangement ⟨90TL4849⟩ (Equation (107)).

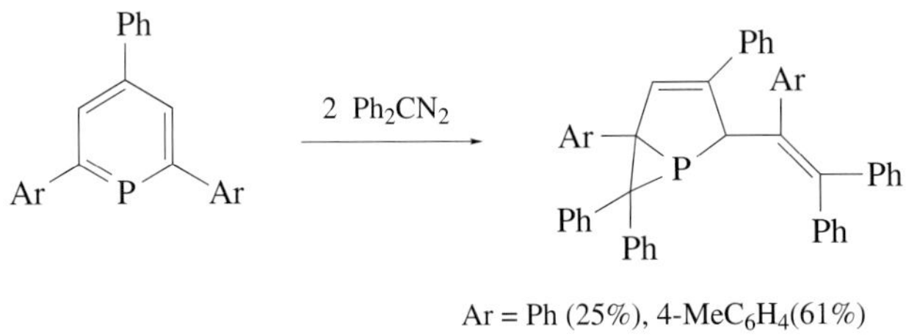

(107)

The structure of one bis-adduct has been confirmed by X-ray analysis. The proposed mechanism involves the isomerisation of the initial 1-phosphanorcaradiene into a 2*H*-phosphepine (Equation (56)), followed by a second addition of carbene onto the P=C double bond of the 7-membered ring. A ring-contraction via a formal [1,3] shift then affords the observed product. With 3,5-diphenylphosphinine, the reaction goes one step further to give the tris-adduct (Equation (108)).

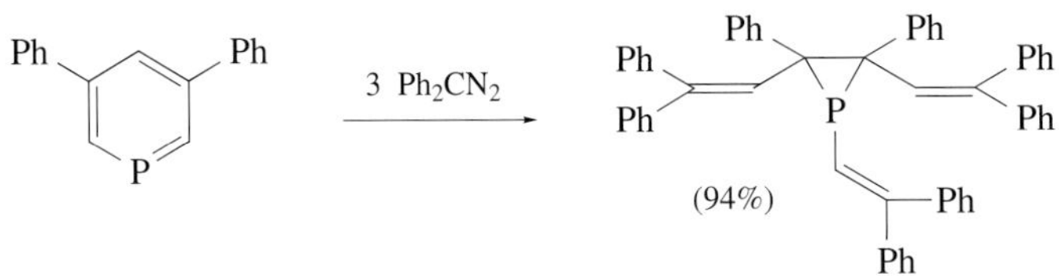

(108)

A series of rearrangements leading to polycyclic phosphiranes has been described ⟨77JA8511, 78H(9)98, 80JOC4683, 87AG(E)85, 87AG(E)1011, 95BSF652⟩ (Equations (109)–(111)).

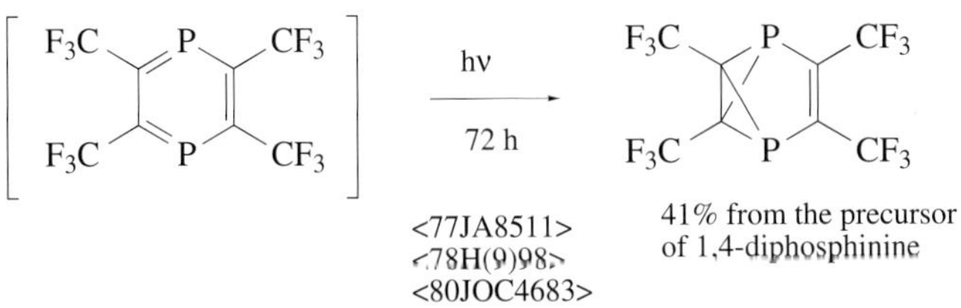

(109)

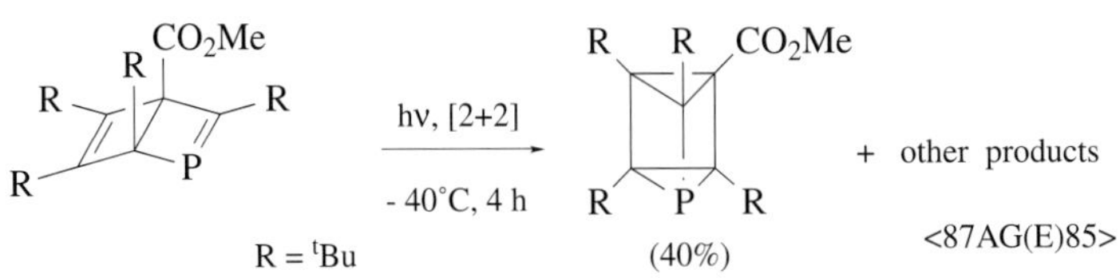

(110)

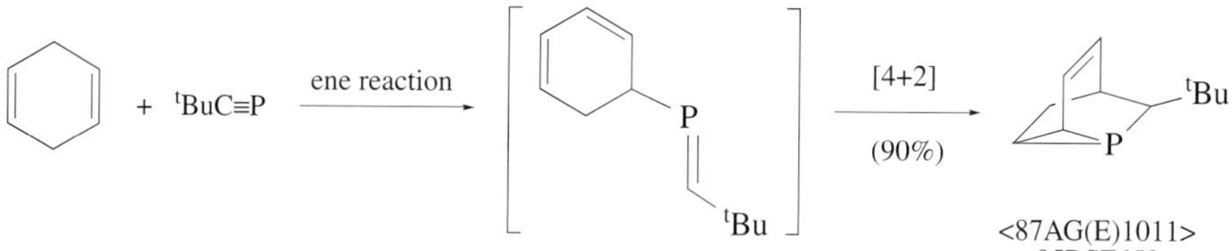

(111)

<87AG(E)1011>
<95BSF652>

The well-known synthesis of zirconacyclopentadienes from zirconocene and alkynes has been applied to bis-alkynyl-phosphines ⟨96PS(109)185, 97OM3086⟩ (Equation (112)).

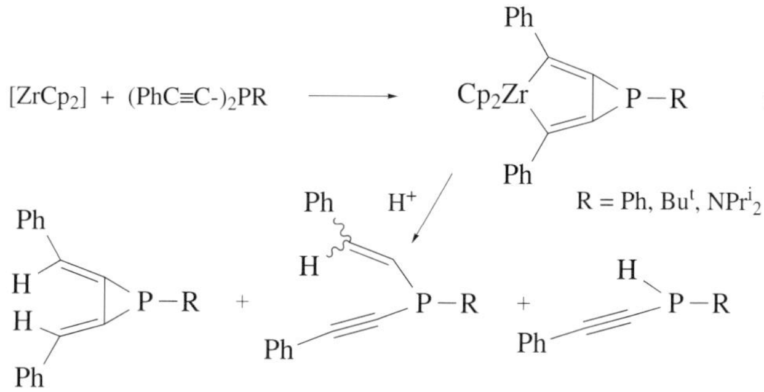

(112)

The bicyclic zirconacyclopentadiene-phosphiranes thus formed can lead to a variety of products via further reactions of the Zr–C bonds. Protonation gives, inter alia, bis-alkylidene-phosphiranes.

Finally, the very interesting *BABAR*-Phos has been synthesised via a formal intramolecular condensation of a phosphinidene with a proximal C=C double bond ⟨99AG(E)1623⟩ (Equation (113)).

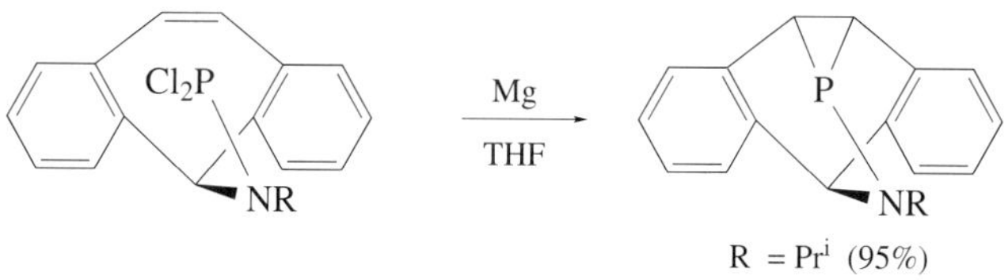

(113)

This polycyclic phosphirane displays outstanding stability and can be used in the platinum-catalysed hydrosilylation of olefins.

2.1.5 REFERENCES

63USP3086053 R.I. Wagner; *US Pat.* 3 086 053 (1963) (*Chem. Abstr.*, 1963, **59**, 10 124).
63USP3086056 R.I. Wagner; *US Pat.* 3 086 053 (1963) (*Chem. Abstr.*, 1964, **60**, 559).
66JA3832 T.J. Katz, C.R. Nicholson and C.A. Reilly; *J. Am. Chem. Soc.*, 1966, **88**, 3832.
67JA1102 R.I. Wagner, L.V.D. Freeman, H. Goldwhite and D.G. Rowsell; *J. Am. Chem. Soc.*, 1967, **89**, 1102.
69JA17 M.T. Bowers, R.A. Beaudet, H. Goldwhite and R. Tang; *J. Am. Chem. Soc.*, 1969, **91**, 17.
69SA(A)819 R.W. Mitchell, L.J. Kuzma, R.J. Pirkle and J.A. Merritt; *Spectrochim. Acta, Part A*, 1969, **25**, 819.
69T1097 S. Chan, H. Goldwhite, H. Keyzer, D.G. Rowsell and R. Tang; *Tetrahedron*, 1969, **25**, 1097.
70JCP2831 M.T. Bowers, R.A. Beaudet, H. Goldwhite and S. Chan; *J. Chem. Phys.*, 1970, **52**, 2831.
70SA(A)249 S. Chan, H. Goldwhite, H. Keyzer and R. Tang; *Spectrochim. Acta, Part A*, 1970, **26**, 249.
71AG(E)125 R. Bausch, E.A.V. Ebsworth and D.W.H. Rankin; *Angew. Chem., Int. Ed. Engl.*, 1971, **10**, 125.
72CC1085 E.W. Koos, J.P. Vander Kooi, E.E. Green and J.K. Stille; *J. Chem. Soc., Chem. Commun.*, 1972, 1085.
73JA4292 E.W. Turnblom and T.J. Katz; *J. Am. Chem. Soc.*, 1973, **95**, 4292.
73JST(18)75 J. Bragin and L.W. Dennis; *J. Mol. Struct.*, 1973, **18**, 75.
74JA317 D.B. Denney and L.S. Shih; *J. Am. Chem. Soc.*, 1974, **96**, 317.
75AG(E)363 E. Niecke and W. Flick; *Angew. Chem., Int. Ed. Engl.*, 1975, **14**, 363.
75ZN(A)1470 Z.-C. Profous, K.-P. Wanczek and H. Hartmann; *Z. Naturforsch., A*, 1975, **30**, 1470.
76JA2109 G.A. Gray, S.E. Cremer and K.L. Marsi; *J. Am. Chem. Soc.*, 1976, **98**, 2109.
77IZV1453 B.A. Arbuzov and E.N. Dianova; *Izv. Akad. Nauk. SSSR, Ser. Khim.*, 1977, 1453.
77JA8511 Y. Kobayashi, S. Fujino, H. Hamana, I. Kumadaki and Y. Hanzawa; *J. Am. Chem. Soc.*, 1977, **99**, 8511.
78AG(E)867 H. Quast and M. Heuschmann; *Angew. Chem., Int. Ed. Engl.*, 1978, **17**, 867.
78CC854 B.C. Campbell, D.B. Denney, D.Z. Denney and L.S. Shih; *J. Chem. Soc., Chem. Commun.*, 1978, 854.

78H(9)98 Y. Kobayashi, S. Fujino, H. Hamana, I. Kumadaki and Y. Hanzawa; *Heterocycles,* 1978, **9**, 98.
79CC390 H. Quast and M. Heuschmann; *J. Chem. Soc., Chem. Commun.,* 1979, 390.
79DOK(244)117 B.A. Arbuzov, E.N. Dianova and Y.Y. Samitov; *Dokl. Akad. Nauk SSSR,* 1979, **244**, 117.
80JOC4683 Y. Kobayashi, S. Fujino, H. Hamana, Y. Hanzawa, S. Morita and I. Kumadaki; *J. Org. Chem.,* 1980, **45**, 4683.
81AG(E)131 E. Niecke, W.W. Schoeller and D.-A. Wildbredt; *Angew. Chem., Int. Ed. Engl.,* 1981, **20**, 131.
81IZV1113 B.A. Arbuzov, E.N. Dianova and S.M. Sharipova; *Izv. Akad. Nauk. SSSR, Ser. Khim.,* 1981, 1113.
81LA977 H. Quast and M. Heuschmann; *Liebigs Ann. Chem.,* 1981, 977.
82AG(E)292 W.J. Richter; *Angew. Chem., Int. Ed. Engl.,* 1982, **21**, 292.
82CB901 H. Quast and M. Heuschmann; *Chem. Ber.,* 1982, **115**, 901.
82CC667 A. Marinetti, F. Mathey, J. Fischer and A. Mitschler; 1982, 667.
82IZV2730 B.A. Arbuzov, E.N. Dianova and Y.Y. Samitov; *Izv. Akad. Nauk. SSSR, Ser. Khim.,* 1982, 2730.
82JA4484 A. Marinetti, F. Mathey, J. Fischer and A. Mitschler; *J. Am. Chem. Soc.,* 1982, **104**, 4484.
82TL4915 G. Märkl and B. Alig; *Tetrahedron Lett.,* 1982, **23**, 4915.
83CB3293 W.J. Richter; *Chem. Ber.,* 1983, **116**, 3293.
83CC1171 E. Niecke, M. Leuer, D.-A. Wildbredt and W.W. Schoeller; *J. Chem. Soc., Chem. Commun.,* 1983, 1171.
83DOK(273)613 B.A. Arbuzov, E.N. Dianova and E. Ya. Zabotina; *Dokl. Akad. Nauk. SSSR,* 1983, **273**, 613.
83IZV2395 B.A. Arbuzov, A.F. Lisin and E.N. Dianova; *Izv. Akad. Nauk. SSSR, Ser. Khim.,* 1983, 2395.
83JA5960 N.S. Rao and L.D. Quin; *J. Am. Chem. Soc.,* 1983, **105**, 5960.
83OMR494 H. Goldwhite, D. Rowsell, L.E. Vertal, M.T. Bowers, M.A. Cooper and S.L. Manatt; *Org. Magn. Reson.,* 1983, **21**, 494.
83TL643 K.S. Fongers, H. Hogeveen and R.F. Kingma; *Tetrahedron Lett.,* 1983, **24**, 643.
83TL3981 G. Märkl and B. Alig; *Tetrahedron Lett.,* 1983, **24**, 3981.
84CC45 A. Marinetti, F. Mathey, J. Fischer and A. Mitschler; *J. Chem. Soc., Chem. Commun.,* 1984, 45.
84OM456 A. Marinetti and F. Mathey; *Organometallics,* 1984, **3**, 456.
84OM1492 A. Marinetti and F. Mathey; *Organometallics,* 1984, **3**, 1492.
84T3273 R. Benn, R. Mynott, W.J. Richter and G. Schroth; *Tetrahedron,* 1984, **40**, 3273.
84TL1345 R. Breslow and L.A. Deuring; *Tetrahedron Lett.,* 1984, **25**, 1345.
85CB97 W.J. Richter; *Chem. Ber.,* 1985, **118**, 97.
85CB1575 W.J. Richter; *Chem. Ber.,* 1985, **118**, 1575.
85CB3419 R. Appel and C. Casser; *Chem. Ber.,* 1985, **118**, 3419.
85CB4285 M. Baudler and J. Germeshausen; *Chem. Ber.,* 1985, **118**, 4285.
85CL441 M. Yoshifuji, K. Toyota and N. Inamoto; *Chem. Lett.,* 1985, 441.
85IC4133 D. Gonbeau, G. Pfister-Guillouzo, A. Marinetti and F. Mathey; *Inorg. Chem.,* 1985, **24**, 4133.
85JA4700 A. Marinetti and F. Mathey; *J. Am. Chem. Soc.,* 1985, **107**, 4700.
85JA5001 A. Marinetti, J. Fischer and F. Mathey; *J. Am. Chem. Soc.,* 1985, **107**, 5001.
85NJC71 D. Gonbeau and G. Pfister-Guillouzo; *Nouv. J. Chim.,* 1985, **9**, 71.
85SYN290 T. Oshikawa and M. Yamashita; *Synthesis,* 1985, 290.
85TL4595 B. Deschamps and F. Mathey; *Tetrahedron Lett.,* 1985, **26**, 4595.
86BCJ3293 T. Oshikawa and M. Yamashita; *Bull. Chem. Soc. Jpn.,* 1986, **59**, 3293.
86CB1977 R. Appel, T. Gaitzsch, F. Knoch and G. Lenz; *Chem. Ber.,* 1986, **119**, 1977.
86CC762 D. Carmichael, P.B. Hitchcock, J.F. Nixon, F. Mathey and A. Pidcock; *J. Chem. Soc., Chem. Commun.,* 1986, 762.
86CZ(110)379 R. Braun and E. Fluck; *Chem.-Ztg.,* 1986, **110**, 379.
86JA4519 L.D. Quin, N.S. Rao, R.J. Topping and A.T. McPhail; *J. Am. Chem. Soc.,* 1986, **108**, 4519.
86JOM(306)29 J. Borm, G. Huttner and O. Orama; *J. Organomet. Chem.,* 1986, **306**, 29.
86NJC321 J.-M. Alcaraz, J. Svara and F. Mathey; *New J. Chem.,* 1986, **10**, 321.
86OM1159 J. Svara and F. Mathey; *Organometallics,* 1986, **5**, 1159.
86OM1161 J. Svara, A. Marinetti and F. Mathey; *Organometallics,* 1986, **5**, 1161.
86PS(27)109 L.D. Quin, N.S. Rao and J. Szewczyk; *Phosphorus Sulfur,* 1986, **27**, 109.
86TL1323 F. Mercier and F. Mathey; *Tetrahedron Lett.,* 1986, **27**, 1323.
86TL2635 S. Lochschmidt, F. Mathey and A. Schmidpeter; *Tetrahedron Lett.,* 1986, **27**, 2635.
87AG(E)85 K. Blatter, W. Rösch, U.-J. Vogelbacher, J. Fink and M. Regitz; *Angew. Chem., Int. Ed. Engl.,* 1987, **26**, 85.
87AG(E)236 G. Märkl, H.J. Beckh, K.K. Mayer, M.L. Ziegler and T. Zahn; *Angew. Chem., Int. Ed. Engl.,* 1987, **26**, 236.
87AG(E)548 L. Ricard, N. Maigrot, C. Charrier and F. Mathey; *Angew. Chem., Int. Ed. Engl.,* 1987, **26**, 548.
87AG(E)1011 E.P.O. Fuchs, W. Rösch and M. Regitz; *Angew. Chem., Int. Ed. Engl.,* 1987, **26**, 1011.
87AG(E)1134 G. Märkl, H.J. Beckh, M.L. Ziegler and B. Nuber; *Angew. Chem., Int. Ed. Engl.,* 1987, **26**, 1134.
87AG(E)1257 O. Wagner, G. Maas and M. Regitz; *Angew. Chem., Int. Ed. Engl.,* 1987, **26**, 1257.
87CB597 B.A. Arbuzov, E.N. Dianova, R.T. Galiaskarova and A. Schmidpeter; *Chem. Ber.,* 1987, **120**, 597.
87CPL(136)224 A.L. Barra and J.B. Robert; *Chem. Phys. Lett.,* 1987, **136**, 224.
87IC1799 D. Gonbeau and G. Pfister-Guillouzo; *Inorg. Chem.,* 1987, **26**, 1799.
87JST(152)69 C.F. Rodriguez and A.C. Hopkinson; *J. Mol. Struct.,* 1987, **152**, 69.
87OM586 C. Charrier, N. Maigrot and F. Mathey; *Organometallics,* 1987, **6**, 586.
87OM2189 A. Marinetti and F. Mathey; *Organometallics,* 1987, **6**, 2189.
87TL1077 L.D. Quin, E.-Y. Yao and J. Szewczyk; *Tetrahedron Lett.,* 1987, **28**, 1077.
87TL2693 G. Märkl, W. Hölzl and I. Trötsch-Schaller; *Tetrahedron Lett.,* 1987, **28**, 2693.

87TL5021 A. Marinetti and F. Mathey; *Tetrahedron Lett.*, 1987, **28**, 5021.
87ZOB2482 M.A. Kazankova, I.L. Rodionov, I.F. Lutsenko, A.N. Chernega, M.Yu. Antipin and Yu.T. Struchkov; *Zh. Obshch. Khim.*, 1987, **57**, 2482.
88CC462 J.M. Vural, S.A. Weissman, S.G. Baxter, A.H. Cowley and C.M. Nunn; *J. Chem. Soc., Chem. Commun.*, 1988, 462.
88CC1156 S.S. Al-Juaid, D. Carmichael, P.B. Hitchcock, S. Lochschmidt, A. Marinetti, F. Mathey and J.F. Nixon; *J. Chem. Soc., Chem. Commun.*, 1988, 1156.
88CPL(149)363 A.L. Barra and J.B. Robert; *Chem. Phys. Lett.*, 1988, **149**, 343.
88JOM(354)83 B. Deschamps and F. Mathey; *J. Organomet. Chem.*, 1988, **354**, 83.
88JPC4886 K.D. Dobbs, J.E. Boggs, A.R. Barron and A.H. Cowley; *J. Phys. Chem.*, 1988, **92**, 4886.
88NJC755 B. Deschamps and F. Mathey; *New J. Chem.*, 1988, **12**,755.
88OM1796 S. Holand and F. Mathey; *Organometallics,* 1988, **7**, 1796.
88OM1875 K. Lammertsma, P. Chand, S.-W. Yang and J.-T. Hung; *Organometallics,* 1988, **7**, 1875.
88TL605 E. Niecke and M. Lysek; *Tetrahedron Lett.*, 1988, **29**, 605.
88ZN(B)1285 W. Schnurr and M. Regitz; *Z. Naturforsch., B,* 1988, **43**, 1285.
89AG(E)225 O. Wagner, M. Ehle and M. Regitz; *Angew. Chem., Int. Ed. Engl.,* 1989, **28**, 225.
89AG(E)315 G. Fritz, T. Vaahs, H. Fleischer and E. Matern; *Angew. Chem., Int. Ed. Engl.,* 1989, **28**, 315.
89AG(E)900 D. Hu, H. Schaüfele, H. Pritzkow and U. Zenneck; *Angew. Chem., Int. Ed. Engl.,* 1989, **28**, 900.
89CB453 E. Niecke, M. Leuer and M. Nieger; *Chem. Ber.,* 1989, **122**, 453.
89CC973 D.S. Hutchings, P.C. Junk, W.C. Patalinghug, C.L. Raston and A.H. White; *J. Chem. Soc., Chem. Commun.,* 1989, 973.
89CC1389 D. Carmichael, P.B. Hitchcock, J.F. Nixon, F. Mathey and L. Ricard; *J. Chem. Soc., Chem. Commun.,* 1989, 1389.
89CPL(163)349 R.G.A.R. Maclagan; *Chem. Phys. Lett.*, 1989, **163**, 349.
89JA9098 F. Mercier, B. Deschamps and F. Mathey; *J. Am. Chem. Soc.,* 1989, **111**, 9098.
89JPC3025 J.A. Boatz and M.S. Gordon; *J. Phys. Chem.,* 1989, **93**, 3025.
89JPC7780 S.M. Bachrach; *J. Phys. Chem.,* 1989, **93**, 7780.
89POL2671 B. Deschamps, L. Ricard and F. Mathey; *Polyhedron,* 1989, **8**, 2671.
89PS(44)291 E. Fluck and R. Braun; *Phosphorus Sulfur Silicon,* 1989, **44**, 291.
89T3061 A. Marinetti and F. Mathey; *Tetrahedron,* 1989, **45**, 3061.
89TL3951 W. Schnurr and M. Regitz; *Tetrahedron Lett.,* 1989, **30**, 3951.
89ZAAC(570)54 G. Fritz, T. Vaahs, H. Fleischer and E. Matern; *Z. Anorg. Allg. Chem.,* 1989, **570**, 54.
90AG(E)1436 S. Haber, R. Boese and M. Regitz; *Angew. Chem., Int. Ed. Engl.,* 1990, **29**, 1436.
90CL827 M. Yoshifuji, H. Yoshimura and K. Toyota; *Chem. Lett.*, 1990, 827.
90CRV997 F. Mathey; *Chem. Rev.,* 1990, **90**, 997.
90PS(47)477 N.H. Tran Huy and F. Mathey; *Phosphorus Sulfur Silicon,* 1990, **47**, 477.
90TL819 S.A. Weissman and S.G. Baxter; *Tetrahedron Lett.,* 1990, **31**, 819.
90TL3429 M. Ehle, O. Wagner, U. Bergsträsser and M. Regitz; *Tetrahedron Lett.,* 1990, **31**, 3429.
90TL4849 G. Märkl, K. Hohenwarter, M.L. Ziegler and B. Nuber; *Tetrahedron Lett.,* 1990, **31**, 4849.
91CB933 A. Maercker and W. Brieden; *Chem. Ber.,* 1991, **124**, 933.
91CB1207 O. Wagner, M. Ehle, M. Birkel, J. Hoffmann and M. Regitz; *Chem. Ber.,* 1991, **124**, 1207.
91CC124 M. Yoshifuji, K. Toyota, H. Yoshimura, K. Hirotsu and A. Okamoto; *J. Chem. Soc., Chem. Commun.,* 1991, 124.
91JCS(D)905 S.S. Al-Juaid, D. Carmichael, P.B. Hitchcock, A. Marinetti, F. Mathey and J.F. Nixon; *J. Chem. Soc., Dalton Trans.,* 1991, 905.
91JOC2205 S.M. Bachrach; *J. Org. Chem.,* 1991, **56**, 2205.
91OM3949 H. Brombach, E. Niecke and M. Nieger; *Organometallics,* 1991, **10**, 3949.
91SL433 H. Memmesheimer, J.R. Al-Dulayymi, M.S. Baird, T. Wettling and M. Regitz; *Synlett,* 1991, 433.
91SL743 P. Le Floch and F. Mathey; *Synlett,* 1991, 743.
91T6673 F. Nief and F. Mathey; *Tetrahedron,* 1991, **47**, 6673.
91TL6537 H.J. Metternich and E. Niecke; *Tetrahedron Lett.,* 1991, **32**, 6537.
91TL6879 K. Toyota, H. Yoshimura, T. Uesugi and M. Yoshifuji; *Tetrahedron Lett.,* 1991, **32**, 6879.
92CB771 P. Becker, H. Brombach, G. David, M. Leuer, H.-J. Metternich and E. Niecke; *Chem. Ber.,* 1992, **125**, 771.
92CB923 M.T. Nguyen, H. Vansweevelt and L.G. Vanquickenborne; *Chem. Ber.,* 1992, **125**, 923.
92CC750 F.A. Ajulu, D. Carmichael, P.B. Hitchcock, F. Mathey, M.F. Meidine, J.F. Nixon, L. Ricard and M.L. Riley; *J. Chem. Soc., Chem. Commun.,* 1992, 750.
92CC1799 S. Haber, P. Le Floch and F. Mathey; *J. Chem. Soc., Chem. Commun.,* 1992, 1799.
92JA8526 X. Li, D. Lei, M.Y. Chiang and P.P. Gaspar; *J. Am. Chem. Soc.,* 1992, **114**, 8526.
92JOC6557 K. Lammertsma, J.-T. Hung, P. Chand and G.M. Gray; *J. Org. Chem.,* 1992, **57**, 6557.
92JST(255)207 S.M. Bachrach; *J. Mol. Struct.,* 1992, **255**, 207.
92OM1438 N.H. Tran Huy, A. Marinetti, L. Ricard and F. Mathey; *Organometallics,* 1992, **11**, 1438.
92OM4365 J.-T. Hung and K. Lammertsma; *Organometallics,* 1992, **11**, 4365.
92POL385 I. Miyahara, A. Hayashi, K. Hirotsu, M. Yoshifuji, H. Yoshimura and K. Toyota; *Polyhedron,* 1992, **11**, 385.
92SYN157 A. Marinetti, L. Ricard and F. Mathey; *Synthesis,* 1992, 157.
92SL635 H. Memmesheimer, U. Bergsträsser, J. Hoffmann, M. Baird and M. Regitz; *Synlett,* 1992, 635.
92TL2335 G. David, E. Niecke and M. Nieger; *Tetrahedron Lett.,* 1992, **33**, 2335.

92ZOB464	O.I. Kolodyazhnyi and S.N. Ustenko; *Zh. Obshch. Khim.*, 1992, **62**, 464.
92ZOB1418	O.I. Kolodyazhnyi and S.N. Ustenko; *Zh. Obshch. Khim.*, 1992, **62**, 1418.
93BSF521	M.L. Sierra, C. Charrier, L. Ricard and F. Mathey; *Bull. Soc. Chim. Fr.*, 1993, **130**, 521.
93BSF636	W.W. Schoeller, W. Haug and J. Strutwolf; *Bull. Soc. Chim. Fr.*, 1993, **130**, 636.
93BSF695	A. Espinosa Ferao, B. Deschamps and F. Mathey; *Bull. Soc. Chim. Fr.*, 1993, **130**, 695.
93CB649	D. Barion, G. David, M. Link, M. Nieger and E. Niecke; *Chem. Ber.*, 1993, **126**, 649.
93CC142	F.A. Ajulu, P.B. Hitchcock, F. Mathey, R.A. Michelin, J.F. Nixon and A.J.L. Pombeiro; *J. Chem. Soc., Chem. Commun.*, 1993, 142.
93JCS(D)1811	D. Carmichael, P.B. Hitchcock, J.F. Nixon, F. Mathey and L. Ricard; *J. Chem. Soc., Dalton Trans.*, 1993, 1811.
93JOC1379	P. Chaquin and A. Gherbi; *J. Org. Chem.*, 1993, **58**, 1379.
93JOC1800	J.-T. Hung and K. Lammertsma; *J. Org. Chem.*, 1993, **58**, 1800.
93JOC6786	J.-T. Hung, S.-W. Yang, G.M. Gray and K. Lammertsma; *J. Org. Chem.*, 1993, **58**, 6786.
93MC68	A.D. Averin, N.V. Lukashev, M.A. Kazankova and I.P. Beletskaya; *Mendeleev Commun.*, 1993, 68.
93OM1207	A. Marinetti, F. Mathey and L. Ricard; *Organometallics,* 1993, **12**, 1207.
93OM1401	J.-T. Hung, P. Chand, F.R. Fronczek, S.F. Watkins and K. Lammertsma; *Organometallics,* 1993, **12**, 1401.
93OM4653	L. Weber, A. Rühlicke, H.-G. Stammler and B. Neumann; *Organometallics,* 1993, **12**, 4653.
93PS(75)225	S. Haber, P. Le Floch and F. Mathey; *Phosphorus Sulfur Silicon,* 1993, **75**, 225.
93TL1331	N.V. Lukashev, P.E. Zhichkin, M.A. Kazankova and I.P. Beletskaya; *Tetrahedron Lett.*, 1993, **34**, 1331.
94AG(E)2330	A.G. Avent, F.G.N. Cloke, K.R. Flower, P.B. Hitchcock, J.F. Nixon and D.M. Vickers; *Angew. Chem. Int. Ed. Engl.*, 1994, **33**, 2330.
94AG(E)2427	R. Streubel, A. Kusenberg, J. Jeske and P.G. Jones; *Angew. Chem. Int. Ed. Engl.*, 1994, **33**, 2427.
94CB313	M. Link, E. Niecke and M. Nieger; *Chem. Ber.*, 1994, **127**, 313.
94CC475	Y.B. Kang, M. Pabel, A.C. Willis and S.B. Wild; *J. Chem. Soc., Chem. Commun.*, 1994, 475.
94IC1153	M.T. Nguyen, E. Van Praet and L.G. Vanquickenborne; *Inorg. Chem.*, 1994, **33**, 1153.
94JA7899	X. Li, S.I. Weissman, T.-S. Lin, P.P. Gaspar, A.H. Cowley and A.I. Smirnov; *J. Am. Chem. Soc.*, 1994, **116**, 7899.
94JA9407	K.K. Laali, B. Geissler, O. Wagner, J. Hoffmann, R. Armbrust, W. Eisfeld and M. Regitz; *J. Am. Chem. Soc.*, 1994, **116**, 9407.
94JA10966	J.-T. Hung, S.-W. Yang, P. Chand, G.M. Gray and K. Lammertsma; *J. Am. Chem. Soc.*, 1994, **116**, 10966.
94JCS(F)1771	M.T. Nguyen, L. Landuyt and L.G. Vanquickenborne; *J. Chem. Soc., Faraday Trans.*, 1994, **90**, 1771.
94JPC6481	E.P.F. Lee, L. Nyulaszi and T. Veszprémi; *J. Phys. Chem.*, 1994, **98**, 6481.
94MI195	M. Yoshifuji, M. Shibata, K. Toyota, I. Miyahara and K. Hirotsu; *Heteroatom Chem.*, 1994, **5**, 195.
94MI567	S. Kobayashi and J.-I. Kadokawa; *Macromol. Rapid Commun.*, 1994, **15**, 567.
94OM640	D.J. Berger, P.P. Gaspar, R.S. Grev and F. Mathey; *Organometallics,* 1994, **13**, 640.
94RHA61	H. Memmesheimer and M. Regitz; *Rev. Heteroatom. Chem.*, 1994, **10**, 61.
94ZN(B)1511	D.J. Brauer, J. Fischer, S. Kucken, K.P. Langhans, O. Stelzer and N. Weferling; *Z. Naturforsch.*, **49b**, 1511.
95AG(E)1849	E. Niecke, P. Becker, M. Nieger, D. Stalke and W.W. Schoeller; *Angew. Chem. Int. Ed. Engl.*, 1995, **34**, 1849.
95BSF652	H. Heydt, U. Bergsträsser, R. Fässler, E. Fuchs, N. Kamel, T. Mackewitz, G. Michels, W. Rösch, M. Regitz, P. Mazerolles, C. Laurent and A. Faucher; *Bull. Soc. Chim. Fr.*, 1995, **132**, 652.
95CC25	B. Manz and G. Maas; *J. Chem. Soc., Chem. Commun.*, 1995, 25.
95CC2507	A. Ostrowski, J. Jeske, P.G. Jones and R. Streubel; *J. Chem. Soc., Chem. Commun.*, 1995, 2507.
95CC257	D.C.R. Hockless, M.A. McDonald, M. Pabel and S.B. Wild; *J. Chem. Soc., Chem. Commun.*, 1995, 257.
95H(40)311	T. Kobayashi, H. Minemura and H. Kato; *Heterocycles,* 1995, **40**, 311.
95JA10785	R. Armbrust, M. Sanchez, R. Réau, U. Bergsträsser, M. Regitz and G. Bertrand; *J. Am. Chem. Soc.*, 1995, **117**, 10785.
95JA11914	T.L. Breen and D.W. Stephan; *J. Am. Chem. Soc.*, 1995, **117**, 11914.
95JOC3723	P. Chaquin and A. Gherbi; *J. Org. Chem.*, 1995, **60**, 3723.
95JOM(489)1	J.-T. Hung and K. Lammertsma; *J. Organomet. Chem.*, 1995, **489**, 1.
95JST(341)13	P.B. Karadakov, J. Gerratt, D.L. Cooper and M. Raimondi; *J. Mol. Struct.*, 1995, **341**, 13.
95JST(343)49	W.W. Schoeller and U. Tubessing; *J. Mol. Struct.*, 1995, **343**, 49.
95OM3592	A.J. Arce, Y. De Sanctis, R. Machado, M.V. Capparelli, J. Manzur and A.J. Deeming; *Organometallics,* 1995, **14**, 3592.
95OM4048	N.H. Tran Huy, L. Ricard and F. Mathey; *Organometallics,* 1995, **14**, 4048.
95SL353	N.H. Tran Huy and F. Mathey; *Synlett,* 1995, 353.
95SYN941	B. Deschamps and F. Mathey; *Synthesis,* 1995, 941.
95ZOR400	A.D. Averin, N.V. Lukashev, A.A. Borisenko, M.A. Kazankova and I.P. Beletskaya; *Zh. Org. Khim.*, 1995, **31**, 400.
95ZOR404	A.D. Averin, N.V. Lukashev, A.A. Borisenko, M.A. Kazankova and I.P. Beletskaya; *Zh. Org. Khim.*, 1995, **31**, 404.
95ZOR495	A.D. Averin, N.V. Lukashev, A.A. Borisenko, M.A. Kazankova and I.P. Beletskaya; *Zh. Org. Khim.*, 1995, **31**, 495.
96JA1690	B. Wang, C.H. Lake and K. Lammertsma; *J. Am. Chem. Soc.*, 1996, **118**, 1690.
96JOC5840	A. Göller, H. Heydt and T. Clark; *J. Org. Chem.*, 1996, **61**, 5840.
96JOC7702	X. Li, K.D. Robinson and P.P. Gaspar; *J. Org. Chem.*, 1996, **61**, 7702.

96JST(371)69 A. Sevin, H. Chevreau and C. Dézarnaud-Dandine; *J. Mol. Struct.*, 1996, **371**, 69.

96OM1301 D.C.R. Hockless, Y.B. Kang, M.A. McDonald, M. Pabel, A.C. Willis and S.B. Wild; *Organometallics*, 1996, **15**, 1301.

96OM1597 K. Waschbüsch, P. Le Floch and F. Mathey; *Organometallics*, 1996, **15**, 1597.

96OM4904 D.J. Berger, P.P. Gaspar, P. Le Floch, F. Mathey and R.S. Grev; *Organometallics*, 1996, **15**, 4904.

96PS(109)185 J.-P. Majoral, A. Igau, M. Zablocka, A. Mahieu, N. Cenac and M.K. Pietrusiewicz; *Phosphorus Sulfur Silicon*, 1996, **109–110**, 185.

96PS(109)609 N.V. Lukashev, A.D. Averin, P.E. Zhichkin, M.A. Kazankova and I.P. Beletskaya; *Phosphorus Sulfur Silicon*, 1996, **109–110**, 609.

96T4709 E.J.P. Malar; *Tetrahedron*, 1996, **52**, 4709.

96ZOR433 A.D. Averin, N.V. Lukashev, A.A. Borisenko, M.A. Kazankova and I.P. Beletskaya; *Zh. Org. Khim.*, 1996, **32**, 433.

97AG(E)1492 R. Streubel, H. Wilkens, A. Ostrowski, C. Neumann, F. Ruthe and P.G. Jones; *Angew. Chem. Int. Ed. Engl.*, 1997, **36**, 1492.

97BSF1033 L. Colombet, A. Sevin and P. Chaquin; *Bull. Soc. Chim. Fr.*, 1997, **134**, 1033.

97CB711 H. Heydt, M. Ehle, S. Haber, J. Hoffmann, O. Wagner, A. Göller, T. Clark and M. Regitz; *Chem. Ber./Recueil*, 1997, **130**, 711.

97CB779 B. Manz, U. Bergsträsser, J. Kerth and G. Maas; *Chem. Ber./Recueil*, 1997, **130**, 779.

97CC1033 A. Göller and T. Clark; *Chem. Commun.*, 1997, 1033.

97JA8432 S. Krill, B. Wang, J.-T. Hung, C.J. Horan, G.M. Gray and K. Lammertsma; *J. Am. Chem. Soc.*, 1997, **119**, 8432.

97JOM(529)127 R. Pietschnig, E. Niecke, M. Nieger and K. Airola; *J. Organomet. Chem.*, 1997, **529**, 127.

97JOM(529)189 D.C.R. Hockless, M.A. McDonald, M. Pabel and S.B. Wild; *J. Organomet. Chem.*, 1997, **529**, 189.

97MI451 D.B. Chesnut, L.D. Quin and S.B. Wild; *Heteroatom Chem.*, 1997, **8**, 451.

97MI(2)255 A. Hinchliffe and H.J.S. Machado; *Electron. J. Theor. Chem.*, 1997, **2**, 255.

97OM1526 N. Mézailles, P.E. Fanwick and C.P. Kubiak; *Organometallics*, 1997, **16**, 1526.

97OM3086 A. Mahieu, Y. Miquel, A. Igau, B. Donnadieu and J.-P. Majoral; *Organometallics*, 1997, **16**, 3086.

97OM4501 N.H. Tran Huy, L. Ricard and F. Mathey; *Organometallics*, 1997, **16**, 4501.

97ZAAC(623)1897 A. Ostrowski, J. Jeske, F. Ruthe, P.G. Jones and R. Streubel; *Z. Anorg. Allg. Chem.*, 1997, **623**, 1897.

98CC867 J. Simon, U. Bergsträsser and M. Regitz; *Chem. Commun.*, 1998, 867.

98CC1177 M. Zablocka, Y. Miquel, A. Igau, J.-P. Majoral and A. Skowronska; *Chem. Commun.*, 1998, 1177.

98CC1529 H. Wilkens, F. Ruthe, P.G. Jones and R. Streubel; *Chem. Commun.*, 1998, 1529.

98CC1761 R. Streubel, H. Wilkens and P.G. Jones; *Chem. Commun.*, 1998, 1761.

98JA7063 T.I. Sølling, M.A. McDonald, S.B. Wild and L. Radom; *J. Am. Chem. Soc.*, 1998, **120**, 7063.

98JOC2814 W. Eisfeld and M. Regitz; *J. Org. Chem.*, 1998, **63**, 2814.

98JST(424)67 C. Dézarnaud-Dandine, H. Chevreau and A. Sevin; *J. Mol. Struct.*, 1998, **424**, 67.

98MI607 K. Tsuji, S. Sasaki and M. Yoshifuji; *Heteroatom. Chem.*, 1998, **9**, 607.

98OM893 J. Foerstner, A. Kakoschke, D. Stellfeldt, H. Butenschön and R. Wartchow; *Organometallics*, 1998, **17**, 893.

98OM2677 N. Mézailles, N. Avarvari, D. Bourissou, F. Mathey and P. Le Floch; *Organometallics*, 1998, **17**, 2677.

98SYN175 H. Heydt, J. Hoffmann, A. Göller, T. Clark and M. Regitz; *Synthesis*, 1998, 175.

99AG(E)1623 J. Liedtke, S. Loss, G. Alcaraz, V. Gramlich and H. Grützmacher; *Angew. Chem. Int. Ed. Engl.*, 1999, **38**, 1623.

99JA11650 K. Lammertsma, B. Wang, J.-T. Hung, A.W. Ehlers and G.M. Gray; *J. Am. Chem. Soc.*, 1999, **121**, 11650.

99JCS(D)2409 N.H. Tran Huy, L. Ricard and F. Mathey; *J. Chem. Soc., Dalton Trans.*, 1999, 2409.

99JOM(580)320 T.I. Sølling, S.B. Wild and L. Radom; *J. Organomet. Chem.*, 1999, **580**, 320.

99JOM(582)53 N.H. Tran Huy, L. Ricard and F. Mathey; *J. Organomet. Chem.*, 1999, **582**, 53.

99MI274 M. Sanchez, R. Réau, C.J. Marsden, M. Regitz and G. Bertrand; *Chem. Eur. J.*, 1999, **5**, 274.

99MI1581 S. Haber, M. Schmitz, U. Bergsträsser, J. Hoffmann and M. Regitz; *Chem. Eur. J.*, 1999, **5**, 1581.

99OM796 B. Wang, K.A. Nguyen, G.N. Srinivas, C.L. Watkins, S. Menzer, A.L. Spek and K. Lammertsma; *Organometallics*, 1999, **18**, 796.

99OM817 J. Simon, U. Bergsträsser, M. Regitz and K. Laali; *Organometallics*, 1999, **18**, 817.

99ZAAC(625)102 R. Streubel, H. Wilkens, F. Ruthe and P.G. Jones; *Z. Anorg. Allg. Chem.*, 1999, **625**, 102.

2000JOC652 N.H. Tran Huy and F. Mathey; *J. Org. Chem.*, 2000, **65**, 652.

2.2
Three-membered Rings: 2. Diphosphiranes and Diphosphirenes

GUITA ETEMAD-MOGHADAM and MAX KOENIG
Université Paul Sabatier, Toulouse Cedex, France

2.2.1 INTRODUCTION

The scope of the three-membered rings with two phosphorus atoms includes essentially the heterocyclopropanes such as diphosphiranes and diphosphirenes. Their chemistry, more recent, is deeply underdeveloped by comparison with its oxygen, nitrogen, and sulfur counterparts. So, in diphosphirane series, the pioneering work of Baudler which allowed the detection of the first diphosphirane in 1977 ⟨77ZN(B)1490⟩, was followed by successful synthesis of various models. During the development of this heterocyclic chemistry, it appeared that they are less stable than their carbon analogues and their thermal stability is highly dependent on the steric requirements of the substituent present. So, various methods were developed during these last decades for the synthesis of stable diphosphiranes and diphosphirenes. Their structural, stereochemical, and thermodynamic parameters were investigated by experimental and theoretical approaches. For the most part, the chemistry of these strained heterocycles has been investigated in the literature as well as in the synthetic approaches than in the field of the reactivity in order to compare their eventual analogy to pure carbon systems. The behavior of the diphosphiranes under various conditions opened a versatile access to tetraphosphacyclohexanes, phosphaalkenes, 1,3-diphosphapropenes, 1,3-diphosphaallenes, but also 1,3-diphosphaallyl anion or cation, and diphosphiranyl radical or 1,3-diphosphapropyl biradical intermediates.

The chemistry of diphosphirenes was recently developed in particular in the field of diphosphirenium and diphosphirenylium salts ⟨99ACR561⟩.

In this review, we have attempted to comprehensively cover the literature until the middle of 1999, including several reviews already written on three-membered heterocycles with two phosphorus atoms ⟨90CRV997, 92CRV1839, B-96MI2, 99ACR561⟩. Taking into account the various synthetic routes, and the different strategies for the stabilization of these strained three-membered heterocycles, we have attempted to divide them into several series:

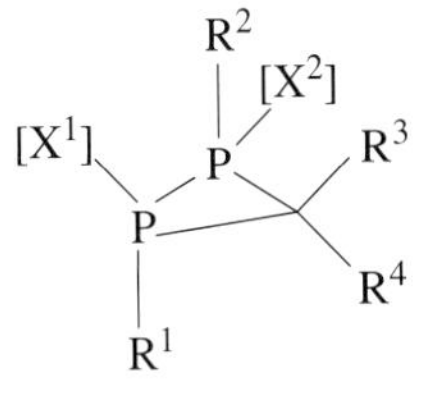

(**A**)

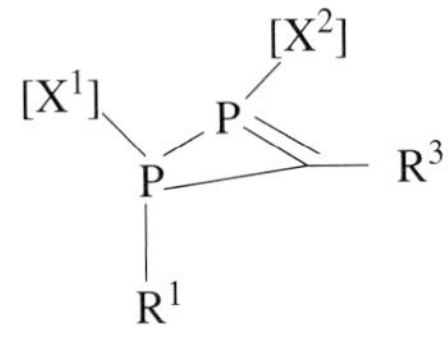

(**B**)

Abbreviations

- Ar = 2,4,6-$Bu^t_3C_6H_2$
- Ar' = 2,4,6-$Pn^t_3C_6H_2$
- Ar_F = 2,4,6-$CF_3C_6H_2$
- Cp^* = C_5Me_5
- TMS = $SiMe_3$
- Tsi = — $C(SiMe_3)_3$
- Mes = 2,4,6-Me_3-C_6H_2
- Mes' = 2,4,6-Pr^i_3-C_6H_2

- X = lone pair l.p.
- X = [M], metal complex or NR_2 (diphosphirenes)

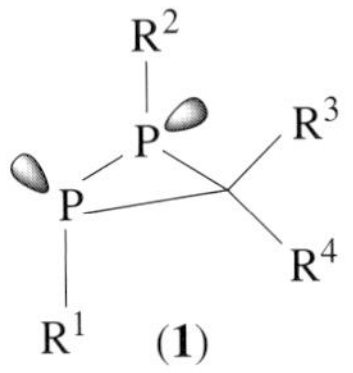

(**1**)

$R^1 = R^2 = Bu^t$

a : $R^3 = R^4 = H$
b : $R^3 = H, R^4 = Me$
c : $R^3 = R^4 = Me$

d : $R^3 = R^4 = Cl$
e : $R^3, R^4 = {=}CH_2$
f : $R^3, R^4 = {=}C(4\text{-}Cl\text{-}C_6H_4)_2$
g : $R^3, R^4 = {=}N\text{-}$ 2,4-$Bu^t_2C_6H_3$
h : $R^3, R^4 = {=}N\text{-}$ 2-$Bu^tC_6H_4$
i : $R^3, R^4 = {=}CMe_2$
j : $R^3, R^4 = {=}P\text{-}$ 2,4-Bu^t_2-6-Me-C_6H_2
k : $R^3, R^4 = $ < (P—Bu^t, P—Bu^t)

$R^1 = R^2 = P(NMe_2)_3$

l : $R^3 = R^4 = H$

Series 1

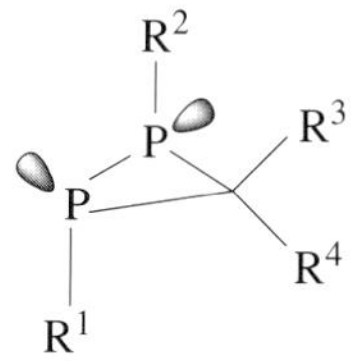

(**2**)

$R^1 = R^2 = Ar$

a : $R^3 = R^4 = H$
b : $R^3 = R^4 = Ph$
c : R^3 , R^4 = fluorenyl
d : $R^3 = R^4 = Cl$
e : $R^3 = R^4 = Br$
f : $R^3 = Ph$, $R^4 = Cl$
g : $R^3 = Me$, $R^4 = Cl$
h : $R^3 = Ph$, $R^4 = Br$
i : R^3, R^4 = $=CMe_2$
j : R^3 , R^4 = ◁
k : $R^3 = R^4 = \ =CCl_2$

$R^1 = Ar$; $R^2 = Ar'$

l : $R^3 = R^4 = Cl$

$R^1 = Ar$; $R^2 = Cp^*$

m : $R^3 = R^4 = H$
n : $R^3 = H$; $R^4 = TMS$

$R^1 = Ar$, $R^2 = Tsi$

o : $R^3 = R^4 = Cl$
p : $R^3 = R^4 = Br$

$R^1 = R^2 = Tsi$

q : $R^3 = R^4 = \ =NCF_3$
r : $R^3 = R^4 = \ =NCH_3$

$R^1 = R^2 = Ar'$

s : $R^3 = R^4 = Cl$

$R^1 = R^2 = Cp^*$

t : $R^3 = R^4 = H$
u : $R^3 = H$; $R^4 = TMS$

$R^1 = R^2 = Ar_F$

v : $R^3 = R^4 = H$

$R^1 = Ar$; $R^2 = OAr$

w : $R^3 = R^4 = Cl$

Series 2

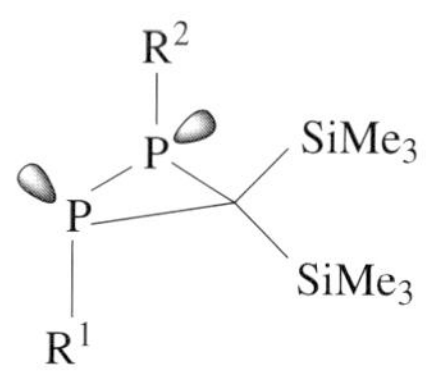

(**3**)

a : $R^1 = Cl$, $R^2 = CH(TMS)_2$
b : $R^1 = Cl$, $R^2 = N(TMS)_2$
c : $R^1 = Cl$, $R^2 = Cp^*$
d : $R^1 = Cl$, $R^2 = Ph$
e : $R^1 = Cl$, $R^2 = CH_2CH_2Cl$
f : $R^1 = Cl$, $R^2 = Bu^t$
g : $R^1 = Cl$, $R^2 = NPr^i_2$
h : $R^1 = Br$, $R^2 = Bu^t$
i : $R^1 = Br$, $R^2 = NPr^i_2$
j : $R^1 = I$, $R^2 = NPr^i_2$
k : $R^1 = NHTMS$, $R^2 = NPr^i_2$
l : $R^1 = Ph_2P$, $R^2 = NPr^i_2$
m : $R^1 = Cp^*$, $R^2 = NPr^i_2$
n : $R^1 = Bu^t$, $R^2 = NPr^i_2$
o : $R^1 = Bu^t$, $R^2 = Ph$
p : $R^1 = R^2 = Ph$
q : $R^1 = -C\equiv C(TMS)$
$R^2 = -C=C=C(TMS)_2$
r : $R^1 = H$, $R^2 = NPr^i_2$
s : $R^1 = Cl$, $R^2 = CH(TMS)_2$
t : $R^1 = N(TMS)_2$, $R^2 = NPr^i_2$
u : $R^1 = PNPr^i_2$, $R^2 = NPr^i_2$
v : $R^1 = AsPh_2$, $R^2 = NPr^i_2$

Series 3

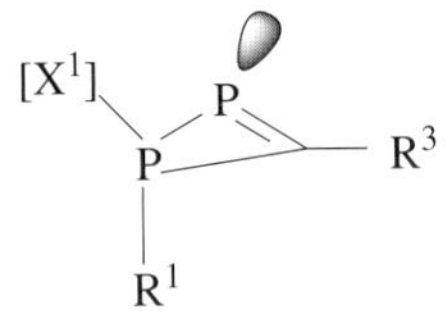

(4)

a : $R^1 = NPr^i_2$, X^1 = l.p., $R^3 = N(SiMe_3)Pr^i$

b : $R^1 = CEt_3$, X^1 = ═ NAr , $R^3 = Bu^t$

c : $R^1 = CEt_3$, X^1 = ═ NBu^t , $R^3 = Bu^t$

d : $R^1 = X_1 = R_3 = NPr^i_2$

e : $R^3 = Bu^t$, R^1,X^1 =

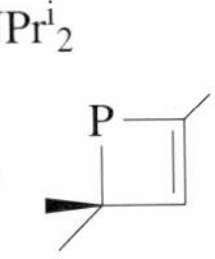

f : $R^1 = NPr^i_2$, X^1 = l.p., $R^3 = NPr^i_2$

Series 4

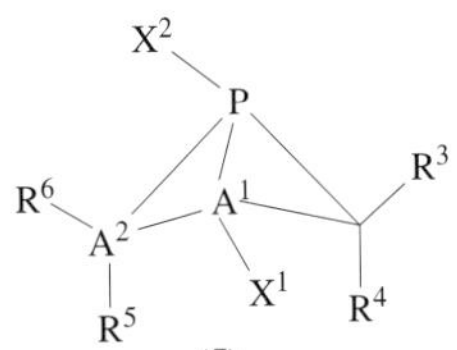

(5)

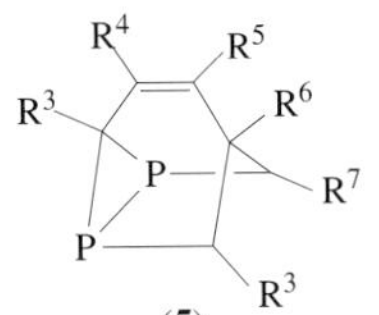

(5)

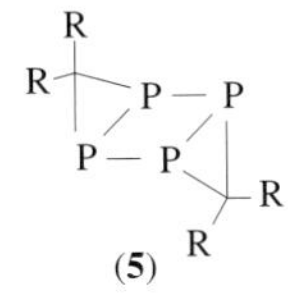

(5)

$A^1 = A^2 = P$ $X^1 = X^2$ = l.p.

a : $R^3 = R^4$ = TMS, $R^5 = N(TMS)[N(TMS)_2]$, R^6 = l.p.

b : $R^3 = R^4$ = TMS, R^5 = Ar, R^6 = l.p.

$A^1 = P$, $A^2 = C$

c : $R^3 = R^6 = Bu^t$, $X^1 = X^2$ = l.p., $R^4 = R^5 = ZrCp_2$

d : $R^3 = R^6 = Bu^t$, X^1 = l.p., $X^2 = Fe(CO)_4$, $R^4 = R^5 = ZrCp_2$

e : $R^3 = R^6 = Bu^t$, $X^1 = X^2$ = l.p., $R^4 = R^5$ = >C(O)

$A^1 = C$, $A^2 = P$

f : $R^3 = R^4 = R^5 = X^1$ = TMS, l.p., $X^2 = R^6$ = l.p.

g : $R^3 = Bu^t$, R^4= H, R^5= H, R^6= H, R^7= H

h : $R^3 = Bu^t$, R^4= H, R^5= Me, R^6= H, R^7= Me

i : $R^3 = Bu^t$, R^4= Me, R^5= H, R^6= H, R^7= Me

j : $R^3 = Bu^t$, R^4= H, R^5= OTMS, R^6= OTMS, R^7= H

k : $R^3 = Bu^t$, R^4= COOMe, R^5= H, R^6= H, R^7= H

l : $R^3 = Bu^t$, $R^4 = P(O)OMe_2$, R^5= H, R^6= H, R^7= H

m : $R^3 = Bu^t$, $R^4 = P(O)OMe_2$, R^5= H, R^6= Me, R^7= H

n : $R^3 = Bu^t$, $R^4 = P(O)OMe_2$, R^5= H, $R^6 = Bu^t$, R^7= H

o : $R^3 = Bu^t$, R^4= H, R^5= COOMe, R^6= H, R^7= H

p : $R^3 = Bu^t$, R^4= H, R^5= OTMS, R^6= OTMS, R^7= H

q : R = TMS

r : 2R = $NR_2^{\cdot}$

Series 5

2.2.1.1 Synopsis of Three-Membered Rings with Two Phosphorus Atoms

2.2.1.1.1 Series 1, structures (1a–l)

Initiated by Baudler ⟨80PAC755⟩, series 1 constitutes the first diphosphirane series, obtained by cyclocondensation of diphosphides with gem-dihalogeno derivatives. The various syntheses and structural determinations (NMR, X-ray diffraction, PES) represented a decisive step in this field.

2.2.1.1.2 Series 2, structures (2a–w)

Series 2 results from cyclopropanation reactions of diphosphenes with various carbenoids ⟨87T1793, 92CRV1839, 92ZAAC(617)53, 95HAC33, 96JOM(514)37, 99UP1⟩. Owing to the presence of bulky substituents on the heteroatoms, these diphosphiranes are stable and their reactivity, in particular their ring opening reactions, have been investigated (cf. Section 2.2.4).

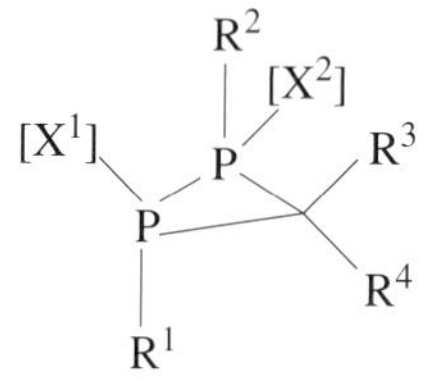

(**6**)

a : $R^1 = CH(TMS)_2$, X^1 = l.p., R^2 = Cl, $X^2 = Fe(CO)_4$, $R^3 = R^4$ = TMS

b : R^1 = Mes, X^1 = l.p., R^2 = Mes, $X^2 = Fe(CO)_4$, $R^3 = R^4$ = H

c : R^1 = Mes', X^1 = l.p., R^2 = Mes', $X^2 = Fe(CO)_4$, $R^3 = R^4$ = H

d : $R^1 = Cp^*$, $X^1 = Cr(CO)_5$, $R^2 = Cp^*$, X^2 = l.p., $R^3 = R^4$ = H

e : R^1 = Cl, X^1 = l.p., R^2 = Ph, $X^2 = W(CO)_5$, $R^3 = R^4$ = TMS

f : R^1 = Mes, $X^1 = Fe(CO)_4$, R^2 = Mes, $X^2 = Fe(CO)_4$, $R^3 = R^4$ = H

g : R^1 = Mes', $X^1 = Fe(CO)_4$, R^2 = Mes', $X^2 = Fe(CO)_4$, $R^3 = R^4$ = H

h : R^1 = Ph, $X^1 = Cr(CO)_5$, R^2 = Ph, $X^2 = Cr(CO)_5$, $R^3 = R^4$ = H

i : R^1 = Ar, X^1 = l.p., $R^2 = FeCp^*(CO)_2$, X^2 = l.p., $R^3 = R^4$ = H

j : R^1 = Ar, X^1 = l.p., $R^2 = RuCp^*(CO)_2$, X^2 = l.p., $R^3 = R^4$ = H

k: R^1 = Ar, X^1 = l.p., $R^2 = FeCp^*(CO)_2$, X^2 = l.p., $R^3 = R^4$ = 

l : R^1 = Ar, X^1 = l.p., $R^2 = RuCp^*(CO)_2$, X^2 = l.p., $R^3 = R^4$ = 

m : R^1 = H, X^1 = l.p., $R^2 = FeCp^*(CO)_2$, X^2 = l.p., $R^3 = R^4$ = TMS

n: R^1 = TMS, X^1 = l.p., $R^2 = FeCp^*(CO)_2$, X^2 = l.p., $R^3 = R^4$ = $=NCF_3$

o: R^1 = TMS, X^1 = l.p., $R^2 = FeCp^*(CO)_2$, X^2 = l.p., $R^3 = R^4$ = $=NPh$

p : R^1 = Ar, X^1 = l.p., $R^2 = FeCp^*(CO)_2$, X^2 = l.p., $R^3 = R^4$ = $=NCF_3$

q : $R^1 = Bu^t$, X^1 = l.p., $R^2 = FeCp^*(CO)_2$, X^2 = l.p., R^3 = Ph, R^4 = TMS

r : R^1 = H, $X^1 = Cr(CO)_5$, $R^2 = FeCp^*(CO)_2$, X^2 = l.p., $R^3 = R^4$ = TMS

s : R^1 = H, $X^1 = Cr(CO)_5$, $R^2 = FeCp^*(CO)_2$, X^2 = l.p., R^3 = TMS, R^4 = Ph

t : R^1 = TMS, $X^1 = Cr(CO)_5$, $R^2 = FeCp^*(CO)_2$, X^2 = l.p., R^3 = TMS, R^4 = Ph

u : R^1 = Ph, X^1 = l.p., $R^2 = CH_2Bu^t$, $X^2 = W(CO)_5$, R^3 = H, R^4 = C(Me) $=CH_2$

v : $R^1 = R^2 = FeCp^*(CO)_2$, $X^1 = X^2$ = l.p., $R^3 = R^4$ = TMS

Series 6

2.2.1.1.3 Series 3, structures (3a–v)

Cyclopropanation reactions with phosphaalkenes and phosphinidenes are involved in this series. The stability of this class of diphosphiranes is related to the presence of intracyclic carbon substituents (trimethylsilyl), various *P*-functionalized diphosphiranes can be obtained in this way ⟨93CB645, 93S763⟩.

2.2.1.1.4 Series 4, structures (4a–f)

Series 4 consists of unsaturated three-membered rings: diphosphirenes and diphosphirenium salts. Their stability can be attributed to the participation of resonance structures ⟨89AG(E)1673, 91AG(E)90, 91JA8160, 92AG(E)1055, 97CC2399, 97PS(124-125)545, 99ACR561⟩.

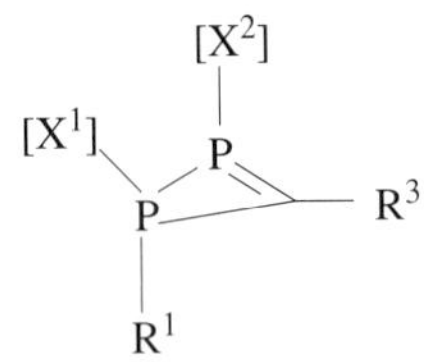

(7)

a : $R^1 = NPr^i_2$, $X^1 = W(CO)_5$, X^2 = l.p., $R^3 = NPr^i_2$

b : $R^1 = NPr^i_2$, $X^1 = Fe(CO)_4$, X^2 = l.p., $R^3 = NPr^i_2$

c : $R^1 = NPr^i_2$, $X^1 = X^2 = W(CO)_5$, $R^3 = NPr^i_2$

d : $R^1 = NPr^i_2$, $X^1 = X^2 = W(CO)_4$, $R^3 = NPr^i_2$

e : $R^1 = Cl$, $X^1 = W(CO)_5$, X^2 = l.p., $R^3 = NPr^i_2$

f : $R^1 = N_3$, $X^1 = W(CO)_5$, X^2 = l.p., $R^3 = NPr^i_2$

g : $R^1 = Ph$, $X^1 = W(CO)_5$, X^2 = l.p., $R^3 = NPr^i_2$

h : $R^1 = OH$, $X^1 = W(CO)_5$, X^2 = l.p., $R^3 = NPr^i_2$

i : $R^1 = F$, $X^1 = W(CO)_5$, X^2 = l.p., $R^3 = NPr^i_2$

j : $R^1 = CH(TMS)_2$, $X^1 = W(CO)_5$, X^2 = l.p., $R^3 = N(TMS)Pr^i$

k : $R^1 = X^1 = NPr^i_2$, $X^2 = W(CO)_5$, $R^3 = NPr^i_2$

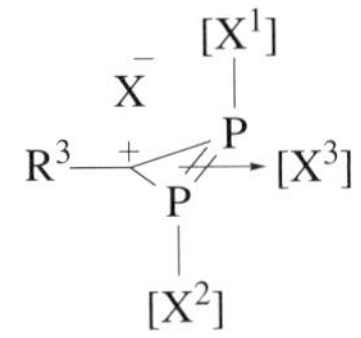

(7)

l : $X^1 = W(CO)_5$, $X^2 = X^3$ = l.p., $R^3 = NPr^i_2$

m : $X^1 = X^2 = W(CO)_5$, X^3 = l.p., $R^3 = NPr^i_2$

n : $X^1 = X^2 = X^3 = W(CO)_5$, $R^3 = NPr^i_2$

o : X^1, X^2, X^3 = 2 $Fe(CO)_3$, $R^3 = NPr^i_2$

Series 7

2.2.1.1.5 *Series 5, structures (5a–r)*

In the series 5 polycyclic models, the diphosphirane structure is inserted into a butterfly (**5a–f**) ⟨87AG(E)1011, 90CB67, 90PS(49-50)297⟩, or more complex environments (**5g–p**) ⟨95BSCF652, 98S125, 98S427⟩, and (**5q, r**) ⟨91AG(E)90, 98SCI2080⟩.

2.2.1.1.6 *Series 6, structures (6a–v)*

The series 6 diphosphiranes are stabilized by transition metal complexes and are prepared by [2 + 1]-cycloaddition reactions of carbenoids to diphosphene complexes.

Four classes of compounds can be distinguished: those with two organic substituents on phosphorus atoms in which one (**6a–e**) or both (**6f–h**) phosphorus lone pairs are metallated, and those in which one substituent is replaced by a metal complex fragment and one (**6r–u**) or both (**6i–q, 6v**) phosphorus lone pairs are non-metallated (Scheme 1) ⟨85JOM(282)53, 90JOM(399)C4, 93CB1963, 93OM731, 94ZK848, 94OM4406⟩.

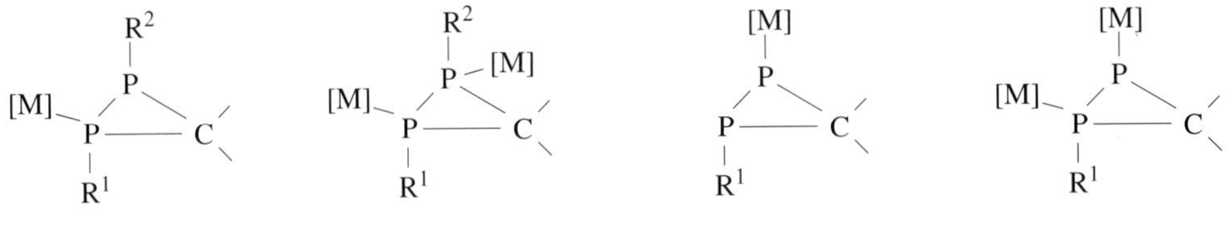

6a-c **6f-h** **6i q,v** **6r-u**

Scheme 1

2.2.1.1.7 *Series 7, structures (7a–o)*

Series 7 includes the η_1 diphosphirene and diphosphirenium complexes (**7a–k**) and the η_2 diphosphirenylium complexes (**7l–o**) in which one or both phosphorus lone pairs are metallated ⟨95CC2113, 97CC2399, 97JA9923, 99ACR561, 99CC1535, 99EJIC1479⟩.

2.2.2 SYNTHESIS

2.2.2.1 Diphosphiranes

2.2.2.1.1 [2 + 1]-Cyclocondensation from P–P and C units

The chemistry of diphosphiranes, phosphorus analogues of cyclopropanes, was first initiated by Baudler *et al.* ⟨77ZN(B)1490, 78ZN(B)1208, 84ZC352⟩ through the [2 + 1]-cyclocondensation of diphosphides with gem-dihalogeno derivatives (alkyl- ⟨82ZAAC(486)39⟩, alkene- ⟨84CB1542⟩, isocyanide- ⟨87CB421⟩, and methylenephosphane- ⟨88CB281⟩ dichlorides) in an apolar medium such as pentane affording a series of Bu^t-*P*-substituted diphosphiranes (**1a–j**) (Series 1) (Equation (1)). The use of carbon tetrachloride led to a first tetraphosphaspiro[2.2]pentane derivative (**1k**) ⟨83AG(E)632⟩.

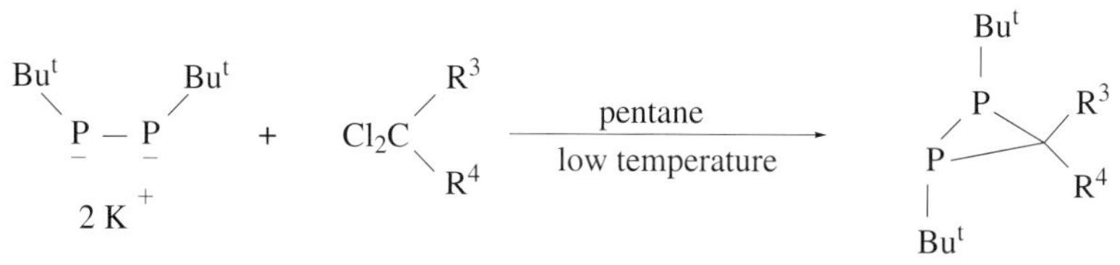

1a-k (1)

2.2.2.1.2 Cyclization from P–C–P units

The intramolecular cyclization of the *P,P*-dihalogenodiphosphinomethane by treatment with a potassium–sodium alloy at low temperature, realized by Lutsenko and co-workers ⟨79ZOB471, 80ZOB689, 84ZC345⟩, constituted another approach to 3-membered rings (Series 1), obtained as a 80 : 20 mixture with its dimer (Equation (2)). Koenig and co-workers have described a similar chemistry ⟨93-TH⟩. The reaction has been carried out at 40°C in the presence of magnesium or at 20°C under sonication in the presence of lithium wire affording the corresponding methylene diphosphirane (Equation (2)) (Series 2).

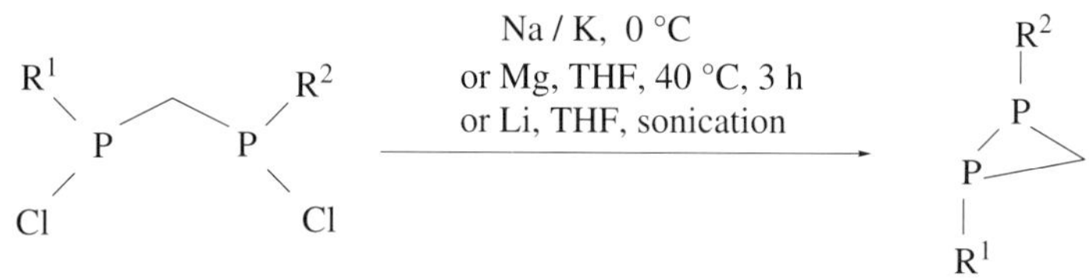

1a : R^1, R^2 = Bu^t
2a : R^1, R^2 = Ar (2)

Another approach was described by Appel where the pyrolysis of a 2,4-diphosphabutene-1 *via* a P–C–P cyclization leads to diphosphirane (**3q**) by means of intramolecular [2 + 2]-cycloaddition ⟨86CB2609⟩.

2.2.2.1.3 [2 + 1]-Cycloaddition from P=P and carbon units

Since 1981, the availability of stable diphosphenes ⟨81JA4587⟩ has made it possible to apply the cyclopropanation reaction of alkenes to these phosphorus–phosphorus double bonds. The [2 + 1]-cycloaddition reactions of carbenoids (diazo derivatives, mono- and dihalogeno- or ethylenic carbenes, and sulfur ylide) with symmetrical and unsymmetrical *trans* diphosphenes were developed by Koenig. In all cases, the cyclopropanation reaction is stereoselective and the corresponding *trans* diphosphiranes are obtained in good yield (Equation (3)) ⟨86TL1145, 87T1793, 87PS(30)425, 89CC593, 89TL177, 95HAC33⟩.

The presence of bulky substituents in diphosphenes has a stabilizing effect, and generally a large excess of reactant is necessary. The synthesis of methylene diphosphirane (**2a**) under sonication by transposition of the Simmons–Smith reaction to diphosphene which, like the synthesis of the *gem*-dihalogenodiphosphiranes (**2d, e**), uses inexpensive KOH pellets (instead of Bu^tOK or Bu^nLi) with a short reaction time has been described ⟨91TL5965⟩.

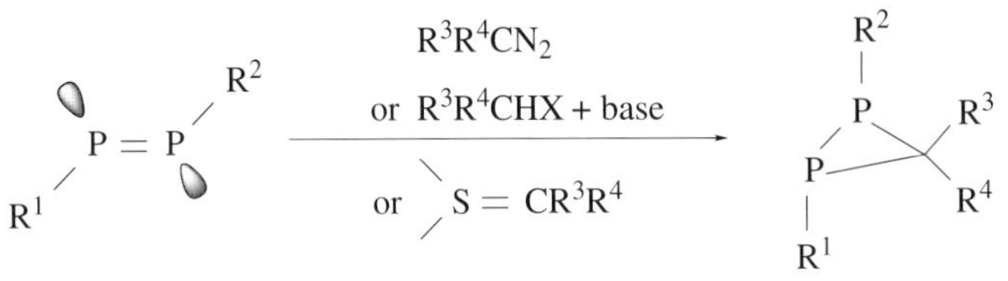

(3)

R^1, R^2 = Ar, Ar', Tsi, Cp*, OAr

2

[2 + 1]-Cycloaddition reactions of alkyl isocyanide with the diphosphene yield the three-membered heterocyclic diphosphirane imines (**2q,r**) (Equation (4)) ⟨92ZAAC(617)53⟩. Whereas the trifluoromethyl compound (**2q**) is thermally very stable as are all the diphosphiranes (**2a–v**), the methyl-substituted derivative (**2r**) slowly loses methyl isocyanide, reforming the diphosphene.

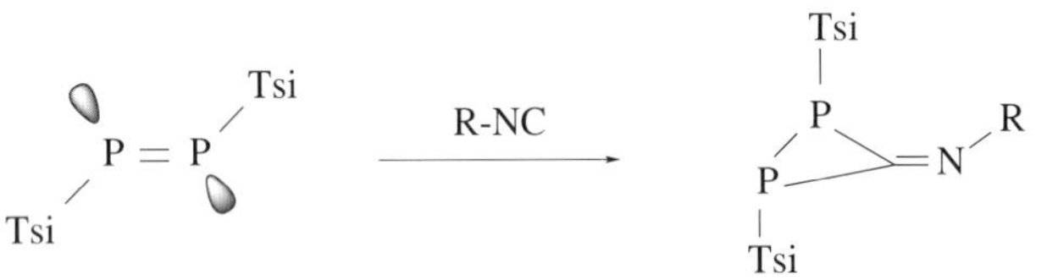

(4)

2q : R = CF_3
2r : R = CH_3

Diphosphiranes were also obtained in the reaction of dichlorocarbene with sterically protected 1,3-diphosphaallene in 34% yield (Equation (5)) ⟨91CC124⟩. The reaction may proceed *via* a phosphinidenephosphirane as an intermediate followed by a rearrangement leading to the thermodynamically stable 3-dichloromethylene–diphosphirane (**2k**).

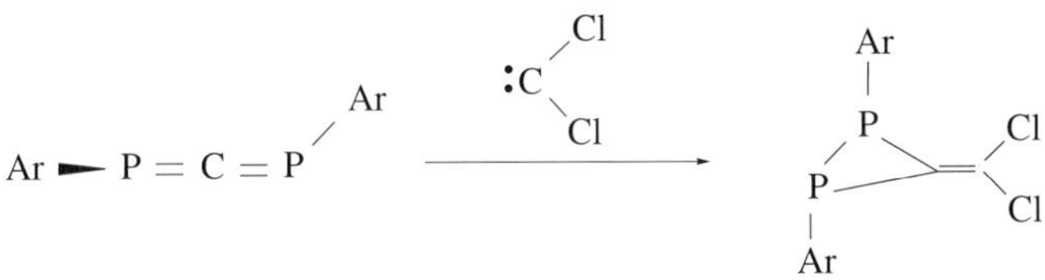

(5)

2k

2.2.2.1.4 Cycloaddition from P=C and phosphorus units

Another important approach, involving the cyclopropanation reaction of phosphinidenes with various phosphaalkenes (Equation (6)) was first mentioned by Niecke and co-workers (Series 3) ⟨83PS(18)35⟩.

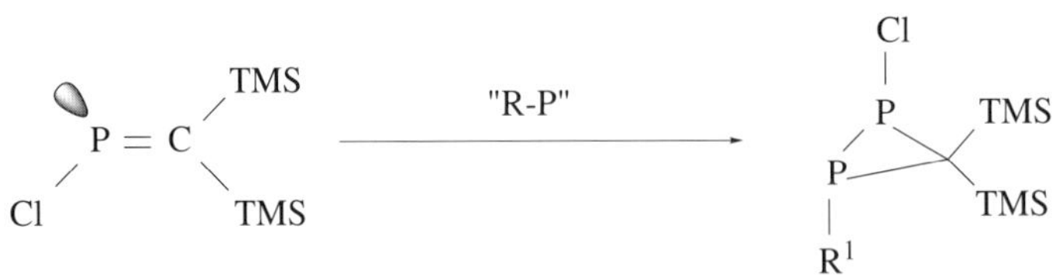

(6)

3

"R-P" = $R^1P(Cl)SiMe_3$, $PhPN_2$, $R^1P = W(CO)_5$

A similar scheme was proposed by Carrié leading to a *P*-chlorodiphosphirane ⟨89NJC891⟩. This kind of chemistry has been extended by Niecke in the reaction of chloro(trimethylsilyl)phosphanes with *P*-chlorophosphaalkenes leading to *P*-chlorodiphosphiranes *via* a 1,2-diphosphapropene-2 intermediate ⟨90CB1245⟩. The *P*-bromo- and *P*-iododiphosphiranes are also accessible by an exchange reaction with the corresponding halosilanes. The reaction of terminal phosphinidene complexes with phosphaalkenes at 120°C in toluene leading to the corresponding uncomplexed diphosphiranes was described by Mathey ⟨93S763⟩. The reduction of the *P*-chlorodiphosphirane by tributylstannane leads to the isolated diphosphirane (**3r**) bearing a P–H bond (Equation (7)) ⟨93CB645⟩.

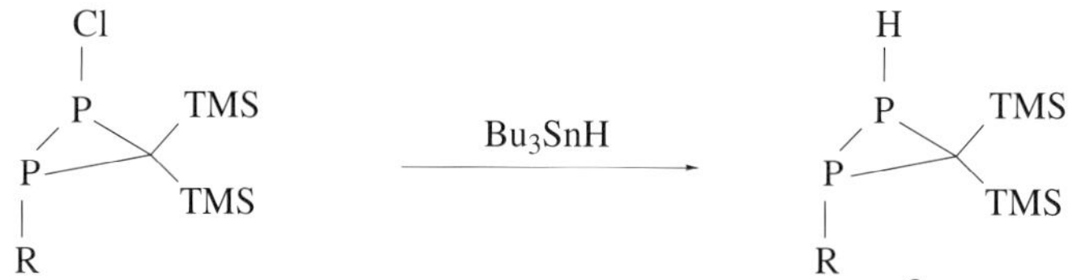

3r (7)

2.2.2.1.5 Miscellaneous

The reaction of triphosphenium salt with dichloromethane, involving the condensation of P_2 and carbon units, was described by Schmidpeter and co-workers ⟨86AG(E)253, 86ZN(B)444⟩. The formation of the diphosphoniodiphosphirane tetrachloroaluminate (**1l**) suggests an initial chloromethylation followed by cleavage of the resulting intermediate and formation of the three-membered ring compound by [2 + 1] addition of a phosphoniophosphanediyl to the P=C system. The unexpected synthesis of free or complexed *P*-chlorodiphosphiranes from anionic hydrido iron complex and *P*-chlorophosphaalkene was also reported ⟨87CC639⟩.

Another synthesis of diphosphiranes was described by Stelzer *via* deprotonation of the bis-secondary phosphines with strong organic bases in the presence of CuCl and pyridine (Equation (8)) ⟨90CB653⟩.

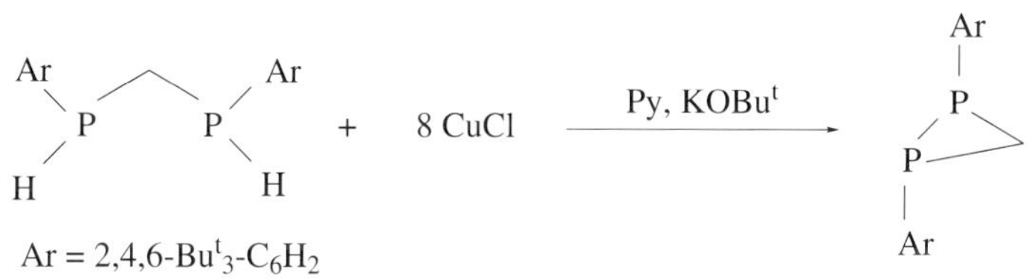

(8)

2.2.2.2 Polycyclic Diphosphiranes

Several specific syntheses of polycyclic diphosphiranes have been reported in the literature. Polycycles containing a diphosphirane ring were described by Regitz from the tandem Diels–Alder and ene reactions of conjugated dienes with phosphaalkynes (Equation (9)) ⟨87AG(E)1011, 95BSCF652, 98S125, 98S427⟩.

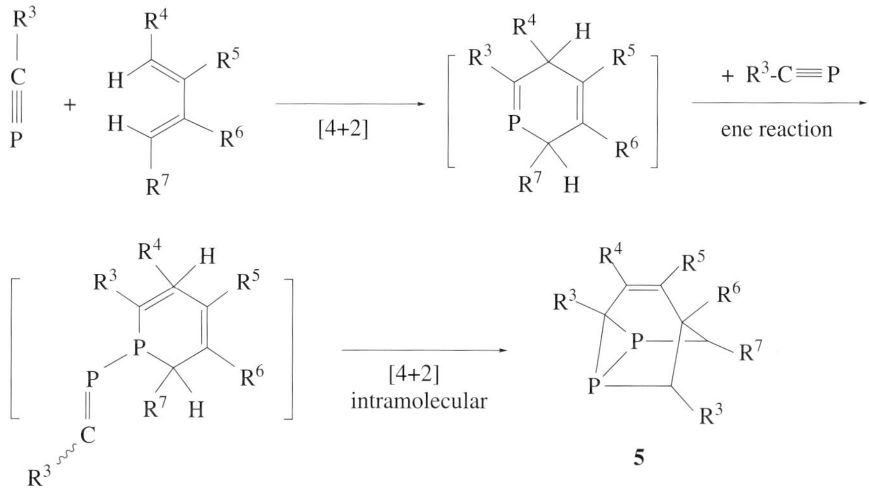

5 (9)

The synthesis and the X-ray structure of tetracyclic triphosphane bearing a diphosphirane system was realized by Märkl from the reaction of 1,3-azaphosphinines and Bu^t-phosphaalkyne on heating for 48 h ⟨88AG(E)389⟩. The decisive step in the formation of the diphosphirane unit is an ene reaction. The syntheses of triphosphabicyclobutanes involving an intramolecular P=P and P=C [2 + 2]-cycloaddition were also reported ⟨88AG(E)957, 88CC945⟩. A complementary access to polycyclic diphosphiranes involves the cyclotelomerization of a phosphaalkyne in the coordination sphere of a transition metal. The reactions of Bu^t-phosphaalkyne with zirconocene ⟨87AG(E)764⟩ and with vanadium complexes ⟨87AG(E)908⟩ were reported (Equation (10)).

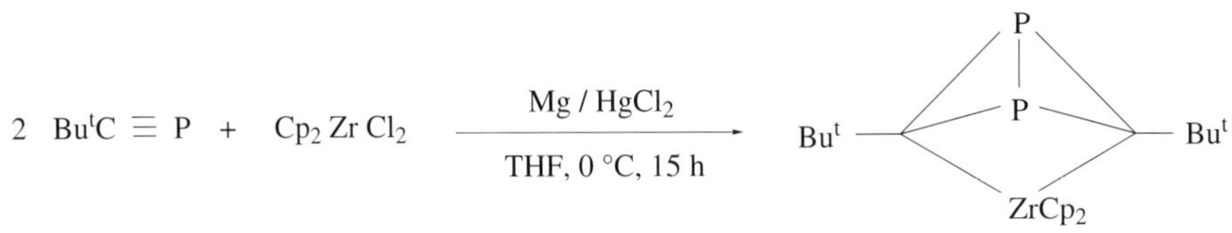

(10)

5c

In both cases, the metal complex fragments were incorporated into the products having the diphosphirane ring and their structures were established by X-ray analysis. An analogous reaction of phosphaalkyne with titanium complex was described by Cowley (Equation (11)), but in that case, the polycyclic diphosphirane spontaneously leaves the coordination sphere of the transition metal ⟨88AG(E)837⟩.

(11)

5e

The unusual formation of the germadiphosphatricyclooctene from a phosphaalkyne and an arylgermene was reported by Escudié and its structure was established by X-ray analysis ⟨93CC569⟩.

2.2.2.3 Diphosphirenes and Diphosphirenium Salts

The synthesis of 1*H*-diphosphirenes was realized by Niecke and co-workers by [2 + 1]-cycloaddition of iminophosphanes to Bu^t-phosphaalkyne at low temperature in hexane ⟨89TL459⟩. These thermally unstable air- and moisture-sensitive diphosphirenes undergo an isomerization to the thermodynamical isomers, azadiphosphetines. The synthesis and the structure of the first stable 1*H*-diphosphirene (**4a**) was reported by the same authors upon reaction of the functionalized aminophosphane with the amino-substituted phosphaalkyne under mild conditions (Equation (12)) ⟨89AG(E)1673⟩.

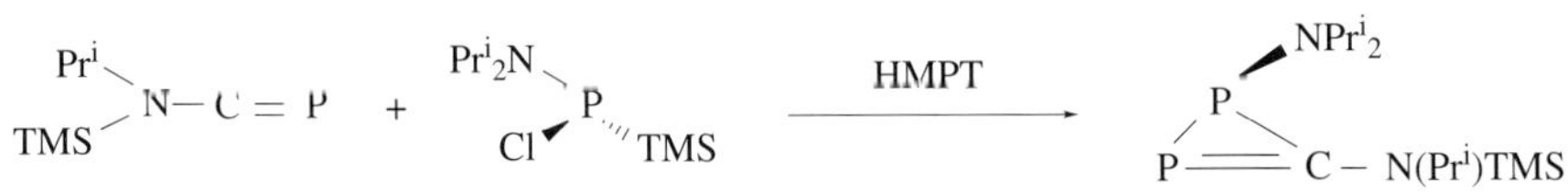

(12)

4a

The stability of the ring has been partly explained by some zwitterionic character and participation of resonance structures due to the amino-substituent.

Whereas diazirines exist only in the 3*H*-form with an intracyclic N=N bond, and are widely used as precursors of electrophilic carbenes, their 3*H*-diphosphirene analogues bearing an intracyclic P=P bond are unstable and are postulated only as intermediates ⟨94CB969⟩. Actually, Niecke and co-workers ⟨91AG(E)90⟩ described the reaction of persilylated 1,2-diphosphapropene-2 with hexachloroethane under mild conditions giving a *trans* 1,2,4,5-tetraphosphatricyclo[3.1.0.0]hexane (**5g**) (Equation (13)). Its structure was established by X-ray analysis (Table 1) and its formation can be attributed to a cyclodimerization of a 3*H*-diphosphirene intermediate. This [2 + 2]-cyclodimerization is typical of diphosphenes with (*Z*) configurations.

On another side, the synthesis of the diphosphirenium salt (**4d**) by treatment of amino-substituted 1,3-diphosphapropene with boron trifluoride–ethylamine complex at room temperature was reported by G. Bertrand and co-workers (Equation (14)) ⟨91JA8160, 94JA6149⟩. Its structure has been established by X-ray diffraction study and shows a substantial participation of resonance structure and a ionic character of the salt.

Table 1. Structural data of diphosphiranes and diphosphirenes [a]

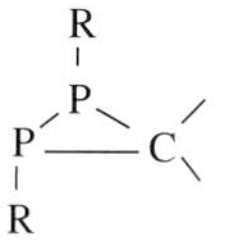 and 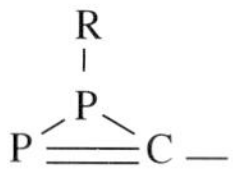

Compound	*No.*	*Bond angles, deg*			*Bond lengths, Å*			*Ref.*
		P–C–	*P–P–*	*C–P–R*	*P–P*	*P–C*	*P–R*	
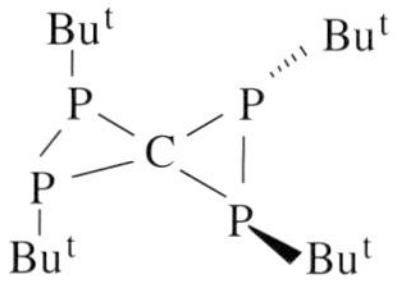	**1k**	72.35	53.83		2.175		1.888	⟨85MI117-01⟩
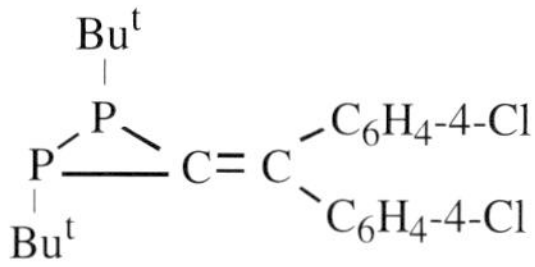	**1f**	75.2	52.2 52.6	107.7 108.2	2.195	1.794 1.804	1.861 1.883	⟨84CB1542⟩
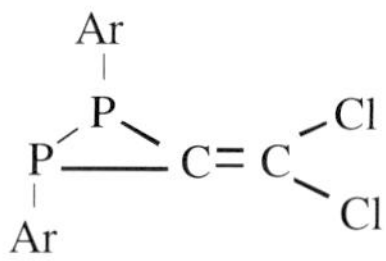	**2k**	78.9	50.4 50.7	128.8 109.8	2.277	1.789 1.795	1.875 1.860	⟨91CC124⟩
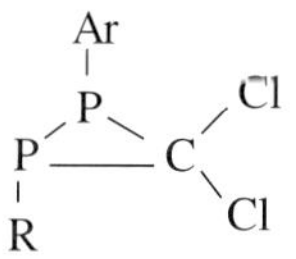 R = Ar	**2d**	74.3	52.7	101.8	2.245	1.863	1.858	⟨90JOC5750⟩
R = Tsi	**2o**	71.4	53.7 54.9	112.3 (Tsi) 97.8 (Ar)	2.117	1.801 1.827	1.803 (Tsi) 1.783 (Ar)	⟨96JOM(514)37⟩
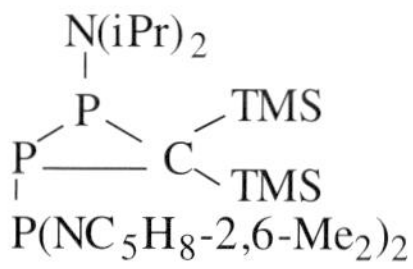	**3u**	69.8	55.0 55.3		2.159	1.184 1.891	2.238 1.693	⟨93CB645⟩
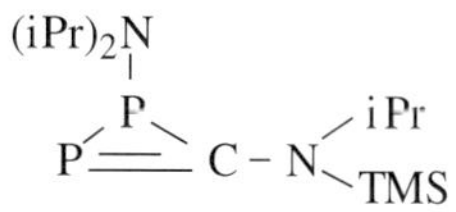	**4a**		52.3 48.9	111	2.218	1.702 1.787	1.702	⟨89AG(E)1673⟩
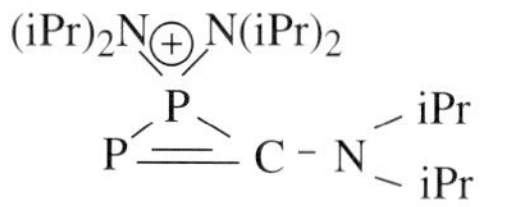	**4d**	73.9	52.8 53.3		2.095	1.747 1.737	1.643 1.637	⟨91JA8160⟩
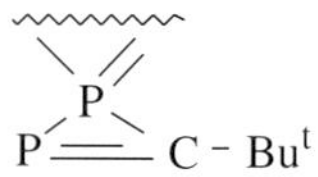	**4e**	75.9	53.3 50.8		2.108	1.74 1.68	1.77 1.80	⟨92AG(E)1055⟩
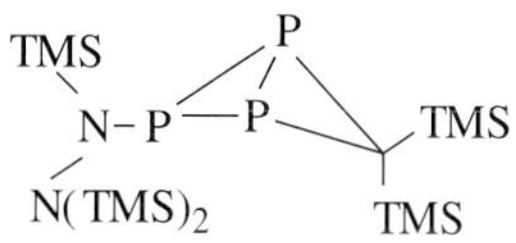	**5a**	70.0	55.1		2.158	1.880 1.883	2.215 2.226	⟨88CC945⟩
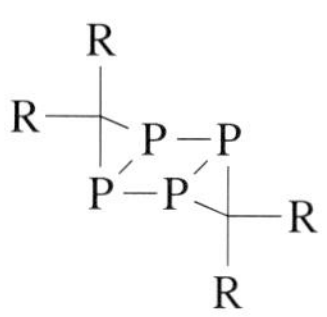 R = TMS	**5q**	72.1	53.9	102.0	2.204	1.871 1.873	2.245	⟨91AG(E)90⟩
2 R = $NR_2^{\cdot}$	**5r**	78.6	50.6	94.9	2.205	1.743 1.738	2.634	⟨98SCI2080⟩

[a] All the data are from X-ray crystal.

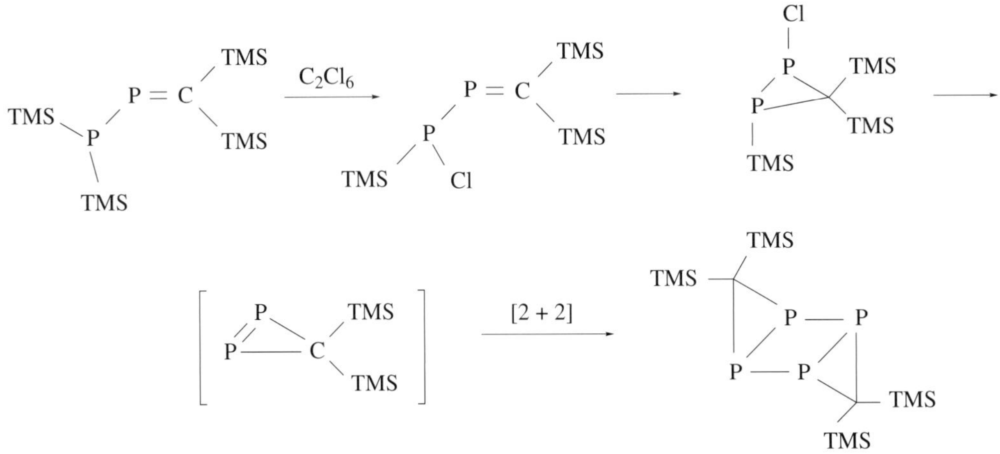

5g (13)

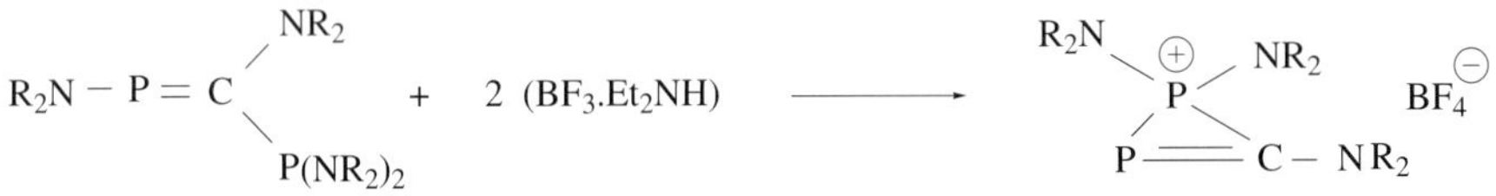

4d (14)

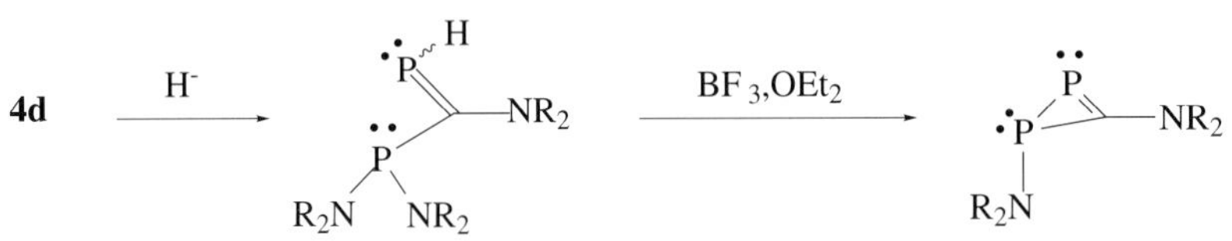

4f (15)

Diphosphirenium salt (**4d**) can be reduced with $LiAlH_4$ leading to *P*-hydrophosphaalkene, which on treatment with a catalytic amount of BF_3, OEt_2 afforded 1*H*-diphosphirene **4f** (Equation (15)) ⟨97CC2399⟩.

Regitz and co-workers have reported a spirocyclotrimerization of a stable phosphaalkyne with aluminum trichloride in a 3 : 1 molar ratio (Scheme 2) ⟨92AG(E)1055⟩. The reaction gives a betaine with the incorporation of the Lewis acid and its structure bearing a spiranic diphosphirene was confirmed by X-ray crystal structure analysis (cf. (**4e**), Table 1).

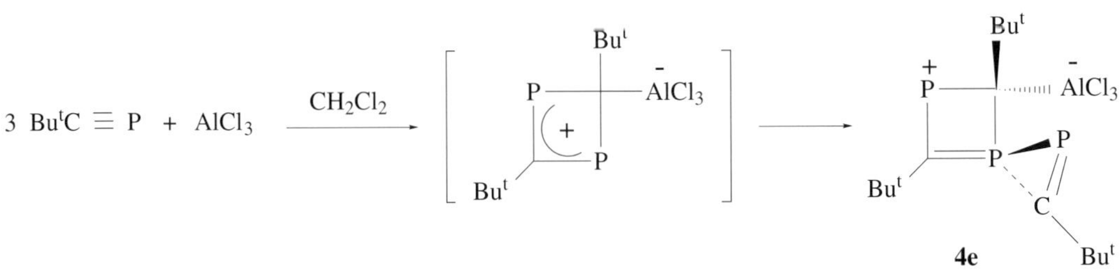

Scheme 2

The addition of the Lewis base (DMSO) to this betaine give the tri-phospha-Dewar benzene derivatives *via* an unstable diphosphirene ring.

2.2.2.4 Complexes of Diphosphiranes

The [2 + 1]-cycloaddition of diazomethane to a diphosphene complex leading to a chromium diphosphirane complex (**6h**) was first realized by Huttner (Equation (16)) ⟨85JOM(282)53⟩.

A related chemistry was reported by Weber with iron- and ruthenium-substituted diphosphenes and sulfur ylides. Thus, the presence of 1,2-diphosphaspiro[2.2]pentane system linked to the metal through

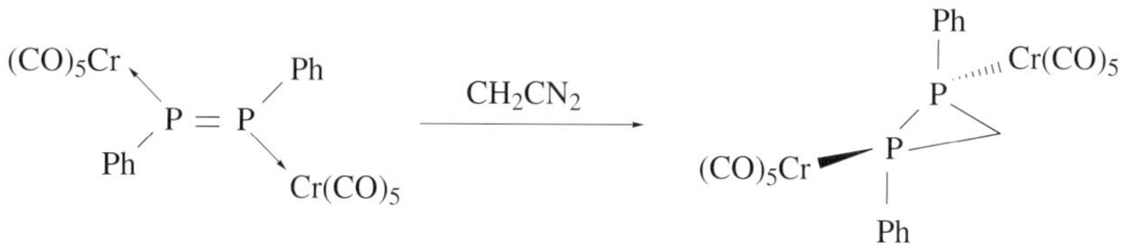

6h

(16)

an Fe–P bond was confirmed by X-ray structure 〈87JOM(327)C63〉〈88OM978〉. The iminodiphosphiranes were also obtained by a similar approach from metallodiphosphenes and trifluoromethyl isocyanide 〈92CRV1839〉.

The bidentate halogenophosphines were dehalogenated by $Fe_2(CO)_9$ to give complexes of diphosphiranes (Equation (17)) 〈90JOM(399)C4, 91ZN(B)1131〉.

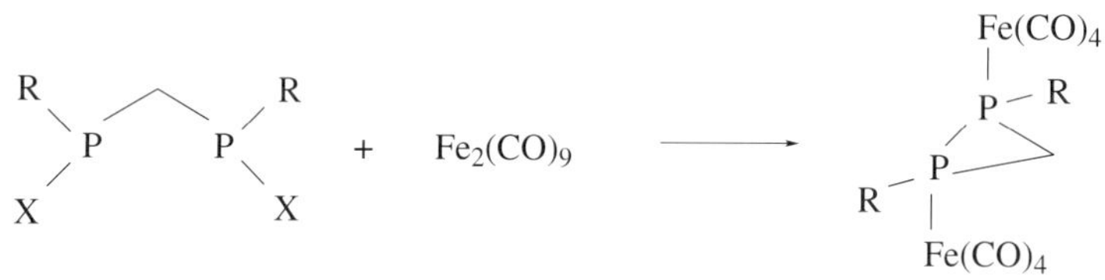

6f,g

(17)

More recently, Weber realized the synthesis of an iron complex of diphosphirane by reaction of the phosphine iron complex ([Fe]–PH_2) with *P*-chlorophosphaalkene (Equation (18)). It was converted into the iron chromium complex, whose structure was elucidated by X-ray analysis 〈92CB1553〉.

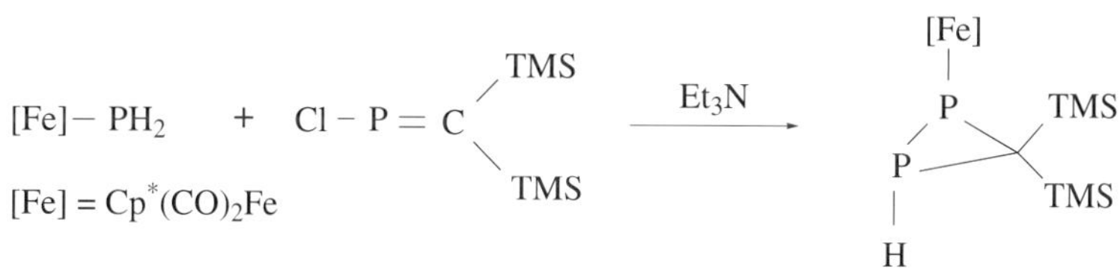

6m

(18)

Another reaction for the synthesis of diphosphirane complexes by thermal rearrangement of metallo-1,2-diphosphapropenes was described by the same authors 〈93CB1963〉. Finally, the synthesis of the $W(CO)_5$–diphosphirane complex was realized by the ene reaction of phosphaalkyne with the corresponding phosphaalkene complex 〈93T10279〉.

2.2.2.5 Diphosphirene, Diphosphirenium and Diphosphirenylium Complexes

Starting from complexed phosphinidene, Mathey and co-workers described the synthesis of the corresponding $W(CO)_5$–diphosphirene complex 〈93PS(77)229〉.

Thermal decomposition of the 2*H*-azaphosphirene complex in the presence of kinetically stabilized phosphaalkynes, affording the corresponding $W(CO)_5$–diphosphirene complex **7j** was reported by Streubel (Equation (19)) 〈95CC2113〉.

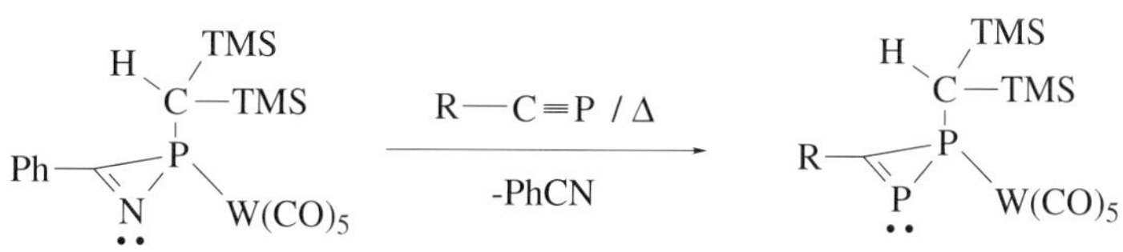

7j

(19)

The several possibilities of η_1 or η_2 complexation of 1*H*-diphosphirenes and diphosphirenium salts were recently studied by Bertrand and co-workers (Equations (20)–(23)) 〈99EJIC1479, 99CC1535〉.

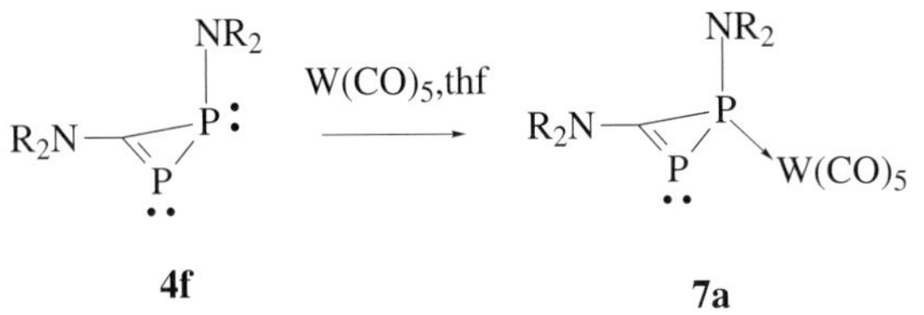

(20)

The 1,3-bis(diisopropylamino)-1*H*-diphosphirene **4f** and the diphosphirenium salt **4d** react with one equivalent of $W(CO)_5$(thf) affording the corresponding η_1 complexes **7a** and **7k**, respectively (Equations (20) and (21)).

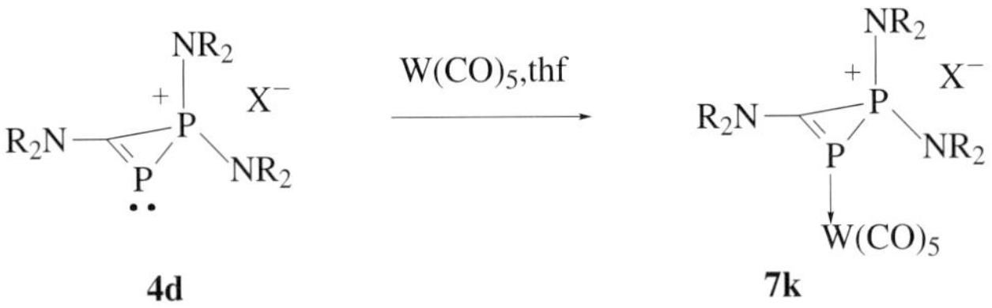

(21)

The dinuclear tungsten complexes **7c** are obtained by treatment of the free 1*H*-diphosphirene **4f** with two equivalents of $W(CO)_5$(thf) or one equivalent of [$W(CO)_4(thf)_2$] (Equation (22)).

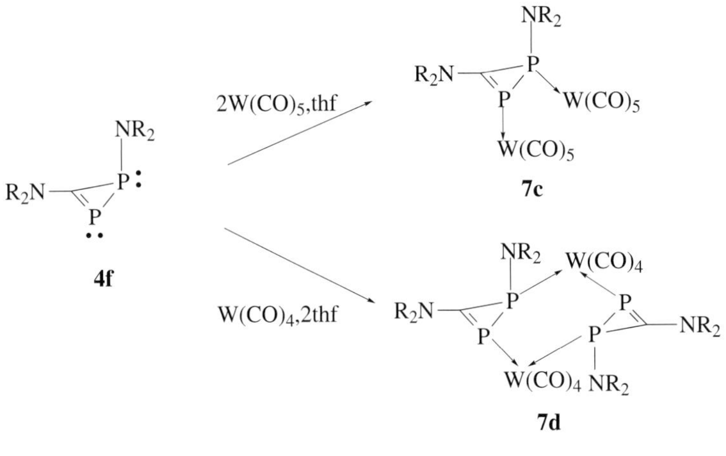

(22)

The η_2 diphosphirenylium salt $W(CO)_5$ complexes **7l** can be prepared by reaction of **7e** with silver trifluoromethane sulfonate (or aluminum or gallium trichloride). This η_2 complex reacts with one or two equivalents of $W(CO)_5$(thf) leading to the di- and tri-nuclear complexes **7m** and **7n** (Equation (23)).

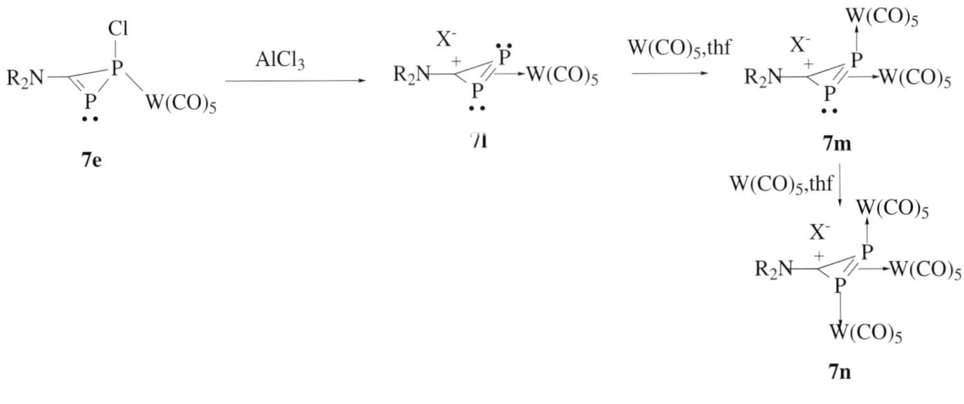

(23)

2.2.3 PHYSICOCHEMICAL DATA AND THEORETICAL CALCULATIONS

2.2.3.1 Stability

Whereas the diphosphiranes exhibit a lower stability than the cyclopropane homologues, they are the stable solids purified by crystallization or column chromatography. Their thermal stability is highly dependent on the steric requirements of the substituents. Thus, whereas the methylene derivative (**1a**) easily dimerizes to six-membered cyclodicarbaphosphane at room temperature, its analogous (**2a**) is thermally stable, owing to the presence of sterically hindered 2,4,6-$Bu_3{}^{t}$-phenyl substituents on the phosphorus atoms (Equation (24)).

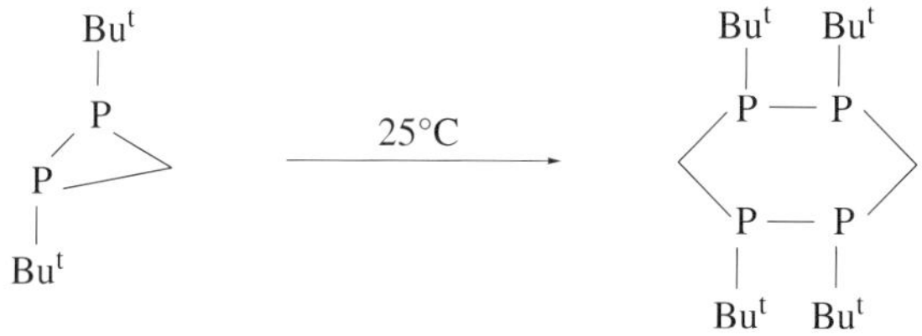

(24)

So, the diphosphiranes and diphosphirenes can be stabilized by the bulky substituents on the phosphorus atoms (**2a–s**) or on the intracyclic carbon atom (Series 3), by aromaticity and resonance character (Series 4), by the presence of a polycyclic or butterfly environment (Series 5), or by complexation (Series 6 and 7).

2.2.3.2 Structural and Spectroscopic Data

2.2.3.2.1 NMR studies

As do all the three-membered carbon–phosphorus rings, the ^{31}P, ^{13}C, and ^{1}H chemical shifts of diphosphiranes appear at high fields. The anisotropy induced by the ring strains is dependent on the steric hindrance and the electronegativity of the ring substituents. Thus, the ^{31}P spectra of non-complexed diphosphiranes (Series 1, 2, and 3) comprise a range from 0 to −200 ppm. For butterfly diphosphiranes (Series 5), the supplementary ring strain involves a shielding of the ^{31}P resonance ($\delta^{31}P = -347$ for (**5e**) ⟨88AG(E)837⟩).

The variations of the ^{31}P chemical shift against the nature of *P*-substituents are reported in Figure 1 for the diphosphiranes (e.g., for the unsymmetrical diphosphirane (**3e**), $\delta_{P-Cl} = -23$ and $\delta_{P-CH_2CH_2Cl} = -155$ ⟨93S763⟩) and in Figure 2 for the complexed diphosphirenes ⟨97JA9923⟩.

The $\delta^{31}P$ of diphosphiranes are also dependent to the nature of the intracyclic carbon substituents (e.g., $\delta = -141$ for (**2a**) ⟨86TL1145⟩, and $\delta = -70$ for (**2d**) ⟨87T1793⟩) (Figure 3). The homocyclic diphosphiranes have a *trans* stereochemistry. The *cis* isomers, obtained after substitution, were detected by ^{31}P NMR whereas the polycyclic homologues were isolated in *cis* form.

A solid-state study of a ^{31}P spin pair using the magic-angle-spinning (MAS) NMR spectroscopy of chloromethylene–diphosphirane (**2k**) was reported ⟨93JMR(A)258⟩. Although the detection of the intracyclic carbon atom in ^{13}C NMR is not easily realized in each case (due to high relaxation times), the

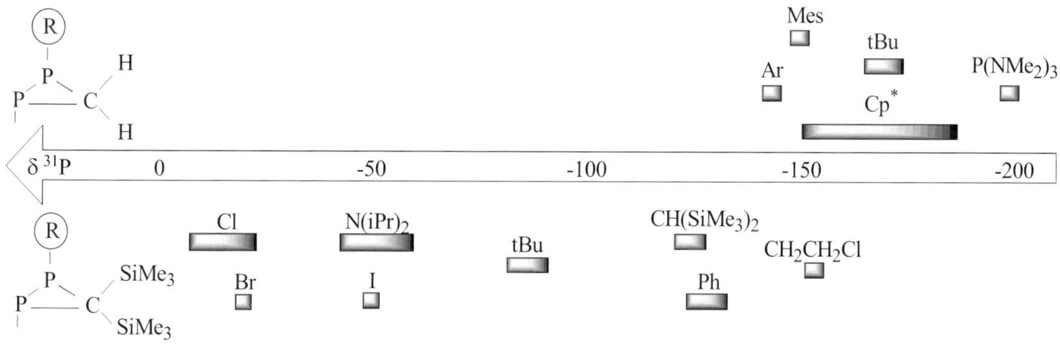

Figure 1. Influence of the *P*-substituents on the ^{31}P chemical shifts of diphosphiranes.

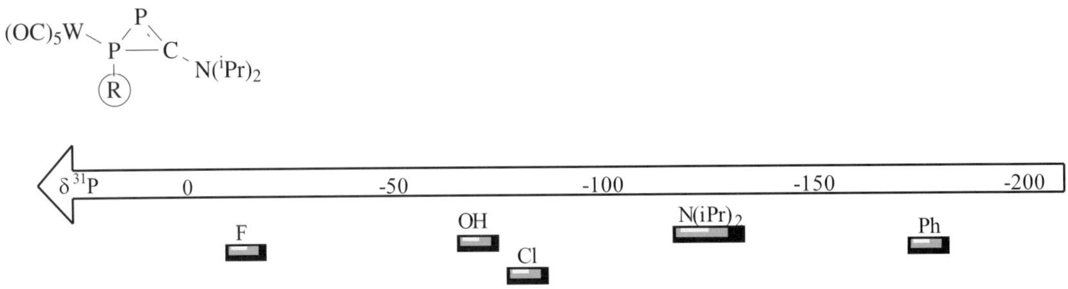

Figure 2. Influence of the *P*-substituents on the ^{31}P chemical shifts of complexed diphosphirenes.

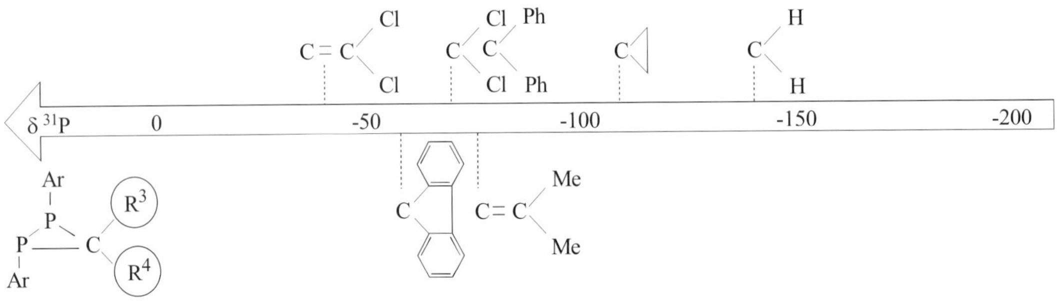

Figure 3. Influence of the *C*-substituents on the ^{31}P chemical shifts of diphosphiranes.

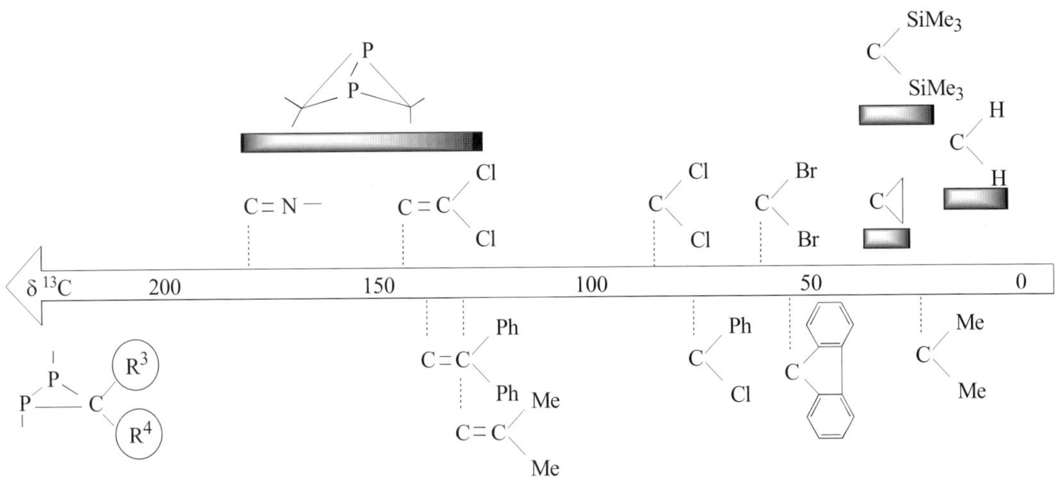

Figure 4. Influence of the *C*-substituents on the ^{13}C chemical shifts of diphosphiranes.

^{13}C chemical shift covers an extremely large range (e.g., $\delta^{13}C = 6$ for (**1a**) ⟨78ZN(B)1208, 84ZC345⟩, 86 for (**2d**) ⟨87T1793⟩, and 180 for (**1g**) ⟨88CB281⟩) (Figure 4).

The $^1J_{PP}$ coupling constants have a negative sign and their absolute values vary widely: for example, 38 Hz (**1g**) $< J_{PP} <$ 276 Hz (**6n**) ⟨92CB1553⟩. The intracyclic $^1J_{CP}$ coupling constants also have a negative sign and their values range from 15 Hz for (**5e**) ⟨88AG(E)837⟩ to 101 Hz in (**2o**) ⟨89CC593⟩. The proton directly bonded to an intracyclic phosphorus atom induces a very low $^1J_{PH}$ coupling constant ($^1J_{PH}$ = 153 Hz for (**3r**) ⟨93CB645⟩ and $^1J_{PH}$ = 149 Hz for (**3s**) ⟨90CC507⟩). These values increase when the diphosphirane is bonded to a metallic complex ($^1J_{PH}$ = 304 Hz in (**6n**) ⟨92CB1553⟩).

2.2.3.2.2 Structural data

About 20 structural studies of the diphosphirane and diphosphirene rings in Series 1 to 6 have been described: (**1a, f, k, l**), (**2d, k, o**), (**3u**), (**4a, d, e**), (**5a, b, c, e**), (**6h, n, o, q**). To these structures were added the polycyclic compounds reported by Regitz ⟨87AG(E)764, 87AG(E)908⟩, Märkl ⟨88AG(E)389⟩, and Escudié ⟨93CC569⟩.

The more significant parameters concerning the non-metallated diphosphirane and diphosphirene rings were collected in Table 1. The X-ray structure of diphosphiranes confirm the *trans* configuration of the substituents. The diphosphirane and diphosphirene rings are very flexible. The ring P–P bond lengths vary between 2.095 Å for (**4d**) and 2.277 Å for (**2k**). At the bottom of the range, the P–P bond seems to have some double bond character despite the observations of Niecke about the instability of 3*H*-diphosphirenes in contrast to 3*H*-diazirines. At the top of the range, the P–P bond length values in diphosphiranes are very close to the P–P bond length values of diphosphanes, despite the considerable strain of these three-membered rings. This is probably due to a strong mixing of the phosphorus lone pairs and the Walsh orbitals, as assigned by theoretical calculations and photoelectron spectroscopy ⟨81CB1004, 92IC2414⟩.

The intracyclic P–C bond length values cover a wide range: 1.794 Å $< d_{P\text{–}C} <$ 1.933 Å in diphosphiranes (**1**), (**2**), (**3**), and (**6**); 1.680 Å $< d_{P=C} <$ 1.737 Å and 1.740 Å $< d_{P\text{–}C} <$ 1.787 Å in diphosphirenes (**4**).

2.2.3.3 Theoretical Calculations

Based on PES spectra and semiempirical calculations, the HOMO of the diphosphirane was assigned to the symmetrical combination of the phosphorus lone pairs which is strongly coupled with the $\sigma_{P\text{–}P}$ orbital ⟨81CB1004, 92IC2414⟩. So, the oxidation of the lone pair destabilizes the cyclic form by weakening both the P–P and P–C bonds, and promotes the cheletropic reactions (see Section 2.2.4.1.5). The *cis/trans* stereochemistry of diphosphiranes is investigated by MNDO calculations of dipole moments in terms of substituent effects on intracyclic phosphorus and carbon atoms. These results compared to experimental dipole moments measurements show the presence of preferential *trans* isomers ⟨92CJC2834⟩.

2.2.3.3.1 Mechanism of the addition of carbenoids to diphosphene: theoretical approach

Although carbene addition to alkenes has been studied extensively, little work has been done on the reactivity of carbenes toward heteroatomic π-bonds; in particular, the problem of the preferential site of addition (π-system or lone pair) has not been fully elucidated ⟨72TL4347, 78JOC2216⟩. The theoretical approaches have been investigated using *ab initio* theory (SCF + CI level) concerning addition of singlet methylene to unsubstituted diphosphene ⟨92IC2414⟩.

A priori, an addition mechanism similar to that observed for ethylene was likely for the diphosphene system; the diffuse character of the P=P π-bond and the energetic positions of $\pi_{P=P}$ and $\pi^*_{P=P}$ orbitals could lead to differences in the interactions governing the approach of the carbene. This heteroatomic system is characterized in its ground state by a quasi-degeneracy of $\pi_{P=P}$ orbital and n_P combination of the phosphorus lone pairs. Among the two possible mechanisms (addition to the π system and addition to the lone pair of phosphorus), the preferential process corresponds to a π addition leading to the diphosphirane ring: the approach of the carbene occurring in the σ plane of the molecule with the displacement of the methylene group toward the middle of the P–P bond, a further pivotal motion of this group leading to ring formation. The change in phosphorus lone pair hybridization only takes place at the end of the reaction at $d <$ 2.2 Å and the reaction occurs without any energy barrier.

Distinguishing between the electrophilic and nucleophilic phases is more difficult than in the case of addition to ethylene. The lower position of the $\pi^*_{P=P}$ orbital with respect to the $\pi^*_{C=C}$ orbital leads to a more important $\pi^*_{P=P} \Leftrightarrow n_{CH_2}$ interaction throughout the reaction pathway. The calculations show that the reaction is stereospecific, giving a *trans* product, the preferential isomer of diphosphirane ($\Delta E_{trans\text{–}cis} = -102.7$ kJ mol^{-1}) ⟨92IC2414⟩.

2.2.3.3.2 Diphosphirenes and diphophirenylium salts

Ab initio MO calculations at the QCISD(T)/6-31G**//MP2/6-31G** + ZPE level on the diphosphaallenes, 1*H*-diphosphirenes, 3*H*-diphosphirenes and phosphanylphosphaalkynes $XX'CP_2$ isomers (X = H, Cl, F) show that the energies of the isomeric species are often reordered following halogen substitution. Generally 1*H*-diphosphirenes are more stable (except for X = H) than the corresponding 3*H*-diphosphirenes ⟨94CB969⟩. So, the transient formation of a 3*H*-diphosphirene or diphosphirenyl radical has been postulated in the formation of **5g** (Equation (13)) or **5r** (Equation (35)), respectively.

Theoretical calculations were performed on diphosphirenylium models HCP_2^+ and $H_2NCP_2^+$ at the

SCF/DZP and B3LYP/DZP levels. The optimized geometric data indicate that both cations adopt perfectly planar geometries with C_{2v} symmetry. Moreover, the frontier orbital system of the amino-substituted cation is in agreement with the observed η_2 coordination mode (Equation (23)) ⟨99EJIC1479⟩.

2.2.4 CHEMICAL PROPERTIES

2.2.4.1 Diphosphiranes

Experimental and theoretical approaches of electrocyclic reactions constitute a well-known field in cyclopropane series. In contrast, in heterocyclic chemistry, and in particular in phosphorus chemistry, with few exceptions, the electrocyclic reactions are rare. The presence of heteroatoms in the molecules makes the application of the Woodward–Hoffmann rules ⟨69AG(E)781⟩ to the ring-opening process of diphosphiranes more difficult. However, the functionalized diphosphiranes constitute the appropriate models for these studies since (i) owing to the presence of the skeleton disymmetry and the enthalpy difference ΔH_{P-P} and ΔH_{P-C}, their ring-opening reactions with preferential P–P bond cleavage can occur under photochemical, thermal, anionic or cationic reaction conditions, and (ii) due to the presence of bulky substituents on phosphorus atoms, the ring-opening products are stable and can be obtained in different conformations ⟨91JOC3438⟩. Moreover, the fragmentation can occur by the cheletropic reactions in oxidative conditions giving rise to a series of phosphaalkenes. Thus, the diphosphiranes can be considered as potential precursors of phosphaalkenes, 1,3-diphosphapropenes, 1,3-diphosphaallenes, but also 1,3-diphosphaallyl anion or cation, and diphosphiranyl radical or 1,3-diphosphapropyl biradical intermediates. Diphosphiranes react with anionic transition metal complexes leading to σ- or π-diphosphaallyl cations.

2.2.4.1.1 *Photochemical reactions*

The irradiation of functionalized symmetrical diphosphiranes **2d–h** gave largely 1,3-diphosphapropenes in the *cis* and *trans* configurations in 50–80% yield ⟨90JOC5750⟩. Their ^{31}P NMR parameters are different and the ratio of *cis/trans* isomers is about 10 : 90 (Equation (25)). The structure of the *trans* isomer has been established by X-ray analysis.

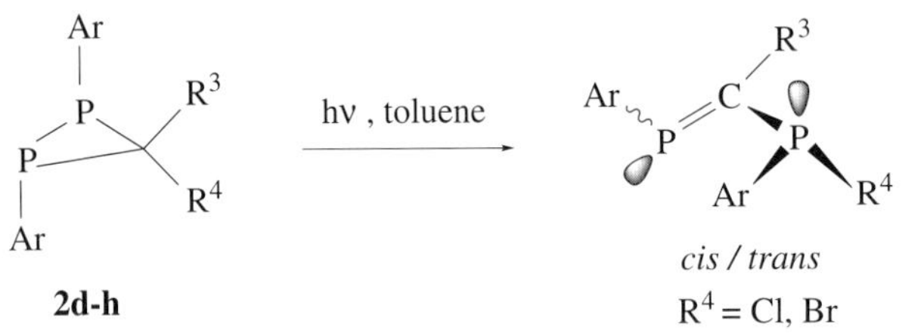

(25)

Under the same conditions as above, the unsymmetrical diphosphiranes (**2o,p**) undergo a regio- and stereoselective ring-opening reaction leading quantitatively to only *trans* 1,3-diphosphapropenes (Equation (26)) ⟨89CC593⟩.

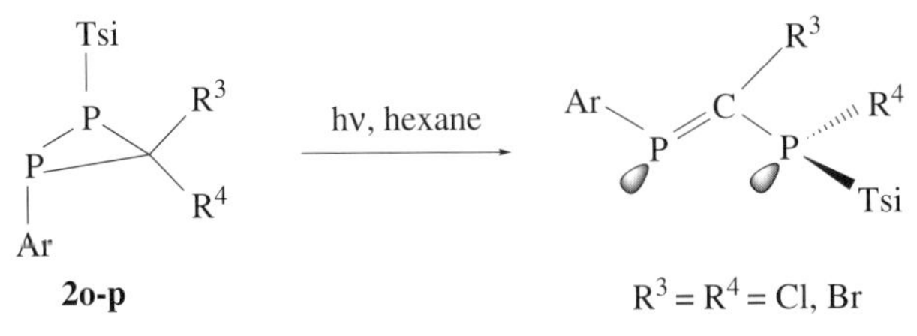

(26)

The mechanism of the photochemical ring opening of functionalized diphosphiranes was studied by the spin-trapping technique using α-phenyl Bu^t-nitrone (PBN) and *in situ* irradiation in the ESR cavity ⟨90JOC5750⟩. The diphosphiranyl and diphosphapropenyl radical intermediates were characterized. The generation of a diphosphiranyl radical intermediate can be the consequence of the homolytic rupture of the C–R^4 bond (R^4 = Cl). This unstable radical, after the preferential P–P bond cleavage, gives the 1,3-diphosphapropenyl radical which undergoes the migration of the halogen radical on the one of the two phosphorus atoms, leading to *P*-chloro-1,3-diphosphapropenes (Scheme 3).

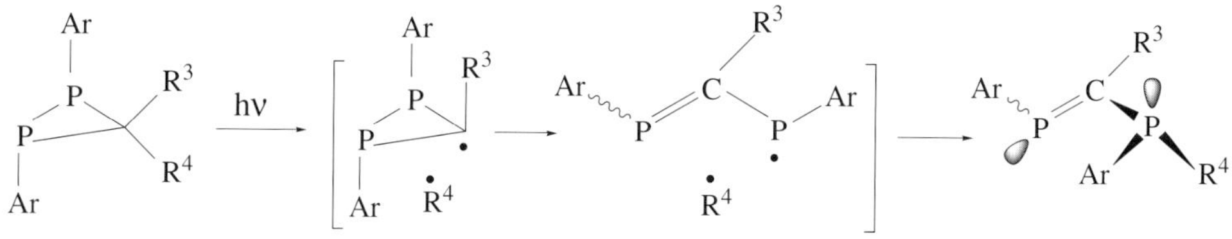

2d-h

Ar = 2,4,6-$Bu^t{}_3C_6H_2$; R^4 = Cl, Br

Scheme 3

The *cis* and *trans* isomers arise neither from the previous *cis–trans* isomerization of the diphosphiranes nor from the *cis–trans* isomerization of the diphosphapropenes, but presumably they are the consequence of a disrotatory process leading to a planar diphosphaallyl intermediate with *endo–exo* substituents on phosphorus. The MNDO and *ab initio* calculations are in agreement with this mechanism ⟨94IC596⟩.

2.2.4.1.2 *Reactions with nucleophiles (anionic ring opening)*

2.2.4.1.2.1 Organolithium reagents

The reaction, at low temperature, of a butyllithium excess on the monohalogenated diphosphiranes (**2f–g**) in diethylether, causes ring-opening by cleavage of the P–P bond and leads to a quantitative formation of 1,3-diphosphapropenes (Scheme 4) ⟨89NJC315, 96JOM(514)37⟩. The reaction is stereospecific and gives only the *trans* isomers.

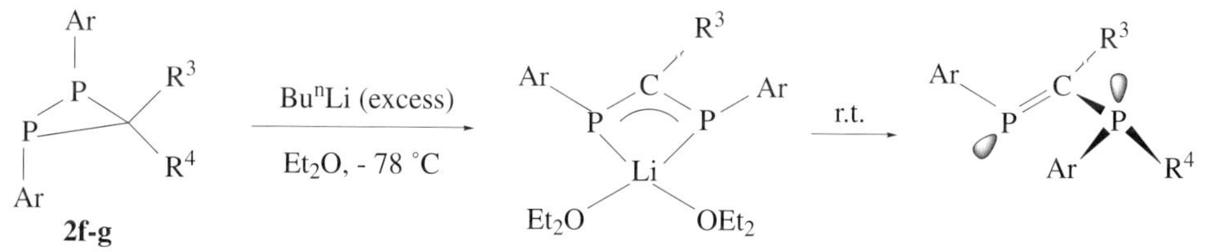

2f-g

Scheme 4

Unlike the cyanocyclopropyl anion ⟨77JA2631⟩, the cyclic diphosphiranyl anion intermediate is not detectable, but the 1,3-diphosphaallyl anion is observed when the progress of the reaction is followed by ^{31}P NMR at low temperature. It appears at $\delta^{31}P = 64.4$ with a $^1J_{P-Li} = 40$ Hz, characteristic of covalent phosphorus–lithium bonds ⟨89NJC315, 96JOM(514)37⟩. The *exo–exo* configuration of this intermediate was confirmed by X-ray diffraction ⟨93PS(77)229⟩. This anion, recrystallized with two molecules of diethylether, has a σ-allyl structure ($d_{P-C} = 1.71$ Å) and the lithium atom is located in the PCP plane ($d_{P-Li} = 2.75$ Å). The *exo–exo* configuration of the diphosphaallyl intermediate is consistent with the conrotatory process of the ring opening of the diphosphirane, as predicted by theoretical calculations ⟨92IC2414⟩.

As with the analogous gem-dihalogenocyclopropanes ⟨63ACS1683, 84T2805⟩, the gem-dihalogeno diphosphiranes (**2d, e**) react with methyllithium at 0°C in diethylether, and lead quantitatively to the 1,3-diphosphaallene (Scheme 5) ⟨89TL177, 89TL187⟩. The reaction does not involve a carbenic insertion in the P–P bond ⟨90PS(49-50)305⟩ but most likely it occurs by the same anionic ring opening as above *via* the unstable similar diphosphaallyl anion.

The *P*-chloro-1,3-diphosphapropenes, obtained by photolysis of the corresponding functionalized diphosphiranes, react with organolithium compounds under the same conditions as their precursors affording the same substituted 1,3-diphosphapropenes or 1,3-diphosphaallene in good yield (Scheme 5) ⟨89TL177⟩. By the same method, the unsymmetrical 1,3-diphosphaallenes were obtained ⟨89CC1732⟩. However, the synthesis of unsymmetrical 1,3-diphosphaallene from the corresponding *P*-chloro-1,3-diphosphapropene (**2o**) (*syn* conformation) does not occur by the organolithium pathway, and it requires a reduction of P–Cl to P–H by $LiAlH_4$, followed by a dehydrohalogenation in the presence of Et_3N or DBU as base (Scheme 6) ⟨91HAC221⟩.

2.2.4.1.2.2 Grignard reagents

The monohalogenated diphosphiranes (**2f–h**) as their photochemical isomers, the 1,3-diphosphapropenes, react with saturated and unsaturated organomagnesium compounds leading to a series of substituted

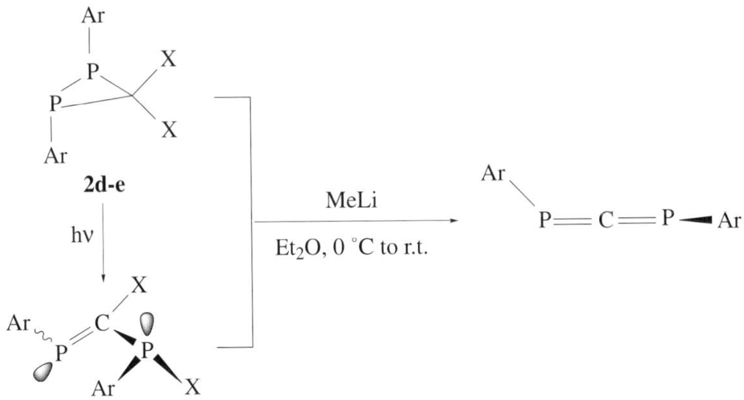

Ar = 2,4,6-$Bu^t_3C_6H_2$; X = Cl, Br

Scheme 5

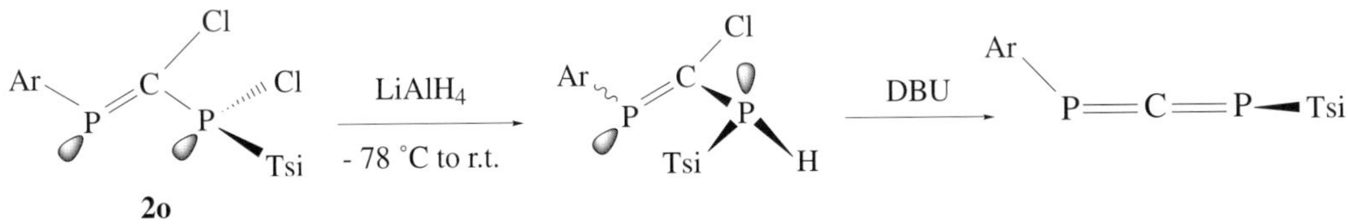

Scheme 6

1,3-diphosphapropenes and 1,3-diphosphadienes. The reaction could also involve an anionic ring opening followed by substitution at phosphorus *via* an allylic anion intermediate detected by ^{31}P NMR spectroscopy ($\delta^{31}P = 54$) (Scheme 7) 〈92JOM(436)169〉. The reaction requires milder conditions with the opened isomers.

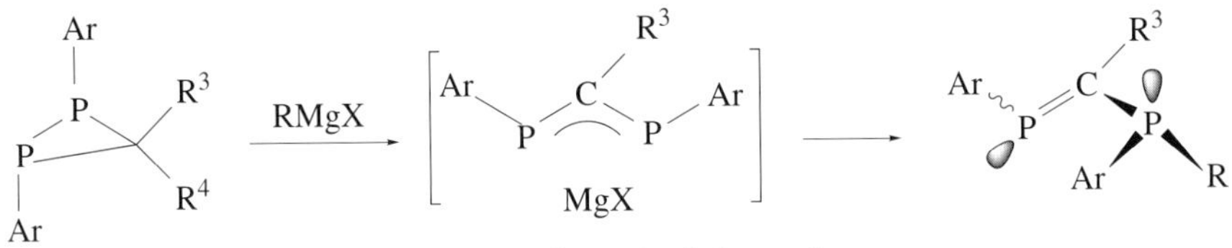

R = Me, Bu^n, vinyl, allyl, crotyl

Scheme 7

With gem-dihalogenodiphosphiranes, the reaction gives quantitative yields of 1,3-diphosphaallene probably involving the same mechanism as Scheme 7 and an unstable allylic anion intermediate 〈92JOM(436)169〉.

2.2.4.1.3 *Reactions with electrophiles (cationic ring opening)*

The reaction of one equivalent of aluminum trichloride to functionalized diphosphiranes (**2f–g**) in an anhydrous solution of dichloromethane at 0°C affords the corresponding phosphonium phosphaalkenes as main products, with the fragmentation derivatives as byproducts (Scheme 8) 〈89TL7411〉. Starting from the corresponding *P*-chloro-1,3-diphosphapropenes, the formation of byproducts can be avoided 〈84TL3687, 86AG(E)454, 89TL7411〉.

The phosphonium phosphaalkenes are formed *via* the diphosphaallyl cation after the cationic rupture of an extracyclic C–Cl bond. Contrary to the photochemical or anionic ring-opening reactions, the cationic intermediates are not detected. The theoretical calculations performed on unsubstituted derivatives reveal for this reaction a P–P bond rupture by a disrotatory process leading directly to an *exo–exo* open chain structure 〈92IC2414〉.

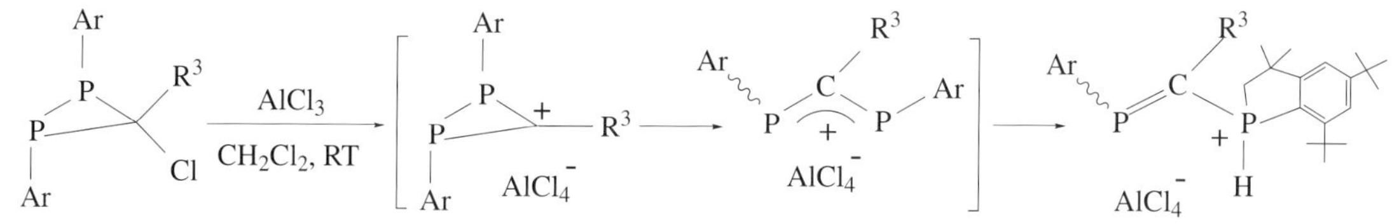

2f-g

R^3 = Ph, Me

Scheme 8

2.2.4.1.4 Reactions with metal carbonyls

The first attempt to complex diphosphirane was by Seyferth who obtained, after the cleavage of the P–P bond, a bridged iron complex in 10% yield (Equation (27)) ⟨87JOM(336)163⟩.

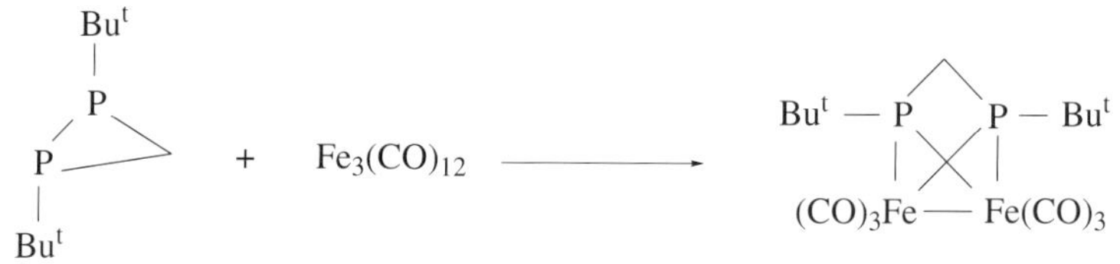

1a (27)

Subsequently, Jutzi allowed a ring diphosphirane to react with [$Cr(CO)_5THF$] and obtained an η^1-diphosphirane pentacarbonyl chromium complex (Equation (28)) ⟨92ZAAC(610)75⟩.

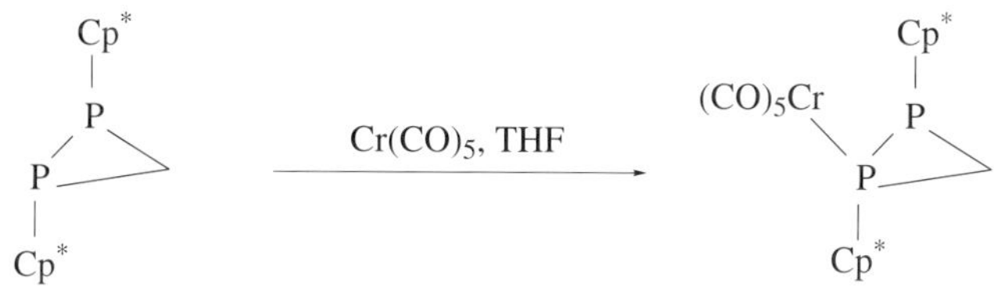

2r 6d (28)

The reaction of anionic transition metal complexes (M = Mo, W, Co) with monohalogenated diphosphiranes (**2f–g**) occurs upon heating and leads to the σ- or π-diphosphaallyl complexes as a result of preferential P–P bond rupture, depending on the reaction conditions, the nature of the intracyclic carbon substituents, and the transition metal complexes (Scheme 9) ⟨93PS(76)41, 93JOM(453)77⟩. π-Donor substituents (R^3 = Ph) promote the η^3-form, whereas σ-donor (R^3 = Me) afford the η^1-complex.

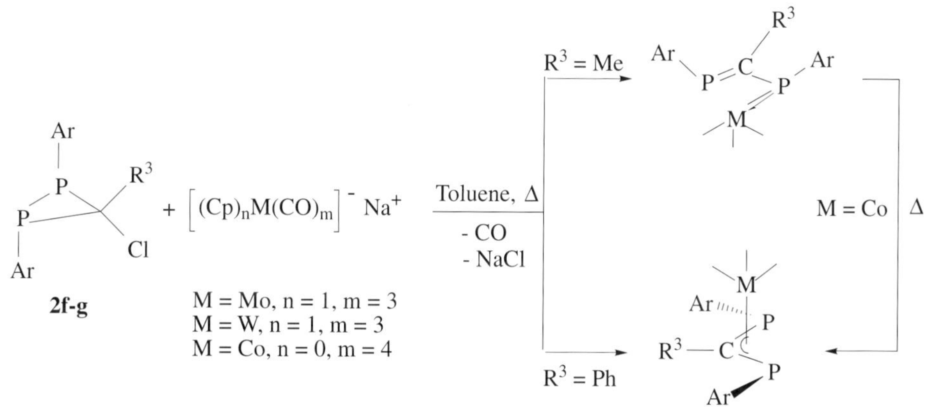

2f-g

M = Mo, n = 1, m = 3
M = W, n = 1, m = 3
M = Co, n = 0, m = 4

Scheme 9

The complexes are thermally labile as shown by variable-temperature NMR spectroscopy. Upon heating in toluene at reflux for 3 h, the σ-diphosphaallyl cobalt complex is irreversibly transformed into the π-allylic complex. This isomerization upon heating constitutes the first example of such interconversion in the diphosphaallyl systems. The synthesis of σ- or π-diphosphaallyl complexes can also be realized from the photochemical isomers of diphosphiranes, the 1,3-diphosphapropenes. In this case, the reaction occurs more readily under milder conditions than above, and the same diphosphaallyl complexes were obtained in 70–90% yield with the same selectivity as their parent diphosphiranes ⟨93JOM(453)77⟩. The π-diphosphaallyl complex obtained by addition of anionic transition metal complexes represents an *efficient stabilization of the corresponding diphosphaallyl cation*. The complexation reactions do not take place with all the anionic transition metal complexes. Treatment of monohalogenated diphosphiranes (**2f–g**) with Na[CpFe(CO)$_2$] in refluxing toluene solution affords the reduced 1,3-diphosphapropenes. Similar results were observed with the *P*-chloro-1,3-diphosphapropenes (Equation (29)).

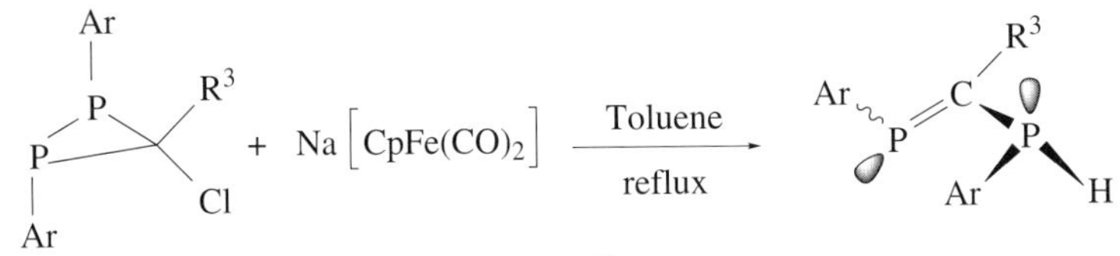

2f-g

(29)

Like their behavior toward the organolithium derivatives or the Grignard reagents, the gem-dihalogeno-diphosphiranes (**2d–e**) or their photochemical opened isomers react with the anionic metal transition complexes leading to 1,3-diphosphaallene in quantitative yield ⟨93JOM(453)77⟩.

2.2.4.1.5 *Oxidation reactions*

Ozonation of (**2d–f**), carried out in dichloromethane at low temperature (−78°C) (Scheme 10) (since ozone is a more suitable reagent for oxidizing the bulky phosphines) ⟨91TL3687⟩, involves the instantaneous oxidation of the phosphorus lone pairs and the formation of the oxidized diphosphirane intermediates detected by ^{31}P NMR. It decomposes at room temperature to phosphaalkene oxide, phosphinic and phosphonic acids.

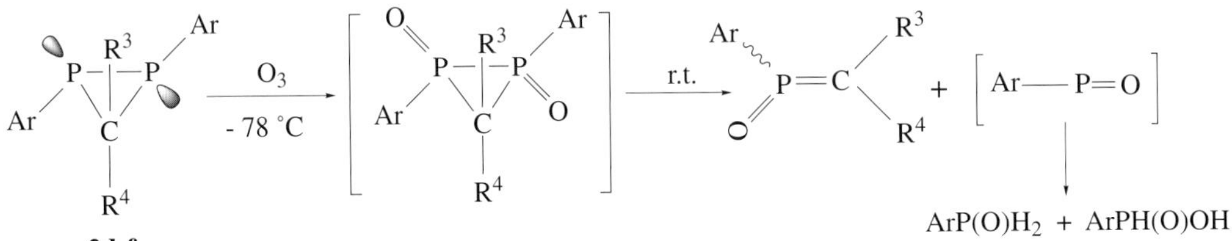

2d-f

Scheme 10

Consequently, the oxidation of the phosphorus lone pairs does not stabilize these strained heterocycles, but allows the production of the phosphaalkene oxides not readily available by other methods ⟨83JA4026, 84AG(E)157⟩.

The reaction of bromine (2 equiv.) with diphosphiranes in dichloromethane at room temperature also involved the fragmentation of the heterocycle leading quantitatively to phosphaalkene and dibromophosphine (Equation (30)) ⟨91TL3687⟩.

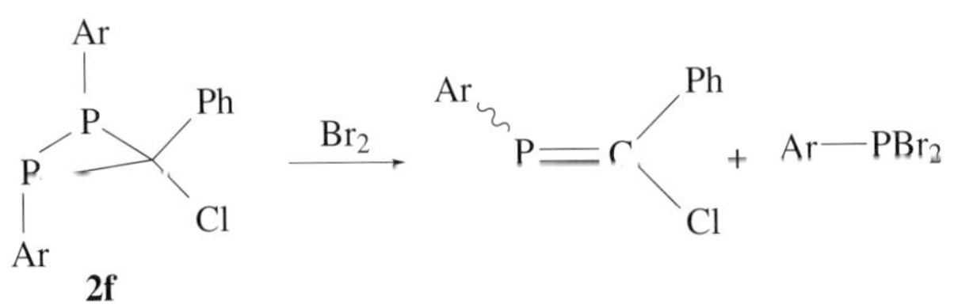

2f

(30)

2.2.4.1.6 *Thermal rearrangements*

The diphosphapropenes obtained after the thermal ring opening of diphosphiranes can exist as the *cis* or *trans* isomer stabilized in a *gauche* or *syn* conformation (Equation (31)) ⟨94IC596⟩. In the absence of nucleophilic or electrophilic assistance, the thermal opening of mono- and dihalogenated diphosphiranes (**2d–h**) leads to *cis* and *trans* 1,3-diphosphapropenes in *gauche* conformation. For the unsymmetrical diphosphiranes (**2o–p**), the thermal ring opening is regio- and stereoselective and leads quantitatively to the *trans* diphosphapropenes in the *syn* conformation.

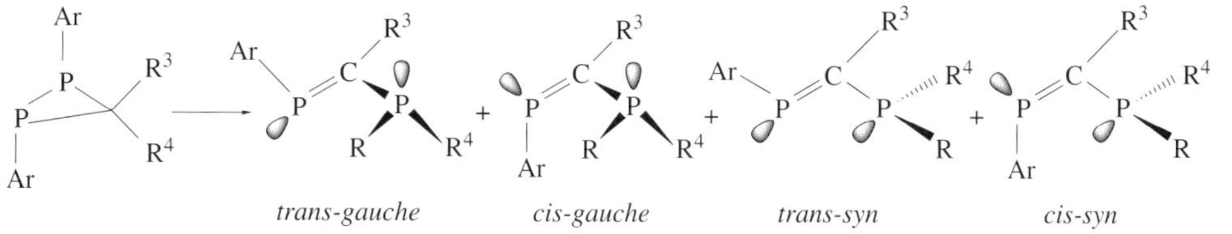

(31)

The theoretical analysis of the possible monomolecular processes for thermal opening of diphosphiranes has revealed two preferred mechanisms, by a cleavage of the P–P bond followed by a halogen migration involving biradical or radical species; either conrotatory opening *via* an *exo–exo* singlet biradical leading to *trans* diphosphapropenes or disrotatory process *via* a radical *exo–endo* species affording *cis* or *trans* diphosphapropenes ⟨94IC596⟩. Only the disrotatory process is in good agreement with the presence of the *cis* isomers. For functionalized diphosphiranes, an initial homolytic cleavage of the exocyclic C–R^4 bond (R^4 = Cl, Br) with formation of cyclic diphosphiranyl radical, which should open relatively easily by P–P bond rupture leading to an *exo–exo* or *exo–endo* radical intermediate, allow the interpretation of all the experimental results. For diphosphiranes bearing no good leaving group substituents (e.g., diphosphaspiropentane (Section 2.2.4.1.7) ⟨92AG(E)625⟩), the biradicalar mechanism appears the most appropriate ⟨94IC596⟩.

More recently, Nguyen and colleagues have carried out an extensive reexploration of (CH_4P_2) potential energy surface related to the transformations of diphosphirane. The authors conclude that the diphosphirane–diphosphapropene ring-chain rearrangements involve unimolecular and concerted processes. In most cases, the ring opening is accompanied by 1,2-hydrogen shift in a single kinetic step and this concerted mechanism proves to be more favorable than pure ring-opening or bond cleavage ⟨99UP2, 99UP3⟩.

In cyclopropane ⟨71AG(E)537, 88JA982⟩ and heterocyclopropane series, the [3 → 2 + 1] cycloelimination reactions have been studied and are conveniently used as a method of generation of silylenes ⟨83JA7776, 87AG(E)776, 88JA3310⟩ or germylenes ⟨88OM2015⟩. In the diphosphirane series where the low P–P bond enthalpy value (50 kcal mol^{-1}) compared to the P–C one (64 kcal mol^{-1}) ⟨B-77MI1⟩ must involve ring opening by a preferential P–P bond rupture, avoiding the fragmentation reactions (The bond enthalpy values do not take into account the strain energies of the ring. In the diphosphirane ring, the calculated strain energy is 14.7 kcal mol^{-1} ⟨90PS(53)7⟩.) However, the cheletropic reactions of diphosphiranes leading to the phosphaalkene and the corresponding phosphinidene can occur under various conditions (Equation (32)):

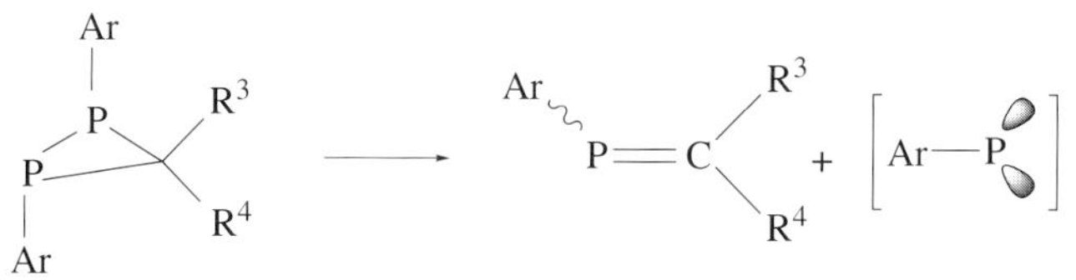

(32)

These cheletropic rearrangements also observed under photochemical irradiation by Koenig and co-workers ⟨87T1793, 90JOC5750⟩ have also been reported by Yoshifuji and co-workers ⟨98HAC607⟩. During the synthesis of diphosphiranes from the symmetrical diphosphene by cyclopropanation, whatever the nature of the carbenoids used (diazo derivatives, carbenes or halogenocarbenes), in all cases the formation as byproducts of the phosphaalkenes and the unstable phosphinidene was observed (5–20%). So, for isopropylidene carbene, the formation of the phosphacumulene, more stable than the parent diphosphirane can reach 30% ⟨91TL3687⟩.

The formation of the phosphaalkenes can be explained either by the addition of excess of carbenes on phosphorus lone pairs of diphosphirane followed by decomposition of the latter, or by the fragmentation of diphosphirane. When an excess of dichloro- or dibromocarbene is added on diphosphirane (**2f**), the formation of the corresponding phosphaalkenes is not detected and the diphosphirane remains unchanged ⟨91TL3687⟩. In contrast, during the sonochemical Simmons–Smith cyclopropanation of diphosphene in THF at 15°C, if the sonication time is prolonged to 4 h, the methylene diphosphirane (**2a**) reacts with the excess of carbenoid, leading to methylene phosphaalkene in quantitative yield ⟨91TL5965⟩. Thus, the addition of methylene carbene on phosphorus lone pairs destabilizes the cyclic form and promotes the cheletropic reaction. It must be noted that diphosphirane (**2a**) is stable without methylene carbene under sonication, and also in the presence of a large excess of diazomethane (Scheme 11).

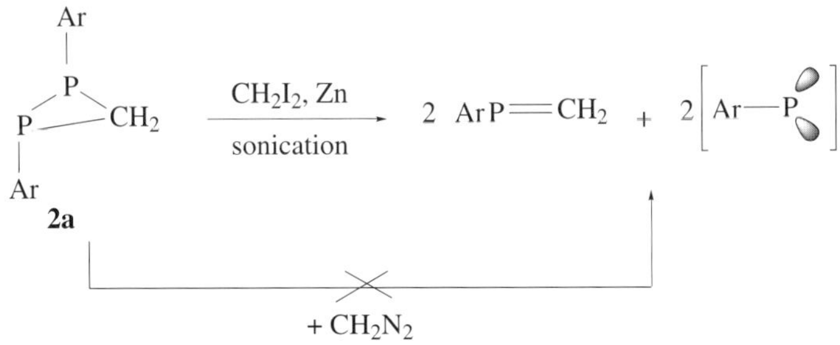

Scheme 11

The fragmentation related to cheletropic reactions of diphosphiranes can also be observed upon heating ⟨94IC596⟩, under photochemical irradiation at 300 nm ⟨90JOC5750⟩ or during reactions with Lewis acids ⟨89TL7411⟩ leading to a variety of substituted phosphaalkenes (minor products). Under these conditions, the fragmentations occur in competition with the electrocyclic ring-opening reactions involving the formation of 1,3-diphosphapropenes as main products.

For unsymmetrical diphosphiranes, whatever the experimental conditions (during the synthesis, under irradiation or heating), the fragmentation products were never observed ⟨91TL3687⟩. This absence of cheletropic reactions is probably due to stronger and shorter P–C bonds for these diphosphiranes than for symmetrical ones (cf. Table 1) ⟨96JOM(514)37⟩.

2.2.4.1.7 *Ring expansion reactions*

Studies on the synthesis and reactivity of oxa- ⟨68JOC3291, 71CC1579⟩ and aza- ⟨73CC33, 74JOC63, 73TL4795⟩ spiropentanes, the heterocyclic analogues of the well-known and highly strained spiropentane system ⟨89JA1501⟩ have led to an improved understanding of chemical bonding, intramolecular interactions, and reaction mechanisms ⟨79AG(E)809, 86AG(E)312⟩. All these spiropentanes rearrange to methylenecyclobutanes ⟨90TL2411⟩. In phosphorus chemistry, the first rearrangement of a diphosphaspiropentane (**2j**) has been observed upon heating at 100°C in degassed toluene leading to the expected diphosphamethylenecyclobutane (70%) but also to the unexpected 1,4-diphosphanorbornadiene (30%) (Equation (33)) ⟨92AG(E)625⟩. This ring expansion reaction was promoted by addition of AIBN.

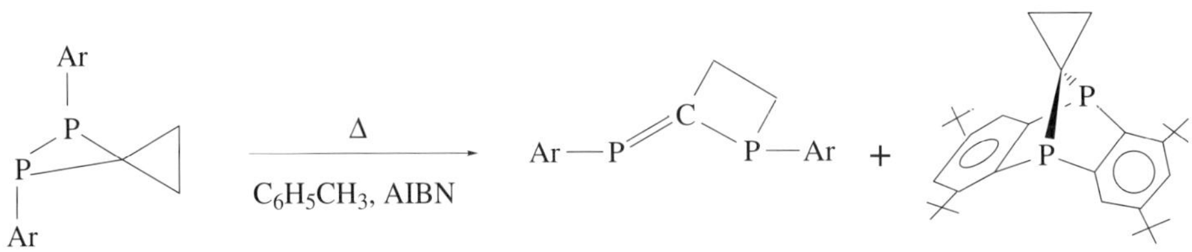

2j (70 %) (30 %) (33)

The diphosphamethylenecyclobutane and the 1,4-diphosphanorbornadiene are formed from the common biradical intermediate after the P–P bond cleavage (**8**):

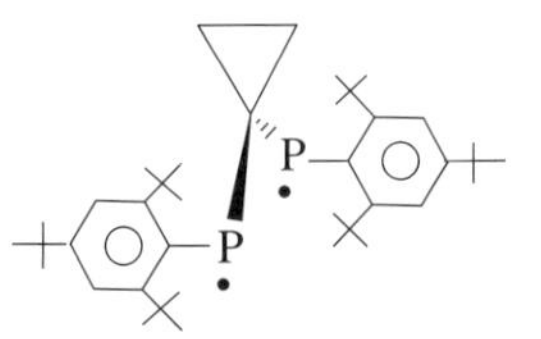

8

The MNDO and *ab initio* calculations performed on unsubstituted derivatives reveal for this reaction a conrotatory process involving an *exo–exo* singlet π-biradical ⟨94IC596⟩.

2.2.4.1.8 *Substitution reactions*

Ring-opening reactions of cyclopropane derivatives are commonly observed; however, there exist only few examples of substitution reactions without ring opening which mainly involve halogen substitutions ⟨83T913, 85JOC3255, 88-MI2⟩. In the diphosphirane series, the ring opening reactions with preferential intracyclic P–P bond rupture were observed. However, the substitution of one aryl group (2,4,6-tri-Bu^t-phenyl) by a chlorine atom leading to the *trans P*-chlorodiphosphirane via the *cis* isomer intermediate has recently been realized in chlorinated solvent under sonication, or in the presence of Brönsted acids or in the presence of gaseous hydrogen chloride in diethylether at room temperature (Scheme 12) ⟨91TL5965, 94-TH⟩. The postulated mechanism probably involved an electrophilic substitution of the aryl group by an acidic proton.

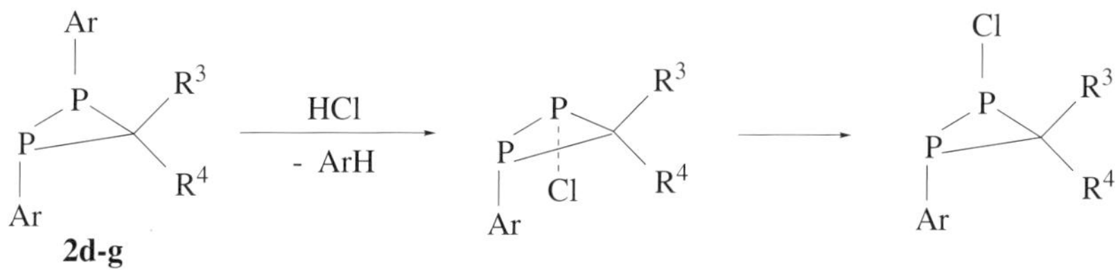

Scheme 12

Whatever the method, starting from a *trans* diphosphirane (**2d–g**), the *trans P*-chlorodiphosphirane is obtained. The substitution of the aryl group by the chlorine atom induces an increase of the ^{31}P NMR chemical shift values (cf. Figure 1). The J_{PP} coupling constant values are slightly higher than in the starting diphosphiranes (e.g., 10–30 Hz). When the reactions were carried out under mild conditions at room temperature, it was possible to detect the formation of cyclic *cis* isomer intermediates ⟨97MGC37⟩.

For diphosphiranes belonging to series 3, the exchange reactions of *P*-chloro derivatives with halosilanes lead to the corresponding *P*-bromo- and *P*-iododiphosphiranes ⟨90CB1245⟩. In the same diphosphirane series, the reduction of the *P*-chlorodiphosphirane to P–H analogous by the tributylstannane method was reported (Equation (7)) ⟨93CB645⟩.

2.2.4.2 Diphosphirenes and Diphosphirenium Salts

The chemical properties of diphosphirenes and diphosphirenium salts were developed recently by Bertrand and co-workers ⟨99ACR561⟩.

2.2.4.2.1 *Reduction reaction*

Electrolytic reduction of the diphosphirenium salt **4d** (or addition of lithium metal) after the ring opening, and migration of amino group, leads to the stable diphosphaallyl radical characterized by ESR spectroscopy (Equation (34)). The splitting pattern has been explained by the presence of two equivalent phosphorus nuclei and one nitrogen atom which exhibit similar hyperfine coupling constants: $a_P = 9.4$ G (2P), $a_N = 9.9$ G (1N) and $a_N = 1.5$ G (2N) ⟨97JA9923⟩. The starting cation **4d** can be regenerated by oxidation.

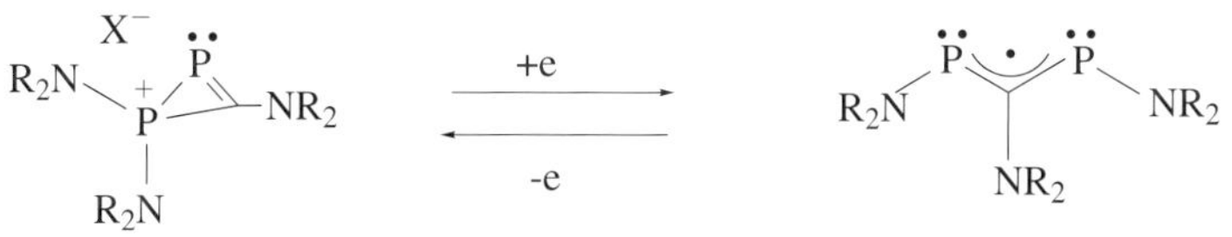

(34)

2.2.4.2.2 *Anionic ring opening*

The ring-opening reactions occur when the diphosphirenium salt was treated with nucleophiles ⟨94CC337⟩ so, **4d** reacts with $LiAlH_4$ leading to the corresponding 1,3-diphosphapropene (as a 50 : 50 mixture of *Z* and *E* isomers) which on treatment with a catalytic amount of BF_3. OEt_2 affords the corresponding diphosphirene **4f** (Equation (15)) ⟨97CC2399⟩.

2.2.4.2.3 Dimerization

Treatment of diphosphirene **4f** with catalytic amounts of BF_3 and triethylamine (5%) in THF at 40°C for 2 h led, after the homolytic cleavage of the P–N bond, to the formation of dimer of the diphosphirenyl radical. Compound **5r**, a tetraphosphabenzene analog of the postulated anti-tricyclohexylene, a singlet biradical valence isomer of benzene, has been isolated (Equation (35)) ⟨98SCI2080⟩.

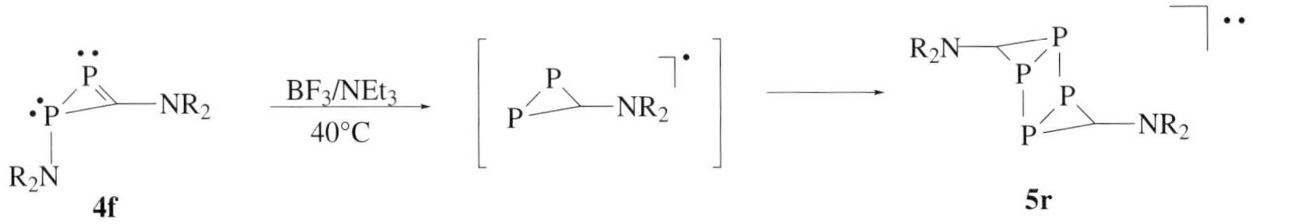

(35)

2.2.4.2.4 Reactions with metal carbonyls

Diphosphirenes, diphosphirenium and diphosphirenylium salts can be stabilized by action of $W(CO)_5$ in the form of η_1 and η_2 complexes (see Section 2.2.2.5; Equations (19)–(23)).

The same η_1 complex **7b** can be obtained by treatment of 1*H*-diphosphirene **4f** with $Fe_2(CO)_9$. Addition of an excess of $Fe_2(CO)_9$ after the P–P bond cleavage, afforded the binuclear and trinuclear complexes (Equation (36)). Treatment of the 1*H*-diphosphirene mononuclear complex **7b** with 2 equivalents of trifluoromethanesulfonic acid led to the diphosphirenylium binuclear complex **7o** ⟨99CC1535⟩.

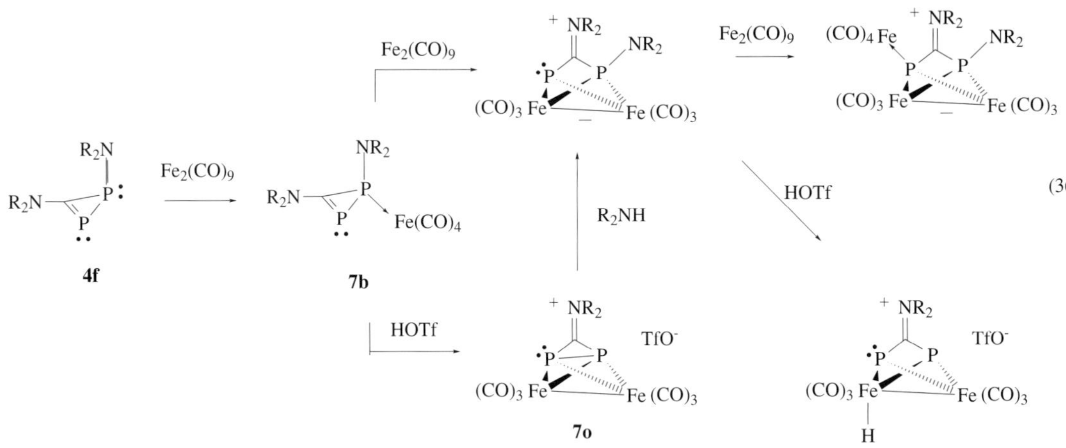

(36)

2.2.4.2.5 Ring expansion reactions

Palladium tetrakis(triphenylphosphane) cleanly reacts at −40°C with one equivalent of diphosphirenium **4d** to yield the 1-pallada(II)-2σ^4,4σ^2-diphosphacyclobutene after P–P bond rupture. When a solution of this complex was refluxed in the presence of the starting diphosphirenium **4d** (0.5 equiv.) a new complex — characterized as pyramidal μ_2-phosphinidene — was formed (Equation (37)) ⟨95AG(E)2677, 95OM3614, 96PS(111)53⟩. The same diphosphirenium **4d** reacts at −50°C with *tert*-butyl isocyanide to give the transient nitrilium salt, which undergoes a ring expansion reaction affording the four-membered heterocycle (Equation (38)) ⟨94CC337⟩.

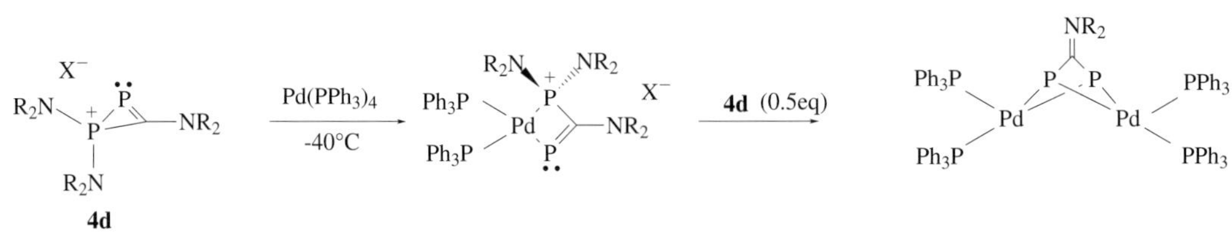

(37)

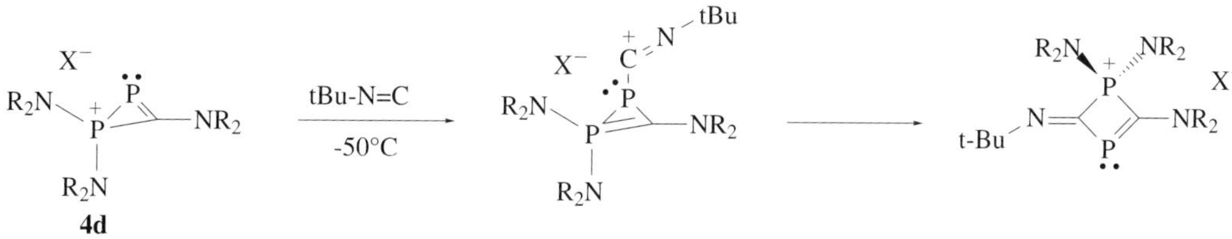

(38)

2.2.4.2.6 Substitution reactions

Various substituents can be introduced in the complexed diphosphirene ring by substitution reaction from **7e** (Equation (39)). So the chlorine atom of the derivative **7e** can easily be substituted for an amino group (**7a**), an azido (**7f**) or a phenyl group (**7g**). The formation of **7g** probably results from the ionization of the P–Cl bond of **7e** leading to a transient diphosphirenylium salt, which abstracts a phenyl group from the tetraphenylborate counterion ⟨97JA9923, 99EJIC1479⟩. The diphosphirenylium salt complex **7l**, treated with various nucleophiles leads to the corresponding 1*H*-diphosphirene complexes **7a–7i**.

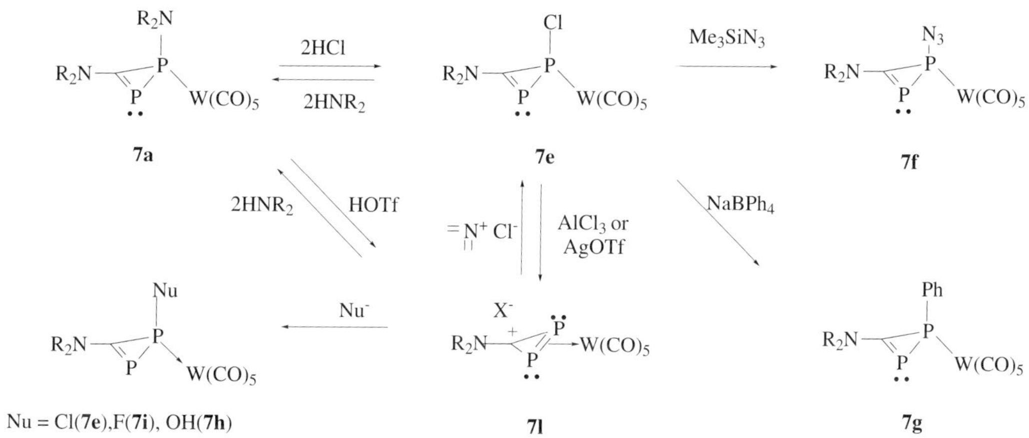

(39)

2.2.5 CONCLUSION

The different synthetic approaches have opened access to various stable or functionalized diphosphiranes and diphosphirenes.

As can be seen from their behavior under various conditions, the reactivity of diphosphiranes is essentially dominated by the ring opening reactions involving a preferential P–P bond rupture. Thus, diphosphiranes can be considered as potential precursors of 1,3-diphosphapropenes, 1,3-diphosphaallenes, 1,3-diphosphaallyl anion or cation, diphosphiranyl or 1,3-diphosphapropanyl biradical intermediates, demonstrating the amazing analogy to pure carbon systems. Moreover, their particular reactivity towards the transition metal complexes has opened original pathways to new complexes allowing, for example, the stabilization of diphosphaallyl cation.

In the diphosphirene series, the stable 1*H*-diphosphirene has been easily obtained, contrary to the corresponding diazirine. The existence of diphosphirenium parent and stable metal carbonyl complexes made it possible to highlight of many examples of opening, cyclic extension and substitution reactions. The particular stabilization of the diphosphirenylium salt (analogous of the cyclopropenium cation) makes it possible to consider in the future a development of the chemistry of new carbenoids or(and) that of complexed P_2 elemental phosphorus.

2.2.6 REFERENCES

63ACS1683 L. Skattebol; *Acta Chem. Scand.*, 1963, **17**, 1683.

68JOC3291 J.K. Crandall and D.R. Paulson; *J. Org. Chem.*, 1968, **33**, 3291.

69AG(E)781 R.B. Woodward and R. Hoffmann; *Angew. Chem., Int. Ed. Engl.*, 1969, **8**, 781.

71AG(E)537 G.W. Griffin; *Angew. Chem., Int. Ed. Engl.*, 1971, **10**, 537.

71CC1579 J. Salaün and J.M. Conia; *J. Chem. Soc., Chem. Commun.*, 1971, 1579.
72TL4347 J.P. Snyder, R.J. Boyd and M.A. Whitehead; *Tetrahedron Lett.*, 1972, 4347.
73CC33 J.K. Crandall and W.W. Conover; *J. Chem. Soc., Chem. Commun.*, 1973, 33.
73TL4795 D.H. Aue, R.B. Lorens and G.S. Helwig; *Tetrahedron Lett.*, 1973, 4795.
74JOC63 J.K. Crandall and W.W. Conover; *J. Org. Chem.*, 1974, **39**, 63.
77JA2631 M.P. Periasamy and H.M. Walborsky; *J. Am. Chem. Soc.*, 1977, **99**, 2631.
B-77MI1 F. Purcell and J.C. Kotz; in "Inorganic Chemistry", W.B. Saunders Co., Philadelphia, 1977, p. 329.
77ZN(B)1490 M. Baudler and B. Carlsohn; *Z. Naturforsch.*, 1977, **32B**, 1490.
78JOC2216 L. Carlsen and J.P. Snyder; *J. Org. Chem.*, 1978, **43**, 2216.
78ZN(B)1208 M. Baudler and F. Saykowski; *Z. Naturforsch.*, 1978, **33B**, 1208.
79AG(E)809 A. de Meijere; *Angew. Chem., Int. Ed. Engl.*, 1979, **18**, 809.
79ZOB471 Z.S. Novikova, A.A. Prishchenko and I.F. Lutsenko; *Zh. Obshch. Khim.*, 1979, **49**, 471.
80PAC755 M. Baudler; *Pure Appl. Chem.*, 1980, **52**, 755.
80ZOB689 A.A. Prishchenko, Z.S. Novikova and I.F. Lutsenko; *Zh. Obshch. Khim.*, 1980, **50**, 689.
81CB1004 R. Gleiter, M.C. Böhm and M. Baudler; *Chem. Ber.*, 1981, **114**, 1004.
81JA4587 M. Yoshifuji, I. Shima, M. Inamoto, K. Hirotsu and T. Higuchi; *J. Am. Chem. Soc.*, 1981, **103**, 4587.
82ZAAC(486)39 M. Baudler and F. Saykowski; *Z. Anorg. Allg. Chem.*, 1982, **486**, 39.
83AG(E)632 M. Baudler and W. Leonhardt; *Angew. Chem., Int. Ed. Engl.*, 1983, **22**, 632.
83JA4026 Th.A. van der Knapp, Th.C. Klebach, R. Lourens, M. Vos and F. Bickelhaupt; *J. Am. Chem. Soc.*, 1983, **105**, 4026.
83JA7776 S. Masamune, S. Murakami, H. Tobita and D.J. Williams; *J. Am. Chem. Soc.*, 1983, **105**, 7776.
83PS(18)35 E. Niecke, R. Rüger, M. Lysek and W.W. Schoeller; *Phosphorus, Sulfur Silicon*, 1983, **18**, 35.
83T913 L.A. Paquette, G.J. Wells, K.A. Horn and T.H. Yan; *Tetrahedron*, 1983, **39**, 913.
84AG(E)157 R. Appel, F. Knoch and H. Kunze; *Angew. Chem., Int. Ed. Engl.*, 1984, **23**, 157.
84CB1542 M. Baudler, F. Saykowski, M. Hintze, K.F. Tebbe, T. Heinlein, A. Vissers and M. Feher; *Chem. Ber.*, 1984, **117**, 1542.
84T2805 D.J. Pasto; *Tetrahedron*, 1984, **40**, 2805.
84TL3687 H.H. Karsch, F.H. Köhler and H.U. Reisacher; *Tetrahedron Lett.*, 1984, **25**, 3687.
84ZC345 I.F. Lutsenko; *Z. Chem.*, 1984, 345.
84ZC352 M. Baudler; *Z. Chem.*, 1984, 352.
85JOC3255 R.C. Gadwood, M.R. Rubino, S.C. Nagarajan and S.T. Michel; *J. Org. Chem.*, 1985, **50**, 3255.
85JOM(282)53 J. Borm, G. Huttner, O. Orama and L. Zsolnai; *J. Organomet. Chem.*, 1985, **282**, 53.
85MI117-01 K.F. Tebbe, T. Heinlein and M. Feher; *Z. Kristallogr.*, 1985, **172**, 89.
86AG(E)253 A. Schmidpeter and S. Lochschmidt; *Angew. Chem., Int. Ed. Engl.*, 1986, **25**, 253.
86AG(E)312 K.B. Wiberg; *Angew. Chem., Int. Ed. Engl.*, 1986, **25**, 312.
86AG(E)454 H.H. Karsch, H.-U. Reisacher and G. Müller; *Angew. Chem., Int. Ed. Engl.*, 1986, **25**, 454.
86CB2609 R. Appel, C. Casser and F. Knoch; *Chem. Ber.*, 1986, **119**, 2609.
86TL1145 J. Bellan, G. Etemad-Moghadam, M. Payard and M. Koenig; *Tetrahedron Lett.*, 1986, **27**, 1145.
86ZN(B)444 S. Lochschmidt, G. Müller, B. Huber and A. Schmidpeter; *Z. Naturforsch.*, 1986, **41B**, 444.
87AG(E)764 P. Binger, B. Biedenbach, C. Krüger and M. Regitz; *Angew. Chem., Int. Ed. Engl.*, 1987, **26**, 764.
87AG(E)776 A. Schäfer, M. Wiedenbruch, W. Saak and S. Pohl; *Angew. Chem., Int. Ed. Engl.*, 1987, **26**, 776.
87AG(E)908 R. Milczarek, W. Rüsseler, P. Binger, K. Jonas, K. Angermund, C. Krüger and M. Regitz; *Angew. Chem., Int. Ed. Engl.*, 1987, **26**, 908.
87AG(E)1011 E.P.O. Fuchs, W. Rösch and M. Regitz; *Angew. Chem., Int. Ed. Engl.*, 1987, **26**, 1011.
87CB421 M. Baudler and J. Simon; *Chem. Ber.*, 1987, **120**, 421.
87CC639 A.M. Caminade, J.P. Majoral, R. Mathieu and Y.Y.C. Yeung Lam Ko; *J. Chem. Soc., Chem. Commun.*, 1987, 639.
87JOM(327)C63 L. Weber and E. Lücke; *J. Organomet. Chem.*, 1987, **327**, C63.
87JOM(336)163 D. Seyferth, T.G. Wood and R.S. Henderson; *J. Organomet. Chem.*, 1987, **336**, 163.
87PS(30)425 M. Koenig, G. Etemad-Moghadam, C. Tachon and J. Bellan; *Phosphorus Sulfur*, 1987, **30**, 425.
87T1793 G. Etemad-Moghadam, J. Bellan, C. Tachon and M. Koenig; *Tetrahedron*, 1987, **43**, 1793.
88AG(E)389 G. Märkl, S. Dietl, M.L. Ziegler and B. Nuber; *Angew. Chem., Int. Ed. Engl.*, 1988, **27**, 389.
88AG(E)837 A.R. Barron, A.H. Cowley, S.W. Hall and C.M. Nunn; *Angew. Chem., Int. Ed. Engl.*, 1988, **27**, 837.
88AG(E)957 R. Appel, B. Niemann and M. Nieger; *Angew. Chem., Int. Ed. Engl.*, 1988, **27**, 957.
88CB281 M. Baudler and J. Simon; *Chem. Ber.*, 1988, **121**, 281.
88CC945 E. Niecke, O. Altmeyer and M. Nieger; *J. Chem. Soc., Chem. Commun.*, 1988, 945
88JA982 K.M. Wilson and K.A. Schnapp; *J. Am. Chem. Soc.*, 1988, **110**, 982.
88JA3310 W. Ando, F. Fujita, H. Yoshida and A. Sekiguchi; *J. Am. Chem. Soc.*, 1988, **110**, 3310.
88OM978 L. Weber, E. Lücke and R. Boese; *Organometallics*, 1988, **7**, 978.
88OM2015 T. Tsumuraya, S. Sato and W. Ando; *Organometallics*, 1988, **7**, 2015.
88-MI2 J. Salaün; in "Topics in Current Chemistry: Small Ring Compounds in Organic Synthesis", ed. A. de Meijere, Springer Verlag, Berlin, 1988.
89AG(E)1673 E. Niecke, R. Streubel, M. Nieger and D. Stalke; *Angew. Chem., Int. Ed. Engl.*, 1989, **28**, 1673.
89CC593 M. Gouygou, J. Bellan, J. Escudié, C. Couret, A. Dubourg, J.P. Declercq and M. Koenig; *J. Chem. Soc., Chem. Commun.*, 1989, 593.
89CC1732 M. Yoshifuji, S. Sasaki and N. Inamoto; *J. Chem. Soc., Chem. Commun.*, 1989, 1732.

89JA1501 R. Boese, D. Bläser, K. Gomann and U.H. Brinker; *J. Am. Chem. Soc.*, 1989, **111**, 1501.

89NJC315 M. Gouygou, C. Tachon, M. Koenig and G. Etemad-Moghadam; *New J. Chem.*, 1989, **13**, 315.

89NJC891 M. Rahmoune, Y.Y.C. Yeung Lam Ko, R. Carrié and F. Tonnard; *New J. Chem.*, 1989, **13**, 891.

89TL177 M. Gouygou, C. Tachon, R. El-Ouatib, O. Ramarijaona, G. Etemad-Moghadam and M. Koenig; *Tetrahedron Lett.*, 1989, **30**, 177.

89TL187 M. Yoshifuji, S. Sasaki, T. Niitsu and N. Inamoto; *Tetrahedron Lett.*, 1989, **30**, 187.

89TL459 E. Niecke and D. Barion; *Tetrahedron Lett.*, 1989, **30**, 459.

89TL7411 M. Gouygou, C. Tachon, G. Etemad-Moghadam and M. Koenig; *Tetrahedron Lett.*, 1989, **30**, 7411.

90CB67 E. Niecke, H.J. Metternich and R. Streubel; *Chem. Ber.*, 1990, **123**, 67.

90CB653 G. Hessler, P.C. Knüppel, O. Stelzer and W.S. Sheldrick; *Chem. Ber.*, 1990, **123**, 653.

90CB1245 R. Streubel and E. Niecke; *Chem. Ber.*, 1990, **123**, 1245.

90CC507 J.P. Majoral, N. Dufour, F. Meyer, A.M. Caminade, R. Choukroun and D. Gervais; *J. Chem. Soc., Chem. Commun.*, 1990, 507.

90CRV997 F. Mathey; *Chem. Rev.*, 1990, **90**, 997.

90JOC5750 M. Gouygou, C. Tachon, M. Koenig, A. Dubourg, J.-P. Declercq, J. Jaud and G. Etemad-Moghadam; *J. Org. Chem.*, 1990, **55**, 5750.

90JOM(399)C4 F. Bitterer, D.J. Brauer, F. Dörrenbach and O. Stelzer; *J. Organomet. Chem.*, 1990, **399**, C4.

90PS(53)7 M. Liu and S.M. Bachrach; *Phosphorus, Sulfur Silicon*, 1990, **53**, 7.

90PS(49-50)297 R. Appel; *Phosphorus, Sulfur Silicon*, 1990, **49–50**, 297.

90PS(49-50)305 M. Koenig, M. Gouygou, C. Tachon, R. El-Ouatib and G. Etemad-Moghadam; *Phosphorus, Sulfur Silicon*, 1990, **49–50**, 305.

90TL2411 T. Miyashi, Y. Takahashi, H. Ohaku, K. Yokogawa, S.I. Morishima and T. Mukai; *Tetrahedron Lett.*, 1990, **31**, 2411.

91AG(E)90 E. Niecke, R. Streubel and M. Nieger; *Angew. Chem., Int. Ed. Engl.*, 1991, **30**, 90.

91CC124 M. Yoshifuji, K. Toyota, H. Yoshimura, K. Hirotsu and A. Okamoto; *J. Chem. Soc., Chem. Commun.*, 1991, 124.

91HAC221 M. Gouygou, M. Koenig, J. Escudié and C. Couret; *Heteroatom Chem.*, 1991, **2**, 221.

91JA8160 F. Castan, A. Baceiredo, J. Fischer, A. De Cian, G. Commenges and G. Bertrand; *J. Am. Chem. Soc.* 1991, **113**, 8160.

91JOC3438 M. Gouygou, M. Koenig, M.-J. Hervé, D. Gonbeau and G. Pfister-Guillouzo; *J. Org. Chem.*, 1991, **56**, 3438.

91TL3687 G. Etemad-Moghadam, C. Tachon, M. Gouygou and M. Koenig; *Tetrahedron Lett.*, 1991, **32**, 3687.

91TL5965 G. Etemad-Moghadam, M. Rifqui, P. Layrolle, J. Berlan and M. Koenig; *Tetrahedron Lett.*, 1991, **32**, 5965.

91ZN(B)1131 F. Bitterer, D.J. Brauer, F. Dörrenbach, F. Gol, P.C. Knüppel, O. Stelzer, C. Krüger and Y.H. Tsay; *Z. Naturforsch.*, 1991, **46B**, 1131.

92AG(E)625 C. Garot, G. Etemad-Moghadam, J.-P. Declercq, A. Dubourg and M. Koenig; *Angew. Chem., Int. Ed. Engl.*, 1992, **31**, 625.

92AG(E)1055 B. Breit, U. Bergsträsser, G. Maas and M. Regitz; *Angew. Chem., Int. Ed. Engl.*, 1992, **31**, 1055.

92CJC2834 J.P. Fayet, D. Mermillod-Blardet, C. Tachon, G. Etemad-Moghadam and M. Koenig; *Can. J. Chem.*, 1992, **70**, 2834.

92CB1553 L. Weber, R. Kirchhoff, H.-G. Stammler and B. Neumann; *Chem. Ber.*, 1992, **125**, 1553.

92CRV1839 L. Weber; *Chem. Rev.* 1992, **92**, 1839.

92IC2414 C. Tachon, M. Gouygou, M. Koenig, M.J. Hervé, D. Gonbeau and G. Pfister-Guillouzo; *Inorg. Chem.*, 1992, **31**, 2414.

92JOM(436)169 R. El-Ouatib, C. Garot, G. Etemad-Moghadam and M. Koenig; *J. Organomet. Chem.*, 1992, **436**, 169.

92ZAAC(610)75 P. Jutzi and S. Opiela; *Z. Anorg. Allg. Chem.*, 1992, **610**, 75.

92ZAAC(617)53 D. Lentz and R. Marschall; *Z. Anorg. Allg. Chem.*, 1992, **617**, 53.

93CB645 R. Streubel, M. Nieger and E. Niecke; *Chem. Ber.*, 1993, **126**, 645.

93CB1963 L. Weber, R. Kirchhoff and R. Boese; *Chem. Ber.*, 1993, **126**, 1963.

93CC569 M. Lazraq, J. Escudié, C. Couret, U. Bergsträsser and M. Regitz; *J. Chem. Soc., Chem. Commun.*, 1993, 569.

93JOM(453)77 R. El-Ouatib, D. Ballivet-Tkatchenko, G. Etemad-Moghadam and M. Koenig; *J. Organomet. Chem.*, 1993, **453**, 77.

93JMR(A)258 R. Challoner, C.A. McDowell, M. Yoshifuji, K. Toyota and J.A. Tossell; *J. Magn. Res., Ser. A*, 1993, **104**, 258.

93OM731 L. Weber, R. Kirchhoff, R. Boese, H.G. Stammler and B. Neumann; *Organometallics,* 1993, **12**, 731.

93PS(76)41 G. Etemad-Moghadam, R. El-Ouatib, D. Ballivet-Tkatchenko and M. Koenig; *Phosphorus, Sulfur Silicon*, 1993, **76**, 41.

93PS(77)229 R. Streubel, N.H. Rran Huy, L. Ricard and F. Mathey; *Phosphorus, Sulfur Silicon*, 1993, **77**, 229.

93S763 R. Streubel, N.H. Tran Huy and F. Mathey; *Synthesis*, 1993, 763.

93T10279 A. Marinetti, L. Ricard, F. Mathey, M. Slany and M. Regitz; *Tetrahedron*, 1993, **49**, 10279.

93-TH C. Garot; *Thesis*, 1993, Toulouse.

94CC337 M. Soleilhavoup, A. Baceiredo, F. Dahan and G. Bertrand; *J. Chem. Soc., Chem. Commun.*, 1994, 337.

94CB969 N.J. Fitzpatrick, D.F. Brougham, P.J. Groazke and M.T. Nguyen; *Chem. Ber.*, 1994, 969.

94IC596 M.J. Hervé, G. Etemad-Moghadam, M. Gouygou, D. Gonbeau, M. Koenig and G. Pfister-Guillouzo; *Inorg. Chem.*, 1994, **33**, 596.

94JA6149	M. Soleilhavoup, Y. Canac, A.M. Polozov, A. Baceiredo and G. Bertrand; *J. Am. Chem. Soc.*, 1994, **116**, 6149.
94OM4406	L. Weber, S. Buchwald, D. Lentz, O. Stamm, D. Preugschat and R. Marschall; *Organometallics*, 1994, **13**, 4406.
94-TH	M. Rifqui; *Thesis*, 1994, Toulouse.
94ZK848	M. Nieger, G. Bruder and R. Appel; *Z. Kristal.*, 1994, **209**, 848.
95AG(E)2677	Y. Canac, A. Baceiredo, H. Gornitzka, D. Stalke and G. Bertrand; *Angew. Chem., Int. Ed. Engl.*, 1995, **34**, 2677.
95CC2113	Streubel, L. Ernst, J. Jeske and P.G. Jones; *J. Chem. Soc., Chem. Commun.*, 1995, 2113.
95HAC33	D.L. An, K. Toyota, M. Yasunami and M. Yoshifuji; *Heteroatom Chem.*, 1995, **6**, 33.
95BSCF652	H. Heydt, U. Bergsträsser, R. Fässler, E. Fuchs, N. Kamel, T. Mackewitz, G. Michels, W. Rösch, M. Regitz, P. Mazerolles, C. Laurent and A. Faucher; *Bull. Soc. Chim. Fr.*, 1995, **132**, 652.
95OM3614	Y. Canac, M. Soleilhavoup, L. Ricard, A. Baceiredo and G. Bertrand; *Organometallics*, 1995, **14**, 3614.
B-96MI2	G. Etemad-Moghadam and M. Koenig; in "Comprehensive Heterocyclic Chemistry II", ed. A.K. Katritzky, Pergamon, 1996, Vol. 1, p. 469.
96JOM(514)37	M. Gouygou, M. Veith, C. Couret, J. Escudié, V. Huch and M. Koenig; *J. Organomet. Chem.*, 1996, **514**, 37.
96PS(111)53	Y. Canac, G. Bouhadir, R. Reau, A. Baceiredo and G. Bertrand; *Phosphorus, Sulfur Silicon*, 1996, **111**, 53.
97CC2399	D. Bourissou, Y. Canac, M.I. Collado, A. Baceiredo and G. Bertrand; *Chem. Commun.*, 1997, 2399.
97JA9923	D. Bourissou, Y. Canac, M.I. Collado, A. Baceiredo and G. Bertrand; *J. Am. Chem. Soc.*, 1997, **119**, 9923.
97MGC37	M. Rifqui, G. Etemad-Moghadam, J. Berlan and M. Koenig; *Main Group Chem.*, 1997, **2**, 37.
97PS(124-125)545	U. Rohde, H. Wilkens and R. Streubel; *Phosphorus, Sulfur Silicon*, 1997, **124–125**, 545.
98HAC607	K. Tsuji, S. Sasaki and M. Yoshifuji; *Heteroatom Chem.*, 1998, 607.
98S125	T.W. Mackewitz and M. Regitz; *Synthesis*, 1998, 125.
98S427	A. Nachbauer, U. Bergsträsser, S. Leininger and M. Regitz; *Synthesis*, 1998, 427.
98SCI2080	Y. Canac, D. Bourissou, A. Baceiredo, H. Gornitzka, W.W. Schoeller and G. Bertrand; *Science*, 1998, **279**, 2080.
99ACR561	D. Bourissou and G. Bertrand; *Acc. Chem. Res.*, 1999, **32**, 561.
99CC1535	D. Bourissou, Y. Canac, H. Gornitzka, A. Baceiredo and G. Bertrand; *Chem. Commun.*, 1999, 1535.
99EJIC1479	D. Bourissou, Y. Canac, H. Gornitzka, C.J. Marsden, A. Baceiredo and G. Bertrand; *Eur. J. Inorg. Chem.*, 1999, 1479.
99UP1	M. Rifqui, K.B. Dillon and M. Koenig; unpublished results.
99UP2	M.T. Nguyen, L. Landuyt, A. Dransfeld and L.G. Vanquickenborne; unpublished results.
99UP3	M.T. Nguyen, S. Creve, L. Landuyt and L.G. Vanquickenborne; unpublished results.

3.1
Phosphetanes — Early Studies

ANGELA MARINETTI and DUNCAN CARMICHAEL
Ecole Nationale Supérieure de Chimie de Paris, Paris, France

3.1.1 INTRODUCTION

This short review covers early studies of the phosphetane ring, up to 1982, and complements the chapter concerning modern aspects of four-membered rings (Chapter 3.2).

The first general synthetic approach to phosphetanes was reported by J.J. McBride and coworkers in 1962 ⟨62JOC606, 62JOC1833⟩ and, for several years, research was essentially limited to a single main series of phosphetane derivatives bearing three-to-five methyl substituents on the ring carbons. These compounds were developed widely. Most studies treated stereochemical aspects, from assignments by NMR spectroscopy to theoretical analyses of stereoselective transformations. Given that the four-membered ring is quite stable under normal conditions (only C-unsubstituted trivalent phosphetanes undergo rapid decomposition and polymerisation reactions) it is surprising that phosphetanes were not used as synthetic intermediates and have only found applications in catalysis very recently. The explanation probably lies in the restricted choice of structures which were available through McBride's method. Recently, synthetic methodology has improved and new developments and applications of phosphetane chemistry are to be expected.

3.1.2 THEORETICAL METHODS

No theoretical treatments appeared prior to 1984 except a single paper of semiempirical (CNDO/2) calculations treating pyramidal inversion barriers in four-membered rings containing first- and second-row elements. It gave an inversion barrier of 44.3 kcal mol^{-1} for 1-methylphosphetane 〈71JA6507〉. The reader is referred to Chapter 3.2 for more recent studies.

3.1.3 EXPERIMENTAL STRUCTURAL METHODS

3.1.3.1 X-Ray Diffraction Studies

Many crystal structures of phosphetane derivatives have been determined, usually to establish stereochemistry. The area should be approached with caution. Many early assignments of *cis–trans* isomers were extrapolated from an X-ray crystal structure of the phosphetanium salt (**2**). Later studies showed that the structurally characterised *cis* isomer (**2**) was the minor reaction product, so assignments based upon correlations with physicochemical data attributed to this compound are incorrect. They should be reversed 〈70CC616〉.

The first crystallographic studies, of the *trans*-phosphetane oxide (**1**) 〈68CC1228, 70JCS(B)938〉, a 2,3,4-trimethylphosphetanic acid 〈68CC1051〉 and the *cis*-phosphetanium bromide (**2**) 〈69JA2255〉, gave bond lengths and angles which are quite representative of these general classes. Similar ring puckerings, conformational preferences, bond angles and distances are found in related series of phosphetane oxides, sulphides and phosphetanium salts 〈78JOC3513, 79AX2601, 82AX2944〉. These studies show that the direction of ring puckering is controlled by the cross-ring interactions between the substituents at C(3) and P. Thus, the solid state conformation of the *trans*-phosphetane oxide (**1**) shows the 3-Me substituent oriented equatorially. The *trans*- and *cis*-1-chloro-2,2,3,4,4-tetramethylphosphetane oxides 〈70JCS(B)934, 71JCS(B)117〉 also have equatorial methyl groups.

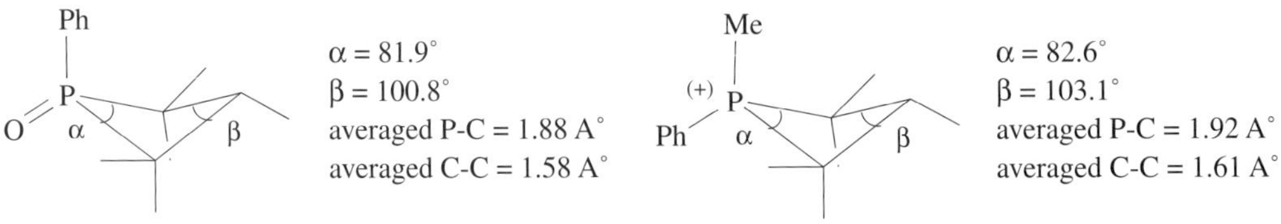

(**1**) (**2**)

dihedral angles C(2)PC(4) / C(2)C(3)C(4) = 23.4° for (**1**) and 24° for (**2**)

Scheme 1

X-ray diffraction studies have also allowed the stereochemistry of phosphetanium salt (**3**) 〈81JCS(P2)1138〉 and the *exo*/*endo* stereochemistry in tetracyclic phosphetane derivatives such as (**4**) to be determined 〈78JCS(P2)1115, 80JCS(P1)2136〉.

Spirocyclic phosphoranes incorporating phosphetane rings have essentially square-pyramidal geometries in the solid state, with only slight distortions towards trigonal bipyramidality 〈78JA433〉. For instance, compounds (**5**) 〈73CC856, 78JA6327〉 and (**6**) 〈78IC3265〉 are located only 16% and 13% of the distance along the Berry coordinate leading towards the trigonal bipyramidal geometry. A trigonal bipyramidal arrangement with the phosphetane adopting the unfavoured diequatorial position has only been found in the very strained compound (**7**) 〈76CC449〉.

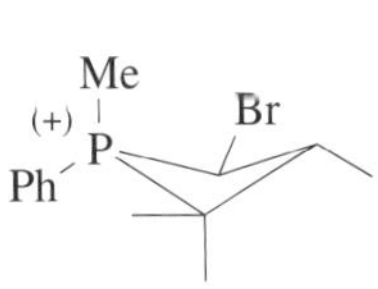

(**3**)

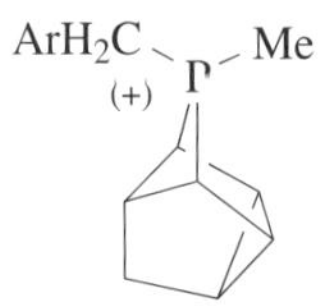

(**4**)

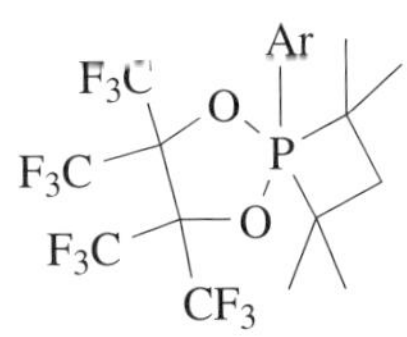

(**5**)

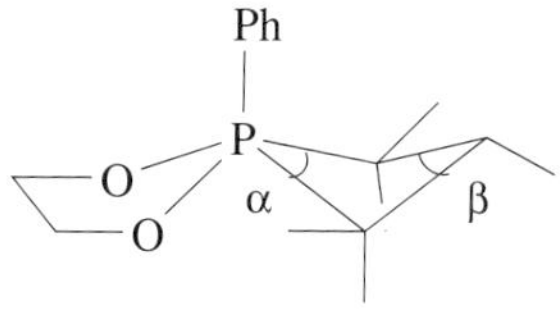

$\alpha = 76.2°$
$\beta = 98.5$ A°
averaged P-C =1.90 A°
averaged C-C = 1.55 A°

(6)

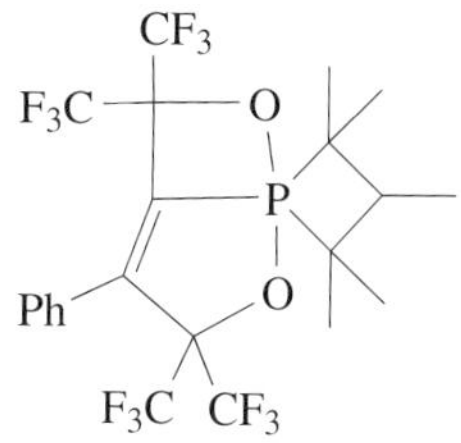

(7)

Scheme 2

3.1.3.2 NMR Studies

The geometrical constraints within the phosphetane ring allow stereochemical assignments to be made straightforwardly on the basis of detailed ^{1}H and ^{13}C NMR studies 〈B-81MI1〉. Selected trends are summarised here. More recent papers giving complete NMR characterisation of new phosphetanes may be found in Chapter 3.2.

3.1.3.2.1 ^{1}H NMR

In trivalent phosphetanes and their oxides, assignment of configuration at the α-carbon can be made from $^2J_{\text{H-P}}$ coupling constants (Scheme 3).

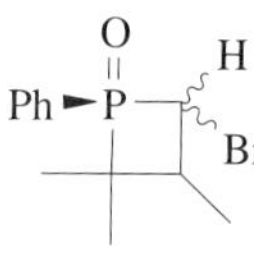

$^2J_{\text{H-P}}$:
trans- isomer 2.5 Hz
cis-isomer 13.5 Hz

(8)

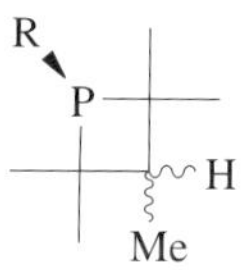

$^3J_{\text{H-P}}$:
cis- isomer 3.5 Hz
trans-isomer 0 Hz

(9)

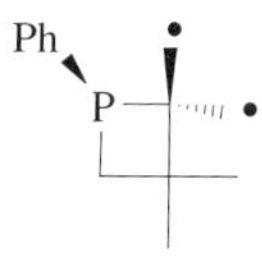

$^3J_{\text{H-P}}$:
cis- Me group 6 Hz
trans-Me group 18 Hz

(10)

Scheme 3

These couplings respect classical Karplus relationships with small numerical values of $^2J_{\text{H-P}}$ identifying protons *trans* to the phosphorus lone pair (or oxygen atom in the corresponding oxides) 〈81JCS(P2)1138〉. In trivalent phosphetanes, large $^3J_{\text{H-P}}$ couplings can also be used diagnostically to identify hydrogen atoms at C(3) lying *cis* to the phosphorus lone-pair. Large couplings are also associated with methyl groups bound to the α-carbon atoms and oriented *cis* to the phosphorus lone-pair 〈67JOC4066, 76JA2109〉. The chemical shift of the hydrogen bound to C-3 in *trans* isomers of (**9**) lies 0.4–0.7 p.p.m. downfield from the *cis* isomer: when R = *p*-BrPh, the chemical shifts are at 2.57 and 2.08 ppm, respectively 〈70CC616, 73JCS(P1)1300〉.

In phosphetane oxides, both *cis*- and *trans*-lying α-methyl groups give large $^3J_{\text{H-P}}$ couplings. However, regular trends in chemical shifts and couplings are observed within a series of homologues such as (**11**) 〈73JCS(D)2641〉. The methyl region of the ^{1}H NMR spectrum also shows distinct differences between *cis* and *trans* diastereoisomers within the related series (**12**) 〈71CC721, 73JA4681〉.

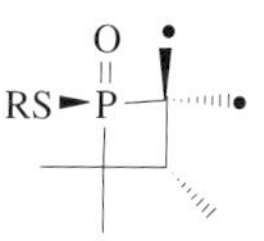

cis- Me group
δ=1.27-1.29 $^3J_{\text{H-P}}$= 20 Hz
trans-Me group
δ=1.33-1.35 $^3J_{\text{H-P}}$= 18 Hz

(**11**)

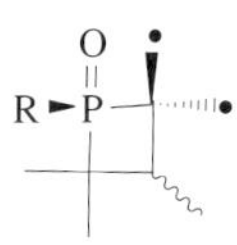

(**12**)

α-Me shifts and couplings (C_6D_6):

	trans-isomer	Δδ	*cis*-isomer	Δδ
R=OMe	1.00[18]; 1.12 [19]	0.12	0.94[18]; 1.22[19]	0.28
NMe_2	1.04[18]; 1.17 [18]	0.13	0.97[17]; 1.21[18]	0.24
Cl	1.03[23]; 1.14 [21]	0.11	0.97[23]; 1.24 [21]	0.27

Scheme 4

3.1.3.2.2 ^{13}C NMR

A tentative estimation of ring contraction effects upon the ^{13}C chemical shifts of the ring carbons has appeared. Generally, shielding seems to occur when going from five- to four-membered rings, but clear-cut trends were not easily established because of superimposed substituent effects 〈76JA2109〉. Within a homologous series, the C(2) and C(4) resonances move to low field upon increasing the electronegativity of the exocyclic phosphorus substituent. The reverse trend is observed for C(3), whose shift is very sensitive to the phosphorus substitution pattern and oxidation state ⟨72JOC3458⟩.

In trivalent phosphetanes, unusually small intracyclic ${}^1J_{C-P}$ couplings are observed. They have been discussed in terms of angular distortions which confer increased *p*-character upon the intracyclic bonds ⟨74CC451, 76JA2109⟩. Ring substituents show characteristic J_{C-P} coupling constants which allow their orientation with respect to the phosphorus substituent to be established: α-methyl groups *trans* to the phosphorus substituent show ${}^2J_{C-P}$ couplings which are much larger in magnitude than the *cis*-methyl groups. The reverse situation applies for β-substituents ⟨72CC367, 72JOC3470, 76JA2109⟩.

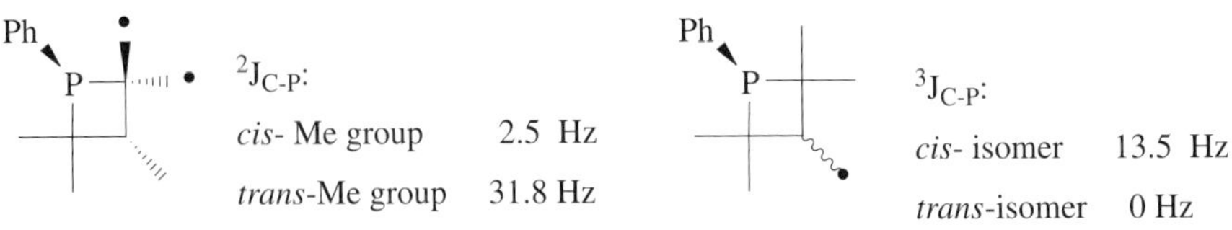

Scheme 5

In phosphetane oxides (**1**) and their analogues, ${}^2J_{C-P}$ couplings to the α-methyl groups are essentially insensitive to isomerism and stereochemistry. Thus, early discriminations between α-methyl groups were based mainly on chemical shifts, with the higher field shifts being associated with pseudoaxial carbons ⟨72JOC3458⟩. The C(3) carbon and its methyl substituent are uniformly found at lower fields in *cis* oxides ⟨71TL3061, 72JOC3458, 72JOC3470, 81JCS(P2)1138⟩. Values for the intracyclic ${}^2J_{C-P}$, exocyclic ${}^1J_{C-P}$ and ${}^3J_{C-P}$ coupling constants for β-methyl groups are also stereodependent: for instance, the *cis*-isomer of (**1**) shows larger ${}^2J_{C-P}$ values for C(3), lower ${}^3J_{C-P}$ values for the methyl group on C(3) and smaller exocyclic ${}^1J_{C-P}$ ($\Delta\delta = 5.0$ Hz) couplings than the *trans* isomer. No general rule can be applied for couplings of *cis* and *trans* methyls found at C(3) because they reflect the other substituents in the ring (cf. (**1**) and (**13**)). These groups cannot always be distinguished by simple means ⟨72JOC3458⟩.

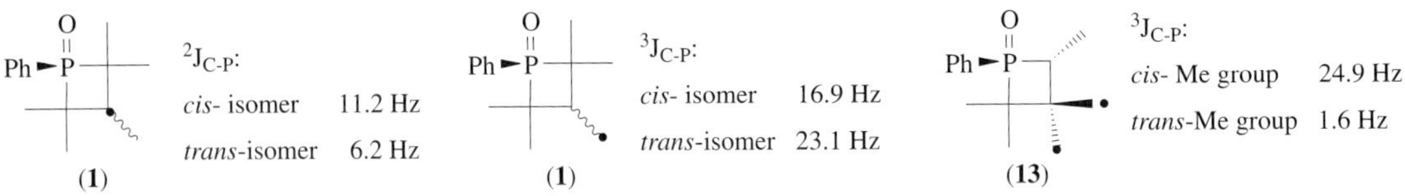

Scheme 6

Representative ^{13}C NMR spectra for phosphetanium salts are available. *Cis–trans* assignments in 1-chloro-1-R-2,2,3,4,4-pentamethylphosphetanium salts have been made on the basis of relative chemical shifts and J_{P-C} couplings to the C(3) carbons: lower field shifts and smaller magnitudes of coupling constant are found in isomers having R-*cis* to the 3-Me group ⟨72JOC3470, 73JOC3199, 76JA2109⟩.

3.1.3.2.3 ³¹P NMR

It became clear from very early studies that the four-membered ring confers large downfield ^{31}P NMR shifts upon trivalent phosphetanes with respect to their cyclic and acyclic analogues. For comparison, 1-phenylphosphetane, 1-phenylphospholane and diethylphenylphosphine show δ^{31}P at +13.9, −15.3 and −15.5 p.p.m., respectively. Representative ^{31}P chemical shifts from the early literature are given in Scheme 7 ⟨73JCS(D)1576, 73JCS(D)2701, 73JCS(P1)1300, 76JA768, 83JOC4466, 84JA7021⟩. These shifts have been rationalised recently in terms of a balance between two opposing effects, a small HOMO–LUMO gap and high *p*-character in the intracyclic P–C bonds ⟨97HAC451⟩.

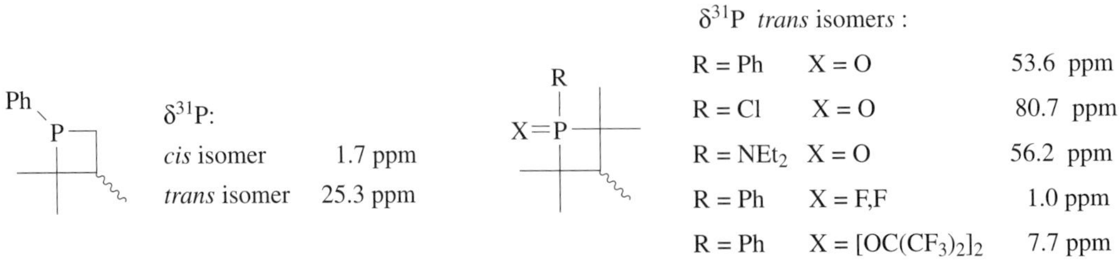

Scheme 7

Within a closely related series of compounds, ^{31}P NMR shifts correlate with the stereochemistry of the C(3) substituent, with trivalent phosphetanes giving more downfield signals for *trans*-isomers than for *cis* ⟨70CC616⟩. However, the situation reverses for isomeric pairs of phosphetane oxides and sulphides where the *trans*-derivatives appear upfield from the *cis* ⟨87SYN265, 92JOC511⟩. Isomeric pairs of phosphetanium salts show only small chemical shift differences.

Compilations of chemical shifts for 1-alkoxy and 1-aryloxy, 1-alkylthio and 1-arylthio, and a series of 1-amino-2,2,3,4,4-pentamethylphosphetane oxides are available in the early literature ⟨74JCS(D)633, 73JCS(D)2641, 73JCS(D)1576⟩.

^{31}P NMR chemical shifts for dioxaphosphoranes have been found to fall in a range from +30 to −5 p.p.m. A tentative correlation between chemical shift and solid state geometry involving the measurement of solid state ^{31}P NMR spectra for several phosphoranes indicated that trigonal bipyramid geometries induce more shielded ^{31}P NMR resonances than square (or rectangular) planar pyramidal arrangements ⟨82JA230⟩.

3.1.3.3 IR Spectroscopy and Mass Spectrometry

Infrared spectra show characteristic phosphetane modes which appear as a series of eight bands between 1260 and 400 cm^{-1}. The highest frequency absorptions, assigned to ν_{as} (CPC), occur at 1230–1236, 1248–1253, 1230–1239, 1257 and 1242 cm^{-1} in phosphetanic amides, esters, thioesters, chloride and acid, respectively. The ν(P=O) bands in these compounds fall close to 1200 cm^{-1}. The corresponding Raman spectra were also reported ⟨73JCS(D)1576, 73JCS(D)2641, 73JCS(D)2701⟩.

1-Ethyl-2,2,3-trimethylphosphetane oxide gives a strong ν(P=O) band at 1205 cm^{-1} and a weaker band at 1241 cm^{-1}. These are shifted to 1187 and 1157 cm^{-1} by exchange with ^{18}O-enriched water. Such "phosphoryl doublets" probably result from Fermi resonance, although the possibility of rapidly equilibrating conformers has also been discussed. The mean ν(P=O) frequency (1223 cm^{-1}) shows a higher phosphoryl force constant than in phospholane analogues (1181 cm^{-1}). This has been attributed to changes in hybridisation of the phosphorus orbitals resulting in greater s-character in the orbitals forming exocyclic bonds ⟨74JA5189⟩.

Electron impact mass spectra of phosphetane oxides, phosphetanic amides, halides and esters have been studied. There appear to be no characteristic fragmentation patterns, so ring cleavage may occur at each ring position ⟨73JCS(D)1576, 73JCS(D)2641, 77OMS624⟩.

3.1.4 THERMODYNAMIC ASPECTS

3.1.4.1 Physical Properties and Physical Constants

This section emphasises some thermodynamic properties in as far as they are specifically related to the four-membered ring structure. These properties are generally dominated by effects resulting from ring strain.

Phosphetanic acids such as (**14**) are stronger acids than their open-chain analogues because of the electron withdrawing effect of the strained ring ⟨57JCS3739, 72JOC3342⟩. Conversely, phosphetane oxides are weaker bases. So, for example, phenol forms weaker hydrogen bonds to phosphetane oxides than to more classical phosphine oxides ⟨68ACS1866⟩.

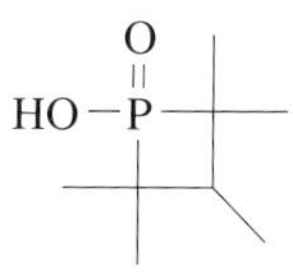

(**14**)

$pK_A = 4.66$ (ethanol)

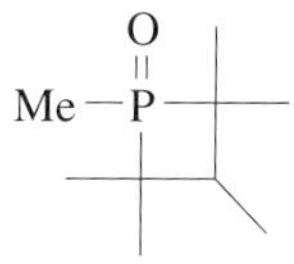

$pK_B = -2.88$

Scheme 8

Strain within the phosphetane ring is relieved upon incorporation into a five-coordinate phosphorane structure. This relief of tension permits the very high hydrolysis rates observed for phosphetanium salts (which proceed through phosphorane intermediates); activation enthalpies ($\Delta H^{\ddagger}$) are reduced by 4–5 kcal mol^{-1} with respect to the corresponding acyclic salts ⟨71JOC3226⟩. Relief of ring strain also explains the highly covalent nature of the P-alkoxy bonds in phosphetane-containing phosphoranes (**15**), which show no evidence for ionisation on the NMR time scale ⟨72JA245⟩.

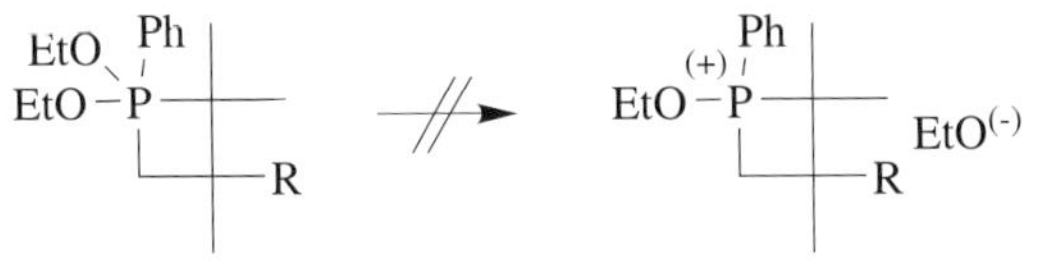

(**15**)

Scheme 9

3.1.4.2 Conformational and Configurational Studies

Conformational flexibility within the phosphetane ring is quite restricted: NMR data for polymethyl-substituted phosphetanes, their oxides and salts show a strong preference for structures which have the C(3) and/or phosphorus substituents lying equatorially, as found in structure (**16a**). X-ray data fully support these analyses. This favoured conformation relieves the 1,3 phenyl-methyl nonbonded interactions found in (**16b**) ⟨76JA2109⟩. If the phosphorus and C(3) substituents are mutually *trans*, the C(3) substituent usually takes the equatorial position.

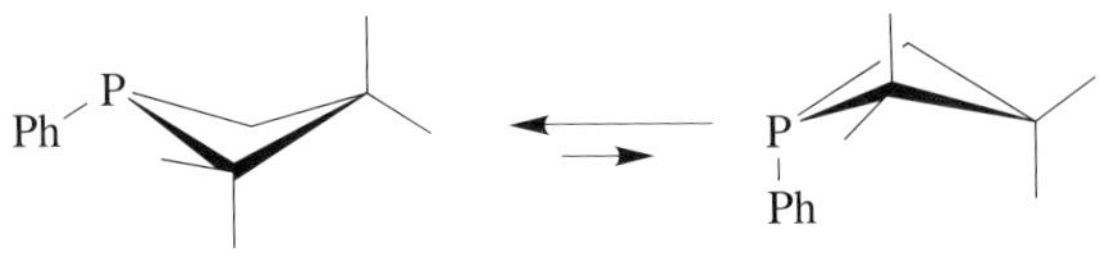

(**16a**) (**16b**)

Scheme 10

As shown in Scheme 11, *cis–trans* isomerisation of phosphetanes involving inversion of configuration at phosphorus usually takes place at very high temperatures ⟨68TL5799⟩. However, in phosphetane (**9**), (R = Ph) isomerisation occurs slowly at 110ºC. This comparatively facile inversion has been attributed to the high degree of substitution of the ring and the phenyl substituent at phosphorus.

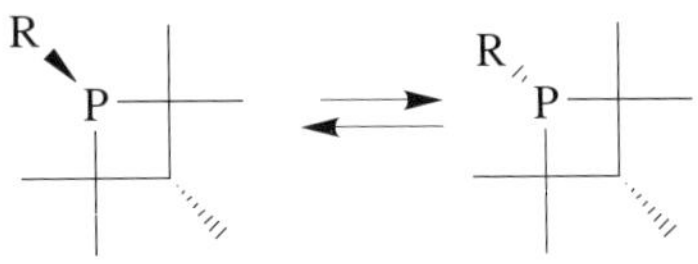

R=*t*-Bu 157°C, 1:1 mixture <73JOC3199>
R=Ph 110°C, 3:2 mixture <67JOC4066>
R=Me 162°C *trans*, unchanged <68TL5799>

trans-(**9**) *cis*-(**9**)

Scheme 11

A second isomerisation pathway for phosphetanes involves stereomutation at phosphorus. This process proceeds through pentacoordinated species which subsequently undergo pseudorotation (see Section 3.1.5.1). For instance, the *cis* ester (**17**) was readily isomerised to its more stable *trans* isomer (Scheme 12). The presumed phosphorane intermediate was intercepted in a trapping reaction with diazomethane which gave the corresponding P-methoxyphosphorane ⟨76TL4381, 79JCS(P1)879⟩.

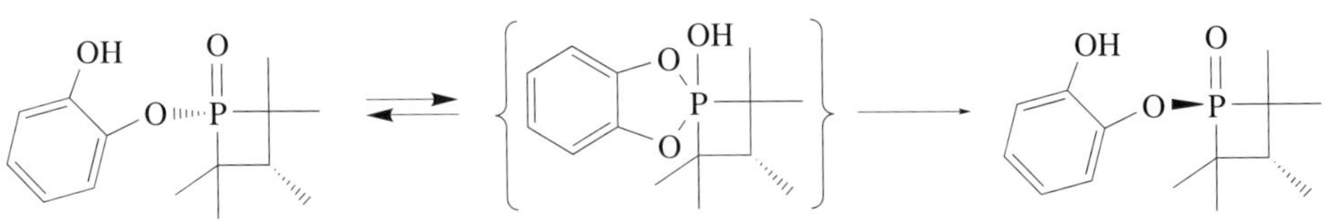

cis-(**17**) *trans*-(**17**)

Scheme 12

The stereomutation of phosphetane oxides may also be provoked by aqueous acids, bases or $SiCl_4$ ⟨69JA7027, 72JA2808⟩. Phosphetanium chlorides (**18**) equilibrate at room temperature to give mixtures favouring the isomer whose larger group at phosphorus lies *cis* to the C(3) substituent. Both of these groups are then able to occupy pseudoequatorial positions ⟨73JOC3199⟩.

The structures of phosphoranes in solution have been discussed extensively. Older NMR data were frequently used to suggest a preference for trigonal–bipyramidal (TBP) geometries over square pyramidal arrangements. However, a more recent paper has shown that there is little difference between ^{31}P NMR chemical shifts in solution and those in the solid state, where most phosphoranes are known to adopt

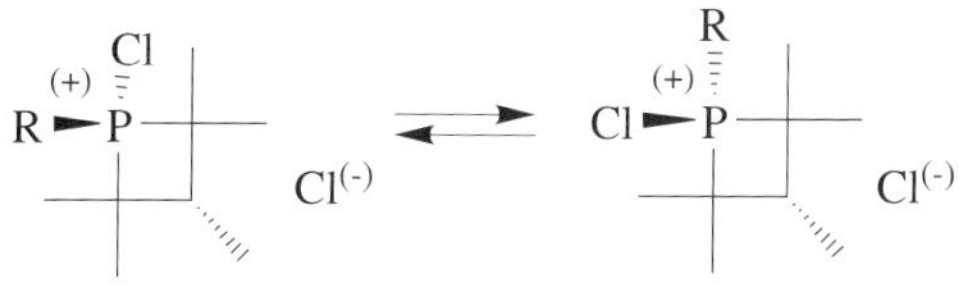

R	*Trans:cis* ratios
R = Me	3:2
Ph	1:2
t-Bu	1:3

trans-(**18**) *cis*-(**18**)

Scheme 13

geometries which are close to square pyramidal ⟨82JA230⟩. The true solution geometries have rarely been established. Difluorophosphoranes and acyclic dioxaphosphoranes such as (**19**) are fluxional and undergo rapid conformational exchange on the ^{1}H NMR scale at room temperature. Analysis of their temperature-dependent NMR spectra allowed the calculation of ΔG^* values (11 kcal mol^{-1} for (**19**)). Berry pseudorotation processes have been suggested in these trigonal–bipyramidal species, with the phosphetane ring preferentially adopting an apical–equatorial configuration ⟨71JA2066, 72CC272, 76JA768⟩. Spirocyclic dioxaphosphoranes such as (**20**) give stable, non-interconverting isomers at room temperature which very probably have square pyramidal geometries ⟨71PS(1)91, 73JCS(P1)1300, 78IC3265⟩. ΔG^* values for pseudorotation in (**20**) were calculated as 16.5 kcal mol^{-1} (R = furyl) and 19.6 kcal mol^{-1} (R = Ph) from NMR coalescence measurements ⟨71CC1011, 72CC554, 82JCS(P1)191⟩. The spirocyclic dithiophosphoranes (**21**) show NMR data which are consistent with rapidly interconverting bipyramidal species or static square pyramids ⟨72CC395, 78JA7300⟩.

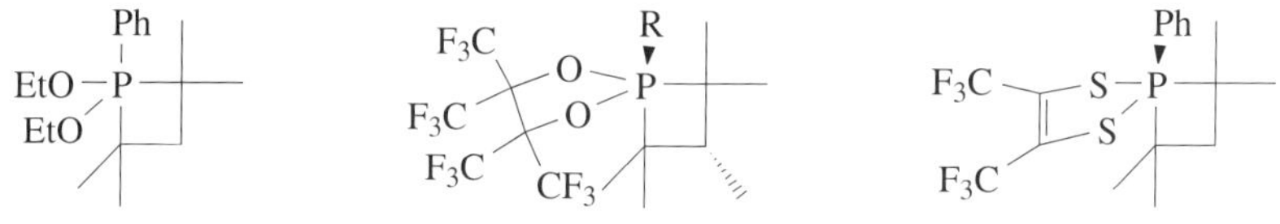

(**19**) (**20**) (**21**)

Scheme 14

A model has been developed which allows estimation of the relative stability of TBP and SP isomers in pentacoordinated phosphorus compounds and the corresponding activation energies for pseudorotation. It incorporates ring strain and steric terms and is in agreement with experimental data ⟨78JA433⟩.

3.1.5 REACTIVITY OF THE PHOSPHORUS ATOM AND SUBSTITUENTS ATTACHED TO THE PHOSPHORUS ATOM

3.1.5.1 Nucleophilic Displacement on Phosphetanium Salts, Phosphetane Oxides and Trivalent Phosphetanes

Nucleophilic displacements have been widely studied from theoretical and synthetic standpoints. Particular effort has been devoted to understanding their stereochemistry.

Hydrolysis of phosphetanium salts may lead either to ring opened products (see Section 3.1.5.5) or phosphetane oxides, with the nature of the product varying as a function of the phosphorus substituent. If the phosphorus atom has good leaving groups such as halides or benzyl functionalities, phosphetane oxides are formed easily ⟨71JOC3226⟩. Whether these oxides form with retention or inversion at phosphorus reflects the nature of the leaving group, experimental conditions and other factors.

Thus, for instance, hydrolysis of benzyl substituted phosphetanium salts (**22**, R = CH_2Ph) occurs with loss of phosphorus stereochemistry ⟨69CC769, 70CC1502⟩ whilst hydrolysis of the alkoxy-derivatives (**22**, R = OEt, OMe) takes place with retention ⟨69JA7027, 73JA8060⟩. When the phosphetane oxide (**1**) is obtained through hydrolysis of the corresponding chloro compound (**22**, R = Cl), the *cis–trans* ratio varies with changes in experimental conditions ⟨67JOC4066, 73JOC3199⟩ (Scheme 15).

Alkaline hydrolysis of 1-dimethylamino-1-methoxyphosphetanium salts occurs with stereospecific methoxy group cleavage ⟨71CC59, 73JA4681⟩ whilst 1-alkoxy-1-methoxyphosphetanium salts give mainly 1-alkoxyphosphetane oxides with partial or complete loss of stereospecificity. *Trans*-1-chloro-2,2,3,4,4-pentamethylphosphetane oxide is obtained stereoselectively upon hydrolysis of the corresponding 1,1-dichlorophosphetanium salt ⟨62JOC1833, 69JA7200, 76JCS(D)979⟩.

Pseudorotational processes, which have been studied using topological methods ⟨69JA7031, 70CC1502, 70ACR321⟩, are involved in determining the product ratios (Scheme 16). The following general principles

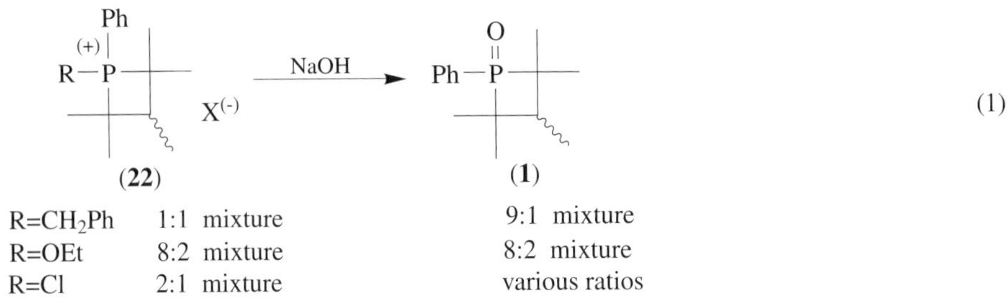

Scheme 15

influence the structure of the preferred pentacoordinate intermediate: (1) the OH group enters at an apical position of the trigonal bipyramid; (2) the four-membered ring preferentially adopts an equatorial–apical arrangement (to minimise strain by permitting a 90° CPC bond angle, cf. 120° for the diequatorial position); (3) the leaving group exits from an apical position; and (4) the most electronegative groups prefer apical positions (the "element effect").

The reaction stereochemistry is then dictated by: (1) the rate of pseudorotation with respect to the rate of R group loss (i.e. the leaving group character of R); and (2) the relative apicophilicity of the two substituents (R, R_1) (a function of their electronegativity and π-donor character ⟨72JA3047, 73JA4681⟩).

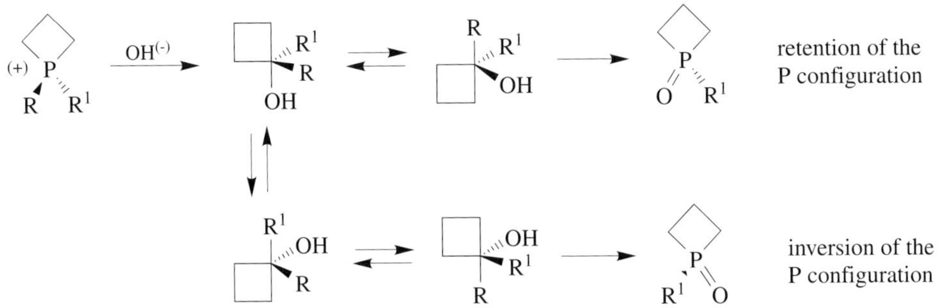

Scheme 16

The above general principles hold for any nucleophilic displacement reaction at phosphorus which involves trigonal bipyramidal intermediates. This class includes the reduction of phosphetane oxides.

In some cases, nucleophilic attack on phosphetanium salts can give stable phosphorane products. These include a simple addition of ethoxide ions ⟨72JA245⟩ (Equation (2)) and the double addition-substitution of catechol at a P-methylthiophosphetanium centre leading to (**23**) (Equation (3)). Compound (**23**) and analogues have also been obtained in reactions of trivalent phosphetanes with $ClNPr^i_2$ and catechol, through the intermediacy of P-amino-phosphetanium salts ⟨77JCS(P1)278⟩. No reaction is observed between catechol and P-alkoxyphosphetanium analogues ⟨78JCS(P1)1326⟩.

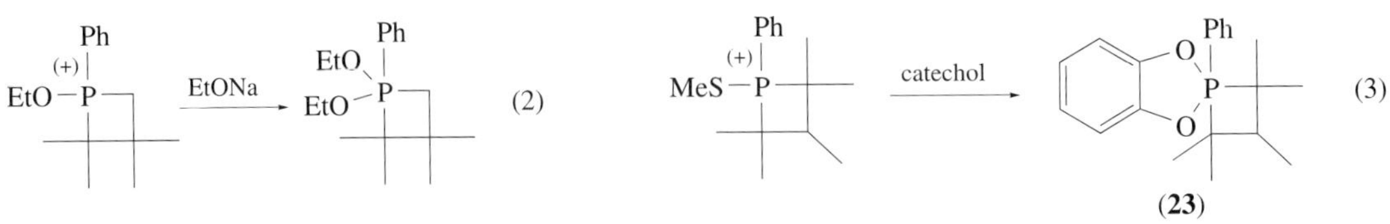

Scheme 17

Nucleophilic displacements at phosphetane oxide centres offer useful syntheses of phosphetane derivatives. Phosphetanic halides are easily accessible and react with alcohols, amines, water, alkyl- and aryllithium derivatives and Grignard reagents to give halogen substitution reactions (Scheme 18). Phosphorus configuration is generally retained: the few exceptions to this rule have again been explained in terms of apicophilicity of the leaving groups and coordinating properties of the reaction medium ⟨74JCS(D)633, 76TL2853⟩.

Substitution reactions with displacement of P-bound alkoxy groups by lithium amides ⟨73JA4681⟩ or hydroxide ions ⟨68CC577⟩ are known and phosphetanic acids are converted into the corresponding chlorides by $SOCl_2$ ⟨79JCS(P1)879⟩. Benzyl groups can equally be displaced from phosphorus through alkaline

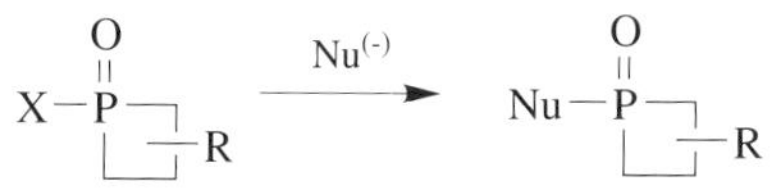

(4)

X = Cl, Nu = MeONa, NaOH, H_2O <69JA7200, 72JOC3458, 73PS(3)45, 73JCS(D)2701>
ROH <74JCS(D)633, 76TL4381>
RLi <67JOC4066, 69JCS(C)1465, 72JOC3458, 73JOC3199, 82JCS(P1)191>
RMgX <69CC769, 70JOC2426, 71JOC3226, 72JCS(P1)713, 73JCS(P1)1300>
R_2NH, RNH_2, NH_3, NaN_3 <69JCS(C)1465, 72JCS(P1)713, 73JCS(D)1576, 76JCS(D)979, 80JCS(P1)705, 71CC442, 74JA4262>
RSNa <73JCS(D)2641, 73JA4681>
X = Br, Nu = $H_2N(CH_2)_3NMe_2$ <73JCS(D)2701>

Scheme 18

hydrolysis ⟨70JOC2426⟩. Conversions of phosphetanic amides to the corresponding esters ⟨76TL2853⟩ or halides ⟨73JCS(D)2701, 72JCS(P1)1184, 73JCS(D)1576⟩ under acidic conditions have also been studied.

Under basic conditions, substitution reactions proceed mainly by an associative mechanism involving pentacoordinated intermediates. Phosphetanic esters undergo alkaline hydrolysis faster than the analogous acyclic dialkylphosphinic esters, with the observed 100-fold rate enhancement reflecting a relief of ring strain in the trigonal bipyramidal intermediate ⟨73JA8088, 73JCS(P2)1798⟩. Similarly, acid or base catalysed ^{18}O exchanges in phosphetane oxides are likely to proceed through phosphorane intermediates and are faster than in other cyclic phosphine oxides ⟨72JA2808, 74JA5199⟩. In cases where reactions are acid catalysed, ring strain induces the opposite effect: slow reaction rates in the acid-catalysed hydrolysis of phosphetanic amides (10^3 times lower than for open chain analogues) suggest an S_N2-like, colinear displacement mechanism rather than formation of a pentacoordinate intermediate ⟨70JA3828, 73JA8073⟩. Kinetic studies also suggest that a dissociative, S_N1 mechanism may operate for the hydrolysis of phosphetanic chlorides ⟨71JA6924⟩ and amides ⟨73JA8073⟩ under acid catalysis.

Nucleophilic substitution at tricoordinate P-chlorophosphetanes is also a useful synthetic tool ⟨69CC855, 73JCS(P1)1300, 84JCS(P2)553⟩. Unlike their phosphetanic chloride counterparts, these reactions usually proceed with complete inversion at phosphorus. However, isomerisation leading to thermodynamically controlled mixtures takes place when electronegative groups such as Cl, OR, or NR_2 are bound to phosphorus in the product. Phenyllithium, when used as the nucleophile, also induces isomerisation ⟨72JCS(P1)713⟩.

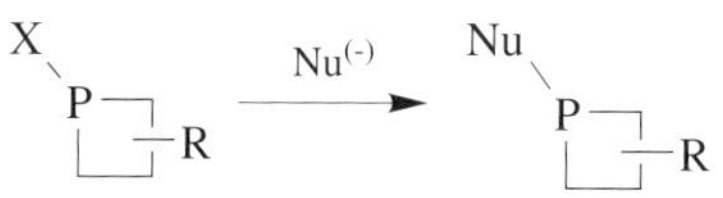

X = Cl
Nu = MeO, PhO, NR_2, NHR, R, Ph

(5)

Scheme 19

3.1.5.2 Electrophilic Additions to the Phosphorus Atom

Electrophilic additions to the phosphorus lone pair of trivalent phosphetanes generally give phosphetanium salts or stable phosphoranes. Alkylation of trivalent phosphetanes to phosphetanium salts with MeBr, MeI ⟨67JOC4066, 81JCS(P2)1138⟩ and $PhCH_2Br$ ⟨69JCS(C)1465, 71JOC3226⟩ has been reported. Chlorination also affords the corresponding phosphonium salts (Equation (6)) but whilst low-temperature chlorination occurs stereospecifically, interconversion of the isomers through pseudorotation occurs at room temperature. The overall *trans–cis* ratio depends upon the nature of the R substituent (1 : 2 for R = Ph, 3 : 2 for R = Me, 1 : 3 for R = *t*Bu) ⟨73JOC3199⟩. The reverse reaction, reduction of P-chlorophosphonium salts to trivalent phosphetanes, may be effected using PBu_3 ⟨78JOC1250⟩.

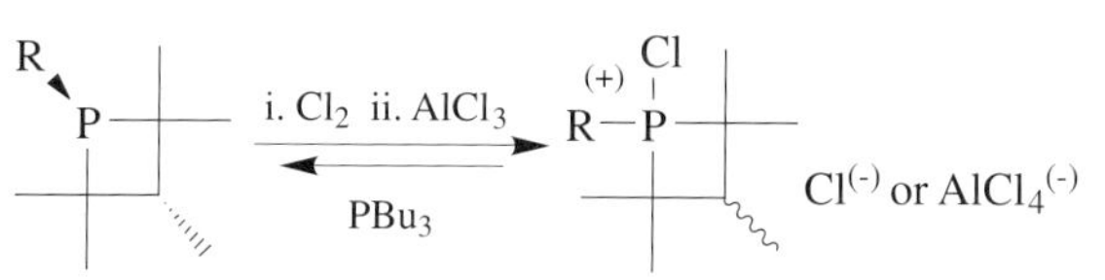

(6)

R = tBu, Me, Ph, Cl

Scheme 20

Substitution reactions of secondary phosphetanes also usually involve electrophilic addition: 1-acetyl-

and 1-chlorophosphetanes have been prepared by reaction of pentamethylphosphetane (**23**) with acetyl chloride and chlorine, respectively ⟨75JCS(P1)1220, 72JCS(P1)713⟩.

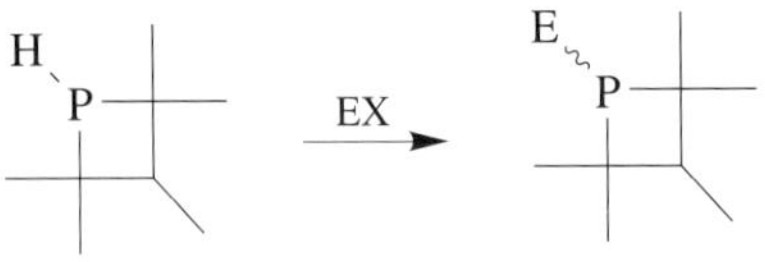

(**23**)

EX = MeCOCl / Et_3N
Cl_2 or ClCOCOCl

(7)

Scheme 21

These 1-chlorophosphetanes react with benzaldehyde and derivatives to give 1-(α-chlorobenzyl)-phosphetane oxides through a multistep transformation involving electrophilic attack of the aldehyde at phosphorus ⟨81JCS(P1)776⟩.

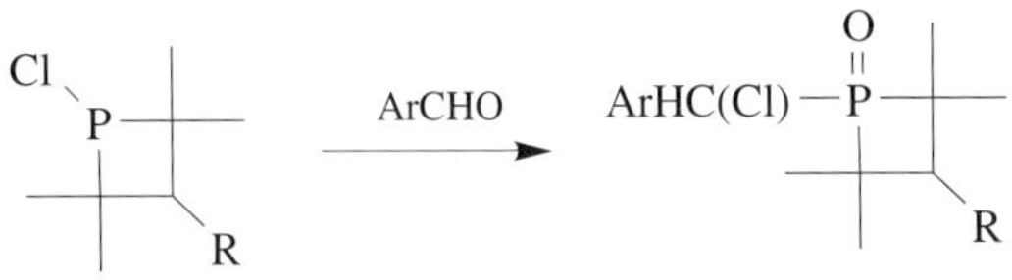

R = H, Me

(8)

Scheme 22

Electrophilic additions to phosphetanes provide a variety of stable phosphoranes. For instance, reactions with diethyl peroxide ⟨71JA2066, 72JA245⟩, 1,2-dioxane ⟨77PS(3)51⟩, hexafluoroacetone ⟨71CC1011, 73JCS(P1)1300, 82JCS(P1)191⟩, 3,4-bis(trifluoromethyl)-1,2-dithietene ⟨72CC395, 78JA7300⟩ and hexafluorobutadiene ⟨74PS(4)217⟩ afford the corresponding phosphoranes. The rate constant for the formation of (**19**) is far greater than for acyclic analogs. A concerted mechanism with insertion of the "biphilic" trivalent phosphorus into the oxygen–oxygen bond rather than a stepwise ionic mechanism has been proposed for this and analogous reactions ⟨72JA245, 77PS(3)51⟩. Bis(trifluoromethyl) peroxide reacts with trivalent phosphetanes to give the difluorophosphoranes (**25**) ⟨72CC272, 76JA768⟩.

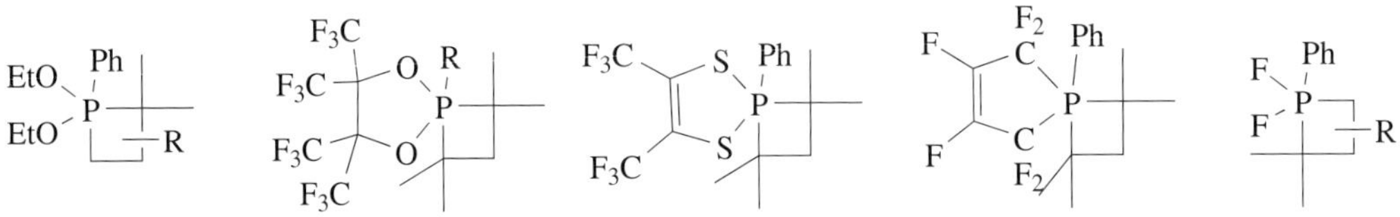

(**19**) (**20**) (**21**) (**24**) (**25**)

Scheme 23

Phosphoranes can be formed in multistep transformations where the phosphorus substituent intervenes: 1-benzyl and 1-(phenylethenyl)phosphetanes afford phosphoranes (**26**) and (**7**), respectively, upon reaction with hexafluoroacetone ⟨73JCS(P1)1300, 76CC449⟩. 1-Acetyl-2,2,3,4,4-pentamethylphosphetane reacts with tetrachloro-*o*-benzoquinone to give phosphorane (**27**), through migration of the acetyl group from phosphorus to the aryloxy moiety ⟨75JCS(P1)1220⟩.

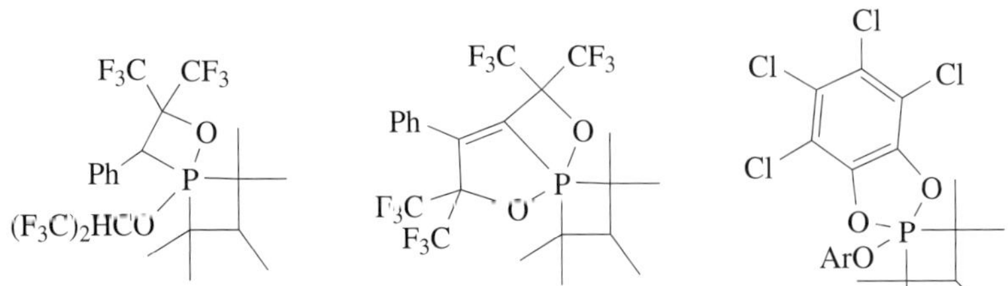

Ar =2-Acetoxy-3,4,5,6-tetra-chlorophenyl

(**26**) (**7**) (**27**)

Scheme 24

3.1.5.3 Oxidation and Reduction Reactions

Early studies showed that reduction of *cis*- or *trans*-2,2,3,4,4-pentamethyl-1-phenylphosphetane oxides with trichlorosilane/triethylamine (or pyridine) gave single isomers of the corresponding trivalent phosphetanes (Equation (9)). Subsequent oxidation of these phosphetanes with H_2O_2 in acetone regenerated the starting oxides. Both reactions were assumed to proceed stereospecifically with retention of configuration at phosphorus ⟨67JOC4066, 69JCS(C)1465⟩. This stereochemistry has been confirmed repeatedly during reductions of many phosphetane oxides. Reductions using Si_2Cl_6 also proceed with complete retention of phosphorus stereochemistry ⟨69JA7027⟩, in sharp contrast to the overall inversion of configuration which is achieved when acyclic phosphine oxides are reduced under analogous conditions. The dichotomy has been rationalised in two ways. One hypothesis invokes special constraints imposed by the four-membered ring and their influence upon the stability and interconversion rates of the intermediate phosphoranes. A second proposition involves a four-centre reaction mechanism, with H transfer from silicon to phosphorus occurring concurrently with oxygen transfer to silicon ⟨69JA7031, 70ACR321, 84JA7021⟩. In the absence of base, trichlorosilane reduces monocyclic phosphetane oxides less stereoselectively (with acid or silane catalysed isomerisation of the product being proposed ⟨81JCS(P2)1138⟩), but stereospecific inversion at phosphorus in the reduction of *exo*-tetracycle (**28**) ⟨84JA7021⟩ was found. Likewise, phenylsilane-promoted reductions are not invariably stereospecific ⟨73JCS(P1)1300, 82JCS(P1)191⟩. A reduction employing $LiAlH_4$ involved mixtures of the isomeric phosphetane oxides (**1**) and led stereoselectively to a 95 : 5 mixture of the corresponding phosphetanes (**9**) ⟨69JA5645⟩.

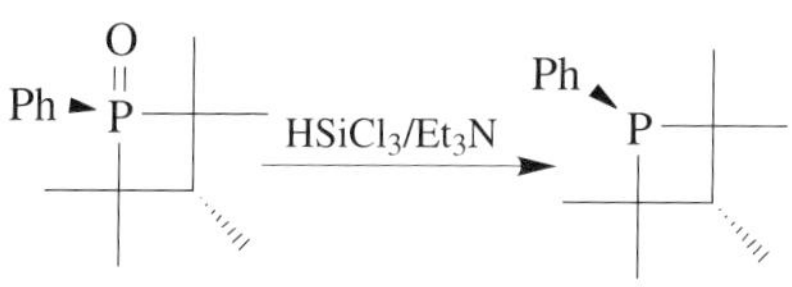

(9)

(**1**) (**9**)

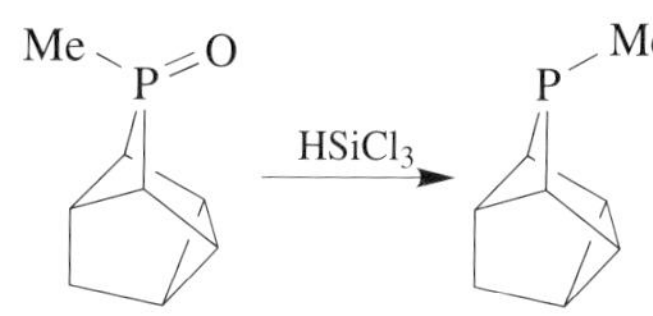

(10)

(**28**) (**29**)

Scheme 25

Reductions of 1-chlorophosphetane oxides using either silanes or $LiAlH_4$ afford the corresponding secondary phosphetanes. Polymethylsiloxane allows selective reduction to 1-chlorophosphetanes ⟨72JCS(P1)713⟩.

Oxidation of trivalent phosphetanes is usually performed with H_2O_2 ⟨67JOC4066, 72JCS(P1)713⟩, phosphetane sulphides and selenides are obtained from trivalent phosphetanes by reaction with elemental sulphur or selenium ⟨71JCS(C)334, 72JCS(P1)713⟩ and tosyl azide converts phosphetanes into the corresponding imides, all with retention of configuration ⟨77JCS(P2)1373⟩.

3.1.5.4 Reactivity at the Phosphorus Substituent

This section comprises a number of essentially unrelated transformations. O-alkylation (or silylation) of phosphetane oxides using trialkyloxonium salts (or trimethylsilyl triflate) yields the corresponding phosphonium salts (Equation (11)) ⟨71CC59, 72JA245, 73JA4681, 78JCS(P1)1326⟩. Retention of stereochemistry is observed when non-coordinating anions are used. S-alkylations of phosphetane sulphides occur analogously ⟨78JCS(P1)1326⟩.

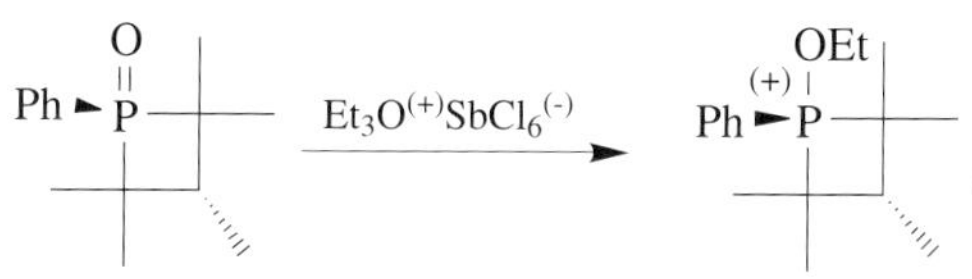

(11)

Scheme 26

Selective demethylation of a 1-methoxyphosphetane oxide to the corresponding acid has been performed using *t*-butylamine ⟨80TL859⟩. N-Chlorination of 1-aminophosphetane oxides with *t*-butyl hypochlorite affords N-monochloroamides and N-dichloroamides which behave as chlorinating agents for anthracene and anisole ⟨80JCS(P1)705, 80JCR(S)49⟩.

Phosphetane oxides can be converted to the corresponding sulphides with boron trisulphide ⟨73CC273⟩. With sulphonyl isocyanates, they give the corresponding imides with retention at phosphorus (Equation (12)) ⟨74TL1693, 77JCS(P2)1373⟩. SF_4 converts 1-phenyl-2,2,3,4,4-pentamethylphosphetane oxide into the difluorophosphorane (**25**) which reacts in turn with $(Me_3SiOCH_2)_2$ to afford the spirophosphorane (**30**) ⟨78IC3265⟩.

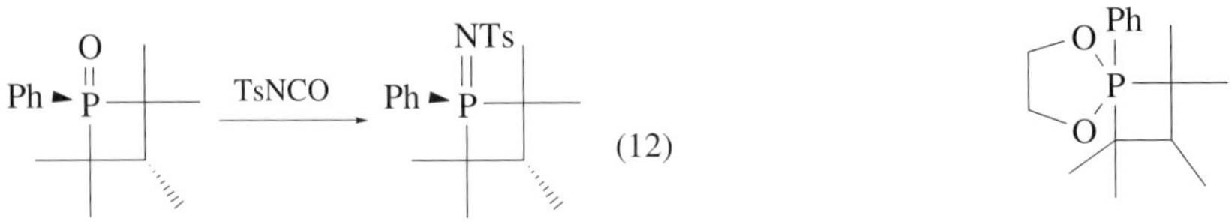

(12)

(**30**)

Scheme 27

α-Deprotonation of the phosphorus substituent in 1-benzyl phosphetanium salts gives ylides which behave as classical Wittig olefination reagents toward aldehydes. Retention of phosphorus configuration is observed in the phosphetane oxide product ⟨69CC789, 69JCS(C)1465⟩.

3.1.5.5 Ring Opening and Ring Expansion Reactions

Very few examples of the ring expansion of trivalent phosphetanes appear in the early literature (for later work, see Chapter 3.2). However, unsaturated electrophiles such as propiolate derivatives, were reacted with phosphetanes to give phospholanes (**31**) or (**32**), predominantly by migration of the bond bearing the CMe_2 group ⟨69JCS(C)1465, 70JCS(C)1855⟩.

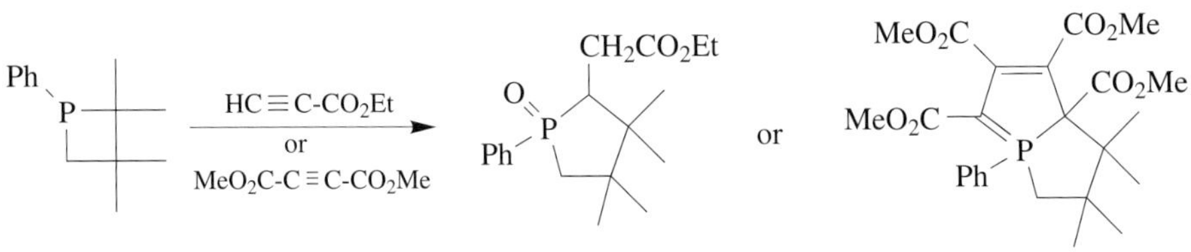

(13)

(**31**) (**32**)

Scheme 28

Other ring openings or ring expansions were observed during reactions of nucleophiles with phosphetanium salts having poor leaving groups at phosphorus. Examples of attacks provoking ring opening reactions are given in Equation (14) ⟨68CC182, 70JCS(C)1855, 70CC1267, 73JA200⟩ and Equation (15) ⟨67JOC4066, 68TL413, 69JCS(C)1465, 70JCS(C)1855⟩. [1,4]-Migration of a P-bound phenyl group occurs during the formation of (**33**). For dissymmetrically substituted phosphetanes, ring cleavage may occur at either of the P–C bonds, with the regiochemistry reflecting the nature of the incoming reagent.

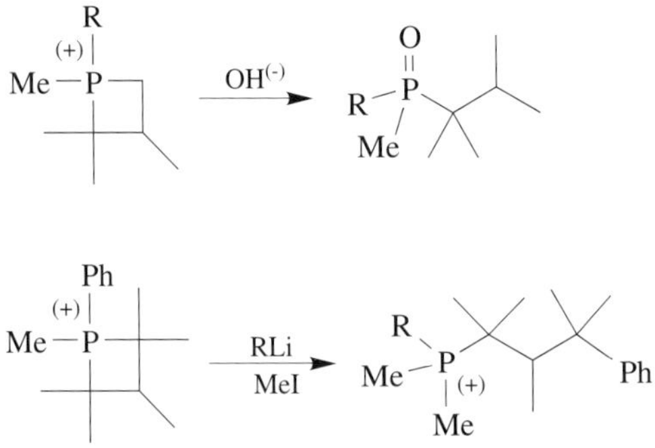

(14)

(15)

(**33**)

Scheme 29

Two further examples of ring expansion occur upon the alkaline hydrolysis of the phosphetanium salts shown in Scheme 30. The phosphorus substituents, a phenyl and a iodomethyl group, respectively, participate in these rearrangements leading to phospholane oxides ⟨67CC1113, 68CC1132, 70JCS(C)1855, 74PS(4)205⟩.

1-Chlorophosphetanium chlorides may decompose at elevated temperatures (90°C) to give acyclic chlorophosphines (Equation (18)) ⟨73JOC3199, 70CC1350⟩.

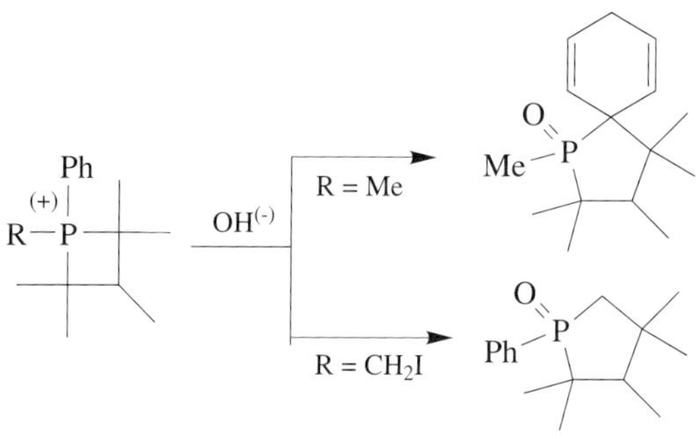

(16)

(17)

Scheme 30

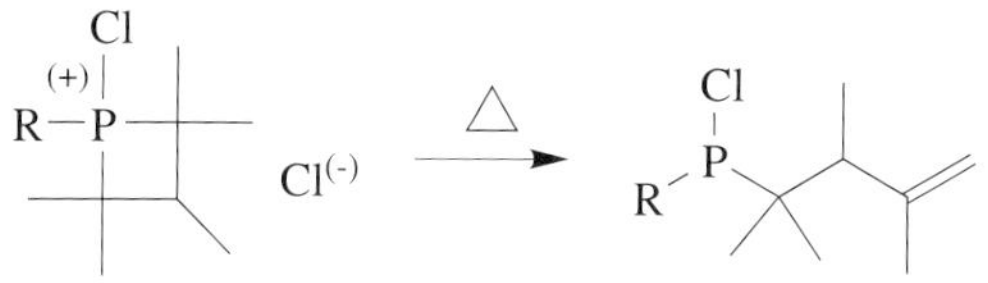

(18)

Scheme 31

Unlike phosphetanium salts, phosphetane oxides are generally stable towards ring cleavage under basic conditions ⟨70JOC2426⟩ but phenyllithium provokes ring opening with [1,4]-phenyl group migration (analogously to the phosphetanium salt opening in Equation (15)) ⟨69JCS(C)1465⟩. Phenyllithium also gives ring opened products with phosphetane sulphides ⟨71JCS(C)334⟩.

A Curtius-like rearrangement of 1-azidophosphetane oxides under photolysis in methanol gives mainly five-membered phosphinamidic esters, along with smaller quantities of ring opened products (Equation (19)) ⟨71CC442, 74JCS(P1)2604, 74JA4262⟩.

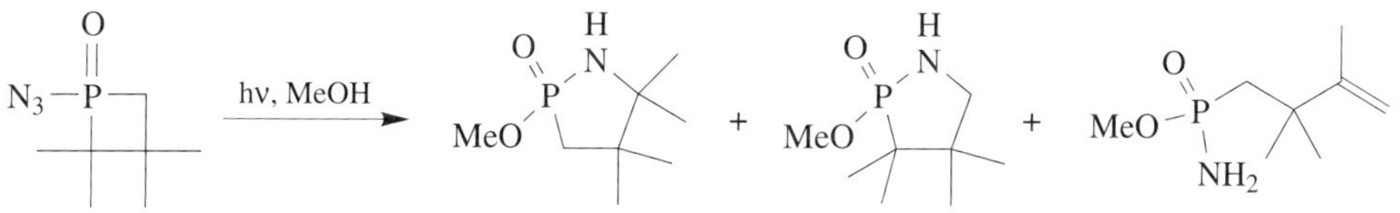

(19)

Scheme 32

3.1.6 RING SYNTHESES

Most synthetic approaches to the phosphetane ring start from "P" and "C_3" fragments. An alternative method, involving "C_2P"/"C" annelation of bis(chloromethyl)phosphinates and diethyl sodium malonate, gave (**34**) in 61% yield (Equation (20)) ⟨69IZV373⟩. This methodology has not been further exploited.

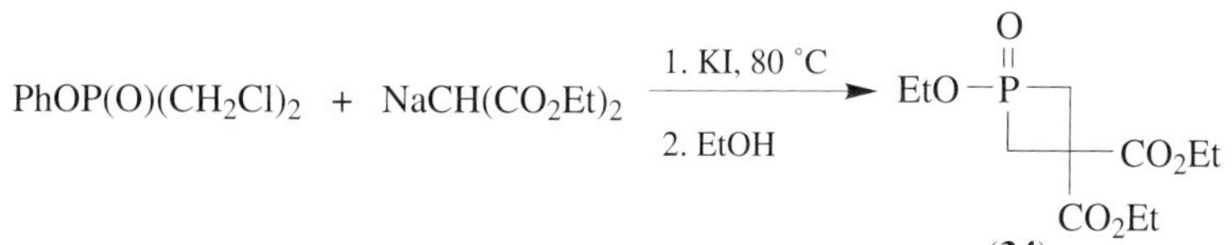

(**34**)

(20)

Scheme 33

3.1.6.1 Addition of Electrophilic Phosphorus Derivatives to Alkenes and Cyclopropanes

The phosphetane ring is most easily synthesised by the formal electrophilic addition of phosphenium cations to appropriately substituted alkenes. The first report, by J.J. McBride *et al.* in 1962, appeared during studies on the use of a $PCl_3/AlCl_3$ adduct for the phosphorylation of alkenes ⟨62JOC606⟩. Reaction of PCl_3 with 2,4,4-trimethyl-2-pentene in the presence of $AlCl_3$ affords 1-chloro-2,2,3,4,4-pentamethylphosphetan-

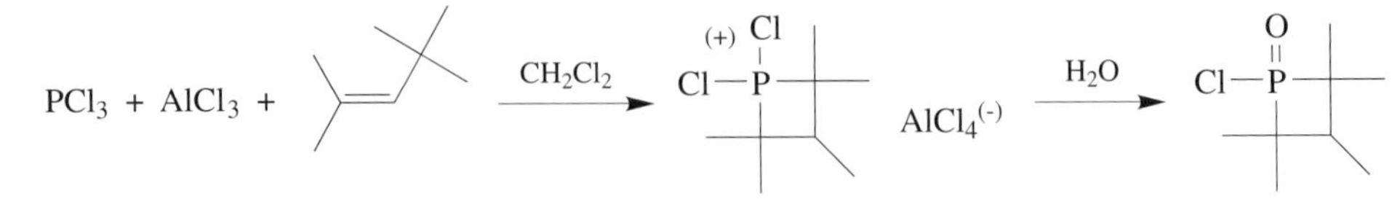

(21)

Scheme 34

1-oxide in 62% yield upon hydrolysis (Equation (21)). The *trans* stereochemistry of the major isomer (>95%) has been established by X-ray crystallography ⟨70JCS(B)934⟩.

It is assumed that a $PCl_2^{(+)}$ $AlCl_4^{(-)}$ phosphenium ion adds to the alkene double bond. Rearrangement of the intermediate carbocation through a [1,2]-methyl shift and a subsequent ring closure provides the four-membered ring (Equation (22)) ⟨62JOC1833⟩.

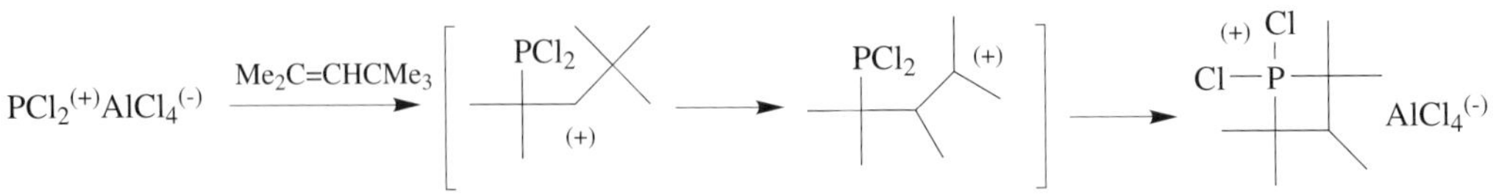

(22)

Scheme 35

This general principle has been applied to the synthesis of phosphetane oxides (**35**) from a wide variety of phosphorus halides (PCl_3, $ArPCl_2$, $MePCl_2$, $PhCH_2PCl_2$ ⟨67JOC4066⟩, PBr_3 ⟨73JCS(D)2701⟩) and alkenes (2,4,4-trimethyl-2-pentene, 2,3,3-trimethyl-1-butene, 3,3-dimethyl-1-butene, 4,4-dimethyl-2-pentene, 3,4,4-trimethyl-2-pentene) ⟨72JOC3458⟩. In some cases where *cis* and *trans* isomerism at phosphorus is possible, the product stereochemistry reflects the experimental procedures employed during the hydrolysis step ⟨67JOC4066, 76JCS(D)979⟩. Amongst others, camphene has been employed in a McBride synthesis to prepare the chiral phosphetane (**36**) ⟨80JCS(P1)2136⟩. The use of 1-bromo-3,3-dimethyl-1-butene allows a potentially reactive bromo substituent to be introduced ⟨81JCS(P2)1138⟩.

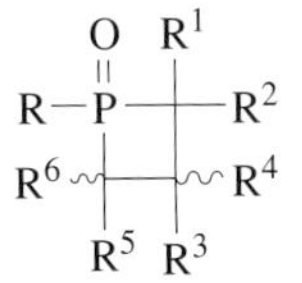

R = Cl, Br, Ar, Me, CH_2Ph
$R^1=R^2=R^3=R^5=R^6=Me$ $R^4=H$
$R^1=R^2=R^3=R^4=Me$, $R^5=R^6=H$
$R^1=R^3=R^4=Me$, $R^2=R^5=R^6=H$
$R^1=R^2=R^3=Me$, $R^4=R^5=H$, $R^6=Br$

(**35**)

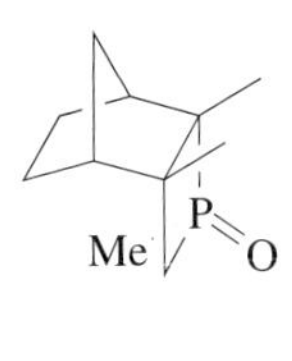

(**36**)

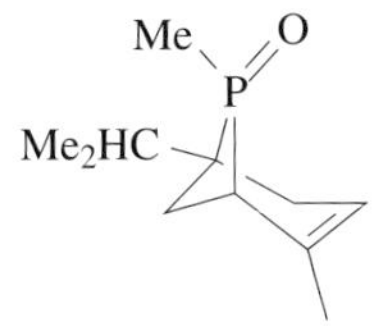

(**37**)

Scheme 36

The rearrangement of the intermediate McBride carbocation is not constrained to [1,2] methyl migrations and [1,2] hydrogen, vinyl or phenyl shifts may also occur. Thus, phosphetanes have also been prepared through reaction of $PhPCl_2/AlCl_3$ mixtures with 2,3-dimethyl-1-butene ⟨67JOC4066⟩, 3,3-dimethyl-1,4-pentadiene and 3-methyl-3-phenyl-1,4-pentadiene ⟨78TL63⟩ (Scheme 37).

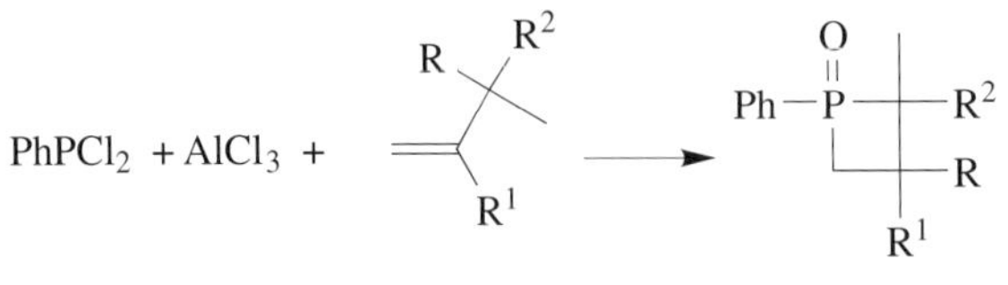

R=H, $R^1=R^2=Me$
R=CH:CH$_2$, $R^1=H$, $R^2=Me$
R=Ph, $R^1=H$, $R^2=CH{:}CH_2$

(23)

Scheme 37

Finally, a multistep reaction of an $MePCl_2/AlCl_3$ adduct with α-pinene furnishes the optically active bicyclic phosphetane oxide (**37**) ⟨78CC125⟩.

3.1.6.2 McCormack Cycloaddition

$MePCl_2$ reacts with norbornadiene to afford the tetracyclic phosphetanium salt (**38**) $R_1 = Me$, $R_2 = Cl$ as a mixture of two equilibrating isomers. Addition of water gives the thermodynamically more stable *exo*-phosphetane oxide (**28**) ⟨65JCS541, 75CC374⟩. Unlike the reactions described in Section 3.1.7.1, Lewis acid activation is not necessary. Attempts to vary the phosphorus substituents failed.

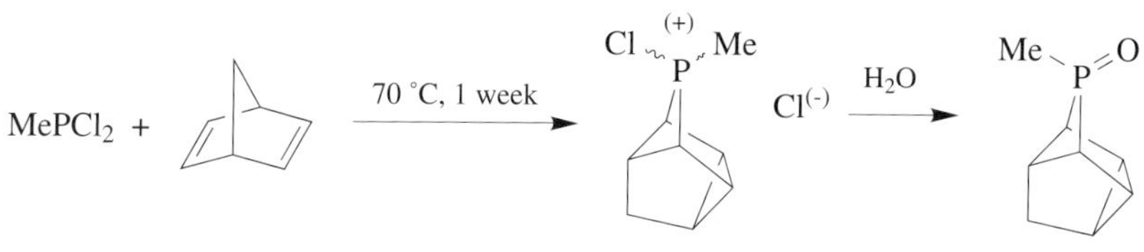

(24)

Scheme 38

3.1.6.3 Electrophilic Cyclisation on Phosphorus Derivatives

Phosphetanium salts were prepared in early studies by reacting 1,3-dielectrophiles with appropriate phosphides. Thus, phosphetanium salts were obtained from lithium or sodium diphenylphosphides through alkylation with substituted 1,3-dihalopropanes and intramolecular cyclisation ⟨68JA518, 76JA2109⟩.

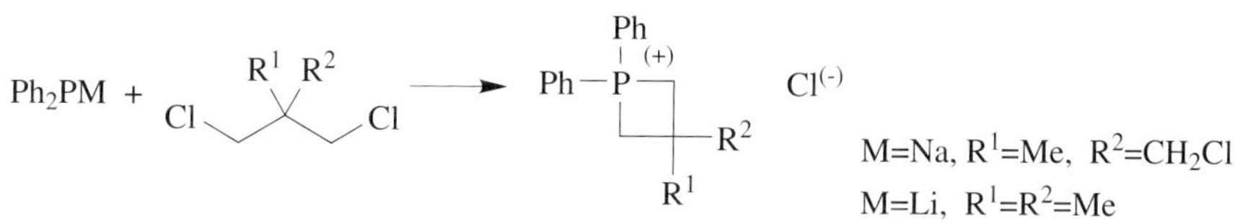

(25)

Scheme 39

3.1.6.4 Nucleophilic Cyclisation on Phosphorus Derivatives

In principle, the reaction of phosphorus halides with difunctional nucleophilic reagents provides an alternative approach to the phosphetane ring. Prior to 1984 it remained largely untested, save for a single report concerning the reaction of magnesium with a bromopropane derivative containing P–Cl functionality (Equation (26)).

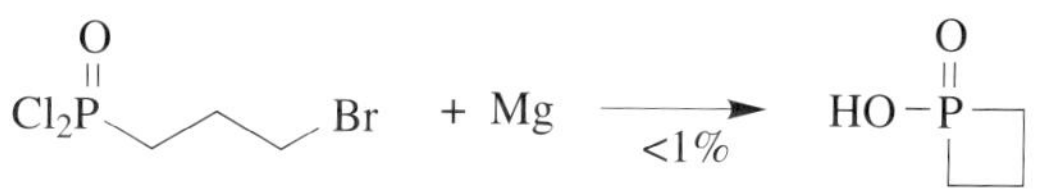

(26)

Scheme 40

The desired phosphetanic acid was obtained in very low yield ⟨57JCS3739⟩.

More recent developments in the synthesis of phosphetanes, which have been mainly driven by the recognition of the potential of cyclic phosphines in homogeneous catalysis, are discussed in Chapters 3.2 and 7.

3.1.7 REFERENCES

57JCS3739 G.M. Kosolapoff and R.F. Struck; *J. Chem. Soc.*, 1957, 3739.
62JOC606 E. Jungermann, J.J. McBride Jr., R. Clutter and A. Mais; *J. Org. Chem.*, 1962, **27**, 606.
62JOC1833 J.J. McBride Jr., E. Jungermann, J.V. Killheffer and R.J. Clutter; *J. Org. Chem.*, 1962, **27**, 1833.
65JCS541 M. Green; *J. Chem. Soc.*, 1965, 541.
67CC1113 S.E. Fishwick, J. Flint, H. Hawes and S. Trippett; *J. Chem. Soc. Chem. Commun.*, 1967, 1113.
67JOC4066 S.E. Cremer and R.J. Chorvat; *J. Org. Chem.*, 1967, **32**, 4066.
68ACS1866 G. Aksnes and P. Albriktsen; *Acta Chem. Scand.*, 1968, **22**, 1866.
68CC182 S.E. Fishwick and J.A. Flint; *J. Chem. Soc. Chem. Commun.*, 1968, 182.
68CC577 W. Hawes and S. Trippett; *J. Chem. Soc. Chem. Commun.*, 1968, 577.
68CC1051 D.D. Swank and C.N. Caughlan; *J. Chem. Soc. Chem. Commun.*, 1968, 1051.
68CC1132 S.E. Cremer; *J. Chem. Soc. Chem. Commun.*, 1968, 1132.
68CC1228 Mazhar-ul-Haque and C.N. Caughlan; *J. Chem. Soc. Chem. Commun.*, 1968, 1228.
68JA518 D. Berglund and D.W. Meek; *J. Am. Chem. Soc.*, 1968, **90**, 518.
68TL413 S.E. Cremer and R.J. Chorvat; *Tetrahedron Lett.*, 1968, 413.
68TL5799 S.E. Cremer, R.J. Chorvat, C.H. Chang and D.W. Davis; *Tetrahedron Lett.*, 1968, 5799.
69CC769 S.E. Cremer, R.J. Chorvat and B.C. Trivedi; *J. Chem. Soc. Chem. Commun.*, 1969, 769.

69CC789 J.R. Corfield, J.R. Shutt and S. Trippett; *J. Chem. Soc. Chem. Commun.*, 1969, 789.

69CC855 D.J.H. Smith and S. Trippett; *J. Chem. Soc. Chem. Commun.*, 1969, 855.

69IZV373 T.A. Zyablikova, A.R. Panteleeva and I.M. Shermergorn; *Izv. Akad. Nauk SSSR, Ser. Khim.*, 1969, 373 (*Chem. Abstr.*, 1969, **70**, 115228t).

69JA2255 C. Moret and L.M. Trefonas; *J. Am. Chem. Soc.*, 1969, **91**, 2255.

69JA5645 P.D. Henson, K. Naumann and K. Mislow; *J. Am. Chem. Soc.*, 1969, **91**, 5645.

69JA7027 K.E. DeBruin, G. Zon, K. Naumann and K. Mislow; *J. Am. Chem. Soc.*, 1969, **91**, 7027.

69JA7031 K.E. DeBruin, K. Naumann, G. Zon and K. Mislow; *J. Am. Chem. Soc.*, 1969, **91**, 7031.

69JA7200 S.E. Cremer and B.C. Trivedi; *J. Am. Chem. Soc.*, 1969, **91**, 7200.

69JCS(C)1465 W. Hawes and S. Trippett; *J. Chem. Soc. C*, 1969, 1465.

70ACR321 K. Mislow; *Acc. Chem. Res.*, 1970, **3**, 321.

70CC616 S.E. Cremer; *J. Chem. Soc. Chem. Commun.*, 1970, 616.

70CC1267 J.R. Corfield and S. Trippett; *J. Chem. Soc. Chem. Commun.*, 1970, 1267.

70CC1350 J.R. Corfield, M.J.P. Harger, R.K. Oram, D.J.H. Smith and S. Trippett; *J. Chem. Soc. Chem. Commun.*, 1970, 1350.

70CC1502 J.R. Corfield, N.J. De'Ath and S. Trippett; *J. Chem. Soc. Chem. Commun.*, 1970, 1502.

70JA3828 P. Haake, R.D. Cook, T. Koizumi, P.S. Ossip, W. Schwarz and D.A. Tyssee; *J. Am. Chem. Soc.*, 1970, **92**, 3828.

70JCS(B)934 Mazhar-ul-Haque; *J. Chem. Soc. (B)*, 1970, 934.

70JCS(B)938 Mazhar-ul-Haque; *J. Chem. Soc. (B)*, 1970, 938.

70JCS(C)1855 J.R. Corfield, M.J.P. Harger, J.R. Shutt and S. Trippett; *J. Chem. Soc. (C)*, 1970, 1855.

70JOC2426 B.R. Ezzell; *J. Org. Chem.*, 1970, **35**, 2426.

71CC59 K.E. DeBruin and M.J. Jacobs; *J. Chem. Soc. Chem. Commun.*, 1971, 59.

71CC442 M.J.P. Harger; *J. Chem. Soc. Chem. Commun.*, 1971, 442.

71CC721 J.R. Corfield and S. Trippett; *J. Chem. Soc. Chem. Commun.*, 1971, 721.

71CC1011 R.E. Duff, R.K. Oram and S. Trippett; *J. Chem. Soc. Chem. Commun.*, 1971, 1011.

71JA2066 D.Z. Denney, D.W. White and D.B. Denney; *J. Am. Chem. Soc.*, 1971, **93**, 2066.

71JA6507 A. Rauk, J.D. Andose, W.G. Frick, R. Tang and K. Mislow; *J. Am. Chem. Soc.*, 1971, **93**, 6507.

71JA6924 P. Haake and P.S. Ossip; *J. Am. Chem. Soc.*, 1971, **93**, 6924.

71JCS(B)117 Mazhar-ul-Haque; *J. Chem. Soc. (B)*, 1971, 117.

71JCS(C)334 J.R. Corfield and S. Trippett; *J. Chem. Soc. (C)*, 1971, 334.

71JOC3226 S.E. Cremer, B.C. Trivedi and F.L. Weitl; *J. Org. Chem.*, 1971, **36**, 3226.

71PS(1)91 D.W. White, N. De'Ath, D.Z. Denney and D.B. Denney; *Phosphorus*, 1971, **1**, 91.

71TL3061 G.A. Gray and S.E. Cremer; *Tetrahedron Lett.*, 1971, 3061.

72CC272 N.J. De'Ath, D.Z. Denney and D.B. Denney; *J. Chem. Soc. Chem. Commun.*, 1972, 272.

72CC367 G.A. Gray and S.E. Cremer; *J. Chem. Soc. Chem. Commun.*, 1972, 367.

72CC395 N.J. De'Ath and D.B. Denney; *J. Chem. Soc. Chem. Commun.*, 1972, 395.

72CC554 R.K. Oram and S. Trippett; *J. Chem. Soc. Chem. Commun.*, 1972, 554.

72JA245 D.B. Denney, D.Z. Denney, C.D. Hall and K.L. Marsi; *J. Am. Chem. Soc.*, 1972, **94**, 245.

72JA2808 D.G. Gorenstein; *J. Am. Chem. Soc.*, 1972, **94**, 2808.

72JA3047 R. Hoffmann, J.M. Howell and E.L. Muetterties; *J. Am. Chem. Soc.*, 1972, **94**, 3047.

72JCS(P1)713 J.R. Corfield, R.K. Oram, D.J.H. Smith and S. Trippett; *J. Chem. Soc., Perkin Trans. 1*, 1972, 713.

72JCS(P1)1184 K. Ellis, D.J.H. Smith and S. Trippett; *J. Chem. Soc., Perkin Trans. 1*, 1972, 1184.

72JOC3342 A.G. Cook and G.W. Mason; *J. Org. Chem.*, 1972, **37**, 3342.

72JOC3458 G.A. Gray and S.E. Cremer; *J. Org. Chem.*, 1972, **37**, 3458.

72JOC3470 G.A. Gray and S.E. Cremer; *J. Org. Chem.*, 1972, **37**, 3470.

73CC273 B.E. Maryanoff, R. Tang and K. Mislow; *J. Chem. Soc. Chem. Commun.*, 1973, 273.

73CC856 J.A. Howard, D.R. Russell and S. Trippett; *J. Chem. Soc. Chem. Commun.*, 1973, 856.

73JA200 K.L. Marsi and J.E. Oberlander; *J. Am. Chem. Soc.*, 1973, **95**, 200.

73JA4681 K.E. DeBruin, A.G. Padilla and M.-T. Campbell; *J. Am. Chem. Soc.*, 1973, **95**, 4681.

73JA8060 D.G. Gorenstein; *J. Am. Chem. Soc.*, 1973, **95**, 8060.

73JA8073 T. Koizumi and P. Haake; *J. Am. Chem. Soc.*, 1973, **95**, 8073.

73JA8088 R.D. Cook, C.E. Diebert, W. Schwarz, P.C. Turley and P. Haake; *J. Am. Chem. Soc.*, 1973, **95**, 8088.

73JCS(D)1576 J. Emsley and J.K. Williams; *J. Chem. Soc., Dalton Trans.*, 1973, 1576.

73JCS(D)2641 R.E. Ardrey, J. Emsley, A.J.B. Robertson and J.K. Williams; *J. Chem. Soc., Dalton Trans.*, 1973, 2641.

73JCS(D)2701 J. Emsley, T.B. Middleton and J.K. Williams; *J. Chem. Soc., Dalton Trans.*, 1973, 2701.

73JCS(P1)1300 R.K. Oram and S. Trippett; *J. Chem. Soc., Perkin Trans. 1*, 1973, 1300.

73JCS(P2)1798 J.I.G. Cadogan, D.T. Eastlick, J.A. Challis and A. Cooper; *J. Chem. Soc., Perkin Trans. 2*, 1973, 1798.

73JOC3199 S.E. Cremer, F.L. Weitl, F.R. Farr, P.W. Kremer, G.A. Gray and H.-O. Hwang; *J. Org. Chem.*, 1973, **38**, 3199.

73PS(3)45 J. Emsley, T.B. Middleton, J.K. Williams and M.F. Crook; *Phosphorus*, 1973, **3**, 45.

74CC451 G.A. Gray and S.E. Cremer; *J. Chem. Soc. Chem. Commun.*, 1974, 451.

74JA4262 J. Wiseman and F.H. Westheimer; *J. Am. Chem. Soc.*, 1974, **96**, 4262.

74JA5189 R.B. Wetzel and G.L. Kenyon; *J. Am. Chem. Soc.*, 1974, **96**, 5189.

74JA5199 R.B. Wetzel and G.L. Kenyon; *J. Am. Chem. Soc.*, 1974, **96**, 5199.

74JCS(D)633 J. Emsley, T.B. Middleton and J.K. Williams; *J. Chem. Soc., Dalton Trans.*, 1974, 633.

74JCS(P1)2604 M.J.P. Harger; *J. Chem. Soc., Perkin Trans. 1*, 1974, 2604.

74PS(4)205 H.A.E.-S. Aly, D.J.H. Smith and S. Trippett; *Phosphorus*, 1974, **4**, 205.
74PS(4)217 D.B. Denney, D.Z. Denney and Y.F. Hsu; *Phosphorus*, 1974, **4**, 217.
74TL1693 C.R. Hall and D.J.H. Smith; *Tetrahedron Lett.*, 1974, 1693.
75CC374 S.E. Cremer, F.R. Farr, P.W. Kremer, H.-O. Hwang, G.A. Gray and M.G. Newton; *J. Chem. Soc. Chem. Commun.*, 1975, 374.
75JCS(P1)1220 S. Trippett and P.J. Whittle; *J. Chem. Soc., Perkin Trans. 1*, 1975, 1220.
76CC449 H.A.E. Aly, J.H. Barlow, D.R. Russell, D.J.H. Smith, M. Swindles and S. Trippett; *J. Chem. Soc. Chem. Commun.*, 1976, 449.
76JA768 N.J. De'Ath, D.B. Denney, D.Z. Denney and Y.F. Hsu; *J. Am. Chem. Soc.*, 1976, **98**, 768.
76JA2109 G.A. Gray, S.E. Cremer and K.L. Marsi; *J. Am. Chem. Soc.*, 1976, **98**, 2109.
76JCS(D)979 J. Emsley, T.B. Middleton and J.K. Williams; *J. Chem. Soc., Dalton Trans.*, 1976, 979.
76TL2853 T. Koizumi, Y. Kobayashi and E. Yoshii; *Tetrahedron Lett.*, 1976, 2853.
76TL4381 G. Kemp and S. Trippett; *Tetrahedron Lett.*, 1976, 4381.
77JCS(P1)278 S. Antczak, S.A. Bone, J. Brierley and S. Trippett; *J. Chem. Soc., Perkin Trans. 1*, 1977, 278.
77JCS(P2)1373 C.R. Hall and D.J.H. Smith; *J. Chem. Soc., Perkin Trans. 2*, 1977, 1373.
77OMS624 C.W. Koch, D.H. Eargle Jr. and G.L. Kenyon; *Org. Mass. Spectrom.*, 1977, **12**, 624.
77PS(3)51 N.J. De'Ath and D.B. Denney; *Phosphorus, Sulfur* 1977, **3**, 51.
78CC125 E. Vilkas, M. Vilkas, D. Joniaux and C. Pascard-Billy; *J. Chem. Soc. Chem. Commun.*, 1978, 125.
78IC3265 W. Althoff, R.O. Day, R.K. Brown and R.R. Holmes; *Inorg. Chem.*, 1978, **17**, 3265.
78JA433 R.R. Holmes; *J. Am. Chem. Soc.*, 1978, **100**, 433.
78JA6327 B.A. Belkind, D.B. Denney, D.Z. Denney, Y.F. Hsu and G.E. Wilson Jr.; *J. Am. Chem. Soc.*, 1978, **100**, 6327.
78JA7300 B.C. Burros, N.J. De'Ath, D.B. Denney, D.Z. Denney and I.J. Kipnis; *J. Am. Chem. Soc.*, 1978, **100**, 7300.
78JCS(P1)1326 S. Antczak and S. Trippett; *J. Chem. Soc., Perkin Trans. 1*, 1978, 1326.
78JCS(P2)1115 Mazhar-ul-Haque, M. Rashid and S.E. Cremer; *J. Chem. Soc., Perkin Trans. 2*, 1978, 1115.
78JOC1250 C. Symmes Jr. and L.D. Quin; *J. Org. Chem.*, 1978, **43**, 1250.
78JOC3513 A. Fitzgerald, J.A. Campbell, G.D. Smith, C.N. Caughlan and S.E. Cremer; *J. Org. Chem.*, 1978, **43**, 3513.
78TL63 M. Rotem and Y. Kashman; *Tetrahedron Lett.*, 1978, 63.
79AX2601 Mazhar-ul-Haque; *Acta Crystallogr.*, 1979, **B35**, 2601.
79JCS(P1)879 G. Kemp and S. Trippett; *J. Chem. Soc., Perkin Trans. 1*, 1979, 879.
80JCR(S)49 M.J.P. Harger and M.A. Stephen; *J. Chem. Res.*, 1980, 49.
80JCS(P1)705 M.J.P. Harger and M.A. Stephen; *J. Chem. Soc., Perkin Trans. 1*, 1980, 705.
80JCS(P1)2136 E. Vilkas, M. Vilkas, J. Sainton, B. Meunier and C. Pascard; *J. Chem. Soc., Perkin Trans. 1*, 1980, 2136.
80TL859 M.D.M. Gray and D.J.H. Smith; *Tetrahedron Lett.*, 1980, **21**, 859.
81JCS(P1)776 N.J. De'Ath, J.A. Miller, M.J. Nunn and D. Stewart; *J. Chem. Soc., Perkin Trans. 1*, 1981, 776.
81JCS(P2)1138 Mazhar-ul-Haque, W. Horne, S.E. Cremer, P.W. Kremer and P.K. Kafarski; *J. Chem. Soc., Perkin Trans. 2*, 1981, 1138.
B-81MI1 L.D. Quin; *The Heterocyclic Chemistry of Phosphorus*, Wiley-Interscience, New York, NY, 1981.
82AX2944 Mazhar-ul-Haque and W. Horne; *Acta Crystallogr.*, 1982, **B38**, 2944.
82JA230 L.W. Dennis, V.J. Bartuska and G.E. Maciel; *J. Am. Chem. Soc.*, 1982, **104**, 230.
82JCS(P1)191 M.P. Johnson and S. Trippett; *J. Chem. Soc., Perkin Trans. 1*, 1982, 191.
83JOC4466 L.D. Quin, J.C. Kisalus and K.A. Mesch; *J. Org. Chem.*, 1983, **48**, 4466.
84JA7021 L.D. Quin, K.C. Caster, J.C. Kisalus and K.A. Mesch; *J. Am. Chem. Soc.*, 1984, **106**, 7021.
84JCS(P2)553 J. Nielsen and O. Dahl; *J. Chem. Soc., Perkin Trans. 2*, 1984, 553.
87SYN265 E. Yao, J. Szewczyk and L.D. Quin; *Synthesis*, 1987, 265.
92JOC511 S.E. Cremer, J.M. Cowles, F.R. Farr, H. Hwang, P.W. Kremer, A.C. Peterson and G.A. Gray; *J. Org. Chem.*, 1992, **57**, 51.
97HAC451 D.B. Chesnut, L.D. Quin and S.B. Wild; *Heteroatom. Chem.*, 1997, **8**, 451.

3.2
Four-membered Rings with 1 Phosphorus Atom

TAKAYUKI KAWASHIMA
The University of Tokyo, Tokyo, Japan

and

RENJI OKAZAKI
Japan Women's University, Tokyo, Japan

3.2.1 INTRODUCTION

3.2.1.1 Historical Perspective

Four-membered rings containing one phosphorus atom are phosphetanes. Although the systematic study has been carried out only using the reaction of phosphorous chlorides with highly substituted olefins in the presence of aluminum trichloride (Scheme 1), some new approaches to this class of compounds have been developed in the last decade. Some four-membered rings with one phosphorus atom appeared in ⟨79COC(2)1127⟩ together with three-membered rings. Since then several interesting compounds including the above ones have also been summarized in ⟨84CHEC(1)(1)523⟩ and ⟨96CHEC(2)(1B)833⟩ covering the literature up to about 1982 and between 1983 and 1994, respectively. In the latter review, a new class of compounds, phosphetenes, which were synthesized for the first time in 1983, were also be described. Here, not only more recent work published until now but also some important earlier work have been collected. Since the first report of four-membered rings containing one arsenic atom ⟨77JCS(D)704⟩ a few new compounds of this class have been synthesized, but four-membered rings with one antimony or bismuth atom are still unknown, to our understanding, and present an open field of research.

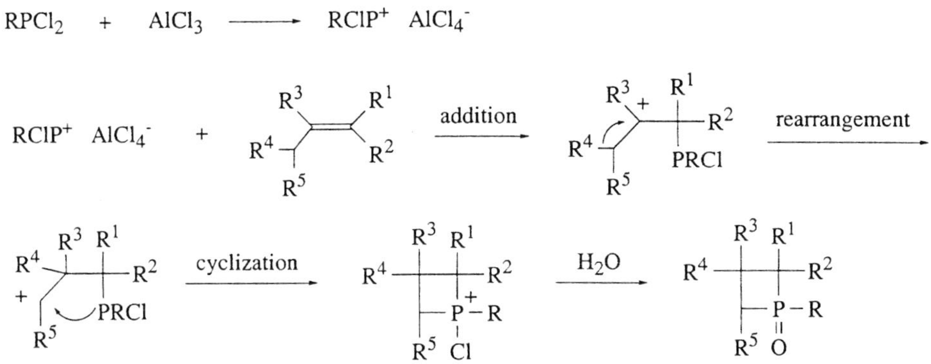

Scheme 1

3.2.1.2 Nomenclature

Four-membered rings containing one phosphorus atom can be classified mainly under three categories.

(1) Phosphetane. A saturated four-membered ring containing one phosphorus atom is named as phosphetane. It can also be named as phosphacyclobutane.

(2) Phosphetene. A four-membered ring with one phosphorus atom and one double bond in the ring is called phosphetene. It can be called phosphacyclobutene. Two isomers based on the position of the double bond may be named, for example, as Δ^1-phosphetene and Δ^2-phosphetene. They can also be named as 3,4-diphydrophosphete and 1,2-dihydrophosphete, respectively.

(3) Phosphete. A four-membered ring with one phosphorus and two double bonds is named as phosphete. It is also named as phosphacyclobutadiene.

3.2.2 EXPERIMENTAL STRUCTURAL METHODS

Structural elucidation of four-membered rings with one phosphorus atom has been mostly carried out by ^{1}H, ^{13}C and ^{31}P NMR spectroscopy. In recent work, X-ray diffraction is also one of the powerful methods for structural determination.

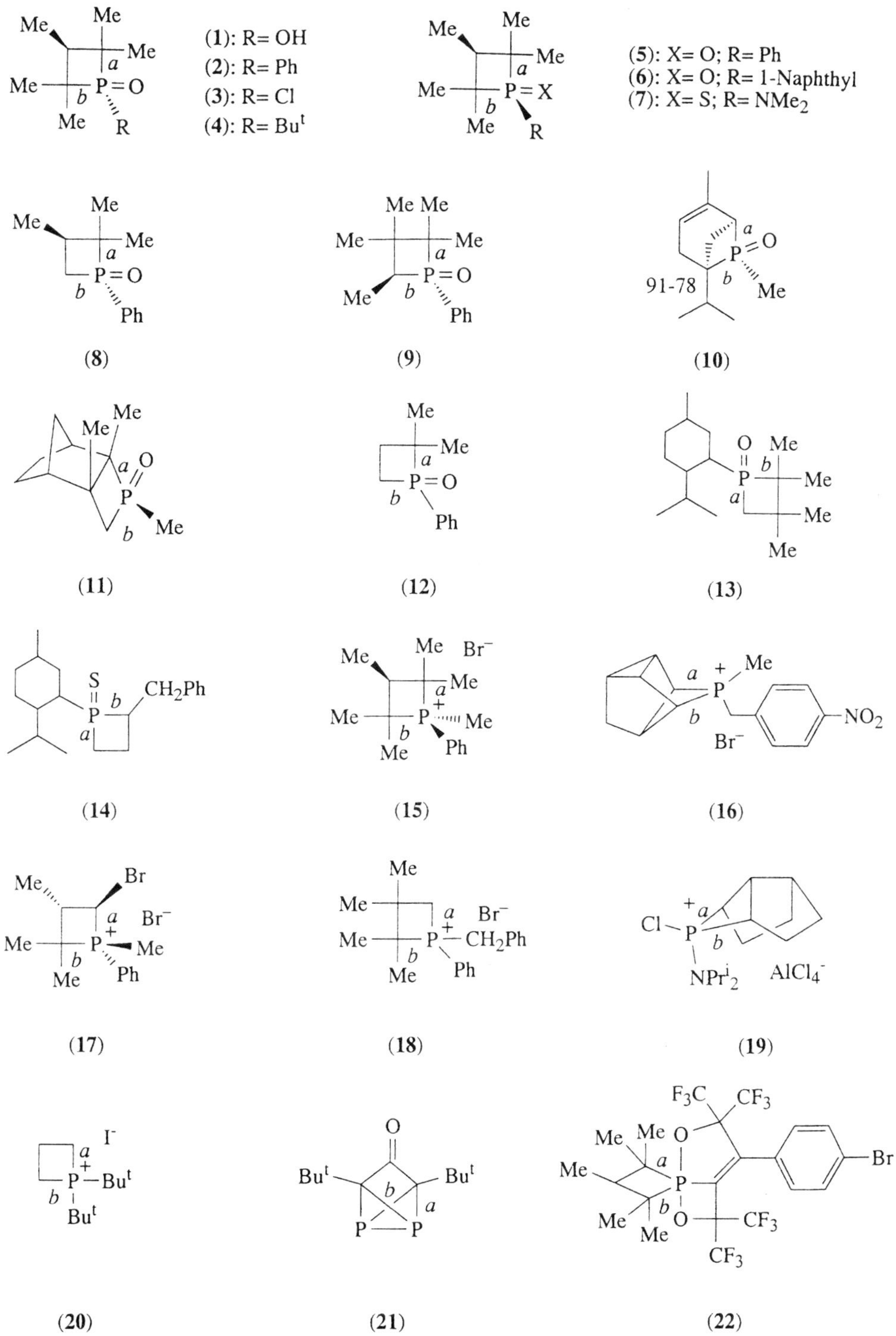

Scheme 2-1

3.2.2.1 X-Ray Diffraction Studies

X-ray crystallographic analysis has been used to determine the crystal structures of phosphetanes having a trivalent, tetravalent, or pentavalent phosphorus atom. The first X-ray crystallographic analysis was reported by Caughlan and co-workers for 1-hydroxy-2,2,3,4,4-pentamethylphosphetane 1-oxide (**1**) <68JCS(CC)1051>. The four-membered ring is significantly puckered, which represents one of structural features of phosphetanes. Selected bond lengths, bond angles, and dihedral angles of the four-membered ring of phosphetanes are summarized in Table 1 (for structural formulae,

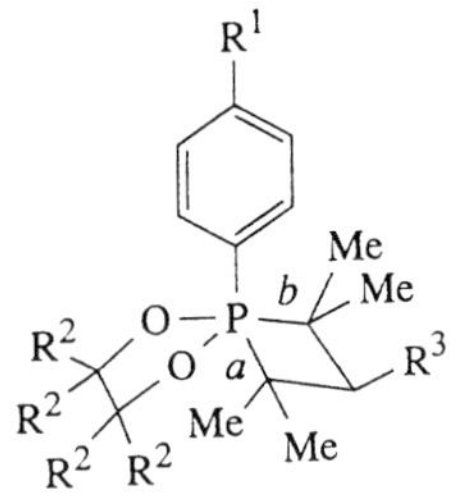

(**23**): R^1= Br; R^2= CF_3; R^3= H
(**24**): R^1, R^2= H; R^3= Me

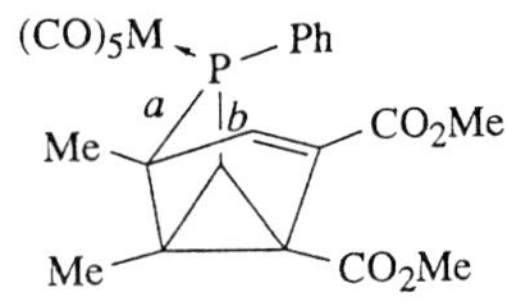

(**25**):M= Cr
(**26**): M= W

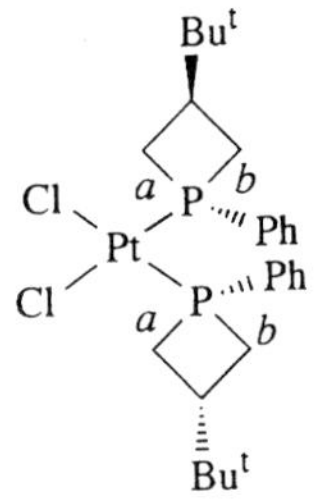

(**27**)

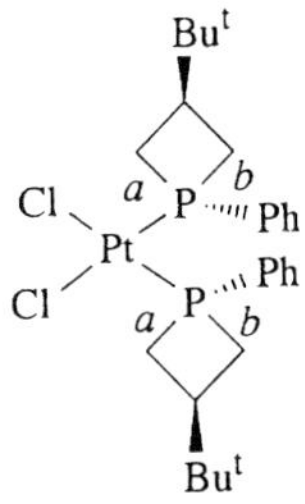

(**28**)

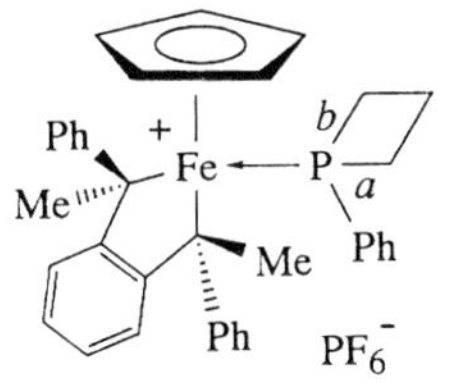

(**29**)

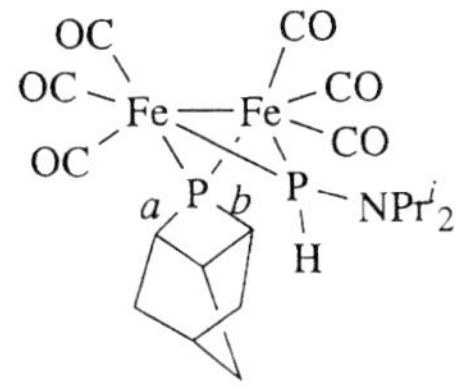

(**30**)

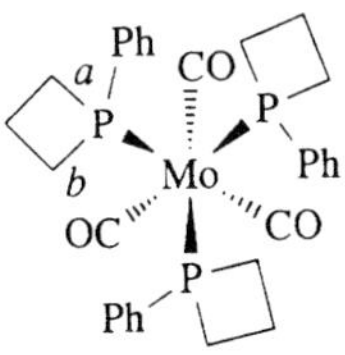

(**31**)

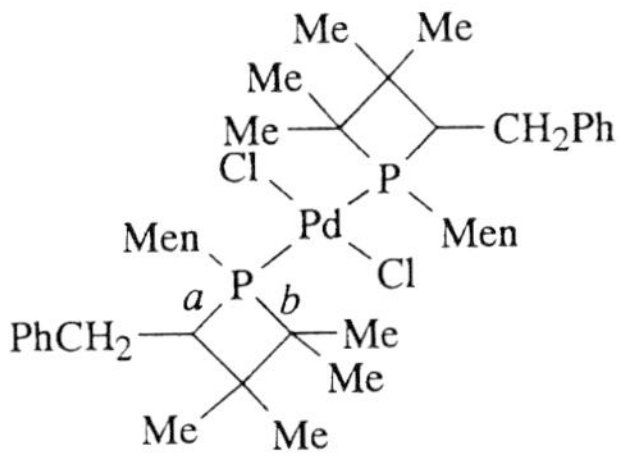

(**32**)

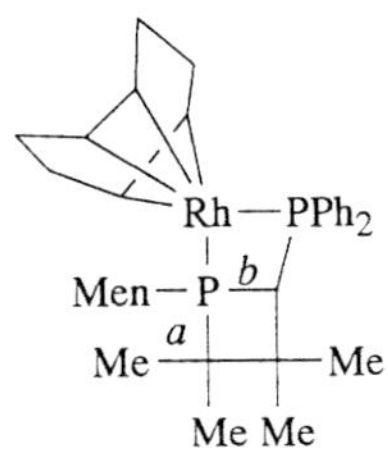

(**33**)

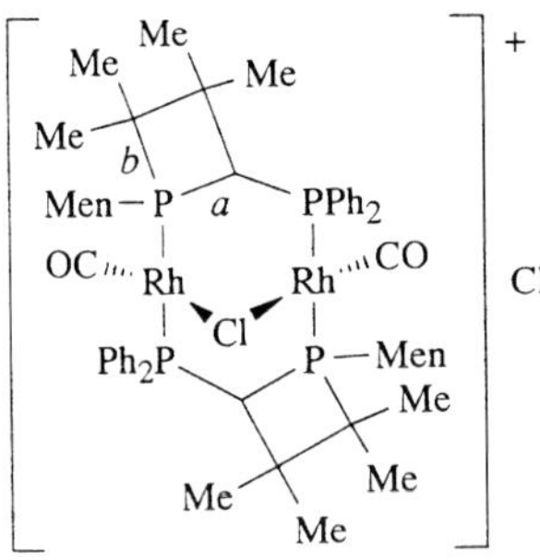

(**34**)

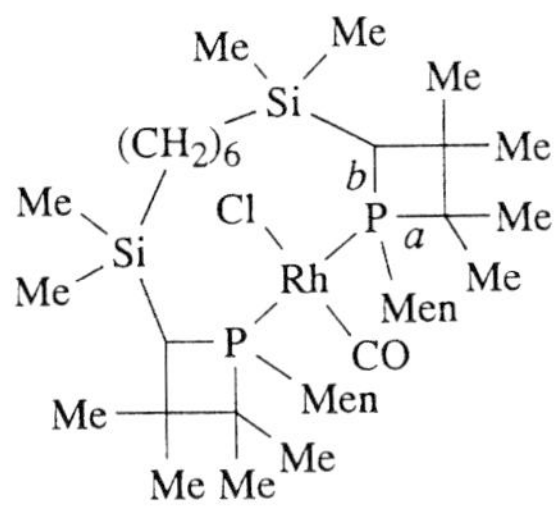

(**35**)

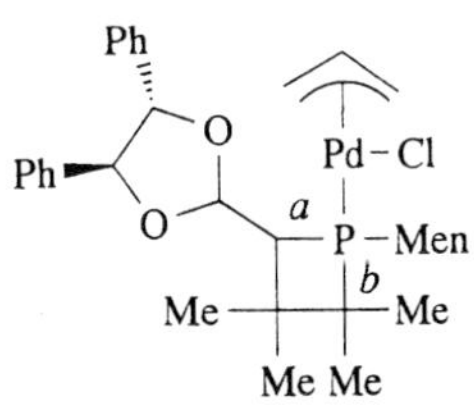

(**36**)

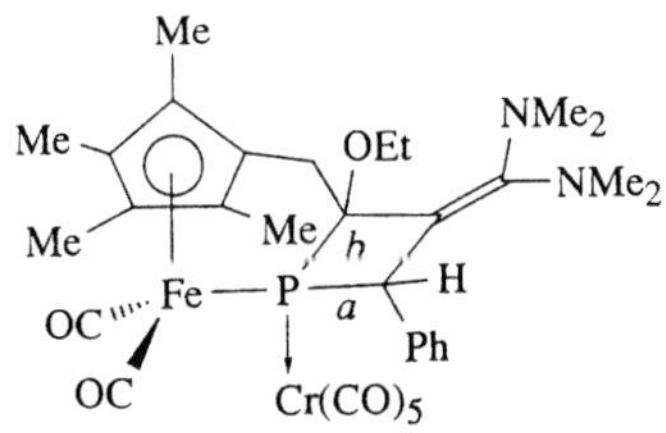

(**37**)

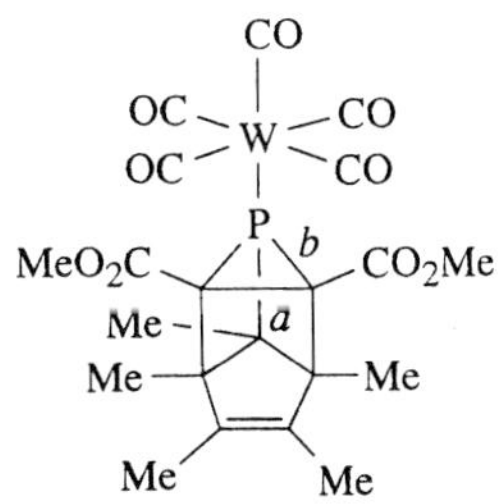

(**38**)

Scheme 2-2

Table 1. Selected bond lengths, bond angles, and dihedral angles of four-membered rings of phosphetanes

Compound	*a* (*Å*)	*b* (*Å*)	*C–P–C* (°)	*Dihedral angle*[a] (°)	*Ref.*
(1)	1.83 (1)	1.83 (2)	82.5 (5)	–	⟨68JCS(CC)1051⟩
(2)	1.86 (1)	1.91 (1)	81.9 (6)	23.4	⟨70JCS(B)938⟩
	1.88 (1)	1.88 (1)	81.8 (6)	19.9	
(3)	1.84 (2)	1.83 (3)	85.9 (11)	26.4	⟨70JCS(B)934⟩
(4)	1.846 (3)	1.845 (4)	80.7 (2)	19.2	⟨84AX(C)1918⟩
(5)	1.836 (6)	1.835 (4)	81.3 (2)	23.8	⟨71JCS(B)117⟩
(6)	1.849 (2)	1.838 (2)	81.1 (1)	30.3	⟨85AX(C)975⟩
(7)	1.8545 (7)	–	81.20 (4)	28.75	⟨85AX(C)935⟩
(8)	1.835 (4)	1.788 (5)	79.4 (2)	16.7	⟨78JOC3513⟩
(9)	1.840 (5)	1.799 (5)	80.8 (2)	29.8	⟨78JOC3513⟩
(10)	1.820 (5)	1.844 (5)	76.9 (2)	41	⟨78JCS(CC)125⟩
(11)[b]	1.831	1.811	81.8	5.0	⟨80JCS(P1)2136⟩
	1.830	1.832	86.0	2.2	
(12)	1.836 (2)	1.807 (3)	80.4 (1)	25.8	⟨82AX(B)2944⟩
(13)	1.806 (5)	1.852 (6)	79.9 (2)	17.6	⟨93T10291⟩
(14)	1.841 (2)	1.848 (2)	79.1 (1)	–	⟨93T10291⟩
(15)	1.905	1.944	82.6	24	⟨69JA2255⟩
(16)	1.82 (1)	1.82 (1)	74.8 (5)	46.6	⟨78JCS(P2)1115⟩
(17)	1.800 (8)	1.829 (8)	80.0 (4)	34.0	⟨81JCS(P2)1138⟩
(18)	1.778 (4)	1.849 (4)	81.1 (2)	28.1	⟨87AX(C)340⟩
(19)	1.820 (15)	1.821 (11)	85.2 (6)	22.4	⟨86JCS(CC)1081⟩
(20)	1.80 (1)	1.79 (1)	81.7 (7)	0	⟨92CB1987⟩
	1.83 (1)	–	82.4 (10)	0	
(21)	1.895 (7)	1.897 (8)	–	–	⟨88AG(E)837⟩
(22)	1.81	1.86	82.9	–	⟨76JCS(CC)449⟩
(23)	1.96 (3)	1.89 (3)	78.1 (9)	–	⟨73JCS(CC)856⟩
(24)	1.894 (5)	1.904 (5)	76.2 (2)	–	⟨78IC3265⟩
(25)	1.950 (3)	1.805 (3)	77.0 (1)	26.3	⟨84NJC453⟩
(26)	1.936 (5)	1.806 (4)	77.6 (2)	26.4	⟨97OM4145⟩
(27) *trans*	1.824 (7)	1.833 (7)	78.7 (3)	7.98	⟨92OM2944⟩
cis	1.835 (7)	1.830 (7)	78.8 (3)	30.67	
(28) *trans*	1.829 (8)	1.836 (8)	79.0 (4)	0.77[c]	
trans	1.815 (9)	1.833 (8)	78.2 (2)	20.01	
(29)	1.85 (2)	1.87 (2)	77.6 (8)	14 (2)[d]	⟨92JCS(D)1751⟩
	1.86 (3)	1.90 (2)	77.0 (1)	18 (2)[d]	
(30)	1.859 (9)	1.83 (1)	73.6 (4)	–	⟨93IC5720⟩
(31)	1.850 (6)	1.861 (6)	77.2 (2)	12.6 (2)[d]	⟨94JCS(CC)475⟩
	1.851 (5)	1.857 (5)	77.1 (2)	–	⟨96OM1301⟩
(32)	1.85 (1)	1.86 (1)	76.2 (6)	–	⟨94OM3956⟩
(33)	1.898 (7)	1.873 (8)	78.3 (3)	–	⟨95OM4983⟩
(34)	1.879 (5)	1.881 (5)	77.1 (2)	–	⟨95OM4983⟩
(35)	1.889 (3)	1.882 (3)	77.6 (1)	–	⟨96JOM(522)223⟩
(36)	1.840 (4)	1.886 (4)	76.7 (2)	–	⟨97JOM(529)465⟩
(37)	1.921 (4)	1.933 (4)	76.7 (2)	–	⟨98CEJ469⟩
(38)	1.924 (3)	1.836 (3)	72.21 (12)	–	⟨99AG(E)215⟩

[a] Dihedral angle between planes C–C–C and C–P–C. [b] Data were obtained from the reported atomic coordinates. [c] 1 ± > 10. [d] Dihedral angle P–C–C–C.

see Schemes 2-1 and 2-2). *trans*-1-Phenyl-, *trans*-1-chloro-, *trans*-1-(*tert*-butyl)-, *cis*-1-phenyl, *cis*-1(1-naphthyl)-2,2,3,4,4-pentamethylphosphetane oxides (**2**) ⟨68JCS(CC)1228, 70JCS(B)938⟩, (**3**) ⟨70JCS(B)934⟩, (**4**) ⟨84AX(C)1918⟩, (**5**) ⟨71JCS(B)117⟩, and (**6**) ⟨85AX(C)975⟩, *cis*-1-(dimethylamino)-2,2,3,4,4-pentamethylphosphetane 1-sulfide (**7**) ⟨85AX(C)975⟩, *trans*-2,2,3-trimethyl-1-phenylphosphetane 1-oxide (**8**) ⟨78JOC3513⟩, *trans*-2,2,3,3,4-pentamethyl-1-phenylphosphetane 1-oxide (**9**) ⟨78JOC3513⟩, 2,6-dimethyl-5-(1-methylethyl)-6-phophabicyclo[3.1.1]pent-2-ene 6-oxide (**10**) ⟨78JCS(CC)125⟩, 2,3,5-trimethyl-3-phosphatricyclo[3.2.1.0^{2,5}]nonane 3-oxide (**11**) ⟨80JCS(P1)2136⟩, 2,2-dimethyl-1-phenylphosphetane 1-oxide (**12**) ⟨82AX(B)2944⟩, optically active 1-menthylphosphetane 1-oxide (**13**) ⟨93T10291⟩ and 1-sulfide (**14**) ⟨93T10291⟩, 1,2,2,3,4,4-hexamethyl-1-phenylphosphetanium bromide (**15**) ⟨69JA2255⟩, tetracyclophosphetanium bromide (**16**) ⟨78JCS(P2)1115⟩, 2-bromo-1,3,4,4-tetramethyl-1-phenylphosphetanium bromide (**17**) ⟨81JCS(P2)1138⟩, 1-benzyl-2,2,3,3-tetramethyl-1-phenylphosphetanium bromide (**18**) ⟨87AX(C)340⟩, 9-chloro-9-diisopropylamino-9-phosphoniatricyclo[5.1.1.0^{4,8}]nonane tetrachloroaluminate (**19**) ⟨86JCS(CC)1081⟩,

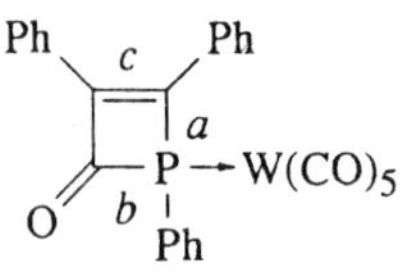

(39)

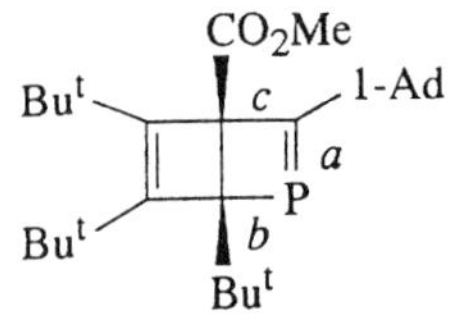

(40)

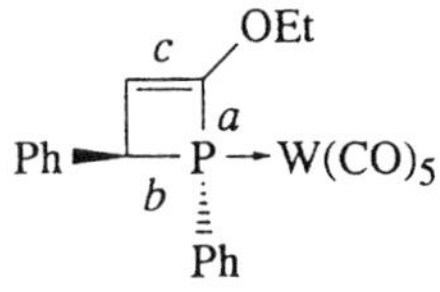

(41)

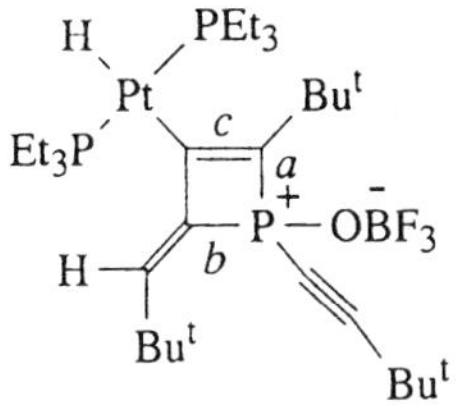

(42)

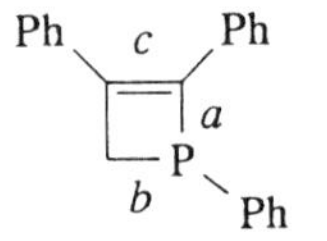

(43)

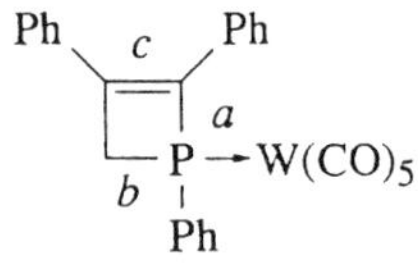

(44)

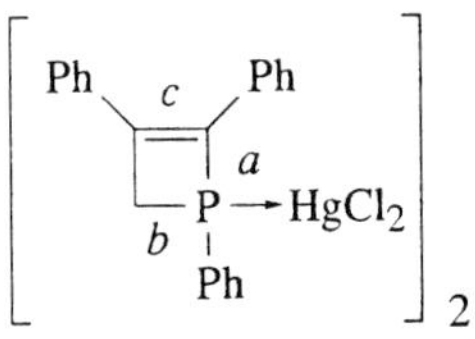

(45)

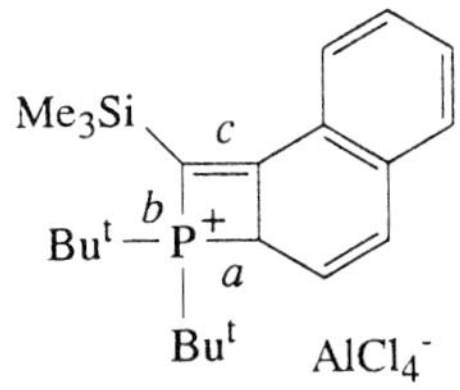

(46)

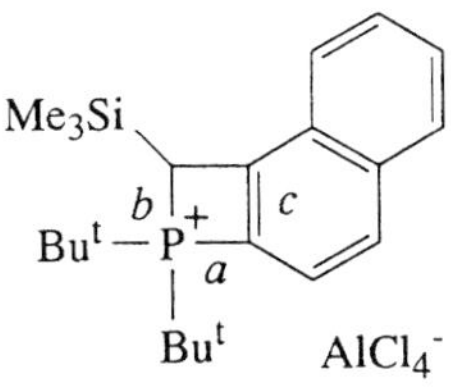

(47)

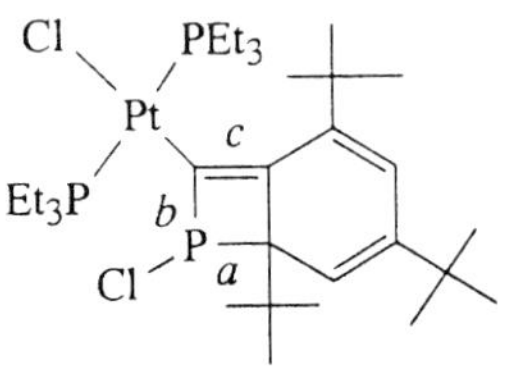

(48)

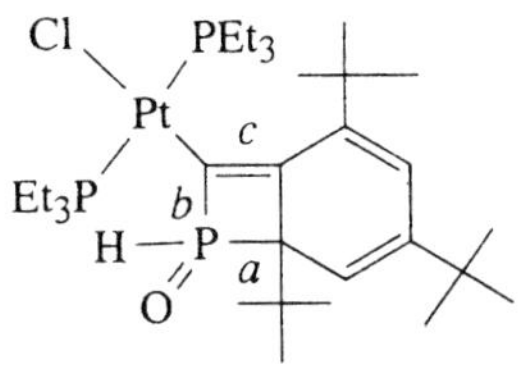

(49)

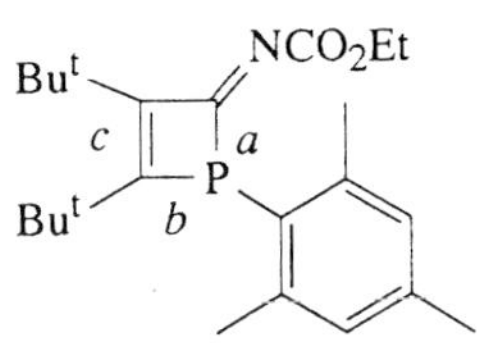

(50)

Scheme 3-1

1,1-di-*tert*-butylphosphetanium iodide (**20**) ⟨92CB1987⟩, diphosphatricyclo[2.1.0.0^{2,5}]-pentane (**21**) ⟨88AG(E)837⟩, pentacoordinate phosphetane with an oxaphosphetane ring (**22**) ⟨76JCS(CC)149⟩, pentacoordinate phosphetanes (**23**) ⟨71JCS(CC)1011, 73JCS(CC)856⟩ and (**24**) ⟨78IC3265⟩ with a 1,3,2-dioxaphospholane ring, (7-phosphatricyclo[3.2.0.0^{4,6}]-2-heptene chromium and tungsten pentacarbonyl complexes (**25**) ⟨84NJC453⟩ and (**26**) ⟨97OM4145⟩, *cis*-dichlorobis(3-*tert*-butyl-1-phenylphosphetane)platinum (II) complexes (**27**) ⟨92OM2944⟩ and (**28**) ⟨92OM2944⟩, phosphetane iron complex (**29**) ⟨92JCS(D)1751⟩, adduct (**30**) ⟨93IC5720⟩ of dinuclear iron phosphinidene complex with norbornadiene, tris(1-phenylphosphetane)moribudenum tricarbonyl complex (**31**) ⟨94JCS(CC)475⟩, *trans*-dicholorobis(2-benzyl-1-menthyl-3,3,4,4-tetramethylphosphetane)palladium (II) complex (**32**) ⟨94OM3956⟩, rhodiumcyclooctadiene complex

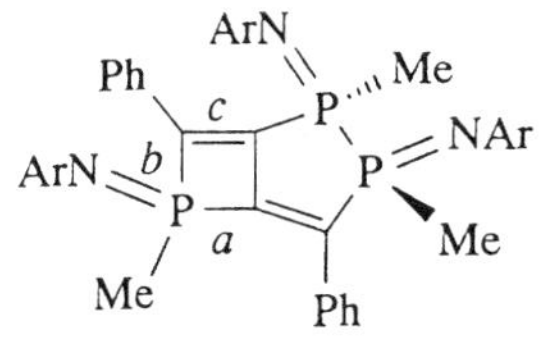

Ar= 2,4,6-$(Bu^t)_3C_6H_2$

(51)

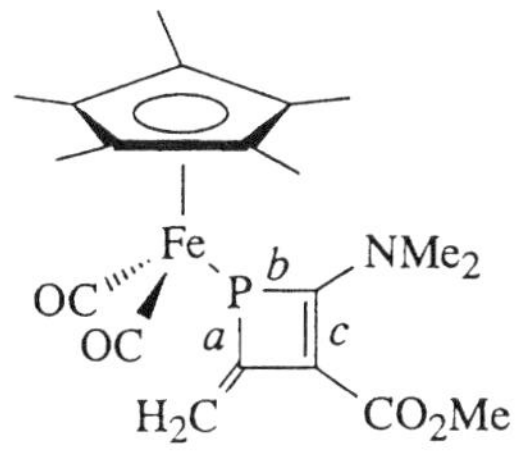

(52)

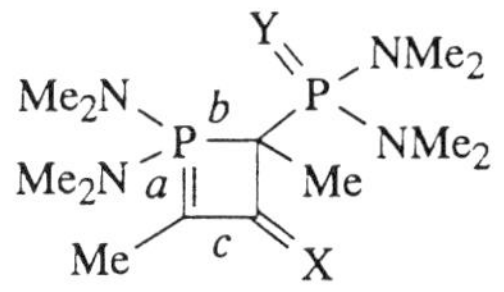

(**53**): X= Y=O
(**54**): X= O; Y= S
(**55**): X= Y= S

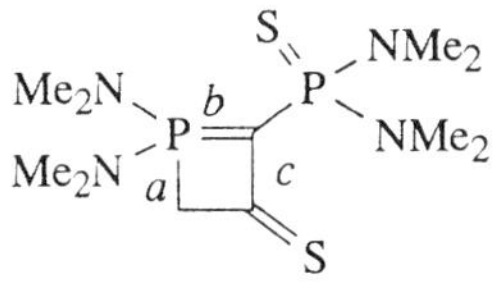

(56)

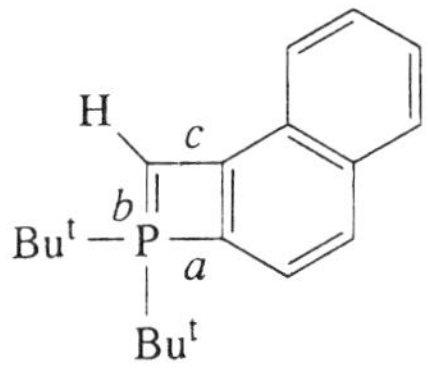

(57)

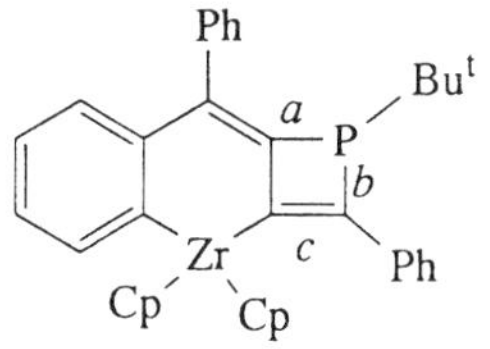

(58)

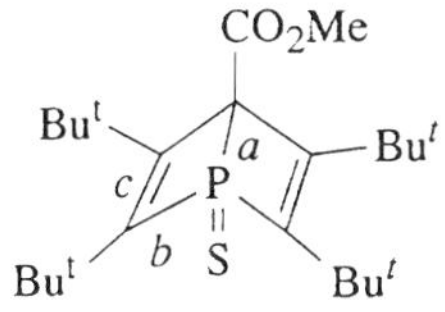

(59)

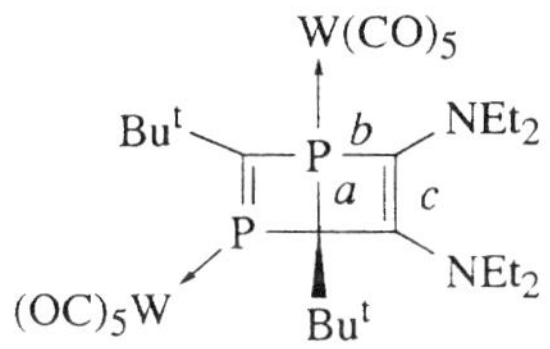

(60)

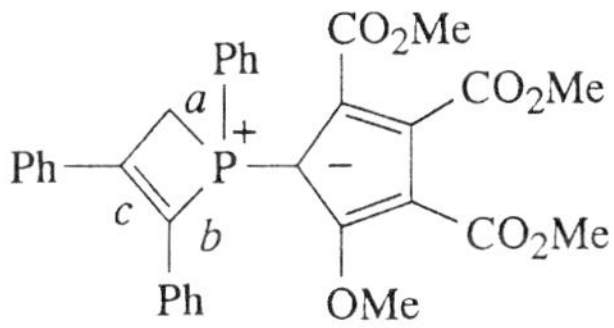

(61)

Scheme 3-2

(**33**) ⟨95OM4983⟩ and dinuclear rhodium complex (**34**) ⟨95OM4983⟩ of 1-menthyl-2-(diphenylphosphino)-3,3,4,4-tetramethylphosphetane, silyl-bridged diphosphine rhodium (I) complex (**35**) ⟨96JOM(522)223⟩, 2-(1,3-dioxolane-2-yl)-1-menthyl-3,3,4,4-tetramethylphosphetane π-allylpalladium complex (**36**) ⟨97JOM(529)465⟩, chromium pentacarbonyl complex (**37**) ⟨98CEJ469⟩ of iron phosphide, tetracyclophosphine tungsten pentacarbonyl complex (**38**) ⟨99AG(E)215⟩ were analyzed to show that all four-membered rings are puckered except for (**19**). The longest P–C bond in the ring is 1.950 (3) Å for (**25**) and the shortest one is 1.778 (4) Å for (**18**). Bond angles C–P–C are scattered in the region of 72.21 (12)° (for **38**) to 86.0° (for **11**). X-ray crystallographic analyses were carried out for 2-oxo-1,2-dihydrophosphete tungsten pentacarbonyl complex (**39**) ⟨85JA5001⟩, Dewar phosphabenzene (**40**) ⟨86AG(E)280⟩, 1,2-dihydrophosphete tungsten pentacarbonyl complex (**41**) ⟨88OM1791⟩, 2-alkylidene-1,2-dihydrophosphete 1-oxide-BF_3 platinum complex (**42**) ⟨89OM1007⟩, free 1,3,4-triphenyl-1,2-dihydrophosphete (**43**) ⟨89JA9129⟩, ⟨90AG(E)75⟩, tungsten pentacarbonyl complex (**44**) and $HgCl_2$ complex (**45**) of (**43**) ⟨91IC3381⟩, phosphetenium salts (**46**) and (**47**) ⟨93AG(E)1359⟩, 1-chloro-1,2-dihydrophosphete-4-yl platinum complex **48** ⟨93OM4265⟩, 1-oxo-1,2-dihydrophosphete-4-yl platinum complex **49** ⟨94OM2444⟩, 2-imino-1,2-dihydrophosphete (**50**) ⟨94TL1527⟩, 2,3,6-triphosphabicyclo[3.2.0]heptan-1,5-diene (**51**) ⟨94CB313⟩, $Cp^*Fe(CO)_2$ 2-methylene-1,2-dihydrophosphete 1-ylide (**52**) ⟨94ZN(B)1693⟩, 4-diaminophosphoryl- and 4-diaminothiophosphoryl-1,1-bis(dimethylamino)-3-oxo-1λ^4-3,4-dihydrophosphetes (**53**) and (**54**) ⟨95ZAAC1539⟩, 4-diaminothiophosphoryl-1,1-bis(dimethylamino)-3-thioxo-1λ^4-3,4-dihydrophosphete (**55**) ⟨96PS(115)3⟩, 2-diaminothiophosphoryl-1,1-bis(dimethylamino)-3-thioxo-1λ^5-3,4-dihydrophosphete (**56**) ⟨96PS(115)3⟩, naphtho-1λ^4-phosphete (**57**) ⟨96CEJ68⟩, 2-alkylidene-3-zirconyl-1,2-dihydrophosphete (**58**) ⟨97AG(E)987⟩, Dewar phosphabenzene sulfide (**59**) and tungsten dinuclear complex (**60**) ⟨98MI1⟩, phosphetanium cyclopentadienylide (**61**) ⟨98HAC21⟩ showed that these four-membered rings are almost planar in sharp contrast to those of phosphetanes. Selected bond lengths and bond angles are summarized in Table 2 (for structural formula,

Table 2. Selected bond lengths and bond angles of four-membered rings of phosphetenes

Compound	*a* (*Å*)	*b* (*Å*)	*c* (*Å*)	*C–P–C* (*°*)	*Ref.*
(39)	1.93 (1)	1.83 (1)	1.36 (1)	71.9 (5)	⟨85JA5001⟩
(40)	1.68 (1)[a]	1.89 (1)	1.54 (1)	79.5 (4)	⟨86AG(E)280⟩
(41)	1.815 (4)	1.902 (5)	1.331 (7)	74.0 (2)	⟨88OM1791⟩
(42)	1.754 (6)	1.766 (6)	1.365 (8)	77.9 (3)	⟨89OM1007⟩
(43)	1.821 (3)	1.886 (4)	1.366 (5)	74.0 (2)	⟨89JA9129⟩
	1.826 (3)	1.872 (4)	1.359 (4)	73.8 (2)	⟨90AG(E)75⟩
(44)	1.816 (19)	1.813 (24)	1.358 (31)	75.4 (10)	⟨91IC3381⟩
(45)	1.80 (1)	1.82 (1)	1.34 (2)	76.8 (6)	⟨91IC3381⟩
	1.80 (1)	1.80 (1)	1.34 (2)	77.9 (6)	
(46)	1.856 (3)	1.798 (3)	1.366 (4)	79.1 (1)	⟨93AG(E)1359⟩
(47)	1.859 (3)	1.781 (3)	1.360 (5)	78.7 (1)	⟨93AG(E)1359⟩
(48)	1.911 (6)	1.678 (5)	1.389 (8)	–	⟨93OM4265⟩
(49)	1.844 (10)	1.776 (3)	1.369 (13)	80.2 (5)	⟨94OM2444⟩
(50)	1.848 (6)	1.850 (6)	1.354 (7)	70.2 (3)	⟨94TL1527⟩
(51)	1.850 (5)	1.840 (7)	1.387 (7)	75.1 (3)	⟨94CB313⟩
(52)	1.862 (7)	1.856 (7)	1.399 (10)	72.2 (3)	⟨94ZN(B)1693⟩
(53)	1.709 (2)[b]	1.849 (2)	1.391 (3)	81.48 (10)	⟨95ZAAC1539⟩
(54)	1.708 (2)[b]	1.863 (2)	1.396 (3)	81.42 (9)	⟨95ZAAC1539⟩
(55)	1.748 (7)[b]	1.848 (7)	1.36 (1)	79.2 (3)	⟨96PS(115)3⟩
	1.733 (7)[b]	1.849 (7)	1.35 (1)	79.9 (3)	
	1.737 (7)[b]	1.873 (7)	1.39 (1)	81.2 (3)	
(56)	1.807 (5)	1.736 (5)[b]	1.395 (6)	79.6 (2)	⟨96PS(115)3⟩
(57)	1.773 (3)	1.772 (8)[b]	1.401 (8)	78.9 (3)	⟨96CEJ68⟩
(58)	1.835 (8)	1.844 (8)	1.53 (1)	72.3 (3)	⟨97AG(E)987⟩
(59)	1.823 (4)	–	1.351 (5)	–	⟨98MI1⟩
(60)	1.874 (8)	1.830 (9)	1.354 (12)	75.7 (4)	⟨98MI1⟩
(61)	1.796 (15)	1.797 (18)	1.354 (22)	78.7 (8)	⟨98HAC21⟩

[a] C=P double bond containing a trivalent phosphorus. [b] C=P double bond containing a pentavalent phosphorus.

see Schemes 3-1 and 3-2). Single bonds for P–C(sp^2) are shorter than those for P–C(sp^3) (see **(41)**, **(42)**, **(43)**, **(44)**, **(45)**, **(48)**, and **(49)**). The double bond between carbon and trivalent phosphorus atoms is significantly shorter than C–P single bonds (0.21 Å for **(40)**), but those between carbon and pentavalent phosphorus atoms are not so much shorter (by 0.21 Å) than C–P single bonds (0.140 Å for **(53)**, 0.155 Å for **(54)**, 0.117 Å (averaged value for three independent molecules) for **(55)**, 0.071 Å for **(56)**), because ylide canonical structure is more important in the latter bonds than in the former bond. It is very interesting to point out that they are equal to each other for **(57)**.

3.2.2.2 NMR Studies

3.2.2.2.1 ^{1}H NMR

Although ^{1}H NMR spectroscopy has been gradually replaced by ^{13}C and ^{31}P NMR in view of usefulness for the determination of all compounds containing one P atom, it still remains one of the important methods. In the case of trivalent phosphetanes **(62)**–**(64)** the stereochemistry of the compounds can be estimated from the coupling constant between the ring proton and ^{31}P nucleus, see Scheme 4. For example, larger coupling constants between proton on carbon-3 and ^{31}P were observed for the *cis*-isomers than the *trans*-isomers as shown in Table 3 ⟨84JCS(P2)553⟩.

Table 3. Selected ^{1}H NMR spectral data of phosphetanes **(62)**–**(64)**

Compounds		δH^a	$^3J_{PH}$ *(Hz)*
(62)	*trans*	2.81	1.3
	cis	2.17	2.4
(63)	*trans*	2.65	<0.5
	cis	1.64	3.4
(64)	*trans*	2.60	0.6
	cis	1.84	2.2

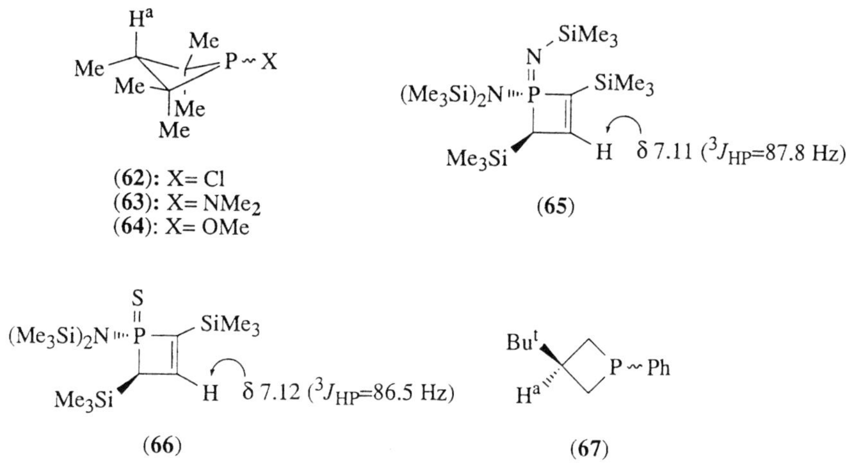

Scheme 4

An unusually large 3-bond P–H coupling constant ($^3J_{HP}$ = 88 and 87 Hz) was observed for a proton of the 3-position of 1,2-dihydrophosphetes (**65**) and (**66**), while the $^3J_{HP}$'s with the *trans*-proton are usually 40–50 Hz, see Scheme 4. The increase of $^3J_{HP}$ by ca. 40 Hz can be reasonably explained by the enlargement of ring strain and a dual pathway for spin–spin coupling in these rings ⟨87TL6121⟩.

NOESY, one of 2D NMR spectroscopy, which has recently been developed, has often contributed to determine such a kind of stereochemistry in solution; for example, multiplets at δ 2.81–2.96 and δ 2.41–2.60, whose assignment is otherwise almost impossible because of their very complex signal, can be assigned to *trans*- and *cis*-(**67**), respectively ⟨92OM2944⟩, see Scheme 4.

3.2.2.2.2 ^{13}C NMR

In ^{13}C NMR spectra not only the chemical shifts but also the coupling constants with ^{31}P nucleus have provided important information about the molecular structure. In 1972 Gray and Cremer collected the ^{13}C NMR data (chemical shifts and coupling constants with ^{31}P) for the ring carbons of polymethylphosphetane, phosphetane oxides, phosphetanium salts, and phosphetane sulfides as shown in Table 4 and Scheme 5 and showed effects of substituents on the phosphorus atom on the chemical shifts ⟨72JOC3458, 72JOC3342⟩. For other phosphetanes and their metal complexes including some compounds shown in Table 4, the data of chemical shifts and coupling constants of the ring carbons along with their ^{31}P chemical shifts are summarized in Table 5 (Scheme 6). In Table 6 the recent data of optically active phosphetane derivatives have been collected (Schemes 7-1 and 7-2). Table 7 summarizes the data for 1,2-dihydrophosphetes and their metal complexes (Schemes 8-1 and 8-2).

The region of ^{13}C NMR chemical shifts of 2-imono-1,2-dihydrophosphetes (**295a–j**) along with that of their ^{31}P NMR chemical shifts are shown in Scheme 9, but those of 1-imono derivatives (**296b–e**) were not reported ⟨94TL1527⟩.

Selected ^{13}C NMR data of Rh π-allyl complex (**297**) and Co π-complex of phosphete (**298**) are given in Scheme 10.

3.2.2.2.3 ^{31}P NMR

^{31}P NMR spectroscopy is the most useful for the phosphorus-containing compounds, because of its ready measurement emerging from 100% natural abundance and high relative intensity. One of noteworthy features of ^{31}P NMR is that the chemical shifts are directly reflecting structural environment of the phosphorus atom, see Tables 4–7. The *trans*-isomers at the 1,3-positions of trivalent phosphetanes (Table 8) resonate more downfield than the *cis*-isomer in ^{31}P NMR, while the reverse situation was found for pentavalent phosphetanes (Table 9) ⟨92JOC511, 84JA7021, 84JCS(P2)553, 87S265⟩.

Recently, the ^{31}P NMR data of 1-amino-2,2,3,4,4-tetramethylphosphetanes (**299**) and their oxides (**300**) and sulfides (**301**) have been reported as shown in Tables 10–12 (Scheme 11), respectively ⟨96PS(115)273⟩. Assignment of *cis*- and *trans*-isomers was carried out only for the trivalent phosphetanes except for one example of sulfide (**301w**).

Table 4. ^{13}C NMR spectral data of polymethylphosphetanes

Compounds	*C-1* δ ($^{1}J_{CP}$/Hz)	*C-2* δ ($^{2}J_{CP}$/Hz)	*C-3* δ ($^{1}J_{CP}$/Hz)	*Ref.*
(1) [a]	48.77 (76.6)	42.90 (11.7)	C-1	⟨72JOC3458⟩
(2) [a]	46.68 (58.7)	44.88 (6.2)	C-1	⟨72JOC3458⟩
(3) [a]	57.35 (56.8)	42.79 (1.6)	C-1	⟨72JOC3458⟩
(4) [a]	47.50 (51.0)	45.71 (5.2)	C-1	⟨72JOC3458⟩
(5) [a]	46.45 (58.4)	48.03 (11.2)	C-1	⟨72JOC3458⟩
(8) [a]	49.96 (69.3)	30.97 (11.9)	37.42 (52.5)	⟨72JOC3458⟩
(9) [b]	49.20 (61.2)	41.74 (15.1)	43.10 (58.4)	⟨72JOC3458⟩
(15) [b]	42.11 (45.3)	50.43 (10.5)	C-1	⟨72JOC3458⟩
trans-**(62)** [c]	31.65 (−7.8)	51.55 (2.1)	C-1	⟨72JOC3470⟩
cis-**(62)** [c]	35.10 (−11.1)	48.04 (7.8)	C-1	⟨72JOC3470⟩
(67) [c]	30.22 (−5.9)	54.02 (2.7)	C-1 [d]	⟨72JOC3470⟩
(69) [c]	34.52 (−2.6)	49.70 (5.9)	C-1	⟨72JOC3470⟩
(70) [c]	26.37 (−7.7)	51.43 (3.3)	C-1	⟨72JOC3470⟩
(71) [c]	28.61 (−6.5)	51.13 (2.9)	C-1	⟨72JOC3470⟩
(72) [a]	44.43 (74.2)	37.10 (10.9)	C-1	⟨72JOC3458⟩
(73) [a]	44.23 (73.5)	37.65 (10.4)	C-1	⟨72JOC3458⟩
(74) [a]	45.41 (59.4)	42.61 (6.3)	C-1	⟨72JOC3458⟩
(75) [a]	44.05 (59.4)	46.89 (10.0)	C-1	⟨72JOC3458⟩
(76) [a]	47.23 (58.3)	45.33 (6.2)	C-1	⟨72JOC3458⟩
(79) [a]	57.05 (55.4)	46.20 (1.8)	C-1	⟨72JOC3458⟩
(80) [e]	48.25 (35.5)	52.89 (9.8)	C-1	⟨72JOC3470⟩
(81) [e]	48.38 (48.1)	50.55 (9.8)	C-1	⟨72JOC3470⟩
(82) [a]	46.96 (56.9)	43.80 (5.7)	C-1	⟨72JOC3458⟩
(83) [a]	46.96 (51.0)	47.39 (11.8)	C-1	⟨72JOC3458⟩
(84) [a]	47.50 (51.0)	47.68 (12.5)	C-1	⟨72JOC3458⟩
(85) [a]	46.62 (47.9)	51.06 (5.4)	C-1	⟨72JOC3458⟩
(86) [a]	45.79 (47.3)	48.73 (6.9)	C-1	⟨72JOC3458⟩
(87) [b]	38.07 (45.2)	51.42 (11.2)	C-1	⟨72JOC3458⟩
(88) [b]	41.41 (45.2)	53.40 (10.9)	C-1	⟨72JOC3458⟩
(89) [b]	41.98 (45.8)	53.44 (19.5)	C-1	⟨72JOC3458⟩
(90) [b]	39.82 (44.0)	51.94 (10.2)	C-1	⟨72JOC3458⟩
(91) [b]	40.40 (44.3)	50.93 (10.2)	C-1	⟨72JOC3458⟩
(92) [f]	42.28 (41.0)	50.82 (9.1)	C-1	⟨72JOC3458⟩
(94) [a]	52.20 (81.8)	30.62 (18.1)	40.74 (67.7)	⟨72JOC3458⟩
(95) [a]	52.93 (79.3)	30.27 (18.3)	39.71 (65.8)	⟨72JOC3458⟩
(96) [a]	52.93 (79.3)	30.27 (18.3)	39.71 (65.8)	⟨72JOC3458⟩
(100) [a]	48.41 (63.1)	34.47 (16.0)	34.56 (52.3)	⟨72JOC3458⟩
(101) [b]	45.53 (48.6)	40.49 (17.1)	25.35 (47.7)	⟨72JOC3458⟩
(102) [b]	44.01 (48.5)	15.9 (15.0)	46.1 (46.1)	⟨72JOC3458⟩
(103) [a]	53.34 (81.8)	32.17 (14.6)	48.19 (67.7)	⟨72JOC3458⟩
(104) [a]	54.47 (78.2)	31.97 (13.8)	47.86 (64.4)	⟨72JOC3458⟩
(105) [a]	50.81 (62.8)	35.04 (11.9)	41.46 (52.0)	⟨72JOC3458⟩
(106) [b]	47.2 (47.2)	30.88 (14.0)	44.00 (46.0)	⟨72JOC3458⟩
(107) [a]	49.33 (49.7)	39.19 (10.5)	42.74 (44.8)	⟨72JOC3458⟩
(108) [a]	52.47 (78.9)	35.57 (106)	51.0 (72.8)	⟨72JOC3458⟩
(109) [a]	53.00 (74.7)	35.64 (10.1)	50.13 (69.7)	⟨72JOC3458⟩
(110) [a]	53.14 (74.2)	33.97 (10.1)	51.72 (67.8)	⟨72JOC3458⟩
(111) [b]	46.24 (45.8)	46.95 (9.7)	36.33 (44.7)	⟨72JOC3458⟩
(112) [a]	60.54 (58.6)	40.60 (6.2)	58.75 (53.5)	⟨72JOC3458⟩
(113) [a]	60.04 (57.8)	35.54 (2.9)	57.74 (51.7)	⟨72JOC3458⟩
(114) [a]	49.52 (80.1)	39.88 (14.0)	47.74 (73.48)	⟨72JOC3458⟩
(115) [a]	50.18 (76.1)	38.84 (13.8)	48.17 (70.4)	⟨72JOC3458⟩
(116) [a]	48.68 (60.4)	36.02 (12.5)	38.38 (56.2)	⟨72JOC3458⟩
(117) [a]	43.77 (46.7)	43.91 13.7)	29.48 (46.5)	⟨72JOC3458⟩
(118) [a]	47.18 (57.8)	58.98 (7.1)	C-1	⟨72JOC3458⟩

[a] In $CHCl_3$. [b] In H_2O. [c] In neat. [d] C-1 and C-3 are equivalent. [e] In CH_2Cl_2–CH_3Cl. [f] In glacial acetic acid.

The substituents on the phosphorus atom significantly affect the chemical shifts of the ^{31}P NMR. For example, in the trivalent 2,2,3,4,4-pentamethylphosphetanes **(62)**–**(64)** the signals may shift downfield in the order NMe_2 < Cl < OMe ⟨84JCS(P2)553⟩, which is simply based on electronegativity of N, Cl, and O atoms.

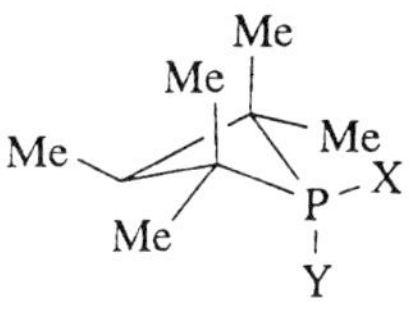

	X	Y
(**68**)	lone pair	Ph
(**69**)	Ph	lone pair
(**70**)	lone pair	Me
(**71**)	Me	lone pair
(**72**)	O	OMe
(**73**)	OMe	O
(**74**)	O	Me
(**75**)	Me	O
(**76**)	O	4-FC_6H_4
(**77**)	O	4-ClC_6H_4
(**78**)	O	3,4-$Cl_2C_6H_3$
(**79**)	Cl	O
(**80**)	Cl	Me
(**81**)	Me	Cl
(**82**)	O	CH_2Ph
(**83**)	CH_2Ph	O
(**84**)	Bu^t	O
(**85**)	S	Ph
(**86**)	Ph	S
(**87**)	Me	Me
(**88**)	Me	Ph
(**89**)	Me	4-FC_6H_4
(**90**)	Me	CH_2Ph
(**91**)	CH_2Ph	Me
(**92**)	CH_2Ph	CH_2Ph
(**93**)	O	Men

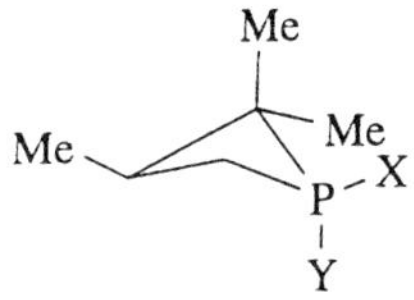

	X	Y
(**94**)	O	OH
(**95**)	O	OMe
(**96**)	OMe	O
(**97**)	O	4-FC_6H_4
(**98**)	O	4-ClC_6H_4
(**99**)	O	2,4-$Cl_2C_6H_3$
(**100**)	Ph	O
(**101**)	Me	Ph
(**102**)	Ph	Me

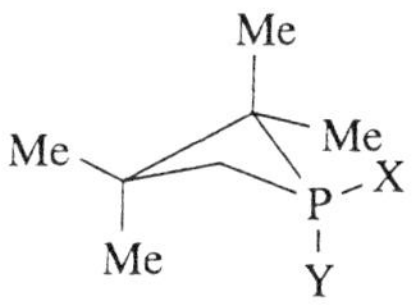

	X	Y
(**103**)	OH	O
(**104**)	OMe	O
(**105**)	Ph	O
(**106**)	Ph	Me
(**107**)	Ph	S

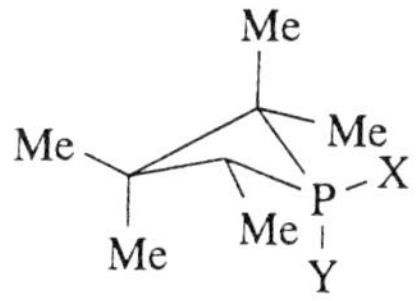

	X	Y
(**108**)	OH	O
(**109**)	O	OMe
(**110**)	OMe	O
(**111**)	Ph	Me
(**112**)	O	Cl
(**113**)	Cl	O

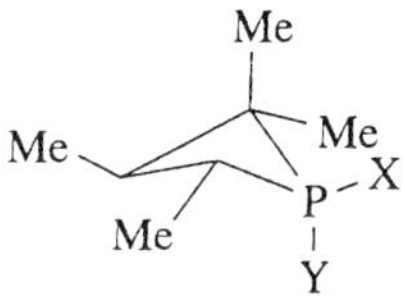

	X	Y
(**114**)	OH	O
(**115**)	OMe	O
(**116**)	Ph	O
(**117**)	Ph	Me

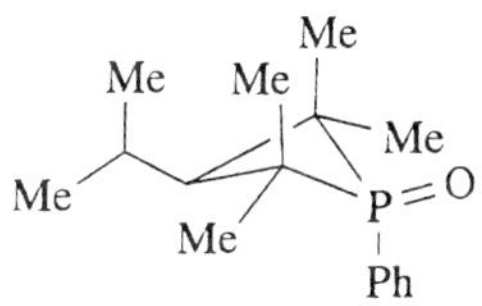

(**118**)

Scheme 5

Table 5. ^{13}C and ^{31}P NMR spectral data of phosphetanes

Compounds	*C-1* δ ($^1J_{CP}$/Hz)	*C-2* δ ($^2J_{CP}$/Hz)	*C-3* δ ($^1J_{CP}$/Hz)	^{31}P δ	*Ref.*
(**11**)	46.1 (56)	42.1 (0)	40.3 (53)		⟨80JCS(P1)2136⟩
(**25**)	61.40 (25.64)	52.22 (9.77)	31.23 (50.05)	15.54 [a,b]	⟨84NJC453⟩
(**146**)	64.8 (19.8)	34.6 (0)	68.8 (0)	−130.2	⟨87AG(E)85⟩
	67.2 (11.8)	46.0 (0)	68.8 (0)		
trans-(**76**)	46.8 (58.1)	45.2 (5.5)	46.8 (58.1)	52.8	⟨87S265⟩
cis-(**76**)	46.1 (59.3)	45.7 (–) [c]	46.1 (59.3)	55.3	⟨87S265⟩
trans-(**77**)	46.9 (58.2)	45.0 (6.6)	46.9 (58.2)	52.3	⟨87S265⟩
cis-(**77**)	46.0 (58.2)	45.0 (6.6)	46.0 (58.2)	55.6	⟨87S265⟩
trans-(**78**)	47.9 (58.2)	45.1 (5.5)	47.9 (58.2)	52.6	⟨87S265⟩
trans-(**97**)	49.3 (63.7)	30.0 (12.1)	36.5 (52.8)	43.4	⟨87S265⟩
trans-(**98**)	50.2 (64.5)	30.8 (12.1)	37.2 (52.4)	43.6	⟨87S265⟩
trans-(**99**)	50.8 (63.1)	29.6 (10.8)	39.9 (55.1)	47.7	⟨87S265⟩

Table 5. (continued)

Compounds	*C-1* δ ($^{1}J_{CP}/Hz$)	*C-2* δ ($^{2}J_{CP}/Hz$)	*C-3* δ ($^{1}J_{CP}/Hz$)	^{31}P δ	*Ref.*
(**300x**)	52.2 (–)[c]	41.4 (–)[c]	C-1	64.56	⟨87PS283⟩
(**300y**)	52.8 (–)[c]	41.7 (–)[c]	C-1	63.4	⟨87PS283⟩
(**141**)[e]	49.1 (72)	36.5 (17)	C-1	74.8[e]	⟨87TL603⟩
(**142**)[e]	55.1 (65)	41.4 (11)	C-1	80.1[e]	⟨87TL603⟩
(**143**)[e]	48.1 (80)	31.9 (12)	C-1	55.8[e]	⟨87TL603⟩
(**144**)[e]	48.0 (79)	32.4 (12)	C-1	53.8[e]	⟨87TL603⟩
(**145**)[e]	48.2 (81)	32.2 (11)	C-1	49.6[e]	⟨87TL603⟩
(**21**)[a]	178.52 (15.1)	–[c]	–[c]	−346.6[a]	⟨88AG(E)837⟩
(**152**)	51.4 (78)	27.9 (18)	31.3 (59)	49.8	⟨88TL1219⟩
(**153**)	52.7 (79)	34.8 (14)	46.4 (69)	46.4	⟨88TL1219⟩
(**154**)	58.7 (67)	14.3 (21)	34.5 (61)	48.6	⟨88TL1219⟩
	56.3 (63)	19.3 (22)	34.0 (67)	46.7	⟨88TL1219⟩
(**155**)	71.9 (68)	26.6 (22)	33.7 (62)	48.4	⟨88TL1219⟩
(**148**)[a]	50.69 (23.1)	65.27 (–)[c]	41.43 (27.2)	37.1[a]	⟨90JCS(CC)153⟩
(**164**)	195.5 (dd) (71.0, 24.6)	30.2 (0)	27.2 (15)	232, 45 (*J*(PP)57)	⟨92AG(E)625⟩
(**165**)	192.6 (dd) (67.5, 61.0)	30.2 (0)	27.2 (14.6)	270, 40 (*J*(PP)76)	⟨92AG(E)625⟩
(**119**)	17.0 (46)	18.6 (23)	C-1	74.7[e]	⟨92CB1987⟩
(**120**)	16.2 (43)	18.4 (23)	C-1	80.3[e]	⟨92CB1987⟩
(**121**)[f]	33.2 (47)	21.3 (22)	C-1	41.9[e]	⟨92CB1987⟩
(**122**)[a]	40.52 (0.0)	47.36 (1.9)	C-1[d]	47.4[a]	⟨92JOC511⟩
(**123**)[a]	40.47 (6.1)	40.17 (14.7)	C-1	39.5[a]	⟨92JOC511⟩
(**124**)	44.12 (52.3)	40.42 (12.4)	C-1	70.4	⟨92JOC511⟩
(**125**)	43.30 (52.8)	40.62 (14.2)	C-1	73.4	⟨92JOC511⟩
(**126**)	45.03 (52.7)	40.35 (12.7)	C-1	70.1[g]	⟨92JOC511⟩
(**127**)	43.95 (52.2)	40.58 (14.2)	C-1	73.3[h]	⟨92JOC511⟩
(**128**)	45.45 (52.2)	41.03 (13.4)	C-1	70.2	⟨92JOC511⟩
(**129**)	44.21 (52.8)	40.85 (14.0)	C-1	73.2	⟨92JOC511⟩
(**130**)	44.04 (56.2)	40.65 (13.6)	C-1	69.2	⟨92JOC511⟩
(**131**)	43.82 (56.3)	40.52 (13.8)	C-1	–	⟨92JOC511⟩
(**132**)[a]	52.27 (47.9)	38.40 (5.2)	C-1	95.7[e]	⟨92JOC511⟩
(**133**)[a]	50.04 (50.2)	42.83 (18.8)	C-1	86.6[e]	⟨92JOC511⟩
(**134**)[a]	48.08 (63.2)	31.03 (7.1)	C-1	55.9	⟨92JOC511⟩
(**135**)[a]	45.60 (63.7)	39.49 (20.1)	C-1	76.7	⟨92JOC511⟩
(**136**)[a]	48.67 (65.0)	31.65 (6.9)	C-1	48.3	⟨92JOC511⟩
(**137**)[a]	49.37 (52.5)	37.11 (8.6)	C-1	74.7	⟨92JOC511⟩
(**138**)[a]	49.47 (52.1)	41.32 (13.5)	C-1	83.6	⟨92JOC511⟩
(**139**)[a]	47.87 (47.3)	39.11 (9.5)	C-1	63.8	⟨92JOC511⟩
(**140**)[a]	48.86 (47.0)	41.88 (11.3)	C-1	67.9	⟨92JOC511⟩
(**29**)[e]	25.79 (31.9)	23.36 (18.5)	C-1	99.47[e]	⟨92JCS(D)1751⟩
(**162**)[a]	39.3 (0)	12.8 (7)	C-1	−23.7[h,j]	⟨92CB1987⟩
(**163**)[a]	18.98 (0)	17.2 (13)	C-1	12.8[i]	⟨92CB1987⟩
(**156**)	46.8 (69.7)	7.52 (27.2)	C-1	53.2	⟨93JCR(S)268⟩
(**157**)	36.8 (69.7)	7.52 (27.2)	C-1	52.6	⟨93JCR(S)268⟩
(**158**)	37.5 (72.0)	7.4 (27.6)	C-1	53.2	⟨93JCR(S)268⟩
(**19**)	49.91 (14.9)	58.67 (58.3)	C-1	79.4[e]	⟨86JCS(CC)1081⟩
(**150**)[a]	69.7 (24.2)	42.4 (3.0)	68.0 (21.3)	11.5[a]	⟨94T759⟩
(**151**)[a]	71.1 (24.9)	42.9 (0)	69.7 (20.2)	7.7[a]	⟨94T759⟩
(**159**)[f]	48.6 (72)	57.9 (17)	C-1[d]	40.0[f]	⟨95JOC6076⟩
(**160**)	44.5 (76)	61.7 (14)	C-1[d]	39.8	⟨95JOC6076⟩
(**161**)	47.3 (73)	28.0 (20)	C-1[d]	46.0	⟨95JOC6076⟩
(**166**)[a]	62.9 (11.3)	39.46 (0)	34.6 (17.3)	113.0[a]	⟨97OM2958⟩
(**167**)[a]	110.7 (19.7)	124.8 (0)	53.0 (17.1)	107.2[a]	⟨98CEJ469⟩

[a] In C_6D_6. [b] $^{1}J_{PW}$ = 239 Hz. [c] Not assigned or not resolved. [d] C-1 and C-3 are equivalent. [e] In CH_2Cl_2 or CD_2Cl_2. [f] In D_2O. [g] $^{5}J_{PF}$ = 6.1 Hz. [h] $^{5}J_{PF}$ = 6.1 Hz. [i] In pentane. [j] $^{1}J_{PF}$ = 663 Hz.

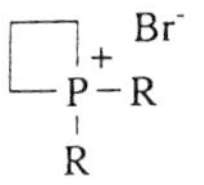

(**119**): R=Pr^i
(**120**): R=Bu^t
(**121**): R=Mes

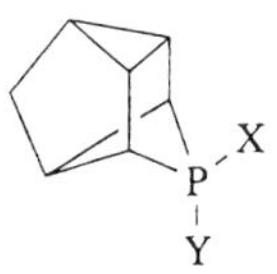

(**122**)-(**145**)

	X	Y		X	Y
(**122**)	lone pair	Me	(**134**)	O	Me
(**123**)	Me	lone pair	(**135**)	Me	O
(**124**)	$PhCH_2$	Me	(**136**)	O	Ph
(**125**)	Me	$PhCH_2$	(**137**)	S	Me
(**126**)	p-$FC_6H_4CH_2$	Me	(**138**)	Me	S
(**127**)	Me	p-$FC_6H_4CH_2$	(**139**)	Se	Me
(**128**)	p-$NO_2C_6H_4CH_2$	Me	(**140**)	Me	Se
(**129**)	Me	p-$NO_2C_6H_4CH_2$	(**141**)	NMe_2	NMe_2
(**130**)	Me	Me	(**142**)	Ni-Pr_2	Cl
(**131**)	Me	CD_3	(**143**)	O	NMe_2
(**132**)	Cl	Me	(**144**)	O	NEt_2
(**133**)	Me	Cl	(**145**)	O	Ni-Pr_2

A counter anion is Br^- for (**124**)-(**131**) and $AlCl_4^-$ for (**132**), (**133**), (**141**), and (**142**).

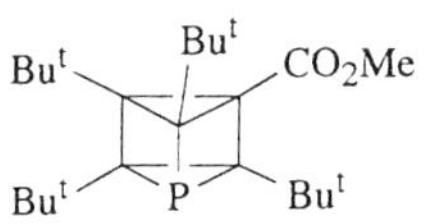

(**146**)

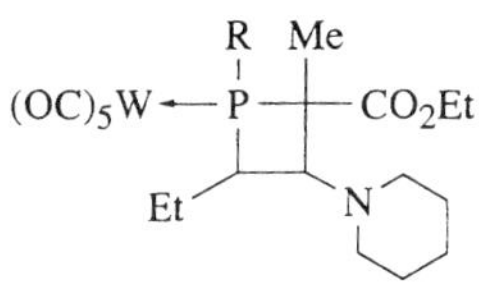

(**147**): R= Me
(**148**): R= Ph
(**149**): R= Bu^t

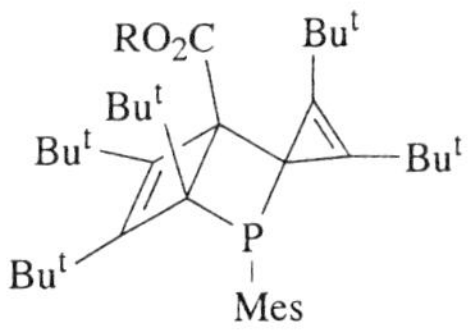

(**150**): R= Me
(**151**): R= Bu^t

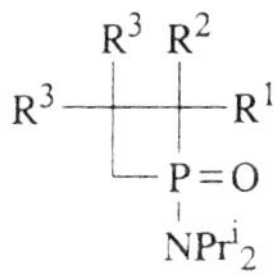

	R^1	R^2	R^3
(**152**):	Me	Me	H
(**153**):	Me	Me	Me
(**154**):	Ph	H	H
(**155**):	Ph	Ph	H

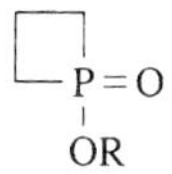

(**156**): R= Et
(**157**): R= Me

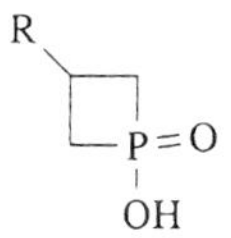

(**158**): R= H
(**159**): R= OH
(**160**): R= OCH_2Ph

(**161**)

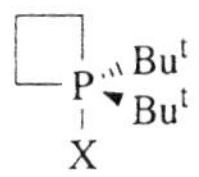

(**162**): X=F
(**163**): X=NH_2

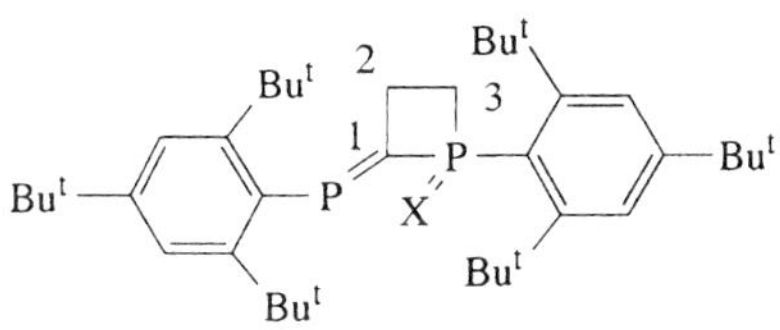

(**164**): X= lone pair
(**165**): X= O

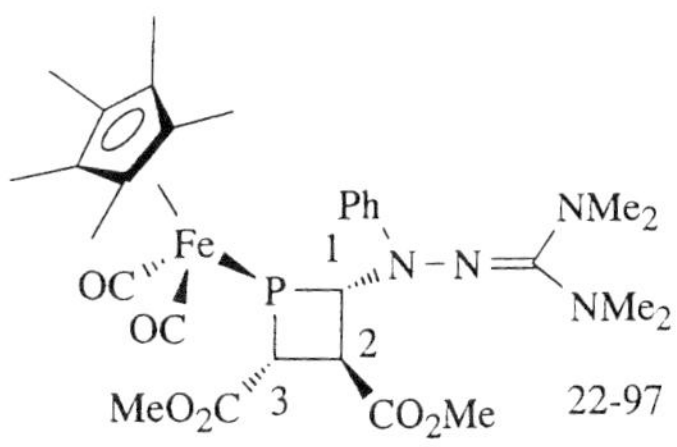

(**166**)

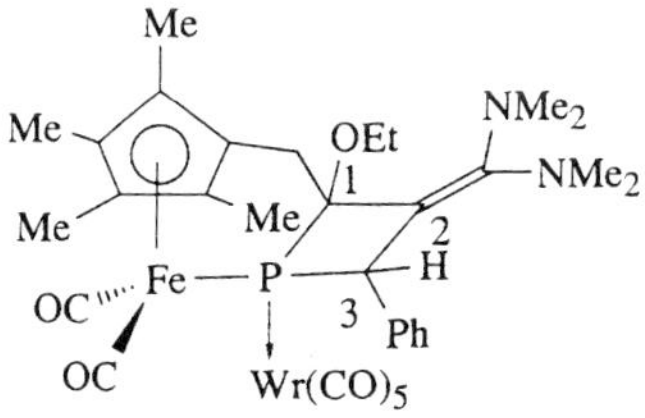

(**167**)

Scheme 6

Table 6. ^{13}C and ^{31}P NMR spectral data of optically active phosphetanes

Compounds	*C-1* δ ($^{1}J_{CP}$/Hz)	*C-2* δ ($^{2}J_{CP}$/Hz)	*C-3* δ ($^{1}J_{CP}$/Hz)	^{31}P δ	*Ref.*
(13)	49.7 (55.1)	36.4 (12.3)	41.7 (51.1)	62.7 [a]	⟨93T10291⟩
	49.2 (58.0)	38.5 (14.0)	46.1 (47.4)	57.7 [a]	⟨93T10291⟩
(93)	46.7 (55.0)	47.4 (12.9)	44.0 (55.1)	67.9	⟨93T10291⟩
(203) [a]	35.4 (4.9)	49.5 (7.2)	33.8 (2.7)	48.6 [a]	⟨93T10291⟩
(168) [a]	39.4 (–)	39.8 (–) [b]	37.0 (9) [b]	17.6 [a]	⟨93T10291⟩
	38.9 (–) [b]	41.3 (–) [b]	39.9 (16.9)	8.7 [a]	⟨93T10291⟩
(169) [a]	36.6 (2.5)	43.0 (5.3)	42.5 (0)	32.1 [a]	⟨93T10291⟩
	36.7 (4.6)	44.1 (5.7)	43.6 (2.9)	26.9 [a]	⟨93T10291⟩
(170) [a]	39.6 (2.9)	46.2 (9.6)	53.4 (9.2)	45.1 [a]	⟨93T10291⟩
(183)	48.5 (53.6)	41.8 (11.6)	52.7 (51.7)	68.8 [a]	⟨93T10291⟩
	48.1 (56.2)	42.8 (15)	53.4 (48.7)	64.5 [a]	⟨93T10291⟩
(184)	48.3 (52.6)	44.5 (6.7)	51.3 (45.6)	64.3 [a]	⟨93T10291⟩
	47.2 (56.3)	45.8 (7.7)	55.2 (40.5)	61.7 [a]	⟨93T10291⟩
(32a) [a]	50.6 (t, 12.9)	45.7 (0)	34.2 (t, 5.1)	62.6 [a]	⟨94TL5861⟩
(32b) [a]	49.0 (t, 14.9)	46.7 (0)	33.9 (t, 5.9) or 52.1 (t, 11.2)	56.9 [a]	⟨94OM3956⟩
(186a) [a]	49.6 (56.0)	42.2 (17.0)	62.4 (39.0)	65.6 [a]	⟨94OM3956⟩
(186b) [a]	48.4 (60.6)	44.3 (16.7)	64.2 (37.0)	59.1 [a]	⟨94OM3956⟩
(204a) [a]	36.8 (3.0) [b]	41.7 (4.4) [b]	43.4 [b]	20.9 [a]	⟨94OM3956⟩
(204b) [a]	36.6 (4.4)	41.7 (4.5)	43.7 (0)	21.0 [a]	⟨94OM3956⟩
(204c) [a]	36.6 (6.2)	42.9 (3.5)	45.1 (6.1)	18.9 [a]	⟨94OM3956⟩
(204d) [a]	36.8 (6.2)	43.0 (4.4)	44.7 (6.3)	18.4 [a]	⟨94OM3956⟩
(206a) [a]	48.5 (55.4)	40.7 (13.6)	53.6 (48.6)	64.8 [a]	⟨94OM3956⟩
(206b) [a]	48.4 (56.0)	40.8 (13.7)	53.9 (49.3)	64.2 [a]	⟨94OM3956⟩
(206c) [a]	48.0 (57.9)	41.7 (13.8)	55.3 (46.7)	60.4 [a]	⟨94OM3956⟩
(206d) [a]	47.9 (57.7)	41.8 (14.8)	55.0 (45.6)	60.1 [a]	⟨94OM3956⟩
(179) [a]	44.4 (dd) [b] (10.7, 2.9)	37.3 (dd) [b] (16.7, 3.2)	38.4 (dd) (26.9, 12.6)	24.4, −18.6 [a] (*J*(PP)87.0)	⟨95OM4983⟩
(196) [a]	51.2 (dd) (57.4, 10.4)	41.5 (dd) (9.2, 4.6)	55.1 (dd) (60.5, 34.8)	64.6, 25.7 [a] (*J*(PP)17.1)	⟨95OM4983⟩
(35)	47.2 (t, 14.9)	45.3 (0)	34.1 (t, 4.8) or 37.4 (t, 8.9)	94.2 [d]	⟨96JOM(522)223⟩
(171a) [a]	43.2 (2.8) [b]	40.6 (0) [b]	27.8 (16.8)	28.2 [a]	⟨96JOM(522)223⟩
(171b) [a]	44.5 (2.7) [b]	41.6 (0) [b]	29.1 (19.6)	22.6 [a]	⟨96JOM(522)223⟩
(172) [a]	43.6 (0) [b]	40.5 (0) [b]	24.4 (18.3)	28.9 [a]	⟨96JOM(522)223⟩
(173) [a]	43.0 (0) [b]	40.5 (0) [b]	25.8 (18.3)	28.8 [a]	⟨96JOM(522)223⟩
(174) [a]	43.6 (0) [b]	40.8 (0) [b]	27.7 (15.7)	29.9 [a]	⟨96JOM(522)223⟩
(175) [a]	44.1 (0) [b]	41.3 (0) [b]	27.0 (16.5)	29.0 [a]	⟨96JOM(522)223⟩
(187a)	50.4 (56.0)	40.7 (15.2)	42.2 (37.6)	68.0	⟨96JOM(522)223⟩
(187b)	51.3 (56.3)	40.4 (13.8)	44.7 (36.0) or 46.8 (40.2)	61.8	⟨96JOM(522)223⟩
(188)	50.3 (55.0)	40.2 (15.1)	39.7 (37.9)	68.7	⟨96JOM(522)223⟩
(189)	50.3 (55.0)	40.3 (15.4)	40.3 (36.5)	–	⟨96JOM(522)223⟩
(190)	50.8 (55.2)	40.8 (15.0)	42.0 (37.5)	72.7	⟨96JOM(522)223⟩
(191)	50.5 (55.1)	40.4 (14.0)	41.4 (38.1)	71.7	⟨96JOM(522)223⟩
(210) [a]	40.3 (0) [b]	43.3 (0) [b]	29.4 (16.6)	28.3 [a]	⟨96JOM(522)223⟩
(211) [a]	40.7 (0) [b]	43.6 (0) [b]	27.0 (17.8)	273 [a]	⟨96JOM(522)223⟩
(212) [a]	40.6 (0) [b]	43.6 (0) [b]	27.0 (18.2)	26.8 [a]	⟨96JOM(522)223⟩
(213)	50.4 (55.8)	40.3 (15.1)	44.4 (36.9)	65.9 [a]	⟨96JOM(522)223⟩
(214)	50.3 (55.1)	40.0 (14.1)	41.3 (38.2)	71.3 [a]	⟨96JOM(522)223⟩
(215)	50.4 (55.4)	40.6 (14)	41.7 (37.6)	67.8	⟨96JOM(522)223⟩
(185a) [a]	48.7 (53.2)	41.7 (12.2)	40.6 (42.5) or 49.3 (51.8)	66.9 [a]	⟨97JOM(529)465⟩
(185b) [a]	47.7 (56.6)	42.7 (12.3)	44.0 (40.8) or 51.8 (47.8)	63.6 [a]	⟨97JOM(529)465⟩
(195a)	48.8 (54.1)	41.3 (12.0)	40.8 (41.3) or 42.3 (52.0)	67.5	⟨97JOM(529)465⟩
(195b)	48.6 (56.6)	42.3 (16.90)	44.0 (41.1) or 44.8 (48.9)	64.5	⟨97JOM(529)465⟩
(205a) [a]	41.9 (3.0)	37.0 (4.0)	42.9 (0)	19.5 [a]	⟨97JOM(529)465⟩
(205b) [a]	41.8 (4.6)	36.9 (5.6)	43.1 (0)	21.6 [a]	⟨97JOM(529)465⟩
(205c) [a]	44.8 (4.3)	37.0 (7.8)	43.1 (2.8)	18.8 [a]	⟨97JOM(529)465⟩
(205d) [a]	43.1 (3.3)	37.2 (6.7)	43.7 (5.2)	18.5 [a]	⟨97JOM(529)465⟩

Table 6. (continued)

Compounds	*C-1* δ ($^{1}J_{CP}/Hz$)	*C-2* δ ($^{2}J_{CP}/Hz$)	*C-3* δ ($^{1}J_{CP}/Hz$)	^{31}P δ	*Ref.*
(**207a**)	53.3 (47.9)	40.7 (13.7)	48.5 (55.2)	64.1	⟨97JOM(529)465⟩
(**207b**)	43.5 (49.8)	40.7 (13.6)	48.7 (55.1)	66.7	⟨97JOM(529)465⟩
(**207c**)	48.1 (58.0)	42.0 (13.7)	55.5 (46.2)	63.3	⟨97JOM(529)465⟩
(**207d**)	48.9 (57.8)	41.7 (13.9)	53.8 (45.7)	64.9	⟨97JOM(529)465⟩
(**208a**)	42.4 (3.9) [b]	37.0 (5.8) [b]	34.7 (3.6) or 36.2 (28.8)	31.0	⟨97JOM(529)465⟩
(**208b**)	42.3 (4.5) [b]	36.7 (4.4) [b]	34.7 (3.5) or 36.3 (28.5)	–	⟨97JOM(529)465⟩
(**208c**)	43.3 (4.3) [b]	36.6 (6) [b]	39.7 (31.8)	25.3	⟨97JOM(529)465⟩
(**208d**)	43.2 (4.6) [b]	36.4 (7.6) [b]	39.6 (30.1)	–	⟨97JOM(529)465⟩
(**209a**)	48.8 (54.6)	41.5 (12.6)	40.3 (40.6) or 44.6 (52.5)	70.5	⟨97JOM(529)465⟩
(**209b**)	48.7 (54.0)	41.4 (12.3)	40.4 (41.3) or 44.5 (53.0)	70.5	⟨97JOM(529)465⟩
(**209c**)	48.1 (56.6)	42.2 (12.5)	44.4 (39.4) or 46.3 (49.5)	–	⟨97JOM(529)465⟩
(**216**)	48.9 (59.6)	–	37.0 (42.8) or 44.7 (50.0)	55.5	⟨97JOC297⟩
(**217a**)	48.9 (58.4)	–	42.0 (49.9)	66.4	⟨97JOC297⟩
(**217b**)	49.3 (54)	–	43.9 (48.7)	59.2	⟨97JOC297⟩
(**218**)	49.2 (55.3)	–	41.8 (50.7)	65.6	⟨97JOC297⟩
(**220**) [a]	24.0 (2.9)	38.8 (0)	24.8 (9.0)	25.8 [a]	⟨97TL2947⟩
(**221**) [a]	25.9 (3.0)	35.9 (3.2)	29.0 (3.0)	21.0 [a]	⟨97TL2947⟩
(**222**)	25.9 (41.1)	35.9 (15.4)	29.1 (39.7)	51.4 (*J*(PB)57.8)	⟨97TL2947⟩
(**223**)	30.3 (39.0)	34.6 (11.8)	31.0 (38.1)	50.5	⟨97TL2947⟩
(**224**)	26.3 (40.5)	19.2 (16.1)	C-1 [c]	38.8 (*J*(PB)53.3)	⟨97TL2947⟩
(**14a**)	45.1 (47.2) [b]	24.0 (18.1)	33.1 (45.7)	77.3 [e]	⟨97T4363⟩
(**14b**)	45.6 (44.4) [b]	22.9 (17.5)	29.9 (48.6)	76.8 [e]	⟨97T4363⟩
(**225**)	21.3 (41.4)	18.8 (17.9)	24.4 (39.3)	51.5 (*J*(PB)46)	⟨97T4363⟩
(**226**)	23.9 (48.1)	15.2 (19.9)	36.6 (45.7)	70.7	⟨97T4363⟩
(**227**)	41.4 (46.9)	31.5 (15.2)	42.3 (48.5)	85.6	⟨97T4363⟩
(**228a**)	43.6 (48.8)	25.0 (18.0)	31.5 (34.1)	78.9	⟨97T4363⟩
(**228b**)	45.0 (43.3) [b]	25.0 (18.0)	34.7 (36.7) [b]	82.5	⟨97T4363⟩
(**228c**)	44.0 (47.8) [b]	26.6 (18.8)	38.1 (30.9) [b]	85.3	⟨97T4363⟩
(**229**) [a]	24.0 (t, 4.6)	25.5 (t, 7.6)	36.3 (0)	18.3 [a]	⟨97TL2947⟩
(**233**)	24.1 (9.0) [b] 31.4 (dd) [b] (40.8, 10.4)	37.8 (0) 34.1 (11.9)	25.4 (7.4) [b] 28.9 (dd) [b] (39.6, 4.4)	49.1, 16.9 (*J*(PP)36.1)	⟨97TL2947⟩
(**176**) [a]	42.0 (4.8)	36.3 (2.7)	42.1 (0)	19.6 [a]	⟨98S1539⟩
(**177**) [a]	42.4 (4.7)	36.7 (2.9)	36.1 (0)	29.3 [a]	⟨98S1539⟩
(**178**) [a]	42.5 (4.7)	36.4 (0)	39.8 (0)	29.5 [a]	⟨98S1539⟩
(**179**) [a]	43.2 (3.8)	36.2 (0)	35.9 (9.2)	33.3 [a] −18.5 (*J*(PP)23))	⟨98S1539⟩
(**181**) [a]	42.2 (4.6)	36.7 (4.5)	48.6 (21.6)	29.2, −17.2 [a]	⟨98S1539⟩
(**182**) [a]	42.5 (5.3)	36.3 (0)	39.8 (0)	30.2, −16.0 [a]	⟨98S1539⟩
(**192a**)	49.0 (53.8)	40.9 (11.9)	40.2 (41.1) or 50.8 (50.5)	73.8	⟨98S1539⟩
(**192b**)	48.2 (57.1)	41.7 (12.4)	43.3 (40.1) or 52.8 (47.6)	73.8	⟨98S1539⟩
(**193**)	48.5 (54.1)	41.4 (11.7)	40.2 (40.8) or 46.1 (52.0)	73.6	⟨98S1539⟩
(**194a**)	48.6 (54.0)	41.2 (13.4)	40.1 (40.0) or 49.2 (52.3)	70.8	⟨98S1539⟩
(**194b**)	47.8 (56.3)	42.0 (12.3)	44.0 (39.0) or 51.2 (49.2)	68.1	⟨98S1539⟩
(**197**)	40.5 (35.2)	42.8 (5.8)	42.5 (34.9)	51.7	⟨98S1539⟩
(**198**)	40.4 (35.2)	43.2 (4.8)	36.7 (36.6)	55.5	⟨98S1539⟩
(**199**)	40.3 (36.3)	43.4 (4.5)	40.8 (37.0)	54.8	⟨98S1539⟩
(**200**)	39.6 (35.1)	43.3 (4.5)	34.6 (37.4)	58.0, 16.0	⟨98S1539⟩
(**201**)	40.2 (35.3)	43.4 (5.0)	42.0 (dd) (46.2, 12.3)	58.6, 16.9	⟨98S1539⟩

Table 6. (continued)

Compounds	*C-1* δ ($^{1}J_{CP}$/Hz)	*C-2* δ ($^{2}J_{CP}$/Hz)	*C-3* δ ($^{1}J_{CP}$/Hz)	^{31}P δ	*Ref.*
(**202**)	40.1 (36.3)	43.1 (4.6)	40.7 (36.9)	57.0, 15.8	⟨98S1539⟩
(**230**) [a]	33.6 (t, 9.7) [b]	30.8 (0)	38.9 (t, 7.9) [b]	6.2 [a]	⟨99CEJ1160⟩
(**231**) [a]	32.1 (5.0) [b]	32.7 (0)	33.9 (t, 7.6) [b]	12.3 [a]	⟨99CEJ1160⟩
(**232**) [a]	39.5 (0)	33.0 (t, 4.0)	43.1 (t, 7.2)	7.8 [a]	⟨99CEJ1160⟩
	37.4 (t, 1.6)	31.9 (t, 4.3)	37.7 (t, 6.8)		
(**234**)	29.3 (4.6) [b]	29.8 (0)	28.4 (dd) [b] (10.5, 8.1)	45.9, −3.4 (*J*(PP)34.6)	⟨99CEJ1160⟩
	41.8 (dd) [b] (39.7, 7.7)	25.4 (11.4)	44.8 (dd) [b] (36.6, 3.0)		
(**235**)	29.3 (4.6) [b]	23.6 (0)	32.5 (dd) [b] (8.4, 0.7)	49.7, 9.4 (*J*(PP)36)	⟨99CEJ1160⟩
	33.6 (dd) [b] (37.4, 3.2)	24.3 (t, 8.0)	37.5 (dd) [b] (40.3, 9.7)		
(**236**)	42.5 (dd) [b] (39.5, 8.1)	34.1 (3.0)	43.8 (dd) [b] (35.5, 3.5)	46.5, −2.3 (*J*(PP)35)	⟨99CEJ1160⟩
(**237**)	28.8 (39.6) [b]	34.0 (9.5)	34.3 (40.1) [b]	52.7	⟨99CEJ1160⟩

[a] In C_6D_6. [b] Not assigned or not resolved. [c] C-1 and C-3 are equivalent. [d] In pentane. [e] In THF.

A similar shift was also observed in the tungsten pentacarbonyl complexes of the following phosphetane and 1,2-dihydrophosphetes. In these cases, the compounds resonate down filed in the order Me < Ph < *t*-Bu, which is attributable to inductive effect (for Me and Ph) and electronic repulsion between bonding electrons (for *t*-Bu), see Scheme 12 ⟨90JCS(CC)153, 90OM788⟩.

It is well known that pentacoordinate phosphorus compounds usually resonate at much higher field than 85% H_3PO_4 as external standard ((**302**) ⟨83JOC2621⟩, (**162**) ⟨92CB1987⟩). As shown in Scheme 13 the chemical shifts are strongly sensitive to the coordination number of the phosphorus atom.

The ^{31}P NMR chemical shifts of the metal complexes are also affected by the metal, the chemical shift increasing in the order W < Cr < Mo in the following cases (Scheme 14) ⟨85JA5001, 84NJC453⟩.

The structures and ^{31}P chemical shifts of other interesting compounds are summarized in Scheme 15. The most upfield shift is δ −346.6 for (**21**), which is a reasonable value in view of the chemical shift of δ −460 for P_4. Two Dewar type benzenes (**240**) and (**244**) and prismane (**146**) which are valence isomers of a phosphabenzene were shown to have very different ^{31}P chemical shifts. Phosphonium ylide (**304**) resonates at slightly higher field than phosphonium salt (**250**), indicating the contribution of ylene structure. In the case of (**305**) the low coordinate λ^3 phosphorus appears at remarkably high field (δ −109), which points to the high contribution of canonical structure B ⟨87AG(E)83, 89NJC335⟩, see Scheme 15.

The coupling constants of ^{31}P with homo and hetero nuclei provide important structural information and there have been reported several examples for multi-bond-through coupling. The coupling constants with ^{19}F (Scheme 16), ^{95}Mo, ^{183}W (see, Schemes 12 and 14), ^{103}Rh (Scheme 18), and ^{185}Pt (see Tables 13 and 14) are demonstrated. The J_{PF} values of fluorine-containing compounds (**162**), (**306**) ⟨78IC3265⟩, (**307**) ⟨93TL4619⟩, and (**308**) ⟨85ZN(B)927⟩ are summarized in Scheme 16. In compound (**308**) the $^{2}J_{PF}$ values are scattered from 8 to 237 Hz.

The *J*(PH) and *J*(PP) coupling constants are also demonstrated for compounds (**30**), (**309**) ⟨83JA2074⟩, (**310**), (**311**), (**312**), (**313**) ⟨96CEJ221⟩, and (**29**) as shown in Scheme 17.

It is interesting to point out that the coupling constants between ^{31}P and ^{103}Rh are 3.4 Hz for the η^3 complex (**297**) and 37.3 and 36.5 Hz for η^4 complexes (**314**) and (**315**), respectively. The coupling constants (^{1}J(PRh)) of diphosphine Rh complexes (**33**), (**34**), and (**35**) were observed between 110 Hz and 126 Hz, which are not so different from each other in sharp contrast to the large difference in their chemical shifts.

The ^{31}P chemical shifts and coupling constants *J*(PP) and *J*(PPt) of 1,2-dihydrophosphete platinum complexes (**48**), (**49**), and (**316**)–(**320**) and 2-alkylidene-1,2-dihydrophosphetes platinum complexes (**321**)–(**324**) are summarized in Tables 13 and 14, respectively.

3.2.2.2.4 ^{17}O NMR

There has been reported only one ^{17}O NMR study on phosphetane 1-oxides (**134**), (**8**) and (**2**) ⟨87MRC271⟩, in which the ^{17}O NMR chemical shifts are extremely sensitive to structural effects such as ring size and the presence of γ-interactions. The data are shown in Scheme 20. When methyl groups are

Table 7. ^{13}C and ^{31}P NMR spectral data of 1,2-dihydrophosphetes

Compounds	*C-1* δ ($^1J_{CP}$/Hz)	*C-2* δ ($^2J_{CP}$/Hz)	*C-3* δ ($^1J_{CP}$/Hz)	^{31}P δ	*Ref.*
(39)	190.72 (34.2)	169.5 (36.6)	151.7 (51.3)	86.9[a]	⟨85JA5001⟩
(257)	202.13 (65.7)	180.35 (70.5)	160.46 (84.3)	63[b]	⟨85JA5001⟩
(266)	191.42 (30.5)	181.30 (33)	163.21 (48.8)	85.8[c]	⟨85JA5001⟩
(40)	241.5 (33.7)	73.4 (6.1)	67.8 (17.6)	314[a]	⟨86AG(E)280⟩
(242)	237.9 (33.4)	73.7 (3.8)	69.9 (19.6)	312[a]	⟨86AG(E)280⟩
(243)	223.6 (36.5)	75.3 (–)[d]	71.8 (19.5)	314[a]	⟨86AG(E)280⟩
(244)	239.8 (33.5)	73.2 (5.9)	67.1 (17.7)	315[a]	⟨86AG(E)280⟩
(245)	242.1 (33.4)	73.6 (5.5)	67.4 (17.9)	315	⟨86AG(E)280⟩
(247)	240.5 (31.7)	75.9 (5.2)	68.0 (17.0)	316	⟨86AG(E)280⟩
(248)	241.7 (32.0)	75.8 (6.3)	68.6 (17.1)	317	⟨86AG(E)280⟩
(249)	242.1 (34.5)	75.9 (6.0)	68.1 (17.3)	316	⟨86AG(E)280⟩
(240)	177.1 (32.6)	139.2 (5.0)	47.5 (47.6)	−19.0	⟨87AG(E)85⟩
	160.2 (21.9)	162.6 (15.0)	47.5 (47.6)		
(59)	158.2 (36.2)	165.8 (15.4)	65.0 (60.1)	39.3	⟨87AG(E)85⟩
(65)	158.73 (51.3)	157.21 (34.9)	56.58 (62.1)	−0.1[b]	⟨87TL6121⟩
(66)	158.34 (36.1)	154.80 (32.7)	58.17 (41.5)	64.4[b]	⟨87TL6121⟩
(251)	109.19 (120.9)	105.56 (15.6)	30.04 (99.2)	−13.0[b]	⟨87TL6121⟩
(241)	168.6 (36.1)	145.7 (6.8)	47.92 (16.1)	51.2	⟨88OM572⟩
(255)[b]	–[d]	–[d]	75.43 (40.3)	42.85[c,e,f]	⟨88OM633⟩
(253)	160.2 (83.8)	113.57 (8.1)	50.82 (58.2)	22.95	⟨88OM1791⟩
(258)	148.28 (50.6)	124.09 (10.1)	39.55 (31.32)	72.72	⟨89JCS(CC)788⟩
(43)[a]	–[d]	–[d]	28.3 (7.1)	−18.0[b]	⟨90AG(E)75⟩
(254)[g]	67.74 (116.2)	–[d]	80.08 (116.8)	19.1[g]	⟨90JCS(CC)153⟩
(259)[a]	95.87 (43.3)	151.35 (4.3)	77.25 (50.8)	−12.5[a]	⟨90JCS(CC)153⟩
(260)[a]	93.34 (54.3)	149.25 (0)	52.14 (26.2)	20.2[a]	⟨90JCS(CC)153⟩
(261)	–[d]	159.41 (10.12)	59.6 (35.16)	97.0	⟨90SL171⟩
(262)[a]	–[d]	158.01 (7.5)	56.67 (32.7)	42.55[a,h]	⟨90OM788⟩
(263)[a]	–[d]	156.32 (7.5)	49.46 (29.7)	74.39[i,j]	⟨90OM788⟩
(43)[b]	–[d]	–[d]	28.1 (7.1)	−16.04[a]	⟨91IC3381⟩
(44)	–[d]	–[d]	35.5 (38.5)	16.9[a,k]	⟨91IC3381⟩
(45)	–[d]	–[d]	33.9 (39.2)	35.45[a]	⟨91IC3381⟩
(252)[a]	–[d]	–[d]	42.7 (59.2)	18.4[a]	⟨91IC3381⟩
(46)	168.6 (36.1)	145.7 (6.8)	42.4 (46)	81.7	⟨93AG(E)1359⟩
(250)	132.6 (26)	172.9 (29)	43.8 (44)	82.3	⟨93AG(E)1359⟩
(268)	43.28 (27.5)	155.61 (5.3)	127.22 (32.3)	78.6[a]	⟨95BSF384⟩
(269)	77.47 (67.5)	154.95 (0)	128.52 (31.7)	83.1[a]	⟨95BSF384⟩
(272)[a]	21.73 (10.8)	145.85 (0)	140.2 (34.7)	−10.6[a]	⟨95AG(E)590⟩
(53)[a]	80.0 (dd)	174.0 (dd)	67.4 (dd)	55.8, 35.6[a]	⟨95ZAAC1539⟩
	(139.6, 13.9)	(7.2, 9.29)	(115.5, 63.6)	(*J*(PP)13.2)	
(54)[a]	80.8 (dd)	174.2 (dd)	75.2 (dd)	55.5, 85.0[a]	⟨95ZAAC1539⟩
	(141.4, 13.5)	(12.4, 9.2)	(92.5, 68.6)	(*J*(PP)0.2)	
(293)[a]	85.2 (dd)	173.2 (dd)	41.3 (dd)	48.6, 63.3[a]	⟨95ZAAC1539⟩
	(135.0, 113.0)	(2.6, 0.9)	(76.3, 30.4)	(*J*(PP)40.2)	
(55)	113.4 (dd)	193.6 (dd)	82.4 (dd)	62.4, 83.5	⟨96PS(115)3⟩
	(113.1, 13.1)	(9.4, 6.1)	(90.9, 55.7)	(*J*(PP)0.4)	
(56)	114.2 (dd)	195.0 (dd)	46.6 (dd)	55.1, 63.7	⟨96PS(115)3⟩
	(137.2, 91.6)	(6.5, 1.5)	(66.0, 29.1)	(*J*(PP)45.5)	
(294)[l]	140.9 (dd)	239.1 (dd)	52.0 (dd)	57.4, 20.4[l]	⟨96ZAAC974⟩
	(170.4, 52.8)	(23.3, 7.4)	(49.3, 33.1)		
(273)[a]	25.8 (9.6)	142.8 (8.2)	135.6 (9.9)	−49.1[a]	⟨96HAC397⟩
(274)[a]	26.6 (10.3)	143.0 (8.5)	136.8 (10.3)	−48.3[a]	⟨96HAC397⟩
(277)[a]	165.80 (dd)	141.25 (dd)	138.50 (dd)	−206.60[a]	⟨96OM1597⟩
	(66.80, 43.40)	(26.00, 13.60)	(18.60, 8.05)	−12.55	
				(*J*(PP)84.10)	
(288)[a]	136.3 (10.2)*	136.6 (10.2)	140.9 (5.5)*	39.8[a]	⟨97AG(E)987⟩
	or 138.1 (6.5)		or 154.6 (5.5)		
(58)	–	214.4 (4.3)	–	59.9	⟨97AG(E)987⟩
(289)	140.9 (13.8)*	214.2 (4.3)	142.7 (10.2)*	61.6[a]	⟨97AG(E)987⟩
	or 142.0 (4.6)		or 143.5 (5.5)		
(290)	–[d]	–[d]	–[d]	73.7	⟨97AG(E)987⟩
(291)[a]	216.3 (23.1)	67.6 (9.7)	31.0 (17.4)	171.4[a]	⟨97AG(E)1095⟩
(292)[a]	210.7 (27.8)	68.0 (11.1)	29.9 (13.0)	114.2[a]	⟨97AG(E)1095⟩
(61)	35.1 (65)	152.7 (9.7)	152.8 (82.9)	15.2	⟨98HAC21⟩
		or 159.3 (7.9)	or 135.2 (71.9)		

Table 7. (continued)

Compounds	*C-1* δ ($^{1}J_{CP}/Hz$)	*C-2* δ ($^{2}J_{CP}/Hz$)	*C-3* δ ($^{1}J_{CP}/Hz$)	^{31}P δ	*Ref.*
(278)[a]	41.4 (43.4)	150.3 (13.8)	104.5 (13.8)	117.4[a]	<98ZOB1449>
(279)[a]	41.3 (36.9)	149.4 (13.5)	112.2 (15.4)	113.1[a]	<98ZOB1449>
(280)[a]	39.7 (38.9)	147.9 (14.5)	118.2 (19.8)	107.0[a]	<98ZOB1449>
(281)[a]	40.8 (42.3)	147.7 (22.9)	119.6 (22.3)	121.0[a]	<98ZOB1449>
(282)[a]	40.8 (37.8)	148.3 (15.0)	109.7 (14.0)	116.8[a]	<98ZOB1449>
(283)[a]	42.3 (42.7)	150.6 (12.2)	104.4 (16.7)	132.8[a]	<98ZOB1449>
(284)[a]	40.5 (42.9)	149.7 (12.3)	111.8 (19.9)	127.3[a]	<98ZOB1449>
(285)[a]	38.6 (40.7)	148.5 (13.8)	117.4 (24.3)	119.1[a]	<98ZOB1449>
(286)[a]	–	147.8 (21.5)	120.0 (26.6)	130.1[a]	<98ZOB1449>
(287)[a]	57.2 (54.6)	156.3 (10.5)	101.6 (90.3)	78.9[a]	<98ZOB1449>

[a] In C_6D_6. [b] In CH_2Cl_2 or CD_2Cl_2. [c] In toluene. [d] Not assigned or not resolved. [e] $^{1}J_{PW}$ = 244 Hz. [f] Minor diastereomer $\delta_P(CH_2Cl_2)$ 47.59. [g] In CD_3OD. [h] $^{1}J_{PW}$ = 239 Hz. [i] In pentane. [j] $^{1}J_{PW}$ = 239 Hz. [k] $^{1}J_{PW}$ = 231.4 Hz. [l] In pyridine-d5.

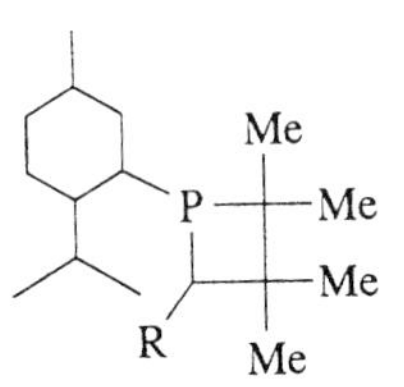

(168): R= H
(169): R= $PhCH_2$
(170): R= Br
(171): R= $SiMe_3$
(172): R= $SiEt_3$
(173): R= $SiMe_2CHMe_2$
(174): R= $SiMe_2Ph$
(175): R= $SiMe_2CH_2CH_2Ph$
(176): R= CH_2OH
(177): R= CH_2CH_2OH
(178): R= $CH_2CH_2CH_2OH$
(179): R= PPh_2
(180): R= CH_2PPh_2
(181): R= $CH_2CH_2PPh_2$
(182): R= $CH_2CH_2CH_2PPh_2$

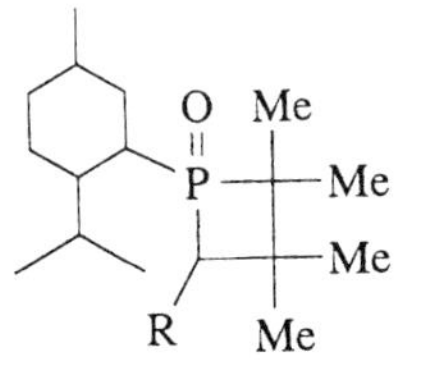

(183) : R= CH_2Ph
(184) : R= Br
(185) : R= $CH_2CH{=}CH_2$
(186) : R= CHO
(187) : R= $SiMe_3$
(188) : R= $SiEt_3$
(189) : R= $SiMe_2CHMe_2$
(190) : R= $SiMe_2Ph$
(191) : R= $SiMe_2CH_2CH_2Ph$
(192) : R= CH_2OH
(193) : R= CH_2CH_2OH
(194) : R= $CH_2CH_2CH_2OH$
(195) : R= CH_2CHO
(196) : R= $P(O)Ph_2$

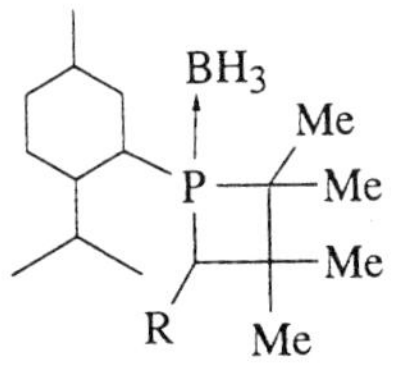

(197): R= CH_2OH
(198): R= CH_2CH_2OH
(199): R= $CH_2CH_2CH_2OH$
(200): R= $CH_2P(BH_3)Ph_2$
(201): R= $CH_2CH_2P(BH_3)Ph_2$
(202): R=$CH_2CH_2CH_2P(BH_3)Ph_2$

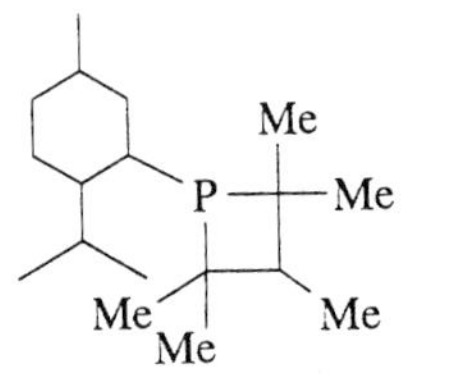

(203)

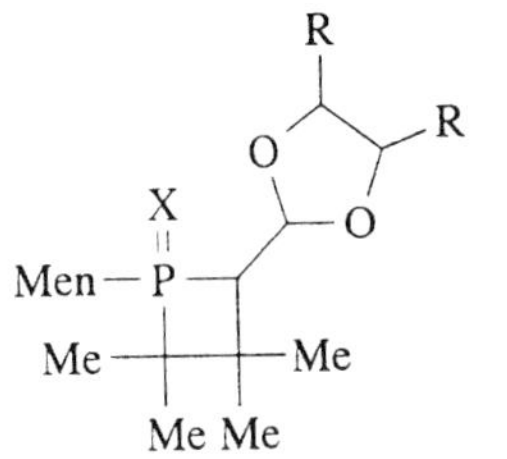

(204): X= lone pair; R= Me
(205): X= lone pair; R= Ph
(206): X= O; R= Me
(207): X= O; R= Ph

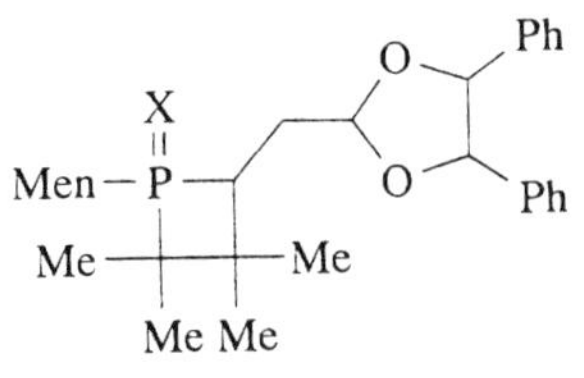

(208): X= lone pair
(209): X= O

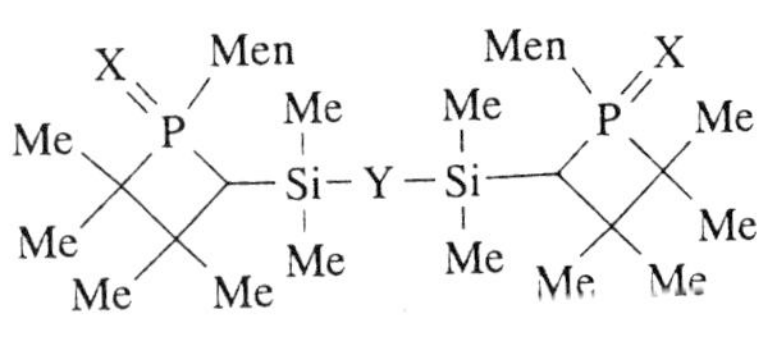

(210): X= lone pair; Y= O
(211): X= lone pair; Y= $(CH_2)_2$
(212): X= lone pair; Y= $(CH_2)_6$
(213): X= Y= O
(214): X= O; Y= $(CH_2)_2$
(215): X= O; Y= $(CH_2)_6$

Scheme 7-1

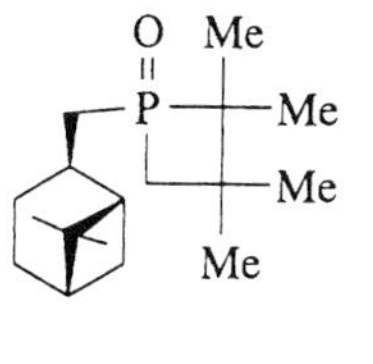

(**216**)

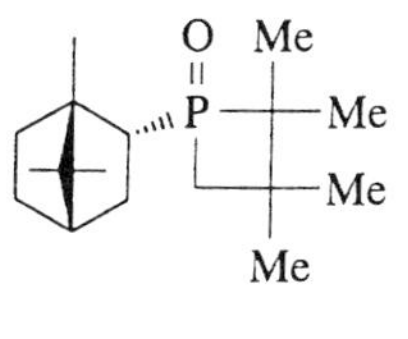

(**217**)

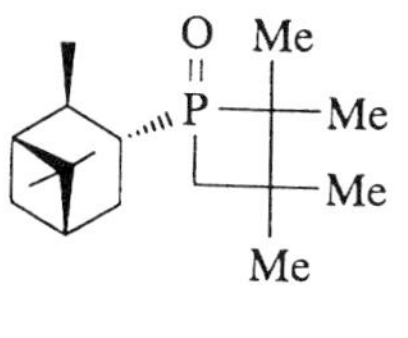

(**218**)

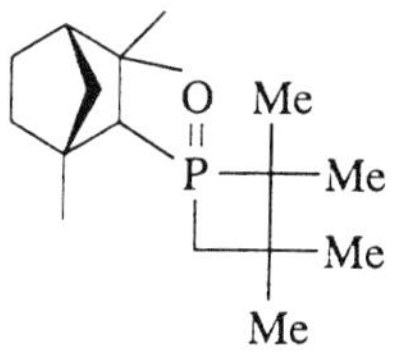

(**219**)

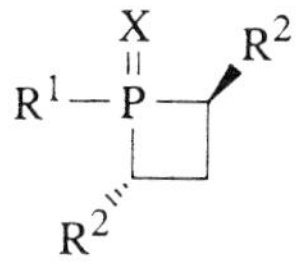

(**220**): X= lone pair; R^1= Ph; R^2= Me
(**221**): X= lone pair; R^1= Mes; R^2= Me
(**222**): X= BH_3; R^1= Ph; R^2= Me
(**223**): X= BH_3; R^1= Mes; R^2= Me
(**224**): X= BH_3; R^1= Mes; R^2= H

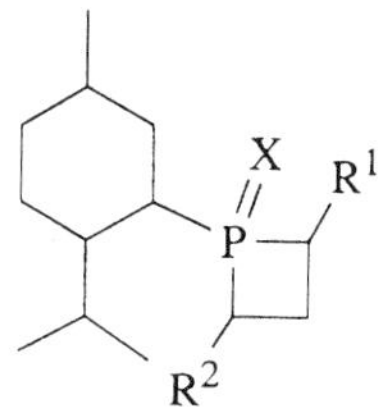

(**225**): X= BH_3; R^1, R^2= H
(**226**): X= S; R^1, R^2= H
(**227**): X= S; R^1, R^2= CH_2Ph
(**228**): X= S; R^1= CH_2Ph; R^2= $SiMe_3$

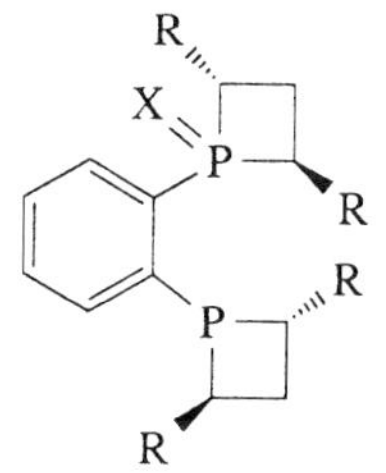

(**229**): X= lone piar; R= Me
(**230**): X= lone piar; R= Pr^i
(**231**): X= lone piar; R= Et
(**232**): X= lone piar; R= Hex^c
(**233**): X= BH_3; R= Me
(**234**): X= BH_3; R= Pr^i
(**235**): X= BH_3; R= Et
(**236**): X= BH_3; R= Hex^c

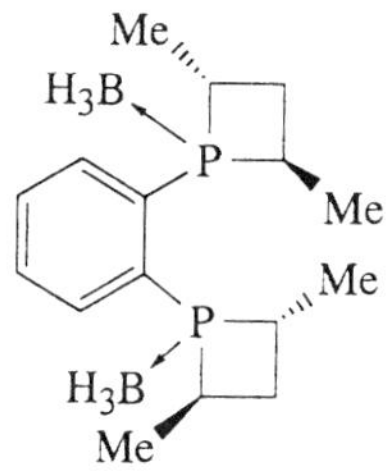

(**237**)

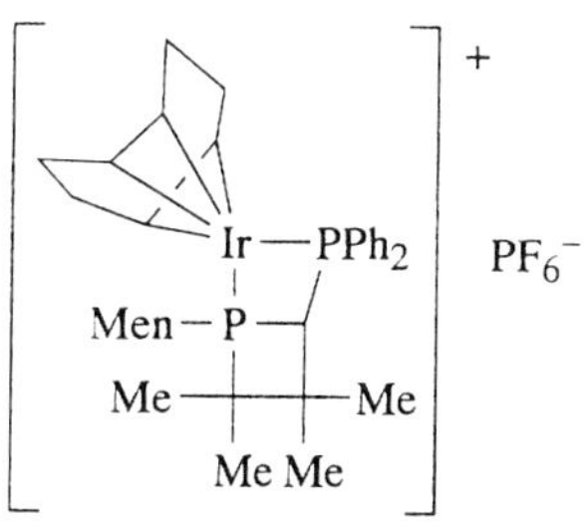

(**238**)

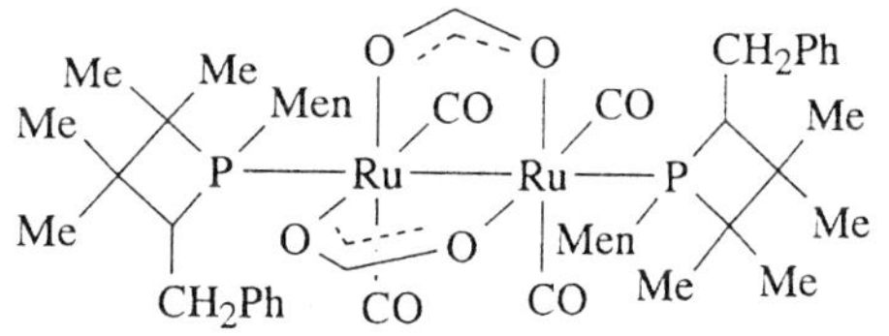

(**239**)

Scheme 7-2

placed on the carbons attached to phosphorus, these groups have a gamma relationship to the phosphoryl oxygen and hence the ^{17}O can experience the familiar γ-shielding effect. Therefore, the upfield shift was observed in (**2**) compared with (**8**). A comparison of the data for these four-membered heterocycles reveals that still another, stronger effect acts on the ^{17}O shift. The bridged phosphetane (**134**) with a larger γ-shielding effect resonates at lower field by 19 ppm and 34.7 ppm than (**8**) and (**2**), respectively. Although the C–P–C angle of (**134**) has never been reported, it can be expected to be smaller than those of (**8**) and (**2**). Therefore, such a large downfield shift is attributed to the contraction of the C–P–C angle, as noted also for cyclic sulfoxides ⟨82JOC3660, 80JOC4807⟩. Very interestingly, it was found that the ^{17}O chemical shifts of the phosphine oxides nicely correlate with the ^{31}P NMR chemical shifts of their phosphines, but not with those of their phosphine oxides ⟨88PS(37)35⟩.

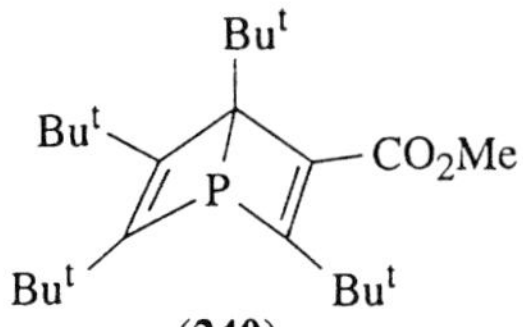

(**240**)

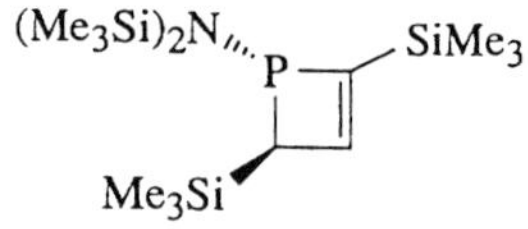

(**241**)

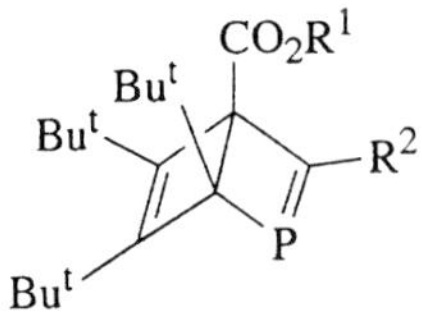

(**242**)-(**249**)

	R^1	R^2
(**242**):	Me	Pr^i
(**243**):	Me	CH_2Bu^t
(**244**):	Me	Bu^t
(**245**):	Me	1-methylcyclohexyl
(**246**):	Me	Mes
(**247**):	Bu^t	Bu^t
(**248**):	Bu^t	1-Ad
(**249**):	Bu^t	1-methylcyclohexyl

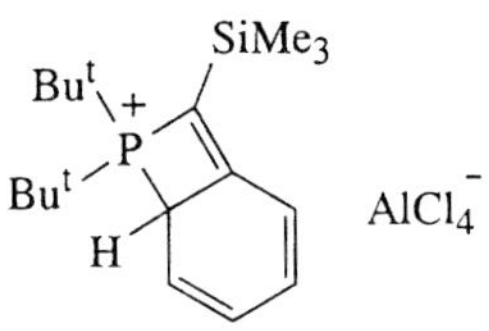

(**250**)

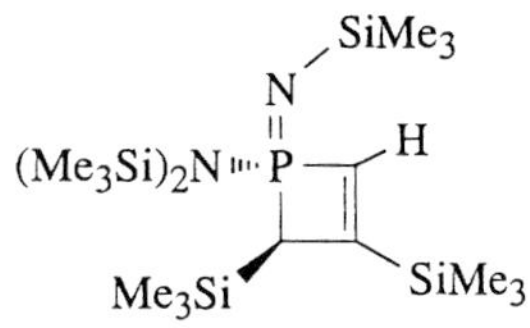

(**251**)

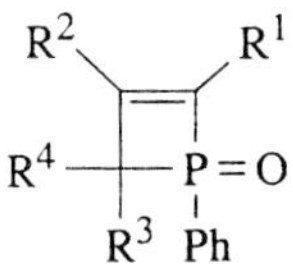

(**252**)-(**256**)

	R^1	R^2	R^3	R^4
(**252**):	Ph	Ph	H	H
(**253**):	OEt	H	Ph	H
(**254**):	H	OEt	Me	CO_2Et
(**255**):	Ph	Ph	OH	H
(**256**):	Et	Et	OH	H

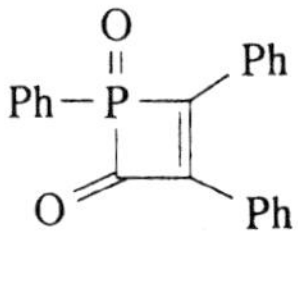

(**257**)

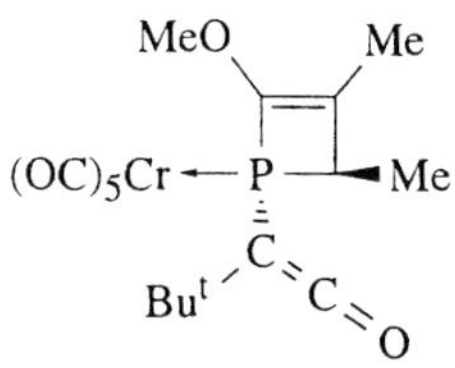

(**258**)

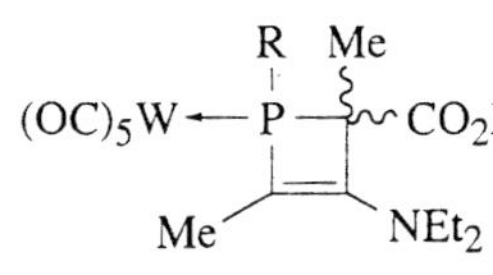

(**259**): R= Ph

(**260**): R= Me

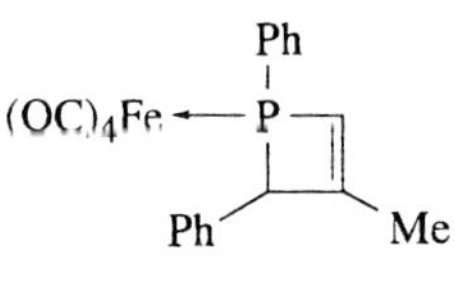

(**261**)

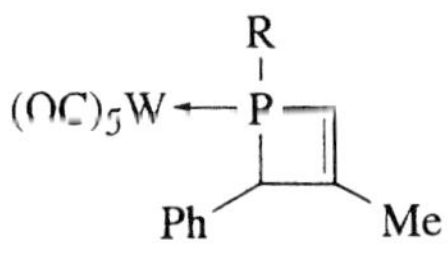

(**262**): R= Ph

(**263**): R= Bu^t

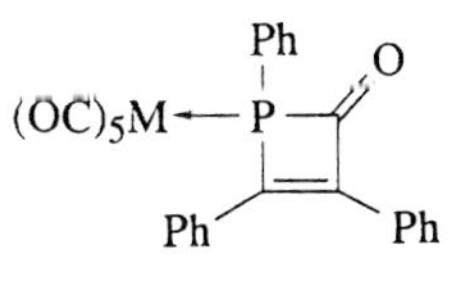

(**264**): M= Mo

(**265**): M= Cr

Scheme 8-1

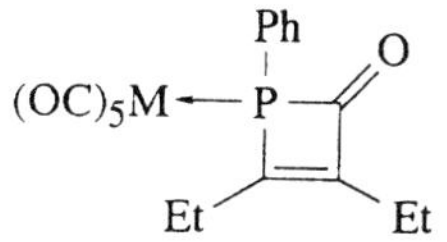

(**266**): M= W
(**267**): M= Cr

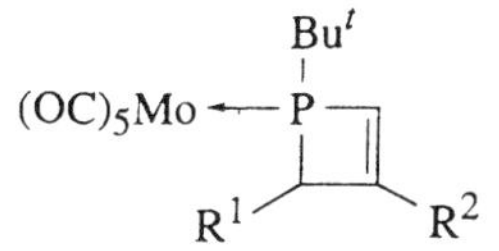

(**268**): R^1= Et; R^2= Ph
(**269**): R^1= OEt; R^2= Me
(**270**): R^1= Ph; R^2= Me

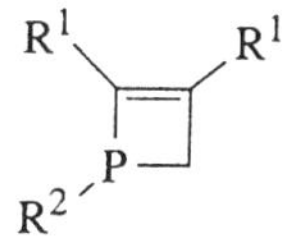

(**271**): R^1= Me; R^2= Ph
(**272**): R^1= Ph; R^2= Et

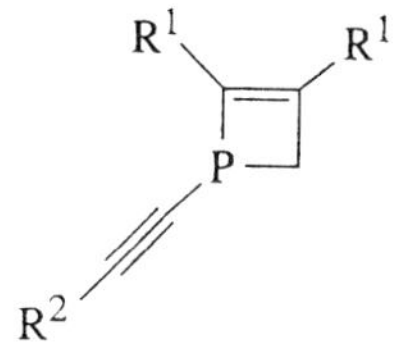

(**273**): R^1= R^2= Ph
(**274**): R^1= Ph; R^2= Hex^n
(**275**): R^1= Et; R^2= Ph
(**276**): R^1= Et; R^2= Hex^n

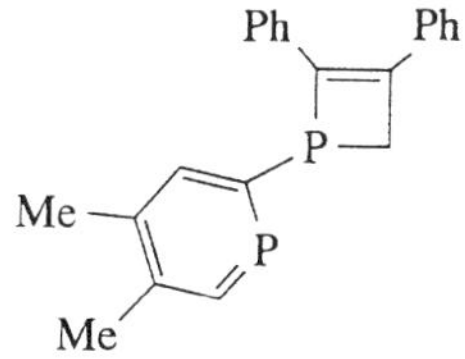

(**277**)

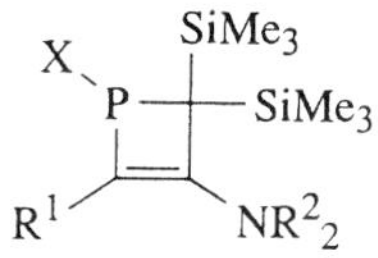

(**278**)-(**286**)

(**278**): X= Cl; R^1= Me; R^2= Et
(**279**): X= Cl; R^1= R^2= Et
(**280**): X= Cl; R^1= Pr^i; R^2= Et
(**281**): X= Cl; R^1= NEt_2; R^2= Et
(**282**): X= Cl; R^1= Et; R^2= Pr^i
(**283**): X= Br; R^1= Me; R^2= Et
(**284**): X= Br; R^1= Et; R^2= Et
(**285**): X= Br; R^1= Pr^i; R^2= Et
(**286**): X= Br; R^1= NEt_2; R^2= Et

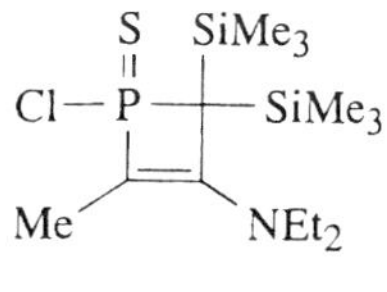

(**287**)

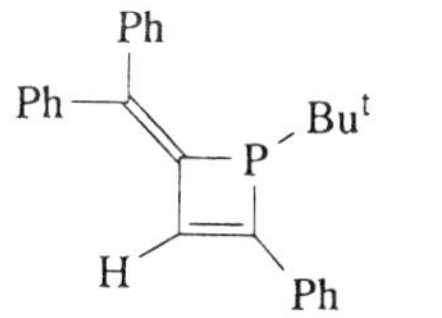

(**288**)

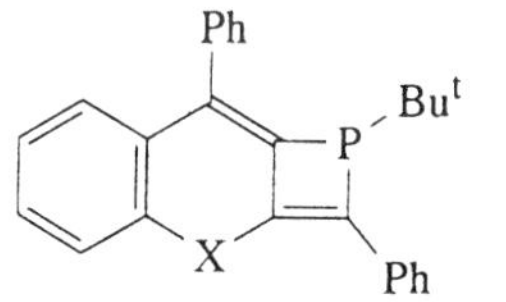

(**289**): X= Zr(Bu^{t}Cp)$_2$
(**290**): X= SbPh

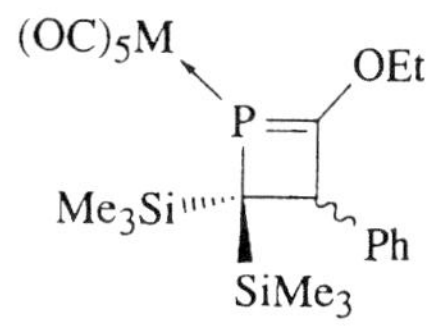

(**291**): M= Cr
(**292**): M= W

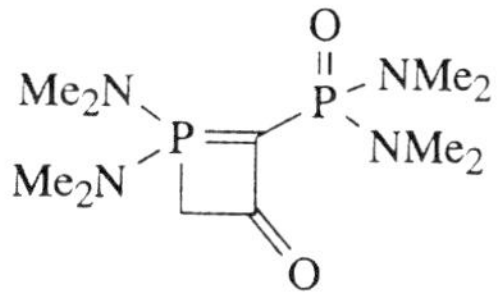

(**293**)

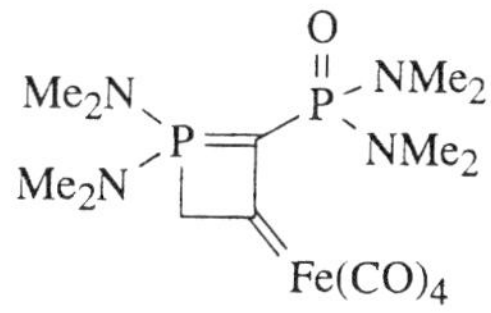

(**294**)

Scheme 8-2

3.2.2.3 Mass Spectrometry

Most of the compounds described here have their molecular ions in electron impact (EI) or fast atom bombardment (FAB) mass mode. High resolution mass spectra are also useful for determination of element

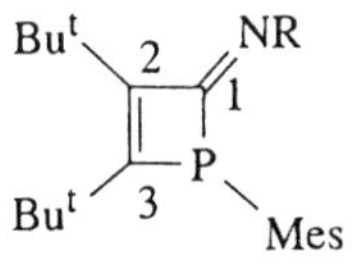

(295a-j) δ_P: -2.3-16.5

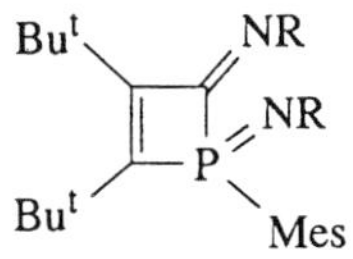

(296b-e)

(no data)

a: R= Ph
b: R= 1-Naphthyl
c: R= m-ClC_6H_4
d: R= p-ClC_6H_4
e: R= m-$NO_2C_6H_4$
f: R= p-$NO_2C_6H_4$
g: R= 2,6-$(Me)_2C_6H_3$
h: R= SO_2Ph
i: R= $PO(OPh)_2$
j: R= CN

C-1: δ_C 175.8-194.0 (^1J(PC)= 9.6-15.0 Hz)
C-2: δ_C 167.1-185.9 (^2J(PC)= 6.6-16.8 Hz)
C-3: δ_C 152.7-154.7 (^1J(PC)= 26.0-28.5 Hz)

Scheme 9

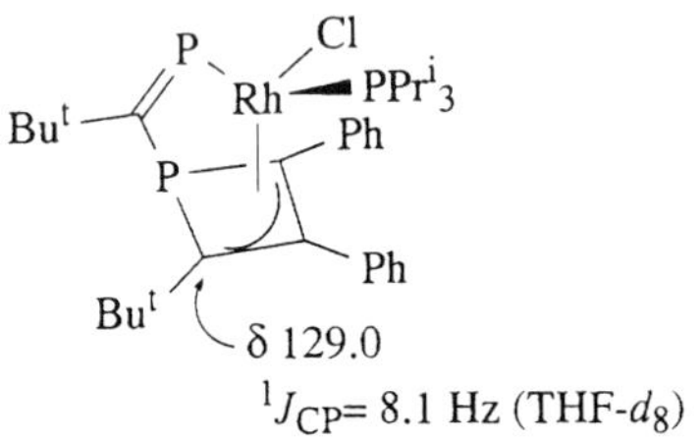

(297)

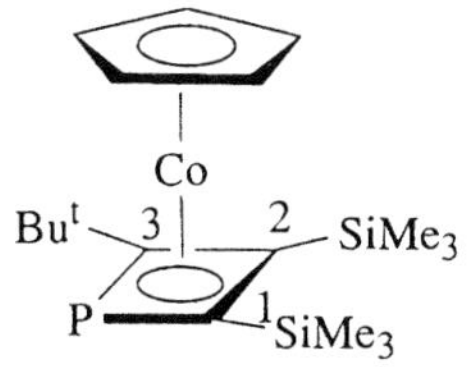

(298)

	δ	J_{CP}/ Hz
C-1	72.6	49.8
C-2	87.3	1.5
C-3	116.2	43.5

Scheme 10

Table 8. ^{31}P NMR spectral data of *trans*- and *cis*-isomers of trivalent phosphetanes

Compounds		$^{31}P/\delta$	*Ref.*
(**62**)	*trans*	169.2	⟨84JCS(P2)553⟩
	cis	149.0	
(**63**)	*trans*	127.5	⟨84JCS(P2)553⟩
	cis	99.6	
(**64**)	*trans*	186.6	⟨84JCS(P2)553⟩
	cis	169.5	
(**65**)	*trans*	−6.8	⟨92OM2944⟩
	cis	−38.3	
(**122**)	*trans*	47.4	⟨92JOC511⟩
(**123**)	*cis*	39.5	⟨92JOC511⟩

composition of the molecules. In the mass spectra of multi carbonyl complexes a stepwise elimination of carbonyl is one of the very important fragmentations, the fragment ion having no carbonyl ligands is often observed as a base peak. The following fragmentation was observed in the mass spectrum of pyrolysis product (**308**) ⟨85ZN(B)927⟩, see Scheme 21. (The authors reported that the fragment at m/z 127 was CF_3P–CCH^+, but it is wrong; the fragments of m/z 127 and 128 should be assigned as shown in the scheme.)

3.2.2.4 IR Spectroscopy

IR spectra are still important for the discussion on functionalities such as carbonyl and phosphoryl groups and so on. Stretching frequencies of the carbonyl ligand (ν_{CO}) and the ring carbonyl group ($\nu_{C=O}$) are summarized in Table 15. The values of $\nu_{C=O}$ scatter in the range of 1710–1730 cm^{-1}, indicating that the ring strain of the four-membered ring is not so large, because P–C bonds are longer than C–C bonds. Stretching frequencies of the phosphoryl group ($\nu_{P=O}$) are also summarized in Table 16. The stretching

Table 9. ^{31}P NMR spectral data of *trans*- and *cis*-isomers of pentavalent phosphetanes

Compounds [a]		$^{31}P/\delta$	*Compounds*		$^{31}P/\delta$
(134)	*trans*	55.9	**(96)**	*trans*	43.8
(135)	*cis*	76.7	**(100)**	*cis*	52.2
(137)	*trans*	74.7	**(76)**	*trans*	52.8
(138)	*cis*	83.6		*cis*	55.3
(139)	*trans*	63.8	**(77)**	*trans*	52.3
(140)	*cis*	67.9		*cis*	55.6
(97)	*trans*	43.4	**(78)**	*trans*	52.6
	cis	51.7		*cis*	54.5
(98)	*trans*	43.6	**(2)**	*trans*	53.6
	cis	48.6	**(5)**	*cis*	–
(99)	*trans*	47.7			
	cis	–			

[a] Compounds except for **(134)**–**(140)**: Ref. 〈87S265〉; Compounds **(134)**–**(140)**: Ref. 〈92JOC511〉.

Table 10. ^{31}P NMR spectral data of *trans*- and *cis*-isomers of trivalent 1-amino-2,2,3,4,4-pentamethylphosphetanes

Compounds		$^{31}P/\delta$	*Compounds*		$^{31}P/\delta$
(62)	*trans*	127.5	**(299j)**	*trans*	77.8
	cis	99.8		*cis*	65.2
(299b)	*trans*	92.76	**(299k)**	*trans*	77.4
	cis	77.55		*cis*	62.3
(299c)	*trans*	128.4	**(299l)**	*trans*	77.8
	cis	97.2		*cis*	65.7
(299d)	*trans*	130.9	**(299m)** [a]		78.45
	cis	94.5			
(3299e)	*trans*	128.0	**(299n)**	*trans*	98.1
	cis	97.2		*cis*	78.5
(299f) [a]		75.4	**(299o)**	*trans*	97.9
				cis	78.4
(299g)	*trans*	94.31	**(299p)**	*trans*	91.3
	cis	73.99		*cis*	70.06
(299h)	*trans*	95.1	**(299q)**	*trans*	97.76
	cis	76.29		*cis*	78.26
(299i)	*trans*	75.22	**(299r)**	*trans*	90.72
	cis	66.1		*cis*	70.39

[a] Only one signal was observed.

Table 11. ^{31}P NMR spectral data of *trans*- and *cis*-isomers of 1-amino-2,2,3,4,4-pentamethylphosphetane 1-oxides

Compounds	$^{31}P/\delta$	*Compounds*	$^{31}P/\delta$
(300a) [a]	53.4	**(300k)** [a]	50.4
(300b) [a]	56.23	**(300l)** [b]	52.1, 50.85
(300c) [a]	57.0	**(300m)** [b]	52.1, 50.89
(300d) [a]	53.7	**(300n)** [b]	52.4, 51.08
(300e) [b]	56.9, 55.2	**(300o)** [b]	52.0, 50.8
(300f) [b]	51.47, 50.19	**(300p)** [b]	50.97, 50.52
(300g) [b]	51.4, 51.03	**(300s)** [b]	56.88, 55.74
(300i) [b]	52.28, 51.02	**(300t)** [b]	51.25, 50.79
(300j) [a]	50.5	**(300u)** [b]	51.2, 50.8

[a] Only one signal was observed. [b] Two signals were observed, but not assigned.

frequency $\nu(PO_2H)$ of the phosphetanic acid (**1**) has been reported to be 1230 cm^{-1}, while its metal salts and ammonium salts have the two frequencies $\nu_{as}(PO_2^-)$ and $\nu_s(PO_2^-)$ around 1160–1130 and 1050–1030 cm^{-1}, respectively.

In some respects the behavior of 2,2,3,4,4-pentamethylphosphetanic acid is expected to parallel that of

Table 12. ^{31}P NMR spectral data of *trans*- and *cis*-isomers of 1-amino-2,2,3,4,4-pentamethylphosphetane 1-sulfides

Compounds	$^{31}P/\delta$	*Compounds*		$^{31}P/\delta$
(301a) [b]	113.27, 108.44	**(301l)** [b]		96.29, 92.45
(301c) [a]	105.0	**(301m)** [b]		96.37, 92.57
(301d) [b]	108.2, 105.56	**(301n)** [b]		96.44, 92.6
(301e) [b]	106.2, 101.8	**(301o)** [b]		96.3, 92.5
(301f) [b]	95.25, 91.49	**(301p)** [b]		95.32, 90.53
(301g) [b]	94.67, 89.76	**(301q)** [b]		94.91, 91.98
(301h) [b]	96.0, 93.0	**(301r)** [b]		94.85, 89.93
(301j) [b]	95.9, 92.3	**(301w)**	*trans*	92.6
			cis	96.4
(301k) [b]	95.79, 92.45			

[a] Only one signal was observed. [b] Two signals were observed, but not assigned.

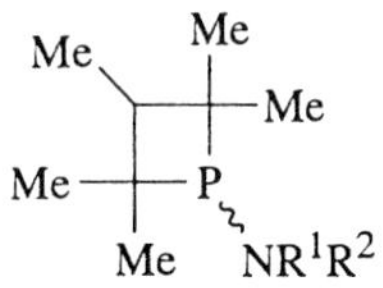

(299a-r)

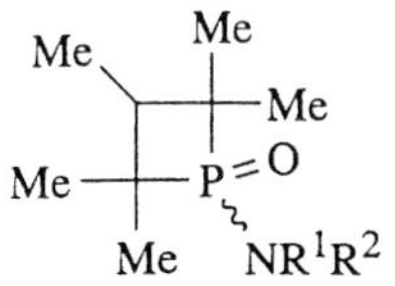

(61)

(300a-g)
(300i-p)
(300s-y)

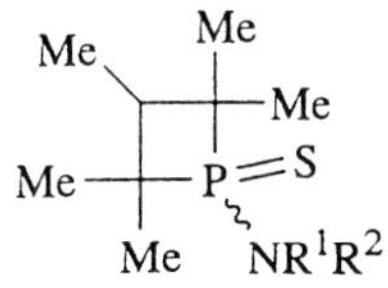

(301a)
(301c-h)
(301j-r)
(301w)

Amino groups:

a: NMe_2

b: NEt_2

c: $NBu^n{}_2$

d: $NBu^i{}_2$

e: 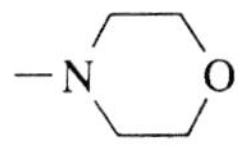

f: $NHOct^n$

g: 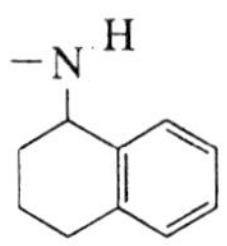

h: 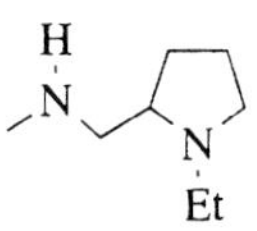

i: 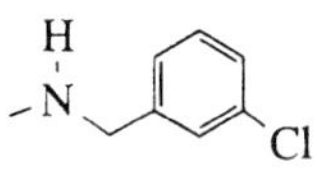

j. 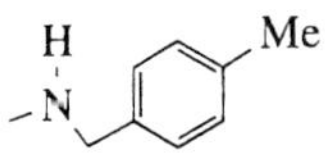

k: 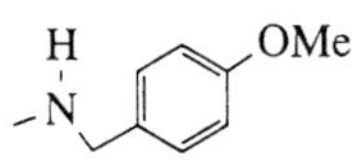

l: 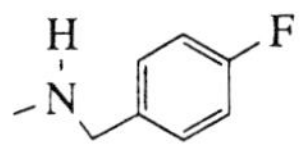

m: 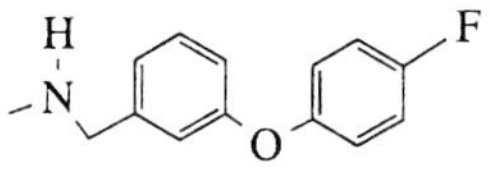

n: 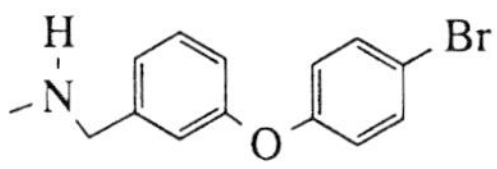

o: 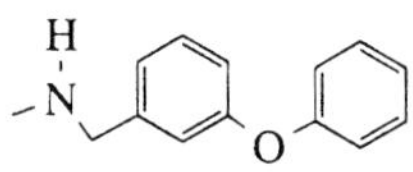

p: 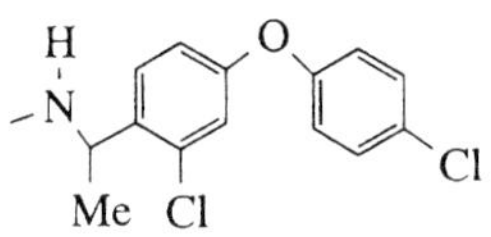

q: 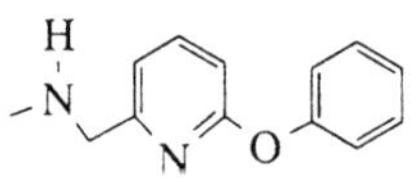

r: 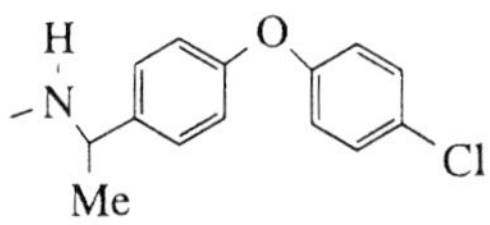

s: 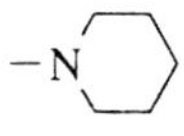

t: 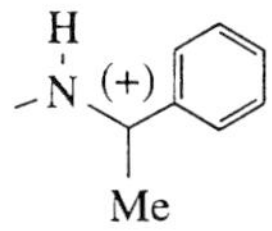

u: 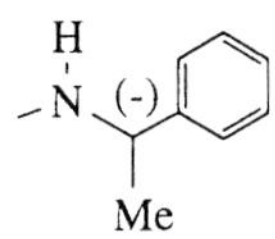

v: 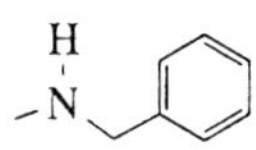

w: 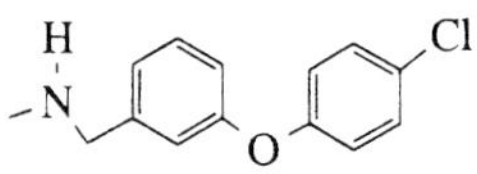

x: 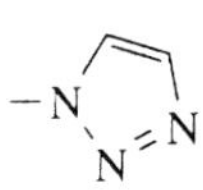

y: 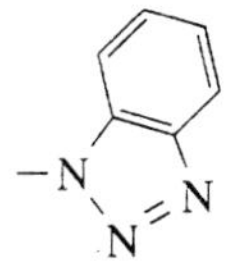

Scheme 11

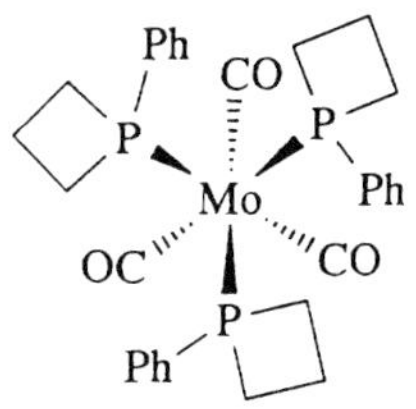

(**31**): δ_P 64.25, $^1J_{MoP}$= 126.7 Hz in CD_2Cl_2)

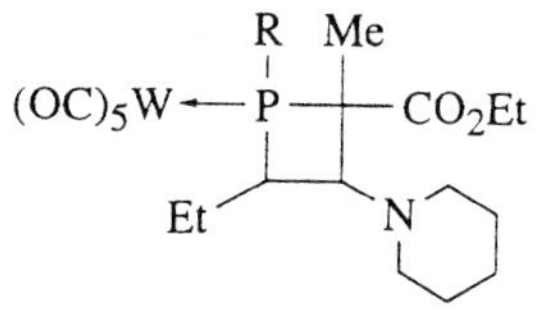

(**146**): R= Me (δ_P 11.8)
(**147**): R= Ph (δ_P 37.1)
(**148**): R= Bu^t (δ_P 59.8) (in C_6D_6)

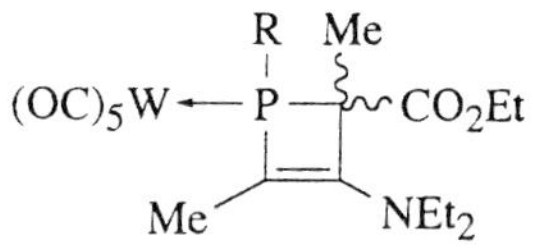

(**258**): R= Me (δ_P -21.5, 20.2)
(**259**): R= Ph (δ_P -0.05, 34.8)
(in C_6D_6)

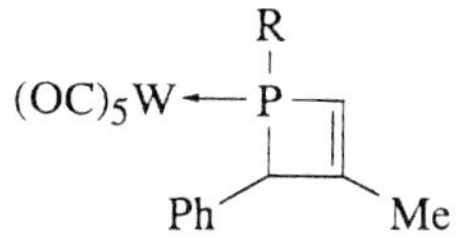

(**261**): R= Ph (δ_P 42.55; $^1J_{WP}$= 239 Hz in C_6D_6)
(**262**): R= Bu^t (δ_P 74.39; $^1J_{WP}$= 239 Hz in pentane)

Scheme 12

Tricoordinate

λ^3

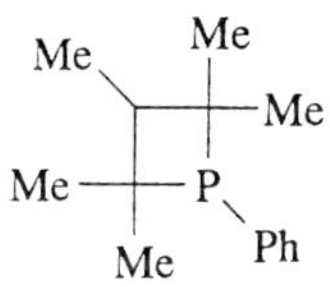

(**66**)
(**67**)

δ_P 1.7, 25.3 in C_6D_6

Tetracoordinate

λ^5

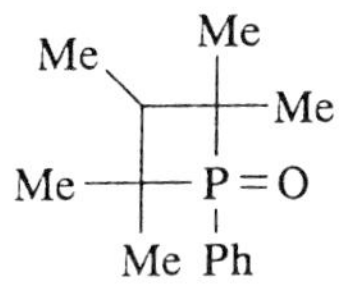

(**2**)

δ_P 53.6 in $CDCl_3$

λ^4

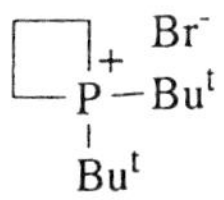

(**119**)

δ_P 80.3

λ^5

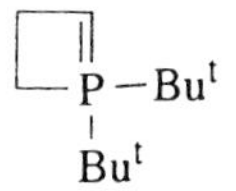

(**303**)

δ_P 53.3

(in CH_2Cl_2)

Pentacoordinate

λ^5

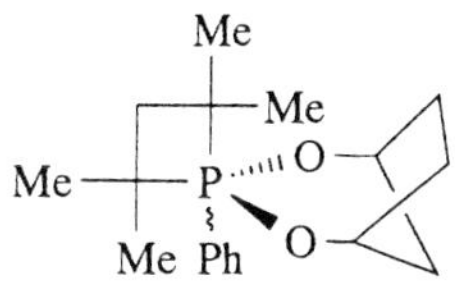

(**302**)

δ_P -42.8, -44.7 in C_6D_6

λ^5

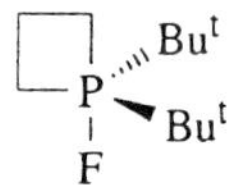

(**162**)

δ_P -23.7 in pentane

Scheme 13

Table 13. ^{31}P chemical shifts and coupling constants *J*(PP) and *J*(PPt) data of platinum complexes of 1,2-dihydrophosphetes

	δ_P				*J (Hz)*						
	P	*L¹*	*L²*	*L³*	*P–Pt*	*L¹–Pt*	*L²–Pt*	*L³–Pt*	*P–L²*	*P–L³*	*L–L*
(**48**)	93.8	12.9	11.2	–	387.7	2737.1	2632.5	–	–	–	–
(**316**)	88.2	–	9.3	−0.1	190.7	–	1793.0	3970.5	18.3	18.3	18.3
(**317**)	92.1	7.2	5.3	–	393.8	2714.6	2588.6	–	–	–	–
(**318**)	88.5	–	13.1	13.8	144.1	–	1762.8	4204.2	19.2	6.4	18.1
(**49**)	25.9	13.0	11.4	–	147.1	2680.0	2575.0	–	9.6	–	–
(**319**)	25.5	10.1	8.2	–	145.7	2537.7	2583.0	–	12.4	–	–
(**320**)	13.1	–	16.4	12.2	20	–	1760	4150	3.2	–	18.1

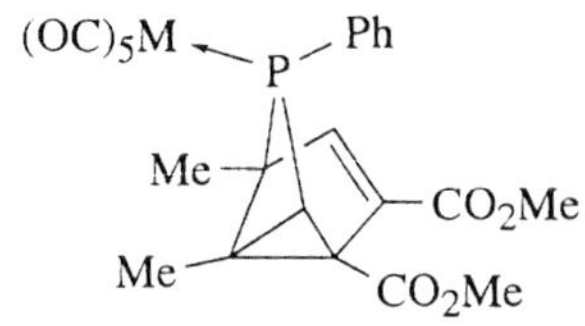

(**26**): M= W (δ_P 15.54; $^1J_{WP}$= 239 Hz in C_6D_6)
(**25**): M= Cr (δ_P 65.36 in toluene)

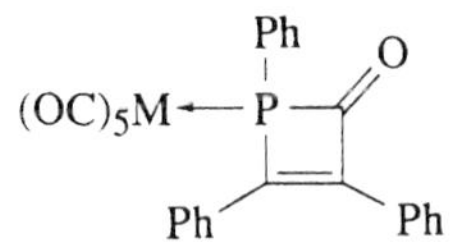

(**39**): M= W (δ_P 86.9; $^1J_{WP}$= 231.9 Hz)
(**263**): M= Mo (δ_P 109.5)
(**264**): M= Cr (δ_P 133.3)

(in C_6D_6)

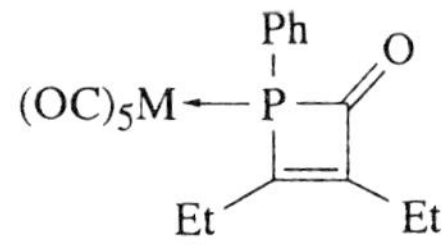

(**265**): M= W (δ_P 85.8; $^1J_{WP}$= 227 Hz in toluene)
(**266**): M= Cr (δ_P 131.8 in CH_2Cl_2)

Scheme 14

Table 14. ^{31}P chemical shifts and coupling constants *J*(PP) and *J*(PPt) data of platinum complexes of 2-alkylidene-1,2-dihydrophosphetes

	Solvent	*δ(P)*	*δ(PEt₃)*	*^3J(PPt) (Hz)*	*^1J(PPt) (Hz)*	*^2J(PP) (Hz)*
(**321**)	CD_3OD	34.2	14.14, 16.02	652	2678	360
(**322**)	$CDCl_3$	34.2	12.1, 12.9	624	2723	375
(**323**)	$CDCl_3$	12.5	2.85	571	2457	–
(**324**)	CD_2Cl_2	49.8	7.2	100	2474	–

Table 15. Stretching frequencies of the ring carbonyl group and the carbonyl ligand of pentacarbonyl metal complexes

Compounds	*$\nu_{C=O}$ (cm^{-1})*	*ν_{CO} (cm^{-1})*	*Ref.*
(**25**)	–	2075, 1985, 1954, 1944	⟨84NJC453⟩
(**26**)	–	2065, 1987, 1958, 1945	⟨84NJC453⟩
(**325**)	–	2070, 1950, 1945	⟨84NJC453⟩
(**39**)	1710	2074, 1955, 1948	⟨85JA5001⟩
(**264**)	1725	2072, 1960, 1955	⟨85JA5001⟩
(**265**)	1710	2067, 1959, 1953	⟨85JA5001⟩
(**266**)	1723	2072, 1988, 1955, 1943, 1910	⟨85JA5001⟩
(**267**)	1730	2065, 1990, 1960, 1946	⟨85JA5001⟩
(**262**)	–	2070, 1950, 1940	⟨90OM788⟩
(**263**)	–	2070, 1948, 1950, 1944	⟨90OM788⟩
(**148**)	–	2075, 1955, 1950, 1940	⟨90JCS(CC)153⟩
(**30**)	–	2041, 2000, 1970, 1962, 1952	⟨93IC5720⟩
(**31**)	–	1950, 1856	⟨94JCS(CC)475⟩
(**52**)	1665	1982, 1940	⟨94ZN(B)1693⟩
(**34**)	–	1951	⟨95OM4983⟩
(**53**)	1625	–	⟨95ZAAC1539⟩
(**166**)	1723	1997, 1939	⟨97OM2958⟩
(**37**)	–	2048, 1995, 1942, 1916, 1890	⟨98CEJ469⟩
(**167**)	–	2058, 1996	⟨98CEJ469⟩

carboxylic acids so that mutual hydrogen bond formation to give cyclic dimers would be anticipated, i.e. (**326**). The ν_{OH} signal at ca. 3500 cm^{-1} is not observed but two very broad and strong peaks at 2250 and 2600 cm^{-1} can be attributed to ν(OH$\cdots$O) of (**326**) by analogy with carboxylic acid dimers. Other broad bands at 955 and 500 cm^{-1} are the corresponding in-plane and out-of-plane OH$\cdots$O bending modes.

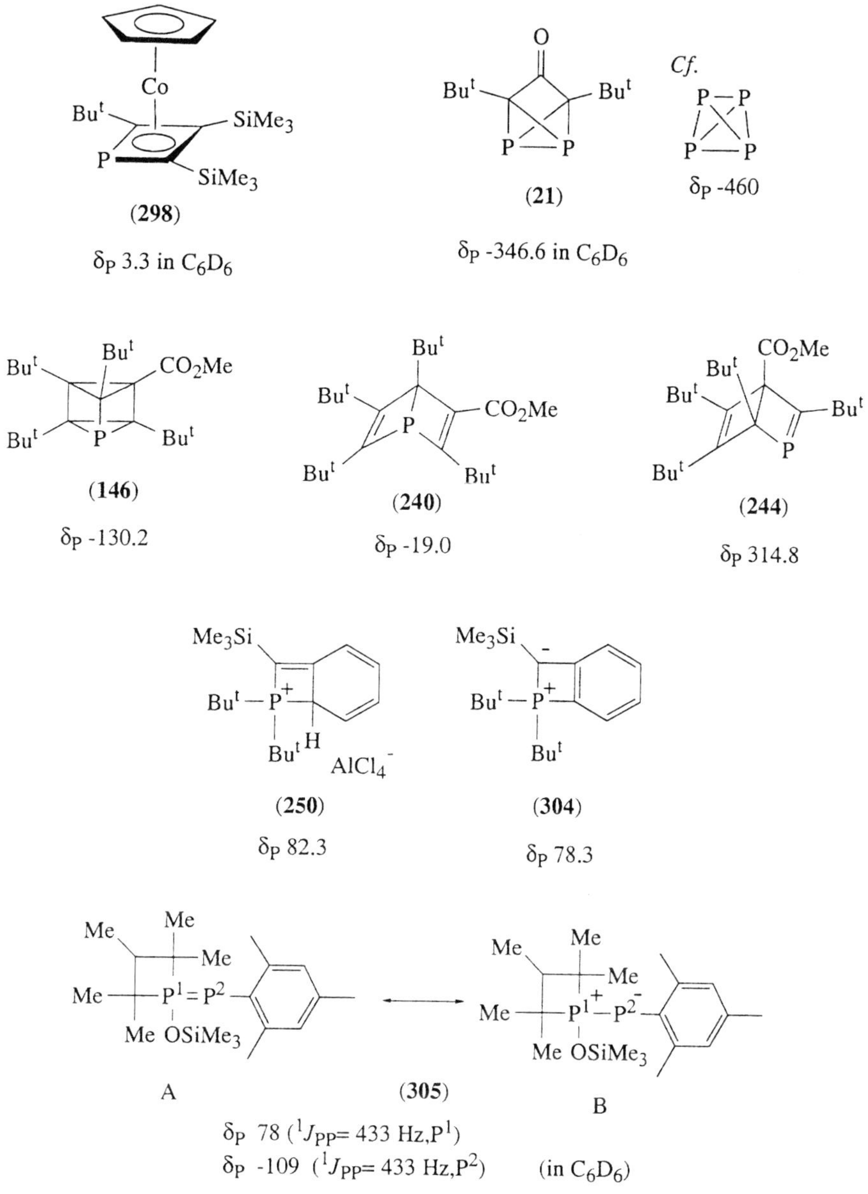

Scheme 15

Table 16. Stretching frequencies of the phosphoryl group of 1,2-dihydrophosphetes

Compounds	$\nu_{P=O}$ (cm^{-1})	*Ref.*	*Compounds*	$\nu_{P=O}$ (cm^{-1})	*Ref.*
(252)	1230	<91IC3381>	**(330)**	1172	<89OM1007>
(257)	1217	<85JA5001>	**(331)**	1150	<89OM1007>
(158)	1205	<93JCR(S)268>	**(332)**	1180	<89OM1007>
(327)	1179, 1037	<84PO325>	**(333)**	1177	<89OM1007>
(328)	1092, 1062	<84PO325>			
(329)	1089, 1056	<84PO325>			

Dilution of the acid in CCl_4 down to 0.001 M concentration failed to produce a signal in the spectrum at ca. 3500 cm^{-1} that could arise from the monomeric acid species. This seems to be evidence that the hydrogen bonding of (**326**) is stronger than that of (RCO_2H) where dilution disturbs the equilibrium towards the monomer sufficiently for the latter's OH bond vibration to be observed. For structural formulas, see Scheme 22.

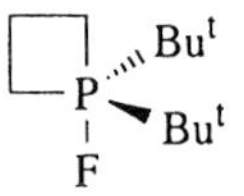

δ_P -23.7 in pentane

$^1J_{PF}$= 663 Hz

(162)

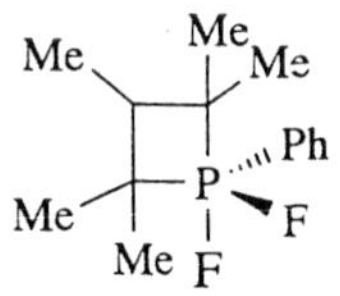

δ_F 58

δ_P -2.87

$^1J_{PF}$= 904 Hz

(306)

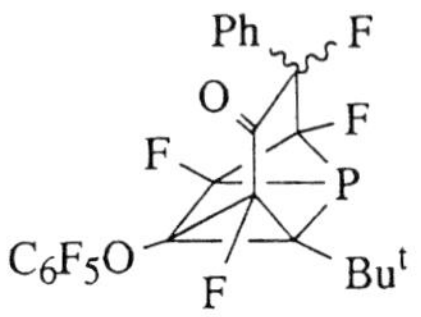

δ_P -86.4 in C_6D_6

$^2J_{PF}$= 174.3 Hz

$^2J_{PF}$= 94.2 Hz

$^3J_{PF}$= 3.2, 1.2 Hz

(307)

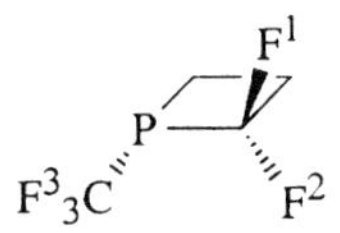

δ_P -6.4 ($^2J_{PP}$= 115 Hz)

$^2J(PF^1)$= 237 Hz

$^2J(PF^2)$= 8 Hz

$^2J(PF^3)$= 71 Hz

(308)

Scheme 16

Table 17. Selected geometric parameters of phosphetanes (**334a**) and (**334b**) and transition states of ring flip (**334c**) and P inversion (**334d**)

Compound	*r(P–C) (Å)*	*r(C–C) (Å)*	*C–P–C (°)*	*δ [a] (°)*	*Basis set*	*Ref.*
(**334a**)	1.9380	1.5618	74.53	156.96	3-21G	⟨89JPC7780⟩
	1.8776	1.5665	76.66	156.62	3-21G*	
	1.8801	1.5444	75.72	155.63	6-31G*	
(**334b**)	1.9375	1.5651	75.31	155.47	3-21G	⟨89JPC7780⟩
	1.8775	1.5711	77.52	154.68	3-21G*	
	1.8789	1.5474	76.21	151.83	6-31G*	
(**334c**)	1.8806	1.5518	77.56	172.22	HF/6-31G*/ /HF/6-31G*	⟨89JPC7780⟩
(**334d**)	1.8365	1.5669	84.23	180.0	HF/6-31G*/ /HF/6-31G*	⟨89JPC7780⟩

[a] Pucker angle of the four-membered ring.

3.2.3 THEORETICAL METHODS

3.2.3.1 *Ab Initio* Studies

The optimized structures of phosphetane (**334**) as well as phosphirane (**335**) have been obtained at the HF–SCF level using the 3-21G, 3-21G*, and 6-31G* basis sets ⟨89JPC7780⟩, see Scheme 23 and Table 17. The lowest energy isomer of phosphetane (**334**) is puckered with the P–H bond being axial (**334a**). The ring strain is estimated as 17.9 kcal mol^{-1}. Optimized transition structures of the ring flip and P inversion are (**334c**) and (**334d**), respectively. The inversion barrier is 42.65 kcal mol^{-1} (calculated at MP2/6-31G*//HF6-31G* with ZPE(3-21G*). The ring flip barrier is calculated to be 2.45 kcal mol^{-1}.

The geometry of 1,2-dihydrophosphete (**336**) are optimized to be (**336a**) at HF/6-31G* and single-point calculations at MP2 are performed along with 1*H*-phosphirene (**338**) and 2*H*-phosphirene (**338**) ⟨91JOC2205⟩, see Scheme 24. The transition structure for inversion at P of (**336**) is also optimized to be (**336b**). The inversion barrier is 46.62 kcal mol^{-1}, see Scheme 24. The allowed conrotatory ring-opening reactions of 1,2-dihydrophosphete (**336**) and 3,4-dihydrophosphete (**337**) are examined together with 1,2-dihydroazete (**340**) and 3,4-dihydroazete (**341**) at the HF/6-31G* and MP2/6-31G* levels ⟨92JOC209⟩. Reasonable activation barriers are obtained only with the inclusion of electron correlation; however, geometry optimization at MP2 did not significantly change the geometry from those obtained at the HF level. The opening of the dihydrophosphetes (**336**) and (**337**) giving phospha-1,3-butadienes (**342**) or (**343**) and (**344**), respectively, is thermoneutral or slightly endothermic, while the opening of the dihydroazetes (**340**) and (**341**) is exothermic, see Scheme 25 and Table 18.

The calculated activation barrier for the opening of (**337**) is 40.76 kcal mol^{-1}. The opening of (**336**) can occur via two diastereomeric pathways. Inward rotation of the heteroatom lone pair is favored for this system; the lower barrier is 24.69 kcal mol^{-1} for (**336**).

Quantum chemical calculations at the MCSCF level have been reported for phosphete (**347**) and

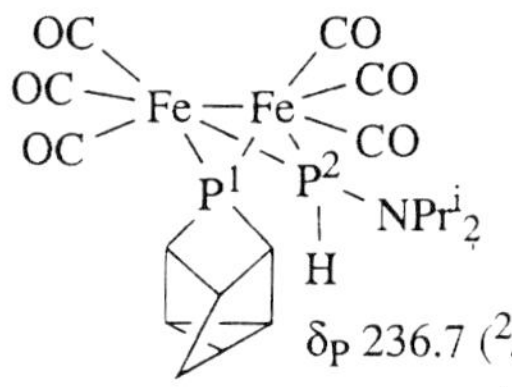

δ_P 236.7 ($^2J_{PP}$= 115 Hz, P^1)

δ_P 133.2 ($^1J_{PH}$= 380 Hz, P^2)

(**30**)

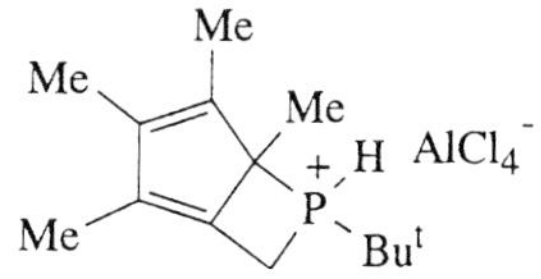

δ_P 87.1 ($^1J_{PH}$= 503.6 Hz)
in CH_2Cl_2

(**309**)

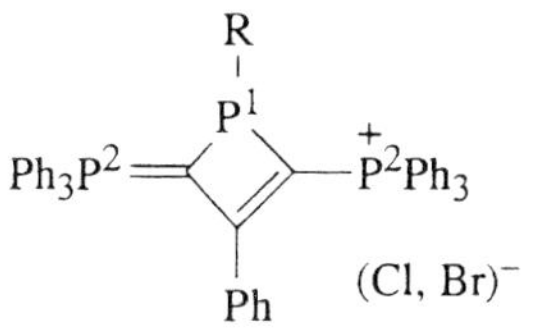

(**310**): R= Ph
(**311**): R= Cl
(**312**): R= Br

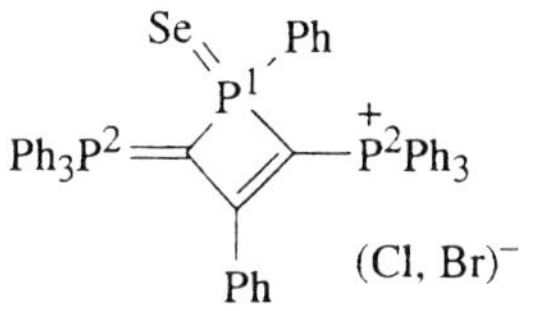

(**313**)

	δ P^1	δ P^2	^2J(PP)/Hz
(**310**)	6.6	20.3	36.6
(**311**)	7.1	97.2	37.4
(**312**)	6.9	96.0	36.6
(**313**)	6.7	46.3	18.3

in C_6D_6

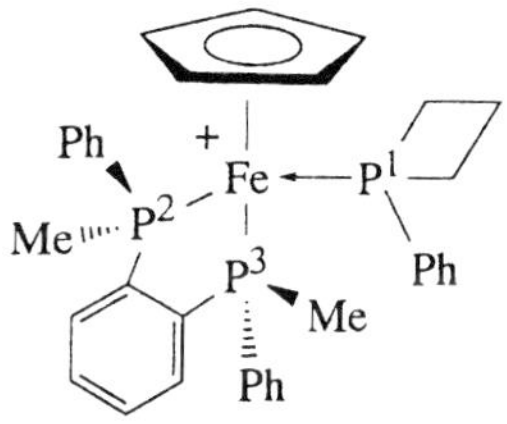

δ_P 99.47 (P^1) ^2J(P^{1}P^2)= 50.7 Hz
δ_P 79.70 (P^2) ^2J(P^{1}P^3)= 49.2 Hz
δ_P 80.73 (P^3) ^2J(P^{2}P^3)= 44.2 Hz

(**29**)

Scheme 17

Table 18. Selected geometric parameters of dihydrophosphetes (**336**) and (**337**)

Compound	*r(P–C1) (Å)*	*r(P–C4) (Å)*	*r(C2–C3) (Å)*	*(C3–C4) (Å)*	*C–P–C (°)*	*Ref.*
(**336a**)	1.892	1.834	1.513	1.325	73.63	⟨91JOC2205⟩ [a]
(**336a**)	1.899	1.832	1.506	1.348	–	⟨92JOC209⟩ [b]
obsd [c]	1.886	1.821	1.517	1.366	74.0	⟨91IC3381⟩
obsd [c]	1.872	1.826	1.510	1.359	73.8	⟨90AG(E)75⟩
(**336b**) [d]	1.856	1.791	1.526	1.330	77.32	⟨91JOC2205⟩ [a]
(**337a**)	1.662	1.883	1.507	1.553	77.9	⟨92JOC209⟩ [a]
(**337a**)	1.691	1.885	1.501	1.551	–	⟨92JOC209⟩ [b]

[a] HF/6-31G* basis set. [b] MP2/6-31G* basis set. [c] X-ray crystal data of 1,3,4-tripheny-1,2-dihydrophosphete. [d] Transition state of inversion at P of (**336**).

its valence isomers, which are compared with those for cyclobutadiene (**345**), aza- (**346**), 1,3-diaza-, 1,3-diphospha-, and tetraphosphacyclobutadienes ⟨93AG(E)617⟩. The relative energy values (in kcal mol^{-1}) are 7.2 (ε_1), 7.4 (ε_2), 13.6 (ε_3), −1.5 (ΔE), where ε_1 and ε_2, and ε_3 are the energies of singlet and triplet transition states of valence isomerization of (**347**), and phosphatetrahedrane, respectively, and ΔE is an energy difference for the fragmentation of phosphahedrane into $CH_2{=}CH_2$ and HC≡P, see Scheme 26 and Table 19.

Relative energies were calculated by RHF/6-31G* for the possible intermediates (phophazene (**352**),

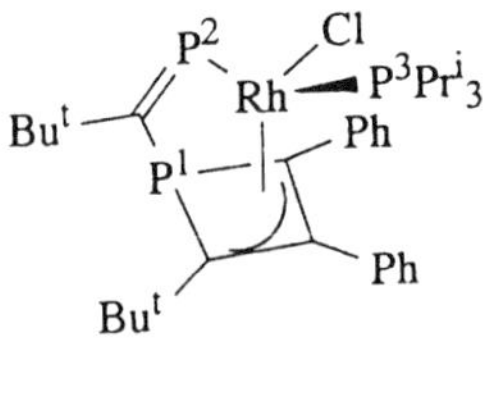

(297)

	δ_P	J_{PP}/ Hz P^1	J_{PP}/ Hz P^2	J_{PRh}/ Hz
P^1	-142.46	-	37.3	3.4
P^2	422.75	37.3	-	30.6
P^3	195.1	3.4	30.6	174.5

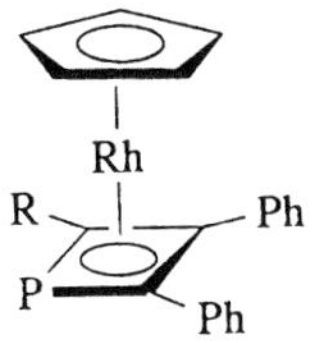

(314): R= Bu^t ; δ_P -48.75 (J_{PRh}= 37.3 Hz)
(315): R=NPr^i(SiMe$_3$); δ_P -26.45 (J_{PRh}= 36.5 Hz)

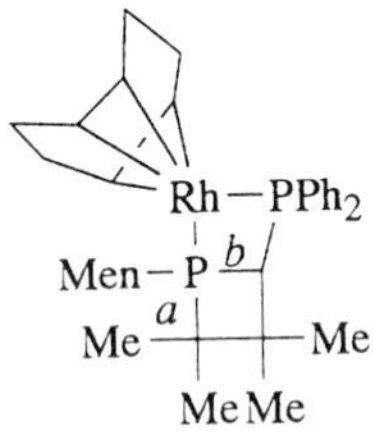

δ_P 4.2 ($^1J_{PRh}$= 121 Hz), -37.3 ($^1J_{PRh}$= 126 Hz) $^2J_{PP}$= 62 Hz

(33)

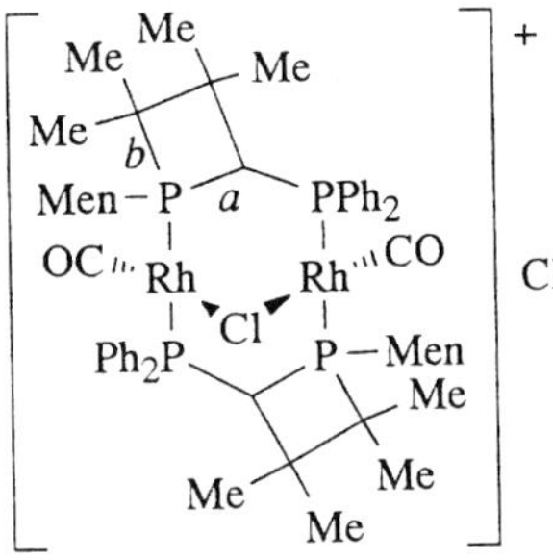

δ_P 84.2 (ddd, $^1J_{PRh}$= 110 Hz, $^2J_{PP}$= 310 Hz, $^2J_{PP}$=26 Hz), 16.5 (ddd, $^1J_{PRh}$= 122 Hz)

(34)

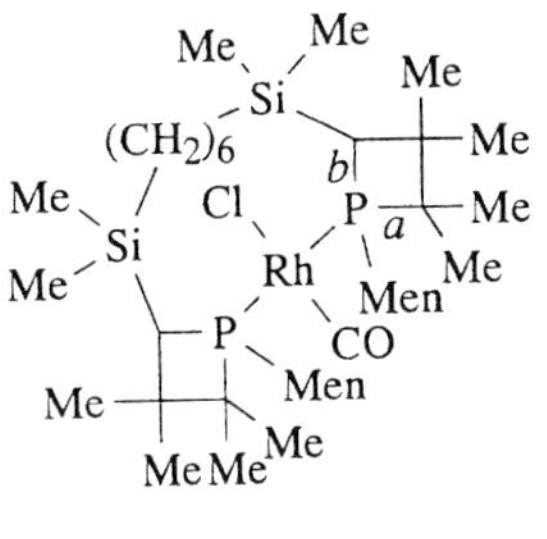

δ_P 94.2 (d, $^1J_{PRh}$= 116 Hz)

(35)

Scheme 18

Table 19. Relative energy values (in kcal mol^{-1}) of (**345**), (**346**) and (**347**)

Compounds	*Symmetry*[a]	ε_1	ε_2	ε_3	ΔE^b
(**345**)	D_{2h} (D_{4h})	8.4	14.8	36.4	−27.2
(**346**)	C_s(C_{2v})	12.1	18.9	30.1	−54.9
(**347**)	C_s(C_{2v})	7.2	7.4	13.6	−1.5

[a] Symmetry in parentheses refers to transition state geometries for valence isomerization. [b] Value for fragmentation of the tetrahedranes into (from top to bottom) 2 C_2H_2, C_2H_2 + HCN, C_2H_2 + HCP, respectively.

iminophosphane (**353**), and spiroazaphosphiridine (**354**)) and the product (iminophosphetene (**355**)) of the reaction of cyclopropenylidenephosphaalkene (**351**) with HN_3, indicating that the product is most stable among them. All steps except for the first step are exothermic and estimated activation energy is in good accordance with the experimental results (Scheme 27) ⟨94TL1527⟩.

Structures of four benzene-ring fused phosphorus heterocycles (**356**)–(**359**) were optimized by *ab initio* SCF calculation (Scheme 28). Selected bond lengths and bond angles are summarized in Table 20.

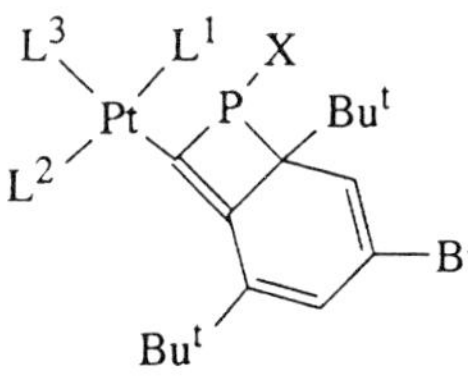

	X	L^1	L^2	L^3
(**48**):	Cl	PEt_3	PEt_3	Cl
(**316**):	Cl	Cl	PEt_3	PEt_3
(**317**):	Br	PEt_3	PEt_3	Br
(**318**):	Br	Br	PPh_3	PPh_3

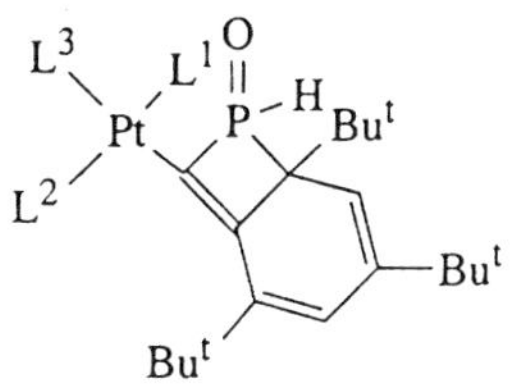

	L^1	L^2	L^3
(**49**):	PEt_3	PEt_3	Cl
(**319**):	PEt_3	PEt_3	Br
(**320**):	Cl	PPh_3	PPh_3

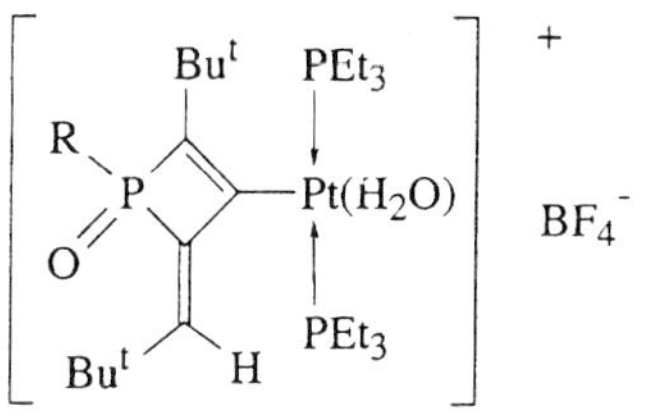

(**321**): R= Ph
(**322**): R= Me

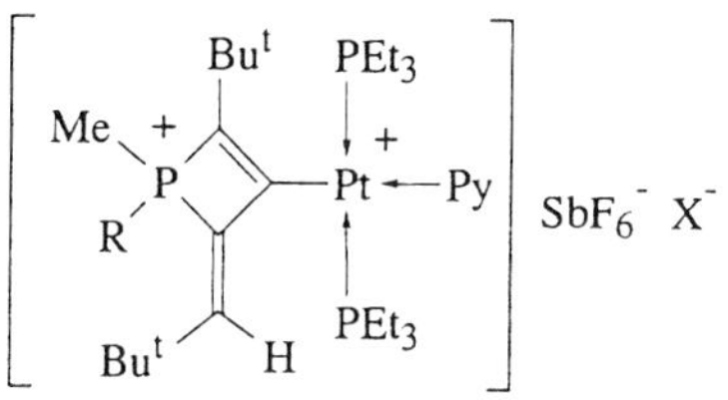

(**323**): R= C≡CBu^t; X =BF_4
(**324**): R= Ph; X= SbF_6

Scheme 19

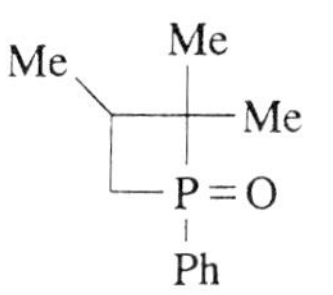

(**8**)

δ_O 74.0
J_{PO}= 175 Hz

C-P-C 79.4°

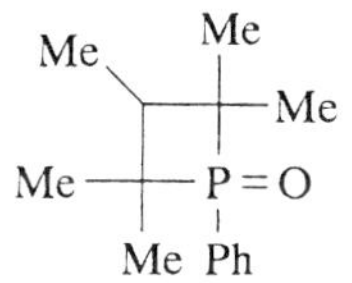

(**2**)

δ_O 58.4
J_{PO}= 203 Hz

C-P-C 80.8°

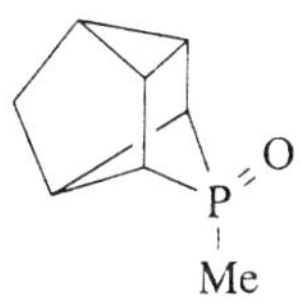

(**134**)

δ_O 93.1
J_{PO}= 179 Hz

C-P-C < 79.4°

Scheme 20

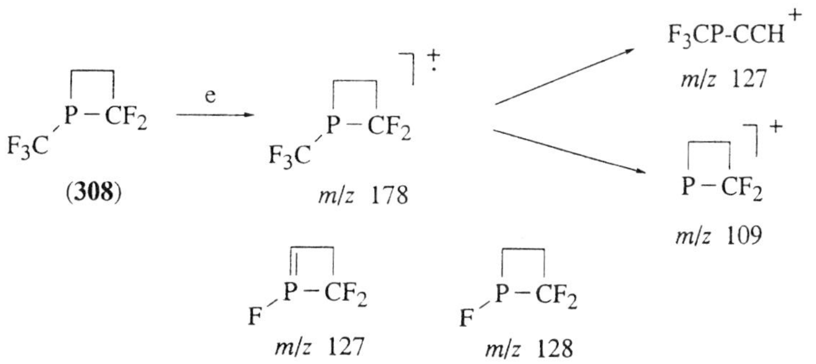

Scheme 21

From these data ^{13}C NMR chemical shifts were calculated by IGLO. These values closely agree with experimental ones 〈96CEJ68〉.

Relative energies of 6 valence isomers (**361**)–(**366**) of 7-phosphanorbornadiene (**360**) were calculated by MP2(fc)/6-31G* and shown in Scheme 29. The selected bond lengths and angle of (**361**) and (**366**)

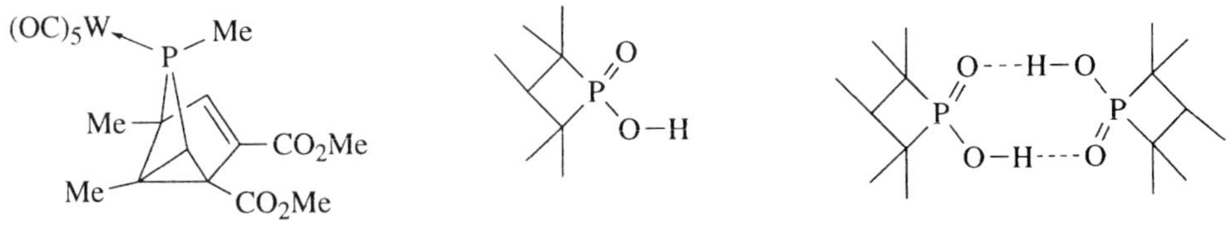

(**325**) (**1**) (**326**)

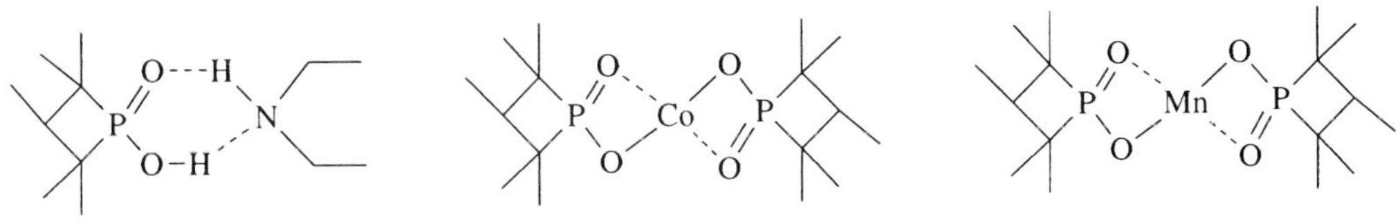

(**327**) (**328**) (**329**)

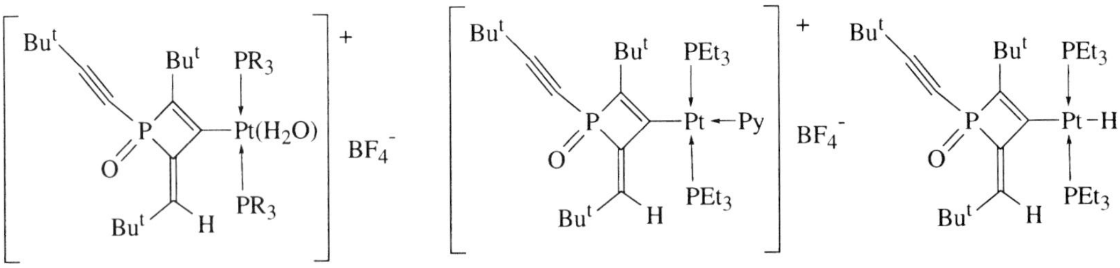

(**330**): R= Et
(**331**): R= Pr^n

(**332**) (**333**)

Scheme 22

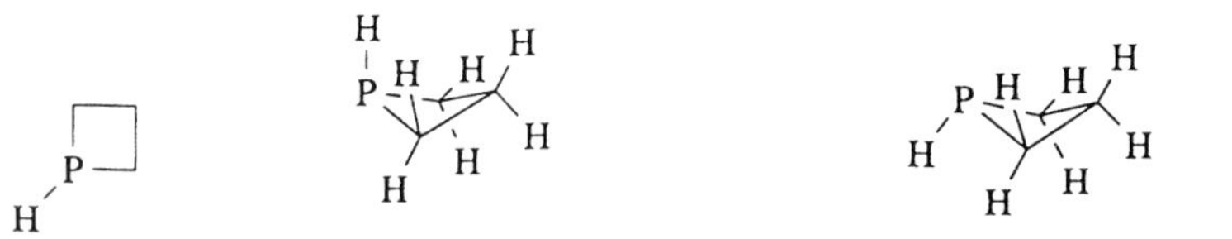

(**334**) (**334a**) (**334b**)

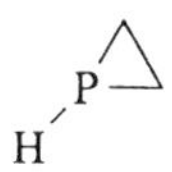

(**335**)

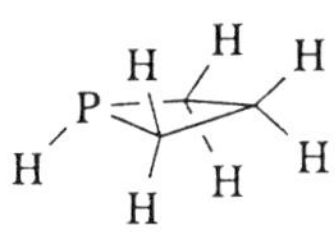

transition state of ring flip

(**334c**)

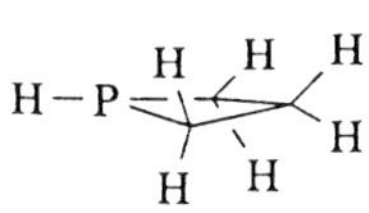

transition state of P inversion

(**334d**)

Scheme 23

Table 20. Selected geometric parameters of benzene-ring fused phosphorus heterocycles (**356**)–(**358**)

Compound	*a* (*Å*)	*b* (*Å*)	*c* (*Å*)	*d* (*Å*)	*C–P–C* (*°*)	*Ref.*
(**356**)	1.898	1.675	1.491	1.415	75.1	<96CEJ68>
(**357**)	1.833	1.650	1.494	1.423	80.8	<96CEJ68>
(**358**)	1.793	1.710	1.452	1.425	80.0	<96CEJ68>
(**359**)	1.774	1.847	1.526	1.391	79.0	<96CEJ68>

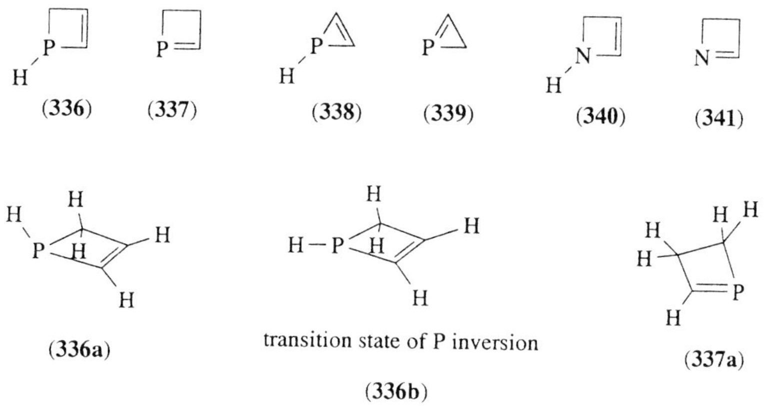

Scheme 24

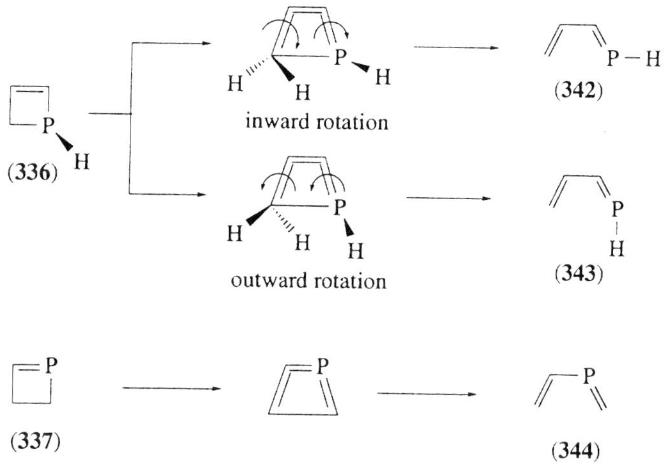

Scheme 25

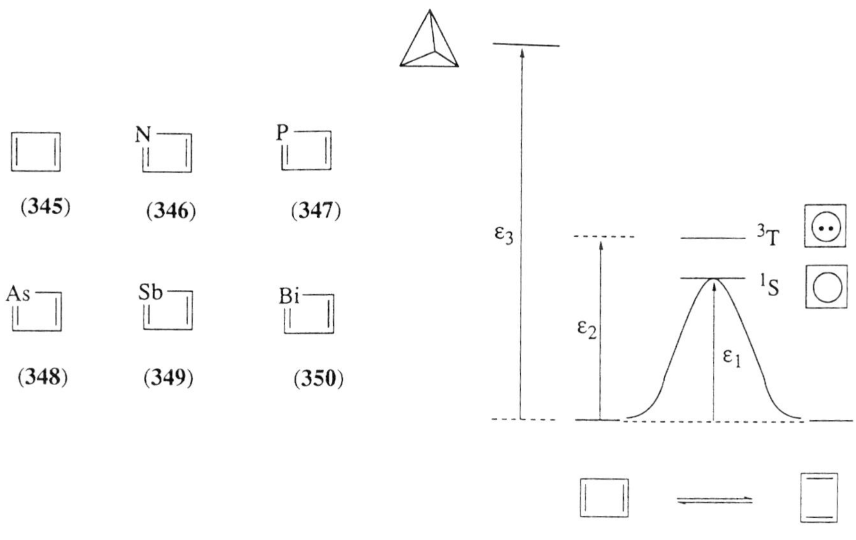

Scheme 26

bearing phosphetane ring(s) are also shown in Scheme 29. The calculated relative energies indicate that a thermal rearrangement is unlikely because of its 6.4 kcal/mol endothermicity. Therefore, the plausible mechanism of rearrangement of $W(CO)_5$-complexed 7-phosphoanorbornadiene (**367**) to the 7-phosphatricyclo[3.2.0^{4,6}]hept-2-ene complex (**26**) was estimated to involve retro[1,4]cycloaddition of phosphinidene metal complex ⟨97OM4145⟩.

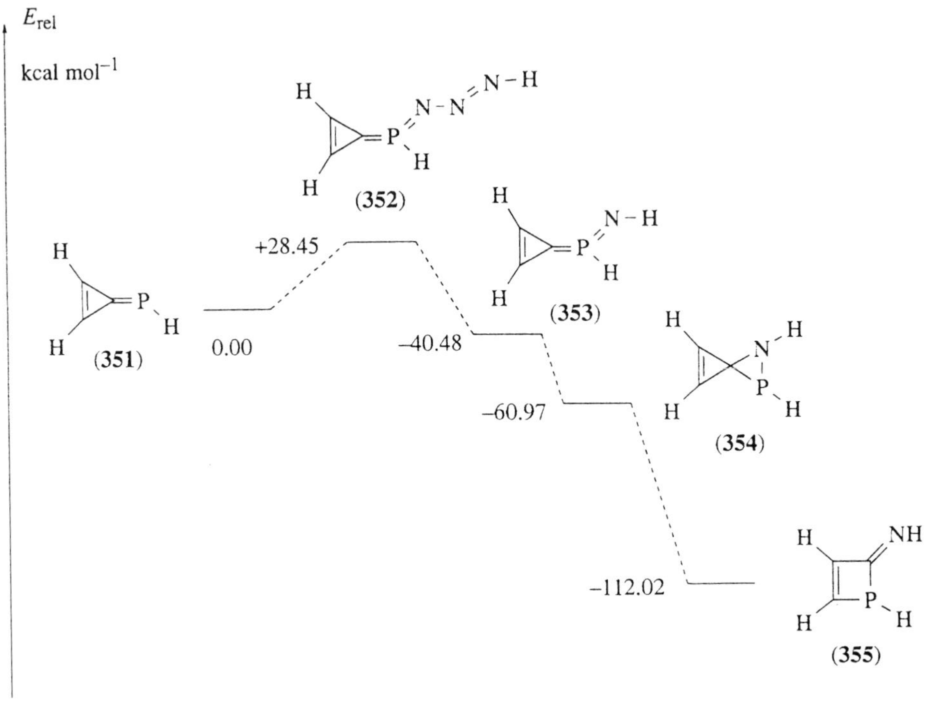

Scheme 27

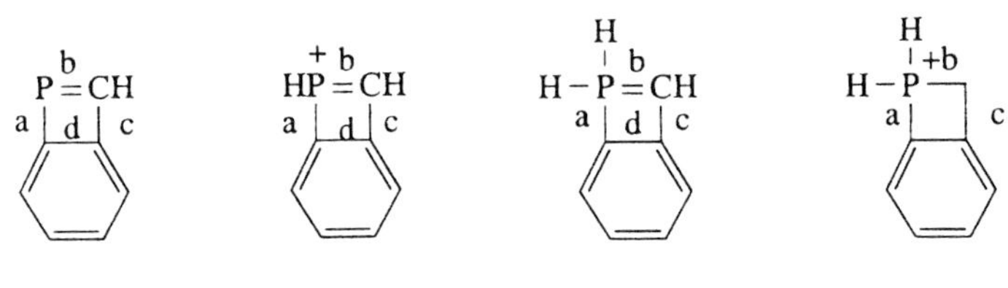

Scheme 28

Table 21. Selected bond lengths and angles of $HP(CH_2)n$ (n = 2–5)

Compound	*n*	*P–C* (*Å*)	*P–H* (*Å*)	*C–P–C* (°)	*H–P–C* (°)
(**335**)	2	1.869	1.417	46.97	95.62
(**334**)	3	1.888	1.422	74.51	96.04
(**368**)	4	1.877 [a]	1.421	91.77	99.05
(**369**)	5	1.860	1.423	98.08	100.74

[a] Mean of 1.869 and 1.886 Å of the two nonequivalent PC bonds in this molecule.

The structures of $HP(CH_2)n$ (n = 2–5) (**335**), (**334**), (**368**), and (**369**) were optimized by MP2/6-31G(d) //MP2/6-311G(d,p) and the selected bond lengths and angles are shown in Table 21 and Scheme 30. Shielding parameters (σ_{iso} (calcd)) in ^{31}P NMR spectra were calculated by GIAO (RHF/6-311G(d,p)) and shown in Scheme 30 together with observed shielding parameters (σ_{iso} (obs)) for 1-phenyl derivatives, $PhP(CH_2)n$ (n = 2–5). It was found that a phosphirane has the highest shielding, while a phosphetane has the lowest shielding, which is in agreement with experimental results ⟨97HAC451⟩.

Ab initio calculations for *o*-, *m*-, and *p*-diphosphabenzenes and their valence isomers, 7 diphosphabenzvalenes, 6 Dewar diphosphabenzenes, 3 diphosphaprismanes, and 5 diphosphabicyclopropenyls, were carried out. Among those optimized structures the selected bond lengths and angles of Dewar diphosphabenzenes (**370**)–(**373**) and diphosphaprismanes (**374**) and (**375**) bearing phosphetane or 1,2-dihydrophosphete ring partially are shown in Table 22 (Scheme 31) ⟨99JA4215⟩.

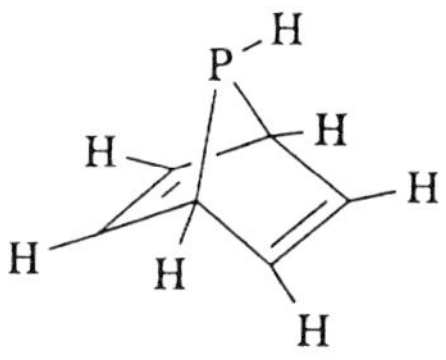

(360)
ΔE= 0.00

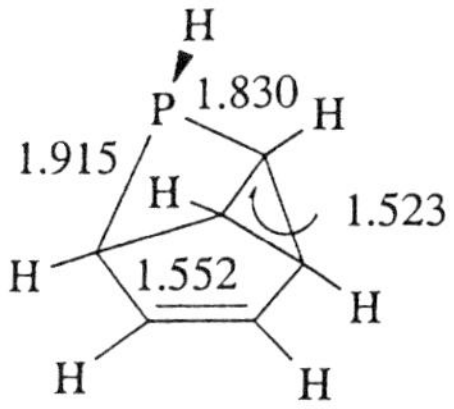

(361)
ΔE= 6.36 kcal/mol

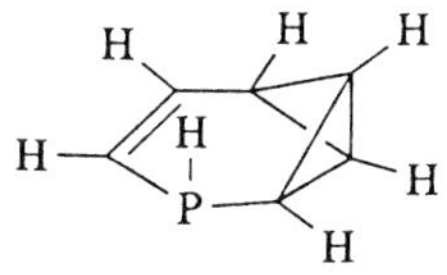

(362)
ΔE= 21.80 kcal/mol

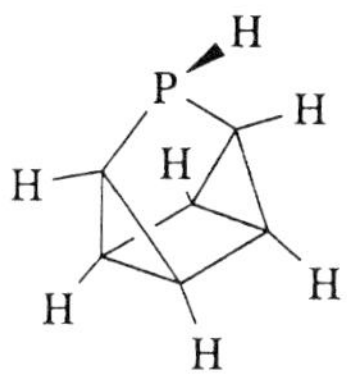

(363)
ΔE=22.54 kcal/mol

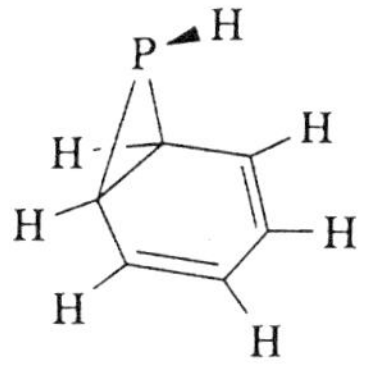

(364)
ΔE=–12.70 kcal/mol

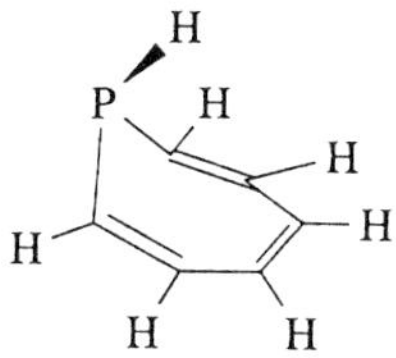

(365)
ΔE=–4.25 kcal/mol

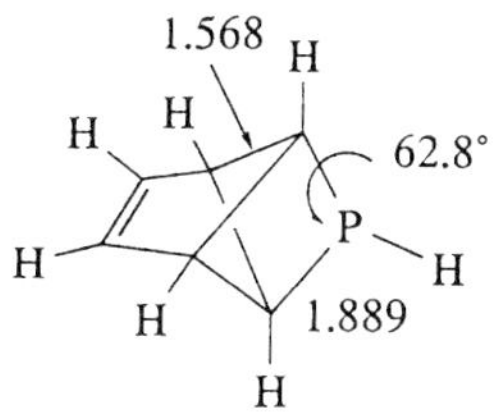

(366)
ΔE=24.74 kcal/mol

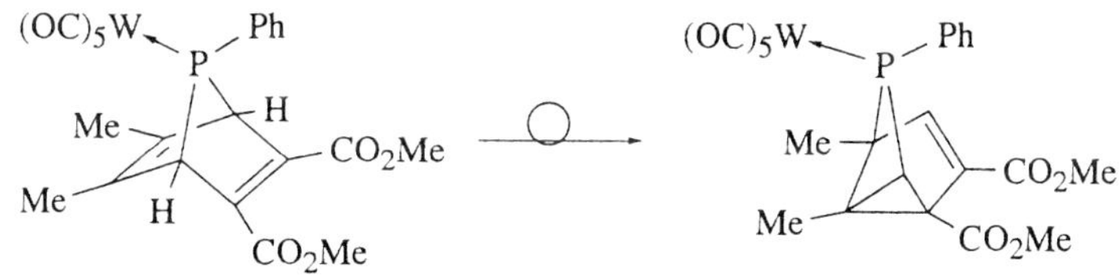

(367) **(26)**

Scheme 29

Table 22. Selected bond lengths and angles of Dewar diphosphabenzenes and diphosphaprismanes

Compound	*a* (*Å*)	*b* (*Å*)	*c* (*Å*)	*d* (*Å*)	*C–P–C* (*°*)
(370)	1.91	1.83	1.35	1.51	73.7
(371)	1.89	1.83	1.35	1.50	74.2
(372)	1.70	1.92	1.55	1.49	76.3
(373)	1.70	1.89	1.56	1.51	77.3
(374)	1.89	1.88	1.54	1.51	74.6
(375)	1.90	1.91	1.53	1.49	73.2

3.2.3.2 Empirical and Semiempirical Methods

Topological resonance energies (TRE) of heterocyclobutadienes containing one N, P, As, Sb, or Bi have been calculated to evaluate their stability or reactivity. TRE is defined as TRE $= \Sigma g_j(x_j - x_j^{\mathrm{ac}})$ from graph

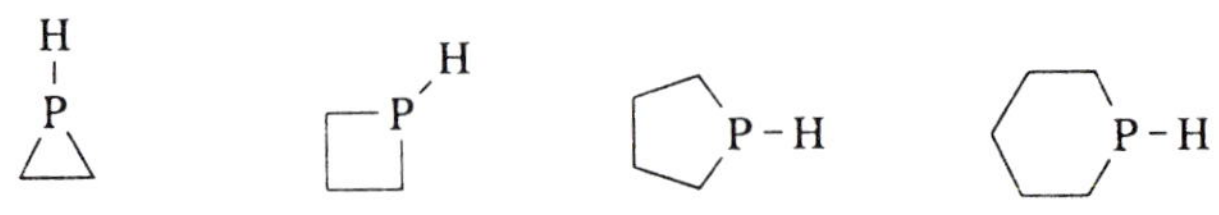

	(335)	(334)	(368)	(369)
σ_{iso}^{calc}	68.3.8	382.4	431.2	420.6
σ_{iso}^{obs}	564.4	314.5	343.7	362.7

Scheme 30

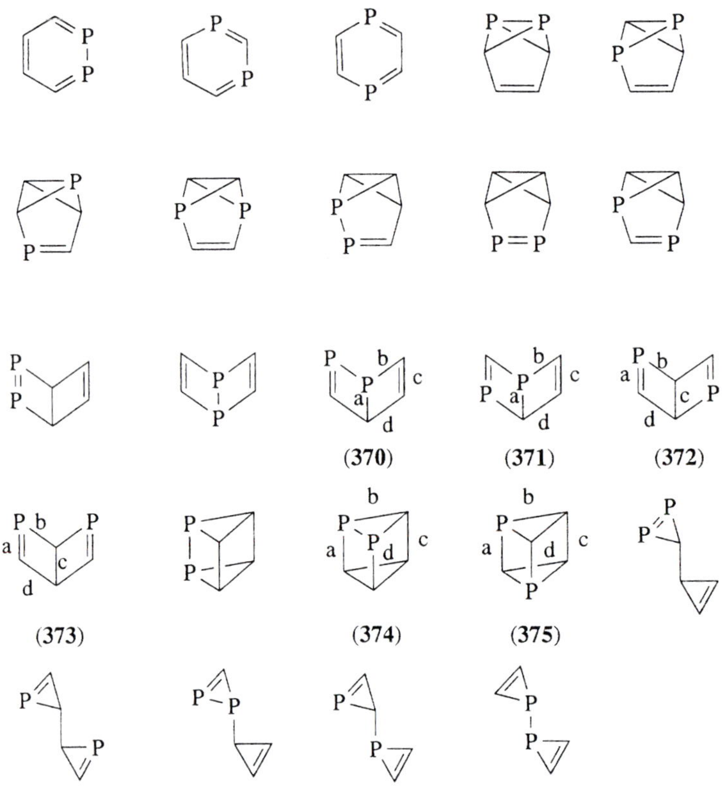

Scheme 31

theory ⟨76JA2750, 77JA1692⟩. However, for open-shell configuration ($g_1 = 2$, $g_2 = g_3 = 1$) TRE values are positive, that is, (**346**): 0.2236, (**347**): 0.2132, (**348**): 0.1856, (**349**): 0.1376, (**350**): 0.0984. Modified TRE*, which can be defined as $TRE^* = \Sigma g_j(x_j - x_j^{ac})$, is necessary to evaluate their stability correctly ⟨88CPL44⟩. Their values become around −0.7524 to −0.5936, indicating that they are antiaromatic and antiaromaticity decreases in the order of As > P > N, but the Sb and Bi derivatives are less antiaromatic. The electron densities for the open-shell configuration are in accordance with ^{13}C NMR values. The mobile bond orders suggest a symmetric four-membered ring with strongly distorted diagonal distances for all the molecules. The molecular stability/reactivity has been discussed in the light of HOMO–LUMO separation (ΔE) in the graph spectrum ⟨90IJC370⟩. The larger gap between HOMO and LUMO makes the molecule more stable. In cyclobutadiene ΔE is zero, and hence the molecule is said to be more reactive and unstable. The values for (**345**)–(**350**) are 0, 0.2523, 0.2057, 0.0735, 0.0806, and 0.08616, respectively, indicating that ΔE values decrease in the order (**346**) > (**347**) > (**349**) > (**350**) > (**348**) > (**345**), which is a similar order to that obtained from the TRE calculation. (There are some mistakes, because ΔE values decrease in the order (**346**) > (**347**) > (**350**) > (**349**) > (**348**) > (**345**) and TRE order is (**350**) > (**349**) > (**346**) > (**347**) > (**348**) > (**345**)).

Hydrolytic reactions of some cyclic esters of alkylphosphates and phosphonate have been studied using molecular mechanics calculations (Allinger 1977 force field, MM2 1985) ⟨90JCS(P2)971⟩. Energy

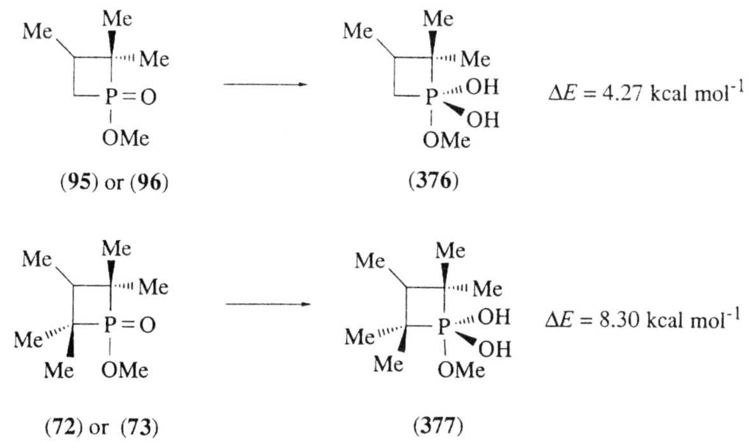

Scheme 32

Table 23. Selected bond lengths and angles of phosphete (**347**) and protonated phosphetes (**378**) and (**379**)

Compound	*a* (*Å*)	*b* (*Å*)	*c* (*Å*)	*d* (*Å*)	*C–P–C* (°)	*C–C–C* (°)	*Ref.*
(**347**) [a]	1.727	1.901	1.351	1.502	73.4	99.1	<93AG(E)617>
(**347**) [b]	1.615	1.951	1.336	1.489	75.5	102.1	<95JST(338)51>
(**347**) [c]	1.576	1.854	1.344	1.495	77.0	97.9	<95JST(338)51>
(**347**) [d]	1.597	1.790	1.350	1.509	77.7	96.2	<95JST(338)51>
(**378**) [b]	1.510	1.782	1.352	1.526	83.0	99.0	<95JST(338)51>
(**378**) [c]	1.513	1.513	1.445	1.445	88.4	93.7	<95JST(338)51>
(**378**) [d]	1.777	1.777	1.407	1.407	73.2	97.8	<95JST(338)51>
(**379**) [b]	1.924	1.924	1.388	1.388	68.8	103.2	<95JST(338)51>
(**379**) [c]	1.776	1.776	1.399	1.399	73.4	98.6	<95JST(338)51>

[a] *Ab initio*. [b] PM3. [c] AM1. [d] MNDO.

differences between the tetracoordinate starting materials ((**95**) or (**96**)) and ((**72**) or (**73**)) and the pentacoordinate intermediates ((**376**) and (**377**)) are 4.27 and 8.30 kcal mol^{-1} for 1-methoxy-2,2,3-trimethyl-1-oxophosphetane ((**95**) or (**96**)) and 1-methoxy-2,2,3,4,4-pentamethyl-1-oxophosphetane ((**72**) or (**73**)), respectively, see Scheme 32. Therefore, the increase in steric energy caused by two methyl groups in the transition state is 4.03 kcal mol^{-1}. The calculated relative rate of hydrolysis of compounds ((**95**) or (**96**)) and ((**72**) or (**73**)) is, therefore, 884 : 1, which is close to the experimental data reported by Hawes <68JCS(CC)577>.

Structures of (CH)*n*EH ($n = 2, 4$; E = N, P, As, Sb), (CH)*n*E ($n = 3, 5$; E = N, P, As, Sb), (CH)*n*EH$^+$ ($n = 3, 5$; E = N, P, As, Sb) were optimized by PM3 semi-empirical calculation and were compared with those previously calculated <95JST(338)51>. The selected bond lengths and angles of four-membered ring compounds with a phosphorus atom are summarized in Table 23 (Scheme 33). The relative energies of protonated species (**378**), (**379**), (**380**) and (**381**) are 13.4, 25.6, 0.0, and 42.9 kcal mol^{-1}, respectively, indicating that C2-protonation is favored over not only C3-protonation, but also favored over P-protonation, in sharp contrast to nitrogen analog where N-protonation is favored over both C-protonations.

For cyclopropane, cyclobutane and cyclobutene containing one Si, N, P, O, or S, ^{13}C–^{13}C-coupling constants were estimated by SCPT INDO calculation. The geometric parameters were optimized by the MNDO method (MOPAC 6.0). The selected bond lengths, angles, and the calculated ^{13}C–^{13}C-coupling constants for phosphetane (**334**) and 1,2-dihydrophosphetane (**336**) are shown in Scheme 34 <97ZOR1703>.

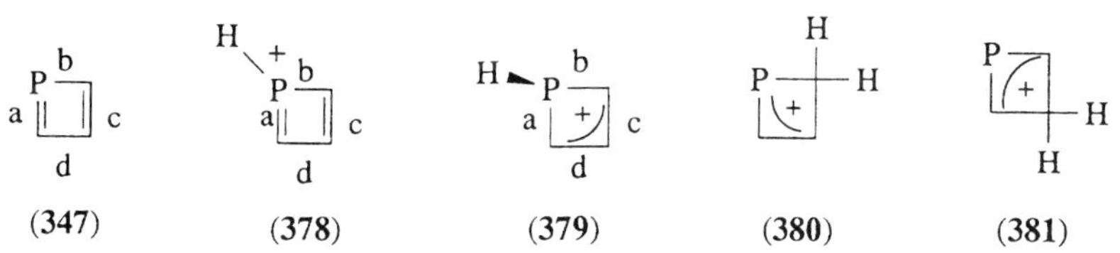

Scheme 33

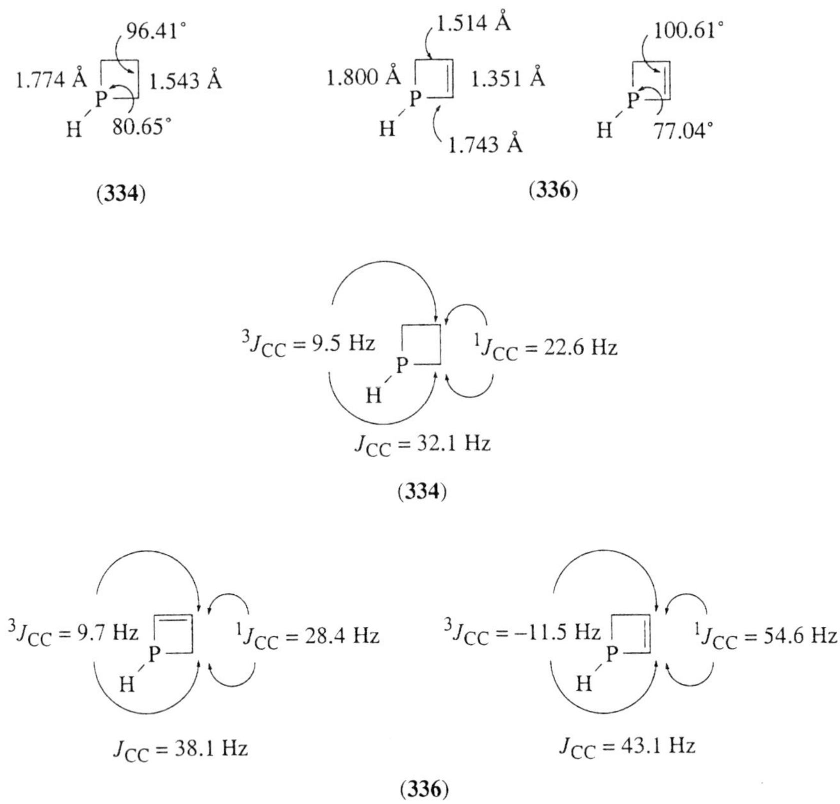

Scheme 34

Table 24. The melting points, μ_{eff}, and conductances

Compounds	*mp* (°C)	μ_{eff} (BM)	*Conductance* (Ω^{-1} cm^2 mol^{-1})	*Ref.*
(**385**)	98	2.80	–	⟨84PO1155⟩
(**386**)	125	3.73	–	⟨84PO1155⟩
(**329** · $2H_2O$)	230	5.87	4.20	⟨84PO1155⟩
(**387**)	290	5.68	2.8	⟨84PO1155⟩
(**382** · $9H_2O$)	>360	3.30	252	⟨84PO1155⟩
(**383**)	>360	1.84	–	⟨84PO1155⟩
(**384**)	>360	–	–	⟨84PO1155⟩

3.2.4 THERMODYNAMIC ASPECTS

3.2.4.1 Physical Properties and Physical Constants

2,2,3,4,4-Pentamethylphosphetanic acid (**1**) is a fairly strong acid, whose pK_a value was reported to be 2.85 ⟨83POL1225⟩. The melting points, μ_{eff}, and conductances of its transition metal salts (**382**)–(**387**) are summarized in Table 24, see Scheme 35. The conductance of compounds (**63**)–(**65**) ⟨84PO1155⟩ are reported to be ca. 100 Ω^{-1} cm^2 mol^{-1} ⟨92CB1987⟩.

Many optically active phosphetanes have been synthesized so far as shown in Schemes 7-1 and 7-2, and their optical rotation data are summarized in Table 25. It was found that the palladium complex of (**205a**) catalyzes asymmetric hydrosilylation of cyclopenadiene by trichlorosilane and asymmetric allylic substitution of 1,3-diphenyl-2-propenyl acetate by sodium dimethyl malonate in 72% ee and 91% ee, respectively ⟨98CCR(178-180)755⟩.

3.2.4.2 Aromaticity

There have been reported three η^4 complexes (**298**), (**314**), and (**315**) of phosphetes, but no discussion has been done with regard to aromaticity.

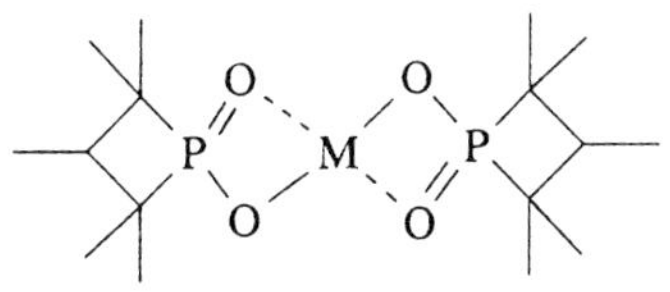

(382): M= Ni
(383): M= Cu
(384): M= Zn

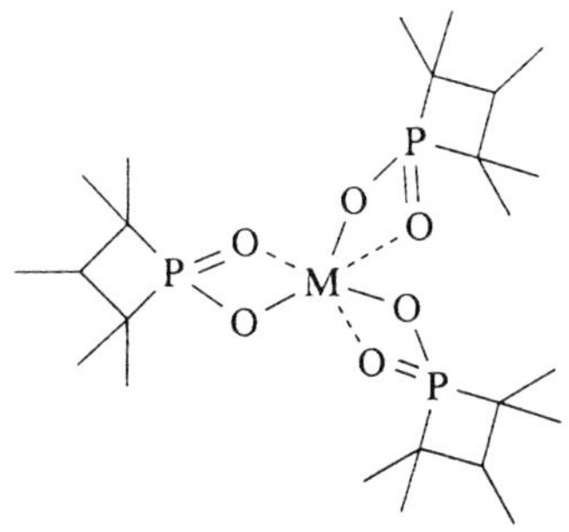

(385): M= V
(386): M=Cr
(387): M=Fe

Scheme 35

Table 25. Optical rotation of chiral phosphetanes

Compound	*[α]D (°)*	*c*	*Solvent*	*Ref.*
(**10**)	−76	2	$CHCl_3$	⟨78JCS(CC)125⟩
(**11**)	−9.4	–	–	⟨80JCS(P1)2136⟩
(**32b**)	−89	0.2	$CHCl_3$	⟨94OM3956⟩
(**186a**)	−218	1	$CHCl_3$	⟨94OM3956⟩
(**186b**)	+92	1	$CHCl_3$	⟨94OM3956⟩
(**204a**)	−189	0.8	C_6H_6	⟨94OM3956⟩
(**204b**)	−148	0.8	C_6H_6	⟨94OM3956⟩
(**204c**)	−53	0.8	C_6H_6	⟨94OM3956⟩
(**204d**)	−15	0.8	C_6H_6	⟨94OM3956⟩
(**206a**)	−81	1	$CHCl_3$	⟨94OM3956⟩
(**206b**)	−52	1	$CHCl_3$	⟨94OM3956⟩
(**206d**)	−7	1	$CHCl_3$	⟨94OM3956⟩
(**33**)	−194	1	$CHCl_3$	⟨95OM4983⟩
(**196**)	−141	1	$CHCl_3$	⟨95OM4983⟩
(**35**)	+44	0.2	$CHCl_3$	⟨96JOM(522)223⟩
(**171a**)	−180	1	C_6H_6	⟨96JOM(522)223⟩
(**175**)	−123	1	C_6H_6	⟨96JOM(522)223⟩
(**174**)	−142	1	C_6H_6	⟨96JOM(522)223⟩
(**19a**)	−47	1	$CHCl_3$	⟨96JOM(522)223⟩
(**184b**)	−33	1	$CHCl_3$	⟨96JOM(522)223⟩
(**188**)	−45	1	$CHCl_3$	⟨96JOM(522)223⟩
(**189**)	−40	1	$CHCl_3$	⟨96JOM(522)223⟩
(**190**)	−45	1	$CHCl_3$	⟨96JOM(522)223⟩
(**213**)	−50	1	$CHCl_3$	⟨96JOM(522)223⟩
(**215**)	−47	1	$CHCl_3$	⟨96JOM(522)223⟩
(**185a**)	−63	1	$CHCl_3$	⟨97JOM(529)465⟩
(**185b**)	−38	1	$CHCl_3$	⟨97JOM(529)465⟩
(**195a**)	−50	1	$CHCl_3$	⟨97JOM(529)465⟩
(**205a**)	−235	0.8	$CHCl_3$	⟨97JOM(529)465⟩
(**207a**)	−126	1	$CHCl_3$	⟨97JOM(529)465⟩
(**209a**)	−26	1	$CHCl_3$	⟨97JOM(529)465⟩
(**217a**)	−57	0.5	$CHCl_3$	⟨97JOC297⟩
(**218a**)	−28	0.5	$CHCl_3$	⟨97JOC297⟩
(**222**)	+103	1	$CHCl_3$	⟨97TL2947⟩
(**223**)	+123	1	$CHCl_3$	⟨97TL2947⟩
(**234**)	+354	1	$CHCl_3$	⟨97TL2947⟩
(**14a**)	−109	0.5	$CHCl_3$	⟨97T4363⟩
(**14b**)	−24	0.5	$CHCl_3$	⟨97T4363⟩
(**180**)	−25	1	CH_2Cl_2	⟨98S1539⟩
(**192a**)	−52	1	$CHCl_3$	⟨98S1539⟩
(**192b**)	−41	1	$CHCl_3$	⟨98S1539⟩
(**194a**)	−45	1	$CHCl_3$	⟨98S1539⟩
(**230**)	−288	0.5	CH_2Cl_2	⟨99CEJ1160⟩
(**234**)	−202	0.5	CH_2Cl_2	⟨99CEJ1160⟩

3.2.4.3 Conformational Studies and Tautomerization

The signals due to diastereotopic protons of the ethoxy groups of diethoxyphosphetane (**388a**) coalesce at 30°C due to the process involves placing the four-membered ring diequatorial as in (**388b**), whose barrier was estimated to be 15 kcal mol^{-1} ⟨71JA2066⟩. The ^{1}H NMR spectrum of 1 : 2-adduct (**389a**) of 2,2,4,4-tetramethylphosphetane with hexafluoroacetone is unchanged from −60 to 160°C. At room temperature and below, the ^{19}F NMR spectrum consists of two signals, but above 120°C these signals broaden and finally coalesce at 155°C. The data imply a pseudorotation, rapid on the NMR time scale down −60°C, between the equivalent structures (**389a**) and (**389b**). At higher temperatures equivalence of the CF_3 groups, but not of the CH_3 groups, is achieved via the high energy trigonal bipyramid (**389c**) having a four-membered ring spanning a diequatorial position. The free energy of activation for this process at 155°C is 20.1 kcal mol^{-1}. The 5 kcal mol^{-1} difference between these free energies is a measure of the difference in the preferences of the phenyl and the ethoxy-group for the apical position.

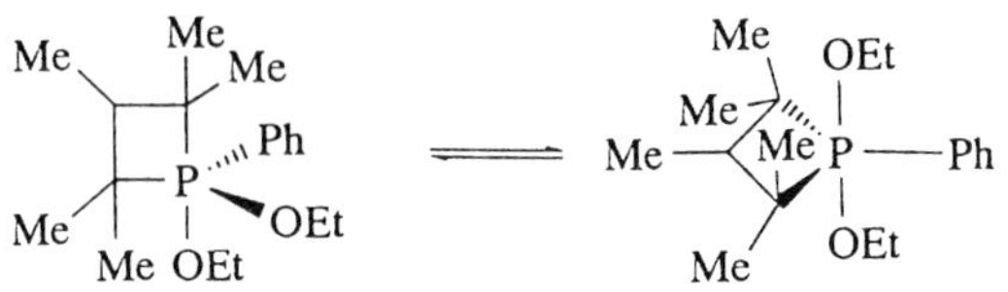

(**388a**) (**388b**)

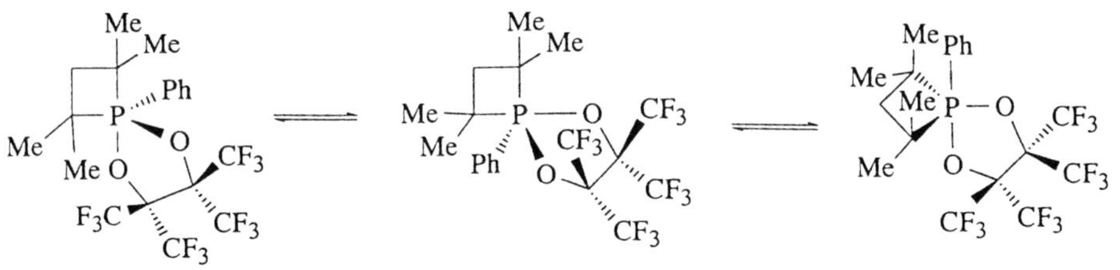

(**389a**) (**389b**) (**389c**)

Scheme 36

Tungstenpentacarbonyl 2-hydroxy-1,2-dihydrophosphetes (**390a,b**), which were prepared by reduction of the corresponding ketone (**266**) with $LiAlH_4$, undergo tautomerization to give open chain formyl-group-containing secondary phosphine complex (**391**). Although (**391**) was not obtained in a pure state, its structure has been strongly supported by the ^{1}H and ^{31}P NMR spectra (δ_P −24.8 ($^{1}J_{PH}$ = 356.4 Hz), δ 9.64 (–C*H*=O, $^{4}J_{HP}$ = 1.46 Hz)).

Water and triethylamine may help the closure of (**391**) back to (**390a,b**) while HCl seems to catalyze the opening of (**390a,b**) giving (**391**) and to simultaneously induce an equilibration between (**390a**) and (**390b**) (88OM633), see Scheme 37.

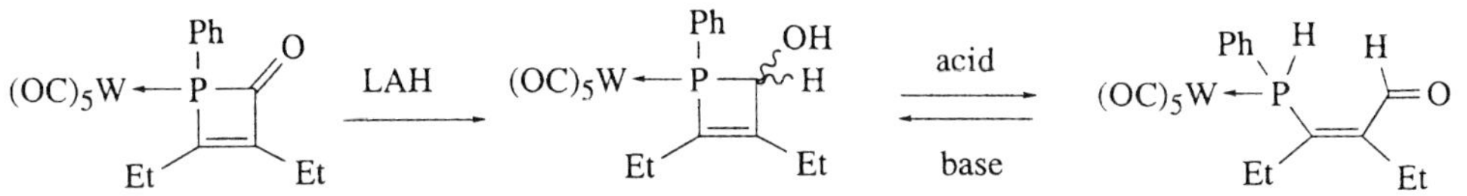

(**266**) (**390a,b**) (**391**)

Scheme 37

Phosphetanium salts (**46**) and (**250**) were prepared by intramolecular electrocyclic ring closure of methylenephosphonium salts (**392**) and (**393**), respectively. This ring closure is reversible, and the methylenephosphonium ion (**393**), which is present at equilibrium in low concentration, reacts with 2,3-dimethylbutadiene (DMB) to give a [2 + 4] cycloadduct (**395**) and an ene reaction product (**396**), though (**393**) could not be spectroscopically detected. On the other hand, (**46**) is stable and does not react with DMB. Upon addition of a tertiary amine base (preferably pyridine) to both (**46**) and (**250**), quantitative tautomerization to phosphetanes (**47**) and (**394**) takes place, but it is irreversible because isomers (**47**) and (**394**) are stabilized by aromatization, see Scheme 38 ⟨93AG(E)1359⟩.

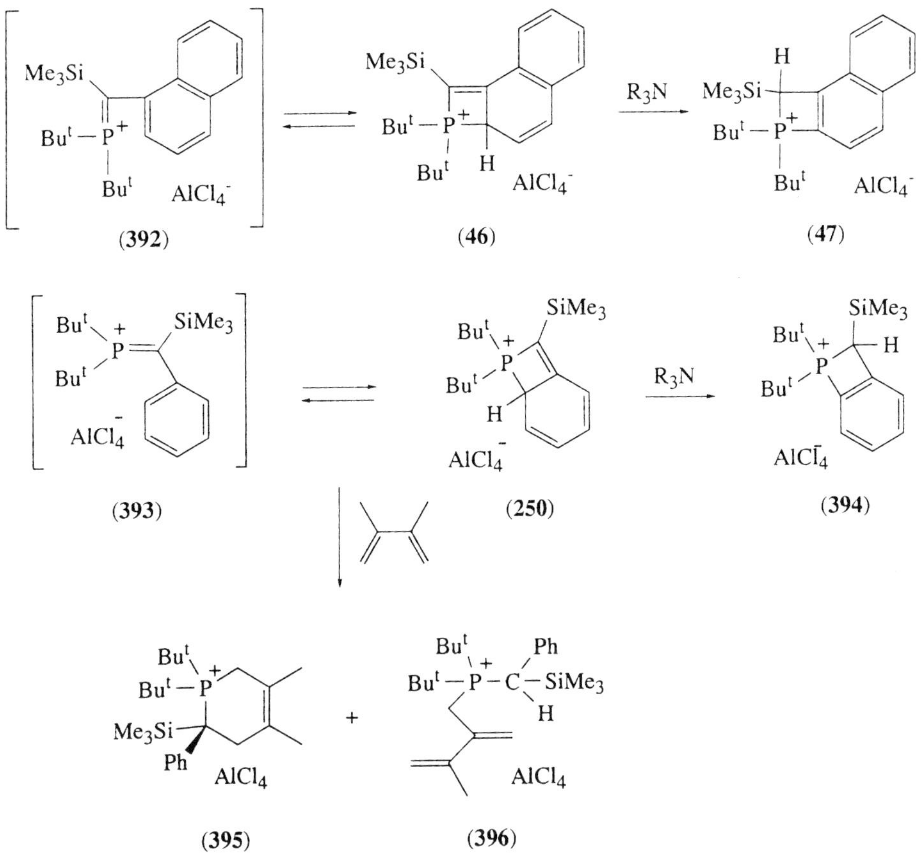

Scheme 38

3.2.5 REACTIVITY OF FULLY CONJUGATED RINGS

As described above only three metal complexes (**298**), (**314**), and (**315**) of phosphetes have been synthesized; *t*-butyl derivative (**314**) of the rhodium complexes is obtained as yellow crystals which melted at 95°C, while cobalt complex (**298**) is red oil. Although all complexes were characterized by ^{1}H, ^{13}C, and ^{31}P NMR spectroscopy and mass spectrometry, their reactivity remains still unknown ⟨87JOMC35, 91AG(E)310⟩.

3.2.6 REACTIVITY OF NONCONJUGATED RINGS

3.2.6.1 Thermal and Photochemical Reactions

Tungstenpentacarbonyl 1,2-dihydrophosphete (**41**) reacts with bis(triphenylphosphine) ethene platinum (0) (**398**) to give five-membered (**399**) and three-membered ring (**400**) complexes containing phosphorus and platinum, indicating that (**41**) undergoes ring opening to form 1-phospha-1,3-butadiene complex (**397**) which has both reactivity as an ene and a diene, see Scheme 39 ⟨91JOMC20⟩.

Dewar phosphabenzene (**244**) is heated at 160°C for 4 d to afford Dewar phosphabenzene (**240**) in 44% yield via phosphabenzene (**401**). On the other hand, Dewar phosphabenzene (**240**) is irradiated ($\lambda > 280$ nm) in $CDCl_3$ to give phosphabenzene (**401**). Irradiation of Dewar phosphabenzene (**244**) for 4 h at −40°C with Pyrex filter results in relatively fast intramolecular [2 + 2] cycloaddition to form, as the main product, the phosphaprismane (**146**) (40%); in addition, phosphabenzene (**401**) formed via phosphabenzbalene (**402**) and phosphabenzvalene (**403**) (each 10%) are formed. The phosphabenzvalene (**403**) is treated with sulfur at 110–120°C for 12 h to afford Dewar phosphabenzene *P*-sulfide (**59**) (65%), see Scheme 40 ⟨87AG(E)85, 99PJC135⟩.

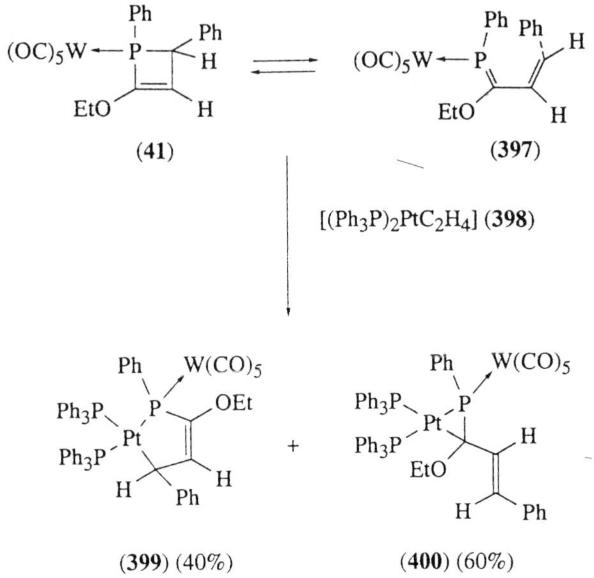

Scheme 39

3.2.6.2 Cycloaddition and Retro-Cycloaddition Reactions

1,3,4-Triphenyl-1,2-dihydrophosphete (**43**) reacts with dimethyl maleate (C_6D_6, 110°C for 14 h) to produce a mixture (3:2) of diastereomeric cycloadducts (**406**) and (**407**), indicating that (**43**) undergoes retro-cycloaddition to give 1-phospha-1,3-butadiene (**405**) under these conditions. Both isomers contain trans ester groups, suggesting a non-concerted pathway via twitter ionic intermediate (**408a**) or (**408b**) for the cycloaddition as shown in Scheme 41. This mechanism is supported by the fact that formation of cycloadducts is faster in $(CD_3)_2CO$ than in C_6D_6.

The same ratio of the diastereomers is obtained from dimethyl fumarate; dimethyl maleate is, however, isomerized to dimethyl fumarate under the reaction conditions. Although concerted mechanism involving intermediate 1-phospha-1,3-butadiene (**405**) cannot be ruled out, the results that the reaction with ethyl acrylate gives a single regioisomer and that no reaction occurred with benzaldehyde suggest that a stepwise mechanism is more plausible ⟨90JCS(CC)1649⟩.

Tungstenpentacarbonyl 1,2-dihydrophosphete (**41**) is also in equilibrium with a retro-cycloaddition product, tungstenpentacarbonyl 1-phospha-1,3-butadiene (**409**), which undergoes Diels–Alder reaction with dienophiles such as maleimide, acetylenedicarboxylate, and benzaldehyde to afford the corresponding cycloadducts (**410**), (**411**), and (**412**), respectively, see Scheme 42 ⟨88TL3077⟩.

Similar reactions of molybdenum pentacarbonyl complex (**269**) of a 1,2-dihydro-2-ethoxyphosphete with *N*-phenylmaleimide and activated acetylenes such as dimethyl acetylenedicarboxylate, ethyl phenylpropiolate, phenylacetylene, and 2-pyridylacetylene give the corresponding [4 + 2] cycloadducts in 30–60% yields. The reaction of 2-phenyl derivative (**270**) with *p*-quinone affords the corresponding [4 + 2] adduct, which seems to be formed by hetero-Diels–Alder reaction with the carbonyl group of the quinone, while the reaction of (**269**) with 1,4-naphthoquinone gives a mixture of products as a result of cycloadditions at either the carbon–carbon double bond or the carbonyl function ⟨95BSF384⟩.

Pt phosphete complex (**48**) in dry benzene undergoes retro-cycloaddition (room temperature for 6 h) to give phosphene Pt complex (**413**), which affords quantitatively 2,4,6-tri-*tert*-butylphenylphosphaalkyne (**414**) via elimination of *trans*-$Cl_2Pt(PEt_3)_2$ as shown in Scheme 43 ⟨94OM2444⟩.

As shown in Scheme 38, the silylated phosphetanium salt (**250**) undergoes retro-cycloaddition to give methylenephosphonium salt (**393**) in low concentration, which reacts with 2,3-dimethylbutadiene (DMB) to afford the [2 + 4] cycloadducts (**395**) and ene reaction product (**396**) ⟨93AG(E)1359⟩.

Dewar phosphabenzenes (**150**) and (**151**) undergo [2 + 2]-retro-cycloaddition to give phosphaalkenes (**415**) and (**416**), respectively. This reaction proceeds in $CHCl_3$ at room temperature (1–2 h), but that in benzene more slowly (>24 h) (Scheme 44) ⟨94T759⟩.

Flash vacuum pyrolysis of dichloroneopentylphosphine over Mg gives the corresponding phosphinidene (**417**), which undergoes intramolecular C–H insertion to afford phosphetane (**418**). The formation of (**418**) was estimated by isolation of isobutene and polymer (**419**) of [CH_2=PH], which are considered as retro-cycloaddition products of (**418**) (Scheme 45) ⟨97TL8417⟩.

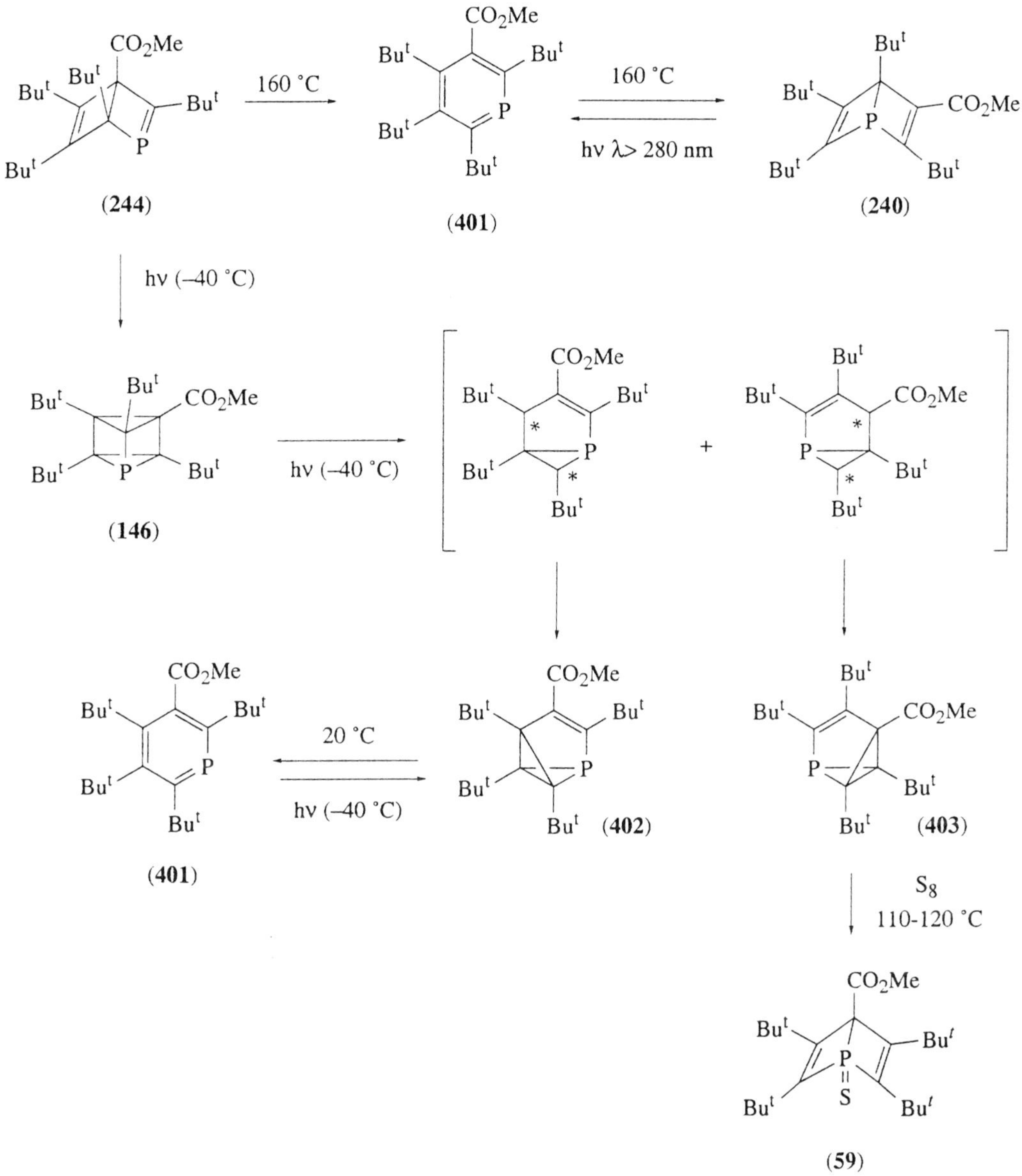

Scheme 40

3.2.6.3 Reactions Involving Rearrangements

1,3,4-Triphenyl-1,2-dihydro-2-oxophosphete tungstenpentacarbonyl (**39**) is reduced with $NaBH_4$ in dioxane at room temperature to give a mixture of two oxaphospholene complexes (**420**) and (**421**) in a ratio of 3 : 1 (55% yield) instead of the expected 2-hydroxy derivative. The oxaphospholenes (**420**) and (**421**) probably result from the spontaneous rearrangement of the oxyanion (**422**) to afford the allylic carbanion (**423**) after initial attack of H^- on the carbonyl group of (**39**) as shown in Scheme 46 ⟨88OM633⟩.

1-Alkynyl-1,2-dihydrophosphetes (**273**), (**274**), and (**275**) are heated at 45°C in benzene to give the corresponding ring enlargement intermediates, which undergo [1,5] hydrogen shift to afford phosphabenzenes (**424**), (**425**), and (**426**), respectively (Scheme 47) ⟨96HAC397⟩.

3.2.7 REACTIVITY OF SUBSTITUENTS ATTACHED TO RING CARBON ATOM

When bicyclophosphetane oxide (**427**) having a C=C double bond on the bridge has reacted with mCPBA, both the epoxidation of the double bond and the Baeyer–Villiger reaction take place to give (**428**)

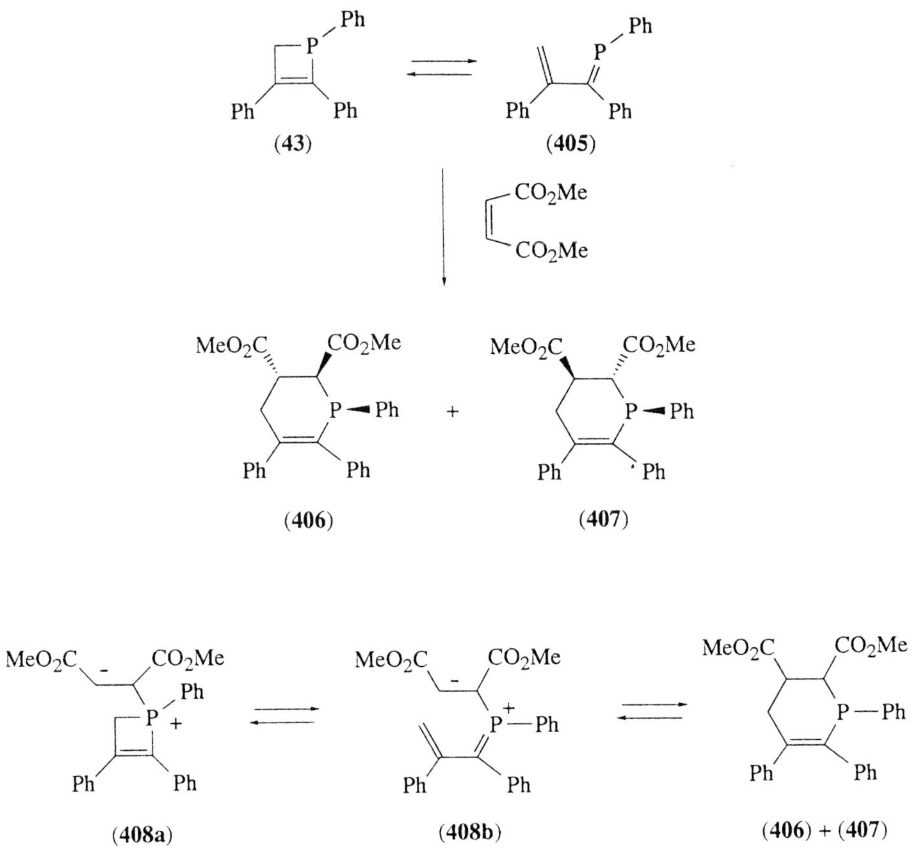

Scheme 41

in 52% yield, as shown in Scheme 48. Probably the former reaction giving (**429**) seems to be faster than the latter reaction ⟨83JOC4466⟩.

The ligand (H_2O) of 2-alkylidene-1,2-dihydrophosphete platinum complex (**330**) can be replaced by pyridine in CH_2Cl_2 to afford (**332**) in 66% yield (Scheme 22) ⟨89OM1007⟩. 1-Mentyl-3,3,4,4-tetramethyl-1,2-dihydrophosphete 1-oxide (**13**) can be converted to the corresponding 2-lithio derivative with *n*-BuLi, which is treated with several electrophiles such as alkyl halides, allyl halides, halosilanes and DMF, etc., to give the corresponding 2-substituted phosphetanes (**183**)–(**191**) (Scheme 49) ⟨94TL5861, 94OM3956, 98CCR(178-180)755⟩.

3.2.8 REACTIVITY OF RING PHOSPHORUS AND SUBSTITUENTS ATTACHED TO RING PHOSPHORUS ATOM

1,3,4-Triphenyl-1,2-dihydrophosphete (**43**) reacts slowly with oxygen at room temperature or rapidly at 150°C to give the corresponding 1-oxide (**252**) ⟨91IC3381⟩.

Oxidative decomplexation of 1,2-dihydrophosphete tungstenpentacarbonyl complex (**430**) with pyridinium perbromide in the presence of 2,2′-bipyridyl gave the corresponding 1-oxide (**254**), see Scheme 50. Similarly, 1,2-dihydrophosphete oxides (**252**) and (**253**) are obtained from (**44**) and (**41**), respectively ⟨90JCS(CC)153, 85JA5001, 88OM1791⟩.

1-Menthylphosphetane 1-oxides (**13**), (**183**), (**184**), and (**93**) and tetracyclic phosphetane oxide (**134**) can be reduced with trichlorosilane to afford the corresponding phosphines (**168**), (**169**), (**170**), (**203**), and (**122**), respectively ⟨84JA7021, 93T10291⟩. Many other phosphetane oxides also can be reduced to give the corresponding phosphines ⟨94OM3956, 95OM4983, 96JOM(522)223, 97JOM(529)465, 98CCR(178-180)755⟩.

Phosphetane boron complexes can be treated with amines such as 1,4-diazabicyclo[2.2.2]octane to give the corresponding phosphines ⟨97TL2947⟩.

Hydrolysis of ethyl phosphetanate (**156**) and methylation of the phosphetanic acid (**158**) with diazomethane have been reported ⟨93JCR(S)268⟩.

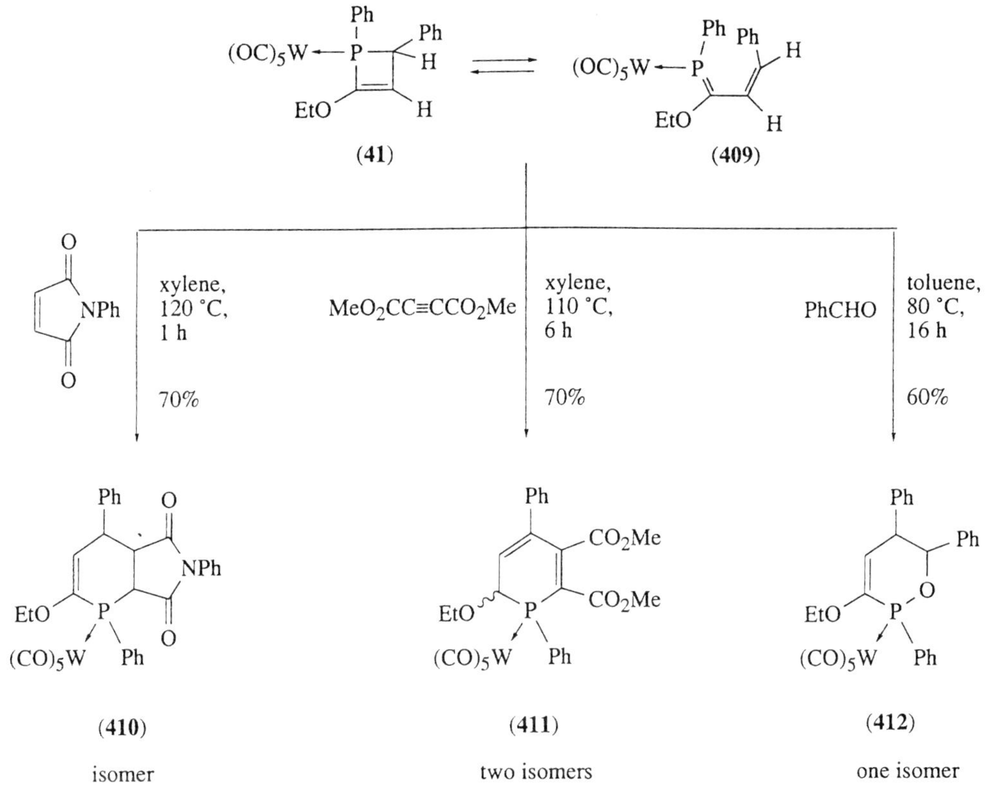

Scheme 42

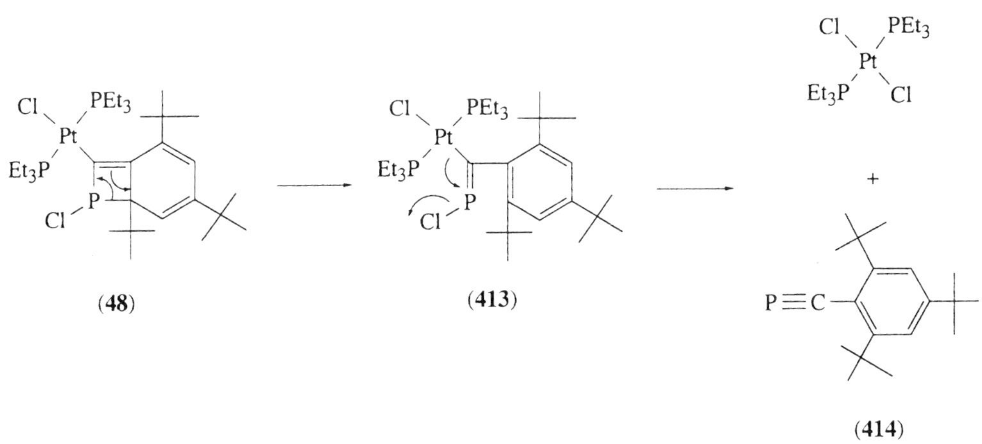

Scheme 43

Phosphetanium salt (**120**) has interesting reactivity in the reaction with nucleophiles. For example, α-proton can be removed with a base such as *n*-BuLi to afford ylide (**303**), which undergoes the Wittig reaction with aldehydes and ketones to give the corresponding 4-butenylphosphine oxide (**432**). Nucleophiles can attack the phosphorus atom of the phosphonium salts to produce sometimes penta-coordinate phosphetanes (**86**) or (**87**), but most of them attack the carbon attached to the phosphorus atom to form functionalized phosphine (**433**) accompanied with the C–P bond cleavage, see Scheme 51 ⟨92CB1987⟩. Phosphide anion can be used as a nucleophile to give the corresponding diphosphine ⟨94CB481, 97JOC2362⟩.

The hydrolysis of 2,3,6-tris((2,4,6-tri-*t*-butylphenyl)imino)-2,4,6-triphosphabicyclo[3.2.0]heptane (**51**) gives phosphinic amide (**434**), which is formed by addition of water to the P=N double bond of a four-membered ring, followed by cleavage of the C–P bond (Scheme 52) ⟨94CB313⟩.

The reactions of 1,2-dihydrophosphetes (**43**), (**271**), and (**272**) with $RPCl_2$ (R = Ph, Me, Cl) give a

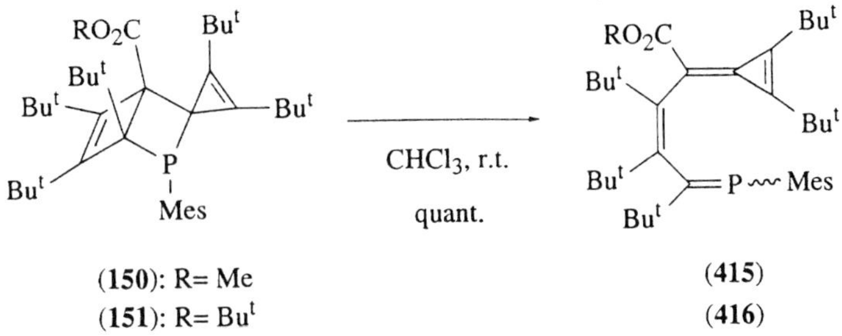

Scheme 44

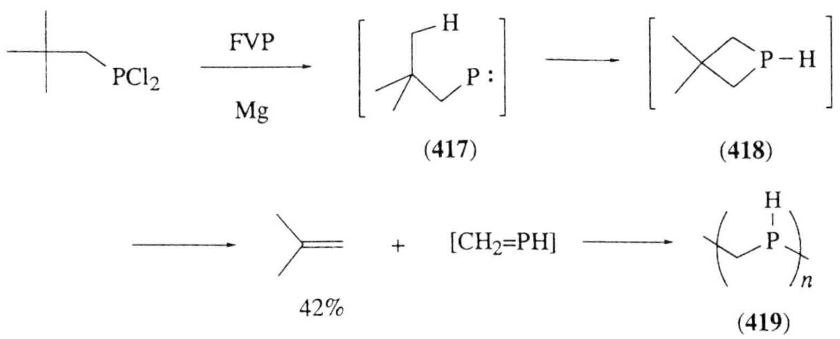

Scheme 45

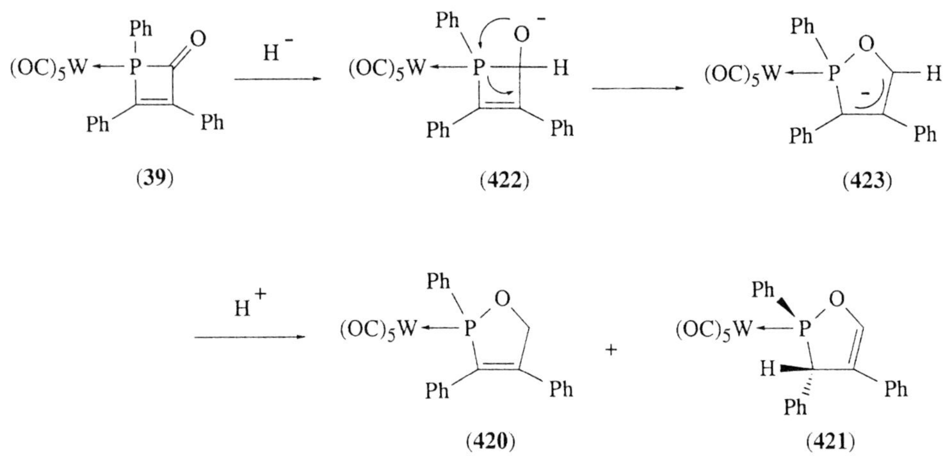

Scheme 46

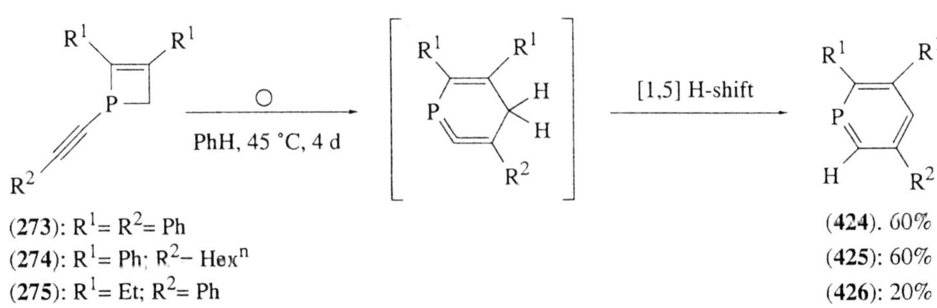

Scheme 47

mixture of 2,3-dihydro-1,2-diphospholes (**435**) and their regio-isomers (**436**). The possible mechanism is shown in Scheme 53 ⟨95AG(E)590⟩.

Treatment of 4-phosphonio-2-phosphoranylidene-1,2-dihydrophosphete (**310**) with dichlorophosphine

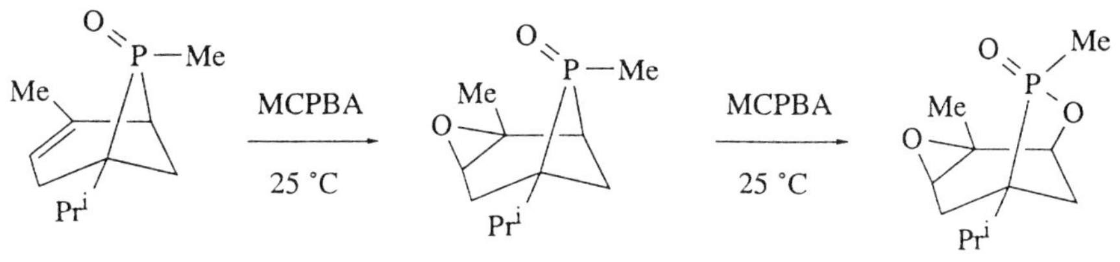

(427) (428) (429)

Scheme 48

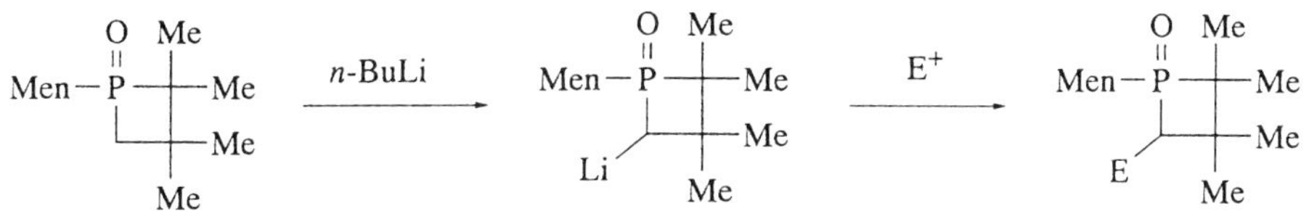

(13) (183)-(191)

Scheme 49

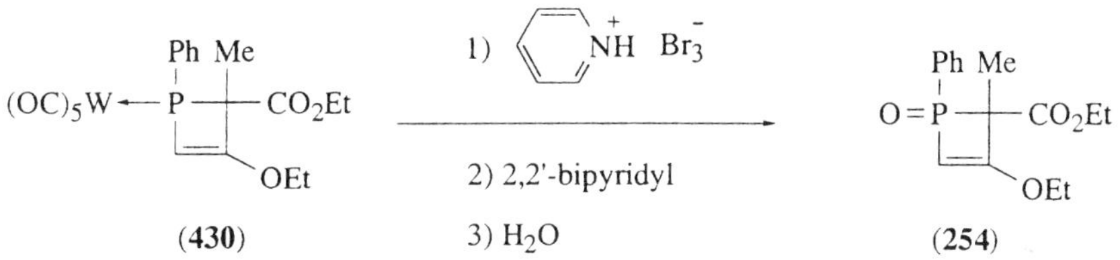

(430) (254)

Scheme 50

in the presence of triphenylphosphine gave 2,3-dihydro-1,2-diphosphole (**437**) and the reaction was induced by nucleophilic attack of triphenylphosphine on the ring and C–P bond cleavage and P–P bond formation followed. Its mechanism is also shown in Scheme 54 ⟨96CEJ221⟩.

3.2.9 SYNTHESIS OF 4-MEMBERED HETEROCYCLES WITH ONE PHOSPHORUS ATOM

3.2.9.1 From Acyclic Compounds by Formation of One Bond to a P Atom

Silylphosphines (**438**) bearing bulky groups such as isopropyl, *t*-butyl, and mesityl groups react with 1,3-dihalopropane to give intermediary 3-halopropylphosphines (**439**), which undergo intramolecular cyclization to afford the corresponding phosphetanium salts (**119**) (23%), (**120**) (76%), and (**121**) (72%), see Scheme 55 ⟨92CB1987⟩.

The phosphenium cations (**440**) and (**441**) which are formed from chlorophosphorus ylides (**442**) and (**443**) by dehalogenation with aluminum chloride, undergo intramolecular cyclization to the benzene ring attached to the ylide carbon to give (**46**) and (**250**) in 80–85% yield, respectively, see Scheme 56.

3.2.9.2 Formation of Two C–P Bonds by Displacement Reactions

As shown in Scheme 57, phenylphosphine iron complex (**444**) reacts with 1,3-dibromopropane in the presence of DBU to afford 1-phenylphosphetane complex (**29**) ⟨92JCS(D)1751⟩. *t*-BuOK can be used as a base instead of DBU ⟨95OM1434⟩.

1,3-Dibromopropane and 1,3-ditosilate derivatives are treated with ammonium hypophosphite in the presence of hexamethyldisilazane (HMDS) to give the corresponding 1-hydroxyphosphetanes (**158**), (**159**), (**160**) and (**161**). Bis(trimethylsilyloxy)phosphine was proposed as active species of this reaction (Scheme 58) ⟨95JOC6076⟩.

Dilithio phosphide (**445**) also reacts doubly with cyclic sulfates to give *o*-(bis(phosphetanyl)benzenes (**229**), (**230**), (**231**), and (**232**), respectively (Scheme 59) ⟨99CEJ1160⟩. A similar reaction of dilithium phenylphosphide with 1,3-dichloropropane gives 1-phenylphosphetane as 80% pure compound after vac-

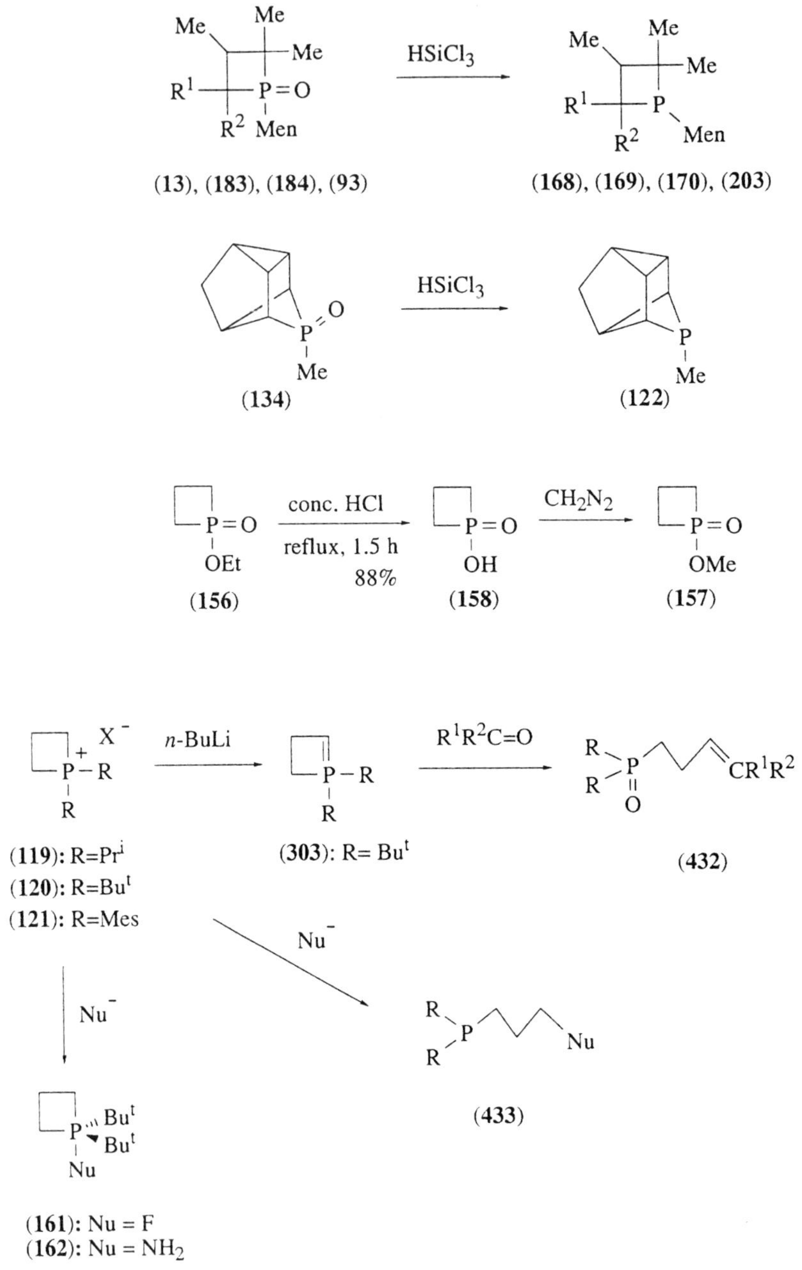

Scheme 51

uum distillation (bp 63°C/0.05 Torr), which was used to synthesize tris(1-phenylphosphetane)moribudenum tricarbonyl complex (**31**) ⟨94JCS(CC)475⟩.

Propane 1,3-bismagnesium bromide reacts with ethyl dichlorophosphate to yield ethyl ester of phosphetanic acid (**156**). The reaction of titanocene metallacyclobutenes (**446**) and (**447**) with dihalophosphine provides another synthetic route to 1,2-dihydrophosphete (**43**) and phosphetane (**67**), respectively, where titanocene is replaced by the phosphine moiety, see Scheme 60 ⟨89JA9129, 90AG(E)75, 92OM2944, 96HAC397, 96OM1597, 98JOM(567)151⟩.

Ylide carbanion of phosphonioylide (**448**) can attack the phosphorus of $RPCl_2$ to form the corresponding diphosphonium salt. Deprotonation with trietylamine followed by intramolecular nucleophilic attack of carbanion on the trivalent phosphorus gives compounds (**310**), (**311**), and (**312**), respectively (Scheme 61) ⟨96CEJ221⟩.

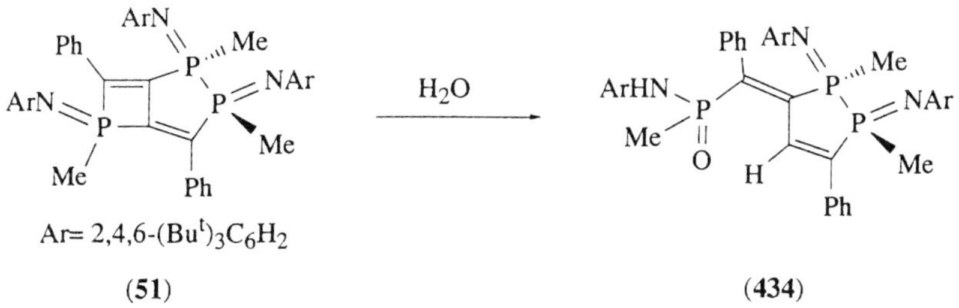

Scheme 52

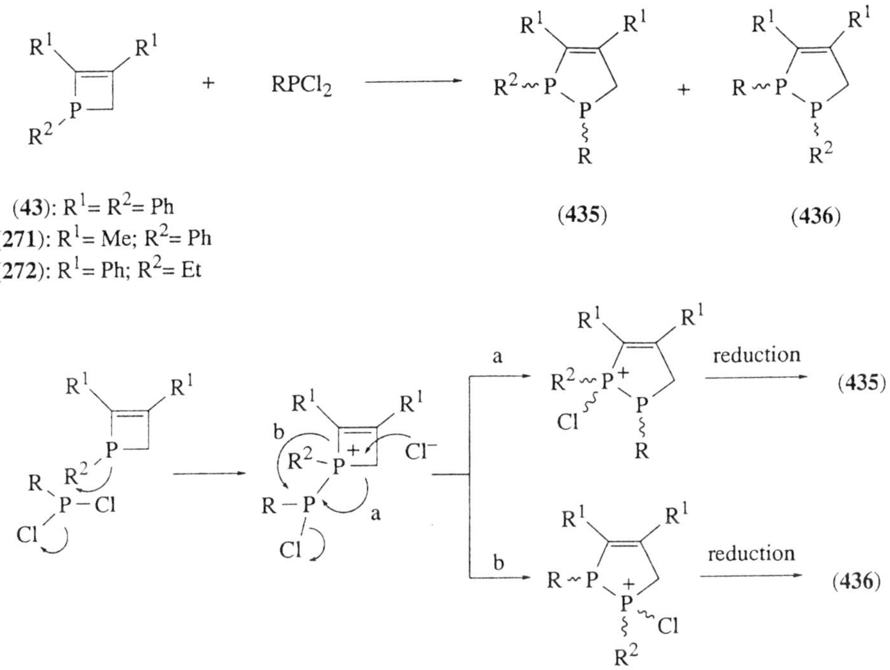

Scheme 53

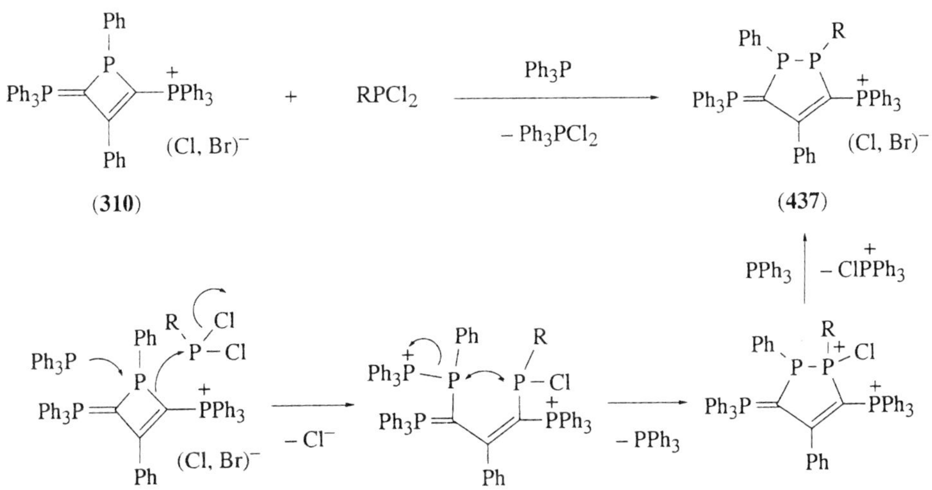

Scheme 54

3.2.9.3 Ring Formation by Cycloaddition Reactions

As shown in Scheme 40, phospha-Dewar benzene (**240**) is obtained via phosphabenzene (**401**) from phospha-Dewar benzene (**244**). Irradiation of (**244**) affords phosphaprismane (**146**), see Scheme

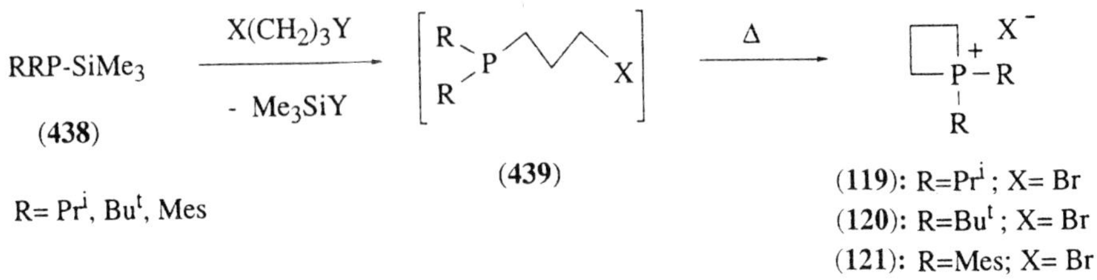

Scheme 55

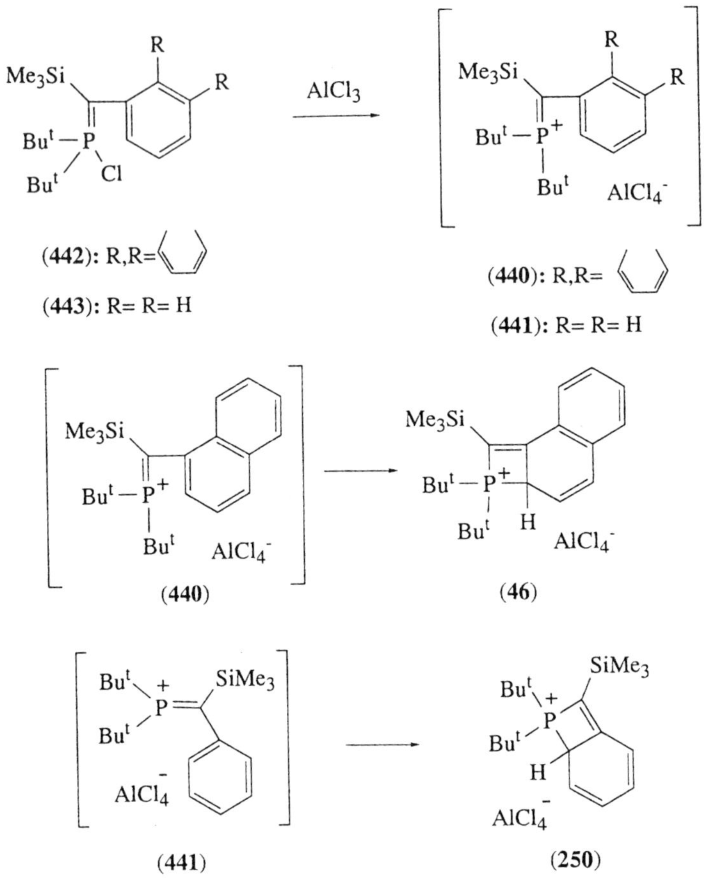

Scheme 56

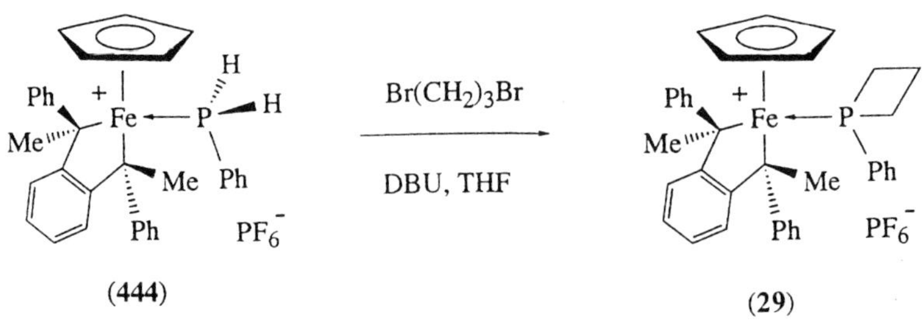

Scheme 57

40 ⟨87AG(E)85⟩. 1-(Bistrimethylsilylamino)-2,4-bis(trimethylsilyl)-1-phospha-1,3-butadiene (**449**) reacted with trimethylsilyl azide to give the corresponding iminodihydrophosphete (**65**) after the formation of the corresponding iminophosphorane (**450**) and its cyclization. 1-Thioxophosphabutadiene (1-sulfide of (**449**)) also readily cyclizes to give 1,2-dihydrophosphete 1-sulfide (**66**), see Scheme 62 ⟨87TL6121⟩.

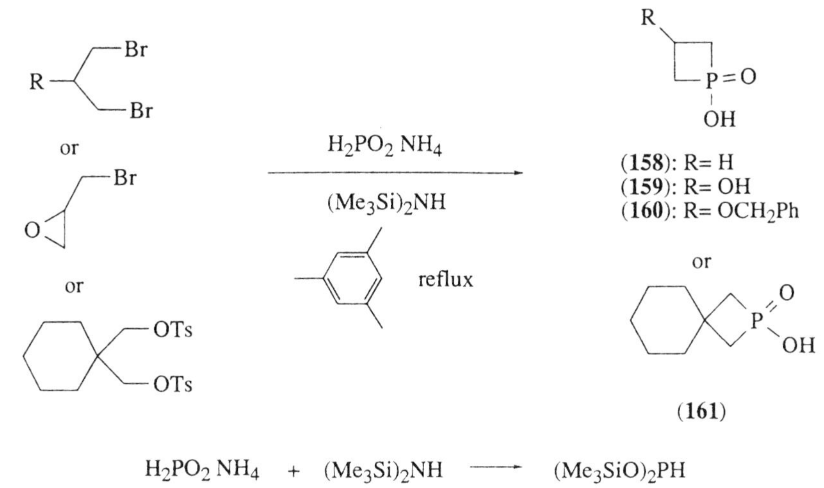

Scheme 58

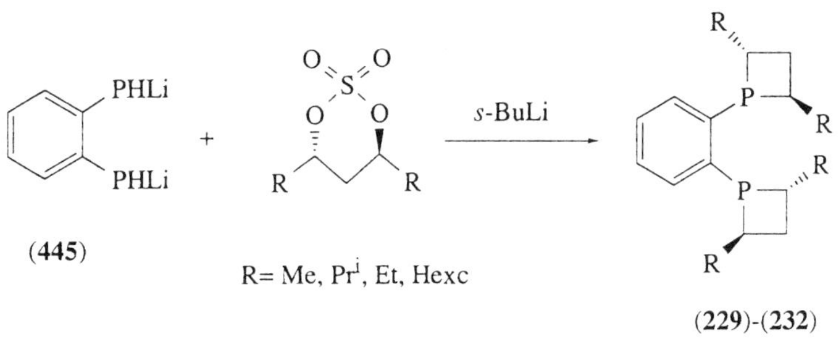

Scheme 59

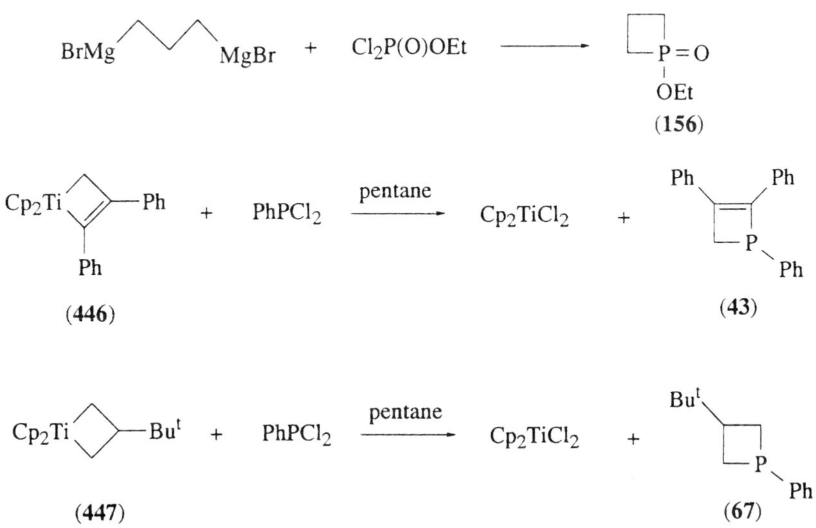

Scheme 60

Similarly, 1-amino-1,2-dihydrophosphete (**241**) is synthesized from the corresponding 1-amino-1-phospha-1,3-butadiene ⟨88OM572⟩.

Metal complexes of 1-phospha-1,3-butadienes give the corresponding metal complexes of 1,2-dihydrophosphetes. The reaction of phosphinidene tungstenpentacarbonyl complex (**451**) formed by retro-cycloaddition of 7-phosphanorbornadiene complex (**367**) with carbene chromium pentacarbonyl complex (**452**) was used as a unique generation method of phosphabutadiene tungsten pentacarbonyl complex (**453**), which cyclizes to give the corresponding 1,2-dihydrophosphete metal complex (**41**) (Scheme 63) ⟨90SL171, 88OM1791⟩. Iron phosphide complex (**52**) is synthesized by a similar electrocyclic reaction followed by

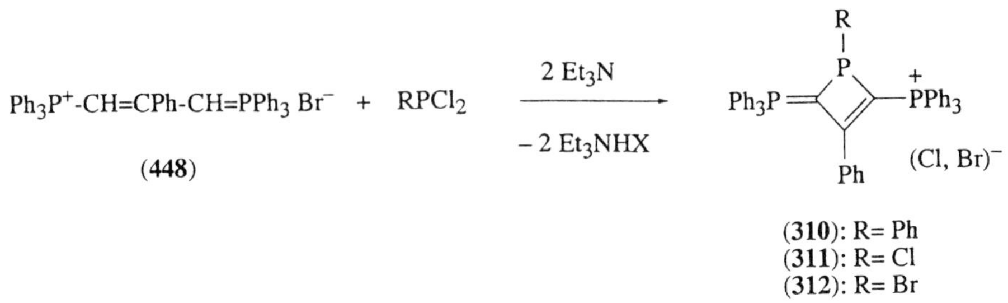

Scheme 61

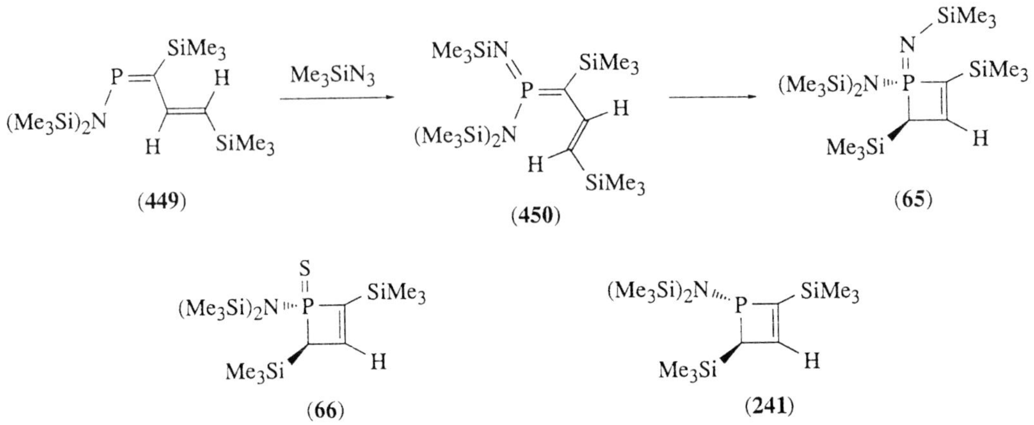

Scheme 62

elimination of dimethylamine ⟨94ZN(B)1693⟩. 2-Phospha-1,3-butadiene metal complexes undergo a similar electrocyclic reaction to give 3,4-dihydrophosphete metal complexes (**291**) and (**292**) ⟨97AG(E)1095⟩. Electrocyclic reaction of *in situ* generated 2-phospha-1,3-butadienes followed by migration of a chlorine from C to P gives 1-chloro-1,2-dihydrophosphetes (**278**)–(**286**) ⟨98ZOB1449⟩.

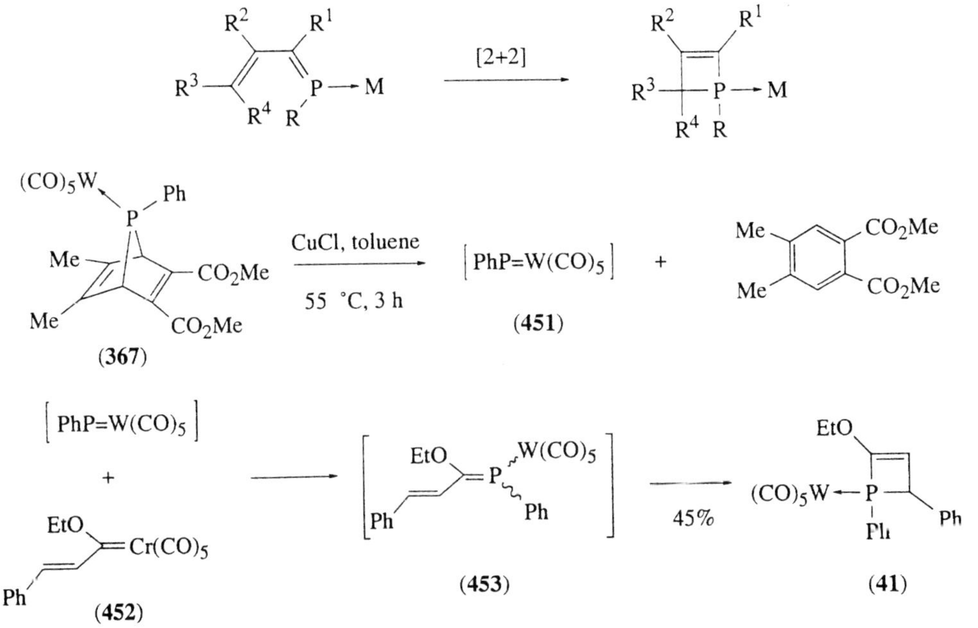

Scheme 63

Dewar benzenes (**242**)–(**249**) are synthesized by Diels–Alder reaction of phosphaalkyne with cyclobutadienes with high regioselectivity, see Scheme 64 ⟨86AG(E)280, 98S1305, 99PJC135⟩. Dewar phosphabenzenes (**150**) and (**151**) with a cyclopropene ring are similarly synthesized ⟨94T759⟩.

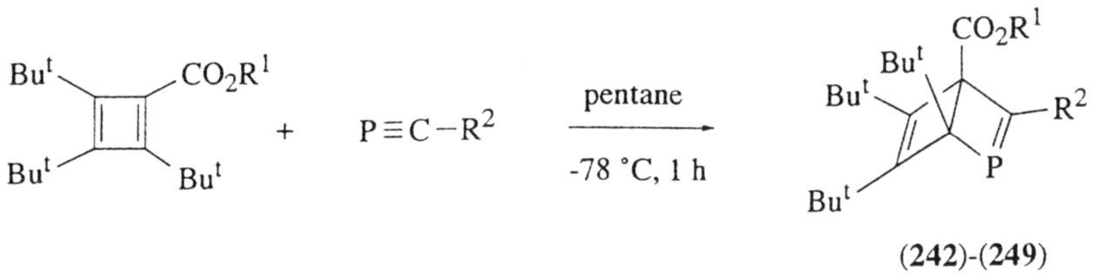

Scheme 64

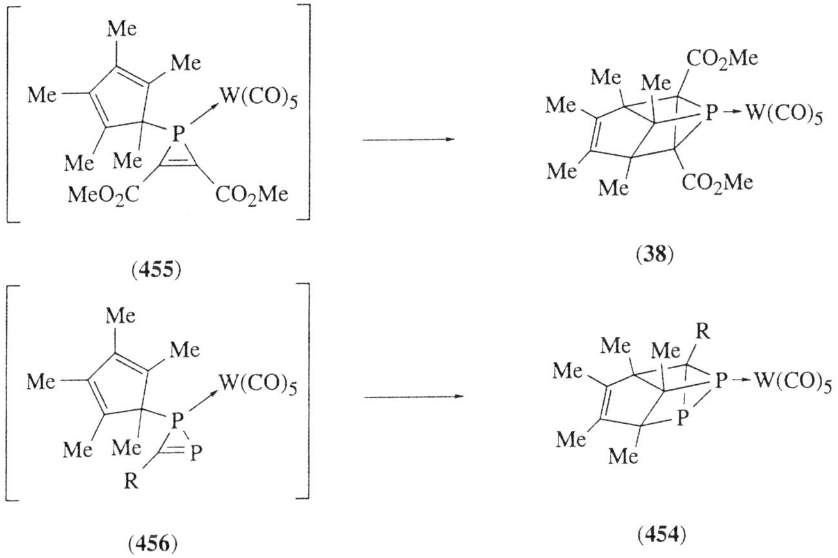

Scheme 65

Tungsten pentacarbonyl tetracyclic phosphine complex (**38**) and diphosphine complex (**454**) are obtained by intramolecular [4 + 2] cycloaddition of *in situ* generated 1-(pentamethylcyclopentadienyl)-phosphirene complex (**455**) and -diphosphirene complex (**456**), respectively (Scheme 65) ⟨99AG(E)215, 97PS(124-125)545⟩.

As shown in Scheme 66 treatment of norbornadiene with dichloromethylphosphine at 60–80°C for 1 week gives a mixture of tetracyclic phosphetanium chlorides in 60–70%. They are readily converted to their tetrachloroaluminates (**132**) and (**133**) which then lead to the corresponding phosphine oxides (**134**) and (**135**) after hydrolysis. These oxides were readily reduced with silanes to give the corresponding trivalent phosphines (**122**) and (**123**), which can be converted into several functionalized derivatives (**124**)–(**131**) and (**136**)–(**145**) ⟨92JOC511⟩.

The [1 + 2 + 2] cycloaddition of divalent cation species $RR'P^+$ with norbornadiene and 1,4-cyclohexadiene gave tetracyclic phosphetanium salts (**142**) and (**19**), respectively, the former of which can be converted to the oxides (**143**), (**144**), and (**145**) ⟨87TL603, 84TL815⟩. The phosphinidene dinuclear iron complex similarly reacts with norbornadiene to afford nortricyclan iron complex (**30**) ⟨93IC5720⟩.

3.2.9.4 Cycloadditions between Two Fragments Containing C=C and C=P

P,N-Di-*t*-butylcarboimidophosphene (**457**) reacts with diphenylketene to afford the corresponding 2-oxo-4-iminophosphetane (**458**) ⟨83PS(18)39⟩.

Phosphabicyclo[2.2.2]octene (**459**) undergoes two retro-Diels–Alder reactions to give a mixture of ethylene and phosphacyclohexadiene (**460**) and a mixture of cyclohexadiene and phosphaalkene (**461**), respectively. The [2 + 2] cycloaddition between phosphaalkene (**461**) and ethylene gives phosphacyclobutane (**308**) ⟨85ZN(B)927⟩. Iron phophetanide complex (**166**) is synthesized by similar [2 + 2] cycloaddition ⟨97OM2958⟩.

The [2 + 2] cycloaddition reaction of bis(trimethylsilyl)amino(trimethylsilylimino)(trimethylsilylmethylene)phosphorane (**462**) with trimethylsilylacetylene affords the regioisomer (**251**) of (**65**) ⟨87TL6121⟩.

Phosphete complexes (**314**) and (**315**) are synthesized by the [2 + 2] cycloaddition of rhodium acetylene complex (**463**) with phosphaalkynes, see Scheme 67 ⟨91AG(E)310⟩.

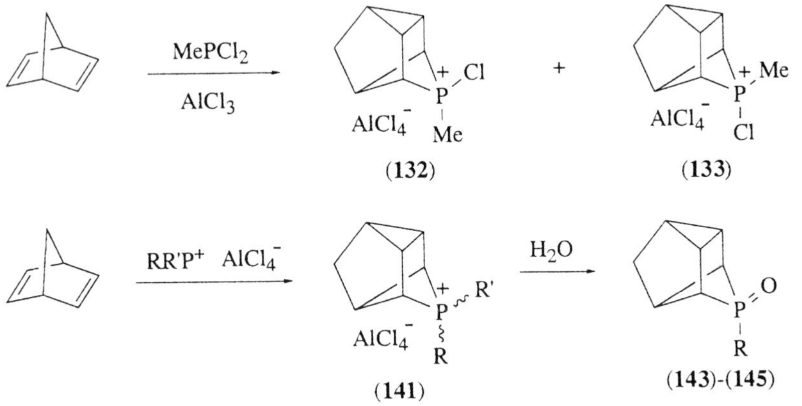

Scheme 66

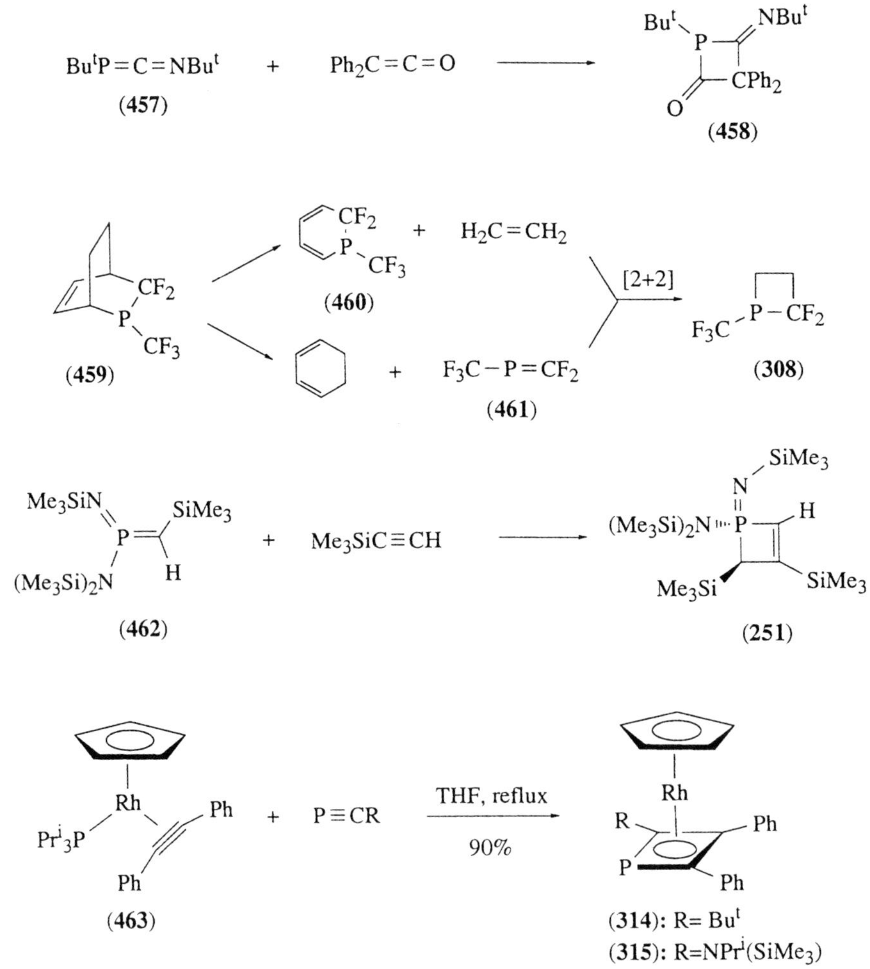

Scheme 67

3.2.9.5 Miscellaneous Routes

Reaction of divalent phosphorus cation species $RR'P^+$ with substituted olefins gives the corresponding phosphetanium salts through rearrangement of the first formed carbocations. This method is also available for 1,5-cyclooctadiene, see Scheme 1 ⟨87S265, 93T10291⟩.

Divalent phosphorus cation species $RR'P^+$ react with cyclopropanes to give insertion products to the

C–C bond between the least and most substituted carbons, which seems to be due to readier electrophilic attack of the phosphorus cation and higher stabilization of the resulting carbocation ⟨86JCS(CC)1081⟩.

Chromium pentacarbonyl carbene complex (**464**) reacts with phosphaalkyne to afford 1,2-dihydrophosphete complex (**258**) with ketene moiety ⟨89JCS(CC)788⟩.

Reaction of di or trialkynylphosphine oxides (**465**) with platinum hydride (**466**) gives either a *Pt,P-μ*-alkenylidene product (**467**) or a 2-alkylidene-1,2-dihydrophosphete 1-oxide product (**468**). The latter type product is favored when a phosphonium salt is used as the substrate instead of a phosphine oxide. Compounds (**330**)–(**333**) can be synthesized by the same method. The mechanism is discussed in detail ⟨91OM1197, 89OM1007⟩, see Scheme 68. Zirconium complexes (**58**) and (**289**) are obtained by sequential reactions involving carbometallation from bis(phenylethynyl)phosphine (**469**) and benzyne zirconium complexes (**470**) ⟨97AG(E)987⟩.

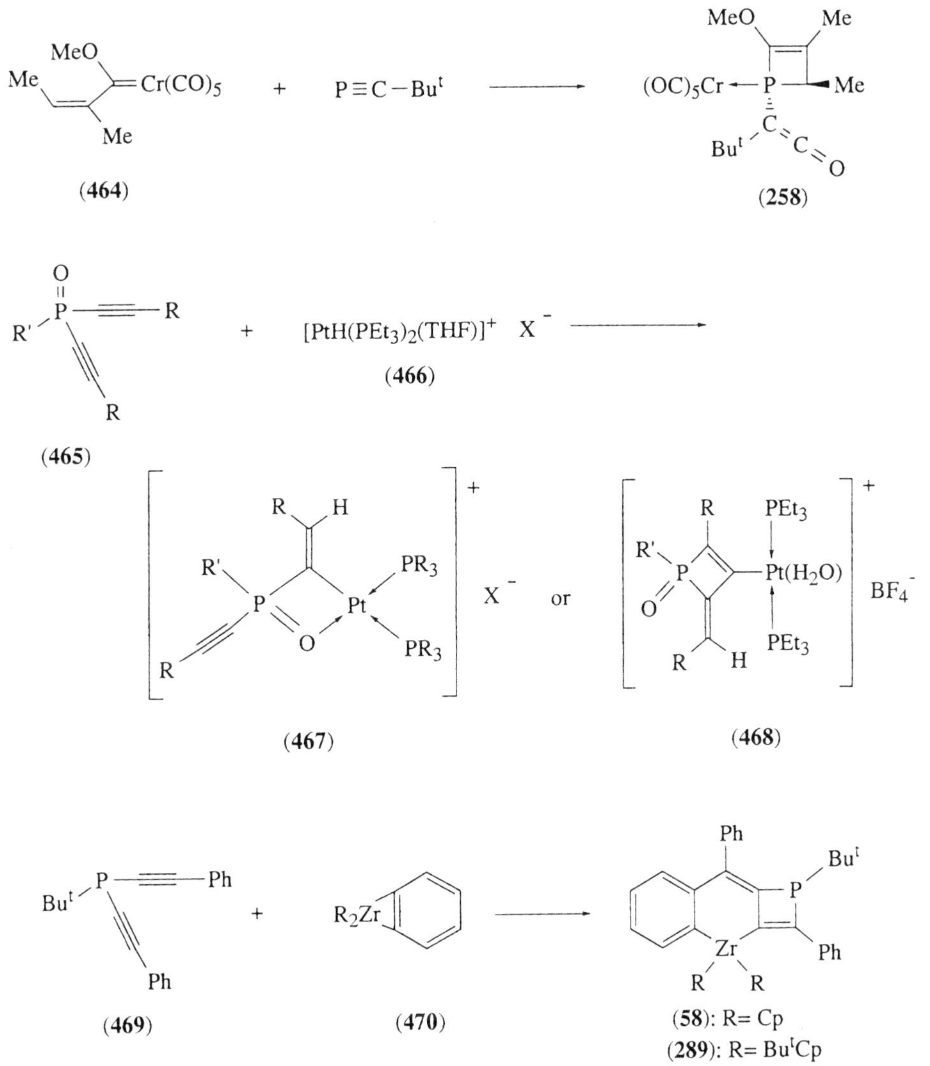

Scheme 68

As shown in Scheme 69, diphosphatricyclo[2.1.0.0^{2,5}]pentane (**21**) is obtained by the reaction of $[(\eta^5\text{-}C_5H_5)_2Ti(CO)_2]$ with *t*-BuC≡P via bisphosphaalkenyl ketone (**471**) ⟨88AG(E)837⟩.

Oxidative addition reactions of $Cl_2C{=}P{-}R$ (R = 2,4,6-tri*t*-butylphenyl) (**472**) with $M(PEt_3)_4$ (M = Pt or Pd) initially give the *trans*-isomer of square planar (**473**), which rearranges to give the corresponding 1,2-dihydrophosphete (**48**) fused benzene ring. Similarly, the corresponding *cis*-complex (**318**) and *trans*-dibromo complex (**317**) are synthesized, see Scheme 70 ⟨94OM2444⟩.

Diphosphaspiro[2.2]pentane (**474**) is heated in the presence of AIBN as a radical initiator to give the corresponding phosphetane (**164**) via diradicals (**475**) and (**476**) ⟨92AG(E)625⟩. Azaphosphaspiro[2.2]pentenes (**477a–j**), which is formed by the reaction of cyclopropenylidenephosphane and azides, undergoes ring

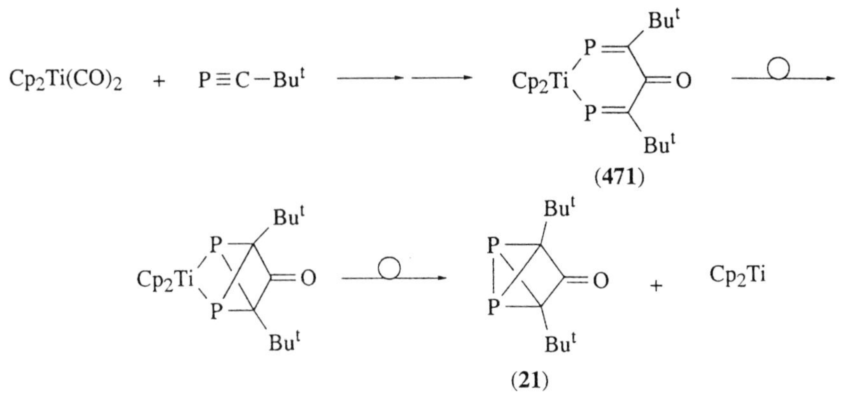

Scheme 69

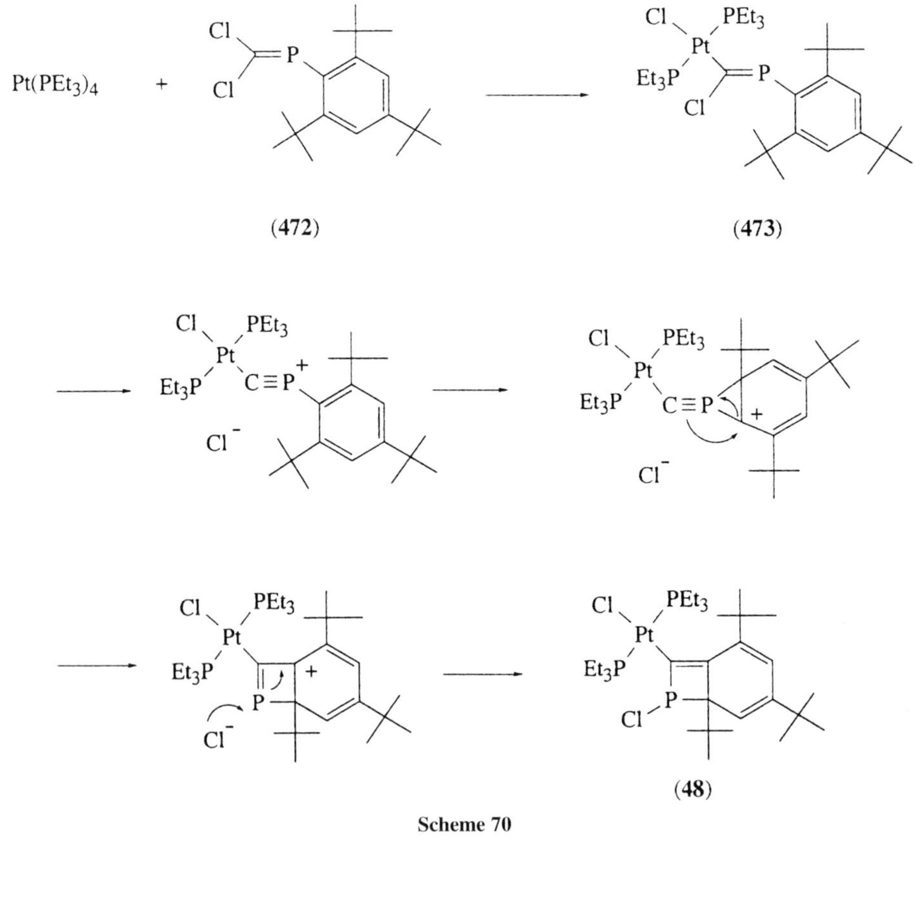

Scheme 70

expansion to give 2-imino-1,2-dihydrophosphetes (**295a–j**) and their over-reaction products (**296b–e**) with azides, see Scheme 9 ⟨94TL1527⟩. Bisphosphirene (**478**) reacts with iminophosphine to afford 1,2-dihydro-1,2-diphosphete (**479**), which undergoes rearrangement to give 1,2-dihydrophosphete (**51**) ⟨94CB313⟩, see Scheme 71.

3.2.10 RING SYNTHESIS BY TRANSFORMATION TO ANOTHER RING

3.2.10.1 By Ring Expansion

Tungstenpentacarbonyl 2-oxo-1,3,4-triphenyl-1,2-dihydrophosphete is treated with $NaBH_4$ in dioxane to afford oxaphospholene complexes (**420**) and (**421**), see Scheme 46 ⟨88OM633⟩.

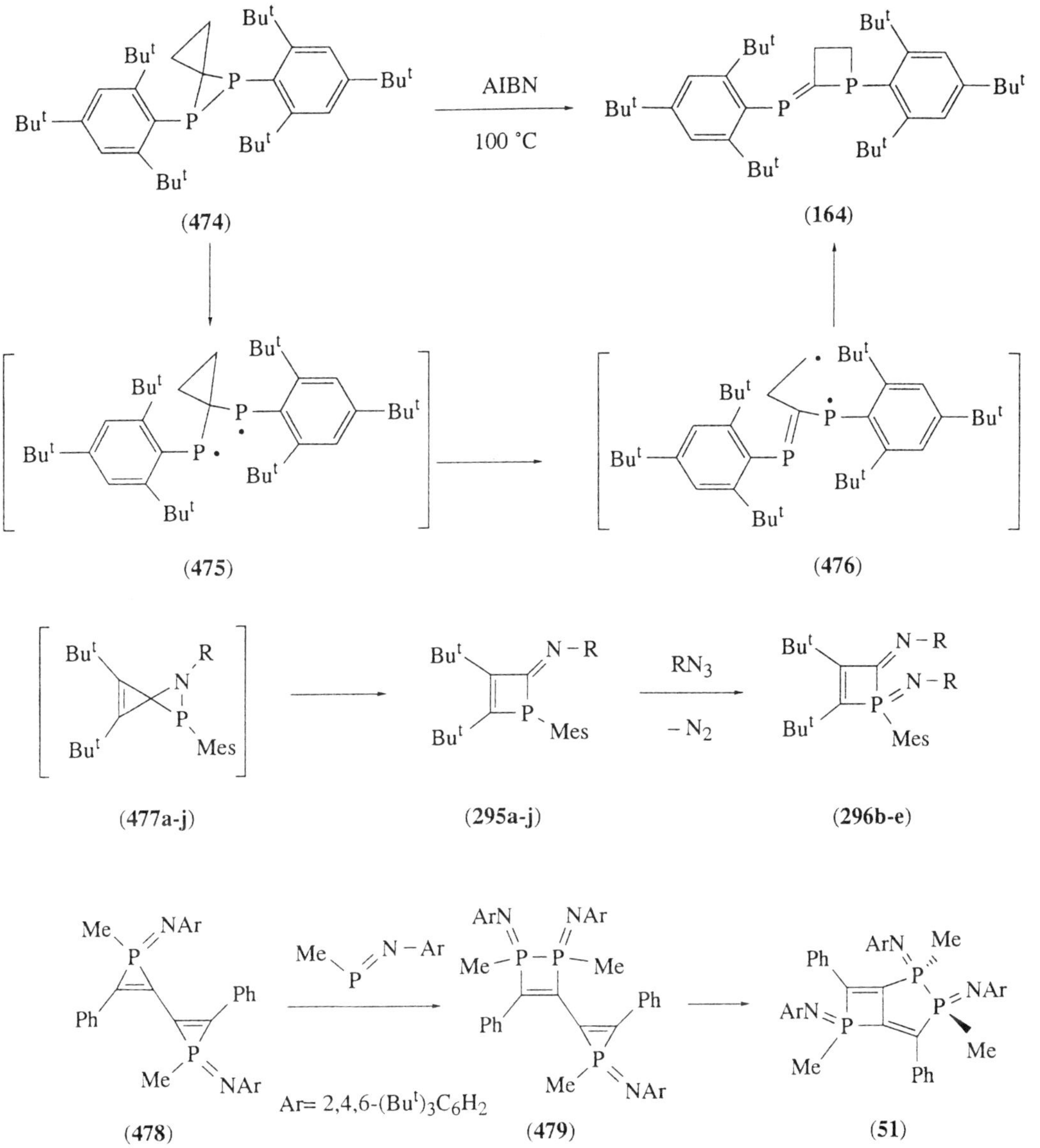

Scheme 71

The reactivity of the strained four-membered phosphorus-containing ring in the tetracyclic system is demonstrated by the insertion of oxygen into the carbon–phosphorus bond in oxide (**134**) with trifluoroperoxyacetic acid to give the tetracyclic oxaphospha derivative (**480**). Use of $NaHCO_3$ in order to remove MCPBA results in hydrolysis of the product giving (**481**), which is attributed to its highly strained structure. *trans*-2,2,3,4,4-Pentamethyl-1-phenylphosphetane 1-oxide (**2**) is oxidized with MCPBA at 25°C to give Baeyer–Villiger product (**482**) in 75% yield. A similar result is obtained for the phosphetane oxide (**8**). Here the reaction can follow two pathways, but attack occurred primarily at the least substituted carbon to give a mixture (isolated yield (56%)) consisting of about 90–95% (**483**) and 5–10% of (**484**). The opposite orientation would normally be encountered in the Baeyer–Villiger reaction of ketones. However, the special steric requirements at phosphorus appear to direct a different course for this and related rearrangement reactions of phosphetanes, see Scheme 72 ⟨83JOC4466, 90PS(54)135⟩.

As shown in Scheme 40, Dewar phosphabenzene (**244**) is heated at 160°C for 4 d to afford Dewar phosphabenzene (**240**) via phosphabenzene (**401**). On the other hand, irradiation of Dewar phosphabenzenes (**240**) and (**244**) gives phosphabenzene (**401**) and the phosphaprismane (**146**) mainly, respectively ⟨87AG(E)85, 99PJC135⟩.

1,3,4-Triphenyl-1,2-dihydrophosphete (**43**) reacts with dimethyl maleate to produce a mixture (3 : 2) of diastereomeric cycloadducts (**406**) and (**407**), see Scheme 41 ⟨90JCS(CC)1649⟩. The phosphole tungsten

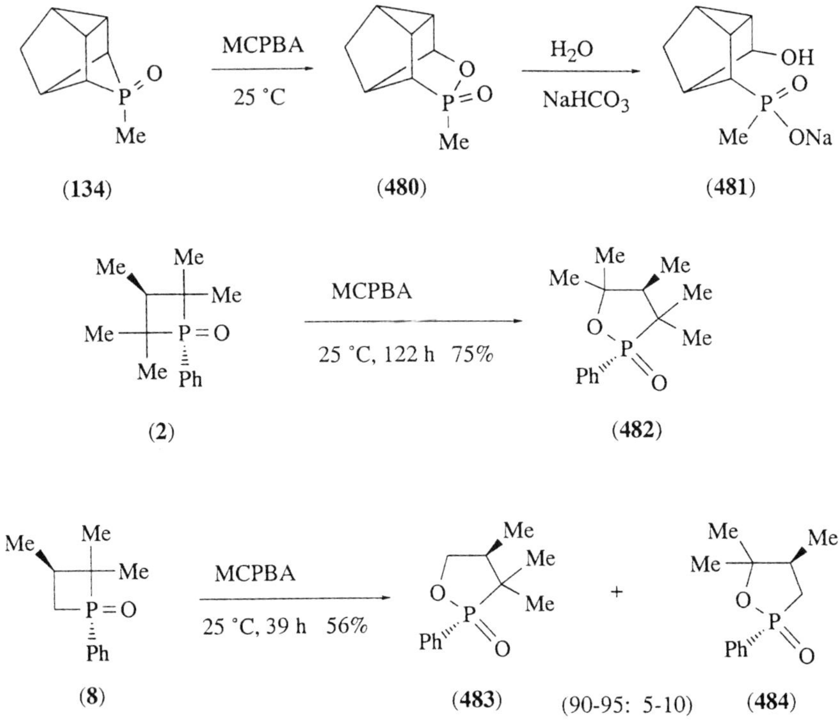

Scheme 72

pentacarbonyl complex (**485**) reacts with dimethyl acetylenedicarboxylate to give Diels–Alder adduct (**367**), which transfers phosphinidene metal complex to diphenylacetylene to afford phosphirene tungsten pentacarbonyl complex (**486**), which is heated at 160ºC for 100–120 h to give the corresponding 2-oxo-1,2-dihydrophosphete complex (**39**). One pot reaction from phosphole metal complex (**485**) in the presence of dimethyl acetylenedicarboxylate and diphenylacetylene gives successfully (**39**), see Scheme 73 ⟨88OM633, 85JA5001⟩.

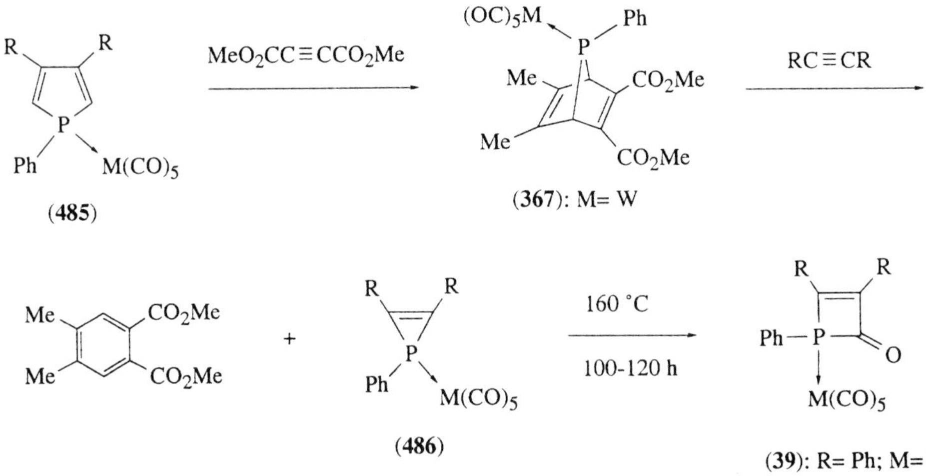

Scheme 73

7-Phosphanorbornadiene metal complexes (**487**), (**367**), and (**488**) rearrange under solar light for 15 d to give the [2 + 2] cycloadducts, quadricyclane derivatives, which undergo disrotatory opening of one of the two cyclopropane rings to give (**25**), (**26**), and (**325**), respectively. The corresponding tricyclophosphetane complex (**489**) is probably formed through 1,3-sigmatropic rearrangement of (**325**), see Scheme 52 ⟨84NJC453, 97OM4145⟩. For ring expansion to 2,3-dihydro-1,2-diphospholes, see Schemes 53 and 54 ⟨95AG(E)590, 96CEJ221⟩, for that to phosphabenzenes, see Scheme 47 ⟨98JOM(567)151, 96HAC397⟩.

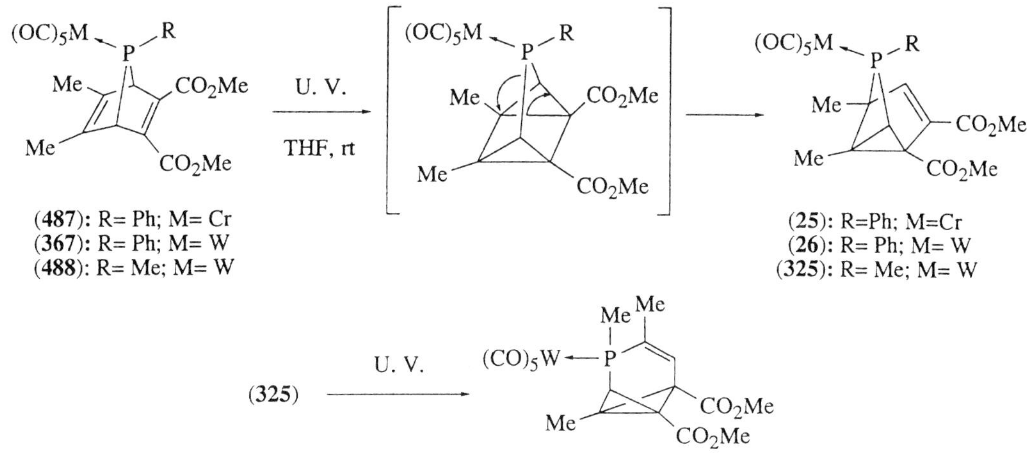

(489)

Scheme 74

3.2.11 REFERENCES

68JCS(CC)577 H. Hawes and S. Trippett; *J. Chem. Soc., Chem. Comm.*, 1968, 577.

68JCS(CC)1051 D.D. Swank and C.N. Caughlan; *J. Chem. Soc., Chem. Commun.*, 1968, 1051.

68JCS(CC)1228 Mazhar-ul-Haque and C.N. Caufhlan; *J. Chem. Soc., Chem. Commun.*, 1968, 1228–1229.

69JA2255 C. Moret and L.M. Trefonas; *J. Am. Chem. Soc.*, 1969, **91**, 2255–2259.

70JCS(B)934 Mazhar-ul-Haque; *J. Chem. Soc. (B)*, 1970, 934–937.

70JCS(B)938 Mazhar-ul-Haque; *J. Chem. Soc. (B)*, 1970, 938–943.

71JCS(B)117 Mazhar-ul-Haque; *J. Chem. Soc. (B)*, 1971, 117–120.

71JCS(CC)1011 R.E. Duff, R.K. Oram and S. Trippett; *J. Chem. Soc., Chem. Commun.*, 1971, 1011–1012.

71JA2066 D.Z. Denney, D.W. White and D.B. Denney; *J. Am. Chem. Soc.*, 1971, **93**, 2066–2067.

72JOC3342 A.G. Cook and G.W. Mason; *J. Org. Chem.*, 1972, **37**, 3342–3345.

72JOC3470 G.A. Gray and S.E. Cremer; *J. Org. Chem.*, 1972, **37**, 3470–3475.

72JOC3458 G.A. Gray and S.E. Cremer; *J. Org. Chem.*, 1972, **37**, 3458–3469.

73JCS(CC)856 J.A. Howard, D.R. Russell and S. Trippett; *J. Chem. Soc., Chem. Commun.*, 1973, 856–857.

76JA2750 J.-i. Aihara; *J. Am. Chem. Soc.*, 1976, **98**, 2750.

76JCS(CC)449 H.A.E. Aly, J.H. Barlow, D.R. Russell, D.J.H. Smith, M. Swindles and S. Trippett; *J. Chem. Soc., Chem. Commun.*, 1976, 449–450.

77JA1692 I. Gutman, M. Milun and N. Trinajstic; *J. Am. Chem. Soc.*, 1977, **99**, 1692.

77JCS(D)704 M. Mickiewicz and S.B. Wild; *J. Chem. Soc. Dalton Trans.*, 1977, 704.

78JCS(CC)125 E. Vilkas, M. Vilkas, D. Joniaux and C. Pascaard-Billy; *J. Chem. Soc., Chem. Commun.*, 1978, 125–126.

78JCS(P2)1115 Mazhar-ul-Haque, M. Rashid and S.E. Cremer; *J. Chem. Soc., Perkin II*, 1978, 1115–1119.

78IC3265 W. Althoff, R.O. Day, R.K. Brown and R.R. Holmes; *Inorg. Chem.*, 1978, **17**, 3265.

78JOC3513 A. Fitzgerald, J.A. Campbell, G.D. Smith, C.N. Caughlan and S.E. Cremer; *J. Org. Chem.*, 1978, **43**, 3513–3517.

79COC(2)1127 D.J.H. Smith; *Compr. Org. Chem.*, 1979, **2**, 1127.

80JCS(P1)2136 E. Vilkas, M. Vilkas, J. Saainton, B. Meunier and C. Pascard; *J. Chem. Soc., Perkin 1*, 1980, 2136–2144.

80JOC4807 E. Block, A.A. Bazzi, J.M. Lambert, S.M. Wharry, K.K. Andersen, D.C. Dittmer, B.H. Pattwardhan and D.H. Smith; *J. Org. Chem.*, 1980, **45**, 4907.

81JCS(P2)1138 Mazhar-ul-Haque, W. Horne, S.E. Cremer, P.W. Kremer and P.K. Kafarski; *J. Chem. Soc., Perkin II*, 1981, 1138–1142.

82AX(B)2944 Mazhar-ul-Haque and W. Horne; *Acta Crystallogr.*, 1982, **B38**, 2944–2946.

82JOC3660 J.C. Dyer, D.L. Harris and S.A. Evans Jr.; *J. Org. Chem.*, 1982, **47**, 3660.

83POL1225 T. Kwon, J.C. Woo and C.S. Chin; *Polyhedron*, 1983, **2**, 1225.

83PS(18)39 O.I. Kolodiazhnyi; *Phosphorus Sulfur*, 1983, **18**, 39.

83JA2074 A.H. Cowley and S.K. Mehretra; *J. Am. Chem. Soc.*, 1983, **105**, 2074.

83JOC2621 E.L. Clennan and P.C. Heah; *J. Org. Chem.*, 1983, **48**, 2621.

83JOC4466 L.D. Quin, J.C. Kisalus and K.A. Mesch; *J. Org. Chem.*, 1983, **48**, 4466.

84AX(C)1918 J.A. Campbell, C.N. Caughlan, A. Fitzgerald, C. Campana and S.E. Cremer; *Acta Cryst.*, 1984, **C40**, 1918.

84CHEC(1)(1)523 K. Dimroth; *Comprehensive Heterocycl. Chem. (1)*, 1984, **1**, 523.

84JA7021 L.D. Quin, K.C. Caster, J.C. Kisalus and K.A. Mesch; *J. Am. Chem.*, 1984, **106**, 7021.

84JCS(P2)553 J. Nielsen and O. Dahl; *J. Chem. Soc., Perkin Trans. 2*, 1984, 553.

84NJC453 A. Marinetti, F. Mathey, J. Fischer and A. Mitschler; *Nouv. J. Chim.*, 1984, **8**, 453.

84PO325	J. Emsley, A.P. Dunning, R.J. Parker, J.K. Williams, S. Brown, S. Earnshaw and D.S. Moore; *Polyhedron*, 1984, **3**, 325.
84PO1155	C.E.J. Dowrick and D.S. Moore; *Polyhedron*, 1984, **3**, 1155.
84TL815	A.H. Cowley, C.A. Stewart, B.R. Whittlesey and T.C. Wright; *Tetrahedron Lett.*, 1984, **25**, 815.
85AX(C)935	M. Gajhede, O. Dahl and J. Nielsen; *Acta Crystallogr.*, 1985, **C41**, 935–937.
85AX(C)975	Mazhar-ul-Haque and J. Ahmed and W. Horne; *Acta Cryst.*, 1985, **C41**, 975.
85JA5001	A. Marinetti, J. Fischer and F. Mathey; *J. Am. Chem. Soc.*, 1985, **107**, 5001.
85ZN(B)927	M. Binnewies, J. Grobe and D. Le Van; *Z. Naturforsch.*, 1985, **40b**, 927.
86AG(E)280	J. Fink, W. Rösch, U.-J. Vogelbacher and M. Regitz; *Angew. Chem., Int. Ed. Engl.*, 1986, **25**, 280.
86JCS(CC)1081	S.A. Weissman, S.G. Baxter, A.M. Arif and A.H. Cowley; *J. Chem. Soc., Chem. Commun.*, 1986, 1081.
87AG(E)83	F. Zurmühlen and M. Regitz; *Angew. Chem., Int. Ed. Engl.*, 1987, **26**, 83.
87AG(E)85	K. Blatter, W. Rösch, U.-J. Vogelbacher, J. Fink and M. Regitz; *Angew. Chem., Int. Ed. Engl.*, 1987, **26**, 85.
87JOMC35	P. Binger, R. Milczarek, R. Mynott and M. Regitz; *J. Organomet. Chem.*, 1987, **323**, C35.
87MRC271	L.D. Quin, J. Szewczyk, K. Linehan and D.L. Harris; *Magn. Reson. Chem.*, 1987, **25**, 271.
87S265	E.-y. Yao, J. Szewczyk and L.D. Quin; *Syntyhesis*, 1987, 265.
87AX(C)340	J.A. Campbell, R. Larsen, C. Campana and S.E. Cremer; *Acta Crystallogr.*, 1987, **C43**, 340–342.
87TL603	S.A. Weissman and S.G. Baxter; *Tetrahedron Lett.*, 1987, **28**, 603.
87TL6121	B.A. Boyd, R.J. Thoma and R.H. Neilson; *Tetrahedron Lett.*, 1987, **28**, 6121.
87PS283	J. Emsley and J.K. Williams; *Phosphorus Sulfur*, 1987, **29**, 283.
88PS(37)35	J. Szewczyk, K. Linehan and L.D. Quin; *Phosphorus Sulfur*, 1988, **37**, 35–40.
88AG(E)837	A.R. Barron, A.H. Cowley, S.W. Hall and C.M. Nunn; *Angew. Chem., Int. Ed. Engl.*, 1988, **27**, 837.
88CPL44	R.K. Mishra and B.K. Mishra; *Chem. Phys. Lett.*, 1988, **151**, 44.
88OM572	B.A. Boyd, R.J. Thoma, W.H. Watson and R.H. Neilson; *Organometallics*, 1988, **7**, 572.
88OM633	A. Marinetti and F. Mathey; *Organometallics*, 1988, **7**, 633.
88OM1791	N.H. Huy, L. Ricard and F. Mathey; *Organometallics*, 1988, **7**, 1791.
88TL1219	S.A. Weissman and S.G. Baxter; *Tetrahedron Lett.*, 1988, **29**, 1219.
88TL3077	N.H. Tran Huy and F. Mathey; *Tetrahedron Lett.*, 1988, **29**, 3077.
89NJC335	F. Zurmühlen and M. Regitz; *New J. Chem.*, 1989, **13**, 335.
89JA9129	K.M. Doxsee, G.S. Shen and C.B. Knobler; *J. Am. Chem. Soc.*, 1989, **111**, 9129.
89JCS(CC)788	K.H. Dötz, A. Tiriliomis and K. Harms; *J. Chem. Soc., Chem. Commun.*, 1989, 788.
89JPC7780	S.M. Bachrach; *J. Phys. Chem.*, 1989, **93**, 7780.
89OM1007	C.M. Lukehart, A.T. McPhail, D.R. McPhail, J.B. Myers Jr. and H.K. Soni; *Organometallics*, 1989, **8**, 1007.
90AG(E)75	W. Tumas, J.A. Suriano and R.L. Harlow; *Angew. Chem., Int. Ed. Engl.*, 1989, **29**, 75.
90PS(54)135	J. Szewczyk, E.-Y. Yao and L.D. Quin; *Phosphorus, Sulfur Silicon*, 1990, **54**, 135–141.
90IJC370	B.K. Mishra, S.C. Varghese, R. Roy and R.K. Mishra; *Indian J. Chem.*, 1990, **29A**, 370.
90JCS(CC)153	A. Marinetti and F. Mathey; *J. Chem. Soc., Chem. Commun.*, 1990, 153.
90JCS(CC)1649	K.M. Doxsee, G.S. Shen and C.B. Knobler; *J. Chem. Soc., Chem. Commun.*, 1990, 1649.
90JCS(P2)971	X. Liao, S. Li and C. Yuan; *J. Chem. Soc., Perkin Trans. 2*, 1990, 971.
90OM788	A. Marinetti, L. Ricard and F. Mathey; *Organometallics*, 1990, **9**, 788.
90SL171	P. Le Floch and F. Mathey; *Synlett.*, 1990, 171.
91AG(E)310	P. Binger, J. Haas, A.T. Herrmann, F. Langhauser and C. Krüger; *Angew. Chem., Int. Ed. Engl.*, 1991, **30**, 310.
91IC3381	K.M. Doxsee, E.M. Hanawalt, G.S. Shen, T.J.R. Weakley, H. Hope and C.B. Knobler; *Inorg. Chem.*, 1991, **30**, 3381.
91JOC2205	S.M. Bachrach; *J. Org. Chem.*, 1991, **56**, 2205.
91JOMC20	P.B. Hitchcock, M.F. Meidine, J.F. Nixon, F. Mathey and N.H. Tran Huy; *J. Organomet. Chem.*, 1991, **406**, C20.
91OM1197	A.C. Dema, X. Li, C.M. Lukehart and M.D. Owen; *Organometallics*, 1991, **10**, 1197.
92AG(E)625	C. Garot, G. Etemad-Moghadam, J.-P. Declercq, A. Dubourg and M. Koenig; *Angew. Chem., Int. Ed. Engl.*, 1992, **31**, 625–626.
92CB1987	D.J. Brauer, A.J. Ciccu, G. Heßler and O. Stelzer; *Chem. Ber.*, 1992, **125**, 1987.
92JCS(D)1751	A. Bader, D.D. Pathak, S.B. Wild and A.C. Willis; *J. Chem. Soc., Dalton Trans.*, 1992, 1751.
92JOC209	S.M. Bachrach and M. Liu; *J. Org. Chem.*, 1992, **57**, 209.
92JOC511	S.E. Cremer, J.M. Cowles, F.R. Farr, H.-O. Hwang, P.W. Kremer, A.C. Peterson and G.A. Gray; *J. Org. Chem.*, 1992, **57**, 511.
92OM2944	W. Tumas, J.C. Huang, P.E. Fanwick and C.P. Kubiak; *Organometallics*, 1992, **11**, 2944.
93AG(E)1359	U. Heim, H. Pritzkow, U. Fleischer and H. Grützmacher; *Angew. Chem., Int. Ed. Engl.*, 1993, **32**, 1359.
93AG(E)617	W.W. Schoeller and T. Busch; *Angew. Chem. Int. Ed. Engl.*, 1993, **32**, 617.
93IC5720	Y.W. Li, M.G. Newton and R.B. King; *Inorg. Chem.*, 1993, **32**, 5720.
93JCR(S)268	S.-C. Wong, N.I. Carruthers and T.-M. Chan; *J. Chem. Res. (S)*, 1993, 268.
93OM4265	H. Jun and R.J. Angelici; *Organometallics*, 1993, **12**, 4265.
93T10291	A. Marinetti and L. Ricard; *Tetrahedron*, 1993, **49**, 10291.
93TL4619	J. Grobe, D. Le Van, B. Broschk and L.S. Kobrina; *Tetrahedron Lett.*, 1993, **34**, 4619.
94T759	E. Fuchs, F. Krebs, H. Hydt and M. Regitz; *Tetrahedron*, 1994, **50**, 59–774.

94TL1527 W. Eisfeld, M. Slany, U. Rergsträsser and M. Regitz; *Tetrahedron Lett.*, 1994, **35**, 1527–1530.

94CB481 A. Hessler, J. Fischer, S. Kucken and O. Stelzer; *Chem. Ber.*, 1994, **127**, 481–488.

94JCS(CC)475 Y.B. Kang, M. Pabel, A.C. Willis and S.B. Wild; *J. Chem. Soc., Chem. Commun.*, 1994, 475–476.

94CB313 M. Link, E. Niecke and M. Nieger; *Chem. Ber.*, 1994, **127**, 313–319.

94TL5861 A. Marinetti; *Tetrahedron Lett.*, 1994, **35**, 5861–5864.

94OM2444 H. Jun, V.G. Young Jr., and R.J. Angelici; *Organometallics*, 1994, **13**, 2444.

94OM3956 A. Marinetti and L. Ricard; *Organometallics*, 1994, **13**, 3956–3962.

94ZN(B)1693 L. Weber, O. Kaminski, H.-G. Stammler, B. Neumann and R. Boese; *Z. Naturforsch. B: Chem. Sci.*, 1994, **49**, 1693–1706.

95JST(338)51 D.J. Berger, P.P. Gaspar and J.F. Liebman; *J. Mol. Struct. (Theochem)*, 1995, **338**, 51.

95BSF384 H. Trauner, E. de la Cuesta, A. Marinetti and F. Mathey; *Bull. Soc. Chim. Fr.*, 1995, **132**, 384–393.

95AG(E)590 N. Maigrot, N. Avarvari, C. Charrier and F. Mathey; *Angew. Chem., Int. Ed. Engl.*, 1995, **34**, 590–592.

95ZAAC1539 E. Fluck, G. Heckmann, F. Rosche, M. Westerhausen and E. Gorhunowa; *Z. Anorg. Allg. Chem.*, 1995, **621**, 1539–1546.

95JOC6076 J.-L. Montchamp, F. Tian and J.W. Frost; *J. Org. Chem.*, 1995, **60**, 6076–6081.

95OM1434 A. Bader, Y.B. Kang, M. Pabel, D.D. Pathak, A.C. Willis and S.B. Wild; *Organometallics*, 1995, **14**, 1434–1441.

95OM4983 A. Marinetti, C. Le Menn and L. Ricard; *Organometallics*, 1995, **14**, 4983–4985.

96CHEC(2)(1B)833 T. Kawashima and T. Okazaki; *Comprehensive Heterocycl. Chem. II*, 1996, **1B**, 833.

96PS(115)3 G. Heckmann, F. Rosche, F. Weller and E. Fluck; *Phosphorus, Sulfur Silicon Relat. Elem.*, 1996, **115**, 3–18.

96CEJ68 U. Heim, H. Pritzkow, U. Fleischer, H. Grützmacher, M. Sanchez, R. Réau and G. Bertrand; *Chem. Eur. J.*, 1996, **2**, 68–74. *Angew. Chem., Int. Ed. Engl.*, 1996, **35**, 1.

96CEJ221 G. Jochem, A. Schmidpeter and H. Nöth; *Chem. Eur. J.*, 1996, **2**, 221–227. *Angew. Chem., Int. Ed. Engl.*, 1996, **35**, 3.

96JOM(522)223 A. Marinetti, V. Kruger, C. Le Menn and L. Ricard; *J. Organomet. Chem.*, 1996, **522**, 223–230.

96PS(115)273 L. Maier and P.J. Diel; *Phosphorus, Sulfur Silicon Relat. Elem.*, 1996, **115**, 273–300.

96HAC397 N. Avarvari, P. Le Floch, C. Charrier and F. Mathey; *Heteroat. Chem.*, 1996, **7**, 397–402.

96OM1301 D.C.R. Hockless, Y.B. Kang, M.A. McDonald, M. Pabel, A.C. Willis and S.B. Wild; *Organometallics*, 1996, **15**, 1301–1306.

96OM1597 K. Waschbüsch, P. Le Floch and F. Mathey; *Organometallics*, 1996, **15**, 1597–1603.

96ZAAC974 F. Rosche, G. Heckmann, E. Fluck and W. Weller; *Z. Anorg. Allg. Chem.*, 1996, **622**, 974–980.

97JOC297 A. Marinetti, F.-X. Buzin and L. Ricard; *J. Org. Chem.*, 1997, **62**, 297–301.

97HAC451 D.B. Chesnut, L.D. Quin and S.R. Wild; *Heteroat. Chem.*, 1997, **8**, 451–457.

97JOM(529)465 A. Marinetti, V. Kruger and L. Ricard; *J. Organomet. Chem.*, 1997, **529**, 465–472.

97PS(124-125)545 U. Rohde, H. Wilkens and R. Streubel; *Phosphorus, Sulfur Silicon Relat. Elem.*, 1997, **124–125**, 545–548.

97AG(E)987 L. Dupuis, N. Pirio, P. Meunier, A. Igau, B. Donnadieu and J.-P. Majoral; *Angew. Chem., Int. Ed. Engl.*, 1997, **36**, 987–989.

97AG(E)1095 R. Streubel, M. Hobbold, J. Jeske and P.G. Jones; *Angew. Chem. Int. Ed. Engl.*, 1997, **36**, 1095.

97ZOR1703 T.A. Kuznetsova, N.V. Istomina and L.R. Krivdin; *Zh. Org. Khim.*, 1997, **33**, 1703.

97JOC2362 A. Hessler, O. Stelzer, H. Dibowski, K. Worm and F.P. Schmidtchen; *J. Org. Chem.*, 1997, **62**, 2362–2369.

97OM2958 L. Weber, M. Scheffer, E. Beckmann, H.-G. Stammler and B. Neumann; *Organometallics*, 1997, **17**, 2958–2962.

97OM4145 R. Wang, C.H. Lake and K. Lammertsma; *Organometallics*, 1997, **16**, 4145–4149.

97TL2947 A. Marinetti, V. Kruger and F.-X. Buzin; *Tetrahedron Lett.*, 1997, **38**, 2947–2950.

97T4363 A. Marinetti, F.-X. Buzin and L. Ricard; *Tetrahedron*, 1997, **53**, 4363–4370.

97TL8417 R.A. Aitken, W. Masamba and N.J. Wilson; *Tetrahedron Lett.*, 1997, **38**, 8417–8420.

98HAC21 E.M. Hanawalt, K.M. Doxcee, G.S. Shen and H. Hope; *Heteroat. Chem.*, 1998, **9**, 21.

98JOM(567)151 N. Avarvari, P. Rosa, F. Mathey and P. Le Floch; *J. Organomet. Chem.*, 1998, **567**, 151–155.

98CEJ469 L. Weber, B. Quasdorff, H.-G. Stammler and B. Neumann; *Chem. Eur. J.*, 1998, **4**, 469–475.

98CCR(178-180)755 A. Marinetti, V. Kruger and F.-X. Buzin; *Coord. Chem. Rev.*, 1998, **178–180**, 755–770.

98S1305 A. Mack, E. Pierron, T. Allspach, U. Bergstrasser and M. Regitz; *Synthesis*, 1998, 1305–1313.

98ZOB1449 A.D. Averin, P. Mukhaiimana, N.V. Lukashev, A.A. Borisenko, M.A. Kazankova and I.P. Beletskaya; *Zh. Obshch. Khim.*, 1998, **68**, 1449.

98S1539 A. Marinetti, V. Kruger and B. Conetoux; *Synthesis*, 1998, 1539–1543.

98MI1 W. Rösch; Ph.D. Thesis, Universität Kaiserslautern, Germany (1998).

99PJC135 M. Regitz and U. Bergsträsser; *Pol. J. Chem.*, 1999, **73**, 135–150.

99AG(E)215 U. Rohde, F. Ruthe, P.G. Jones and R. Streubel; *Angew. Chem., Int. Ed.*, 1999, **38**, 215–217.

99CEJ1160 A. Marinetti, J.-P. Genet, S. Jus, D. Blanc and V. Ratovelomanana-Vidal; *Chem. Eur. J.*, 1999, **5**, 1160–1165.

99JA4215 L. Colombet, F. Volatron, P. Maitre and P.C. Hiberty; *J. Am. Chem. Soc.*, 1999, **121**, 4215–4221.

3.3
Diphosphetanes, Dihydrodiphosphetes, and Diphosphetes

MARTIN HOFMANN and ULRICH ZENNECK
University of Erlangen–Nürnberg, Erlangen, Germany

3.3.1 INTRODUCTION

The first example of a four-membered heterocycle with two phosphorus atoms was reported in 1964 by Mahler ⟨64JA(86)2306⟩, who prepared a 1,2-dihydro-1,2-diphosphete derivative (**1**). The saturated 1,2- and 1,3-diphosphetane (**2, 3**) rings followed in 1976 when Becker started developing the chemistry of phosphaalkenes, as the head-to-head, respectively, head-to-tail cyclodimerization gave access to both rings ⟨76ZA(432)242⟩. A transition metal mediated [2 + 2]-cyclodimerization of phosphaalkynes leads to the formation of the fully unsaturated diphosphetes (**4**), which are stable only as η^4-ligands. Free diphosphetes are highly reactive, but may be used as interesting intermediates for the synthesis of organophosphorus cage compounds.

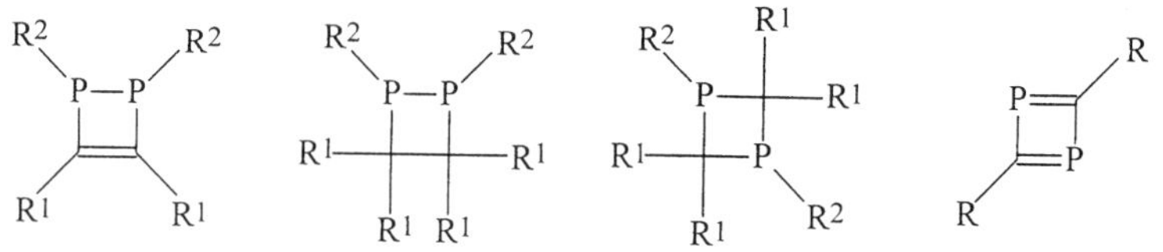

(**1**) (**2**) (**3**) (**4**)

Scheme 1

3.3.1.1 Reviews

Fluck and Neumüller reviewed the complete field covering the literature up to 1988 ⟨B-89MI1⟩. Synthesis, NMR spectroscopic results and structural data are presented mainly. Synthesis and properties of diphosphetanes are summarized by Afarinkia, ⟨92H(34)369⟩ and those of 1,2-dihydro-1,2-diphosphetes by Mathey *et al.* ⟨92MI(13)61⟩. The extraordinary synthetic potential of phosphaalkynes as precursors of 1,3-diphosphetes and of polycyclic P–C cage compounds, which include 1,3-diphosphetanes as building blocks, is described in several reviews by Regitz and Binger ⟨88AG(E)(27)1484, B-93MI2, B-95MI6, 99ASIOM(S1)199, 97CB(130)823⟩, Nixon ⟨88CRV(88)1327, 95CRV(145)201, 95CSR319⟩, and Weber ⟨97AOC(41)1, 97CCR(158)1⟩. Bertrand and Fluck reported about ylidic four π-electron four-membered λ^5-phosphorus heterocycles ⟨98AG(E)(37)271, 98PAC(70)819⟩. A monography by Dillon *et al.* was devoted to organophosphorus chemistry including transition metal complexes of phosphorus heterocycles. It touches the topic of this review at several points ⟨B-98MI7⟩.

3.3.2 THEORETICAL METHODS

Theoretical calculations utilizing *ab-initio* and semi-empirical methods have been applied to all principal ring structures and reveal details on their relative energy ⟨95AG(E)(34)484, 96CB(129)419⟩. According to Gaussian calculations, $1\lambda^5,2\lambda^3$-diphosphetes and 1,2-dihydro-$1\lambda^3$-$3\lambda^3$-diphosphetes are higher in energy, but kinetically stabilized if compared to their isomer 1,2-dihydro-$1\lambda^3$-$2\lambda^3$-diphosphete ⟨97JOM(529)177, 99CEJ(5)274⟩. MCSCF *ab-initio* MO calculations ⟨93AG(E)(32)617⟩ on the highly reactive λ^3,λ^3-1,2- and 1,3-diphosphetes lead to a singlet ground state with a triplet state close by. Their principal structures are a trapezoid (1,2) and a parallelogram (1,3) and the P–C bond lengths are alternating ⟨96JPC(100)6456⟩. 1,2-diphosphete is 44 kJ mol^{-1} more stable than its 1,3-isomer. Protonation of hypothetical diphosphatetrahedrane ($C_2H_2P_2$) in the gas phase leads to hydro-1,3-diphosphete cations ⟨98JPO(11)678⟩. In contrast to the free rings, π-complexed 1,3-diphosphetes have been prepared and the PES spectra of an irontricarbonyl and a cyclopentadienyl cobalt complex were correlated to INDO SCF calculations ⟨92OM(11)177⟩. Both metals interact strongly with this planar π-ligand, which shows no sign of localization of the π-electrons. This is in agreement with the tough reaction conditions, which are necessary for successful decomplexation experiments ⟨96S265⟩. *Ab-initio* calculations on the 1,3-diphosphete dianion $C_2H_2P_2{}^{2-}$ reveal a D_{2h} symmetry, close to that one found for the π-complexed 1,3-diphosphete ligand of cobalt ⟨93JPC(97)6384⟩. Ring opening of 1,2-dihydro-1,2- and -1,3-diphosphetes have been studied by *ab-initio* methods ⟨92JOC(57)2040, 98EJIC951⟩. Ring strain energies of possible isomers of 1,3-diphosphetanes were calculated ⟨95JPO(8)742⟩. The ring strain can be compared to that of tetraphosphacubane which is composed of 1,3-diphosphetane units. Both are less strained than the corresponding carbon hydrides. The ring strain energy of tetraphosphacubane results to 264 kJ mol^{-1}, ⟨93TL(34)6365⟩ upon oxidation of the phosphorus atoms by oxygen or sulphur, the ring strain increases ⟨98TL(39)4211⟩. Tetraphosphacubane is an unexpectedly strong base in the gas phase ⟨95JOC(60)47, 96JOC(61)7813⟩. MP2 calculations give a hint on the possibility to encapsulate small cations like Li^+ and Be^{2+} ⟨96CJC(74)901⟩.

According to *ab-initio* calculations the electronic structure of 1,3-dihydro-1,3-diphosphetes with bulky P-substituents is very complex. A triplet state contributes significantly to the ground state and the P-lone pairs form a cyclic π-system together with the two π-electrons of the C–R-fragments. Thus, the cyclic delocalized π^6-system 1,3-diphosphacyclobutane-2,4-diyl is formed. Because of the triplet contribution, the authors prefer to call the compound a 1,3-diphosphacyclobutane-2,4-diyl derivative ⟨95AG(E)(34)555, 96PS(111)42, 98AG(E)(37)949⟩. 1,3,5-Triphosphabicyclo[2.1.0]pent-2-ene with its 1,2-dihydro-1,3-diphosphete partial structure exhibits a fast 1,3-phosphorus exchange, which has been analysed by *ab-initio* calculations ⟨95CC2395⟩.

NMR data of tetraphosphacubane derivatives have been calculated by utilizing the IGLO-method ⟨94PS(93/94)365⟩. Gaussian 94 allows the calculation the ^{31}P shift tensor of 1,2-dihydro-1,2-diphosphetes ⟨99JPC(A)(103)1029⟩.

3.3.3 EXPERIMENTAL STRUCTURAL METHODS

3.3.3.1 Structural Aspects

$1\lambda^5,3\lambda^5$-Diphosphetes (**5**) are practically planar with intraring P–C bond distances in the range from 172 to 177 pm and P–C–P as well as C–P–C bond angles between 88 and 92° ⟨B-89MI1, 95CB(128)959⟩. $1\lambda^5,2\lambda^3$-diphosphetes (**6**) are also almost planar. The intraring distances are, respectively, P–C 179–183 (λ^5) and 183–192 pm (λ^3), C–C 136–139 pm, and P–P 217–230 pm. The geometric parameters suggest a localized C=C double bond and a strongly polarized P–P single bond ⟨95JA(117)10785, 96AG(E)(35)2228, 97JA(119)9720, 99CEJ(5)274⟩. The 1,2-dihydro-$1\lambda^5,3\lambda^3$-diphosphete ring of (**7**) is almost coplanar with its NR_2 substituents, whereas the PX_2 (X = S, Se) plane is arranged perpendicular to them. The P–C distances are 180–184 (λ^5) and 175–178 pm (λ^3) ⟨00T(56)27, 00T(56)1147⟩.

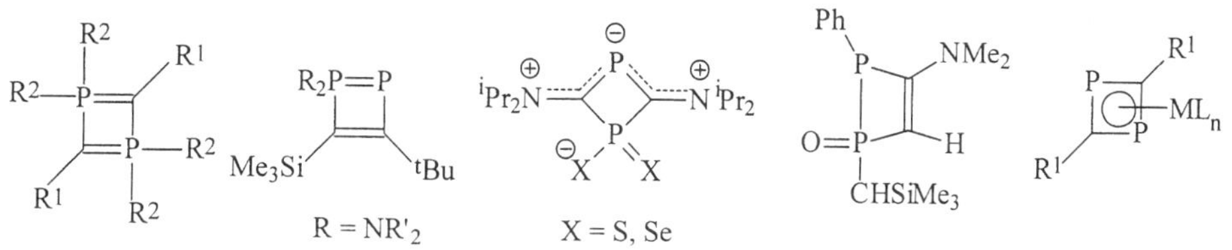

(**5**) (**6**) (**7**) (**8**) (**9**)

Scheme 2

The P–C-bond lengths of 1,2-dihydro-$1\lambda^5,2\lambda^3$-diphosphete (**8**) are not equivalent either (λ^5-P–C = 175 pm; λ^3-P–C = 185 pm) ⟨97JOC(62)7605⟩.

Structural information on $1\lambda^3,3\lambda^3$-diphosphetes is accessible only to π-co-ordinated species (**9**). Structural data are available for complexes with the transition metals Co, Ir, Fe, Mo, Ni, Rh, Ti, and W. First examples have been achieved with cobalt ⟨86AG(E)(25)644, 86JCS(CC)737, 88CB(121)637⟩. This field has been reviewed extensively (*vide supra*). With few exceptions, the co-ordinated rings generally form slightly distorted squares with identical P–C-distances. Depending on the metals, the P–C bond lengths range from 174 to 186 pm, most of them are close to 180 pm. Due to the different sizes of P- and C-atoms, the intraring bond angles are around 84° for C–P–C-, and 96° for P–C–P-angles. An early transition metal complex of Binger, (COT)(1,3-diphosphete)Ti ⟨95CB(128)1261⟩ exhibits a partial localized π-electron system, which causes alternating intraring P–C-bond lengths (174 vs. 183 pm).

Only two $1\lambda^3,2\lambda^3$-diphosphete complexes have been reported to date by Binger (Ti) and Zenneck (Fe) ⟨95CB(128)1261, 99OM(18)2021⟩. As for their 1,3-isomers, the 1,2-diphosphete ligands are planar. The P–C-bond lengths range from 180 to 182 pm, thus indicating a comparable bonding situation for 1,3- and 1,2-diphosphete ligands.

The structural data of the trapezoid 1,2-dihydro-1,2-diphosphetes, including $\lambda^3,\lambda^3,\lambda^3,\lambda^5$, and σ-complexed λ^3,λ^3-species have been summarized by Mathey *et al.* ⟨92MI(13)61⟩. Most λ^3-P–C_{ring} bond distances are 184 ± 1 pm. An example for 1,2-dihydro-$1\lambda^3,3\lambda^3$-diphosphetes is given by (**10**) ⟨83CB(116)938⟩. The ring is slightly puckered (dihedral angle 7.5°) and the P=C double bond is significantly shortened in respect to the P–C single bonds (168 vs. 179, 189, and 191 pm). 1,2-Dihydro-$1\lambda^3,3\lambda^5$-diphosphetes (**11**), too, have a localized λ^5-P=C double bond ⟨86CB(119)535⟩.

3,4-Diisopropylidene-1,2-diphosphetane (**12**) is puckered as well with a dihedral angle of 15.3° and the P–C single bonds within the ring are 184 pm long ⟨93CB(126)845⟩. In contrast, 1,3-dihydro-1,3-diphosphete (**13**) is planar with $P–C_{ring}$ distances of 175 pm ⟨95AG(E)(34)555⟩. In principle (**13**) has the same π-electron count within the ring as the dihydrodiphosphetes and (**12**); however, the shortened P–C-bond distances and *ab-initio* MO calculations clearly indicate a strong interaction of the two carbon π-electrons and the phosphorus lone pairs; (**13**) is thus better described as a delocalized π^6-electron system.

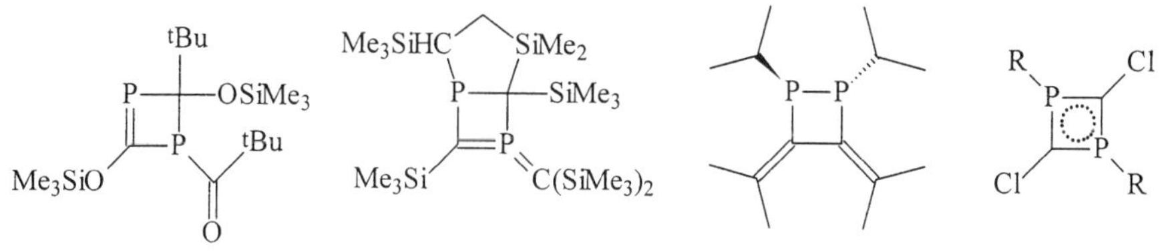

(**10**) (**11**) (**12**) (**13**)

R = 1,3,5-$^tBu_3C_6H_2$, NR_2

Scheme 3

2,4-Diallenyl-1,3-diphosphetane (**14**), a representative of the 2,4-dimethylene-1,3-diphosphetanes, is a nearly planar (bent by 3.1°) parallelogram with P–C single bond distances in the normal range (181–186 pm) ⟨90TL(31)4429⟩. The same principal structure features are found for 2,4-diphosphoranediyl-1,3-diphosphetane (**15**), but the intraring P–C-distances are slightly shorter (176–178 pm) ⟨95AG(E)(34)1853⟩.

1,2-Diphosphetanes as well as 1,3-diphosphetanes are folded molecules with regular to slightly elongated P–C bonding distances (187–195 pm) ⟨79AG(E)(18)872, 90CB(123)733, 82ZAAC(485)23⟩. However, if the 1,3-diphosphetane unit is part of a polycyclic system, as for example in tetraphosphacubane (**16**) ⟨89AG(E)(28)1013⟩ or in hexaphosphahomopentaprismane (**17**) ⟨89AG(E)(28)900⟩, the rings might be close to planarity with P–C distances from 186 to 191 pm. A gas phase electron diffraction study on the perfluorinated 1,3-diphosphetane derivative (**18**) confirms the X-ray results on this class of compounds. The ring is puckered (dihedral angle 35.4°) and the C–P–C angles within the ring are very small (77.6°) ⟨86IC(25)3177⟩.

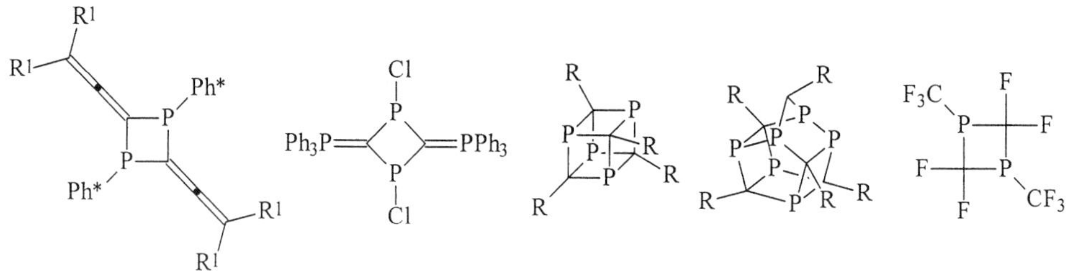

(**14**) (**15**) (**16**) (**17**) (**18**)

Ph* = 1,3,5-$^tBu_3C_6H_2$; $R^1 = C_6H_4F$ R = tBu

Scheme 4

3.3.3.2 Spectroscopic Aspects

^1H-, ^{13}C- and ^{31}P-NMR spectroscopy play a very important role for the characterization of the compounds. A survey on the ^{31}P-NMR spectroscopy of low co-ordinated phosphorus compounds with P-element multiple bonds was given by Karaghiosoff ⟨B-90MI3⟩. ^1H-NMR data are of special interest only if ring protons are present; however, only few of the many known compounds fulfil this condition. Therefore no clear trends for the proton chemical shifts or coupling constants can be deduced from the data for different bonding situations within the rings. Saturated C–H units in diphosphetanes which are not influenced by polar groups or nearby transition metals, absorb in the range 1–2.2 ppm ⟨84TL(25)1963, 91AG(E)(30)207⟩. Unluckily the C–H protons of the unsaturated compounds (**19**) ⟨92ZAAC(612)56⟩ and (**20**) ⟨89ZAAC(576)81⟩ are found in the same region (Scheme 5). ^{13}C-NMR is plagued by the same problem.

Table 1. Selected ^{13}C- and ^{31}P-NMR data

Compound	$\delta(^{31}P)$	J_{PP} *[Hz]*	$\delta(^{13}C)$	J_{PC}	*Reference*
10	279.1 (P1)	16.2	223.2 (C1)	53.2 (J_{P1C})	⟨83CB(116)938⟩
				11.6 (J_{P2C})	
	104.1 (P2)		79.9 (C2)	27.2 (J_{P1C})	
				1.3 (J_{P2C})	
16	257.4	8.52	−29.1	13.43	⟨89AG(E)(28)1013⟩
20a	43.4		19.8		⟨89ZAAC(576)81⟩
21	−35.9 (P1)	64	159.1		⟨86OM(5)623⟩
	−35.7 (P2)		144.0		
22	−21.3		43.3	23.7	⟨92HAC(3)395⟩
	−14.7		42.9	25.7	
23	24.1 (P2)	22.8	46.4	31.9	⟨91AG(E)(30)207⟩
	−133.6 (P1)			12.5	
24	56.29 (P1)	41.8	157.7		⟨86OM(5)623⟩
	54.26 (P2)		157.45		
25	21.7		163.15	13.7	⟨96S265⟩
26	157.6		212.19	5.38	⟨99OM(18)2021⟩
			164.28	16.15	
27	−84.63		134.89	27.66	⟨99OM(18)2021⟩

The chemical shift ranges of sp^3- and sp^2-carbons partly overlap. This is mainly due to charges and substituents, the phosphorus atoms play a minor role. ^{13}C- and ^{31}P-NMR data for 1λ^5,3λ^5-diphosphetes of the type (**5**) (R = H, Me, Ph; R′ = NMe$_2$, NEt$_2$) have been summarized ⟨89ZAAC(576)81⟩. Some selected NMR-data are presented in Table 1.

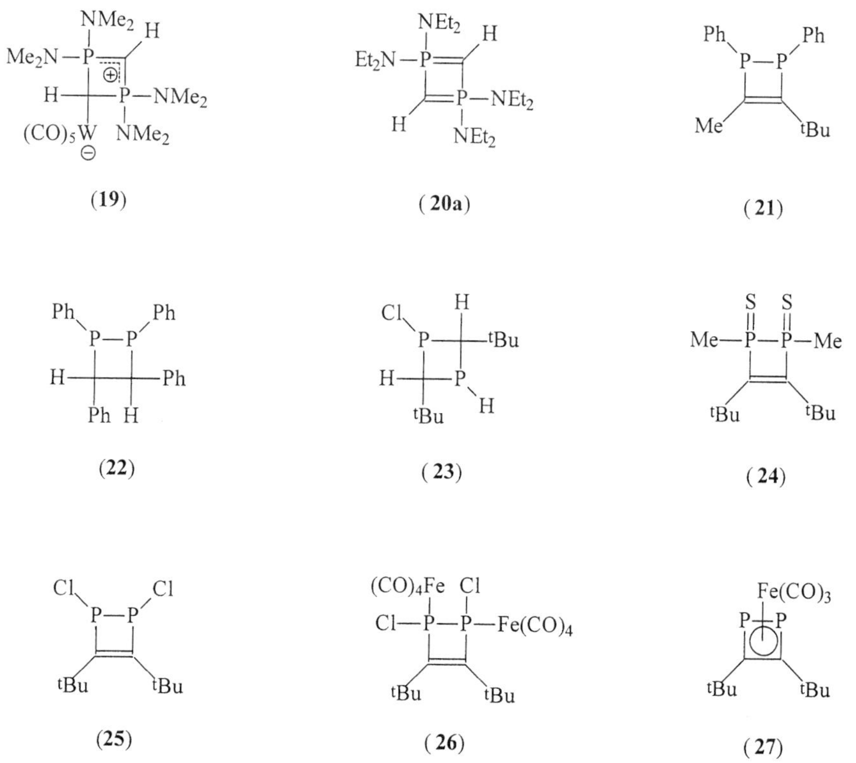

Scheme 5

Even for ^{31}P-NMR an overlap of chemical shift and coupling constant ranges is observed. Not surprisingly $^2J_{PP}$ values in 1,3-ring systems are found to be smaller (5.5–35 Hz) than the $^1J_{PP}$ values for 1,2-rings (60–130 Hz) in most cases. But some rings definitely violate this rule. Extraordinary small PP coupling is found for the 1,2-dihydro-1,2-diphosphete (**28**) ($^1J_{PP}$ = 29.3 Hz) ⟨86OM(5)623⟩ and (**29**) ($^1J_{PP}$ = 36.5 Hz) ⟨92BSF(129)76⟩. Because of the metal co-ordination of one of the P-atoms in (**29**), a reduction of the coupling constant from the 61.0 Hz found for the free ring is not difficult to understand, but such arguments are not available for (**28**). Remarkably large $^2J_{PP}$ values are found for a derivative of (**14**) (45.7 Hz) ⟨90TL(31)4429⟩, (**30**) (54.0 Hz) ⟨93AG(E)(32)1424⟩ and (**31**) (89.0 Hz) ⟨86CB(119)535⟩.

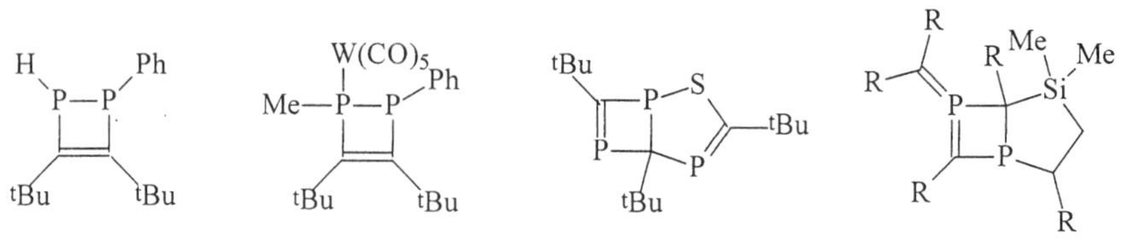

(**28**) (**29**) (**30**) (**31**) R = $SiMe_3$

Scheme 6

Under the conditions of the mass spectroscopy experiment, diphosphetanes may undergo a [2 + 2]-cycloreversion reaction. Examples are given by ⟨85PS(21)349, 85PS(25)345, 95ZAAC(621)1407⟩.

IR spectroscopy is predominantly used to generate finger print spectra. Decisive information is obtained, however, if stretching modes of polar groups connected to the rings are observable, e.g. C=N or C=O. He(I) PE spectra have been interpreted with the help of semi-empirical or *ab-initio* MO calculations ⟨76PS(2)283⟩. The two highest occupied MOs of $1\lambda^5,3\lambda^5$-diphosphetes of the type (**5**) (R = H, Ph; R′ = NEt_2) are π-ring orbitals ⟨88CB(121)2071⟩. Metal complexed $1\lambda^3,3\lambda^3$-diphosphetes interact significantly stronger with cobalt(I) and iron(0) than their carbacyclic analogue cyclobutadiene ⟨92OM(11)177⟩. According to its high symmetry, tetraphosphacubane (**16**) exhibits a triply degenerate delocalized state as HOMO. The first ionization takes place predominantly in the vicinity of the P-atoms and the second one is due to high-lying σ-MOs ⟨90CB(123)757⟩. EPR-data are only available from the redox products of (η^6-toluene)(η^4-di-*t*-butyl-1,3-diphosphete)iron ⟨87JOM(334)C35⟩. Evidence is found for a LUMO which consists of a combination of iron d-functions and π-orbitals of the diphosphete ligand, whereas the HOMO is located at the iron atom.

3.3.3.3 Thermodynamic Aspects

1,2-Dihydro-1,2-diphosphetes are thermodynamically favoured over other $C_2P_2R_4$-derivatives. Therefore 1,3-diphosphetane-2,4-diyl (**13**) as well as 1,2-dihydro-1,3-diphosphete (**32**), both rearrange quantitatively to form corresponding 1,2-dihydro-1,2-diphosphetes ⟨97JOM(529)177, 98AG(E)(37)949⟩.

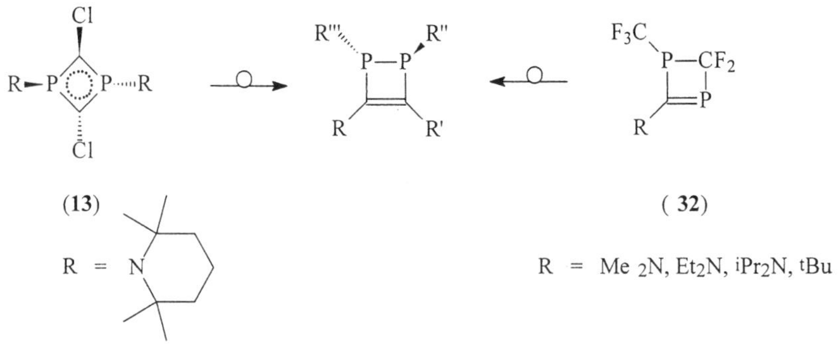

(**13**) (**32**)

R = Me $_2$N, Et_2N, iPr_2N, tBu

Scheme 7

The existence of an equilibrium between 1,2-dihydro-1,2-diphosphete and its 1,4-diphosphabutadiene open-chain isomer has been proved only for the bistungsten complex (**33**) ⟨90PO(9)1363⟩. The free ring shows no activity of that sort ⟨92JOC(57)2040⟩. Evidence for a dimerization equilibrium between phosphaalkene and 1,3-diphosphetane was found for the ethynyl compounds (**34**) and (**35**). The dimerization of (**35**) is a slow reaction at room temperature, daylight regenerates the monomer ⟨84CB(117)2693⟩. Flash vacuum pyrolysis of the iminodiphosphetane (**36**) gave access to the open-chain monomer ⟨83JA(105)7194⟩. In contrast to this, already slight heating of λ^5,λ^5-diphosphetane (**37**) yields the bis(methylene)phosphorane (**38**) ⟨92CB(125)771⟩.

Stereoisomerism in respect to the heteroelement is frequently reported. Two σ^3,λ^3 phosphorus atoms in a four ring principally lead to *cis* and *trans* isomers. 1,2-Dihydro-1,2-diphosphetes ⟨91AG(E)(30)207⟩ strongly favour the *trans* configuration. This situation changes and may be reversed, if a transition

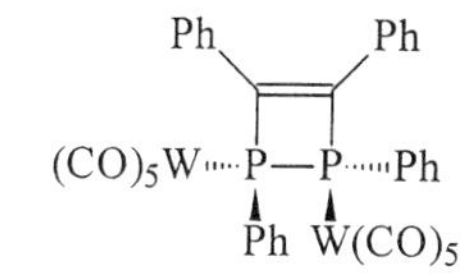

(**33**)

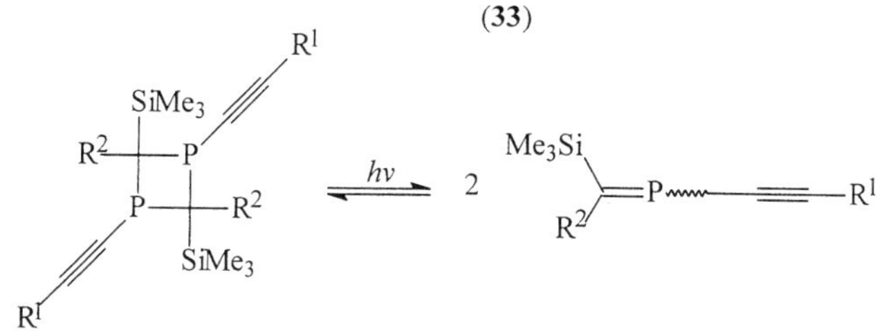

(**34**) R^1, R^2 = Ph, $SiMe_3$ (**35**)

(**36**)

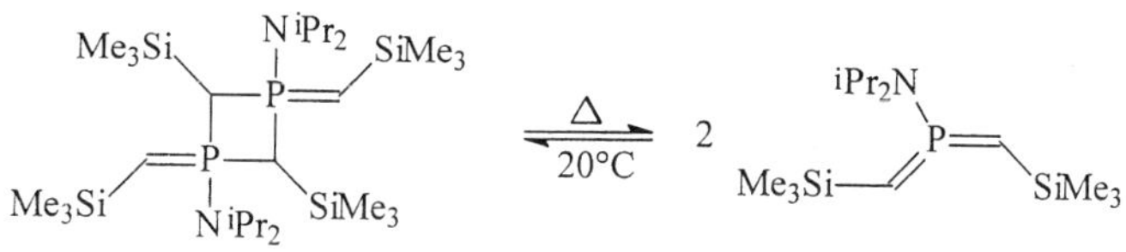

(**37**) (**38**)

Scheme 8

metal fragment like $W(CO)_5$ is linked to the phosphorus atom(s) of compound (**33**) ⟨97OM(16)2506, 97OM(16)4501⟩. 1,3-Rings do not discriminate as clearly between the two possibilities, thus (**39**) forms a *cis/trans* mixture (1 : 2) whereas (**40**) consists of the *cis* isomer only ⟨91AG(E)(30)207⟩. The same situation is created in λ^5,λ^5-rings if different substituents are attached to the phosphorus atoms. Surprisingly the *cis* isomer is found only in (**41**) ⟨92ZAAC(609)99⟩.

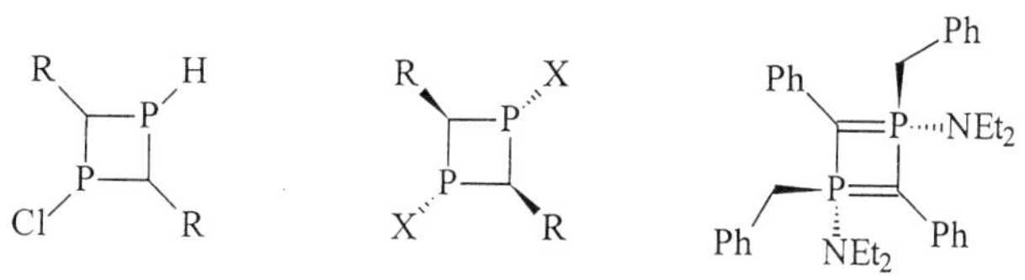

(**39**) (**40**) (**41**)

R = tBu; X = Cl, Br

Scheme 9

3.3.4 REACTIVITY STUDIES

3.3.4.1 Reactivity of $1\lambda^5,3\lambda^5$-Diphosphetes

The reactivity of $1\lambda^5,3\lambda^5$-diphosphetes (**20b**) (Scheme 10) was reviewed ⟨94RHA(11)65⟩. Electrophiles like proton active compounds, AlR_3, BF_3, InR_3, $GeCl_2$, $SnCl_2$, $Mo(CO)_5$, $W(CO)_5$, and $Fe(CO)_4$ add to one of the ring carbon atoms of the diphosphetes and salts are formed ⟨97ZAAC(623)919,

96ZAAC(622)942, 93ZAAC(619)1551, 92ZAAC(612)56, 86PS(29)23⟩. $Fe(CO)_5$ or CS_2 add to one carbon atom of (**20b**) and form inner salts, too (**42, 43**). A successive valence isomerization of the products, however, results in the formation of a dihydrophosphete. The second phosphorus atom moves to an exocyclic position ⟨96PS(115)3, 96ZAAC(622)974⟩. COS and CO_2 react in a similar way, but the zwitterionic intermediates are unobservable ⟨95ZAAC(621)1539⟩. Carbodiimides and alkynes cause a ring expansion and 2,4-diphosphapyridine ⟨98ZN(B)(53)443⟩ or 1,3-diphosphabenzene ⟨97JOM(529)223, 94ZAAC(620)483⟩ derivatives are obtained, respectively.

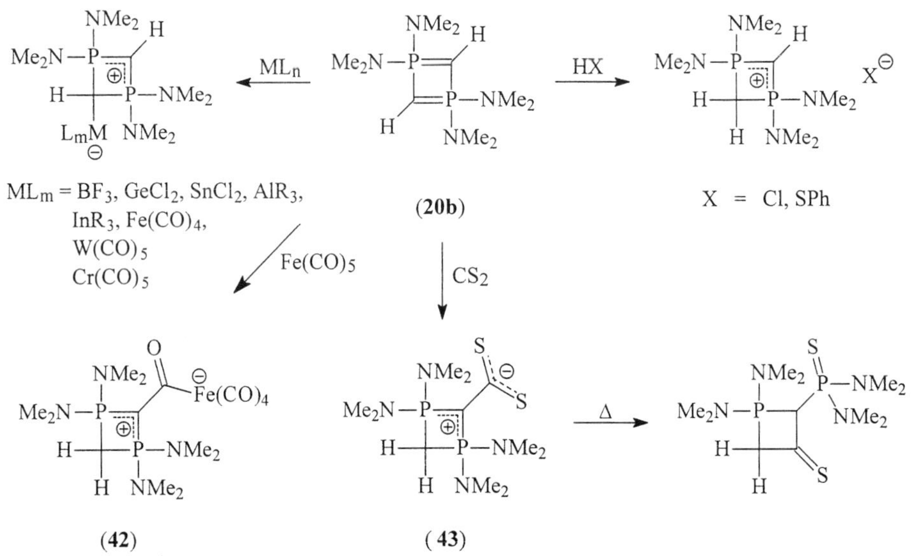

Scheme 10

3.3.4.2 Reactivity of $1\lambda^5,2\lambda^3$-Diphosphetes

The λ^3-phosphorus atom of $1\lambda^5,2\lambda^3$-diphosphetes (**6**) adds one or two electrophiles of different kinds ⟨97JA(119)9720, 99CEJ(5)274⟩. Photolysis of (**6**) gives rise to the isomeric 1,2-dihydro-1,2-diphosphete (**44**) ⟨99CEJ(5)274⟩.

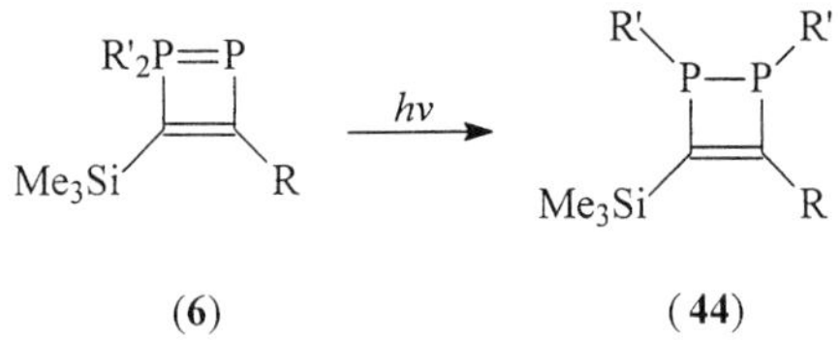

R = tBu, R' = NiPr$_2$

Scheme 11

3.3.4.3 Cyclic Addition Reactions of λ^3,λ^3-Diphosphetes

Oxidative decomplexation of a Hf-co-ordinated 1,3-diphoshete ⟨95AG(E)(34)2227⟩ or the diphosphabicyclobutanediyl ligand of the Zr complex (**45**) yield 1,3- or 1,2 diphosphetes as reactive intermediates, depending on the agents used. The 1,3-isomer effectively forms tetraphosphacubanes (**16**) upon double [2 + 2]-cyclic addition via tricyclic (**46**) ⟨92AG(E)(31)758⟩. Reactive intermediate 1,2-diphosphete (**47**) shows one [2 + 2]-cyclic addition step only and a tricyclic compound (**48**) is formed as the stable product ⟨95AG(E)(34)484⟩. Both, highly reactive (**46**), and stable (**48**), have a saturated central ring connecting two 1,2-dihydro-1,2-diphosphetes. The second cyclic addition step, however, is impossible for (**48**), as the unsaturated rings occupy different sides of the central one. A corresponding structure to (**48**) for the dimer of 1,3-diphosphete (**4**) is strongly disfavoured, because of sterical interactions of the bulky substituents. In contrast, (**47**) cannot form a dimer of type (**46**) as the main product, again because of strong repulsion by the substituents.

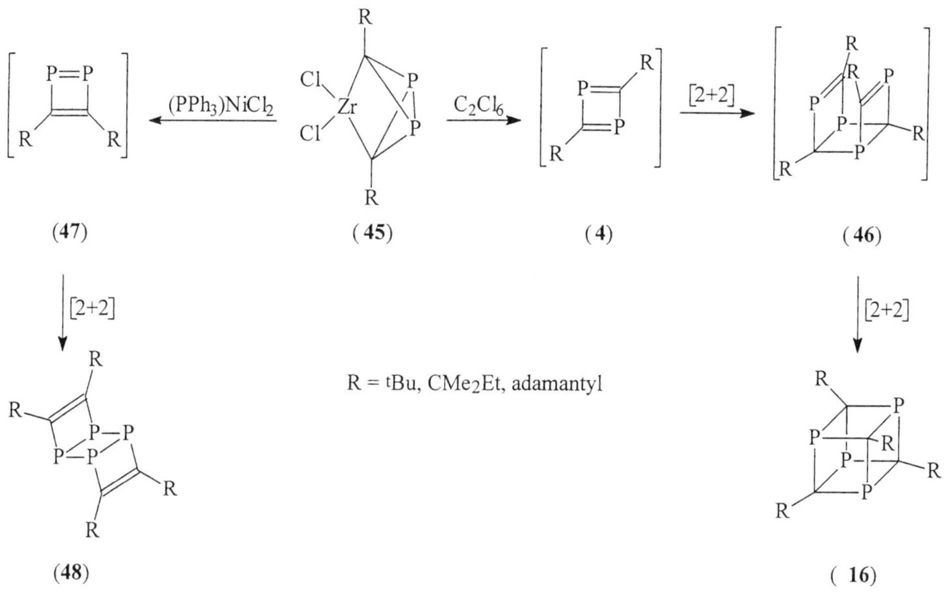

Scheme 12

3.3.4.4 Addition Reactions of λ^3,λ^3-Diphosphetes

If (**45**) is degraded in the presence of transition metal carbonyl complexes or halogenating agents, dimerization of (**4**) is prevented and complexes of the type (**9**) (ML_n = $Fe(CO)_3$) and P-halogenated 1,2-dihydro-1,2-diphosphetes (**25**) or 1,3-diphosphetanes (**39**, **40**) are formed ⟨91AG(E)(30)207⟩. The only successful decomplexation experiment of a 1,3-diphosphete ligand by C_2Cl_6 results in an 1,3 → 1,2-isomerization and the formation of a P,P-dichlorinated 1,2-dihydro-1,2-diphosphete derivative (**25**) ⟨96S265⟩.

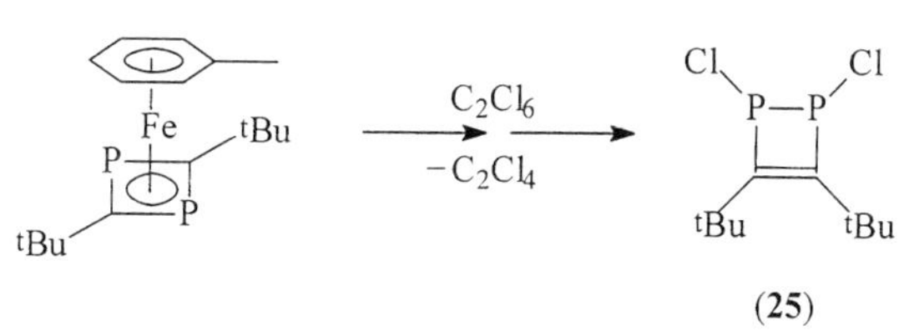

Scheme 13

3.3.4.5 Reactivity of Nonconjugated Rings

Treatment of 1,2-dihydro-3-alkynyl-1λ^5,2λ^5-diphosphetes with iminophosphanes leads to bicyclic ring expansion products ⟨94CB(127)313⟩. The 1,2-dihydro-1,2-diphosphete ring is transformed into a 1,2-diphosphacyclopent-3-ene which is condensed to a dihydrophosphete. P-imino substituted 1,2-dihydro-1λ^5,2λ^5-diphosphetes lose an iminophosphane unit upon heating and rearrange quantitatively to form a 1,2-dihydro-2-aza-λ^3-phosphete derivative ⟨96RJOC(32)407⟩.

Bis(tungstenpentacarbonyl) complexes of 1,2-dihydro-1,2-diphosphetes exhibit a formal ring expansion reaction with electron-poor dienophiles. 1,4-Diphosphacyclohexenes or cyclohexadienes are formed. This reaction was interpreted in terms of a [4 + 2]-cyclic addition reaction of the open-chain diphosphabutadienes ⟨90PO(9)1363⟩. 1,3-Dihydro-1,3-diphosphete (**13**) exhibits a unique reactivity. Thermolysis leads to the corresponding 1,2-dihydro-1,2-diphosphete. A phosphanylcarbene is believed to be the decisive intermediate ⟨98AG(E)(37)949⟩. An alternative valence isomerization process utilizes a photochemical ring closure of a derivative (**13a**) to yield diphospha bicyclobutane (**49**) and its thermal ring opening product *gauche*-1,4-diphosphabutadiene (**50**) ⟨99AG(E)(38)3028⟩; (**13a**) may be deprotonated to yield the anionic diphosphacarbene (**51**). Addition of $AlMe_3$ to (**51**) results in the formation of the stable alane adduct (**52**) ⟨99AG(E)(38)3031⟩.

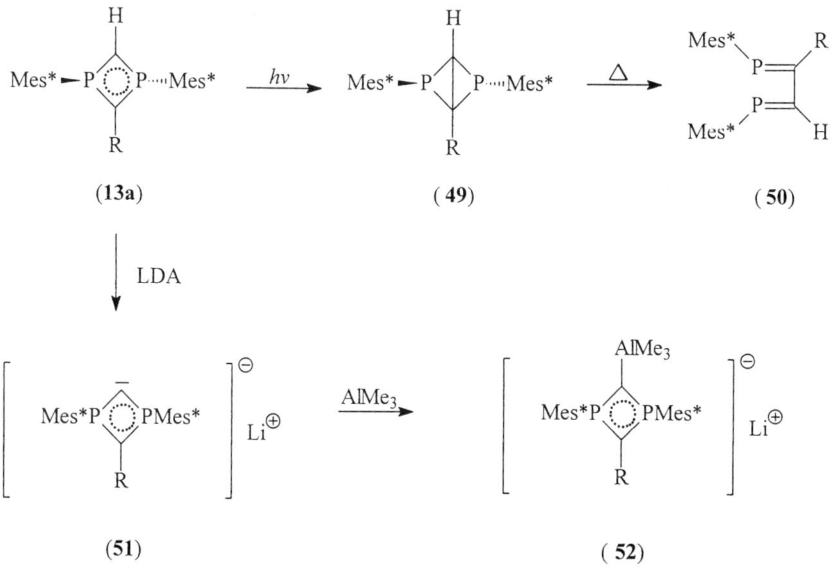

Scheme 14

Dienophiles like phosphaalkyne (**54a**) or alkynes and triphosphadewarbenzene derivative (**53**) undergo a [2 + 2 + 2] homo Diels–Alder reaction which leads to tetracyclic products like (**55**). Cyclic dienophiles yield pentacycles ⟨97JOM(529)215, 99S1363⟩.

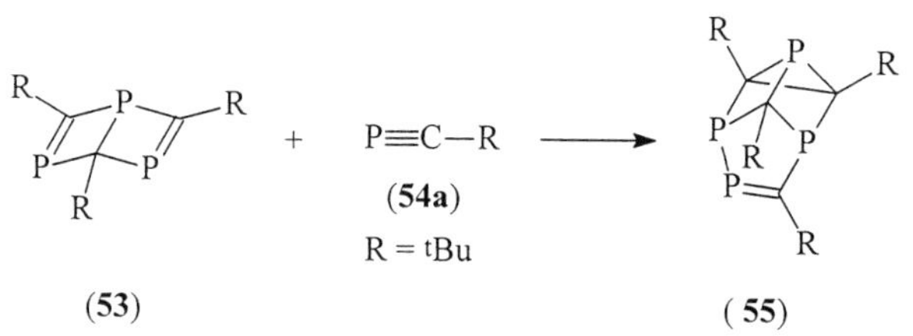

Scheme 15

The P–P-bond of a 1,2-dihydro-1,2-diphosphete derivative may be opened selectively by controlled hydrolysis and a phosphaalkene is formed ⟨97JOC(62)7605⟩.

3.3.4.6 Reactivity of Ring Heteroatoms and Substituents Attached to Ring Heteroatoms

Mono- or diprotonation of tetraphosphacubane or tricyclic 1,2-dihydro-1,2-diphosphetes with superacids and monoalkylation or alkynylation with strong electrophiles leads to stable phosphonium salts ⟨93JOC(58)4105, 95JOC(60)47, 95JOC(60)3149⟩. One or both chloro substituents of 2,4-diphosphoranediyl-1,3-diphosphetane (**56**) may be removed as chloride anions by $AlCl_3$. The resulting electronic deficiency of one phosphorus atom causes an electronic rearrangement, which imports the exocyclic double bond into the ring and 1,2-dihydro-1,3-diphosphete salts like (**57**) are formed ⟨95AG(E)(34)1853⟩; (**57**) itself gives access to several derivatives ⟨97CB(130)1519⟩.

3.3.4.7 σ-Complexation of Transition Metals

λ^3-Phosphorus atoms as parts of rings may function as σ-donor ligands of transition metals in many examples. The above-mentioned reviews present the details. Such metal-ligand units may remain intact over several reaction steps, including open-chain starting materials and products ⟨88IC(27)2269, 88OM(7)1791,

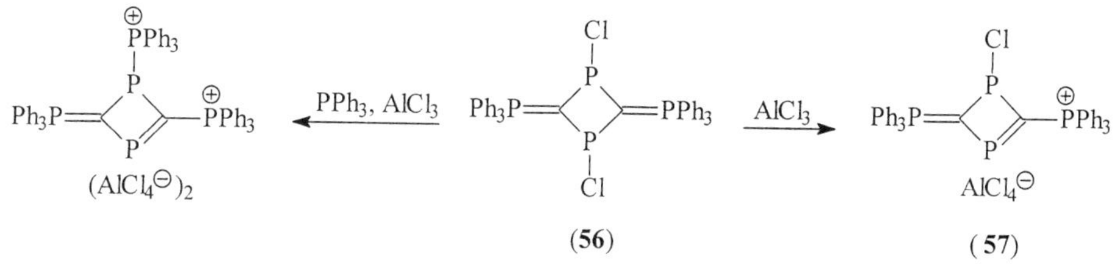

Scheme 16

92BSF(129)76, 97AX(C)(53)42, 98CC1537⟩. π-Co-ordinated diphosphetes still exhibit ligand properties through the lone pairs of the phosphorus atoms. One or two transition metals may be bonded this way ⟨86AG(E)(25)644, 91HAC(2)253⟩. 1,2-Dichloro-1,2-diphosphete (**25**) adds two $Fe(CO)_4$-fragments at room temperature. The resulting diiron complex (**26**) allows the first targeted preparation of an 1,2-diphosphete derivative (*vide infra*) ⟨99OM(18)2021⟩.

3.3.4.8 Oxidation of Ring Heteroatoms

The oxidation of λ^3-phosphorus atoms of stable heterocycles is a general reaction without disruption of the basic structures. It includes diphosphetanes, especially tetraphosphacubane (**16**) or other related P–C-cage compounds, and monocyclic or fused 1,2-dihydro-1,2-diphosphete derivatives like (**48**). Oxidation agents used are: tetrachloroorthochinone, RN_3, diazoalkanes, S_8, Se_8, $PSCl_3$, and $[(CH_3)_3Si]_2O_2$ ⟨86OM(5)623, 92AG(E)(31)879, 94S1337, 94ZN(B)(49)1361, 95HAC(6)1, 95JOC(60)3149, 97CB(130)1519, 97JOC(62)7605⟩.

3.3.4.9 Reactions with Reducing Agents

Alkali metals, Li alkyls, and Pt(0)-complexes are able to cleave the P–P bond of 1,2-dihydro-1,2-diphosphetes or one substituent of a phosphorus may be removed as well ⟨84JCS(CC)1707, 92JA(114)9453, 90AG(E)(29)534⟩. The resulting mono and dianion lithium salts (**58**) and (**59**), respectively, give access to many new 1,2-dihydro-1,2-diphosphetes (**60**) and open-chain *cis*-diphosphaethene derivatives (**61**). If the incoming group R^1 of (**60**) is an ω-chloroalkyl, self quaternization leads to bicyclic (**62**) and related compounds ⟨86OM(5)623⟩. The reaction of dianion (**59**) with organo dihalides leads to diphosphacyclopentene derivatives: Me_2SiCl_2 yields siladiphosphacyclopentenes ⟨92OM(11)459, 94BSF(131)397⟩, methylene dichloride diphosphacyclopentenes ⟨92PO(11)601⟩, and C_2Cl_4 tetraphosphafulvalene, ⟨88AG(E)(27)950⟩, respectively.

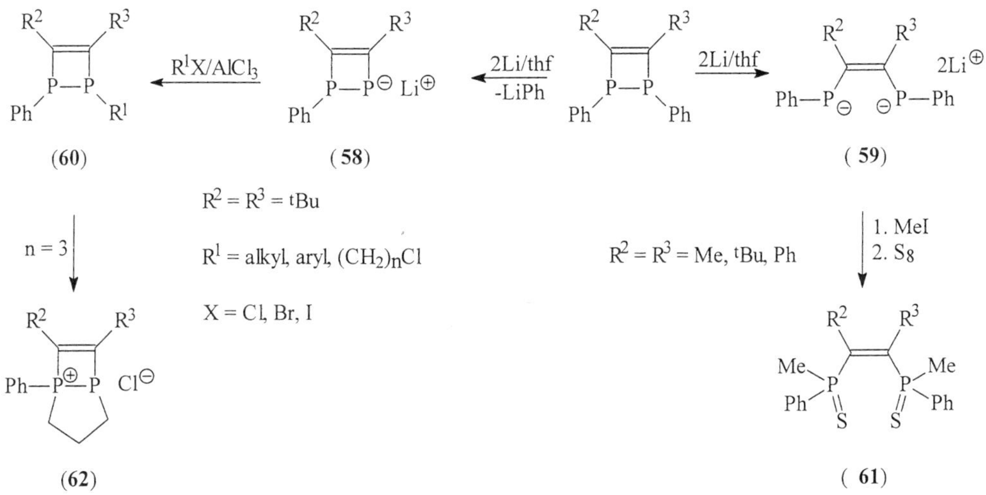

Scheme 17

3.3.4.10 Other Reactions

$1\lambda^5,2\lambda^3$-diphosphete (**6**) reacts with a variety of substrates including transition metal complex fragments, alkylating, and strongly oxidizing agents. The λ^3-phosphorus atom is the reactive part of (**6**), and the P–P-bond persists in all cases ⟨97JA(119)9720⟩.

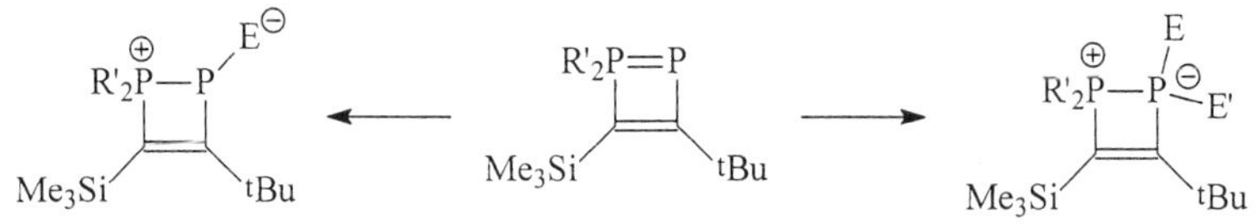

(**6**)

R' = iPr$_2$N; E = BH$_3$, Se, Fe(CO)$_4$, W(CO)$_5$; E, E' = O, O; S, S; Se, Se; W(CO)$_5$, O.

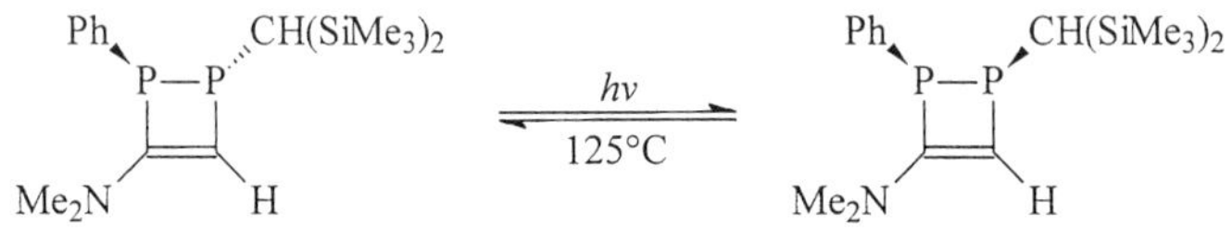

(**63a**) (**63b**)

Scheme 18

Chlorine iodine exchange reaction on 1-chloro-2-hydro-1,3-diphosphete was furnished with $ISiMe_3$ ⟨93BSF(130)642⟩. The reactions of $Mn_2(CO)_{10}$ and 1,2-diphenyl-1,2-diphosphete derivatives result in the formation of a binuclear Mn_2 complex. It contains a ring expanded $Mn(PPh)_2C_2R_2$ unit, which is π-complexed to a $Mn(CO)_3$ fragment ⟨95IC(34)5070⟩. Photolysis of *trans*-1,2-dihydro-1,2-diphosphete (**63a**) converts it partly into its *cis* isomer (**63b**). The mixture of isomers can be reconverted completely into the *trans* isomer (**63a**) ⟨97JOC(62)7605⟩.

3.3.5 SYNTHESIS OF FOUR-MEMBERED HETEROCYCLES WITH TWO P-ATOMS

3.3.5.1 From Acyclic Compounds by Formation of one P–C- or P–P-Bond

One of the rare examples of P–C-bond formation was discovered by Niecke *et al.*, when they prepared 1,2-dihydro-$1\lambda^5,3\lambda^5$-diphosphete (**65**). The final step starts with a 1,3-H-shift from a methyl group to the negatively charged carbon of (**64**) and the ring is closed. The authors prefer a polar formulation for (**65**) ⟨94AG(E)(33)982⟩.

A high-yield route to *trans*-3,4-diisopropyliden-1,2-diphosphetane (**12**) utilizes reductive P–P-coupling of a bis(chlorophosphane) ⟨93CB(126)845⟩.

3.3.5.2 Formation of Two C–P Bonds

An Arbusov reaction of diphospha species (**66**) and methylene dichloride leads to λ^5,λ^5-1,3-diphosphetane (**67**) ⟨77JGCU(47)2409⟩.

Dipotassium diphosphide (**68**) and 1-chloro-2-bromoethane form 1,2-diphosphetanes in a nucleophilic substitution reaction ⟨84TL(25)1963⟩. Hydrostannylation of a P-chloro phosphaalkene and subsequent R_3SnCl elimination is a low-yield route to specific 1,3-diphosphetanes ⟨98HAC(9)453⟩. Hydrostannylation of (**54**) leads in moderate to good yield to stannyl-1,2-dihydro-1,3-diphosphete derivatives (**69**) ⟨98EJIC227⟩. In a 1,3-cyclosubstitution reaction of phosphaalkene (**70**) the lone pairs of the λ^3-P-atoms attack the carbon atoms of the other molecule and chloride is eliminated. The λ^5,λ^5-diphosphete diphosphonium salt (**71**) is very stable and exhibits no ylid reactivity ⟨77AG(E)(16)402⟩. $(NMe_2)_3P{=}CCl_2$ is an alternative educt for salts of the type (**71**) (R = NMe_2) ⟨83CB(116)2275⟩.

Bis(trimethylsilyl)ylide (**72**) reacts with phosphorus trichloride or tribromide to give dihalophosphino (trimethylsilyl)ylide (**73**). In a two-step self condensation of (**73**) two equivalents of halogeno(trimeth-

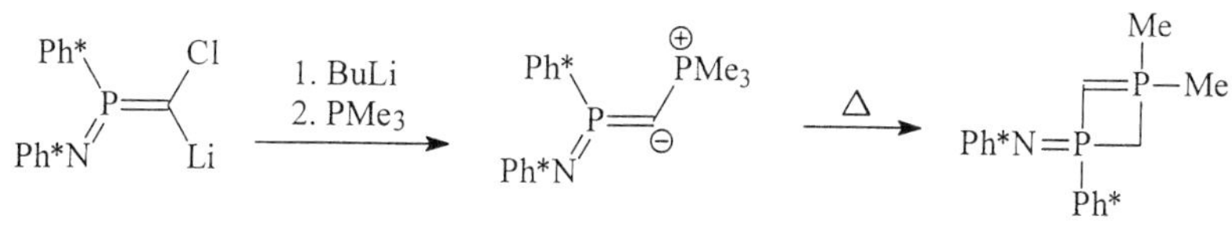

Ph* = 1,3,5-$^{t}Bu_3C_6H_2$ (**64**) (**65**)

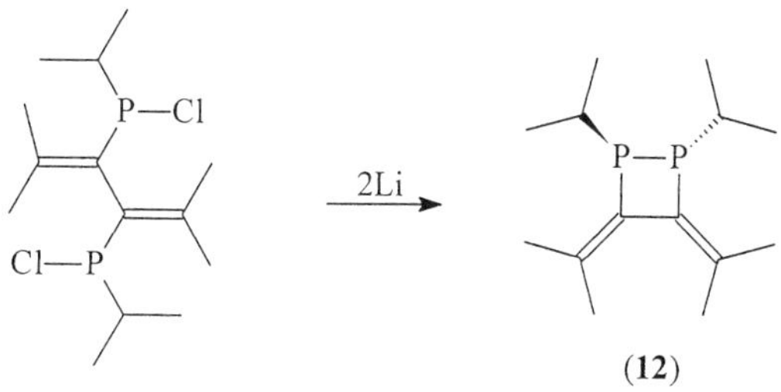

(**12**)

Scheme 19

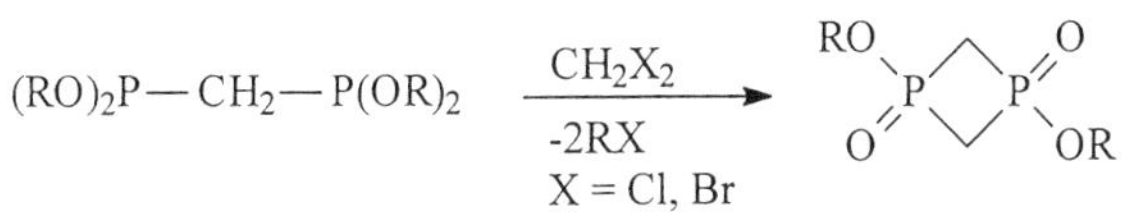

(**66**) (**67**)

Scheme 20

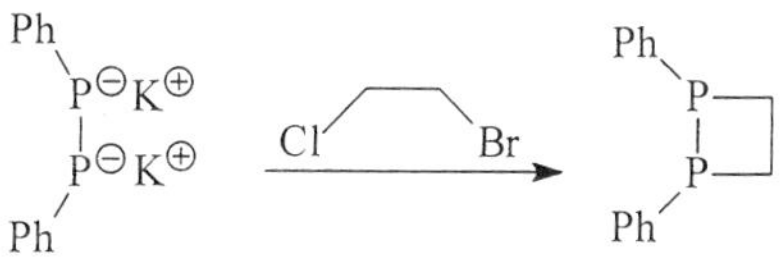

(**68**)

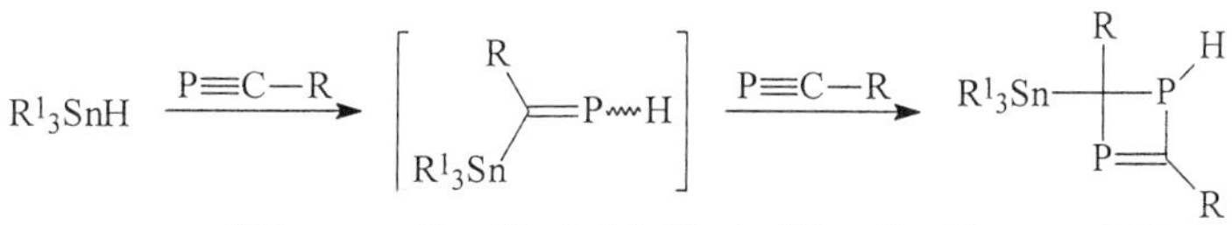

(**54**) R = ^{t}Bu, 1-Ad, ^{t}Pent; R^1 = ^{n}Bu, Ph (**69**)

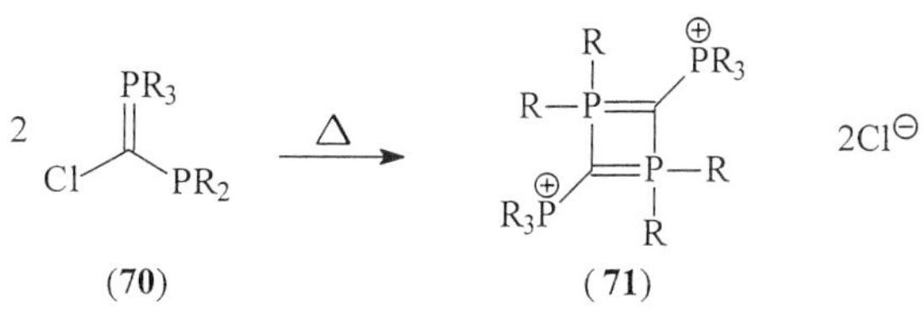

(**70**) (**71**)

R = NMe_2, Ph

Scheme 21

yl)silane are eliminated and P,P-dihalogeno-2,4-diphosphoranediyl-1,3-diphosphetanes (**74**) are formed ⟨95AG(E)(34)1853, 97CB(130)1519⟩.

3.3.5.3 Ring Formation by Organic Cyclic Addition Reactions

Cyclic addition reactions of unsaturated organophosphorus compounds are the most prominent routes to 4-ring heterocycles containing two phosphorus atoms.

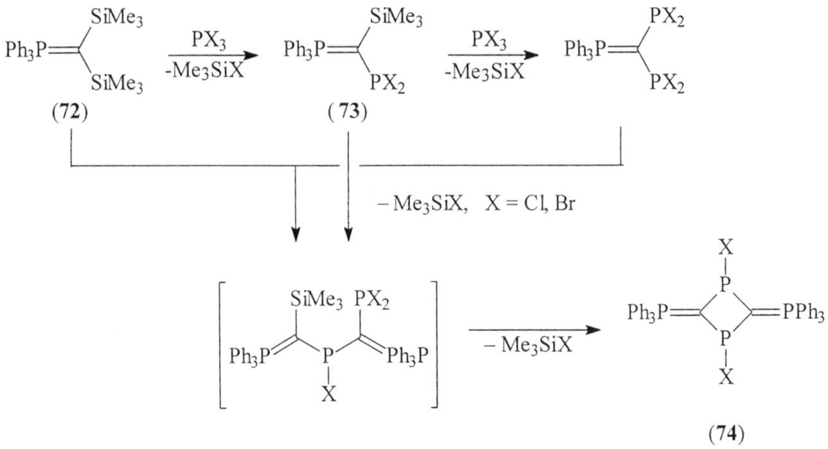

Scheme 22

λ^5-Phosphaalkyne is formed as a reactive intermediate by photolysis of diazo compound (**75**). It dimerizes spontaneously in a head-to-tail manner, yielding $1\lambda^5,3\lambda^5$-diphosphete (**76**) ⟨90AG(E)(29)1429⟩. In the presence of *t*-butyl phospaalkyne (**54a**) a head-to-head cross-coupling occurs and $1\lambda^5,2\lambda^3$-diphosphete (**6**) is formed in high yield ⟨95JA(117)10785⟩. A partial oxidation of amino substituted phosphaalkyne (**54b**) by sulphur or selenium leads to the quantitative formation of 1,2-dihydro-$1\lambda^5,3\lambda^3$-diphosphetes (**7**) ⟨00T(56)27, 00T(56)1147⟩.

$1\lambda^5,3\lambda^5$-Diphosphetanes (**37**) and (**77**) (*cis* and *trans* isomers) are formed by a head-to-tail [2 + 2]-cyclic addition reaction of the formal C=P-double bonds of bis(methylene)- or iminomethylenephosphoranes (Section 3.3.3.3) ⟨80CB(113)1549, 92CB(125)771⟩.

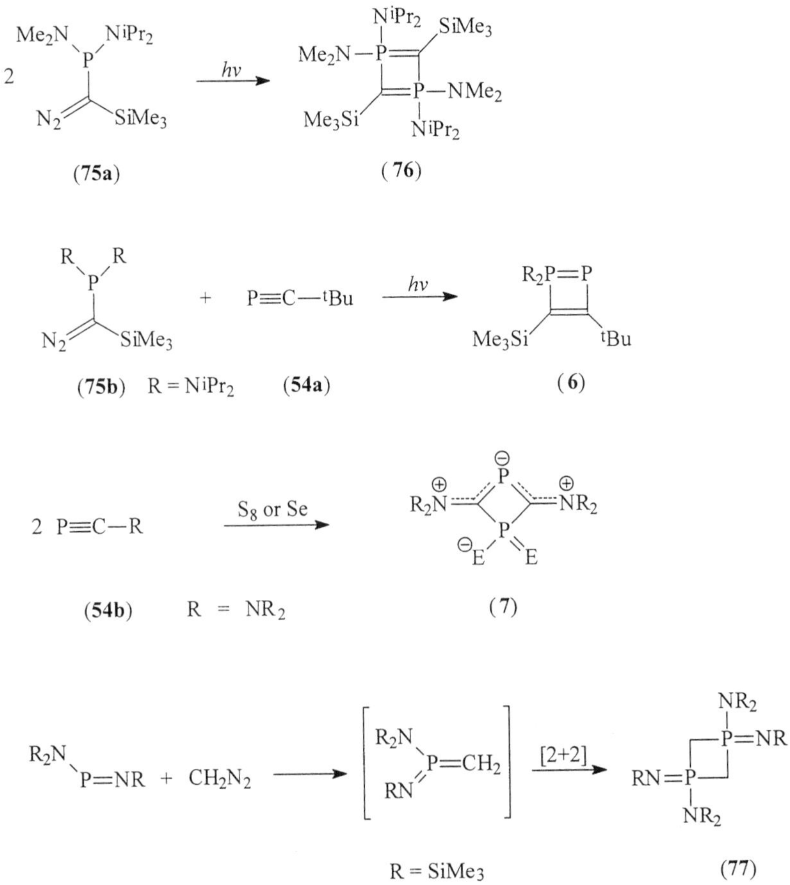

Scheme 23

The preparative potential of phosphaalkenes, including *in-situ* prepared species, was explored soon after the first access to this class of compounds and an early review of Appel *et al.* outlined the most important facts ⟨81AG(E)(20)731⟩. The preferred route of cyclodimerization is head-to-tail, which yields 1,3-diphosphetanes ⟨76ZOK(46)2268, 84ZAAC(517)75, 80ZAAC(462)130, 81ZAAC(479)41, 82JGCU(52)2465, 85PS(25)345, 91CB(124)2843⟩. The same is observed for the fluorinated species RP=CF_2 (R = alkyl, aryl, CF_3) ⟨86IC(25)3177, 92ZN(B)(47)321, 94ZN(B)(49)1361, 96AX(C)(52)1919⟩. Phosphacummulenes RP=C=X (X = CR_2, NR, O) are often postulated to be reactive intermediates, exhibiting the same reactivity as phosphaalkenes. They lead to 2,4-dimethylene-1,3-diphosphetane derivatives ⟨80TL(21)2497, 83TL(24)2639, 86CB(119)2466, 86ZAAC(534)31, 90TL(31)4429⟩. The alternative head-to-head cyclodimerization leads to 1,2-diphosphetanes ⟨84CB(117)3151, 86CB(119)2915⟩. An interesting example was given by Grützmacher and Pritzkow. Reactive phosphaalkene (**79**) is formed by catalytic alkanechloride abstraction of the P-chlorinated ylid (**78**). At room temperature it dimerizes, yielding the *trans*-1,2-diphosphetane (**80**) ⟨92AG(E)(31)99⟩. [2 + 2]-Cyclic addition of an alkene (**82**) with a diphosphene (**81**) results in the formation of 1,2-diphosphetanes (**83**) exclusively. Dimetallated diphosphene and electron-poor alkenes exhibit similar reactivity ⟨90CB(123)733⟩. Monometallated diphosphene and nitriles react in a [2 + 1 + 1]-manner to form 1-ferrio-2,4-diimino-1,3-diphosphetanes ⟨94OM(13)4406⟩. Lithoxyphosphaalkyne $(DME)_2LiOC{\equiv}P$ (**54c**) reacts smoothly with $Cp^{**}(CO)_2Fe$ (Cp^{**} = C_5Me_5 or other bulky substituted Cp derivatives) to afford 1,3-diferrio-1,3-diphosphetane-2,4-diones (**84**) by a [2 + 2]-cyclic addition reaction of a phosphaketene intermediate ⟨96OM(15)128⟩. Corresponding 2,4-diimine derivatives of (**84**) are accessible as well ⟨95ZAAC(621)1407⟩.

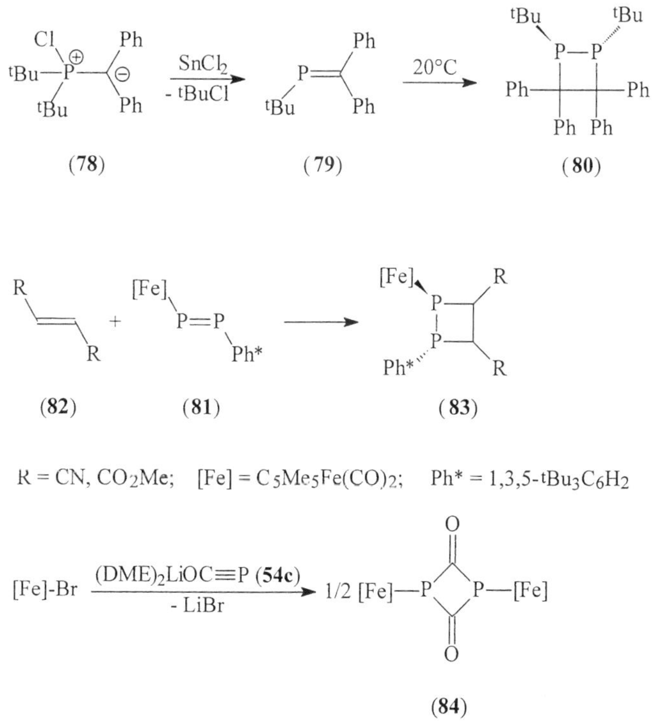

Scheme 24

Tetraphosphacubane (**16**) is formed in small amounts by the thermal cyclooligomerization of pure phosphaalkyne (**54**) ⟨89AG(E)(28)1013⟩. Much better yields of (**16**) are obtained by using 1,3-diphosphete as a reactive intermediate (Section 3.3.4.3).

The reaction of electrophiles E = AlX_3, $B(OTf)_3$, and GaX_3 with three phosphaalkyne molecules leads almost quantitatively to the spirocyclic inner salt (**85**) ⟨95JOC(60)6362, 96CB(129)489⟩. Bases like DMSO remove $AlCl_3$ from (**85**). Repetitive cyclic addition reactions of the intermediates and an additional phosphaalkyne generates the tetraphospha tetracycles (**55**) or (**86**) ⟨92AG(E)(31)1055⟩. Photochemical activation of (**86**) causes a complex rearrangement process, which gives access to other polycyclic tetramers of (**54**) ⟨97AG(E)(36)1337⟩. $GaEt_3$ add three, and $AlEt_3$ four molecules of (**54**), respectively, to form a metal containing polycycles with 1,3-diphosphetane substructure units ⟨98EJIC1597⟩. Product (**87**) loses its AlEt-fragment, when reacted with $Fe_2(CO)_9$, to yield (**88**) ⟨97JOM(539)61⟩.

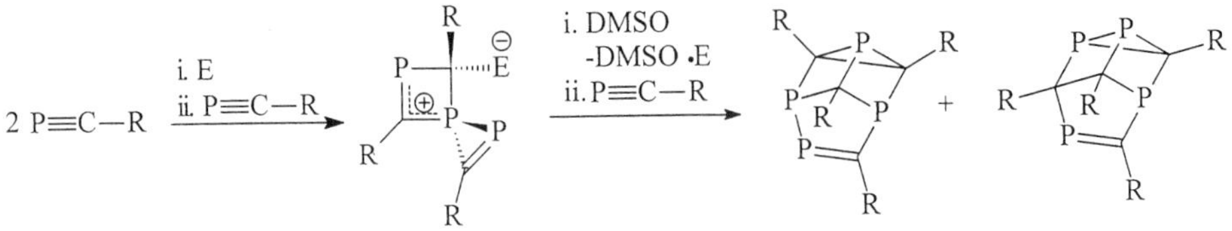

(54a) E = AlCl3 **(85)** **(55)** **(86)**

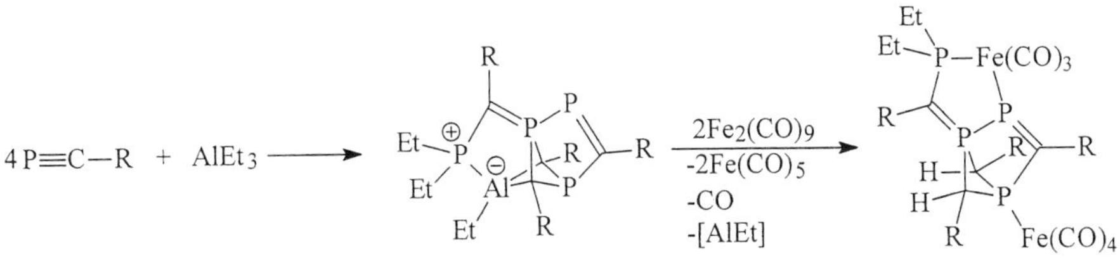

(54a) **(87)** **(88)**

Scheme 25

Complex trapping reactions of in situ generated silene or phosphinidene by (**54**) results in the formation of tricyclic frameworks (**89, 90**), which contain 1,2-diphosphetane subunits ⟨94JOM(464)41, 95CB(128)991⟩.

A 1,2-diphosphetane as part of a hexacyclic diphospha hydrocarbon was made photochemically ⟨83JA(105)6871⟩. Bisphosphaallene (**91**) is a spectroscopically observable intermediate, which undergoes an intramolecular [2 + 2]-cyclic addition reaction at room temperature forming 3-naphtho-1,2-dihydro-1,2-diphosphete, (**92**) effectively ⟨97LA121⟩.

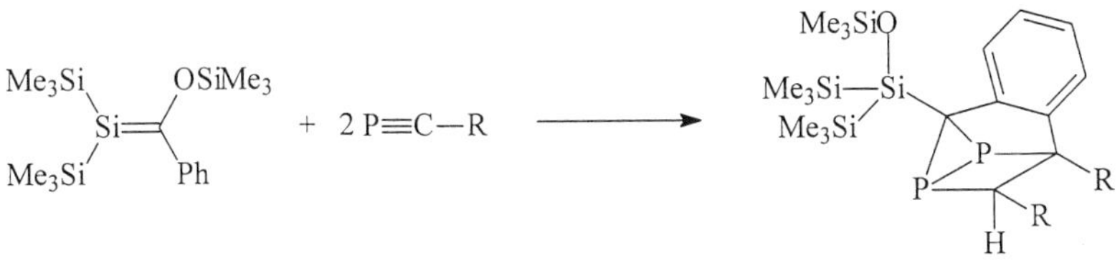

(54a) R = tBu **(89)**

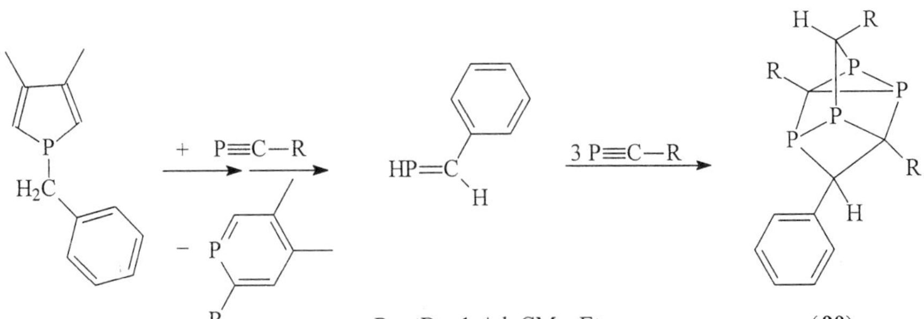

R = tBu, 1-Ad, CMe2Et **(90)**

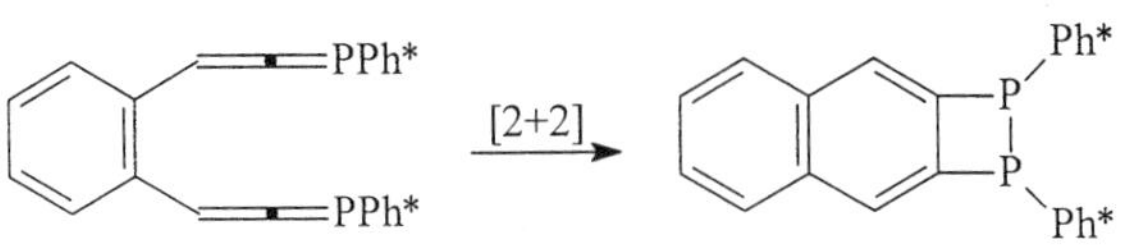

(91) Ph* = 1,3,5-tBu3C6H2 **(92)**

Scheme 26

The head-to-tail [2 + 2]-cyclic addition of perfluorophosphaalkene and phosphaalkyne derivatives yield 1,2-dihydro-1,3-diphosphetes (**32**) ⟨97JOM(529)177⟩.

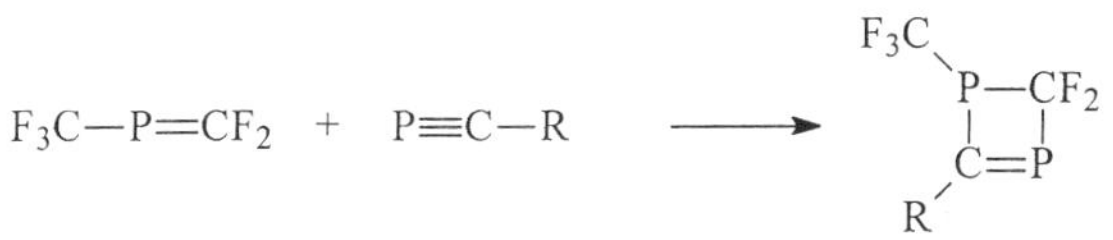

R = Me_2N, Et_2N, $^iPr_2N$, ^tBu; (**32**)

Scheme 27

3.3.5.4 Ring Formation by Cyclic Addition Reactions in the Co-ordination Sphere of a Transition Metal

Phosphaalkynes (**54**) can be used very successfully for the synthesis of transition metal co-ordinated 1,3-diphosphetes in complexes of type (**9**). This field was opened by Binger, Nixon, Regitz, *et al.* ⟨86JCS(CC)737, 86AG(E)(25)644⟩; some reviews reveal the most important facts ⟨88AG(E)(27)1484, 88CRV(88)1327, B-93MI5⟩. Examples for transition metal complexes used for this purpose are $CpM(C_2H_4)_2$ (Cp = cyclopentadienyl derivatives including indenyl, dicarbollide; M = Co, Rh, Ir), $M(CO)_5L$ (M = Cr, Mo, W; L = THF) ⟨95CC1671⟩, $M(CO)_3L_3$ (M = Mo, W; L = CH_3CN) ⟨98ZAAC(624)399, MI3⟩, $Fe_2(CO)_9$ (η^6-toluene)FeL_2 (L = C_2H_4, L_2 = η^4-1-methylnaphthalene) ⟨89JCS(CC)1409, 95AG(E)(34)198⟩, and (1,5-COD)$_2$Ni (cyclodimerization of diisopropylaminophosphaalkyne) ⟨96ZAAC(622)24⟩. From the reaction of $[(C_2H_4)_2RhCl]_2$ with (**54**) some mechanistic insight into these cyclic addition reactions was acquired. The two intermediates (**93**) and (**94**) have been fully characterized ⟨88AG(E)(27)1157⟩. A comparable co-ordinated 1-hydro-2,4-diphosphete substructure has been found for an iron complex of a heptamer of (**54a**) ⟨89AG(E)(28)900⟩ and a few other complexes ⟨94AG(E)(33)463, 94JOM(466)153⟩. Homoleptic 2,4-di-*tert*-butyl-1,3-diphosphete complexes are formed in the reaction of (acac)$_2$Ni with ^{n}BuLi and (**54a**), ⟨90JCS(CC)1541⟩ and reactions of Mo-vapour ⟨94JCS(CC)489⟩ or [*fac*-$(CO)_3(MeCN)_3Mo$] ⟨MI3⟩ with (**54a**) result in the formation of complexes (**95**) and (**96**). Tris(1,3-diphosphete)Mo complex (**96**) exhibits a unique temperature-dependent, reversible P–C bond formation between two of the heterocycles in the solid state ⟨MI3⟩.

Two routes have been reported for the preparation of 1,2-diphosphete complexes. Cyclodimerization of (**54**) at a cyclooctatetraene titanium fragment results in the formation of a mixture of 1,2- and 1,3-diphosphete (COT)Ti complexes ⟨95CB(128)1261⟩. A specific preparation of a 1,2-diphosphete iron complex is based on a thermal dechlorination and rearrangement of *trans*-binuclear 1,2-dichloro-1,2-diphosphete complex (**26**) ⟨99OM(18)2021⟩.

The first cyclotrimer of (**54**), characterized unambiguously, was formed at a V(I) centre. The triphospha ligand of (**98**) is a triphospha Dewar benzene derivative ⟨87AG(E)(26)908⟩. Free triphospha Dewar benzene (**53**) is formed by oxidative decomplexation of the triphospha cyclohexa-1,4-diyl-2,5-diene ligand of Hf complex (**99**) ⟨95AG(E)(34)2227, 97CB(130)1491⟩.

Compound (**54**) may add to complex cluster structures under formation of cyclic C_2P_2-substructures as well. A good example was given by Huttner and Imhof ⟨93JOM(447)31⟩. An organometallic [2 + 1 + 1]-cyclic addition reaction between the phosphinidene complex (**100**), (**54**) and a co-ordinated CO gives access to 1,2-diphosphacyclobutenone complex (**101**) ⟨98CEJ(4)1917⟩.

Cyclotetramerization of (**54**) at a vanadium(V) centre, metal to phosphorus transfer of a ^{t}BuN-fragment, and subsequent elimination of an alkyne are the main steps of the high-yield formation of aza tetraphospha quadricyclane (**102**) with its 1,3-diphosphetane basic ring ⟨98AG(E)(37)1233⟩.

3.3.5.5 Various Routes

Dehalogenation of P-fluorinated λ^5-phosphaalkenes is a useful route for the preparation of 1λ^5,3λ^5-diphosphetes ⟨90ZAAC(588)147, B-89MI4, 98ZAAC(624)1116⟩. 1,3-Dihydro-1,3-diphosphete (**13**) was found to be the product of the chlorine abstraction of phosphaalkene (**103a**) with ^{n}BuLi ⟨95AG(E)(34)555⟩. As reported above, rearrangement reactions lead to some 1,2-dihydro-1,2-diphosphete derivatives ⟨97JOM(529)177, 98AG(E)(37)949⟩. However, if the P-amino phosphaalkene derivative (**103b**) is dehalogenated by a group-4 metallocene, (**104**) is formed directly ⟨97OM(16)4551⟩. A complex rearrangement and oxidation process takes place in the case of the first successful ligand displacement of a 1,3-diphosphete ligand. It results

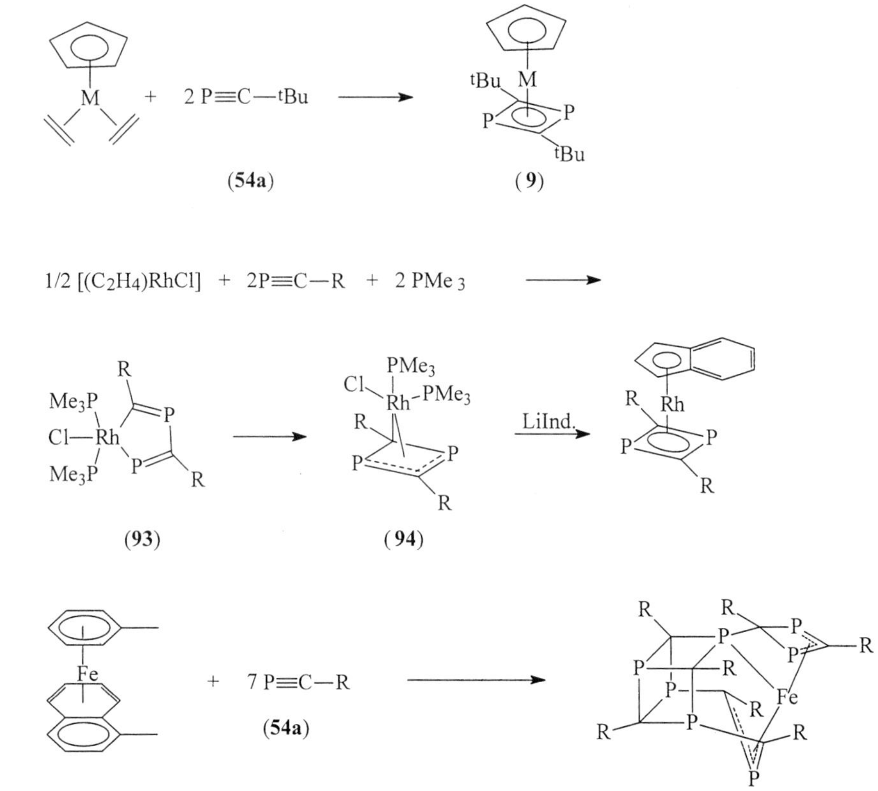

Scheme 28a

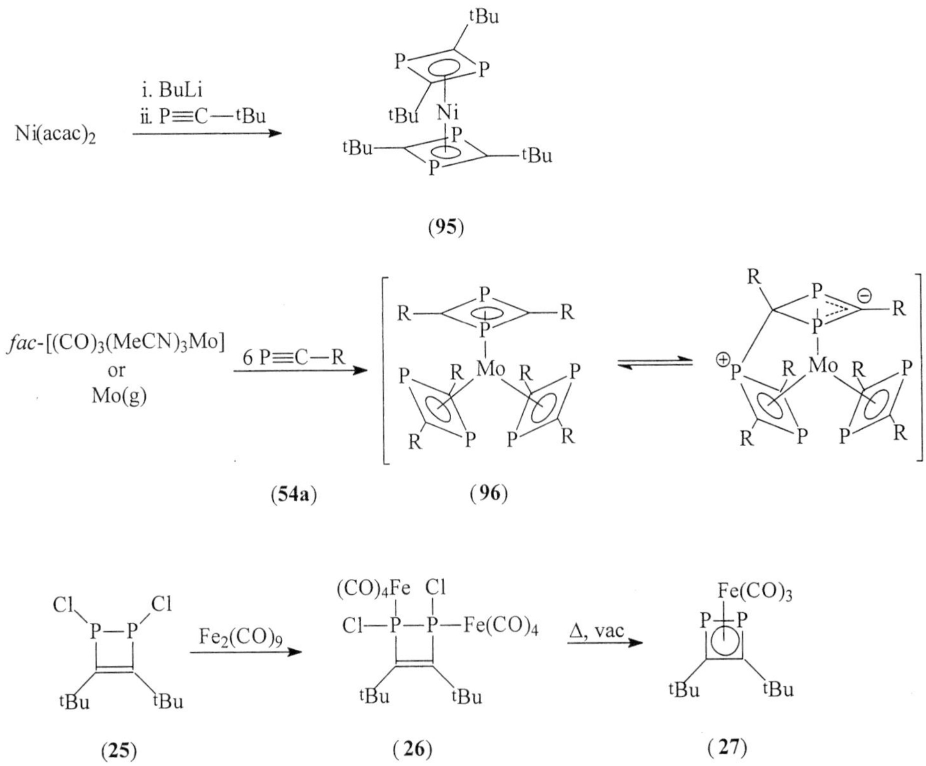

Scheme 28b

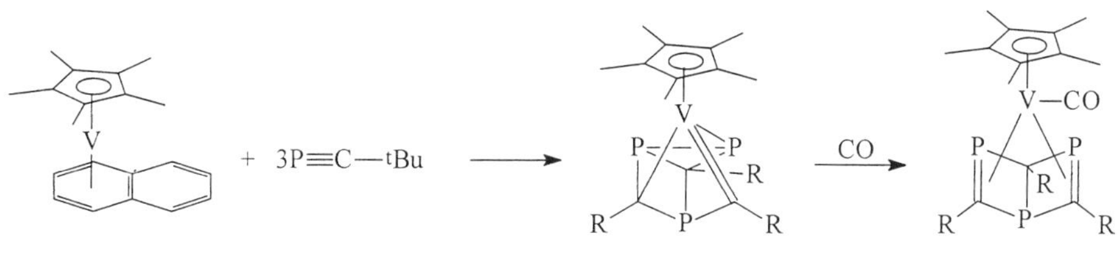

(54a) (97) (98)

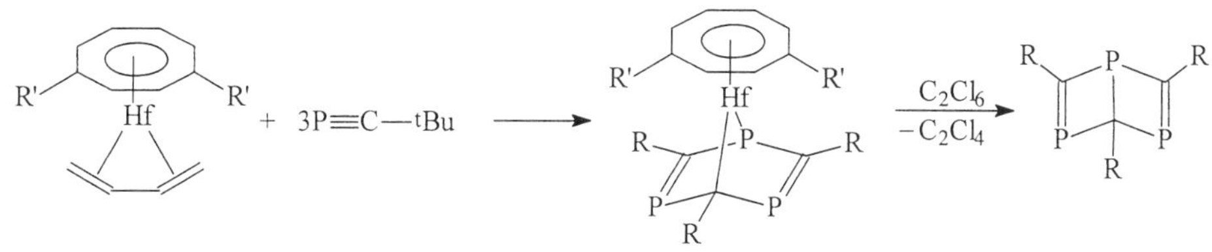

R' = H, SiMe3 (54a) (99) R = tBu (53)

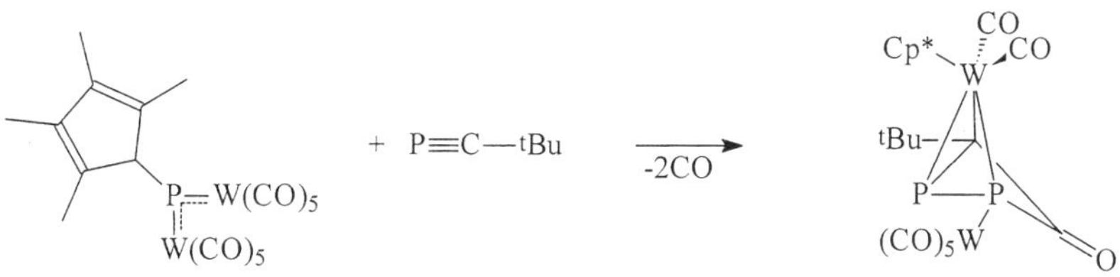

(100) (54a) (101)

Scheme 29

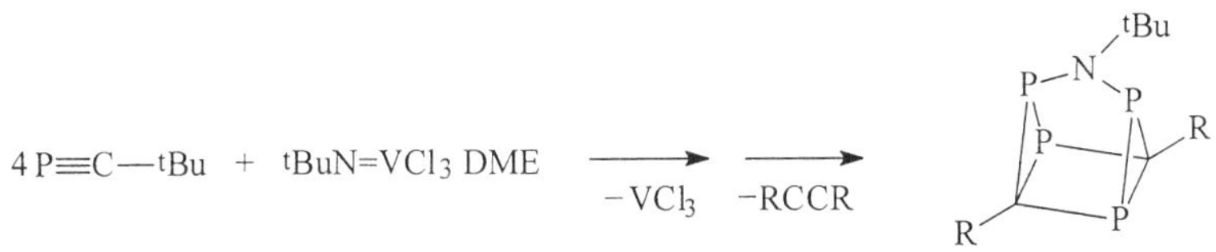

(54a) (102)

Scheme 30

in the formation of 1,2-dichloro-1,2-diphosphetane (**25**) most probably via a tetracyclic intermediate ⟨96S265⟩.

The first and simplest route to 1,2-dihydro-1,2-diphosphetes (**1**) utilizes cyclopolyphosphines and alkynes (max. 55% yield) ⟨64JA(86)2306, 73CB(106)1453, 86OM(5)623⟩. Somewhat related to this method are the photocleavage of a P-phenylphosphaalkene in the presence of tolane ⟨83TL(24)1975⟩ and the ring contraction of the readily available 1,2,3-triphospholenes ⟨81JOC(46)3⟩.

Alkylbis(trimethylsilyl)phosphanes and oxalylchloride yield 1,2-dihydro-1,2-diphosphetes as well ⟨80TL(21)1923⟩. The only access known to a 2,3-dihydro-1,2-diphosphete like (**105**) is based on a complex reaction sequence, starting with a phosphole derivative ⟨91OM(10)3958⟩.

Phosphaalkene (**106**) is desilylated by phosgene. A proposed bisphosphaalkene intermediate undergoes intramolecular cyclic addition and bicyclic 1,2-diphosphetane (**107**) is formed as the stable product ⟨79AG(E)(18)872⟩. A head-to-head dimerization of a C-dichloro-phosphaalkene under reducing conditions leads to a 3,4-dichloro-1,2-diphosphetane ⟨MI2⟩. Oxidation of phosphaalkyne (**54c**) with grey selenium or sulphur leads to $1\lambda^3,3\lambda^5$-heterocycle (**108**) ⟨MI1⟩. Compound (**108**) is isolobal with the transition metal stabilized species (**109**), which is accessible via an aminophosphaalkyne (**54b**) ⟨92AG(E)(31)95⟩. The same aminophosphaalkyne leads to salt (**110**) in good yields upon reacting with methyl iodide or triflate ⟨94HAC(5)337⟩. Compounds (**109**) and (**110**) are planar rings, their π-electrons and charges are delocalized. Addition of PBr_3 to phosphaalkyne (**54a**) leads to bicyclic triphospha[1.1.1]pentane (**111**) in good yield ⟨85MI(59)55⟩. An access to a 1,2-diphosphetane not utilizing a [2 + 2]-cyclic addition

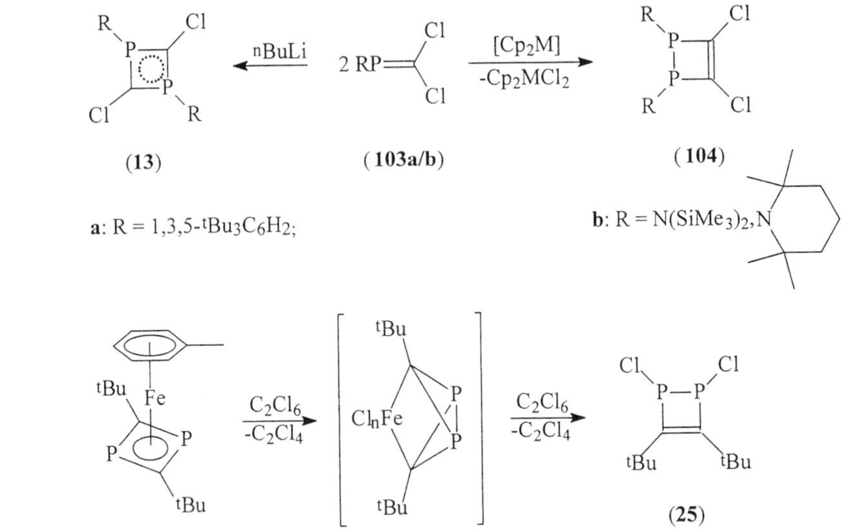

Scheme 31

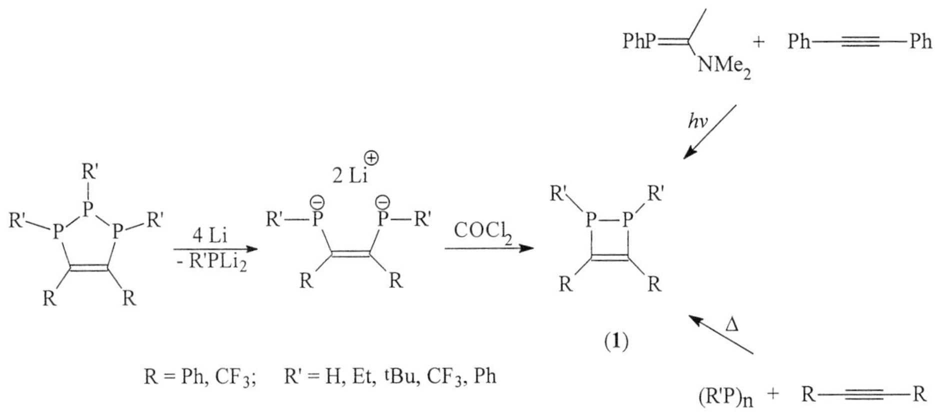

Scheme 32

reaction of phosphaalkenes was discovered by Regitz *et al.* ⟨94JOM(464)41⟩. 1,3-Diphosphetanes as part of polycyclic polyphosphorus hydrocarbons were observed for the first time by reacting (**54**) with the Lewis acid $TaCl_5$. An ionic product was isolated: $(Me_3CCP)_6P^+TaCl_6^-$ ⟨85MI(59)55⟩. Several different metal-containing or metal-free cages can be made in small yields by reacting (**54a**) with reactive arene iron(0) complexes. Examples are the pentamer (**112**) and the hexaphosphahomopentaprismane (**17**). Better yields for (**112**) are attained by two alternative routes: reduction of (**54a**) at a potassium mirror and oxidation of the mixture by air ⟨93PS(77)5⟩ or oxidation of mixtures of di- and triphospholyl anions ⟨89JOM(375)C31⟩. The K/air route, based on (**54**), gives better yields; however, they differ from run to run.

3.3.5.6 Ring Synthesis by Transformation to Another Ring

1,2-Dihydro-1λ^5,2λ^5-diphosphetes (**114**) are formed upon insertion of phosphanes RPX (X = NR, $W(CO)_5$) into the P–C-bond of λ^5-phosphirenes (**113**) in good yield ⟨87OM(6)207, 94CB(127)313, 97OM(16)4501, 99OM(18)796⟩.

Phosphinidene insertion into a phosphirene ring is a general route to 1,2-dihydro-1,2-diphosphetes (**1**) ⟨87(AG(E)(26)548⟩. Ring opening and dimerization of a 1H-diphosphirene tungsten complex under mild conditions yields a tricyclic species (**115**) with a diphosphetane substructure ⟨91JCS(CC)1305⟩.

The carbon atom of *tert*-butyl isonitrile inserts into the P–P-bond of diphosphirenium salt (**116**), to furnish salt (**117**), which may be alkylated to yield neutral (**118**) ⟨94CC337⟩.

A metallacycle transfer into a 1,2-dihydro-1λ^3,2λ^3-diphosphete (**120**) with two different P-substituents is achieved by reacting ZrP-heterocycle (**119**) with $PhPCl_2$ ⟨97OM(16)365⟩.

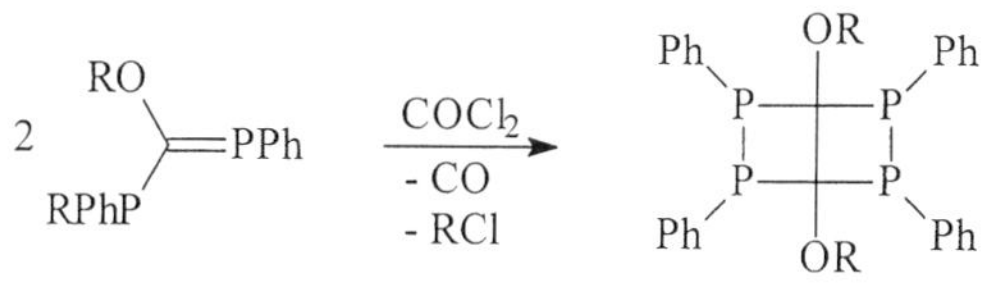

(**106**) R = $SiMe_3$ (**107**)

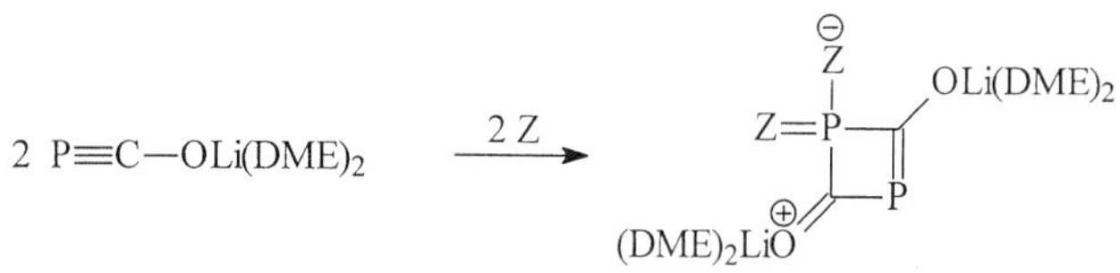

(**54c**) Z = S, Se (**108**)

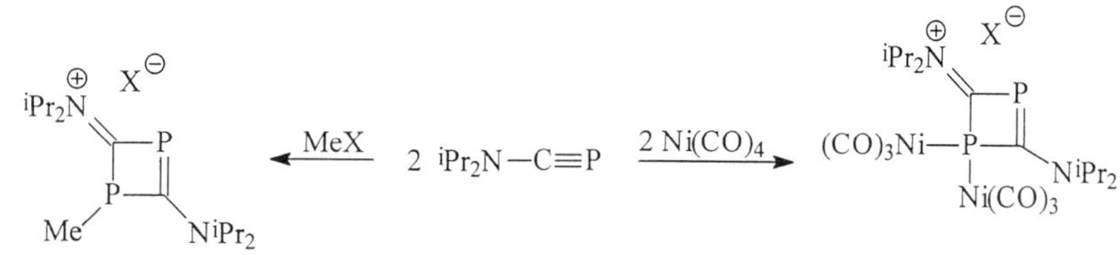

(**110**) X = I, O_3SCF_3; (**54b**) (**109**)

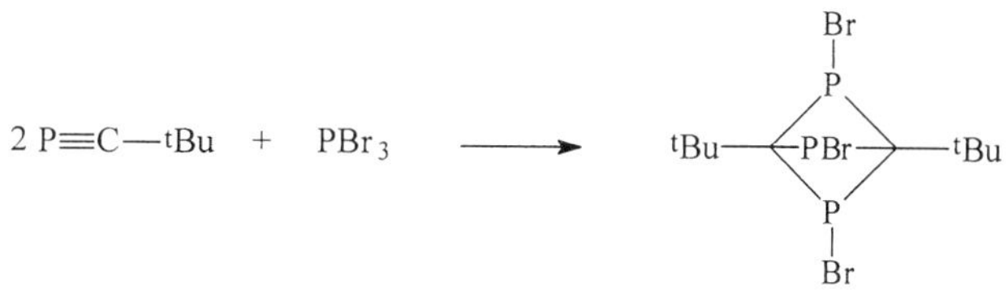

(**54a**) (**111**)

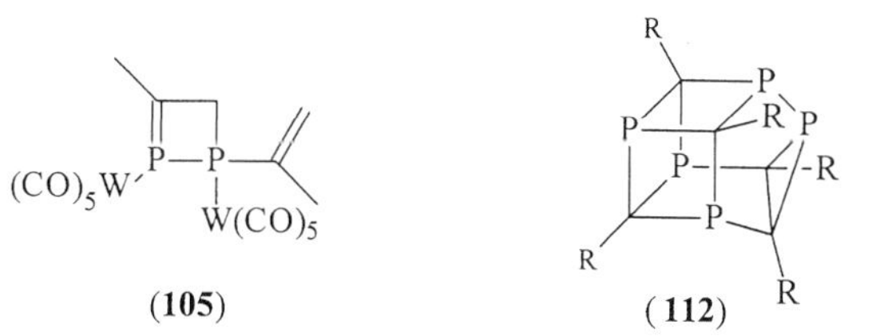

(**105**) (**112**) R = ${}^tBu$

Scheme 33

A photochemical electrocyclization of triphosphol (**121**) affords triphospha bicyclopentene (**122**) with a 1,2-dihydro-1,3-diphosphete substructure ⟨98CC1537⟩.

3.3.6 SYNTHESIS OF PARTICULAR CLASSES OF COMPOUNDS

All relevant reactions leading to the heterocycles are described in Section 3.3.5. Section 3.3.6 points out only the preparative most useful routes to the ring compounds.

3.3.6.1 1,2- and 1,3-Diphosphetes

$1\lambda^5,3\lambda^5$-Diphosphetes can be made best from C–Cl-phosphaalkenes (**70**) (Section 3.3.5.2) or C-diazo diaminophosphane precursors (**75**). *In-situ* prepared (**75**) and *t*-butyl phospaalkyne (**54a**) cyclodimerize

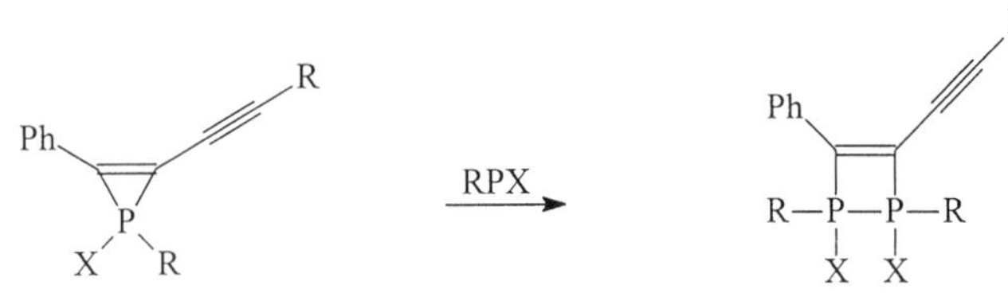

(**113**) R = alkyl, aryl; X = NR, W(CO)$_5$ (**114**)

Scheme 34

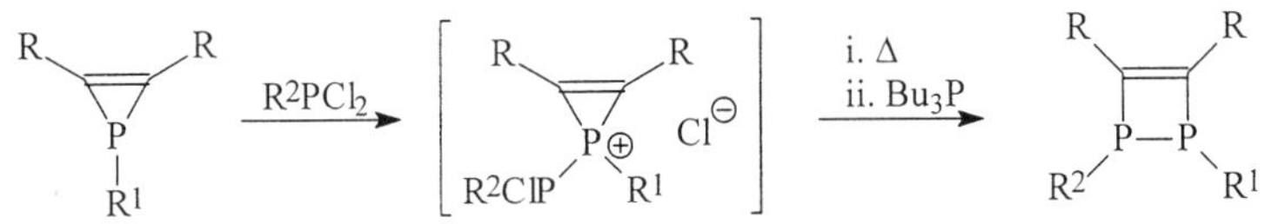

R = alkyl, Ph; R^1 = Me, ^{t}Bu, Ph; R^2 = Me, ^{t}Bu, Ph; (**1**)

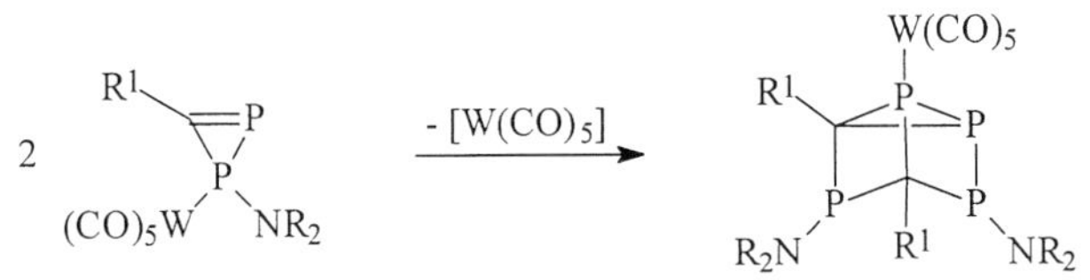

R = C$_6$H$_{11}$; R^1 = ^{t}Bu (**115**)

Scheme 35

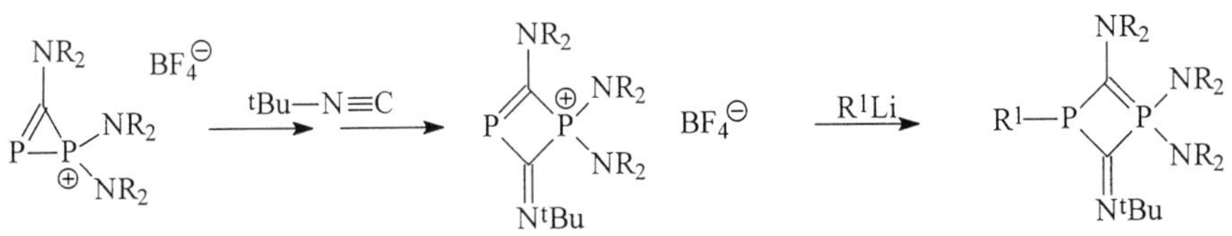

(**116**) (**117**) R = ^{i}Pr; R^1 = Bu, Me (**118**)

Scheme 36

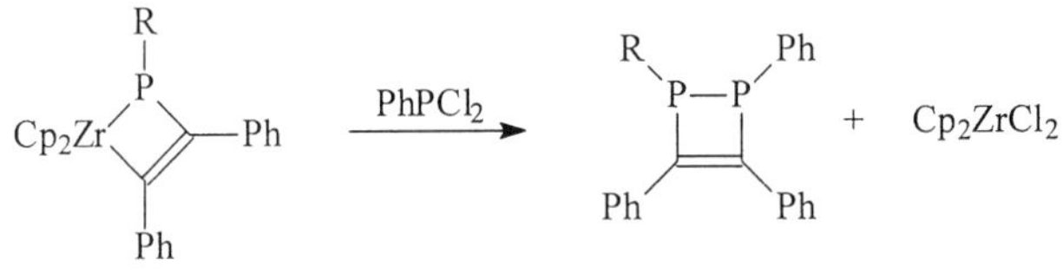

(**119**) R = C$_6$H$_2$-2,4,6-^{t}Bu$_3$ (**120**)

Scheme 37

in a head-to-head manner to form 1λ^5,2λ^3-diphosphete (**6**) (Section 3.3.5.3). Free 1λ^3,3λ^3-diphosphetes (**4**) as well as 1λ^3,2λ^3-diphosphetes (**47**) are accessible as reactive intermediates only by oxidative degradation of Zr-complex (**45**) (Section 3.3.4.3). η^4-Complexed 1λ^3,3λ^3-diphosphetes are made easily from phosphaalkynes (**54**) (Section 3.3.5.4).

3.3.6.2 1,3-Dihydro-1,3-Diphosphetes

1,3-Dihydro-1,3-diphosphete (**13**) can be made by chlorine abstraction of C–Cl$_2$-phosphaalkene (**103a**) (Section 3.3.5.5).

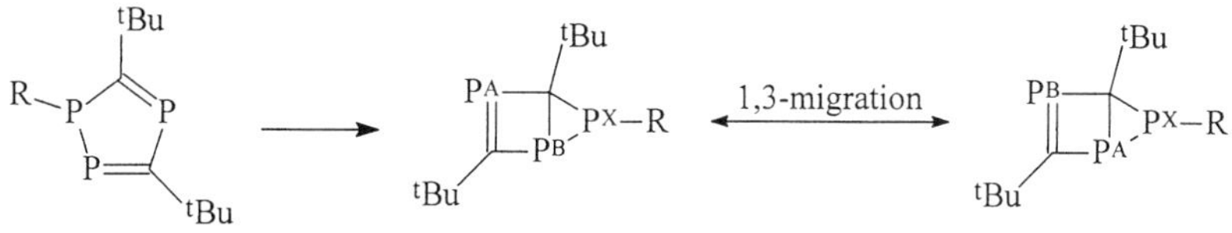

(121) (122)

Scheme 38

3.3.6.3 1,2-Dihydro-1,2-Diphosphetes

1,2-Dihydro-1λ^5,2λ^5-diphosphete (**114**) are made from λ^5-phosphirenes in good yield (Section 3.3.5.6). The chemistry of 1,2-dihydro-1λ^3,2λ^3-diphosphete (**1**) was reviewed including all relevant syntheses ⟨92MI(13)61⟩. Favourable reactions yielding (**1**) are: ring contraction of 1,2,3-triphospholenes (15–43%), cyclopolyphosphanes plus alkynes (max. 55%, difficult workup) (Section 3.3.5.5), and phosphiniden insertion into phosphirenes (30–80%) (Section 3.3.5.6). Tricyclic (**48**) includes two 1,2-dihydro-1,2-diphosphete rings. It is formed upon cyclodimerization of reactive 1,2-diphosphete (**47**) (Section 3.3.4.3). The only access known to a 2,3-dihydro-1,2-diphosphete (**105**) is based on a phosphole derivative (Section 3.3.5.5).

3.3.6.4 1,2-Dihydro-1,3-Diphosphetes

An iminophosphane leads to 1,2-dihydro-1λ^5,3λ^5-diphosphete (**65**) (Section 3.3.5.1). An oxidative cyclodimerization of amino-substituted phosphaalkyne (**54b**) by sulphur or selenium leads to the quantitative formation of 1,2-dihydro-1λ^5,3λ^3-diphosphetes (**7**) ⟨00T(56)27, 00T(56)1147⟩. 1,2-Dihydro-1λ^5,3λ^3-diphosphete (**108**) and 1,2-dihydro-1λ^3,3λ^3-diphosphetes (**109, 110**) are made in good yield from N- and O-substituted phosphaalkynes (**54a–c**) (Section 3.3.5.5). 1,3-Dihydro-1λ^3,3λ^3-diphosphete (**10**) is synthesized in 75% yield by reacting a phosphaalkene with phosgene ⟨83CB(116)938⟩.

3.3.6.5 1,2- and 1,3-Diphosphetanes

1λ^5,3λ^5-Diphosphetanes (**105**) are formed in moderate yield by a cyclic addition reaction of bis(methylene)- or imino-methylenephosphoranes (Section 3.3.5.3). λ^3-Phosphaalkenes are excellent precursors for many different λ^3,λ^3-diphosphetanes. Head-to-head cyclic addition leads to 1,2-diphosphetanes, the preferred head-to-tail dimerization to 1,3-diphosphetanes (Section 3.3.5.3). This principal route has been reviewed recently ⟨92H(34)369⟩. *trans*-3,4-Dipropyliden-1,2-diphosphetane is accessible in 83% yield by a reductive coupling reaction (Section 3.3.5.1).

3.3.6.6 Diphosphetanes as Parts of Polycyclic Compounds

Tetraphosphacubanes (**16**) are preferably made (30–75%) by using Zr complexes (**45**) as educt (Section 3.3.4.3). Tetracyclic tetramers of phosphaalkyne (**54**), which include 1,3- as well as 1,2-diphosphetane subunits, are produced best in a reaction of (**54**) with $AlCl_3$ (Section 3.3.5.3). Pentamers of (**54**) and other P–C-cage compounds can be made in poor to moderate yields by reductive routes (Section 3.3.5.5).

3.3.7 REFERENCES

64JA(86)2306	W. Mahler; *J. Am. Chem. Soc.*, 1964, **86**, 2306.
73CB(106)1453	A. Ecker and U. Schmidt; *Chem. Ber.*, 1973, **106**, 1453.
76PS(2)283	A.H. Cowley; *Phosphorus Sulfur*, 1976, **2**, 283.
76ZA(432)242	G. Becker; *Z. Anorg. Allg. Chem.*, 1976, **432**, 242.
76ZOK(46)2268	A.A. Prishchenko and I.F. Lutsenko; *Zh. Obshch. Khim.*, 1976, **46**, 2268.
77AG(E)(16)402	R. Appel, F. Knoll and H.-D. Wihler; *Angew. Chem., Int. Ed. Engl.*, 1977, **16**, 402.
77JGCU(47)2409	Z.S. Novikova, A.A. Prishchenko and I.F. Lutsenko; *J. Gen. Chem. USSR*, 1977, **47**, 2409.
79AG(E)(18)872	R. Appel, V. Barth, M. Halstenberg, G. Huttner and J.V. Seyerl; *Angew. Chem., Int. Ed. Engl.*, 1979, **18**, 872.
80CB(113)1549	E. Niecke and D.-A. Wildbredt; *Chem. Ber.*, 1980, **113**, 1549.

80TL(21)1923	R. Appel and V. Barth; *Tetrahedron Lett.*, 1980, **21**, 1923.
80TL(21)2497	R. Appel and B. Laubach; *Tetrahedron Lett.*, 1980, 2497.
80ZAAC(470)144	G. Becker and G. Gutekunst; *Z. Anorg. Allg. Chem.*, 1980, **470**, 144.
81AG(E)(20)731	R. Appel, B. Niemann and M. Nieger; *Angew. Chem., Int. Ed. Engl.*, 1981, **20**, 731.
81JOC(46)3	C. Charriere, J. Guilhem and F. Mathey; *J. Org. Chem.*, 1981, **46**, 3.
81ZAAC(479)41	G. Becker, W. Uhl and H.-J. Wessely; *Z. Anorg. Allg. Chem.*, 1981, **479**, 41.
82JGCU(52)2465	L.N. Markovskii, V.D. Romanenko, A.V. Ruban and S.V. Iksanova; *J. Gen. Chem. USSR*, 1982, **52**, 2465.
82ZAAC(485)23	G. Becker, W. Massa, O. Mundt and R. Schmidt; *Z. Anorg. Allg. Chem.*, 1982, **485**, 23.
83CB(116)938	R. Appel, V. Barth and F. Knoch; *Chem. Ber.*, 1983, **116**, 938.
83CB(116)2275	R. Appel, U. Baumeister and F. Knoch; *Chem. Ber.*, 1983, **116**, 2275.
83JA(105)6871	C. Charrier, H. Bonnard, G. de Lauzon and F. Mathey; *J. Am. Chem. Soc.*, 1983, **105**, 6871.
83JA(105)7194	C. Wentrup, H. Briehl, G. Becker, G. Uhl, H.-J. Wessely, A. Maquestiau and R. Flammang; *J. Am. Chem. Soc.*, 1983, **105**, 7194.
83TL(24)1975	A. Meriem, J.P. Majoral, M. Revell and J. Navech; *Tetrahedron Lett.*, 1983, **24**, 1975.
83TL(24)2639	R. Appel and W. Paulen; *Tetrahedron Lett.*, 1983, **24**, 2639.
84CB(117)2693	R. Appel, C. Casser and F. Knoch; *Chem. Ber.*, 1984, **117**, 2693.
84CB(117)3151	R. Appel, F. Knoch and H. Kunze; *Chem. Ber.*, 1984, **117**, 3151.
84JCS(CC)1707	C. Charriere, F. Mathey, F. Robert and Y. Jeannin; *J. Chem. Soc. Chem. Commun.*, 1984, 1707.
84TL(25)1963	T. Kauffmann, E. Antfang and J. Olbrich; *Tetrahydron Lett.*, 1984, **25**, 1963.
84ZAAC(517)75	G. Becker, W. Massa, R.E. Schmidt and G. Uhl; *Z. Anorg. Allg. Chem.*, 1984, **517**, 75.
85MI(59)55	G. Becker, W. Becker, R. Knebl, H. Schmidt, U. Weeber and M. Westerhausen; *Nova Acta Leopold.*, 1985, **59**, 55.
85PS(21)349	M. Binnewies, J. Grobe and D. Le Van; *Phosphorus Sulfur*, 1985, **21**, 349.
85PS(25)345	J. Heinicke and A. Tzschach; *Phosphorus Sulfur*, 1985, **25**, 345.
86AG(E)(25)644	P. Binger, R. Milczarek, R. Mynott, M. Regitz and W. Rösch; *Angew. Chem., Int. Ed. Engl.*, 1986, **25**, 644.
86CB(119)535	R. Appel, E. Gaitzsch, K.-H. Dunker and F. Knoch; *Chem. Ber.*, 1986, **119**, 535.
86CB(119)2466	R. Appel, V. Winkhaus and F. Knoch; *Chem. Ber.*, 1986, **119**, 2466.
86CB(119)2915	R. Appel, C. Casser, F. Knoch and B. Niemann; *Chem. Ber.*, 1986, **119**, 2915.
86IC(25)3177	B. Steger, H. Oberhammer, J. Grobe and D. Le Van; *Inorg. Chem.*, 1986, **25**, 3177.
86JCS(CC)737	P.B. Hitchcock, M.J. Maah and J.F. Nixon; *J. Chem. Soc., Chem. Commun.*, 1986, 737.
86OM(5)623	C. Charriere, N. Maigrot, F. Mathey, F. Robert and Y. Jeannien; *Organometallics*, 1986, **5**, 623.
86PS(29)23	B. Neumüller and E. Fluck; *Phosphorus Sulfur*, 1986, **29**, 23.
86ZAAC(534)31	G. Becker, H. Riffel, W. Uhl and H.-J. Wessely; *Z. Anorg. Allg. Chem.*, 1986, **534**, 31.
87(AG(E)(26)548	L. Ricard, N. Maigrot, C. Charriere and F. Mathey; *Angew. Chem., Int. Ed. Engl.*, 1987, **26**, 548.
87AG(E)(26)908	R. Milczarek, W. Rüsseler, P. Binger, K. Jonas, K. Angermund, C. Krüger and M. Regitz; *Angew. Chem., Int. Ed. Engl.*, 1987, **26**, 908.
87JOM(334)C35	M. Driess, D. Hu, H. Pritzkow, H. Schäufele, U. Zenneck, M. Regitz and W. Rösch; *J. Organomet. Chem.*, 1987, **334**, C35.
87OM(6)207	N. Hoa Tran Huy and F. Mathey; *Organometallics*, 1987, **6**, 207.
88AG(E)(27)950	N. Maigrot, L. Ricard, C. Charrier and F. Mathey; *Angew. Chem., Int. Ed. Engl.*, 1988, **27**, 950.
88AG(E)(27)1157	P. Binger, B. Biedenbach, R. Mynott, C. Krüger, P. Betz and M. Regitz; *Angew. Chem., Int. Ed. Engl.*, 1988, **27**, 1157.
88AG(E)(27)1484	M. Regitz and P. Binger; *Angew. Chem., Int. Ed. Engl.*, 1988, **27**, 1484.
88CB(121)637	P. Binger, R. Milczarek, R. Mynott, C. Krüger, Y.-H. Tsay, E. Raabe and M. Regitz; *Chem. Ber.*, 1988, **121**, 637.
88CB(121)2071	T. Veszpremi, R. Geiter, E. Fluck, J. Svara and B. Neumüller; *Chem. Ber.*, 1988, **121**, 2071.
88CRV(88)1327	J.F. Nixon; *Chem. Rev.*, 1988, **88**, 1327.
88IC(27)2269	I.G. Phillips, R.G. Ball and R.G. Cavell; *Inorg. Chem.*, 1988, **27**, 2269.
88OM(7)1791	N.H.T. Huy, L. Ricard and F. Mathey; *Organometallics*, 1988, **7**, 1791.
89AG(E)(28)900	D. Hu, H. Schäufele, H. Pritzkow and U. Zenneck; *Angew. Chem., Int. Ed. Engl.*, 1989, **28**, 900.
89AG(E)(28)1013	T. Wettling, J. Schneider, O. Wagner, C.G. Kreiter and M. Regitz; *Angew. Chem., Int. Ed. Engl.*, 1989, **28**, 1013.
89JCS(CC)1409	H.F. Dare, J.A.K. Howard, M.U. Pilotti, F.G.A. Stone and J. Szameitat; *J. Chem. Soc., Chem. Commun.*, 1989, 1409.
89JOM(375)C31	R. Bartsch, P.B. Hitchcock and J.F. Nixon; *J. Organomet. Chem.*, 1989, **375**, C31.
89ZAAC(576)81	E. Fluck, B. Neumüller and H. Riffel; *Z. Anorg. Allg. Chem.*, 1989, **576**, 81.
90AG(E)(29)534	N. Meigrot, L. Ricard, C. Charrier and F. Mathey; *Angew. Chem., Int. Ed. Engl.*, 1990, **29**, 534.
90AG(E)(29)1429	G.R. Gillette, A. Baceiredo and G. Bertrand; *Angew. Chem., Int. Ed. Engl.*, 1990, **29**, 1429.
90CB(123)733	L. Weber, M. Frebel and R. Boese; *Chem. Ber.*, 1990, **123**, 733.
90CB(123)757	R. Gleiter, K.-H. Pfeifer, M. Baudler, G. Scholz, T. Wettling and M. Regitz; *Chem. Ber.*, 1990, **123**, 757.
90JCS(CC)1541	T. Wettling, G. Wolmershäuser, P. Binger and M. Regitz; *J. Chem. Soc., Chem. Commun.*, 1990, 1541.
90PO(9)1363	N. Maigrot, C. Charriere, L. Ricard and F. Mathey; *Polyhedron*, 1990, **9**, 1363.
90TL(31)4429	G. Märkl, P. Kreitmeyer, H. Nöth and K. Polborn; *Tetrahydron Lett.*, 1990, **31**, 4429.
90ZAAC(588)147	B. Neumüller, H. Riffel and E. Fluck; *Z. Anorg. Allg. Chem.*, 1990, **588**, 147.

91AG(E)(30)207 P. Binger, T. Wettling, R. Schneider, F. Zurmühlen, U. Bergsträsser, J. Hoffmann, G. Maas and M. Regitz; *Angew. Chem., Int. Ed. Engl.*, 1991, **30**, 207.

91CB(124)2843 E. Fuchs, B. Breit, H. Heydt, W. Schoeller, T. Busch, C. Krüger, P. Betz and M. Regitz; *Chem. Ber.*, 1991, **124**, 2843.

91HAC(2)253 P.B. Hitchcock, M.J. Maah and J.F. Nixon; *Heteroatom. Chem.*, 1991, **2**, 253.

91JCS(CC)1305 F. Mercier, L. Ricard, F. Mathey and M. Regitz; *J. Chem. Soc., Chem. Commun.*, 1991, 1305.

91OM(10)3958 N.H.T. Huy, L. Ricard and F. Mathey; *Organometallics*, 1991, **10**, 3958.

92AG(E)(31)95 J. Grobe, D. Le Van, M. Hegemann, B. Krebs and M. Läge; *Angew. Chem., Int. Ed. Engl.*, 1992, **31**, 95.

92AG(E)(31)99 H. Grützmacher and H. Pritzkow; *Angew. Chem., Int. Ed. Engl.*, 1992, **31**, 99. According to a personal communication, the diphosphetane is a 1,2-species, not a 1,3-one.

92AG(E)(31)758 T. Wettling, B. Geissler, R. Schneider, S. Barth, P. Binger and M. Regitz; *Angew. Chem., Int. Ed. Engl.*, 1992, **31**, 758.

92AG(E)(31)879 M. Birkel, J. Schulz, U. Bergsträsser and M. Regitz; *Angew. Chem., Int. Ed. Engl.*, 1992, **31**, 879.

92AG(E)(31)1055 B. Breit, U. Bergsträsser, G. Maas and M. Regitz; *Angew. Chem., Int. Ed. Engl.*, 1992, **31**, 1055.

92BSF(129)76 N. Maigrot, L. Ricard, C. Charrier, P. Le Goff and F. Mathey; *Bull. Soc. Chim. Fr.*, 1992, **129**, 76.

92CB(125)771 P. Becker, H. Brombach, G. David, M. Leuer, H.-J. Metternich and E. Niecke; *Chem. Ber.*, 1992, **125**, 771.

92H(34)369 K. Afarinkia; *Heterocycles*, 1992, **34**, 369.

92HAC(3)395 G. Martin, E. Ocando-Mavarez, A. Osorio and M. Laya; *Heteroatom. Chem.*, 1992, **3**, 395.

92JA(114)9453 G. Sillett, L. Ricard, C. Patois and F. Mathey; *J. Am. Chem. Soc.*, 1992, **114**, 9453.

92JOC(57)2040 S.M. Bachrach and M. Liu; *J. Org. Chem.*, 1992, **57**, 2040.

92MI(13)61 F. Mathey, C. Charrier, N. Maigrot, A. Marinetti, L. Ricard and N.H.T. Huy; *Comments Inorg. Chem.*, 1992, **13**, 61.

92OM(11)177 R. Gleiter, I. Hyla-Kryspin, P. Binger and M. Regitz; *Organometallics*, 1992, **11**, 177.

92OM(11)459 M.L. Sierra, N. Maigrot, C. Charrier, L. Ricard and F. Mathey; *Organometallics*, 1992, **11**, 459.

92PO(11)601 N. Maigrot, M.L. Sierra, C. Charrier, L. Ricard and F. Mathey; *Polyhedron*, 1992, **11**, 601.

92ZAAC(609)99 E. Fluck and R. Braun; *Z. Anorg. Allg. Chem.*, 1992, **609**, 99.

92ZAAC(612)56 E. Fluck, M. Spahn, G. Heckmann and H. Borrmann; *Z. Anorg. Allg. Chem.*, 1992, **612**, 56.

92ZN(B)(47)321 J. Grobe, D. Le Van, S. Martin and J. Szameitat; *Z. Naturforsch.*, B, 1992, **47**, 321.

93AG(E)(32)617 W.W. Schoeller and T. Busch; *Angew. Chem., Int. Ed. Engl.*, 1993, **32**, 617.

93AG(E)(32)1424 E. Lindner, C. Haase, H.A. Mayer, M. Kemmler, R. Fawzi and M. Steinmann; *Angew. Chem., Int. Ed. Engl.*, 1993, **32**, 1424.

93BSF(130)642 R. Streubel, M. Frost, N. Nieger and E. Niecke; *Bull. Soc. Chim. Fr.*, 1993, **130**, 642.

93CB(126)845 W. Brieden and T. Kellerson; *Chem. Ber.*, 1993, **126**, 845.

93JOC(58)4105 K.K. Laali, M. Regitz, M. Birkel, P.J. Stang and C.M. Crittell; *J. Org. Chem.*, 1993, **58**, 4105.

93JOM(447)31 W. Imhof and G. Huttner; *J. Organomet. Chem.*, 1993, **447**, 31.

93JPC(97)6384 R.A.J. Janssen; *J. Phys. Chem.*, 1993, **97**, 6384.

93PS(77)5 D. Böhm, D. Hu and U. Zenneck; *Phosphorus Sulfur*, 1993, **77**, 5.

93TL(34)6365 S.M. Bachrach and L.M. Perriott; *Tetrahedron Lett.*, 1993, **34**, 6365.

93ZAAC(619)1551 E. Fluck, G. Heckmann and K. Lange; *Z. Anorg. Allg. Chem.*, 1992, **619**, 1551.

94AG(E)(33)463 P.B. Hitchcock, C. Jones and J.F. Nixon; *Angew. Chem., Int. Ed. Engl.*, 1994, **33**, 463.

94AG(E)(33)982 W. Schilbach, V.V.D. Gönna, D. Gudat, M. Nieger and E. Niecke; *Angew. Chem., Int. Ed. Engl.*, 1994, **33**, 982.

94BSF(131)397 N. Maigrot, M. Sierra, C. Charrier and F. Mathey; *Bull. Soc. Chim. Fr.*, 1994, **131**, 397.

94CB(127)313 M. Link, E. Niecke and M. Nieger; *Chem. Ber.*, 1994, **127**, 313.

94CC337 M. Soleilhavoup, A. Baceiredo, F. Dahan and G. Bertrand; *J. Chem. Soc., Chem. Commun.*, 1994, 337.

94JCS(CC)489 F.G.N. Cloke, F.R. Flower, P.B. Hitchcock and J.F. Nixon; *J. Chem. Soc., Chem. Commun.*, 1994, 489.

94JOM(464)41 B. Breit, R. Boese and M. Regitz; *J. Organomet. Chem.*, 1994, **464**, 41.

94JOM(466)153 P.B. Hitchcock, M.J. Maah and J.F. Nixon; *J. Organomet. Chem.*, 1994, **466**, 153.

94HAC(5)337 J. Grobe, D. Le Van, M. Hegemann, B. Krebs and M. Läge; *Heteroatom. Chem.*, 1994, **5**, 337.

94OM(13)4406 L. Weber, S. Buchwald, D. Lentz, O. Stamm, D. Preugschat and R. Marschall; *Organometallics*, 1994, **13**, 4406.

94PS(93/94)365 U. Fleischer, C.V. Wüllen and W. Kutzelnigg; *Phosphorus Sulfur*, 1994, **93/94**, 365.

94RHA(11)65 G. Heckmann and E. Fluck; *Rev. Heteroatom. Chem.*, 1994,**11**, 65.

94S1337 B. Geißler, T. Wettling, S. Barth, P. Binger and M. Regitz; *Synthesis*, 1994, 1337.

94ZAAC(620)483 E. Fluck, K. Bieger, G. Heckmann and B. Neumüller; *Z. Anorg. Allg. Chem.*, 1994, **620**, 483.

94ZN(B)(49)1361 M. Fild, P.G. Jones, K. Ruhnau and C. Thöne; *Z. Naturforsch. Teil B*, 1994, **49**, 1361.

95AG(E)(34)198 D. Böhm, F. Knoch, S. Kummer, U. Schmidt and U. Zenneck; *Angew. Chem., Int. Ed. Engl.*, 1995, **34**, 198.

95AG(E)(34)484 B. Geißler, S. Barth, U. Bergsträßer, M. Slany, J. Durkin, P.B. Hitchcock, M. Hofmann, P. Binger, J.F. Nixon, P.V.R. Schleyer and M. Regitz; *Angew. Chem., Int. Ed. Engl.*, 1995, **34**, 484.

95AG(E)(34)555 E. Niecke, A. Fuchs, F. Baumeister, M. Nieger and W.W. Schoeller; *Angew. Chem., Int. Ed. Engl.*, 1995, **34**, 555.

95AG(E)(34)1853 H.-P. Schrödel, G. Jochem, A. Schmidpeter and H. Nöth; *Angew. Chem., Int. Ed. Engl.*, 1995, **34**, 1853.

95AG(E)(34)2227 P. Binger, S. Leininger, J. Stannek, B. Gabor, R. Mynott, J. Bruckmann and C. Krüger; *Angew. Chem., Int. Ed. Engl.*, 1995, **34**, 2227.

95CB(128)959 H.H. Karsch, T. Rupprich and M. Heckel; *Chem. Ber.*, 1995, **128**, 959.

95CB(128)991 M. Julino, M. Slany, U. Bergsträßer, F. Merçier, F. Mathey and M. Regitz; *Chem. Ber.*, 1995, **128**, 991.

95CB(128)1261 P. Binger, G. Glaser, S. Albus and C. Krüger; *Chem. Ber.*, 1995, **128**, 1261.

95CC1671 M. Scheer, K. Schuster, T.A. Budzichowski, M.H. Chrisholm and W.E. Streib; *J. Chem. Soc., Chem. Commun.*, 1995, 1671.

95CC2395 S.M. Bachrach, V. Caliman and J.F. Nixon; *J. Chem. Soc., Chem. Commun.*, 1995, 2395.

95CRV(145)201 J.F. Nixon; *Chem. Rev.*, 1995, **145**, 201.

95CSR319 J.F. Nixon; *Chem. Soc. Rev.*, 1995, 319.

95HAC(6)1 X.-B. Ma, M. Birkel, T. Wettling and M. Regitz; *Heteroatom. Chem.*, 1995, **6**, 1.

95IC(34)5070 P. Le Floch, N. Maigrot, L. Ricard, C. Charrier and F. Mathey; *Inorg. Chem.*, 1995, **34**, 5070.

95JA(117)10785 R. Armbrust, M. Sanchez, R. Réau, U. Bergsträsser, M. Regitz and G. Bertrand; *J. Am. Chem. Soc.*, 1995, **117**, 10785.

95JOC(60)47 K.K. Laali, B. Geissler, M. Regitz and J.J. Houser; *J. Org. Chem.*, 1995, **60**, 47.

95JOC(60)3149 K.K. Laali, B. Geissler and M. Regitz; *J. Org. Chem.*, 1995, **60**, 3149.

95JOC(60)6362 K.K. Laali, B. Geissler, M. Regitz and J.J. Houser; *J. Org. Chem.*, 1995, **60**, 6362.

95JPO(8)742 W.W. Schoeller, B. Goette and U. Tubbesing; *J. Phys. Org. Chem.*, 1995, **8**, 742.

95ZAAC(621)1407 L. Weber, E. Dobbert, S. Buchwald, H.-G. Stammler and B. Neumann; *Z. Anorg. Allg. Chem.*, 1995, **621**, 1407.

95ZAAC(621)1539 E. Fluck, G. Heckmann, F. Rosche, M. Westerhausen and E. Gorbunowa; *Z. Anorg. Allg. Chem.*, 1995, **621**, 1539.

96AG(E)(35)2228 M. Sanchez, R. Réau, F. Dahan, M. Regitz and G. Bertrand; *Angew. Chem., Int. Ed. Engl.*, 1996, **35**, 2228.

96AX(C)(52)1919 P.G. Jones; *Acta Cryst.*, 1996, **C52**, 1919.

96CB(129)419 W.W. Schoeller and U. Tubbesing; *Chem. Ber.*, 1996, **129**, 419.

96CB(129)489 B. Breit and M. Regitz; *Chem. Ber.*, 1996, **129**, 489.

96CJC(74)901 O. Mó and M. Yáñez; *Can. J. Chem.*, 1996, **74**, 901.

96JOC(61)7813 J.-L.M. Abboud, M. Herreros, R. Notario, O. Mó, M. Yáñez, M. Regitz and J. Elguero; *J. Org. Chem.*, 1996, **61**, 7813.

96JPC(100)6456 L. Nyulászi and T. Veszprémi; *J. Phys. Chem.*, 1996, **100**, 6456.

96OM(15)128 L. Weber, B. Torwiehe, G. Bassmann, H.-G. Stammler and B. Neumann; *Organometallics*, 1996, **15**, 128.

96PS(111)42 W.W. Schoeller, J. Strutwolf and E. Niecke; *Phosphorus Sulfur*, 1996, **111**, 42.

96PS(115)3 G. Heckmann, F. Rosche, F. Weller and E. Fluck; *Phosphorus Sulfur*, 1996, **115**, 3.

96RJOC(32)407 A.D. Averin, N.V. Lukashev, A.A. Borisenko, M.A. Kazankova and I.P. Beletskaya; *Russ. J. Org. Chem.*, 1996, **32**, 407.

96S265 F. Knoch, S. Kummer and U. Zenneck; *Synthesis*, 1996, 265.

96ZAAC(622)24 J. Grobe, D. Le Van, F. Immel, M. Hegemann, B. Krebs and M. Läge; *Z. Anorg. Allg. Chem.*, 1996, **622**, 24.

96ZAAC(622)942 F. Rosche, G. Heckmann, B. Neumüller and E. Fluck; *Z. Anorg. Allg. Chem.*, 1996, **622**, 942.

96ZAAC(622)974 F. Rosche, G. Heckmann, E. Fluck and F. Weller; *Z. Anorg. Allg. Chem.*, 1996, **622**, 974.

97AG(E)(36)1337 A. Mack, B. Breit, T. Wettling, U. Bergsträßer, S. Leininger and M. Regitz; *Angew. Chem., Int. Ed. Engl.*, 1997, **36**, 1337.

97AOC(41)1 L. Weber; *Adv. Metalorg. Chem.*, 1997, **41**, 1.

97AX(C)(53)42 P.G. Jones and C. Thöne; *Acta Cryst. Part C*, 1997, **53**, 42.

97CB(130)823 A. Mack and M. Regitz; *Chem. Ber./Recueil*, 1997, **130**, 823.

97CB(130)1491 P. Binger, S. Leininger, K. Günther and U. Bergsträßer; *Chem. Ber./Recueil*, 1997, **130**, 1491.

97CB(130)1519 H.-P. Schrödel and A. Schmidpeter; *Chem. Ber./Recueil*, 1997, **130**, 1519.

97CCR(158)1 L. Weber; *Coord. Chem. Rev.*, 1997, **158**, 1.

97JA(119)9720 M. Sanchez, R. Réau, H. Gornitzka, F. Dahan, M. Regitz and G. Bertrand; *J. Am. Chem. Soc.*, 1997, **119**, 9720.

97JOC(62)7605 T.W. Mackewitz, C. Peters, U. Bergsträsser, S. Leininger and M. Regitz; *J. Org. Chem.*, 1997, **62**, 7605.

97JOM(529)177 J. Grobe, D. Le Van, B. Broschk, M. Hegemann, B. Lüth, G. Becker, M. Böhringer and E.-U. Würthwein; *J. Organometal. Chem.*, 1997, **529**, 177.

97JOM(529)215 P. Binger, S. Leininger, M. Regitz, U. Bergsträßer, J. Bruckmann and C. Krüger; *J. Organometal. Chem.*, 1997, **529**, 215.

97JOM(529)223 E. Fluck, G. Heckmann, E. Gorbunowa, M. Westerhausen and F. Weller; *J. Organometal. Chem.*, 1997, **529**, 223.

97JOM(539)61 A. Hoffmann, S. Leininger and M. Regitz; *J. Organometal. Chem.*, 1997, **539**, 61.

97LA121 G. Märkl, R. Hennig and H. Nöth; *Liebigs Ann. Recueil*, 1997, 121.

97OM(16)365 T.L. Breen and D.W. Stephan; *Organometallics*, 1997, **16**, 365.

97OM(16)2506 N.H. Tran Huy, Y. Inubushi, L. Ricard, F. Mathey; *Organometallics*, 1997, **16**, 2506.

97OM(16)4501 N.H. Tran Huy, L. Ricard and F. Mathey; *Organometallics*, 1997, **16**, 4501.

97OM(16)4551 N. Cénac, A. Chrostowska, J.-M. Sotiropoulos, B. Donnadieu, A. Igau, G. Pfister-Guillouzo and J.-P. Majoral; *Organometallics*, 1997, **16**, 4551.

97ZAAC(623)919 E. Gorbunowa, G. Heckmann, M. Westerhausen and E. Fluck; *Z. Anorg. Allg. Chem.*, 1997, **623**, 919.

98AG(E)(37)271 G. Bertrand; *Angew. Chem., Int. Ed. Engl.*, 1998, **37**, 271.

98AG(E)(37)949	O. Schmidt, A. Fuchs, D. Gudat, M. Nieger, W. Hoffbauer, E. Niecke and W.W. Schoeller; *Angew. Chem., Int. Ed. Engl.*, 1998, **37**, 949.
98AG(E)(37)1233	F. Tabellion, A. Nachbauer, S. Leininger, C. Peters, M. Regitz and F. Preuss; *Angew. Chem., Int. Ed. Engl.*, 1998, **37**, 1233.
98CC1537	V. Caliman, P.B. Hitchcock and J.F. Nixon; *Chem. Commun.*, 1998, 1537.
98CEJ(4)1917	M. Scheer, E. Leiner, P. Kramkowski, M. Schiffer and G. Baum; *Chem. Eur. J.*, 1998, **4**, 1917.
98EJIC951	W.W. Schoeller, U. Tubbesing and A.B. Rozhenko; *Eur. J. Inorg. Chem.*, 1998, 951.
98EJIC1597	A. Hoffmann, A. Mack, R. Goddard, P. Binger and M. Regitz; *Eur. J. Inorg. Chem.*, 1998, 1597.
98EJIC227	M. Schmitz, R. Göller, U. Bergsträßer, S. Leininger and M. Regitz; *Eur. J. Inorg. Chem.*, 1998, 227.
98HAC(9)453	M. Schmitz, S. Leininger, U. Bergsträßer and M. Regitz; *Heteroatom. Chem.*, 1998, **9**, 453.
98JPO(11)678	M. Yáñez; *J. Phys. Org. Chem.*, 1998, **11**, 678.
98PAC(70)819	E. Fluck, G. Heckmann and E. Gorbunowa-Jonas; *Pure Appl. Chem.*, 1998, **70**, 819.
98TL(39)4211	S.M. Bachrach and B.D. Gailbreath; *Tetrahedron Lett.*, 1998, **39**, 4211.
98ZAAC(624)399	M. Scheer and J. Krug; *Z. Anorg. Allg. Chem.*, 1998, **624**, 399.
98ZAAC(624)1116	G. Heckmann, S. Plank, H. Borrmann and E. Fluck; *Z. Anorg. Allg. Chem.*, 1998, **624**, 1116.
98ZN(B)(53)443	G. Heckmann, E. Jonas, E. Fluck and B. Neumüller; *Z. Naturforsch. Teil B*, 1998, **53**, 443.
99AG(E)(38)3028	E. Niecke, A. Fuchs and M. Nieger; *Angew. Chem., Int. Ed. Engl.*, 1999, **38**, 3028.
99AG(E)(38)3031	E. Niecke, A. Fuchs, M. Nieger, O. Schmidt and W.W. Schoeller; *Angew. Chem., Int. Ed. Engl.*, 1999, **38**, 3031.
99ASIOM(S1)199	A. Mack and M. Regitz; *Adv. Strain. Inter. Mol.*, 1999, Suppl. 1, 199.
99CEJ(5)274	M. Sanchez, R. Réau, C.J. Marsden, M. Regitz and G. Bertrand; *Chem. Eur. J.*, 1999, **5**, 274.
99JPC(A)(103)1029	G.M. Bernard, G. Wu, M.D. Lumsden, R.E. Wasylishen, N. Maigrot, C. Charrier and F. Mathey; *J. Phys. Chem. A*, 1999, **103**, 1029.
99OM(18)796	B. Wang, K.A. Nguyen, G.N. Srinivas, C.L. Watkins, S. Menzer, A.L. Spek, K. Lammertsma; *Organometallics*, 1999, **18**, 796.
99OM(18)2021	F.W. Heinemann, S. Kummer, U. Seiss-Brandl and U. Zenneck; *Organometallics*, 1999, **18**, 2021.
99S1363	P. Binger, K. Günther and M. Regitz; *Synthesis*, 1999, 1363.
00T(56)27	J. Grobe, D. Le Van, T. Pohlmeyer, F. Immel, H. Pucknat, B. Krebs, J. Kuchinke and M. Läge; *Tetrahedron*, 2000, **56**, 27.
00T(56)1147	J. Grobe, D. Le Van, T. Pohlmeyer, F. Immel, H. Pucknat, B. Krebs, J. Kuchinke and M. Läge; *Tetrahedron*, 2000, **56**, 1147.
B-89MI1	E. Fluck, B. Neumüller; in "Rings, Clusters, and Polymers of Main Group and Transition Elements", ed. H.W. Roesky, Elsevier, Amsterdam, 1989, p. 201.
B-93MI2	M. Regitz; in "Oganic Synthesis via Organometallics (OSM4)", eds. D. Enders, H.-J. Gais and W. Keim, Viehweg, Braunschweig, 1993, p. 93.
B-90MI3	K. Karaghiosoff; in "Multiple Bonds and Low Coordination in Phosphorus Chemistry", eds. M. Regitz and O.J. Scherer, Thieme, Stuttgart, 1990, p. 463.
B-89MI4	E. Fluck and B. Neumüller; in "Rings, Clusters, and Polymers of Main Group and Transition Elements", ed. H.W. Roesky, Elsevier, Amsterdam, 1989, p. 201.
B-93MI5	P. Binger; in "Multiple Bonds and Low Coordination in Phosphorus Chemistry", eds. M. Regitz and O.J. Scherer, Thieme, Stuttgart, 1990, p. 90.
B-95MI6	M. Regitz, A. Hoffmann and U. Bergsträßer; in "Modern Acetylene Chemistry", eds. P.J. Stang and F. Diederich, VCH, New York, 1995, p. 73.
B-98MI7	K.B. Dillon, F. Mathey and J.F. Nixon; *Phosphorus the Carbon Copy*, Wiley, Chichester, 1998.
MI1	G. Becker, K. Hübler and H. Richard; to be published.
MI2	N. Lavaud-Dufour; *Dissertation*, Toulouse, 1991.
MI3	H. Geiger, F.W. Heinemann, H. Pritzkow, C. Topf and U. Zenneck; to be published.

4.1
Phospholanes and Phospholenes

T.P. KEE
University of Leeds, Leeds, UK

4.1.1 INTRODUCTION

Five-membered heterocyclic ring systems, both saturated and unsaturated, are of enormous importance in chemistry and biology. For example, the saturated furan ring (**1**) is a fundamental component of nucleosides and nucleotides whilst the nitrogen analogue, pyrrolidine (**2**), as a component of the indole ring system (**3**), is also hugely important in many natural products ⟨92MI⟩.

Indeed, phosphorus-containing saturated five-membered heterocycles such as (**4**) and (**5**) have been known, characterised and used in organic synthesis for some years. However, the parent phospha-carbocycles (**6**) and (**7**), phospholanes and phospholenes, respectively, have only more recently become the focus of attention, mainly due to their applications as ligands in catalysis and organic transformations.

This chapter summarises most of the major developments in synthesis, structure and synthetic applications of both phospholanes and phospholenes. Indeed, some aspects of the synthesis and application of phospholanes and phospholenes have been reviewed recently ⟨99RHC227⟩. The older material (<1980) has been extensively described by Quin ⟨81MI⟩.

4.1.2 SYNTHESIS OF PHOSPHOLANES AND PHOSPHOLENES

4.1.2.1 Classical Approaches

Phospholanes and phospholenes are never made directly by hydrogenation of phospholes; most commonly, the ring is created while phosphorus is in the oxide or sulphide form, and the process is completed by removal of these atoms. For the synthesis of 2- and 3-phospholenes, the McCormack cycloaddition of trivalent phosphorus halides with dienes, followed by hydrolysis, is the method of choice for obtaining the requisite *P*-oxides. This process, first described in US patents ⟨53USP2663737-38⟩ by W.B. McCormack of the duPont Co., has proved over the years to be of monumental importance in heterocyclic phosphorus chemistry. Some simple examples of typical syntheses are shown in Scheme 1.

The McCormack reaction has very great versatility, and a considerable variety of phospholene oxides, monocyclic as well as multicyclic, have been prepared with it. Not only do phosphonous dihalides participate, but also PCl_3 and PBr_3 ⟨63T1563⟩, and some ester derivatives of these ⟨62IZV65, 63ZOB783⟩, are

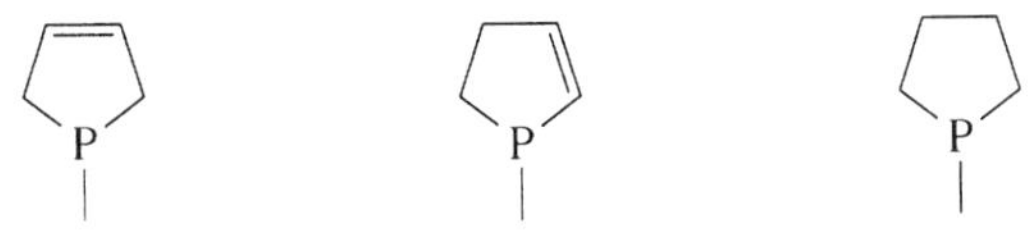

3-phospholene 2-phospholene phospholane

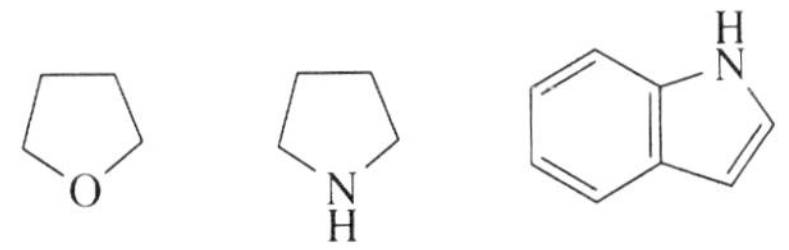

(**1**) (**2**) (**3**)

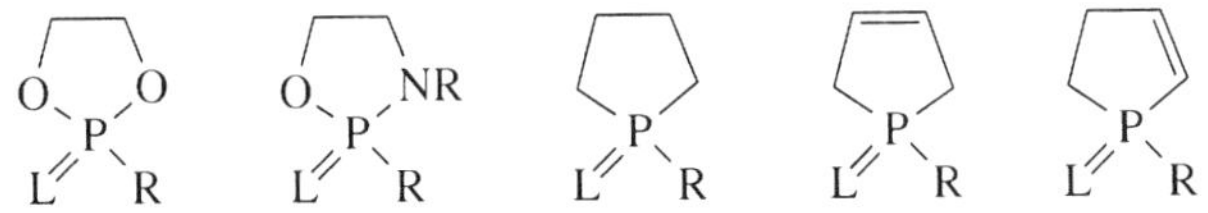

(**4**) (**5**) (**6**) (**7**)

L = lone pair, O, S; R = organic group

useful reactants. As the examples show, with some combinations of reactants the 3-phospholene system is favoured, but the intermediate cycloadduct and the 3-phospholene oxide can enter into rearrangement processes that lead to the 2-phospholene as the major product. Reaction mixtures are always dominated by one of the two regioisomers, and usually can be separated with ease. Some structural limitations in the McCormack cycloaddition do exist, however. Thus, branched-chain phosphonous dihalides react very slowly (e.g. isopropylphosphonous dihalides) or not at all (e.g. *t*-butylphosphonous dichloride), and dienes that contain two substituents on a terminal carbon of the diene fail to react. Branched-chain substituents can, however, be placed indirectly on phosphorus by reaction of *P*-halophospholene derivatives with organometallic reagents ⟨73JOC1858⟩. Both of these limitations appear to have a steric origin, which is especially evident from the behaviour displayed by the 1,3-pentadienes (piperylenes); the *trans*-isomer reacts readily but the more crowded *cis*-isomer is virtually inert ⟨70JA4303⟩. *Trans,trans*-2,4-hexadiene is of special interest in that a mechanistic feature of the McCormack reaction is revealed by the exclusive formation of a 3-phospholene oxide with the two methyls *cis*-oriented ⟨70JA4303⟩. This implies that the cycloaddition is concerted and disrotatory, and thus may be considered as a cheletropic process, a mechanism known also for the addition of SO_2 to dienes. The phosphorus atom constitutes a third chiral center and isomers are possible as in Equation (1).

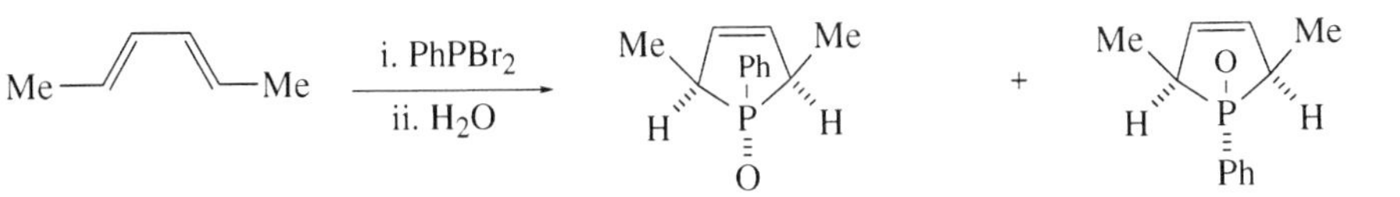

(1)

cis, cis (73%) *cis, trans (27%)*

Isomers are also formed from the use of *trans*-piperylene; in Equation (2) the *cis* : *trans* ratio was 2 : 1.

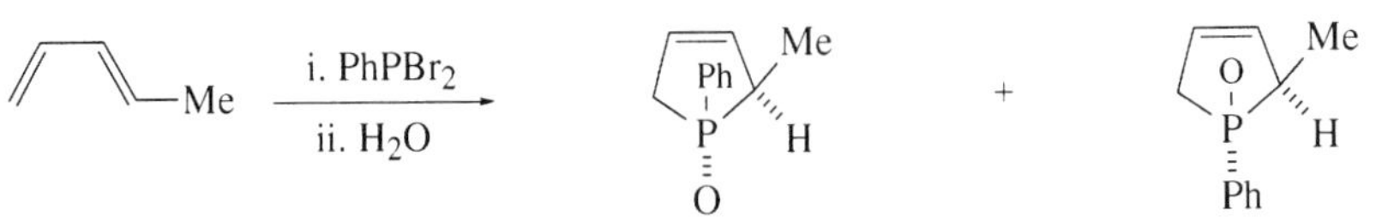

(2)

cis *trans*

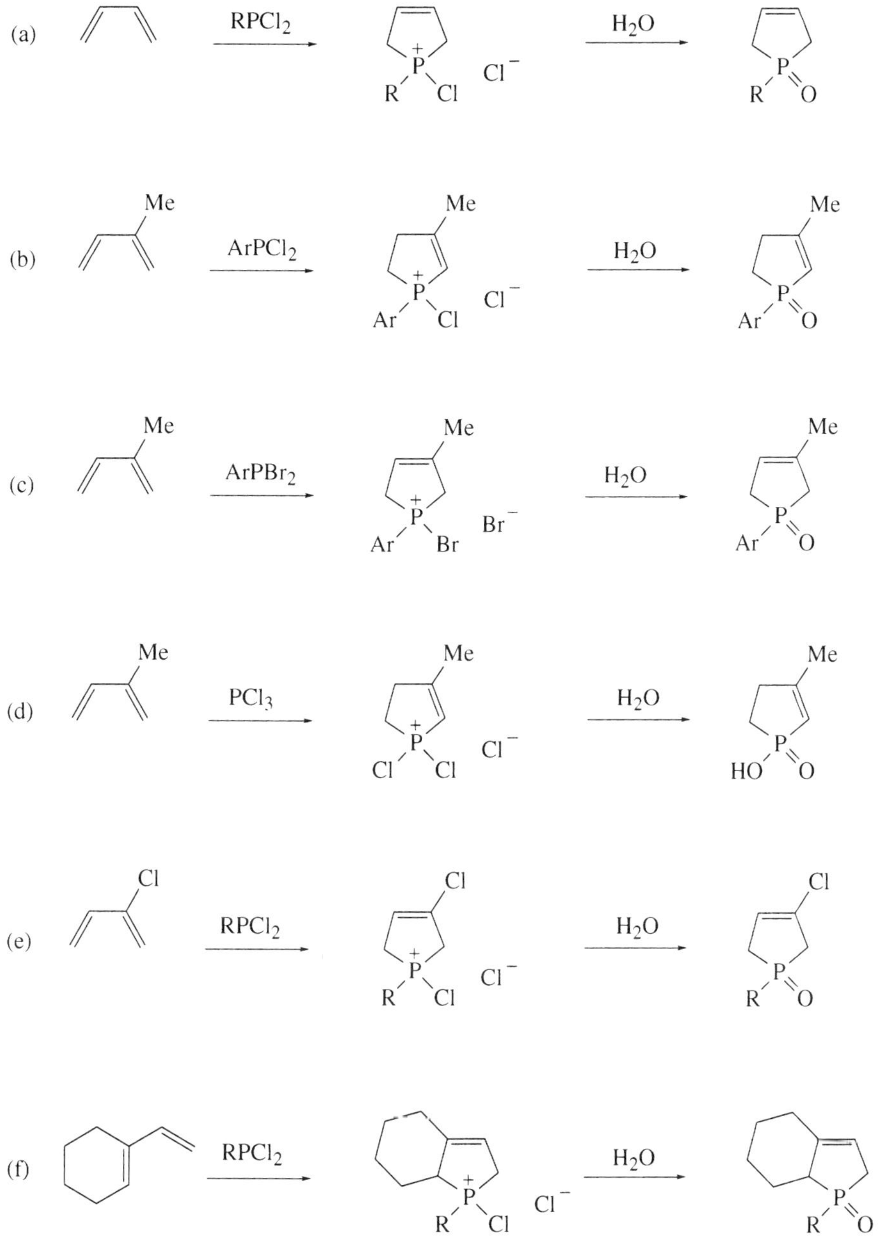

Scheme 1

The phospholene oxides resulting from the McCormack reaction also serve as highly useful starting materials for the synthesis of phospholane oxides by simple hydrogenation techniques (Equation (3)) ⟨72JOC238, 76JOC1155⟩.

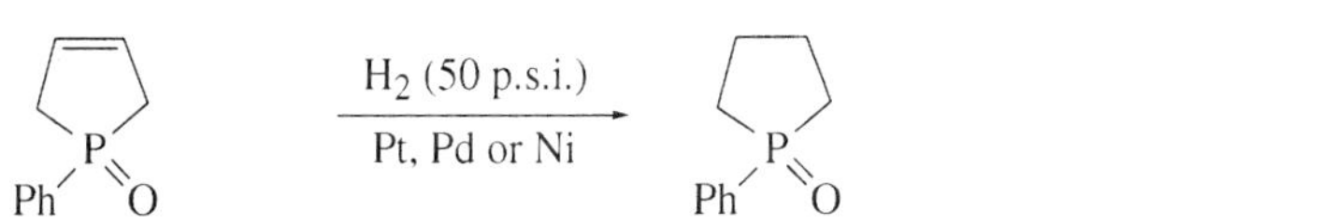

(3)

Many of the known phospholene oxides have been reduced to phosphine form, and this constitutes the best method at present for obtaining both 2- and 3-phospholenes. The latter are of special interest when the ring bears 2- or 2,5-substituents; with $HSiCl_3$ as the reducing agent (and some others), the stereochemistry at phosphorus in the oxide is preserved, and because of the high barrier to pyramidal inversion characteristic of phosphines there can be stable *cis,trans*-isomers where the tricovalent phosphorus acts as a chiral center (Equation (4)) ⟨65CB171, 69JA2788, 74JOC265⟩.

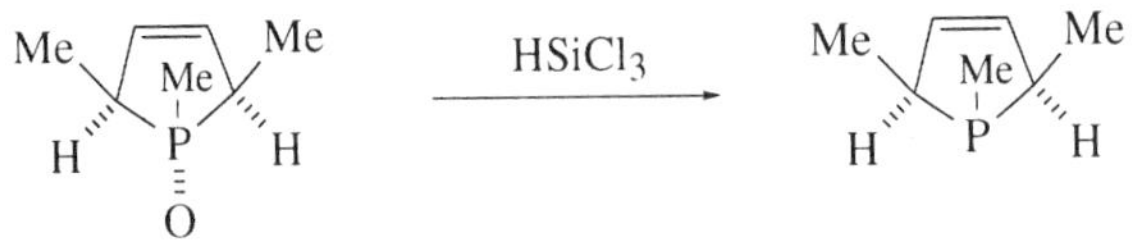

(4)

Examples with several substitution patterns are known. There have been some significant developments in the McCormack reaction since the 1981 review ⟨81MI⟩; one of some value is the discovery that performing the reaction under pressure ⟨89CC967⟩, greatly speeds up the cycloaddition and gives a more tractable solid product, and another is the application of the multivariate optimisation approach to improving the yield of 1-methoxy-3-methyl-2-phospholene oxide from workup of the reaction product of isoprene and PCl_3, which resulted in an overall yield of 90% ⟨93JCS(P2)629⟩. Other advances in the McCormack process include the introduction of two new types of dienophiles, phosphenium ions ⟨83JA7443, 83JA7444, 86IC740, 92JOC5189⟩ which react quite rapidly with dienes and alkyl (methylthio) chlorophosphine ⟨88ZN(B)257⟩ (Scheme 2).

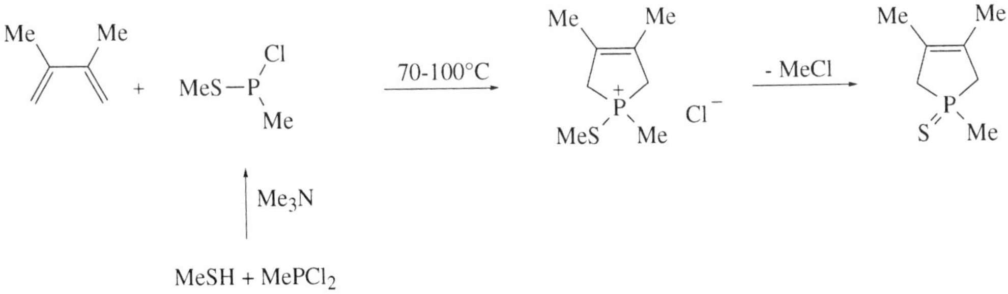

Scheme 2

No other synthetic method for phospholene oxides approaches the McCormack reaction in simplicity and versatility. Some specialised methods do exist, however. One is shown here that allows 2,2-disubstitution in the phospholene ring, which cannot be obtained with the McCormack method (Scheme 3) ⟨75T1907⟩.

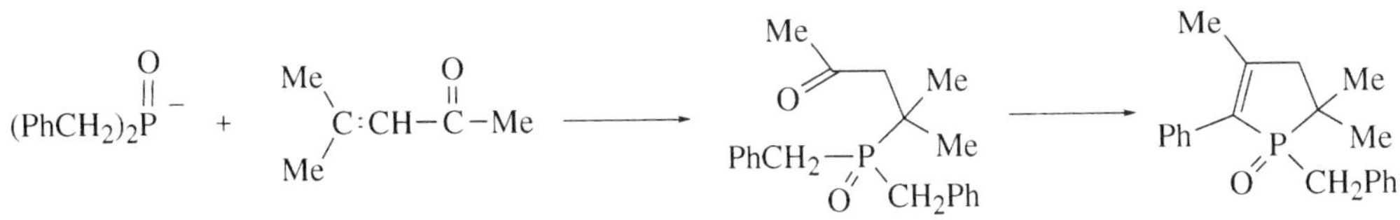

Scheme 3

Phospholene oxides from the McCormack reaction also stand as useful precursors of phospholanes; as noted, the double bond can be easily hydrogenated, and then the oxygen removed from phosphorus with silanes, etc. (Scheme 4).

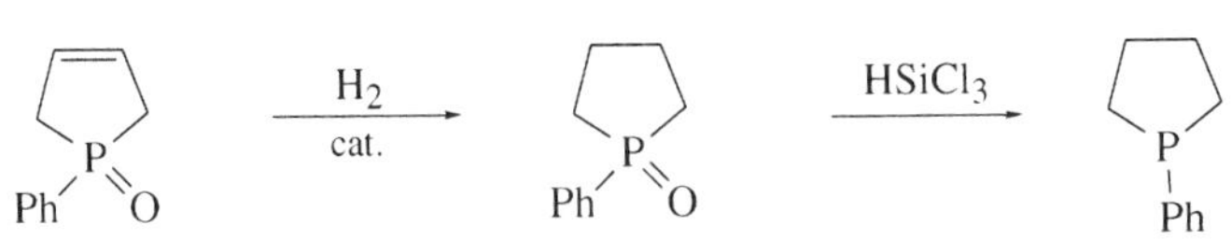

Scheme 4

Phospholanes can be prepared by other cyclisation methods, however, some of which are outlined below (Equations (5) ⟨16CB437⟩, (6) ⟨65HCA133⟩ and (7) ⟨61CB113⟩; Scheme (5) ⟨74JOC1531⟩).
Phospholanes and their oxides are known with a variety of functional groups on ring carbons (e.g. 2- and 3-keto, 2- and 3-hydroxy, 2,5- and 2,3-dihydroxy, 3-carboalkoxy, 3,4-epoxy, 3,4-dinitro, 3-CN, 3-R_2N, 3-OR, etc.).

4.1.2.2 Recent Developments

The synthesis and resolution of *trans*-2,5-diphenylphospholanic acid (**8**) has been described (Scheme 6). Compound (**8**) was subsequently converted into the optically active 1,2,5-triphenylphospholane (**9**) which

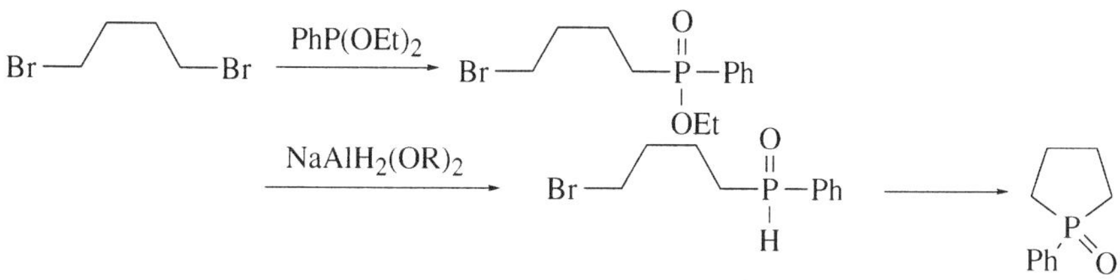

Scheme 5

was then used as a chiral ligand in the rhodium-catalysed hydrogenation of *N*-acetyldehydrophenylalanine methyl ester to afford methyl *N*-acetylphenylalaninate quantitatively in 82% e.e. ⟨99TL2939⟩.

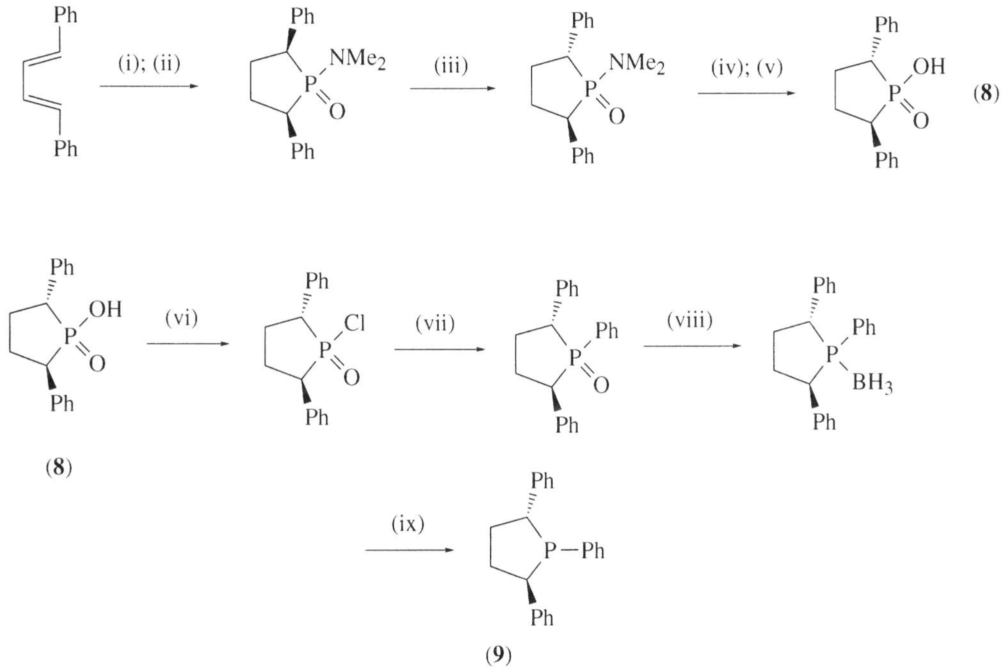

Scheme 6. (i) Cl_2PNMe_2, $AlCl_3$, CH_2Cl_2, −10°C then $NaHCO_3$/EDTA, 0°C; (ii) 5% Pd/C; H_2 (30–50 atm),CH_2Cl_2, 16 h, 65%; (iii) NaOMe (5 equiv.), MeOH, 25°C, 16 h, 100%; (iv) HCl_{aq}, MeOH; (v) quinine (1 equiv.) diastereoisomeric separation 40%; (vi) $(COCl)_2$, THF 100%; (vii) Ph_2CuLi, −78°C, THF, 65%; (viii) $LiAlH_4$, $CeCl_3$, then BH_3-THF, 73%; (ix) DABCO, toluene, 100%.

Novel chiral bisphosphine ligands bearing the (2*S*,5*S*)-dimethylphospholano group on a chiral biphenyl backbone have been designed and prepared. Asymmetric hydrogenations of a ketone and an olefin with the rhodium(I) complex catalysts prepared *in situ* from $[Rh(COD)Cl]_2$ and (*R*)-2,2′-bis-[(2″*S*,5″*S*)-2″,5″-dimethylphospholano]-5,5′-dimethoxy-4,4′,6,6′-tetramethyl-1,1′-biphenyl and (*S*)-2,2′-bis-[(2″*S*,5″*S*)-2″,5″-dimethylphospholano]-5,5′-dimethoxy-4,4′,6,6′-tetramethyl-1,1″-biphenyl gave the corresponding alcohol and alkane in high yields but with different degrees of selectivity ⟨99CPB436⟩.

C_2-Symmetrical ferrocenyl diamino diphosphines (diamino FERRIPHOS ligands) have proved to be excellent ligands for the rhodium-catalysed enantioselective reduction of methyl-α-acetamidoacrylates. The fact that their synthesis is straightforward, they are both air-stable and easy to modify, makes these ligands especially interesting for transition metal-catalysed hydrogenations. α-Amino acids with enantioselectivities greater than 95% (and in certain cases up to 99.3% e.e.) have been obtained without the need for further re-crystallisation ⟨99TA375⟩.

A family of diphosphine ligands has been prepared by Michael addition of *o*-anisylphenyl phosphide to diethyl vinylphosphonate and elaboration to phospholanes based on hexane-2,5-diol or mannitol (Scheme 7) ⟨99CC261⟩. Related work has appeared from Holz *et al.* (Scheme 8) ⟨98JOC8031⟩.

The phosphorylation reaction of 1,4- and 1,5-diketones with dimethylphosphites, hypophosphorous acid and its esters, and phosphines leading predominantly to cyclic compounds has been reviewed ⟨98RCB1851⟩.

Phospholanes have appeared as part of a recent review on optically active phosphine ligands where the emphasis is laid on the utility of these ligands in transition metal-catalysed asymmetric reactions ⟨98JSO511⟩.

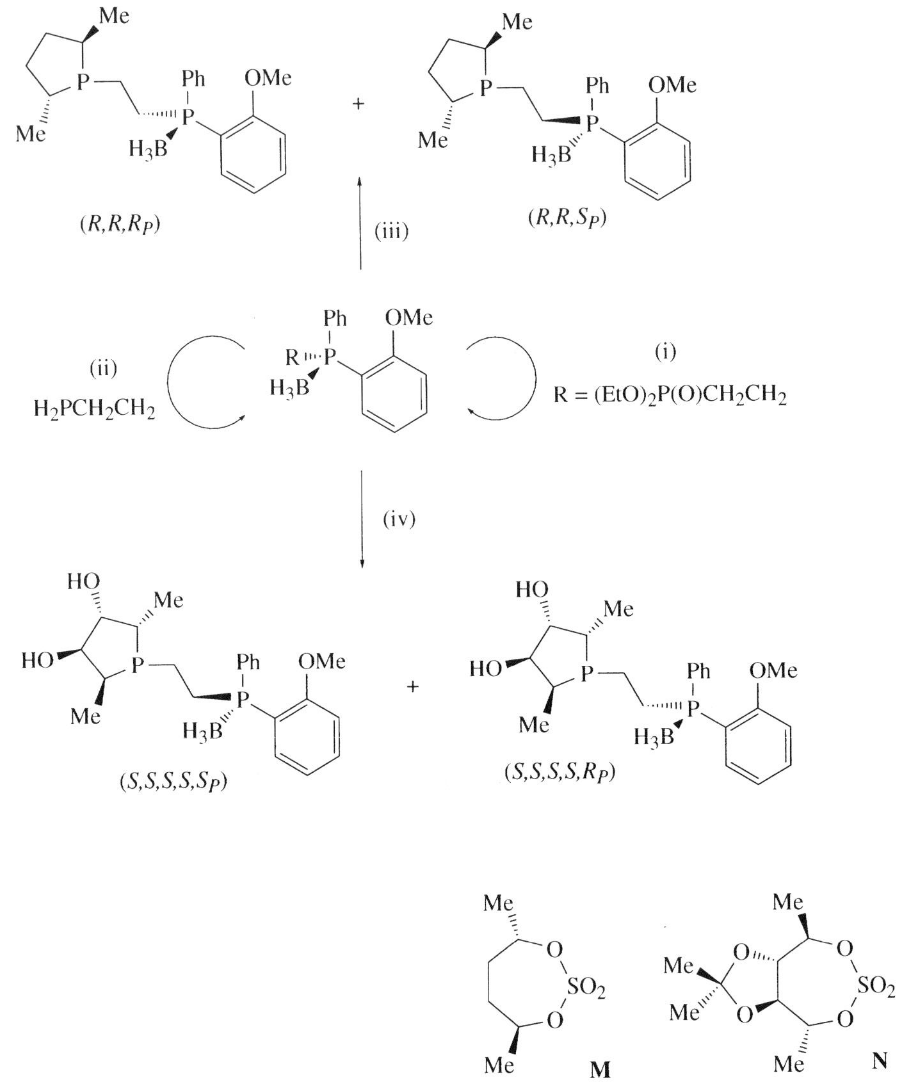

Scheme 7. (i) $CH_2{=}CHP(O)(OEt)_2$, KO^{t}Bu, THF, 95%; (ii) DABCO, C_7H_8; AlH_3, Et_2O; H_2O then CaH_2; (iii) BuLi, THF, −78°C then **M** then further BuLi; $Me_2S.BH_3$, 45% overall; (iv) BuLi, −78°C then **N**; then repeat; $Me_2S.BH_3$, 30% isolated yield; (v) NaH, MeI, THF, >80%; (vi) $HBF_4.OMe_2$ then $NaHCO_3$.

A novel enantioselective route to enantiomerically pure phospholanes has been developed exploiting the chiral lithium amide base approach (Scheme 9) ⟨98JOC912⟩.

Cyclo-phosphorylation of 1-hexadecylglycerol with phosphorus oxychloride afforded 4-hexadecyloxymethyl-2-oxo-2-chloro-1,3 2-dioxaphospholane in high yield. By means of ^1H, ^{13}C and ^{31}P-MR, this was shown to be a 1 : 9 mixture of two stable epimers with equatorial and axial configuration of the chlorine atom, respectively. This phospholane is susceptible to nucleophilic reactions with $KHCO_3$ and $^iPr_2NH$ to afford the chlorine substitution products in high (80–85%) yields with the phospholane cycle remaining intact. Interaction of the phospholane with water or choline tosylate also initially produces substituted phospholanes by NMR. Subsequent hydrolysis results in cleavage of the phospholane ring and formation of isomeric 1-hexadecylglycero-2- and 3-phosphates and 1-hexadecylglycero-2- and 3-phosphocholines, respectively. A method for the separation of isomeric phosphates has been developed as part of this work thus making it possible to isolate 1-hexadecylglycero-2-phosphate and 1-hexadecylglycero-3-phosphate, 4-alkyloxymethyl-2-oxo-2-chloro-1,3,2-dioxaphospholane, 1-alkylglycero-2-phosphate, 1-alkylglycero-3-phosphate ⟨97BK747⟩.

The zirconaindane phospholane (**10**) reacts either with $PhSbCl_2$ or Me_2SnCl_2 to give tricyclic systems incorporating phosphorus, antimony or tin; the same reaction involving (**10**) and R_2PCl (R = Ph, Et) leads to acyclic–cyclic diphosphanes (Scheme 10) ⟨97CC1239⟩.

(Trimethylsilyl)phosphine (Me_3SiPH_2) undergoes radical P–H bond addition to vinylphosphines and

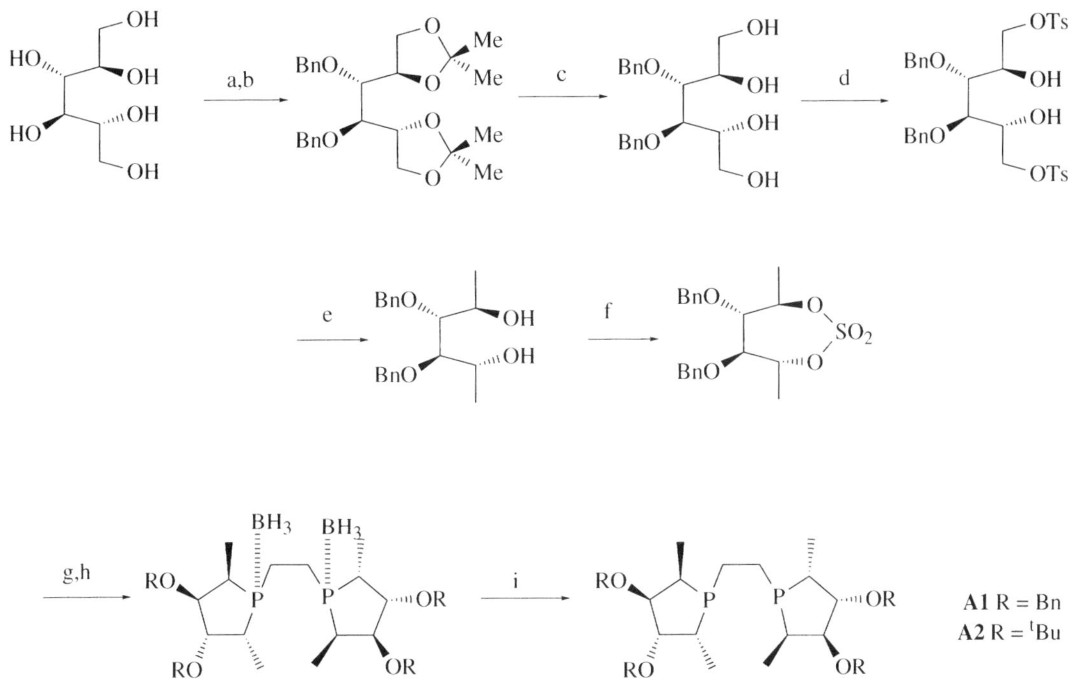

Scheme 8. (a) acetone, $ZnCl_2$, rt; (b) BnBr, NaOH, THF; (c) 70% aq. HOAc, rt; (d) TsCl, pyridine, 0°C; (e) $LiAlH_4$, THF; (f) $SOCl_2$, $RUCl_3$, $NaIO_4$, MeCN, H_2O, CCl_4; (g) BuLi, 1,2-$H_2P(CH_2)_2PH_2$, THF; (h) BH_3.THF; (i) DABCO, toluene, 40°C, 48 h.

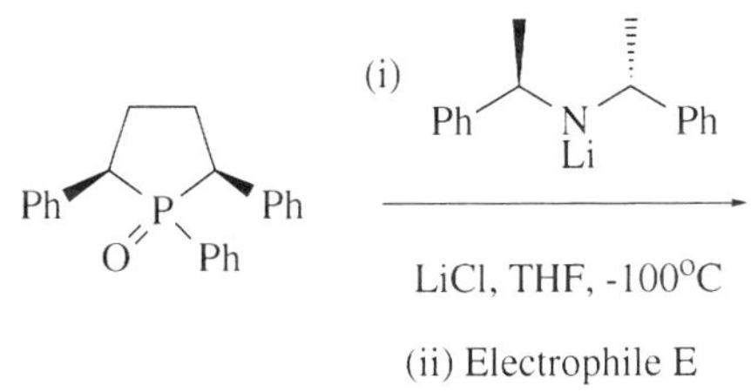

Scheme 9

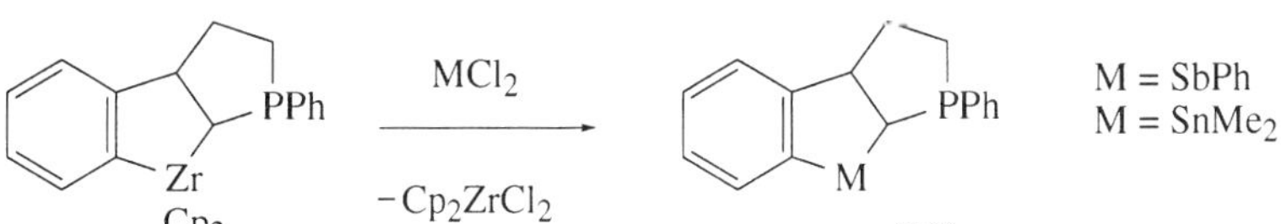

(**10**)

Scheme 10

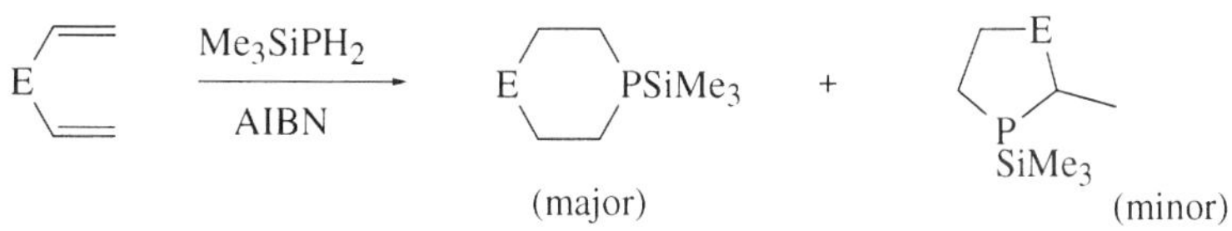

E = PhP, Me_2Si

Scheme 11

-silanes to form new di-phospha- and 4-silaphosphinanes (Scheme 11). Reactions are accompanied by formation of only small quantities of the Markovnikov addition product phospholanes. Methanolysis of these new silylphosphines affords new phosphines ⟨97IC1867⟩.

Phosphorylation reactions of 1,4-diketones, 1,5-diketones, and methylene-1,5-diketones with bidentate phosphorus nucleophiles (phosphines and phosphonous acid derivatives) have been shown to be convenient methods of synthesis of saturated functionalised phospholanes and phosphinanes ⟨97HAC217⟩.

The synthesis of two ethylene-bridged bidentate phosphines has been described. 1,2-bis-(1-phospholano)ethane (**11**) was synthesised by the stepwise addition of 2,2-dioxo-1,3,2-dioxathiepane to $H_2PCH_2CH_2PH_2$ (Scheme 12). This bis(phosphine) forms 2 : 1 complexes of the form $Fe(PP)_2Cl_2$ when added to iron(II) chloride ⟨96IC2546⟩.

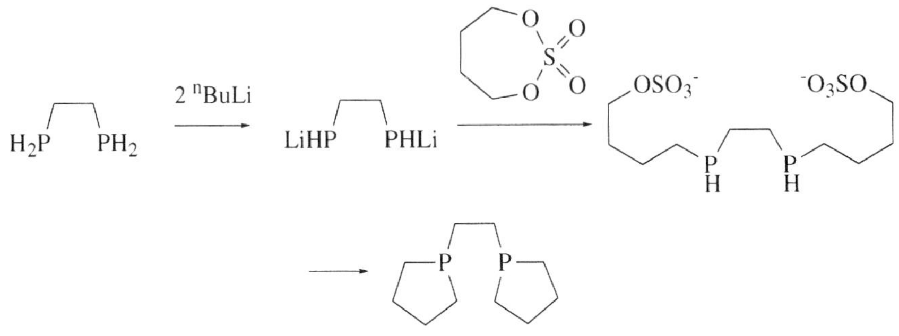

(**11**)

Scheme 12

Phospholanes have been prepared from red phosphorus and α,ω-dihaloalkanes in a single step in super-basic media ⟨95ZOB1096⟩.

Diastereomerically pure 5′-*O*-DMT-nucleoside 3′-*O*-(2-thio-1,3,2-oxathiaphospholanes) (**12**) and their oxathiaphospholane ring-substituted analogues (**13**) have been used for the synthesis of stereoregular oligo(nucleoside phosphorothioates) (S-oligos). The oxathiaphospholane ring-opening condensation requires the presence of strong organic base, preferably DBU. The yield of a single coupling step is ca. 95% and the resulting S-oligos are free of nucleobase- and sugar-phosphorothioate backbone modifications. The diastereomeric purity of products has been estimated on the basis of diastereoselective degradation with Nuclease P1 and a mixture of snake venom phosphodiesterase and *serratia marcescens* endonuclease. Thermal dissociation studies of heteroduplexes S-oligos/DNA and S-oligos/RNA showed that their stability is stereochemistry- and sequence-dependent ⟨95JA12019⟩.

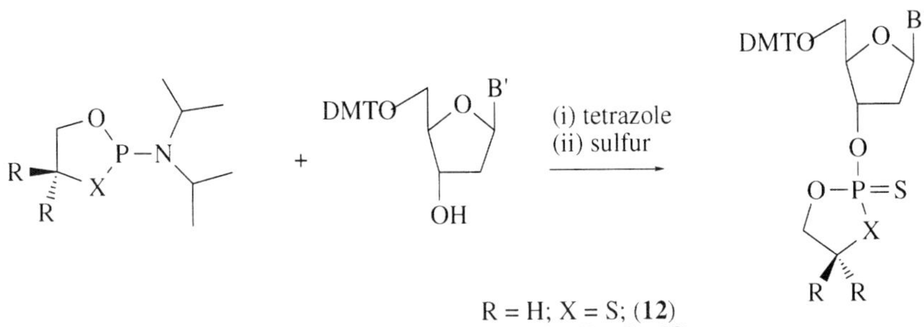

R = H; X = S; (**12**)
R = Me; X = S; (**13**)
DMT = dimethoxytrityl

(1*S*,2*S*)-Trans-cyclopentane-1,2-diyl-bis(phosphine) has been obtained by reduction, with $LiAlH_4$, of the cyclopentane-1,2-diyl-bis(1′,3′,2′)-dioxaphospholane (1*S*,2*S*)-*trans*-$C_5H_8[P(OCH(CO_2Pr)R_2]_2$, which itself was isolated in diastereomerically pure form from the reaction between, *trans*-$C_5H_8[PCl_2]_2$ and (2*R*,3*R*)-(+)-disopropyl tartrate ⟨95ZN(B)1004⟩.

The formation of the *N*-phosphanyl derivatives of oxaza- and diazaphosphinanes and -phospholanes has been reported *via* reaction of PX_3 (X = Cl, Br) with alcoholamines and diamines ⟨94PS(88)75⟩.

1,4-*Bis*-(1-phospholano)butane has been prepared in high yield by reacting trimethylphosphite with 1,4-di-lithiumbutane. Monodentate 1-alkylated phospholanes can be prepared in an analogous manner from alkyldimethoxyphosphine derivatives (Scheme 13) ⟨93S39⟩.

New, optically active phospholanes have been derived from tartaric acid (**14**) and (**15**) ⟨87JOM(328)71⟩.

Metallo-derivatives of 3-oxaphospholanes, 3-thiaphospholanes and 3-azaphospholanes have been described along with phospholanes containing hydrazones ⟨81ZOB514⟩.

Phosphanyl sugar derivatives, analogs of sugars having phosphorus in place of the ring oxygen, have been synthesised from 2- and 3-phospholenes as starting materials. Catalytic *cis*-dihydroxylation of 2-phospholene or 3-phospholene 1-oxide derivatives with osmium(VIII) oxide in the presence of a co-oxidant was shown to afford 3-deoxy- or 1-deoxy-tetrofuranose-type phosphanyl sugar derivatives,

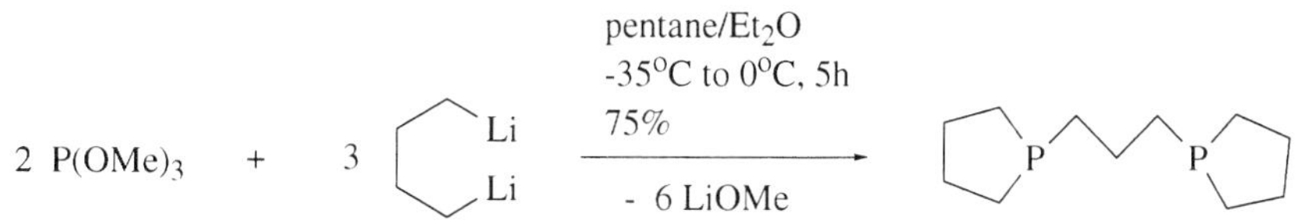

Scheme 13

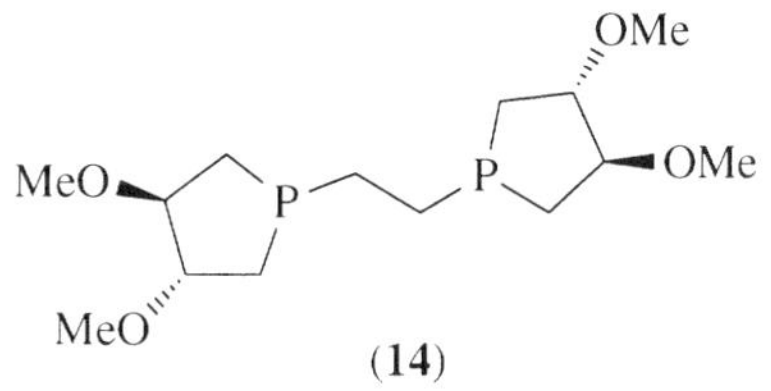

(**14**)

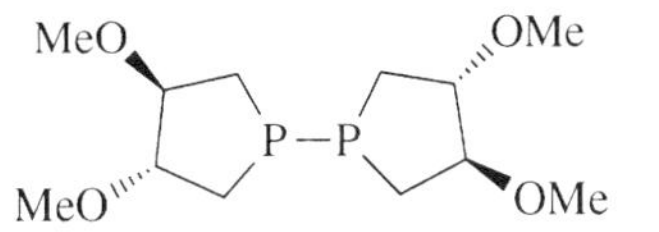

(**15**)

respectively. *cis*-Dihydroxylation of 4-acyloxy-2-phospholene 1-oxide derivatives gave tetrofuranose type phosphanyl sugar derivatives ⟨97JCC499⟩.

2- and 3-Phospholenes have been used as starting materials in the synthesis of sugar derivatives having a phosphorus atom in the hemiacetal ring of the sugar (phosphono sugars). Alkylation or benzyloxymethylation of 3-phospholene 1-oxides afforded phospholene derivatives with more than five carbon atoms. Epoxidation followed by epoxide ring opening and *cis*-dihydroxylation of 5-benzyloxymethyl-2-phospholene derivatives afforded four pentofuranose type phosphono sugars (arabinose, lyxose, ribose, and xylose). Nucleoside and isonucleoside derivatives of phosphono sugars have also been reported ⟨96PS(110)405⟩.

The first 3-silylated phosphol-3-enes have been synthesised using a modified McCormack reaction sequence. The complete protodesilylation observed in standard conditions is suppressed when reaction between 2-silylated 1,3-butadiene and dichlorophenylphosphine is carried out in the presence of a small amount of copper stearate in acetic anhydride at 50ºC. The new phospholenes contain a reactive vinylsilane function which has been epoxidised by *meta*-chloroperbenzoic acid. The bicyclic compound has a *cis* structure as established by classical spectroscopic methods and corroborated by molecular modelling ⟨95PS(105)101⟩.

The reaction of the *bis* (2-chloroethyl)amino-substituted benzoxazaphosphinone (**16**) with hexafluoroacetone (HFA) proceeds with insertion of the carbonyl group of HFA into the heterocycle of (**16**) to form the oxazaphosphepinedione (**17**) (Scheme 14).

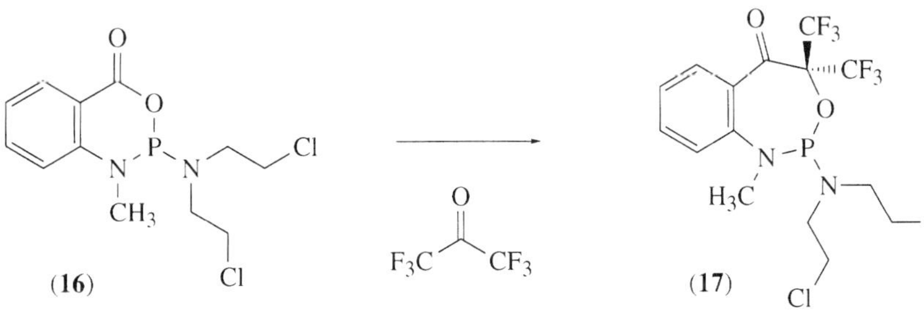

Scheme 14

Triethyl phosphite and triphenyl phosphite react with perfluorinated diketones with formation of the pentaoxyphosphoranes containing a dioxaphospholene ring system (**18**) (Scheme 15).

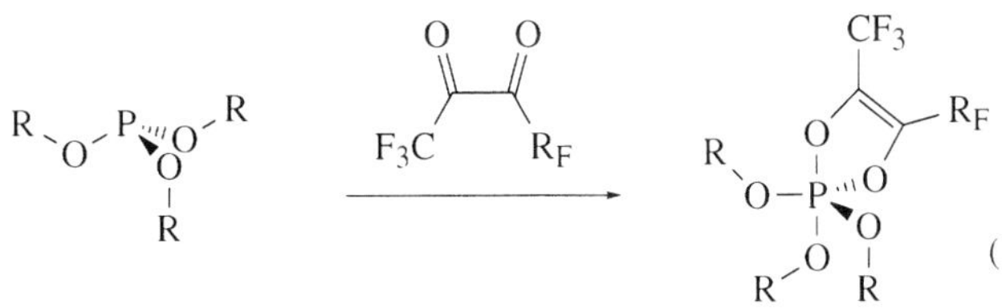

(**18**)

R = Et, Ph; $R_F = CF(CF_3)_2$; $(CF_2)_2CF_3$

Scheme 15

The reaction of (2-chloroethyl)amino- and bis(2-chloroethyl)amino-substituted 1,3,5,2-triazaphosphorinanediones with the perfluorinated diketone $C_2F_5C(O)C(O)CF_3$ affords spirophosphorane derivatives with the dioxaphospholene ring system in two cases only (Scheme 10). 2-[*Bis*-(2-chloroethyl) amino]-4*H*-1,3,2-benzodioxaphosphin-4-one reacts with perfluorinated alpha-diketones with insertion to afford dioxa- and oxazaphosphepinone ring systems (Scheme 16). The expected oxidative addition of the diketones to λ^3 phosphorus with formation of spirophosphoranes was not observed ⟨96CB725⟩.

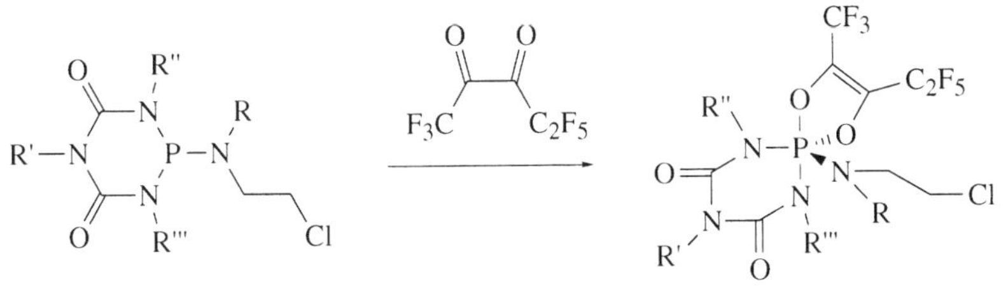

R' = R'' = R''' = Me; R = CH_2CH_2Cl
R' = R'' = R''' = Me; R = H
R' = R'' = Me; R''' = Ph; R = CH_2CH_2Cl
R' = R'' = R''' = Ph; R = CH_2CH_2Cl

Scheme 16

Novel phosphaheterocycles and arsa-heterocycles have been prepared from cyclophosphanes or cycloarsanes and 1,3-dienes ⟨95ZN(B)189⟩.

It is possible to prepare a range of main-group heterocycles *via* zirconium metallacycles, including phospholes and phospholenes. The scope and potential for this metallacycle transfer reaction have also been examined ⟨94JA1880⟩.

1-Phenylphosphol-2-en-1-oxide reacts with aryl iodides in the presence of ammonium formate and a catalytic amount of $Pd(OAc)_2$ to give 3-aryl-1-phenylphospholane-1-oxides, whereas 4-aryl-1-phenylphosphol-2-en-1-oxides are obtained from 1-phenylphosphol-3-en-1-oxide under modified Heck reaction conditions employing aryldiazonium salts and catalytic amounts of $Pd(OAc)_2$ (Scheme 17) ⟨98TL6707⟩.

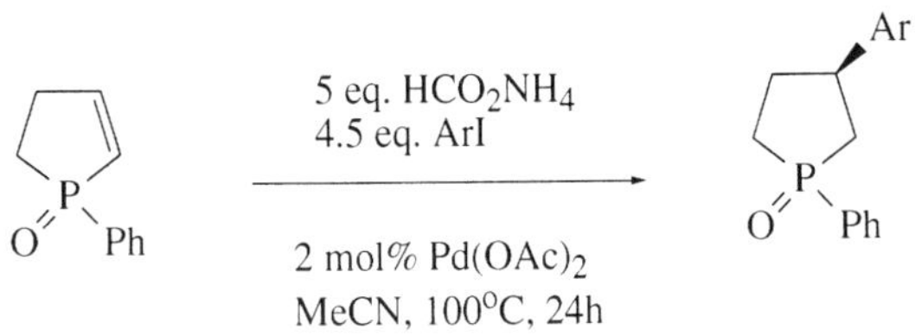

Ar = Ph; 67% yield
Ar = 4-MeC_6H_4; 69% yield
Ar = 4-MeOC_6H_4; 91% yield

Scheme 17

Borane complexes of several trialkyl-type C_2-symmetric diphosphines bearing two chiral phospholanes have been prepared by the reaction employing a lithium salt of a (2*S*,5*S*)-2,5-dimethyl-phospholane-borane complex which was formed *in situ* from a borane complex of (2*S*,5*S*)-2,5-dimethyl-1-phenylphospholane ⟨96SL1211⟩.

The triphenylphosphine deoxygenation of 2,3-dimethyllbenzofuran, 2,3-dimethylindole- and 2,3-dimethylindenedioxetanes leads to the respective epoxides through the intermediary phospholanes, of which the indene and indole derivatives were detected by low-temperature NMR spectroscopy; with excess Ph_3P the benzofuran dioxetane affords a benzofuran derivative, a novel process in which the resulting benzofuran epoxide is deoxygenated *in situ* ⟨94TL6063⟩.

The base-catalysed addition of the optically active [P–H]-containing phospholanes tarpholane and diopholane to olefins activated by diphenylphosphino or diphenylphosphinoyl groups gives unsymmetrical chelate phosphines (**19**)–(**26**) which are used as cocatalysts in the hydrogenation of (*Z*)-α-*N*-acetamidocinnamic acid, in the cross-coupling of 1-phenylethyl Grignard with vinylbromide, and in the hydrosilylation of acetophenone with diphenylsilane ⟨91JOM(413)55⟩.

Reactions of 1-phenyl- and 1-methoxy-2-phospholene 1-oxides with bromine in aqueous organic solvents or in a protic medium, such as methanol, have been shown to afford the corresponding 2-bromo-3-hydroxy- or 2-bromo-3-methoxyphospholane 1-oxide derivatives in high yields ⟨93HAC553⟩.

The cheletropic cycloaddition of $[ClP(NPr_2^{i})]^+AlCl_4^-$ with 1-substituted dienes at 0°C has been found to afford 1-(*N*,*N*-diisopropylamino)-1-chloro-2-alkyl-δ^3-phospholenium tetrachloroaluminates. The

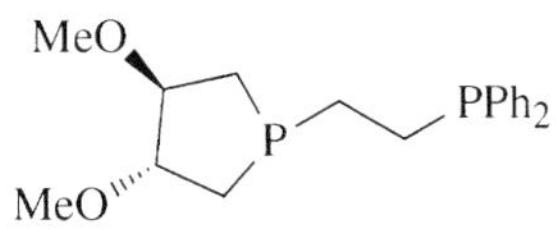

(19)

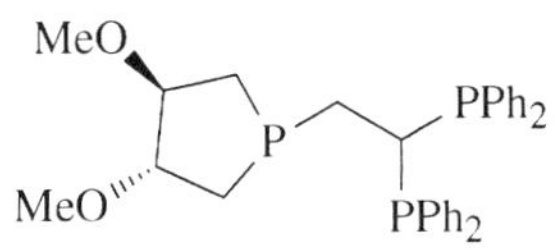

(20)

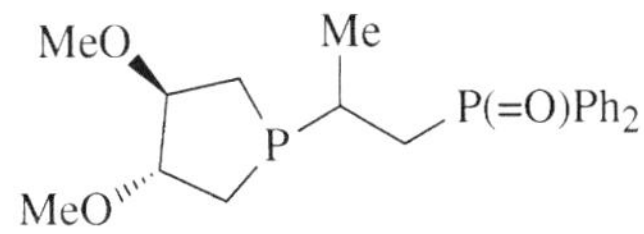

(21)

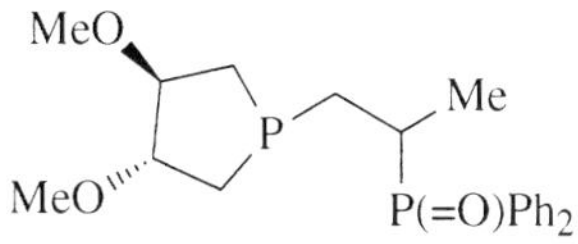

(22)

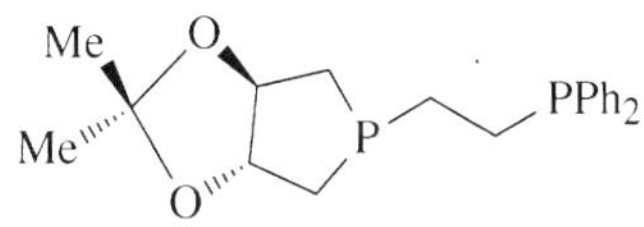

(23)

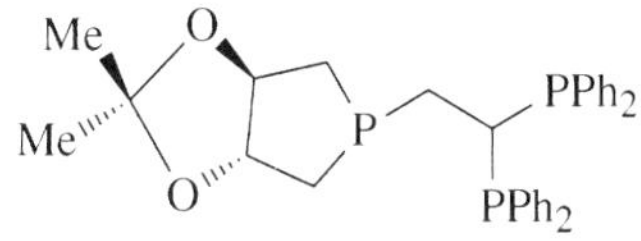

(24)

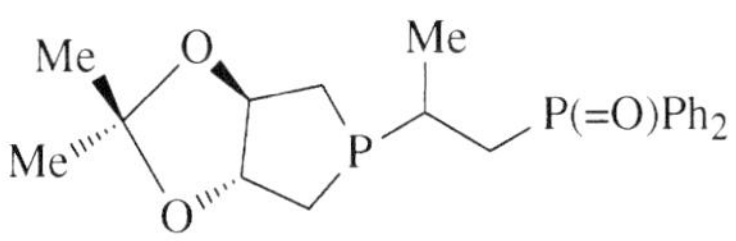

(25)

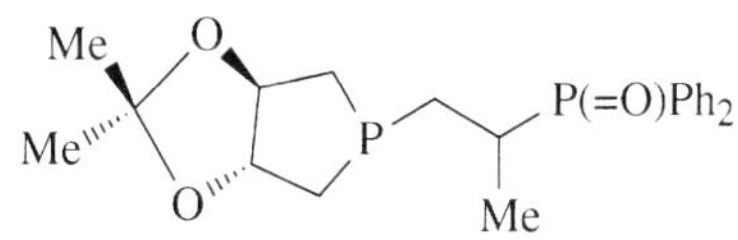

(26)

stereoselectivity of these reactions was found to range from 5 : 1 to 100 : 1. Hydrolysis of these cycloadducts afforded a diastereomeric mixture of 1-(*N*,*N*-diisopropylamino)-1-oxo-2-alkyl-δ^3 phospholenes. The ratio of the δ^3-phospholene amides differed significantly from the ratio of the intermediate δ^3-phospholenium ions, implying that the hydrolysis reactions occurred *via* five-coordinate phosphoranes which underwent pseudorotation prior to elimination of HCl. Hydrogenation of the δ^3-phospholene amides afforded saturated phospholane amides which underwent regioselective deprotonation and subsequent stereospecific alkylation reactions with alkyl halides. 1-(*N*,*N*-Diisopropylamino)-1-oxo-2,5-dimethyl- and -2,5-dibenzylphospholanes were converted by acid-promoted hydrolysis to (2*R**,5*R**)-2,5-dimethyl- and (2*S**,5*S**)-2,5-dibenzylphospholanic acid respectively (Scheme 18) ⟨92JOC5189⟩.

Phospholenes have been exploited in a novel conversion route to phospho-sugar derivatives (Scheme 19) ⟨88CC569⟩.

Phosphorane fragmentation reactions have been examined by Hall and co-workers in the reaction of δ-phospholenes with phenanthraquinone ⟨81PS(10)17⟩. Similarly, the reactivity of δ-phospholenes with sulfenate and sulfenamides has been examined ⟨81PS(10)67⟩ along with the isomerisation of δ^3 to δ^2 phospholenes promoted by ethyl acrylate ⟨81PS(10)185⟩.

4.1.3 CHEMICAL REACTIVITY

The chemistry of phospholanes and phospholenes, with P in the three-coordinate state, is dominated by the very high sensitivity of the phosphorus atom to various oxidising and electrophilic reagents that result in conversion to higher coordination states (usually C.N.4). In this sense their chemistry is not exceptionally different from common tertiary phosphines. The attack of electrophiles normally occurs first at phosphorus, and the product may react further at the double bond.

There is a rich chemistry known for phospholene oxides, too extensive to be included in this review. Most of the background information has been summarised ⟨81MI⟩. In general, the common reactions of alkene chemistry occur at the double bond, although there are some notable failures (hydration with sulphuric acid, hydroboration, oxymercuration, etc.). Some of the failures may be correctable with different conditions, as was demonstrated in the case of addition of dichlorocarbene from an NaOH–$CHCl_3$ mixture, which is only successful under phase-transfer conditions ⟨93S931⟩. The resulting dichlorocarbene adducts undergo ring expansion under heating (Scheme 20).

Another example is the failure of 2-phospholene oxides to add bromine; when the bromination is conducted in water or alcohols, bromohydrins or bromo alkoxy derivatives can be easily obtained ⟨73TL1249⟩. The 2,5 positions of the 3-phospholene oxide are activated and can be deprotonated by strong bases such as *n*-butyllithium; the resulting carbanions act in conventional manner and can be used

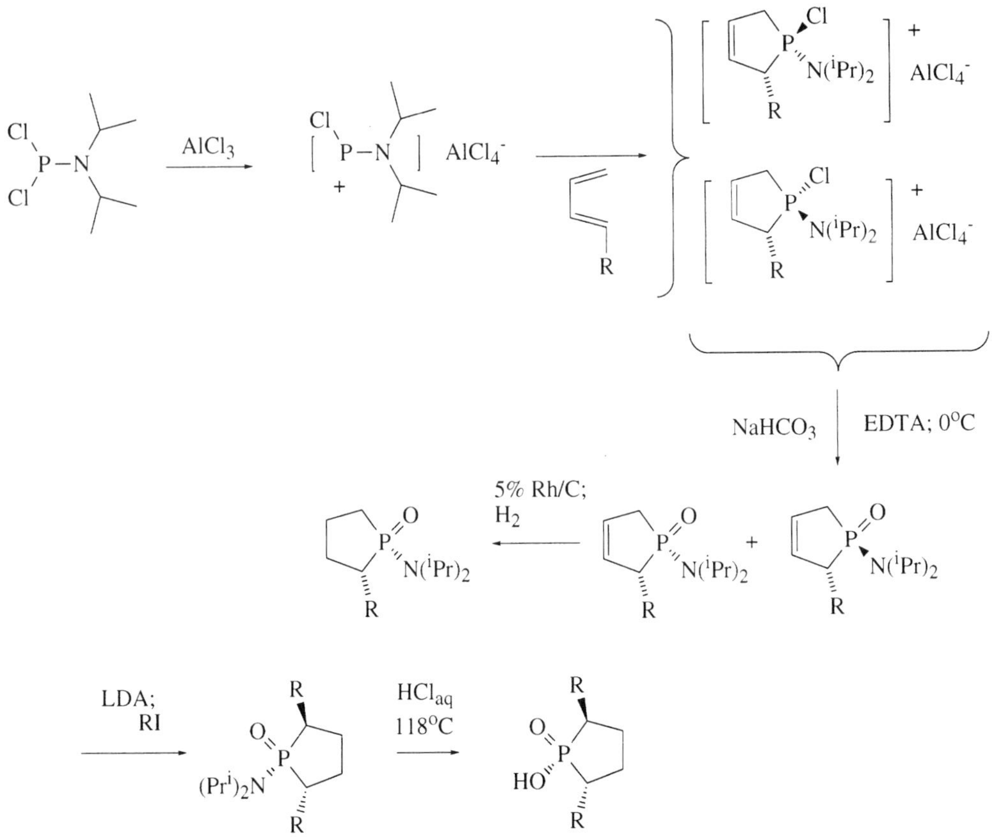

Scheme 18

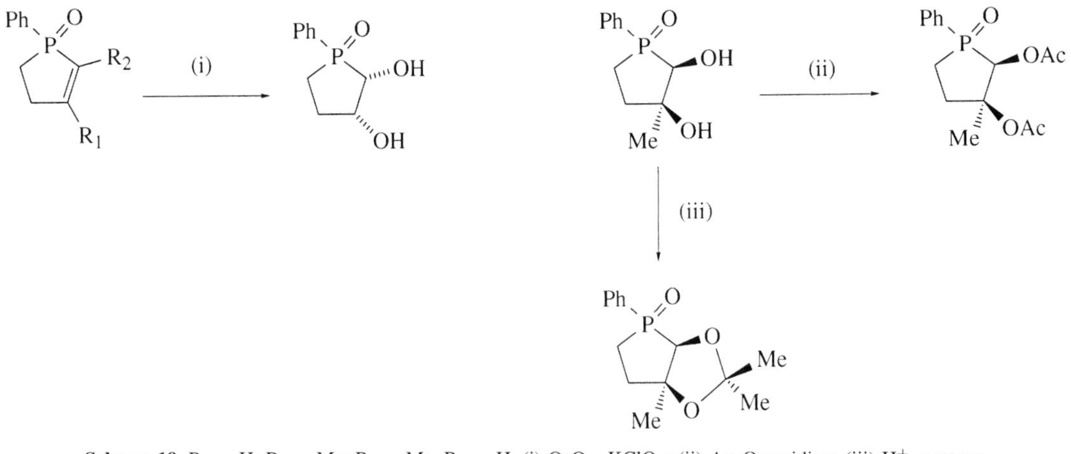

Scheme 19. $R_1 = H$; $R_2 = Me$; $R_1 = Me$, $R_2 = H$. (i) OsO_4; $KClO_4$; (ii) Ac_2O-pyridine; (iii) H^+, acetone.

in condensation reactions to prepare functionalised phospholene oxides ⟨71BSF317, 72BSF3494⟩. A novel type of compound is formed when the carbanion is reacted with Ph_2PCl; a phosphino group is installed at the 2-position of the 2-phospholene ring ⟨71BSF317⟩. Upon reaction with aromatic nitriles, unusual 7-membered rings are produced (Scheme 21) ⟨73T2909⟩.

The double bond in 2-phospholene oxides is activated by the phosphoryl group, and nucleophilic additions then are possible with such reactants as OH^-, RO^-, CN^-, RNH_2, $(RO)_2P(O)H$, etc. ⟨77GEP2541355, 77GEP2552319⟩. The activating effect also makes it possible to perform Diels–Alder ⟨71CJC530⟩ and 1,3-dipolar cycloadditions at the double bond, and permits the ready displacement of leaving groups from the 3-position by attack of nucleophiles (e.g. of Cl^- by $NaOCH_3$, H_2O, R_2NH, etc. ⟨74JOC686, 77JOC552, 78JOC108⟩).

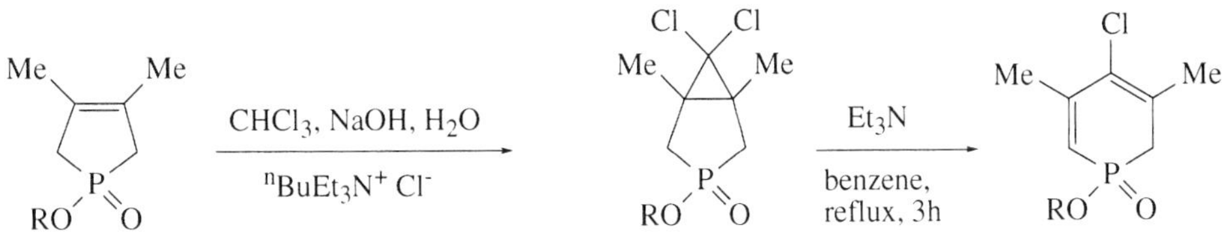

Scheme 20

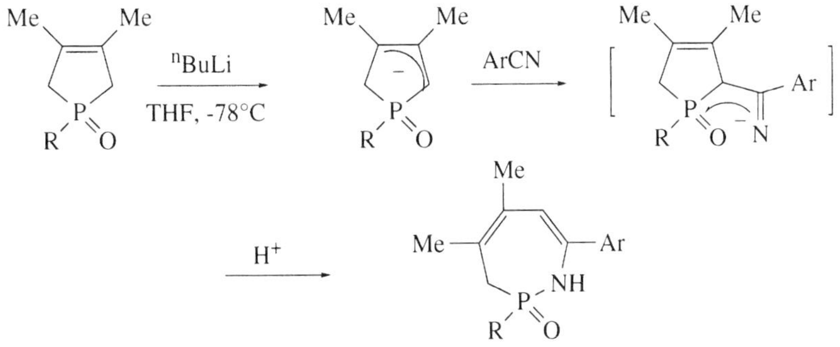

Scheme 21

4.1.4 APPLICATIONS OF PHOSPHOLANES AND PHOSPHOLENES IN CATALYSIS

The family of diphosphine ligands prepared by Michael addition of *o*-anisylphenyl phosphide to diethyl vinylphosphonate and elaboration to phospholanes as illustrated in Scheme 7 have been used in some rhodium-catalysed hydrogenations of enamides (Table 1) ⟨99CC261⟩. Related work has appeared from Holz *et al.* (Scheme 8 and Table 2) ⟨98JOC8031⟩.

New novel chiral monophosphines, 2,5-dialkyl-7-phenyl-7-phosphabicyclo-[2.2.1]-heptanes have been prepared according to the route outlined in Scheme 22 and exploited in the asymmetric palladium-catalysed allylic alkylation reaction (Table 3) ⟨97JOC4521⟩.

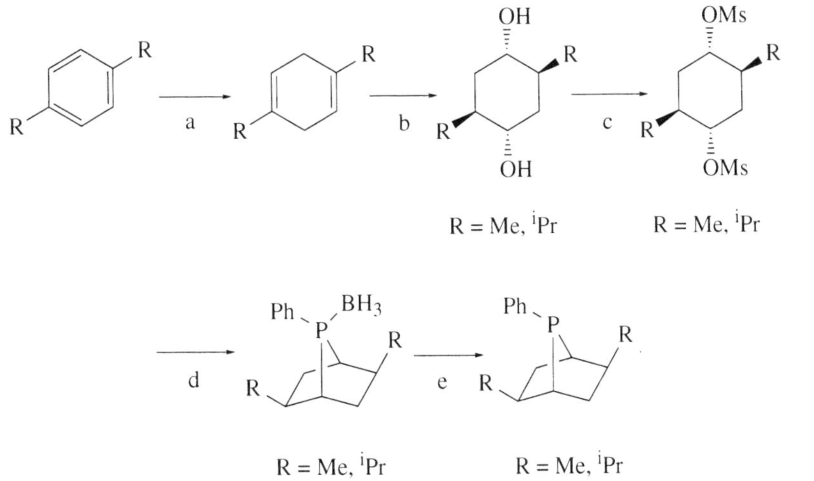

A, B

Scheme 22. (a) Li, $EtNH_2$; (b) $IpcBH_2$, Et_2O; H_2O_2, NaOH; (c) MsCl, Et_3N, CH_2Cl_2; (d) Li_2PPh, THF; BH_3.THF; (e) $HBF_4.OMe_2$, CH_2Cl_2; $NaHCO_3$.

Two 1,1′-bis(phospholano)ferrocene ligands have been described by the Burk group along with preliminary catalytic studies which revealed that enantioselectivities as high as 83% could be achieved in the hydrogenation of several model olefin, ketone, and hydrazone substrates ⟨94TL9363⟩ (Scheme 23 and Table 4).

Chiral C_2-symmetric bis(phospholane) ligands have been prepared and exploited in rhodium-catalysed asymmetric hydrogenation reactions. A practical, one-pot procedure which utilises enantiomerically pure 1,4-diol cyclic sulphates for the preparation of 1,2-bis(phospholano)ethanes and

Table 1. Conditions: Substrate : catalyst 100 : 1, $(COD)_2RhBF_4$ as precursor ($HBF_4.OMe_2$ deboronation *in situ*), 1.3 bar, MeOH

Substrate	*Catalyst precursor*	*e.e.* (%)
1	C2	19 *S*
1	D2	85 *S*
1	C1	43 *S*
1	D1	92 *S*
1	A	38 *S*
1	B	60 *R*
2	C2	58 *S*
2	D2	67 *S*
2	C1	82 *S*
2	D1	88 *S*
2	A	5 *R*
2	B	36 *R*
3	D2	77 *S*
3	C1	72 *S*
3	D1	90 *S*

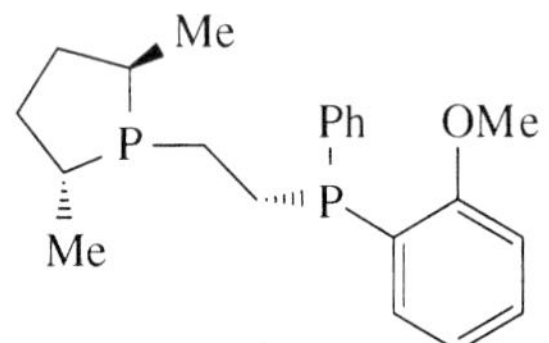

A

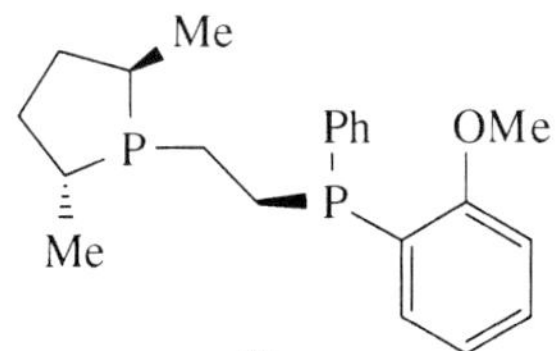

B

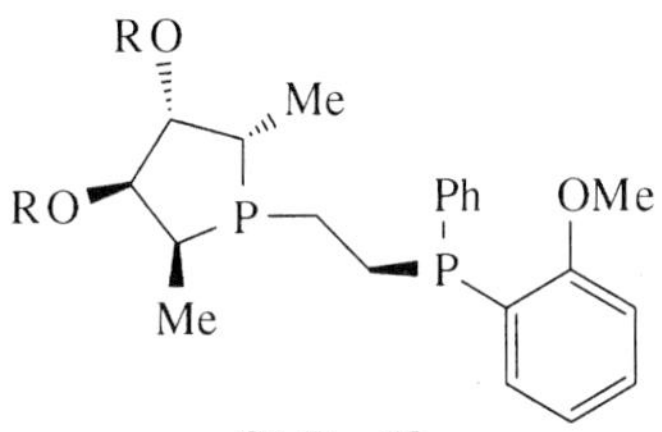

C1 R = H
C2 R = Me

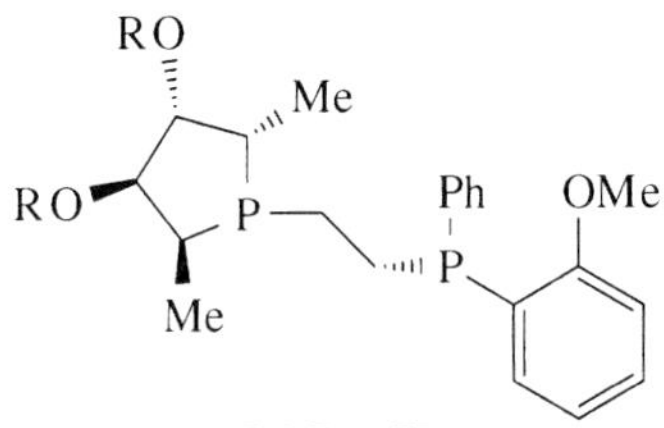

D1 R = H
C2 R = Me

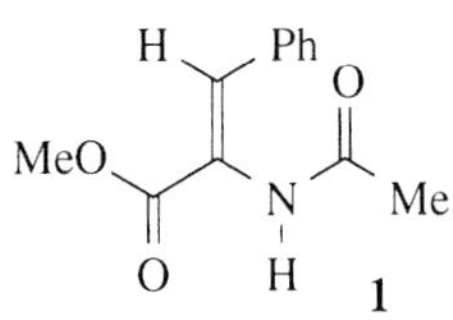

1

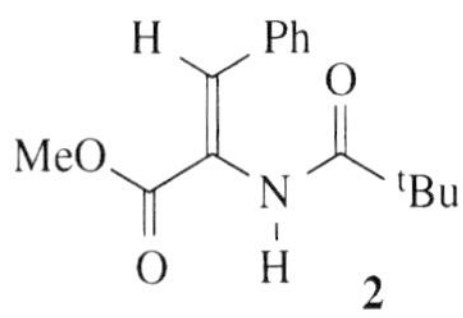

2

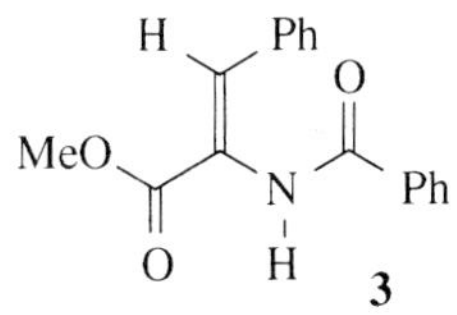

3

1,2-bis(phospholano)benzenes (DuPHOS) is also described (Scheme 24). Cationic rhodium complexes bearing these new ligands have been found to behave as very efficient catalyst precursors for the asymmetric hydrogenation of a broad range of α-(*N*-acylamino)acrylate (enamide) substrates (Table 5). Significantly, a variety of unnatural and non proteinaceous α-amino acid derivatives were obtained directly with enantioselectivities approaching 100% when using the DuPHOS ligands of Scheme 24. Substrate-to-catalyst ratios of 10,000 : 1 were routinely used, and ratios as high as 50,000 : 1 were demonstrated in these reactions ⟨93JA10125⟩.

Enantiomerically pure *trans*-2,5-disubstituted-1-phenylphospholanes have been described and employed in the preparation of a new series of C_3-symmetric tris(phospholane) ligands (Scheme 25). Rhodium complexes bearing the new phosphine ligands were prepared and shown to act as efficient catalyst precursors for the enantioselective hydrogenation of various unsaturated substrates (Table 6) ⟨91TA569⟩.

A greatly improved process has been developed for synthesis of the glutarate derivative (**27**), a key

Table 2. Conditions: 1.0 atm overall pressure over the solution; standard conditions of 0.01 mmol pre-catalyst and 1.0 mmol prochiral olefin in 15 ml solvent

Ligand	*R^1*	*R^2*	*R^3*	*Time for 50% conversion (min)*	*e.e. (%)*
A1	Ph	NHAc	CO_2H	62	93.5 *S*
A1	Ph	NHAc	CO_2Me	52	97.5 *S*
A1	H	CH_2CO_2H	CO_2H	24	98.0 *S*
A1	H	CH_2CO_2Me	CO_2Me	9	98.9 *S*
A1	Ph	NHBz	$P(O)(OMe)_2$	780	95.4 *S*
A2	Ph	NHAc	CO_2H	115	96.8 *R*
A2	Ph	NHAc	CO_2Me	125	98.4 *S*
A2	H	CH_2CO_2H	CO_2H	11	96.7 *S*
A2	H	CH_2CO_2Me	CO_2Me	8	99.1 *S*
A2	Ph	NHBz	$P(O)(OMe)_2$	500	98.8 *S*

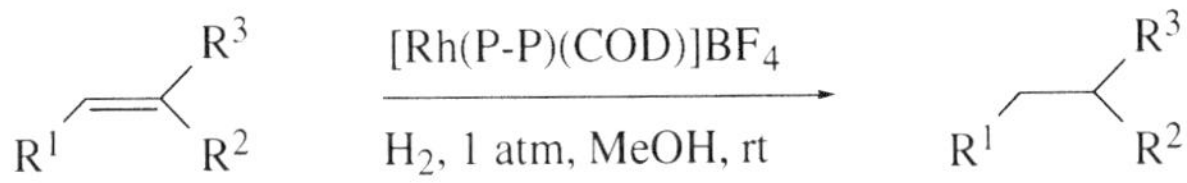

Table 3. Reaction performed under N_2 using 1,3-diphenyl-2-propenyl acetate as nucleophile Nu (300 mol%), BSA [bis(trimethylsilyl)acetamide] (300 mol%), KOAc (2 mol%), toluene, [Pd] 1.4 mol% and L^*; e.e. values measured by HPLC using Chiralcel OD column and absolute configuration determined by comparison of optical rotation values to literature

L^*	*[Pd]*	*[Pd] : L^**	*Nu*	*Yield (%)*	*Reaction time (h)*	*e.e. (%)*
A	$Pd_2(dba)_3$	1 : 2.2	$CH_2(CO_2Me)_2$	96	1.5	74 *R*
A	$Pd(OAc)_2$	1 : 2.2	$CH_2(CO_2Me)_2$	98	4.0	72 *R*
A	$[Pd(C_3H_5)Cl]_2$	1 : 1.1	$CH_2(CO_2Me)_2$	97	5.0	60 *R*
A	$[Pd(C_3H_5)Cl]_2$	1 : 2.2	$CH_2(CO_2Me)_2$	93	2.0	95 *R*
A	$[Pd(C_3H_5)Cl]_2$	1 : 3.3	$CH_2(CO_2Me)_2$	96	1.5	96 *R*
A	$[Pd(C_3H_5)Cl]_2$	1 : 2.2	$CH_2(CO_2Me)_2$	80	1.0	97 *R*
A	$[Pd(C_3H_5)Cl]_2$	1 : 2.2	$CH_2(CO_2Me)_2$	95	2.0	96 *R*
A	$[Pd(C_3H_5)Cl]_2$	1 : 2.2	$CH_2(CO_2Me)_2$	99	2.0	>97 *R*
A	$[Pd(C_3H_5)Cl]_2$	1 : 2.2	$CH(NHAc)(CO_2Et)_2$	95	2.0	>99 *S*
B	$[Pd(C_3H_5)Cl]_2$	1 : 2.2	$CH_2(CO_2Me)_2$	99	3.5	78 *R*

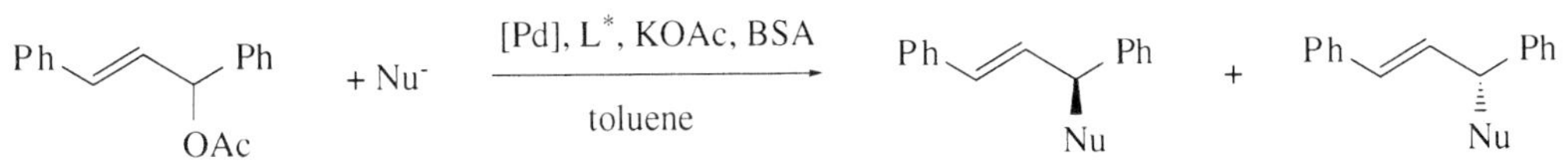

Table 4. Reaction conditions: 0.05–0.1 M substrate, methanol solvent, 60 psi H_2, 25°C, catalyst precursor $[(COD)Rh(P_2)]^+OTf^-$ (0.2 mol%; P_2 = **L1; L2**); reactions were stirred for 6–24 h and in all cases 100% conversion was observed; e.e. values were determined by chiral capillary GC using Astec's Chiraldex G-PN column for reactions involving B and C or Chrompack's Chiralsil-L-Val column for reactions involving A and D

Substrate	*Ligand*	*e.e. (%)*
A	**L1**	64
A	**L2**	83
B	**L1**	72
B	**L2**	83
C	**L1**	33
C	**L2**	58
D	**L2**	31

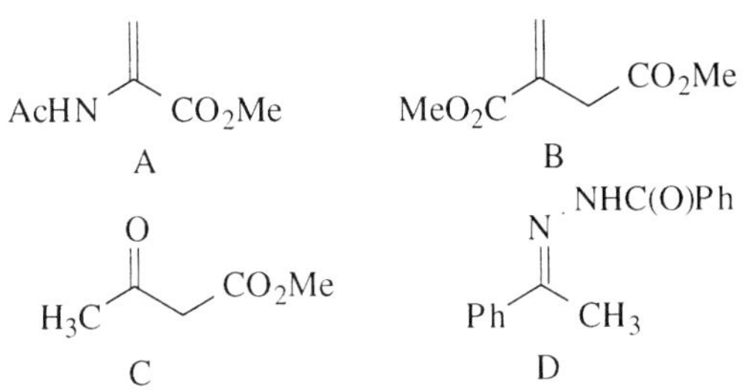

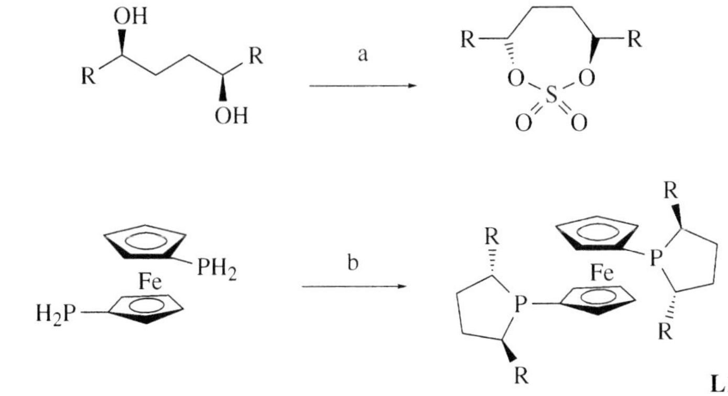

L1 R = Me
L2 R = Et

Scheme 23. (a) $SOCl_2$, $RuCl_3$, $NaIO_4$; (b) ^{n}BuLi (2 eq.), cyclic sulphate X (2 eq.), ^{n}BuLi (2.2 eq.).

Table 5. Reactions were carried out at 20–25°C and an initial H_2 pressure of 30 psi (2 atm) with 0.25–0.35 M methanol solutions of substrate and the catalyst precursors $[(COD)Rh(P_2)]^+OTf^-$ (0.1 mol%; P_2 = **L3–L10**); reaction time for complete conversion was 1 h

Ligand	*Substrate E e.e. (%)*	*Substrate F e.e. (%)*
L3	91.4	85
L4	98.1	93
L5	97.7	92
L6	96.4	93
L7	99.0	98
L8	99.4	99
L9	99.8	>99
L10	95.4	87

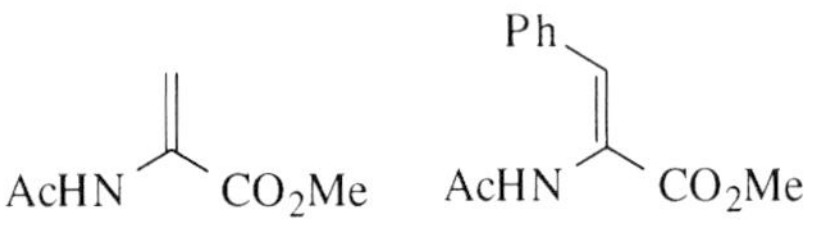

E F

Table 6. Reactions were carried out in Fisher–Porter tubes at 20–25°C and an initial H_2 pressure of 30 psi (2 atm) with 0.25–0.35 M methanol solutions of substrate and the catalyst precursors $[(COD)Rh(P)]^+SbF_6^-$ (0.1 mol%; P = **L11**)

Substrate	*Ligand*	*e.e.* [a] *(%)*	*Time* [b] *(h)*
B	**L11**	94 *S*	20
F	**L11**	89 *S*	72
F	**L11**	40 *S*	48

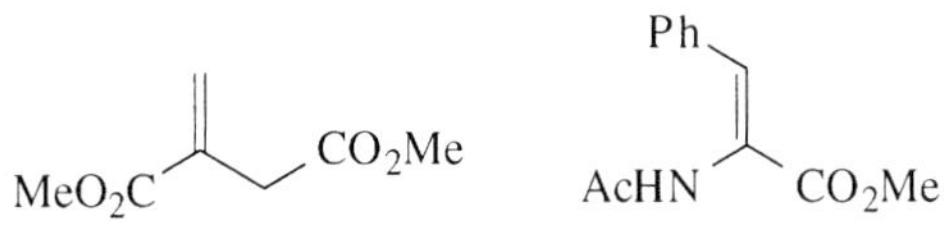

B F

[a] Absolute configurations determined by optical rotation. [b] Time required for 100% conversion of substrate

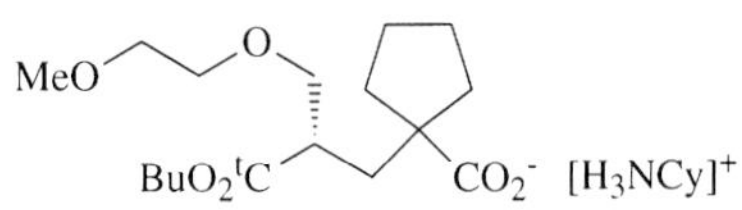

(**27**)

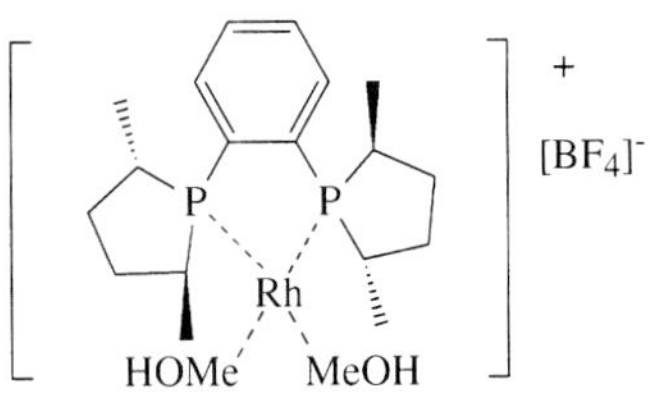

(**28**)

Table 7. Reactions were carried out at 25°C under an initial H_2 pressure of 60 psi and as 0.5 M solutions in methanol of substrate using the catalyst precursor [(COD)Rh(*S,S*)-Et-DuPHOS]OTf (0.2 mol%) for 48 h unless stated otherwise

R′	*R*	*R″*	*E/Z*	*e.e.* [a] (%)
Ac	H	Et	–	>99 (*S*)
Bz	Me	Et	3	96 (*S*)
Bz	nPr	Me	3	98 (*S*)
Ac	^{i}Pr	Et	6	96 (*S*) [b]
Bz	^{i}Pr	Et	6	97 (*S*) [b]
Bz	^{c}Pr	Me	9	98 (*S*)
Bz	$CH_2{}^i$Pr	Et	2.5	>99 (*S*)
Ac	$^nC_5H_{11}$	Et	3.5	>99 (*R*) [c]
Bz	$^nC_5H_{11}$	Et	3.5	>99 (*S*)
Bz	$^cC_6H_{11}$	Me	3	95 (*S*)
Ac	Ph	Et	9	96 (*S*) [d]
Bz	Ph	Me	10	98 (*S*)
Bz	α-naphthyl	Et	3	93 (*S*) [e]
Bz	2-thienyl	Me	4	98 (*S*)

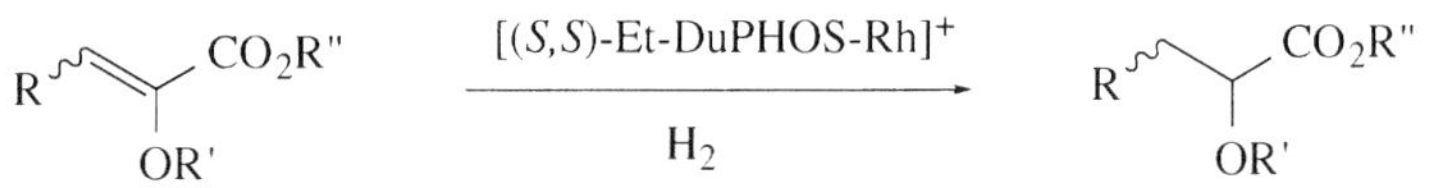

[a] E.e. values determined by chiral HPLC or chiral capillary GC. Absolute configurations were assigned by comparing the sign of optical rotation of derived α-hydroxyesters or 1,2-diols with that of known compounds. [b] Initial pressure 90 psig. [c] Reaction catalysed by (*R,R*)-Et-DuPHOS-Rh. [d] Reaction performed with the (*S,S*)-Me-DuPHOS-Rh catalyst (0.4 mol%). [e] Reaction performed in CH_2Cl_2.

intermediate required for Pfizer's drug candoxatril. Cationic (*R,R*)-Me-DuPHOS-Rh catalyst (**28**) was found to allow highly efficient and enantioselective hydrogenation of a unique carboxylate substrate to afford the desired product in >99% ee and high yield (95%). The robust nature of the process was validated on a 12 kg reaction scale. A novel mechanism for the hydrogenation process is proposed. Through use of a labile η^6-benzene-Rh-Me-DuPHOS complex, the postulated catalytic intermediates have been independently synthesised and interconversion of these key catalytic intermediates demonstrated ⟨99JOC3290⟩.

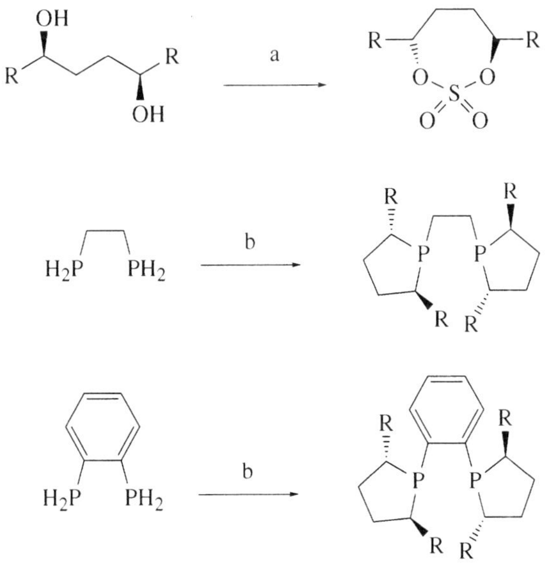

(*S,S*)-**L3** R = Me
(*S,S*)-**L4** R = Et
(*S,S*)-**L5** R = ^{n}Pr
(*R,R*)-**L6** R = ^{i}Pr

(*S,S*)-**L7** R = Me
(*S,S*)-**L8** R = Et
(*S,S*)-**L9** R = ^{n}Pr
(*R,R*)-**L10** R = ^{i}Pr

Scheme 24. (a) $SOCl_2$, $RuCl_3$, $NaIO_4$; (b) ^{n}BuLi (2 eq.), cyclic sulphate X (2 eq.), ^{n}BuLi (2.2 eq.).

Asymmetric catalytic hydrogenation of Stobbe-derived itaconates has provided practical access to 2-alkylsuccinates (Scheme 26) ⟨98AG(E)1931⟩.

The asymmetric hydrogenation of α-(acetyloxy)- and α-(benzoyloxy)acrylates catalysed by cationic rhodium–DuPHOS complexes has been examined by the Burk group (Table 7). A wide range of substrates

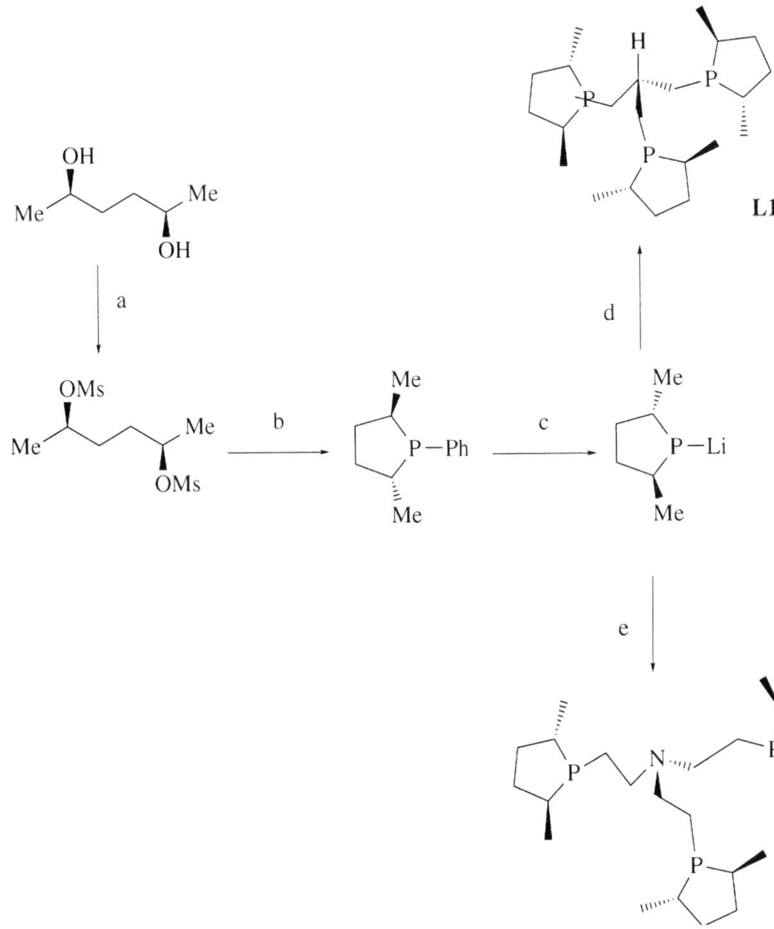

L11

L12

Scheme 25. (a) MsCl, NEt_3; (b) Li_2PPh; (c) Li, THF; (d) $HC(CH_2Cl)_3$, THF; (e) $N(CH_2CH_2Cl)_3$.

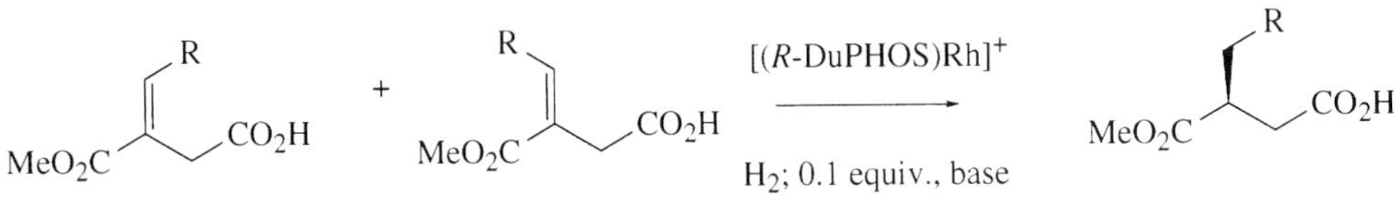

Scheme 26

were prepared *via* Horner–Emmons condensation protocol and hydrogenated subsequently under mild conditions (60 psi H_2) at substrate-to-catalyst ratios of 500. Overall, enol ester substrates were reduced by cationic Et-DuPHOS-Rh catalysts with very high levels of enantioselectivity (93–99% e.e.). Importantly, substrates bearing β-substituents could be employed as E/Z isomeric mixtures with no detrimental effect on the selectivity. Labelling studies indicated that no significant E/Z isomerisation of the substrates occurs during the course of these reactions ⟨98JA4345⟩.

Asymmetric [3 + 2] cycloadditions of 2,3-butadienoates with electron-deficient olefins have been found to be catalysed by novel chiral 2,5-dialkyl-7-phenyl-7-phosphabicyclo-[2.2.1]-heptanes (Scheme 27) ⟨97JA3836⟩.

Highly enantioselective rhodium-catalysed hydrogenations have been described with a new chiral 1,4-bisphosphine (**29**) containing a cyclic backbone (Table 8) ⟨97JA1799⟩.

Convenient asymmetric synthesis of α-1-arylalkylamines through the enantioselective hydrogenation of enamides *via* [Me-DuPHOS]Rh-complexes has been reported by the Burk group ⟨96JA5142⟩. The same group have demonstrated how asymmetric hydrogenation reactions can provide practical access to a diverse array of chiral building blocks. The Burk group has also developed a variety of highly selective rhodium and ruthenium hydrogenation catalysts based on chiral 1,2-bis(phospholano)benzene (DuPHOS) and 1,2-bis(phospholano)ethane (BPE) ligands. The expanding utility of these catalyst systems is revealed through highly enantioselective syntheses of α-amino acids, β-branched amino acids, and β-hydroxy

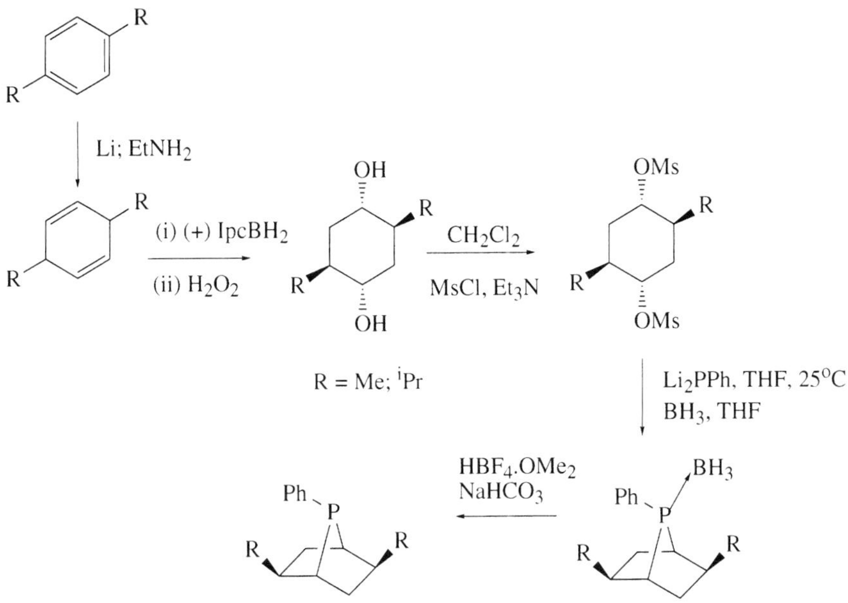

Scheme 27

Table 8. Reactions were performed at 25°C under 1 atm H_2 for 24 h [substrate (0.5 mmol, 0.125 M)/$[Rh(COD)_2]BF_4$/ligand (**21**) = 1 : 0.01 : 0.011] (reactions are quantitative)

Entry	*Solvent*	*NEt₃* (%)	*e.e.* [a] (%)
1	EtOH	0	89.2
2	EtOH	50	93.3
3	EtOH	50	83.6
4	$ClCH_2CH_2Cl$	50	93.4
5	THF	50	96.8
6	THF	5	95.1

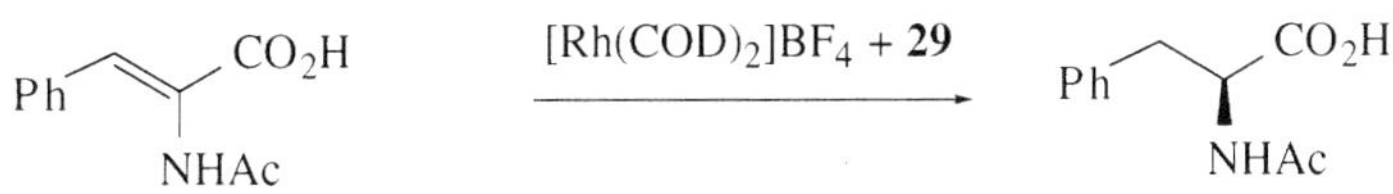

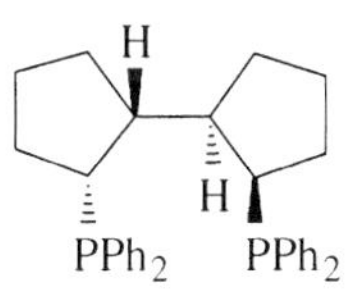

(**29**)

[a] Determined by GC using a Chirasil-VAL III FSOT column on the corresponding methyl ester. The *S* absolute configuration was determined by comparing the optical rotation with the reported value. [b] $[Rh(COD)Cl]_2$ (0.5 mol%) was used as the catalyst precursor. [c] A ratio of $[Rh(COD)]_2BF_4$ (0.1 mol%)/ligand (**21**) (0.11 mol%)/Et_3N (5 mol%) was used

esters. The versatility of the catalysts derives from careful ligand design which allows variation of the steric environment imposed by the phospholane ligands such that they can accommodate different steric demands of each substrate class of interest ⟨96PAC37, 95JA9375, 95JA4423, 94JA10847, 91JA8518⟩ (Tables 9–11).

A practical enantioselective synthesis of α-aminodicarboxylates using ruthenium-bis(phospholanes) as catalysts has been reported ⟨94JOC3676⟩.

4.1.5 STRUCTURAL ANALYSES OF PHOSPHOLANES AND PHOSPHOLENES

Bicyclophosphites based on linear 1,2,3-triols with terminal substituents have been found to be stable. Thus a series of hitherto unknown phospholane esters, including optically active derivatives, have been synthesised and examined by single crystal X-ray diffraction ⟨96PS(113)1⟩.

Table 9. Reactions were performed at 25°C under 60 psig. H_2 using 0.25 M methanol solutions of substrate and the catalyst precursors [DuPHOS-Rh(COD)]$^+$OTf$^-$ or [BPE-Rh(COD)]$^+$OTf$^-$ (0.2 mol%); reaction time was 24 h, reactions being 100% complete within that time (reactions are quantitative)

Catalyst	*e.e.* [a] (%)
(*S,S*)-Me-DuPHOS-Rh	91.9 (*S*)
(*S,S*)-Et-DuPHOS-Rh	74.4 (*S*)
(*R,R*)-Pr-DuPHOS-Rh	45.2 (*R*)
(*R,R*)-^{i}Pr-DuPHOS-Rh	13.6 (*S*)
(*R,R*)-Me-BPE-Rh	96.9 (*R*)
(*S,S*)-Et-BPE-Rh	89.3 (*S*)
(*R,R*)-^{i}Pr-BPE-Rh	11.8 (*S*)

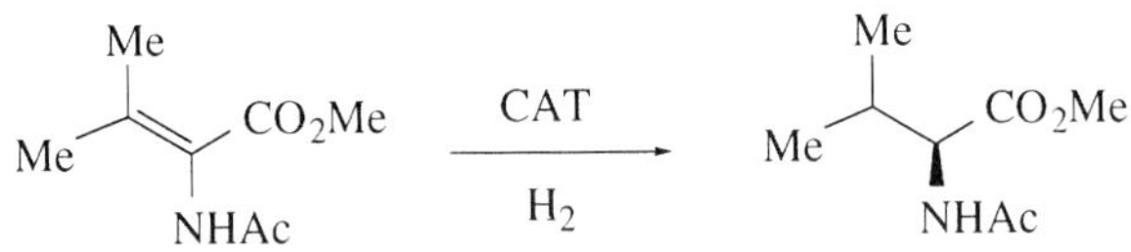

R-DuPHOS =
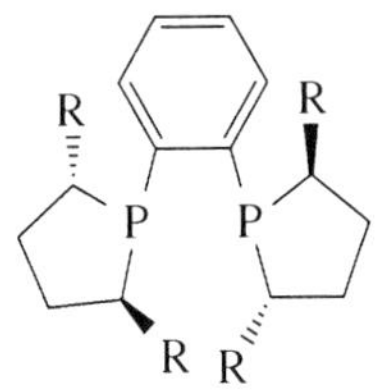

R-BPE =
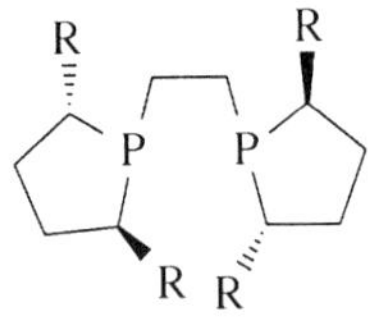

[a] Determined by chiral capillary GC using a Chirasil-L-VAL column (25 m). Absolute configurations were assigned by comparing the sign of optical rotation with that of authentic *N*-acetylvaline methyl ester.

Photoelectronic spectra and the electronic structure of organophosphorus compounds, including oxathiaphospholenes and phospholes, have been reported ⟨92ZOB2693⟩.

Conformational and stereodynamic investigations of 2-dialkylamino-1,3-dioxa-2-phospholanes and 3-methyl-1,3-oxaza-2-phospholanes have been examined using ^{13}C-NMR and theoretical MNDO analyses (Scheme 28) ⟨84JCS(P2)1207⟩. Analyses reveal that phosphorus-bound dialkylamino functions prefer the conformational arrangement illustrated in (i) and tend not to possess planar hetero-cyclopentane rings but have puckered envelope-type conformations as shown in (iv) of Scheme 28.

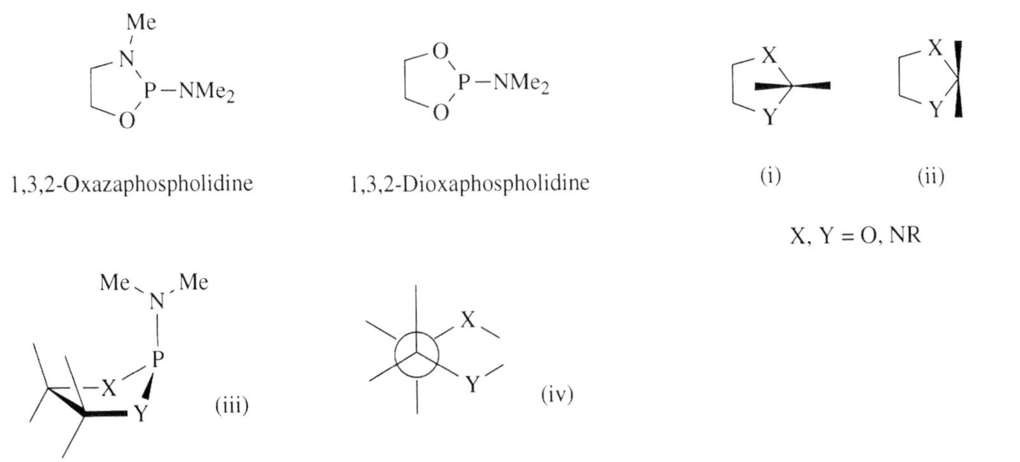

Scheme 28

Table 10. Reactions were performed at 35°C under 60 psig. H_2 using 0.25 M–0.1 M methanol/H_2O (9/1 v/v), using the catalyst (*R*,*R*)-[^{i}Pr-BPE-$RhBr_2$] (0.2 mol%) as precursor; reaction time was 20 h, reactions being 100% complete within that time (reactions are quantitative)

R	*R′*	*e.e.* [a] (%)	*Config.* [b]
CH_3	CH_3	99.3	*S*
CH_3	*t*-Bu	99.4	*S*
C_2H_5	CH_3	98.6	*S*
C_3H_7	C_2H_5	98.8	*S*
$(CH_3)_2CH$	C_2H_5	99.0	*R*
$C_{11}H_{23}$	CH_3	98.7	*S*
CH_3OCH_2	CH_3	95.5	*R*
$ClCH_2$	CH_3	76	*R*
c-C_6H_{11}	CH_3	99.1	*R*

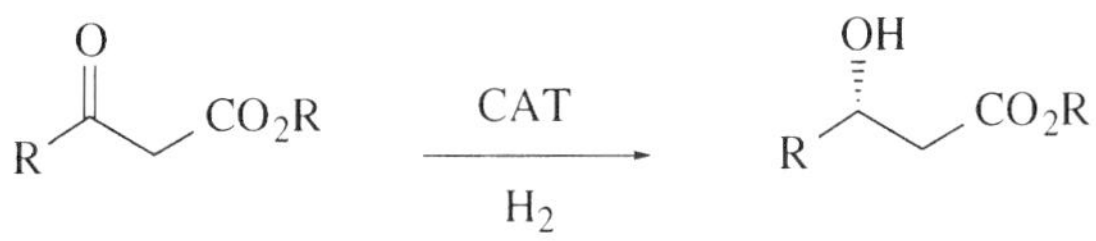

(*R*,*R*)^{i}Pr-BPE =
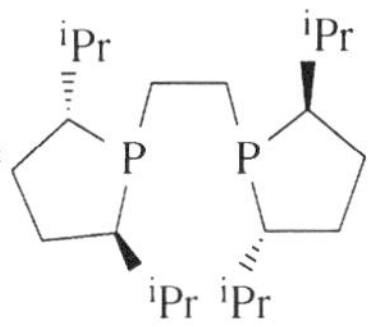

[a] Determined by chiral capillary GC using a Chirasil-L-VAL column (25 m). [b] Absolute configurations were assigned by comparing the sign of optical rotation with that of known alcohols.

Configurational inversion at phosphorus has been explored in dimetalla-phospholanes (Scheme 29 and Table 12) ⟨82JOM(224)247⟩.

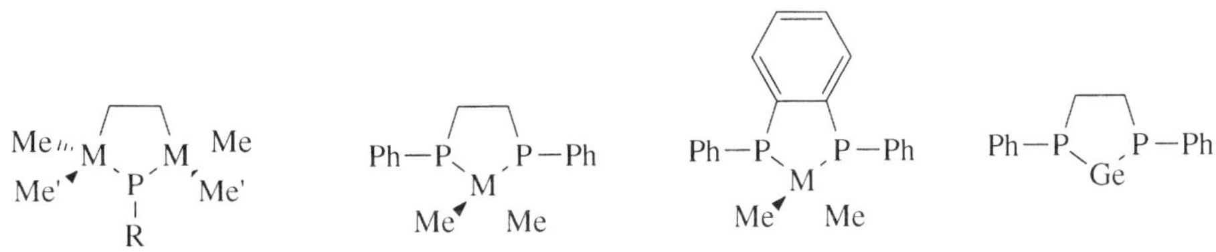

(I)-(IV) (V)-(VII) (VIII)-(X) (XI)

(I) M = Si, R = Ph
(II) M = Si, R = ^{t}Bu
(III) M = Ge, R = Ph
(IV) M = Ge, R = ^{t}Bu

(V) M = Si
(VI) M = Ge
(VII) M = Sn

(VIII) M = Si
(IX) M = Ge
(X) M = Sn

Scheme 29

Pure diastereoisomers of 2-{[(R_c)-1-(α-naphthyl)ethyl]amino}-2-thiono-1,3,2-oxathiaphospholane and *O*,*S*-dimethyl-*N*-[(R_c)-1-(α-naphthyl)ethyl]phosphoramido-thioate have been obtained, and both "slow"-migrating isomers studied by X-ray crystallography which demonstrated their (R_P,R_c) absolute configuration. Therefore, the absolute configuration of both "fast"-migrating isomers must be (S_P,R_c). In (R_P,R_c)-1 oxathiaphospholane, the ring adopts the open-envelope conformation with the C2 atom in the flap position; the S1-P-O1 angle is 97°. DBU-assisted methanolysis of (R_P,R_c)-1 ("slow") followed by *S*-methylation, gave (S_P,R_c)-2 ("fast"). This result has been interpreted in terms of an "adjacent" type mechanism on the regio- and stereoselective 1,3,2-oxathiaphospholane ring-opening process. Suggestions are presented regarding the absolute configuration at the phosphorus atom in diastereoisomers of 5′-*O*-protected nucleoside 3′-*O*-(2-thiono-1,3,2-oxathiaphospholanes), which are synthons for stereocontrolled synthesis of oligo(nucleoside phosphorothioate)s ⟨92JA10197⟩.

Table 11. Reactions were performed at 20–25°C under 30 psig. H_2 using 0.25 M–0.35 M methanol solutions of substrate using the catalyst $[(COD)RhL_2]^+OTf^-$ (0.1 mol%) as precursor; reaction time for complete conversion was 1–2 h

R′	*Ligand*	*e.e.* [a] (%)	*Config.* [b]
C_6H_5	**L3**	89	*S*
$4\text{-}NO_2C_6H_4$	**L1**	90	*R*
$3\text{-}ClC_6H_4$	**L4**	91	–
1-naphthyl	**L2**	94	*R*
1-naphthyl	**L3**	93	*S*
CO_2Et	**L3**	99	*S*
CO_2Et	**L4**	>99	*R*
CF_3	**L3**	94	*S*
CF_3	**L2**	>95	*R*

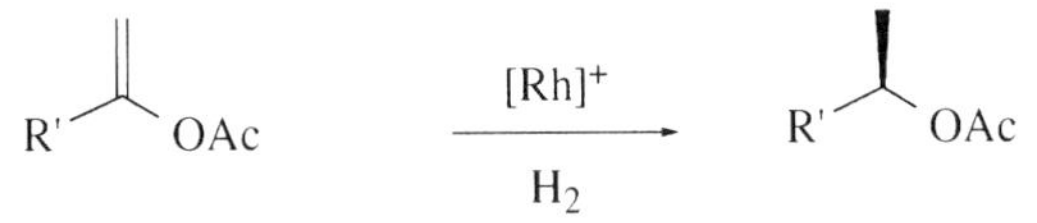

R-BPE =

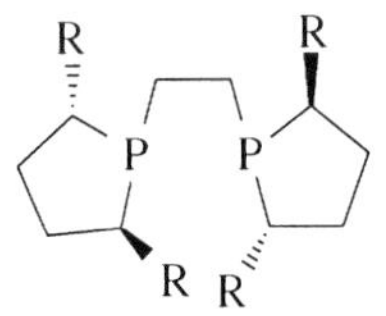

L1 R = Me
L2 R = Et

R-DuPHOS =

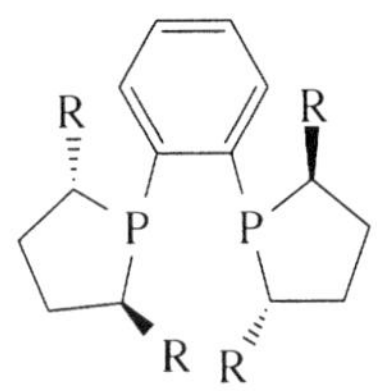

L3 R = Me
L4 R = Et

[a] Determined by chiral capillary GC. [b] Product absolute configurations were assigned by optical rotations.

Table 12. Inversion barriers at phosphorus for compounds (I)–(XI) calculated from the Eyring equation

Compound	t_c [a] (°C)	K_c [b] (s^{-1})	ΔG_c [c]
(I)	−17	58	12.8
(II)	−57	20	11.2
(III)	78	270	18.3
(IV)	57	80	16.5
(V)	123	400	18.6
(VI)	195	910	21.4
(VII)	169	1510	19.0
(VIII)	113	370	18.2
(IX)	175	630	20.8
(X)	125	780	18.2
(XI)	145	963	19.0

[a] t_c = coalescence temperature. [b] K_c = rate constant at temperature t_c. [c] In kcal/mol.

ACKNOWLEDGEMENTS

Grateful thanks are extended to Professor L.D. Quin for his contributions on some of the earlier published work on phospholenes and phospholanes and to Professor F. Mathey for his valuable advice throughout the process of completing this chapter.

4.1.6 REFERENCES

16CB437 G. Grüttner and E. Krause; *Chem. Ber.,* 1916, **49**, 437.
53USP2663737-38 W.B. McCormack; *US Pat.* 2 663 737-38 (1953) (*Chem. Abstr.,* 1955, **49**, 7602).
61CB113 K. Issleib and S. Häusler; *Chem. Ber.,* 1961, **94**, 113.
62IZV65 B.A. Arbuzov and L.A. Shapshinskaya; *Izv. Akad. Nauk. SSSR, Ser. Khim.,* 1962, 65.
63T1563 U. Hasserodt, K. Hunger and F. Korte; *Tetrahedron,* 1963, **19**, 1563.
63ZOB783 N.A. Razumova and A.A. Petrov; *Zh. Obshch. Khim.,* 1963, **33**, 783.
65CB171 H. Fritzsche, U. Hasserodt and F. Korte; *Chem. Ber.,* 1965, **98**, 171.
65HCA133 L. Maier; *Helv. Chim. Acta,* 1965, **48**, 133.
69JA2788 K. Naumann, G. Zon and K. Mislow; *J. Am. Chem. Soc.,* 1969, **91**, 2788.
70JA4303 L.D. Quin and T.P. Barket; *J. Am. Chem. Soc.,* 1970, **92**, 4303.
71BSF317 J.P. Lampin, F. Mathey and B. Bartet; *Bull. Soc. Chim. Fr.,* 1971, 317.
71CJC530 T.H. Chan and L.T.L. Wong; *Can. J. Chem.,* 1971, **49**, 530.
72JOC238 K.L. Marsi, F.B. Burns and R.T. Clark; *J. Org. Chem.,* 1972, **37**, 238.
72BSF3494 J.P. Lampin, L. Eberhard and F. Mathey; *Bull. Soc. Chim. Fr.,* 1972, 3494.
73JOC1858 L.D. Quin, S.G. Borleske and J.F. Engel; *J. Org. Chem.,* 1973, **38**, 1858.
74JOC265 K.L. Marsi; *J. Org. Chem.,* 1974, **39**, 265.
73T2909 L. Eberhard, J.P. Lampin and F. Mathey; *Tetrahedron,* 1973, **29**, 2909.
73TL1249 D.G. Smith and D.J.H. Smith; *Tetrahedron Lett.,* 1973, 1249.
74JOC686 L.D. Quin and R.C. Stocks; *J. Org. Chem.,* 1974, **39**, 686.
74JOC1531 R.B. Wetzel and G.L. Kenyon; *J. Org. Chem.,* 1974, **39**, 1531.
75T1907 R. Bodalski, K.M. Pietrusiewicz and J. Koszuk; *Tetrahedron,* 1975, **31**, 1907.
76JOC1155 A. Fitzgerald, G.D. Smith, C.N. Caughlan, K.L. Marsi and F.B. Burns; *J. Org. Chem.,* 1976, **41**, 1155.
77GEP2541355 R. Schliebs and H.D. Block; *Ger. Pat. 2541355* (1977) (*Chem. Abstr.,* 1977, **87**, 68494).
77GEP2552319 H.D. Block; *Ger. Pat. 2552319* (1977) (*Chem. Abstr.,* 1977, **87**, 102433).
77JOC552 H. Tomioka, N. Toriyama and Y. Izawa; *J. Org. Chem.,* 1977, **42**, 552.
78JOC108 R.S. Jain, H.F. Lawson and L.D. Quin; *J. Org. Chem.,* 1978, **43**, 108.
81MI L.D. Quin; "The Heterocyclic Chemistry of Phosphorus", Wiley-Interscience, New York, 1981.
81PS(10)47 P.J. Hammond, J.R. Lloyd and C.D. Hall; *Phosphorus Sulfur,* 1981, **10**, 47.
81PS(10)67 P.J. Hammond, J.R. Lloyd and C.D. Hall; *Phosphorus Sulfur,* 1981, **10**, 67.
81PS(10)185 P.D. Beer, P.J. Hammond and C.D. Hall; *Phosphorus Sulfur,* 1981, **10**, 185.
81ZOB514 M.A. Pudovik, G.Y. Gadzhiev, E.Y. Dzhalilov, M.D. Medvedeva and A.N. Pudovik; *Zh. Obshch. Khim.,* 1981, **51**, 514.
82JOM(224)247 C. Couret, J. Escudie, B. Saintroch, J.D. Andriamizaka and J. Satge; *J. Organomet. Chem.,* 1982, **224**, 247.
83JA7443 C.K. SooHoo and S.G. Baxter; *J. Am. Chem. Soc.,* 1983, **105**, 7443.
83JA7444 A.H. Cowley, R.A. Kemp, J.G. Lasch, N.C. Norman and C.A. Stewart; *J. Am. Chem. Soc.,* 1983, **105**, 7444.
84JCS(P2)1207 W.B. Jennings, M.S. Tolley, J.H. Hargis, S.D. Worley and L. Chang; *J. Chem. Soc., Perkin Trans. 2,* 1984, 1207.
86IC740 A.H. Cowley, R.A. Kemp, J.G. Lasch, N.C. Norman, C.A. Stewart, B.R. Whittlesey and T.C. Wright; *Inorg. Chem.,* 1986, **25**, 740.
87JOM(328)71 H. Brunner and R. Sievi; *J. Organomet. Chem.,* 1987, **328**, 71.
88CC569 M. Yamashita, M. Uchimura, A. Iida, L. Parkanayi and J. Clardy; *J. Chem. Soc., Chem. Commun.,* 1988, 569.
88ZN(B)257 J. Apitz, J. Grobe and D. LeVan; *Z. Naturforsch., B,* 1988, 257.
89CC967 N.S. Isaacs and G.N. El-Din; *J. Chem. Soc., Chem. Commun.,* 1989, 967.
91JA8518 M.J. Burk; *J. Am. Chem. Soc.,* 1991, **113**, 8518.
91JOM(413)55 H. Brunner and S. Limmer; *J. Organomet. Chem.,* 1991, **413**, 55.
91TA569 M.J. Burk, J.E. Feaster and R.L. Harlow; *Tetrahedron: Asymmetry,* 1991, **2**, 569.
92JA10197 W.J. Stec, M.W. Wieczorek and J. Blaszcyk; *J. Am. Chem. Soc.,* 1992, **114**, 10197.
92JOC5189 R.P. Polniaszek; *J. Org. Chem.,* 1992, **57**, 5189.
92MI J. March, "Advanced Organic Chemistry. Reactions, Mechanisms and Structure", John Wiley and Sons, New York, 1992.
92ZOB2693 V.V. Zverev, Z.G. Bazhanova and A.R. Burilov; *Zh. Obshch. Khim.,* 1992, **62**, 2693.
93HAC553 M. Yamashita, A. Iida, H. Mizuno, Y. Miyamoto, T. Morishita, N. Sata, K. Kiguchi, A. Yabui and T. Oshikawa; *Heteroatom Chem.,* 1993, **4**, 553.
93JA10125 M.J. Burk, J.E. Feaster, W.A. Nugent and R.L. Harlow; *J. Am. Chem. Soc.,* 1993, **115**, 10125.
93JCS(P2)629 G.V. Coleman, D. Price, A.R. Harrocks and J.E. Stephenson; *J. Chem. Soc., Perkin Trans. II,* 1993, 629.
93S39 R. Emrich and P.W. Jolly; *Synthesis,* 1993, 39.
93S931 G. Keglevich; *Synthesis,* 1993, 931.
94JA1880 P.J. Fagan, W.A. Nugent and J.C. Calabrese; *J. Am. Chem. Soc.,* 1994, **116**, 1880.
94JA10847 M.J. Burk, J.R. Lee and J.P. Martinez; *J. Am. Chem. Soc.,* 1994, **116**, 10847.
94JOC3676 T. Pham and W.D. Lubell; *J. Org. Chem.,* 1994, **59**, 3676.
94PS(88)75 C. Mundt and L. Riesel; *Phosphorus Sulfur,* 1994, **88**, 75.
94TL6063 W. Adam, M. Ahrweiler, D. Reinhardt and M. Sauter; *Tetrahedron Lett.,* 1994, **35**, 6063.
94TL9363 M.J. Burk and M.F. Gross; *Tetrahedron Lett.,* 1994, **35**, 9363.

95JA4423	M.J. Burk, T.G.P. Harper and C.S. Kalberg; *J. Am. Chem. Soc.*, 1995, **117**, 4423.
95JA9375	M.J. Burk, M.F. Gross and J.P. Martinez; *J. Am. Chem. Soc.*, 1995, **117**, 9375.
95JA12019	W.J. Stec, A. Grajkowski, A. Kobylanska and B. Karwowski; *J. Am. Chem. Soc.*, 1995, **117**, 12019.
95PS(105)101	R. Boukherroub, E. Garrigues and G. Manuel; *Phosphorus Sulfur*, 1995, **105**, 101.
95ZN(B)189	A. Karst, B. Broschk, J. Grobe and D. Levan; *Z. Naturforsch. Teil B*, 1995, **50**, 189.
95ZN(B)1004	C. Eckert, L. Dahlenburg and A. Wolski; *Z. Naturforsch. Teil B*, 1995, **50**, 1004.
95ZOB1096	N.K. Gusarova, S.I. Shaikhudinova, V.I. Dimitiev and S.F. Malysheva; *Zh. Obshch. Khim.*, 1995, **65**, 1096.
96CB725	A. Kadyrov, I. Neda, T. Kaukorat, R. Sonnenburg and A. Fischer; *Chem. Ber.*, 1996, **129**, 725.
96IC2546	L.D. Field, I.P. Thomas; *Inorg. Chem.*, 1996, **35**, 2546.
96JA5142	M.J. Burk, Y.M. Wang and J.R. Lee; *J. Am. Chem. Soc.*, 1996, **118**, 5142.
96PAC37	M.J. Burk, M.F. Gross, T.G.P. Harper, C.S. Kalberg, J.R. Lee and J.P. Martinez; *Pure and Applied Chem.*, 1996, **68**, 37.
96PS(110)405	M. Yamashita, A. Yabui, K. Suzuki, S. Kumagai and T. Oshikawa; *Phosphorus Sulfur*, 1996, **110**, 405.
96PS(113)1	E.E. Nifantyev, A.M. Koroteev, M.P. Koroteev, S.V. Meshkov, V.K. Belsky and A.R. Bekker; *Phosphorus Sulfur*, 1996, **113**, 1.
96SL1211	T. Morimoto, N. Ando and K. Achiwa; *Synlett*, 1996, 1211.
97BK747	S.I. Malekin, N.Y. Khromova, A.V. Kisin, Y.L. Kruglyak and V.K. Kurochkin; *Bioorg. Khim.*, 1997, **23**, 747.
97CC1239	M. Zablocka, A. Igau, B. Donnadieu, J.P. Majoral and A. Skowronska; *Chem. Commun.*, 1997, 1239.
97HAC217	V.I. Vysotskii; *Heteroatom Chem.*, 1997, **8**, 217.
97IC1867	M.L.J. Hackney, D.M. Schubert, P.F. Brandt, R.C. Haltiwanger and A.D. Norman; *Inorg. Chem.*, 1997, **36**, 1867.
97JA1799	G.X. Zhu, P. Cao, Q.Z. Jiang and X.M. Zhang; *J. Am. Chem. Soc.*, 1997, **119**, 1799.
97JA3836	G.X. Zhu, Z.G. Chen, Q.Z. Jiang, D.M. Xiao, P. Cao and X.M. Zhang; *J. Am. Chem. Soc.*, 1997, **119**, 3836.
97JCC499	M. Yamashita, A. Yabui, K. Suzuki, Y. Kato, M. Uchimura, A. Iida, H. Mizuno, K. Ikai, T. Oshikawa, L. Parkanayi and J. Clardy; *J. Carbohydrate Chem.*, 1997, **16**, 499.
97JOC4521	Z.G. Chen, Q.Z. Jiang, G.X. Zhu, D.M. Xiao, P. Cao, C. Guo and X.M. Zhang; *J. Org. Chem.*, 1997, **62**, 4521.
98AG(E)1931	M.J. Burk, F. Bienewald, M. Harris and A. Zanotti-Gerosa; *Angew. Chem., Int. Ed. Engl.*, 1998, **37**, 1931.
98JA4345	M.J. Burk, C.S. Kalberg and A. Pizzano; *J. Am. Chem. Soc.*, 1998, **120**, 4345.
98JOC912	S.C. Hume and R.S. Simpkins; *J. Org. Chem.*, 1998, **63**, 912.
98JOC8031	J. Holz, M. Quirmbach, U. Schmidt, D. Heller, R. Sturmer and A. Borner; *J. Org. Chem.*, 1998, **63**, 8031.
98JSO511	T. Imamoto; *J. Synth. Org. Chem.*, 1998, **56**, 511.
98RCB1851	V.I. Vysotskii; *Russ. Chem. Bull.*, 1998, **47**, 1851.
98TL6707	P. Desmazeau, J.Y. Legros and J.C. Fiand; *Tetrahedron Lett.*, 1998, **39**, 6707.
99CC261	D. Carmichael, H. Doucet and J.M. Brown; *Chem. Commun.*, 1999, 261.
99CPB436	N. Fujie, M. Matsui and K. Achiwa; *Chem. Pharm. Bull.*, 1999, **47**, 436.
99JOC3290	M.J. Burk, F. Bienewald, S. Challenger, A. Derrick and J.A. Ramsden; *J. Org. Chem.*, 1999, **64**, 3290.
99RHC227	Y. Yamanoi and T. Imamoto; *Rev. Heteroatom Chem.*, 1999, **20**, 227.
99TA375	J.J.A. Perea, M. Lotz and P. Knochel; *Terahedron: Asymmetry*, 1999, **10**, 375.
99TL2939	F. Guillen and J.C. Fiaud; *Tetrahedron Lett.*, 1999, **40**, 2939.

4.2.1
Five-membered rings. Phospholes: Early Literature 1953–1994

LOUIS D. QUIN
University of Massachusetts at Amherst, Amherst, USA

4.2.1.1 INTRODUCTION

The fully unsaturated five-membered ring system containing a phosphine unit (phosphole) did not make an appearance in the chemical literature until 1953, when Wittig and Geissler announced the synthesis of the dibenzophosphole derivative (**1**) ⟨53LA(580)44⟩. Another six years were to pass before two groups independently reported the construction of the pentaphenyl derivative (**2**) of the monocyclic parent ⟨59CI(L)1250, 59JA3163⟩. The complexity of these derivatives prevented detailed study of the ring system, and the immediate question that arose of the presence of cyclic delocalization could not be addressed. The few reports that appeared in the next 8–10 years showed that simpler derivatives, notably (**3**) ⟨62CI(L)359⟩, (**4**) ⟨67AG(E)86⟩, and (**5**) ⟨67JA5984⟩ could be prepared; the character of the phosphole ring then began to emerge and the study of the chemistry of the phosphole system commenced in earnest.

Interest has remained high, and many phospholes are now known, including some with phosphorus in

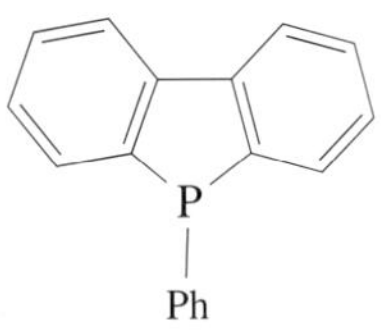

(**1**)

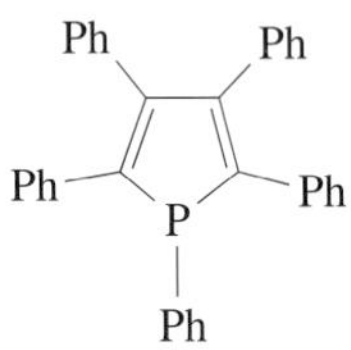

(**2**)

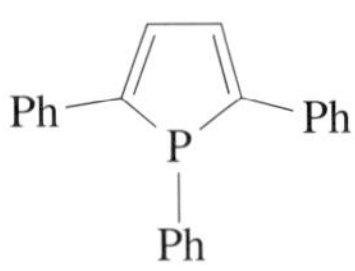

(**3**)

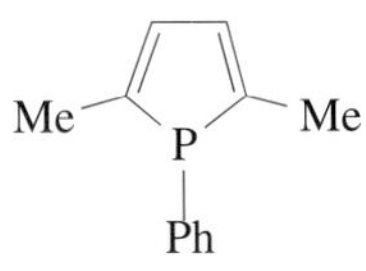

(**4**)

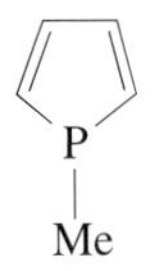

(**5**)

other states of coordination (number of atoms directly attached to phosphorus) and with various rings fused to the parent. A major development has been the synthesis and study of a fascinating array of metallic derivatives and coordination complexes, some of which exhibit unique and valuable properties. Research in phosphole chemistry continues unabated, and it is a major part of contemporary phosphorus chemistry. Recognition of phospholes as having importance in heterocyclic chemistry has been slower to develop, however, and in the first edition of *Comprehensive Heterocyclic Chemistry* (CHEC-I) ⟨84CHEC-I(1)518⟩ only four pages were devoted to the field. Even some later monographs on heterocyclic chemistry fail to mention the existence of this field. It is hoped that this chapter will aid in heightening the interest of general heterocyclic chemists in this area. To this end, the chapter is written insofar as possible with that audience in mind, and some of the intricacies of phosphorus chemistry are not allowed to dominate the presentation. There are, in fact, many reviews of phosphole chemistry that provide great detail on the subject, and are recommended to those wishing further insight. For a complete coverage of the early work (to about 1976) on phospholes, three reviews are available ⟨70JHC1, B-70MI01, 76JHC1⟩. Later reviews were published by Hughes in 1979⟨B-79MI01⟩, by Mathey in 1980 ⟨80MI01⟩ and Quin in 1981 ⟨B-81MI01⟩. Reviews from the late 1980s and early 1990s include those of Mathey in 1988 ⟨88CRV429⟩, Quin and Hughes in 1990 ⟨B-90MI01⟩, and lastly Hughes in 1992 ⟨B-92MI01⟩.

A complexity, yet at the same time a source of fascinating chemistry, of systems based on phosphorus is that the phosphorus may appear in functional groups where this atom is attached to anywhere from one to six other atoms. This has led to a description of the group by its "coordination number" (C.N.), which refers to this number of attached atoms. Thus simple phospholes are designated as having C.N.-3, but the coordination number may range from two to six. However, some of these coordination numbers are rather rare, and most of phosphole chemistry is based on C.N.-3 and C.N.-4, which includes especially the oxides, sulfides, and salts of the phosphines. Structures for the various coordination numbers are given below.

C.N.-3. This is considered to be the parent ring system, with analogy to pyrroles, and accorded the most attention as possibly possessing the cyclic delocalization found in the N, S, and O counterparts.

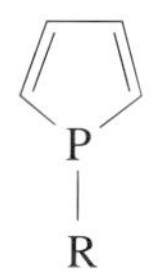

C.N.-4. Salts, oxides, sulfides, imines, and ylides of the parent. Here cyclic delocalization is not possible since the lone pair on phosphorus is used in a fourth bond. Most of the oxides and some of the other functions are unstable at or slightly above room temperature, forming Diels–Alder dimers in the manner of thiophene oxides.

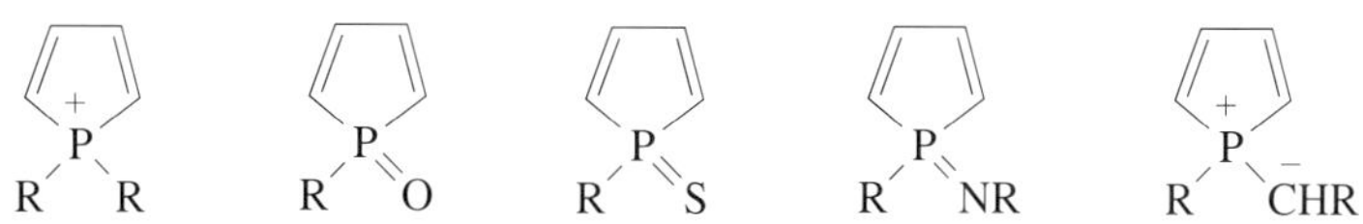

C.N.-2. Phosphole anions (**6**) *and 2H-phospholes* (**7**). Very different species, the former is quite stable and possesses significant cyclic delocalization, while the latter readily undergoes dimerization. Phosphole anions can take the place of cyclopentadienide ion in metallocene complexes, forming another family of delocalized heterocycles, such as the phosphacymantrene (**8**) and phosphaferrocene (**9**) shown below.

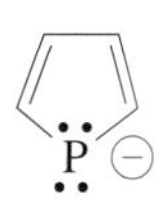

(**6**)

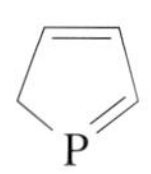

(**7**)

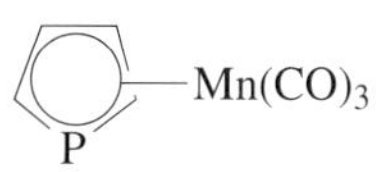

(**8**)

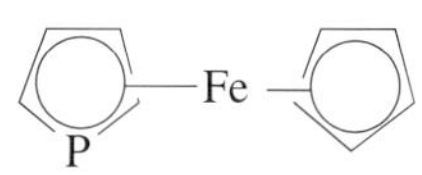

(**9**)

C.N.-5. Phosphoranes. These are uncommon derivatives, following the usual rules of phosphoranes that stability requires the presence of multiple electronegative substituents. Unlike the lower coordination numbers, this condition develops two sets of bond angles around phosphorus, 120° in the equatorial plane and 90° in the two apical positions.

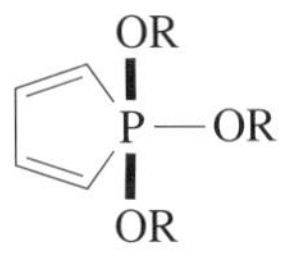

C.N.-6. Hexacoordinate phosphorus. This is the rarest of the coordination numbers, and generally exists in anionic form. The geometry at phosphorus is square pyramidal.

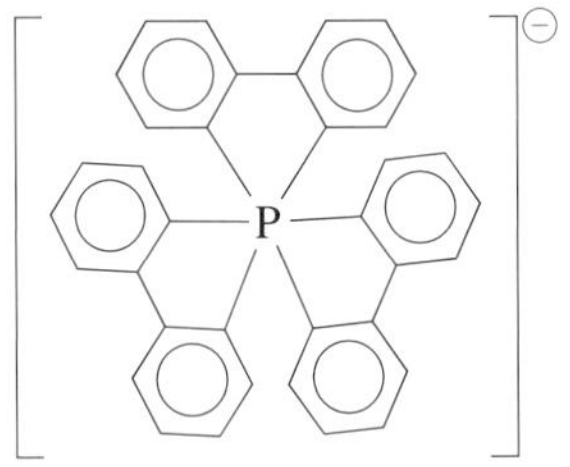

The parent phosphole ring can be fused to a number of other cyclic structures; for many of these compounds the coordination number can be varied, but most of the known compounds are found in phosphole form, or as the oxide derivative. The ring systems that are the most common are shown below. By far the most widely studied is the dibenzo derivative, because it is easily synthesized and worked with. However, the system displays few properties of a phosphole and is more like a diphenylphosphine. This system is accorded abbreviated attention; for current interest, a series of papers by Cornforth et al. starting in 1982 should be consulted ⟨82JCS(P1)2289, 82JCS(P1)2317⟩.

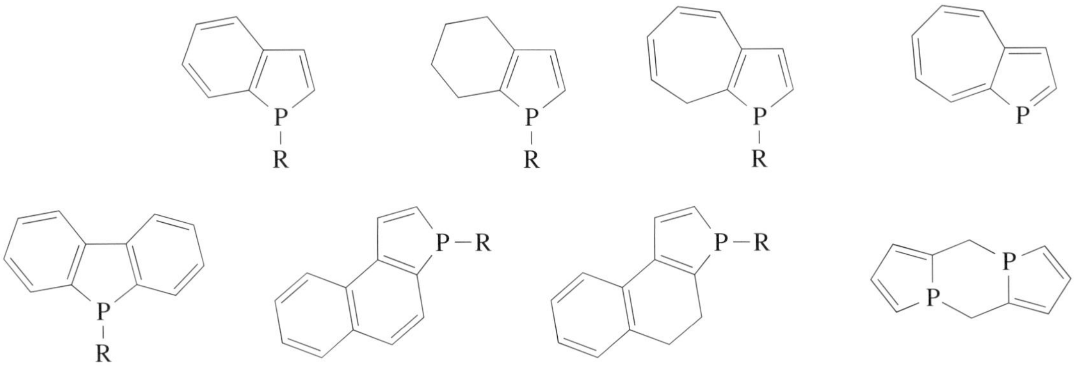

From the above introductory survey of structural possibilities, it is easily seen that a very great variety of phosphole derivatives is known. However, many remain to be synthesized, and even the information on some of those that exist is sketchy. Unanswered questions abound, and it is a field that can be expected to thrive and grow in the years ahead, as it slowly approaches the degree of maturity that exists for the related N, S, and O heterocyclic systems.

4.2.1.2 THEORETICAL STUDIES OF PHOSPHOLES

Knowledge of the electronic structure of the phosphole ring is far less advanced than it is for the related heterocyclopentadienes, and in spite of the years of study of this ring system the degree of cyclic electron delocalization remains unclear. No thermochemical measurement of the resonance energy has ever been made with appropriate compounds, and the chemical and physical properties, while sometimes suggestive of a degree of delocalization, are not compelling. Nor have purely theoretical treatments given much insight into the electronic structure; indeed, many have merely contributed uncertainty, with some studies speaking to a significant degree of delocalization, others denying it almost completely. The last theoretical treatment of the electronic structure was reported in 1976, before the advent of the supercomputer. It is hoped that a detailed reexamination of the electronic structure with the advanced computational techniques of today might at last lead to a clearer understanding of this matter. Indeed steps in this direction are already being taken; beginning in 1988, there have been a series of publications on calculations of the geometry of pyramidal and planar phosphole, and of the energy difference between the two forms. Also, for the first time, theoretical calculations of the ^{31}P NMR shift of phospholes (see Section 4.2.1.3.2.1), and of the relative stabilities and energy of interconversion of 1*H*-, 2*H*-, and 3*H*-phosphole, have been made. Some of these theoretical treatments are discussed in more detail later in this section. It does seem obvious that the new techniques can make a major contribution to our understanding of the true nature of phospholes.

Three years after phospholes first became available (as the pentaphenyl derivative ⟨59CI(L)1250, 59JA3163⟩, Brown published a theoretical treatment of the unsubstituted system that was based on the relatively simple molecular orbital theory of the time ⟨62JCS929⟩. With the assumption that the system was planar, it was concluded that the conjugation energy for the phosphole ring was considerable, 1.49β in a calculation giving 1.27β for pyrrole. Also in 1962 appeared thermochemical measurements of the oxidation of pentaphenylphosphole, showing that, relative to the *P*-oxide, the phosphole was more stable by 39 ± 13 kcal mol^{-1}, and this was taken as a measure of the conjugation energy ⟨62JCS2932⟩. However, studies with such a highly substituted derivative cannot give much information on the true nature of the ring, and it was not until synthetic work was initiated by three groups ⟨67AG(E)86, 67JA5984, 69CR(C)1066⟩ that simple phospholes without the interfering ring phenyls became available and made it possible to begin serious experimental examination of the question of cyclic delocalization in the ring. Some spectral properties pointed to a degree of delocalization (see Section 4.2.1.3), and an X-ray diffraction analysis of the solid 1-benzylphosphole showed bond order changes that supported the concept ⟨70JA5779⟩. That phosphorus had retained the pyramidal character of a phosphine was first revealed by stereochemical studies with phospholes lacking symmetry ⟨70JA1442, 71JA6205⟩, and confirmed by the X-ray analysis of 1-benzylphosphole ⟨73JCS(D)1888⟩. The latter analysis showed that the phosphorus was slightly out of the plane of the four ring carbons, but more importantly that the angle at P between the benzylic C and the ring C was 106°. These observations voided the early treatment of Brown, who had assumed full planarity of the system. The stereochemical studies showed that the inversion barrier in 1-isopropyl-2-methyl-5-phenylphosphole (16 kcal mol^{-1}) was some 20 kcal mol^{-1} below that of a phosphine model, and this was interpreted as meaning the planar transition state for the inversion was stabilized by cyclic electron delocalization ⟨71JA6205⟩. Some of the theoretical studies that followed were greatly concerned with this point, and questioned the extent of electron delocalization that could occur from pyramidal phosphorus. Several such studies were made over the period 1970–1976 but gave conflicting results, some stating that conjugation was virtually absent in the system ⟨73AG(E)145, 73T469, 75JCS(P2)974⟩, others that it could be considerable ⟨71JA6507, 74JA6904, 74PS(4)231, 76JA407, 76JA4365⟩. Advances in computational techniques were made during this period, and are reflected in the approaches taken by the authors. Thus the Pople CNDO/2 semiempirical MO theory was applied by Rauk et al. ⟨71JA6507⟩, and an advanced semiempirical method was also used by Andose et al. ⟨74JA6904⟩, that supported the concept of delocalization in the planar transition state. Kaufmann and Mathey ⟨74PS(4)231⟩ employed LCAO–MO–CNDO/2 calculations and found that 2*p*–3*d* π interaction played a role both in the pyramidal state and in the planar transition state of inversion. However, Palmer and Findlay ⟨75JCS(P2)974⟩ performed a treatment by linear combination of Gaussian orbitals (LCGO) which showed no delocalization, and all of the known experimental data were reinterpreted on that basis. However, within a year Epiotis and Cherry ⟨76JA4365⟩ published the results of SCF–MO–CNDO/2 calculations that found considerable "aromaticity" to be present. With the exception of the last paper, the salient points of all of these theoretical evaluations have been reviewed in 1976 by Hughes and Kleemola ⟨76JHC1⟩. The theoretical inquiry rested at this inconclusive stage for 12 years. Probably the consensus opinion of the phosphorus community during that period was that expressed by Mathey in a 1988 review: the phosphole ring may have only weak delocalization, less than that in furan, and ought not to be considered as an aromatic system ⟨88CRV429⟩. This position relative to furan was emphasized by Bird who devised an "index of aromaticity" in heterocyclic rings ⟨85T1409⟩ that was applied to some phosphorus systems ⟨90T5697⟩. The index is based on published experimental bond lengths. Using a scale of 0 (none) to 100 (high), the heterocyclopentadiene series had the indices (I_5) S, 66; Se, 59; NH, 59; O, 43; $PhCH_2P$, 35.5; 1,2,5-Ph_3 phosphole, 20. This accords with the view that some phospholes may have a measure of delocalization but less than that in furan.

Great advances in computational techniques and in computer power were made in the years following the 1976 theoretical treatment of Epiotis and Cherry, and now the phosphole problem is again a subject for theoretical investigation. In the first of the new round of publications, Baldridge and Gordon reported the optimized geometries for pyramidal 1*H*-phosphole, using the 3-21G* basis set at the SCF level of computation, employing GAUSSIAN 82 ⟨88JA4204⟩. Hughes and Edgecombe showed in 1992 that with an *ab initio* method using the 6-311G* basis set and the GAUSSIAN 86 program series it was possible to compute optimized geometries for both the pyramidal and the planar state of phosphole, as well as their total energies ⟨92H(33)563⟩. From the latter, the energy difference, and thus the inversion barrier, was calculated to be 24.4 kcal mol^{-1}. There is no experimental value for the inversion barrier in phosphole itself, but as noted a value of 16 kcal mol^{-1} was found for a 1,2,5-trisubstituted phosphole ⟨71JA6205⟩.

Table 1. Calculated and experimental molecular parameters for phosphole

	Phosphole [a]							*Thiophene* [b]
	Pyramidal					*Planar*		
	A	*B*	*C*	*D*	*E*	*B*	*D*	
P–C_2(C_5)	1.815	1.817	1.7998	1.817	1.783 [c]	1.716	1.716	1.714
C_2(C_5)–C_3(C_4)	1.334	1.331	1.3618	1.332	1.343	1.371	1.373	1.370
C_3–C_4	1.475	1.473	1.4506	1.471	1.438	1.423	1.420	1.423
<C_2–P–C_5	81.6	89.5	90.8	89.68	90.7	98.2	98.2	92.17
<P–C–C	110.9	111.2	110.6	110.63	110 [c]	105.6	105.5	111.47
<C–C–C	113.9		114.1	114.19	114.1		115.4	112.45
<X–P–C_2(C_5) [d]	99.2	99.3			106 [c]	130.9		

[a] A = Baldridge and Gordon ⟨88JA4204⟩; B = Hughes and Edgecombe ⟨92H(33)563⟩; C = Bachrach ⟨93JOC5414⟩; D = Chesnut and Quin ⟨94JA9638⟩; E = experimental for 1-benzylphosphole ⟨70JA5779⟩.
[b] ⟨94JA9638⟩. [c] Averaged value. [d] X=H or CH_2.

The authors conclude that this reduced inversion barrier is consistent with some electron delocalization in the planar transition state. In 1993 Bachrach also reported geometry calculations for pyramidal phosphole using *ab initio* calculations at the MP2/6-31G* level ⟨93JOC5414⟩, and Chesnut and Quin in 1994 reported the geometries of both pyramidal and planar 1*H*-phosphole ⟨94JA9638⟩. In the latter two studies, P was displaced from the four-carbon planar segment by 0.18 and 0.167 Å, respectively. Bachrach and Perriott have calculated parameters for all possible isomers from methyl substitution on the 1*H*-, 2*H*-, and 3*H*-phosphole isomers and on some vinyl derivatives as well ⟨94JOC3394⟩. Steric interaction causes a preference for P- over C-substitution.

The good agreement between the geometries calculated by the various groups, and with the experimental values for 1-benzylphosphole, is shown in Table 1. It is seen that the experimental bond lengthening at C_2–C_3 in 1-benzylphosphole is between the values for planar and pyramidal phosphole, as it is also for C_3–C_4. If it can be concluded that no delocalization is present in the calculated pyramidal 1*H*-phosphole, then these experimental values are in the correct direction to indicate some delocalization in 1-benzylphosphole, as of course was concluded by Bird in his treatment of this compound. There is a striking agreement for the lengths of these same carbons between the experimental (microwave) values for thiophene and the planar form of 1*H*-phosphole. The thiophene parameters are also included in Table 1 for comparison. This might suggest that the degree of delocalization in planar phosphole would be the same as that in thiophene, which is generally accepted to be considerable. This is supported by a calculation of the Bird indices ⟨94PC01⟩ for planar 1*H*-phosphole and 1-methylphosphole using the computed bond-length values of Chesnut and Quin. The indices (I_5) are 75 and 83, respectively, compared to the experimentally based value for thiophene of 66. It follows that a natural phosphole which for steric reasons adopts a shape with an angle intermediate between the extreme values of pyramidal and planar might partake of an intermediate degree of delocalization, which need not be trivial. The view that natural phospholes may vary in their degree of delocalization depending on their structure was early emphasized by Kaufmann and Mathey ⟨74PS(4)231⟩, thus making a flat statement that, as a family, phospholes lack delocalization seem incautious. Qualitative variation in the degree of delocalization is a point that has already appeared in the literature concerned with the chemical character of phospholes, as is noted in Section 4.2.1.5.

Bachrach has also considered the relative stabilities of the isomeric forms resulting from placement of the "extra hydrogen" at ring positions 1, 2, and 3 of phosphole (Scheme 1) ⟨93JOC5414⟩.

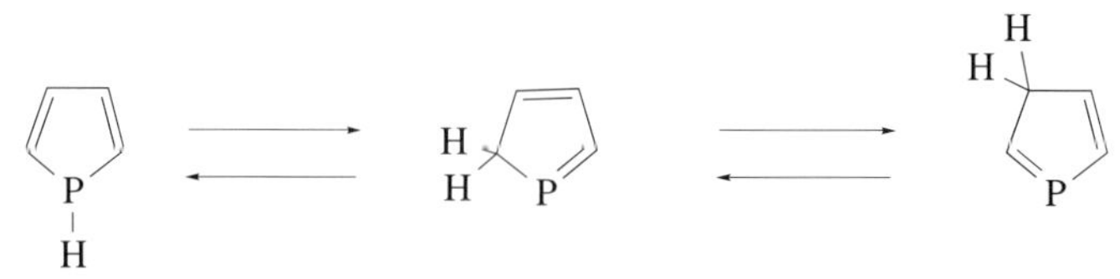

1*H* - 2*H* - 3*H* -

Scheme 1

These forms are related through [1,5] sigmatropic hydrogen shifts. An interesting discovery is that 2*H*-phosphole is the most stable of the three, in particular some 6.53 kcal mol^{-1} more stable than the 1*H*-isomer. The activation energy for the 1*H* to 2*H* conversion is only 16.0 kcal mol^{-1}. Indeed, it is seen in Section 4.2.1.5.8 that 1*H*-phosphole exists only at low temperature and rearranges to 2*H*-phosphole as

Table 2. Calculated and experimental molecular parameters for phospholide ion

	A [a]	*B* [a]	*Experiment* [b]
P–C2(C5)	1.749	1.764	1.715 [c]
C2(C5)–C3(C4)	1.405	1.376	1.396
C3–C4	1.402	1.416	1.424
<C2–P–C5	94.2	89.71	90.5
<P–C–C	108.5	112.23	112.3 [c]
<C–C–C	114.4	112.92	112.6

[a] A = Malar ⟨92JOC3694⟩; B = Chesnut and Quin ⟨94JA9638⟩. [b] For LiC_4H_4P. TMEDA ⟨89AG(E)1367⟩. [c] Averaged.

the temperature rises, then undergoing dimerization. Activation energies for the cycloaddition of ethyne to the phosphole isomers were calculated in later work; the 2*H*-isomer had the lowest energy in this process which was confirmed to be of the concerted [4+2] type ⟨94JOC5027⟩.

The situation regarding delocalization in phospholide ions is quite different and not controversial. The problem of pyramidal structure at P is of course absent, and the parent phospholide ion is isoelectronic with thiophene. Some early calculations of Kaufmann and Mathey ⟨74PS(4)231⟩ suggested that the ion would be highly delocalized. Calculations by Malar ⟨92JOC3694⟩ using the all-valence SCF-MO method SINDO1, which includes some *d*-orbital contribution, suggest that the delocalization is better described as only "moderate"; they concluded that the ion possesses 90.7% of the aromaticity of cyclopentadienide ion. Optimized geometries for the system are given in Table 2, where it will be seen that the C_2–C_3 and C_3–C_4 bonds are nearly equal (1.402 and 1.405 Å, respectively) from the delocalization. The P–C bond is shortened to 1.749 Å. The C–P–C bond angle is 94.2° and thus not greatly different from that in phospholes. Chesnut and Quin ⟨94JA9638⟩ included phospholide ion in their theoretical calculations and found C–C bond lengths that are not quite so equalized (1.376 and 1.424 Å, respectively), with C–P at 1.764 Å. Their results are, in fact, in very close agreement with the experimental structural parameters reported for lithium phospholide as a TMEDA solvate ⟨89AG(E)1367⟩; the data are compared in Table 2. This paper makes no conclusions on the extent of delocalization, except to note that it is present and important.

Chesnut and Quin have extended their calculations to include a treatment of the cause of the very strong deshielding of ^{31}P that characterizes all known phospholide ions. This theoretical work is presented in Section 4.2.1.3.2.1. Relevant to the present discussion is the conclusion that, although delocalization is extensive in the phospholide ring, this is not the direct cause of the deshielding, and there is no connection with the increased P=C double bond character from the delocalization as was first suggested when the strong deshielding was discovered ⟨79CC401⟩. The deshielding arises from an orbital effect of the lone pair on phosphorus that lies in the plane of the ring, placed there by the conjugation in the ring.

4.2.1.3 EXPERIMENTAL STRUCTURAL METHODS

4.2.1.3.1 X-ray Diffraction Analysis

The structures of six phospholes (two with a metal σ-bonded to P) have been determined by X-ray diffraction analysis (Table 3). In five of these, the C–P–C internal angle was determined, and found to be quite small (89.2–91.8°), as it is in thiophene (92.2°) ⟨61JSP(7)58⟩. The phosphorus atom is slightly out of the plane of the ring carbons, and retains its pyramidal configuration, a point of relevance in discussion of the degree of delocalization in the phosphole ring (Section 4.2.1.2). Thus, in 1-benzylphosphole, the angle at phosphorus between C_2 or C_5 and the benzyl substituent is 106°, which places the substituent below the plane of the ring atoms by 66.9°. This bond angle is greater with the tungsten substituents, and the pyramidal shape at P is diminished.

Delocalization is characteristically associated with increases in double bond lengths and decreases in single bond lengths, and in one structure, that of 1-benzylphosphole, these effects are clearly apparent. C-substitution, especially where leading to steric crowding, appears by other criteria to be associated with diminution of delocalization, and to some extent this can be detected in the crowded structures of Table 3. Thus, the C–C single bond is smallest in 1-benzylphosphole (1.438 Å) and increases to as much as 1.472 Å with substituents at positions 2, 3, and 4. Unfortunately, obtaining structural information by X-ray analysis in the phosphole field has been hampered by the low melting points of most of the simple compounds, and further exploration of the presence of delocalization effects by structural analysis has not yet been possible. (The bonding in the tungsten derivatives is not clear and is not included in this discussion.)

Table 3. Structural parameters for phospholes

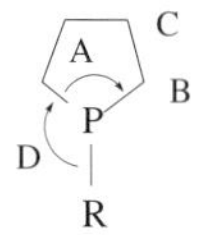

Compound

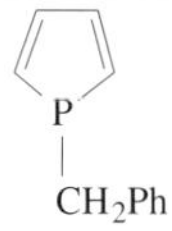

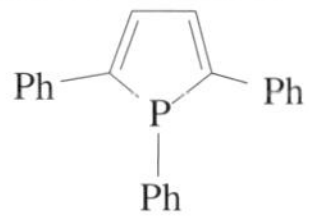

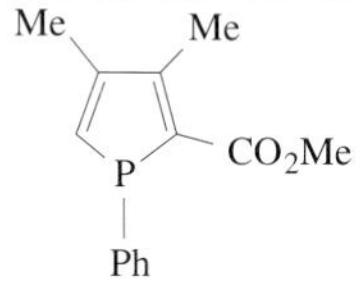

Bond angles [a]	*A*	*B*	*C*	*D*	*A*	*B*	*C*	*D*	*A*	*B*	*C*	*D*
	90.7	110.0	114.1	106.0	-	-	-	-	89.2	111.5 112.9	112.8 113.1	102.9
Bond lengths [a]	P-C(R)	$P-C_1$	C_1-C_2	C_3-C_4	P-C(R)	$P-C_1$	C_1-C_2	C_3-C_4	P-C(R)	$P-C_1$	C_1-C_2	C_3-C_4
	1.858	1.783	1.343	1.438	1.822	1.822	1.349	1.440	1.840	1.804 1.789	1.360 1.332	1.472
Ref	73JCS(D)1888				71CC1062				80CSC905			

Compound

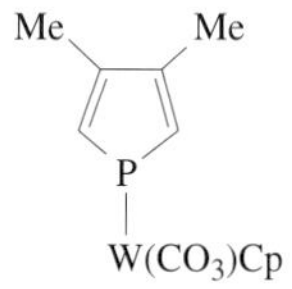

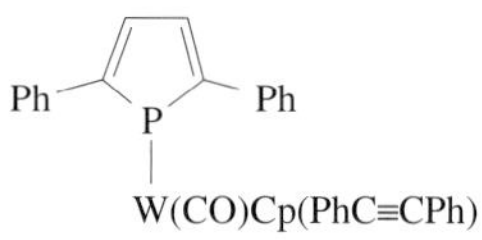

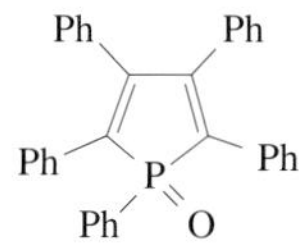

Bond angles [a]	*A*	*B*	*C*	*D*	*A*	*B*	*C*	*D*	*A*	*B*	*C*	*D*
	90.7	111.3	113.0	116.2 112.7	91.4	109.7	114.5	110.8 112.9	93.3	108.7	114.6	108.1
Bond lengths [a]	P-C(R)	$P-C_1$	C_1-C_2	C_3-C_4	P-C(R)	$P-C_1$	C_1-C_2	C_3-C_4	P-C(R)	$P-C_1$	C_1-C_2	C_3-C_4
	2.571	1.771	1.37	1.447	2.448	1.795	1.363	1.439	1.796	1.810	1.355	1.505
Ref	93OM98				93OM98				76CB877			

Compound

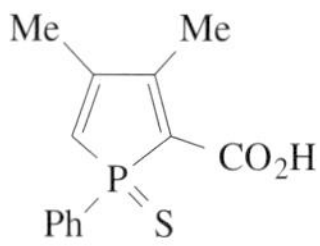

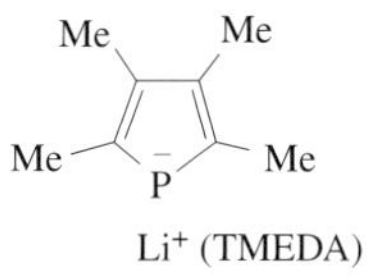

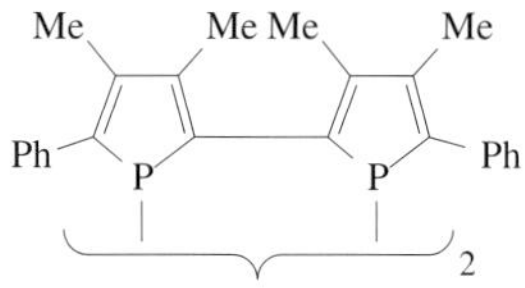

Bond angles [a]	*A*	*B*	*C*	*D*	*A*	*B*	*C*	*D*	*A*	*B*	*C*	*D*
	91.4	110.5	113.9	108.1	90.5	112.3	112.4	-	-	-	-	-
Bond lengths [a]	P-C(R)	$P-C_1$	C_1-C_2	C_3-C_4	P-C(R)	$P-C_1$	C_1-C_2	C_3-C_4	P-C(R)	$P-C_1$	C_1-C_2	C_3-C_4
	1.807	1.813 1.786	1.353 1.336	1.494	-	1.751	1.396	1.424	-	1.785	1.374	1.442
Ref	80CSC901				89AG(E)1367				83JCS(D)841			

Structures for several phospholes with higher coordination numbers at phosphorus have also been reported and are included in Table 3, as are some reports on benzo derivatives. A structure for the 2-coordinate phospholide ion (as a lithium salt) is also included; the effects of delocalization are pronounced in this case, consistent with other observations on such ions. The C–C single bond is decreased to 1.424 Å, and the C=C double bond is increased to 1.396 Å.

A number of structures of phosphole coordination complexes have been determined but are not included in this review. Many of them are described elsewhere ⟨B-83MI01⟩.

Table 3. *Continued*

Compound	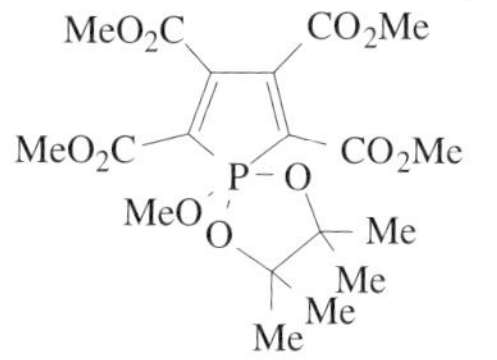				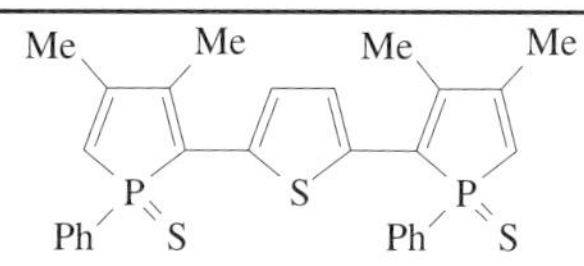				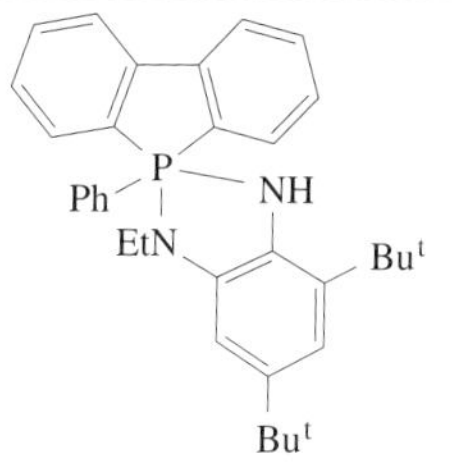			
Bond angles [a]	*A*	*B*	*C*	*D*	*A*	*B*	*C*	*D*	*A*	*B*	*C*	*D*
	87.9	111.1	115.5	-		92.6[b] 92.2			86.3	113.7	112.2	-
Bond lengths [a]	P-C(R)	P-C_1	C_1-C_2	C_3-C_4	P-C(R)	P-C_1	C_1-C_2	C_3-C_4	P-C(R)	P-C_1	C_1-C_2	C_3-C_4
		1.837 1.885	1.334 1.340	1.462	-	-	-	-		1.85	1.37	1.47
Ref	82AX(B)3024				91HAC377				83CB1914			

Compound	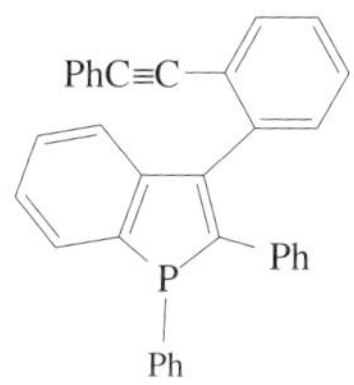			
Bond angles [a]	*A*	*B*	*C*	*D*
	90.1	111.4	114.2	102.7
Bond lengths [a]	P-C(R)	P-C_1	C_1-C_2	C_3-C_4
	1.833	1.826 1.805	1.352 1.402	1.466
Ref	83CB3271			

[a] Averaged values reported for two bonds in a molecule in some cases.
[b] See ref. for additional data.

4.2.1.3.2 NMR Spectroscopy

4.2.1.3.2.1 ^{31}P NMR Spectroscopy

The immediate determination of the ^{31}P NMR shift is virtually routine in the characterization of new phosphorus compounds or analysis of reaction mixtures, and this is certainly the case in the subfield of phosphole chemistry. The ^{31}P NMR shift of phospholes is significantly displaced in the downfield direction relative to model tertiary or secondary phosphines, and substitutions can produce additional profound modifications of the shifts. With a proper understanding of the structure–shift relations, the technique can be of enormous importance in structure assignments, and is given special emphasis in this section. Wherever possible, the considerable NMR data available are organized around the shift changes that take place when structural modifications are made on the simplest phospholes with tertiary structure at phosphorus, 1-methylphosphole or 1-phenylphosphole. Much less is known about 1*H*-phospholes, but effects there are treated as well. All ^{31}P shifts are referenced to 85% H_3PO_4 as 0 ppm; downfield shifts bear a positive sign. Literature before about 1978–1980 used the opposite sign convention, but these have been converted to the current convention. The shifts for phospholes with C.N.-3 are listed in Table 4; data for phosphole oxides, phosphole sulfides, phospholium salts, and ylides are given in Table 5. References are given in the tables, and are not repeated in the discussion which follows.

Table 4. ^{31}P NMR shifts of phospholes

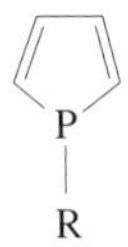

$\delta\,^{31}P$	Ref
R = H : -49.2(C_4D_8O)	87TL5025
Me: -8.7(neat)	69JA3308
Bu^n: +5.5	72OMR(4)171
Ph: +6.4	81S983
$PhCH_2$: +8.0(neat)	73JOC1858
$PhCH_2CH_2$: +5.8 (neat)	73JOC1858
4-Cl-$C_6H_4CH_2$: +5.9(neat)	73JOC1858
$SnMe_3$: -50.09	88CC770

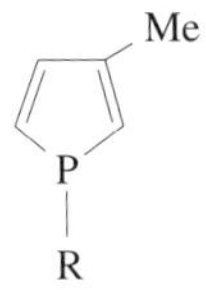

$\delta\,^{31}P$	Ref
R = Me: -6.9(neat)	73JOC1858
Bu^n: +8.5	72OMR(4)171
Bu^t :+40.5 (neat)	72T4171
$PhCH_2$: +11.5(neat)	73JOC1858
Ph: +7.5	81S983

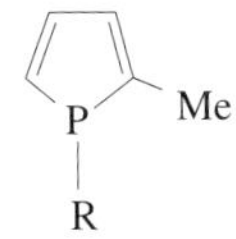

$\delta\,^{31}P$	Ref
R = Me: -7.3 (neat)	73JOC1858

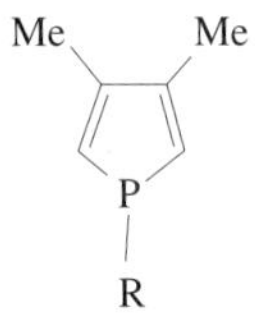

$\delta\,^{31}P$	Ref
R = H: -59.4(C_4D_8O)	87TL5025
Me: -20.2(neat)	73JOC1858
Bu^n: -6.5	76T2395
Bu^t:+27.5	72T4171
$PhCH_2$: -3.0(neat)	73JOC1858
CN: -54.7	88OM1796
$W(CO_3)Cp$: -105.2(CD_2Cl_2)	93OM98
Me_3CO: +82.6	88OM1796
9-fluorenyl: +1.0	88OM1796
Ph: -4.4	81S983
$SnMe_3$: -65.09	88CC770

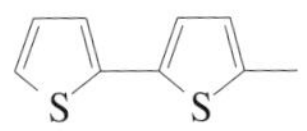

: -25.2 (C_6D_6) 90AG(E)655

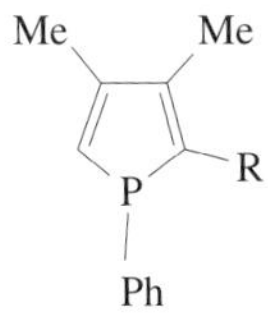

$\delta\,^{31}P$	Ref
R = Me: +6.6	94OM925
COOH: + 5.7(CH_2Cl_2)	92BSF486
COOEt: +5.0	76T2395
CHO: +1.3(CH_2Cl_2)	92BSF486
COMe: +5.0	76T2395
COPh: +11.3(THF)	92BSF486
Br: +10.2(toluene)	92BSF486
$SiMe_3$: +15.3(THF)	92BSF486

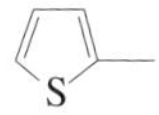

: + 6.6(CH_2Cl_2) 90JOC2494

4-$MeOC_6H_4$: +4.8(CH_2Cl_2) 90JOC2494

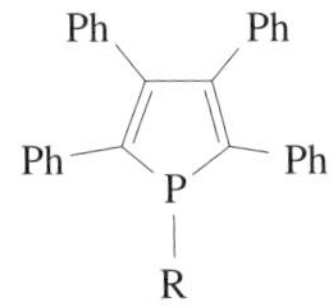

$\delta\,^{31}P$	Ref
R = H: -40.9(C_4D_8O)	83JA6871
Ph: +13.9	82TL511

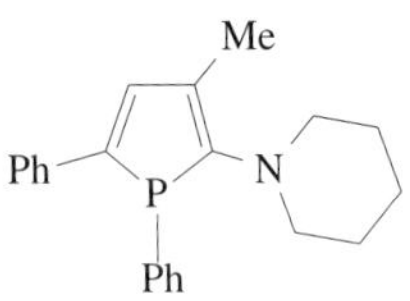

$\delta\,^{31}P$	Ref
-6.3	88TL4581

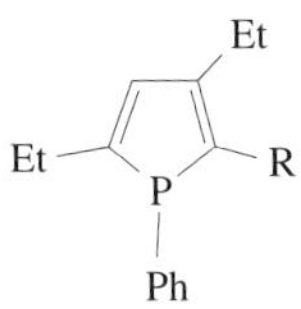

$\delta\,^{31}P$ Ref

R = 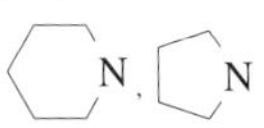

88TL4581

-9.6 -1.7

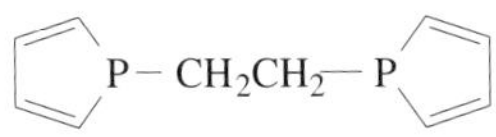

$\delta\,^{31}P$	Ref
+9.96(CD_2Cl_2)	87TL5025

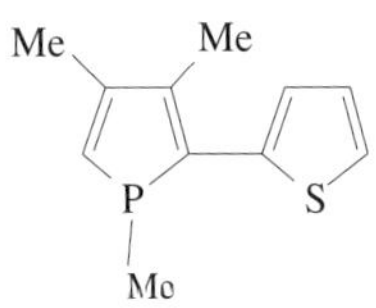

$\delta\,^{31}P$	Ref
-8.4	90JOC2494

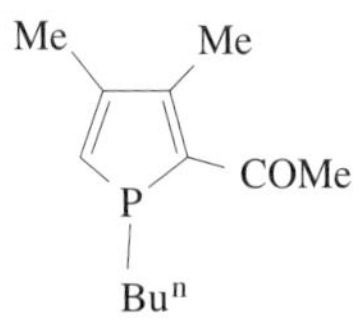

$\delta\,^{31}P$	Ref
+6.8	76T2395

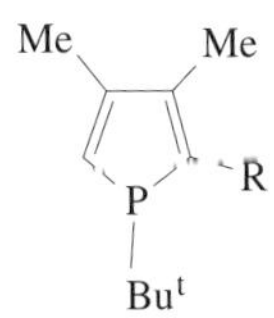

$\delta\,^{31}P$	Ref
R = COMe: +35.1	76T2395
COOEt: +34.3	76T2395

Table 4. *Continued*

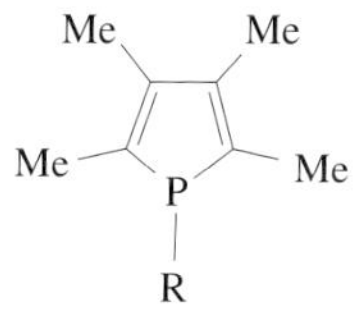

$\delta\,^{31}P$	*Ref*
R = Ph: 12.71	88OM921
Cl: +80.69(C_6D_6)	88OM921
H: -37.0 (C_4D_8O)	87TL5025
$SnMe_3$: -45.96	88CC770

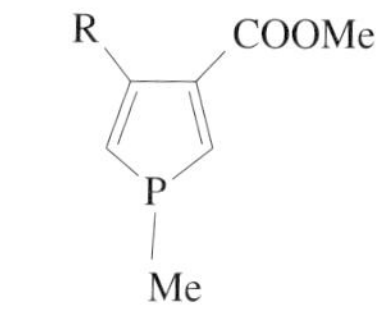

$\delta\,^{31}P$	*Ref*
R = H: +3.0	73JOC1858
Me: -12.6	73JOC1858

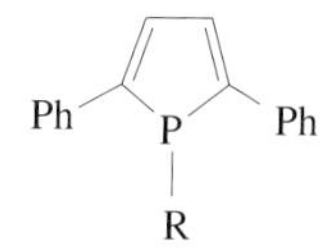

$\delta\,^{31}P$	*Ref*
R = H: -54.2 (C_4D_8O)	83JA6871
Ph: +0.32	87TL5025
Br: +31.1	82JOM(231)361
MeO: +112.9	82JOM(231)361
EtO: +107.4	82JOM(231)361
$Cp(CO)_3W$: -98.8(C_6D_6)	93OM98

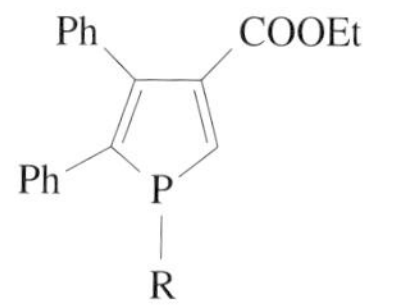

$\delta\,^{31}P$	*Ref*
R = Me: -0.4(CH_2Cl_2)	93BSF695
$EtO_2CCH_2CH_2$: +11.9(CH_2Cl_2)	93BSF695

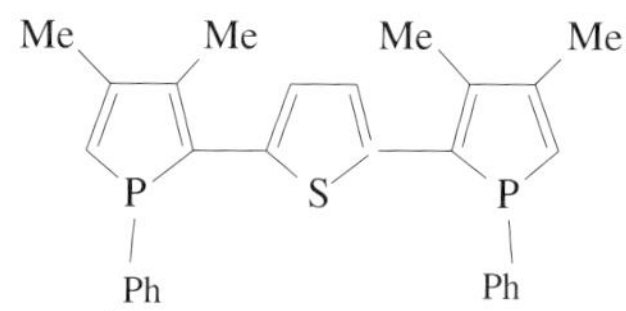

$\delta\,^{31}P$	*Ref*
+4.0	91HAC377

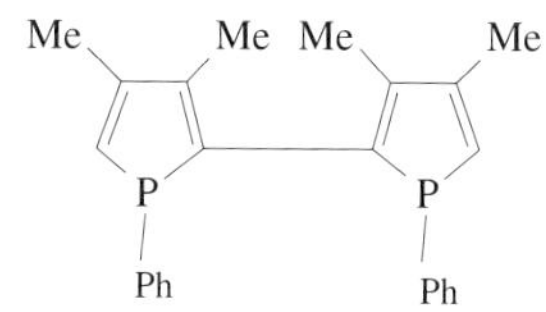

$\delta\,^{31}P$	*Ref*
+12.5	86JOM(316)271

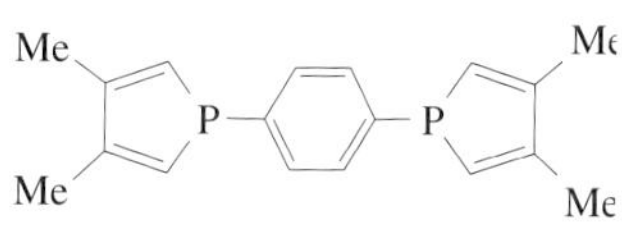

$\delta\,^{31}P$	*Ref*
-3.4(CD_2Cl_2)	90HAC187

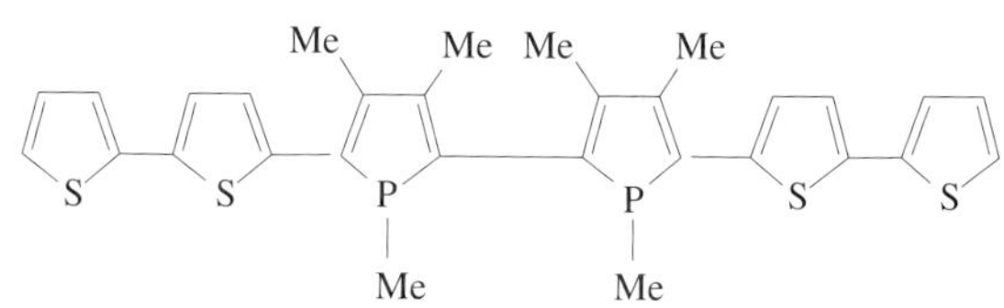

$\delta\,^{31}P$	*Ref*
+7.5	90AG(E)655

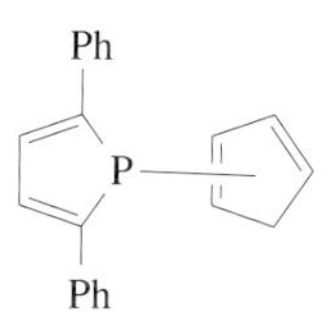

$\delta\,^{31}P$	*Ref*
isomers: -13.3 (CH_2Cl_2) and -15.7 (CH_2Cl_2)	87JOM(332)141

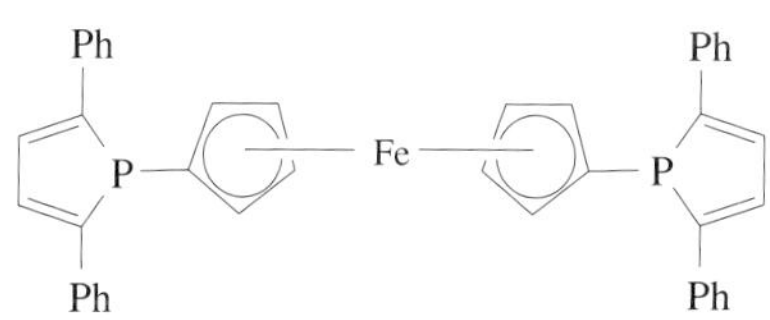

$\delta\,^{31}P$	*Ref*
-5.5(CH2Cl2)	87JOM(332)141

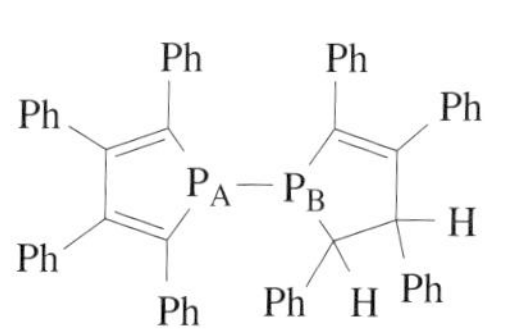

$\delta\,^{31}P$	*Ref*
P_A +28.7, P_B -1.7 (^1J = 370 Hz)	83JA6871

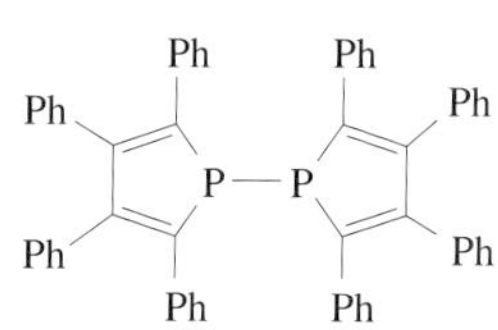

$\delta\,^{31}P$	*Ref*
-14.6	83JA6871

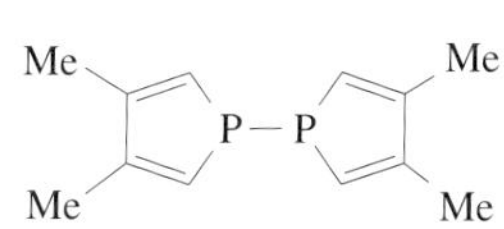

$\delta\,^{31}P$	*Ref*
-25.3	83OM1234

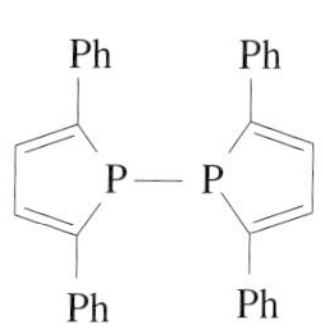

$\delta\,^{31}P$	*Ref*
-27.9	82JOM(231)361

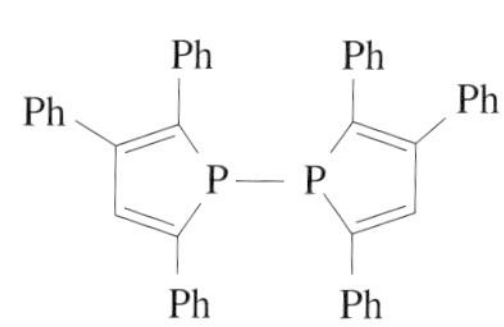

$\delta\,^{31}P$	*Ref*
-11.8	82JOC2376

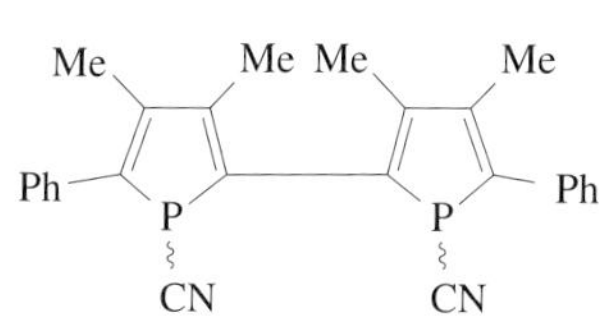

$\delta\,^{31}P$	*Ref*
isomers: -29.06, -33.32	94JA3306

Table 4. *Continued*

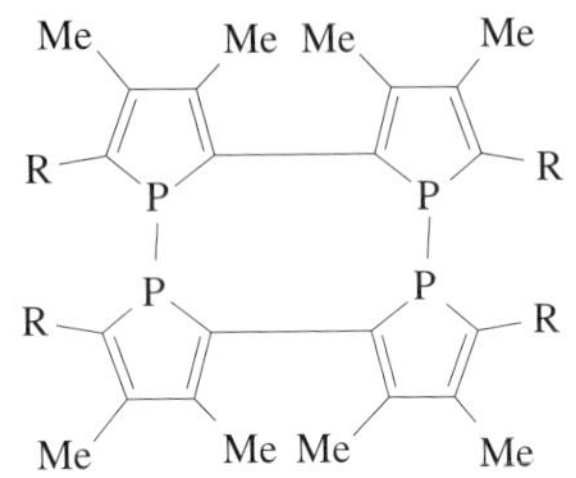

$\delta^{31}P$ — *Ref*
R = Ph: -11.6 — 82JA2077

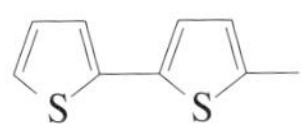

:-2.1 — 90AG(E)655

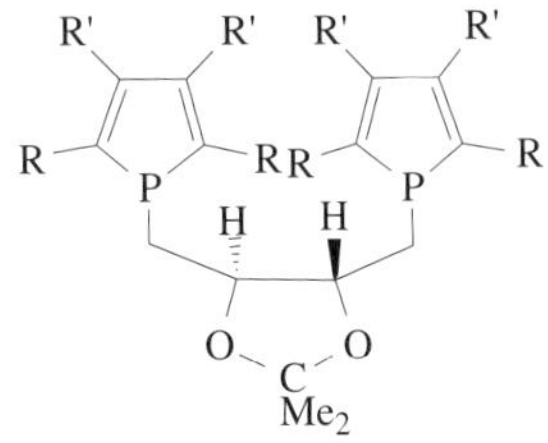

$\delta^{31}P$ — *Ref*
R = Ph, R' = H: -8.3 — 93PS(85)207
R = H, R' = Me: -13.3 — 93PS(85)207

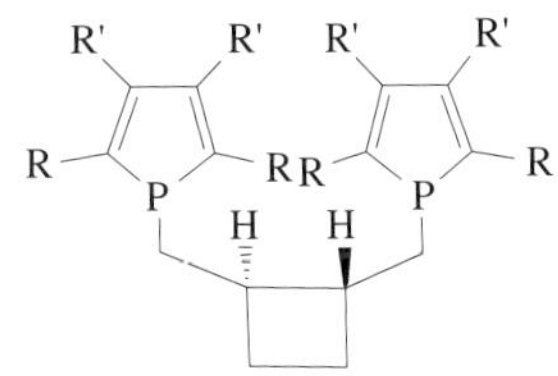

$\delta^{31}P$ — *Ref*
R = Ph, R' = H: -0.8 — 93PS(85)207
R = H, R' = Me: -9.0 — 93PS(85)207

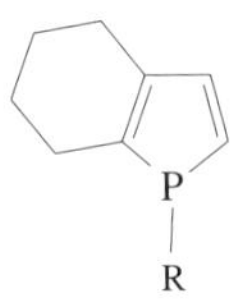

$\delta^{31}P$ — *Ref*
R = Me: -7.6 — 82PS(12)161
Ph: +10.5 — 82PS(12)161

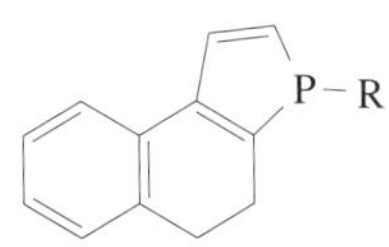

$\delta^{31}P$ — *Ref*
R = Me: -4.1 — 82PS(12)161
Ph: +12.8 — 82PS(12)161

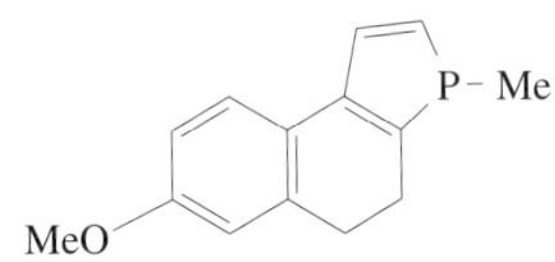

$\delta^{31}P$ — *Ref*
-4.9 — 82PS(12)161

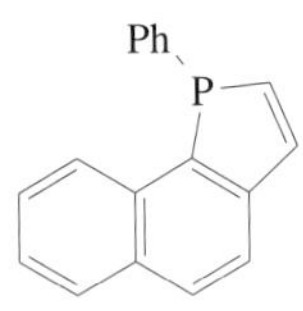

$\delta^{31}P$ — *Ref*
+0.1 — 82PS(12)161

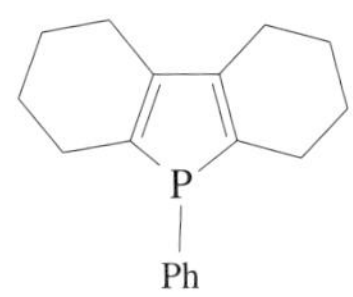

$\delta^{31}P$ — *Ref*
+7.6 — 78CJC1952

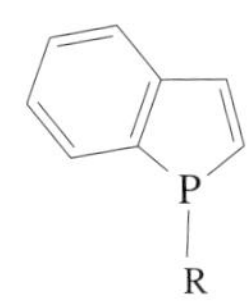

$\delta^{31}P$ — *Ref*
R = Ph: -2 — 82PS(13)259
Bu^t:+23.5 — 82PS(13)259

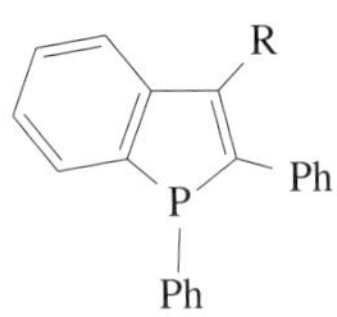

$\delta^{31}P$ — *Ref*
R = H: -1.1 — 84CB990
Cl: +0.6 — 84CB990

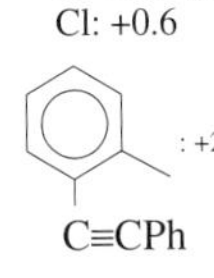

: +2.4($CHCl_3$) — 77CB2168

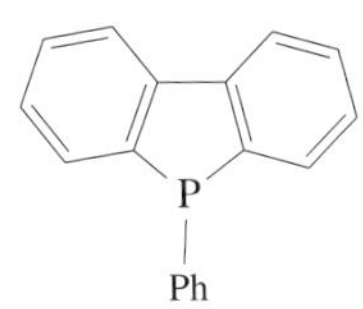

$\delta^{31}P$ — *Ref*
-10.2 — 65CB576

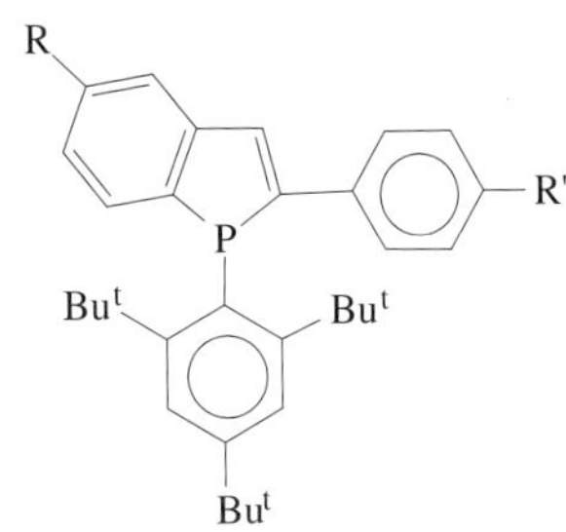

$\delta^{31}P$ — *Ref*
R = R' = H: -13.92 — 93TL3103
R = OMe, R' = H: -15.74 — 93TL3103
R = H, R' = OMe: -14.22 — 93TL3103
R = R' = OMe: -15.67 — 93TL3103

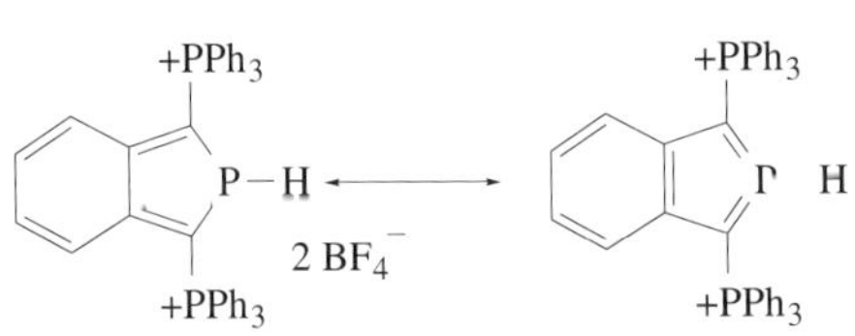

$\delta^{31}P$ — *Ref*
+41.7 (also +23.1, ^2J=12.3) — 91AG(E)308

[a] In $CDCl_3$ unless otherwise noted.

4.2.1.3.2.1.1 Effect of substitution at the P-methyl group

Tertiary phosphines (and other classes of phosphorus compounds with carbon substituents) respond to the effect well known in ^{13}C NMR that replacement of H of a methyl group by a methyl group causes a significant downfield shift (the β-deshielding effect), and replacement of H on the methyl group of ethyl causes a smaller upfield shift (the γ-shielding effect). Average values for saturated tertiary phosphines are +13.5 and −4 ppm, respectively, and are additive ⟨73OMR(5)17⟩. An examination of the data available for phospholes clearly reveals these effects to be operating, but data are insufficient to calculate appropriate constants. Thus, 1,3-dimethylphosphole has a shift of *δ* −6.9; 3-methyl-1-*t*-butylphosphole has a shift of *δ* +40.5, which implies a quite reasonable β-deshielding of 15.8 ppm per methyl in the *t*-butyl group. An identical value is found in the replacement of the 1-methyl group of 1,3,4-trimethylphosphole by *t*-butyl (*δ* +27.5). However, the crowded *t*-butyl is not the best model for making this calculation and the β-constant of 15.8 ppm must be considered as a tentative value. Nevertheless, this reasoning allows an interpretation of the shift of *δ* +5.5 reported for 1-*n*-butylphosphole; here there is a combination of the large β-deshielding and small γ-shielding, and if the former value is taken as +15.8 ppm from the *t*-butyl group calculation, then the γ-shielding is seen to be −1.6 ppm. Similar values for γ-shielding are found in the 1,3,4-trisubstituted compounds (1,3,4-trimethyl, *δ* −20.2 (neat); 3,4-dimethyl-1-*n*-butyl, *δ* −6.5: γ-shielding of −2.1 ppm) and in the 1,3-disubstituted compounds (1,3-dimethyl-, *δ* −6.9; 3-methyl-1-*n*-butyl-, *δ* +8.5: γ-shielding of −0.4 ppm). Using this logic, the values of *δ* +5.8 for 2-phenethylphosphole, and *δ* +9.96 for 1,2-bis-(2-phospholyl)-ethane are understandable.

Replacement of a methyl H in 1-methylphosphole by phenyl causes a substantial downfield shift (1-benzylphosphole, *δ* +8.9) implying that a β-phenyl has a constant of +17.6 ppm. The constancy of this value is seen in the comparison of, for example, 1,3-dimethylphosphole (*δ* −6.9) to 1-benzyl-3-methylphosphole (*δ* −3.0: Δ*δ* 18.4 ppm).

4.2.1.3.2.1.2 Effect of replacement of methyl by other substituents

In tertiary phosphines, replacing methyl by phenyl causes a substantial downfield shift, and this is found also among phospholes (1-phenyl, +6.4 from −8.7, Δ*δ* +15.1 ppm; 3-methyl-1-phenyl, +7.5 from −6.9, Δ*δ* +14.4, etc.). Replacing methyl by Cl, Br, or OR causes much more pronounced downfield shifts ((**10**), *δ* +80.69; (**11**), *δ* +31.1; (**12**), *δ* +82.6). On the other hand, a very strong shielding takes place when methyl is replaced by cyano ((**13**), *δ* −54.7). These effects of P-substituents are common among phosphines in general.

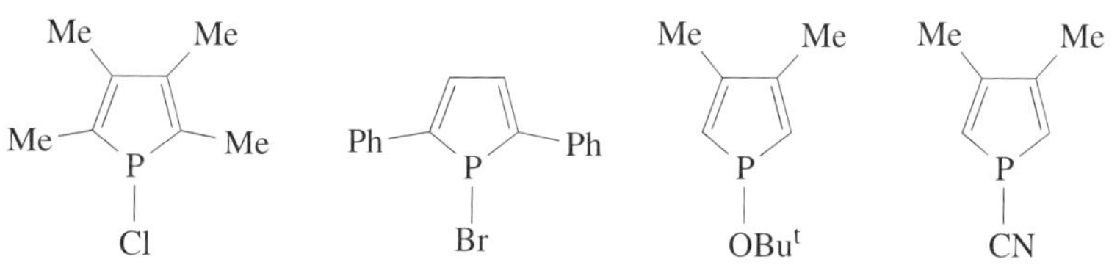

(**10**) (**11**) (**12**) (**13**)

4.2.1.3.2.1.3 Effect of alkyl or aryl substitution on ring carbons

A peculiarity noted in ^{31}P values of 3-phospholenes ⟨72PS(2)55⟩ is very evident also in the phosphole family: replacement of a 3H by methyl causes a small downfield shift (to *δ* −6.9) from the value for the 1-methyl derivative (*δ* −8.7), but a second methyl placed at the 4-position causes a very strong *upfield shift* (to *δ* −20, Δ*δ* −11.5 ppm). The same effect of 3,4-dimethyl substitution is seen in the *n*-butyl- (Δ*δ* −15.0 ppm), P-phenyl (Δ*δ* −11.9 ppm) and P-benzyl series (Δ*δ* −11.9 ppm), and elsewhere. No satisfactory explanation of this effect seems to have been published but one intuitively thinks of the steric crowding accompanying *vic*-dimethyl substitution as being involved. The existence of this effect needs to be remembered when examining the literature, since many known phospholes in fact have 3,4-dimethyl substitution, and the reported shifts are no doubt some 10 ppm more upfield than less crowded, more representative derivatives. As will be noted, the 3,4-dimethyl effect is found in phospholes with higher coordination numbers, and even in phospholide ions.

It might be expected that 2,3,4-trimethyl substitution would lead to even further upfield shifting if simple steric crowding were involved, but this does not seem to occur. 1-Phenyl-2,3,4-trimethylphosphole

Table 5. ^{31}P NMR shifts of phosphole derivatives with C.N. -4 and C.N. -5 [a]

A. Oxides

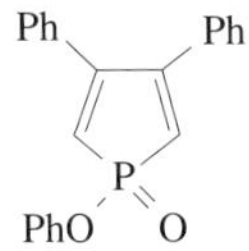

$\delta\,^{31}P$	*Ref*
+ 46.6	71JA4541

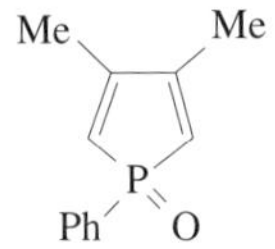

$\delta\,^{31}P$	*Ref*
+42.3	91HAC359

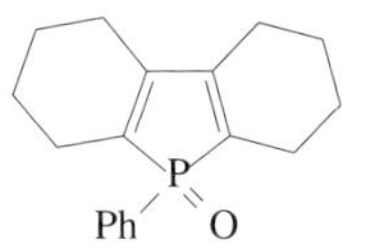

$\delta\,^{31}P$	*Ref*
+ 55.7	78CJC1952

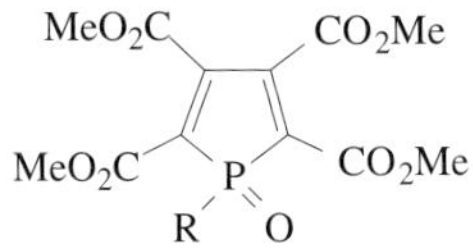

$\delta\,^{31}P$	*Ref*
R = Me : + 46.3	88JCS(P1)175
Ph : + 38.7	88JCS(P1)175
MeO : +36.88	84JCS(P1)1627
EtO : +35.28	84JCS(P1)1627
Pr^{i}O : +34.0	88JCS(P1)175

B. Sulfides and selenides

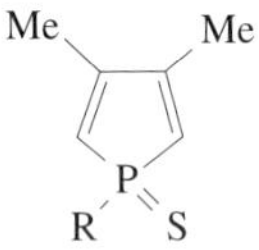

$\delta\,^{31}P$ R =	*Ref*
Bu^n : +52.2	76T2395
Bu^t : +70.5	76T2395
Ph : +46.1	76T2395

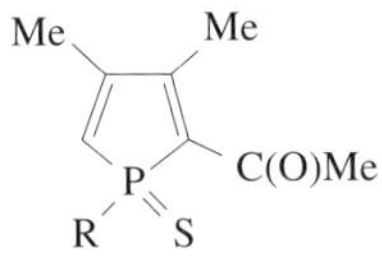

$\delta\,^{31}P$ R =	*Ref*
Bu^n : +55.3	76T2395
Bu^t : +73.8	76T2395
Ph : +48.0	76T2395

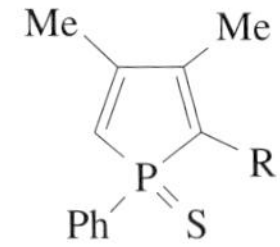

$\delta\,^{31}P$ R =	*Ref*
EtOOC : +48.8	76T2395
HOOC : +53.2	76T2395

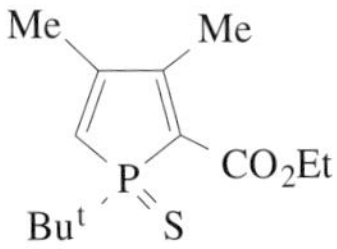

$\delta\,^{31}P$	*Ref*
+73.7	76T2395

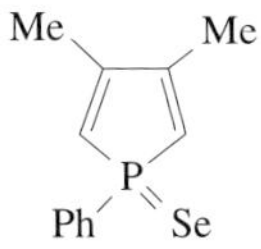

$\delta\,^{31}P$	*Ref*
+28.5	72T4171

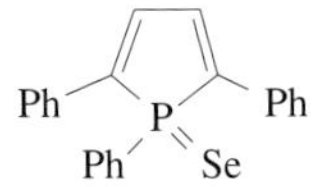

$\delta\,^{31}P$	*Ref*
+35.7	86JCR(S)392

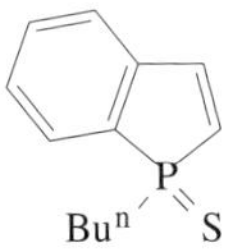

$\delta\,^{31}P$	*Ref*
+51	82PS(13)259

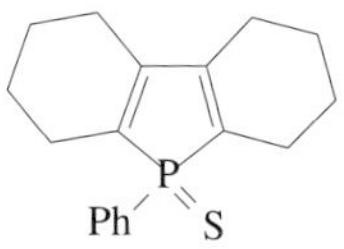

$\delta\,^{31}P$	*Ref*
+50.9	78CJC1952

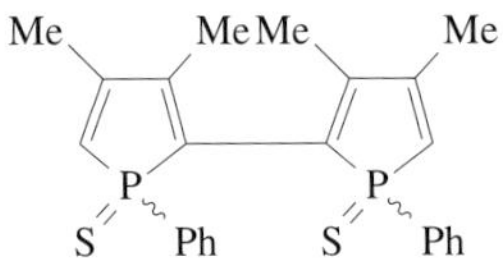

$\delta\,^{31}P$	*Ref*
+47.1 and +47.7	86JOM(316)271

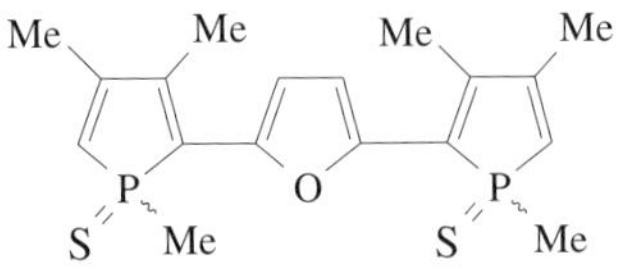

$\delta\,^{31}P$	*Ref*
+44.73 and +45.3	91HAC377

has δ +6.6, whereas the 3,4-dimethyl derivative has δ −4.4. Thus the third methyl has caused a *downfield shift* of 11.0 ppm.

Along the same line, placement of substituents at the 2,3 positions does not cause an upfield shift; 1,2-dimethylphosphole has almost the same shift (δ −7.3) as 1-methylphosphole, and 1-methylcyclohexano[*b*]phosphole (**14**) has a shift of δ −7.6. Similar values are found for the tricyclic phospholes (**15**) and (**16**). Even 2,3,4,5-tetramethyl substitution does not cause pronounced shift effects. The 1-phenyl derivative has δ +12.71, compared to δ +6.4 for 1-phenylphosphole.

Data are insufficient to evaluate the effect of replacing ring H by aryl groups. Compared to 1-phenylphosphole (δ +6.4), the replacement of both 2 and 5 hydrogens by phenyl leads to a small net shielding (to δ +3). Just as did a methyl substituent, an aryl group placed at the 2-position of a 3,4-dimethyl derivative switches off the shielding caused by this arrangement; relative to (**17**), phospholes (**18**) and (**19**) have significantly more downfield shifts. A 2-thienyl group has a similar effect ((**20**), δ

Table 5. *Continued*

C. Salts

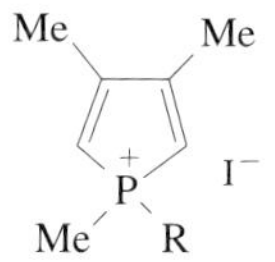

$\delta\, ^{31}P$ *Ref*
R=Me : + 34.0 (D_2O) 73JOC1954
Bu^t : +52.8 72T4171

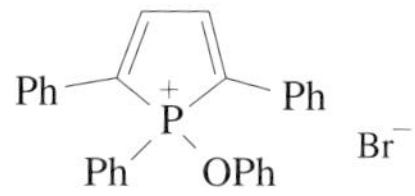

$\delta\, ^{31}P$ *Ref*
+74.6 77JCS(P1)1044

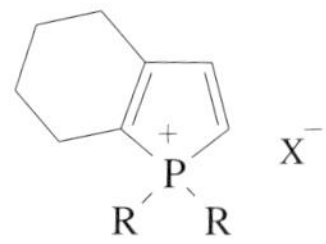

$\delta\, ^{31}P$ *Ref*
R = Me (I^-): +44.1 79OMR(12)442
$PhCH_2$(Br^-) : +50.179OMR(12)442

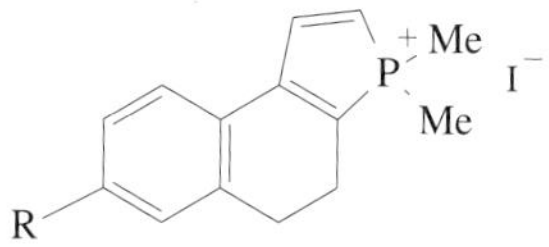

$\delta\, ^{31}P$ *Ref*
R = H : +48.3 79OMR(12)442
MeO : +46.4 79OMR(12)442

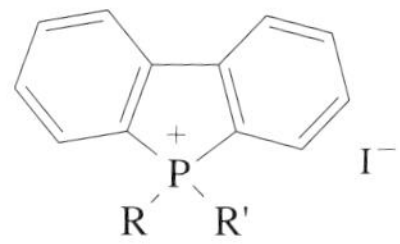

$\delta\, ^{31}P$ *Ref*
R = R' =
Me : +27 (TFA) 70JCS(B)1527
R = Me, R' = Ph :
+ 21.5 (?) 72 JCS(P1)2713

D. Ylides

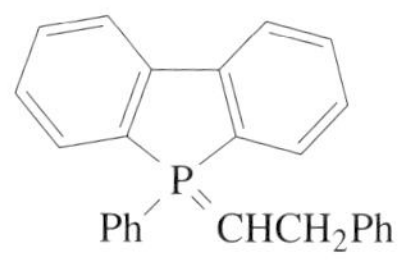

$\delta\, ^{31}P$ *Ref*
+17 72JCS(P1)2713

E. C.N.-5

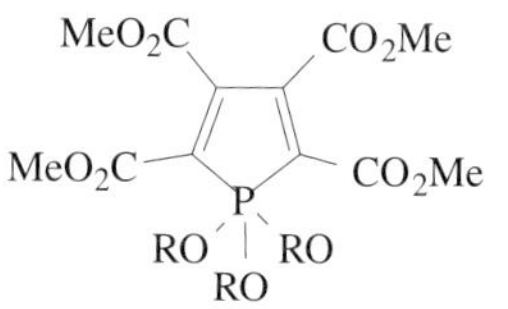

$\delta\, ^{31}P$ *Ref*
R =
Me : -37.4 (toluene) 84JCS(P1)1627
Et : -41.84 84JCS(P1)1627

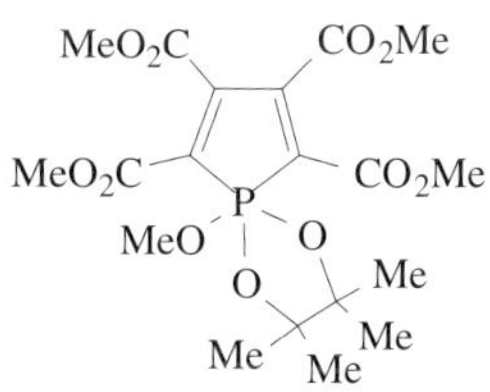

$\delta\, ^{31}P$ *Ref*
-35.5 81TL3533

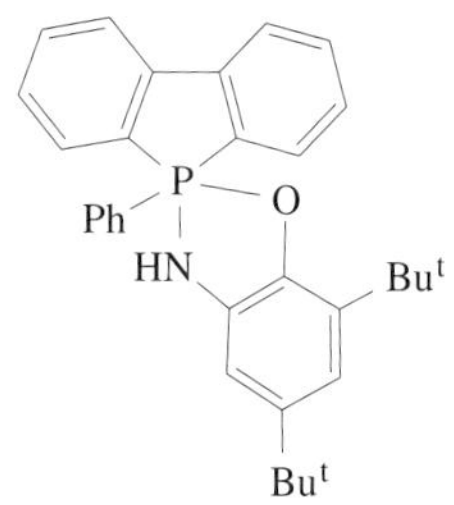

$\delta\, ^{31}P$ *Ref*
-57.5(acetone) 83CB1914

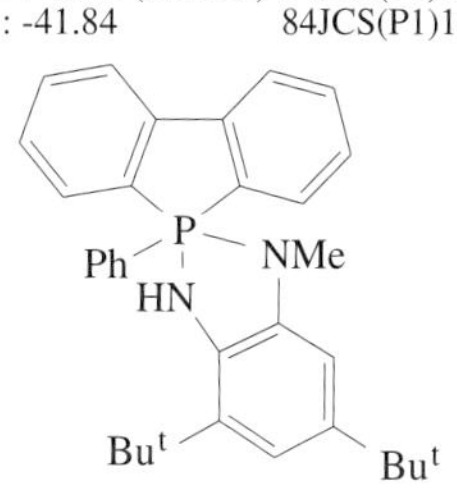

$\delta\, ^{31}P$ *Ref*
-68.94 (toluene) 83CB1914

[a] In $CDCl_3$ unless otherwise noted.

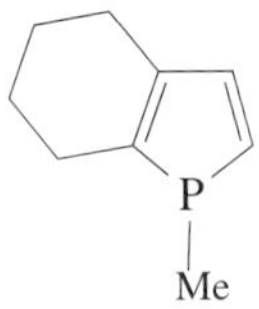
(**14**), δ −7.6

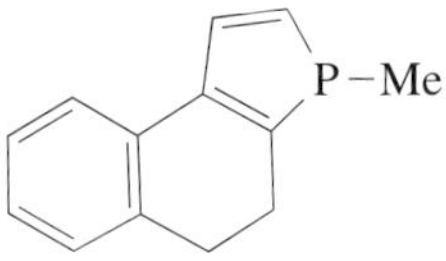
(**15**), δ −4.1

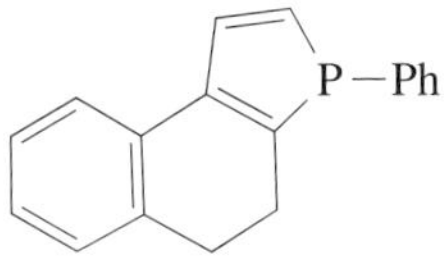
(**16**), δ +12.8

+6.6), and does not seem to introduce any significant shift influence from its electron-releasing character. Thus, (**20**) has exactly the same shift as the 2,3,4-trimethyl derivative (**19**).

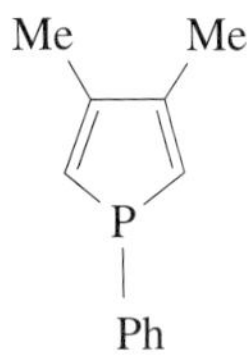
(**17**), δ −4.4

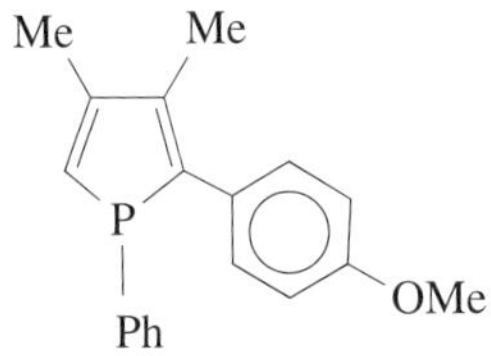
(**18**), δ +4.8

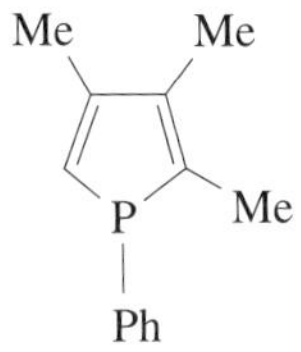
(**19**), δ +6.6

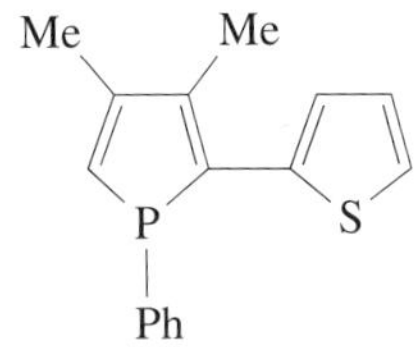
(**20**), δ +6.6

4.2.1.3.2.1.4 Replacement of ring H by conjugating groups

The 3-carbomethoxy derivative of 1-methylphosphole (**21**) was the first functionalized phosphole to be prepared (Scheme 2), and it was discovered that the substituent had a very pronounced deshielding effect (of +11.7 ppm) on the ^{31}P NMR shift (*δ* +3.0). In some way, the conjugation of the ester with a ring double bond (resonance form (**22**)) is strongly influencing the phosphine group, and the explanation offered was that the phosphorus lone pair was also involved in the conjugation (resonance form (**23**)).

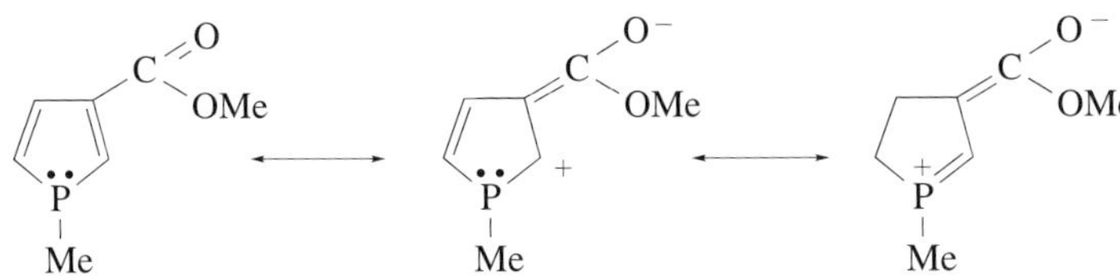
(**21**) (**22**) (**23**)

Scheme 2

As will be seen, the ^{1}H NMR and UV spectra are in accord with the conjugation of form (**22**), but do not speak to form (**23**). Substitution of 4-H by methyl caused an upfield shift (*δ* −12.6), of the same magnitude (Δ*δ* −15.6) as seen on converting 1,3-dimethylphosphole to 1,3,4-trimethylphosphole (Δ*δ* −11.5 ppm), and thus providing a further extension of the 3,4-disubstitution effect.

The effect of a conjugating group at the 2-position can only be analyzed by examining compounds with 3,4-dimethyl substitution, since simpler models are not yet known. A 2-carboethoxy group placed on 3,4-dimethyl-1-phenylphosphole caused a downfield shift of 9.4 ppm (*δ* +5.0), an effect like that of a 2-methyl group (Δ*δ* +11.0 ppm). Similarly, a 2-acetyl group led to a shift also of *δ* +5.0. The acetyl group had the same effect when placed on 1-*n*-butyl-3,4-dimethylphosphole (*δ* +6.8, Δ*δ* +13.5 ppm). Other conjugating groups (–CHO, –COPh, –COOH) show similar effects. These results suggest that the net effect is again simply switching off the shielding of 3,4-dimethyl substitution, with no obvious conjugating effect being transmitted to phosphorus as might be associated with resonance forms such as (**24**) (Scheme 3).

Substitution on 1-*t*-butyl-3,4-dimethylphosphole by 2-carboethoxy caused a net upfield shift of −6.8 ppm, and the same effect was observed for 2-acetyl (Δ*δ* −7.6 ppm). The upfield shift may be more a consequence of steric influences from the bulky *t*-butyl group, and not indicative of a conjugative effect. Other ester derivatives are known ((**25**) and (**26**)) but are not useful in ascertaining conjugative effects without related model compounds.

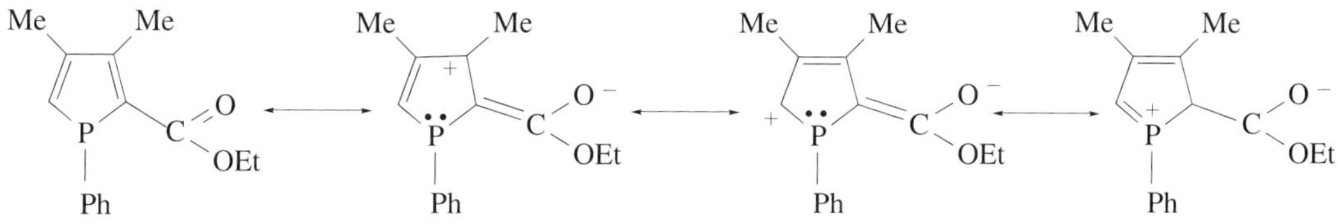

(24)

Scheme 3

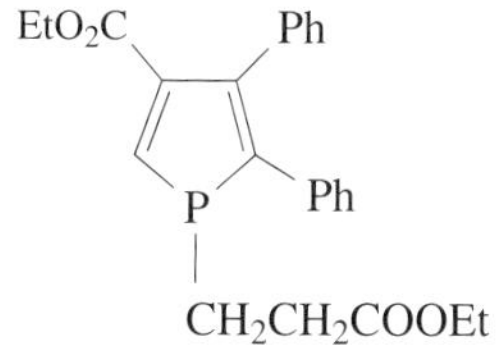

(25), δ +11.9

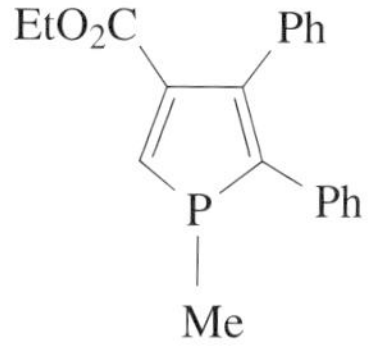

(26), δ -0.4

4.2.1.3.2.1.5 Effects from halo, silyl, and amino substituents

Only one C-halophosphole (2-bromo-3,4-dimethyl-1-phenyl) is known; its shift of +10.2 suggests a stronger deshielding than the switching off of the 3,4-dimethyl effect. This was true also when TMS replaced Br (*δ* +15.3).

There are three 2-amino-substituted phospholes ((**27**), (**28**), and (**29**)), but with an unusual 2,3,5-substitution pattern that makes it difficult to assess the effect of the amino groups. There is a suggestion, however, of upfield shifting coming from this substituent; if one ignores the 5-methyl substituent of (**27**), then a comparison can be made to 3-methyl-1-phenylphosphole (*δ* +7.5). The shift of (**27**) is *δ* −6.3, thus suggesting an upfield shift of −13.8 ppm due to the amino group. It will be of interest to confirm this value with other aminophospholes, since a shift of this magnitude is of considerable importance; it suggests that there is strong electron release into the ring (resonance forms (**30**) and (**31**)) that in some way has an influence on phosphorus (Scheme 4). This point clearly deserves further study, as it may bear on the nature of electron delocalization in the phosphole ring (see Section 4.2.1.2).

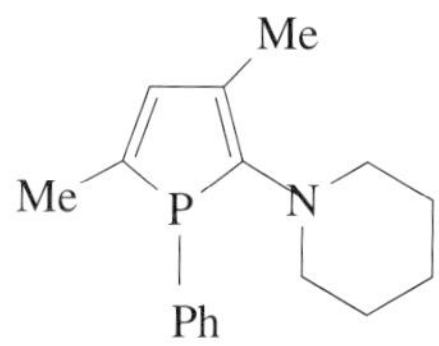

(**27**), δ -6.3

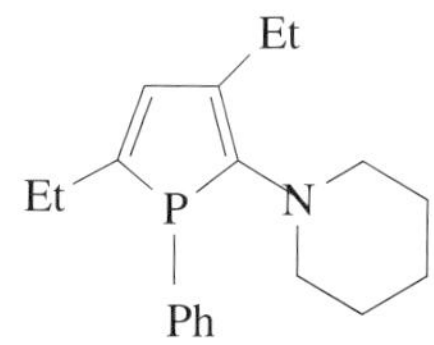

(**28**), δ -9.6

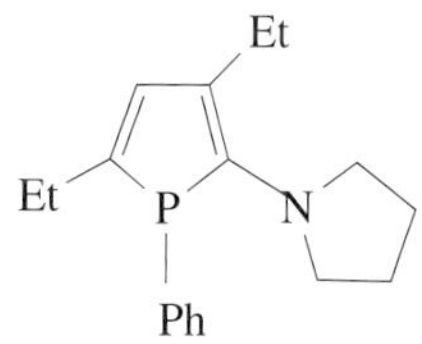

(**29**), δ -1.7

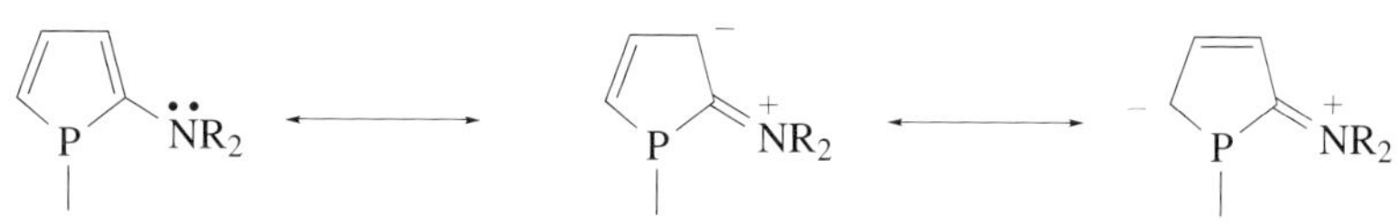

(**30**) (**31**)

Scheme 4

4.2.1.3.2.1.6 Replacement of methyl on P by hydrogen

In simple tertiary phosphines, replacement of one carbon substituent by hydrogen causes a very strong upfield shift, and this is seen also among phospholes. The shifts of 1*H*-phospholes have been recorded at low temperatures (−90°C), since they rapidly rearrange by [1,5]H-shifts and then dimerize (see Section 4.2.1.5.8). Structures (**32**) and (**33**) also clearly show the operation of the 3,4-dimethyl shielding effect (Δδ −10.2 ppm) in the 1*H*-phospholes.

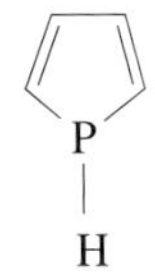

(**32**), δ -49.2

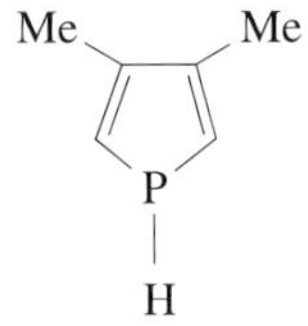

(**33**), δ -59.4

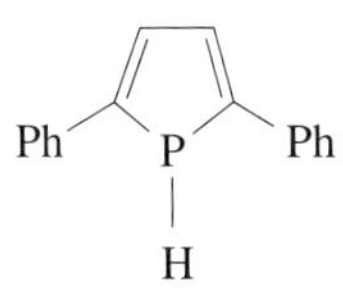

δ -54.16

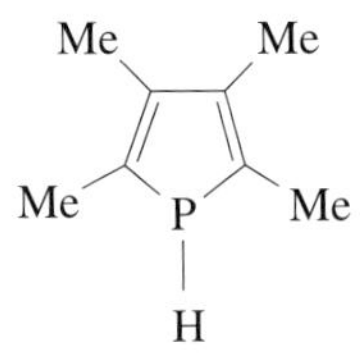

δ -37.0

The ^{31}P–^{1}H coupling constants for the 1*H*-phospholes range from 207.3 to 233.1 Hz, which is within the range for typical secondary phosphines, although on the high side. This is suggestive of low *s*-character in the P–H bond <83JA6871>.

4.2.1.3.2.1.7 Effects of benzo fusion

The fusion of a benzo group on the 2,3-positions of 1-phenylphosphole causes an upfield shift of 8.4 ppm, to δ −2. The effect seems to be additive; the shift for dibenzophosphole (δ −10.2) is an additional 8.2 ppm upfield. The shift for 1-*t*-butylphosphindole is also found in a relatively shielded position (δ +23.5; compare to δ +40.5 for 3-methyl-1-*t*-butylphosphole).

The only phosphole with a naphtho group fused to the *b*-face (**34**) has a shift of δ +0.1, not surprisingly quite close to the value for 1-phenylphosphindole (**35**).

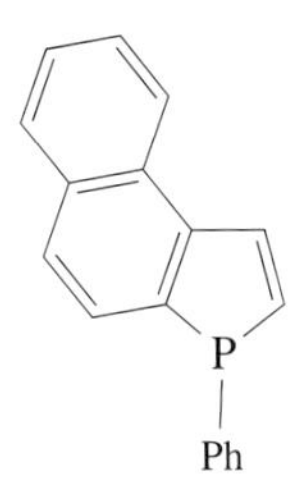

(**34**), δ +0.1

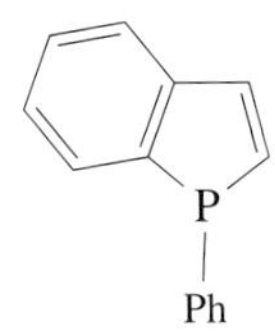

(**35**), δ -2

4.2.1.3.2.1.8 Conversion to C.N.-4 and C.N.-5 species

All conversions of phosphines to derivatives with C.N. 4 are accompanied by sizeable downfield shifts, and this is true of phospholes. Phosphole oxides are rare species, since they tend to dimerize unless sterically protected (see Section 4.2.1.6.1). Nevertheless, some data are available and collected in Table 5. The magnitude of the downfield shifting that takes place is not unlike that known for simple phosphine to oxide conversions. Phosphole sulfides and phospholium ions (both Table 5) also give downfield shifts of expected size. The 3,4-dimethyl shielding effect is evident in all of these classes on examining the tables.

Phospholes with C.N.-5 (phosphoranes) are quite rare, but give the highly characteristic upfield shifts always associated with this coordination state. Trioxyphosphoranes with the phosphole nucleus, for example, have shifts around δ −37 (Table 5).

Table 6. ^{31}P NMR shifts of phospholide ions [a]

Compound	δ ^{31}P	*Ref*
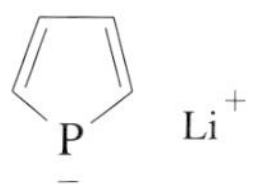	+77.2	87TL5025
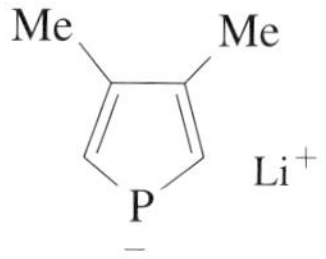	+55.8	87TL5025
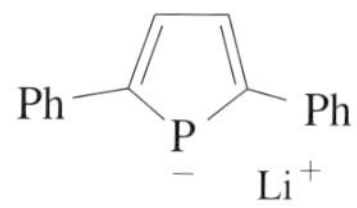	+78.8	87TL5025
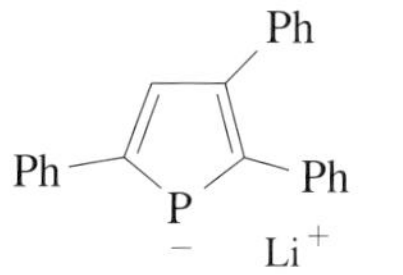	+102.5	82JOM(231)361
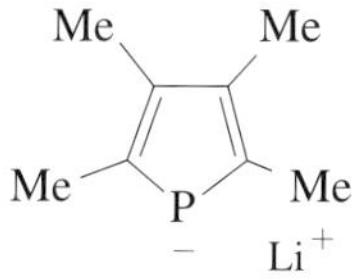	+63.02	87TL5025
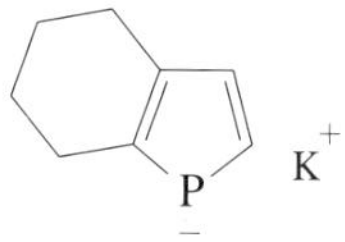	+63.02	87TL5025
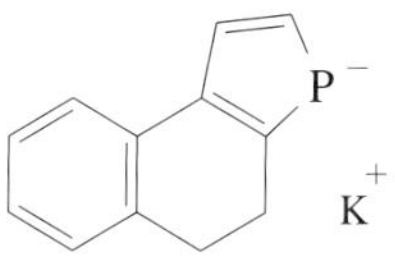	+73.3	79CC401
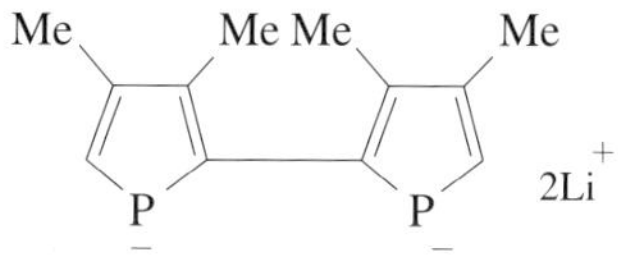	+56.7	86JOM(316)271
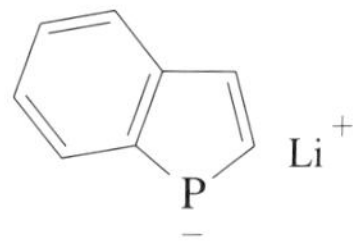	+40	82PS(13)259
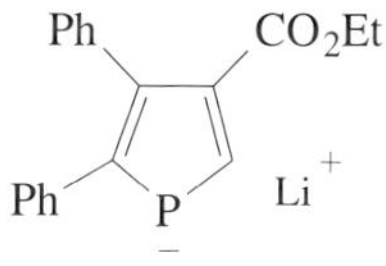	+116.2	93BSF695

[a] In THF.

4.2.1.3.2.1.9 Phospholide ions

Dramatic downfield shifts accompany the cleavage of P-substituents with formation of phospholide ions. These shifts are considerably further downfield than those for simpler phosphide ions, and have been taken to be a reflection of strong electron delocalization in the phospholide ion (see Section 4.2.1.2). Data are collected in Table 6, where it will be seen that typical shifts fall in the region δ +70 to +100. Once again the 3,4-dimethyl shielding effect is present and indeed is of the largest size ($\Delta\delta$ 21.4 ppm) of all families examined. Phospholide ions with this substitution have shifts in the range δ +50 to +60.

4.2.1.3.2.1.10 Theoretical aspects

It was noted in Section 4.2.1.3.2.1 that the ^{31}P NMR signal for phospholes is generally downfield-shifted relative to model (unsaturated) phosphines. This effect was early attributed to cyclic delocalization in the phosphole ring, which could reduce the shielding effect by diminishing the contribution of the lone pair on phosphorus and give it more 4-coordinate character. However, later work has emphasized the low level of delocalization in phospholes that is occasioned by its pyramidal shape. New computational techniques have now made it possible to explore this point by calculating the ^{31}P NMR shifts for 1*H*-phosphole in its natural pyramidal form and in its forced planar form, where delocalization can be appreciable (see Section 4.2.1.2) ⟨94JA9638⟩. The surprising result was obtained that the ^{31}P shifts were quite similar, differing by only 9.0 ppm, and thus speaking against any important role for delocalization in establishing the chemical shift. Furthermore, the calculations showed that a large downfield shift of 110.9 ppm should occur on removing the proton on phosphorus, just as is observed experimentally (126.4 ppm). However, this shift does not arise directly from the delocalization in the phospholide ion, but from the presence of the *second* lone pair on phosphorus which is oriented in the plane of the ring. It is this lone pair then that is the cause of the deshielding; it has an orbital energy value significantly closer to that of the HOMO–LUMO gap than does the lone pair in planar 1*H*-phosphole, which is perpendicular to the plane of the ring. This was shown to lead to a significant paramagnetic contribution to the net shielding. This strong deshielding effect was also calculated to be present (147.7 ppm) when pyrrole was deprotonated to form the pyrrolide ion; the reverse effect, shielding by 100 ppm, was found when a lone pair on thiophene was protonated to form the thiophenium ion.

4.2.1.3.2.2 ^{1}H NMR spectroscopy

Most of the compounds listed in Table 4 have also been characterized by ^{1}H NMR spectroscopy, and this table therefore stands as a source of references for obtaining detailed spectral information for particular phospholes.

The ^{1}H NMR spectra of phospholes without C-substituents are of the AA′BB′X type and are quite complex. The first measurement was made on 1-methylphosphole at 100 MHz; the spectrum had to be interpreted with the aid of computer simulation ⟨69JA3308⟩. Later measurements at higher field gave spectra that agreed well with that for 1-methylphosphole (Table 7). The chemical shifts of the ring protons are of considerable interest, for they exhibit strong deshielding relative to model phosphines and appear in the same region as do the signals for thiophene. Also pointing to the strength of the downfield shifting is the observation that, in 1-benzylphosphole, the phosphole and phenyl proton signals actually overlap in the region δ 6.31–7.29 ⟨73JOC1858⟩. In thiophene, this downfield shifting is taken to imply the existence of an aromatic ring current, and this interpretation was applied to 1-methylphosphole in the 1969 study. However, later studies of phospholes revealed that the cyclic delocalization cannot be very extensive (see Section 4.2.1.2), and was considered to be *less* than that in thiophene. This apparent contradiction has not yet been satisfactorily explained; it cannot be attributed to experimental differences in the measurements of the two compounds as suggested by Palmer and Findlay ⟨75JCS(P2)974⟩.

Spectra for C-methyl phospholes are much easier to interpret since the differences in the chemical shifts of the ring protons are of significant size, and observed chemical shifts for the α- and β-protons ⟨73JOC1858⟩ are quite consistent with the values seen in Table 7. When the conjugating ester group is placed at the 3-position, the expected additional deshielding takes place (numerals are δ values) ⟨73JOC1858, 93BSF695⟩:

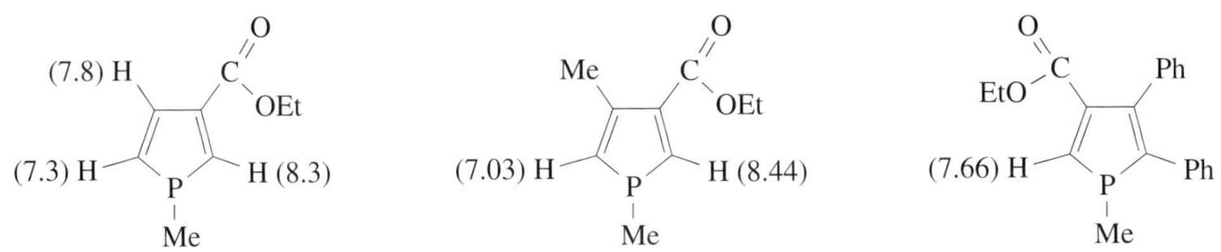

The effects of a conjugating group at the 2-position are not yet fully known, since all compounds of this type so far studied have methyls also at positions 3 and 4. Any shift effect at H-5 is not pronounced; typical shifts for 2-acetyl and 2-carboethoxy are in the range δ 6.5–6.75 ⟨76T2395⟩.

The effect of electron-releasing groups on the ring protons cannot yet be interpreted, due to the scarcity of structures available. There is a strong suggestion, however, that a 2-amino group introduces significant shielding at H-4. Assuming that the 3,5-dialkyl substitution in compounds (**27**)–(**29**) introduces no appreciable effect (not yet confirmable), the signal for H-4 in these amino derivatives falls in the range δ 6.32–6.47 ⟨88TL4581⟩, whereas the proton in this position in the unsubstituted phospholes has a signal around δ 7.0.

2-(2-Thienyl) derivatives of 3,4-dimethylphospholes (e.g. (**37**)) also may exhibit small shielding of H-5 relative to the 2,3,4-trimethyl compound (**36**). A 2-(2-phospholyl) substituent (**38**) ⟨86JOM(316)271⟩ shows the opposite effect.

The ^{31}P–^{1}H coupling constants for the 2- and 3-protons are typical of cyclic vinyl phosphines and while they are rather large they are not reflective of any special property of the phosphole ring system. For all of the monocyclic phospholes listed in Table 7, the $^{2}J_{PH}$ values for H-2 fall in the range 34–42 Hz; the $^{3}J_{PH}$

Table 7. Proton NMR spectra of P-substituted phospholes

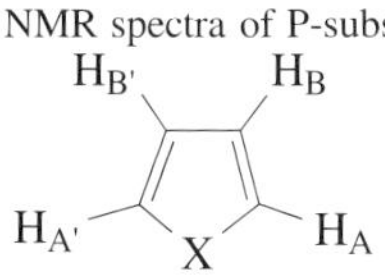

	P–CH₃ [a]	*P–Ph* [b]	*P–H* [b]	*(P–CH₂)₂* [b]	*P⁻ Li⁺* [b]
δ H_A	(6.5–7.5)	6.94	7.10	6.76	6.78
δ H_B	(6.5–7.5)	7.00	7.33	7.03	6.62
$^2J_{P\text{-}HA}$	38.49	38.7	39.8	37.1	40.76
$^3J_{P\text{-}HB}$	13.77	14.5	14.33	15.3	6.42
$^3J_{HA\text{-}HB}$	7.24	7.2	7.05	7.2	5.61
$^4J_{HA\text{-}HB'}$	1.12	1.1	1.04	1.1	1.23
$^4J_{HA\text{-}HA'}$	3.01	2.7	2.78	2.8	3.56
$^3J_{HB\text{-}HB'}$	1.96	2.2	2.3	2.3	2.43

[a] At 100 MHz. From calculated spectrum; error limits not shown <69JA3308>.
[b] <87TL5025>.

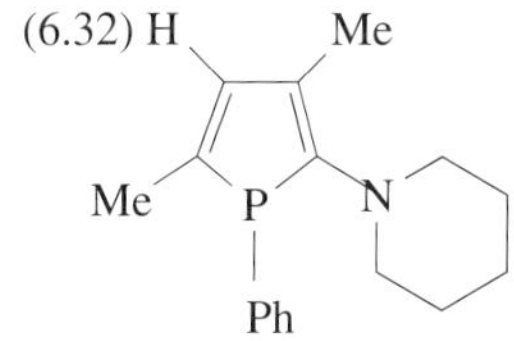

(27)

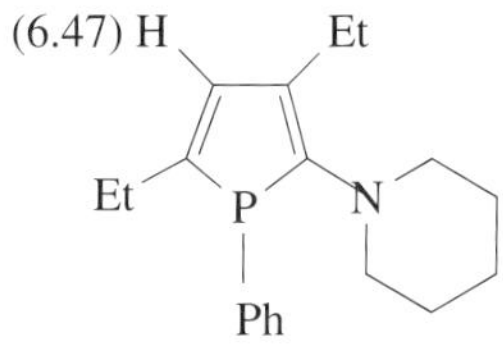

(28)

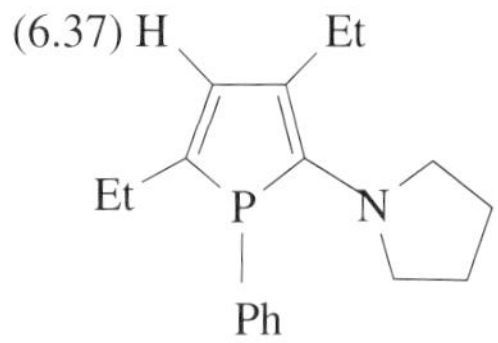

(29)

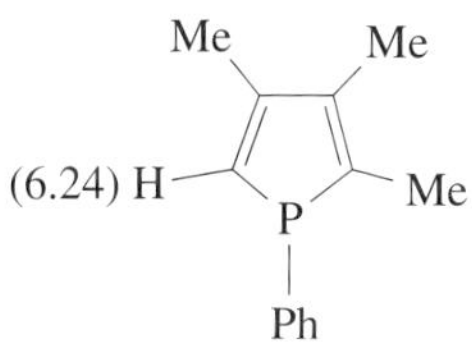

(36)

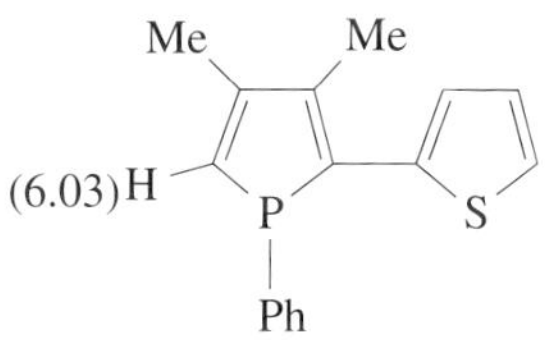

(37)

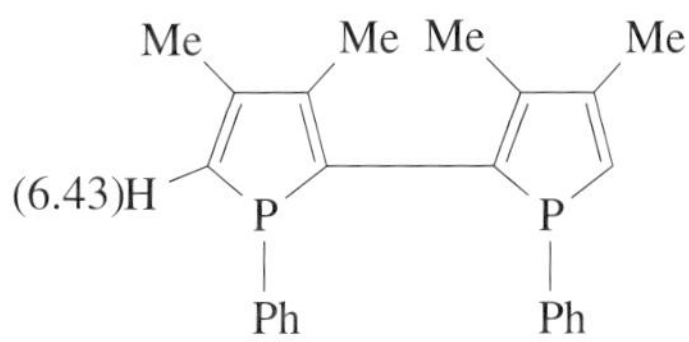

(38)

values for H-3 are around 12–17 Hz. These should be compared to the 2-phospholene model compounds **(39)** <71JOC1297> and **(40)** <B-81MI03>, which have constants in the same ranges.

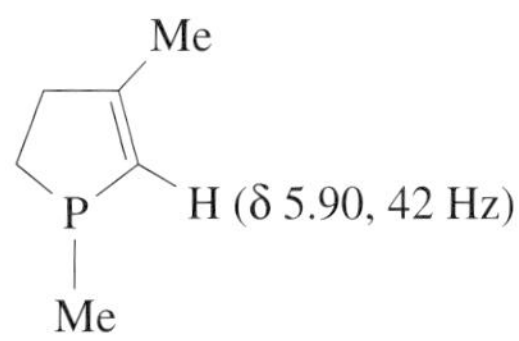

(39)

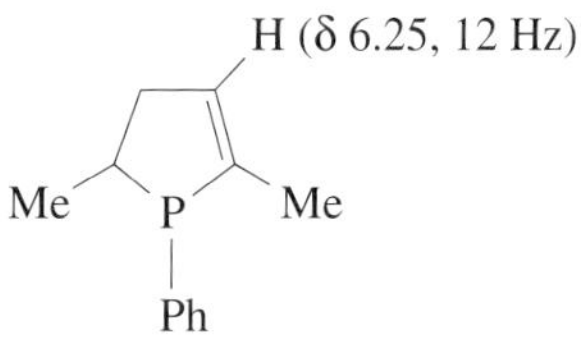

(40)

Methyl groups attached to the phosphorus atom **(34)** <B-81MI03> typically have signals around δ 1.3–1.4, values noticeably more downfield than found for simple methylphosphines (generally δ 0.7–1.0) but not dissimilar to those for cyclic vinyl phosphines (e.g. 1,3-dimethyl-2-phospholene, δ 1.23). However, the introduction of a conjugating carbonyl substituent in the 3-position clearly leads to considerable additional deshielding at a P-methyl group; placement of a 3-carboethoxy group on 1-methylphosphole and on 1,3-dimethylphosphole leads to shifts of δ 1.92 and 1.96 respectively, for the P–Me group. The coupling constant for the ^{31}P interaction with the methyl group is frequently zero, and does not exceed 2 Hz. Such small $^2J_{PH}$ values are not unique to the phosphole ring, however; benzylic protons have signals around δ 3.0, and generally show no detectable coupling with ^{31}P.

It has been difficult to establish the effects on ^1H NMR spectra when phospholes are converted to derivatives with higher coordination numbers, since these species (especially oxides) are prone to dimerize

Table 8. ^{13}C NMR of some P-phenylphospholes [a,b]

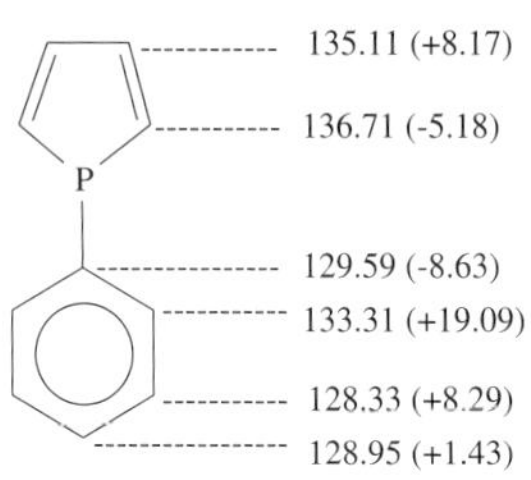

(**41**)

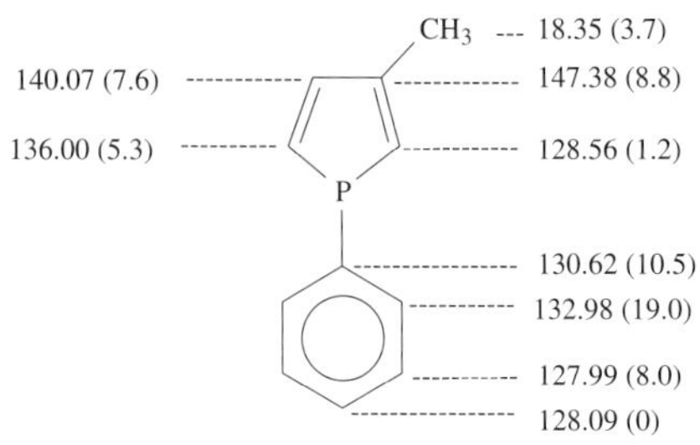

(**42**)

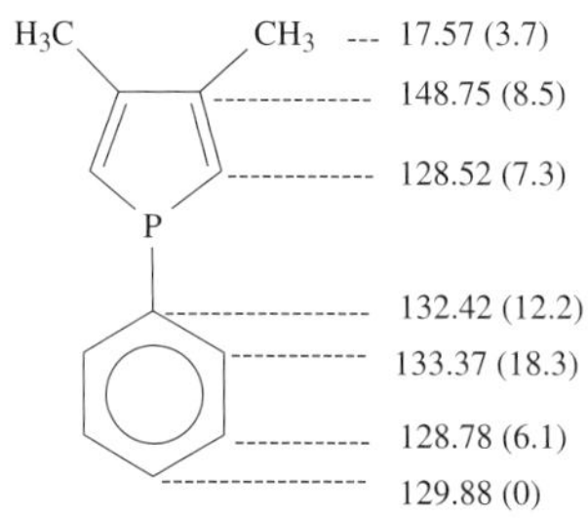

(**43**)

[a] Values in parentheses are J_{PC}, in Hertz.
[b] Refs: (**41**) ⟨72TL3353⟩; (**42**) and (**43**) ⟨80OMR(14)14⟩.

(see Section 4.2.1.6). Those that are stable are multiply substituted, and the chemical shifts will be subject to modification by the substituents. Some structures that show the effects at the α-position are given below. The effects are in fact not great, both with regard to shifts and coupling to ^{31}P, although there is some diminution of the latter value ⟨76T2395⟩.

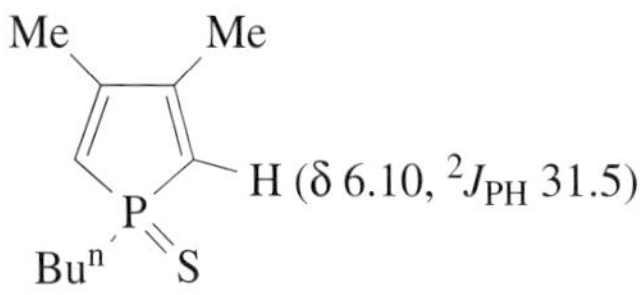

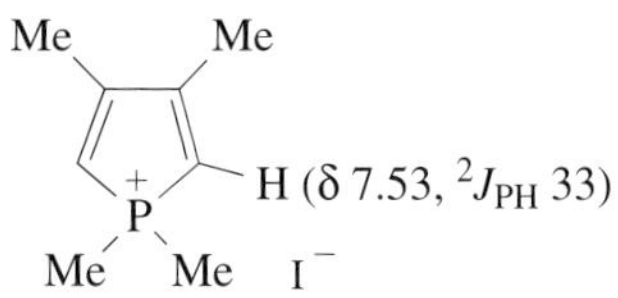

4.2.1.3.2.3 ^{13}C NMR spectroscopy

^{13}C NMR has been of great value in characterizing phospholes, and data are available for several of the compounds recorded in Table 4. Not only are the chemical shifts of value but so also are the ^{13}C–^{31}P coupling constants. Influences on the magnitude of the coupling constant are frequently of steric origin, where the relative disposition of the carbon and the lone pair on phosphorus seems to be important ⟨B-87MI01⟩.

Detailed considerations of the ^{13}C NMR spectra of phospholes ⟨72TL3353, 80OMR(14)14⟩ and of dibenzophospholes ⟨80OMR(14)14, 87MRC774⟩ have been published, and some correlations with the extent of cyclic delocalization in the ring system have been suggested. Thus, the 1972 paper proposed that, in a series of P-phenylphospholes, the relative extent of cyclic delocalization could be revealed by the chemical shifts of the *ipso* carbon of the phenyl ring, with delocalization increases associated with increased shielding of this carbon. Another indication of the degree of delocalization was said to be provided by the magnitude of $^{1}J_{CP}$ for C-2 of the ring, where an increase in this value correlated with an increase in delocalization. The data in Tables 8–10 show differences in the two parameters, but since the degree of electron delocalization in the compounds is not a measurable quantity by other techniques, the interpretation of these relations remains subjective. A useful observation of structural importance that came from the 1980 paper was the increase in shielding of a 3-methyl carbon when a second methyl was placed at C-4. The increase, while small (Tables 8 and 9; 0.78 ppm in the P-phenyl compounds (**42**)

Table 9. ^{13}C NMR of some P-alkylphospholes

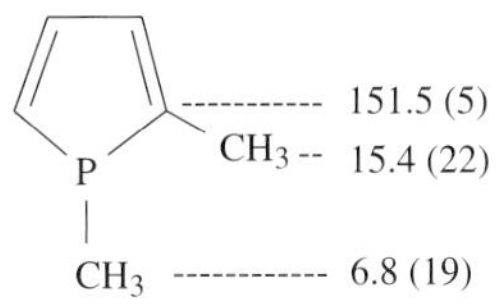

(**44**)[c]

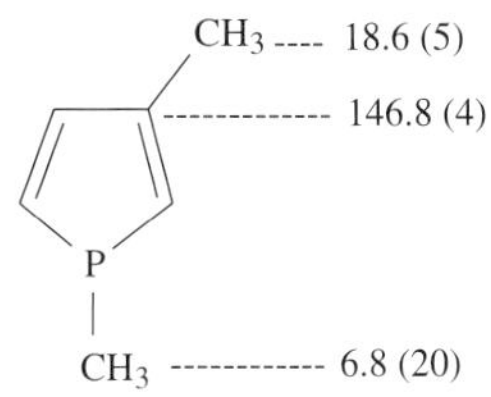

(**45**)[c]

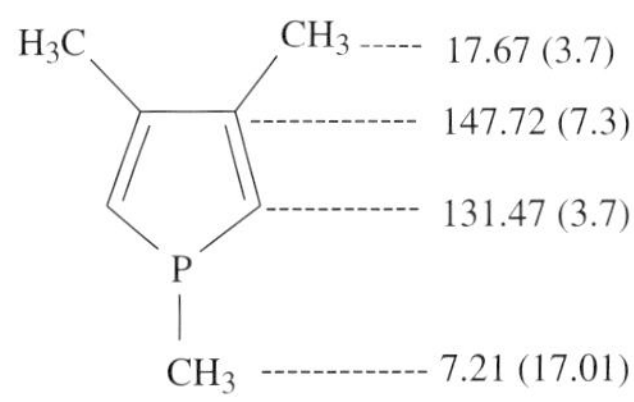

(**46**)[c]

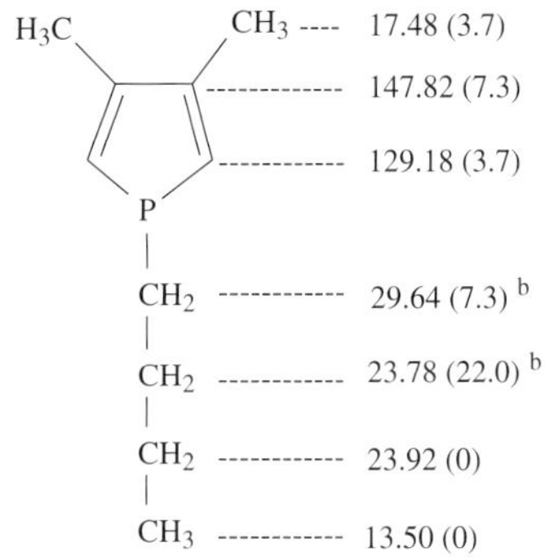

(**47**)[c]

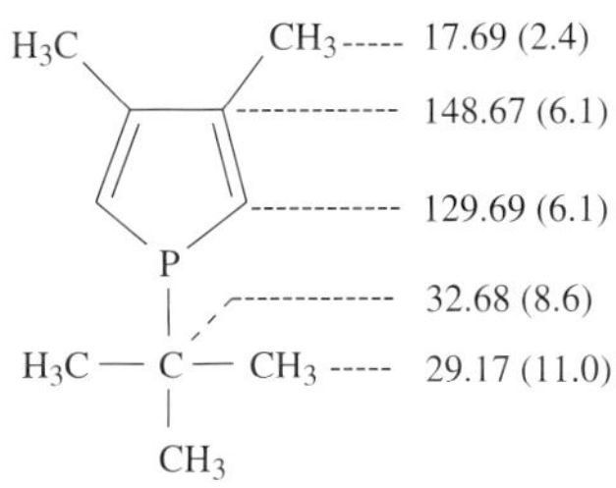

(**48**)[c]

[a] Values in parentheses are J_{PC}, in Hertz.
[b] These assignments require confirmation.
[c] Refs: (**44**) and (**45**) ⟨73OMR(5)161⟩; (**46**), (**47**) and (**48**) ⟨80OMR(14)14⟩.

and (**43**); 0.93 in the P-methyl compounds (**45**) and (**46**) may be associated with the operation of the γ-shielding effect which arises from steric proximity of the groups.

The representative spectra presented in Tables 8–10 reveal some other characteristics of phosphole spectra:

(i) In C-unsubstituted compounds, there is little difference in the shifts of the α- and β-carbons; coupling constants are larger for the latter, which aids in assignments.

(ii) Coupling constants (both ^1J and ^2J) are much larger for exocyclic carbons than for endocyclic carbons. The ^2J values are controlled by the relation of the lone pair with respect to the coupled carbon ⟨B-87MI01⟩. The pyramidal shape at phosphorus has the effect of placing the lone pair close to a 2-substituent, a relation giving strong coupling (compound (**44**), 22 Hz for 2-Me). This effect operates on substituents attached to P also (compound (**47**), 22 Hz for C-2 of *n*-butyl; compounds (**41**), (**42**), and (**43**), 18–19 Hz for the *ortho*-phenyl carbon).

(iii) ^1J to an exocyclic carbon seems to be quite variable, and not obviously related to the structure. Thus, P-methyls have values consistently of 17–20 Hz, but the value for a *t*-butyl substituent is 8.6 Hz. For phenyl carbons, values range from −8.63 to 12.2 Hz. A very large value was observed for a P-cyano substituent (83.0 Hz).

Table 10. ^{13}C NMR of some phospholes with functional groups [a]

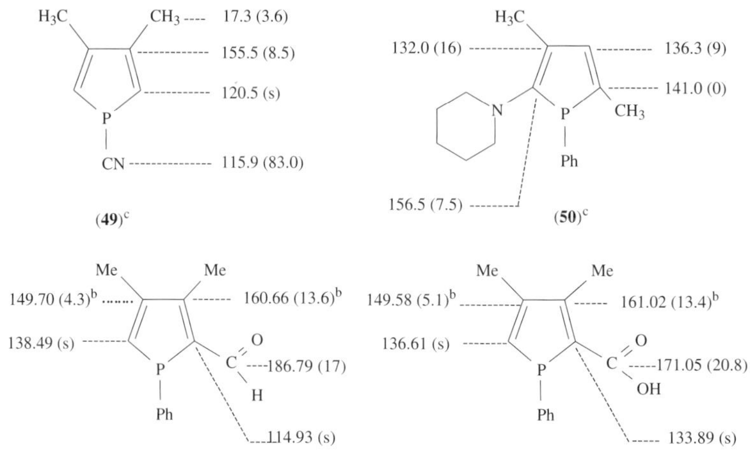

(49)[c] **(50)**[c] (51)[c] **(52)**[c]

[a] Values in parentheses are J_{PC}, in Hertz.
[b] Could be reversed.
[c] Refs: (**49**) ⟨88OM1796⟩; (**50**) ⟨88TL4581⟩; (**51**) and (**52**) ⟨92BSF486⟩.

As in the case of ^{1}H NMR of phospholes, few model compounds are available to assess the effect of a change in coordination number on ^{13}C NMR spectra. However, some striking changes do take place; the β-carbon in sulfides and salts is strongly deshielded, and the coupling constants to ^{31}P are greatly increased, especially at the α-carbon. None of these effects are unique to phospholes, however, and are known also for derivatives of other cyclic, α,β-unsaturated phosphorus compounds in these coordination states ⟨B-81MI02⟩. Compound (**53**) illustrates these spectral effects ⟨78CJC1952⟩.

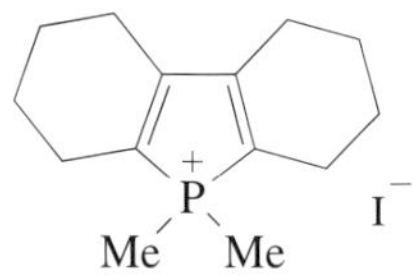

β-C: δ 157.1, $^{2}J_{PC}$ 22 Hz
α-C: δ 122.9, $^{1}J_{PC}$ 79.2 Hz

(53)

4.2.1.3.3 Mass Spectrometry

Following the various NMR techniques in value for structural characterization is mass spectrometry. It was early recognized ⟨69JA3308⟩ that the phosphole ring was of good stability under electron impact; the molecular ion (m/z 98) of 1-methylphosphole was the most abundant peak (R.A. 100), followed closely by M^{+}-1 (R.A. 98). The ring was retained in some other fragments (m/z 83, M^{+}–Me, R.A. 47; m/z 96, M^{+}-2, R.A. 26). Ring fragmentation with loss of ethyne accounted for a peak at m/z 72 ($C_3H_5P^+$, R.A. 11). Numerous later studies have confirmed that the base peak is usually the molecular ion in a variety of compounds (e.g. in methyl 1-methylphosphole-3-carboxylate ⟨73JOC1858⟩; in 1-cyano-3,4-dimethylphosphole ⟨88OM1796⟩; in 1-phenyl-2,3,4-trimethylphosphole ⟨94OM925⟩; in ethyl 2,3-diphenyl-1-methylphosphole-4-carboxylate ⟨93BSF695⟩; in 2-acetyl-3,4-dimethyl-1-phenylphosphole ⟨76T2395⟩; in 3,4-dimethyl-1-phenyl-2(2-thienyl)phosphole ⟨90JOC2494⟩; in bis(3,4-dimethyl-1-phenyl-2-phospholyl) ⟨86JOM(316)271⟩; in 3,5-dimethyl-1-phenyl-2-piperidinophosphole ⟨88TL4581⟩; in 2-bromo-3,4-dimethyl-1-phenylphosphole ⟨92BSF486⟩; and in others). A notable exception occurs with P-benzyl compounds, where the familiar cleavage of benzyl with formation of the $C_7H_7^+$ ion accounted for the base peak, and the molecular ion had R.A. 39% ⟨70JA5779⟩. Certain carbonyl derivatives are also exceptional in giving base peaks from fragmentation of this function: 3,4-dimethyl-2-ethoxycarboxyl-1-phenylphosphole, M^+ (65%), M^+ $-C_2H_4$ (100%), M^+ $-C_2H_4$–CO (53%), M^+ $-C_2H_4$–CHO (75%) ⟨92BSF486⟩; 2-benzoyl-3,4-dimethyl-1-phenylphosphole, M^+ (6%), M^+ –PhCO (100%).

Table 11. UV spectra of some phospholes

Phosphole	*Solvent*	λ_{max}	log ϵ	*Ref.*
1-Me	isooctane	286	3.89	⟨69JA3308⟩
1-$PhCH_2$	95% EtOH	286	3.56	⟨73JOC1858⟩
1-$PhCH_2CH_2$	95% EtOH	284	3.93	⟨73JOC1858⟩
1-$PhCH_2$-3,4-diMe	95% EtOH	280	3.26	⟨73JOC1858⟩
1-Me-3-COOMe	C_6H_{12}	307		⟨73JOC1858⟩
1,4-diMe-3-COOMe	C_6H_{12}	300		⟨73JOC1858⟩
1,2,5-triPh	EtOH	372, 221	4.26, 4.37	⟨74PS(4)279⟩
2-MeCO-3,4-diMe-1-Ph	MeOH	211.5	4.12	⟨76T2395⟩

4.2.1.3.4 Ultraviolet Spectroscopy

Simple P-alkylphospholes possess a unique absorption band, of moderate intensity, in the region 280–290 nm. First noted with 1-methylphosphole (λ_{max} 286 nm in isooctane, log ϵ 3.89, ⟨69JA3308⟩), this band cannot be associated either with the diene component of the ring, or with the phosphine unit, considered separately, since neither have absorption bands near this region. The band is clearly descriptive of the phosphole ring system, but its electronic origin, and relevance to delocalization phenomena, remain to be evaluated. Other phospholes have also given this absorption; some representative data are given in Table 11.

As noted in this table, P-phenylphospholes also give characteristic absorption, and the 1,2,5-triphenyl derivative absorbs at such long wavelength (372 nm) that it is yellow. It also has the rare property of exhibiting triboluminescence ⟨89MI01⟩, and a reversible color change (yellow to red) on boiling in CCl_4. This has been attributed to a reversible formation of a phosphole trichloromethide salt ⟨82CJC138⟩.

4.2.1.3.5 Infrared Spectroscopy

Little use has been made of the IR spectra of phospholes. A few reports exist on such spectral measurements, but primarily they have been made for the purpose of characterizing functional groups on the ring. From the spectrum reported for methyl 1-methylphosphole-3-carboxylate (ν_{CH}3110, $\nu_{C=C}$ 1535, $\nu_{C\text{-}O}$ 1250 and 1186, unassigned at 1075, 1062, 888, 793, 740, 700 cm^{-1} ⟨73JOC1858⟩), it is deduced that the ring double bonds absorb around 1535 cm^{-1}, although in this model compound the conjugating group should introduce as usual a shift of the double bonds to lower frequency. In 1-cyano-3,4-dimethylphosphole ⟨88OM1796⟩, ν_{CN} has been reported at 2160 cm^{-1}, which is at the low frequency end of the C–CN range but like that of S–CN (2170 cm^{-1}). In 2-acetyl-3,4-dimethyl-1-phenylphosphole $\nu_{C=O}$ appeared at 1655 cm^{-1} ⟨76T2395⟩.

4.2.1.3.6 Photoelectron Spectroscopy

The photoelectron spectrum of 1-phenylphosphole contains bands of equal intensity at 8.45 and 9.25 eV ⟨73AG(E)145⟩; other phospholes give similar bands ⟨76JA407⟩. The theory of these signals has been discussed and interpreted in various ways, and they have figured in discussions favoring the view that phospholes can be considered as "aromatic" ⟨76JA4365⟩, as being non-aromatic ⟨75JCS(P2)974⟩, and as having a form of hyperconjugative delocalization involving the σ-orbital of the exocyclic P-substituent ⟨76JA407⟩. A reinterpretation of the data would be timely, since there is fresh interest in the nature of bonding in phospholes (see Section 4.2.1.2).

4.2.1.3.7 Dipole Moments

The dipole moments (in D units) of a few phospholes have been determined ⟨72CR(C)100⟩, with the following results: 1-*n*-butyl, 1.23; 1-*n*-butyl-3,4-dimethyl, 2.00; 3-methyl-1-phenyl, 1.39; 3,4-dimethyl-1-phenyl, 1.52; 1,2,5-triphenyl, 1.45. These values are close to those found for saturated phosphines (e.g. 1-*n*-butylphospholane, 1.93; 1-phenylphospholane, 1.52). The negative end of the dipole in the phospholes is the phosphorus atom, which is the same as found for thiophene and the opposite of the case for pyrrole ⟨72JHC817⟩.

4.2.1.3.8 Electron Spin Resonance Spectroscopy

The only application of this technique in phosphole chemistry has been in the detection of radical intermediates in the cleavage of P-substituents with alkali metals at −80°C. Radical anions at phosphorus were formed as the initial products from 1-methyl-2,5-diphenylphosphole, 1,2,5-triphenylphosphole, and pentaphenylphosphole. The ^{31}P nucleus exhibited strong coupling with the electron ⟨71T5705⟩. Radical anions were also detected in the reaction of 1,2,5-triphenylphosphole oxide and pentaphenylphosphole oxide with alkali metals ⟨71CC782⟩.

4.2.1.4 THERMODYNAMIC ASPECTS

4.2.1.4.1 Physical Properties

In general, the simple phospholes are malodorous, colorless materials that require great care in handling because of their ease of oxidation. Those with more complex structures, especially where conjugating substituents are present, are less volatile and oxidizable, and can be relatively inoffensive. These characteristics figure prominently in the design of experiments to determine physical and other properties of phospholes.

Most of the simple phospholes are liquids that can be distilled. Distillations must be conducted under air-free conditions; generally reduced pressure is used to avoid thermal decomposition. Some typical b.p. (°C) are: 1-Me, 83–85 ⟨69JA3308⟩; 1,3,4-triMe, 60 at 15 mm Hg ⟨72OMR(4)171⟩; 1-benzyl, 71–72 at 0.2 mm Hg, m.p. 34–34.5°C ⟨73JOC1858⟩; 1-Ph, 55 at 0.1 mm Hg ⟨81S983⟩; 3,4-diMe-1-Ph, 78 at 0.1 mm Hg ⟨81S983⟩; 1-cyanoethyl-3,4-diMe, 80 at 0.1 mm Hg ⟨76CJC2402⟩; Me 1,4-diMe-phosphole-3-carboxylate, 67 at 0.3 mm Hg ⟨73JOC1858⟩; 1-Me-cyclohexano[b]-phosphole, 60 at 0.03 mm Hg ⟨82PS(12)161⟩; 1-Ph-2,3,4,5-tetraMe, 105–110 at 0.5 mm Hg ⟨88OM921⟩; 1,2-diPh-5-isopropyl, 100–105 at 0.025 mm Hg, m.p. 73–76°C ⟨71JA6205⟩; 1-cyano-3,4-diMe, 80 at 20 mm Hg, m.p. 52°C ⟨88OM1796⟩.

As noted in the above listing, a few simple phospholes are liquids at room temperature, but those with higher molecular weight are generally solids (m.p. °C): 1,2,5-triPh, 184–186 ⟨67AG(E)86⟩; 2,2′-bis(3,4-diMe-1-Ph-phospholyl) ⟨86JOM(316)271⟩; Me 3,4-diMe-1-Ph-phosphole-2-carboxylate, 61–63 ⟨80CSC905⟩; 1-Ph-phosphindole, 66–68 ⟨72CC496⟩; 1,2-diPh-3-*n*-Bu-benzophosphindole, 84–84.5 ⟨67JA5732⟩; 1-Ph-dibenzophosphole, 93–94 ⟨53LA(580)44⟩.

In general, unless polar substituents are present, phospholes are to be viewed as rather non-polar substances, and show little or no solubility in water. They do have good solubility in most common organic solvents. Frequently used for extraction purposes are chloroform, methylene chloride, benzene, toluene, and others. Reactions such as cleavage of P-substituents with metals are usually performed in THF. Simple P-alkyl and P-phenylphospholes are soluble in alkanes, and some 2-aminophospholes are soluble in ether-petroleum ether. Examples of systems used for the crystallization of phospholes include: CH_2Cl_2 for 3,4-dimethyl-1-phenyl-2-carboxylic acid ⟨92BSF486⟩; aqueous MeOH for methyl 3,4-dimethyl-1-phenyl-1-phosphole-2-carboxylate ⟨80CSC905⟩; chloroform for 1,2,5-triphenylphosphole ⟨65JCS2184⟩.

Liquid chromatography on silica gel has been employed in the isolation and purification of a number of less volatile and less easily oxidized phospholes. Some solvent systems employed include: toluene for 3,4-dimethyl-1-phenyl-2-carbaldehyde ⟨92BSF486⟩; hexane–CH_2Cl_2 for a 2,2′-biphospholyl ⟨86JOM(316)271⟩; ether–petroleum ether for some 2-aminophospholes ⟨88TL4581⟩; toluene–hexane (1 : 4) for 1-methyl-2,3,5-triphenylphosphole ⟨82JOC2376⟩; hexane–ether (95 : 5) for 1-phenyl-2-(2-thienyl)-3,4-dimethylphosphole ⟨90JOC2494⟩, 2,5-bis(2-phospholyl)thiophene ⟨91HAC377⟩, and 2-bromo-3,4-dimethyl-1-phenylphosphole ⟨92BSF486⟩; pentane–CH_2Cl_2 for 1-phenyl-2,3,4-trimethylphosphole ⟨94OM925⟩. 3-Methyl-4,5-dihydro-(3*H*)-benzo[*e*]phosphindole was chromatographed on alumina (neutral) with benzene as eluent ⟨82PS(12)161⟩; Florisil with pentane as eluent was used for 1-isopropyl-2-methyl-5-phenylphosphole, and also for a phosphindole ⟨71JA6205⟩.

The high-pressure liquid chromatography technique does not appear to have been useful in phosphole isolation or purification. However, the more volatile phospholes can safely be subjected to gas chromatography. For 1-methylphosphole, a column of silicone oil SE-30 at 50°C was used ⟨69JA3308⟩; the 3-carboxymethyl derivative required a temperature of 150°C on this column ⟨73JOC1858⟩. In this paper, C-methyl phospholes were chromatographed on OV-17 at 100–125°.

Conversion of phospholes to derivatives with coordination number 4 completely changes their properties. As noted in Section 4.2.1.6, such derivatives frequently dimerize, even on formation. Those that remain monomeric are invariably more polar than the phosphole, and are non-distillable but chromatographable solids, recrystallizable from common solvents when thermally stable.

4.2.1.4.2 Thermal Stability

1*H*-Phosphole and some of its C-substituted derivatives are stable only at low temperatures (−70 to −90°C), rapidly isomerizing at higher temperatures and then undergoing dimerization (see Section 4.2.1.5.8). Replacement of H by methyl or phenyl gives the common phospholes that are stable for considerable periods at room temperature, and as noted above are distillable. Placement of a carboalkoxy group at the 3-position decreases the stability, and a distilled sample turned into a solid brown mass after a few hours at room temperature ⟨73JOC1858⟩. Prolonged thermal treatment of phospholes produces profound chemical change. Some (e.g., 1-benzyl-3,4-dimethylphosphole in refluxing xylene for 25 h) form insoluble, unidentifiable solids ⟨73JOC1858⟩; 1,3-dimethylphosphole, on the other hand, in part undergoes Diels–Alder dimerization on heating at 120°C for 17 h in xylene ⟨85PS(25)117⟩. Heating 3,4-dimethyl-1-phenylphosphole at temperatures around 150–170°C leads to rearrangement to a 2*H*-isomer, as in the case of 1*H*-phospholes, which then dimerizes (see Section 4.2.1.5.9). When the phosphole is heated at 170°C for a prolonged period, a tetramer (**54**) is formed along with other products ⟨82JA2077⟩. A different reaction course is followed by 1,2,5-triphenylphosphole; at 230°C for 10 days, a 1,1′-biphospholyl (**55**) is formed, probably via a 2*H*-phosphole intermediate ⟨82JOC2376⟩.

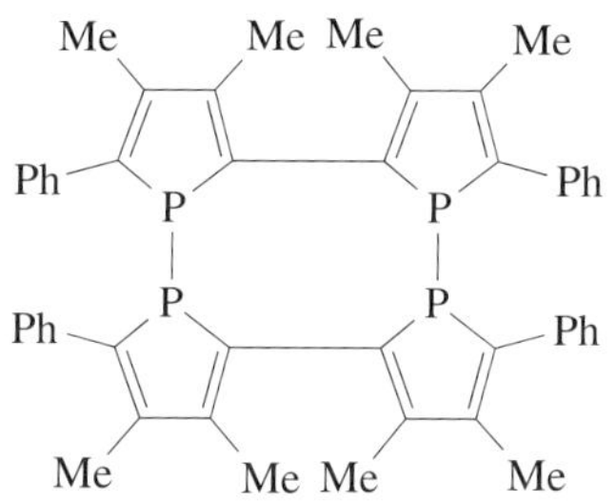

(**54**)

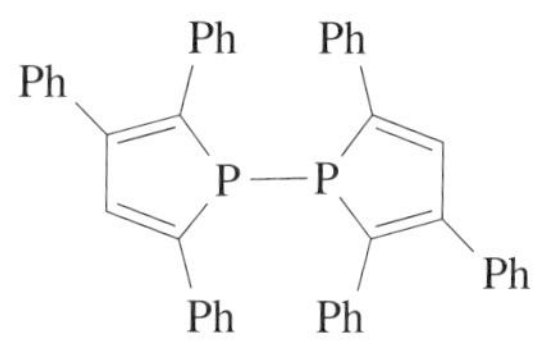

(**55**)

Evident in the above discussion is the great instability of the 2*H*-phosphole ring; this structure lacks any stabilization by cyclic delocalization, and as true of the C=P double bond in general (except those with special stabilizing substituents) is prone to dimerize upon formation. The 2*H*-phosphole ring system can be stabilized, however, in some specialized forms that are described in Section 4.2.1.7.1.

4.2.1.4.3 Basicity of Phospholes

A significant property of phospholes is their reduced basicity relative to other types of cyclic phosphines. This was first noticed in work on 1-methylphosphole, where it was found that some phospholene impurities could be selectively extracted from a pentane solution of a phosphole with 2 N HCl ⟨69JA3308⟩. With 6 N HCl, the phosphole was extracted from the pentane, but could not be recovered on neutralization; the products were unrecognizable. The pK_a of 1-methylphosphole was then determined to be approximately 0.5 by following the changes in the strength of the UV maximum at 284 nm in various concentrations of HCl. This pK_a is approximately 5 units lower than might be expected for a divinyl phosphine, implying that the ring structure has a profound effect on basicity at phosphorus. This property is in common with that of pyrrole and the first interpretation given to this property was that the phosphole system was similarly highly delocalized. As other evidence accumulated over the years that the system probably has rather weak delocalization (see Section 4.2.1.2), the low basicity seems to require another explanation. A possible one is that the small internal C–P–C bond angle (about 90°C) imposed by the ring structure is probably retained in the protonated form, which however is sp^3-hybridized, thus creating an energy barrier to the protonation ⟨78JCS(P1)98⟩. It has also been suggested that the low basicity arises from the possible degeneracy of the *n*- and highest π-orbitals ⟨B-80MI01⟩.

The site of protonation of phospholes in nonaqueous solvents by dry HCl or CF_3SO_3H has been found to be the phosphorus atom when the reaction is conducted at −90°C ⟨86JCS(P2)629⟩; in pyrroles it is a ring carbon that undergoes protonation. The 1*H*-phospholium ions are not stable with Cl^- as counterion, and consequences of the P-protonation by HCl are discussed in Section 4.2.1.5.3. When the counterion is nonnucleophilic, the 1*H*-phospholium ions are sufficiently stable to be directly observable ⟨71JHC877, 72CJC1008⟩.

4.2.1.4.4 Resonance Energies

Missing in phosphole chemistry are adequate thermochemical measurements that would provide experimental determinations of resonance energies. The only thermochemical measurements so far reported were made on pentaphenylphosphole and its oxide; heats of combustion were -4349.17 ± 3.4 and -4308.26 ± 5.2 kcal mol^{-1}, respectively. Heats of formation were calculated to be $+93.7 \pm 3.4$ and $+52.8 \pm 5.2$, respectively <62JCS932>. It was possible to determine the bond dissociation energy D (P=O) for the oxide, which was found to be 100.4 ± 9.5 kcal mol^{-1}. This value is significantly smaller than those for triphenylphosphine oxide (128.4 ± 5.5 kcal mol^{-1}) and for trimethylphosphine oxide (139.3 ± 3.0 kcal mol^{-1}). It was concluded that the difference of 39 ± 13 kcal mol^{-1} between D (P=O) for trimethylphosphine oxide and pentaphenylphosphole oxide gave an indirect measure of the conjugation energy of the system. At the time, this value seemed acceptable as an indicator of quite high resonance energy in the phosphole ring. Unfortunately, pentaphenylphosphole is a very poor model for studies of the properties of the parent phosphole ring, and as noted in Section 4.2.1.2, over the years the view has developed that the resonance energy is probably rather small, perhaps less than that in furan. This is now being substantiated by application of modern sophisticated computational techniques to phospholes. Some calculated values for 1*H*-phosphole with pyramidal phosphorus are: 12.9 kcal mol^{-1} ⟨75JCS(P2)974⟩, 12.23 kcal mol^{-1} ⟨88JA4204⟩, and 13.18 kcal mol^{-1} ⟨94JA9638⟩. The latter two values were derived from isodesmic bond separation methods, used for the first time in studies of phosphole.

4.2.1.4.5 Basicity of Phospholide Ions

Compared to noncyclic phosphide ions, phospholide ions show a definite decrease in basicity that is consistent with the high degree of electron delocalization evident from other properties (see Sections 4.2.1.2 and 4.2.1.3.2). The ions (**56**) and (**57**) in THF solution were found not to be protonated by ethanol, implying an acidity for the conjugate acid (the 1*H*-phosphole) greater than that of ethanol, pK_a 15.9 in water ⟨79CC401⟩.

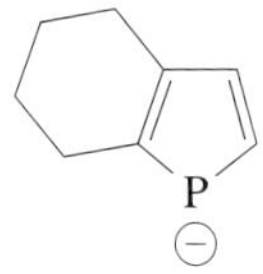

(**56**)

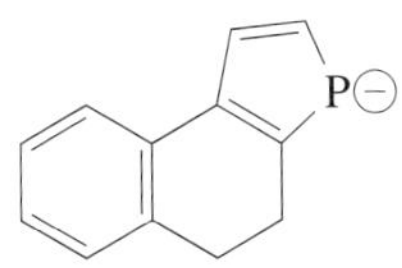

(**57**)

No other measure of the basicity of phospholide ions has yet been published. For comparison, the pK_a of diphenylphosphine is 21.7.

4.2.1.5 REACTIVITY OF C.N.-3 PHOSPHOLES

4.2.1.5.1 General

Studies of the reactivity of phospholes have revealed that the system behaves very differently from the related S, N, and O ring systems. In the first place, all attempts at performing electrophilic substitution of the ring protons, a commonplace process for the other systems, have so far resulted in failure; attack occurs instead at the nucleophilic phosphorus atom. This is also a site of reactivity when Diels–Alder reactions are attempted with certain alkynic dienophiles, although the expected cycloaddition to a diene moiety can be realized. In fact, it is quite common for the phosphole ring to be destroyed in its reactions rather than preserved in the characteristic way of the related heterocyclic systems. However, these structural modifications are of considerable value as a source of other phosphorus heterocyclic systems. Another differentiating property of phospholes, and of the related phosphole anion (phospholide), is that an extraordinarily rich chemistry based on metal coordination complexes of various types has developed. In short, the phosphole ring, while disappointing in not giving some of the expected reactions of a truly cyclically delocalized system, has nevertheless displayed its own unique reactivity that has not only led to fascinating results for the researcher but also to the creation of some quite novel heterocyclic phosphorus compounds.

A point of caution needs to be made in interpreting phosphole chemistry: there are distinct differences in reactivity that are imposed by the pattern of substituents, so much so that a study made with one phosphole as a model may give specific results that are not extendable to other phospholes. In the early days of phosphole chemistry, 1,2,5-triphenylphosphole was frequently the model used because it was readily available. It is, in fact, a rather poor model. Even the placement of methyl groups on the 3 and 4 positions can have significant effects on reactivity at phosphorus. An excellent example of the influence of structure on the outcome of a reaction is seen in Section 4.2.1.5.5.1 for the case of addition of diethyl acetylenedicarboxylate.

4.2.1.5.2 Reactions on Phosphorus with Conversion to C.N.-4

Some of the expected simple reactions of phosphines occur also with phospholes. Phosphorus can be oxidized with common oxidizing agents to form P-oxides, but with few exceptions these are unstable, usually spontaneously so, and give crystalline dimers (see Section 4.2.1.6.1). Intermediate stability was exhibited by 3,4-dimethyl-1-phenylphosphole (**58**) which when oxidized with *t*-butyl hydrogen peroxide at room temperature gave an oxide (**59**) detectable by ^{31}P NMR (δ +42.3). However, dimer (**60**) begins to form in a few hours and after several days is the only product that can be observed (Scheme 5) ⟨91HAC359⟩. Pentaphenylphosphole ⟨61JA4406⟩ and 1,2,3-triphenylphosphole ⟨65JCS2184⟩ give oxides that can be isolated and preserved. Sulfides are more stable; they are prepared by direct sulfuration of phospholes (see Section 4.2.1.6.2).

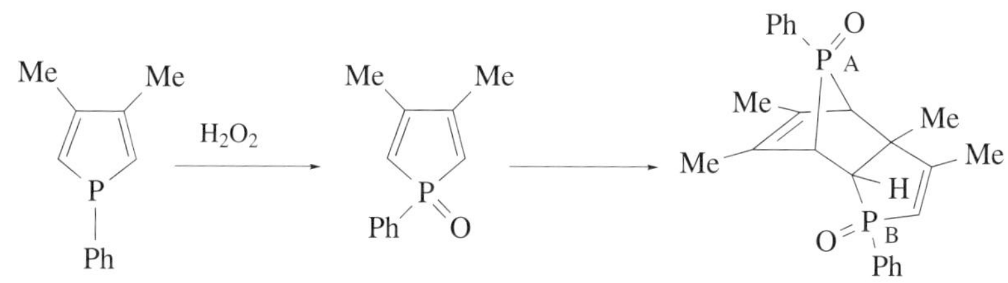

(**58**), δ ^{31}P −4.4 (**59**) δ ^{31}P +42.3 (**60**), δ ^{31}P +74.3 (A), +53.6 (B); $^{3}J_{PP}$ = 38.8 Hz

Scheme 5

Simple alkyl halides give crystalline quaternary salts, although these seem to form more slowly than those of other cyclic phosphines ⟨69JA3308⟩. They too will dimerize on formation or on heating ⟨73JOC1954⟩ but less readily so than oxides, so that some simple phospholium salts (see Section 4.2.1.6.3) are known (e.g. (**61**), ⟨73JOC1858⟩).

Methylene iodide gives an iodomethyl salt (**62**) with 1,2,5-triphenylphosphole which is of interest because it reacts with base to undergo rearrangement to the homophosphole oxide (**63**) (Equation 1) ⟨71CJC879⟩. The mechanism was not confirmed.

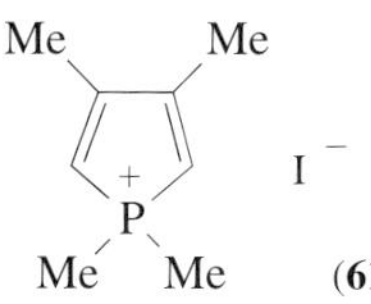

(**61**)

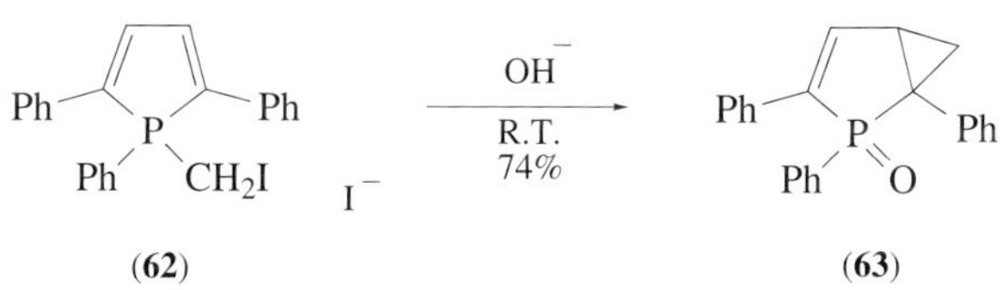

(**62**) (**63**) (1)

It has also been noted that, while stable at room temperature, a solution of triphenylphosphole in carbon tetrachloride undergoes a distinct color change (yellow to red) which is reversible on cooling ⟨82CJC138⟩. Benzotrichloride also reacts on heating but the color change is not reversible. The phosphorus species formed in these reactions are not fully understood, but may stem from an initially formed phospholium salt, for example (**64**) (Equation 2).

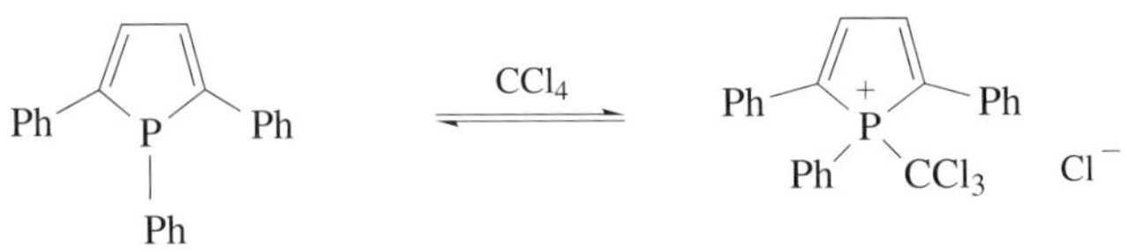

(2)

(**64**)

Azides have been found to attack the phosphorus atom of 1,2,5-triphenylphosphole to give imino derivatives in a version of the well-known Staudinger reaction of phosphines ⟨72CC1242⟩. Yields of the crystalline imines (see also Section 4.2.1.6.4) are in the range 40–95%.

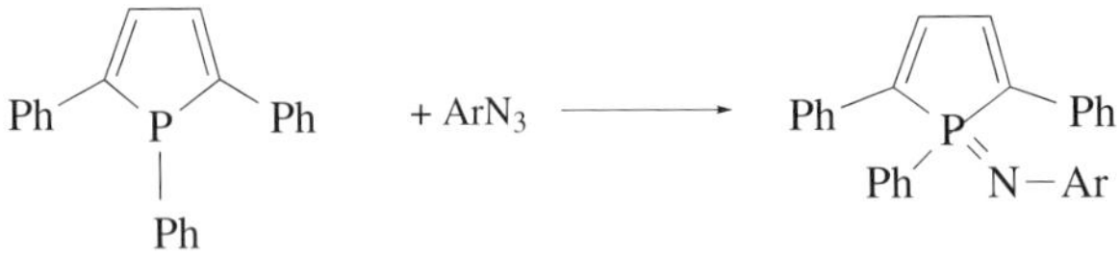

(3)

4.2.1.5.3 Protonation of Phospholes

1-Alkylphospholes have been shown to be much weaker bases than other tertiary phosphines; a pK_a of 0.5 was recorded for the 1-methyl derivative ⟨69JA3308⟩. The site of proton attachment in nonaqueous systems has been investigated ⟨86JCS(P2)629⟩ and it was conclusively shown that if the protonation is allowed to occur at low temperature (−70°C) and with an acid whose anionic part is nonnucleophilic (e.g. CF_3SO_2OH) the proton attaches to phosphorus to form a 1*H*-phospholium ion. These ions are quite stable up to about −30°C, but decomposition occurs at room temperature (Equation 4).

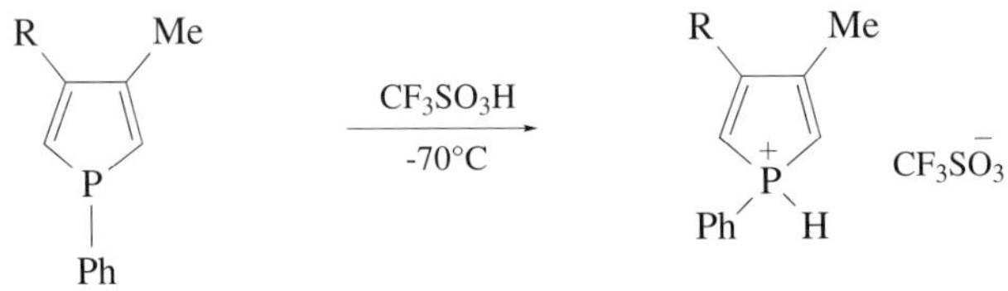

(4)

R = H, δ ^{31}P +18.9, $^{3}J_{PH}$ = 530 Hz
R = Me, δ ^{31}P +9.5, $^{3}J_{PH}$ = 547 Hz

The ions were found to react with phosphole in the solution to form a dimer-like structure (**65**) (Equation 5).

(5)

(**65**)

With an acid whose anionic part has some nucleophilic character, such as HCl, a secondary reaction occurs. The P-protonation in CD_2Cl_2 at −90°C can be detected by ^{31}P NMR but on warming to −70°C the P–H form disappears and a signal for the 1-chloro-2,5-dihydrophospholium ion (**68**) appears. This substance is the same as formed in the McCormack cycloaddition of dienes and $RPCl_2$ ⟨53USP2663737⟩, and is easily recognized (Scheme 6).

Apparently the nucleophilic chloride ion first added to the phospholium ion to form a five-coordinate species (**66**), which undergoes a [1,5] sigmatropic hydrogen shift to give a 2*H*-phosphole derivative (**67**), then adding HCl. Such [1,5] shifts have been encountered in other types of phosphole reactions and are

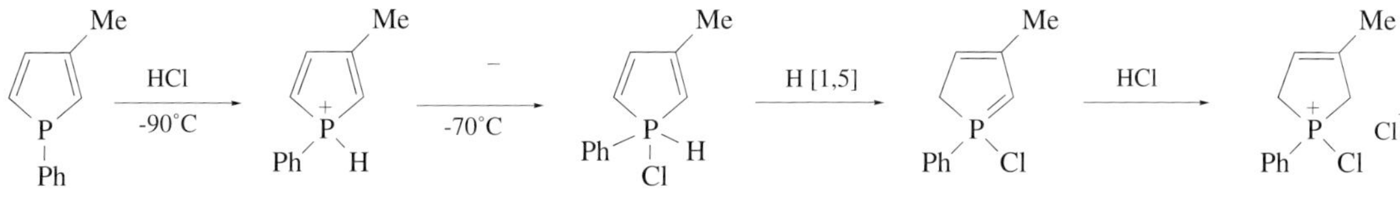

(**66**) (**67**) (**68**)

Scheme 6

discussed in Section 4.2.1.5.9. In an earlier study ⟨82PS(12)161⟩ protonation of phospholes with dry HCl had been performed at room temperature; this formed immediately the chlorodihydrophospholium ion, and was considered at that time to indicate that carbon was the site of protonation. On the other hand, 1,2,5-triphenylphosphole is not protonated when dry HCl is passed into a benzene solution, although a stabilized 1*H*-phospholium $TaCl_6^-$ salt can be obtained when HCl is passed into a solution of the $TaCl_5$ complex of the phosphole ⟨71JHC877⟩. Protonation of 1,2,5-triphenylphosphole by triflic acid also occurs on P ⟨83JOM(256)103⟩.

4.2.1.5.4 Cycloadditions with the Diene Moiety

The first successful Diels–Alder reaction of a phosphole was accomplished with the 3,4-dimethyl-1-phenyl derivative, which gave the crystalline cycloadduct (**69**) with *N*-phenylmaleimide (50%) (Equation 6) ⟨81TL319⟩.

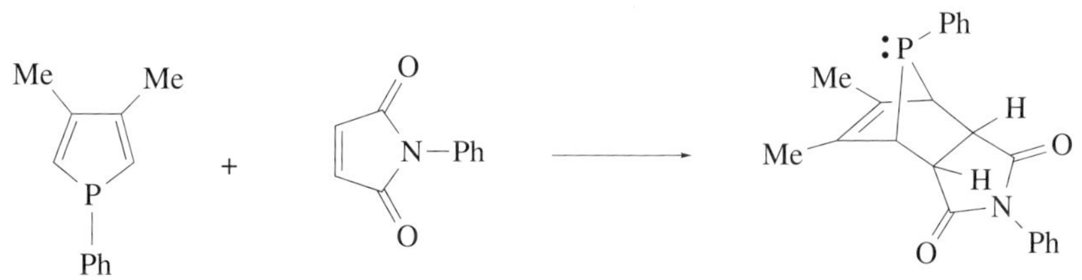

(6)

(**69**)

The structure at phosphorus is of interest in this adduct; the phenyl substituent is *anti* to the double bond, while in cycloadditions of the corresponding phosphole oxide with the same dienophile the structure (**70**) with *syn* phenyl is the exclusive product (Equation 7) ⟨91HAC359⟩. In reactions of both systems, only the *endo* fused product is formed; these stereochemical points have been discussed fully ⟨90RHA39⟩.

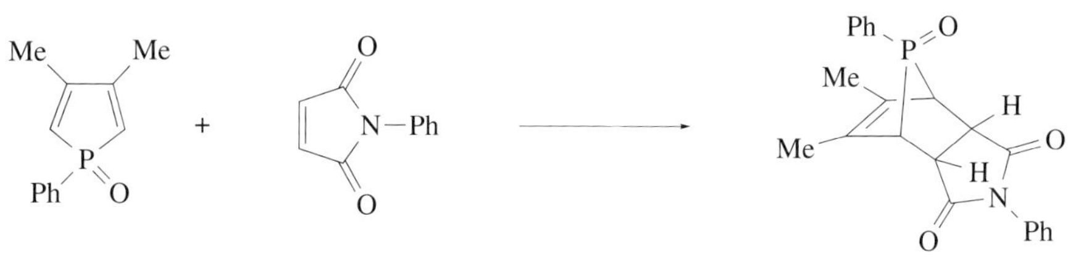

(7)

(**70**)

Another example of a Diels–Alder reaction giving the *anti* isomer is seen in the dimerization of phosphole (**71**) ⟨85PS(25)117⟩. This occurred on heating in xylene solution at 120ºC for 17 h. The dimer (**72**) was observed by ^{31}P NMR spectroscopy; it was not isolated but has been made indirectly by isomerization at P of the *syn* dimer (**73**), made from the oxide dimer (**74**) (Scheme 7).

As noted in Section 4.2.1.4, however, thermal treatment of phospholes can also lead to other products, and the simple dimerization process is probably of little value. The dimer framework is much more easily formed from phosphole oxides.

An important new development is the demonstration that Diels–Alder reactions with phospholes and fumaronitriles are greatly accelerated under pressure, and this has led to the synthesis of a number of new 7-phosphanorbornene derivatives. Thus, 3,4-dimethyl-1-phenylphosphole reacted with fumaronitrile

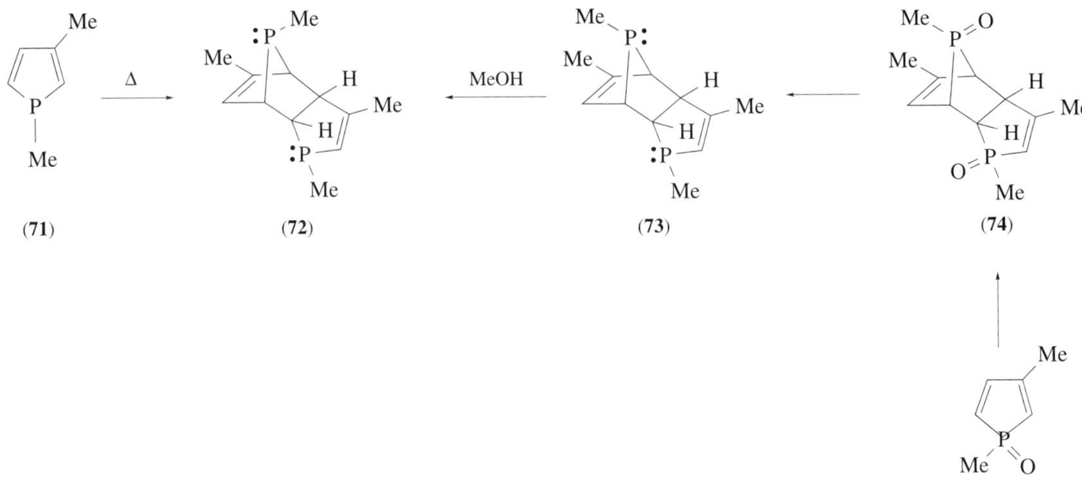

Scheme 7

at 30°C under 9 kbar for 24 h to give an adduct in 67% yield (Equation 8) ⟨89T7083⟩. The stereochemistry at phosphorus was confirmed by X-ray analysis.

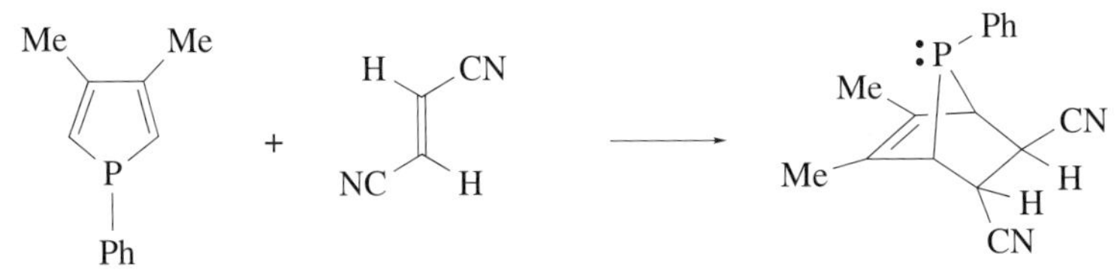

(8)

1-Phenylphosphole and 3-methyl-1-phenylphosphole also gave such adducts, but they were difficult to separate from some accompanying P-oxides. Other dienophiles (maleic anhydride, tetracyanoethylene, benzoquinone) were not useful in this process.

A fascinating aspect of phosphole chemistry is revealed in consideration of a different form of Diels–Alder cycloadditions. It is feasible to conduct reactions with a phosphole while it is coordinated to a metal center, and this has made possible the performance of several Diels–Alder reactions with dienophiles that similarly are held in a state of coordination or are bonded to phosphorus. In the absence of the metal, no reaction occurs between the phosphole and the dienophile. An excellent example is the intramolecular cycloaddition of a 3-butenyl group with the phosphole diene in a complex with $W(CO)_5$ (Equation 9) ⟨86OM1161⟩.

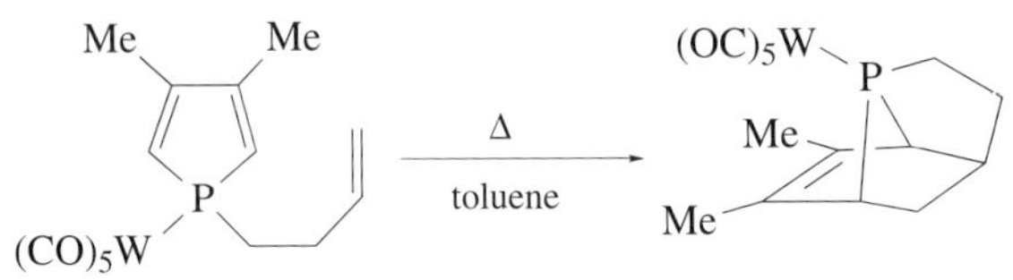

(9)

Another metal-promoted cycloaddition is found when, in the presence of $NiBr_2$, diphenylvinylphosphine acts as a dienophile towards 3,4-dimethyl-1-phenylphosphole (Equation 10) ⟨89IC3453⟩.

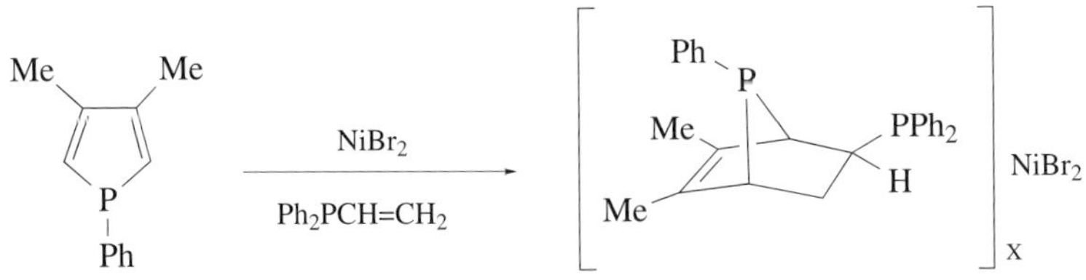

(10)

Since the cycloaddition occurs within the coordination sphere of a $NiBr_2$ complex of the two ligands, it was considered to be an intramolecular process. A similar reaction occurred with phenyldivinylphosphine to form triphosphine (**75**), isolated in free form by treating the complex with NaCN.

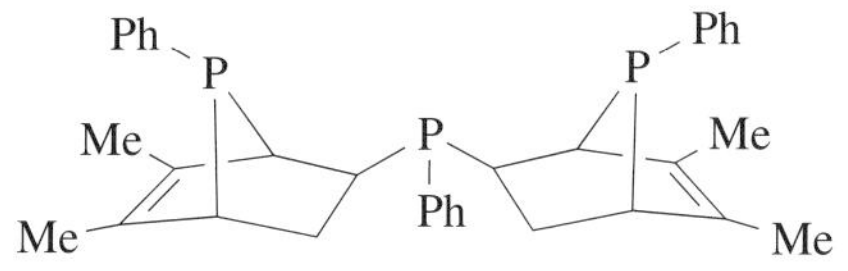

(75)

Similar intramolecular Diels–Alder reactions between phospholes and vinylphosphines were also effected with complexes of ruthenium ⟨89IC4132, 92OM1840⟩, palladium ⟨85JA6396⟩, molybdenum ⟨89IC4536⟩, platinum ⟨89IC217⟩, and iron ⟨92OM4069⟩, with varying levels of reactivity and diastereoselectivity to various vinylphosphine derivatives.

A vinylcarbene complex (**76**) also is capable of acting as a dienophile towards 3,4-dimethyl-1-phenylphosphole, again considered to be an intramolecular cycloaddition within the coordination sphere ⟨88OM2233⟩. A novel 7-phosphanorbornene derivative (**77**) was formed (Equation 11).

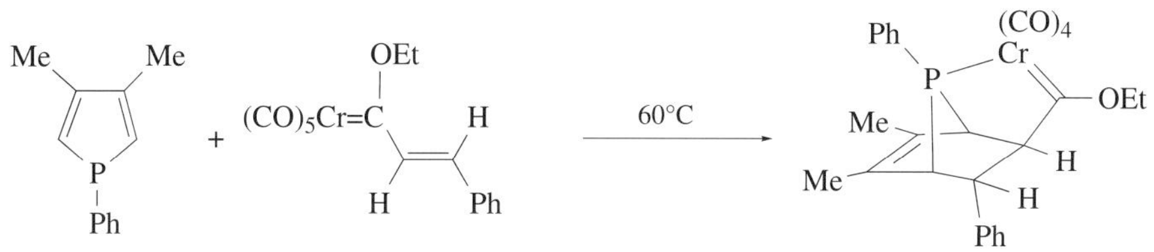

(76) (77) (11)

4.2.1.5.5 Reactions on Phosphorus Followed by Attack on Ring Carbon

4.2.1.5.5.1 Acetylene dicarboxylates

One of the earliest reactions performed on a phosphole is that with diethyl acetylenedicarboxylate, carried out with 1,2,5-triphenylphosphole ⟨68T3437⟩. This reaction is best accomplished by exposing the phosphole to excess carboxylate for two days at room temperature. It was obvious to the investigators that it was not a Diels–Alder cycloadduct; rather it resulted from a 1 : 2 reaction with the ester. Its true structure was later found to be that of a novel tricycle with phosphorus in the ylidic condition ((**78**) ⟨70JCS(C)386⟩). This compound is of some interest as on refluxing in chloroform it gave a rearrangement product believed to be the phosphonin derivative (**79**) (Scheme 8); this still constitutes the only report of a monocyclic phosphonin with P in the C.N.-3 condition.

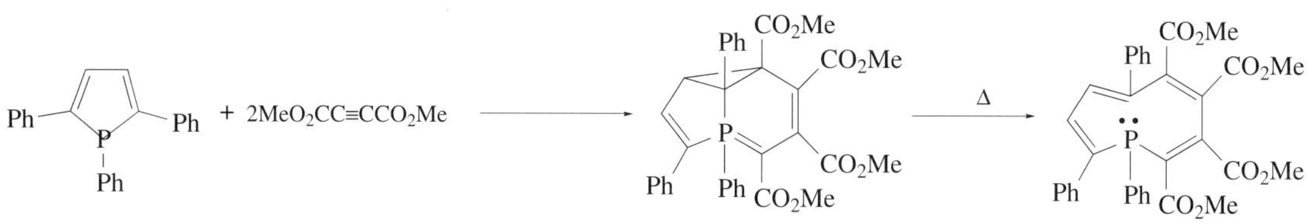

(78) (79)

Scheme 8

The importance of the substitution pattern on the phosphole in this reaction was obvious from continued studies ⟨78JHC1319⟩ that showed that a much higher order of reactivity and the formation of a quite different 1 : 2 adduct (**82**) could be experienced with phospholes of structure (**80**) and (**81**). Structure (**82**) also is novel in heterocyclic phosphorus chemistry and does not seem to have been approached by any other method (Equation 12).

1-Phenylphosphole also reacts rapidly with the ester, but gave a mixture of products in low yield. One product is probably of the same type as (**82**); the other is tentatively assigned structure (**83**). The reaction of a phosphindole has also been investigated ⟨76JHC65⟩. All of these early results with acetylenedicarboxylate have been summarized in a review ⟨81H(15)637⟩. In terms of mechanism, in every case it appears that the

(**80**), R= H
(**81**), R = Me

(**82**)

initial step is the action of the phosphorus atom as a typical nucleophilic phosphine in attacking on the triple bond (Equation 13). Subsequent reactions involve attack on the ring carbons and on a second alkynic group, but details of these processes have not been fully established.

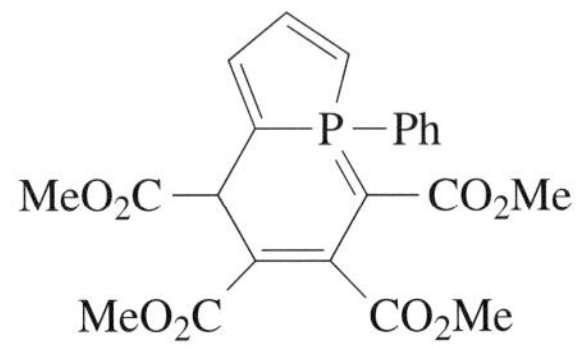

(**83**)

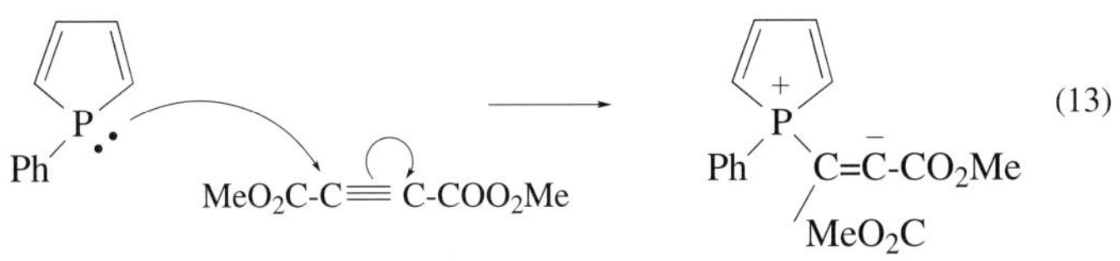

A report that includes structural proofs by X-ray crystallography has served to clarify some of the complex chemistry of these reactions ⟨93CJC1873⟩. In particular, structure (**78**) was confirmed, and a new structure (**84**) was determined as a reaction product from 1,2,5-triphenylphosphole. The phosphonin structure (**79**) was not confirmed, and structure (**85**) was found instead.

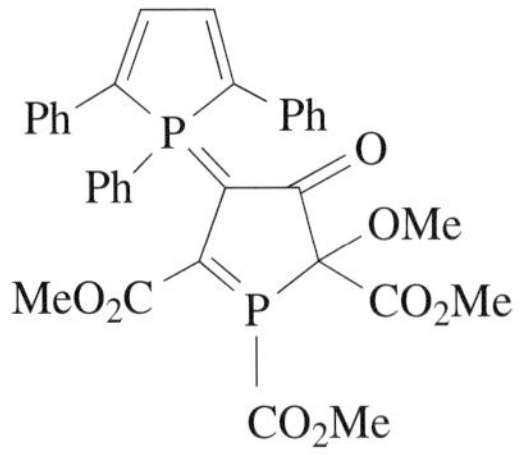

(**84**)

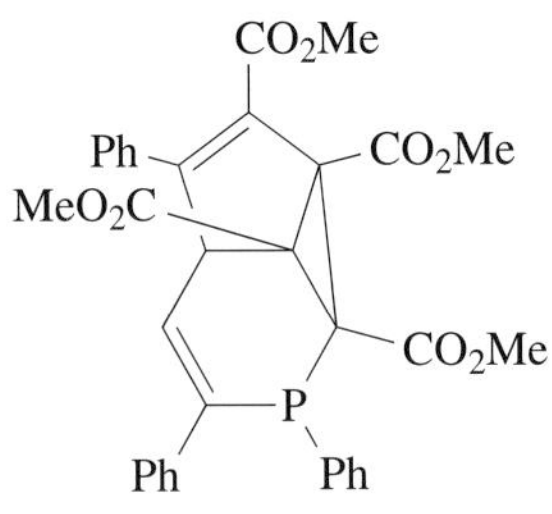

(**85**)

When 3,4-dimethylphenylphosphole is heated in xylene at 140°C with propiolic esters, quite different products are formed. These are Diels–Alder adducts of 2*H*-phospholes, which are formed from the phospholes by [1,5] phenyl migrations. A similar result has been obtained with 1,2,5-triphenylphosphole and alkynes at higher temperatures. These reactions are described in Section 4.2.1.7.1.

4.2.1.5.5.2 Aromatic acid chlorides

A remarkable and valuable reaction takes place when aromatic acid chlorides are reacted with phospholes and the mixture then treated with water. The products possess the 1,2-dihydrophosphinine ring structure. Apparently the nucleophilic phosphorus displaces chlorine from the acid chloride; water then adds to the phosphonium ion to form a pentacoordinate intermediate (**86**) which undergoes the ring expansion to the phosphinine derivative (**87**) ⟨73T707⟩ (Scheme 9).

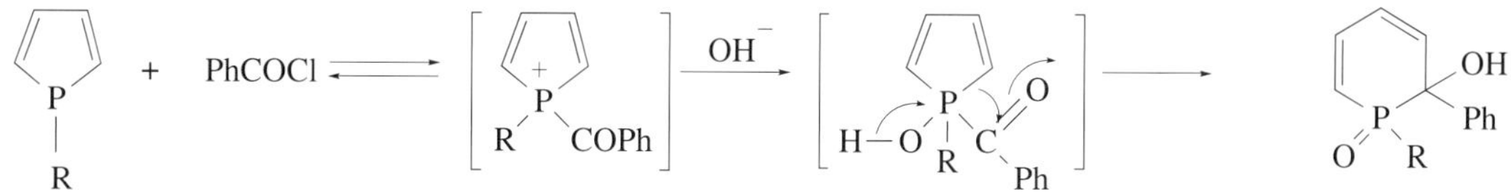

(**86**) (**87**)

Scheme 9

In the second paper on this reaction ⟨75CJC855⟩, some eight aryl acid chlorides were reacted with 3,4-dimethyl-1-phenylphosphole, giving yields generally in the range 60–80%. Thenoyl chloride also reacted to give low yields of the same type of product, but nicotinoyl chloride gave no product. It was later found that picolinoyl chloride could be used and provided access to the 2-(2-pyridyl)phosphinine derivative ⟨82TL1565⟩. Various other phospholes, including 1-phenyl and 1-benzyl ⟨84PS(19)45⟩, and 1-benzyl-3,4-dimethyl ⟨79TL1753⟩ could also be used in the process. Application of the ring expansion process to dibenzophospholes and phosphindoles is also possible (Equation 14), and has provided dihydrophosphaphenanthrenes and dihydroisophosphinolines ⟨80TL1441, 81NJC187⟩.

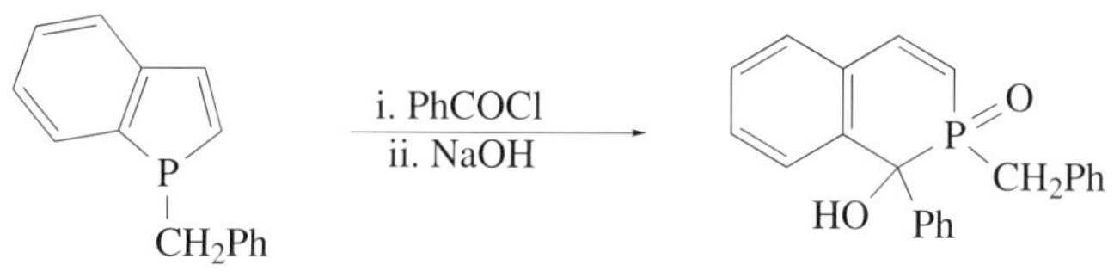

(14)

The dihydrophosphinines can be converted to the phosphinines, and this has provided a route to some valuable members of this family as well as the phosphaphenanthrene and isophosphinoline systems ⟨79TL1753, 84PS(19)45, 80TL1441, 81NJC187⟩.

4.2.1.5.5.3 Diphenylketene

In a study of the effect of pressure on phosphole reactions, the combination of 3,4-dimethyl-1-phenylphosphole with diphenylketene was investigated ⟨89RTC120, 89T7083⟩. The reaction was conducted at 25ºC at 9–10 kbar for 10 h and yielded a yellow crystalline product. The best yield was about 30%. The solid was proved by X-ray diffraction analysis to have the unusual spirodioxaphosphorane structure (**89**) (Scheme 10) and thus the Ph-P moiety has been completely removed from the carbon atoms of the phosphole ring. The product contains four moles of the ketene relative to the Ph-P unit. A plausible explanation for the course of the reaction is shown in Scheme 10. A critical step is seen to be the expulsion of the bridging phosphorus from (**88**), proposed by analogy to the 7-phosphanorbornenes with five-coordinate phosphorus, which undergo such fragmentations ⟨83TL5831⟩. This still remains as the only example of the complete destruction of the phosphole ring, but others may be encountered. The influence of pressure conditions on other phosphole reactions certainly deserves continued exploration.

4.2.1.5.6 Additive Reactions to a Double Bond

Relatively few reactions are known where only the equivalent of one double bond is involved in a reaction. Carbanion addition to the phosphole ring is of this type. *n*-Butyllithium adds to 1-*n*-butyl-3,4-dimethylphosphole to give 3-phospholene (**90**) (Equation 15) ⟨72OMR(4)171⟩.

An extension of this reaction type was the intramolecular addition of a carbanion to the ring, which provided a rare type of bicyclic compound (**91**) with phosphorus at a bridgehead position (Equation 16) ⟨87OM586⟩.

A photochemical dimerization is known for phospholes that can be described as a [2+2] cycloaddition to a double bond (Equation 17) ⟨69TL5037⟩.

A [2+2] head-to-head adduct also has been detected when 1-phenylphosphole is photolyzed in the presence of $Mo(CO)_6$. The reaction takes place in the coordination sphere of Mo, and the product appears in the form of a Mo complex (Equation 18) ⟨80JA5809⟩.

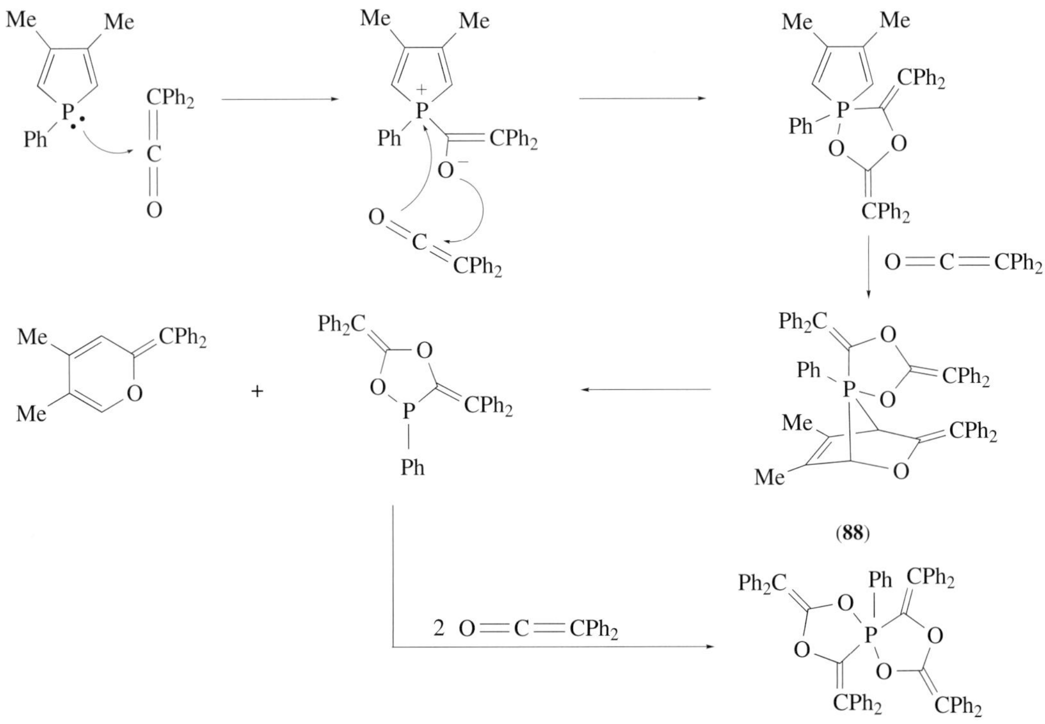

(88)

(89)

Scheme 10

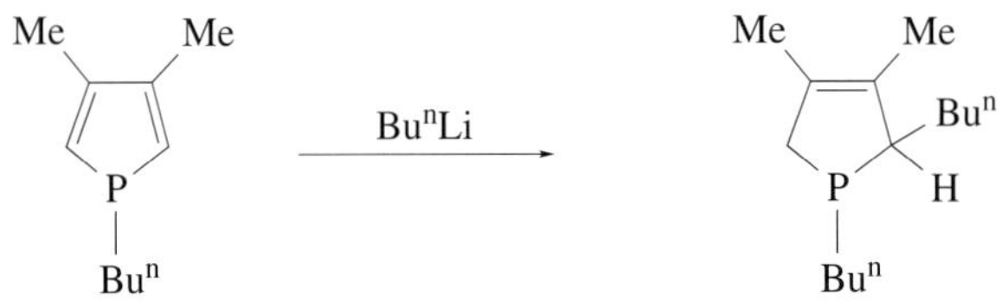

(90)

(15)

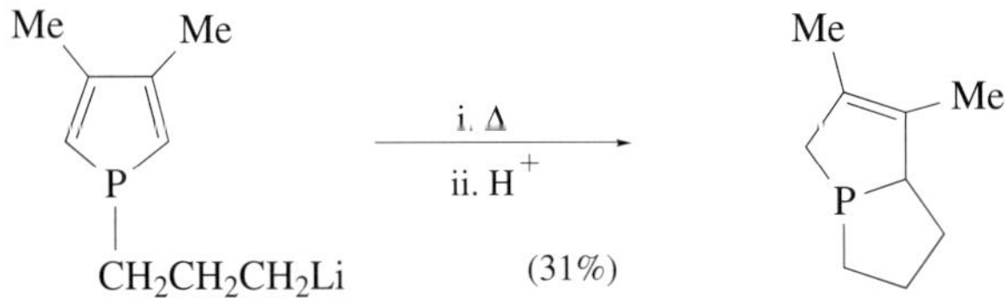

(91)

(16)

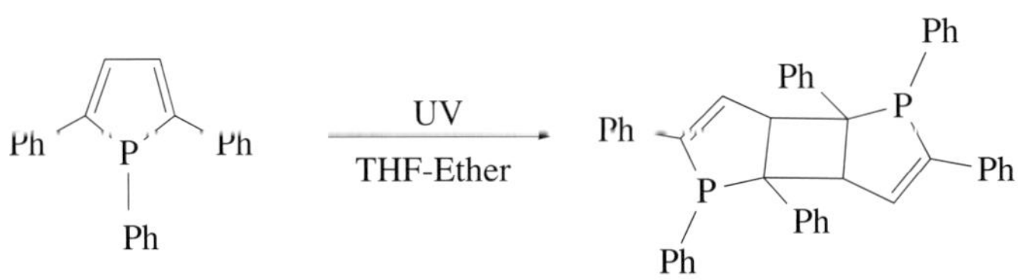

(17)

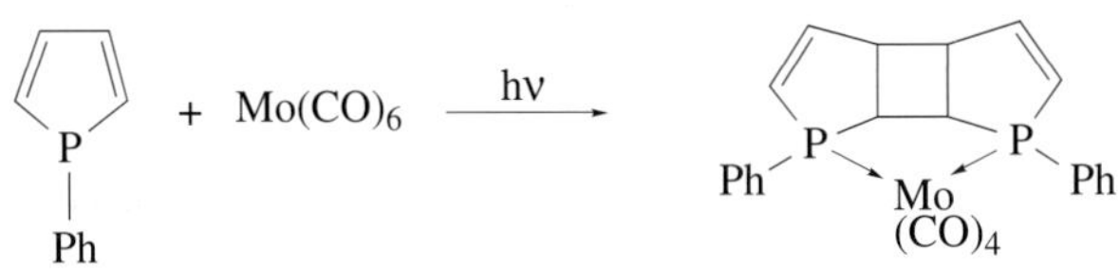

(18)

4.2.1.5.7 Substitution and Cleavage Reactions on Phosphorus; Phospholide Ions

1-Chlorotetraphenylphosphole was synthesized as early as 1961 from the reaction of 1,4-dilithiotetraphenylbutadiene with PCl_3 ⟨61JA4406⟩, but only in the late 1970s have simpler 1-halophospholes been synthesized (from cleavage of the P–P bond in biphospholyls; see below). These halides undergo the expected replacement reactions with simple nucleophiles (see Section 4.2.1.11.8). Of special interest are the formation of a biphosphole with a phospholide ion, and a σ-bonded metal derivative from, e.g. the anion of $M(CO)_5^-$ where M = Mn or Re ⟨79JCS(D)814⟩ (Scheme 11).

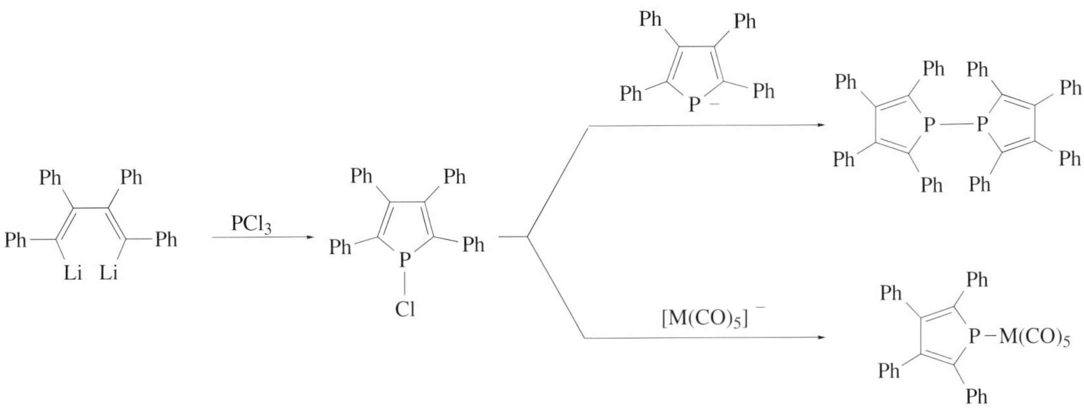

Scheme 11

More surprising is the replacement of a carbon substituent on P, as has been observed in the reaction of 3,4-dimethyl-1-phenylphosphole with *t*-butyllithium (Equation 19).

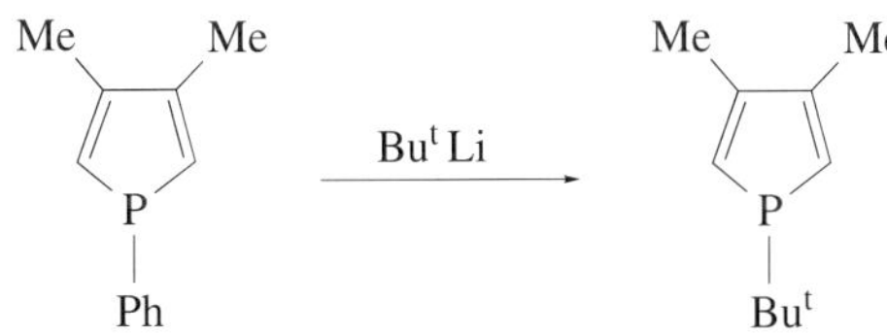

(19)

This is an excellent method for the formation of *t*-butyl phospholes ⟨72OMR(4)171, 72T4171⟩. Carbon substituents (phenyl, alkyl, benzyl) on phosphorus can be cleaved by direct reaction with metals (potassium, lithium, sodium) and this constitutes an outstanding method for the generation of phospholide ions. The reactions are easily conducted in inert solvents such as THF or DME and proceed with nearly quantitative formation of the phospholide ion. This reaction was first observed by Braye with pentaphenyl- and 1,2,5-triphenylphospholes, to give (**92**) and (**93**), but has since been extended to include the formation of a great variety of phospholides, such as (**94**)–(**97**); dibenzophosphole also readily forms a phospholide ion (**98**): (**92**) and (**93**) ⟨71T5523⟩; (**94**) and (**95**) ⟨82TL511⟩; (**96**) and (**97**) ⟨79CC401⟩; (**98**) ⟨74PS(4)199⟩.

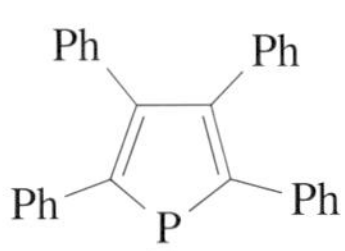

(**92**)

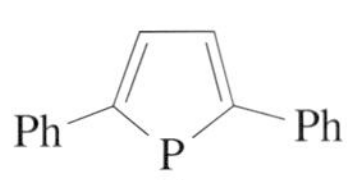

(**93**)

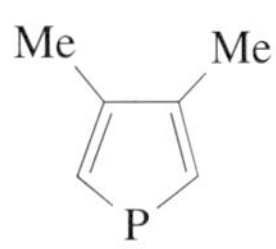

(**94**)

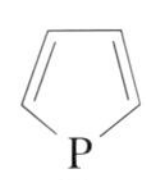

(**95**)

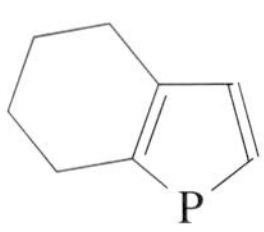

(**96**)

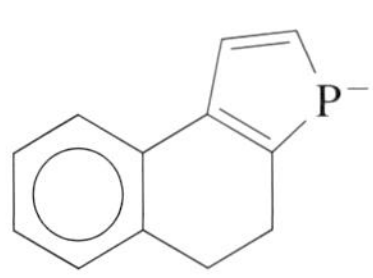

(**97**)

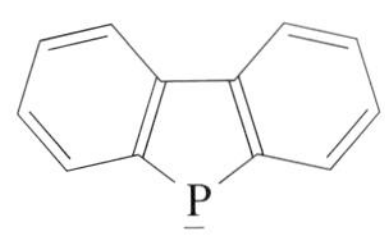

(**98**)

To avoid the presence in the reaction medium of the cleaved carbon fragment, the bis(phospholyl)-ethane (**99**) is useful, since no such fragment remains ⟨87OM586⟩ (Equation 20).

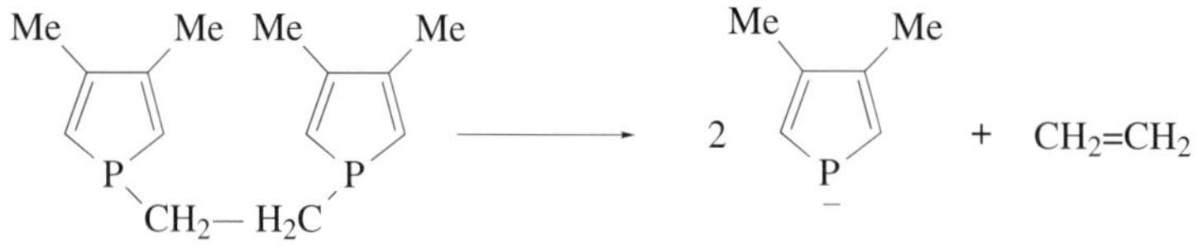

(99)

(20)

Phospholides are extremely important in synthesis; they act as highly nucleophilic species that can be alkylated with simple and functional alkyl halides, and other electrophiles, to produce new phosphole derivatives. An excellent example is provided by the early synthesis of a *P*-vinylphosphole (**100**) (Scheme 12) ⟨71T5523⟩.

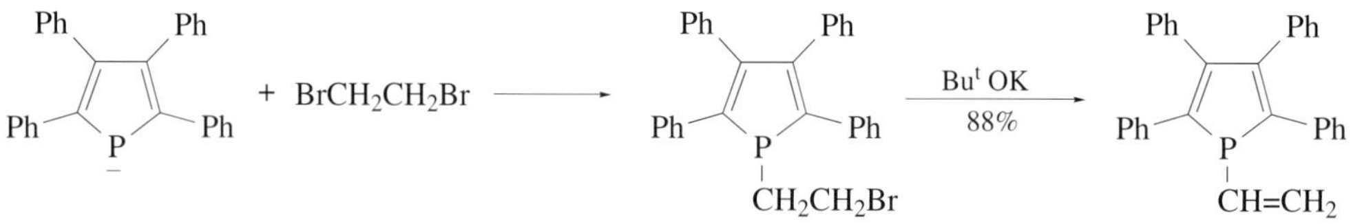

(**100**)

Scheme 12

Phospholide ions are also of value in the formation of 1,1′-biphospholyls; the coupling can be effected with phosgene or iodine ⟨82JOM(231)361⟩. The latter reaction, in the case of the 3,4-dimethyl derivative, is best accomplished in a medium free of the phenyl anion that accompanies the P–C cleavage. This anion was removed by conversion to triphenylaluminum with $AlCl_3$ ⟨83OM1234⟩. Some biphospholyls synthesized in this way are shown as (**101**) ⟨82JOM(231)361⟩, (**102**) and (**103**) ⟨79JCS(D)814, 82JOC2376⟩, and (**104**) ⟨83OM1234⟩.

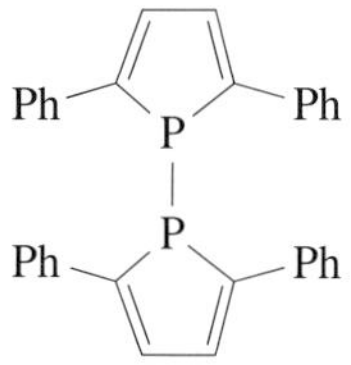

(**101**)

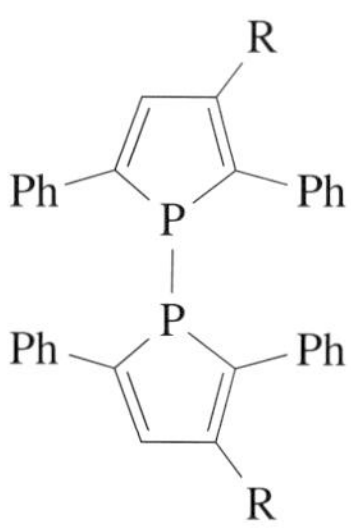

(**102**), R = Ph
(**103**), R = H

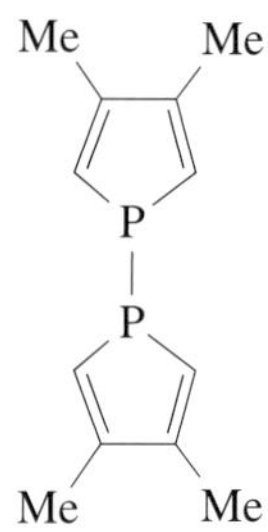

(**104**)

Another extremely important property of phospholide ions is their behavior as ligands in coordination chemistry. This subject has been reviewed in detail ⟨B-83MI01, 87NJC585⟩, and for the most part is beyond the scope of this chapter. However, complexes of the metallocene type deserve comment here, since in them the phospholide ion displays some chemistry of interest from the heterocyclic chemist's view. The first phosphametallocenes (phosphacymantrenes, (**105**)) were formed when 1-phenylphospholes were heated with $[Mn_2(CO)_{10}]$ ⟨75JOM(93)377, 78JA5748⟩. In this reaction, the P-phenyl bond is cleaved to form the phospholide ion, which proceeds to complex with manganese. Phosphaferrocene (**106**) is formed similarly (Scheme 13) ⟨77JA3537⟩.

Phospholide ions can also be made separately by the cleavage reaction with metals, and then complexed with the proper metallic compounds. Diphosphaferrocene (**107**) was first made this way, by reacting the phospholide ion with $FeCl_2$ (Equation 21) ⟨78JOM(156)C33⟩.

These novel complexes are of interest because they possess some of the chemistry of a highly delocalized system, specifically in their giving C-substitution products with acyl halides in the presence of $AlCl_3$ (e.g. Mn ⟨76TL4155, 78JA5748⟩; Fe ⟨77JOM(139)77, 80NJC683, 87ICA(130)93⟩). The high nucleophilicity at phosphorus is reduced in these complexes, and it no longer is the site of attack by electrophilic reagents. In addition to the Friedel–Crafts type of reaction, the Vilsmeier formylation has been observed. These reactions are of synthetic value for obtaining C-functional phospholes and are presented from that standpoint in Section 4.2.1.11.

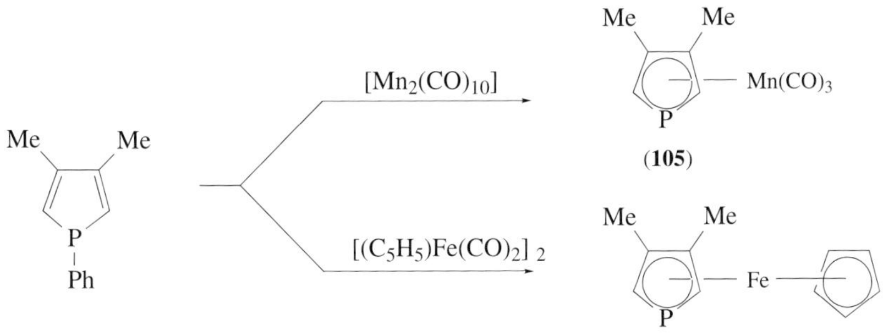

(105)

(106)

Scheme 13

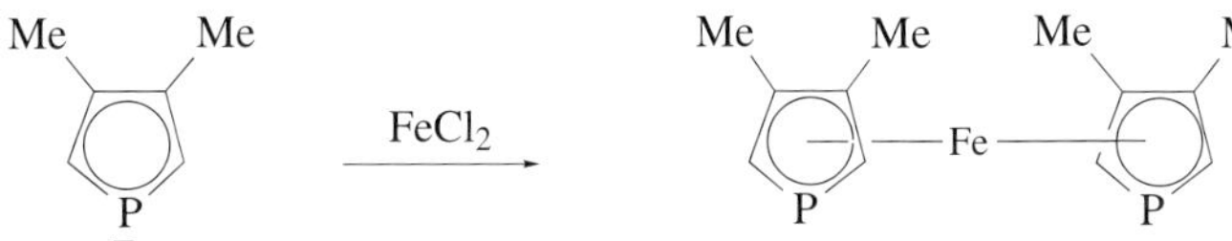

(107) (21)

4.2.1.5.8 1*H*-Phospholes from Protonation of Phospholide Ions

The phospholide ions are also used in the synthesis of 1*H*-phospholes. However, in a paper that is of great importance in phosphole chemistry, 1*H*-phospholes were found to be quite unstable, undergoing a [1,5] sigmatropic hydrogen shift to form 2*H*-phospholes (e.g. (**108**)) which then dimerize ⟨82TL511⟩ by a [4+2] cycloaddition (e.g. to (**109**)). This reaction can be so rapid that the parent 1*H*-phosphole and some substituted derivatives can only be observed at low temperatures, and dimerization is complete at room temperature (Scheme 14) ⟨83PS(18)51⟩.

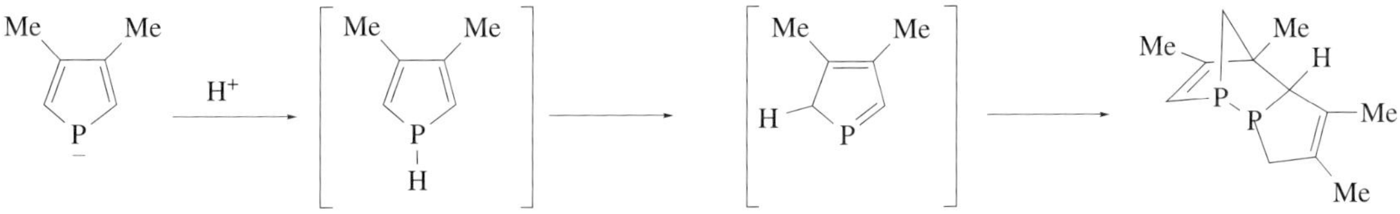

(108) (109)

Scheme 14

The W(CO)$_5$ complex of the 1*H*-phosphole is more stable, lasting for 4 h at room temperature ⟨86POL1413⟩. Even the polyphenylphosphole anions give 1*H*-derivatives that undergo ready rearrangement (in contrast to the earlier reports of Braye and of Freedman); thus, 1,2,5-triphenylphosphole forms the phospholide ion in 100% yield in THF at room temperature with potassium, and when the solution is quenched with acetic acid, the product observed is the dimer (**110**) (Scheme 15).

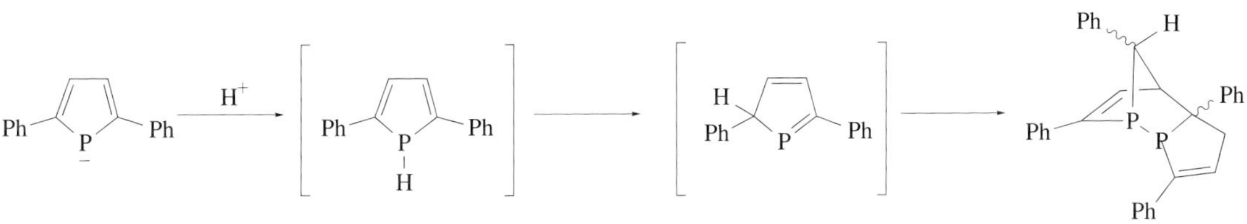

(110)

Scheme 15

Pentaphenylphosphole also gives the anion quantitatively; on acetic acid quenching at room temperature, the initially formed 1*H*-phosphole (**111**) can be observed spectroscopically (δ ^{31}P −40.9, J_{PH} = 218 Hz) but this also undergoes the rearrangement. In this case, the 2*H*-phosphole forms more slowly, which allows a reaction between the unreacted 1*H*-form and the 2*H*-form to take place. The product is again a dimer, but of different structure ((**112**), ⟨83JA6871⟩) (Scheme 16).

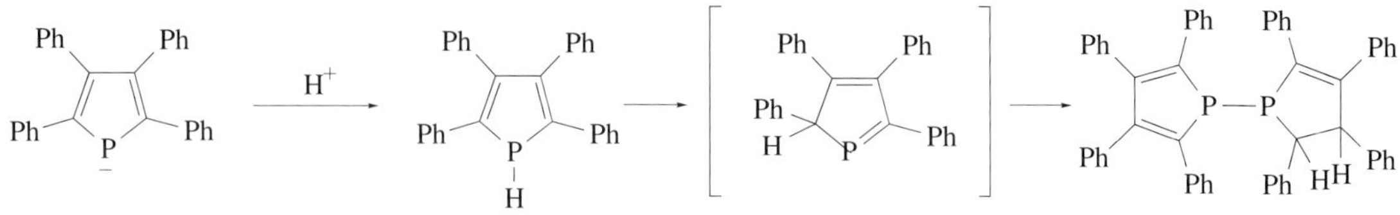

(111) (112)

Scheme 16

The Diels–Alder [4+2] structure of the 2*H*-dimers was first elucidated from spectroscopic data. The P–P bond was evident from the ^{31}P NMR data, which gave a doublet of doublets with a large P–P coupling constant (e.g. for (**110**), δ −25.86 and +17.11, J = 200 Hz). Other NMR data were interpreted on the basis of the dimeric structure. Conclusive proof of the [4+2] structure, with the endo fusion of the rings, came from the X-ray diffraction analysis of the product (**109**) from protonation of 3,4-dimethylphospholide ion ⟨82CC1272⟩. Another significant event was the stabilization of the 2*H*-phosphole (**108**) formed from this ion as a coordination complex; the anion was first coordinated to either tungsten or chromium carbonyl and then protonated, whereupon the [1,5] H-shift occurred to the 2*H*-phosphole (e.g. (**113**)), which was unable to undergo the dimerization (Scheme 17) ⟨84JA826⟩.

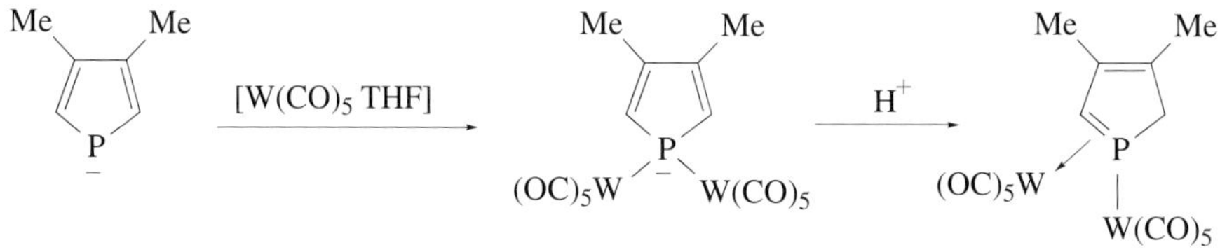

(113)

Scheme 17

The discovery that 1*H*-phospholes are unstable relative to 2*H*-phospholes is of theoretical significance, and represents a unique difference from pyrrole chemistry. The subject has received comment in Section 4.2.1.7.1.

4.2.1.5.9 [1,5] Sigmatropic Rearrangement of P-Substituents

The important observation was made in 1981 that a P-phenyl substituent on a phosphole could undergo migration to carbon, creating thereby the highly reactive 2*H*-phosphole system ⟨81JA4595⟩. This reaction, a [1,5] sigmatropic rearrangement, occurred in 100% yield when conducted for 60 h at 170ºC with

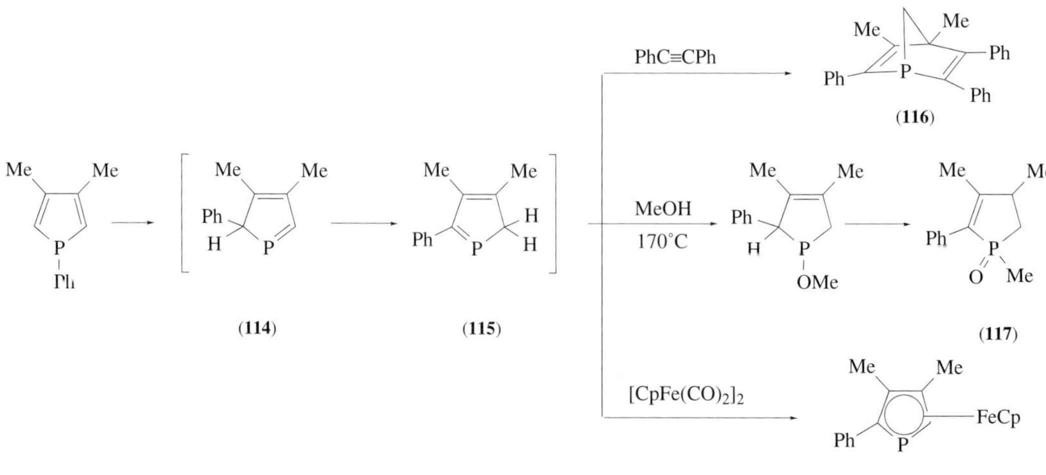

(116) (114) (115) (117) (118)

Scheme 18

3,4-dimethyl-1-phenylphosphole. The initially-formed 2*H*-phosphole (**114**) rearranged to (**115**), which was detected by various trapping reactions (Scheme 18). Additions occur to the diene unit (with alkynes, forming (**116**); see also ⟨89NJC369⟩) or at the P=C bond (to (**117**)). Trapping as a coordination complex (**118**) was also possible.

The intermediate 2*H*-phosphole (**115**) could also be trapped with alkenes to yield isolatable 1-phosphanorbornene derivatives (e.g. (**119**) major and (**120**) minor; Equation (22). ⟨89JOC4754⟩)

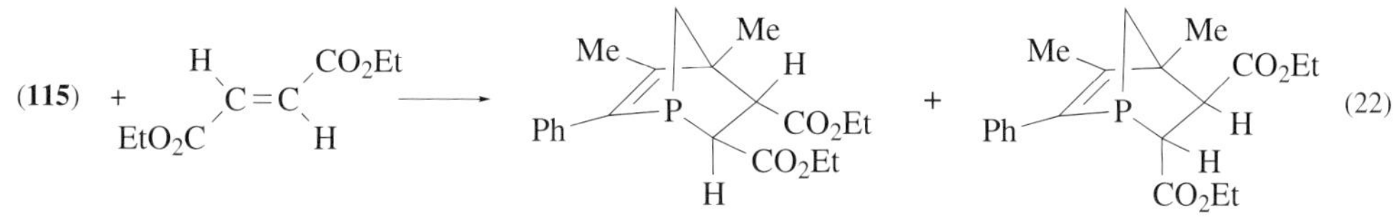

(**119**) (**120**)

1,2,5-Triphenylphosphole also gave a 2*H*-phosphole rearrangement product (**121**), which was trapped with tolane in 60% yield. These alkyne reaction products are of interest in possessing the rare bicyclo[2.2.1] framework with phosphorus at a bridgehead position. Also of interest was the observation that the adduct (**122**) was thermally unstable; the bridging Ph_2C unit was eliminated at 230ºC to form a phosphinine, and this constitutes a new and valuable entry to this ring system (Scheme 19). Other phosphinines have also been prepared by strong heating of the triphenylphosphole in the presence of an alkyne ⟨82JOC2376⟩.

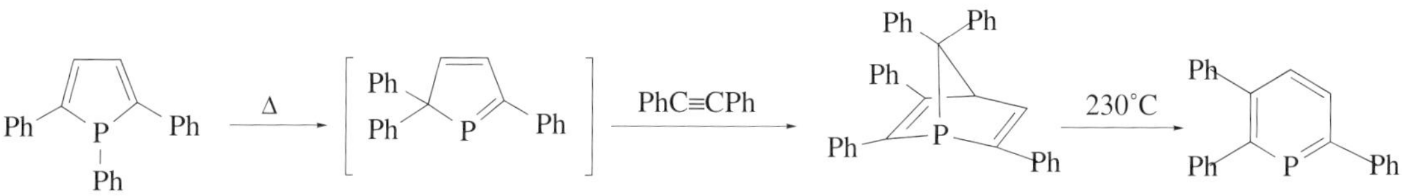

(**121**) (**122**)

Scheme 19

When the pyrolysis of 1,2,5-triphenylphosphole was conducted at a higher temperature (230ºC) in the absence of a trapping agent, the product was found to have the rare biphospholyl structure (**125**) ⟨82JOC2376⟩. This was proposed to result from a second [1,5] phenyl shift in (**121**) to form 3*H*-phosphole (**123**) followed by hydrogen shifts to a 1*H*-phosphole structure (**124**) which then lost hydrogen (Scheme 20).

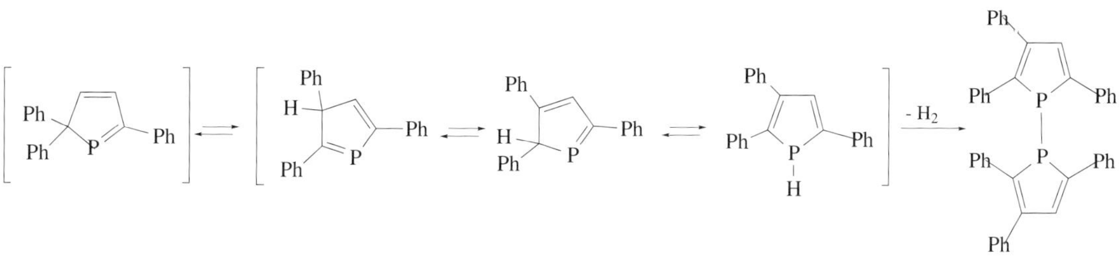

(**121**) (**123**) (**124**) (**125**)

Scheme 20

2,2′-Diphospholes are themselves subject to dimerization, thereby forming cyclic tetramers containing four phosphole rings ⟨82JA2077, 90AG(E)655⟩. The reaction with 3,4-dimethyl-1-phenylphosphole can be conducted at 170–180ºC; *N,N*-dimethyl-4-bromoaniline greatly enhances the reaction rate, and leads to yields of the tetramer of 40–45% after 18 h (Equation 23) ⟨94JA3306⟩.

The P–P bond in the tetramers can be cleaved by sodium in THF, giving a diphospholide which can be used for elaboration to macrocyclic systems on reaction with dihalides. Cleavage of only one P–P bond is also possible. The macrocycles formed from the two types of diphospholide ions are of considerable interest as novel ligands in coordination chemistry.

A bisphosphole (**126**) has also been subjected to the rearrangement; the connecting *p*-phenylene group migrated in both rings to give the novel bis-2*H*-phosphole (**127**) trapped with tolane (giving (**128**) ⟨90HAC187⟩) as well as with metal complexes (Scheme 21).

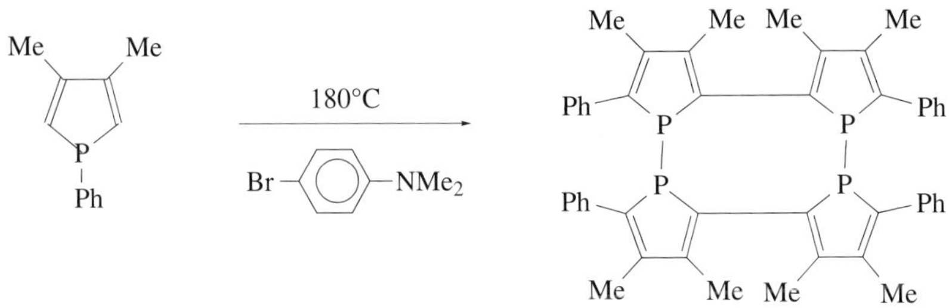

(23)

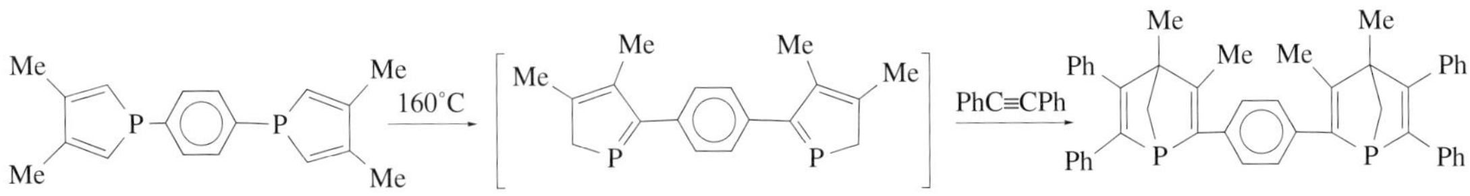

(126) (127) (128)

Scheme 21

The mechanism of [1,5] sigmatropic rearrangements of P-substituents has been treated by Mathey ⟨83JA6871⟩. The reaction is also known for P in the four- and five-coordinate states, and in fact occurs more readily in the latter state. An important feature is hyperconjugative overlap of the σ-orbital of an exocyclic P-substituent with the π* orbital of the diene unit, which is geometrically most favorable in the trigonal bipyramidal structure of C.N.-5.

4.2.1.5.10 Thermal Dimerization of Phospholes

Novel 2,2′-bisphospholenes can be obtained when certain phospholes are heated at 140–170°C in the presence of nickel(II) chloride ⟨84JA425, 85IC4141⟩. The products are obtained in the form of a nickel complex, but the free bisphospholene can be liberated with sodium cyanide. This product has the exclusive stereochemistry shown in structure (**129**) (Equation 24).

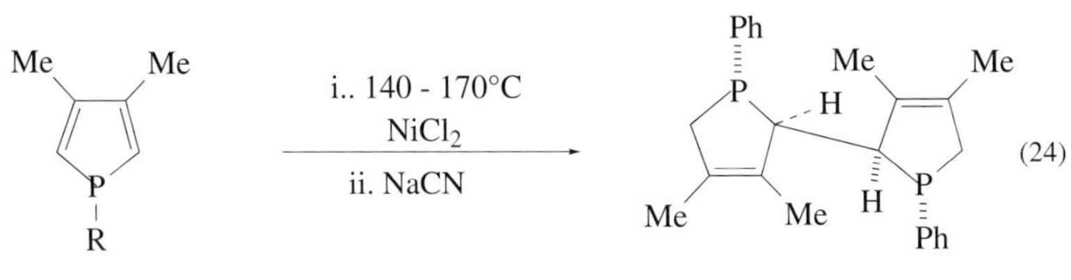

R = Ph or Bu^t (129)

(24)

In the presence of the nickel chloride, the thermal [1,5] migration of the P-substituent described in Section 4.2.1.5.9 is completely avoided. The bisphospholenes are of considerable value as ligands in coordination chemistry.

4.2.1.5.11 Coordination with Metals

There is a vast and increasing literature on the formation, characterization, and practical applications of metallic complexes of phospholes. An excellent summary of the various types of complexes that can be prepared is provided in a review ⟨B-83MI01⟩. Phospholes have unusually great variety in the way bonding to metals can occur, and are known to form many different types of complexes through the use of electrons on phosphorus and/or those in the diene unit. In general phospholes are better ligands than noncyclic tertiary phosphines such as the commonly used triphenylphosphine, but the substitution pattern on the phosphole ring can play a very important role in the reactivity to metals. It was noted in quite early studies of complexation that P-methyl- and P-benzylphospholes showed no tendency to form complexes with $NiCl_2$, but the 3,4-dimethyl derivatives of these phospholes readily formed stable complexes ⟨73PS(2)205⟩. A similar effect was noted in a series of P-phenyl phospholes ⟨B-83MI01⟩, which appear to be more reactive than P-methyl counterparts. In general, complexing ability is maximized in the compound 3,4-dimethyl-1-phenylphosphole, and this is the phosphole most frequently used in complexes prepared as possible catalysts. Dibenzophosphole is also frequently used as a ligand, but its mode of bonding is restricted to the two-electron type typical of a tertiary phosphine. Complexes containing 1-phenylphosphole, 1,2,5-triphenylphosphole, pentaphenylphosphole, and some 1-alkyl-3,4-dimethylphospholes have also been included in studies on potential catalyst systems. In this section, only special reactions that occur with the phosphole moiety while in coordinated form are considered. Examples have, of course, already been encountered, as in the intramolecular form of Diels–Alder cycloadditions described in Section 4.2.1.5.4. Other important processes include:

(i) Diels–Alder reaction of σ-complexes of phospholes with $Cr(CO)_5$, $Mo(CO)_5$, or $W(CO)_5$ (Equation 25) ⟨82CC667⟩.

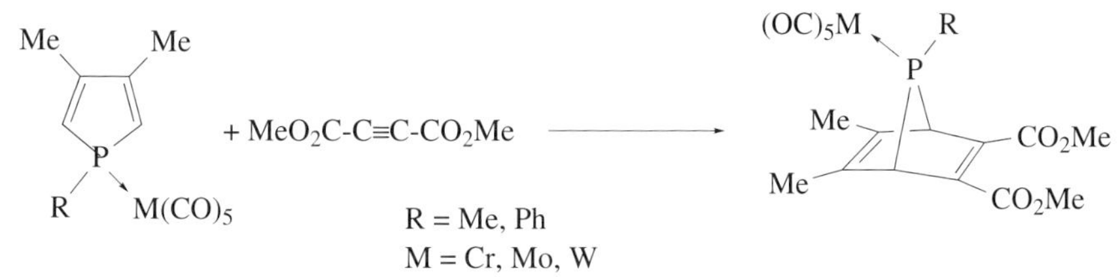

(25)

This reaction permitted the successful construction of a stable 7-phosphanorbornadiene derivative. While it was not possible to release the free phosphine from the complex, the complexes acted, on thermolysis, as precursors of the synthetically valuable phosphinidene complex, $[RPM(CO)_5]$.

(ii) [2+2] and [4+2] dimerization of σ-complexes. UV irradiation of some phosphole • $M(CO)_5$ complexes (but not of the free phosphole) causes the formation of a Diels–Alder dimer, which acts as a bidentate ligand to the metal center. The dimer has the unique *exo* fusion of the two rings, and is the only dimeric product known to date to have this configuration proved by X-ray analysis ⟨80JA5809⟩. The complex where R=Ph was broken on heating with sulfur to form the novel *exo*-fused dimeric disulfide (**130**) (Scheme 22).

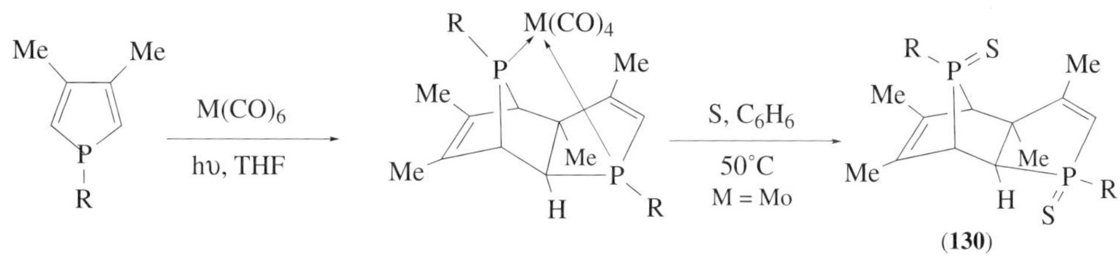

(**130**)

Scheme 22

(iii) Rearrangement of P-β-chloroethylphospholes to dihydrophosphepins. β-Chloroethyl-3,4-dimethylphosphole forms a complex with $W(CO)_5$ which on treatment with $AlCl_3$ at room temperature rearranges with the insertion of the ethylene unit into the ring. The product is a 1-chlorodihydrophosphepin (**131**) (Scheme 23a) ⟨85TL3461⟩. The chlorine is reactive, and can be displaced on reaction with methanol at room temperature with retention of the metal group.

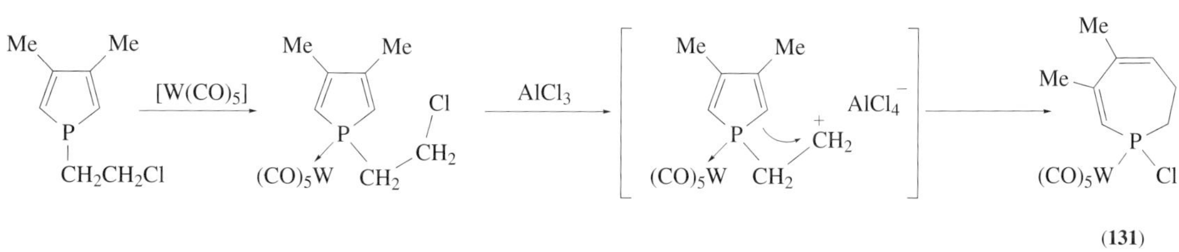

(**131**)

Scheme 23a

(iv) Isomerization to complexes of 2*H*-phospholes. The anionic complex (**132**), formed by addition of 3,4-dimethylphospholide ion to $Cr(CO)_5$, can be alkylated on P to form the stable 1,3,4-trimethylphosphole complex (**133**), but on reaction with water the protonated product undergoes a [1,5] sigmatropic H-shift to form the complex (**134**) of the 2*H*-phosphole ⟨84JA826⟩. The same phospholide ion reacts with two moles of $W(CO)_5$ to give the anionic complex (**135**), which also gives a stable 2*H*-phosphole complex (**136**) on protonation (Scheme 23b).

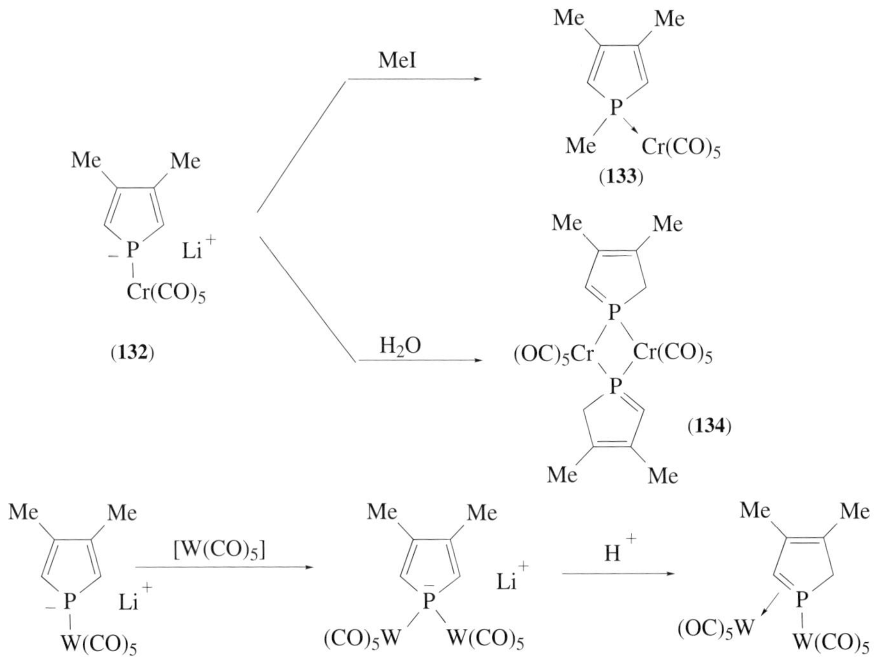

(132) (133) (134) (135) (136)

Scheme 23b

4.2.1.6 REACTIVITY AND SYNTHESIS OF PHOSPHOLES WITH C.N.-4

4.2.1.6.1 Phosphole Oxides

4.2.1.6.1.1 Synthesis

The great majority of phosphole oxides are unstable and rapidly undergo Diels–Alder dimerization upon formation (Equation 26). This is true for all types of phosphoryl-containing functionalities studied to date (phosphine oxides, phosphinates, phosphinamides).

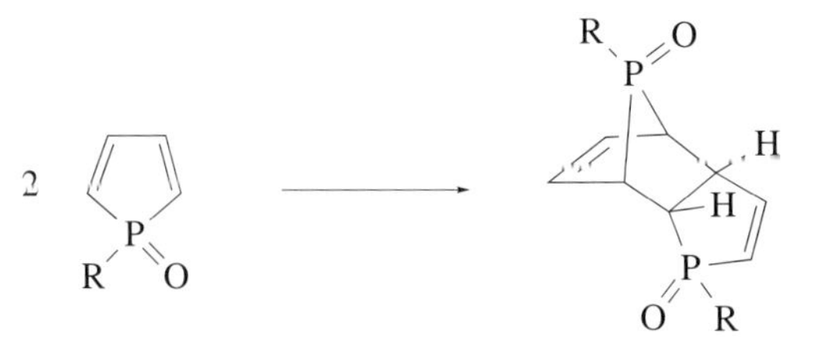

(26)

There are, however, some exceptions, which appear to be associated with steric crowding that may be experienced around the phosphorus atom by substitutions at C-2 and C-5 (as in (**138**) and (**142**)), or at C-3 and C-4 by phenyl as in (**140**). All known pentasubstituted oxides (**137**), (**139**), and (**141**) are stable. Compound (**143**) cannot be isolated but has reasonable stability in solution, no doubt associated with the presence of the 3,4-dimethyl substitution ((**137**) 〈61JA4406〉; (**138**) 〈65JCS2184〉; (**139**) 〈83TL1423〉; (**140**) 〈71JA4541〉; (**141**) 〈74PS(4)199〉; (**142**) 〈74PS(4)199〉; (**143**) 〈91HAC359〉).

There are two general methods for the creation of phosphole oxides. The most obvious is the direct oxidation of phospholes, which is easily accomplished with hydrogen peroxide, *t*-butyl hydroperoxide, *m*-chloroperbenzoic acid, and the like. The stable compounds (**137**), (**138**), and (**143**) were prepared this way. The second method is indirect, and involves the elimination of hydrogen halide from halophospholane oxides. These substances are generally obtained by addition of halogen (usually bromine) to the double bond of 3-phospholene oxides, themselves readily available from the McCormack reaction (Chapter 4.1), or by bromination at the 2-position of phospholane oxides. The elimination of hydrogen halide occurs at room temperature in the presence of a base, commonly triethylamine. This method is used for the

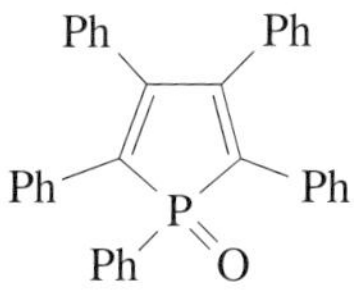

(**137**)

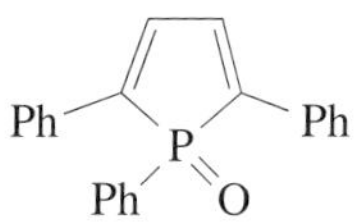

(**138**)

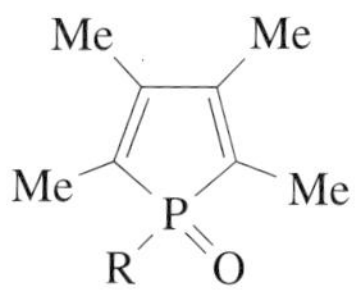

(**139**), R = Me, Ph, or *t*-Bu

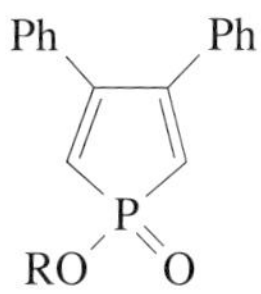

(**140**), R = H or Ph

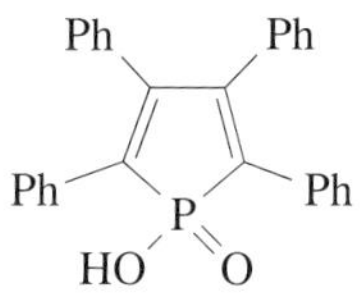

(**141**)

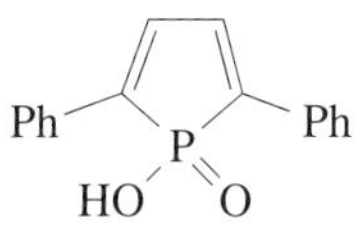

(**142**)

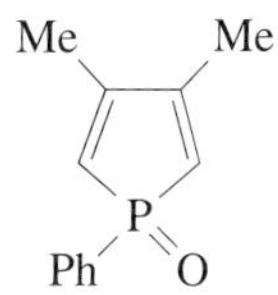

(**143**)

synthesis of phosphole oxides that are easily dimerized or are to be trapped with dienophiles present in the reaction medium (see Section 4.2.1.12.2). The considerable literature on the generation of unstable phosphole oxides has been extensively reviewed by Mathey ⟨88CRV429⟩ and Quin ⟨90RHA39⟩.

A more direct method, but of limited scope, consists of the direct formation of the phosphole ring in the chlorophosphonium condition, which with aqueous base gives the phosphole oxide ⟨83TL1423⟩ in 60–78% overall yield. The reactants are phenyl-, methyl- or *t*-butylphosphonous dichloride and the aluminum chloride σ-complex of cyclobutadienes (Scheme 24).

The starting complexes were formed from reaction of 2-butyne with aluminum chloride, and accordingly the phosphole product is a tetramethyl derivative. With this degree of substitution, the phosphole oxides are stable, isolable solids.

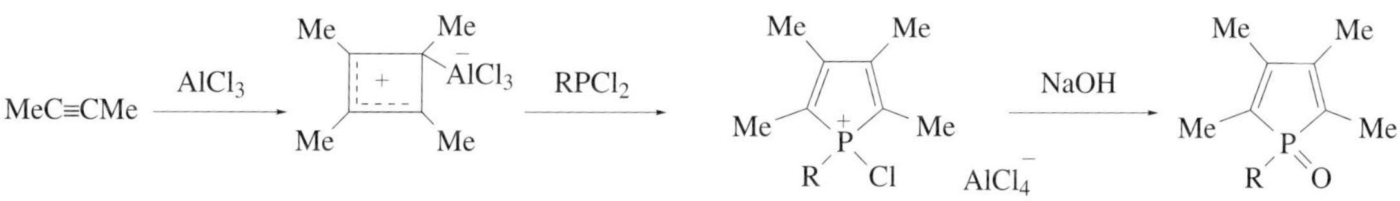

Scheme 24

This method has also been used to synthesize the bicyclic phosphole oxide ((**144**), obtained as a 3 : 1 mixture with the 2,3-cyclohexano isomer, but isolable) (Equation 27).

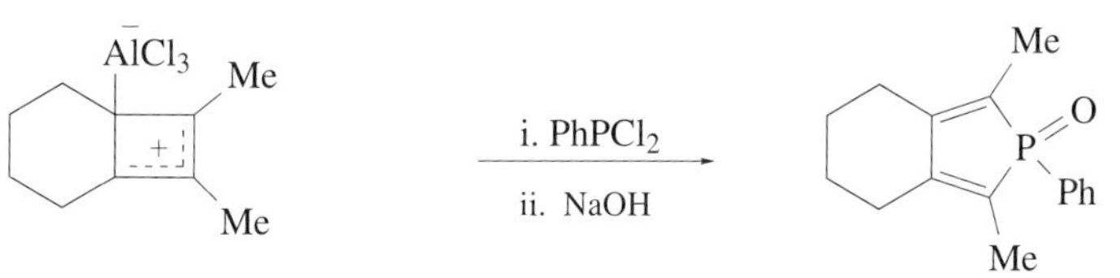

(**144**)

(27)

Benzo[*b*]phosphole oxides (phosphindole oxides) and dibenzo[*b,d*]phosphole oxides are stable as the monomer, since Diels–Alder dimerization would require disruption of the aromatic ring system. The former are prepared by the dehydrohalogenation method (R=Ph ⟨71CJC530⟩; R=MeO ⟨74AJC831, 82PS(13)259, 81NJC187⟩) (Scheme 25).

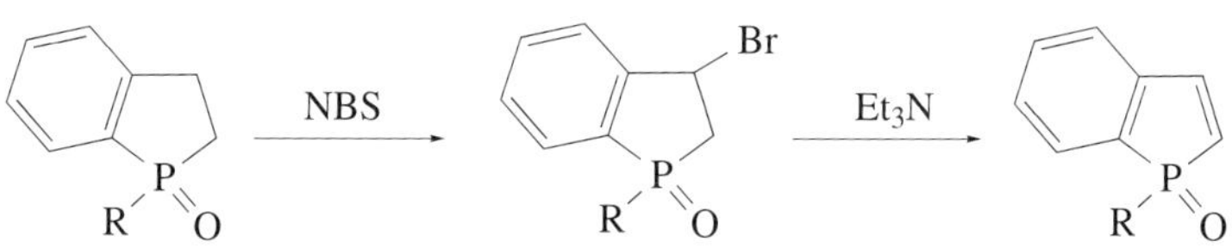

Scheme 25

Dibenzophosphole oxides are prepared by the direct oxidation (H_2O_2) of the phospholes ⟨69JOC1777⟩; a diarylphosphinic acid can also be used for generating the phosphole ring ⟨56JOC238, 59JOC638⟩ (Equation 28).

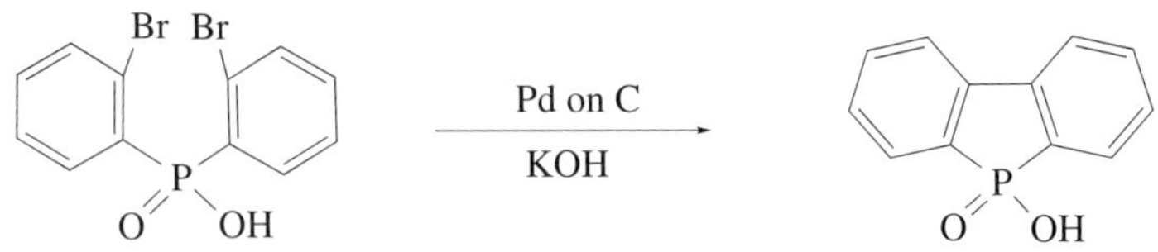

(28)

Benzo[*c*]phosphole oxides (isophosphindole oxides) have remained elusive; all attempts to generate them have led to dimeric products. The first approach to an isophosphindole oxide involved debromination of the dibromide (**145**) with copper powder and yielded a yellow solid with the composition of a dimer of the expected isophosphindole oxide (**146**) to which structure (**147**) was assigned ⟨70CC122⟩. This was supported by the spectral properties, and the yellow color indicated the presence of the phosphole oxide ring. To account for this structure, the intermediate isophosphindole oxide acts as a diene in condensation with a double bond of the benzo group. When *N*-phenylmaleimide was present in the reaction medium, (**146**) was trapped as the Diels–Alder adduct (**148**), in the same way that monocyclic phosphole oxides can be trapped (see Scheme 26).

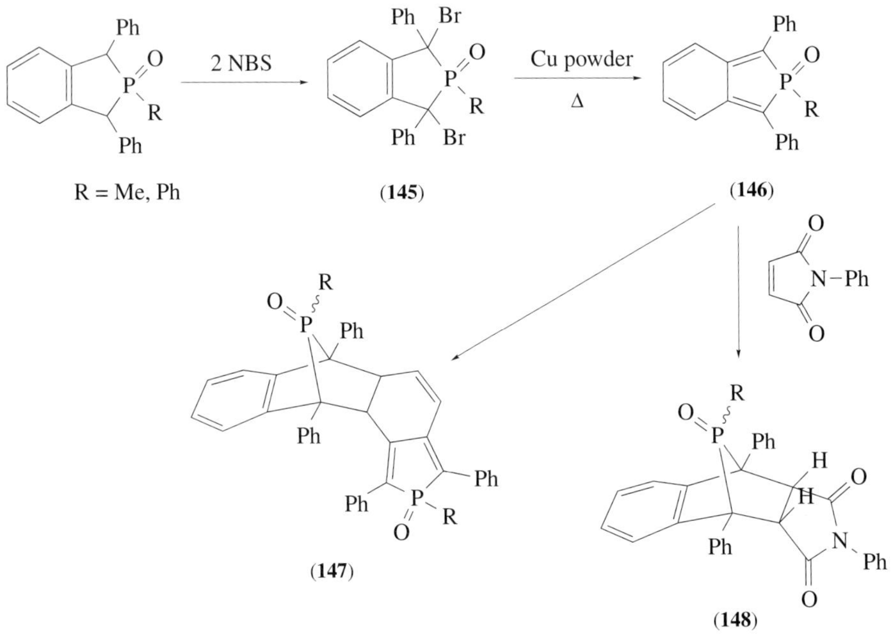

Scheme 26

Another approach involved the dehydrobromination of bromide (**149**), which also led to a dimeric product, (**151**) (Scheme 27) ⟨73TL4815, 75T2537⟩.

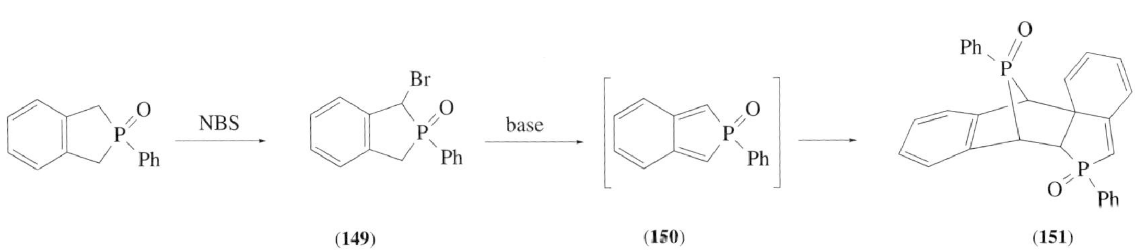

Scheme 27

In this case, the evidence supported a dimer structure of the same Diels–Alder type as observed from monocyclic phosphole oxides, and this was supported by later studies ⟨85MRC929⟩. Dienophiles present as trapping agents gave the expected adducts of (**150**).

Functionalized phosphole oxides are quite rare; the only stable type that is known has four carboxylate groups, with phosphorus in phosphinate form ⟨84JCS(P1)1627, 88JCS(P1)175⟩ or in phosphine oxide form ⟨88JCS(P1)175⟩. A five-coordinate adduct (**152**) from a phosphite or a phosphinite is converted with HBr to the phosphole oxide (**153**) or is first allowed to rearrange to a 2*H*-phosphole (**154**) which is then converted to (**153**) (Scheme 28). Since the starting phosphorus materials are the common trialkyl phosphites or dialkyl phosphonites, respectively, and the yields are quite high, the method is of considerable synthetic value for this type of phosphole derivative.

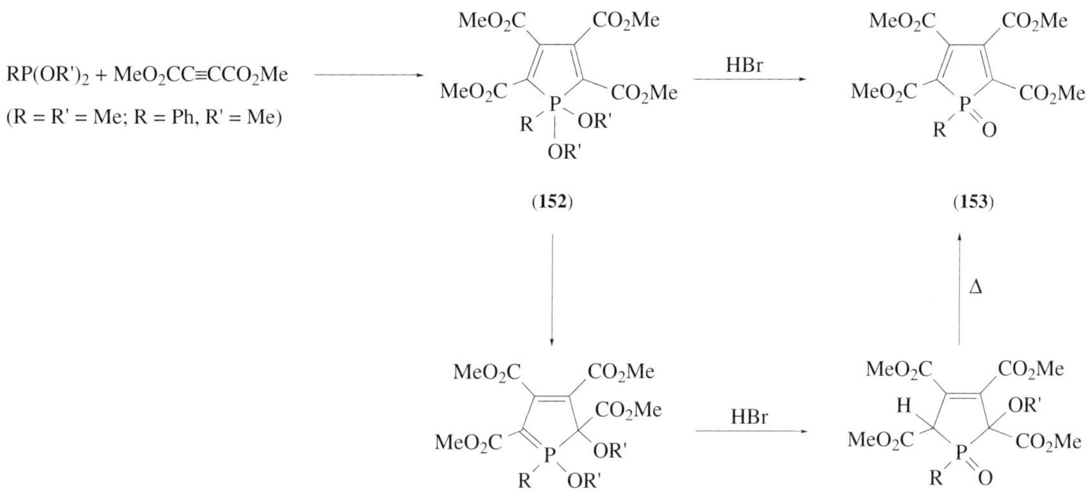

(**152**) (**153**) (**154**)

Scheme 28

4.2.1.6.1.2 Reactions

Because the few stable phosphole oxides that are known have special steric features that can modify reactivity, relatively little is known about the reaction chemistry of the ring systems. The unstable structures are of interest, however, for the dimers that form spontaneously are of value as derivatives of the 7-phosphanorbornene ring system, a subject reviewed in some detail elsewhere ⟨88CRV429, 90RHA39⟩. Of more importance is the reaction of the unstable phosphole oxides with dienophiles when present as trapping agents in the reaction medium, which also gives 7-phosphanorbornene derivatives. Numerous examples of such compounds are known and are described in the reviews cited. The choice of dienophile is somewhat restricted, however. The most commonly used type has the maleimide or maleate structure; some examples are given in Schemes 29 ⟨91HAC359⟩ and 30 ⟨85JA3389⟩.

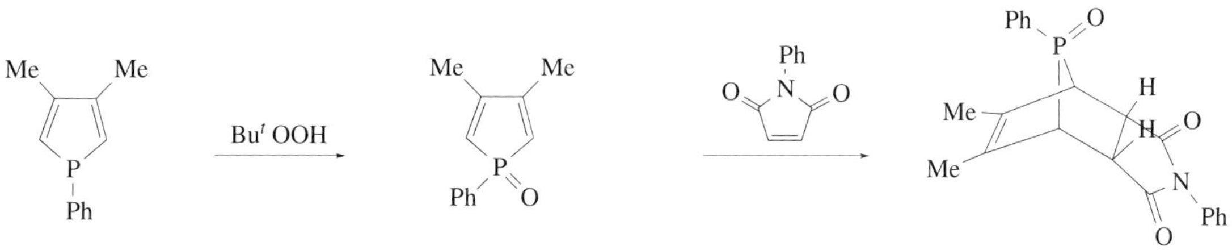

Scheme 29

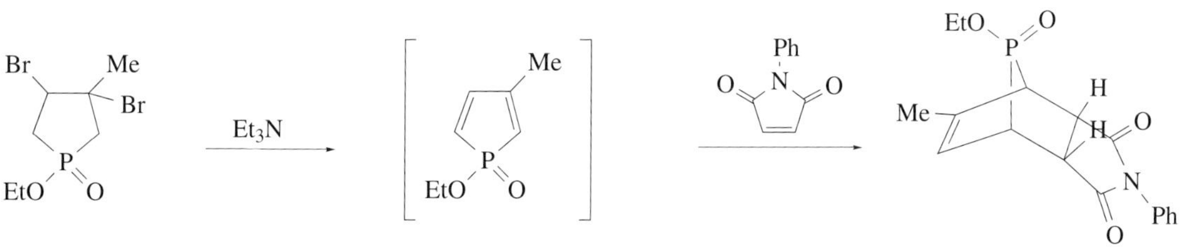

Scheme 30

Ethyne dicarboxylate has given cycloaddition reactions with some stabilized, highly substituted phosphole oxides, but the cycloadducts (7-phosphanorbornadiene derivatives) readily lose the phosphorus bridge under the reaction conditions and have never been characterized ⟨65JCS2184, 72JA4761⟩. However, diaroyl ethynes have been reported to give stable 7-phosphanorbornadiene derivatives (**155**) with pentaphenylphosphole oxide when the reaction was conducted under high pressure (Equation 29) ⟨90H(30)201⟩.

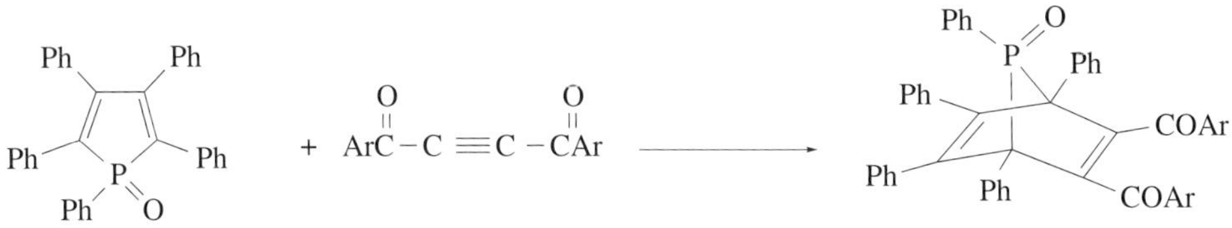

(29)

(**155**)

Attempts to trap the less stable phosphole oxides by having ethyne derivatives present in the medium of their generation have not been successful, since the ethynes used so far have been sensitive to the reagents in the medium.

Phosphole oxides also participate in a cycloaddition reaction with oxyallyl cations ⟨75T53⟩, a process that gives rise to novel phosphatropolone structures. Preformed 1,2,5-triphenylphosphole oxide gave a derivative of the 8-phosphabicyclo[3.2.1]octan-6-ene-3-one ring system ((**157**), 91%). The oxyallyl cation (**156**) was prepared from α,α′-dibromodibenzylketone and NaI in acetonitrile (Scheme 31).

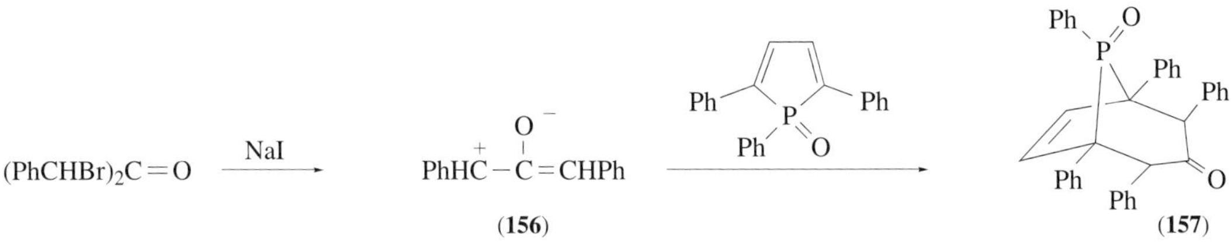

Scheme 31

The reaction was also attempted on 3,4-dimethyl-1-phenylphosphole oxide generated in the reaction medium; the expected cycloadduct was obtained in only 6% yield, dimerization of the oxide being the major reaction pathway.

Additions to only one double bond of a phosphole oxide are also possible. Diazomethane gives a pyrazoline, thermally converted to a cyclopropano derivative, as has been observed with 1,2,5-triphenylphosphole oxide ⟨65JCS2184, 71CJC874⟩ and 3,4-dimethyl-1-phenylphosphole oxide ⟨89T7083⟩ (Scheme 32).

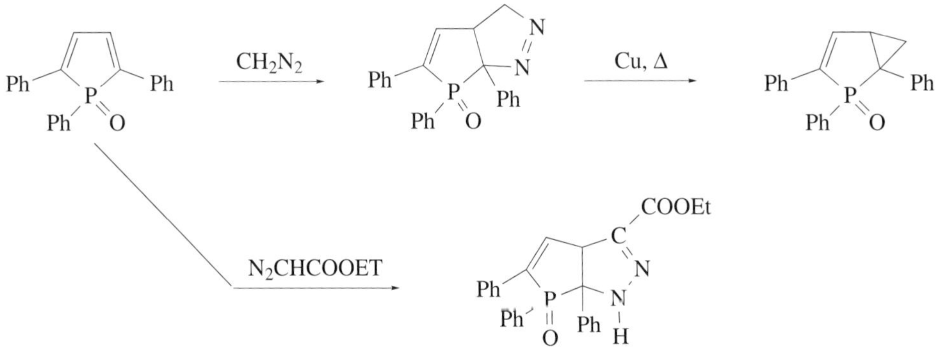

Scheme 32

A phosphole oxide may be an intermediate when diazomethane reacts with a phosphole in the presence of water, since the reaction product is the pyrazoline that would be expected from addition to the double bond of a phosphole oxide. The technique was applied in high yield to several phospholes (Scheme 33) ⟨89TL3525, 93CB473⟩.

The orientation of addition of diazomethane is the opposite of that proposed in the earlier work ⟨65JCS2184⟩. The ^{1}H NMR spectrum of (**158**) does support this structure, since large $^3J_{PH}$ values (14–19

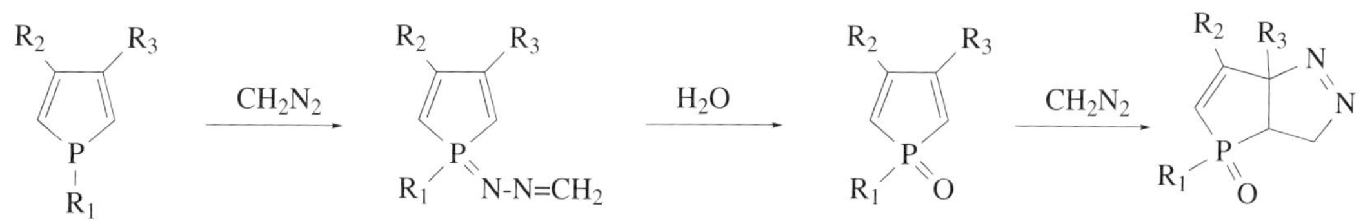

(**158**)

Scheme 33

Hz) were observed for the CH_2 group. It was also found possible to epoxidize one of the double bonds of 3,4-dimethyl-1-phenylphosphole oxide with *m*-chloroperbenzoic acid ⟨91HAC359⟩. The product (**159**) was a stable solid, a mixture of diastereoisomers (Equation 30).

(**159**)

(30)

Phosphine oxides are normally reducible to the phosphines with trichlorosilane or phenylsilane (Equation 31), and some examples with cyclic systems have recently been reviewed ⟨B-90MI01⟩. This reaction probably can be applied to the stable phosphole oxides, and would have practical value as a phosphole synthesis when applied to phosphole oxides that are prepared by methods other than the direct oxidation method.

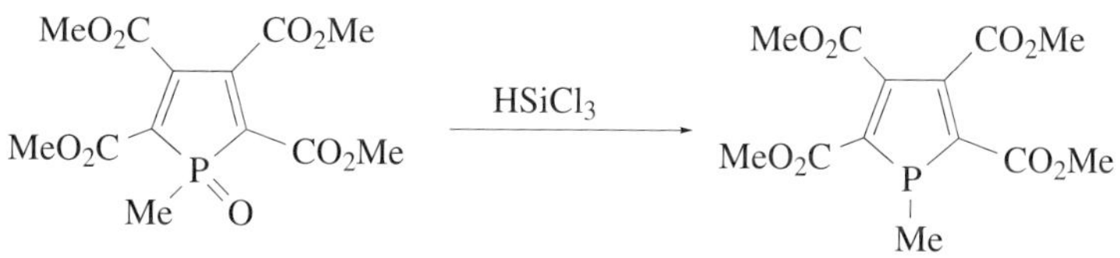

(31)

4.2.1.6.2 Phosphole Sulfides and Selenides

4.2.1.6.2.1 Synthesis

The only approach to phosphole sulfides involves direct addition of sulfur to preformed phospholes (e.g. ⟨65JCS2184, 68TL1755, 69CR(C)1066, 70BSF4433⟩). The most commonly studied compound is 3,4-dimethyl-1 phenylphosphole sulfide, easily prepared in high yield from the phosphole on refluxing with sulfur in benzene. Unlike the corresponding oxide, the sulfide is a stable, isolable solid; stability of the phosphole sulfides is in fact one of their notable characteristics, and makes them quite valuable as starting materials in processes that are denied to most phosphole oxides. The known stable phosphole sulfides that are available as starting materials are shown below (**160–163**): (**160**) ⟨65JCS2184⟩; (**161**) ⟨69CR(C)1066⟩; (**162**) ⟨70BSF4433⟩; (**163**) ⟨61JA4406⟩. Many other phosphole sulfides have been derived from them as will be seen in the next section.

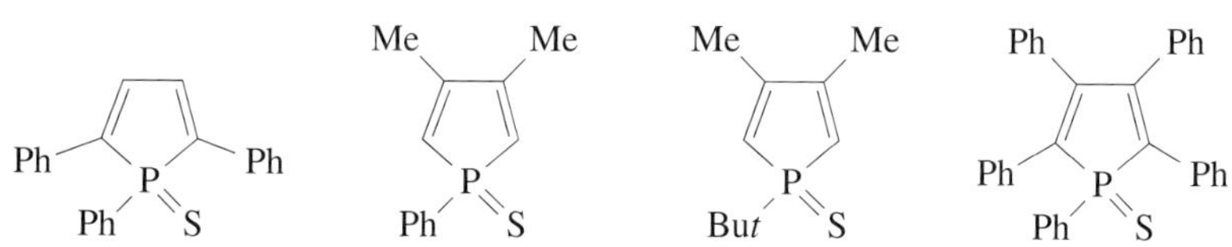

(**160**) (**161**) (**162**) (**163**)

Selenides are also prepared from the phospholes by reaction with selenium ⟨65JCS2184, 72T4171⟩. Stable compounds analogous to (**160**), (**161**) and (**163**) are known, but have been little studied.

4.2.1.6.2.2 Reactions

Phosphole sulfides are generally not prone to dimerization by a Diels–Alder reaction, although exceptions are known (e.g. ⟨81PS(10)175⟩), but they do participate in cycloaddition reactions with various dienophiles ⟨76T2427⟩. Thus, the 3,4-dimethyl-1-phenyl derivative (**161**) gave an adduct with *N*-methyl-1,2,4-triazoline-3,5-dione consisting of two isomers, the major being obtained in pure form by recrystallization (Equation 32). The isomerism was attributed to configurational differences at phosphorus. Yields were not reported in the original work, but a later preparation of the same adduct gave a yield after recrystallization of 34% of the isomer mixture (**164**)–(**165**) ⟨85PS(25)201⟩. The *N*-phenyl derivative was prepared in 49% yield, also as an isomer mixture. The major isomer from these reactions appears from NMR studies to have the P-substituent *syn* to the C=C bridge, as in (**164**).

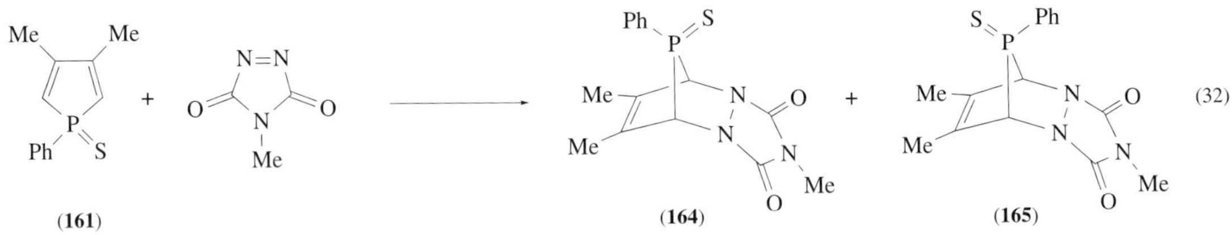

(**161**) (**164**) (**165**)

Diels–Alder additions (*endo*) were also reported to occur with maleic anhydride to give a single isomer (**166**), and with dimethyl ethynedicarboxylate, although in the latter case the adduct was unstable and lost the P bridge (Equation 33) ⟨76T2427⟩. Yields were not reported.

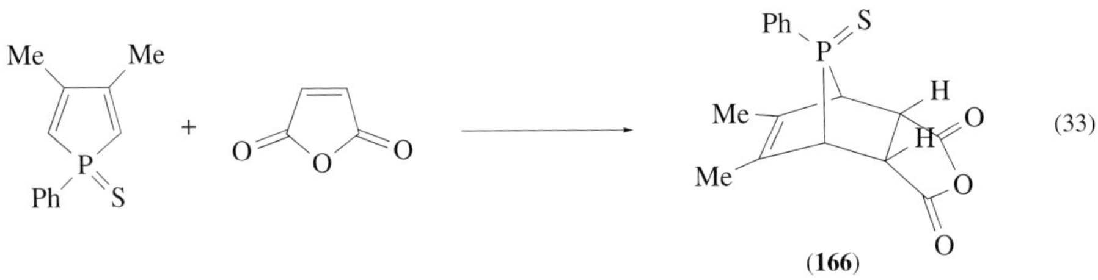

(**166**)

The triple bond of phosphaalkynes also acts as a dienophile with phosphole sulfides, giving a cycloadduct (**167**) that readily loses the bridging phosphine sulfide group and is converted to a phosphinine derivative, (**168**) (Scheme 34) ⟨86ZN(B)931⟩. The reaction has practical value as a phosphinine synthesis.

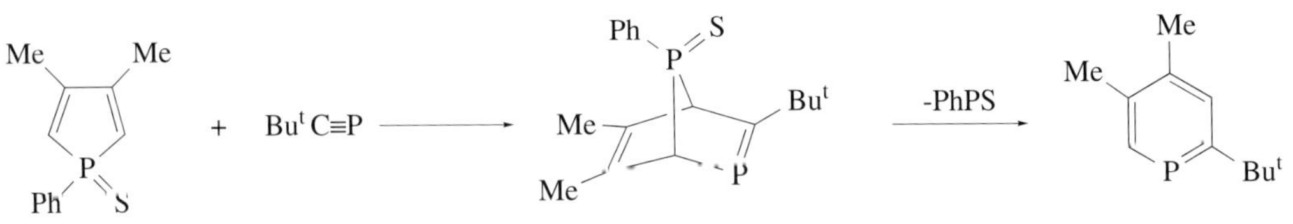

(**167**) (**168**)

Scheme 34

3,4-Dimethyl-1-phenylphosphole selenide gave adducts with maleic anhydride (49%) and N-phenylmaleimide (72%; ⟨85PS(25)201⟩). With the triazolinediones, the addition products fragmented easily.

In behavior unlike that known for phosphole oxides, phosphole sulfides also act as dienophiles in reaction with dienes, thereby giving valuable cycloadducts such as (**169**) and (**170**) (Scheme 35) ⟨76T2427⟩. The latter is a member of a rare class of phosphorus heterocycles describable as phosphasteroids.

Reaction at one of the double bonds of phosphole sulfides also takes place with diazo compounds. Thus, ethyl diazoacetate condensed over a period of 45 days with phosphole sulfide (**161**) to give the cycloaddition product (**171**) in 54% yield ⟨91JOC4031⟩. A single isomer was obtained; the regiochemistry was established by ^{1}H and ^{13}C NMR techniques to be the same as that proposed in the work in 1989 on phosphole oxides ⟨89TL3525⟩. When the reaction was attempted at a higher temperature (150°C), nitrogen was lost and the cyclopropano derivative (**172**) resulted, in 69% yield. The same product was formed when the initial adduct (**171**) was heated under similar conditions (Scheme 36).

The cyclopropano derivative (**172**) can be viewed as a sulfide of a homophosphole, and was converted to the homophosphole (**174**) in 74% yield by a reduction procedure that involved first conversion to the

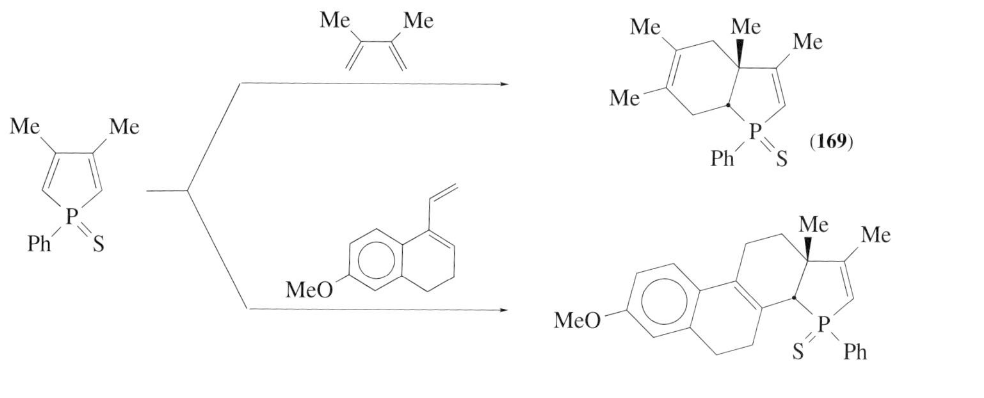

(169)

(170)

Scheme 35

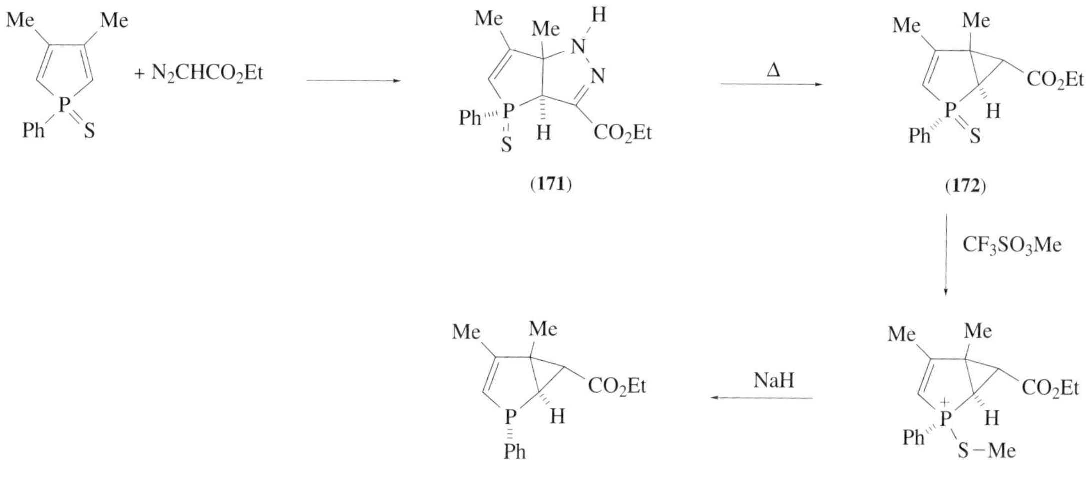

(171) (172) (173) (174)

Scheme 36

S-methyl derivative (**173**) and reaction with NaH. Reduction of (**172**) to (**174**) was also achieved with Bu_3P or Zn/Hg. A similar series of reactions was applied to the dithiophosphinate (**175**), giving homophosphole (**176**) (Scheme 37). The homophospholes are useful as precursors on thermolysis of phosphinines.

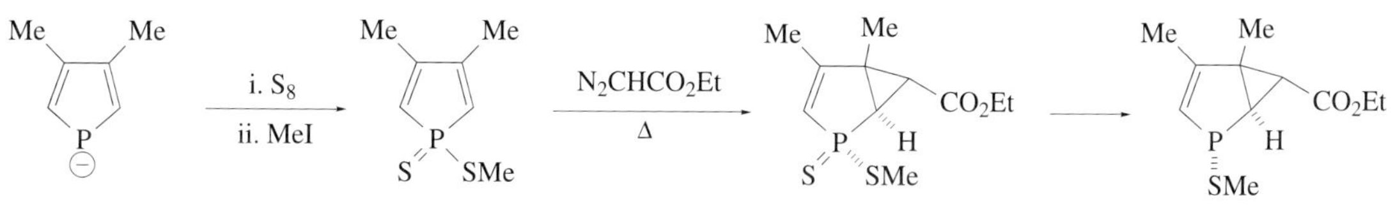

(175) (176)

Scheme 37

The sensitivity of S to electrophilic reagents proved to be a complication when an attempt was made to perform cycloaddition with the oxyallyl cation, which had worked well with phosphole oxides. The reagent attacked on S, presumably to give a spirophosphorane, such as (**177**), which rearranged to leave oxygen on phosphorus as in (**178**) (Scheme 38) (31%; ⟨75T53⟩). Other products observed were the dimer of the phosphole oxide and the cycloadduct of the oxide with the oxyallyl cation.

The reactivity of the sulfur atom to attack by electrophiles is also revealed by the intramolecular quaternization of the bromoalkyl derivative (**179**) (Equation 34) ⟨81JOC4386⟩.

In addition to the simple removal of sulfur to generate the phosphole (Bu_3P or Zn/Hg), reaction at sulfur occurs with other reagents, and of special value is that taking place with certain coordination complexes.

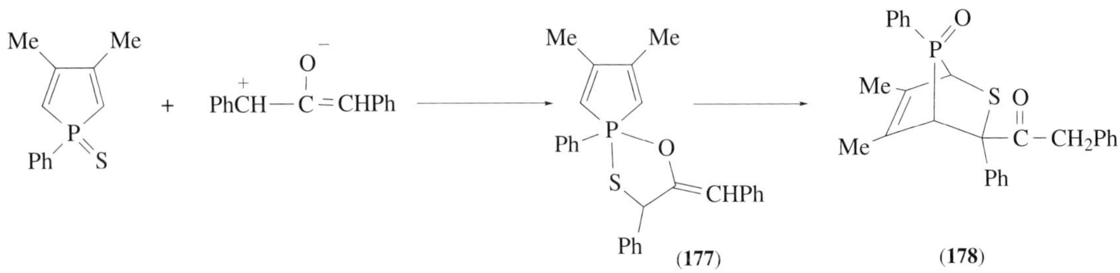

(177) (178)

Scheme 38

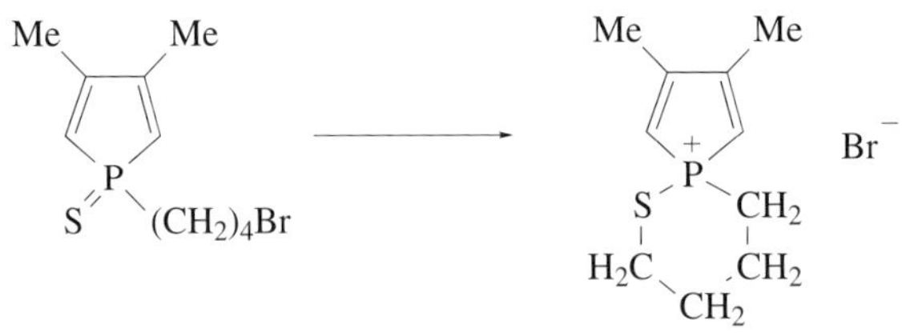

(179) (34)

With $Fe_2(CO)_9$, phosphole sulfide (**161**) underwent a reduction-complexation process that resulted in the formation of the σ-complex (**180**) ⟨77JOM(136)241⟩. With $Mn_2(CO)_{10}$, phosphole sulfide (**161**) underwent both loss of sulfur and of phenyl, giving rise to the phosphacymantrene derivative (**181**) ⟨79JOM(165)129⟩.

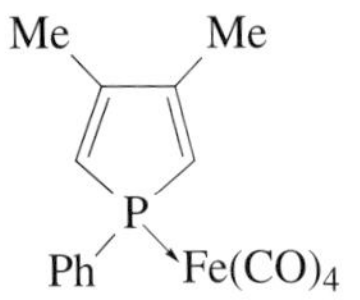

(180)

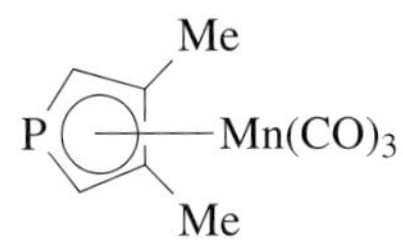

(181)

One of the most valuable reactions of 3,4-dimethyl-1-phenylphosphole sulfide is that with strong bases such as the butyllithiums ⟨73TL3255, 74T3127⟩. A proton is abstracted from a ring methyl group, and the resulting anion is resonance-delocalized. Attack of electrophiles can then occur either on the exocyclic carbon (as with $Ph_2C{=}O$ to form (**182**)) or at the 2-position of the ring to form (**183**) (Scheme 39). It is of particular importance when attack occurs on the ring (as with CO_2, ethyl acetate, and certain ketones; a later report ⟨92OM1411⟩ showed that Ph_2PCl also reacted at this position) and a proton migration follows to regenerate the methyl group, for this constitutes a technique for the direct introduction of functional substituents on the phosphole sulfide ring. Removal of sulfur then gives the functionalized phosphole. The full scope of this process is described in Sections 4.2.1.11.2 and 4.2.1.11.3.

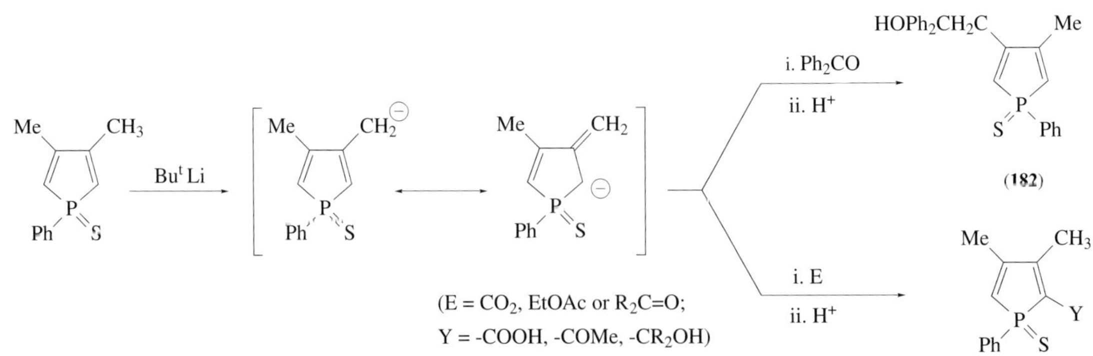

(182) (183)

Scheme 39

4.2.1.6.3 Phospholium Salts

Alkyl halides react with phospholes to give crystalline phospholium salts; many such derivatives have been prepared. While some of these are stable as monomers, others proceed to dimerize during formation or on heating of the phospholium salts ⟨73JOC1954, 79OMR(12)442⟩. Some examples of monomeric phospholium salts are shown below ⟨65JCS2184, 73JOC1954⟩.

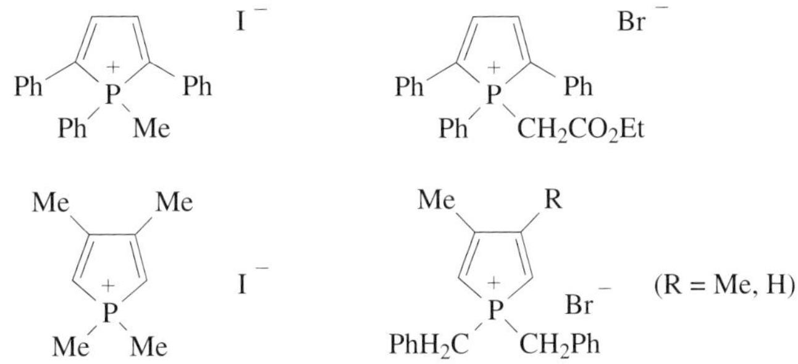

The dimeric products have the typical Diels–Alder structure as in (**184**); the bridging P atom possesses quite distinctive ^{31}P NMR characteristics.

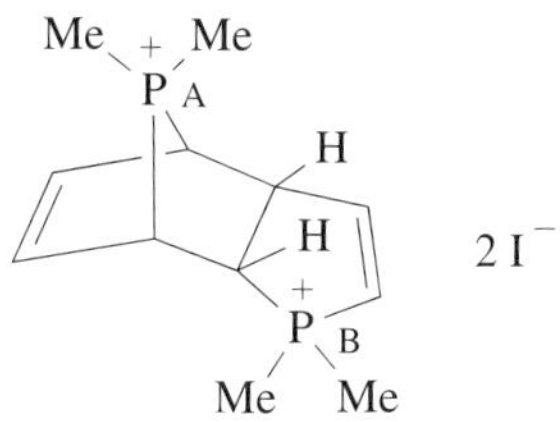

$\delta\ ^{31}P_A = +98.3$
$\delta\ ^{31}P_B = +58.8$
$^3J_{PP} = 35.5$ Hz

(**184**)

As is discussed in Section 4.2.1.6.5, P-alkylphospholium ions react with strong bases to give ylides which participate in Wittig reactions. Other reactivity of the ring system is for the most part unexplored, an exception being the reaction that occurs when iodomethyl salts are treated with aqueous base (Equation 35) ⟨71CJC879⟩.

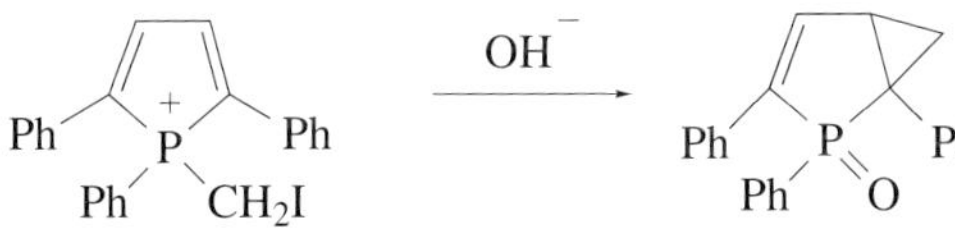

(35)

In the dibenzophosphole system, ring expansion to dihydrophosphinines occurs (Equation 36) ⟨69JCS(C)252⟩. A novel spirocyclic ion (**185**) has been formed by an intramolecular quaternization (Scheme 40) ⟨71T5523⟩.

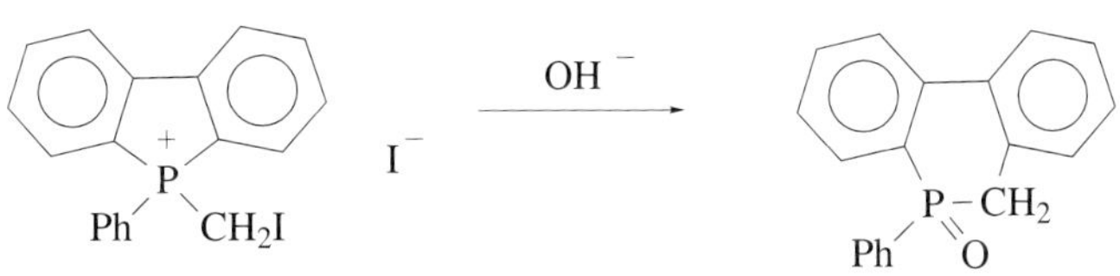

(36)

(**135**)

4.2.1.6.4 Iminophospholes

The imino group may be attached to phosphorus of the phosphole ring by conventional methods of phosphine chemistry to form the family of iminophospholes, which may be considered to be nitrogen analogues of ylides (Equation 37).

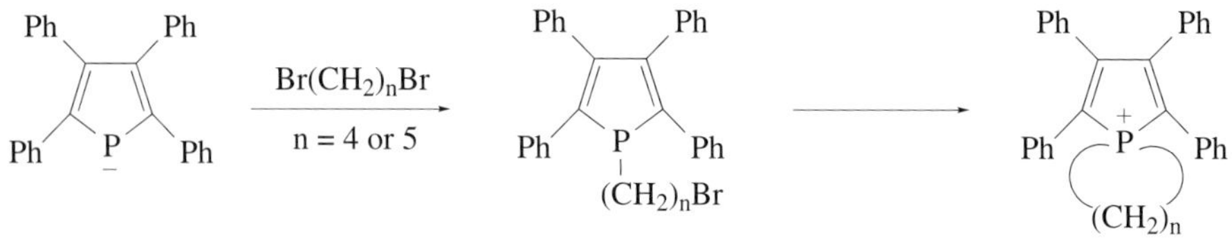

Scheme 40

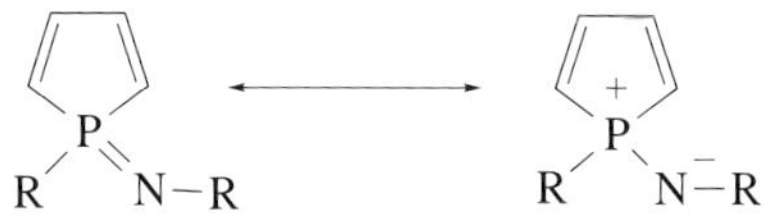

(37)

The best known of these methods is the reaction of azides with phosphines (the Staudinger reaction), and this has been applied both to dibenzophosphole ⟨64CB741⟩ and to several 1,2,5-triarylphospholes ⟨72CC1242, 74JCS(P1)1694⟩. The reactions with the phospholes were conducted at 100°C in toluene and gave generally good yields of the iminophospholes. The triarylphospholes also reacted with nitrenes from various sources to give iminophospholes.

The iminophospholes are crystalline compounds, of widely varied color. They have good stability except on exposure to water. It is notable that the color differences disappear in solution, and in general the iminophospholes have strong UV absorption maxima at 210 and 380–405 nm.

Iminophospholes have high electrophilicity at the phosphorus atom, and are easily hydrolyzed to phosphole oxides. The electrophilic activity also is evident in an unusual reaction discovered for iminophospholes with *o*-nitroaryl substituents on nitrogen. These compounds on being heated at 160°C in mesitylene gave benzofurazan derivatives (32–65% yields) along with the phosphole oxide ⟨74JCS(P1)1694⟩. The mechanism proposed involved attack of an oxygen of the nitro group on the electrophilic phosphorus to form a five-coordinate intermediate which collapsed to products. The reaction seems to be peculiar to iminophospholes, since some noncyclic iminophosphines with *o*-nitroaryl N-substituents did not give these products. Ring strain in the phosphole derivatives, which is diminished in the five-coordinate state, may account for their higher reactivity (Scheme 41).

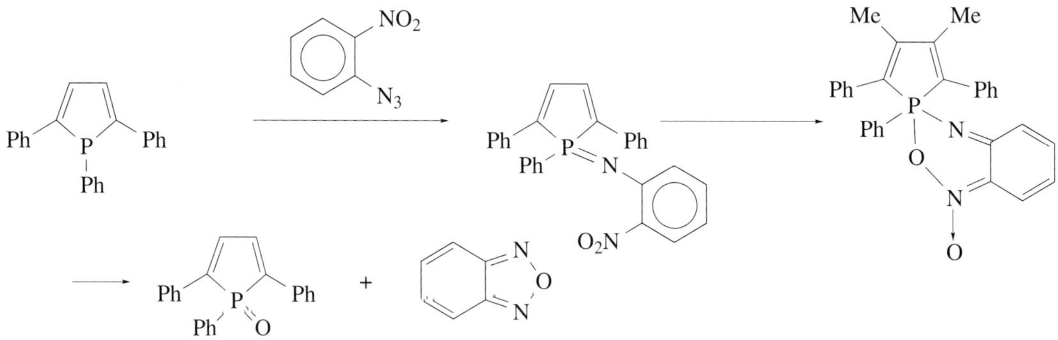

Scheme 41

4.2.1.6.5 Phosphole Ylides

Phosphole ylides may be represented by the resonance structures of Equation (38).

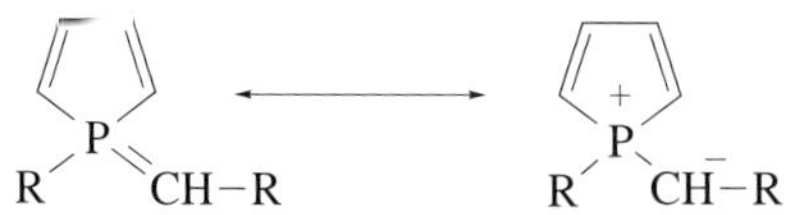

(38)

It was early recognized ⟨72CC1242, 74JCS(P1)1694⟩ that carbenes, generated from diazo compounds, added to phosphorus to produce the ylide structure. 1,2,5-Triphenylphosphole has been the usual participant in this reaction (Schemes 42 and 43). When some multicyclic diazo compounds were used (e.g. (**186**) and (**188**)) as sources of carbenes, the ylides were thermally unstable. Compound (**187**), formed from (**186**), decomposed with loss of phosphorus to give 7,10-diphenylfluoranthene ⟨74CC902, 78LA74⟩; ylide (**189**) from (**188**) rearranged by a novel Stevens-like process to spirophosphinine (**190**) ⟨84CC1217⟩.

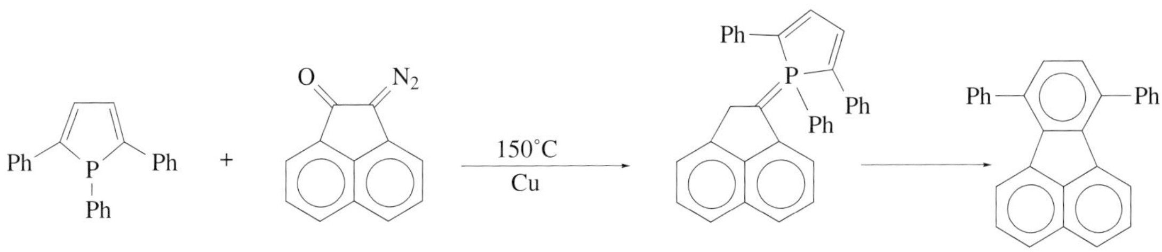

(**186**) (**187**)

Scheme 42

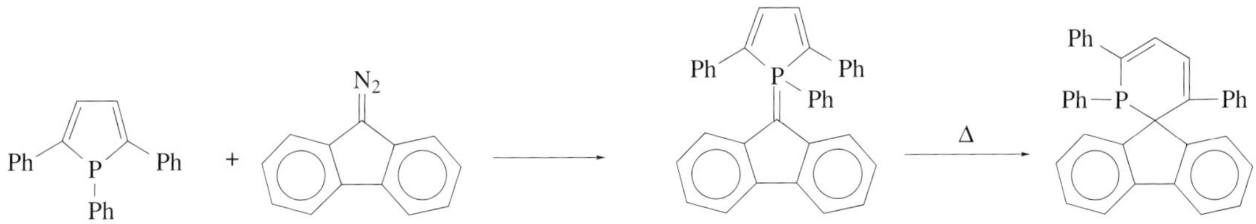

(**188**) (**189**) (**190**)

Scheme 43

Phosphole ylides have also been prepared by the common procedure of abstraction of an α-proton from P-alkyl phospholium salts (Equation 39) ⟨65JCS2184⟩.

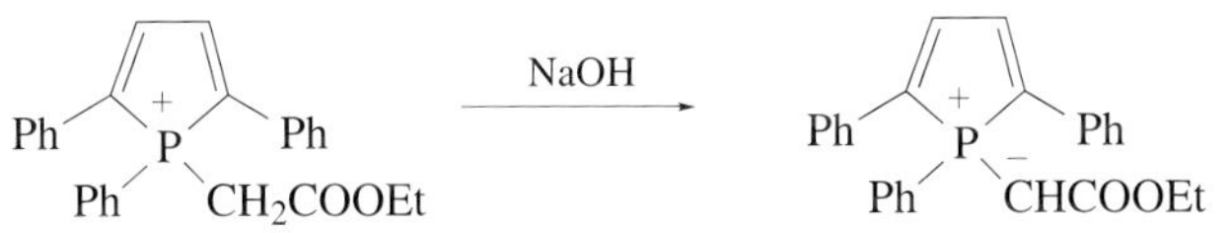

(39)

Ylides derived from benzophospholes have attracted interest as participants in the Wittig reaction, where they give exceptionally high ratios of *E*- to *Z*-alkenes (90 : 10 or better) ⟨87TL3445, 93JOC1985⟩. They are also of interest in giving rather stable oxaphosphetane intermediates (**191**) in this reaction (Equation 40); these five-coordinate phosphole derivatives are discussed in Section 4.2.1.7.3.

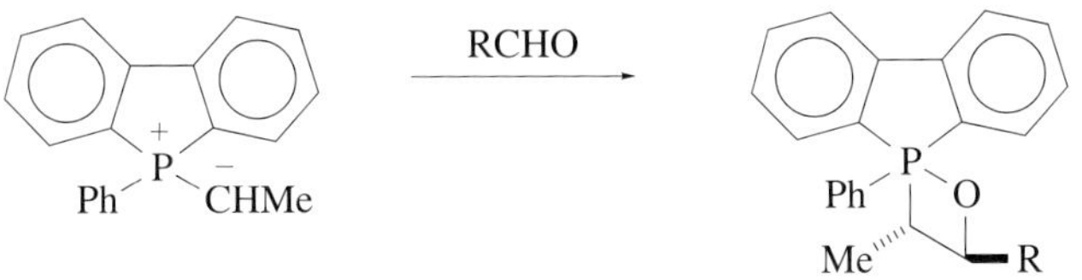

(40)

(**191**)

Most of the interest in phosphole ylides has rested in the reactivity external to the ring, and little is known about the character of the phosphole ring itself.

Ylides of another type are also known, where the functionality is incorporated within the ring. These endocyclic ylides, frequently the result of complex reactions, are highly reactive to water, and virtually nothing is known about the chemistry of the ring system. The structures of this type that are known are summarized in structures (i) to (vi), the leading references being: (i) ⟨69JCS(C)1100⟩; (ii) ⟨86CC1574⟩; (iii) and (iv) ⟨72CJC3625⟩; (v) ⟨78JHC1319⟩; (vi) ⟨70JCS(C)1855⟩.

4.2.1.6.6 Phosphole-Boranes

A Lewis salt (**192**) of 3,4-dimethyl-1-phenylphosphole with BH_3 is formed by reaction with $H_3B \bullet SMe_2$. The reaction occurs rapidly at 0ºC and is near quantitative ⟨94OM925⟩. The salt can be deprotonated

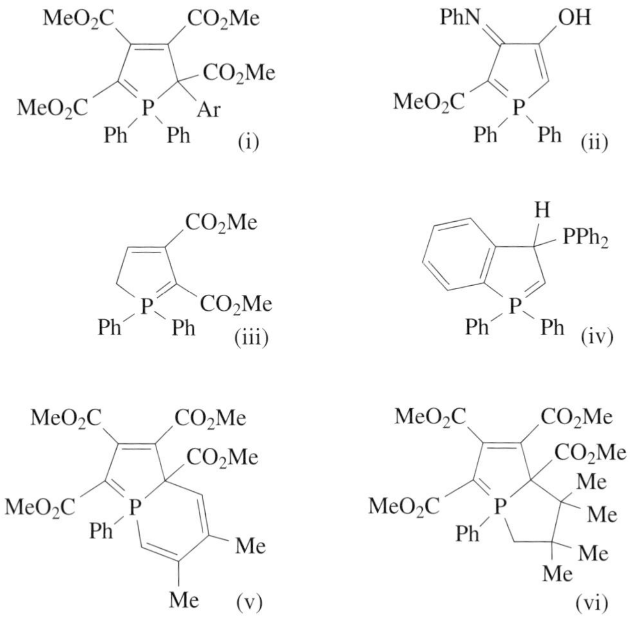

at a methyl group at −80°C with *sec*-butyllithium; the allylic ion (**193**) so produced reacts at the 2-position on protonation or methylation, giving the novel phosphine complexes (**194**) and (**196**) that are isomeric with phosphole complexes and can generate the free phosphole isomers (**195**) and (**197**) on decomplexation (Scheme 44).

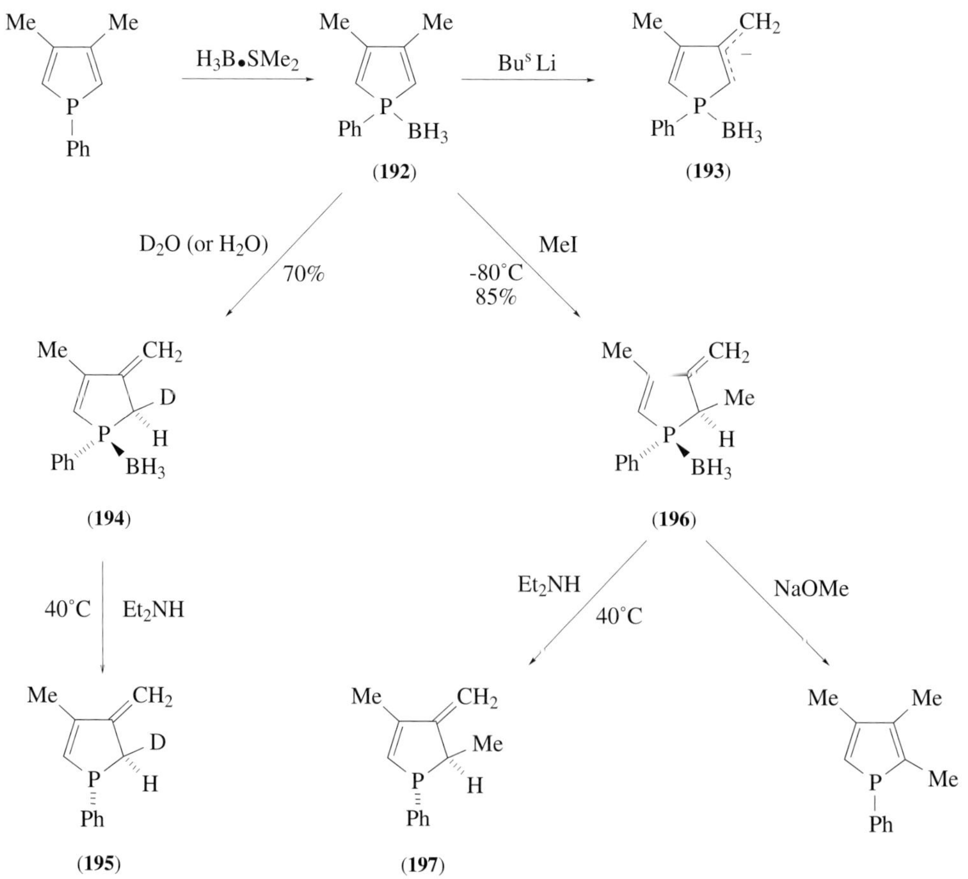

Scheme 44

Silylation and stannylation also occurred at the 2-position, but quenching allylic anion (**193**) with diethyl oxalate gave the product of γ-attack (**198**). With ethyl acetate, attack occurred at the 2-position, but the product underwent reduction to give the 2-phospholene complex (**199**).

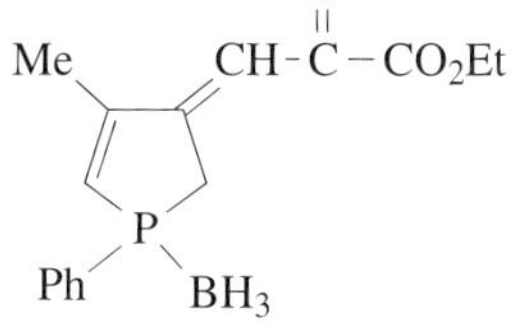

(**198**) (60%)

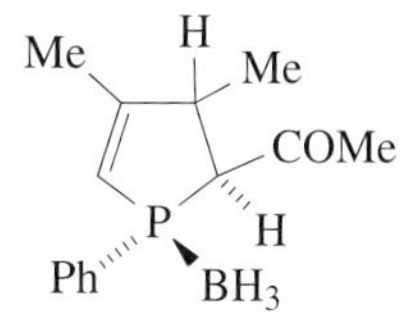

(**199**)

4.2.1.7 REACTIVITY AND SYNTHESIS OF PHOSPHOLES WITH C.N.-2 AND C.N.-5

4.2.1.7.1 2*H*-Phospholes

The C=P double bond was unknown until the discovery of the delocalized phosphinines in 1966 ⟨66AG(E)846⟩, and it was not until 1978 that the first stable compound (**200**) with an isolated double bond was discovered ⟨78JA4886⟩.

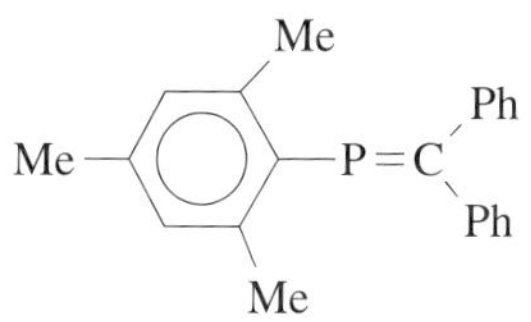

(**200**)

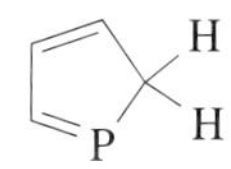

(**201**)

The 2*H*-phosphole structure (**201**) requires such a bond, and although compelling evidence for the fleeting existence of 2*H*-phospholes accumulated after 1981 (see below), it was not until 1987 that the first example of a stable monocyclic 2*H*-phosphole (**203**) was reported ⟨87JOM(332)C1⟩. As is common in studies of low-coordination phosphorus species, the ring was heavily substituted by very large substituents to endow the system with kinetic stability. The synthesis is outlined in Scheme 45; the migration of TMS from P to O in (**202**) is a known technique for creating the C=P double bond.

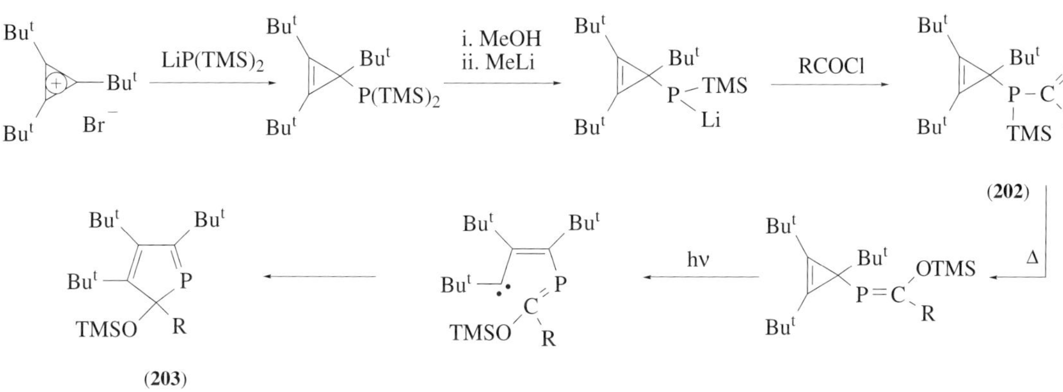

Scheme 45

2*H*-phosphole (**203**) was a yellow oil that could be distilled by the Kugelrohr technique at 170–180ºC (10^{-3} mmHg). In C_6D_6, the ^{31}P NMR shift was δ +159, entirely consistent with the presence of a C=P double bond, and this was confirmed by the ^{13}C NMR spectrum, which showed phosphorus to be directly coupled to an sp^2 carbon (δ 194.7, $^1J_{PC}$ 68.7 Hz) and to an sp^3 carbon (δ 37.1, $^1J_{PC}$ 51.4 Hz). The 2*H*-phospholes where the 2-phenyl was replaced by 2,4,6-trimethylphenyl, 2,4,6-trimethoxyphenyl, and methyl were also made in good yield by this method. These 2*H*-phospholes are thermally stable but very sensitive to water and were hydrated on attempted chromatography on "Kieselgel" to form the phospholanone (**204**). Other reactions observed (Scheme 46) included a Diels–Alder cycloaddition with methyl propiolate (giving (**205**)), and a 1,3-dipolar cycloaddition with mesityl nitrile oxide that formed (**206**).

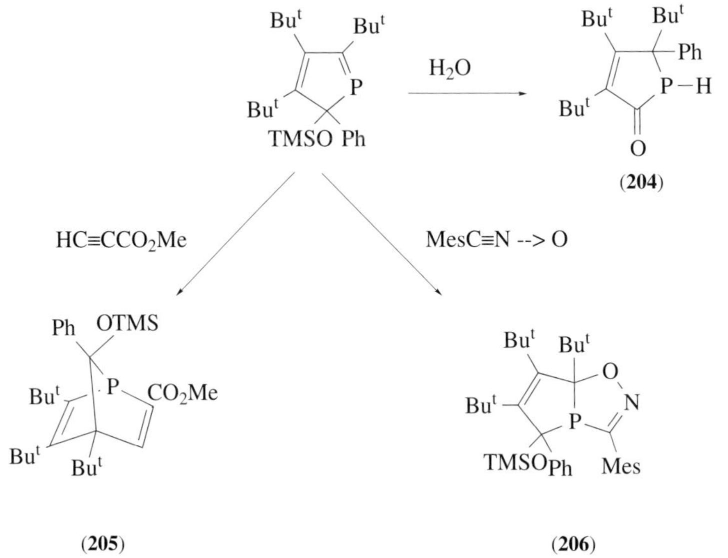

Scheme 46

The 2*H*-phosphole moiety has also been stabilized by incorporation into some phosphaazulene derivatives synthesized in the Märkl laboratory. The structures ((**208**), (**209**), and (**210**)) are shown in Equation (41) and Scheme 47.

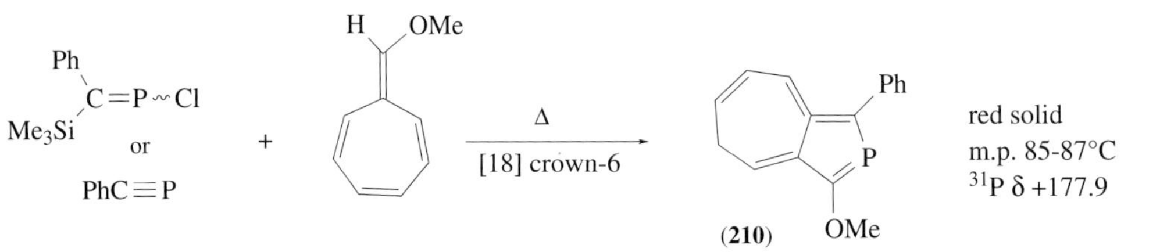

(41)

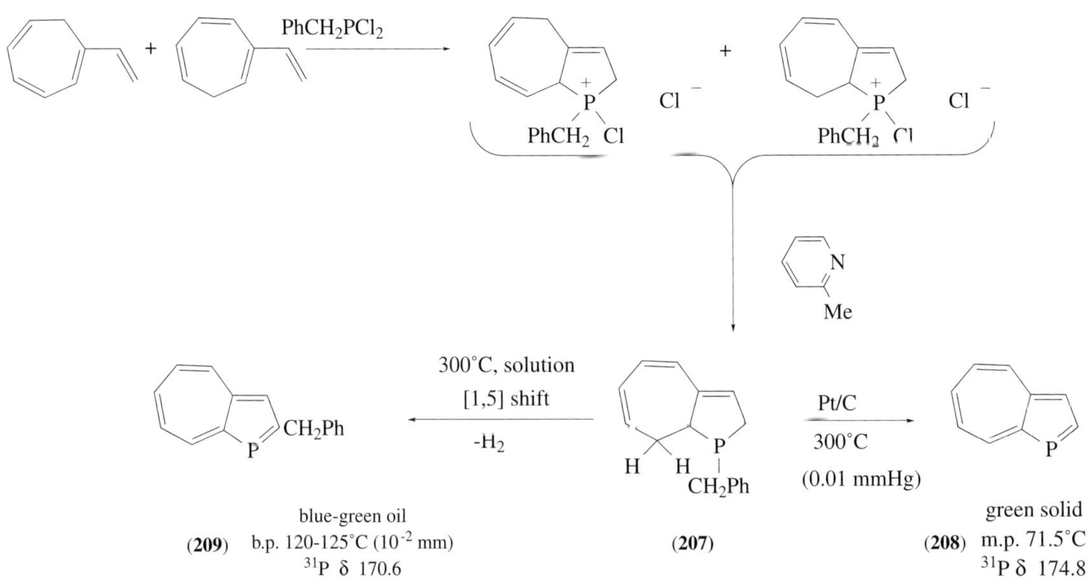

Scheme 47

These remarkable compounds were of good stability; in fact the molecular ion was the base peak in the mass spectrum of (**209**) ⟨83AG(E)57⟩. All showed strong UV absorption that was appropriate for an azulene derivative, and strongly deshielded ^{31}P ⟨92TL5495⟩ due to the double bond. Some other properties are noted by the structures. The methods of synthesis of the phosphazulenes are shown in Scheme 47. In the synthesis of phosphole (**207**), conventional reactions (McCormack cycloaddition, dehydrohalogenation with α-picoline) were remarkably applicable. When (**207**) was heated at 300°C (0.01 mm Hg) in the gas phase over Pt/C, the expected elimination of toluene occurred forming phosphaazulene (**208**) ⟨92TL5495⟩. Earlier it had been noted that when (**207**) was heated at 300°C in diphenyl ether or xylene, the familiar [1,5] sigmatropic shift of the P-benzyl group occurred; aromatization to phosphaazulene (**209**) by elimination of H_2 followed. Phosphaazulene (**210**) ⟨83AG(E)879⟩ was prepared more directly in a simple condensation of a phosphanitrile (or equivalent) with a heptafulvene. Another type of bicyclic 2*H*-phosphole has the 2-keto structure (**211**). It was synthesized by two groups ⟨83ZOB2796, 86CB2915⟩, and found to be stable and spectroscopically observable up to about −20°C. A novel [2+2] dimerization ensued; the structure of this product (**212**) was elucidated by X-ray analysis in the later study.

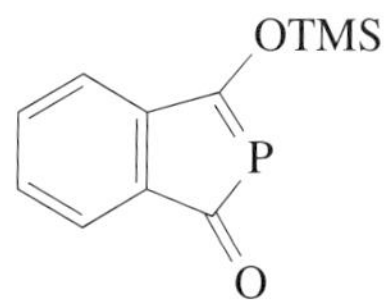

(**211**)

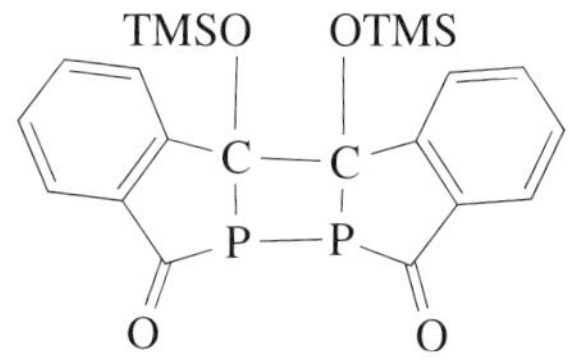

(**212**)

As noted, the existence of 2*H*-phospholes as fleeting intermediates had been placed on firm ground (by work of the Mathey laboratory) that preceded the actual isolation of a stable monocyclic 2*H*-phosphole. Some of this work is described here; to summarize, it was discovered that at high temperatures a phenyl group on phosphorus of a phosphole migrated in a [1,5] sigmatropic shift to an adjacent carbon, giving the highly reactive, non-observable 2*H*-phosphole (e.g. (**213**) and (**215**); a proton shift ensued in (**213**) to give (**214**)) (Scheme 48) ⟨81JA4595⟩. The 2*H*-phospholes could be trapped as noted in Scheme 49. Computational studies by Bachrach support the notion that 1*H*-phospholes rearrange to the 2*H*-isomers, which are of lower energy ⟨94JOC5027⟩.

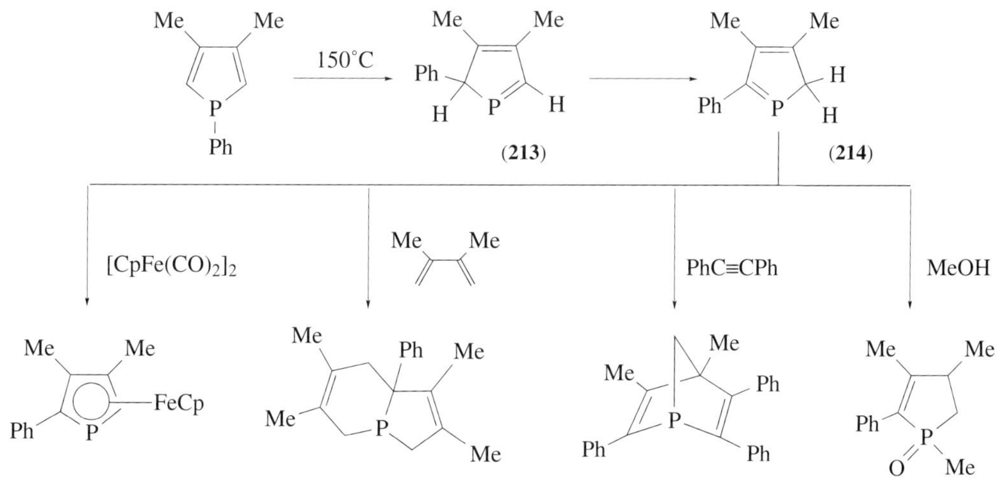

Scheme 48

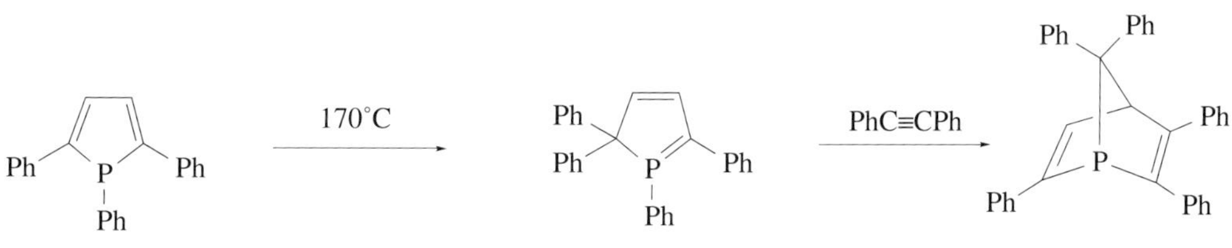

Scheme 49

It was later noted ⟨89JOC4754⟩ that activated alkenes reacted with the 2*H*-phospholes (faster than with the 1*H*-isomer) to form the very valuable 1-phosphanorbornene system (Equation 42).

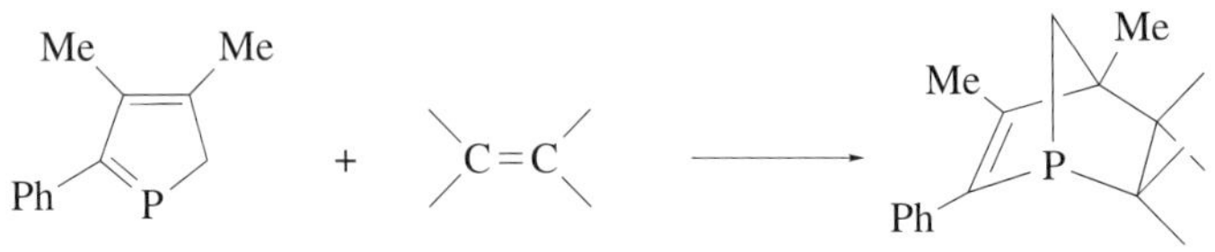

(42)

1*H*-Phospholes, prepared by protonation of phospholide ions at −90°C, behave in a similar manner; their rearrangement is so fast that it occurs on gradually raising the temperature of the protonation mixture ⟨82TL511, 83JA6871⟩. The 2*H*-phospholes rapidly form dimers as discussed in Section 4.2.1.5.8.

If the 1*H*-phosphole is heavily substituted, as in (**216**), stability and lifetime of the materials are increased, and an addition of the 1*H*-phosphole to the 2*H*-phosphole occurs (Scheme 50).

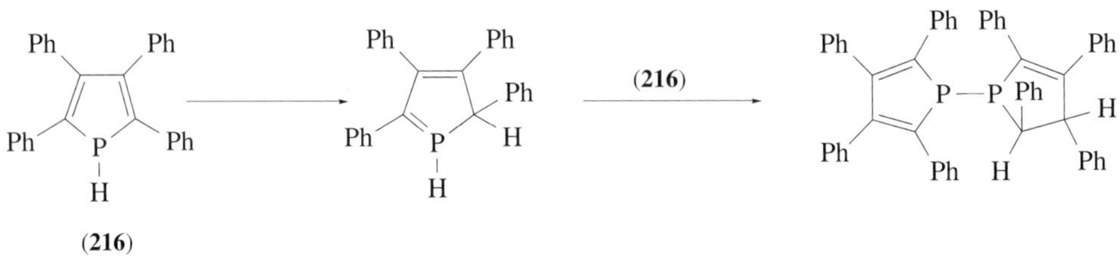

Scheme 50

The important observation was made that the dimer (e.g. (**218**) from 2*H*-phosphole (**217**)), whose structure was proved by X-ray analysis, thermally rearranged to the exo isomer (**219**), implying reversibility in the dimerization process ⟨82CC1272, 83JA6871⟩. It was found possible to conduct the retro-dimerization in refluxing toluene, and when dienes or ethynes were present, the 2*H*-phosphole could be trapped (Scheme 51). Adducts of alkenes with 2*H*-phospholes showed similar reversibility ⟨89JOC4754⟩.

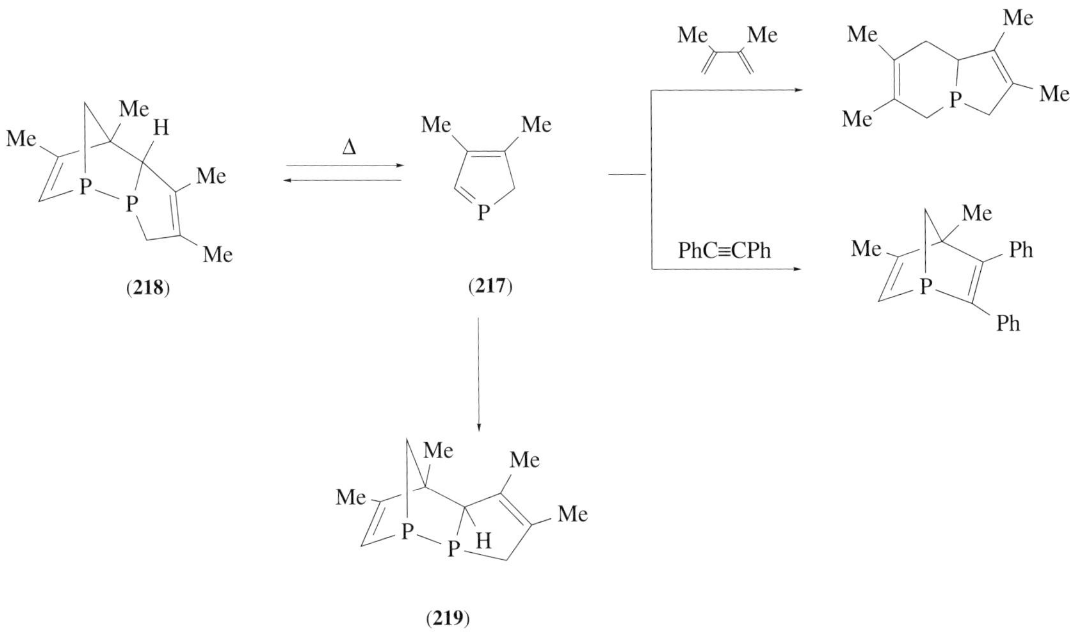

Scheme 51

It was also shown by the Mathey laboratory that the 2*H*-phosphole moiety could be preserved in the form of a chromium or tungsten (e.g. (**221**)) complex ⟨84JA826⟩. This was accomplished by preparing a phospholide ion complex (e.g. complex (**220**)) and treating it with acid at room temperature; P-protonation occurred and the usual rearrangement to the 2*H*-phosphole followed (Equation 43). The nature of the complex then prevented the usual dimerization from taking place. This is an excellent example of the power of conducting reactions in the coordination sphere of a metal.

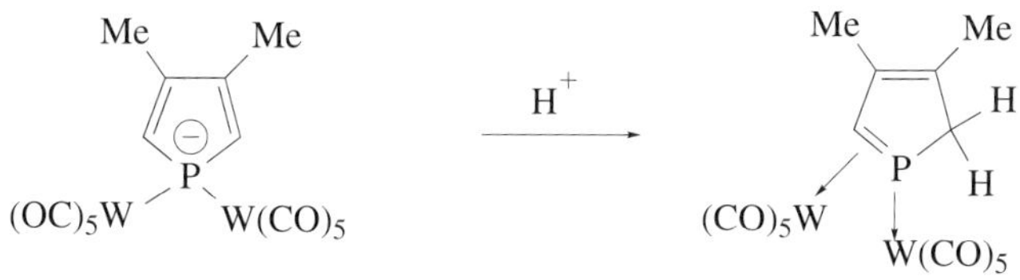

(43)

(**220**) (**221**)

4.2.1.7.2 3*H*-Phospholes

There are no examples yet known of 3*H*-phospholes; in fact, the only stable structure based on a 3*H*-phosphole seems to have the ylidic structure (**222**) 〈86CC1574〉.

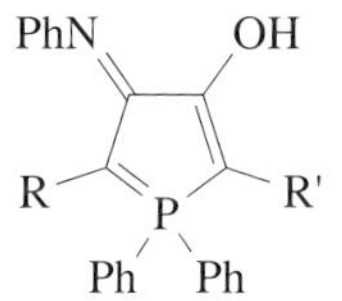

(**222**)

4.2.1.7.3 Five-Coordinate Phospholes (Phosphoranes)

The phosphole moiety has been incorporated in the trioxyphosphorane structures (**223**)–(**225**). In this coordination state, the ^{31}P shift is typically strongly negative and the experimental values for these compounds support the structures, as do other spectral data. Compounds (**223**)–(**225**) are stable in solution at −10°C but decompose at room temperature ((**223**) 〈81CC420, 81TL3533〉; (**224**) and (**225**) 〈81CC420〉). However, the bicyclic phosphorane (**226**) was a stable crystalline solid, and its structure was confirmed by X-ray analysis 〈82AX(B)3024〉.

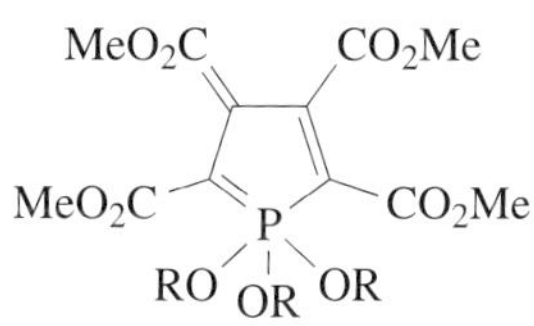

(**223**), R = Me
(**224**), R = Et
(**225**), R = Pr^i

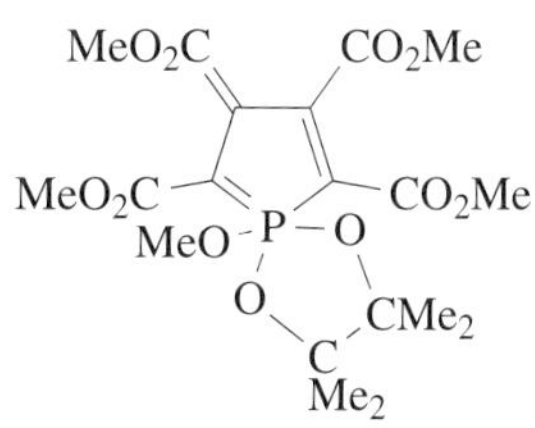

(**226**)

Such structures are not made from the phospholes as reactants but are prepared indirectly by condensation reactions of alkynic derivatives with trialkyl phosphites 〈81CC420, 81TL3533〉, whereupon a cyclization process occurs that gives the phosphole ring. In the reaction of dimethyl ethynedicarboxylate with trialkyl phosphites, an intermediate has been detected at −50°C that has structure (**227**) 〈84JCS(P1)1627〉.

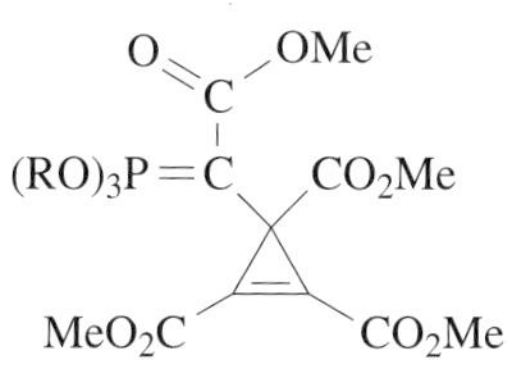

(**227**)

At room temperature, the monocyclic phosphoranes undergo rearrangement by a [1,5] sigmatropic migration of an alkoxy group from phosphorus to an α-ring carbon, thus creating an ylidic structure (**228**) ⟨81TL3533, 84JCS(P1)1627⟩.

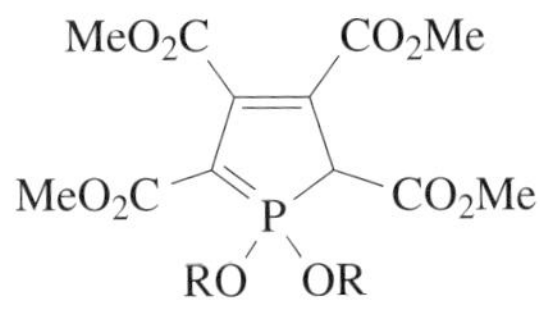

(**228**)

It has also been shown that dialkyl phosphonites give similar phosphoranes with dimethyl acetylene dicarboxylate ⟨87PS(29)123⟩. These products are less stable than those from phosphites, and undergo the [1,5] rearrangement at −50°C.

Transient intermediates with five carbon substituents attached to phosphorus have been proposed in related reactions between phosphines and acetylene dicarboxylates ⟨69JCS(C)1100, 71PS(1)139⟩ but are not directly detectable ⟨93CJC1873⟩, such reactions giving the ylide from the [1,5] rearrangement process (Scheme 52).

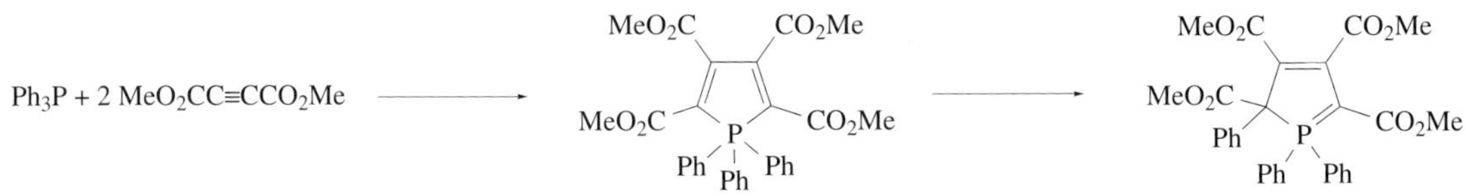

Scheme 52

However, a C.N.-5 phosphole with all-carbon substituents has been prepared by another method ⟨72JCS(D)102⟩ (Equation 44).

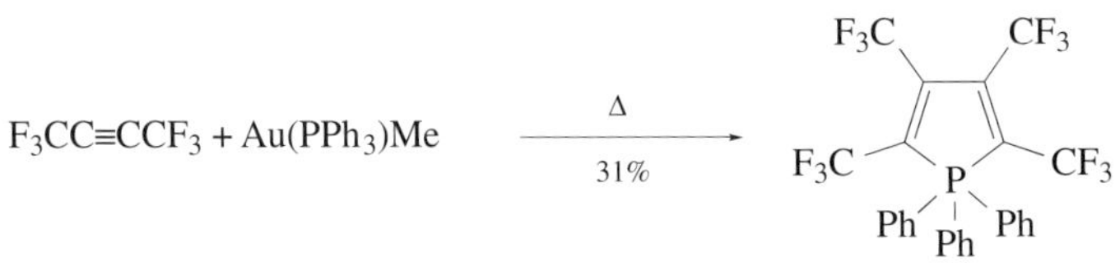

(44)

4.2.1.8 REACTIVITY OF RING SUBSTITUENTS

4.2.1.8.1 Reactivity of P-Substituents

There are no examples of reactions occurring on a P-substituent that comprise retention of the substituent in a modified form. The only type of reactivity known consists of the displacement or cleavage of some P-substituents.

The phenyl group on phosphorus can be replaced by an alkyl group on reaction with alkyllithium reagents ⟨72T4171⟩. This has proved to be a process of considerable value for the introduction of the *t*-butyl group on phosphorus; 3,4-dimethyl-1-*t*-butylphosphole was prepared for the first time by this reaction. The reaction is conducted in hexane solution at room temperature in the presence of TMEDA; it is presumed to proceed by expansion of the coordination number of phosphorus with formation of a phosphoranyl anion (**229**). This anion returns to the three-coordinate phosphole by loss of phenyl anion. The phenyl group may also be displaced by *n*-butyl in similar fashion ⟨72OMR(4)171⟩ (Scheme 53).

Another type of reactivity displayed by P-phenyl substituents is their ready cleavage by alkali metals. This is a reaction of great importance in phosphole chemistry as the products are phospholide ions, widely used as precursors of other P-substituted phospholes (see Section 4.2.1.5.7). The P-methyl group can also be cleaved by metals, as can an ethylene linkage between two phosphole nuclei (Equation 45).

Halogen on phosphorus may be displaced by alcohols (Equation 46) ⟨82JOM(231)361⟩.

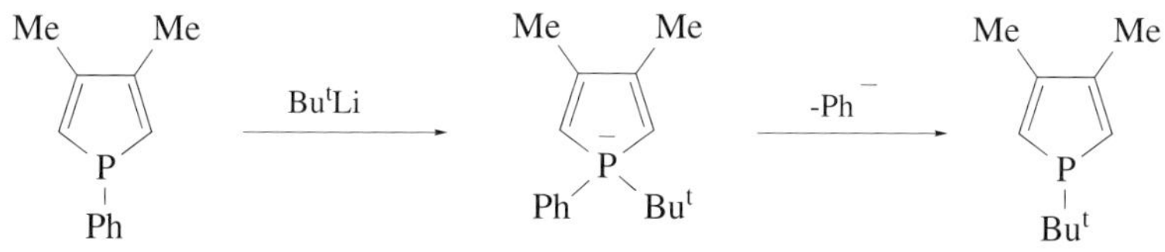

(**229**)

Scheme 53

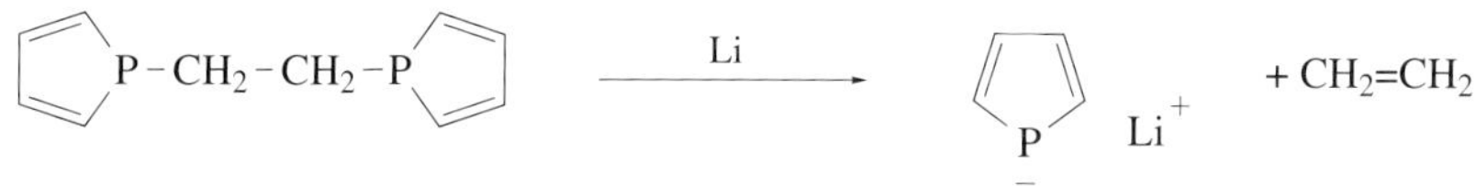

(45)

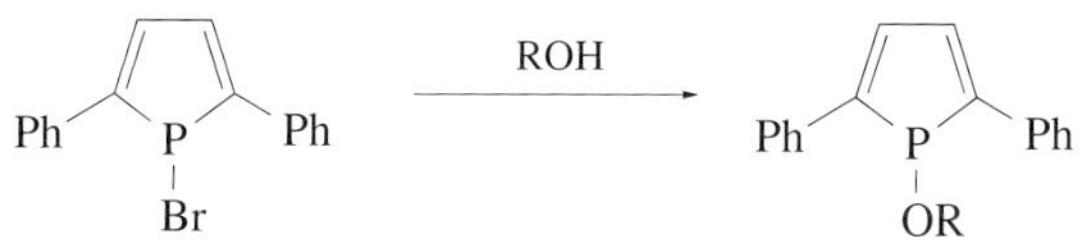

(46)

4.2.1.8.2 Reactivity of Substituents on Carbon

The phosphole literature in its present state is devoid of information on the reactivity of carbon-based substituents of any type attached to ring carbons. Phospholes are π-excessive heterocyclic systems, and C-alkyl substituents will not be activated; substituents such as ester and carbonyl should exhibit normal reactivity. At present there is only one known C-halo derivative of a phosphole. In this compound (2-bromo-3,4-dimethyl-4-phenylphosphole), the bromine may be replaced with lithium, and the lithio derivative used in conventional reactions with electrophiles ⟨92BSF486⟩ (see Section 4.2.1.11).

4.2.1.9 RING SYNTHESES

4.2.1.9.1 General

Construction of the phosphole ring, with one useful exception, is accomplished by methods that are quite different from those employed for the N, S, O ring systems. Because of fundamental differences in the chemistry of phosphines versus amines, none of the familiar carbonyl condensation processes are known to be applicable to phosphole synthesis. Thus an attempt to use the Paal–Knorr condensation of 1,4-dicarbonyl compounds with $PhPH_2$ ⟨65JCS2184⟩ and with PH_3 ⟨88ZOB783⟩ failed to give phospholes. The only successful method known to the present that is common to both the phosphole and pyrrole ring systems is the condensation of 1,3-diynes with primary phosphines or amines. This method has some generality that makes it of value in phosphole synthesis, but the other methods that are used for direct ring synthesis are of limited scope. In fact, the best methods for obtaining phospholes are based not on direct ring closures but on the synthesis first of the 3-phospholene ring system, with subsequent conversion to phospholes. These methods are treated as ring transformations and are discussed in Section 4.2.1.10.

4.2.1.9.2 Condensation of Alkynic Compounds with Phosphines

The first example of this method was discovered by G. Märkl in 1967 ⟨67AG(E)86⟩, and was one of the earliest that provided variety in the substitution pattern on the phosphole ring; the others known at that time, to be discussed, were limited to the formation of C-phenyl derivatives. Märkl employed diynes with both 1- and 4-substituents, and thus created phospholes with 1,2,5-trisubstitution (Equation 47).

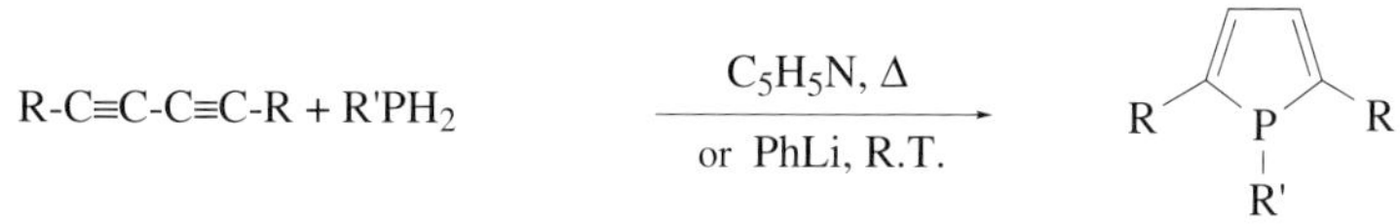

(47)

The primary phosphine needed for this process may be replaced by the dimethylol derivative ($RP(CH_2OH)_2$), a material that is far easier to handle but gives significantly lower yields than the

Table 12. Phospholes from phenylphosphine with diynes [a]

$R(C{\equiv}C)_2R'$ *R*	*R'*	*Phosphole*	*Yield (%)*	*Ref.*
Ph	Ph	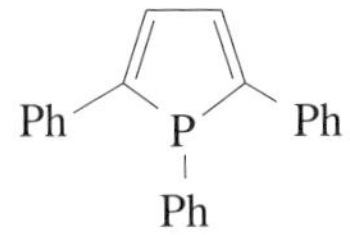	60	67AG(E)86
2-Naph	2-Naph	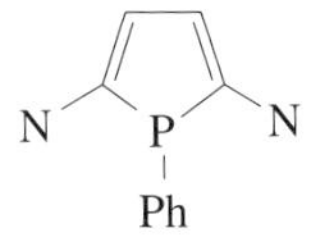	89.5	67AG(E)86
p-Tol	*p*-Tol	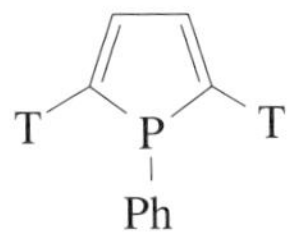	59	67AG(E)86
2-BrC_6H_4	2-BrC_6H_4	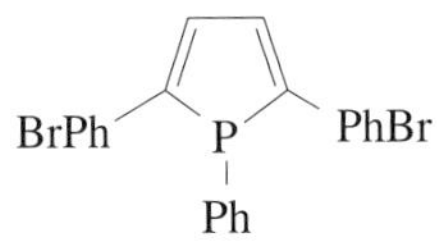	-	67AG(E)86
Me	Me	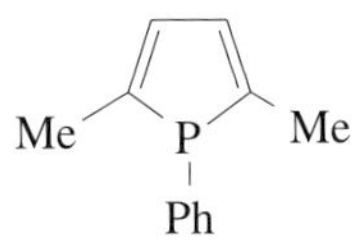	51	67AG(E)86
Ph	Me	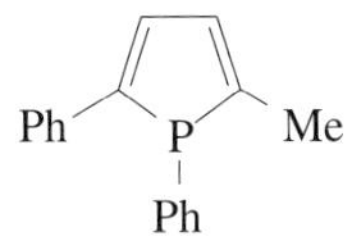	30	71JA6205
Me	$PhCH_2CH_2$	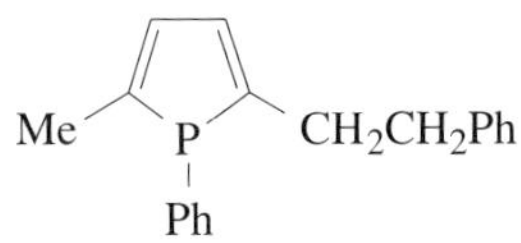	70	71JA6205
Ph	Pr^i	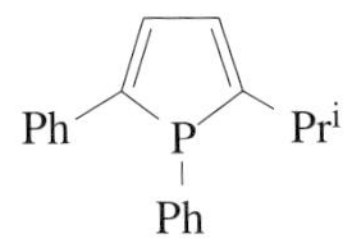	86	71JA6205

[a] N = 2-naphthyl, T = p-tolyl, BrPh = 2-bromophenyl.

free phosphine. Thus, 1,2,5-triphenylphosphole was formed in 60% yield from phenylphosphine and 1,4-diphenylacetylene, but only 20.5% from the dimethylol derivative.

Phospholes prepared in the original report as well as by later users of this method are shown in Table 12. In addition to bases, the reaction may be promoted by metal salts and by free radicals ⟨71T5523⟩. Märkl had employed only symmetrical diynes in the reaction, but unsymmetrical diynes work quite well ⟨71JA6205⟩.

The alkyne-phosphine condensation can be performed in intramolecular fashion, allowing the synthesis of the benzo[*b*]phosphole (phosphindole) ring system, as seen in Scheme 54. The phosphindoles were obtained in 90–96% yield ⟨93TL3103⟩.

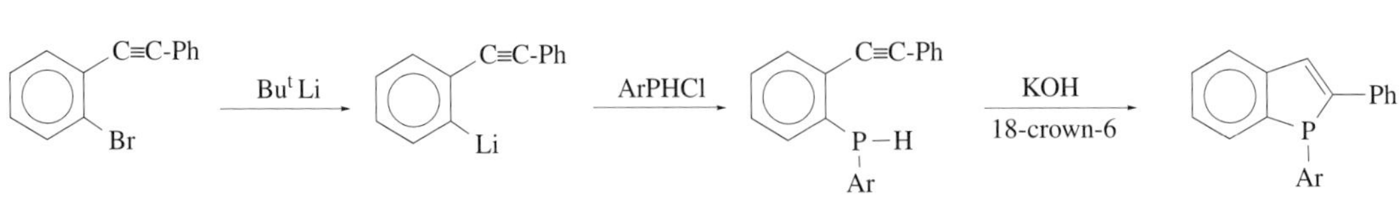

Ar = 2,4,6-tri-Bu^t C_6H_2

Scheme 54

4.2.1.9.3 Reaction of 1,4-Dimetallic Derivatives of Dienic Systems with Phosphonous Dichlorides

This was the first approach used for the synthesis of a monocyclic phosphole, the pentaphenyl derivative, and stands as a landmark in heterocyclic phosphorus chemistry. In 1959, two laboratories reported the synthesis of this compound. In one synthesis, the dilithio derivative of tetraphenylbutadiene was reacted with phenylphosphonous dichloride (Scheme 55); the reaction proceeded in excellent yield (84% ⟨59JA3163, 60JA5099⟩). Since the starting diene is easily prepared from diphenylethyne and lithium, it remains the best way to prepare this derivative, which has found use in many subsequent studies. In addition to the P-phenyl derivative, the P-benzyl compound was also reported (38% ⟨61JA4406⟩).

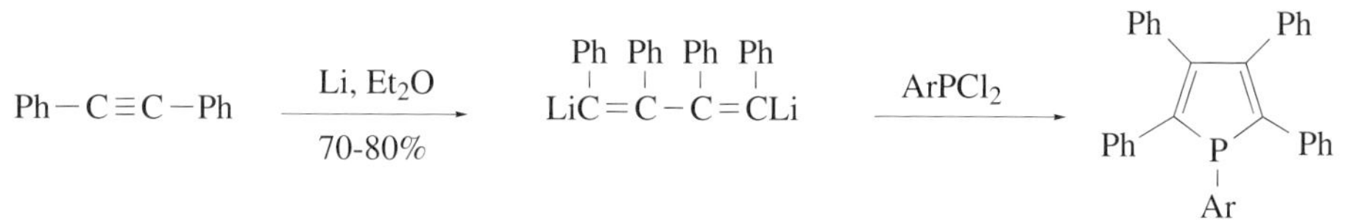

Scheme 55

The other synthesis of pentaphenylphosphole ⟨59CI(L)1250⟩ employs an iron carbonyl complex (**230**) as the metallic reactant with phenylphosphonous dichloride (Scheme 56).

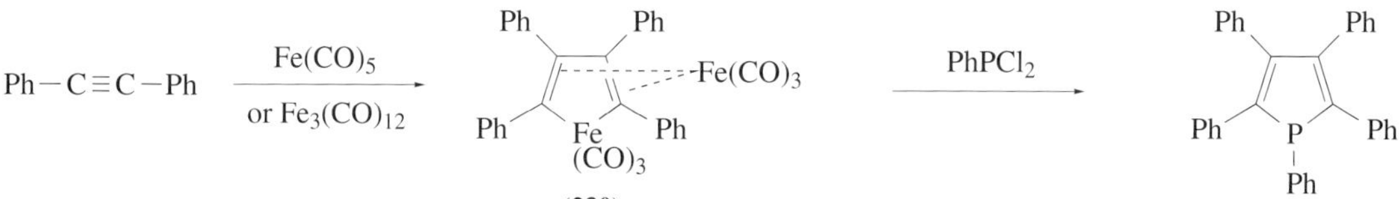

(**230**)

Scheme 56

A 1,4-dilithio system was used in the synthesis of the first phosphindole ⟨67JA5732⟩. *n*-Butyllithium added to diphenylethyne to give the dilithium derivative (**231**), which was then treated with phenylphosphonous dichloride to form crystalline 3-*n*-butyl-1,2-diphenylphosphindole (**232**) in 40% yield (Scheme 57). This method remains of value, and has been employed by later workers to prepare the phosphindole as a model in other studies ⟨71JA6205, 76JHC65⟩.

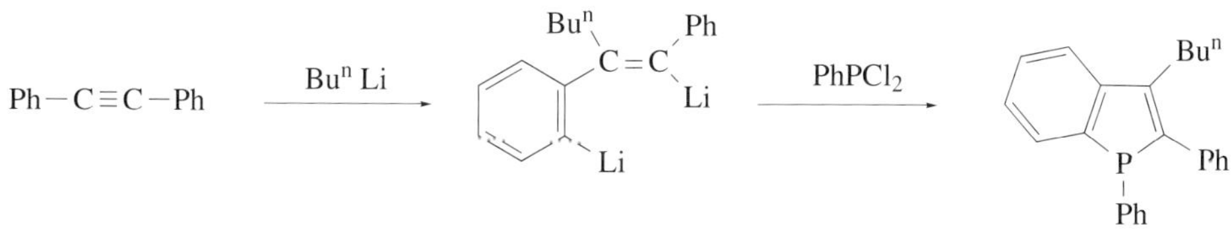

(**231**) (**232**)

Scheme 57

A modification of this approach has been used to synthesize a C-unsubstituted phosphindole (as the oxide) (Scheme 58) ⟨93CC1309⟩.

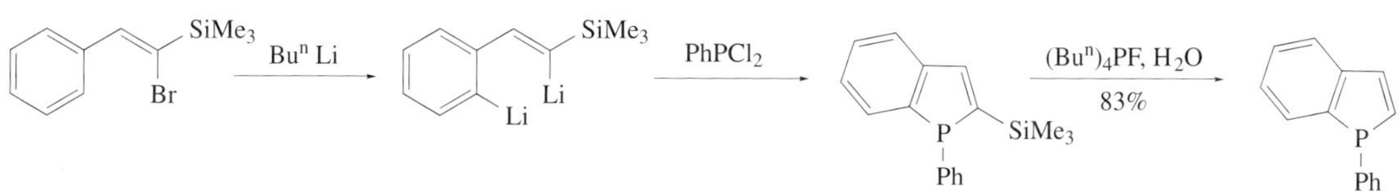

Scheme 58

The *o,o′*-dilithio derivative of biphenyl was used in the first synthesis of a dibenzophosphole, the P-phenyl derivative ⟨53LA(580)44⟩. While this compound is now better prepared by other methods (87% from Ph_4P^+ and MeLi ⟨87SR1307⟩), the reaction principle found early use ⟨64CB747⟩ in the preparation of substituted dibenzophospholes. It remains of synthetic value; examples include the synthesis (63%) of 1-(2-thienyl)dibenzophosphole from reaction of dilithiodiphenyl with 2-thienylphosphonous dichloride

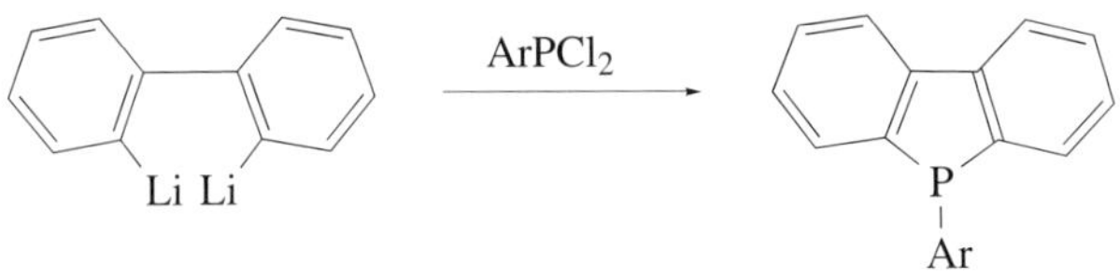

(48)

⟨79ZN(B)1116⟩, and of the P-Cl derivative (not isolated) of some 4,6-diarylbenzophospholes (Equation 48) ⟨82JCS(P1)2317⟩.

A new one-step approach to main group heterocycles involves the reaction of the easily prepared zirconacyclopentadienes with halides (Scheme 59). Yields of heterocycles from zircona derivatives are high. The method has so far only been used for the synthesis of two phospholes ((**233**) and (**234**)), both fully substituted, but appears capable of extension to other types: (**233**) ⟨88JA2310, 92OS(70)272⟩; (**234**) ⟨89AG(E)1367⟩.

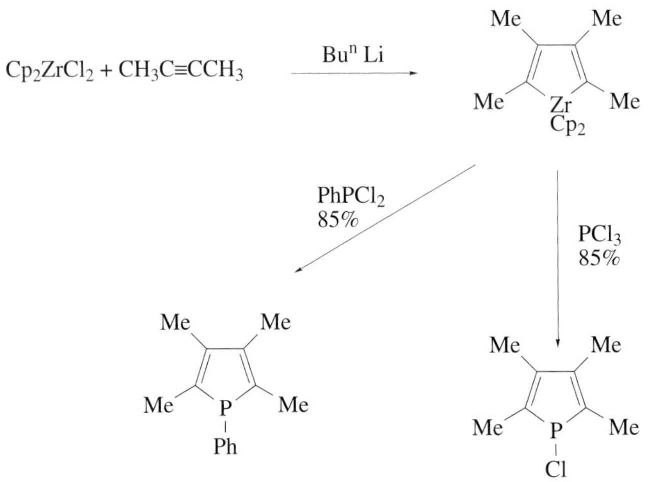

(**233**) (**234**)

Scheme 59

4.2.1.9.4 Reaction of 1,4-Dihalobutadienes with Metallic Phosphides

The first example of this reaction type involved the conversion of the 1,4-dilithio derivative of tetraphenylbutadiene to the 1,4-diiodo derivative (70.5%) and then reaction with the disodium salt of phenylphosphine to form pentaphenylphosphole (88%) (Scheme 60) ⟨61JA4406⟩.

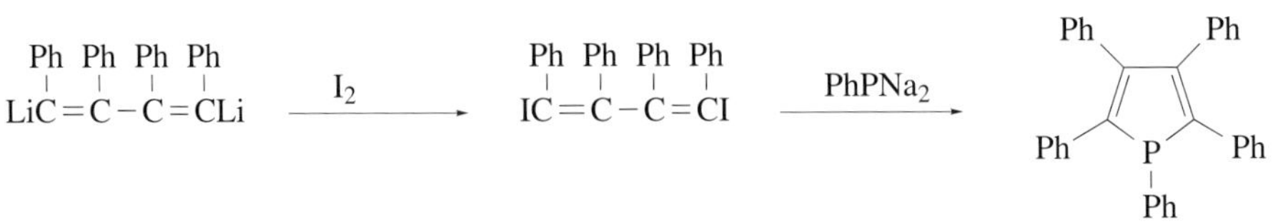

Scheme 60

1-Phenylphosphole has been prepared by the related reaction of 1,4-dichlorobutadiene with $PhPLi_2$ (25%) ⟨87AG(E)229⟩, and other extensions of the general method seem possible.

4.2.1.9.5 Reaction of Enamines with Phosphonous Dichlorides

A discovery in phosphole synthesis in 1988 is that reacting a mixture of two moles of an enamine with a phosphonous dichloride at room temperature results in a one-step construction of the phosphole ring, bearing an amino substituent ⟨88TL4581⟩. This reaction is considered in more detail in Section 4.2.1.11.5 treating functional phospholes; it constitutes an important advance in this area, and is the only source at present of aminophospholes. Further development of the method could make it of major significance in phosphole chemistry.

4.2.1.9.6 Creation of the Phosphole Ring in the Coordination Sphere of Metallic Derivatives

An entirely different approach for the creation of the phosphole ring is found in the field of coordination chemistry. A coordinating metal group can significantly modify the chemistry of a phosphine ligand, allowing novel and useful reactivity to develop without loss of the stabilizing metallic group. The principle has been exploited for phosphole synthesis in brilliant fashion by the laboratory of F. Mathey. Thus, a palladium(0) complex catalyzed the insertion of an alkyne into a phosphirene when the latter was in the coordination sphere of a $W(CO)_5$ group ⟨87TL5021⟩, thereby creating a phosphole. Here also the method is of special value as a source of C-functional phospholes and is discussed from that standpoint in Section 4.2.1.11. In the synthesis of six different phosphole complexes by this method, moderate to excellent yields were obtained (33–85%). After the phosphole complex has been created, the phosphole must be displaced from the coordinating metal; a useful procedure involves reaction with I_2 and 1-methylimidazole ⟨84CC45⟩.

Tricyclic compounds containing two phosphole rings have been synthesized by the reaction of trishydroxymethylphosphine with certain 3-hydroxypenta-1,4-diynes (Equation 49) ⟨74PS(4)279⟩. The products possess the novel diphosphadihydroindazene ring system. Derivatives with R=H, Me, and Et have been prepared in low yield (10–37%).

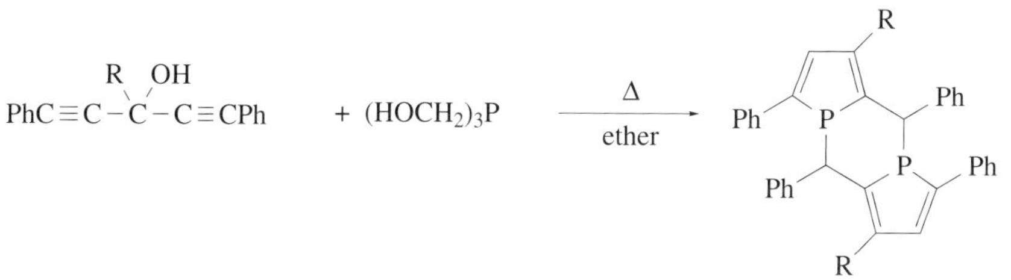

(49)

4.2.1.10 PHOSPHOLE SYNTHESIS BY RING TRANSFORMATION

4.2.1.10.1 General

The McCormack cycloaddition of dienes and phosphorus(III) halides (see Chapter 4.1) has made readily available a great variety of 3-phospholene derivatives, and these have figured heavily in the development of synthetic approaches to phospholes. Numerous techniques have been devised for transformations to the fully unsaturated ring system; they continue to be of great synthetic value, and in some cases these methods are clearly superior to those based on direct formation of phospholes by ring-closure methods. There are also some syntheses based on transformation of phosphole derivatives, but these methods are generally limited to modification of the substitution at phosphorus rather than at ring carbons. A useful exception is the formation of carbanions from phosphole sulfides which can then react with electrophiles; this method is of interest in forming C-functional phospholes and is described in Section 4.2.1.11.2.

4.2.1.10.2 Dehydrohalogenation of McCormack Cycloadducts (Halophospholenium Halides)

The initially formed product in the McCormack cycloaddition can be described as a halophospholenium halide (see Chapter 4.1). Depending upon the reactant combination, the double bond can be found in the 2- or 3-position. When it is preserved in the 3-position, these adducts can undergo dehydrohalogenation to form phospholes (Scheme 61).

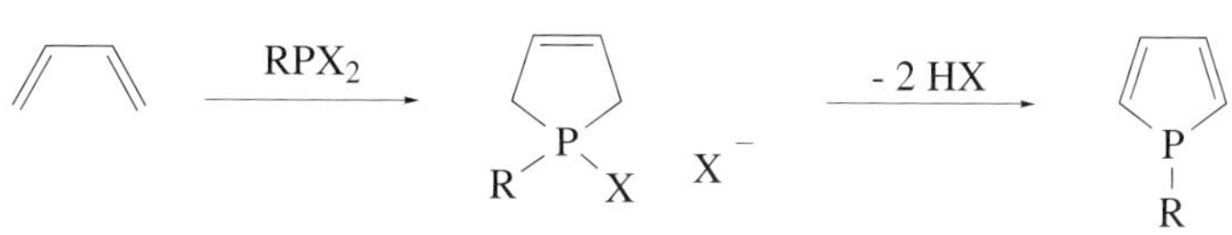

Scheme 61

One of the earliest phosphole syntheses can be considered to be of this type, although the mechanism has not been confirmed: heating at high temperature (226–230°C) of a mixture of 1,4-diphenylbutadiene with phenylphosphonous dichloride gave 1,2,5-triphenylphosphole (**236**), a yellow solid, in 50–66% yield ⟨62CI(L)359⟩. The yield can be improved to 70% by conducting the reaction in excess phenylphosphonous dichloride as solvent ⟨83JOM(256)103⟩, and this remains the best method to obtain this compound. Similarly, treating 1,2,3,4-tetraphenylbutadiene with phenylphosphonous dichloride gave pentaphenylphosphole ⟨65JCS2184⟩, and reacting 1,4-di-*p*-tolyl butadiene with *p*-tolyl PCl_2 gave a 13% yield of 1,2,5-tritolylphosphole ⟨74JCS(P1)1694, 78LA74⟩. The process can be expressed as proceeding through the McCormack cycloadduct (**235**) which then loses HCl (Scheme 62). However, an attempt to synthesize the McCormack cycloadduct from the same reactants at 100°C was unsuccessful, and the mechanism remains unconfirmed. An attempt to use this method for the synthesis of 1,3,4-triphenylphosphole was not successful ⟨74PS(4)199⟩.

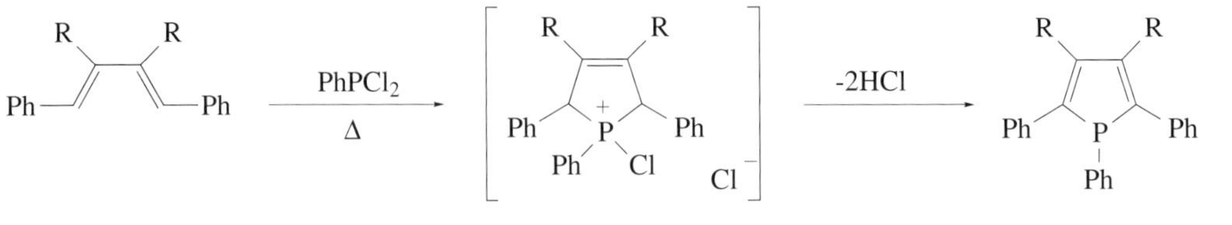

(**235**)

(**236**), R = H

(**237**), R = Ph

Scheme 62

A more clear-cut case of the dehydrohalogenation of McCormack cycloadducts was discovered by Mathey, who accomplished the synthesis of phospholes by the action of DBU as a base in benzene solution on the isolated cycloadduct (Equation 50) ⟨69CR(C)1066⟩.

(50)

This was a major event in phosphole chemistry, as it allowed a number of phospholes, inaccessible by other methods, to be synthesized. This remains the best and most general of all phosphole syntheses, and has been widely used in this field. A review ⟨88CRV429⟩ provides a summary of the 25 phospholes prepared by dehydrohalogenation of halophospholenium halides (some prepared as discussed in Section 4.2.1.10.3). Yields are improved by the use of benzene–CH_2Cl_2 as solvent ⟨70BSF4433⟩ and of α-picoline as the base ⟨81S983⟩. In the latter paper, it was reported that phenylphosphonous dichloride could be used instead of the less readily available dibromide in those cases where it gives a 3-phospholenium adduct rather than a 2-phospholenium adduct. These improvements in this "one-pot" method led to yields of 85% for 3,4-dimethyl-1-phenylphosphole, 75% for 3-methyl-1-phenylphosphole, and 50% for phenylphosphole ⟨81EUP41447⟩. Some of the low yields reported in earlier work (many around 15–30%) could probably be improved by the newer techniques. Another valuable discovery is that the McCormack reaction can be greatly accelerated by the use of pressure in the range of 4–7 kbar ⟨89CC967⟩. Thus, the reaction of 2,3-dimethylbutadiene with phenylphosphonous dibromide normally requires 7–10 days at room temperature for completion, but at 7 kbar it is complete in 2 h. Pressure also has the effect of giving a granular, more tractable form of the cycloadduct. The Mathey dehydrohalogenation is then performed directly on the reaction mixture with α-picoline to give a yield of 77%. Similarly were prepared 1-phenylphosphole from butadiene (75%) and 3-methyl-1-phenylphosphole from isoprene (75%).

The scope of the Mathey method is quite broad and can be extended to the synthesis of multicyclic phospholes (types (**238**) and (**239**) ⟨82PS(12)161⟩; (**240**) ⟨92TL5495⟩; (**241**) ⟨78CJC1952⟩). Also, adducts from some phosphonous dichlorides fail to form phospholes, and the less accessible bromides must be used.

The mechanism of the dehydrohalogenation process has not been directly studied. A reasonable explanation is shown in Scheme 63; it is based on observations made on the course of events of the reverse process, the reaction of phospholes with anhydrous HCl ⟨86JCS(P2)629⟩ (see Section 4.2.1.4.3). A key feature is the formation of a transient 2*H*-phosphole derivative which undergoes a [1,5]-sigmatropic shift to the phosphole.

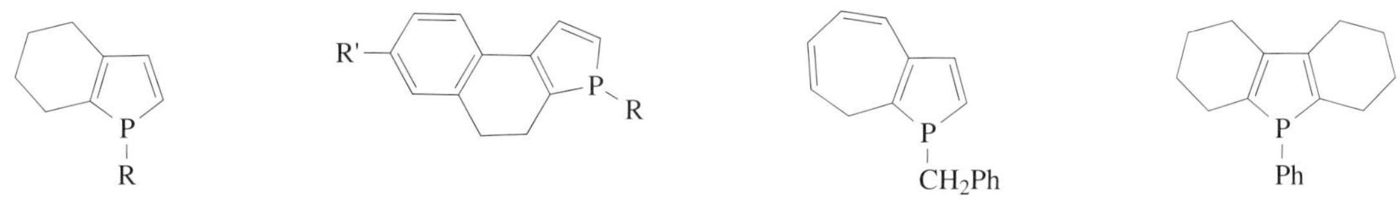

(**238**) R: Me, 18%; Ph, 25%

(**239**) R, R': Me, H, 20%; Ph, H, 40%; Me, MeO, 47%

(**240**), 34%

(**241**), 38%

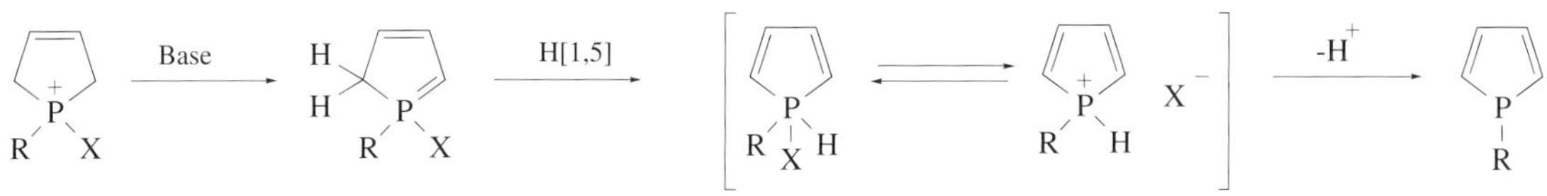

Scheme 63

4.2.1.10.3 Dehydrohalogenation of Halophospholenium Halides from Other Sources

Since the addition to the phosphorus atom of a 3-phospholene gives the same structure as formed in the McCormack cycloaddition, 3-phospholenes can be viewed as starting materials for the Mathey dehydrohalogenation procedure to form phospholes. Normally, the 3-phospholenes are themselves prepared from McCormack cycloadducts, either directly by dehalogenation ⟨64JOC836⟩ or by reduction of the phospholene oxide formed on hydrolysis of the adducts ⟨65CB171⟩. There are, however, occasions when the 3-phospholene to phosphole conversion can be useful. Thus, 3-phospholene oxides or sulfides can undergo C-substitution reactions that are forbidden to phospholes; after such a reaction, the oxygen or sulfur can be removed, for example with trichlorosilane or phenylsilane, and the phosphine converted to the phosphole. This approach was used ⟨73JOC1858⟩ in the first synthesis of a C-functional phosphole ((**244**), see Section 4.2.1.11.2). The acid (**242**) was recovered after conversion of 1-methyl-3-phospholene oxide to its anion with *n*-BuLi, followed by carbonation. The phosphine oxide function in the methyl ester was reduced to the phosphine; bromine addition gave the bromophospholenium bromide (**243**) needed for the Mathey conversion to the phosphole (**244**) (Scheme 64).

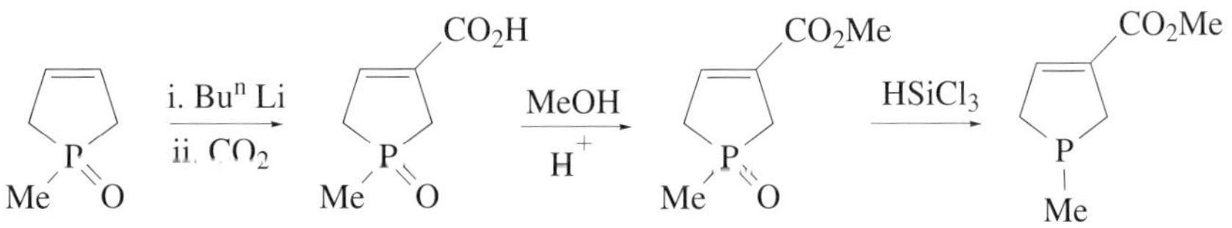

(**242**)

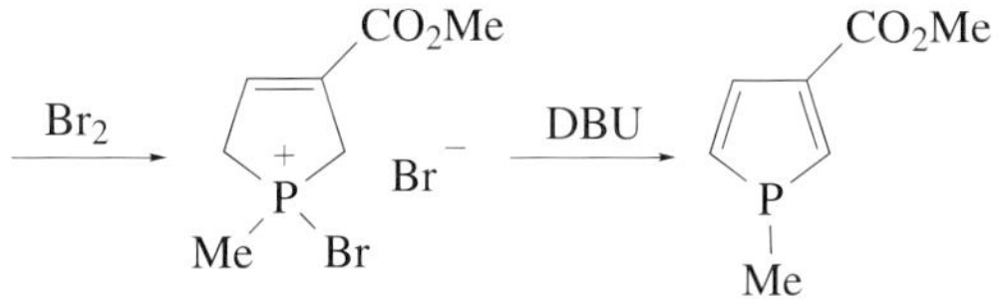

(**243**) (**244**)

Scheme 64

The reaction has been extended to the synthesis of a 2,2′-biphosphole ⟨86JOM(316)271⟩. The biphospholene derivative (**245**) is directly available from the reductive dimerization of phospholes ⟨84JA425⟩. Addition of bromine gave the salt (**246**) which on dehydrohalogenation with α-picoline gave the biphosphole (**247**) (Scheme 65). Phospholes (**248**)–(**250**) were also prepared by this method ⟨90JOC2494, 91HAC377⟩.

It is also possible to place a variety of alkyl substituents on phosphorus of the phospholene system by preparing the P-halo derivative and reacting it with organometallic derivatives. Several phospholes have been prepared by this approach (Scheme 66) ⟨73JOC1858⟩. An obvious advantage is that the need for the

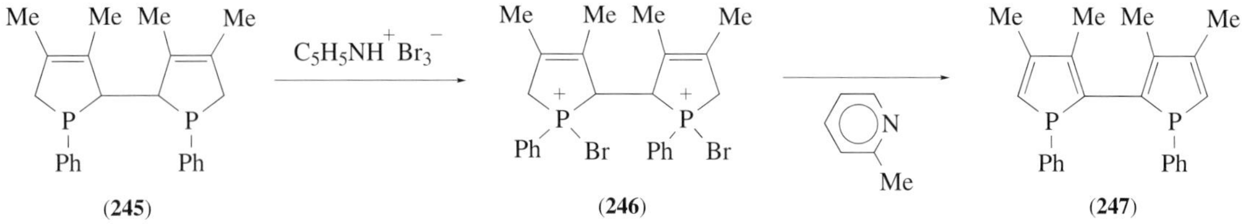

Scheme 65

alkylphosphonous dichloride required in a McCormack reaction is eliminated. The dichlorides are difficult to prepare and handle.

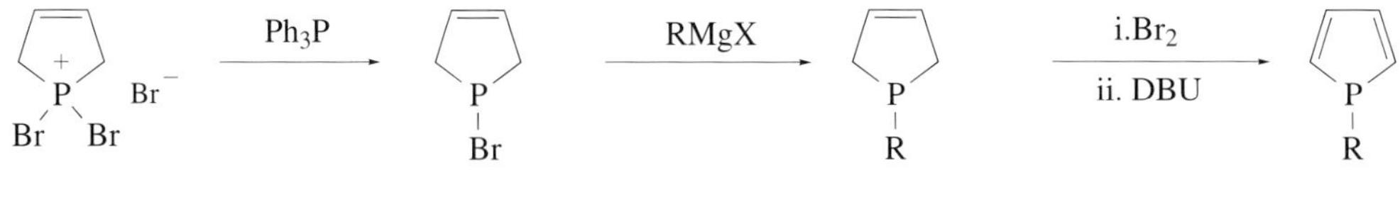

R = $PhCH_2$, $PhCH_2CH_2$, p-$ClC_6H_4CH_2CH_2$

Scheme 66

McCormack type compounds can be generated by another approach, involving quaternization of P-bromophospholenes. Benzyl bromide is especially effective as the quaternizing agent, although methyl bromide may also be employed ⟨73JOC1858⟩. The bromophospholenium salts can then be treated in the usual way to form phospholes. By this method, 1-benzyl-3,4-dimethylphosphole was first synthesized. The yield in the final step was 28.6% with DBU as the base, but this was later improved to 75% with α-picoline (Scheme 67) ⟨81S983⟩. The method has also been used to prepare a series with XCH_2– or XCH_2CH_2– as P-substituents (X = OR, SR, C(O)R, $CH(OR)_2$, COOR, $CONR_2$, and CN; ⟨76CJC2402⟩).

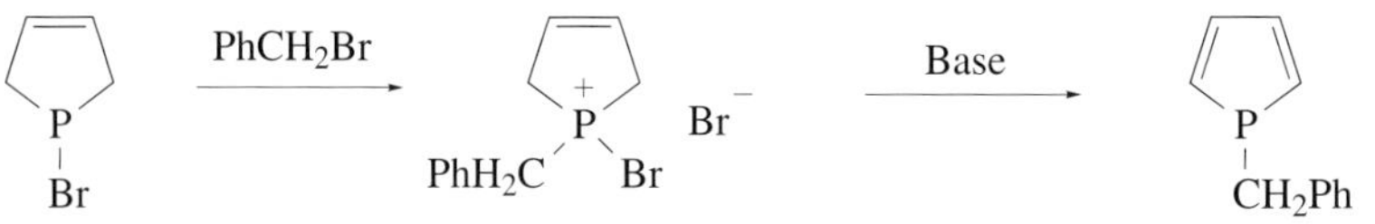

Scheme 67

4.2.1.10.4 Dehydrohalogenation of C-Halo Phospholanes

The first synthesis of methylphosphole ⟨67JA5984, 69JA3308⟩ and of phenylphosphole ⟨68TL1755⟩ employed 3-phospholene oxides from McCormack cycloadducts as the starting materials. Bromine was added to the double bond to form the dibromophospholane oxides; these were deoxygenated with trichlorosilane and phenylsilane, respectively, to form the dibromophospholanes. Potassium *t*-butoxide was used for the dehydrohalogenation of the methyl derivative (13–15% from the dibromo oxide), and DBU for the phenyl derivative (47%) (Scheme 68). These two studies were important in providing the first examples of C-unsubstituted phospholes, which allowed the beginning of serious inquiry into the matter of cyclic delocalization in the phosphole ring.

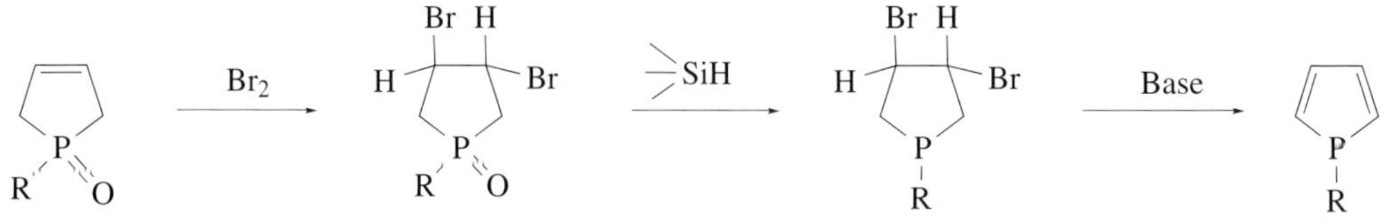

Scheme 68

This method is still of value and has been used to prepare 1-benzylphosphole (26%), 1-(2-phenylethyl)phosphole (10.1%) and 1-[2-(*p*-chlorophenyl)ethyl]phosphole (17.3%) ⟨73JOC1858⟩.

The concept of preparing a C-bromo P-oxide as an intermediate for conversion to a phosphole has appeared in modified form in the synthesis of fused-ring phospholes, for example (**251**) (Scheme 69) ⟨79PS(5)349⟩.

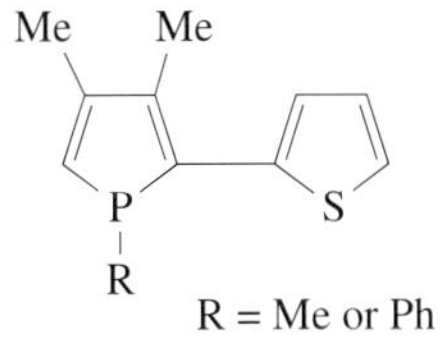

(**248**)

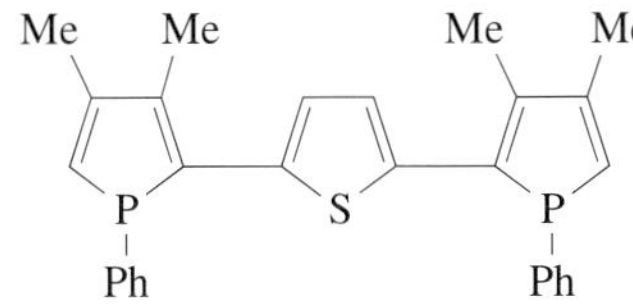

(**249**)

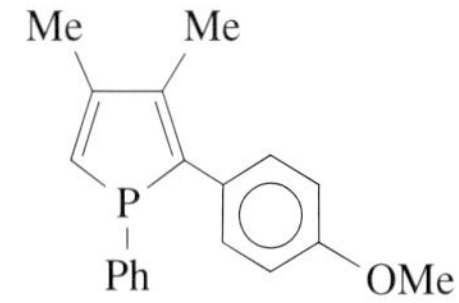

(**250**)

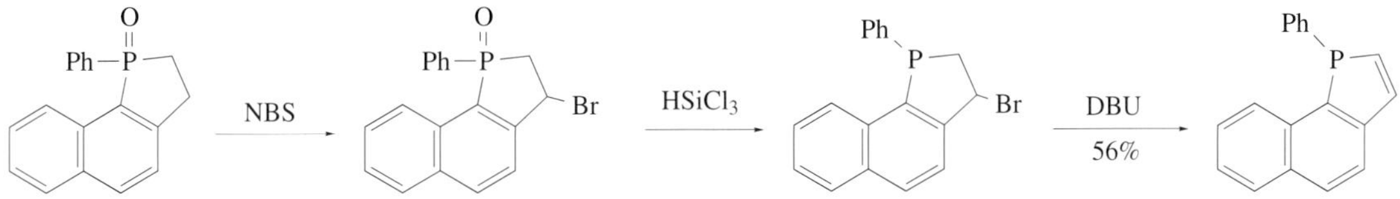

(**251**)

Scheme 69

In the case of the phosphindole system, a C-bromo P-oxide (**252**) may be generated as a precursor to the phosphole oxide, which is then reduced to the phosphole group by a silane ⟨71CJC530⟩. In this fused ring system, the phosphole oxide moiety is stable and does not undergo the rapid dimerization characteristic of most monocyclic phospholes. The entire synthesis is shown in Scheme 70.

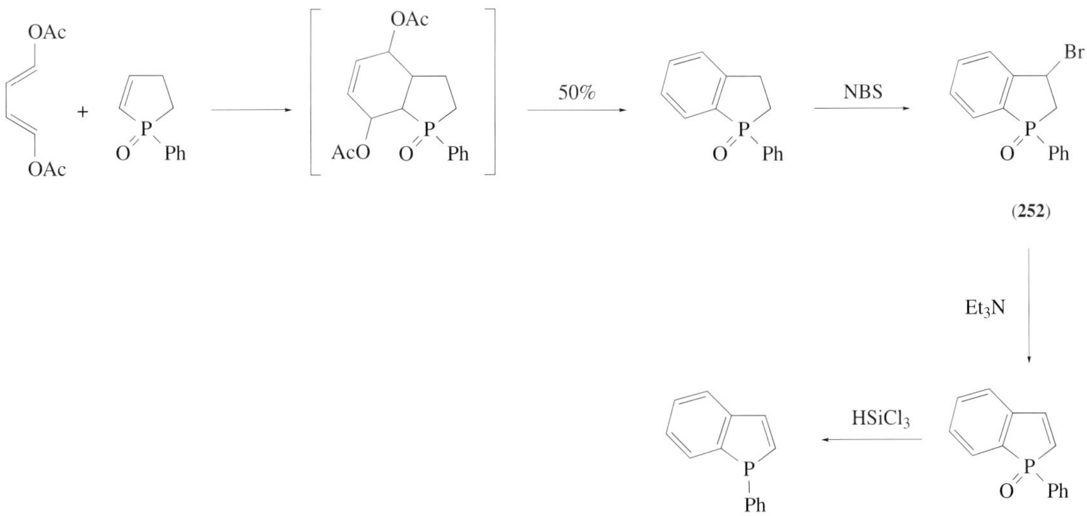

Scheme 70

This was not the case for a benzo[*g*]phosphindole oxide ((**253**) ⟨82PS(12)161⟩); application of the above reaction sequence gave a dimer, rather than the desired monomer, of the benzophosphindole oxide (**253**).

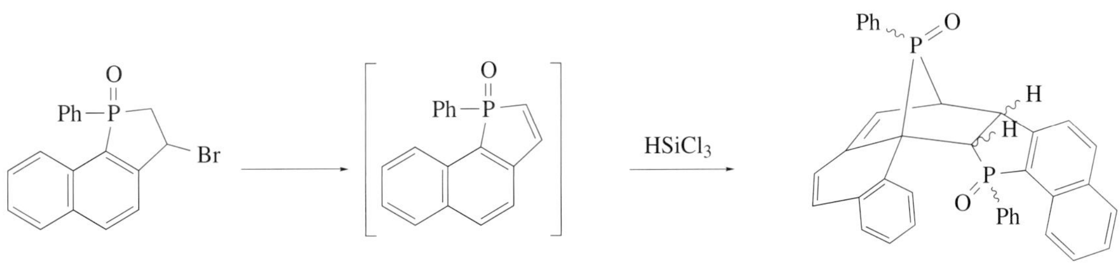

(**253**) (**254**)

Scheme 71

It was necessary to reverse the reaction sequence so as to prepare the bromophosphine for the dehydrohalogenation step as in Scheme 69. The structure of the dimer was proposed, but not fully established, as having the usual Diels–Alder adduct framework.

4.2.1.10.5 Dehalogenation of Halophospholium Tetrachloroaluminates

A new approach to tetrasubstituted phospholes is based on the observation that the tetrachloroaluminate salt of a tetrasubstituted cyclobutadiene–$AlCl_3$ σ-complex undergoes insertion of phosphorus into the ring when reacted with a phosphonous dichloride. A chlorophospholium salt is formed ⟨83TL1423⟩ and hydrolysis gives a stable phosphole oxide. Later workers ⟨88OM921⟩ showed that the salt (**255**) could be dehalogenated to the phosphole (**256**), and thereby accomplished the first synthesis of 1-phenyl-2,3,4,5-tetramethylphosphole (Scheme 72).

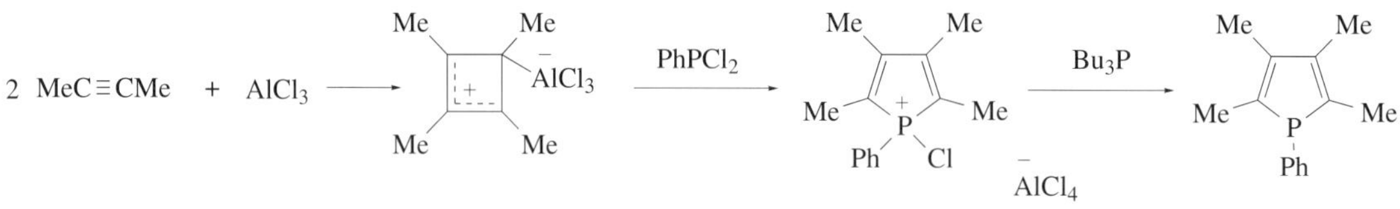

(**255**) (**256**)

Scheme 72

The method is particularly simple, and consists of the formation first of the cyclobutadiene salt from 2-butyne and $AlCl_3$, addition of the phosphonous dichloride, and finally tri-*n*-butylphosphine as the reducing agent, without isolation of any intermediates. The yield was 68%. The method will have limited scope, however, due to restrictions in the alkyne that will form the cyclobutadiene complex.

4.2.1.10.6 Displacement of P-Substituents

The P-phenyl group of a phosphole can be directly displaced by reaction with alkyl lithium reagents in TMEDA. Both *t*-butyl ⟨72T4171⟩ and *n*-butyl ⟨72OMR(4)171⟩ have been placed on P by this method. With 3,4-dimethyl-1-phenylphosphole, the former reaction occurred in 70% yield, the latter in 31.5% (with some oxide) (Scheme 73).

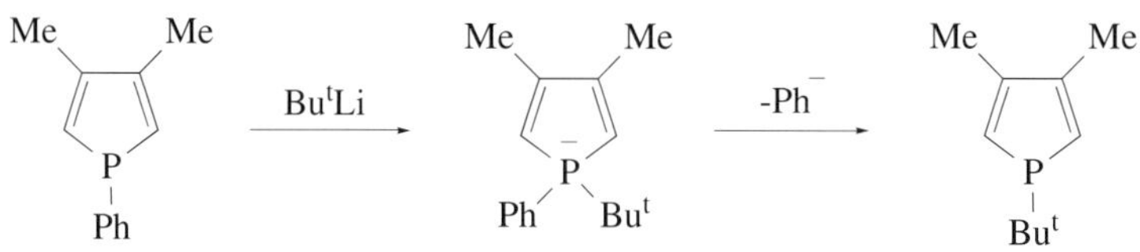

Scheme 73

This method is of considerable value for the introduction of the *t*-butyl group on phosphorus, as this group cannot be used as the P-substituent in a phosphonous dihalide in the McCormack reaction because of steric restrictions.

A more versatile method for a group interchange at P consists of cleaving a P-phenyl substituent with a metal (lithium, sodium, or potassium) to form the phospholide ion, which can then be reacted with electrophiles. In every case known the electrophile has attacked exclusively at P, even though electron delocalization in the phospholide ion is extensive (see Section 4.2.1.2). This reaction was discovered early in the development of phosphole chemistry ⟨71T5523⟩, and has been widely used over the following years (Scheme 74).

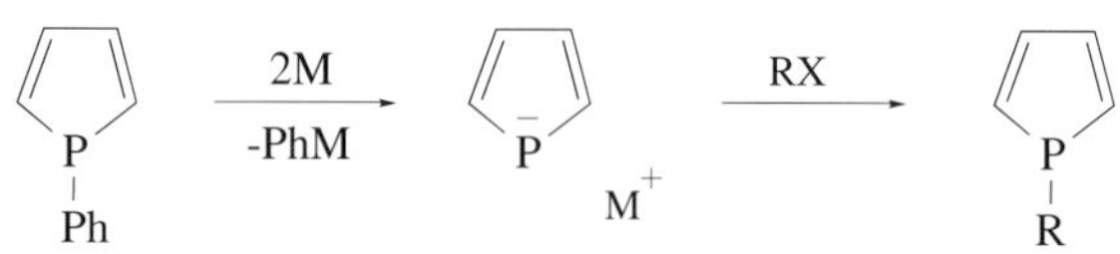

Scheme 74

A considerable number of alkyl halides, dihaloalkanes, and halides with other functional groups has been used for this purpose (for a summary, see ⟨B-90MI01⟩). Biphospholes (**257**) are created by reaction of the anion with phosgene ⟨82JOM(231)361⟩ or iodine ⟨83OM1234⟩. The biphospholes serve as precursors of other P-functional phospholes; reaction with bromine gives 1-bromophospholes, which with alcohols provide 1-alkoxyphospholes (Scheme 75). This is the only method presently available for synthesizing these derivatives ⟨82JOM(231)361⟩.

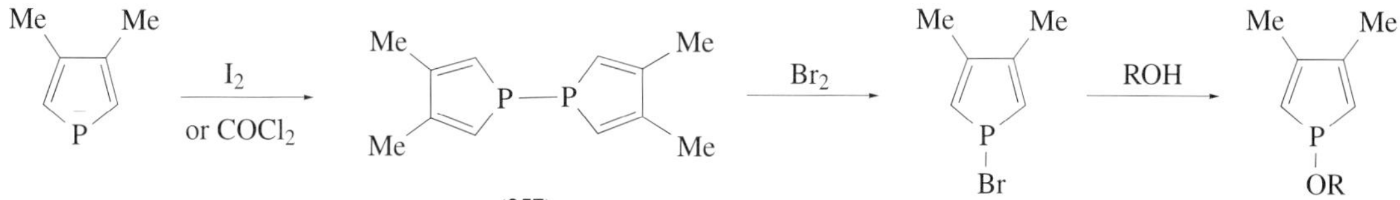

(**257**)

Scheme 75

4.2.1.11 SYNTHESIS OF PARTICULAR TYPES OF PHOSPHOLES

4.2.1.11.1 General

With the exception of the ring synthesis that provides aminophospholes (see Section 4.2.1.9.5), the reactions used for the construction of the phosphole ring are not amenable to the direct synthesis of phospholes with C-substituents other than alkyl or aryl groups. This is a drastic difference with regard to the related heterocycles with N, S, or O in the ring. Additionally, there are no known direct electrophilic substitution reactions that take place on phosphole ring carbons. Indeed, the first functionalized phosphole (a carboxylate ⟨72TL299⟩) was not prepared until 1972. To address the need for functional phospholes, techniques have been devised that are based on appropriate substitutions on dihydro derivatives, followed by conversion to phospholes, or on phospholes where phosphorus bears additionally a blocking function such as O, S, or a metallic group, which can be removed at the end of the synthesis. For the most part, these synthetic procedures are multistep with low overall yields. For these reasons, functionalized phospholes remain rare. There remains much to be done in this area, both in the creation of practical new methods and in the generation of new structures.

4.2.1.11.2 Carboxylic Acids and Derivatives

The first synthesis of a carboxylate was based on the use of a phospholene oxide as starting material ⟨72TL299⟩. Carbanion formation occurs relatively readily with these compounds; reaction of the resonating ion (**258**) with carbon dioxide installed the carboxy group on the ring at positions 2 and 4 with subsequent double bond rearrangement to (**259**) and (**260**). The mixture was then esterified; isomer (**261**) was unstable at room temperature and formed an insoluble polymer, making it possible to isolate (**262**). The oxygen on P in (**262**) was then removed, and the phosphine treated with bromine to provide a bromophospholenium ion (**263**). This is the same species as formed in the McCormack diene-phosphonous dibromide cycloaddition reaction, widely used as an immediate precursor of phospholes (see Section 4.2.1.10.2). By this approach (Scheme 76), two phosphole-3-carboxylates ((**264**) and (**265**)) have been synthesized ⟨72TL299, 73JOC1858⟩. These phosphole carboxylates are not stable at room temperature; they can be distilled if this is performed rapidly at reduced pressures, but they require temperatures around −75°C for storage.

Phosphole sulfides are frequently stable, unlike oxides which rapidly dimerize, and can be used as precursors of phospholes by simple reduction methods. In another approach ⟨76T2395⟩, this stability was taken advantage of for the introduction of the carboxylate group on the 3,4-dimethylphosphole ring (Scheme 77). The sulfide forms a carbanion (**266**) which can be quenched with diethyl carbonate to form a phosphole sulfide 2-carboxylate. This was converted to the phosphole by removing the sulfur with a more nucleophilic phosphine (tri-*n*-butylphosphine or tris-(2-cyanoethyl)phosphine). The phosphole-2-carboxylates prepared this way ((**267**) and (**268**)) were significantly more stable than the 3-carboxylates and could be stored at room temperature. 3,4-Dimethyl substitution is well known to increase the stability of phospholes, and at present it is unknown if this substitution pattern or the position of the carboxylate is responsible for the increased stability.

Until 1992, no C-metal derivative of a phosphole was known, and some of the conventional routes to carbonyl derivatives from this type of precursor were not practical. However, 2-bromo-3,4-dimethyl-1-phenylphosphole (**269**) has now been synthesized (see Section 4.2.1.11.8 ⟨92BSF486⟩) and this allowed the synthesis of the 2-lithio derivative (**270**) and its use in various processes (Scheme 78). Thus, the 2-COOH

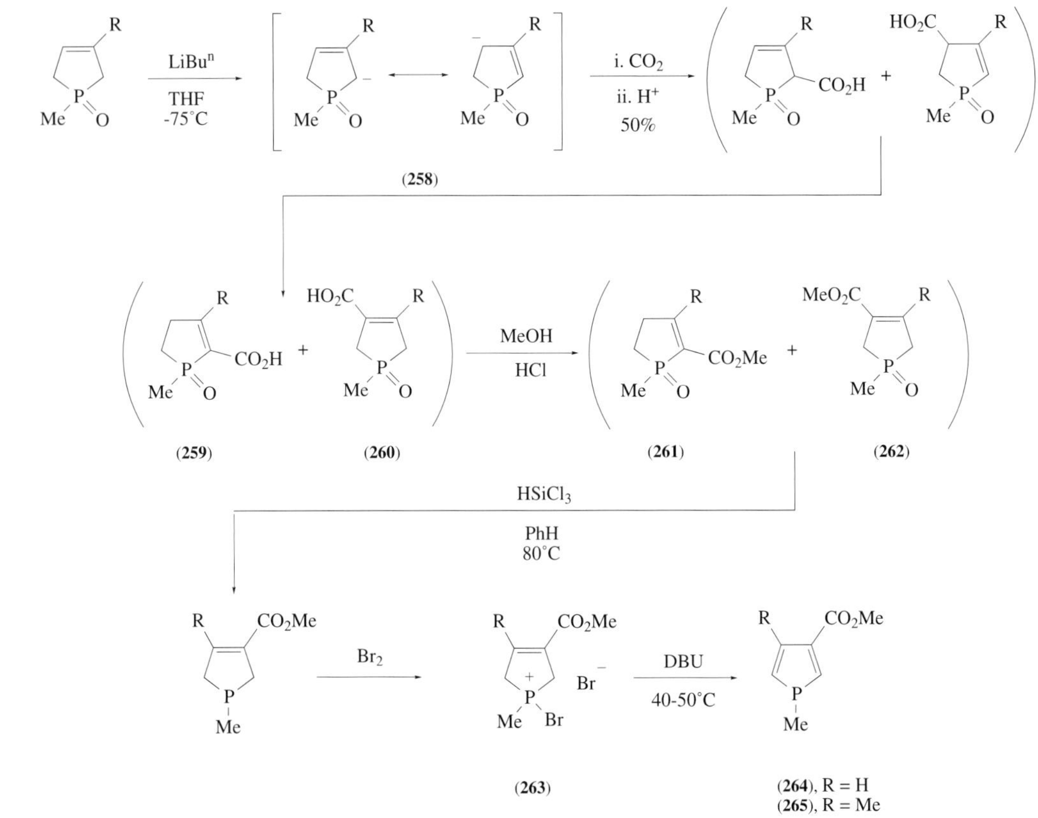

Scheme 76

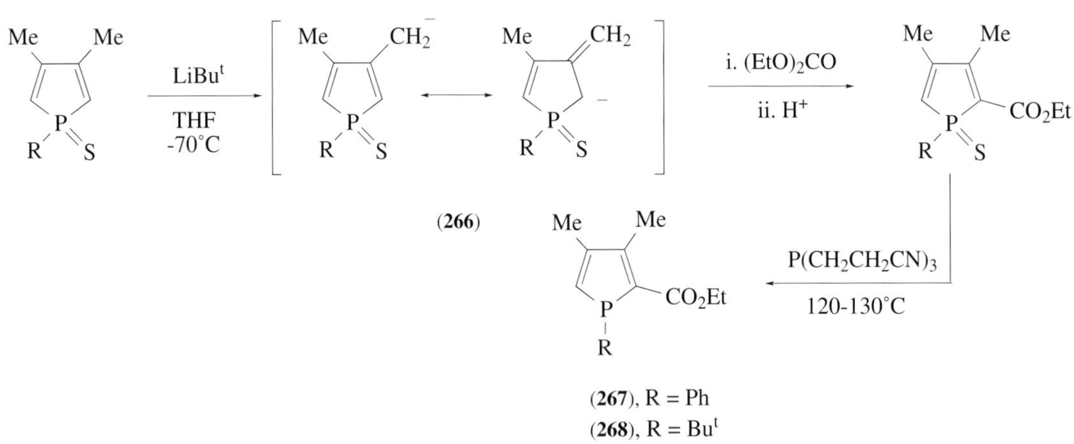

Scheme 77

(**271**) and 2-COOEt (**267**) derivatives were prepared from this organometallic. Compound (**271**) is the first example of a phosphole carboxylic acid; its properties are yet to be evaluated.

Two phosphole-3-carboxylates with 4,5-diphenyl substituents ((**272**) and (**273**)) have been prepared by a rather specialized approach ⟨93BSF695⟩, outlined in Scheme 79. This method was designed to generate phospholide ions with a C-functional group (e.g. (**274**)), but the ions can be alkylated on P to form phospholes. The ions are formed by a base-induced retro-Michael reaction. While of value in producing an otherwise inaccessible substitution pattern on the ring, the method is less versatile than those discussed above. Unlike the 3-carboxylates (**264**) and (**265**), these more highly substituted compounds were quite stable.

A property of certain metallic complexes with the phosphole ring may allow them to become useful precursors of carboxylates (as indeed they have in the synthesis of ketones, see Section 4.2.1.11.3). In the phosphaferrocene ring, where the phosphole moiety is in the form of the phospholide ion, direct substitution reactions are possible, carboxyethylation among them (Equation 51) ⟨80NJC683⟩. There is, however, no report on the conversion of the complexes to the free phospholes.

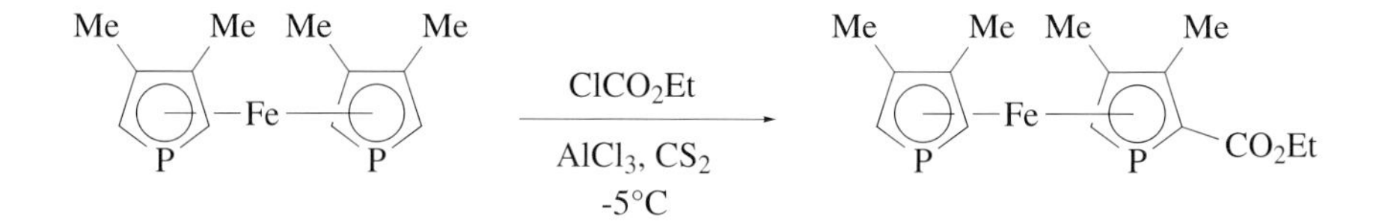

(51)

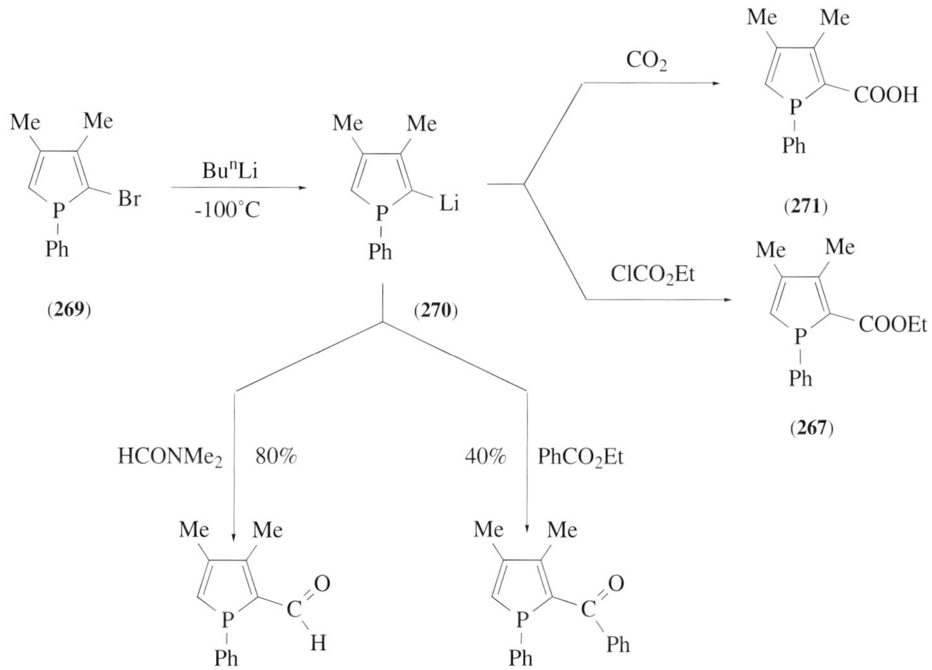

Scheme 78

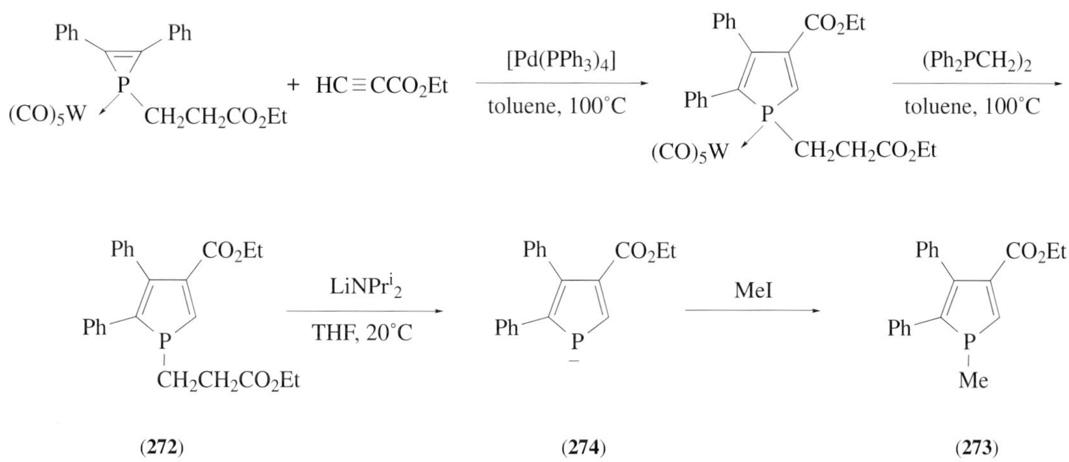

Scheme 79

4.2.1.11.3 Aldehydes and Ketones

A method similar to that used for phosphole carboxylate synthesis has been employed for the formation of ketones (**275**), (**276**), and (**277**) (Scheme 80); here the carbanion from a phosphole sulfide is reacted with ethyl acetate and the sulfur then removed with a phosphine ⟨76T2395⟩. These ketones are of good stability and can be purified by distillation. The method appears extendable to the introduction of other acyl groups.

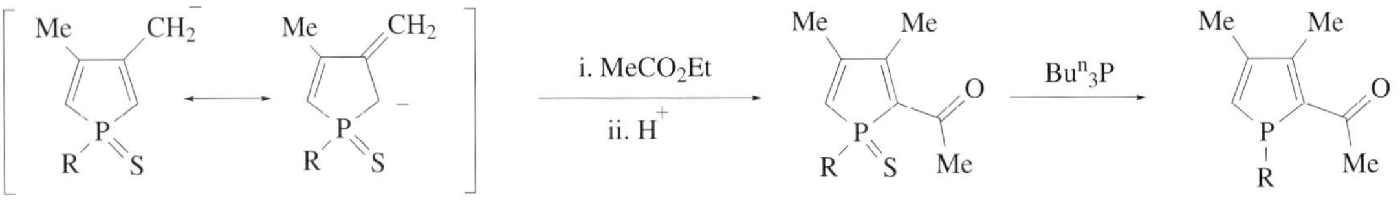

(266)

(275), R = Bu^n
(276), R = Bu^t
(277), R = Ph

Scheme 80

With the synthesis of the first organometallic derivative of a phosphole (**270**), a new approach to ketones (and also to the first phosphole carbaldehyde) became possible (see Scheme 78; ⟨92BSF486⟩).

The sensitivity of metal complexes of phospholide ions to electrophilic substitution has been employed in another approach to phosphole ketones ⟨85JOC467⟩. The molybdenum complex (**278**) was acetylated at the 2-position under Friedel–Crafts conditions, and the metallic group then removed with carbon monoxide under pressure. Trimethyl phosphite also can be used for this purpose (Scheme 81).

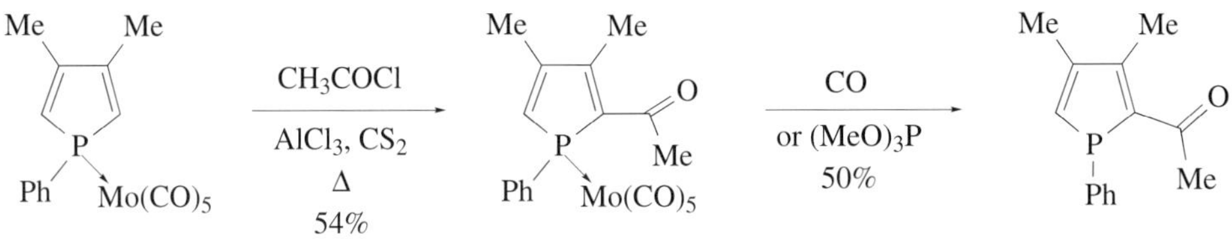

(278) (279)

Scheme 81

Some other metallic complexes were also found in this study to be acetylated readily under the same conditions, but the metallic group has not yet been removed to give the free phosphole ketones. These reactions are summarized in Scheme 82 and Eqs. (52) and (53).

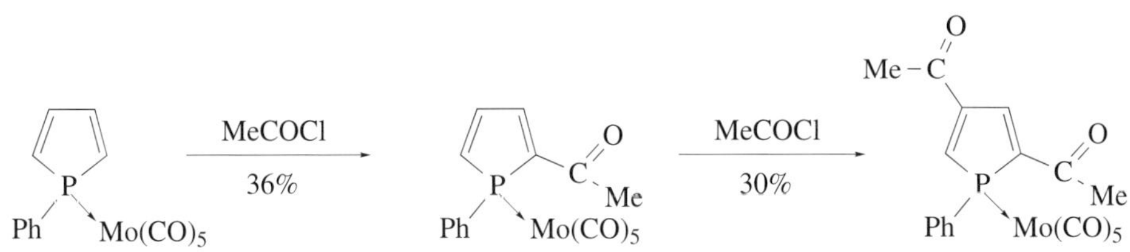

(280)

Scheme 82

The first example of a diacetylated product (**280**) was also encountered in this study, as was the formation of a 3-benzoyl product (**281**). This orientation presumably arose from a steric effect associated with the use of the more bulky reagent benzoyl chloride for attack on the ring.

The phospholide moiety of a monophosphaferrocene was acetylated by treatment with acetic anhydride in the presence of trifluoromethylsulfonic acid ⟨87ICA(130)93⟩. Again the acetylated phosphole was not released from the complex. Benzoyl chloride also was used to benzoylate the phospholide moiety of (**282**) (Eqs. 54a and 54b). Diphosphaferrocenes can also be acetylated under similar conditions ⟨80JA994⟩.

The aldehyde function has been placed on the phospholide ring of phosphaferrocenes by a conventional Vilsmeier reaction ⟨80JA994⟩ (Eqs. 54c and 54d).

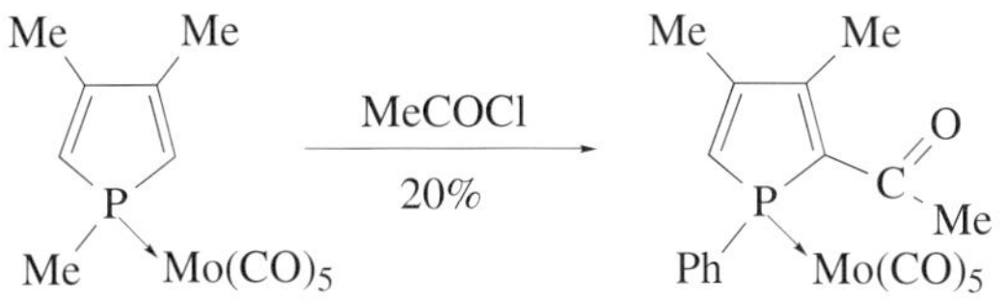

(52)

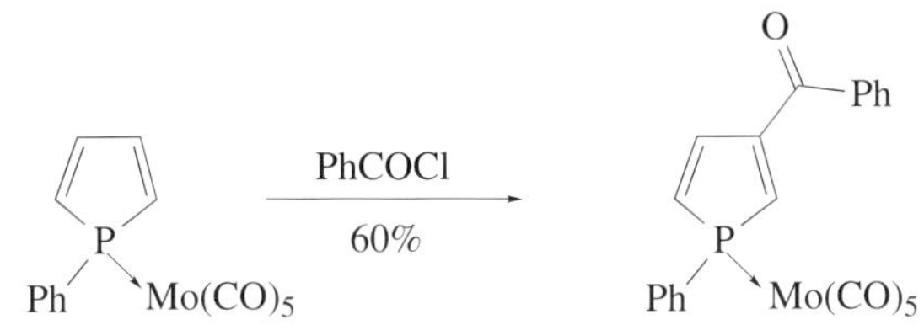

(53)

(**281**)

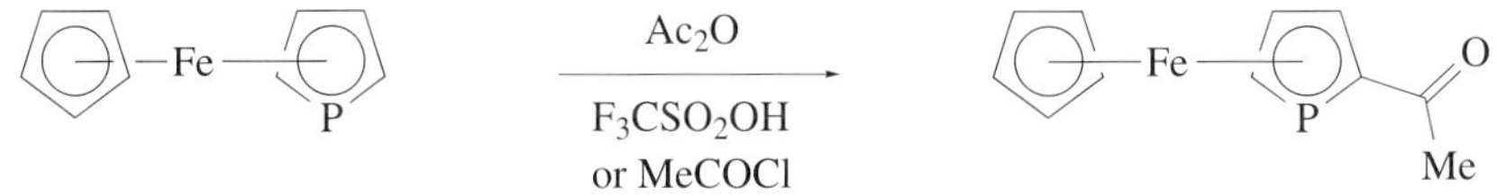

(54a)

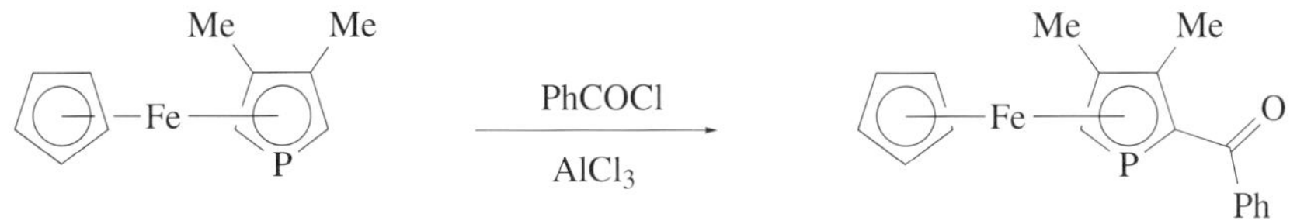

(54b)

(**282**)

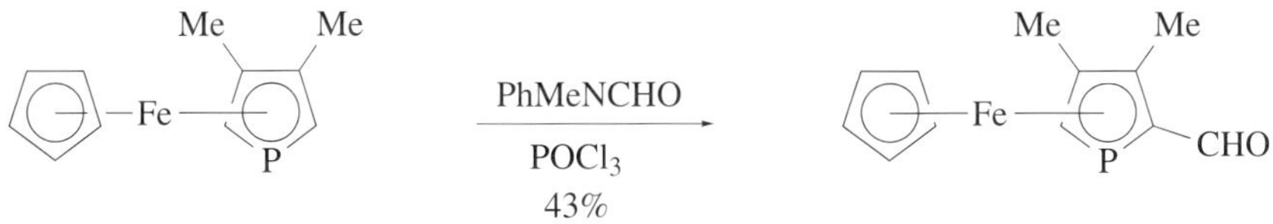

(54c)

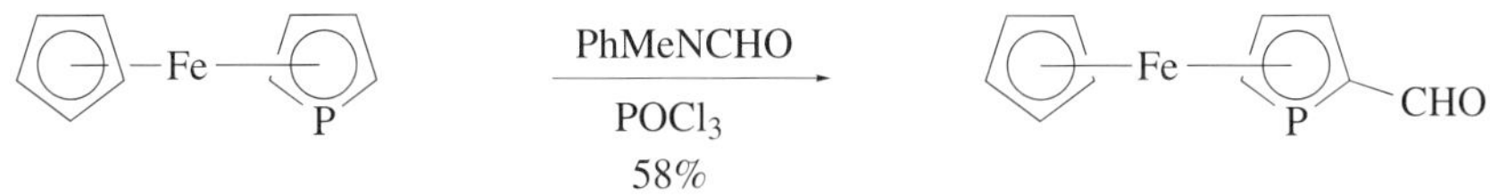

(54d)

(**282**)

4.2.1.11.4 Nitriles

Phospholes with a C-nitrile group remain unknown at this time. A P-nitrile has been prepared and has considerable value for the introduction of other types of substituents on phosphorus (Scheme 83) ⟨88OM1796⟩.

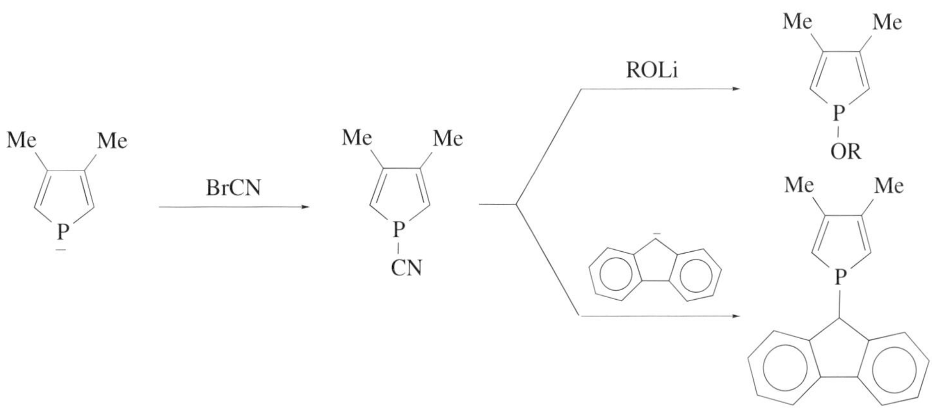

Scheme 83

A tungsten carbonyl complex of the 1-cyanophosphole has also been used with success in the synthesis of other P-substituted complexes (e.g. the P–OH complex (**292**) see below).

4.2.1.11.5 Amines

The only phospholes with amino substituents are compounds (**283**), (**284**), and (**285**) (Eqs. 55). Unlike the C-functional phospholes discussed to the present, these compounds resulted directly from the ring-forming process ⟨88TL4581⟩. The starting materials were simple enamines and phenylphosphonous dichloride, which were condensed in a single step. Because of the simplicity of the method, it could become the most important one for providing a phosphole with a C-function. The yields for the single step are 30–35%, surpassing the yields of all of the multistep functionalizations discussed in this section.

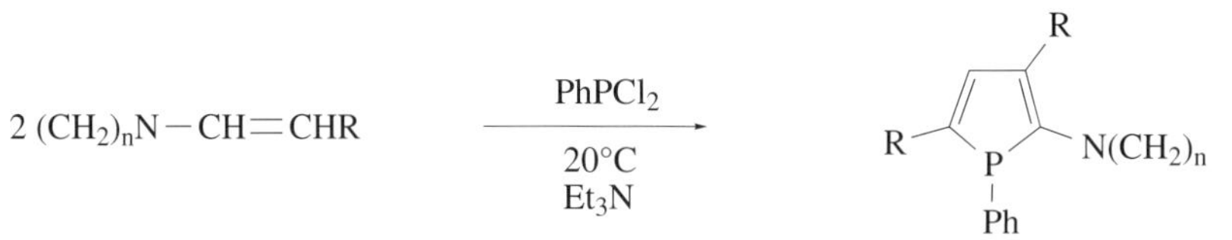

(55)

(**283**), R = Me, n = 5
(**284**), R = Et, n = 5
(**285**), R = Et, n = 4

The mechanism of this process is not obvious and has received no experimental work. The phosphole ring was indicated by the ^{31}P NMR spectra, and the compounds gave the correct molecular weight by mass spectral measurements. ^{13}C NMR spectral features were used to assign the position of the substituents.

Aminophospholes in the form of tungsten carbonyl complexes have been synthesized as follows (Equation 56) ⟨90PS(47)477⟩.

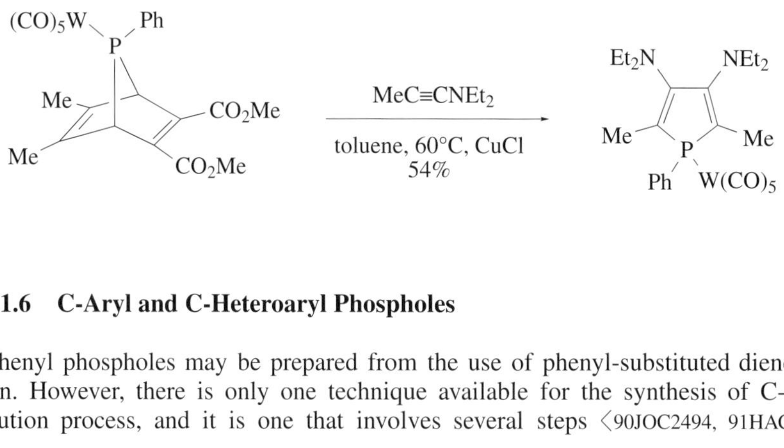

(56)

4.2.1.11.6 C-Aryl and C-Heteroaryl Phospholes

C-Phenyl phospholes may be prepared from the use of phenyl-substituted dienes in the McCormack reaction. However, there is only one technique available for the synthesis of C-aryl phospholes by a substitution process, and it is one that involves several steps ⟨90JOC2494, 91HAC377⟩. Again a metal

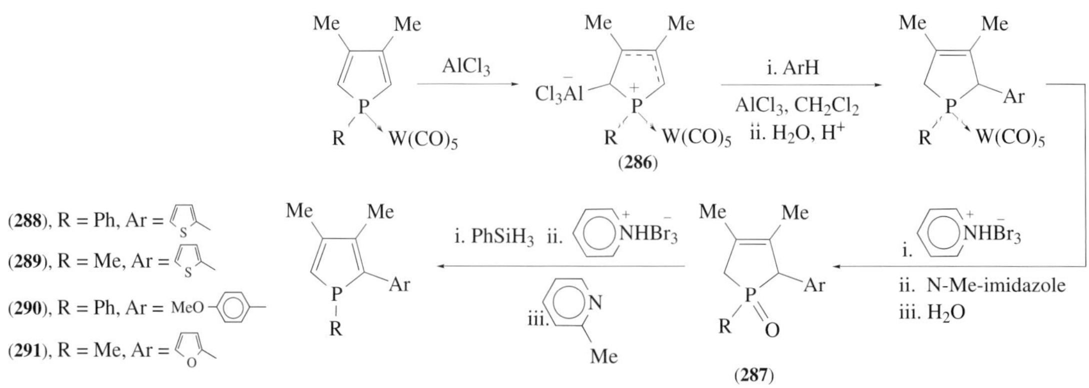

Scheme 84

complex (with $W(CO)_5$) of a phosphole is the starting material; in the presence of aluminum chloride, electrophilic character in the ring is increased, expressed by complex (**286**), and attack on electron-rich aromatic systems becomes possible. The complex after the arylation reaction is broken to generate a phospholene oxide (**287**), which is then converted to a phosphole in the same way (reduction to the phospholene, followed by bromine addition and dehydrohalogenation with base) as used in the first synthesis of phosphole-3-carboxylates ⟨72TL299⟩. These reactions are summarized in Scheme 84. The compounds formed have structures (**288**)–(**291**).

4.2.1.11.7 Hydroxy and Alkoxy Derivatives

C-alkoxy derivatives of a phosphole have resulted from the insertion of ethoxyethyne into a phosphirene complex (Equation 57) ⟨87TL5021⟩.

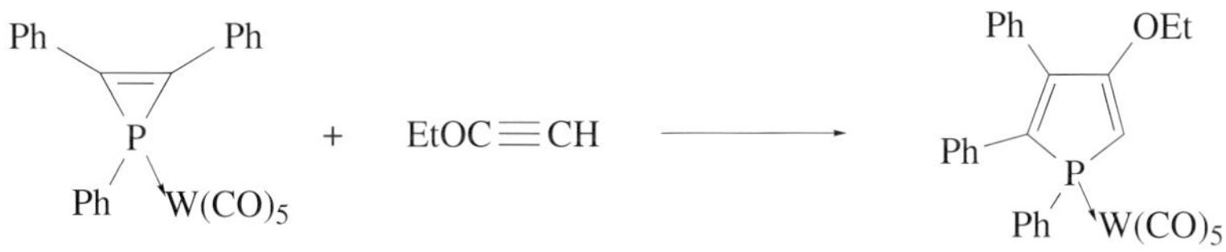

(57)

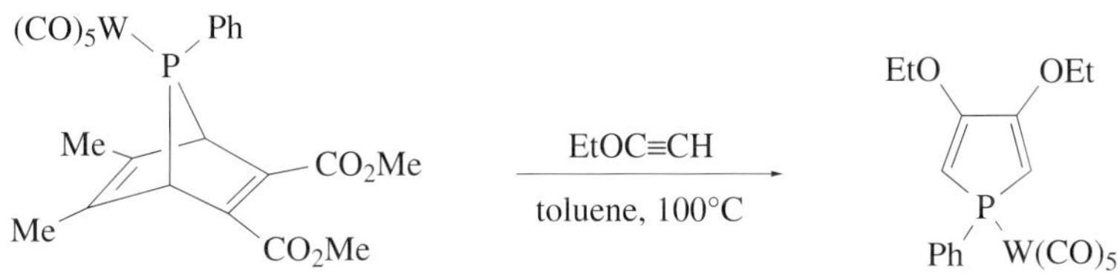

(58)

Related complexes have been formed from a norbornadiene complex (Equation 58) ⟨90PS(47)477⟩.

In neither case has the decomplexed phosphole yet been reported, however. A C-hydroxy derivative of a phosphole would of course be involved in tautomerism with the keto isomer, with the latter probably the observable form (as with O, N, S cycles). However, no model compounds have yet been prepared to test this point.

A P-hydroxy phosphole has been reported in the form of a tungsten carbonyl complex (**292**) (Equation 59) ⟨88OM1796⟩. Normally P–OH groups in phosphines undergo rearrangement to –P(O)H forms, but the complex prevents this from occurring.

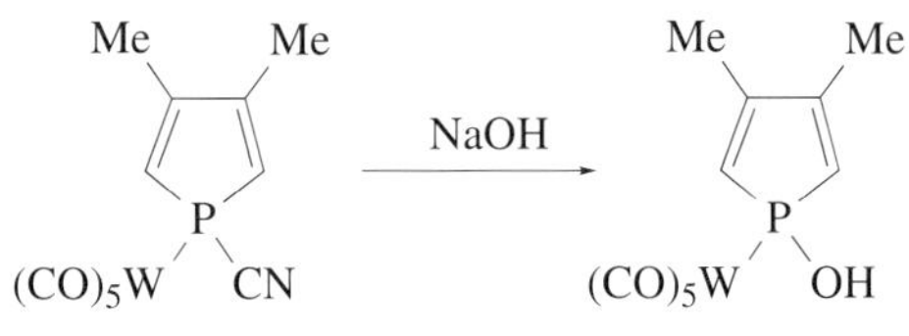

(59)

(**292**)

4.2.1.11.8 Halo Substituents

The only known C-halophosphole is 2-bromo-3,4-dimethyl-1-phenylphosphole (**293**) prepared as shown in Scheme 85 ⟨92BSF486⟩. As noted in Section 4.2.1.11.3, this compound is important as a precursor of the 2-lithio derivative.

4.2.1.11.9 Silyl Derivatives

The only C-silylated derivative of a phosphole was prepared from the lithio compound (**294**) (Equation 60) ⟨92BSF486⟩.

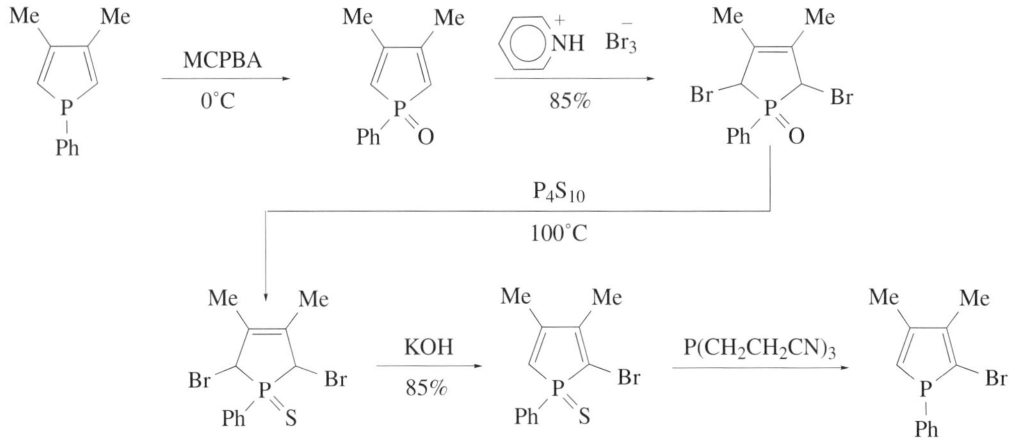

(293)

Scheme 85

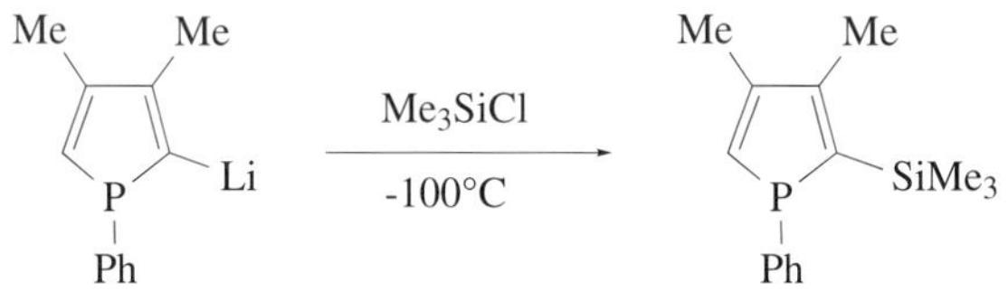

(294) (60)

4.2.1.12 PHOSPHOLES AS PRECURSORS FOR OTHER PHOSPHORUS HETEROCYCLES

4.2.1.12.1 7-Phosphanorbornenes and 7-Phosphanorbornadienes

The only synthetic approach to the valuable 7-phosphanorbornene system is through Diels–Alder chemistry on phospholes, especially in the four-coordinate state (P-oxides and sulfides). Some of this chemistry has been described in Sections 4.2.1.6.1 and 4.2.1.6.2, and it has been reviewed elsewhere ⟨90RHA39⟩. The 7-phosphanorbornenes are used as precursors for the 2,3-oxaphosphabicyclo[2.2.2]octene ring system, which finds value as a source, on thermal or photochemical fragmentation, of 3-coordinate phosphoryl species (metaphosphates and dioxophosphoranes, ⟨B-92MI02⟩). An example of the synthesis of ethyl metaphosphate is shown in Scheme 86.

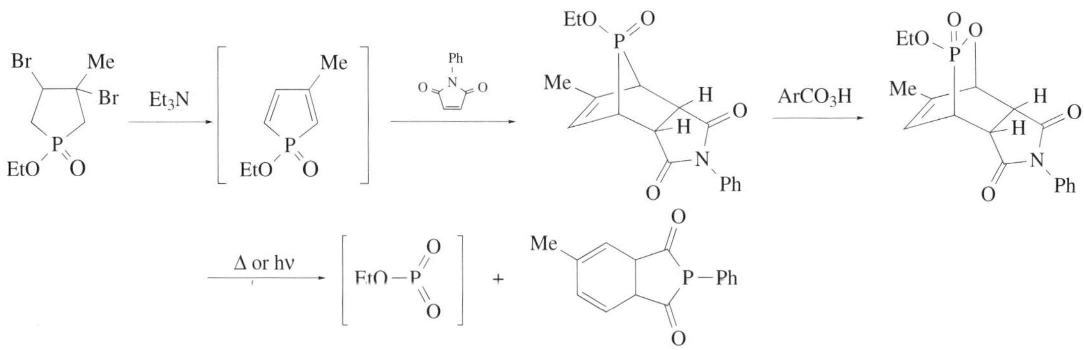

Scheme 86

Phenyl dioxophosphorane is especially easily generated in a sequence that starts from 3,4-dimethyl-1-phenylphosphole ⟨91HAC359⟩ (Scheme 87).

There are no stable derivatives of the 7-phosphanorbornadiene ring system with phosphorus in the phosphine state, but one example of an observable, though rather unstable, oxide (**295**) has been reported ⟨84H(22)2713⟩, and the ring system is also known in the form of metal complexes (**297**) ⟨82CC667⟩ (Eqs. 61a–c).

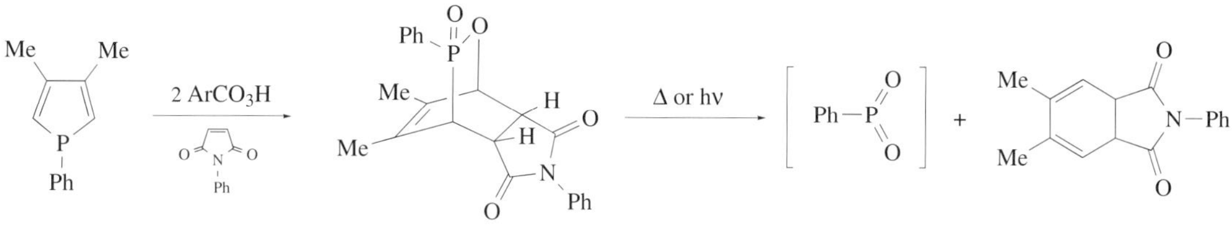

Scheme 87

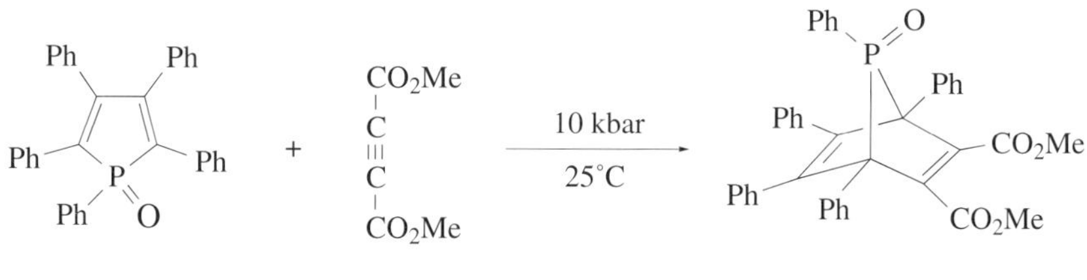

(61a)

(**295**)

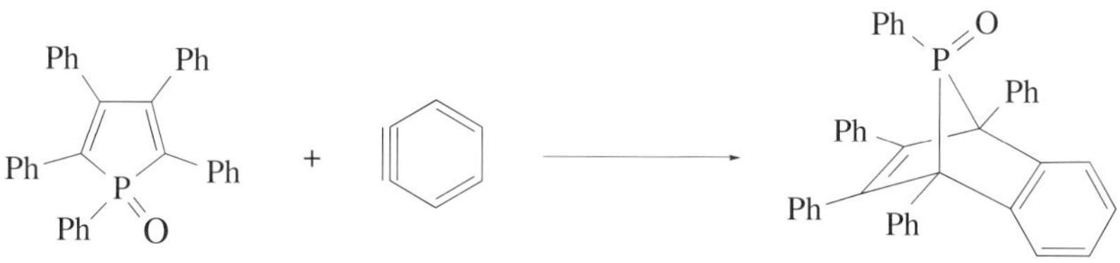

(61b)

(**296**)

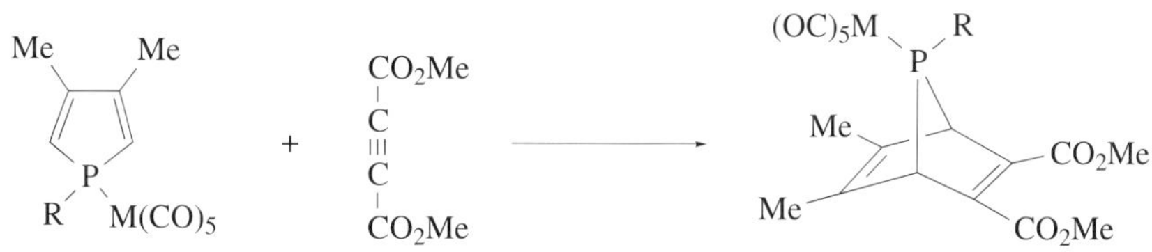

(61c)

M = Cr, Mo, W

(**297**)

A stable benzo derivative (**296**) is also known ⟨72JA4761⟩. The complexes (**297**) have synthetic utility also; when heated in inert solvents in the presence of an alkene ⟨84OM456⟩ or an ethyne ⟨82JA4484⟩, the P-containing fragment is released from the bicyclic ring and transferred to the unsaturated reactant (Scheme 88).

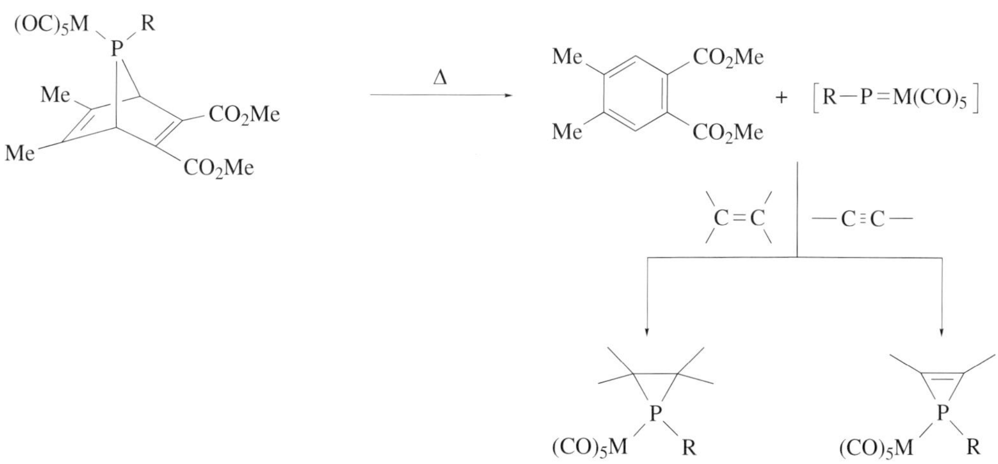

Scheme 88

This results in the formation of complexes of phosphiranes and phosphirenes, respectively, which can be decomplexed to give the free phosphines. Phosphirenes were synthesized for the first time by this technique. A one-pot synthesis of phosphirene complexes from a 3,4-dimethyl-1-phenylphosphole complex with tungsten has even been developed ⟨85JA4700⟩. These applications based on phosphole chemistry are included in a review of the three-membered ring systems ⟨87AG(E)275⟩.

4.2.1.12.2 1-Phosphanorbornadienes

The sole source of the valuable 1-phosphanorbornadiene ring system is based on a novel feature of phosphole chemistry; as seen in Section 4.2.1.5.9, at high temperatures, some phospholes undergo [1,5] sigmatropic rearrangements to give 2*H*-phospholes, and these can participate in Diels–Alder reactions that serve to generate the bicyclic system having phosphorus at the bridgehead position. Again 3,4-dimethyl-1-phenylphosphole is a particularly useful reactant ⟨81JA4595, 89NJC369⟩, and has been reacted with several ethyne derivatives to form 1-phosphanorbornadienes (Scheme 89); the other regioisomer is formed in lesser amount.

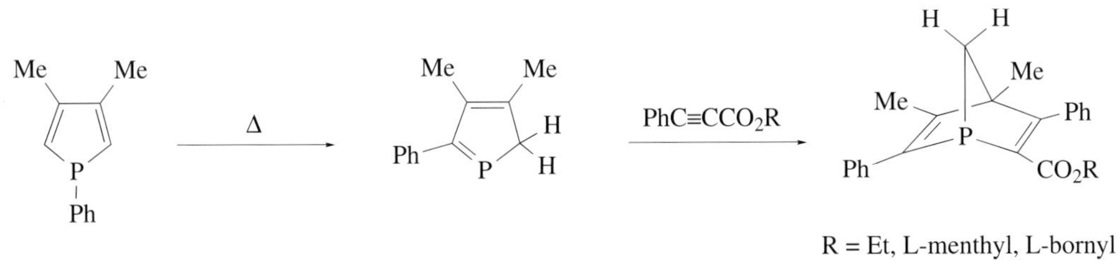

Scheme 89

4.2.1.12.3 Phosphinines

Several processes have already been noted that amount to conversion of the phosphole ring to the phosphinine ring. These are summarized in Schemes 90 to 92 and are included in a 1992 review of some work in phosphinine chemistry ⟨92RHA1⟩. The methods are of considerable synthetic value and are evaluated along with other methods in Section 5.2. Various groups R and Ar may be present in the methods in Scheme 90 and Scheme 91, making some novel substitution patterns possible in the phosphinines.

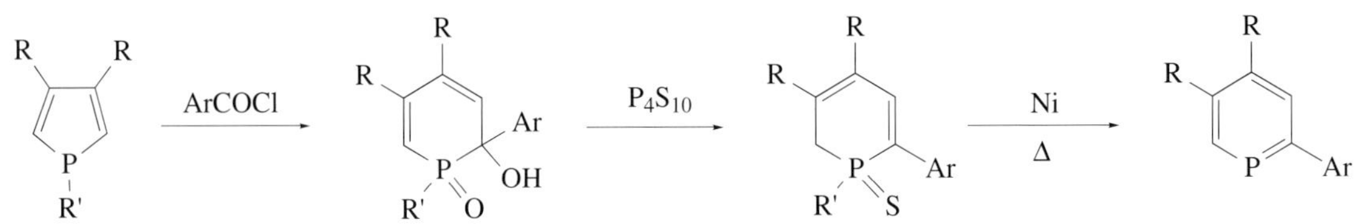

Scheme 90

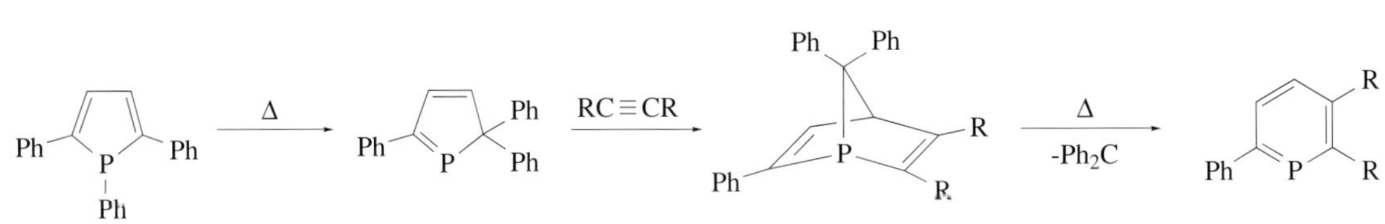

Scheme 91

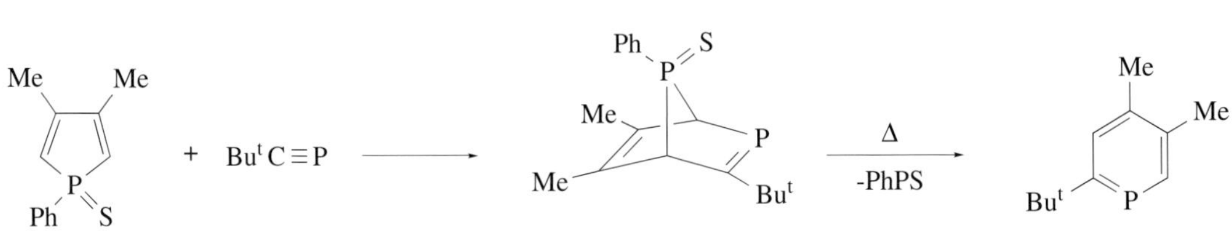

Scheme 92

4.2.1.13 REFERENCES

53LA(580)44	G. Wittig and G. Geissler; *Liebigs Ann. Chem.*, 1953, **580**, 44.
53USP2663738	W.B. McCormack; *US Pat.* 2 663 738 (1953) (*Chem. Abstr.*, 1955, **49**, 7601).
56JOC238	L.D. Freedman and G.O. Doak; *J. Org. Chem.*, 1956, **21**, 238.
59CI(L)1250	E.H. Braye and W. Hübel; *Chem. Ind. (London)* 1959, 1250.
59JA3163	F.C. Leavitt, T.A. Manuel and F. Johnson; *J. Am. Chem. Soc.* 1959, **81**, 3163.
59JOC638	L.D. Freedman and G.O. Doak; *J. Org. Chem.*, 1959, **24**, 638.
60JA5099	F.C. Leavitt, T.A. Manuel, F. Johnson, L.U. Matternas and D.S. Lehman; *J. Am. Chem. Soc.*, 1960, **82**, 5099.
61JA4406	E.H. Braye, W. Hübel and I. Caplier; *J. Am. Chem. Soc.*, 1961, **83**, 4406.
61JSP(7)58	B. Bak, D. Christensen, L. Hansen-Nygaard and J. Rastrup-Andersen; *J. Mol. Spectrosc.*, 1961, **7**, 58.
62CI(L)359	I.G.M. Campbell, R.C. Cookson and M.B. Hocking; *Chem. Ind. (London)*, 1962, 359.
62JCS929	D.A. Brown; *J. Chem. Soc.*, 1962, 929.
62JCS2932	A.F. Bedford, D.M. Heinekey, I.T. Millar and C.T. Mortimer; *J. Chem. Soc.*, 1962, 2932.
64CB741	G. Wittig and E. Kochendoerfer; *Chem. Ber.*, 1964, **97**, 741.
64CB747	G. Wittig and A. Maercker; *Chem. Ber.*, 1964, **97**, 747.
64JOC836	L.D. Quin and D.A. Mathewes; *J. Org. Chem.*, 1964, **29**, 836.
65CB171	H. Fritzsche, U. Hasserodt and F. Korte; *Chem. Ber.*, 1965, **98**, 171.
65JCS2184	I.G.M. Campbell, R.C. Cookson, M.B. Hocking and A.N. Hughes; *J. Chem. Soc.*, 1965, 2184.
66AG(E)846	G. Märkl; *Angew. Chem., Int. Ed. Engl.*, 1966, **5**, 846.
67AG(E)86	G. Märkl and R. Potthast; *Angew. Chem., Int. Ed. Engl.*, 1967, **6**, 86.
67JA5732	M.D. Rausch and L.P. Klemann; *J. Am. Chem. Soc.*, 1967, **89**, 5732.
67JA5984	L.D. Quin and J.G. Bryson; *J. Am. Chem. Soc.*, 1967, **89**, 5984.
68T3437	A.N. Hughes and S. Uaboonkul; *Tetrahedron*, 1968, **24**, 3437.
68TL1755	G. Märkl and R. Potthast; *Tetrahedron Lett.*, 1968, **9**, 1755.
69CR(C)1066	F. Mathey; *C. R. Hebd. Seances Acad. Sci., Ser. C*, 1969, **269**, 1066.
69JA3308	L.D. Quin, J.G. Bryson and C.G. Moreland; *J. Am. Chem. Soc.*, 1969, **91**, 3308.
69JCS(C)252	D.W. Allen and I.T. Millar; *J. Chem. Soc. (C)*, 1969, 252.
69JCS(C)1100	N.E. Waite, J.C. Tebby, R.S. Ward and D.H. Williams; *J. Chem. Soc. (C)*, 1969, 1100.
69JOC1777	B.R. Ezzell and L.D. Freedman; *J. Org. Chem.*, 1969, **34**, 1777.
69TL5037	T.J. Barton and A.J. Nelson; *Tetrahedron Lett.*, 1969, **10**, 5037.
70BSF4433	F. Mathey and R. Mankowski-Favelier; *Bull. Soc. Chim. Fr.*, 1970, 4433.
70CC122	J.M. Holland and D.W. Jones; *J. Chem. Soc., Chem. Commun.*, 1970, 122.
70JA1442	W. Egan, R. Tang, G. Zon and K. Mislow; *J. Am. Chem. Soc.*, 1970, **92**, 1442.
70JA5779	P. Coggon, J.F. Engel, A.T. McPhail and L.D. Quin; *J. Am. Chem. Soc.*, 1970, **92**, 5779.
70JCS(C)386	N.E. Waite and J.C. Tebby; *J. Chem. Soc. (C)*, 1970, 386.
70JCS(C)1855	J.R. Corfield, M.J.P. Harger, J.R. Shutt and S. Trippett; *J. Chem. Soc. (C)*, 1970, 1855.
70JHC1	A.N. Hughes and C. Srivanavit; *J. Heterocycl. Chem.*, 1970, **7**, 1.
71CC782	C. Thomson and D. Kilcast; *J. Chem. Soc., Chem. Commun.*, 1971, 782.
71CJC530	T.H. Chan and L.T.L. Wong; *Can. J. Chem.*, 1971, **49**, 530.
71CJC874	A.N. Hughes and C. Srivanavit; *Can. J. Chem.*, 1971, **49**, 874.
71CJC879	A.N. Hughes and C. Srivanavit; *Can. J. Chem.*, 1971, **49**, 879.
71JA4541	F.B. Clarke, III and F.H.Westheimer; *J. Am. Chem. Soc.*, 1971, **93**, 4541.
71JA6205	W. Egan, R. Tang, G. Zon and K. Mislow; *J. Am. Chem. Soc.*, 1971, **93**, 6205.
71JA6507	A. Rauk, J.D. Andose, W.G. Frick, R. Tang and K. Mislow; *J. Am. Chem. Soc.*, 1971, **93**, 6507.
71JHC877	R. Chuchman, D.G. Holah, A.N. Hughes and B.C Hui; *J. Heterocycl. Chem.*, 1971, **8**, 877.
71JOC1297	L.D. Quin, J.J. Breen and D.K. Myers; *J. Org. Chem.*, 1971, **36**, 1297.
71PS(1)139	N.E. Waite, D.W. Allen and J.C. Tebby; *Phosphorus*, 1971, **1**, 139.
71T5523	E.H. Braye, I. Caplier and R. Saussez; *Tetrahedron*, 1971, **27**, 5523.
71T5705	D. Kilcast and C. Thomson; *Tetrahedron*, 1971, **27**, 5705.
72CC496	W.B. Farnham and K. Mislow; *J. Chem. Soc., Chem. Commun.*, 1972, 496.
72CC1242	J.I.G. Cadogan, R. Gee and R.J. Scott; *J. Chem. Soc., Chem. Commun.*, 1972, 1242.
72CJC1008	D. Budd, R. Chuchman, D.G. Holah, A.N. Hughes and B.C. Hui; *Can. J. Chem.*, 1972, **50**, 1008.
72CJC3625	M. Davies, A.N. Hughes and S.W.S. Jafry; *Can. J. Chem.*, 1972, **50**, 3625.
72CR(C)100	H. Lumbroso, D.M. Bertin and F. Mathey; *C. R. Hebd. Seances Acad. Sci., Ser. C*, 1972, **274**, 100.
72JA4761	J.K. Stille, J.L. Eichelberger, J. Higgins and M.E. Freeburger; *J. Am. Chem. Soc.*, 1972, **94**, 4761.
72JCS(D)102	C.M. Mitchell and F.G.A. Stone; *J. Chem. Soc., Dalton Trans.*, 1972, 102.
72JHC817	G. Marino; *J. Heterocycl. Chem.*, 1972, **9**, 817.
72OMR(4)171	F. Mathey and R. Mankowski-Favelier; *Org. Magn. Reson.*, 1972, **4**, 171.
72PS(2)55	J.J. Breen, J.F. Engel, D.K. Myers and L.D. Quin; *Phosphorus*, 1972, **2**, 55.
72T4171	F. Mathey; *Tetrahedron*, 1972, **28**, 4171.
72TL299	L.D. Quin and S.G. Borleske; *Tetrahedron Lett.*, 1972, **13**, 299.
72TL3353	T. Bundgaard and H.J. Jakobsen; *Tetrahedron Lett.*, 1972, **13**, 3353.
73AG(E)145	W. Schäfer, A. Schweig, G. Märkl, H. Hauptmann and F. Mathey; *Angew. Chem., Int. Ed. Engl.*, 1973, **12**, 145.
73JCS(D)1888	P. Coggon and A.T. McPhail; *J. Chem. Soc., Dalton Trans.*, 1973, 1888.

73JOC1858 L.D. Quin, S.G. Borleske and J.F. Engel; *J. Org. Chem.*, 1973, **38**, 1858.
73JOC1954 L.D. Quin, S.G. Borleske and J.F. Engel; *J. Org. Chem.*, 1973, **38**, 1954.
73OMR(5)17 L.D. Quin and J.J. Breen; *Org. Magn. Reson.*, 1973, **5**, 17.
73PS(2)205 L.D. Quin, J.G. Bryson and J.F. Engel; *Phosphorus*, 1973, **2**, 205.
73T469 H.L. Hase, A. Schweig, H. Hahn and J. Radloff; *Tetrahedron*, 1973, **29**, 469.
73T707 F. Mathey; *Tetrahedron,* 1973, **29**, 707.
73TL3255 F. Mathey; *Tetrahedron Lett.*, 1973, **14**, 3255.
73TL4815 T.H. Chan and K.T. Nwe; *Tetrahedron Lett.*, 1973, **14**, 4815.
74AJC831 D.J. Collins, L.E. Rowley and J.M. Swan; *Aust. J. Chem.*, 1974, **27**, 831.
74CC902 J.I.G. Cadogan, R.J. Scott and N.H. Wilson; *J. Chem. Soc., Chem. Commun.*, 1974, 902.
74JA6904 J.D. Andose, A. Rauk and K. Mislow; *J. Am. Chem. Soc.*, 1974, **96**, 6904.
74JCS(P1)1694 J.I.G. Cadogan, R.J. Scott, R.D. Gee and I. Gosney; *J. Chem. Soc., Perkin Trans 1*, 1974, 1694.
74PS(4)199 L.D. Freedman, B.R. Ezzell, R.N. Jenkins and R.M. Harris; *Phosphorus*, 1974, **4**, 199.
74PS(4)231 G. Kaufmann and F. Mathey; *Phosphorus*, 1974, **4**, 231.
74PS(4)279 G. Märkl, H. Hauptmann and F. Lieb; *Phosphorus*, 1974, **4**, 279.
74T3127 F. Mathey; *Tetrahedron*, 1974, **30**, 3127.
75CJC855 F. Mathey, D. Thavard and B. Bartet; *Can. J. Chem.*, 1975, **53**, 855.
75JCS(P2)974 M.H. Palmer and R.H. Findlay; *J. Chem. Soc., Perkin Trans. 2*, 1975, 974.
75JOM(93)377 F. Mathey; *J. Organomet. Chem.*, 1975, **93**, 377.
75T53 Y. Kashman and O. Awerbouch; *Tetrahedron*, 1975, **31**, 53.
75T2537 T.H. Chan and K.T. Nwe; *Tetrahedron*, 1975, **31**, 2537.
76CJC2402 F. Mathey, J.-P. Lampin and D. Thavard; *Can. J. Chem.*, 1976, **54**, 2402.
76JA407 W. Schäfer, A. Schweig and F. Mathey; *J. Am. Chem. Soc.*, 1976, **98**, 407.
76JA4365 N.D. Epiotis and W. Cherry; *J. Am. Chem. Soc.*, 1976, **98**, 4365.
76JHC1 A.N. Hughes and D. Kleemola; *J. Heterocycl. Chem.*, 1976, **13**, 1.
76JHC65 A.N. Hughes, K. Amornraksa, S. Phisithkul and V. Reutrakul; *J. Heterocycl. Chem.*, 1976, **13**, 65.
76T2395 F. Mathey; *Tetrahedron*, 1976, **32**, 2395.
76T2427 Y. Kashman, I. Wagenstein and A. Rudi; *Tetrahedron*, 1976, **32**, 2427.
76TL4155 F. Mathey; *Tetrahedron Lett.*, 1976, **17**, 4155.
77JA3537 F. Mathey, A. Mitschler and R. Weiss; *J. Am. Chem. Soc.*, 1977, **99**, 3537.
77JOM(136)241 F. Mathey and G. Muller; *J. Organomet. Chem.*, 1977, **136**, 241.
77JOM(139)77 F. Mathey; *J. Organomet. Chem.*, 1977, **139**, 77.
78CJC1952 F. Mathey and D. Thavard; *Can. J. Chem.*, 1978, **56**, 1952.
78JA4886 Th.C. Klebach, R. Lourens and F. Bickelhaupt; *J. Am. Chem. Soc.*, 1978, **100**, 4886.
78JA5748 F. Mathey, A. Mitschler and R. Weiss; *J. Am. Chem. Soc.*, 1978, **100**, 5748.
78JCS(P1)98 D.W. Allen, I. Nowell, A.C. Oades and P.E. Walker; *J. Chem. Soc., Perkin Trans. 1*, 1978, 98.
78JHC1319 D.G. Holah, A.N. Hughes and D.. Kleemola; *J. Heterocycl. Chem.*, 1978, **15**, 1319.
78JOM(156)C33 G. de Lauzon, F. Mathey and M. Simalty; *J. Organomet. Chem.*, 1978, **156**, C33.
78LA74 J.I.G. Cadogan, A.G. Rowley and N.H. Wilson; *Liebigs Ann. Chem.*, 1978, 74.
79CC401 L.D. Quin and W.L. Orton; *J. Chem. Soc., Chem. Commun.*, 1979, 401.
79JCS(D)814 E.W. Abel and C. Towers; *J. Chem. Soc., Dalton Trans.*, 1979, 814.
79JOM(165)129 A. Brèque, F. Mathey and C Santini; *J. Organomet. Chem.*, 1979, **165**, 129.
79PS(5)349 W.L. Orton, K.A. Mesch and L.D. Quin; *Phosphorus Sulfur*, 1979, **5**, 349.
79TL1753 F. Mathey; *Tetrahedron Lett.*, 1979, **20**, 1753.
79ZN(B)1116 D.W. Allen and B.G. Hutley; *Z. Naturforsch., Teil B*, 1979, **34**, 1116.
80CSC905 R.B. Knott, H. Honneger, A.D. Rae, F. Mathey and G. de Lauzon; *Cryst. Struct. Commun.*, 1980, **9**, 905.
80JA994 G. de Lauzon, B. Deschamps, J. Fischer, F. Mathey and A. Mitschler; *J. Am. Chem. Soc.*, 1980, **102**, 994.
80JA5809 C.C. Santini, J. Fischer, F. Mathey and A. Mitschler; *J. Am. Chem. Soc.*, 1980, **102**, 5809.
80MI01 F. Mathey; *Top. Phosphorus Chem.*, 1980, **10**, 1.
80NJC683 G. de Lauzon, B. Deschamps and F. Mathey; *Nouv. J. Chim.*, 1980, **4**, 683.
80OMR(14)14 G.A. Gray and J.H. Nelson; *Org. Magn. Reson.*, 1980, **14**, 14.
80TL1441 F. Nief, C. Charrier, F. Mathey and M. Simalty; *Tetrahedron Lett.*, 1980, **21**, 1441.
81CC420 J.C. Tebby, S.E. Willetts and D.V. Griffiths; *J. Chem. Soc., Chem. Commun.*, 1981, 420.
81EUP41447 A. Breque, G. Muller, H. Bonnard, F. Mathey and P. Savignac; *Eur. Pat.*, 41 447 (1981) (*Chem Abstr.*, 1982, **96**, 143 076x).
81H(15)637 A.N. Hughes; *Heterocycles*, 1981, **15**, 637.
81JA4595 F. Mathey, F. Mercier, C. Charrier, J. Fischer, and A. Mitschler; *J. Am. Chem. Soc.*, 1981, **103**, 4595.
81JOC4386 S. Holand and F. Mathey; *J. Org. Chem.*, 1981, **46**, 4386.
81NJC187 F. Nief, C. Charrier, F. Mathey and M. Simalty; *Nouv. J. Chim.*, 1981, **5**, 187.
81PS(10)175 G. Muller, H. Bonnard and F. Mathey; *Phosphorus Sulfur*, 1981, **10**, 175.
81S983 A.Brèque, F. Mathey and P. Savignac; *Synthesis*, 1981, 983.
81TL319 F. Mathey and F. Mercier; *Tetrahedron Lett.*, 1981, **22**, 319.
81TL3533 R. Burgada, Y. Leroux and Y.O. El Khoshnieh; *Tetrahedron Lett.*, 1981, **22**, 3533.
82AX(B)3024 I. Bkouche-Waksman, P. L'Haridon, Y. Leroux and R. Burgada; *Acta Crystallogr., Sect. B*, 1982, **38**, 3024.
82CC667 A. Marinetti, F. Mathey, J. Fischer and A. Mitschler; *J. Chem. Soc., Chem. Commun.*, 1982, 667.

82CC1272 G. de Lauzon, C. Charrier, H. Bonnard, F. Mathey, J. Fischer and A. Mitschler; *J. Chem. Soc., Chem. Commun.*, 1982, 1272.
82CJC138 M.B. Hocking and T.A. Smyth; *Can. J. Chem.*, 1982, **60**, 138.
82JA2077 F. Mathey, F. Mercier, F. Nief, J. Fischer and A. Mitschler; *J. Am. Chem. Soc.*, 1982, **104**, 2077.
82JA4484 A. Marinetti, F. Mathey, J. Fischer and A. Mitschler; *J. Am. Chem. Soc.*, 1982, **104**, 4484.
82JCS(P1)2289 J. Cornforth, R.H. Cornforth and R.T. Gray; *J. Chem. Soc., Perkin Trans. 1*, 1982, 2289.
82JCS(P1)2317 J. Cornforth, D.D. Ridley, A.F. Sierakowski, D. Uguen and T.W. Wallace; *J. Chem. Soc., Perkin Trans. 1*, 1982, 2317.
82JOC2376 C. Charrier, H. Bonnard and F. Mathey; *J. Org. Chem.*, 1982, **47**, 2376.
82JOM(231)361 C. Charrier, H. Bonnard, F. Mathey and D Neibecker; *J. Organomet. Chem.*, 1982, **231**, 361.
82PS(12)161 L.D. Quin, K.A. Mesch and W.L. Orton; *Phosphorus Sulfur*, 1982, **12**, 161.
82PS(13)259 F. Nief, C. Charrier, F. Mathey and M. Simalty; *Phosphorus Sulfur*, 1982, **13**, 259.
82TL511 G. de Lauzon, C. Charrier, H. Bonnard and F. Mathey; *Tetrahedron Lett.*, 1982, **23**, 511.
82TL1565 J.-M. Alcaraz, A. Breque and F. Mathey; *Tetrahedron Lett.*, 1982, **23**, 1565.
83AG(E)57 G. Märkl and E. Seidel; *Angew. Chem., Int. Ed. Engl.*, 1983, **22**, 57.
83AG(E)879 G. Märkl, E. Seidel and I. Trötsch, I. *Angew. Chem., Int. Ed. Engl.*, 1983, **22**, 879.
83JA6871 C. Charrier, H. Bonnard, G. de Lauzon and F. Mathey; *J. Am. Chem. Soc.*, 1983, **105**, 6871.
83JOM(256)103 B. Lukas, R.M.G. Roberts, J. Silver and A.S. Wells; *J. Organomet. Chem.*, 1983, **256**, 103.
83OM1234 S. Holand, F. Mathey, J. Fischer and A. Mitschler; *Organometallics*, 1983, **2**, 1234.
83PS(18)51 C. Charrier, H. Bonnard, G. de Lauzon, S. Holand and F. Mathey; *Phosphorus Sulfur*, 1983, **18**, 51.
83TL1423 K.S. Fongers, H. Hogeveen and R.F. Kingma; *Tetrahedron Lett.*, 1983, **24**, 1423.
83TL5831 K.C. Caster and L.D. Quin; *Tetrahedron Lett.*, 1983, **24**, 5831.
83ZOB2796 L.N. Markovskii, V.D. Romanenko, A.V. Ruban, S.V. Iksanova; *Zh. Obshch. Khim.*, 1983, **52**, 2796.
84CC45 A. Marinetti, F. Mathey, J. Fischer and A. Mitschler; *J. Chem. Soc., Chem. Commun.*, 1984, 45.
84CC1217 D.G. Gilheany, D.A. Kennedy, J.F. Malone and B.J. Walker; *J. Chem. Soc., Chem. Commun.*, 1984, 1217.
84CHEC-I(1)518 K. Dimroth; *Comp. Heterocycl. Chem., 1st edn.*, 1984, **1**, 518–522.
84H(22)2713 K. Matsumoto, S. Hashimoto, S. Otani and T. Uchida; *Heterocycles*, 1984, **22**, 2713.
84JA425 F. Mercier, F. Mathey, J. Fischer and J.H. Nelson; *J. Am. Chem. Soc.*, 1984, **106**, 425.
84JA826 S. Holand, C. Charrier, F. Mathey, J. Fischer and A. Mitschler; *J. Am. Chem. Soc.*, 1984, **106**, 826.
84JCS(P1)1627 J.C. Caesar, D.V. Griffiths, J.C. Tebby and S.E. Willetts; *J. Chem. Soc., Perkin Trans. 1*, 1984, 1627.
84OM456 A. Marinetti and F. Mathey; *Organometallics*, 1984, **3**, 456.
84PS(19)45 J.-M. Alcaraz, E. Deschamps and F. Mathey; *Phosphorus Sulfur*, 1984, **19**, 45.
85IC4141 F. Mercier, F. Mathey, J. Fischer and J.H. Nelson; *Inorg. Chem.*, 1985, **24**, 4141.
85JA3389 L.D. Quin and B.G. Marsi; *J. Am. Chem. Soc.*, 1985, **107**, 3389.
85JA4700 A. Marinetti and F. Mathey; *J. Am. Chem. Soc.*, 1985, **107**, 4700.
85JA6396 M.S. Holt, J.H. Nelson, P. Savignac and N.W. Alcock; *J. Am. Chem. Soc.*, 1985, **107**, 6396.
85JOC467 C.C. Santini and F. Mathey; *J. Org. Chem.*, 1985, **50**, 467.
85MRC929 L.D. Quin and F.C. Bernhardt; *Magn. Reson. Chem.*, 1985, **23**, 929.
85PS(25)117 L.D. Quin and K.C. Caster; *Phosphorus Sulfur*, 1985, **25**, 117.
85PS(25)201 R. Hussong, H. Heydt and M. Regitz; *Phosphorus Sulfur*, 1985, **25**, 201.
85T1409 C.W. Bird; *Tetrahedron*, 1985, **41**, 1409.
85TL3461 B. Deschamps and F. Mathey; *Tetrahedron Lett.*, 1985, **26**, 3461.
86CB2915 R. Appel, C. Casser, F. Knoch and B. Niemann; *Chem. Ber.*, 1986, **119**, 2915.
86CC1574 J. Barluenga, F. Lopez and F. Palacios; *J. Chem. Soc., Chem. Commun.*, 1986, 1574.
86JCS(P2)629 L.D. Quin, S.E. Belmont, F. Mathey and C. Charrier; *J. Chem. Soc., Perkin Trans. 2*, 1986, 629.
86JOM(316)271 F. Mercier, S. Holand and F. Mathey; *J. Organomet. Chem.*, 1986, **316**, 271.
86OM1161 J. Svara, A. Marinetti and F. Mathey; *Organometallics*, 1986, **5**, 1161.
86POL1413 S. Holand, F. Mathey and J. Fischer; *Polyhedron*, 1986, **5**, 1413.
86ZN(B)931 W. Rösch and M. Regitz; *Z. Naturforsch., Teil B*, 1986, **41**, 931.
87AG(E)229 A.J Ashe III, S. Mahmoud, C. Elschenbroich and M. Wunsch; *Angew. Chem., Int. Ed. Engl.*, 1987, **26**, 229.
87AG(E)275 F. Mathey; *Angew. Chem., Int. Ed. Engl.*, 1987, **26**, 275.
87ICA(130)93 R.M.G. Roberts and A.S. Wells; *Inorg. Chim. Acta*, 1987, **130**, 93.
87JOM(332)C1 F. Zurmühlen and M. Regitz; *J. Organomet. Chem.*, 1987, **332**, C1.
87MRC774 J.H. Nelson, S. Affandi, G.A. Gray and E.C. Alyea; *Magn. Reson. Chem.*, 1987, **25**, 774.
87NJC585 F. Mathey; *New J. Chem.*, 1987, **11**, 585.
87OM586 C. Charrier, N. Maigrot and F. Mathey; *Organometallics*, 1987, **6**, 586.
87PS(29)123 J.C. Caesar, D.V. Griffiths and J.C. Tebby; *Phosphorus Sulfur*, 1987, **29**, 123.
87SR1307 S. Affandi, R.L. Green, B.T. Hsieh, M.S. Holt, J.H. Nelson and E.C. Alyea; *Synth. React. Inorg. Metal-Org. Chem.*, 1987, **17**, 307.
87TL3445 E. Vedejs and C. Marth; *Tetrahedron Lett.*, 1987, **28**, 3445.
87TL5021 A. Marinetti and F. Mathey; *Tetrahedron Lett.*, 1987, **28**, 5021.
88CRV429 F. Mathey; *Chem. Rev.*, 1988, **88**, 429.
88JA2310 P.J. Fagan and W.A. Nugent; *J. Am. Chem. Soc.*, 1988, **110**, 2310.
88JA4204 K.K. Baldridge and M.S. Gordon; *J. Am. Chem. Soc.*, 1988, **110**, 4204.
88JCS(P1)175 J.C. Caesar, D.V. Griffiths and J.C. Tebby; *J. Chem. Soc., Perkin Trans. 1*, 1988, 175.
88OM921 F. Nief, F. Mathey, L. Ricard and F. Robert; *Organometallics*, 1988, **7**, 921.

88OM1796	S. Holand and F. Mathey; *Organometallics*, 1988, **7**, 1796.
88OM2233	N.H. Tran Huy and F. Mathey; *Organometallics*, 1988, **7**, 2233.
88TL4581	W.H.-L.W.Tan and A. Foucaud; *Tetrahedron Lett.*, 1988, **29**, 4581.
88ZOB783	S.M. Kalinov, M.F. Rostovskaya and V.I. Vysotskii; *Zh. Obshch. Khim.*, 1988, **58**, 783.
89AG(E)1367	T. Douglas and K.H. Theopold; *Angew. Chem., Int. Ed. Engl.*, 1989, **28**, 1367.
89CC967	N.S. Isaacs and G.N. El-Din; *J. Chem. Soc., Chem. Commun.*, 1989, 967.
89IC217	J.A. Rahn, M.S. Holt, G.A. Gray, N.W. Alcock and J.H. Nelson; *Inorg. Chem.*, 1989, **28**, 217.
89IC3453	L. Solujic, E.B. Milosavljević, J.H. Nelson, N.W. Alcock and J. Fischer; *Inorg. Chem.*, 1989, **28**, 3453.
89IC4132	R. Vac, J.H. Nelson, E.B. Milosavljević, L. Solujić and J. Fischer; *Inorg. Chem.*, 1989, **28**, 4132.
89IC4536	S. Affandi, J.H. Nelson and J. Fischer; *Inorg. Chem.*, 1989, **28**, 4536.
89JOC4754	P. Le Goff, F. Mathey and L. Ricard; *J. Org. Chem.*, 1989, **54**, 4754.
89MI01	M.B. Hocking, D.M. Preston and J. Zink; *J. Lumin.*, 1989, **43**, 309.
89NJC369	A. Brèque, J.-M. Alcaraz, L. Ricard, F. Mathey, A. Tambuté and P. Macaudière, P. *New J. Chem.*, 1989, **13**, 369.
89RTC120	N.S. Isaacs, G.N. El-Din and M.G.B. Drew; *Recl. Trav. Chim. Pays-Bas*, 1989, **108**, 120.
89T7083	N.S. Isaacs and G.N. El-Din; *Tetrahedron*, 1989, **45**, 7083.
89TL3525	D. Oebels and F.-G. Klärner; *Tetrahedron Lett.*, 1989, **30**, 3525.
90AG(E)655	M.-O. Bevierre, F. Mercier, L. Ricard and F. Mathey; *Angew. Chem., Int. Ed. Engl.*, 1990, **29**, 655.
90H(30)201	K. Matsumoto, S. Hashimoto and T. Uchida; *Heterocycles*, 1990, **30**, 201.
90HAC187	F. Mercier and F. Mathey; *Heteroatom Chem.*, 1990, **1**, 187.
90JOC2494	E. Deschamps and F. Mathey; *J. Org. Chem.*, 1990, **55**, 2494.
90PS(47)477	N.H. Tran Huy and F. Mathey; *Phosphorus Sulfur Silicon*, 1990, **47**, 477.
90RHA39	L.D. Quin; *Rev. Heteroatom Chem.*, 1990, **3**, 39.
90T5697	C.W. Bird; *Tetrahedron*, 1990, **46**, 5697.
91HAC359	L.D. Quin and X.-P. Wu; *Heteroatom Chem.*, 1991, **2**, 359.
91HAC377	E. Deschamps, L. Ricard and F. Mathey; *Heteroatom Chem.*, 1991, **2**, 377.
91JOC4031	S. Holand, L. Ricard and F. Mathey; *J. Org. Chem.*, 1991, **56**, 4031.
92BSF486	E. Deschamps and F. Mathey; *Bull. Soc. Chim. Fr.*, 1992, **129**, 486.
92H(33)563	A.N. Hughes and K.E. Edgecombe; *Heterocycles*, 1992, **33**, 563.
92JOC3694	E.J.P. Malar; *J. Org. Chem.*, 1992, **57**, 3694.
92OM1411	B. Deschamps and F. Mathey; *Organometallics*, 1992, **11**, 1411.
92OM1840	H.-L. Ji, J.H. Nelson, A. De Cian, J. Fischer, L. Solujić and E.B. Milosavljević; *Organometallics*, 1992, **11**, 1840.
92OM4069	D. Bhaduri, J.H. Nelson, C.L. Day, R.A. Jacobson, L. Solujić and E.B. Milosavljević; *Organometallics*, 1992, **11**, 4069.
92OS(70)272	P.J. Fagan and W.A. Nugent; *Org. Synth.*, 1992, **70**, 272.
92RHA1	F. Mathey; *Rev. Heteroatom Chem.*, 1992, **6**, 1.
92TL5495	G. Märkl and J. Reindel; *Tetrahedron Lett.*, 1992, **33**, 5495.
93BSF695	A.E. Ferao, B. Deschamps and F. Mathey; *Bull. Soc. Chim. Fr.*, 1993, **130**, 695.
93CB473	F.-G. Klärner, D. Oebels and W.S. Sheldrick; *Chem. Ber.*, 1993, **126**, 473.
93CC1309	J. Kurita, M. Ishii, S. Yasuike and T. Tsuchiya; *J. Chem. Soc., Chem. Commun.*, 1993, 1309.
93CJC1873	M.B. Hocking and F.W. van der Voort Maarschalk; *Can. J. Chem.*, 1993, **71**, 1873.
93JOC1985	E. Vedejs and M.J. Peterson; *J. Org. Chem.*, 1993, **58**, 1985.
93JOC5414	S.M. Bachrach; *J. Org. Chem.*, 1993, **58**, 5414.
93TL3103	G. Märkl, G.Y. Jin and K.P. Berr; *Tetrahedron Lett.*, 1993, **34**, 3103.
94JA3306	F. Laporte, F. Mercier, L. Ricard and F. Mathey; *J. Am. Chem. Soc.*, 1994, **116**, 3306.
94JA9638	D.B. Chesnut and L.D. Quin; *J. Am. Chem. Soc.*, 1994, **116**, 9638.
94JOC3394	S.M. Bachrach and L. Perriott; *J. Org. Chem.*, 1994, **59**, 3394.
94JOC5027	S.M. Bachrach; *J. Org. Chem.*, 1994, **59**, 5027.
94OM925	N.H. Tran Huy and F. Mathey; *Organometallics*, 1994, **13**, 925.
94PC01	C.W. Bird, Personal communication, 1994. *J. Organomet. Chem.*, 1997, **529**, 395.
B-70MI01	F.G. Mann, *The Heterocyclic Derivatives of Phosphorus, Arsenic, Antimony, and Bismuth*, Wiley-Interscience, New York, 2nd. ed., 1970, ch. 1.
B-79MI01	A.N. Hughes; in *New Trends in Heterocyclic Chemistry*, eds. R.B. Mitra, N.R. Ayyangar, V.N. Gogte, R.M. Acheson and N. Cromwell, Elsevier, Amsterdam, 1979, p. 216.
B-80MI01	I. Jardine, *Anticancer Agents based on Natural Products*, M. Cassidy, New York, 1980, p. 319.
B-81MI01	L.D. Quin; *The Heterocyclic Chemistry of Phosphorus*, Wiley-Interscience, New York, 1981, ch. 2.
B-81MI02	L.D. Quin; *The Heterocyclic Chemistry of Phosphorus*, Wiley-Interscience, New York, 1981, ch. 6.
B-81MI03	L.D. Quin; *The Heterocyclic Chemistry of Phosphorus*, Wiley-Interscience, New York, 1981, ch. 7.
B-83MI01	F. Mathey, J. Fischer and J.H. Nelson; *Structure and Bonding*, Springer-Verlag, Berlin, 1983, vol. 53, p. 153.
B-87MI01	L.D. Quin; in *Phosphorus-31 NMR Spectroscopy in Stereochemical Analysis*, eds. J.G. Verkade and L.D. Quin, VCH, Deerfield Beach, FL, 1987, ch. 12.
B-90MI01	L.D. Quin and A.N. Hughes; in *The Chemistry of Organophosphorus Compounds*, ed. F.R. Hartley, Wiley, Chichester, 1990, vol. 1, ch. 10.

B-92MI01 A.N. Hughes, in *Handbook of Organophosphorus Chemistry*, ed. R. Engel, Marcel Dekker, New York, 1992, ch. 10.

B-92MI02 L.D. Quin, S. Jankowski, G.S. Quin, A. Sommese, J.-S. Tang and X.-P. Wu; *ACS Symposium Series 486*, 1992, p. 115.

4.2.2
Five-membered rings. Phospholes: Recent literature 1994–mid-1999

LOUIS D. QUIN and GYÖNGYI S. QUIN
University of North Carolina at Wilmington, Wilmington, NC, USA

4.2.2.1 INTRODUCTION

The phosphole ring system did not appear in the literature until 1953, when Wittig and Geissler reported it in the form of the dibenzo derivative **1** ⟨53LA(580)44⟩. This compound largely has the chemistry of a triarylphosphine, and is not suited for the study of the parent heterocycle. A few years later, the first monocyclic phosphole derivative (**2**) was reported by two groups ⟨59CI(L)1250, 59JA3163⟩. Again the heavy substitution retarded the development of the chemistry of the ring system, and to some extent this was true of the next phosphole to be synthesized, the triphenyl derivative **3** ⟨62CI(L)359⟩. It was not until simpler derivatives such as **4** ⟨66AG(E)846⟩ and **5** ⟨67JA5984⟩ became available that the true chemistry of the phosphole ring system began to emerge.

The field has grown rapidly since these early discoveries. A table of phospholes in Chapter 4.2.1 with phosphorus having coordination number 3 (C.N.3) that were known up to 1994–1995 and for which ^{31}P NMR data were reported can be used to reveal the scope of the field; some 95 structures have been listed in this table. The number of known phospholes is actually greater than this, since ^{31}P NMR data were not reported for all compounds. The phosphole ring is primarily found in the simple monocyclic form with C.N.3 (the 1H-phosphole ring system), but many variations are possible, including (1) fused-ring systems, such as the benzo derivative **6**, (2) the 2-coordinate phospholide anion **7**, (3) systems with C.N.4, 5 and 6

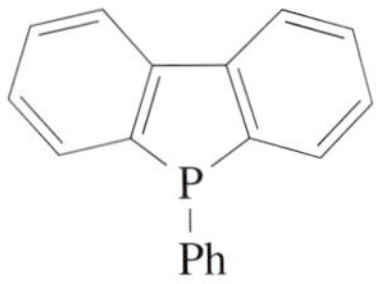
(1)

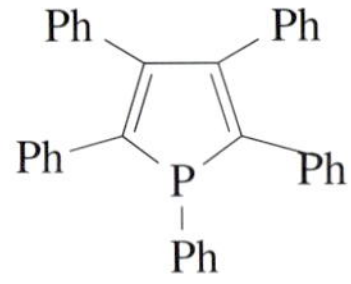
(2)

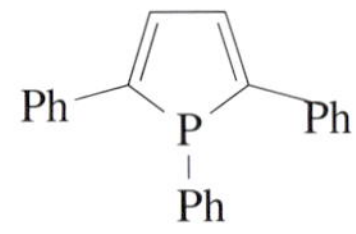
(3)

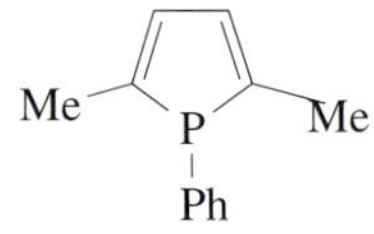
(4)

(5)

(**8, 9** and **10**, respectively), and (4) the rare 2H-position isomer (**11**). The 3H-phosphole structure is known, but so far in only one compound (**12**).

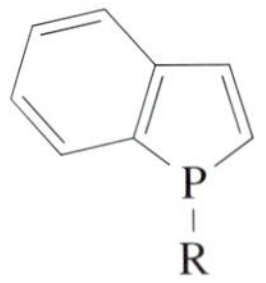
(6)

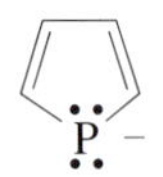
(7)

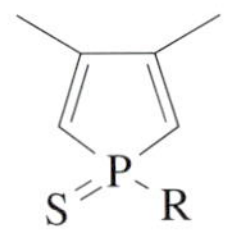
(8)

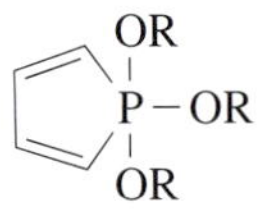
(9)

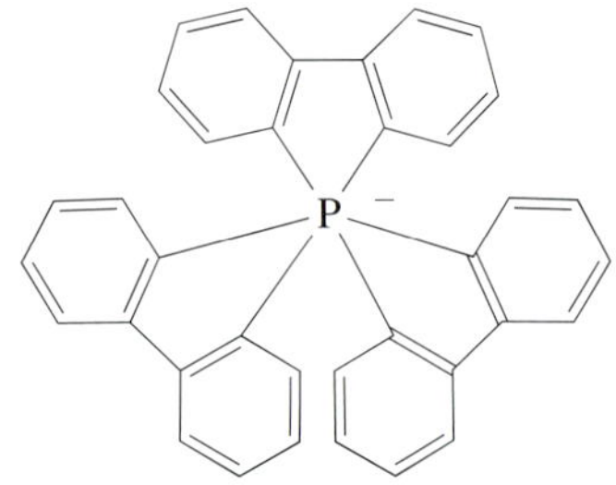
(10)

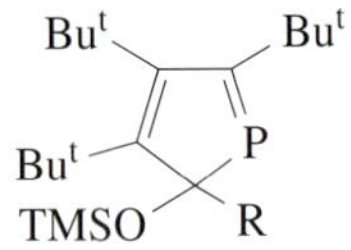
(11)

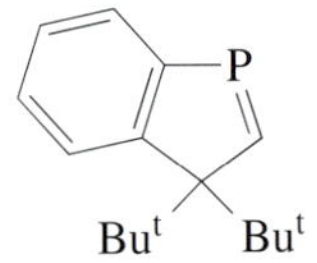
(12)

A very large field has also developed around the metal coordination compounds of 1H-phospholes, which can exhibit various complexing modes. These complexes are of interest as synthetic intermediates, since valuable reactions of the phosphole moiety may be conducted within the coordination sphere of the metal. More recently, certain types of phosphole complexes are finding use in homogeneous catalysis, a point to be considered in Chapter 7.

Although simple 1H-phospholes with hydrogen on phosphorus are unstable at room temperature ⟨82TL511⟩, most other phosphole derivatives have good stability and the ring system can be constructed in a variety of ways. Some of the more common ways are illustrated in Equations 1–3.

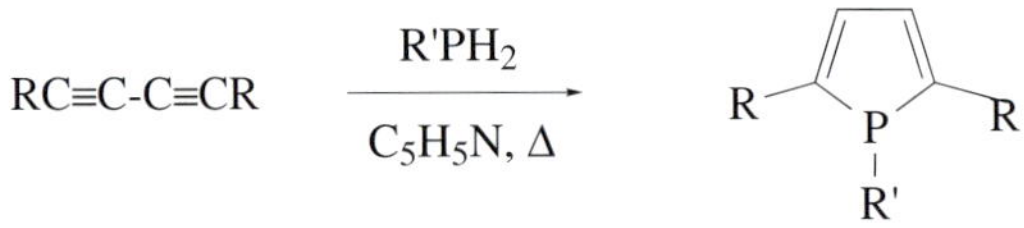
(1)

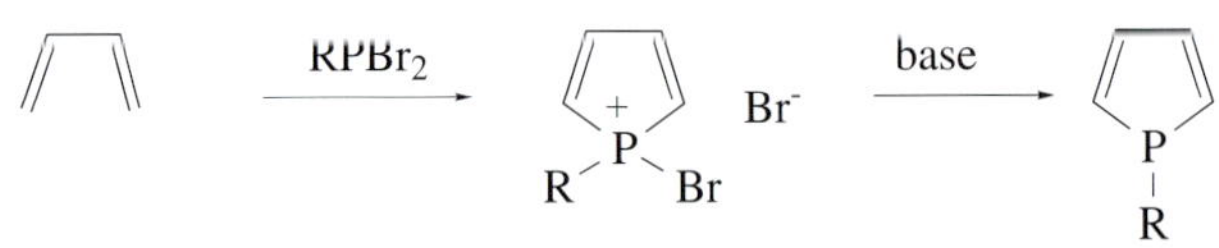
(2)

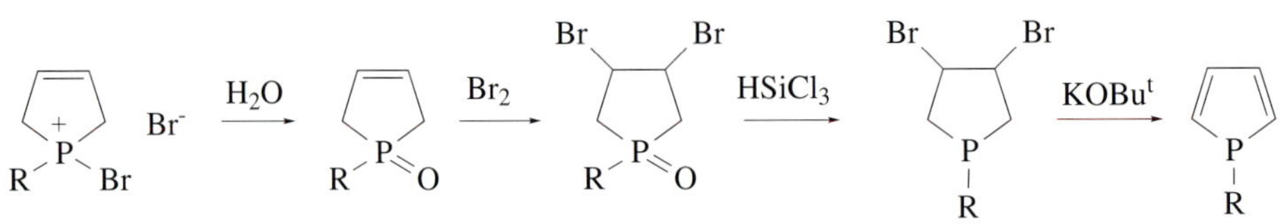
(3)

Once the phosphole ring has been constructed, some synthetically useful chemical changes can be performed on phospholes with preservation of the ring system. This does not include electrophilic substitution, however; this process fails, since the phosphorus atom is the most nucleophilic site in a normal phosphole (an exception was reported in 1997, and is described in Section 4.2.2.8). A few useful processes are seen in Equations 4–6.

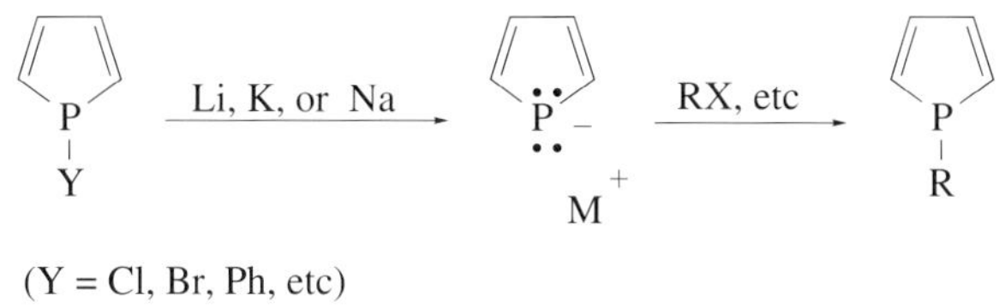

(4)

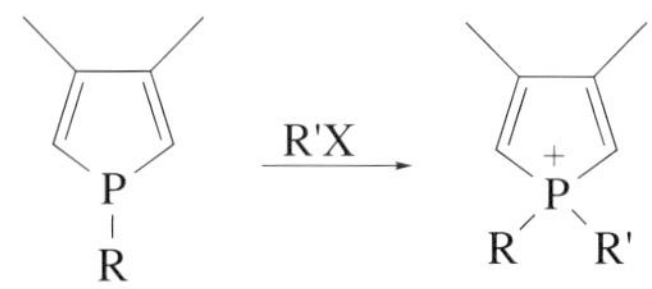

(5)

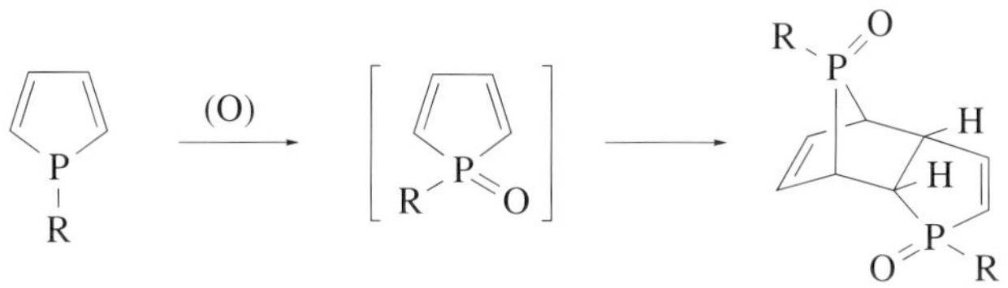

(6)

At elevated temperatures, phospholes can exhibit various forms of chemical change, resulting in modification of the ring system. Phospholes with P-alkyl substituents undergo a form of Diels–Alder dimerization such as that seen in Equation 6, but this process is much more common with 4-coordinate phospholes such as the P-oxides, P-sulfides and quaternary salts, where the dimerization can, in most cases, occur even on formation of the monocyclic phosphole derivative. Phospholes can also undergo Diels–Alder reactions with strong dienophiles (Equation 7).

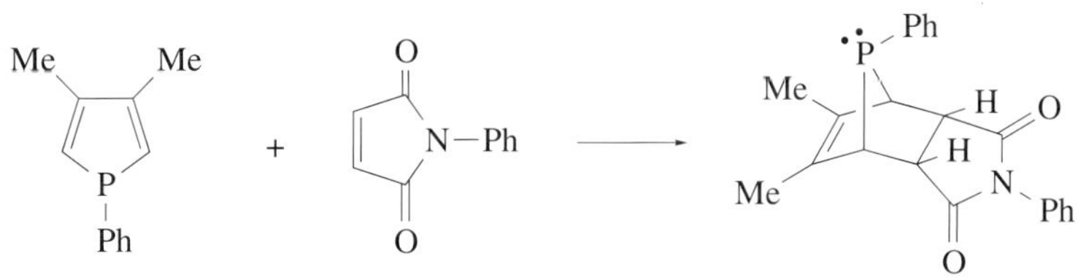

(7)

A process that has proved to be of considerable synthetic value involves the thermally induced [1,5] sigmatropic rearrangement of a P-aryl substituent to a ring carbon, as in Equation 8. The initial product from 3,4-dimethyl-1-phenylphosphole is seen to be a 2H-phosphole (**13**), which immediately undergoes isomerization to 2H-phosphole **14**. This phosphole is also not isolatable, and undergoes dimerization as shown.

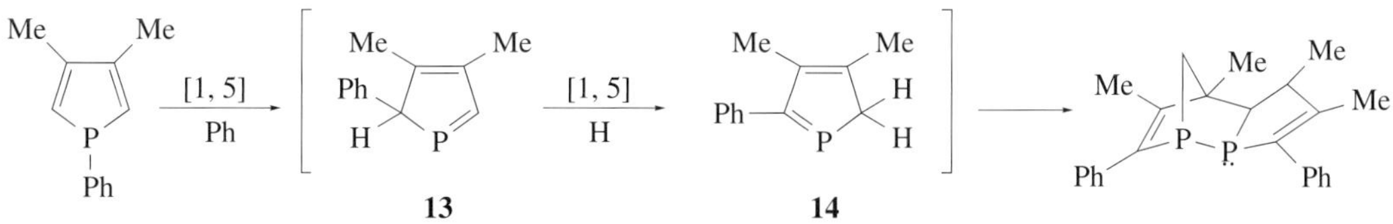

(8)

The 2H-phosphole can be trapped by various reagents as seen in Scheme 1.

Yet another thermal reaction is illustrated with 1,2,5-triphenylphosphole, which at 230ºC is converted to a biphospholyl (Equation 9).

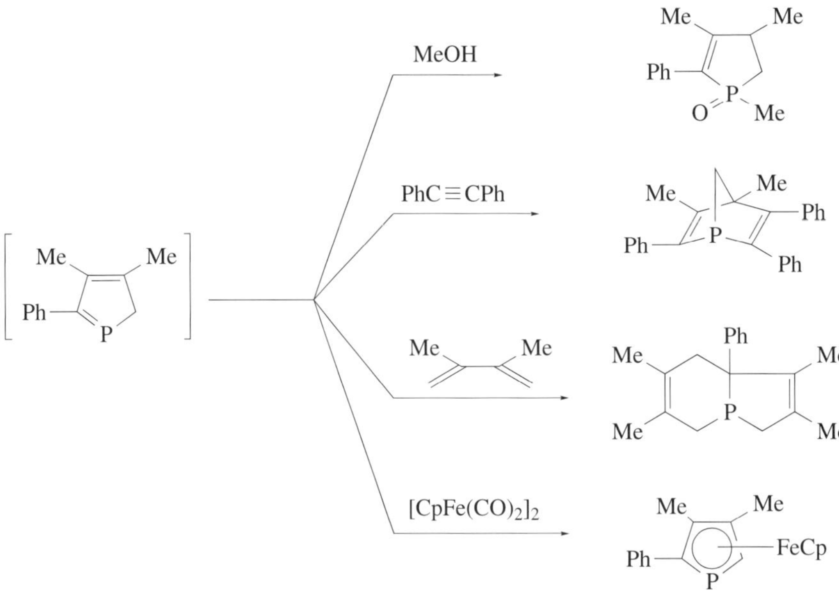

Scheme 1

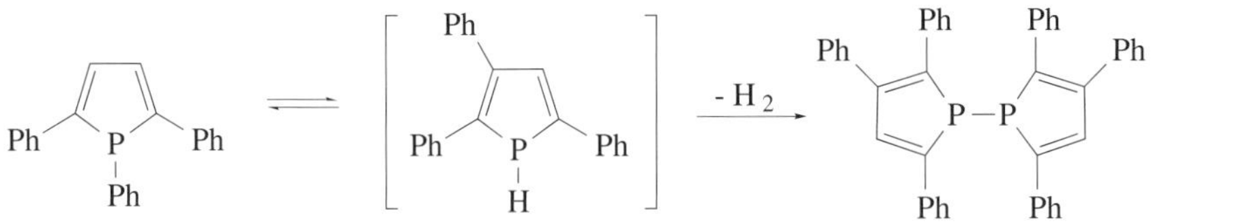

(9)

Much of the early interest in phosphole chemistry was centered on the possibility that the system might be cyclically delocalized with 6 π electrons and thus have the properties of aromaticity. The resonance picture of such delocalization is shown with Equation 10, which is the expression widely used in connection with the truly delocalized related 5-membered heterocycles pyrrole, thiophene and furan.

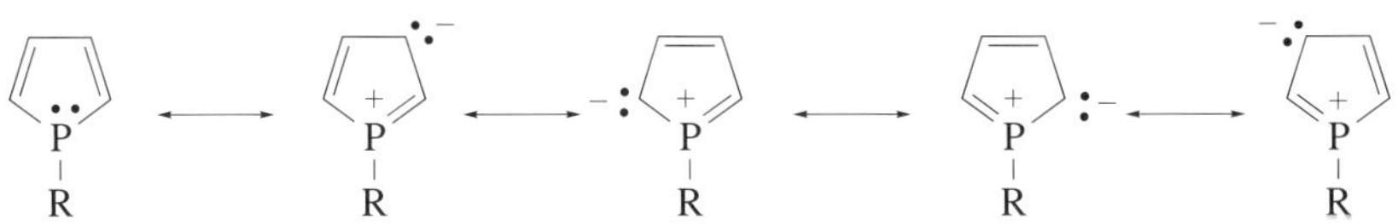

(10)

However, just as is true of any 3-coordinate phosphorus compound, the phosphorus atom has pyramidal structure. Although the inversion barrier (16 kcal/mol ⟨71JA6205⟩) is only about half that of simple phosphines, it nevertheless is significant, and the pyramidal structure is a deterrent to the efficient overlap of orbitals on carbon and phosphorus. As a result, phospholes are of a low order of delocalization, less than that of furan, and are in fact generally considered to have little aromatic character, as has been expressed in recent reviews ⟨88CRV429, B-92MI01⟩. One of the simplest of the early phospholes, the 1-benzyl derivative, provides informative data in this regard. This compound is a low-melting crystalline solid, and X-ray diffraction analysis was performed with it ⟨70JA5779⟩. This was the first report that established firmly the pyramidal character at phosphorus, and showed that the effects of bond shortening at the P–C bonds and of the C_3–C_4 bond, and lengthening of the C=C bonds, as called for by the resonance hybrid of Equation 10, were present but of small magnitude. Indeed, a calculation of the Bird Index of aromaticity, which is based on bond parameters, showed that 1-benzylphosphole with an Index of 35.5 was the least aromatic of the family of 5-membered ring heterocycles (thiophene, 66; selenophene, 59; pyrrole, 59; furan, 43). Some other properties of the compounds are consistent with this view of little delocalization, although others are ambiguous (such as the pronounced deshielding of the protons on carbon, a property generally accepted to be a result of an aromatic ring current). Numerous theoretical considerations of the bonding in phospholes have been made; the most revealing are the more recent ones, summarized in Chapter 4.2.1, where the power of super-computing has been employed. The picture of little delocalization has been confirmed and directly related to the pyramidal structure at phosphorus. Indeed, calculations show that as the pyramid is flattened, delocalization increases, to the point that in fully planar 1H-phosphole the delocalization would exceed that of the other 5-membered heterocycles. As will be seen in Section 4.2.2.3.2, this point has very recently been tested experimentally by constructing phospholes with greatly reduced pyramidal structure, whereupon experimental indications were obtained of a system with a significant level of delocalization.

Phospholes have been extensively characterized by various spectroscopic techniques. Proton NMR attracted early interest, where it was found, as already noted, that the shifts for both the α and β protons were strikingly downfield (e.g. in 1-benzylphosphole, these signals overlap with those of the phenyl group at δ 6.31–7.29 ⟨73JOC1858⟩). The ^{31}P coupling to the α protons is especially strong (typically 34–42 Hz). The ^{31}P chemical shifts for phospholes are significantly downfield of non-cyclic divinylphosphines or of dihydrophospholes, but this effect has recently been shown by theoretical computations not to be a consequence of electron delocalization. Thus, calculations of the ^{31}P NMR shift of pyramidal and planar 1H-phosphole (where delocalization is strong) showed only a deshielding of 9 ppm in the latter ⟨94JA9638⟩. Removal of the proton to give the phospholide ion is accompanied by a strong downfield shift, reproduced by calculation, where it has been attributed to a paramagnetic contribution of the lone pair orbital in the plane of the ring and not to a conjugative interaction. Another unique phosphole spectral feature is found in their mass spectra; even in the simplest phospholes, the molecular ion is the strongest signal. Simple phospholes also display a characteristic absorption band in the ultraviolet spectrum, not found in acyclic models (e.g. 1-methylphosphole, λ_{max} 286 nm in isooctane, log ϵ 3.89 ⟨69JA3308⟩).

The literature on phosphole chemistry is extensive, but has been reviewed frequently over the years. The very recent CHEC-II chapter on phospholes ⟨96CHECII(2)757⟩ probably provides the most complete coverage of the field. It is reproduced with minor corrections in Chapter 4.2.1. Some of the other reviews that are available are those by Mann ⟨B-70MI01⟩, Mathey ⟨80MI01, 88CRV429⟩ and ⟨94PS(87)139⟩, Quin ⟨B-81MI01⟩, Quin and Hughes ⟨B-90MI01⟩, and Hughes ⟨B-92MI01⟩. The recent discussion of phospholes in the monograph by Dillon, Mathey and Nixon ⟨B-98MI01⟩ is also of importance. Reviews by Mathey have also appeared on the chemistry of phospholide ions ⟨94CCR(137)1, 94JOM(475)25⟩

In viewing the literature of the period 1995–mid-1999, these authors note increased emphasis on the creation of properly substituted phospholes for use as ligands in metal complexes that might function as homogeneous catalysts, especially to effect asymmetric syntheses. This has led to renewed interest in dibenzophosphole derivatives as ligands. Attention continues to be given to theoretical aspects of the phosphole ring system. The creation of novel derivatives and of new synthetic procedures remain an active part of the field of phosphole chemistry. The discussion to follow in this chapter will emphasize these and other aspects of the continuing development of this branch of phosphorus chemistry.

4.2.2.2 THEORETICAL STUDIES OF PHOSPHOLES

Along with the great increase in experimental work on the phosphole ring system in recent years has been heightened interest in theoretical aspects of the system. As discussed in Chapter 4.2.1, modern super-computing power makes possible some very sophisticated and reliable calculations of geometrical parameters, as well as evaluations of the aromatic stabilization energy (ASE). Theoretical analysis of the reactions of phospholes also is of current interest.

The calculation of geometrical parameters at various levels of theory was outlined in Chapter 4.2.1. To these may be added calculations at the HF/6-31G** level ⟨95JST(331)109⟩, the MP2(fc)/6-31G* level ⟨95AG(E)337⟩ and MP2/6-31G* level ⟨95JPC586⟩. Semi-empirical calculations by the AM1 method also have been reported ⟨97JHC1387⟩. Calculations of molecular parameters and energies for 1H-, 2H-, and 3H-phospholes were consistent with an earlier report ⟨93JOC5414⟩ in showing the greater thermodynamic stability of 2H-phosphole ⟨94OM4732⟩. In general the results of all of the newly reported calculations of molecular parameters are in accord with those of the previous studies.

The calculation of other molecular properties of phospholes is also being undertaken. Joining an earlier report on the calculation of ^{31}P NMR shifts ⟨94JA9638⟩, reports have appeared on calculations of dipole polarizabilty and hyperpolarizability ⟨95JST(331)109, 97JA6575⟩ (which are of interest in pointing out the molecular features desirable for non-linear optical materials, suggesting potential value of conjugated phospholes in this application), and magnetic susceptibilities ⟨95AG(E)337, 98IC4413, 98JPC(A)9912⟩ which are useful in assessing aromatic stabilization. The heat of hydrogenation of 1H-phosphole to form phospholane has been calculated by the PM3 method to be 51.5 kcal/mol, a reasonable value when compared to that (39.2 kcal/mol) of the much more aromatic pyrrole ⟨95JST(338)51⟩. The gas-phase basicity of the phospholide ion for protonation at the 2-position was calculated to be 341.4 kcal/mol, in good agreement with an experimentally determined value of 338 ± 3 kcal/mol ⟨94OM4732⟩.

Calculations have shown (Chapter 4.2.1) that the pyramidal condition of phosphorus in phospholes, with a substantial barrier (about 16–17 kcal/mol) to inversion, prevents the attainment of conjugation and the achievement of a significant level of ASE, but that when the phosphorus is planarized the ASE becomes quite high. Planarization of phospholes continues to receive theoretical attention ⟨95JPC586, 96JPC6194, 98IC4413⟩. Molecular modelling shows that reduction of the pyramidality at phosphorus can be achieved by creating steric interactions of the substituent at the P atom with the ring ⟨96JOC7801⟩. Placement of π-electron acceptor substituents (such as BH_2 and AlH_2 ⟨95JPC586⟩) at the α-position significantly reduce the inversion barrier and the pyramidality. In explaining these effects, it was postulated that some pentacoordinate character for P is indicated. This is a concept for phosphines that has been used occasionally over the years but strong arguments against the inclusion of d-orbitals to explain the bonding at phosphorus have been presented (see, e.g. ⟨93JA1051⟩), and the proposal for its presence in phospholes has not been accepted by others ⟨96JPC13447, 98IC4413⟩. Planarization at phosphorus is increased by installing additional P atoms in the ring, and in pentaphosphole full planarity with high ASE is achieved ⟨96JPC13447⟩. Planarity is also developed at phosphorus by embedding this atom at a bridgehead position in the fully unsaturated phosphindolizine ring system (**15**) ⟨98NJC651⟩.

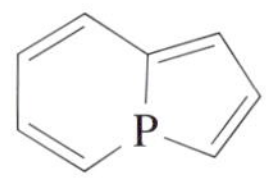

(**15**)

A recent calculation of the aromatic stabilization energy in the parent pyramidal phosphole gives a value of 7.0 kcal/mol ⟨95AG(E)337⟩, to be compared to a value of 25.5 kcal/mol for pyrrole. However, calculations indicate that planar phosphole would have higher ASE values than pyrrole and thiophene ⟨98IC4413⟩. This paper presents a valuable discussion of the various criteria presently in use in assessing aromaticity, including the Bird index and the Bond Shortening (BDSHRT) index, both of which are based on bond parameters, and the nucleus-independent chemical shifts (NICS), which are derived from calculations at the center of aromatic systems ⟨96JA6317⟩. The Julg Index, again based on bond parameters, also finds use in evaluating relative aromaticities ⟨96JPC13447⟩. A new proposal is to employ average bond order deviation from an ideal system, using semi-empirical AM1 calculations ⟨97JHC1387⟩. By any of these criteria, pyramidal phosphole is seen to be a system of only low ASE. On the other hand, the phospholide ion, where the problem of pyramidality is absent, has a very high ASE by calculation (29.0 kcal/mol ⟨96OM1756⟩).

The concept from calculations that increasing the steric bulk at phosphorus would reduce the pyramidality and increase delocalization has been tested experimentally. Phospholes with very large aryl substituents were synthesized and their structural parameters determined by X-ray diffraction analysis. The data are discussed more fully in Section 4.2.2.3.2, but do confirm the expected flattening and increased delocalization. The compound giving the best indication of increased aromaticity had structure **16a**; the shortening of the P–C and C_3–C_4 bonds, and lengthening of the C_α–C_β bonds, was pronounced and led to a Bird Index of 56.5 ⟨97JA5095⟩, a great increase over the value (35.5) for 1-benzylphosphole, and close to that of pyrrole (59). The compound had significant flattening of the pyramid; the sum of the bond angles at P was 331.7° (compared to 302.7° for 1-benzylphosphole), and the angle by which the exocyclic C atom at P was out of the plane of the ring C_2–P–C_5 atoms was reduced to 45° from 66.9° in the benzyl derivative.

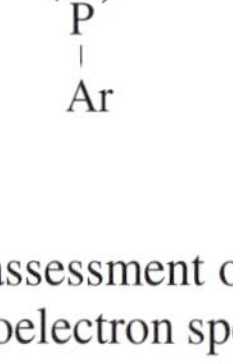

(**16**)

a, Ar = 2,4,6-$Bu^t_3C_6H_2$
b, Ar = 2,4,6-$Pr^iC_6H_2$
c, Ar = 6-Me-2,4-$Bu^tC_6H_2$
d, Ar = 2,4,6-$Me_3C_6H_2$
e, Ar = Ph

Another experimental assessment of the aromaticity in compound **16a**, as well as in related compounds **16b–e**, was made by photoelectron spectroscopy studies. The first to be studied (**16c**) gave the lowest value ever recorded up to that time for the lone pair ionization energy of a phosphole (7.9 eV). Of significance also was the fact that this value was higher than that measured for the corresponding tetrahydro derivative (7.55 eV) ⟨96JOC7808⟩. In all other PES studies of phospholes ⟨96CHECII(2)757⟩, no difference was observed from the value of the tetrahydro compound. Calculations supported an explanation for these facts based on partial flattening of the phosphorus pyramid to an out-of-plane value of 55.7°. PES studies later were extended to the more crowded phosphole **16a**, and this led as expected to an even lower value for the lone pair ionization potential (7.5 eV) ⟨98JOM(566)29⟩. In this paper, a comparison of experimental and calculated PE spectra for all of the compounds of structure **16a–e** was presented.

Just as a decrease in pyramidality should increase delocalization, an *increase* in pyramidality should further reduce the already low level of delocalization in a phosphole. This appears to have been demonstrated with two phospholes (**17** and **18**) where strongly conjugating unsaturated groups are attached to the 2,5-positions. X-ray data are presented in Section 4.2.2.3.2 that show definite increases in the pyramidality (bond angle sums of 292.4° ⟨96BSF33⟩ and 299.3° ⟨99CC345⟩, respectively) and length changes in the direction of reduced delocalization. In both papers, it was suggested that the conjugating α-substituents interacted with the diene unit of the ring, virtually eliminating any interaction of the phosphorus lone pair with the diene. This point has not yet been examined with the aid of theoretical calculations. In both compounds, the length of the bond between the substituent and the phosphole ring shows some shortening (Section 4.2.2.3.2) which does support the proposal of conjugation.

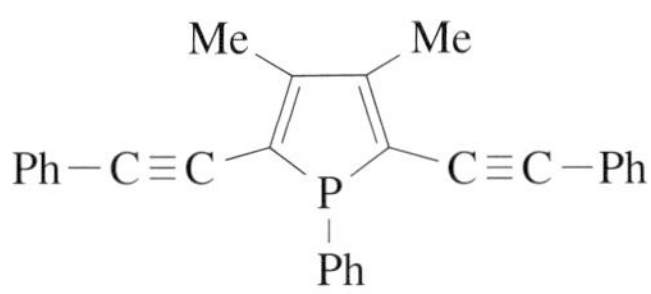

(**17**)

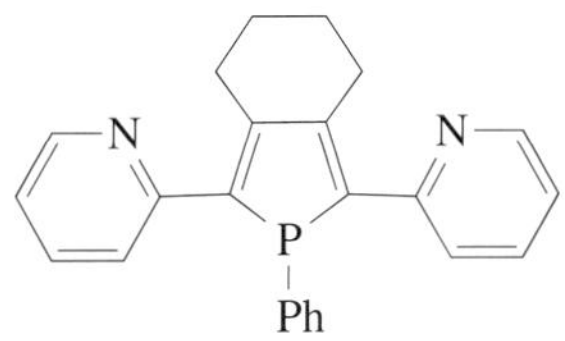

(**18**)

The application of computational methods to interpret reaction characteristics of phospholes was introduced by Bachrach in three papers discussed in CHEC-II ⟨93JOC5414, 94JOC3394, 94JOC5027⟩. These studies have been continued with the treatment of cycloaddition reactions of 2H-phospholes. That the P=C bond in a 2H-phosphole can act as a dienophile toward a diene in the Diels–Alder reaction was noted in Scheme 1 ⟨81JA4595⟩ to give the phosphabicyclo[4.3.0]nonane framework. Calculations have now shown that this is indeed the expected thermodynamic product for the reaction of 2H-phosphole with butadiene ⟨96CJC839⟩. The alternative of the 2H-phosphole acting as the diene toward a double bond of butadiene to form a 1-phosphanorbornene was found to be the kinetically favored path. The results are strikingly similar to those obtained in calculations with cyclopentadiene taking the place of 2H-phosphole. In another study ⟨97JCO198⟩, the reaction of 2H-phosphole with phosphaethene was studied as model for the dimerization of the phosphole. From the results it could be concluded that the dimerization would give the P–P bonded product with endo ring fusion, as observed experimentally. The reaction is kinetically controlled and is concerted and synchronous, with a very low activation energy (5–7 kcal/mol). This low activation energy accounts for the failure to observe the 2H-phospholes. With phosphaketene (HP=C=O) as the reactant ⟨97JOM(529)15⟩, the Diels–Alder [4 + 2] path was favored both thermodynamically and kinetically over the alternative of a [2 + 2] cycloaddition. Preferred products from these cycloadditions are shown in Equation 11. The opposite prediction was made for the reaction of cyclopentadiene with ketene.

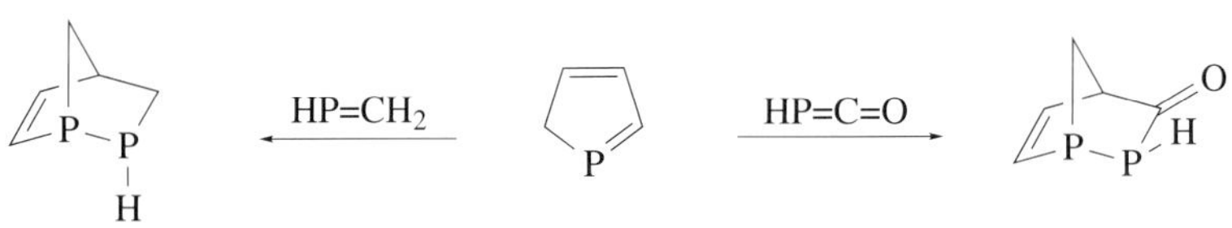

(11)

There are two theoretical studies on the structure and properties of polyphospholes, where the linkage is at the α-positions and strong conjugation could exist ⟨94CPL(224)213, 98SM(96)177⟩. Such structures could be of interest as electrically conducting polymers. Experimentally, the largest polyphosphole consists of four phosphole units ⟨94AG(E)1158⟩. The X-ray analysis of this compound (Section 4.2.2.3.2) supplied the information that the bond length between the central phosphole units was 1.431 Å, and that between the terminal units was 1.451 Å. This suggested to the authors that there was little bond shortening, and therefore it was unlikely that the rings were conjugated. However, the paper by Salzner et al. on theoretical grounds predicted the presence of conjugation, and suggested that these bond lengths are not necessarily indicative of its absence ⟨98SM(96)177⟩.

4.2.2.3 EXPERIMENTAL STRUCTURAL METHODS

4.2.2.3.1 General

Virtually all of the new phospholes reported in Table 1 were characterized by ^{1}H, ^{31}P and ^{13}C NMR spectroscopy, and much new data have been published. In general, the ^{1}H NMR data fit into the patterns outlined in Chapter 4.2.1. They will receive no special mention. Some new results for ^{31}P NMR and ^{13}C NMR spectra do call for discussion in this section, however, as do some structural studies made with X-ray diffraction analysis. Mass spectra are reported for most of the compounds in Table 1, but call for no special comment. This is true also for the less commonly measured infrared and ultraviolet spectra. An application of photoelectron spectroscopy in phosphole chemistry was discussed in Section 4.2.2.2.

4.2.2.3.2 X-ray Diffraction Analysis

Bond angle and length data for the phosphole ring in recently synthesized compounds are presented in Table 2. Numerous metal complexes of phospholes have also been subjected to X-ray analysis, but the data are not included in this table. While most of the structures are complex, some new data provide information on the effect of ring substituents on the degree of pyramidality of the phosphorus atom, and the relation of pyramidality and electron delocalization in the ring. This is revealed by the classical effect in 5-membered rings of delocalization causing shortening of the X–C_1, X–C_5 and C_3–C_4 bonds and lengthening of the C_2=C_3 and C_4=C_5 bonds. To assess these effects in phospholes, data for new compounds are compared to those of the simplest phosphole for which data are available, namely, 1-benzylphosphole ⟨73JCS(D)1888⟩. This compound is free of any conjugating substituents on the ring, and of steric crowding from multiple substitution. For reference purposes, parameters for this compound are shown in Table 3. The sum of the bond angles around phosphorus can be taken as the deviation from planarity (360°; a simple trialkylphosphine has a sum of about 295°(e.g. Me_3P, 296°) ⟨96CB1083⟩. Compared to 1-benzylphosphole (302.7°), new compound **17** with a sum of 292.4° can be said to have increased pyramidality ⟨96BSF33⟩. A similar but less pronounced result (299.3°) has been obtained for compound **18** (Section 4.2.2.2). The pyridyl rings are nearly co-planar (dihedral angles 25.6° and 7.0°), which would allow for conjugation between the rings. The authors of both papers suggest that conjugation with C-substituents would greatly reduce conjugation with the phosphorus lone pair.

If no other effects were operative, the increased pyramidality would suggest diminished electron release from phosphorus to the ring, a relation clearly established by the theoretical studies cited in Chapter 4.2.1. Since the heavy C-substitution in the new compounds compared to 1-benzylphosphole may introduce perturbing steric interactions along with conjugative effects, a comparison of carbon–carbon bond lengths may not be very meaningful, but the average length of the P–C_1 and P–C_5 bonds do vary in the expected direction for diminished electron delocalization. Thus, compared to 1-benzylphosphole with 1.783 Å, both compound **17** and **18** have longer C–P bonds (1.818 Å and 1.806 Å, respectively). In compound **17**, conjugation between the phosphole ring and the acetylenic groups would be indicated by a reduction in the normal length (1.43 Å ⟨B-85MI01⟩) of the sp^2–sp bond, and indeed there is a modest change (to 1.420 Å) in this direction. A similar small effect (to 1.466 Å from the sp^2–sp^2 length of 1.48 Å) was seen for the phosphole–pyridyl bond in compound **18**. It would be helpful to have data for other such phospholes to confirm if the apparent reduced electron release from phosphorus can be more firmly attributed to the proposed conjugative electronic effect of the C-substituents.

X-ray analyses have also been performed on the two phospholes (**16a** ⟨97JA5095⟩ and **16b** ⟨97JOM(532)109⟩) that have been constructed to test the possibility that increased steric bulk at phosphorus might cause flattening of the pyramid, thus permitting greater electron delocalization.

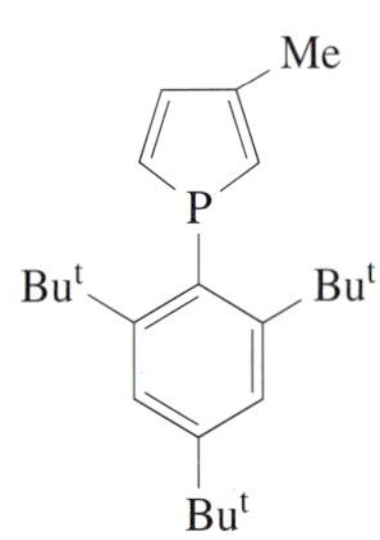

(**16a**)

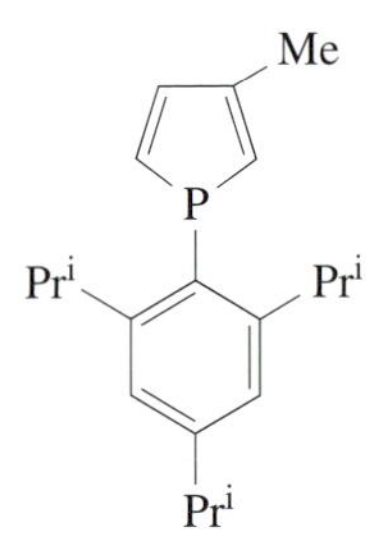

(**16b**)

Table 1. Recently reported phospholes and ^{31}P NMR spectra

Compound		$\delta\ ^{31}P$	Ref.
A. Monocyclic Phospholes			
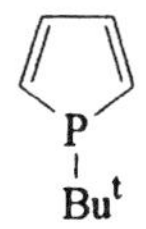		+ 40.0	94OM4732
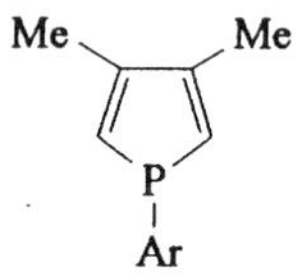	Ar =		
	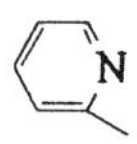 :	+1.8	97AG(E)98
	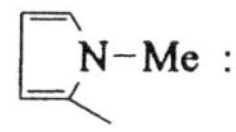 :	- 32	98OM2996
	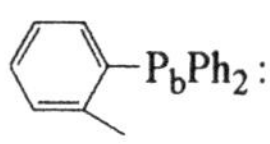 :	- 7.0 (CH_2Cl_2) (P_b -10.5, J_{PP} = 110Hz)	98OM2996
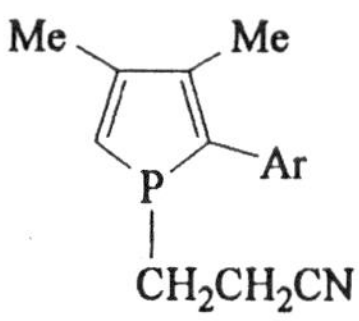	Ar =		
	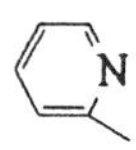 :	+1.8	97AG(E)98
	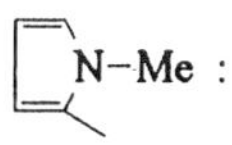 :	- 0.2(C_6D_6)	98OM2996
	Ph :	-1.4	97AG(E)98
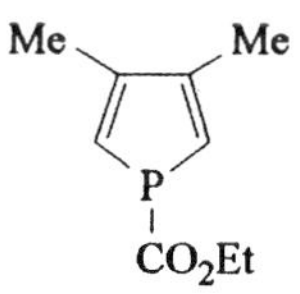		-0.6	97AG(E)98
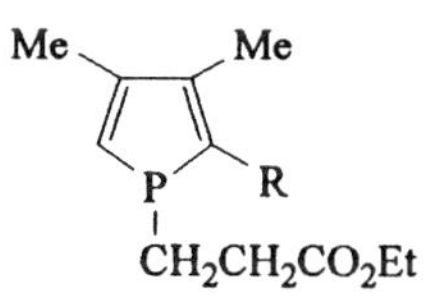	R=		
	CO_2Et :	+3.9	97AG(E)98
	CN :	+12.11	96BSF541
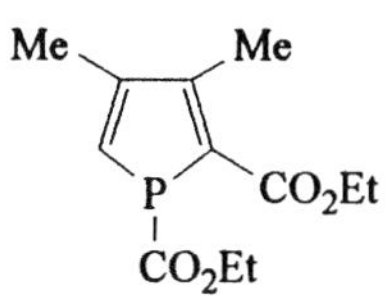		+5.1	97AG(E)98

Table 1. *Continued*

Compound	$\delta\ ^{31}P$	Ref.
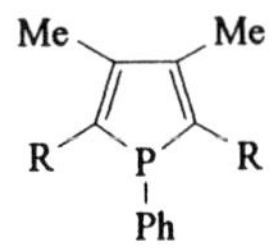	R= Me_3Si : +35.6 MeOOC : +11.9(CD_2Cl_2) Br : +17.04(CH_2Cl_2)	 98CR(C)715 98CR(C)715 94AG(E)1158
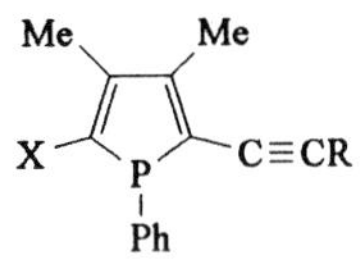	X = H R = Ph : +12.4 R = Me_3Si : +13.7 X = Br R = Ph : +20.2 R = Me_3Si : +21.2	96BSF33
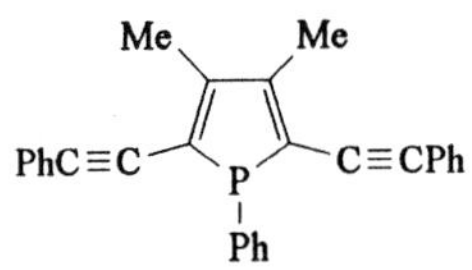	+22.6	96BSF33
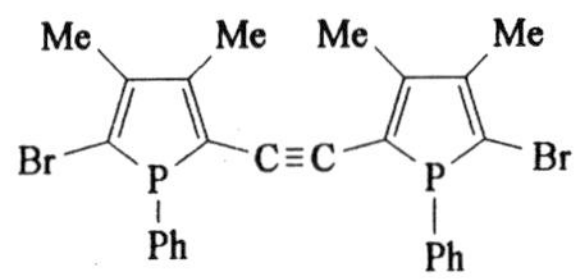	+20.4, +20.5	96BSF33
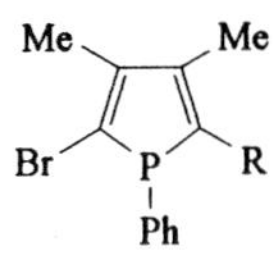	R= COOH : +14.96(DMSO) CHO : +12.07(TMF)	94AG(E)1158
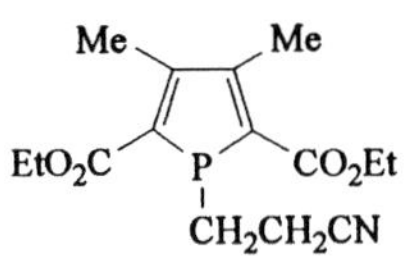	+10.0	97AG(E)98
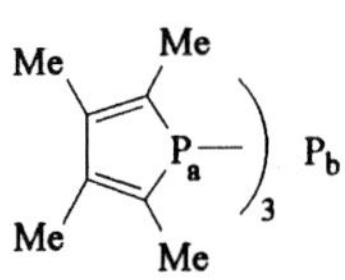	-9.55 (P_b -51.06, J_{PP} = 292.3 Hz)	97JOM(529)75

Table 1. *Continued*

Compound	$\delta\ ^{31}P$	Ref.
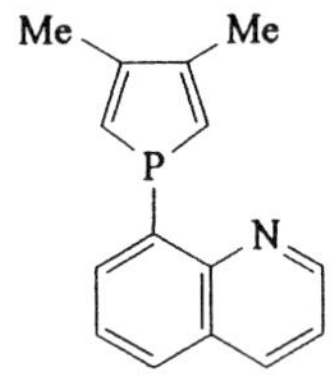	-4.5	97JOM(548)17
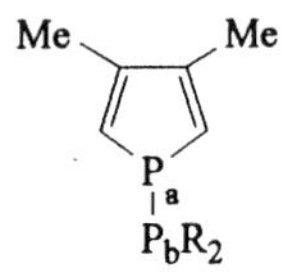	R =	97JOM(529)197
	Bu^t: -18.5 (P_b + 34.8, J_{PP} = 388)	
	Pr^i : -23.8 (P_b + 8.7, J_{PP} = 327.9)	
	Et : -17.3 (P_b -20.3, J_{PP} = 281.1)	
	Ph : -21.7 (P_b -8.7, J_{PP} = 230.4)	
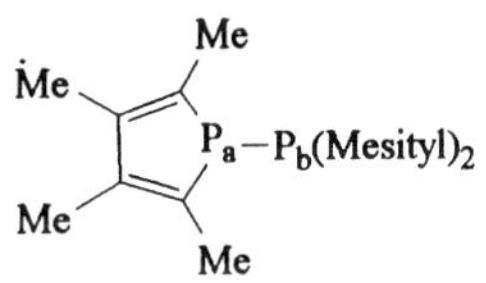	-2.36(P_b -24.49, J_{PP} = 256Hz)	95BSF649
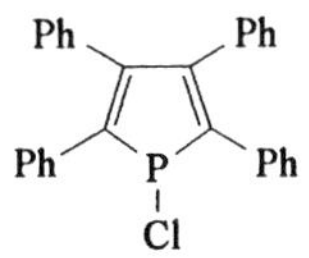	+77.0(C_6D_6)	96PS(118)309
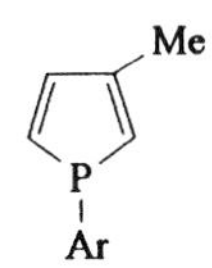	Ar =	
	2,4- Bu^t_2-6-MeC_6H_2 : +1.8	96JOC7801
	2,4,6- Bu^t_3-C_6H_2 : -0.40	97JA5095
	2,4,6- Pr^i_3-C_6H_2 : -8.6	97JOM(532)109
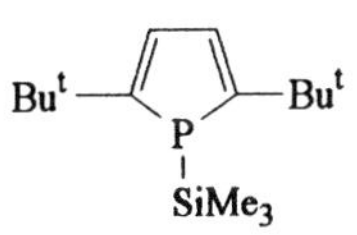	-41.3(THF)	94CC1167

Table 1. *Continued*

Compound	$\delta\,^{31}P$	Ref.
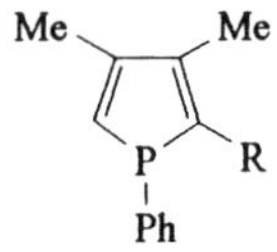	R = COCl : (no data) $CONH_2$: +4.4 CN : (no data) Me_3Sn : +15.2(THF)	96BSF541 96BSF541 96BSF541 96BSF541
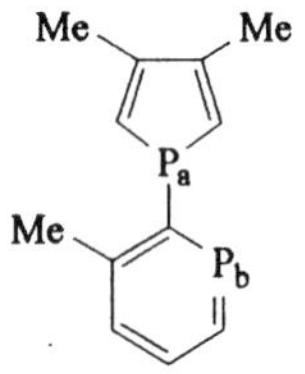	+0.6 (P_b +214.3, $^2J_{PP}$ = 32.20Hz)	95BSF910
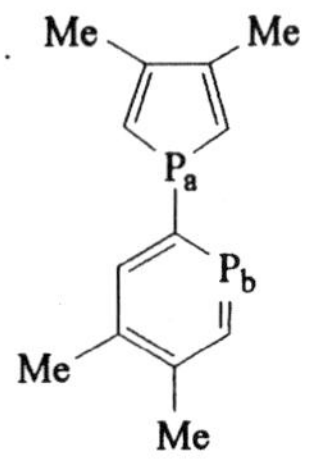	+2.90 (P_b +202.70, $^2J_{PP}$ = 81.65Hz)	95BSF910
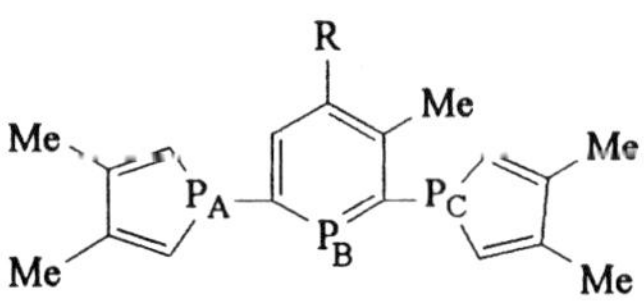	R = Br : P_A +0.85 (P_B +217.95, $^2J_{P_AP_B}$= 91.30 Hz); P_C +3.15 ($^2J_{P_CP_B}$=32.30 Hz) Me : P_A +1.40 (P_B +211.40, $^2J_{P_AP_B}$= 91.65); P_C +1.55 ($^2J_{P_CP_B}$= 36.0 Hz)	95BSF910
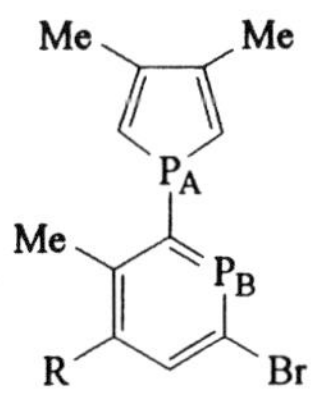	R = Br : +1.87 (P_B+201.13, $^2J_{P_AP_B}$=30.1 Hz) Me : -0.50 (P_B+193.60, $^2J_{P_AP_B}$=31.7 Hz)	95BSF910

Table 1. *Continued*

Compound	$\delta\ ^{31}P$	Ref.
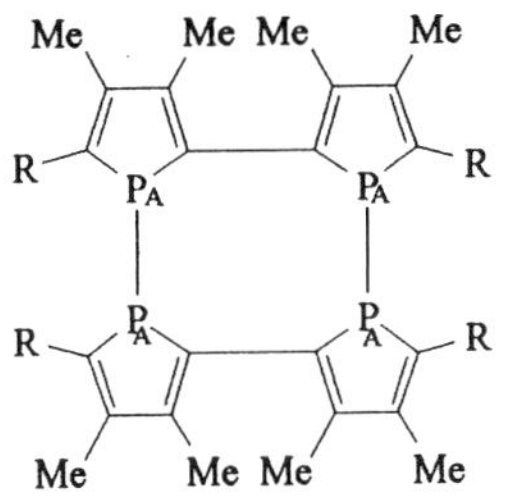	R = 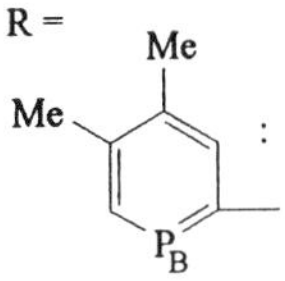 : -5.85 (P_B -186.55, ΣJ_{PP} = 66.65 Hz)	95BSF910
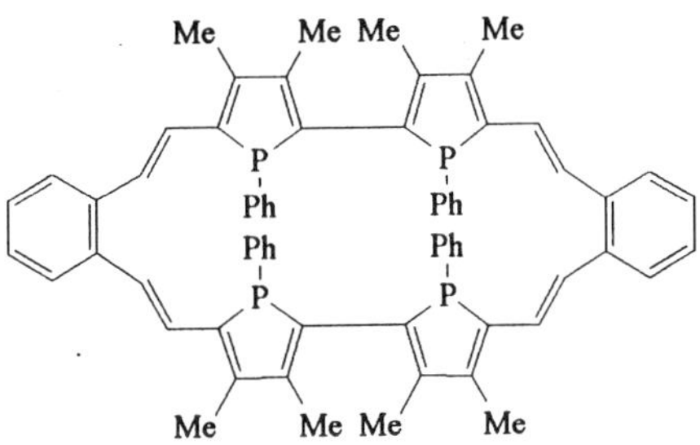	+15.9(CH_2Cl_2)	95CC1561
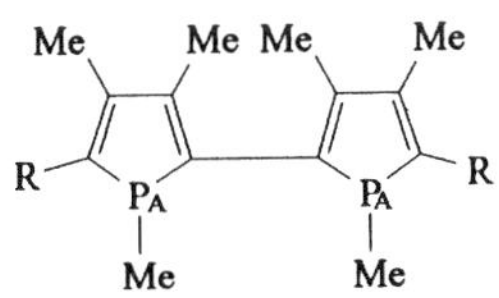	R = 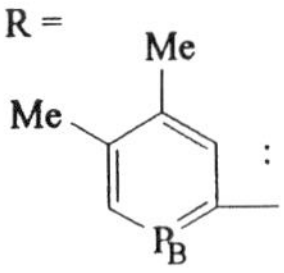 : +7.80 (C_6D_6) (P_B +186.0; AA'XX' spectrum, $^9J_{AA'}$ = 0, $^3J_{XX'}$ =55.5, $^3J_{AX}$ =43.0, $^6J_{AX'}$ = 1.0 Hz)	95BSF910
	Br : +23.5(THF)	96AG(E)1158
	COOH : +23.3(THF)	96AG(E)1158
	CHO : +20.5	96AG(E)1158
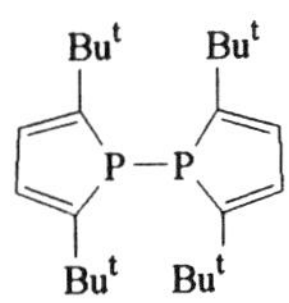	-28.0	94CC1167
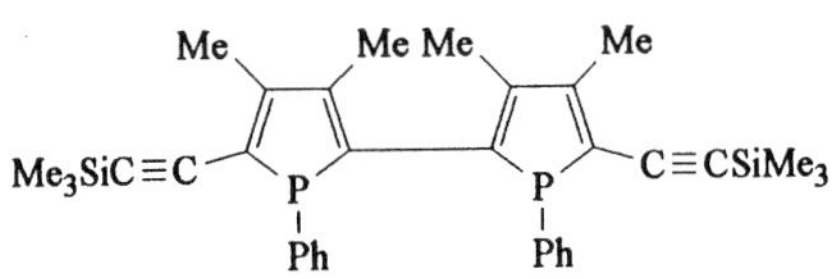	+23.9, +24.2, +24.5 (pseudo-t)	96BSF33

Table 1. *Continued*

Compound	$\delta\ ^{31}P$	Ref.
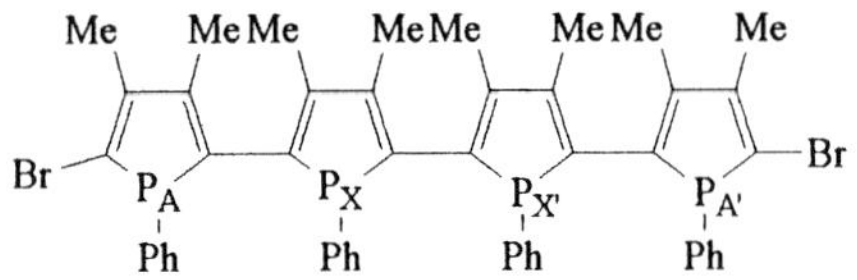	AA'XX', δ (AA') = +29.69, δ (XX') = +21.29 $^3J_{AX}$ = 61.4Hz, $^3J_{XX'}$ = 55.5Hz, $^6J_{AX'}$ = -0.36Hz	94AG(E)1158
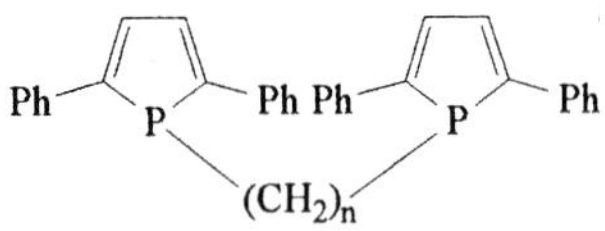	n = 2, +6 3, +2.1 4, +2.5 5, +3.3 6, +3.5	98PS(142)117
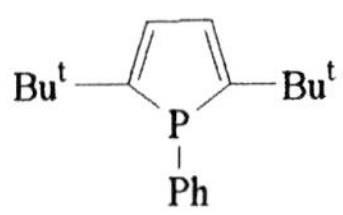	+0.7	94CC1167
B. Multicyclic Phospholes		
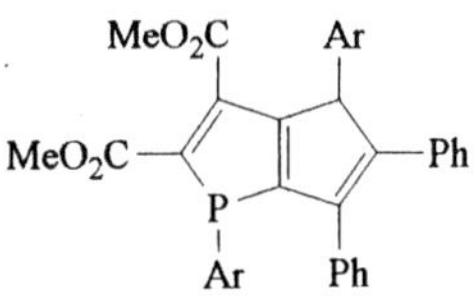	Ar = (2,4,6 - tri-t-butyl) phenyl : -1.90, 5.90 ($^3J_{PP}$ = 105 Hz)	95TL6655
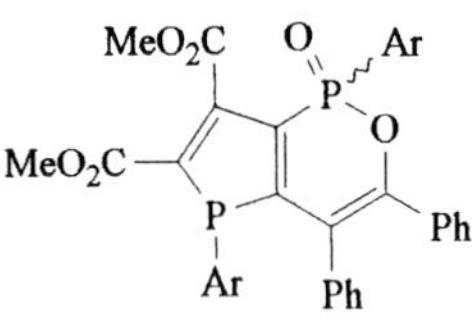	(no data)	95TL6655
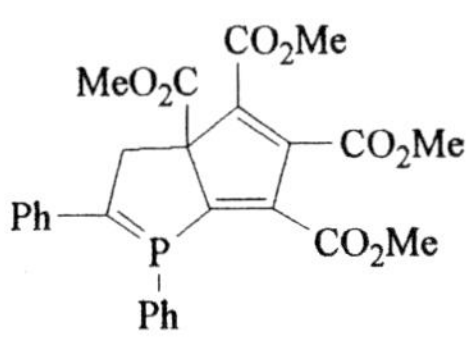	+72.8	98HAC9
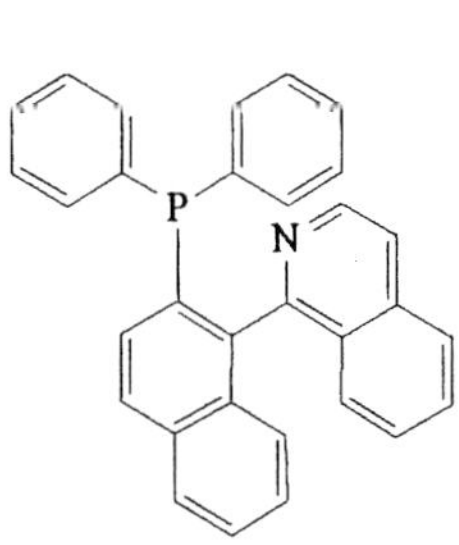	-20.0	97TA3775

Table 1. *Continued*

Compound	$\delta\ ^{31}P$	Ref.
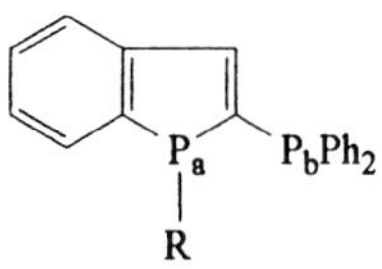	R = Ph : +10.6(THF), (P_b -14, $^2J_{P_aP_b}$= 50 Hz) Bu^t: + 3.7 (P_b -12, $^2J_{P_aP_b}$-52 Hz)	97CC279
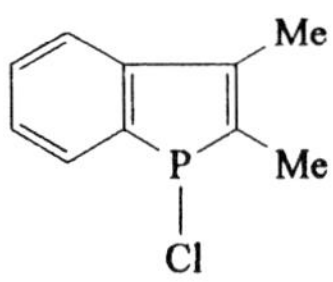	+78.0(CD_2Cl_2)	94JOM(464)149
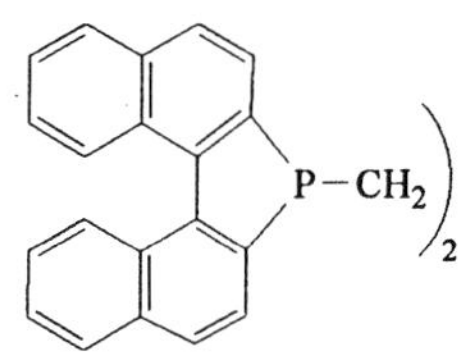	-3.78 (DMSO-d_6)	95JOM(491)91
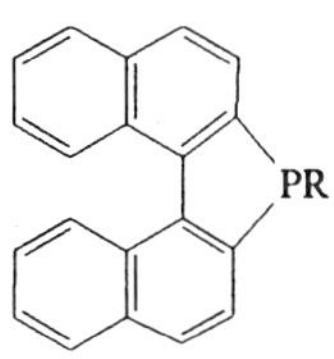	R = Ph : +4.69 $EtMeCHCH_2$: -11.2	93CC1124
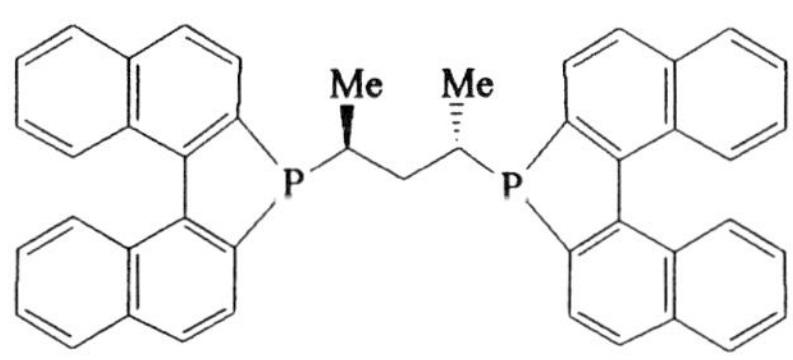	+7.52 (DMSO-d_6)	95JOM(491)91
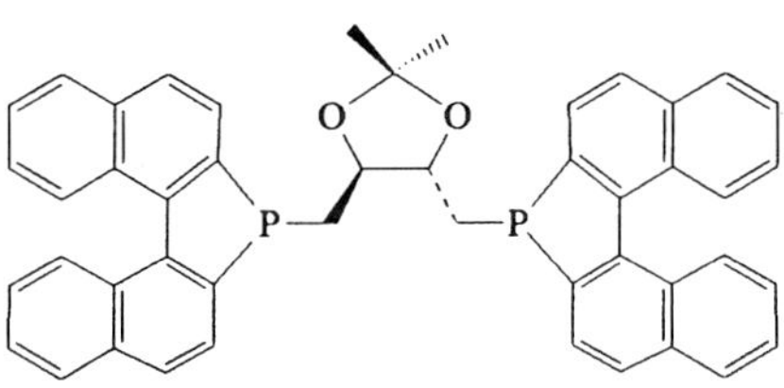	+6.28 (DMSO-d_6)	95JOM(491)91
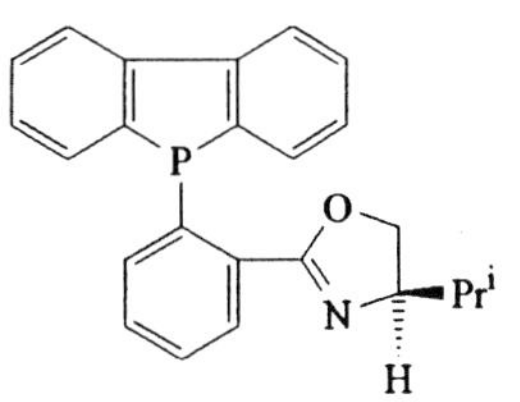	-17.6	96T7547

Table 1. *Continued*

Compound	$\delta\ ^{31}P$	Ref.
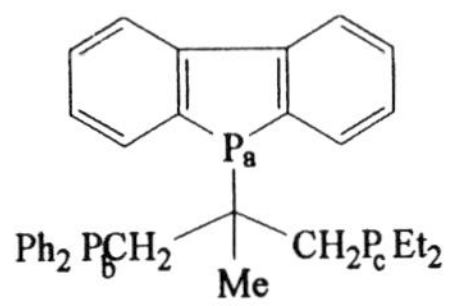	-23.55 (P_b, -26.64, P_c -34.89)	95CB293
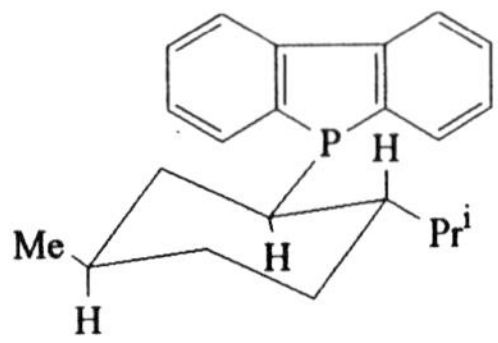	-7.5	98S45
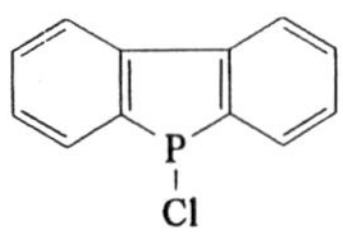	+69.0 (C_6D_6) +58.1 (CD_2Cl_2)	96PS(118)309 96PS(115)227
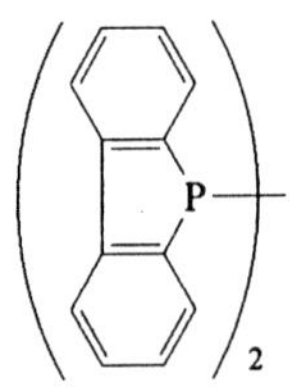	-21.5(C_4D_8O)	94JOM(464)149
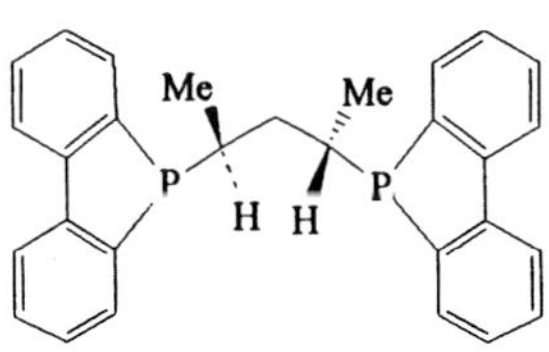	+1.07	97JOM(540)15

Table 1. *Continued*

C. Phospholides and Phospholide Complexes

Compound	$\delta\ ^{31}P$	Ref.
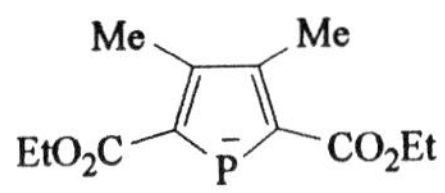	+141.5 [C_4D_8O + HC(O)NMe_2]	97AG(E)98
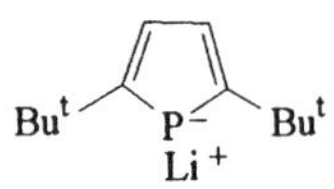	+60.4	94CC1167
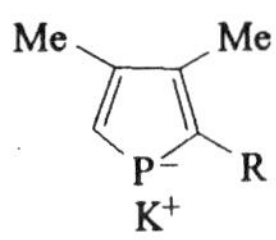	R =	
	CN : + 111.86 (THF-d_8)	96BSF541
	MeS : + 67 (THF-d_8)	96BSF541
	Me_3Si : + 93 (THF-d_8)	96BSF541
	Me_3Sn : + 89 (THF-d_8)	96BSF541
	EtO_2C : + 99.7 (THF-d_8)	97AG(E)98
	2-pyridyl : + 79.5 (THF-d_8)	97AG(E)98
	Ph : + 72.0 (THF-d_8)	97AG(E)98
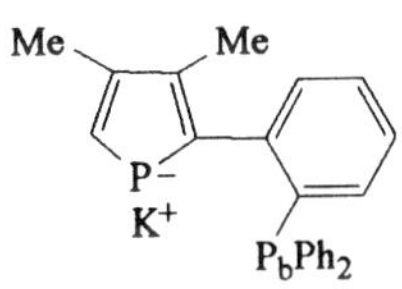	+86.2(THF) (P_b -14.7, J_{PP} = 25Hz)	98OM2996
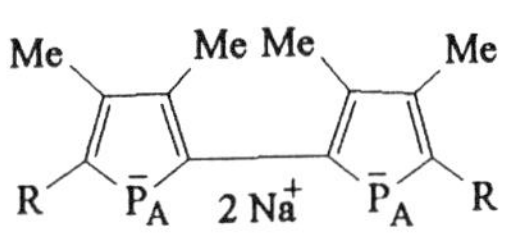	R = 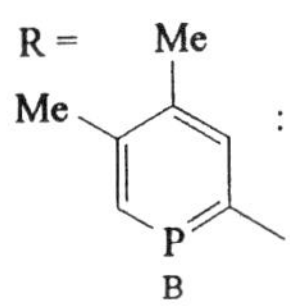 : + 64.90 (THF) (P_B +186.0; AA'XX' spectrum, $^9J_{AA'}$ = 0, $^3J_{AX}$ = 55.0, $^6J_{AX'}$ = 1.0, $^3J_{XX'}$ = 72.0 Hz)	95BSF910
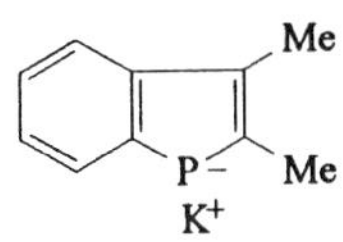	+46.2 (C_4D_8O)	94JOM(464)149
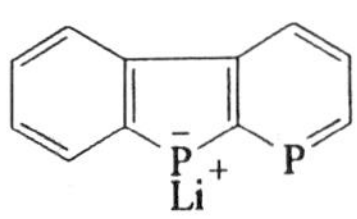	+ 14.3 (C_4D_8O)	94JOM(464)149
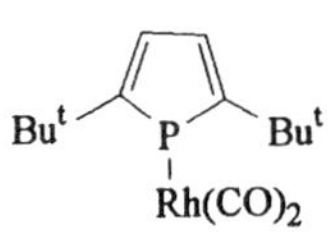	+1.0	94CC1167
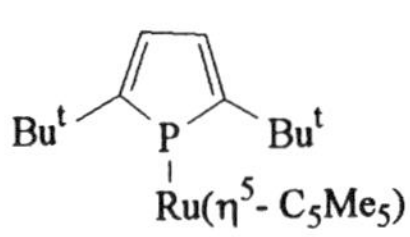	-63.0	94CC1167

Table 1. *Continued*

D. Phospholes with Coordination Number 4

Compound	δ ^{31}P	Ref.
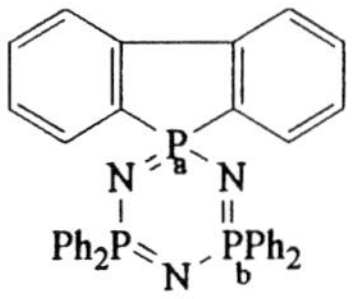	+ 24.4 (P_b +17.0)	97JCS(P2)15
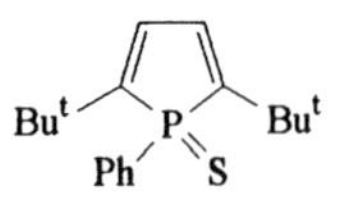	+48.4	94CC1167
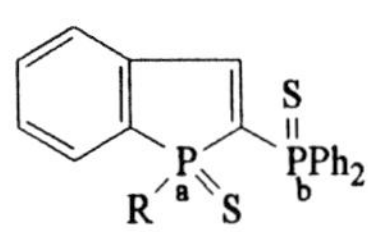	R = Ph : + 51.4(C_6D_6) (P_b +39.2, $^2J_{P_aP_b}$ = 24.7 Hz) Bu^t : + 78.6 (C_6D_6) (P_b+ 38.9, $^2J_{P_aP_b}$ = 24.9 Hz)	97CC279
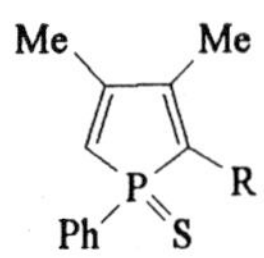	R : MeS : + 50.4 Br : +50.2 (CH_2Cl_2)	 96BSF(133)541 92BSF(129)486
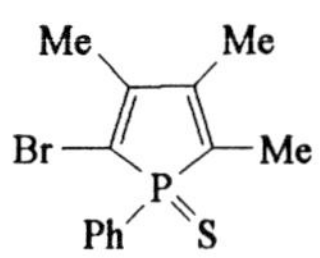	+54.4 (CH_2Cl_2)	94AG(E)1158
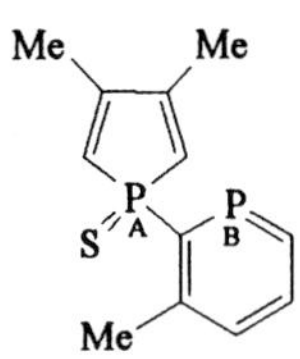	+ 45.85 (P_B +212.55, $^2J_{P_AP_B}$ = 104.85 Hz)	95BSF910
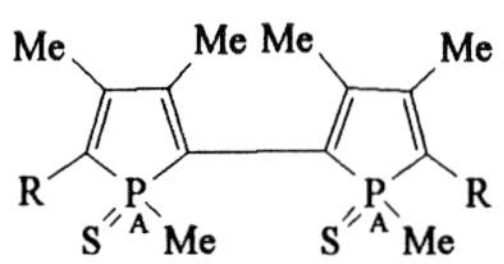	R = 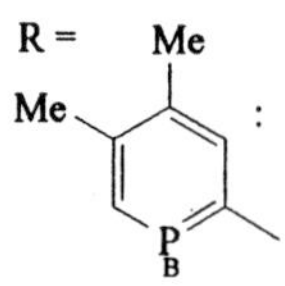 : + 50.55 (P_B + 190.10; pseudo-d, Σ J_{PP} = 11.25 Hz)	95BSF910

Table 1. *Continued*

Compound	$\delta\ ^{31}P$	Ref.
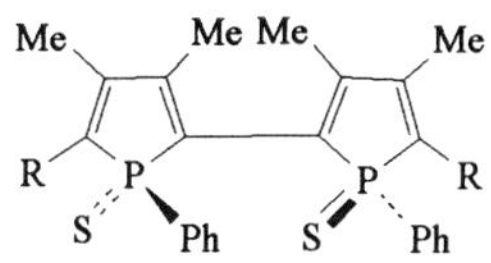	+ 50.75 (diastereomer + 51.37)	96CC2287
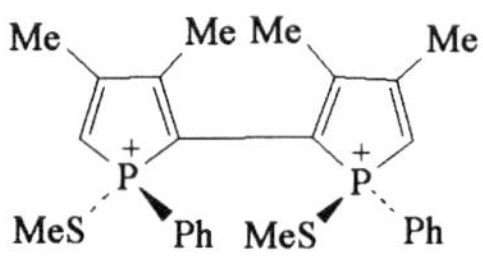 2 $CF_3SO_3^-$	+ 52.95 (diastereomer + 54.27)	96CC2287
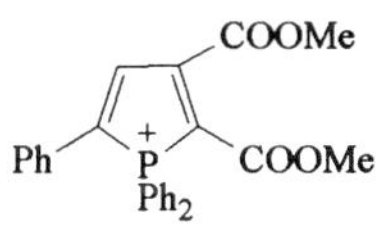	+37.2 (MeOH)	94ZOB1566
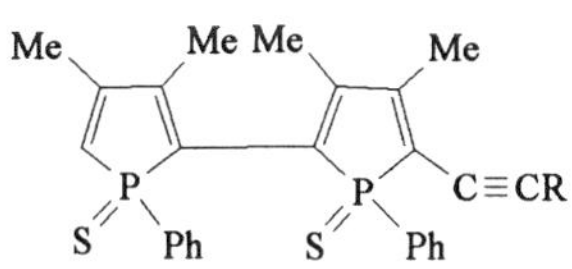	R = Ph: + 51.0 Me_3Si: + 50.3	96BSF33
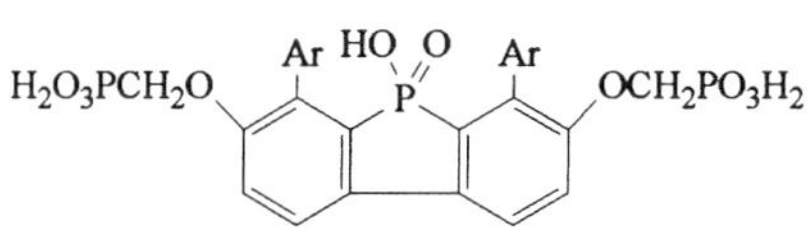	Ar = 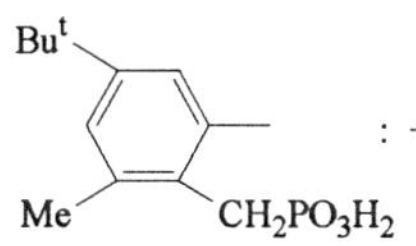 : +34.1 (also +24.1, +14.0) (and related compounds)	96JCS(P1)2889

Phosphole **16a**, with the largest substituent, indeed has a lower degree of pyramidality (angle sum 331.7°), and has significant shortening of the C_3–C_4 bond (1.402 Å) and the P–C bonds (ave. 1.744 Å); some lengthening of the C=C bonds also occurs (ave. 1.350 Å). Another useful measure of the effect of the P-substituent is provided by the angle by which the P-substituent is deflected from the plane of ring atoms C_2–P–C_5. For **16a**, this value is 45.0°, while it is much larger in 1-benzylphosphole (66.9°). Phosphole **16b** has parameters that show a smaller degree of flattening of the pyramid (angle sum 314.4°, deflection angle 58.0°). The Bird Index of aromaticity (discussed in Chapter 4.2.1) can be derived from bond parameters, and phosphole **16a** has an index (56.5) well above that of furan (43) and quite close to that of pyrrole (59). 1-Benzylphosphole has an index of only 35.5. Other aspects of the increased delocalization in phosphole **16a** are discussed in Sections 4.2.2.2 and 4.2.2.8.

The phospholes **16a** and **16b** exhibit another effect due to the steric crowding; the plane of the aryl substituent is orthogonal to the plane of the phosphole ring, as will be seen in a discussion (Section 4.2.2.3.4) of the ^{13}C NMR spectra of phosphole **16a.** This conformation is maintained in solution, but the less hindered **16b** undergoes free rotation around the P–C bond in solution. The steric crowding at the P-substituent does not influence greatly the degree to which the phosphorus atom is displaced from the plane of C_2–C_3–C_4–C_5; this displacement is 0.294 Å in **16a**, and 0.21 Å in 1-benzylphosphole.

4.2.2.3.3 P-31 NMR Spectroscopy

Tables of ^{31}P NMR data for phospholes were provided in Chapter 4.2.1, along with a discussion of the effects of substituents on the chemical shifts. A considerable amount of new data has been collected, and is recorded in Table 1. Some unusual substituents effects have been noted among the new data, and are discussed below. In their interpretation, it is important to remember that making a connection between ^{31}P NMR shift and aromaticity is hazardous; previously reported theoretical results ⟨94JA9638⟩ suggest that there is only a modest downfield shift (10.9 ppm) on converting pyramidal, non-aromatic 1-methylphosphole to the planar, strongly aromatic form, and that the very strong deshielding in phospholide ions, which are highly aromatic, arises not directly from the delocalization but from the delocalization placing the second lone pair on phosphorus into the plane of the ring. This has the effect of reducing the HOMO–LUMO gap. A further complication in assessing substituent effects is that substitution of the 3,4-hydrogens in phospholes by the innocuous methyl group causes substantial upfield shifting, for which no satisfactory explanation other than relating the effect to steric interactions has been presented. There are many examples of this effect, but important to the discussion to follow are data for the P-phenyl derivatives, **19** and **20**.

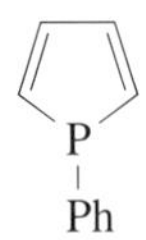

(**19**), $\delta^{31}P$ +6.4

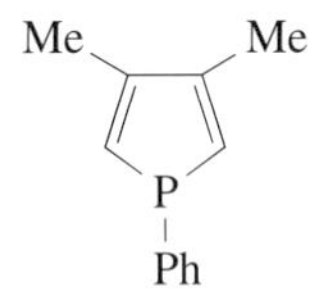

(**20**), $\delta^{31}P$ -4.4

For reliable information on the effect of particular ring substituents, it would be much more desirable to deal with data for derivatives of the parent compound **19**. In practice, 3,4-dimethyl-1-phenylphosphole is by far the most common starting material for elaboration of other phospholes, since it is particularly easily synthesized. The newly observed effects are:

(a) Replacement of phenyl on P by electron releasing phosphino and 2-pyrryl groups causes substantial upfield shifting as seen in compounds **21**–**23**.

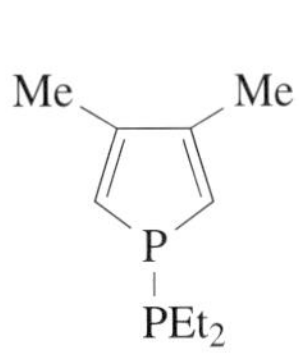

(**21**), $\delta^{31}P$ -17.3

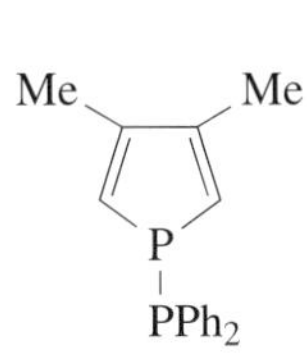

(**22**), $\delta^{31}P$ -21.7

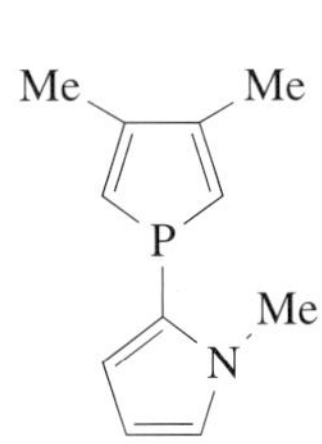

(**23**), $\delta^{31}P$ -32

Replacing the phenyl groups in **22** by mesityl should not have a large effect on the ^{31}P shift, suggesting that the assignments for this compound of δ −2.36 to the ring P and δ −24.49 to the phosphine P seen in Table 1 ⟨95BSF649⟩ may need confirmation.

On the other hand, electron-withdrawing groups on P have surprisingly little effect, as seen for the 2-pyridyl (**24**), 2-phosphininyl (**25**), and carbethoxy (**26**) compounds.

(b) More substantial downfield shifting accompanies the placement of a substituent at the 2-position, as is seen with compounds **27**, **28** ⟨90JOC2494⟩ and **29** ⟨92BSF486⟩. From the disparate nature of the electronic characteristics of the substituting groups, the shifting does not appear to arise solely from conjugative or inductive effects, and is possibly better associated with steric congestion from the presence of four adjacent ring substituents.

Table 2. Structural parameters for phospholes and phospholide complexes [a,b]

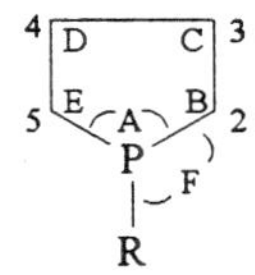

Compound	Bond Angles[a]						Bond Lengths[a]						Ref.
	A	B	C	D	E	F	P-C(R)	P-C_2	C_2-C_3	C_3-C_4	C_4-C_5	C_5-P	
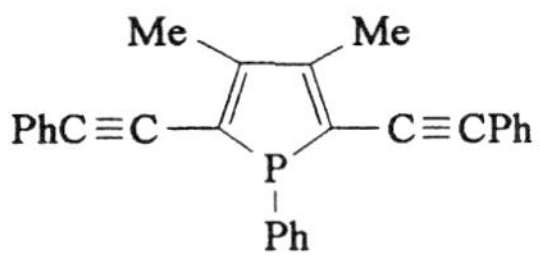	89.19	111.5	113.9	113.3	111.6	102.74 100.42	1.839	1.815	1.367	1.458	1.365	1.821	96BSF33
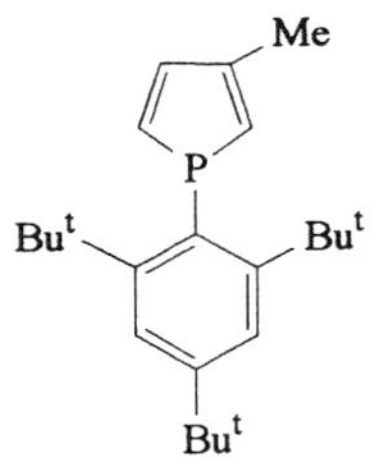	91.7	109.9	112.9	115.2	108.7	118.3 121.7	1.812	1.741	1.347	1.402	1.352	1.746	97JA5095
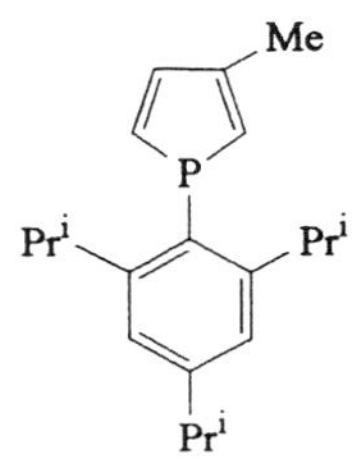	91.1	-	-	-	-	109.6 113.7	1.835	1.782	1.340	1.436	1.366	1.778	97JOM(532)109

Table 2. *Continued*

	A	B	C	D	E	F	P-C(R)	$P-C_2$	C_2-C_3	C_3-C_4	C_4-C_5	C_5-P	Ref.
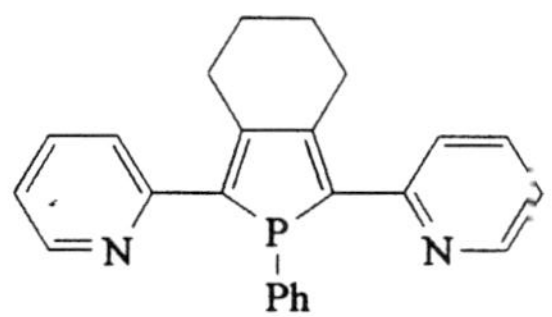	-	-	-	-	-	-	-	1.806	1.354	1.478	1.356	1.806	99CC345
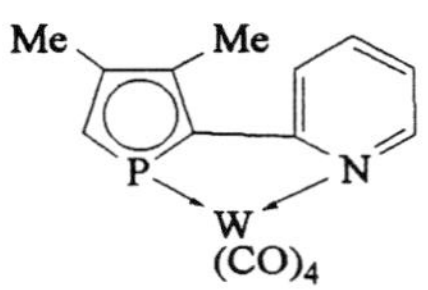	91.62	-	-	-	-	-	-	1.746	1.421	1.444	1.443	1.761	97JOM(548)17
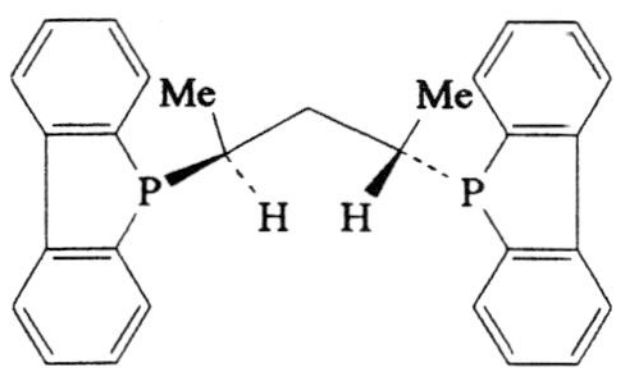	104.8	-	-	-	-	-	1.857	1.794	1.390	1.434	1.407	1.804	97JOM(540)15
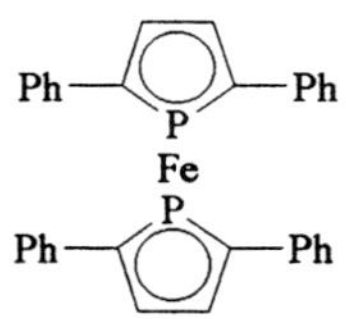	90.2	111.5	113.0	113.1	112.1	-	-	1.783	1.405	1.416	1.414	1.776	97JOM(527)305

Table 2. *Continued*

	A	B	C	D	E	F	P-C(R)	P-C_2	C_2-C_3	C_3-C_4	C_4-C_5	C_5-P	Ref.
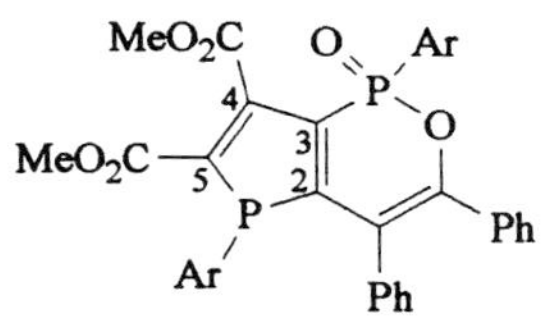 Ar = 2,4,6-$Bu^t_3C_6H_2$	92.8	-	114.1	-	-	-	1.837	1.782	1.399	1.436	1.367	1.755	95TL6655
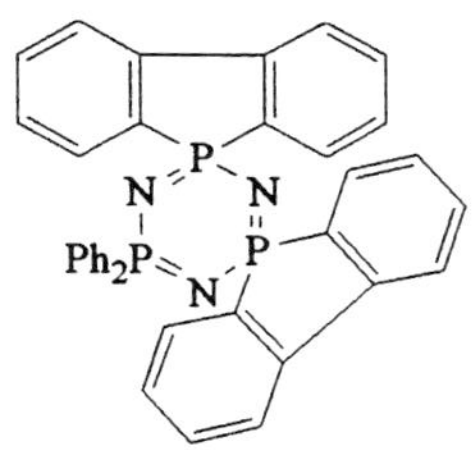	91.04	-	-	-	-	-	-	1.816	.	.	.	1.807	97JCS(P2)15
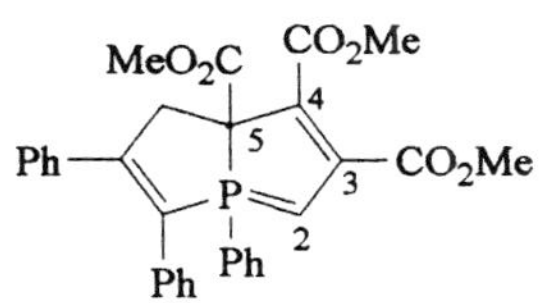	96.5	107.3	118.1	115.1	101.6	117.3	1.781	1.737	1.414	1.368	1.518	1.853	98HAC9
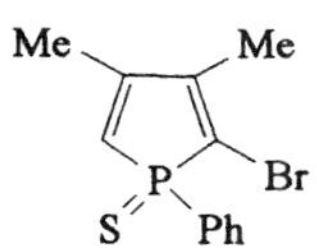	90.6	.	112.1	.	.	108.3 105.6	.	1.805	1.331	1.499	1.336	1.785	98AX(C)676

Table 2. *Continued*

	A	B	C	D	E	F	P-C(R)	P-C_2	C_2-C_3	C_3-C_4	C_4-C_5	C_5-P	Ref.
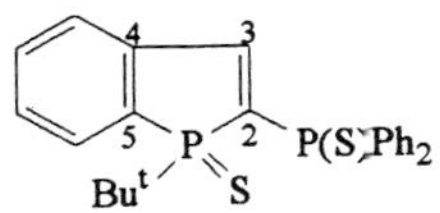	91.4	109.2	-	-	-	-	-	1.831	1.340	1.468	1.391	1.814	97CC279
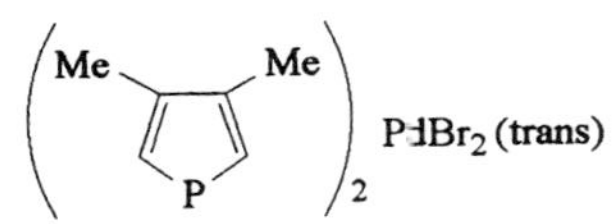	91.9	-	-	-	-	-	-	1.793	.	.	.	1.782	96IC1486
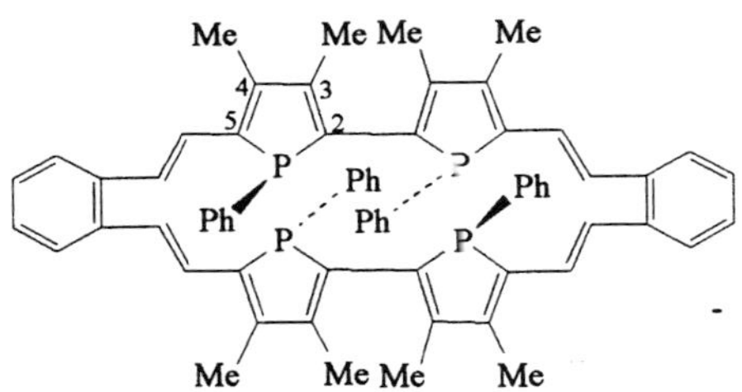	-	-	-	-	-	-	1.830	1.807	1.356	1.469	1.364	1.822	95CC1561
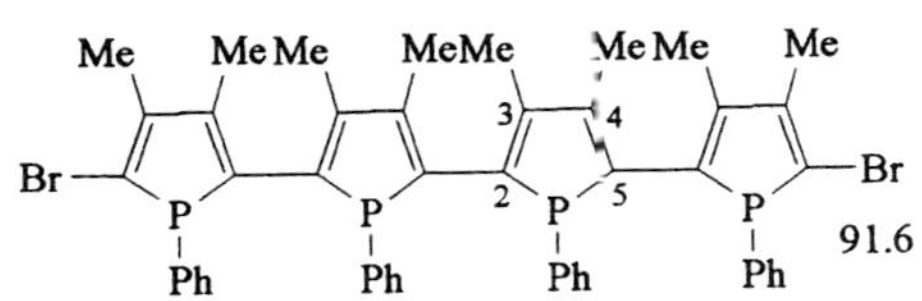	91.6	-	-	-	-	106.7 106.3	1.814	1.822	1.374	1.445	1.383	1.790	94AG(E)1158

Table 2. *Continued*

	A	B	C	D	E	F	P-C(R)	P-C_2	C_2-C_3	C_3-C_4	C_4-C_5	C_5-P	Ref.
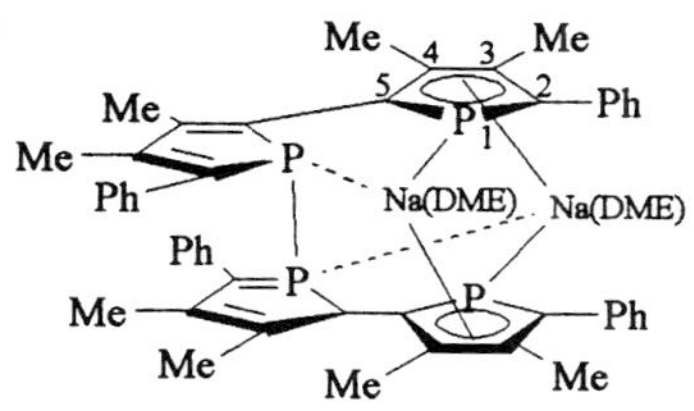	91.01	-	-	-	-	-	-	1.765	1.413	1.423	1.406	1.766	96AG(E)1125
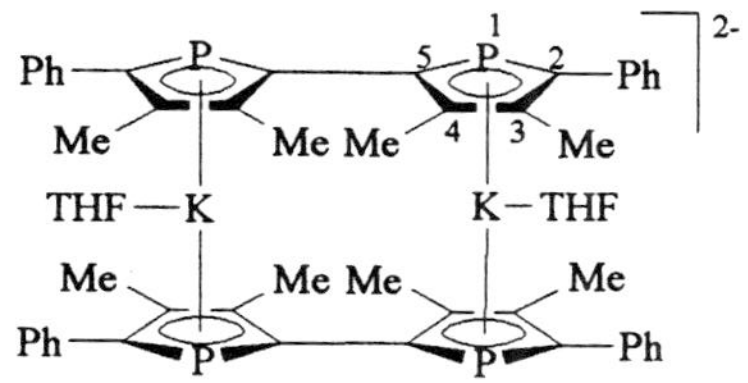	90.8	-	-	-	-	-	-	1.779	1.399	1.415	1.411	1.759	96AG(E)1125

[a] Deviations omitted

[b] Data from the following structure determinations were not reported: 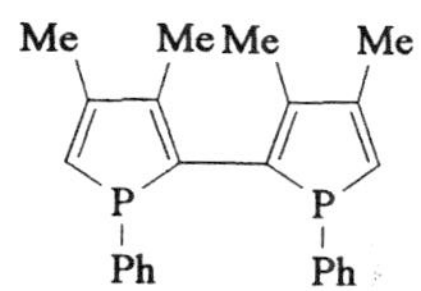 and bis-sulfide: 96CC2287 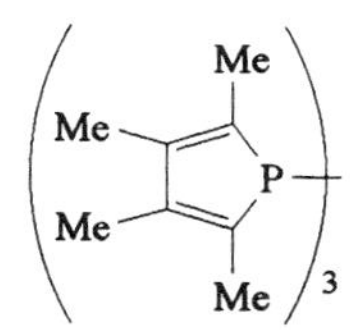 P : 97JOM(529)75

Table 3. Pyramidal shapes and the effect on bond lengths in monocyclic phospholes

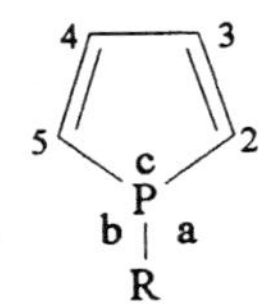

Compound	Bond Angles					Bond Lengths					Ref.
	a	b	c	Σ (a,b,c)	α[a]	P-C_2	P-C_5	C_2-C_3	C_4-C_5	C_3-C_4	
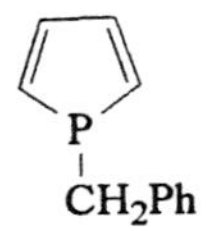	106.1	105.9	90.7	302.7	66.9	1.780	1.786	1.343	1.343	1.438	73JCS(D)1888
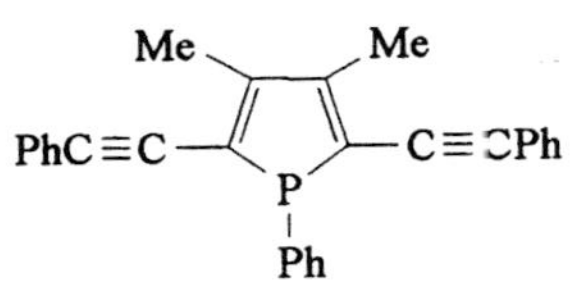	102.74	100.42	89.19	292.4	-	1.815	1.821	1.367	1.365	1.458	96BSF33
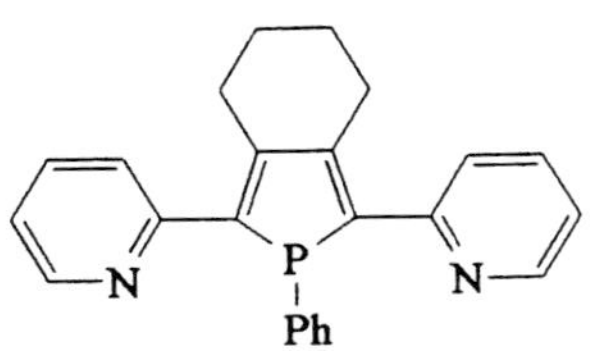	-	-	-	299.3	-	1.806	1.806	1.354	1.356	1.478	99AG(E)345

Table 3. *Continued*

Compound	Bond Angles					Bond Lengths					Ref.
	a	b	c	Σ (a,b,c)	α[a]	P-C_2	P-C_5	C_2-C_3	C_4-C_5	C_3-C_4	
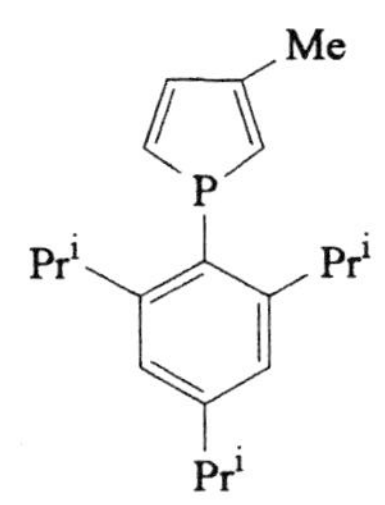	109.6	113.3	99.1	314.4	58.0	1.782	1.778	1.340	1.366	1.436	97JOM(532)109
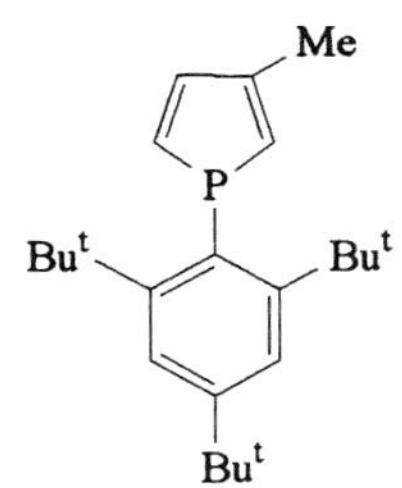	121.74	118.3	91.7	331.74	45.0	1.746	1.741	1.347	1.352	1.402	97JA5095

[a] Angle by which R is below the plane of the C_2–P–C_5 segment.

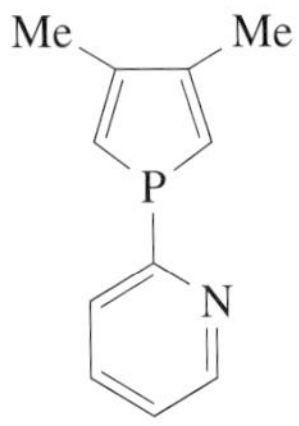

(**24**), δ^{31}P +1.8

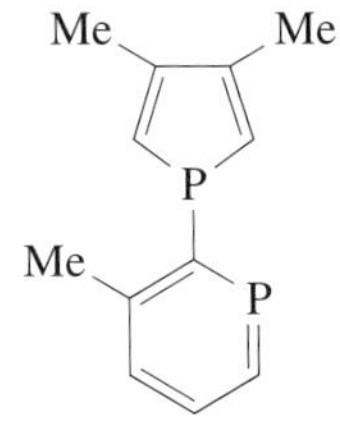

(**25**), δ^{31}P +0.60

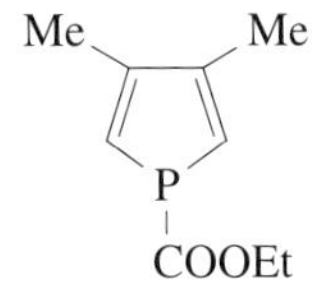

(**26**), δ^{31}P +5.1

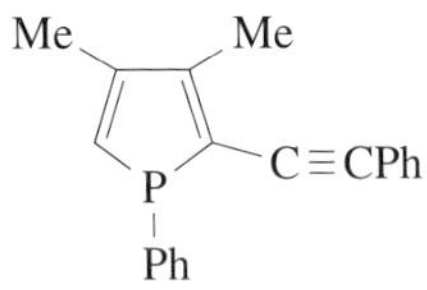

(**27**), δ^{31}P +12.4

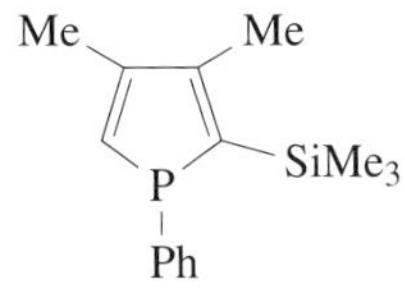

(**28**), δ^{31}P +15.3

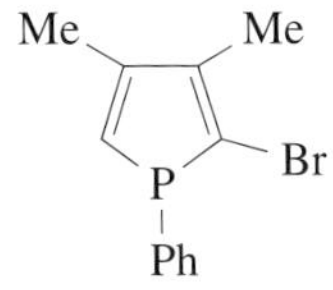

(**29**), δ^{31}P +10.2

The deshielding is intensified with additional substitution at the 5-position as in **30** and **31**. However, pronounced deshielding (17 ppm) also occurs with methyl substitution as in compound **32**, which is suggestive of the importance of steric effects. Consistent with this is the effect of replacing the 2,5-methyl groups by trimethylsilyl, which gave a shift to δ +35.6.

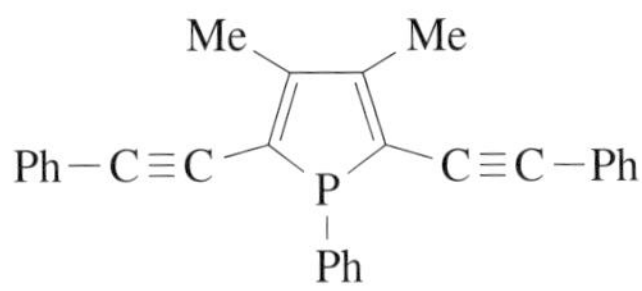

(**30**), δ^{31}P +22.6

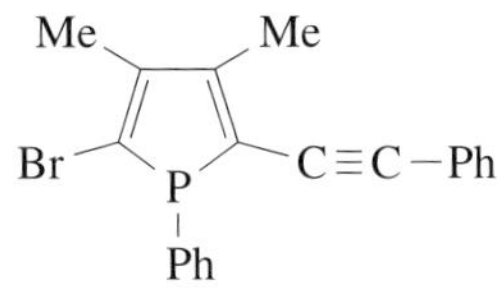

(**31**), δ^{31}P +20.2

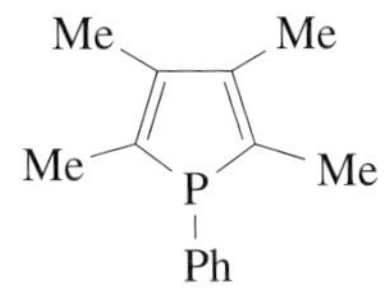

(**32**), δ^{31}P +12.7

(c) It was noted in Section 4.2.2.3.2 on X-ray diffraction data that new compounds showing alteration of the pyramidal shape at phosphorus have been synthesized. Thus, two acetylenic and 2-pyridyl substituents at the α-positions tend to increase the pyramidality, which would be in the direction of diminished delocalization. The downfield shifting of the ^{31}P signal seen in item (b) would not be consistent with decreased aromaticity; according to the theoretical calculations, planar phospholes are downfield of pyramidal ones. Phospholes that have been made to be more planar through the presence of large aryl substituents on phosphorus do not reveal much of an effect on the ^{31}P shift (**16a,** δ −0.40; **16c**, δ +1.8). That steric effects may be operative in **16a** is revealed by the upfield shifting of 7.8 ppm as the two *ortho* *i*-propyl groups of **16b** are replaced by *t*-butyl. The latter is clearly more planar as seen from the X-ray results.

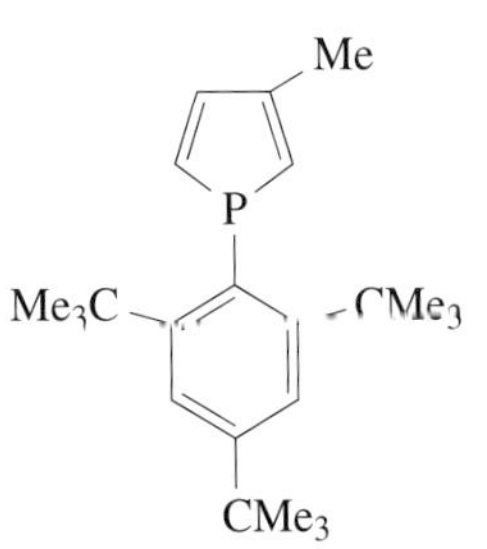

(**16a**)

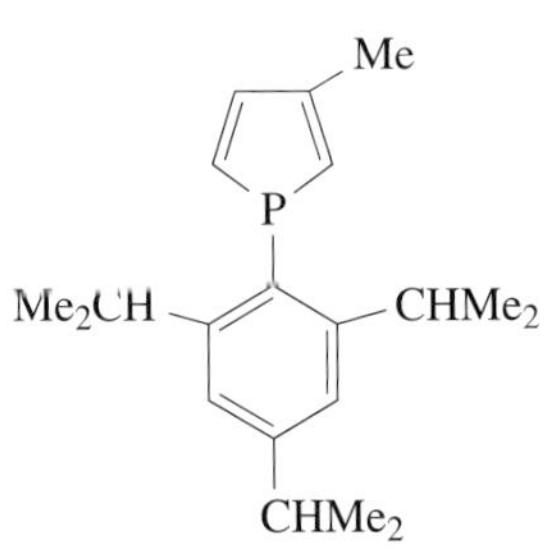

(**16b**)

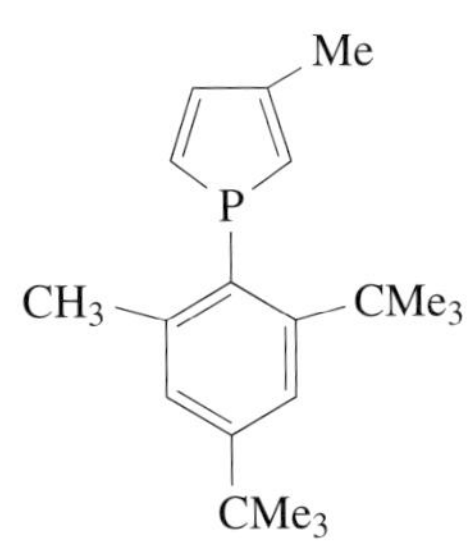

(**16c**)

(d) As noted above, monocyclic phospholide ions show striking downfield shifting relative to acyclic phosphide ions and typically have shifts at δ +70 to +100. However, benzo fusion at the α,β positions greatly diminishes the magnitude of the deshielding, as seen in the phosphindolide **33** with δ +46.2 and the dibenzophospholide ion **34** with δ +15. The reasonable suggestion has been made ⟨94JOM(464)149⟩ that these shifts indicate diminished delocalization of the phospholide charge. However, the theoretical study of monocyclic phospholides ⟨94JA9638⟩ shows that deshielding in the ion is not directly related to delocalization but to a modification of HOMO–LUMO energies, and it is possible that in the dibenzophospholide ion the ^{31}P shifts are not controlled by delocalization effects. The point clearly needs further study. That extensive electron release is present from the phospholide moiety in **33** to the benzo group may be seen in the very strong upfield shifting reported for two of the CH carbons in the ^{13}C NMR spectrum (δ 115.1 and 116.1).

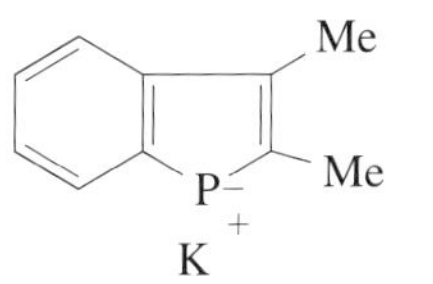

(**33**)

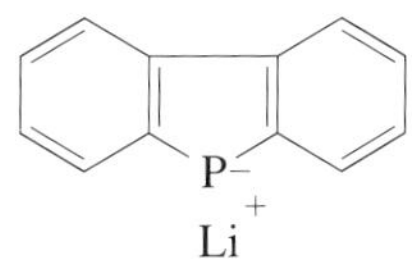

(**34**)

Relatively little has been published on solid-state ^{31}P NMR spectra of phospholes. Two recent papers do point out the value of obtaining such spectra, which can provide specialized and fundamentally important data on physical chemical aspects of phosphole structure and behavior. The papers are on (a) the shift tensors of dibenzophosphole and its chalcogenides ⟨96IC3904⟩, (b) the spectrum of the cyclic "tetramer" (**35**) from 3,4-dimethyl-1-phenylphosphole, which shows complex spinning sidebands under certain conditions, giving spectra that are useful in providing information, *inter alia*, on the sign of P–P coupling ⟨97JMR(124)366⟩ in this complex molecule.

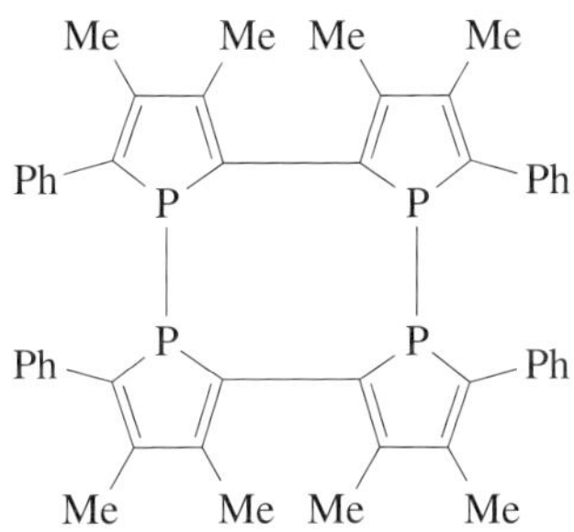

(**35**)

4.2.2.3.4 C-13 NMR Spectroscopy

The characteristics of ^{13}C NMR of phospholes were described in Chapter 4.2.1, and most of the newly published spectral data can be interpreted along the lines cited there. Some recently synthesized structures are of interest, however, in giving spectra which reveal the presence of restricted rotation around the exocyclic C–P bond, leading to a preferred conformation. This is a newly observed property of phospholes. The conformational effect was seen in cases where the P-substituent has cyclic structure. Phospholes with sterically demanding substituted-phenyl groups on P give both chemical shift and coupling constant indications of the restricted rotation resulting in a preferred conformation with the phosphole and benzene rings being orthogonal. As an example, the data for compound **16a** ⟨97JA5095⟩ may be considered. The two *ortho* carbons have different chemical shifts and $^{2}J_{PC}$ values (δ 151.9, J = 2.2 Hz, and δ 158.6, J = 12.3 Hz). The larger coupling constant would arise from a conformation with a dihedral angle of 0º relating the lone pair and the *ortho* carbon, a well-known effect of stereospecificity in two-bond coupling in phosphines. Similarly, the *meta* carbons give different signals (δ 119.6, $^{3}J_{PC}$ = 0, and δ 122.9, $^{3}J_{PC}$ = 9.1 Hz); the larger coupling is again to be associated with the carbon closer to the lone pair. The orthogonal relation of the two rings was in fact confirmed by the X-ray diffraction analysis. Small effects were observed also on the phosphole ring carbons arising from the steric crowding of the P-substituent.

In the related phosphole with the smaller isopropyl groups replacing the *t*-butyl groups, no indications of the restricted rotation were present. Consistent with the need for the bulk of the *t*-butyl group, the

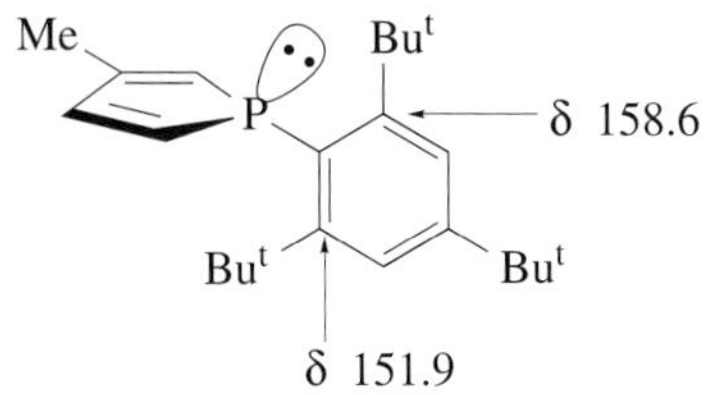

(**16a**)

compound **16c** with *t*-butyl and methyl in the *ortho* positions also gave ^{13}C spectra with both the two-bond and the three-bond coupling constants having different values.

Another case of apparent restricted rotation being revealed by ^{13}C NMR features is provided by 1-(1-*N*-methyl-2-pyrrolyl)-3,4-dimethylphosphole ⟨98OM2996⟩. Again employing the argument of stereospecificity in two-bond coupling, the very large value of 37.8 Hz for the coupling to the β-carbon of the pyrrole ring implies a conformation (**36a**) with orthogonal rings, where the *N*-methyl group must lie under the phosphole ring to place the β-carbon in close proximity to the lone pair. It is not clear at this time why this conformation would be of low energy relative to the rotamer **36b** with the *N*-methyl group in the less crowded position above the phosphole ring.

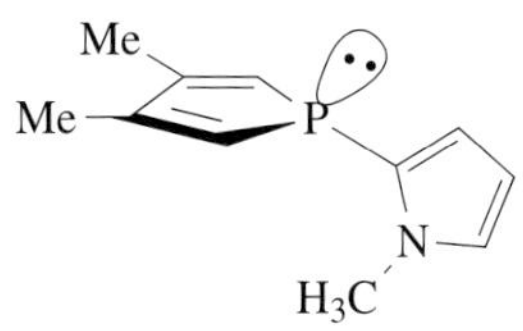

(**36a**)

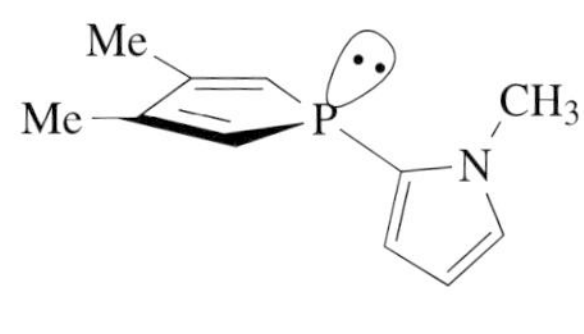

(**36b**)

4.2.2.4 NEW TYPES OF 1H-PHOSPHOLES

Table 1 lists the new phospholes prepared during the subject period; it contains entries for a number of compounds where new functional groups have been attached at one or both of the α-positions. Included are:

(a) Acetylenic groups. Reacting α-lithio derivatives of phospholes with acetylenic sulfones resulted in displacement of the sulfonyl group with attachment of the acetylene group to the ring ⟨96BSF33⟩. A number of compounds were prepared by this method (Equation 12).

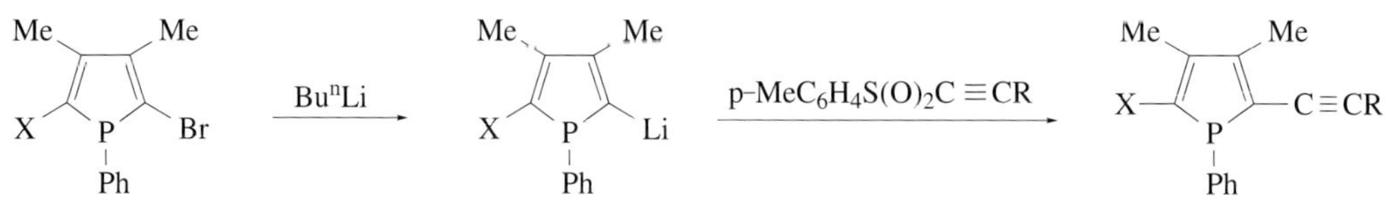

(12)

R = Ph, X = H
R = TMS, X = H
R = Ph, X = Br
R = TMS, X = Br

The process was extended to include the synthesis of derivatives with two acetylenic groups on the ring (Equation 13).

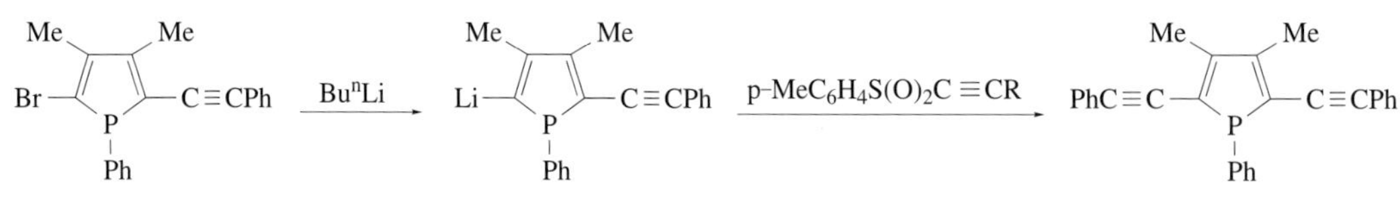

(13)

(b) 2-Phenyl, 2-Pyridyl and 2-Carboethoxy Substituents by Rearrangement of the 1-Substituted Phospholes. Holand, Jeanjean and Mathey ⟨97AG(E)98⟩ have developed a technique for the rearrangement of a P-substituent to the α-position. The process is performed with potassium *t*-butoxide at high temperature in a sealed tube. The phosphole undergoes its normal high-temperature [1,5] sigmatropic rearrangements to the 2H-phosphole structure, which is converted by the base to the phospholide ion before the normal dimerization can take place. The phospholide ion is then quenched with an electrophile (typically $BrCH_2CH_2CN$ or ClCOOEt) to give the 2-substituted 1H-phosphole. The phospholide can also be considered a synthetic target of this approach, and can be used in other reactions. The process was applied to 3,4-dimethyl-1-phenylphosphole and also to the 1-(2-pyridyl) derivative **37** with which the process is illustrated in Equation 14.

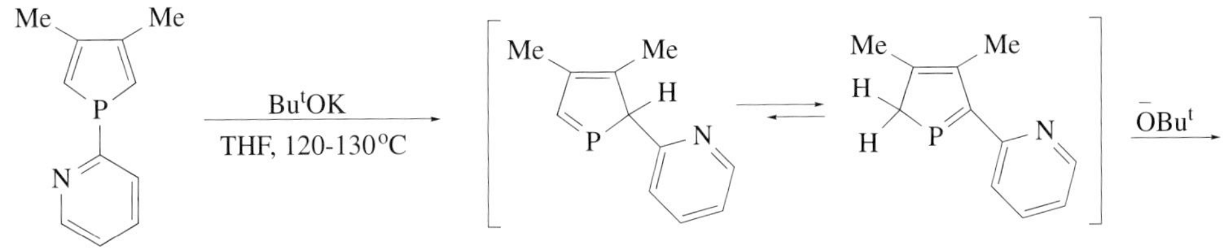

(14)

(**37**)

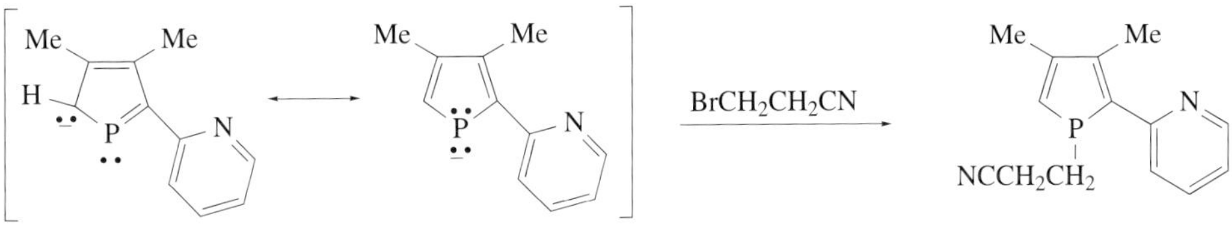

It was also possible to prepare the dimer of the 2H-phosphole from the rearrangement of the 1-carboethoxy derivative, and then subject it to de-dimerization to the phospholide ion with $KOBu^t$ (Equation 15).

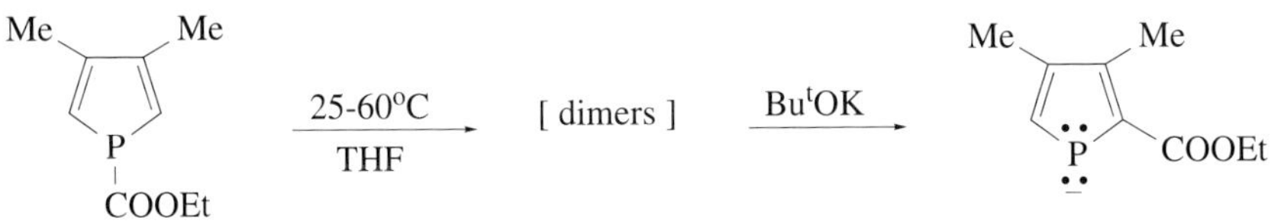

(15)

The method was later extended to the synthesis of the 2-(1-methyl-2 pyrrolyl) derivative **36** (Section 4.2.2.3.3) and the 2-(diphenylphosphinophenyl derivative **38**. The synthesis of the precursor of phospholide **38** is shown in Equation 16.

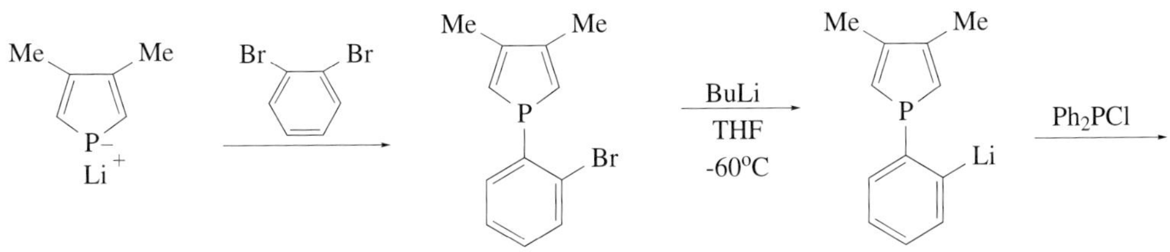

(16)

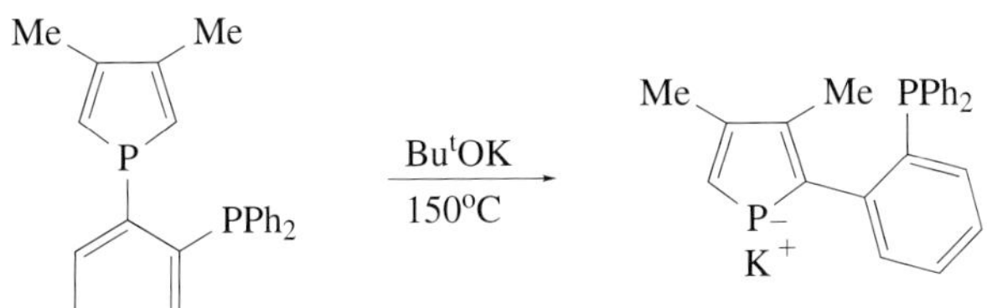

(**38**)

(c) Carboxamide and Cyano Groups. A procedure has been reported for converting an α-COOH group to the amide and then to the nitrile (Equation 17) ⟨96BSF541⟩. The P-phenyl group in nitrile **39** can be cleaved in the usual way to give the phospholide ion with the nitrile substituent, of value in other syntheses.

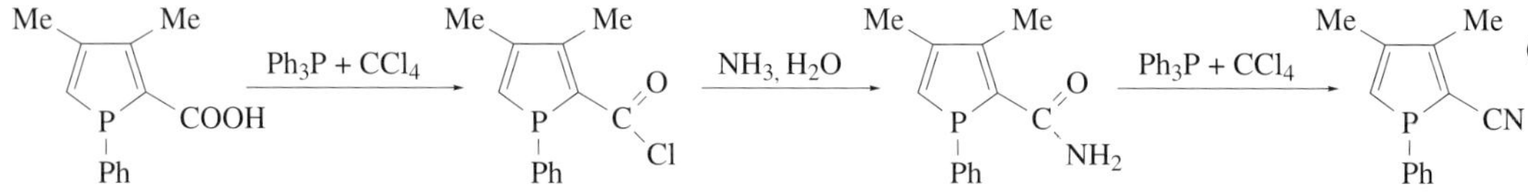

(17)

(**39**)

(d) Methylthio Substituent. Here the starting material is the 4-coordinate phosphole sulfide **40**; it is first lithiated and the anion reacted with elemental sulfur (presumably to form the sulfide anion, not detected) which is then methylated. On reaction with lithium, the P=S group was reduced to the 3-coordinate state, and the P-phenyl group underwent cleavage (Equation 18). The product is a phospholide with the methylthio substituent; this was characterized as the 1,1′-diphosphaferrocene derivative (**41**), but presumably could be used as a precursor of other methylthiophospholes ⟨96BSF541⟩.

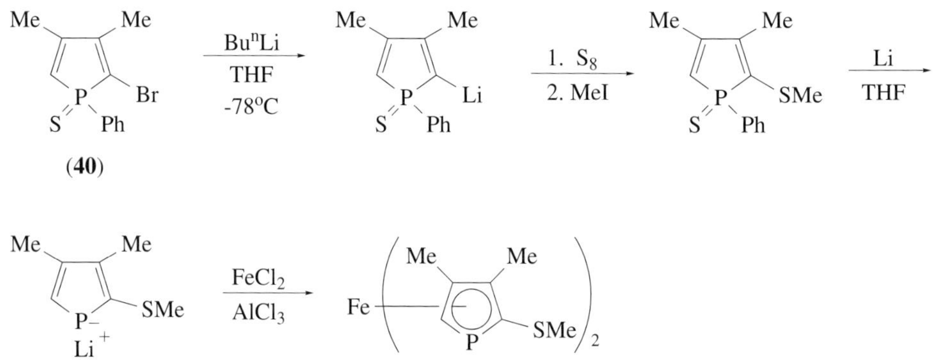

(18)

(**40**)

(**41**)

(e) The Trimethylstannyl Group. A coupling reaction between a bromophosphole and trimethylstannyl chloride provided a phosphole with an α-trimethylstannyl substituent (Equation 19) ⟨96BSF541⟩. Here again the P-phenyl substituent was cleaved to give the functionalized phospholide anion.

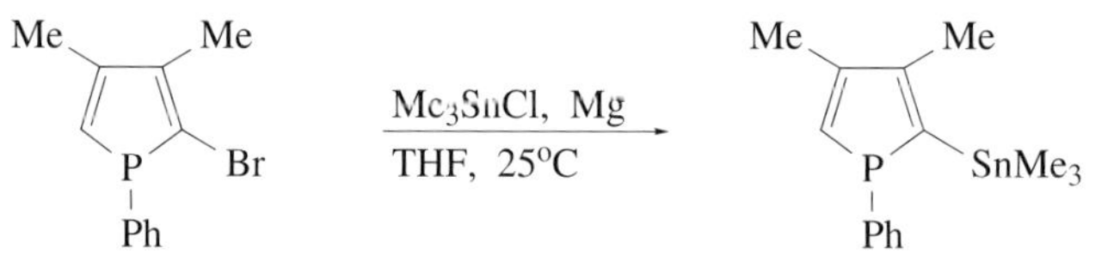

(19)

(f) 2,5-Difunctional Phospholes. The valuable 2,5-dibromo derivative can be prepared from 3,4-dimethyl-1-phenylphosphole-1-oxide. This is first brominated with pyridinium tribromide to give the previously reported dibromophospholene **42** ⟨92BSF486⟩. The steps of Equation 20 are then used to obtain the dibromophosphole **43** ⟨94AG(E)1158⟩. This compound is a key intermediate in other syntheses of new phospholes with 2,5 substituents, as will be discussed.

Treatment of the dibromophosphole **43** with one equivalent of butyllithium gave the monolithio derivative **44**, which was carbonated to give the carboxylic acid **45** or formylated with dimethylformamide to give aldehyde **46** (Scheme 2) ⟨94AG(E)1158⟩. The lithio derivative could also be coupled on treatment with $CuCl_2$ to give the biphosphole **47**, in which the P atoms are chiral. That a single product was observed implies that either a *meso* form or a racemate had been synthesized.

Similar reactions were used to prepare the first known α-linked quaterphosphole **50** (Scheme 3) ⟨94AG(E)1158⟩, starting with diphosphole **47**. The quaterphosphole was a mixture of diastereoisomers. The structure and conformation of the predominant isomer was established by X-ray crystallography (Section 4.2.2.3.2).

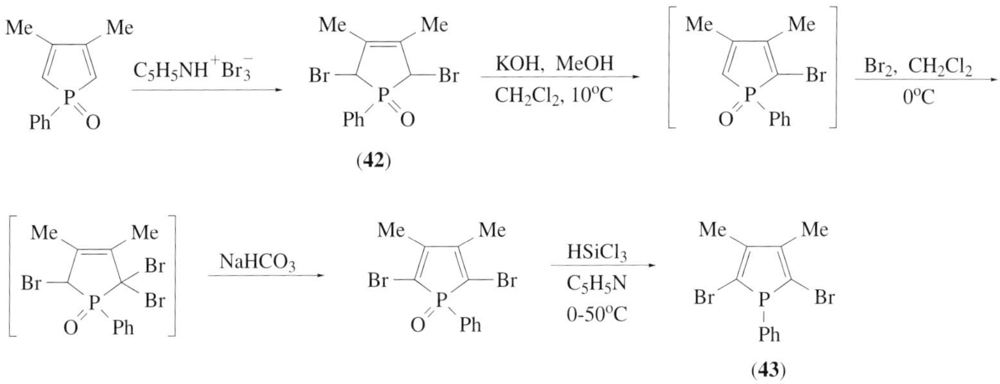

(20)

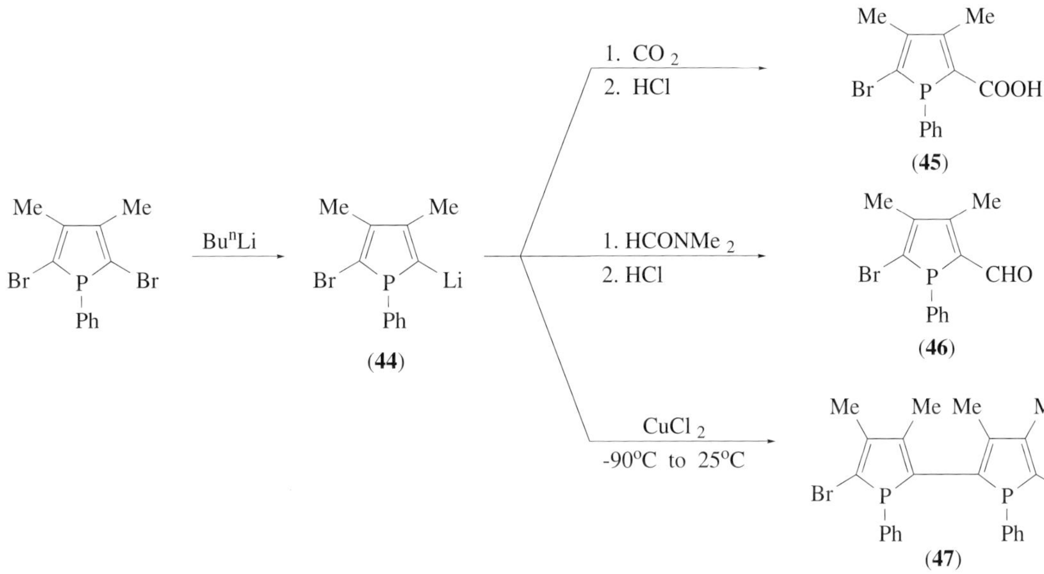

Scheme 2

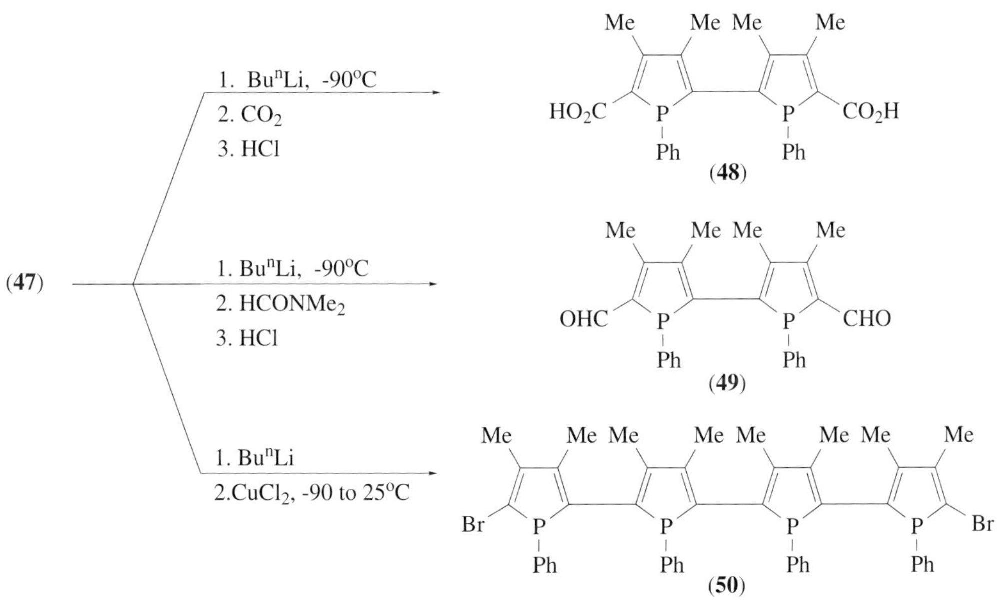

Scheme 3

When the dibromophosphole **43** was reacted with excess butyllithium in hexane at 50°C, a high yield of the 2,5-dilithio derivative was obtained without any cleavage of the P-phenyl bond ⟨98CR(C)715⟩. The dilithio compound gave phosphole **51** with chlorotrimethylsilane, and acid **52** on carbonation. The latter was esterified to give the phosphole dicarboxylate (**53**) (Equation 21).

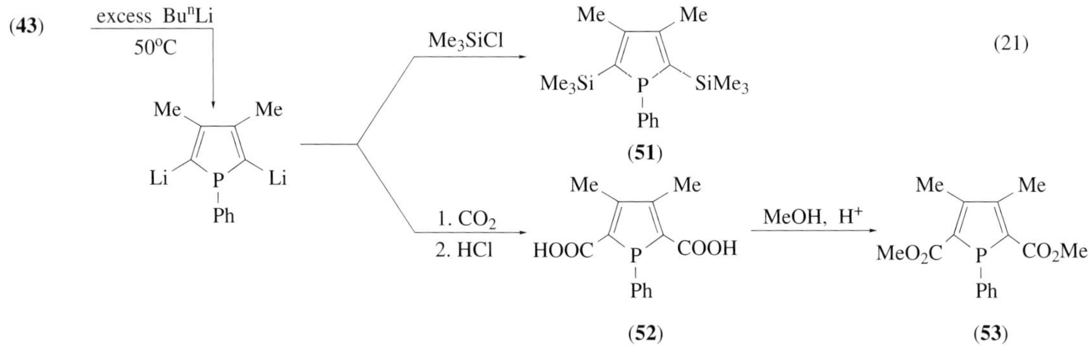

The dialdehyde **49** reacts with a bis-ylide to form in 60% yield a macrocycle with 24 ring members (Equation 22) ⟨95CC1561⟩. The macrocycle was shown by X-ray analysis to have the staggered arrangement of P-phenyl substituents as shown in structure **54**.

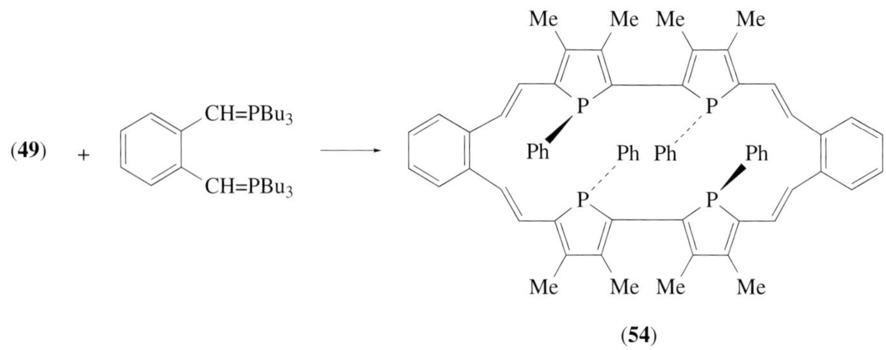

(22)

(g) 2,5-Di-*t*-butyl Substitution. One of the earliest phosphole ring syntheses has now been employed to construct a phosphole with *t*-butyl substituents at the 2- and 5-positions ⟨94CC1167⟩. This involved the addition of phenylphosphine to the di-acetylene **55**. Several derivatives (Scheme 4) were made by conventional reactions.

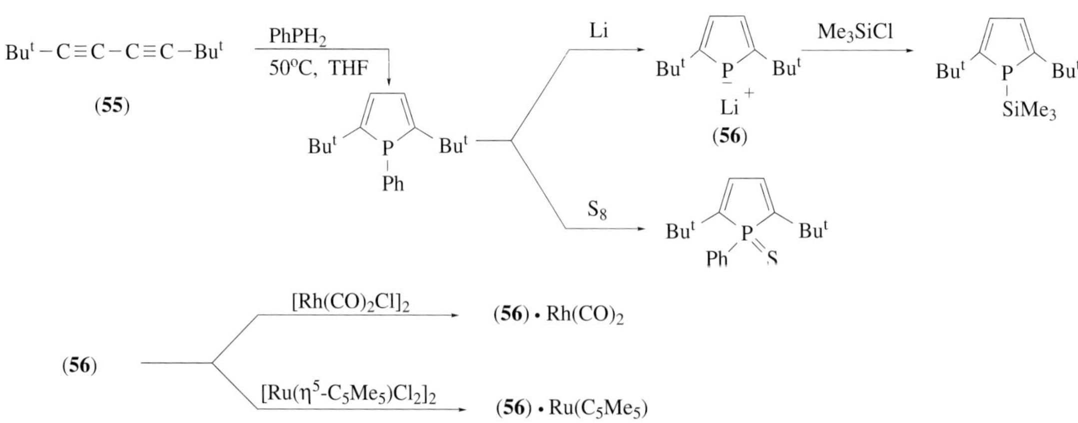

Scheme 4

Several phospholes with new types of P-substituents that have recently been reported include the following:

(a) The Phosphinine Group. The 3,4-dimethylphospholide ion was reacted with α-bromophosphinines to effect displacement of bromine and attachment of a phosphinine group to the phosphole ring (Equation 23–25) ⟨95BSF910⟩. The reaction is catalyzed by 5% Pd(dibenzylideneacetone)$_2$, Pd(dba)$_2$. The 1-trimethylstannyl derivative prepared from **57** with Me_3SnCl also gave similar results with bromophosphinines.

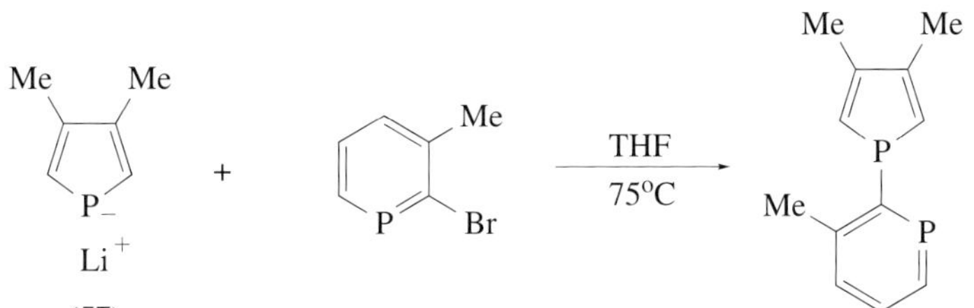

(**57**)

(23)

(**57**)

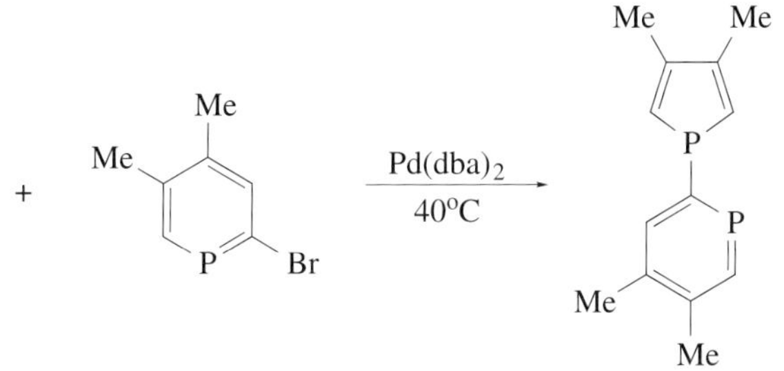

(24)

(**57**)

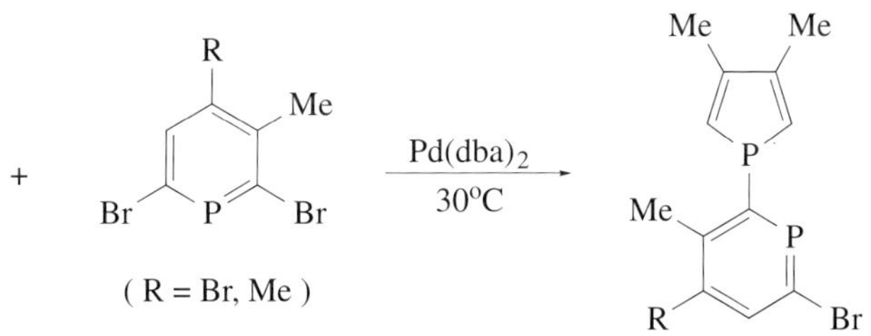

(25)

These novel phospholes underwent the usual [1,5] sigmatropic shift of the P-aryl group to give the 2-H phosphole which then dimerized.

(b) Dialkyl- and Diarylphosphino Groups. It has been demonstrated that phospholide anions react with phosphinous chlorides to establish the P–P bond. The phosphinophospholes, formed in good yield (Equation 26), are described as pale yellow solids, stable for several weeks if stored at −30°C ⟨97JOM(529)197⟩. These new phospholes formed η^1 complexes (**58**) at the phosphole P on reaction with $W(CO)_5(MeCN)$, but with formation of side products.

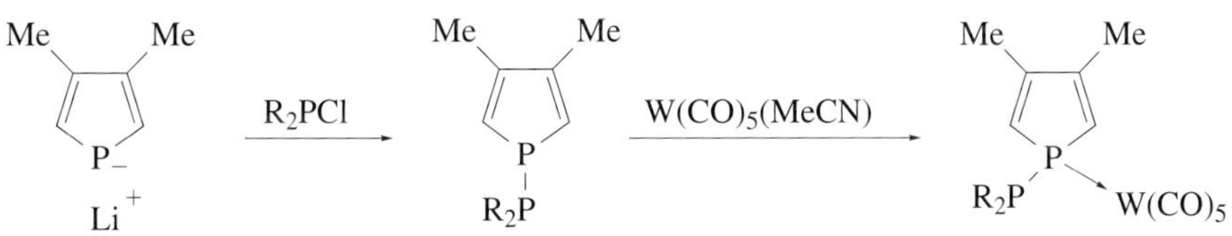

(**58**)

(26)

The same complexes could be formed by reacting the phosphinous chlorides with the pre-formed phospholide-$W(CO)_5$ derivative **59** (Equation 27).

Dimesitylphosphinous bromide also reacted with a phospholide ion (2,3,4,5-tetramethyl) to give a phosphinophosphole **60**. This compound was used in a reaction with a lanthanide-series metal, ytterbium,

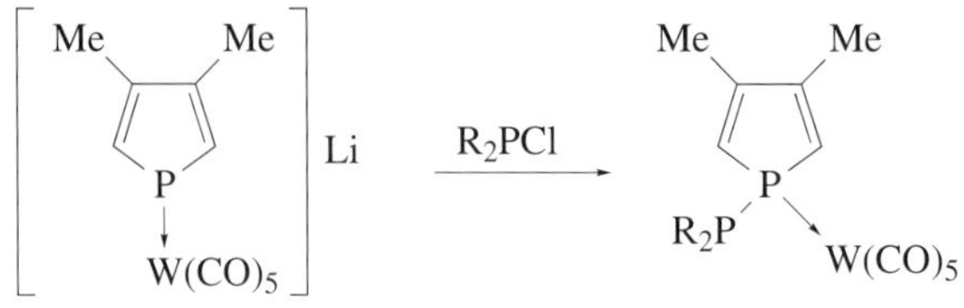

(27)

(**59**)

with the expectation that the metal would insert into the P–P bond ⟨95BSF649⟩. This product (**61**) could not be observed; apparently it underwent disproportionation to give Yb(II)bis(dimesitylphosphide), which was isolated and subjected to X-ray analysis. The fate of the phosphole moiety was not established.

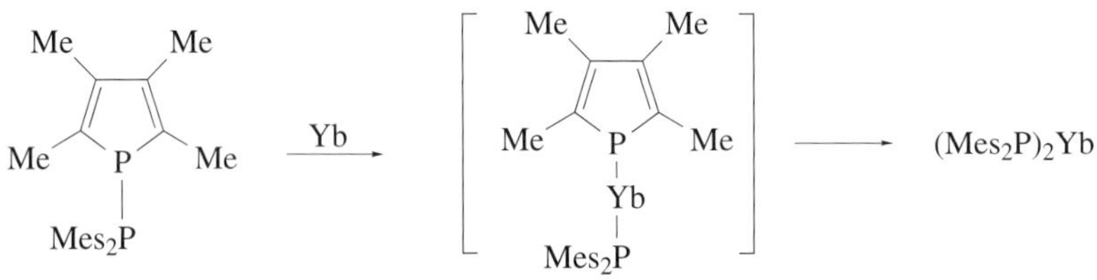

(28)

(**60**) (**61**)

(c) The Diphospholylphosphino Group. The novel structure **62**, described as a triphospholylphosphine, was prepared by the reaction of a lithium phospholide with phosphorus trichloride (Equation 29) ⟨97JOM(529)75⟩. The product (23% yield) was a yellow crystalline solid, stable when stored at -30°C.

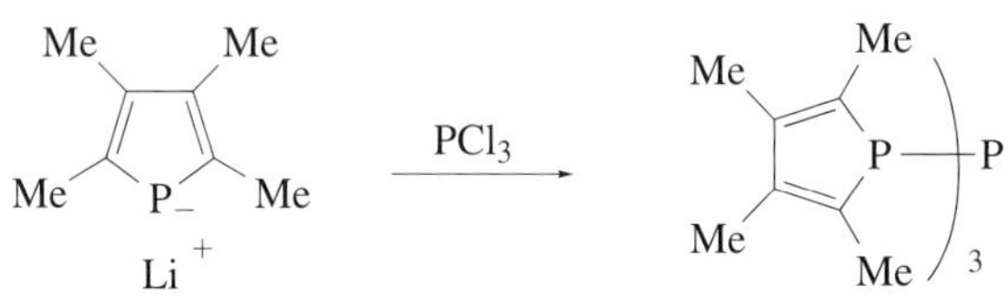

(29)

(**62**)

(d) Pyridine and Quinoline Substituents. The reactive bromine at the 2-position of pyridine can be displaced by attack of the 3,4-dimethylphosphole anion in the presence of CuI as catalyst, to give 1-(2-pyridyl)phospholes (Equation 30) ⟨97AG(E)98⟩. In similar fashion, the 8-quinolinyl group was attached to the phosphole ring ⟨97JOM(548)17⟩.

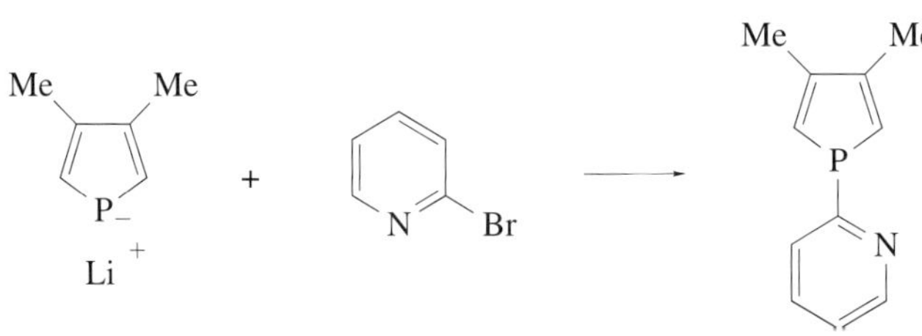

(30)

(e) Phospholes with Very Large 1-Substituents. In connection with studies on the effect of flattening the phosphorus pyramid on aromaticity, several phospholes with very bulky 2,4,6-trisubstituted aryl substituents on phosphorus have been prepared. Steric hindrance prevented the use of the McCormack cycloaddition of a diene with $ArPX_2$ to form the 5-membered ring with the desired P-substituent. The substituent was attached after the ring had been constructed with a P-halo group, and conventional reactions then used to create the phosphole system (Equation 31). Phospholes **16b** ⟨96JOC7801⟩ and **16c** ⟨97JOM(532)109⟩ were prepared as shown in Equation 31. The larger 2,4,6-tri-*t*-butylphenyl substituent required a modification of this procedure; the deoxygenation of the phospholene oxide intermediate in

Equation 31 was prevented by the steric hindrance, and an alternate path (Equation 32), again used in earlier phosphole syntheses, was employed to synthesize the phosphole (**16a**) with this P-substituent ⟨97JA5095⟩.

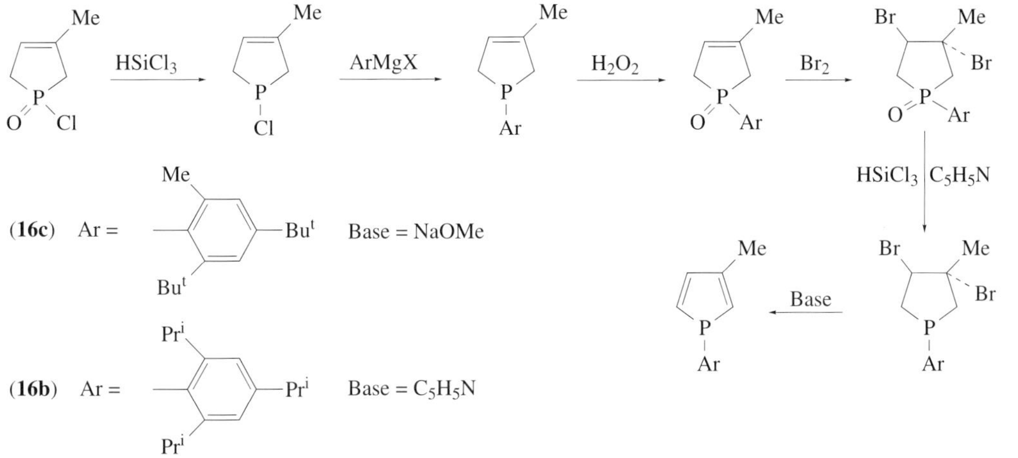

(31)

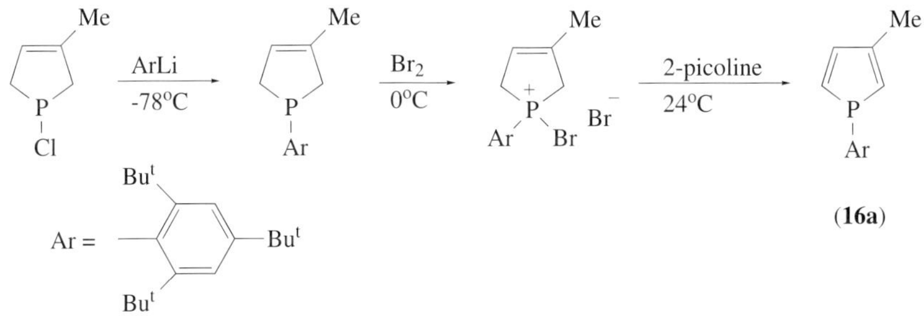

(32)

(**16a**)

4.2.2.5 NEW RING SYNTHESES

A phospholium ion with novel substitution (**63**) has been claimed to result from a new ring-synthesis method ⟨94ZOB1566⟩. Here a cyclization of dimethyl acetylenedicarboxylate with a β-bromovinylphosphine takes place, possibly through an ylide adduct that eliminates Br^- (Equation 33). The ^{31}P NMR shift of δ +37.2 is consistent with this structure, as is the presence of a downfield ^{1}H signal (δ 8.20) with a large coupling constant to ^{31}P of 36 Hz.

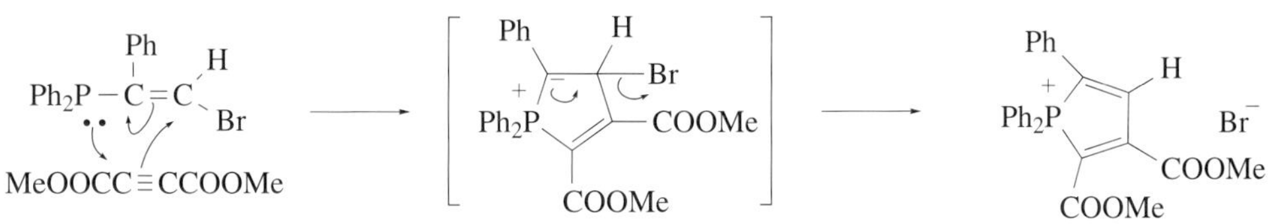

(33)

(**63**)

Zirconacyclopentadienes, easily synthesized in a reductive reaction of zirconocene dichloride with an alkyne, can react with phosphonous dihalides to form phosphole derivatives. The original example ⟨94JA1880⟩ is expressed by Equation 34.

A zirconacyclopentadiene was later used in a synthesis of a bicyclic phosphole and a phosphindole (see Section 4.2.2.6) and to the synthesis of a polymer containing phosphole units ⟨97MM5566⟩ (Equation 35). These polymers exhibit photoluminescence, with the emission of blue light.

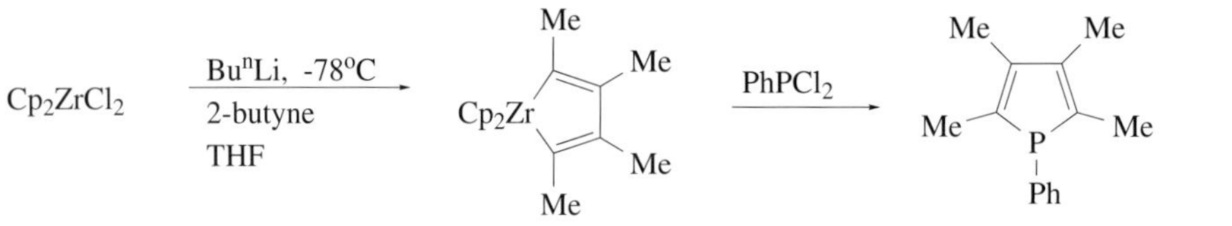

(34)

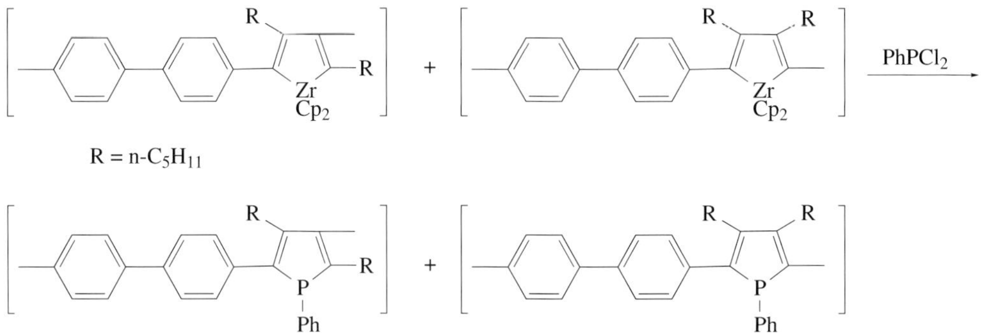

$R = n\text{-}C_5H_{11}$

(35)

4.2.2.6 NEW SYNTHESES AND DERIVATIVES OF BICYCLIC PHOSPHOLES AND DIBENZOPHOSPHOLES

(a) Phosphindoles. A new approach to the phosphindole ring system consists of the synthesis of zirconaindene derivatives (**65**) and their condensation with phosphorus chlorides. This is a variant of the phosphole synthesis from zirconacyclopentadienes noted above. The required zirconaindenes were prepared from a transient benzynezirconocene (**64**), as is illustrated in Equation 36 for the synthesis of a novel P-chlorophosphindole derivative 〈94JOM(464)149〉.

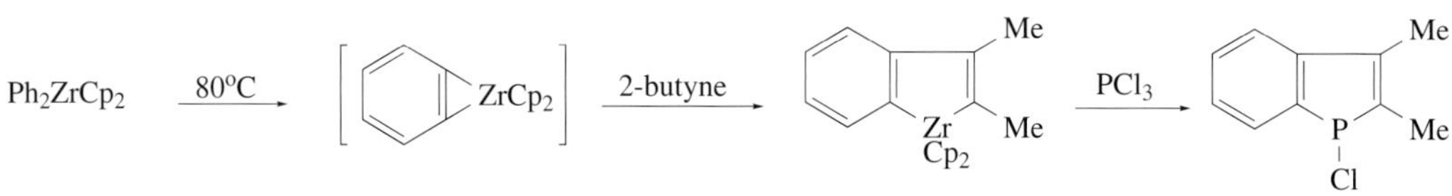

(**64**) (**65**)

(36)

By using a phosphinoacetylene derivative in the reaction with the benzynezirconocene to form the zirconaindene, the first phosphindoles with a phosphino substituent were prepared (Equation 37) 〈97CC279〉. Sulfur could be added to both phosphorus atoms to produce the phosphindole sulfide derivatives.

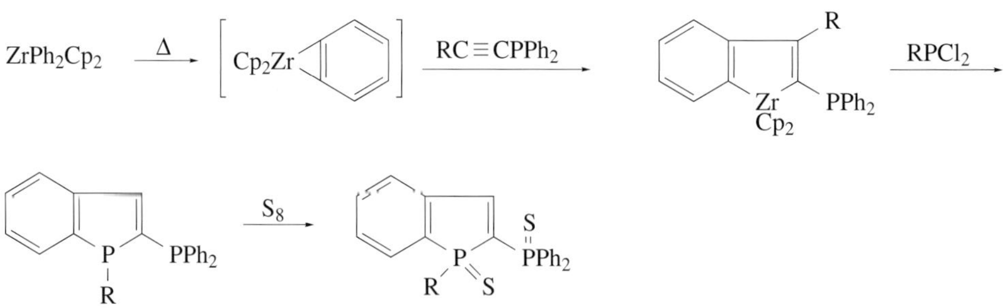

(37)

The phosphindole ring system has been created in an unusual reaction of a lithium arylphosphide with diphenylacetylene 〈94AG(E)353〉. As expressed by the example of Equation 38, an initial solvent-stabilized adduct undergoes ring-closure by loss of an *ortho* group on the P-aryl substituent. The structure of the product was confirmed to be that of the lithium phosphindolide **66**.

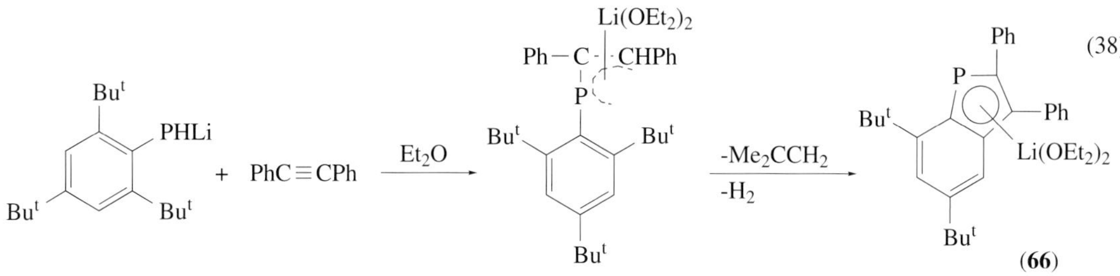

(38)

(**66**)

(b) The Isophosphindole Anion. Schmidpeter and Thiele ⟨91AG(E)308⟩ reported several years ago on the synthesis and delocalization phenomena of novel α,α'-bis(triphenylphosphonio) derivatives of the isophosphindole anion. A related structure **67**, where vinyl replaces one phenyl, has now been synthesized (Equation 39) ⟨95BSF280⟩, and the theory of the bonding in such systems developed in detail. This compound is extensively delocalized and may be considered as an example of a "heteronaphthalenic" π-system.

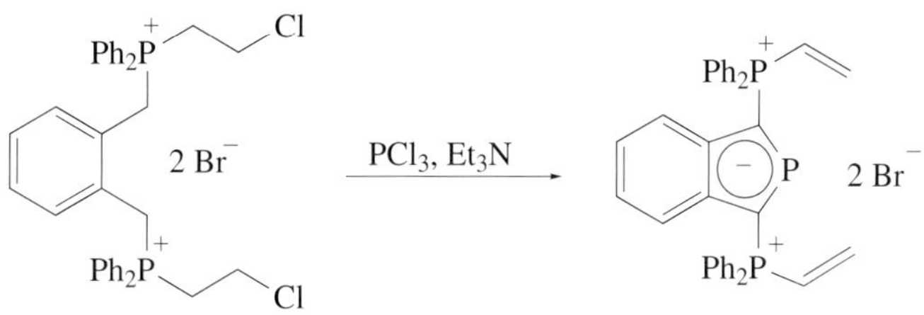

(39)

(**67**)

When **67** was reacted with HBF_4, protonation occurred on carbon to form a novel tri-cation **68**, and not on phosphorus as had been observed with the triphenylphosphonio derivative (Equation 40).

(**67**)

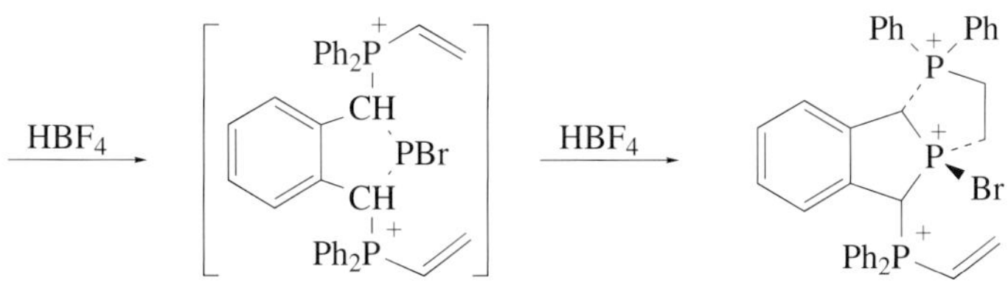

(40)

(**68**)

(c) Cycloalkano[c]phospholes. In another process based on zirconium chemistry, the method of Fagan et al. ⟨94JA1880⟩ was used to prepare zirconacyclopentadienes from the reaction of a diyne with Cp_2Zr, which were then reacted with phenylphosphonous dichloride (Equation 41). This effected ring closure to the phosphole with a cycloalkano group fused at the 3,4-position ⟨99CC345⟩. The particular phospholes synthesized contained 2,5-di(2-pyridyl) substituents (**69** and **70**).

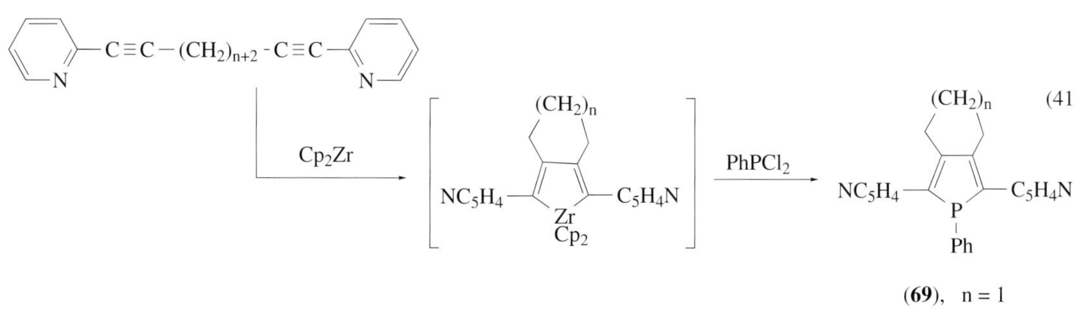

(41)

(**69**), n = 1

(**70**), n = 2

Phosphole (**71**) with 2,5-diphenyl substitution was also prepared in this manner.

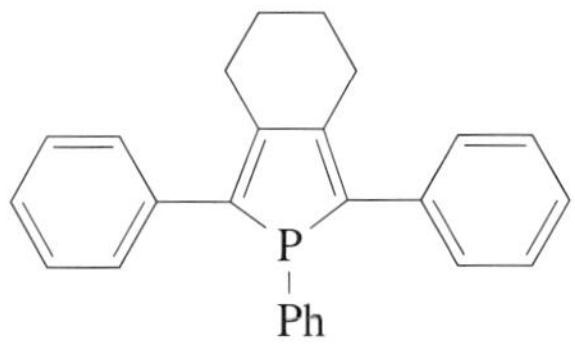

(**71**)

(d) The Phospholo[3,2-b]phosphole (1,4-Diphosphadihydropentalene) System. This bicyclic system consists of two fused 1H-phosphole rings, shown in structure **73**. It is of interest because the cleavage of the two P-phenyl substituents would give the diphosphapentalene dianion, a 10-π cyclic system that is potentially aromatic and not previously synthesized. Märkl, Hennig and Nöth ⟨95TL6655⟩ accomplished the synthesis of **73** by reacting E,E-3,4-(diphosphamethylene)cyclobutene **72** (which undergoes ring-opening by Cope rearrangement) with dimethyl acetylenedicarboxylate (Equation 42). It was proposed that the reaction proceeded by a radical mechanism. The structure was well established by spectral analysis.

(42)

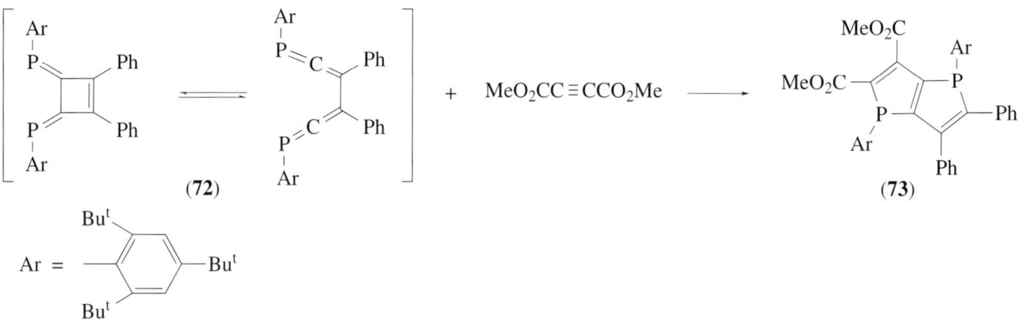

The phospholophosphole has strong UV absorption at 249 and 394 nm; it is sensitive to light-promoted air oxidation, thereby adding a molecule of oxygen to give a presumed adduct **74**, which rearranges to the new fused phosphole system **75** (Equation 43). The structure of this unusual phosphole derivative was established by X-ray diffraction analysis, which served also to confirm the structure of the phospholophosphole.

(43)

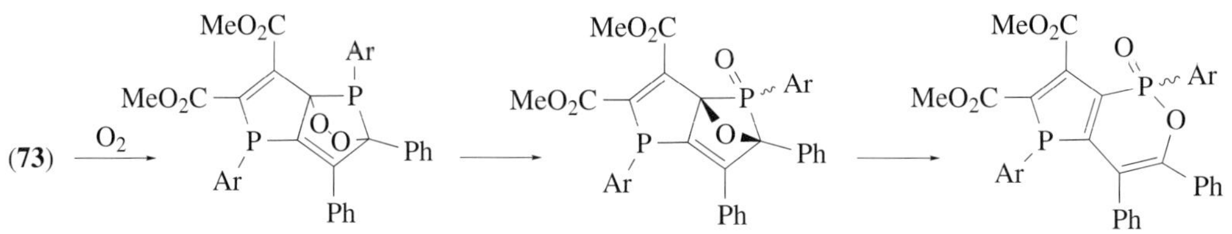

(e) Dibenzophospholes. Much of the recent work on dibenzophosphole derivatives has been concerned with developing their potential as ligands in coordination chemistry. Two types of chirality can be created around the dibenzophosphole nucleus, and chiral derivatives are being exploited in new catalyst systems for asymmetric synthesis. In creating one form of chiral dibenzophosphole derivatives, groups with asymmetric carbon atoms are attached to the phosphorus atom. In the example of Equation 44 ⟨96T7547⟩, a chiral dibenzophosphole is formed by reacting the lithium derivative of 5H-dibenzophosphole with a chiral oxazoline derivative (**76**) bearing a 2-fluorophenyl substituent at the 2-position. The halo substituent is activated by the oxazoline group and is replaceable by nucleophiles. The (+)–(*S*) stereoisomer of compound **77** was synthesized in this fashion.

A complex of this phosphine with Pd has been considered for use as a catalyst for the enantioselective dicarbonylation of styrene ⟨98HCA764⟩.

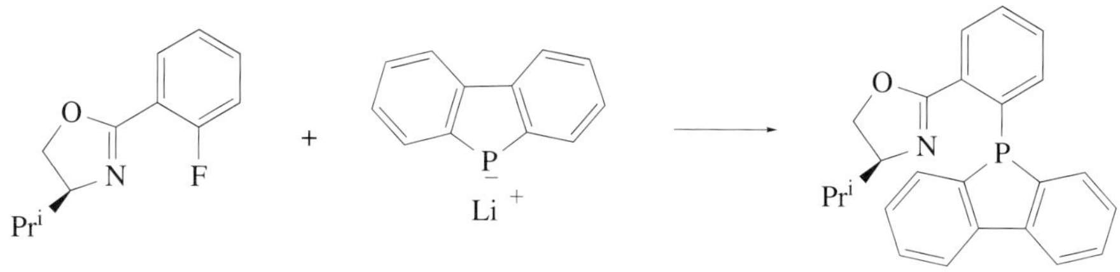

(76) (77)

(44)

Another example (Equation 45) ⟨97JOM(540)15⟩ uses lithiodibenzophosphole in a double displacement of the tosyl groups from the (2*S*,4*S*) stereoisomer of the ditosylate of 2,4-pentanediol. The product (**78**) is known as BDBPP from the trivial name 2,4-bis(dibenzophospholyl)pentane; platinum complexes have been found to be effective catalysts in the asymmetric hydroformylation of styrene derivatives with H_2 and CO.

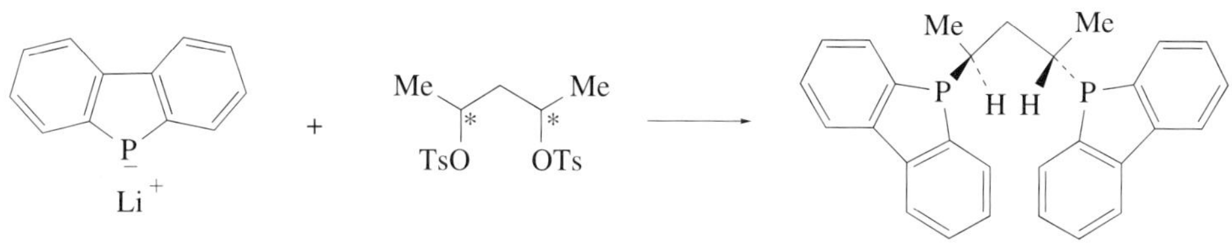

(**78**) (S, S)

(45)

A related compound has the structure **79**, and is prepared similarly ⟨98JOM(560)257⟩. It is related to the well-known ligand CHIRAPHOS (**80**), but differs in having rigidity around the substituents on phosphorus.

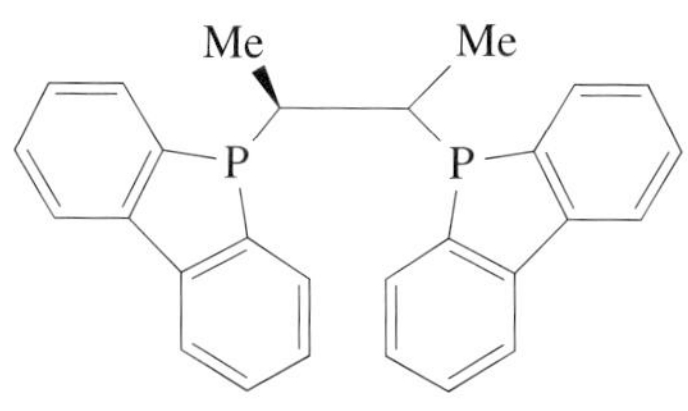

(**79**)

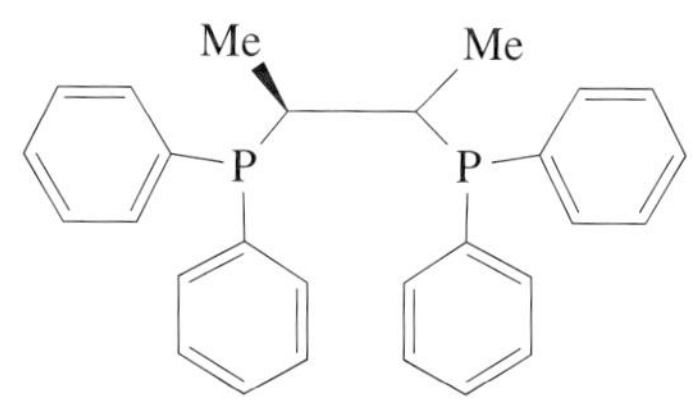

(**80**)

An unusual reaction has resulted in the formation of dibenzophosphole bearing the optically active menthyl group ⟨98S45⟩. When the 2,2′-dilithio derivative prepared from 2,2′-dibromobiphenyl was reacted with chlorodi-(−)-menthylphosphine, the expected bis(dimenthylphosphino)biphenyl was not obtained; only one dimenthylphosphino group became attached, after which the product underwent a cyclization reaction with the remaining lithio function. There was obtained a 26% yield of the 5-(−)-menthyldibenzophosphole (**81**, Equation 46).

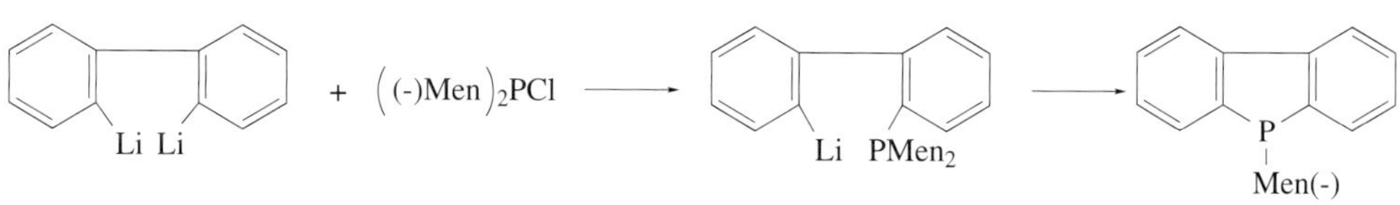

(**81**)

(46)

It is most uncommon to experience a cleavage of an alkyl C–P bond by a lithioaryl group, although cleavage of aryl-P groups are known. However, intramolecular ring closure to a dibenzophosphole derivative, attributable to the same process, was noted when 2-lithio-2′-diphenylphosphinyl-1,1′-biphenyl was quenched with chlorodimenthylphosphine (Equation 47); the lithio function apparently had attacked on the diphenylphosphino group to displace a phenyl anion, which was then quenched with chlorodimenthylphosphine. In earlier work ⟨96JOM(507)257⟩, it had been observed that quenching of 2,2′-dilithiobiphenyl with

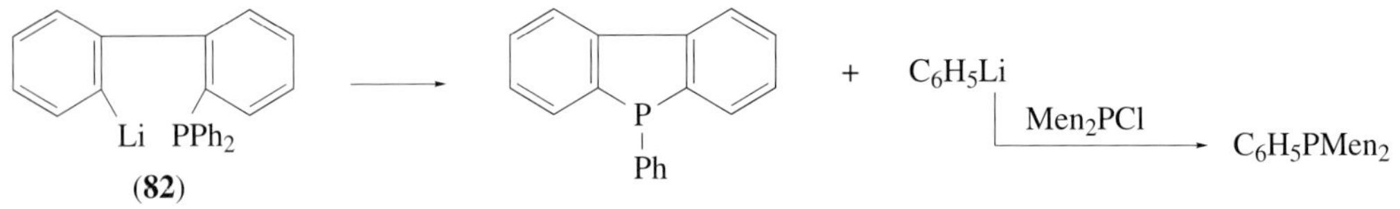

(82) (47)

chlorodiphenylphosphine led to a similar reaction, with formation of 5-phenyldibenzophosphole and triphenylphosphine in equal amounts. An initially formed mono lithio derivative (**82**) presumably attacks the tertiary phosphine group with elimination of phenyllithium (Equation 48). This type of reaction was also observed ⟨96TL5347⟩ when the (*R*)-stereoisomer of biphenyl derivative **83** (which exhibits atropisomerism from restricted rotation around the biphenyl bond) was reacted with butyllithium at −70°C and the mixture allowed to warm to −20°C (Equation 48). The dibenzophosphole **84** proved to be racemic.

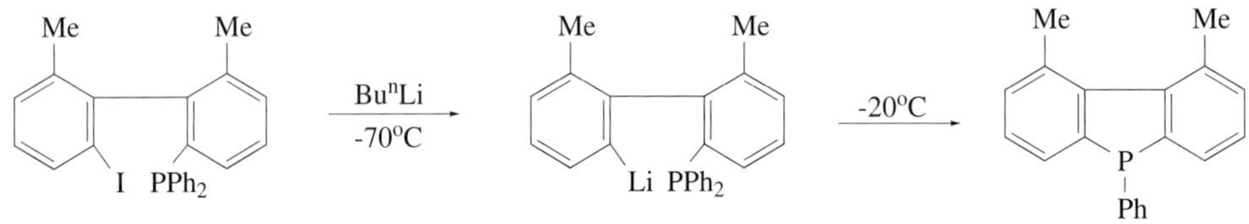

(R) (**83**) (**84**) (48)

Dibenzophosphole was a component of a racemic chiral tripodal ligand (**85**) synthesized several years ago ⟨93ZN(B)1681⟩. It remains of interest, and a new complex with $Fe(NCMe)_3^{2+}$ has been reported and thoroughly studied by NMR and X-ray diffraction methods ⟨95CB293⟩. This ligand has now been obtained in optically active form ⟨95ZN(B)729⟩.

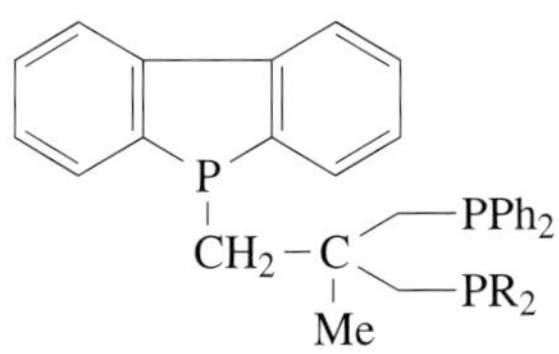

(**85**)

Other tripodal ligands containing dibenzophosphole (DBP) groups are **86** ⟨94CB1837⟩, **87** ⟨94JOM(468)149⟩, **88** ⟨95ZN(B)1045⟩, **89** ⟨95ZN(B)1287⟩, **90** ⟨97JOM(539)67, 96CB697⟩, **91** ⟨98JOM(553)433⟩ and **92** ⟨99EJICI405⟩.

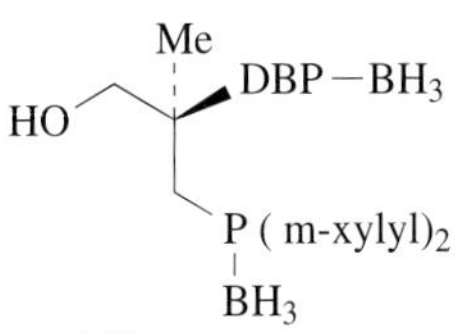

(**86**)

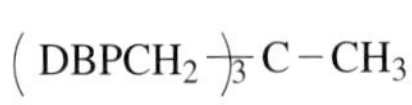

(**87**)

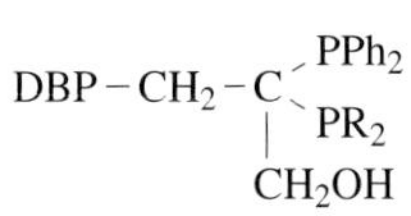

(**88**)

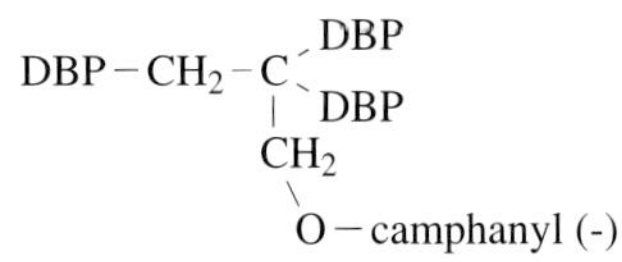

(**89**)

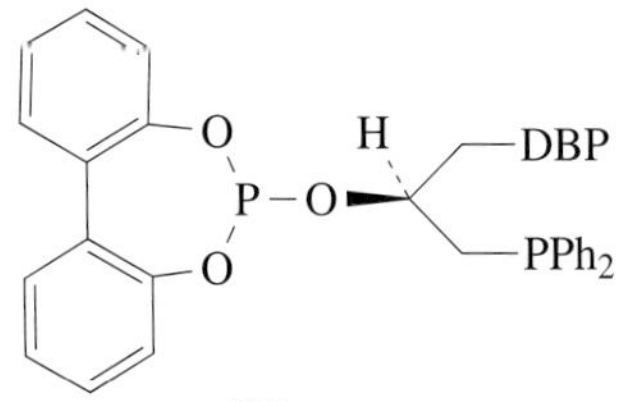

(**90**)

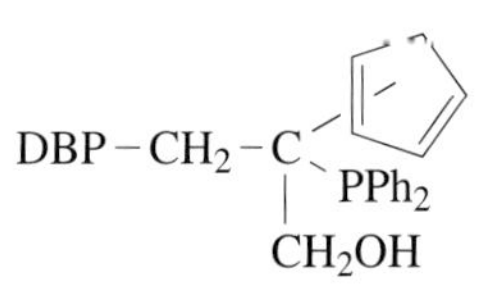

(**91**)

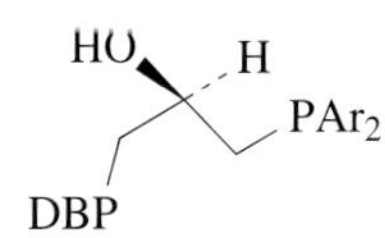

(**92**)

The other form of chirality of current interest in dibenzophosphole chemistry is that of atropisomerism, where dissymmetry develops from intramolecular crowding and axial chirality is present. The fusion of two benzo groups to dibenzophosphole, to provide the dinaphtho[2,1-b;1′,2′-d]phosphole system **93**, leads to this type of chirality, first detected in 1993 ⟨93CC1124, 93JOM(445)71⟩. Here the interference of H_1 and H_{13} prevents the molecule from becoming planar ⟨94AX(C)769⟩; for the P-phenyl derivative, spontaneous resolution from ethanol provides optically active isomers, but they undergo fast racemization in solution ⟨94JOC6363⟩. The binaphthophospholes are useful ligands in metal complexes (e.g. ⟨94JOM(469)229, 94JOM(475)307⟩). Energy barriers to atropisomerization for a number of derivatives fall in the 55–56 kJ mol^{-1} range. Oxides and salts of binaphthophospholes also display atropisomerism ⟨94SC1271⟩, with slightly higher barriers for isomerization (60 kJ mol^{-1}, compared to 55–60 kJ mol^{-1}). The subject has very recently been reviewed ⟨97CB543⟩.

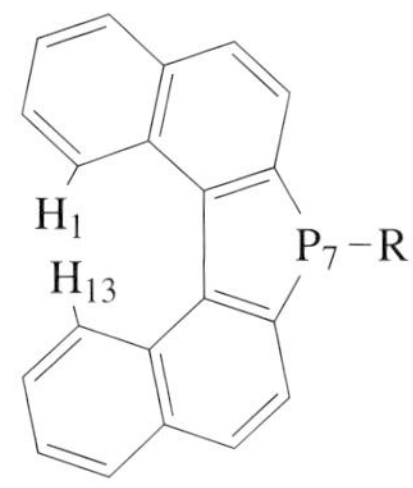

(**93**)

Three main methods of synthesis are used to obtain the binaphthophospholes ⟨94JOC6363⟩. The first method (Equation 49) employs a McCormack reaction to form the central phosphole ring, under high temperature conditions that effect dehydrohalogenation of the initial adduct. These conditions have been known for many years to be useful for the synthesis of 1,2,5-triphenylphosphole ⟨62CI(L)359⟩. Aromatization by loss of H_2 gave the naphthalene rings.

(49)

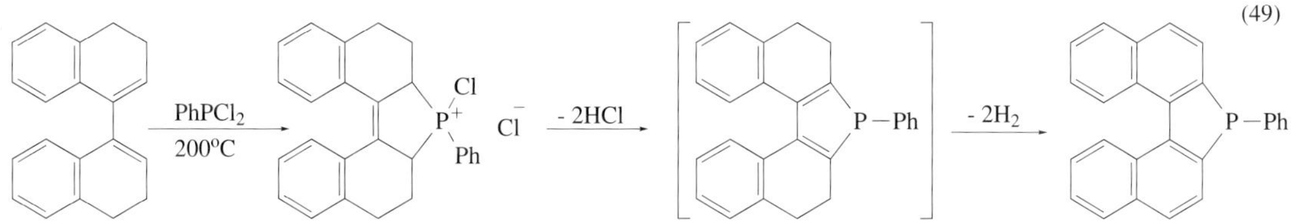

The second method, preferred to the first, employs a typical dibenzophosphole synthesis of a dilithio biphenyl with a phosphonous dichloride (Equation 50).

(50)

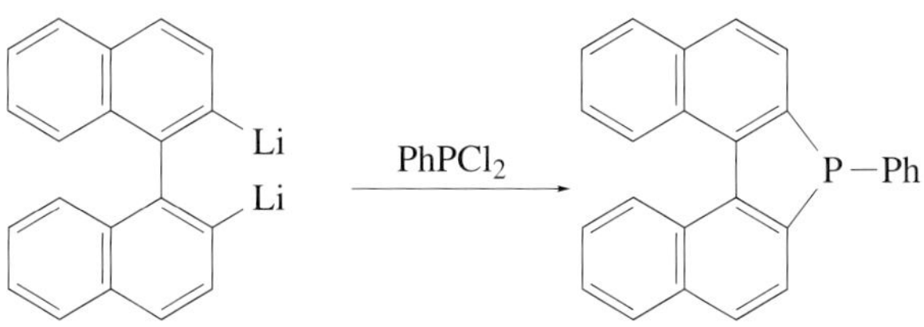

The third method makes use of the P-phenyl derivative of the binaphthophosphole, prepared by one of the preceding methods. Phenyl is cleaved with lithium, and the anion alkylated with an alkyl halide or tosylate.

The attachment of groups to phosphorus that have chiral carbon functions is made possible by the third method ⟨93CC1124⟩, and this has led to some interesting optically active ligands for use in coordination complexes. An example is the synthesis (Equation 51) of bis(binaphthophospholes) (BNP) such as compound **94**, where chirality is due both to atropisomerism and to the two asymmetric carbon atoms ⟨95JOM(491)91⟩. Certain platinum complexes exhibit remarkable activity in the hydroformylation of styrene, with high regioselectivity and moderate enantioselectivity.

The dibenzophosphole group may also be a substituent on a system possessing atropisomerism. Thus, in the novel compound **95**, synthesized as shown in Equation 52 ⟨97TA3775⟩, there is restricted rotation around the bond connecting the naphthalene group to the isoquinoline group. This dibenzophosphole was resolved into the enantiomers by making a mixture of diastereomers with an optically active palladium complex.

(51)

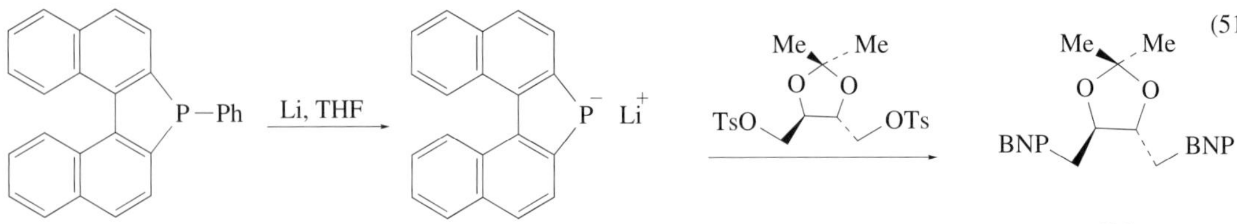

(**94**)

(52)

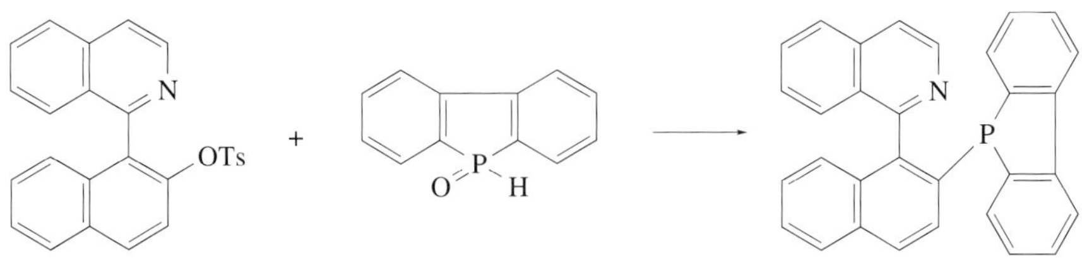

(**95**)

Ts = CF_3SO_2

Two examples of dibenzophosphole derivatives with tetracoordinate phosphorus have been recently described. In one of these reports 〈96JCS(P1)2889〉, which follows up on an earlier description of such compounds 〈87JCS(P1)871〉, the phosphorus atom was in the state of a phosphinic acid (e.g. **96**); compounds of this type are unusual in having complex aryl substituents at the 1,6-positions, each of which bears a phosphonic acid group. Thus these molecules have five ionizable hydrogens, and have some water solubility. Catalysis of the hydration of 2-methylpropene to *tert*-butyl alcohol was effected by these compounds, with efficiency somewhat exceeding that of a solution of *p*-toluenesulfonic acid of comparable acidity.

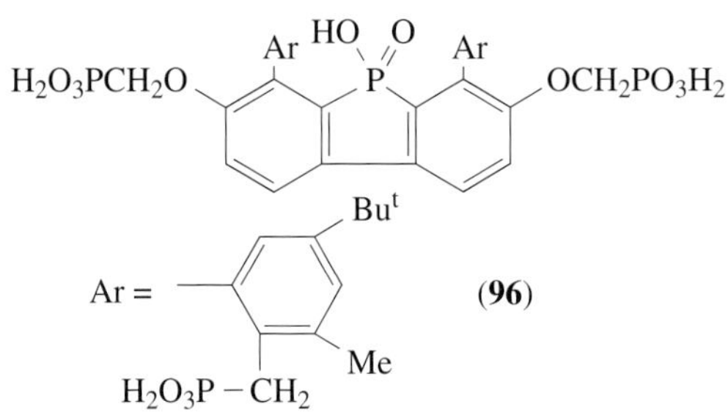

(**96**)

The other type of C.N.4 dibenzophosphole derivative has phosphorus in the state of a phosphazene with spirocyclic structure 〈97JCS(P2)15〉. The synthetic method exemplified in Equation 53 gave a separable mixture of **97** and **98**. In earlier work the trispirocyclotriphosphazene had been prepared 〈94JCR(M)887〉.

(53)

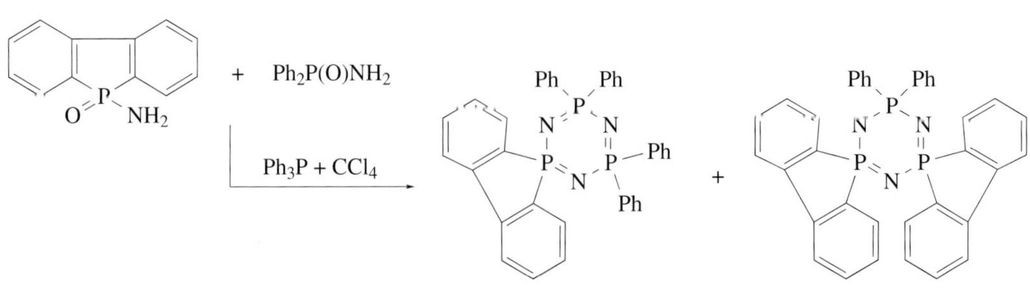

(**97**) (**98**)

4.2.2.7 2H- AND 3H-PHOSPHOLES

2H-phospholes can be generated by the thermal [1,5] sigmatropic rearrangement of a P-substituent to the 2-position. While the 2H-phospholes so produced are not stable, they nevertheless can give rise to valuable new heterocyclic compounds through dimerization or trapping with dienophiles or dienes. A variety of C- and P-substituted 1H-phospholes have been found to undergo the equilibration with the 2H-phosphole isomer ⟨93BSF843⟩. Recent reviews ⟨97CR(B)701, 98JOM(557)117⟩ summarize the known chemistry of the 2H-phospholes. Trapping reactions of the highly reactive 2H-phospholes are now being used to generate novel 1-phosphanorbornadienes, such as those with water-solubilizing phosphonate groups at the α or β positions ⟨96JOC3531⟩. This is accomplished by using acetylenic phosphonic acid derivatives as trapping agents for the 2H-phosphole. In Equation 54 is illustrated the reaction with derivatives of phenylethynylphosphonic acid, which leads primarily to 1-phosphanorbornadiene with α-phosphonic acid substituents.

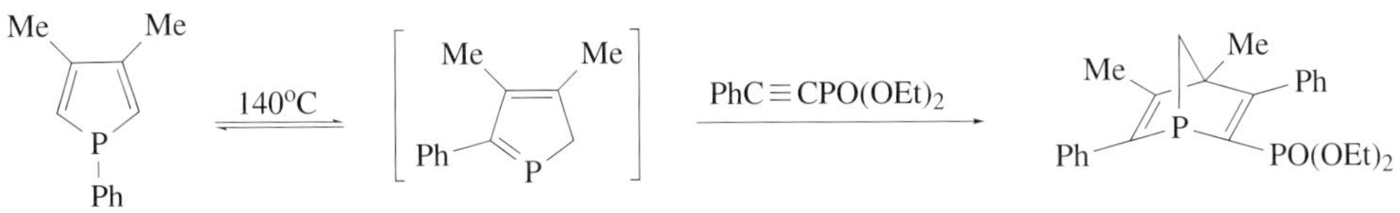

(54)

With ethynylphosphonamide derivative **99**, a 3 : 1 mixture of α- and β-substituted adducts was obtained (Equation 55). These were easily separated on silica gel and hydrolyzed to the phosphonic acids; a rhodium complex of the sodium salt of the β-isomer was found to be an effective catalyst for the hydrogenation of (Z)-α-(*N*-acetamido)cinnamic acid.

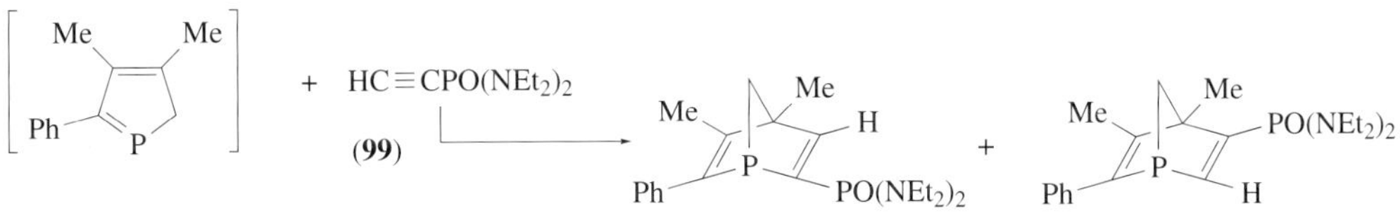

(55)

Water soluble 1-phosphanorbornadiene derivatives of structure **101** have also been synthesized ⟨93JOM(462)103⟩. The reaction of 2H-phosphole **100** with maleic anhydride gives the anhydride adduct, which can be hydrolyzed to give the sodium salt **101** (Equation 56).

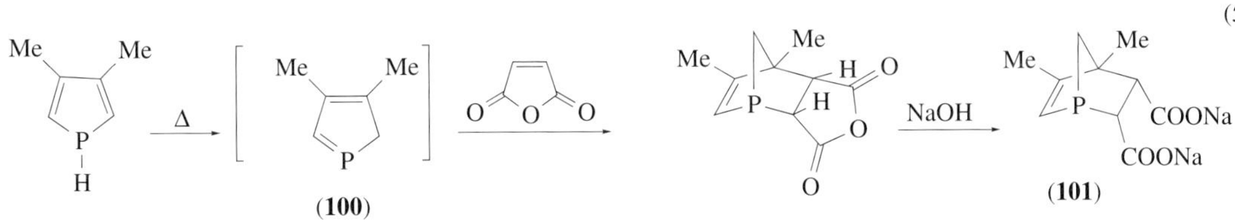

(56)

A 1-phospholo substituent can also participate in the [1,5] sigmatropic rearrangement ⟨92BSF1⟩; each phosphole moiety undergoes the rearrangement to a 2H-phosphole, and reaction with diphenylacetylene then results in the creation of a structure with two 1-phosphanorbornadiene groups attached at their α-positions (Equation 57). Both *meso* and racemic forms are obtained; these have been separated as their $PdCl_2$ complexes, and the racemic form then resolved with an optically active palladium complex. The recovered di-phosphine is known as (*R*,*R*)- or (*S*,*S*)-BIPNOR. The complex Rh(cod)(BIPNOR)$^+$ PF_6^- has proved to be highly effective for the asymmetric hydrogenation of C=C and C=O groups ⟨98JOM(557)117⟩.

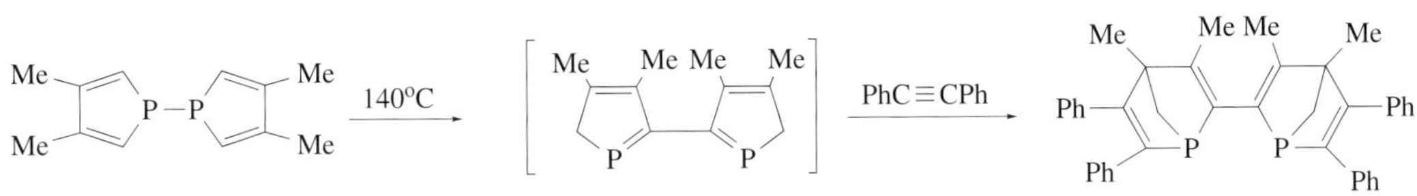

(57)

2H-Phospholes have been shown to be formed within the coordination sphere of a transition metal. This has been demonstrated by using a special type of complex that contains both 3,4-dimethyl-1-phenylphosphole and a second ligand with the capability of trapping the 2H-phosphole as formed in a Diels–Alder reaction ⟨97JA12560⟩. With *trans*-crotyldiphenylphosphine as the trapping ligand, as in complex **102**, the [1,5] rearrangement was conducted at ~150°C in diglyme to give a product **103** that was the $Mo(CO)_4$ complex of a phosphinomethyl-1-phosphanorbornene (Equation 58).

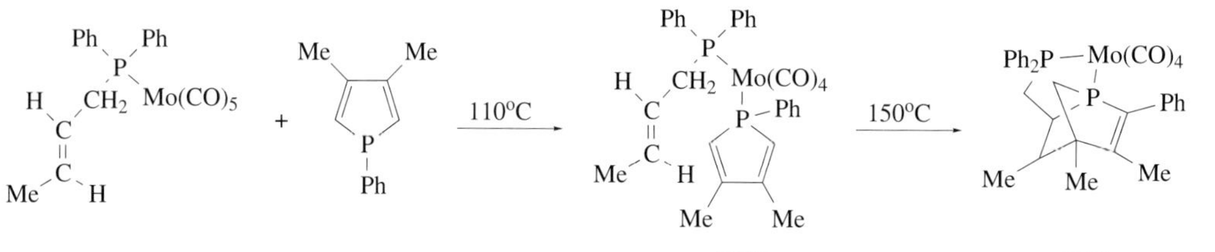

(58)

(**102**) (**103**)

The same process was performed with related complexes; of interest is the case where dibenzophosphole served as the phosphine on the crotyl group (i.e., in **103** where DBP replaces Ph_2P).

A novel 2H-phosphole based on the phospholo[1,2-a]phosphole ring system, with P in the 4-coordinate ylide structure, resulted from the condensation of two moles of dimethyl acetylenedicarboxylate with 1,3,4-triphenyl-1,2-dihydrophosphete ⟨98HAC9⟩. The reaction (Equation 59) occurred simply on mixing at room temperature, and within an hour a yellow solid identified as ylide **104** precipitated. Its unusual structure was confirmed by X-ray diffraction analysis; it was also well characterized spectroscopically.

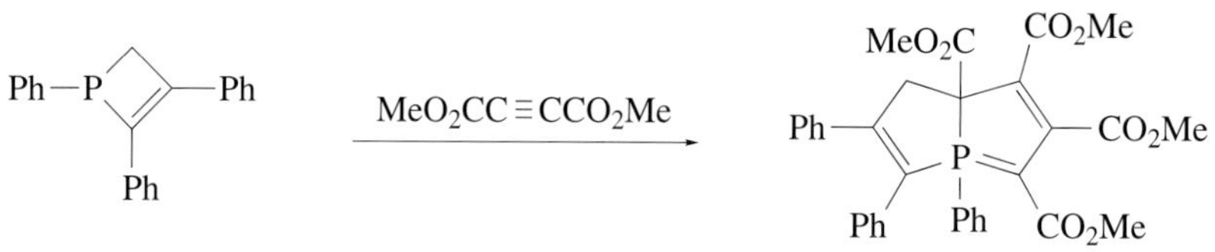

(59)

(**104**)

The monocyclic 3H-phosphole ring system remains unknown, but a benzo fused derivative (thus a 3H-phosphindole) appears to have been formed as a minor product on reaction of (2,4,6-tri-*t*-butylphenyl)phosphonous dichloride under flash vacuum pyrolysis conditions over Mg (Equation 60) ⟨96PS(111)814⟩. A phosphinidene is presumed to be the initial product, which must undergo cyclization with an *ortho*-*t*-butyl group. Loss of hydrogen must also occur.

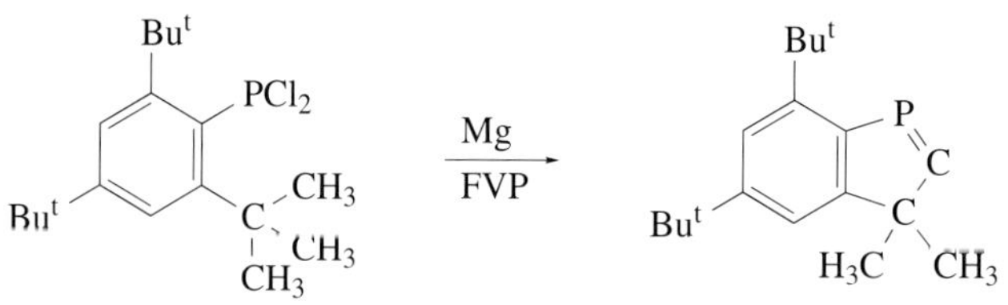

(60)

4.2.2.8 NEW REACTIONS OF PHOSPHOLES

(1) Thermal Coupling within the Coordination Sphere in Palladium (II) Halide Complexes. In addition to the useful thermal [1,5] sigmatropic rearrangements of 1H-phospholes to 2H-phospholes, thermally induced reactions of other types are possible with phospholes. As reported in Chapter 4.2.1, thermal dimerization of $NiCl_2$ complexes of phospholes to bis-phospholenes (Equation 61) can take place on heating at 140–170°C ⟨84JA425, 85IC4141⟩.

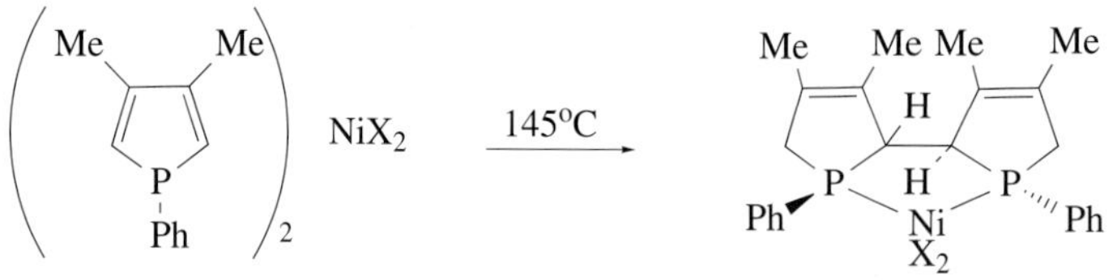

(61)

More recently, thermolysis of phospholes in the coordination sphere of other metals has been studied. It has been found ⟨94IC109⟩ that platinum (II) complexes give three different dimerization products (Equation 62): a [4 + 2] cycloaddition dimer (**105**) with exo ring fusion, a [2 + 2] cycloaddition dimer (**106**), and a dimeric structure with an exo-methylene group (**107**). Continued heating converts dimers **105** and **106** into the exo-methylene compound. All structures were proved by X-ray diffraction analysis. The dimerizations are intramolecular and occur only with the *cis*, not the *trans*, coordination complex.

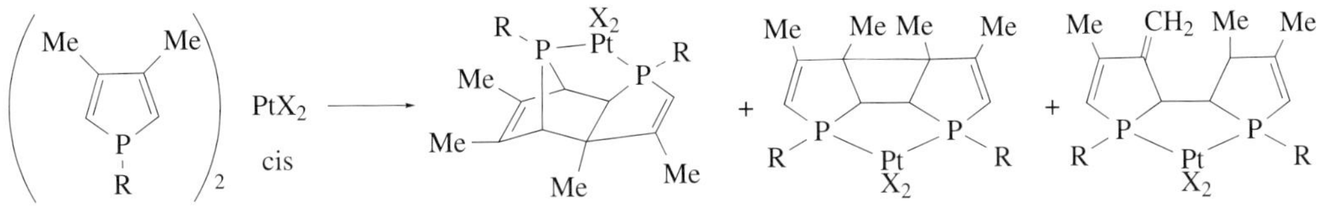

(62)

(**105**), [4+2] (**106**), [2+2] (**107**), "exo-methylene"

Complexes with Pd (II) have also been studied ⟨96IC1486⟩. With the *cis* or *trans* complexes of $PdCl_2$ or $PdBr_2$, four products are obtained on thermolysis in $ClCH_2CH_2Cl$ at 145°C in a pressure tube, as well as in the solid state at 140°C (Equation 63). Three of the four products have the same framework as do those from the platinum complexes, and are referred to in Equation 63 as [4 + 2], [2 + 2] and "exo-methylene" products. Again the reactions are intramolecular and occur with the *cis* isomer, but the *trans* isomer can rearrange to the cis isomer under the reaction conditions, and thus gives the same products. The fourth compound **108** is formed in a proposed intermolecular process. With the larger I^- anion, the *trans* isomer is more stable; rearrangement to the *cis* isomer does not occur, and thus no intramolecular coupling reactions take place.

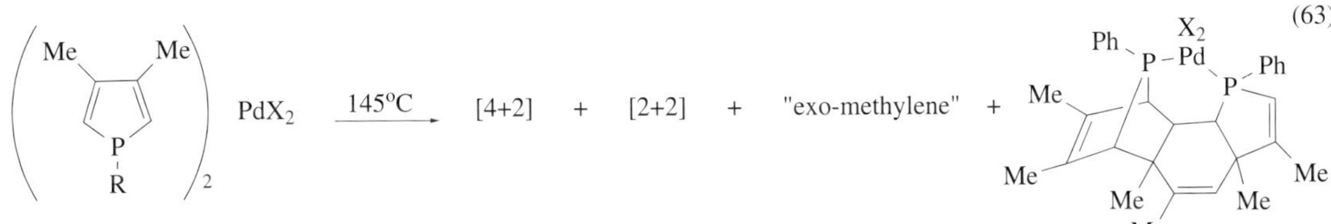

(63)

(**108**)

(2) Photocycloadditions within the Coordination Sphere of Metals. The [2 + 2] dimer **106** of DMPP was formed on sunlight irradiation of the complex $[(\eta^5\text{-}C_5H_5)Ru(DMPP)_2L]PF_6$, where L is a simple ligand such as MeCN, Ph_3P, $PhSO_2CH{=}CH_2$, etc., and DMPP is 3,4-dimethyl-1-phenylphosphole ⟨93PS(77)268, 97JOM(529)395⟩. Some [4 + 2] dimer also accompanied the main product. Similar results were obtained earlier on DMPP carbonyl complexes of Cr(0), Mo(0) and W(0). The [2 + 2] adducts appear to be formed by an intramolecular reaction within the coordination sphere of the metal. Uncomplexed phospholes do not give these [2 + 2] dimers. The dimerization presumably occurs by a biradical mechanism ⟨97JOM(529)395⟩.

(3) Stereoselective Diels–Alder Reaction within the Coordination Sphere of Metals. Several examples of Diels–Alder reactions of simple phospholes were described in Chapter 4.2.1. The reactions are regiospecific and give one (endo-anti, **109**) of the several possible diastereomeric forms of the resulting 7-phosphanorbornene derivatives (Equation 64).

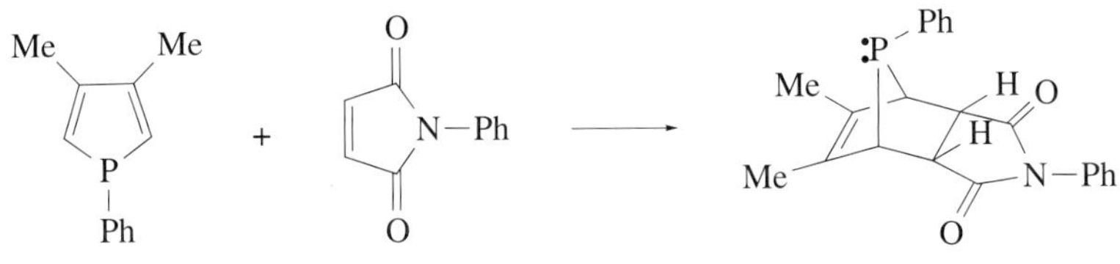

(64)

(**109**)

Diels–Alder reactions can also be performed with the phosphole while held within the coordination sphere of metals. The phosphole is seen to be more reactive in the complex than in free form. This process has been extensively studied by Nelson and co-workers, and reviewed recently ⟨B-94MI01⟩. These reactions are also regiospecific, but give the exo adduct with the P-substituent in the syn position. The metal complex appears to act as a template, holding both the diene and dienophile in position, so that the cycloaddition is intramolecular. In an extensive series of new studies by other workers, this process has been used to prepare optically active complexes of the 7-phosphanorbornene system, from which optically active 7-phosphanorbornenes are released on reaction with KCN or 1,2-bis(diphenylphosphino)ethane. In carrying out the process, a complex is first formed with either the *R* or *S* isomer of the Pd compound **110** and 3,4-dimethyl-1-phenylphosphole. Chlorine on Pd is replaced by perchlorate, and the optically active complex **112** (confirmed by X-ray analysis ⟨96TA45⟩) is then reacted with a dienophile (C=C–Y in Equation 65); several have been used, of structures new to Diels–Alder reactions with the free phosphole. The 7-phosphanorbornene system with exo-syn geometry is formed from: divinyl sulfoxide ⟨95CC1747, 98JCS(D)893⟩, phenyl vinyl sulfide ⟨96JCS(D)4443⟩, diphenylvinylarsine ⟨96OM3640⟩, *N,N*-dimethylacrylamide ⟨96CC591⟩, ethyl vinyl ketone ⟨97CC1987⟩, methyl *trans*-4-oxo-2-pentenoate ⟨98TA2961⟩, 1-methyl-2-vinylpyrrole ⟨98TA423⟩, 2-vinylpyridine ⟨98OM3931⟩, and 2-methylene-3-quinuclidinone ⟨99OM650⟩. When chloro complex **111** is used in the Diels–Alder reaction, the product is the endo-syn isomer ⟨97CC1987, 98OM3931⟩, although the exo-syn isomer is formed in the case of divinyl sulfide. All structures were confirmed by X-ray analysis as well as by extensive NMR studies.

(65)

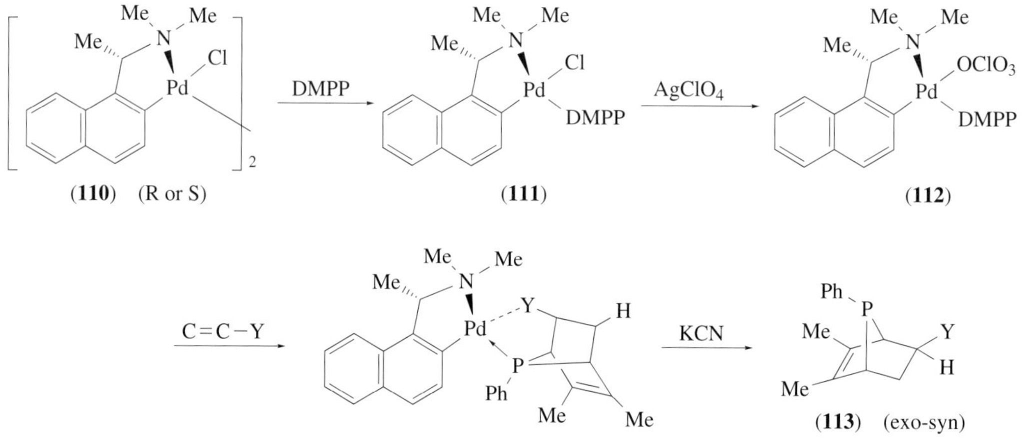

Dienophiles bearing phosphino substituents are also used, but are first complexed with the palladium reagent and then the complex is reacted with the phosphole. Phosphines that have been used include diphenylvinylphosphine and diphenyl-1-propenylphosphine (E and Z) ⟨97IC2138⟩, divinylphenylphosphine ⟨96IC4798⟩, phenyldi[(Z)-prop-1-enyl]phosphine ⟨97CC751⟩, and racemic methylphenylvinylphosphine ⟨97CC2397⟩. The adduct from the latter phosphine is of special interest because the phosphine group in the dienophile is chiral. A technique was developed to obtain the exo-syn phosphanorbornene in individual diasteromeric forms with the exocyclic P in either *R* or *S* configuration, and with the ring P in both having the *R* configuration.

Another example of an intramolecular Diels–Alder reaction in a complex is shown in Equation 66 ⟨99SRI395⟩. Here, however, the reaction was not diastereospecific and a 1 : 1 mixture of adducts (**113** and **114**) was obtained.

(66)

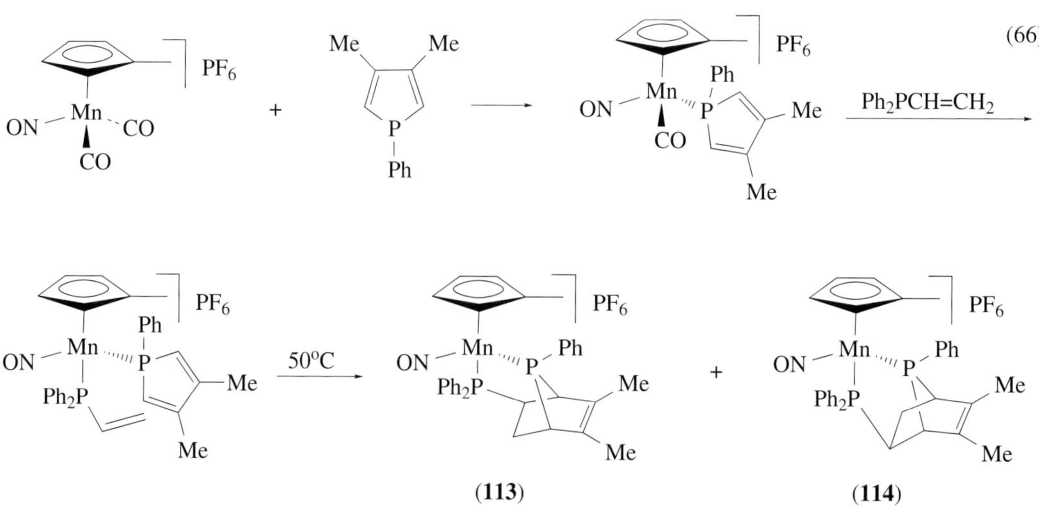

(4) Dimer Formation from a Phosphole-Borane. Phospholes and C.N.4-phosphole derivatives are well known to undergo Diels–Alder dimerization to form adducts containing the 7-phosphanorbornene moiety, as in Equation 67 (a recent report ⟨96ZOR446⟩ of a different dimeric structure arising from the oxide of 3,4-dimethyl-1-phenylphosphole conflicts with the earlier literature ⟨91HAC359⟩ and does not provide adequate experimental justification).

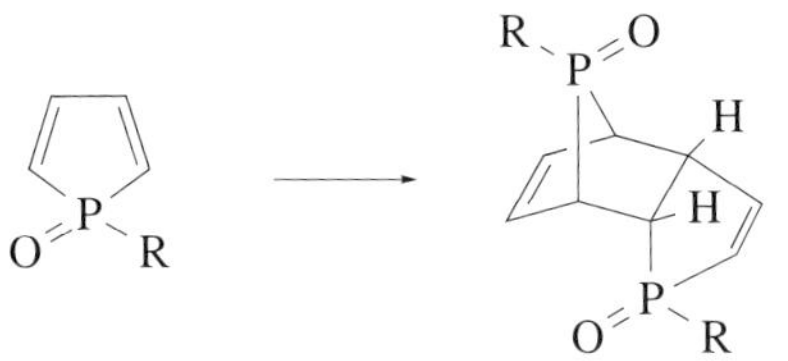

(67)

It has now been found that a quite different dimeric structure containing a 7-membered ring component can be formed during reactions applied to the adduct **115** of 3,4-dimethyl-1-phenylphosphole and borane ⟨98CR(C)53⟩. This adduct was first converted to the anion **116** with sec.-butyllithium at −78°C, and then one equivalent of *N*-bromosuccinimide was added to form the bromomethyl derivative. After overnight standing, a new compound was recovered by silica gel chromatography in 30% yield and found to have the constitution of a dimer of the original adduct. Structure **119** was assigned by NMR and X-ray diffraction analyses. The mechanism proposed for the formation of **119** assumes that the negative site of **116b** is attacked by the intermediate allylic bromo derivative **117** to couple the rings (Equation 68). Nucleophilic attack of hydride (from a borane) leads to the final product (Equation 69).

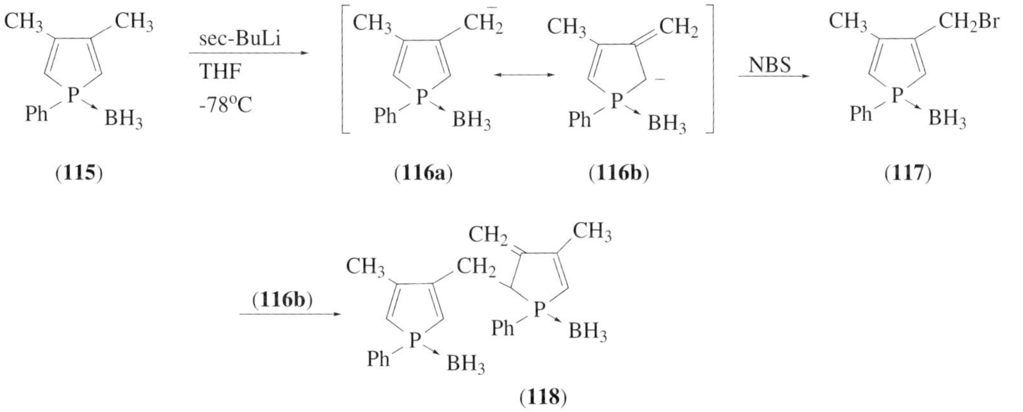

(68)

(**115**) (**116a**) (**116b**) (**117**)

(**118**)

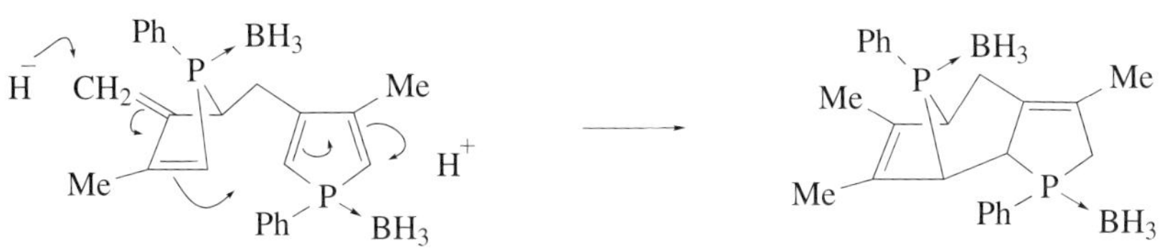

(69)

(**119**)

(5) Electrophilic Substitution of a Ring Proton. The more extensive electron delocalization of phosphole **16a**, seen in its modified bond parameters (Section 4.2.2.3.2), prompted an attempt at electrophilic substitution on the phosphole ring ⟨97JA5095⟩. No previous case of such a reaction has been reported for a free phosphole. The Friedel–Crafts acylation with acetyl chloride-$AlCl_3$ indeed gave a mixture of the three possible mono-substitution products in the ratio 49 : 22 : 22 (Equation 70). The major isomer was shown to have the 2-acetyl structure (**120**), which is consistent with the chemistry of pyrroles.

(6) Electrophilic Substitution on a Benzene Ring of Dibenzophosphole. Electrophilic substitutions at aromatic C–H of 3-coordinate phospholes are generally complicated by oxidative attack on the phosphorus atom. A technique has been developed that avoids this problem in the sulfonation of 5-phenyldibenzophosphole ⟨95AG(E)811⟩, which under customary sulfonation procedures is converted to the phosphole oxide. The procedure calls for the dissolution of orthoboric acid in 96% H_2SO_4, followed by the addition of oleum (65% SO_3). Excess SO_3 is removed *in vacuo*. The result is the

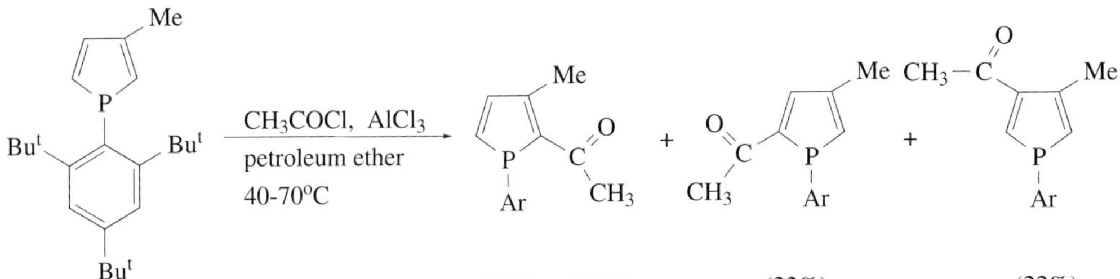

(70)

(**16a**) (**120**) (49%) (22%) (22%)

formation of a superacid $[B(OSO_3H)_4]^-$ which does not have oxidizing ability. To this residue is added the dibenzophosphole. The mixture is heated at 145°C for 15 hr, resulting in a 89% yield of the disulfonated compound **121** (Equation 71) ⟨95AG(E)811⟩. That *meta*-substitution occurs suggests that the species undergoing the reaction is the protonated phosphine.This technique does not appear to have been applied to monocyclic phospholes, but certain of them, such as those with increased aromaticity through large substituents on P, may well respond favorably to these conditions, as suggested by the successful demonstration of Friedel–Crafts acylation as noted in Equation 70 above.

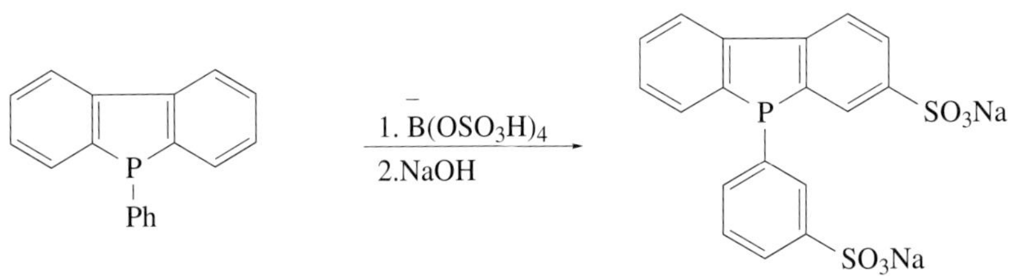

(71)

(**121**)

(7) Reaction of the Tetramer of Dimethyl Acetylenedicarboxylate with 1,2,5-Triphenylphosphole. An unusual product (**123**) was obtained by the reaction of the tetramer of dimethyl acetylenedicarboxylate (**122**) with 1,2,5-triphenylphosphole (Equation 72) ⟨94CJC2428⟩. This compound, a red crystalline solid, was examined by X-ray crystallography and its structure confirmed. The bridgehead phosphorus, which is incorporated in a 2-phospholene ring and also a dihydrophosphinine ring, is in the ylide state.

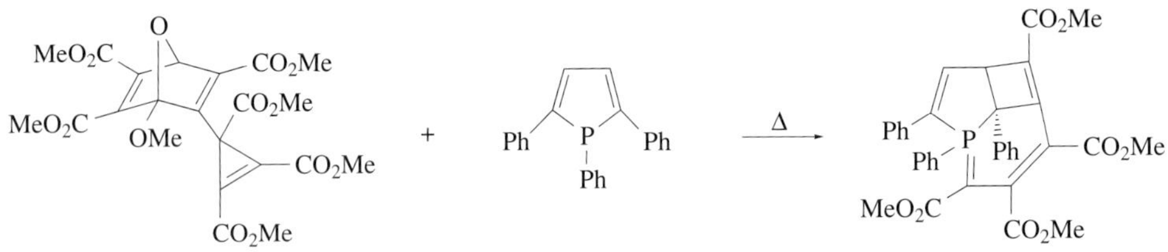

(72)

(**122**) (**123**)

4.2.2.9 NEW METALLIC COMPLEXES OF PHOSPHOLES AND PHOSPHOLIDE IONS

Numerous complexes of phospholes, dibenzophospholes and their derived phospholide ions have been prepared, many in connection with the search for efficient catalysts for homogeneous reactions. The more effective of these will be treated in Chapter 7. Described briefly here will be some of the new complexes formed with phospholes and phospholide ions. There are also numerous complexes formed with the dibenzophosphole framework; this ligand has the properties of a tertiary phosphine and these complexes are more of interest in coordination chemistry than in phosphole chemistry.

(1) 1-Phenylphosphole and 3,4-dimethyl-1-phenylphosphole react with the clusters $[Os_3(CO)_{11}(MeCN)]$ or $[Os_3(CO)_{10}(MeCN)_2]$ by replacement of MeCN. The coordination is at the phosphorus atom ⟨93PS(77)33⟩. The latter phosphole, as well as the biphosphole (3,4-dimethylphospholyl)$_2$, were used in the synthesis of novel square planar complexes of general formula $Pd(R_3P)_4$ ⟨98BCJ2885⟩.

(2) The sterically crowded 1-aryl-3-methylphospholes **16a–c** displace one or both PhCN groups from $PtCl_2(PhCN)_2$, with the most crowded **16a** (with reduced pyramidality) being the less reactive. Both *cis*

and *trans* isomers at Pt were formed. The phosphole ligands lack symmetry from the presence of the 3-methyl group, and the complexes are obtained as mixtures of diastereoisomers ⟨99IC831⟩.

(3) 3,4-Dimethyl-1-phenylphosphole-1-sulfide acts as a 6-electron (η^4–C_4, η^1–S) chelating ligand to Cr or Mo, forming complexes of structure **124** ⟨97PS(123)119⟩.

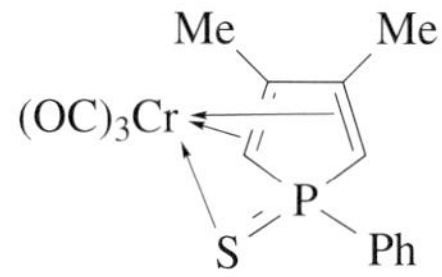

(**124**)

(4) Complexes of 3,4-dimethyl-1-phenylphosphole (DMPP) with Ru of the formula [CpRu(DMPP)$_2$L] PF_6, where L represents a variety of molecular ligands, were found to undergo [2 + 2] photoinitiated (sunlight) dimerization of the phosphole ligands. The dimer is of the same structure as formed thermally from other complexes (Equation 61, Section 4.2.2.8) ⟨93PS(77)268⟩.

(5) The phospholide ion from cleavage of phenyl from 2,5-di-*t*-butyl-1-phenylphosphole has been used to form a complex with Ru and Rh in which it acts as an η^5 ligand through use of the delocalized 6π aromatic system. The large *t*-butyl substituents block the formation of an η^1 complex at the phosphorus lone pair. This leads to the formation of complexes **125** and **126** ⟨94CC1167⟩. The ion has also formed a complex with Mo, having the structure [Mo_2(phospholide)$_2$(CO)$_4$] in which there is a Mo to Mo triple bond ⟨94CC2459⟩.

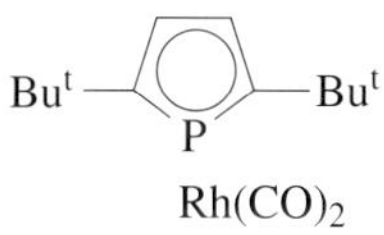

(**125**)

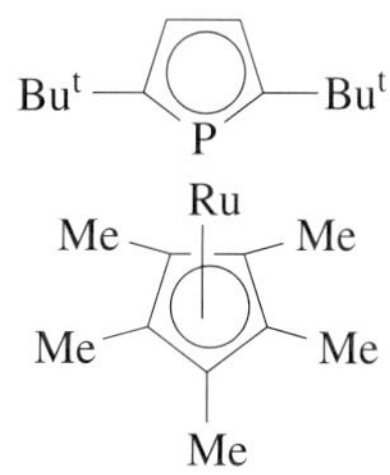

(**126**)

(6) Both potassium 1,2,3,4-tetramethylphospholide (Tmp) and potassium 3,4-dimethyl-1-phenyl phospholide (Dmp) form complexes on reaction with $SmCl_3$ in toluene ⟨95IC1306⟩. The solids were shown by X-ray analysis to have the compositions [(Tmp)$_6$Sm$_2$(KCl)$_2$(toluene)$_3$] in a polymer, and [(η^5-Dmp)$_4${μ-(η^5,η^1)-Dmp}$_2$Sm$_2$], respectively. Samarium complexes were also obtained on reacting a phosphindolide (**127**) with $SmI_2(THF)_2$ and by cleavage of the P–P bond of **128** with Sm in THF ⟨94JOM(464)149⟩.

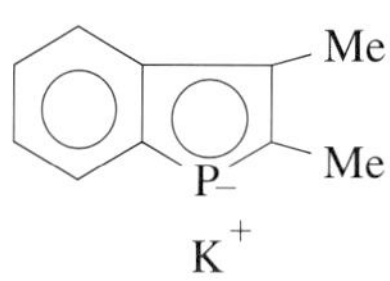

(**127**)

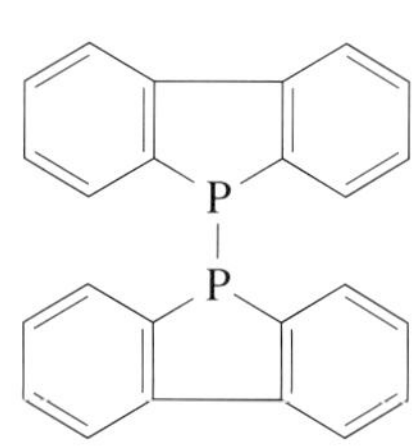

(**128**)

(7) The tetraphosphole macrocycle **35** has been cleaved at one P–P bond by Na in dimethoxyethane to gave the complex **129** in which coordination to the sodium ion is present (by X-ray analysis) ⟨96AG(E)1125⟩. With potassium, both P–P bonds are cleaved to give the potassium ion complex **130**.

(8) A bisphosphonio isophosphindolium cation (cf.**67**, Section 4.2.2.6) with 2-coordinate phosphorus forms the stable gold complex **131** ⟨95CB259⟩. With CuBr, an unusual binuclear complex **132**, proved by X-ray analysis, was formed ⟨95CC1541⟩.

(9) It was shown many years ago ⟨80JA994⟩ that the phospholide moiety of a phosphaferrocene can be formylated; conversion of the aldehyde group to other substituents that bear a second P or N coordinating

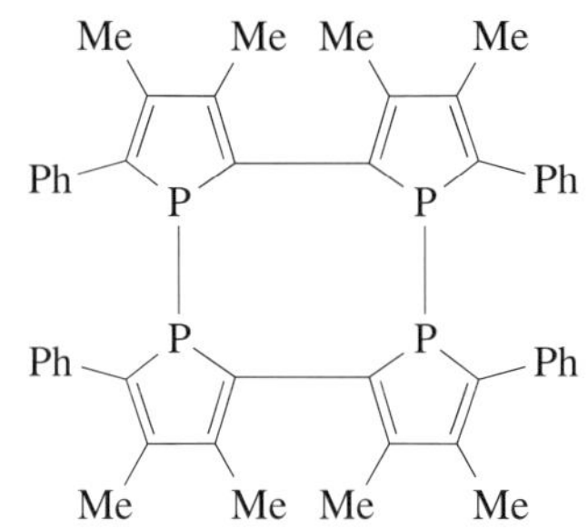

(**35**)

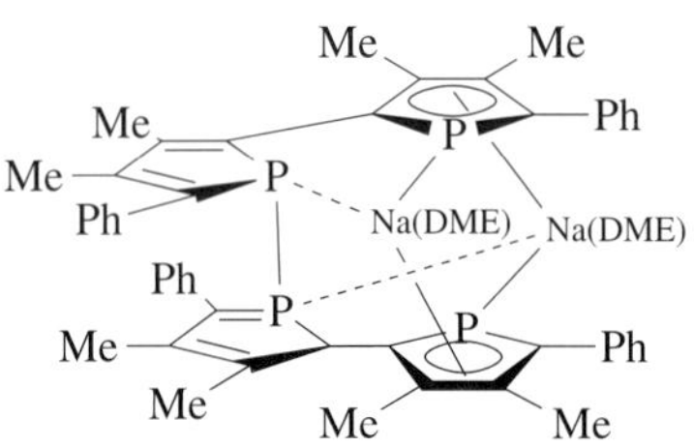

(**129**)

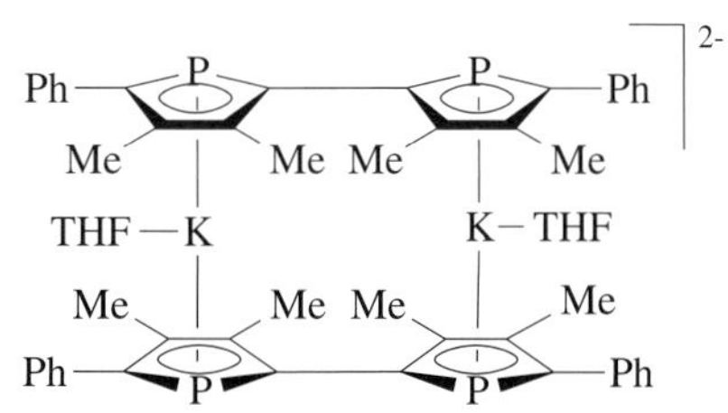

(**130**)

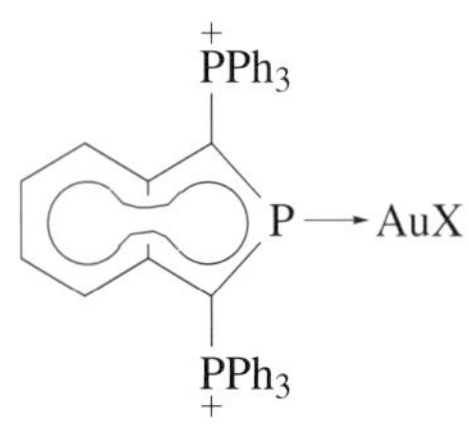

(**131**)

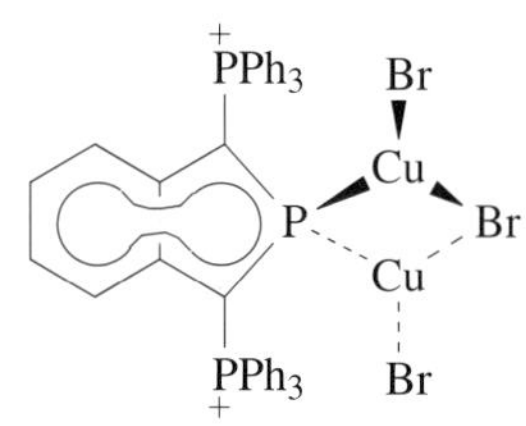

(**132**)

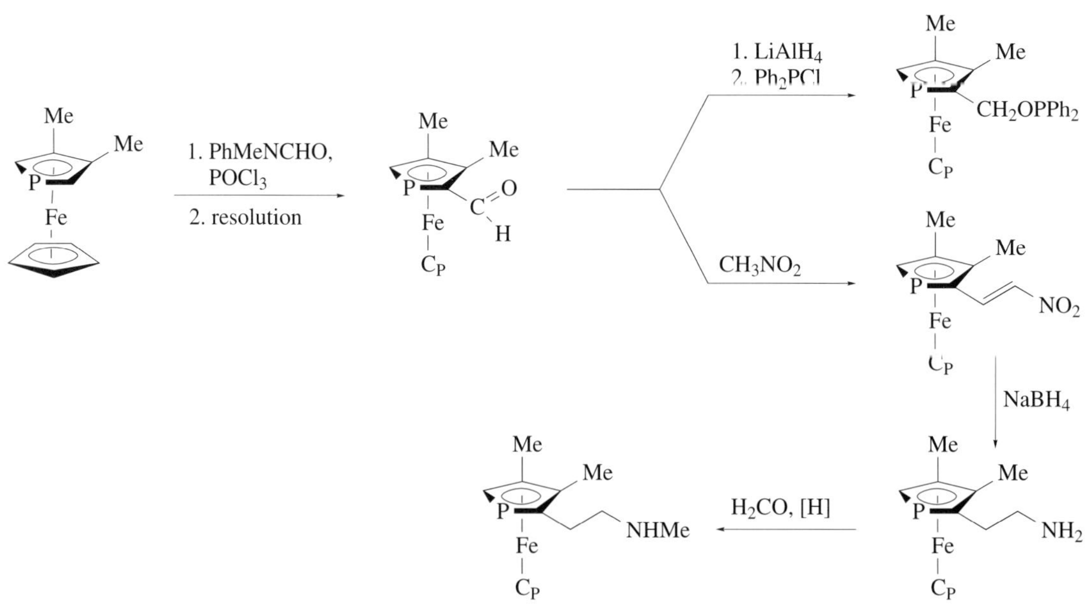

Scheme 5

center gives a potential chelating ligand where one center is planar phosphorus ⟨97OM2862⟩. Some examples of these new derivatives that have been synthesized are given in Scheme 5.

(10) 2-Substituted phosphaferrocenes are chiral molecules, and a resolution has been performed on compound **133** (Equation 73) using a chiral HPLC column (Chiralcel OD) ⟨98JOC4168⟩. The structure of the (+) enantiomer was proved by X-ray crystallography. The OH group of this enantiomer was replaced by Cl, and this in turn by Ph_2P (Equation 73), giving the (−) enantiomer of **134**. This compound functions as a chiral bisphosphine on coordination; a rhodium complex gave good enantioselectivity in olefin hydrogenation experiments.

(133) (134) (73)

(11) 1,2,5-Triphenylphosphole (TPP) formed the complex $Rh(CO)Cl(TPP)_2$ which was used to catalyze the hydroformylation of styrene ⟨95BSF815⟩. Zerovalent palladium complexes of this phosphole have been formed by reacting it with bis(dibenzylideneacetone)Pd(0) ⟨94JOM(481)253⟩.

4.2.2.10 REFERENCES

53LA(580)44 G. Wittig and G. Geissler; *Liebigs Ann. Chem.*, 1953, **580**, 44.
59CI(L)1250 E.H. Braye and W. Hübel; *Chem. Ind. (London)* 1959, 1250.
59JA3163 F.C. Leavitt, T.A. Manuel and F. Johnson; *J. Am. Chem. Soc.* 1959, **81**, 3163.
62CI(L)359 I.G.M. Campbell, R.C. Cookson and M.B. Hocking; *Chem. Ind. (London)*, 1962, 359.
66AG(E)846 G. Märkl; *Angew. Chem., Int. Ed. Engl.*, 1966, **5**, 846.
67JA5984 L.D. Quin and J.G. Bryson; *J. Am. Chem. Soc.*, 1967, **89**, 5984.
69JA3308 L.D. Quin, J.G. Bryson and C.G. Moreland; *J. Am. Chem. Soc.*, 1969, **91**, 3308.
70JA5779 P. Coggon, J.F. Engel, A.T. McPhail and L.D. Quin; *J. Am. Chem. Soc.*, 1970, **92**, 5779.
71JA6205 W. Egan, R. Tang, G. Zon and K. Mislow; *J. Am. Chem. Soc.*, 1971, **93**, 6205.
73JCS(D)1888 P. Coggon and A.T. McPhail; *J. Chem. Soc., Dalton Trans.*, 1973, 1888.
73JOC1858 L.D. Quin, S.G. Borleske and J.F. Engel; *J. Org. Chem.*, 1973, **38**, 1858.
80JA994 G. de Lauzon, B. Deschamps, J. Fischer, F. Mathey and A. Mitschler; *J. Am. Chem. Soc.*, 1980, **102**, 994.
80MI01 F. Mathey; *Top. Phosphorus Chem.*, 1980, **10**, 1.
81JA4595 F. Mathey, F. Mercier, C. Charrier, J. Fischer, and A. Mitschler; *J. Am. Chem. Soc.*, 1981, **103**, 4595.
82TL511 G. de Lauzon, C. Charrier, H. Bonnard and F. Mathey; *Tetrahedron Lett.*, 1982, **23**, 511.
84JA425 F. Mercier, F. Mathey, J. Fischer and J.H. Nelson; *J. Am. Chem. Soc.*, 1984, **106**, 425.
85IC4141 F. Mercier, F. Mathey, J. Fischer and J.H. Nelson; *Inorg. Chem.*, 1985, **24**, 4141.
87JCS(P1)871 J. Cornforth; *J. Chem. Soc., Perkin Trans. 1*, 1987, 1897.
88CRV429 F. Mathey; *Chem. Rev.*, 1988, **88**, 429.
90JOC2494 E. Deschamps and F. Mathey; *J. Org. Chem.*, 1990, **55**, 2494.
91AG(E)308 A. Schmidpeter and M. Thiele; *Angew. Chem., Int. Ed. Engl.*, 1991, **30**, 308.
91HAC359 L.D. Quin and X.-P. Wu; *Heteroatom Chem.*, 1991, **2**, 359.
92BSF1 M.-O. Bevierre, F. Mercier, L. Ricard and F. Mathey; *Bull. Soc. Chim. Fr.*, 1992, **129**, 1.
92BSF486 E. Deschamps and F. Mathey; *Bull. Soc. Chim. Fr.*, 1992, **129**, 486.
93BSF843 F. Laporte, F. Mercier, L. Ricard and F. Mathey; *Bull. Soc. Chim. Fr.*, 1993, **130**, 843.
93CC1124 A. Dore, D. Fabbri, S. Gladiali and O. De Lucchi; *J. Chem. Soc., Chem. Commun.*, 1993, 1124.
93JA1051 E. Magnusson; *J. Am. Chem. Soc.*, 1993, **115**, 1051.
93JOC5414 S.M. Bachrach; *J. Org. Chem.*, 1993, **58**, 5414.
93JOM(445)71 A.A. Watson, A.C. Willis and S.B. Wild; *J. Organomet. Chem.*, 1993, **445**, 71.
93JOM(462)103 F. Mercier and F. Mathey; *J. Organomet. Chem.*, 1993, **462**, 103.
93PS(77)33 A.J. Arce, Y. de Sanctis, A.J. Deeming, J. Mazur and N.I. Powell; *Phosphorus Sulfur Silicon*, 1993, **77**, 33.
93PS(77)268 H.-L. Ji, J.H. Nelson and J. Fischer; *Phosphorus Sulfur Silicon*, 1993, **77**, 268.
93ZN(B)1681 H. Heidel, G. Huttner and G. Helmchen; *Z. Naturforsch.,Teil B*, 1993, **48**, 1681.
94AG(E)353 E. Niecke, M. Nieger and P. Wenderoth; *Angew. Chem., Int. Ed. Engl.*, 1994, **33**, 353.
94AG(E)1158 E. Deschamps, L. Ricard and F. Mathey; *Angew. Chem., Int. Ed.Engl.*, 1994, **33**, 1158.
94AX(C)769 K. Tani, T. Yamagata and H. Tashiro; *Acta Crystallogr., Part C*, 1994, **50**, 769.

94CB1837 T. Seitz, A. Muth and G. Huttner; *Chem. Ber.*, 1994, **127**, 1837.
94CC1167 D. Carmichael, L. Ricard and F. Mathey; *J. Chem. Soc., Chem. Commun.*, 1994, 1167.
94CC2459 D. Carmichael, L. Ricard and F. Mathey; *J. Chem. Soc., Chem. Commun.*, 1994, 2459.
94CCR(137)1 F. Mathey; *Coord. Chem. Rev.*, 1994, **137**, 1.
94CJC2428 M.B. Hocking and F.W. van der Voort Maarschalk; *Can. J. Chem.*, 1994, **72**, 2428.
94CPL(224)213 M. Karelson and M.C. Zerner; *Chem. Phys. Lett.*, 1994, **224**, 213.
94IC109 W.L. Wilson, J.A. Rahn, N.W. Alcock, J. Fischer, J.H. Frederick and J.H. Nelson; *Inorg. Chem.*, 1994, **33**, 109.
94JA1880 P.J. Fagan, W.A. Nugent and J.C. Calabrese; *J. Am. Chem. Soc.*, 1994, **116**, 1880.
94JA9638 D.B. Chesnut and L.D. Quin; *J. Am. Chem. Soc.*, 1994, **116**, 9638.
94JCR(M)887 C. Combes, H.-J. Cristau, W.-S. Li, M. McPartlin, F. Plénat and I.J. Scowen; *J. Chem. Res.(M)*, 1994, 887.
94JOC3394 S.M. Bachrach and L. Perriott; *J. Org. Chem.*, 1994, **59**, 3394.
94JOC5027 S.M. Bachrach; *J. Org. Chem.*, 1994, **59**, 5027.
94JOC6363 S. Gladiali, A. Dore, D. Fabbri, O. De Lucchi and G. Valle; *J. Org. Chem.*, 1994, **59**, 6363.
94JOM(464)149 F. Nief and L. Ricard; *J. Organomet. Chem.*, 1994, **464**, 149.
94JOM(468)149 A. Muth, O. Walter, G. Huttner, A. Asam, L. Zsolnai and C. Ammerich; *J. Organomet. Chem.*, 1994, **468**, 149.
94JOM(469)229 K. Tani, H. Tashiro, M. Yoshida and T. Yamagata; *J. Organomet. Chem.*, 1994, **469**, 229.
94JOM(475)25 F. Mathey; *J. Organomet. Chem.*, 1994, **475**, 25.
94JOM(475)307 S. Gladiali, D. Fabbri, G. Banditelli, M. Manassero and M. Sansoni; *J. Organomet. Chem.*, 1994, **475**, 307.
94JOM(481)253 W.A. Herrmann, W.R. Thiel, T. Priermeir and E. Herdtweck; *J. Organomet. Chem.*, 1994, **481**, 253.
94OM4732 L.S. Sunderlin, D. Panu, D.B. Puranik, A.J. Ashe, III, and R.B. Squires; *Organometallics*, 1994, **13**, 4732.
94PS(87)139 F. Mathey; *Phosphorus Sulfur Silicon*, 1994, **87**, 139.
94SC1271 D. Fabbri, S. Gladiali and O. De Lucchi; *Synth. Commun.*, 1994, **24**, 1271.
94ZOB1566 V.N. Chistokletov, E.D. Maiorova and A.Yu. Platonov; *Zhur. Obshch. Khim.*, 1994, **64**, 1566.
95AG(E)337 P.v.R. Schleyer, P.K. Freeman, H. Jiao and B. Goldfuss; *Angew. Chem., Int. Ed. Engl.*, 1995, **34**, 337.
95AG(E)811 W.A. Herrmann, G.P. Albanese, R.B. Manetsberger, P. Lappe and H. Bahrmann; *Angew. Chem., Int. Ed. Engl.*, 1995, **34**, 811.
95BSF280 D. Gudat, V. Bajorat and M. Nieger; *Bull. Soc. Chim. Fr.*, 1995, **132**, 280.
95BSF649 S. Atlan, F. Nief and L. Ricard; *Bull. Soc. Chim. Fr.*, 1995, **132**, 649.
95BSF815 C. Bergounhou, D. Neibecker and R. Réau; *Bull. Soc. Chim. Fr.*, 1995, **132**, 815.
95BSF910 K. Waschbusch, P. Le Floch and F. Mathey; *Bull. Soc. Chim. Fr.*, 1995, **132**, 910.
95CB259 D. Gudat, M. Nieger and M. Schrott; *Chem. Ber.*, 1995, **128**, 259.
95CB293 H. Heidel, J. Scherer, A. Asam, G. Huttner, O. Walter and L. Zsolnai; *Chem. Ber.*, 1995, **128**, 293.
95CC1541 D. Gudat, M. Schrott and M. Nieger; *J. Chem. Soc., Chem. Commun.*, 1995, 1541.
95CC1561 E. Deschamps, L. Ricard and F. Mathey; *J. Chem. Soc., Chem Commun.*, 1995, 1561.
95CC1747 S.-Y. Siah, P.-H. Leung and K.F. Mok; *J. Chem. Soc., Chem. Commun.*, 1995, 1747.
95IC1306 H.-J. Gosink, F. Nief, L.Ricard and F. Mathey; *Inorg. Chem.*, 1995, **34**, 1306.
95JOM(491)91 S. Gladiali, D. Fabbri and L. Kollàr; *J. Organomet. Chem.*, 1995, **491**, 91.
95JPC586 L. Nyulászi; *J. Phys. Chem.*, 1995, **99**, 586.
95JST(331)109 A. Hinchliffe and H.J. Soscún M.; *J. Mol. Struct.*, 1995, **331**,109.
95JST(338)51 D.J. Berger, P.P. Gaspar and J.F. Liebman; *J. Mol. Struct.*, 1995, **338**, 51.
95TL6655 G. Märkl, R. Hennig and H. Nöth; *Tetrahedron Lett.*, 1995, **36**, 6655.
95ZN(B)729 H. Heidel, G. Huttner and L. Zsolnai; *Z. Naturforsch., Teil B*, 1995, **50**, 729.
95ZN(B)1045 T. Seitz, A. Muth and G. Huttner; *Z. Naturforsch., Teil B*, 1995, **50**, 1045.
95ZN(B)1287 T. Seitz, A. Asam, G. Huttner, O. Walter and L. Zsolnai; *Z. Naturforsch., Teil B.*, 1995, **50**, 1287.
96AG(E)1125 F. Paul, D. Carmichael, L. Ricard and F. Mathey; *Angew. Chem., Int. Ed. Engl.*, 1996, **35**, 1125.
96BSF33 S. Holand, F. Gandolfo, L. Ricard and F. Mathey; *Bull. Soc. Chim. Fr.*, 1996, **133**, 33.
96BSF541 B. Deschamps and F. Mathey; *Bull. Soc. Chim. Fr.*, 1996, **133**, 541.
96CB697 J. Scherer, G. Huttner and M. Büchner; *Chem. Ber.*, 1996, **129**, 697.
96CB1083 G. Jochem, H. Nöth and A. Schmidpeter; *Chem. Ber.*, 1996, **129**, 1083.
96CC591 P.-K. Leung, S.-K. Loh, K.F. Mok, A.J.P. White and D.J. Williams; *Chem. Commun.*, 1996, 591.
96CC2287 O. Tissot, M. Gouygou, J.-C. Daran and G.G.A. Balavoine; *Chem. Commun.*, 1996, 2287.
96CHECII(2)757 L.D. Quin; *Comp. Heterocycl. Chem., 2nd edn.*, 1996, **2**, 757.
96CJC839 S.M. Bachrach and L.M. Perriott; *Can. J. Chem.*, 1996, **74**, 839.
96IC1486 W.L. Wilson, J. Fischer, R.E. Wasylishen, K. Eichele, V.J. Catalano, J.H. Frederick and J.H. Nelson; *Inorg. Chem.*, 1996, **35**, 1486.
96IC3904 K. Eichele, R.E. Wasylishen, J.M. Kessler, L. Solujic and J.H. Nelson; *Inorg. Chem.*, 1996, **35**, 3904.
96IC4798 S. Selvaratnam, K.F. Mok, P.-H. Leung, A.J.P. White and D.J. Williams; *Inorg. Chem.*, 1996, **35**, 4798.
96JA6317 P.v.R. Schleyer, C. Maerker, A. Dransfeld, H. Jiao and N.J.R. van Eikema Hommes; *J. Am. Chem. Soc.*, 1996, **118**, 6317.
96JCS(D)4443 P.-H. Leung, S.-K. Loh, K.F. Mok, A.J.P. White and D.J. Williams; *J. Chem. Soc., Dalton Trans.*, 1996, 4443.
96JCS(P1)2889 J. Cornforth; *J. Chem. Soc., Perkin Trans. 1*, 1996, 2889.
96JOC3531 S. Lelièvre, F. Mercier and F. Mathey; *J. Org. Chem.*, 1996, **61**, 3531.
96JOC7801 L.D. Quin, G. Keglevich, A.S. Ionkin, R. Kalgutkar and G. Szalontai; *J. Org. Chem.*, 1996, **61**, 7801.
96JOC7808 L. Nyulászi, G. Keglevich and L.D. Quin; *J. Org. Chem.*, 1996, **61**, 7808.

96JOM(507)257	O. Desponds and M. Schlosser; *J. Organomet. Chem.*, 1996, **507**, 257.
96JPC6194	L. Nyulászi; *J. Phys. Chem.*, 1996, **100**, 6194.
96JPC13447	M.N. Glukhovtsev, A. Dransfeld and P.v.R. Schleyer; *J. Phys. Chem.*, 1996, **100**, 13447.
96OM1756	B. Goldfuss, P.v.R. Schleyer and F. Hampel; *Organometallics*, 1996, **15**, 1756.
96OM3640	B.-H. Aw, P.-H. Leung, A.J.P. White and D.J. Williams; *Organometallics*, 1996, **15**, 3640.
96PS(111)814	R.A. Aitken, A.H. Cowley, F.P. Gabbaï and W. Masamba; *Phosphorus Sulfur Silicon*, 1996, **111**, 814.
96PS(115)227	S.E. Johnson and C.B. Knobler; *Phosphorus Sulfur Silicon*, 1996, **115**, 227.
96PS(118)309	H.T. Teunissen, C.B. Hansen and F. Bickelhaupt; *Phosphorus Sulfur Silicon*, 1996, **118**, 309.
96T7547	M. Peer, J.C. de Jong, M. Kiefer, T. Langer, H. Rieck, H. Scheli, P. Sennhen, J. Sprinz, H. Steinhagen, B. Wiese and G. Helmchen; *Tetrahedron*, 1996, **52**, 7547.
96TA45	S.-K. Loh, K.F. Mok, P.-H. Leung, A.J.P. White and D.J. Williams; *Tetrahedron Asymmetry*, 1996, **7**, 45.
96TL5347	M. Cereghetti, W. Arnold, E.A. Broger and A. Rageot; *Tetrahedron Lett.*, 1996, **37**, 5347.
96ZOR446	J. Leis, K. Pihlaja and M. Karelson; *Zhur. Org. Khim.*, 1996, **32**, 446.
97AG(E)98	S. Holand, M. Jeanjean and F. Mathey; *Angew. Chem., Int. Ed. Engl.*, 1997, **36**, 98.
97CB543	S. Gladiali and D. Fabbri; *Chem. Ber. Rec.*, 1997, **130**, 543.
97CC279	Y. Miquel, A. Igau, B. Donnadieu, J.-P. Majoral, L. Dupuis, N. Pirio and P. Meunier; *Chem. Commun.*, 1997, 279.
97CC751	P.-H. Leung, S. Selvaratnam, C.R. Cheng, K.F. Mok, N.H. Rees and W. McFarlane; *Chem. Commun.*, 1997, 751.
97CC1987	P.-H. Leung, S.-K. Loh, J.J. Vittal, A.J.P. White and D.J. Williams; *Chem. Commun.*, 1997, 1987.
97CC2397	A. Liu, K.F. Mok and P.-H. Leung; *Chem. Commun.*, 1997, 2397.
97CR(B)701	F. Mathey and F. Mercier; *C. R. Hebd. Seances Acad. Sci., Ser. IIb*, 1997, **324**, 701.
97IC2138	B.-H. Aw, T.S.A. Hor, S. Selvaratnam, K.F. Mok, A.J.P. White, D.J. Williams, N.H. Rees, W. McFarlane and P.-H. Leung; *Inorg. Chem.*, 1997, **36**, 2138.
97JA5095	G. Keglevich, Z. Böcskei, G.M. Keserü, K. Ujszászy and L.D. Quin; *J. Am. Chem. Soc.*, 1997, **119**, 5095.
97JA6575	I.D.L. Albert, T.J. Marks and M.A. Ratner; *J. Am. Chem. Soc.*, 1997, **119**, 6575.
97JA12560	K. Maitra, V.J. Catalano and J.H. Nelson; *J. Am. Chem. Soc.*, 1997, **119**, 12560.
97JCO198	U. Salzner, S.M. Bachrach and D.C. Mulhearn; *J. Comput. Chem.*, 1997, **18**, 198.
97JCS(P2)15	C. Combes-Chamalet, H.-J. Cristau, M. McPartlin, F. Plénat, I.J. Scowen and T.M. Woodroffe; *J. Chem. Soc., Perkin Trans. 2*, 1997, 15.
97JHC1387	B.S. Jursic; *J. Heterocycl. Chem.*, 1997, **34**, 1387.
97JMR(124)366	G. Wu, B. Sun, R.E. Wasylishen and R.G. Griffin; *J. Magn. Reson.*, 1997, **124**, 366.
97JOM(527)305	P.B. Hitchcock, G.A. Lawless and I. Marziano; *J. Organomet. Chem.*, 1997, **527**, 305.
97JOM(529)15	U. Salzner and S.M. Bachrach; *J. Organomet. Chem.*, 1997, **529**, 15.
97JOM(529)75	A.H. Cowley, S.M. Dennis, S. Kamepalli, C.J. Carrano and M.R. Bond; *J. Organomet. Chem.*, 1997, **529**, 75.
97JOM(529)197	D. Schmidt, S. Krill, B. Wang, F.R. Fronczek and K. Lammertsma; *J. Organomet. Chem.*, 1997, **529**, 197.
97JOM(529)395	H.-L. Ji, J.H. Nelson, A. DeCian, J. Fischer, B. Li, C. Wang, B. McCarty, Y. Aoki, J.W. Kenney, III, L. Solujic and E.B. Milosavljevic'; *J. Organomet. Chem.*, 1997, **529**, 395.
97JOM(532)109	G. Keglevich, L.D. Quin, Z. Böcskei, G.M. Keserü, R. Kalgutkar and P.M. Lahti; *J. Organomet. Chem.*, 1997, **532**, 109.
97JOM(539)67	J. Scherer, G. Huttner and H. Heidel; *J. Organomet. Chem.*, 1997, **539**, 67.
97JOM(540)15	I. Tóth, C.J. Elsevier, J.G. de Vries, J. Bakos, W.J.J. Smeets and A.L. Spek; *J. Organomet. Chem.*, 1997, **540**, 15.
97JOM(548)17	B. Deschamps, L. Ricard and F. Mathey; *J. Organomet. Chem.*, 1997, **548**, 17.
97MM5566	S.S.H. Mao and T.D. Tilley; *Macromolecules*, 1997, **30**, 5566.
97OM2862	C. Ganter, L. Brassat, C. Glinsböckel and B. Ganter; *Organometallics*, 1997, **16**, 2862.
97PS(123)119	B. Deschamps, L. Ricard and F. Mathey; *Phosphorus Sulfur Silicon*, 1997, **123**, 119.
97TA3775	H. Doucet and J.M. Brown; *Tetrahedron Asymmetry*, 1997, **8**, 3775.
98AX(C)676	O. Tissot, M. Gouygou, J.-C. Daran and G. Balavoine; *Acta Crystallogr., Part C*, 1998, **54**, 676.
98BCJ2885	T. Kojima, K. Saeki, K. Ono and Y. Matsuda; *Bull. Chem. Soc. Jpn.*, 1997, **71**, 2885.
98CR(C)53	N.H. Tran Huy, L. Ricard and F. Mathey; *C. R. Hebd. Seances Acad. Sci., Ser. IIc*, 1998, **1**, 53.
98CR(C)715	E. Deschamps and F. Mathey; *C. R. Hebd. Seances Acad. Sci., Ser.IIc*, 1998, **1**, 715.
98HAC9	E.M. Hanawalt, K.M. Doxsee, G.S. Shen, T.J.R. Weakley, C.B. Knobler and H. Hope; *Heteroatom Chem.*, 1998, **9**, 9.
98HCA764	A. Aeby, F. Bangerter and G. Consiglio; *Helv. Chim. Acta*, 1998, **81**, 764.
98IC4413	A. Dransfeld, L. Nyulaszi and P.v.R. Schleyer; *Inorg. Chem.*, 1998, **37**, 4413.
98JCS(D)893	P.-H. Leung, S.-Y. Siah, A.J.P. White and D.J. Williams; *J. Chem. Soc., Dalton Trans.*, 1998, 893.
98JOC4168	S. Qiao and G.C. Fu; *J. Org. Chem.*, 1998, **63**, 4168.
98JOM(553)433	B. Antelmann, G. Huttner and U. Winterhalter; *J. Organomet. Chem.*, 1998, **553**, 433.
98JOM(557)117	F. Mathey, F. Mercier, F. Robin and L. Ricard; *J. Organomet. Chem.*, 1998, **557**, 117.
98JOM(560)257	W.A. Schenk, M. Stubbe and M. Hagel; *J. Organomet. Chem.*, 1998, **560**, 257.
98JOM(566)29	L. Nyulászi, L. Soós and G. Keglevich; *J. Organomet. Chem.*, 1998, **566**, 29.
98JPC(A)9912	R. Balawender, L. Komorowski, F. De Proft and P. Geerlings; *J. Phys. Chem. A*, 1998, **102**, 9912.
98NJC651	L. Nyulászi, U. Bergsträsser, M. Regitz and P.v.R. Schleyer; *New J. Chem.*, 1998, 651.
98OM2996	S. Holand, N. Maigrot, C. Charrier and F. Mathey; *Organometallics*, 1998, **17**, 2996.
98OM3931	G. He, S.-K. Loh, J.J. Vittal, K.F. Mok and P.-H. Leung; *Organometallics*, 1998, **17**, 3931.

98PS(142)117	S. Bousquet, J.-J. Brunet, T. Courcet and D. Neibecker; *Phosphorus Sulfur Silicon*, 1998, **142**, 117.
98S45	H. Brunner and M. Janura; *Synthesis*, 1998, 45.
98SM(96)177	U. Salzner, J.B. Lagowski, P.G. Pickup and R.A. Poirier; *Synth. Met.*, 1998, **96**, 177.
98TA423	S.-K. Loh, J.J. Vittal and P.-H. Leung; *Tetrahedron Asymmetry*, 1998, **9**, 423.
98TA2961	P.-H. Leung, H. Lang, A.J.P. White and D.J. Williams; *Tetrahedron Asymmetry*, 1998, **9**, 2961.
99CC345	C. Hay, D. Le Vilain, V. Deborde, L. Toupet and R. Réau; *Chem. Commun.*, 1999, 345.
99EJICI405	J. Karas, G. Huttner, K. Heinze, P. Rutsch and L. Zsolnai; *Eur. J. Inorg. Chem.*, 1999, 405.
99IC831	Z. Csók, G. Keglevich, G. Petöcz and L. Kollár; *Inorg. Chem.*, 1999, **38**, 831.
99OM650	Y. Song, J.J. Vittal, S.-H. Chan and P.-H. Leung; *Organometallics*, 1999, **18**, 650.
99SRI395	K.D. Redwine, V.J. Catalano and J.H. Nelson; *Synth. React. Inorg. Metal-Org. Chem.*, 1999, **29**, 395.
B-70MI01	F.G. Mann, *The Heterocyclic Derivatives of Phosphorus, Arsenic, Antimony, and Bismuth*, Wiley-Interscience, New York, 2nd. ed., 1970, ch. 1.
B-79MI01	A.N. Hughes; in *New Trends in Heterocyclic Chemistry*, eds. R.B. Mitra, N.R. Ayyangar, V.N. Gogte, R.M. Acheson and N. Cromwell, Elsevier, Amsterdam, 1979, p. 216.
B-81MI01	L.D. Quin; *The Heterocyclic Chemistry of Phosphorus*, Wiley-Interscience, New York, 1981, ch. 2.
B-85MI01	J. March, *Advanced Organic Chemistry*, 3rd ed., John Wiley and Sons, Inc., New York, 1985, p. 19.
B-90MI01	L.D. Quin and A.N. Hughes; in *The Chemistry of Organophosphorus Compounds*, ed. F.R. Hartley, Wiley, Chichester, 1990, vol. 1, ch. 10.
B-92MI01	A.N. Hughes, in *Handbook of Organophosphorus Chemistry*, ed. R. Engel, Marcel Dekker, New York, 1992, ch. 10.
B-94MI01	J.H. Nelson, in *Phosphorus-31 NMR Spectral Properties in Compound Characterization and Structural Analysis*, ed. L.D. Quin and J.G. Verkade, VCH Publishers, Inc., New York, 1994, ch. 16.
B-98MI01	K.B. Dillon, F. Mathey and J.F. Nixon, *Phosphorus: The Carbon Copy*, Wiley, Chichester, 1998, ch. 8.

4.3
Heterophospholes

ALFRED SCHMIDPETER
Ludwig-Maximilians-Universität, München, Germany

4.3.1 SCOPE, CONTENTS AND REVIEWS

Heterophospholes are five-membered 6π-systems. To allow full cyclic conjugation their phosphorus ring member as a rule must be two-coordinate and is then contributing one π-electron. The ring, if neutral, must hence in addition to the phosphorus atom contain one member such as NR, O, S (or Se, Te) contributing two π-electrons. Furthermore, the ring may contain one or more two-coordinate nitrogen atoms. Accordingly, two aza-, oxa-, thiaphospholes, six diaza-, oxaza-, thiazaphospholes, six triaza-, oxadiaza-, thiadiazaphospholes, and two tetraza-, oxatriaza-, thiatriazaphospholes can be anticipated, all in all 48 ring systems ⟨B-89MI307, B-90MI258, 90PS(49/50)349⟩. The six ring systems with just one heteroatom in addition to phosphorus and adjacent or non-adjacent to it, the 1,2- and 1,3-azaphospholes, the 1,2- and 1,3-oxaphospholes and the 1,2- and 1,3-thiaphospholes are covered in Sections 4.3.2 and 4.3.3. The remaining 42 systems with two and more heteroatoms besides phosphorus — as far as they are known — are the subject of Sections 4.3.4 and 4.3.5. Of the azaphospholes, only those isomers with a three-coordinate nitrogen atom are included, i.e. of 1,2-azaphospholes only the 1*H*-isomers and not the 2*H*-isomers. In the latter the phosphorus atom is three-coordinate and due to its inherent pyramidal preference ⟨96JPC6194, 98IC4413⟩ will not allow aromaticity.

A heterophosphole may also contain a second two-coordinate phosphorus atom (or a two-coordinate arsenic or antimony atom) in the ring. The resulting aza- and thiadiphospholes, diazadiphospholes, azaphosphaarsoles and azaphosphastiboles are covered in Section 4.3.6. If phosphorus atoms are the only heteroatoms (possibly together with an arsenic or antimony atom) the five-membered ring as a rule must be negatively charged to establish a 6π-system. 1,2- and 1,3-Diphospholides are covered in Sections 4.3.2 and 4.3.3, respectively, together with the other ring systems with two adjacent and non-adjacent heteroatoms mentioned above. In phospholes that contain more than one two-coordinate phosphorus atoms in the ring, the three-coordinate phosphorus atom attains planarity more easily ⟨98IC4413⟩ and consequently even neutral oligophospholes may be aromatic. Oligophospholes and oligophospholides with three, four and five phosphorus atoms (or in part arsenic or antimony atoms) are the subject of Section 4.3.7.

For many heterophospholes benzo derivatives and other *CC*-fused bicyclic systems are possible. With a nitrogen atom in the bridgehead position azaphospholes also can *CN*-anellate to heterocycles of pyridine or azole type. Even the *CP*-anellation of heterophospholes seems possible as the three-coordinate phosphorus in a bridgehead position also reaches planarity more easily than usual ⟨98NJC651⟩.

A great variety of systems is thus feasible and although heterophospholes are only a recent branch of heterocyclic chemistry, quite a number of them has been synthesized during the last third of the century. The first examples, i.e. 2*H*-1,2,3-diazaphospholes, were found not earlier than 1967 ⟨67CHE601⟩ and remained the only ones for seven more years! The first fully unsaturated phospholes with a non-adjacent second heteroatom were 1,3-benzazaphospholes. They were first described in 1978 ⟨78TL441⟩ and are consequently already mentioned in the first edition of *Comprehensive Heterocyclic Chemistry* ⟨84CHEC-I(1)523⟩, together with 1,3-benzothiaphospholes and 1,3-benzodiphospholes ⟨80TL3483⟩. 1,3-Benzazaarsoles were also first studied at that time ⟨78JOM(154)1⟩.

An early review ⟨78S557⟩ covers heterocycles containing a PNN-linkage and includes 1,2,3 diazaphospholes and 1,2,3,4-triazaphospholes. Azaphospholes and, more general, heterophospholes were reviewed in 1989 ⟨B-89MI307⟩ and 1990 ⟨B-90MI258⟩, anellated heterophospholes in 1994 ⟨94T7675⟩. Another review gives special emphasis to 1,3-heterophospholes ⟨94MI307⟩. Azaphospholes are also contained in a review on cyclic mono- and diphosphazenes ⟨90UK931⟩.

Oligophospholide anions are found in a review on phosphanes ⟨93CRV1623⟩, in reviews of their own ⟨94CCR(137)1, 94JOM(475)25⟩ and in a book comparing phosphorus to carbon ⟨B-98MI1⟩. Syntheses of 1,3-diphospholide anions are mentioned in a review on the chemistry of 1,2-dihydro-1,2-diphosphetes ⟨92CIC61⟩.

Synthetic routes to heterophospholes have been surveyed ⟨86PS(28)71⟩ and also the use of α-aminocarbonyl compounds for the synthesis of 1,3,2-oxazaphospholes ⟨92RHA25⟩ and the use of 2-substituted cycloiminium salts for the synthesis of anellated azaphospholes ⟨95S361⟩. Reviews on phosphaalkynes ⟨90CRV191, B-90MI58, B-90MI295⟩ and more general on compounds containing phosphorus–carbon multiple bonds ⟨94CRV1413⟩ demonstrate their potential for heterophosphole synthesis. A review on the preparative use of metallophosphaalkenes ⟨96AG292⟩ also includes the synthesis of 1,2,3-diazaphospholes. A review on the transition-metal assisted synthesis of rings from phosphaalkenes and phosphaalkynes includes some heterophosphole syntheses ⟨97AOC(41)1⟩. The new Houben–Weyl series *Science of Synthesis* will in its Volumes 12 and 13 on hetarenes also cover the synthesis of heterophospholes.

Cycloaddition reactions of heterophospholes have been reviewed along with those of other compounds with P=C or P=N bonds ⟨86PS(26)203⟩. A review on cationic low coordinated phosphorus compounds as ligands also deals with a number of heterophospholium ions ⟨97CCR(163)71⟩.

A list of compounds, complete up to mid 1987 is provided by a compilation of ^{31}P NMR shifts of heterocycles of two-coordinate phosphorus ⟨88PS(36)217⟩ and a comprehensive list of anellated heterophospholes is contained in the review mentioned above ⟨94T7675⟩. *Comprehensive Heterocyclic Chemistry II* contains chapters about all the then known heterophospholes with two ⟨96CHEC-II(3)709, 96CHEC-II(3)715⟩ or more ⟨96CHEC-II(4)771⟩ heteroatoms. The following sections present an updated version thereof. The chapters of CHEC-II in addition cover the analogous arsenic and antimony compounds ⟨96CHEC-II(3)709, 96CHEC-II(4)819⟩.

I wish to thank Raj Kumar Bansal, Jaipur, and Konstantin Karaghiosoff, München, for their help and advice.

4.3.2 PHOSPHOLES WITH ONE ADDITIONAL HETEROATOM ADJACENT TO P

4.3.2.1 Introduction

Fully unsaturated five-membered ring systems with one other heteroatom adjacent to phosphorus have been known only since 1986 ⟨86CB410⟩ and therefore no example is contained in the first edition of *Comprehensive Heterocyclic Chemistry* ⟨84CHEC-I(1)523⟩. These systems include now 1,2-azaphospholes (**1**), 1,2-oxaphospholes (**2**), 1,2-thiaphospholes and -selenaphospholes (**3**) and are mentioned in several reviews ⟨B-89MI307, B-90MI258, 96CHEC-II(3)709⟩. The then known compounds are listed in a compilation of phosphorus-31 chemical shifts ⟨88PS(36)217⟩ and in a review on anellated heterophospholes ⟨94T7675⟩. Rings with two adjacent phosphorus atoms, i.e. 1,2-diphospholides (**4**), are known as salt, or zwitterion or complex.

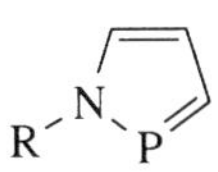

(**1**)

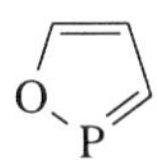

(**2**)

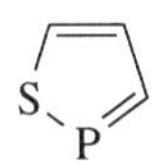

(**3**)

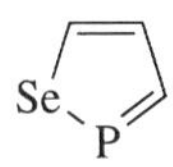

(**4**)

4.3.2.2 Spectra, Structure, and Bonding

4.3.2.2.1 *Quantum chemical calculations*

Among other azaphospholes, 1,2-azaphosphole (**1**) has been studied for an isodesmic ring fragmentation reaction and compared to pyrrole and pyrazole. The calculated heats suggest that in conjugative interaction P=C is more comparable to C=C than to N=C ⟨93JPC4011⟩. The phosphorus analogues of the cyclopentadienyl anion $P_n(CH)^-_{5-n}$ have been examined by SINDO1 for their aromaticity. Of the two diphospholides ($n = 2$), the 1,3-isomer (88%) has a marginally higher relative aromaticity than the 1,2-isomer (**4**, 86% of the aromaticity of the cyclopentadienyl anion) ⟨92JOC3694⟩. A more recent treatise of oligophosphole aromaticity ⟨98IC4413⟩ also includes the diphospholes and diphospholides.

4.3.2.2.2 *X-ray structure analyses*

The structures of the five-membered ring of the 1,2-azaphospholes (**6, 8**) ⟨98ICA273, 00S417⟩ and of the diphosphonio-1,2-diphospholide (**40**) (Section 4.3.2.6) ⟨96CEJ221⟩ are in accord with a cyclic delocalized system. In the latter, the P–P bond of 207.7(1) pm is only somewhat longer than that of acyclic *trans*-diphosphenes (202–204 pm). On bromination it widens to 221.6(2) pm. Methylation of a ring phosphorus atom in (**40**) apparently does not prevent the 6π delocalization as in the case of other phospholides and leaves the P–P bond length unchanged ⟨96CB1083⟩.

4.3.2.2.3 *NMR spectra*

A low-field chemical shift is characteristic for heterophospholes in general with most of the shifts between $\delta^{31}P = 50$ and 300 ⟨88PS(36)217, B-90MI258, B-90MI463, 00S417⟩. Within this range, the shift depends mainly on the ring members adjacent to phosphorus ⟨B-90MI463⟩ and is found for 1,2-azaphospholes at $\delta^{31}P = 169$ to 238, for 1,2-oxaphospholes at $\delta^{31}P = 303$ ⟨99EJO587⟩ and for 1,2-thiaphospholes at $\delta^{31}P =$ 198 to 211.

1,2-Diphospholide anions (**4**) are characterized by $\delta^{31}P = 152$ to 206 and by a large coupling constant $^1J_{PP}$ of roughly 440 Hz. The signals of their mono- and bis(triphenylphosphonio) derivatives (**44**) and (**40**) are found much more downfield at $\delta^{31}P = 218$ to 317 and show a still increased coupling $^1J_{PP} = 465$ to 480 Hz ⟨94AG708, 96CEJ221⟩.

Proton and carbon-13 NMR data are available for 1,2-azaphospholes ⟨86CB410, 98ICA273, 00S417⟩, 1,2-benzazaphospholes ⟨89PS(46)31⟩, 1,2-thiaphospholes ⟨94BCJ2785⟩, and for the 1,2-diphosphaindenide (**44**) ⟨94AG708⟩. The signal for C-3 in the 1,2-azaphospholes (**8**) appears at $\delta^{13}C = 174$ to 181 as a doublet with $^1J_{PC} = 41$ to 44 Hz.

4.3.2.2.4 *Mass spectra*

The mass spectra of the 1,2-azaphospholes (**8**) often show M^+ as the base peak ⟨00S417⟩. In the mass spectra of 1,2-thiaphospholes (**19**) and (**25**), M^+ forms the base peak and M^+–PS a characteristic fragment peak ⟨88TL4535, 94BCJ2785⟩. The base peak of lithium diphospholides (**35**; R = Et, Ph) corresponds to the mass of the anion ⟨95AG623⟩.

4.3.2.3 1,2-Azaphospholes (1)

1,2-Azaphospholes such as (**5**) with a cyano, nitroaryl, or other electron-withdrawing substituent in 3-position can be prepared by condensating the appropriate β-methylene ketimine with PCl_3 or PBr_3 and a base ⟨89EGP268704⟩. The yield can be improved if DABCO is used in place of triethylamine (Equation (1)). In a related synthesis, at first a 2-methyl 1,2-azaphospholium bromide is formed, which is subsequently demethylated with sodium in tetrahydrofuran to yield the neutral diazaphosphole (**6**) ⟨98ICA273⟩.

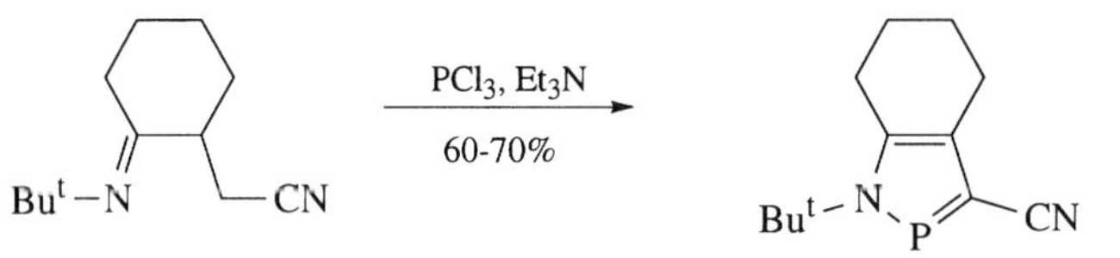

(**5**)

(1)

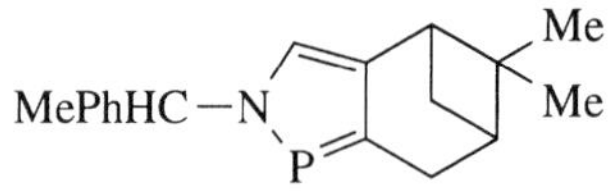

(**6**)

Another synthetic pathway to a variety of alkyl/aryl-substituted 1,2-azaphospholes (**8**) has been developed involving reactions of the cycloadducts (**7**), generated *in situ* from imidovanadium trichlorides R-$NVCl_3$ (R = Pr^i, Bu^t, 1-Ad) and phosphaalkynes R^1-CP (R^1 = Bu^t or other tertiary alkyl substituent), with acetylenes (R^2/R^3 = H/Pr, H/Bz, H/Ph, Me/Me, Me/Ph, Ph/Ph). The reaction is regiospecific (Equation (2)).

4,5-Dicyano-1,3,2-diazaphospholes (**9**; R^1 = Me, $SiMe_3$) are converted to 5-cyano-1,2-azaphospholes (1,2-azaphosphole-5-carbonitriles) (**10**) by reaction with alkynes R^2–C≡C–R^3 (R^2/R^3 = H/Pr^n, H/Bu^n, H/CH_2NMe_2, H/Ph, H/CO_2Et, Me/Me, Me/CO_2Me, Ph/CO_2Et, Ph/OCOPh, CO_2Me/CO_2Me). In some cases of asymmetrical alkynes, the other isomer (**11**) also forms. The preference for position 3 increases

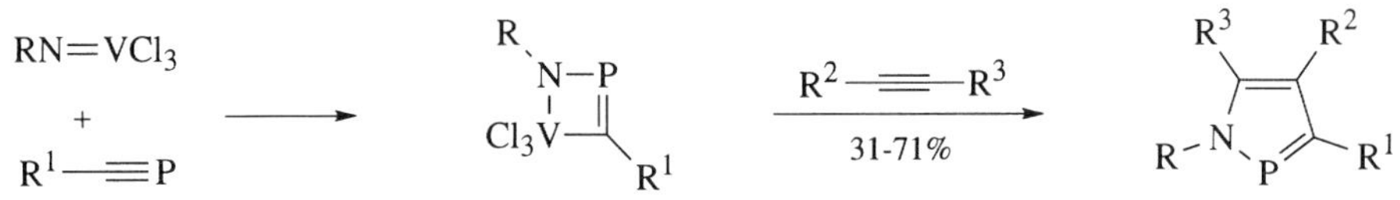

(7) (8)

(2)

in the order H < CO_2Et < Ph < alkyl (Equation (3)). The products are remarkably stable both to oxidation and hydrolysis. Compound (**10**), R^1 = Me, R^2 = H, R^3 = Ph, taken as an example, does not react with elemental sulfur or methanol or both of them (typically a reactive combination towards other azaphospholes) 〈86CB410〉.

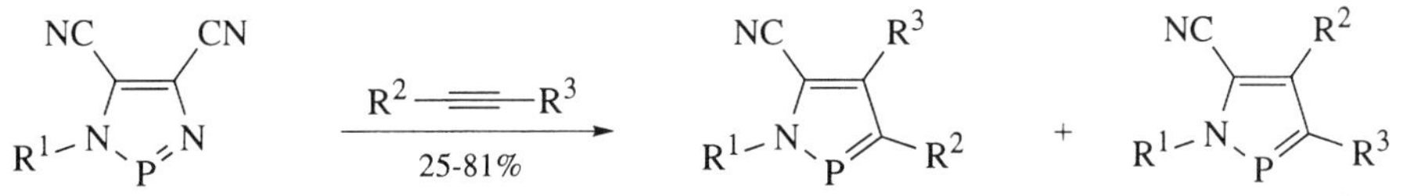

(9) (10) (11)

(3)

3-Phenyl-1,2,3,4-triazaphosphole (**12**; R = Bu^t) prepared from the [3 + 2]-cycloaddition of the phosphaalkyne with phenyl azide, on flash vacuum pyrolysis splits off dinitrogen. The intermediate azaphosphirene (**13**) may open the ring to give either a carbene or a phosphinidene. These species subsequently undergo [1,5]-electrocyclization followed by proton migrations to form anellated azaphospholes: the 1,2-benzazaphosphole (**14**), and the 1,3-benzazaphosphole (**15**) in a ratio 4 : 1. The isomer mixture can be separated by medium pressure chromatography. An analogous reaction is observed for the 3-naphthyl-1,2,3,4-triazaphosphole and gives the two isomers in a ratio of 3 : 1 (Equation (4)) 〈87T3247, 89PS(46)31〉.

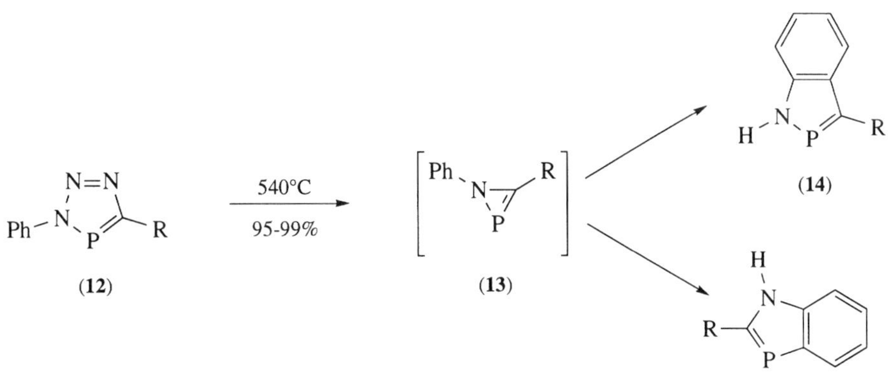

(12) (13) (14) (15)

(4)

The chiral 1,2-azaphosphole (**6**) gave a 1 : 1-adduct with C_3H_5NiCl and a 4 : 1-adduct with Ni(0). In both cases the phosphorus atom is coordinating to nickel. The complexes of P-coordinated 1,2-azaphospholes with $Fe(CO)_4$ and $W(CO)_5$ have, respectively, been isolated 〈00S417〉 or postulated as an intermediate 〈99AG158〉.

Alkyl trifluoromethane-sulfonates alkylate 1,2-azaphospholes of type (**8**) selectively and quantitatively at the phosphorus atom. The oxidative addition of two equivalents of diethyl azodicarboxylate to the phosphorus atom affords a zwitterionic product 〈00S417〉. Azaphospholes (**8**) undergo Diels–Alder reactions with electron-poor acetylenes to give in 63–73% yield azaphosphanorbornadienes 〈00S417〉. The same reaction is observed for the transient $W(CO)_5$ complex mentioned above 〈99AG158〉.

4.3.2.4 1,2-Oxaphospholes (2)

The dimesityl-1,2,4-oxadiphosphole (**16**) reacts at −20°C with dimethyl acetylenedicarboxylate in a [4 + 2] cycloaddition/retro [4 + 2] sequence to yield selectively the 1,2-oxaphosphole (**17**) which, however, is too labile to be isolated by chromatography or distillation (Equation (5)) 〈99EJO587〉.

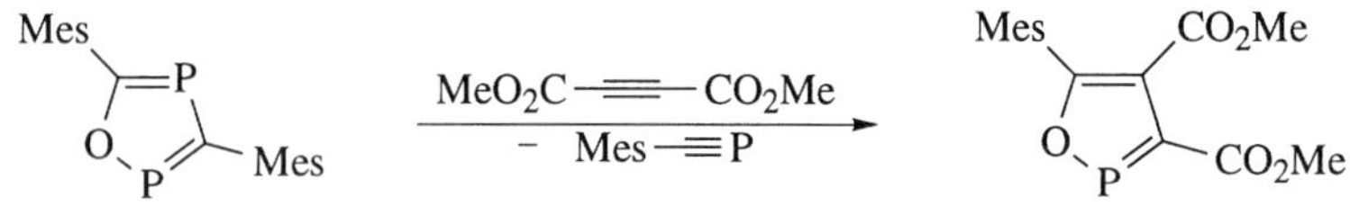

(16) (17) (5)

4.3.2.5 1,2-Thiaphospholes, 1,2-Selenaphospholes (3)

The first representative, 3,4,5-triphenyl-1,2-thiaphosphole (**19**) was observed as a side product and isolated in low yields (1% or below) along with 3,5-diphenyl-1,2,4-thiadiphosphole from the reaction of 1,3,4-oxathiazol-2-ones (**18**), Ar = 2-MeC_6H_4, 2,6-$Cl_2C_6H_3$, and a 1-chloro-2-trimethylsilyl-phosphaalkene (Equation (6)) ⟨88TL4535⟩. The corresponding reaction of the methyl derivative (**20**) yields *E*/*Z*-mixtures of 2-chloro-3-trimethylsilyl-thiaphosphiranes (**21**). Reaction with dimethyl acetylenedicarboxylate in the presence of KF (18-crown-6) gives the 1,2-thiaphosphole (**22**, R = CO_2Me) in poor yield (Equation (7)) ⟨89TL4501⟩. 1,2-Thiaphospholes (**22**, R = CN and a tertiary alkyl group in place of Ph) can be obtained in yields up to 81% from the reaction of 1,2,4-thiadiphospholes (Section 4.3.6.8) and dicyano acetylene ⟨00TH3⟩.

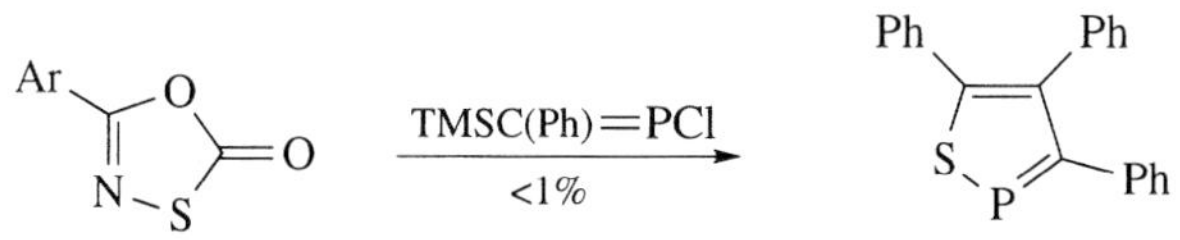

(18) (19) (6)

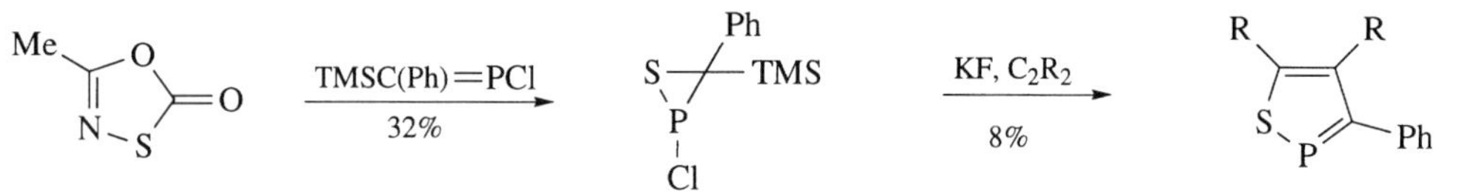

(20) (21) (22) (7)

3,5-Diaryl-substituted 1,2-thiaphospholes (**25**; R^1, R^2 = Ph, 4-MeC_6H_4, 4-$MeOC_6H_4$, 2-thienyl, styryl) can efficiently be prepared from α,β-unsaturated ketones (**23**). Reaction with excess P_4S_{10} in the presence of triethylamine gives 2,9-dithia-1-phosphabicyclo[4.3.0]nona-3,7-diene-1-sulfides (**24**) ⟨82TL1263⟩. Their treatment with $Bu^n{}_3P$ affords the thiaphospholes (Scheme 1) ⟨93CL1563, 94BCJ2785⟩. A mixture of 1,2- and 1,3-selenaphospholes is obtained from the reaction of 3,5-di-*t*-butyl-1,2,4-selenadiphosphole with dimethyl acetylenedicarboxylate (Section 4.3.6.8) ⟨99TH1⟩.

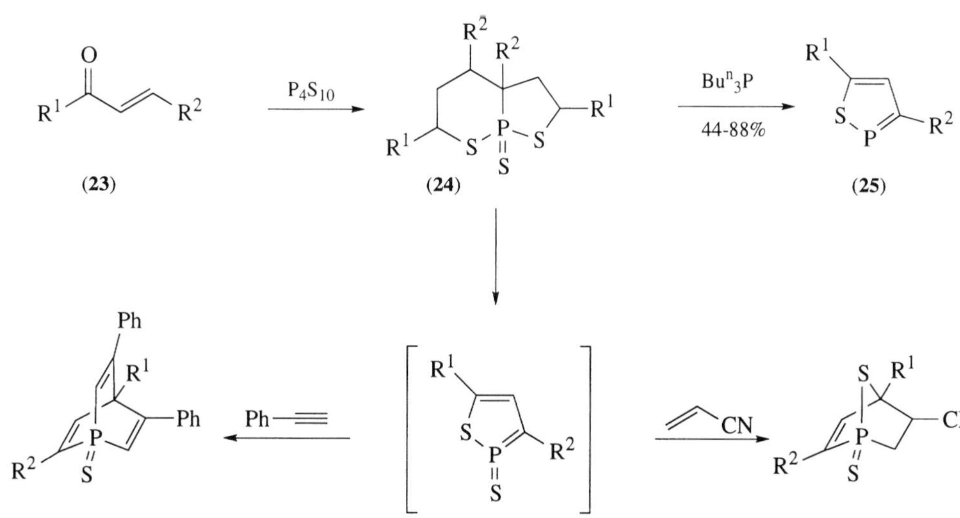

(23) (24) (25)

(28) (26) (27)

Scheme 1

Thermolysis of (**24**) in absence of a phosphine generates 1,2-thiaphosphole 2-sulfides (**26**) which can be trapped with suitable dienophiles such as acrylonitrile, styrene, butyl vinyl ether, and norbornene. The [4 + 2] cycloadducts such as (**27**) are formed regiospecifically ⟨85BCJ667⟩. The reaction with dimethyl acetylenedicarboxylate or phenylethyne affords 1-phosphabarrelene-1-sulfides (**28**) (Scheme 1) ⟨86BCJ2047⟩.

The 1,2-thiaphospholes (**25**) react thermally with norbornadiene, norbornene, or diethyl azodicarboxylate to produce double Diels–Alder cycloadducts (**29**) and (**30**). With mixtures of norbornadiene and methyl acrylate or acrylonitrile, crossed double Diels–Alder cycloadducts are obtained. In the presence of $AlCl_3$, the 1,2-thiaphospholes (**25**) react with acrylic esters, acrylonitrile or methyl vinyl ketone to give tandem Diels–Alder/Michael adducts such as (**31**). The $AlCl_3$-promoted reaction of (**25**) with ethyl propiolate affords with the elimination of a sulfur atom the stable phosphinine (**32**) (Scheme 2) ⟨94BCJ2785⟩.

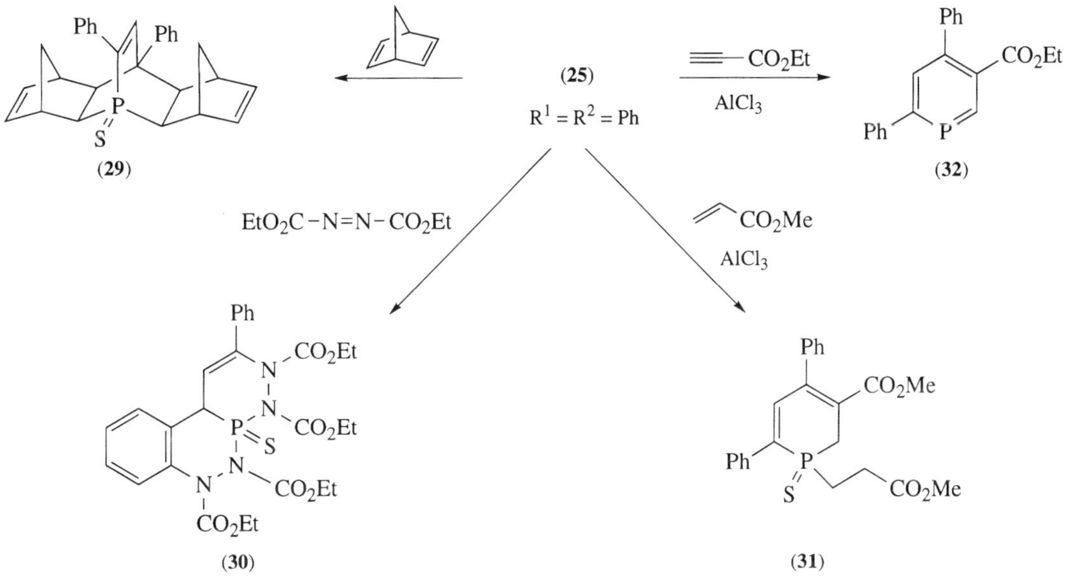

Scheme 2

An attempt to synthesize 1,2-thiaphospholes and 1,2-selenaphospholes from triphenylphosphoniumylidyl thioxophosphines and selenoxophosphines, respectively, and phenacyl halides failed ⟨95ZN(B)1543⟩.

4.3.2.6 1,2-Diphospholides (4)

1,2-Dihydrophosphetes (**33**), accessible from titanacyclobutenes and dihalophosphines undergo a ring enlargement reaction with PCl_3 and at the same time act as reducing agent. The resulting 1,2-diphosphacyclopent-3-enes (**34**; R = Et, Ph) are further reduced by lithium in THF to yield the lithium 1,2-diphospholides (**35**) (Equation (8)). They can be used for the preparation of 1,2-diphosphaferrocenes ⟨95AG623⟩.

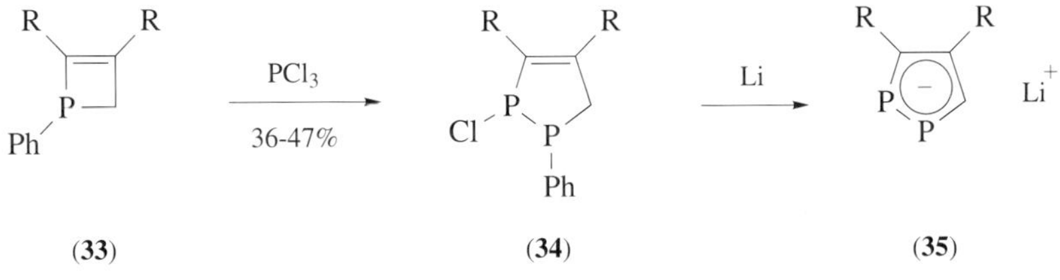

(8)

Other 1,2-diphosphaferrocenes have been obtained by direct synthesis ⟨91CC1293, 91CC1659, 93OM731, 95CB665, 96ZAAC(622)543, 97AOC(41)1⟩. The 1,2-diphosphaferrocene (**37**) for example results from the persilylated phosphinomethylenephosphine (**36**) by complex formation and thermal rearrangement of the primary complex (Equation (9)).

Condensation of the 1,3-diphosphonio-propenide cation (**38**) with two equivalents each of phosphorus trichloride, triphenylphosphine and triethylamine yields the 1,2-dichloro-1,2-dihydro-1,2-diphospholide

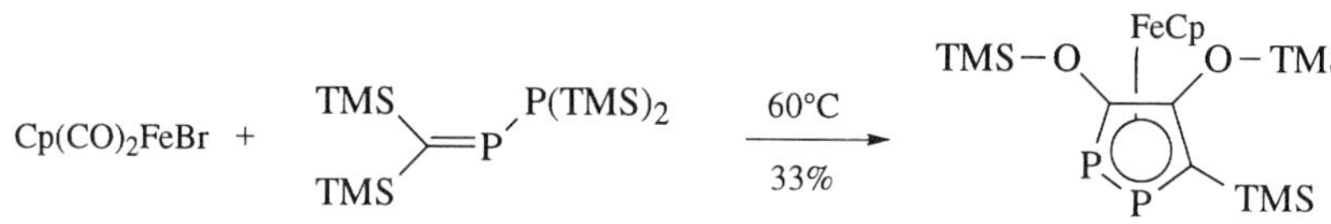

(36) (37)

(9)

cation (**39**). Its reduction by tributylphosphine gives the 3,5-bis(triphenylphosphonio)-1,2-diphospholide cation (**40**) in the form of a mixed chloride/bromide ⟨96CEJ221⟩. The cation is stable to hydrolysis, adds chlorine, bromine or iodine to the P–P bond and can be regained from these adducts (Equation (10)).

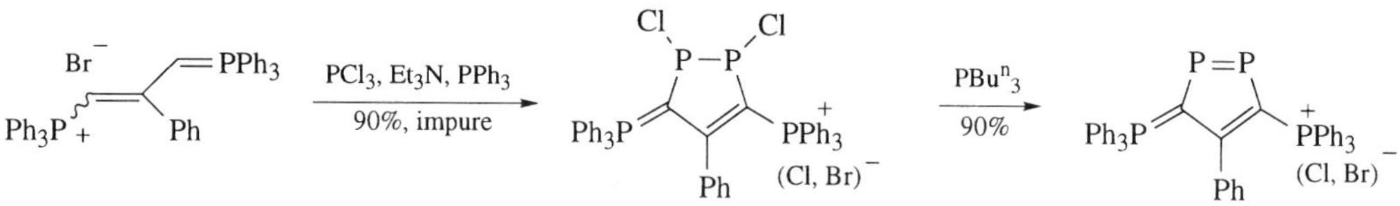

(38) (39) (40)

(10)

The reaction of benzyltriphenylphosphonium bromide (**41**) with phosphorus trichloride or tribromide and triethylamine leads *via* the *C*-dihalophosphino-*C*-phenyl ylides (**42**) to the 1,2-dihalo-3-triphenylphosphoranediyl-1,2-diphosphaindanes (**43**). Their yield increases with reaction time, and for X = Br the product (**43**) can account for up to 50% of the reacted trihalide. The corresponding reactions of (3-methylbenzyl)triphenylphosphonium bromide yield mixtures of the 5-Me and 7-Me derivatives. The bromides (**43**) have been reduced with magnesium in THF to yield the 3-triphenylphosphonio-1,2-diphosphaindenide (**44**) (Equation (11)) ⟨94AG708⟩.

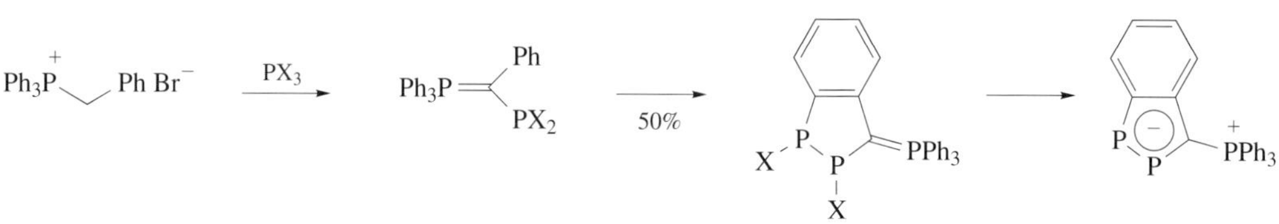

(41) (42) (43) (44)

(11)

4.3.3 PHOSPHOLES WITH ONE ADDITIONAL HETEROATOM NON-ADJACENT TO P

4.3.3.1 Introduction

The known fully unsaturated systems containing 1,3-heterophospholes are listed in Table 1. In addition some 1,3-heterophospholes with a partly saturated anellated ring are known.

The family embraces monocyclic 1,3-azaphospholes (**1**), 1,3-oxaphospholes (**2**), 1,3-thiaphospholes (**3**), and 1,3-selenaphospholes (**4**) as well as benzo derivatives (**5**), (**6**), (**7**) of the former three and a naphtho derivative (**8**) of (**1**). While the benzo derivatives (**5**) result from the 4,5-anellation to (**1**), the 1,5- and 1,2-anellations lead to two isomeric bicyclic systems such as (**9**) and (**14**) with a bridgehead nitrogen atom. The 1,5- and 1,2-anellated 1,3-azaphospholes embrace the pyrido derivatives, i.e. the 2-phosphaindolizines (**9**) and 1-phosphaindolizines (**14**), the isoquinolo derivatives (**10**) and (**15**), the pyridazo derivatives (**11**) and (**16**), the phthalazo derivative (**17**), the pyrazo derivatives (**12**) and (**18**), and the benzothiazolo derivative (**13**). In the final example to be mentioned here the 1,3-azaphosphole is fused both with its 1,2- and its 1,5-bond to further rings and becomes part of the tetracyclic system (**19**). In further examples which are not shown here, the fused ring is not fully unsaturated. According to fusion nomenclature compounds (**9**), (**14**) and (**13**) are [1,3]azaphospholo[1,5-*a*]pyridines, -[1,2-*a*]pyridines, and -[5,1-*b*]benzothiazoles.

In neutral 1,3-diphospholes (**20**) the second heteroatom is a three-coordinate pyramidal phosphorus atom which is only very weakly conjugated to the remaining π-system. Full cyclic conjugation occurs only in their anions, i.e. in the 1,3-diphospholide anions (**21**). The same applies to the 1,3-benzodiphospholes (**23**) and their anions (**24**). Isomeric to a 1*H*-1,3-diphosphole in the sense of a substituent shift from C-2 to P-3 would be the analogue of an Arduengo-carbene with phosphorus in place of nitrogen. Some cyclic delocalization has been concluded for it ⟨00EJI369⟩, but so far only a dimer has been obtained

Table 1. The known monocyclic and fused 1,3-heterophospholes and -diphospholes

Monocyclic 1,3-heterophospholes

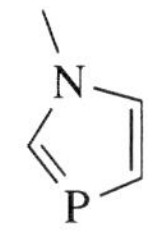

(**1**)

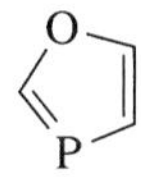

(**2**)

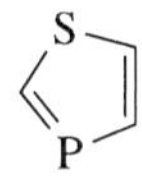

(**3**)

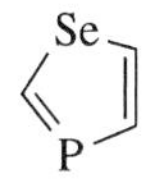

(**4**)

1,3-Benzoheterophospholes and 1,3-naphthoheterophospholes

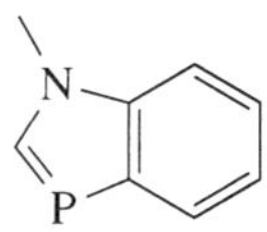

(**5**)

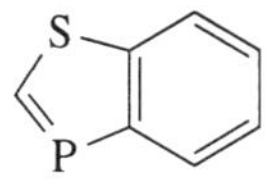

(**6**)

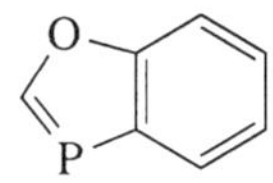

(**7**)

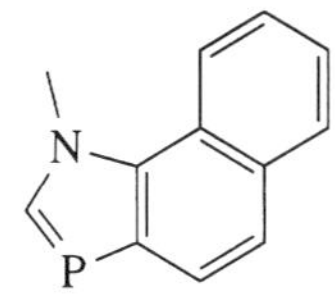

(**8**)

1,5-Anellated 1,3-azaphospholes

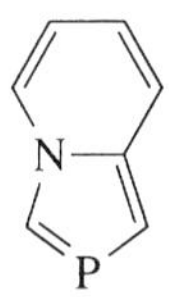

(**9**)

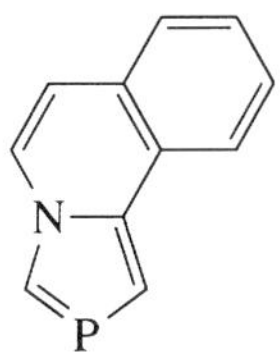

(**10**)

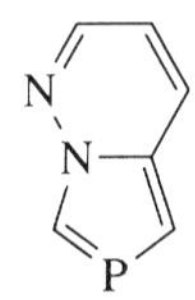

(**11**)

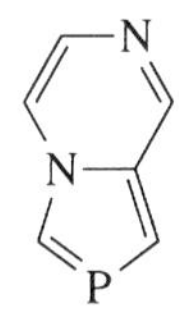

(**12**)

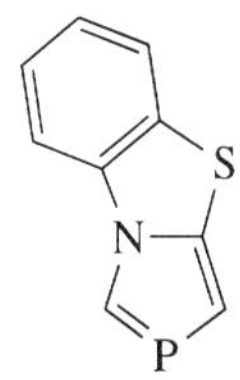

(**13**)

1,2-Anellated 1,3-azaphospholes

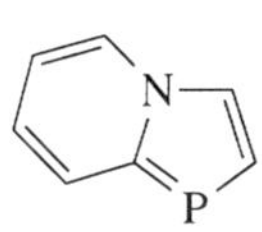

(**14**)

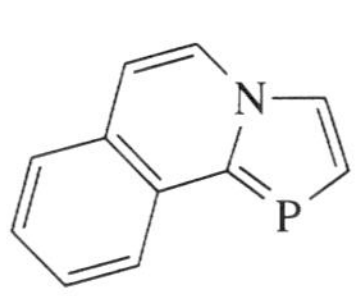

(**15**)

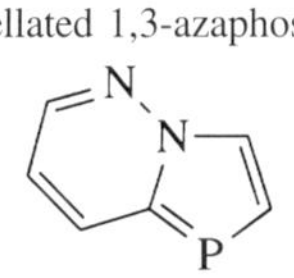

(**16**)

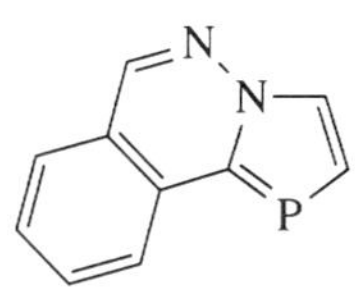

(**17**)

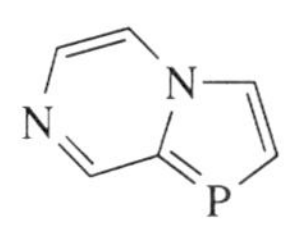

(**18**)

At the same time 1,2- and 1,5-anellated 1,3-azaphospholes

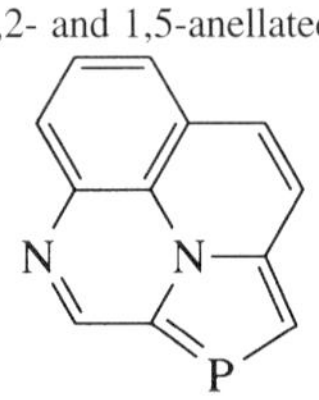

(**19**)

1,3-Diphospholes and 1,3-diphospholides/phosphaarsolides

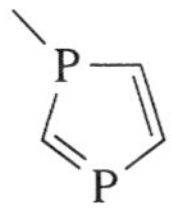

(**20**)

(**21**)

(**22**)

1,3-Benzodiphospholes and 1,3-diphospholides

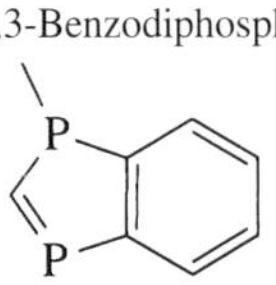

(**23**)

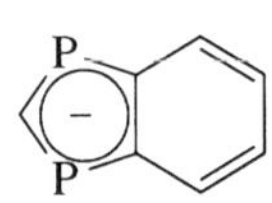

(**24**)

⟨97BSF853⟩. The 1,3-phosphaarsolide ion (**22**) contains both a two-coordinate phosphorus and arsenic atom as non-adjacent ring members.

Reviews dealing with the synthesis and reactivity of 1,3-heterophospholes are mentioned in Section 4.3.1. A review ⟨94MI307⟩ on 1,3-heterophospholes compares their reactivity to that of di- and triazaphospholes which are covered in Sections 4.3.4 and 4.3.5. In contrast to 1,3-azaphospholes the ring of the latter contains one or two two-coordinate nitrogen atoms from a formal RC/N exchange, lowering the charge density at the phosphorus atom and facilitating the attack by hard nucleophiles.

4.3.3.2 Spectra, Structure, and Bonding

4.3.3.2.1 Theoretical methods and PE-spectra

The electronic structure of 1,3-azaphospholes (**1**) (and of other azaphospholes, see following sections) has been investigated by means of *ab initio* quantum chemical calculations and HeI photoelectron spectroscopy ⟨92JPC623, 92JA9080, 96PS(109-110)109⟩. The calculated total energy of the 3*H*-isomer (**1′**) is significantly higher than that of the 1*H*-isomer (**1**). The aromatic stabilization — as measured by the heats of bond separation reactions — of (**1′**) is smaller than that of (**1**). While the aromaticity of the 1*H*-isomer (**1**) is found to be similar to that of imidazole, the 3*H*-isomer (**1′**) is much less aromatic and generally even considered nonaromatic and nonconjugated. While the 1*H*-isomer (**1**) is planar with calculated bond lengths between those of isolated single and double bonds, the 3*H*-isomer (**1′**) is nonplanar, the arrangement around the phosphorus atom pyramidal, and the C–P bond length close to the single bond value. UV and PE spectra of 3-alkyl 1,3-benzazaphospholes are consequently completely different from those of their 1*H*-isomers ⟨94MI307, 96PS(109-110)109⟩. For (**1**) a NICS of −14.7 ppm has been calculated ⟨97MI1⟩, close to that of pyrrole ⟨96JA6317⟩. Structure and aromaticity of lithium 1,3-azadiphospholide have been investigated by ab initio methods ⟨01MI(538)189⟩.

(1)

(1) **(1′)**

Earlier MNDO calculations are in accord with the above results ⟨91PS(56)189⟩. Their main emphasis is on 1,3-benzazaphospholes (**5**) and 1,3-benzodiphospholes (**23**) and their anions. The remarkable reactivity differences caused by the homologous N/P-exchange is readily explained. Other MNDO and CNDO/2 calculations and HeI photoelectron spectra ⟨89JOM(373)49, 89JOM(373)57, 94MI307⟩ include 1,3-benzoxaphospholes (**7**). Their HOMO has predominantly P=C character although this unit strongly interacts with the benzo π-system. As compared to phosphinine, the results indicate for them a lower electrophilicity and a lower reactivity in addition reactions.

Comparison of π-ionization energies of a series of compounds including 1,3-azaphospholes (**1**), 1,3-benzazaphospholes (**5**) and 1,3-benzoxaphospholes (**7**) and of ring fragmentation reaction heats suggest that in conjugative interactions P=C is more comparable to C=C than to N=C π-systems ⟨93JPC4011⟩. Charge densities calculated by MNDO for indolizine, its 2-phospha derivative (**9**) and its 1-phospha derivative (**14**) show the π-electron excess in the five-membered ring to increase in this order ⟨91CB475⟩. As a dicoordinate phosphorus atom lowers the π-electron density at the adjacent carbon atoms, 2-phosphaindolizines are less reactive towards electrophiles in positions 1 and 3 than indolizines.

PE spectra of a 1,3-thiaphosphole (**3**) and a 1,3-selenaphosphole (**4**) were recorded. The positions of the π-ionization energies have been found to be similar to those of thiophene and selenophene ⟨95JCS(P2)315⟩.

As mentioned in Section 4.3.2.2.1, aromaticity of the 1,3-diphospholide anion (**21**) was found somewhat higher than that of the 1,2-isomer ⟨92JOC3694⟩.

4.3.3.2.2 X-ray structure analyses

Several structures of 1,3-heterophospholes and 1,3-diphospholes have been determined by X-ray diffraction techniques. The bond distances within the ring and the bond angle at the two-coordinate phosphorus atom are given in Table 2. In part they are also contained in a review ⟨88ZSK122⟩. In general the structures show a planar five-membered ring with averaged bond distances. The angle at the phosphorus atom clearly depends on the size of the second heteroatom: in case of O, N it ranges from 88 to 92°, in

Table 2. Bond lengths (pm) in 1,3-heterophosphole, 1,3-diphosphole and 1,3-diphospholide rings and the angle at the two-coordinate phosphorus atom from X-ray structure analyses

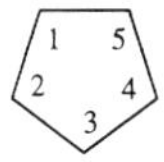	*Bonds*					*Angle*	*Ref.*
	1.2	*2.3*	*3.4*	*4.5*	*1.5*		
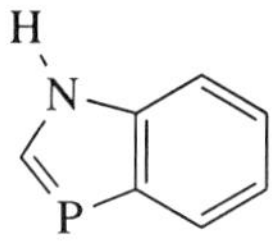	135.1	169.5	180.7	142.2	137.1	88.2°	<86ZAAC(540/541)336>
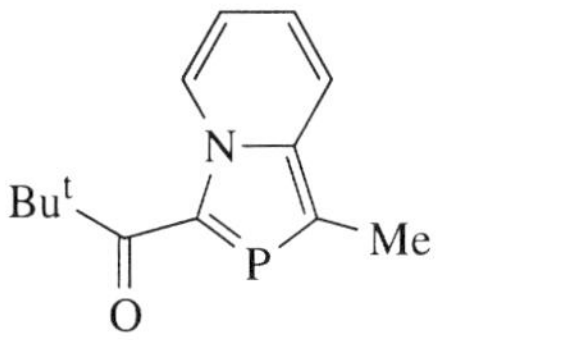	140.0	174.7	170.8	139.5	139.5	91.9°	<98HAC333>
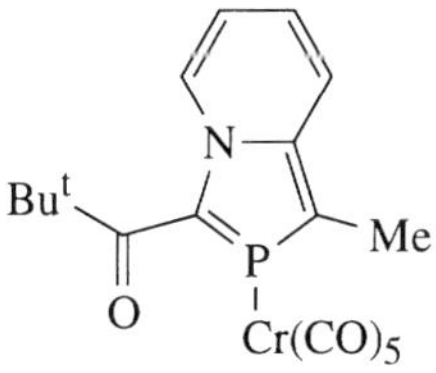	138.2	172.4	171.7	138.7	140.4	93.5°	<98EJI1079>
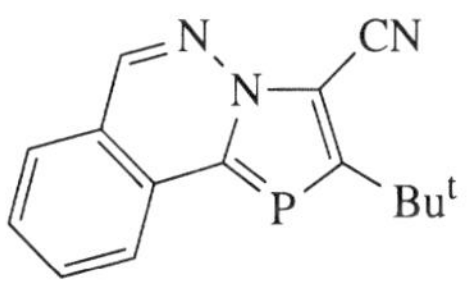	136.5	173.4	175.4	137.1	137.2	89.9°	<92TL1049>
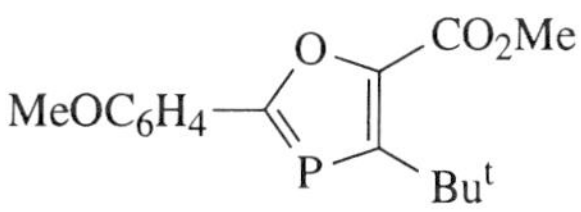	135.2	171.8	178.5	137.1	137.2	89.0°	<00T63>
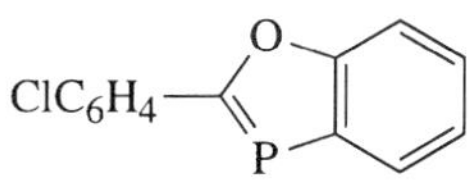	136.9	171.2	178.8	139.3	137.4	88.1°	<85ZAAC(520)107>
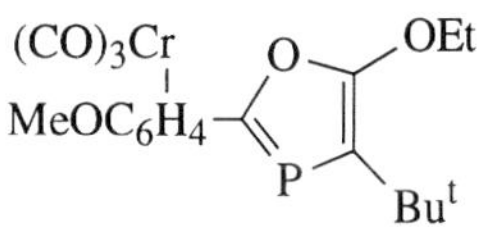	136.9	170.5	177.2	133.9	135.5	89.5°	<89CC788> <93T5577>
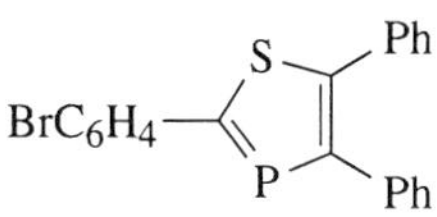	172.3	169.1	171.9	136.3	167.7	96.3°	<87TL2119>
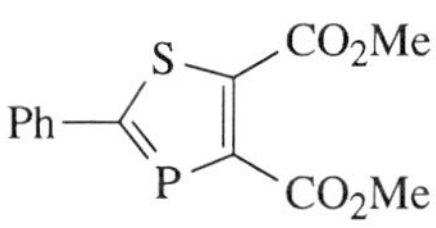	169.9	171.5	176.1	136.5	171.3	93.8°	<92TL4421>
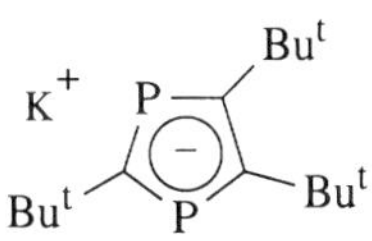	171.9	174.0	177.7		177.5	97.2°	<00OM219>
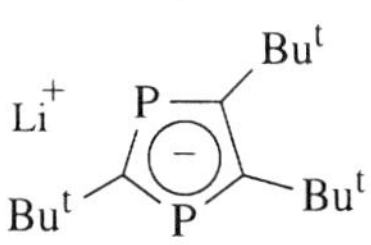	174.8	172.7	179.0	140.9	180.3	96.0° 96.7°	<96ZK(211)341>
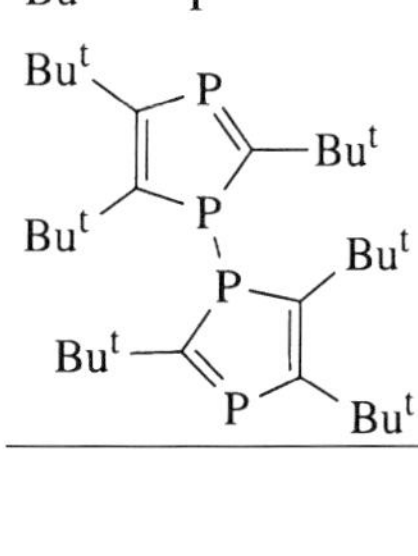	178.4	167.3	181.6	139.7	181.9	100.5°	<93CC267>

case of S, P from 94 to 100°. Coordination of the phosphorus atom of course widens its endocyclic angle, though only slightly ⟨98EJI1079⟩.

The PC bond lengths in Table 2 cover almost the whole range between the average values for a localized single (185 pm) and for a localized double (167 pm) bond ⟨88ZSK122, B-90MI157, 98HAC333⟩. Of the two bonds connecting the two-coordinate phosphorus to the carbon atoms in most cases the bond 2,3 is shorter than the bond 3,4. The difference may serve as a straightforward indicator of cyclic delocalization: the more the two lengths deviate from each other, the less pronounced it is. The difference is small (2–5 pm) for 1,3-thiaphospholes and 1,2- and 1,5-anellated 1,3-azaphospholes, larger (7–11 pm) for 1,3-benzazaphospholes and 1,3-benzoxaphospholes and largest (14 pm) for the 1,3-diphosphole. In the latter case, the phosphorus atom in 1-position is pyramidal and does not participate in the π-system which therefore is of diene character. Consequently the two PC bond lengths are close to those of localized single and double bonds. The calculated and experimental structures of 1,2-, 1,5-, and 4,5-anellated 1,3-azaphospholes have been compared ⟨98HAC333⟩.

For two of the compounds of Table 2, aromaticity indices were derived from the molecular dimensions (1,3-oxaphosphole: 38; 1,3-thiaphosphole: 55). According to them, aromaticity is not greatly reduced if dicoordinate nitrogen in a five-membered heterocycle is replaced by dicoordinate phosphorus ⟨90T5697⟩.

The structure of lithium and potassium salts, of an indium(I) complex and of some transition metal η^5-complexes of 1,3-diphospholide ions have been determined by X-ray diffraction ⟨87CC1146, 90AG575, 90CC472, 92AG1082, 92CIC61, 92JOM(430)C10, 93JOM(462)319, 93PS(77)254, 97JOM(529)69, 00JCS(D)1715, 00OM219⟩. They all demonstrate a fully cyclic delocalization. In the structure of the $W(CO)_5$ complex of a 1,3,6-triphosphafulvene all atoms of the skeleton are coplanar ⟨00AG2892⟩.

4.3.3.2.3 NMR spectra

^{31}P NMR data of 1,3-heterophospholes have been compiled ⟨88PS(36)217, 94T7675⟩ and discussed several times ⟨B-89MI307, B-90MI258, B-90MI463, 94MI307⟩. The typical shift range is $\delta^{31}P = 50$ to 180 with a high-field extreme of $\delta = -5$ and a low-field extreme of $\delta = 235$. Table 3 lists characteristic monocyclic examples and their benzo derivatives. The low-field chemical shifts observed for monocyclic 1,3-thiaphospholes (and 1,3-selenaphospholes) is ascribed to a transannular electron repulsion ⟨94MI307⟩. Benzo anellation is generally accompanied by a significant upfield shift.

In 1,3-azaphospholes electron donating 2- or 4-substituents cause an upfield shift, while electron withdrawing 2-substituents induce a downfield shift, increasing e.g. in the order $C_6H_5 < C_6H_4NO_2 < CO_2Et$, CN < COPh ⟨91CB475, 94T7675⟩. The effects correspond to those of α-substituents in acyclic two-coordinate phosphorus compounds ⟨86PS(29)73, B-90MI463⟩. For 1,5-anellated 1,3-azaphospholes, $\delta^{31}P$ depends on the nature of the fused ring and decreases in the order benzothiazole > oxazoline > thiazoline > pyridine over a range of $\Delta\delta = 20$. The shifts of 1-phosphaindolizines (**14**) and 2-phosphaindolizines (**9**) are found well separated in different regions of the above range, the former at the high-field end, $\delta = 68$ to 76 ⟨87TL1511⟩, the latter at the low-field end, $\delta = 121$ to 184 ⟨91CB475, 92IJC254, 98HAC333⟩. They reflect thus the difference in charge density as calculated (Section 4.3.3.2.1).

The two-coordinate phosphorus signals of 1,3-diphospholes are found further downfield than those of 1,3-heterophospholes (Table 3) and are particularly sensitive to substituent effects. This reflects the absent interaction of the three-coordinate phosphorus atom with the π-system. In accord the chemical shift of the two-coordinate phosphorus does hardly change if the three-coordinate phosphorus atom is oxidized ⟨90JOM(382)53⟩. It compares well with $\delta^{31}P$ of the electron-poor phosphaalkenes ⟨86PS(29)73⟩.

The signals of monocyclic 1,3-diphospholide anions are observed at lower field than those of their benzo derivatives (Table 3) ⟨89PO849, 90AG575, 92PO601, 97JOM(529)69, 00OM219⟩. The effects of benzo anellation and of electron donating substituents are similar to those in 1,3-heterophospholes. Coupling constants $^2J_{PP}$ in 1,3-diphospholes ⟨93CC267⟩, 1,3-benzodiphospholes ⟨90JOM(382)53⟩ and 1,3-diphospholides ⟨92PO601, 97JOM(529)69⟩ range from 9 to 39 Hz.

For most of the newly synthesized 1,3-heterophospholes and 1,3-diphospholes ^{1}H and ^{13}C NMR spectra are given and discussed. A characteristic feature is the low-field signals due to protons which are adjacent to the phosphorus atom ⟨99CC1283⟩. Coupling constants $^2J_{PH}$ are typically large, ranging from 30 to 50 Hz. The coupling constants $^1J_{PC}$ vary between 20 and 60 Hz.

4.3.3.2.4 Mass spectra

Mass spectra have been reported of 1,3-azaphospholes (**1**) ⟨87TL1089, 87CB1809, 88JOM(338)329, 98S1305⟩, 1,3-benzazaphospholes (**5**) ⟨78TL441, 81ZAAC(481)22, 82TL3643⟩, 2-phosphaindolizines (**9**) ⟨98HAC333⟩, 1-

Table 3. Characteristic chemical shifts $\delta^{31}P$ of 1,3-heterophospholes, 1,3-diphospholes, 1,3-diphospholides and their benzo derivatives

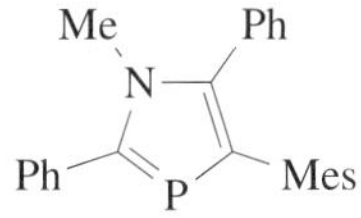

106

98S1305

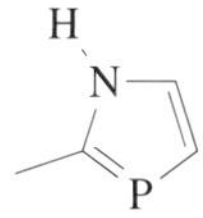

92

86TL5699

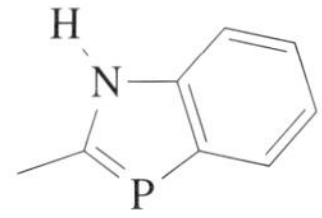

70

78TL441

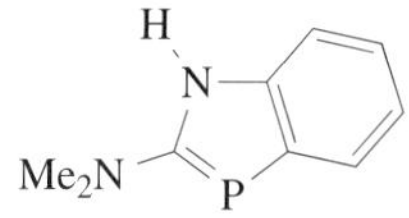

-5

81ZAAC(481)22

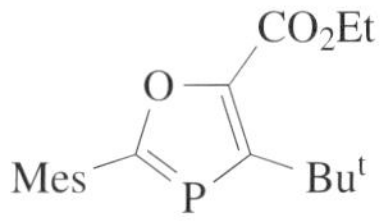

129

00T63

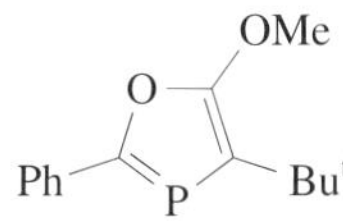

118

93T5577

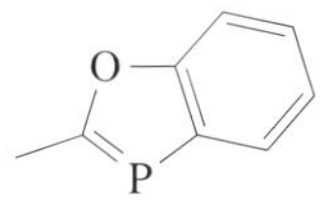

85

85PS(25)345

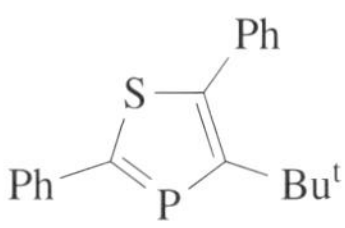

221

98S1305

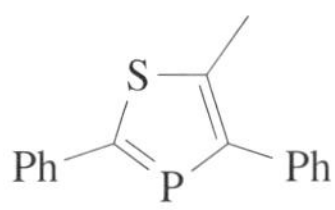

217

87TL2119

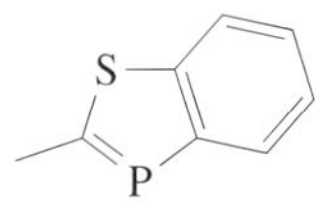

71

80TL3483

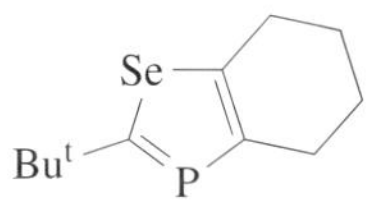

223

91SL356

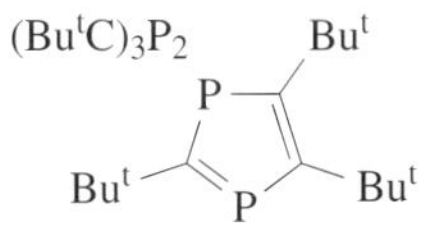

303

25, $^2J_{pp} = 24$ Hz

93CC267

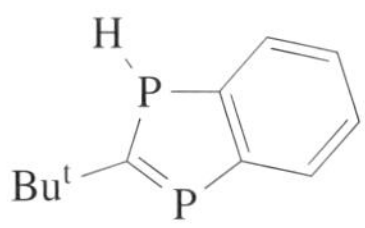

263

-85, $^2J_{pp} = 23$ Hz

90JOM(382)53

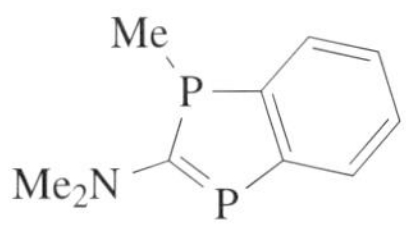

104

2, $^2J_{pp} = 27$ Hz

81TL4475

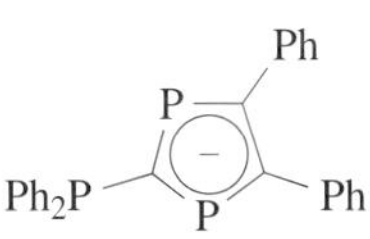

225

97JOM(529)69

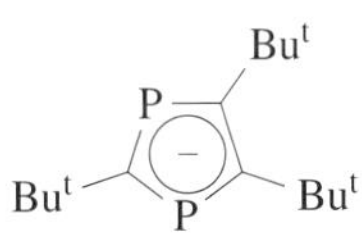

188

89POL849

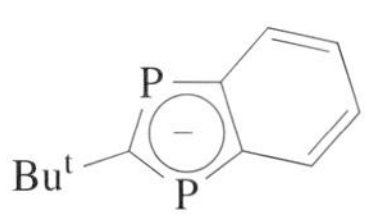

145

90JOM(382)53

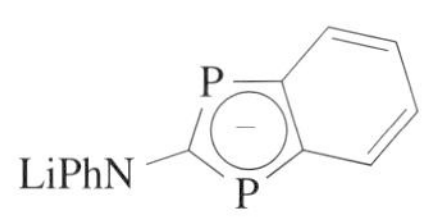

84

88ZC255

phosphaindolizines (**14**) ⟨87TL1511⟩, 1,3-azaphospholes with 1,5-anellated oxazoline ring ⟨94PS(86)139⟩, 1,3-benzoxaphospholes (**7**) ⟨80ZC342⟩, 1,3-thiaphospholes (**3**) (81TL2119, ⟨98S1305, 99CC1283⟩ and 1,3-benzodiphospholes (**23**) (87TL4475, ⟨90JOM(382)53⟩. While 1,3-benzazaphospholes (**5**) give mass spectra with few lines and the loss of nitrile as important fragmentation step, the fragmentation of 1,3-benzoxaphospholes (**7**) is complex. The mass spectrum of the unsubstituted thiaphosphole (**3**) shows the molecular ion as the base peak and a major daughter peak due to loss of acetylene ⟨99CC1283⟩. For a 1,3-thiaphosphole (**3**) the loss of a sulfur atom is observed ⟨98S1305⟩. Mass spectra of 1,3-azaphosphole complexes have also been reported ⟨99JOM(577)337⟩. The mass spectral fragmentation patterns of a 2-phosphaindolizine (**9**) and of a 1,3-azaphospholo-isoquinoline (**10**) have been discussed ⟨98HAC333, 99HAC598⟩.

4.3.3.2.5 UV spectra

For 1,3-heterophospholes ⟨78TL441, 80ZC342, 81ZAAC(481)22, 82TL3643, 83JOM(258)257, 85PS(25)345, 86TL5699, 89ZC410, 99CC1283⟩ and their benzo derivatives, as well as for a 1,3-benzodiphosphole ⟨90JOM(382)53⟩ UV spectra have been reported and discussed. By comparison, the electronic similarity of 1,3-azaphospholes (**1**) and 1,3-azarsoles with pyrroles and imidazoles, that of 1,3-benzazaphospholes (**5**) and 1,3-benzazarsoles with indoles and benzimidazoles or that of 1,3-benzoxaphospholes (**7**) and 1,3-benzoxarsoles with benzofuranes and benzoxazoles were shown ⟨94MI307⟩. Replacing a two-coordinate nitrogen atom in an azole for phosphorus, then replacing this by arsenic, is accompanied by a bathochromic shift of the absorption maximum. As compared to pyrrole, imidazole, and oxazole, 1,3-azaphospholes (**1**) and 1,3-azarsoles like thiazoles show an additional $\pi \rightarrow \pi^*$ band ⟨86TL5699⟩. A red shift of the corresponding bands is observed for 1,3-thiaphosphole (**3**) and 1,3-thiaarsole ⟨99CC1283⟩.

2-Aryl-1,3-benzoxaphospholes exhibit a strong blue fluorescence ($\lambda = 416$ nm) in the concentration range 10^{-2} to 10^{-4} M ⟨85PS(25)345⟩.

4.3.3.3 1,3-Azaphospholes (1), 1,3-Benzazaphospholes (5, 8) and other 4,5-Anellated 1,3-Azaphospholes

4.3.3.3.1 Reactivity

The title compounds (**1**) and (**5**) are colorless to yellow oils or solids which are generally stable in air for some time ⟨78JOM(154)1, 78TL441, 81ZAAC(481)22, 82TL3643, 83JOM(258)257, 84PS(20)347, 86TL5699, 89ZC71⟩. In contrast to most of the azaphospholes with more than two heteroatoms (which are covered in Section 4.3.4) they are not (or only slowly) attacked by water, dilute acids or bases, alcohols and amines. They neither add hard nor soft nucleophiles such as PhSH or $PhAsH_2$ ⟨94MI307⟩. 2-Methyl- and 2-phenyl-1,3-benzazaphosphole are not attacked by dilute cold sulfuric acid, 2-*t*-butyl-1,3-benzazaphosphole (**25**) under these conditions adds water to give (**26**) selectively in *E*-configuration ⟨99S264⟩. 1,2-Dimethyl-1,3-benzazaphosphole (**27**) is slowly oxidized in methanol solution to give the phosphinic ester (**28**) ⟨94MI307⟩.

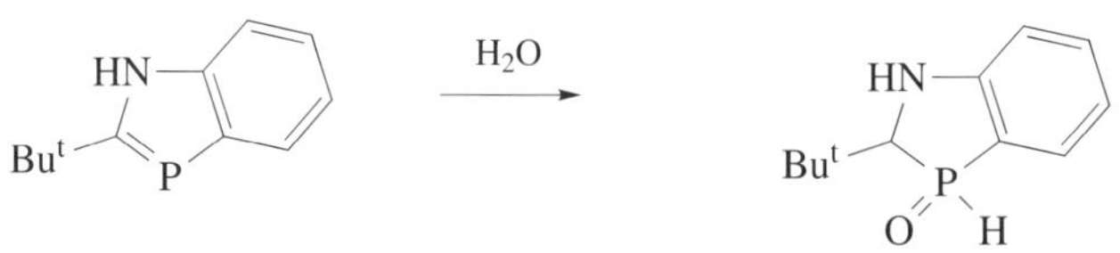

(1)

(**25**) (**26**)

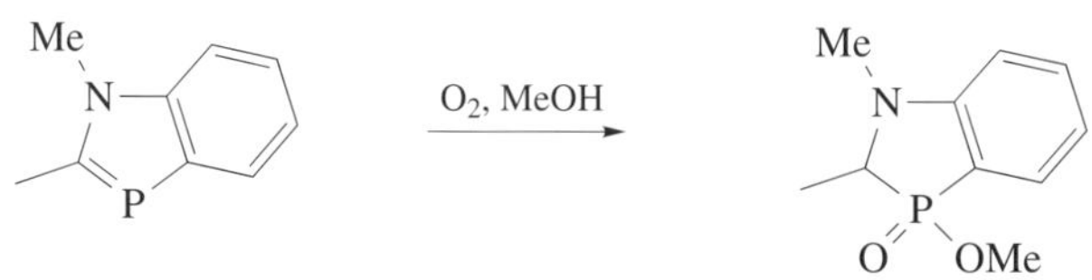

(2)

(**27**) (**28**)

4.3.3.3.1.1 Tautomerism

In accordance with the calculated energies (Section 4.3.3.2.1) for *N*-unsubstituted representatives only the 1*H*-isomers are observed spectroscopically and no 3*H*-isomer can be detected. 4-Hydroxy-1,3-azaphospholes do not form the keto tautomer ⟨87TL1089, 88TL3387⟩.

4.3.3.3.1.2 Reactivity towards electrophiles

N-Unsubstituted 1,3-benzazaphospholes and 1-methyl-1,3-benzazaphospholes are not basic and are not extracted from an organic phase by dilute sulfuric acid. They do not react with methyl iodide but are *P*-alkylated by $Me_3O^+BF_4^-$ ⟨83JOM(258)257, 94MI307⟩. A *P*–$Cr(CO)_5$ complex of 2-phenyl-1,3-benzazaphosphole is known ⟨81ZAAC(481)22⟩. 1,3-Benzazaphospholes undergo *N*-substitution with PCl_3 and they are *N*-acylated and *N*-trimethylsilylated after deprotonation with LiBu or more conveniently with $LiNEt_2$. 1,3-Azaphospholes react in the same way ⟨81ZAAC(481)22, 83JOM(258)257, 86TL5699⟩. The lithium 1,3-azaphospholides and 1,3-benzazaphospholides are intensely colored and can be isolated as mono-etherates. 2-Substituents ($SiMe_3$, $SnMe_3$, CO_2SiMe_3, SMe, $Fe(CO)_2Cp$) are introduced in 1-methyl-1,3-benzazaphospholes by 2-lithiation and subsequent reaction with the electrophile ⟨82TL3643, 83JOM(258)257, 86TL5699, 88ZC255⟩. Addition of $LiBu^t$ to the P=C bond is a side reaction to the metallation, however.

1,2-Dimethyl-1,3-benzazaphosphole (**27**) adds *o*-chloranil in a 1 : 2 ratio to the phosphorus atom ⟨83TL5481⟩.

4.3.3.3.1.3 Substituent reactivity

4-Trimethylsiloxy-1,3-azaphospholes (Section 4.3.3.3.2.4) on treatment with methanol or sodium methoxide and acyl chloride are converted to the hydroxy or acyloxy derivatives ⟨87TL1089, 88TL3387⟩. The 1-methyl-1,3-benzazaphospholyl-2-carboxylic silylester (Section 4.3.3.3.1.2) is quickly hydrolyzed to the acid ⟨82TL3643⟩.

4.3.3.3.2 Synthesis

4.3.3.3.2.1 Survey of the synthetic routes

Several synthetic routes to the title compounds may be distinguished in view of the fragments from which the ring is formed.

NCCP + C (Section 4.3.3.3.2.2): The procedure by which the first representatives were obtained ⟨78TL441, 78JOM(154)1⟩ proved very effective. It starts with the four-membered chain of a 2-aminoalkylphosphine or 2-aminophenylphosphine and completes the ring by condensation with a carboxylic acid derivative. In the former case full unsaturation is reached by thermolysis of the condensation product ⟨94MI307⟩.

CNCCP (Section 4.3.3.3.2.3): This variant of the above 1,3-benzazaphosphole synthesis starts from *N*-acylated 2-bromoanilines. After arylation of triethylphosphite to give the CNCCP chain, the ring is closed by reduction ⟨99S264⟩.

CNCC + electrophilic P (Section 4.3.3.3.2.4): This procedure was first developed for the synthesis of 1,5-anellated 1,3-azaphospholes like 2-phosphaindolizines ⟨91CB475⟩. It needs a four-membered chain with reactive methylene groups at both ends for condensation with phosphorus trichloride to form the ring. For 1,5-anellated 1,3-azaphospholes, 1,2-dialkyl cycloiminium ions are used, for monocyclic 1,3-azaphospholes a ketene *N*-methylaminal. To yield 1,3-benzazaphosphole according to this pattern, phenyl isonitrile can serve as the starting material.

CNCC + nucleophilic P (Section 4.3.3.3.2.5): Here the fragment pattern is the same as before but the condensation reaction is of inverse polarity. Oxazolium ions or zwitterionic münchnones contribute the CNCC chain and tris(trimethylsilyl)phosphine contributes the fifth ring member.

CNC + CP (Section 4.3.3.3.2.6): This route involves the cycloaddition of phosphaalkynes RC≡P or of phosphaalkenes $R(Me_3Si)C{=}PCl$ which lose Me_3SiCl in the course of the reaction. Reaction partners are 1,3-dipoles such as nitrile ylides or precursors thereof.

CCNCP: As mentioned in Section 4.3.2.3, 1,3-benzazaphospholes and the corresponding naphtho derivatives result from the electrocyclization of intermediate phenyl- or naphthyliminoacyl phosphinidenes ⟨87T3247, 89PS(46)31⟩.

4.3.3.3.2.2 Synthesis from 2-aminoorganylphosphines

(2-Aminoethyl)benzylphosphine (**29**) is condensed with iminoester hydrochlorides, R = H, Me, Et, Ph. Flash vacuum pyrolysis of the products (**30**) yields 1,3-azaphospholes (**31**) ⟨76JPR(318)409, 86TL5699, 86ZC407, 89ZC71, 91CB493⟩.

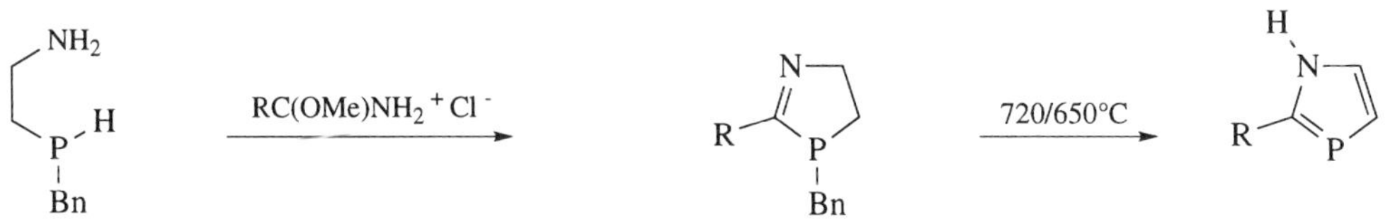

(3)

(**29**) (**30**) (**31**)

1,3-Benzazaphospholes (**33**) are obtained from the condensation of 2-aminophenylphosphines (**32**), R^1 = H, Me, with iminoester hydrochlorides, R^2 = H, Me, Bu^t, Ph. A wide variety of carboxylic acid derivatives can be used for the condensation ⟨78JOM(154)1, 78TL441, 81ZAAC(481)22, 82TL3643, 83JOM(258)257⟩. This synthesis has also been used for specially substituted compounds ⟨95BSF306⟩ and has recently been improved ⟨99S264⟩.

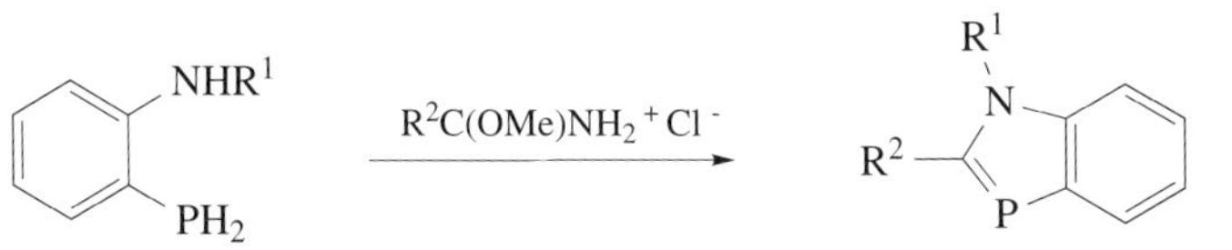

(4)

(**32**) (**33**)

In another version of this route, 2-aminophenylphosphine is first persilylated. The product on reacting with diphenylcarbodiimide followed by the action of methanol affords 2-anilino-1,3-benzazaphosphole (**33**), R^1 = H, R^2 = NHPh ⟨89ZC410⟩.

4.3.3.3.2.3 Synthesis from triethylphosphite

The reaction of *N*-acyl-2-bromoanilines with triethylphosphite in the presence of $NiCl_2$ furnishes high yields of *N*-acyl-2-aminobenzenephosphonic esters (**34**), R = Me, Bu^t, Ph. Their reduction with excess $LiAlH_4$ provides the 1,3-benzazaphospholes (**35**) as the main products, which thus are available in a three-step procedure ⟨99S264⟩.

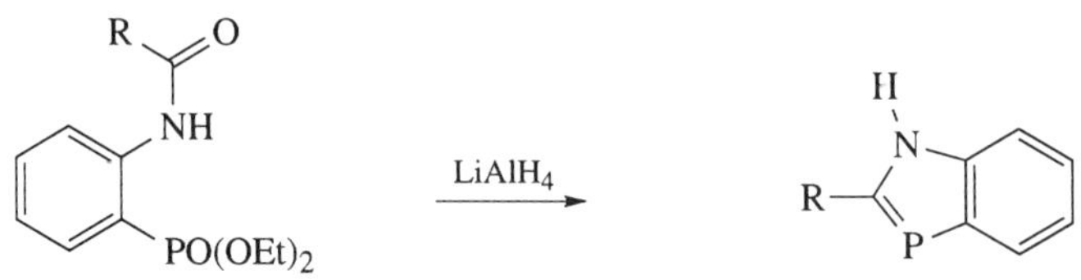

(5)

(**34**) (**35**)

4.3.3.3.2.4 Synthesis from chlorophosphines

Condensation of ketene aminals (**36**), R^1 = H, alkyl, aryl, R^2 = CN, $C_6H_4NO_2$-4 with phosphorus trichloride or tribromide gives 1,3-azaphospholes (**37**) ⟨89EGP268702⟩.

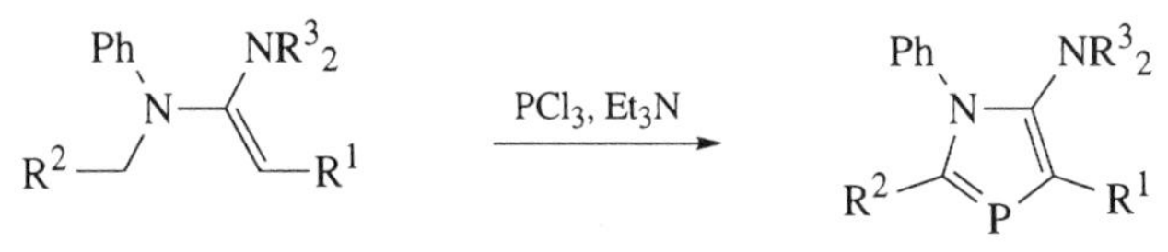

(6)

(**36**) (**37**)

The reagent (**38**) generated *in situ* from the action of LiBu^t on phenyl isonitrile and *t*-butyl dichlorophosphine give the 3*H*-1,3-benzazaphosphole (**39**) which is converted to the 1*H*-1,3-benzazaphosphole (**40**) by flash vacuum pyrolysis ⟨89JOM(364)C17⟩.

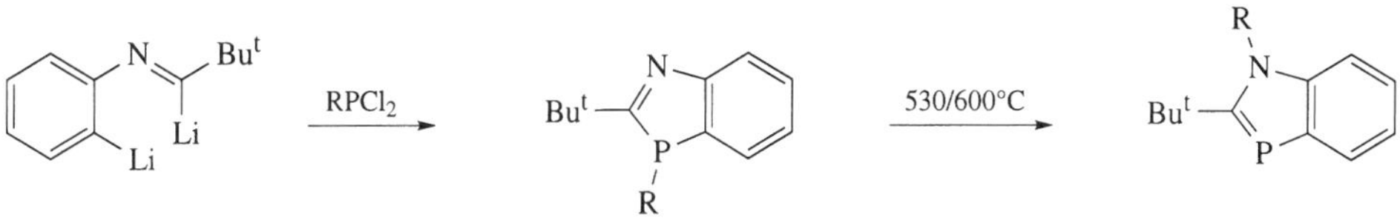

(**38**) (**39**) (**40**) (7)

4.3.3.3.2.5 Synthesis from tris(trimethylsilyl)phosphine

Oxazolium perchlorates (**41**) enter a KF assisted condensation with tris(trimethylsilyl)phosphine to give 1,3-azaphospholes (**42**), R = Me, Et, Bu^n, R^1 = H, Me, Et, Ph, R^2 = H, Me, Et, Ph. Suitably substituted chloromethyleneiminium chlorides can also be used for the condensation ⟨86TL4419⟩.

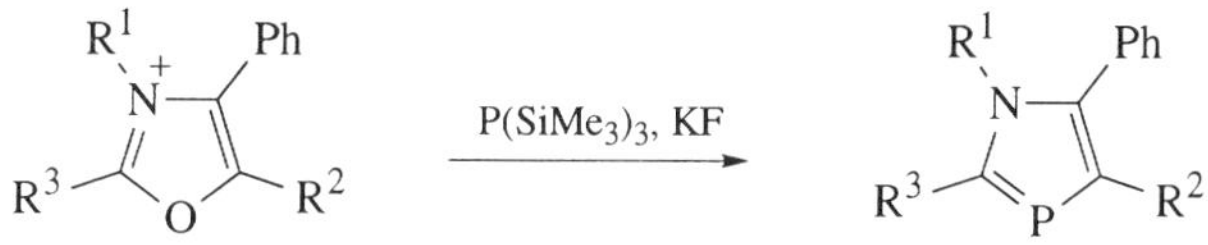

(**41**) (**42**) (8)

Münchnones (**43**) and tris(trimethylsilyl)phosphine give 4-trimethylsiloxy-1,3-azaphospholes (**44**), Ar = Ph, C_6H_4OMe-4 ⟨87TL1089, 88TL3387⟩ and after methanolysis, hydroxy-1,3-azaphospholes (**45**).

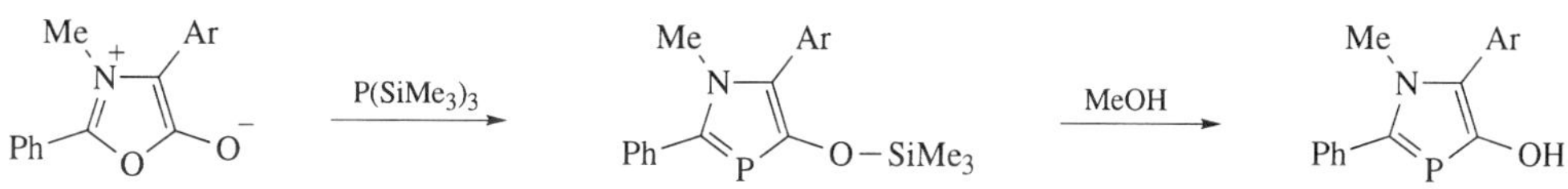

(**43**) (**44**) (**45**) (9)

4.3.3.3.2.6 Synthesis from phosphaalkynes/alkenes

The reaction of münchnones (**43**), Ar = Ph, C_6H_4OMe-4, and a (trimethylsilyl)chlorophosphaalkene, R = Ph, ⟨86TL4419⟩ or a phosphaalkyne, R = H, Bu^t, Mes ⟨87CB1809, 88AG1541, 88JOM(338)329, 98S1305⟩ gives 1,3-azaphospholes (**46**).

(**43**) 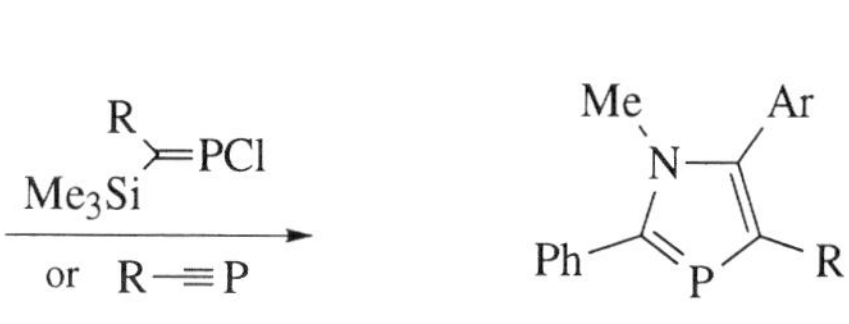 (10)

(**46**)

Cycloaddition of diphenylnitrilium ylide (**47**), Ar = Ph, and a (trimetylsilyl)chlorophosphaalkene, R = Ph ⟨88TL785⟩, or of (**47**), Ar = $C_6H_4NO_2$-4, and a phosphaalkyne, R = Bu^t, CMe_2Et, 1-adamantyl, yields 1-unsubstituted 1,3-azaphospholes (**48**). The cycloaddition of the phosphaalkynes is regiospecific, the constitution of (**48**) could not be determined with certainty, however ⟨90CRV191⟩.

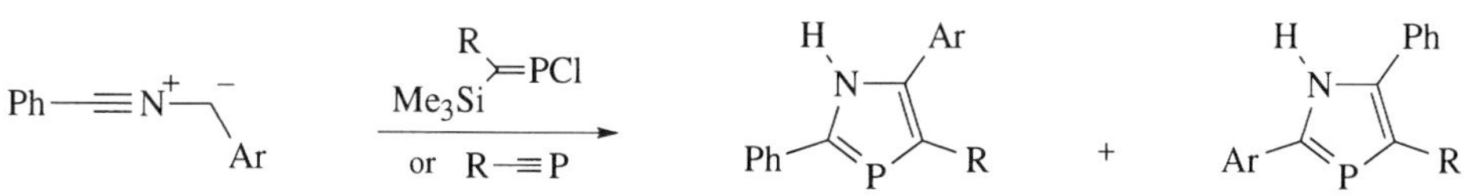

(**47**) (**48**) (11)

4.3.3.3.3 Application

7-Amino-1*H*-1,3-azaphospholo[4,5-*d*]-pyrimidine and several of its derivatives exhibit superior activity against TNFα and IL-1β production in the lipopolysaccharide or a mixture of phorbol myristate acetate and interferon gamma stimulated human monocytic THP-1 cells ⟨97MI2⟩.

4.3.3.4 1,2- and 1,5-Anellated 1,3-Azaphospholes (9–19)

4.3.3.4.1 Reactivity

Depending on their substituents 2-phosphaindolizines (**9**) are hydrolysed more or less readily. The ring first opens at the P=C bond ⟨91CB475⟩. 1,3-Azaphospholo[1,5-*b*]benzothiazoles (**13**) hydrolyse rapidly ⟨92HAC359⟩.

In contrast to 2-phosphaindolizines (**9**) which are stable in dry air and which cannot be alkylated ⟨91CB475⟩, 1-phosphaindolizines (**14**) readily decompose in air and form phosphaindolizinium salts with methyl iodide ⟨87TL1511⟩. This is in accord with the higher positive charge at the phosphorus atom in (**9**) than in (**14**) (Section 4.3.3.2).

4.3.3.4.1.1 Reactivity towards electrophiles and nucleophiles

As shown for the representative (**49**), 2-phosphaindolizines (**9**) can be brominated in 1-position to (**50**) with bromine in presence of a base or with NBS, and are substituted by PCl_3 or $PhPCl_2$ (but not by Ph_2PCl, MeCOCl or PhCOCl) or by perselenophosphonic anhydride ⟨92HAC359, 93PS(77)235, 98HAC445⟩. Hydrolysis of the bromo derivative (**50**) interestingly involves the debromination of C-1 coupled to the oxidation of the phosphorus atom to (**51**).

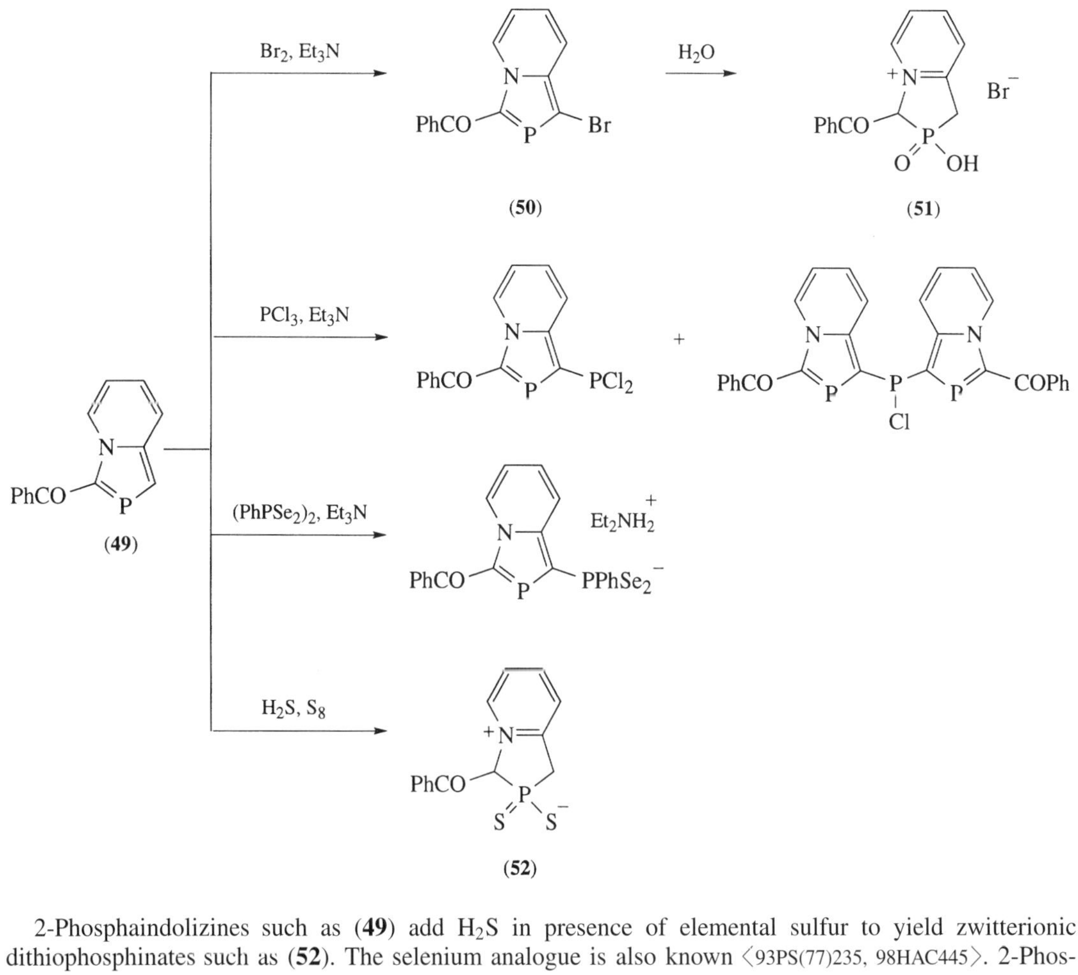

(**50**) (**51**) (**52**)

2-Phosphaindolizines such as (**49**) add H_2S in presence of elemental sulfur to yield zwitterionic dithiophosphinates such as (**52**). The selenium analogue is also known ⟨93PS(77)235, 98HAC445⟩. 2-Phosphaindolizines act as *P*-coordinated ligands L in forming complexes of the type $LM(CO)_5$, *cis*-$L_2M(CO)_4$ and $L_3M(CO)_3$, M = Cr, Mo, W ⟨98EJI1079⟩. The first type of complexes is also reported with 2,3-dihydro-1,3-azaphospholo[5,1-*b*]thiazole and -oxazole as L ⟨99JOM(577)337⟩.

4.3.3.4.1.2 Cycloaddition reactions

1,3-Azaphospholo[5,1-*b*]benzothiazoles (**13**), as well as part of the 2-phosphaindolizines (**9**) and 1,3-azaphospholo[5,1-*a*]isoquinolines (**10**) do not react with 2,3-dimethylbutadiene or isoprene alone. However, if the reaction is carried out in the presence of O_2, S_8 or Se_n, from (**53**) as an example the [4 + 2]-cycloadducts (**54**), X = O, S, Se with pentavalent phosphorus are obtained ⟨93PS(77)235, 99HAC598⟩. The reaction with isoprene occurs regioselectively ⟨93PS(77)235, 94T7675⟩. This reaction is analogous to that of 2-phenyl-4,5-dimethyl-phosphinine ⟨84CC508, 90HAC37⟩.

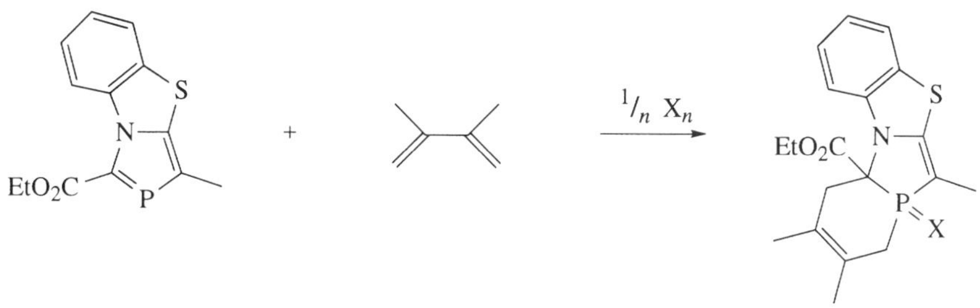

(12)

(**53**) (**54**)

4.3.3.4.2 Synthesis

4.3.3.4.2.1 Survey of the synthetic routes

A variety of anellated 1,3-azaphospholes (**9–19**) have been synthesized by variants of three procedures also used for monocyclic 1,3-azaphospholes ⟨94T7675⟩. The analogy of the syntheses of indolizines and phosphaindolizines has been pointed out comprehensively ⟨98JIC648⟩.

NCCC + electrophilic P (Section 4.3.3.4.2.2): This procedure parallels the classical indolizine synthesis ⟨69CB669⟩. A 2-methylcycloiminium ion provides the four-membered chain which is condensed with phosphorus trichloride to the five-membered ring ⟨95S361⟩.

CNCPC (Section 4.3.3.4.2.2): Here the 1,3-azaphosphole is anellated to the pyridine ring by the 1,5-electrocyclization of an intermediate bis(pyridiniumylidyl)phosphenium chloride ⟨99TL1565⟩.

CNCC + nucleophilic P (Section 4.3.3.4.2.3): As in the case of monocyclic1,3-azaphospholes (Section 4.3.3.3.2.5) tris(trimethylsilyl)phosphine provides the phosphorus ring member by replacing the oxygen atom of an oxazolium ring.

CNC + CP (Section 4.3.3.4.2.4): Azomethine ylides of pyridine, quinoline, isoquinoline and related nitrogen heterocycles undergo 3 + 2 cycloaddition with a variety of dipolarophiles leading to fused heterocycles ⟨73S469⟩. This method has been extended to the use of a phosphaalkyne as dipolarophile.

4.3.3.4.2.2 Synthesis from phosphorus trichloride

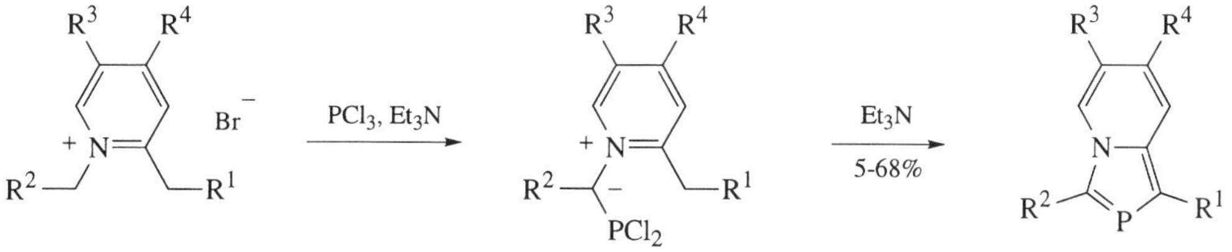

(13)

(**55**) (**56**) (**57**)

Salts resulting from the *N*-alkylation (R^2 = $C_6H_4NO_2$-4, COPh, CO_2Et, CN) of 2-alkylpyridines (**55**),

R^1 = H, Me, Ph, $R^{3,4}$ = H, Me, Bu, (**58**), of 2,5-dimethylpyrazine (**60**), of 2-alkylbenzothiazoles (**62**), of 2-methylthiazoline (**64**), and of 2-ethyloxazoline (**66**) undergo cyclocondensation with PCl_3 to give 1,3-azaphospholo[1,5-*a*]pyridines (2-phosphaindolizines) (**57**, **59**), the 1,3-azaphospholo[1,5-*a*]pyrazine (**61**) 〈90PS(49/50)349, 91CB475, 92IJC254, 98HAC333〉, 1,3-azaphospholo[5,1-*b*]benzothiazoles (**63**), 1,3-azaphospholo[5,1-*b*]thiazolines (**65**), and 1,3-azaphospholo[5,1-*b*]oxazoline (**67**) 〈92HAC351, 94PS(86)139〉, respectively. A PCl_2-substituted pyridinium ylide (**56**) and the corresponding benzothiazolium ylide have been isolated as intermediates 〈96PS(112)121〉.

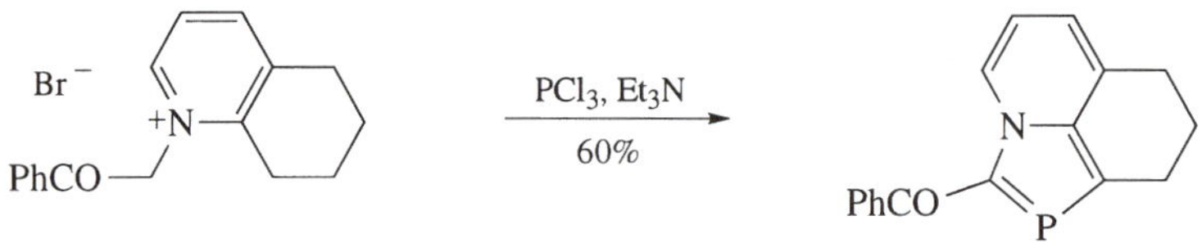

(**58**) (**59**) (14)

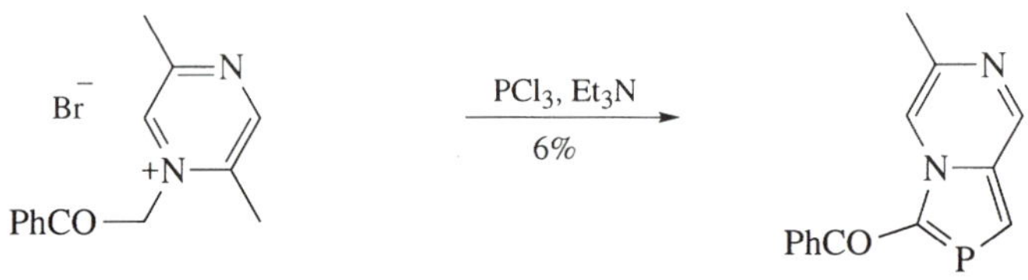

(**60**) (**61**) (15)

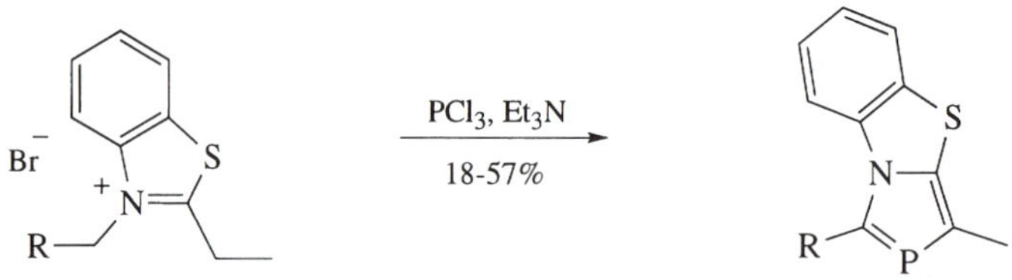

(**62**) (**63**) (16)

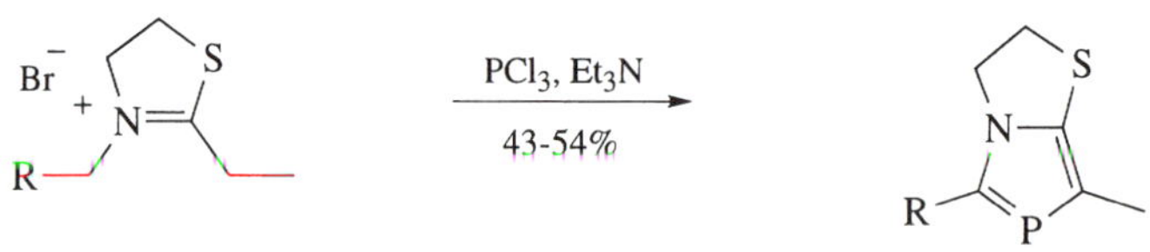

(**64**) (**65**) (17)

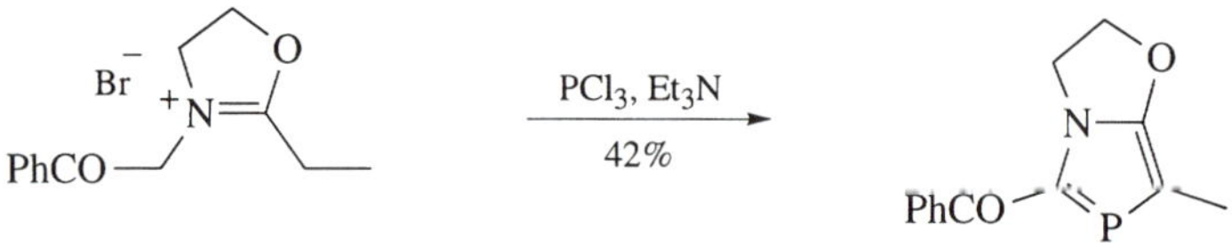

(**66**) (**67**) (18)

In a one pot synthesis, 2-phosphaindolizines (**70**) can be prepared from *N*-alkoxycarbonylmethyl pyridinium bromides (**68**) (R^1 = Me, Et; $R^{2,3,4}$ = H, Me), phosphorus trichloride, and triethyl amine in a molar ratio of 2 : 1 : 4. In an initial condensation reaction the phosphenium chloride (**69**) seems to be formed which undergoes 1,5-electrocyclization on standing at room temperature 〈99TL1565〉. A crossed reaction leads to a mixture of compounds (**70**) including those with mixed substituents R. 1,3-Azaphospholo[5,1-*a*]isoquinolines of type (**10**) are obtained in the same way 〈99HAC598〉.

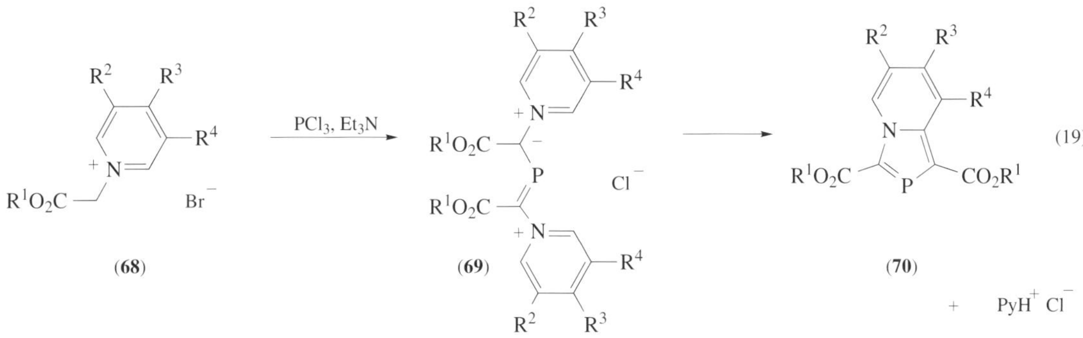

(19)

4.3.3.4.2.3 *Synthesis from tris(trimethylsilyl)phosphine*

1,3-Oxazolo[3,2-*a*]pyridinium salts (**71**), R = Me, Pr^i, Bu^t, Ph, react with $P(SiMe_3)_3$ or $LiP(SiMe_3)_2$ to afford 1-phosphaindolizines (**72**) ⟨87TL1511⟩.

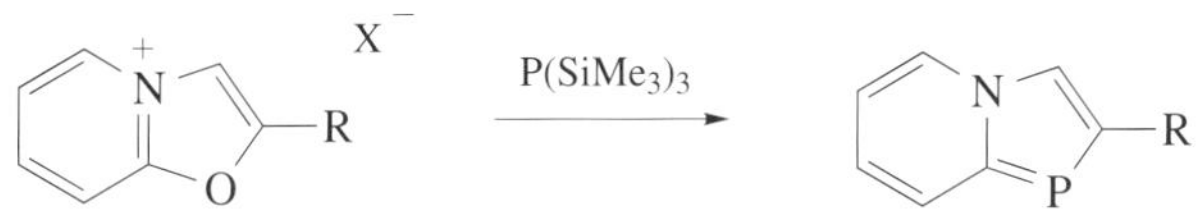

(71) (72) (20)

4.3.3.4.2.4 *Synthesis from phosphaalkynes*

The pyridinium ylide (**73**) undergoes regiospecific [3 + 2] cycloaddition with *t*-butyl-phosphaacetylene to give the 2-phosphaindolizine (**74**). Cycloadditions of pyridinium (**75**), X = CH, Y = CH, pyridazinium (**75**), X = N, Y = CH, and pyrazinium (**75**), X = CH, Y = N, dicyanomethylides are not regioselective and a mixture of the two isomers and is obtained in each case. However, regioselectivity is observed when *t*-butyl and isopropoxy groups are introduced in 4-position of (**75**), X = CH, Y = CH. With isoquinolinium and phthalazinium ylides the reaction is again regiospecific, though the dipole orientation is reversed as compared to the formation of (**74**), yielding tricyclic systems of type (**15**) and (**17**) ⟨92TL1049⟩.

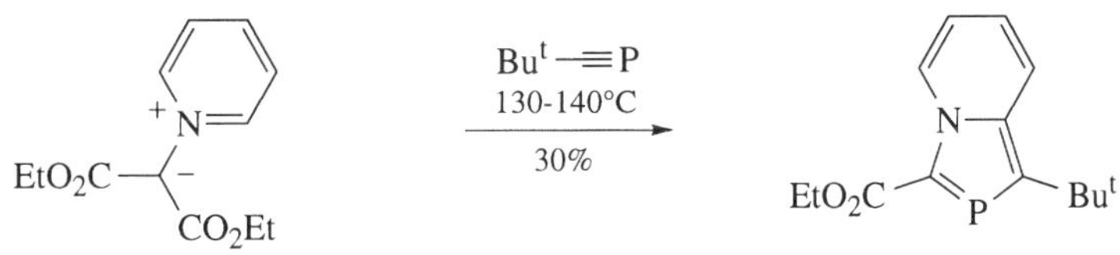

(73) (74) (21)

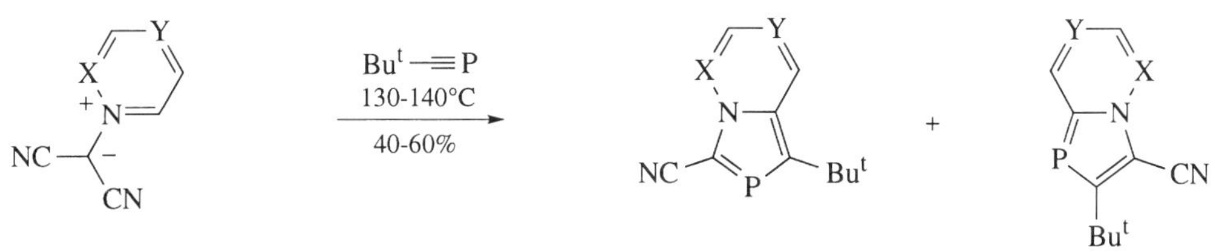

(75) (76) (77) (22)

1,3-Dipoles generated from 3*H*-pyrido[1,2,3-*de*]quinoxalinium bromides react analogously with *t*-butyl phosphaacetylene and yield regiospecifically azaphosphaullazines of type (**19**) ⟨94CC1121⟩.

4.3.3.4.3 ***Application***

A catalyst from the 2-phosphaindolizine (**57**), $R^2 = CO_2Et$, $R^{1,3,4} = H$, and $Rh(CO)_2acac$ has been used for the hydroformylation of styrene ⟨96CC2071, 99JMOC(A143)143⟩.

4.3.3.5 1,3-Oxaphospholes (2) and 1,3-Benzoxaphospholes (7)

4.3.3.5.1 Reactivity

2-Alkyl-1,3-benzoxaphospholes are colorless liquids, thermally stable up to 130°C; the 2-aryl derivatives are yellow crystalline solids ⟨80JOM(186)39, 80ZC342, 83ZC439, 85PS(25)345⟩. Air oxidation of the alcoholic solution is fast for the 2-methyl, medium for the 2-*t*-butyl and slow for the 2-phenyl derivatives. It proceeds probably *via* cycloaddition of triplet oxygen to the P=C bond and subsequent fission of it.

2-*t*-Butyl-1,3-benzoxaphosphole (**78**) has mostly been used to investigate the reactivity. While it does not react with MeI, MeCOCl, Me_3SiCl, PCl_3, water, alcohols and amines, it adds HCl, PhSH, $PhAsH_2$ and in presence of AIBN also Me_3SnH. Adduct (**79**) undergoes dismutation to yield the dihydro derivative (**80**), which also results from (**78**) and Me_3SnH. At −60°C (**78**) and $LiBu^t$ rapidly give the adduct (**81**) which on hydrolysis furnishes (**82**) as 1 : 9 mixtures of *cis*/*trans* isomers ⟨84PS(20)347⟩.

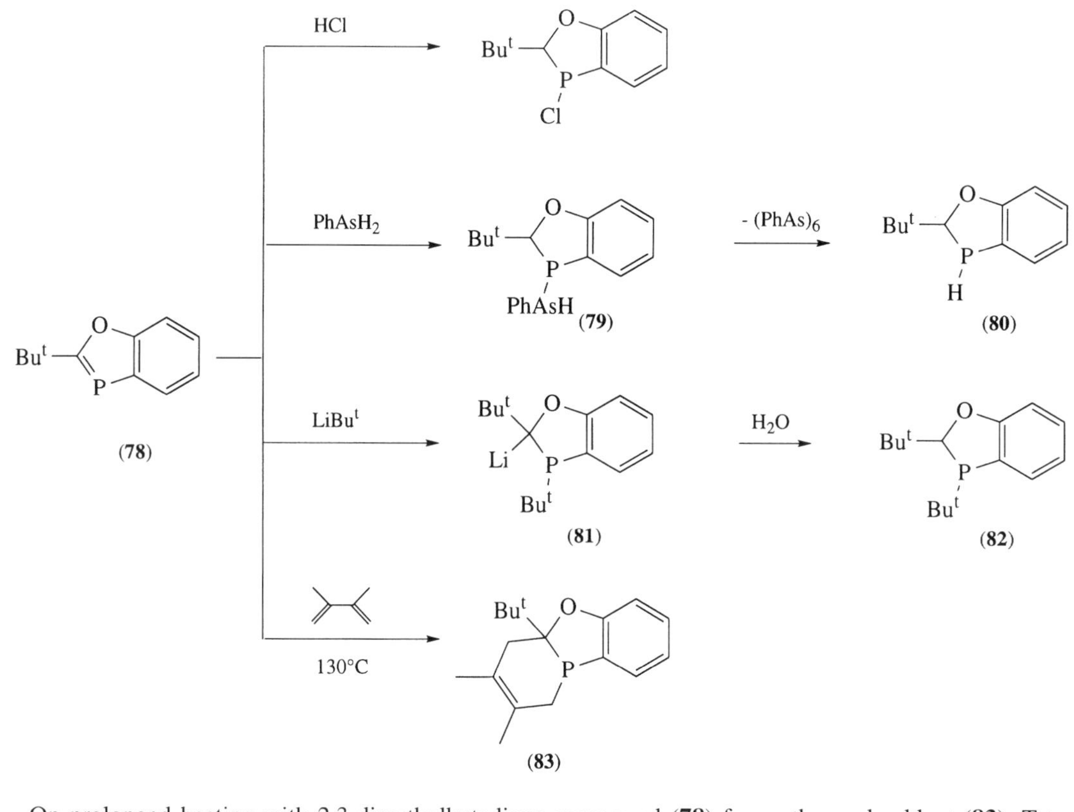

On prolonged heating with 2,3-dimethylbutadiene compound (**78**) forms the cycloadduct (**83**). Tetrachloro-*o*-benzoquinone adds in the same way but much more quickly; in a subsequent reaction a second molecule adds to the phosphorus atom ⟨83TL5481⟩.

4.3.3.5.2 Synthesis

4.3.3.5.2.1 Survey of the synthetic routes ⟨00T63⟩

OCCP + C (Section 4.3.3.5.2.2): This synthesis of 1,3-benzoxaphospholes (**7**) is analogous to that of 1,3-benzazaphospholes (**33**) in Section 4.3.3.3.2.2 using 2-phosphinophenols to start with.

COCC + P (Section 4.3.3.5.2.3): Monocyclic 1,3-oxaphospholes (**2**) can be prepared like 1,3-azaphospholes (**1**) from ketene aminals (Section 4.3.3.3.2.4).

CCO or COC + CP (Section 4.3.3.5.2.4): The addition of phosphaalkynes to a diazoketone with elimination of dinitrogen or to an isomünchnone with elimination of phenyl isocyanate provides a simple and selective access to 1,3-oxaphospholes.

CO + C + CP (Section 4.3.3.5.2.4): In this synthesis of (**2**) a phosphaalkyne reacts with a chromium pentacarbonyl carbene complex which provides a carbon monoxide and a carbene unit as the other ring constituents.

4.3.3.5.2.2 Synthesis from 2-phosphinophenols

Condensation of 2-phosphinophenols (**84**) with *N*-arylimidoyl chlorides give 1,3-benzoxaphospholes (**85**) ⟨80JOM(186)39, 80ZC342, 85PS(25)345⟩. They are also obtained by acylation of (**84**) and dehydration of the intermediate.

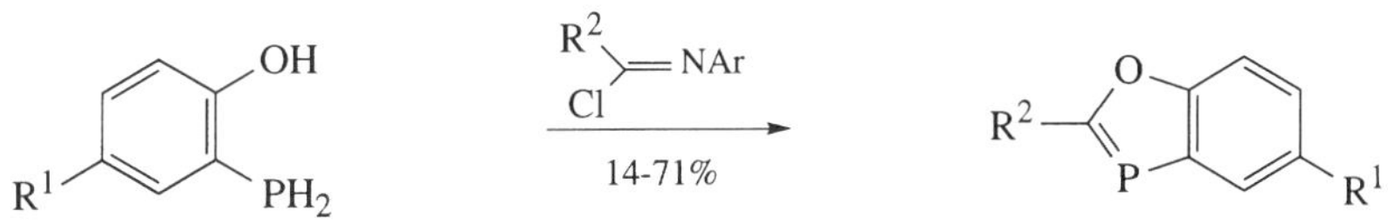

(**84**) (**85**) (23)

4.3.3.5.2.3 Synthesis from phosphorus trihalides

Condensation of α-(4-nitrobenzyl)oxy- or α-cyanomethoxy-substituted enamines (**86**) with PCl_3 or PBr_3 and a tertiary amine leads to 1,3-oxaphospholes (**87**) ⟨89EGP268700⟩. 1,3-Thiaphospholes and 1,3-selenophospholes are obtained analogously (Section 4.3.3.6.2.4).

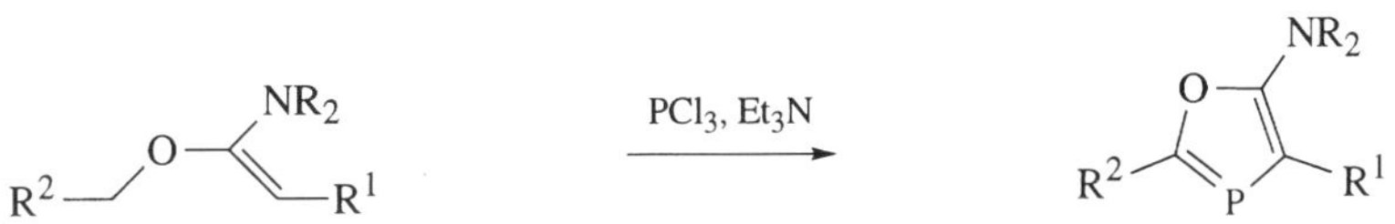

(**86**) (**87**) (24)

4.3.3.5.2.4 Synthesis from phosphaalkynes

Chromiumpentacarbonyl arylmethoxycarbene complexes (**88**), Ar = Ph, 2-naphthyl, 2-furyl, 2-thienyl, 3,4-dimethoxyphenyl, 2-(1,4-dimethoxy)naphthyl, react with *t*-butyl phosphaacetylene to give 1,3-oxaphospholes (**89**) together with phosphahydroquinone derivatives. Depending on Ar the ratio varies from 1 : 1 to 1 : 0 ⟨89CC788, 90NJC433, 93T5577⟩.

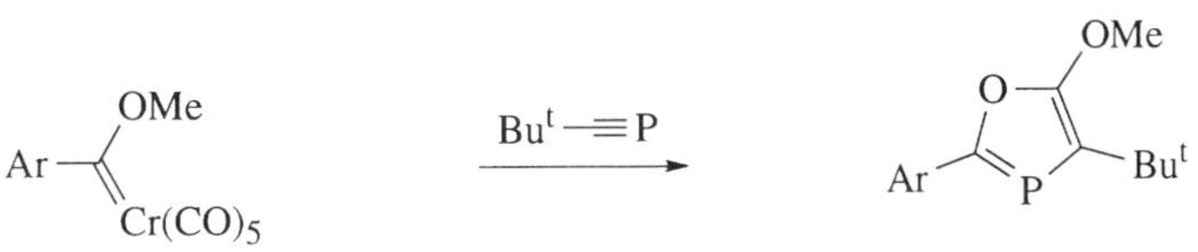

(**88**) (**89**) (25)

Catalytic elimination of nitrogen from the diazo compound (**90**) in the presence of a phosphaalkyne yields the bicyclic 1,3-oxaphosphole (**91**) ⟨00S360⟩. The result can be explained in terms of a rearrangement of a phosphirene formed as an intermediate ⟨00T63⟩.

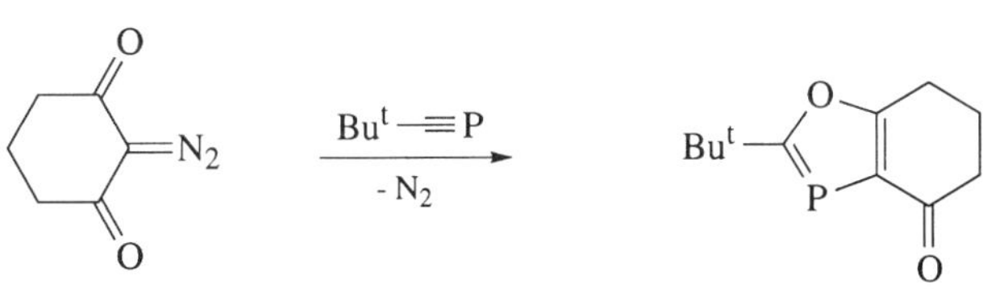

(**90**) (**91**) (26)

Phosphaalkynes react with the isomünchnones (**92**) (R = Me, Et; Ar = a variety of aryl substituents) in a regiospecific process to furnish the monocyclic 1,3-oxaphospholes (**93**). The bicyclic intermediates from the 1,3-dipolar cycloaddition reaction of the two components cannot be detected ⟨00T63⟩.

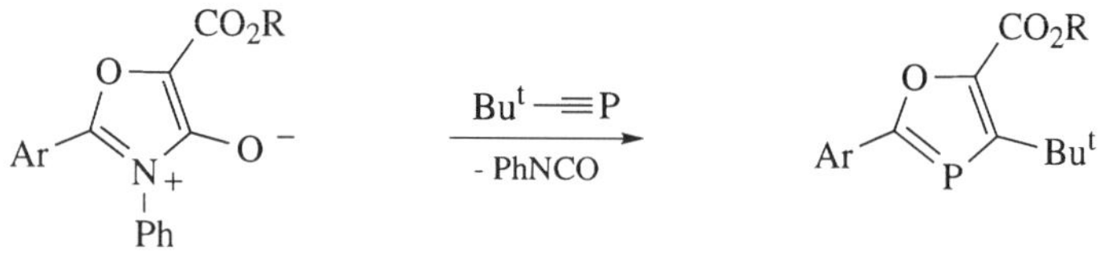

(92) (93) (27)

4.3.3.6 1,3-Thia/selenaphospholes (3, 4) and 1,3-Benzothiaphospholes (6)

4.3.3.6.1 Reactivity

Unsubstituted 1,3-thiaphosphole (**3**) is an air sensitive liquid ⟨99CC1283⟩; substituted derivatives are colorless to pale yellow air-stable crystalline solids. The addition of $LiBu^t$ to the 2,4,5-triphenyl derivative seems the only reaction described ⟨87TL2119⟩. No H/D exchange is observed for (**3**) in CH_3CO_2D ⟨99CC1283⟩.

4.3.3.6.2 Synthesis

4.3.3.6.2.1 Survey of the synthetic routes

SCCP + C (Section 4.3.3.6.2.2): 1,3-Benzothiaphospholes (**6**) are prepared from 2-phosphinothiophenol as the 1,3-benzoxaphosphole (**7**) from 2-phosphinophenol (Section 4.3.3.5.2.2).

SCP and SeCP + CC (Section 4.3.3.6.2.3): This synthesis of (**3**) involves the cycloaddition of an acetylene to an intermediate thioacylphosphinidene fragment and is related to the electrocyclization of the phenyliminoacylphosphinidene mentioned in Section 4.3.2.3. A selenadiphosphole provides the CSeP fragment for the synthesis of (**4**).

CSCC and CSeCC + electrophilic P (Section 4.3.3.6.2.4): The unsubstituted prototype (**3**) becomes accessible by the Sn/P replacement in a dihydro-1,3-thiostannole. Other syntheses of (**3**) and (**4**) according to this pattern are analogous to that of (**2**) in Section 4.3.3.5.2.3.

CSCC + nucleophilic P (Section 4.3.3.6.2.5): This synthesis follows the same pattern but with inverse polarity; it involves an S/P exchange in the ring.

CSC + CP (Section 4.3.3.6.2.6): This route involves the cycloaddition of a phosphaalkyne or phosphaalkene to a mesoionic 1,3-dithiolium-4-olate and the elimination of COS.

CCSe + CP (Section 4.3.3.6.2.6): The cycloaddition of a phosphaalkyne to an intermediate selenocarbonylcarbene yields a 1,3-selenaphosphole (**4**).

4.3.3.6.2.2 Synthesis from 2-phosphinothiophenol

2-Phosphinothiophenol (**94**) undergoes condensation with amide acetals or with benzaldehyde to give 1,3-benzothiaphospholes (**95**) ⟨80TL3483⟩:

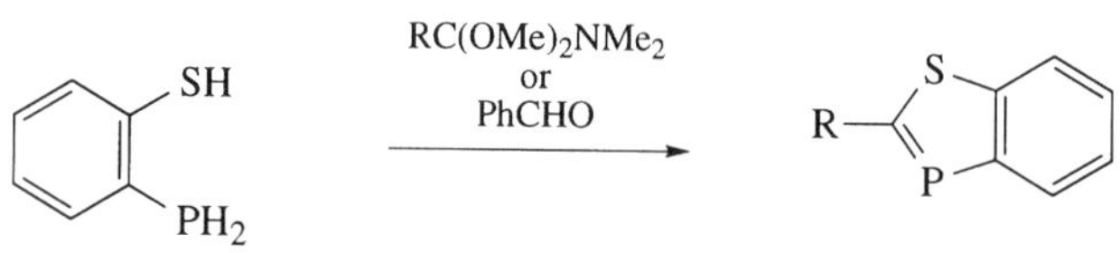

(94) (95) (28)

4.3.3.6.2.3 Synthesis from a thiaphosphirane or a selenadiphosphole

The thiaphosphirane (**96**) reacts with dimethyl acetylenedicarboxylate in presence of KF(18-crown-6) to the 1,3-thiaphosphole (**97**) ⟨92TL4421⟩. A mixture of 1,2- and 1,3-selenaphospholes is obtained from the reaction of 3,5-di-*t*-butyl-1,2,4-selenadiphosphole with dimethyl acetylenedicarboxylate (Section 4.3.6.8) ⟨99TH1⟩.

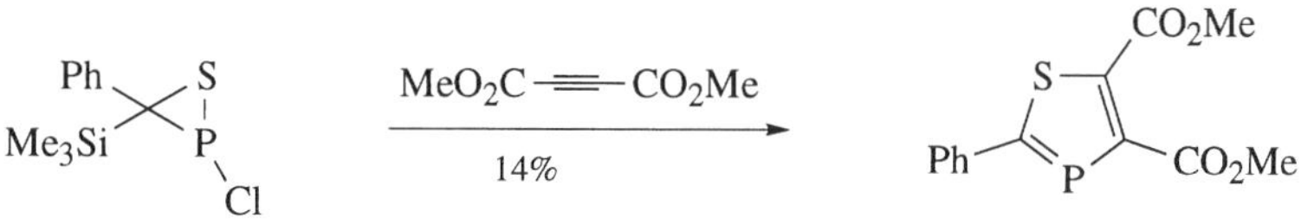

(96) (97) (29)

4.3.3.6.2.4 *Synthesis from phosphorus trihalides*

Treatment of the 2,3-dihydro-1,3-thiastannole (**98**) with phosphorus tribromide followed by DBU affords the unsubstituted 1,3-thiaphosphole (**99**) ⟨99CC1283⟩.

(98) (99) (30)

Substituted 1,3-thiaphospholes ⟨89EGP268701⟩ and 1,3-selenaphospholes ⟨89EGP268703⟩ are obtained in analogy to the synthesis of 1,3-oxaphospholes in Equation (24).

4.3.3.6.2.5 *Synthesis from tris(trimethylsilyl)phosphine*

Tris(trimethylsilyl)phosphine effects an S/P-exchange in a 1,3-dithiolium-4-olate (**100**), $R^{1,2}$ = Ph, and gives the respective 1,3-thiaphosphole (**101**), $R^3 = OSiMe_3$ ⟨87TL1089⟩.

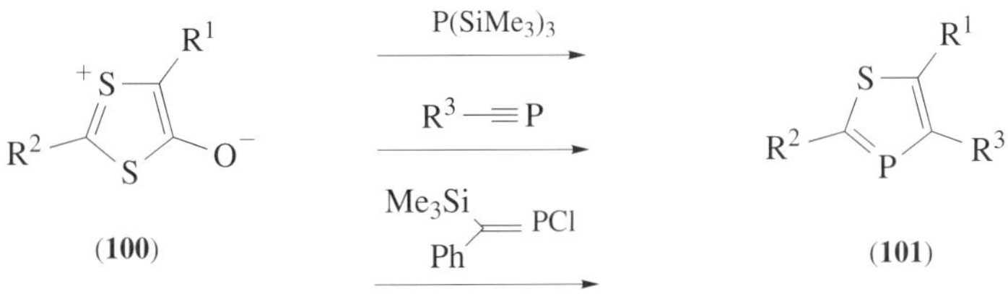

(100) (101) (31)

4.3.3.6.2.6 *Synthesis from phosphaalkynes/alkenes*

The 1,3-dipolar cycloaddition of (**100**) and a phosphaalkyne ⟨87CB1809, 88AG1541, 98S1305⟩ or a (trimethylsilyl)chlorophosphaalkene ⟨87TL2119⟩ yields 1,3-thiaphospholes (**101**), $R^{1,2}$ = Ph, R^3 = Bu^t, Mes, and R^1 = Me, Ph, 1-piperidyl, 1-morpholyl, R^2 = Ph, C_6H_4Me-4, C_6H_4Br-4, C_6H_4OMe-4, R^3 = Ph, respectively (Equation (31)).

Photolysis or thermolysis of 1,2,3-selenadiazoles (**102**) in presence of phosphaalkynes, R^2 = Bu^t, CMe_2Et, 1-adamantyl, gives rise to 1,3-selenaphospholes (**103**) ⟨91SL356⟩.

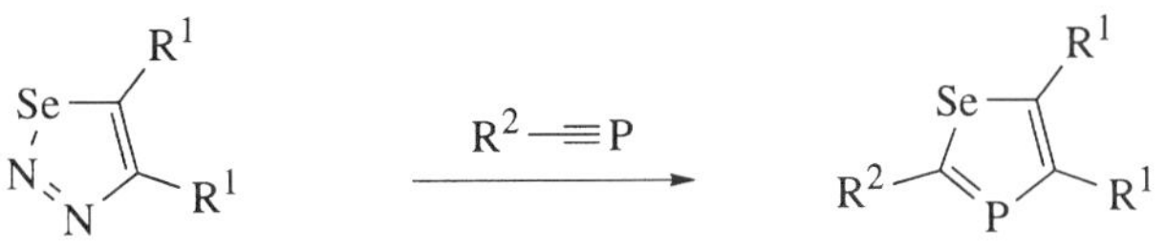

(102) (103) (32)

4.3.3.7 1,3-Diphospholides (21), 1,3-Phosphaarsolides (22) and 1,3-Benzodiphospholides (24)

4.3.3.7.1 *Reactivity*

The reactivities of 1,3-benzazaphospholes (**5**) and 1,3-benzodiphospholes (**23**) as well as those of their anions have been compared ⟨91PS(56)189⟩.

4.3.3.7.1.1 Complex formation, reactions with electrophiles and oxidation of 1,3-diphospholides

A variety of transition metal complexes has been prepared from 1,3-diphospholide anions ⟨94CCR(137)1, 95CCR(145)201, 95CSR319, 96BSB675, B-98MI1⟩. In these complexes the ligand may be η^1(*P*)-bonded ⟨91OM2835⟩, η^3(*PCP*)-bonded ⟨89JOM(373)C17, 90CC472⟩ or η^5-bonded ⟨87CC1146, 87JOM(334)C35, 90AG575, 90CC472, 91JOM(415)C15, 92JOM(430)C10, 92OM459, 92PO601, 93CC311, 94JOM(479)C28, 95CC1659, 96JCS(D)739, 96JOM(512)141, 97JOM(529)69, 98CC797, 01JCS(D)1013⟩. An example of fluxional η^3/η^5-bonding is also known ⟨95JOM(487)C21⟩. The phosphorus atoms in the ring exert a greater electron-withdrawing effect on the metal atom than the carbon atoms ⟨97JOM(529)375, B-98MI1⟩. The nido-structured indium diphospholide $(Bu^tC)_3P_2In$ is formed along with the analogously structured indium triphospholide (Section 4.3.7.2.2) ⟨99OM793, 00JCS(D)1715⟩. Of the 1,3-benzodiphospholide anion (**106**) η^1(*P*)- and η^5-complexes are known and an η^1-complex of the 1,3-benzodiphosphole (**107**) ⟨90JOM(382)53, 90PS(49/50)355⟩.

Protonation of the 1,3-diphospholide anion $(PhC)_2P_2CH^-$ generates the transient 2*H*-1,3-diphosphole $(PhC)_2P_2CH_2$ which can be trapped ⟨91OM2835⟩. Oxidation of the anion (**104**) gives the 1,1′-biphospholyl (**105**) ⟨93CC267⟩. Protonation or oxidation of a mixture of the diphospholide and triphospholide anions (**104**) and $(Bu^tC)_2P_3^-$ give the tricyclic systems $H_2(Bu^tC)_6P_4$ and $H_2(Bu^tC)_5P_5$ ⟨90CC1307, 96BSB675, 97JOM(536-537)273⟩ or the cage compound $(Bu^tC)_5P_5$ ⟨89JOM(375)C31⟩, respectively.

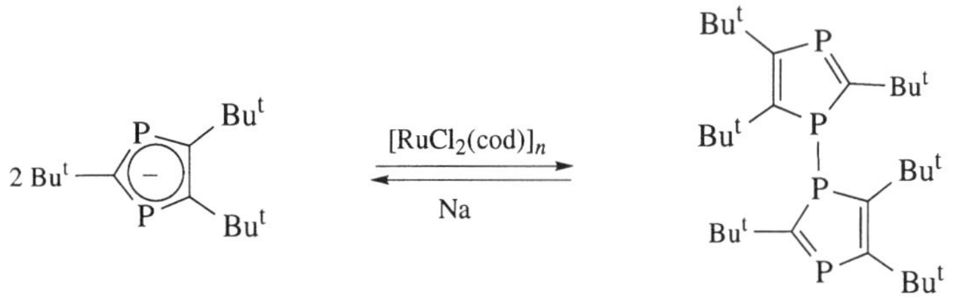

(33)

(**104**) (**105**)

Lithium 2-*t*-butyl-1,3-benzodiphospholide (**106**) reacts with ethyl bromide and pivaloyl chloride to give the 1-substituted products, such as (**107**), which can be oxidized to (**108**). Hydrogen chloride (but not methanol) yields the 1,3-benzodiphosphole (**109**), which immediately rearranges to the isomer (**110**). The same rearrangement is observed for the 1-SiMe$_3$ and 1-PPh$_2$ derivatives ⟨90JOM(382)53, 90PS(49/50)355⟩. Lithium 2-trimethylsiloxy-1,3-benzodiphospholide is 1-trimethylsilylated (88ZC250), lithium 2-anilido-1,3-benzodiphospholide is both *N*- and 1-trimethylsilylated ⟨85ZAAC(529)216⟩.

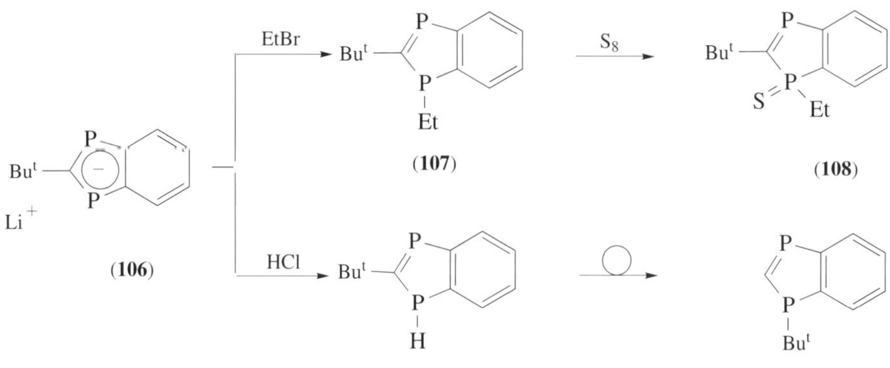

(**107**) (**108**)

(**106**)

(**109**) (**110**)

4.3.3.7.1.2 Addition and cycloaddition reactions of 1,3-dipospholes

The 1,3-benzodiphosphole (**107**) adds HCl and MeOH as well as LiBu to its P=C bond ⟨90JOM(382)53, 90PS(49/50)355⟩. The addition of HCl and ROH is also known of a 2,4,5-tri-*t*-butyl-1,3-diphosphole ⟨88TL5867⟩.

Compound (**107**) gives a Diels–Alder addition of 1,4-diphenylbutadiene to its P=C bond ⟨90JOM(382)53, 90PS(49/50)355⟩. As mentioned in Section 4.3.3.7.1.1, cycloaddition reactions are induced by the protonation of (**104**). Phosphorus-carbon cage compounds derived from (**104**) have been reviewed ⟨99MI257⟩.

4.3.3.7.2 *Synthesis*

As pointed out in the introduction (Section 4.3.3.1), neutral 1,3-diphospholes (**20**) and 1,3-benzodiphospholes (**23**) have a pyramidal phosphorus atom and are not fully conjugated; their anions, however, are. Syntheses therefore generally aim to these anions, i.e. to alkali metal 1,3-diphospholides (**21**) and 1,3-benzodiphospholides (**24**). In some cases they can be converted to neutral diphospholes by protonation, alkylation, silylation, acylation or oxidation (Section 4.3.3.7.1.1). In some other cases 1,3-diphospholes and 1,3-benzodiphospholes have also been prepared by direct synthesis.

4.3.3.7.2.1 *Synthesis from 1,2-diphosphinobenzenes*

Condensation of sodium 1,2-diphosphino benzene (**111**) with carbonic or carboxylic acid derived carbenium salts give 1,3-benzodiphospholes (**112**), $R^2 = H$, Me, NMe_2 ⟨81TL4475, 83PS(18)15, 87JOM(330)17⟩.

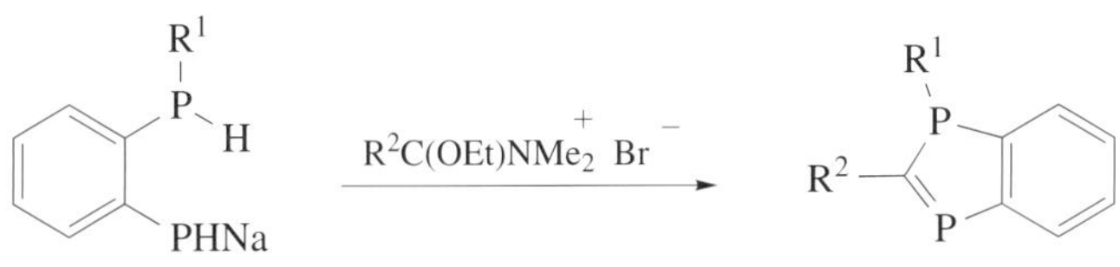

(34)

(**111**) (**112**)

Tetrakis(trimethylsilyl)-1,2-diphosphinobenzene (**113**) after lithiation reacts with carbon dioxide, diphenylcarbodiimide and acyl chlorides to give lithium 1,3-benzodiphospholides (**114**), R = $OSiMe_3$ ⟨88ZC255⟩, NPhLi ⟨85ZAAC(529)216⟩, Me, Bu^t, Ph, Mes ⟨90JOM(382)53, 90PS(49/50)355⟩. Dihydro-1,3-benzodiphospholes have also been prepared from 1,2-diphosphinobenzenes ⟨84ZAAC(519)75⟩. The parent 1,3-diphosphindolide (**114**), R = H, has been obtained starting from the reaction of 1,2-bis(lithiophenylphosphino)benzene with dichloromethane ⟨97BSF853⟩.

(35)

(**114**)

(**113**)

4.3.3.7.2.2 *Synthesis from dihydro-1,2-diphosphetes*

This route to monocyclic 1,3-diphospholides is related to the 1,3-benzodiphosphole synthesis in Section 4.3.3.7.2.1 as the reductive opening of a dihydro-1,2-diphosphete ⟨92CIC61⟩ results in a 1,2-diphosphinoalkene. Reaction of dihydro-1,2-diphosphetes (**115**) and (**120a**) with lithium metal and carbon disulfide, methyl iodide ⟨90AG575⟩, or dimethyldichlorosilane, acyl chloride, methanol ⟨92OM459⟩, or dichloromethane ⟨92PO601⟩ gives dihydro-1,3-diphospholes (**116**), (**118**) and (**121a**), respectively. Reaction with excess lithium converts them into diphosphole dianions of unknown structure which are monoprotonated by dry ammonium chloride to (**117**), (**119**), R = $(CH_2)_{3/2}$, Bn, Ph, (**122a**), R = Me, Et, Bu^t. Related to the second step is the reduction of the octaphenyltetraphosphafulvalene to give the 2,2′-bis(1,3-diphospholyl) dianion ⟨92AG1082⟩.

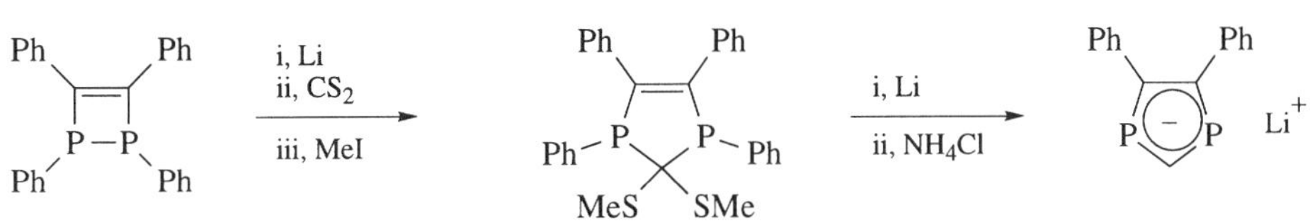

(36)

(**115**) (**116**) (**117**)

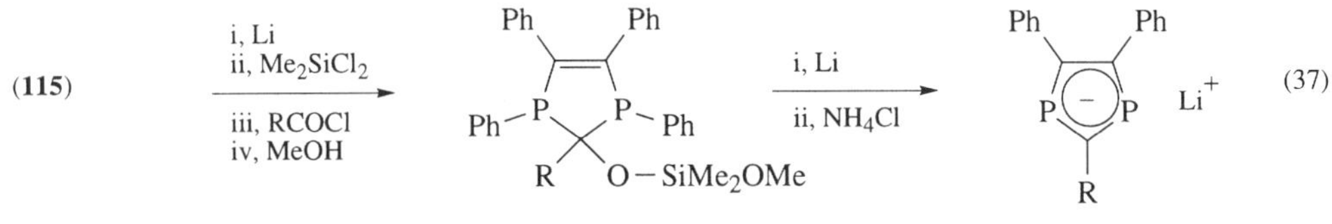

(**118**) (**119**)

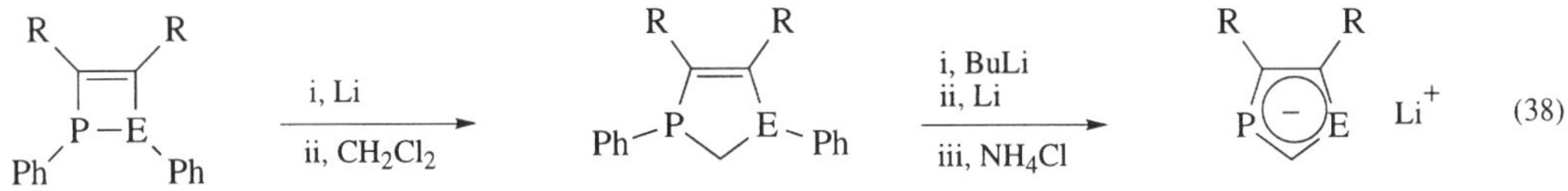

(**120a**) E = P
(**120b**) E = As

(**121a**) E = P
(**121b**) E = As

(**122a**) E = P
(**122b**) E = As

The 1,3-phosphaarsolide (**122b**), R = Et is accessible from the dihydro-1,2-phospharsete (**120b**) ⟨93BSF521, 93PS(77)254⟩. 2-Diphenylphosphino substituted 1,3-diphospholides such as (**119**), R = PPh_2 are obtained in the same way *via* the respective dihydro-1,3-diphospholes ⟨97JOM(529)69⟩.

4.3.3.7.2.3 Synthesis from phosphaalkynes

The 1,3-diphospholide anion (**123**) is formed in the reaction of $LiP(SiMe_3)_2$ or $LiSi(SiMe_3)_3$ and $Bu^tC{\equiv}P$ ⟨85MI55, 87CC1146, 96MI1⟩. Reduction of *t*-butyl-phosphaacetylene with sodium amalgam ⟨89PO2407⟩ or a low-valent tantalum species leads to a 1 : 1 mixture of the sodium 1,3-diphospholide (**123**) and 1,3,4-triphospholide (**124**) which can be separated by fractional crystallization ⟨89PO849⟩. Mixtures of $Pr^iC{\equiv}P$ and $Bu^tC{\equiv}P$ have also been reduced ⟨91JOM(415)C15⟩. From the reduction of the biphospholyl (**105**) with sodium (Equation (33)) (**123**) is obtained free from (**124**).

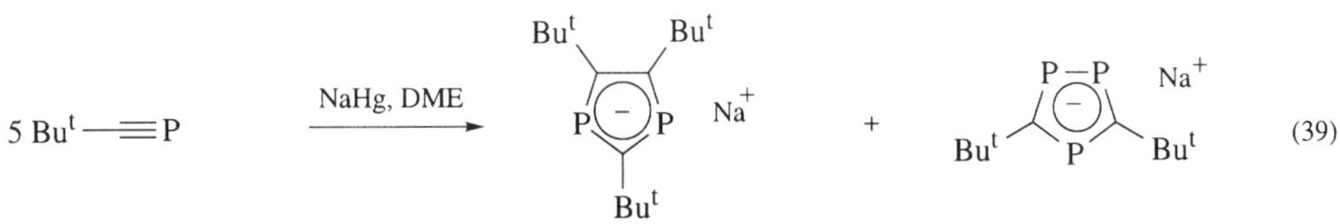

(**123**) (**124**)

Heating 1,2,4,5-tetrazine-dicarboxylates (**125**) with *t*-butyl-phosphaacetylene yields a 1,2,3-azadiphospholyl 1,3-diphosphole (**126**), R = Me, Et ⟨88TL5867⟩.

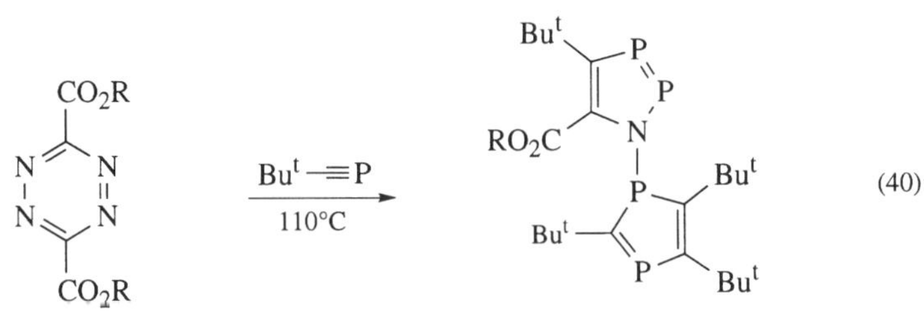

(**125**) (**126**)

The synthesis of the 1,3,6-triphosphafulvene (**127**), R = 2,4,6-$Bu^t_3C_6H_2$, (Equation (41)) may also involve a phosphaalkyne as an intermediate ⟨00AG2892⟩.

4.3.3.7.2.4 Synthesis from a 1,3,5-triphosphinine

The reduction of 2,4,6-tri-*t*-butyl-1,3,5-triphosphinine with a potassium mirror generated the 2,4,5-tri-*t*-butyl-1,3-diphospholide anion (**123**) ⟨00OM219⟩.

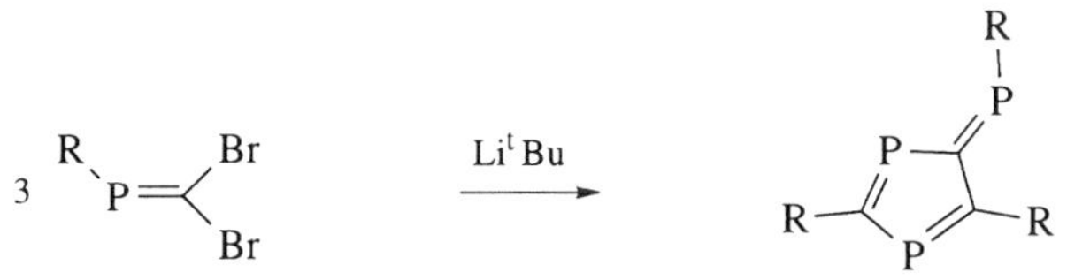

(41)

(127)

4.3.4 PHOSPHOLES WITH TWO ADDITIONAL HETEROATOMS

4.3.4.1 Introduction

This section covers the heterocycles with one two-coordinate phosphorus atom and two other heteroatoms in the ring (Table 4). Of diaza-, oxaza- and thiazaphospholes six isomers each are possible (see Section 4.3.1). Monocyclic and fused examples of all the diazaphosphole isomers (**1**, **4**, **5**, **7**, **8**, **9**) are known; however, only two of the oxazaphosphole (**2**, **10**) and three of the thiazaphosphole isomers (**3**, **6**, **11**) have become available. In addition, 1,3,2-dithia- and diselenaphospholium cations (**12**) and methylidene-1,4,2-dithiaphospholes come under this heading.

Only for 1,2,3- and 1,3,2-diheterophospholes (**7–12**) benzo derivatives can be conceived and to a greater part they are in fact known (Table 4). In every other case of anellation a nitrogen atom must occupy a ring junction position and becomes thus also a member of the anellated ring. There is one example, however, of a phosphorus atom in this position (of a fused 1,2,4-diazaphosphole, Table 4), where it seems to gain planarity and to allow cyclic delocalization.

In Table 4 the known systems are ordered according to (1) the number of heteroatoms adjacent to phosphorus, (2) the number of heteroatoms adjacent to each other, (3) the nature of the heteroatoms: N, O, S, Se, and (4) whether they are monocyclic, 4,5-anellated or otherwise anellated. This order is essentially also used to arrange the compounds in the following sections.

4.3.4.2 Spectra, Structure and Bonding

For better comparison, spectra, structure and bonding of the compounds of Sections 4.3.5 and 4.3.6 are here presented and discussed together with those of Section 4.3.4.

4.3.4.2.1 Theoretical methods

Quantum chemical calculations on different levels have been reported for 2*H*-1,2,3-diazaphospholes (**7**) ⟨83T1507⟩, for 4,6-diamino-1,3,5-triaza-2-phosphapentalenes derived from 1,3,2-diazaphospholes (**9**) ⟨86CB3213⟩, for the tetrazaphospholium ion (Section 4.3.5) ⟨93CB1513⟩, for 1*H*-1,2,3-azadiphospholes (Section 4.3.6) ⟨89NJC309⟩ and for 1*H*-1,2,3-benzazadiphospholes (Section 4.3.6) ⟨93PS(76)45⟩. Planar and nonplanar structures of 2*H*-1,3,2-diazaphosphole and its 2-boryl derivative as well as of 4*H*-1,2,4-diazaphosphole and its 4-boryl and 3,5-diboryl derivatives are compared by correlated *ab initio* calculations ⟨95JPC586⟩. For the parent 4*H*-1,4,2-diazaphosphole (**5**), 1*H*-1,2,3-diazaphosphole (**8**), and 1*H*-1,2,4,3-triazaphosphole (Section 4.3.5) NICS of -13.5, -14.7 and -17.4 ppm have been calculated ⟨97MI1⟩ and may be compared to -15.1 ppm for pyrrole ⟨96JA6317⟩.

The electronic structure of 1,3-azaphospholes (Section 4.3.3.3), of 1*H*- and 4*H*-1,2,4-diazaphospholes, and of 1*H*-, 2*H*-, 3*H*- and 4*H*-1,2,3,4-triazaphospholes (Section 4.3.5.3) has been investigated by *ab initio* quantum chemical calculations and HeI ultraviolet photoelectron spectroscopy. The total energy of the tautomers containing a three-coordinate phosphorus atom is significantly higher (by 25–35 kcal mol^{-1}) than the total energy of those containing two-coordinate phosphorus ⟨92JA9080, 92JPC623⟩. The aromatic stabilization — as measured by the heats of bond separation reactions — changes in a similar way. The calculated optimized geometries reveal that the rings containing two-coordinate phosphorus can be characterized by bond lengths between those of isolated single and double bonds. In the tautomers containing three-coordinate phosphorus, however, P–C and P–N bond lengths are much closer to the single-bond value, indicating that conjugation is in force only on the rest of the molecule. In order to assess the conjugative interaction between the P=C moiety and the rest of the molecule, ring fragmentation

Table 4. Known phospholes with two additional heteroatoms other than phosphorus, as well as their oxo and anellated derivatives

1,2,4-diazaphospholes (1)

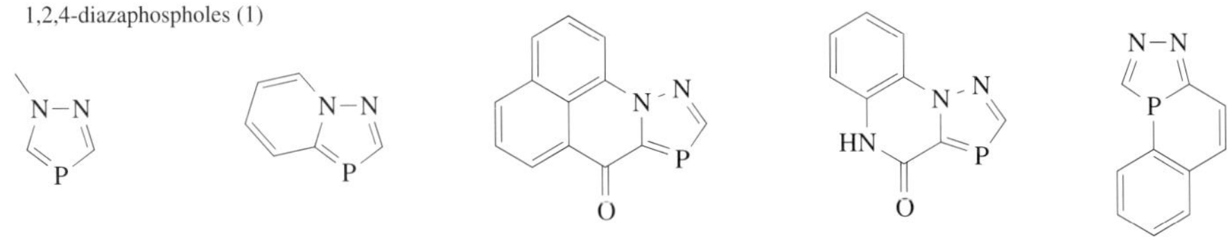

1,2,4-oxazaphospholes (2)

1,2,4-thiazaphospholes (3)

1*H*-1,4,2-diazaphospholes (4)

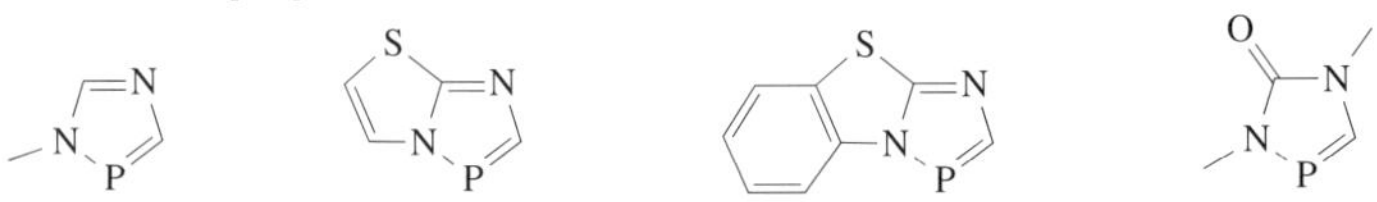

4*H*-1,4,2-diazaphospholes (5)

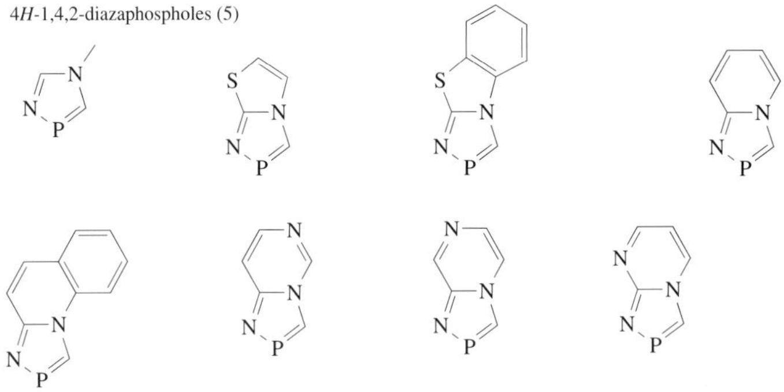

1,3,4-thiazaphospholes (6)

2*H*-1,2,3-diazaphospholes (7)

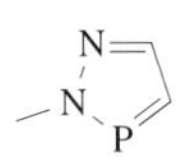

1*H*-1,2,3-diazaphospholes (8)

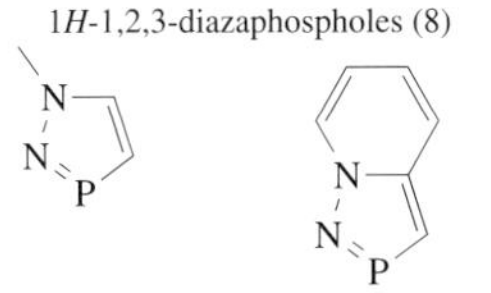

1,3,2-diazaphospholes (9)

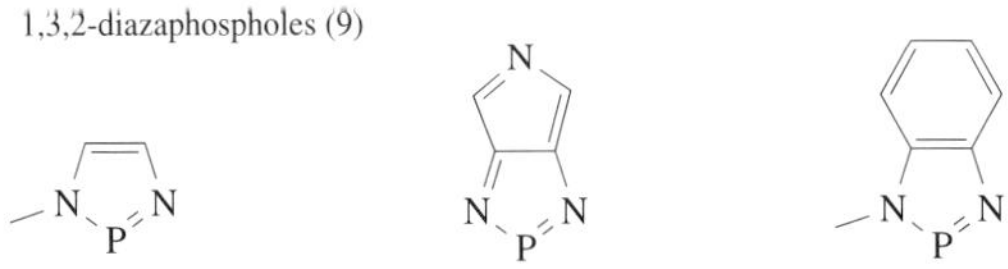

1,3,2-oxazaphospholes (10)

1,3,2-thiazaphospholes (11)

1,3,2-dithia/selenaphospholes (12)

1,4,2-dithiaphospholes (13)

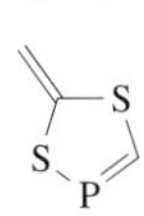

reactions have been investigated. The heats of these reactions indicate that the conjugation of the P=C moiety (and thus the aromatic stability) increases with the increasing number of nitrogen atoms in the ring. The photoelectron spectra of the investigated compounds are in accordance with the electronic structure of the compounds containing two-coordinate phosphorus. When the electron is removed from π-orbitals, ionization energies of azaphospholes are similar to those of the corresponding azoles ⟨92JA9080⟩.

The comparison of π-ionization energies of compounds containing C=C, C=N, and C=P double bonds reveals a more pronounced similarity between the carbon- and phosphorus-containing species than between the nitrogen- and phosphorus-containing species. Ring fragmentation reaction heats of the corresponding azaphospholes and azoles show a similar behavior. All these facts suggest that, although the strength of the two-coordinate P=C bond is known to be significantly smaller than that of the C=C or C=N bond, in conjugative interactions two-coordinate P=C is more comparable to a C=C than to a N=C π-system. By using perturbation arguments, earlier observations about the similarities in chemical reactions of the two-coordinate P=C and C=C bonds become understandable ⟨93JPC4011⟩.

An aspect which has attracted much interest and which was debated controversially, is the stabilization of the cyclic phosphenium cations by π-electron delocalization. Such cations derive from 1,3,2-diaza-, 1,3,2-thiaza- and 1,3,2-dithiaphospholes (**9**, **11**, **12**) ⟨90IC4938, 96ZOB386, 97CC2095⟩ and from 1,3,4,2-thiadiaza- and tetrazaphospholes (Sections 4.3.5.3 and 4.3.5.2) ⟨95MI239, 95MI(357)9, 93CB1513⟩. The calculated enthalpies of isodesmic reactions suggest a weak aromatic character of these cations ⟨97CCR(163)71⟩. For 1,3,2-diazaphospholium ions aromatic stabilization energies (24–26 kcal mol^{-1}) comparable to that of pyrrole have been found ⟨97T2357, 99EJI41⟩. Despite the isoelectronic relationship to carbenes the phosphorus atom in these cations behaves purely electrophilic ⟨98EJI1087⟩.

4.3.4.2.2 *X-ray and electron diffraction*

The known molecular structures of the heterophospholes of Sections 4.3.4 and 4.3.5, and in part of Section 4.3.6 are compiled in Tables 5–10. In part they have been reviewed earlier ⟨88ZSK122, B-89MI307, B-90MI258⟩.

All structures show a planar ring with averaged bond distances in accord with a delocalized π-bond system. The angle at the two-coordinate phosphorus atom in azaphospholes increases with the number n of two-coordinate nitrogen atoms adjacent to phosphorus, $n = 0$: 85–87°, $n = 1$: 87–93°, $n = 2$: 90–100°. A comparatively large angle at a phosphorus atom is found in heterophospholes with a sulfur or with a second phosphorus ring member. The endocyclic angle of course increases with the coordination of the phosphorus atom. The PC bond lengths in heterophospholes generally range from 170 to 175 pm as compared to mean values of 167 pm and 185 pm for the localized double and single PC bonds in phosphaalkenes ⟨86ZOB2433, 88ZSK122, B-90MI157⟩. PN bonds in azaphospholes are 162 to 170 pm long as compared to mean values of 156 pm and 167 pm for the more or less localized double and single bonds in amino(imino)phosphines ⟨88ZSK122, 88ZOB2045, B-90MI293⟩. PP bond lengths in 1,2,3-azadiphospholes are only slightly longer than in diphosphenes (200–204 pm) ⟨88ZSK122, B-90MI321⟩. The PS bond length in the 1,3,2-benzodithiaphospholium cation is the shortest non–terminal phosphorus(III)–sulfur bond reported so far ⟨89IC144⟩. From the thiaza and diazaphosphole molecular dimensions aromaticity indices were derived ⟨90T5697⟩.

In *P*-coordinate azaphosphole complexes (Tables 7 and 9) the metal atom lies strictly in the plane of the ring and the ring structure is not significantly different from that of the free azaphosphole.

Some azaphospholes are not stable as monomers but oligomerize. In other cases the complex formation or oxidation of monomeric azaphospholes is accompanied by an oligomerization. For the oligomerization, the monomers add to each other through their P=N bonds forming a cyclophosphazane ring. Obviously for steric reasons, the eight-membered cyclophosphazane ring and thus the tetramer is generally preferred. X-ray structure investigations have been reported of 1,2,4,3-triazaphosphole tetramers ⟨84CJC43, 92JCS(P2)671⟩ and of a $[Cr(CO)_5]_2$ complex thereof ⟨81AG420⟩ as well as of the $[Fe(CO)_4]_4$ complex ⟨84ZOB271⟩ and of the tetroxide ⟨81ZSK98, 92RHA25⟩ of a 1,3,2-oxazaphosphole tetramer. For the sulfide of a 1,2,4,3-triazaphosphole a trimeric structure was shown by ^{31}P NMR spectroscopy ⟨94HAC385⟩.

The molecules of *N*-unsubstituted 1,2,4-diazaphospholes — like those of pyrazoles — are connected to each other by NH...N hydrogen bonds forming rings or helices in the crystal ⟨99ZN(B)187⟩.

4.3.4.2.3 *NMR spectra*

A compilation of ^{31}P NMR data is available for heterophospholes complete up to mid 1987 ⟨88PS(36)217⟩ and for anellated heterophospholes up to 1993 ⟨94T7675⟩. Some characteristic data are also found where

Table 5. Bond lengths (pm) in 1,2,4-diazaphosphole rings (**1**) and angle(s) at the phosphorus atom

	Bonds					*Angle*	*Ref.*
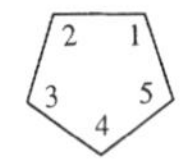	*1,2 NN*	*2,3 CN*	*3,4 PC*	*4,5 PC*	*1,5 CN*		
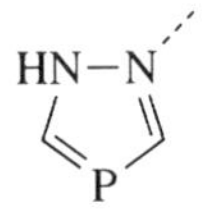	132.3	130.5	171.0	170.7	130.4	85.1°	⟨99ZN(B)187⟩
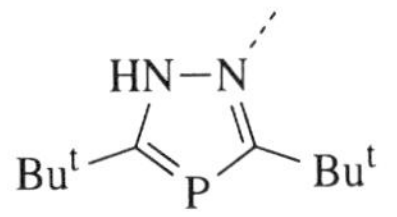	136.0	133.0	173.7	175.2	131.8	87.3°	⟨99ZN(B)187⟩
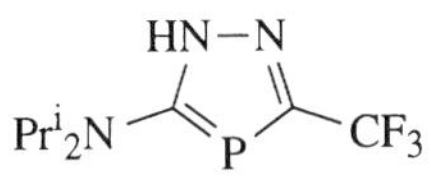	136.6	135.8	177.3	175.6	131.8	85.7°	⟨92CB411⟩
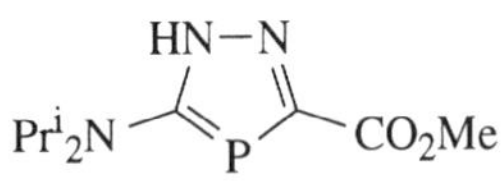	135.2	136.2	177.1	174.2	133.8	86.2°	⟨92CB411⟩
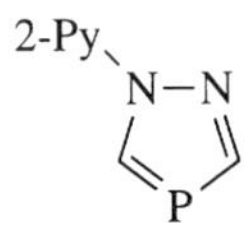	135.3	134.2	170.8	174.8	130.5	85.3°	⟨99ZN(B)187⟩
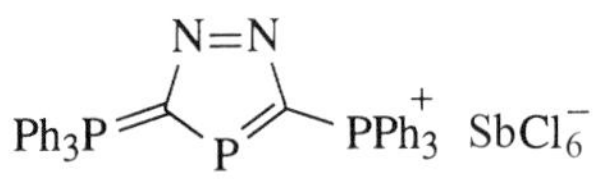	131.4	135.1	174.0	173.6	134.8	84.7°	⟨97CB89, 01MI1⟩
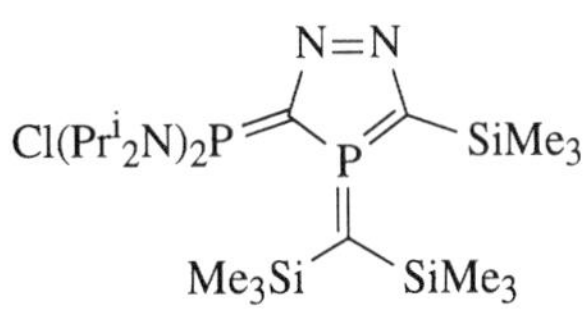	134.8	134.9	173.3	171.1	135.5	89.9° 135.7° 133.7°	⟨93BSF757⟩

In an earlier investigation, similar data was obtained for 1,2,4-diazaphosphole ⟨89DIS(B)3773⟩.
In addition the structures of two 1-ribofuranosyl-1,2,4-diazaphospholes have been mentioned ⟨90JMC572⟩.

more generally the ^{31}P NMR spectra of two-coordinate phosphorus compounds are treated ⟨B-87MI61, B-90MI295, B-90MI463, B-91MI29⟩. A theoretical study correlates the structures of dicoordinated phosphenium cations and their phosphorus-31 chemical shifts ⟨95MI(357)9⟩, ⟨97PS(126)89⟩.

A low-field chemical shift is characteristic for heterophosphole phosphorus atoms. Its magnitude depends mainly on the ring members adjacent to phosphorus, but is also influenced by other factors. Even neglecting single extreme values for the more populated systems, only broad and overlapping ranges result.

Ring members adjacent to P	*δ^{31}P Ranges in heterophospholes*
C, C	50–180
C, N	160–270
C, P	240–360
C, S	180–310
N, N	220–300
N, P	300–350
N, O	230–270
N, S	310–330
S, S	~410

Electron-withdrawing substituents in 4-position of 1,2,3-diazaphospholes cause a downfield ^{31}P shift. *N*-Deprotonation as well as *N*-protonation or alkylation in different heterophosphole systems give rise to shifts in opposite directions ⟨88PS(36)217⟩. Correlations of the available data demonstrate that δ^{31}P in a heterophosphole is influenced in very much the same way as δ^{13}C or δ^{15}N in the corresponding position

Table 6. Bond lengths (pm) of a 1,4,2-diazaphosphole (**4**) and a 1,3,4-thiazaphosphole ring (**6**) and angle at the phosphorus atom

	Bonds					*Angle*	*Ref.*
	4,5 CN	*1,5 CN*	*1,2 PN*	*2,3 PC*	*3,4 CN*	*CPN*	
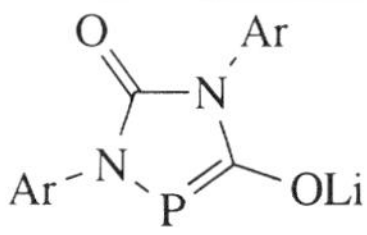 Ar = 2,6-$Pr^i_2C_6H_3$	138.2	133.8	177.4	175.9	141.4	87.9°	⟨98MI1⟩

	Bonds					*Angle*	*Ref.*
	1,2 CS	*2,3 CN*	*3,4 PN*	*4,5 PC*	*1,5 CN*	*CPN*	
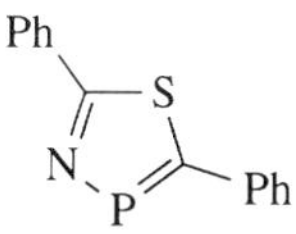	172.8	132.4	168.4	166.9	172.8	98.0°	⟨85AG125⟩

of the heterocycle with CH or N replacing P ⟨86ZN(B)553, 88PS(36)217⟩. This suggests a rather congruent bonding situation in heterophospholes and in the better known non-phosphorus containing five-membered heterocycles.

For many heterophospholes, δ^{13}C values are given and discussed, some $\delta^{14,15}$N values are also available ⟨77IZV532, 86ZN(B)553, 87JA627, 87JOM(319)49, 89JCR(S)250, 91MRC1045⟩.

Characteristic one bond ^{31}P,^{13}C coupling constants are $^{1}J_{PC}$ = 30-70 Hz. They are found to be negative ⟨71DOK1081, 78ZSK64, 91MRC1045⟩. The observed coupling constants are larger than in phosphines and they extend to considerably higher values than in phosphinines ⟨76JA5451, 79JMR199⟩. The two bond coupling constant $^{2}J_{PC}$ to C-5 in 1,2,3-diazaphospholes ranges from 0 to −12 Hz ⟨91MRC1045⟩.

One bond ^{31}P,^{15}N coupling constant in heterophospholes are generally positive ⟨91MRC1045⟩. Ranges and individual values are: $^{1}J_{PN}$ = 24 to 83 Hz for 2*H*-1,2,3-diazaphospholes ⟨77IZV532, 86ZN(B)553, 91MRC1045⟩, 92–97 Hz for 1*H*-1,2,3-diazaphospholes ⟨86ZN(B)553⟩, 87 Hz for 1,2,3-diazaphospholium ions ⟨91MRC1045⟩, 87 and 89 Hz for 1,4,2-diazaphospholo[4,5-*a*]-pyridines ⟨91MRC1045⟩, 93 Hz for 1,3,2-oxazaphospholium ions ⟨87JA627⟩, 79 Hz for 3*H*-1,2,3,4-triazaphospholes ⟨91MRC1045⟩, and 79–89 Hz for 1*H*- and 2*H*-1,2,4,3-triazaphospholes as well as for 4-BF_3 complexes thereof ⟨89JCR(S)250⟩; $^{2}J_{PN}$ = 2 to 7 Hz for 1,2,4-diazaphospholes ⟨86ZN(B)553, 87JOM(319)49⟩, 6 to 9 Hz for 1*H*- and 2*H*-1,2,3-diazaphospholes ⟨86ZN(B)553, 91MRC1045⟩, 0.5 and 2 Hz for 1,4,2-diazaphospholo[4,5-*a*]pyridines ⟨B-90MI258⟩, 0 Hz and 4–8 Hz for 1*H*- and 2*H*-1,2,4,3-triazaphospholes and for 4-BF_3 complexes thereof ⟨89JCR(S)250⟩.

As in α-unsubstituted phosphinines ⟨66AG907, 83TL2645, B-90MI220⟩ and phosphaalkenes with a hydrogen atom *trans* to the phosphorus substituent ⟨84ZOB1520, B-90MI463⟩, the two bond ^{31}P,^{1}H coupling is large, $^{2}J_{PCH}$ = 29 to 57 Hz. For 1,2,3-diazaphospholes it is found to be positive ⟨71DOK1081, 71ZSK798⟩. A three bond phosphorus coupling to a carbon bonded ring hydrogen is in the range of $^{3}J_{PH}$ = 4 to 9 Hz.

One bond ^{31}P,^{31}P coupling in 1,2,3-azadiphospholes, $^{1}J_{PP}$ = 465-481 Hz, is somewhat smaller than in diphosphenes, $^{1}J_{PP}$ = 525 to 670 Hz ⟨86PS(29)73, B-90MI321⟩. $^{2}J_{PP}$ is found to be 47–59 Hz in 1,2,4-thiadiphospholes ⟨86AG570, 88TL4535⟩ and 23–25 Hz in 1,2,3,5-diazadiphospholes ⟨85AG127⟩.

^{1}H- and in some cases ^{13}C-, ^{15}N- and ^{31}P-NMR spectroscopy have been used to study proton exchange in *N*-unsubstituted di- and triazaphospholes. The exchange is found to be rapid for 1,2,3-diazaphospholes ⟨77IZV532, 77ZOB576, 84PS(21)335⟩ and 1,2,4-diazaphospholes ⟨86ZN(B)553, 87CB1645, 89JOM(338)329⟩ but slow enough for 1,4,2-diazaphospholes ⟨86PS(28)289⟩ and 1,2,4,3-triazaphospholes ⟨84CC183⟩ to allow the individual tautomers to be observed. The 2*H*-tautomers are favored in the case of 5-*t*-butyl-1,2,3,4-triazaphosphole ⟨88AG1541⟩ and 5-dimethylamino-1,2,4,3-triazaphosphole ⟨80ZN(B)1222⟩. The two-coordinate phosphorus atom in no case takes part in the proton exchange ⟨80ZN(B)1222, 86ZN(B)553⟩.

^{31}P Solid state NMR has been studied of 1,3,2-benzodithiaphospholium and -thiazaphospholium salts ⟨92IC3386⟩.

Table 7. Bond lengths (pm) in 2*H*- and 1*H*-1,2,3-diazaphosphole rings (**7, 8**) and angle at the phosphorus atom

	*Bonds* 1,5 CN	1,2 NN	2,3 PN	3,4 PC	4,5 CC	*Angle*	*Ref.*
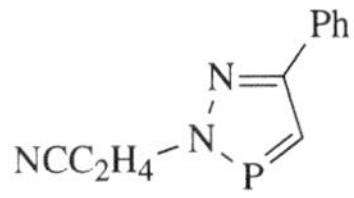	134.3	134.4	167.6	170.2	140.7	88.8°	⟨73DOK1101⟩
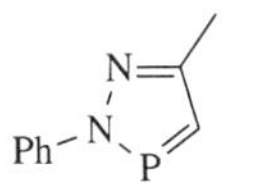	132	132	164	175	142	92°	⟨68DOK95⟩
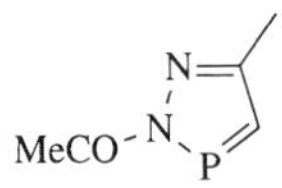	134	134	168	175	144	89°	⟨71DOK1081⟩
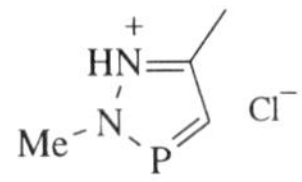	132.6	134.3	166.1	174.6	141.5	88.2°	⟨78CB1558⟩
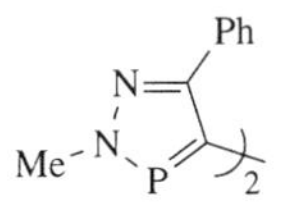	132.5	135.4	168.5	172.6	142.6	89.5°	⟨88IZV150⟩
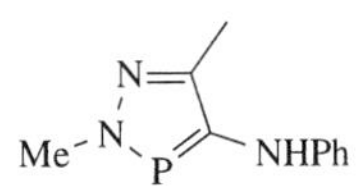	132.9	134.7	167.4	168.8	141.6	90.4°	⟨83CB549⟩
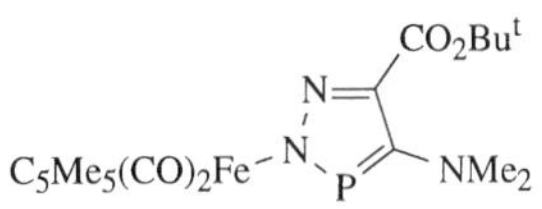	134.6	134.2	170.3	174.6	142.5	91.0°	⟨95OM581⟩
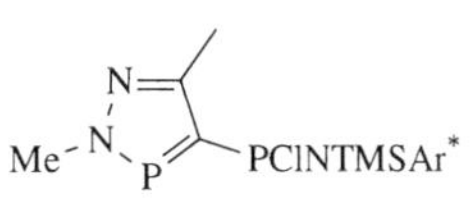	132.5	135.2	168.9	171.8	141.9	89.2°	⟨93BSF726⟩
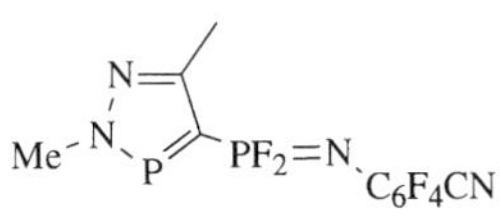	132.2	135.7	166.4	171.7	141.7	88.3°	⟨99IC2791⟩
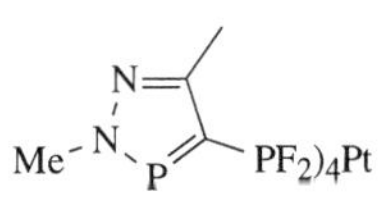	133	138	169	172	145	91.7°	⟨99IC4056⟩
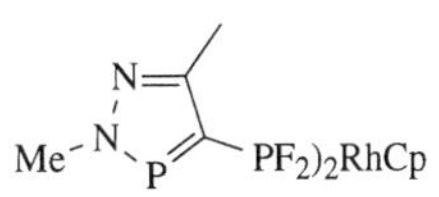	131.6 132.2	135.5 135.4	167.0 167.2	171.3 171.3		88.3° 88.6°	⟨99OM3306⟩
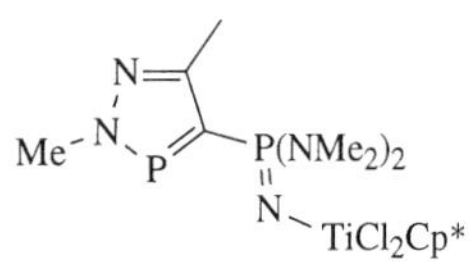	133.0	136.3	166.9	172.1	141.1	89.3°	⟨99IC2791⟩
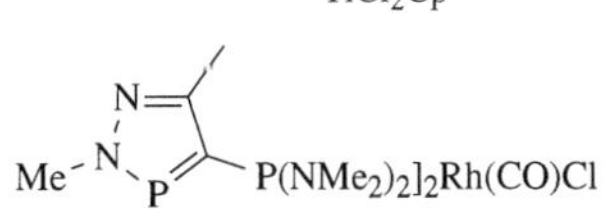	132.5 134.4	135.6 135.0	167.9 167.8	171.2 170.4	141.3 140.9	88.9° 89.1°	⟨99OM3306⟩
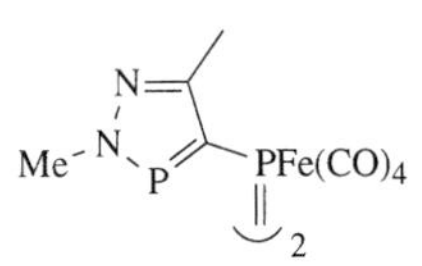	132.0 132.2	133.7 135.1		171.3 171.4	142.3 141.0	88.7° 89.3°	⟨88CC574⟩
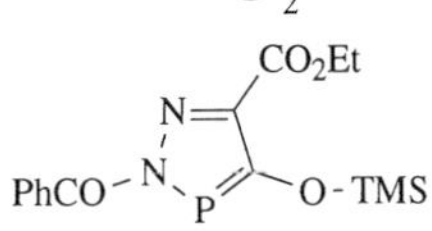	132.2	134.5	171.9	172.0	143.3	86.6°	⟨96TH1⟩

Table 7. (continued)

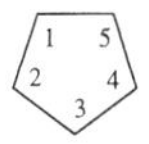	*Bonds*					*Angle*	*Ref.*
	1,5 CN	*1,2 NN*	*2,3 PN*	*3,4 PC*	*4,5 CC*		
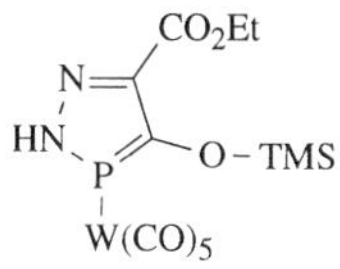	134.3	134.0	166.0	169.5	141.4	91.7°	⟨96TH1⟩
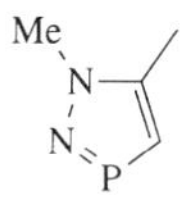	135.6	133.0	165.4	172.3	138.1	92.9°	⟨79AG442⟩
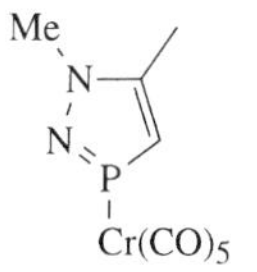	136.2	135.4	163.7	172.8	138.2	95.2°	⟨80JOM(185)53⟩

* Ar = 2,4,6-$Bu^t{}_3C_6H_2$.

4.3.4.2.4 Other spectroscopic methods

For many heterophospholes UV spectra have been reported ⟨81JA2465, 81ZN(B)1273, 82TL3643, 83JOM(258)257, 86CB3213, 88TL5867, 89NJC309⟩. The replacement of a pyridinic N atom in an azole for a P atom in a heterophosphole is generally accompanied by a bathochromic shift of λ_{max} ⟨68DOK95, 78TL441, 80ZC342, 85PS(25)345, 86TL5699⟩. As compared to pyrrole, imidazole, and oxazole, 1,3-azaphospholes like thiazoles show an additional $\pi \rightarrow \pi^*$ band ⟨86TL5699⟩. 2-Aryl-1,3-benzoxaphospholes exhibit a strong blue fluorescence ($\lambda = 416$ nm) in the concentration range 10^{-2}–10^{-4} M ⟨85PS(25)345⟩. The deep red color of the 4,6-diamino-1,3,5-triaza-2-phosphapentalenes is due to a CT band at 500 nm ⟨86CB3213⟩. Electronic absorption and magnetic circular dichroism of 1,3,2-benzothiazaphospholium and -dithiaphospholium ions are compatible with a delocalized π-electron system over the heterocyclic moiety ⟨90OM1085⟩.

Raman ⟨68DOK95, 71DOK1081, 72JPR806⟩ and IR spectra ⟨68DOK95, 71DOK1081, 72JPR806, 84IZV1182, 84S591⟩ of 1,2,3-diazaphospholes, IR spectra of 1,2,4-diazaphospholes ⟨85ZN(B)1077, 86JOM(306)39⟩, 1,2,4-oxazaphospholes ⟨85ZN(B)1077, 86JOM(306)39, 86S31⟩, 1,2,3,4-triazaphospholes ⟨85ZN(B)1077, 86JOM(306)39⟩, and 1,3,2-benzodithiaphospholium salts ⟨88CC842, 89IC144⟩ as well as mass spectra of various heterophospholes ⟨71IZV2611, 78TL441, 80ZN(B)1222, 81ZN(B)1273, 86TL4415, 87CB1809, 89JOM(338)329, 95S173, 95ZN(B)558⟩ have been reported.

4.3.4.3 1,2,4-Diaza-, 1,2,4-Oxaza-, and 1,2,4-Thiazaphospholes (1, 2, 3)

1,2,4-Diazaphospholes resemble in their chemical behavior pyrazoles from which they derive formally by a CH/P-exchange in 4-position. They differ for example in their hydrolytic stability from 1,2,3-diazaphospholes (next section) which also derive from pyrazoles by this exchange, however in 3-position.

4.3.4.3.1 Reactivity

1,2,4-Diazaphospholes are colorless to pale yellow, distillable oils or crystalline solids which may exhibit a characteristic unpleasant penetrating smell ⟨87TH1⟩.

Monocyclic 1,2,4-diazaphospholes are rather unreactive; neither 1,2-additions to the P=C bond, nor 1,1-additions to the phosphorus atom are known. They are stable to hydrolysis even by concentrated hydrochloric acid ⟨84AG901⟩. They are degraded, however, by 30% aqueous hydrogen peroxide and bromine in dry media to phosphoric acid and PBr_3, respectively ⟨87TH1⟩. 1,2,4-Diazaphospholo[1,5-*a*]pyridines are rapidly oxidized in air ⟨87TL1511⟩. 1,2,4-Diazaphospholes (**13**) (R = H, Me, Ph) are methylated and protonated at the pyridinic nitrogen atom and they add Lewis acids to this nitrogen as well ⟨87TH1⟩.

Table 8. Bond lengths (pm) in 1,3,2-diaza-, -thiaza- and -dithiaphosphole rings (**9**, **11**, **12**) and bond angle at the phosphorus atom

	*Bonds*					*Angle*	*Ref.*
	4,5 *CC*	*1,5* *CN/S*	*1,2* *PN/S*	*2,3* *PN/S*	*3,4* *CN/S*		
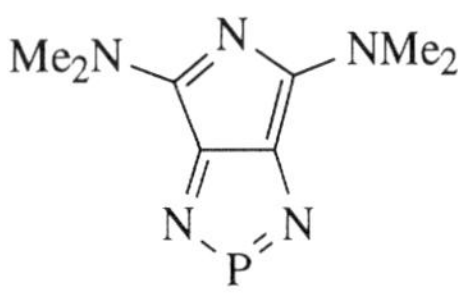	137.6	134.5	166.5	166.1	134.9	99.8°	⟨86CB3213⟩
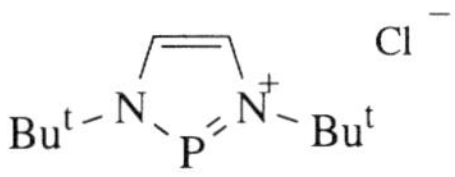	134.7	138.9	166.0	166.2	137.9	90.4°	⟨99EJI41⟩
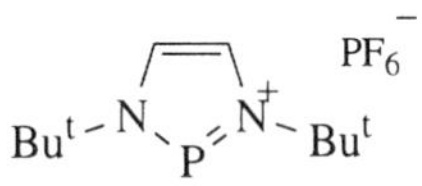	134.1	137.7	166.1	166.2	137.6	90.7°	⟨99EJI41⟩
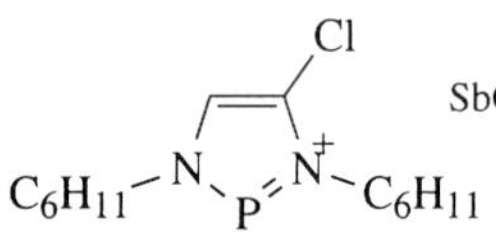	135.1	137.0	165.6	167.3	36.3	90.0°	⟨90DOK(312)623⟩
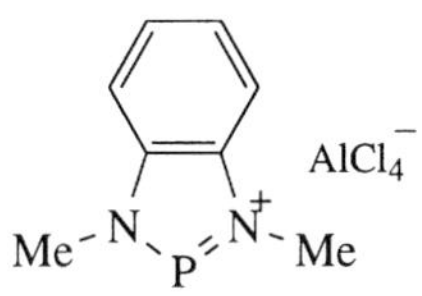	137.8	137.8	163.8	164.8	138.0	91.3°	⟨89AX(C)1485⟩
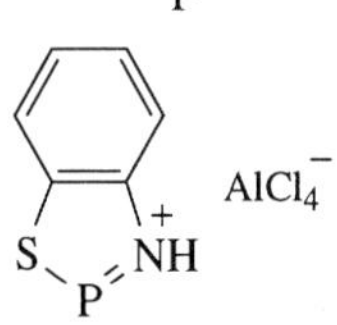	139.4	173.2	203.4	164.3	138.7	93.7°	⟨90IC4938⟩
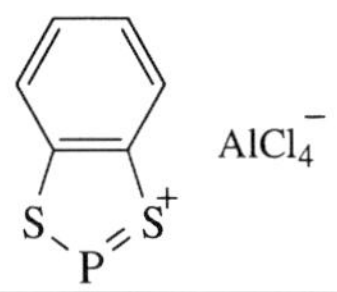	138.3	171.1	201.5	201.6	172.8	97.6°	⟨89IC144⟩

In addition the structures of 1,3,2-diazaphosphole Mo- and W-complexes have been investigated by X-ray diffraction ⟨01JOM(617-618)383⟩.

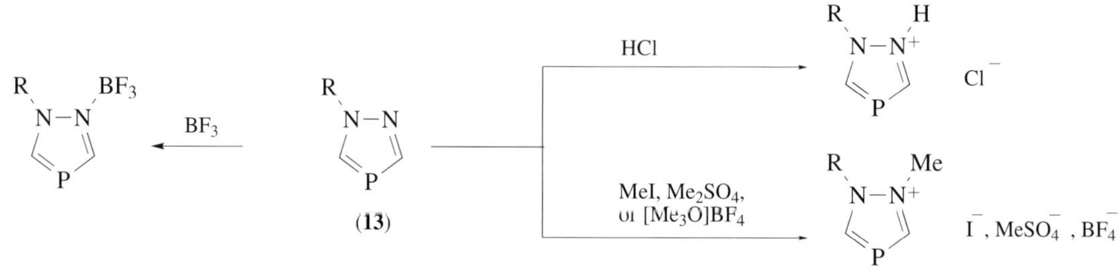

(**13**)

N-Unsubstituted 1,2,4-diazaphospholes (**14**) undergo *N*-alkylation on reaction with alkyl vinyl ether, sulfur ylides and diazo compounds ⟨95HAC403⟩. They react with acyl chlorides in a 2 : 1 molar ratio to give a mixture of the *N*-acylated diazaphosphole and the diazaphosphole hydrochloride. Preparative *N*-acylation is achieved in presence of a tertiary amine. Sulfonyl chlorides and phosphorus trichloride also give *N*-substitution reactions ⟨87TH1⟩. Conversely, 1-acyl ⟨87CB1645, 87TH1⟩, 1-phosphoryl ⟨87CB1645⟩, and 1-(thio)phosphinyl ⟨87CB1645, 87TH1⟩ groups are lost on methanolysis, hydrolysis, or aminolysis.

Table 9. Bond lengths (pm) in 2*H*-, 1*H*-, 4*H*-1,2,4,3-triazaphosphole rings and a tetrazaphosphole ring as well as the angle at the phosphorus atom

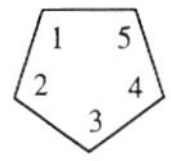	*Bonds*					*Angle*	*Ref.*
	1,5 CN	*1,2 NN*	*2,3 PN*	*3,4 PN*	*4,5 CN*		
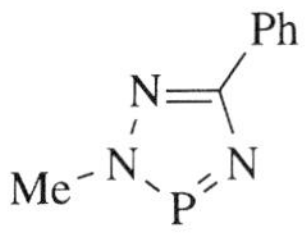	132.8	135.4	166.8	161.6	135.7	92.4°	⟨80CR(C)271⟩
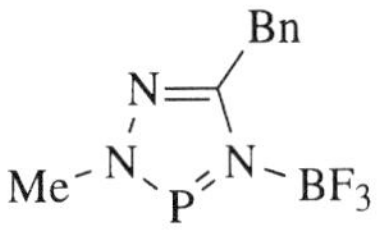	131.6 130.4	136.2 136.4	164.4 163.1	164.4 163.2	135.6 136.1	86.6° 89.5°	⟨89JCR(S)250⟩
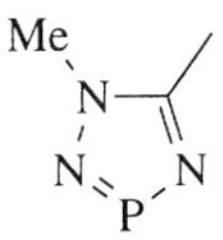	134.2	133.2	163.8	163.6	134.1	96.9°	⟨79CB3159⟩
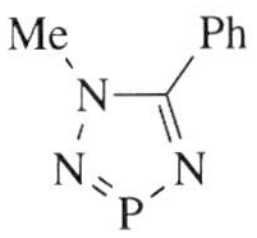	135.8	135.1	162.7	165.4	132.0	97.0°	⟨78CR(C)319⟩
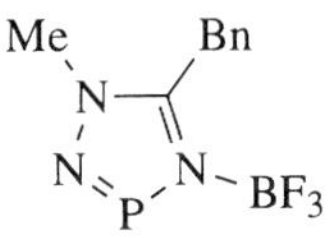	134.3	135.7	162.2	170.1	133.5	93.1°	⟨92PS(69)189⟩
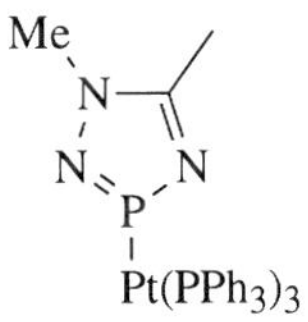	132	139	164	163	131	96.5°	⟨83JOM(256)375⟩
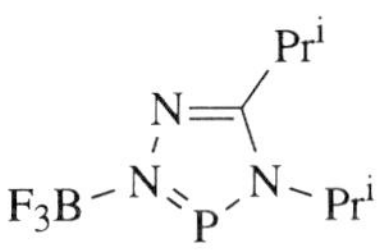	131.5	136.3	163.7	166.9	137.4	89.5°	⟨92JCS(P2)671⟩
	Bonds					*Angle*	*Ref.*
	2.3 NN	*1,2 NN*	*1,5 PN*	*4,5 PN*	*3,4 NN*		
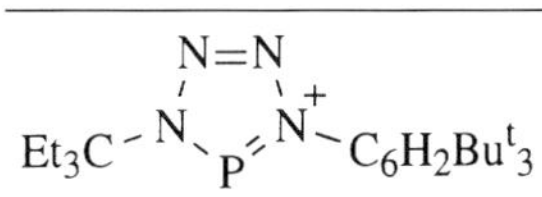	126.5	136.8	165.4	165.3	136.4	86.7°	⟨93CB1513⟩

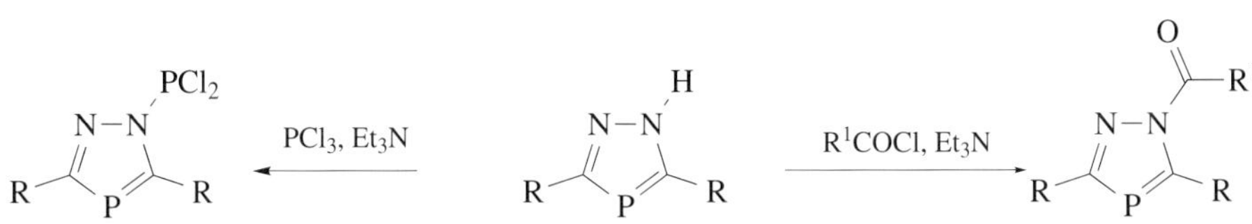

(**14**)

The *N*-trimethylsilyl groups of a 1,2,4-diazaphosphole can be cleaved by hydrolysis or methanolysis to give the *N*-unsubstituted compound ⟨87CB1645, 88ZN(B)1285, 89NJC359, 89S511⟩. $P(SiMe_3)_3$ converts *vice versa* *N*-unsubstituted 1,2,4-diazaphospholes into the *N*-trimethylsilyl derivatives ⟨87TH1⟩.

Table 10. Bond lengths (pm) in 1,2,4- and 1,2,3-azadiphosphole and 1,2,3,4-diazadiphosphole rings as well as angles at the two-coordinate phosphorus atoms

	Bonds					*Angles*		*Ref.*
	1,2 PN	*2,3 PC*	*3,4 PC*	*4,5 PC*	*1,5 CN*	*CPN*	*CPC*	
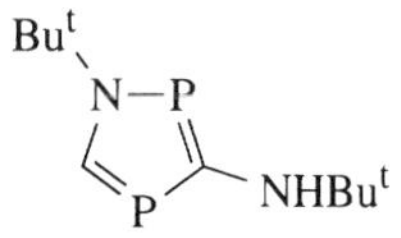	181.4 174.1	169.9 172.1	177.8 176.2	175.4 172.9	126.5 133.5	99.7° 94.7°	94.3° 93.9°	⟨91HAC577⟩
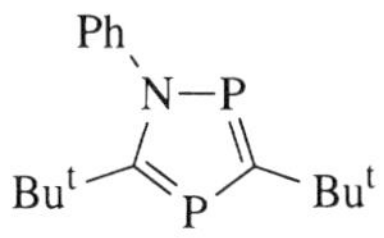	171.5	169.7	174.6	174.0	136.6	95.0°	95.8°	⟨99CC2363⟩

	Bonds					*Angles*		*Ref.*
	1,2 PN	*2,3 PP*	*3,4 PC*	*4,5 CC*	*1,5 CN*	*PPN*	*PPC*	
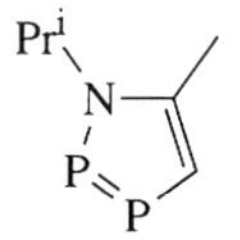	170.6	206.3	174.0	136.4	137.3	95.4°	93.0°	⟨89NJC309⟩
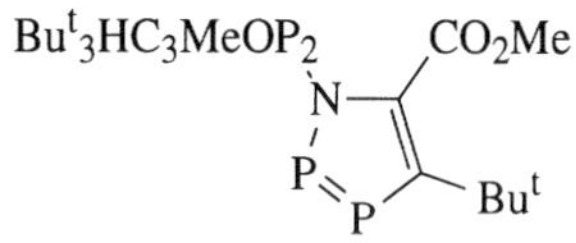	168.7	205.5	183	130.9	142.0	95.8°	94.0°	⟨88TL5867⟩

	Bonds					*Angles*		*Ref.*
	2,3 PN	*3,4 PP*	*4,5 PC*	*1,5 CN*	*1,2 NN*	*PPN*	*PPC*	
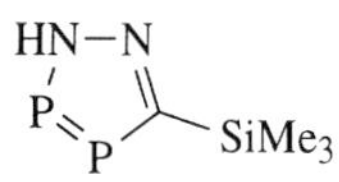	167.0	207.1	176.8	132.0	135.7	93.5°	92.1°	⟨96AG2282⟩

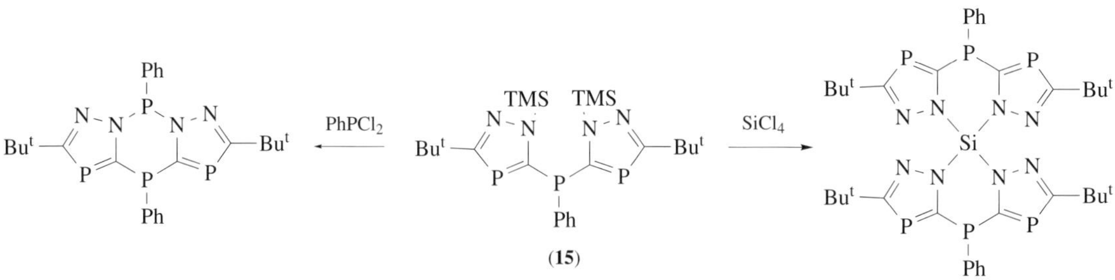

(**15**)

Scheme 3

By substitution of the *N*-trimethylsilyl group other silyl, phosphino, arsino and stibino derivatives have been prepared and starting from a bis(1,2,4-diazaphospholyl)phosphine polycyclic structures have been built in this way (Scheme 3) ⟨88TL925⟩.

1-Dimethylphosphoryl-1,2,4-diazaphospholes eliminate $MeOPO_2$ on heating to give the 1-methyl derivatives ⟨87CB1645⟩.

N-Unsubstituted 1,2,4-diazaphospholes (**14**) are deprotonated by aqueous sodium hydroxide ⟨87TH1⟩ or butyllithium ⟨84AG899, 87CB1809⟩; reprotonation and methylation ⟨84AG899, 87CB1809, 87TH1⟩ take place at nitrogen again.

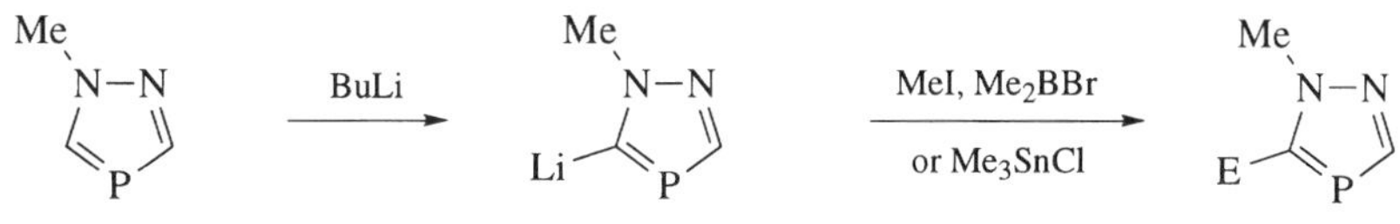

(16) E = Me, Me$_2$B, Me$_3$Sn (17)

1-Methyl-1,2,4-diazaphosphole (**16**) is metallated by butyllithium selectively in the 5-position ⟨87JOM(319)49⟩. From the subsequent reaction with an electrophilic halide, 5-methyl-, 5-stannyl-, and 5-boryldiazaphospholes (**17**) are obtained ⟨87JOM(319)49, 87TH1⟩.

Catalytic amounts of potassium *t*-butoxide in acetonitrile-d_3 effect an H/D-exchange, fast in the 5- and slow in the 3-position of 1-phenyl-1,2,4-diazaphosphole ⟨87JOM(319)49⟩.

1,2,4-Oxaza- and 1,2,4-thiazaphospholes are colorless to pale yellow, air-stable oils or solids ⟨88AG1541⟩. 2-Phenyl-5-*t*-butyl-1,2,4-thiazaphosphole is reported to be thermally stable up to 200°C. It does not hydrolyze in aqueous methanol ⟨87S689⟩. The [3 + 2] dipolar cycloaddition of nitrile oxides to the P=C bond in 1,2,4-oxazaphospholes ⟨84CC1634, 87S689, 90PS(49/50)317⟩ as mentioned in Section 4.3.4.3.2 is the only reaction described so far.

4.3.4.3.2 Synthesis

Condensation of 1,3-bis(dimethylamino)-2-phosphaallylic cations (**18**) (R^1 = H, Bu^t, Ph) with hydrazines and hydroxylamine gives 1,2,4-diazaphospholes (**19**), R = H, Me, Ph ⟨84AG901⟩, and 1,2,4-oxazaphospholes (**20**) ⟨86PS(28)71⟩. The unsubstituted 1,2,4-diazaphosphole is preparatively accessible only through this route so far.

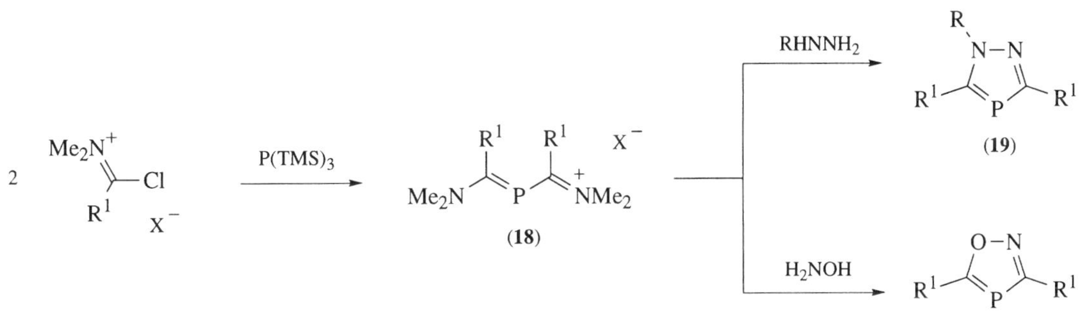

(18) (19) (20)

Monocyclic (**22**), R = Me, Et, Ph; R^1, R^2 = H, Me, Pr^i, Bu^t, Ph ⟨84AG901⟩, and bicyclic 1,2,4-diazaphospholes (**24**), R = Me, Pr^i, Bu^t, Ph, (**26**) ⟨87TL1511⟩, are obtained from oxadiazolium cations (**21**), (**23**), and zwitterions (**25**) by condensation with tris(trimethylsilyl)phosphine ⟨88TL3387⟩.

An important third route to 1,2,4-diazaphospholes is the [3 + 2] cycloaddition of phosphaalkynes and diazoalkanes. The necessary phosphaalkynes are prepared separately (R^1 = neopentyl, Pr^i, Bu^t, other tertiary alkyl groups, mesityl, trimethylsilyl) or *in situ* (R^1 = Me, by flash vacuum pyrolysis of dichloro ethylphosphine) ⟨89PS(46)31⟩. In most cases diazoalkanes add to phosphaalkynes regiospecifically (except R^1 = H, NPr$^i_2$, see Section 4.3.4.5.2) ⟨88JOM(338)329, 92CB411, 97BSB455, 98S1305⟩ with the formation of a PC bond. The stereochemistry of the [3 + 2] cycloaddition reaction of phosphaacetylene with diazomethane has been investigated by *ab initio* calculations. The regioisomeric intermediate leading to the 1,2,4-diazaphosphole was found to be more stable than the one leading to the 1,2,3-diazaphosphole ⟨97JCPC609⟩.

The cycloaddition products of phosphaalkynes and primary (R^3 = H) ⟨84AG898, 84CC1634, 86JOM(306)39, 86S31, 87CB1645, 88JOM(338)329⟩, silyl- (R^3 = SiMe$_3$, SiPr$^i_3$, SiMe$_2$Bu^t) ⟨88TL925, 96T1053⟩, phosphino- (R^3 = PPh$_2$, P(OMe)$_2$, P(NPr$^i_2$)$_2$) ⟨90AG316, 99CEJ274⟩, acyl- (R^3 = COMe, COPh, etc.) and phosphoryl-diazoalkanes (R^3 = POPh$_2$, PO(OMe)$_2$) ⟨84AG898, 87CB1645, 87PS(30)479⟩ rearrange to give 1,2,4-diazaphospholes. The isomer resulting from a 1,5-shift of R^3 is always formed; for R = aryl a minor amount of the other isomer is formed in addition. (For the reaction products of diazomethanes with $R^{2,3}$ = alkyl, aryl see ⟨87TL2693, 90T3859⟩.)

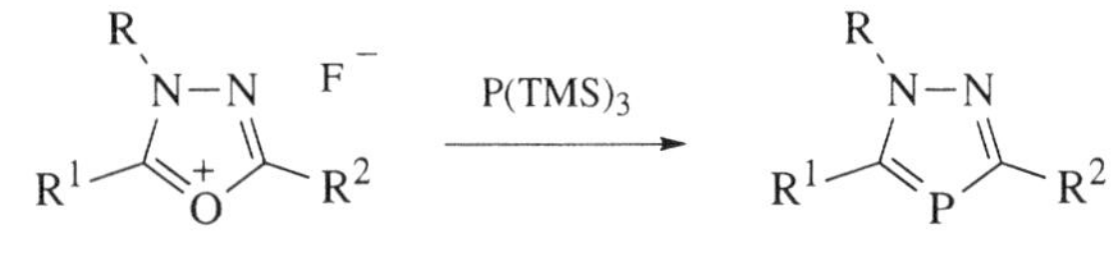

(21) (22)

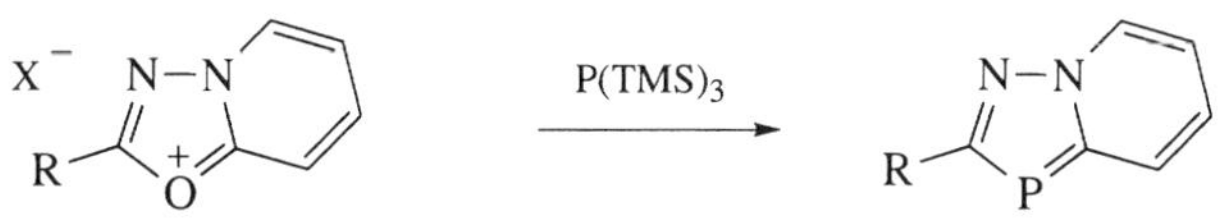

(23) (24)

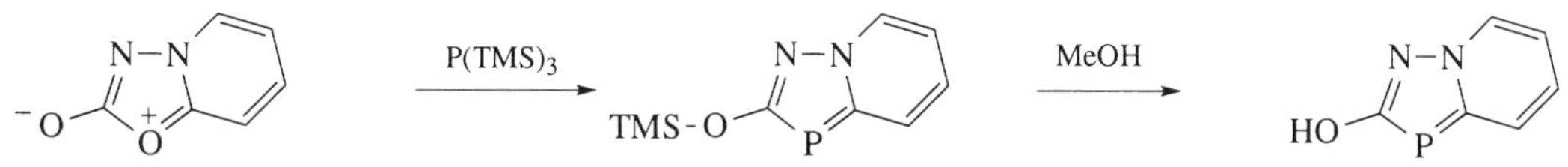

(25) (26)

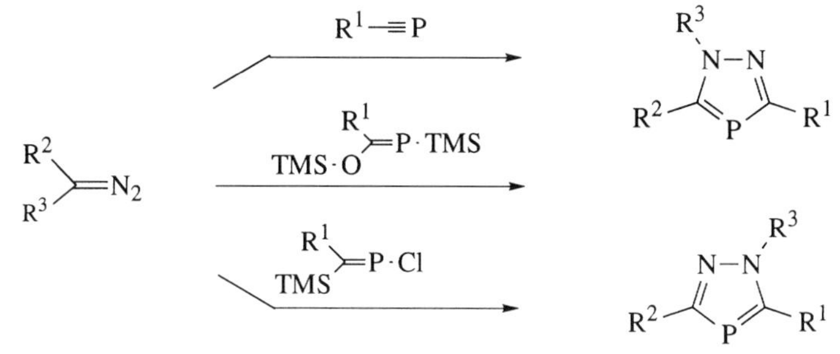

Bis- and tris(trimethylsilyl-diazomethyl)phosphines undergo double and triple regiospecific cycloadditions with a phosphaalkyne ⟨88TL925⟩.

More 1,2,4-diazaphospholes (R^3 = H) result from the reaction of primary diazoalkanes with a phosphaalkene of the type $R^1(Me_3SiO)C{=}PSiMe_3$, R^1 = Bu^t ⟨85ZN(B)1077, 87CB1809⟩, 1-Ad ⟨86S31⟩ or a phosphaalkene of type $R^1(Me_3Si)C{=}PCl$, R^1 = Ph ⟨84AG899, 88TL785⟩, $SiMe_3$ ⟨84CC1640, 88ZN(B)1285, 88JOM(338)329⟩, CO_2Et ⟨86TL5611⟩. In the reaction of lithio or other metallated diazomethanes (**27**) with *P*-chloro trimethylsilylphosphaethenes (**28**) the phosphavinyl diazoalkanes (**29**) can be isolated at −78°C as intermediates. At higher temperatures they undergo cyclization to 1,2,4-diazaphospholes (**30**), including the tris(trimethylsilyl) derivative ⟨88ZN(B)1285, 89NJC359, 89S511⟩. The intermediate with R^2 = Me_3Si may react with Bu^{t}C≡P in a cycloaddition as described above to produce a phosphaalkenyl substituted 1,2,4-diazaphosphole (**31**) (Scheme 4) ⟨89S511⟩.

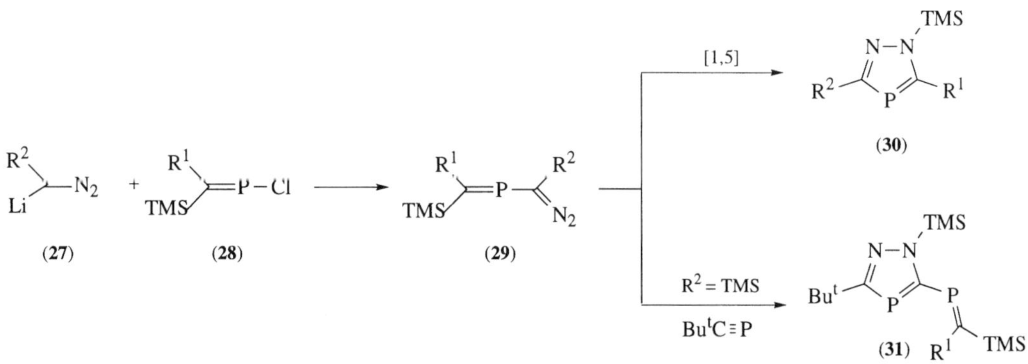

(27) (28) (29) (30) (31)

Scheme 4

From the cycloaddition of trialkylsilyl diazoketones and phosphaalkenes result 3-acyl-1,2,4-diazaphospholes ⟨96T1053, 97CB779⟩.

Trimethylsilyl- and diaminophosphinosubstituted lithiodiazomethanes (**27**) react with the triphosphenium salt $(Me_2N)_3PPP(NMe_2)_3^+BPh_4^-$ to give the correspondingly 3,5-substituted 1,2,4-diazaphospholes probably via the trimethylsilyl- or diaminophosphino-phosphaethyne ⟨99CR(C)351⟩. Thermolysis of a diaminophosphino-phosphiranyl-diazomethane leads to a 1,3-bis(diaminophosphino)-5-phosphiranyl-1,2,4-diazaphosphole via a transient diaminophosphino-phosphaethyne ⟨99CR(C)351⟩.

Phosphaalkyne equivalents for the reaction with ethyl diazoacetate and for the synthesis of 1,2,4-diazaphospholes are 1-chloroalkylphosphines (**32**), R^1 = Me ⟨96IC6667⟩, 1,1-dichloroalkylphosphines (**33**), R^1 = Me, Pr^n, *n*-pentyl, allyl, $PhCH_2$, Ph, PhS, Me_3Si, ⟨90PS(49/50)317, 92HAC337⟩ in the presence of pyridine or triethylamine or allyl dichlorophosphines (**34**), R^1 = CMe=CH-CO_2Me, in the presence of the equimolar amount of DABCO (Scheme 5) ⟨90TL1147⟩. In the first case the transient 1-phosphapropene forms an adduct, the aromatization of which to the 1,2,4-diazaphosphole is achieved by chlorination and HCl abstraction (Scheme 5).

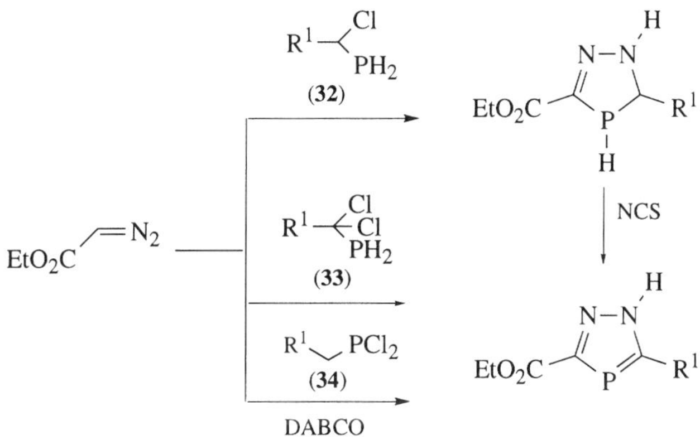

Scheme 5

Reactions of diphenylnitrile imine with the phosphaalkyne $Bu^tC{\equiv}P$ ⟨84AG898, 87S689, 88AG1541⟩ or with the phosphaalkene $Ph(Me_3Si)C{=}PCl$ ⟨88TL785⟩ also give 1,2,4-diazaphospholes. If the nitrile imine has a methyl group in place of one of the phenyl groups, 2% of the isomeric 2*H*-1,2,3-diazaphosphole is formed along with the 1,2,4-diazaphosphole ⟨88AG1541⟩ (see Section 4.3.4.5.2).

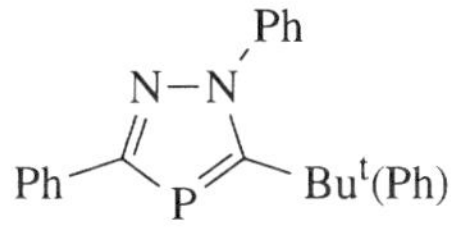

(35)

Sydnones (**36**) provide another source for the CNN fragment. With a phosphaalkyne ⟨87CB1809⟩ and with two types of phosphaalkenes ⟨86TL4415, 87CB1809⟩ they also give 1,2,4-diazaphospholes (R = H, Me, CH_2Ph, Ph; R^1 = H, Me, Ph) (Scheme 6). Only in one case there is a minor amount of the 1*H*-1,2,3-diazaphosphole formed (Section 4.3.4.5.2).

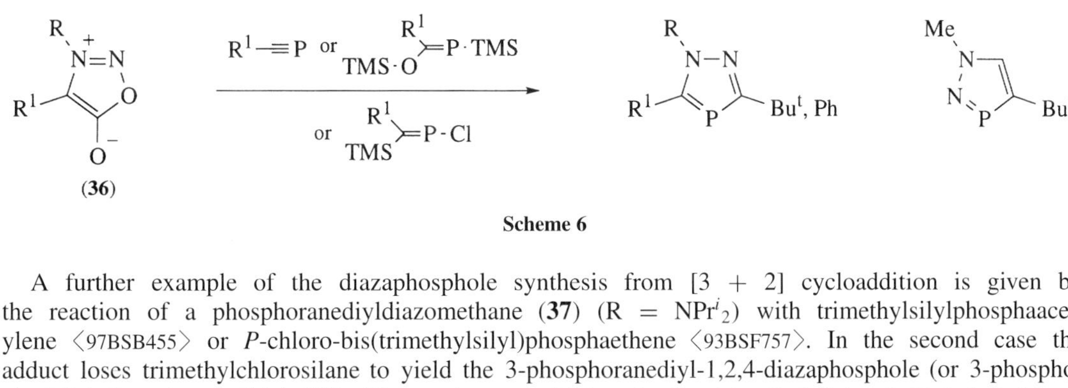

Scheme 6

A further example of the diazaphosphole synthesis from [3 + 2] cycloaddition is given by the reaction of a phosphoranediyldiazomethane (**37**) (R = NPr^i_2) with trimethylsilylphosphaacetylene ⟨97BSB455⟩ or *P*-chloro-bis(trimethylsilyl)phosphaethene ⟨93BSF757⟩. In the second case the adduct loses trimethylchlorosilane to yield the 3-phosphoranediyl-1,2,4-diazaphosphole (or 3-phosphonio-1,2,4-diazaphospholide) (**38**). The analogous addition to the trimethylsilyl substituted *P*-chlorobis(methylene)phosphorane yields a 4-methylene derivative of this diazaphosphole (molecular structure in Table 5). It provides the only fully characterized example of this type so far. Methyl triflate methylates the compound at N-1 and gives a phosphoniomethylene-diazaphosphole cation.

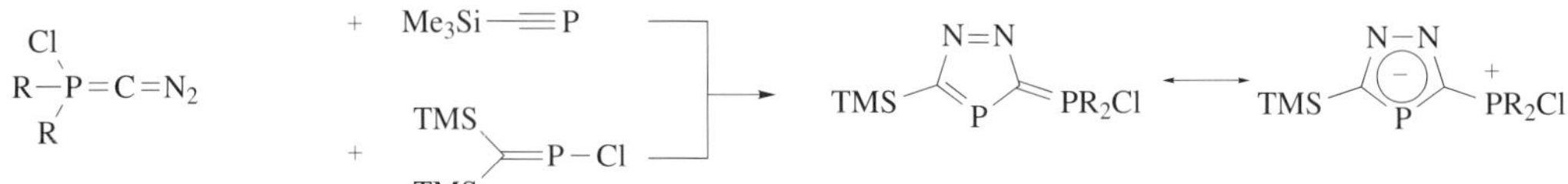

(37) (38)

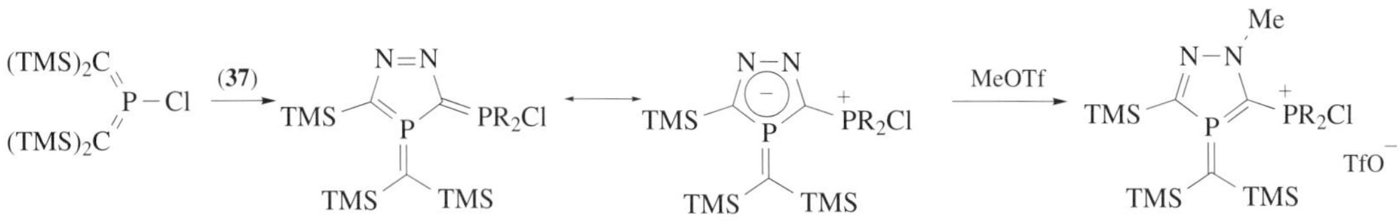

(39) (40) (41)

Another cationic 3-phosphonio-1,2,4-diazaphosphole derivative (molecular structure in Table 5) results from the reaction of $(Ph_3PCPCl)_2$ with trimethylsilylazide ⟨97CB89⟩ (see Section 4.3.5.2.2). The synthesis of dimeric 4-imino-1,2,4-diazaphospholes ⟨00T35⟩ is described in Section 4.3.5.2.1.

Nitrile oxides (**42**) (R = Bu^t, Ph, and other aryl groups) and phosphaalkynes (R^1 = neopentyl, Pr^i, Bu^t, other tertiary alkyl groups, mesityl, trimethylsilyl, and from *in situ* preparation, also R^1 = Me) give 1,2,4-oxazaphospholes (**43**) ⟨84AG898, 84CC1634, 86JOM(306)39, 86S31, 87S689, 88AG1541, 89PS(46)31, 97BSB455, 98S1305, 00TH2⟩ in a regiospecific addition. In case of sterically less demanding substituents the oxazaphosphole adds a second molecule of nitrile oxide ⟨87S689, 88JOM(338)329, 89PS(46)31⟩ yielding a symmetrical bicyclic compound (**44**) (see Section 4.3.4.3.1). In place of phosphaalkynes, certain phosphaalkenes may also be used ⟨84CC1640, 85ZN(B)1077, 88AG1541, 88TL785⟩. From the reaction of mesityl nitrile oxide with 3,5-di-*t*-butyl-1,2,4-thiadiphosphole (Section 4.3.6.8) or with the tetracyclic adduct from *t*-butylphosphaethyne and selenium (Section 4.3.6.8) the respective 1,2,4-oxazaphosphole (**43**) has been isolated ⟨99TH1, 00TH3⟩.

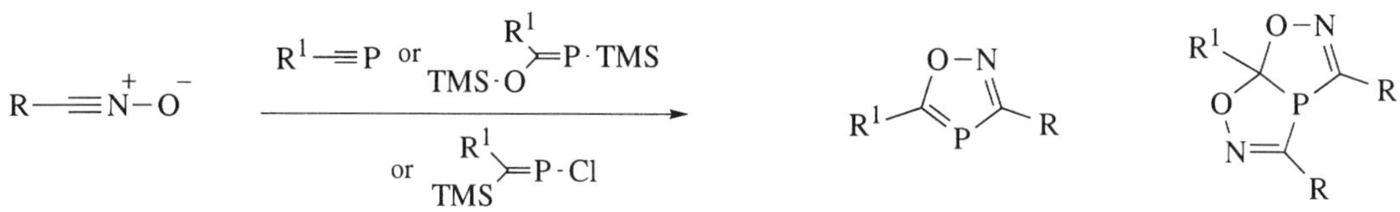

(42) (43) (44)

Dichloromethylphosphine $Cl_2CH\text{-}PH_2$ can also serve as synthetic equivalent of phosphaacetylene. Its reaction with benzonitrile oxide in pyridine leads to a transient 1,2,4-oxazaphosphole which, as in the above examples, adds immediately a second molecule of nitrile oxide ⟨89CC988, 90PS(49/50)317⟩.

Phenylnitrile sulfide, from the thermolysis of an oxathiazolone (**45**) (R = Ph), adds with the same orientation as the nitrile oxides to $Bu^tC{\equiv}P$ ⟨87S689, 88AG1541⟩. More 1,2,4-thiazaphospholes (**46**) are produced with phosphaalkenes, R = Ph, 2- and 4-tolyl, R^1 = Ph, $SiMe_3$ ⟨88TL4535⟩; for side products of this reaction, see Sections 4.3.2.5 and 4.3.6.7.

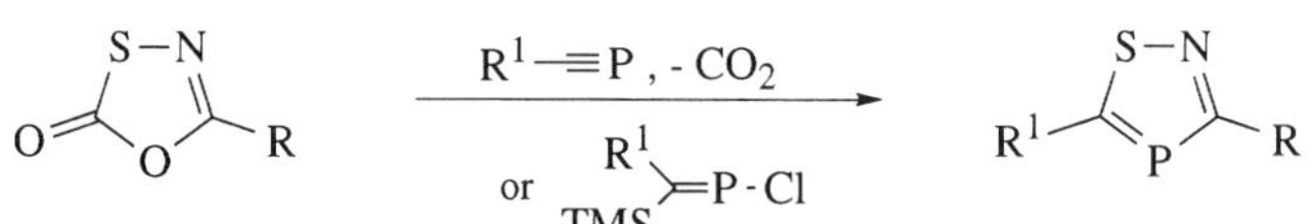

(45) (46)

4-Phenyloxathiazolium-olate (**47**) and the phosphaalkyne $Bu^tC{\equiv}P$ react to yield two isomeric 1,2,4-thiazaphospholes: besides the main product (**48**) with the expected order of substituents, 3% of the product (**49**) with the reversed order is formed ⟨87CB1809, 88AG1541⟩.

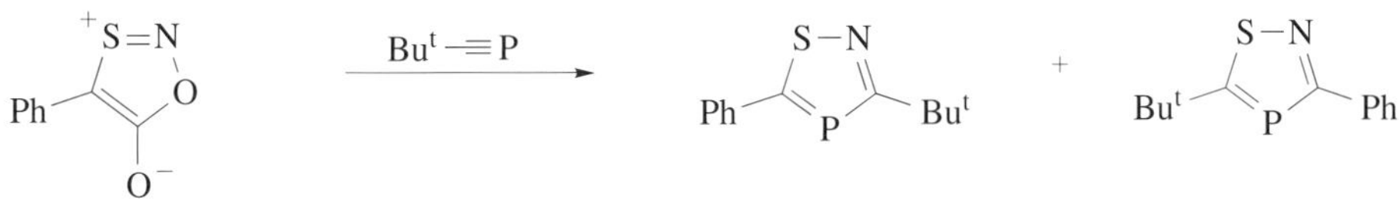

(47) (48) (49)

4.3.4.3.3 Application

1-α-D-Ribofuranosyl-1,2,4-diazaphosphole, the antiviral agent Ribavirin analogue 1-β-D-ribofuranosyl-1,2,4-diazaphosphole-3-carboxamide and 1-β-D-ribofuranosyl-1,2,4-diazaphosphole-3-thiocarboxamide have been prepared. The latter gave a 25% increase in life span against L1210 in female BDF_1 mice ⟨90JMC572, 91CA62507b⟩.

4.3.4.4 1,4,2-Diazaphospholes (4, 5) and 1,3,4-Thiazaphospholes (6)

1,4,2-Diazaphospholes have sometimes been misnumbered as 1,3,4-diazaphospholes.

4.3.4.4.1 Reactivity

The 1,4,2-diazaphospholes are colorless to pale yellow crystalline solids, stable towards atmospheric oxygen. For reactions, mainly the 1,4,2-diazaphospholo[4,5-*a*]pyridines (**50**) were used. They have a characteristic naphthalene-like smell ⟨86PS(28)289⟩. They do not react with sulfur, selenium, or methyl iodide. With dimethyl sulfate, alkylation at the two-coordinate nitrogen occurs. Hydrolysis opens the diazaphosphole ring to give a zwitterionic 2-aminopyridinio-methylphosphonite (**51**). Methanol adds to the P=C bond. The adduct with diethylamine (**52**) is largely dissociated but its product with concomitant oxidation with sulfur or selenium (**53**) can be isolated.

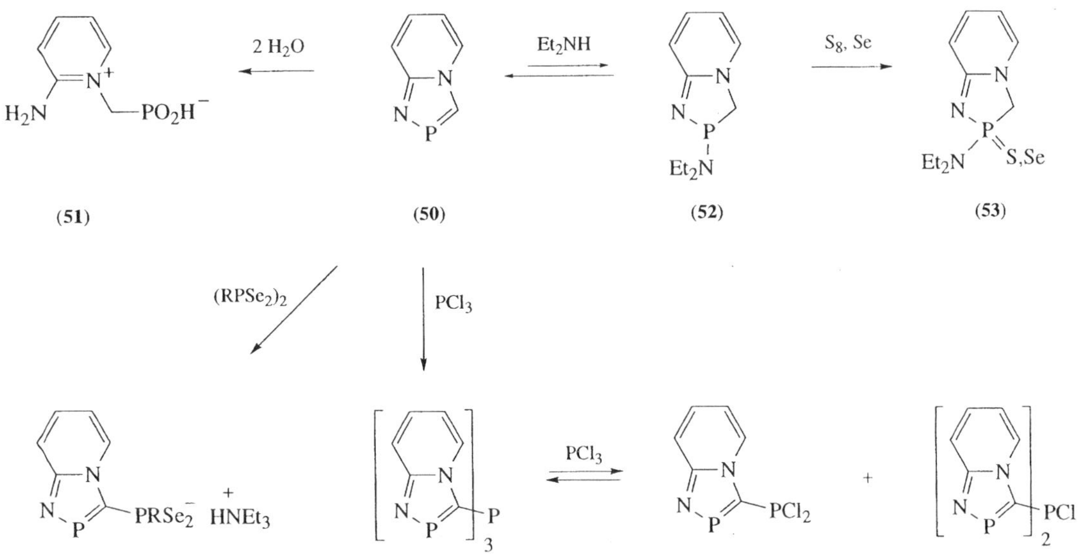

(51) (50) (52) (53)

(55) (54)

With Ph_2PCl, $MePCl_2$, $ClCH_2PCl_2$ and PCl_3 substitution at the carbon atom next to phosphorus takes place. The tris(1,4,2-diazaphospholyl)phosphine (**54**), resulting in the latter case, and phosphorus trichloride take part in a substituent exchange equilibrium ⟨86PS(28)289, 95CB581⟩. The reaction of (**50**) with selenophosphonic acid anhydrides $(RPSe_2)_2$ (R = Ph, 4-$MeOC_6H_4$, 4-$Me_2NC_6H_4$) in the presence of NEt_3 leads to triethylammonium selenophosphinates (**55**) ⟨93PS(77)235⟩.

2,3-Dimethylbutadiene and isoprene undergo a [4 + 2] cycloaddition to the P=C bond of 1,4,2-diazaphospholo[4,5-*a*]pyridines (**50**) and [5,4-*b*]benzothiazoles (**56**) ⟨86PS(28)289, 91HAC369⟩. 1,4,2-Diazaphospholo[5,4-*b*]thiazolines (**57**) react with 1,3-butadienes (R = H, Me) and in addition with elemental sulfur or selenium (Scheme 7). The cycloaddition of isoprene proceeds regioselectively ⟨94T7675⟩.

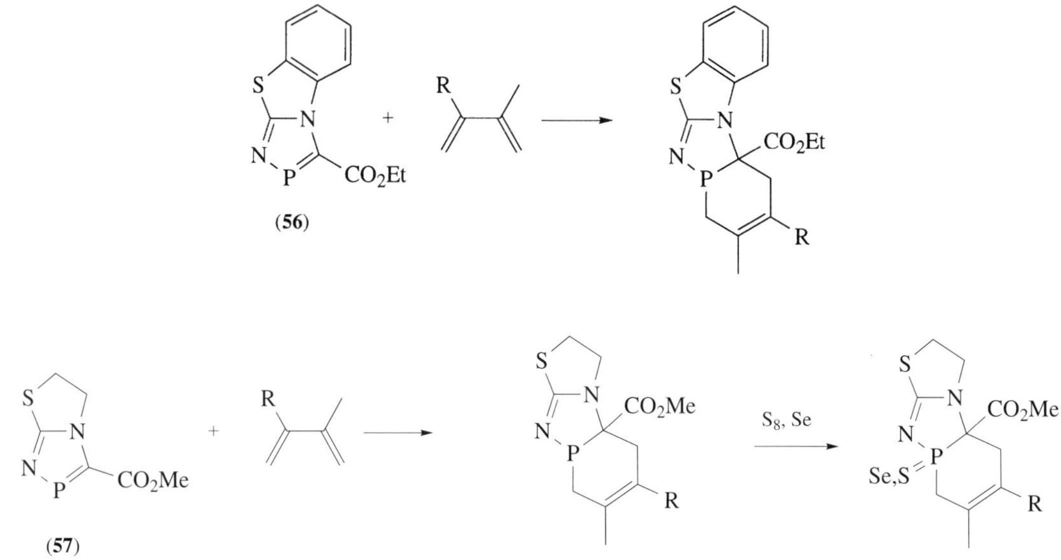

Scheme 7

1,4,2-Diazaphospholo[5,4-*b*]thiazoles do not react with methyl iodide but are alkylated by dimethylsulfate at the nitrogen atom next to phosphorus ⟨92S267⟩.

The 1,3,4-thiazaphospholes (**58**) are colorless, yellow, or red crystalline solids ⟨85AG125⟩; the 2-phenyl derivative is a pale yellow liquid. They are not oxidized by atmospheric oxygen or sulfur ⟨86PS(28)71⟩. Dimethyl sulfate or trimethyloxonium tetrafluoroborate methylates the nitrogen atom ⟨85AG125, 86PS(28)71⟩ but methyl iodide does not react. Borontrifluoride coordinates to nitrogen ⟨B-90MI258⟩. Alcohols, thiols, water, and hydrogen chloride add to the P=C bond to give 4,5-dihydro-1,3,4-thiazaphospholes but diethylamine does not react ⟨85AG125, 86PS(28)71⟩. Phenols (**59**) with a heterodiene moiety in the *ortho*-position react to yield tricyclophosphoranes (**60**) ⟨B-90MI258⟩.

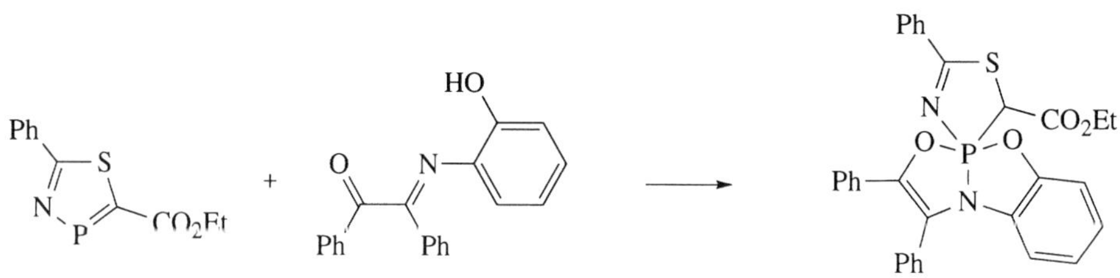

(58) **(59)** **(60)**

Azodicarboxylic esters (R = Me, Et, Pr) and glyoxal-bis(cyclohexyl)imine (R = C_6H_{11}) undergo [4 + 1] cycloaddition reactions to the phosphorus atom of (**58**). Butadienes (R = H, Me) and 1,4-diphenyl-1-azabutadiene give [4 + 2] cycloaddition reactions to the P=C bond (Scheme 8) ⟨B-90MI258⟩. With benzaldazine a criss-cross-type cycloaddition takes place ⟨B-90MI258⟩ and the reaction with diphenyldiazomethane yields a bicyclic phosphirane (**61**) among other products ⟨87CB597⟩.

4.3.4.4.2 Synthesis

4,5-Anellated 1,4,2-diazaphospholes (**61**) ⟨92S267, 92ZN(B)373, 91IJC(B)992, 94PS(97)141, 95S361, 98JIC648⟩ and 1,3,4-thiazaphospholes (**62**) ⟨85AG125⟩ can be prepared in a condensation reaction from PCl_3 with partners having a four-membered chain and a suitable terminal reactivity (Scheme 9). They have been obtained from 2-aminothiazoline, 2-aminothiazoles, 2-aminobenzothiazoles or 2-aminopyridines and ethyl bromoacetate or phenacyl bromide. 2-Amino-1-(ethoxycarbonylmethyl)pyridinium bromides (R = H, Me)

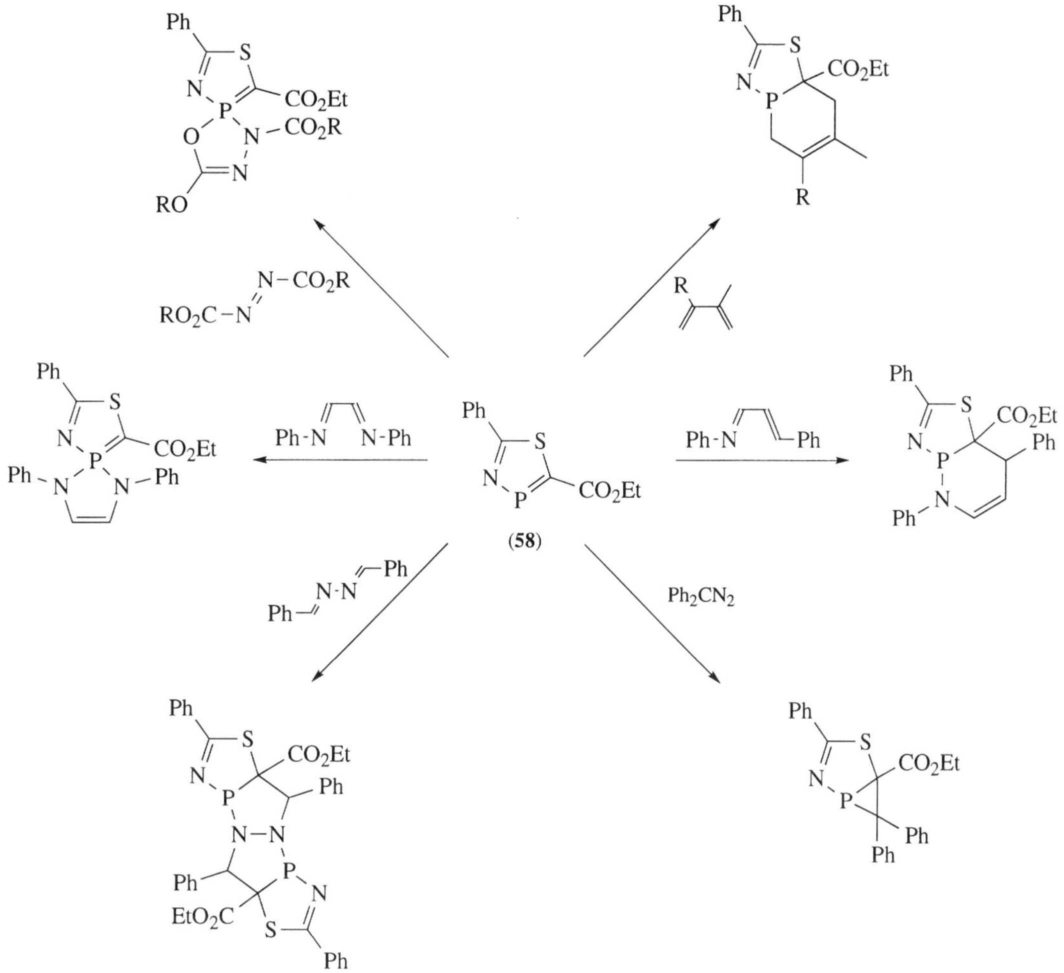

Scheme 8

condense with phosphorus trichloride in the presence of triethylamine in acetonitrile to give 1-aza-2-phosphaindolizines, i.e. [1,4,2]diazaphospholo[4,5-*a*]pyridines. If the reaction is carried out in toluene, the intermediate (1-alkyl-1,2-dihydro-2-pyridylideneamino)dichlorophosphines can be observed and in some cases isolated. In the same way 3-substituted 5,6-dihydrothiazolo[3,2-*d*][1,4,2]diazaphospholes (R = CO_2Me, CO_2Et, CO_2Bu, COPh), thiazolo[3,2-*d*][1,4,2]diazaphosphole and its benzo derivatives (R = CO_2Me, CO_2Et, COPh; R^1 = H, OMe) are obtained. In this case the intermediate dichlorophosphines also have been isolated ⟨91TL6433, 92ZN(B)373, 91IJC(B)992⟩.

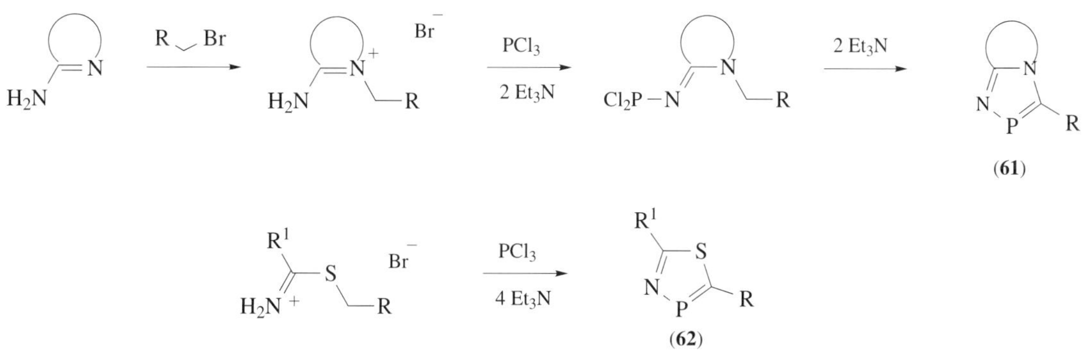

Scheme 9

The salts from thioamides or thiourea (R^1 = Me, Ph, NMe_2) with ethyl bromoacetate or benzyl bromides (R = CO_2Et, Ph, 4-substituted phenyl) yield the corresponding 1,3,4-thiazaphospholes (**62**) on condensation with PCl_3 (Scheme 9) ⟨85AG125⟩.

5-Unsubstituted monocyclic and fused 1,4,2-diazaphospholes are synthesized by using chloromethyl-dichlorophosphine (**64**) for condensation (Scheme 10). With amidines (**63**) (R^1/R^2 = Me/H, Ph/H, Ph/Ph, SMe/Me) mixtures of 3*H*- and 1*H*-1,4,2-diazaphospholes (**65**) and (**66**) are obtained; for R^2 = H they equilibrate by proton exchange ⟨86PS(28)289, 90PS(49/50)349⟩.

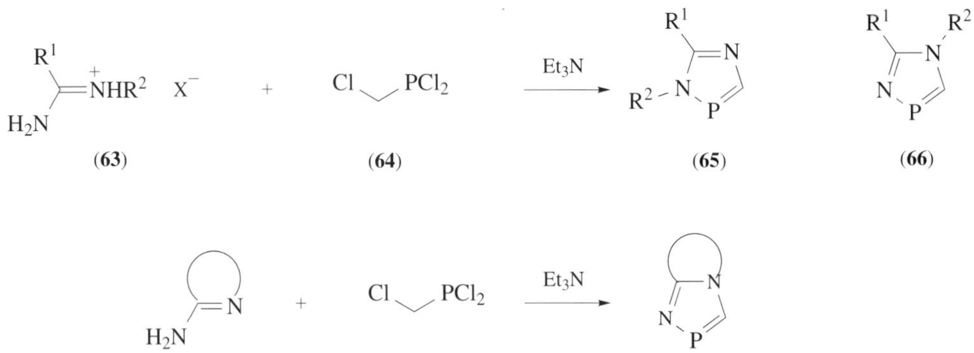

Scheme 10

Condensation of 2-aminopyridines (R = H, Me), 2- and 4-aminopyrimidines (R = H, Me), 2-aminopyrazine, and 2-aminoquinoline with chloromethyl-dichlorophosphine yields regiospecifically 1,4,2-diazaphospholo[4,5-*a*]pyridines, 1,4,2-diazaphospholo[4,5-*a*]pyrimidines, 1,4,2-diazaphospholo[4,5-*a*]pyrazine, and [1,4,2]diazaphospholo[4,5-*a*]quinoline, respectively (Scheme 10). In the case of 2-aminopyrimidines, it has been possible to isolate a diazadiphosphetidine as an intermediate. On treatment with NEt_3 it changes into the final product, presumably through intramolecular nucleophilic displacement of the chlorine atom by the pyridinic nitrogen atom accompanied by the rupture of the diazadiphosphetidine ring and loss of hydrogen chloride ⟨86PS(28)71, 86PS(28)289, 91HAC369, 94T7675, 95CB581⟩.

In the same way 2-aminothiazoline (**67**) can be condensed with chloromethyl-dichlorophosphine. The reaction is again regiospecific, but the orientation is reversed and the 5,6-dihydro-thiazolo[3,2-*e*][1,2,4]diazaphosphole (**68**) is obtained (Scheme 11). In this case the reaction probably initiates at the ring nitrogen atom ⟨86PS(28)289, 94T7675⟩.

When 2-amino-1,3-thiazole (**69**) or 2-amino-1,3-benzothiazoles are used for the cyclocondensation, mixtures of isomeric 1,5- and 4,5-anellated 1,4,2-diazaphospholes, resulting from both orientations of the CP unit with respect to the NCN fragment in the aminothiazole, are generally obtained. Thus, the reaction of 2-amino-1,3-thiazole leads to thiazolo[2,3-*d*]-1,4,2-diazaphosphole (**70**) (70%) and thiazolo[3,2-*d*]-1,4,2-diazaphosphole (**71**) (10%). The ratio of the two isomers obtained from 2-aminobenzothiazoles strongly depends on the substituents in the benzo ring (R^1/R^2 = H/H, H/OMe, Me/H) ⟨94T7675, 95CB581⟩

The analogous condensation of thiobenzamide (**72**) with chloromethyl-dichlorophosphine yields 2-phenyl-1,3,4-thiazaphosphole (**73**) ⟨86PS(28)289⟩.

Lithiumphosphide adds to two moles of 2,6-diisopropylphenylisocyanate accompanied by the loss of hydrogen to give 5-oxo-1,4,2-diazaphospholide (**74**) ⟨98MI1⟩.

(**74**)

4.3.4.5 1,2,3-Diazaphospholes (7, 8)

4.3.4.5.1 Reactivity

1,2,3-Diazaphospholes are colorless to pale yellow distillable liquids or crystalline solids ⟨72JPR806, 80CB2278⟩. They are stable to oxidation by air and they do not react with elemental sulfur ⟨67CHE601,

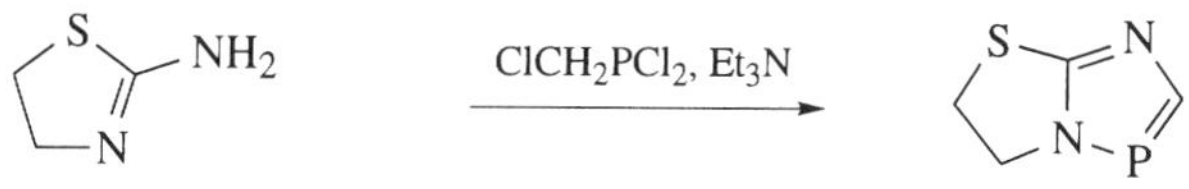

(67) (68)

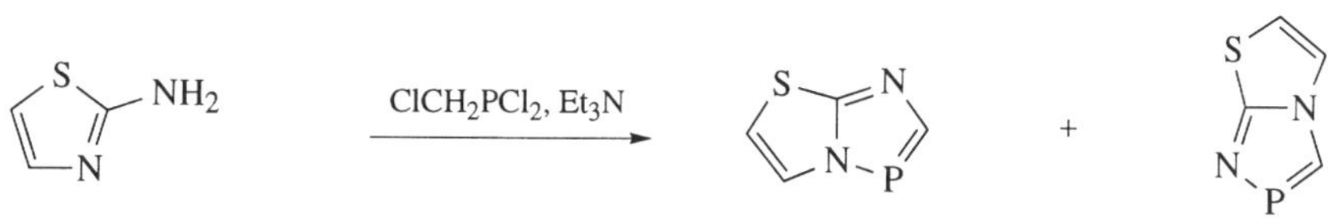

(69) (70) (71)

(72) (73)

Scheme 11

72JPR806, 81CR(C)187, 83CB549⟩. They are readily hydrolyzed ⟨67CHE601, 72JPR806⟩ to give the hydrazone from which they originate and phosphorous acid. The flash pyrolysis of some 2-methyl-1,2,3-diazaphospholes has been investigated ⟨90PS(49/50)363⟩. While only a few reactions of 1*H*-1,2,3-diazaphospholes are reported, the chemistry of the 2*H*-isomers is well studied.

4.3.4.5.1.1 N-coordination and N-substitution

Hydrogen halides give *N*-protonation of 2*H*-1,2,3-diazaphospholes (molecular structure of a hydrochloride in Table 7) but also add to the P=C bond (see below). Hydrolysis opens the P=N bond of 1,2,3-diazaphospholo[1,5-*a*]pyridines ⟨95S173⟩.

While alkyl halides do not react with 1*H*- and 2*H*-1,2,3-diazaphospholes, alkylation with dimethyl sulfate or trimethyloxonium tetrafluoroborate and protonation take place at the two-coordinate nitrogen atom and leave the phosphorus atom two-coordinate ⟨67CHE601, 80CB2278, 83CB549, 95S173⟩. This is also true for the addition of BF_3 ⟨80CB2278⟩ and in some cases even for the coordination of transition metals (see below). Methylation of both 2,5- and 1,5-dimethyl-1,2,3-diazaphosphole gives the same cation (**75**) (X = $MeSO_4$, BF_4), which on demethylation yields only the 2,5-isomer ⟨80CB2278⟩.

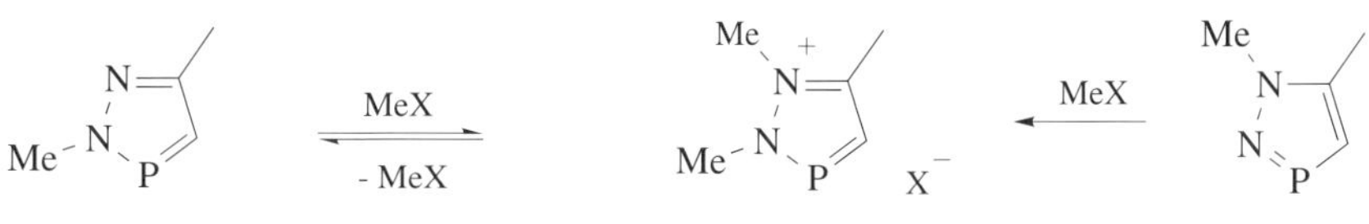

(**75**)

The *N*-unsubstituted 5-methyl-1,2,3-diazaphosphole adds aryl isocyanates to N-2 ⟨77IZV532⟩ and is acylated reversibly ⟨77IZV532, 77ZOB576⟩.

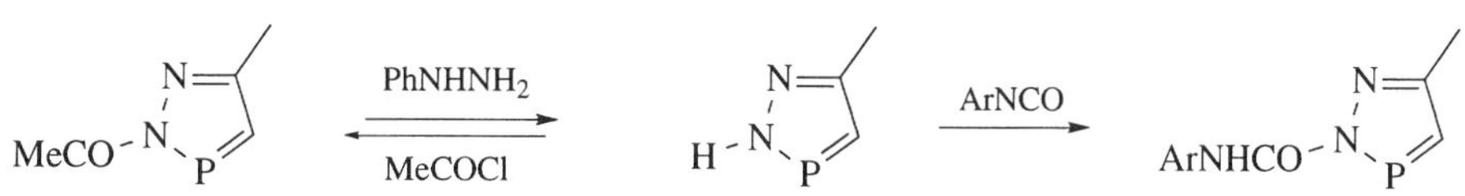

(**76**)

Alkyl triflates react with 2-trimethylsilyl-1,2,3-diazaphospholes (**77**) exclusively in a 2 : 1 molar ratio and yield 1,2-dialkyl-1,2,3-diazaphospholium triflates ⟨96TH1⟩. Methanolysis converts (**77**) to the

N-unsubstituted compounds. In the latter the proton is bonded to N-2 and forms an intermolecular hydrogen bond to N-1 which allows a rapid exchange between the 2*H*- and the 1*H*-isomer ⟨77ZOB576, 96TH1⟩. 2-Acyl-1,2,3-diazaphospholes (molecular structure see Table 7) are readily obtained from the trimethylsilyl derivatives and carbonic acid chlorides ⟨96TH1⟩. More substitution reactions are achieved with chlorophosphines, dichlorophosphines and phosphorus trichloride (Scheme 12) ⟨83ZN(B)1484, 96TH1⟩. The trimethylsilyl group is also replaced in the reaction with cyclopropenylium bromides or triphenylchloromethane. Alkyl halides with less carbenium character do not react ⟨96TH1⟩.

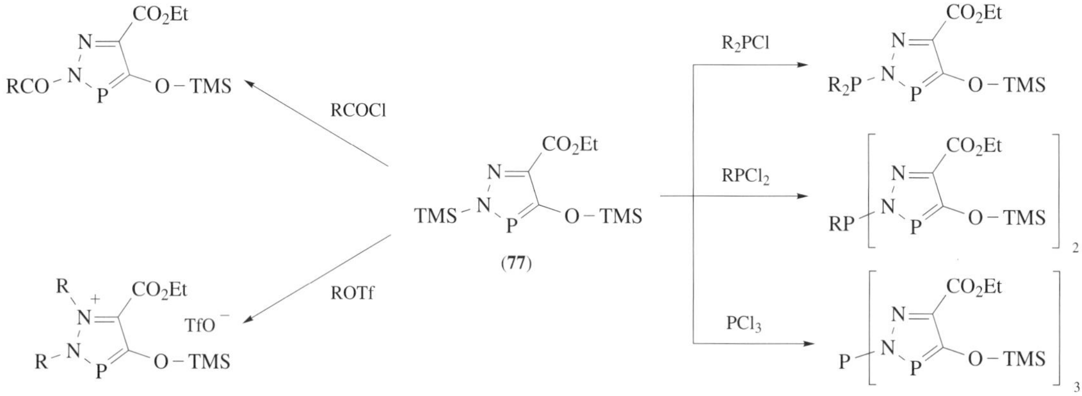

Scheme 12

The opening of the PN bond by methanolysis is observed for some 1,2,3-diazaphosphole complexes ⟨97ICA(256)47⟩ (see Section 4.3.4.5.1.6).

4.3.4.5.1.2 Polar addition to the P=C bond and substitution at C-4

When a polar reagent adds to the P=C bond of a heterophosphole the more electronegative part generally becomes bonded to the phosphorus. The addition reactions are often reversible and the addition equilibria are more or less balanced.

The equilibria of the reactions of 2*H*-1,2,3-diazaphospholes (**78**) with alcohols ROH depend on the nature of the *N*-substituent R^1. With R^1 = Me ⟨76ZOB590, 77ZOB576, 81CR(C)187, 84PS(21)335⟩, Et ⟨67CHE601, 72JPR806, 76ZOB590⟩, CH_2Ph ⟨76ZOB590⟩, the 3-alkoxy-3,4-dihydro-1,2,3-diazaphospholes (**79**) have been obtained. The addition is reversible ⟨76ZOB590, 77ZOB576⟩; an excess of alcohol degrades the diazaphosphole to the hydrazone and $P(OR)_3$ ⟨67CHE601, 72JPR806, 90ZOB35⟩. Addition of propynol to the 2-acetyl-5-methyl-1,2,3-diazaphosphole is followed by a rearrangement to the 3-allenyl-dihydrodiazaphosphole-3-oxide (**80**) ⟨91PS(56)129⟩ (Scheme 13).

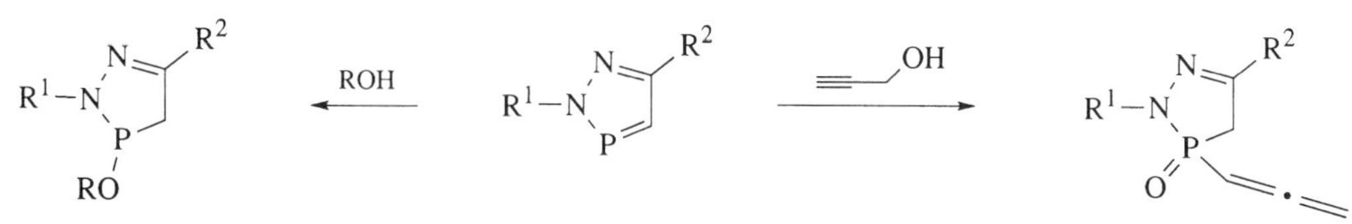

(79) **(78)** **(80)**

Scheme 13

Reaction with ethylene glycol gives a spirophosphorane ⟨81CR(C)187, 84PS(21)335⟩; catechol degrades the ring completely ⟨91PS(57)83⟩. Ammonia as well as primary and secondary amines do not react ⟨67CHE601, 72JPR806, 85CB1621⟩ except with 2-acetyl-5-methyl-1,2,3-diazaphosphole ⟨72ZOB1939⟩. To 1,2,3-diazaphospholo[1,5-*a*]pyridines (as representatives of 1*H*-1,2,3-diazaphospholes) water adds with rupture of the P=N bond ⟨95S173⟩. Dialkylthiophosphoric acids also add to the PC bond of (**78**) ⟨96MI94⟩.

The addition of hydrogen halides HX (X = Cl, Br, I) to a 2*H*-1,2,3-diazaphosphole yields the 3-halodihydrodiazaphosphole in equilibrium with the diazaphospholium halide (Scheme 14). This equilibrium depends strongly on the substituent R^1 at nitrogen ⟨80CB2278⟩ and in particular on the substituent R^2 at C-4 ⟨84S591⟩. For R^1/R^2 = Ph/H ⟨72JPR806, 80CB2278⟩, COMe/H ⟨72JPR806⟩, Me/Cl, Ph/Cl ⟨84PS(21)97⟩, Me/Br, Ph/Br ⟨84S591⟩, Me/I, Ph/I ⟨84PS(21)97⟩$^1/R^2$ = Me/H ⟨80CB2278⟩, Me/Me, Me/SMe, Me/SPh ⟨84S591⟩ the ionic form is favored. In the latter cases the dihydrodiazaphosphole is present albeit in low equilibrium concentration, as shown by the reaction with sulfur ⟨80CB2278⟩. Phosphorylsulfenyl chlorides add to the PC bond of (**78**) and also oxidize the phosphorus atom ⟨98ZOB1479⟩.

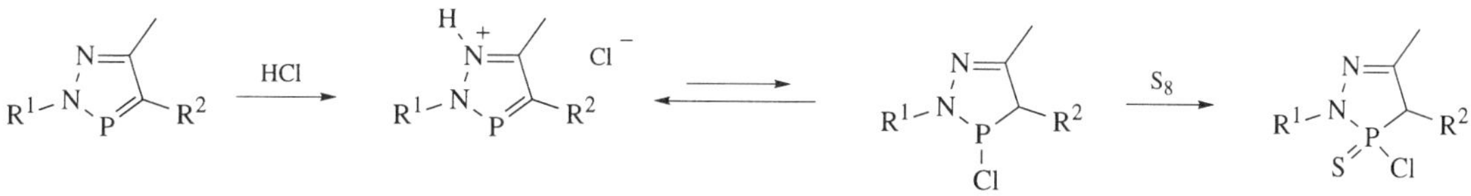

Scheme 14

The 4-aminomethylene-1,2,3-diazaphospholium chlorides and bromides (**114**), which exist in equilibrium with their covalent form (see Section 4.3.4.5.2), react with water, alcohol, amines and sulfur as expected for halophosphines. In contrast, the respective tetraphenylborates and triflates are not susceptible to oxidation ⟨97PS(123)125⟩.

The 1,2-addition to the PC bond of a 2*H*-1,2,3-diazaphosphole may be followed by a 1,2-elimination, thus restoring the diazaphosphole system. In the case of 4-unsubstituted 1,2,3-diazaphospholes this combination of addition and elimination will result in a substitution at C-4. The substitution was first shown for 2*H*-1,2,3-diazaphospholes with phosphorus(III) halides and thiophosphoryl halides ⟨76CC887, 79AG442, 80CB2278⟩. The method has recently been improved ⟨99IC1971⟩ for the preparation of the 4-dichlorophosphino-2,5-dimethyl-1,2,3-diazaphosphole and many derivatives thereof (see Section 4.3.4.5.1.5). The reaction rate increases in the order Ph_2PCl < $PhPCl_2$ < PCl_3. Likewise, with $P(CN)_3$ the 4-dicyanophosphinodiazaphosphole is obtained ⟨80CB2278⟩. 1,2,3-Diazaphospholo[1,5-*a*]pyridines also undergo substitution in 4-position with PCl_3 ⟨95S173⟩.

The reaction of 2*H*-1,2,3-diazaphospholes with bromine involves a sequence of 1,2-addition/elimination equilibria, R = Me, Ph ⟨85CB1621⟩. 4-Bromo-1,2,3-diazaphospholes (**81**) become accessible this way (Scheme 15).

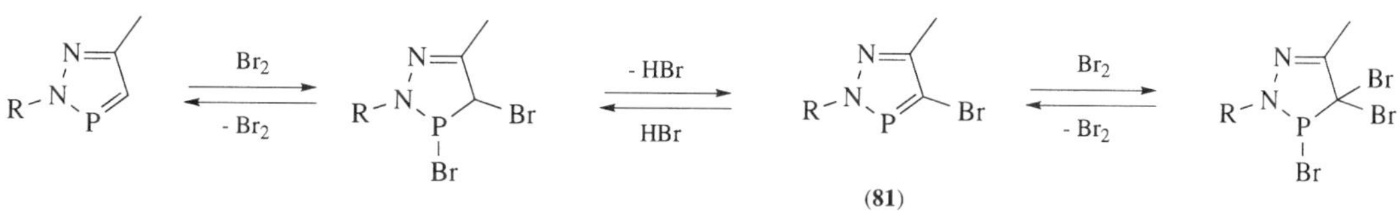

Scheme 15

Similarly, 4-phenylthio- and 4-methylthio-1,2,3-diazaphospholes result from reactions with sulfenyl halides ⟨84S591, 85CB1621⟩. Iodine chloride reacts to give both the 4-chloro- and the 4-iododiazaphosphole ⟨84S591⟩. Chlorination and bromination of diazaphospholes at C-4 has been achieved also with a chlorophosphorane or SO_2Cl_2 and BrCN or NBS, respectively ⟨84S591, 00ZOB1048⟩.

Lithiumalkyls (R = Me, Bu, Bu^t) as well as lithium diisopropylamide (R = NPr^i_2) add readily to the P=C bond of 2,5-dimethyl-1,2,3-diazaphosphole (**82**) ⟨84PS(21)97, 87JOM(319)49⟩. With methyl iodide and organotin chlorides the adduct yields the 4-substituted dihydro-1,2,3-diazaphospholes (**83**). Only in case of reaction of methyllithium (R = Me) lithiation at C-4 is observed as a minor side reaction (Scheme 16).

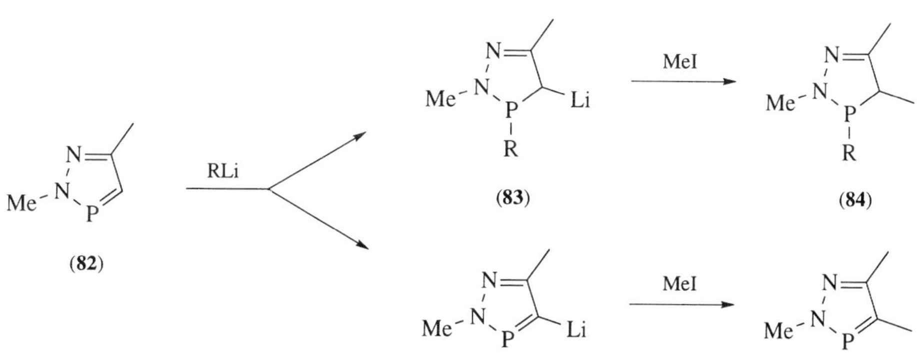

Scheme 16

1,2,3-Diazaphospholes generally do not react with aliphatic and aromatic aldehydes or ketones ⟨67CHE601, 72JPR806⟩. A 4,5-tetramethylene 1,2,3-diazaphosphole (**85**) (R = cyclohexyl) adds cyclohexanone and water in a slow reaction ⟨90ZOB35⟩.

2-Phenyl-5-methyl-1,2,3-diazaphosphole and aliphatic cyclic ketones condense to give the 4-alkenyl diazaphospholes and 4-hydroxyalkyl diazaphospholes ⟨85ZOB1464, 91ZOB1063⟩. 2,5-Dimethyldiazaphosphole (**86**) adds hexafluoroacetone giving the 4-hydroxyalkyl derivative (**87**) ⟨B-90MI258⟩. 2-Acetyl-5-methyl-1,2,3-diazaphosphole (**88**) and an electron-poor isocyanate likewise yield the 4-carbamoyl derivative ⟨84IZV1182⟩ (Scheme 17).

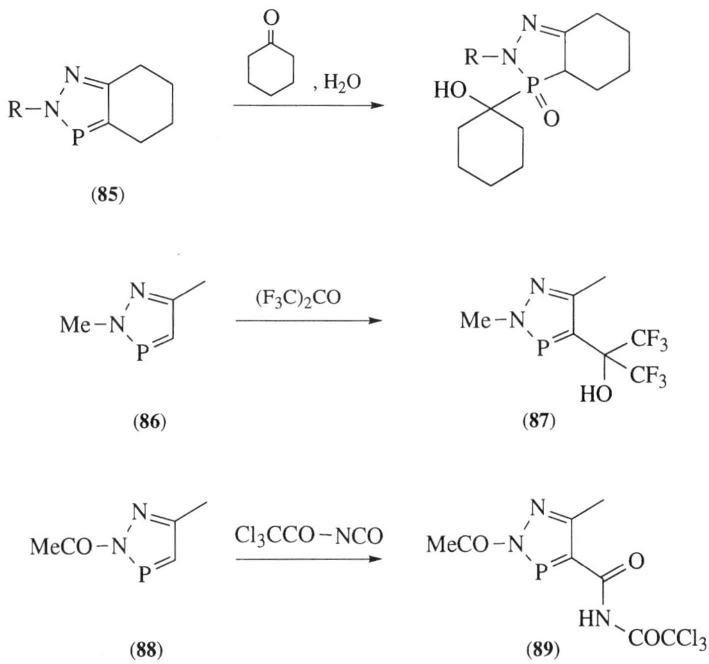

Scheme 17

2-Phenyl-5-methyl-1,2,3-diazaphosphole (**90**) on irradiation loses hydrogen and yields the 4,4′-bis(diazaphospholyl) (**91**) (molecular structure in Table 7) ⟨84IZV1182, 85ZOB1464, 88IZV150⟩.

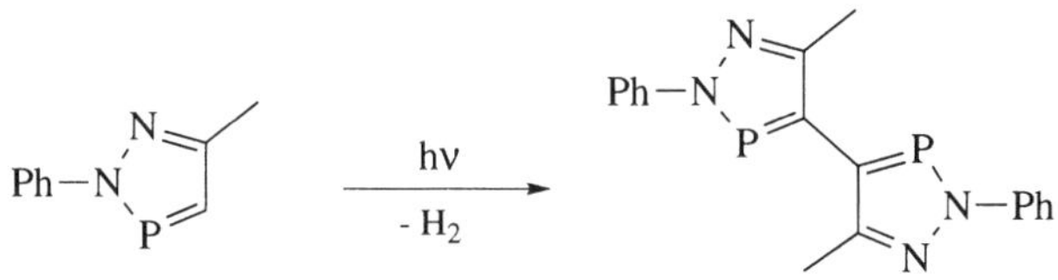

(**90**) (**91**)

4.3.4.5.1.3 Reactions with dienes and heterodienes

2,5-Dimethyl-1,2,3-diazaphosphole (**86**) does not react with 2,3-dimethylbutadiene up to +90°C. [4 + 2] Cycloadditions to the P=C bond of 2-acetyl- (**92**) and 2-phenyl-5-methyl-1,2,3-diazaphosphole are observed with isoprene ⟨91S783⟩, 2,3-dimethylbutadiene ⟨86BSF115⟩, cyclopentadiene, and pentachloro-cyclopentadiene ⟨79DOK1117⟩ as well as with 1-vinyl-3,3-dimethylallenyl diethylphosphonate ⟨92ZOB1505⟩. The addition of isoprene is regioselective, that of the vinylallene is not. The cycloaddition with cyclopentadiene is reversible. Under kinetic control, the *endo*-cycloadduct (**93**) is formed, which on heating changes to the more stable *exo*-isomer (Scheme 18) ⟨86BSF115⟩.

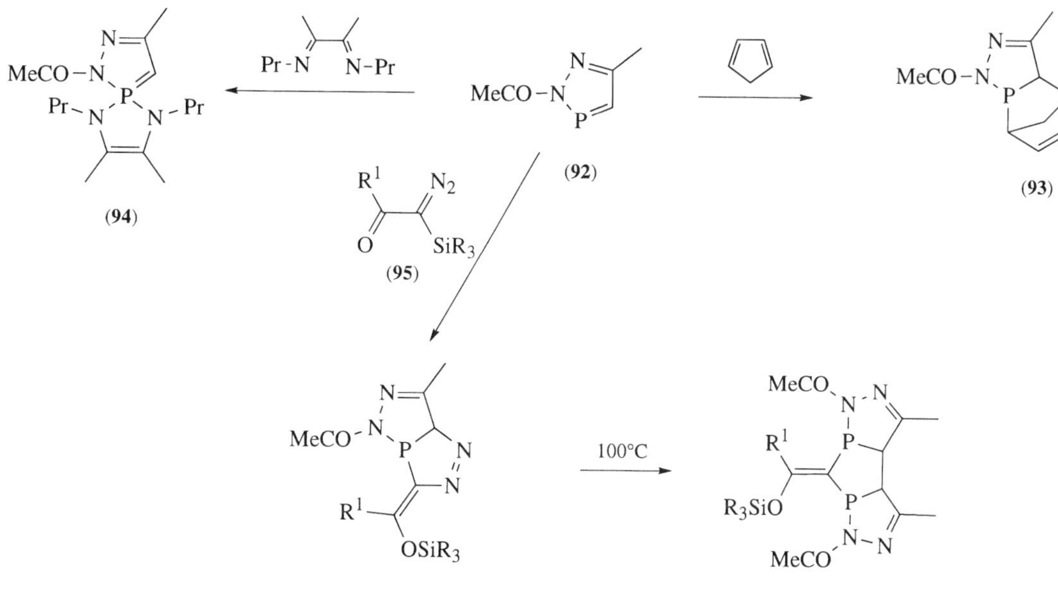

Scheme 18

With diacetyl dipropyldiimine, [4 + 1] cycloaddition to the phosphorus of 2-acetyl-5-methyl-1,2,3-diazaphosphole takes place with formation of a spirocyclic ylide (**94**) ⟨86TL2971⟩.

(1-Diazo-2-oxoalkyl)silanes (**95**) react with 2-acyl-1,2,3-diazaphospholes such as (**92**) to form cycloadducts (**96**). Thermal impact on them generates the tricyclic compounds (**97**) (Scheme 18) ⟨99EJO2633⟩.

Azodicarbonic acid dimethyl ester adds to 5-ethoxycarbonyl-2-trimethylsilyl-4-trimethylsiloxy-1,2,3-triazaphosphole (**77**) in a 2 : 1 molar ratio to give a spirotricyclic zwitterion ⟨96TH1⟩.

Tetrahalobenzoquinones add selectively to the P=C bond of (**77**) ⟨96TH1⟩.

4.3.4.5.1.4 Reactions with 1,3-dipoles

Dipolar [3 + 2] cycloaddition reactions are covered in a review ⟨86PS(26)203⟩. Nitrile ylides ⟨83CJC1105, 93CJC1105⟩, nitrile imines ⟨83CB549⟩, and nitrile oxides ⟨81CC1131, 83T1507⟩ add to the P=C bond of 2*H*-1,2,3-diazaphospholes with the nitrile carbon atom becoming bonded to the phosphorus atom, as has also been found for the cycloaddition of these dipoles to phosphaalkenes ⟨86PS(26)203, B-90MI157⟩ and phosphaalkynes ⟨88AG1541, B-90MI58⟩. The nitrile imine cycloadduct (**98**), X = NPh, on heating loses benzonitrile and yields the 4-anilinodiazaphosphole (**99**) ⟨83CB549⟩.

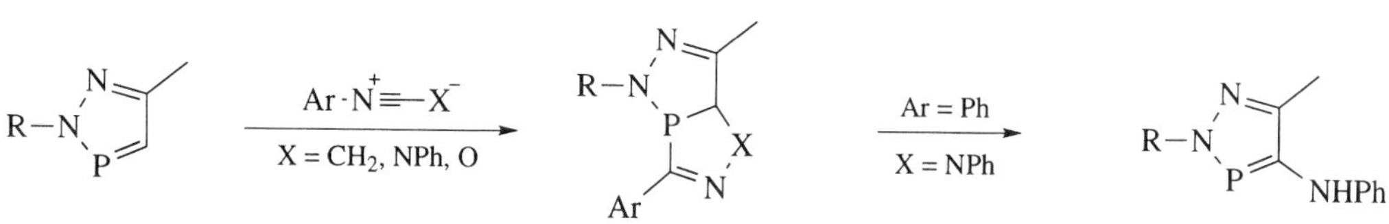

(**98**) (**99**)

In the reactions of diazoalkanes $R^4R^3CN_2$ with 2*H*-1,2,3-diazaphospholes (Scheme 19), various products are formed depending on the diazoalkane substituents ⟨86PS(26)203⟩. In the cases of CR^3R^4 = CH_2 ⟨81IZV1113⟩, CMe_2 ⟨82IZV1196, 84ZOB1511, 85ZOB3, 90ZOB35⟩, CMePh ⟨83DOK613, 90ZOB35⟩, CPh_2 ⟨87CB597, 90ZOB35⟩, fluorenediyl ⟨81IZV1113⟩, and $CHPO(OEt)_2$ ⟨85ZOB1471⟩ only the C–N bonded cycloadduct (**100**) is formed but in the cases of CR^3R^4 = $CHPr^i$ ⟨85ZOB1471⟩, $CHBu^t$ ⟨85ZOB1471⟩, CHCOMe ⟨87ZOB1699, 93PS(77)246⟩, $CHCO_2Et$ ⟨83IZV2395⟩ the P–N bonded isomer (**101**) is also obtained. The reaction rate increases with increasing nucleophilicity of the diazoalkane in the order Ph_2CN_2 < $PhMeCN_2$ < Me_2CN_2 ⟨85ZOB3⟩. The cycloadducts of both types with CR^3R^4 = $CHPO(OEt)_2$, CHCOMe, and $CHCO_2Et$ stabilize themselves by two subsequent 1,3-proton shifts from carbon to nitrogen to give (**102**) and (**103**). Otherwise the cycloadducts lose nitrogen.

From the P–N bonded cycloadducts (**101**) with CR^3R^4 = $CHPr^i$, $CHBu^t$, 4-alkyldiazaphospholes (**104**)

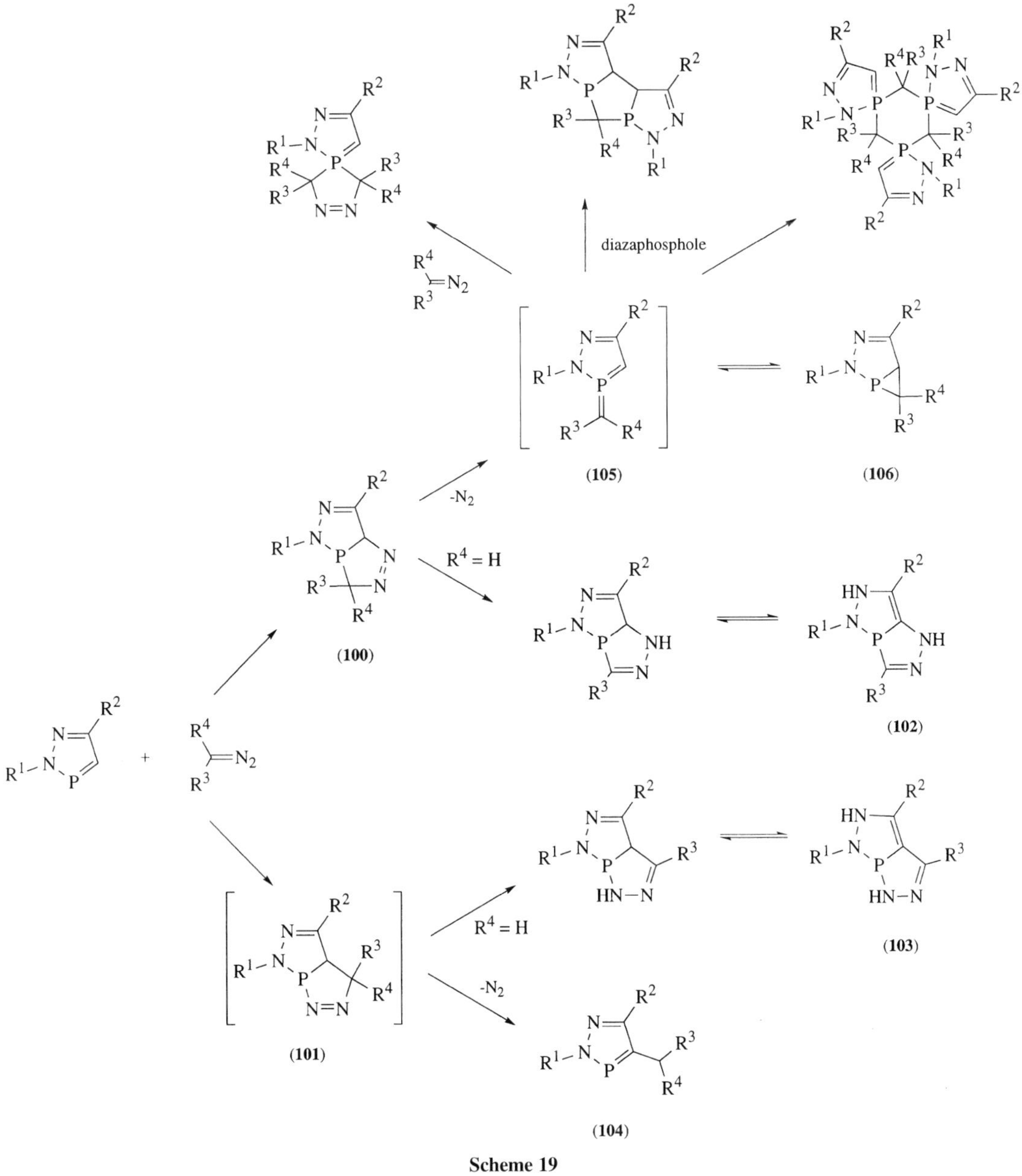
(100) (101) (102) (103) (104) (105) (106)

Scheme 19

are obtained (which are products formally of a carbene insertion into the diazaphosphole C–H bond). From the C–N bonded cycloadducts (**100**) result diazaphosphole *P*-ylides (**105**) as intermediates ⟨79DOK1150, 81IZV1113, 82IZV1196, 83DOK613, 84ZOB1511, 85ZOB3, 85ZOB1471, 87ZOB1699⟩. They either isomerize to phosphiranes (**106**), CR^3R^4 = CMePh ⟨83DOK613⟩, CPh_2 ⟨78DOK358, 79DOK1117, 79DOK1150, 82IZV2730, 87CB597⟩, fluorenediyl ⟨87ZOB1949⟩, or trimerize, CR^3R^4 = CH_2, fluorenediyl ⟨81IZV1113⟩, or add another diazaphosphole molecule, CR^3R^4 = CMe_2 ⟨82IZV1196, 82IZV2718, 84IZV2023, 84ZOB1511, 85ZOB3⟩, $CHPr^i$ ⟨85ZOB1471⟩, or add another diazoalkane molecule, CR^3R^4 = CMe_2 ⟨83ZOB3⟩.

Phenylazide reacts with 2*H*-1,2,3-diazaphospholes with loss of nitrogen to give a cyclic trimer of the diazaphosphole *P*-imide which is suggested as an intermediate ⟨81IZV1600, 90ZOB35⟩.

4.3.4.5.1.5 Substituent reactions

The PCl_2 group of the 4-dichlorophosphino-1,2,3-diazaphospholes can be converted to PF_2 and can be transformed by substitution and oxidation reactions as usual for dichlorophosphines ⟨80CB2278⟩. All reactions selectively occur at the exocyclic phosphorus atom. A large variety of derivatives have been prepared this way ⟨99IC1971, 99IC2791⟩. A diazaphospholyl aryliminophosphine that offers the combination

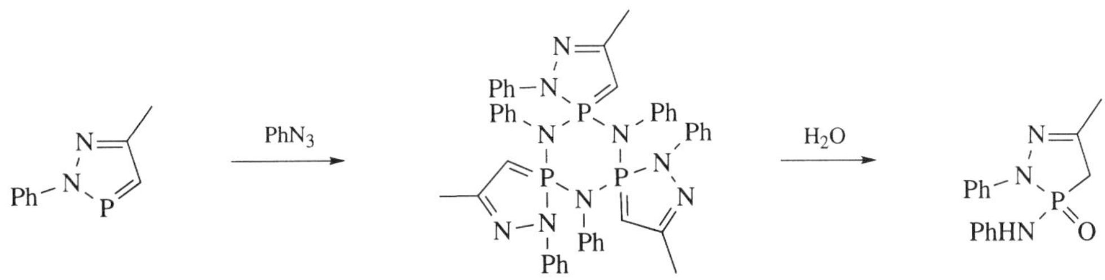

(**107**)

of a cyclic and acyclic two-coordinate phosphorus atom is also accessible ⟨93BSF726⟩. (For the molecular structure of an intermediate compound see Table 7.)

4.3.4.5.1.6 Formation of transition metal complexes

Low valent transition metal centers coordinate preferentially to the phosphorus atom of azaphospholes. *P*-Coordinated complexes of 1,5-dimethyl-1,2,3-diazaphosphole with $Cr(CO)_5$ (molecular structure in Table 7), $W(CO)_5$, $Fe(CO)_4$, $MnC_5H_4Me(CO)_2$ ⟨80JOM(185)53⟩, and $Pt(PPh_3)_3$ ⟨83JOM(256)375⟩, of 2,5-dimethyl-4-(dimethylthiophosphoryl)- ⟨80CB2278⟩ and 1,5-dimethyl-1,2,3-diazaphosphole ⟨80JOM(185)53⟩ with $Cr(CO)_5$ and of 5-ethoxycarbonyl-4-trimethylsiloxy-1,2,3-diazaphosphole with $W(CO)_5$ (molecular structure in Table 7) ⟨96TH1⟩ have been prepared. The 1,5-dimethyl-1,2,3-diazaphosphole can be displaced from its $Cr(CO)_5$ complex by tetraalkylammonium iodide or bromide or by phosphorus trichloride ⟨80JOM(185)53⟩.

With Pd(II), Pt(II), and Au(III), *P*- and *N*-coordination is observed: 2,5-dimethyl-1,2,3-diazaphosphole with $PtCl_2PEt_3$ gives only the *cis-P*-coordinated complex, with $PdCl_2PEt_3$ only the *trans-N*-coordinated complex ⟨84ICA(85)L33, 88IC2612⟩, and with $PtBr_2PEt_3$ both, *cis* and *trans* in a ratio 1 : 2 ⟨88IC2612⟩. Methanolysis of these complexes opens the PN bond and yields five-membered chelates with a *P*-bonded phosphonite and an *N*-coordinated hydrazone function ⟨97ICA(256)47⟩*cis-N*- or *cis-P*-coordinated $AuMe_2Cl$-complexes ⟨80ICA(46)167⟩. Stable, insoluble 1 : 1 complexes with AuClCO, $AuCl_3$ ⟨80ICA(46)167⟩, and CuI ⟨67CHE601⟩ have also been reported.

2,5-Dimethyl-1,2,3-diazaphospholes with a 4-phosphino group coordinate to transition metals only through this substituent, as shown by the reaction of the 4-bis(dimethylamino)phosphino and the bis(trifluorethoxy)phosphino derivative with $[Rh(CO)_2Cl]_2$ ⟨99OM3306⟩, of the 4-difluorophosphino derivative with $CrCO)_5THF$, $Mo(CO)_4nbd$, $Mo(CO)_3(MeCN)_3$ ⟨99IC1971⟩, $CpRu(PPh_3)_2Cl$, $CpRh(CO)_2$ ⟨99OM3306⟩, $Pd(cod)Cl_2$, $Pt(cod)Cl_2$, MePt(cod)Cl ⟨99IC4056⟩ and of the 4-dichlorophosphino derivative with $Cr(CO)_6$ ⟨80CB2278⟩, $HFeW(CO)_9^-$ ⟨89OM244⟩, and $HFe(CO)_4^-$ ⟨88CC574⟩; in the latter cases, in addition to Cl/H exchange, and reduction to the diphosphene is observed. The molecular structures of some of the resulting complexes are given in Table 7.

The diazaphospholyl iminophosphine (see above) uses alternatively N-1, P-3 or the side group phosphorus atom for transition metal complexation ⟨93BSF726⟩.

4.3.4.5.2 Synthesis

2*H*-1,2,3-Diazaphospholes (**111**) are easily accessible from the PCl_3 condensation of the alkylketone hydrazones (**108**) (Scheme 20). A wide variety of hydrazones has been used: R^1, R^2, R^3 = H, alkyl, aryl and in addition R^1 = acyl, R^3 = Ph_2PO ⟨67CHE601, 70ZOB1501, 72JPR806, 72ZOB1939, 76AG91, 76CC887, 77ZOB576, 78CB1558, 79AG442, 80CB2278, 81CR(C)187, 83ZAAC(501)57, 84PS(21)335, 91S783⟩. Condensation may be achieved by refluxing in benzene. Depending on the nature of R^1, the intermediate reached after the loss of two moles HCl is covalent (**109**), R^1 = MeCO, Ph, pyridyl ⟨70ZOB1501, 72JPR806, 72ZOB1939, 74P197, 76AG91, 76CC887, 77ZOB576, 78CB1558, 80CB2278⟩, or ionic (**110**), R^1 = Me ⟨78CB1558, 80CB2278⟩. The third mole HCl is lost on further refluxing or with the help of a base. From ketone hydrazones with two different alkyl groups of similar reactivity (e.g. methyl ethyl ketone) two isomeric products (e.g. R^2, R^3 = Me and R^2 = Et, R^3 = H) are obtained ⟨72ZOB1939, 72JPR806⟩. In some cases (e.g. R^1, R^2 = Me, R^3 = H) besides the 2*H*-, the 1*H*-1,2,3-diazaphosphole (**112**) is found as a second product ⟨79AG442⟩.

Hydrazones of formyl-1,3,3-trimethyl-2-methylene-indoline (**113**) condense with PCl_3 or PBr_3 to

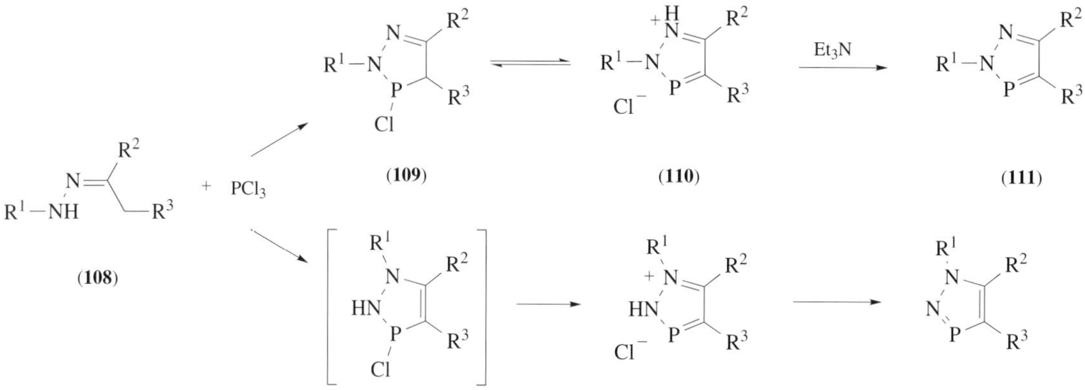

(108) (109) (110) (111) (112)

Scheme 20

methylene-3-halo-1,2,3-diazaphospholines which exist in equilibrium with their ionic isomers (**114**) and which can be converted into fully ionic tetraphenylborates or triflates 〈97PS(123)125〉.

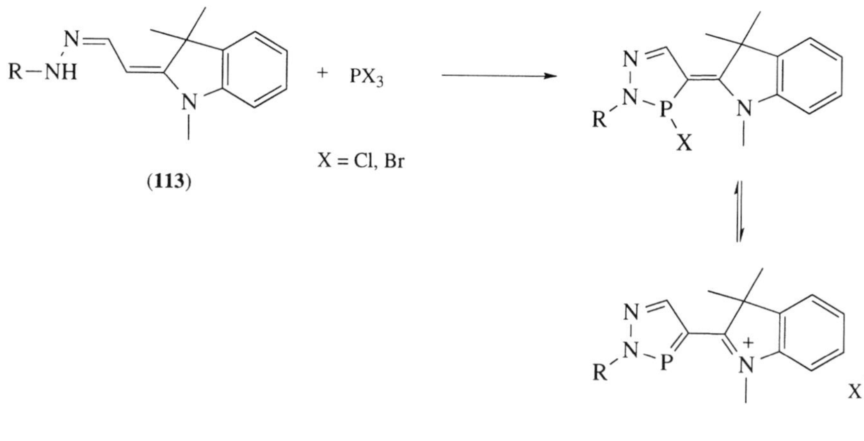

X = Cl, Br

(**113**) (**114**)

A cyclic silyl hydrazone derivative 〈83ZAAC(501)57, 74P197〉 have also been used for the synthesis of 2*H*-1,2,3-diazaphospholes by the PCl_3 condensation. 2,4,5-Triphenyl-1,2,3-diazaphosphole is best prepared from the respective azoalkene and a fused benzothiadiphosphole as phosphorus donating reagent 〈95TL447〉.

Treatment of a 2-acetyl-3-methoxy-3,4-dihydro-1,2,3-diazaphosphole with phenylhydrazine gives the 2-phenyl derivative in addition to the deacylated 1,2,3-diazaphosphole 〈77ZOB576〉.

2-Alkyl-1-aminopyridinium iodides (**115**) condense with PCl_3 in presence of NEt_3 to give 1,2,3-diazaphospholo[1,5-*a*]pyridines (**116**). 2-Methyl-1-aminopyridinium iodide on reacting with two equivalents PCl_3 under these conditions forms 1-dichlorophosphino-1,2,3-diazaphospholo[1,5-*a*]pyridine, $R^1 = PCl_2$ 〈95S173〉.

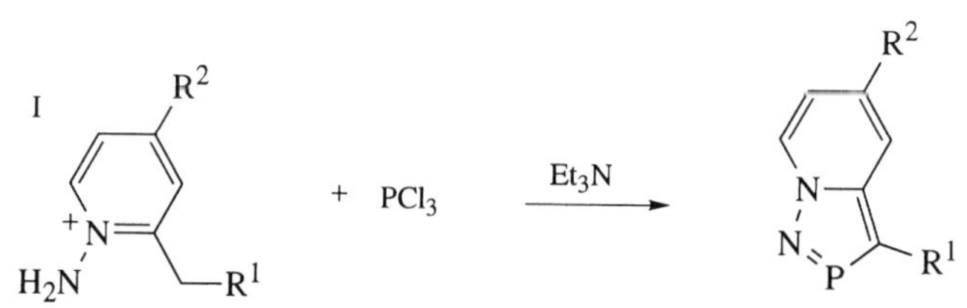

(**115**) (**116**)

1,2,4-Diazaphospholes (**117**) are generally preferred over the 1,2,3-isomers (**118**) as products of the 1,3-dipolar cycloaddition of diazo compounds to phosphaalkynes and often the reaction is regiospecific

(Section 4.3.4.3.2). 1,2,3-Diazaphospholes (**118**) are obtained as minor products, however, from the cycloaddition of HC≡P to a diazo ester (15%) ⟨88JOM(338)329⟩, as well as of $Bu^tC{\equiv}P$ to nitrile imines (2%) ⟨87S689⟩ and to a sydnone (13%) ⟨88AG1541⟩.

4-Aminosubstituted 1,2,3-diazaphospholes (**118**), R^1 = $Pr^i{}_2N$, are obtained from $Pr^i{}_2NC{\equiv}P$ and methyl diazoacetate or trifluoromethyldiazomethane as 1 : 4 and 1 : 2 mixture with the corresponding 1,2,4-diazaphospholes (**117**) ⟨92CB411, 93PS(76)5⟩. Treatment of the metallophosphaalkene $(Me_2N)_2C{=}P\text{-}Fe(CO)_2Cp^*$ with ethyl or *t*-butyl diazoacetate affords the 2-metallated 1,2,3-diazaphosphole ⟨95OM581⟩.

(**117**) (**118**)

Acetylenedicarboxylic esters and (less easily) acetylenemonocarboxylic esters replace nitriles from the 4,5-positions of 1,2,4,3-triazaphosphole rings (**119**) to give carboxyl derivatives of 1,2,3-diazaphospholes (**120**), presumably by a cycloaddition/cycloreversion mechanism, R/R^1 = Me/Ph, $SiMe_3/NMe_2$ ⟨83ZN(B)1484⟩.

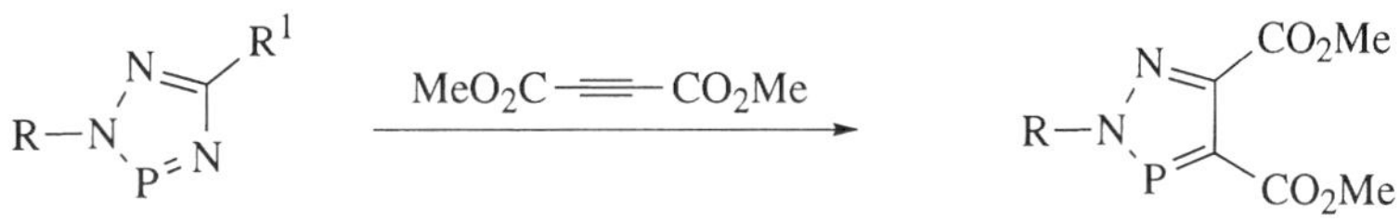

(**119**) (**120**)

α-Diazocarboxylic chlorides (**121**) react with tris(trimethylsilyl)phosphine to give good yields of the 2-trimethylsilyl-1,2,3-diazaphospholes (**122**), R = CO_2Me, CO_2Et, CO_2Bu^t, $CONEt_2$, PO_3Me_2 ⟨96TH1⟩.

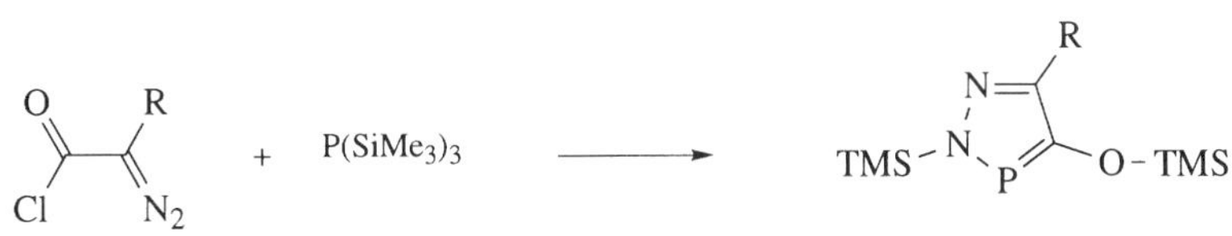

(**121**) (**122**)

4.3.4.5.3 Application

Hydrochlorides of 2*H*-1,2,3-diazaphospholes (Section 4.3.4.5.1.2), i.e. 3-chloro-3,4-dihydro-2*H*-1,2,3-diazaphospholes generated *in situ* from a ketone hydrazone and PCl_3 (see Section 4.3.4.5.2), are useful intermediates for the synthesis of indoles ⟨87T2775, 88JCS(P1)971⟩, pyrroles ⟨89JCS(P1)1053⟩, pyrrolylacetates and 1,2-dihydro-2-alkenyl-3*H*-pyrazol-3-ones ⟨91CC34, 92JCS(P1)1729, 93HAC497⟩. The role of 1,2,3-diazaphosphole intermediates in the room temperature synthesis of aza- and 1,2-diaza-heterocycles has been reviewed ⟨96MI103⟩.

The usual synthesis from a hydrazone and PCl_3 condensation (Section 4.3.4.5.2) has been used to prepare *p*-(1,2,3-diazaphosphol-5-yl)benzylphosphonates, analogues to Fostedil in which the benzothiazole unit is replaced by a diazaphosphole ring. Their activity to inhibit calcium transfer through biological membranes is weak, however ⟨93PS(85)225⟩.

4.3.4.6 1,3,2-Diaza-, 1,3,2-Oxaza-, and 1,3,2-Thiazaphospholes (9, 10, 11)

1,3,5-Triazaphosphapentalenes, 1,3,2-benzodiazaphospholes, 1,3,2-benzoxazaphospholes, 1,3,2-benzothiazaphospholes are contained in a review on anellated heterophospholes ⟨94T7675⟩. The synthesis of 1,3,2-oxazaphospholes from α-aminocarbonyl compounds has also been reviewed ⟨92RHA25⟩. Being isoelectronic to "Arduengo carbenes" 1,3,2-diazaphospholium (and 1,3,2-diazaarsolium) ions have attracted much attention ⟨89IC144, 90IC4938, 92DOK(324)330, 92PS(65)131, 96TL9025, 96ZOB386, 97CC2095, 97T2357, 98EJI1087, 99EJI41⟩. Depending on the argument used, cyclic delocalization or aromaticity seems to contribute to the stability of these phosphenium ions in a rather different degree ⟨97CCR(163)71⟩.

4.3.4.6.1 Reactivity

The salts of the 4,5-dicyano-1,3,2-diazaphospholyl anion are colorless crystalline solids ⟨81ZN(B)1273, 83TL2137⟩. The anion can be alkylated ⟨81ZN(B)1273, 84CC183, 86CB410⟩ and silylated ⟨86CB410⟩ at nitrogen. Borontrifluoride becomes *N*-coordinated as well. Solutions of the dimethylammonium salt (**123**) contain a low equilibrium concentration of the isomeric 2-dimethylamino-2,3-dihydrodiazaphosphole (**124**), as shown by the oxidation with sulfur or selenium to (**125**) ⟨B-90MI258⟩.

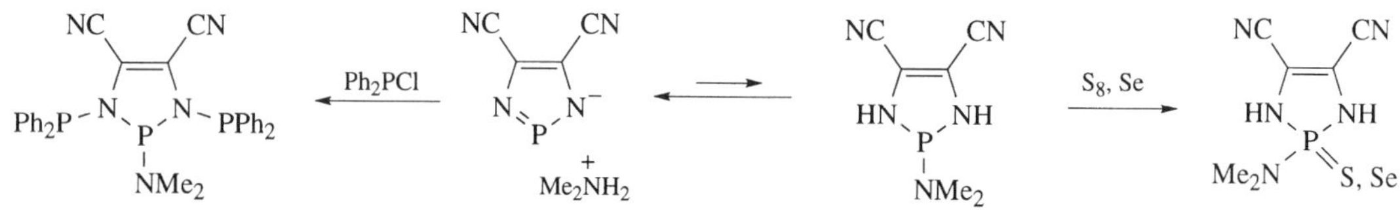

(**126**) (**123**) (**124**) (**125**)

Reactions with methanol and hydrogen chloride yield the 2-methoxy- and 2-chloro-2,3-dihydro derivatives; an excess, however, degrades the products to DAMN and $P(OMe)_3$ or PCl_3, respectively ⟨83TL2137⟩. With tetramethylethylene glycol a spirophosphorane is formed ⟨83TL2137⟩. Diphenylchlorophosphine reacts in a 1 : 2 ratio and gives the poorly soluble 1,3-bis(diphenylphosphino)-2-dimethylamino-2,3-dihydrodiazaphosphole (**126**) ⟨B-90MI258⟩. Dialkylamines react at the cyano groups giving 4,6-diamino-1,3,5-triaza-2-phosphapentalenes (see Section 4.3.4.6.2).

The 1,3,2-diazaphospholes are colorless to pale yellow distillable liquids or crystalline solids ⟨81ZN(B)1273, 86CB410, 86PS(28)71, 96ZOB1455⟩. They are stable to oxidation by air and do not react with sulfur. For the reactions mostly the 1-methyl-4,5-dicyano derivative was used. Bromine degrades it to PBr_3. It readily adds water, alcohols, or phenols to the P=N bond yielding the 2,3-dihydrodiazaphospholes ⟨B-90MI258⟩. The BF_3-adduct of 1-benzyl-1,3,2-benzodiazaphosphole (**127**) adds diethylamine or alcohols in the same way; with butanol a 1,1-addition to the phosphorus atom is observed ⟨84CJC43⟩. In the reaction with diols, 1,2-addition of one hydroxy group is followed by 1,1-addition of the other to yield a spirocyclic phosphorane (**128**) ⟨84CJC43⟩. The addition of carboxyl chlorides to the P=N bond of 1,3,2-benzodiazaphospholes (**130**), R^1 = H, Bn, present in equilibrium with their tetramers, opens the way to novel complexing agents ⟨00PS(159)215⟩.

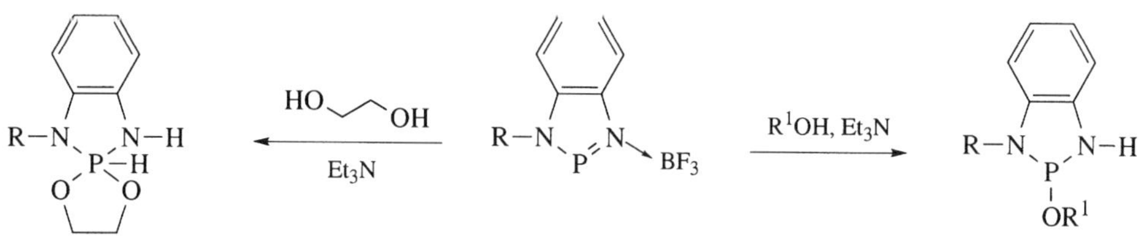

(**128**) (**127**)

The [4 + 2] cycloaddition of 4,5-dicyano-1,3,2-diazaphospholes, such as (**129**), with acetylenes yields 1,2-azaphospholes (Section 4.3.2.3). Azodicarboxylic esters (R = Et, Pr^i) give 1,1-addition accompanied by dimerization of the product (Scheme 21) ⟨B-90MI258⟩.

1,3,2-Benzodiazaphospholes (**130**) (thermally generated from their tetramers) as well as 1-phenyl-1,3,2-benzodiazaphosphole and 1,3,2-benzoxazaphosphole $AlCl_3$ adducts undergo [4 + 1] cycloadditions with α-diimines or benzil; here the products remain monomeric (Scheme 22) ⟨84TL5521, 86TL2971⟩. The same cycloadditions were observed with the monocyclic 1,3,2-diazaphosphole (**148**) (Section 4.3.4.6.2) ⟨96ZOB1455⟩.

The 4,6-diamino-1,3,5-triaza-2-phosphapentalenes are deep red crystalline solids. For reactions, the 4,6-bis(diethylamino) derivative (**131**) was used (Scheme 23). The compound is not oxidized by air and

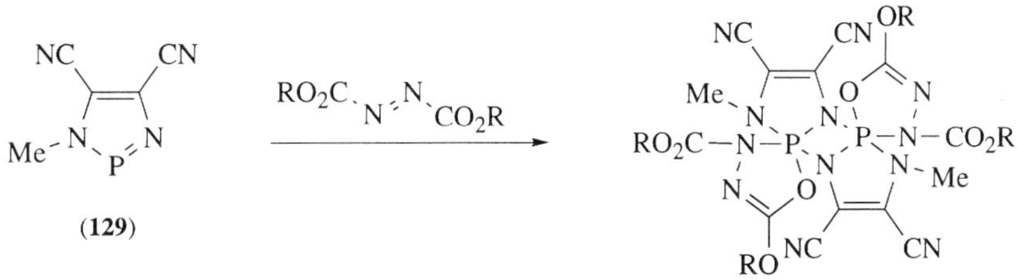

(129)

Scheme 21

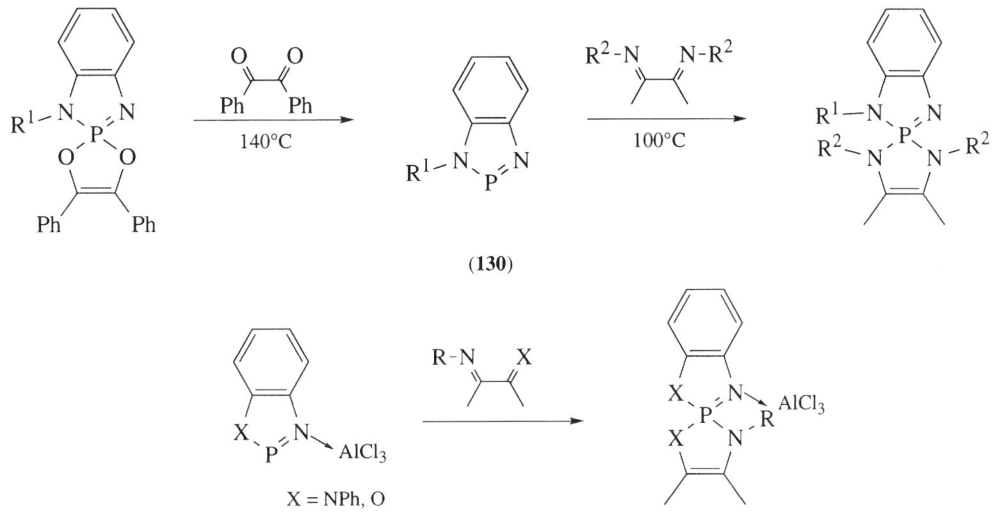

(130)

X = NPh, O

Scheme 22

does not react with sulfur or selenium. Alkylation takes place at a nitrogen atom adjacent to phosphorus. With diethylamine and alcohols no reaction is observed ⟨86CB3213, 87PS(30)780⟩. On simultaneous oxidation with sulfur or selenium, however, alcohols add to the P=N bond yielding 1,2-dihydropentalene 2-sulfides or 2-selenides (**132**) ⟨86PS(28)71⟩.

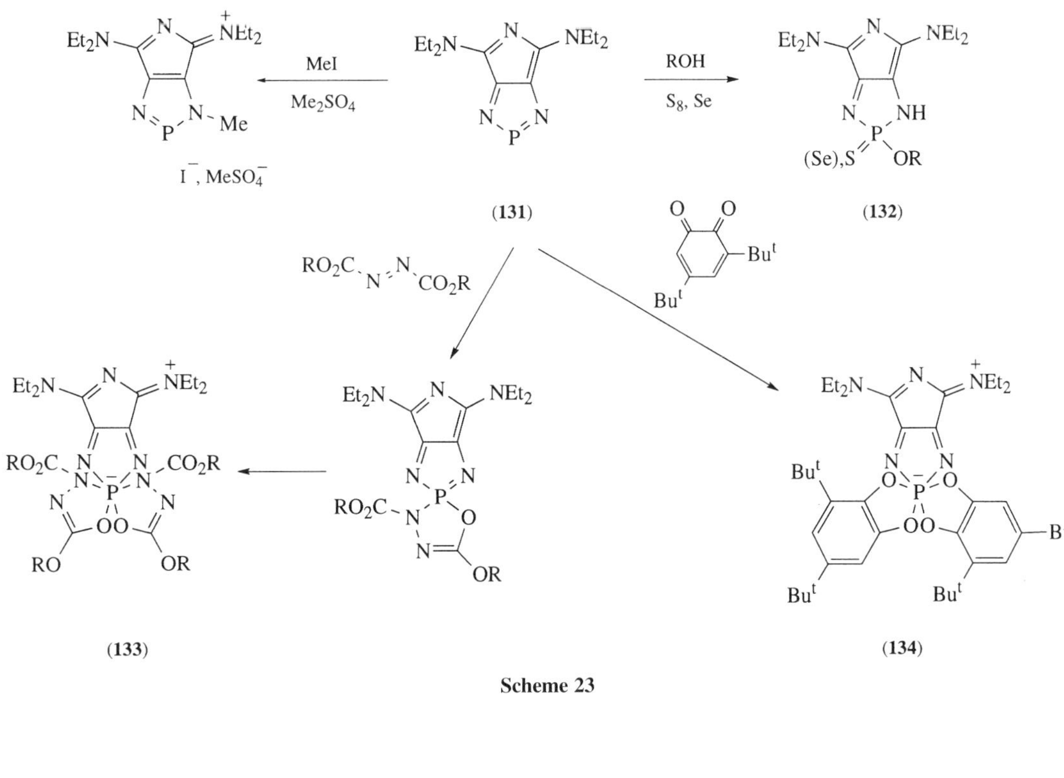

(131) (132) (133) (134)

Scheme 23

Azodicarboxylic esters, R = Et, Pr^i, and 2,4-di(*t*-butyl)benzoquinone undergo a double [4 + 1] cycloaddition to the phosphorus atom resulting in zwitterionic products (**133**) and (**134**) with six-coordinate ⟨87PS(30)780⟩ phosphorus.

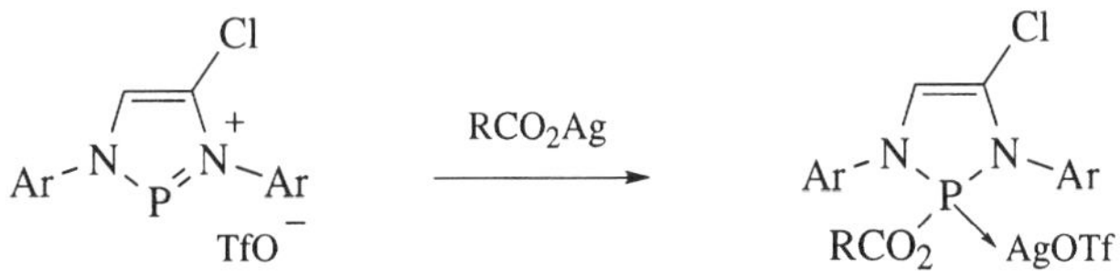

(**135**)

1,3,2-Diazaphospholium salts are reported as colorless or orange crystals ⟨96TL9025, 97CC2095⟩. In case of (**135**), Ar = 2,6-$Me_2C_6H_3$, the addition of silver acetates (R = Me, CF_3) has been observed ⟨96CB465⟩ and the formation of *P*-coordinated Mo(0)- and W(0)-complexes has been reported ⟨01JOM(617-618)383⟩.

4.3.4.6.2 Synthesis

Diaminomaleodinitrile and PCl_3 in refluxing acetonitrile yield a cyclic chlorophosphine (**136**) which gives the diazaphospholes by treating with a tertiary base and then with an alkylating agent, R = Me, Et, $CH_2C_2H_3$, CH_2Ph, CH_2COPh, CH_2CO_2Et ⟨83TL2137, 86CB410⟩. Condensation of the diaminomaleodinitrile with $P(NMe_2)_3$ gives a product which is not the cyclic aminophosphine (**137**) but the isomeric dimethylammonium diazaphospholate (**123**) ⟨81ZN(B)1273, 83TL2137⟩. Further heating converts it to the 4,6-bis(dimethylamino)-1,3,5-triaza-2-phosphapentalene (**138**) (molecular structure in Table 8). Other secondary amines react analogously ⟨86CB3213⟩. The benzaldehyde derivative of DAMN (**139**) may also be used for condensation (Scheme 24) ⟨84CC183⟩.

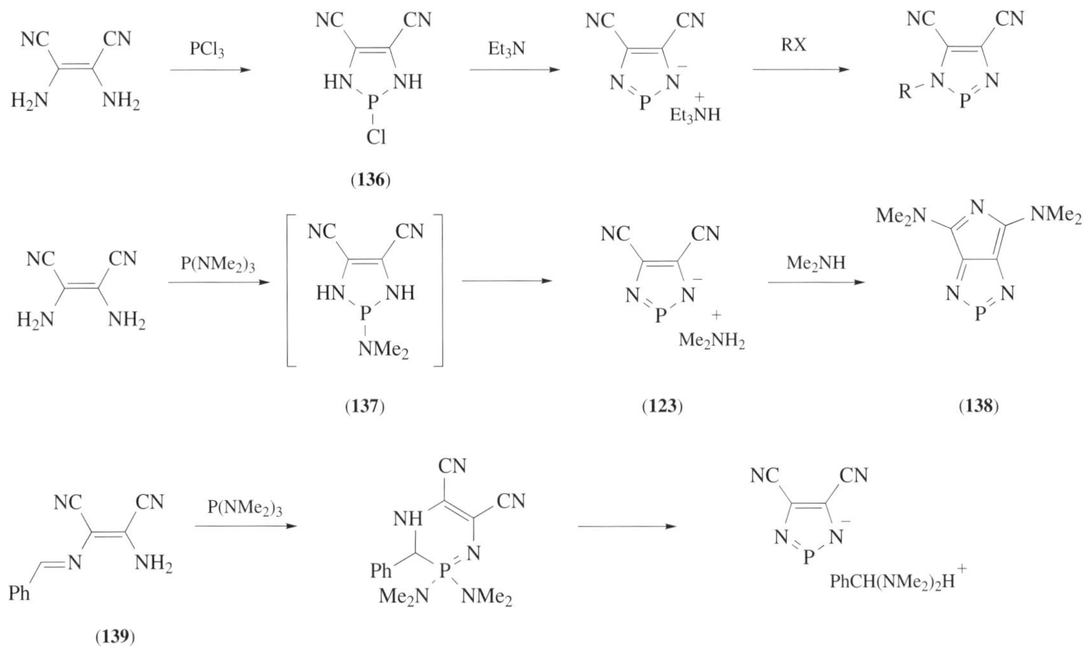

Scheme 24

The transamination product of sarcosine nitrile (**140**) and $P(NMe_2)_3$ rearranges to a cyclic aminophosphine (**141**) which on thermolysis (120°C/20 mbar) yields the diazaphosphole (**142**) ⟨86PS(28)71⟩.

Condensation of a 1,2-diiminoethane (**143**), R = cyclohexyl, Mes, with PCl_3 yields a 2,4-dichloro-dihydro-1,3,2-diazaphosphole (**144**) ⟨88DOK(298)369, 00CEJ3414⟩ which is converted to the 1,3,2-diazaphospholium salt (**145**) by antimony pentachloride ⟨90DOK(312)623⟩. The cyclic chlorophosphine (**146**), X = Cl, enters an equilibrium with the ionic form (**147**) in solution ⟨92DOK(324)330⟩ and yields on thermolysis the 1,3,2-diazaphosphole (**148**) ⟨96ZOB1455⟩. The dissociation has been questioned in regard to more recent investigations, however ⟨00CEJ3414⟩.

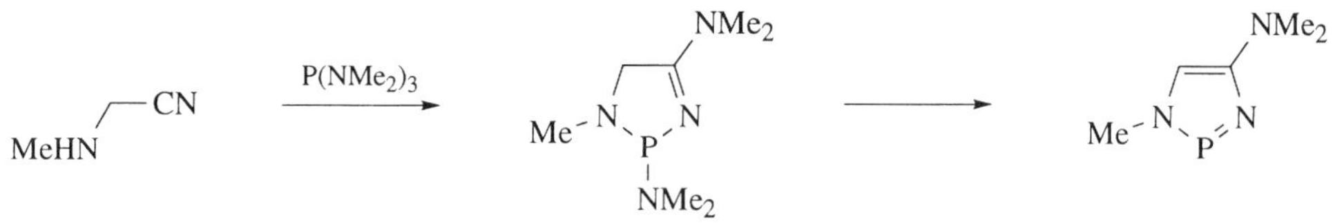

(**140**) (**141**) (**142**)

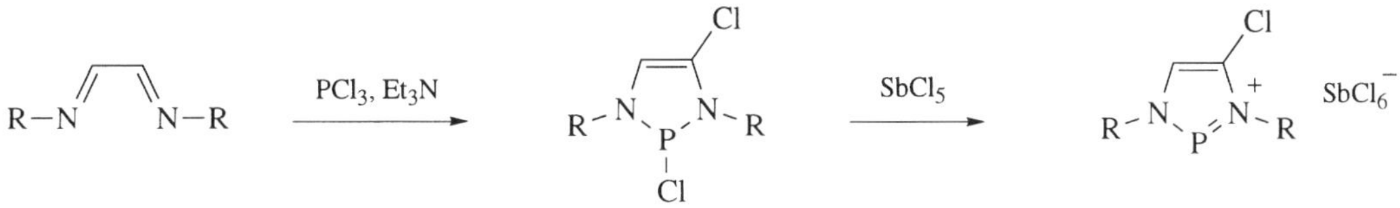

(**143**) (**144**) (**145**)

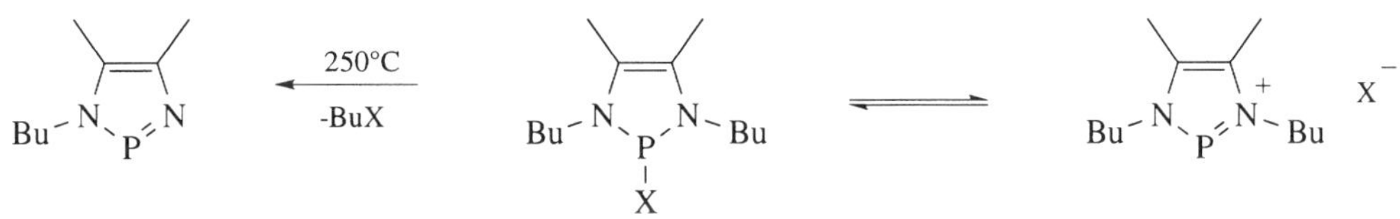

(**148**) (**146**) (**147**)

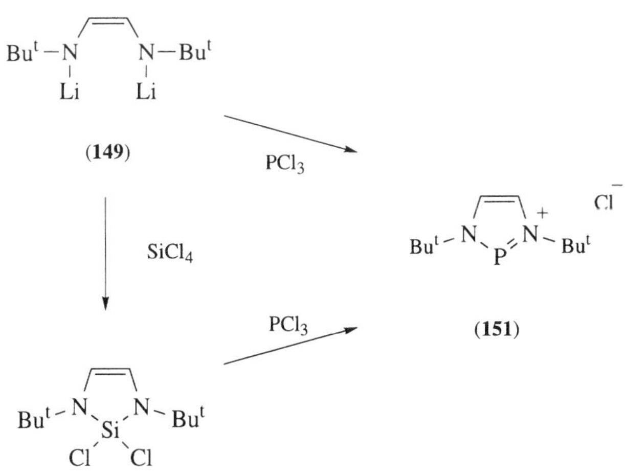

(**149**) (**150**) (**151**)

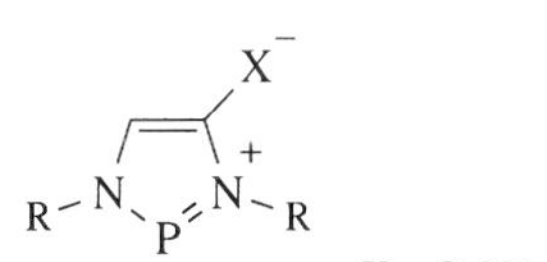

(**152**)

The synthesis of the 4,5-unsubstituted compound (**151**) starts with the reduction of the diimine to the dilithio compound (**149**) and perhaps its conversion to the dichlorosilane (**150**). (**151**) is found to be ionic in solution and in the solid state ⟨96TL9025, 97CC2095, 99EJI41⟩ and the chloride anion has been exchanged for $GaCl_4^-$ and PF_6^- (for molecular structures see Table 8). Mesoionic 1,3,2-diazaphospholium-4-olates (**152**) have been considered ⟨96PS(111)197⟩.

Chlorophosphines of the type (**144**) are readily reduced by $LiBEt_3H$ or $LiAlH_4$ to the corresponding phosphines, which can be converted to phospholium salts (**145**), BF_4^-, TfO^- in place of $SbCl_6^-$, by reaction with trityl tetrafluoroborate or reaction with triflic acid with evolution of hydrogen ⟨00AG3211⟩.

1,3,2-Benzodiazaphospholes (**155**) from the condensation of *o*-phenylenediamines (**153**), R = H, alkyl, Ph, with $P(NMe_2)_3$, prove to be tetrameric or polymeric (**154**). They monomerize, however, on heating

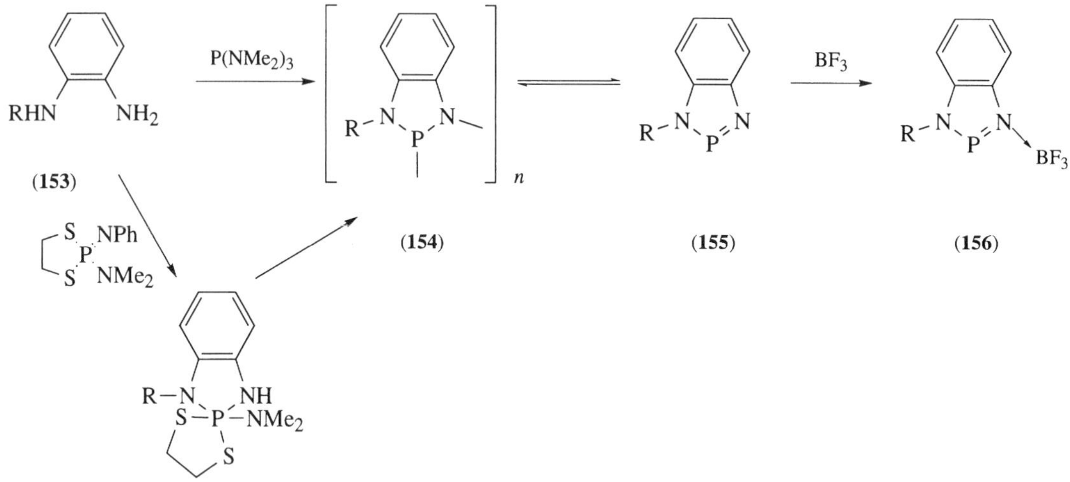

(153) (154) (155) (156)

Scheme 25

or as BF_3- (**156**) or $AlCl_3$-complexes ⟨79TL447, 84CJC43, 93PS(75)245, 93T4651⟩. One representative (R = CH_2Ph) has been prepared using an iminodithiophosphate for condensation by a subsequent reductive elimination of disulfide (Scheme 25) ⟨80TL1307⟩.

4,5-Diaminopyrimidine is reported to react with triphenylphosphite to give 1,3,2-diazaphospholo[1,5-*d*]pyrimidine. Formation of 1,3,2-diazaphospholopyridines from diaminopyridines under similar conditions is also claimed ⟨79NAR251⟩. In both cases, however, the physical properties of the isolated materials are in accord with oligomers rather than with monomeric azaphospholes.

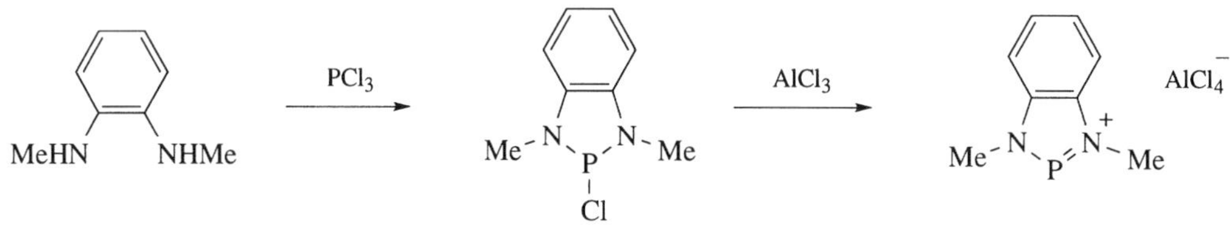

(157) (158)

Scheme 26

The 1,3-dimethyl-1,3,2-benzodiazaphospholium cation can be prepared as tetrachloroaluminate (**158**) (molecular structure in Table 8) ⟨89AX(C)1485⟩ from the 2-chloro-1,3-dimethyl-2,3-dihydro-1*H*-1,3,2-diazaphosphole (**157**); the latter is obtained from the condensation of PCl_3 with *N,N′*-dimethyl-*o*-phenylenediamine (Scheme 26) ⟨81JCS(P2)1411⟩.

2-2′-Dipyridyl chelates the phosphorus atom of the cation 2,4,6-$Me_3C_6H_2$-NP^+. The product may be understood as an imide of the dipyrido-1,3,2-diazaphospholium ion ⟨00CC2087⟩. PCl_3 condensation of a 2-aminoketone hydrobromide (**159**) yields the tetramer of a 1,3,2-oxazaphosphole (**160**) ⟨81ZOB475, 84ZOB271⟩. Boiling with Et_2OBF_3 in benzene gives a complex of the monomer with the borontrifluoride (**161**) most probably coordinated to the nitrogen. Starting from a diketo amine (**162**) (R = Bu^t), a bicyclic product (**163**) (with a planar phosphorus coordination) is obtained. Protonation opens one ring and gives a 1,3,2-oxazaphospholium cation (**164**) (Scheme 27) ⟨87JA627⟩.

Condensation of *o*-aminophenol and *o*-aminothiophenol with PCl_3, $P(NMe_2)_3$ or $P(NEt_2)_3$ gives covalent hydrochlorides and tetramers, respectively, of 1,3,2-benzoxazaphosphole and 1,3,2,-benzothiazaphosphole (Scheme 28) ⟨B-81MI413⟩. Action of a Lewis acid converts them in each case to the heterophospholium ion ⟨88CC842, 89IC144, 90IC4938⟩ (molecular structure of the 1,3,2-benzothiazaphospholium ion in Table 8). Their aromatic character is best shown for the 1,3,2-benzothiazaphospholium ions ⟨90OM1085, 92IC3386⟩.

The reported 2,2-dichloro-1,3,2λ^5-benzodiazaphospholium, -benzoxazaphospholium, and -benzothiazaphospholium hexachloroantimonates are formally the chlorination products of the aromatic systems ⟨91PS(55)255, 93PS(82)211⟩. 1,3,2λ^5-Benzoxazaphospholes are dimeric. They have been prepared from quite a number of different routes ⟨B-87MI617⟩.

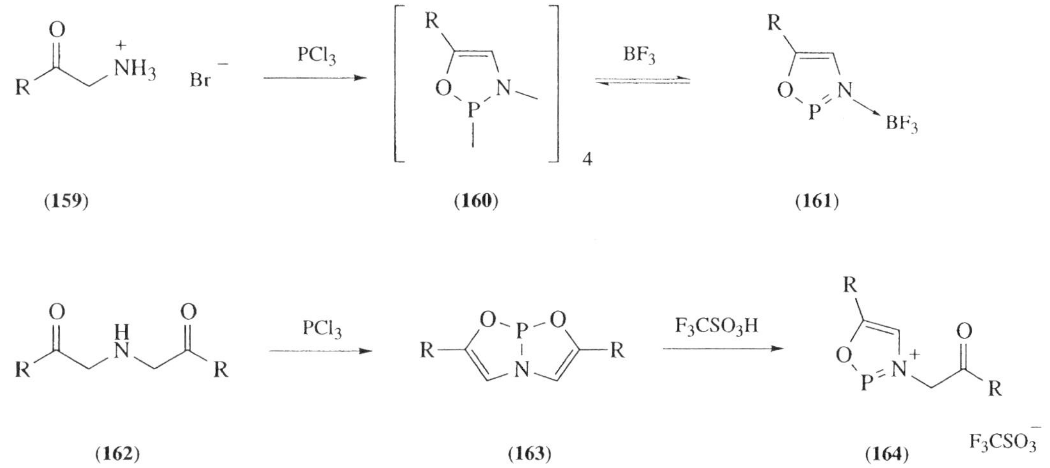

Scheme 27

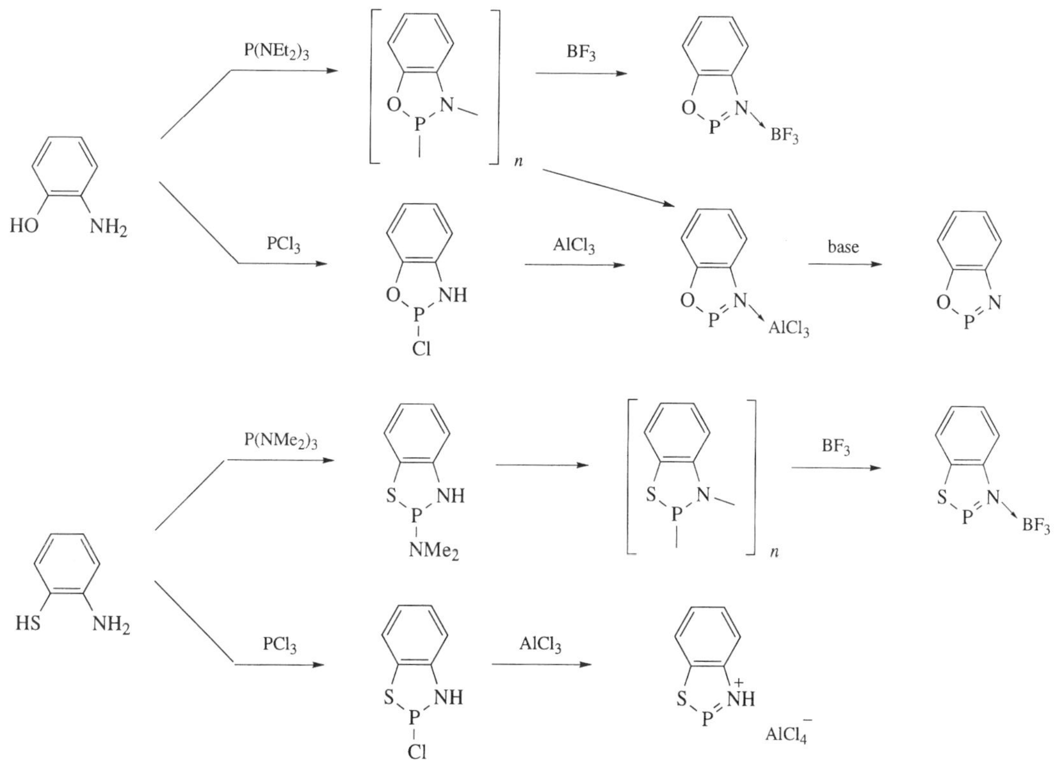

Scheme 28

4.3.4.7 1,3,2-Dithia- and 1,3,2-Diselenaphospholium Salts (12)

Only benzo derivatives of 1,3,2-dithia- and 1,3,2-diselenaphospholium salts (**12**) are known. They are contained in a review on anellated heterophospholes ⟨94T7675⟩. The method used for the generation of 1,3,2-benzodiaza- and -thiazaphospholium cations has been extended to benzodithiaphospholium cations (**166**) ⟨88CC842, 89IC144, 89PS(41)37⟩ (molecular structure of the tetrachloroaluminate in Table 8). Starting compounds are the 2-chloro-1,3,2-benzodithiaphospholes (**165**), which are obtained from the condensation of *o*-phenylenedithiols with PCl_3 (Scheme 29) ⟨73ZN(B)363⟩. The Lewis acidity of the cation towards pyridine has been determined using $\delta^{13}C$ of C-4 in the pyridine ring as a probe for the complexation equilibrium ⟨92HAC157⟩.

The 1,3,2-benzodiselenophospholium ion has been mentioned as an unpublished result ⟨90IC4938⟩. 2,2-Dichloro-1,3,2λ^5-benzodioxaphospholium and -benzodithiaphospholium hexachloroantimonates have been described ⟨78JCS(D)1465, 90ZAAC(585)197, 93PS(82)211⟩.

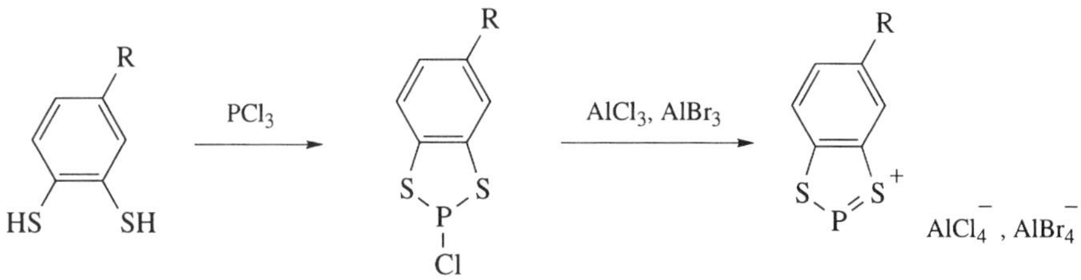

(165) (166)

Scheme 29

4.3.4.8 1,4,2-Dithiaphospholes

The ylide type CS_2 complexes (**167**), R = piperidyl, morpholyl, react with a large excess of *t*-butylphosphaacetylene in a [3 + 2] cycloaddition affording the 5-methylidene-1,4,2-dithiaphospholes (**168**). At higher temperatures the reaction proceeds to a mixture of the 1,2,4- and 1,3,4-thiadiphosphole (**169**) and (**170**) (see Sections 4.3.6.8 and 4.3.6.10) ⟨00JA4557⟩.

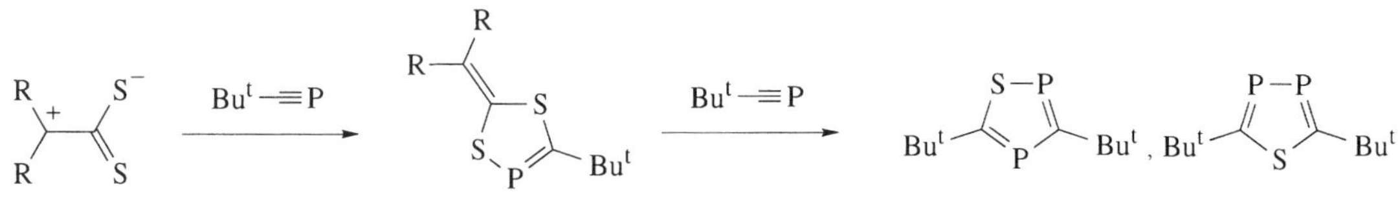

(167) (168) (169) (170)

4.3.5 PHOSPHOLES WITH THREE OR FOUR ADDITIONAL HETEROATOMS

4.3.5.1 Introduction

Five-membered rings with one phosphorus and three nitrogen atoms include the 1,2,3,4- and the 1,2,4,3-alternative. Depending on the position of the *N*-substituent, for each of them three isomers are possible. In the first case only 3*H*-1,2,3,4-triazaphospholes (**1**) and 2*H*-1,2,3,4,-triazaphospholes (**2**) are known, however, and no anellated 1,2,3,4-triazaphospholes. Of the 1,2,4,3-triazaphospholes examples of the 2*H*-, 1*H*- and 4*H*-isomer (**3**), (**4**) and (**5**), as well as 1,5-anellated examples have been reported. Table 11 shows the known systems. Their spectra, structure (Table 9), and bonding are covered in Section 4.3.4.2.

4.3.5.2 1,2,3,4-Triazaphospholes (1, 2) and Tetrazaphospholium Salts (7)

4.3.5.2.1 Reactivity

The 1,2,3,4-triazaphospholes are colorless distillable liquids or crystalline solids ⟨84CC1634, 86JOM(306)39, 87T3247, 88TL785⟩. The hydrolysis of 2-trimethylsilyl-5-*t*-butyl-1,2,3,4-triazaphosphole gives the corresponding NH-compound. The flash vacuum pyrolysis of the 3-phenyl derivative (**8**) ⟨87T3247, 89PS(46)31⟩ leads to a mixture of 1,2- and 1,3-benzazaphospholes (R = Me, Bu^t) as the major and minor reaction products (see Sections 4.3.2.3 and 4.3.3.3).

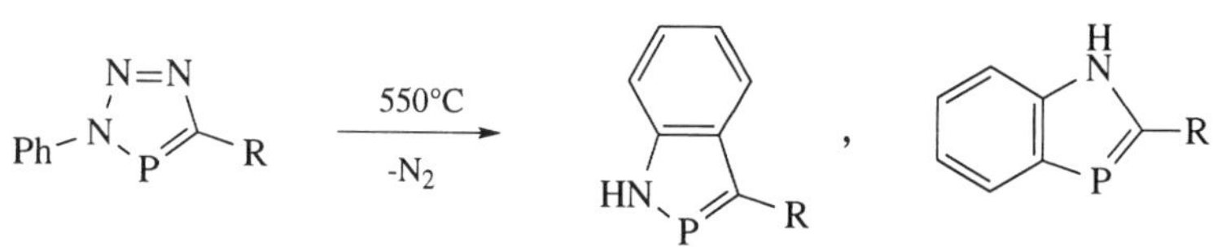

(8)

Table 11. Known 1,2,3,4- and 1,2,4,3-triazaphospholes and oxo and anellated derivatives of the latter, as well as 1,3,4,2-thiadiaza- and tetraazaphospholium ions

3*H*-1,2,3,4-triazaphospholes (**1**)

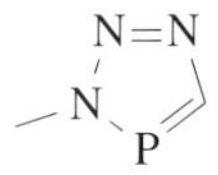

2*H*-1,2,3,4-triazaphospholes (**2**)

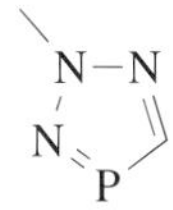

2*H*-1,2,4,3-triazaphospholes (**3**)

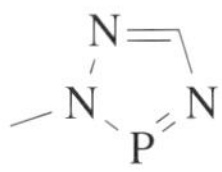

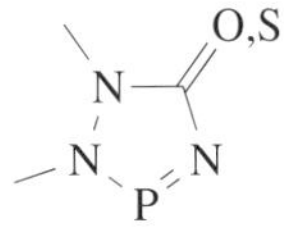

1*H*-1,2,4,3-triazaphospholes (**4**)

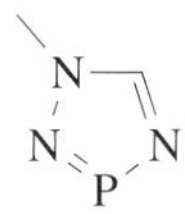

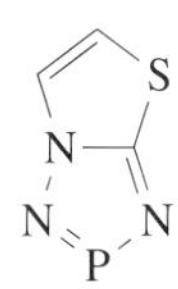

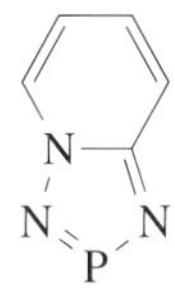

4*H*-1,2,4,3-triazaphospholes (**5**)

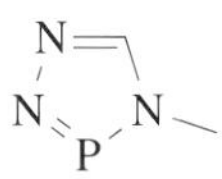

3*H*-1,3,4,2-thiadiazaphospholium (**6**)

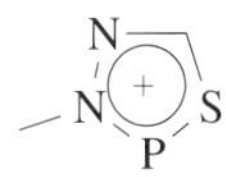

1,4*H*-tetraazaphospholium (**7**)

The 3-phenyl-1,2,3,4-triazaphosphole (**9**) reacts with α-trialkylsilyldiazoketones (**10**) to form dimers of 4-imino-1,2,4-diazaphospholes (**11**). With 3-alkoxycarbonyl-1,2,3,4-triazaphospholes (**12**) from the same reaction 4-alkoxy-4-isocyanato-1,2,4-diazaphospholes (**13**), R^1, R^2 = Bu^t, 1-adamantyl, are obtained. A monomeric 4- imino-1,2,4-diazaphosphole seems to be the precursor of both types of products (Scheme 30) ⟨00T35⟩.

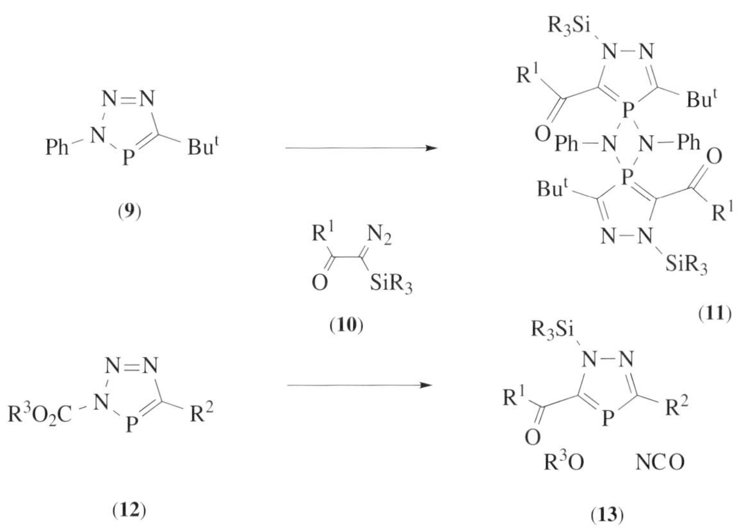

Scheme 30

4.3.5.2.2 Synthesis

The 1,2,3,4-triazaphosphole ring may be generated from the combination of PC + NNN or of PN + CNN fragments. According to the first pattern the regiospecific cycloaddition of phosphaalkynes, R^1 = H, neopentyl, secondary and tertiary alkyl, Mes ⟨86JOM(306)39, 86S31, 87T3247, 88AG1541, 98S1305, 00TH2⟩ or Me, if prepared *in situ* ⟨89PS(46)31⟩ and of phosphaalkenes $R^1(Me_3Si)C{=}PCl$, R^1 = Me, Pr^n, *n*-pentyl, allyl, $PhCH_2$ ⟨92HAC337⟩, Ph ⟨88TL785⟩, $SiMe_3$ ⟨84CC1640, 92HAC337, 97BSB455⟩, to azides (R = H and a wide variety of substituents) gives 3*H*-1,2,3,4-triazaphospholes (**14**). In case of R = H ⟨88AG1541⟩ and $SiMe_3$ ⟨87T3247⟩ they rearrange to give the 2*H*-isomers (**15**). The synthesis from phosphaalkenes $R^1(Me_3SiO)C{=}PSiMe_3$ is not successful here ⟨85ZN(B)1077⟩.

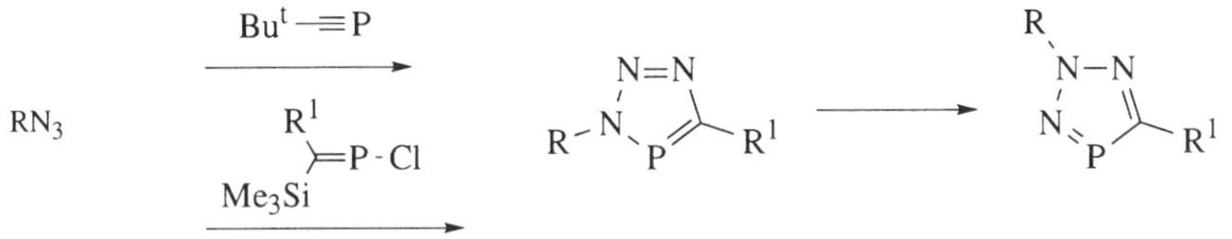

(**14**) (**15**)

With 2,4,6-triazidopyridines a cycloaddition to three moles of Bu^tCP has been accomplished ⟨96IZV252, 96KGS67⟩ and the regioselectivity of this reaction has been explained by a quantum-chemical study ⟨97CHE587⟩.

A 5-phosphonio-substituted 1,2,3,4-triazaphosphole (**17**) results from the reaction of a dichlorophosphinomethyl-phosphonium salt (**16**) with an excess of DABCO and mesityl azide. The 2-phosphonio-1-phosphaethyne is suggested to be the key intermediate ⟨91CC302⟩.

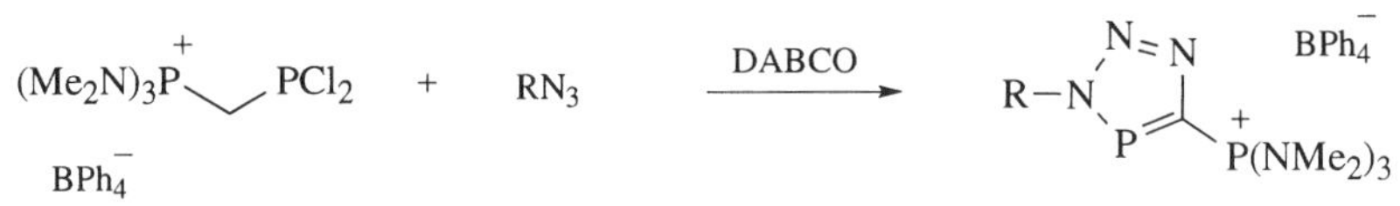

(**16**) (**17**)

Products from the condensation of a bis(trimethylsilyl)ylide and PCl_3 such as the bis(triphenylphosphoranediyl)-dichloro-1,3-diphosphetane (**18**) serve as synthetic equivalents of a phosphonio-phosphaethyne. Cycloaddition reactions with phenyl- or trimethylsilylazide lead to 5-phosphonio-1,2,3,4-triazaphospholes (**19**). In case of R = Me_3Si, the product loses trimethylchlorosilane and gives the triphenylphosphoranediyl-1,2,3,4-triazaphosphole (**20**) (or triphenylphosphonio-1,2,3,4-triazaphospholide) which finally converts to a 1,2,4-diazaphosphole (**21**) (see Section 4.3.4.3.2) ⟨97CB89⟩.

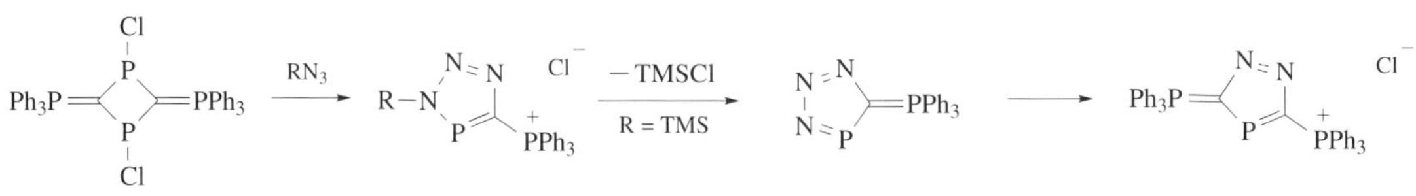

(**18**) (**19**) (**20**) (**21**)

According to the PN + CNN pattern the reaction of a phosphino-diazomethane (**22**) (R = dicyclohexylamino) and a *P*-iodo iminophosphine (**23**) ⟨94BSF748⟩ and also the reaction of a phosphoranediyl-diazomethane (**25**) (R = Pr^i_2N) and a *P*-chloro-iminophosphine (**26**) ⟨93BSF757⟩ give the corresponding phosphino and phosphonio 1,2,3,4-triazaphospholes (**24**) and (**27**) (Ar = 2,4,5-$Bu^t_3C_6H_2$), respectively. The chlorophosphonio substituent of (**27**) is converted to a phosphoryl substituent in (**28**) by hydrolysis.

The cycloaddition of an iminophosphenium ion (**29**) (Ar = 2,4,6-$Bu^t_3C_6H_2$) to azides (R = Bu^t, CEt_3) yields tetrazaphospholium ions (**30**) ⟨93CB1513⟩.

4.3.5.3 1,2,4,3-Triazaphospholes (3, 4, 5) and 1,3,4,2-Thiadiazaphospholium Salts (6)

1,2,4,3-Triazaphospholes are related to 1,2,4-triazoles by a CH/P exchange and to tetrazoles by a N/P exchange; they have been compared to these analogues with respect to their electronic structure and NMR

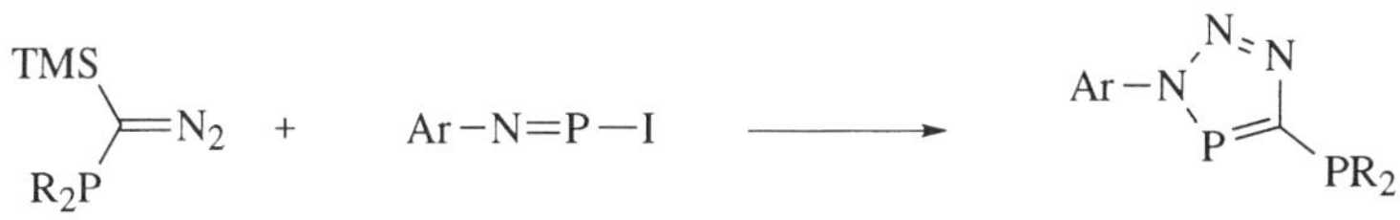

(22) (23) (24)

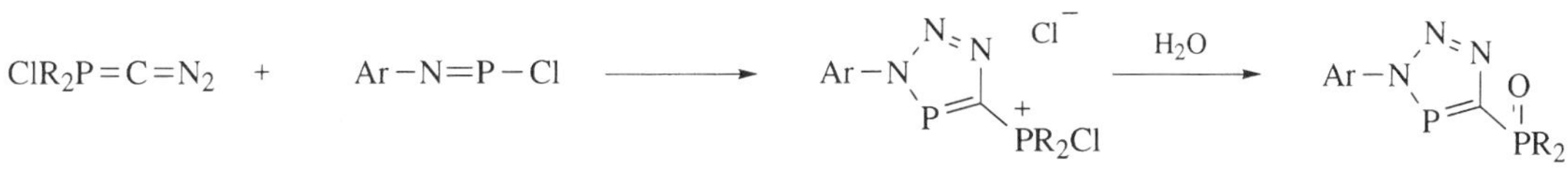

(25) (26) (27) (28)

(29) (30)

spectra ⟨89JCR(S)250⟩. 1*H*-, 2*H*- and 4*H*-1,2,4,3-triazaphospholes differ considerably in their behavior. 1,2,4,3λ^5-Triazaphospholes, -oxadiazaphospholes and -thiadiazaphospholes are covered in a review on cyclophosphazanes with five- and six-coordinate phosphorus ⟨B-87MI617⟩.

4.3.5.3.1 *Reactivity of 1,2,4,3-triazaphospholes*

The 1,2,4,3-triazaphospholes are colorless or pale yellow distillable liquids or crystalline solids ⟨76T2039, 80ZN(B)1222⟩. They are not oxidized by air and are reluctant to react with sulfur. The flash thermolysis of two 1,2,4,3-triazaphospholes has been investigated ⟨90PS(49/50)363⟩.

4.3.5.3.1.1 Reaction at the ring nitrogen

The *N*-unsubstituted 5-dimethylamino-1,2,4,3-triazaphosphole is converted to its anion by sodium, potassium, or their hydrides ⟨80ZN(B)1222⟩. Boryl ⟨86PS(28)71⟩, silyl ⟨80ZN(B)1222⟩, stannyl, phosphino, and (thio)phosphinyl ⟨80ZN(B)1222⟩ derivatives have been prepared mostly from reaction of the triazaphosphole with the corresponding chloride; the substituents become attached to N-2.

1*H*- and 2*H*-1,2,4,3-triazaphospholes are not methylated by methyl iodide and not protonated by acetic or trichloroacetic acid; they are, however, decomposed by trifluoroacetic acid ⟨74CR(C278)355, 76T2039⟩. They are protonated by triflic or tetrafluoroboric acid and methylated by methyl triflate or trimethyloxonium tetrafluoroborate at N-4 ⟨91TL501, 92TH1⟩. Borontrifluoride coordinates to the same nitrogen ⟨76T2039, 87TL6049, 89JCR(S)250⟩; in a 4*H*-1,2,4,3-triazaphosphole it coordinates to N-2. The molecular structure of the BF_3-complex of an example of each of the three isomeric 1,2,4,3-triazaphospholes is shown in Table 9.

4.3.5.3.1.2 Addition to the P=N bond

Hydrogen chloride adds to the P=N bond to give 3-chloro-dihydro-1,2,4,3-triazaphospholes (**31**) which may be converted to the thiochloride (**32**) ⟨94HAC385⟩.

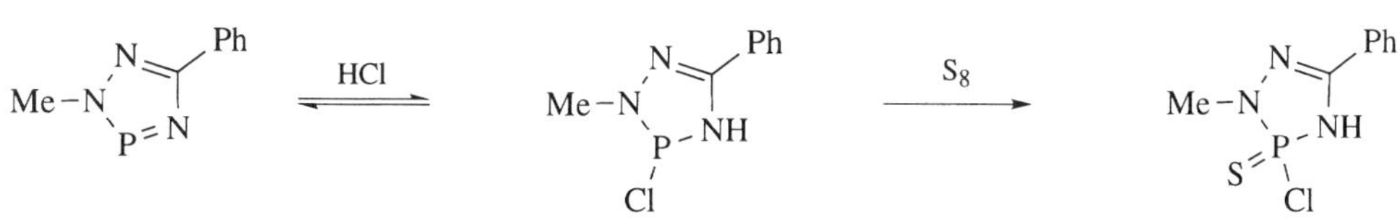

(31) (32)

The tendency of the three 1,2,4,3-triazaphosphole isomers to give 1,2-additions is quite different and decreases in the order 4*H* > 2*H* > 1*H*. Thus the 2*H*-isomers add readily water ⟨84PS(20)333⟩, alcohols and phenols ⟨74CR(C278)355, 76T2039, 84PS(20)333⟩, or amines ⟨74CR(C278)355, 76T2039, 84CC183, 87TL6049⟩ and yield 3,4-dihydrotriazaphospholes. An excess of water or alcohol degrades them to the amidrazone and H_3PO_3 or $P(OR)_3$, respectively ⟨84PS(20)333⟩. In contrast, the 1*H*-isomers are reported not to react with alcohols or amines ⟨76T2039⟩. In fact, they give a balanced equilibrium with alcohols and are degraded to $P(OR)_3$ by an excess ⟨92TH1⟩.

The said tendency also holds for the oligomerization: 1*H*-1,2,4,3-triazaphospholes are monomeric and the 2*H*-isomers are so too, although for some of them (mainly with RO or RS as 5-substituents) tetramers (or other oligomers) are reported to exist in a temperature-controlled equilibrium with the monomers ⟨87TL6049, B-90MI258, 91CC23, 93CJC1200⟩. 4*H*-1,2,4,3-Triazaphospholes on the other hand are not known at all as such but only in form of their 2,3-dihydro derivatives, i.e. their tetramers and adducts (see Section 4.3.5.3.2).

Regardless of the differences in the addition equilibria the combined action of an alcohol, ammonia or an amine and of sulfur or selenium on 1*H*- or 2*H*-1,2,4,3-triazaphospholes gives the dihydro-triazaphosphole-3-chalcogenides (Scheme 31) ⟨92TH1, 93CJC1200, 94HAC385⟩. This applies also to [1,5]-anellated 1,2,4,3-triazaphospholes ⟨92TH1, 94PS(93/94)381, 95ZN(B)558⟩.

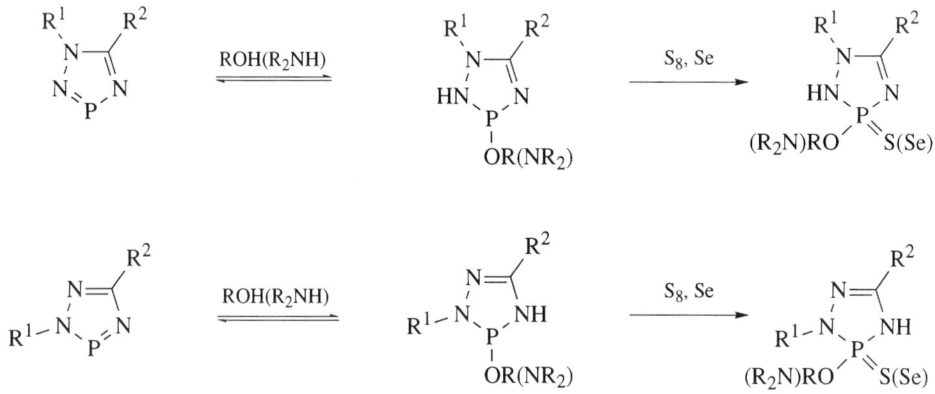

Scheme 31

Hydrogen sulfide and sulfur together react with 2-methyl-5-phenyl-1,2,4,3-triazaphosphole (**33**) to give the dihydro triazaphosphole sesquisulfide (**34**). Sulfur alone gives in a slow reaction the trimeric triazaphosphole sulfide (**35**) (Scheme 32) ⟨94HAC385, 94PS(93/94)383⟩.

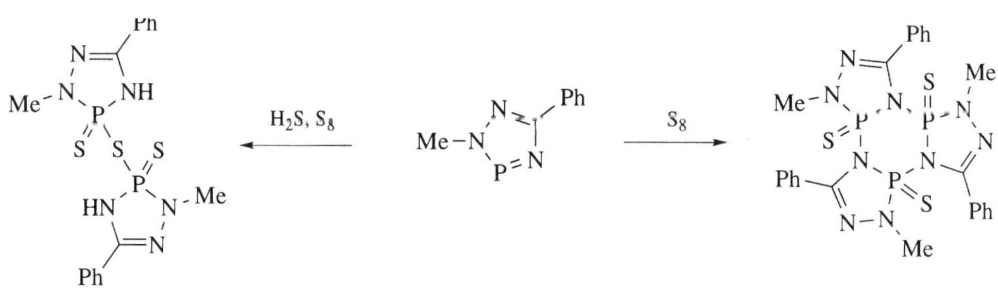

(**34**) (**33**) (**35**)

Scheme 32

Grignard reagents add to the P=N bond as well. Protonation or alkylation of the resulting anion yields 3,4-dihydro-triazaphospholes ⟨84PS(20)333⟩.

Some further reactions with difunctional partners obviously involve the addition of the protic part to the P=N bond as the initial step (Scheme 33). The reaction of 2*H*-triazaphospholes with glycols ⟨74CR(C278)355, 76T2039, 84CC183, 87TL6049⟩ gives hydrido-spirophosphoranes (**36**). Another example of this type is provided by the 2 : 1 condensation products from N^3-amidrazones and $P(NMe_2)_3$ mentioned in Section 4.3.5.3.2. The spirocyclic phosphazenes (**37**) resulting from the reaction with 2-azidoalcohols dimerize to (**38**) ⟨82TL853⟩. In the adducts of aromatic alcohols with a heterodiene function in the 2-position ⟨77ZN(B)841⟩ the phosphorus atom also becomes the center of a tricyclic system (**39**).

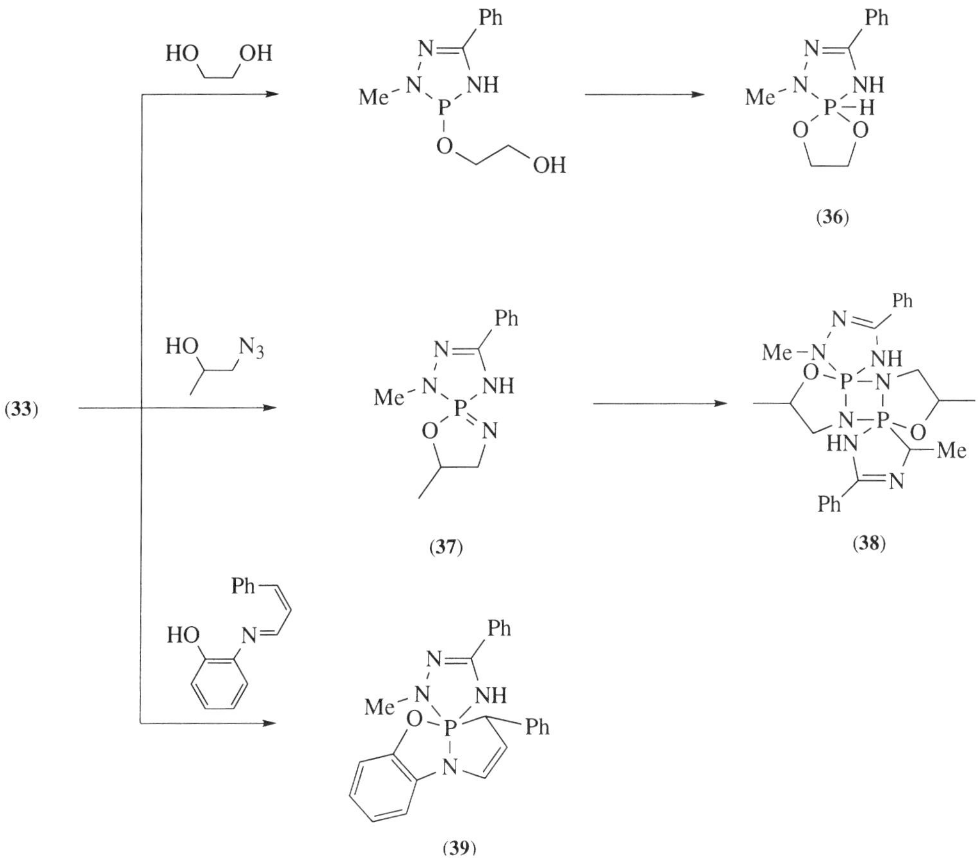

Scheme 33

In the reaction with 2-hydroxy derivatives of benzaldehyde, acetophenone and benzophenone (**40**) the phenol group adds to the P=N bond, the carbon atom of the carbonyl group inserts into the P–NMe bond and the phosphorus atom becomes oxidized to give (**41**) ⟨96CB1493⟩.

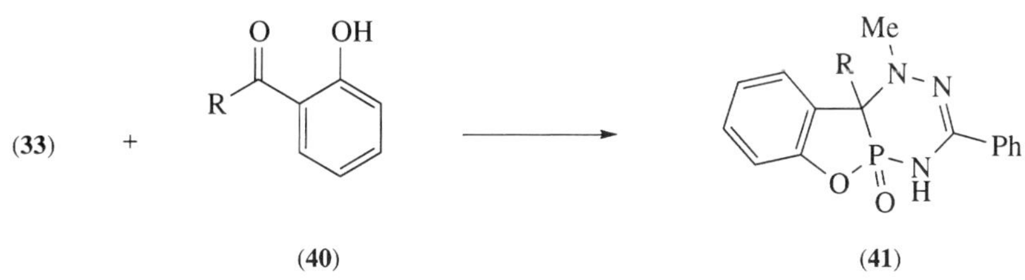

4.3.5.3.1.3 Oxidative addition to phosphorus

On chlorination or bromination of 2*H*-1,2,4,3-triazaphospholes, a 1,1-addition of the halogen to the phosphorus atom takes place, accompanied by a dimerization ⟨77AG554, 81ZAAC(475)211, 96PS(118)129⟩. The dichloride can also be obtained directly from the amidrazone and PCl_5 (see Section 4.3.5.3.2). It can be converted to the difluoride by SbF_3 ⟨81IC1222, 81ZAAC(475)211⟩. With half a mole of bromine ⟨86PS(28)71, 96PS(118)129⟩ or if the triazaphosphole is reacted with its dibromide ⟨83IC3696⟩ a dimer (**42**), trimer and tetramer with mixed phosphorus oxidation states are formed. In part they can be converted to the respective dialkylamino derivatives, e.g. (**42**), X = NMe_2 ⟨96PS(118)129⟩. An equimolar mixture of the triazaphosphole and the dimer of its dihalide (**43**) slowly loses phosphorus trihalide and results in a product (**44**) in which three triazaphosphole rings are fused ⟨82IC3916⟩. Even more complex polycyclic triazaphospholes are formed from such mixtures (X = Cl, Br) ⟨B-87MI617⟩.

2*H*-1,2,4,3-Triazaphospholes do not react with iodine alone but yield the 3,3-diamino-1,2,4,3λ^5-triazaphospholes (**45**) on the combined action of iodine and a primary or secondary amine (Scheme 34) ⟨91PS(56)17, 92TH1⟩.

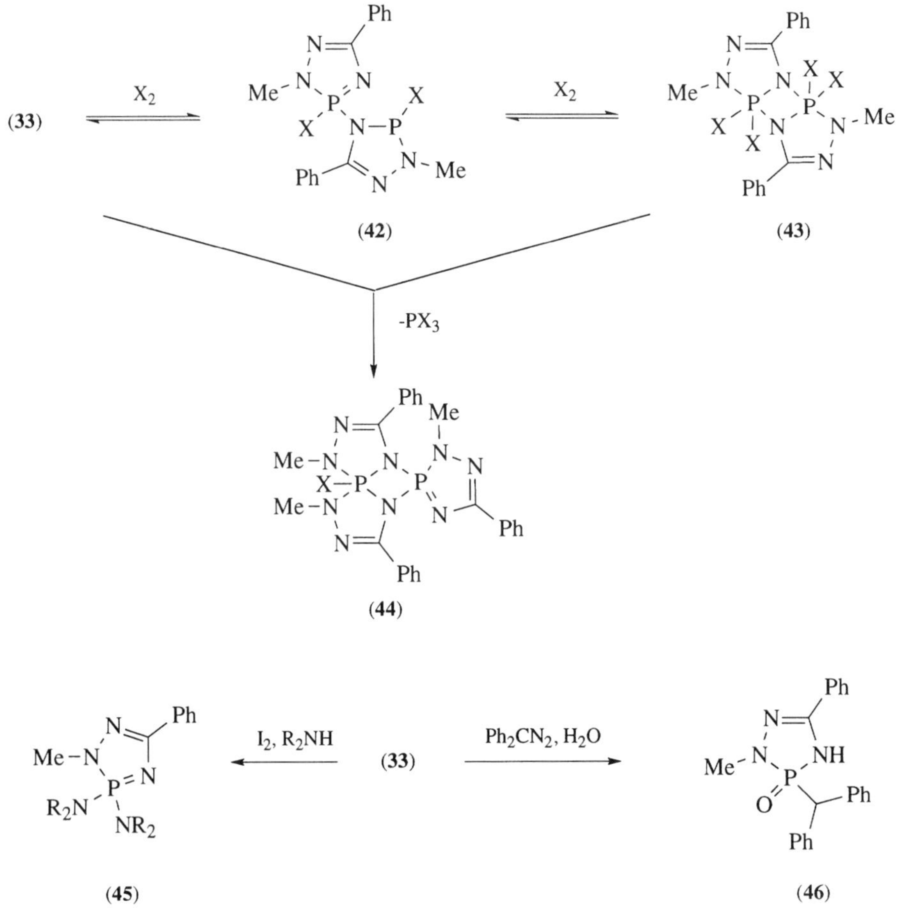

Scheme 34

4.3.5.3.1.4 Cycloaddition reactions

2*H*-1,2,4,3-Triazaphospholes are reported not to react with phenylazide, phenyl-benzoyldiazomethane, maleic anhydride, and tetracyanoethylene ⟨76T2039⟩. From the [4 + 2] cycloaddition with electron-poor acetylenes, 1,2,3-diazaphospholes are obtained ⟨83ZN(B)1484⟩ (see Section 4.3.4.5.2). [2 + 3] Cycloaddition of 2-methyl-5-phenyltriazaphosphole with diphenyldiazomethane yields, after hydrolysis, the 3,4-dihydrotriazaphosphole-3-oxide (**46**), presumably *via* the *P*-ylide as an intermediate (Scheme 34) ⟨86PS(26)203⟩.

With α-diimines, R = Pr^n, Ph [4 + 1] cycloadducts (**47**) are formed (Scheme 35) ⟨86TL2971, 87TL6049, 93CJC1200⟩, which are similar in structure and bonding to the products of a 2 : 1 reaction of amidrazone hydrochloride and phosphorus pentachloride; see Section 4.3.5.3.2 ⟨83CB1468, 91PS(56)17⟩. A comparable spirocyclic derivative results from the addition of 1,2-diaminopyridines under simultaneous oxidation ⟨93JPR458⟩. The [4 + 1] cycloadducts with azodicarboxylic esters (**48**), R = Me, Et, Pr^i, CH_2Ph, ⟨81CB825⟩ and 1,2-diketones (**49**), (R = Me, Ph, ⟨84TL5521⟩ are dimeric (**50**).

The double [4 + 1] cycloaddition with tetrachloro- and 3,5-di(*t*-butyl)-*o*-benzoquinone results in zwitterionic products (**51**) with six-coordinate phosphorus ⟨84TL5521⟩.

4.3.5.3.1.5 Formation of transition metal complexes

Triazaphospholes yield solid products with Cu(I), Hg(II) and Cd(II) salts ⟨76T2039⟩.

P-Coordinated complexes of 2-methyl-5-phenyl- and 1,5-dimethyl-triazaphosphole with $Cr(CO)_5$ and $Mo(CO)_5$ ⟨80JOM(185)53⟩ and of the latter with $Pt(PPh_3)_3$ ⟨83JOM(256)375⟩ have been prepared. The molecular structure of this complex is shown in Table 9. With $PtCl_2(PEt_3)$ only the *trans*-N-4-, with $PtBr_2(PEt_3)$ both the *cis*-P and the *trans*-N-4-coordinated complexes are obtained ⟨84ICA(85)L33, 88IC2612⟩; their alcoholysis yields novel chelates ⟨97ICA(256)47⟩. A *cis*-N-2 or -N-4 coordinated complex of 1,5-dimethyltriazaphosphole with Me_2AuCl and a poorly soluble 1 : 1 complex with ClAuCO have also

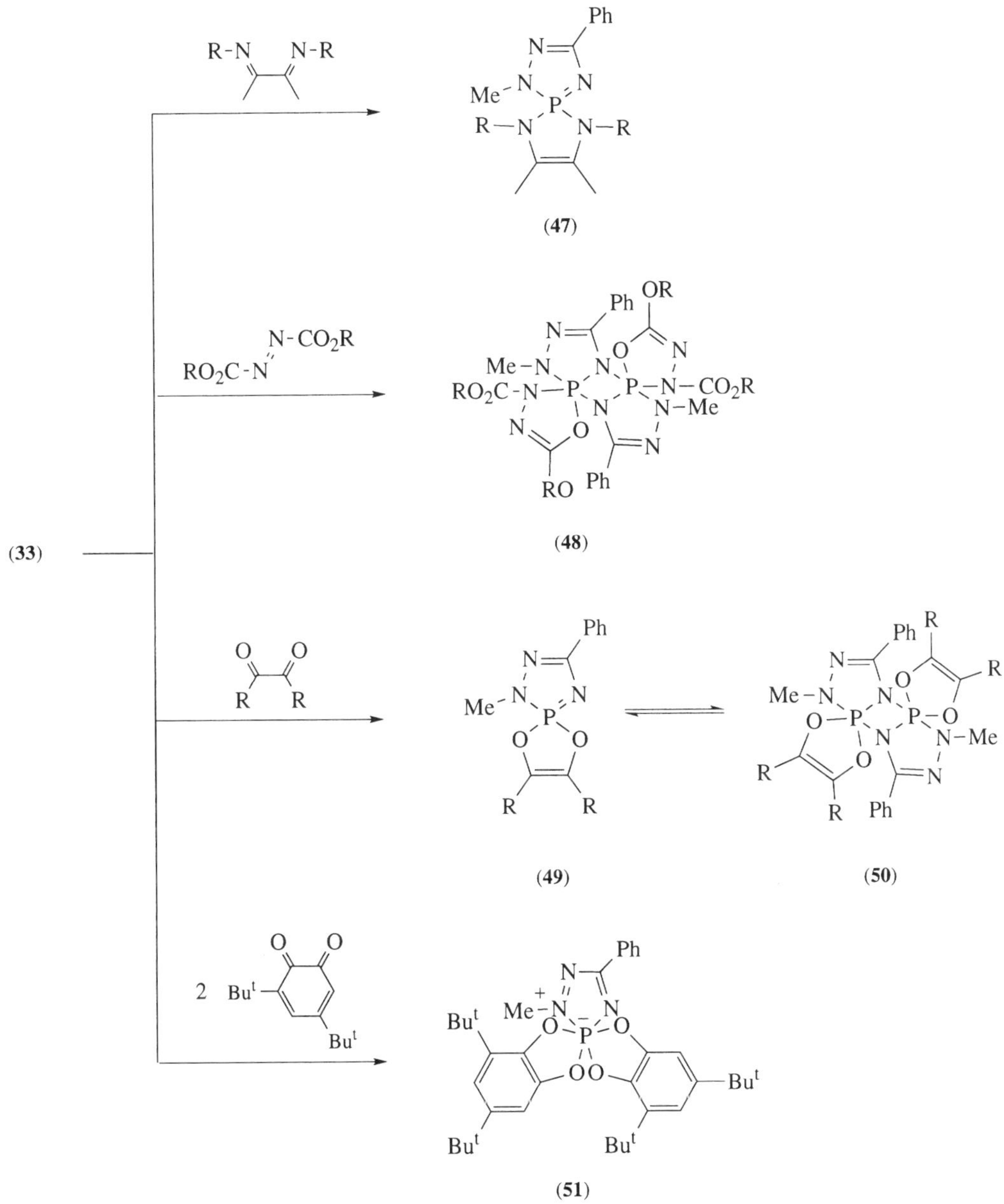

Scheme 35

been reported ⟨80ICA(46)167⟩. *P*-Coordination of the transition metal may induce an addition to the P=N bond ⟨81AG420, 84ICA(85)L33, 88IC2612⟩.

4.3.5.3.2 *Reactivity of 1,3,4,2-thiadiazaphospholium ions*

The ring phosphorus has phosphenium character. It adds tetrachloro-*ortho*-benzoquinone and 2,3-butanedione and it also reacts with dimethylaniline and anisole in a *para*-substitution ⟨93PS(76)79⟩.

4.3.5.3.3 *Synthesis*

1,2,4,3-Triazaphospholes are always synthesized from condensation reactions. The first examples were obtained according to the pattern PNC + NN, most of them according to NCNN + P from amidrazones or their heterocyclic equivalents and a phosphorus(III) partner such as PCl_3 or $P(NMe_2)_3$ or a phosphorus(V) partner and subsequent reduction.

PCl_3 condensation of N^1- and N^2-methylamidrazones gives 2- and 1-methyl-1,2,4,3-triazaphospholes (**52**) and (**53**) (R = Me, Ph). The first step is achieved in refluxing benzene, the second with the help of NEt_3 (Scheme 36) ⟨86PS(28)71⟩.

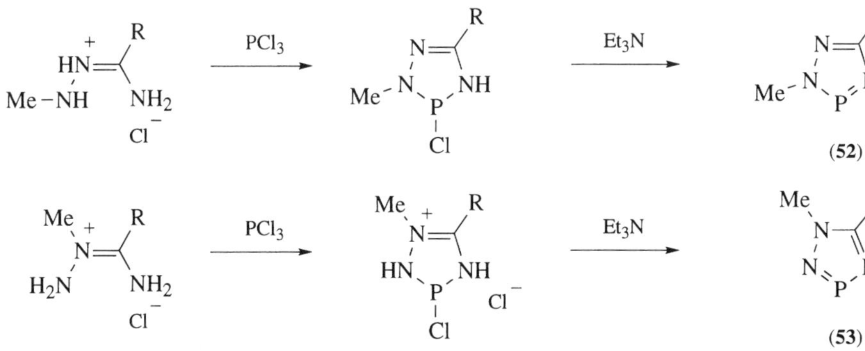

Scheme 36

From condensation of an amidrazone hydrochloride or hydroiodide or an amidrazone benzaldehyde derivative with $P(NMe_2)_3$, *N*-unsubstituted 1,2,4,3-triazaphospholes (**54**), (**55**), as well as 1- and 2-substituted triazaphospholes are obtained (Scheme 37) (74(CR278)355, ⟨76T2039, 80ZN(B)1222, 84CC183, 92PS(69)189⟩. 5.5′-Bis(1,2,4,3-triazaphospholyls) (**56**) have also been obtained this way ⟨87TL6049⟩.

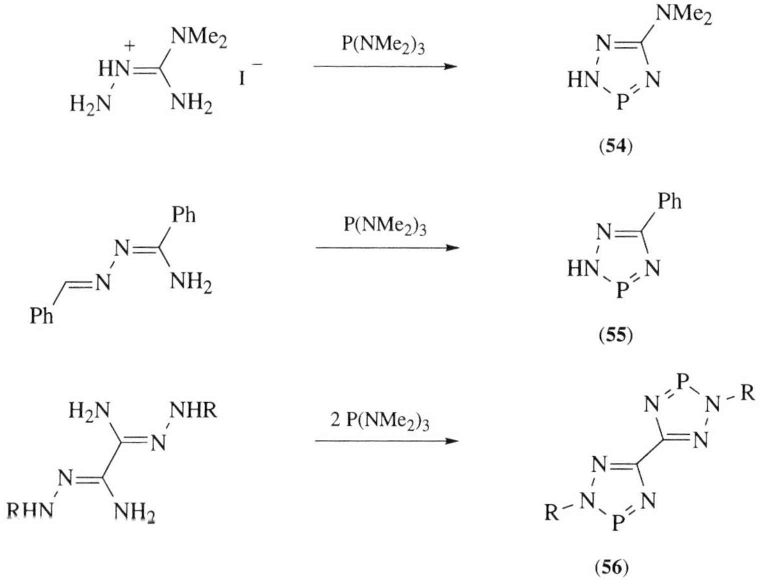

Scheme 37

4*H*-1,2,4.3-Triazaphospholes (**58**) are obtained from the reaction of N^3-substituted amidrazones with PCl_3, R^1/R^2 = Ph/Me ⟨86TL2971⟩, Ph/Ph ⟨92TH1⟩, or with tris(dimethylamino)phosphine, R^1 = Me, Pr^i, CH_2Ph, Ph; R^2 = Me, Pr^i, Bu^s, neopentyl, CH_2Ph, Ph ⟨92JCS(P2)671⟩. They are oligomeric, however (see Section 4.3.5.3.1.2), and no monomeric representative has yet been prepared ⟨80ZN(B)1222, 84CC183, 92JCS(P2)671⟩. Primary products of the reaction of N^3-substituted amidrazones with PCl_3 are the 3-chloro-dihydro-1,2,4,3-triazaphospholes (**57**) which are converted to the 1,2,4,3-triazaphospholium tetrachloroaluminates (**59**) by aluminum trichloride ⟨92TH1⟩. The molecular structure of a 4,5-disubstituted 1,2,4,3-triazaphosphole tetramer has been elucidated by X-ray analysis ⟨92JCS(P2)671⟩. Treating a tetramer with BF_3OEt_2 or $AlCl_3$ affords the 2-BF_3 complex (**60**) or 2-$AlCl_3$ complexes of the monomer (Scheme 38) ⟨86TL2971, 92JCS(P2)671, 92TH1⟩ (molecular structure of a BF_3 complex, $R^1 = R^2 = Pr^i$, in Table 9).

Spirocyclic (**62**) and phosphino substituted dihydro-1,2,4,3-triazaphospholes (**63**) result from the reaction of N^3-substituted amidrazones (**61**) and $P(NMe_2)_3$ in a 2 : 1 and 1 : 2 molar ratio (Scheme 39) ⟨80ZN(B)1222, 92JCS(P2)671, 92TH1⟩.

The semicarbazide (**64**) reacts with $P(NMe_2)$ in boiling acetonitrile to give a 3-dimethylamino-5-oxo-1,2,4,3-triazaphospholane (**65**) (Scheme 40). It is converted by trimethylchlorosilane to the

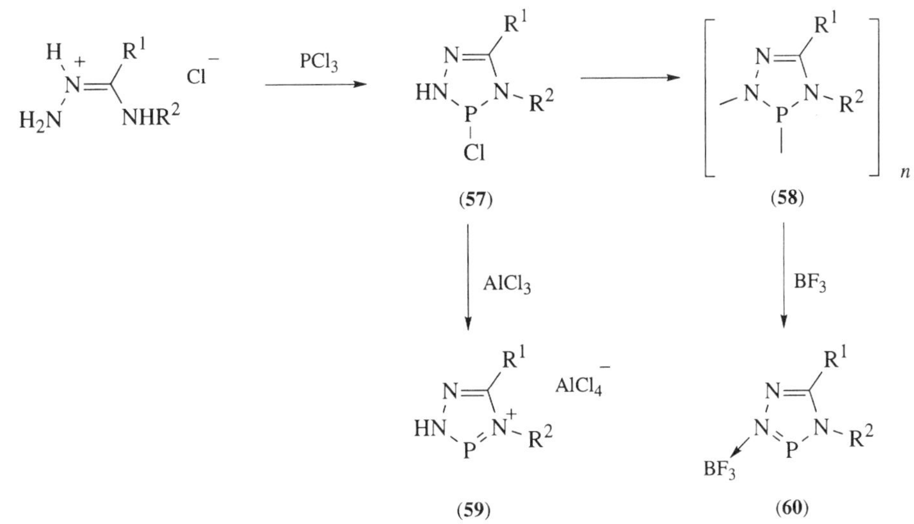

Scheme 38

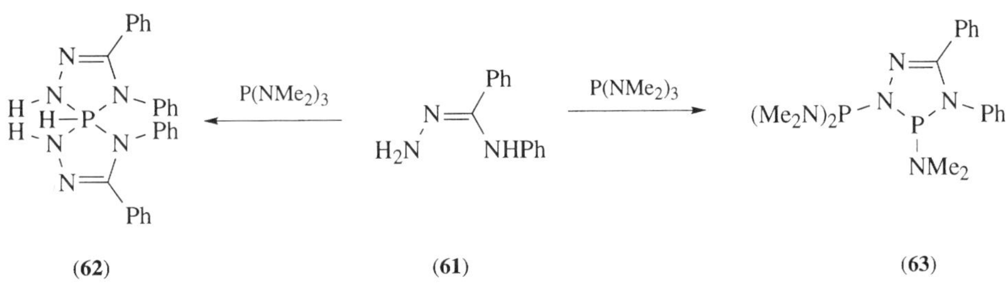

Scheme 39

5-trimethylsiloxy-triazaphosphole (**66**) which exists in a temperature controlled equilibrium with its tetramer. At higher temperatures the condensation affords the 5-oxo-dihydro-1,2,4,3-triazaphosphole (**67**) ⟨91CC23⟩.

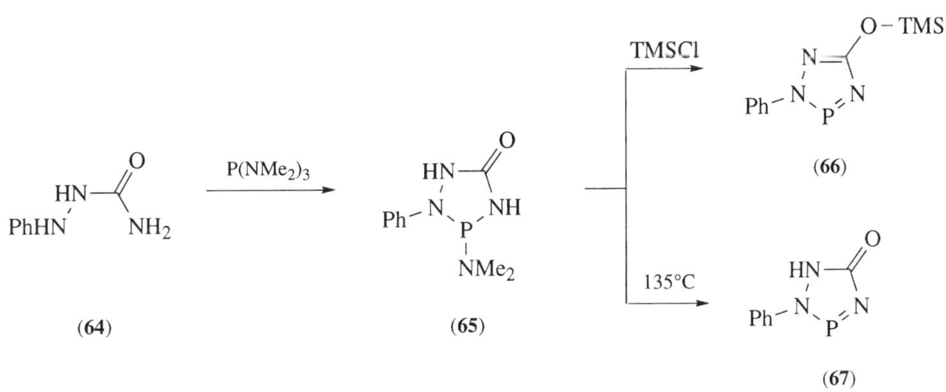

Scheme 40

The 5-thioxo-dihydro-1,2,4,3-triazaphosphole (**69**) is obtained from a thiosemicarbazide (**68**) in the same way or its hydrobromide with PBr_3. With an amine they both give the respective ammonium 1,2,4,3-triazaphosphole-5-thiolate (**70**) which is readily methylated to the 1-methyl-5-methylthio-1,2,4,3-triazaphosphole (**71**) ⟨91CC23, 93CJC1200⟩.

The *N*-unsubstituted *S*-methylthiosemicarbazide hydroiodide (**72**) affords the 2-trimethylsilyl-5-methylthio-1,2,4,3-triazaphosphole (**73**) from its reaction with PCl_3 and Me_3SiCl (B90MI258) or with $P(NMe_2)_3$ and $(Me_3Si)_2NH$ ⟨93CJC1200⟩. The N^1-phenyl-, N^2-methyl-, N^3-methyl- and N^3-ethyl-substituted *S*-methylthiosemicarbazide hydroiodides have also been used for condensation with $P(NMe_2)_3$. The results correspond to those obtained with other amidrazones ⟨93CJC1200⟩.

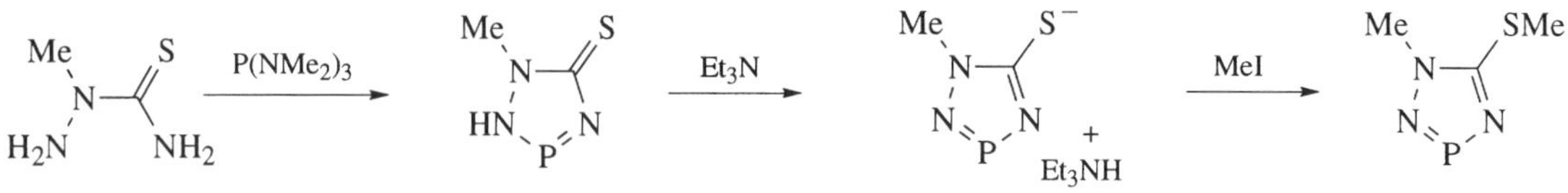

(68) (69) (70) (71)

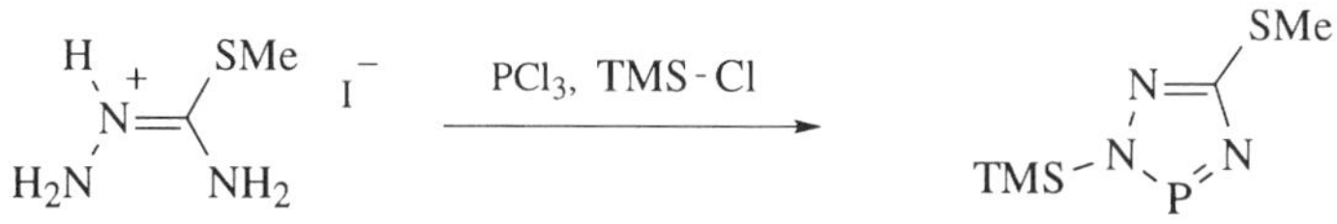

(72) (73)

[1,5]Anellated 1,2,4,3-triazaphospholes (**75**), (**77**), (**79**) are also prepared from condensation with PCl_3 or $P(NMe_2)_3$ (Scheme 41) ⟨95S361⟩. Starting compounds are 1-amino-2-iminopyridines (**74**) ⟨93JPR458⟩ or their hydroiodides, i.e. 1,2-diaminopyridinium iodides (**76**), R = H, Me ⟨94PS(93/94)381, 95ZN(B)558⟩ and 2,3-diaminothiazolium chloride (**78**) ⟨93JPR458⟩.

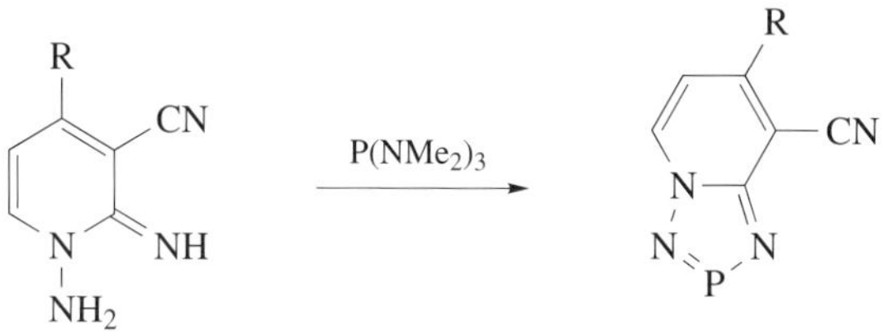

(74) (75)

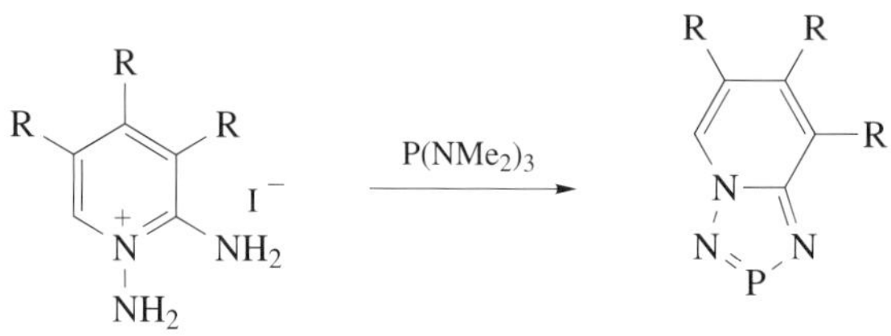

(76) (77)

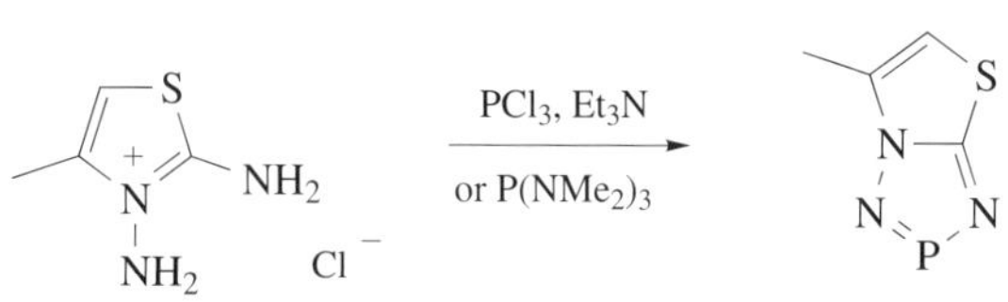

(78) (79)

Scheme 41

[4,5]Anellated 1,2,4,3-triazaphospholes (which are of 4*H*-type) cannot be obtained this way, however (Scheme 42). The condensation of a 2-hydrazinothiazolium chloride (**80**) ⟨90PS(47)173⟩ or of the 2-hydrazinobenzothiazole ⟨85S1067⟩ with tris(dialkylamino)phosphines lead instead to dihydro derivatives (**81**). The condensation of the 2-hydrazinoquinoline (**82**) under severe conditions yields a product assumed to be a polymer of the triazaphospholoquinoline (**83**) ⟨96MI96⟩.

Similar to 1,3,2-benzodiazaphospholes (Section 4.3.4.6.2), 1,2,4,3-triazaphospholes can be prepared *via* a pentavalent phosphorus compound which eliminates a disulfide (Scheme 43) ⟨80TL1307⟩.

The equimolar condensation of N^1- and N^2-methylamidrazones with phosphorus pentachloride yields dimers and hydrochlorides of 3,3-dichloro-1,2,4,3λ^5-triazaphospholes ⟨77AG554, 81IC1222, 81ZAAC(475)211⟩.

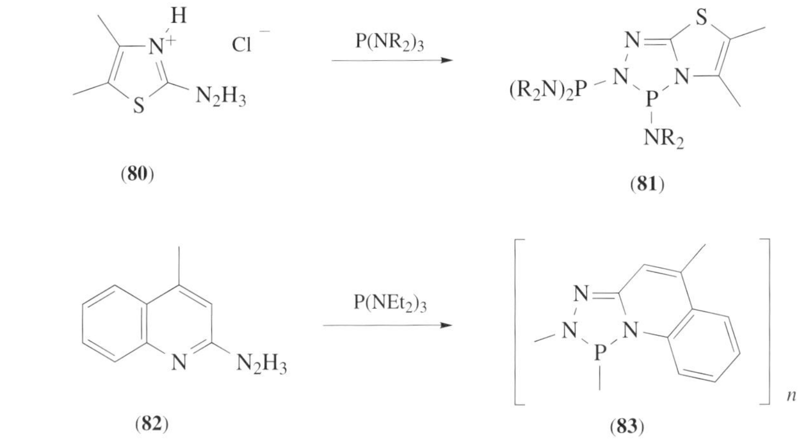

Scheme 42

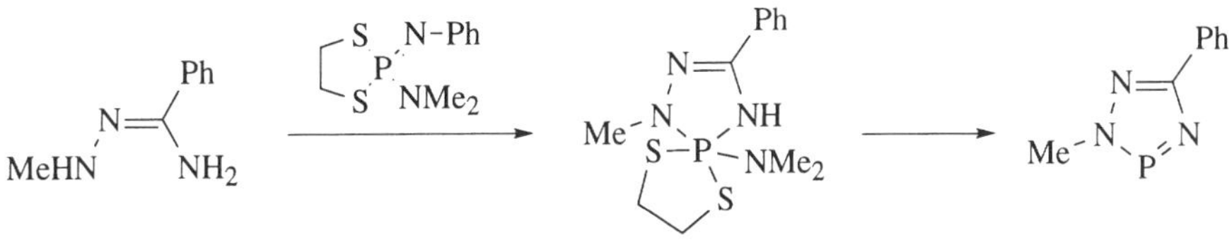

Scheme 43

The former also result from the chlorination of the triazaphospholes (see Section 4.3.5.3.1.3) and can be reduced to them again by reaction with propanedithiol ⟨77AG554⟩. The 2 : 1 condensation of amidrazones and phosphorus pentachloride gives spirocyclic 1,2,4,3λ^5-triazaphospholes ⟨83CB1468⟩.

1,2,4,3-Triazaphospholes (**85**, **86**), R^1 = Me, Ph; R^2 = Bu^t, CH_2Ph, Ph, are also synthesized from imidato-chlorophosphines (**84**) from reaction with hydrazines (Scheme 44) ⟨74CR(C278)355, 76T2039⟩. Rather different conditions are necessary to eliminate ethanol from the cyclic ethoxyphosphine intermediates for the different isomers: 1*H*-1,2,4,3-triazaphospholes (**86**) form easily at room temperature, while the 2*H*-isomers (**85**) need 100°C or an OEt/Cl-exchange and subsequent elimination of HCl by base.

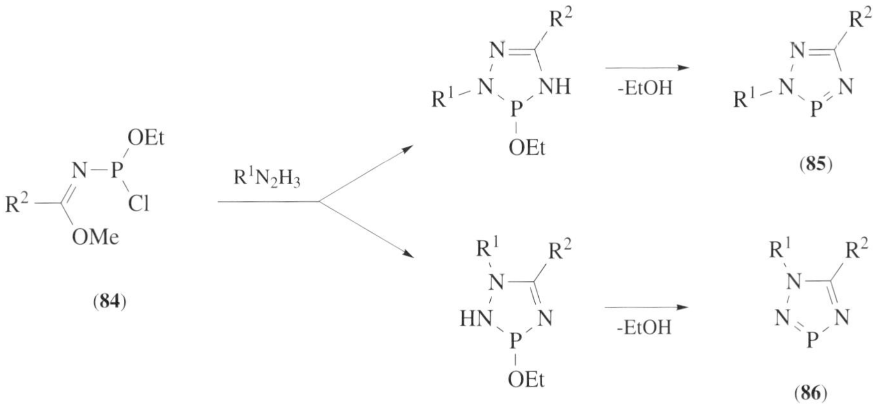

Scheme 44

Methyl esters of dithiocarbazic acids (**87**) and PCl_3 react in boiling benzene to give 2-chloro-dihydro-1,3,4,2-thiadiazaphospholes (**88**). Aluminum trichloride abstracts chloride from them and yields the 1,3,4,2-thiadiazaphospholium salts (**89**), R = H, Ph, Ac ⟨93MI641, 93PS(76)79, 93PS(76)339, 94ZOB149, 95PS(104)33, 98PS(140)183⟩. The geometry of the cation has been calculated ⟨95MI239⟩. Elimination of HCl in case of R = H does not give the monomeric 1,3,4,2-thiadiazaphosphole but an oligomer thereof (**90**) for which $n = 2$ is suggested (but seems rather doubtful) ⟨95PS(104)33⟩ (Scheme 45).

1,3,4,2λ^5-Oxadiazaphospholes, 1,3,4,2λ^5-thiadiazaphospholes and 1,3,4,2λ^5-selenadiazaphospholes are known with a variety of phosphorus substituents and are in every case dimeric ⟨B-87MI617⟩. They

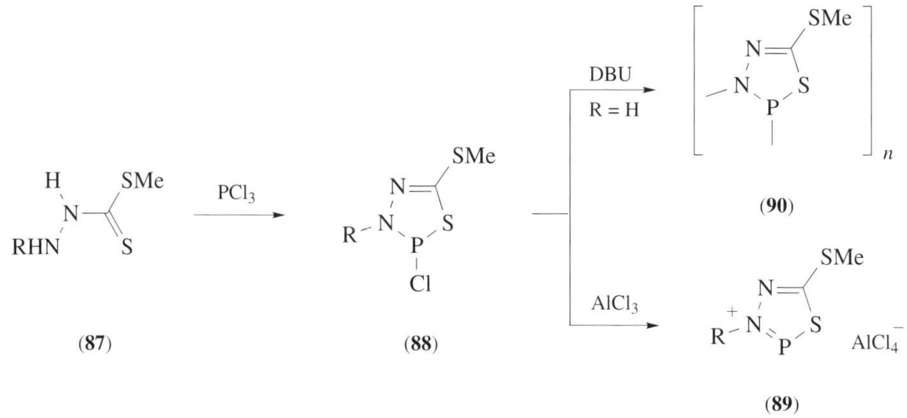

Scheme 45

are prepared starting from carbonic and thiocarbonic acid hydrazides ⟨69AG707, 75CB820, 77CB1124, 77ZN(B)845, 81IC1222, 82PS(14)49⟩. Alternatively, 2-phosphino-tetrazoles are converted to thiophosphinoyl and selenophosphinoyl tetrazoles and the latter are thermolyzed at 100°C ⟨93ZN(B)93⟩.

4.3.6 DIPHOSPHOLES AND TRIPHOSPHOLES WITH ONE OR TWO ADDITIONAL HETEROATOMS

4.3.6.1 Introduction

This section covers the heterocycles with two non-adjacent or adjacent two-coordinate phosphorus atoms and one or two nitrogen atoms or one chalcogen atom in the ring (Table 12). Fused examples are known only where a *CC*-anellation is possible. Section 4.3.6.4 also includes rings with an arsenic or antimony atom in place of a phosphorus atom and Section 4.3.6.3 includes a ring with a third phosphorus atom. Oxa- and thiatetraphospholes P_4E, E = O, S are not included. *Ab initio* calculations indicate that the cyclic planar form is the most stable conformation of P_4O and P_4S ⟨91CPH(149)325⟩. The latter was evidenced as a secondary product from P_4 and S_8 in an argon discharge ⟨91JPC75⟩.

Spectra, structure and bonding of the compounds of Table 12 are discussed in Section 4.3.4.2. According to several different criteria, 1,2,4-aza-, 1,2,4-oxa- and 1,2,4-thiadiphospholes are somewhat less aromatic than the analogous ring systems, pyrrole, furan and thiophene ⟨99CC2363⟩. More calculations deal with thia- and selenadiphospholes ⟨95JCS(P2)315, 00JA4557⟩.

Structural data of azadiphospholes and diazadiphospholes from X-ray investigations are given in Table 10. Their PP bonds (206–207 pm) are only slightly longer than in diphosphenes (200–204 pm) ⟨88ZSK122, B-90MI321⟩.

For 1,2,4-aza-, 1,2,4-oxa- and 1,2,4-thiadiphospholes the difference in the two ^{31}P shifts decreases significantly as the electronegativity of the 1-atom becomes smaller ⟨99CC2363⟩. One bond ^{31}P,^{31}P coupling in 1,2,3-azadiphospholes, $^1J_{PP}$ = 465–481 Hz, is somewhat smaller than in diphosphenes, $^1J_{PP}$ = 525–670 Hz ⟨86PS(29)73, B-90MI321⟩. The cross ring coupling constant $^2J_{PP}$ is found to be 30–35 Hz in 1,2,4-azadiphospholes ⟨99CC2363⟩, 18 Hz in 1,2,4-oxadiphospholes ⟨99EJO587⟩, 47–59 Hz in 1,2,4-thiadiphospholes ⟨86AG570, 88TL4535⟩ and 23–25 Hz in 1,2,3,5-diazadiphospholes ⟨85AG127⟩.

4.3.6.2 1,2,4-Azadiphospholes (1)

1,2,4-Azadiphospholes (**1**) can be made up of two phosphaalkyne molecules and an imino group. For the synthesis of (**18**) the *t*-butyl-phosphaacetylene is heated with the imide (**16**), R = Et, Ph, *p*-Tol, (yield 15–31%) or (**17**), R = Pr, Pr^i, CH$_2$Bu^t, Cy (yield 55–87%) ⟨99CC2363⟩. The molecular structure of (**18**), R = Ph, is contained in Table 10.

Aminophosphaethynes (**20**), R^1/R^2 = Bu^t/H, Bu^i/SiMe$_3$, cyclohexyl/SiMe$_3$, obtained from the adducts of alkyl isocyanates and tris(trimethylsilyl)phosphine (**19**), dimerize in alkaline medium to give 3-amino-1,2,4-azadiphospholes (**21**) ⟨91HAC577, 92HAC435⟩. The molecular structure for R^1 = Bu^t, R^2 = H is given in Table 10.

Table 12. Known di- and triphospholes with one or two additional heteroatoms as well as their anellated derivatives

1,2,4-azadiphospholes (**1**)

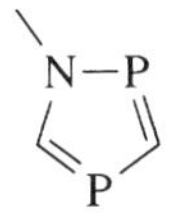

1,2,5-azadiphospholes (**2**)

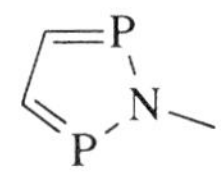

1,2,3-azadiphospholes (-phosphaarsoles, -phosphastiboles) (**3**)

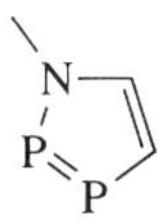

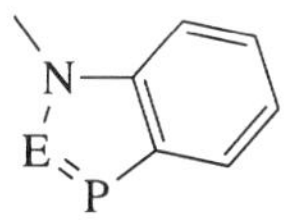

E = P, As, Sb

1,2,3,5-diazadiphospholes (**4**)

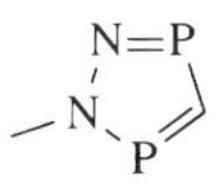

1,2,3,4-diazadiphospholes (**5**)

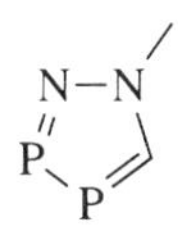

1,4,2,3-diazadiphospholes (**6**)

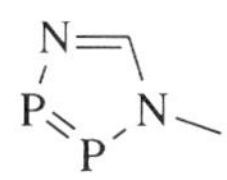

1,2,3,5-azatriphospholes (**7**)

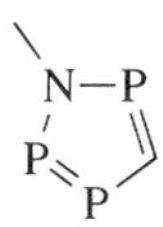

1,2,4-oxadiphospholes (**8**)

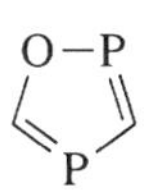

1,2,4-thia/selena/telluradiphospholes (**9**), (**10**), (**11**)

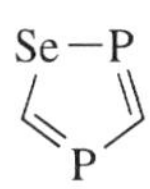

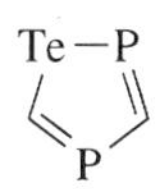

1,2,5-thiadiphospholes (**12**)

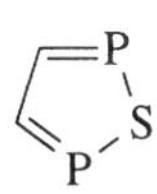

1,3,4-thia/selenadiphospholes (**13**), (**14**)

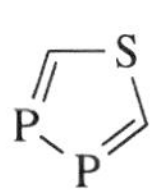

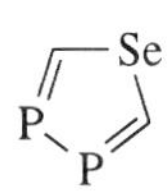

1,2,3-oxadiphospholes (**15**)

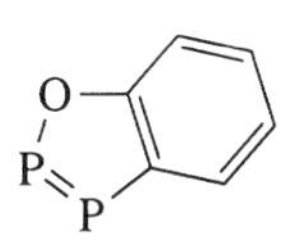

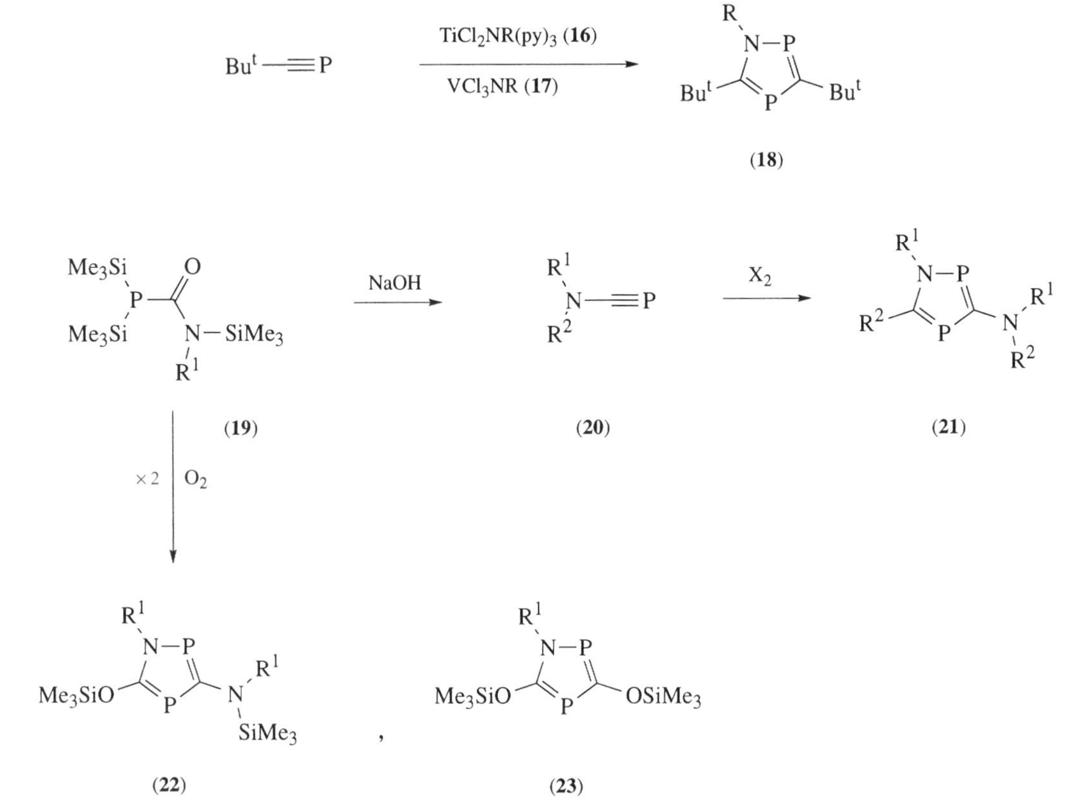

Scheme 46

Flash pyrolysis of (**19**), $R^1 = Pr^i$, Bu^i, Cy, or R^1 = Ph, $PhCH_2$, under controlled air oxidation leads to the 5-trimethylsiloxy (**22**) and 3,5-bis(trimethylsiloxy) derivatives (**23**) (Scheme 46) ⟨92TH2⟩.

4.3.6.3 1,2,5-Azadiphospholes (2) and 1,2,3,5-Azatriphospholes (7)

Like the 1,2,4-azadiphospholes (**1**), the 1,2,5-azadiphospholes (**2**) too can be made up of two phosphaalkyne molecules and an imino group. Examples (**26**) have in fact been synthesized in a similar way from the imine (**25**) and the phosphaalkynes (**24**), R = Bu^t, other tertiary alkyl ⟨98AG1318, 99PS(144-146)261⟩. With more (**24**), the azatetraphosphaquadricyclanes (**28**) are formed and the 1,2,3-azadiphospholes (**26**) are obtained as mixtures with (**28**). As an intermediate, resulting from the addition of a phosphaalkyne and the elimination of an alkyne molecule, the 1,2,3,5-azatriphosphole (**27**) is assumed (Scheme 47) ⟨98AG1318⟩.

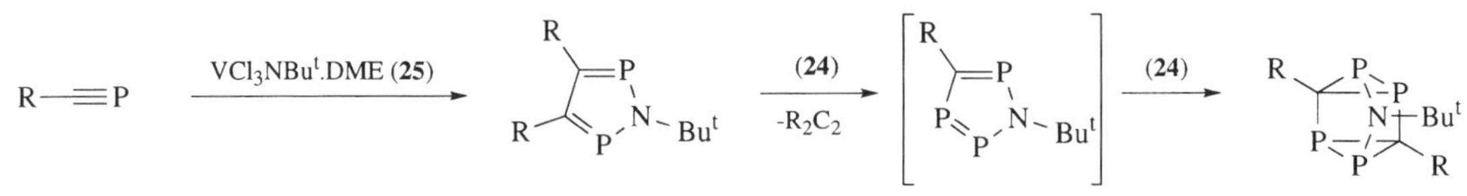

(24) (26) (27) (28)

Scheme 47

4.3.6.4 1,2,3-Azadiphospholes, 1,2,3-Benzazadiphospholes and 1,3,2-Benzazaphosphaarsoles and -stiboles (3)

Pyrolysis of crude bis(diisopropylamino)phosphine or of diallyl(diisopropylamino)phosphine yields among other products 1-isopropyl-5-methyl-1,2,3-azadiphosphole, a colorless pyrophoric solid ⟨89NJC309, 97HAC97⟩. Its molecular structure is given in Table 10. As compared to acyclic diphosphenes which adopt the *trans*-configuration, the 1,2,3-azadiphospholes contain a *cis*-diphosphene unit and as a consequence of the cyclic π-conjugation the P=P bond is slightly elongated.

1,3-Diphospholyl substituted 1,2,3-azadiphospholes (**30**) are formed by reaction of 1,2,4,5-tetrazine-dicarboxylates (**29**) with four moles of *t*-butylphosphaethyne in an unusual reaction sequence ⟨88TL5867⟩. They are yellow liquids and they readily add hydrogen chloride or methanol to the P=C bond of the diphospholyl substituent to yield pale yellow crystalline products (**31**) ⟨88TL5867⟩. The molecular structure with R = Me is given in Table 10.

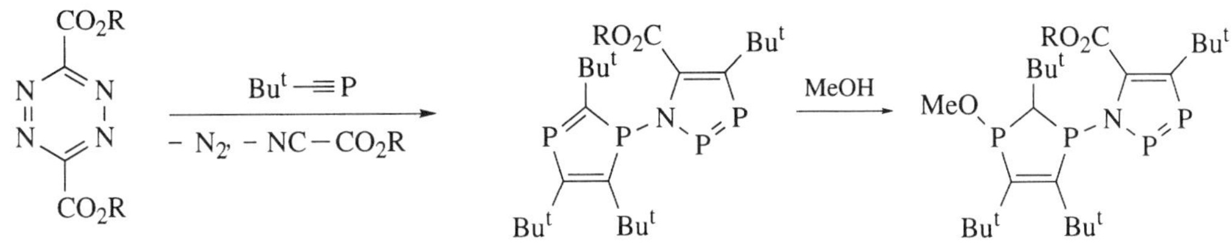

(**29**) (**30**) (**31**)

N-Unsubstituted 1,2,3-benzazadiphospholes (**33**), R^1, R^2 = H, Me, Cl, were prepared from a cyclocondensation with 2-aminophenylphosphines (**32**) in a synthesis analogous to that used for 1,3-benzazaphospholes (Section 4.3.3.3.2.2) ⟨90TL4463, 93PS(76)45⟩.

By a strong base they are converted to the anions. 1,3-Dienes such as 2,3-dimethylbutadiene give a [4 + 2] cycloaddition to the P=P bond ⟨90TL4463⟩.

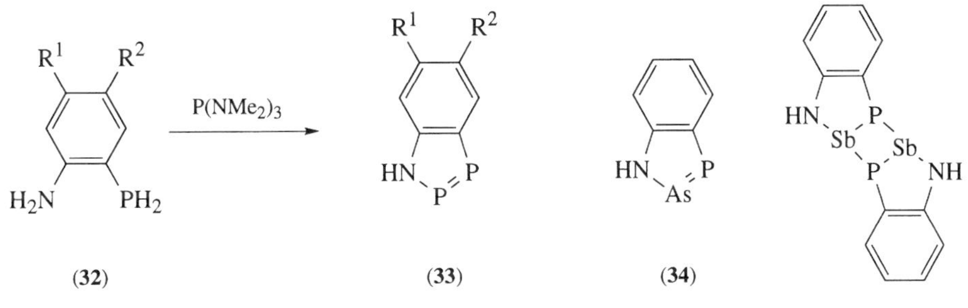

(**32**) (**33**) (**34**) (**35**)

Analogues to (**33**) with As or Sb in 2-position (**34**) and (**35**) were prepared in a corresponding manner. The 1,3,2-benzazaphosphastibole is dimeric (**35**) ⟨93PS(76)45⟩.

4.3.6.5 1,2,3,5-Diazadiphospholes (4)

1,2,3,5-Diazadiphospholes (**38**), R = Me, Ph, are obtained from the condensation of methylene bis(dichlorophosphine) (**36**) and a primary hydrazine with the diazadiphospholanes (**37**) as intermediates ⟨85AG127⟩.

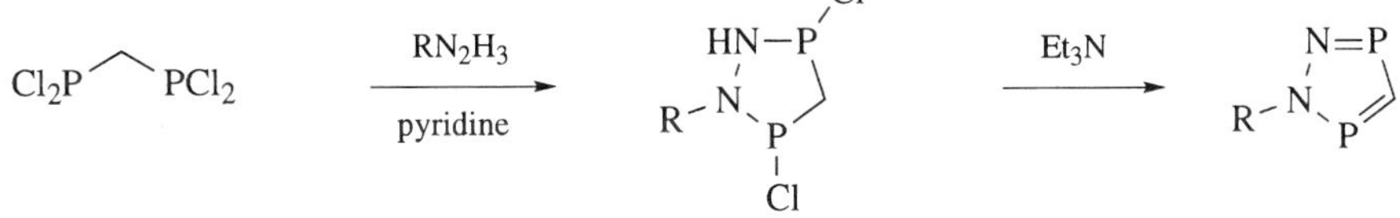

(**36**) (**37**) (**38**)

1-Phenyl-1,2,3,5-diazadiphosphole is a colorless, air-sensitive liquid. It is not alkylated by methyl iodide or trimethyloxonium tetrafluoroborate and it does not react with biacetyl and *o*-naphthoquinone. Additions of hydrogen chloride, methanol, and diethylamine always take place at both the P=C and P=N bond and yield the tetrahydro derivatives ⟨85AG127⟩.

4.3.6.6 1,2,3,4-Diazadiphospholes (5) and 1,4,2,3-Diazadiphospholes (6)

Lithium 5-trimethylsilyl-1,2,3,4-diazadiphospholide (**40**), M = Li, is obtained from the lithiated trimethylsilyl-diazomethane (**39**) and white phosphorus ⟨96AG2282⟩. It can be protonated to the 2*H*-1,2,3,4-diazadiphosphole (**41**) and the latter can be deprotonated to (**40**), M = Na. The molecular structure of (**41**) is contained in Table 10.

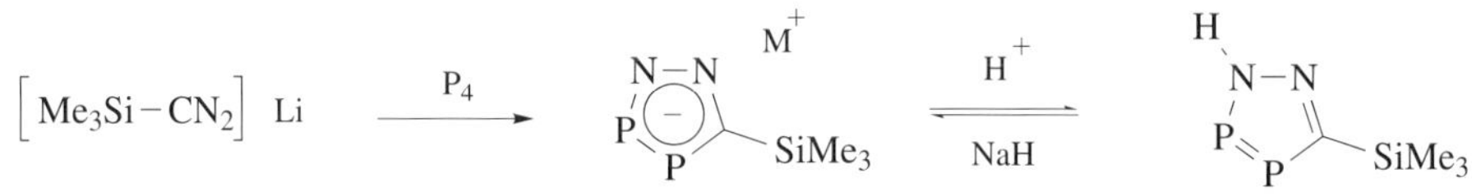

(**39**) (**40**) (**41**)

The electronic structures of diphosphorus compounds bridged by urea and also that of a 1,4,2,3-diazadiphosphole derivative were calculated ⟨99MI485⟩.

4.3.6.7 1,2,4-Oxadiphospholes (8)

The 1,2,4-oxadiphosphole (**43**) has been obtained from vacuum thermolysis of the phosphaalkene (**42**). It readily gives [4 + 2] cycloadditions with phosphaalkynes and acetylenedicarboxylate. The latter reaction provides an access to 1,2-oxaphospholes (Section 4.3.2.4) ⟨99EJO587⟩.

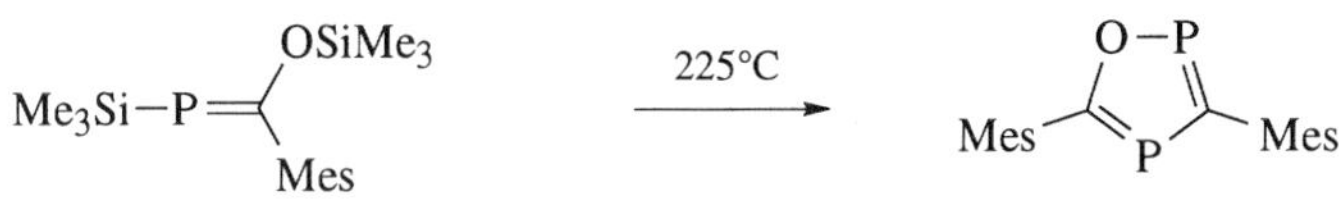

(**42**) (**43**)

4.3.6.8 1,2,4-Thia/Selena/Telluradiphospholes (9), (10), (11)

The chalcogenadiphospholes (**9**), (**10**) and (**11**) can formally be made up of two phosphaethyne molecules and a chalcogen atom. In this sense thiophosphinato manganese complexes (**44**), R = Me, cyclohexyl, Ph, or the cobalt complex (**45**) and the phosphaethynes (**46**), also with 1-adamantyl in place of Bu^t, give the 1,2,4-thiadiphospholes (**47**). In case of R = Ph, the bicyclic product (**48**), containing one more molecule of the phosphaethyne, is also formed ⟨93AG1521, 94JOM(464)C31, 96JOM(524)67⟩. The thiadiphosphole (**47**) also forms in the reaction of the phosphaethyne (**46**) with elemental sulfur in boiling DME ⟨00TH1⟩. With $TaCl_3$ as catalyst this reaction can be performed at room temperature giving a 66% yield of (**47**) ⟨00TH3⟩.

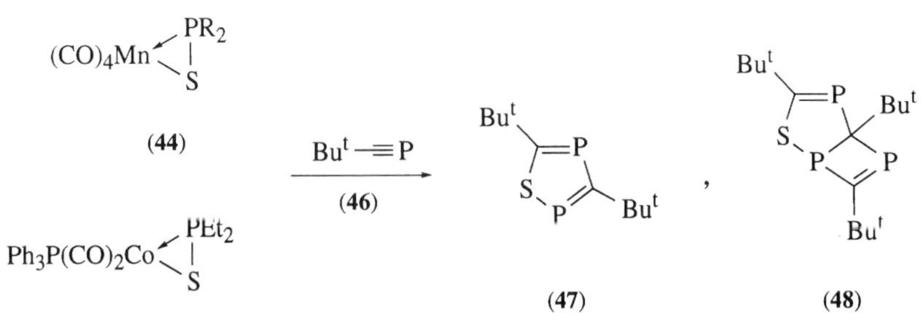

(**44**) (**45**) (**47**) (**48**)

Carbon disulfide also can supply the sulfur atom necessary for the formation of (**47**) ⟨00T27, 00JA4557⟩ and the thiadiphosphole (**47**) is obtained from the room temperature reaction of (**46**) and CS_2 in 92% yield ⟨00TH1⟩. As mentioned in Section 4.3.4.8, the reaction of (**46**) with ylide type CS_2-complexes such as (**49**), R = piperidyl, morpholyl, leads to a mixture of the 1,2,4-thiadiphosphole (**47**) and the

1,3,4-thiadiphosphole (**50**) (see Section 4.3.6.9). (**47**) Gives η^1-complexes with $Cr(CO)_5$ and $W(CO)_5$ as well as both a *cis* and a *trans* complex with $PtCl_2PEt_3$ in which the metal is attached to P-2 ⟨00JA4557, 00TH3⟩. η^5-Complexes with $Cr(CO)_3$ and $W(CO)_3$ are also accessible ⟨00TH3⟩.

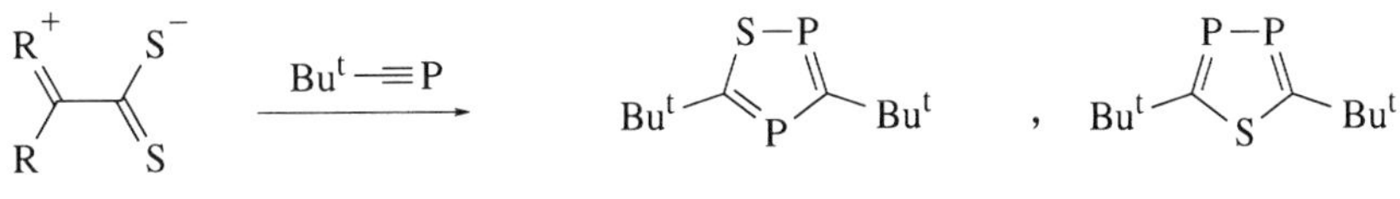

(**49**) (**47**) (**50**)

Scheme 48

From the condensation of lithium bis(trimethylsilyl)phosphide with carbon disulfide 3-trimethylsilylthio-5-trimethylsilyl-1,2,4-thiadiphosphole (**51**) is obtained as a distillable yellow liquid ⟨86AG570⟩. As mentioned in Section 4.3.4.3.2 the diphenyl-1,2,4-thiadiphosphole (**52**) is found in 2–4% yield as crystalline side product from the reaction of the phosphaalkene $Ph(Me_3Si)C{=}PCl$ with an oxathiazolone ⟨88TL4535⟩.

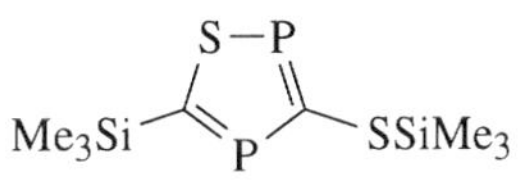

(**51**)

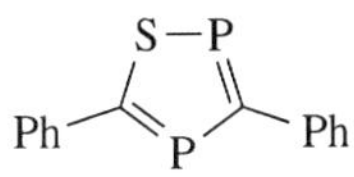

(**52**)

The 1,2,4-selenadiphospholes (**53**) and 1,2,4-telluradiphospholes (**54**) can be obtained from the reaction of (**46**), or phosphaethynes with other tertiary alkyl groups in place of Bu^t, with elemental selenium in toluene at 70ºC in the presence of NEt_3 (with up to 89% yield) or in boiling DME (79%) and with elemental tellurium at 120ºC (15–20%) or in boiling DME (36%) ⟨99S1642, 00TH1⟩ (Scheme 49).

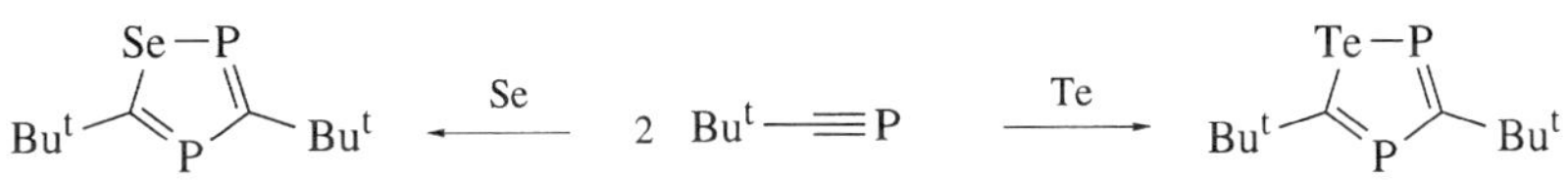

(**53**) (**46**) (**54**)

Scheme 49

Access to the selenadiphosphole (**53**) is also provided by the thermal reaction of the 1,2,3-selenadiazole (**56**) with the phosphaalkyne (**46**) (17% yield) ⟨96PS(115)99⟩, the reaction of elemental selenium with the 1-bis(trimethylsilyl)methyl-1,2,4-triphosphole (**55**) (see Section 4.3.7) ⟨96BSB675⟩ and the reaction of $(Et_2NCS_2)_2Se$ (**58**), E = Se, with the lithium 1,4,2-diphosphastibolide (**57**) (see Section 4.3.7) (60% yield). The analogous reaction of (**58**), E = Te, gives the 1,2,4-telluradiphosphole (**54**) (32% yield) ⟨99TL3815⟩ (Scheme 50).

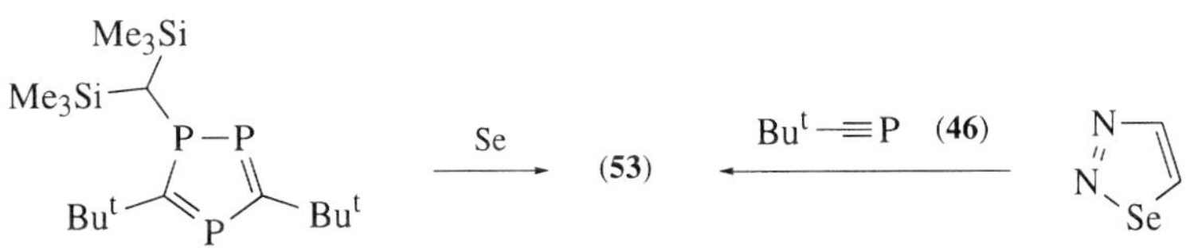

(**55**) (**56**)

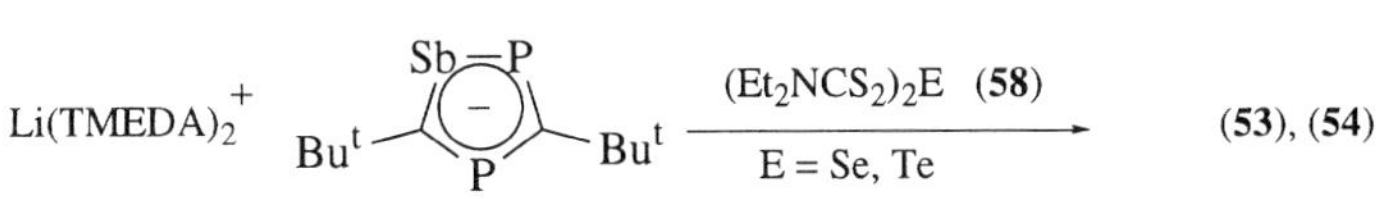

(**57**)

Scheme 50

1,2,4-Thiadiphospholes (**47**, also with other tertiary alkyl groups besides Bu^t) add dicyano-acetylene and subsequently eliminate phosphaacetylenes to yield 1,2-thiaphospholes (Section 4.3.2.5) ⟨00TH3⟩. They also add two equivalents of a phosphaacetylene to a tetracyclic product and two equivalents of mesitylnitrile oxide to a tricyclic product, the decomposition of which leads to 1,2,4-oxazaphospholes (Section 4.3.4.3.2) ⟨00TH3⟩. Tetrahalo-*o*-benzoquinones give a regioselective [4 + 2] cycloaddition to the P-2, C-3 double bond of (**47**) ⟨00TH3⟩.

Although the 1,2,4-telluradiphosphole (**54**) does not exhibit any selective cycloaddition reactivity, two equivalents of the phosphaalkyne (**46**) do undergo cycloaddition to the 1,2,4-selenadiphosphole (**53**) to afford a tetracyclic product ⟨99S1642, 99TH1⟩. With dimethyl acetylenedicarboxylate, (**53**) gives a mixture of 1,2- and 1,3-selenaphospholes (Sections 4.3.2.5 and 4.3.3.6) ⟨99TH1⟩. Of the 1,2,4-selenadiphosphole (**53**), $Cr(CO)_5$ and $W(CO)_5$ complexes with the metal attached to P-2 ⟨99JOM(580)156⟩ as well as $Cr(CO)_3$ and $Mo(CO)_3$ complexes ⟨96BSB675⟩ are reported, of the 1,2,4-telluradiphosphole (**54**), the $W(CO)_5$ complex, also with the metal coordinated to P-2 ⟨99JOM(580)156⟩.

4.3.6.9 1,2,5-Thiadiphospholes (12)

The reaction of tetrakis(trifluoromethyl)-1,4-diphosphinine (**59**) with elemental sulfur results in the correspondingly substituted 1,4-diphosphabarrelene and 1,2,5-thiadiphosphole (**60**) ⟨81JA2465⟩. (**60**) behaves as a diene and undergoes [4 + 2] cycloadditions with bis(trifluoromethyl)acetylene and maleic anhydride (Scheme 51) ⟨81JA2465⟩.

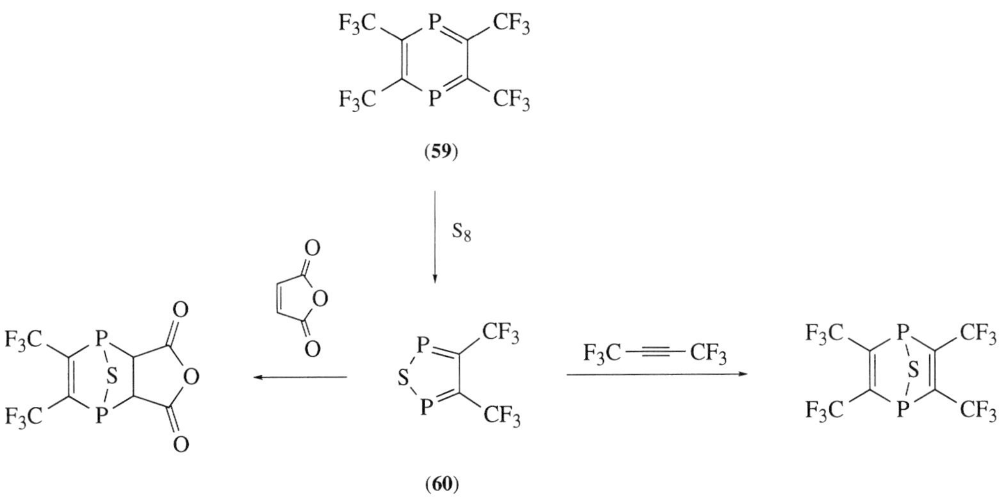

Scheme 51

4.3.6.10 1,3,4-Thia/Selenadiphospholes (13), (14)

The 1,3,4-thiadiphospholes and 1,3,4-selenadiphospholes (**62**), E = S, Se; R = Bu^t, 1-adamantyl, are obtained from the cycloadduct (**61**) of the *t*-butylimidovanadium trichloride and the respective phosphaalkyne ⟨99TH2⟩. The synthesis of the 2,5-di-*t*-butyl-1,3,4-thiadiphosphole (**50**) from CS_2 and *t*-butyl-phosphaacetylene ⟨00JA4557⟩ is mentioned in Sections 4.3.4.8 and 4.3.6.8 (Scheme 48). (**50**) Was found to be much more reactive than the 1,2,4-isomer (**47**) in forming a $PtCl_2PEt_3$ complex. This complex is symmetric with both phosphorus atoms *cis* coordinated to the metal ⟨00JA557⟩

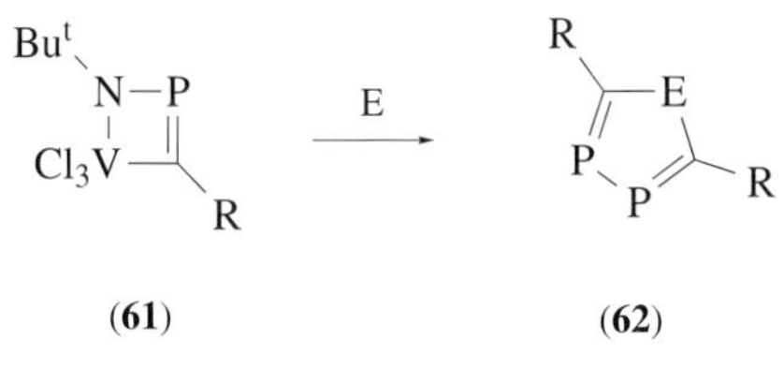

Reaction of the transition metal substituted bis(trimethylsilyl)phosphine (**63**), M = Fe, Ru, with carbon disulfide afforded the unstable phosphaalkene (**64**), which decomposed to the 1,3,4-thiadiphosphole (**65**) by extrusion of bis(trimetylsilyl)sulfide (Scheme 52) ⟨97OM3188, 97AOC(41)1⟩.

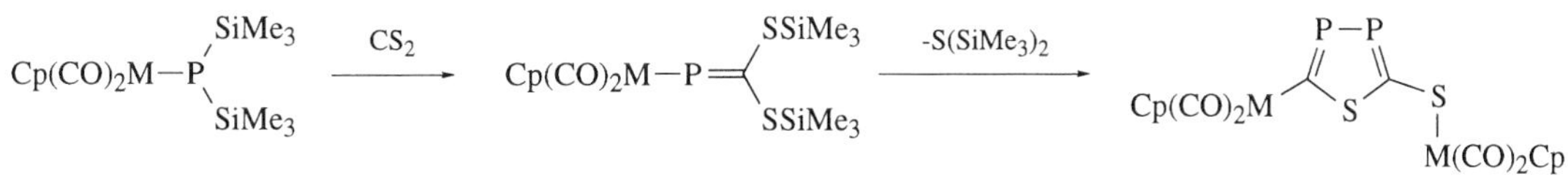

(**63**) (**64**) (**65**)

Scheme 52

4.3.6.11 1,2,3-Oxadiphospholes (15)

The dimer of 1,2,3-benzoxadiphosphole (**67**) is formed as one of the products from the thermal decomposition of the 2-phosphinophenol (**66**). Its dimeric structure is shown by an X-ray analysis ⟨98HAC183⟩.

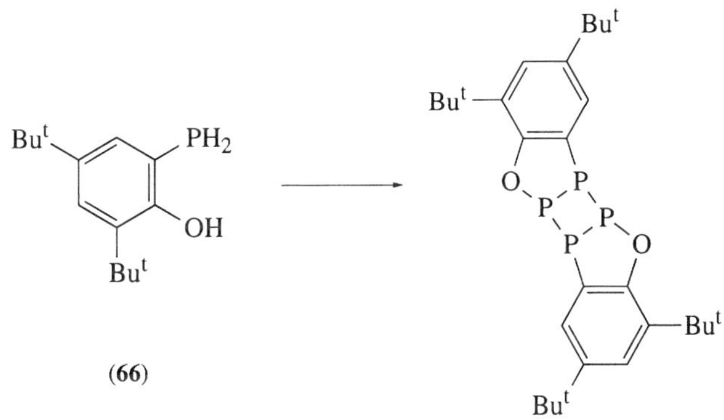

(**66**) (**67**)

4.3.7 OLIGOPHOSPHOLES, DIPHOSPHAARSOLES, AND DIPHOSPHASTIBOLES

4.3.7.1 Introduction

This section covers five-membered rings with three, four and five phosphorus atoms but no additional heteroatoms such as nitrogen or chalcogen. These rings of the general formula $RP_n(CH)_{5-n}$ are strictly speaking therefore no heterophospholes, but they are closely related to them; in particular they continue the series which had been started for $n = 2$ with the 1,2-diphospholes in Section 4.3.2.6 and the 1,3-diphospholes in Section 4.3.3.7. Furthermore, one of the phosphorus atoms of the ring may be replaced by an arsenic or antimony atom. The known systems are shown in Table 13. They attain full cyclic conjugation if all phosphorus atoms (and may be the arsenic or antimony atom) are two-coordinate. The ring consequently must therefore be negatively charged as in the 1,2,4- and 1,2,3-triphospholides (**1**), (**5**), 1,2,4- and 1,4,2-diphosphaarsolides (**2**), (**3**), 1,4,2-diphosphastibolides (**4**), tetraphospholides (**6**) and pentaphospholide (**7**). Neutral oligophospholes contain a tricoordinate phosphorus atom; its pyramidality decreases and hence the aromaticity of the oligophosphole increases with the number of dicoordinate phosphorus atoms in the ring ⟨98IC4413⟩. As a consequence, a variety of stable 1*H*-1,2,4-triphospholes could be prepared (Section 4.3.7.3).

The chemistry of polyphospholide anions is covered in recent reviews ⟨93CRV1623, 94CCR(137)1, 95CCR(145)201, 95CSR319, B-98MI1, 99MI257⟩. A review on "Rings with Phosphorus Carbon Multiple Bonds" ⟨B-89MI195⟩ also includes tri- and tetraphospholes. A recent review contains synthetic routes to and structures of cage compounds derived from triphosphole rings ⟨99MI257⟩.

Pentaphospholide complexes are contained in further reviews ⟨87CIC1, 90AG1137, 90ZC41, 92CCR259, 99ACR751, 99PS(144-146)141, 00CIUZ374⟩.

Table 13. Known oligophospholides/oligophospholes and diphosphaarsolides/stibolides

1,2,4-triphospholides / phospholes

(**1**)

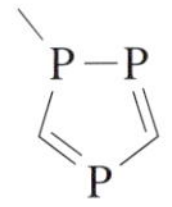

1,2,4-diphosphaarsolides

(**2**)

1,4,2-diphosphaarsolides

(**3**)

1,4,2diphosphastibolides

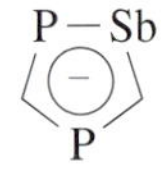

(**4**)

1,2,3-triphospholides

(**5**)

tetraphospholides

(**6**)

pentaphospholide

(**7**)

Phosphorus analogues of the cyclopentadienide anion, i.e. phospholide, 1,2-diphospholide, 1,3-diphospholide, 1,2,4-triphospholide (**1**), 1,2,3-triphospholide (**5**), tetraphospholide (**6**) and pentaphospholide (**7**) ions were examined by complete structural optimization using the all-valence electron SCF MO method SINDO1. Optimized geometries obtained with inclusion of d-orbitals on phosphorus show that the ring bond lengths CC, CP, and PP are in the ranges 137–142 pm, 174–177 pm, and 217–220 pm, respectively. These values are between those of the bond lengths of the respective single and double bonds and are in close agreement with the observed values of the ligated rings in complexes. The bond-order definition of aromaticity, based on the ring-current concept, reveals that the phosphorus analogues of cyclopentadienide anion in the above order retain 91, 86, 88, 96, 86, 95 and 100% aromaticity relative to the carbocyclic system $C_5H_5^-$. A rather dramatic result is that the pentaphosphorus analogue P_5^- is as aromatic as the carbocyclic anion ⟨92JOC3694⟩. Its structure and bonding has been discussed earlier ⟨84CJC341, 89AG500, 89CB2121, 90PAC423⟩.

As has been mentioned above, the aromaticity of oligophospholes becomes higher as the number of dicoordinate phosphorus atoms in the ring increases. The change is correlated with the bond angle sum about the tricoordinate phosphorus atom ⟨98IC4413⟩. The inversion barrier about this phosphorus atom is strongly reduced by incorporating two dicoordinate phosphorus atoms into the ring ⟨96JPC6194⟩.

4.3.7.2 1,2,4-Triphospholides (1)

4.3.7.2.1 Synthesis

The phosphaalkyne Bu^{t}C≡P or its precursor, the phosphaalkene Bu^t(Me$_3$SiO)C=PSiMe$_3$, reacts in DME with lithium bis(trimethylsilyl)phosphide, LiP(SiMe$_3$)$_2$, to give the 3,5-di-*t*-butyl-1,2-bis(trimethylsilyl)-dihydro-1,2,4-triphospholide (**8**) which spontaneously loses the trimethylsilyl groups. The resulting DME solvated lithium 1,2,4-triphospholide (**9**) is isolated in yellow crystals ⟨85MI55, 87CC1146, B-96MI161⟩. With phenyl in place of *t*-butyl groups as 3,5-substituents the dihydro-1,2,4-triphospholide is stable and its DME solvated lithium salt can be isolated in dark green crystals ⟨D-96MI161⟩.

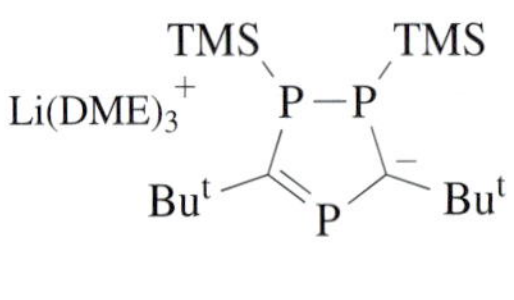

(**8**)

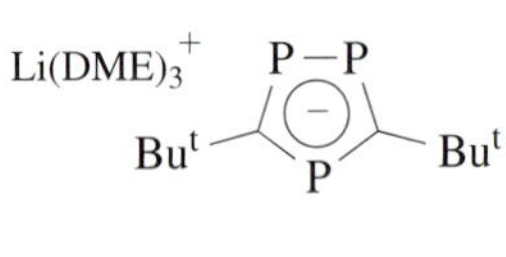

(**9**)

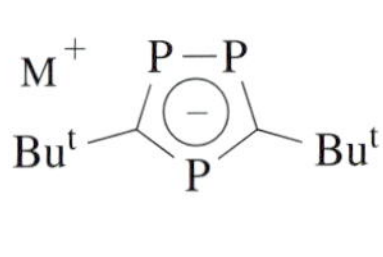

(**10**)

Reduction of the *t*-butyl-phosphaalkyne directly with sodium amalgam ⟨89PO2407⟩ or mediated by a reduced tantalum compound ⟨89PO849⟩ leads to an equimolar mixture of the sodium salt of the

1,3-diphospholide (Section 4.3.3.7) and of the 1,2,4-triphospholide (**10**), M = Na. The two salts can be separated by fractional crystallization. Starting with a mixture of $Pr^iC{\equiv}P$ and $Bu^tC{\equiv}P$ a complex mixture of differently substituted di- and triphospholides results ⟨91JOM(415)C15⟩. In place of the phosphaalkyne $Bu^tC{\equiv}P$, the phosphaalkene $Bu^t(Me_3SiO)C{=}PSiMe_3$ can also be used and reduced with sodium metal in DME ⟨96BSB675⟩. The potassium salt (**10**), M = K, can also be prepared ⟨99OM793⟩.

The lithium salt of the 1,1,5,5-tetrakis(trimethylsilyl)-2,3,4-triphospha-pentadienide anion (**11**) rearranges in DME above 100ºC (presumably *via* the loss and [2 + 3]cycloaddition of $Me_3SiC{\equiv}P$) to form the lithium 1,2,3,5-tetrakis(trimethylsilyl)-dihydro-1,2,4-triphospholide (**12**). In air it is oxidized to give the lithium salt of the 3,5-bis(trimethylsilyl)-1,2,4-triphospholide anion (**13**) which is isolated as yellow microcrystalline solid ⟨96AG354⟩.

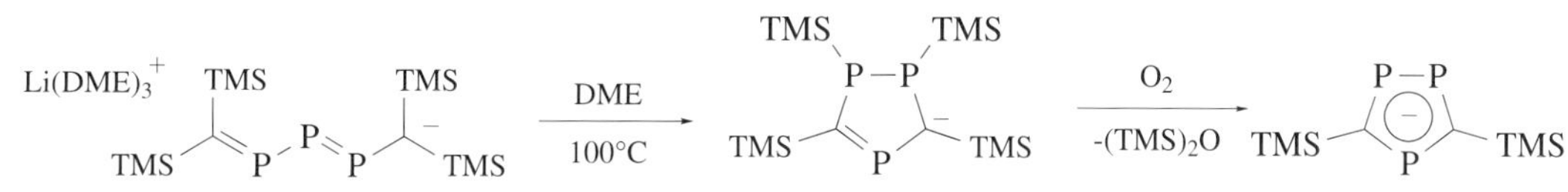

(**11**) (**12**) (**13**)

A bis(triphenylphosphonio)-substituted and consequently cationic 1,2,4-triphospholide (**15**) results from the reduction of the respective trihalo-1,2,4-triphospholane (**14**) which in turn is obtained from the condensation of the triphenylphosphonium bis(trimethylsilyl)methylide with phosphorus trichloride or tribromide. The chloride, bromide and triflate of the triphospholide (**15**) are isolated in yellow crystals ⟨96ZN(B)1022⟩.

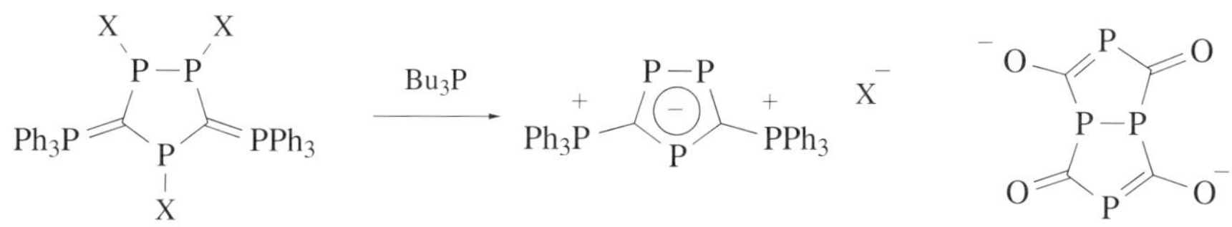

(**14**) (**15**) (**16**)

The phosphorus analogue PCO^- of the cyanate ion on oxidation by SO_2 or I_2 forms a tetrameric dianion (**16**) in which two dihydro-1,2,4-triphospholonolate rings are fused. It is isolated as DME and THF solvated lithium salt. The molecule is folded along the P–P bond ⟨95ZAAC(621)34, B-96MI161⟩.

4.3.7.2.2 Reactivity

The protonation of the 3,5-di-*t*-butyl-1,2,4-triphospholide ion (**10**) takes place at a carbon ring member. The resulting 1,2,4-triphospha-cyclopenta-1,3-diene (**17**) dimerizes *via* successive [4 + 2] and [2 + 2] cycloadditions to give a dihomocubane-structured cage molecule (**18**) containing five directly bonded phosphorus atoms and no C–C bond (Scheme 53) ⟨89CC1046, 90PS(49/50)463, 97JOM(536-537)273⟩. The triphospha-cyclopentadiene (**17**) is stable as η^4-ligand in a cobalt complex ⟨88CC819⟩. If a mixture of the 1,3-diphospholide (Section 4.3.3) and 1,2,4-triphospholide is protonated, the subsequent combination of the two phospha-cyclopentadienes ends with the [4 + 2] cycloaddition ⟨97JOM(536-537)273⟩.

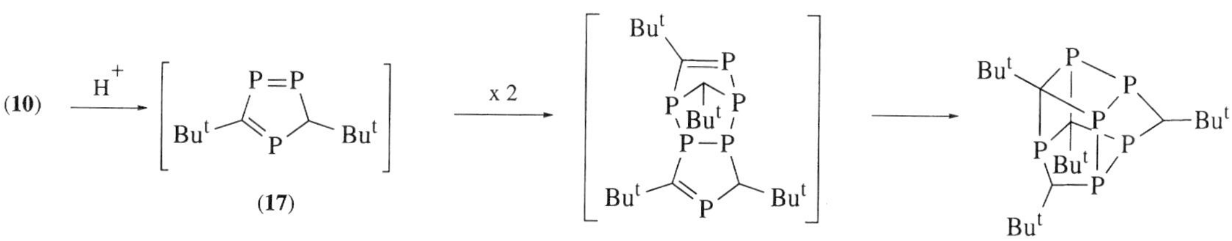

(**18**)

Scheme 53

In contrast to the protonation, alkylation with bis(trimethylsilyl)bromomethane and stannylation of the 1,2,4-triphospholide ion (**10**) takes place at P-1 and affords stable 1*H*-1,2,4-triphospholes (**19**) and (**24**) (Section 4.3.7.3.1).

The oxidative coupling of the 1,2,4-triphospholide ion (**10**) with the 1,3-diphospholide ion $(Bu^tC)_3P_2^-$ (Section 4.3.3.7) by $FeCl_3$ or $CoBr_2$ results in a hexacyclic neutral cage product $(Bu^tC)_5P_5$ ⟨89JOM(375)C31, 90PS(49/50)463⟩.

Several types of 1,2,4-triphospholide transition metal complexes are known, with η^1-coordination of P-1 (in rare cases of P-4) to one or two metal centers and sometimes additional η^4-coordination of the remaining four ring members, with η^2-coordination of P-1,2, as well as with η^5-coordination and sometimes additional η^1-coordination of P-1 and P-2 or P-4, or additional η^2-coordination of P-1,2 ⟨94CCR(137)1, 95CCR(145)201, 96CC441, 96JA7630, 96JCS(D)739, 96JOM(512)141, 97CC1739, 98ZAAC(624)399, 99JOM(584)87, 00AG2174, 00MI(11)405, 00OM1713⟩. In a cationic Fe–Fe triple-decker complex, the 1,2,4-triphospholide also serves as the central $\eta^5:\eta^5$ ligand of ⟨97MI309⟩. η^1-Complexes are fluxional in solution as the metal atom undergoes an intramolecular shift between P-1 and P-2. In most of the η^1-complexes of the 3,5-di-*t*-butyl-1,2,4-triphospholide the metal atom is coplanar with the ring and the sum of bond angles around the ligating phosphorus atom P-1 is 360° ⟨88CC1615, 89NJC353, 93JOM(462)319, 95CC1661⟩.

Sometimes, however, the metal atom deviates from the plane of the ring and the arrangement about P-1 is pyramidal with a sum of angles equal to 348° ⟨89NJC353⟩.

η^5-Complexes of the di-*t*-butyl-1,2,4-triphospholide encompass Ti, Yb, V, Cr, Mo, W, Mn, Fe, Ru, Co, Rh, Ir and Ni, as well as Sr, In, Tl, Sn and Pb derivatives. Most of the complexes are prepared from an alkali metal 1,2,4-triphospholide (**10**) ⟨89JOM(375)C31, 90CC472, 90CC1307, 94CCR(137)1, 95CCR(145)201, 97CC1739, 99CC1731, 99JCS(D)4057, 99OM793, 00CC2027, 00JCS(D)1715, 00JCS(D)3534, 01JCS(D)1013⟩ or from a 1-stannyl-1,2,4-triphosphole (**24**) (Section 4.3.7.3) ⟨99CEJ3143, 00AG2174, 00OM4283⟩, but some have also been obtained from the phosphaalkyne $Bu^tC{\equiv}P$ by metal vapor synthesis ⟨87JOM(334)C35, 94MI369, 94MI385, 95CCR(145)201, 96JA7630, 99CC1731, 99OM793⟩ or from reaction with a metal complex ⟨97MI309, 98ZAAC(624)399⟩$^tC{\equiv}P$ ⟨99CC1731⟩.

4.3.7.3 1,2,4-Triphospholes

4.3.7.3.1 Synthesis

The alkylation of the 1,2,4-triphospholides (**10**) and (**13**) with bis(trimethylsilyl)bromomethane gives the triphospholes (**19**) and (**20**) which are isolated as orange and yellow crystals ⟨95CC1661, 98AG1139⟩. The arrangement about P-1 is pyramidal in (**19**) with a sum of angles of 342° ⟨95CC1661⟩ and planar in (**20**) with a sum of angles of 359° ⟨98AG1139⟩. The triphosphole (**19**) with a tri-*t*-butylphenyl in place of the bis(trimethylsilyl)methyl substituent is also predicted to possess a planar P-1 coordination ⟨99JOM(588)28⟩.

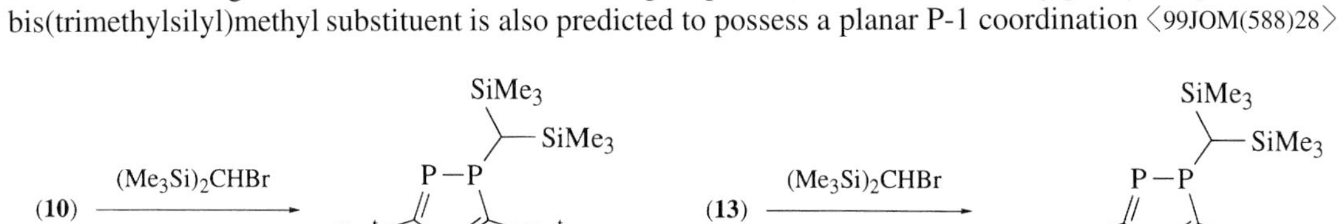

(**19**) (**20**)

The reaction of the carbene (**21**) with *t*-butyl-phosphaethyne gives the 1*H*-1,2,4-triphosphole (**22**) for which also a planar coordination of the tricoordinate P-1 is found ⟨00AG2393⟩. A product analogous to (**22**) results from the ring contraction of the 2,4,6-tri-tert.butyl-1,3,5-triphosphinine induced by a

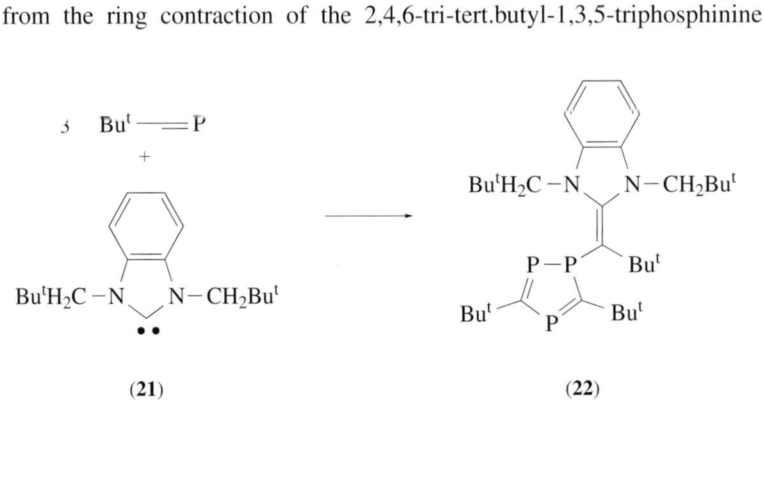

(**21**) (**22**)

monocyclic carbene. The three-coordinate phosphorus atom in its η^5-ligated Mo(CO)3 complex is also planar ⟨00CC1305⟩.

The reaction of (**10**), M = Na, with the chlorostannanes (**23**) affords the 1-triorganylstannyl-1,2,4-triphospholes (**24**), R = Me, Bu^n, Ph, in yellow crystals. In case of R = Ph the tricoordinate P-1 deviates only a little from the plane of the ring, but shows a pyramidal coordination with a sum of bond angles of 310°. In solution, the $SnPh_3$ group rapidly shifts between P-1 and P-2 ⟨99CEJ3143, 99PS(144-146)725⟩.

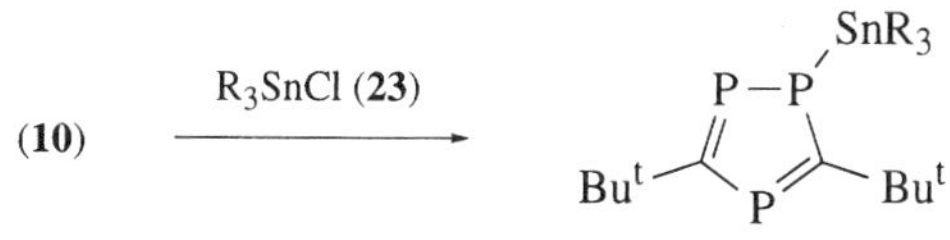

(**24**)

4.3.7.3.2 *Reactivity*

On standing in sunlight at room temperature the 1,2,4-triphosphole (**19**) slowly electrocyclizes to give the triphosphabicyclo[2.1.0]pentene (**25**), R = $CH(SiMe_3)_2$, which undergoes a rapid [1,3]-phosphorus migration ⟨95CC2395, 98CC1537⟩. The process can be stopped in (**26**) by ligation with tungsten pentacarbonyl (Scheme 54).

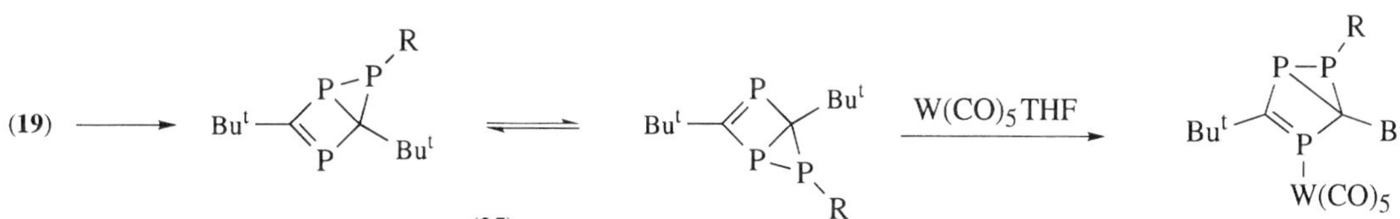

(**25**)

(**26**)

Scheme 54

The triphosphole (**19**) reacts with elemental sulfur and selenium in different ways. From the reaction with sulfur the pentasulfide (**27**) has been isolated, while the reaction with selenium gave the 1,2,4-selenadiphosphole (**28**) (Section 4.3.6.8) ⟨96BSB675⟩. Addition of methyl lithium gives the triphosphacyclopentenyl lithium (**29**) (Scheme 55).

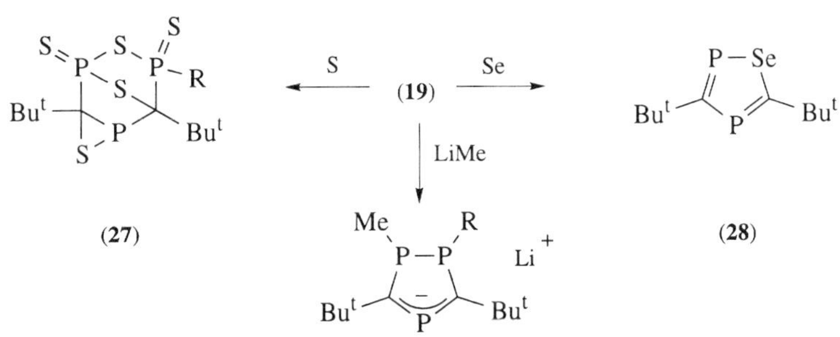

(**27**)

(**28**)

(**29**)

Scheme 55

The 1,2,4-triphosphole (**19**) forms a variety of transition and main group metal complexes. It gives 1 : 1- and 1 : 2-*trans*-$PtCl_2PEt_3$ complexes in which the metal is attached to P-4 in the plane of the ring and in the second case also to P-1 ⟨95CC1661, 98HAC1⟩. η^5-Complexes of (**19**) are known with Cr, Mo, W ⟨97CC1305⟩, Fe ⟨95CC1661⟩, Ru ⟨96CC2751, 97PS(124-125)567, 99JOM(584)58⟩. In the binuclear complex (**19**)$Co_2(CO)_5$ the triphosphole acts as a six-electron donor ⟨00JOM(601)335⟩.

The 1-triphenylstannyl-triphosphole (**24**), R = Ph, adds two moles of the *t*-butyl-phosphaethyne to give the tetracyclic compound (**30**) in a chemoselective and diastereospecific reaction. With CpCo it forms an η^4-complex not involving P-1 ⟨99CEJ3143⟩. More general, however, the 1-triorganylstannyl-1,2,4-triphopholes (**24**) can serve as starting compounds for η^5-triphospholyl complexes as has been demonstrated by the synthesis of hexaphospha-stannocene ⟨99CEJ3143⟩ and hexaphospha-manganocene ⟨00AG2174⟩ as well as of η^5-complexes of Co and Ni ⟨00OM4283⟩.

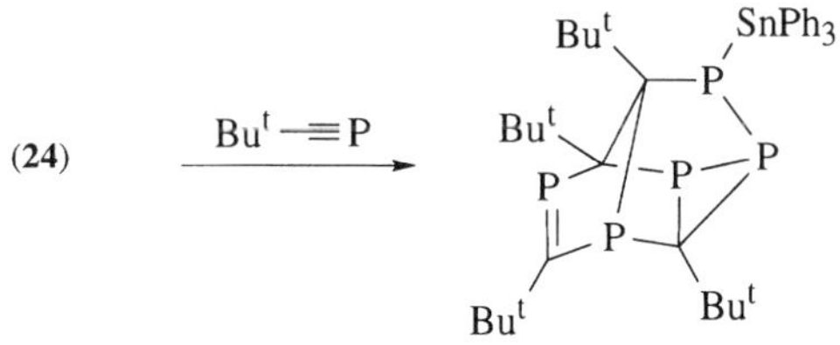

(30)

4.3.7.4 1,2,4- and 1,4,2-Diphosphaarsolides (2), (3)

The reaction of the lithium bis(trimethylsilyl)arsenide (**33**) with *t*-butyl-phosphaacetylene (**31**) ⟨93PS(77)45, 94JOM(480)45⟩ or a phosphaalkene precursor thereof (**32**) leads to a mixture of lithium 1,2,4-triphospholide (**9**), 1,2,4-diphosphaarsolide (**34**) and 1,4,2-diphosphaarsolide (**35**) (Scheme 56). Attempts to extend this reaction to the preparation of a phosphadiarsolide ring system led to the tetraarsacubane instead ⟨93AG86⟩.

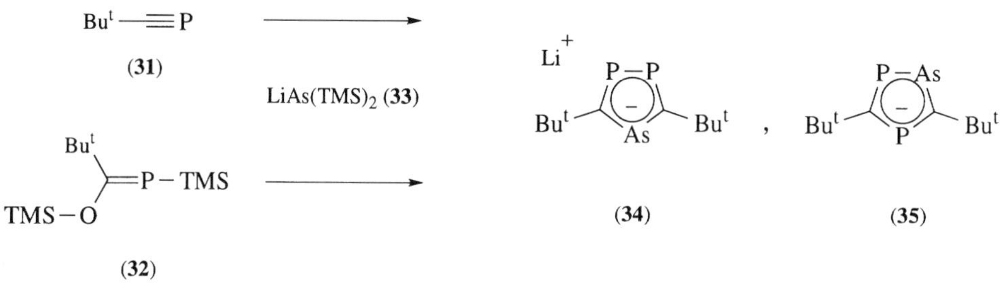

Scheme 56

The 1,2,4-diphosphaarsolide ion (**34**) has been used as η^5-ligand in Mn-, Fe-, Ru and Mo-complexes ⟨93PS(77)45, 93PS(77)275, 94JOM(480)45⟩ and as middle deck of a triple-decker ruthenium complex ⟨95OM4382⟩. η^5-Sandwich and triple-decker complexes of the 1,4,2-diazaphospholide ion (**35**) are also known ⟨94JOM(480)45, 95CCR(145)201, 95OM4382⟩.

4.3.7.5 1,4,2-Diphosphastibolides (4)

Lithium 3,5-di-*t*-butyl-1,4,2-diphosphastibolide (**36**) has been prepared in analogy to (**35**) (Scheme 56) using $LiSb(SiMe_3)_2$. In this case the reaction is regioselective, yielding only the 1,4,2-isomer (**36**) ⟨97JOM(527)291, 00OM1713⟩. In its *trans*-complex with $PtCl(PEt_3)_2$ (**36**) is η^1 ligated through P-1. Treatment of *cis*-$PtCl_2(PEt_3)_2$ with two equivalents initiates a ring-coupling reaction of (**36**) ⟨97JCS(D)4321⟩. From (**36**) sandwich and triple-decker complexes of Fe and Ru have been obtained ⟨96CC1591, 97JOM(527)291⟩. In a secondary co-ordination $W(CO)_5$ is η^1 ligated to P-1 of such complexes ⟨97JCS(D)2183⟩. More η^5 complexes of (**36**) have been prepared of Co, Rh and Ir. The Co complexes also contain an η^4 coordinated diphosphastibacyclopentadiene ligand originating from the anion (**36**) by protonation at C-3 ⟨97JOM(534)89⟩. An anionic Yb complex contains one η^2 (P-1, Sb-2) coordinated and two η^5 coordinated diphosphastibolide ligands ⟨00OM1713⟩. The reaction of (**36**) with TlCl affords a polymeric product with $\eta^5 : \eta^5$ coordination ⟨98OM3826⟩.

Two equivalents of the diphosphastibolide (**36**) react with $FeCl_3$ or $PbCl_2$ to produce the tetracyclic compound (**37**) which rearranges when forming the $W(CO)_5$ complex (**38**) (Scheme 57) ⟨97CC305, 97JCS(D)4321, 99JCS(D)4057⟩.

The reaction of (**36**) with $(Et_2NCS_2)_2E$, E = Se, Te, leads to a 1,2,4-selena and 1,2,4-telluradiphosphole ⟨99TL3815⟩ (Section 4.3.6.8).

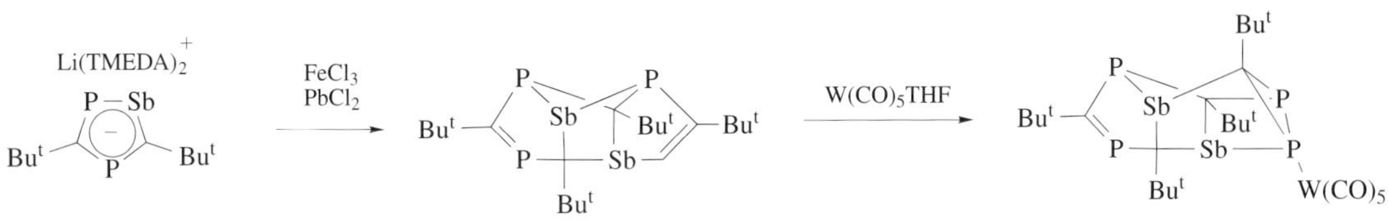

(36) (37) (38)

Scheme 57

4.3.7.6 1,2,3-Triphospholides (5)

The parent 1,2,3-triphospholide was first detected as a trace product in the reaction of P_4 with sodium in diglyme ⟨90ZN(B)1139⟩. A specific synthesis of the 4,5-diphenyl-1,2,3-triphospholide (**40**) starts from the tetraphenyl-dihydro-1,2-diphosphet (**39**) and involves several steps (Scheme 58). In the last step the central P–P bond of the bis(dihydrotriphospholyl) is cleaved followed by subsequent cleavage of the P–Ph bonds. The potassium salt is solvated by DME and isolated as a white powder ⟨93CC267, 93PS(77)254, 94BSF397⟩. The attempted synthesis of the 1,2,3-triphosphindolide ion failed ⟨97BSF853⟩.

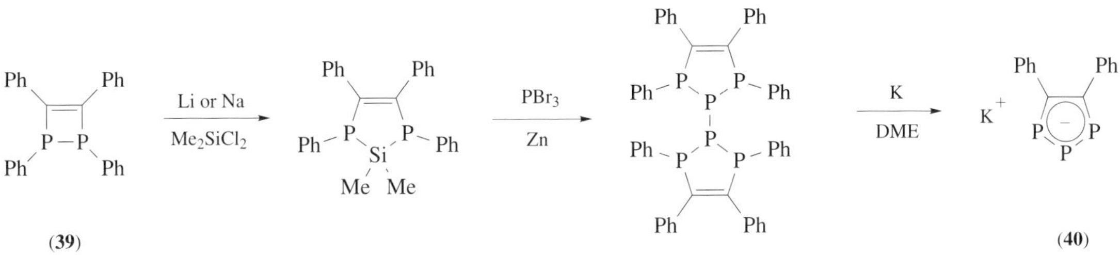

(39) (40)

Scheme 58

The 4,5-diphenyl-1,2,3-triphospholide ion coordinates molybdenum and tungsten pentacarbonyl in two steps to P-1,3 and to P-2 ⟨94BSF397⟩. With $FeCl_2$ a sandwich complex has been obtained ⟨93PS(77)254⟩. In the reaction of a P_4 iron complex with tolane the 1,2,3-triphospholyl ring is formed as part of a sandwich complex ⟨00AG1484⟩.

4.3.7.7 Tetraphospholides and Pentaphospholide (6), (7)

The sodium 1,2,3-triphospholide (Section 4.3.7.6), tetraphospholide (**6**) and pentaphospholide (**7**) are formed together with polycyclic sodium phosphides as the minor products from elemental white phosphorus with sodium (P/Na ratio = 3) in boiling diglyme; the carbon ring members originate from the solvent ⟨87ZAAC(544)87, 90ZN(B)1139⟩. The pentaphospholide (but not the tri- or tetraphospholide) is also accessible in tetrahydrofuran from the reaction of white phosphorus with lithium phosphide ⟨87ZAAC(544)87⟩ or with sodium metal or sodium phosphide in the presence of 18-crown-6 ⟨88AG288, 89ZN(B)381⟩ or in boiling dimethylformamide from red phosphorus with potassium phosphide ⟨91CB1159⟩. The solutions of alkali pentaphospholides MP_5 are gold-orange to red in color and are extremely sensitive to oxidation.

Calculations show that the pentaphosphole P_5H resulting from the protonation of (**7**) is planar. In spite of its significant aromatic stabilization, it is, however, unstable relative to dimeric and polymeric products ⟨B-95MI1, 96IC4690, 96JPC13447⟩. The same is true for alkylpentaphospholes. The pentaphospholide ion (**7**) reacts with alkyl halides RBr and RI, but the alkylpentaphospholes RP_5 which are to be assumed as the primary products rearrange to alkylpolyphosphines such as R_3P_7 and R_3P_9 ⟨88AG288⟩. The trimethylsilyl pentaphosphole is known as a complex ligand (see below).

A series of anionic and neutral η^5-pentaphospholyl metal tricarbonyl complexes $[\eta^5\text{-}P_5M(CO)_3]^-$, M = Cr, Mo, W, and $\eta^5\text{-}P_5Mn(CO)_3$ have been prepared from alkali pentaphospholides. $K(DMF)_3[\eta^5\text{-}P_5W(CO)_3]$ is silylated with Me_3SiCl to give the $W(CO)_3$ complex of trimethylsilyl-pentaphosphole Me_3SiP_5 ⟨91AG590⟩. The decaphosphatitanocene dianion has been obtained in $[K(18\text{-crown-}6)_2]_2Ti(P_5)_2$ from $TiCl_4$ and P_4 by reduction ⟨01MI2⟩.

Of the pentaphospholide ion, only η^5-coordination is known, sometimes with additional η^1- or η^2-coordination ⟨90AG1137, 95AG1454⟩. A characteristic feature is the formation of triple-decker complexes with P_5 as the middle deck ⟨89CB2049, 94CCR(137)1, 97CB1299, 98RCB1583, 99OM2833, 99RCB1362⟩. Their electronic structure has been explored ⟨89JA2030⟩.

Pentaphospholide sandwich complexes of Fe, Ru and Cr are obtained from white phosphorus and the appropriate substrate complexes ⟨87AG59, 87CIC1, 88CB935, 90AG1137, 95AG1454⟩ and also directly from LiP_5, LiC_5Me_5 and $FeCl_2$ ⟨88AG288⟩. A third approach involves a PCl_3 complex as the building block for the η^5-P_5 unit ⟨93AG641, 97CB1299⟩. The triple-decker complexes of the type $(\eta^5\text{-CpCr})_2P_5$ have been prepared from white phosphorus ⟨86AG349, 90JCS(D)977⟩, from a P_2 chromium complex ⟨99OM2833⟩ and from a PCl_3 chromium complex ⟨97CB1299⟩. The complex Cp^*FeP_5 can be used to transfer the P_5 ring or fragments thereof ⟨98CEJ1910, 99PS(144-146)141⟩.

4.3.8 REFERENCES

66AG907 G. Märkl; *Angew. Chem.*, 1966, **78**, 907; *Angew. Chem. Int. Ed. Engl.*, 1966, **5**, 846.

67CHE601 N.P. Ignatova, N.N. Melnikov and N.I. Shvetsov-Shilovskii; *Chem. Heterocycl. Compd.*, 1967, **3**, 601.

68DOK95 A.F. Vasilev, L.V. Vilkov, N.P. Ignatova, N.N. Melnikov, V.V. Negrebetskii, N.I. Shvetsov-Shilovskii and L.S. Khaikin; *Dokl. Akad. Nauk SSSR*, 1968, **183**, 95.

69AG707 J. Ebeling and A. Schmidpeter; *Angew. Chem.*, 1969, **81**, 707; *Angew. Chem. Int. Ed. Engl.*, 1969, **8**, 674.

69CB669 F.W. Kröck and F. Kröhnke; *Chem. Ber.*, 1969, **102**, 669.

70ZOB1501 N.I. Shvetsov-Shilovskii, N.P. Ignatova and N.N. Melnikov; *Zh. Obshch. Khim.*, 1970, **40**, 1501; *J. Gen. Chem. USSR*, 1970, **40**, 1488.

71DOK1081 L.V. Vilkov, L.S. Melnikov, V.V. Negrebetskii and N.I. Shvetsov-Shilovskii; *Dokl. Akad. Nauk SSSR*, 1971, **197**, 1081.

71IZV2611 R.G. Kostyanovskii, V.G. Plekhanov, N.P. Ignatova, R.G. Bobkova and N.I. Shvetsov-Shilovskii; *Izv. Akad. Nauk SSSR, Ser. Khim.*, 1971, 2611; *Bull. Acad. Sci. USSR, Div. Chim. Sci.*, 1971, 2486.

71ZSK798 V.V. Negrebetskii, A.V. Kessenikh, A.F. Vasilev, N.P. Ignatova, N.I. Shvetsov-Shilovskii and N.N. Melnikov; *Zh. Strukt. Khim.*, 1971, **12**, 798.

72JPR806 A.F. Vasilev, L.V. Vilkov, N.P. Ignatova, N.N. Melnikov, V.V. Negrebetskii, N.I. Shvetsov-Shilovskii and L.S. Khaikin; *J. Prakt. Chem.*, 1972, **314**, 806.

72ZOB1939 N.I. Shvetsov-Shilovskii, N.P. Ignatova, R.G. Bobkova, V.Ya. Manyukhina and N.N. Melnikov; *Zh. Obshch. Khim.*, 1972, **42**, 1939; *J. Gen. Chem. USSR*, 1972, **42**, 1932.

73DOK1101 G. Andrianov, Yu.T. Struchkov, N.I. Shvetsov-Shilovskii, N.P.V. Ignatova, R.G. Bobkova and N.N. Melnikov; *Dokl. Akad. Nauk SSSR*, 1973, **211**, 1101.

73S469 C.G. Stuckwisch; *Synthesis*, 1973, 469.

73ZN(B)363 M. Baudler, A. Moog, K. Glinka and U. Kelsch; *Z. Naturforsch. B*, 1973, **28**, 363.

74CR(C278)355 Y. Charbonnel and J. Barrans; *C.R. Acad. Sci. Paris, Ser. C*, 1974, **278**, 355.

74P197 L. Dulog, F. Nierlich and A. Verhelst; *Phosphorus*, 1974, **4**, 197.

75CB820 A. Schmidpeter and J. Luber; *Chem. Ber.*, 1975, **108**, 820.

76AG91 J. Luber and A. Schmidpeter; *Angew. Chem.*, 1976, **88**, 91; *Angew. Chem. Int. Ed. Engl.*, 1976, **15**, 111.

76CC887 J. Luber and A. Schmidpeter; *Chem. Commun.*, 1976, 887.

76JA5451 A.J. Ashe, R.R. Sharp and J.W. Tolan; *J. Am. Chem. Soc.*, 1976, **98**, 5451.

76JPR(318)409 A. Tzschach and J. Heinicke; *J. Prakt. Chem.*, 1976, **318**, 409.

76T2039 Y. Charbonnel and J. Barrans; *Tetrahedron*, 1976, **32**, 2039.

76ZOB590 R.G. Bobkova, N.P. Ignatova, N.I. Shvetsov-Shilovskii, V.V. Negrebetskii and A.F. Vasilev; *Zh. Obshch. Khim.*, 1976, **46** 590; *J. Gen. Chem. USSR*, 1976, **46**, 588.

77AG554 A. Schmidpeter, J. Luber and H. Tautz; *Angew. Chem.*, 1977, **89**, 554; *Angew. Chem. Int. Ed. Engl.*, 1977, **16**, 546.

77CB1124 A. Schmidpeter and J. Luber; *Chem. Ber.*, 1977, **110**, 1124.

77IZV532 V.V. Negrebetskii, L.Ya. Bogelfer, A.V. Vasilev, R.G. Bobkova, N.P. Ignatova, N.I. Shvetsov-Shilovskii and N.N. Melnikov; *Izv. Akad. Nauk SSSR, Ser. Khim.*, 1977, 532.

77ZN(B)841 A. Schmidpeter, M. Junius, J.H. Weinmaier, J. Barrans and Y. Charbonnel; *Z. Naturforsch. B*, 1977, **32**, 841.

77ZN(B)845 A. Schmidpeter, J. Luber and Th. von Criegern; *Z. Naturforsch. B*, 1977, **32**, 845.

77ZOB576 R.G. Bobkova, N.P. Ignatova, N.I. Shvetsov-Shilovskii, N.N. Melnikov, V.V. Negrebetskii, L.Ya. Bogelfer, S.F. Dymova and A.F. Vasilev; *Zh. Obshch. Khim.*, 1977, **47**, 576; *J. Gen. Chem. USSR*, 1977, **47**, 527.

78CB1558 P. Friedrich, G. Huttner, J. Luber and A. Schmidpeter; *Chem. Ber.*, 1978, **111**, 1558.

78CR(C)319 J.-P. Legros, Y. Charbonnel, J. Barrans and J. Galy; *C.R. Acad. Sci. Paris, Ser. C*, 1978, **286**, 319.

78DOK358 B.A. Arbuzov, V.D. Cherepinskii-Malov, E.N. Dianova, A.I. Gusev and V.A. Sharapov; *Dokl. Akad. Nauk SSSR*, 1978, **243**, 358.

78JCS(D)1465 K.B. Dillon, R.N. Reeve and T.C. Waddington; *J. Chem. Soc., Dalton Trans.*, 1978, 1465.

78JOM(154)1 J. Heinicke and A. Tzschach; *J. Organomet. Chem.*, 1978, **154**, 1.

78S557 J.-P. Majoral; *Synthesis*, 1978, 557.

78TL441 K. Issleib, R. Vollmer, H. Oehme and H. Meyer; *Tetrahedron Lett.*, 1978, **19**, 441.

78ZSK64 V.V. Negrebetskii, L.Ya. Bogelfer, R.G. Bobkova, N.P. Ignatova and N.I. Shvetsov-Shilovskii; *Zh. Strukt. Khim.*, 1978, **19**, 64; *J. Struct. Chem. USSR*, 1978, **19**, 52.
79AG442 J.H. Weinmaier, J. Luber, A. Schmidpeter and S. Pohl; *Angew. Chem.*, 1979, **91**, 442; *Angew. Chem. Int. Ed. Engl.*, 1979, **18**, 411.
79CB3159 S. Pohl; *Chem. Ber.*, 1979, **102**, 3159.
79DOK1117 B.A. Arbuzov, E.N. Dianova and Yu.Yu. Samitov; *Dokl. Akad. Nauk SSSR*, 1979, **244**, 1117.
79DOK1150 B.A. Arbuzov, V.D. Cherepinskii-Malov, E.N. Dianova, A.I. Gusev and V.A. Sharapov; *Dokl. Akad. Nauk SSSR*, 1979, **247**, 1150.
79JMR199 V. Galasso; *J. Magn. Reson.*, 1979, **34**, 199.
79NAR251 T.A. Khwaja and H. Pande; *Nucleic Acid Res.*, 1979, **7**, 251.
79TL447 C. Malavaud, M.T. Boisdon, Y. Charbonnel and J. Barrans; *Tetrahedron Lett.*, 1979, **20**, 447.
80CB2278 J.H. Weinmaier, G. Brunnhuber and A. Schmidpeter; *Chem. Ber.*, 1980, **113**, 2278.
80CR(C)271 J.-P. Legros, Y. Charbonnel and J. Barrans; *C.R. Acad. Sci. Paris, Ser. C*, 1980, **291**, 271.
80ICA(46)167 K.C. Dash, H. Schmidbaur and A. Schmidpeter; *Inorg. Chim. Acta*, 1980, **46**, 167.
80JOM(185)53 J.H. Weinmaier, H. Tautz, A. Schmidpeter and S. Pohl; *J. Organomet. Chem.*, 1980, **185**, 53.
80JOM(186)39 J. Heinicke, B. Raap and A. Tzschach; *J. Organomet. Chem.*, 1980, **186**, 39.
80TL1307 J.-P. Majoral, R. Kraemer, Th. N'Gando M'Pondo and J. Navech; *Tetrahedron Lett.*, 1980, **21**, 1307.
80TL3483 K. Issleib and R. Vollmer; *Tetrahedron Lett.*, 1980, **21**, 3483.
80ZC342 J. Heinicke and A. Tzschach; *Z. Chem.*, 1980, **20**, 342.
80ZN(B)1222 A. Schmidpeter and H. Tautz; *Z. Naturforsch. B*, 1980, **35**, 1222.
81AG420 A. Schmidpeter, H. Tautz, J. von Seyerl and G. Huttner; *Angew. Chem.*, 1981, **93**, 420; *Angew. Chem. Int. Ed. Engl.*, 1981, **20**, 408.
81CB825 H. Tautz and A. Schmidpeter; *Chem. Ber.*, 1981, **114**, 825.
81CC1131 R. Carrié, Y.Y.C. Yeung Lam Ko, F. de Sarlo and A. Brandi; *J. Chem. Soc., Chem. Commun.*, 1981, 1131.
81CR(C)187 N. Ayed, R. Mathis, F. Mathis and B. Baccar; *C.R. Acad. Sci. Paris, Ser. C*, 1981, **292**, 187.
81IC1222 R.O. Day, R.R. Holmes, H. Tautz, J.H. Weinmaier and A. Schmidpeter; *Inorg. Chem.*, 1981, **20**, 1222.
81IZV1113 B.A. Arbuzov, E.N. Dianova and S.M. Sharipova; *Izv. Akad. Nauk SSSR, Ser. Khim.*, 1981, 1113.
81IZV1600 B.A. Arbuzov, E.N. Dianova, and S.M. Sharipova; *Izv. Akad. Nauk SSSR, Ser. Khim.*, 1981, 1600.
81JA2465 Y. Kobayashi, S. Fujino and I. Kumadaki; *J. Am. Chem. Soc.*, 1981, **103**, 2465.
81JCS(P2)1411 W.B. Jennings, D. Randall, S.D. Worley and J.H. Hargis; *J. Chem. Soc., Perkin Trans. 2*, 1981, 1411.
B-81MI413 C. Malavaud, L. Lopez, Th. N'Gando M'Pondo, M.T. Boisdon, Y. Charbonnel and J. Barrans; *ACS Symp. Ser.*, 1981, **171**, 413.
81TL4475 K. Issleib, E. Leissring and H. Meyer; *Tetrahedron Lett.*, 1981, **22**, 4475.
81ZAAC(475)211 A. Schmidpeter, H. Tautz and F. Schreiber; *Z. Anorg. Allg. Chem.*, 1981, **475**, 211.
81ZAAC(481)22 K. Issleib and R. Vollmer; *Z. Anorg. Allg. Chem.*, 1981, **481**, 22.
81ZN(B)1273 A. Schmidpeter and K. Karaghiosoff; *Z. Naturforsch. B*, 1981, **36**, 1273.
81ZOB475 Yu.V. Balitskii, V.V. Negrebetskii and Yu.G. Gololobov; *Zh. Obshch. Khim.*, 1981, **51**, 475.
81ZSK98 M.Yu. Antipin, Yu.T. Struchkov, Yu.V. Balitskii and Yu.G. Gololobov; *Zh. Strukt. Khim.*, 1981, **22**, 98.
82IC3916 R.O. Day, A. Schmidpeter and R.R. Holmes; *Inorg. Chem.*, 1982, **21**, 3916.
82IZV1196 B.A. Arbuzov, E.N. Dianova and I.E. Galeeva; *Izv. Akad. Nauk SSSR, Ser. Khim.*, 1982, 1196.
82IZV2718 I.A. Litvinov, Yu.T. Struchkov, B.A. Arbuzov, E.N. Dianova and I.E. Galeeva; *Izv. Akad. Nauk SSSR, Ser. Khim.*, 1982, 2718.
82IZV2730 B.A. Arbuzov, E.N. Dianova and Yu.Yu. Samitov; *Izv. Akad. Nauk SSSR, Ser. Khim.*, 1982, 2730.
82PS(14)49 A. Schmidpeter, J. Gross, E. Schrenk and W.S. Sheldrick; *Phosphorus Sulfur*, 1982, **14**, 49.
82TL853 M.R. Marre, M.T. Boisdon and M. Sanchez; *Tetrahedron Lett.*, 1982, **23**, 853.
82TL1263 H. Yamaguchi, S. Kametani, T. Karakasa, T. Saito and S. Motoki; *Tetrahedron Lett.*, 1982, **23**, 1263.
82TL3643 J. Heinicke and A. Tzschach; *Tetrahedron Lett.*, 1982, **23**, 3643.
83CB549 J. Högel, A. Schmidpeter and W.S. Sheldrick; *Chem. Ber.*, 1983, **116**, 549.
83CB1468 A. Schmidpeter, M. Nayibi, P. Mayer and H. Tautz; *Chem. Ber.*, 1983, **116**, 1468.
83CJC1105 Y.Y.C. Yeung Lam Ko, R. Carrié, F. de Sarlo and A. Brandi; *Can. J. Chem.*, 1983, **61**, 1105.
83DOK613 B.A. Arbuzov, E.N. Dianova and E.Ya. Zabotina; *Dokl. Akad. Nauk SSSR*, 1983, **273**, 613.
83IC3696 R.O. Day, A. Schmidpeter and R.R. Holmes; *Inorg. Chem.*, 1983, **22**, 3696.
83IZV2395 B.A. Arbuzov, A.F. Lisin and E.N. Dianova; *Izv. Akad. Nauk SSSR, Ser. Khim.*, 1983, 2395.
83JOM(256)375 J.G. Kraaijkamp, G. van Koten, K. Vrieze, D.M. Grove, E.A. Klop, A.L. Spek and A. Schmidpeter; *J. Organomet. Chem.*, 1983, **256**, 375.
83JOM(258)257 J. Heinicke, A. Petrasch and A. Tzschach; *J. Organomet. Chem.*, 1983, **258**, 257.
83PS(18)15 K. Issleib, E. Leissring and H. Schmidt; *Phosphorus Sulfur*, 1983, **18**, 15.
83T1507 Y.Y.C. Yeung Lam Ko, F. Tonnard and R. Carrié; *Tetrahedron*, 1983, **39**, 1507.
83TL2137 K. Karaghiosoff, J.-P. Majoral, A. Meriem, J. Navech and A. Schmidpeter; *Tetrahedron Lett.*, 1983, **24**, 2137.
83TL2645 G. Märkl and K. Hock; *Tetrahedron Lett.*, 1983, **24**, 2645.
83TL5481 J. Heinicke and A. Tzschach; *Tetrahedron Lett.*, 1983, **24**, 5481.
83ZAAC(501)57 M. Hesse and U. Klingebiel; *Z. Anorg. Allg. Chem.*, 1983, **501**, 57.
83ZC439 J. Heinicke and A. Tzschach; *Z. Chem.*, 1983, **23**, 439.
83ZN(B)1484 A. Schmidpeter and H. Klehr; *Z. Naturforsch. B*, 1983, **38**, 1484.

84AG898	W. Rösch and M. Regitz; *Angew. Chem.*, 1984, **96**, 898; *Angew. Chem. Int. Ed. Engl.*, 1984, **23**, 900.
84AG899	G. Märkl and I. Trötsch; *Angew. Chem.*, 1984, **96**, 899; *Angew. Chem. Int. Ed. Engl.*, 1984, **23**, 901.
84AG901	A. Schmidpeter and A. Willhalm; *Angew. Chem.*, 1984, **96**, 901; *Angew. Chem. Int. Ed. Engl.*, 1984, **23**, 903.
84CC183	L. Lopez, J.-P. Majoral, A. Meriem, Th. N'Gando M'Pondo, J. Navech and J. Barrans; *J. Chem. Soc., Chem. Commun.*, 1984, 183.
84CC508	J.M. Alcaraz and F. Mathey; *J. Chem. Soc., Chem. Commun.*, 1984, 508.
84CC1634	Y.Y.C. Yeung Lam Ko, R. Carrié, A. Muench and G. Becker; *J. Chem. Soc. Chem., Chem. Commun.*, 1984, 1634.
84CC1640	Y.Y.C. Yeung Lam Ko and R. Carrié; *J. Chem. Soc., Chem. Commun.*, 1984, 1640.
84CHEC-I(1)523	K. Dimroth; in "Comprehensive Heterocyclic Chemistry", eds. A.R. Katritzky and C.W. Rees, Pergamon, 1984, **Vol. 1**, p. 523.
84CJC43	C. Malavaud, Th. N'Gando M'Pondo, L. Lopez, J. Barrans and J.-P. Legros; *Can. J. Chem.*, 1984, **62**, 43.
84CJC341	N.C. Baird; *Can. J. Chem.*, 1984, **62**, 341.
84ICA(85)L33	J.G. Kraaijkamp, G. van Koten, K. Vrieze, D.M. Grove, G. Abbel, C.H. Stam and A. Schmidpeter; *Inorg. Chim. Acta*, 1984, **85**, L33.
84IZV1182	B.A. Arbuzov, E.N. Dianova and E.Ya. Zabotina; *Izv. Akad. Nauk SSSR, Ser. Khim.*, 1984, 1182.
84IZV2023	I.A. Litvinov, Yu.T. Struchkov, B.A. Arbuzov, E.N. Dianova and I.E. Galeeva; *Izv. Akad. Nauk SSSR, Ser. Khim.*, 1984, 2023.
84PS(20)333	M. Haddad, Th. N'Gando M'Pondo, C. Malavaud, L. Lopez and J. Barrans; *Phosphorus Sulfur*, 1984, **20**, 333.
84PS(20)347	J. Heinicke and A. Tzschach; *Phosphorus Sulfur*, 1984, **20**, 347.
84PS(21)97	W. Rösch and M. Regitz; *Phosphorus Sulfur*, 1984, **21**, 97.
84PS(21)335	N. Ayed, B. Baccar, F. Mathis and R. Mathis; *Phosphorus Sulfur*, 1984, **21**, 335.
84S591	W. Rösch and M. Regitz; *Synthesis*, 1984, 591.
84TL5521	O. Diallo, M.T. Boisdon, C. Malavaud, L. Lopez, M. Haddad and J. Barrans; *Tetrahedron Lett.*, 1984, **25**, 5521.
84ZAAC(519)75	K. Issleib, E. Leissring and M. Riemer; *Z. Anorg. Allg. Chem.*, 1984, **519**, 75.
84ZOB271	A.N. Chernega, M.Yu. Antipin, Yu.T. Struchkov, Yu.V. Balitskii, Yu.G. Gololobov and I.E. Boldeskul; *Zh. Obshch. Khim.*, 1984, **54**, 271; *J. Gen. Chem. USSR*, 1984, **54**, 238.
84ZOB1511	A.V. Ilyasov, A.N. Chernega, A.A. Nafikova, I.E. Galeeva, E.N. Dianova and B.A. Arbuzov; *Zh. Obshch. Khim.*, 1984, **54**, 1511; *J. Gen. Chem. USSR*, 1984, **54**, 1346.
84ZOB1520	A.A. Prishchenko, A.V. Gromov, Yu.N. Luzikov, A.A. Borisenko, E.I. Lazhko, K. Klaus and I.F. Lutsenko; *Zh. Obshch. Khim.*, 1984, **54**, 1520; *J. Gen. Chem. USSR*, 1984, **54**, 1354.
85AG125	A. Schmidpeter, K. Karaghiosoff, C. Cleve and D. Schomburg; *Angew. Chem.*, 1985, **97**, 125; *Angew. Chem. Int. Ed. Engl.*, 1985, **24**, 123.
85AG127	A. Schmidpeter, Ch. Leyh and K. Karaghiosoff; *Angew. Chem.*, 1985, **97**, 127; *Angew. Chem. Int. Ed. Engl.*, 1985, **24**, 124.
85BCJ667	H. Tanaka, S. Kametani, T. Saito, S. Motoki; *Bull. Chem. Soc. Jpn.*, 1985, **58**, 667.
85CB1621	J. Högel and A. Schmidpeter; *Chem. Ber.*, 1985, **118**, 1621.
85MI55	G. Becker, W. Becker, R. Knebl, H. Schmidt, U. Weeber and M. Westerhausen; *Nova Acta Leopoldina*, 1985, **59**, 55.
85PS(25)345	J. Heinicke and A. Tzschach; *Phosphorus Sulfur*, 1985, **25**, 345.
85S1067	J.L. Zhang and Z.S. Cao; *Synthesis*, 1985, **11**, 1067.
85ZAAC(520)107	H.-D. Hausen and G. Weckler; *Z. Anorg. Allg. Chem.*, 1985, **520**, 107.
85ZAAC(529)216	K. Issleib, H. Schmidt and P. Bergmann; *Z. Anorg. Allg. Chem.*, 1985, **529**, 216.
85ZN(B)1077	F. Zurmühlen, W. Rösch and M. Regitz; *Z. Naturforsch. B*, 1985, **40**, 1077.
85ZOB3	B.A. Arbuzov, E.N. Dianova, I.Z. Galeeva, I.A. Litvinov, Yu.T. Struchkov, A.N. Chernov and A.V. Ilyasov; *Zh. Obshch. Khim.*, 1985, **55**, 3; *J. Gen. Chem. USSR*, 1985, **55**, 1.
85ZOB1464	B.A. Arbuzov, E.N. Dianova, E.Ya. Zabotina, Yu.Ya. Efremov, R.L. Korshunov and R.Z. Musin; *Zh. Obshch. Khim.*, 1985, **55**, 1464; *J. Gen. Chem. USSR*, 1985, **55**, 1304.
85ZOB1471	B.A. Arbuzov, E.N. Dianova and E.Ya. Zabotina; *Zh. Obshch. Khim.*, 1985, **55**, 1471; *J. Gen. Chem. USSR*, 1985, **55**, 1310.
86AG349	O.J. Scherer, J. Schwalb, G. Wolmershäuser, W. Kaim and R. Gross; *Angew. Chem.*, 1986, **98**, 349; *Angew. Chem. Int. Ed. Engl.*, 1986, **25**, 363.
86AG570	R. Appel and R. Moors; *Angew. Chem.*, 1986, **98**, 570; *Angew. Chem. Int. Ed. Engl.*, 1986, **25**, 567.
86BCJ2047	H. Tanaka and S. Motoki; *Bull. Chem. Soc. Jpn.*, 1986, **59**, 2047.
86BSF115	Y.Y.C. Yeung Lam Ko, R. Carrié, L. Toupet and F. de Sarlo; *Bull. Soc. Chim. Fr.*, 1986, 115.
86CB410	K. Karaghiosoff, H. Klehr and A. Schmidpeter; *Chem. Ber.*, 1986, **119**, 410.
86CB3213	K. Karaghiosoff, W.S. Sheldrick and A. Schmidpeter; *Chem. Ber.*, 1986, **119**, 3213.
86JOM(306)39	W. Rösch, U. Vogelbacher, T. Allspach and M. Regitz; *J. Organomet. Chem.*, 1986, **306**, 39.
86PS(26)203	B.A. Arbuzov and E.N. Dianova; *Phosphorus Sulfur*, 1986, **26**, 203.
86PS(28)71	A. Schmidpeter; *Phosphorus Sulfur*, 1986, **28**, 71.
86PS(28)289	K. Karaghiosoff, C. Cleve and A. Schmidpeter; *Phosphorus Sulfur*, 1986, **28**, 289.
86PS(29)73	S. Lochschmidt and A. Schmidpeter; *Phosphorus Sulfur*, 1986, **29**, 73.

86S31 T. Allspach, M. Regitz, G. Becker and W. Becker; *Synthesis*, 1986, 31.
86TL2971 O. Diallo, M.T. Boisdon, L. Lopez, C. Malavaud and J. Barrans; *Tetrahedron Lett.*, 1986, **27**, 2971.
86TL4415 G. Märkl and S. Pflaum; *Tetrahedron Lett.*, 1986, **27**, 4415.
86TL4419 G. Märkl and G. Dorfmeister; *Tetrahedron Lett.*, 1986, **27**.
86TL5611 P. Pellon and J. Hamelin; *Tetrahedron Lett.*, 1986, **27**, 5611.
86TL5699 J. Heinicke; *Tetrahedron Lett.*, 1986, **27**, 5699.
86ZAAC(540/541)336 G. Becker, W. Massa, O. Mundt, R.E. Schmidt and C. Witthauer; *Z. Anorg. Allg. Chem.*, 1986, **540/541**, 336.
86ZC407 J. Heinicke and A. Tzschach; *Z. Chem.*, 1986, **26**, 407.
86ZN(B)553 A. Schmidpeter and B. Wrackmeyer; *Z. Naturforsch. B*, 1986, **41**, 553.
86ZOB2433 L.N. Markovskii and V.D. Romanenko; *Zh. Obshch. Khim.*, 1986, **56**, 2433; *J. Gen. Chem.*, 1986, **56**, 2153.
87AG59 O.J. Scherer and T. Brück; *Angew. Chem.*, 1987, **99**, 59; Angew. *Chem. Int. Ed. Engl.*, 1987, **26**, 59.
87CB597 B.A. Arbuzov, E.N. Dianova, R.T. Galiaskarova and A. Schmidpeter; *Chem. Ber.*, 1987, **120**, 597.
87CB1645 W. Rösch, U. Hees and M. Regitz; *Chem. Ber.*, 1987, **120**, 1645.
87CB1809 W. Rösch, H. Richter and M. Regitz; *Chem. Ber.*, 1987, **120**, 1809.
87CC1146 R. Bartsch, P.B. Hitchcock and J.F. Nixon; *J. Chem. Soc., Chem. Commun.*, 1987, 1146.
87CIC1 O.J. Scherer; *Comments Inorg. Chem.*, 1987, **6**, 1.
87JA627 A.J. Arduengo III, C.A. Stewart, F. Davidson, D.A. Dixon, J.Y. Becker, S.A. Culley and M.B. Mizen; *J. Am. Chem. Soc.*, 1987 **109**, 627.
87JOM(319)49 S. Kerschl, B. Wrackmeyer, A. Willhalm and A. Schmidpeter; *J. Organomet. Chem.*, 1987, **319**, 49.
87JOM(330)17 K. Issleib and H. Schmidt; *J. Organomet. Chem.*, 1987, **330**, 17.
87JOM(334)C35 M. Driess, D. Hu, H. Pritzkow, H. Schäufele, U. Zenneck, M. Regitz and W. Rösch; *J. Organomet. Chem.*, 1987, **334**, C35.
B-87MI61 E. Fluck and G. Heckmann; in "^{31}P NMR Spectroscopy in Stereochemical Analysis", eds. J.G. Verkade and L.D. Quin, VCH Publishers, Deerfield Beach, 1987, p. 61.
B-87MI617 A. Schmidpeter; in "The Chemistry of Inorganic Homo- and Heterocycles", eds. I. Haiduc and D.B. Sowerby, Vol. 2, Academic Press, London 1987, p. 617.
87PS(30)479 M. Regitz, W. Rösch, T. Allspach, U. Annen, K. Blatter, J. Fink, M. Hermesdorf, H. Heydt, U. Vogelbacher and O. Wagner; *Phosphorus Sulfur*, 1987, **30**, 479.
87PS(30)780 K. Karaghiosoff, W.S. Sheldrick and A. Schmidpeter; *Phosphorus Sulfur*, 1987, **30**, 780.
87S689 W. Rösch and M. Regitz; *Synthesis*, 1987, 689.
87T2775 G. Baccolini, R. Dalpozzo and E. Errani; *Tetrahedron*, 1987, **43**, 2775.
87T3247 W. Rösch, T. Facklam and M. Regitz; *Tetrahedron*, 1987, **43**, 3247.
87TH1 A. Willhalm; *Ph.D. Thesis*, Universität München, 1987.
87TL1089 G. Märkl and G. Dorfmeister; *Tetrahedron Lett.*, 1987, **28**, 1089.
87TL1511 G. Märkl and S. Pflaum; *Tetrahedron Lett.*, 1987, **28**, 1511.
87TL2119 G. Märkl, E. Eckl, U. Jakobs, M.L. Ziegler and B. Nuber; *Tetrahedron Lett.*, 1987, **28**, 2119.
87TL2693 G. Märkl, W. Hölzl and I. Trötsch-Schaller; *Tetrahedron Lett.*, 1987, **28**, 2693.
87TL6049 T. N'Gando M'Pondo, C. Malavaud, L. Lopez and J. Barrans; *Tetrahedron Lett.*, 1987, **28**, 6049.
87ZAAC(544)87 M. Baudler, D. Düster and D. Ouzounis; *Z. Anorg. Allg. Chem.*, 1987, **544**, 87.
87ZOB1699 B.A. Arbuzov, E.N. Dianova and E.Ya. Zabotina; *Zh. Obshch. Khim.*, 1987, **57**, 1699; *J. Gen. Chem. USSR*, 1987, **57**, 1515.
87ZOB1949 B.A. Arbuzov, E.N. Dianova, R.T. Galiaskarova, P.P. Chernov, I.A. Litvinov and V.A. Naumov; *Zh. Obshch. Khim.*, 1987, **57**, 1949; *J. Gen. Chem. USSR*, 1987, **57**, 1745.
88AG288 M. Baudler, S. Akpapoglou, D. Ouzounis, F. Wasgestian, B. Meinigke, H. Budzikiewicz and H. Münster; *Angew. Chem.*, 1988, **100**, 288; *Angew. Chem. Int. Ed. Engl.*, 1988, **27**, 280.
88AG1541 M. Regitz and P. Binger; *Angew. Chem.*, 1988, **100**, 1541; *Angew. Chem. Int. Ed. Engl.*, 1988, **27**, 1484.
88CC574 A.G. del Pozo, A.-M. Caminade, F. Dahan, J.-P. Majoral and R. Mathieu; *J. Chem. Soc., Chem. Commun.*, 1988, 574.
88CC819 R. Bartsch, P.B. Hitchcock and J.F. Nixon; *J. Chem. Soc., Chem. Commun.*, 1988, 819.
88CC842 N. Burford, B.W. Royan, A. Linden and T.S. Cameron; *J. Chem. Soc., Chem. Commun.*, 1988, 842.
88CC1615 R. Bartsch, D. Carmichael, P.B. Hitchcock, M.F. Meidine, J.F. Nixon and G.J.D. Sillett; *J. Chem. Soc., Chem. Commun.*, 1988, 1615.
88CB935 O.J. Scherer, T. Brück and G. Wolmershäuser; *Chem. Ber.*, 1988, **121**, 935.
88DOK(298)369 A.M. Kibardin, I.A. Litvinov, V.A. Naumov, Yu.T. Struchkov, T.V. Gryaznova and Yu.B. Michailov; *Dokl. Akad. Nauk SSSR*, 1988, **298**, 369.
88IC2612 J.G. Kraaijkamp, D.M. Grove, G. van Koten and A. Schmidpeter; *Inorg. Chem.*, 1988, **27**, 2612.
88IZV150 B.A. Arbuzov, E.N. Dianova, E.Ya. Zabotina, I.A. Litvinov and V.A. Naumov; *Izv. Akad. Nauk SSSR, Ser. Khim.*, 1988, 150.
88JCS(P1)971 G. Baccolini, R. Dalpozzo and P.E. Todesco; *J. Chem. Soc., Perkin Trans. 1*, 1988, 971.
88JOM(338)329 E.P.O. Fuchs, M. Hermesdorf, W. Schnurr, W. Rösch, H. Heydt, M. Regitz and P. Binger; *J. Organomet. Chem.*, 1988, **338**, 329.
88PS(36)217 K. Karaghiosoff and A. Schmidpeter; *Phosphorus Sulfur*, 1988, **36**, 217.
88TL785 G. Märkl, I. Troetsch-Schaller and W. Hölzl; *Tetrahedron Lett.*, 1988, **29**, 785.
88TL925 H. Keller and M. Regitz; *Tetrahedron Lett.*, 1988, **29**, 925.

88TL3387	G. Märkl and S. Pflaum; *Tetrahedron Lett.*, 1988, **29**, 3387.
88TL4535	G. Märkl and W. Hölzl; *Tetrahedron Lett.*, 1988, **29**, 4535.
88TL5867	G. Märkl, S. Dietl, M.L. Ziegler and B. Nuber; *Tetrahedron Lett.*, 1988, **29**, 5867.
88ZC255	H. Schmidt, E. Leissring and K. Issleib; *Z. Chem.*, 1988, **28** 255.
88ZN(B)1285	W. Schnurr and M. Regitz; *Z. Naturforsch. B*, 1988, **43**, 1285.
88ZOB2045	A.N. Chernega, A.A. Korkin, M.Yu. Antipin, Yu.T. Struchkov; *Zh. Obshch. Khim.*, 1988, **58**, 2045; *J. Gen. Chem. USSR*, 1988, **58**, 1823.
88ZSK122	A.N. Chernega, M.Yu. Antipin and Yu.T. Struchkov; *Zh. Strukt. Khim.*, 1988, **29**, 122.
89AG500	T.P. Hamilton and H.F. Schaefer III; *Angew. Chem.*, 1989, **101**, 500; *Angew. Chem., Int. Ed. Engl.*, 1989, **28**, 485.
89AX(C)1485	N. Burford, A.I. Dipchand, B.W. Royan and P.S. White; *Acta Cryst. C*, 1989, **45**, 1345.
89CB2049	O.J. Scherer, T. Brück and G. Wolmershäuser; *Chem. Ber.*, 1989, **122**, 2049.
89CB2121	R. Janoscheck; *Chem. Ber.*, 1989 **122**, 2121.
89CC788	K.H. Dötz, A. Tiriliomis and K. Harms; *J. Chem. Soc., Chem. Commun.*, 1989, 788.
89CC988	J.C. Guillemin, M. le Guennec and J.M. Denis; *J. Chem. Soc., Chem. Commun.*, 1989, 988.
89CC1046	R. Bartsch, P.B. Hitchcock and J.F. Nixon; *J. Chem. Soc., Chem. Commun.*, 1989, 1046.
89DIS(B)3773	Y. Younghee; *Diss. Abstr. Int. B*, 1989, **49**, 3773.
89EGP268700	H. Hartmann and M. Weber; (Technische Hochschule "Carl Schorlemmer", Leuna-Merseburg) Ger. (East) DD 268,700 (7 June 1989) (*Chem. Abstr.*, 1990, **112**, P 98839j).
89EGP268701	H. Hartmann and M. Weber; (Technische Hochschule "Carl Schorlemmer", Leuna-Merseburg) Ger. (East) DD 268,701 (7 June 1989) (*Chem. Abstr.*, 1990, **112**, P 98840c).
89EGP268702	H. Hartmann and M. Weber; (Technische Hochschule "Carl Schorlemmer", Leuna-Merseburg) Ger. (East) DD 268,702 (7 June 1989) (*Chem. Abstr.*, 1990, **112**, P 98841d).
89EGP268703	H. Hartmann and M. Weber; (Technische Hochschule "Carl Schorlemmer", Leuna-Merseburg) Ger. (East) DD 268,703 (7 June 1989) (*Chem. Abstr.*, 1990, **112**, P 98842e).
89EGP268704	H. Hartmann and M. Weber; (Technische Hochschule "Carl Schorlemmer", Leuna-Merseburg) Ger. (East) DD 268,704 (7 June 1989) (*Chem. Abstr.*, 1990, **112**, P 98843f).
89IC144	N. Burford, B.W. Royan, A. Linden and T.S. Cameron; *Inorg. Chem.*, 1989, **28**, 144.
89JA2030	W. Tremel, R. Hoffmann and M. Kertesz; *J. Am. Chem. Soc.*, 1989, **111**, 2030.
89JCR(S)250	M. Haddad, M.-T. Boisdon, L. Lopez, C. Malavaud, G. Pelletier, J. Barrans, G. Pfister-Guillouzo and J.P. Legros; *J. Chem. Res. (S)*, 1989, 250; *J. Chem. Res. (M)*, 1989, 1815.
89JCS(P1)1053	G. Baccolini; *J. Chem. Soc. Perkin Trans. 1*, 1989, 1053.
89JOM(338)329	E.P.O. Fuchs, M. Hermesdorf, W. Schnurr, W. Rösch, H. Heydt, M. Regitz and P. Binger; *J. Organomet. Chem.*, 1988, **338**, 329.
89JOM(364)C17	J. Heinicke; *J. Organomet. Chem.*, 1989, **364**, C17.
89JOM(373)49	J. Heinicke, G. Csonka, J. Réffy and T. Veszprémi; *J. Organomet. Chem.*, 1989, **373**, 49.
89JOM(373)57	J. Heinicke, L. Nyulászi, G. Csonka, J. Réffy and T. Veszpremi; *J. Organomet. Chem.*, 1989, **373**, 57.
89JOM(373)C17	R. Bartsch, P.B. Hitchcock and J.F. Nixon; *J. Organomet. Chem.*, 1989, **373**, C17.
89JOM(375)C31	R. Bartsch, P.B. Hitchcock and J.F. Nixon; *J. Organomet. Chem.*, 1989, **375**, C31.
B-89MI195	E. Fluck and B. Neumüller; in "Rings, Clusters and Polymers of Main Group and Transition Elements", ed. H.W. Roesky, Elsevier, Amsterdam, 1989, p. 195.
B-89MI307	A. Schmidpeter and K. Karaghiosoff; in "Rings, Clusters and Polymers of Main Group and Transition Elements", ed. H.W. Roesky, Elsevier, Amsterdam, 1989, p. 307.
89NJC309	W. Güth, T. Busch, W.W. Schoeller, E. Niecke, B. Krebs, M. Dartmann and P. Rademacher; *New J. Chem.*, 1989, **13**, 309.
89NJC353	P.B. Hitchcock, J.F. Nixon and G.J.D. Sillett; *New J. Chem.*, 1989, **13**, 353.
89NJC359	A. Igau, A. Baceiredo, G. Bertrand, K. Kühnel-Lysek and E. Niecke; *New J. Chem.*, 1989, **13**, 359.
89OM244	H. Westermann, M. Nieger and E. Niecke; *Organometallics*, 1989, **8**, 244.
89PO849	A.H. Cowley and S.W. Hall; *Polyhedron*, 1989, **8**, 849.
89PO2407	R. Bartsch and J.F. Nixon; *Polyhedron*, 1989, **8**, 2407.
89PS(41)37	N. Burford and B.W. Royan; *Phosphorus Sulfur Silicon*, 1989, **41**, 37.
89PS(46)31	M. Hermersdorf, M. Birkel, H. Heydt, M. Regitz and P. Binger; *Phosphorus Sulfur Silicon*, 1989, **46**, 31.
89S511	W. Schnurr and M. Regitz; *Synthesis*, 1989, 511.
89TL4501	G. Märkl and W. Hölzl, *Tetrahedron Lett.*, 1989, **30**, 4501.
89ZC71	J. Heinicke, *Z. Chem.*, 1989, **29**, 71.
89ZC410	H. Schmidt, E. Leissring and C. Wirkner; *Z. Chem.*, 1989, **29**, 410.
89ZN(B)381	M. Baudler and D. Ouzounis; *Z. Naturforsch. B*, 1989, **44**, 381.
90AG316	T. Facklam, O. Wagner, H. Heyth and M. Regitz; *Angew. Chem.*, 1990, **102**, 316; *Angew. Chem. Int. Ed. Engl.*, 1990, **29**, 314.
90AG575	N. Maigrot, L. Ricard, C. Charrier and F. Mathey; *Angew. Chem.*, 1990, **102**, 575; *Angew. Chem. Int. Ed. Engl.*, 1990, **29**, 534.
90AG1137	O.J. Scherer; *Angew. Chem.*, 1990, **102**, 1137; *Angew. Chem. Int. Ed. Engl.*, 1990, **29**, 1104.
90CC472	R. Bartsch, P.B. Hitchcock and J.F. Nixon; *J. Chem. Soc., Chem. Commun.*, 1990, 472.
90CC1307	R. Bartsch, P.B. Hitchcock and J.F. Nixon; *J. Chem. Soc., Chem. Commun.*, 1990, 1307.
90CRV191	M. Regitz; *Chem. Rev.*, 1990, **90**, 191.

90DOK(312)623 I.A. Litvinov, V.A. Naumov, T.V. Gryaznova, A.N. Pudovik and A.M. Kibardin; *Dokl. Akad. Nauk USSR*, 1990, **312**, 623.

90HAC37 S. Holand, J.M. Alcaraz, L. Ricard and F. Mathey; *Heteroatom Chem.*, 1990, **1**, 37.

90IC4938 N. Burford, A.I. Dipchand, B.W. Royan and P.S. White; *Inorg. Chem.*, 1990, **29**, 4938.

90JCS(D)977 L.Y. Goh, R.C.S. Wong, C.K. Chu and T.W. Hambley; *J. Chem. Soc., Dalton Trans.*, 1990, 977.

90JMC572 T.A. Riley, S.B. Larson, T.L. Avery, R.A. Finch and R.K. Robins; *J. Med. Chem.*, 1990, **33**, 572.

90JOM(382)53 K. Issleib, H. Schmidt and E. Leissring; *J. Organomet. Chem.*, 1990, **382**, 53.

B-90MI58 M. Regitz; in "Multiple Bonds and Low Coordination in Phosphorus Chemistry", eds. M. Regitz and O.J. Scherer, Georg Thieme, Stuttgart, 1990, p. 58.

B-90MI157 R. Appel; in "Multiple Bonds and Low Coordination in Phosphorus Chemistry", eds. M. Regitz and O.J. Scherer, Georg Thieme, Stuttgart, 1990, p. 157.

B-90MI220 G. Märkl; in "Multiple Bonds and Low Coordination in Phosphorus Chemistry", eds. M. Regitz and O.J. Scherer, Georg Thieme, Stuttgart, 1990, p. 220.

B-90MI258 A. Schmidpeter and K. Karaghiosoff; in "Multiple Bonds and Low Coordination in Phosphorus Chemistry", eds. M. Regitz and O.J. Scherer, Georg Thieme, Stuttgart, 1990, p. 258.

B-90MI293 E. Niecke; in "Multiple Bonds and Low Coordination in Phosphorus Chemistry", eds. M. Regitz and O.J. Scherer, Georg Thieme, Stuttgart, 1990, p. 293.

B-90MI295 M. Regitz; in "Heteroatom Chemistry", ed. E. Block, VCH Publishers, New York, 1990, p. 295.

B-90MI321 M. Yoshifuji; in "Multiple Bonds and Low Coordination in Phosphorus Chemistry", eds. M. Regitz and O.J. Scherer, Georg Thieme, Stuttgart, 1990, p. 321.

B-90MI463 K. Karaghiosoff; in "Multiple Bonds and Low Coordination in Phosphorus Chemistry", eds. M. Regitz and O.J. Scherer, Georg Thieme, Stuttgart, 1990, p. 463.

90NJC433 K.H. Dötz; *New J. Chem.*, 1990, **14**, 433.

90OM1085 B.W. Royan, N. Burford, J. Waluk and J. Michl; *Organometallics*, 1990, **9**, 1085.

90PAC423 B.M. Gimarc; *Pure Appl. Chem.*, 1990, **62**, 423.

90PS(47)173 J.L. Zhang and W. Xu; *Phosphorus Sulfur Silicon*, 1990, **47**, 173.

90PS(49/50)317 J.M. Denis, J.C. Guillemin and M. le Guennec; *Phosphorus Sulfur Silicon*, 1990, **49**/**50**, 317.

90PS(49/50)349 A. Schmidpeter, R.K. Bansal, K. Karaghiosoff, F. Steinmüller and C. Spindler; *Phosphorus Sulfur Silicon*, 1990, **49**/**50**, 349.

90PS(49/50)355 H. Schmidt, K. Issleib and E. Leissring; *Phosphorus Sulfur Silicon*, 1990, **49**/**50**, 355.

90PS(49/50)363 O.S. Diallo, L. Lopez and J. Barrans; *Phosphorus Sulfur Silicon*, 1990, **49**/**50**, 363.

90PS(49/50)463 J.F. Nixon; *Phosphorus Sulfur Silicon*, 1990, **49**/**50**, 463.

90T3859 J.Y. Thorval, W. Nagai, Y.Y.C. Yeung Lam Ko and R. Carrié; *Tetrahedron*, 1990, **46**, 3859.

90T5697 C.W. Bird; *Tetrahedron*, 1990, **46**, 5697.

90TL1147 P. Pellon, S. Himdi-Kabbab, I. Rault, F. Tonnard and J. Hamelin; *Tetrahedron Lett.*, 1990, **31**, 1147.

90TL4463 K. Rauzy, M.R. Mazieres, P. Page, M. Sanchez and J. Bellan; *Tetrahedron Lett.*, 1990, **31**, 4463.

90UK931 R.I. Tarasova and V.V. Moskva; *Usp. Khim.*, 1990, **59**, 931; *Russ. Chem. Rev.*, 1990, **59**, 538.

90ZAAC(585)197 J. Gloede and R. Waschke; *Z. Anorg. Allg. Chem.*, 1990, **585**, 197.

90ZC41 M. Scheer and E. Herrmann; *Z. Chem.*, 1990, **30**, 42.

90ZN(B)1139 M. Baudler and J. Hahn; *Z. Naturforsch. B*, 1990, **45**, 1139.

90ZOB35 B.A. Arbuzov, E.N. Dianova, E.Ya. Zabotina and I.Z. Akhmetkhanova; *Zh. Obshch. Khim.*, 1990, **60**, 35; *J. Gen. Chem. USSR*, 1990, **60**, 29.

91AG590 M. Baudler and Th. Etzbach; *Angew. Chem.*, 1991, **103**, 590; *Angew. Chem. Int. Ed. Engl.*, 1991, **30**, 580.

91CA62507b H.D.H. Showalter; *Chem. Abstr.*, 1991, **114**, 62507b; *Chemtracts. Org. Chem.*, 1990, **3**, 311.

91CB475 R.K. Bansal, K. Karaghiosoff, N. Gupta, A. Schmidpeter and C. Spindler; *Chem. Ber.*, 1991, **124**, 475.

91CB493 J. Heinicke, I. Kovacs and L. Nyulászi; *Chem. Ber.*, 1991, **124**, 493.

91CB1159 M. Baudler and Th. Etzbach; *Chem. Ber.*, 1991, **124**, 1159.

91CC23 Y. Kandri Rodi, L. Lopez, C. Malavau, M.T. Boisdon and J. Barrans; *J. Chem. Soc., Chem. Commun.*, 1991, 23.

91CC34 G. Baccolini and P. Sgarabotto; *J. Chem. Soc., Chem. Commun.*, 1991, 34.

91CC302 U. Fleischer, H. Grützmacher and U. Krüger; *J. Chem. Soc., Chem. Commun.*, 1991, 302.

91CC1293 L. Weber, R. Kirchhoff, R. Boese and H.-G. Stammler; *J. Chem. Soc., Chem. Commun*, 1991, 1293.

91CC1659 E. Niecke and D. Schmidt; *J. Chem. Soc., Chem. Commun.*, 1991, 1659.

91CPH(149)325 P.J. Bruna and S.D. Peyerimhoff; *Chem. Phys.*, 1991, **149**, 325.

91HAC369 I.A. Litvinov, K. Karaghiosoff, A. Schmidpeter, E.Ya. Zabotina and E.N. Dianova; *Heteroatom Chem.*, 1991, **2**, 369.

91HAC577 A.S. Ionkin, S.N. Ignateva, I.A. Litvinov, V.A. Naumov and B.A. Arbuzov; *Heteroatom Chem.*, 1991, **2**, 577.

91IJC(B)992 R.K. Bansal, V. Kabra, R. Munjal and N. Gupta; *Indian J. Chem. B.*, 1991, **33**, 992.

91JOM(415)C15 R. Batsch and J.F. Nixon; *J. Organomet. Chem.*, 1991, **415**, C15.

91JPC75 Z. Mielke, G. Brabson and L. Andrews; *J. Phys. Chem.*, 1991, **95**, 75.

B-91MI29 A. Schmidpeter and K. Karaghiosoff; in "Handbook of ^{31}P NMR Data", ed. J.C. Tebby, CRC Press, Boca Raton, 1991, p. 29.

91MRC1045 B. Wrackmeyer, E. Kupce and A. Schmidpeter; *Magn. Reson. Chem.*, 1991, **29**, 1045.

91OM2835 M.L. Sierra, N. Maigrot, C. Charrier, L. Ricard and F. Mathey; *Organometallics*, 1991, **10**, 2835.

91PS(55)255	J. Gloede; *Phosphorus Sulfur Silicon*, 1991, **55**, 255.
91PS(56)17	O.S. Diallo, L. Lopez, Y. Kandri Rodi and J. Barrans; *Phosphorus Sulfur Silicon*, 1991, **56**, 17.
91PS(56)129	R. Chen, B. Cai, L. Liu, G. Li and L. He; *Phosphorus Sulfur Silicon*, 1991, **56**, 129.
91PS(56)189	H.-D. Schädler, H. Schmidt and M. Frenzel; *Phosphorus Sulfur Silicon*, 1991, **56**, 189.
91PS(57)83	R. Chen and B. Cai; *Phosphorus Sulfur Silicon*, 1991, **57**, 83.
91S783	R. Chen, B. Cai and G. Li; *Synthesis*, 1991, 783.
91SL356	B. Burkhart, S. Krill, Y. Okano, W. Ando and M. Regitz; *Synlett*, 1991, **5**, 356.
91TL501	O.S. Diallo, L. Lopez and J. Barrans; *Tetrahedron Lett.*, 1991, **32**, 501.
91TL6433	R.K. Bansal, D.C. Sharma and R. Mahnot; *Tetrahedron Lett.*, 1991, **32**, 6433.
91ZOB1063	E.N. Dianova, E.Ya. Zabotina, I.Z. Akhmetkhanova and Ya.D. Samuilov; *Zh. Obshch. Khim.*, 1990, **61**, 1063; *J. Gen. Chem. USSR*, 1990, **61**, 965.
92AG1082	N. Maigrot, L. Ricard, C. Charrier and F. Mathey; *Angew. Chem.*, 1992, **104**, 1082; *Angew. Chem., Int. Ed. Engl.*, 1992, **31**, 1031.
92CB411	J. Grobe, D. Le Van, M. Hegemann, B. Krebs and M. Läge; *Chem. Ber.*, 1992, **125**, 411.
92CCR259	M. Di Vaira and P. Stoppioni; *Coord. Chem. Rev.*, 1992, **120**, 259.
92CIC61	F. Mathey, C. Charrier, N. Maigrot, A. Marinetti, L. Ricard and N.H. Tran Huy; *Comments Inorg. Chem.*, 1992, **13**, 61.
92DOK(324)330	A.M. Kibardin, T.V. Gryaznova, N.N. Sarvarova, L.V. Ermolaeva, S.G. Vulfson and A.N. Pudovik; *Dokl. Akad. Nauk SSSR*, 1992, **324**, 330.
92HAC157	C. Payrastre, Y. Madaule, J.G. Wolf, T.C. Kim, M.-R. Mazières, R. Wolf and M. Sanchez; *Heteroatom Chem.*, 1992, **3**, 157.
92HAC337	C. Grandin, E. About-Jaudet, N. Collignon, J.-M. Denis and P. Savignac; *Heteroatom Chem.*, 1992, **3**, 337.
92HAC351	R.K. Bansal, R. Mahnot, D.C. Sharma, K. Karaghiosoff and A. Schmidpeter; *Heteroatom Chem.*, 1992, **3**, 351.
92HAC359	R.K. Bansal, N. Gupta, V. Kabra, C. Spindler, K. Karaghiosoff and A. Schmidpeter; *Heteroatom Chem.*, 1992, **3**, 359.
92HAC435	L. Ermolaeva and A. Ionkin; *Heteroatom Chem.*, 1992, **3**, 435.
92IC3386	R.D. Curtis, B.W. Royan, R.E. Wasylishen, M.D. Lumsden and N. Burford; *Inorg. Chem.*, 1992, **31**, 3386.
92IJC254	R.K. Bansal, V. Kabra, N. Gupta and K. Karaghiosoff; *Indian J. Chem.*, 1992, **31b**, 254.
92JA9080	L. Nyulászi, T. Veszprémi, J. Réffy, B. Burkhardt and M. Regitz; *J. Am. Chem. Soc.*, 1992, **114**, 9080.
92JCS(P1)1729	G. Baccolini, D. Evangelisti, C. Rizzoli and P. Sgarabotto; *J. Chem. Soc., Perkin Trans. 1*, 1992, 1729.
92JCS(P2)671	M. Haddad, F. Dahan, J.-P. Legros, L. Lopez, M.-T. Boisdon and J. Barrans; *J. Chem. Soc., Perkin. Trans. 2*, 1992, 671.
92JOC3694	E.J. Padma Malar; *J. Org. Chem.*, 1992, **57**, 3694.
92JOM(430)C10	R. Bartsch, A. Gelessus, P.B. Hitchcock and J.F. Nixon; *J. Organomet. Chem.*, 1992, **430**, C10.
92JPC623	T. Veszprémi, L. Nyulászi, J. Réffy and J. Heinicke; *J. Phys. Chem.*, 1992, **96**, 623.
92OM459	M.L. Sierra, N. Maigrot, C. Charrier, L. Ricard and F. Mathey; *Organometallics*, 1992, **11**, 459.
92PO601	N. Maigrot, M.L. Sierra, C. Charrier, L. Ricard and F. Mathey; *Polyhedron*, 1992, **11**, 601.
92PS(65)131	T.S. Cameron, B. Borecka and A. Linden; *Phosphorus Sulfur Silicon*, 1992, **65**, 131.
92PS(69)189	M. Haddad, J.P. Legros, L. Lopez, M.T. Boisdon and J. Barrans; *Phosphorus Sulfur Silicon*, 1992, **69**, 189.
92RHA25	Yu.G. Gololobov and Yu.V. Balitskii; *Rev. Heteroatom Chem.*, 1992, **6**, 25.
92S267	R.K. Bansal, R. Mahnot, D.C. Sharma and K. Karaghiosoff; *Synthesis*, 1992, 267.
92TH1	F. Steinmüller; *Ph.D. Thesis*, Universität München, 1992.
92TH2	A.S. Ionkin; *Ph.D. Thesis*, Kazan Scientific Centre of the Russian Academy of Sciences, 1992.
92TL1049	U. Bergsträßer, A. Hoffmann and M. Regitz; *Tetrahedron Lett.*, 1992, **33**, 1049.
92TL4421	G. Märkl, W. Hölzl, A. Kallmünzer, M.L. Ziegler and B. Nuber; *Tetrahedron Lett.*, 1992, **33**, 4421.
92ZN(B)373	K. Karaghiosoff, R.K. Bansal and N. Gupta; *Z. Naturforsch. B*, 1992, **47**, 373.
92ZOB1505	N.G. Khusainova, L.V. Smirnova, E.A. Irtuganova and Ya. Zabotina; *Zh. Obshch. Khim.*, 1992, **62**, 1505; *J. Gen. Chem. USSR*, 1992, **62**, 1234.
93AG641	M. Scheer, G. Friedrich and K. Schuster; *Angew. Chem.*, 1993, **105**, 641; *Angew. Chem. Int. Ed. Engl.*, 1993, **32**, 593.
93AG86	P.B. Hitchcock, J.A. Johnson and J.F. Nixon; *Angew. Chem.*, 1993, **105**, 86; *Angew. Chem. Int. Ed. Engl.*, 1993, **32**, 103.
93AG1521	E. Lindner, C. Haase, H.A. Mayer, M. Kemmler, R. Fawzi and M. Steimann; *Angew. Chem.*, 1993, **105**, 1521; *Angew. Chem. Int. Ed. Engl.*, 1993, **32**, 1424.
93BSF521	M.L. Sierra, C. Charrier, L. Ricard and F. Mathey; *Bull. Soc. Chim. Fr.*, 1993, **130**, 521.
93BSF726	V.D. Romanenko, V.L. Rudzevich, A.O. Gudima, M. Sanchez, A.B. Rozhenko, A.N. Chernega and M.R. Mazières; *Bull. Soc. Chim. Fr.*, 1993, **130**, 726.
93BSF757	A. Baceiredo, M. Nieger, E. Niecke and G. Bertrand; *Bull. Soc. Chim. Fr.*, 1993, **130**, 757.
93CB1513	G. David, E. Niecke, M. Nieger, V. von der Gönna and W.W. Schoeller; *Chem. Ber.*, 1993, **126**, 1513.
93CC267	S.S. Al-Juaid, P.B. Hitchcock, R.M. Matos and J.F. Nixon; *J. Chem. Soc., Chem. Commun.*, 1993, 267.
93CC311	R. Bartsch, P.B. Hitchcock and J.F. Nixon; *J. Chem. Soc., Chem. Commun.*, 1993, 311.
93CJC1105	Y.Y.C. Yeung Lam Ko, R. Carrié, F. De Sarlo and A. Brandi; *Can. J. Chem.*, 1983, **61**, 1105.

93CJC1200 Y. Kandri Rodi, L. Lopez, C. Malavaud and M.-T. Boisdon; *Can. J. Chem.*, 1993, **71**, 1200.
93CL1563 S. Motoki, T. Sakai, I. Shinoda, T. Saito and T. Uchida; *Chem. Lett.*, 1993, 1563.
93CRV1623 M. Baudler and K. Glinka; *Chem. Rev.*, 1993, **93**, 1623.
93HAC497 G. Baccolini and D. Biondi; *Heteroatom Chem.*, 1993, **4**, 497.
93JOM(462)319 P.B. Hitchcock, R.M. Matos and J.F. Nixon; *J. Organomet. Chem.*, 1993, **462**, 319.
93JPC4011 L. Nyulászi, T. Veszprémi and J. Réffy; *J. Phys. Chem.*, 1993, **97**, 4011.
93JPR458 A. Schmidpeter, F. Steinmüller and E.Ya. Zabotina; *J. Prakt. Chem.*, 1993, **335**, 458.
93MI641 T.-B. Huang, J.-L. Zhang, C.-G. Zhan and Y.-J. Zhang; *Sci. China, Ser. B*, 1993, **36**, 641.
93OM731 L. Weber, R. Kirchhoff, R. Boese, H.-G. Stammler and B. Neumann; *Organometallics*, 1993, **12**, 731.
93PS(75)245 S. Kaji Nitcheu T., C. Malavaud, L. Lopez, Y. Kandri Rodi and E.M. Essassi; *Phosphorus Sulfur Silicon*, 1993, **75**, 245.
93PS(76)5 J. Grobe, D. Le Van, M. Hegemann, B. Lüth and J. Nientiedt; *Phosphorus Sulfur Silicon*, 1993, **76**, 5.
93PS(76)45 M.R. Mazieres, K. Rauzy, J. Bellan, M. Sanchez, G. Pfister-Guillouzo and A. Senio; *Phosphorus Sulfur Silicon*, 1993, **76**, 45.
93PS(76)79 J.-L. Zhang, T-B. Huang and H.-Y. Lu; *Phosphorus Sulfur Silicon*, 1993, **76**, 79.
93PS(76)339 J.-L. Zhang, T.-B. Huang and H.-Y. Lu; *Phosphorus Sulfur Silicon*, 1993, **76**, 339.
93PS(77)45 R. Bartsch, P.B. Hitchcock, J.A. Johnson, R.M. Matos and J.F. Nixon; *Phosphorus Sulfur*, 1993, **77**, 45.
93PS(77)235 K. Karaghiosoff, G. Hackenbracht, A. Schmidpeter, R.K. Bansal, N. Gupta, D.C. Sharma, R. Mahnot and V. Kabra; *Phosphorus Sulfur Silicon*, 1993, **77**, 235.
93PS(77)246 E. Zabotina and E. Dianova; *Phosphorus Sulfur Silicon*, 1993, **77**, 246.
93PS(77)254 C. Charrier, M. Sierra, N. Maigrot, L. Ricard and F. Mathey; *Phosphorus Sulfur Silicon*, 1993, **77**, 254.
93PS(77)275 R. Bartsch, J.A. Johnson and J.F. Nixon; *Phosphorus Sulfur*, 1993, **77**, 275.
93PS(82)211 J. Gloede; *Phosphorus Sulfur Silicon*, 1993, **82**, 211.
93PS(85)225 Y. Kandri Rodi, L. Lopez, J. Bellan, J. Barrans and E.M. Essassi; *Phosphorus Sulfur Silicon*, 1983, **85**, 225.
93T4651 S. Kaji Nicheu and C. Malavaud; *Tetrahedron*, 1993, **49**, 4651.
93T5577 K.H. Dötz, A. Tiriliomis and K. Harms; *Tetrahedron*, 1993, **49**, 5577.
93ZN(B)93 A. Schmidpeter and G. Jochem; *Z. Naturforsch. B*, 1993, **48**, 93.
94AG708 G. Jochem, A. Schmidpeter, M. Thomann and H. Nöth; *Angew. Chem.*, 1994, **106**, 708; *Angew. Chem. Int. Ed. Engl.*, 1994, **33**, 663.
94BCJ2785 I. Shinoda, A. Takahashi, T. Saito and T. Uchida; *Bull. Chem. Soc. Jpn.*, 1994, **67**, 2785.
94BSF397 N. Maigrot, M.L. Sierra, C. Charrier and F. Mathey; *Bull. Soc. Chim. Fr.*, 1994, **131**, 397.
94BSF748 V. Romanenko, A.O. Gudima, A.N. Chernega, J.-M. Sotriopoulos, G. Alcaraz and G. Bertrand; *Bull. Soc. Chim. Fr.*, 1994, **131**, 748.
94CC1121 U. Bergsträßer, J. Stannek and M. Regitz; *J. Chem. Soc., Chem. Commun.*, 1994, 1121.
94CCR(137)1 F. Mathey; *Coord. Chem. Rev.*, 1994, **137**, 1.
94CRV1413 A.C. Gaumont and J.M. Denis; *Chem. Rev.*, 1994, **94**, 1413.
94HAC385 A. Schmidpeter, F. Steinmüller and K. Karaghiosoff; *Heteroatom Chem.*, 1994, **5**, 385.
94JOM(464)C31 E. Lindner, T. Schlenker and C. Haase; *J. Organomet. Chem.*, 1994, **464**, C31.
94JOM(475)25 F. Mathey; *J. Organomet. Chem.*, 1994, **475**, 25.
94JOM(479)C28 P. Binger and G. Glaser; *J. Organomet. Chem.*, 1994, **479**, C28.
94JOM(480)45 S.I. Al-Juaid, P.B. Hitchcock, J.A. Johnson and J.F. Nixon; *J. Organomet. Chem.*, 1994, **480**, 45.
94MI307 J. Heinicke; *Trends Organomet. Chem.*, 1994, **1**, 307.
94MI369 A.G. Avent, F.G.N. Cloke, K.R. Flower, P.B. Hitchcock, J.F. Nixon and D.M. Vickers; Abstracts XVIth International Conference on Organometallic Chemistry, 1994, p. 369.
94MI385 F.G.N. Cloke, K.R. Flower, P.B. Hitchcock and J.F. Nixon; Abstracts XVIth International Conference on Organometallic Chemistry, 1994, p. 385.
94PS(86)139 R.K. Bansal, C.B. Jain, N. Gupta, V. Kabra, K. Karaghiosoff and A. Schmidpeter; *Phosphorus Sulfur Silicon*, 1994, **86**, 139.
94PS(93/94)381 R.K. Bansal, N. Gandhi, A. Schmidpeter and K. Karaghiosoff; *Phosphorus Sulfur Silicon*, 1994, **93/94**, 381.
94PS(93/94)383 A. Schmidpeter and F. Steinmüller; *Phosphorus Sulfur Silicon*, 1994, **93/94**, 383.
94PS(97)141 R.K. Bansal, V. Kabra, R. Munjal and N. Gupta; *Phosphorus Sulfur Silicon*, 1994, **97**, 141.
94T7675 R.K. Bansal, K. Karaghiosoff and A. Schmidpeter; *Tetrahedron*, 1994, **50**, 7675.
94ZOB149 T.-B. Huang, J.-L. Zhang, X-L. Xu, W.-F. Huang, Y.-J. Zhang and X.-P. Liu; *Zh. Obshch. Khim.*, 1994, **64**, 149; *Russian J. Gen. Chem.*, 1994, **64**, 138.
95AG623 N. Maigrot, N. Avarvari, C. Charrier and F. Mathey; *Angew. Chem.*, 1995, **107**, 623; *Angew. Chem. Int. Ed. Engl.*, 1995, **34**, 590.
95AG1454 M. Detzel, G. Friedrich, O.J. Scherer and G. Wolmershäuser; *Angew. Chem.*, 1995, **107**, 1454; *Angew. Chem. Int. Ed. Engl.*, 1995, **34**, 1321.
95BSF306 R. Beugelmans and M. Chbani; *Bull. Soc. Chim. Fr.*, 1995, **132**, 306.
95CB665 L. Weber, O. Sommer, H.-G. Stammler, B. Neumann and U. Kölle; *Chem. Ber.*, 1995, **128**, 665.
95CB581 K. Karaghiosoff, R. Mahnot, C. Cleve, N. Gandhi, R.K. Bansal and A. Schmidpeter; *Chem. Ber.*, 1995, **128**, 581.
95CC1659 F.G.N. Cloke, K.R. Flower, P.B. Hitchcock and J.F. Nixon; *J. Chem. Soc., Chem. Commun.*, 1995, 1659.
95CC1661 V. Caliman, P.B. Hitchcock and J.F. Nixon; *J. Chem. Soc., Chem. Commun.*, 1995, 1661.

95CC2395 S.M. Bachrach, V. Caliman and J.F. Nixon; *J. Chem. Soc., Chem. Commun.*, 1995, 2395.

95CCR(145)201 J.F. Nixon; *Coord. Chem. Rev.*, 1995, **145**, 201.

95CSR319 J.F. Nixon; *Chem. Soc. Rev.*, 1995, 319.

95HAC403 P. de Dianous, S. Himdi-Kabbab and J. Hamelin; *Heteroatom Chem.*, 1995, **6**, 403.

95JCS(P2)315 L. Nyulászi, P. Várnai, S. Krill and M. Regitz; *J. Chem. Soc., Perkin Trans. 2*, 1995, 315.

95JPC586 L. Nyulászi; *J. Phys. Chem.*, 1995, **99**, 586.

95JOM(487)C21 F.G.N. Cloke, K.R. Flower, C. Jones, R.M. Matos and J.F. Nixon; *J. Organomet. Chem.*, 1995, **480**, C21.

B-95MI1 V.I. Minkin, M.N. Glukhovtsev and B.Y. Simkin; "Aromaticity and Antiaromaticity", Wiley, New York, 1995.

95MI239 C.-G. Zhan; *Chinese J. Org. Chem.* (Youji Huaxue), 1995, **15**, 239.

95MI(357)9 C.-J. Zhang, C.-G. Zhan and T.-B. Huang; *J. Mol. Struct. (Theochem)*, 1995, **357**, 9.

95OM581 L. Weber, O. Kaminski, H.-G. Stammler and B. Neumann; *Organometallics*, 1995, **14**, 581.

95OM4382 P.B. Hitchcock, J.A. Johnson and J.F. Nixon; *Organometallics*, 1995, **14**, 4382.

95PS(104)33 T.-B. Huang and J.-L. Zhang; *Phosphorus Sulfur Silicon*, 1995, **104**, 33.

95S173 R.K. Bansal, G. Pandey, R. Gupta, K. Karaghiosoff and A. Schmidpeter; *Synthesis*, 1995, 173.

95S361 R.K. Bansal, K. Karaghiosoff, N. Gandhi and A. Schmidpeter; *Synthesis*, 1995, 361.

95TL447 G. Baccolini, G. Orsolan and E. Mezzina; *Tetrahedron Lett.*, 1995, **36**, 447.

95ZAAC(621)34 G. Becker, G. Heckmann, K. Hübler and W. Schwarz; *Z. Anorg. Allg. Chem.*, 1995, **621**, 34.

95ZN(B)558 R.K. Bansal, N. Ghandi, A. Schmidpeter and K. Karaghiosoff; *Z. Naturforsch. B*, 1995, **50**, 558.

95ZN(B)1543 A. Schmidpeter, S. Plank and K. Polborn; *Z. Naturforsch. B*, 1995, **50**, 1543.

96AG292 L. Weber; *Angew. Chem.*, 1996, **108**, 292; *Angew. Chem. Int. Ed. Engl.*, 1996, **35**, 271.

96AG354 V. Thelen, D. Schmidt, M. Nieger, E. Niecke and W.W. Schoeller; *Angew. Chem.*, 1996, **108**, 354; *Angew. Chem. Int. Ed. Engl.*, 1996, **35**, 313.

96AG2282 C. Charrier, N. Maigrot, L. Ricard, P. le Floch and F. Mathey; *Angew. Chem.*, 1996, **108**, 2282; *Angew. Chem. Int. Ed. Engl.*, 1996, **35**, 2133.

96BSB675 V. Caliman, P.B. Hitchcock, J.F. Nixon and N. Sakarya; *Bull. Soc. Chim. Belg.*, 1996, **105**, 675.

96CB465 D. Gudat, A.W. Holderberg, S. Kotila and M. Nieger; *Chem. Ber.*, 1996, **129**, 465.

96CB1083 G. Jochem, H. Nöth and A. Schmidpeter; *Chem. Ber.*, 1996, **129**, 1083.

96CB1493 A. Schmidpeter, F. Steinmüller and H. Nöth; *Chem. Ber.*, 1996, **129**, 1493.

96CC441 M.H.A. Benvenutti, P.B. Hitchcock, J.F. Nixon and M.D. Vargas; *J. Chem. Soc., Chem. Commun.*, 1996, 441.

96CC1591 M.D. Francis, D.E. Hibbs, M.B. Hursthouse, C. Jones and K.M.A. Malik; *J. Chem. Soc., Chem. Commun.*, 1996, 1591.

96CC2071 B. Breit; *J. Chem. Soc., Chem. Commun.*, 1996, 2071.

96CC2751 P.B. Hitchcock, J.F. Nixon and N. Sakarya; *J. Chem. Soc., Chem. Commun.*, 1996, 2751.

96CEJ221 G. Jochem, A. Schmidpeter and H. Nöth; *Chem. Eur. J*, 1996, **2**, 221.

96CHEC-II(3)709 A. Schmidpeter; "Comp. Heterocycl. Chem.", 2nd ed., Pergamon, 1996, **3**, 709.

96CHEH-II(3)715 A. Schmidpeter; "Comp. Heterocycl. Chem.", 2nd ed., Pergamon, 1996, **3**, 715.

96CHEC-II(4)771 A. Schmidpeter; "Comp. Heterocycl. Chem.", 2nd ed., Pergamon, 1996, **4**, 771.

96CHEC-II(4)819 A. Schmidpeter; "Comp. Heterocycl. Chem.", 2nd ed., Pergamon, 1996, **4**, 819.

96IC4690 L. Nyulászi; *Inorg. Chem.*, 1996, **35**, 4690.

96IC6667 A.-C. Gaumont, B. Pellerin, J.-L. Cabioch, X. Morise, M. Lesvier, P. Savignac, P. Guenot and J.-M. Denis; *Inorg. Chem.*, 1996, **35**, 6667.

96IZV252 S.V. Chapyshev, U. Bergsträßer and M. Regitz; *Izv. Akad. Nauk SSSR, Ser. Khim.*, 1996, **45**, 252.

96JA6317 P. von R. Schleyer, C. Maerker, A. Dransfeld, H. Jiao and N.J.R.V. Eikema Hommes; *J. Am. Chem. Soc.*, 1996, **118**, 6317.

96JA7630 P.L. Arnold, F.G.N. Cloke, P.B. Hitchcock and J.F. Nixon; *J. Am. Chem. Soc.*, 1996, **118**, 7630.

96JCS(D)739 M.H.A. Benvenutti, P.B. Hitchcock, J.F. Nixon and M.D. Vargas; *J. Chem. Soc., Dalton Trans.*, 1996, 739.

96JOM(512)141 C. Müller, R. Bartsch, A. Fischer, P.G. Jones and R. Schmutzler; *J. Organomet. Chem.*, 1996, **512**, 141.

96JOM(524)67 E. Lindner, E. Bosch, C. Maichle-Mößmer and U. Abram; *J. Organomet. Chem.*, 1996, **524**, 67.

96JPC6194 L. Nyulászi; *J. Phys. Chem.*, 1996, **100**, 6194.

96JPC13447 M.N. Glukhovtsev, A. Dransfeld and P. von R. Schlcycr; *J. Phys. Chem.*,1996, **100**, 13447.

96KGS67 S.V. Chapyshev, U. Bergsträßer and M. Regitz; *Khim. Geterotsikl. Soedin.*, 1996, 67.

96MI1 G. Becker, J. Weidlein, S. Abele, M. Böhringer, K.W. Klinkhammer, F. Schaller and H. Schmidt; Abstracts, XIth International Symposium on Organosilicon Chemistry, Montpellier 1996.

96MI94 N.G. Khusainova, T.A. Zyablikova, G.R. Reshetkova, R.A. Cherkasov and A.N. Pudovik; Abstracts, XIth International Conference on Chemistry of Phosphorus Compounds, Kazan, 1996, p. 94.

96MI96 B.J Busykin, R.M. Eliseenkova; Abstracts, XIth International Conference on Chemistry of Phosphorus Compounds, Kazan, 1996, p. 96.

96MI103 G. Baccolini; *Topics Heterocycl. Syst.*, 1996, **1**, 103.

B-96MI161 G. Becker, G. Ditten, K. Hübler, U. Hübler, K. Merz, M. Niemeyer, N. Seidler, M. Westerhausen and Z. Zheng; in "Organosilicon Chemistry II", eds. N. Auner and J. Weis, VCH, Weinheim, 1996, p. 161.

96PS(109-110)109 L. Nyulászi and T. Veszprémi; *Phosphorus Sulfur Silicon*, 1996, **109–110**, 109.

96PS(111)197 E.E. Korshin, R.M. Eliseyenkova, L.G. Zakharova, L.I. Sabirova and Y.A. Levin; *Phosphorus Sulfur Silicon*, 1996, **111**, 197.

96PS(112)121	R.K. Bansal, N. Gupta, R. Gupta, G. Pandey and M. Agarwal; *Phosphorus Sulfur Silicon*, 1996, **112**, 121.
96PS(115)99	M. Regitz and S. Krill; *Phosphorus Sulfur Silicon*, 1996, **115**, 99.
96PS(118)129	A. Schmidpeter, H. Tautz and F. Steinmüller; *Phosphorus Sulfur Silicon*, 1996, **118**, 129.
96T1053	B. Manz and G. Maas; *Tetrahedron*, 1996, **52**, 10053.
96TH1	W. Nießen; *Ph.D. Thesis*, Universität Kaiserslautern, 1996.
96TL9025	M.K. Denk, S. Gupta and R. Ramachandran; *Tetrahedron Lett.*, 1996, **37**, 9025.
96ZAAC(622)543	L. Weber, O. Sommer, H.-G. Stammler and B. Neumann; *Z. Anorg. Allg. Chem.*, 1996, **622**, 543.
96ZK(211)341	R. Ficker, W. Hiller, M. Böhringer and G. Becker; *Z. Kristallogr.*, 1996, **211**, 341.
96ZN(B)1022	H.-P. Schrödel, A. Schmidpeter, H. Nöth and M. Schmidt; *Z. Naturforsch. B*, 1996, **51**, 1022.
96ZOB386	L.V. Ermolaeva, A.M. Kibardin, A.N. Pudovik and A.I. Konovalov; *Zh. Obshch. Khim.*, 1996, **66**, 386; *Russ. J. Gen. Chem.*, 1996, **66**, 375.
96ZOB1455	A.M. Kibardin, T.V. Gryaznova, A.N. Pudovik and V.A. Naumov; *Zh. Obshch. Khim.*, 1996, **66**, 1455; *Russ. J. Gen. Chem.*, 1996, **66**, 1418.
97AOC(41)1	L. Weber; *Adv. Organomet. Chem.*, 1997, **41**, 1.
97BSB455	W. Fiedler, M. Regitz and G. Bertrand; *Bull. Soc. Chim. Belg.*, 1997, **106**, 455.
97BSF853	N. Maigrot, C. Charrier, L. Ricard and F. Mathey; *Bull. Soc. Chim. Fr.*, 1997, **134**, 853.
97CB89	H.-P. Schrödel and A. Schmidpeter; *Chem. Ber./Recueil*, 1997, **130**, 89.
97CB779	B. Manz, U. Bergsträßer, J. Kerth and G. Maas; *Chem. Ber./Recueil*, 1997, **130**, 779.
97CB1299	M. Scheer, K. Schuster, A. Krug and H. Hartung; *Chem. Ber./Recueil*, 1997, **130**, 1299.
97CC305	S.J. Black, M.D. Francis and C. Jones; *Chem. Commun.*, 1997, 305.
97CC1305	V. Caliman, P.B. Hitchcock, J.F. Nixon, L. Nyulászi and N. Sakarya; *J. Chem. Soc., Chem. Commun.*, 1997, 1305.
97CC1739	V. Caliman, P.B. Hitchcock and J.F. Nixon; *J. Chem. Soc., Chem. Commun.*, 1997, 1739.
97CC2095	C.J. Carmalt, V. Lomeli, B.G. McBurnett and A.H. Cowley; *J. Chem. Soc., Chem. Commun.*, 1997, 2095.
97CCR(163)71	D. Gudat; *Coord. Chem. Rev.*, 1997, **163**, 71.
97CHE587	S.V. Chapyshef and V.M. Anisimov; *Chem. Heterocycl. Compd.*, 1997, **33**, 587.
97HAC97	E. Ocando-Mavarez, G. Martin, L. Rodriguez and W. Munoz; *Heteroatom Chem.*, 1997, **8**, 97.
97ICA(256)47	J.G. Kraaijkamp, D.M. Grove, G. van Koten, J.M. Ernsting, A. Schmidpeter, K. Goubitz, C.H. Stam and H. Schenk; *Inorg. Chim. Acta*, 1997, **256**, 47.
97JCPC609	L. Nyulászi, P. Várnai, W. Eisfeld and M. Regitz; *J. Comput. Chem.*, 1997, **18**, 609.
97JCS(D)2183	S.J. Black, M.D. Francis and C. Jones; *J. Chem. Soc., Dalton Trans.*, 1997, 2183.
97JCS(D)4321	S.J. Black, D.E. Hibbs, M.B. Hursthouse, C. Jones, K.M.A. Malik and R.C. Thomas; *J. Chem. Soc., Dalton Trans.*, 1997, 4321.
97JOM(527)291	M.D. Francis, D.E. Hibbs, M.B. Hursthouse, C. Jones and K.M.A. Malik; *J. Organomet. Chem.*, 1997, **527**, 291.
97JOM(529)69	C. Charrier, N. Maigrot and F. Mathey; *J. Organomet. Chem.*, 1997, **529**, 69.
97JOM(529)375	R. Bartsch, S. Datsenko, N.V. Ignatiev, C. Muller, J.F. Nixon and C.J. Pickett; *J. Organomet. Chem.*, 1997, **529**, 375.
97JOM(534)89	S.J. Black and C. Jones; *J. Organomet. Chem.*, 1997, **534**, 89.
97JOM(536-537)273	V. Caliman, P.B. Hitchcock and J.F. Nixon; *J. Organomet. Chem.*, 1997, **536–537**, 273.
97MI1	R.K. Bansal and P. von R. Schleyer; personal communication.
97MI2	G.R. Revankar, T.S. Rao, J.O. Ojwang, R.F. Rando and T.L. Wallace; Book of Abstracts, 213th ACS National Meeting, San Francisco 1997.
97MI309	D. Böhm, F. Heinemann, D. Hu, S. Kummer and U. Zenneck; *Collect. Czech. Chem. Commun.*, 1997, **62**, 309.
97OM3188	L. Weber, S. Uthmann, B. Torwiehe, R. Kirchhoff, R. Boese and D. Bläser; *Organometallics*, 1997, **16**, 3188.
97PS(123)125	A.A. Tolmachev, S.I. Dovgopoly, A.N. Kostyuk, I.V. Komarov and A.M. Pinchuk; *Phosphorus Sulfur Silicon*, 1997, **123**, 125.
97PS(124-125)567	P.B. Hitchcock, J.F. Nixon and N. Sakarya; *Phosphorus Sulfur Silicon*, 1997, **124–125**, 567.
97PS(126)89	C.-J. Zhang and C.-G. Zhan; *Phosphorus Sulfur Silicon*, 1997, **126**, 89.
97T2357	R.R. Sauers; *Tetrahedron*, 1997, **53**, 2357.
98AG1139	F.G.N. Cloke, P.B. Hitchcock, P. Hunnable, J.F. Nixon, L. Nyulászi, E. Niecke and V. Thelen; *Angew. Chem.*, 1998, **110**, 1139; *Angew. Chem. Int. Ed. Engl.*, 1998, **37**, 1083.
98AG1318	F. Tabellion, A. Nachbauer, S. Leininger, C. Peters, M. Regitz and F. Preuss; *Angew. Chem.*, 1998, **110**, 1318; *Angew. Chem. Int. Ed. Engl.*, 1998, **37**, 1083.
98CC797	P.L. Arnold, F.G.N. Cloke and J.F. Nixon; *J. Chem. Soc., Chem. Commun.*, 1998, 1537.
98CC1537	V. Caliman, P.B. Hitchcock and J.F. Nixon; *J. Chem. Soc., Chem. Commun.*, 1998, 1537.
98CEJ1910	O.J. Scherer, S. Weigel and G. Wolmershäuser; *Chem. Eur. J.*, 1998, **4**, 1910.
98EJI1079	N. Gupta, C.B. Jain, J. Heinicke, R.K. Bansal and P.G. Jones; *Eur. J. Inorg. Chem.*, 1998, 1079.
98EJI1087	D. Gudat; *Eur. J. Inorg. Chem.*, 1998, 1087.
98HAC1	V. Caliman, P.B. Hitchcock and J.F. Nixon; *Heteroatom Chem.*, 1998, **9**, 1.
98HAC183	J. Heinicke, M. He, R. Kadyrov and P.G. Jones; *Heteroatom Chem.*, 1998, **9**, 183.

98HAC333	N. Gupta, C.B. Jain, J. Heinicke, N. Bharatiya, R.K. Bansal and P.G. Jones; *Heteroatom Chem.*, 1998, **9**, 333.
98HAC445	R.K. Bansal, N. Gupta, N. Bharatiya, G. Gupta, A. Surana, G. Hackenbracht and K. Karaghiosoff; *Heteroatom Chem.*, 1998, **9**, 445.
98IC4413	A. Dransfeld, L. Nyulászi and P. von R. Schleyer; *Inorg. Chem.*, 1998, **37**, 4413.
98ICA273	K. Angermund, A. Eckerle, J. Monkiewicz, C. Krüger and G. Wilke; *Inorg. Chim. Acta*, 1998, **270**, 273.
98JIC648	R.K. Bansal, N. Gupta and A. Surana; *J. Indian Chem. Soc.*, 1998, **75**, 648.
B-98MI1	K.B. Dillon, F. Mathey and J.F. Nixon; "Phosphorus: The Carbon Copy", Wiley, Baffins Lane, 1998.
98MI1	G. Becker, M. Bader, U. Hübler and W. Schwarz; "Poster A3, XIII Internat. Conf. Organomet. Chem.", München, 1998.
98NJC651	L. Nyulászi, U. Bergsträßer, M. Regitz, P. von R. Schleyer; *New J. Chem.*, 1998, 651.
98OM3826	M.D. Francis, C. Jones, G.B. Deacon, E.E. Delbridge and P.C. Junk; *Organometallics*, 1998, **17**, 3826.
98PS(140)183	T.-B. Huang, L.-F. Liu, J.-L. Zhang, X.-L. Xu, W.-F. Huang, R.-Y. Chen, K. Wang, X.M. Yu and X.P. Liu; *Phosphorus Sulfur Silicon*, 1998, **140**, 183.
98RCB1583	A.R. Kudinov, D.A. Loginov, P.V. Petrovskii and M.I. Rubinskaya; *Russ. Chem. Bull.*, 1998, **47**, 1583.
98S1305	A. Mack, E. Pierron, T. Allspach, U. Bergsträßer and M. Regitz; *Synthesis*, 1998, 1305.
98ZAAC(624)399	M. Scheer and J. Krug; *Z. Anorg. Allg. Chem.*, 1998, **624**, 399.
98ZOB1479	N.G. Khusianova, T.A. Zyablikova, G.R. Reshetkova, E.A. Lamm and R.A.; *Zh. Obshch. Khim.*, 1998, **68**, 1479; *Russ. J. Gen. Chem.*, 1998, **68**, 1411.
99ACR751	O.J. Scherer; *Acc. Chem. Res.*, 1999, **32**, 751.
99AG158	U. Rohde, F. Ruthe. P.G. Jones and R. Streubel; *Angew. Chem.*, 1999, **111**, 158; *Angew. Chem. Int. Ed. Engl.*, 1999, **38**, 218.
99CC1283	A.J. Ashe III and X. Fang; *J. Chem. Soc., Chem. Commun.*, 1999, 1283.
99CC1731	F.G.N. Cloke, J.R. Hanks, P.B. Hitchcock and J.F. Nixon; *J. Chem. Soc., Chem. Commun.*, 1999, 1731.
99CC2363	F.G.N. Cloke, P.B. Hitchcock, J.F. Nixon, D.J. Wilson, F. Tabellion, U. Fischbeck, F. Preuss, M. Regitz and L. Nyulászi; *J. Chem. Soc., Chem. Commun.*, 1999, 2363.
99CEJ3143	A. Elvers, F.W. Heinemann, B. Wrackmeyer and U. Zenneck; *Chem. Eur. J.*, 1999, **5**, 3143.
99CEJ274	M. Sanchez, R. Reau, C.J. Marsden, M. Regitz and G. Bertrand; *Chem. Eur. J.*, 1999, **5**, 274.
99CR(C)351	D. Bourissou, P. Le Floch, F. Mathey and G. Bertrand; *C.R. Acad. Sci., Ser. IIc: Chim.*, 1999, 351.
99EJI41	M.K. Denk, S. Gupta and A.J. Lough; *Eur. J. Inorg. Chem.*, 1999, 41.
99EJO587	A. Mack, U. Bergsträßer, G.J. Reiß and M. Regitz; *Eur. J. Org. Chem.*, 1999, 587.
99EJO2633	J. Kerth and G. Maas; *Eur. J. Org. Chem.*, 1999, 2633.
99HAC598	R.K. Bansal, L. Hemrajani and N. Gupta; *Heteroatom Chem.*, 1999, **10**, 598.
99IC1971	M.D. Mikoluk and R.G. Cavell; *Inorg. Chem.*, 1999, **38**, 1971.
99IC2791	M.D. Mikoluk, R. McDonald and R.G. Cavell; *Inorg. Chem.*, 1999, **38**, 2791.
99IC4056	M.D. Mikoluk, R. McDonald and R.G. Cavell; *Inorg. Chem.*, 1999, **38**, 4056.
99JCS(D)4057	J.J. Durkin, M.D. Francis, P.B. Hitchcock, C. Jones and J.F. Nixon; *J. Chem. Soc., Dalton Trans.*, 1999, 4057.
99JMOC(A143)143	B. Breit; *J. Mol. Catal. A*, 1999, **143**, 143.
99JOM(577)337	C.B. Jain, D.C. Sharma, N. Gupta, J. Heinicke and R.K. Bansal; *J. Organomet. Chem.*, 1999, **577**, 337.
99JOM(580)156	M.D. Francis, D.E. Hibbs, P.B. Hitchcock, M.B. Hursthouse, C. Jones, T. Mackewitz, J.F. Nixon, L. Nyulászi, M. Regitz and N. Sakarya; *J. Organomet. Chem.*, 1999, **580**, 156.
99JOM(584)58	G.K.B. Clentsmith, P.B. Hitchcock, M.B. Hursthouse, J.F. Nixon, L. and Sakarya; *J. Organomet. Chem.*, 1999, **584**, 58.
99JOM(584)87	C.S.J. Callaghan, P.B. Hitchcock, M.B. Hursthouse and J.F. Nixon; *J. Organomet. Chem.*, 1999, **580**, 156.
99JOM(588)28	L. Nyulászi and J.F. Nixon; *J. Organomet. Chem.*, 1999, **588**, 28.
99MI257	J.F. Nixon; *Adv. Strained Interesting Org. Mol.*, 1999, 257.
99MI485	L.L. Lohr and W.M. Ahrens; *J. Mol. Struct.*, 1999, **485–486**, 57.
99OM793	C. Callaghan, G.K.B. Clentsmith, F.G.N. Cloke, P.B. Hitchcock, J.F. Nixon and D.M. Vickers; *Organometallics*, 1999, **18**, 793.
99OM2833	P. Sekar, M. Scheer, A. Voigt and R. Kirmse; *Organometallics*, 1999, **18**, 2833.
99OM3306	M.D. Mikoluk, R. McDonald and R.G. Cavell; *Organometallics*, 1999, **18**, 3306.
99PS(144-146)141	O.J. Scherer, C. Hofmann, M. Ehses, S. Weigel and G. Wolmershäuser; *Phosphorus Sulfur Silicon*, 1999, **144–146**, 141.
99PS(144-146)261	A. Mack, F. Tabellion, C. Peters, A. Nachbauer, U. Bergstrasser, F. Preuss and M. Regitz; *Phosphorus Silicon Sulfur*, 1999, **144–146**, 261.
99PS(144-146)725	A. Elvers, F. Heinemann, S. Kummer, B. Wrackmeyer, M. Zeller and U. Zenneck; *Phosphorus Sulfur Silicon*, 1999, **144–146**, 725.
99RCB1362	A.R. Kudinov, P.V. Petrovskii and M.I. Rubinskaya; *Russ. Chem. Bull.*, 1999, **48**, 1362.
99S264	R.K. Bansal, N. Gupta, J. Heinicke, G.N. Nikonov, F. Saguitova and D.C. Sharma; *Synthesis*, 1999, 264.
99S1642	S.F.M. Asmus, U. Bergsträßer and M. Regitz; *Synthesis*, 1999, **9**, 1642.
99TH1	S.F.M. Asmus; *Ph.D. Thesis*, Universität Kaiserslautern, 1999.
99TH2	C. Peters; *Ph.D. Thesis*, Universität Kaiserslautern, 1999.
99TL1565	R.K. Bansal, A. Surana and N. Gupta; *Tetrahedron Lett.*, 1999, **40**, 1565.
99TL3815	M.D. Francis, C. Jones and C.P. Morley; *Tetrahedron Lett.*, 1999, **40**, 3815.

99ZN(B)187	K. Polborn, A. Schmidpeter, G. Märkl and A.Willhalm; *Z. Naturforsch. B*, 1999, **54**, 187.
00AG1484	O.J. Scherer, T. Hilt and G. Wolmershäuser; *Angew. Chem.*, 2000, **112**, 1484; *Angew. Chem. Int. Ed. Engl.*, 2000, 39, 1426.
00AG2174	T. Clark, A. Elvers, F.W. Heinemann, M. Hennemannn, M. Zeller and U. Zenneck; *Angew. Chem.*, 2000, **112**, 2174; *Angew. Chem. Int. Ed. Engl.*, 2000, **39**, 2087.
00AG2393	F.E. Hahn, L. Wittenbecher, D. Le Van, R. Fröhlich and B. Wibbeling; *Angew. Chem.*, 2000, **112**, 2393; *Angew. Chem. Int. Ed. Engl.*, 2000, **39**, 2307.
00AG2892	S. Ito, H. Sugiyama and M. Yoshifuji; *Angew. Chem.*, 2000, **112**, 2892; *Angew. Chem., Int. Ed. Engl.*, 2000, **39**, 2781.
00AG3211	D. Gudat, A. Haghverdi and M. Nieger; *Angew. Chem.*, 2000, **112**, 3211; *Angew. Chem. Int. Ed. Engl.*, 2000, **39**, 3084.
00CC1305	S.C. Clendenning, P.B. Hitchcock, J.F. Nixon and L. Nyulászi; *J. Chem. Soc., Chem. Commun.*, 2000, 1305.
00CC2027	M.D. Francis, P.B. Hitchcock and J.F. Nixon; *Chem. Commun.*, 2000, 2027.
00CC2087	N. Burford, T.S. Cameron, K.N. Robertson, A.D. Phillips and H.A. Jenkins; *J. Chem. Soc. Chem. Comm.*, 2000, 2087.
00CIUZ374	O.J. Scherer; *Chem. uns. Zeit*, 2000, **34**, 374.
00CEJ3414	D. Gudat, A. Haghverdi and M. Nieger; *Chem. Eur. J.*, 2000, **6**, 3414.
00EJI369	W.W. Schoeller; *Eur. J. Inorg. Chem.*, 2000, 369.
00JA4557	S.E. d'Arbeloff-Wilson, P.B. Hitchcock, S. Krill, J.F. Nixon, L. Nyulászi and M. Regitz; *J. Am. Chem. Soc.*, 2000, **122**, 4557.
00JCS(D)1715	G.K.B. Clentsmith, F.G.N. Cloke, M.D. Francis, J.C. Green, P.B. Hitchcock, J.F. Nixon, J.L. Suter and D.M. Vickers; *J. Chem. Soc., Dalton Trans.*, 2000, 1715.
00JCS(D)3534	F.G.N. Cloke, J.C. Green, J.R. Hanks, J.F. Nixon and J.L. Suter; *J. Chem. Soc., Dalton Trans.*, 2000, 3534.
00JOM(601)335	R.B. Hitchcock, J.F. Nixon and N. Sakarya; *J. Organomet. Chem.*, 2000, **601**, 335.
00MI1	J. Heinicke, K. Steinhauser, N. Peulecke and A. Spannenberg; to be published.
00MI(11)405	M.H. Araujo, P.B. Hitchcock, J.F. Nixon and M.D. Vargas; *J. Braz. Chem. Soc.*, 2000, **11**, 405.
00OM219	F.G.N. Cloke, P.B. Hitchcock, J.F. Nixon and D.J. Wilson; *Organometallics*, 2000, **19**, 219.
00OM1713	G.B. Deacon, E.E. Delbridge, G.D. Fallon, C. Jones, D.E. Hibbs, M.B. Hursthouse, B.W. Skelton and A.H. White; *Organometallics*, 2000, **19**, 1713.
00OM4283	F.W. Heinemann, H. Pritzkow, M. Zeller and U. Zenneck; *Organometallics*, 2000, **19**, 4283.
00PS(159)215	Y. Kandri, A. El Halloui, E.M. Essassi, C. Dargelas, L.L. Lopez and J. Bellan; *Phosphorus Sulfur Silicon*, 2000, **159**, 215.
00S360	S.G. Ruf, A. Mack, J. Steinbach, U. Bergsträßer and M. Regitz; *Synthesis*, 2000, 360.
00S417	C. Peters, F. Tabellion, M. Schröder, U. Bergsträßer, F. Preuss and M. Regitz; *Synthesis*, 2000, 417.
00T27	J. Grobe, D. Le Van, T. Pohlmeyer, F. Immel, H. Pucknat, B. Krebs, J. Kuchinke and M. Läge; *Tetrahedron*, 2000, **56**, 27.
00T35	J. Kerth, U. Werz and G. Maas; *Tetrahedron*, 2000, **56**, 35.
00T63	S.G. Ruf, U. Bergsträßer and M. Regitz; *Tetrahedron*, 2000, **56**, 63.
00TH1	R. Ruf, *Ph.D. Thesis*, Universität Stuttgart, 2000.
00TH2	E. Pierron; *Ph.D. Thesis*, Universität Kaiserslautern, 2000.
00TH3	T. Schmidt; *Ph.D. Thesis*, Universität Kaiserslautern, 2000.
00ZOB1048	V.F. Mironov, N.G. Khusainova, G.R. Reshetkova, T.A. Zyablikova and R.A. Cherkasov; *Zh. Obshch. Khim.*, 2000, **70**, 1048; *Russ. J. Gen. Chem.*, 2000, **70**, 984.
01JCS(D)1013	R. Bartsch, F.G.N. Cloke, J.C. Green, R.M. Matos, J.F. Nixon, R.J. Suffolk, J.L. Suter and D.J. Wilson; *J. Chem. Soc., Dalton Trans.*, 2001, 1305.
01JOM(617-618)383	D. Gudat, A. Haghverdi and M. Nieger; *J. Organomet. Chem.*, 2001, **617–618**, 383.
01MI1	A. Schmidpeter, F. Breitsameter, P. Mayer and G. Kramer, to be published.
01MI2	E. Urnezius, W. Brennessel and J. Ellis; Book of Abstracts, 221st ACS National Meeting, San Diego 2001, Abstract 650.
01MI(538)189	T. Veszpremi, J. Matrai, J. Heinicke, M.K. Kindermann; *J. Mol. Struct. (Theochem.)*, 2001, **538**, 189.

5.1
Six-membered rings: Phosphinanes, Dihydro- and Tetrahydro-phosphinines

MICHAEL J. GALLAGHER
University of New South Wales, Sydney, NSW, Australia

5.1.1 INTRODUCTION

This chapter deals with six-membered rings containing five carbons and one phosphorus and none, one or two double bonds and in which the smallest ring containing the heteroatom is six membered. Not surprisingly, such structures have been known for a long time and probably the first phosphorus–carbon heterocycle ever made was 1-phenylphosphinane ⟨15CB1473⟩. As in all areas of phosphorus chemistry consistency of nomenclature leaves much to be desired. The IUPAC name for the parent ring, phosphinane, is used here though the older alternative, phosphorinane, is still used, and is the preferred name in Chemical Abstracts (CA). Unsaturated derivatives are named as reduced forms of the fully unsaturated phosphinine (CA phosphorin). Polycyclic systems are named in CA as analogues of the corresponding nitrogen heterocycles, e.g., phosphinoline, etc., though, curiously, acridophosphine, not phosphacridine.

Earlier reviews of a more general nature deal with some of the topics covered here, notably books by Mann ⟨70MI1⟩, Quin ⟨81MI1⟩ and the first and second editions of CHEC Other Reviews containing much useful data deal with phosphasugars, ⟨83JCC(2)351, 84ACC(42)135⟩, cyclic phosphines ⟨90MI2(10)⟩, stereochemistry ⟨87MI1⟩ ^{31}P NMR data, including coupling constant data, ⟨91MI1, 94MI1⟩; ring enlargement reactions ⟨93S931⟩ and cycloadditions. ⟨86PS(26)203⟩ Generally speaking the emphasis is on work done in the last twenty years.

In this chapter there will be no separation on the basis of the oxidation state of the phosphorus atom since the two common levels (P(III) and P(V)) are interconvertible by standard chemical procedures and other oxidation levels are rare except as reactive intermediates which, even if visible (usually by ^{31}P NMR), are not isolated.

5.1.2 STRUCTURE AND STEREOCHEMISTRY

Structural information is principally obtained by X-ray and NMR spectroscopy. In phosphinanes, ^{1}H spectra will often be too complex to be useful as a result of the extra $^{2,3}J_{PH}$ and, in general, ^{13}C data will be more informative, particularly if the ring carbons are substituted, since angular dependence of $^{2,3}J_{PC}$ is well established and a large body of data is available ⟨87MI1, 94MI1⟩ though caution is necessary with P(III). ^{31}P NMR spectra will, unless ^{1}H decoupled, often be too broad to be of value, though if the fully coupled spectra can be resolved they are very useful in assigning both structure and stereochemistry. Diastereoisomers normally have different ^{31}P chemical shifts but $\Delta\delta$ may be quite small and assignments are usually insecure unless supported by other data. Many early studies did not report ^{31}P shifts. Unsaturated rings and those carrying several substituents are generally more easily analysed.

5.1.2.1 Phosphinanes

The ring is normally a conformationally mobile chair, flattened at the phosphorus end. There is no evidence of strain; ^{31}P NMR shifts, after allowing for substituent effects elsewhere in the ring, are not usually significantly different from their acyclic analogues and activation energies for inversion at P(III) are similar for ring and chain molecules ⟨81JA7602, 69JA5686⟩. The greater length of the P–C vs C–C bonds means that the substituents at phosphorus are less affected by the conformational effects so well established in cyclohexanes, e.g., where there are two substituents at phosphorus, i.e., P(V), there is often little difference in conformational preference of the groups unless they are very bulky, a fact which can make assignment of stereochemistry at phosphorus difficult. Quin *et al.* have observed that both conformers may occur in the same crystal ⟨74TL3473, 72PS(1)1255⟩. P(III) systems possess a preference for axial orientation of the substituent at phosphorus including H ⟨87MI1, 81MI1, 71T4245⟩. The parent compound, (**1**), was shown in an early ^{1}H NMR study to contain essentially only the axial isomer and temperature lowering failed to reveal the equatorial form. A subsequent study of (**2**) found only the *cis* isomer with axial P–H. The 2,6-methyls of (**2**) effectively lock the 4-substituent in the equatorial position but the *trans* isomer with P–H equatorial should be perfectly reasonable energetically. Despite the failure to detect the other conformer, it still seems most likely that the ring is in mobile equilibrium and the axial isomer is strongly preferred, since P-alkyl phosphinanes exhibit both isomers interconverting, though still with an axial preference unless very bulky, e.g., *t*-butyl. Equally surprisingly, phenyl substituents at phosphorus in both P(III) and P(V) prefer the axial position in both solid state and solution much more so than alkyl groups.

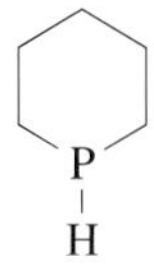

(**1**)

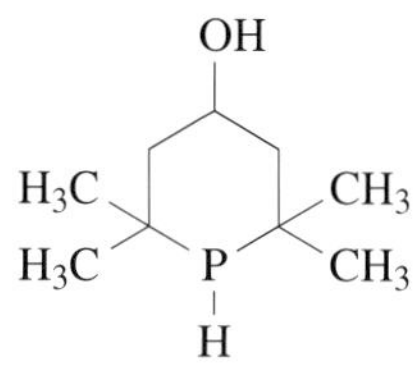

(**2**)

Where the substituent at phosphorus is a chalcogen, the doubly bound atom normally prefers the equatorial position, possibly to minimise electron repulsion. In general, however, conformational effects of substituents at the ring carbons energetically outweigh those at phosphorus. This may also reasonably be attributed to the greater length of the P–C bond though it does not account for the axial preference of the P–H and P–Ph groups, the origins of which remain obscure though, at least in the case of hydrogen, the axial preference is observed with other heterocycles, e.g., sulphur ⟨71T4245⟩.

The difference in ^{31}P chemical shift ($\Delta\delta$) between diastereoisomers arising from the ring structure is often large (2–9 ppm) in P(III) but small (ca. 2 ppm) in P(IV) at least in 1,4-disubstituted systems for which most data are available. Limited data suggest smaller differences for other substitution patterns. P(III) molecules with an axial substituent at P will usually absorb upfield of their equatorial isomers but the data base is too small for complete confidence and the reverse seems true for phenyl substituents ⟨83OMR345⟩. Chalcogenides on the other hand, though showing smaller differences, show consistently an upfield shift of the axial isomer and there are sufficient data for this to be a good indicator where both isomers are visible ⟨87MI1, 95JOC6281⟩. Many preparations have afforded only one isomer including phosphinane itself and its sulphide, in both of which only the H_{ax} isomer was observed.

Conformational analyses of many functionalised phosphinanes have been carried out using ^{1}H and ^{13}C NMR ⟨87MI1, 81MI1, 88IZV(K)59, 86ZOB817, 86ZOB1973, 83OMR457, 83OMR345⟩ and many are supported by X-ray structures ⟨95JOC6281, 95PS(103)137, 95PS(104)223, 92ZOB762, 87IZV(K)79, 86IZV(K)69, 85ZSK(26)78, 84ZSK(25)79, 80JA4379⟩. With only occasional minor distortions the ring shapes are predictable and straightforward.

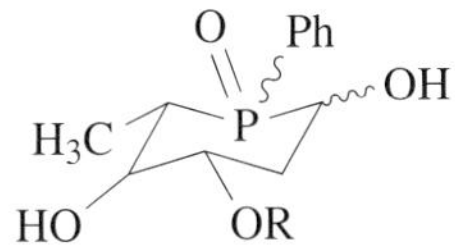

(**3**)

Probably the largest single class of phosphinanes is the phosphapyranoses in which the phosphorus replaces the ring oxygen of a pyranose, e.g., (**3**). Since they are invariably prepared from carbohydrate precursors of known stereochemistry, the ^{1}H, ^{13}C and X-ray data allow the determination of absolute stereochemistry at phosphorus, a unique situation in phosphinanes. There appears to be a useful correlation between $^2J_{PCH}$ and the dihedral angle in the system O=P–C–H which, if supported by further data, could well be a most useful stereochemical tool for phosphorus(V) heterocycles in general ⟨82JOC191, 96HAC391, 91S1099⟩.

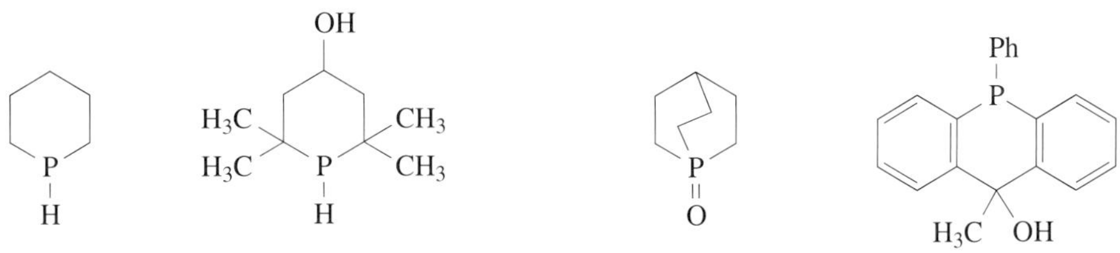

(**4**) (**5**) (**6**) (**7**)

Bi- and polycyclic phosphinanes have been less studied though several rigid systems have been prepared, mainly for spectroscopic studies, e.g., (**6**, **7**). X-ray structures confirm that the phosphinane ring is a chair in the more complex structures also and there does not seem to be an example of a well defined boat or twist boat structure except in structurally constrained systems, e.g., (**6**). ^{1}H spectra of non-rigid systems with more than one ring are often too complex to be useful. ^{17}O, ^{18}O and ^{15}N NMR spectra of P heterocycles show considerable promise ⟨77JA3243, 80TL1361, 91MI2, 95JOC6281⟩ particularly in conformational analysis, but have not been extensively used, probably because of difficulties in measurement, though improving instrumentation may overcome this. δ^{17}O in particular, shows large axial-equatorial differences (8–33 ppm) and the data once again support an axial upfield of equatorial relationship ⟨86JA6651⟩.

5.1.2.2 Di- and Tetrahydrophosphinines

Increasing the unsaturation in the ring reduces the possibilities for structural and stereochemical complexity and simplifies spectroscopic analysis. Where the unsaturation arises from ring fusion to a benzenoid system then the phosphorus ring adopts a flexible half chair conformation amenable to ^{1}H and ^{13}C NMR analysis which will usually be straightforward depending on the degree of complexity elsewhere in the molecule. Non-fused systems can be obtained by dehydration of phosphorinanols and from cycloaddition of P(II) systems to conjugated dienes, a field which has increased rapidly in recent times ⟨90MI1⟩. Dihydrophosphinines, most commonly the 1,4-isomers both with and without benzene ring annelation, are folded about the longitudinal axis, e.g., (**7**), about which it can also flex and does so in the *trans* isomer. The *cis* isomer, depending on the size of the substituents, is normally constrained to one conformer to reduce non-bonded interactions between the 1,4-substituents. The 1,2-isomers are less common though known for some time. Both types routinely occur as intermediates in phosphinine syntheses.

5.1.3 SYNTHESIS AND REACTIVITY

5.1.3.1 Phosphinanes

The first report of the parent compound, (**1**) and some P-alkyl derivatives, was by Wagner in a 1963 patent and involved the bisalkylation of sodium phosphide with 1,5-dichloropentane but no data were reported ⟨63USP3086053⟩. The method does not seem to have been repeated, probably because of the difficulties and dangers associated with working with phosphine gas on a preparative scale. However, the confirmation of this phosphine alkylation route in the synthesis of other ring sizes, notably phosphiranes, supports its validity here also. The first published method of preparing 1-phenylphosphinane, from a bis-Grignard and a phosphorus dihalide, ⟨15CB1473⟩ is little used as it is inflexible and does not give good yields though it has been applied in making 1-phospha(2,2,2)bicyclooctane, (**6**) ⟨74JA5189⟩. Variants on this reaction use lithium instead of magnesium, which generally gives a cleaner reaction, or the dipotassium derivative of the phosphine, with an appropriate dihalide. A somewhat better yielding reaction involves a one-pot bisalkylation of a primary phosphine with 1,5-dibromopentane. Air-sensitive materials can be avoided by preparing diethyl 5-bromopentylphosphonate from the dibromide and triethyl phosphite and cyclising this in a Grignard procedure, or by reducing it to the H-phosphinate and cyclising to the phosphinate. The ester was then converted to phosphinane, (**1**), by reduction with diphenyl silane. The reduction method is important since it was subsequently extensively used in the synthesis of phosphapyranoses by Inokawa and Yamamoto and their coworkers where the cyclisation step occurs by addition of the phosphinate anion to a sugar carbonyl.

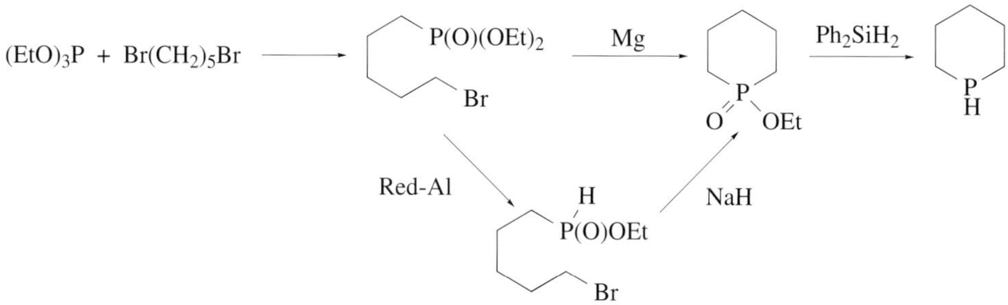

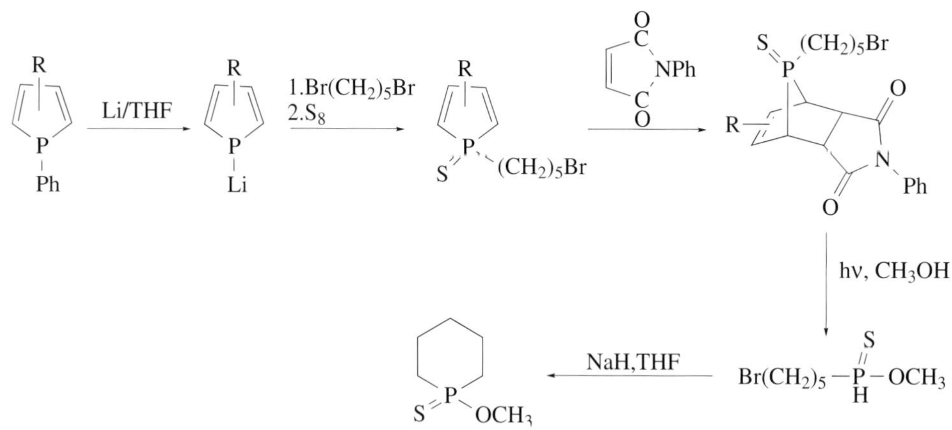

An alternative method of forming the P–H phosphinate is by photolysis of the N-phenylmaleimide adduct of the phosphole sulphide ⟨81JOC4386⟩ and could also be made by alkylation of *i*-PrOP(O)H_2 ⟨96PS(115)255⟩. Märkl developed a useful one-pot synthesis by reacting 1,5-dibromopentane with tetraphenyl diphosphine, itself generated *in situ* ⟨63AG(E)479⟩. This is still a convenient route to the unsubstituted (at carbon) ring and the cyclic phosphonium salt obtained can be conveniently converted to other useful compounds.

Though wasteful in consuming only half of the available phosphorus this may be overcome by replacing Ph_2PPPh_2 with Ph_2PSiMe_3 ⟨95JOC6281⟩. Another variant on cyclising 1,5-dihalides involves

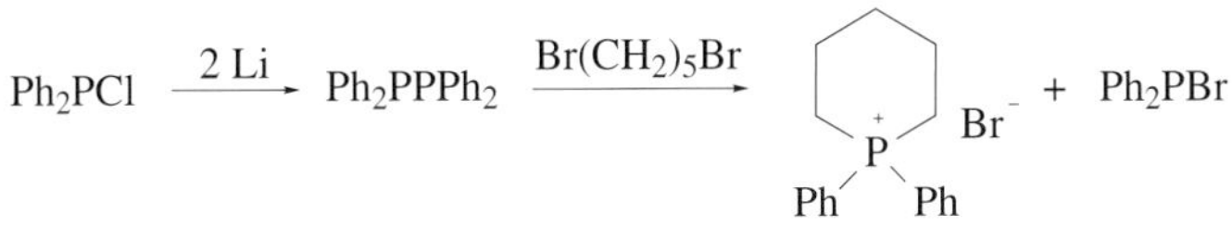

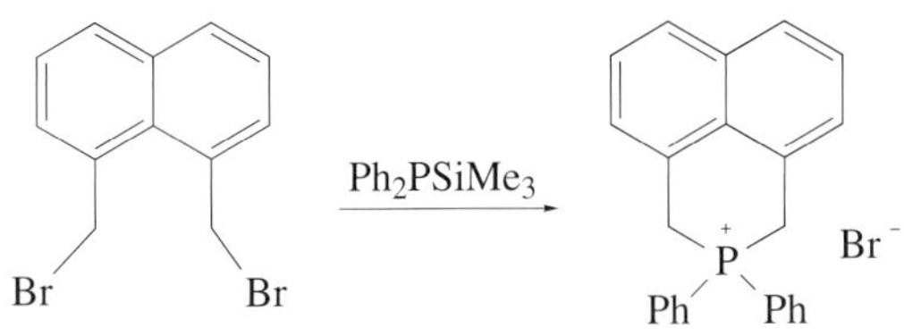

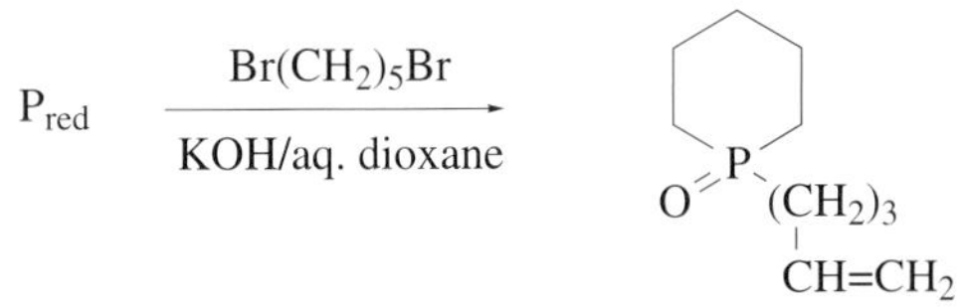

using $(Me_3SiO)_2PH$, obtained by silylating $H_2P(O)OH$ or its salts ⟨95JOC6076, 97IC1867⟩; hydrolysis of the product then gives the cyclic phosphinic acid, 1-oxo-1-hydroxyphosphinane, in good yield. Perhaps the simplest of all methods is the reaction of 1,5-dibromopentane with red phosphorus in strongly alkaline conditions in the presence of a phase transfer catalyst to give 1-oxo-1-(4-pentenyl)phosphinane in low yield; presumably these conditions generate phosphide ions ⟨95ZOB1096⟩. However, all methods using 1,5-dihalides will be limited since the terminal carbons must be primary for reasonable yields.

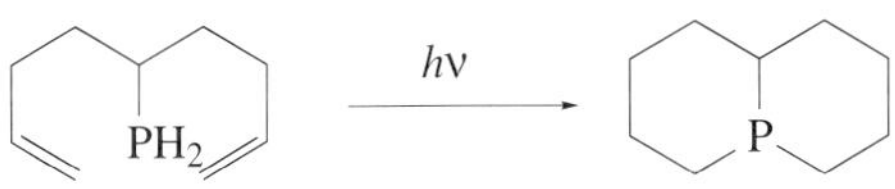

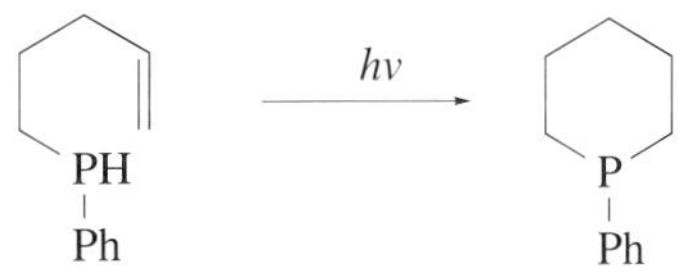

A potentially more general route involves the radical addition of primary phosphines to appropriate dienes but though this reaction has been extensively applied to P-heterocycle synthesis it seems to have found little use in phosphinanes, possibly because other methods are available. Cyclisations are initiated either photochemically ⟨66JCS(C)245⟩ or with AIBN since peroxides would rapidly be consumed in oxidation of P(III) to P(V). Phenyl phosphinane has been made this way and an interesting example is the double cyclisation to give *cis*- and *trans*-1-phosphabicyclo(4,4,0)decane ⟨87ZAAC(553)136⟩.

Relatively little effort has been directed towards synthesising monocyclic phosphinanes by modification of readily available derivatives but Mathey and his collaborators have converted 4-phosphinanones into halogenated (Br, F) and alkylated derivatives ⟨72BSF4021, 73CR(C)(277)45⟩. Schmutzler has prepared 1,1,1-trifluorophosphinane, a rare example of a $\lambda^5\sigma^5$-phosphinane and Schmidbaur and Moertl have converted 1-methylphosphinane into an interesting series of bicyclic ylides ($\lambda^5\sigma^4$) ⟨64IC421, 80ZN(B)990⟩.

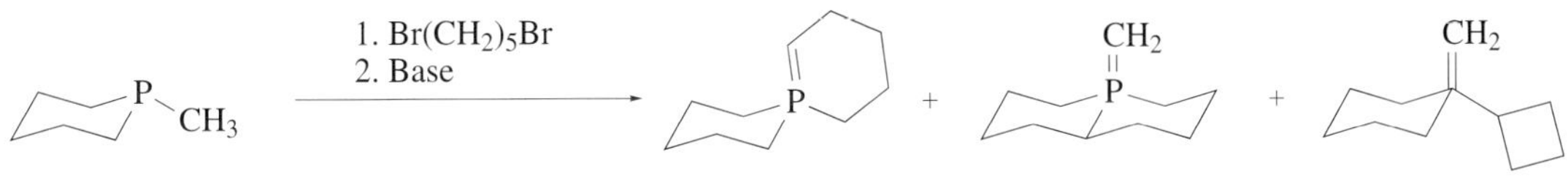

Functionalised derivatives of phosphinanes are more readily accessed by Michael-type additions of P(III)–H compounds to α,β-unsaturated nitriles or carbonyl compounds. This was initially done with simple acrylate derivatives followed by standard Dieckmann or aldol cyclisations. The products may be hydrolysed to phosphinanones or elaborated to more complex systems ⟨81JA2032⟩.

Replacing the acrylates with substituted divinyl ketones gives a 4-phosphinanone directly, a reaction which proceeds well even with heavily substituted ketones and affords a convenient route to bicyclic systems. Though basic conditions are commonly used in the addition reactions of phosphines, they frequently proceed as well in the absence of a catalyst. An obvious limitation is that only the 4-position is functionalised and this is reflected in the number of such compounds known. A further difficulty is the requirement for a primary phosphine, a class of compounds notorious for their air sensitivity and foul smell, but this may often be circumvented by masking the P–H bond either as P–CH_2OH or P–$SiMe_3$.

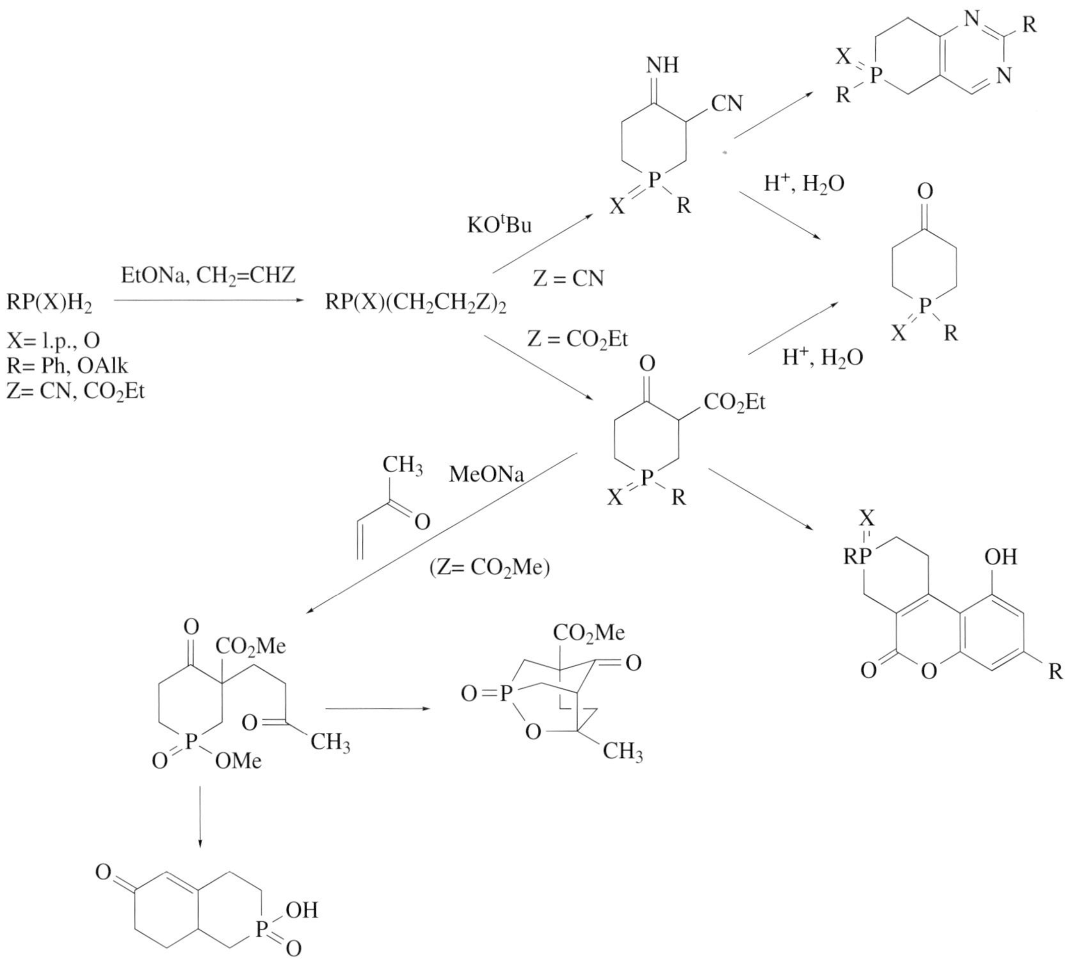

Phosphinate esters, AlkOP(O)H_2, are readily accessible and may be used instead ⟨73HCA489, 75PS(5)91⟩, but are somewhat less reactive as nucleophiles and will usually only add twice to primary carbons, whether halides or Michael acceptors.

Exceptionally, addition of methyl 2-methoxycarbonylethylphosphinate to methyl cyclohexeneoate proceeds sluggishly and in modest yield: subsequent cyclisation affords, unexpectedly, the *cis*-octahydrophosphinolin-4-one system ⟨87AJC1353⟩.

Addition of P(III)–H to aldehydes or ketones is a general reaction, normally acid catalysed and has also been used ⟨61JA168⟩ as have mixed systems ⟨83ZOB2645⟩. Though 1,5-dialdehydes are not generally available, Vysotskii and colleagues have studied 1,5-diketones and related systems in a series of papers ⟨97HAC217⟩ Cyclic conjugated dieneones afford the expected bicyclic systems ⟨72T4091, 76JOC589, 88ZOB1931⟩ and eneamines and Mannich bases have also been used ⟨89ZOB476, 83ZOB1050⟩. Bickelhaupt's group have elegantly applied eneamine alkylation to the synthesis of 1-phosphaadamantane ⟨83T4225⟩.

Phosphinanes may also be prepared by reduction of their unsaturated congeners and recent work has shown that direct catalytic hydrogenation of 3-oxo-3-phosphabicyclo (3,1,0) adducts of carbenes and dihydrophospholes is practical and convenient ⟨97MOL(2)43⟩.

Much chemical effort has been expended on the synthesis of pyranoses in which the ring oxygen is replaced by phosphorus in the hope of finding pharmaceutically valuable structures and they probably now constitute the largest single group of phosphinanes. Their development is due almost entirely to

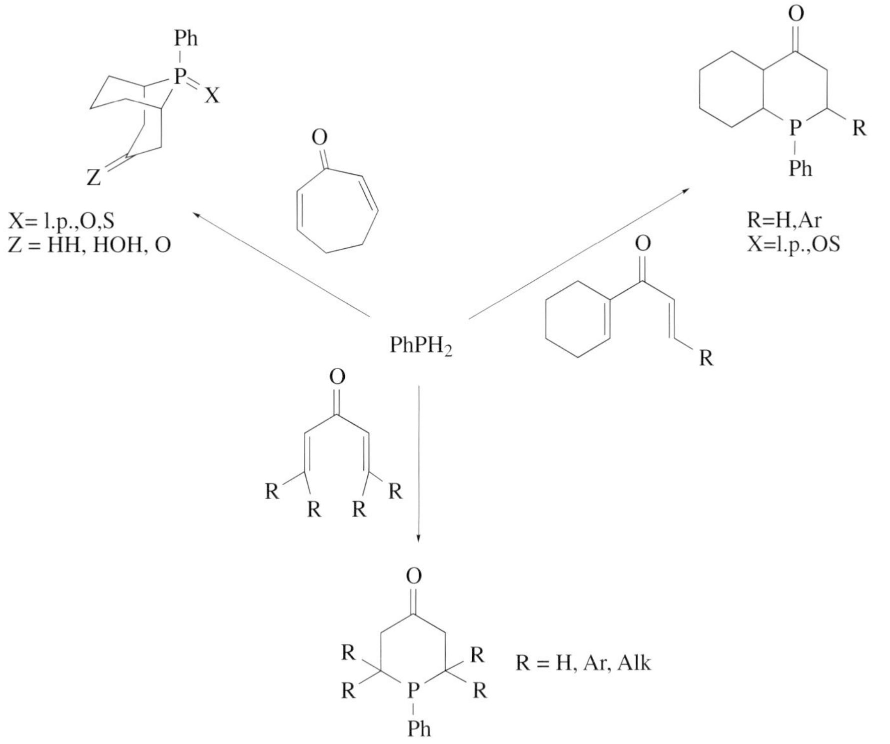
X= l.p.,O,S
Z = HH, HOH, O
R=H,Ar
X=l.p.,OS
$PhPH_2$
R = H, Ar, Alk

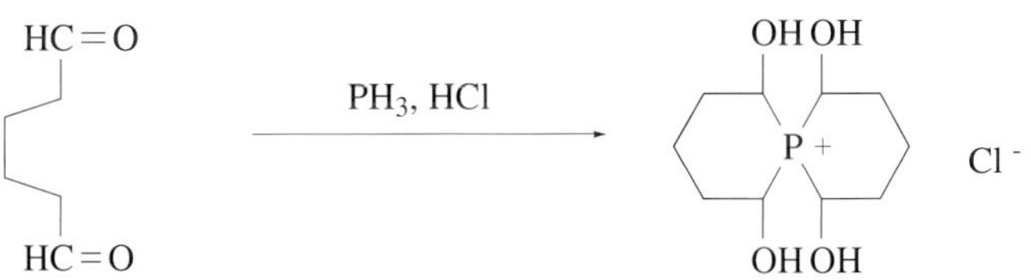
PH_3, HCl

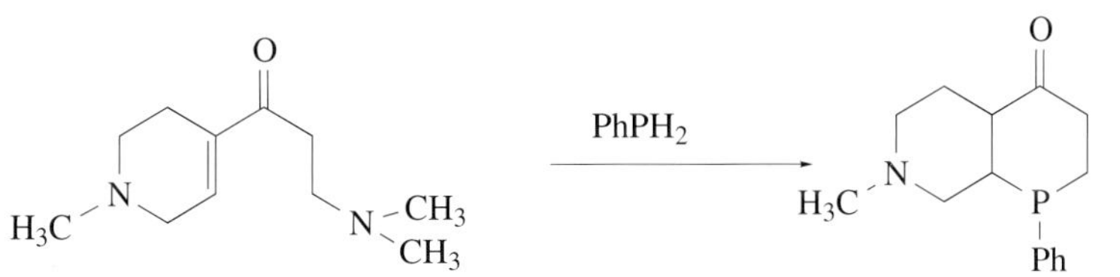
$PhPH_2$

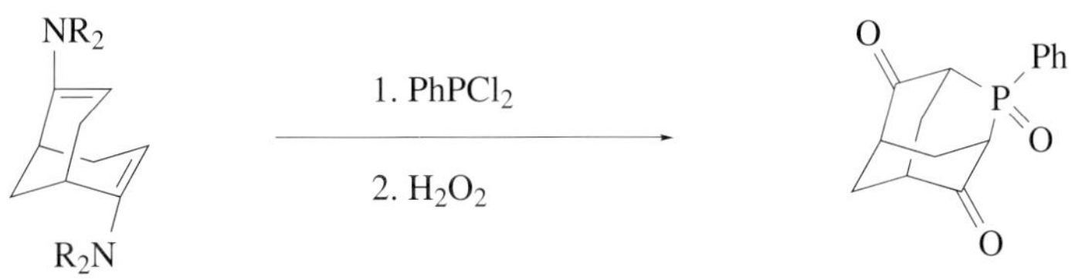
1. $PhPCl_2$
2. H_2O_2

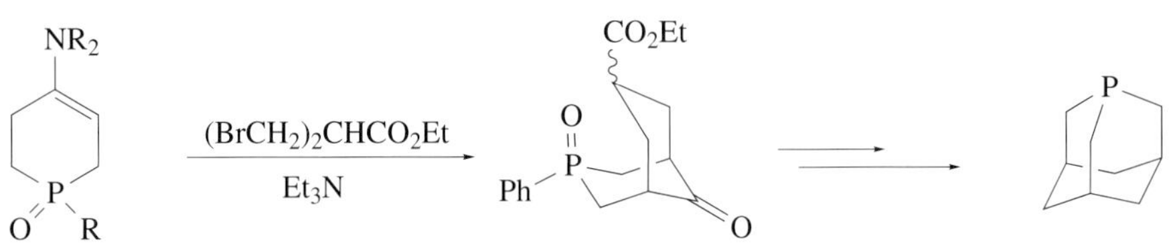
$(BrCH_2)_2CHCO_2Et$
Et_3N

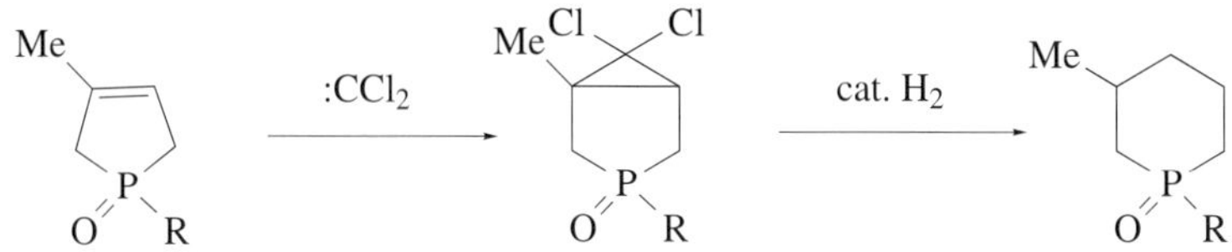

the work of Inokawa and Yamamoto and their coworkers ⟨83JCC(2)351, 84ACC(42)135, 94BCJ2596, 91BCJ869, 88BCJ2499, 87CL2081, 85CAR(141)335⟩. The major route, a variant on the first preparation, involves using a carbohydrate precursor with a protected carbonyl group, insertion of phosphorus, usually by an Arbuzov or Michaelis–Becker reaction, partial reduction, deprotection and closing the ring usually by the addition of the P–H bond to the carbonyl:

$$\text{P(O)H} + \text{C=O} \rightleftharpoons \text{(O)P-CH(OH)}$$

This reaction is reversible and preserves the anomeric character at the α-carbon. The Japanese workers have made extensive and elegant use of partial reduction of P(O)OAlk to P(O)H in this work. Initially $LiAlH_4$ was the reducing agent but most work has been done using $NaAlH_2(OCH_2CH_2OCH_3)_2$ to reduce the phosphonate ester to a H-phosphinate. The nature of the intermediate is unclear ⟨75PS(6)61⟩ but the overall cyclisation is not in doubt and the products of this path have been authenticated by detailed X-ray and NMR means.

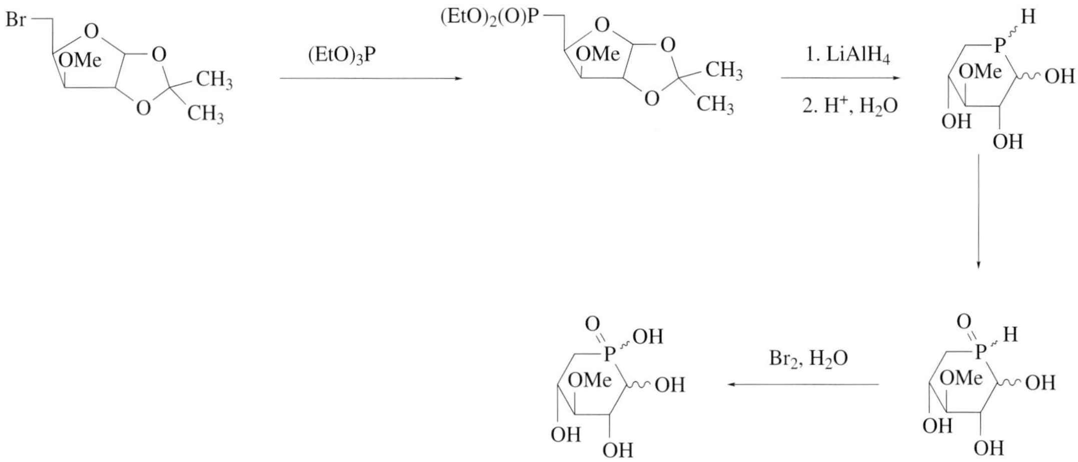

No information concerning biological properties has been forthcoming which may reflect the presence of a bulky alkyl or aryl substituent at phosphorus since, from the isosteric point of view, it would seem that the grouping P(O)H is closer to CH(OH) but relatively few such compounds have been made though the first phosphasugar was of this structure ⟨68JOC4455⟩.

5.1.3.2 Di- and Tetrahydrophosphinines

Initial preparations were either of fused ring systems, intermediates *en route* to phosphinines or, in the tetrahydro case, dehydration products, both intentional and inadvertent, of 4-phosphinanols obtained from 4-phosphinanones by reduction or Grignard reactions though the first unsaturated phosphinane appears to have arisen as a byproduct of an early attempt to make phosphininium salts ⟨66JA1034⟩.

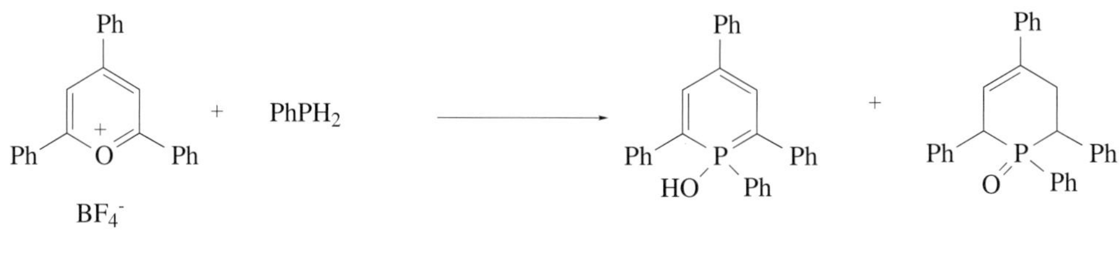

In more recent times the rapid growth of understanding in mechanistic aspects of phosphorus chemistry has led to completely new methods and it is convenient to separate the synthetic routes into those which, by and large, use the classical methods of organic synthesis, and those making extensive use of reactive organophosphorus intermediates. Much of the interest in these studies stems from Märkl's preparation of a stable phosphinine and the consequent surge of interest in novel heteroaromatic systems involving second and higher row elements.

5.1.3.2.1 Classical routes

Mann and his collaborators in very early work obtained tetrahydro phosphinoline and isophosphinoline systems by intramolecular alkylation at P(III) including the first examples of resolved chiral phosphonium salts ⟨55JCS4107, 51JCS411, 47JCS1634⟩.

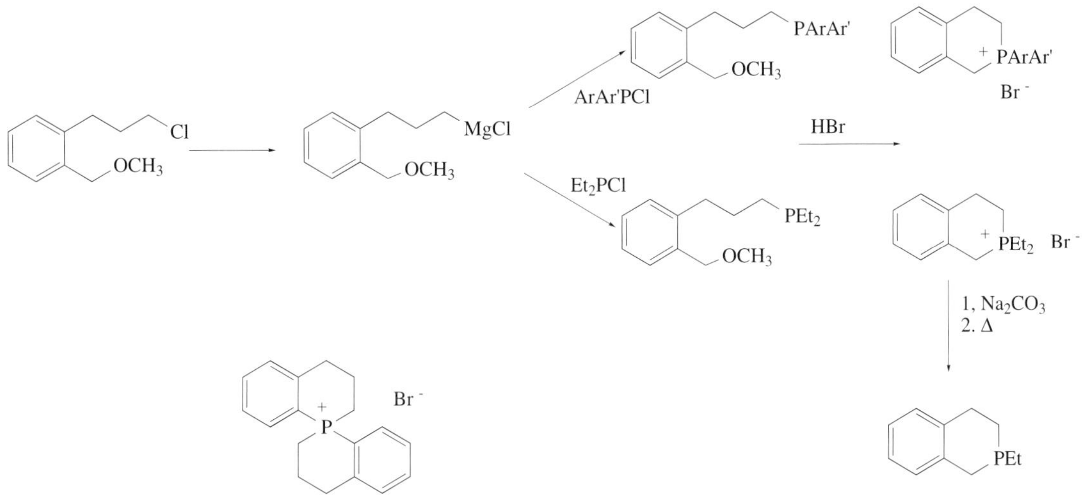

Märkl and coworkers prepared similar systems in the same way as part of synthetic routes to λ^5-phosphinines and λ^5-phosphinolines ⟨72AG(E)1016,1017, 63AG(E)153,179⟩;

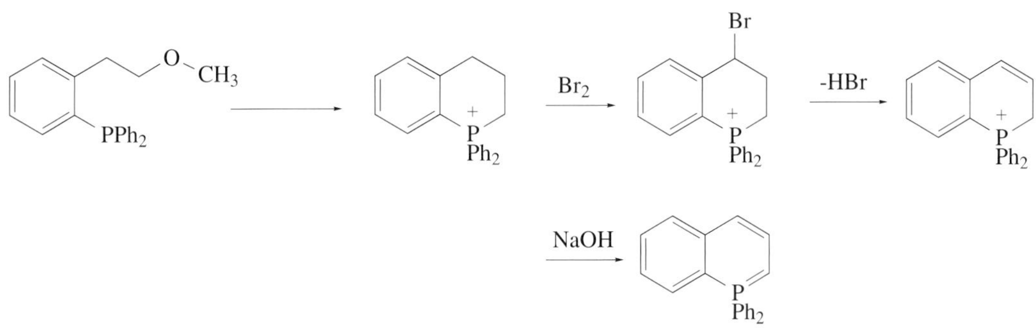

They also showed that phosphines add to 1,4-diynes giving 1,4-dihydrophosphinines directly ⟨68TL3813, 77S842⟩. Phosphonous halide–aluminium chloride complexes add to vinyl cyclopropanes to give 1,2,5,6-tetrahydrophosphinines after hydrolysis ⟨79TL1077⟩.

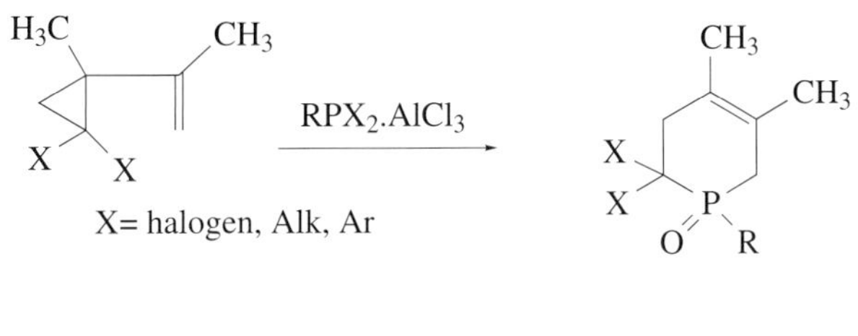

Friedel–Crafts and related cyclisations have been used extensively in the preparation of mono- and bis-benzannelated phosphinanes (reduced derivatives of phosphinolines, isophosphinolines, phosphanthridines, etc.) ⟨84T4731, 82JA3107,3114, 77JOC1170, 75JOC1150, 74CC501, 73USSRP371237, 72TL4977, 72PS(2)49, 64JOC2382, 62JCS3729, 72PS(2)49⟩.

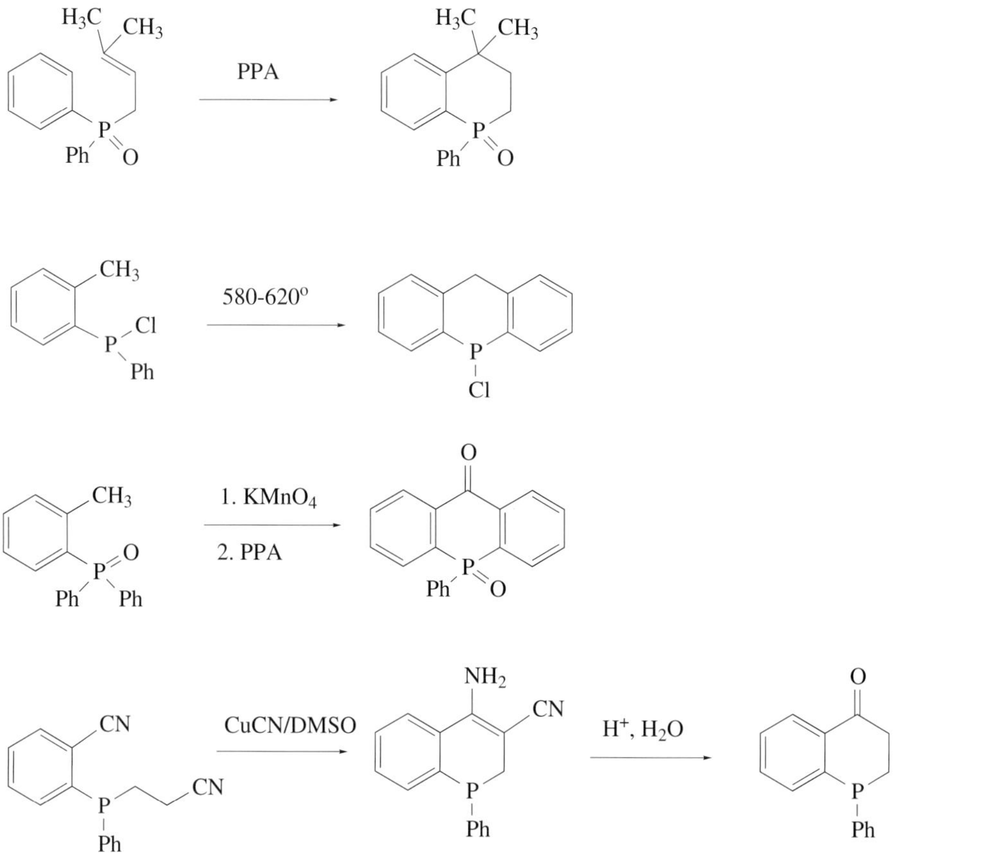

Yields are often not good since conditions need to be forcing as the ring carrying the phosphorus is usually deactivated; similar effects are found in Fischer indolisation reactions of phosphinanone phenylhydrazones ⟨72JOC4487⟩. Where appropriate precursors are available, better results are obtained using Dieckmann and related cyclisations. A novel approach to the phosphajulolidine system involves

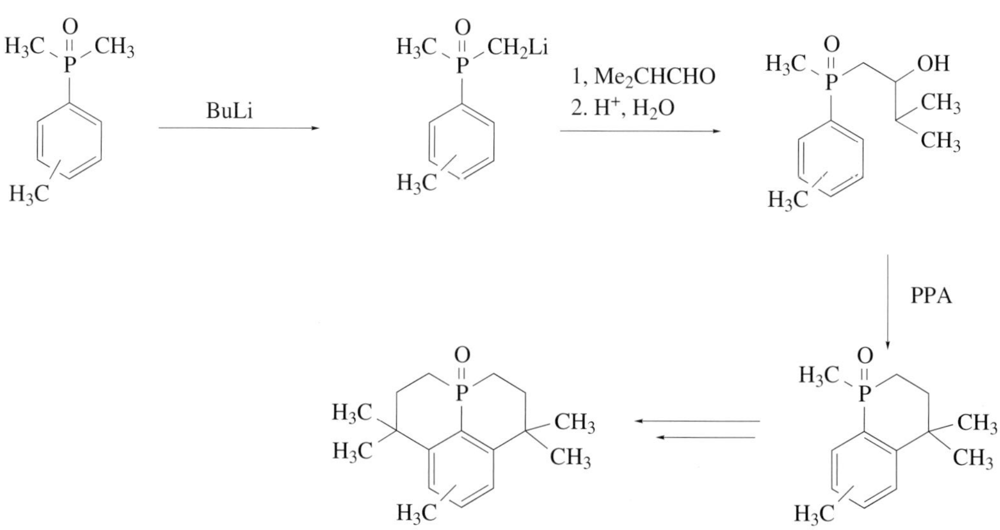

PPA cyclisation of the alcohols formed from aryldimethylphosphine oxide anions and *iso*butyraldehyde ⟨81JOC361⟩, a nice variation on Berlin's allylic cyclisation which he and his collaborators have successfully used to obtain phosphorus analogues of steroids ⟨90HAC93, 75JOC1150⟩.

5.1.3.2.2 *Non-classical routes*

The burgeoning interest in novel heteroaromatic systems and λ^3-phosphaaromatics in particular ⟨71JA3293, 73AG(E)931, 90MI1(220), 98MI1⟩ has led to the preparation of many di- and tetrahydro systems as intermediates. Two major routes have emerged: (a) ring expansions of five-membered rings ⟨93S931⟩; (b) cycloadditions involving reactive intermediates in which the phosphorus atom is multiply bound to carbon. ⟨98MI1, 90MI1(220), 85PS(22)35⟩.

5.1.3.2.3 *Ring enlargements*

Early examples ⟨71JCS(C)1064, 69JCS(C)252⟩ extended the known tendency for phenyl group migration from phosphorus to an adjacent electrophilic carbon in the alkaline hydrolysis of phosphonium salts ⟨68JCS(C)2760⟩, a reaction generally accepted to proceed by a phosphorane intermediate. An improved reaction involves using P-aroyl phosphonium salts of phospholes ⟨73T707, 75CJC855, 78CJC1952, 78TL133, 79TL1753, 80TL1441, 81NJC187, 82TL1565, 84PS(19)45⟩ though this reaction fails if the P-substituent of the phosphole is *t*-butyl or a dihydrophosphole is used, presumably because the phosphorane intermediate no longer has the geometry with hydroxyl and migrating group apical as the preferred conformation.

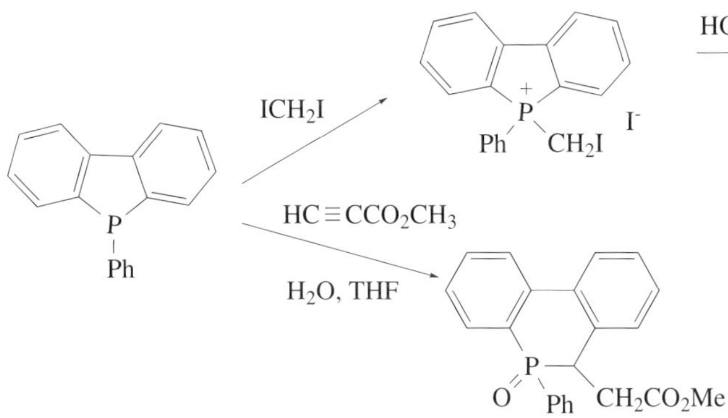

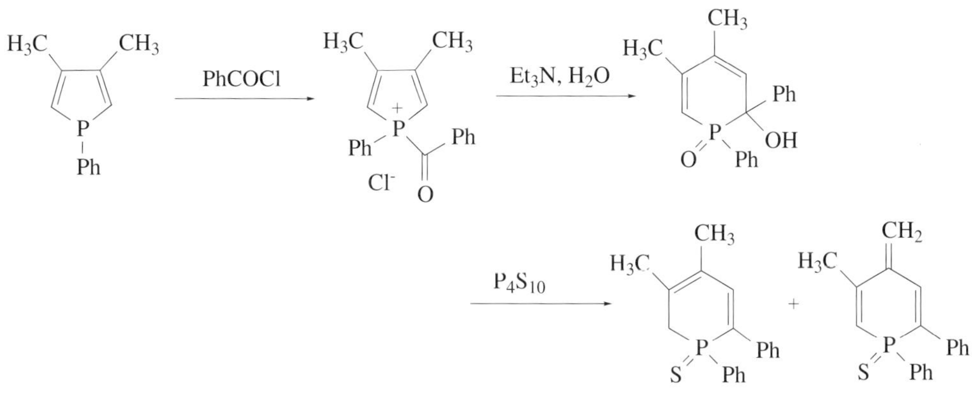

The 1,2-dihydro-2-hydroxyphosphinines can be transformed into 1,2-dihydrophosphinines or to phosphinines. Aldol condensations of dicarbonyl compounds obtained by ozonolysis of appropriately substituted phospholenes ⟨90HAC93, 85PS(22)35⟩ or by other means ⟨TL944381,4385⟩ afford mono-, di- and polycylic systems including phosphasteroid structures.

Of greater generality is the use of diazo compounds or carbenes to convert five-membered to six-membered rings, the latter being more generally useful particularly if phosphinines are not the desired

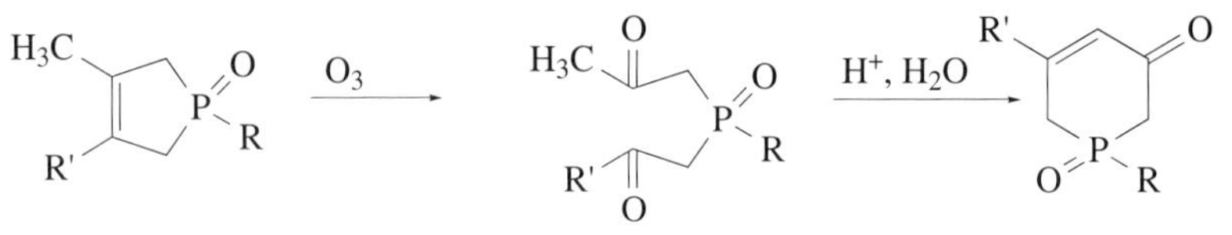

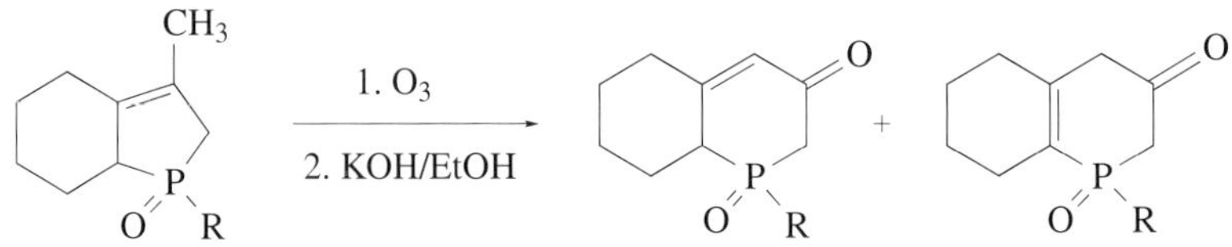

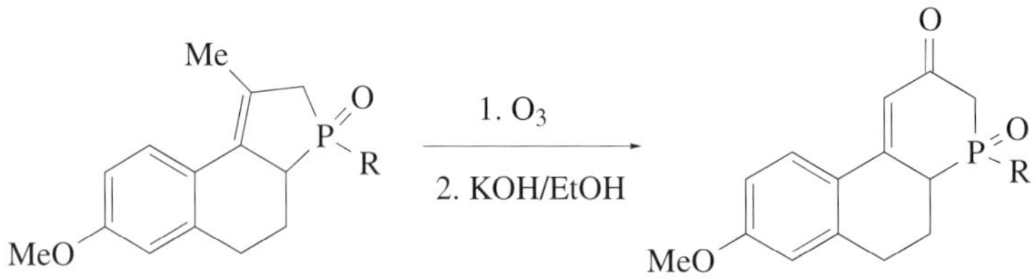

end product (see however ⟨91JOC4031⟩). The 2-phosphabicyclo(3,1,0) hexanes formed by formal addition of a carbene to a phosphole double bond, when heated or photolysed afford the bis-4,4′-dihydrophosphinine and the dihydro-4,4′-bisphosphinine, presumably by radical coupling ⟨71CJC874,879⟩. Similar products are obtained by SeO_2 oxidation of 4-phosphinanones.

Addition of dichlorocarbene, generated from chloroform under phase transfer conditions, to 1-substituted-1-oxo-2,5-dihydrophospholes provides the 3-phosphabicyclo(3,1,0) hexanes which undergo ring-opening reactions either thermally in the presence of base to give dihydrophosphinines or by metal-catalysed (Ag^+, Hg^{2+}) solvolysis to tetrahydrophosphinines.

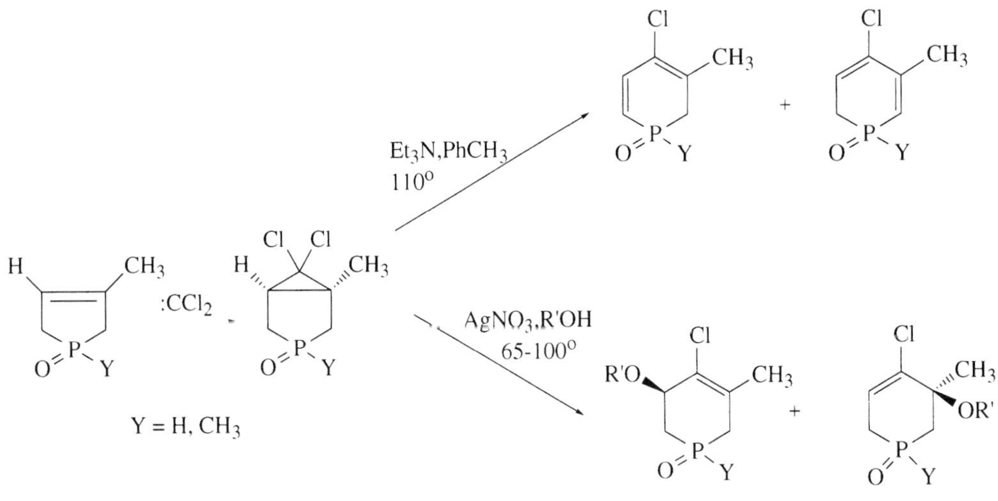

This approach has been developed by Keglevich and his collaborators into a flexible and general method for this class of compounds ⟨99JCS(P1)1801, 93S931⟩. Addition of the carbene is normally stereoselective, the carbene entering from the side remote from the P-substituent in the oxophospholene starting material unless this substituent is alkoxy when both diastereoisomers are obtained, presumably because the two substituents are then roughly equivalent. The ring-opening process is not regioselective and when the dihydrophosphole is not symmetrically substituted regioisomers are obtained. Yields are fair to very good and a wide range of starting materials is available. However, with a bulky substituent at phosphorus (mesityl) the reaction changes: stereoselectivity in formation of the bicyclic intermediate depends on the carbene source and regiospecific ring opening occurs both during the reaction or on subsequent pyrolysis of the adducts ⟨99JCS(P1)1801⟩, a different isomer being formed in each process. More unusual ring expansions include the reaction of P-mesityl-2,3-di-*t*-butylphosphatriafulvene acting as a cyclopropenium phosphide with DMAD and water to give a 1,2-dihydrophosphinine ⟨91S1099⟩ and the reactions of

1,2,3-triphenyl-2-phosphete, presumably in the tautomeric 1-phosphabutadiene form, with dienophiles ⟨90CC1649⟩.

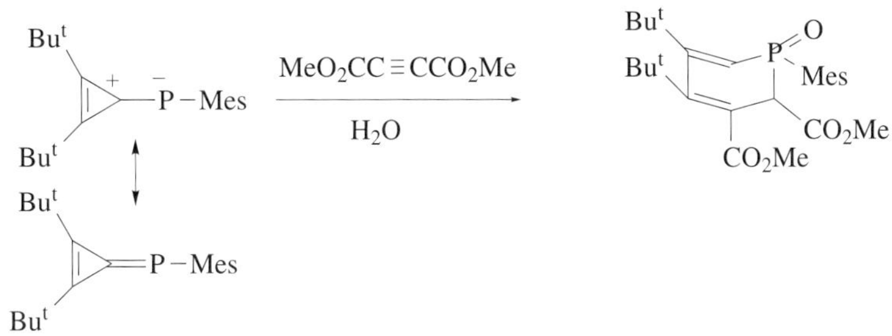

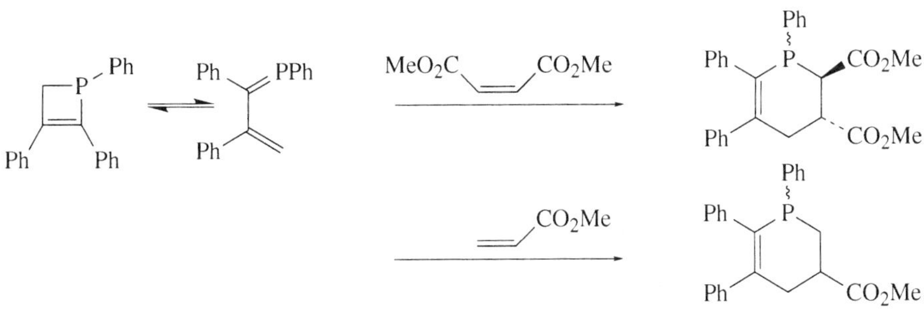

The latter reactions proceed non-stereospecifically, suggesting a stepwise pathway, i.e., despite their similarity with carbocyclic analogues they are not concerted processes.

5.1.3.2.4 Cycloadditions

Despite the characterisation of HCP (phosphaethyne) in 1961, Dimroth's synthesis of phosphacyanines in 1964 and Markl's demonstration of the stability of λ^3-phosphinines in 1966, it was not until 1976 when Becker obtained ⟨76ZAAC(423)242⟩ the first stable compound with an isolated phosphorus–carbon double bond that chemists really believed that compounds with multiple bonds between carbon and phosphorus could be generated, used in reactions and, in many instances, isolated. This scepticism was probably not for want of trying. F.G. Mann told the writer in the early 60s that he had unsuccessfully tried catalytic dehydrogenation of phosphorus heterocycles in the 1930s. The discovery of stable species with P=C bonds has led to the development of phosphorus analogues of the most useful of ring forming reactions, the Diels–Alder and related cycloaddition processes, the first example of a [4 + 2] addition being reported by Appel and coworkers in 1983 ⟨83TL3591, 85CB814⟩.

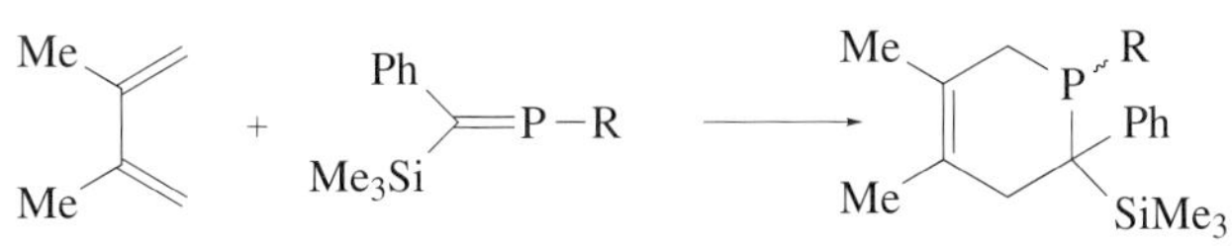

As in the ring expansion of phosphetes, these reactions were not stereospecific and the mixtures of stereoisomers obtained could be thermally equilibrated suggesting a stepwise rather than a concerted process, an observation that has been repeated in other cycloadditions of phospha alkenes. Less than a decade later a review ⟨90MI1(220)⟩ listed more than 200 acyclic phosphaalkenes, so capability for cycloadditions is extensive but most of these fall outside the scope of this chapter. Often the products proceeded directly to phosphinines or were designed to be easily transformed to aromatic systems.

The simplest phospha alkenes ($RP(X)=CH_2$; R = Alkyl, aryl; X = lone pair, S) are accessible by thermal extrusion under mild conditions from 2-phosphabicyclo(2,2,2)octa-5,7-dienes, which are themselves readily available by Diels–Alder reaction of 1,2-dihydrophosphinines with an alkyne, usually DMAD, the driving force for the fragmentation being the formation of the aromatic ring ⟨84CC1214, 87TL5783, 88JOC1722⟩.

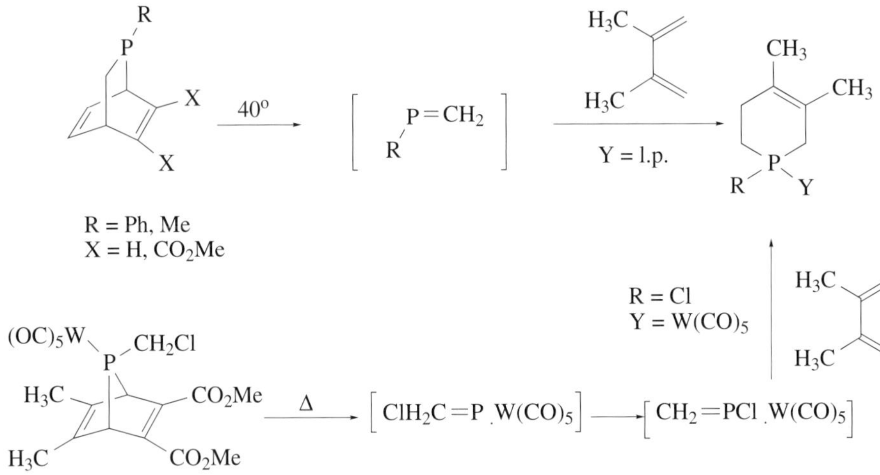

An interesting variant involves an apparent rearrangement in chloromethyl phosphinidene expelled in the same way. The mild conditions which suffice for phosphaalkenes and their sulphides are unsuccessful with the oxides and attempts to prepare them this way fail. An even simpler method, dehydrohalogenation of $ClCH_2PH_2$ with DABCO, proceeds at −50°C. All these species add readily to dienes to give the corresponding tetrahydrophosphinines ⟨84CC1214, 87TL5783, 87TL5811, 91HAC283⟩. α,α-Dichlorophosphines react in the same way with organic bases and undergo intramolecular [4 + 2] cyclisations if appropriately designed. The resulting tetrahydrophosphinines may be dehydrochlorinated again to give 1,4-dihydrophosphinines and subsequent products, promising a rich chemistry for these species ⟨96PS(109-110)461⟩.

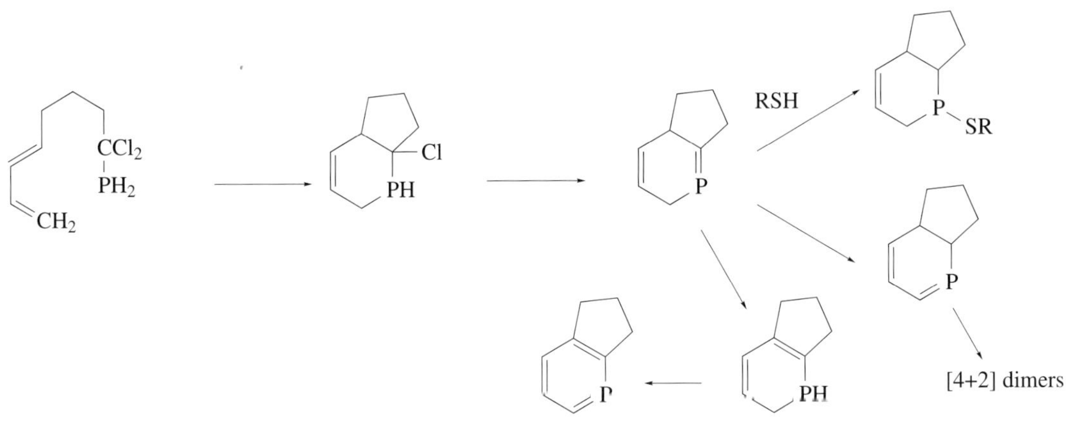

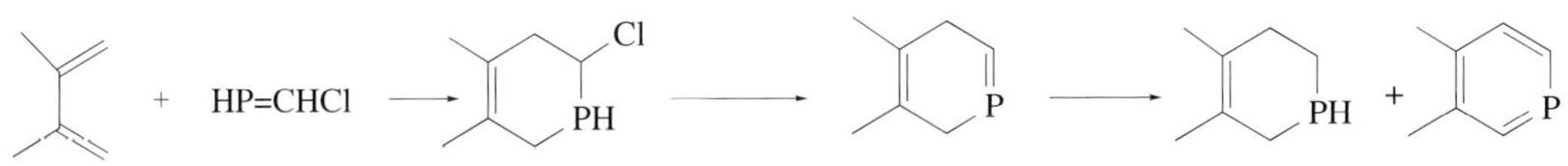

Even more remarkably, in the presence of the strong organic base, DBU, secondary vinylphosphines rearrange to phosphaalkenes which again may be cyclised ⟨94CC945⟩.

Mathey and coworkers, utilising an elegant and ingenious "phospha-Wittig" reaction ⟨88AG(E)1382, 90JA2407, 90OM788, 90OM793⟩, have made a range of simple phosphaalkenes as their metal carbonyl complexes and reacted them with electron-rich acyclic dienes, obtaining fair to reasonable yields of substituted 1,2,5,6-tetrahydrophosphinines with high regio- and stereoselectivity. Regioselectivity is probably controlled by the P=C bond which is polarised in the sense $P^{+}=C^{-}$; stereoselectivity suggests that the reactions have a high degree of concerted character. 1-Phospha-1,3-dienes, also made by this Wittig reaction and in equilibrium with the corresponding phosphetes, may be trapped by good dienophiles such as N-phenylmaleimide or 1,4-naphthoquinone and also react with alkynes to give 1,4-dihydrophosphinines ⟨95BSF384⟩.

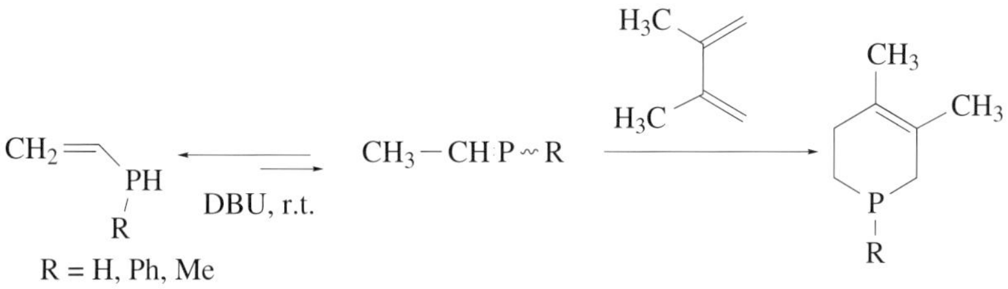

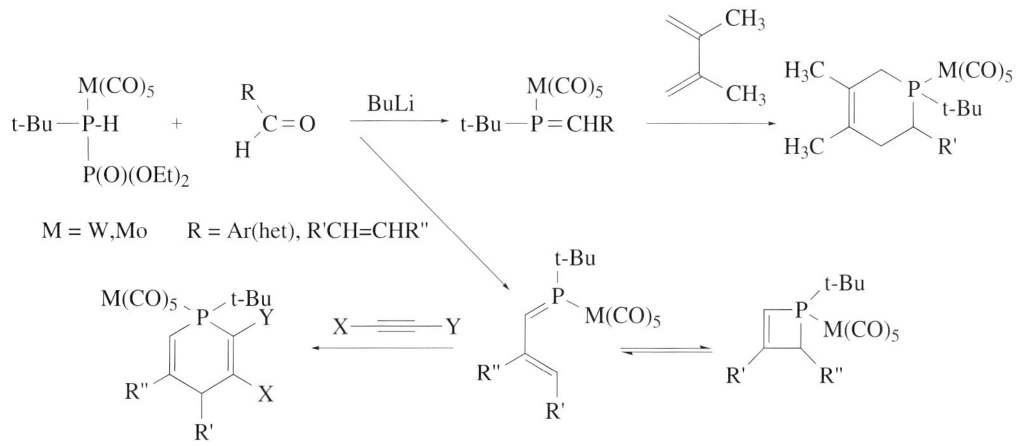

It is interesting to note that in reactions with quinones addition to C=O occurs also. Mathey's group has also shown that formation of 1-phospha-1,3,5-triene by a novel phosphinidene–carbene dimerisation occurs with spontaneous cyclisation to a 1,2-dihydrophosphinine ⟨88TL4289⟩.

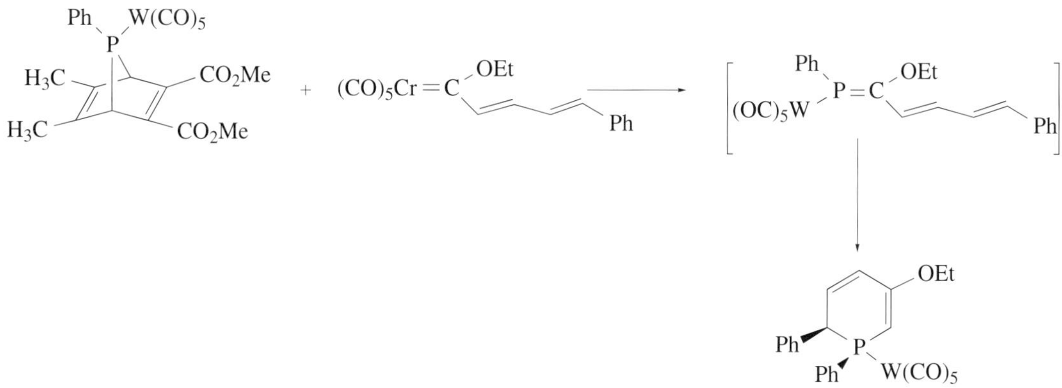

[4 + 2] Cycloadditions of phosphaalkynes has been markedly less useful since it is difficult to stop the reactions at the equimolar level and the product often arises from attack of a second molecule of the phosphaalkyne on the initial adduct, regardless of the proportions of reagents used. Quantum chemical calculations of the model reaction between HCP and butadiene suggest that the first step of these sequential reactions, the [4 + 2] cycloaddition, has the highest activation energy so that the subsequent steps occur more rapidly and the initial product is not seen ⟨99EJOC363⟩. However, 2-mesitylphosphaethyne reacts with buta-1,3-diene to give 2-mesityl-3,6-dihydrophosphinine, the first example of a stable, characterisable, reduced phosphinine with a P=C double bond; its structure was established unequivocally by X-ray of the $W(CO)_5$ complex. Similarly, cyclohexa-1,3-diene affords 3-*t*-butyl-2-phosphabicyclo(2,2,2)octa-1,4-diene by reaction with *t*-butyl phosphaethyne at 120°C. If the temperature is allowed to rise further reactions occur. In the same vein, the P–$W(CO)_5$ complex of a 3,4,5,6-tetrahydrophosphinine had been previously synthesised using the phospha-Wittig reaction ⟨91OM1190⟩.

It is significant that relatively forcing conditions or very reactive dieneophiles are necessary to bring these reactions about and the chemistry of these reactive enes and ynes is a constant balancing act between

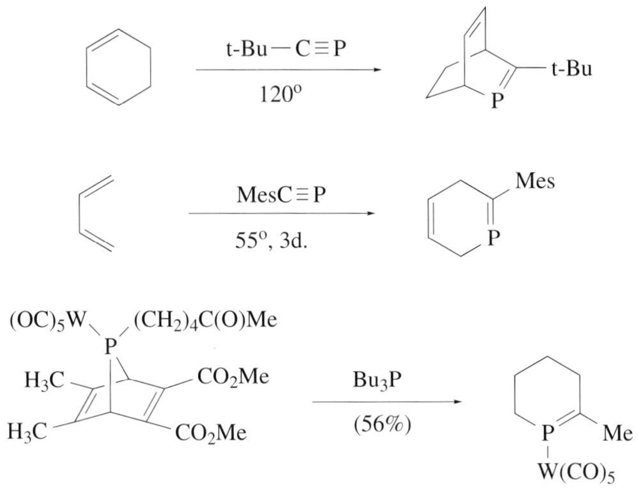

kinetic stabilisation by the attachment of bulky groups or effectively blocking their useful chemistry by overprotection ⟨88TL1681, 86ZN(B)931, 98S1305⟩.

5.1.3.2.5 Additions to phosphaaromatics

Unlike their nitrogen analogues phosphaaromatics will undergo cycloaddition reactions with electrophilic dienophiles, though usually under reasonably forcing conditions giving derivatives of the 1-phosphabicyclo(2,2,2)octatriene system (phosphabarrelenes) ⟨68AG(E)702,733, 71TL1249, 72JA7596, 74TL4369⟩. Transition metal complexes of phosphinines, presumably because of reduced aromatic character, undergo similar cycloadditions but with dienes and 1,3-dipoles as well as dienophiles and usually under much milder conditions ⟨84TL207, 87TL3475⟩. With dienes and 1,3-dipoles, it is the P=C bond which reacts as a dienophile in [4 + 2] and [3 + 2] cyclisations whereas electrophilic dieneophiles add 1,4.

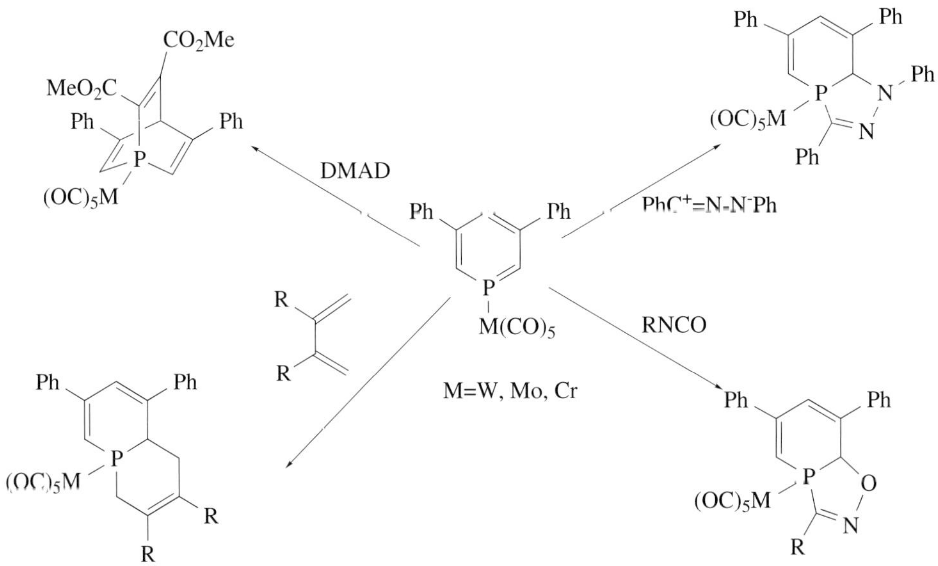

Addition of sulphur to phosphinines affords the sulphide as a transient reactive intermediate which reacts readily with dienes at the P=C bond and with dienophiles in a 1,4 sense ⟨84CC508, 84TL4659⟩.

The elegant work of Motoki and his colleagues has lead to an alternative generation of these interesting $\lambda^5\sigma^3$ reactive intermediates starting from readily accessible bicyclic compounds obtained from the reaction of P_4S_{10} with chalcones ⟨82TL1263, 86BCJ2047⟩. Protic solvents will add to these intermediates to give 1,2-dihydrophosphinines ⟨87BCJ1558⟩.

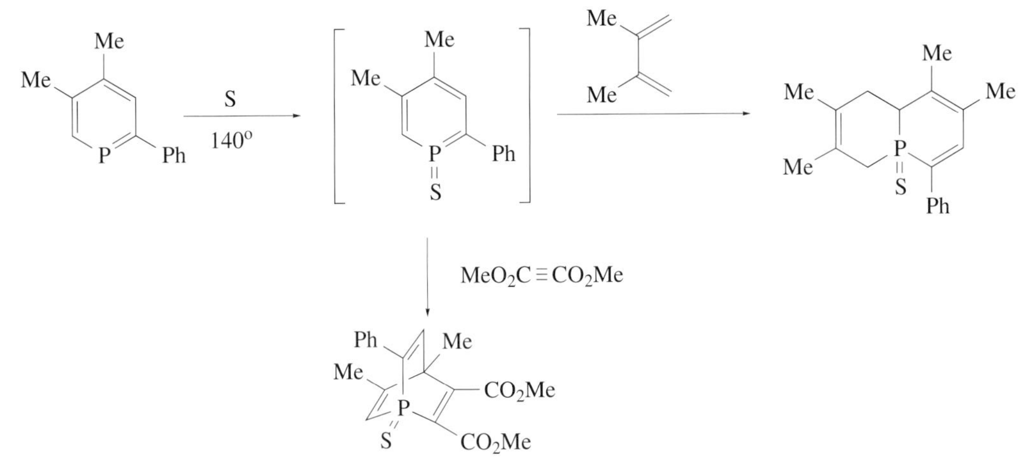

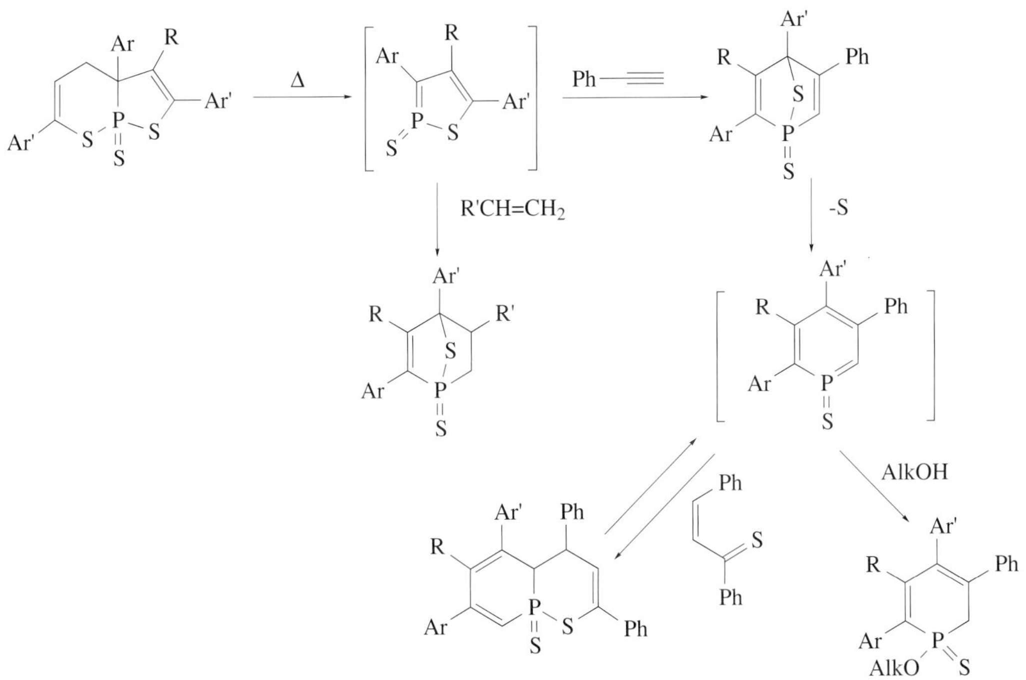

A novel, general and high yielding approach to cyclic phospha-olefin synthesis involves ring closing metathesis of appropriately substituted acyclic phosphinate esters with a transition metal catalyst ⟨99JOC2119⟩.

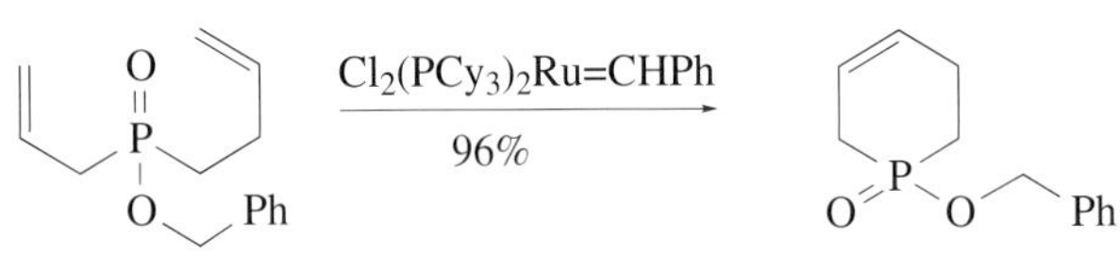

5.1.3.2.6 *Miscellaneous*

These are reactions of some interest but which would not normally be considered as routes to di- and tetrahydrophosphinines. Hydrolysis of 1,1-dihalo-$\lambda^5\sigma^4$-2,4,6-triphenyl-phosphinines affords 1-hydroxy-1-oxo-1,2-dihydrophosphinines ⟨72AG(E)1090⟩ and pyrolysis of the 1-acetoxy-1-alkoxy derivative of the same system gives 1-alkoxy-1-oxo-2-acyl derivative. The 1-allyloxy-phosphinines of this system undergo an irreversible, intramolecular, sterospecific, apparently non-Woodward–Hoffmann, Claisen-type rearrangement yielding a 4-allyl-1,4-dihydrophosphinine which in turn rearranges to the 2-isomer which then undergoes an intramolecular Diels–Alder to give a novel tricyclic compound ⟨81CB1752⟩.

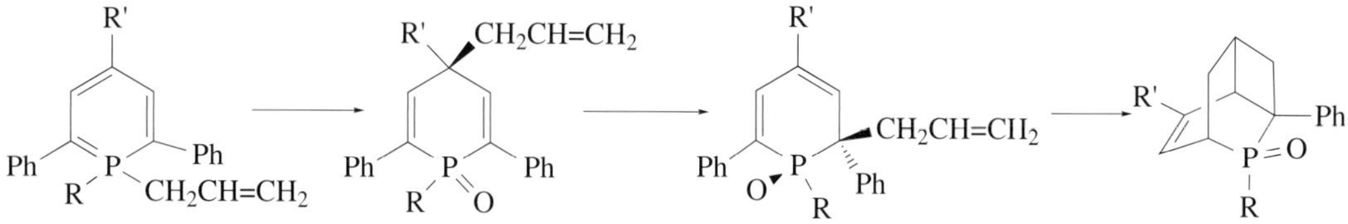

The 9-phosphabarbaralene system, which displays some of the spectroscopic features expected of a degenerate Claisen compound, is readily accessible from cyclooctatetraene or from heterocycles obtained by reacting the cyclotetraenide dianion with $PhPCl_2$ ⟨70JA734, 83TL3981, 86JA529⟩.

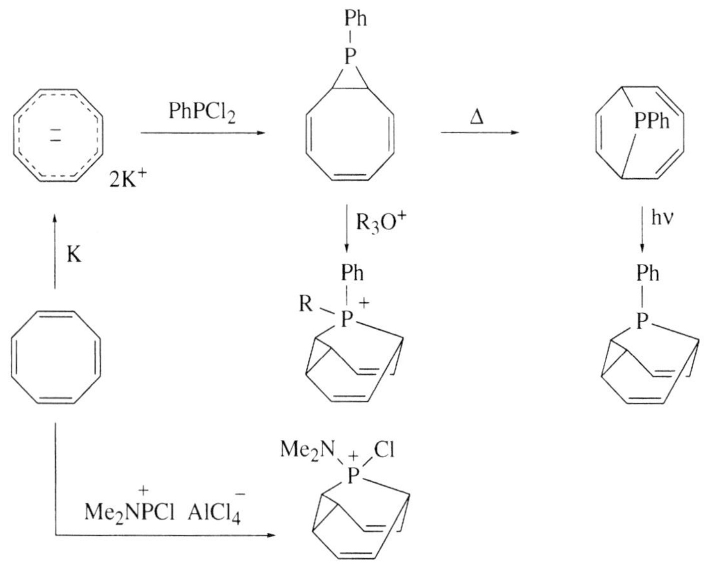

5.1.4 REFERENCES

15CB1473	G. Grüttner and M. Wiernik; *Chem. Ber.*, 1915, **48**, 1473.
47JCS1634	F.G. Holliman and F.G. Mann; *J. Chem. Soc.*, 1947, 1634.
51JCS411	M.H. Beeby and F.G. Mann; *J. Chem. Soc.*, 1951, 411.
55JCS4107	M.H. Beeby and F.G. Mann; *J. Chem. Soc.*, 1955, 4107.
61JA168	S.A. Buckler and V.P. Wystrach; *J. Am. Chem. Soc.*, 1961, **83**, 168.
62JCS3729	E.R. Lynch; *J. Chem. Soc.*, 1962, 3729.
63AG(E)153,179	G. Märkl; *Angew. Chem. Int. Ed. Engl.*, 1963, **2**, 153, 179.
63AG(E)479	G. Märkl; *Angew. Chem. Int. Ed. Engl.*, 1963, **2**, 479.
63USP3086053	R.I. Wagner; *US Pat.* 3086053; (*Chem. Abstr.*, 1963, **59**, 10124).
64IC421	R. Schmutzler; *Inorg. Chem.*, 1964, **3**, 421.
64JOC2382	G.O. Doak, L.D. Freedman, and J.B. Levy; *J. Org. Chem.*, 1964, **29**, 2382.
66JA1034	C.C. Price, T. Parasan and T.V. Lakshminarayan; *J. Am. Chem. Soc.*, 1966, **88**, 1034.
66JCS(C)245	J.H. Davies, J.D. Downer, and P. Kirby; *J. Chem. Soc. (C)*, 1966, 245.
68JOC4455	R.L. Whistler and C.C. Wang; *J. Org. Chem.*, 1968, **33**, 4455.
68AG(E)702,733	G. Märkl and F. Lieb; *Angew. Chem. Int. Ed. Engl.*, 1968, **8**, 702, 733.
68JCS(C)2760	J.J. Brophy and M.J. Gallagher; *J. Chem. Soc. (C)*, 1968, 2760.
68TL3813	G. Märkl and H. Ollbrich; *Tetrahedron Lett.*, 1968, 3813.
69JA5686	R. Baechler, W.B. Farnham and K. Mislow; *J. Am. Chem. Soc.*, 1969, **91**, 5686.
69JCS(C)252	D.W. Allen and I.T. Millar; *J. Chem. Soc.*, 1969, 252.
70JA734	T.J. Katz, J.C. Carnahan, G.M. Clarke and N. Acton; *J. Am. Chem. Soc.*, 1970, **92**, 734.
70MI1	F.G. Mann; "The Heterocyclic Derivatives of Phosphorus, Arsenic, Antimony and Bismuth", Wiley-Interscience, New York, 1970.
71CJC874,879	A.N. Hughes and C. Srinanavit; *Can. J. Chem.*, 1971, **49**, 874, 879.

71JA3293 A.J. Ashe III; *J. Am. Chem. Soc.*, 1971, **93**, 3293.
71JCS(C)1064 E.M. Richards and J.C. Tebby; *J. Chem. Soc.*, 1971, 1064.
71T4245 J.B. Lambert and W.L. Oliver; *Tetrahedron*, 1971, **27**, 4245.
71TL1249 G. Märkl, F. Lieb and C. Martin; *Tetrahedron Lett.*, 1971, 1249.
72AG(E)1016,1017 G. Märkl and K. Heier; *Angew. Chem. Int. Ed. Engl.*, 1972, **11**, 1016, 1017.
72AG(E)1090 H. Kanter and K. Dimroth; *Angew. Chem. Int. Ed. Engl.*, 1972, **11**, 1090.
72BSF4021 F. Mathey and G. Muller; *Bull. Soc. Chim. Fr.*, 1972, 4021.
72JA7596 A.J. Ashe III and M.D. Gordon; *J. Am. Chem. Soc.*, 1972, **94**, 7596.
72JOC4487 K.C. Srivastava and K.D. Berlin; *J. Org. Chem.*, 1972, **37**, 4487.
72PS(1)1255 A.T. McPhail, J.J. Breen, J.C.H. Steele and L.D. Quin; *Phosphorus Sulfur*, 1972, **1**, 255.
72PS(2)49 C.E. Griffin and W.S. Bryant; *Phosphorus Sulfur*, 1972, **2**, 49.
72T4091 Y. Kashman and E. Benary; *Tetrahedron*, 1972, **28**, 4091.
72TL4977 C.N. Robinson and W.A. Pettit; *Tetrahedron Lett.*, 1972, 4977.
73AG(E)931 G. Märkl and F. Kneidl; *Angew. Chem. Int. Ed. Engl.*, 1973, **12**, 931.
73CR(C)(277)45 F. Mathey and G. Muller; *C.R. Acad. Sci., Ser. C*, 1973, **277**, 45.
73HCA489 L. Maier; *Helv. Chim. Acta*, 1973, **56**, 489.
73T707 F. Mathey; *Tetrahedron*, 1973, **29**, 707.
73USSRP371237 E.A. Chernyshov, E.F. Bugerenko and G.P. Matriicheva; *USSR Pat.* 3711237(1973); (*Chem. Abstr.*, 1973, **79**, 53543).
74CC501 Y. Segall, A. Granoth and A. Kalir; *Chem. Commun.*, 1974, 501.
74JA5189 R.B. Wetzel and G.L. Kenyon; *J. Am. Chem. Soc.*, 1974, **96**, 5189.
74TL3473 L.D. Quin, A.T. McPhail, S.O. Lee and K.D. Onan; *Tetrahedron Lett.*, 1974, 3473.
74TL4369 G. Märkl and K.H. Meier; *Tetrahedron Lett.*, 1974, 4369.
74TL4381,4385 G. Märkl and D. Matthes; *Tetrahedron Lett.*, 1974, 4381, 4385.
75CJC855 F. Mathey, D. Thavard and B. Bartet; *Can. J. Chem.*, 1975, **53**, 855.
75JOC1150 G.A. Dilbeck, D.L. Morris and K. Berlin; *J. Org. Chem.*, 1975, **40**, 1150.
75PS(5)91 M.J. Gallagher and J. Sussmann; *Phosphorus Sulfur*, 1975, **5**, 91.
75PS(6)61 M.J. Gallagher and G. Pollard; *Phosphorus Sulfur*, 1975, **6**, 61.
76JOC589 J.R. Wiseman and H.O. Krabbenhoft; *J. Org. Chem.*, 1976, **41**, 589.
76ZAAC(423)242 G. Becker; *Z. Anorg. Allg. Chem.*, 1976, **423**, 242.
77JA3243 G.A. Gray and T.A. Albright; *J. Am. Chem. Soc.*, 1977, **99**, 3243.
77JOC1170 K.C. Chen, S.E. Ealick, D. van der Helm, J. Barycki and K.D. Berlin; *J. Org. Chem.*, 1977, **42**, 1170.
77S842 G. Märkl, H. Baier and R. Liebl; *Synthesis*, 1977, 842.
78CJC1952 F. Mathey and D. Thavard; *Can. J. Chem.*, 1978, **56**, 1952.
78TL133 F. Mathey; *Tetrahedron Lett.*, 1978, 133.
79TL1077 Y. Kashman and A. Rudi; *Tetrahedron Lett.*, 1979, **20**, 1077.
79TL1753 F. Mathey; *Tetrahedron Lett.*, 1979, **20**, 1753.
80JA4379 J.C. Galluci and R.R. Holmes; *J. Am. Chem. Soc.*, 1980, **102**, 4379.
80TL1361 W. Stec and W.S. Zielinski; *Tetrahedron Lett.*, 1980, 1361.
80TL1441 F. Nief, C. Charrier, F. Mathey and M. Simalty; *Tetrahedron Lett.*, 1980, **21**, 1441.
80ZN(B)990 H. Schmidbaur and A. Moertl; *Z. Naturforsch.*, 1980, **35b**, 990.
81CB1752 K. Dimroth, O. Schaffer and G. Weierhaeuser; *Chem. Ber.*, 1981, **114**, 1752.
81JA2032 J.B. Rampal, K.D. Berlin, N.S. Pantales, A.C. McGuffy and D. van der Helm; *J. Am. Chem. Soc.*, 1981, **103**, 2032.
81JA7602 J.B. Rampal, N. Satyamurthy, J.M. Bowen, N. Purdie and K.D. Berlin; *J. Am. Chem. Soc.*, 1981, **103**, 7602.
81JOC361 C.H. Chen, K.E. Brightly and F.M. Michaejs; *J. Org. Chem.*, 1981, **46**, 361.
81JOC4386 S. Holand and F. Mathey; *J. Org. Chem.*, 1981, **46**, 4386.
81MI1 L.D. Quin; "The Heterocyclic Chemistry of Phosphorus", Wiley-Interscience, New York, 1981.
81NJC187 F. Nief, C. Charrier, F. Mathey and M. Simalty; *Nouv. J. Chim.*, 1981, **5**, 187.
82JA3107,3114 N. Gurusamy, K.D. Berlin and M.B. Hossain; *J. Am. Chem. Soc.*, 1982, **104**, 3107, 3114.
82JOC191 H. Yamomoto, K. Yamomoto, S. Inokawa, M. Armour and P. Lugar; *J. Org. Chem.*, 1982, **47**, 191.
82TL1263 H. Yamaguchi, S. Kametani, T. Karahaasa, T. Saito and S.Motoki; *Tetrahedron Lett.*, 1982, **23**, 1263.
82TL1565 J.-M. Alcaraz, A. Brèque and F. Mathey; *Tetrahedron Lett.*, 1982, **23**, 1565.
83JCC(2)351 Z.J. Witczak and R.L. Whistler; *J. Carbohydr. Chem.*, 1983, **2**, 351.
83OMR345 K.M. Pietrusiewcz; *Org. Magn. Res.*, 1983, **21**, 345.
83OMR457 J.A. Hirsch and K. Banasiak; *Org. Magn. Res.*, 1983, **21**, 457.
83T4225 H.J. Meeuwissen, T.H. van der Knaap and F. Bickelhaupt; *Tetrahedron*, 1983, **39**, 4225.
83TL3591 R. Appel and R. Zimmerman; *Tetrahedron Lett.*, 1983, **24**, 3591.
83TL3981 G. Märkl and B. Alig; *Tetrahedron Lett.*, 1983, **24**, 3981.
83ZOB1050 Y.G. Bosyakov, D.G. Kim, A.P. Logunov, G.P. Revenko, T.I. Shiganakova and O.V. Rabetskaya; *Zh. Obsch. Khim.*, 1983, **53**, 1050 (*Chem. Abstr.*, 1983, **99**, 195085).
83ZOB2645 V.I. Visotski, Y.T. Prikhodko, G.V. Pavel and M.N. Tilichenko; *Zh. Obsch. Khim.*, 1983, **53**, 2645; (*Chem. Abstr.*, 1983, **100**, 191953).
84ACC(42)135 H. Yamomoto and S. Inokawa; *Adv. Carbohydr., Chem. Biochem.*, 1984, **42**, 135.
84CC508 J.M. Alcaraz and F. Mathey; *Chem. Commun.*, 1984, 508.

84CC1214 E. Deschamps and F. Mathey; *Chem. Commun.*, 1984, 1214.

84PS(19)45 J.-M. Alcaraz, E. Deschamps and F. Mathey; *Phosphorus Sulfur*, 1984, **19**, 45.

84T4731 C. Laurenço, L. Villien and G. Kaufmann; *Tetrahedron*, 1984, **40**, 2731.

84TL207 J.-M. Alcaraz and F. Mathey; *Tetrahedron Lett.*, 1984, **25**, 207.

84TL4659 J.M. Alcaraz and F. Mathey; *Tetrahedron Lett.*, 1984, **25**, 4659.

84ZSK(25)79 A.I. Yanovski, A.E. Shalamov, Y.T. Struchkov, A.P. Loguunov, Y.G. Bosyakov and O.V. Agashkin; *Zh. Strukt. Khim.*, 1985, **25**, 79; (*Chem. Abstr.*, 1984, **102**, 62330).

85CB814 R. Appel, F. Knoch and R. Zimmerman; *Chem. Ber.*, 1985, **118**, 814.

85CAR(141)335 K. Seo and M. Yamashita; *Carbohydr. Res.*, 1985, **141**, 335.

85PS(22)35 L.D. Quin and J.C. Kisalus; *Phosphorus Sulfur*, 1985, **22**, 35.

85ZSK(26)78 A.E. Shalamov, A.I. Yanovskii, Y.T. Struchkov, A.P. Logunov, Y.G. Bosyakov and O.V. Agashkin; *Zh. Strukt. Khim.*, 1985, **26**, 78; (*Chem. Abstr.*, 1985, **103**, 105064).

86BCJ2047 H. Tanaka and S. Motoki; *Bull. Chem. Soc. Jpn.*, 1987, **59**, 2047.

86IZV(K)69 A.E. Shalamov, A.I. Yanovskii, Y.T. Struchkov, O.V. Agashkin, G.P. Revenko, A.P. Logunov and Y.G. Bosyakov; *Izv. Akad. Nauk. Kaz. SSR, Ser. Khim.*, 1986, 69; (*Chem. Abstr.*, 1987, **106**, 84742).

86JA529 S.A. Weissmann, S.G. Baxter, A.M. Arif and A.H. Cowley; *J. Am. Chem. Soc.*, 1986, **108**, 529.

86JA6651 E.L. Eliel, S. Chandrasekaran, L.E. Carpenter and J.G. Verkade; *J. Am. Chem. Soc.*, 1986, **108**, 6651.

86PS(26)203 B.A. Arbuzov and E.N. Dianova; *Phosphorus Sulfur*, 1986, **26**, 203.

86ZN(B)931 W. Rösch and M. Regitz; *Z. Naturforsch.*, 1986, **41b**, 931.

86ZOB817 L.P. Krasnomolova, O.V. Agashkin, A.P. Logunov and Y.G. Bosyakov; *Zh. Obsch. Khim.*, 1986, **56**, 817; (*Chem. Abstr.*, 1987, **106**, 84729).

86ZOB1973 Y.G. Bosyakov, G.P. Revenko and A.P. Logunov,*Zh. Obsch. Khim.*, 1986, **56**, 1973, 1978, 1983.

87AJC1353 Nur Franisal and M.J. Gallagher; *Aust. J. Chem.*, 1987, **40**, 1353.

87BCJ1558 H. Tanaka and S. Motoki; *Bull. Chem. Soc. Jpn.*, 1987, **60**, 1558.

87CL2081 H. Yamamoto, T. Hanaya, N. Shigetoh, H. Kawamoto and S. Inokawa; *Chem. Lett.*, 1987, 2081.

87IZV(K)79 A.E. Shalamov, A.I. Yanovskii, Y.T. Struchkov, S.K. Shishkin, Y.G. Bosyakov and O.V. Agashkin; *Izv. Akad. Nauk. Kaz. SSR, Ser. Khim.*, 1987, 79; (*Chem. Abstr.*, 1987, **107**, 217739).

87MI1 J.G. Verkade and L.D. Quin (eds.); "Phosphorus-31 Spectroscopy in Stereochemical Analysis", VCH Publishers, Florida, 1987.

87TL3475 G. Märkl and H.-J. Beckh; *Tetrahedron Lett.*, 1987, **28**, 3475.

87TL5783 L.D. Quin, A.N. Hughes and B. Pete; *Tetrahedron Lett.*, 1987, **28**, 5783.

87TL5811 B. Pellerin, P. Guenot and J.-M. Denis; *Tetrahedron Lett.*, 1987, **28**, 5811.

87ZAAC(553)136 F. Krech, K. Issleib, A. Zschunke and H. Meyer; *Z. Anorg. Allg. Chem.*, 1987, **553**, 136.

88AG(E)1382 A. Marinetti and F. Mathey; *Angew. Chem. Int. Ed. Engl.*, 1988, **27**, 1382.

88BCJ2499 T. Hanaya, N. Shigitoh and H. Yamamoto; *Bull. Chem. Soc. Jpn.*, 1988, **61**, 2499.

88IZV(K)59 A.E. Shalamov, O.V. Agashkin, A.I. Yanovskii, Y.T. Struchkov, A.P. Logunov and Y.G. Bosyakov; *Izv. Akad. Nauk. Kaz. SSR, Ser. Khim.*, 1988, 59; (*Chem. Abstr.*, 1989, **110**, 135369).

88JOC1722 L.D. Quin, A.N. Hughes, J. Kisalus and B. Pete; *J. Org. Chem.*, 1988, **53**, 1722.

88TL1681 U. Annen and M. Regitz; *Tetrahedron Lett.*, 1988, **29**, 1681.

88TL4289 N.H. Tran Huy, F. Mathey and L. Ricard; *Tetrahedron Lett.*, 1988, **29**, 4289.

88ZOB1931 G.M. Akhmetova, G.M. Isaeva and B.M. Butin; *Zh. Obshch. Khim.*, 1988, **58**, 1931; (*Chem. Abstr.*, 1989, **111**, 194867).

89ZOB476 V.N. Zemlyanoi, O.L. Kolodyazhnyi and V.P. Kukhar; *Zh. Obshch. Khim.*, 1989, **59**, 476; (*Chem. Abstr.*, 1989, **111**, 134336).

90CC1649 K.M. Doxsee, G.S. Shen and C.B. Knobler; *Chem. Commun.*, 1990, 1649.

90HAC93 L.D. Quin and B.G. Marsi; *Heteroatom. Chem.*, 1990, **1**, 93.

90JA2407 P. Le Floch, A. Marinetti, L. Ricard and F. Mathey; *J. Am. Chem. Soc.*, 1990, **112**, 2407.

90MI1(220) M. Regitz and O.J. Scherer (eds.); "Multiple Bonds and Low Coordination in Phosphorus Chemistry", George Thieme Verlag, Stuttgart, 1990.

90MI2(10) F. Hartley (ed.); "The Chemistry of Organophosphorus Compounds", Vol 1, John Wiley, Chichester, 1990.

90OM788 A. Marinetti, L. Ricard and F. Mathey; *Organometallics*, 1990, **9**, 788.

90OM793 A. Marinetti, S. Bauer, L. Ricard and F. Mathey; *Organometallics*, 1990, **9**, 793.

91BCJ869 T. Hanaya, H. Yamamoto and T. Ohmse; *Bull. Chem. Soc. Jpn.*, 1991, **64**, 869.

91HAC283 L.D. Quin, J.-S. Tang and G. Keglevich; *Heteroatom Chem.*, 1991, **2**, 283.

91JOC4031 S. Holand, L. Ricard and F. Mathey; *J. Org. Chem.*, 1991, **56**, 4031.

91MI1 J.C. Tebby (ed.); "Phosphorus-31 NMR Data", CRC Press, Boca Raton, FL, 1991.

91MI2 D.W. Boykin (ed.); "^{17}-O NMR Spectroscopy in Organic Chemistry", CRC Press, Boca Raton, FL, 1991.

91OM1190 A. Marinetti, P. Le Floch and F. Mathey; *Organometallics*, 1991, **10**, 1190.

91S1099 E. Fuchs, B. Breit, U. Bergstrasser, J. Hoffmann, H. Heydt and M. Regitz; *Synthesis*, 1991, 1099.

92ZOB762 Y.G. Bosyakov and G.P. Revenko; *Zh. Obshch. Khim.*, 1992, **62**, 762, 767; (*Chem. Abstr.*, 1993, **118**, 191804).

93S931 G. Keglevich; *Synthesis*, 1993, 931.

94BCJ2596 T. Hanaya, Y. Fujii and H. Yamamoto; *Bull. Chem. Soc. Jpn.*, 1994, **67**, 2596.

94CC945 A.-C. Gaumont, J.-C. Guillemin and J.-M. Denis; *Chem. Commun.*, 1994, 945.

94MI1 L.D. Quin and J.G. Verkade (eds.); "Phosphorus-31 NMR Spectral Properties in Compound Characterisation and Structural Analysis", VCH Publishers, Florida, 1994.

95BSF384 H. Trauner, E. de la Cuesta, A. Marinetti and F. Mathey; *Bull. Soc. Chim. Fr.*, 1995, **132**, 384.

95JOC6076 J.-L. Montchamp, F. Tian and J.W. Frost; *J. Org. Chem.*, 1995, **60**, 6076.

95JOC6281 M.K. Tasz, A. Gamliel, O.P. Rodriguez, T.M. Lane, S.E. Cremer and D.W. Bennet; *J. Org. Chem.*, 1995, **60**, 6281.

95PS(103)137 S.E. Cremer and P.E. Fanwick; *Phosphorus Sulfur Silicon*, 1995, **103**, 137.

95PS(104)223 M.K. Tasz, S.E. Cremer and P.E. Fanwick; *Phosphorus Sulfur Silicon*, 1995, **104**, 223.

95ZOB1096 N.K. Gusarova, S.I. Shaikhudinova, D.I. Dmitriev, S.F. Malasheva, S.N. Arbusova and B.A. Trofimov; *Zh. Obshch. Khim.*, 1995, **65**, 1096; (*Chem. Abstr.*, 1996, **124**, 87133).

96HAC391 C.J.R. Fookes and M.J. Gallagher; *Heteroatom Chem.*, 1996, **7**, 391.

96PS(109-110)461 A.C. Gaumont, J.-F. Pilard and J.M. Denis; *Phosphorus Sulfur Silicon*, 1996, **109-110**, 461.

96PS(115)255 M.J. Gallagher, M.G. Ranasinghe and I.D. Jenkins; *Phosphorus Sulfur Silicon*, 1996, **115**, 255.

97HAC217 V.I. Vysotski; *Heteroatom Chem.*, 1997, **8**, 217.

97IC1867 M.L.J. Hackney, D.M. Schubert, P.F. Brandt, R.C. Halliwenger and A.D. Norman; *Inorg. Chem.*, 1997, **36**, 1867.

97MOL(2)43 G. Keglevich, A. Tungler, K. Ujszaszy, T. Novak and L. Toke; *http://mdpc.org/molecules/papers/feb97p3.pdf.*

98MI1 K.B. Dillon, F. Mathey and J.F. Nixon; "Phosphorus: The Carbon Copy", Wiley, Chichester, 1998.

98S1305 A. Mack, T. Allspan, E. Pierronand and M. Regitz; *Synthesis*, 1998, 1305.

99EJOC363 W. Fiedler, O. Lieber, U. Bergstrasser and M. Regitz; *Eur. J. Org. Chem*, 1999, 363.

99JCS(P1)1801 G. Keglevich, G.M. Keseru, H. Fornitos, A. Szollosy, C. Ludanye and L. Toke; *J. Chem. Soc. (P1)*, 1999, 1801.

99JOC2119 M. Bujard, V. Gouverneur and C. Mioskowski; *J. Org. Chem.*, 1999, **64**, 2119.

5.2
Six-membered Rings: Phosphinines

PASCAL LE FLOCH
Ecole Polytechnique, Palaiseau, France

5.2.1 INTRODUCTION

Undoubtedly, a real landmark of the chemistry of low-coordinated phosphorus compounds was set when Märkl reported in 1966 the successful preparation of the λ^3-2,4,6-triphenylphosphinine from the reaction between a source of PH_3 and the corresponding pyrylium salt ⟨66AG(E)846⟩. Beyond the simple

1 λ^3-phosphinine

2 λ^5-phosphinine

Figure 1. Numbering scheme of phosphinines.

incorporation of a P=C double bond in a six-membered ring, this discovery pointed out that the concept of aromaticity could be extended to higher main group elements.

Several reviews on λ^3-phosphinines and their λ^5-derivatives, whose reading is recommended for further details, have appeared in periodicals or books. For a complete coverage of the early work (up to 1983) six reviews on λ^3-phosphinines (four by Märkl ⟨72MIV2-01, 77PS(3)77, 82MIV2-01, 82HOU(E1)72⟩, one by Bickelhaupt and Jongsma ⟨77MIV2-01⟩ and one by Ashe ⟨78ACR153⟩), and four reviews on λ^5-phosphinines ⟨73TCC(38)1, 82HOU(E1)783, 82ACR58, 84CHEC-I(1)493⟩ by Dimroth are available. More recent work has been reviewed by Märkl ⟨B-90MIV2-01⟩, Hewitt ⟨96CHEC-II(5)639⟩ and Mathey ⟨92MIV2-01, B-98MIV2-01⟩. The present review which aims to be comprehensive will cover all topics of the chemistry of phosphinines including theoretical and experimental methods, syntheses and reactivity of the ring and perspectives. Specific topics dealing with the coordination chemistry of these heterocycles have been for the most part deliberately excluded.

Various names have been given to these heterocycles. IUPAC ⟨83PAC410⟩ proposed that six-membered saturated rings be named as derivatives of phosphinane and the unsaturated rings as phosphinines. For a long time Chemical Abstracts ignored IUPAC rules and used the term "phosphorin" to call both saturated and unsaturated species. Another frequently encountered terminology is "phosphabenzene". Throughout this chapter, the IUPAC recommendations will be followed. The valence state of phosphorus will be designed by the term λ. Thus aromatic derivatives such as (**1**) will be named as λ^3-phosphinines (a trivalent phosphorus atom with a coordination number of 2) and compounds such as (**2**) are λ^5-phosphinines (a pentavalent phosphorus atom with a coordination number of 4). The numbering of the ring carbons is given below (Figure 1). It will be used systematically throughout this review.

5.2.2 THEORETICAL METHODS

Phosphinines, like other group 15 heterobenzenes (As, Sb), have been widely studied from a theoretical standpoint. A first important subject of investigation concerned the determination of their degree of aromaticity. Each method led to the conclusion that λ^3-phosphinines are aromatic whereas λ^5-phosphinines may be viewed as phosphonium ylides. Calculations (SCF/3-21G*) of bond separation (BS) and superhomodesmic reactions (SHR) for a series of hetero analogues of benzene predicted that pyridine is as aromatic as benzene and that λ^3-phosphinine possesses 88% (BS) to 90% (SHR) of benzene aromatic stabilization ⟨88JA4204⟩. Other calculations (MP2/6-31G*) using isodesmic reaction energies have confirmed these results ⟨90MIV2-01, 92JA9080, 96JPC6456⟩. The resonance energy of λ^3-phosphinine (26.0 kcal mol^{-1}) has been calculated using the AM1 method and compared to that of benzene (28.3 kcal mol^{-1}), pyridine (25.6 kcal mol^{-1}) and to that of other known or yet unknown heterocycles ⟨89H1135⟩. Another important criterion to evaluate aromaticity is the geometry of the ring. Recently, the structure of the parent compound has been investigated at the Hartree–Fock and correlated levels of theory using various basis sets ⟨89MIV2-01, 93JPC4011⟩. Small discrepancies between theoretical and experimental data were found (electron diffraction and microwave spectroscopy ⟨74JCP2840, 78ACR153⟩) concerning C–C bond lengths, theoretical values being closer to an ideal aromatic structure (P–C2, 1.737 Å; C2–C3, 1.396 Å; C4–C4, 1.398 Å). To be fully comprehensive, the theoretical structure of 2-chlorophosphinine has been calculated ⟨94HAC131⟩. λ^5-phosphinines have been less studied. Isodesmic reactions and the calculated structure of the 1,1-dihydro compound (MP2/6-31G*) have been used to compare the conjugative abilities of λ^5–P=C and λ^3–P=C bonds in unsaturated systems ⟨96JPC6456⟩. From this study it was concluded that these systems cannot be considered as simple ylides. Other criteria have been used to estimate the degree of aromaticity of C_5H_5P, such as absolute hardness (HOMO–LUMO gap) ⟨93OM5005⟩, statistical evaluation of the deviations in peripheral bond orders (Bird's index) ⟨86T89, 90T5697⟩, and nics values (nuclear independent chemical shift) ⟨99JOC5524⟩. In the latter work, the effects of CH replacement by nitrogen atoms on the aromaticity of λ^3-phosphinines have also been discussed. Electron correlation calculations (MP2) of the dipole moments and static dipole polarizabilities have been reported for

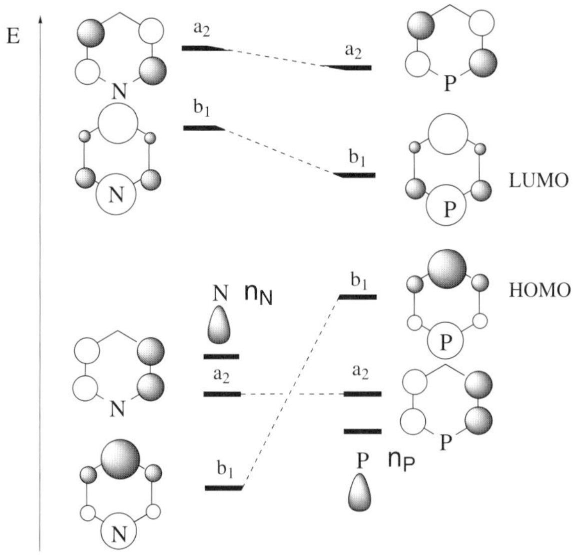

Figure 2. MO scheme of pyridine and phosphinine.

benzene and 10 aromatics including λ^3-phosphinine (C_5H_5P) ⟨94MP557⟩. Electronic spectra of the parent compound and pyridine have been calculated (CASSCF/CASPT2 method). Both spectra compare with experimental values and confirm that the lowest exited state corresponds to a π to π^* transition in phosphinine and that excitation energies are shifted to lower values and separations are smaller than in pyridine ⟨95TCA(92)67⟩. Theoretical studies (HF/4-31G, AM1 and PM3) of the gas phase proton affinities of molecules containing phosphorus–carbon multiple bonds, including C_5H_5P, have been carried out ⟨84JPC1981, 93IJQ343, 95JST(338)51⟩. Although there is a poor agreement between values obtained by *ab initio* (HF/4-31G, 2.4 kcal mol^{-1}) ⟨84JPC1981⟩ and semi-empirical methods (PM3, 27.8 kcal mol^{-1}) ⟨95JST(338)51⟩, both studies show that protonation is favored at the phosphorus atom over protonation at the C2-site.

Molecular orbital calculations have been widely used to illustrate the difference between benzene, pyridine and other group *15* heterobenzenes. Early calculations demonstrated that both HOMO and LUMO of phosphinines are of π-type ⟨70TL4941, 71AG(E)656, 73JA928, 75JCS(P2)841, 76JA4410, 82JA425⟩. More recent investigations, taking into account results obtained from photoelectron and electron transmission spectroscopies, have shown that the orbital which describes the lone pair in phosphinine is only the third highest occupied level (HOMO-2) ⟨89OM2804, 94HAC131, 96JPC6456, 99JOC5524⟩ (in pyridine the HOMO describes the lone pair). In the C_{2v} symmetry of heterobenzenes, the degenerate HOMO (e_{1g}) and LUMO (e_{2u}) of benzene both split into non-degenerate a_2 and b_1 orbitals. It was shown that relative positions of these orbitals depend both on the resonance integral ($\beta_{CX} < \beta_{CC}$) (X = heteroatom) and on the π-electronegativities (χ_X) of the heteroatom (inductive effect of the p_z orbital). Whereas the a_2 orbital is almost unaffected by the replacement of CH by N or P, the b_1 orbital which has a non-zero coefficient at the heteroatom experiences a dramatic change. Thus the ordering of π-HOMOs is inverted between pyridine and phosphinine (Figure 2). In both compounds the ordering of LUMOs is maintained although that of phosphinines lie at lower energies, comparatively. The analysis of the orbital splitting in phosphinines (INDO/S and PPP methods) has been used to establish that the effective π-electronegativity of phosphorus (as well as that of As and Sb) is higher than that of carbon ⟨89OM2804⟩ and to demonstrate why phosphinine displays a comparable π-donor but a better π-acceptor strength than benzene ⟨91AG(E)547⟩.

The MO scheme of λ^5-phosphinine has been constructed by combining that of λ^3-phosphinine with that of H_2 (MP2/6-31G*). This results in a strong destabilization of the HOMO which has the b_1 symmetry ⟨96JPC6456⟩. Finally, calculations of charge distributions have been used to illustrate the difference of reactivity between λ^3-phosphinines and pyridines. In phosphinines, there is a substantial electronic transfer from phosphorus to the carbocyclic C_5 system. Thus, the phosphorus atom bears a positive charge whereas the α carbon atoms (C2 and C6) are negatively charged. The opposite is found in pyridine where these two carbons are positively charged. This specific electronic distribution explains why nucleophiles will react at phosphorus rather than at adjacent carbon atoms as in pyridine. Additionally, it was shown that this charge distribution is almost insensitive to the presence of a chlorine substituent at the C2, C3 or C4 positions ⟨94HAC131⟩. The charge distribution of λ^5-phosphinines has been compared to that of the pentadienyl anion ⟨76JA4410, 96JPC6456⟩ (Figure 3). The phosphorus atom is positively charged whereas all

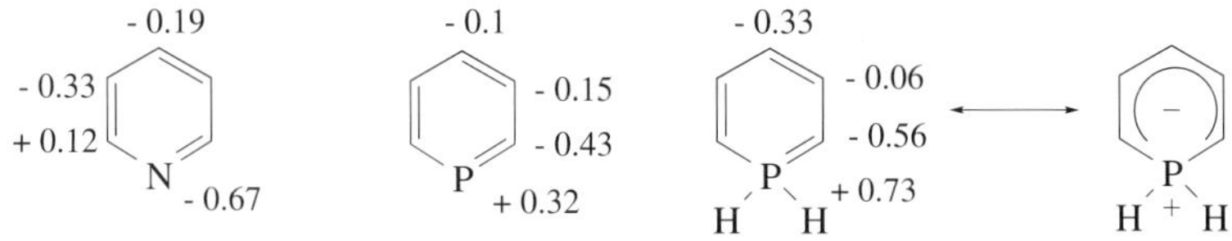

Figure 3. Mulliken charge distribution of pyridine, λ^3-phosphinine and λ^5-phosphinine.

the carbon atoms of the ring bear a negative charge thus confirming the partial ylidic character of these compounds.

5.2.3 EXPERIMENTAL STRUCTURAL METHODS

5.2.3.1 X-Ray Diffraction Analysis

The first X-ray structures reported were those of the 2,6-dimethyl-4-phenylphosphinine ⟨68AG(E)811⟩ and of a dihydro-λ^3-phosphaphenantrene derivative ⟨68TL6227⟩. Since then, a number of free phosphinines or their complexes have been structurally characterized. Concerning uncomplexed species, the most significant features concern the opening of the internal angle (C–P–C) and the P=C bond lengths. The opening of the angle varies from 100 to 106º depending on the substitution scheme. These values which are relatively small compared to that reported for pyridine (up to 117º) are characteristic of heavier group *15* heterobenzenes. They reflect both the lengthening of the heteroatom-carbon bond and the difficulty of the heteroatom to achieve sp^2-hybridization. For stibabenzene derivative, the internal angle (93º) is close to the angle formed by 5p orbitals. In general P=C bond distances fall in the range from 1.73 to 1.75 Å. These distances are shorter than single P–C bonds (1.83 Å) but longer than P=C bonds in phosphaalkenes (1.66–1.68 Å) and clearly show the electronic delocalization in phosphinines. It has been estimated that the CP bond in C_5H_5P is 7.5% shorter than the sum of covalent radii of C and P ⟨78ACR153⟩. To accommodate this distortion, P–C–C and C–C–C angles slightly exceed the ideal value of 120º. Carbon–carbon distances which generally vary from 1.39 to 1.42 Å are close to "normal aromatic bond length" (1.395 Å). Some selected examples are listed in Figure 4 (bond lengths are expressed in Å and bond angles in degrees). In most cases, phosphinines are strictly planar, but bulky substituents

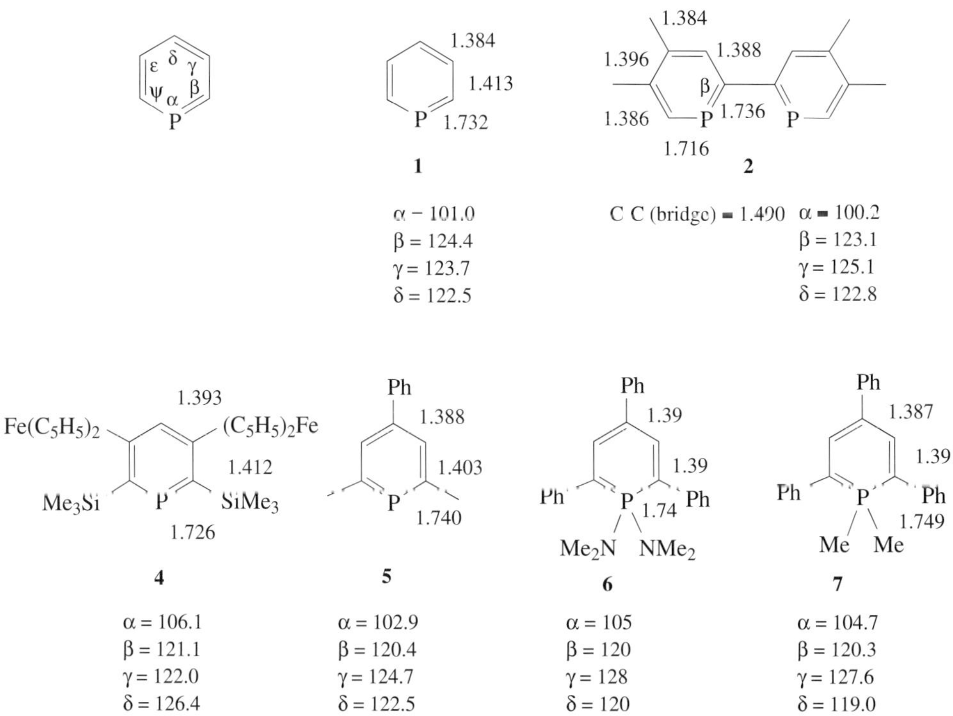

Figure 4. X-ray structures of some λ^3- and λ^5-phosphinines: (**1**) ⟨74JCP2840⟩; (**3**) ⟨92OM2475⟩; (**4**); ⟨97OM4089⟩; (**5**) ⟨68AG(E)811⟩; (**6**) ⟨70AG(E)898⟩; (**7**) ⟨70JCS(A)1832⟩.

can induce some severe distortions. This has been illustrated with the highly distorted structure of the 2,4,5,6-tetrakis(tertiobutyl)-3-methylester derivative which shows a twisted-boat form (up to 33.7° out of planarity) ⟨87CB819⟩. A number of X-ray structures of λ^3-phosphinine complexes have also been reported (see Section 5.2.5.1.3). Three structures of λ^5-phosphinines are known ⟨69CC1057, 69AG(E)769, 70AG(E)898, 70JCS(A)1832⟩.

5.2.3.2 NMR Spectroscopy

5.2.3.2.1 ^{31}P NMR spectroscopy

^{31}P NMR chemical shifts of λ^3-phosphinines appear at significantly lower field (between 178 and 293 ppm, downfield from 85% H_3PO_4) compared to classical tertiary phosphanes. Nevertheless, this deshielding is less than what is routinely observed for other unsaturated species such as phosphaalkenes. The parent compound C_5H_5P resonates at 211 ppm ⟨71JA3293, 76JA5451⟩. Chemical shifts are obviously highly dependent on the substitution pattern of the ring. Some examples are listed in Figure 5 (literature before about 1978–1980 used the opposite sign convention, but these have been converted to the current convention). The substitution of the adjacent carbon atoms (C2 and C6) seems to play a crucial role.

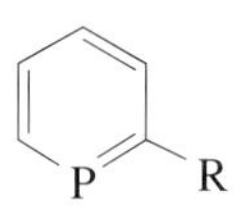

1 : R = H; 211
8 : R = Bu^t; 202
9 : R = Ph; 201
10 : R = Cl; 200.7
11 : R = Br; 210.4
12 : R = I; 233.1

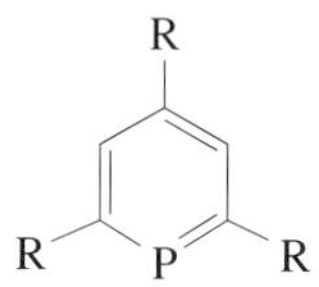

13 : R = Ph; 178
14 : R = Bu^t; 178.2

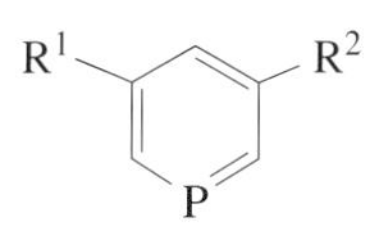

15 : R^1 = R^2 = Ph; 207.8
16 : R^1 = Ph, R^2 = Cl; 215
17 : R^1 = Ph, R^2 = Br; 217.1

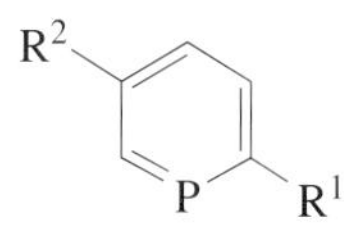

18 : R^1 = R^2 = Ph; 198
19 : R^1 = Br, R^2 = Me; 206.4
20 : R^1 = CO_2Et, R^2 = OH; 217

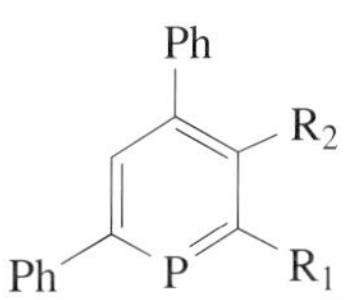

21 : R^1 = R^2 = CO_2Me; 208
22 : R^1 = PPh_2, R^2 = Me; 212.7

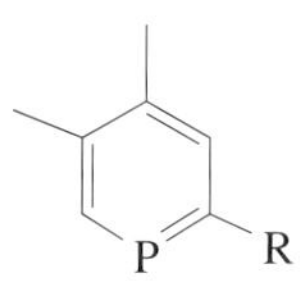

23: R = H; 195
24 : R = Ph; 189
25 : R = Cl; 179.8
26 : R = Br; 193
27 : R = I; 223
28 : R = $SiMe_3$; 219
29 : R = Zr(Me)Cp_2; 224.4
30 : R = 2-thienyl; 177.7
31 : R = 2-furyl; 172.8

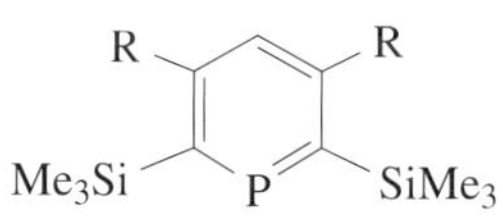

32 : R = H; 254.6
33 : R = Ph; 269.40
34 : R = $SiMe_3$; 266.5

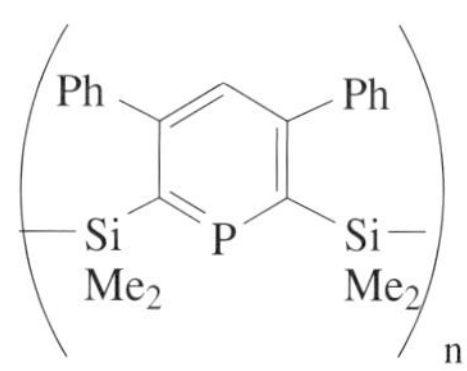

35 : n = 3; 293.1
36 : n = 4; 277.5

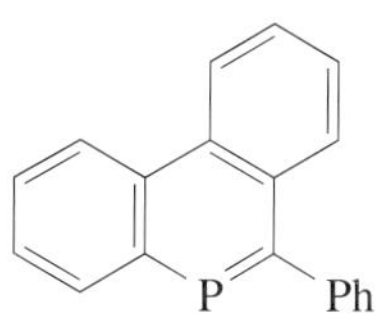

37 : 186

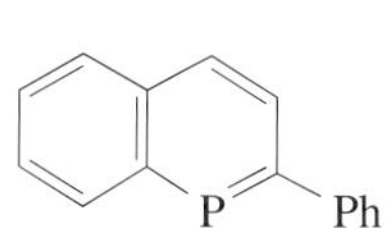

38 : 197

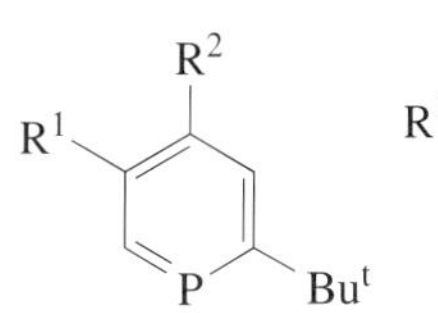

39 : R^1 = Et_2B, R^2 = Et; 182.2
40 : R^1 = Et, R^2 = BEt_2; 193.6

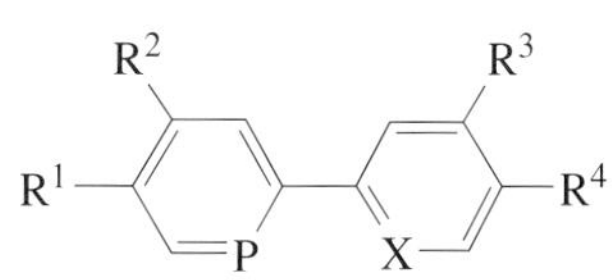

41 : R^1 = R^2 = Me,
R^3 = R^4 = H, X = N; 184.9
3 : R^1 = R^2 = Me,
R^3 = R^4 = Me, X = P; 178.32
42 : R^1 = R^2 = H,
R^3 = R^4 = H, X = P; 196.9

Figure 5. ^{31}P NMR chemical shifts of some λ^3-phosphinines: (**1**) ⟨76JA5451⟩; (**8**) ⟨86ZN(B)931⟩; (**9**) ⟨84PS(19)45⟩; (**10**), (**25**) ⟨89TL817⟩; (**11**) ⟨93JA10665⟩; (**12**) ⟨96OM794⟩; (**13**) ⟨72MIV2-01⟩; (**14**) ⟨74TL3179⟩; (**15**) ⟨B-90MIV2-01⟩; (**16**), (**17**) ⟨83TL2645⟩; (**18**) ⟨82JOC2376⟩; (**19**) ⟨91OM2432⟩; (**20**) ⟨86TL4299⟩; (**21**) ⟨87TL1093⟩; (**22**) ⟨90TL4589⟩; (**23**) ⟨84TL4659⟩; (**24**), (**30**), (**31**) ⟨84PS(19)45⟩; (**26**) ⟨90POL991⟩; (**27**) ⟨92TL3537⟩; (**28**) ⟨91OM2432⟩; (**29**) ⟨97JA9417⟩; (**32**), (**33**) ⟨96JA11978⟩; (**34**) ⟨97OM4089⟩; (**35**), (**36**) ⟨98SCI1587⟩; (**37**) ⟨80TL1441⟩; (**38**) ⟨72AG(E)1017⟩; (**39**), (**40**) ⟨96JOM(520)211⟩; (**41**) ⟨82TL1565⟩; (**3**) ⟨91JA667⟩; (**42**) ⟨98JOC4826⟩.

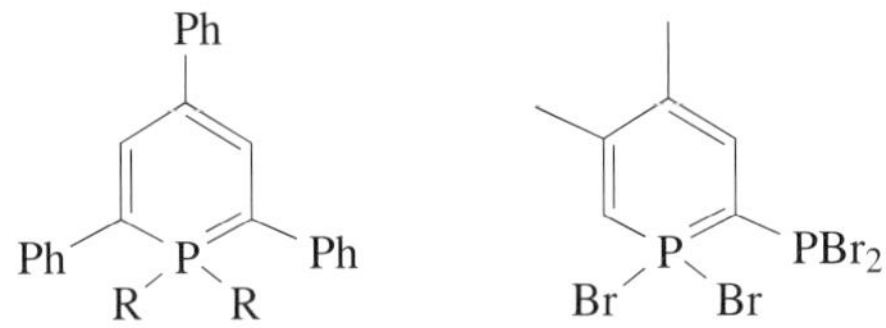

7 : R = Me; 8
43 : R = Cl; 17
6 : R = N(Me)$_2$; 42.5
44 : R = SMe$_2$; 42
45 : R = OMe; 65
46 : R = F; 73.3

47 : -47.5

Figure 6. ^{31}P NMR chemical shifts of some λ^5-phosphinines: (**7**), (**43**), ⟨81PS(10)305⟩; (**6**), (**44**), (**45**) ⟨72TL829⟩; (**46**) ⟨77CB395⟩; (**47**) ⟨96OM1597⟩.

Thus, the presence of electropositive atoms such as silicon, zirconium or phosphorus induces a significant deshielding due to the increase of s character in the hybrid sp^2 orbital at phosphorus (decrease of s character in the lone pair). On the other hand, whereas the nature of functional groups at C3 position only induces small variations, substitution at C4 has a significant effect. Additionally, the IGLO-method (Individual Gauge for Localized Orbitals) has been used to predict the chemical shift of C_5H_5P ⟨93PS(77)105⟩ and MP2 calculations have been carried out to include correlation in the determination of the ^{31}P NMR shielding ⟨96HAC307⟩. Absolute signs of ^nJ(P–H), ^2J(P–Si) and ^2J(P–Sn) coupling constants have been determined ⟨96JOM(520)211⟩.

^{31}P NMR chemical shifts of λ^5-phosphinines are very different from that of their λ^3-congeners. Their values depend strongly on the nature of substituents at phosphorus. Compounds bearing aryl or alkyl substituents generally show signals in the range of triphenyl or trimethylmethylylides. Some derivatives have been listed in Figure 6. Low-field values are generally encountered in the case of electron-attracting substituents which tend to increase the positive charge at phosphorus.

5.2.3.2.2 *^{1}H NMR spectroscopy*

The most significant feature in ^{1}H NMR spectra of heterobenzenes (P, As, Sb) is the downfield shift of the proton resonances (Figure 7). In particular, protons located at adjacent carbons (C2 and C6) are strongly deshielded (δ = 8.61 ppm in C_5H_5P). The same trend is observed for H3,5 and H4 but the magnitude of the shift is considerably smaller. This effect has been explained in terms of magnetic anisotropy of group 15 atoms. It was also demonstrated that, after correction for the local anisotropic effect, these values are consistent with smaller ring currents than in benzene (δ = 7.00 ppm for C_5H_5P) ⟨76JA5451⟩. The proton NMR spectrum of C_5H_5P displays a very characteristic AB_2X_2 spin pattern which was successfully simulated. In general, like for other phosphaaromatic heterocycles, the magnitude of ^2J(P–H) coupling constants are relatively large because of the lone pair effect (Θ = 0) and the magnitude of ^nJ(H–H) coupling constants compare with those recorded for benzene and its functionalized derivatives. The ^{1}H NMR spectra of λ^5-phosphinines markedly differ from their λ^3-counterparts for obvious reasons. The lack of downfield for H3 and H5 as well as the large upfield shift of H2,6,4 is consistent with the loss of ring currents effects.

5.2.3.2.3 *^{13}C NMR spectroscopy*

As observed for protons, adjacent carbon atoms (C2,6) are significantly deshielded ($\Delta\delta$ = 25.4 ppm) whereas C3,5 only undergo a small deshielding ($\Delta\delta$ = 4.9 ppm). The chemical shift of C4 remains nearly unchanged compared to benzene ($\Delta\delta$ = 0.1 ppm). It has been proposed that the source of these large downfield shifts results from variations of the interatomic terms (Karplus and Pople's theory) associated with diamagnetic currents centered at phosphorus ⟨76JA5451⟩. Whereas downfield signals characterizing C2,6 can be easily ascribed, those corresponding to C3,5 and C4 require a careful examination since their relative positions can be readily exchanged with minor modifications of the substitution scheme of the ring. Some examples are listed in Figure 8. The magnitude of ^nJ(P–C) coupling constants in phosphinines has been the subject of theoretical investigation (SCF-INDO). It was shown that the orbital–dipole and spin–dipole terms outweigh the Fermi-contact term. The magnitude of the ^1J(P–C) is controlled by the

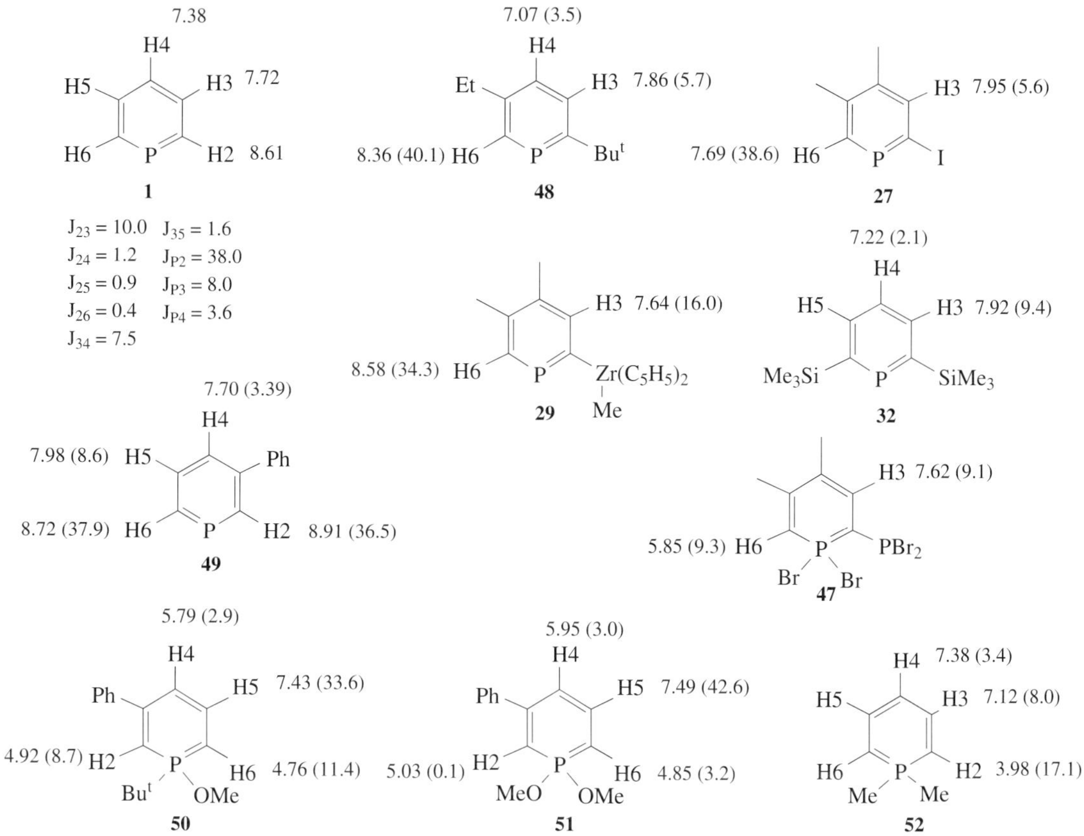

Figure 7. ^{1}H NMR chemical shifts of some λ^3- and λ^5-phosphinines (values in parentheses are ^nJ(P=H) coupling constants): (**1**), ⟨76JA5451⟩; (**48**), ⟨96JOM(520)211⟩; (**27**), ⟨92TL3537⟩; (**49**), (**50**), (**51**), ⟨83CB445⟩; (**29**), ⟨94CC2065⟩; (**32**), ⟨96JA11978⟩; (**47**), ⟨96OM1597⟩, (**52**), ⟨76JA7861, 77TL407⟩.

orbital term whereas that of ^2J and ^3J(P–C) depend on the spin–dipole interactions ⟨79JMR(34)199⟩. In 2-substituted silylphosphinines, the magnitude of ^1J(P–C) are relatively large due to the increase of s character in the P=C bond ⟨99MIV2-01⟩. Heteronuclear correlation experiments ⟨96JOM(520)211⟩ confirmed the absolute signs of ^nJ(P–C) coupling constants proposed earlier ⟨74TL3179, 79JMR(36)181, 79JMR(34)199⟩. In general, the magnitude of ^1J(C–H) coupling constants compare with those of benzene because of similar hybridization of carbon atoms ⟨76JA5451⟩. In λ^5-phosphinines, both chemical shifts and magnitude of coupling constants compare in general with those recorded for phosphorus ylides ⟨74TL3179⟩.

5.2.3.3 Mass Spectrometry

There are no systematic studies of the mass spectra of phosphinines. However, gathered pieces of data clearly show that group *15* heterobenzenes present mass spectra typical of aromatic compounds. These compounds generally show intense molecular peaks. The most significant fragmentations of C_5H_5P are the loss of C_2H_2 or HCP from the molecular ion ⟨71JA3293⟩.

5.2.3.4 Photoelectron (PE) and Electron Transmission Spectroscopies (ETS)

The UV–PE spectra of all heterobenzenes (P, As, Sb, Bi) have been recorded. Their analysis revealed that the ordering of the MOs is inverted (see Section 5.2.2) when compared to those of pyridine. Thus, the two first ionization potentials (IP) (9.2 and 9.8 eV) of C_5H_5P are respectively ascribed to $3b_1$ and $1a_2$ π-orbitals whereas the third (10.0 eV) corresponds to the ionization of the lone pair σ n_p (a_1) ⟨71AG(E)656, 72AG(E)631, 73JA928, 89OM2804⟩. Theoretical calculations also support the assignment of the third IP to the n_P MO (see Section 5.2.2). The analysis of angular and energy dependence of band intensities confirmed that the $1a_2$ and a_1 orbitals are close in energy ⟨76HCA1944⟩. Measurements on the 3-methyl-4-chlorophosphinine reveal striking similarities between values recorded and those of the o-chlorotoluene thus showing

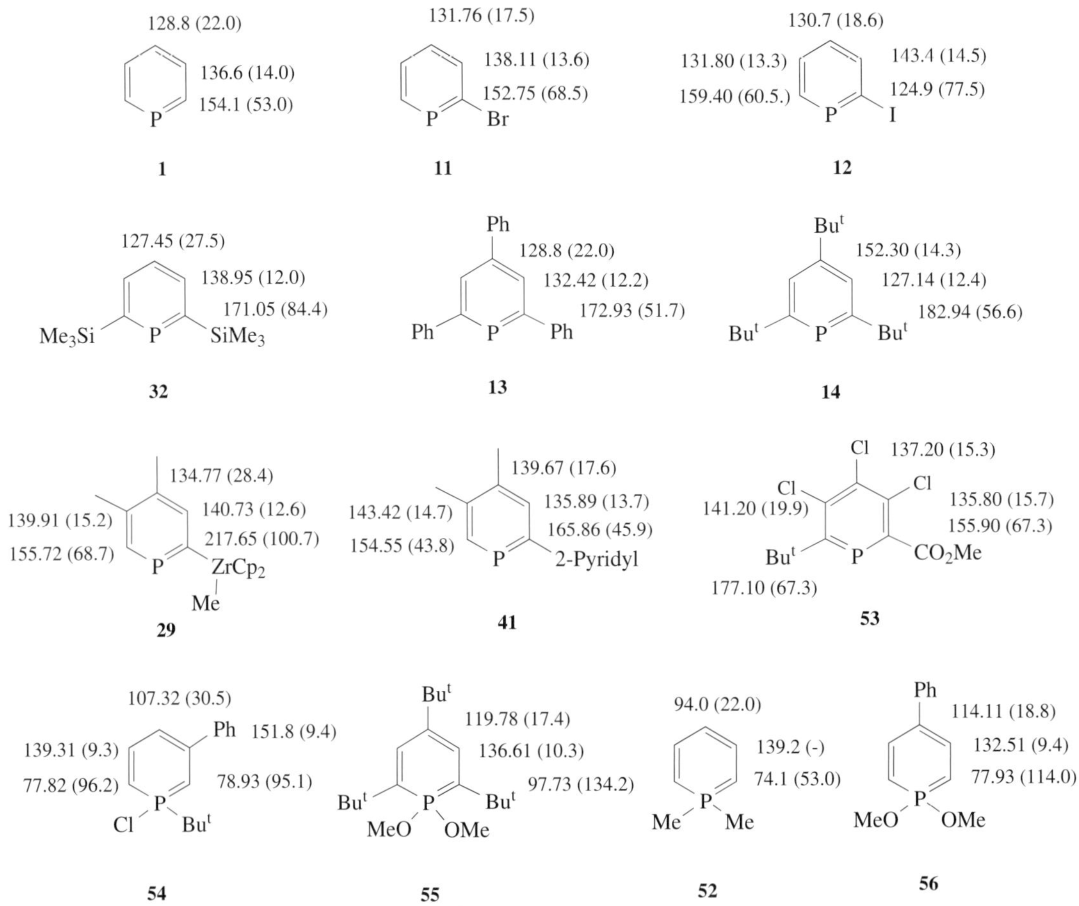

Figure 8. ^{13}C NMR chemical shifts of some λ^3- and λ^5-phosphinines (values in parentheses are ^nJ(P–C) coupling constants): (**1**), ⟨76JA5451⟩; (**11**), ⟨93JA10665⟩; (**12**), ⟨96OM794⟩; (**32**), ⟨96JA11978⟩; (**13**), (**14**), ⟨74TL3179⟩; (**29**), ⟨94CC2065⟩; (**41**), ⟨82TL1565⟩; (**53**), ⟨90CB935⟩; (**54**), (**55**), ⟨74TL3179⟩; (**52**) ⟨76JA7861⟩; (**56**), ⟨83CB445⟩.

that the conjugative abilities of P=C and C=C systems are quite similar ⟨94HAC131, 95JST(347)57⟩. Electron Transmission Spectroscopy (ETS) has been employed to study temporary negative ion formation in some heterobenzenes (P, As, Sb) in the gas phase and to determine Electron Affinities (EA). The main conclusion of this study was that the LUMO of these molecules is a b_1 orbital ⟨82JA425⟩. As quoted above, it must be added that a combination of UV spectroscopy and magnetic circular dichroism experiments established the relative π electronegativity of P in C_5H_5P ⟨89OM2804⟩. A study on PE spectra of seven λ^5-phosphinines has confirmed the idea that their π-electronic structure involves some back-bonding from the pentadienyl anion unit to the positively charged phosphorus atom ⟨76JA4410⟩.

5.2.3.5 Ultraviolet and Infrared Spectroscopies

UV spectra of λ^3-phosphinines are very similar to those of benzene and pyridine derivatives. The UV spectrum of C_5H_5P (**1**) shows intense bands at 213 (ε = 19000) and 246 nm (ε = 8500) ⟨71JA3293⟩, the tris(tertiobutyl) compound (**14**) at 262 nm (ε = 8500) ⟨68AG(E)460⟩, and the 2,4,6-triphenylphosphinine (**13**) at 278 (ε = 41000) and 314 (ε = 12600) ⟨66AG(E)846⟩. The IR spectra of phosphinines are also similar to those of corresponding benzene derivatives. Only in the region 1200–1400 cm^{-1} are additional bands detected. Going from λ^3- to λ^5-phosphinines, UV spectra are profoundly modified and new bands appear in the visible region since these compounds are colored, ranging from yellow to red. Thus, for example, the spectrum of the 1,1-dimethoxy-2,4,6-triphenyl derivative (**45**) shows bands at 417 (ε = 13700), 305 (ε = 11900), 278 (ε = 12700) and 220 (ε = 16600) ⟨73TCC(38)1⟩. It must also be noted that some derivatives show fluorescence. The IR spectra of λ^5-phosphinines show additional bands between 1180 and 1220 cm^{-1} and others in the region between 718 and 1160 cm^{-1} which characterize the P–X (X = O, S, N) bonds ⟨73TCC(38)1⟩.

5.2.3.6 Dipole Moments

The dipole moments of C_5H_5P (1.46 ± 0.04 D) and other heterobenzenes have been determined in cyclohexane from standard measurements of dielectric constants and calculations of polarization at infinite dilution ⟨73T475, 74JCS(F2)1222, 75TL2749, 75JCS(P2)841, 75CPH345⟩. This result correlates with the more precise gas phase data (1.54 ± 0.02 D) ⟨72MIV2-02⟩. Also, the negative end of the dipole is shown to be at phosphorus. Additional experiments carried out on 4-methylphosphinine (1.77 ± 0.12 D) have confirmed this result since the value recorded exceeds the one obtained from the parent compound.

5.2.3.7 Cyclic Voltammetry

Only two studies report on the electrochemical behavior of free λ^3-phosphinines. Whereas the oxidation process appears to be irreversible, C_5H_5P is reversibly reduced at $E_{1/2} = -2.27$ V *vs* SCE) (SCE = standard calomel electrode) to give the mono radical anion C_5H_5P ⟨91AG(E)547, 94JA6217⟩. For the sake of comparison, benzene is reduced under the same conditions at a more negative potential ($E_{1/2} = -3.42$ V *vs* SCE). The second study deals with the reduction of the a 2,2′-biphosphinine (**3**). In this case, the first monoelectronic reduction ($E_{1/2} = -1.85$ V *vs* SCE) is chemically reversible whereas the second monoelectronic reduction ($E_{1/2} = -2.42$ V *vs* SCE) becomes reversible only under forced conditions. A comparison with the 4,4′-dimethyl-2,2′-bipyridine, which is reduced at more negative potentials, suggests that biphosphinines should behave as better π-acceptor ligands than bipyridine towards transition metals ⟨92OM2475⟩.

5.2.4 THERMODYNAMIC ASPECTS

5.2.4.1 Physical Properties and Thermal Stability

λ^3-phosphinines exist as liquids or solids which can be malodorous. Their resistance towards air oxidation is strongly dependent on their substitution scheme, especially at C2,6 positions. As a whole, unsubstituted compounds, liquid at room temperature, are easily oxidized but 2,6-disubstituted phosphinines turn out to be very air-stable. Usually, λ^3-phosphinines are soluble in common organic solvents such as alkanes, benzene, toluene, ethers (Et_2O, THF), DMF and moderately soluble in acetonitrile and alcohols. When no polar groups are present, λ^3-phosphinines can be considered as non-polar substances. They can be purified by simple distillation under inert atmosphere or by column chromatography using silica gel or alumina as support provided that separations are quickly carried out. Solvents for crystallization are relatively diverse and mainly depend on the solubility of the compound. Due to the absence of lone pair at phosphorus, λ^5-derivatives appear to be more stable towards air oxidation. They usually exist as colored solids soluble in ethers and in polar organic solvents. In general, λ^3-phosphinines withstand high temperatures without apparent decomposition. A large number of derivatives have been synthesized by thermolysis from diverse precursors. The parent compound has been produced by flash-vacuum-thermolysis (FVT) of the bis(allyl)-vinylphosphine at 700ºC under 1.3×10^{-3} mbar ⟨93CC1295, B-96MIV2-01⟩. On the contrary, λ^5-phosphinines are less stable under thermolytic conditions and decompose to yield λ^3-derivatives.

5.2.4.2 Basicity of λ^3-Phosphinines

Unlike pyridines and common phosphines, λ^3-phosphinines show no basic nor nucleophilic character. Indeed, strong alkylating or protonating reagents such as R_3OBF_4 ROTf, CF_3CO_2H do not react with phosphinines to give the corresponding phosphininium salts. Preliminary studies suggested a pK_a difference of 15 units between protonated pyridine (pK_a = 5) and protonated phosphinine (pK_a = −10) ⟨71MIV2-01⟩. Gas phase proton affinities of C_5H_5X (X = N, P, As) have been determined by ion-cyclotron resonance techniques to be 219.4, 195.8 and 189.3 kcal mol^{-1}, respectively ⟨85OM457⟩. The proton affinity of the parent compound C_5H_5P is 30 kcal mol^{-1} less than that of trimethylphosphine and only slightly greater than that of PH_3 (188.7 kcal mol^{-1}). In the same study, deuterium labeling experiments have shown that the protonation takes place at the phosphorus atom. Correlations between core-ionization energies and proton affinities suggest that the low basicity of heterobenzenes mainly results from the inability of the ring to undergo geometrical change upon protonation ⟨79JA1764⟩. Indeed, contrary to pyridine, the lone pair at phosphorus is largely s-hybridized and a protonation would imply that it should gain a significant p-character to take part effectively in bonding (calculations have shown that the lone pair in phosphinines bears a significant s-character: 29.1% in C_5H_5N compared to 63.8% in C_5H_5P ⟨99JOC5524⟩). Such rehybridization is hampered by the rigidity of the C–P–C angle which cannot expand because of the resulting strain placed on the C–C–C angles ⟨95JST(338)51⟩.

5.2.5 REACTIVITY OF λ^3-PHOSPHININES

5.2.5.1 Reactions at the P Atom

5.2.5.1.1 Reduction and oxidation processes, reactions with radicals

λ^3-Phosphinines are easily reduced by alkali metals (Li, Na, K) to give paramagnetic mono- and diamagnetic dianion radicals ⟨67AG(E)446, 74MP601, 81JPC1202⟩. The formation of a paramagnetic trianion radical has also been reported in the case of the 2,4,6-triphenylphosphinine. ESR (Electron Spin Resonance) experiments demonstrate that the mono anion radical can be regarded as a "P-centered radical" ($g = 2.0045$) whereas trianionic species (**58**) have to be classified as "hydrocarbon radicals" ($g = 2.0027$) (Scheme 1). In general these reductions take place in THF or DME at room temperature using alkali or K/Na-alloy. As their carbon counterparts these highly colored (color depending on the charge) solutions of these anion radicals are extremely oxygen-sensitive.

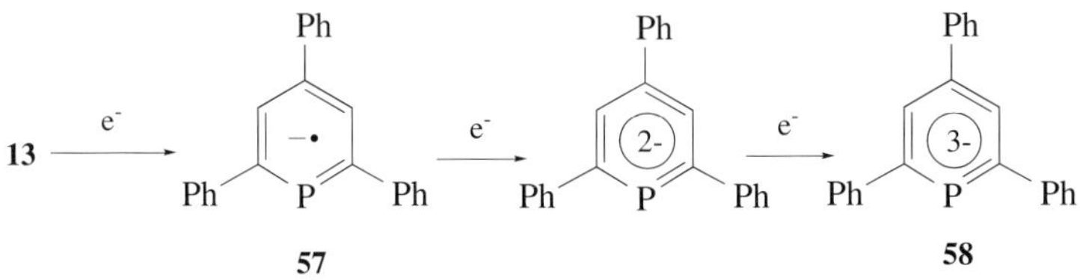

Scheme 1

Studies have not been limited to the triphenyl derivative (**13**). Radical anions of phosphanaphthalene (**59**) ⟨74TL1267⟩, 4,4′- and 2,2′-biphosphinine derivatives such as (**60**), (**61**), (**62**) and (**3**) have also been reported ⟨97MRC384, 99OM3348⟩ (Figure 9). EPR spectra of (**60**), (**61**) and (**62**) show that the unpaired electron is delocalized on the two rings. Formation of trianions with these latter 4,4′-derivatives have also been evidenced. In this case hyperfine patterns show that the π-spin population is only restricted to one phosphinine ring and its phenyl substituents, probably as a result of a perturbation caused by the association with three alkali metal counterions. On the other hand, information relative to diamagnetic dianions are rare. A trianionic derivative of a 1,4-diphosphaindene ($\mathbf{63}^{3-}$) including a dianionic phosphinine unit and a 2,2′-biphosphinine species whose NMR data have been recorded ⟨99OM3348⟩ are the only two examples known.

Oxidation of λ^3-phosphinines has been widely studied since it provides a straightforward entry to λ^5-derivatives. This part will be discussed in Section 5.2.7.2.2. These reactions proceed through

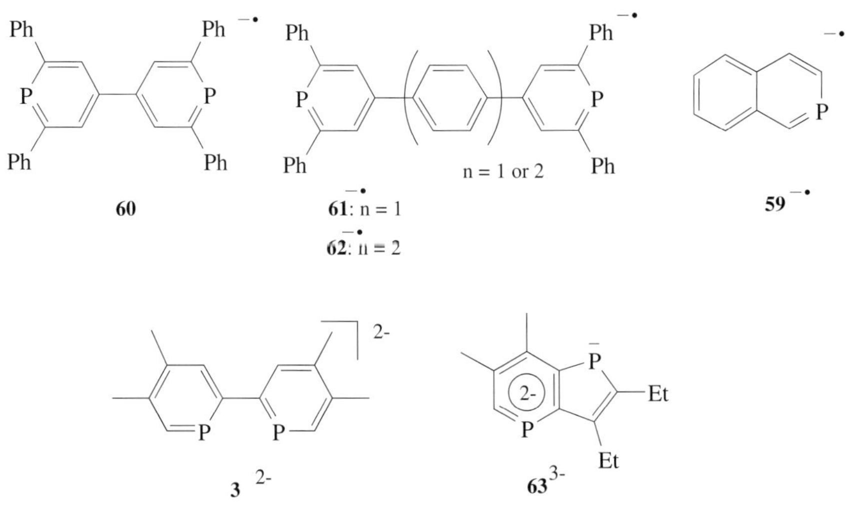

Figure 9. Some mono radical anions, dianions and trianions of phosphinines.

the formation of phosphinine radical cations such as (**64**) though there is no definite evidence that ESR-signals observed are not related to radicals formed upon reactions with alkoxy, aryloxy and carboxy anions ⟨81CB3004⟩. Oxidations can be carried out with various reagents such as 2,4,6 triphenylphenoxyl ⟨67AG(E)85⟩, $Hg(OAc)_2$ ⟨68AG(E)881⟩, $Pb(OCOPh)_4$ ⟨67AG(E)711⟩ and tetrachloro-p-benzoquinone.

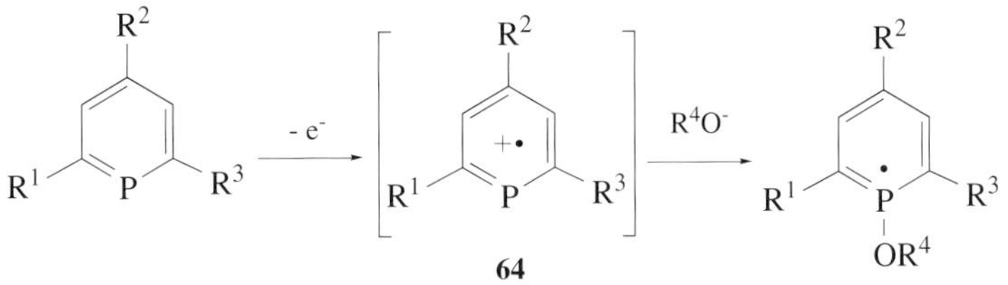

Scheme 2

Another important point concerns the oxidation of the P lone pair with oxygen and sulfur. Although phosphinine-oxides have never been detected, it is likely that they exist (**65**) as highly reactive species. They have been trapped by water to give compounds such as (**66**) ⟨73CB1001⟩. Molecular oxygen was observed to react at P and C_4 to yield, depending on the substitution scheme, a 4,4′-peroxy-bis-phosphininic acid ⟨68AG(E)371⟩ or a dioxa-phosphabarrelene adduct resulting from a [4 + 2] cycloaddition ⟨72AG(E)506⟩. Phosphinine sulfides are more stable than the oxides and compounds such as (**67**) and (**68**) have been detected by ^{31}P NMR spectroscopy ⟨84CC508, 88CC493⟩. It was also noted that reaction of sulfur in excess leads to a tricyclic compound (**69**) (Scheme 3) ⟨90HAC37⟩.

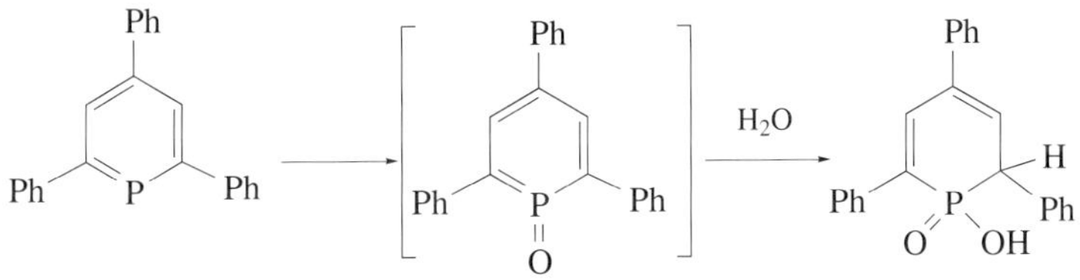

13 **65** **66**

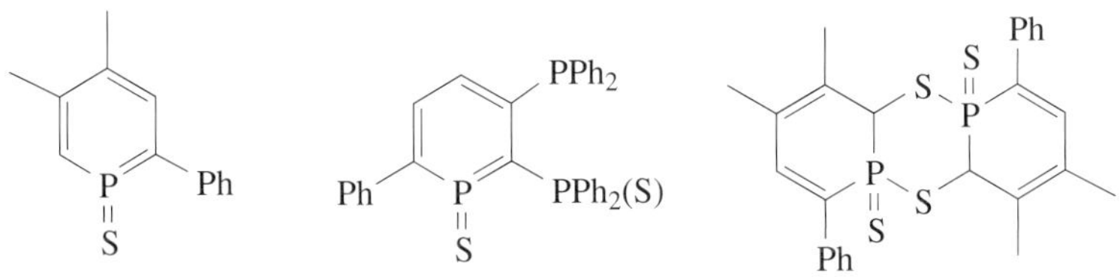

67 **68** **69**

Scheme 3

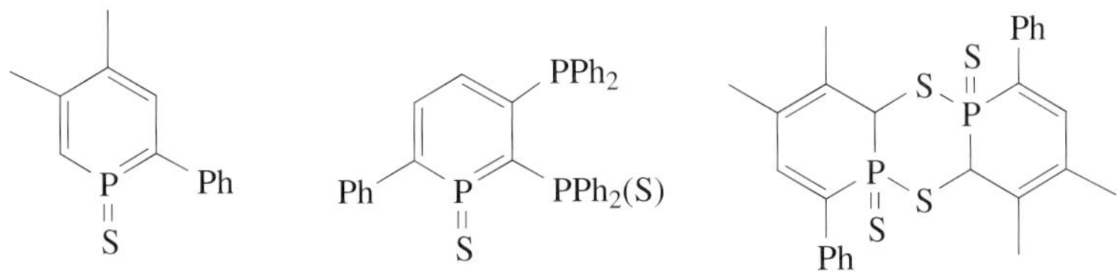

67 **68** **69**

Reactions with radicals give 1R-phosphininyl radicals which in turn may be used as reactive intermediates for the synthesis of λ^5-phosphinines (see Section 5.2.7.2.2) (Scheme 4). The transient formation of these radicals has been ascertained by complementary experiments with 1,2-dihydrophosphinines ⟨69TL1231⟩. The synthetic utility of these radicals will be addressed in Section 5.2.7.2.2.

5.2.5.1.2 Reactions with nucleophiles

The reactivity of phosphinines towards nucleophiles has been extensively studied. According to the charge distribution diagram, it is not surprising that reactions exclusively take place at the phosphorus

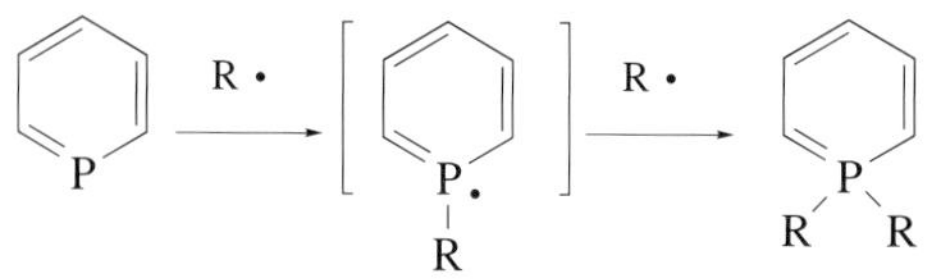

Scheme 4

atom to give delocalized dihydrophosphinine carbanions such as (**70**). These show an ambident character ⟨67AG(E)87, 68TL3611⟩ and can be used to synthesize either 1,n-dihydrophosphinines (n = 2 or 4) or λ^5-phosphinines (see Section 5.2.7.2.2) depending on the nature of the electrophile. Thus, whereas "soft" alkylating (S_N2) agents react at the phosphorus atom lone pair to give λ^5-phosphinines (**7**), hard electrophiles (S_N1) react at C2 or C4 to give 1,2- (**71**) or 1,4-dihydro (**72**) derivatives, respectively (Scheme 5) ⟨69TL1231, 71TL1215, 74TL4501, 74AG(E)408, 77TL407, 81TL1207⟩. In general, the formation of the anion occurs in aprotic polar solvents (THF, Et_2O, DME, toluene) at low temperature using a wide range of Grignard or lithio compounds (R = Ph, Bu, Bu^t, heteroaryl, ferrocenyl). This reactivity is not limited to (**13**), as every λ^3-phosphinine undergoes these nucleophilic attacks whatever their substitution scheme. A different outcome is observed for 2-halogeno derivatives. Even though nucleophiles still attack at the P-atom, the advancement of the reaction seems to be dependent on the ability of the anion formed to release or not the halogen atom. Thus, a 2-iodophosphinine (**27**) was shown to react with EtONa to give the 1,1-diethoxy compound (**13**) (Scheme 6) ⟨90BSB4849⟩.

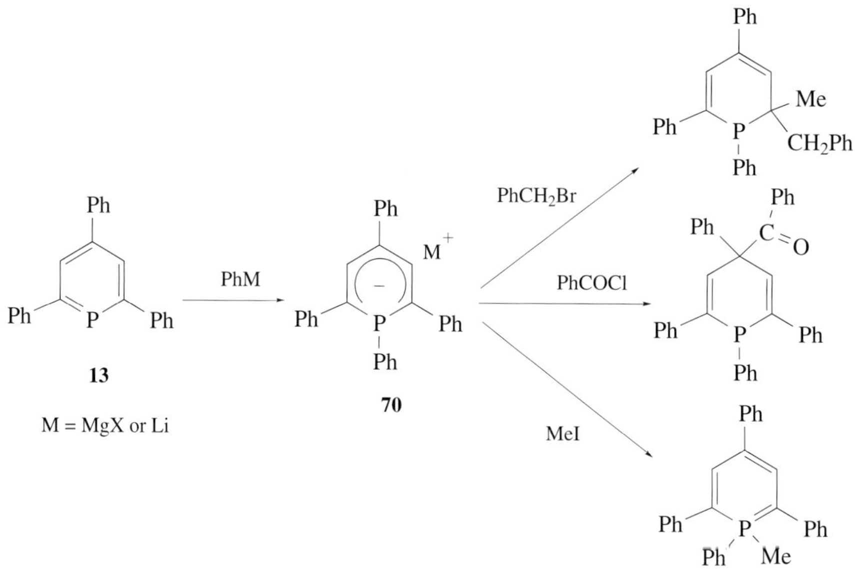

Scheme 5

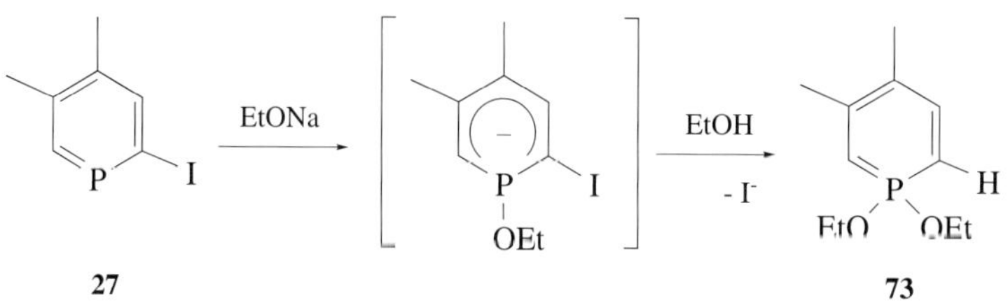

Scheme 6

Phosphinine complexes of transition metals also undergo nucleophilic attacks to yield dihydrophosphinine anion complexes such as (**75**) which can be subsequently transformed to 1,2- and 1,4-dihydrophosphinines complexes such as (**76**) and (**77**) (Scheme 7). Many examples illustrate this reactivity, mostly with pentacarbonyl metals of group 6. In particular, this approach has been used to alkylate a phosphinine–$W(CO)_5$ complex (**74**) at the C2 position to give (**79**) via the P–Cl derivative (**78**) ⟨90POL991, 91OM2432⟩. Also, neutral and cationic platinum complexes of 2,2′-pyridylphosphinine ⟨91IC4693⟩ and

2,2′-biphosphinine ⟨93PS(77)255⟩ can be converted to the corresponding dihydrophosphinine complexes via reaction with alcohols. Finally, η^6-phosphinine–$Cr(CO)_3$ complexes were shown to react with nucleophiles to give delocalized anion complexes, subsequently transformed into λ^5-phosphinines ⟨83JOM(247)271⟩.

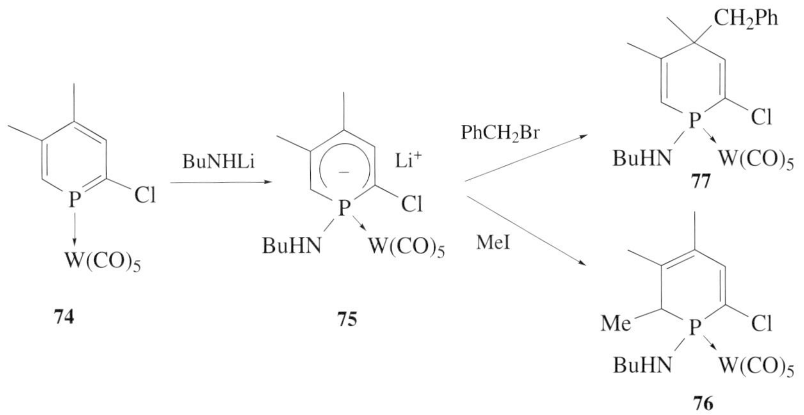

Scheme 7

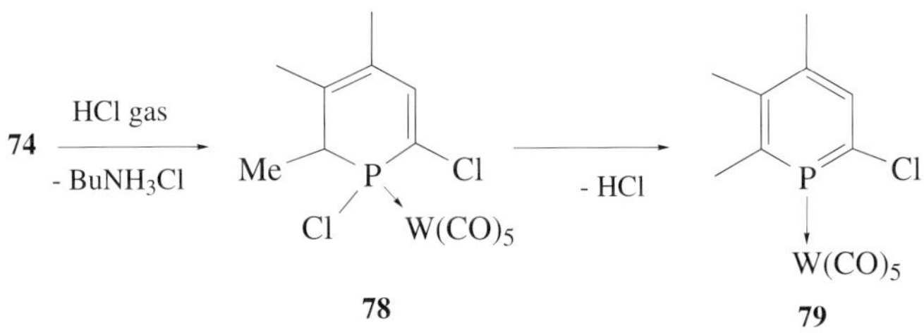

Scheme 8

Diazoalkanes, as soft nucleophiles, also react at the P atom. The zwitterions formed evolve differently, depending on the nature of the group and on the substitution scheme of the ring. In some cases, this reactivity has been used to prepare λ^5-phosphinines (see Section 5.2.7.2.2). In aprotic solvents such as CH_2Cl_2, hexacyclic diphosphachiropteradienes (**80–82**) resulting from the formal reaction of two molecules of (**15**) with the carbene are formed (Scheme 9) ⟨87AG(E)236⟩. Other aspects of the reactivity of carbenes involving the reactivity of the P=C double bond will be addressed in Section 5.2.5.2.

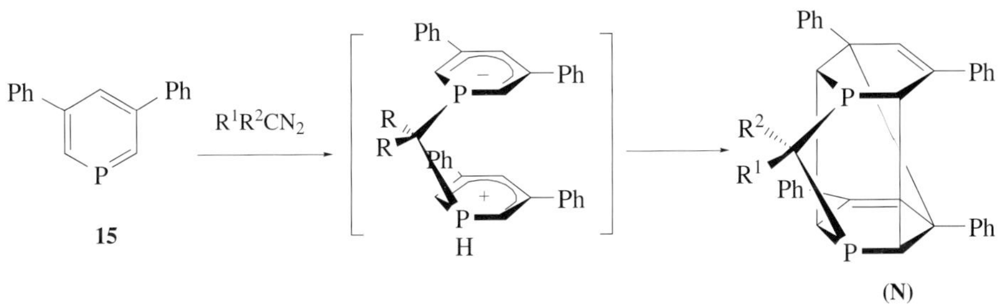

80 : $R^1 = R^2 = H$
81 : $R^1 = R^2 = Me$
82 : $R^1 = Me$; $R^2 = Ph$

Scheme 9

5.2.5.1.3 Complexation of phosphinines

A wide range of λ^3-phosphinines complexes have been synthesized with various metallic centers. All these developments have been briefly reviewed ⟨98CCR(179-180)771⟩. Eight modes of coordination have been counted for λ^3-phosphinines. The mode I in which the ligand behaves as a 2e donor is by far the most

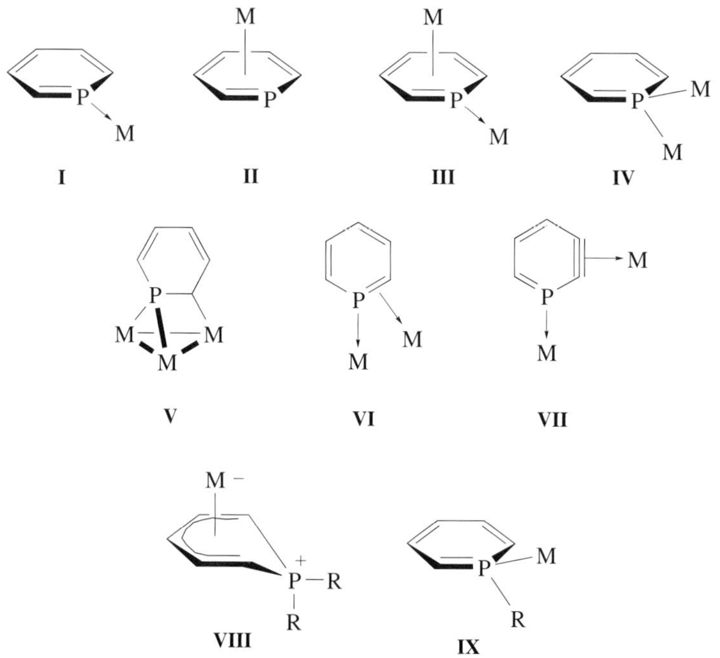

Figure 10. Bridging modes of phosphinines.

widespread. In general, this type of coordination is favored when no substituents are present at C2,6. As phosphinines are as good σ-donors and better π-acceptor than arenes, η^6-complexes (mode II, 6e donor and mode III, 8e donor) are also found. Other modes such as IV–VII are more anecdotal. Whereas in modes IV (μ-P, 2e donor) and V (μ3, 4e donor), the phosphorus atom lone pair bridges a dimetallic or a trimetallic unit, mode VI (η^1-P, η^2-PC, 4e donor) shows that the P=C double bond can act as a second binding site. As we will see below, mode VII (η^1-P, η^2-C2C3, 4e donor) is only encountered in phosphabenzyne complexes (see Section 5.2.5.3.2). With λ^5-phosphinines, only two modes are known. Whereas the first one involves the coordination of the unsaturated system (mode VIII, η^5-, 6e donor) ⟨76AG(E)503, 81PS(10)285, 81PS(10)305, 83JOM(247)271⟩, in mode IX, the phosphorus atom is connected both to a substituent and a metal (Figure 10) ⟨75TL545, 74AG(E)408⟩.

Early studies took advantage of the availability of the 2,4,6-triphenyl derivative (**13**) whose coordinating properties have thus been studied in depth. A number of η^1-complexes (mode I) such as M(**13**)(CO)$_5$ (M = Cr, Mo, W) ⟨73JOM(49)453, 73CB2222, 78JOM(148)C31⟩, Ni(**13**)(PR$_3$)$_2$ ⟨81LA1139⟩, Rh(**13**)$_3$Cl, Ru(**13**)$_2$Cl$_2$ ⟨72JHC1457⟩, Au(**13**)I ⟨73SRT375⟩, Mn(**13**)(CO)$_2$(η^5C$_5$H$_5$) ⟨80JOM(187)277⟩ have been described. As noted in Section 5.2.5.2, carbonyl complexes of group 6 metals have been utilized to promote various reactions at the P=C double bond or the 1,4-phosphadiene moiety. It is now clearly established that phosphinines favor the coordination of electron-rich centers because of their strong π-acceptor ability. This property was illustrated by the synthesis of several homoleptic complexes of C_5H_5P (Ni ⟨92AG(E)1343⟩, Cr ⟨93JOM(459)157⟩, W ⟨98OM4417⟩) such as (**83**) ⟨94JA6217⟩. Bidentate ligands such as 2-phosphinophosphinines can bind two metal centers (**84**) ⟨97CB843, 97JOM(541)277⟩, and an interesting complex of 2-(2′-pyridyl)phosphinine (**85**) was also characterized ⟨91IC4693, 92IC5117⟩. But mostly, the work focused on 2,2′-biphosphinines ⟨96CB263⟩. These ligands, as their bipyridine counterparts, are powerful chelates of various metal centers such as group 6 M(CO)$_4$ ⟨91JA667, 92OM2475⟩, Ni^0 ⟨95IC11⟩, RuCl$_2$ ⟨92ICA(199-200)437, B-96MIV2-02⟩, Ru(η^5-C$_5$Me$_5$)Cl ⟨96OM3267⟩, CuI ⟨96BSF691⟩. A series of homoleptic complexes of group 6 metals such as (**86**) have also been characterized ⟨98IC3154⟩ and the reaction with Mn$_2$(CO)$_{10}$ yielded the bimetallic complex (**87**) in which the ligand behaves as a 8e donor ⟨95IC5070⟩. Electrochemical studies revealed that these ligands can also stabilize highly reduced centers, such as Ni^{2-} ⟨95IC11⟩. Cationic Rh^{I} and Ir^{I} complexes of a phosphinine-based tridentate ligand ⟨98IC5313⟩ and the silacalix-[4]-macrocycle (**88**) have also been prepared ⟨98SCI1587⟩. Finally, many studies focused on η^6-coordination mode of phosphinines with various centers such as Cr^0 ⟨70CB2541, 72CB1148, 78JOM(148)C31, 91AG(E)547⟩, Mn^{I} ⟨80JOM(187)277⟩, V^0 ⟨91AG(E)547, 93OM3373⟩ (**89**), and Fe^0 (**90**) (Figure 11). This latter complex shows a catalytic activity for the formation of pyridines from nitriles and alkynes ⟨96PS(109)173, 96OM2713, 98MIV2-02⟩.

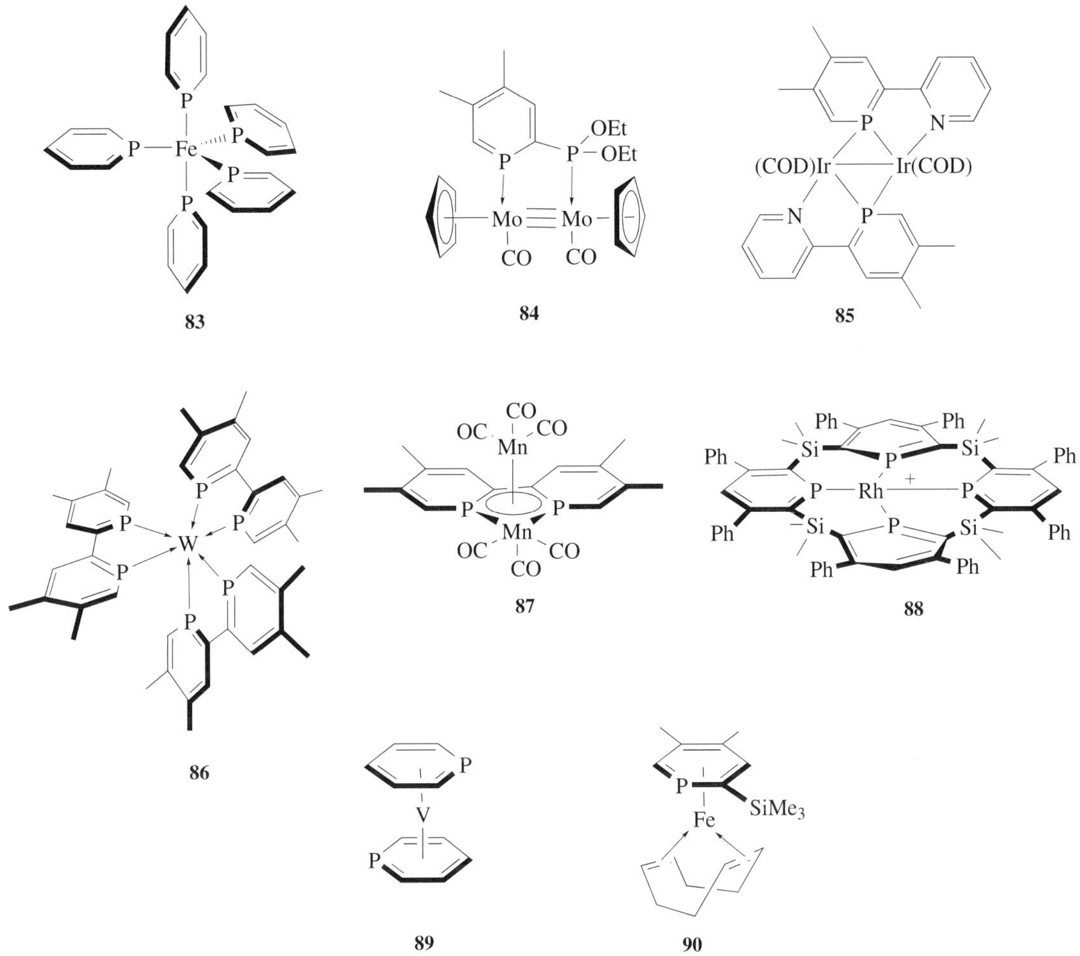

Figure 11. Some complexes of phosphinines.

5.2.5.2 Reactions at the P=C Double-Bond (Carbenes, Cycloadditions)

These reactions have been studied in depth. As we will see below, this bond can be engaged in various addition and cycloaddition processes including [2 + 1], [3 + 2] and [4 + 2] (depending on the fact that the lone pair is protected or not). We will also consider reactions in which the P=C bond participates as part of a conjugated system. Reactions with carbenes have been extensively studied, and it is fairly difficult to distinguish in some cases whether the initial reaction takes place at the P atom or at the P=C double-bond in a concerted fashion. A first interesting example is provided by the reaction of chlorocarbenoids [:CRCl] with (**13**) and (**14**). The chloro-P-phosphepines (**91**) and (**92**) transiently formed rearrange to yield 1,3,5-triphenyl-2-R-benzenes (**93**) and (**94**) via extrusion of chlorophosphinidene (Scheme 10) ⟨71TL1269, 84AG(E)894⟩. This reaction raises the problem of the phosphanorcaradiene/phosphatropylidene rearrangement. Experiments with $W(CO)_5$ complexes shed some light on this valence tautomerism. Indeed, the condensation of diazoalkanes with complex (**95**) leads to the formation of bicyclic compounds such as (**96**). It is not known whereas the initial step results from a [2 + 1] or from a [3 + 2] addition with loss of N_2, but the formation of (**96**) cannot be rationalized without postulating that phosphanorcaradiene (**97**) equilibrates with phosphatropylidene (**98**). Then the diazoalkane reacts with the P=C bond of (**98**) via a [3 + 2] addition, followed by a 1,3 H-shift (Scheme 11) ⟨87AG(E)1134⟩. Diphenyl diazomethane was also reacted with (**95**) to produce a 1-chlorodehydrophosphepine after HCl treatment ⟨87AG(E)1134⟩. A bicyclic derivative and a phosphirane were obtained when the reaction was carried out with the free ligand ⟨90TL4849⟩.

Phosphinine complexes can react as dipolarophiles. Diphenylnitrile imine and nitrile oxides react at the P=C system to form bicyclic compounds (**99–102**) and (**103**) (Scheme 12) ⟨87TL3475⟩.

As shown above, coordination of the lone pair generally increases the reactivity of the P=C double-bond system in partially desaromatizing the ring. This effect was used to promote various

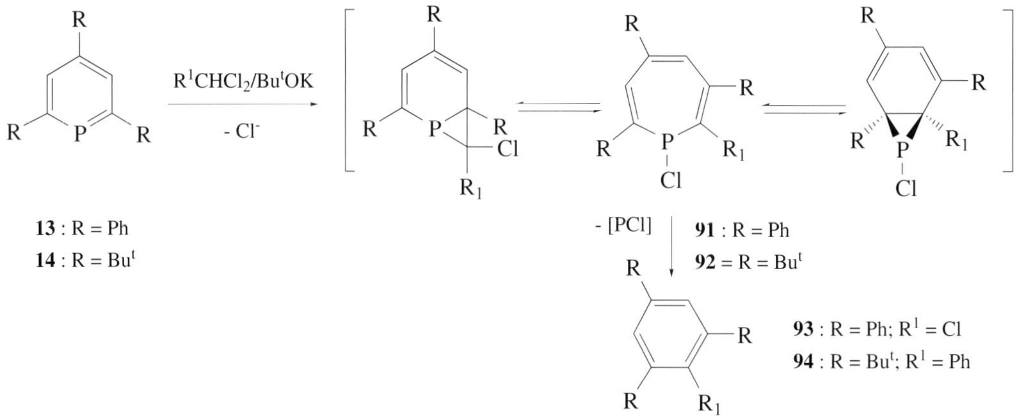

Scheme 10

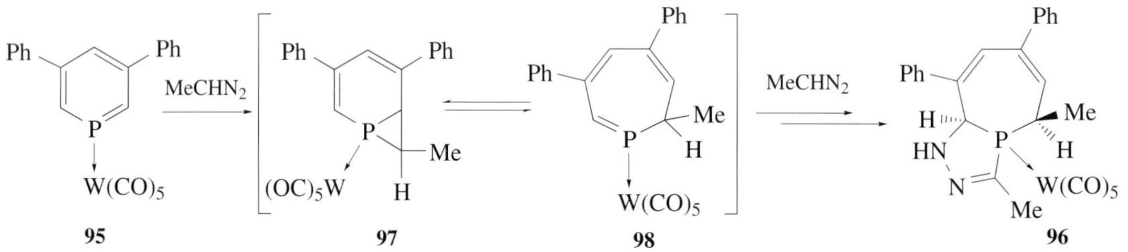

Scheme 11

[4 + 2] cycloadditions with phosphinines and first experiments were carried out with sulfides (see Section 5.2.5.1.1). Heating (90–110°C) a phosphinine (**24**) with sulfur in the presence of a 1,3-diene results in the formation of the bicyclic derivative (**104**) (Equation (1)) ⟨84CC508, 84TL4659, 91JA667, 92OM2475, 93PS(76)33, B-96MIV2-02⟩. This method was successfully employed to the functionalization of 2-bromophosphinines (see Section 5.2.7.1.4). Noteworthy, the reaction takes place at the less hindered carbon. Pentacarbonyl complexes such as (**95**) display a similar reactivity but at higher temperature (140°C, 24 h) since the desaromatization is less pronounced in this case ⟨84TL207, 87TL3475⟩. Various bicyclic compounds (**104–106**) have thus been synthesized (Equation (2)).

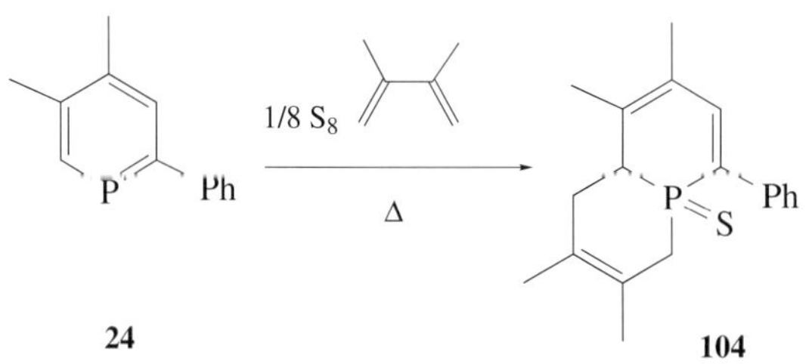

24 **104**

(1)

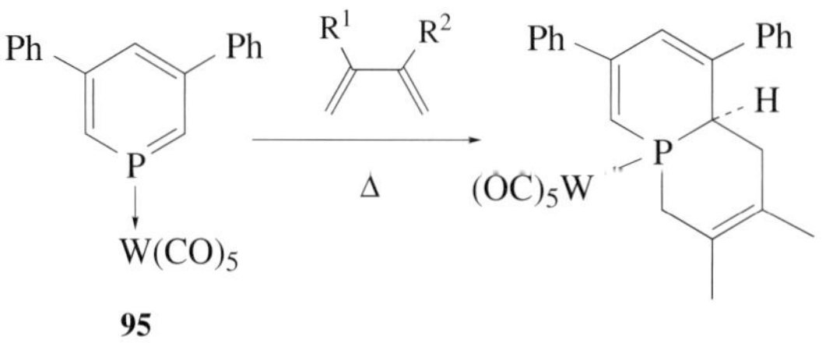

95

(2)

104 : R^1 = R^2 = Me
105 : R^1 = R^2 = OSiMe$_3$
106 : R^1 = R^2 = H or Me

A last important point concerns reactions in which phosphinines behave as 1-phospha-1,3-butadienes. Due to their strong aromatic character phosphinines are not usually good partners in these processes.

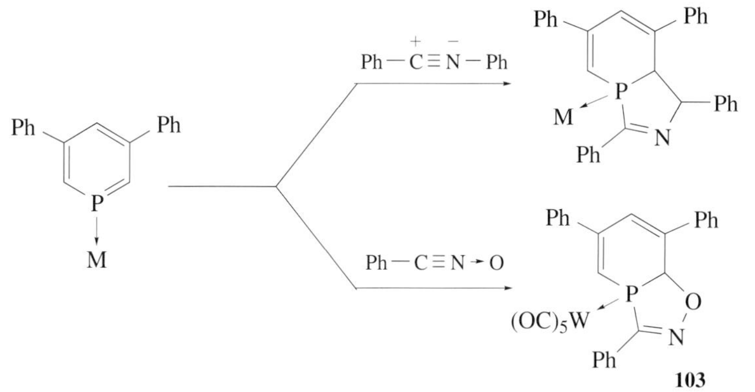

99 : M = $Cr(CO)_5$
100 : M = $Mo(CO)_5$
101 : M = $W(CO)_5$
102 : M = lone pair

Scheme 12

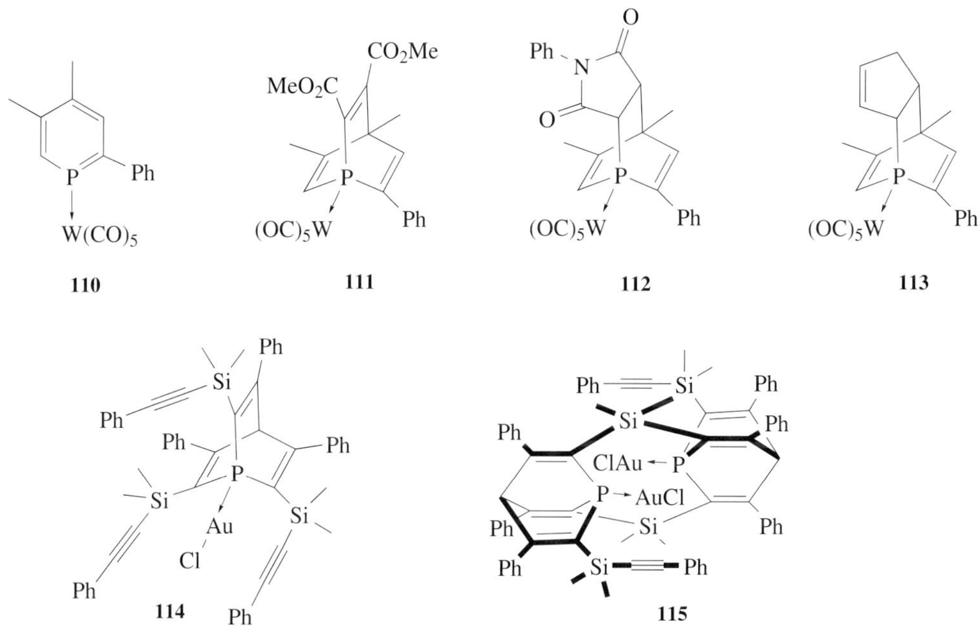

Figure 12. Trapping compounds illustrating the reactivity of phosphinines as 1-phospha-1,3-butadienes.

Nevertheless, in the case of electron deficient alkynes such as hexafluorobutyne, dicyanoacetylene and benzyne, [4 + 2] cycloadditions yield 1-phosphabarrelenes (**107–109**) (Equation (3)) ⟨68AG(E)733, 72JA7596⟩. Sulfides ⟨84CC508⟩ and complexes such as (**110**) display a similar behavior with dimethylacetylenedicarboxylate, N-phenylmaleimide and even cyclopentadiene under forcing conditions to give (**111**), (**112**) and (**113**), respectively ⟨84TL207, 87TL3475⟩. Phosphinine gold(I) chloride complexes undergo cycloadditions with alkynes at room temperature to give 1-phosphabarrelene derivatives such as (**114**) and (**115**) (Figure 12) ⟨99MIV2-01⟩.

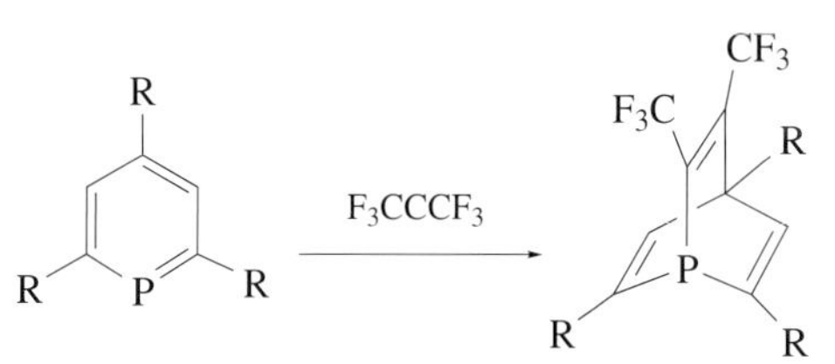

(3)

107 : R = Ph
108 : R = Bu^t
109 : R = H

5.2.5.3 Reactions at Ring Carbon Atoms

5.2.5.3.1 Nucleophilic aromatic substitutions

Nucleophilic aromatic substitutions with phosphinines are rare. The first reported example describes the substitution of a 3-chloro derivative (**16**) by amido groups to give (**116**) and (**117**) (Equation (4)). A different mechanism involving the transient formation of a phosphabenzyne was also proposed ⟨83TL5051⟩. A 2-bromophosphinine (**26**) has been reacted with lithium tetramethylpiperidine (LiTMP) to yield the corresponding 2,2′-biphosphinine (**3**) (Scheme 13) ⟨94BSF330⟩. A possible mechanism involves the preliminary attack of the amide at phosphorus and the substitution of the bromine atom of a second molecule by the 1-amino-phosphacyclohexadienyl anion (**118**) thus formed.

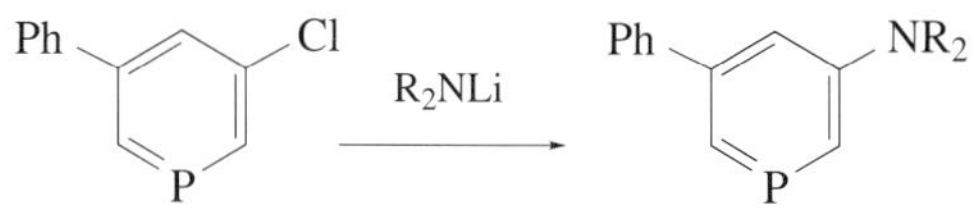

(4)

16

116 : R = Pr^i

117 : R-R = $(CH_2)_5$

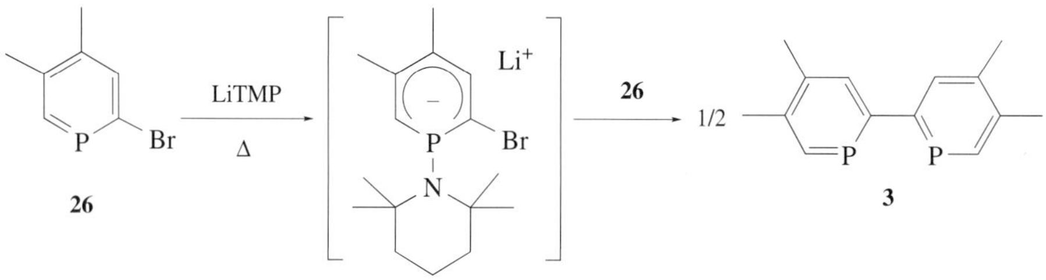

26 **118** **3**

Scheme 13

5.2.5.3.2 2-Metallophosphinines

Due to the high electrophilicity of the phosphorus atom, halogen–metal exchanges allowing to derivatize halogenobenzenes or halogeno heterocycles cannot be simply duplicated with halogenophosphinines. Accordingly, reactions with strongly electropositive metals such as Li, Na or K only lead to unidentified compounds and polymers. A common way to take advantage of the reactivity of these non-isolable organometallic intermediates is the use of trapping procedures (Barbier's method). Magnesium compounds transiently formed react with $R_nSiCl_{(4-n)}$ and R_3SnCl to yield the corresponding silyl and tin derivatives of phosphinines ⟨92BSF291⟩. An example describing the synthesis of the 2,6-disilylphosphinines (**120**) from 2,4,6-tribromophosphinine (**119**) is presented (Equation (5)). The first reported stable and isolable organometallic derivative was the zinc complex (**121**), obtained by reacting a 2-iodophosphinine (**27**) with activated zinc ⟨92TL3537⟩. This method allowed the synthesis of 2-functionalized phosphinines in modest yields including a silver salt obtained by metathesis (Scheme 14) ⟨96OM794, 96OM802⟩. Other examples will be presented in Section 5.2.8. The method was extended to P–W(CO)$_5$ complexes and successfully applied to the synthesis of a 2,2′-biphosphinine W(CO)$_4$ complex ⟨92BSB609⟩.

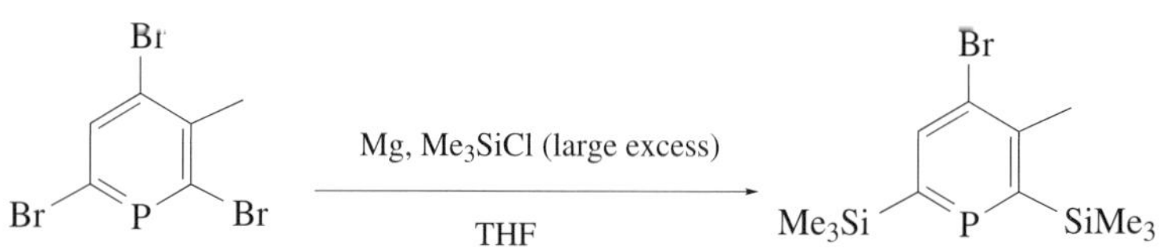

(5)

119 **120**

2-Lithiophosphinines such as (**122**) are only available when the lone pair is complexed to a metal fragment. In these species, the metallic fragment not only activates the C–X (X = Br or I) bond in

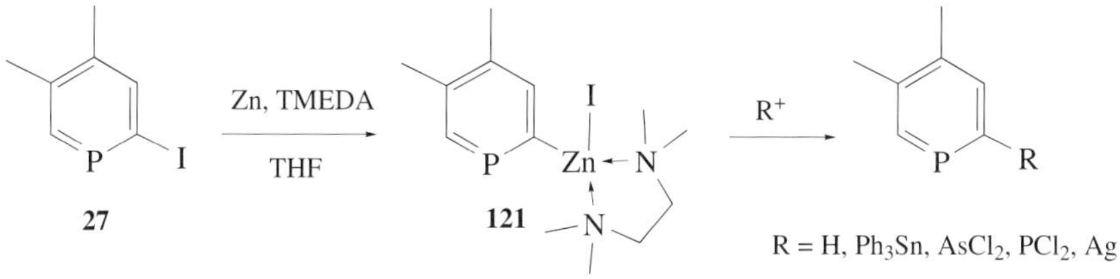

27 **121**

Scheme 14

decreasing the aromaticity within the ring and but also protects the P=C bond from nucleophilic attack. This method was employed to synthesize λ^3-phosphinines (**123–126**) since the complexing group is readily removed at the end of the reaction by heating with 1,2-bis(diphenylphosphino)-1,2-ethane (dppe) (Scheme 15) ⟨91OM2432, 92TL3537⟩. A 4,4′,5,5′-tetramethyl-2,2′-biphosphinine $Mo(CO)_4$ complex has been obtained via the $NiBr_2$.DME coupling of the corresponding 2-lithio derivative ⟨92OM2475⟩.

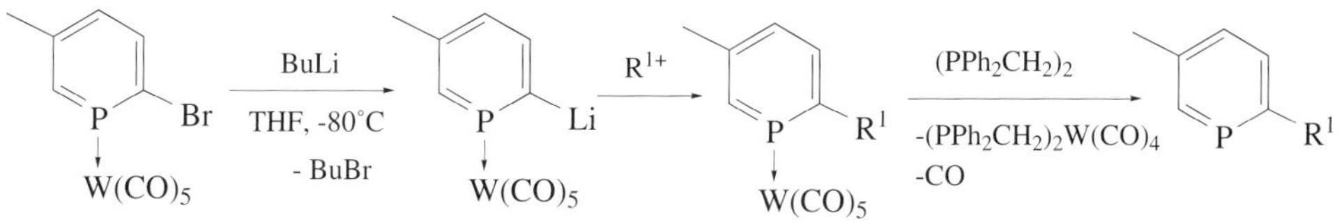

122

123 : R^1 = PPh_2
124 : R^1 = I
125 : R^1 = $SiMe_3$
126 : R^1 = CO_2Et

Scheme 15

A second important functionalization pathway relies on the reactivity of zirconocene. The "$ZrCp_2$" fragment efficiently inserts into the C–Br bonds of phosphinines (**26**), (**11**), (**127**) and (**128**) to give zirconium IV derivatives such as (**129**). One of these complexes has been structurally characterized ⟨93CC789⟩. The cleavage of the Zr–C2 bond with electrophiles furnishes an interesting entry to 2-functional compounds (see Section 5.2.8). These zirconium complexes were shown to be synthetically valuable as precursors of 2,2′-biphosphinines (**3**, **130–132**) through a coupling reaction with [Ni(dppe)Cl_2] (Scheme 16) ⟨98JOC4826⟩. Finally, the most striking illustration of their reactivity was given by the synthesis of η^2-phosphabenzyne–zirconocene complexes which are the first heterobenzyne complexes known to date. A thermally promoted α-elimination of methane or benzene from the corresponding Zr–Me (**133**) or Zr–Ph (**134**) compounds gives the benzyne complexes which can be isolated either as the PMe_3 adduct (**135**) or as the dimer (**136**) if the reaction is carried out without additional phosphine (Scheme 17) ⟨94CC2065, 97JA9417, 98JOM(567)151⟩. The reactivity of these complexes will be addressed in Section 5.2.8.

Cross-coupling reactions catalyzed by palladium and nickel(0) complexes which are commonly used in the chemistry of arenes and heterocycles can be successfully extended to the chemistry of phosphinines. Indeed, Pd or Ni(0) complexes insert into the C2–Br bond under classical conditions. Interestingly, in the case of disymmetrically substituted 3-methyl-2,6 dibromophosphinines (**19**), the insertion takes place at the more hindered position. One of these palladium derivatives (**137**) has been structurally characterized ⟨93JA10665⟩. Calculations have revealed that a through-space destabilizing interaction between the methyl group located at C3 and the lone pair at bromine facilitates the insertion at this position (Equation (6)).

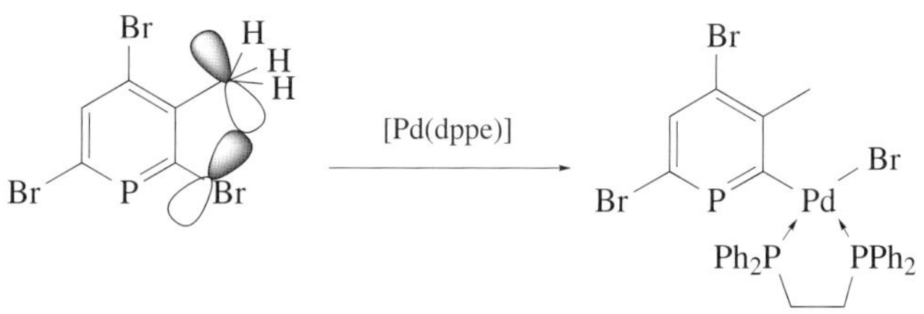

119 **137** (6)

Reaction of organotin compounds (Stille coupling) give access to a wide range of 2,6-difunctionalized phosphinines such as (**140**) which was obtained via a two step sequence from (**138**) via (**139**) taking into profit the regioselectivity of the first insertion (Scheme 18) ⟨95S717⟩. Nickel(0) complexes also

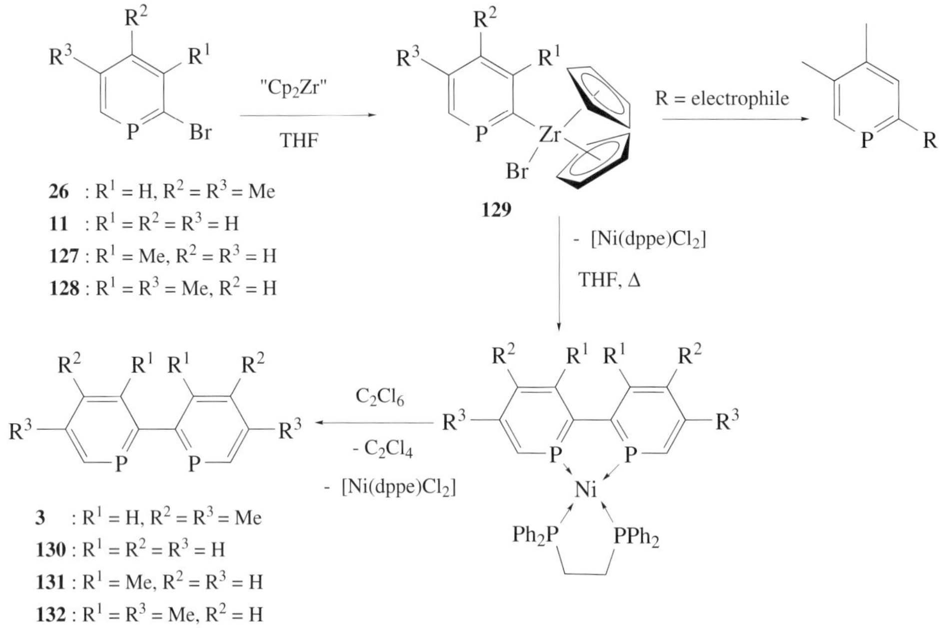

Scheme 16

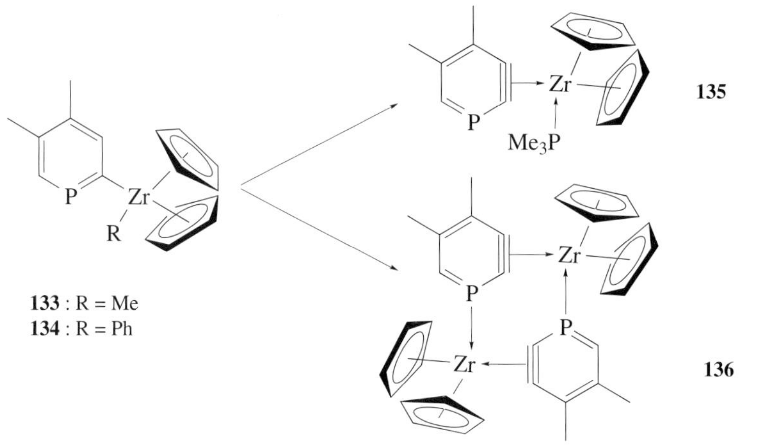

Scheme 17

catalyze the coupling between phospholide anions (**141**) and bromo (**26**) or polybromophosphinines to give 2-(phospholyl)-phosphinines such as (**142**) (Equation (7)) ⟨95BSF910⟩. Some developments of this chemistry will be presented and discussed in Section 5.2.8.

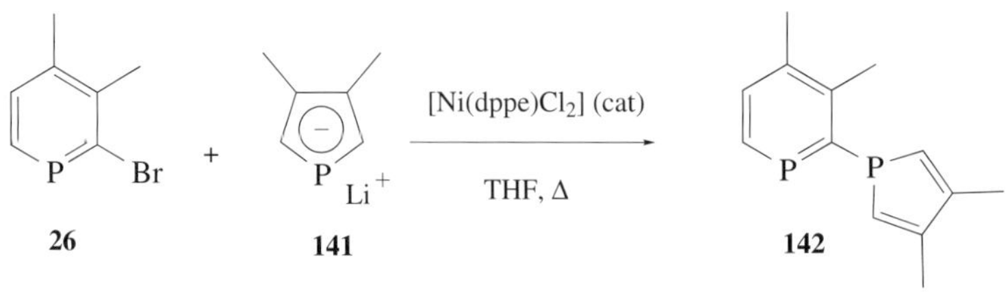

(7)

Finally, there is only one example describing the substitution of the bromine atom by a boron group. "Super-hydride" reacts with 2-bromophosphinines such as (**26**) to give the lithium salt (**143**) which in turn can be used as a precursor of 2-ethylphosphinines (**144**) through reaction with iodine (Scheme 19) ⟨96CC971⟩.

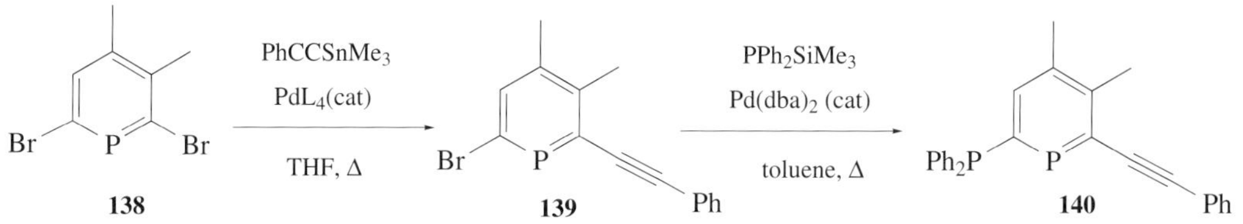

Scheme 18

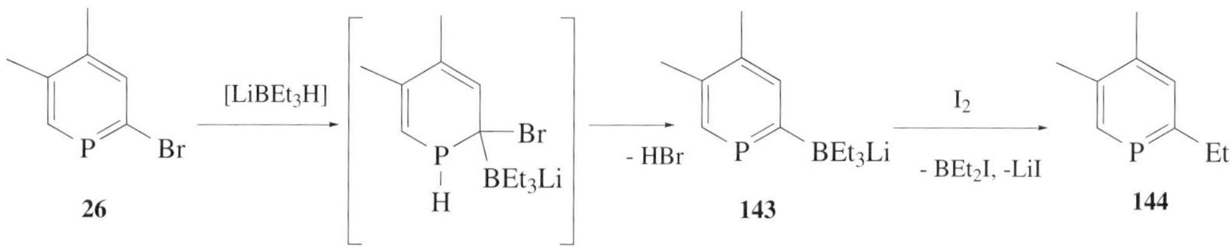

Scheme 19

5.2.6 REACTIVITY OF λ^5-PHOSPHININES

5.2.6.1 Reactions at the P Atom

5.2.6.1.1 Reduction and oxidation processes, reactions with radicals

λ^5-Phosphinines can be oxidized chemically or electrochemically to form cation radicals. The ESR spectrum of (**145**) has been recorded ⟨67AG(E)711, 69AG(E)769, 75JA5526⟩. These cation radicals readily react with water to give very stable neutral radicals such as (**146**) (Scheme 20) ⟨81CB3004, 81CB3019⟩. Similar experiments have also been carried out with λ^5-phosphinines bearing R_2N and F groups at P. ESR experiments have shown that the corresponding radicals are mainly localized on the carbocyclic part of the ring, the P atom playing no role in the delocalization of the unpaired electron. When oxidation is carried out with hydrogen peroxide in the presence of LiBr or with $KMnO_4$, 1,4-dihydrophosphinine oxides such as (**147**) and (**148**) are formed, respectively ⟨73TCC(38)1⟩. Only one report refers to the reduction of λ^5-phosphinines. Reaction of Na/K alloy with the 1,1-diphenyl derivative of the 2,4,6-triphenylphosphinine gives the corresponding 1-Ph-phosphininyl anion (**149**) ⟨69TL1231⟩.

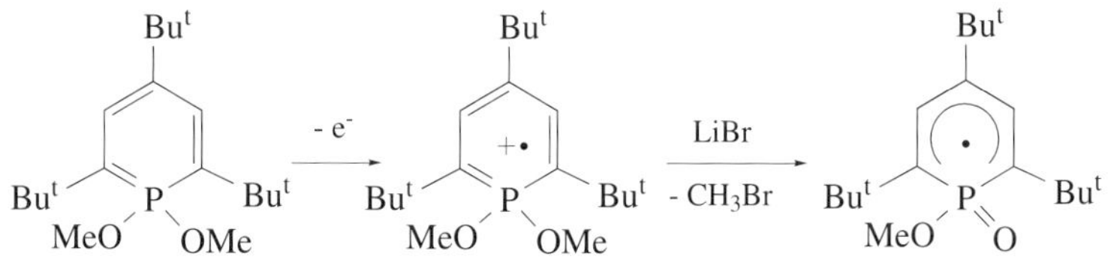

Scheme 20

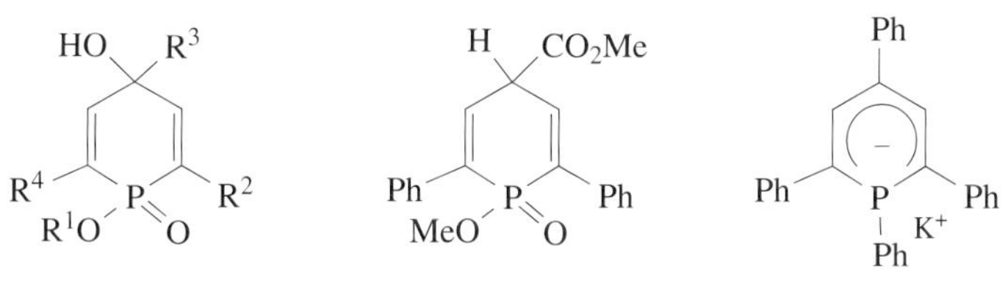

5.2.6.1.2 Reactions at the P atom and of substituents attached to P

The reactivity at P-substituents has been thoroughly explored. These studies mainly focused on substitutions of halogen substituents. Thus 1,1-dichloro or dibromo-λ^5-phosphinines serve as precursor for the synthesis of a wide range of P-functionalized compounds (Scheme 21). Most of these reactions have been carried out on 2,4,6-triphenyl or tris(tertiobutyl) derivatives. Nucleophilic or electrophilic derivatization pathways as well as halogen metathesis are possible ⟨72TL835, 77CB395, 81CB3004, 87CB1249⟩. These reactions are not restricted to dihalogeno compounds and 1,1-diamino and in some cases 1,1′-dialkoxy compounds also undergo exchange with nucleophiles in acidic medium. Finally, 1-methyl-1-phenyl derivative (**150**) of 2,4,6-triphenylphosphinine reacts with a base to form a carbanion (**151**) which subsequently reacts with electrophiles such as MeI or PhCHO to give compounds (**152**) or (**153**) (Scheme 22) ⟨81AG(E)871⟩.

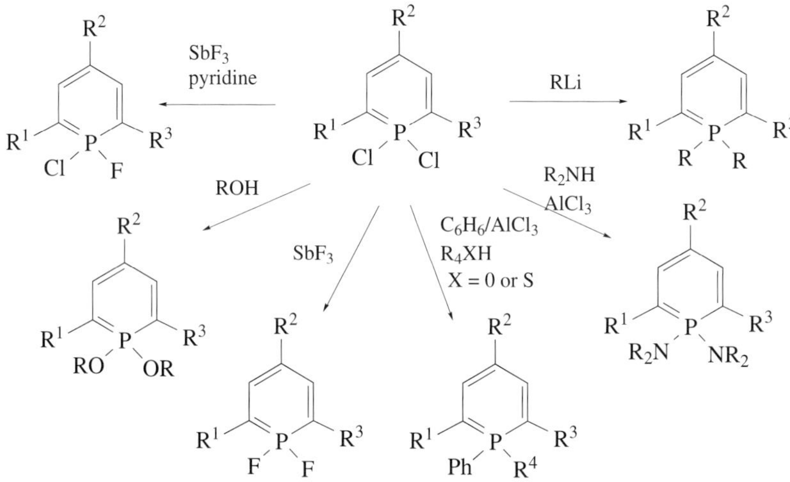

Scheme 21

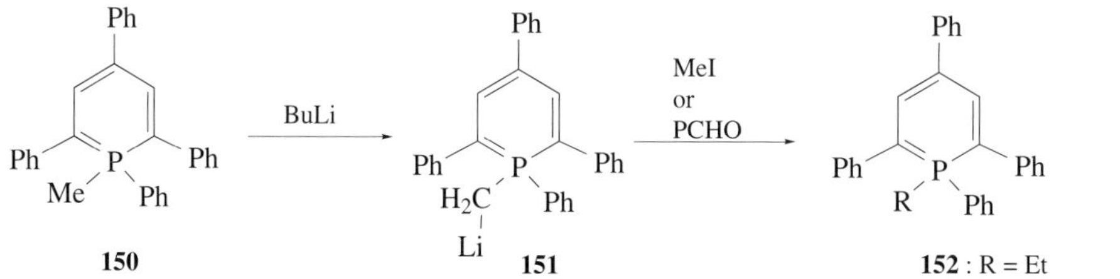

150 **151** **152** : R = Et
153: R = CHCHPh

Scheme 22

Thermal and photochemical rearrangements have been reported but from a synthetic point of view, thermolysis is more interesting as, in some cases, it yields λ^3-phosphinines (see Section 5.2.7.1.5). Two different types of rearrangements have been reported. The first one, similar to the Fries and Claisen transpositions, is based on the reactivity of 1-acetoxy substituted phosphinines. With compounds such as (**154**), a migration of the acetyl group to the C2 position has been observed leading to (**155**). This rearrangement can also be promoted by daylight. Prolonged irradiation of (**154**) leads to 2,4,6-triphenyl toluene via (**156**) (Scheme 23) ⟨72TL1045, 74CB3501, 76CB3099⟩. The second rearrangement involves the reactivity of allyloxy groups. With compound (**157**), a migration of the allyl moiety is observed upon heating, leading to (**159**) which equilibrates with (**158**) via a Cope rearrangement. Under more drastic conditions this reaction leads to tricyclic derivative (**160**) ⟨75AG(E)112, 76AG(E)238, 81CB1752⟩.

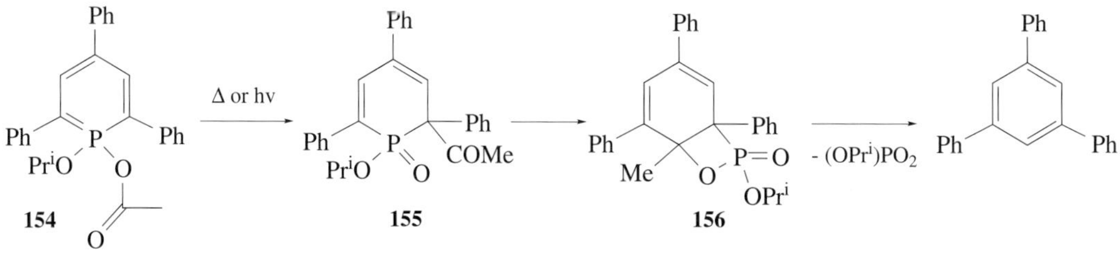

154 **155** **156**

Scheme 23

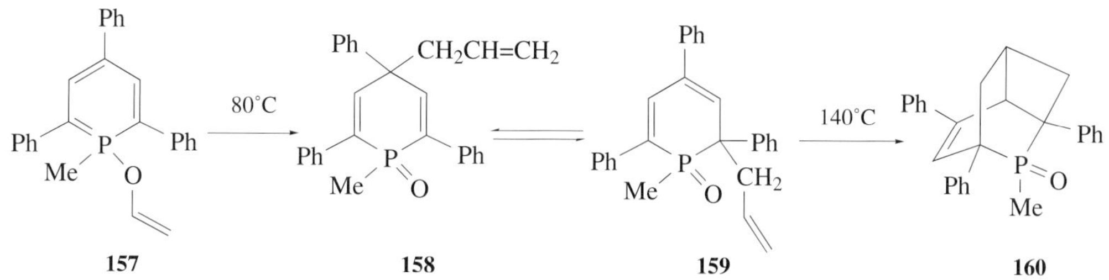

157 **158** **159** **160**

Scheme 24

5.2.6.2 Reactions of Ring Carbon Atoms and of their Substituents

5.2.6.2.1 Ring carbon atoms

According to the charge distribution observed in λ^5-phosphinines, reactions with electrophiles are expected to occur at C2 or C4 positions. The most significant results were obtained with compounds bearing no groups at the C4 position 〈72AG(E)1016, 83TL5051〉. Phosphanaphthalene (**161**) reacts with MeI and RCOCl to yield the C4 functionalized λ^5-phosphanaphthalene (**162**) or (**163**) (Equation (8)). This approach can be used to circumvent the lack of Friedel–Craft reactions with λ^3-phosphinines.

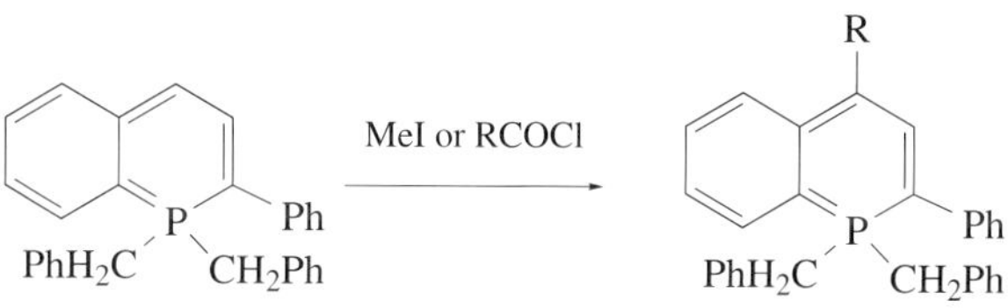

(8)

161

162 : R = Me
162: R = PhCO

5.2.6.2.2 Ring carbon substituents

Each transformation reported involves the chemistry of C4-alkyl substituted λ^5-1,1-dimethoxy-2,6-diphenyl or ditertiobutylphosphinines. Treatment with $Ph_3C^+BF_4^-$ affords in both cases the corresponding stable carbenium salts (Equation (9)) 〈72TL839, 75CB1384〉. These salts have been used to produce a wide range of C4-functionalized compounds. Nucleophilic addition of CN^- give the cyano derivative (**164**) and PPh_3 reacts to yield the Wittig salt (**165**) 〈79CB1273〉 subsequently transformed into (**166**) through reaction with benzaldehyde. Aldehyde (**168**) obtained from (**167**) by oxidation with $KMnO_4$ is also an important synthon for the preparation of cyano (**169**) and nitro (**170**) derivatives 〈75AG(E)11〉. The reaction of (**171**) with arene diazonium salts in aqueous acetonitrile solutions leads to the 4-arylazo (**172**) 〈73AG(E)753〉 and the acetamino (**173**) 〈75AG(E)112, 80CB3313〉 derivatives. In the absence of electrophiles (**171**) dimerizes to yield, after treatment with an equivalent of $Ph_3C^+BF_4^-$, phosphacyanamine (**174**) (Scheme 25) 〈72TL843〉. Finally, the reaction of carbenium salts towards diisopropyl-ethyl-amine was also investigated 〈77CB1497〉.

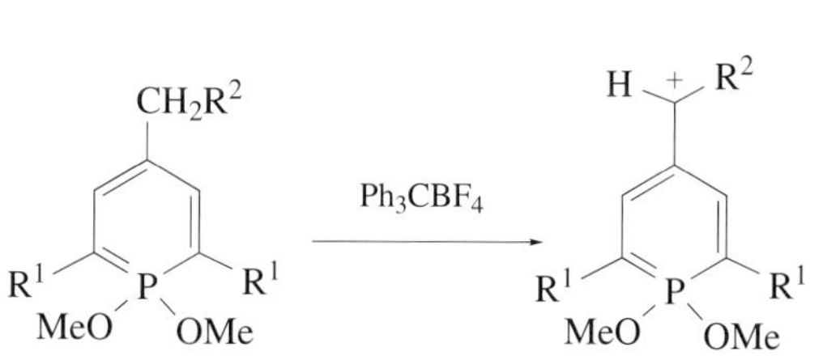

(9)

R^1 = Ph, Bu^t

R^2 = H, $CH(CH_3)_2$, CH_2CH_3

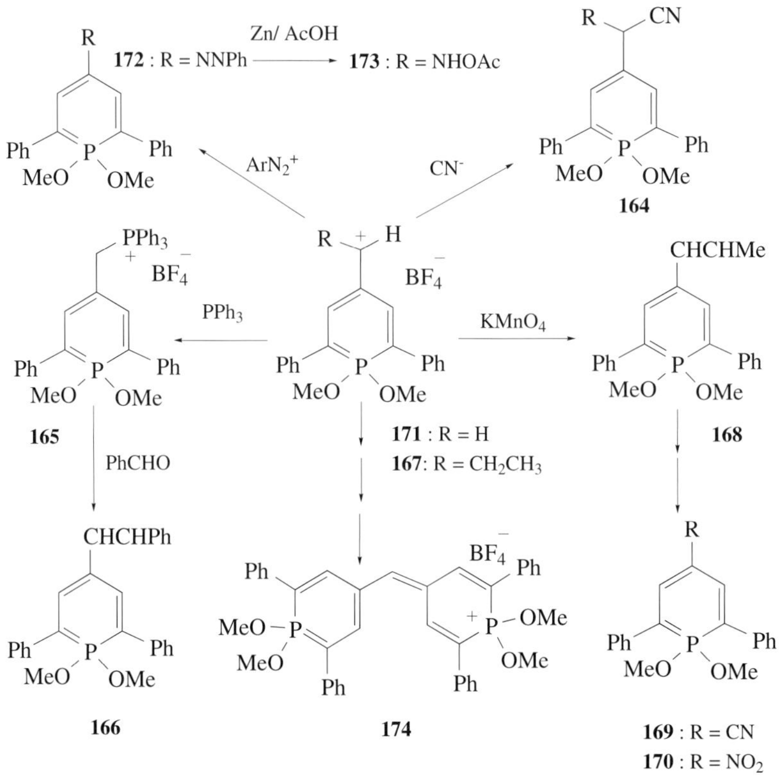

Scheme 25

5.2.7 RING SYNTHESES

5.2.7.1 λ^3-Phosphinines

5.2.7.1.1 O, P exchange

As stated in the introduction, the O^+/P exchange in pyrylium salts using a source of PH_3 was the first known synthetic approach to aryl-substituted phosphinines. Although many other routes have been devised since this discovery, its usefulness remains large because of its simplicity and of the availability of numerous pyrylium salts. Several sources of PH_3 have been used such as $P(CH_2OH)_3$ 〈66AG(E)846, 67AG(E)711〉, $P(SiMe_3)_3$ 〈67AG(E)458〉 and PH_4I 〈67AG(E)944〉. The first compound to be described was 2,4,6-triphenylphosphinine (**13**) which remained for a long time the most readily available derivative. In any case, the mechanism proceeds via an initial attack of the phosphine at C_2. After elimination of formaldehyde from the phosphonium salt (**175**) thus formed, an intramolecular nucleophilic addition of the phosphine to the carbonyl group of the open chain valence isomer (**176**) takes place leading to the ring closure. The last step is the aromatization which yields (**13**) (Scheme 26). Interestingly, when using PH_3 in the presence of strong acids, not only aryl but also 2- and 4-alkyl-, 2,4-dialkyl- and 2,4,6-trialkylpyridinium salts are converted to the corresponding phosphinines. An interesting illustration of this reactivity was provided by the synthesis of the bulky 2,4,6-tris(tertiobutyl)phosphinine (**14**) 〈68AG(E)460〉. Shortly after, this method was successfully extended to the synthesis of fused systems such as (**61**) 〈70TL645〉, (**177**) 〈73TCC(38)1〉 and (**178**) 〈71CB2984〉.

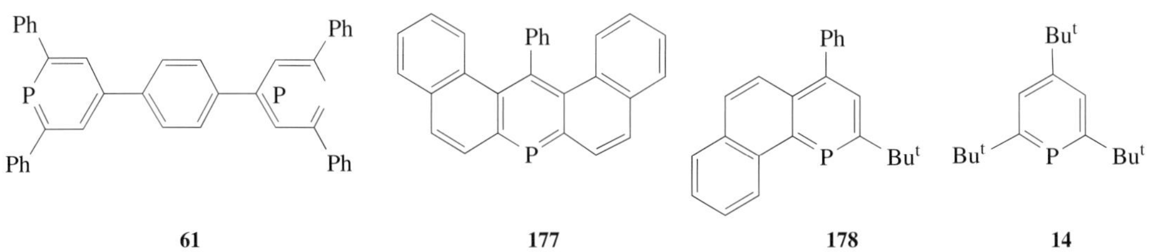

61 177 178 14

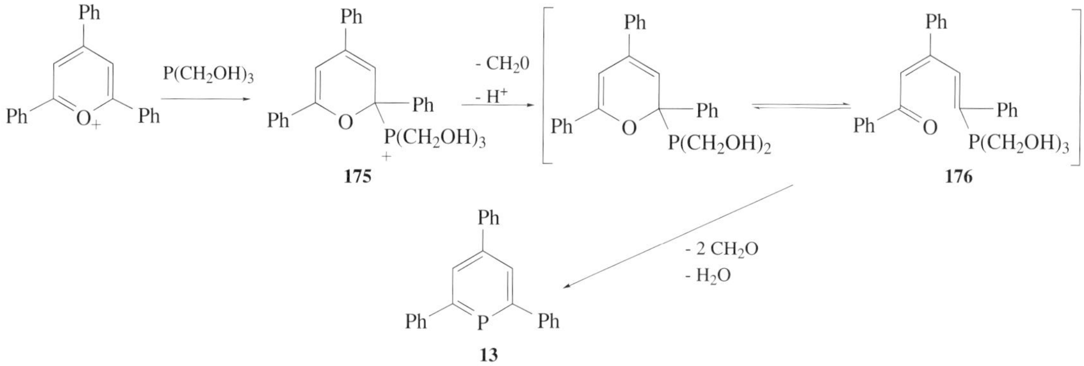

Scheme 26

5.2.7.1.2 *From cycloaddition processes with phosphaalkenes and phosphaalkynes*

Progresses in the chemistry of low coordinated phosphorus compounds such as phosphaalkenes and phosphaalkynes have made cycloaddition processes with conjugated 1,3-dienes and other unsaturated precursors an attractive approach for the synthesis of phosphinines. As a wide range of functional phosphaalkenes are now available, the substitution scheme of the ring can be modified at will. With phosphaalkenes, this approach generally involves two steps: the cycloaddition in which the ring is formed followed by an aromatization process. Some examples are listed below. P-chloro phosphaalkenes of the type $ClP{=}CR(SiMe_3)$ ($R = Ph, SiMe_3, CO_2Et$) have been extensively used in these cycloadditions ⟨89TL5245, 86TL5611, 86TL4299, 91HAC439⟩. An illustration of their utility is provided by the synthesis of a 3-hydroxyphosphinines (**179**) and (**180**) (Scheme 27). Note that they also react as dienophiles with pyrones to give phosphinines via a mechanism which involves the concomitant loss of CO_2 (Equation (10)) ⟨82AG(E)370⟩. The drastic conditions used strongly suggest the transient formation of the phosphaalkyne PhCP.

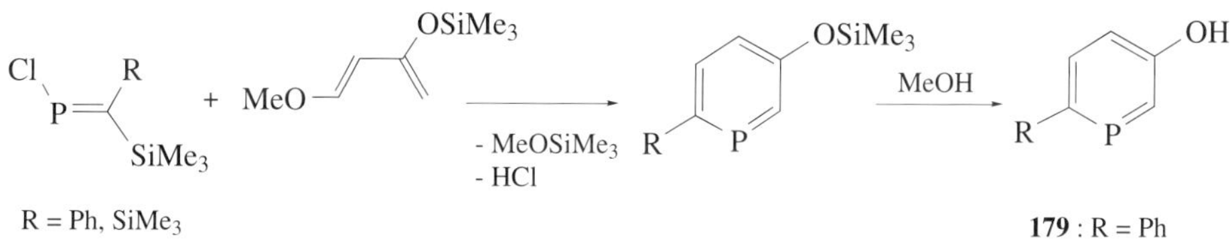

179 : R = Ph
180 : R = SiMe$_3$

Scheme 27

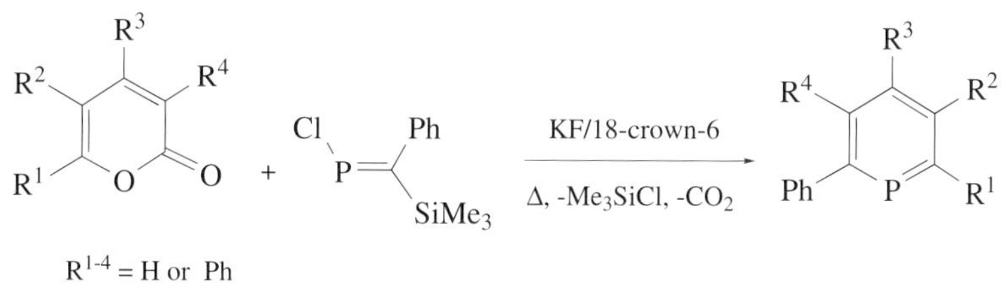

(10)

Trihalogenophosphaalkenes $XP{=}CX_2$, generated by the dehydrohalogenation of the tetrahalogenophosphanes X_2PCHX_2 have been employed to prepare 2-halogenophosphinines in large quantities ⟨89TL817, 90POL991, 91OM2432, 93PS(76)33, 92TL3537, 92BSB609, 93PAC621, 93PS(76)75, 95JOC7439⟩. Some of them are listed in Figure 13.

Tertiobutylphosphaethyne, due to its good kinetic stability, has been widely used in cycloaddition reactions. It displays a very rich chemistry towards various unsaturated cyclic compounds such as pyrones and 1,3-cyclohexadienes ⟨88AG(E)1541, 99PJC(73)135⟩, cyclopentadienones ⟨96ZN931, 87PS(30)479, 89TL5245⟩, activated cyclopentadienes ⟨90CB935⟩ and stannoles ⟨96JOM(520)211⟩. All these [4 + 2] cycloadditions give phosphinines in moderate to good yields. Some examples are listed in the following equations (Equations (11–15)). It must be also noted that cycloaddition of Bu^tCP with a kinetically stabilized cyclobutadiene affords a 2-P Dewar benzene derivative which isomerizes at 160°C to give the 1-P isomer ⟨87AG(E)85, 87CB819⟩.

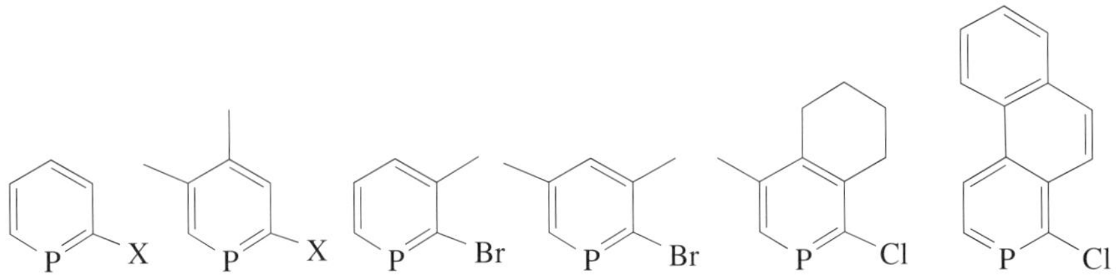

10 : X = Cl **25** : X = Cl **127** **128** **181** **182**
11 : X = Br **26** : X = Br
12 : X = I **27** : X = I

Figure 13. 2-Halogenophosphinines obtained from cycloaddition of XP=CX$_2$ with 1,3-dienes.

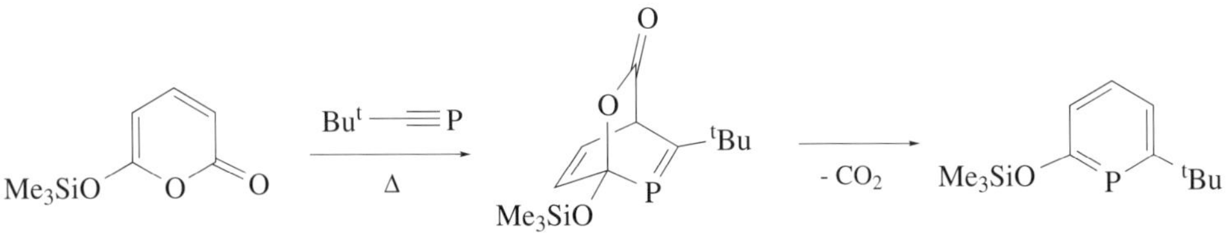

(11)

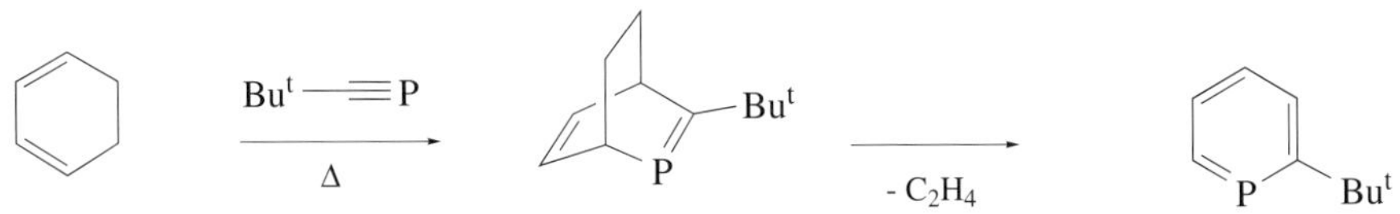

(12)

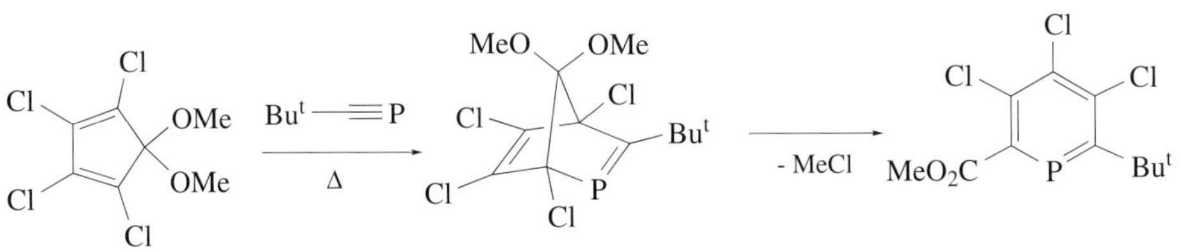

(13)

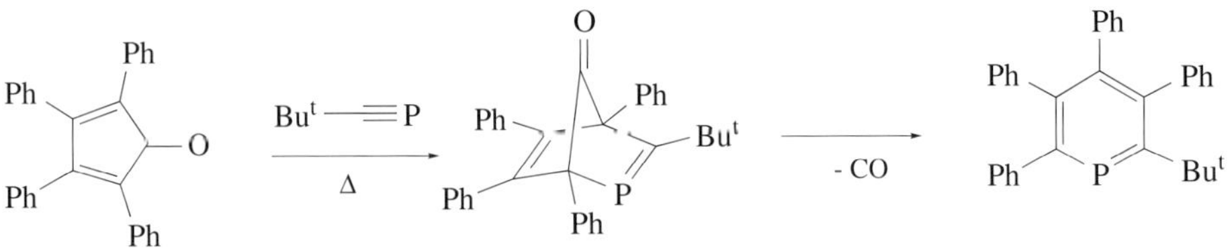

(14)

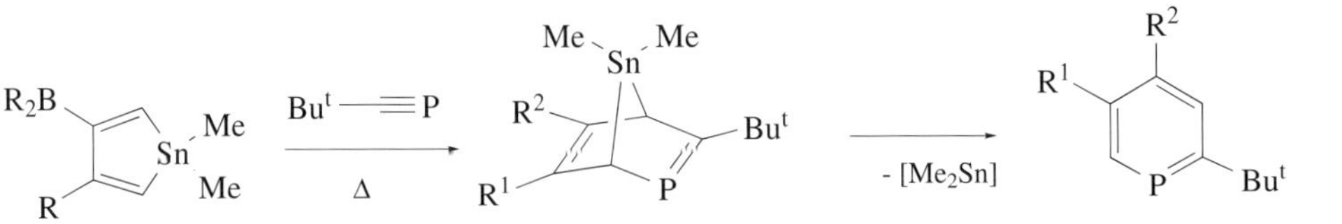

(15)

R^1 = substituent, R^2 = BR$_2$ R^1 = substituent, R^2 = BR$_2$
R^1 = BR$_2$, R^2 = substituent R^1 = BR$_2$, R^2 = substituent

A 3-hydroxyphosphanaphthalene (**183**) has also been obtained from the reaction of Bu^{t}CP with diphenylketene. Two mechanistic pathways were proposed: a [4 + 2] cycloaddition followed by a [1,5]-shift of hydrogen or a ene-reaction followed by a 6π-electrocyclization (Scheme 28) ⟨92TL1597⟩. Finally, the reaction of Fischer carbenes with Bu^{t}CP emphasizes the striking analogy between phosphorus and carbon. Like alkynes, this phosphaalkyne reacts with chromium carbene complexes providing an original access to benzannelated phosphinines such as (**184**) ⟨88AG(E)713⟩ or some phosphaphenols when vinylic carbenes complexes are used (Scheme 29) ⟨93T5577⟩.

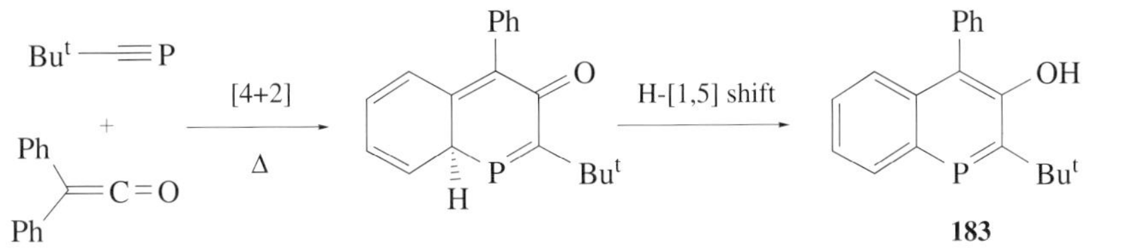

183

Scheme 28

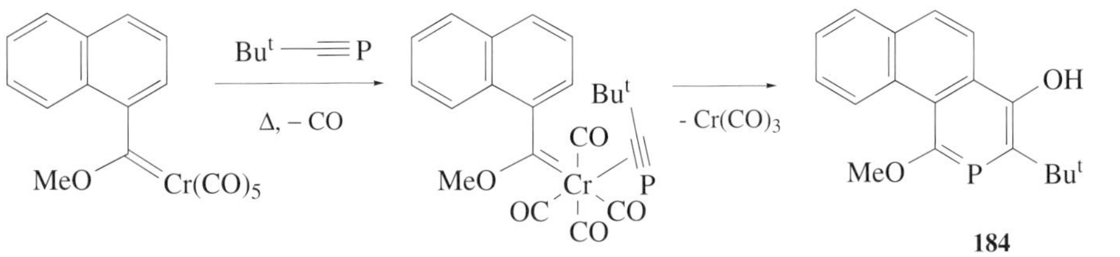

184

Scheme 29

5.2.7.1.3 From 4- and 5-membered rings

Ring expansions are powerful synthetic approaches since they usually give access to functionalized phosphinines. Concerning four-membered rings, only two strategies, both relying on the use of 1,2-dihydrophosphetes as precursors, have been devised. The first one takes advantage of the equilibrium between 1,2-dihydrophosphetes and 1,4-phosphabutadienes. 1-Alkynyl-1,2-dihydrophosphetes rearrange to phosphinines upon heating in benzene for several days (Scheme 30). The mechanism of this transformation which has not been totally rationalized would involve a 6π-electrocyclization followed by a [1,5] shift of hydrogen ⟨96HAC397⟩. The second approach also takes advantage of the 1,3-phosphadiene character of 1,2-dihydrophosphete complexes. At high temperature, the phosphadiene reacts with an activated alkyne to give dihydrophosphinine complexes such as (**185**) via a classical [4 + 2] cycloaddition. After removal of the complexing group, phosphinine (**186**) is obtained by thermal elimination of isobutane (Scheme 31) ⟨95BSF384⟩.

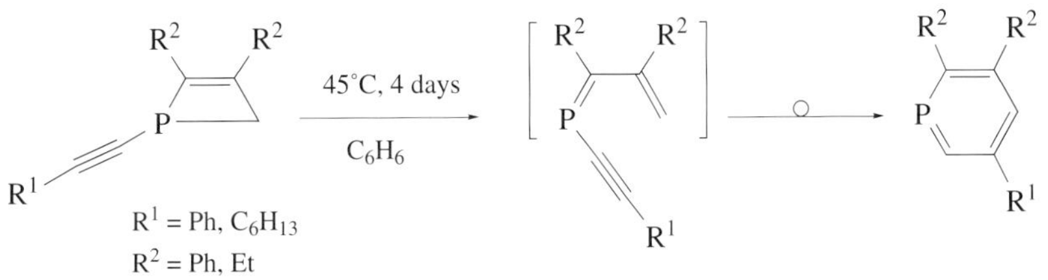

Scheme 30

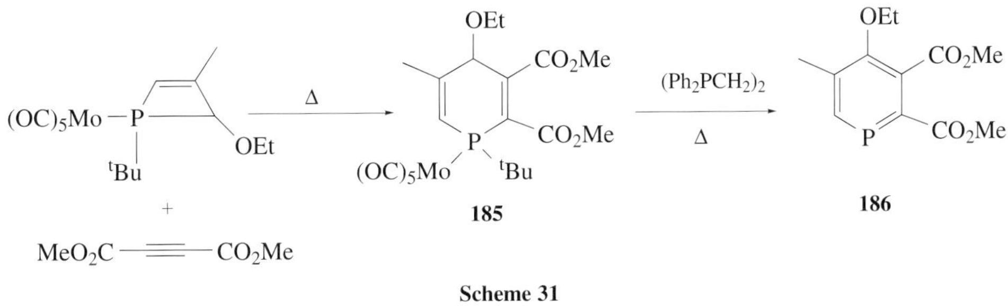

185

186

Scheme 31

Ring expansions from phospholes, which are much more documented, have constituted for a long time one of the most straightforward access to large quantities of phosphinines. These reactions can be classified in two groups. Reactions involving the phosphole skeleton as the diene component in [4 + 2] cycloadditions and reactions in which a source of "CR_2" inserts into one bond of the ring. The former relies on the equilibrium between phosphole and 2*H*-phosphole in the case of the 1,2,5-triphenyl derivative ⟨81JA4595, 82JOC2376⟩. This approach was extended to the synthesis of a bis(diphenylphosphino)-substituted phosphinine (**187**) (Scheme 32) ⟨88JHC155⟩. Also, 1,2-thiaphospholes have been used as a source of masked 1,4-phosphadienes (Equation (16)) ⟨87BCJ1558, 94BCJ2785⟩. Finally, reaction of Bu^tCP with a phosphole-sulfide, used as an activated diene, led to phosphinine (**189**). In this process, the loss of a phosphinidene sulfide promotes the aromatization (Scheme 33) ⟨86ZN(B)931⟩.

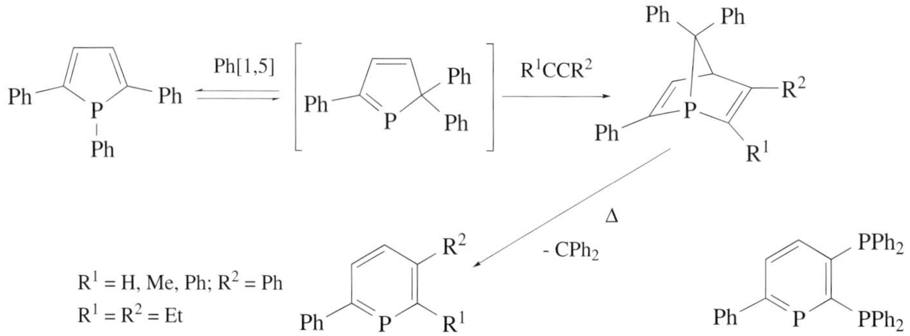

187

Scheme 32

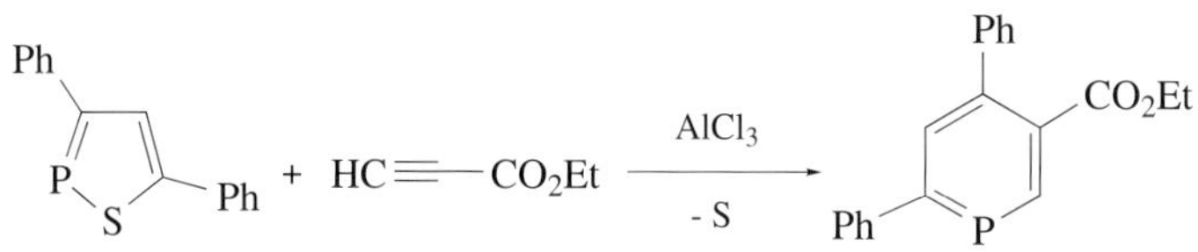

(16)

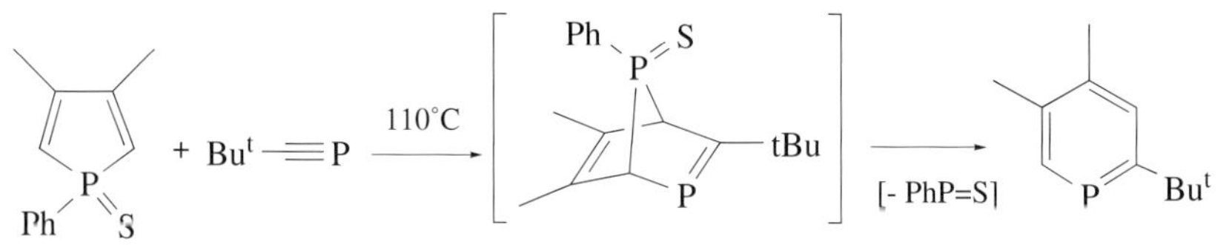

189

Scheme 33

Ring expansion of phospholes to phosphinines by formal insertion of "CR_2" in one bond of the ring can be easily achieved by reaction with acid chlorides in basic medium. This transformation involves the preliminary formation of a 1,2-dihydrophosphinine oxide (**190**) which is subsequently transformed into the corresponding sulfide (**191**) upon treatment with P_4S_{10} or the Laweson reagent $[ArPS_2]_2$. Phosphinines are then obtained via thermolysis in the presence of nickel powder in toluene. This method, limited to the use of aromatic acid chlorides, has thus been used to synthesize some 2-functionalized phosphinines such as the 2-phenyl compound (**24**) (Scheme 34) ⟨79TL1753, 84PS(19)45, 82TL1565, 84IC3463⟩. Accordingly, 2-phosphanaphthalene (**37**) (Equation (17)) and phosphaphenanthrene (**192**) (Equation (18)) were obtained from the corresponding benzo and dibenzophospholes, respectively ⟨80TL1441, 81NJC187⟩.

The reaction of ethyldiazocetate with a phosphole sulfide yields a bicyclic derivative (**193**) resulting from the addition of the carbene at the C2–C3 bond. It subsequently rearranges under thermal conditions in the presence $P(OPh)_3$ to give an ester derivative (**194**) (Scheme 35) ⟨91JOC4031⟩. Finally, the reaction of a dichlorocarbene with a phospholene oxide turns out to be an original way to produce 4-chlorophosphinines such as (**197**). This three-step process involves the successive formation of 1,2-dihydrophosphinines oxides (**195**) and (**196**) (Scheme 36) ⟨93JOC977⟩.

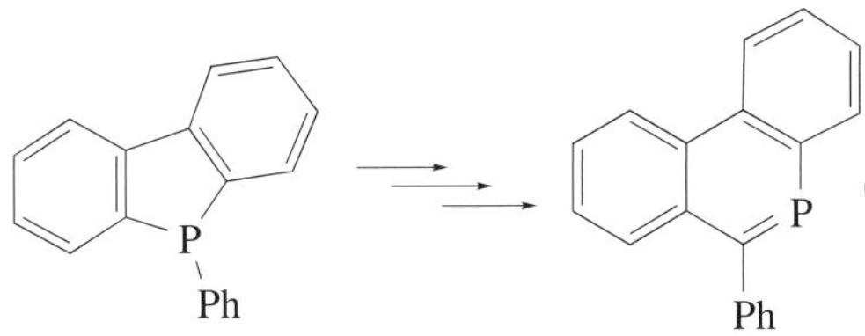

37 (17)

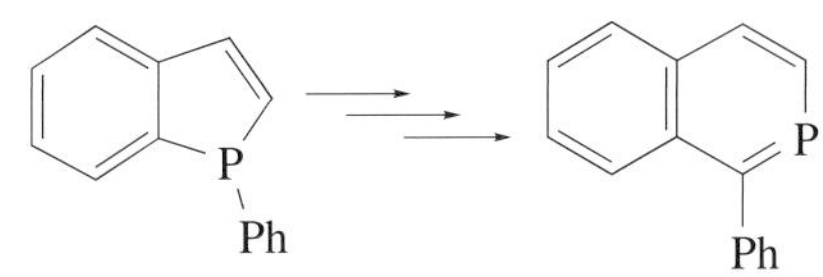

192 (18)

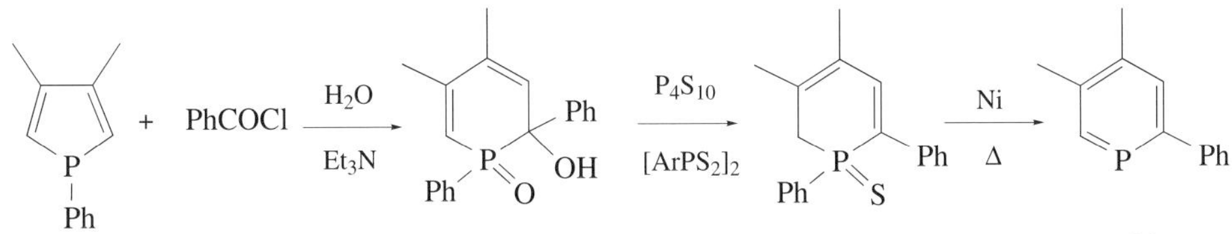

190 191 24

Scheme 34

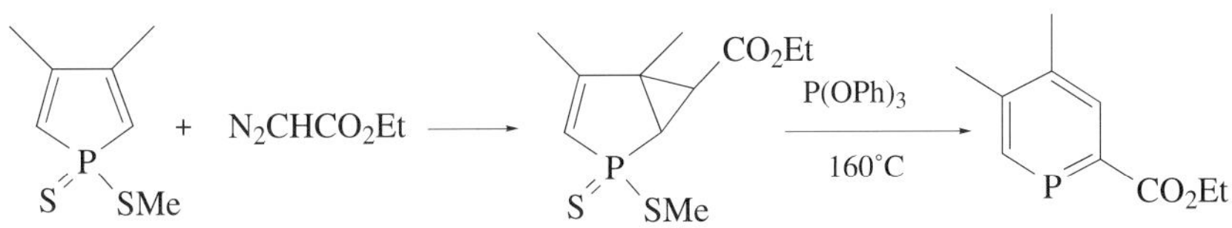

193 194

Scheme 35

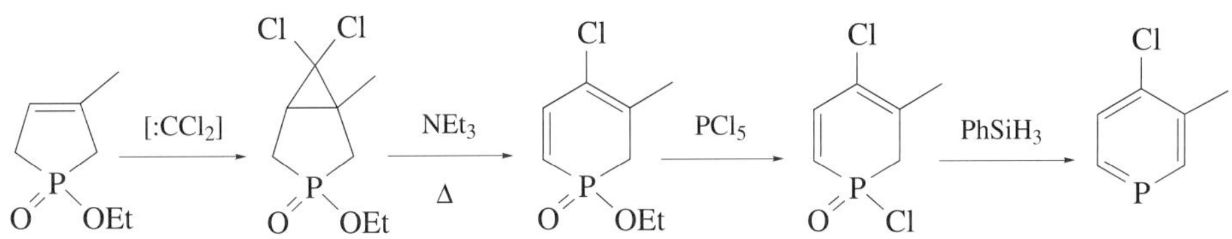

195 196 197

Scheme 36

5.2.7.1.4 From dihydrophosphinines

There are numerous reactions describing the synthesis of phosphinines from dihydrophosphinines. This aromatization step is, in general, encountered as the final step of a multi-step sequence. It can be either promoted by amines when the substituent at phosphorus is a halogen atom or thermally when the substituent is hydrogen or a carbon atom. Elimination of HX (X = Cl, Br) occurs with 1,2- and 1,4-dihydrophosphinines. One of the most representative example is provided by the synthesis of the parent derivative C_5H_5P (**1**) which relies on a Sn/P exchange from stannacyclohexadienes ⟨71JA3293⟩. After reaction of this tin derivative with phosphorus tribromide, bromophosphacyclohexadiene (**198**) is aromatized to (**1**) with a base. One can note that this methodology has also been successfully extended to the synthesis of arsenic, boron ⟨71JA1804⟩, and antimony ⟨71JA6690⟩ compounds. Some 2,6-disubstituted phosphinines ⟨75TL1083⟩ and fused ring systems such as (**199**) (Equation (19)) ⟨67AG(E)567, 68AG(E)889, 75T2931, 75RTC7⟩, (**200**) ⟨68AG(E)465⟩ and (**201**) ⟨73TL2397, 75T1097⟩ were also synthesized from the corresponding P–Cl precursors. Other examples of such eliminations have already been presented in Section 5.2.7.1.2.

There are only four examples of aromatization involving a P–H bond. Two examples, leading to (**1**) and its 4-substituted derivatives, rely on a modified Sn/P exchange procedure which involves, in the final step, the elimination of MeOH (Scheme 38) ⟨73AG(E)931, 79JOMC(173)125⟩. The third example is a thermal process in which the formation of the 3,4-dihydrophosphinine occurs via a 6π-electrocyclization. However, it was not proved whether the elimination of H_2 proceeds from the 2-phosphacyclohexadiene (**202**) or from its PH isomer ⟨93CC1295, B-96MIV2-01⟩. Lastly, a dimethylphosphinine (**23**) was obtained from the

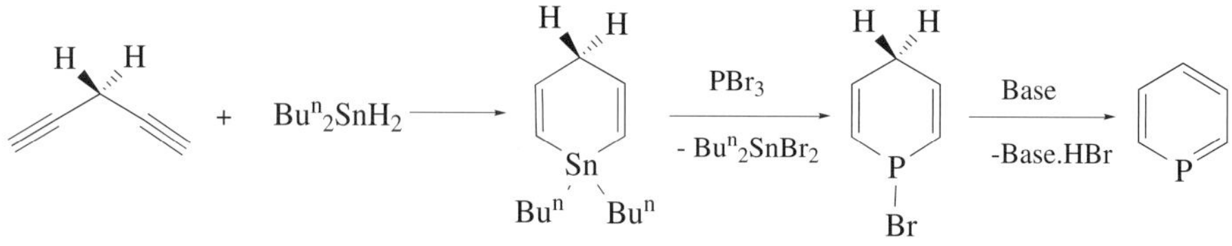

198 1

Scheme 37

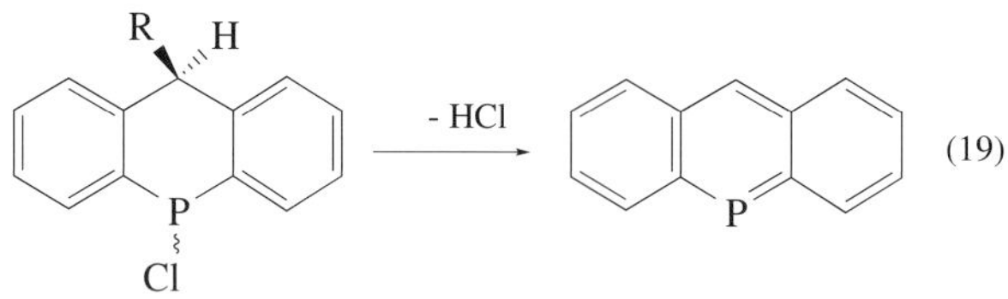

199

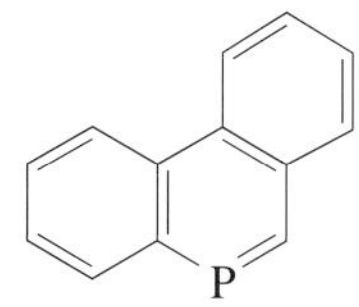

200

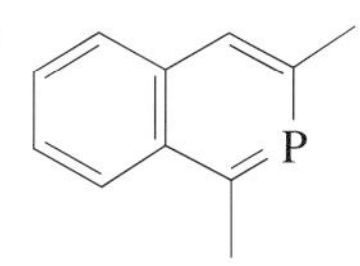

201

dehydrohalogenation of the 1,2,5,6-tetrahydro derivative (**203**). This sequence involves the formation of the transient 2,5-dihydrophosphinine (**204**) which isomerizes to give the 1,2-dihydro intermediate (**205**) (Scheme 39). This approach has been extended to the synthesis of bicyclic phosphinine (**206**) 〈96PS(109)461〉.

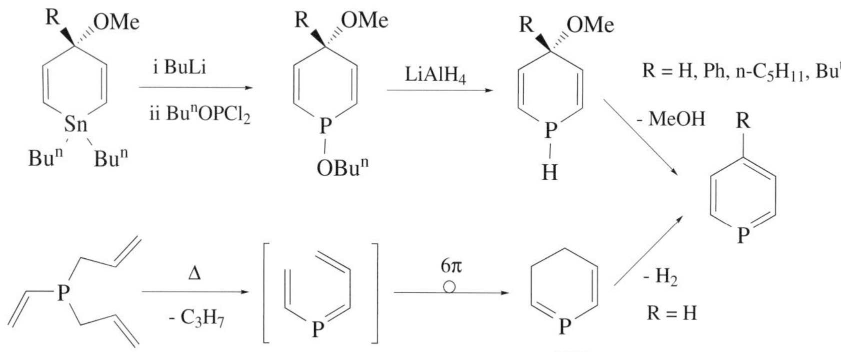

202

Scheme 38

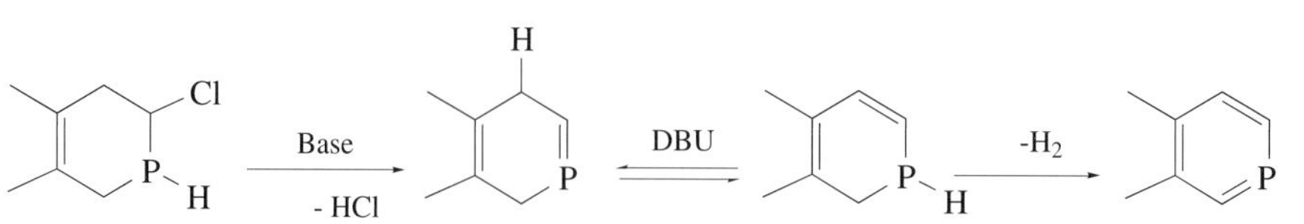

203 204 205 23

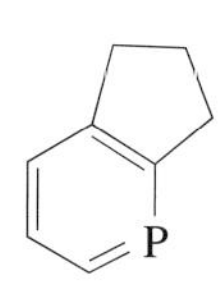

206

Scheme 39

Thermal elimination from dihydrophosphinines bearing carbon substituents at P is one of the most important approach to functional phosphinines. These processes, which often require high temperatures (200–300°C) are favored by the presence of good radical leaving groups at phosphorus such as Bu^t

or CH_2Ph ⟨70TL645, 71TL1215, 72TL4925⟩. The first 1-phosphanaphthalene (**38**) was synthesized from the bicyclic 1-benzyl-1,2-dihydrophophanaphthalene (**207**) by elimination of toluene ⟨72AG(E)1017⟩. A synthesis of 4-functionalized phosphanaphthalenes (**209**) and (**210**), requiring the aromatization of 1,4-dihydrophosphanaphthalenes (**208**), was also reported. In this approach, compound (**38**) is used as precursor for (**208**) *via* a sequence which involves a classical attack at phosphorus by a Grignard derivative followed by a C-alkylation (see Section 5.2.5.1.2) (Scheme 40) ⟨72AG(E)1017, 74TL4501⟩. Finally, another example of aromatization from the 1,4-dihydrophosphinine skeleton is provided by the preparation of a 4,4′-biphosphinine (**60**) (Equation (20)) ⟨70TL645⟩.

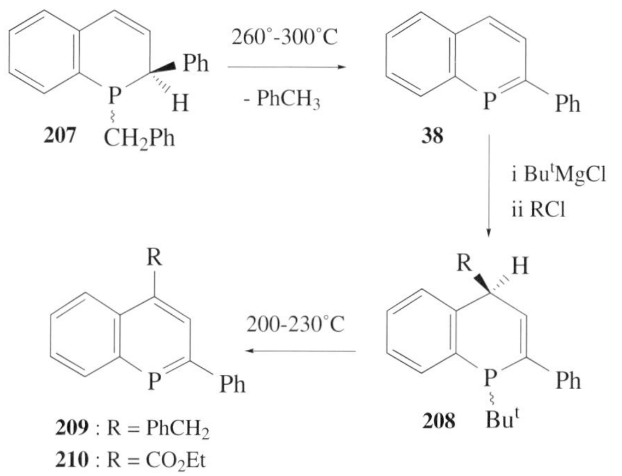

207, **38**, **208**

209 : R = $PhCH_2$
210 : R = CO_2Et

Scheme 40

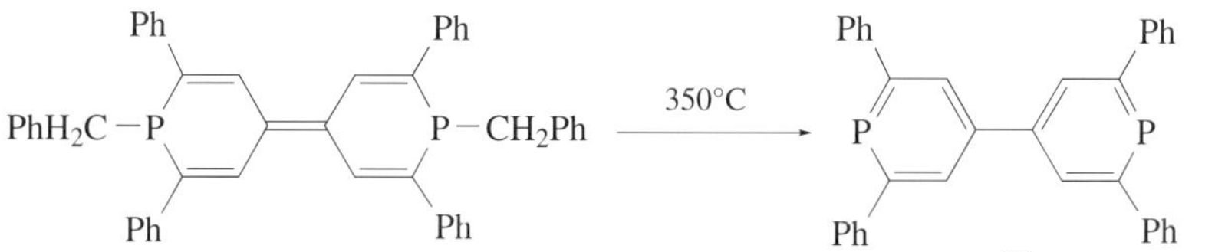

60 (20)

A great number of transformations have been achieved using phosphacyclohexenones such as (**211**) as starting precursors. These heterocycles are easily converted to 1,2-dihydrophosphinines using Grignard or lithium reagents. Conveniently, the aromatization step can be carried out without further purification of the 1,2-dihydrophosphinine. This approach is useful for the synthesis of 3,5-disubstituted derivatives ⟨84CB763⟩. A modified approach involving the preliminary preparation of (**212**) from (**211**) allowed the synthesis of the first 3-phosphaphenol (**213**) (Scheme 41) ⟨77TL3445⟩.

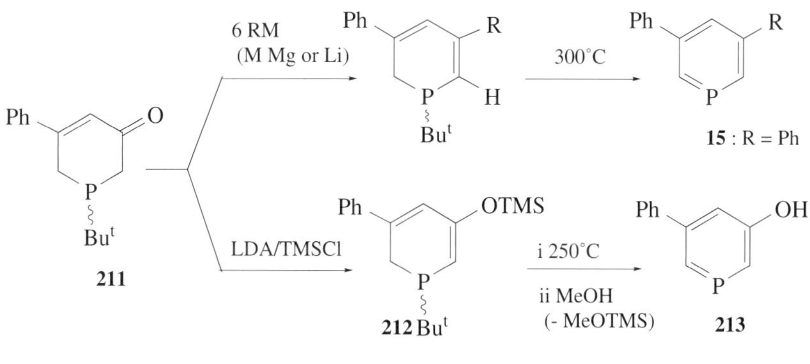

211, **212**, **213**

15 : R = Ph

Scheme 41

Phosphinines were also synthesized from 1,2-dihydrophosphinines sulfides. There, the reduction is carried either with Ni powder or a tertiary phosphine (see Section 5.2.7.1.4). In the same vein, bicyclic sulfides were converted into phosphinines via a retro Diels–Alder reaction. Most interestingly, this method allowed for the first synthesis of a 2,2′-biphosphinine (**3**) (Scheme 42) ⟨91JA667⟩. In a earlier development, 2-phenyl-3,4-dimethylphosphinine (**24**) was converted into the 3,4-dimethyl derivative (**23**) via the tricyclic sulfide (**214**) (Scheme 43) ⟨84TL4659⟩. In this process (**24**) can be regarded as a synthetic equivalent of HCP.

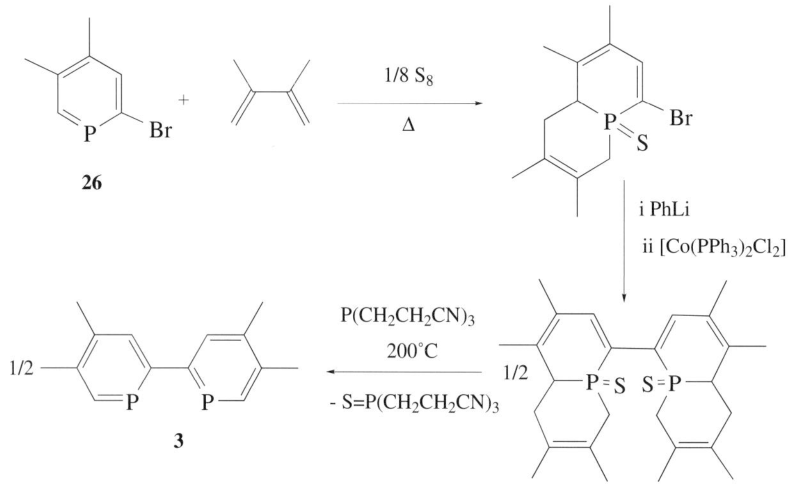

Scheme 42

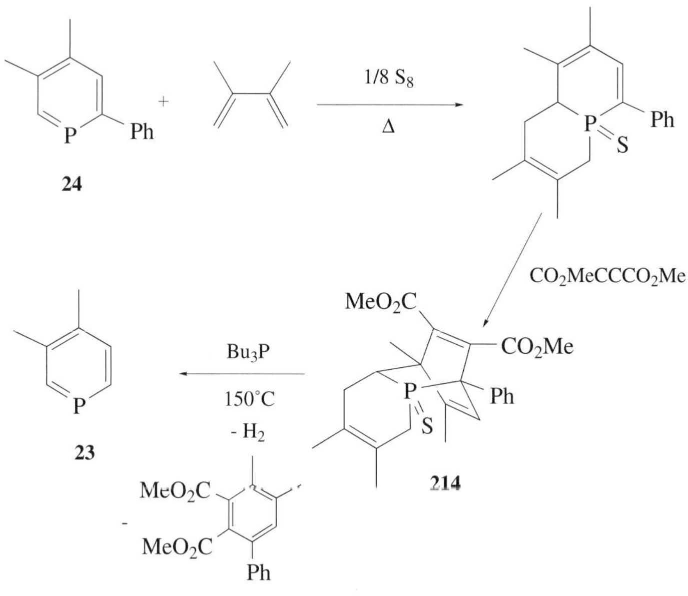

Scheme 43

5.2.7.1.5 *From λ^5-phosphinines*

Although of interest from a mechanistic point of view, transformation of λ^5-phosphinines into λ^3-compounds bears less synthetic value than the other known procedures since λ^5-phosphinines are generally produced directly from λ^3-compounds ⟨71TL1215⟩. Nevertheless, in some cases, this approach gives access to functional derivatives which are not available by other synthetic routes. These transformations require the presence of good leaving groups at phosphorus such as $PhCH_2$, Bu^t or allyl. Eliminations from 1,1-diarylamino- and 1,1-di(alkylthio)-λ^5-phosphinines were also reported (Equation (21)) ⟨68AG(E)981, 69AG(E)776, 72TL829, 72TL835, 77CB395, 83TL5051⟩. One of the most interesting application is provided by C4-functionalized derivatives of the 1,1-methoxy-λ^5-2,6-diphenylphosphinine which are readily converted to phosphinines (**215–217**). The aromatization is achieved by treatment with LiI and $HSiCl_3$ or with HCl and Ph_2SiH_2 (Equation (22)) ⟨87CB1245⟩.

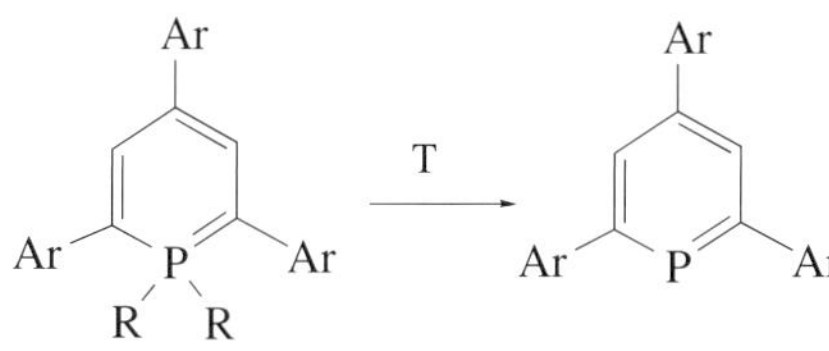

(21)

R = $PhCH_2$, T = 250°C
R = Ar_2N, T = 150-180°C
R = SR, T = 200°C; R-R = $-(CH_2)_3$, T = 80°C

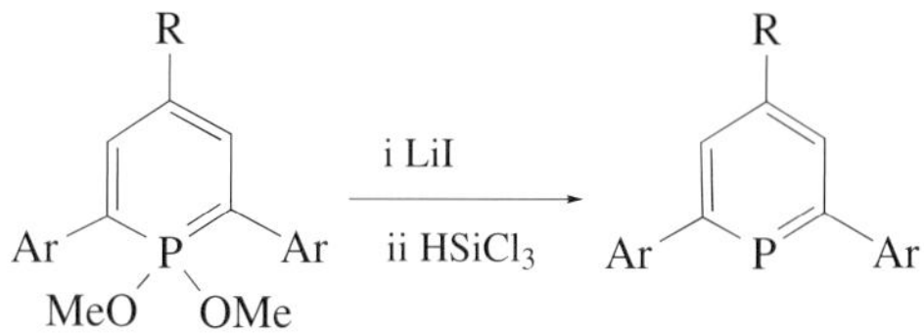

(22)

215 : R = CO_2Me
216: R = CH_2CN
217: R = CN

Reactions involving phosphacyclohexenones as starting materials are of particular interest since phosphinines can be obtained *via* a two steps sequence. Reaction of $HSiCl_3$ with (**211**) yields a 1-Bu^t-1-chloro-λ^5-1,2-dihydrophosphinine (**54**) which is subsequently thermolysed into (**49**) in rather good yields ⟨74TL4381, 83CB445, 83TL1756⟩. The first halogenophosphinines were synthesized following a modified procedure. Reaction of (**211**) with one equivalent of PCl_5 or PBr_3 gives the λ^5-phosphinine (**218**) which is aromatized to (**16**) or (**17**) (Scheme 44). Tri- and tetrachlorophosphinines, obtained as a mixture, have been prepared when an excess of PCl_5 is used. The reduction was achieved with triphenylphosphine (Scheme 45) ⟨83TL2645⟩.

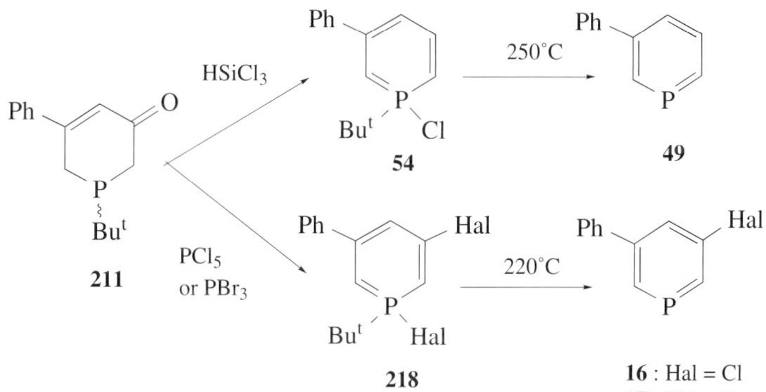

16 : Hal = Cl
17 : Hal = Br

Scheme 44

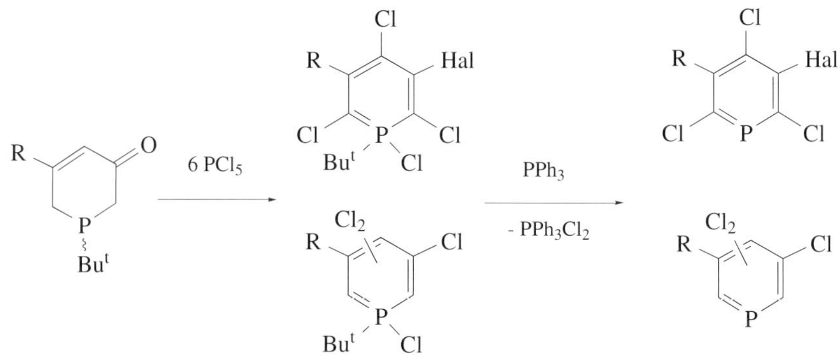

Scheme 45

More recently, 2,6 and 2,4,6-tribromophosphinines have been synthesized in good yields via a bromination of the phosphinine ring. Indeed, 2-bromophosphinine (**11**) is easily converted into (**220**) with three equivalents of bromine. Though the nature of intermediates were not determined with accuracy, 1,1-dibromo-λ^5-tribromophosphinines, such as (**219**), are very likely formed. In this reaction, aromatization is achieved with triethylamine (Scheme 46) ⟨92BSF291, 93JA10665⟩. Finally 2-dibromophosphinophosphinine (**221**) has been prepared by elimination of bromine from the corresponding 1,1-dibromo-λ^5-compound (**47**), which is obtained by reaction of PBr_3 with 2-bromophosphinine (**26**) (Scheme 47) ⟨96OM1597⟩.

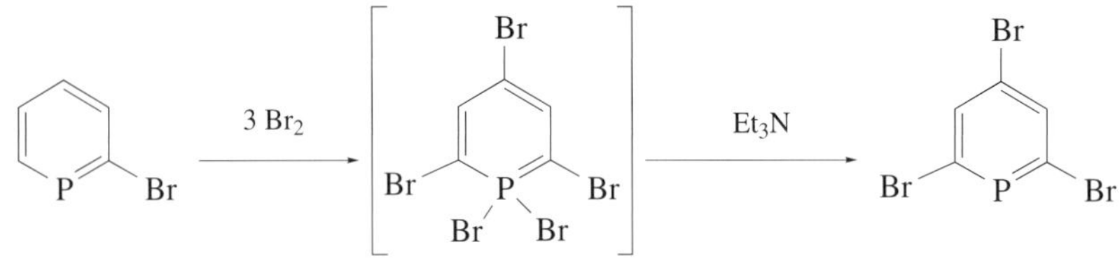

11 **219** **220**

Scheme 46

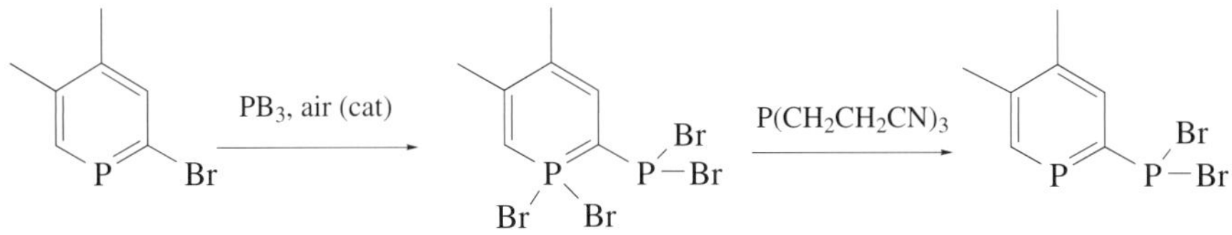

26 **47** **221**

Scheme 47

5.2.7.1.6 From aza- and diazaphosphinines

Compounds such as 1,3-aza- (**222**) ⟨87TL1093, 90TL4589, 95TL3839⟩ and 3,5-diazaphosphinines (**226**) ⟨91AG(E)106⟩ were first synthesized by Märkl. Due to the replacement of one or two CH units in the ring by more electronegative N atoms, these azaphosphinines behave as formal 1,4-[P–CR] dipoles. Their reactivity towards alkynes provides an easy entry to polyfunctional phosphinines. These transformations rely on a [4 + 2] cycloaddition/cycloreversion sequence involving the transient formation of an azaphosphabarrelene (**223**) which eliminates one molecule of nitrile to yield the phosphinine ring. Several disymmetrical phosphinines such as (**224**) and (**225**) have been prepared (Scheme 48). On the other hand, 3,5-diazaphosphinines have been used to prepare symmetrically tetrasubstituted phosphinines such as (**227**) (Scheme 49) ⟨91AG(E)106⟩.

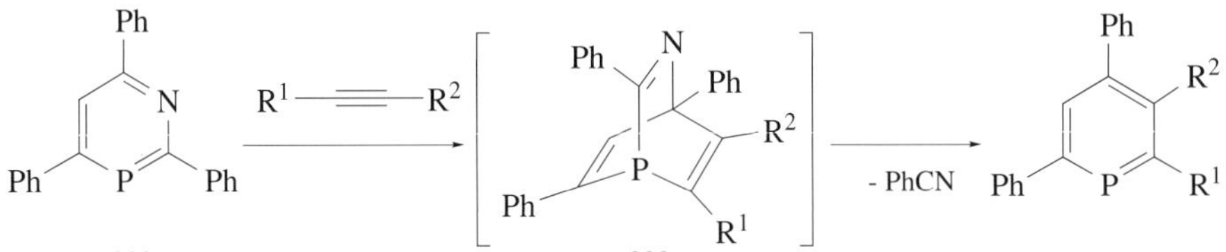

222 **223**

22 : $R^1 = PPh_2$; $R^2 = Me$
224 : $R^1 = P(O)Ph_2$; $R^2 = Me$
225 : $R^1 = P(O)Ph_2$; $R^2 = Ph$

Scheme 48

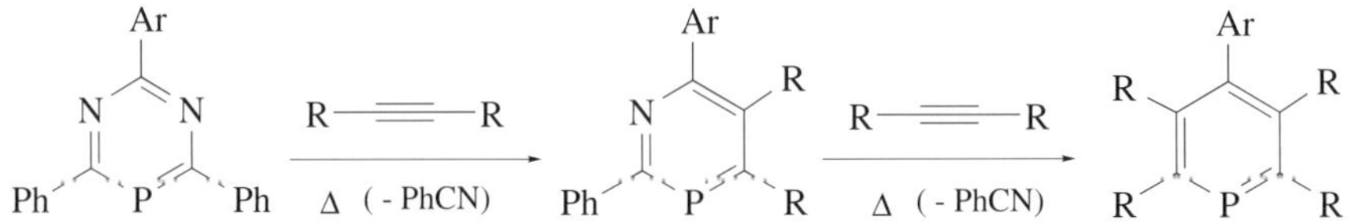

226 **227** **228** : R = COPh

Scheme 49

1,2-Aza- (**231**) and 1,3,2-diazaphosphinines (**229**) are significantly more reactive towards alkynes than their 1,3- and 1,3,5-congeners ⟨99JOC5524⟩. In general, these diazaphosphinines undergo a first cycloaddition with alkynes at 80–90°C without any additional pressure. 1,2-Azaphosphinines (**231**) thus formed are still reactive enough to give a second cycloaddition with a second alkyne to yield the desired phosphinine. Although no intermediates have been isolated, these transformations likely involve the transient formation of azabarrelenes such as (**230**) and (**232**). Using this approach, a number of 2,3,5,6-tetrafunctional phosphinines such as (**233–235**) have been synthesized (Scheme 50) ⟨96JA11978⟩. This approach proved to be powerful and allowed the preparation of polydentate ligands including phosphinines as subunits ⟨97OM4089, 98JOM(567)151, 98SCI1587⟩. Nevertheless and contrary to 1,3- and 1,3,5-diaza derivatives, no functional groups have been introduced at the C4 position. Major developments of these two complementary strategies will be presented in Section 5.2.8.

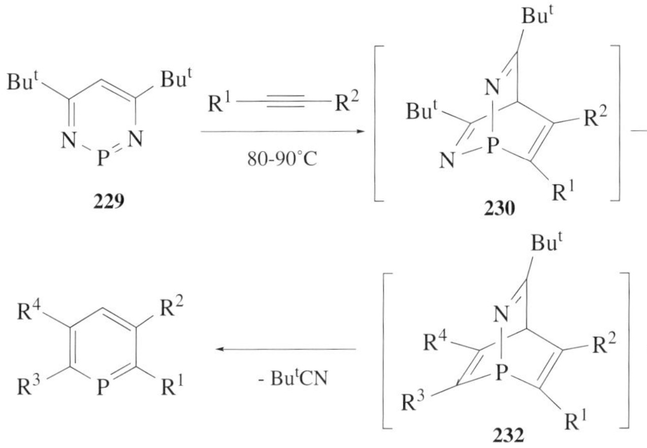

4 : $R^1 = R^3 = SiMe_3$; $R^2 = R^4 = C_{10}H_9Fe$

233 : $R^1 = CH(OEt)_2$; $R^2 = Me$; $R^3 = SiMe_3$; $R^4 = H$

234 : $R^1 = Me$, $R^2 = CCMe$; $R^3 = SiMe_3$; $R^4 = Ph$

235 : $R^1 = R^2 = PPh_2$; $R^3 = SiMe_3$; $R^4 = H$

Scheme 50

5.2.7.2 λ^5-Phosphinines

5.2.7.2.1 λ^5-Phosphinines from O, P exchange and from phosphonium salts

Unlike their λ^3-counterparts, syntheses leading to the direct formation of λ^5-phosphinines are rather rare. Only two approaches have been studied so far. The first one relies on the classical O^+/P exchange from pyrylium salts with bis(hydroxymethyl)phosphine. This transformation presumably involves the transient formation of a 1R-phosphininyl cation (**236**) ⟨66JA1034⟩. Some important compounds have thus been synthesized (Scheme 51) ⟨69TL1231, 81TL2889⟩.

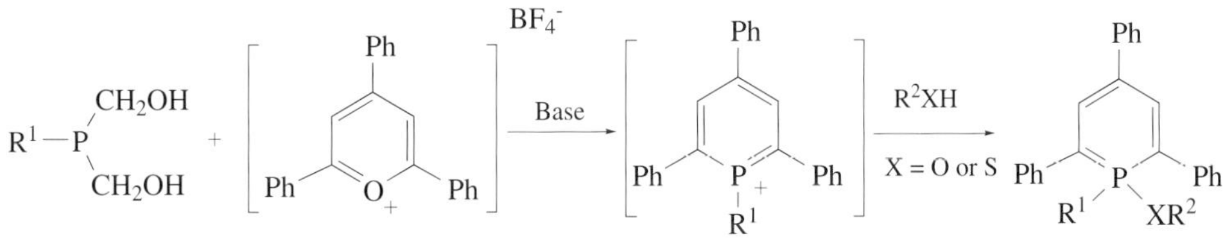

236

$R^1 = PhCH_2$, Ph

$XR^2 = OMe$, OEt, OPh, SEt

Scheme 51

The second method, involving the cyclization of phosphonium salts, is well adapted to the preparation phosphanaphthalene derivatives such as (**237**) (Equation (23)) ⟨63AG(E)168, 63AG(E)479, 65AG(E)1023, 66AG(E)588, 72AG(E)1016⟩.

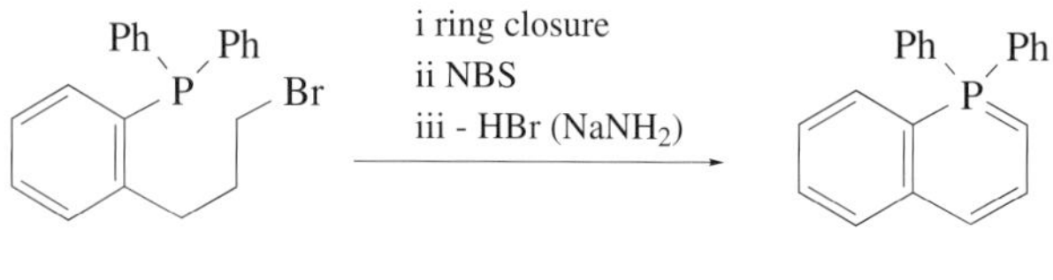

237

(23)

5.2.7.2.2 *From λ³-phosphinines*

Various approaches have been devised to synthesize λ^5-phosphinines and some of these have been briefly covered over the precedent paragraphs (see Section 5.2.5.1). One of the most common ways relies on the reaction of nucleophiles at the P atom of phosphinines. As shown previously, the anions formed show an ambident character and can be alkylated at P- or C- depending on the nature of the electrophile used. When the reaction proceeds at P, λ^5-phosphinines are thus formed in a straightforward fashion. Various derivatives (**150–152**) of the 2,4,6-triphenylphosphinine (**13**) were thus synthesized (Scheme 52) ⟨67AG(E)87, 68TL3611, 71TL1215, 74AG(E)408⟩. The addition of diazoalkanes at phosphorus is also an efficient route. Although the mechanism of this transformation has not been completely elucidated, it is believed that the first step involves an initial attack of the nucleophilic carbon at the phosphorus atom to give a zwitterion (**238**) which evolves to ylide (**239**) by loss of N_2. Subsequent reaction with alcohols yields λ^5-phosphinines (Scheme 53) ⟨74AG(E)148, 75CB3671⟩. The use of ethyldiazoacetate ⟨75CB3656⟩ leads to similar results. The reaction of azide is of particular interest. The imino derivative (**240**) formed after loss of nitrogen dimerizes to give spirotricyclic diphosphazane (**241**) (Scheme 54). This last reaction has been extended to various substituted phosphinines ⟨93TL3107⟩.

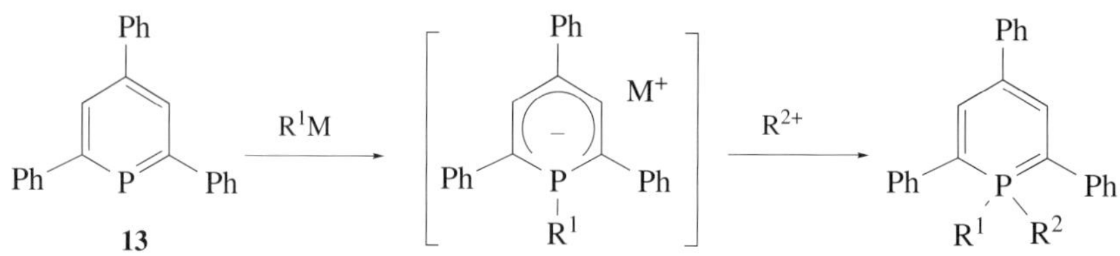

13

7 : $R^1 = R^2 = Me$
150 : $R^1 = Ph$; $R^2 = Me$
152 : $R^1 = Ph$; $R^2 = Et$

Scheme 52

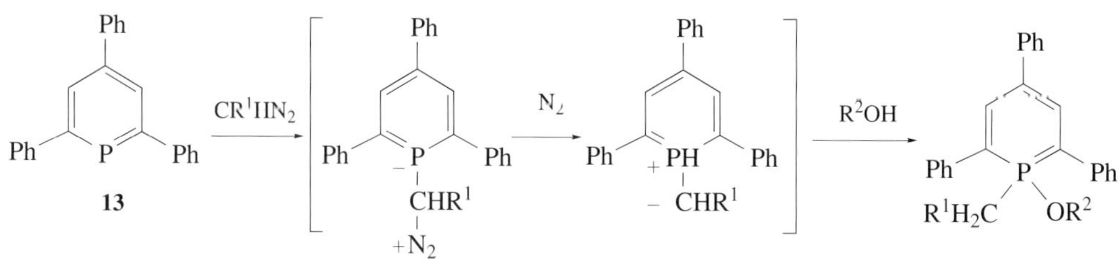

13 238 239

$R^1 = H$; $R^2 =$ alkyl or aryl
$R^1 = CO_2C_2H_5$; $R^2 =$ alkyl

Scheme 53

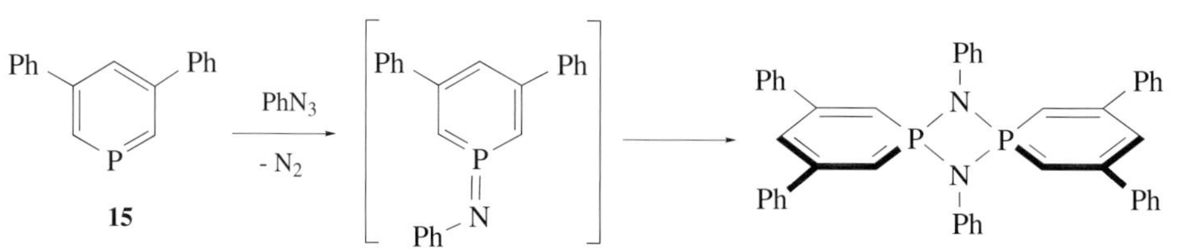

15 240 241

Scheme 54

The addition of radicals such as diphenylaminyl, triphenyl-phenoxyl, diarylmercury ⟨69AG(E)770, 69TL1231⟩, usually provides λ^5-phosphinines in high yields. The synthesis of 1,1-Bis-[diphenylamino]-2,4,6-triphenylphosphinine (**242**) is described (Equation (24)). The most interesting and efficient method is probably the addition of chlorine or bromine which gives access to 1,1-dihalo-λ^5-phosphinines which in turn are powerful precursors via nucleophilic substitution (see Section 5.2.6.1.2). This transformation requires additional irradiation (daylight for Br_2 and high pressure Hg-lamp for Cl_2) (Equation (25)) ⟨72AG(E)1090, 77CB395⟩. Finally, aryldiazonium salts react with phosphinines in the presence of water, alcohol phenols or thiols to give 1-aryl-XR-λ^5-phosphinines (X = O or S) with the concomitant release of nitrogen ⟨72AG(E)1091, 75CB3281, 75CB3671⟩. Although no mechanistic studies have been undertaken, this transformation is thought to involve in a first step a donor–acceptor type complex between the diazonium and the ring. An addition of aryl radicals then ends the sequence.

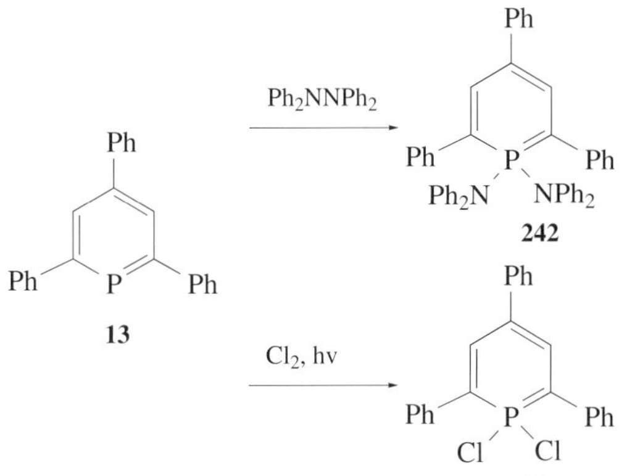

13 242 43

(24)

(25)

Oxidative addition of protic nucleophile also constitutes a valuable approach. These transformations, which are promoted by silver ($AgOCOCF_3$) or mercury salts $(HgOCOCH_3)_2$, give access to 1,1-dialkoxy, diaryloxy, crown ethers derivatives, dithio, diamino and amino compounds (Equation (26)) ⟨68AG(E)881, 72TL829, 74CB3501, 75TL541, 81LA919⟩. Two mechanistic pathways have been proposed to explain their formation. The first mechanism postulates successive oxidations of the phosphorus followed by nucleophilic attack of the alcohol. The second implies the transient formation of a phosphinine Hg^{2+} complex which would add one molecule of alcohol at the P=C bond. Then one OAc group is transferred from the metal to phosphorus. In the presence of excess nucleophile, this OAc group is displaced from phosphorus.

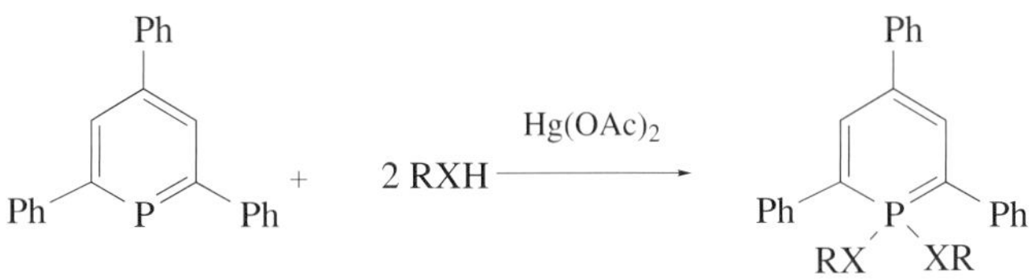

13

X = O, S, N

(26)

5.2.7.2.3 *From dihydro and tetrahydrophosphinine oxides*

This method relies on the dihydrophosphinines oxides/1-hydroxy-1-organo-λ^5-phosphinines tautomery (see Scheme 55) ⟨72TL829⟩. In the presence of alkylating agents such as oxoniums, $(MeO)_2SO_2$ or alkyl iodides, and a base, dihydrophosphinines oxides are converted in 1-alkoxy-1-organo-λ^5-phosphinines in fair to good yields ⟨65AG(E)914⟩. An elegant illustration was given by the synthesis of indolo-λ^5-phosphinine (**243**) ⟨78PS(5)257⟩. This sequence has been successfully applied to 1,4-dihydro and 1,2,3,4-tetrahydro derivatives. Thus, oxide (**244**) is converted to the 1,4-dimethoxy-1,2,6-triphenyl compound (**245**) (Equation (27)) ⟨80CB3313⟩. Interestingly, this conversion can also be achieved intramolecularly. When an alkoxy group is present at the C2 position, a thermally promoted migration at the oxygen takes place to yield a 1-acyloxy-1-alkoxy derivative ⟨72TL1045⟩.

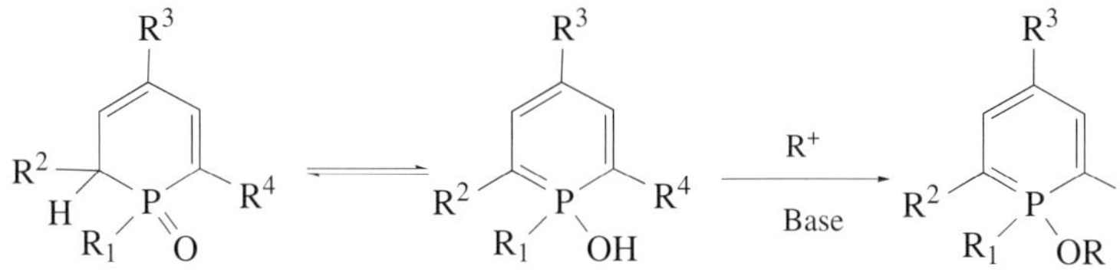

Scheme 55

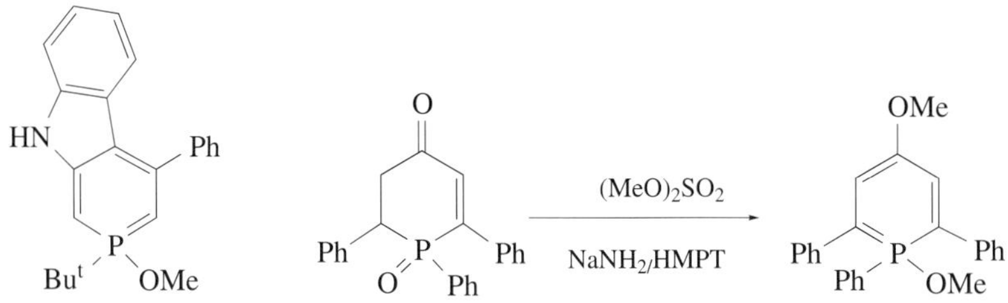

243 **244** **245** (27)

5.2.7.2.4 *Synthesis by ring contraction*

Ring contractions leading to phosphinines are rare. To the best of our knowledge, Quin *et al.* reported the only example ⟨84JOC3157⟩. This quite complicated transformation relies on the chemistry of a 3,8-phosphanedione. Silylation of this compound with $(Me_3Si)_2COCF_3$ provides the syliloxy derivative (**246**) which reacts with triethylamine to give a dihydrophosphonine derivative (**247**). Heating this compound results in an intramolecular ring closure with a concomitant sylil migration from C–O to P–O leading to the λ^5-cyclopentaphosphinine-7-one (**248**) (Scheme 56).

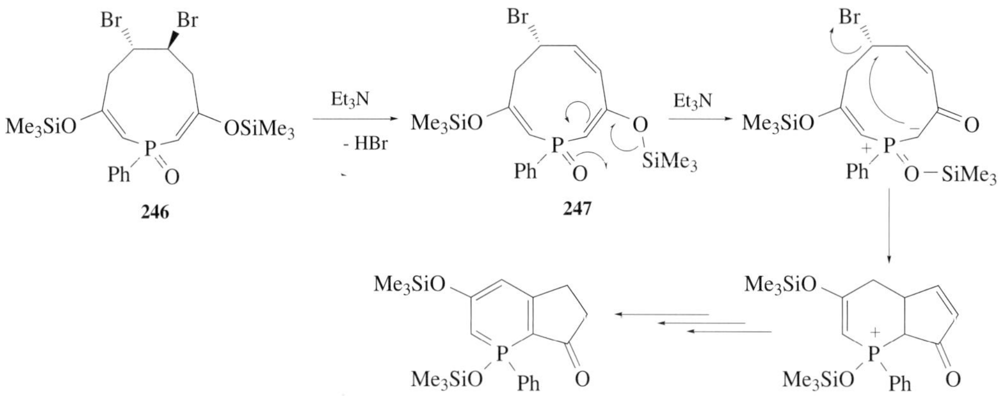

248

Scheme 56

5.2.8 SYNTHESIS OF PARTICULAR TYPES OF PHOSPHININES

A great deal of effort has been devoted to the synthesis of functional phosphinines and various approaches have been studied. For a long time, no general methods allowing to graft functional groups onto a preformed phosphinine ring, like in O, S, and N heterocyclic chemistry, were available. This mainly results from the difficulty to transpose with phosphinines some conventional synthetic methodologies such as proton or halogen abstraction, involving nucleophilic or basic reagents, or Friedel–Craft reactions. Over the last decade, specific methods taking into account the reactivity of the phosphorus atom have been devised and significant progresses have been made. The most important developments are presented in this section.

5.2.8.1 Functional Phosphinines

As stated previously (see Section 5.2.7.1) ring transformations of other phosphorus heterocycles into phosphinines have been widely studied. Ring expansion from phospholes was used to prepare 2-thienyl (**30**), 2-furyl (**31**) and the important 2-(2′-pyridyl)-phosphinine (**41**) ⟨84PS(19)45, 82TL1565⟩. A 3-vinyl-derivative (**249**) has been obtained using the reactivity of dimethyltitanocene towards the diyne $(PhCC)_2$ ⟨96HAC397⟩. The most important developments were obtained with transformation involving the reactivity of aza- and diazaphosphinines towards alkynes. Thus, the reaction of 1,3-aza and 1,3,5-diaza derivatives with diethylacetylene dicarboxylate respectively yielded compounds (**21**) ⟨87TL1093, 90TL4589⟩ and (**256**) ⟨91AG(E)106⟩. Symmetrical and unsymmetrical tetrafunctional derivatives such as (**251–254**) were readily synthesized from the reaction of 1,3,2-diazaphosphinines with alkynes and diynes (Figure 14) ⟨97OM4089, 96JA11978⟩.

Although of more limited synthetic value, cycloadditions of 1,3-dienes or their precursors with phosphaalkenes and phosphaalkynes also provide an interesting entry to polyfunctional phosphinines. Ester (**255**) and acidic derivatives (**256**) of 3-phosphaphenol ⟨86TL4299⟩, 2-phosphaphenol (**257**) ⟨89TL5245⟩ and C3-substituted ester (**258**) and ether (**259**) derivatives ⟨91HAC439⟩ have been characterized. A wide range of 2-tertiobutyl substituted phosphinines, (**260–263**), have been synthesized from the reaction of $PCBu^t$ with functional pyrones (Figure 15) ⟨86ZN(B)931, B-90MIV2-02⟩.

The reactivity of 2-halogenophosphinines allowed for significant advances. Once the phosphorus lone pair is protected by $W(CO)_5$, the 2-lithio or 2-zinc compounds can be prepared prior to functionalization (see Section 5.2.5.3). Tin (**264**), silver (**265**) and dichloroarsine (**266**) derivatives were prepared by this way ⟨92OM2475, 92TL3537⟩. Insertion of zirconocene into the C–Br bond of 2-bromophosphinines led to

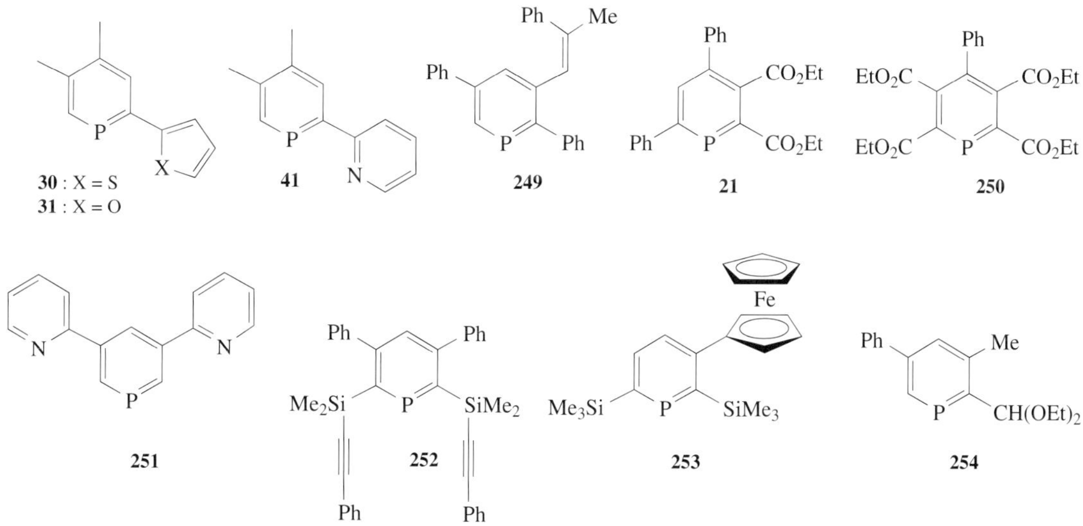

Figure 14. Some examples of 2-functional phosphinines.

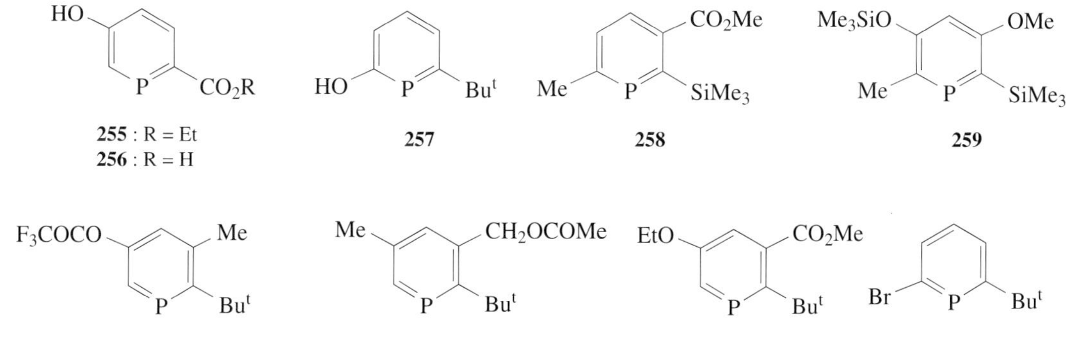

Figure 15. Functional phosphinines prepared from cycloadditions of phosphaalkenes and phosphaalkynes with 1,3-dienes and pyrones.

2-cyano (**267**) and a 2-deutero (**268**) phosphinines after treatment with BrCN and D_2O, respectively (see Section 5.2.5.3) ⟨93CC789⟩.

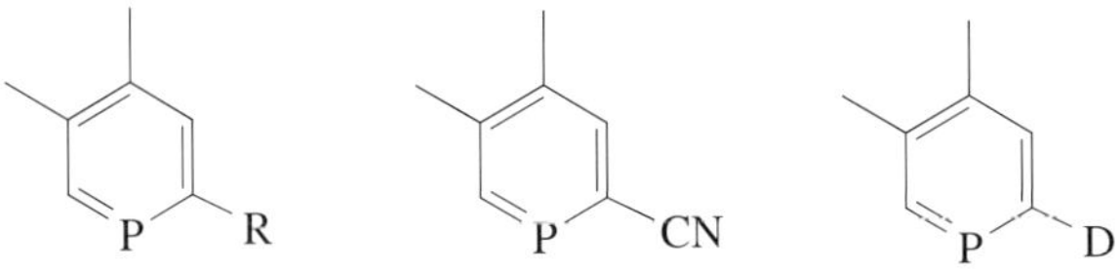

264 : R = $SnPh_3$ **267** **268**
265 : R = Ag
266 : R = $AsCl_2$

Phosphabenzyne–zirconocene complexes also display a very rich chemistry. Their reactivity towards alkynes, ketones, aldehydes and $P(S)Ph_3$ respectively gives access to vinyl (**272**), secondary (**270**), tertiary (**279**) alcohols and thio (**71**) compounds *via* hydrolysis of the corresponding metallacycles ⟨94CC2065, 97JA9417⟩. Interestingly, an enantiomerically pure phosphinine (**273**) was obtained by reaction with (+)-camphor.

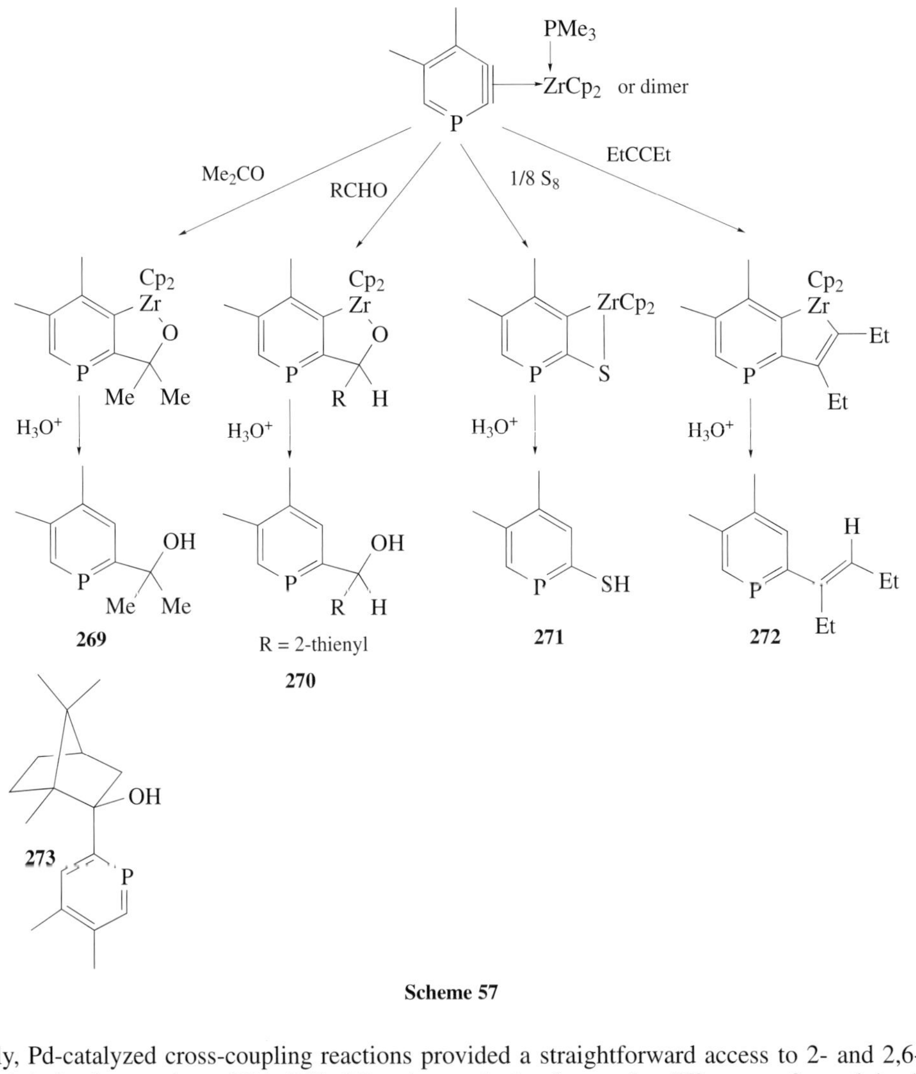

Scheme 57

Finally, Pd-catalyzed cross-coupling reactions provided a straightforward access to 2- and 2,6-difunctionalized derivatives such as (**274–279**). Most interestingly, due to the difference of reactivity between the bromine atoms, some unsymmetrically substituted compounds could be synthesized (Figure 16) ⟨93JA10665, 95S717⟩.

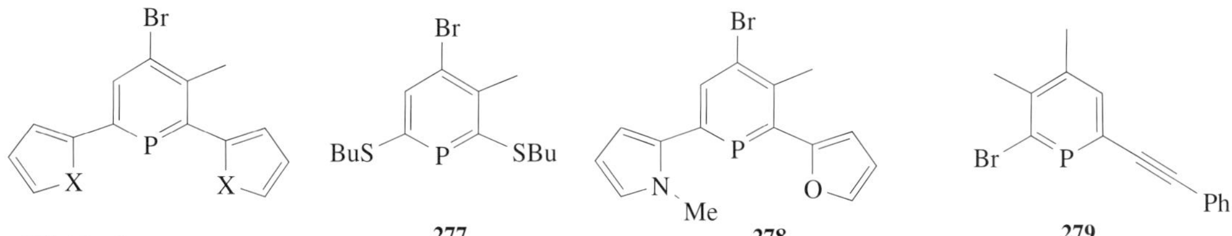

274 : X = S
275 : X = O
276 : X = NMe

Figure 16. 2,6-Difunctional phosphinines obtained from Pd-catalyzed cross-coupling reactions.

5.2.8.2 Polydentate Phosphorus Ligands and Macrocycles including Phosphinines

A great attention has been paid to the synthesis of biphosphinines. As seen previously, a 4,4′-(Section 5.2.7.1.4) and some 2,2′-derivatives (Sections 5.2.5.3.2 and 5.2.7.1.4) are known. Only two functional 2,2′-biphosphinines have been synthesized so far, a 6,6′-diester (**280**) obtained via the ring expansion from the corresponding biphosphole ⟨91JOC4031⟩ and two dibromo derivatives, (**281**) and (**282**), obtained as a mixture in a cross-coupling reaction ⟨95S717⟩. Two 3,3′-biphosphinines, (**283**) and (**284**), have been synthesized by reaction of a 1,3-azaphosphinine with 1,3-diynes. The same approach was used to prepare some bis(phosphinines) (**285**–**289**). Syntheses of silyl substituted bis(phosphinines) (**290**–**292**) rely on the reactivity of a diazaphosphinine (**229**) with diynes (see Section 5.2.7.1.6) ⟨97OM4089⟩. Finally, the carbenium salt of a 1,1-dimethoxy-λ^5-derivative was coupled to the corresponding bis(phosphinine) (**293**) (Figure 17) ⟨87CB1245⟩.

A number of 2-phosphino substituted phosphinines are known. The two most convenient approaches

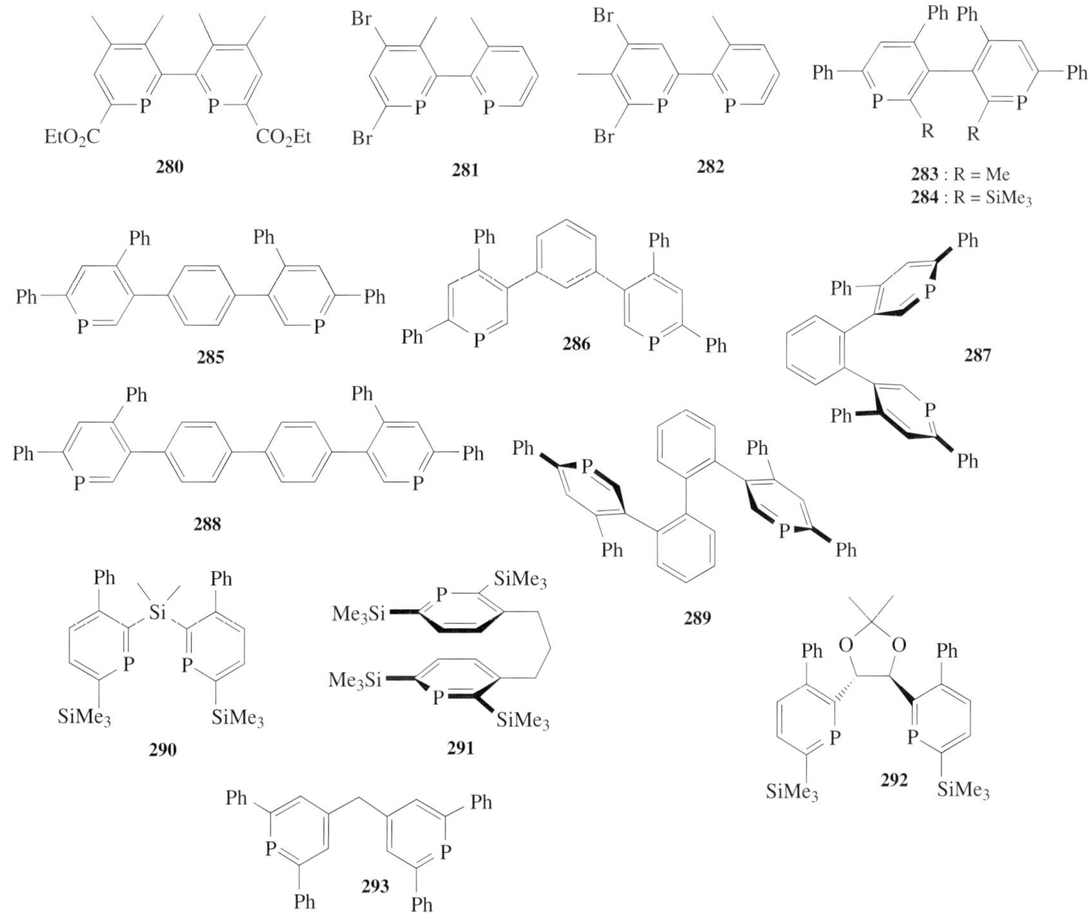

Figure 17. Selected examples of bis(phosphinines).

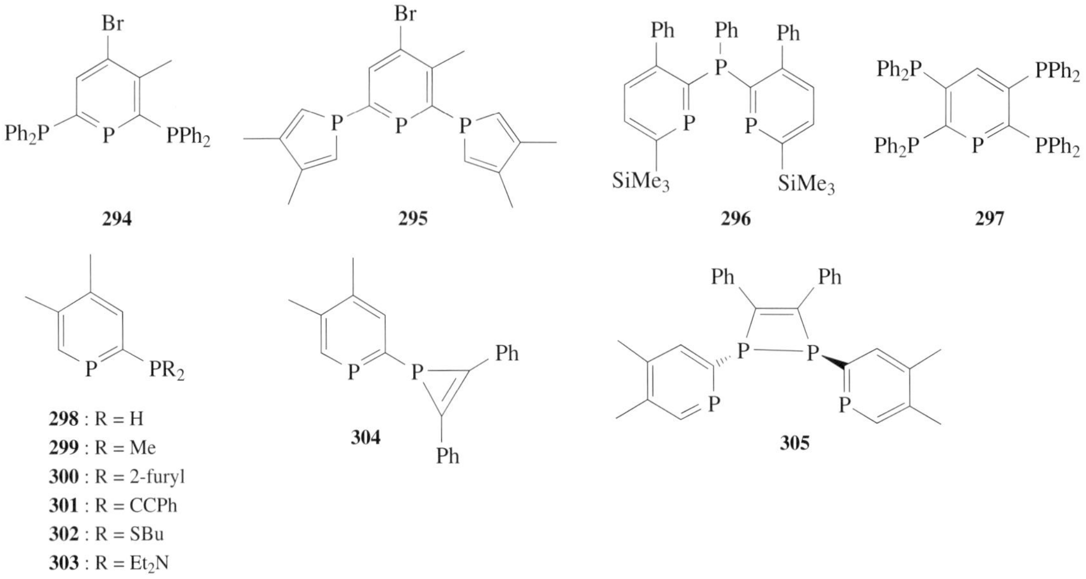

Figure 18. Some phosphinophosphinines.

are, cross-coupling reactions of bromophosphinines with the corresponding anions or their silyl or tin derivatives (syntheses of **294** and **295**) and cycloadditions of 1,3,2-diazaphosphinines with phosphino-substituted alkynes (syntheses of **296** and **297**). The reactivity of 2-dibromophosphino-phosphinine (**221**) has also been explored. Compounds (**298–305**) have been prepared using either nucleophilic or electrophilic conditions (Figure 18) ⟨96OM1597⟩.

Interestingly, 2-phospholylphosphinines have been used as precursors in the synthesis of polydentate ligands. The strategy is based on the characteristic [1,5] sigmatropic shifts of P-substituents of phospholes. Phosphaferrocenyl- (**307**) ⟨97CB843⟩ and phosphanorbornadienyl-phosphinine bidentate ligands (**308**) (Scheme 58) ⟨95BSF910⟩ were readily synthesized by heating phosphinine (**142**), precursor of the 2H-phosphole (**306**) with $[Fe(\eta^5\text{-}C_5H_5)CO_2]_2$ and tolane, respectively. Thermolysis of (**142**) at 180°C for 60 h without trapping reagent yields tetramer (**309**) which is subsequently transformed into dianion (**310**) and tetradentate ligand (**311**).

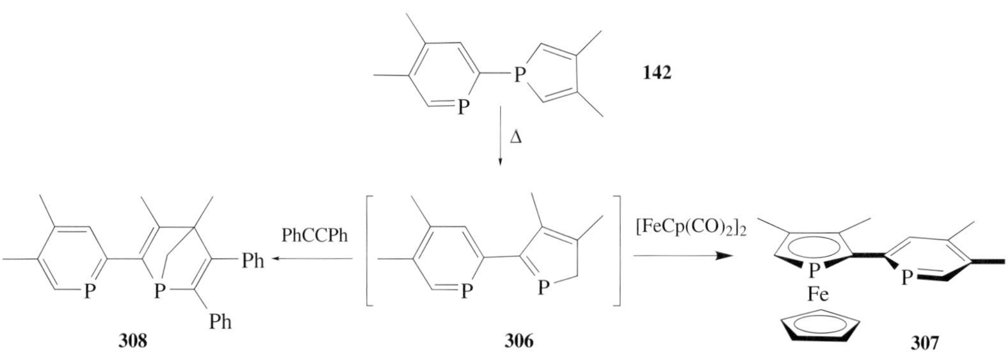

Scheme 58

There exist four types of phosphinine based-tridentate ligand. All were built following the same strategy which is detailed in scheme 60 for ligand (**313**). Cycloaddition of phosphinine (**252**) with two equivalents of diazaphosphinine (**229**) yields intermediate (**312**) which subsequently reacts with trimethylsilyacetylene to give (**313**). Phosphole (**314**), phospholide (**315**) and phosphaferrocene (**316**) based-tridentates were assembled using the corresponding bis(alkynyl) substituted compounds ⟨99OM4205⟩. A similar strategy was employed to prepare the tetrakis(phosphinine) (**317**) and the silacalix-phosphinines (**35**) (**36**) (**318**) (**319**) which are the first examples of macrocycles bearing sp^2-hybridized phosphorus atoms (Figure 19). All these macrocycles are flexible in solution but adopt a 1,3-anti conformation in the solid state ⟨98SCI1587, 99MIV2-02⟩.

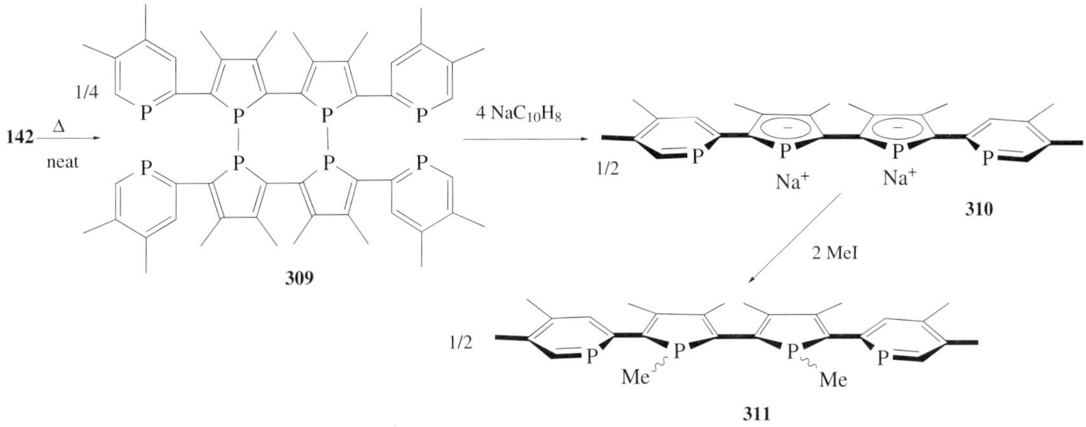

Scheme 59

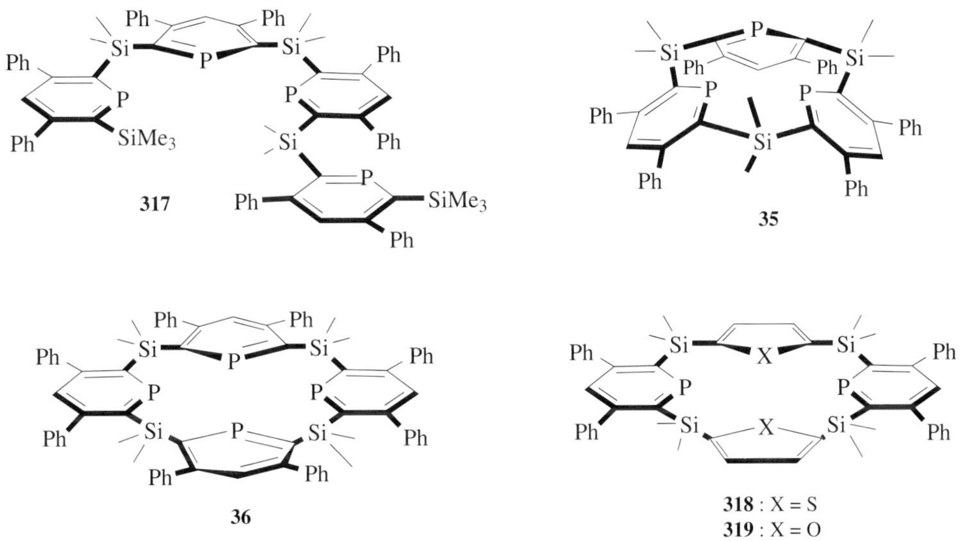

Figure 19. Phosphinine-based polydentate ligands and macrocycles.

5.2.9 PERSPECTIVES IN PHOSPHININE CHEMISTRY

For a long time, phosphinines, like other low-coordinated phosphorus compounds, have been considered as curiosities for phosphorus chemists. This time is now gone and it is obvious that this chemistry has come to maturity. Although some improvements have still to be achieved in their synthesis to diversify the availability of functional groups (aldehydes, amines, imines), their use in more applied projects can now be seriously envisaged. Three possible fields of investigations can be thought of. First, in coordination chemistry, the suitable balance between a poor σ-donating and a very strong π-accepting ability make phosphinines attractive ligands for the stabilization of electron-rich metallic centers, usually stable only when CO is present as ancillary ligands. In this perspective the use of phosphinine-based polydentate and macrocyclic ligands should lead to significant results. A second exciting field of investigation is homogeneous catalysis, where these ligands should be tested in processes usually involving strong π-acceptor tertiary phosphines. With this aim in view, recent results obtained in the Rh-catalyzed hydroformylation process are particularly encouraging since they show that phosphinine complexes withstand high H_2 pressure ⟨96CC2071, 97JCS(P1)2681, 99GEP19621967, 99GEP19743197⟩. A third area of development concerns the building of phosphinine-based edifices and polymeric materials and studies of their specific properties. At the present time, this direction has not been investigated yet. Although some valuable synthetic approaches exist and should allow the synthesis of supramolecular edifices, no simple method allowing the preparation of defined oligomers or polymeric materials has been devised so far. Theoretical calculations (HF crystal orbital calculations) have shown that poly(2,5-phosphinines) should

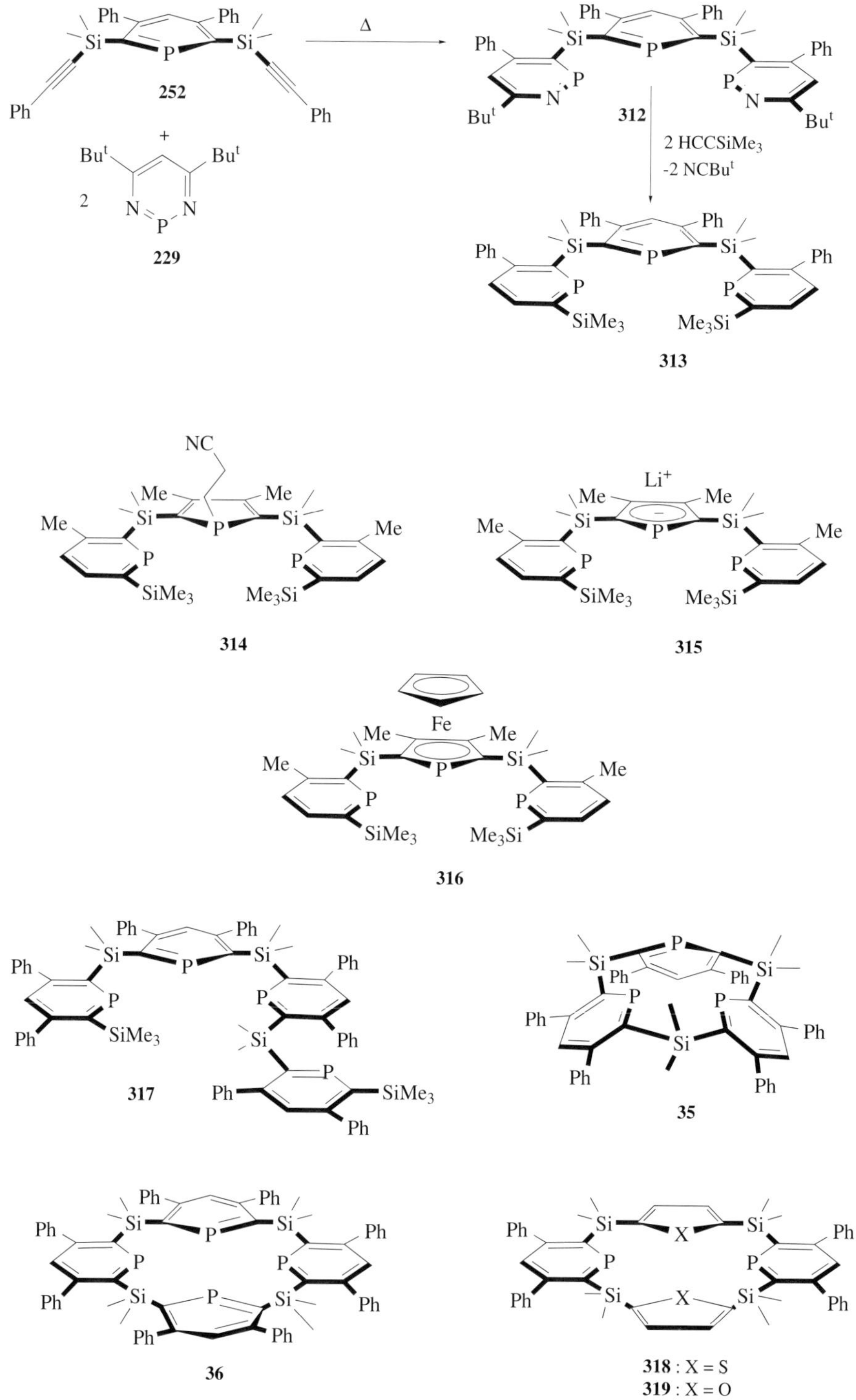

Scheme 60

display interesting conductive abilities ⟨97MIV2-01⟩. Finally, the synthesis of substances of biological interest incorporating phosphinines has not been reported.

5.2.10 REFERENCES

63AG(E)168 G. Märkl; *Angew. Chem. Int. Ed. Engl.*, 1963, **2**, 168.
63AG(E)479 G. Märkl; *Angew. Chem. Int. Ed. Engl.*, 1963, **2**, 479.

65AG(E)914 G. Hilgetag and H. Teichmann; *Angew. Chem. Int. Ed. Engl.*, 1965, **4**, 914.
65AG(E)1023 G. Märkl; *Angew. Chem. Int. Ed. Engl.*, 1965, 4, 1023.
66AG(E)588 G. Märkl and H. Olbrich; *Angew. Chem. Int. Ed. Engl.*, 1966, **5**, 588.
66AG(E)846 G. Märkl; *Angew. Chem. Int. Ed. Engl.*, 1966, **5**, 846.
66JA1034 C.C. Price, T. Parasavan and T. Lakshminarayan; *J. Am. Chem. Soc.*, 1966, **88**, 1034.
67AG(E)87 G. Märkl, F. Lieb and A. Merz; *Angew. Chem. Int. Ed. Engl.*, 1967, **6**, 87.
67AG(E)85 K. Dimroth, N. Greig, H. Perst and F.W. Steuber; *Angew. Chem. Int. Ed. Engl.*, 1967, **6**, 85.
67AG(E)446 K. Dimroth and W. Steuber; *Angew. Chem. Int. Ed. Engl.*, 1967, **6**, 446.
67AG(E)458 G. Märkl, F. Lieb and A. Merz; *Angew. Chem. Int. Ed. Engl.*, 1967, **6**, 458.
67AG(E)567 P. de Koe and F. Bickelhaupt; *Angew. Chem. Int. Ed. Engl.*, 1967, **6**, 567.
67AG(E)711 K. Dimroth, N. Greif, W. Städe and F.W. Steuber; *Angew. Chem. Int. Ed. Engl.*, 1967, **6**, 711.
67AG(E)944 G. Märkl, F. Lieb and A. Merz; *Angew. Chem. Int. Ed. Engl.*, 1967, **6**, 944.
68AG(E)371 K. Dimroth, K. Vogel, W. Mach and U. Schoeler; *Angew. Chem. Int. Ed. Engl.*, 1968, **7**, 371.
68AG(E)460 K. Dimroth and W. Mach; *Angew. Chem. Int. Ed. Engl.*, 1968, **7**, 460.
68AG(E)465 P. de Koe, R. van Veen and F. Bickelhaupt; *Angew. Chem. Int. Ed. Engl.*, 1968, **7**, 465.
68AG(E)733 G. Märkl and F. Lieb; *Angew. Chem. Int. Ed. Engl.*, 1968, **7**, 733.
68AG(E)811 J.C. Bart and J.J. Daly; *Angew. Chem. Int. Ed. Engl.*, 1968, **7**, 811.
68AG(E)881 K. Dimroth and W. Städe; *Angew. Chem. Int. Ed. Engl.*, 1968, **7**, 881.
68AG(E)889 P. de Koe and F. Bickelhaupt; *Angew. Chem. Int. Ed. Engl.*, 1968, **7**, 889.
68AG(E)981 K. Dimroth and W. Städe; *Angew. Chem. Int. Ed. Engl.*, 1968, **7**, 981.
68TL3611 G. Märkl and A. Merz; *Tetrahedron Lett.*, 1968, 3611.
68TL6227 W. Fischer, E. Hellner, A. Chatzidakis and K. Dimroth; *Tetrahedron Lett.*, 1968, 6227.
69AG(E)769 U. Thewalt; *Angew. Chem. Int. Ed. Engl.*, 1969, **8**, 769.
69AG(E)770 K. Dimroth, A. Hettche, W. Städe and F.W. Steuber; *Angew. Chem. Int. Ed. Engl.*, 1969, **8**, 770.
69AG(E)776 K. Dimroth, A. Hettche, W. Städe and F.W. Steuber; *Angew. Chem. Int. Ed. Engl.*, 1969, **8**, 776.
69CC1057 J.J. Daly and G. Märkl; *J. Chem. Soc. Chem. Commun.*, 1969, 1057.
69TL1231 G. Märkl and A. Merz; *Tetrahedron Lett.*, 1969, 1231.
70AG(E)898 U. Thewalt, Ch. Bugg and A. Hettche; *Angew. Chem. Int. Ed. Engl.*, 1970, **9**, 898.
70CB2541 J. Deberitz and H. Nöth; *Chem. Ber.*, 1970, **103**, 2541.
70JCS(A)1832 J.J. Daly; *J. Chem. Soc. (A)*, 1970, 1832.
70TL645 G. Märkl, D.E. Fischer and H. Olbrich; *Tetrahedron Lett.*, 1970, 645.
70TL4941 H. Oehling and A. Schweig; *Tetrahedron Lett.*, 1970, 4941.
71AG(E)656 H. Oehling, W. Schäfer and A. Schweig; *Angew. Chem. Int. Ed. Engl.*, 1971, **9**, 656.
71CB2984 K. Dimroth and H. Odenwälder; *Chem. Ber.*, 1971, 2984.
71JA1804 A.J. Ashe III and P. Shu; *J. Am. Chem. Soc.*, 1971, **93**, 1804.
71JA3293 A.J. Ashe III; *J. Am. Chem. Soc.*, 1971, **93**, 3293.
71JA6690 A.J. Ashe III; *J. Am. Chem. Soc.*, 1971, **93**, 6690.
71MIV2-01 H. Oehling and A. Schweig; *Phosphorus*, 1971, 203.
71TL1215 G. Märkl and A. Merz; *Tetrahedron Lett.*, 1971, 1215.
71TL1269 G. Märkl and A. Merz; *Tetrahedron Lett.*, 1971, 1269.
72AG(E)506 K. Dimroth, A. Chatzidakis and O. Schaffer; *Angew. Chem. Int. Ed. Engl.*, 1972, **11**, 506.
72AG(E)631 A. Schweig, W. Schäfer and K. Dimroth; *Angew. Chem. Int. Ed. Engl.*, 1972, **11**, 631.
72AG(E)1016 G. Märkl and K.H. Heier; *Angew. Chem. Int. Ed. Engl.*, 1972, **11**, 1016.
72AG(E)1017 G. Märkl and K.H. Heier; *Angew. Chem. Int. Ed. Engl.*, 1972, **11**, 1017.
72AG(E)1090 H. Kanter and K. Dimroth; *Angew. Chem. Int. Ed. Engl.*, 1972, **11**, 1090.
72AG(E)1091 O. Schaffer and K. Dimroth; *Angew. Chem. Int. Ed. Engl.*, 1972, **11**, 1091.
72CB1148 H. Vahrenkamp and H. Nöth; *Chem. Ber.*, 1972, **105**, 1148.
72JA7596 A.J. Ashe III and M.D. Gordon; *J. Am. Chem. Soc.*, 1972, **94**, 7596.
72JHC1457 M. Fraser, D.G. Holah, A.N. Hughes and B.C. Hui; *J. Heterocycl. Chem.*, 1972, 1457.
72MIV2-01 G. Märkl; *Lect. Heterocycl. Chem.*, 1972, **1**, S-69.
72MIV2-02 R.L. Kuczkowski and A.J. Ashe III; *J. Mol. Spectrosc.*, 1972, **42**, 457.
72TL829 A. Hettche and K. Dimroth; *Tetrahedron Lett.*, 1972, 829.
72TL835 K. Dimroth, A. Hettche, H. Kanter and W. Städe; *Tetrahedron Lett.*, 1972, 835.
72TL839 K. Dimroth, W. Schäfer and H.H. Pohl; *Tetrahedron Lett.*, 1972, 839.
72TL843 W. Schäfer and K. Dimroth; *Tetrahedron Lett.*, 1972, 843.
72TL1045 A. Hettche and K. Dimroth; *Tetrahedron Lett.*, 1972, 1045.
72TL4925 G. Märkl and D.E. Fischer; *Tetrahedron Lett.*, 1972, 4925.
73AG(E)753 W. Schäfer and K. Dimroth; *Angew. Chem. Int. Ed. Engl.*, 1973, **12**, 753.
73AG(E)931 G. Märkl and F. Kneidl; *Angew. Chem. Int. Ed. Engl.*, 1973, **12**, 931.
73CB1001 A. Hettche and K. Dimroth; *Chem. Ber.*, 1973, **106**, 1001.
73CB2222 H. Vahrenkamp and H. Nöth; *Chem. Ber.*, 1973, **106**, 2222.
73JA928 C. Batich, E. Heilbronner, V. Hornung, A.J. Ashe III, D.T. Clark, U.T. Cobley, D. Kilcast and I. Scanlan; *J. Am. Chem. Soc.*, 1973, **95**, 928.
73JOM(49)453 J. Deberitz and H. Nöth; *J. Organomet. Chem.*, 1973, **49**, 453.
73SRT375 K.C. Dash, J. Eberlein and H. Schmidbaur; *Synth. React. Inorg. Metal-Org. Chem.*, 1973, **3**, 375.
73T475 H.L. Hase, A. Schweig, H. Hahn and J. Radloff; *Tetrahedron*, 1973, **29**, 475.

73TCC(38)1	K. Dimroth; *Top. Curr. Chem.*, 1973, **38**, 1.
73TL2397	H.G. de Graaf, J. Dubbledam, H. Vermeer and F. Bickelhaupt; *Tetrahedron Lett.*, 1973, 2397.
74AG(E)148	P. Kieselack and K. Dimroth; *Angew. Chem. Int. Ed. Engl.*, 1974, **13**, 148.
74AG(E)408	G. Märkl and C. Martin; *Angew. Chem. Int. Ed. Engl.*, 1974, **13**, 408.
74CB3501	M. Constentla and K. Dimroth; *Chem. Ber.*, 1974, **107**, 3501.
74JCS(F2)1222	D.T. Clark and I.W. Scanlan; *J. Chem. Soc., Faraday Trans. 2*, 1974, **70**, 1222.
74JCP2840	T.C. Wong and L.S. Bartell; *J. Chem. Phys.*, 1974, **61**, 2840.
74MP601	F. Gerson, G. Plattner, A.J. Ashe III and G. Märkl; *Mol. Phys.*, 1974, **28**, 601.
74TL1267	C. Jongsma, H.G. de Graaf and F. Bickelhaupt; *Tetrahedron Lett.*, 1974, 1267.
74TL3179	T. Bundgaard, H.J. Jakobsen, K. Dimroth and H.H. Pohl; *Tetrahedron Lett.*, 1974, 3179.
74TL4381	G. Märkl and D. Matthes; *Tetrahedron Lett.*, 1974, 4381.
74TL4501	G. Märkl and K.H. Heier; *Tetrahedron Lett.*, 1974, 4501.
75AG(E)11	H.H. Pohl and K. Dimroth; *Angew. Chem. Int. Ed. Engl.*, 1975, **14**, 11.
75AG(E)112	K. Dimroth and M. Lückoff; *Angew. Chem. Int. Ed. Engl.*, 1975, **14**, 112.
75CB1384	H.H. Pohl and K. Dimroth; *Chem. Ber.*, 1975, **108**, 1384.
75CB3281	O. Schaffer and K. Dimroth; *Chem. Ber.*, 1975, **108**, 3281.
75CB3656	P. Kieselhack, C. Helland and K. Dimroth; *Chem. Ber.*, 1975, **108**, 3656.
75CB3671	P. Kieselhack and K. Dimroth; *Chem. Ber.*, 1975, **108**, 3671.
75CPH345	W. von Niessen, G.H.F. Diercksen and L.S. Cederbaum; *Chem. Phys.*, 1975, **10**, 345.
75JA5526	D. Griller, K. Dimroth, T.M. Fyles and K.U. Ingold; *J. Am. Chem. Soc.*, 1975, **97**, 5526.
75JCS(P2)841	M.H. Palmer, R.H. Findlay, W. Moyes and A.J. Gaskell; *J. Chem. Soc. Perkin Trans. 2*, 1975, 841.
75RTC7	F. Bickelhaupt and H. Vermeer; *Recl. Trav. Chim. Pays-Bas*, 1975, **93**, 7.
75T1097	H.G. de Graaf and F. Bickelhaupt; *Tetrahedron*, 1975, **31**, 1097.
75T2931	C. Jongsma, H. Vermeer, F. Bickelhaupt, W. Schäfer and A. Schweig; *Tetrahedron*, 1975, **31**, 2931.
75TL541	H. Kanter and K. Dimroth; *Tetrahedron Lett.*, 1975, 541.
75TL545	H. Kanter and K. Dimroth; *Tetrahedron Lett.*, 1975, 545.
75TL1083	A.J. Ashe III, W.-T. Chan and E. Perozzi; *Tetrahedron Lett.*, 1975, 1083.
75TL2749	A.J. Ashe III and W.-T.Chan; *Tetrahedron Lett.*, 1975, 2749.
76AG(E)238	W.J. Seifert, O. Schaffer and K. Dimroth; *Angew. Chem. Int. Ed. Engl.*, 1976, **15**, 238.
76AG(E)503	M. Lückhoff and K. Dimroth; *Angew. Chem. Int. Ed. Engl.*, 1976, **15**, 503.
76CB3099	M. Constentla and K. Dimroth; *Chem. Ber.*, 1976, **109**, 3099.
76HCA1944	A.J. Ashe III, F. Burger, M.Y. El-Sheik, E. Heilbronner, J.P. Maier and J.-F. Müller; *Helv. Chim. Acta*, 1976, **59**, 1944.
76JA4410	W. Schäfer, A. Schweig, K. Dimroth and H. Kanter; *J. Am. Chem. Soc.*, 1976, **98**, 4410.
76JA5451	A.J. Ashe III, R.R. Sharp and J.W. Tolan; *J. Am. Chem. Soc.*, 1976, **98**, 5451.
76JA7861	A.J. Ashe III and T.W. Smith; *J. Am. Chem. Soc.*, 1976, **98**, 7861.
77CB395	H. Kanter, W. Mach and K. Dimroth; *Chem. Ber.*, 1977, **110**, 395.
77CB1497	T. Debaerdemaeker, H.H. Pohl and K. Dimroth; *Chem. Ber.*, 1977, **110**, 1497.
77MIV2-01	C. Jongsma and F. Bickelhaupt; in "Topics in Non-Benzenoid Aromatic Chemistry", **Vol. II**, Hirokowa Publ., Tokyo, 1977, p. 139.
77PS(3)77	G. Märkl; *Phosphorus Sulfur*, 1977, **3**, 77.
77TL407	A.J. Ashe III and T.W. Smith; *Tetrahedron Lett.*, 1977, 407.
77TL3445	G. Märkl, G. Adolin, F. Kees and G. Zander; *Tetrahedron Lett.*, 1977, 3445.
78ACR153	A.J. Ashe III; *Acc. Chem. Res.*, 1978, **11**, 153.
78JOM(148)C31	K.C. Nainan and C.T. Sears; *J. Organomet. Chem.*, 1978, **148**, C31.
78PS(5)257	G. Märkl, G. Habel and H. Baier; *Phosphorus and Sulfur*, 1978, **5**, 257.
79CB1273	K. Dimroth, H.H. Pohl and K.-W. Wichmann; *Chem. Ber.*, 1979, **112**, 1273.
79JA1764	A.J. Ashe III, M.K. Bahl, K.D. Bomben, W.-T. Chan, J.K. Gimzewski, P.G. Sitton and T.D. Thomas; *J. Am. Chem. Soc.*, 1979, **101**, 1764.
79JMR(34)199	V. Galasso; *J. Magn. Resonance*, 1979, **34**, 199.
79JMR(36)181	V. Galasso; *J. Magn. Resonance*, 1979, **36**, 181.
79JOMC(173)125	G. Märkl, F. Kees, P. Hofmeister and C. Soper; *J. Organomet. Chem.*, 1973, **173**, 125.
79TL1753	F. Mathey; *Tetrahedron Lett.*, 1979, 1753.
80CB3313	K. Dimroth and M. Lückoff; *Angew. Chem. Int. Ed. Engl.*, 1980, **113**, 3313.
80JOM(187)277	F. Nief, C. Charrier, F. Mathey and M. Simalty; *J. Organomet. Chem.*, 1980, **187**, 277.
80TL1441	F. Nief, C. Charrier, F. Mathey and M. Simalty; *Tetrahedron Lett.*, 1980, **21**, 1441.
81AG(E)871	K. Dimroth and H. Kaletsch; *Angew. Chem. Int. Ed. Engl.*, 1981, **20**, 871.
81CB1752	K. Dimroth, O. Schaffer and G. Weiershäuser; *Chem. Ber.*, 1981, **114**, 1752.
81CB3004	K. Dimroth and W. Heide; *Chem. Ber.*, 1981, **114**, 3004.
81CB3019	K. Dimroth and W. Heide; *Chem. Ber.*, 1981, **114**, 3019.
81JA4595	F. Mathey, F. Mercier and C. Charrier; *J. Am. Chem. Soc.*, 1981, **103**, 4595.
81JPC1202	H. Plato, W. Lubitz and K. Möbius; *J. Phys. Chem.*, 1981, **85**, 1202.
81LA919	G. Märkl, H. Baier and R. Lieb; *Liebigs Ann. Chem.*, 1981, 919.
81LA1139	H. Lehmkuhl, R. Paul and R. Mynott; *Liebigs Ann. Chem.*, 1981, 1139.
81NJC187	F. Nief, C. Charrier, F. Mathey and M. Simalty; *Nouv. J. Chim.*, 1981, **5**, 187.
81PS(10)285	K. Dimroth, M. Lückoff and J. Kaletsch; *Phosphorus Sulfur*, 1981, **10**, 285.

81PS(10)305 K. Dimroth, S. Berger and J. Kaletsch; *Phosphorus Sulfur*, 1981, **10**, 305.

81TL1207 G. Märkl, C. Martin and W. Weber; *Tetrahedron Lett.*, 1981, **22**, 1207.

81TL2889 G. Märkl, A. Merz and H. Rausch; *Tetrahedron Lett.*, 1981, 2889.

82ACR58 K. Dimroth; *Acc. Chem. Res.*, 1982, **15**, 58.

82AG(E)370 G. Märkl, G.Y. Jin and E. Silbereisen; *Angew. Chem., Int. Ed. Engl.*, 1982, **21**, 370.

82HOU(E1)72 G. Märkl; in "Methoden Org. Chem. (Houben-Weyl)", ed. M. Regitz, 1982, E1, 72.

82HOU(E1)783 K. Dimroth; in "Methoden Org. Chem. (Houben-Weyl)", ed. M. Regitz, 1982, E1, 783.

82JA425 P.D. Burrow, A.J. Ashe III, D.J. Belville and K.D. Jordan; *J. Am. Chem. Soc.*, 1982, **104**, 425.

82JOC2376 C. Charrier, H. Bonnard and F. Mathey; *J. Org. Chem.*, 1982, **47**, 2376.

82MIV2-01 G. Märkl; *Chem. Zeit*, 1982, **16**, 139.

82TL1565 J.-M. Alcaraz, A. Brèque and F. Mathey; *Tetrahedron Lett.*, 1982, **23**, 1565.

83CB445 G. Märkl, K. Hock and D. Matthes; *Chem. Ber.*, 1983, **116**, 445.

83JOM(247)271 K. Dimroth and H. Kaletsch; *J. Organomet. Chem.*, 1983, **247**, 271.

83PAC410 W.H. Powel; *Pure Appl. Chem.*, 1983, **55**(2), 409.

83TL1756 G. Märkl and K. Hock; *Chem. Ber.*, 1983, **116**, 1756.

83TL2645 G. Märkl and K. Hock; *Tetrahedron Lett.*, 1983, **24**, 2645.

83TL5051 G. Märkl and K. Hock; *Tetrahedron Lett.*, 1983, **24**, 5051.

84AG(E)894 G. Märkl and W. Burger; *Angew. Chem. Int. Ed. Engl.*, 1984, **23**, 894.

84CB763 G. Märkl, K. Hock and L. Merz; *Chem. Ber.*, 1984, **117**, 763.

84CC508 J.-M. Alcaraz and F. Mathey; *J. Chem. Soc., Chem. Commun.*, 1984, 508.

84CHEC-I(1)493 K. Dimroth; in "Comp. Heterocycl. Chem.", 1st ed., 1984, **1**, 493.

84IC3463 A. Brèque, C. Santini, F. Mathey, J. Fischer and A. Mitschler; *Inorg. Chem.*, 1984, **23**, 3463.

84JOC3157 N.S. Rao and L.D. Quin; *J. Org. Chem.*, 1984, **49**, 3157.

84JPC1981 L.L. Lohr, H.B. Schlegel and K. Morokuma; *J. Phys. Chem.*, 1984, **88**, 1981.

84PS(19)45 J.-M. Alcaraz, E. Deschamps and F. Mathey; *Phosphorus Sulfur*, 1984, **19**, 45.

84TL207 J.-M. Alcaraz and F. Mathey; *Tetrahedron Lett.*, 1984, **25**, 207.

84TL4659 J.-M. Alcaraz and F. Mathey; *Tetrahedron Lett.*, 1984, **25**, 4659.

85OM457 R.V. Hodges, J.L. Beauchamp, A.J. Ashe III and W.-T. Chan; *Organometallics*, 1985, **4**, 457.

86T89 C.W. Bird; *Tetrahedron*, 1986, **42**, 89.

86TL4299 P. Pellon, Y.Y.C. Yeung Lam Ko, P. Cosquer, J. Hamelin and R. Carrié; *Tetrahedron Lett.*, 1986, **27**, 4299.

86TL5611 P. Pellon and J. Hamelin; *Tetrahedron Lett.*, 1986, **27**, 5611.

86ZN(B)931 W. Rösch and M. Regitz; *Z. Naturforsch.*, 1986, **41b**, 931.

87AG(E)85 K. Blatter, W. Rösch, U.-J. Vogelbacher, J. Fink and M. Regitz; *Angew. Chem. Int. Ed. Engl.*, 1987, **26**, 85.

87AG(E)236 G. Märkl, H.J. Beckh, K.K. Mayer, M.L. Ziegler and T. Zahn; *Angew. Chem. Int. Ed. Engl.*, 1987, **26**, 236.

87AG(E)1134 G. Märkl, H.-J. Beckh, M.L. Ziegler and B. Nuber; *Angew. Chem. Int. Ed. Engl.*, 1987, **26**, 1134.

87BCJ1558 H. Tanaka and S. Motoki; *Bull. Chem. Soc. Jpn.*, 1987, **60**, 1558.

87CB819 G. Maas, J. Fink, H. Wingert, K. Blatter and M. Regitz; *Chem. Ber.*, 1987, **120**, 819.

87CB1245 K. Dimroth and A. Kaletsch; *Chem. Ber.*, 1987, **120**, 1245.

87CB1249 K. Dimroth and A. Kaletsch; *Chem. Ber.*, 1987, **120**, 1249.

87PS(30)479 U. Vogelbacher and O. Wagner; *Phosphorus Sulfur*, 1987, **30**, 479.

87TL1093 G. Märkl and G. Dorfmeister; *Tetrahedron Lett.*, 1987, **28**, 1093.

87TL3475 G. Märkl and H.-J. Beckh; *Tetrahedron Lett.*, 1987, **28**, 3475.

88AG(E)713 K.H. Dötz, A. Tiriliomis, K. Harms, M. Regitz and U. Annen; *Angew. Chem. Int. Ed. Engl.*, 1988, **27**, 713.

88AG(E)1541 M. Regitz and P. Binger; *Angew. Chem. Int. Ed. Engl.*, 1988, **27**, 1541.

88CC493 D.G. Holah, A.N. Hugues and K.L. Knudsen; *J. Chem. Soc., Chem. Commun.*, 1988, 493.

88JA4204 K.K. Baldridge and M.S. Gordon; *J. Am. Chem. Soc.*, 1988, **110**, 4204.

88JHC155 D.G. Holah, A.N. Hughes, K.L. Knudsen and R. Perrier; *J. Heterocycl. Chem.*, 1988, **25**, 155.

89H1135 M.J.S. Dewar and A. Holder; *Heterocycles*, 1989, **28**, 1135.

89MIV2-01 S.M. Bachrach; *J. Comput. Chem.*, 1989, **10**, 392.

89OM2804 J. Waluk, H.-P. Klein, A.J. Ashe III and J. Michl; *Organometallics*, 1989, **8**, 2804.

89TL817 P. Le Floch and F. Mathey; *Tetrahedron Lett.*, 1989, **30**, 817.

89TL5245 G. Märkl and A. Kallmünzer; *Tetrahedron Lett.*, 1989, **30**, 5245.

90BSB4849 G. Märkl, K. Hohenwarter, M.L. Ziegler and B. Nuber; *Tetrahedron Lett.*, 1990, **31**, 4849.

90CB935 U. Annen, M. Regitz and H. Kluge; *Chem. Ber.*, 1990, **123**, 935.

90HAC37 S. Holand, J.-M. Alcaraz, L. Ricard and F. Mathey; *Heteroatom Chem.*, 1990, **1**, 37.

B-90MIV2-01 G. Märkl; in "Multiple Bonds and Low Coordination in Phosphorus Chemistry", eds. M. Regitz and O.J. Scherer, Georg Thieme Verlag, New York, 1990, p. 220.

B-90MIV2-02 M. Regitz; in "Multiple Bonds and Low Coordination in Phosphorus Chemistry", eds. M. Regitz and O.J. Scherer, Georg Thieme Verlag, New York, 1990, p. 58.

90MIV2-01 C.W. Bock, M. Trachtman and P. George; *Struct. Chem.*, 1990, **1**, 345.

90POL991 P. Le Floch, L. Ricard and F. Mathey; *Polyhedron*, 1990, **9**, 991.

90T5697 C.W. Bird; *Tetrahedron*, 1990, **46**, 5697.

90TL4589 G. Märkl, C. Dörges, T. Riedl, F.-G. Klärner and C. Lodwig; *Tetrahedron Lett.*, 1990, **31**, 4589.

90TL4849	G. Märkl, K. Hohenwarter, M.L. Ziegler and B. Nuber; *Tetrahedron Lett.*, 1990, **31**, 4849.
91AG(E)106	G. Märkl and C. Dörges; *Angew. Chem. Int. Ed. Engl.*, 1991, **30**, 106.
91AG(E)547	C. Elschenbroich, M. Nowotny, B. Metz, W. Massa, J. Graulich, K. Biehler and W. Sauer; *Angew. Chem. Int. Ed. Engl.*, 1991, **30**, 547.
91IC4693	B. Schmid, L.M. Venanzi, A. Albinati and F. Mathey; *Inorg. Chem.*, 1991, **30**, 4693.
92ICA(199-200)437	D. Carmichael, P. Le Floch, L. Ricard and F. Mathey; *Inorganica Chim. Acta*, 1992, **199–200**, 437.
91JA667	P. Le Floch, D. Carmichael, L. Ricard and F. Mathey; *J. Am. Chem. Soc.*, 1991, **113**, 667.
91JOC4031	S. Holand, L. Ricard and F. Mathey; *J. Org. Chem.*, 1991, **56**, 4031.
91HAC439	M. Abbari, Y.Y.C. Yeung Lam Ko and R. Carrié; *Heteroatom Chem.*, 1991, **2**, 439.
91OM2432	P. Le Floch, D. Carmichael and F. Mathey; *Organometallics*, 1991, **10**, 2432.
92AG(E)1343	C. Elschenbroich, M. Nowotny, A. Behrendt, W. Massa and S. Wocaldo; *Angew. Chem. Int. Ed. Engl.*, 1992, **31**, 1343.
92BSB609	H.T. Teunissen and F. Bickelhaupt; *Bull. Soc. Chim. Belg.*, 1992, **101**, 609.
92BSF291	P. Le Floch, D. Carmichael and F. Mathey; *Bull. Soc. Chim. Fr.*, 1992, **129**, 291.
92IC5117	B. Schmid, L.M. Venanzi, T. Gerfin, V. Gramlich and F. Mathey; *Inorg. Chem.*, 1992, **31**, 5117.
92JA9080	L. Nyulaszi, T. Veszpremi, J. Réffy, B. Burkhardt and M. Regitz; *J. Am. Chem. Soc.*, 1992, **114**, 9080.
92MIV2-01	F. Mathey; *Rev. Heteroatom Chem.*, 1992, **6**, 1.
92OM2475	P. Le Floch, D. Carmichael, L. Ricard, F. Mathey, A. Jutand and C. Amatore; *Organometallics*, 1992, **11**, 2475.
92TL1597	G. Märkl, A. Kallmünzer, H. Nöth and K. Pohlmann; *Tetrahedron Lett.*, 1992, **33**, 1597.
92TL3537	H.T. Teunissen and F. Bickelhaupt; *Tetrahedron Lett.*, 1992, **33**, 3537.
93CC789	P. Le Floch, L. Ricard and F. Mathey; *J. Chem. Soc., Chem. Commun.*, 1993, 789.
93CC1295	P. Le Floch and F. Mathey; *J. Chem. Soc. Chem., Commun.*, 1993, 1295.
93IJQ343	A.B. Pierini and J.S. Duca Jr.; *Int. J. Quant. Chem.*, 1993, **48**, 343.
93JA10665	P. Le Floch, D. Carmichael, L. Ricard and F. Mathey; *J. Am. Chem. Soc.*, 1993, **115**, 10665.
93JOC977	G. Keglevich, K. Ujszaszi, A. Kovacs and L. Töke; *J. Org. Chem.*, 1993, **58**, 977.
93JOM(459)157	C. Elschenbroich, M. Nowotny, J. Kroker, A. Behrendt, W. Massa and S. Wocaldo; *J. Organomet. Chem.*, 1993, **459**, 157.
93JPC4011	L. Nyulaszi, T. Veszprémi and J. Réffy; *J. Phys. Chem.*, 1993, **97**, 4011.
93OM3373	C. Elschenbroich, F. Bär, E. Bilger, D. Mahrwald, M. Nowotny and B. Metz; *Organometallics*, 1993, **12**, 3373.
93OM5005	J.A. Chamizo, J. Morgado and P. Sosa; *Organometallics*, 1993, **12**, 5005.
93PAC621	F. Bickelhaupt; *Pure Appl. Chem.*, 1993, **65**, 621.
93PS(76)33	P. Le Floch, D. Carmichael and F. Mathey; *Phosphorus, Sulfur Silicon*, 1993, **76**, 33.
93PS(76)75	H.T. Teunissen and F. Bickelhaupt; *Phosphorus, Sulfur Silicon*, 1993, **76**, 75.
93PS(77)105	U. Fleisher and W. Kutzelnigg; *Phosphorus, Sulfur Silicon*, 1993, **77**, 105.
93PS(77)255	D. Carmichael, P. le Floch and F. Mathey; *Phosphorus, Sulfur Silicon*, 1993, **77**, 255.
93T5577	K.H. Dötz, A. Tiriliomis and K. Harms; *Tetrahedron*, 1993, **49**, 5577.
93TL3107	G. Märkl, H. Sommer and H. Nöth; *Tetrahedron Lett.*, 1993, **34**, 3107.
94BCJ2785	I. Shinoda, A. Takahashi, T. Saito and T. Uchida; *Bull. Chem. Soc. Jpn.*, 1994, **67**, 2785.
94BSF330	P. Le Floch, L. Ricard and F. Mathey; *Bull. Soc. Chim. Fr.*, 1994, **131**, 330.
94CC2065	P. Le Floch, A. Kolb and F. Mathey; *J. Chem. Soc., Chem. Commun.*, 1994, 2065.
94HAC131	L. Nyulaszi and G. Keglevich; *Heteroatom Chem.*, 1994, **5**, 131.
94JA6217	C. Elschenbroich, M. Nowotny, A. Behrendt, K. Harms, S. Wocaldo and J. Pebler; *J. Am. Chem. Soc.*, 1994, **116**, 6217.
94MP557	E.F. Archibong and A.J. Thakkar; *Mol. Phys.*, 1994, **81**, 557.
95BSF384	H.-G. Trauner, E. de la Cuesta, A. Marinetti and F. Mathey; *Bull. Soc. Chim. Fr.*, 1995, **132**, 384.
95BSF910	K. Waschbüsch, P. Le Floch and F. Mathey; *Bull. Soc. Chim. Fr.*, 1995, **132**, 910.
95JOC7439	H.T. Teunissen, J. Hollebeek, P.J. Nieuwenhuizen, B.L.M. van Baar, F.J.J. de Kanter and F. Bickelhaupt; *J. Org. Chem.*, 1995, **60**, 7439.
95IC11	P. Le Floch, L. Ricard, F. Mathey, A. Jutand and C. Amatore; *Inorg. Chem.*, 1995, **34**, 11.
95IC5070	P. le Floch, N. Maigrot, L. Ricard, C. Charrier and F. Mathey; *Inorg. Chem.*, 1995, **34**, 5070.
95JST(338)51	D.J. Berger, P.P. Gaspar and J.F. Liebman; *J. Mol. Struct. (Theochem)*, 1995, **338**, 51.
95JST(347)57	L. Nyulaszi and T. Vespremi; *J. Mol. Struct. (Theochem)*, 1995, **347**, 57.
95S717	H.-G. Trauner, P. Le Floch, J.-M. Lefour, L. Ricard and F. Mathey; *Synthesis*, 1995, 717.
95TCA(92)67	J. Lorentzon, M.P. Fülscher and B.O. Ross; *Theor. Chim. Acta*, 1995, **92**, 67.
95TL3839	G. Märkl and S. Dorsch; *Tetrahedron Lett.*, 1995, **36**, 3839.
B-96MIV2-01	P. Le Floch and F. Mathey; in "Synthetic Methods of Organometallic and Inorganic Chemistry, Hermann/Brauer", ed. W.A. Hermann, Georg Thieme Verlag, New York, 1996, **Vol. 3**, p. 8.
B-96MIV2-02	P. Le Floch, D. Carmichael and F. Mathey; in "Synthetic Methods of Organometallic and Inorganic Chemistry, Hermann/Brauer", ed. W.A. Hermann, Georg Thieme Verlag, New York, 1996, **Vol. 3**, p. 167.
96BSF691	P. Le Floch, L. Ricard and F. Mathey; *Bull. Soc. Chim. Fr.*, 1996, **133**, 691.
96CB263	F. Mathey and P. Le Floch; *Chem. Ber.*, 1996, **129**, 263.
96CC971	D. Carmichael, P. Le Floch, H.-G. Trauner and F. Mathey; *J. Chem. Soc., Chem. Commun.*, 1996, 971.
96CC2071	B. Breit; *J. Chem. Soc., Chem. Commun.*, 1996, 2071.

96CHEC-II(5)639	G.G. Hewitt; in "Comprehensive Heterocyclic Chemistry II", eds. A.R. Katritzky, C.W. Rees and E.F.V. Scriven, Pergamon, New York, 1996, **Vol. 5**, p. 639.
96HAC307	D.B. Chesnut and E.F.C. Bird; *Heteroatom Chem.*, 1996, **7**, 307.
96HAC397	N. Avarvari, P. Le Floch, C. Charrier and F. Mathey; *Heteroatom Chem.*, 1996, **7**, 397.
96JA11978	N. Avarvari, P. Le Floch and F. Mathey; *J. Am. Chem. Soc.*, 1996, **118**, 11978.
96JOM(520)211	B. Wrackmeyer and U. Klaus; *J. Organomet. Chem.*, 1996, **520**, 211.
96JPC6456	L. Nyulaszi and T. Veszpremi; *J. Phys. Chem.*, 1996, **100**, 6456.
96OM794	H.T. Teunissen and F. Bickelhaupt; *Organometallics*, 1996, **15**, 794.
96OM802	H.T. Teunissen and F. Bickelhaupt; *Organometallics*, 1996, **15**, 802.
96OM1597	K. Waschbüsch, P. Le Floch and F. Mathey; *Organometallics*, 1996, **15**, 1597.
96OM2713	F. Knoch, F. Kremer, U. Schmidt, U. Zenneck, P. Le Floch and F. Mathey; *Organometallics*, 1996, **15**, 2713.
96OM3267	P. Le Floch, S. Mansuy, L. Ricard, F. Mathey, A. Jutand and C. Amatore; *Organometallics*, 1996, **15**, 3267.
96PS(109)173	D. Böhm, H. Geiger, F. Knoch, S. Kummer, P. Le Floch, F. Mathey, U. Schmidt and U. Zenneck; *Phosphorus, Sulfur Silicon*, 1996, **109**, 173.
96PS(109)461	A.-C. Gaumont, J.-F. Pilard and J.-M. Denis; *Phosphorus, Sulfur Silicon*, 1996, **109**, 461.
96ZN931	G. Märkl and F. Kneidl; *Angew. Chem. Int. Ed. Engl.*, 1973, **12**, 931.
97CB843	K. Waschbüsch, P. Le Floch, L. Ricard and F. Mathey; *Chem. Ber.*, 1997, **130**, 843.
97JA9417	P. Rosa, P. Le Floch, L. Ricard and F. Mathey; *J. Am. Chem. Soc.*, 1997, **119**, 9417.
97JCS(P1)2681	B. Breit, R. Winde and K. Harms; *J. Chem. Soc., Perkin Trans.*, 1997, 2681.
97JOM(541)277	N. Mézailles, P. Le Floch, K. Waschbüsch, L. Ricard, F. Mathey and C.P. Kubiak; *J. Organomet. Chem.*, 1997, **541**, 277.
97OM4089	N. Avarvari, P. Le Floch, L. Ricard and F. Mathey; *Organometallics*, 1997, **16**, 4089.
97MIV2-01	C.-S. Lin, J. Li and C.-W. Liu; *Chin. J. Chem.*, 1997, **15**, 289.
97MRC384	F. Gerson, P. Merstetter, S. Pfenninger and G. Märkl; *Magn. Res. Chem.*, 1997, **35**, 384.
98CCR(179-180)771	P. Le Floch and F. Mathey; *Coord. Chem. Rev.*, 1998, **179–180**, 771.
98IC3154	P. Rosa, L. Ricard, P. Le Floch, F. Mathey, G. Sini and O. Eisenstein; *Inorg. Chem.*, 1998, **37**, 3154.
98IC5313	N. Mézailles, N. Avarvari, L. Ricard, F. Mathey and P. Le Floch; *Inorg. Chem.*, 1998, **37**, 5313.
98JOC4826	P. Rosa, N. Mézailles, F. Mathey and P. Le Floch; *J. Org. Chem.*, 1998, **63**, 4826.
98JOM(567)151	N. Avarvari, P. Rosa, F. Mathey and P. Le Floch; *J. Organomet. Chem.*, 1998, **567**, 151.
B-98MIV2-01	K.B. Dillon, F. Mathey and J.F. Nixon; in "Phosphorus: The Carbon Copy", eds. K.B. Dillon, F. Mathey and J.F. Nixon, Wiley, Chichester, 1998, p. 235.
98MIV2-02	P. Le Floch, F. Knoch, F. Kremer, F. Mathey, J. Scholz, W. Scholz, K.-H. Thiele and U. Zenneck; *Eur. J. Inorg. Chem.*, 1998, 119.
98OM4417	C. Elschenbroich, S. Voss, O. Schiemann, A. Lippek, K. Harms; *Organometallics*, 1998, **17**, 4417.
98SCI1587	N. Avarvari, N. Mézailles, L. Ricard, P. le Floch and F. Mathey; *Science*, 1998, **8**, 1587.
99GEP19621967	B. Breit, R. Paciello, B. Geisler and M. Röper (BASF A.-G.); *Ger. Pat.* 19621967, (1999) (*Chem. Abstr.*, 1999, 128:129496).
99GEP19743197	R. Paciello, E. Zeller, B. Breit and M. Röper (BASF A.-G.); *Ger. Pat.* 19743197, (1999) (*Chem. Abstr.*, 1999, 130:129496).
99JOC5524	G. Frison, N. Avarvari, A. Sevin, F. Mathey and P. Le Floch; *J. Org. Chem.*, 1999, **64**, 5524.
99MIV2-01	N. Mézailles, L. Ricard, F. Mathey and P. Le Floch; *Eur. J. Inorg. Chem.*, 1999, 2233.
99MIV2-02	N. Avarvari, N. Mézailles, N. Maigrot, L. Ricard, F. Mathey and P. Le Floch; *Chem. Eur. J.*, 1999, **5**, 2109.
99PJC(73)135	M. Regitz and U. Bergsträsser; *Pol. J. Chem.*, 1999, **73**, 135.
99OM3348	P. Rosa, L. Ricard, F. Mathey and P. Le Floch; *Organometallics*, 1999, **18**, 3348.
99OM4205	X. Sava, N. Mézailles, N. Maigrot, F. Nief, L. Ricard, F. Mathey and P. Le Floch; *Organometallics*, 1999, **18**, 4205.

5.3
Six-membered Rings with Two or More Heteroatoms with at least One Phosphorus Atom

GOTTFRIED MAERKL and PETER KREITMEIER
University of Regensburg, Regensburg, Germany

5.3.1 INTRODUCTION

A. Michaelis (1847–1916) and A.E. Arbuzow (1877–1968) in the last decades of the nineteenth century and the first few decades of the twentieth century, laid the foundation of phosphorus chemistry, which soon acquired a reputation for evil smells and, later, for drastic — often lethal — physiological effects.

The first comprehensive compilation of organic phosphorus compounds was that by Kosolapoff ⟨B-50MI625-01⟩. "Kosolapoff" was the standard book in the field for many years, and the second edition, "Kosolapoff–Maier" ⟨B-72MI625-01⟩, had expanded to eight volumes, indicating the enormous development in phosphorus chemistry in the period since 1950.

F.G. Mann, working in the field since 1935, is undoubtedly the father of phosphorus as well as arsenic heterocyclic chemistry. Most of his work, mainly on saturated and benzo-annelated ring systems, is summarized in his monograph ⟨B-70MI625-01⟩. Other important early contributions to the synthesis of phosphorus heterocycles were made by K. Issleib and by B.A. Arbuzov, the son of A.E. Arbuzov.

After the first synthesis of a λ^3-phosphinin derivative, the 2,4,6-triphenyl compound (**A**) by G. Maerkl ⟨66AG(E)846⟩, and the isolation of the unsubstituted phosphinin (**B**) by A. Ashe III ⟨71JA3293⟩, a dramatic development and an avalanche of publications followed.

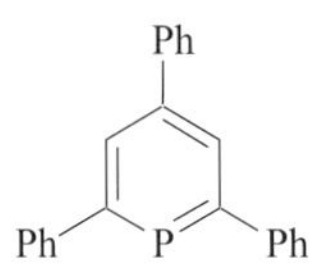

(**A**)

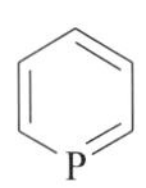

(**B**)

The question of the possible existence of λ^3-phosphinines containing more than one heteroatom, e.g. N, P, with a Hückel aromatic character, has provided the impetus for much experimental and theoretical activity in the field since 1966. Early reviews, by Berlin and Hellwege ⟨B-69MI625-01⟩ and Quin ⟨B-81MI625-01⟩ took into account these new developments, but they dealt only with rings containing P and C. The most comprehensive survey of heterocyclic phosphorus chemistry is that in Houben-Weyl ⟨63HOU(12)203, 82HOU(E1/2) ⟩. The six-membered heterocycles are treated by Maerkl (λ^3-phosphorus) ⟨82HOU(E1)72⟩ and Dimroth (λ^5-phosphorus) ⟨82HOU(E1)783⟩. In the first edition of *Comprehensive Heterocyclic Chemistry*, the entire field of phosphorus heterocycles was covered by Dimroth ⟨82CHEC-I(1)493⟩. More recent work was reviewed in a 1990 monograph ⟨B-90MI625-01⟩.

The 1,4-diphosphinin (**277**) was described as early as 1976 ⟨76TL3715⟩, but it took another 29 years for the first 1,3-diphosphinines (**218**) to be obtained ⟨95AG(E)81⟩ (Chapter 5.3.6.2). The key building block for the synthesis of (**218**) was *tert*-butylphosphaethyne ⟨81ZN(B)16⟩, the preparation of which was certainly another milestone in phosphorus heterocyclic chemistry, as it also enabled the synthesis of 2,4,6-tri-*tert*-butyl-1,3,5-triphosphinin (**354**) in 1995.

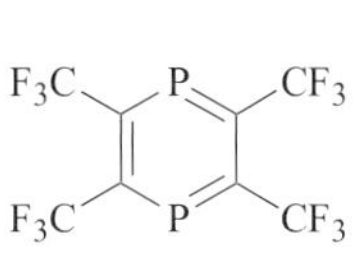

(**277**)

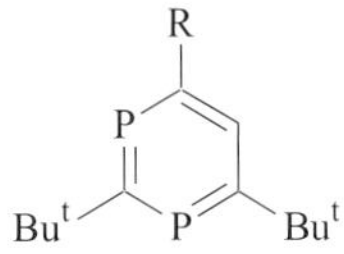

(**218**)

Bu^t—C≡P

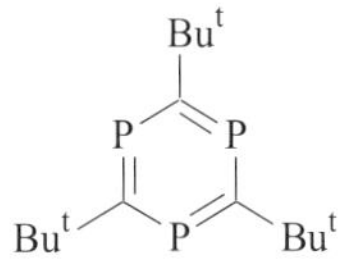

(**354**)

The behavior of phosphaethynes, mainly *tert*-butylphosphaethynes in the coordination sphere of transition metals open up an entirely new field of phosphorus and polyphosphorus heterocycles with unusual structures ⟨88AG(E)1485, 88CRV1327⟩.

Aromatic six-membered phosphorus heterocycles containing nitrogen include derivatives of all the three isomeric azaphosphinines (**C**)–(**E**) and the diazaphosphinines, e.g. (**F**) and (**G**). They are 6π-aromatic Hückel systems, although the 1,2- and the 1,4-azaphosphinines are highly reactive towards nucleophiles as a result of the N/P electronegativity difference.

(**C**)

(**D**)

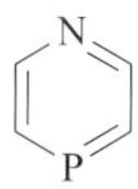

(**E**)

(**F**)

(**G**)

Also in this case transition metal complexes, e.g. like (**110**), are valuable intermediates for the general synthesis of 1,3,5-diazaphosphinines (**111**) in excellent yields ⟨96JA11978⟩.

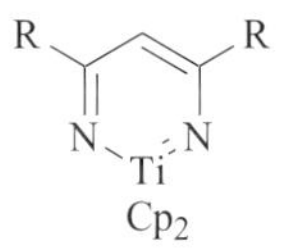

(**110**)

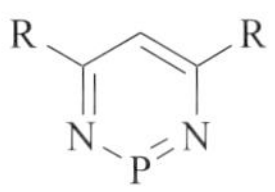

(**111**)

Cycloaddition reactions with R–C≡P also open several approaches to azaphosphinines.

Another large chapter of six-membered phosphorus heterocycles with more than one heteroatom is represented by those with $\lambda^5\sigma^4$-phosphorus. The first synthesis of a λ^5-phosphinin, the 1,1-diphenyl compound (**H**) was described already in 1963 by Maerkl ⟨63AG(E)479⟩. Because of the ready availability of λ^3-phosphinines, and the realization that these can be transformed into λ^5-phosphinines via anionic, cationic and radical intermediates, the chemistry of λ^5-phosphinines has expanded rapidly ⟨82HOU(E1)783⟩.

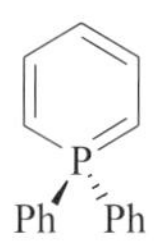

(**H**)

Several independent synthetic approaches to 1,2λ^5- and 1,4λ^5-azaphosphinines (**I**)–(**K**), as well as 1λ^5,3λ^5-diphosphinines (**L**) and 1λ^5,3λ^5,5λ^5-triphosphinines (**M**), have been published. Synthetic work on polyhalogenated aza-λ^5-phosphinines, e.g. (**N**), has been carried out, mainly by Russian research groups ⟨66ZOB735, 70ZAAC(372)233⟩.

The chemistry of λ^5-phosphinines culminates in the six-membered cyclotriphosphazenes, for instance the so-called phosphonitrilic halides (**O**), which are not discussed in this review, except in connection with the polyheteroatom systems to which they may give rise.

Although a detailed review on λ^3-phosphinines appears separately, it is not possible to discuss the chemistry of the systems with more than one heteroatom without occasional reference to the monoheteroatom systems.

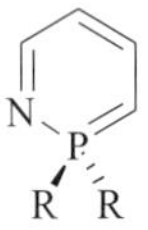
(I)

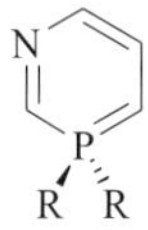
(J)

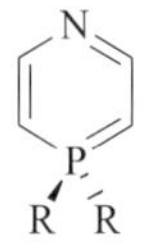
(K)

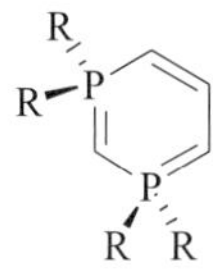
(L)

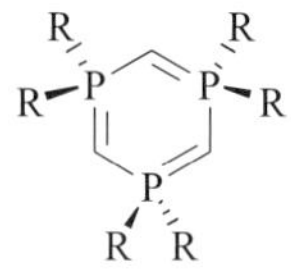
(M)

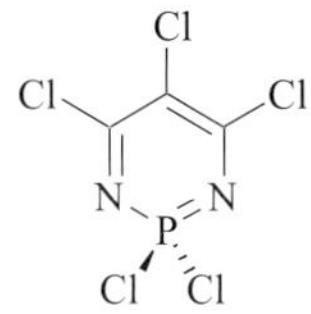
(N)

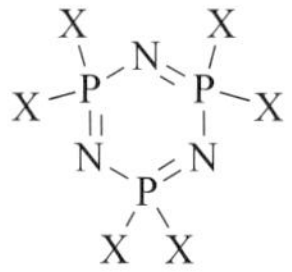
(O)

5.3.2 NOMENCLATURE

Nomenclature in phosphorus chemistry is in a somewhat confused state, and for this reason information retrieval is not a straightforward matter. The saturated heterocyclic systems, particularly the oxygen-containing compounds, are for the most part cyclic phosphonate, phosphate or phosphite esters or amides, and the names may be formed from the open-chain structures from which they are derived. In this chapter we follow the IUPAC recommendations of 1983 ⟨83PAC410⟩. The fully unsaturated six-membered ring (**A**), with phosphorus in its normal (trivalent) state, is therefore named "phosphinine". When a ring contains "hypervalent" phosphorus, this is denoted using the λ convention: (**B**) is λ^5-phosphinine. Optionally, λ^3 may be used, when it is wished to emphasize that it is the P(III) valence state of phosphorus coordination number 2 that is being discussed, but in compound names this is understood unless specified otherwise. *Chemical Abstracts* perseveres with the rules which applied before 1983 ⟨B-69MI625-02⟩, whereby the ring in (**A**) is named as "phosphorin", and derivatives of (**B**) are called 1,1-dihydrophosphorins. The saturated ring (**C**) is "phosphinane" (formerly phosphorinane). "Replacement nomenclature" is also used widely in the primary literature: (**A**) by this method is called "phosphabenzene".

(A)

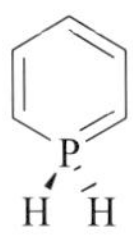
(B)

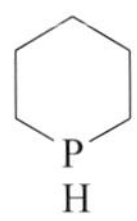
(C)

The ring systems described in the present chapter contain other heteroatoms, and the normal Hantzsch–Widman rules are applied. Compounds like (**111**) (see Introduction) are therefore "diazaphosphinines". The order of precedence in heteroatoms numbering should be noted: oxygen > sulfur > nitrogen > phosphorus. In this chapter we normally place the P-atom at the bottom of the ring, but this is seldom position 1, when other heteroatoms are present.

The three tricyclic ring systems, e.g. (**D**)–(**F**), are usually trivially named, as indicated below the structures.

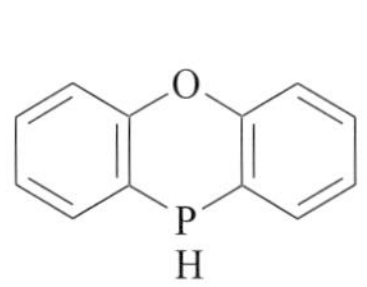
Phenoxaphosphine
(D)

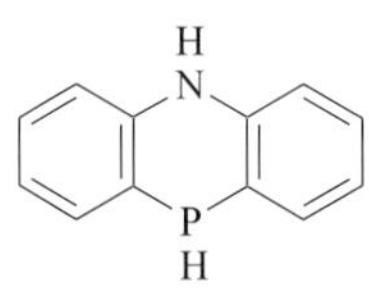
5,10-Dihydrophenophosphazine
(E)

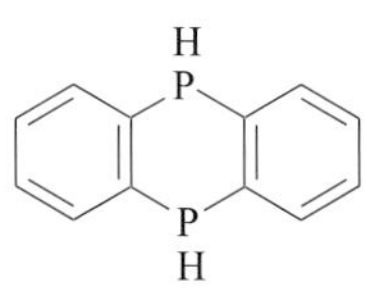
5,10-Dihydrophosphanthrene
(F)

5.3.3 RING SYSTEMS WITH ONE PHOSPHORUS AND ONE NITROGEN ATOM

All of the fully conjugated six-membered ring systems with one nitrogen and one phosphorus atom — the 1,2-, 1,3- and 1,4-azaphosphinines — are known, and all may have λ^3- or λ^5-phosphorus. The 1,3λ^3-azaphosphinines are quite stable while the 1,2λ^3- and 1,4λ^3-azaphosphinines are — because of the electronegativity difference between N and P — are highly reactive and add even weak nucleophiles.

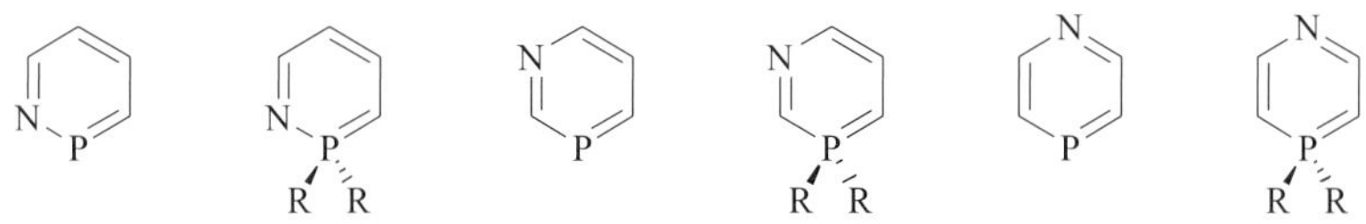

(I) (II) (III) (IV) (V) (VI)

5.3.3.1 1,2-Azaphosphinines

The first example of a fully unsaturated 1,2λ^3-azaphosphinine was reported in 1987 by Foucaud and co-workers ⟨87TL4673⟩. Two moles of an aliphatic imine (**1**) with a bulky substituent R react with one mole of dichlorophenylphosphine to give mainly the 1,2-dihydro-1,2-azaphosphinines (**2**) ⟨86TL4725, 90T6715⟩. Pyrolysis results in loss of benzene and an alkene, and formation of the azaphosphinine (**3**). Nucleophiles add readily to compound (**3**), reforming the 1,2-dihydro system, with λ^5-phosphorus in the case of the water adduct (**6**) ⟨87TL4673⟩. The dihydro-azaphosphinines (**2**) are converted into the 2-oxides (**4**), (O_2) and 2-sulfides (**5**) (S_8/*N*-methylimidazole) (Scheme 1) ⟨90T6715⟩.

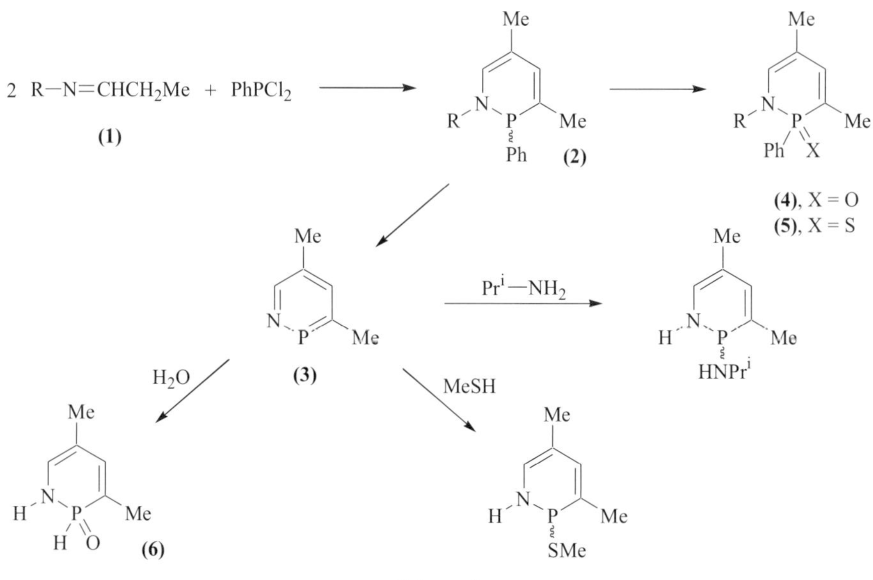

Scheme 1

The excellent and unbelievable simple synthesis of 1,3,2-diazaphosphinines (**7**) by reaction of the easily available 1,3,2-diazatitana-cyclohexa-3,6-dienes with PCl_3 by Mathey *et al.* was the beginning of an explosive development of the synthesis and chemistry of 1,2-azaphosphinines (**8**) (Scheme 2) ⟨96JA11978⟩. The 1,2-azaphosphinines (**8**) are prepared by reaction of (**7**) with various alkynes.

Smooth [4 + 2]-cycloaddition reactions of ^{t}BuC≡P with the 1,2-azaphosphinines (**8**) proceed already at room temperature to yield the corresponding azadiphosphabarrelenes (**9**) (Scheme 2).

Pyrolysis of (**9c**), R^1 = $SiMe_3$, R^2 = H furnishes 2-*tert*-butyl-4-(trimethylsilyl)-1,3-diphosphinine (**10**), pyrolysis of the other barrelenes leads exclusively back to the starting 1,2-azaphosphinines ⟨98EJOC2039⟩.

This elegant synthesis of 1,2-azaphosphinines was expanded by reaction of (**7**) with further alkynes furnishing a whole series of new 1,2-azaphosphinines (**8**) (Scheme 3) ⟨97OM4089⟩.

2,5-Bis(1-propynyldimethylsilyl)-3,4-dimethyl-1R-phospholes (**11**) react with 2 mole 4,6-di-*tert*-butyl-1,3,2-diazaphosphinine (**7**) to give the bis-Diels–Alder adducts, which eliminate 2 mole ^{t}BuC≡N in a retro-Diels–Alder reaction to give (**12**) with two 1,2-azaphosphinine units (Scheme 4) ⟨99EJOC2039⟩.

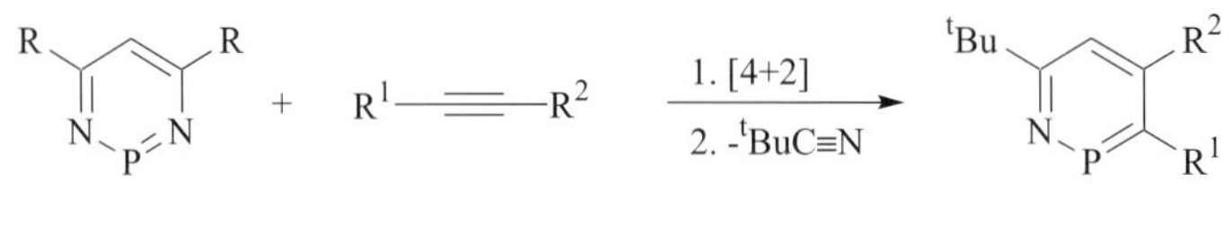

(7), R = ^{t}Bu, Ph **(8)**

(a) $R^1 = R^2 = SiMe_3$, **(b)** $R^1 = R^2 = Et$, **(c)** $R^1 = SiMe_3$, $R^2 = H$,
(d) $R^1 = SiMe_3$, R^2 = ferrocenyl <96JA11978> <98EJOC2039>.

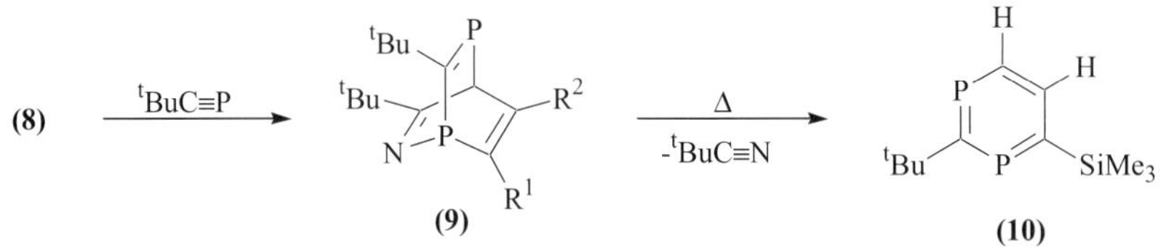

(a) $R^1 = R^2 = SiMe_3$, **(b)** $R^1 = R^2 = Et$, **(c)** $R^1 = SiMe_3$, $R^2 = H$, **(d)** $R^1 = SiMe_3$, R^2 = ferrocenyl

Scheme 2

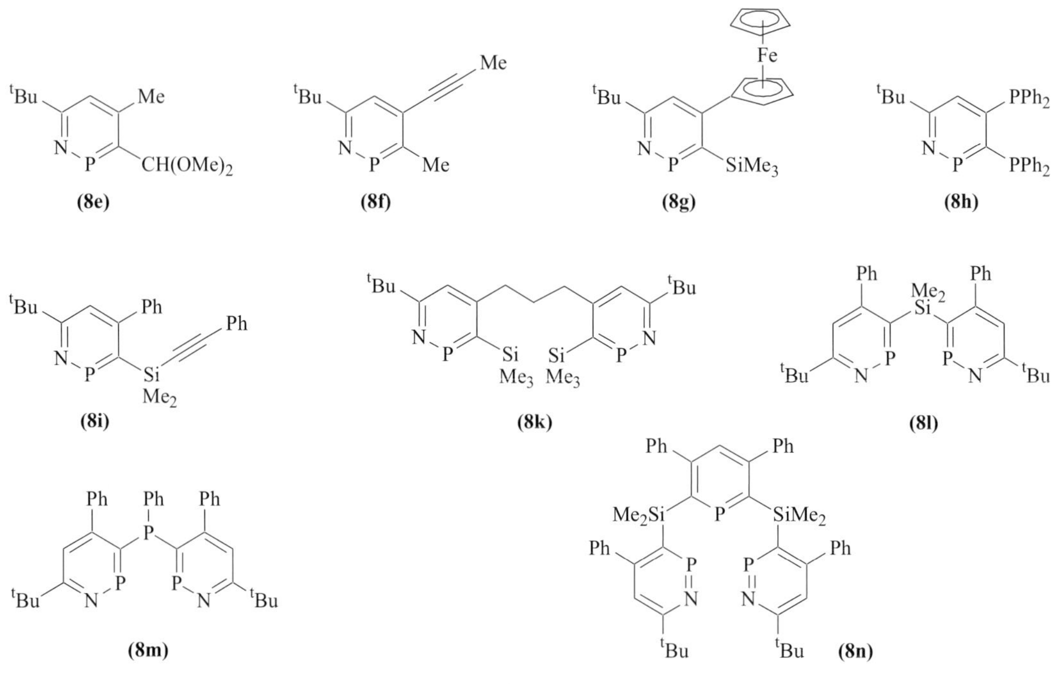

Scheme 3

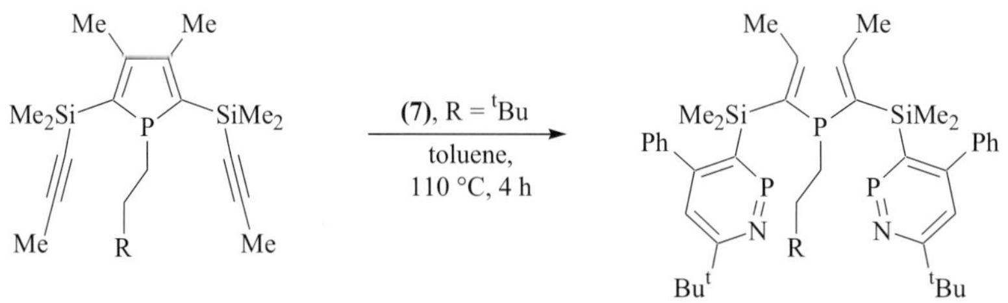

(11) **(12)**, R = Cl, CN, COOEt

Scheme 4

The 1,2-benzazaphosphinine system (**16**) arose in an unusual rearrangement of the phosphirenes (**14**) ⟨92TL2335⟩, which are formed by cycloaddition of the phosphanetriylammonium ion (**13**) ⟨88AG(E)1715⟩ to phenyl- or diphenyl-acetylene (Scheme 5). Triethylphosphine and 2,3-dimethylbutadiene were found to add to the tetrachloroaluminate salt (**15**) at the P-atom, but, apart from spectroscopic data, no chemistry of (**16**) is yet reported.

The synthesis of the 5,6-dihydrodibenz[*c,e*][1,2]azaphosphinines (9,10-dihydro-9-aza-10-phosphaphenanthrenes (**17**) was published in 1960 ⟨60JA5685, 60JCS5034⟩; for an account of their chemistry the reviews cited should be consulted.

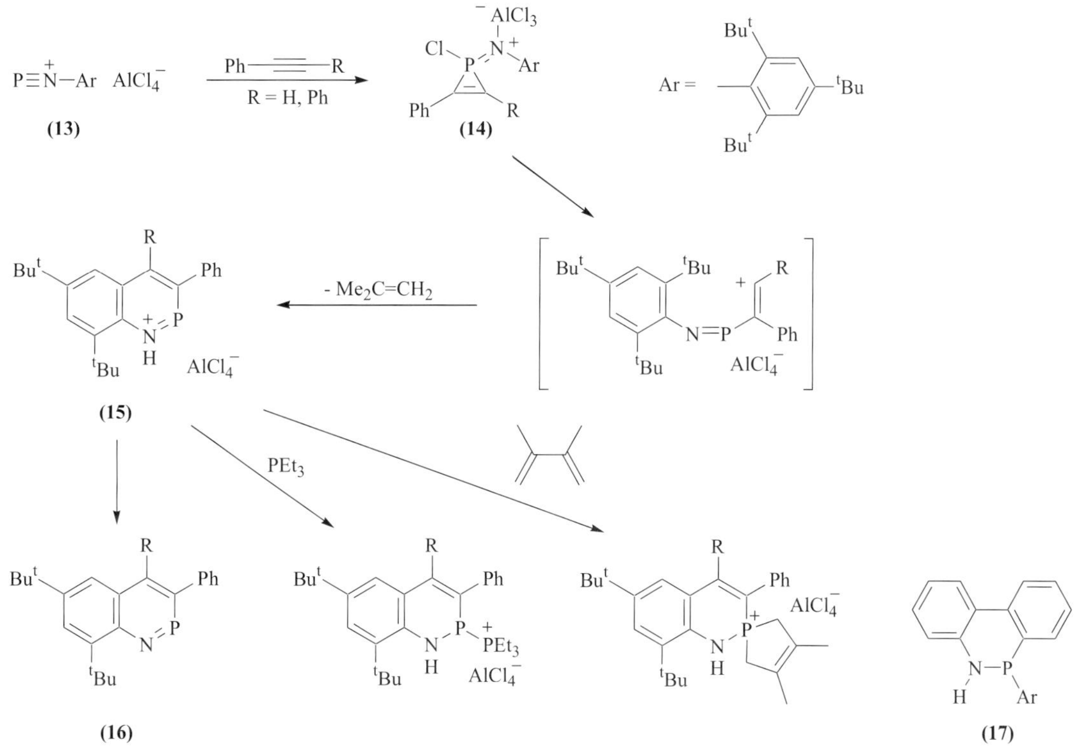

Scheme 5

Several methods for the synthesis of 1,2λ^5-azaphosphinines are described in the literature.

(i) Thermolysis of the methiodides (**18**) of 1-*t*-butyl-1,2-dihydro-1,2-azaphosphinines, which are available from aliphatic imines as shown in Scheme 6 ⟨86TL4725, 90T6715⟩. Basification forms (**19**) (Scheme 6). On exposure to moisture and air they are degraded to the phosphine oxides (**20**) ⟨91TL2619⟩. The full paper appeared in 1995 ⟨95T9625⟩.

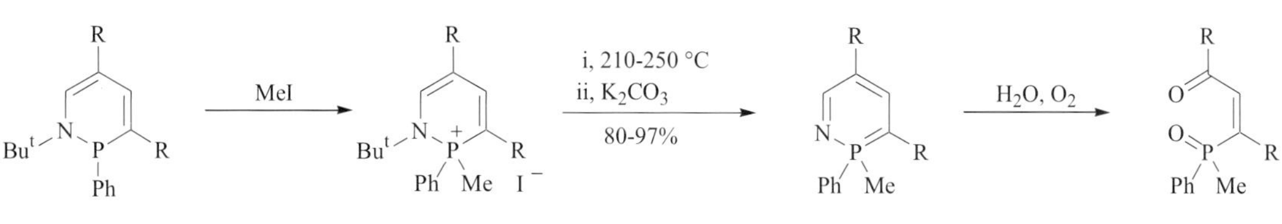

Scheme 6

(ii) *N*-Vinyliminotrimethoxyphosphorane (**21**), easily prepared by the Staudinger reaction of α-azidostyrene with trimethyl phosphite at room temperature, reacts with electron-deficient acetylenes in a formal (4 + 2)-cycloaddition reaction to give the cyclic phosphoranes (**22**). The 1,2λ^5-azaphosphinines (**23**) are formed by the spontaneous elimination of methanol (Scheme 7) ⟨85CL1459⟩.

In the same manner methoxy diphenyl (1-phenylvinylimino)phosphorane (**24**) react with electron-deficient acetylenes to give the 1,2λ^5-azaphosphinines (**25**) ⟨98H1903⟩.

The 1,1-dimethoxy-1,2λ^5-azaphosphinines (**23**) rearrange at 180°C by 1,2-methylmigration to give the *N*-methyl-derivatives 1,2-λ^5-azaphosphinines (**26**), (**23**) as well as (**25**) react with dimethylacetylendicarboxylate and form via the azaphosphabarrelenes (**27**) the λ^5-phosphinines (**28**) (Scheme 8) ⟨98H1903⟩.

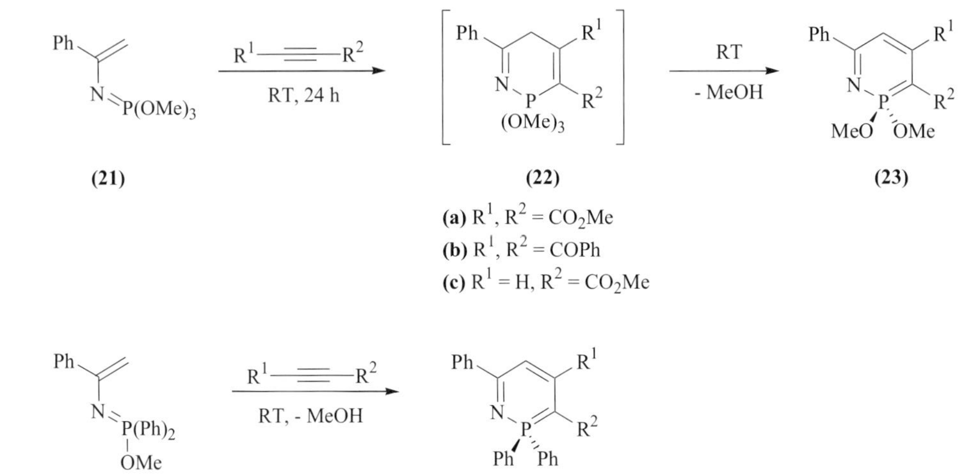

(21) (22) (23)

(a) $R^1, R^2 = CO_2Me$
(b) $R^1, R^2 = COPh$
(c) $R^1 = H, R^2 = CO_2Me$

(24) (25)

Scheme 7

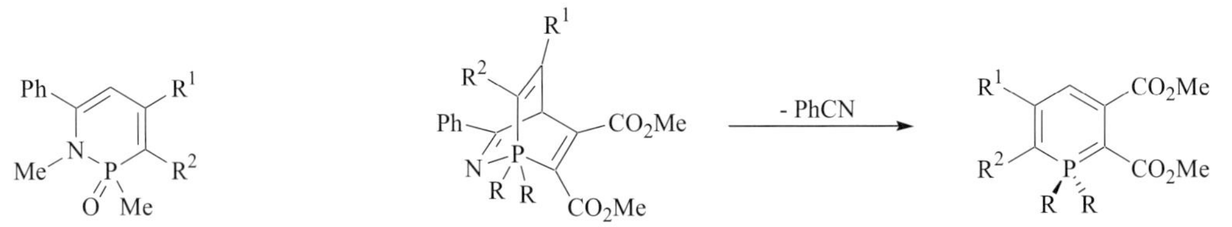

(26) (27) (28)

$R^1, R^2 = CO_2Me$
$R^1 = H, R^2 = CO_2Me$

R = OMe;
$R^1, R^2 = CO_2Me$;
$R^1 = H, R^2 = CO_2Me$
R = Ph;
$R^1 = H, R^2 = CO_2Me$

Scheme 8

(iii) Dialkyl (3-methyl-1-vinyl-1,2-butadienyl)phosphonates (**29**) undergo a cyclocondensation with the phosphoramidite (**30**) with formation of the substituted 1,2λ^5-azaphosphinines (**31**) (Scheme 9) ⟨85ZOB1710⟩. Similar reactions had been described earlier ⟨76ZOB51, 81ZOB767, 82ZOB789, 83ZOB930⟩.
The structure of (**31**) was confirmed by ^{1}H, ^{13}C NMR and MS data.

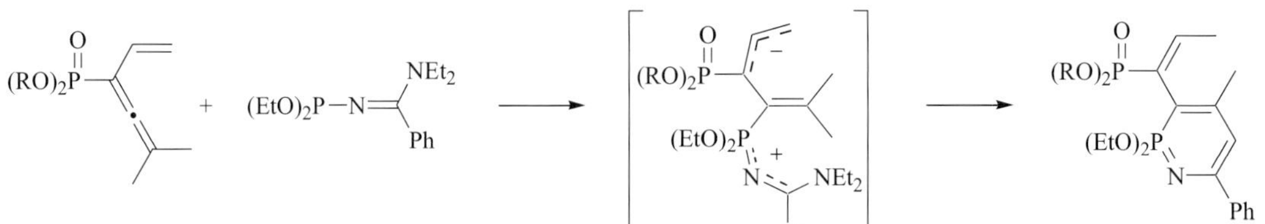

(29) (30) (31)

Scheme 9

(iv) A fourth route to 1,2λ^5-azaphosphinines was reported by Bertrand's group ⟨94JOC5927⟩. The *N,N'*-bis[(diisopropylamino)phosphino]carbodiimide (**32**) adds two moles of DMAD in a stepwise fashion as shown, to form the intermediate (**33**), which rearranges on heating by methoxycarbonyl group migration to form the fused compound (**34**) (Scheme 10).

1*H*-1,2λ^5-Azaphosphinin-6-ones (**37**) are prepared by reaction of DMAD with metallated *N*-alkoxy-carbonyl-alkyldiphenylphosphazenes (**35**) via the DMAD-adduct (**36**) (Scheme 11) ⟨96TL2841⟩.

The 1,2-azaphosphoniumsalt (**38**) ⟨94JOC7144, 96HCC233⟩ forms the azaylide (phosphineimine) (**39**) by deprotonation with NaHDMS in THF. By a [2 + 2]-cycloaddition of (**39**) with dimethylacetylenedicar-

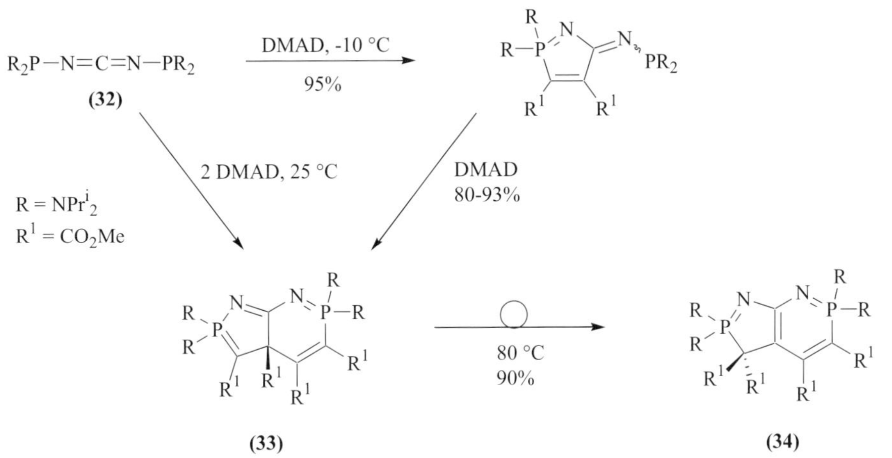

Scheme 10

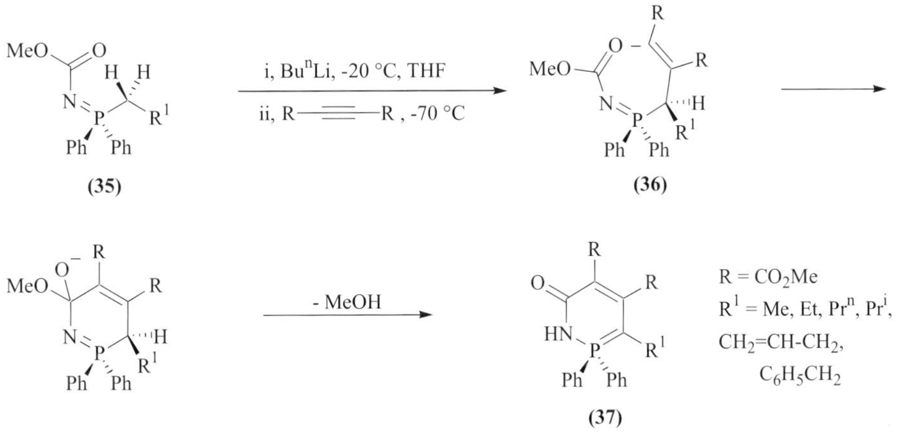

Scheme 11

boxylate (DMAD) at 0ºC the bicyclic intermediate (**40**) is formed followed by a valence isomerization to give the 8-membered azaylide (**41**). The structure of (**41**) is confirmed by an X-ray structural analysis (Scheme 12) 〈99JCS(P1)1577〉.

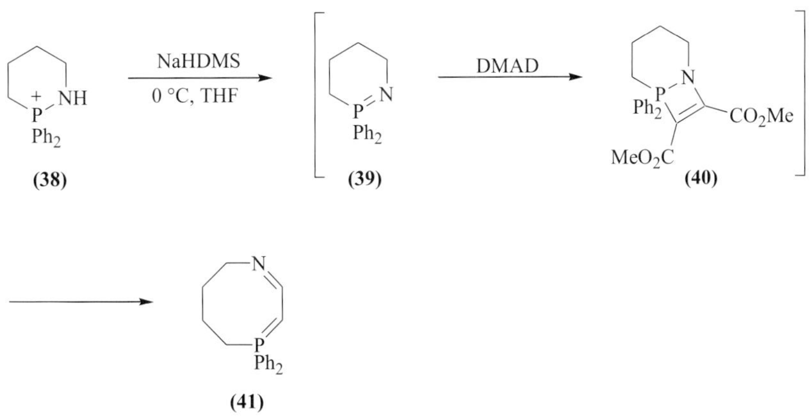

Scheme 12

The phosphiranes (**44**) contain a tetrahydro-1,2-aza-phosphinine structure. Starting out from dibenzotropone the amines (**42**) are prepared according to known methods. Lithiation and reaction with PCl_3 yields the dichloroaminophosphane (**43**). By dehalogenation with Mg the phosphiranes (**44**) are easily available. The formation of the three-membered ring formally is an intramolecular [2 + 1]-cycloaddition of a R_2*N*-P-phosphinidene to the CC-double bond. According to the similarity of (**44**) with the tetracyclic C_9H_{10}-barbaralane (replacement of two carbon atoms by P-*N*) the authors call the phosphiranes (**44**) phosphaazabarbaralanes (BABAR-Phos).

Transition metal complexes (Cu, Pt) of (**44**) are described (Scheme 13) ⟨99AG(111)1724⟩.

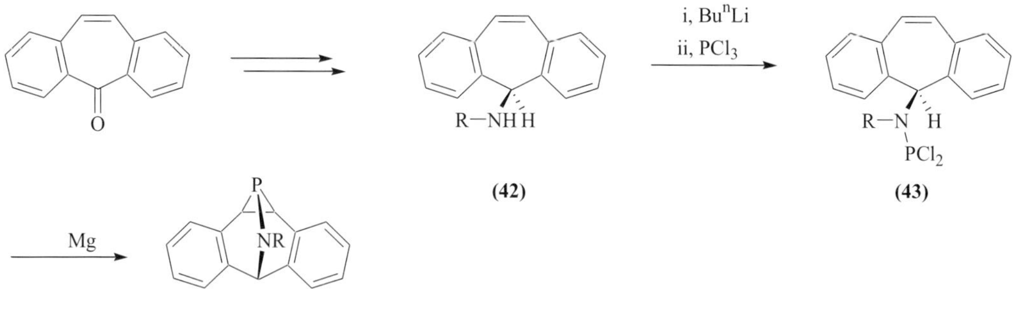

(**42**) (**43**) (**44**)

Scheme 13

5.3.3.2 1,3-Azaphosphinines

A very straightforward synthesis of 2,4,6-triaryl-1,3λ^3-azaphosphinines was published by Maerkl and Dorfmeister ⟨87TL1093⟩. The readily available triaryl-1,3-oxazinium salts (**45**) undergo a smooth O/P exchange reaction with tris(trimethylsilyl)phosphine to give the 1,3-azaphosphinines (**46**) in yields of 20–30%.

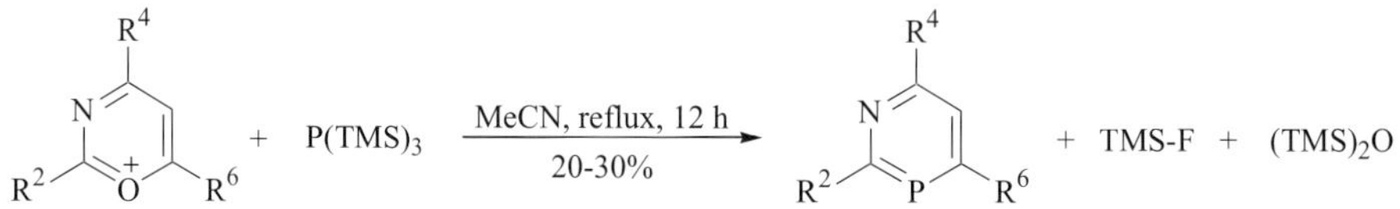

(**45**) (**46**)

Appel and Poppe prepared the 4-amino-1,3-azaphosphinine (**49**) as a yellow oil by cycloaddition of the phospha-ynamine (**47**) to the 1,3-oxazin-6-one (**48**), followed by loss of carbon dioxide and hydrolysis (Scheme 14) ⟨89AG(E)53⟩.

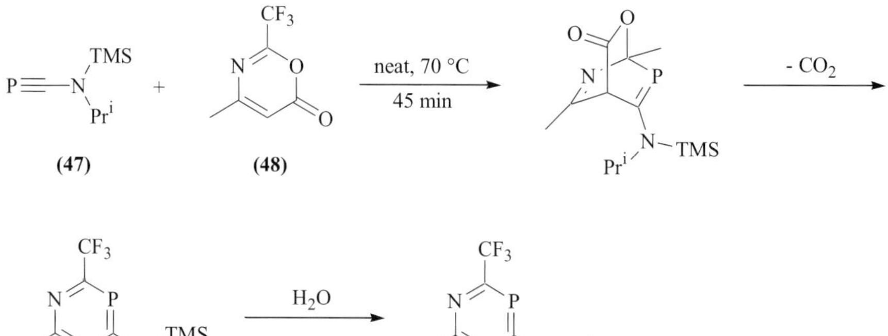

(**47**) (**48**) (**49**)

Scheme 14

A fascinating entry to the 1,3-azaphosphinines starts from tri-*t*-butylazete (**50**), which undergoes facile cycloaddition to *t*-butylphosphaethyne in a sterically controlled reaction to yield the 1,3-Dewar azaphosphabenzene (**51**) [$\delta(^{31}P) = 202$] ⟨88AG(E)272⟩. This substance undergoes thermal valence isomerization to the 1,3-azaphosphinine (**52**) [$\delta(^{31}P)$ = 147]. An alternative route to (**52**) is the photochemical valence isomerization of (**51**) to the prismane (**53**) [$\delta(^{31}P)$ = −111], which at room temperature slowly rearranges to the 1,4-azaphosphinine (**54**) [$\delta(^{31}P)$ = 228]. On irradiation, (**54**) is converted into the azaphosphabenzvalene (**55**) [$\delta(^{31}P$ = −50], which undergoes aromatization to (**52**) at 140ºC (Scheme 15) 〈B-90MI625-01〉.

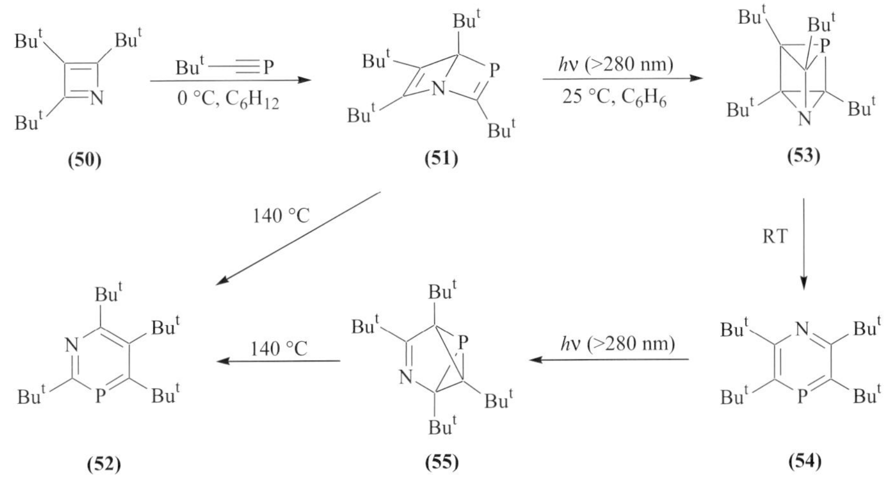

Scheme 15

A 1,3λ^5-azaphosphinine was prepared by condensation of diethyl phenylethynyl-phosphonite (**56**) with the nitrile ylide formed by deprotonation and chloride ion loss from (**57**). Cyclization and prototropic isomerization of an intermediate ethynyl-1,4-dipole (**58**) led to the product (**59**) (Scheme 16), the structure of which was confirmed by X-ray, ^{31}P NMR and MS data 〈86ZOB1661〉.

1,3λ^5-Azaphosphinines can also be prepared from the 1,3λ^3 compounds. Addition of an alkyllithium, followed by alkylation, which works in the case of the phosphinines, fails with the triaryl-1,3-azaphosphinines because the initial RLi adduct at P extrudes R–P and forms a triarylpyrrolylanion 〈92TH625-01〉. However, the reaction with diazoalkanes in the presence of alcohols forms the 1-alkoxy-1-alkyl adducts (**60**) 〈92TH625-01〉. In the absence of alcohol, diazomethane forms unusual 1 : 2 compounds (**61**) known as diaza-diphospha-chiropteradienes (Scheme 17) 〈90TL6999〉 (see also 〈87AG(E)236〉).

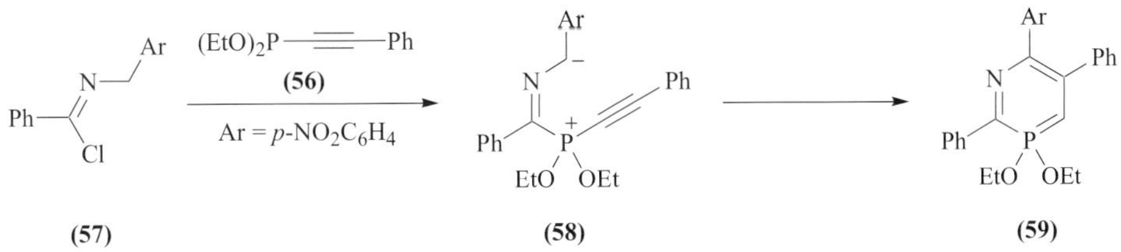

Scheme 16

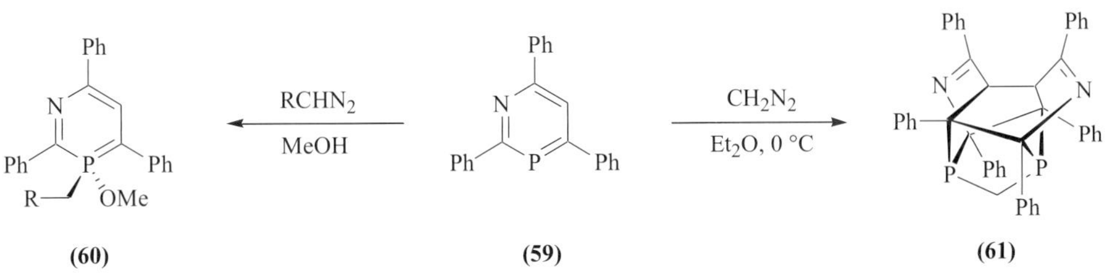

Scheme 17

The cycloadditions of alkynes to triaryl-1,3-azaphosphinines has been extensively investigated. Addition occurs across P/C-6, to form an unstable azaphosphaarrelene (**62**) which decomposes under the reaction conditions, expelling aryl carbonitrile and formation of the phosphinine (**63**) (Scheme 18). Acetylenic esters, alkynyldiphenylphosphine oxides, and alkynyldiphenylphosphines all undergo the reaction, the last requiring heating under pressure ⟨87TL1093, 90TL4589, 92TH625-01, 93TH625-01⟩.

t-Butylphosphaethyne reacts in a completely different way, forming a 2 : 1 adduct (**64**) with far-reaching rearrangement of the ring skeleton ⟨88AG(E)389⟩.

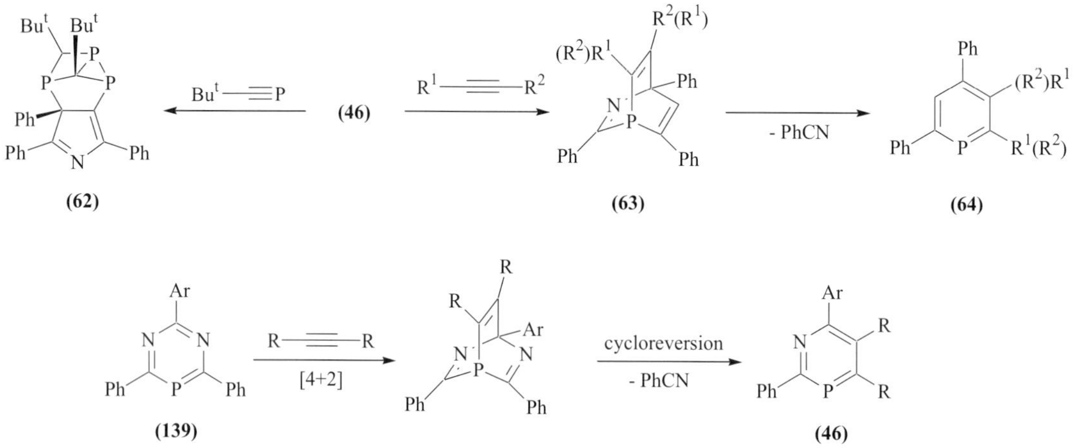

Scheme 18

By reaction of 1,3,5λ^3-diazaphosphinines (**139**) with alkynes in a (4 + 2)-cycloaddition- and cycloreversion-sequence 1,3λ^3-azaphosphinines (**46**) are formed, further reaction of (**46**) with a second mole alkyne yields λ^3-phosphinines ⟨91AG(E)106, 92TH625-01⟩ (Scheme 18) (see also Scheme 43).

5.3.3.3 1,4-Azaphosphinines

The 1,4-azaphosphinines include the long-known phenophosphazine system, the chemistry of which is somewhat different from that of the monocyclic compounds. Therefore phenophosphazines will be discussed in the following section.

The first, and for a long time the only 1,4λ^3-azaphosphinine derivatives (**66a,b**) were synthesized in 1972 by thermolysis of 4-*t*-butyl-1,4-dihydro-1,4-azaphosphinines (**65**) and the 4-*t*-butyl-4-chloro-1,4λ^5-azaphosphinine (**67**) respectively (Scheme 19) ⟨72AG(E)1019, 79TH625-01⟩. The spectral data, and particularly the very low-field ^{31}P chemical shift are consistent with the Hückel aromatic structure (**66**).

In Scheme 15 Regitz's preparation of the tetra-*t*-butyl-1,4-azaphosphinine (**54**) from the azaphosphaprismane (**53**) was mentioned.

The electronegativity difference between nitrogen and phosphorus renders the 1,4-azaphosphinines (**66**) very reactive towards nucleophiles. Protic reagents H-X add the X to phosphorus, and the proton to nitrogen. Water, followed by peroxide oxidation gives the phosphinic acid (**68**).

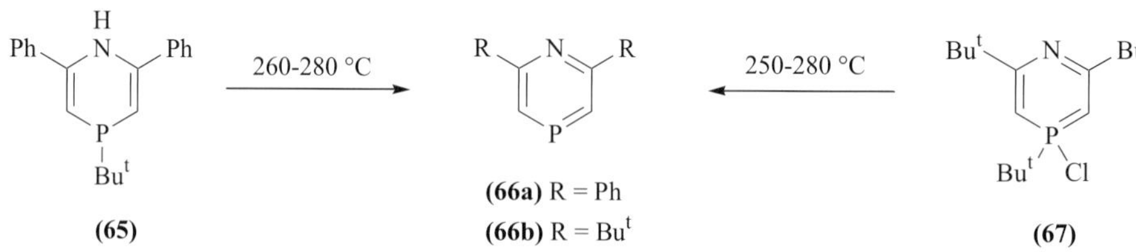

Scheme 19

Aprotic species (PhLi) attack the P; on neutralization the N is protonated (**69**). Scheme 20 summarizes these reactions.

4,5-Dihydro-1*H*-pyrazolo[3,4*b*][1,4]azaphosphinines (**71**) are accessible by phosphorylation of 1,3,3-trimethyl-2-(1-R-3-methyl-5-pyrazolyl-iminoethylidene) indolines (**70**) with $PHal_3$ at the two nucleophilic carbon centers to give the fused 1,2-dihydro-1,4-azaphosphinines (**71**) (Scheme 21) ⟨99HAC391⟩.

Nucleophilic substitution reactions at the P–Hal bond are described with morpholine, aniline, methanol and water. Oxidation and reaction with sulfur yields the pentavalent derivates (Scheme 21).

The 1,4λ^5-azaphosphinines are available by a variety of methods. Cyclocondensation of the λ^5-phosphazenes (phosphinimines, (**72**) with DMAD in acetonitrile proceeds, probably via the dihydrophosphazete

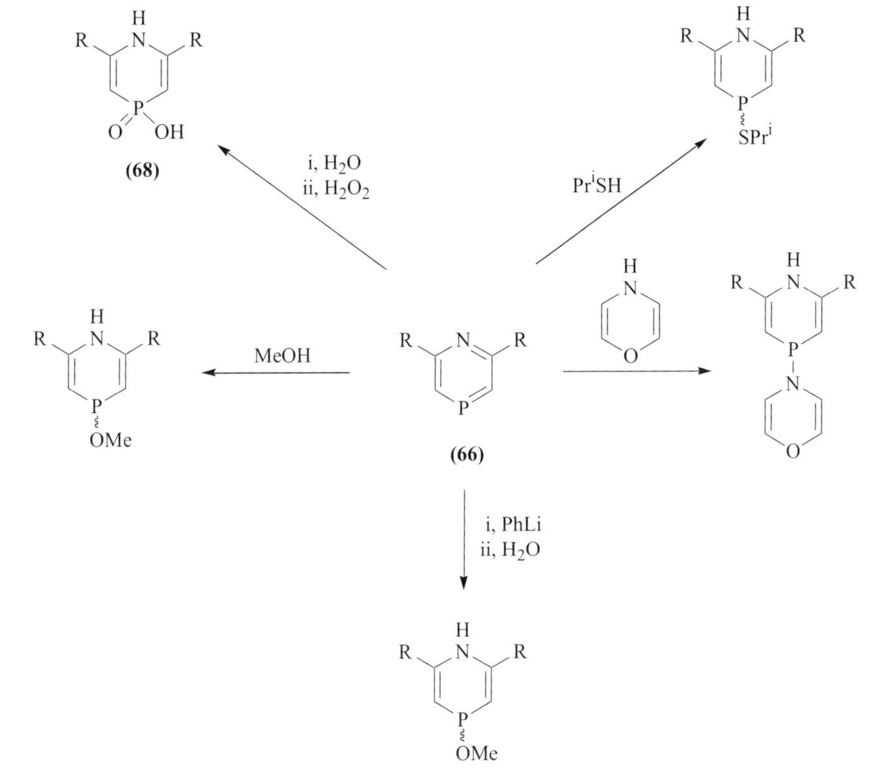

Scheme 20

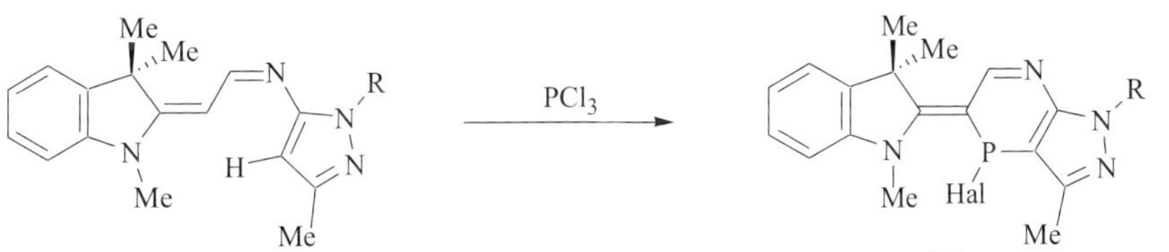

(70) **(71)**

R = Ph, CH_2CH_2CN
Hal = Cl, Br

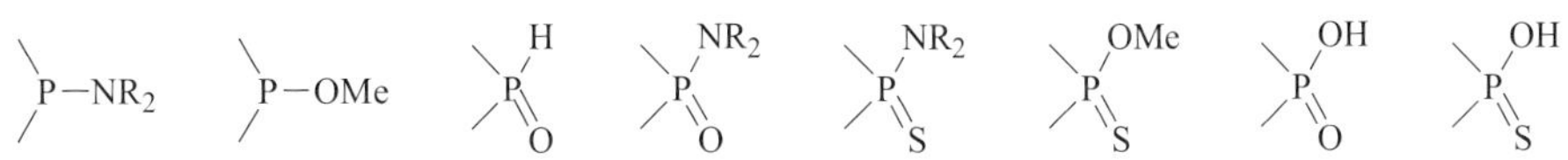

Scheme 21

(**73**), with insertion of the acetylenic moiety into the P–N bond. The enamino-phosphorane products (**74**) cyclize on warming, losing aniline and forming the azaphosphinines (**75**) ⟨85CC1681⟩. With butyllithium the enamino-phosphoranes cyclize to give 1,4λ^5-azaphosphinin(1*H*)-2-ones (**76**) (Scheme 22) ⟨89JCS(P1)2273⟩. Azaphosphinin-2-one (**79**) are also the outcome of the condensation of the phosphinimines (**77**) with acetylenic esters, followed by cyclization of the ylides (**78**) with KH (Scheme 22) ⟨90JOM(382)61⟩.

A quite extraordinary cycloaddition route to a 1,4λ^5-azaphosphinine was described by Bertrand and co-workers ⟨94JA8087⟩. The 1,2λ^5-azaphosphete (**80**) reacts with DMAD to give the azaphosphinine (**80**), as shown by X-ray crystallography (Scheme 23).

By a number of related methods Simalty and others have produced di-(β-oxoalkyl)phosphonium salts (**81**) (Scheme 24). These are readily converted into the cyclic dienimines — the 1,4λ^5-azaphosphininium salts (**82**) — by reaction with ammonium acetate in refluxing acetic acid ⟨84PS(19)159⟩. Deprotonation (NaH/DMSO ⟨72TL4363⟩ or KOH/EtOH) gives the 1,4λ^5-azaphosphinines (**83**) ⟨70TL4371, 71TL4749,

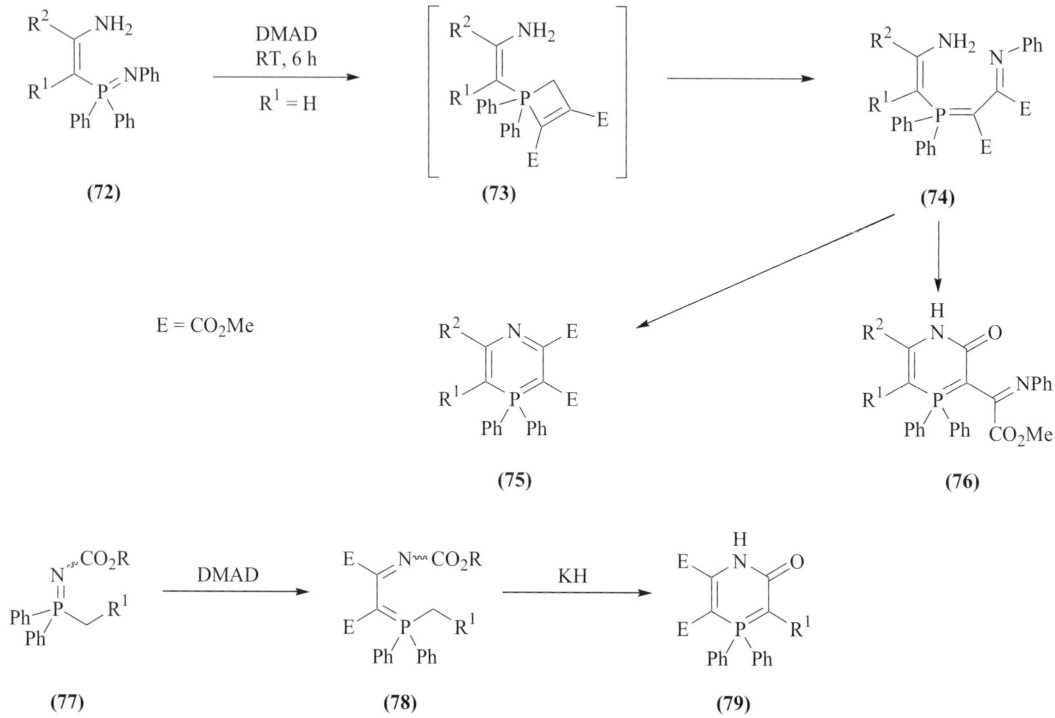

Scheme 22

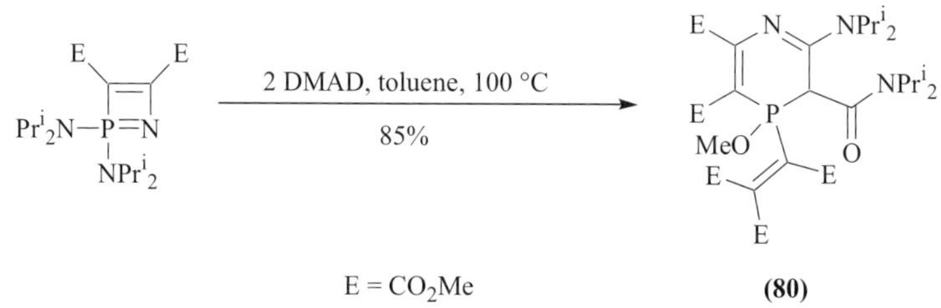

Scheme 23

72CR(C)2206, 72JHC347, 72T3343, 79S109〉. Quin *et al.* have shown that di-(β-oxoalkyl)phosphine oxides (**85**) can be prepared from phospholene oxides (**84**) by ozonolysis and reductive work-up 〈84S1074〉. These cyclize under similar conditions ($NH_4OAc/AcOH$) to form 1,4-dihydro-1,4-azaphosphinine 4-oxides (**86**). Silylation (bistrimethylsilyl trifluoroacetamide or Et_3SiBr) occurs to form the 4-silyloxy-1,4λ^5-azaphosphinines (**87**) (Scheme 25) (silylation at the nitrogen atom could not completely be ruled out) 〈90PS(54)1〉.

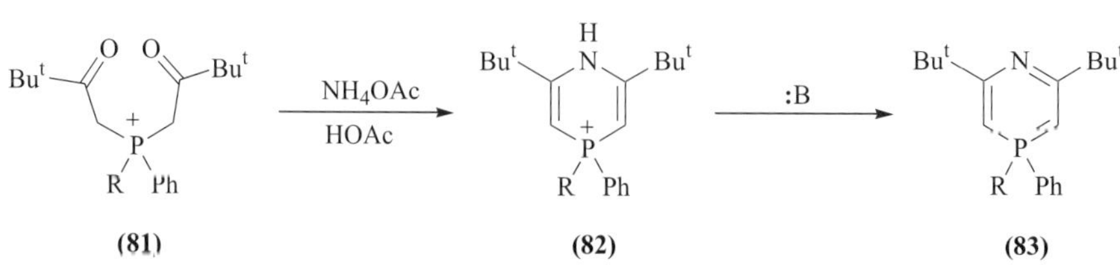

Scheme 24

Novel pyrazolo[5,4-b]azaphosphinines (**91**) have been synthesized from 4-phosphorylated 5-formamidinopyrazoles.

Into *N*,*N*-dimethyl-N-5-pyrazolylformamidine (**88**) a $-PCl_2-$ moiety can be introduced in 3-position of the pyrazole ring to give (**89a**), reaction with secondary amines (**89b**) followed by phosphoniumsalt-formation (**90**) and deprotonation with base yields the ylid, which undergoes ring closure to (**91**) by an intramolecular alkylation reaction (Scheme 26) 〈99MC161〉.

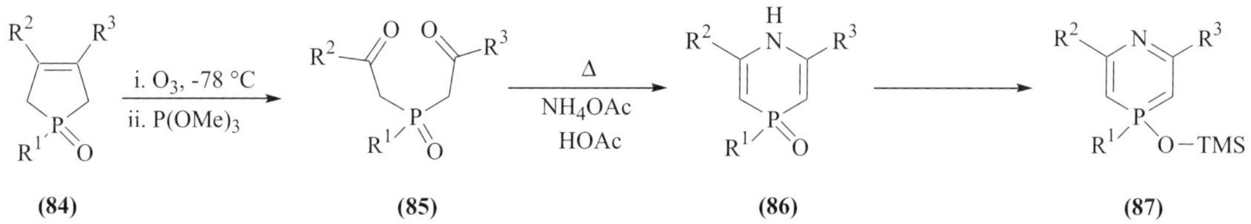

R^1, R^2, R_3 = Ph, Me; R^1 = Ph, R^2 = Me, R^3 = H; R^1, R^2 = Me, R^3 = H

Scheme 25

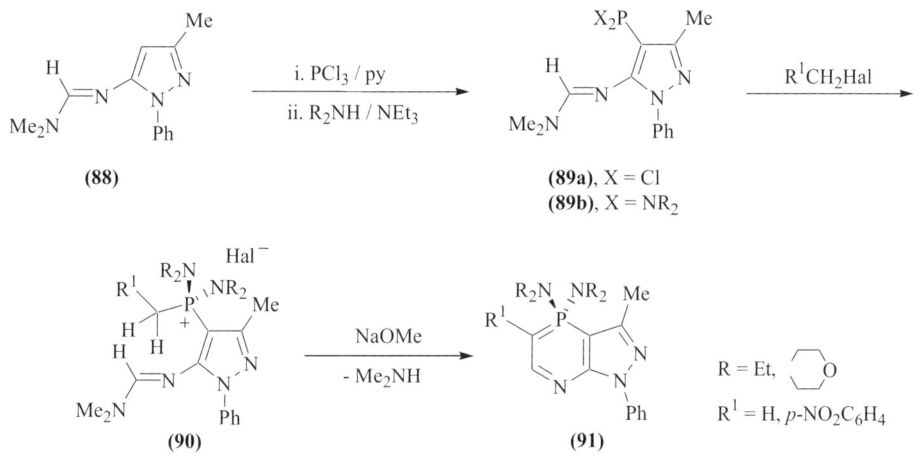

Scheme 26

Acetaldehyde-1,1-dimethylhydrazone reacts with acetylchloride in the presence of base to form (**92**) which on treatment with PCl_5 yields the 1,4-dihydro-1,4-azaphosphininiumsalt (**94**) via the intermediate (**93**) (Scheme 27) ⟨99JGU833⟩.

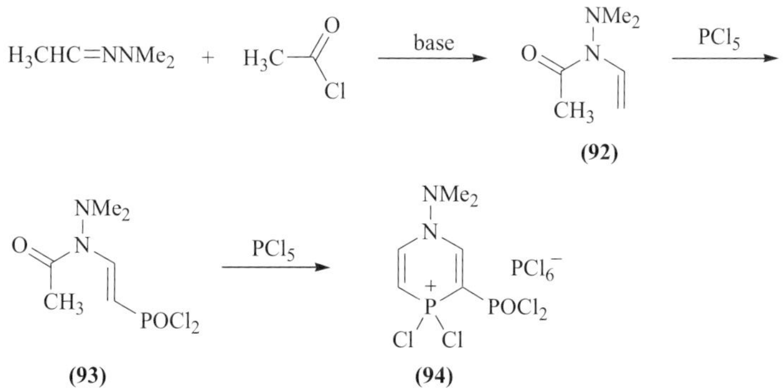

Scheme 27

4-Mercapto-6-oxo-1,4-azaphosphinane-2-carboxylicacid-4-oxide (**100**) as an inhibitor of mammalian dihydroorotase was synthesized starting out with the amino acid (**95**), which was cyclisized by reaction with a 1-fold excess of 1-(3-dimethyl-aminopropyl)-3-ethylcarbodiimide (DEPC) to give 4-hydroxy-6-oxo-1,4-azaphosphinane-2-carboxylicacid-4-oxide (**96**). Protection of the functional groups by reaction with Ph_2CN_2 (**97**) and deprotection of the $P(O)OCHPh_2$-group with ethanol yields >P(O)OH which can be transformed into (**99**) by reaction with Ph_2CHSH/DEPC/NEt_3, removal of the protecting groups with TFA yields (**100**) (Scheme 28) ⟨98JMC4550⟩.

The enzymic evaluation of the *cis*- and *trans*-isomers of (**100**) against mammalian dihydroorotase is discussed.

1,4-Azaphosporinanes (**101**) are accessible by alkylation of PH_3 or primary phosphanes with bis(chloroethyl)amines in the superbasic medium DMSO/KOH beside diprimary and disecondary bis(phosphinoethyl)amines (**102**) (Scheme 29) ⟨98JOM39⟩.

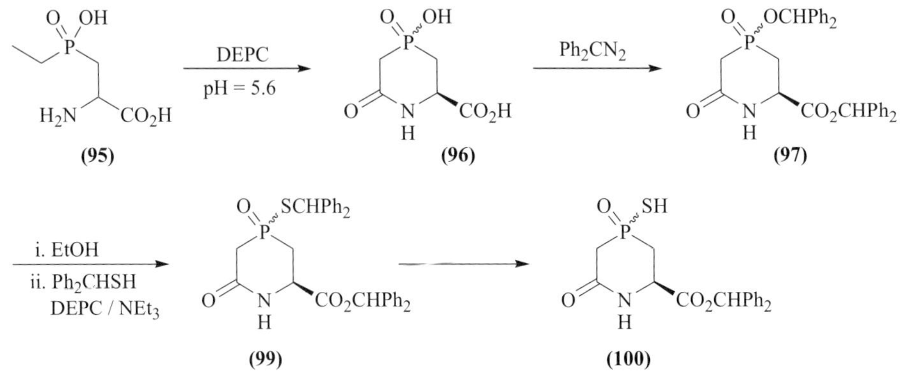

Scheme 28

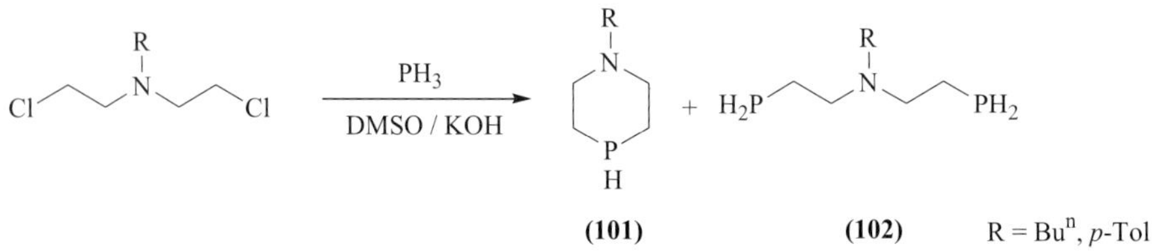

Scheme 29

5.3.3.4 Phenophosphazines

Fully unsaturated phenophosphazines (dibenzo[1,4]azaphosphinines) are unknown, but the 5,10-dihydro compounds are very well known, having first been prepared well over a century ago. Diphenylamines are the starting materials for all their syntheses. The reactivity of the 1,5-dihydrophenophosphazines is mainly classical phosphorus chemistry; it has been reviewed by Freedman and Freeman ⟨87CRV289⟩.

When diphenylamine is strongly heated with PCl_3 the chlorophenophosphazine (**103**) is formed. Aqueous work-up produces the dihydrophenophosphazine oxide (**104**) (Scheme 30) ⟨1888CB1497⟩. The structure of (**104**) was finally confirmed in 1960 ⟨60HCA1826⟩. The reaction was extended by Freedman and co-workers to substituted arylamines ⟨75JOC766, 81PS(9)269⟩. The spiro compounds (**105**) have also been detected in the reaction (Scheme 30) ⟨71JCS(A)1213⟩.

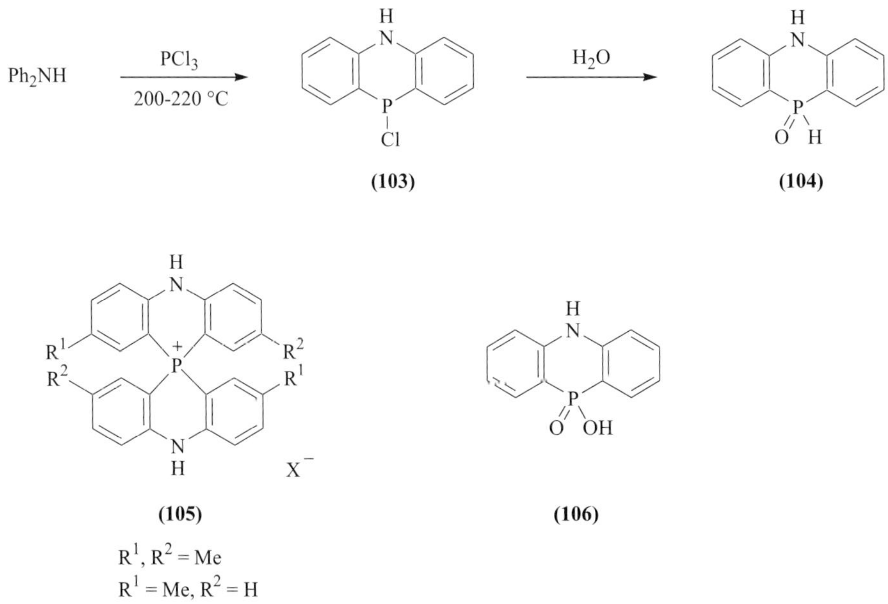

Scheme 30

The practically planar tricyclic structure of 5,10-dihydro-10-hydroxy-phenophazine-10-oxide (**106**) is in sharp contrast to the structure of other compounds in this class ⟨95ZOB793⟩.

When diphenylamines are heated with $PSCl_3$ at 200°C in an autoclave the 10-sulfides (**107**) are formed ⟨61BRP860629, 81ZOB2222⟩. These products are available by other routes ⟨60HCA1826⟩.

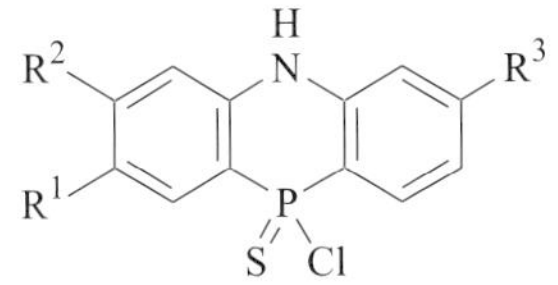

(107a) R^1, R^2, R^3 = H
(107b) R^1 = H, R^2, R^3 = Me
(107c) R^1 = Me, R^2, R^3 = H
(107c) R^1 = Cl, R^2, R^3 = H

A second, versatile method for the synthesis of dihydrophenophosphazines (**108**) involves the reaction of 2,2′-dilithiodiphenylamines with phenyldichlorophosphine or diaryl phenylphosphonites ⟨64JOC3410⟩. When trilithiotriphenylamine reacts with triaryl phosphites, the azaphosphatripycene (**109**) is formed ⟨69AG(E)987, 78CB1798⟩ (Scheme 31).

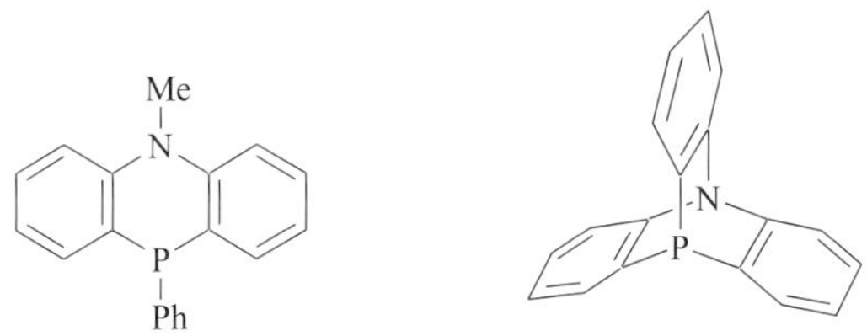

(108) **(109)**

Scheme 31

5.3.4 RING SYSTEMS WITH ONE PHOSPHORUS AND TWO NITROGEN ATOMS

Of the six possible diaza-phosphinines 1,3,2-(I), 1,5,3-(III) and 1,5,2-diazaphosphinines (IV) are known as λ^3- as well as λ^5-derivatives. 1,2,3- and 1,2,4-diazaphosphinines are known only as λ^5-systems (**X**, **XII**). 1,4,2-Diazaphosphinines (**VII**) have not yet been described in the literature.

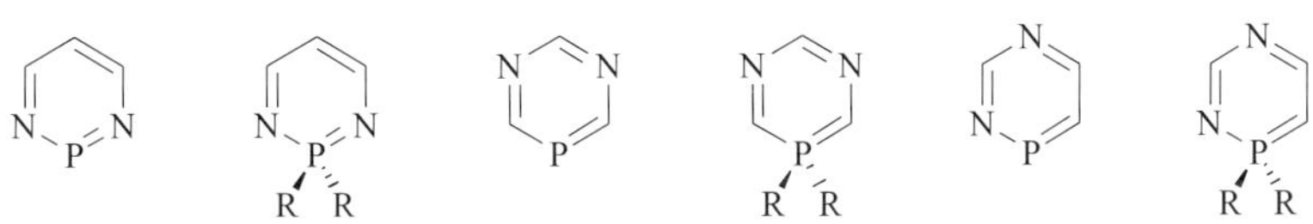

(I) **(II)** **(III)** **(IV)** **(V)** **(VI)**

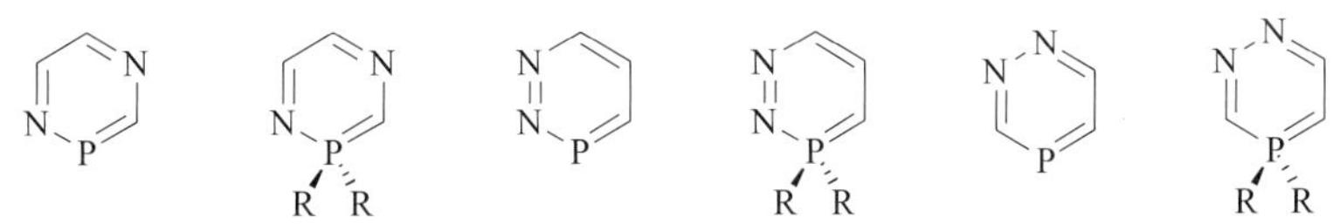

(VII) **(VIII)** **(IX)** **(X)** **(XI)** **(XII)**

The 1,3,2λ^3-diazaphosphinines are highly reactive already with weak nucleophiles similar to the behavior of the 1,2λ^3- and 1,4λ^3-azaphosphinines.

5.3.4.1 1,3,2-Diazaphosphinines

Since 1996 the synthesis as well as the chemistry of 1,3,2λ^3-diazaphosphinines has been well established. In 1988 Doxsee *et al.* for the first time have described the synthesis of 1,3,2-diazatitana-cyclohexa-3,6-dienes (**110**) by reaction of nitriles with $Cp_2Ti{=}CH_2$ or Cp_2TiMe_2 ⟨93OM3776⟩. The reaction of (**110**) with PCl_3 and triethylamine directly affords 1,3,2-diazaphosphinines (**111**) (Scheme 32) ⟨96JA11978⟩.

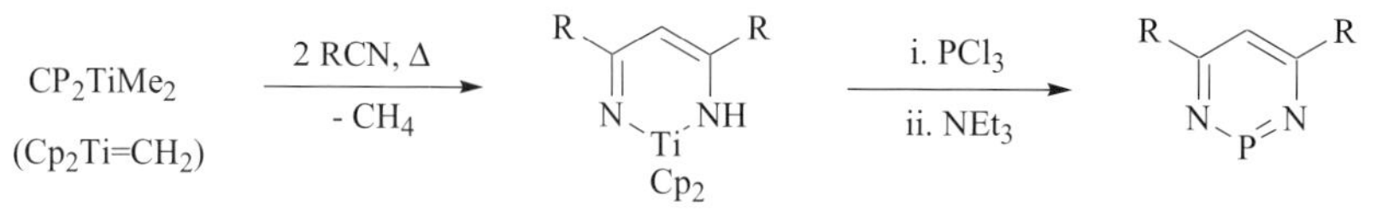

(110) **(111)**

R = Bu^t (45%, $\delta(^{31}P)$ = 267.5, orange oil)
R = Ph ($\delta(^{31}P)$ = 269.2)

Scheme 32

The 1,3,2-diazaphosphinines (**111**) are very sensitive to air and moisture, similar to the P=N-bond in (**8**) and (**66**). The P–N-bond already at room temperature reacts with protic reagents R–OH, R–SH to give the 1,2-dihydro-1,3,2-diazaphosphinines (**112**) (Scheme 33) ⟨97OM4089⟩.

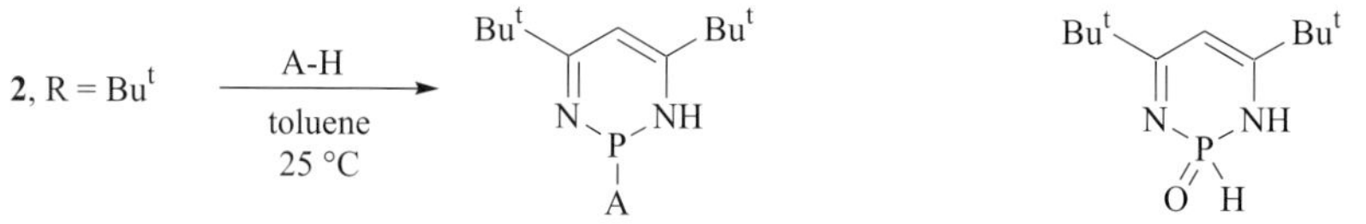

(112), A = OMe, SEt, OH **(112)**, A = OH

Scheme 33

The most interesting feature of the 1,3,2-diazaphosphinines (**111**) is their (4 + 2)-cycloaddition-chemistry with alkynes. Depending on the reaction conditions and the stoichiometry in a cycloaddition–cycloreversion sequence 1,2-azaphosphinines (**8**) and in the following λ^3-phosphinines (**113**) are formed (Scheme 34) (see also 1,2-azaphosphinines) ⟨97OM4089⟩.

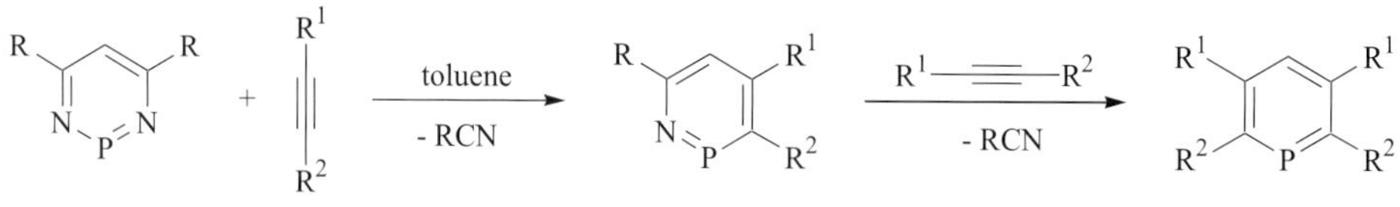

(111) **(8)** **(113)**

(**8**), R^1 = Me, R^2 = CH(OEt)$_2$, R = Bu^t (2h, 90 °C)
R^1 = –≡–Me, R^2 = Me, R = Bu^t (3h, 100 °C)
R^1 = $C_{10}H_9Fe$, R^2 = $SiMe_3$ (xylene, 15h, 150 °C)
R^1, R^2 = PPh_2, R = Bu^t (4h, 120 °C)
(further 1,2-azaphosphinines with other substituents R^1, R^2 are listed in the chapter 1,2-azaphosphinines)

Scheme 34

The [4 + 2]-cycloaddition of the 1,3,2-diazaphosphinine (**111**) with ^{t}BuC≡P in toluene at 25°C yields the heterobarrelene (**114**). At 90–100°C in toluene a cycloreversion–cycloaddition sequence yields the triphosphabarrelene (**115**), which finally reacts with a third mole ^{t}BuC≡P to give the (2 + 2 + 2)homo-Diels–Alder-adduct (**116**) (Scheme 35) ⟨98EJOC2039⟩.

A large number of papers from Russian laboratories dealing with the synthesis of 2,2-dichloro-1,3,2-λ^5-diazaphosphinines by reaction of malononitrile, derivatives of molononitrile, and other monosubstituted nitriles with PCl_5 were published between 1966 and 1980. A brief overview of this work is given below.

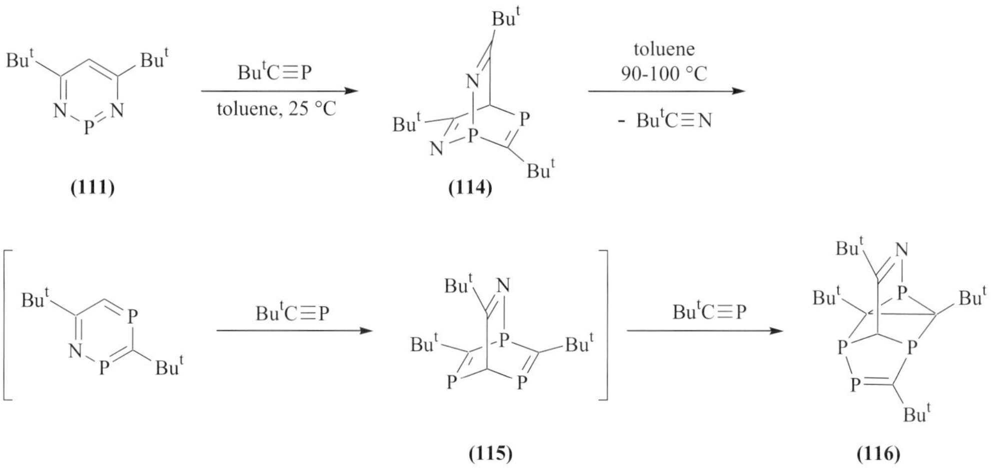

Scheme 35

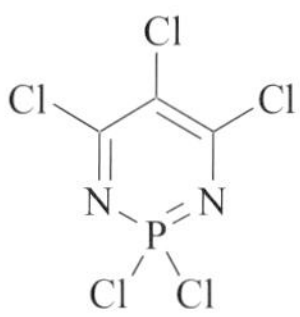

66ZOB735

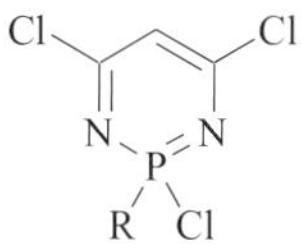

66ZOB1254

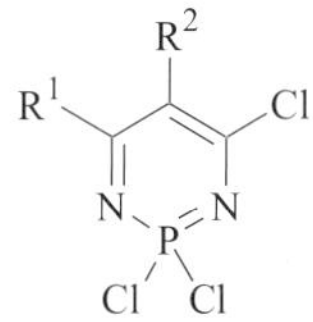

71ZOB988, 71ZOB2390, 73ZOB261, 74ZOB1247, 76ZOB2217, 81ZOB517

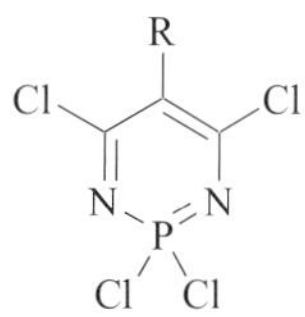

R = H, alkyl 77ZOB2522

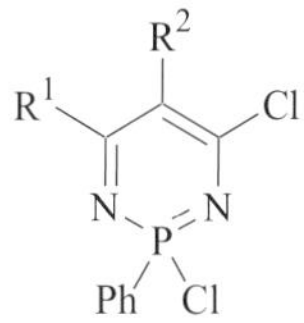

78ZOB799

Reactions at the PCl_2 group, such as aminolysis and alcoholysis, have also been described ⟨70ZOB1466, 73ZOB16, 76ZOB1482, 76ZOB2199, 80ZOB1313⟩. Attempts to reduce the chlorine atoms away, to produce a 1,3,2λ^3-diazaphosphinine, have so far met with failure ⟨94UP625-01⟩.

1,2-Dihydro-1,3,2-diazaphosphinines (**117**), easily obtained by treatment of 4-amino-1-azadienes with dichloro(diisopropylamino)phosphane undergo a cycloaddition in a [5 + 2]-fashion to dimethylacetylenedicarboxylate (DMAD) to furnish the bicyclic iminophosphoranes (**118**).

The formation of the polycyclic phosphoniumsalts (**120**) by reaction of (**118**) with $ClCF_2COOH$ and MeI makes the existence of an equilibrium between the valence isomers (**118**) and (**119**) with a phosphinalkylene unit obvious (Scheme 36) ⟨97JOM233⟩.

The nature of the products formed by the reaction of the 1,2-dihydro-1,3,2-diazaphosphinine (**117**), $R^1 = c\text{-}C_6H_{11}$; $R^2 = C_6H_4\text{-}p\text{-}OMe$; $R^3 = Bu^t$ ⟨97JOM233⟩, with DMAD depends dramatically on the reaction conditions.

With excess DMAD in ether at −20 to 25°C the 1 : 2 adduct (**121**) with a ylid structure is obtained, the bicyclic 1 : 1 adduct (**118**) cannot be proved as an intermediate.

Treatment of a dilute solution of (**117**) (0.05 M) in ether with dimethyl- or diethylacetylenedicarboxylate (molar ratio 1 : 2,4) for 72 h and then slowly raising the temperature to room temperature results in the formation of the λ^5-dihydrodiazaphosphaazulene derivatives (**122a**) (89% yield) and (**122b**) (85%). In this case the bicyclic 1 : 1 adduct (**118**) could be transformed into (**122**) by reaction with a further mole DMAD at −20°C.

Possible reaction mechanisms for the formation of (**121**) and (**122**) are discussed (Scheme 37) ⟨98EJOC1425⟩.

A variety of benzo-annelated 1,3,2-diazaphosphinines are known. The reaction of anthranilamide (**123**; X = O) or anthranilthioamide (X = S) with P_2S_5 in refluxing pyridine gave the pyridinium salt of (**124**), from which by treatment with base or SiO_2 (**124**) could be isolated (Scheme 38) ⟨85HCA900⟩.

Treatment of (**124**) with aqueous alkali and dimethyl sulfate gave the trimethyl derivative (**125**). The structures of (**124**) and (**125**) were proved by X-ray analysis ⟨85JCS(P2)1913⟩; see also ⟨87JOM(325)153, 90IJC(B)711⟩. These products and others prepared from substituted anthranilamides were synthesized with a view to the development of antiparasitic compounds ⟨82MI625-01⟩.

Schmutzler's group has investigated P(III) analogues of these compounds ⟨88ZN(B)937, 88IC1911, 93PS(80)141, 93ZN(B)333, 93PS(80)173, 93PS(78)271, 94JFC51, 95PS(106)65, 95PS(107)69⟩.

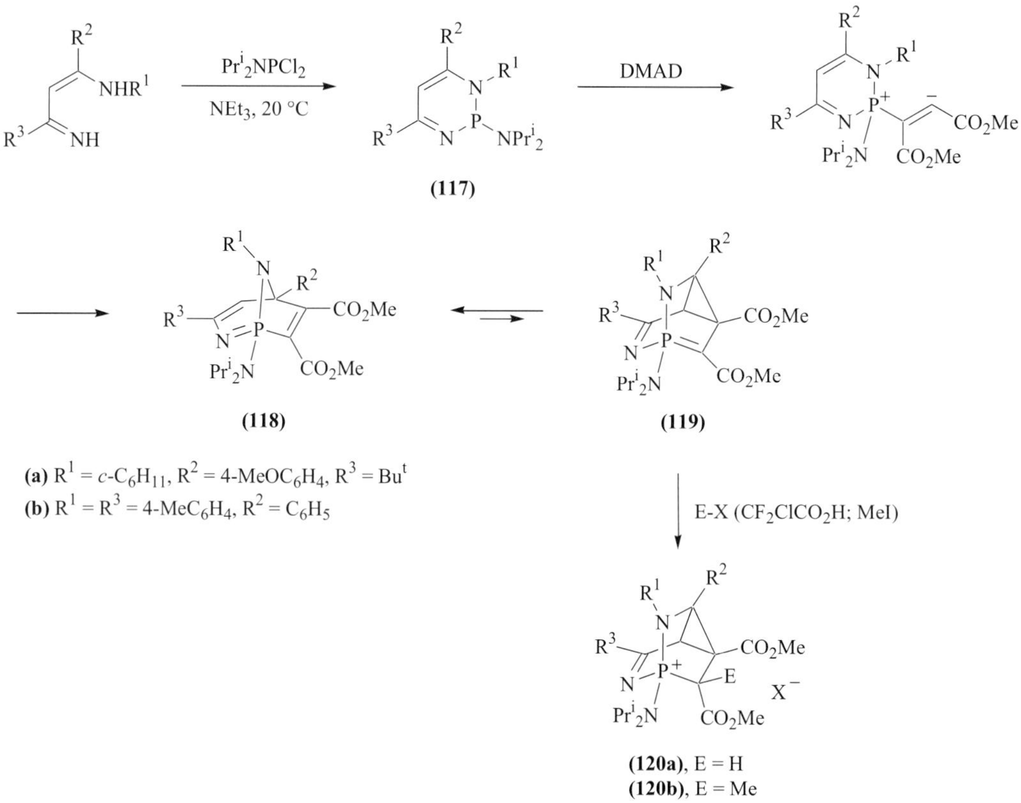

Scheme 36

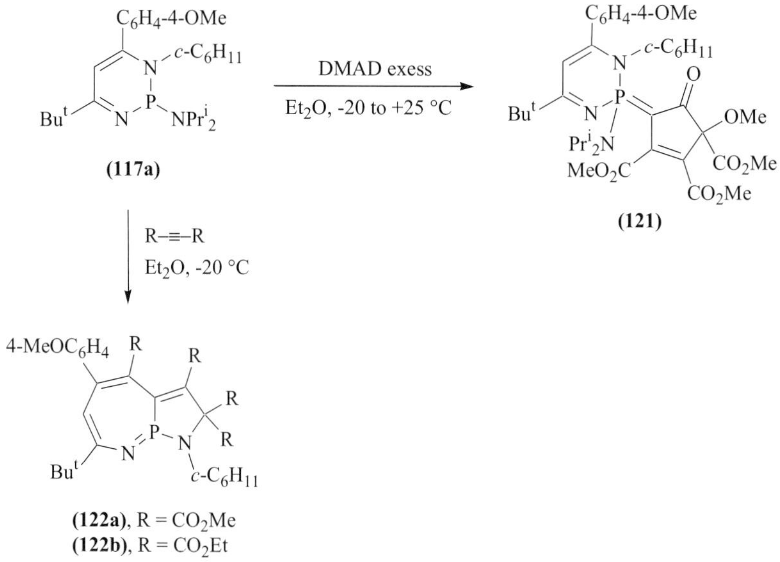

Scheme 37

Starting out from *N*-methyl-isatinicacidanhydride (**126**) the amides (**127**) are prepared, which react with PCl_3 to give (**128**). Numerous substitution and oxidation reactions of (**128**) were studied (Scheme 39) ⟨96ZN(B)1627⟩.

Reaction of (**128**) with secondary amines yields the amino phosphines (**130**), hydrolysis the secondary phosphinoxide (**129**). With hexafluoroacetone (**130**) reacts in a well known manner to give the λ^5-phosphoranes (**131**) ⟨96ZN(B)1627⟩.

In a stereoselective *Pudovic*-reaction (**129**) adds to the 3-thiazoline (2,2-dimethyl-1-thia-4-aza-

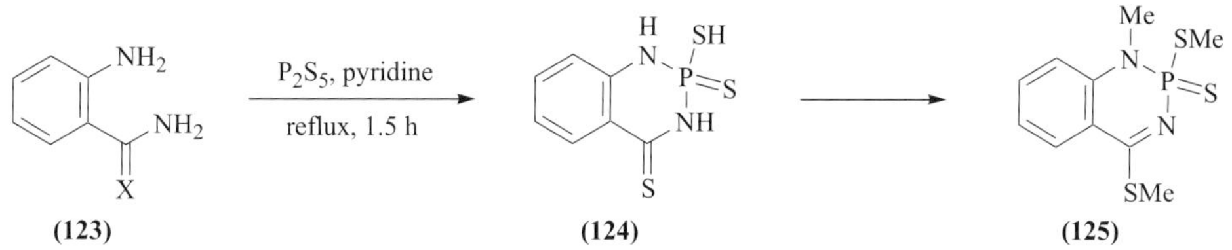

Scheme 38

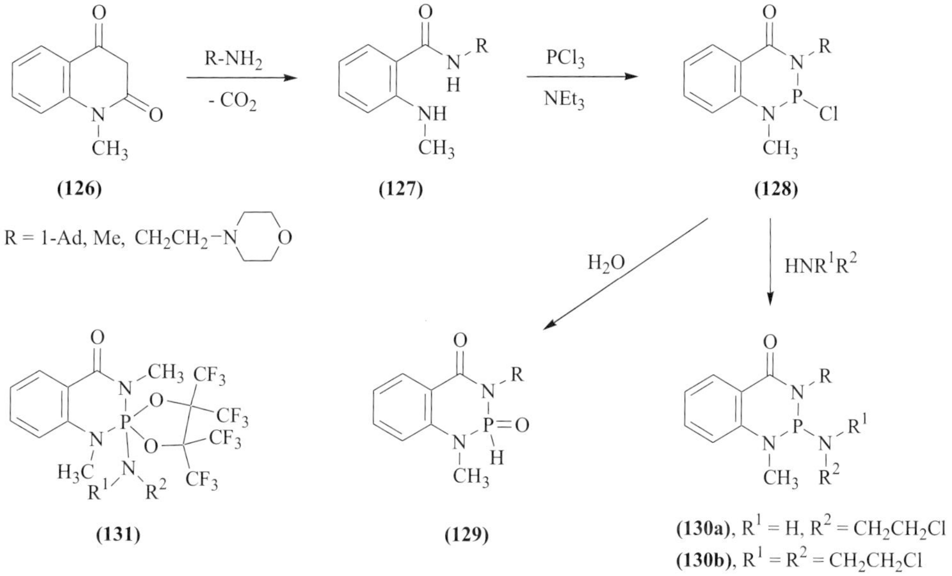

Scheme 39

spiro[4,4]non-3-en) by formation of the diastereomeric adducts (**132a**) and (**132b**) (Scheme 40) ⟨96ZN(B)1305⟩.

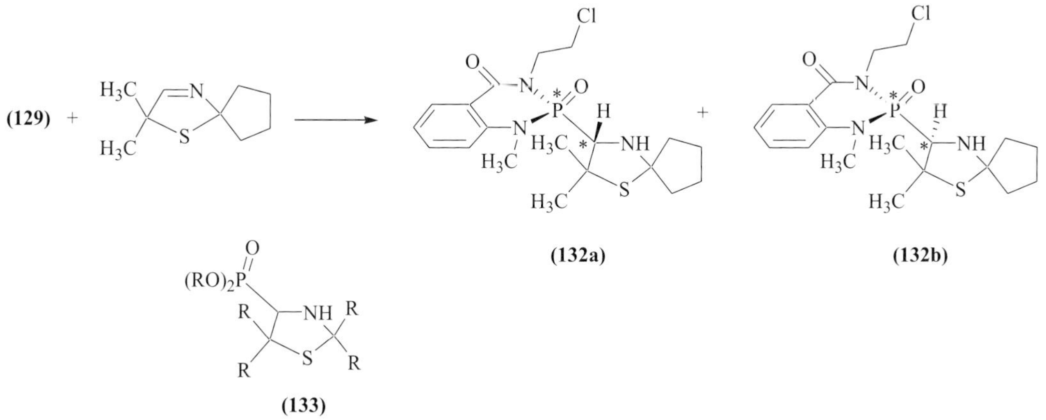

Scheme 40

The adducts (**132**) are of pharmacological interest since they are related to (**133**), which may be considered as phosphonic acid analogues of penicillamine.

An unusual intramolecular N-alkylation reaction was observed during the oxidative addition of hexafluoroacetone and tetrachloro-benzoquinone to (**130b**), the products are the tricyclic ammonium salts (**134**) and (**135**) (Scheme 41) ⟨96S473⟩.

If the ring opening of the *N*-methyl-isatinicacidanhydride (**126**) with amines is carried out with α,ω-diamines, the bis-amides (**136**) are obtained, reaction of (**136**) with PCl_3 yields the 5,6-benzo-2-chloro-1-methyl-1,3,2-diaza-phosphorin-4-one (**137**) connected by $-(CH_2)_n-$ links (Scheme 42) ⟨99ZAAC1732⟩.

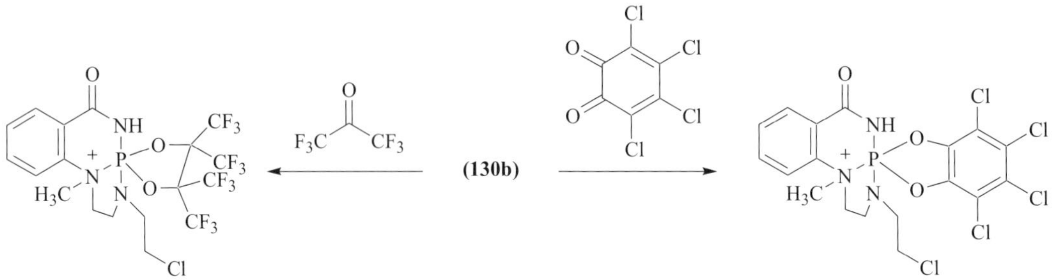

(134) (130b) (135)

Scheme 41

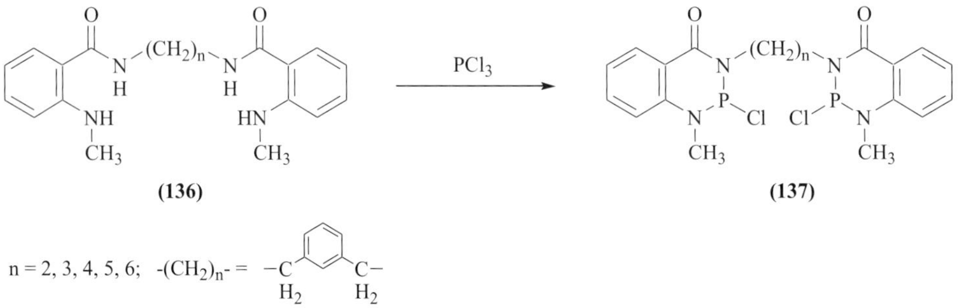

(136) (137)

Scheme 42

1,6-Trimethylene-5-cyano-2-mercapto-1,3,2-diazaphosphorine-2,4-dithione and pyrazolo-[3,4-d]-1,3,2-diazaphosphorines with anti-viral activities were described by Russian authors ⟨96MC191, 98KFZ16⟩.

5.3.4.2 1,3,5-Diazaphosphinines

The triaryl-1,3,5λ^3-diazaphosphinines (**139**) are prepared in a way similar to the 1,3λ^3-azaphosphinines. It is necessary to use the oxadiazinium tetrafluoroborates (**138**) and even then the yields are poor (3–5%) ⟨91AG(E)106⟩.

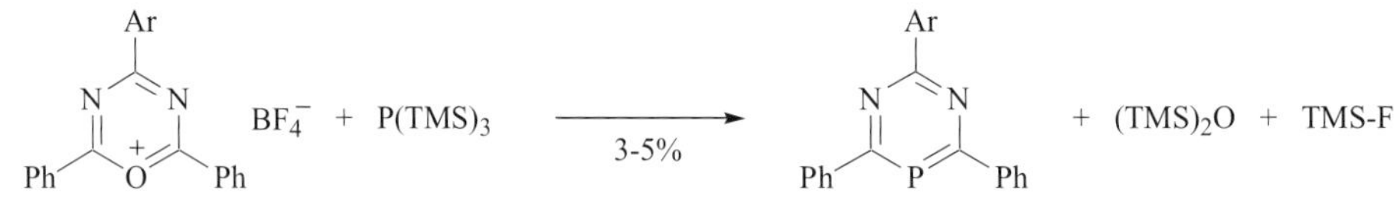

(138) (139)

^{31}P Chemical shifts of compounds (**139**) are in the region δ 135–141 ppm, depending upon the substituent group Ar.

The 1,3λ^3,5λ^3-diazaphosphinines react with alkynes in a cascade of (4 + 2)-cycloaddition and cycloreversion reactions, giving the substituted λ^3-phosphinines (**149**) (Scheme 43) ⟨91AG(E)106, 92TH625-01⟩. With diynes the cycloadditions yield oligomeric phosphinines, and finally polymers with λ^3-phosphinine units ⟨96TH625-01⟩.

The fully reduced 1,3,5-diazaphosphinanes (**141**) have been established in the literature since the early 1980s. Their synthesis is as outlined in Scheme 44. Their *P*-oxides and *P*-sulfides are the only formal representatives of the 1,3,5λ^5-diazaphosphinines that have been described. The following references are the main ones to cover the area ⟨80IZV721, 80IZV952, 80IZV1438, 81IZV1539, 81PS(10)207, 81TL229, 82IZV127⟩.

The electrochemical oxidation of 5-*p*-toluidino-methyl-1,3-di(*p*-tolyl)-1,3,5-diazaphosphorinane (**142**) to give 1,3-di-*p*-tolyl-5-toluidinomethyl-1,3-diaza-5-phosphorinan-2-ylium perchlorate (**143**) depending on the anode material, the electrolyte and the anode potential (Scheme 45) ⟨97RCB1154, 98RJE313, 99RJGC886⟩.

A series of transition metal complexes of 1,3,5-diazaphosphorinanes (**141**) has been described by Russian authors: Synthesis, crystal and molecular structure of conformers of *cis*-bis(1,3,5-triphenyl-1,3,5-diazaphosphorinane)dichloroplatinum(II)complex ⟨93IZV1650⟩.

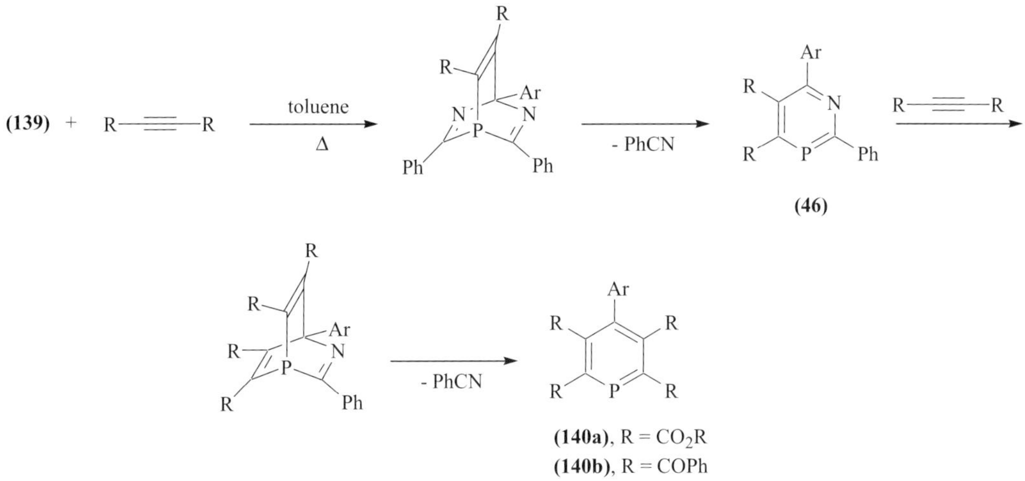

Scheme 43

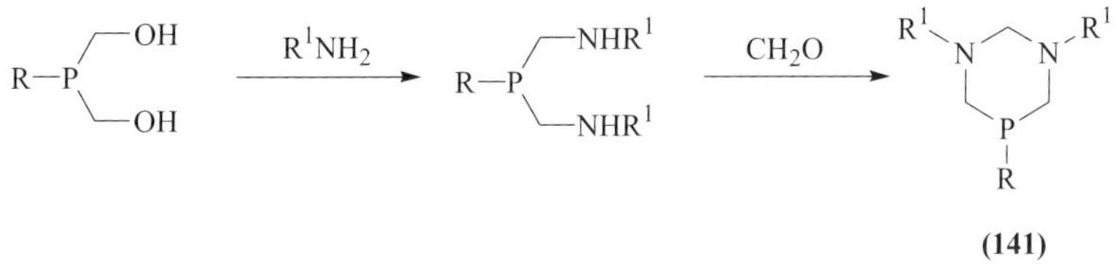

Scheme 44

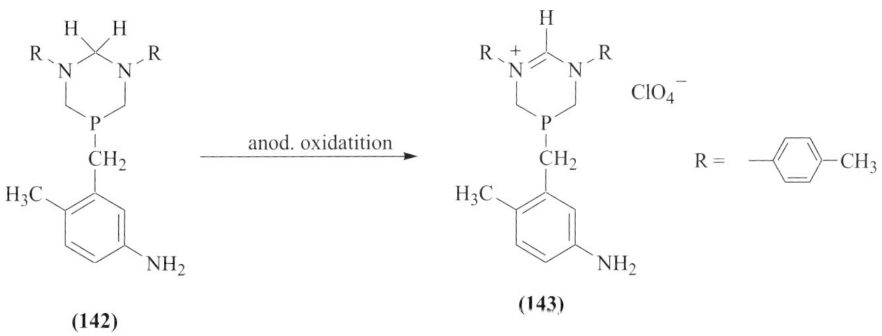

Scheme 45

Molybdenum (0) and tungsten (0) carbonyl-complexes, molecular and crystal structure of *cis*-bis(1,3,5-triphenyl-1,3,5-diazaphosphorinane)tetracarbonylmolybdenum ⟨96RJCC35⟩.

Synthesis of an unusual dichlorotris(1,3,5-triphenyl-1,3,5-diazaphosphorinane)platinum-(II)monohydrate complex and its behavior in solution ⟨96RJCC514⟩.

1,3,5-Triaza-7-phosphaadamantane (TPA) (**144**) may be mentioned here, as a bridged 1,3,5-diazaphosphinane. Its synthesis has been described by several groups [$\delta(^{31}P) = -102$ ppm] (Scheme 46) ⟨74JHC407, 75CZ246, 75JHC579, 77ZN(B)499, 90PS(48)37⟩.

The synthesis of 7-alkyl/aryl-1,3,5-triazaphosphoniaadamantane salts (**145**) by reaction of the hydroxymethylphosphoniumsalts with NH_4OAc has been described (Scheme 46).

Treatment of (**2**) with sodium in liquid ammonia results in the reductive cleavage to give (**146**) ⟨95CB891⟩.

The interest in TPA arises from its capacity to form metal complexes with mercury (TPA•HgX_2, X = Cl, Br, I, SCN, CN, NO_3) ⟨89CJC1319⟩, ruthenium and rhodium ($RuCl_2(TPA)_4$, $RuCl_2(TPA)_4$•HCl, $RhCl(TPA)_3$) ⟨92OM1990⟩ salts. The water-soluble complex $RuCl_2(TPA)_4$ is catalytically active in the reduction of unsaturated aldehydes to unsaturated alcohols. $RhCl(TPA)_3$ catalyzes olefin hydrogenation ⟨95JOM(488)99⟩. There is considerable recent interest in the gold(I) complexes of TPA ⟨95IC75⟩. TPA with $(Me_2S)AuCl$ in $CHCl_3$ affords the complex (TPA)AuCl in high yield. TPA•HCl in MeOH/MeCN

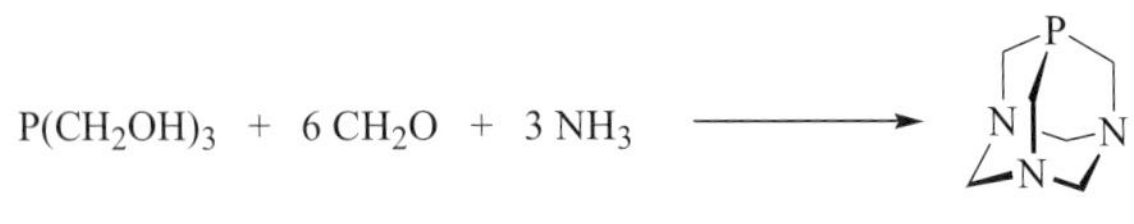

(144)

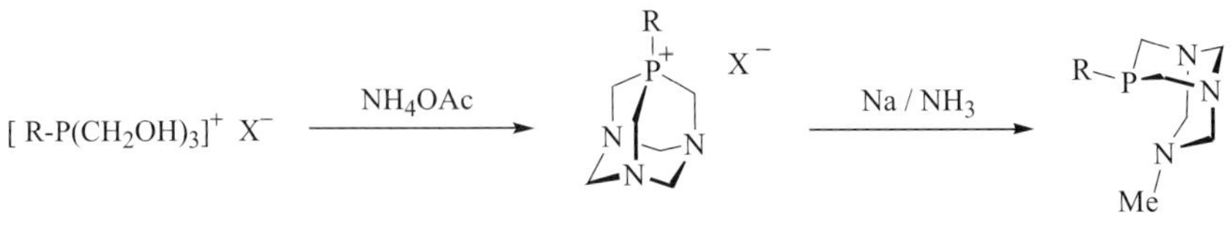

R = Me, Et, *c*-C_6H_{11}, Bz, Ph
X = Cl, I, PF_6, BPh_4

(145) **(146)**

Scheme 46

with (tetrahydrothiophene)AuCl gives the protonated complex (TPA•HCl)AuCl. Both complexes have photoluminescence properties. The HCl-free complex in the solid state luminesces intensely red (674 nm at 78 K), the HCl complex yellow (596 nm at 78 K). This different behavior is clearly a consequence of the change in the aureophilic Au• • •Au interaction (3.092 Å vs. 3.322 Å). It should be pointed out that the aggregation of Au(I) complexes through the forces originating from Au• • •Au contacts is emerging as an element in the design of supramolecular frameworks. The Au• • •Au forces are similar in energy to those of hydrogen bonds. Another interesting luminescent gold(I) complex with TPA is [(MeTPA)$_3$AuI]I$_3$ ⟨95OM4194⟩; see also ⟨94CC431⟩.

5.3.4.3 1,4,2-Diazaphosphinines

The only known example of this ring system is the 3,4-dihydro-2λ^5-compound (**147**) which was reported to be formed by the action of tris(dimethylamino)phosphine on monobenzylidene diaminomaleonitrile. On standing in acetonitrile it undergoes an unusual ring contraction: benzylidene bisdimethylaminoaminal and a proton are eliminated and the anion of dicyano-1,3,2-diazaphosphole is formed (Scheme 47) ⟨84CC183⟩.

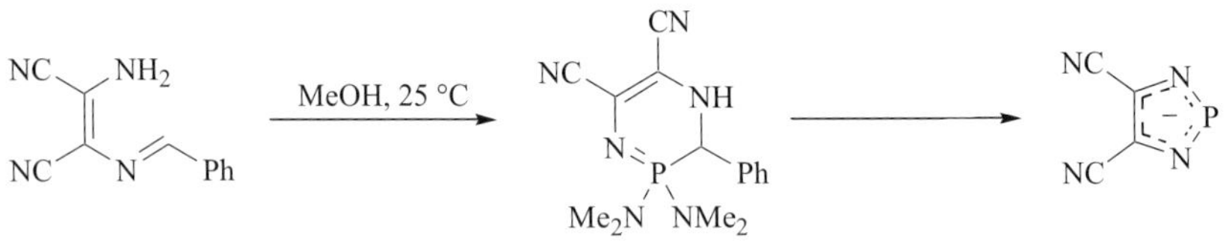

(147)

Scheme 47

5.3.4.4 1,5,2-Diazaphosphinines

The reaction of acetylurea with PCl_5 leads to 2,2,4,6-tetrachloro-1,5,2λ^5-diazaphosphinine (**148**) (m.p. 79–80°C) by a sequence of phosphorylation and chlorination steps ⟨87ZOB228, 91ZOB643, 93ZOB1976⟩. Chloroacetylurea with PCl_5 produces 2,2,3,4,6-pentachloro-1,5,2λ^5-diazaphosphinine (**149**) (m.p. 91–93°C) in a similar reaction sequence ⟨91ZOB643⟩. *N*-Acetyl-*N*′-methylurea with PCl_5 forms the salt (**150**), which can be reduced to (**151**), and thermally decomposed with loss of methylchloride to form (**148**) (Scheme 48) ⟨90ZOB462, 92ZOB306⟩.

The spectroscopic properties of all of the compounds described in this section have been thoroughly examined, including ^{35}Cl NQR and ^{31}P, ^{1}H and ^{13}C NMR.

The halogenated diaza- and particularly the triaza-λ^5-phosphinines and -diphosphinines can be considered as chimeras of cyanuric chloride, for the carbon-containing portion of the ring, and cyclotriphosphazene hexachloride — phosphonitrilic chloride — for the phosphorus-containing part. Their chemistry is understandable when this is born in mind. With only two nitrogen atoms present, as in the compounds

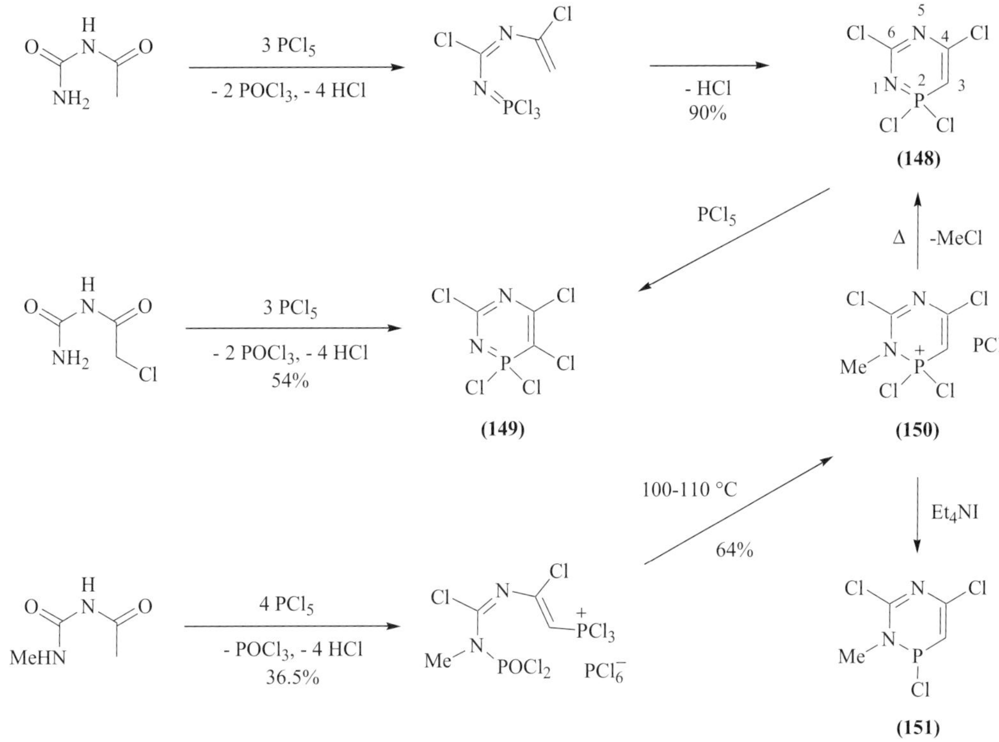

Scheme 48

discussed in the present section, the *P*-halogen atoms are the more reactive, and their replacement by nucleophiles, hydrolysis and so on, can occur while the *C*-halogens are retained. When three nitrogen atoms are present the *C*-halogens are more reactive. The references cited, and ⟨94ZN(B)171⟩, cover the principal features of the chemistry of the polyhalogenated 1,5,2λ^5-diazaphosphinines.

Barluenga and co-workers have reported the synthesis of 1,5,2λ^5-diazaphosphinin-6-ones (**154**) by cyclization of the enamino-phosphinimines (**153**). These are prepared from the alkyldiphenylphosphinimines (**152**) as indicated in Scheme 49 ⟨87TL2785, 89JCS(P1)615⟩.

Another very unusual synthesis of the 6-ones and the corresponding thiones was published by Bertrand and co-workers ⟨93JA7880, 94JA8087⟩. The 1,2λ^5-azaphosphetes (**155**) react with isocyanates and isothiocyanates forming the 6 oxo(thioxo)-1,5,2λ^5-diazaphosphinines (**156**) and (**157**) (Scheme 50).

Other 1,5,2λ^5-diazaphosphinines and 1,5,2λ^5-diazaphosphinin-6-ones have been reported, including some imidazo-fused derivatives which are analogous to the purine nucleosides ⟨78MI625-01⟩, and some with arylazo substituents at C-6 ⟨90HCA1888⟩.

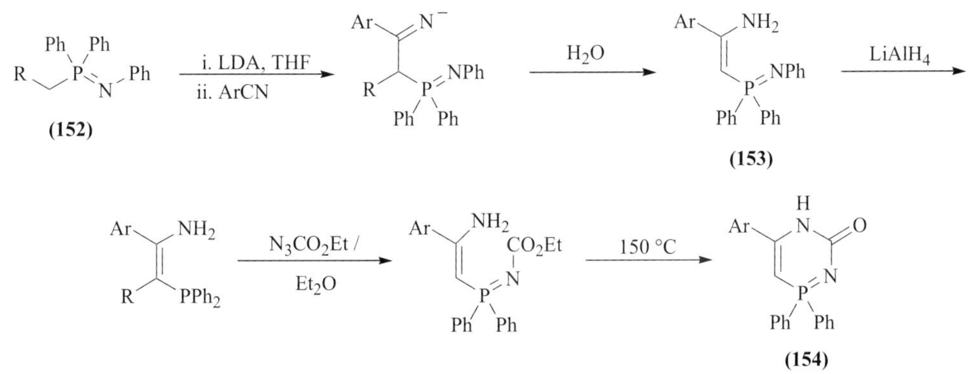

Scheme 49

Pyrazolo[4,5-c]diazaphosphinines (**161**) have been synthesized from 4-phosphorylated 5-formamidinopyrazoles (**158**). The dichlorophosphinine was transformed into bis(dialkylamino)phosphines (**159**). The imidophosphonicdiamide (**160**), which was prepared from (**159**) by chlorination with C_2Cl_6 followed

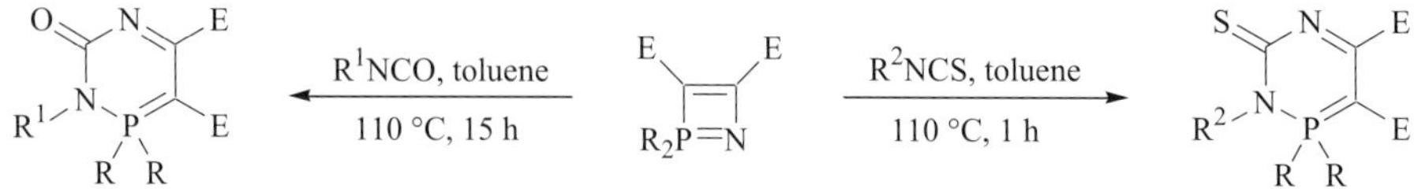

(156) (155) (157)

R = NPr$^i_2$

Scheme 50

by reaction with NH_3 undergoes cyclization *in situ* to give the novel ring system (**161**) (Scheme 51) ⟨99MC161⟩.

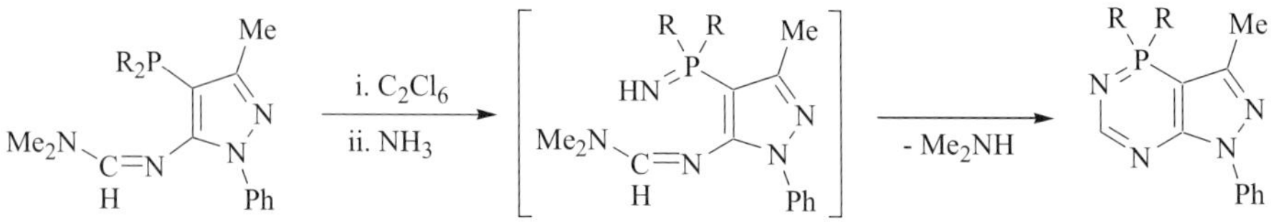

(158), R = Cl
(159), R = NEt$_2$, —N(morpholino)O

(160) (161)

Scheme 51

5.3.4.5 1,2,3-Diazaphosphinines

The *N*-phosphino-*C*-thiophosphinoyl-nitrilimine (**162**) is a surprisingly stable compound (m.p. 100°C). In cycloaddition reactions it can behave as a formal 1,4-dipole as well as a 1,3-dipole. With DMAD the 1,4-dipolar mode operates; cycloaddition takes place across C and P to give the 1,2,3λ^5-diazaphosphinine (**163**) as the first example of this ring system (orange crystals, m.p. 147°C) (Scheme 52). The spectroscopic data, especially the $\delta(^{31}P)$-values (+68.6, +14.5 ppm) and the high field shift of C-4 in the ^{13}C NMR [δ = 92.2 (dd, J((PC) = 107.5 and 11.7 Hz)], attest the ylide-like character of (**163**). An X-ray crystal structure determination has been carried out ⟨90AG(E)1123, 94JA8087⟩.

The synthesis is restricted to very electron-poor alkynes; even with propiolic acid 1,3-dipolar addition forms the expected pyrazole.

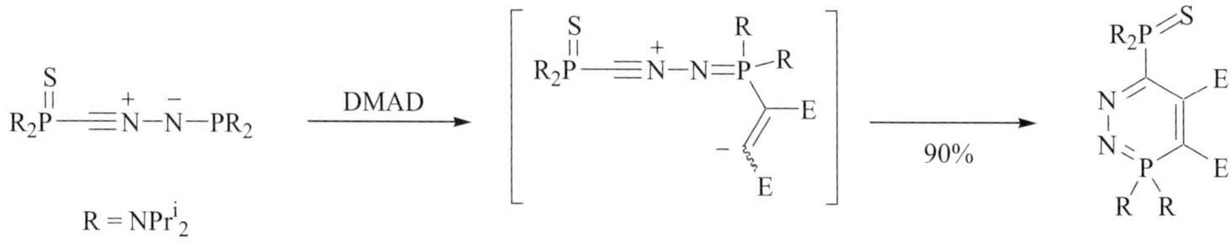

R = NPr$^i_2$

(162) (163)

Scheme 52

5.3.4.6 1,2,4-Diazaphosphinines

(4+2)-Cycloaddition of phosphinodiazoalkanes (**164**) to electrophilic alkynes, probably by the two-step process indicated in Scheme 53, produces the 1,2,4λ^5-diazaphosphinines (**165**). The NMR spectra of compounds (**165**) are consistent with their ylide character ⟨90AG(E)314⟩. The 1,4-dihydro-1,2,4λ^5-diazaphosphinine oxides (**166**) are formed in a similar cycloaddition of nitrilimines to acetylenic phosphonites ⟨73ZOB2178⟩ (Scheme 53).

This synthesis is analogous to the 1,3-dipolar cycloaddition of diphenylnitrilimine to diphenylvinylphosphane published already in 1965 ⟨65AG(E)1124⟩.

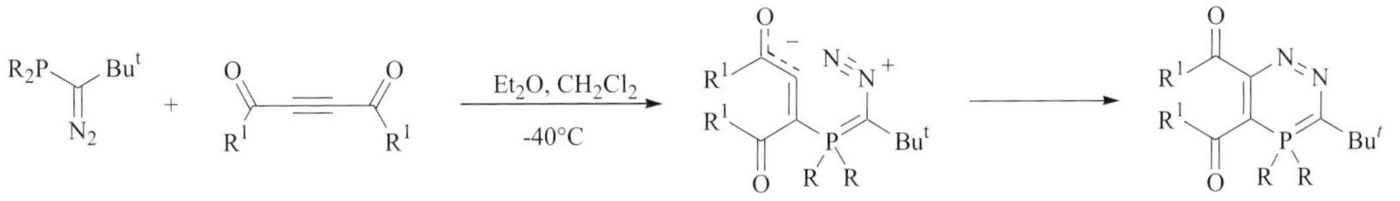

(164)

(165a), R = R^1 = OMe [$\delta(^{31}P)$ 1.5]
(165b), R = OMe, R^1 = Ph [0.3]
(165c), R = Ph, R^1 = OMe [-27.4]
(165d), R = R^1 = Ph [-32.0]

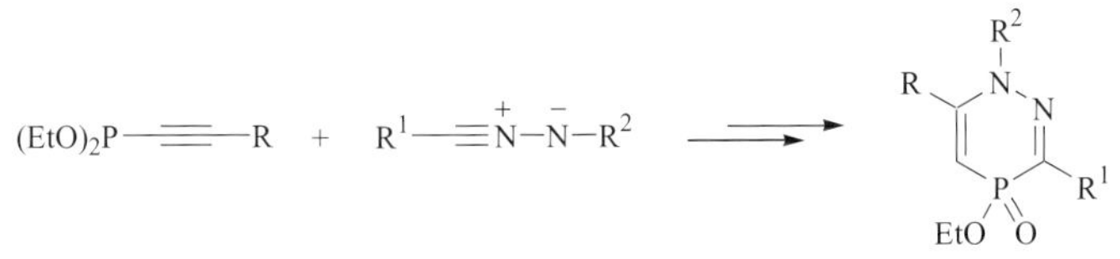

(166a)

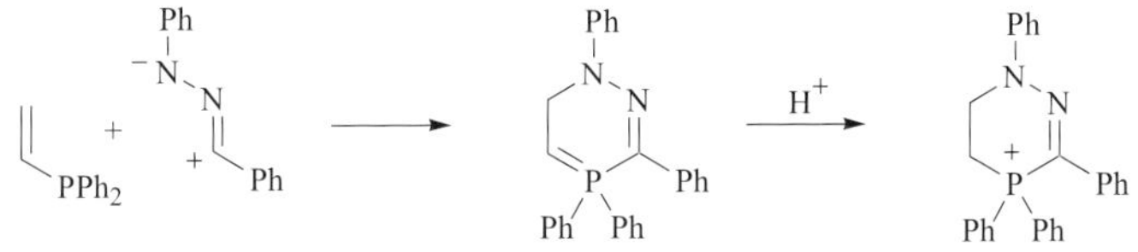

(166b)

Scheme 53

9-Oxa-3,4-diaza-1-phosphabicyclo[4.3.0]-1,1-diethoxy-nona-3,5-dienes (**168**) (1,4-dihydro-1,2,4-diazaphosphinines) with a λ^5-phosphorane structure are formed by the cycloaddition of diarylnitrilimines to 3-diethoxy-phosphino-4,5-dihydrofurane (Scheme 54) ⟨99RJGC736⟩.

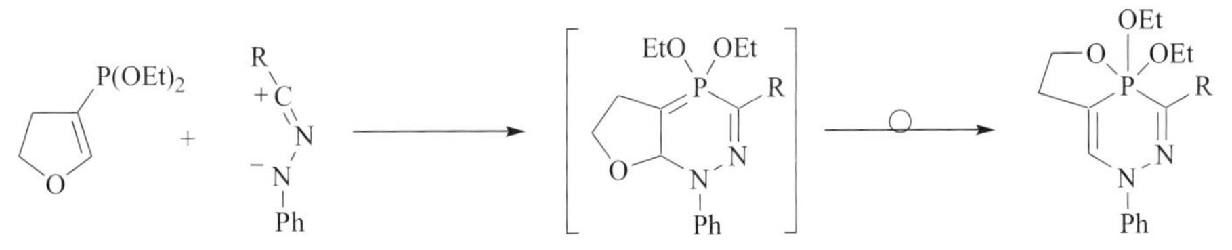

(167) **(168)**

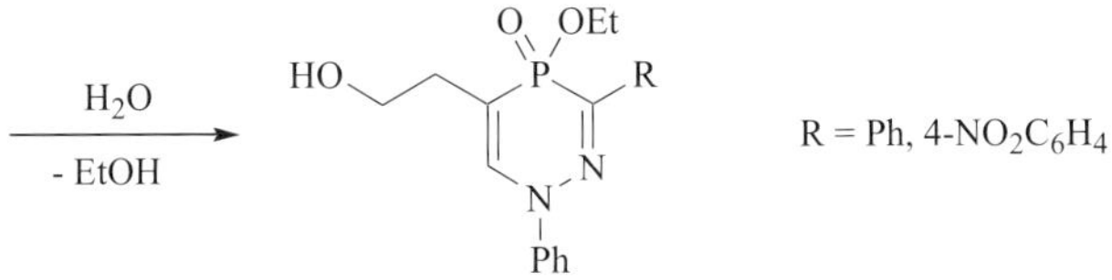

(169)

Scheme 54

The possible intermediate is (**167**) (see also ⟨65AG(E)1124⟩), which rearranges to give the five coordinate phosphorus atom in (**168**).

The crystal structure confirms a planar structure of the bicyclic core. Hydrolysis is accompanied by cleavage of the endocyclic P–O bond and elimination of an alkoxy group to give 1,4-dihydro-5-(2-hydroxyethyl)-1,2,4-diazaphosphinine-4-oxide derivatives (**169**).

5.3.5 RING SYSTEMS WITH ONE PHOSPHORUS AND THREE NITROGEN ATOMS

Of the possible six triazaphosphinine isomers onyl derivatives of the 1,2,3,4λ^5(I)- 1,2,4,3λ^5(II)- and 1,3,5,2λ^5(III)-triazaphosphinines have been reported. Triazaphosphinines with $\lambda^3\sigma^2$-phosphorus are not yet known.

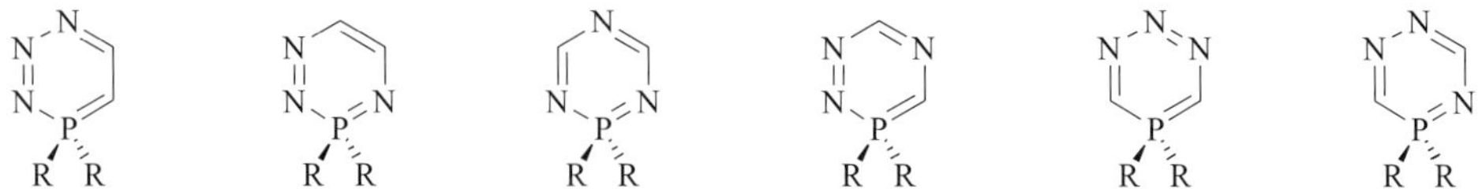

(I) **(II)** **(III)** **(IV)** **(V)** **(VI)**

5.3.5.1 1,2,3,4-Triazaphosphinines

Bertrand and co-workers studied the reaction of the azidophosphines (**188a,b**) with DMAD, and found that, as with other instances of phosphino-1,3-dipoles, the (4 + 2) mode of cycloaddition was favored. In this way the 1,2,3,4λ^5-triazaphosphinine diesters (**189**) were obtained as yellow crystalline materials (Scheme 61a). The NMR spectra [(**189**): $\delta(^{31}P)$ 5.74; $\delta(^{13}C)$ 81.77 d, $^1J(PC)$ 113.2 Hz, C-5; 150.4 d, $^2J(PC)$ 2.4 Hz, C-6] are consistent with the P ylide/phosphinimine character of the ring, and the X-ray data are in agreement with this. The products (**189**) heating in toluene lose nitrogen and form the 1,2-azaphosphetes (**80**), which undergo a versatile selection of cycloaddition reactions, providing phosphorus-containing six-membered rings ⟨93JA7880, 94JA8087⟩.

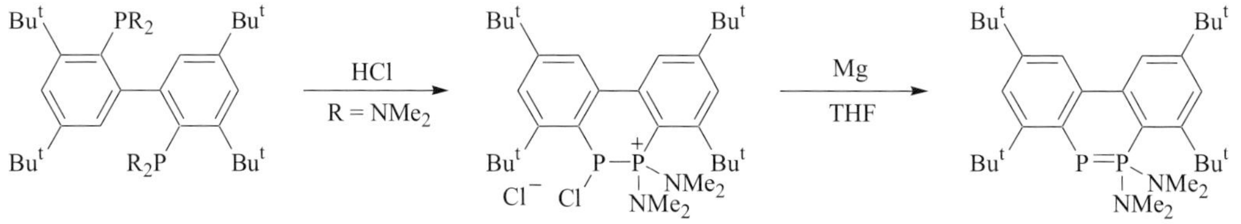

(188) **(189)** **(190)**

Scheme 61a

The ligand behavior of 1,2,3,4λ^5-triazaphosphinine (**189**) (R^1 = $N(^iPr)_2$, R^2 = CO_2Me) had been studied. With $PdCl_2(PhCN)_2$ resp. $W(CO)_5(THF)$ at r.t. the formation of the triazaphosphinine complexes (**190**) and (**191**) can be proved, which at r.t. smoothly loses N_2 to give the 1,2λ^5-azaphosphet-complexes (**192**) and (**193**) (Scheme 62a) ⟨96JA1038⟩.

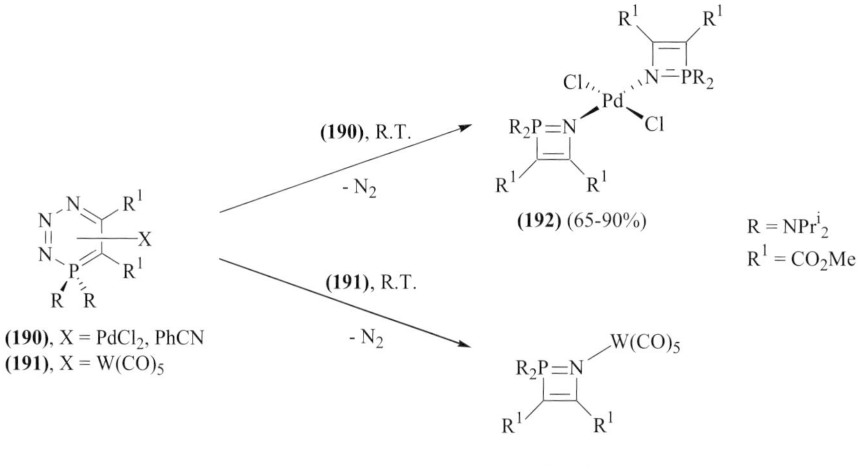

Scheme 62a

Addition of *cis*-$Mo(CO)_4$ $(pip)_2$ leads to the isolated (η^1-triazaphosphinine) $Mo(CO)_4$ (pip) complex (70%) which within two days at r.t. transforms into the η^1-(five-membered cyclophosphazene)-$Mo(CO)_5$ complex (**194**). With two equiv. of $W(CO)_5$ (pip) the complex (**195**) featuring a five-membered cyclophosphazene bonded to $W(CO)_5$ (pip) via a hydrogen bond is isolated (80%). $Cp_2Zr{\bullet}HCl$ reacts with (**189**) to give the five-membered zirkonacyclophosphazene (**196**) (45%) probably via the seven-membered zircona-ring system (Scheme 63a).

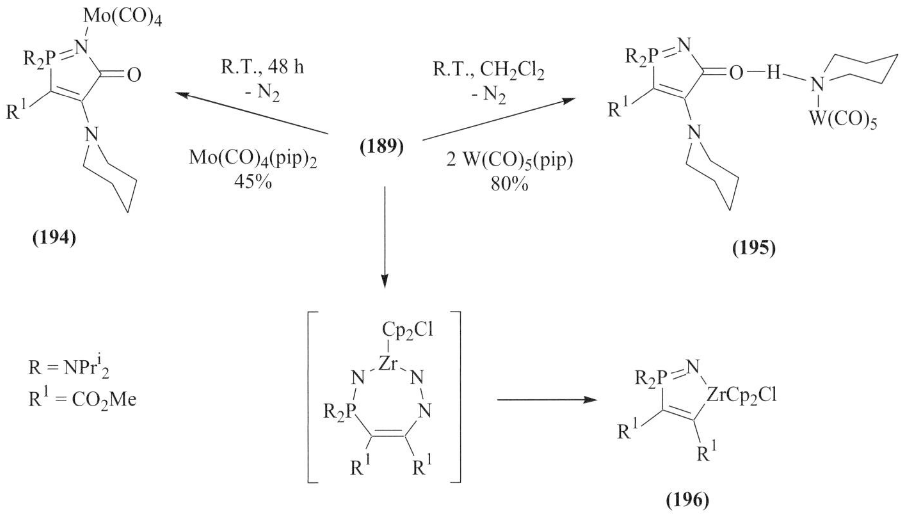

Scheme 63a

5.3.5.2 1,3,5,2-Triazaphosphinines

1,3,5,2λ^5-Triazaphosphinines with halogen and amino substituents have been investigated extensively, mainly between 1970 and 1980. The principal workers in the area were Glemser, Schmidpeter, Kornuta and Kukhar. The figure below provides references to six important compounds or groups of compounds studied at that time. Their syntheses are mostly based on the treatment of dicyanimides, *N*-cyanobenzamidines, *N*-cyanocarboxamides or imidoguanidines with PCl_5.

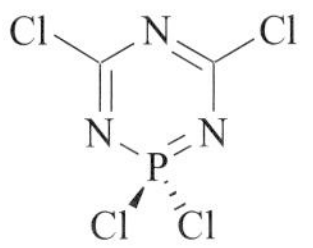

70ZAAC(372)233

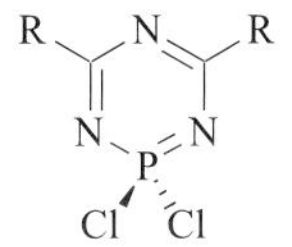

R = CCl_3, CF_3, CCl_2R^1, etc.

76CB2960, 76ZOB243, 76ZOB1482, 77CB1148, 77CB3231, 77ZN(B)117, 78ZOB1181, 79ZOB1671, 80ZOB311

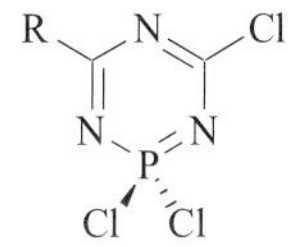

R = aryl, alkyl

77ZOB342, 78ZOB2218, 79ZOB1777, 81ZC137, 81ZC143

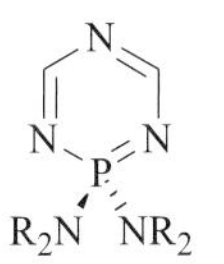

71AG(E)397

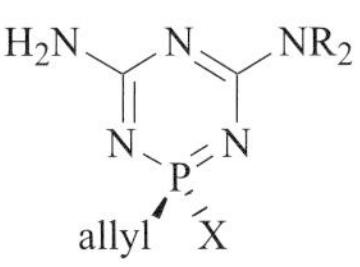

X = halogen, H

71ZN(B)650, 77ZAAC(436)136

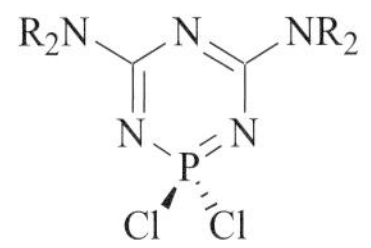

77ZAAC(436)136

N,N′-Diphenylformamidine provided the C-6 for cyclization of the diphenylphospho-guanidine (**170**) in Schmidpeter and Ebeling's synthesis of the 2,2-diphenyl-1,3,5,2λ^5-triazaphosphinine (**171**) ⟨67AG(E)565⟩.

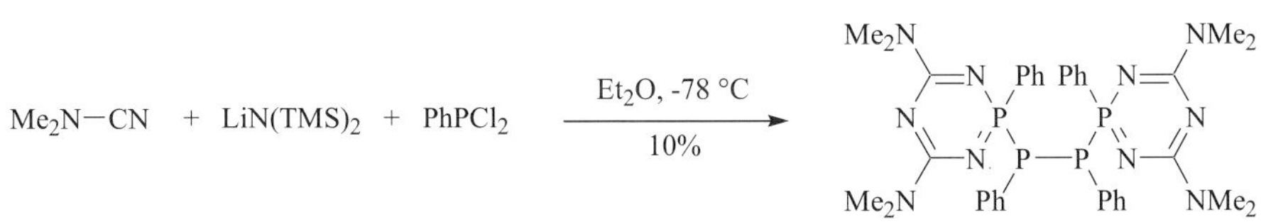

(172)

Only one synthesis in this area has been reported since 1983. When Me_2NCN was reacted with $LiN(SiMe_3)_2$ and then with $PhPCl_2$ in CH_2Cl_2 at −78°C, an unusual product (**172**) (m.p. 228°C) can be isolated, in which two 4,6-bis(dimethylamino)-2-phenyl-1,3,5,2λ^5-triazaphosphinine rings are connected by a $-P_2Ph_2-$ bridge ⟨88CB1681⟩.

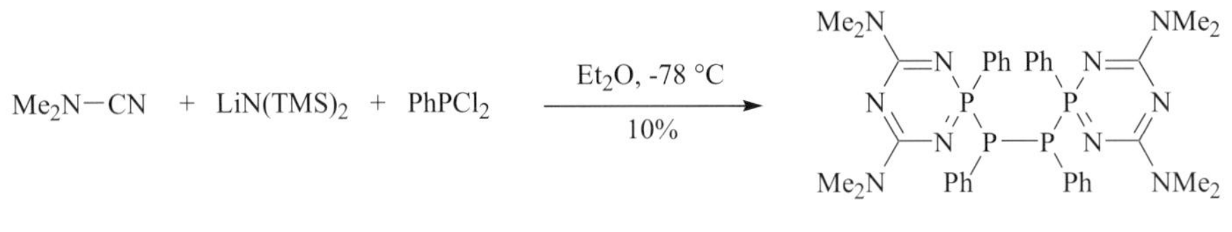

(172)

The presence of $\lambda^5\sigma^4$- and $\lambda^3\sigma^3$-phosphorus atoms is evident from the ^{31}P NMR spectrum of (**172**) [$\delta(^{31}P) = 46.3$ and -46.6, AA′AA′ spin system].

X-ray crystal structure data are available for two 1,3,5,2λ^5-triazaphosphinines with P-amino substitution. Dipole moment measurements ⟨83ZOB559⟩ and conformational studies ⟨85ZOB162⟩ have also been carried out.

Bromophenyldiazine (**173**) is an unusual versatile reagent, which enables the synthesis of neutral, mono- and dicationic 4–16-membered phospha-, diphospha- and tetraphospha-heterocycles. In a special case (**173**) reacts with R_2P–$SnMe_3$ to give the following reaction mixture (Scheme 55):

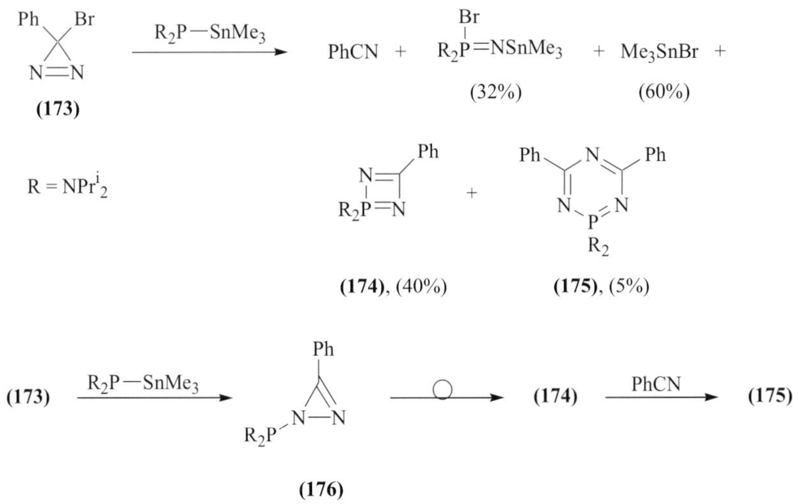

Scheme 55

The possible reaction mechanism starts with the formation of the phosphino-1*H*-diazirine intermediate (**176**), which rearranges to give (**174**). The 1,3,5,2λ^5-triazaphosphinine is the result of a (4 + 2)-cycloaddition of RCN to (**174**) (Scheme 55) ⟨96JA1060⟩.

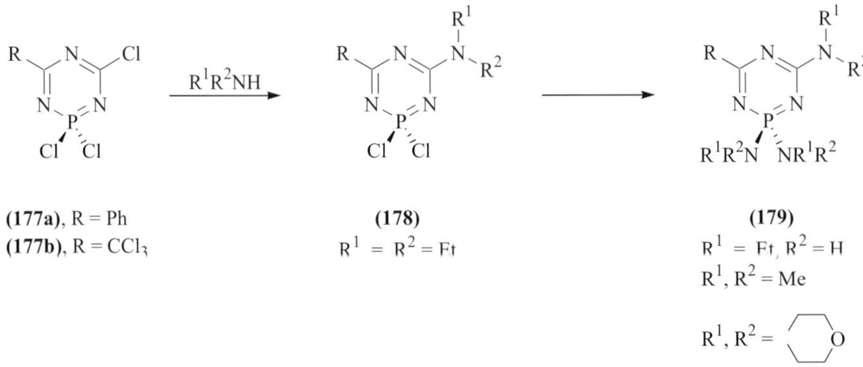

Scheme 56

The chlorine atoms in the 1,3,5,2λ^5-triazaphosphinines have acid chloride character and are readily replaced by nucleophiles. In the 2,2,4-trichlorocompounds (**177**) the reaction with amines occurs selectively. At a molar ratio of (**177**) to amine of 1 : 2 only the C-4 chlorine is replaced. With excess of amine (dimethylamine, morpholine, ethylamine) the triamino derivatives (**179**) are formed, even at 20°C. The

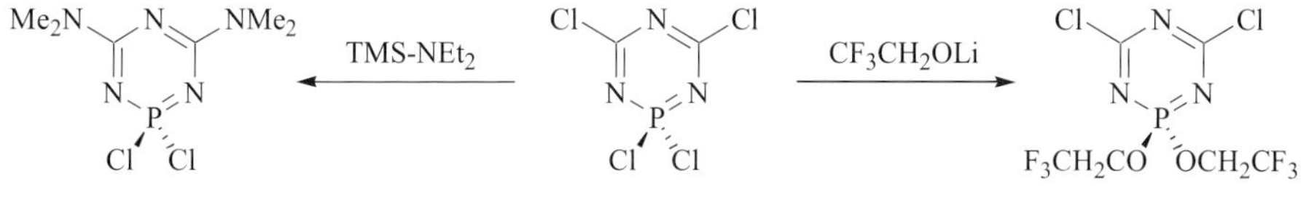

(180) **(179)**

Scheme 57

reaction with diethylamine stops at the stage of the monoamino derivative (**178**) for steric reasons (Scheme 56) ⟨79ZOB1777⟩.

In (**177a**) with sodium benzothiolate (ratio 1 : 1) only the halogen at C-4 is replaced. With sodium methoxide and sodium phenoxide (ratio 1 : 1) a mixture of monosubstitution products at C-6 and at the phosphorus atom is formed ⟨82ZOB590⟩. Roesky and Mainz confirmed this substitution pattern ⟨86ZAAC(540/1)212⟩. In the tetrachloro-triazaphosphinine (**180**), with Me_3SiNMe_2 the halogens in the 4- and 6-positions, and with CF_3CH_2OLi only the *P*-halogen atoms are substituted (Scheme 57). The tetrakis(dimethylamino) substitution product can be prepared from (**180**).

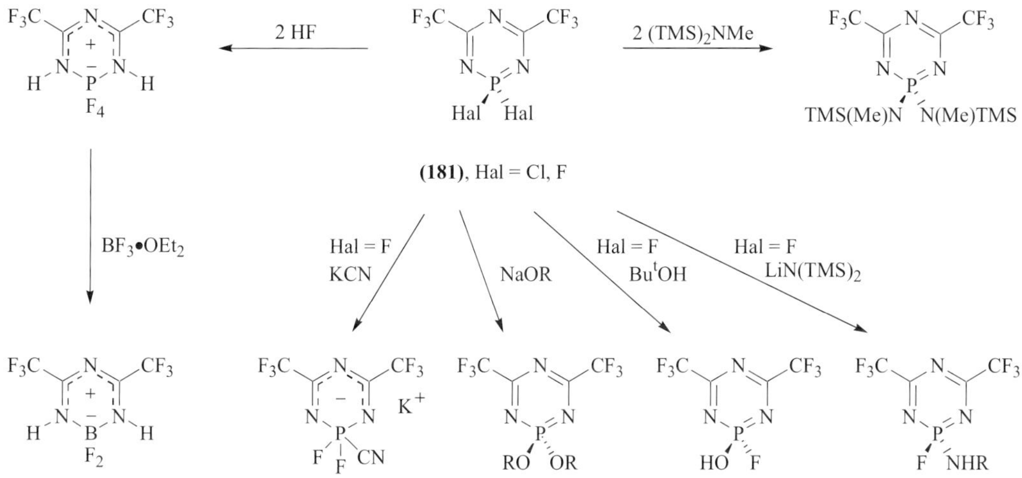

Scheme 58

The 2,2-dihalo-4,6-bis(trifluoromethyl) compounds (**181**) undergo an interesting variety of substitution and addition reactions (Scheme 58) which have been studied by Klingebiel and co-workers ⟨88CB627, 88CB1119, 88ZN(B)1010⟩ and others ⟨74ZOB2104, 76CB2960, 82ZOB2742, 87ZOB1980⟩.

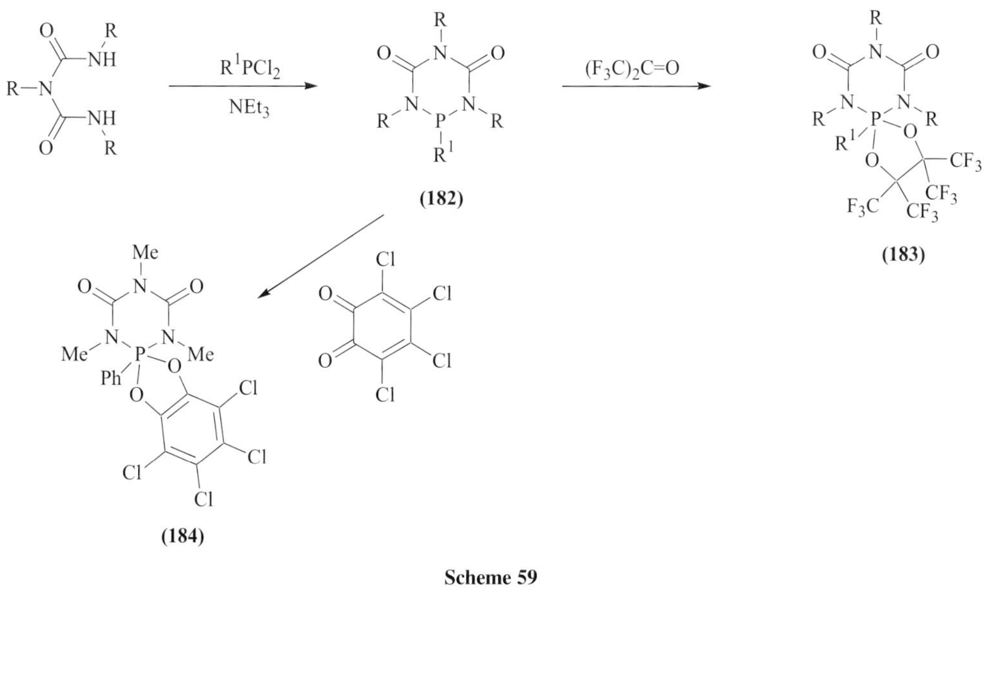

Scheme 59

The synthesis of 1,3,5-triaza-2-phosphorin-4,6-diones (**182**) by reaction of biuret derivatives with $RPCl_2$ has seen studied thoroughly by Schmutzler *et al.* (Scheme 59) ⟨92ZN(B)760, 93ZN(B)443, 93ZN(B)853, 94ZN(B)145, 95ZN(B)1833, 94CB1395, 95ZAAC2001⟩.

With hexafluoroacetone the λ^5-phosphoranes (**183**) are formed ⟨95ZN(B)1785⟩, with tetrachloroorthobenzoquinone the adducts (**184**) with pentacoordinated phosphorus are obtained ⟨96ZN(B)1494⟩.

Addition of aldehydes to the secondary phosphinoxide (**185**) yield the carbinols (**186**) ⟨95ZN(B)1833⟩. In the so-called *Pudowik*-reaction (**185**) adds to a series of 3-thiazolines in the presence of base to give the adducts (**187**) (Scheme 60) ⟨96ZN(B)1486⟩.

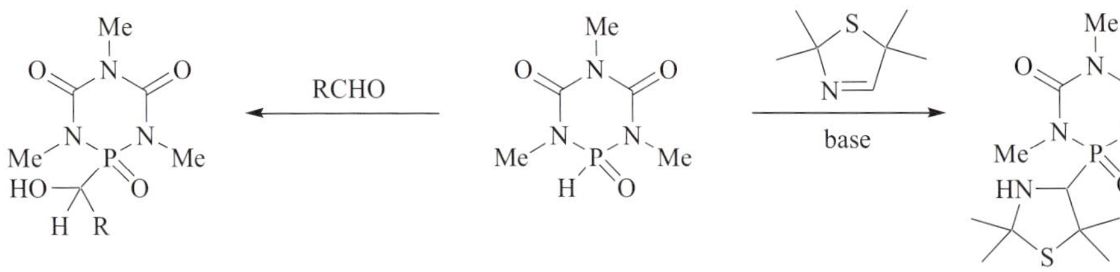

(**186**) (**185**) (**187**)

Scheme 60

An X-ray of the 2-trichloromethylderivative of (**182**) (R = Me, R^1 = CCl_3) shows an envelope conformation with phosphorus out of the plane of the other atoms ⟨96ZN(B)1501⟩.

5.3.6 RING SYSTEMS WITH TWO PHOSPHORUS ATOMS

Not all of the possible diphosphinines (permuting λ^3-P and λ^5-P over the 1,2-, 1,3- and 1,4-orientations gives a total of nine), are known, and some of those are too unstable to isolate. Representatives of the depicted λ^3- and λ^5-diphosphinines I–VI have been published. In addition, the 1,4-dihydro-1,4-diphosphinines have been synthesized with their dibenzo compounds (5,10-dihydrophosphanthrenes), since the early 1960s.

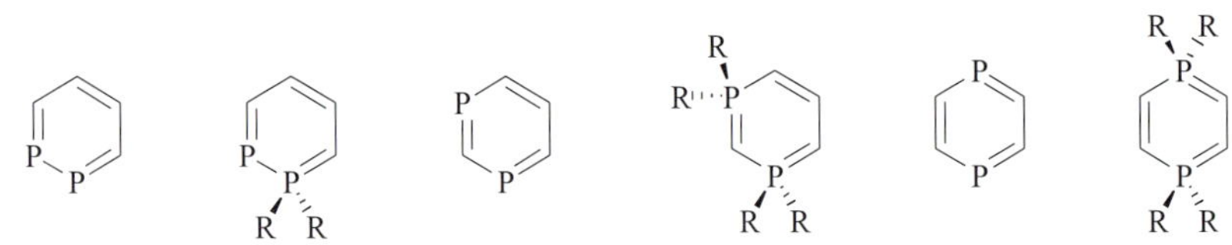

(**I**) (**II**) (**III**) (**IV**) (**V**) (**VI**)

5.3.6.1 1,2-Diphosphinines

Although diphosphenes (R–P=P–R) ⟨81JA4587⟩, and also five-membered rings with P=P bonds ⟨86TL5367⟩, are known, until 1991 only one attempt, by Bickelhaupt and co-workers, at the synthesis of a 1,2-diphosphinine has been reported. 2,2′-Bis(bis(dimethylamino)-phosphino)-3,3′,5,5′-tetra-*t*-butylbiphenyl (**188**; R = NMe_2) on treatment with hydrogen chloride was not converted into the expected bis-dichlorophosphine derivative (**188**; R = Cl) but formed (**189**) by cyclization. Dehalogenation of (**189**) with Mg produced the orange λ^3,λ^5-diphosphenanthrene (**190**), which could not be purified. However, the

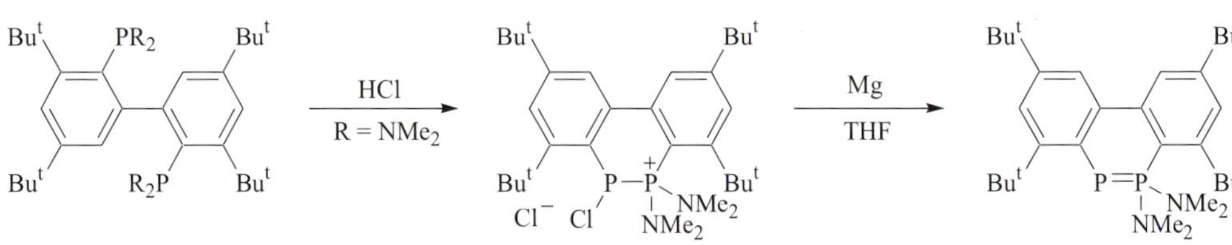

(**188**) (**189**) (**190**)

Scheme 61b

^{31}P NMR data confirmed the structure [$\delta(^{31}P) = 94.5$ (d), -129.0 (d), $^{1}J_{(pp)} = 469$ Hz] (Scheme 61b) ⟨91HAC17⟩.

In a highly sophisticated reaction sequence the 7-phosphanorbornadiene precursor (**191**) at 60ºC generates the transient [*cis*-2-butene-1,4-diyl-1,4-diphosphinidene]decacarbonylditungsten (**192**). In the presence of CuCl (**192**) reacts with 3-hexyne to yield the cycloadduct (**194**) which probably is the result of the [2 + 2]-cycloaddition with the valence isomer 3,6-dihydro-1,2-diphosphine (**193**) (Scheme 62b).

$\delta(^{31}P)$ ($CDCl_3$) $= -43.94$ ppm; $^{1}J(^{31}P, ^{183}W) = 162.9$ Hz; $^{2}J(^{31}P, ^{183}W) = 85.7$ Hz.

With 2,3-dimethylbutadiene and (**192**, **193**) [3,4-dimethyl-1,6-diphosphabicyclo[4.4.0]deca-3,8-di-ene]decacarbonylditungsten (**195**) is generated ⟨99JOM53⟩.

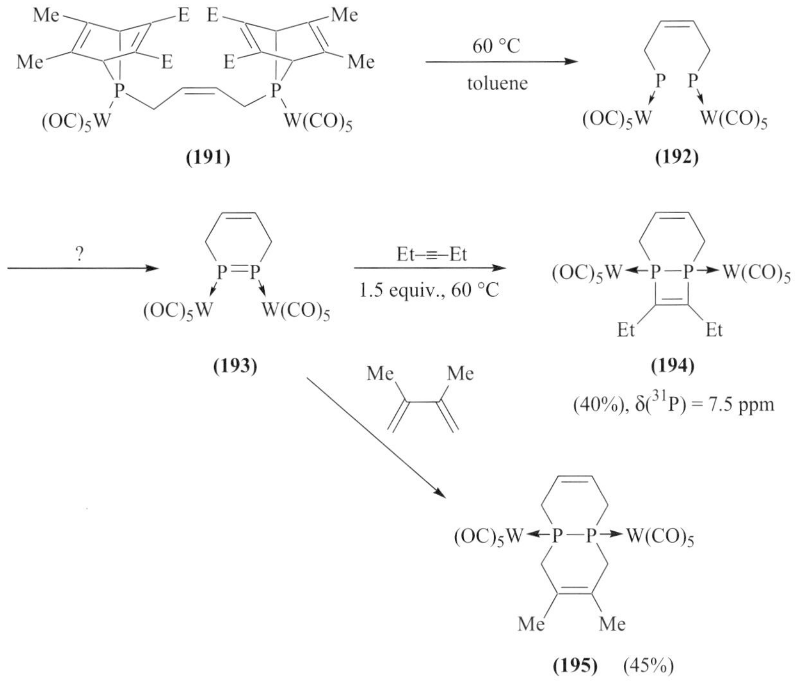

Scheme 62b

The thermal decomposition of the diphosphene $Bu^t_2P–P{=}P(Me)Bu^t_2$ (**196**) at 70ºC (72 h) in the presence of 2,3-dimethyl-1,3-butadiene as trapping reagent yields Bu^t_2PMe (52%) as main product beside $(Bu^t_2P)_2PH$ (12%) and 1,2-bis(di-*tert*-butylphosphino)-4,5-dimethyl-1,2-diphosphacyclohexene-4 (**197**) ⟨97ZAAC1769⟩ (Scheme 63b).

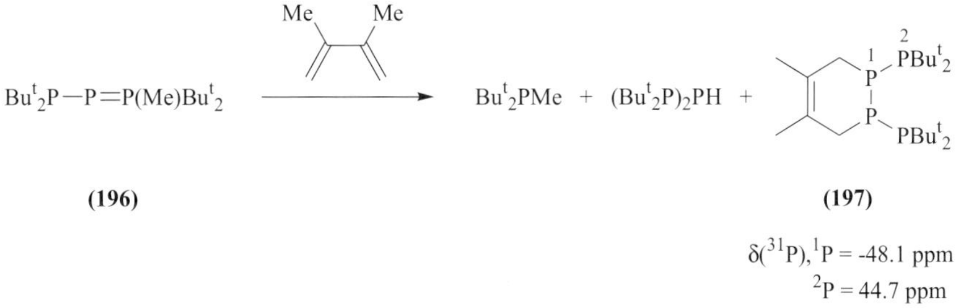

Scheme 63b

The diallyldiisopropylaminophosphine (**198**) was pyrolysed in a stirred flow reactor at 400–450ºC to give the intramolecular (4 + 2)-cycloadduct (**200**) of the 1-phosphabutadiene (**199**) ⟨97HAC97⟩ (Scheme 64).

The thermal conversion of the 1-alkynyl-1,2-dihydrophosphetes (**201**) yields the 2,3,5-trisubstituted λ^3-phosphinine (**202**) beside a trace of the tetrahydro-1,2-phosphinine (**203**) (Scheme 65).

Intermediate obviously is the phosphadiene (**204**), the intramolecular cycloaddition leads to (**202**). (**203**) is the result of a (4 + 2)-cycloaddition of the phosphadiene (**204**) with itself.

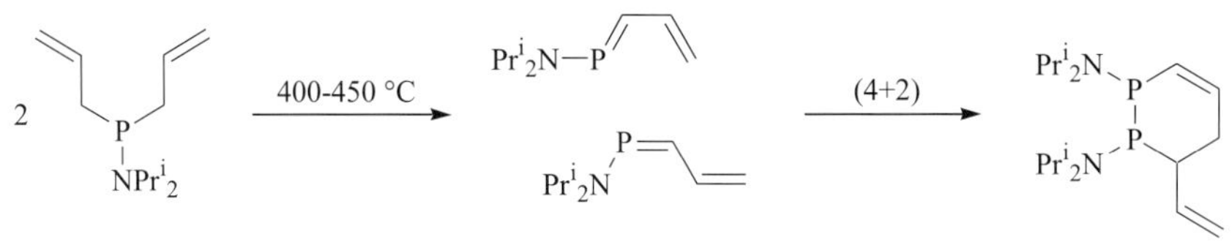

(198) (199) (200)

Scheme 64

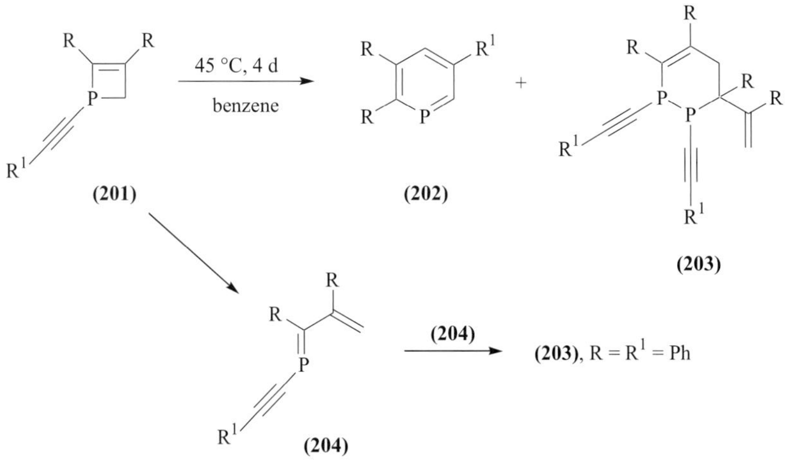

(201) (202) (203) (204)

Scheme 65

(**203**), MS, $m/z = 648$; ^{31}P NMR (C_6D_6), $\delta = -27.3, -49.3$; $^1J(P/P) = 200$ Hz 〈96HAC397〉.

Mathey *et al.* described the formation of a 2-(3,4-dimethylphospholyl)phosphinine (**205**) by substitution of 2-bromo-3-methyl-λ^3-phosphinine with lithium-3,4-dimethylphospholide (in the presence of Pd (0) or Ni (0) catalysts). Thermolysis of (**205**) at 180ºC in xylene leads to the formation of the dimer (**206**) (60%) which may be considered as a tetrahydro-1,2-diphosphinine. The presence of the P–P-bond is demonstrated by the existence of a huge ^1J(P/P) coupling constant of 221.0 Hz (Scheme 66) 〈95BSF(129)910〉.

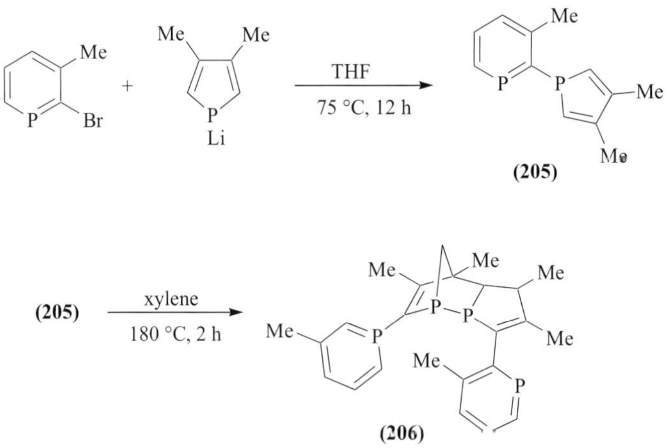

(205) (206)

Scheme 66

Another rather unusual approach to tetrahydro-1,2-diphosphinines also was described by Mathey. A metastable Ga^I-bromide in toluene/THF reacts with lithium-2,5-bis(*tert*-butyl)phospholide, the pentane extract contains the η^5-phospholylgallium (**207**) ($\delta(^{31}P) = 67$ ppm) beside (**208**) ($\delta(^{31}P) = -16{,}7, -30.2$ ppm, $J(^1P,^2P) = 214$ Hz) (Scheme 67) 〈99AG(111)1757〉.

The unusual dienophilicity of trimethylsilylphosphaacetylene (**209**) opens an unusual series of cycloaddition reactions. With butadienes-1,3 the reaction sequence: [4 + 2]-cycloaddition, ene- and intramolecular [4 + 2]-cycloaddition affords the diphosphatricyclooctenes (**210**) (Scheme 68) 〈99EJOC363〉.

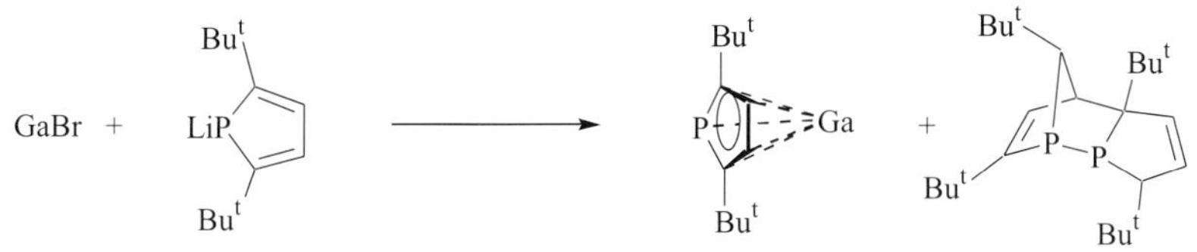

(207) (208)

Scheme 67

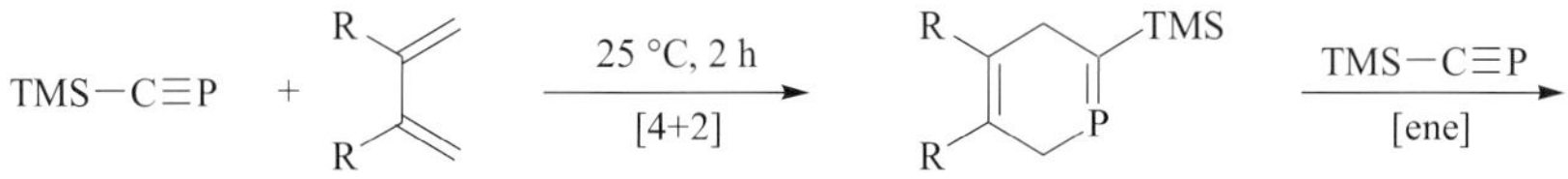

(209)

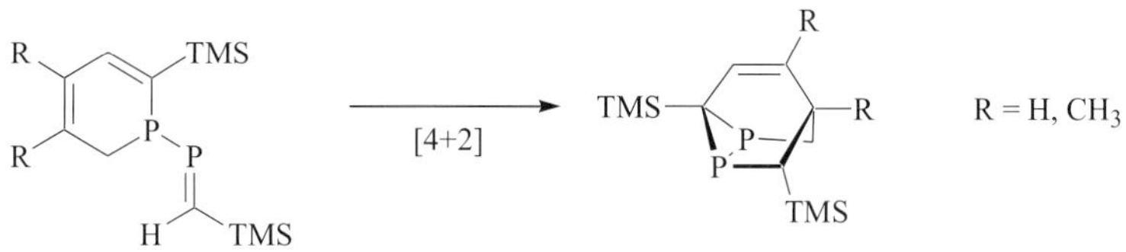

(210)

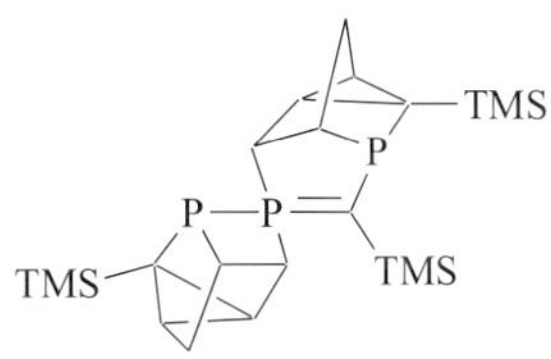

(211)

Scheme 68

An *ab initio* MO study of the HOMO/LUMO situation for the [4 + 2]-cycloaddition of (**209**) in comparison with Bu^{t}C≡P has been carried out. The result of the reaction of (**209**) with cyclopentadiene forming (**211**) is completely unexpected ⟨99EJOC363⟩. Here by the sequence Diels–Alder, homo-Diels–Alder and homo-[4 + 1]-cycloadditions the formation of (**211**) can be rationalized. The system (**211**) represents a 1,2-diphosphorane (Scheme 68).

The extremely noxius and pyrophoric 1,4-diphosphinobutane (**212**) reacts with 1 molequiv. Bu^{n}Li in hexane (24 h) to give the 1-lithium-1,2-diphosphorane (**213**), which on treatment with $Br(CH_2)_4Cl$, followed by addition of a second mole Bu^{n}Li finally leads to the formation of the bicyclic-1,2-diphosphorane (**214**) (65%, ^{31}P NMR, $\delta = -77.5$ ppm) (Scheme 69) ⟨96CC537, 96CC539, 98JCS(P1)1657⟩.

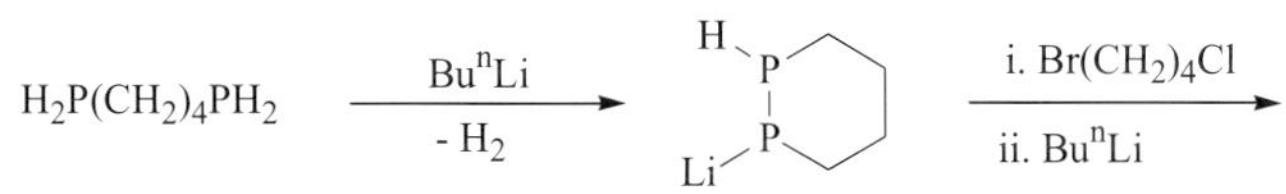

(212) (213)

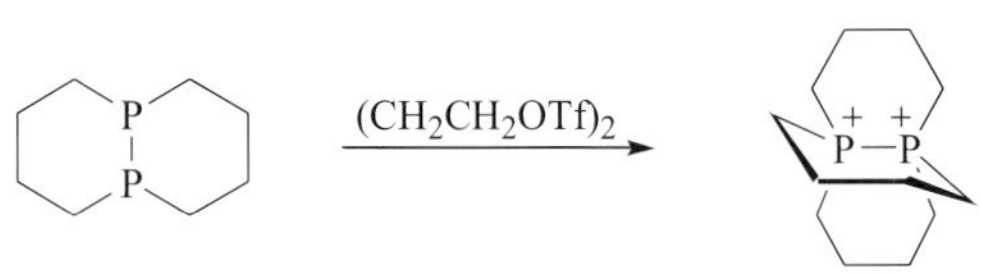

(214) (215)

Scheme 69

The diphosphabicyclo[4.4.0]decane (**214**) can be alkylated with $(CH_2CH_2OTf)_2$ to give the bisphosphoniumsalt (**215**).

5.3.6.2 1,3-Diphosphinines

The behavior of *t*-butylphosphaethyne in the coordination sphere of transition metals opens up an entirely new field of phosphorus compounds with unusual structures ⟨88AG(E)1485, 88CRV1327⟩.

When bis(η^2-ethene)(η^6-toluene)iron (0) (**216**) ⟨86AG(E)831⟩ was reacted at −40°C with acetylene or at −30°C with terminal alkynes and subsequently with excess of *t*-butylphosphaethyne, the stable sandwich complexes (**217**) with diphosphinine and diphosphet ligands could be isolated in 35–55% yield. The (2 + 2 + 2)-cycloaddition of two moles Bu^{t}C≡P with R–≡–H is regiospecific. Only the 1,3- and not the 1,4-diphosphinine is formed and the substituent R is exclusively beside the P-atom (Scheme 70). The stable free diphosphinines are obtained by oxidation of (**217**) with $FeCl_3$ or with CCl_4 at 80°C ⟨95AG(E)81⟩. The spectroscopic data of (**218a**) are in agreement with the expected Hückel aromatic 6π system: (**218**), $\delta(^{31}P) = 220.2$ (d) 237.0 (d), $J(P/P) = 24.7$ Hz; ^1H NMR, δ(H-5) = 8.78 (ddd, J(HP) = 41.8, J(HP) = 4.3, J(HH) = 11.1 Hz), δ(H-4) = 8.06 (ddd, J(HP) = 11.1, J(HP) = 6.8, J(HH) = 11.1 Hz). The structure of (**217**) was confirmed by an X-ray analysis ⟨95AG(E)198⟩.

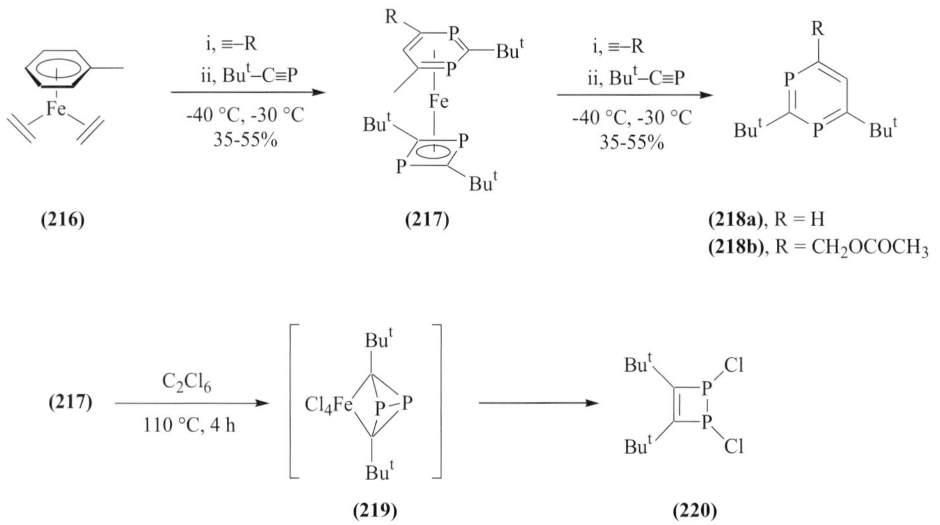

Scheme 70

Also the crystal structure of the sandwich-complex (**217b**) R = $-CH_2O-C(O)CH_3$ has been published ⟨95ZK801⟩.

The oxidation of (**217**) with hexachloroethane C_6Cl_6 results in the formation of the *trans*-1,2-dichloro-3,4-di-*tert*-butyl-1,2-diphosphetene (**220**) probably via the tricyclic intermediate (**219**) ⟨96S265⟩. The structure of (**220**) has been confirmed by a X-ray analysis (Scheme 70).

Triphenylphosphonium-1-trimethylsilylbenzylides (**221**) are condensed with the bis(dichlorophosphino)-substituted ylid (**222**) to give the 1,3-diphosphatetraline derivatives (**224**). By dechlorination of (**224**) with $GaCl_3$ the novel dicationic 1λ^3,3λ^3-diphosphanaphthalene ring system (**225**) is obtained, intermediate is the open-chained bis-ylid (**223**). $\delta(^{31}P)$ NMR, (**225a**): P(A) = 269.3 ppm; P(B) = 348.7; (**225b**): P(A) = 262.4 ppm; P(B) = 344.8; (**225c**): P(A) = 268.4 ppm; P(B) = 344.5 (Scheme 71) ⟨96ZN(B)773⟩.

Allylidentriphenylphosphorane (**226**), which may be considered as a phosphonio-substituted allylanion reacts with PCl_3 in pyridine at 0°C, the ^{31}P NMR spectra indicate the formation of complex mixtures of reaction products. With a 3 : 1 initial molar ratio of (**226**) and PCl_3 the 1,4,7,10-tetraphosphaphenalene derivative (**228**) is the main product after 2d. No intermediate on the way to (**228**) can be identified, probably (**226**) with PCl_3 forms the 3 : 1 condensation product (**227**), which is an tris-ylid, able to condensate with PCl_3 by cyclization to give (**228**). X-ray analysis and the ^{31}P NMR spectroscopy prove the structure (Scheme 72) ⟨98EJIC1907⟩.

The stable phosphaalkynes undergo cycloaddition reactions with tropone, to give cycloadducts, which may be considered as derivatives of the 1,3-diphosphinines. At 95°C without solvent in a Schlenk pressure tube in the presence of an excess of the phosphaalkyne tropone reacts with a first mole R–C≡P to give the

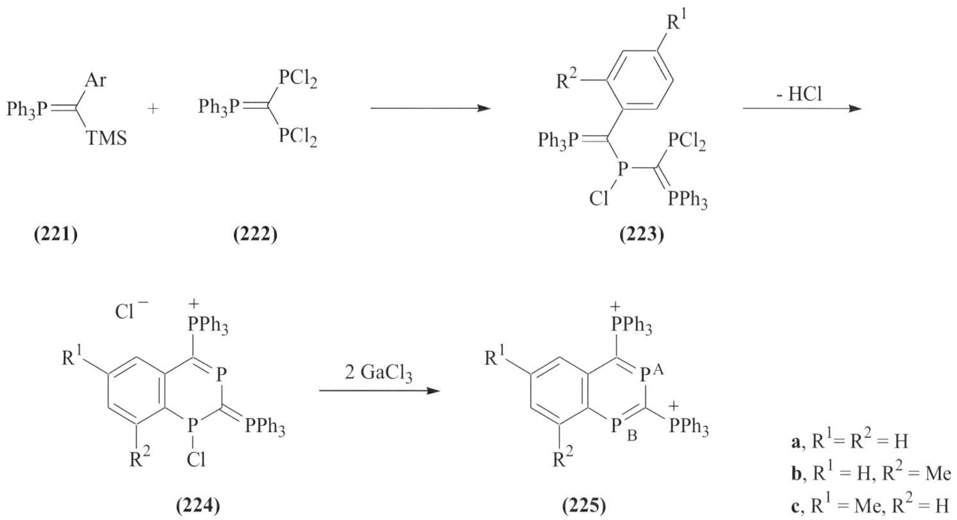

Scheme 71

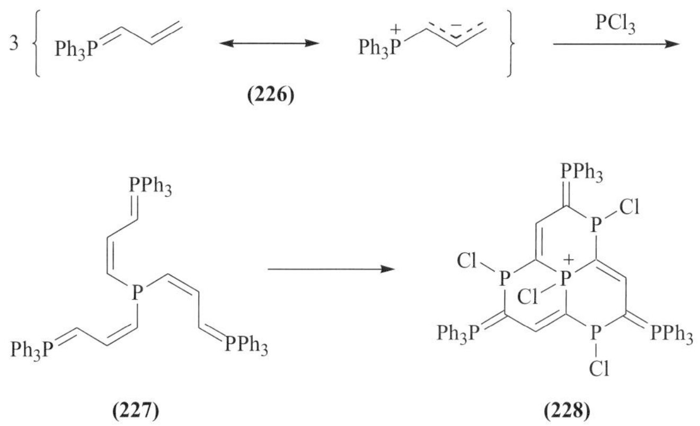

Scheme 72

Diels–Alder adduct (**229**) followed by a [2 + 2 + 2]-cycloaddition with a second mole R–C≡P to yield the tetracyclic phosphorus-cage compound (**230**) (Scheme 73).

When equimolar amounts of R–C≡P in toluene are heated with tropone the adduct (**230**) reacts subsequently with a second mole tropone in a [8 + 2]-cycloaddition to form the pentacyclic adducts (**231**) (Scheme 73).

The tetracycle (**230**) (R = Bu^t) in a (4 + 2)-cycloaddition reacts with 2,3-dimethylbutadiene in toluene at 125°C after cyclopropyl-allyl rearrangement to yield (**232**). With mesitylnitriloxide the (3 + 2)-cycloadduct (**233**) is formed (Scheme 74) ⟨95JOC5884⟩.

1λ^5,3λ^5-Diphosphinines have been studied in the last years mainly by Fluck *et al.* However, already Märkl in 1963 described with the synthesis of 1,1,3,3-tetraphenyl-1,3-diphosphinine (**236**) the first derivative ⟨63ZN(B)1136⟩. 1.1.3.3-Tetraphenyl-1,3-diphosphoniocyclohexen-4 (**234**) is easily deprotonated to give (**235**), which can be considered as a λ^5,λ^5-phosphamethinecyanine. Further deprotonation with aqueous base affords (**236**) as an orange, noncrystalline powder (Scheme 75).

With the synthesis of 1,1,3,3-tetrakis(dimethylamino)-1λ^5,3λ^5-diphosphete (**239**) by treatment of methyl(bisdimethylamino)difluorophosphorane (**237**) with *n*-butyllithium via the transient λ^5-phospha-alkyne (**238**) ⟨85ZN(B)1258, 86PS(29)23⟩, Fluck and co-workers obtained a highly reactive compound which undergoes at least formal Diels–Alder reactions with alkynes at room temperature, forming 1λ^5,3λ^5-diphosphinines (**241**) in fair to good yields. The transient Dewar isomer intermediates (**240**) were not detected. Symmetrical and some unsymmetrical alkynes give single regioisomers; other unsymmetrical alkynes form mixtures of regioisomeric products (Scheme 76) ⟨87CZ309, 89NJC383, 90ZAAC(588)181, 91ZN(B)202, 94ZAAC(620)483⟩. Photoelectron spectra and MO (MNDO) calculations on some of the products

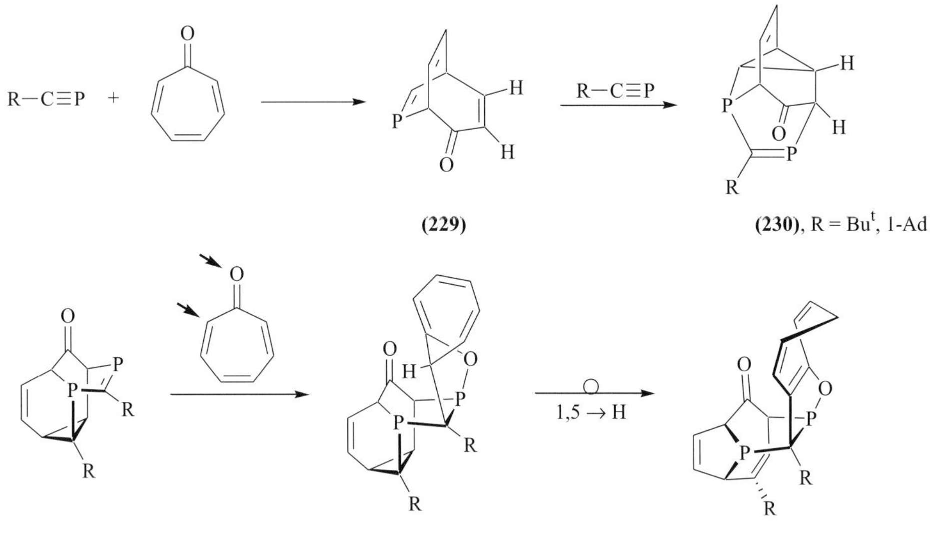

(229) (230), R = Bu^t, 1-Ad

(230), R = Bu^t, 1-Ad, CMe_2Et (231)

Scheme 73

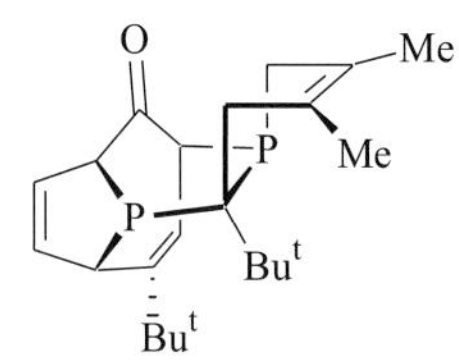

(232)

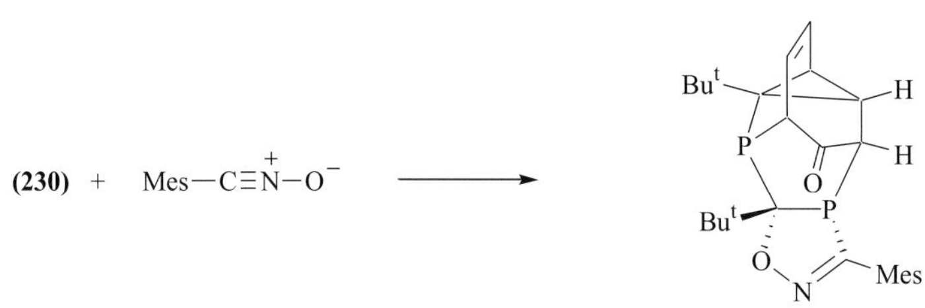

(233)

Scheme 74

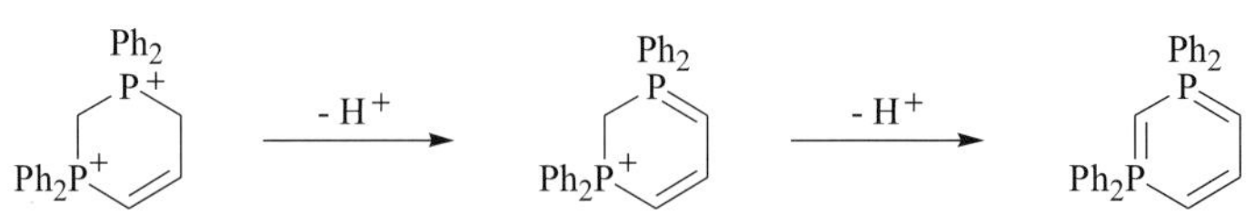

(234) (235) (236)

Scheme 75

from this study have been reported ⟨91CB2071, 97JOM223⟩. With HC≡C–$(CH_2)_n$–C≡H the derivatives (**242**) are obtained.

The ^{31}P NMR data of all new compounds are presented, also the X-rays of (**241b**), $R^1 = C_6H_5$, $R^2 =$ –C≡C–C_6H_5, (**241d**) and (**241e**) ⟨97JOM223⟩.

The alkinylphosphanes react with the 1λ^5,3λ^5-diphosphete (**239**) to yield the 1λ^5,3λ^5-diphosphinine derivatives (**241f**) (Scheme 77).

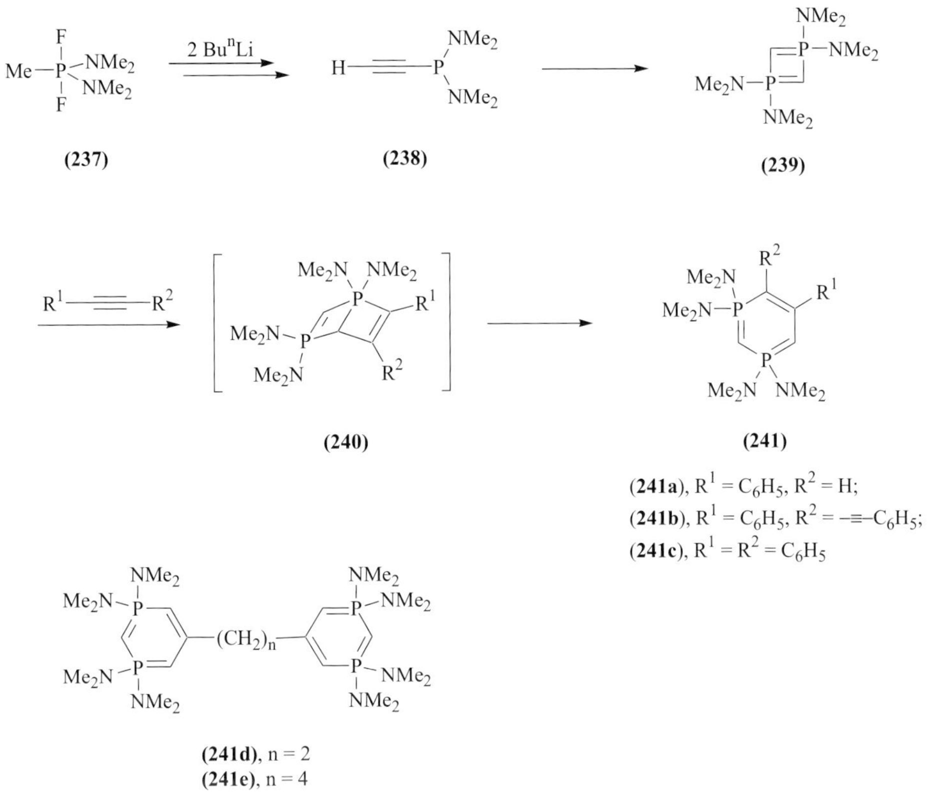

Scheme 76

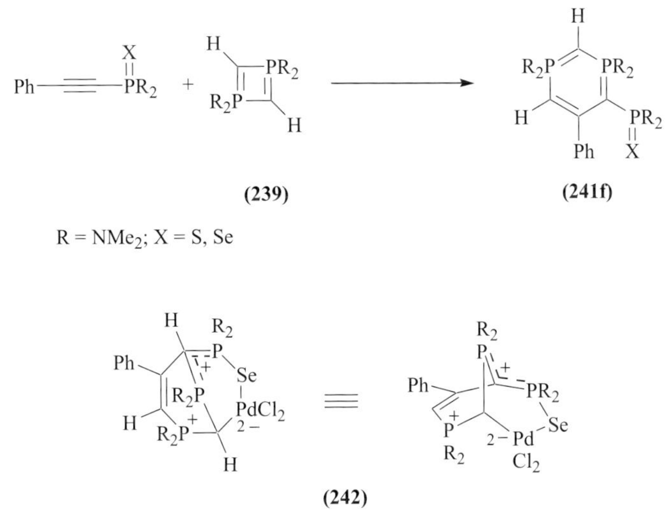

Scheme 77

The selenide (**241f**), X = Se with (benzonitrile)$_2$ $PdCl_2$ forms the red crystalline complex (**242**). The NMR (^{31}P, ^{13}C, ^{1}H) and the X-ray analysis confirm the bicyclic structure (**242**) ⟨99ZAAC1712⟩.

The 1,3-diphosphinine (**241g**) can be isolated from a mixture resulting from the reaction of diphosphet (**239**) and Et–N=C=S (Scheme 78).

The possible mechanism for the formation of (**241g**) as well as the spectroscopic data and the X-ray are described and discussed ⟨95HAC355⟩.

Schmidbaur's synthesis of 5-methyl-1,13,3-tetraphenyl-1λ^5,3λ^5-diphosphinine (**243**) (Scheme 79) ⟨90AG(E)569⟩ is reminiscent of an early approach by Maerkl to the same ring system ⟨63ZN(B)1136⟩.

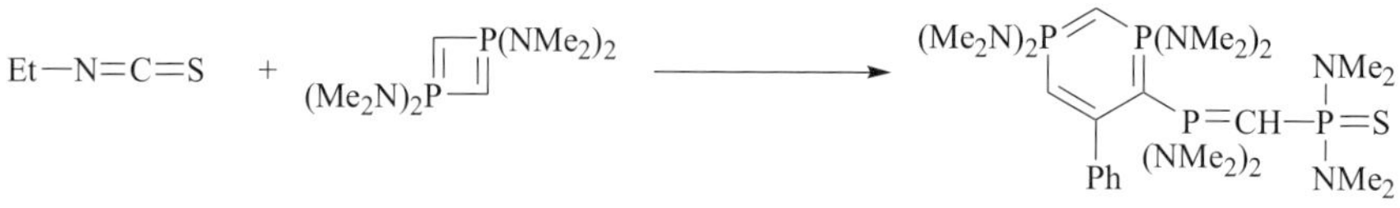

(239) **(241g)**

Scheme 78

The exocyclic double bond moves into the ring on ylide formation thus providing full cyclic conjugation. The saturated six-membered cyclic diphosphonium salt (**244**) behaves differently on deprotonation; a 1,2-didehydro-1,4,5,6-tetrahydro-$1\lambda^5,3\lambda^5$-diphosphinine (**245**) with a carbodiphosphorane structural element is produced (Scheme 80) ⟨81CB3063⟩.

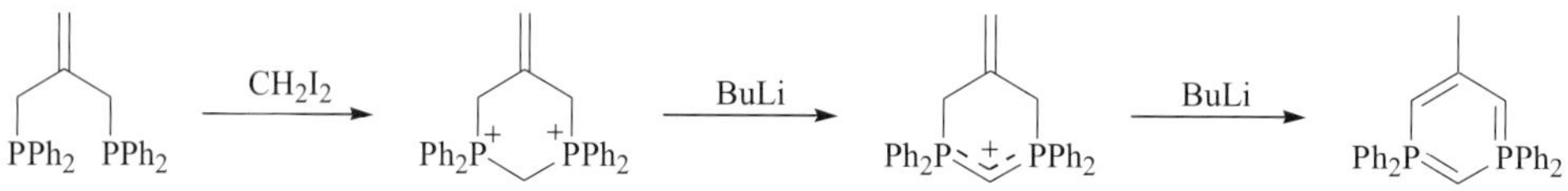

(243)

Scheme 79

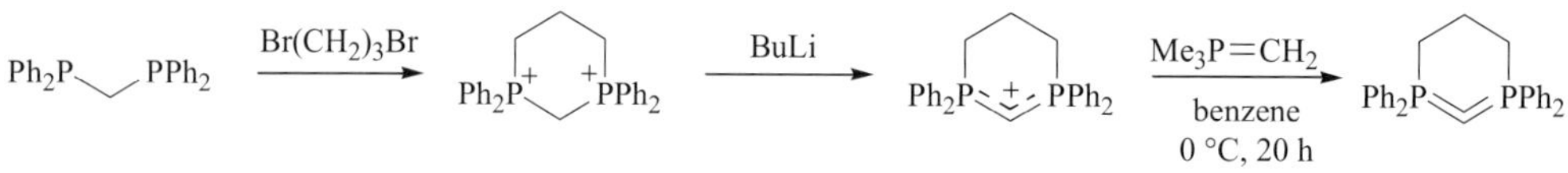

(244) **(245)**

Scheme 80

The chemistry of the $1\lambda^5,3\lambda^5$-diphosphinines known up to now is practically restricted to the various ways in which they can form metal complexes. Structures (**246**)–(**248**) illustrate bridging across ylide carbon atoms ⟨93JOM(459)73⟩, between two substituents ⟨94ZAAC(620)483⟩, and apparent η^6-arene bonding to a transition metal ⟨92OM2863⟩, respectively. The ligand of (**246**) is produced by air oxidation of that of (**247**) (Scheme 81).

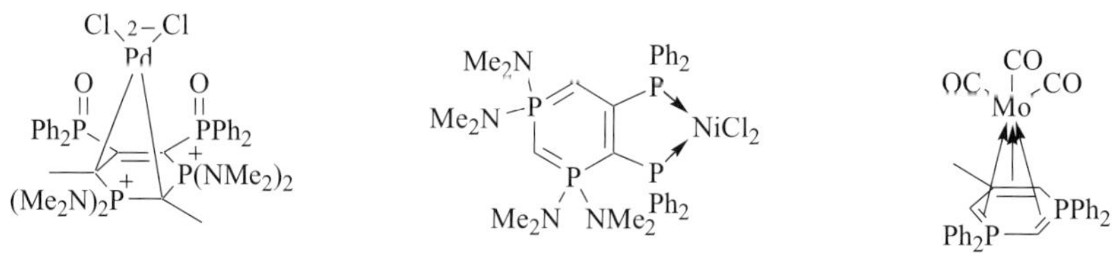

(246) **(247)** **(248)**

Scheme 81

A large paper deals with the synthesis of polycyclic phosphorus cage compounds containing 1,3-diphosphinine and phosphinine units by tandem Diels–Alder and ene reactions ⟨95BSF(132)652⟩.

1,3-Dienes react irrespective of the employed stoichiometry with two moles of phosphaalkyne to give the diphosphatricyclic compounds (**253**) and (**254**) with a diphosphirane unit (Scheme 82) ⟨87AG(E)1011⟩ (see also Scheme 68).

The formation of (**253**) and (**254**) is the result of a sequence of [4 + 2]-, ene- and intramolecular [4 + 2]reactions (R^1–R^4 = H; R^2 = Me; R^2, R^3 = Me; R^4 = Me; R^1, R^4 = Me, R^3 = CO_2Me; R^4 = CO_2Me; R^4 = $PO(OMe)_2$; R^2 = Me, R^4 = $PO(OMe)_2$; R^2 = Bu^t, R^4 = $PO(OMe)_2$; R^2, R^3 = $OSiMe_3$, all substituents not mentioned are H).

This reaction sequence has been extended by cycloadditions with (**255**) to give (**258**), with (**256**) to give (**259**) and with (**257**) to give (**260**) and (**261**) (Scheme 83) ⟨95BSF(132)652⟩.

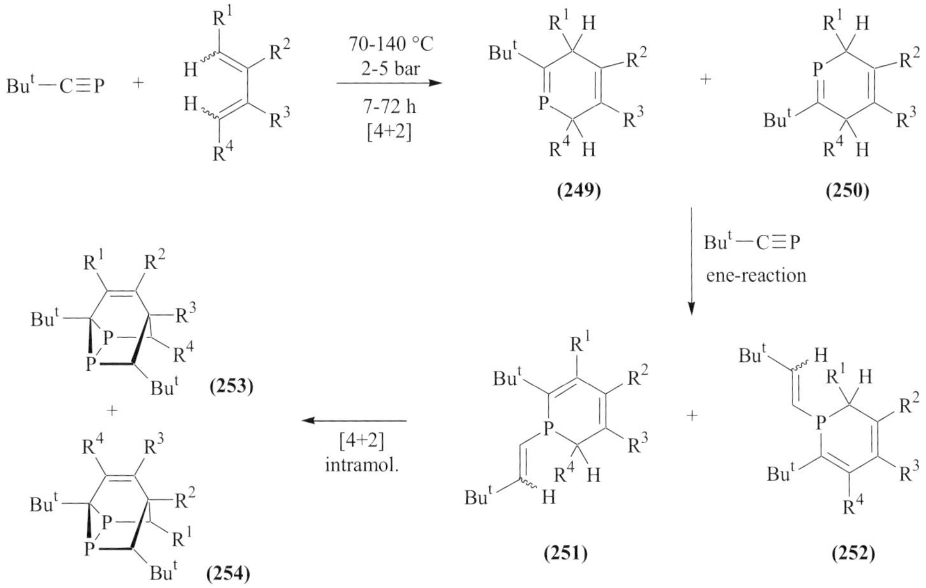

Scheme 82

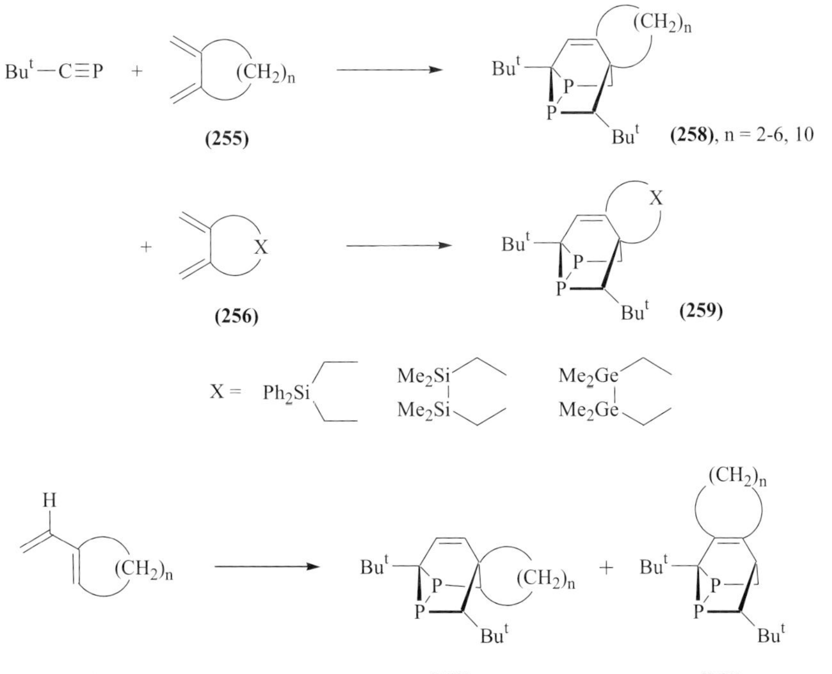

Scheme 83

In a closely related reaction 5,8-bis(trimethylsilyl)cycloocta-1,3,6-triene adds alkynes in toluene at 90ºC, the product (**262**) is the result of a [4 + 2]-cycloaddition at the bicyclic valence isomer.

When the tricyclic compound (**262**), R = CO_2Me is subjected to thermolysis at 140ºC for seven days in the presence of $Bu^tC{\equiv}P$, a new, phosphorus-containing product, the diphosphatricyclooctene (**263**) can be isolated in good yield (55%). The first step of the reaction is most certainly a retro-Diels–Alder reaction of (**262**) (Scheme 84).

The cyclobutene or the valence isomeric 1,3-dienes by reaction with $Bu^tC{\equiv}P$ form (**263**) ⟨99EJOC1041⟩.

The Diels–Alder reaction of 7,8-dichlorocycloacta-1,3,5-triene with 1 mole $Bu^tC{\equiv}P$ in toluene at 120ºC for 12 h yields the adduct (**264**) via the bicyclic valence isomer (Scheme 85).

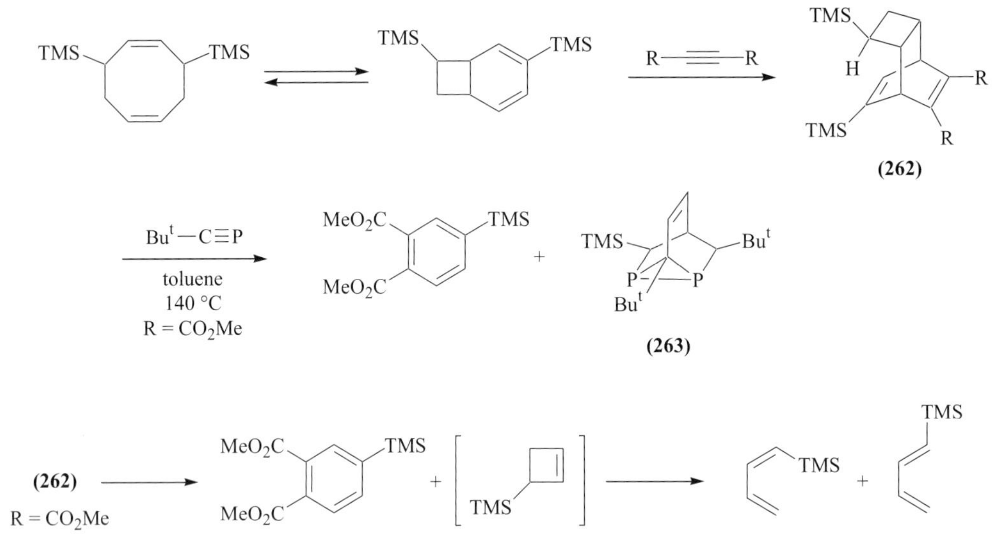

Scheme 84

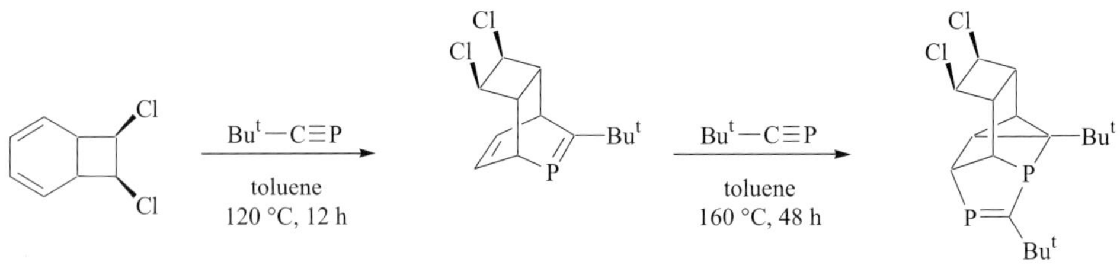

(264) **(265)**

Scheme 85

Under more drastic reaction conditions and prolonged reaction times (**264**) adds a second mole Bu^{t}C≡P to form the polycyclic compound (**265**) with a 1,3-diphosphorane structure ⟨99S644⟩.

The further chemistry of the diphosphatricyclooctenes (**266**) obtained by reaction of 1,3-dienes with Bu^{t}C≡P ⟨87AG(E)1011⟩ ⟨95BSF(132)652⟩ is described ⟨98S427⟩ (Scheme 86).

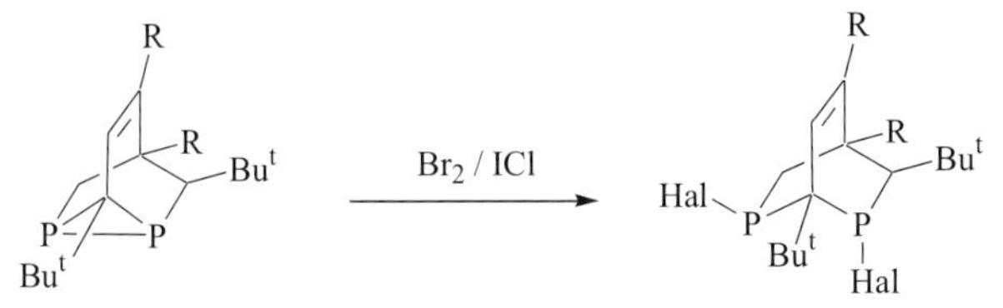

(266) **(267)**, Hal = Br, I

Scheme 86

With Br_2 or ICl the diphosphacyclopropane ring is opened to give the bicyclic halophosphines (**267**).The stereochemistry of (**267**) with the *endo* and *exo*-position of the halogen atoms has been proven by a X-ray analysis ⟨98S427⟩.

The chemistry of the pentacyclic system (**231**) obtained by reaction of two moles of tropone with two moles Bu^{t}C≡P ⟨95JOC5884⟩, has been studied in more detail. With alkyl halides (RHal, R = Me, Et, $PhCH_2$, Hal = Br, I) quaternization at P^{12} occurs. Azides (R–N_3, R = Tos, Ph, Tms) undergo the Staudinger reaction at P^2 to give the phosphinimines. Depending on the stoichiometry Bu^{t}OOH, S_2 and Se form the monoxides, sulfides and selenides at P^{12} as well as at P^2, respectively the dioxides, disulfides and diselenides (Scheme 87). Tetrachloro- and tetrabromo-*o*-quinone yield mixtures of the monophosphoranes with pentavalent phosphorus either at P^{12} or at P^2, f.i. (**268**) ⟨96S87⟩.

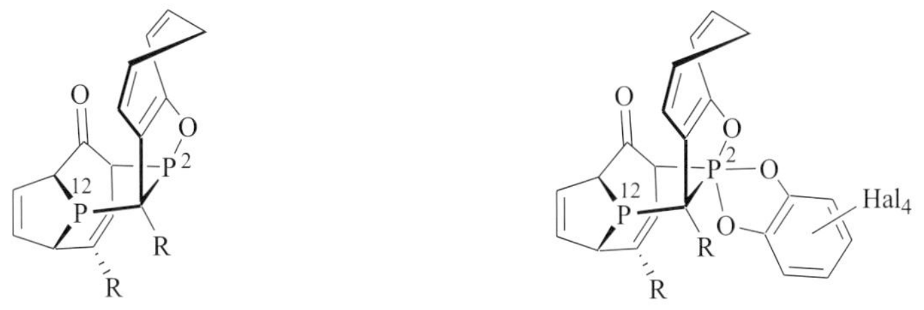

(**231**), R = Bu^t (**268**), R = Bu^t

Scheme 87

The fully saturated system (1,3-diphosphinanes) have also received attention. A double Arbuzov reaction on the bis-phosphonite (**269**) forms the cyclic diphosphinate (**270**), which can be reduced to the diphosphine (**271**) (Scheme 88) ⟨89PS(41)113⟩. The synthesis and stereochemistry of the 1,3-diphenyl analogue (**272**) have also been reported ⟨89PS(44)235⟩.

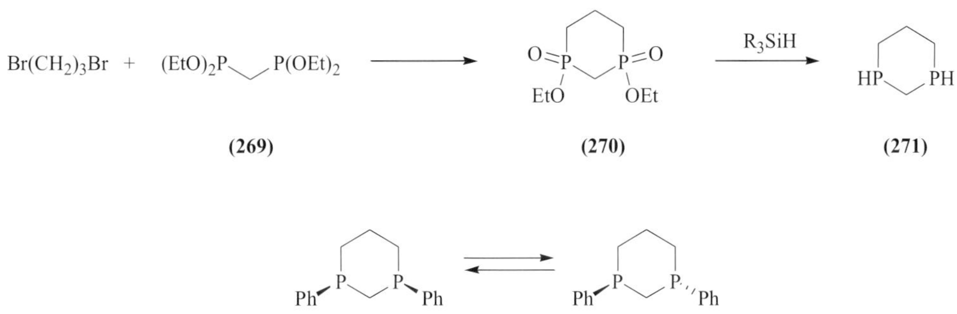

(**269**) (**270**) (**271**)

(**272**)

Scheme 88

5.3.6.3 1,4-Diphosphinines, Phosphanthrenes

The tetrakistrifluoromethyl compound (**277**) is the only known fully unsaturated derivative with λ^3-phosphorus. It has not been isolated, being handled only in hexane solution, and even there it is somewhat unstable. It is formed from the diphosphabarrelene (**273**) ⟨60JA1515, 61JA3432⟩, which forms adducts with methanol (**274**), diazomethane (**275**) and phenyl azide (**276**). These all thermally decompose by cycloreversion to produce the diphosphinine (**277**), along with the appropriate enol ether, pyrazole and triazole fragments, respectively (Scheme 89) ⟨76TL3715, 77TL867⟩.

The chemistry of (**277**) is mainly characterized by (4 + 2)- and (4 + 1)-cycloadditions. It reacts with alkynes to form the 1,4-diphosphabarrelenes (**278**) ⟨76TL3715⟩. By heating with CCl_4 in a sealed tube 7,7-dichloro-1,4-diphosphabicyclo[2.2.1]heptadiene (**279**) is isolated, besides the 1,4-dichloro-1,4-dihydro-1,4-diphosphinine (**280**), which is observed only in solution ⟨68CC1350, 77TL867, 77TL3057⟩. Elemental sulfur adds to (**277**), 7-thia-1,4-diphosphabicyclo[2.2.1]heptadiene (**281**) is obtained, and this is also formed in the reaction with 7-thiabicyclo[4.1.0]heptane along with the cyclohexene (4 + 2)-cycloadduct (**282**) ⟨81JA2465⟩. The epithio derivative (**281**) reacts with (**277**) further to give the noninterceptible (4 + 2)-cycloadduct (**289**) which suffers a (4 + 2)-cycloreversion reaction, forming the barrelene (**273**) and the thiadiphosphole (**284**) (Scheme 90) ⟨81JA2465⟩.

Irradiation of (**277**) in octafluoropentane with a high-pressure Hg lamp in a sealed tube causes valence isomerization to the 1,4-diphosphabenzvalene (**285**) (Scheme 90), which readily forms (4 + 2)-cycloadducts across the remaining double bond, with, e.g. 2,3-dimethylbutadiene or furan ⟨80JOC4683⟩.

The existence of 1λ^5,4λ^5-diphosphinines has not been firmly established. In 1967 Aguiar *et al.* published the acid-catalyzed dimerization of alkynyldiphenylphosphines (**286**) to the bisphosphonium salts (**287**) ⟨67JA3067, 67JA4235, 69JOC2681⟩. However, according to electrochemical studies only the formation of the radical cations (**288**) by one-electron transfer can be observed, and the neutral 1λ^5,4λ^5-diphosphinine species is not obtained (Scheme 91) ⟨77JA6656⟩.

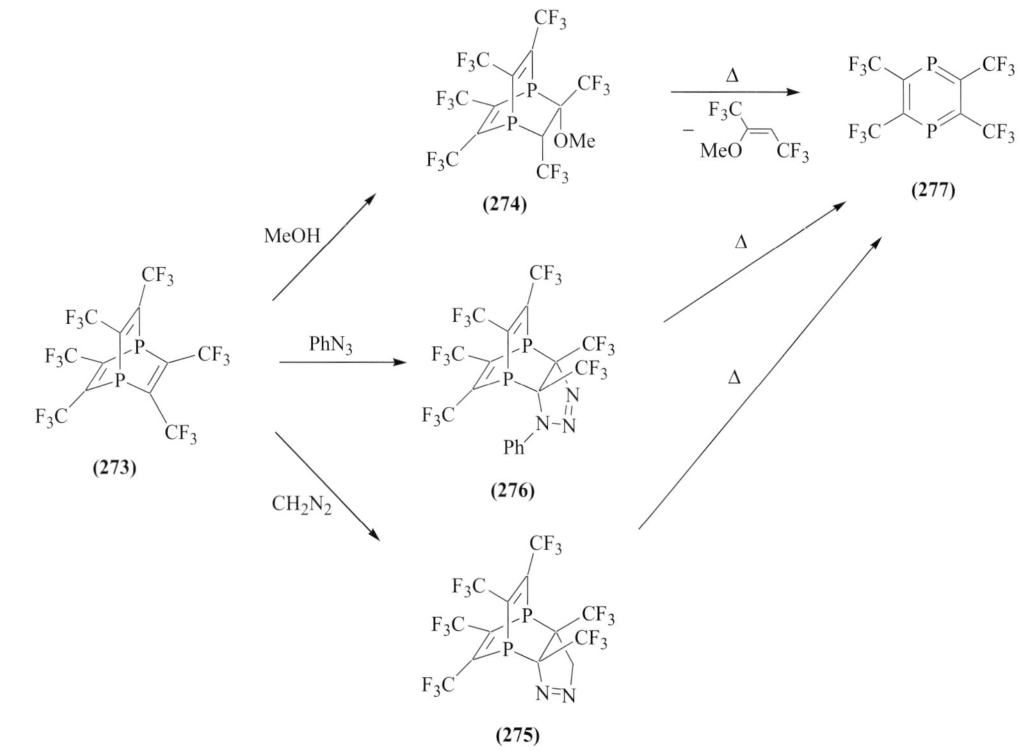

Scheme 89

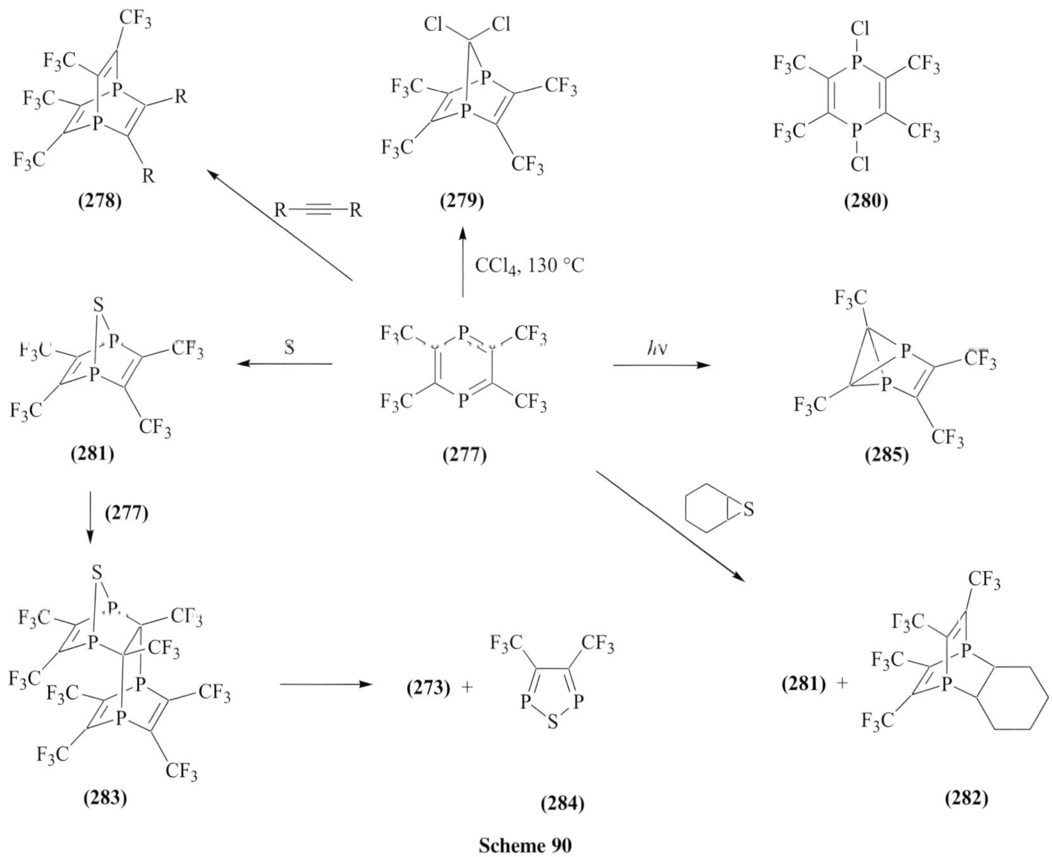

Scheme 90

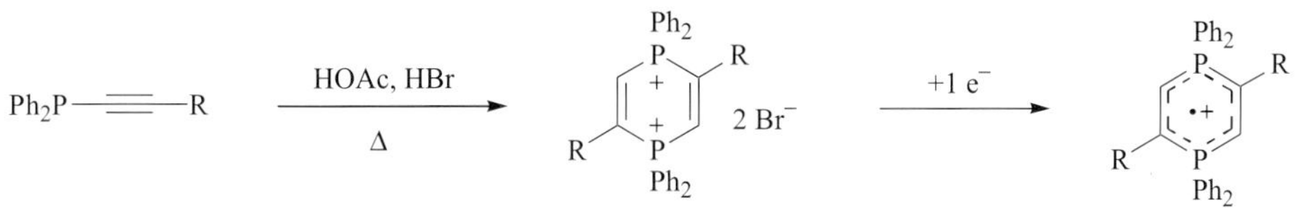

(286) **(287)** **(288)**

Scheme 91

In another attempt the bis-phosphonium salt (**289**) was treated with butyl and phenyl lithium, but ring opening rather than simple di-deprotonation was observed (Scheme 92) ⟨66JA4090⟩.

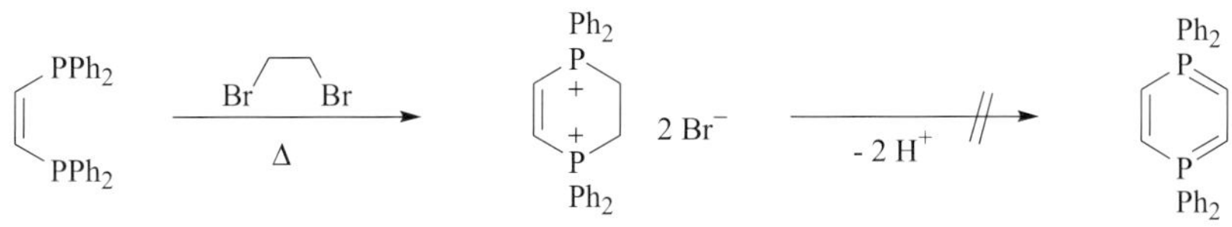

(289)

Scheme 92

1,4-Dihydro-1,4-diphosphinines (**291**) are simply obtained by phase-transfer KOH-catalyzed addition of primary phosphines to dialkynyl phosphines (**290**) (Scheme 93) ⟨75TL3171, 77S842, 85CB2365⟩. Because of the configurational stability of phosphorus, the products (**291**) are formed as mixtures of *cis* and *trans* isomers, the *cis* predominant (Scheme 93).

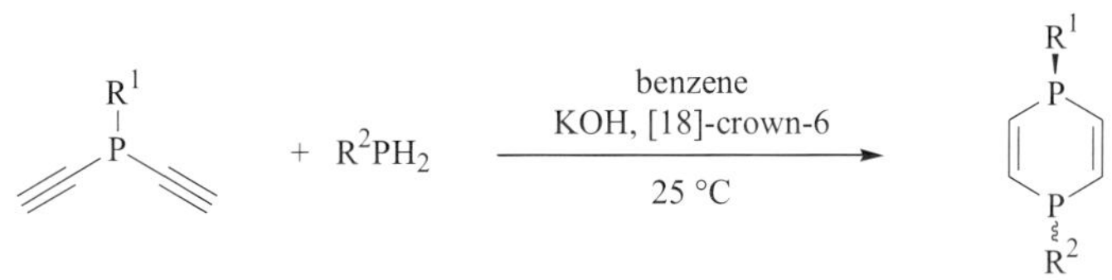

(290) **(291)**

Scheme 93

Fluck's group has reported the synthesis, X-ray crystal structure, and NMR spectrum of the 1,4-disulfide (**292**) ⟨92PS(71)245, 95MRC553⟩ (Scheme 94).

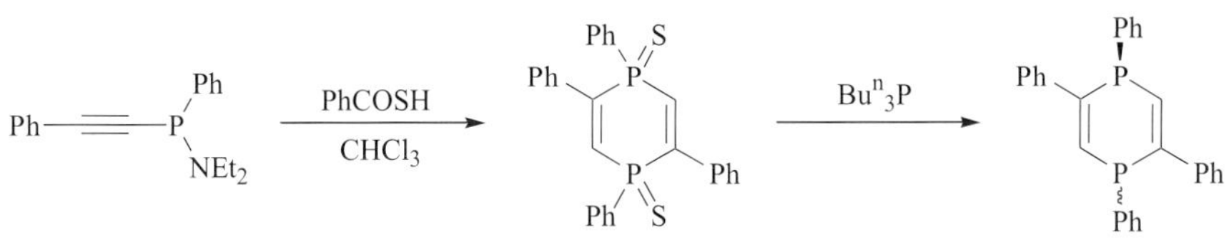

(292) **(293)**

Scheme 94

The reduction of the 1,4-disulfide was achieved with tri-*n*-butylphosphane, the resulting 1,2,4,5-tetraphenyl-1,4-dihydro-1,4-diphosphinine, yellow crystals, exists as a mixture of *cis*/*trans*-isomers (^{31}P NMR, δ(*trans*-**293**) = −29.9 ppm, δ(*cis*-**293**) = −21.6 ppm).

In an analogous manner the following alkynes have been cyclized: $CH_3(CH_2)_2$–≡–P(Ph)NEt_2, Me_3Si–≡–P(Ph)NEt_2 ⟨95ZAAC1981⟩.

The reaction of 1,2-bis(lithiophenylphosphino)benzene (**294**) with tetrachloroethylene leads to a rather complex mixture of dibenzotetraphosphafulvalene (**295**) and its six-membered ring isomer (**296**) in an overall yield of 10%.

Of the five possible isomers for (**295**) as well as (**296**) in case of (**295**) four isomers, in case of (**296**) three main isomers could be observed. The X-ray crystal structure analysis of one isomer of

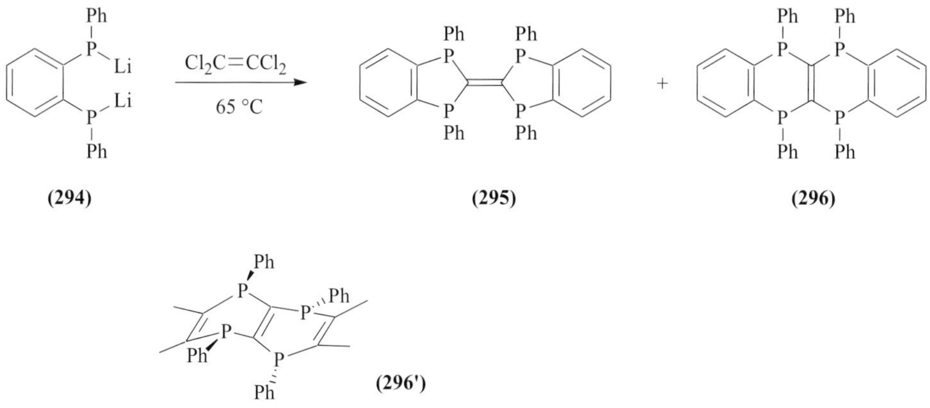

Scheme 95

(**296**) (m.p. > 250°C, ^{31}P NMR, $\delta = -2.5$ ppm) confirms the expected boat-conformation for the 1,4-dihydro-1,4-diphosphinines and a *trans*-orientation to each other (**296′**) (Scheme 95) ⟨97BSF853⟩.

An interesting ring system with a formal 1,4-diphosphinine structure is the 1,3,5,7-tetra-*tert*-butyl-4,8-diphospha-s-indacene (**299**). The rather simple synthesis starts out from lithium-1,3-di-*tert*-butylcyclopentadien by reaction with PCl_3. The possible intermediate 6-chloro-6-phospha-pentafulvalene (**297**) dimerizes, in the presence of TMEDA the dimer (**298**) becomes spontaneously dehydrochlorinated to form (**299**) in 8% yield (Scheme 96).

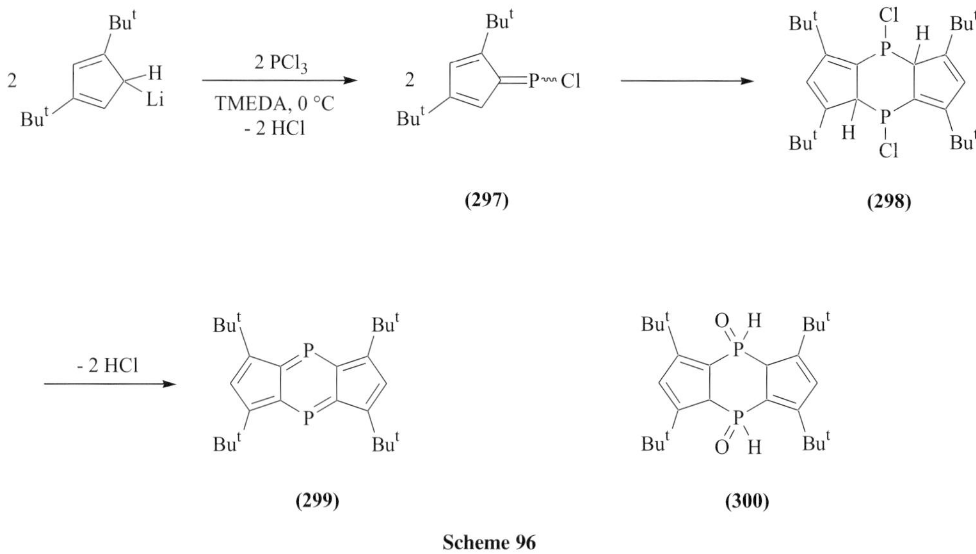

Scheme 96

Compound (**299**) is a blue black fine crystalline solid, m.p. 223–225°C (decomp.), ^{31}P NMR, $\delta = 275.6$ ppm, UV/VIS (CH_2Cl_2): $\lambda_{max}(\log \varepsilon) = 281$ (4.06), 320 (3.84), 384 (4.66), 433 (3.63), 615 (4.25), 760 (2.48) (sh), 917 (2.26) (sh).

The diphosphaindacene adds readily two mole equivalents of water to give the colorless bis-phosphinoxide (**300**) ⟨96TL3829⟩.

1,4-Diphenyl-1,4-diphosphinanes (**301**) have also been prepared and studied ⟨89PS(44)235⟩.

(Divinyl)phosphines react with (trimethylsilyl)phosphine in a radical mechanism to yield the 1.4-diphosphorinanes (**302**) as a diastereomeric mixture (Scheme 97).

In case of R = NEt_2 only one isomer is observed, methanolysis yields the very air sensitive diphosphane (**303**) ⟨97IC1867⟩.

5,10-Dihydrophosphanthrenes (**304**) were synthesized by Davis and Mann via the route shown in Scheme 98 ⟨62CI(L)1539, 64JCS3770⟩. Like the monocyclic compounds (**291**), they show *cis*/*trans* isomerism of the substituents on P. The central ring is folded, as with the phenoxaphosphines and the dihydrophenophosphazines.

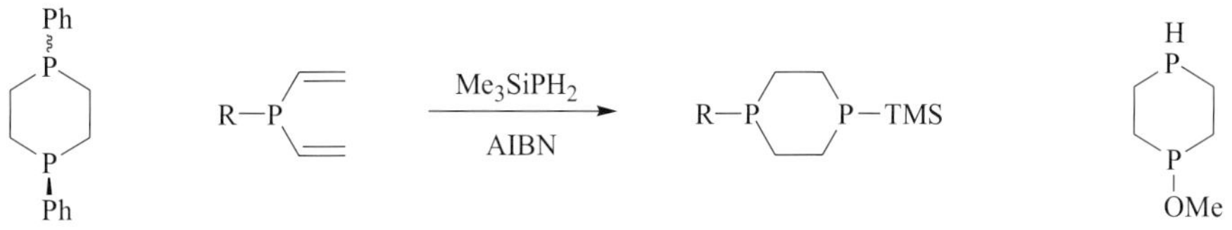

(301) R = Ph, NEt_2 **(302)** **(303)**

Scheme 97

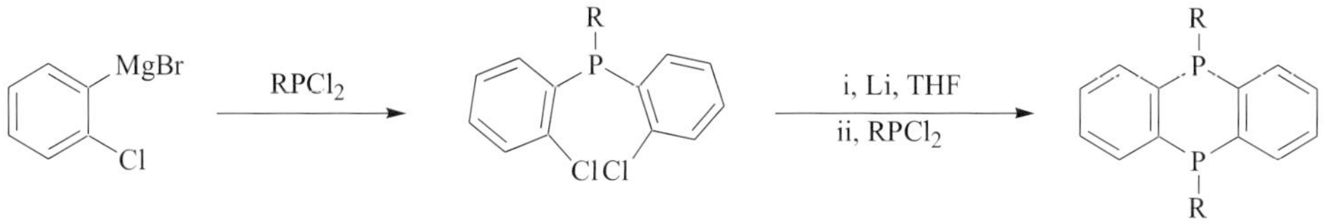

(304), R = Et, Ph

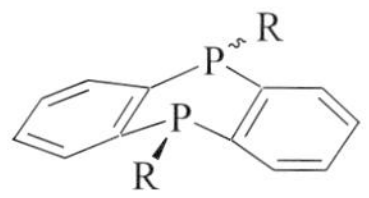

(304)

Scheme 98

5.3.7 RING SYSTEMS WITH TWO PHOSPHORUS ATOMS AND ONE NITROGEN ATOM

Of the 12 possible azadiphosphinines with λ^3,λ^3 or λ^5,λ^5-phosphorus only derivatives of the $1,2\lambda^5,6\lambda^5$-azadiphosphinines (II) and $1,2\lambda^5,4\lambda^5$-azadiphosphinines (IV) are known. No aza-λ^3,λ^3- or aza-λ^3,λ^5-diphosphinines have been described yet.

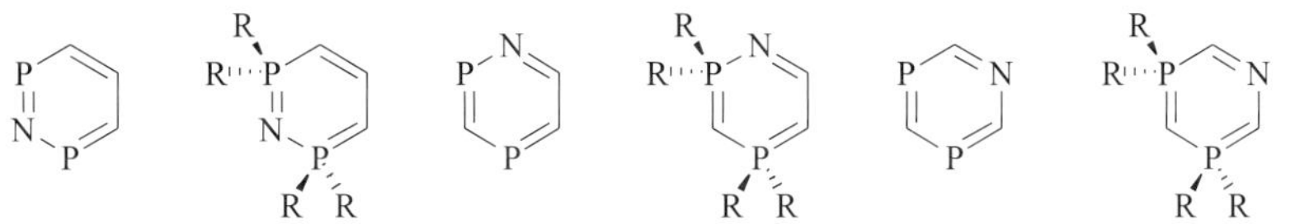

(I) **(II)** **(III)** **(IV)** **(V)** **(VI)**

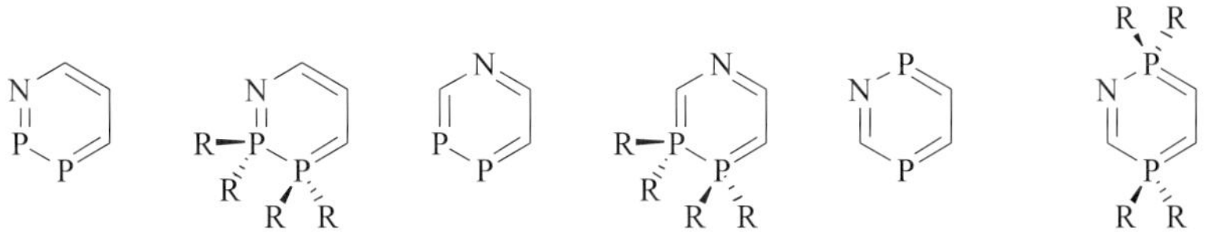

(VII) **(VIII)** **(IX)** **(X)** **(XI)** **(XII)**

5.3.7.1 1,2,6-Azadiphosphinines

The preparation of these compounds is similar to that of the 1,3-diphosphinines (**243**) by Schmidbauer (Scheme 79). The initial quaternization is carried out on the diphosphazane (**305**) ⟨76JA4410⟩, to give (**306**). Deprotonation and tautomerization are brought about by triethyl(ethylidene)phosphorane (Scheme 99) ⟨91CB1525⟩.

The ^{31}P and ^{13}C NMR [δ(C-3/5) = 53.5, $^1J_{(PC)}$ = 111.7 Hz] demonstrate the ylide character of (**307**). An X-ray crystal structure determination has been carried out.

The tetrahydro-1,2,6-azadiphosphinine (**310**) is accessible by condensation of (**309**) (which may be considered as λ^5,λ^5-bisphospha-trimethine-cyanine) with *N,N,*bis(dichlorophosphanyl)aniline (**308**) (Scheme 100).

In (**310**) the phosphacyanin structure is retained. Compound (**310**) reacts with elemental sulfur to give the 2,6-dichloro-1,4-diphenyl-2,6-dithioxo-3,5-bis(triphenyl-phosphorandiyl)-1,2,6-azadiphosphi-

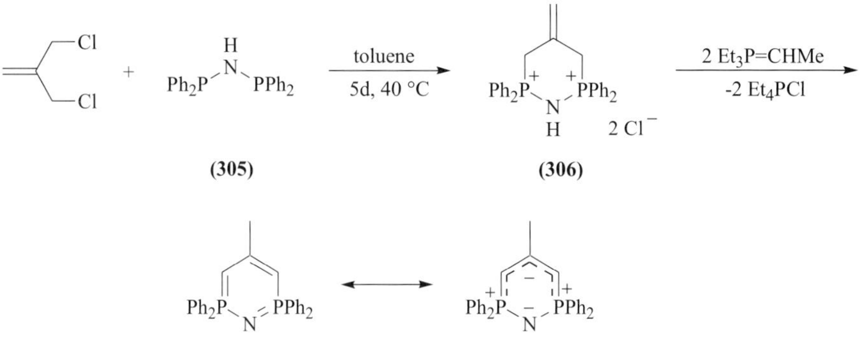

(305) (306)

(307)

Scheme 99

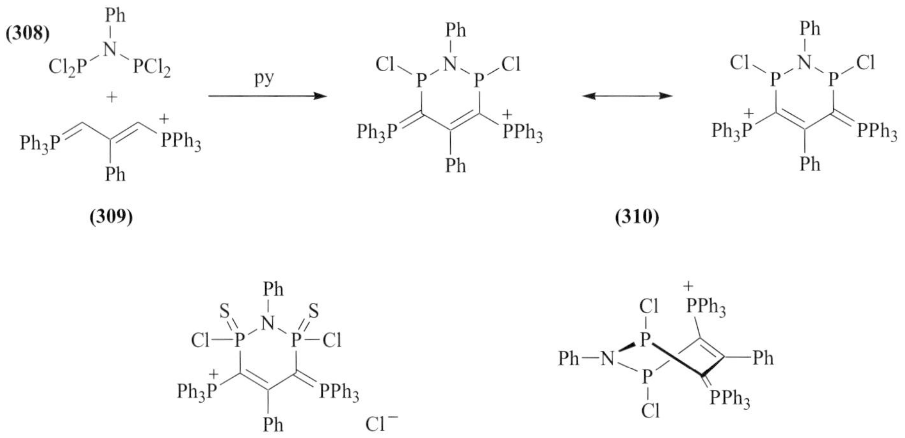

(308) (309) (310) (311) (310)

Scheme 100

nenium-bromide (**311**). According to the X-ray (**310**) possesses a twist structure with *trans*-position of the chorine atoms (Scheme 100) 〈96ZN(B)1761〉.

5.3.7.2 1,2,4-Azadiphosphinines

The diphosphete (**239**), which added alkynes in a synthesis of 1λ^5,3λ^5-diphosphinines (Scheme 76), also smoothly adds to a wide variety of carbonitriles, to form 1,2λ^5,4λ^5-azadiphosphinines (**312**) 〈91PS(56)49, 91ZN(B)762〉 〈91ZN(B)1130, 91PS(56)49, 94RHA65, 95RHA6121〉. NMR and an X-ray structure determination on (**312**), R = 4–F_3C–C_6H_4) confirmed the structures of the products which are very likely formed by valence isomerization via the Dewar intermediate as shown (Scheme 101).

The correctness of this reaction mechanism is proved by cycloaddition of 2-cyanopyridine to 1,1,3,3-tetrakis(dimethylamino)-1λ^5,3λ^5-diphosphet (**239**) (Scheme 102). The (4 + 2)- resp. (2 + 2)-cycloaddition of the nitrile group is followed by a spontaneous valence-isomerization to give 6-pyridyl-1-aza-2λ^5,4λ^5-diphosphinine (**313**) 〈99ZN(B)1478〉.

The X-ray of (**313**) also confirms the planar structure of both 6-membered rings. The torsion angle between both rings is only 15º.

2,2,4,4-Tetrakis-(dimethylamino)-6-amino-1-aza-2λ^5,4λ^5-diphosphinine (**317**) has been prepared by reacting 1,1,3,3-tetrakis(dimethylamino)-1λ^5,3λ^5-diphosphete (**239**) with bis(tri-methylsilyl)carbodiimide (**314**) in a 2 : 1 molar ratio (Scheme 103).

The mechanism can be rationalized by silylation of (**239**) to give (**315**), the silylation reagent (**314**) forms cyanamid which undergoes a (4 + 2)-cycloaddition to give (**316**). By valence-isomerization of the bicyclic adduct the azadiphosphinine (**317**) is obtained. The X-ray confirms the planar structure of the ring skeleton 〈98ZN(B)443〉.

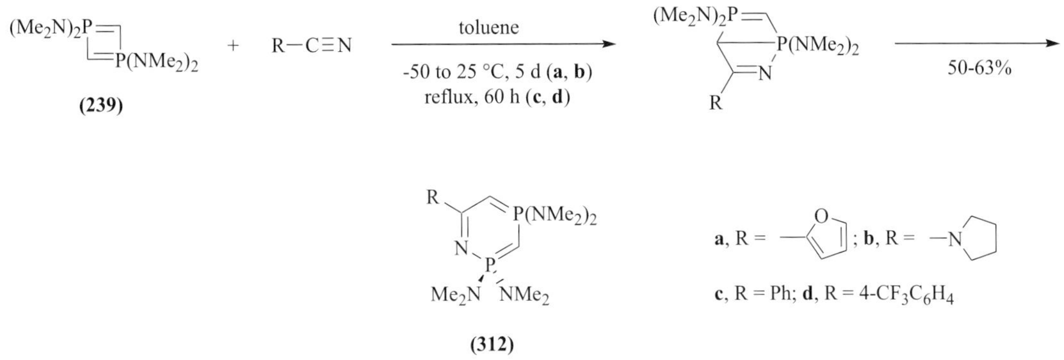

Scheme 101

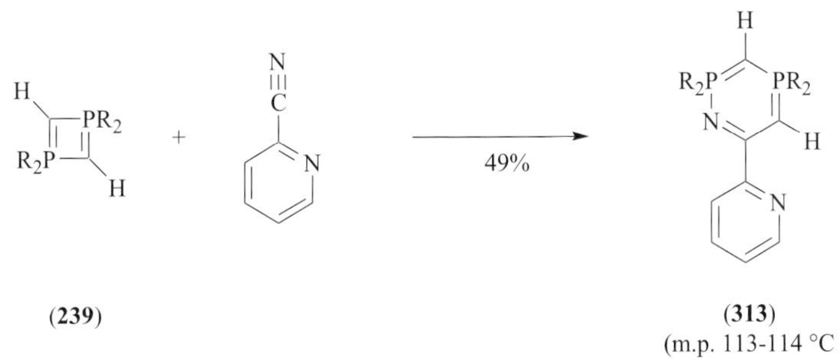

Scheme 102

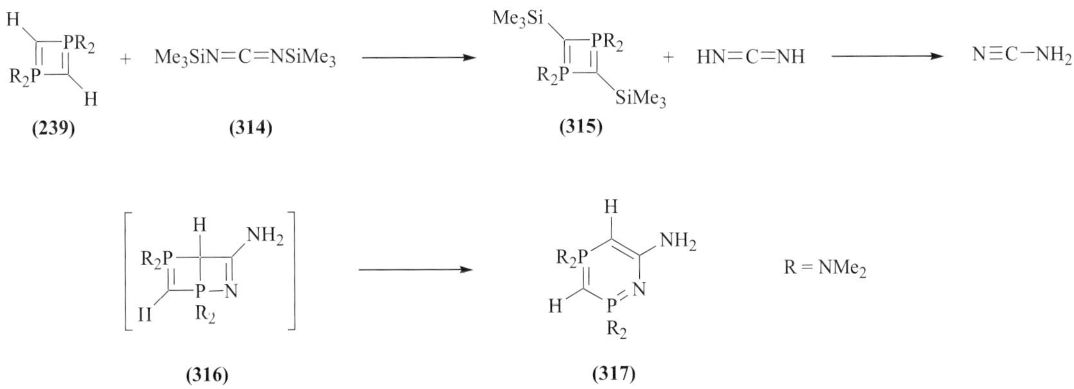

Scheme 103

5.3.8 RING SYSTEMS WITH TWO PHOSPHORUS AND TWO NITROGEN ATOMS

Of the 18 possible diazadiphosphinines with λ^3,λ^3- and λ^5,λ^5-phosphorus only derivatives of the 1,3,2λ^5,4λ^5(II)-, 1,3,2λ^5,5λ^5(XVIII)- and the 1,4,2λ^5,5λ^5(XVI)-diazadiphosphinines are yet known.

Fused ring systems containing the 2,5-dihydro-1,3,2,5-diazadiphosphinine moiety are also known.

5.3.8.1 1,3,2,4-Diazadiphosphinines

Hexachlorodiphosphazenium chloride (**318**) reacts with a variety of carbonitriles to form 2,2,4,4,6-pentachloro-1,3,2λ^5,4λ^5-diazadiphosphinines (**319**) (Scheme 104) 〈77PS(3)209, 77ZAAC(433)229, 77ZAAC(434)95〉.

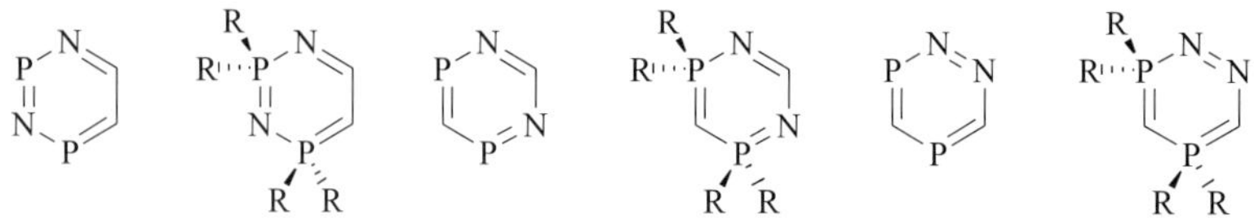

(I) (II) (III) (IV) (V) (VI)

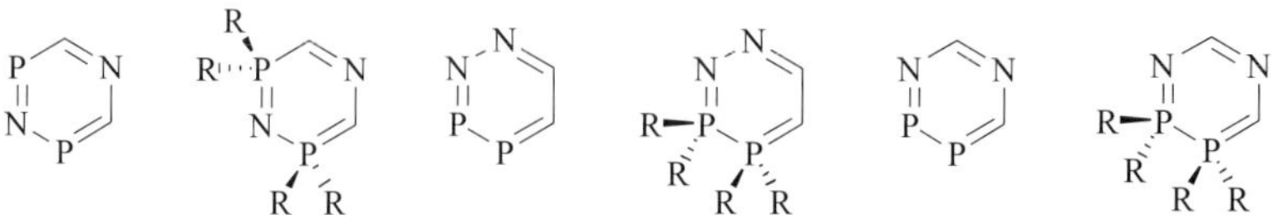

(VII) (VIII) (IX) (X) (XI) (XII)

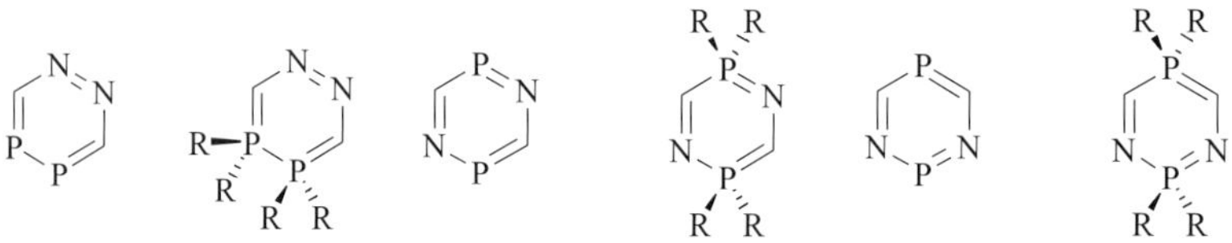

(XIII) (XIV) (XV) (XVI) (XVII) (XVIII)

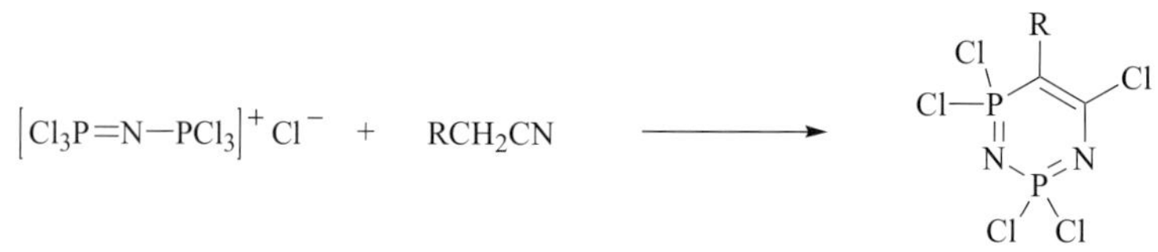

(318) (319)

a, R = Cl; **b**, R = CH_2Cl ($CHCl_2$, CCl_3); **c**, R = H; **d**, R = Ph; **e**, R = CN

Scheme 104

5.3.8.2 1,3,2,5-Diazadiphosphinines

The regiospecific (3 + 2)-cycloaddition of the bis(diazomethyl)phosphane (**320**) with *t*-butylphospha-alkyne gives the phosphine (**321**), substituted by 1,2,4-diazaphosphole units. With dichlorophenylphosphine (**321**) condenses smoothly to furnish compound (**322**), with a 1,3,2,5-diazadiphosphinine unit. An analogous condensation with $SiCl_4$ results in the spiro-linked silicon heterocycle (**323**) with a 1,3,2,5-diazaphosphasilabenzene structural unit (Scheme 105) ⟨88TL925⟩.

A comparable reaction sequence gives access to the tris(1,2,4-diazaphospholyl)phosphane (**324**), which was converted into the heavily heterosubstituted triptycene analogues (**325a–d**) (Scheme 106) ⟨88TL925⟩.

Treatment of dicyanophosphine in CH_2Cl_2 with 2 molequiv. Schwartz reagent $[Cp_2ZrHCl]_n$ at 0°C for 1 h gives quantitatively the dizirconated compound (**326**). Treatment of (**326**) with $PhPBr_2$ in CH_2Cl_2 at room temperature affords the 1,4-diphospha-2,6-diazines (2,6-dihydro-1,3,2,5-diazadiphosphinines) (**327**) (yield 21%), with $PhP(CN)_2$ the yield can be increased up to 52% (Scheme 107).

The condensation of (**326**) with ClP=NAr yields (**328**) as an orange powder. An X-ray crystal structure confirms (**328**) with the low coordinated phosphorus ⟨99OM3138⟩.

5.3.8.3 1,4,2,5-Diazadiphosphinines

Starting out from *N,N′*-diarylformamidines the reaction with PCl_3 in the presence of base leads to the *N*-phosphorylated amidines, which dimerize in a mixture of pyridine/NEt_3 to give (**329**). The chlorophosphane units react with R_2NH/NEt_3 and then with sulfur forming (**330**) ⟨99MC1⟩ (Scheme 108).

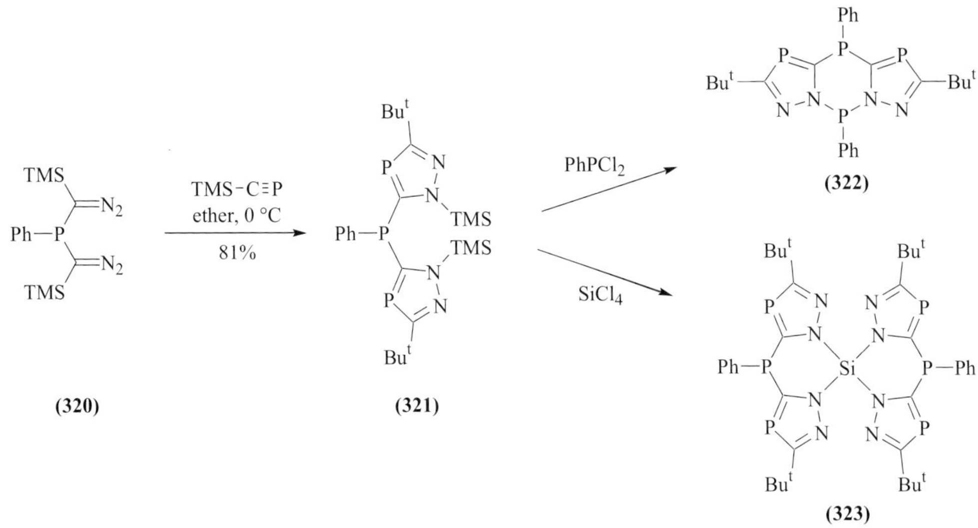

Scheme 105

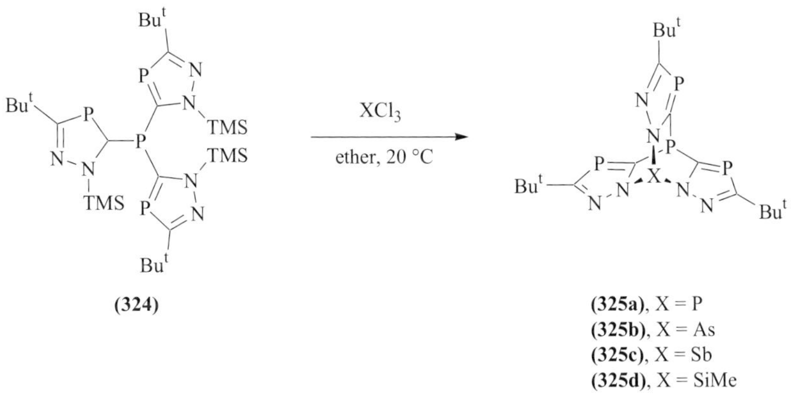

Scheme 106

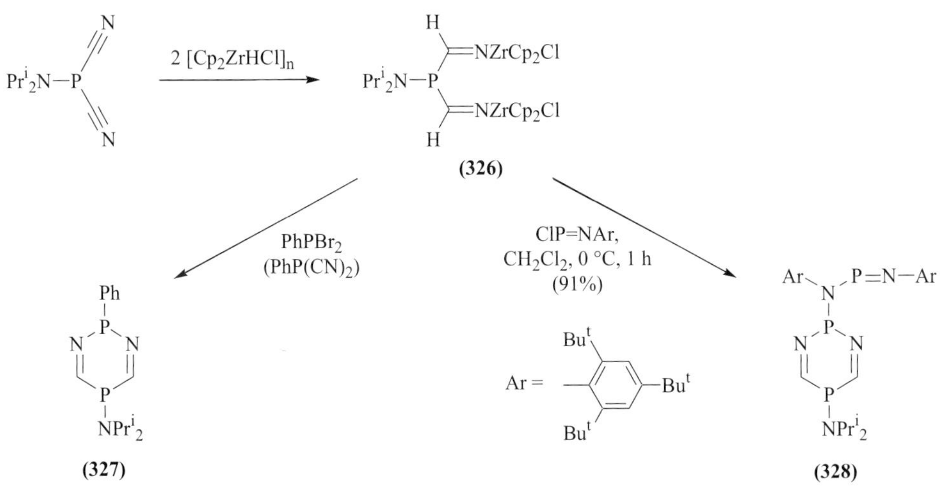

Scheme 107

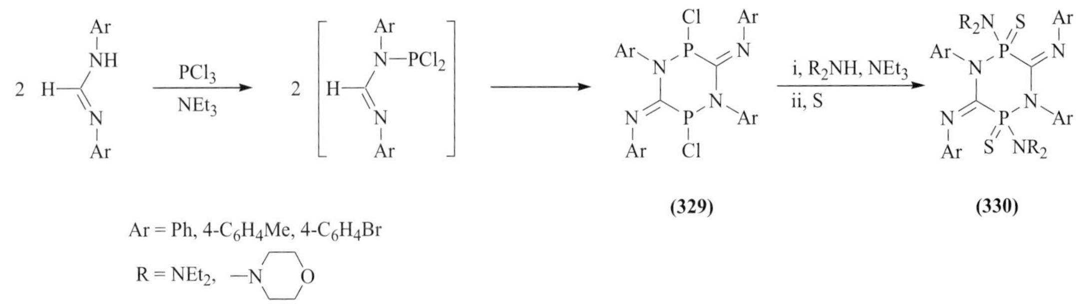

Scheme 108

5.3.9 RING SYSTEMS WITH TWO PHOSPHORUS AND THREE NITROGEN ATOMS

Of the λ^3,λ^3- and λ^5,λ^5-triazadiphosphinines only derivatives of the 1,3,5,2λ^5,4λ^5-triazadiphosphinines (X) are known, which have been prepared by a variety of methods.

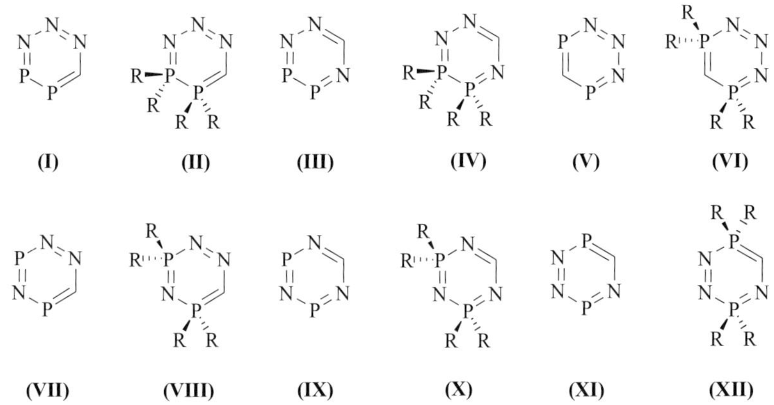

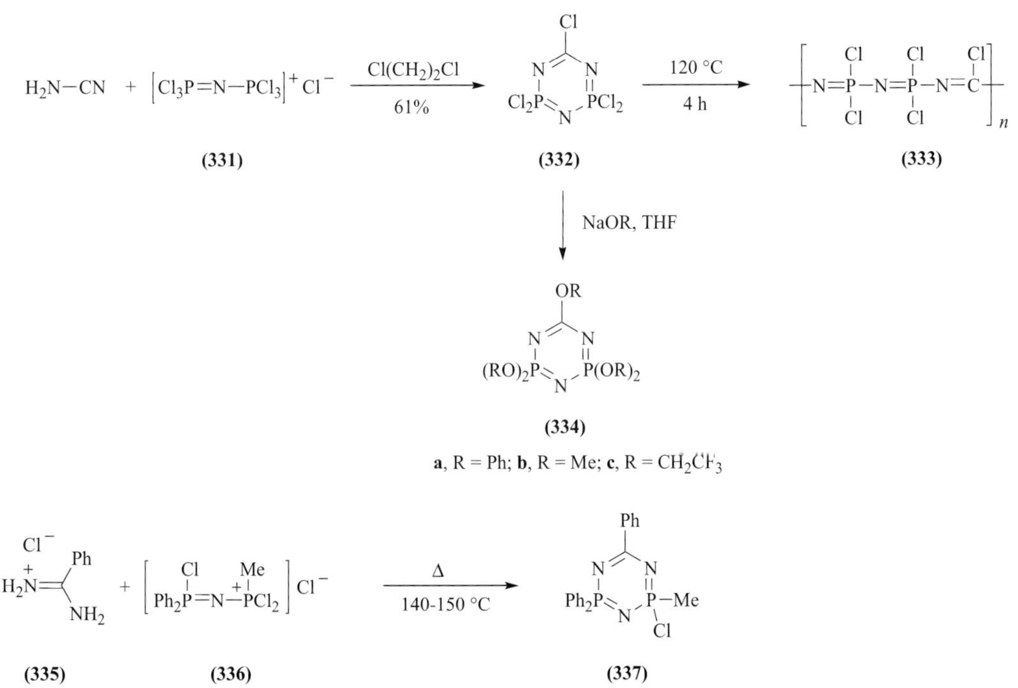

Scheme 109

5.3.9.1 1,3,5,2,4-Triazadiphosphinine

The pentachloro compound (**332**) was reported in 1975 to be formed by the action of cyanamide on hexachlorodiphosphazenium chloride (**331**) (Scheme 109) 〈75ZN(B)808〉, and its reactions have been well studied since that date. The chlorine atoms are easily replaced by amines 〈77ZN(B)254〉, alkoxides including trifluoroethoxide, and aryloxides to give (**334**) 〈93IC5088〉. On heating to 120°C a polychloropolycarbophosphazene (**333**) is formed, which, like its monomer, readily undergoes nucleophilic displacement of the chlorine atoms 〈93IC5088〉.

A detailed six-step procedure for the synthesis of 2-chloro-2-methyl-4,4,6-triphenyl-1,3,5,2λ^5,4λ^5-triazadiphosphinine (**337**) by thermal cyclocondensation of benzamidinium chloride (**335**) with the diphosphazene (**336**) (Scheme 109) under 10^{-2} Torr vacuum at 140–150°C has been described 〈89IS24〉.

Derivatives with the 6-position unsubstituted were described by Schmidpeter and Ebeling. The 2,2,4,4-tetraphenyl and -tetramethyl compounds (**339**) were prepared using *N,N'*-diphenylformamidine and the aminodiphosphadiazenes (**338**) 〈67AG(E)565〉. The tetrakisdialkylamino compounds (**341**) were formed from the corresponding aminodiphosphazenes (**340**) using 1,3,5-triazine as the one-carbon source (Scheme 110) 〈71AG(E)397〉.

(338) R = Ph, Me **(339)**

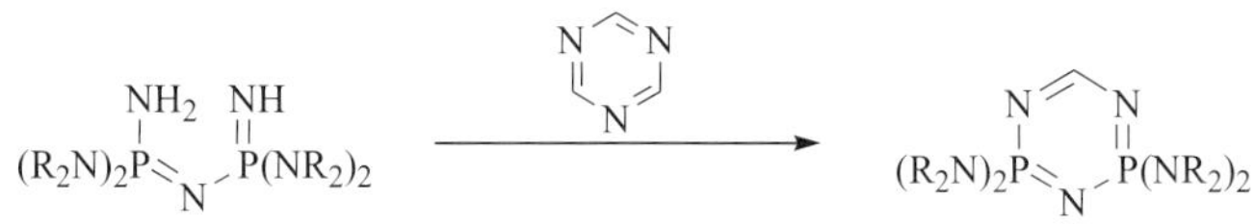

(340) R = Me, Et, Bu **(341)**

Scheme 110

The lithium disilylbenzamidine salt (**342**) with Ph_2PCl_3 forms the pentaphenyl compound (**343**) and in addition the eight-membered ring system (**344**) (Scheme 111) 〈91IC3402〉.

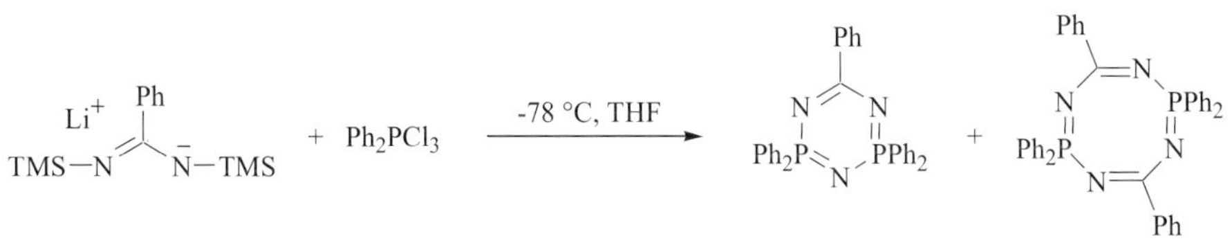

(342) **(343)** **(344)**

Scheme 111

A series of dumb-bell compounds with two tetraphenyl-1,3,5,2λ^5,4λ^5-triazadiphosphinine rings connected by a polyfluorinated chain has been described 〈79JFC(13)189, 79JFC(13)199, 80JFC(15)327〉.

5.3.10 RING SYSTEMS WITH THREE PHOSPHORUS ATOMS

The 1,3,5-triphosphinines are known as all-λ^3; all-λ^5 and mixed λ^3,λ^5-derivatives (III, IV). The 1,2,3-triphosphinines (I, II) are only known as 1λ^5,2λ^3,3λ^5-system (II). The 1,2,4-triphosphinines (V, VI) hitherto are described only as di- and tetrahydro systems.

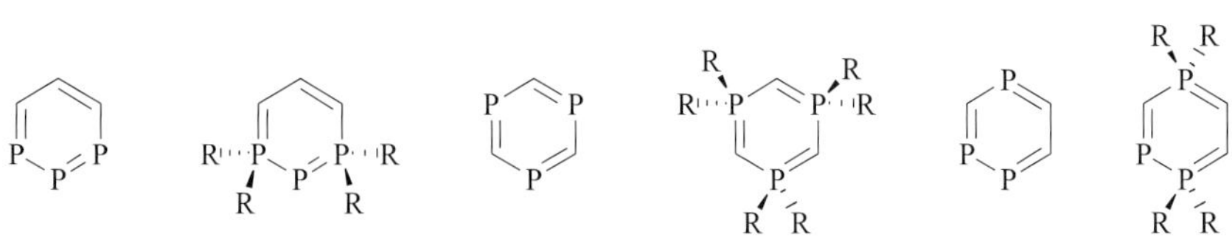

(I) **(II)** **(III)** **(IV)** **(V)** **(VI)**

5.3.10.1 1,2,3-Triphosphinines

Treatment of 1,1-bis[(diphenylphosphonyl)methyl]ethane (**345**) with an equimolar quantity of PCl_3 in the presence of $SnCl_2$ affords the cyclic triphosphenium salt (**346**). The $SnCl_2$ acts as a reducing agent, being oxidized to $SnCl_4$ which acts as a Cl^- acceptor to give the $(SnCl_6)^{2-}$ counterion (Scheme 112) ⟨93CB601⟩. The Ph_2P–P–PPh_2 structural element in (**346**) is unusual; it is to some extent analogous to the carbodiphosphoranes $R_3P{=}C{=}PR_3$. Deprotonation accompanied by tautomerization to (**347**) could be achieved by DBU as base in the two-phase system acetonitrile/hexane, giving the product as a yellow, air-sensitive microcrystalline solid.

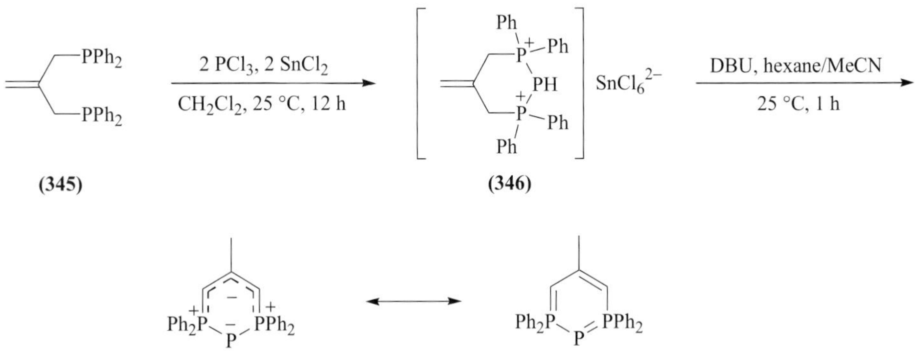

(347)

Scheme 112

The spectral data of (**347**) (^{31}P NMR (C_6D_6) $\delta = +16.1$ (d, P-1, P-3), −232.94 (t, P-2), ^1J(PP) = 374.5 Hz) are consistent with the structure.

This synthesis is closely related to that of the $1\lambda^5,3\lambda^5$-diphosphinine (**245**) and the $1,2\lambda^5,6\lambda^5$-azadiphosphinine (**307**).

5.3.10.2 1,3,5-Triphosphinines

The synthesis of tri-*t*-butyl-$1\lambda^3,3\lambda^3,5\lambda^3$-triphosphinine arose from studies of the behavior of Bu^t–C≡P in the coordination sphere of transition metals. The first indication of the transition metal template catalyzed trimerization of Bu^t–C≡P was the isolation of the vanadium-complexed 1,3,5-triphospha-Dewar benzene (**348**) ⟨87AG(E)908⟩. The formation of the triphosphabarrelene complexed to zirconium (**349**) implied the intermediate formation of the complexed tri-*t*-butyltriphosphinine derivative. The free barrelene could be obtained by oxidative decomplexation using C_2Cl_6 ⟨95AG(E)81⟩.

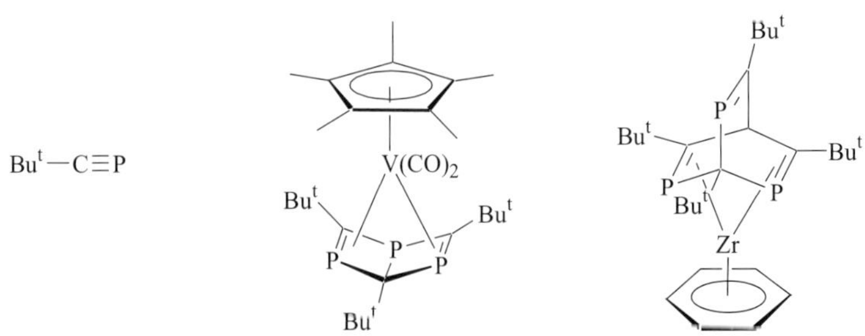

(348) **(349)**

Shortly afterwards, Binger in a fascinating study, reported that the reaction of the hafnium complex (**350**) with (Bu^t–C≡P) is temperature dependent. At −78 to 0°C the complex (**351**) is formed, in which three molecules of (Bu^t–C≡P) are bound to the metal atom. The spectroscopic data of (**351**) (^{31}P NMR, ^{13}C NMR) are discussed without assignment of a definitive structure. At 0–20°C the reaction with (Bu^t–C≡P) remains at the stage of the cyclotrimer complex (**352**). By reaction of the complex (**351**) with hexachloroethane at 25°C the first derivative of a $1\lambda^3,3\lambda^3,5\lambda^3$-triphosphinine (**354**) was obtained as a yellow microcrystalline powder (Scheme 113). The 1,3,5-triphospha-Dewar benzene hafnium complex (**352**) releases the Dewar benzene (**355**) by oxidation with C_2Cl_6. The valence-isomerization (**355**) → (**354**) can be achieved by heating to 90ºC ⟨95AG(E)2227⟩. The complex (**352**) is unstable in solution, at 70ºC it rearranges quantitatively to give (**412**).

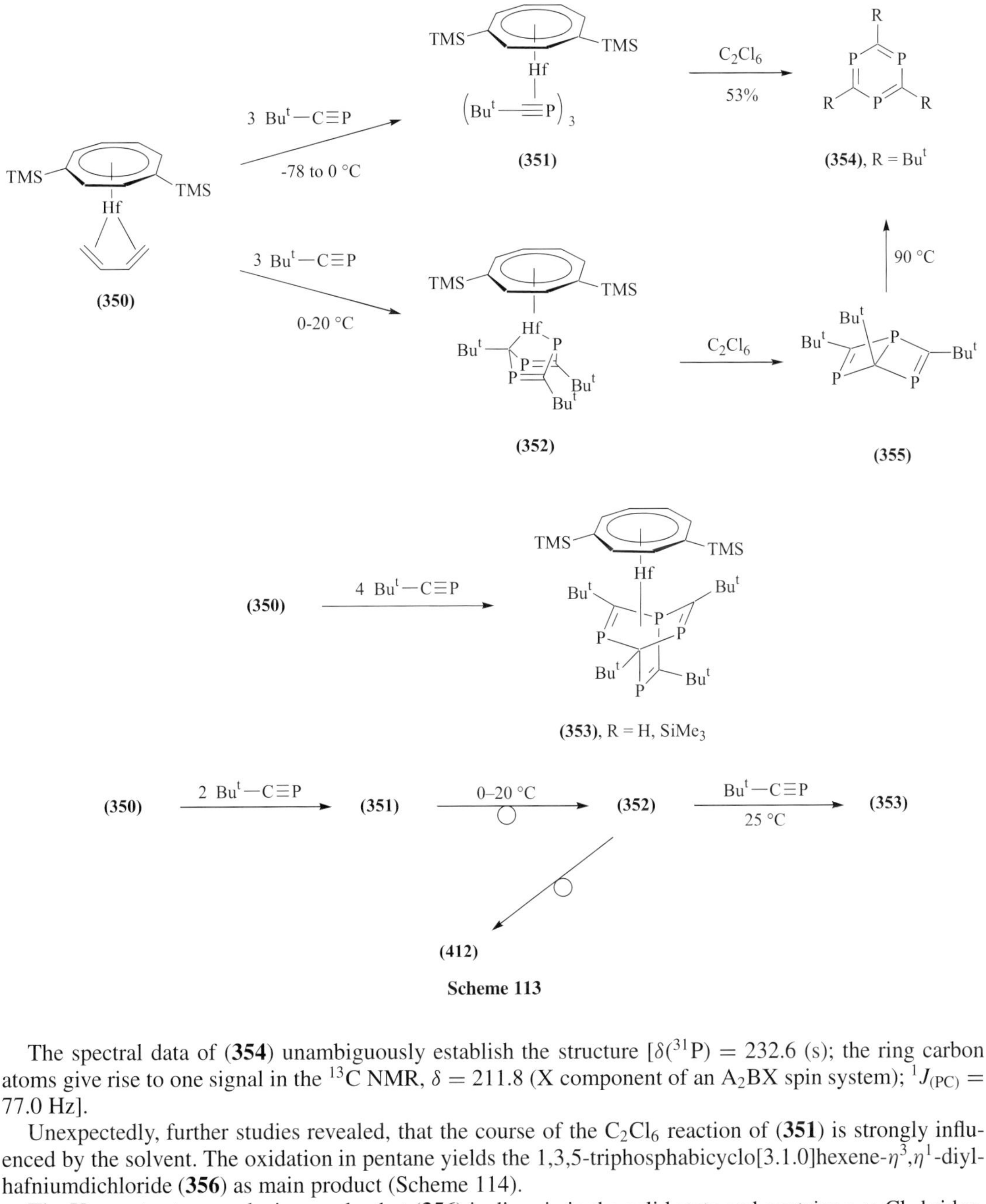

Scheme 113

The spectral data of (**354**) unambiguously establish the structure [$\delta(^{31}P) = 232.6$ (s); the ring carbon atoms give rise to one signal in the ^{13}C NMR, $\delta = 211.8$ (X component of an A_2BX spin system); $^1J_{(PC)} =$ 77.0 Hz].

Unexpectedly, further studies revealed, that the course of the C_2Cl_6 reaction of (**351**) is strongly influenced by the solvent. The oxidation in pentane yields the 1,3,5-triphosphabicyclo[3.1.0]hexene-η^3,η^1-diyl-hafniumdichloride (**356**) as main product (Scheme 114).

The X-ray structure analysis reveals, that (**356**) is dimeric in the solid state and contains a μ-Cl_2 bridge. In solution (**356**) is monomeric and is therefore electronically and coordinatively unsaturated. It reacts with electron-donating ligands such as isonitriles and phosphanes to give the new Hf-complexes (**357**) and (**358**) (Scheme 114).

Hydrolysis of the complexes (**352**), (**357**), (**358**) yields, depending on the reaction conditions either the secondary phosphinoxide (**359**) or (**360**), (**361**) with 1,3,5-triphosphabenzene-skeletons ⟨99EJIC763⟩ (Scheme 114).

New chemoselective cyclooligomerizations of R–C≡P with Bu^{t}N=VCl$_3$•DME (**A**) and Bu^{t}N=VCl$_3$ (**B**) are described ⟨98AG(110)1318⟩.

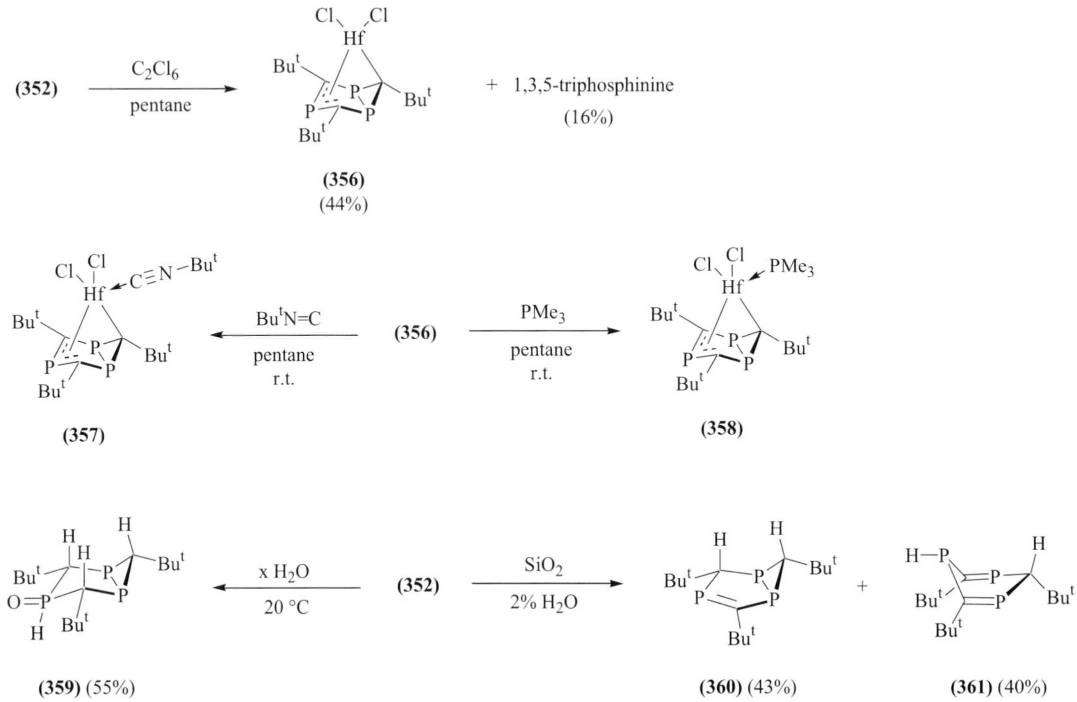

Scheme 114

The reaction of the V-complex (**B**) with R–C≡P totally unexpected and hitherto unknown yields the metal-free 1,3,5-triphosphinines (**354**).

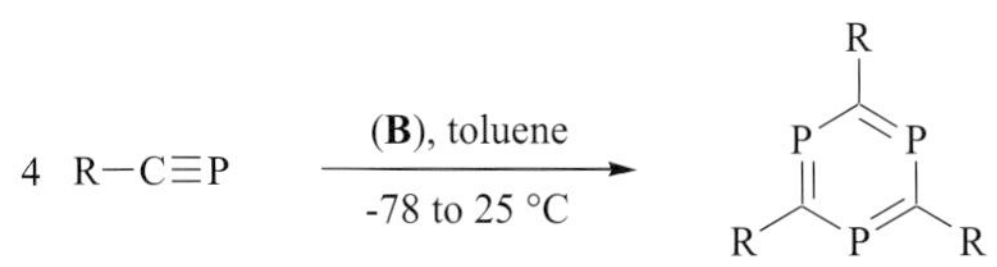

(**354**), R = Bu^t (68%), $Pent^t$ (59%); Ad (36%); C_7H_{13} (44%). The ^{31}P NMR data are in the range of $\delta = 238.1$–242.8 ppm, the ^{13}C NMR data in the range of $\delta = 208.8$–212.2 ppm, the ^{13}C value of the already known (**354**), R = Bu^t is $\delta = 211.8$⟨98AG(110)1318⟩.

The question of the stability and reactivity of 1,3,5-triphospha-Dewar-benzenes is continued to be studied.

In contrast to Dewar-benzenes, which are much less stable than the corresponding benzene-derivatives the Dewar-phosphabenzenes are the more stable isomers ⟨86AG(E)(25)280⟩.

Accordingly the 2-Dewar-phosphabenzene (**362**) undergoes rearrangement to the 1-Dewar-phosphabenzene (**363**) via the λ^3-phosphinine on being heated to 160°C ⟨87AG(E)(26)85⟩.

Homo-Diels–Alder cycloaddition reactions take place under mild conditions, when 2,4,6-tri-*tert*-butyl-1,3,5-triphospha-Dewar-benzene (**364**) is allowed to react with electron-deficient alkynes. The resultant [2 + 2 + 2]-cycloadducts (**365**), (**366**) can be isolated in excellent yields (Scheme 115).

The reaction of (**364**) with tri-*tert*-butylazacyclobutadiene yields the *exo*-cycloadducts (**365a**) (25%, m.p. 68°C) and (**365b**) (Scheme 115) ⟨97JOM215⟩.

The tetraphosphasemibullvalene (**367**) was detected first by the reaction of Bu^t–C≡P with tropone (4.9%) at 95°C ⟨95JOC5884⟩, whereas (**230**) is the main product.

Thermolysis of Bu^tC≡P itself at 95°C yields tetraphospha-tetracyclooctene (**368**), (**367**) and the cage compound (**369**) in a ratio of 55 : 25 : 20. Mesitylnitrileoxide in toluene adds to the PC-double bond in (**367**) furnishing (**372**) (92%, m.p. 123°C). The X-ray structure analysis of (**373**) confirms the exo-adduct ⟨97AG(109)1396⟩.

Contrary to semibullvalene itself (**367**) at room temperature does not show a Cope rearrangement. The tetraphosphacages (**370**) and (**371**) ⟨92AG(E)1055, 95CB(129)489⟩ — connected by a photochemical equilibrium (1 : 1) at 10–15°C — by heating up to 150°C undergo a valence isomerization to give

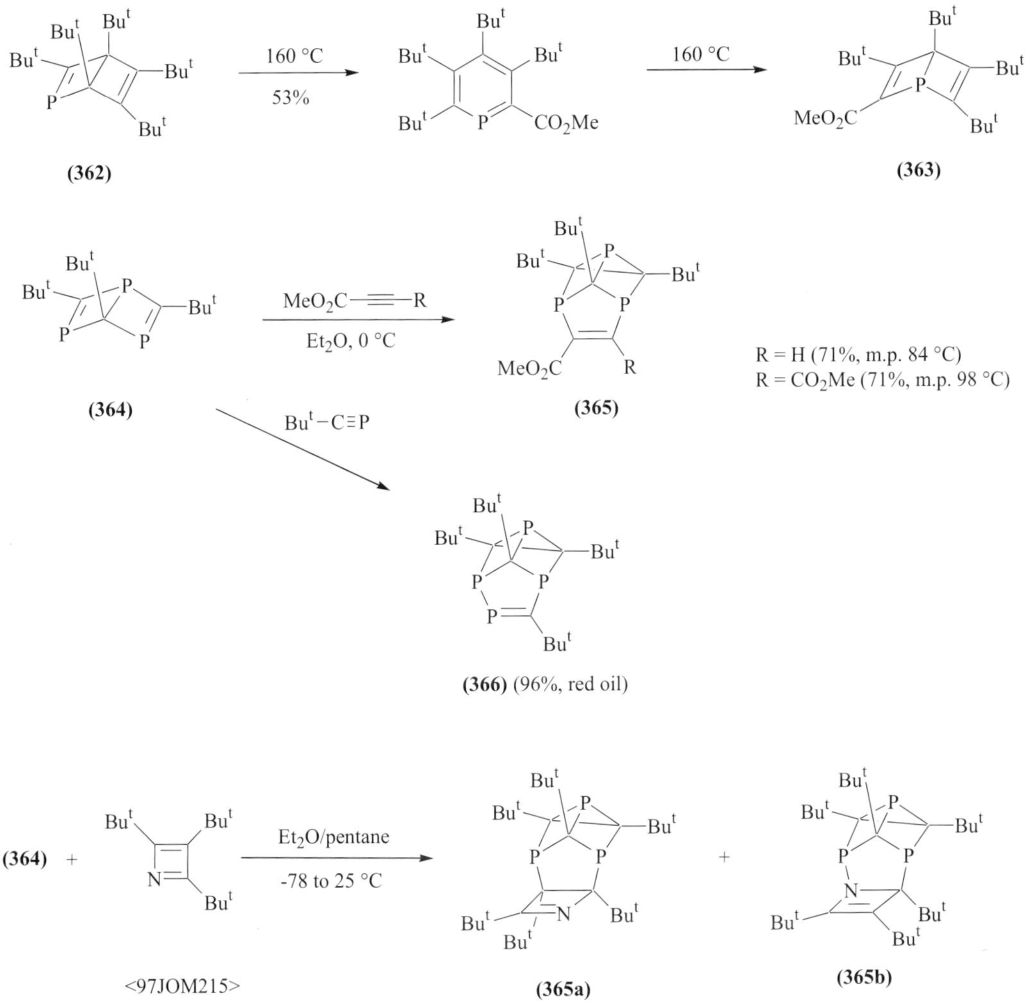

Scheme 115

(**369**). Photolysis of (**369**) in C_6D_6 (10–15ºC) yields the semibullvalene (**367**) (75%). The photochemical isomerization of (**368**) also yields the semibullvalene (**367**) (Scheme 116) ⟨97AG(109)1396⟩.

The homo-Diels–Alder reactions of 2,4,6-tri-*tert*-butyl-1,3,5-triphospha-Dewar benzene (**355**) with alkynes furnishing triphosphabishomoprismanes (**373**) have been studied thoroughly. These cycloadditions are regioselective in the sense that the two P-atoms interact with the CC-triple-bond (Scheme 117). In solution (**373**) rearranges slowly, by irradiation faster, to give the triphosphasemibullvalenes (**367**); the yields are in the range of 70–88%.

The structures of (**373**) were confirmed mainly by ^{31}P and ^{13}C NMR spectroscopy ⟨99S1363⟩.

The 2,4,6-tri-*tert*-butyl-1,3,5-triphosphinine (**354**), R = Bu^t reacts under mild conditions (25ºC, 12 h) with a further equiv. Bu^{t}C≡P to give the barrelene (**374**). The analogous reaction of (**354**) with di(isopropylamino)phosphaethyne leads to an unexpected result. According to ^{31}P NMR control experiments already at −20ºC within a few minutes the tetraphosphasemibullvalene derivative (**375**) is formed quantitatively, orange crystals (Scheme 118) (^{31}P NMR, $\delta(P^1) = -19.6$; $\delta(P^7)$, −92.1; $\delta(P^4) = 88.7$), $\delta(P^3) = 347.5$). The structure elucidation of (**375**) was possible by the X-ray analysis ⟨98EJIC2071⟩.

The chemistry of the $1\lambda^3,3\lambda^3,5\lambda^3$-triphosphinines offers several unexpected results.

The η^6-1,3,5-triphosphabenzene transition metal complexes (**376**)–(**379**) have been synthesized for the first time by the displacement of η^6-arene ligands with the 1,3,5-triphosphinine (Scheme 119).

All new complexes have been characterized mainly by ^{1}H and ^{31}P NMR spectroscopy. (^{31}P NMR, (**376**), $\delta = 57.6$; (**377**), $\delta = 38.6$; (**378**), $\delta = 59$; (**379**), $\delta = -4.7$ ppm) ⟨99EJOC83⟩.

The first examples of η^1-complexes of 2,4,6-tri-*tert*-butyl-1,3,5-triphosphinines (**354**) are described, their structure was established by ^{31}P and ^{195}Pt NMR spectroscopy. When (**380**), $R^1 = R^2$ = Me was recrystallized from CH_2Cl_2 in the air at room temperature, addition of three moles H_2O to the coordinated

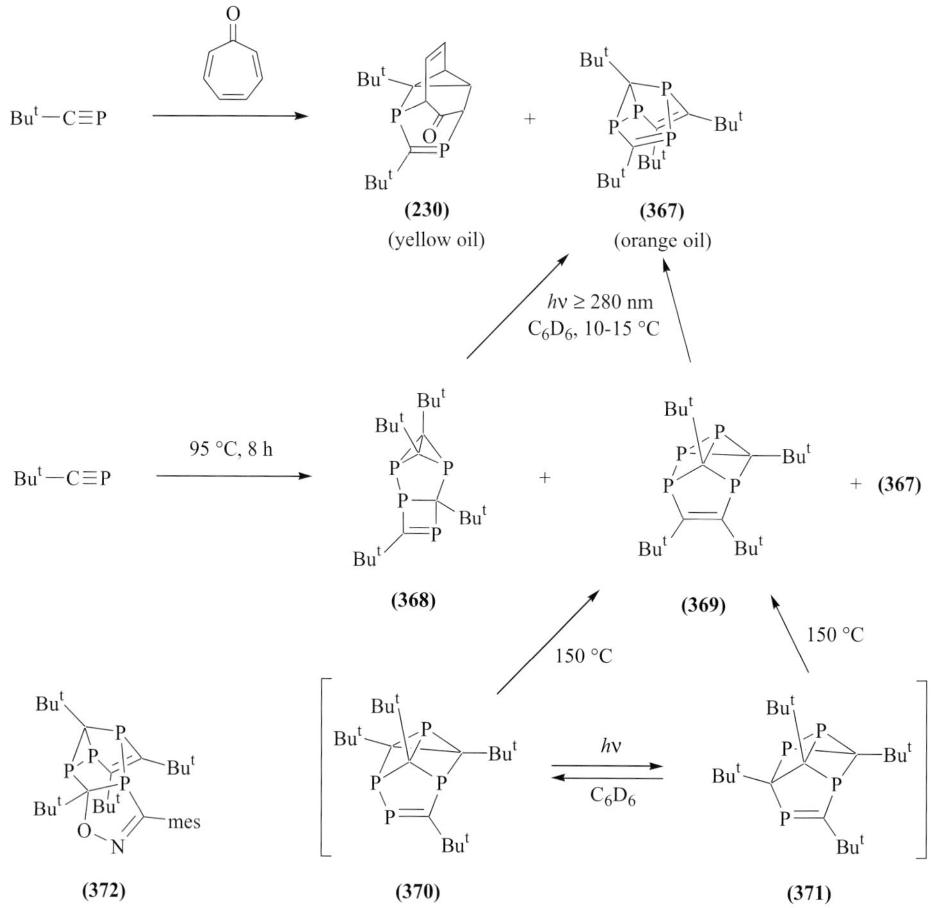

Scheme 116

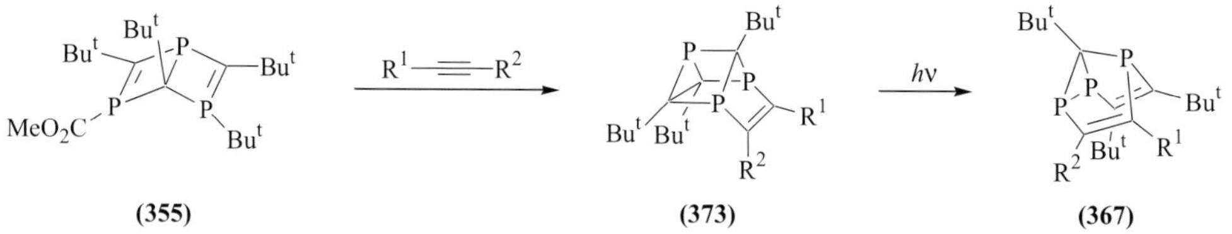

$R^1 = R^2 = H$; $R^1 = R^2 = MePh$, Tms; $R^1 = H$, $R^2 = Ph$; $R^1 = Tms$, $R^2 = -(CH_2)_2-{\equiv}-Tms$

Scheme 117

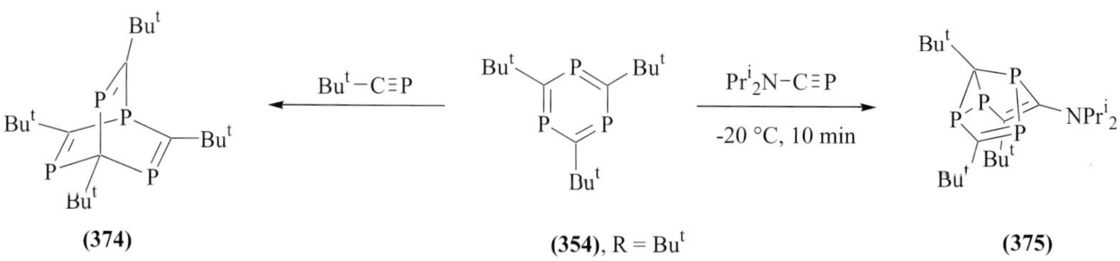

Scheme 118

triphosphinine ring occurs to give the tris-phosphinoxide (**381**). The structure was established by a single crystal X-ray diffraction study ⟨99CC1377⟩.

Cocondensation of electron beam vaporized scandium with an excess of Bu^{t}C≡P at 77 K affords the air sensitive forrest-green 22-electron scandium triple decker complex (**382**) (5–10%). The molecular structure

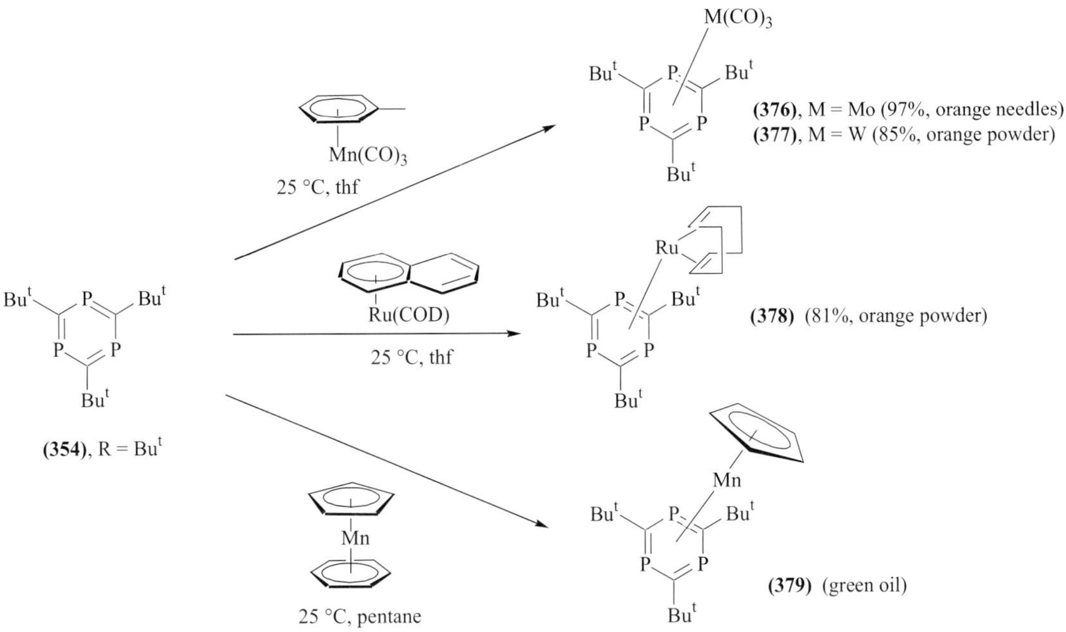

Scheme 119

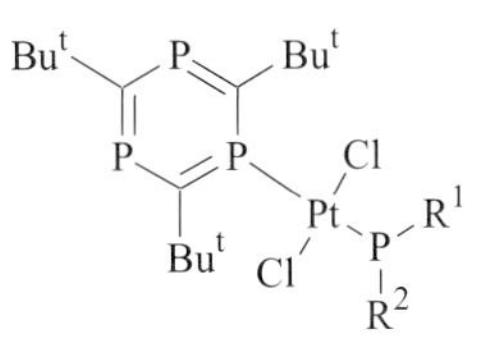

(380)

$R^1 = R^2 =$ Me, Et;
$R^1 =$ Me, $R^2 =$ Ph;

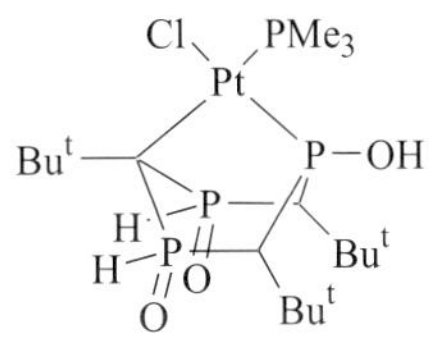

(381)

has been determined by X-ray crystallography, the central 2,4,6-tri-*tert*-butyl-1,3,5-triphosphabenzene ring in the triple decker structure lies on a crystallographic mirror plane and is planar with no significant variation in the P–C bond lengths ⟨96JA7630⟩.

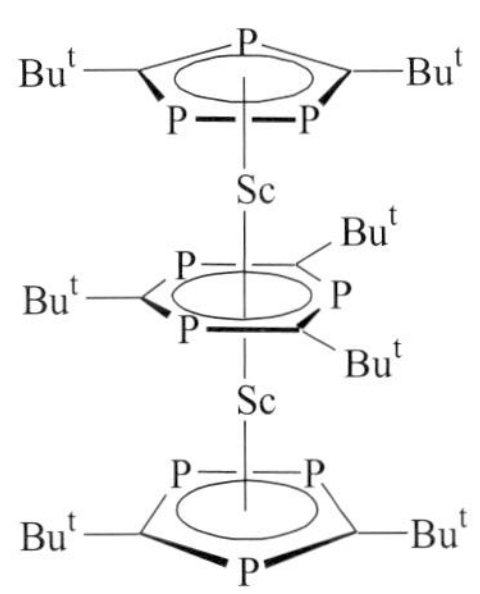

(382)

In a rather unusual manner 2,4,6-tri-*tert*-butyl-1,3,5-triphosphabenzene (**354**) reacts with a potassium mirror in tetrahydrofurane at −78°C which results in the abstraction of one P^{3-} and generation of the 2,4,5-tri-*tert*-butyl-1,3-diphosphacyclopentadienide anion (**383**) (Scheme 120). The potassium salt (**383**) forms an air sensitive white powder (40% yield). The X-ray analysis of the potassium salt (**383**) confirms the structure, it consists of chains of alternating K(THF) cations and $P_2C_3Bu^t{}_3$ anions, the alkali metal is interacting in an η^5-bonding mode with the ring ⟨00OM219⟩.

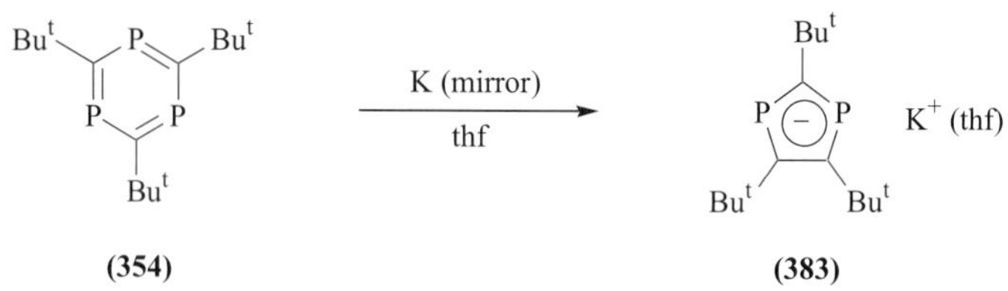

Scheme 120

A series of heteroatom bridged 1,3,5-triphosphinines has been published in the last year.

An 1,2,4-oxadiphosphole (**385**) was synthesized for the first time by vacuum thermolysis of the mesitylphosphaalkene (**384**). The intermediate probably is Mes–C≡P ⟨99EJOC587⟩ (Scheme 121).

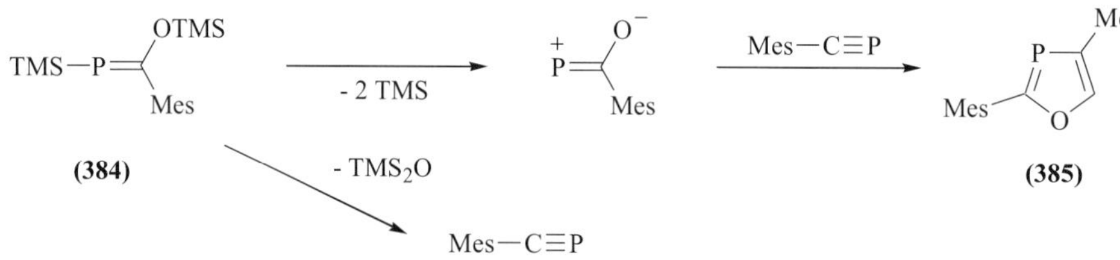

Scheme 121

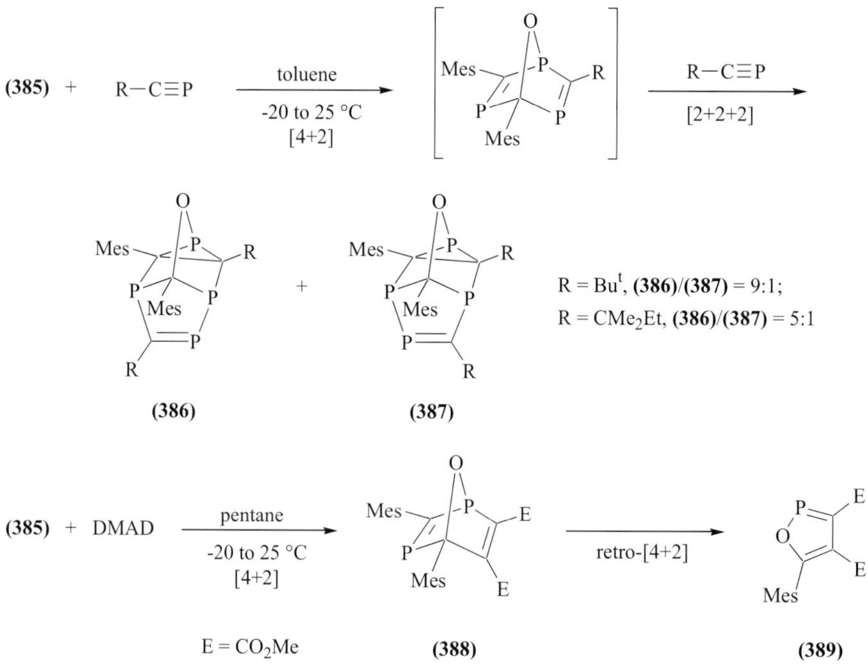

Scheme 122

The new phosphole (**385**) offers an enormous [4 + 2]-cycloaddition reactivity. With phosphaalkynes R–C≡P a sequence of [4 + 1] and [2 + 2 + 2]-cycloaddition reactions yields the oxatetraphosphadeltacyclenes, which can be isolated by column chromatography as a mixture of (**386**) and (**387**) (95–98%) (Scheme 122).

With DMAD (**385**) reacts at −20ºC to 25ºC via the [4 + 2] intermediate (**388**), which undergoes a retro-[4 + 2]-reaction to furnish the labile 1,2-oxaphosphole (**389**) ⟨99EJOC587⟩ (Scheme 122).

The phosphaalkynes react with elemental selenium or tellurium to furnish the 1,2,4-selena- and 1,2,4-tellura-diphospholes (**390**) and (**391**). With the exception of (**390**), R = 1-Ad the phospholes are distillable oils (Scheme 123) ⟨99S1642⟩.

The phosphaalkynes react with a deficit of selenium furnishing the tetracyclic products (**392**) in very good yields (Scheme 123).

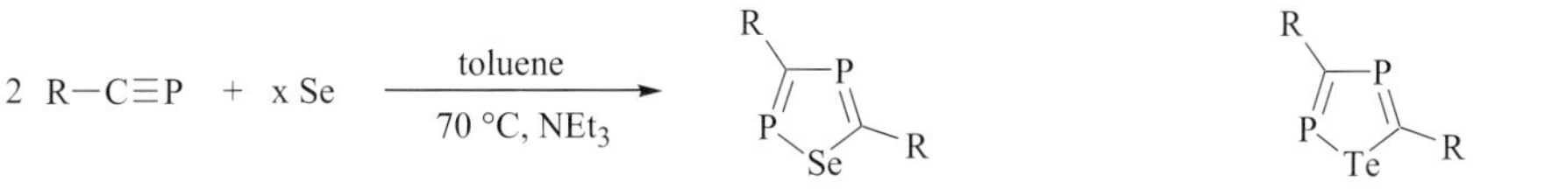

(390)
R = Bu[t] (89%);
Pent[t] (81%);
1-Ad (80%, m.p. 163 °C);
C_6H_{10}-CH_3 (74%)

(391)
R = Bu[t] (15%);
Pent[t] (20%);
1-Ad (15%)

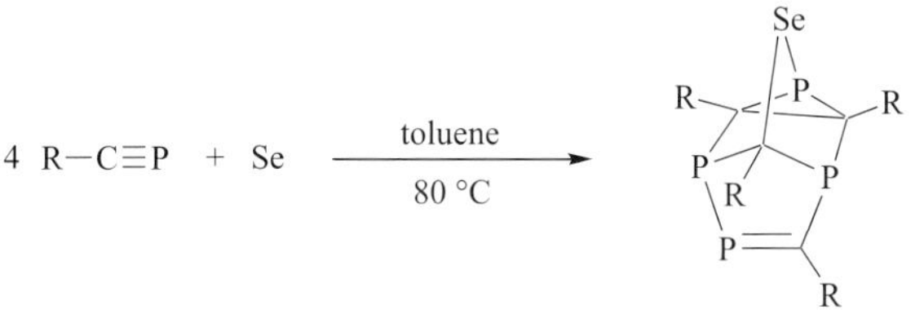

R = Bu[t] (95%, m.p. 138 °C);
Pent[t] (18%, m.p. 151 °C);
1-Ad (49%, m.p. 236 °C)

(392)

Scheme 123

The structure of (**392**), R = Bu^t was confirmed by an X-ray crystallographic analysis ⟨99S1642⟩.

The stable bis(amino)silylene (**393**) undergoes a [1 + 4]-cycloaddition with 2,4,6-tri-*tert*-butyl-1,3,5-triphosphinine (**354**) to afford the rather exotic product (**394**) with a norbornadiene skeleton (Scheme 124). The molecular structure is supported by the NMR data and confirmed by X-ray analysis ⟨99CC2451⟩.

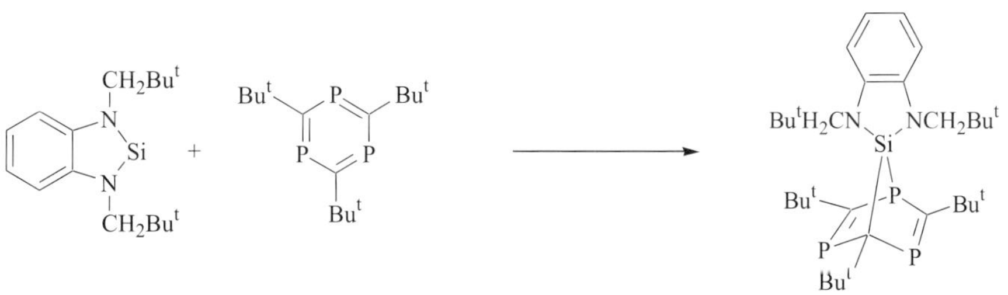

(393) **(354)** **(394)**

orange-red crystals,
m.p. 224-225 °C, 66%

Scheme 124

An unusual tris-triphenylphosphonio-substituted 1,3,5-triphosphinine was prepared by Schmidpeter *et al.* The condensation of triphenylphosphonium-bis(trimethylsilyl)methylide (**395**) with $PHal_3$ (Hal = Cl, Br) gives the cyclic trimers (**396**). In solution they dissociate to give the ionic species. (**396**), Hal = Cl reacts with Ph_2P–$SiMe_3$ to give (**397**) and with trimethylsilylmorpholin (**398**) is formed (as dication) (Scheme 125). With $GaCl_3$ (**396**), Hal = Cl forms the tris(triphenylphosphonio)-1,3,5-triphosphinine-trication (**399**) ⟨97ZN(B)162⟩.

Difluoromethylene(trifluoromethyl)phosphane (**401**), available most conveniently by thermolysis of the stannylphosphane (**400**), dimerizes and trimerizes on warming from a cold trap to give the 1,2- and 1,3-diphosphetanes (**402**) and (**403**) and, in yields of up to 12%, the 1,3,5-triphosphinane (**404**) (Scheme 126) ⟨83IC2573, 84AG(E)710⟩. The structures of the products were mainly elucidated by ^{19}F NMR spectroscopy.

The hexamer of Bu^{t}C≡P (**407**) can be considered as a dimer of a 1,3,5-triphosphinane ⟨94AG(106)2284⟩. The reaction of a mixture of $P_3C_2(Bu^t)_2^-$ (**405**) and $P_2C_3(Bu^t)_3^-$ (**406**) (formed by treatment of Me_3Si–P=C($OSiMe_3$)Bu^t with Na in dimethoxyethane) with the labile complex $PtCl_2$ (cod) (cod≡1,5-cyclooctadiene) at −30ºC yields a mixture of organophosphorus compounds. By chromatography with silica gel/petrolether a colorless compound was isolated, which by X-ray analysis could be identified as the first hexamer of Bu^{t}C≡P (**407**) (Scheme 127).

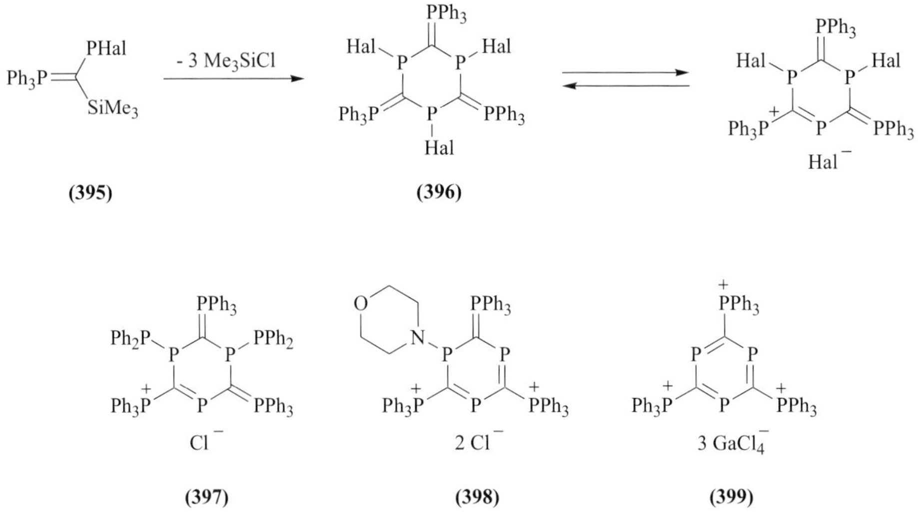

Scheme 125

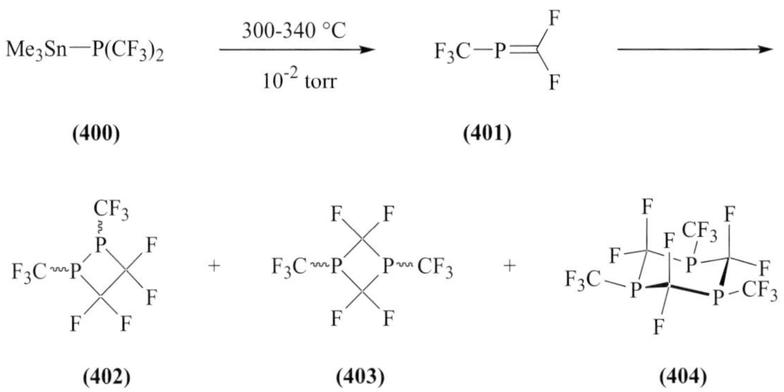

Scheme 126

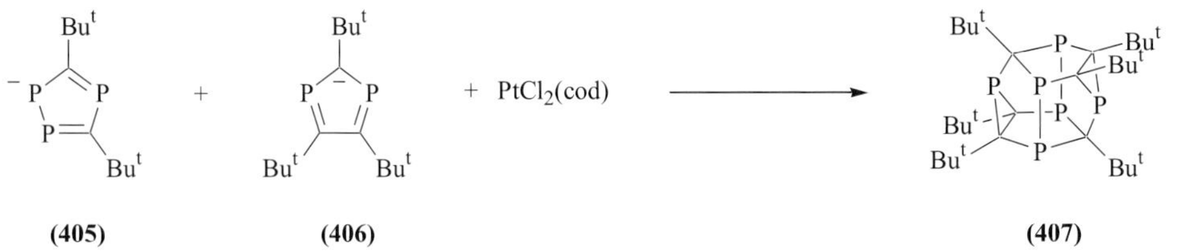

Scheme 127

The hexaphosphorus cage could be considered as dimer of a valence isomeric 1,3,5-triphosphinane. *Ab initio* calculations for the formation of HC≡P oligomers are discussed ⟨94AG(106)2284⟩.

5.3.10.3 1,2,4-Triphosphinines

Dihydro-1,2,4-triphosphinines as well as valence isomers of 1,2,4-triphosphinines are connected with 1,3,5-triphosphinines and its valence isomers by rearrangement reactions.

In the presence of $AlCl_3$($GaCl_3$) phosphaalkynes RC≡P undergo spirocyclotrimerization with incorporation of the Lewis-acid to form (**408**) (Scheme 128) ⟨96CB(129)489⟩.

The formation of (**408**) probably proceeds via the addition of the Lewis acid to the alkyne-C-atom. Already with dimethylsulfoxide as base the $AlCl_3$ in (**408**), R = Bu^t is removed generating the highly reactive spirocyclic diphosphete (**409**), which at −45°C undergoes ring enlargement to the 1,3,5-triphospha-Dewar benzene (**355**). (**355**) could not be isolated, however with a further mole R–C≡P it is trapped as the tetraphosphatetracyclic cage (**356**). If (**409**), R = Bu^t in the presence of a

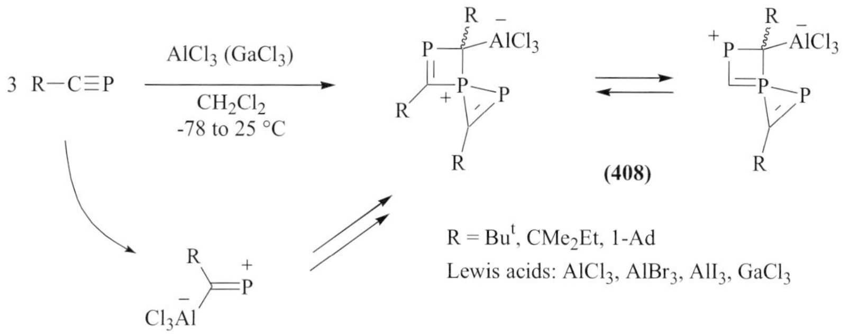

Scheme 128

suspension of excess $AlCl_3$ in dichloromethane was treated with DMSO at −78°C, the unsymmetrical 1,2,4-triphospha-Dewar benzene (**410**) was generated, which also adds further Bu^{t}C≡P to give (**411**), an isomer of (**366**) (Scheme 129).

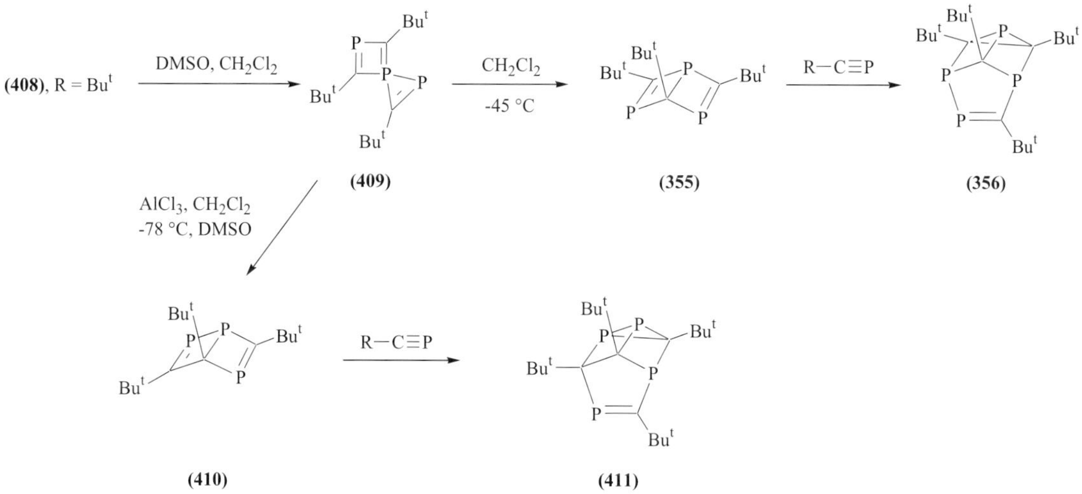

Scheme 129

The ^{31}P NMR spectra prove the structure of (**408**)–(**411**) ⟨96CB(129)489⟩.

(η^4-Butadien)[η^8-1,4-bis(trimethylsilyl)cyclooctatetraen]hafnium (**350**) reacts at 0°C within 2 h with three moles of Bu^{t}C≡P to give the Hf-complex (**353**) of 1,3,5-triphosphinine. Cleavage of (**353**) with C_2Cl_6 yields 1,3,5-triphospha-Dewar benzene (**355**). Already at room temperature, faster at 70°C (1 h) in toluene, (**353**) rearranges, leading to the formation of the stable Hf-complex of the unsymmetrical 1,2,4-triphosphinine (**412**). With the mild chlorination reagent C_2Cl_6 the 1,2,4-triphospha-Dewar benzene (**413**) is displaced from (**412**), a pale yellow oil, ^{31}P NMR, $\delta = 350.4$ (P^2), $\delta = -28.4$ (P^4), $\delta = -154.9$ (P^1) with characteristic coupling constants ($J(P^1,P^2) = 191.8$; $J(P^1,P^4) = 4.2$, $J(P^2,P^4) = 87.2$ Hz). A mechanism for the rearrangement of (**353**) into (**412**) is discussed (Scheme 130) ⟨97CB1491⟩.

In the presence of PMe_3 at 50°C (**412**) is transformed into the Hf-complex (**414**) with 1,2,3-triphosphete, proved by an X-ray analysis ⟨97CB1491⟩.

The 1,2,5-triphosphinine-complex (**415**) can be obtained by the reaction of MesC≡P with [$W(CO)_5$(thf)] at −50°C in THF/*n*-hexane. From the complex reaction mixture a pentamer of MesC≡P (**415**) can be isolated by column chromatography in 10% yield, orange red crystals. The formation of (**415**) was confirmed by a X-ray structure analysis and the ^{31}P NMR data. (**415**) is the first pentamerization product of a phosphaalkyne. The triphosphinine ligand in (**415**) is also a derivative of the unsymmetrical 1,2,4-triphosphinine (Scheme 131) ⟨99AG(111)3384⟩.

When triphenylstannyl-3,5-di(*tert*-butyl)-1,2,4-triphosphole (**416**) in hexane was reacted with Bu^{t}C≡P at 70°C (12 h) the phosphorus–carbon cage (**418**) was obtained in 80% yield as a pure, yellow crystalline compound (Scheme 132) ⟨99CEJ(5)3143⟩.

The formation of (**418**) starts with a [4 + 2]-cycloaddition of Bu^{t}C≡P to (**416**), followed by a (2 + 2 + 2) homo-Diels–Alder reaction of (**417**) with a second mole Bu^{t}C≡P. The cage compound is asymmetric and contains seven stereogenic centers. Surprisingly only one pair of enantiomers has been formed ⟨99CEJ(5)3143⟩.

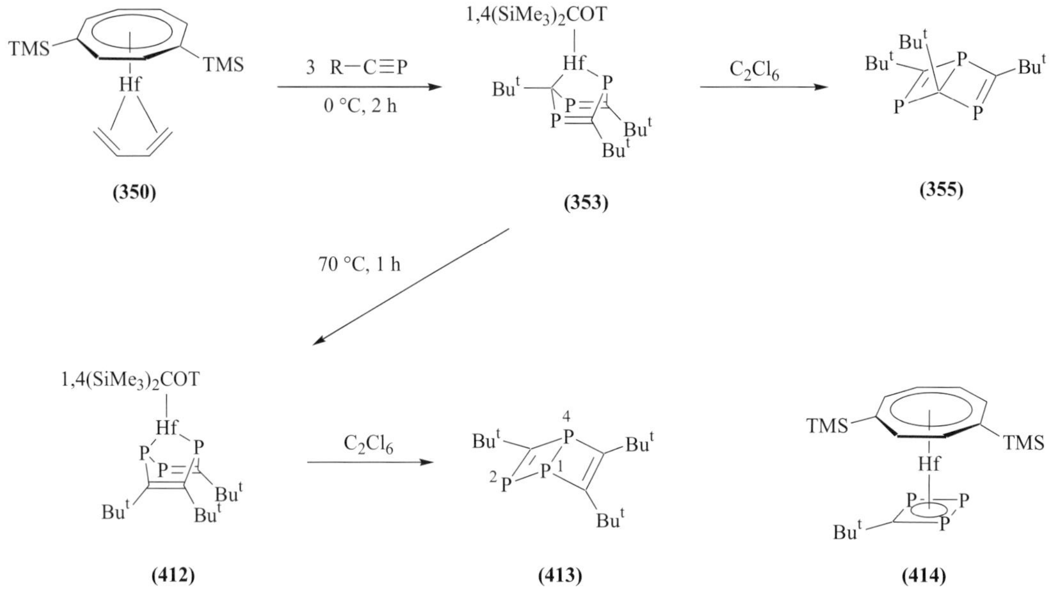

Scheme 130

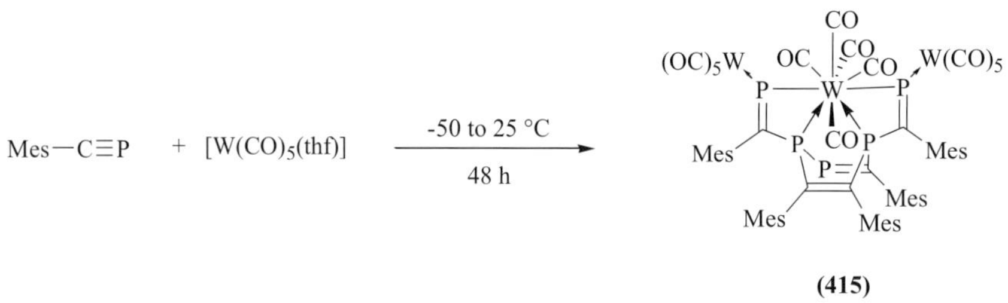

Scheme 131

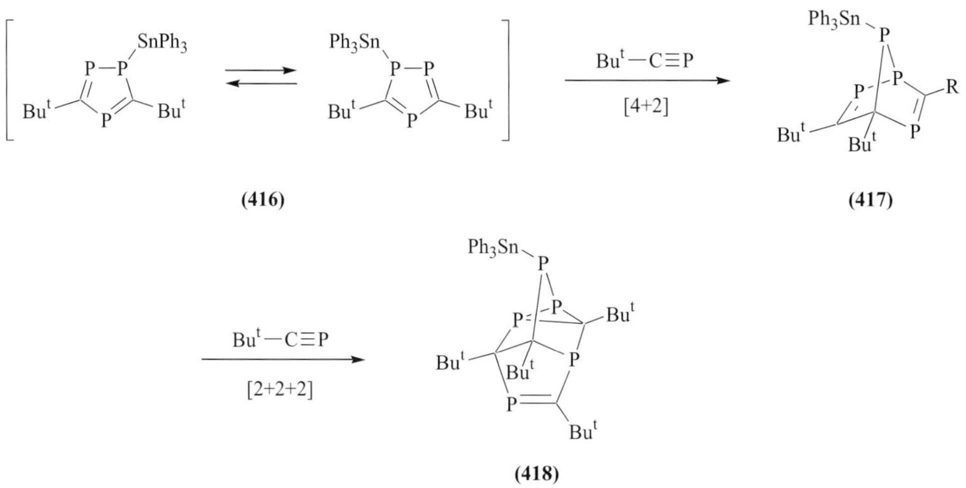

Scheme 132

The chemoselective cyclooligomerizations of R–C≡P with Bu^{t}N=VCl_3•DMF (**A**) and Bu^{t}N=VCl_3 (**B**) ⟨98AG(110)1318⟩ have been already mentioned. The V-complex (**B**) catalyzes the trimerization of R–C≡P to give 1,3,5-triphosphinines. By reaction of R–C≡P with the V-complex (**A**) the hitherto also unknown azatetraphosphaquadricyclanes (**419**) are obtained (Scheme 133) ⟨98AG(110)1318⟩.

For (**419**), R = Bu^t the X-ray structure analysis confirms the cage. The intermediates are the azadiphosphole (**420**), the azatriphosphanorbornadiene (**421**) and the azatriphosphole (**422**).

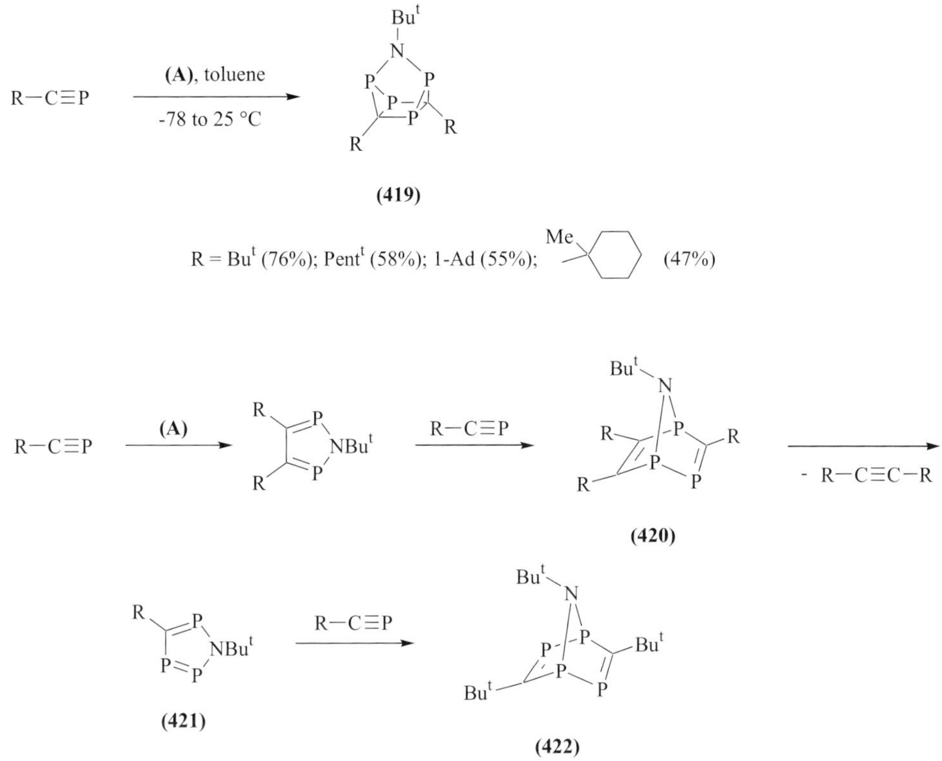

Scheme 133

The tetrakis(dimethylamino)-1,3-diphosphete (**239**) provides access to 1,3,5-triphosphinines with λ^5-phosphorus. Its synthesis is given in Scheme 134. In 1990 one of the by-products of this reaction (3–8% yield) was identified as 1,1,3,3,5,5-hexakis(dime-thylamino)-1λ^5,3λ^5,5λ^5-triphosphinine (**423**) 〈90JCS(P1)1223〉. The spectroscopic and X-ray structural data confirm the structure. The PE spectrum of (**423**) has been reported 〈91CB2071〉, 〈98ZAAC1116〉.

Cycloaddition of (**239**) to *tert*-butylphosphaethyne forms the mixed P-valence triphosphinine (**425**), probably via the triphospha-Dewar benzene (**424**), which was not isolated (Scheme 134) 〈86AG(E)1002, 87ZN(B)1213〉.

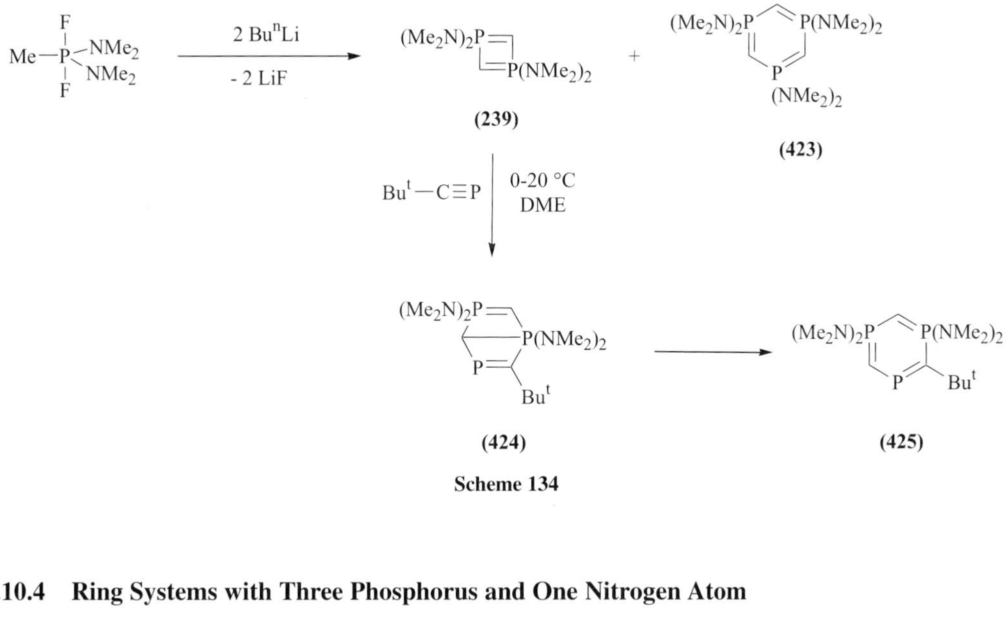

Scheme 134

5.3.10.4 Ring Systems with Three Phosphorus and One Nitrogen Atom

Of the possible azatriphosphinines with all-λ^3, all-λ^5 and mixed valencies none is known yet.

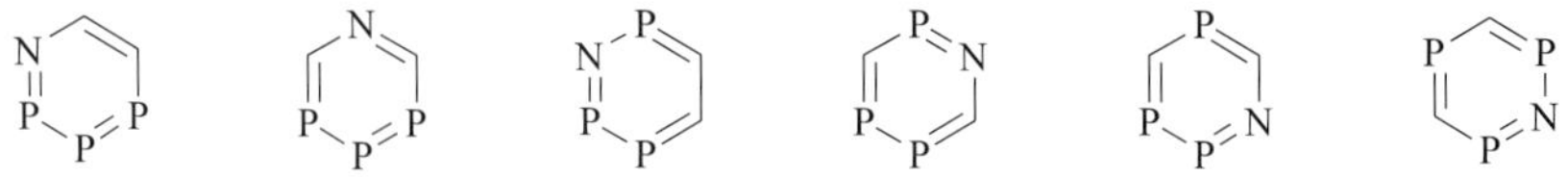

(I) (II) (III) (IV) (V) (VI)

5.3.11 RING SYSTEMS WITH THREE PHOSPHORUS AND TWO NITROGEN ATOMS

Of the possible diazatriphosphinines with all-λ^3, all-λ^5- and mixed valencies only the 1,3,2λ^5,4λ^5,6λ^5-diazatriphosphinine-system(VII) is described in the literature.

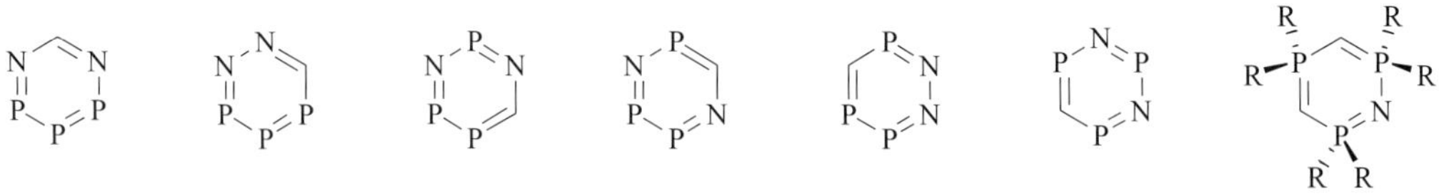

(I) (II) (III) (IV) (V) (VI) (VII)

5.3.11.1 1,3,2,4,6-Diazatriphosphinines

Of the possible isomers only derivatives of the 1,3,2λ^5,4λ^5,6λ^5-diazatriphosphinines(VII) are described in the literature. These were prepared by rearrangement of the heptamethylcyclotriphosphazenium salt (**426**) under the action of base (sodium hexamethyldisilazane). The authors propose a four-center mechanism, with the ylide carbon at P-2 attacking P-6 and displacing the ring N-atom carrying the methyl group. The ring carbon in the product (**427**) has a strong ylide character; hydrolysis and protonation proceed as shown, and in the NMR the *C*-H signal is exchange-broadened and detectable only under favorable conditions. Benzoylation occurs both at C and exocyclic N, forming the derivative (**428**), the structure of which was solved by X-ray crystallography (Scheme 135) ⟨75CC454, 77CJC2534, 77CJC3651⟩.

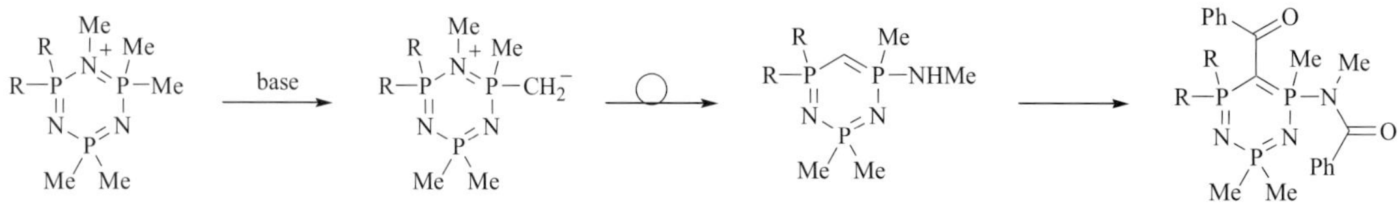

(426) (427) (428)

Scheme 135

5.3.12 RING SYSTEMS WITH FOUR PHOSPHORUS ATOMS

Of the possible tetraphosphinines with all-λ^3, all-λ^5 and mixed valencies only valence isomeric tetraphosphinanes with 1,2,3,5- and 1,2,4,5-tetraphosphaskeletons are known.

5.3.12.1 1,2,3,5-Tetraphosphinines

The polycyclic compounds with valence isomeric 1,3,5- and 1,2,4-triphosphinane-structures (**370**) and (**371**) ⟨92AG(E)1033, 95CB(129)489⟩ on irradiation with a high-pressure Hg lamp (Philips HPK 125) in C_6D_6 form an 1 : 1 equilibrium. By heating (**370**) and (**371**) without solvent up to 150ºC a skeleton rearrangement occurs yielding (**369**) (yield 100% from (**370**), 48% from (**371**), which may be considered as a valence isomer of 1,2,3,5-tetraphosphinane (Scheme 136), see also (Scheme 116) ⟨97AG(109)1396⟩.

Irradiation of (**369**) under the same conditions yields the tetraphosphasemibullvalene (**367**) (see also Scheme 116).

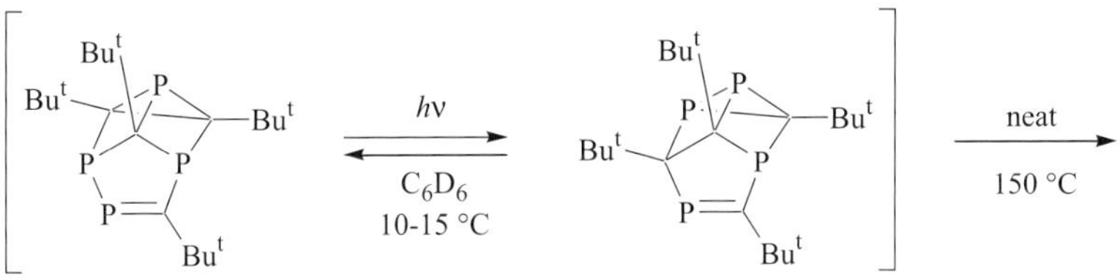

(370) **(371)**

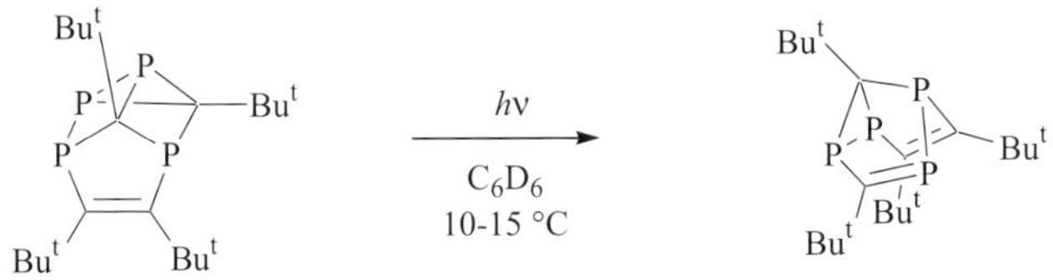

(369) **(367)**

Scheme 136

5.3.12.2 1,2,4,5-Tetraphosphinines

The reaction of the V-complex $Bu^tN{=}VCl_3$ DME (**A**) with $Bu^tC{\equiv}P$ (toluene, $-78°C \rightarrow 25°C$) (see also Scheme 133) yields as an intermediate the 1-aza-2,3,5-triphosphole (**421**), which undergoes a [4 + 2]-cycloaddition with a third mole $Bu^tC{\equiv}P$ to give (**422**), which can be considered as a 1,4-dihydro-1,2,4,5-tetraphosphinine. This norbornadiene-derivative is not stable, it undergoes spontaneously an intramolecular [2 + 2]-cycloaddition to the aza-tetraphosphaquadricyclan (**423**) with a valence isomeric 1,2,4,5-tetraphosphinane structure (Scheme 137) ⟨98AG(110)1318⟩.

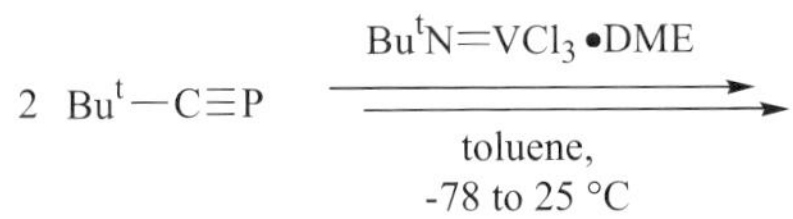

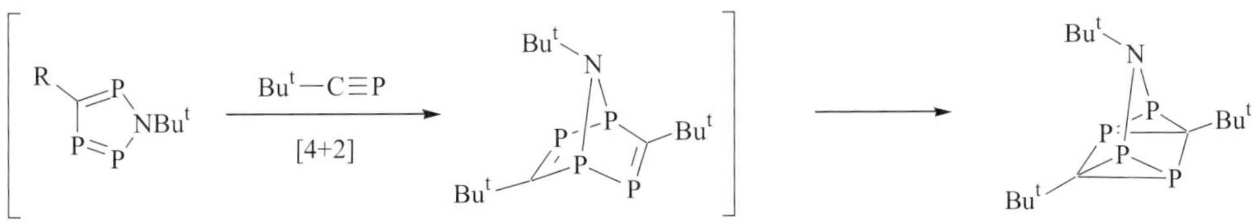

(421) **(422)** **(423)**

Scheme 137

The fully saturated cyclo-3,6-dicarba-1,2,4,5-tetraphosphane was obtained by Baudler *et al.* ⟨78ZN(B)1208⟩ and Stelzer *et al.* ⟨91ZN(B)1131⟩ by spontaneous dimerization of the corresponding diphosphirane at room temperature (Scheme 138).

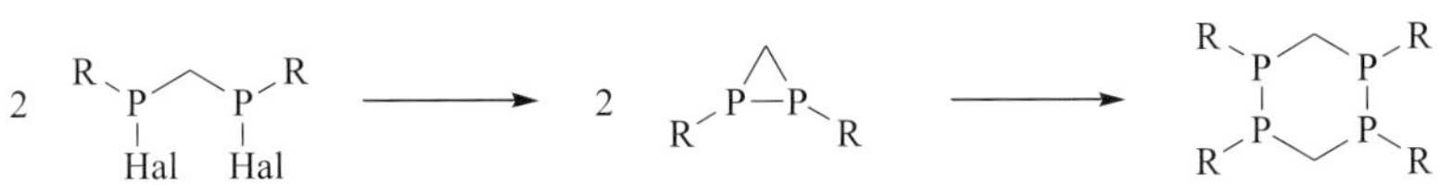

R = Bu^t, 2,4,6-$Me_3C_6H_2$; Hal = Cl, Br

Scheme 138

5.3.13 RING SYSTEMS WITH ONE PHOSPHORUS AND ONE OXYGEN ATOM

λ^3-Oxaphosphinines, f.i. (I) with pyrylium salt structures have not yet been described.

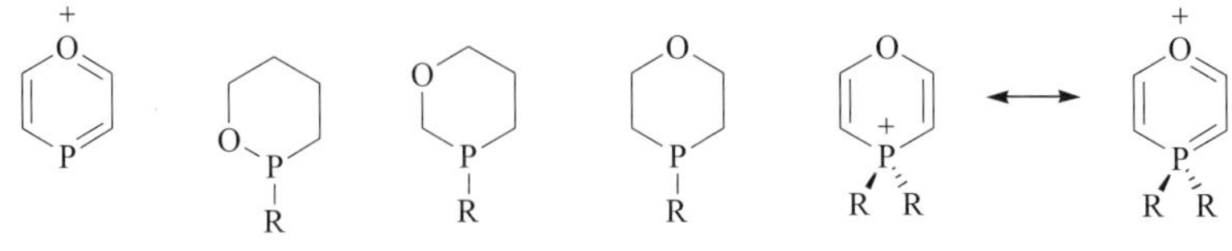

(I) **(II)** **(III)** **(IV)** **(V)**

However, derivatives of the saturated oxaphosphinanes (II)–(IV) are well known. The 4,4-diaryl-1,4-oxaphosphininium-salts (V) can be drawn as mesomeric 1,4-λ^5-oxaphosphinines, emphasizing their relationship to the 1,4λ^5-azaphosphinines ⟨72CR(C)2206⟩.

Phosphorus heterocycles with phosphorus in 2-position (II) as well as 1,3,2-dioxaphosphinanes, means cyclic esters R_2P–OR (phosphinites) and RP $(OR)_2$ (phosphonites) are not discussed here.

5.3.13.1 1,3-Oxaphosphinines

The 1,3-oxaphosphinines are represented by the dihydro derivatives (**431**), which are formed from 3-hydroxypropylphosphine (**429**) by reaction with *N*-arylpivalimidoyl chlorides (**430**). The P=C bond in (**431**) is highly reactive; it adds HCl to give the chlorophosphine (**432**), and is reduced by methylphenylphosphine forming 2-*tert*-butyl-1,3-oxaphosphinane (**433**) (Scheme 139) ⟨91CB493⟩.

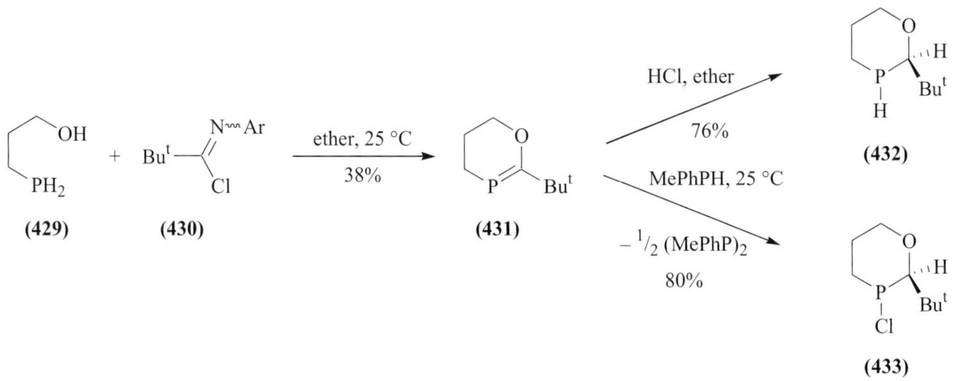

Scheme 139

1,3-Oxaphosphinanes (**435**) have been prepared by acid-catalyzed condensation of 3-hydroxypropylphenylphosphine (**434**) with aldehydes and ketones. The secondary phosphine is easily available by radical addition of phenylphosphine to allylic alcohols ⟨59DOK(127)1217, 63IZV502, 72SRI223⟩. Primary hydroxypropylphosphines condense similarly, forming 3-unsubstituted 1,3-oxaphosphinanes (**436**) ⟨73ZC291, 78PS(5)81⟩ (Scheme 140).

The 1,3-oxaphosphinan-2-ones (**437**) are accessible by AIBN-initiated intramolecular radical cyclization of the P–H bond to the double bond of allyloxycarbonyl phosphines (Scheme 140) ⟨82ZN(B)965⟩. The tervalent phosphorus in (**437**) reacts as in normal phosphines; sulfides and phosphonium salts can be prepared.

The configuration and conformation of 1,3-oxaphosphinanes using NMR measurements has been discussed ⟨86DOK(291)154⟩.

Interesting 1,3-oxaphosphinanes with bi- and tricyclic structures had been described during the last years.

The primary bis-phosphines (**438**) react with acetylacetone or trifluoroacetylacetone to give the bis(phospha-adamantyl)alkanes (**439**) as a new class of very bulky diphosphines (Scheme 141) ⟨99CC901⟩.

With 1,2-diphosphinebenzene (**440**) acetylacetone forms the benzannelated diphosphaadamantane (**441**). The X-ray crystal structures of (**439**) and (**440**) as $PdCl_2$-complexes have been determined.

The bulky diphosphines of the type (**439**) are patented as cocatalysts ⟨98PCT Int. Appl. WO9842717A119881001⟩.

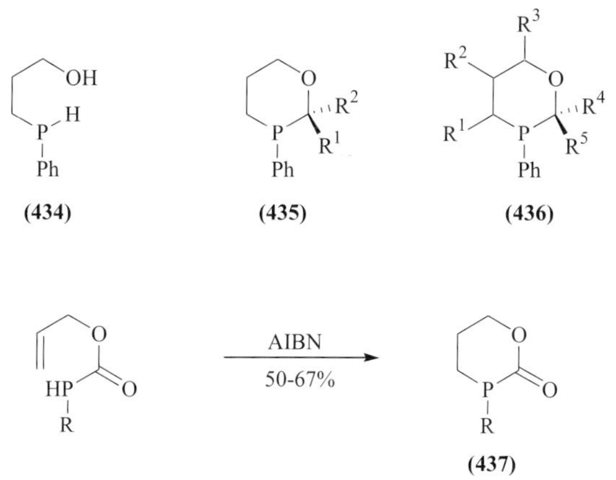

Scheme 140

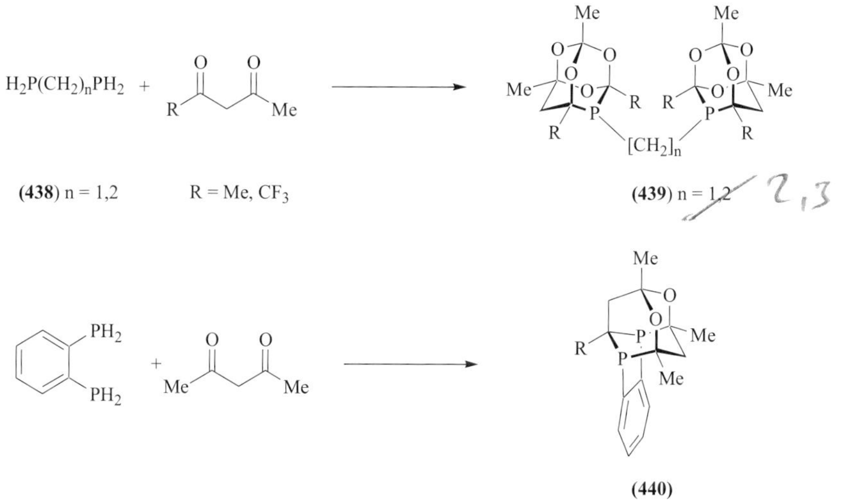

Scheme 141

In a similar reaction PH_3 adds to hexafluoropentane-2,4-dione and trifluoropentane-2,4-dione to give primarily the primary α-hydroxyphosphines (**441**), precursors for the resulting secondary phosphines 6,9-dioxa-2-phosphabicyclo[3.3.1]nonane-3β,7α-diol (**442**) and 2,4,8-trioxa-6-phosphaadamantane (**443**) (Scheme 142).

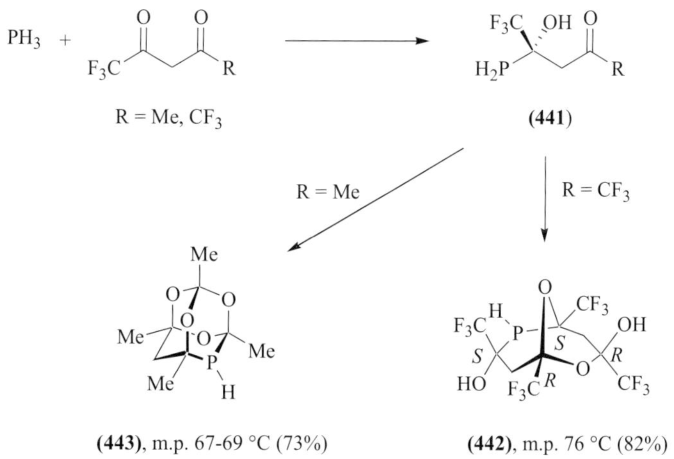

(**443**), m.p. 67-69 °C (73%)

(**442**), m.p. 76 °C (82%)

Scheme 142

The structures of (**442**) and (**443**) were established by single-crystal X-ray structure analysis. The phosphine (**443**) reveals a characteristic heteroadamantane geometry ⟨97CB1547⟩.

1-Oxa-1,2-dihydro-4*H*-3,1-benzoxaphosphinines (**443**) were prepared by the reaction of *o*-(butoxymethoxymethyl)phenyl-MgBr with derivatives of chloromethyl-phosphonic and -phosphinic acids (Scheme 143).

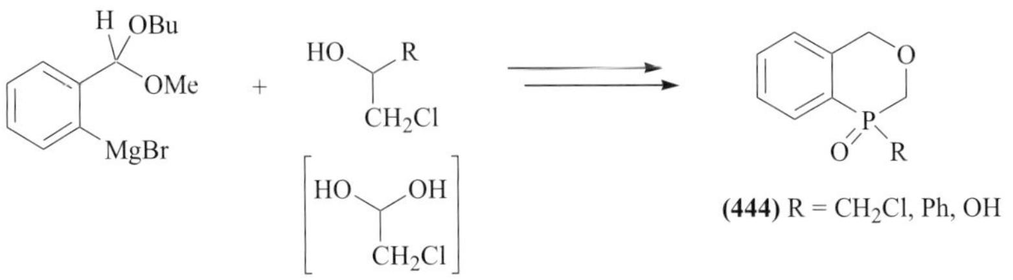

Scheme 143

The structure of (**444**) was determined by X-ray crystallography ⟨92IZV2174⟩.

5.3.13.2 1,4-Oxaphosphinines

The 1,4-oxaphosphinine system is represented by the 4*H*-1,4-oxaphosphininium salts, described mainly by Simalty and co-workers, and the 4-alkyl-, arlyl-, alkoxy- and hydroxy-1,4-oxaphosphinine 4-oxides, studied by several of Russian groups. The dibenzo analogues — the phenoxaphosphines — are discussed in the next section.

A very general method for the synthesis of 1,4-oxaphosphininium salts (**447**) is by reaction of acetylenic phosphines (**445**) with bromomethyl ketones to give the intermediate phosphonium salts (**446**), which readily cyclize (Scheme 144) ⟨68BSF4938, 72BSF3532, 72T3343, 84PS(19)159⟩. Alternatively, dialkynyl phosphonium salts can be cyclized, using ammonium acetate in acetic acid ⟨72TL4363, 80TL3037, 84PS(19)159⟩.

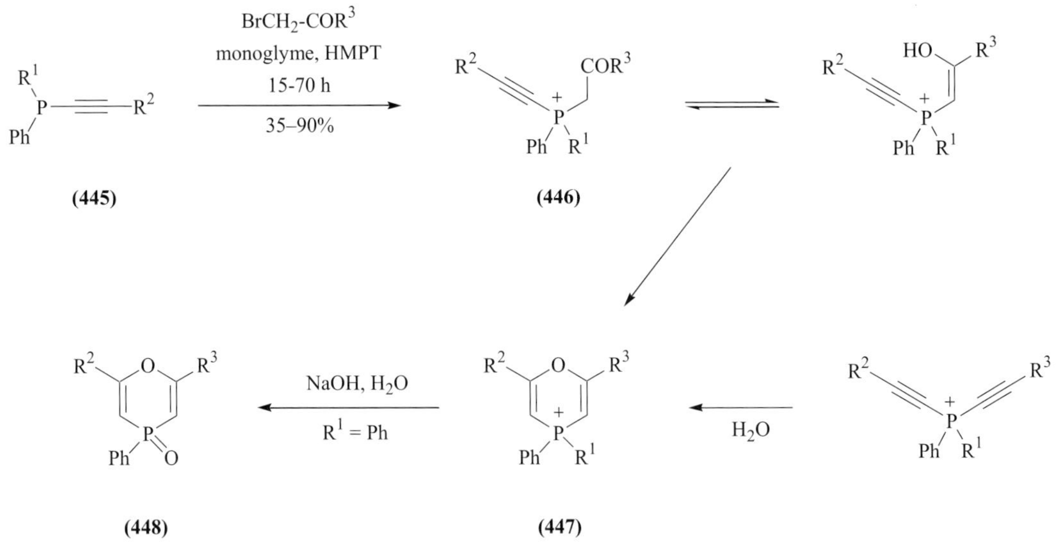

Scheme 144

A *P*-phenyl group can be removed from the phosphonium salts by the action of aqueous NaOH, to form the oxaphosphinine 4-oxides (**448**) ⟨72T3343, 84PS(19)159⟩. The oxides are also formed directly, by cyclization of dialkynyl phosphine oxides ⟨71TL4749⟩.

The cyclic phosphinic acids, the 4-hydroxy-1,4-oxaphosphinine 4-oxides (**449**), may be made by acid-promoted cyclization of bis-(β-chlorostyryl)-phosphinic acids ⟨84ZOB1481, 85ZOB780⟩, and their esters (**450**) result from the reaction of bromomethyl ketones with dialkyl alkynylphosphonites (Scheme 145) ⟨89DOK(304)625⟩.

The action of PCl_5 on divinyl ether followed by partial hydrolysis of the product, forms the 4-chloro-1,4-oxaphosphinine 4-oxide (**451**) ⟨87ZOB954, 87ZOB1518⟩. The 4-*t*-butyl-1,4-oxaphosphinine

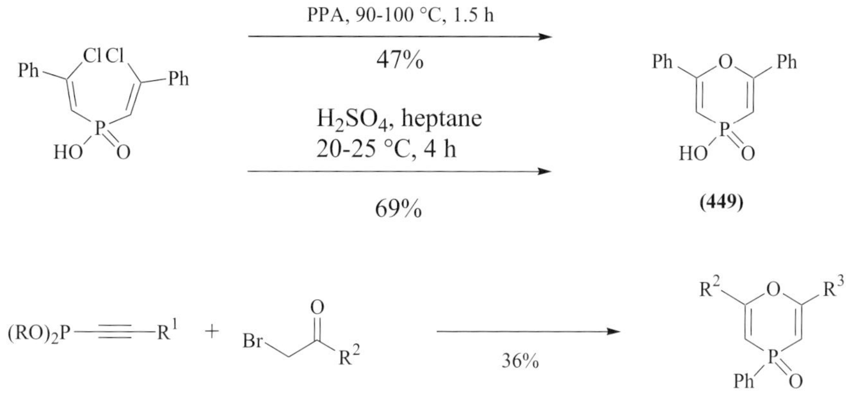

(449) (450)

Scheme 145

oxides (**452**) are of considerable synthetic value. Their acid-catalyzed cleavage yields the 1,5-diketones (**453**), which cannot be isolated; intramolecular aldol condensation gives rise to the enones (**454**). Reduction of (**454**) with $SiHCl_3$ and thermolysis at 200–250°C opens a useful route to the λ^3-phosphinines (**455**) (R^1 = H, Me; R^2 = Me, Bu^t, Ph, 4-MeC_6H_4, 4-ClC_6H_4, α-furyl) (Scheme 146) ⟨74TL4385, 83CB445, 83CB1756⟩. The synthesis of 3-hydroxyphosphinines also starts from the phosphaenones (**454**) ⟨77TL3445⟩.

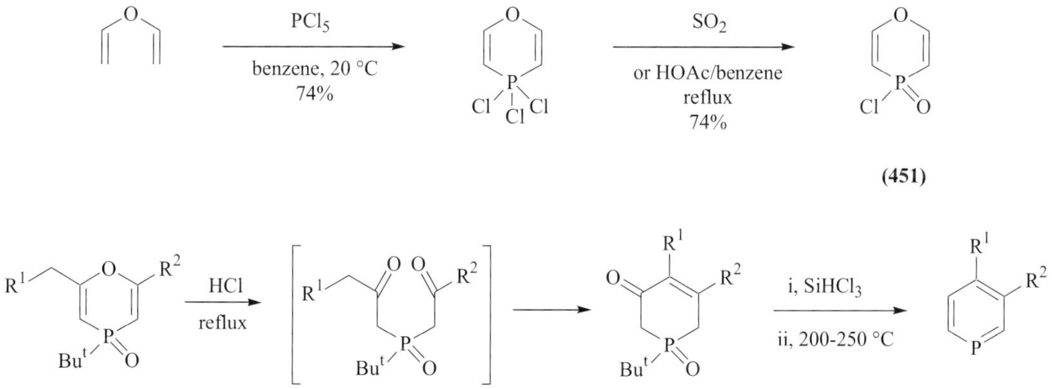

(451)

(452) (453) (454) (455)

Scheme 146

The 1,4-oxaphosphinane system was not known until 1984, when Samaan reported the synthesis of the quaternary 1,4-oxaphosphinanes (e.g. (**458**)), 1,4-thiaphosphinanes and 1,4-azaphosphinanes by base-induced addition of H_2O, H_2S and $MeNH_2$ to the diallyldiphenylphosphonium salts (**457**) (Scheme 147) ⟨74TL3927, 78CB579⟩.

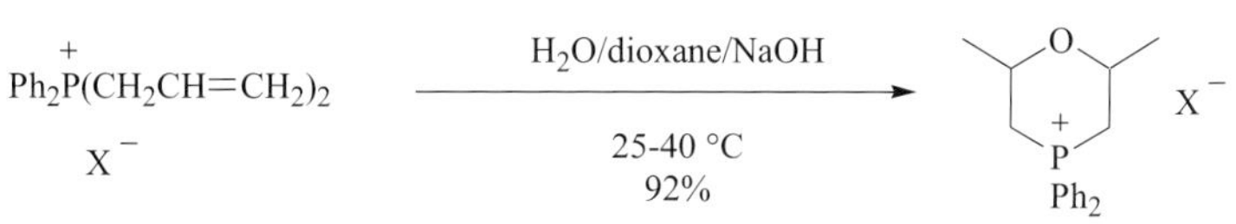

(457) (458)

Scheme 147

Since that time a number of other preparations have appeared, relying both on C–O and C–P bond formation for the cyclization ⟨87IZV2397, 89ZOB1047⟩. Sugar analogues have also been reported in this series.

5.3.13.3 Phenoxaphosphinines

The reaction of diphenyl ether with $PCl_3/AlCl_3$ occurs exclusively in the 4-position. However, di-*p*-tolyl ether (**459**) reacts with refluxing $PCl_3/AlCl_3$ (4 h), and after hydrolysis the 2,8-dimethylphenoxaphosphinic acid (**460**) can be isolated (Scheme 148) ⟨61JOC284, 65CB1681, 68JOC474, 74URP515754⟩. The directive effects of substituents in this reaction have been studied ⟨83JHC1601, 88PS(35)215⟩.

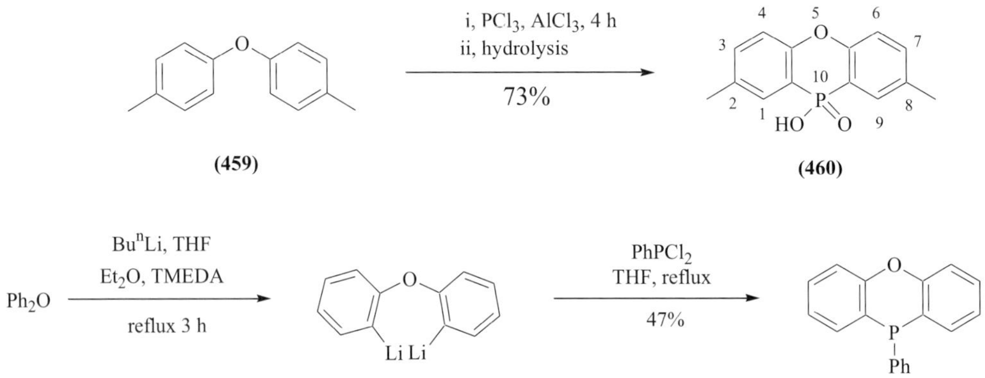

(459) **(460)** **(461)**

Scheme 148

An elegant synthesis of phenoxyphosphines is by the dilithiation of diphenyl ether in the *o,o'*-positions and reaction with $PhPCl_2$ to give 10-phenylphenoxaphosphine (**461**) (Scheme 148) ⟨57JA339, 86USP4698447, 87ZOB716, 90ZOB1257⟩.

In a third method the *o*-phenoxybenzenediazonium salt (**462**) is refluxed with PCl_3/CuBr/EtOAc and then with $AlCl_3$; distillation of the reaction mixture yields (**463**) (Scheme 149). Use of $PhPCl_2$ instead of PCl_3 provides the phenoxaphosphine (**461**) in 26% yield ⟨51JA5658, 52JA2884, 74IJ779⟩.

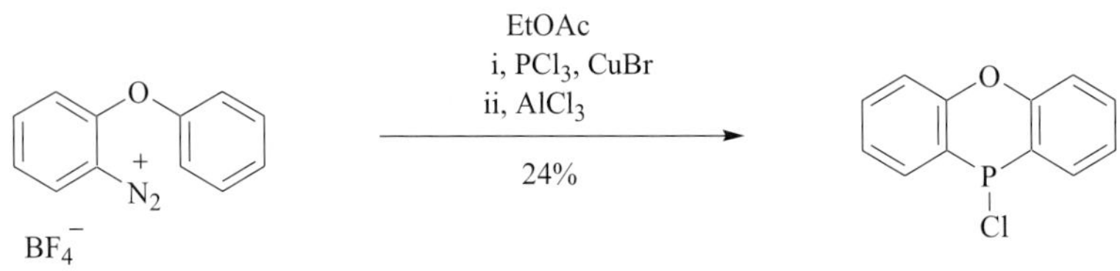

(462) **(463)**

Scheme 149

The stability of the phenoxaphosphine ring system is demonstrated by the fact that $KMnO_4$ only oxidizes methyl groups on the flanking rings to give the carboxylic acids in methyl-substituted phenoxaphosphinic acids and 10-phenylphenoxaphosphine 10-oxides; the heterocyclic nucleus is unattacked ⟨65CB1681, 80PP160, 81JAP(K)8157796⟩. Under vigorous reduction conditions (Ph_2SiH_2, 250°C) 2,8-dimethylphenoxaphosphinic acid (**460**) forms the secondary phosphine (**464**) ⟨65CB1681⟩. Esterification of phenoxaphosphinic acids can be carried out, e.g. by treating the potassium salts with alkyl halides ⟨68IJ651⟩.

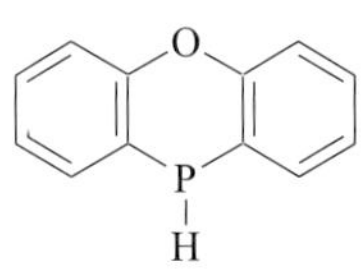

(464)

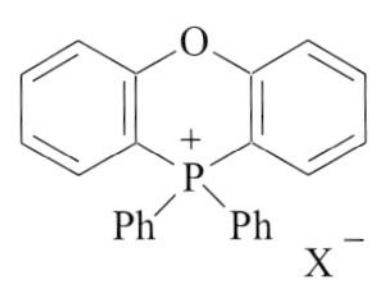

(465)

Phenoxaphosphines with P(III), e.g. the 10-phenyl compound (**461**), are much slower in their reaction with alkylating agents than triphenylphosphine ⟨79PS(7)309⟩. On the other hand, arylation can be achieved ($PhBr/AlCl_3$, 210°C, 2 h), giving the 10,10-diphenyl phosphonium salt (**465**) (isolated as the iodide) ⟨40JCS1192, 53JCS3746⟩.

Transition metal complexes ML_2Cl_2 have been described for (**461**) with M = Ni, Pd and Pt ⟨91ZOB1641⟩.

5.3.14 RING SYSTEMS WITH ONE PHOSPHORUS AND TWO OXYGEN ATOMS

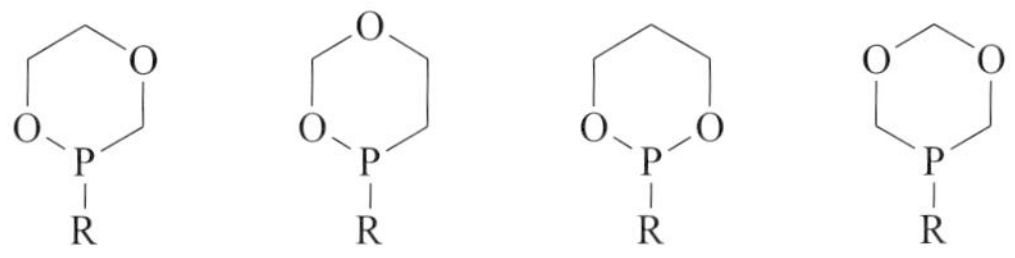

(I) **(II)** **(III)** **(IV)**

5.3.14.1 1,3,5-Dioxaphosphinanes

Epstein and Buckler discovered the acid-catalyzed reaction of phenylphosphine with benzaldehyde, giving 2,4,6-tetraphenyl-1,3,5-dioxaphosphinane. The 1,3-glycol (**466**), formed by the addition of two moles of benzaldehyde to the phosphine, is a probable intermediate, and the ring closure is an acetalation of the third mole of benzaldehyde by (**466**) (Scheme 150) ⟨62T1231⟩.

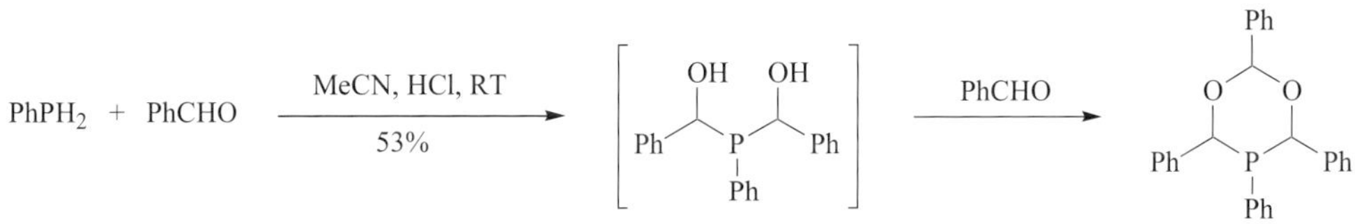

Scheme 150

This reaction proved to be very versatile. Most of the more recent work has been done by B.A. Arbuzov's group.

Extension to aliphatic aldehydes leads to the 2,4,6-trialkyl-5-phenyl derivatives (**467**). When H_3PO_2 is refluxed in toluene with benzaldehyde or isobutyraldehyde the phosphinic acids (**468**); R = Ph, Pr^i) are formed ⟨68IZV397, 92EUP463994⟩. Di(hydroxymethyl)phosphines (**469**) react with aldehydes, ketones and acetals to form the 2-mono- or 2,2-disubstituted 1,3,5-dioxaphosphinanes (**470**) ⟨86IZV2506, 86ZOB2256⟩. With triethyl orthoformate the 2-ethoxy derivatives (**471**) are obtained (Scheme 151) ⟨86IZV418, 86IZV640⟩.

(467) **(468)**

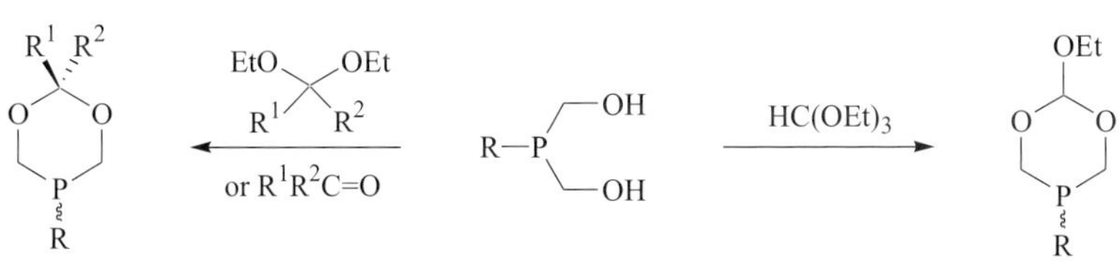

(470) **(469)** **(471)**

Scheme 151

When tri(hydroxymethyl)phosphine (**472**) reacts at 180–200°C with formaldehyde 5-methyl-1,3,5-dioxaphosphane 5-oxide (**473**) is formed. The initial 5-hydroxymethyl compound undergoes a thermal

Buckler–Trippett reaction under the reaction conditions (Scheme 152) <78ZOB97, 78ZOB2653>. 5-Oxides are also formed directly from α,α'-dihydroxydialkylphosphine oxides (Scheme 152) <75JOC2056>. The 5-sulfides are formed in a similar way (Scheme 152) <86IZV640, 86IZV418>.

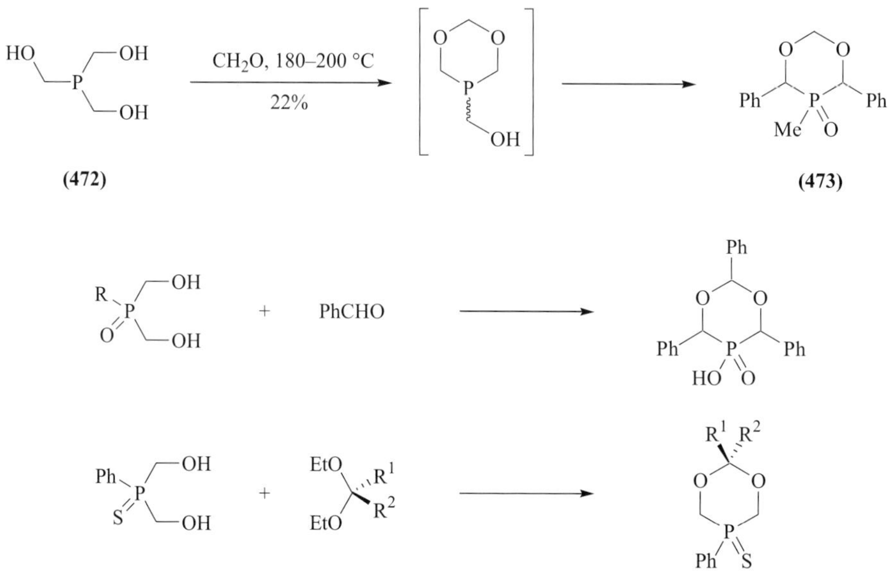

Scheme 152

The synthesis of the unsubstituted 1,3,5-dioxaphosphinane was achieved simply by thermolysis of the phosphonium salt $[PhP(CH_2OH)_3]^+Cl^-$ <80IZV1626>.

5-Amino-substituted 1,3,5-dioxaphosphinanes (**475**) could be synthesized by the reaction of aminophosphines (**474**) with aldehydes, e.g. isobutyraldehyde (Scheme 153) <83IZV2550>.

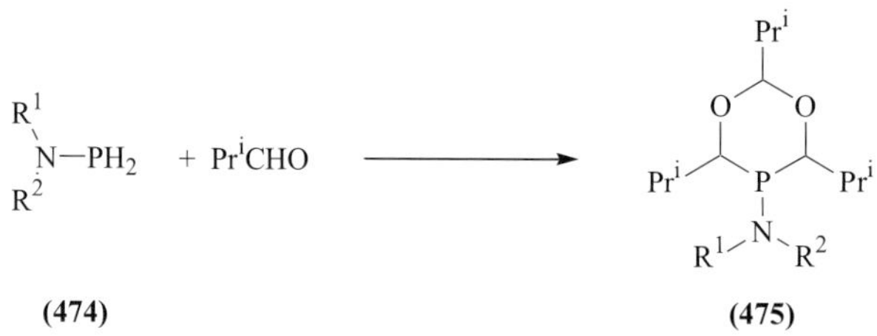

$R^1 = H, R^2 = Ph; R^1, R^2 = Et_2, -(CH_2)_5-, -(CH_2)_2O(CH_2)_2-$

Scheme 153

The *P*-oxides, -sulfides and -selenides are also produced from the P(III) compounds by action of H_2O_2 or ozone, sulfur and selenium <80IZV1626, 86IZV640, 86ZOB2256>. Imines (phenylimines, for instance) are readily obtained by Staudinger reactions using azides <83IZV2550>.

The conformations of 1,3,5-dioxaphosphinines have been studied extensively, normally by a combination of NMR and dipole moment measurements, but X-ray evidence is also available. Ringchair conformations are the general rule. With P(III) at the 5-position, substituents (Ph has been most fully investigated) tend to occupy the axial position, but this does not hold universally. 5-Phenyl-1,3,5-dioxaphosphinane has predominantly axial Ph, as does 5-oxide, but both stereoisomers of the 2-ethoxy-5-phenyl compound have the Ph predominantly equatorial <86IZV418>. The 5-sulfides and 5-selenides of these compounds have the Ph axial, not withstanding the anomeric effect which should apply at position 2 and favor OEt-axial conformers <86IZV640>. References to conformational studies on 1,3,5-dioxaphosphinanes, widely scattered through the Russian literature, include those above and the following <78DOK(240)331, 80IZV1626, 82IZV127, 82IZV594, 84IZV1056, 85IZV1296, 86ZOB2256>. Calculations by the CNDO/S method relating to the conformations of 1,3,5-dioxaphosphinanes have been reported <84ZOB2532>. The Se-77 and P-31 spin–spin interaction in several 1,3,5-dioxaphosphinane-selenides has been studied <90IZV776>.

5.3.15 APPLICATIONS

The uses claimed for the P–N heterocyclic systems described in this review have been few. The phosphorylated anthranilamide derivatives described were developed in a search for antiparasitic agents ⟨82MI625-01⟩. The catalytic activity of complexes of 1,3,5-triaza-7-phosphaadamantane ⟨95JOM(488)99⟩ and the interesting luminescence properties of its gold complexes ⟨95OM4194, 94CC431, 95IC75⟩ have been mentioned.

Derivatives of 1,3,2-dioxaphosphinane- and 1,4,2-dioxaphosphinanes, not discussed in this review, play an important role because of their antiviral activities.

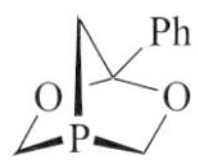

(476)

The insecticidal properties of the 1,3,5-dioxaphosphinane derivative (**476**), against rice weevils, houseflies and aphids, have been reported ⟨77MI625-01⟩. 2,4,6-Triisopropyl-1,3,5-trioxaphosphinane (TTP) has been incorporated into a number of catalytic formulations, such as the ternary system $Co_2(CO)_8/R_3N/TTP$ for hydroformylation of olefins ⟨72USP3644529⟩, and in other combinations for diene alkylation and alkenylation ⟨69GEP1300695, 70GEP(O)1965047, 70USP3524896, 71USP3567792⟩.

The dibenzo[1,2]dioxaphosphinines have been described in the patent literature as having applications for stabilization of organic polymers, as flame retardants, and as polymerization promoters ⟨72USP3702878, 89EUP304782, 94EUP582957, 94JAP(K)6145182, 95EUP632050⟩. Phenoxaphosphines (chapter 13.3) are also effective flame retardants. Polyesters formed from the di-acid chloride (**477**) by interfacial copolymerization with bisphenols have been described in numerous reports and patents, e.g. ⟨81JAP(K)81150098, 84MI625-01⟩. Good thermal stability is reported for benzimidazole and benzoxazole polymers containing phenoxaphosphine rings ⟨80PP160⟩.

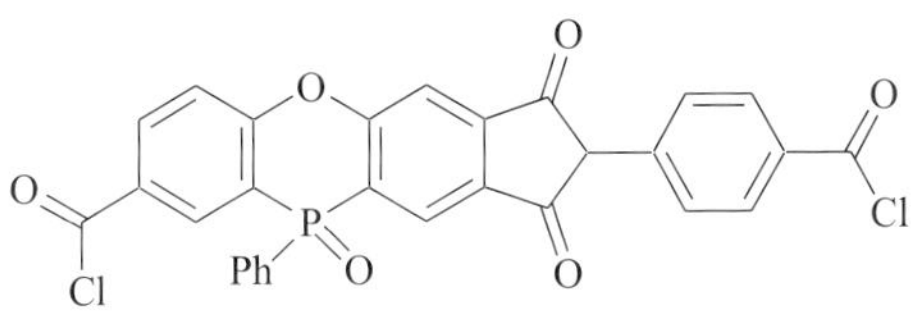

(477)

5.3.16 THEORETICAL STUDIES AND AROMATICITY

For the most part, theoretical methods have been applied to the question of the aromaticity of the various phosphorus-containing ring systems, and this question is relevant only to systems with complete cyclic π-conjugation. Applications to other areas have been rare.

The Hückel aromaticity of the λ^3-phosphinines was supported by X-ray crystal structure analysis combined with: (1) PE spectral studies of λ^3-phosphinine ⟨73JA928, 76HCA1944, 78ACR153⟩, 2,4,6-tri-*t*-butyl-λ^3-phosphinine ⟨71AG(E)656⟩, 3-methyl-2-phosphanaphthalene ⟨75JES91⟩, 10-methyl-9-phosphaanthracene ⟨75T2931⟩, 2,4,6-tri-*t*-butyl-1,3,5-triphosphinine and 2,4-di-*t*-butyl-1,3-diphosphinine ⟨98EJIC1619⟩; (2) *ab initio* calculations using an STO-3G basis set ⟨70JCP(52)2769⟩, CNDO/2 calculations ⟨67JCP(47)158⟩ and a CNDO/S-method ⟨74TCA19⟩; (3) molecular orbital calculations for 2,4,6-tri-*t*-butyl-1,3,5-triphosphinine with Gaussian 94 ⟨98EJIC1619⟩ or MP2/6-31G* ⟨96JMS368⟩; (4) the derived resonance energies using the perturbation graph theory and PE experimental vertical π ionization potentials ⟨78T3419⟩.

Geometries, aromatic character, Mulliken charge distribution, and MO diagrams of 1,2-aza-, 1,3,2-diaza-, 1,3-aza-, and 1,3,5-diazaphosphinines have been calculated and compared to those of phosphinine and pyridine. This study reveals that the introduction of nitrogen atoms at the position adjacent to phosphorus significantly reduces the aromatic delocalization and induces a [1,4] dipolar character through an increase of the positive charge on the P-atom. This phenomenon does not occur in 1,3-aza- and 1,3,5-diazaphosphinines, which exhibit a poor dipolar character ⟨99JOC5524⟩.

MO-Diagrams of 1,2-aza-, 1,3,2-diaza-, 1,3-aza- and 1,3,5-diazaphosphinines were calculated at the B3LYP/6-31G* level. The inversion of the highest occupied π-orbitals sequence between pyridine and phosphinine has been discussed already in the past ⟨78ACR153, 70TL4941⟩.

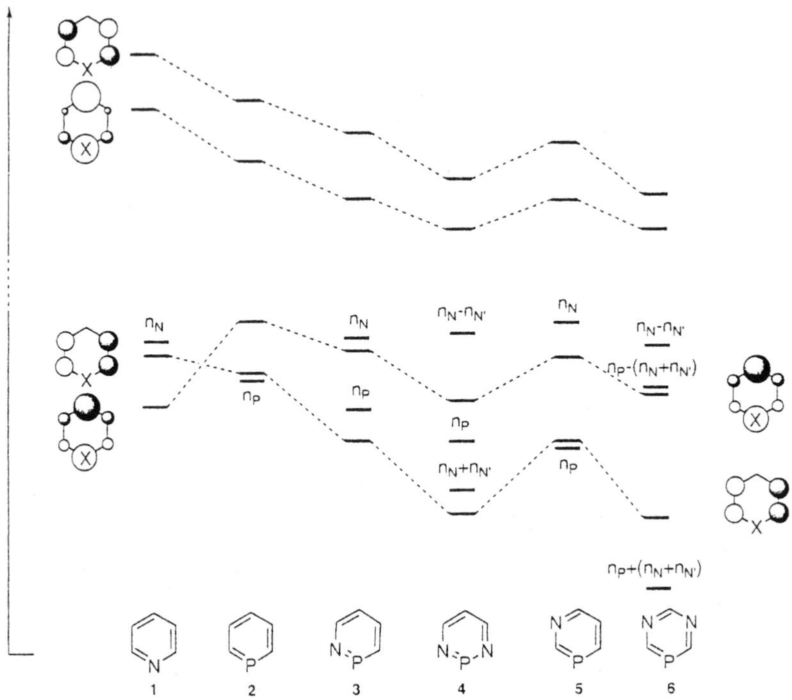

Figure 1. Mos for (**1–6**) and their relative levels, at the B3LYP/6-31G* level. n_N, $n_{N'}$, and n_P mean lone pair at nitrogen and/or phosphorus.

Table 1. Coefficients of the highest π and lowest π* MO of **2–6**

πMO	*2*	*3*	*4*	*5*	*6*
πLUMO					
$P(3p_z)$	0.45	0.47	0.52	0.44	0.45
$C_4(2p_z)$	0.31	0.29	0.28	0.33	0.35
πHOMO					
$P(3p_z)$	0.42	0.40	0.38	0.42	0.45
$C_4(2p_z)$	−0.31	−0.31	−0.35	−0.29	−0.28

A first consequence of the replacement of CH by N concerns the position of the LUMO and LUMO + 1, which both decrease in energy as the number of nitrogen atoms increases. In all azaphosphinines studied, the HOMO now mostly describes the lone pair at N or an anti-symmetric combination of the two associated AOs in the case of the diazaphosphinines. Though the localization of the N-atoms does not notably influence the relative level of the HOMOs, it significantly modifies the coefficients of the π-orbitals, especially at P and C-4. When one ore two N-atoms are incorporated at the α-position at P, the LUMO is more clearly localized on the P-atom whereas the weight of C-4 increases in the highest π-HOMO. The opposite phenomenon is observed when the nitrogen atoms are in β-position. These trends are supported by the evolution of Mulliken charges and by the analysis of the NBO π-population ⟨99JOC5524⟩ (Figure 1; Table 1).

The 24-valence isomers of diphosphinine were studied at an electron-correlated *ab initio* level (MP2, MP4SDTQ, CCSD, CCSD(T) and by d-functional theory (B3LYP) ⟨99JA4215⟩.

While all cyclic, conjugated six-membered phosphorus heterocycles with $\lambda^3\sigma^2$-phosphorus are clearly Hückel aromatic systems, the bonding situation in the cyclic conjugated six-membered systems containing $\lambda^5\sigma^4$-phosphorus has been discussed rather controversially. Schweig and co-workers, for example, discussed a mixture of Hückel and Möbius aromaticity and ylide character ⟨76JA4410⟩. Bird has used the bond lengths reported for 14 phosphorus-containing rings, including seven with more than one heteroatom, to provide estimates of the bond order variation about the ring, and thence an index (I_6) of the aromaticity of the ring ⟨90T5697⟩. Perhaps the most surprising feature of the figures he derives is the fact that rings with λ^5-phosphorus appear to be, on this criterion, scarcely less aromatic than those with λ^3.

The correlation of the bonding situation with the PE spectra has been discussed ⟨95JST(347)57⟩. Dewar and Holder compared the aromatic energies of 1,2-, 1,3- and 1,4-diphosphinines according to their heats of formation obtained from AM1 semiempirical methods ⟨89H(28)1135⟩. HMO calculations of phosphorus heterocycles with $\lambda^5\sigma^4$-phosphorus were reported as early as 1968 ⟨68RRC1623⟩.

Although the molecular geometries of λ^3- and λ^5-phosphinines are nearly identical (no significant deviations from planarity or bond alternation are observed with either system), the λ^5-phosphinines as well as the mono-, di-, tri- and tetra-aza-λ^5-phosphinines, the λ^5,λ^5-diphosphinines and the λ^5,λ^5,λ^5-triphosphinines are certainly not Hückel aromatic systems. Their description as cyclic phosphorus ylides with ionic resonance structures is probably close to the truth. The ^{1}H and ^{13}C NMR evidence is particularly strong on this point.

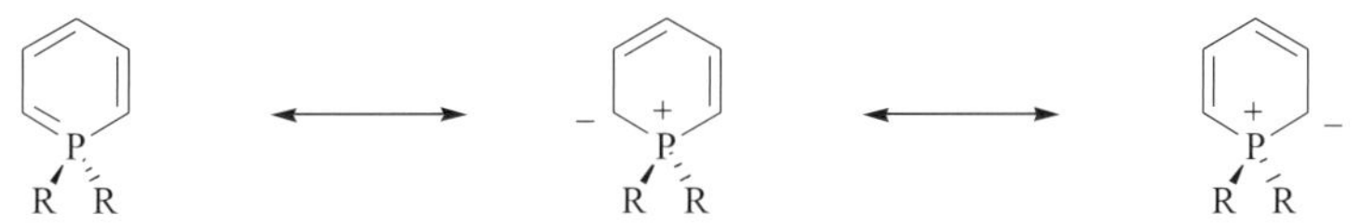

(26)

The PE spectra of the 1λ^5,3λ^5-diphosphinines of type (**241**) as well as the MO calculations (MNDO) have been reported ⟨91CB2071⟩. The calculations predict, in agreement with the P-ylide character of the 1λ^5,3λ^5-diphosphinines, a strong negative charge for the ring carbon atoms and a strong positive net charge for the P-atoms ⟨91CB2071⟩.

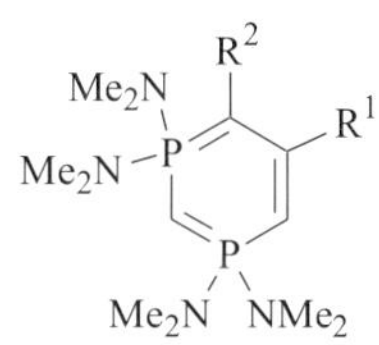

(241)

5.3.17 EXPERIMENTAL STRUCTURAL METHODS

5.3.17.1 X-ray Crystallography

X-ray crystallography has become a routine tool for structure determination since the 1970s, and many instances of its application to phosphorus-containing six-membered heterocyclic compounds have been reported. The data from a number of determinations carried out before 1990 have been collected and discussed ⟨90T5697⟩.

The crystal structures of the sandwich-complex (**217**) ⟨95AG(E)198⟩ as well as the 2,4-di-*tert*-butyl-1λ^3,3λ^3-diphosphinine (**218a**) have been reported ⟨95ZK801⟩. The X-ray structural analysis of 2,4,6-tri-*tert*-butyl-1λ^3,3λ^3,5λ^3-triphosphinine (**354**) confirms its complete planarity.

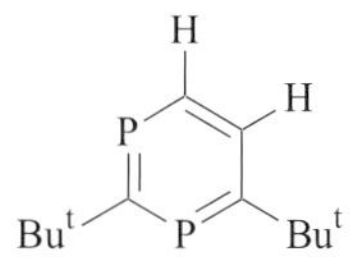

(218a)

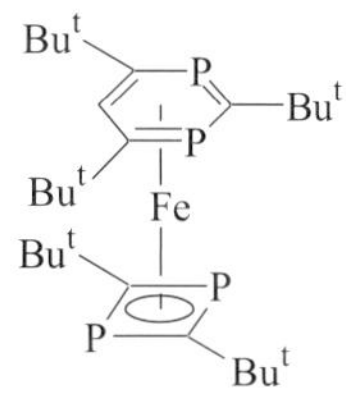

(217)

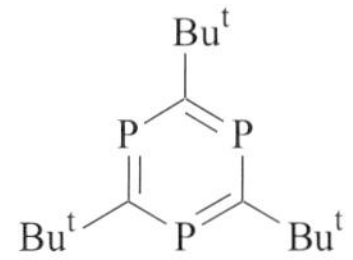

(354)

In contrast to benzene, but in accordance with λ^3-phosphinines, (**354**) is far from being regular hexagonal. The P–C bond lengths are equal with average bond lengths of 1.727 (±0.008)Å, they are in the range of those found for phosphinines (1.73–1.75 Å) but considerably longer than P=C bond lengths in phosphaalkenes (1.66–1.68 Å). The average C–P–C angles are 109.3° (±0.3°) (C–C–C angle in benzene 120°, C–N–C angle in pyridine 117°, C–P–C angle in phosphines 101–103°). In order to accommodate this angular change the P–C–P angle has to exceed 120°. In (**354**) the P–C–P angles are 130.7° (±0.4°).

Some X-ray structure determinations of π- and σ-complexes of 1,3,5-triphosphines (**354**) have been published, however most of them are characterized by ^1H and ^{31}P NMR spectroscopy.

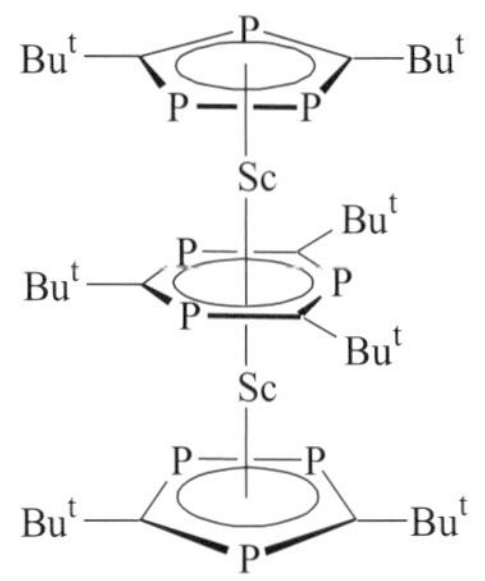

(382)

The molecular structure of the triple decker complex (**382**) has been determined by X-ray crystallography. The 1,3,5-triphosphabenzene rings lie on a crystallographic mirror plane and are planar with no significant variation in the P–C bond length (1.793(8)–1.807(8)) 〈96JA7630〉. In comparison to the free ligand (**354**) the P–C bonds are lengthened by 0.064–0.089 Å, the expansion of the P–C–P angles compared to (**354**) is 3.3–3.4 Å.

The crystal structures of a number of $1\lambda^5,3\lambda^5$-diphosphinines have been measured, e.g. (**243**) 〈90AG(E)516〉, (**241**) 〈91ZN(B)202〉 and the $NiCl_2$ complex (**248**) 〈94ZAAC(620)483〉. The triphosphinine (**425**), with two λ^5 and one λ^3 P-atoms, is an interesting case. According to the X-ray analysis the ring system is slightly puckered, the P(5)–C(4) and P(5)–C(6) distances (170.9 pm (2), 170.0 pm (2)) are very close to the values of normal λ^3-phosphinines, and P(1) and P(3) are more or less ylide phosphorus atoms with polar P–C bonds 〈86AG(E)1002, 87ZN(B)1213〉. The triphosphinine (**423**), with all P-atoms λ^5, is strongly ylidic in character, with an almost planar ring system, C–P bond lengths 168.7–169.7 ppm, P–C–P angles 126.6–127.0° and C–P–C angles 113.0–113.4° 〈90JCS(P1)1223, 98ZAAC1116〉.

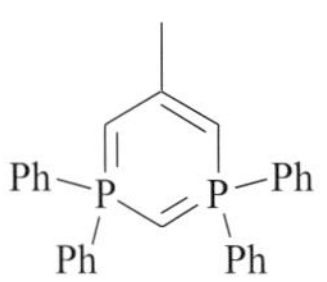

(243)

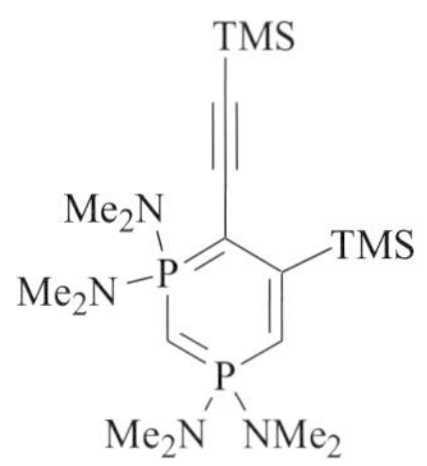

(241)

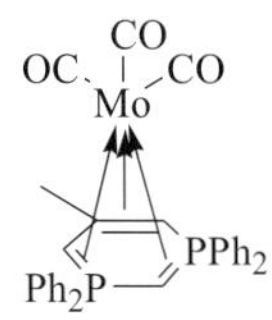

(248)

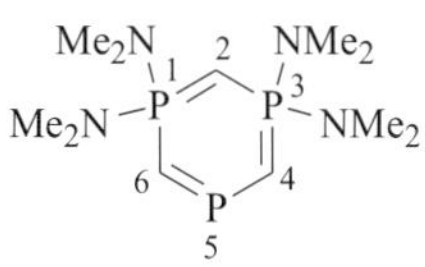

(425)

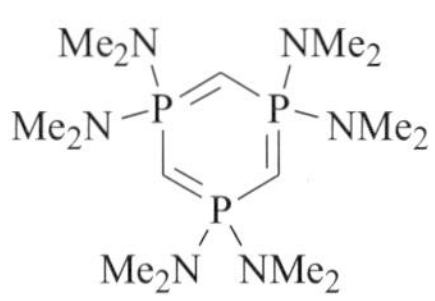

(423)

A series of structure determinations by X-ray analysis for valence isomers of 1,3,5-triphosphinines was reported.

1,3,5-triphosphabicyclo[3.1.0]hexene-η^3,η^1-diylhafnium-dichloride (**356**) 〈99EJIC763〉 is dimeric in the solid state and contains a μ-Cl_2 bridge.

The X-ray of tetraphosphasemibullvalene (**375**) elucidated the structure 〈98EJIC2071〉.

The structure determination of several polycyclic derivatives of tri- and tetraphosphinine valence isomers was possible only by X-ray analysis. The structure of the phosphorus–carbon cage (**418**) was published in 1999〈99CEJ(5)3143〉. The P–C single bonds are in the range of 1.831(3) to 1.913(3) Å, the P–P distances are 215.5 and 220.9 Å, the phosphaalkene bond is indicated by the PC distance of 167.8(3) Å.

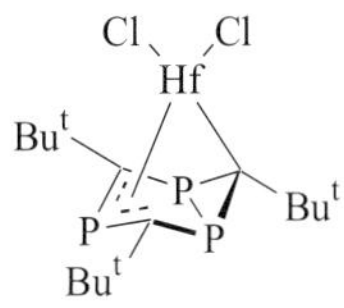

(356)

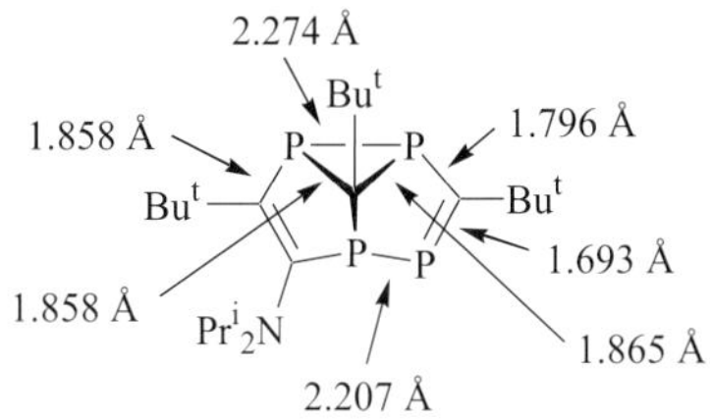

(375)

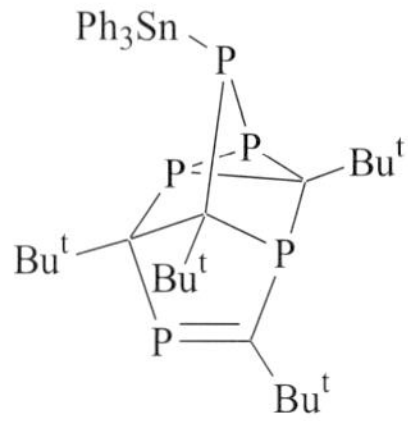

(418)

X-rays of several bridged 1,3,5-triphosphinines and valence isomers have been published.

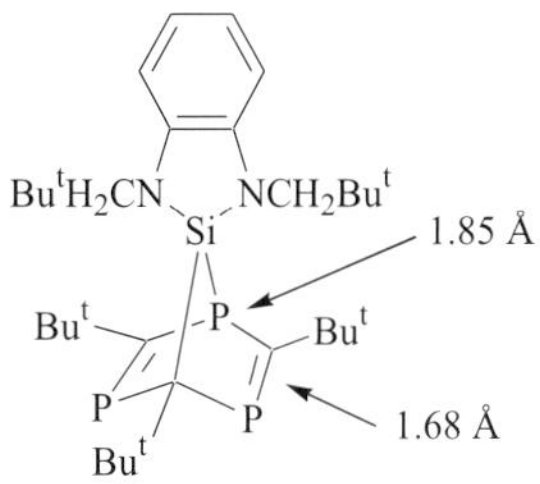

(394)

<99CC2451>

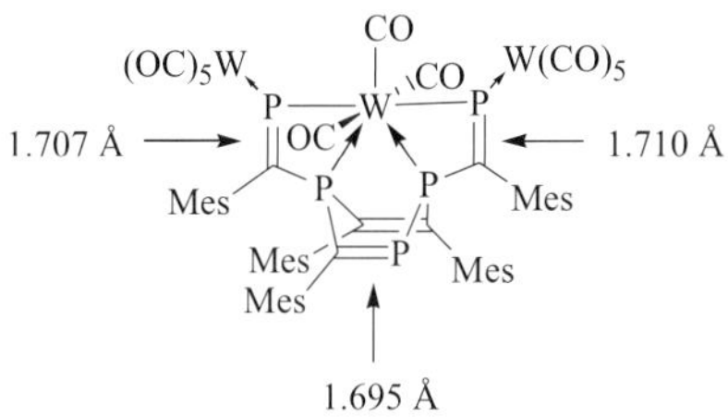

(353)

<99AG3384>

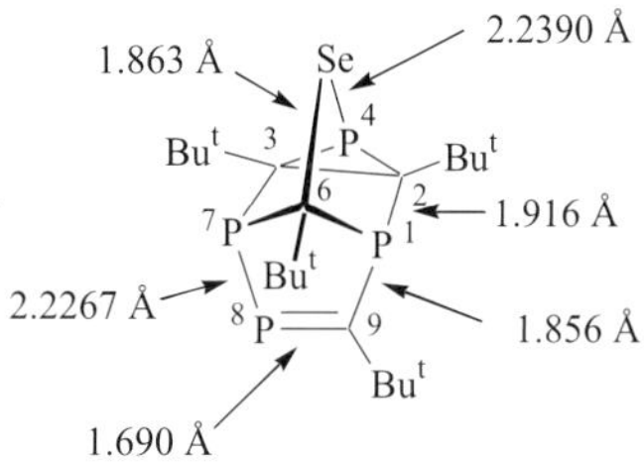

(392)

<99S1642>

X-ray data are available for the 1,3λ^5- and 1,4λ^5-azaphospinines (**59**) ⟨88ZOB526⟩ and (**80**) ⟨94JA8087⟩. P–C-bond lengths are fairly similar to those obtained with the λ^3-P systems: in (**80**) the mean distance is 172.6 pm; the C–P–C angle is 105.8°.

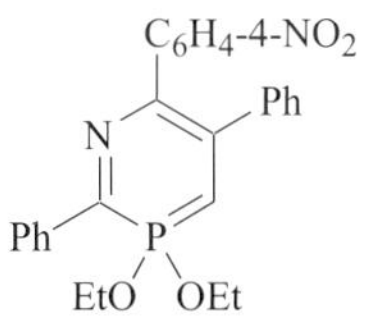

(59)

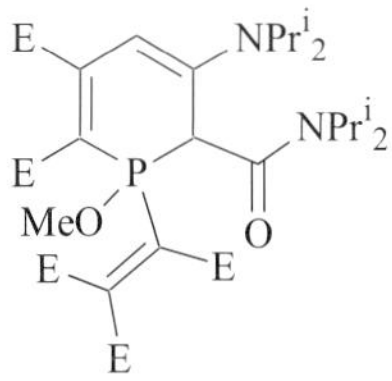

(80) E = CO_2Me

In the 1,2,3λ^5-diazaphosphinine (**163**) an ylide-like structure is consistent with the bond length data, and with the NMR measurements ⟨90AG(E)1123⟩. A later paper reports a number of crystal structures of other di- and tri-aza-λ^5-phosphinines ⟨94JA8087⟩.

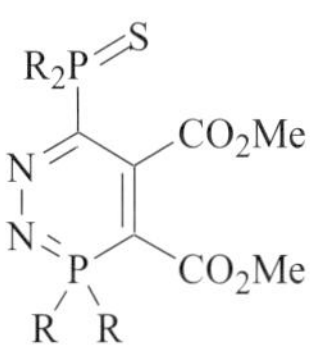

(163) R = NPr^i2

Data on two fused 1,3,2λ^5-diazaphosphinines have been reported. The structures of the molybdenum complex ⟨88ZN(B)937⟩ and the spiro-fused cyclotriphosphazine ⟨88IC1911⟩ were confirmed by X-ray crystallography.

Table 2 lists (not comprehensively) references to crystal structure data of monocyclic ring systems covered in this review chapter. Ring carbon substituents are not indicated in the columns.

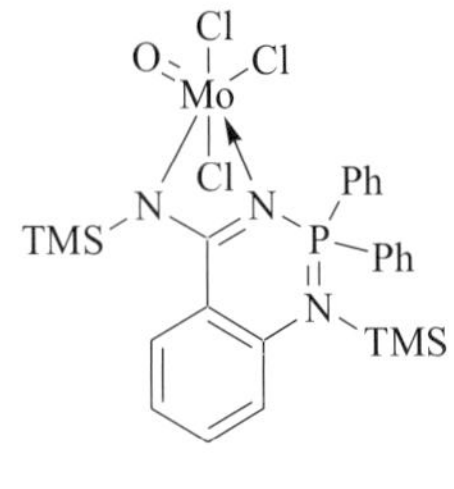

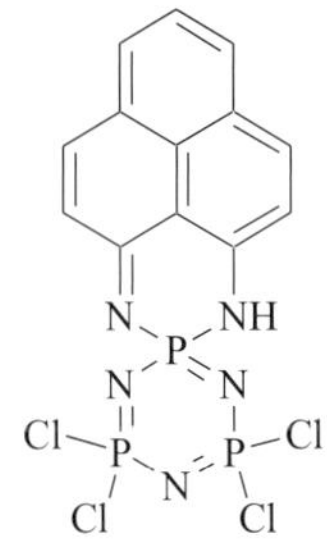

Table 2. X-ray crystal structures for ring systems with nitrogen and phosphorus

1	*2*	*3*	*4*	*5*	*6*	*Ref.*
N		λ^5P(OEt)$_2$				⟨88ZOB526⟩
λ^3P		λ^3P				⟨95AG(E)198⟩
λ^5P(NR$_2$)$_2$		λ^5P(NR$_2$)$_2$				⟨91ZN(B)202⟩
λ^5PPh$_2$		λ^5PPh$_2$				⟨90AG(E)516, 91CB1525⟩
N			λ^5P(OMe)R			⟨94JA8087⟩
NH			λ^5P(=O)Ph			⟨71TL4749⟩
$\lambda^3\sigma^3$PPH			$\lambda^3\sigma^3$Pt-Bu			⟨85CB2365⟩
λ^5P(=S)Ph			λ^5P(–S)Ph			⟨92PS(71)245⟩
N	N	λ^5P(NR)$_2$)$_2$				⟨90AG(E)1123⟩
N	N		λ^5PPh$_2$			⟨90AG(E)314⟩
N	λ^5PMe$_2$			N		⟨90HCA1888⟩
N	λ^5PMe$_2$			NH	C=O	⟨94JA8087⟩
N	λ^5PPh$_2$				λ^5PPh$_2$	⟨91CB1525⟩
N	λ^5P(NR$_2$)$_2$		λ^5P(NR$_2$)$_2$			⟨91PS(56)49⟩
λ^5P(NR$_2$)$_2$		λ^5P(NR$_2$)$_2$		λ^3P		⟨87ZN(B)1213⟩
λ^5P(NR$_2$)$_2$		λ^5P(NR$_2$)$_2$		λ^5P(NR$_2$)$_2$		⟨90JCS(P1)1223⟩
N	N	N	λ^5P(NR$_2$)$_2$			⟨94JA8087⟩
N	λ^5P(NR$_2$)$_2$	N		N		⟨87ZOB1980, 90ZOB1975⟩
N	λ^5PMe$_2$	N	λ^5PMe(R)		λ^5PMe$_2$	⟨77CJC2534⟩

5.3.17.2 NMR Spectroscopy

5.3.17.2.1 ^{1}H NMR spectroscopy

The ^{1}H NMR spectra of the λ^3-phosphinines are very characteristic for this ring system. The δ-values of the ring hydrogen atoms are in the typical aromatic region, especially the α-hydrogens (H-2) are shifted fairly downfield compared to the signals of the aromatic hydrocarbons. Very characteristic furthermore are the P-(H-2) coupling constants in the region of 30–40 Hz. In Scheme 154 the δ- and J-values of several λ^3-P-heterocycles, discussed in this review, are presented.

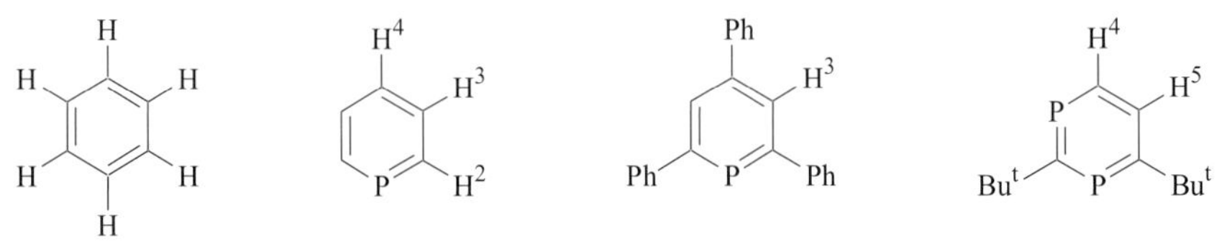

δ(H) = 7.26

δ(H-2) = 8.61
^2J(P,H) = 40.0 Hz
δ(H-3) = 7.72
^3J(P,H) = 8.0 Hz
δ(H-4) = 7.38
^4J(P,H) = 3.0 Hz

δ(H-3) = 8.10
^3J(P,H) = 6.0 Hz

δ(H-4) = 8.78 (ddd)
^2J(P,H) = 41.8 Hz
^3J(P,H) = 4.3 Hz
^3J(H,H) = 11.1 Hz
δ(H-5) = 8.06 (ddd)
^3J(P,H) = 11.1 Hz
^3J(P,H) = 8.06 Hz
^3J(H,H) = 11.1 Hz

Scheme 154

The ^{1}H NMR spectra of the azaphosphinines with ring hydrogens are certainly influenced from the electronegative nitrogen atoms. The introduction of N-atoms at the α- or γ-position relative to P reduces the aromatic delocalization and induces a dipolar character through an increase of the positive charge on the P-atom, this phenomenon does not occur in 1,3-aza- and 1,3,5-diazaphosphinines (Scheme 155).

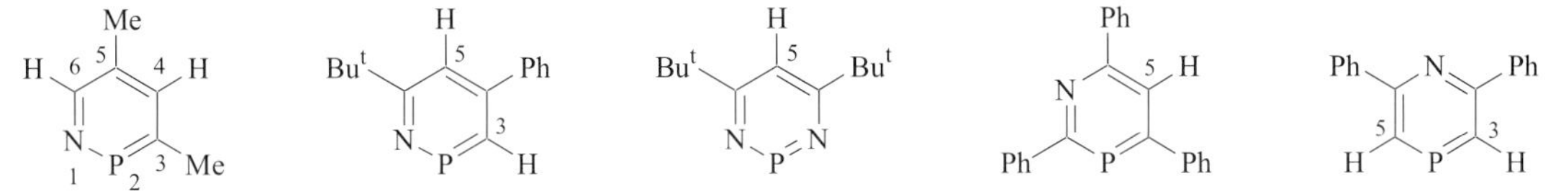

(3)	**(8)**	**(111)**	**(46)**	**(66)**
δ(H-4) = 7.58 (d)	δ(H-3) = 8.64 (d)	δ(H-5) = 7.22 (d)	δ(H-5) = 8.14 (d)	δ(H-3) = 8.66
^{3}J(P,H) = 9.0 Hz	^{2}J(P,H) = 31.1 Hz	^{4}J(P,H) = 4.4 Hz	^{3}J(P,H) = 5.25 Hz	^{2}J(P,H) = 36.0 Hz
δ(H-6) = 9.11 (d)	δ(H-5) = 6.98-7.45			
^{3}J(P,H) = 21.0 Hz	(+C_6H_5)			

Scheme 155

Since most of the known λ^3-di- and triphosphinines as well as the λ^3-azaphosphinines are highly substituted, the characteristic features of the H-shifts and P,H-coupling constants are therefore only of limited value.

The ^{1}H NMR spectra of λ^5-phosphinines and λ^5-azaphosphinines are less characteristic; however, the signals of hydrogen atoms between two λ^5P-atom are shifted extremely highfield (Scheme 156) because of the shielding effect of P(V).

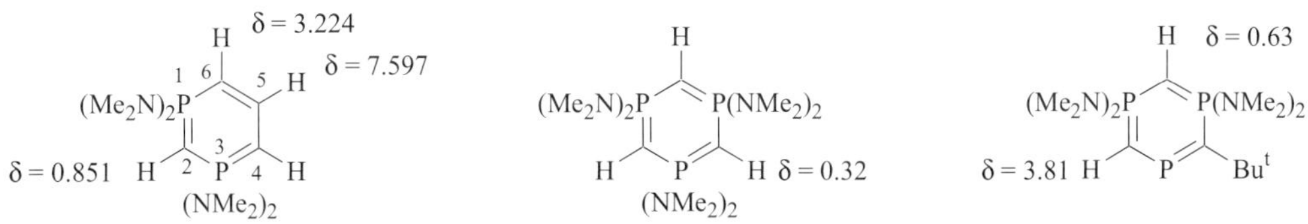

(241) **(423)** **(425)**

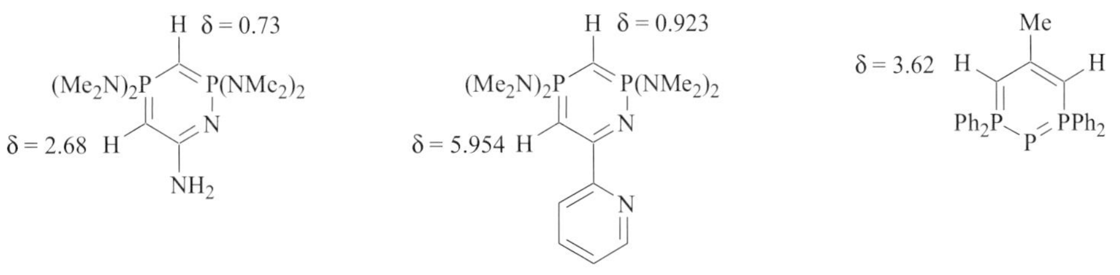

(317) **(313)**

Scheme 156

5.3.17.2.2 $^{31}P\{^{1}H\}$ *NMR spectroscopy*

The ^{31}P NMR spectroscopic δ-values as well as the P/H and the P/C-coupling constants are the most reliable data for the elucidation of the bonding situation in phosphorus compounds. In the $\lambda^3\sigma^2$-phosphinines the phosphorus is strongly deshielded and therefore the δ-values are considerably shifted downfield. In Scheme 157 several $1\lambda^3$-phosphinines, $1\lambda^3,3\lambda^3$-diphosphinines and $1\lambda^3,3\lambda^3,5\lambda^3$-triphosphinines and their $\delta(^{31}P)$-values are listed.

The nature of the transition metal σ- or π-complexes becomes obvious by the ^{31}P NMR data (Scheme 158).

The smaller coupling of (P_A,Pt) compared to (PMe_3, Pt) is interpreted with the smaller s-character of the P lone pair in the triphosphinine.

The ^{31}P NMR as well as the ^{13}C NMR signals appear as single signals indicating a delocalized 6π-system. The η^6-coordination is indicated by strong upfield shifts ($\Delta\delta = 84.8$–77.8).

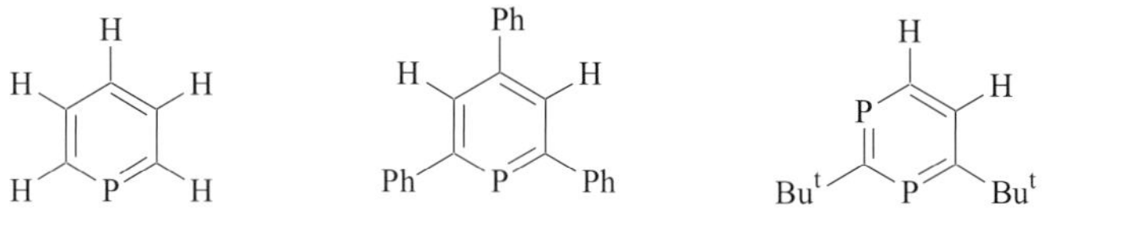

(218a)

$\delta(^{31}P) = 211.0$
$^{2}J(P,H) = 40.0$ Hz
$^{3}J(P,H) = 8.0$ Hz
$^{4}J(P,H) = 3.0$ Hz

$\delta(^{31}P) = 178.2$
$^{3}J(P,H) = 6.0$ Hz

$\delta(^{31}P) = 220.0; 237.0$
$^{2}J(P,P) = 24.7$ Hz
$^{2}J(P,H) = 41.8$ Hz
$^{3}J(P,H) = 4.3$ Hz; 11.1 Hz
$^{4}J(P,H) = 8.06$ Hz

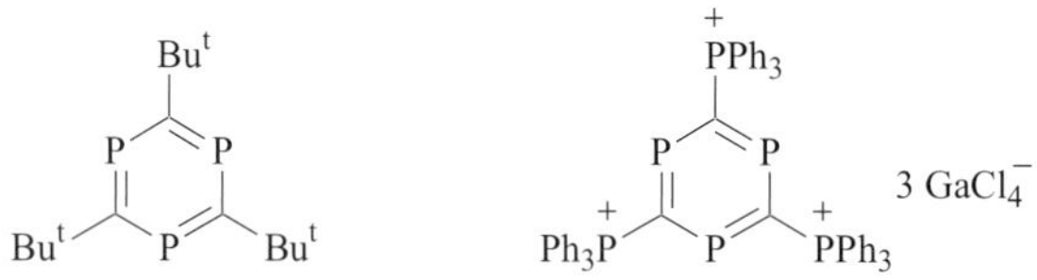

(354)

$\delta(^{31}P) = 232.6$

(399)

$\delta(^{31}P) = 310.0$

Scheme 157

(380)

(376–379)

$\delta(^{31}P_A) = 203.0$
$\delta(^{31}P_B) = 264.8$
$^{1}J(P_A,Pt) = 2418$ Hz
$^{2}J(P_A,P_B) = 36.3$ Hz
$\delta(^{31}PMe_3) = -18.3$
$^{1}J(P,PMe_3) = 2886$ Hz

M = $Mo(CO)_3$, $\delta(^{31}P) = 57.6$ **(376)**
M = $W(CO)_3$, $\delta(^{31}P) = 38.6$ **(377)**

M = Ru  , $\delta(^{31}P) = 59.0$ **(378)**

M = Mo  , $\delta(^{31}P) = -4.7$ **(379)**

Scheme 158

In the aza-λ^3-phosphinines the electronegativity of the nitrogen in *o*- and *p*-position increases the strong deshielding influence on the $\lambda^3\sigma^2$-phosphorus which results in a further downfield shift of the $\delta(^{31}P)$-values (Scheme 159).

As the examples in Scheme 159 show, the influence of the ring substituents on the $\delta(^{31}P)$-values is nearly negligible, furthermore in the 1,3λ^3-aza-phosphinines the $\delta(^{31}P)$-values are not shifted downfield, since its dipolarity is not as pronounced as it is in the 1,2- and 1,4-aza-λ^3-phosphinines.

The P^{31} *NMR spectroscopic data* of the λ^5-phosphinines differ entirely from that of the λ^3-phosphinines. In Scheme 160 the $\delta(^{31}P)$ data of several λ^5-phosphinines are listed.

The λ^5-phosphinines can be considered as cyclic phosphinealkenes and because of the nesomeric phosphorus-ylid structure the electron density in 2-, 4- and 6-position is high with the consequence of a distinct high field shift in the ^{1}H NMR spectra of the hydrogens in these positions (Scheme 160).

The importance of the ^{31}P NMR spectroscopy for the identification of the 6π-aromatic λ^3-phosphinine structures can be estimated by discussion of the ^{31}P NMR spectra of valence-isomeric structures of λ^3-phosphinines and related structures (Scheme 161).

The structure elucidation of these polycyclic systems closely related to di- and triphosphinine is distinctly supported by the ^{31}P NMR data. The δ-values of the 3-membered phosphirane and diphosphirane rings are shifted fairly upfield, while phosphaalkene double bonds are clearly identified by $\delta(^{31}P)$-values further downfield ($\delta = 350$–600) than the $\delta(^{31}P)$-values of λ^3-phosphinines ($\delta = 170$–250).

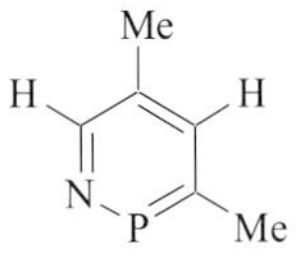

(3)

$\delta(^{31}P) = 263.3$

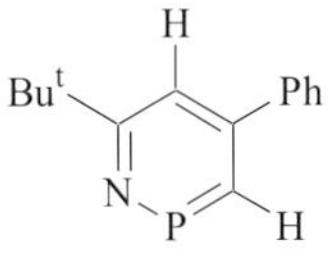

(8)

$\delta(^{31}P) = 273.1$

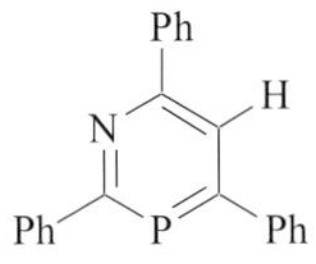

(46)

$\delta(^{31}P) = 163.0$

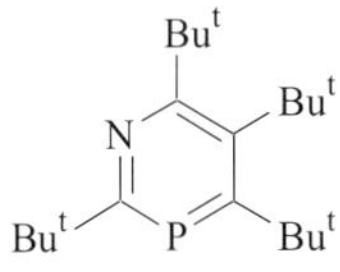

(52)

$\delta(^{31}P) = 147.0$

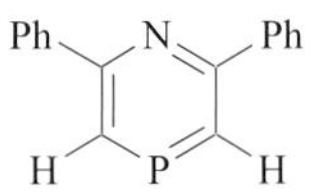

(66)

$\delta(^{31}P) = 243.4$

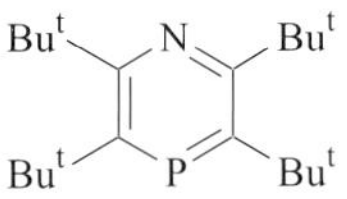

(54)

$\delta(^{31}P) = 228.0$

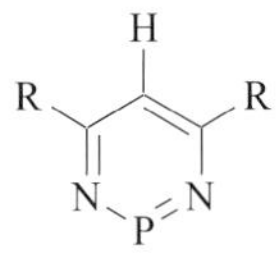

(111)

R = Bu^t, $\delta(^{31}P) = 267.5$
R = Ph, $\delta(^{31}P) = 269.2$

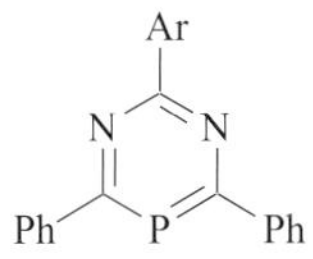

(139)

$\delta(^{31}P) = 135–141$

Scheme 159

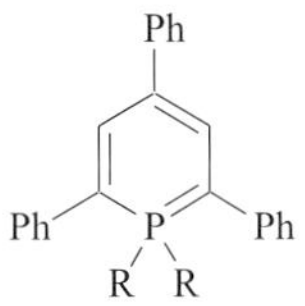

R = OMe, $\delta(^{31}P) = 65.2$
R = NMe$_2$, $\delta(^{31}P) = 42.5$

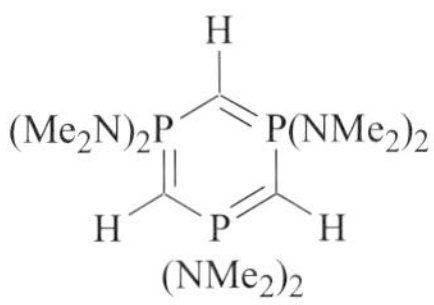

(423)

$\delta(^{31}P) = 65.4$
$^2J(P,P) = 73$ Hz

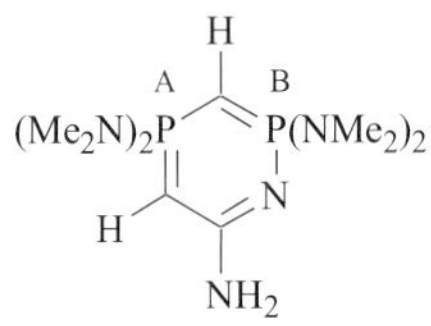

(317)

$\delta(^{31}P_A) = 62.82$
$\delta(^{31}P_B) = 52.24$
$^2J(P,P) = 67.9$ Hz

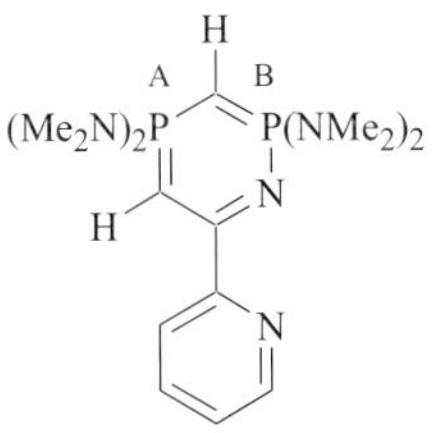

(313)

$\delta(^{31}P_A) = 59.8$
$\delta(^{31}P_B) = 52.5$
$^2J(P,P) = 58.7$ Hz

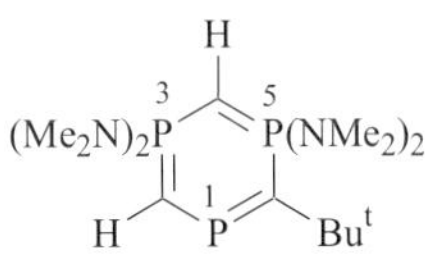

(425)

$\delta(^{31}P_1) = 295.5$
$\delta(^{31}P_3) = 57.5$
$\delta(^{31}P_5) = 59.0$
$^2J(P,P) = 45.0; 49.2$ Hz
$^2J(P,P) = 30.0; 49.2$ Hz

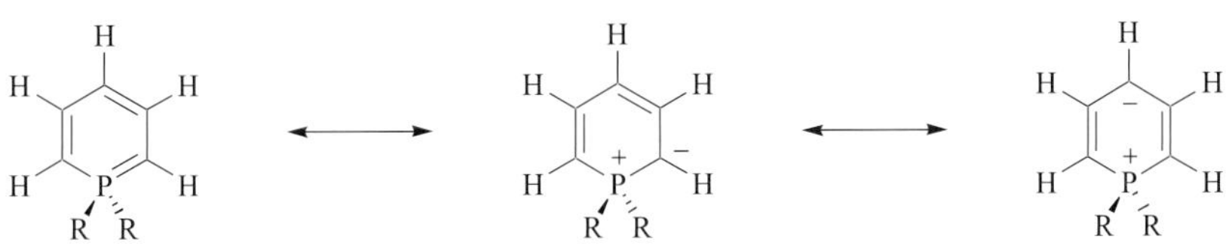

Scheme 160

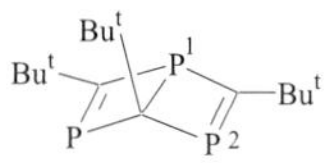

(355)

$\delta(^{31}P_1) = 93.6$
$\delta(^{31}P_2) = 336.8$
$J(P,P) = 23.5$ Hz

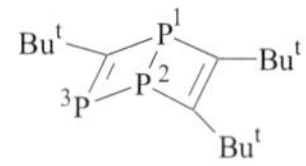

(363)

$\delta(^{31}P_1) = -28.4$
$\delta(^{31}P_2) = -154.9$
$\delta(^{31}P_3) = 350.4$
$J(P_1,P_2) = 4.2$ Hz
$J(P_2,P_3) = 191.8$ Hz
$J(P_1,P_3) = 87.2$ Hz

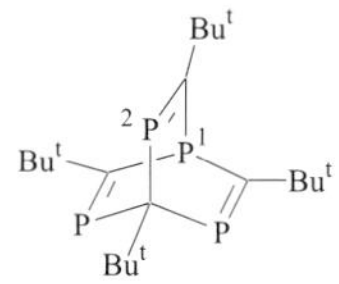

(374)

$\delta(^{31}P_1) = -87.0$
$\delta(^{31}P_2) = 324.1$

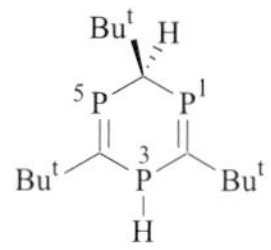

(361)

$\delta(^{31}P_1) = 241.3$
$\delta(^{31}P_2) = -249.0$

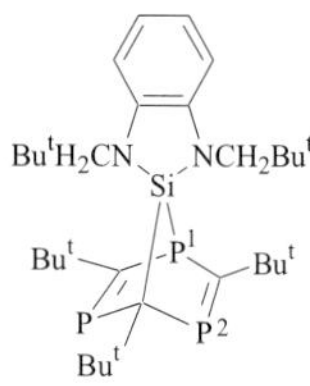

(394)

$\delta(^{31}P_1) = -63.8$
$\delta(^{31}P_2) = 320.4$

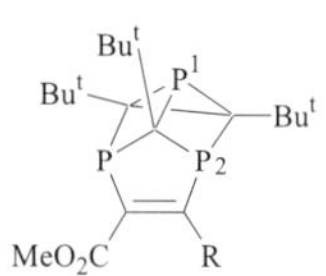

(365)

R = H **(265a)**
$\delta(^{31}P_1) = -160.0$
$\delta(^{31}P_2) = 82.0$
R = CO_2Me **(265b)**
$\delta(^{31}P_1) = -158.0$
$\delta(^{31}P_2) = 104.2$

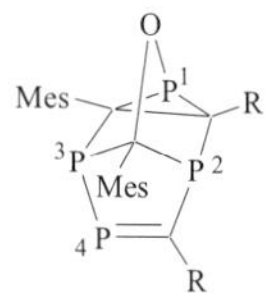

(387)

$\delta(^{31}P_1) = -7.2$
$\delta(^{31}P_2) = 67.1$
$\delta(^{31}P_2) = 98.7$
$\delta(^{31}P_2) = 356.0$

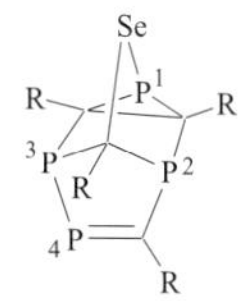

(392)

$\delta(^{31}P_1) = -100.5$
$\delta(^{31}P_2) = 123.5$
$\delta(^{31}P_2) = 136.6$
$\delta(^{31}P_2) = 411.1$

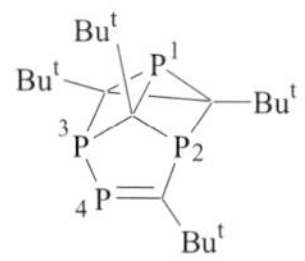

(366)

$\delta(^{31}P_1) = -160.0$
$\delta(^{31}P_2) = 111.5$
$\delta(^{31}P_2) = 134.3$
$\delta(^{31}P_2) = 417.1$

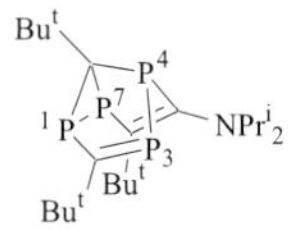

(375)

$\delta(^{31}P_1) = -19.6$
$\delta(^{31}P_3) = 347.5$
$\delta(^{31}P_2) = 88.7$
$\delta(^{31}P_2) = -92.1$
$J(P_1,P_7) = 174.6$ Hz
$J(P_1,P_3)$, $J(P_1,P_4) = 23.5$ Hz
$J(P_3,P_4) = 270.0$ Hz
$J(P_1,P_4) = 23.5$ Hz
$J(P_4,P_7) = 16.4$ Hz

Scheme 161

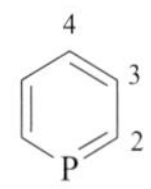

$\delta(^{13}C\text{-}2) = 154.1$
$^1J(P,C) = 53.0$ Hz
$\delta(^{13}C\text{-}3) = 133.6$
$^2J(P,C) = 14.0$ Hz
$\delta(^{13}C\text{-}4) = 128.8$
$^3J(P,C) = 22.0$ Hz

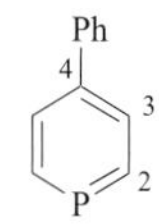

$\delta(^{13}C\text{-}2) = 154.6$
$^1J(P,C) = 55.3$ Hz
$\delta(^{13}C\text{-}3) = 133.5$
$^2J(P,C) = 15.3$ Hz
$\delta(^{13}C\text{-}4) = 142.3$
$^3J(P,C) = 21.0$ Hz

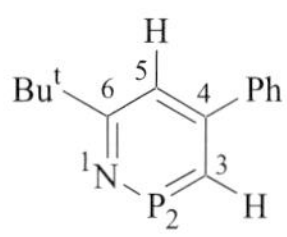

(8)

$\delta(^{13}C\text{-}3) = 153.5$
$^1J(P,C) = 74.9$ Hz
$\delta(^{13}C\text{-}4) = 152.8$
$^2J(P,C) = 10.6$ Hz
$\delta(^{13}C\text{-}5) = 117.4$
$^3J(P,C) = 31.9$ Hz
$\delta(^{13}C\text{-}6) = 177.0$
$^2J(P,C) = 29.0$ Hz

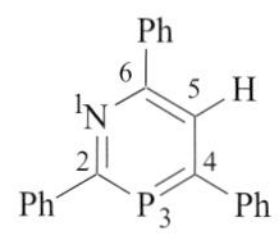

(46)

$\delta(^{13}C\text{-}2) = 191.5$
$^1J(P,C) = 69.3$ Hz
$\delta(^{13}C\text{-}4) = 157.1$
$^1J(P,C) = 18.3$ Hz
$\delta(^{13}C\text{-}5) = 119.8$
$^2J(P,C) = 10.1$ Hz
$\delta(^{13}C\text{-}6) = 180.5$
$^3J(P,C) = 52.6$ Hz

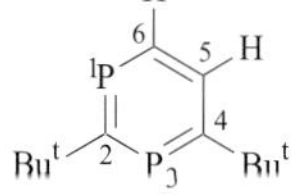

(218a)

$\delta(^{13}C\text{-}2) = 217.1$
$J(P,C) = 76.9$; 70.8 Hz
$\delta(^{13}C\text{-}4) = 150.7$
$J(P,C) = 51.8$; 15.2 Hz
$\delta(^{13}C\text{-}5) = 129.9$
$J(P,C) = 13.5$; 9.9 Hz
$\delta(^{13}C\text{-}6) = 181.2$
$J(P,C) = 63.9$; 18.3 Hz

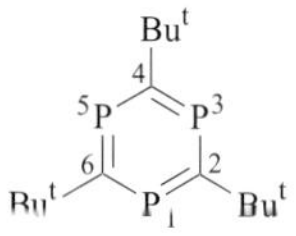

(254)

$\delta(^{13}C\text{-}2,4,6) = 211.8$
$J(P,C) = 77.0$ Hz

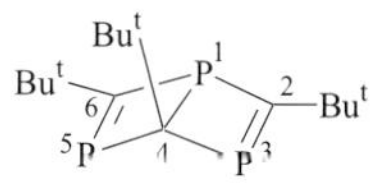

(355)

$\delta(^{13}C\text{-}2,6) = 236.7$
$^1J(P,C) = 55.9$; 47.5 Hz
$\delta(^{13}C\text{-}4) = 46.6$
$^1J(P,C) = 39.8$; 28.0 Hz

Scheme 162

5.3.17.2.3 *$^{13}C\{^1H\}$ NMR spectroscopy*

The $\delta(^{13}C)$-values and the P,C-coupling constants are valuable data and characteristic for λ^3-phosphinines (Scheme 162). Especially the d-values of carbon atoms adjacent to $\lambda^3\sigma^2$-phosphorus are shifted fairly downfield with large P,C-coupling constants.

The ^{13}C NMR δ-values of λ^5-phosphinines also have very characteristic features different from that of λ^3-phosphinines.

For the same reason as in the case of the δ-values for the α-H in λ^5-phosphinines also the δ-values of the α-carbons in λ^5-phosphinines are shifted upfield. Especially in the $(P\lambda^5)$=C–$P(\lambda^5)$ constellation, the δ(C)-values are very often beyond the TMS-signal (Scheme 163).

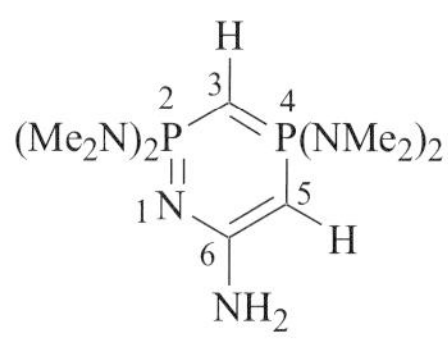

(317)

$\delta(^{13}C\text{-}3) = -2.97$
$\delta(^{13}C\text{-}5) = 39.72$
$\delta(^{13}C\text{-}6) = 167.44$
$^1J(P,C) = 133.4;\ 138.2$ Hz
$^2J(P,C) = 153.5$ Hz

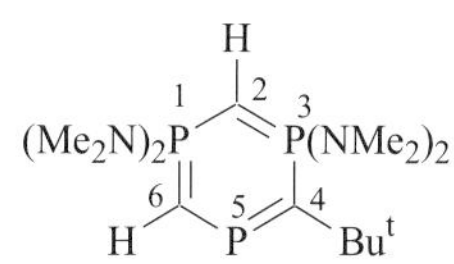

(425)

$\delta(^{13}C\text{-}2) = 8.3$
$\delta(^{13}C\text{-}4) = 100.8$
$\delta(^{13}C\text{-}6) = 62.8$

Scheme 163

Two useful reviews by Fluck and Heckmann on the NMR spectra of cyclic phosphorus compounds should be noted ⟨94RHA65, 95RHA121⟩.

5.3.18 REFERENCES

1888CB1497 A. Schenk and A. Michaelis; *Ber. Dtsch. Chem. Ges.*, 1888, **21**, 1497.
40JCS1192 M. Chatt and F.G. Mann; *J. Chem. Soc.*, 1940, 1192.
B-50MI625-01 G.M. Kosolapoff; *Organophosphorus Compounds*, Wiley, New York, Chapman and Hall, London, 1950.
51JA5658 G.O. Doak and L.D. Freedman; *J. Am. Chem. Soc.*, 1951, **73**, 5658.
52JA2884 L.D. Freedman and G.O. Doak; *J. Am. Chem. Soc.*, 1952, **74**, 2884.
53JCS3746 F.G. Mann and I.T. Miller; *J. Chem. Soc.*, 1953, 3746.
57JA339 J.M. Davidson and C.M. French; *J. Am. Chem. Soc.*, 1957, **79**, 339.
59DOK(127)1217 B.A. Arbuzov, G.M. Vinokurova and J.A. Perfil'eva; *Dokl. Akad. Nauk SSSR*, 1959, **127**, 1217 (*Chem. Abstr.*, 1950, **54**, 1377).
60HCA1826 M. Häring; *Helv. Chim. Acta*, 1960, **43**, 1826.
60JA1515 C.G. Krespan, B.C. McKusick and T.L. Cairns; *J. Am. Chem. Soc.*, 1960, **82**, 1515.
60JA5685 M.J.S. Dewar and V.P. Kubba; *J. Am. Chem. Soc.*, 1960, **82**, 5685.
60JCS5034 I.G.M. Campbell and J.K. Way; *J. Chem. Soc.*, 1960, 5034.
61BRP860629 G.V. McHattie (I.C.I.); *Br. Pat.* 860269 (1961) (*Chem. Abstr.*, 1962, **57**, 2256).
61JA3432 C.G. Krespan; *J. Am. Chem. Soc.*, 1961, **83**, 3432.
61JOC284 L.D. Freedman, G.O. Doak and J.R. Edminston; *J. Org. Chem.*, 1961, **26**, 284.
62CI(L)1539 M. Davis and F.G. Mann; *Chem. Ind. (London)*, 1962, 1539.
62T1231 M. Epstein and S.A. Buckler; *Tetrahedron*, 1962, **18**, 1231.
63AG(E)479 G. Maerkl; *Angew. Chemie., Int. Ed. Engl.*, 1963, **2**, 479.
63HOU(12)203 K. Sasse; *Methoden Org. Chem. (Houben-Weyl)*, 1963, **12**.
63IZV502 B. Arbuzov and G.M. Vinokurova; *Izv. Akad. Nauk SSSR, Ser. Khim.*, 1963, 502 (*Chem. Abstr.*, 1963, **59**, 3947).
63ZN(B)1136 G. Märkl; *Z. Naturforsch.*, 1963, **18b**, 1136.
64JCS3770 M. Davis and F.G. Mann; *J. Chem. Soc.*, 1964, 3770.
64JOC3410 G. Baum, H.A. Lloyd and C. Tamborski; *J. Org. Chem.*, 1964, **29**, 3410.
65AG(E)1124 G. Märkl; *Angew. Chem. Int. Ed. Engl.*, 1965, **4**, 1124.
65CB1681 H. Fritzsche, U. Hasserodt and F. Korte; *Chem. Ber.*, 1965, **98**, 1681.
66AG(E)846 G. Maerkl; *Angew. Chemie., Int. Ed. Engl.*, 1966, **5**, 846.

66JA4090 A.M. Aguiar and H. Aguiar; *J. Am. Chem. Soc*, 1966, **88**, 4090.
66ZOB735 V.I. Shevchenko and V.P. Kukhar; *Zh. Obshch. Khim.*, 1966, 735.
66ZOB1254 V.I. Shevchenko and P.P. Kornuta; *Zh. Obshch. Khim.*, 1966, **36**, 1254 (*Chem. Abstr.*, 1966, **65**, 15381).
67AG(E)565 A. Schmidpeter and J. Ebeling; *Angew. Chemie., Int. Ed. Engl.*, 1967, **6**, 565.
67JA3067 A.M. Aguiar, K.C. Hansen and G.S. Reddy; *J. Am. Chem. Soc.*, 1967, **89**, 3067.
67JA4235 A.M. Aguiar and K.C. Hansen; *J. Am. Chem. Soc.*, 1967, **89**, 4235.
67JCP(47)158 D.P. Santry and G.A. Segal; *J. Chem., Phys.*, 1967, **47**, 158.
68BSF4938 M. Simalty and H. Chahine; *Bull. Soc. Chim. Fr.*, 1968, 4938.
68CC1350 I.M. Downie, J.B. Lee and M.F.S. Matough; *J. Chem. Soc., Chem. Commun.*, 1968, 1350.
68IJ651 I. Granoth, A. Kalir and Z. Pelah; *Isr. J. Chem.*, 1968, 6, 651 (*Chem. Abstr.*, 1969, **70**, 68471).
68IZV397 T.A. Zyablikova, I.M. Magdeev and I.M. Shermergorn; *Izv. Akad. Nauk SSSR, Ser. Khim.*, 1968, 397 (*Chem. Abstr.*, 1969, **69**, 59345).
68JOC474 J.B. Levy, L.D. Freedman and G.O. Doak; *J. Org. Chem.*, 1968, **33**, 474.
68RRC1623 R. Vilceanu, A. Balint, Z. Simra, C. Rentia and G. Unterweger; *Rev. Roumaine Chim.*, 1968, **13**, 1623 (*Chem. Abstr.*, 1969, **71**, 6656).
69AG(E)987 D. Hellwinkel and W. Schenk; *Angew. Chemie., Int. Ed. Engl.*, 1969, **8**, 987.
69GEP1300695 J. Galinke, M. Dohr and G. Klement; *Ger. Pat.* 1300695 (1969) (*Chem. Abstr.*, 1969, **71**, 82061).
69JOC2681 A.M. Aguiar, J.R.S. Irelan, G.W. Prejean, J.P. John and C.J. Morrow; *J. Org. Chem.*, 1969, **34**, 2681.
B-69MI625-01 K.D. Berlin and D.M. Hellwege; in "Topics of Phosphorus Chemistry," eds. M. Grayson and E.J. Griffith, Interscience, New York, 1969, vol. 6.
B-69MI625-02 "Nomenclature of Organic Chemistry", IUPAC, Butterworths, London, 1969.
70GEP(O)1965047 J.E. Bozik, Ch.Y. Wu and H.E. Swift; *Ger. Pat.* 1965047 (1970) (*Chem. Abstr.*, 1970, **73**, 67042).
70JCP(52)2769 W.J. Hehre, R. Ditchfield, R.F. Stewart and A.J. Pople; *J. Chem. Phys.*, 1970, **52**, 2769.
B-70MI625-01 F.G. Mann; "The Heterocyclic Derivatives of Phosphorus, Arsenic, Antimony and Bismuth", 2nd ed., Interscience, New York, 1970.
70TL4371 M. Simalty and L. van Chau; *Tetrahedron Lett.*, 1970, **11**, 4371.
70TL4941 H. Oehlig and A. Schweig; *Tetrahedron Lett.*, 1970, **56**, 4941.
70USP3524896 J.E. Bozik, H.E. Swift and Ch.Y. Wu; *US Pat.* 3524896 (1970) (*Chem. Abstr.*, 1970, **73**, 87392).
70ZAAC(372)233 M. Becke-Goehring and D. Jung; *Z. Anorg. Allg. Chem.*, 1970, **372**, 233.
70ZOB1466 N.D. Bodnarchuk, B.B. Gavrilenko and G.I. Derkach; *Zh. Obshch. Khim.*, 1970, **40**, 1466 (*Chem. Abstr.*, 1971, **74**, 125661).
71AG(E)397 A. Schimdpeter and C. Weingand; *Angew. Chem., Int. Ed. Engl.*, 1971, **10**, 397.
71AG(E)656 H. Oehling, W. Schäfer and A. Schweig, *Angew. Chem. Int. Ed. Engl.*, 1971, **10**, 656.
71JA3293 A.I. Ashe III; *J. Am. Chem. Soc. (A)*, 1971, **93**, 3293.
71JCS(A)1213 T. Tarantelli and C. Furlani; *J. Chem. Soc. (A)*, 1971, 1213.
71TL4749 J.C. Williams Jr., J.A. Kuczkowski, N.A. Portnoy, K.S. Yong, J.D. Wander and A.M. Aguiar; *Tetrahedron Lett.*, 1971, **12**, 4749.
71USP3567792 J.E. Bozik, H.E. Swift and C.Y. Wu; *US Pat.* 3567792 (1971) (*Chem. Abstr.*, 1971, **75**, 48391).
71ZOB988 P.P. Kornuta, A.J. Kalenskaya and V.I. Shevkenko; *Zh. Obshch. Khim.*, 1971, **41**, 988 (*Chem. Abstr.*, 1971, **75**, 140805).
71ZOB2390 P.P. Kornuta, A.J. Kalenskaya and V.I. Shevkenko; *Zh. Obshch. Khim.*, 1971, **41**, 2390 (*Chem. Abstr.*, 1972, **76**, 153718).
71ZN(B)650 J. Ebeling, M.A. Leva, H. Stary and A. Schmidpeter; *Z. Naturforsch., Teil B*, 1971, 26, 650.
72AG(E)1019 G. Märkl and D. Mathes; *Angew. Chem., Int. Ed. Engl.*, 1972, **11**, 1019.
72BSF3532 M. Simalty and M.H. Mebazaa; *Bull. Soc. Chim. Fr.*, 1972, 3532.
72CR(C)2206 R. Fugnitto, M.H. Mebazaa and M. Simalty; *C.R. Hebd. Seances Acad. Sci., Ser. C*, 1972, **274C**, 2206.
72JHC347 L. Hungeford and L.M. Trefonas; *J. Heterocycl. Chem.*, 1972, **9**, 347.
B-72MI625-01 G.M. Kosolapoff and L. Maier, "Organic Phosphorus Compounds", Wily Interscience, New York, 1972.
72SRI223 K. Isslieb, H. Oehme and M. Scheibe; *Synth. React. Inorg. Metal-Org. Chem.*, 1972, **2**, 223 (*Chem. Abstr.*, 1972, **77**, 152272).
72T3343 M. Simalty and M.H. Mebazaa; *Tetrahedron*, 1972, **28**, 3343.
72TL4363 M.H. Mebazaa and M. Simalty; *Tetrahedron Lett.*, 1972, **13**, 4363.
72USP3702878 T. Saito; *US Pat.* 3702878 (1972) (*Chem. Abstr.*, 1973, **78**, 43708).

72USP3644529	E.R. Tucci, H.I. Thayer and J.V. Ward; *US Pat.* 3644529 (1972) (*Chem. Abstr.*, 1972, **76**, 99116).
73JA928	C. Batich, E. Heilbronner, V. Hornung, A.J. Ashe III, D.T. Clarke, U.T. Cobley, D. Kilcast and I. Scanlan; *J. Am. Chem. Soc.*, 1973, **95**, 928.
73ZC291	H. Oehme and E. Leissring; *Z. Chem.*, 1973, **13**, 291.
73ZOB16	V.I. Shevkenko, A.I. Kalenskaya and P.P. Kornuta; *Zh. Obshch. Khim.*, 1973, **43**, 16 (*Chem. Abstr.*, 1973, **78**, 111265).
73ZOB261	P.P. Kornuta, A.I. Kalenskaya, O.B. Lobanov and V.I. Shevchenko; *Zh. Obshch. Khim.*, 1973, **43**, 261 (*Chem. Abstr.*, 1973, **78**, 147926).
73ZOB2178	L.A. Tamm, V.N. Chistokletov and A.A. Petrov; *Zh. Obshch. Khim.*, 1973, **43**, 2178 (*Chem. Abstr.*, 1974, **80**, 48108).
74IJ779	J.B. Levy; *Isr. J. Chem.*, 1974, **12**, 779.
74JHC407	D.J. Daigle, A.B. Pepperman Jr. and S.L. Vail; *J. Heterocycl. Chem.*, 1974, **11**, 407.
74TCA19	K.W. Schulte and A. Schweig; *Theor. Chim. Acta*, 1974, **33**, 19 (*Chem. Abstr.*, 1974, **80**, 145150).
74TL3927	S. Samaan; *Tetrahedron Lett.*, 1974, **15**, 3927.
74TL4385	G. Maerkl and D. Matthes; *Tetrahedron Lett.*, 1974, **15**, 4385.
74URP515754	B.I. Stepanov, N.N. Bychkov and A.I. Bokanov; *USSR Pat.* 515574 (1974) (*Chem. Abstr.*, 1976, **95**, 124150).
74ZOB1247	V.P. Kukhar' and A.P. Boiko; *Zh. Obshch. Khim.*, 1974, **44**, 1247 (*Chem. Abstr.*, 1974, **81**, 77885).
74ZOB2104	V.P. Kukhar' and T.N. Kasheva; *Zh. Obshch. Khim.*, 1974, **44**, 2103 (*Chem. Abstr.*, 1975, **82**, 4216).
75CC454	H.P. Kalhoun, R.T. Oakley and N.L. Paddock; *J. Chem. Soc., Chem. Commun.*, 1975, 454.
75CZ246	E. Fluck and J.E. Förster; *Chem.-Ztg.*, 1975, **99**, 246.
75JES91	W. Schäfer, A. Schweig, H. Vermeer, F. Bickelhaupt and H. de Graaf; *J. Electron Spectrosc.*, 1975, **6**, 91 (*Chem. Abstr.*, 1975, **83**, 8612).
75JHC579	D.J. Daigle and A.B. Pepperman; *J. Heterocycl. Chem.*, 1975, **12**, 579.
75JOC766	R.N. Jenkins and L.D. Freedman; *J. Org. Chem.*, 1975, **40**, 766.
75JOC2056	A.B. Pepperman Jr., G.J. Boudreaux and T.H. Siddatt; *J. Org. Chem.*, 1975, **40**, 2056.
75T2931	C. Jongsma, H. Vermeer, F. Bickelhaupt, W. Schäfer and A. Schweig; *Tetrahedron*, 1975, **31**, 2931.
75TL3171	G. Maerkl, D. Matthes, A. Donaubauer and H. Baier; *Tetrahedron Lett.*, 1975, **16**, 3171.
75ZN(B)808	E. Fluck, E. Schmid and W. Haubold; *Z. Naturforsch., Teil B*, 1975, **30**, 808.
76CB2960	G. Schöning and O. Glemser; *Chem. Ber.*, 1976, **109**, 2960.
76HCA1944	A.J. Ashe III, P. Burger, M.Y. Eh. Sheik. E. Heilbronner, J.P. Maier and J.-F. Müller; *Helv. Chim. Acta*, 1976, **59**, 1944.
76JA4410	W. Schäfer, A. Schweig, K. Dimroth and H. Kanter; *J. Am. Chem. Soc.*, 1976, **98**, 4410.
76TL3715	Y. Kobayashi, I. Kumadaki, A. Ohsawa and H. Hamana; *Tetrahedron Lett.*, 1976, **17**, 3715.
76ZOB51	L.I. Molyavko and V.A. Shokol; *Zh. Obshch. Khim.*, 1976, **46**, 72.
76ZOB1482	P.P. Kornuta, L.I. Derii, A.I. Kalenskaya and V.I. Shevchenko; *Zh. Obshch. Khim.*, 1976, **46**, 1482 (*Chem. Abstr.*, 1965, **85**, 192682).
76ZOB243	V.P. Kukahr' and T.N. Kaheva; *Zh. Obshch. Khim.*, 1976, **46**, 243 (*Chem. Abstr.*, 1976, **84**, 180176).
76ZOB2199	F.V. Dyakiv, A.M. Zelizny and E.F. Shevchenko; *Zh. Obshch. Khim.*, 1976, 2199.
76ZOB2217	V.P. Kukhar' and A.P. Boiko; *Zh. Obshch. Khim.*, 1976, 46, 2217 (*Chem. Abstr.*, 1977, 86, 72591).
77CB1148	G. Schöning and O. Glemser; *Chem. Ber.*, 1977, **110**, 1148.
77CB3231	G. Schöning and O. Glemser; *Chem. Ber.*, 1977, **110**, 3231.
77CJC2534	R.T. Oakley, N.L. Paddock, S.J. Rettig and J. Trotter; *Can. J. Chem.*, 1977, **55**, 2534.
77CJC3651	R.T. Oakley and N.L. Paddock; *Can. J. Chem.*, 1977, **55**, 3651.
77JA6656	R.D. Rieke, R.A. Copenhafer, C.K. White, A. Aguiar, J.C. Williams Jr. and M.S. Chattha; *J. Am. Chem. Soc.*, 1977, **99**, 6656.
77MI625-01	N.S. Lagoshina, E.S. Koslov and A.I. Sedlov; *Fizol. Akt. Veshchestra*, 1977, **9**, 30 (*Chem. Abstr.*, 1977, **88**, 17270).
77PS(3)209	E. Fluck and E. Scmid; *Phosphorus Sulfur*, 1977, **3**, 209.
77S842	G. Maerkl, H. Baier and R. Liebl; *Synthesis*, 1977, 842.
77TL867	Y. Kobayashi, I. Kumadaki, A. Ohsawa and H. Hamana; *Tetrahedron Lett.*, 1977, **18**, 867.

77TL3057 Y. Kobayashi, I. Kumadaki, A. Ohsawa and H. Hamana; *Tetrahedron Lett.*, 1977, **18**, 3057.

77TL3445 G. Maerkl, G. Adolin, F. Kees and G. Zander; *Tetrahedron Lett.*, 1977, **18**, 3445.

77ZAAC(433)229 E. Fluck, E. Schmid and H. Haubold; *Z. Anorg. Allg. Chem.*, 1977, **433**, 229.

77ZAAC(434)95 E. Fluck, E. Schmid and H. Haubold; *Z. Anorg. Allg. Chem.*, 1977, **434**, 95.

77ZAAC(436)136 W. Pinkert, G. Schöning and O. Glemser; *Z. Anorg. Allg. Chem.*, 1977, **436**, 136.

77ZOB342 P.P. Kornuta, N.V. Kolotilo and L.N. Markovskii; *Zh. Obshch. Khim.*, 1977, **42**, 342 (*Chem. Abstr.*, 1977, **87**, 23223).

77ZOB2522 P.P. Kornuta, L.S. Kuz'menko, M.M. Vovk, E.A. Romanenko, S.V. Iksanova and L.N. Markovskii; *Zh. Obshch. Khim.*, 1977, **42**, 2522 (*Chem. Abstr.*, 1978, 88, 105276).

77ZN(B)117 G. Schöning and O. Glemser; *Z. Naturforsch., Teil B*, 1977, **32**, 117.

77ZN(B)254 E. Fluck and E. Schmid; *Z. Naturforsch., Teil B*, 1977, **32**, 254.

77ZN(B)499 E. Fluck, J.E. Förster, J. Weidlein and E. Häddicke; *Z. Naturforsch., Teil B*, 1997, **32**, 499.

78ACR153 A.J. Ashe III; *Acc. Chem. Res.*, 1978, **11**, 153.

78CB579 S. Samaan; *Chem. Ber.*, 1978, **111**, 579.

78CB1798 D. Hellwinkel, W. Schenk and W. Blaicher; *Chem. Ber.*, 1978, **111**, 1798.

78DOK(240)331 B.A. Arbuzov, O.A. Erastov, S.N. Ignat'eva, T.A. Zyablikova, R.P. Arshinova and I.K.H. Shakirov; *Dokl. Akad. Nauk SSSR*, 1978, **240**, 331 (*Chem. Abstr.*, 1978, **89**, 107816).

78MI625-01 P.A. Bartlett, J.T. Hunt, J.L. Adams and J.C.E. Gehret; *Bioorg. Chem.*, 1978, **7**, 421 (*Chem. Abstr.*, 1979, **90**, 168675).

78PS(5)81 A. Zschunke, H. Meyer, E. Leissring, H. Oehme and K. Isslieb; *Phosphorus Sulfur*, 1978, **5**, 81 (*Chem. Abstr.*, 1979, **90**, 86608).

78T3419 W.C. Hendorn and C. Parkyani; *Tetrahedron*, 1978, **34**, 3419.

78ZN(B)1208 M. Baudler and F. Saykowski; *Z. Naturforsch.*, 1978, **33B**, 1208.

78ZOB97 R.K. Valetdinov, A.N. Zuikova., T.A. Zyablikova and A.V. Il'yasov; *Zh. Obshch. Khim.*, 1978, **48**, 97 (*Chem. Abstr.*, 1997, **90**, 5649).

78ZOB799 P.P. Kornuta and L.S. Kuz'menko; *Zh. Obshch. Khim.*, 1978, **48**, 799 (*Chem. Abstr.*, 1978, **89**, 109733).

78ZOB1181 P.P. Kornuta and N.V. Kolotilo; *Zh. Obshch. Khim.*, 1978, **48**, 1181 (*Chem. Abstr.*, 1978, **89**, 109392).

78ZOB2218 P.P. Kornuta, N.V. Kolotilo and L.N. Markovskii; *Zh. Obshch. Khim.*, 1978, **48**, 2218 (*Chem. Abstr.*, 1979, **90**, 72154).

78ZOB2653 R.K. Valetdikov, A.N. Zuikova, N.Sh. Yaminskaya, T.A. Zyablikova, E.N. Ofitserov and A.V. Il'yasov; *Zh. Obshch. Khim.*, 1978, **48**, 2653 (*Chem. Abstr.*, 1979, **90**, 152298).

79JFC(13)189 R.H. Kratzer, K.J.L. Paciorek, J. Kaufman, T.I. Ito and J.H. Nakahar; *J. Flourine Chem.*, 1979, **13**, 189 (*Chem. Abstr.*, 1979, **90**, 204187).

79JFC(13)199 R.H. Kratzer, K.J.L. Paciorek, J. Kaufman, T.I. Ito and J.H. Nakahar; *J. Flourine Chem.*, 1979, **13**, 199 (*Chem. Abstr.*, 1979, **90**, 204188).

79PS(7)309 D.W. Allen, I.W. Nowell and P.E. Walker; *Phosphorus Sulfur*, 1979, **7**, 309 (*Chem. Abstr.*, 1980, **93**, 45454).

79S109 J. Skolimowski and M. Simalty; *Synthesis*, 1979, 109.

79ZOB1671 P.P. Kornuta and N.V. Kolotilo; *Zh. Obshch. Khim.*, 1979, **49**, 1671 (*Chem. Abstr.*, **91**, 175309).

79ZOB1777 P.P. Kornuta, N.V. Kolotilo and V.N. Kalinin; *Zh. Obshch. Khim.*, 1979, **49**, 1777 (*Chem. Abstr.*, 1980, **92**, 6616).

80IZV721 B.A. Arbuzov, O.A. Erastov, G.A. Nikonov, R.P. Arshinova and R.A. Kadyrov; *Izv. Akad. Nauk SSSR, Ser. Khim.*, 1980, **92**, 721 (*Chem. Abstr.*, 1980, **93**, 95334).

80IZV952 B.A. Arbuzov, O.A. Erastov and G.A. Nokonov; *Izv. Akad. Nauk SSSR, Ser. Khim.*, 1980, 952 (*Chem. Abstr.*, 1980, **93**, 95346).

80IZV1438 B.A. Arbuzov, O.A. Erastov and G.A. Nokonov; *Izv. Akad. Nauk SSSR, Ser. Khim.*, 1980, 1428 (*Chem. Abstr.*, 1980, **93**, 204761).

80IZV1626 B.A. Arbuzov, O.A. Erastov, S.Sh. Khetkurova, T.A. Zyablikova, R.P. Arshinova and R.A. Kadyrov; *Izv. Akad. Nauk SSSR, Ser. Khim.*, 1980, **92**, 1626 (*Chem. Abstr.*, 1980, **93**, 220857).

80JFC(15)327 K.J.L. Paciorek, T.I. Ito and R.U. Kratzer; *J. Flourine Chem.*, 1980, **15**, 327 (*Chem. Abstr.*, 1980, **93**, 132558).

80JOC4683 Y. Kobayashi, S. Fujino, H. Hamana, Y. Hanzawa, S. Morita and I. Kumadaki; *J. Org. Chem.*, 1980, **45**, 4683.

80PP160 M. Sato and M. Yokoyama; *Polym. Prepr.*, 1980, **21**, 160 (*Chem. Abstr.*, 1982, **96**, 200304).

80TL3037 J. Skolimowski and M. Simalty; *Tetrahedron Lett.*, 1090, **21**, 3037.

80ZOB311	A.P. Boiko, V.P. Kukhar' and A.A. Kisilenko; *Zh. Obshch. Khim.*, 1980, **50**, 311 (*Chem. Abstr.*, 1980, **93**, 25879).
80ZOB1313	P.P. Kornuta, L.S. Kuz'menko and V.N. Kalinin; *Zh. Obshch. Khim.*, 1980, **50**, 1313 (*Chem. Abstr.*, 1980, **93**, 204498).
81CB3063	H. Schmidbauer and T. Costa; *Chem. Ber.*, 1981, **114**, 3063.
81IZV1539	B.A. Arvuzov, O.A. Erastov, G.N. Nikonov, T.A. Zyablikova, D.S. Yufit and Yu.T. Struchkov; *Izv. Akad. Nauk SSSR, Ser. Khim.*, 1981, **93**, 1539 (*Chem., Abstr.*, 1981, **95**, 186521).
81JA2465	Y. Kobayashi, S. Fujino and J. Kumadaki; *J. Am. Chem. Soc.*, 1981, **103**, 2465.
81JA4587	M. Yoshifuji, I. Shima, N. Inamato, K. Hirotsu and T. Higuchi; *J. Am. Chem. Soc.*, 1981, **103**, 4587.
81JAP(K)8157796	M. Yokoyama; Jpn. Kokai 8157796 (1981) (*Chem. Abstr.*, 1981, **95**, 204155).
81JAP(K)81150098	Osuka Pharmacutical Co.; *Jpn. Kokai* 81150098 (1981) (*Chem. Abstr.*, 1982, **97**, 6518).
B-81MI625-01	L.D. Quin; "The Heterocyclic Chemistry of Phosphorus", Wiley, New York, 1981.
81PS(9)269	J.R. Butler, H.S. Freeman and L.D. Freedman; *Phosphorus Sulfur*, 1981, **9**, 269 (*Chem. Abstr.*, 1981, **95**, 80896).
81PS(10)207	A.W. Frank; *Phosphorus Sulfur*, 1981, **10**, 207 (*Chem. Abstr.*, 1981, **95**, 169290).
81TL229	G. Maerkl and G. Yu Jin; *Tetrahedron Lett.*, 1981, **22**, 229.
81ZC137	E. Fischer, H. Weber and M. Michalik; *Z. Chem.*, 1981, **21**, 137.
81ZC143	E. Fischer, H. Weber, M. Michalik and B. Thomas; *Z. Chem.*, 1981, **21**, 143.
81ZOB517	A.P. Boiko and V.P. Kukhar'; *Zh. Obshch. Khim.*, 1981, **51**, 517 (*Chem. Abstr.*, 1981, **95**, 7234).
81ZOB767	Yu.M. Dangyan, G.A. Panosyan, M.G. Voskanyan and Sh.O. Badanyan; *Zh. Obshch. Khim.*, 1981, **51**, 767.
81ZOB2222	N.A. Rozanel'skaya, A.I. Bokanov, V.V. Negebetskii and B.I. Stepanov; *Zh. Obshch. Khim.*, 1981, **51**, 2222 (*Chem. Abstr.*, 1982, **96**, 20193).
81ZN(B)16	G. Becker, G. Gresser and W. Uhl; *Z. Naturforsch., Teil B*, 1981, **36**, 16.
82CHEC-I(1)493	K. Dimroth; *Comp. Heterocycl. Chem.*, 1982, **1**, 493.
82HOU(E1/2)	M. Regitz (ed.); *Methoden Org. Chem. (Houben-Weyl)*, 1982, **E1**, **E2**.
82HOU(E1)72	G. Maerkl; *Methoden Org. Chem. (Houben-Weyl)*, 1982, **E1**, 72.
82HOU(E1)783	K. Dimroth; *Methoden Org. Chem. (Houben-Weyl)*, 1982, **E1**, 783.
82IZV127	B.A. Arbuzov, O.A. Erastov, T.A. Zyablikova, S.N. Igna'eva, G.N. Nokonov, R.P. Arshinova and R.A. Kadyrov; *Izv. Akad. Nauk SSSR, Ser. Khim.*, 1982, **94**, 127 (*Chem. Abstr.*, 1982, **96**, 162833).
82IZV594	R.A. Kadyrov, R.P. Arshinova and B.A. Arbuzov; *Izv. Akad. Nauk SSSR, Ser. Khim.*, 1982, **94**, 594 (*Chem. Abstr.*, 1982, **97a**, 39013).
82MI625-01	K. Nagarajan and V.P. Arya; *J. Sci. Ind. Res.*, 1982, **41**, 232 (*Chem. Abstr.*, 1982, **97**, 16421).
82ZOB590	P.P. Kornuta, N.V. Kilotilo and L.N. Markovskii; *Zh. Obshch. Khim.*, 1982, **52**, 590 (*Chem. Abstr.*, 1982, **97**, 23896).
82ZOB789	N.G. Khusainova, Z.A. Bredikhiga, A.D. Sinitsa, V.I. Kal'chenko and A.N. Pudovik; *Zh. Obshch. Khim.*, 1982, **52**, 789 (*Chem. Abstr.*, 1982, **97**, 92406).
82ZOB2742	P.P. Kornuta, N.V. Kolotilo and L.N. Markovskii; *Zh. Obshch. Khim.*, 1982, **52**, 2742 (*Chem. Abstr.*, 1983, **98**, 126252).
82ZN(B)965	T. Thamm and E. Fluck; *Z. Naturforsch., Teil B*, 1982, **37**, 965.
83CB445	G. Maerkl, K. Hock and D. Matthes; *Chem. Ber.*, 1983, **116**, 445.
83CB1756	G. Maerkl and K. Hock; *Chem. Ber.*, 1983, **116**, 1756.
83IC2573	A.B. Burg; *Inorg. Chem.*, 1983, **22**, 2573.
83IZV2550	B.A. Arbuzov, O.E. Erastov, S.N. Ignat'eva, I.P. Romanov, O.V. Ovodova and R.P. Arshinova; *Izv. Akad. Nauk SSSR, Ser. Khim.*, 1983, **95**, 2550 (*Chem. Abstr.*, 1984, **100**, 121212).
83JHC1601	B. Renger; *J. Heterocycl. Chem.*, 1983, **20**, 1601.
83PAC410	IUPAC Commission of the Nomenclature of Organic Compounds; *Pure Appl. Chem.*, 1983, **55**, 410.
83ZOB559	S.B. Bulgarevich, P.P. Kornuta, N.V. Kolotilo, Y.A. Yusman, N.A. Ivanova, D.Yu. Movshovich, V.A. Kogan and O.A. Osipov; *Zh. Obshch. Khim.*, 1983, **53**, 559 (*Chem. Abstr.*, 1983, **99**, 5711).
83ZOB930	Yu.M. Dangyan, G.M. Panosyan, M.G. Voskanyan and Sh.O. Badanyan; *Zh. Obshch. Khim.*, 1982, **53**, 930 (*Chem. Abstr.*, 1983, **99**, 70855).
84AG(E)710	J. Grobe and D. Le Van; *Angew. Chemie., Int. Ed. Engl.*, 1984, **23**, 710.
84CC183	L. López, J.-P. Majoral, A. Meriem, T. N'Gando M'Pondo, J. Navech and J. Barrans; *J. Chem. Soc., Chem. Commun.*, 1984, 183.
84IZV1056	B.A. Arbuzov, I.A. Litvinov, Yu.T. Struchkov, S.Sh. Khetagurova and O.A. Erastov; *Izv. Akad. Nauk SSSR, Ser; Khim.*, 1984, **96**, 1056 (*Chem. Abstr.*, 1984, **101**, 91093).

84MI625-01 M. Sato, H. Kondo and M. Yokoyama; *J. Appl. Polym, Sci.*, 1984, **29**, 299 (*Chem. Abstr.*, 1984, **100**, 68856).

84PS(19)159 J. Skolimowski and R. Skowronski; *Phosphorus Sulfur*, 1984, **19**, 159 (*Chem. Abstr.*, 1984, **101**, 91062).

84S1074 L.D. Quin, C.C. Henderson, N.S. Rao and J.C. Kisalus; *Synthesis*, 1984, 1074.

84ZOB1481 G.K. Federova, L.G. Anan'eva and O.A. Yakovchuk; *Zh. Obshch. Khim.*, 1984, **54**, 1481 (*Chem. Abstr.*, 1985, **102**, 6654).

84ZOB2532 L.V. Ermolaeva and R.P. Arshinova; *Zh. Obshch. Khim.*, 1984, **54**, 2532 (*Chem. Abstr.*, 1985, **102**, 132159).

85CB2365 G. Maerkl, W. Weber and W. Weiss; *Chem. Ber.*, 1985, **118**, 2365.

85CC1681 J. Barluenga, F. López and F. Palacios; *J. Chem. Soc., Chem. Commun.*, 1985, 1681.

85CL1459 Y. Nakadaira, R. Sato and H. Sakurai; *Chem. Lett.*, 1985, 1459.

85HCA900 K. Nagarajan, S.J. Shenoy, H. Fritz, O. Hosang and W.J. Richter; *Helv. Chim. Acta*, 1985, **68**, 900.

85IZV1296 R.M. Aminova, T.A. Zyablikova, M.B. Zuev and Yu.Yu. Samitov; *Izv. Akad. Nauk SSSR, Ser. Khim.*, 1985, **97**, 1296 (*Chem. Abstr.*, 1986, **104**, 34142).

85JCS(P2)1913 R.M. Acheson, C.T. Lines, M.R. Bryce, Z. Dauter, C.D. Reynold and A. Schmidpeter; *J. Chem. Soc., Perkin Trans. 2*, 1985, 1913.

85ZN(B)1258 J. Svaram E. Fluck and H. Riffel; *Z. Naturforsch., Teil B*, 1985, **40B**, 1258.

85ZOB162 Yu.A. Nevskaya, E.V. Kryuchkova and T.N. Sumarokova; *Zh. Obshch. Khim.*, 1985, **55**, 165.

85ZOB780 G.K. Federova, L.G. Anan'eva, N.G. Feshchenko, O.A. Yakovchuk and T.M. Zhovtyak; *Zh. Obshch. Khim.*, 1985, **55**, 780 (*Chem. Abstr.*, **103**, 1966161).

85ZOB1710 Z.A. Bredikhina, N.G. Khusainova, Yu.Ya. Efremov, R.L. Korshunov and A.N. Pudovik; *Zh. Obshch. Khim.*, 1985, **55**, 1710 (*Chem. Abstr.*, 1986, **104**, 34056.

86AG(E)(25)280 J. Fink, W. Rösch, U.-J. Vogelbacher and M. Regitz; *Angew. Chem. Int. Ed. Engl.*, 1986, **25**, 280.

86AG(E)831 U. Zenneck and W. Franck; *Angew. Chem., Int. Ed. Engl.*, 1986, **25**, 831.

86AG(E)1002 E. Fluck, G. Becker, B. Neumüller, R. Knebl, G. Heckmann and H. Riffel; *Angew. Chem., Int. Ed. Engl.*, 1986, **25**, 1002.

86DOK(291)154 Yu.Yu. Samitov; *Dokl. Akad. Nauk SSSR*, 1986, **291**, 154 (*Chem. Abstr.*, 1987, **107**, 198480).

86IZV418 B.A. Arbuzov, O.A. Erastov and A.S. Ionkin; *Izv. Akad. Nauk SSSR, Ser. Khim.*, 1996, **98**, 418 (*Chem. Abstr.*, 1987, **106**, 18695).

86IZV640 B.A. Arbuzov, O.A. Erastov, A.S. Ionkin and S.N. Ignat'eva; *Izv. Akad. Nauk SSSR, Ser. Khim.*, 1996, **98**, 640 (*Chem. Abstr.*, 1987, **106**, 119941).

86IZV2506 B.A. Arbuzov, O.A. Erastov, A.S. Ionkin and S.N. Ignat'eva; *Izv. Akad. Nauk SSSR, Ser. Khim.*, 1996, **98**, 2506 (*Chem. Abstr.*, 1987, **107** 96797).

86PS(29)23 B. Neumüller and E. Fluck; *Phosphorus Sulfur*, 1986, **23**, 23 (*Chem. Abstr.*, 1988, **108**, 56210).

86TL4725 C. Bourdieu and A. Foucaud; *Tetrahedron Lett.*, 1986, **27**, 4725.

86TL5367 F.A. Neugebauer and H. Fischer; *Tetrahedron Lett.*, 1986, 5367.

86USP4698447 D.M. Blum; *US Pat.* 4698447 (1986) (*Chem. Abstr.*, **108**, 38135).

86ZAAC(540/1)212 H.W. Roesky and B. Mainz.; *Z. Anorg. Allg. Chem.*, 1986, **540/541**, 212.

86ZOB1661 V.A. Galishev, V.N. Sakharov, T.S. Dolgushina, L. Tamm and A.A. Petrov; *Zh. Obshch. Khim.*, 1986, **56**, 1661 (*Chem. Abstr.*, 1987, **106**, 198473).

86ZOB2256 I.I. Patsanovskii, N.M. Kudyakov, T.A. Zyablikova, I.K. Kovrizhnokova, V.Y. Votkovskii, E.A. Ishmaeva, A.V. Il'yasov, M.G. Voronkov and A.N. Pudovik; *Zh. Obshch. Khim.*, 1996, **56**, 2256 (*Chem. Abstr.*, 1987, **107**, 198473).

87AG(E)(26)85 K. Blatter, W. Rösch, U.-J. Vogelbacher, J. Fink and M. Regitz; *Angew. Chem. Int. Ed.*, 1987, **26**, 85.

87AG(E)236 G. Märkl, H.J. Beckh, K.K. Mayer, M.L. Ziegler and T. Zahn; *Angew. Chem., Int. Ed. Engl.*, 1987, **26**, 236.

87AG(E)908 R. Milczarek, W. Rüsseler, P. Binger, K. Jonas, K. Angermund, C. Krüger and M. Regitz; *Angew. Chem., Int. Ed. Engl.*, 1987, **26**, 908.

87AG(E)1011 E.P.O. Fuchs, W. Rösch and M. Regitz; *Angew. Chem., Int. Ed., Engl.*, 1987, **26**, 1011.

87CRV289 L.D. Freedman and H.S. Freeman; *Chem. Rev.*, 1987, **87**, 289.

87CZ309 E. Fluck, B. Neumüller and G. Heckmann; *Chem. Ztg.*, 1987, **111**, 309.

87IZV2397 N.A. Bondarenko, M.V. Raitarskaya, M.V. Rudomino and E.N. Tsetkov; *Izv. Akad. Nauk SSSR, Ser. Khim.*, 1987, **99**, 2397 (*Chem. Abstr.*, 1988, **109**, 149627).

87JOM(325)153 M.R. Bryce and R.S. Mathews; *J. Organomet. Chem.*, 1987, **325**, 153.

87TL1093 G. Maerkl and F. Dorfmeister; *Tetrahedron Lett.*, 1987, **28**, 1093.

87TL2785 J. Barluenga, F. López and F. Palacios; *Tetrahedron Lett.*, 1987, **28**, 2875.

87TL4673 C. Bourdieu and A. Foucaud; *Tetrahedron Lett.*, 1987, **28**, 4673.

87ZOB228 V.G. Rozinov, M.Yu. Dmitrichenko, V.I. Donskikh, G.V. Dolgushin and V.P. Feshin; *Zh. Obshch. Khim.*, 1987, **57**, 228 (*Chem. Abstr.*, 1987, **107**, 217733).

87ZOB716 N.L. Kostenko, S.V. Nesterova, S.V. toldov, N.K. Skortsov and V.O. Reikhsfel'd; *Zh. Onshch. Khim.*, 1987, 57, 716 (*Chem. Abstr.*, 1987, *107*, 217834).

87ZOB954 V.G. Rozinov, V.E. Kolbina and V.I. Dokskikh; *Zh. Obshch. Khim.*, 1987, **57**, 954 (*Chem. Abstr.*, 1988, **108**, 186836).

87ZOB1518 S.V. Fridland, M.N. Miftakhov and V.P. Arkhipov; *Zh. Obshch. Khim.*, 1987, **57**, 1518 (*Chem. Abstr.*, 1988, **109**, 6166).

87ZOB1980 A.N. Chernega, I.E. Boldeskul, M.Yu. Antipin, Yu.T. Struchkov, P.P. Kornuta and N.V. Kolotilo; *Zh. Obshch. Khim.*, 1987, **57**, 1980 (*Chem. Abstr.*, 1988, **109**, 129144).

87ZN(B)1213 E. Fluck, G. Becker, B. Neumüller, R. Knebl, G. Heckmann and H. Riffel; *Z. Naturforsch., Teil B*, 1987, **42B**, 1213.

88AG(E)272 U.-J. Vogelbacher, M. Ledderman, T. Schach, G. Michels, U. Hees and M. Regitz; *Angew. Chem., Int. Ed. Engl.*, 1988, **27**, 272.

88AG(E)389 G. Maerkl, S. Dietl, M.L. Ziegler and B. Nuber; *Angew. Chem., Int. Ed. Engl.*, 1988, **27**, 389.

88AG(E)1485 M. Regitz and P. Binger; *Angew. Chem., Int. Ed. Engl.*, 1988, **27**, 1485.

88AG(E)1715 E. Niecke, N. Nieger and F. Reichert; *Angew. Chem., Int. Ed. Engl.*, 1988, **27**, 1715.

88CB627 M. Meyer and U. Klingbiel; *Chem. Ber.*, 1988, **121**, 627.

88CB1119 M. Meyer and U. Klingbiel; *Chem. Ber.*, 1988, **121**, 1119.

88CB1681 H.W. Roesky, U. Scholtz, A. Schmidpeter, K. Karaghiosoff and W.S. Sheldrick; *Chem. Ber.*, 1988, **121**, 1681.

88CRV1327 J.F. Nixon; *Chem. Rev.*, 1988, **88**, 1327.

88IC1911 R.C. Haddon, S.V. Chichester-Hicks and S.L. Mayo; *Inorg. Chem.*, 1988, **27**, 1911.

88PS(35)215 G. Reuger; *Phosphorus Sulfur*, **35**, 215.

88TL925 H. Keller and M. Regitz; *Tetrahedron Lett.*, 1988, **29**, 925.

88ZN(B)937 U. Scholtz, M. Noltemeyer and H.W. Roesky; *Z. Naturforsch., Teil B*, 1988, **43**, 937.

88ZN(B)1010 M. Meyer, U. Klingbiel, J. Kadel and H. Oberhammer; *Z. Naturforsch., Teil B*, 1988, **43**, 1010.

88ZOB526 A.N. Chekhlov, V.A. Galishev, T.S. Dolgushina, V.K. Brel, A.A. Petrov and I.V. Martynov; *Zh. Obshch. Khim.*, 1988, **58**, 526 (*Chem. Abstr.*, 1989, **110**, 24003).

89AG(E)53 R. Appel and M. Poppe; *Angew. Chem., Int. Ed. Engl.*, 1989, **28**, 53.

89CJC1319 E.C. Alyea, K.J. Fisher and S. Johnson; *Can. J. Chem.*, 1989, **67**, 1319.

89DOK(304)625 G.Yu. Trishin, R.N. Burangulova, I.V. Konovalova, L.A. Burnaeva, G.S. Akimova, V.N. Chistokletov and A.N. Pudovik; *Dokl. Akad. Nauk SSSR*, 1989, **304**, 625 (*Chem. Abstr.*, 1989, **112**, 35988).

89EUP304782 H.J. Kleiner, J. Gersdorf and U. Bastian; *Eur. Pat.* 304782 (1989) (*Chem. Abstr.*, 1989, **111**, 174807).

89H(28)1135 M.J.S. Dewar and A.J. Holder; *Heterocycles*, 1989, **28**, 1135.

89IS24 A. Schmidpeter, P.P. Power and R. Das; *Inorg. Synth.*, 1989, **25**, 24.

89JCS(P1)615 J. Barluenga, M. Ferrero, F. López and F. Palacios; *J. Chem. Soc., Perkin Trans. 1*, 1989, 615.

89JCS(P1)2273 J. Barluenga, F. López and F. Palacios; *J. Chem. Soc., Perkin Trans. 1*, 1989, 2273.

89NJC383 E. Fluck, B. Neumüller, G. Heckmann, W. Plass and P.G. Jones; *New J. Chem.*, 1989, **13**, 383 (*Chem. Abstr.*, 1990, **112**, 158375).

89PS(41)113 M. Fild and T. Oehimigen; *Phosphorus Sulfur*, 1989, **41**, 113 (*Chem. Abstr.*, 1989, **111**, 153941).

89PS(44)235 P.J. Brooks and M.J. Gallagher; *Phosphorus Sulfur*, 1989, **44**, 235 (*Chem. Abstr.*, 1990, **112**, 118972).

89TH625-01 S. Dietel; *Ph.D. Thesis.*, University of Regensberg, 1989.

89ZOB1047 V.I. Vyotskii, S.M. Kalnov, A.I. Kuz'min and A.V. Nasedkin; *Zh. Obshch. Khim.*, 1989, **59**, 1047 (*Chem. Abstr.*, 1990, **112**, 77349).

90AG(E)314 T. Facklam, O. Wagner, H. Heydt and M. Regitz; *Angew. Chem., Int. Ed. Engl.*, 1990, **29**, 314.

90AG(E)516 H. Schmidbauer, C. Paschalidis, O. Steigelman and G. Müller; *Angew. Chem. Int. Ed., Engl.*, 1990, **29**, 516.

90AG(E)569 H. Schmidbauer, C. Paschalidis, O. Steigelmann and G. Müller; *Angew. Chem., Int. Ed. Engl.*, 1990, **29**, 569.

90AG(E)1123 M. Granier, A. Baceiredo, M. Nieger and G. Bertrand; *Angew. Chem., Int. Ed. Engl.*, 190, **29**, 1123.

90HCA1888 W. Ried, M. Fulde and J.W. Bats; *Helv. Chim. Acta*, 1990, **73**, 1888.

90IJC(B)711 B.R. Shinde, S.J. Shenoy and N.R. Pai; *Indian J. Chem., Sect. B*, 1990, **29**, 711 (*Chem. Abstr.*, 1990, **113**, 231504).

90IZV776 T.A. Zyablikova, A.V. Il'yasov, S.N. Ignat'eva, S.F. Malatsion and O.A. Erastor; *Izv. Akad. Nauk. SSSR, Ser. Khim*, 1990, (4) 776-80.

90JCS(P1)1223 E. Fluck, G. Heckmann, W. Plaass, M. Spahn and H. Borrmann; *J. Chem. Soc., Perkin Trans. 1*, 1990, 1223.

90JOM(382)61 J. Barluenga, F. López and F. Palacios; *J. Organomet. Chem.*, 1990, **382**, 61 (*Chem. Abstr.*, 1990, **113**, 172151).

B-90MI625-01 G. Maerkl; in "Multiple Bonds and Low Coordination in Phosphorus Chemistry", eds. M. Regitz and O.J. Scherer, Georg Thieme Verlag, Stuttgart, 1990.

90PS(48)37 K.J. Fisher, E.C. Alyea and N. Shahnazarian; *Phosphorus Sulfur*, 1990, **48**, 37 (*Chem. Abstr.*, 1990, **113**, 132321).

90PS(54)1 L.D. Quin, J.C. Kisalus J.J. Skolimowski and N.S. Rao; *Phosphorus Sulfur*, 1990, **54**, 1 (*Chem. Abstr.*, 1994, **114**, 42909).

90T5697 C.W. Bird; *Tetrahedron*, 1990, **46**, 5697.

90T6715 W.H.-L. Wai Tan, C. Bourdieu and A. Foucaud; *Tetrahedron*, 1990, **46**, 6715.

90TL4589 G. Markl. C. Dörges, T. Reidl, F.-G. Klärner and C. Lodwig; *Tetrahedron Lett.*, 1990, **31**, 4589.

90TL6999 G. Maerkl, C. Dörges, H. Nöth and K. Polborn; *Tetrahedron Lett.*, 1990, **31**, 6999.

90ZAAC(588)181 E. Fluck, W. Plass and G. Heckmann; *Z. Anorg. Allg. Chem.*, 1990, **588**, 181.

90ZOB462 M.Y. Dmitrichenko, G.V. Dolgushin, V.G. Rozinov and V.I. Donskikh; *Zh. Obshch. Khim.*, 190, **60**, 462 (*Chem. Abstr.*, 1990, **113**, 59335).

90ZOB1257 S.V. Toldov, N.L. Kostenko, V.K. Bel'skii, N.K. Skvortsov and V.O. Reikhsfeld; *Zh. Obshch. Khim.*, 1990, **60**, 1257 (*Chem. Abstr.*, 1990, **113**, 213521).

90ZOB1975 I.E. Boldeskul, A.S. Tarasevich, A.F. Obodovskaya, Z.A. Skarikova and P.P. Kornuta; *Zh. Obshch. Khim.*, 1990, **60**, 1975 (*Chem. Abstr.*, 1991, **114**, 42962).

91AG(E)106 G. Maerkl and C. Dörges; *Angew. Chem., Int. Ed. Engl.*, 1991, **30**, 106.

91CB493 J. Heinicke, I. Kovacs and L. Nyulászi; *Chem. Ber.*, 1991, **124**, 493.

91CB1525 H. Schmidbauer, S. Gamper, C. Paschalidis, O. Steigelmann and G. Müller; *Chem. Ber.*, 1991, **124**, 1525.

91CB2071 R. Gleiter, T. Veprémi and E. Fluck; *Chem. Ber.*, 1991, **124**, 2071.

91HAC17 Y. van den Winkel, J. van der Laarse, F.J.J. De Kanter, T. van der Does, F. Bickelhaupt, W.J.J. Smeets and A.L. Spek; *Heteroatom Chem.*, 1991, **2**, 17 (*Chem. Abstr.*, 1991, **114**, 229033).

91IC3402 V. Chandrasekhar, T. Chivers, S.S. Kumaravel, A. Meetsma and J.C. van de Grampel; *Inorg. Chem.*, 1991, **30**, 3402.

91PS(56)49 E. Fluck, G. Heckmann, W. Plass, H. Bögge and A. Müller; *Phosphorus Sulfur, Silicon Relat. Elements*, 1991, **56**, 49 (*Chem. Abstr.*, 1991, **114**, 247391).

91PSS49 E. Fluck, G. Heckmann, W. Plass, H. Bögge and A. Müller; *Phosphorus, Sulfur, Silicon*, 1991, **56**, 49.

91TL2619 C. Bedel and A. Foucaud; *Tetrahedron Lett.*, 1991, **32**, 2619.

91ZN(B)202 E. Fluck, W. Plass, G. Heckmann, H. Bögge and A. Müller; *Z. Naturforsch.*, 1991, **46B**, 202.

91ZN(B)762 E. Fluck, M. Spahn and G. Heckmann; *Z. Naturforsch.*, 1991, **46B**, 762.

91ZN(B)1130 E. Fluck, M. Spahn and G. Heckmann; *Z. Naturforsch.*, 1991, **46B**, 1130.

91ZN(B)1131 F. Bitterer, D.J. Brauer, F. Dörrenbach, F. Gol, P.C. Knüppel, O. Stelzer, C. Krüger and Y.H. Tsay; *Z. Naturforsch.*, 1991, **46B**, 1131.

91ZOB643 M.Y. Dmitrichenko, V.G. Rozinov, V.I. Donskikh, G.V. Ratovskii, V.E. Efremov, G.V. Dolgushin, V.Y. Vitkovskii and S.V. Shilin; *Zh. Obshch. Khim.*, 1991, **61**, 643 (*Chem. Abstr.*, 1991, **115**, 136226).

91ZOB1641 N.K. Svorttsov, S.V. Toldov, V.K. Bel'skii and V.E. Konovalov; *Zh. Obshch. Khim.*, 1991, **61**, 1641 (*Chem. Abstr.*, 1992, **116**, 119720).

92AG(E)1033 T. Arnold, B. Orschel and H.-U. Reissig; *Angew. Chem., Int. Ed. Engl.*, 1992, **31**, 1033.

92AG(E)1055 B. Breit, U. Bergsträßer, G. Maas and M. Regitz; *Angew. Chem. Int. Ed. Engl.*, 1992, **31**, 1055.

92EUP463994 P. Nesvedba; *Eur. Pat.* 463994 (1992) (*Chem. Abstr.*, 1992, **116**, 153125).

92IZV2174 A.N. Bovin, A.N. Chekhlov and E.N. Tsvetkov; *Izv. Akad. Ser. Khim.*, 1992, (9), 2174.

92OM1990 D.J. Darensbourg, F. Jóo, M. Kannisto, A. Katho and J.H. Reibenspies; *Organometallics*, 1992, **11**, 1990 (*Chem. Abstr.*, 1992, **117**, 39021).

92OM2863 S.F. Gamper and H. Schmidbauer; *Organometallics*, 1992, **11**, 2863 (*Chem. Abstr.*, 1992, **117**, 90443).

92PS(71)245 E. Fluck, K. Bieger, G. Heckmann and F. Weller; *Phosphorus Sulfur Silcon Relat. Elem.*, 1992, **71**, 245.

92TH625-01 C. Dörges; *Ph.D. Thesis*, University of Regensburg, 1992.

92TL2335	G. David, E. Niecke and M. Nieger; *Tetrahedron Lett.*, 1992, **33**, 2335.
92ZOB306	M.Y. Dmitrichenko, V.G. Rozinov, V.I. Donskikh, G.V. Ratovskii, V.G. Efremov and G.V. Dolgushin; *Zh. Obshch. Kkim.*, 1992, **62**, 306 (*Chem. Abstr.*, 1993, **118**, 169178).
92ZN(B)760	C. Müler, T.G. Meyer, M. Farkens, R. Sonnenburg and R. Schmutzler; *Z. Naturforsch.*, 1992, **47B**, 760.
93CB601	S.F. Gamper and A. Schmidbauer; *Chem. Ber.*, 1993, **126**, 601.
93IC5088	H.R. Alcock, S.M. Coley, I. Manners, K.B. Visscher, M. Parvez, O. Nuyken and G. Renner; *Inorg. Chem.*, 1993, **32**, 5088.
93IZV1650	A.A. Karasik, S.V. Bobrov, G.N. Nikonor, I.A. Litvinov and A.G. Akhmadullin; *Izv. Akad. Nauk, Serv. Khim*, 1993, 1650.
93JA7880	J. Tejeda, R. Réau, F. Dahan and G. Bertrand; *J. Am. Chem. Soc.*, 1993, **115**, 7880.
93JOM(459)73	E. Fluck, K. Bieger, G. Heckmann and B. Neumüller; *J. Organomet. Chem.*, 1993, **459**, 73 (*Chem. Abstr.*, 1994, **120**, 270734).
93OM3776	N.A. Petasis and D.-K. Fu; *Organometallics*, 1993, **12**, 3776.
93PS(78)271	I. Neda, A.K. Fischer, P.G. Jones and R. Schmutzler; *Phosphorus Sulfur Silicon*, 1993, **78**, 271.
93PS(80)141	A. Chaturvedi, P.N. Nagar and G. Srivastava; *Phosphorus Sulfur Silicon*, 1993, **80**, 141 (*Chem. Abstr.*, 1994, **20**, 134711).
93PSS(80)173	I. Neda, A. Fischer, P.G. Jones, R. Schmutzler; *Phosphorus Sulfur Silicon*, 1993, **80**, 173.
93TH625-01	R. Dedio; *Ph.D. Thesis*, University of Regensberg, 1993.
93ZOB1976	M.Yu. Dmitrichenko, V.G. Rozinov, G.V. Ratovskii, G.V. Dolgushin, V.I. Donskikhand and L.M. Sergienko; *Zh. Obshch. Khim.*, 1993, **63**, 1976 (*Chem. Abstr.*, 1994, **121**, 57571).
93ZN(B)333	I. Neda, H.J. Pinta and R. Schmutzler; *Z. Naturforsch.*, 1993, **48B**, 333.
93ZN(B)443	I. Neda, M. Farkens, A.K. Fischer, P.G. Jones and R. Schmutzler; *Z. Naturforsch.*, 1993, **48B**, 443.
93ZN(B)853	M. Farkens, I. Neda, A.K. Fischer, P.G. Jones and R. Schmutzler; *Z. Naturforsch.*, 1993, **48B**, 853.
94AG(106)2284	V. Caliman, P.B. Hitchcock, J.F. Nixon, M. Hofmann and P.v.R. Schleyer; *Angew. Chem.*, 1994, **106**, 2284.
94CB1395	V.A. Pinchuk, I. Neda, C. Müller, A. Fischer, Y.G. Shermolovich, P.G. Jones and R. Schmutzler; *Chem. Ber.*, 1994, **127**, 1395.
94CC431	J.P. Fackler Jr., R.J. Staples and Z. Assefa; *J. Chem. Soc., Chem. Commun.*, 1994, 431.
94EUP582957	H.J. Kleiner; *Eur. Pat.* 582957 (1994) (*Chem. Abstr.*, 1994, **120**, 270819).
94JA8087	K. Bieger, J. Tejeda, R. Reau, F. Dahan and G. Bertrand; *J. Am. Chem. Soc.*, 1994, **116**, 8087.
94JAP(K)6145182	T. Haruna, E. Hida and R. Hamada; *Jpn. Kokai* 06145182 (1994) (*Chem. Abstr.*, 1994, **121**, 256037).
94JFC51	I. Neda, H.-J. Plinta, A. Fischer, P.G. Jones and R. Schmutzler; *J. Fluorine Chem.*, 1994, **69**, 51.
94JOC5927	G. Veneziani, R. Reau, F. Dahan and G. Bertrand; *J. Org. Chem.*, 1994, **59**, 5927.
94JOC7144	T. Sakai, T. Kodoma, T. Fujimoto, K. Ohta, I. Yamamoto and K. Kakehi; *J. Org. Chem.*, 1994, **59**, 7144.
94RHA65	G. Heckmann and E. Fluck; *Rev. Heteroatom Chem.*, 1994, **11**, 65.
94ZAAC(620)483	E. Fluck, K. Bieger and G. Heckmann; *Z. Anorg. Allg. Chem.*, 1994, **620**, 483.
94ZN(B)145	M. Farkens, T.G. Meyer, I. Neda, R. Sonnonburg, C. Muller, A.K. Fischer, P.G. Jones and R. Schmutzler; *Z. Naturforsch.*, 1994, **49B**, 145.
94ZN(B)171	G.V. Dolgushin, P.A. Nitikin, Yu.L. Sapozhnikov, M.Yu. Dmitrichenko, V.G. Rozonov and M.G. Voronkov; *Z. Naturforsch.*, 1994, **49B**, 171.
94UP625-01	G. Maerkl; *Unpublished Results*, University of Regensberg, 1994.
95AG(E)81	P. Binger, G. Glaser, B. Gabor and R. Mynott; *Angew. Chem., Int. Ed. Engl.*, 1995, **34**, 81.
95AG(E)198	D. Böhm, F. Koch, S. Kummer, U. Schmidt and U. Zenneck; *Angew. Chem. Int. Ed. Engl.*, 1995, **34**, 198.
95AG(E)2227	P. Binger, S. Leininger, J. Stannek, B. Gabor, R. Mynott, J. Bruckmann and C. Krüger; *Angew. Chem., Int. Ed. Engl.*, 1995, **34**, 2227.
95BSF(129)910	K. Waschbüsch, P. Le Floch and F. Mathey; *Bull. Soc. Chim. Fr.*, 1995, **132**, 910.
95BSF(132)652	H. Heydt, U. Bergsträßer, R. Fäßler, E. Fuchs, N. Kamel, Th. Mackewitz, G. Michels, W. Rösch, M. Regitz, P. Mazerolles, Ch. Laurent and A. Faucher; *Bull. Soc. Chem. Fr.*, 1995, **132**, 652.
95CB(129)489	B. Breit and M. Regitz; *Chem. Ber.*, 1995, **129**, 489.

95CB891	B. Assmann, K. Angermaier, M. Paul, J. Riede and H. Schmidbaur; *Chem. Ber.*, 1995, 891.
95EUP632050	H.J. Kleiner; *Eur. Pat.* 632050 (1995) (*Chem. Abstr.*, 1995, **122**, 160936).
95HAC355	E. Fluck, F. Roshe, G. Heckmann and F. Weller; *Heteroatom Chem.*, 1995, **6**, 355.
95IC75	Z. Assefa, B.G. McBurnett, R.J. Staples and P.J. Fackler Jr.; *Inorg. Chem.*, 1995, **34**, 75.
95JOC5884	M. Julino, U. Bergsträßer and M. Regitz; *J. Org. Chem.*, 1995, **60**, 5884.
95JOM(488)99	D.J. Darensbourg, N. White Stafford, F. Jóo and J.H. Reibenspies; *J. Organomet. Chem.*, 1995, **488**, 99 (*Chem. Abstr.*, 1995, **122**, 314219).
95JST(347)57	L. Nyulászi and T. Veszpremi; *J. Mol. Struct.*, 1995, **347**, 57.
95MRC553	G. Heckmann and E. Fluck; *Magn. Reson. Chem.*, 1995, **33**, 553 (*Chem. Abstr.*, 1995, **123**, 228306).
95OM4194	J.M. Forward, J.P. Fackler Jr., and R.J. Staples; *Organometallics*, 1995, **14**, 4194.
95PS(106)65	C. Melnicky, I. Neda and R. Schmutzler; *Phosphorus Sulfur Silicon Relat. Elements*, 1995, **106**, 65 (*Chem. Abstr.*, 1996, **124**, 176287).
95PS(107)69	A. Vollbrecht, I. Neda, A. Fischer, P.G. Jones and R. Schmutzler; *Phosphorus Sulfur Silicon Relat. Elements*, 1995, **107**, 69 (*Chem. Abstr.*, 1996, **124**, 343452).
95RHA6121	E. Fluck and G. Heckmann; *Rev. Heteroatom Chem.*, 1995, **12**, 6121.
95T9625	A. Foucaud and Ch. Bedel; *Tetrahedron*, 1995, **51**, 9625.
95ZAAC1981	K. Bieger, G. Heckmann, E. Fluck, F. Weller, K. Peters and E.M. Peters; *Z. Anorg. Allg. Chem.*, 1995, **621(12)**, 1981.
95ZAAC2001	V.A. Pinchuk, C. Müller, A. Fischer, H. Thönnesen, P.G. Jones, R. Schmutzler, L.N. Markovsky, Y.G. Shermolovich and A.M. Pinchuk; *Z. Anorg. Allg. Chem.*, 1995, **621**, 2001.
95ZK801	F. Knoch, D. Boehm, S. Kummer and U. Zenneck; *Z.F. Kristallographie*, 1995, **210**, 801.
95ZN(B)1785	I. Neda, M. Farkens, A.K. Fischer, H. Thönnesen, P.G. Jones and R. Schmutzler; *Z. Naturforsch.*, 1995, **50B**, 1785.
95ZN(B)1833	I. Neda, M. Farkens, H. Thönnesen, P.G. Jones and R. Schmutzler; *Z. Naturforsch.*, 1995, **50B**, 1833.
95ZOB793	N.K. Skvortsow, S.V. Toldov, A.I. Stash and V.K. Bel'skii; *Z. Obshch. Khim.*, 1995, 793.
96CB(129)489	B. Breit and M. Regitz; *Chem. Ber.*, 1996, **129**, 489.
96CC537	R.W. Alder, D.D. Ellis, J.K. Hogg, A. Martin, A. Guy Orpen and P.N. Taylor; *J. Chem. Soc. Chem. Commun.*, 1996, 537.
96CC539	R.W. Alder, D.D. Ellis, A. Guy Orpen and P.N. Taylor; *J. Chem. Soc. Chem. Commun.*, 1996, 539.
96HAC397	N. Avarvari, P. Le Floch, C. Charrier and F. Mathey; *Heteroatom Chem.*, 1996, **7**, 397.
96HCC233	T. Sakai, K. Jda, T. Uchiyama, T. Fujimoto, K. Ohta and I. Yamamoto; *Heterocycl. Commun.*, 1996, 233.
96JA1038	K. Bieger, G. Bouhadir, R. Réau, F. Dahan and G. Bertrand; *J. Am. Chem. Soc.*, 1996, **118**, 1038.
96JA1060	G. Alcaraz, V. Piquet, A. Baceiredo, F. Dahan, W.W. Schoeller and G. Bertrand; *J. Am. Chem. Soc.*, 1996, 1060.
96JA7630	P.L. Arnold, F.G.N. Cloke, P.B. Hitchcock and J.F. Nixon; *J. Am. Chem. Soc.*, 1996, **118,** 7630.
96JA11978	N. Avarvari, P. Le Floch and F. Mathey; *J. Am. Chem. Soc.*, 1996, **118**, 11978.
96MC191	D.B. Nilov, A.V. Kadushkin, N. Solov'era, A.L. Sedov and V.G. Granik; *Mendeleev Chem. J.*, 1996, 191.
96JMS368	S.R. Bachrach and P. Magdalinos; *J. Molec. Struct. (Theochem.)*, 1996, **368**, 1.
96RJCC35	A.A. Karasik, S.V. Bobrov, G.N. Nikonov, I.A. Litvinov and A.S. Dokuchaev; *Russ. J. Coord. Chem.*, 1996, **22**, 35.
96RJCC514	A.A. Karasik, S.V. Bobrov, G.N. Nikonow and K.M. Enikeev; *Russ. J. Coord. Chem.*, 1996, **22**, 514.
96S87	M. Julino, U. Bergsträßer and M. Regitz; *Synthesis*, 1996, 87.
96S265	F. Knoch, S. Kummer and U. Zenneck; *Synthesis*, 1996, 265.
96S473	I. Neda, C. Melnicky, A. Vollbrecht and R. Schmutzler; *Synthesis*, 1996, 473.
96TH625-01	U. Schairer; *Ph.D. Thesis*, University of Regensburg, 1996.
96TL2841	J.M. Álvarez-Gutiérrez and F. López-Ortiz; *Tetrahedron Lett.*, 1996, 2841.
96TL3829	S. Schardt and K. Hafner; *Tetrahedron Lett.*, 1996, **37**, 3829.
96ZN(B)773	G. Jorchem and A. Schmidpeter; *Z. Naturforsch.*, 1996, **51B**, 773.
96ZN(B)1305	H. Gröger, J. Wilken, J. Martens, I. Neda, V. Pinchuk, H. Thönnessen, P.G. Jones, R. Schmutzler; *Z. Naturforsch.*, 1996, **51B**, 1305.
96ZN(B)1494	V.A. Pinchuk, I. Neda, C. Müller, H. Thönnesen, P.G. Jones and R. Schmutzler; *Z. Naturforsch.*, 1996, **51B**, 1494.

96ZN(B)1486	I. Neda, Th. Kaukorat, P.G. Jones, R. Schmutzler; *Z. Naturforsch.*, 1996, **56B**, 1486.
96ZN(B)1501	TH. Kaukorat, I. Neda, H. Thönnessen, P.G. Jones and R. Schmutzler; *Z. Naturforsch.*, 1996, **56B**, 1501.
96ZN(B)1627	F. Borkenhagen, I. Neda, H. Thönnessen, P.G. Jones and R. Schmutzler; *Z. Naturforsch.*, 1996, **51B**, 1627.
96ZN(B)1761	G. Jochen, M. Schmidt, H. Nöth and A. Schmidpeter; *Z. Naturforsch.*, 1996, **56B**, 1761.
97AG(109)1396	A. Maak, B. Breit, Th. Wettling, U. Bergsträßer, St. Leininger and M. Regitz; *Angew. Chem.*, 1997, **109**, 1396.
97BSF853	N. Maigrot, C. Charrier, L. Ricard and F. Mathey; *Bull. Soc. Chem. Fr.*, 1997, **134**, 853.
97CB1491	P. Binger, St. Leininger, K. Günther and U. Bergsträßer; *Chem. Ber./Recueil*, 1997, 1491.
97CB1547	G. Bekiaris, E. Lork and W. Offermann; *Chem. Ber./Recueil*, 1997, **130**, 1547.
97HAC97	E. Malvarenz, G. Martin, L. Rodoiguez and W. Munoz; *Heteroatom Chem.*, 1997, **8**, 97.
97IC1867	M.L.J. Hackney, D.M. Schubert, P.F. Brandt, R.C. Haltiwanger and A.D. Normann; *Inorg. Chem.*, 1997, **36**, 1867.
97JOM215	P. Binger, St. Leininger, M. Regitz, U. Bergsträßer, J. Bruckmann and C. Krüger; *J. Organomet. Chem.*, 1997, **529**, 215.
97JOM223	E. Fluck, G. Heckmann, E. Gorbunowa, M. Westerhausen and F. Weller; *J. Organomet. Chem.*, 1997, **529**, 223.
97JOM233	J. Barluenga, M. Thomás, K. Bieger, S. Garcia-Granda, R. Santiago-Garcia; *J. Organomet. Chem.*, 1997, **529**, 233.
97OM4089	N. Avarvari, P. Le Floch, L. Richard and F. Mathey; *Organometallics*, 1997, **16(19)**, 4089.
97RCB1154	V.A. Zagumennov, A.A. Karasik, E.V. Nikitin and G.N. Nikonow; *Russ. Chem. Bull. (Transl. Izv. Akad. Nauk, Ser. Khim.)*, 1997, **46**, 1154.
97ZAAC1769	E. Matern, G. Fritz and J. Pikies; *Z. Anorg. Allg. Chem.*, 1997, **623**, 1769.
97ZN(B)162	H.-P. Schrödel and A. Schmidpeter; *Z. Naturforsch.*, 1997, **52B**, 162.
98AG(110)1318	F. Tabellion, A. Nachbauer, St. Leininger, Ch. Peters, M. Regitz and F. Preuss; *Angew. Chem.*, 1998, **110**, 1318.
98EJIC1619	R. Gleiter, H. Lange, P. Binger, J. Stannek, C. Krüger, J. Bruckmann, U. Zenneck and S. Kummer; *Eur. J. Inorg. Chem.*, 1998, 1619.
98EJIC1907	F. Breitsameter, K. Polborn and A. Schmidpeter; *Eur. J. Inorg. Chem.*, 1998, 1907.
98EJIC2071	P. Binger, St. Stutzmann, J. Bruckmann, C. Krüger, J. Grobe, D. Le Van and T. Pohlmeyer; *Eur. J. Inorg. Chem.*, 1998, 2071.
98EJOC1425	J. Barluenga, M. Tomás, K. Bieger, S. Gareia-Granda and R. Santiago-Garcia; *Eur. J. Org. Chem.*, 1998, 1425.
98EJOC2039	N. Avarvari, L. Ricard, F. Mathey, P. Le Floch, O. Löber and M. Regitz; *Eur. J. Org. Chem.*, 1998, 2039.
98H1903	H. Yamamoto, T. Kobayashi and M. Nitta; *Heterocycles*, 1998, 1903.
98JCS(P1)1657	R.W. Alder, D. Ellis, D. Dianne, R. Gleiter, Ch.J. Harris, H. Lange, A. Guy Orpen, D. Read and P.N. Taylor; *J. Chem. Soc. Perkin Trans. 1*, 1998, 1657.
98JMC4550	M.K. Manthey, D.T.C. Huang, W.A. Bubb and R.I. Christopherson; *J. Med. Chem.*, 1998, **41(23)** 4550.
98JOM39	A. Haßler, S. Kucken, O. Stelzer and W.S. Sheldrick; *J. Organomet. Chem.*, 1998, 39.
98KFZ16	N.B. Nilov, N.P. Solov'eva, I.S. Nikolaeva, V.V. Peters, L.Y. Krylova, T.A. Gus'kova and Y.G. Granik; *Khim.-Farm. Zh.*, 1998, **32**, 16.
98PCT Int. Appl. WO9842717A119881001	J.C.L.J. Suykerbuyk and P.G. Pringle; Shell Internationale Research Maatschappij B.V., 25 March 1998. Title: Diphosphines.
98RJE313	V.A. Zagumennov, A.N. Belousov, V.V. Yanilkin, A.A. Kavasik and E.V. Nikitin; *Russ. J. Electrochem. (Eng. Transl.)*, 1998, **34**, 313.
98S427	A. Nachbauer, U. Bergsträßer, S. Leininger and M. Regitz; *Synthesis*, 1998, 427.
98ZAAC1116	G. Heckmann, S. Plank, H. Borrmann and E. Fluck; *Z. Anorg. Allg. Chem.*, 1998, **624**, 1116.
98ZN(B)443	G. Heckmann, S. Plank, E. Fluck, A. Dashti-Mommertz and B. Neumüller; *Z. Naturforsch.*, 1998, **53B**, 443.
99AG(111)1724	J. Liedtke, S. Loss, G. Alcaraz, V. Gramlich and H. Grützmacher; *Angew. Chem.*, 1999, **111**, 1724.
99AG(111)1757	A. Schnepf, G. Stößer, D. Carmichael, F. Mathey and H. Schnöckl; *Angew. Chem.*, 1999, **111**, 1757.
99AG(111)3384	P. Kramkowski and M. Scheer; *Angew. Chem.*, 1999, **111**, 3384.

99CC901 V. Gee, A.G. Orpen, H. Phetmung, P.G. Pringle and R.I. Pugh; *J. Chem. Soc. Chem. Commun.*, 1999, 901.

99CC1377 S.B. Clendenning, P.B. Hitchcock and J.F. Nixon; *J. Chem. Soc. Chem. Commun.*, 1999, 1377.

99CC2451 S.B. Clendenning, B. Gehrhus, P.B. Hitchcock and J.F. Nixon; *J. Chem. Soc. Chem. Commun.*, 1999, 2451.

99CEJ(5)3143 A. Elvers, F.W. Heinemann, B. Wrackmeyer and U. Zenneck; *Chem. Eur. J.*, 1999, **5**, 3143.

99EJIC763 P. Binger, St. Stutzmann, J. Stannek, K. Günther, P. Phillips, R. Mynott, J. Bruckmann and C. Krüger; *Eur. J. Inorg. Chem.*, 1999, 763.

99EJOC83 P. Binger, St. Stutzmann, J. Stannek, B. Gabor and R. Mynolt; *Eur. J. Org. Chem.*, 1999, 83.

99EJOC363 W. Fiedler, O. Löber, U. Bergsträßler and M. Regitz; *Eur. J. Org. Chem.*, 1999, 363.

99EJOC587 A. Mack, U. Bergsträßler, G.J. Reiß and M. Regitz; *Eur. J. Org. Chem.*, 1999, 587.

99EJOC1041 M.A. Hofmann, A. Nachbauer, U. Bergsträßler and M. Regitz; *Eur. J. Org. Chem.*, 1999, 1041.

99HAC391 A.A. Tolmachev, S.I. Dorgopoly, A.N. Kostyuk, E.S. Kozlov, A.O. Pushechnikow and W. Holzer; *Heteroatom Chem.*, 1999, **10**, 391.

99EJOC2039 X. Sava, N. Mézailles, N. Maigrot, F. Nief, L. Ricard, F. Mathey and P. Le Floch; *Eur. J. Org. Chem.*, 1999, 2039.

99JA4215 L. Colombet, F. Volatron, P. Maitre and P.C. Hiberty; *J. Am. Chem. Soc.*, 1999, **121**, 4215.

99JCS(P1)1577 I. Uchiyama, T. Fujimoto, A. Kakehi and I. Yamamoto; *J. Chem. Soc., Perkin Trans. 1*, 1999, 1577.

99JGU833 V.G. Rozinov, V.E. Kolbina, L.G. Rozinova and V.A. Lopyrev; *J. Gen. Chem. USSR (Engl. Transl.)*, 1999, 833.

99JGU886 V.A. Zagumennov, A.A. Karasik, E.V. Nikitin and G.N. Nikonov; *Russ. J. Gen. Chem.*, 1999, **9**, 886.

99JOC5524 G. Frison, A. Sevin, N. Avarvari, F. Mathey and P. Le Floch; *J. Org. Chem.*, 1999, **64**, 5524.

99JOM53 N.H. Tran Huy, L. Ricard and F. Mathey; *J. Organomet. Chem.*, 1999, **582**, 53.

99MC1 G.V. Oshovsky, A.M. Pinchuk, A.N. Chernega, I.I. Pervac and A.A. Tolmacher, *Mendeleev Chem. J.*, 1999, 1.

99MC161 G.V. Oshovsky, A.M. Pinchuk and A.A. Tolmacher; *Mendeleev Chem. J.*, 1999, 161.

99OM3138 A. Maraval, B. Donnadieu, A. Igau and J.-P. Majoral; *Organometallics*, 1999, **18**, 3138.

99RJGC736 Yu.G. Trishin, V.I. Namestnikow and V.K. Bel'skii; *Russ. J. Gen. Chem. (Engl. Transl.)*, 1999, **69(5)**, 736.

99RJGC886 V.A. Zagumemov, A.A. Karasik, E.V. Nikitin and G.N. Nikonov; *Russ. J. Gen. Chem. (Engl. Transl.)*, 1999, **69**, 886.

99S1363 P. Binger, K. Günther and M. Regitz; *Synthesis*, 1999, 1363.

99S1642 S.M.F. Asmus, U. Bergsträßer and M. Regitz; *Synthesis*, 1999, **9**, 1642.

99S644 O. Löber, U. Bergsträßer and M. Regitz; *Synthesis*, 1999, 644

99ZAAC1712 S. Plank, G. Heckmann, W. Schwarz, B. Neumüller and E. Fluck; *Z. Anorg. Allg. Chem.*, 1999, **625**, 1712.

99ZAAC1732 Z. Fei, H. Thönnessen, P.G. Jones and R. Schmutzler; *Z. Anorg. Allg. Chem.*, 1999, **625**, 1732.

99ZN(B)1478 G. Heckmann, S. Plank, E. Fluck, A. Dashti-Mommertz and B. Neumüller; *Z. Naturforsch.*, 1999, **54B**, 1478.

00OM219 F.G.N. Cloke, P.B. Hitchcock, J.F. Nixon and D.J. Wilson; *Organometallics*, 2000, **19**, 219.

6.1
Macro- and Spiro-heterocycles

MICHAEL PABEL and S. BRUCE WILD
Australian National University, Canberra, ACT, Australia

6.1.1 INTRODUCTION

The eclectic nature of molecules containing phosphorus in 7- and larger-membered rings has necessitated the use of broad subject headings for this chapter. The inclusion of numerous structural diagrams in the text, however, will assist the reader in locating material of interest. A comprehensive review by Caminade and Majoral of phosphorus-containing macrocycles and cryptands has been published ⟨94CRV1183⟩ and there are two important earlier reviews by Gallagher ⟨87MI934-01⟩ and Quin ⟨B-88MI934-01⟩ concerning the stereochemical aspects of heterocyclic phosphorus compounds.

6.1.2 PARENT RINGS WITH ONE PHOSPHORUS ATOM

6.1.2.1 Seven-Membered Rings

6.1.2.1.1 Monocyclic phosphepanes

The slow addition of a solution of potassium diphenylphosphide in dioxane-tetrahydrofuran to a boiling solution of 1,6-dibromohexane gives (6-bromohexyl)diphenylphosphine, which immediately undergoes intramolecular quaternization to 1,1-diphenylphosphepanium bromide (**1**; R^1 = H, R^2 = Ph, X = Br), isolated as the perchlorate (**1**; R^1 = H, R^2 = Ph, X = ClO_4) in 36% yield ⟨63AG(E)479⟩. The cyclic bromide can also be isolated in 54% yield from the reaction between 1,6-dibromohexane and tetraphenyldiphosphine in boiling 1,2-dichlorobenzene ⟨63AG(E)620⟩. The 4-methyl-substituted salt (**1**; R^1 = Me, R^2 = Ph, X = Br), however, was obtained in 85% yield by the diphosphine method ⟨72JHC331⟩. Heating under reflux of a solution of (**1**; R^1 = H, R^2 = Ph, X = Br) in 2 N NaOH led to the isolation of 1-phenylphosphepane 1-oxide (**2**; R = H, X = O), which was converted into the free phosphine (**3**) by reduction with triphenylsilane ⟨76JA2109⟩. Quaternization of (**3**) with benzyl bromide afforded the phosphepanium salt (**1**; R^1 = H, R^2 = Bn, X = Br). ^{13}C NMR data for (**1**)–(**3**) and related compounds having ring sizes of three to eight members have been reported ⟨76JA2109⟩. The UV irradiation of (5-hexenyl)phenylphosphine in 5% petroleum ether solution affords 1-phenylphosphepane (**3**) in 41% yield, which was purified as the phosphine sulfide (**2**; R = H, X = S) ⟨66JCS(C)245⟩. Phenylphosphine, in the complex (**4**), reacts with 1,6-dibromohexane in the presence of KOBu^t to give the 1-phenylphosphepane complex (**5**) in 63% yield ⟨95OM1434⟩. Pyrophoric bis(trimethylsiloxy)phosphine, the isolation of which was avoided by *in situ* generation from ammonium hypophosphite and hexamethylsilazane, reacts with 1,6-dibromohexane in boiling mesitylene in a double Arbusov reaction to give phosphepanic acid (**6**) in 58% yield after hydrolysis ⟨95JOC6076⟩. A two-step synthesis of 1-methyl β,β-diketophosphepane 1-oxide is available ⟨77JA8370⟩. Thus, the 1,2-dimethylenecyclobutane (**7**) is employed as the diene in the McCormack cycloaddition with methyldichlorophosphine to give the bicyclic phospholene derivative (**8**). Ring-opening ozonolysis of (**8**) at −78°C in methanol, followed by reduction of the ozonide with trimethylphosphite, gives (**9**).

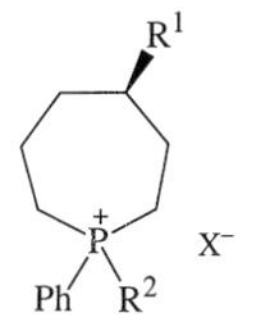

(1)

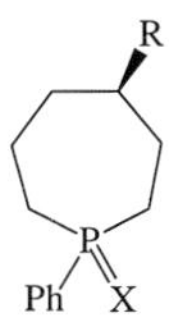

(2)

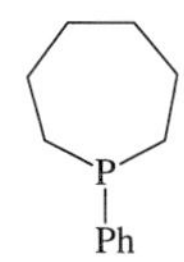

(3)

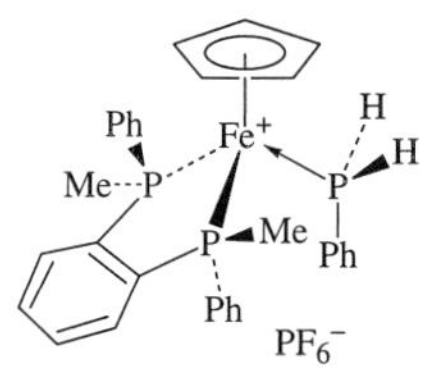

(4)

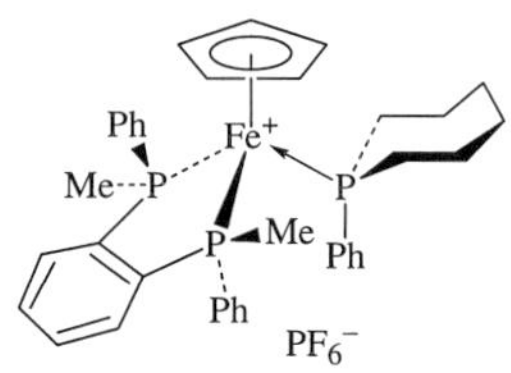

(5)

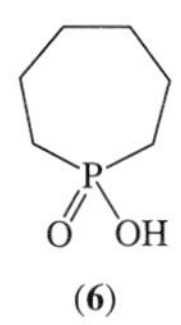

(6)

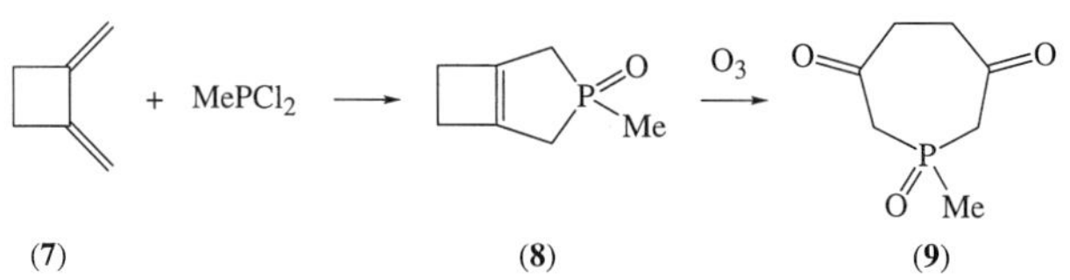

(7) **(8)** **(9)**

6.1.2.1.2 Monocyclic phosphepins

The parent 1*H*-phosphepin (**10**) is not known, although there is evidence for the phosphatropylium ion (**11**) from collision activation experiments between phosphorus and benzene in the gas phase ⟨92JA7582⟩. The chemical valence-isomerization of (**12**; R = Me) in boiling diphenyl ether affords the 2,7-dihydrophosphepin oxide (**13**); oxidative decarboxylation of (**12**; R = H) in aqueous pyridine at a platinum electrode furnishes in 64% yield (**14**), which undergoes a similar rearrangement into the 2,7-dihydrophosphepin oxide (**15**) upon heating in the high-boiling ether. Bromination of (**15**) gives (**16**), which, when dehydrobrominated with triethylamine at room temperature, affords 1-phenylphosphepin 1-oxide (**17**) ⟨70TL1273⟩. The azadihydrophosphepin oxide (**18**; R = H) was prepared by ring-enlargement of 2-lithio-oxo-phospholenene with benzonitrile. Chlorination of the phosphine oxide with PCl_5 gives (**18**; R = Cl), whereas UV irradiation in methanol causes rearrangement into (**19**) ⟨72TL1949, 72T5367⟩.

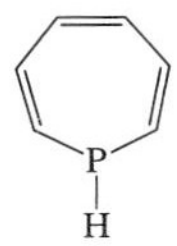

(**10**)

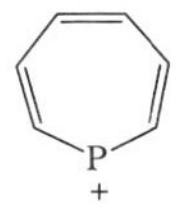

(**11**)

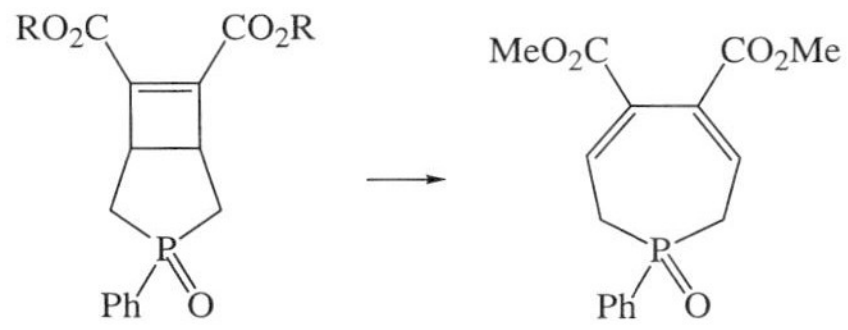

(**12**) (**13**)

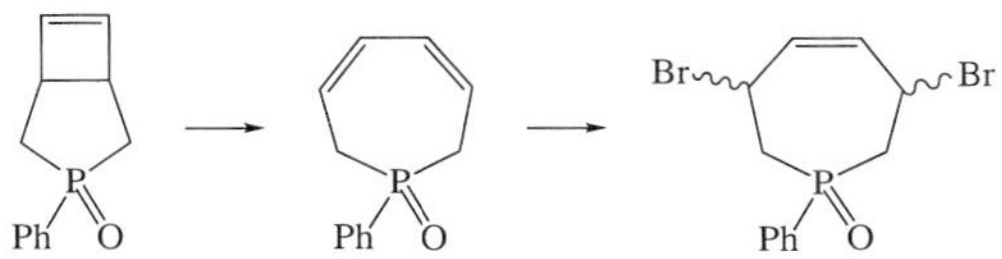

(**14**) (**15**) (**16**)

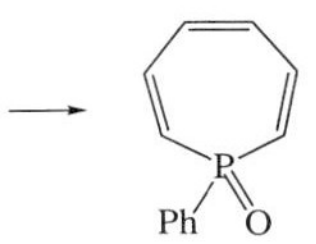

(**17**)

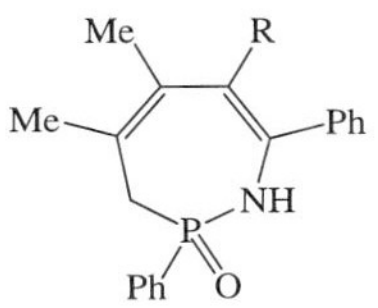

(**18**)

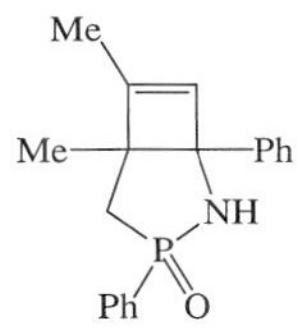

(**19**)

Cycloaddition of phenylphosphine to 1,5-hexadiyne in boiling benzene in the presence of AIBN gives 1-phenyl-4,5-dihydrophosphepin (**20**) in 33% yield; (**20**) was characterized by alkylation and oxidation at phosphorus. The 4,5-dihydrophosphepin oxide (**21**) quantitatively isomerizes into the 2,7-dihydro compound (**22**) when boiled in benzene in the presence of triethylamine; the latter, upon reduction with trichlorosilane in boiling benzene, affords 2,7-dihydrophosphepin (**23**) in 32% yield. The radical cycloaddition of phenylphosphine to 2,6-octadiyne and 3,4-dimethyl-1,5-hexadiyne similarly affords the substituted 4,5-dihydrophosphepins (**24**) and (**25**), respectively ⟨73TL1455⟩. The reductive coupling of (**26**) by the McMurray reaction gave (**27**) as a distillable oil that crystallized in 8–10% yield after purification by HPLC ⟨84AG(E)894⟩.

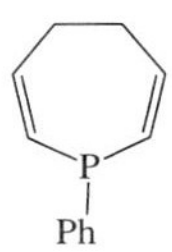

(**20**)

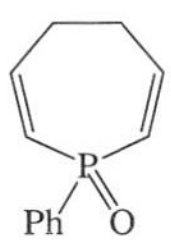

(**21**)

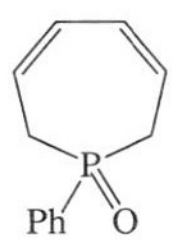

(**22**)

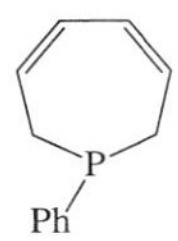

(**23**)

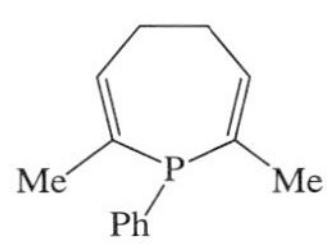

(**24**)

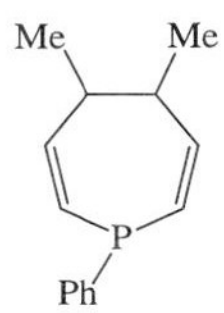

(**25**)

The addition of dichlorocarbene to the 1,2-dihydrophosphinine-1-oxides (**28**; Y = MeO, Ph) under phase transfer conditions, followed by alkaline dehydrochlorination of the intermediate (**29**; Y = MeO, Ph), afforded the phosphepin-1-oxides (**30**; Y = MeO, Ph) in 3–5% yields ⟨87JOC3983⟩. A number of related compounds has been prepared similarly ⟨93S931⟩.

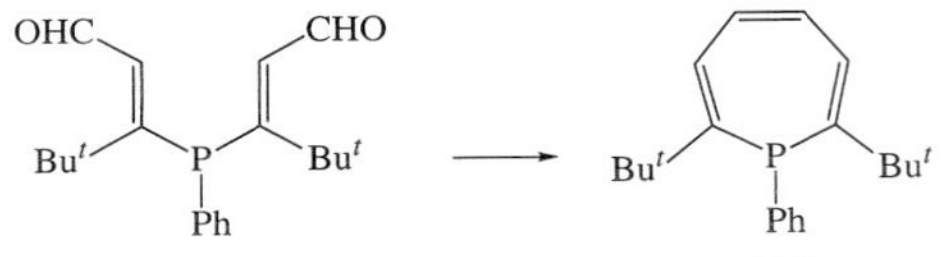

(26) (27)

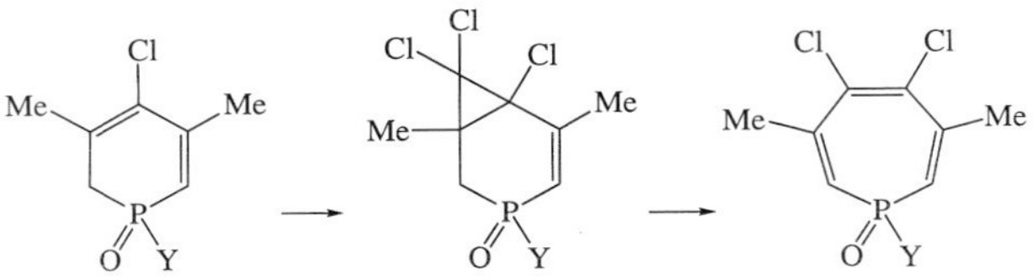

(28) (29) (30)

3,5-Diphenyl-λ^3-phosphinine (**31**) reacts with diphenyldiazomethane to give the 1-phosphanocaradiene derivative (**32**); the latter, when treated with hydrogen chloride, affords (**33**). When co-ordinated to zerovalent metals in the complexes [M(CO)$_5$(**31**)] (M = Cr, Mo, W), diazomethane and monosubstituted diazomethanes react to give (**34**; M = Cr, Mo, W); the latter, depending upon R and the solvent from which they are recrystallized, undergo 1,3-H shifts producing the isomeric hydrazones (**35**; M = W) and (**36**; M = Cr, Mo, W). The reaction of [M(CO)$_5$(**31**)] (M = Cr, Mo) with diphenyldiazomethane produces [M(CO)$_5$(**32**)] (M = Cr, Mo), which are in equilibrium with the complexes (**37**; M = Cr, Mo); (**37**; M = Cr, Mo), react reversibly with hydrogen chloride to give [M(CO)$_5$(**33**)] (M = Cr, Mo) <87AG(E)1134>.

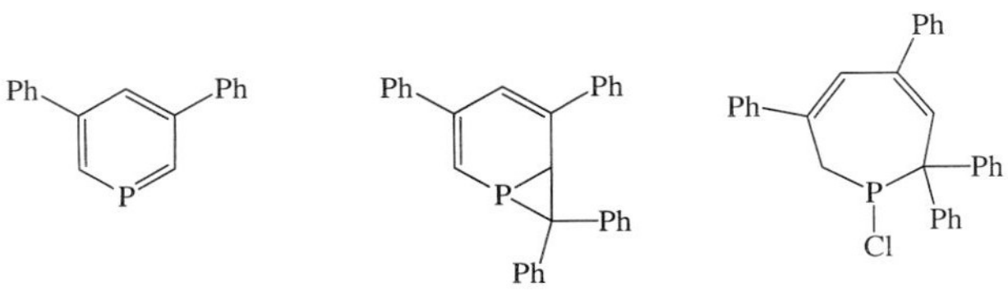

(31) (32) (33)

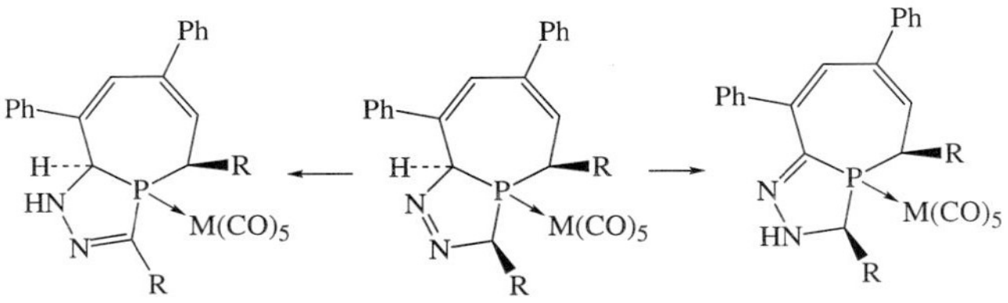

(35) (34) (36)

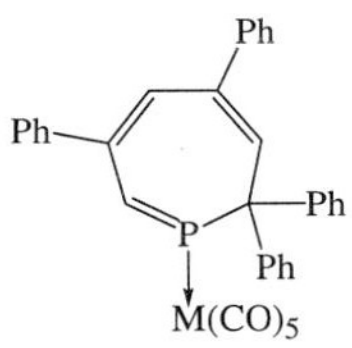

(37)

In the presence of aluminum trichloride, the 1-β-chloroethylphosphole complex (**38**) rearranges into the 6,7-dihydrophosphepin derivative (**39**; X = Cl), which, upon methanolysis, yields the phosphinite complex (**39**; X = OMe) <85TL3461, 92SL363>.

(38) (39)

Substitution of the bromine atoms in (**40**) by disodium phenylphosphide, followed by oxidation of the intermediate phosphine with hydrogen peroxide, gave (**41**) in 10% yield, along with two other unidentified products ⟨94CC67⟩. Dichlorocarbene ring enlargement of the tetrahydrophosphinine oxides (**42**; R = H, Cl) affords the 2,7-dihydrophosphepin oxide (**43**; R = H, Cl) ⟨98JCR(S)210⟩. Compounds (**44**) and (**45**) react with $NaOH–H_2O–CHCl_3$ under phase transfer conditions to give the substituted 2,3,6,7-tetrahydrophosphepin oxides (**46**) and (**47**) by an unexpected path involving isomerization of the starting materials and cyclopropanation via Michael addition of CCl_3^- ⟨98JCS(P1)3323⟩. A review is available concerning the methods of preparing phosphepine derivatives from 5- and 6-membered starting materials ⟨93S931⟩.

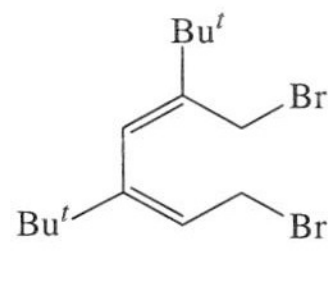

(**40**)

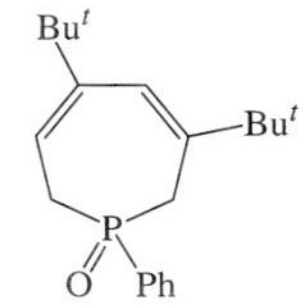

(**41**)

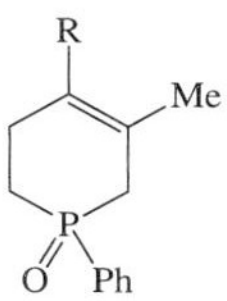

(**42**)

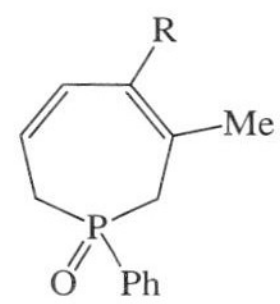

(**43**)

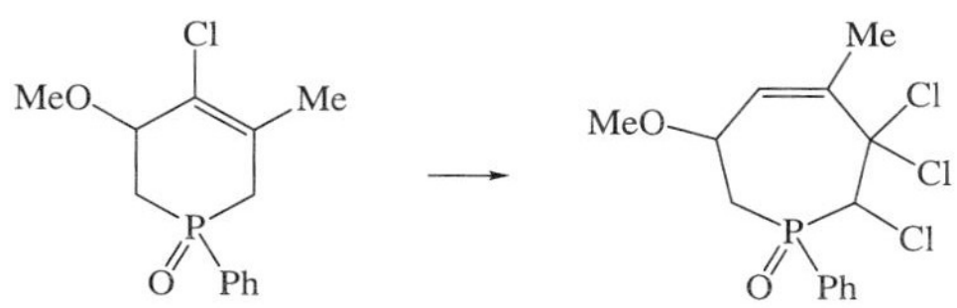

(**44**) (**46**)

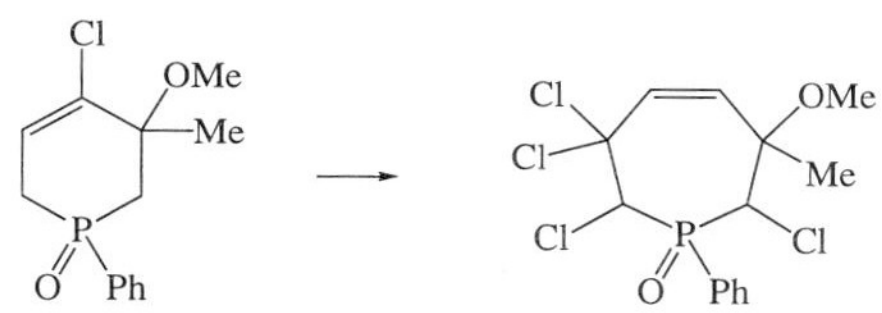

(**45**) (**47**)

6.1.2.1.3 *Fused benzo and dibenzo derivatives*

1-Phenyl-3-benzophosphepin (**48**) was the first phosphepin derivate to be isolated. Thermally labile (**48**) ($t_{1/2}$ ca. 120 min in decalin at 80°C) was synthesized by the base-catalyzed addition of phenylphosphine to 1,2-diethynylbenzene. The stable oxide (**49**) was also prepared ⟨83TL2545⟩.

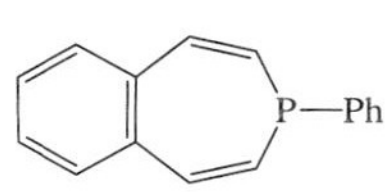

(**48**)

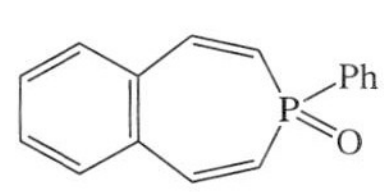

(**49**)

Flash vacuum pyrolysis of (**50**) results in valence isomerization and ring-opening to give the 1-benzophosphepin oxide (**51**) in 85% yield; (**51**) was reduced with trichlorosilane to 1-phenyl-1-benzophosphepin (**52**) in 80% yield ⟨91CC1227⟩. The phosphepin (**52**) (but not the phosphepin oxide (**51**)) is thermally labile in solution, decomposing into naphthalene and, presumably, cyclopoly(phenylphosphine) ($t_{1/2}$ ca. 90 min in toluene at 80°C). Compound (**52**) has also been prepared from (**53**) by lithiation, treatment with dichlorophenylphosphine to give (**54**), and desilylation of (**54**) with tetra-*n*-butylammonium fluoride. The dibenzophosphepins (**55**) and (**56**) were prepared similarly and were found to be stable for 20 h in toluene at 60°C ⟨93CC1817⟩.

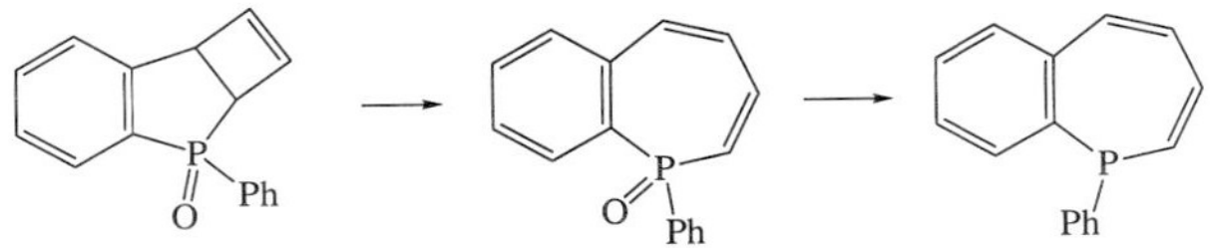

(50) **(51)** **(52)**

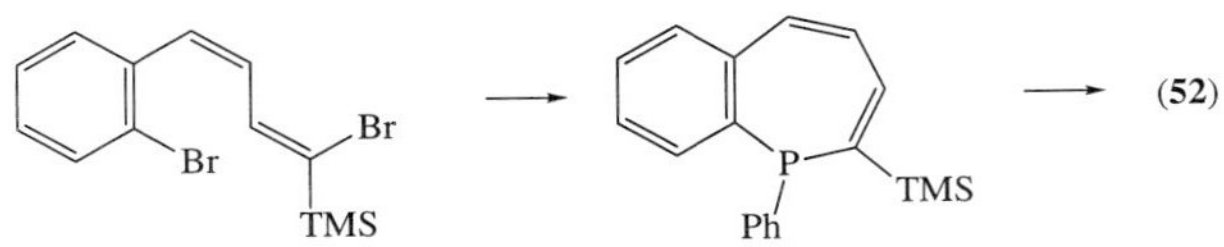

(53) **(54)**

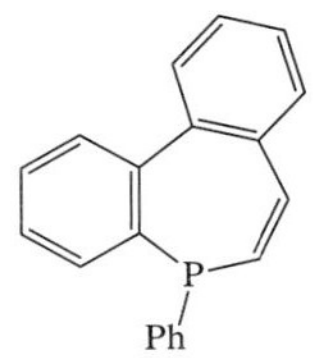

(55)

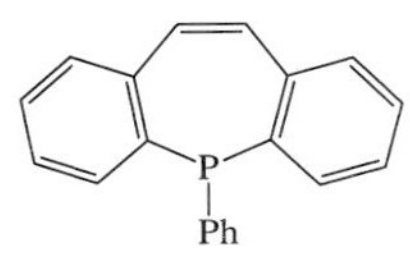

(56)

Photolysis of the phosphine (**57**) with iron pentacarbonyl in THF afforded (**58**) in 40% yield ⟨83CB1923⟩. The structure of (**58**) was determined by X-ray crystallography. Oxidative decomposition of (**58**) with cerium(IV) or oxygen gave the free phosphepin oxide (**59**).

Dimethylacetylene dicarboxylate inserts into a P–C bond of the diphosphapentalene (**60**) to give the diphosphaazulene (**61**), which was characterized by X-ray crystallography ⟨91ZN(B)197⟩.

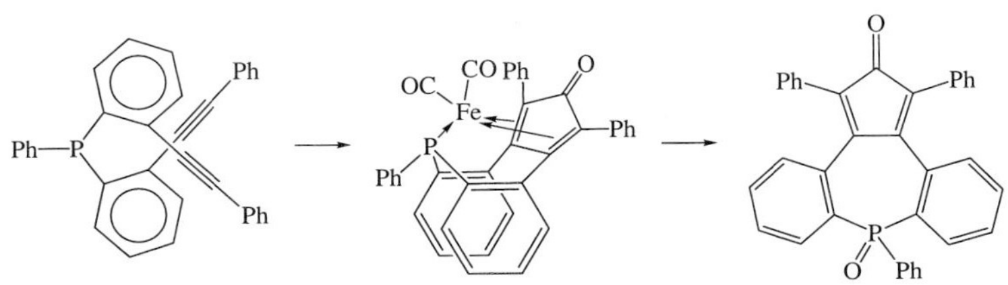

(57) **(58)** **(59)**

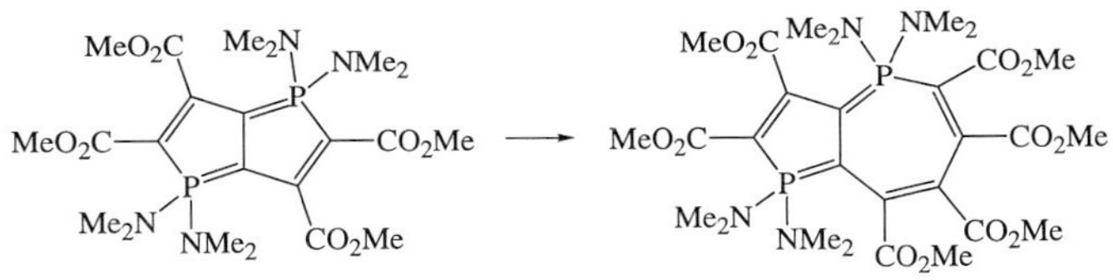

(60) **(61)**

2,2′-Dilithiobibenzyl reacts with phenyldichlorophosphine to give 1-phenyl-4,5-dihydrodibenzo[*b*,*f*]phosphepin (**62**), which exhibits isodimorphism, that is, it exists in low- and high-melting forms; (**62**) was

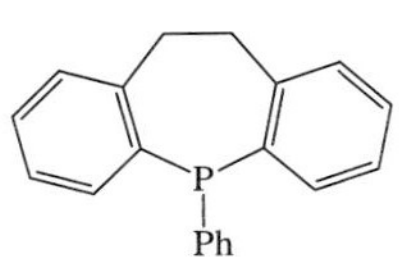

(62)

converted into an oxide, a methiodide, and a palladium(II) derivative ⟨53JCS1130⟩. The atropisomeric dinaphthophosphepins (±)-(**63**) (R = Me, Ph) can be prepared from 2,2′-dimethyl-1,1′-dinaphthyl by lithiation and addition of the respective dichlorophosphine ⟨94TA511, 94JOC6363⟩. The resolutions of (±)-(**63**) (R = Me, Ph) were accomplished by the method of metal complexation with use of an optically active ortho-metallated palladium(II)–amine complex. The absolute configuration of (*S*)-(−)-(**63**) (R = Ph) was determined from the X-ray crystal structure of (**64**). The optically active phosphepin in conjunction with rhodium(I) is an effective auxiliary for the asymmetric hydroformylation of styrene. The secondary phosphine (±)-(**63**; R = H) has been isolated in 75% yield following the two-phase phosphination of 2,2′-bis(chloromethyl)-1,1′-binaphthyl with phosphine in the presence of potassium hydroxide ⟨98IC6408⟩. Functionalization of the secondary phosphine or its borane adduct at phosphorus gave a variety of axially disymmetric chelating tertiary phosphines.

The bis(trimethylsilyl)phosphines (**65**; R = Ph, Bu^t) react with the diacid dichloride (**66**) to give the dibenzophosphepin derivatives (**67**) ⟨87CC1753⟩.

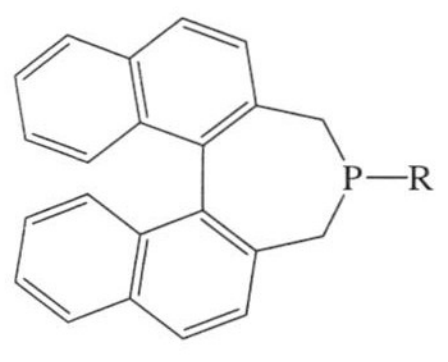

(**63**)

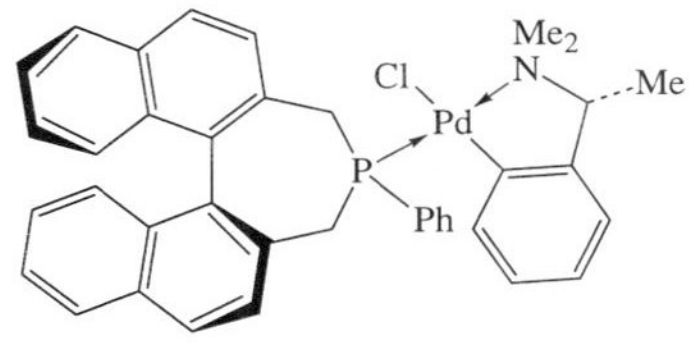

(**64**)

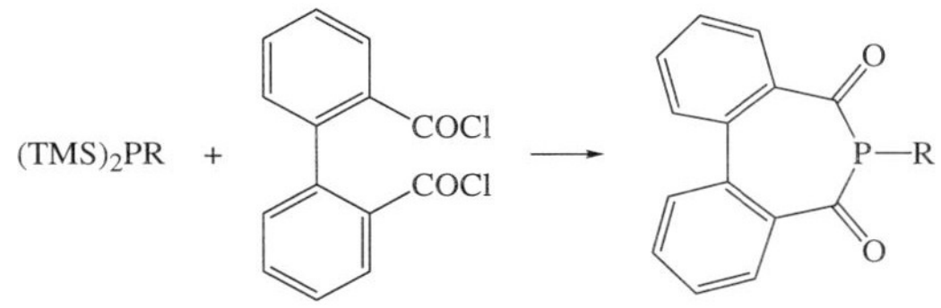

(**65**) (**66**) (**67**)

6.1.2.1.4 Bicyclic systems

Dipotassium cyclooctatetraenide in THF reacts with dichlorophenylphosphine to give (**68**; R = Ph), which upon warming stereoselectively isomerizes into (**69a**) ⟨66JA3832, 73JA4292⟩. Rapid gas phase pyrolysis of (**69a**) at 480°C, or heating at 100°C overnight in chloroform containing hydrogen chloride, causes inversion at phosphorus and formation of (**69b**). Pure (**69b**) was obtained via its palladium(II) chloride derivative. Diastereomers (**69a**) and (**69b**) react with retention of the configuration at phosphorus with methyl iodide and hydrogen peroxide to give the characteristic methiodides and oxides, respectively, but both (**69a**) and (**69b**) give the same oxide with air, i.e. the oxide of (**69b**).

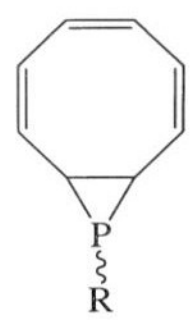

(**68**)

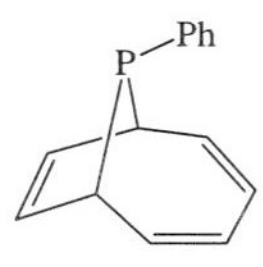

(**69a**)

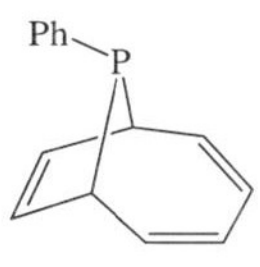

(**69b**)

Conformational barriers in cations of the type $[Rh(COD)L_2]^+$ containing (**69a**) or (**69b**) have been investigated by ^{1}H, ^{13}C, and ^{31}P NMR spectroscopy ⟨90JOM(385)131⟩. A correlation exists between the diastereomer of the complex and the Rh–P rotational barrier and the catalytic activity and selectivity of isomerization and hydrogenation of 1-hexene. In subsequent work, the unusual ^{31}P NMR chemical shifts of 7-phosphanorbornene compounds were rationalized by *ab initio* calculations ⟨93JA11984⟩.

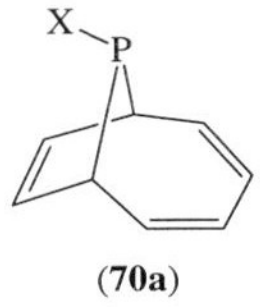

(**70a**)

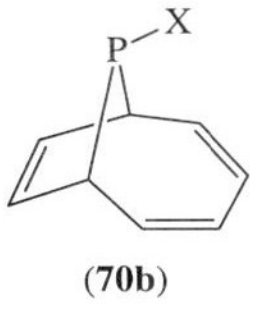

(**70b**)

Diethylaminodichlorophosphine reacts with cyclooctatetraenyldilithium to give diastereomers (**70a**; X = NEt_2) and (**70b**; X = NEt_2) ⟨84JOM(273)1⟩. The *syn* diastereomer (**70a**) is halogenated by phosphorus(III)-bromide or -chloride or KF/[18]-crown-6 to give the corresponding halophosphines as unequal mixtures of diastereomers. The stereochemistry of the substitution of the chloro compound with lithium alkyls, phenolates, thiophenolates, and alcoholates was investigated. The compound (**70b**; X = Bu^t) substitutes carbon monoxide in $Fe_2(CO)_9$ with retention of configuration at phosphorus, whereas (**70a**; X = Bu^t) co-ordinates with inversion ⟨85TL1039⟩.

The mixed isomers (**71a**) (**71b**) react with aqueous formaldehyde and sulfuric acid to give the corresponding bis(hydroxymethyl)phosphonium salts that were individually converted into the corresponding methylphosphine oxides (**72a**) and (**72b**) by heating at 150°C for 13 h in 36% sulfuric acid ⟨93JCS(D)2563⟩. Although 90% of (**71b**) was converted into (**72b**) under these conditions, only 10% of (**71a**) was converted into (**72a**).

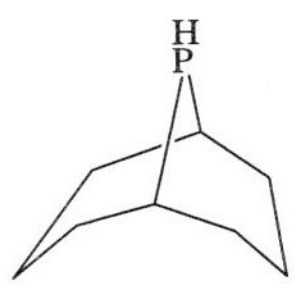

(**71a**)

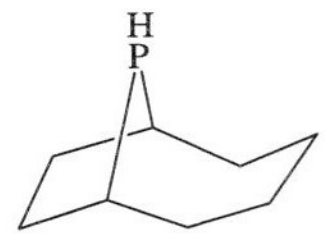

(**71b**)

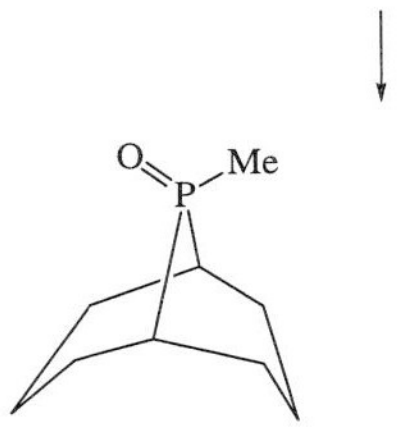

(**72a**)

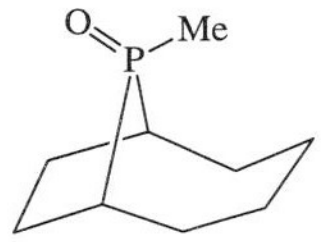

(**72b**)

The crystallographically characterized salt (**73**) was converted into the corresponding ethylidene derivative by deprotonation; the ylide reacted with straight-chain or α-mono-substituted aliphatic aldehydes with formation of *E*-alkenes with 94–96% selectivity ⟨93JOC1985⟩. More general applications of this synthesis will require an inexpensive source of the 9-phosphabicyclo[4.2.1]nonane skeleton ⟨82TL4915, 85CB97⟩.

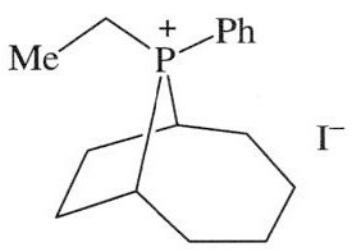

(**73**)

An easy route to 7-membered rings containing phosphorus has been discovered ⟨92OM1438⟩. Terminal phosphinidine complexes react with conjugated dienes to give 3,4,7-trihydrophosphepin complexes via the [3,3]-phospha-Cope rearrangement of intermediate 1,2-divinylphosphirane complexes. The P=C bond in the 7-membered ring thus generated reacts further with the diene to give the corresponding [2 + 4] cycloadduct. Thus, (**74**) reacts with an excess of 2,3-dimethyl-butadiene to give (**75**) in 55% yield.

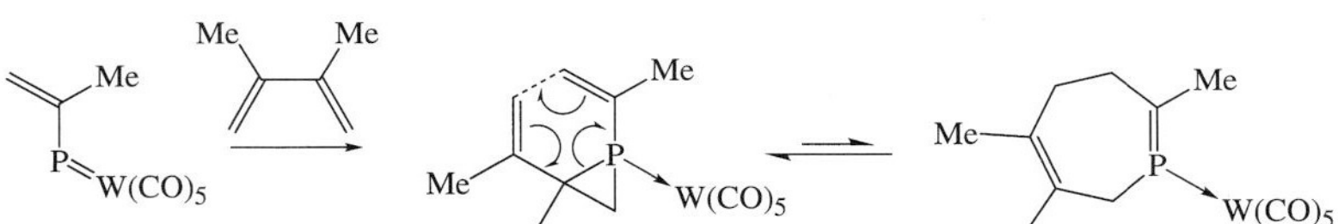

(**74**)

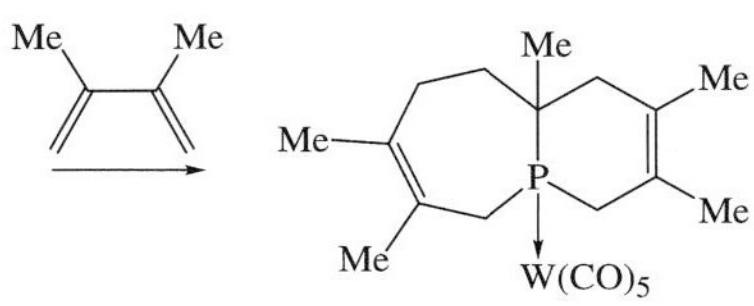

(**75**)

6.1.2.2 Eight-Membered Rings

The first authenticated synthesis of a phosphocane, the phosphocane oxide (**76**), was reported in 1972 ⟨72JHC331⟩ by Märkl's method of cleavage of the appropriate tetraphenylalkyldiphosphonium salt ⟨63AG(E)620⟩. 1,2-Dimethylenecyclopentane as the diene in the McCormack cycloaddition with methyldichlorophosphine gave the bicyclic phospholene (**77**), which, when subjected to ring-opening ozonolysis followed by reduction of the ozonide with trimethylphosphite, furnishes the 3,7-dioxophosphocane 1-oxide (**78**) ⟨77JA8370⟩. Phosphocanic acid (**79**) was prepared in 43% yield from the double Arbusov reaction between bis(trimethylsiloxy)phosphine and 1,6-dibromoheptane in the presence of ammonium hypophosphite followed by hydrolysis of the intermediate trimethylsilylated cyclic intermediates with aqueous sodium chloride ⟨95JOC6076⟩.

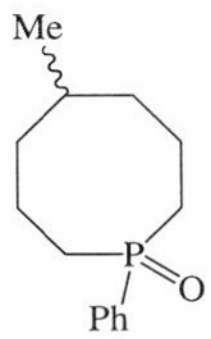

(**76**)

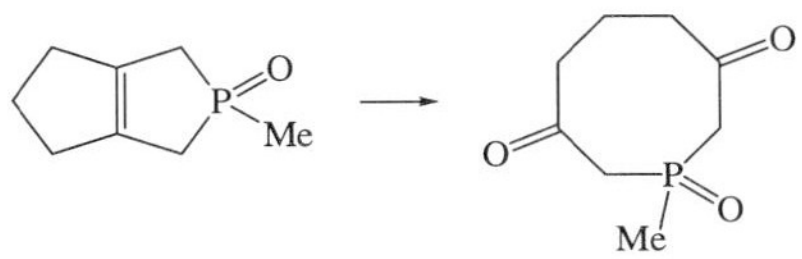

(**77**) (**78**)

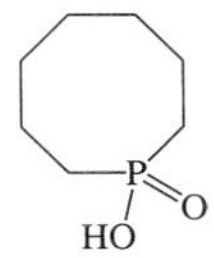

(**79**)

6.1.2.3 Nine-Membered Rings

5-Phenyl-5,5′-spirobi[dibenzophosphole] (**80**; R^1 = Ph, R^2 = H) when briefly heated at 210°C gives the isomeric 17-phenyl-17*H*-tetrabenzo[*b*,*d*,*f*,*h*]phosphonin (**81**; R^1 = Ph, R^2 = H), which readily forms a

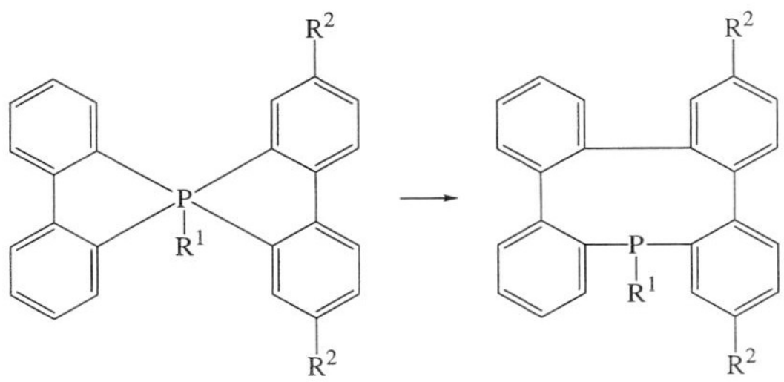

(**80**) (**81**)

methiodide ⟨64CB747⟩. The *p*-dimethylaminophenyl phosphorane (**80**; R^1 = $C_6H_4NMe_2$-*p*, R^2 = H) gives (**81**; R^1 = $C_6H_4NMe_2$-*p*, R^2 = H) and (**80**; R^1 = Ph, R^2 = NMe_2) gives (**81**; R^1 = Ph, R^2 = NMe_2).

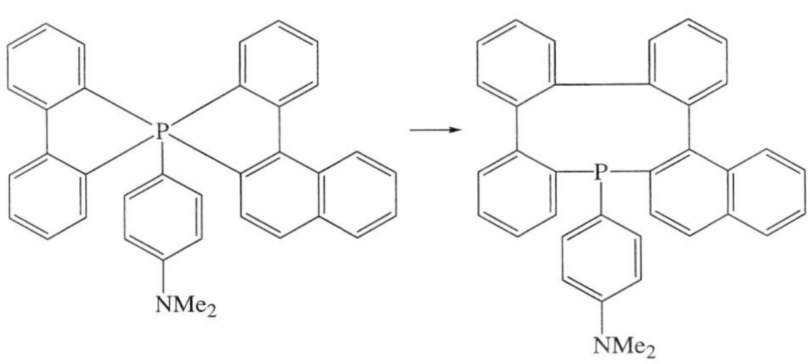

(**82**) (**83**)

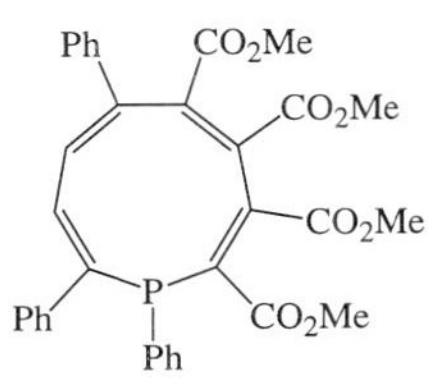

(**84**)

The phosphorane (**82**) undergoes a similar isomerization, to give (**83**) ⟨64CB747⟩. The phosphorane (**80**; R^1 = Me, R^2 = H) can also be converted by heating for 10 min at 260°C into the 17-methyl isomer (**81**; R = Me, R^1 = H) and the biphenylyl-phosphole (**80**; R^1 = 2-C_6H_4•C_6H_5, R^2 = H) gives (**81**; R^1 = 2-C_6H_4•C_6H_5, R^2 = H) under similar conditions ⟨65CB576⟩ Dimethyl acetylenedicarboxylate reacts with 1,2,5-triphenylphosphole to give (**84**) ⟨70JCS(C)386⟩.

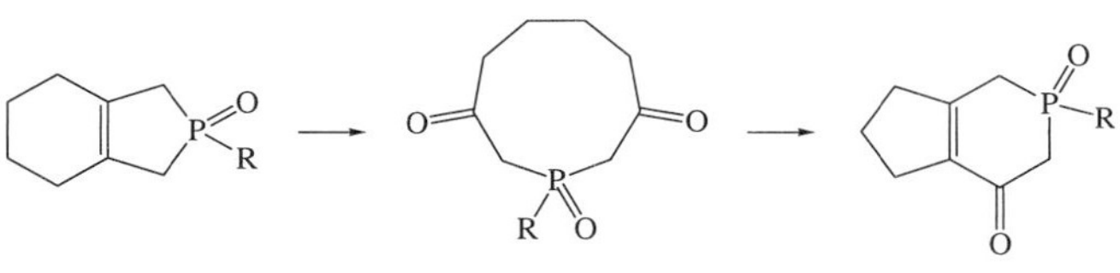

(**85**) (**86**) (**87**)

Oxidative ring-opening of the bicyclic phospholenes (**85**; R = OH, Me, Ph) by ozonolysis affords the phosphonane diones (**86**; R = OH, Me, Ph), which are valuable intermediates for the synthesis of other 9-membered rings ⟨77JA8370, 82JOC905, 82JA1893⟩. Under acidic or basic conditions (**86**; R = Me, OH) undergo intramolecular aldol condensations to give (**87**; R = Me, OH). Similarly, (**88**; R = Me, Ph) react with triethylamine to give (**89**; R = Me, Ph) rather than the desired phosphonen diones. Debromination of (**88**; R = OH) with zinc in acetic acid gave crystalline (**90**) in 89% yield. The crystal structure of (**86**; R = Me) revealed a twist-chair conformation of the 9-membered ring in the solid state, but rapid conformational interconversion was observed in solution for this and other phosphonanes ⟨82JA1893⟩. The bis(tosylhydrazone) (**91**), upon borohydride reduction, regenerated (**85**; R = Me), but the reduction of (**86**; R = Me, Ph) with H_2/Raney nickel gave the phosphonane diols (**92**; R = Me, Ph) as diastereomeric mixtures. The diones (**86**; R = Me, Ph) reacted with 1,2-ethylenedithiol in the presence of BF_3•OEt_2 as catalyst to give the bis(dithioketals) (**93**; R = Me, Ph) and with bis(trimethylsilyl)acetamide under neutral conditions at room temperature to give the enol silyl ethers (**94**; R = Me, Ph). Raney nickel in boiling ethanol desulfurizes (**93**; R = Me) to give the phosphonane oxide (**95**; R = Me), which can be reduced to the free phosphonane (**96**) with diphenylsilane; quaternization of (**96**) with methyl iodide gave the phosphonanium iodide (**97**). Spectroscopic evidence for the single product of dehydration of the diol (**92**; R = Me) with $POCl_3$-pyridine supports the formulation of (**98**).

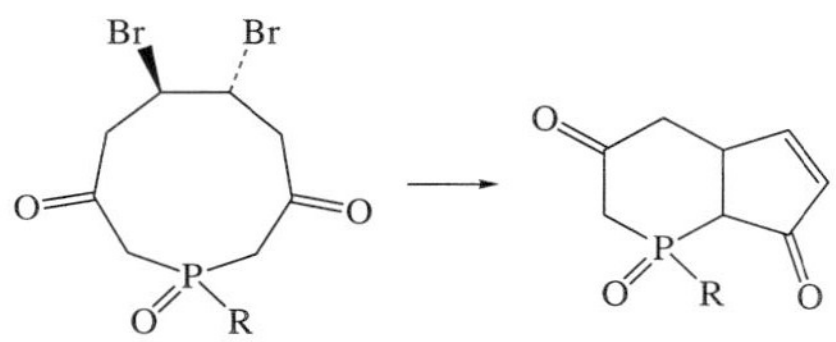

(**88**) (**89**)

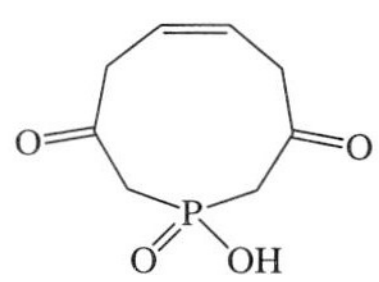

(**90**)

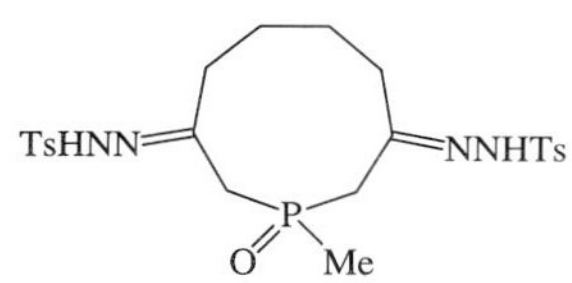

(**91**)

Silylation of (**88**; R = Ph) led quantitatively to (**99**), which, when reacted with triethylamine (1 equiv.) gave (**100**) as a diastereomeric mixture ⟨84JOC3157⟩. Full debromination of (**99**) was achieved by lengthy exposure to triethylamine, which resulted in the formation of the unstable all-*cis* phosphonin oxide (**101**) and bicyclic (**102**) in almost equal amounts.

The dibenzophosphonane oxide (**103**) smoothly dehydrates to (**104**) (23% yield); the latter can be reduced to the free dibenzophosphonin (**105**) with $HSiCl_3$-pyridine. Both (**104**) and (**105**) display the *cis–trans* configuration and severe twisting from planarity prevents attainment of electronic delocalization ⟨82JOC905⟩.

The phosphonin oxides (**106**; R = Bu^t, Ph) are the first recognizable products of the reaction of (**68**; R = Bu^t, Ph) with peroxides ⟨83JA5960, 86PS(27)109⟩. The oxides are stable below −15°C, but undergo intramolecular cycloaddition to give *trans*-3*a*,7*a*-dihydrophosphindoles on warming to room temperature. A similar rearrangement occurs for (**68**; R = *c*-C_6H_{11}, Me) ⟨85CB97⟩.

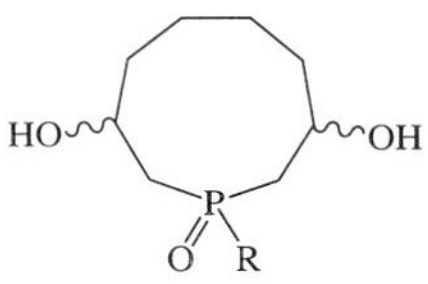

(**92**)

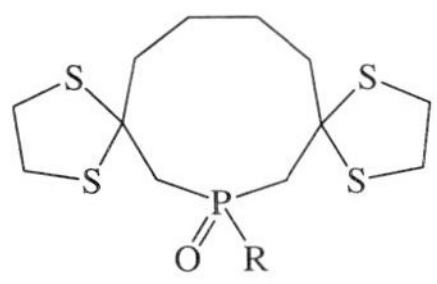

(**93**)

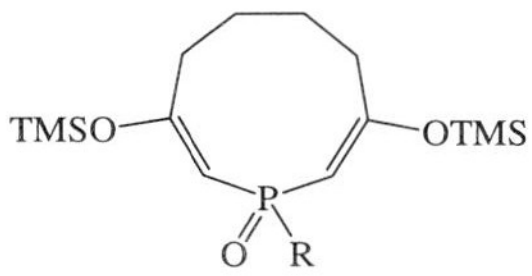

(**94**)

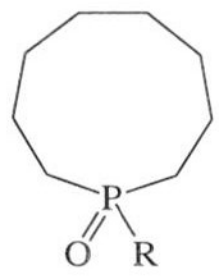

(**95**)

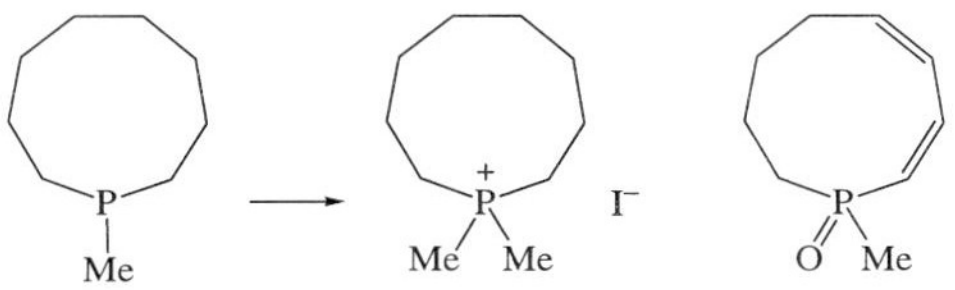

(**96**) (**97**) (**98**)

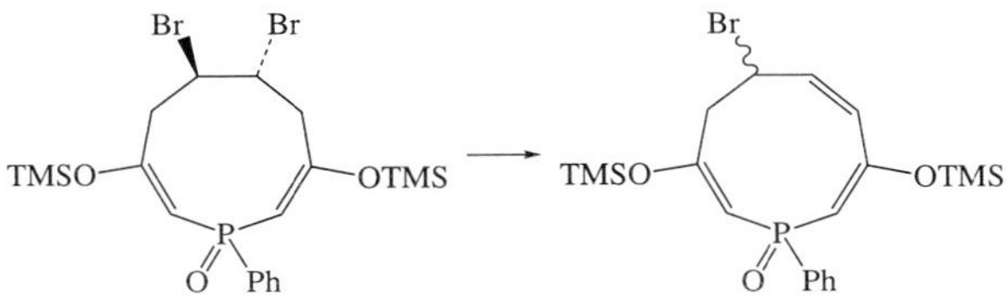

(**99**) (**100**)

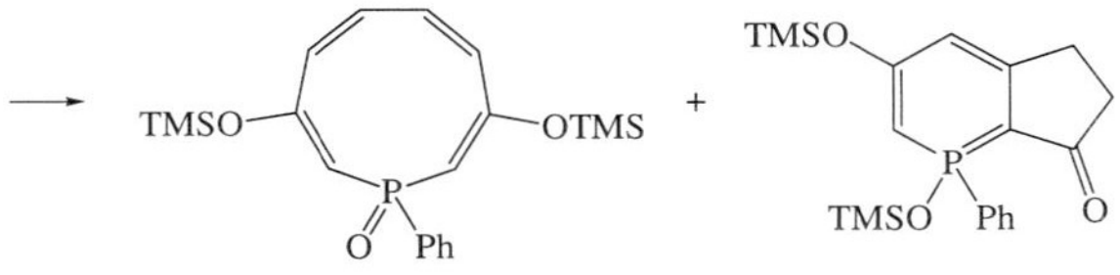

(**101**) (**102**)

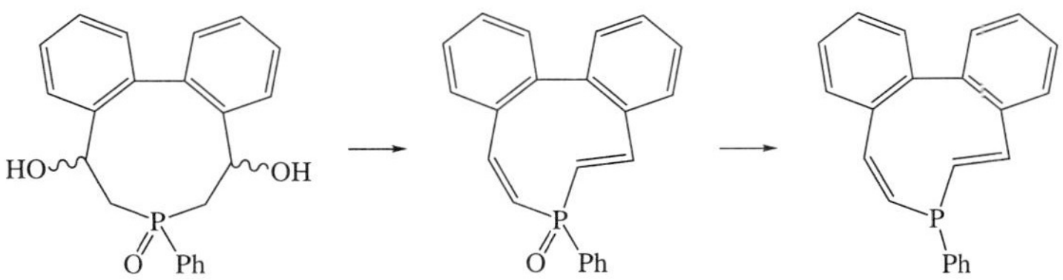

(**103**) (**104**) (**105**)

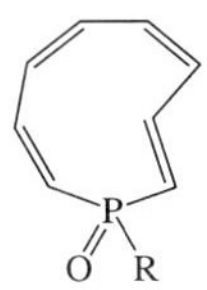

(**106**)

6.1.3 RINGS WITH TWO OR MORE HETEROATOMS

6.1.3.1 Cyclic Quaternizations at Phosphorus

The biphenylyldiphosphines (**107**; R = Me, Et) are important precursors of 7-, 8-, 9- and 10-membered rings by cyclic diquaternization with alkylene dibromides. Thus, (**107**; R = Et), when heated in a sealed tube at 100°C with methylene dibromide, gives the 1,3-diphosphepinium dibromide (**108**; R = Et) ⟨66CI196, 67JCS(C)1869⟩. In a similar manner, the reaction of (**107**; R = Me) with methylene dibromide affords (**108**; R = Me) ⟨82CB1367⟩. The dimethyldiphosphonium salt was converted into (**109**) by deprotonation with methyllithium. Salts (**108**) and (**109**) are dissymmetric. Further deprotonation of (**109**) with KH gave the 9λ^5-phosphaphenanthrene (**110**). 8-Membered (**111**) was obtained when (**107**; R = Et) was heated with 1,2-dibromoethane for 12 h under the conditions described above ⟨66CI196, 67JCS(C)1869⟩; the 9-membered 6,7,8,9-tetrahydro-5*H*-dibenzo[*f*,*h*][1,5]diphosphoninium bromide (**112**) and the 10-membered [1,6]diphosphecinium dibromides (**113**) and (**114**) were obtained under similar conditions by cyclic diquaternizations of (**107**; R = Et) with 1,3-dibromopropane, 1,4-dibromobutane, and 1,2-xylylene dibromide, respectively. The reaction of 1,2-bis(diphenylphosphino)ethene with 1,2-xylylene dibromide in boiling benzene afforded in high yield the 2,5-benzodiphosphocinium dibromide (**115**); with 1,2-bis(diphenylphosphino)ethane the reaction yielded (**116**) ⟨68JOC579⟩. The dibromide (**115**) was converted into the dibromide (**116**) by hydrogenation over a rhodium–alumina catalyst in methanol. The deprotonation of (**115**) in an effort to obtain the diylide was unsuccessful. The diphosphines $R_2PCH_2PR_2$ (R = Me, Ph) react with 1,2-xylylene dibromide to give the bis(phosphonium) salts (**117**; R = Me, Ph) ⟨82OM1266⟩. By stepwise deprotonation of the salts, the mono-, bis- ⟨81CB1428⟩, and tris-ylides ⟨82OM1266⟩ (**118**)–(**120**) were prepared; (**120**) has been used as a ligand for transition metal ions ⟨82OM1266, 83AG(E)907⟩. The 16-membered macrocycle (**121**) was synthesized in four steps from tribenzylphosphine by successive quaternizations with the appropriate alkylating agents and debenzylations with LAH ⟨95ICA(231)225⟩. The diastereomer of (**121**) that was isolated (δ_P −13.9) formed moderately air-stable complexes of the type [MnX_2(**121**)] (X = Cl, Br, I, NCS).

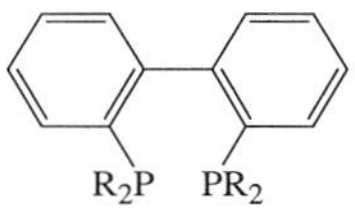

(**107**)

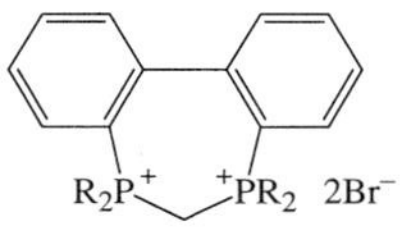

(**108**)

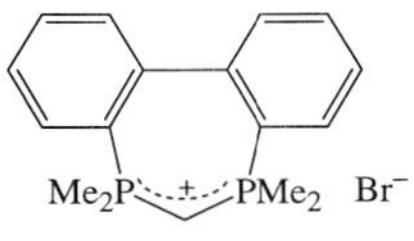

(**109**)

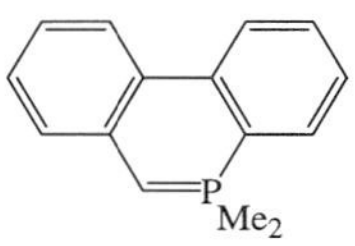

(**110**)

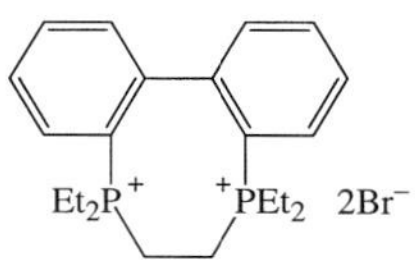

(**111**)

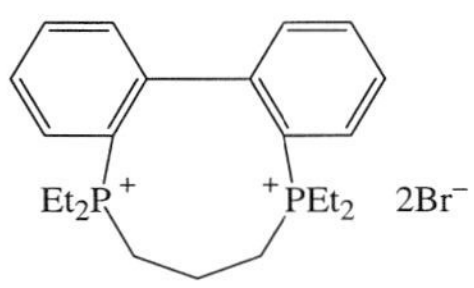

(**112**)

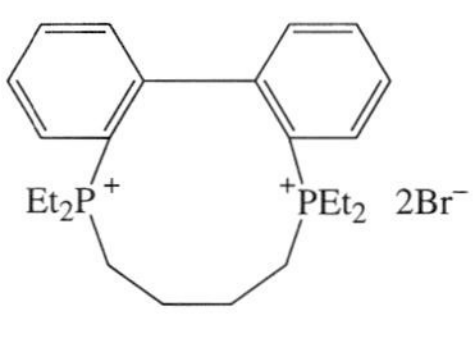

(**113**)

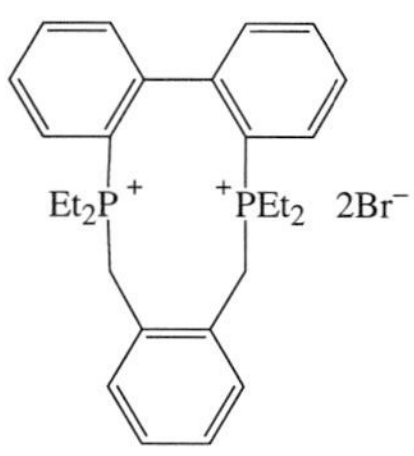

(**114**)

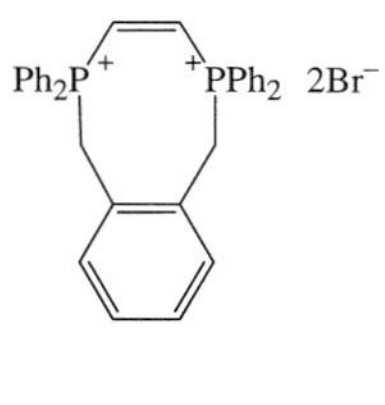

(**115**)

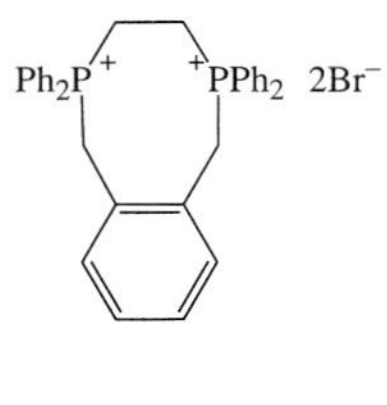

(**116**)

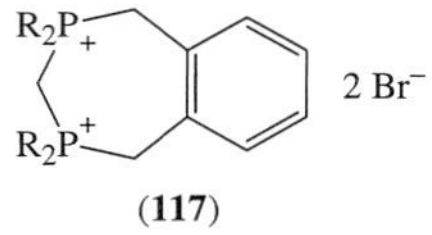

(**117**)

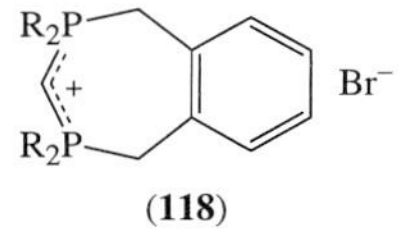

(**118**)

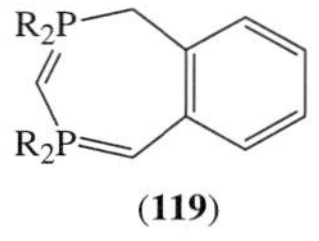

(**119**)

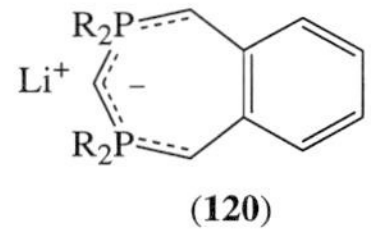

(**120**)

Double deprotonation of (**122**) with strong base gave 3,7-dimethyl-1,1,5,5-tetraphenyl-1λ^5,5λ^5-diphosphocin (**123**) ⟨92OM986⟩. The crystal structure of (**123**)•0.5Et_2O revealed a puckered tub-like conformation of the 8-membered ring with two conjugatively isolated methallylic fragments.

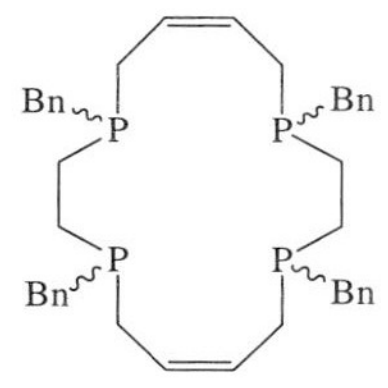

(**121**)

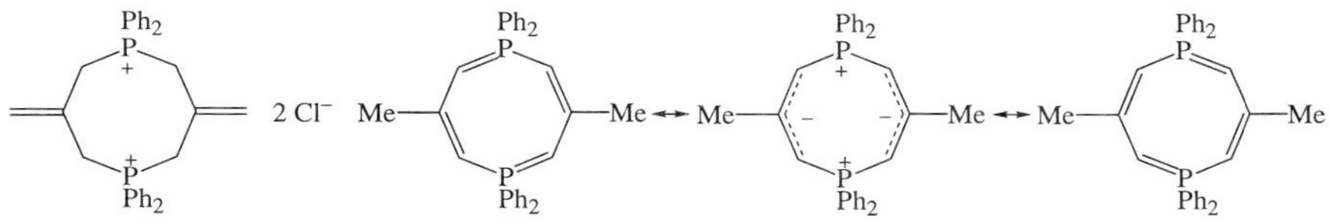

(**122**) (**123**)

The bis(phosphonium) salts (**124**; X = Br, I) ⟨89PS(44)235, 91PS(56)27⟩ were prepared from 1,3-bis(diphenylphosphino)propane and the appropriate 1,3-dihalopropane in boiling DMF. Alkaline hydrolysis of (**124**) afforded the *cis* and *trans* bis(phosphine oxides) (**125a**) (**125b**), which were separated by chromatography and individually reduced to the corresponding bis(tertiary phosphines) (**126a**) (**126b**) with $HSiCl_3$ ⟨89PS(44)235⟩ or LAH ⟨91PS(56)27⟩. The phosphine-thioether (**127**) was also prepared and platinum(II) derivatives isolated ⟨90IC691⟩. A review of the coordination chemistry of medium-sized ring bidentates is available ⟨92CCR(117)133⟩. The crystal structures of (**125a**) and (**125b**), and the thia compound (**128**), revealed boat-chair conformations for the three compounds in the solid state. The relative energies of all possible conformations of (**125b**) and (**127**) have been calculated; their appropriately weighted contributions led to good simulations of the ^1H NMR spectra ⟨97T7461⟩.

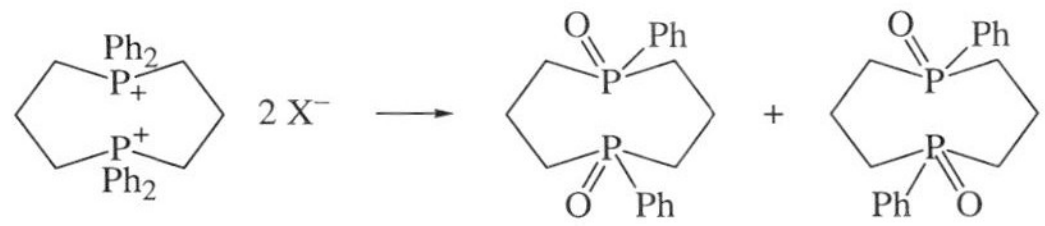

(**124**) (**125a**) (**125b**)

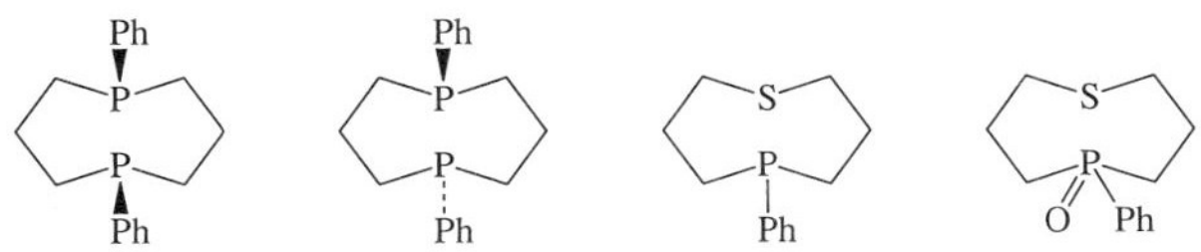

(**126a**) (**126b**) (**127**) (**128**)

As part of an investigation of the intra-bridgehead chemistry of diphosphines, novel syntheses of medium-sized rings having 8, 9, and 10 members and a range of substituents at phosphorus have been developed by an approach that gives the potentially chelating *syn* diastereomers of the diphosphines ⟨96CC537, 98JCS(P1)1643⟩. The synthetic route is summarized by the conversions (**129**)–(**132**). The propellane salts (**133**; n = 0, 1) have also been prepared; their properties are markedly dependent upon ring size ⟨98JCS(P1)1657⟩. Quaternization of (**134**) with 1,3-bis(trifluoromethanesulfonyloxy)propane gave (**135**) in 83% yield, which in turn was reduced with LAH to (**136**) in 85% yield. The crystal and molecular structures of (**135**) and (**136**) were determined.

An 80% yield of (**137**) was obtained by the condensation of 1,2-ethylenebis(diphenylphosphine) with *cis*- or *trans*-1,4-dichlorobut-2-ene ⟨87TL1259⟩. Catalytic hydrogenation of this compound over Raney nickel afforded the saturated macrocycle (**138**).

In related work, the macrocyclic bis(phosphonium) salts (**139**)–(**141**) were prepared in 16–40% yield ⟨84NJC191⟩. Numerous macrocyclic poly(phosphonium) salts have been synthesized in remarkably high yields in similar condensations ⟨75MI934-01⟩. Alkaline hydrolysis of (**139**) afforded the *cis* and *trans* diastereomers of the corresponding bis(phosphine oxide), which were separated by chromatography, individually reduced with hexachlorodisilane to the corresponding bis(tertiary phosphines), and quaternized with methyl iodide. The reaction of 1,3-bis(diphenylphosphino)butane with $Me_2S{\bullet}BH_3/I_2$ in benzene gave the boronium iodide (**142**) in 74% yield ⟨85ICA293⟩.

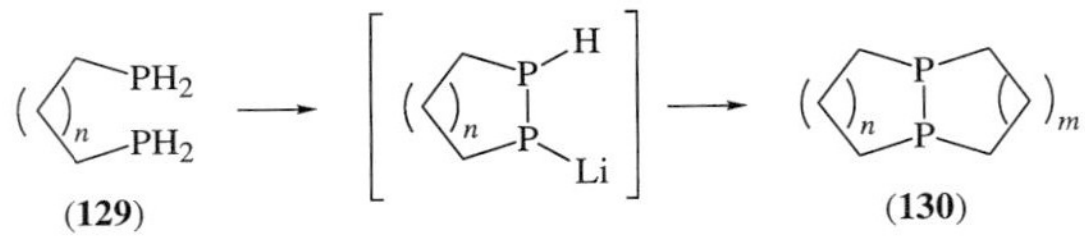

(129) (130)

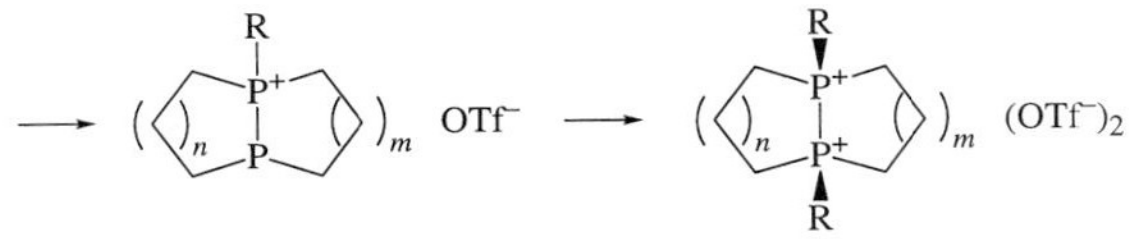

(131) (132)

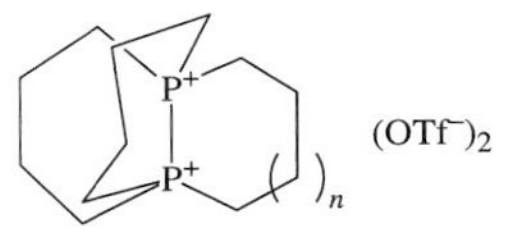

(133)

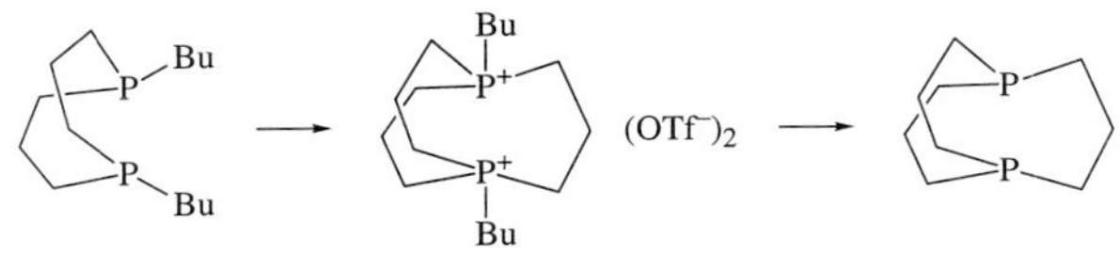

(134) (135) (136)

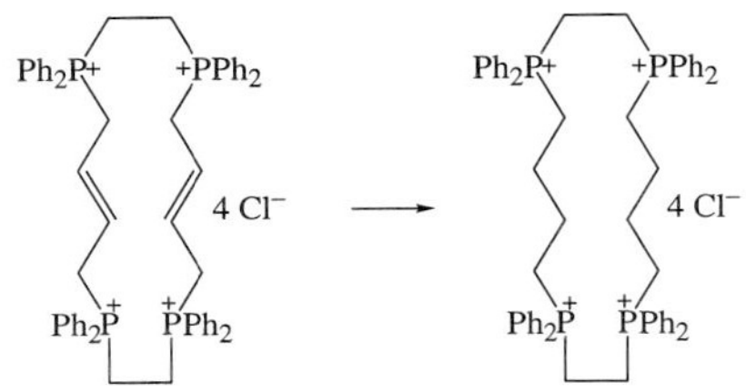

(137) (138)

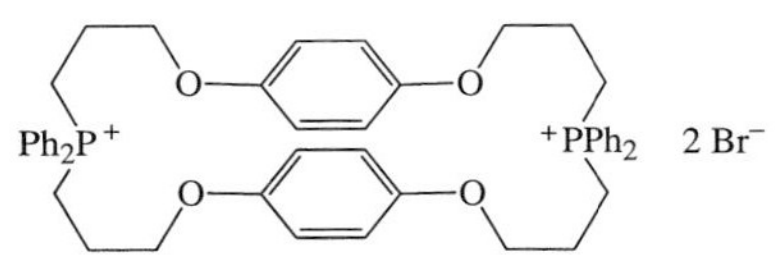

(139)

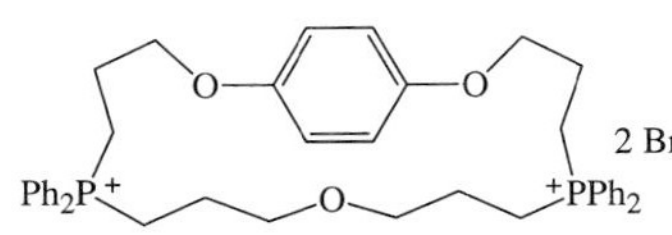

(140)

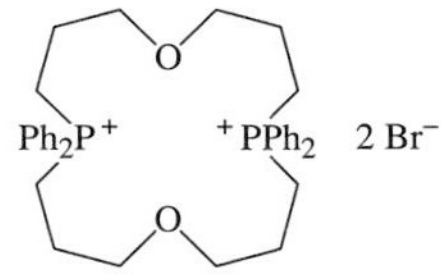

(141)

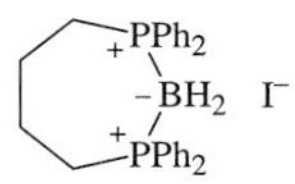

(142)

6.1.3.2 Cyclizations Involving Organophosphides and Other Nucleophiles

The bis(tertiary phosphine) (**143**) was isolated in 35% yield following treatment of dilithium methylenebis(phenylphosphide) with 1,4-dichlorobutane: a single diastereomer was observed in the ^{31}P NMR spectrum of the compound ⟨86CB2832⟩. Similar reactions between dilithium 1,2-ethylene- and 1,3-propylenebis(phenylphosphide) and the appropriate α,ω-dihaloalkane yielded (**144**) as a mixture of two diastereomers ⟨87PS(30)389⟩.

An interesting and effective strategy was used to prepare the heterocycle (**145**) ⟨85IC1613⟩. Treatment of α-naphthylmethylphosphine with two equivalents of 1-bromo-3-chloropropane afforded, after sulfurization, (**146**), which was condensed under high-dilution with the bisphosphide (**147**) to give (**148**); removal of the α-naphthylmethyl group from (**148**) with potassium naphthalenide, followed by the sulfur with LAH/TMS-Cl, afforded (**145**) as a mixture of two diastereomers. A number of transition metal complexes of (**145**) were isolated. A similar P_2N macrocycle was reacted with $CpFe(CO)_2I$ and $Cp^*Fe(CO)_2I$ under photochemical conditions ⟨94JOM(468)199⟩.

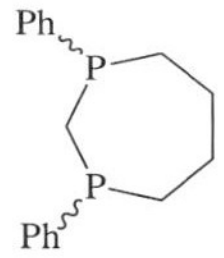
(**143**)

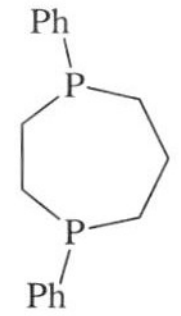
(**144**)

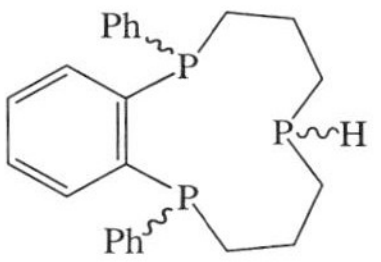
(**145**)

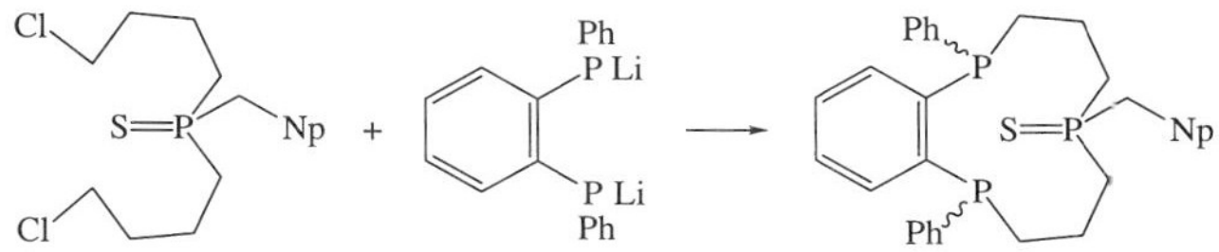
(**146**) (**147**) (**148**)

The *cis*-P_2S_2 macrocycle (**149**) was synthesized as a mixture of (R^*,R^*) and (R^*,S^*) diastereomers from 1-mercapto-2-(phenylphosphino)benzene by deprotonation and treatment with 1,3-dihalopropanes: depending on the base employed, strapping at phosphorus or sulfur was achieved in the first step ⟨85JA2141⟩. The P_3S macrocycle (**150**), together with a trace of (**151**), was prepared by the high-dilution condensation of (**152**) with (**147**). On the other hand, the conversion of (**153**) into (**154**), followed by deprotection at sulfur and ring closure with 1,3-dichloropropane, afforded the *trans*-P_2S_2 macrocycle (**155**). The structures of (R^*,S^*)-(**149**) and $[Pt\{(R^*,S^*)\text{-}(\mathbf{155})\}](PF_6)_2 \bullet C_6H_6$ were determined by X-ray crystallography. Because of the highly stereoselective formation of R^*,S^* (meso) diastereomers of products of (**147**) with electrophiles, (**150**) was also considered to be predominantly of the R^*,S^* stereochemistry. In later work, the synthesis and co-ordination chemistry of other 14-membered P_2S_2-macrocycles were reported ⟨94JCS(D)3031⟩. The reaction of dipotassium 1,2-ethylenebis(phenylphosphide) with bis(2-chloroethyl)amine in THF at room temperature led to the 18-membered P_4N_2 macrocycle (**156**; X = NH), for which four of the five possible diastereomers were isolated ⟨83ICAL17⟩. The crystal structure of one of the diastereomers in the complex $[Ni(\mathbf{156})](BPh_4)_2 \bullet 2Me_2CO$ was determined: the two nitrogen atoms of the macrocycle were not coordinated to the nickel. The synthesis and characterization of the diastereomers (**156**; X = S) and several cobalt(II) and nickel(II) complexes have been described ⟨84JCS(D)575, 84JCS(D)995, 85JCS(D)1425⟩. This work has been reviewed ⟨B-90MI934-01, 94CRV1183, 86PAC1429⟩. The phosphino-crown ethers and -thioethers (**157**; R = Bn, c-C_6H_{11}, Ph) and (**158**) were prepared in one step from the appropriate dilithium organophosphides and 1,2-bis(2-choroethoxy)ethane or 1,2-bis(2-chloroethylthio)ethane, respectively ⟨92TL3621⟩.

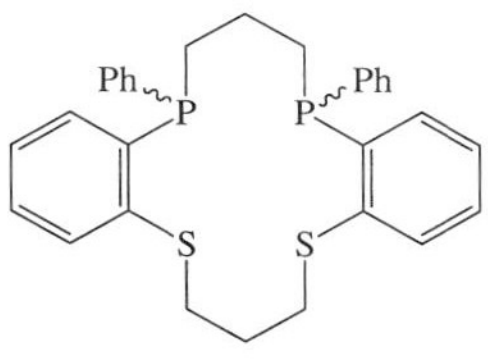
(**149**)

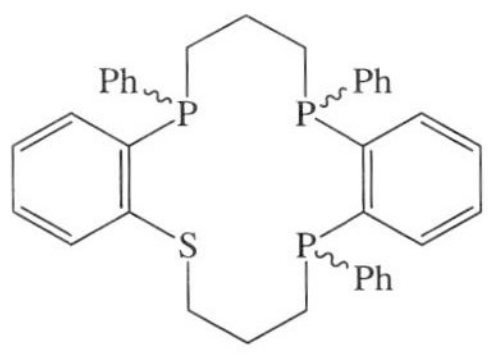
(**150**)

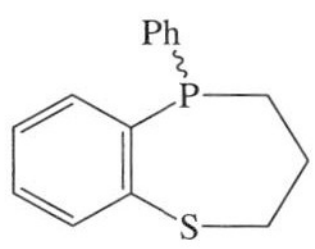
(**151**)

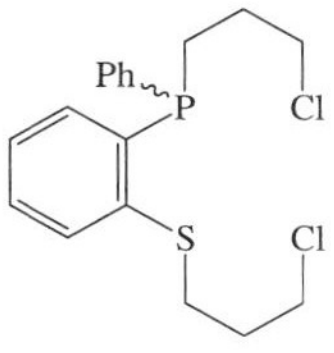
(**152**)

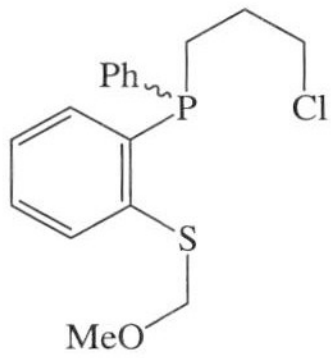
(**153**)

High-dilution syntheses of (**159**) and (**160**) were achieved by reactions of dilithium 1,3-propylenebis-(phenylphosphide) with the appropriate terminal dichloro-heterosubstituted-alkanes ⟨86JA8302⟩. The crystal structures of $(\pm)$-$[PtCl_2\{(R^*,R^*)\text{-}(\mathbf{159})\}]$ and one enantiomer of $(\pm)$-$[Ni(NCS)_2\{(R^*,R^*)\text{-}(\mathbf{160})\}](NCS)_2$

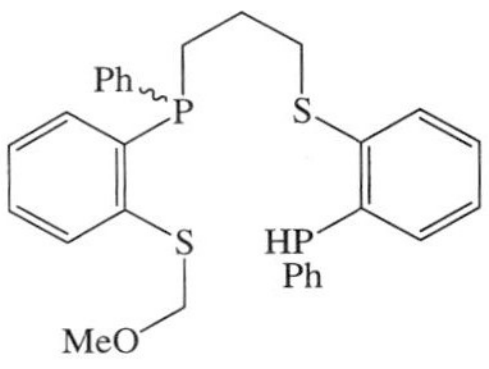

(154)

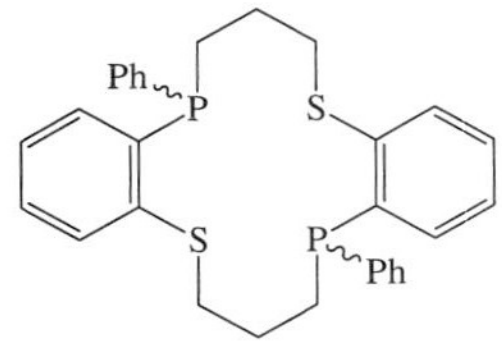

(155)

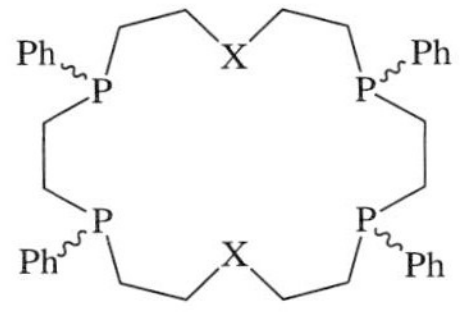

(156)

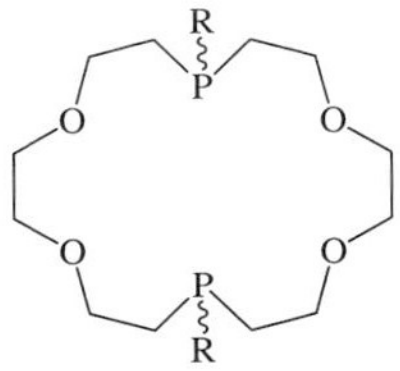

(157)

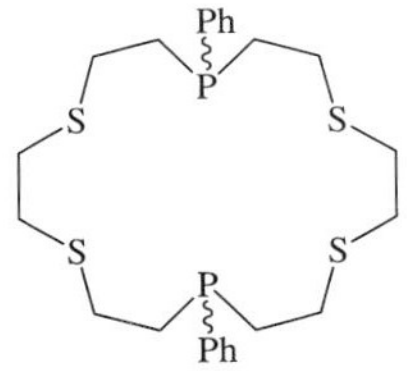

(158)

(spontaneous resolution) were determined. The 16- and 18-membered P_2 macrocycles (**161**; $n = 2$ and R = Me or OMe; $n = 3$, R = OMe) were synthesized from the corresponding terminal *p*-toluenesulfonate esters and the bis(phosphide) (**147**) under high-dilution conditions ⟨92CB679⟩. The compound (**161**; $n = 3$, R = OMe) was obtained in 87% yield. X-ray crystal structure analyses of the C_1 and C_2 diastereomers of (**161**; $n = 2$, R = Me) and the C_1 diastereomer of (**161**; $n = 2$, R = OMe) and its nickel(II) chloride adduct were performed. In later work, the compounds (**162**; $n = 2$ to 6) were prepared in optically active form from (*S*)-1,1′-binaphthyl-2,2′-diol and square-planar complexes of the type [MCl_2(**162**; $n = 2$ to 5)] isolated and characterized by NMR, CD, and selected X-ray crystal structure determinations ⟨94HCA409, 94CB1411, 94TA189⟩.

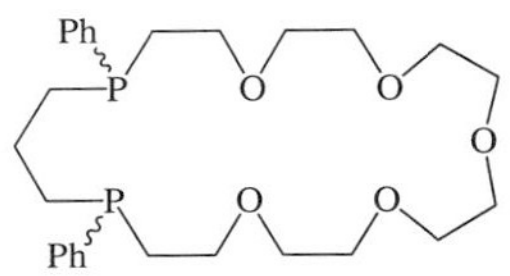

(159)

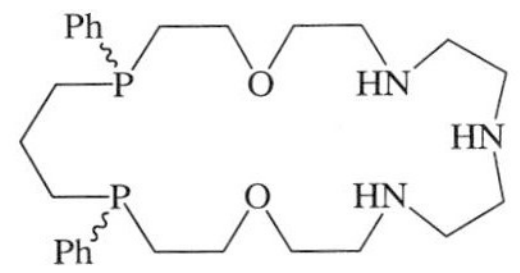

(160)

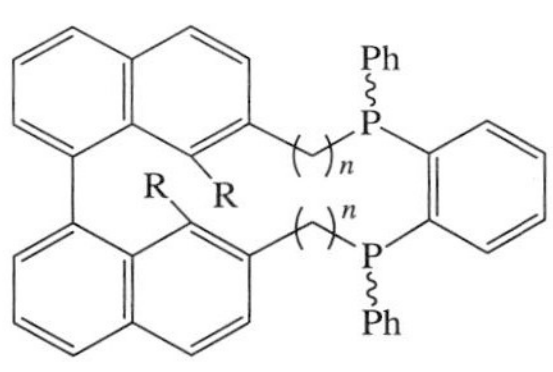

(161)

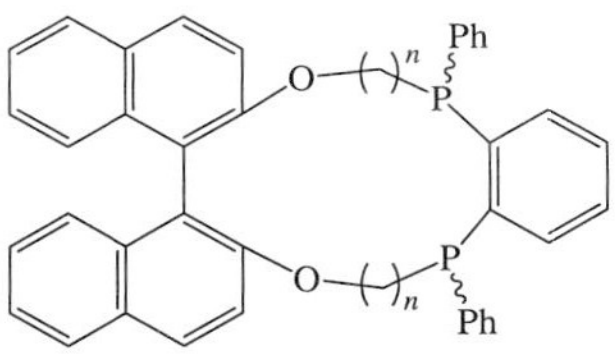

(162)

Treatment of (**163**) with (*R*,*R*)-(**164**) gave with complete stereoselectivity (**165**), which was isolated as colorless crystals ⟨95JCS(D)2095⟩.

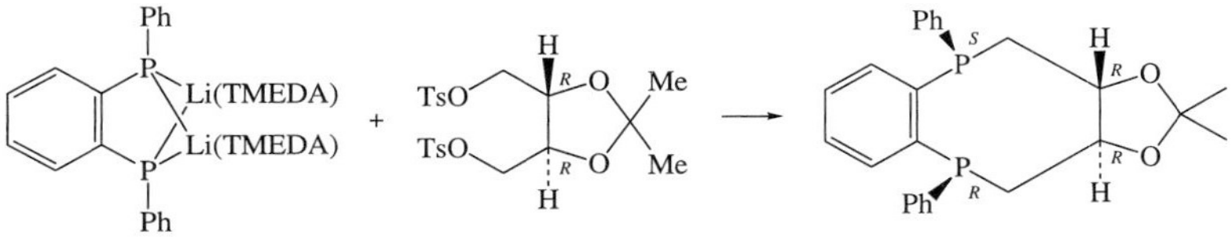

(163) (164) (165)

The two families of macrocyclic phosphine oxide disulfoxides (**166**; X = $(CH_2)_{2-4}$, 1,2-xylyl, 1,3-xylyl) and (**167**; X = $(CH_2)_{3-5}$, 1,2-xylyl, 1,3-xylyl) were prepared stereoselectively for comparisons of three-point binding abilities with monoalkylammonium ions ⟨92TL2107, 93PAC461, 93JA7900, 97T12249⟩. Compound (**167**; X = 1,2-xylyl) also binds halide ions ⟨94JA4069⟩.

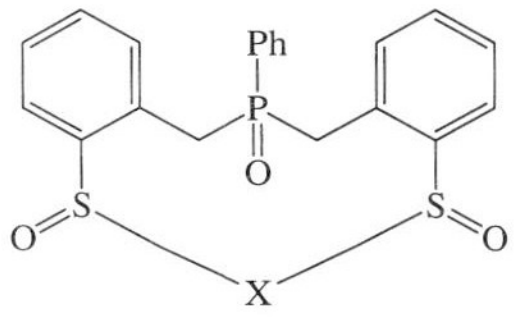

(**166**)

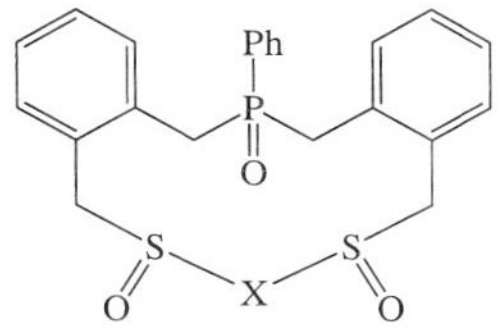

(**167**)

The PS_2 macrocycle (**168**) was synthesized by reacting tetramethylammonium bis(2-thiolatoethyl)phenylphosphine with molybdenum hexacarbonyl and treating the resulting complex with 1,2-dichloroethane, but the ligand could not be liberated from the metal ⟨93CC54⟩. Accordingly, the 9-membered ring was prepared directly in 44% yield from the phosphine-thiol and 1,2-dichloroethane in the presence of caesium carbonate.

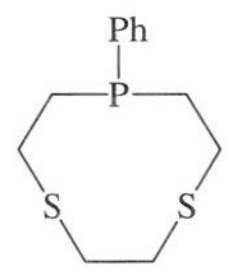

(**168**)

The condensation of (**169**; X = Li) with 2,6-bis(bromomethyl)pyridine in THF gave (**170**) in 38% yield ⟨97ICA99⟩. Fourteen-membered *syn*/*anti*-(**171**) was isolated in 26% yield as a 1 : 1 mixture following the Cs_2CO_3-mediated coupling of bis(3-chloropropyl)sulfoxide with (**169**; X = H) in hot DMF ⟨99JCS(D)1277⟩. The 1 : 1 mixture of *syn*/*anti*-(**171**) transformed over 7 days in chloroform into (**172**); in the presence of a trace of iodine, the rearrangement took 30 min.

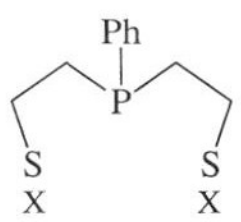

(**169**)

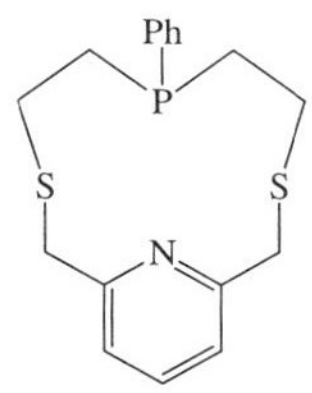

(**170**)

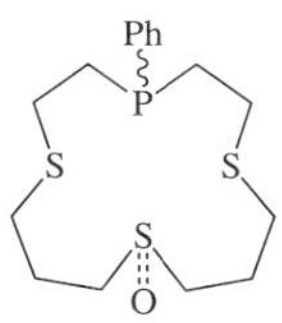

(**171**)

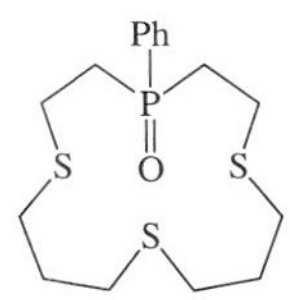

(**172**)

Treatment of (**173**; R = Bu^t, Ph) with sodium hydride followed by the appropriate alkylene ditosylate afforded the dibenzophospha-crown ethers (**174**; R = Bu^t, Ph) and (**175**; R = Ph) ⟨83RTC326⟩. These compounds are capable of solubilizing lithium ions in chloroform. The macrocycle (**176**; X = P(O)OMe) was prepared in 34% yield from (**177**) and 1,4-dibromobutane in the presence of potassium carbonate in DMF ⟨93JA6995⟩. Treatment of (**176**; X = P(O)OMe) with TMS-I and sodium hydroxide gave (**176**; X = P(O)ONa) and reduction of (**176**; X = P(O)OMe) with LAH followed by quaternization with methyl iodide afforded the bis(dimethylphosphonium) salt (**176**; X = $PMe_2{}^+I^-$). The binding abilities of (**176**; X = P(O)ONa and $PMe_2{}^+I^-$) towards substituted naphthalenes were investigated.

3,4-Dimethyl-1-phenylphosphole (**178**), when heated at 160°C for 170 h, produced red crystals of (**179**) in ca. 20% yield ⟨82JA2077⟩ and prolonged heating of (**180**) at 180°C gave (**181**) in 35% yield ⟨95BSF910⟩. The crystal structures of (**179**) and the pentacarbonylmolybdenum(0) derivative have been determined ⟨83JCS(D)841⟩. Six-unit mixed chains containing four thiophene and two phospholyl rings have been prepared ⟨90AG(E)655⟩.

One or both of the P–P bonds in (**179**) can be cleaved by sodium naphthalenide in THF to yield the salts (**182**) and (**183**), which react with dibromomethane to give (**184**) or (**185**) in 45 and 80% yields, respectively (abbreviated structures are given here and henceforth) ⟨94JA3306⟩. The crystal and molecular structures of (**183**)•2DME have been determined ⟨96AG(E)1125⟩. The reaction of (**182**) with tetrachloroethylene unexpectedly gave (**186**) in 25% yield, which was structurally characterized. The dianion (**183**) can be converted by treatment with 1,4-dibromobutane into the 12-membered compound

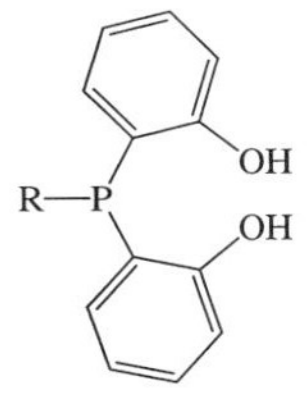

(173)

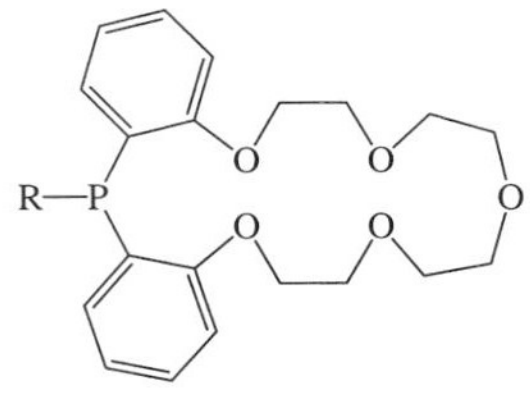

(174)

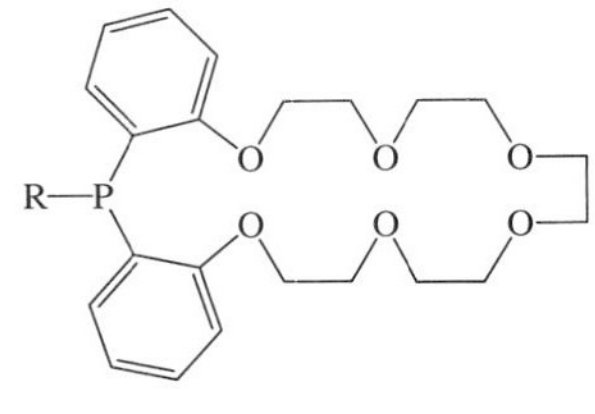

(175)

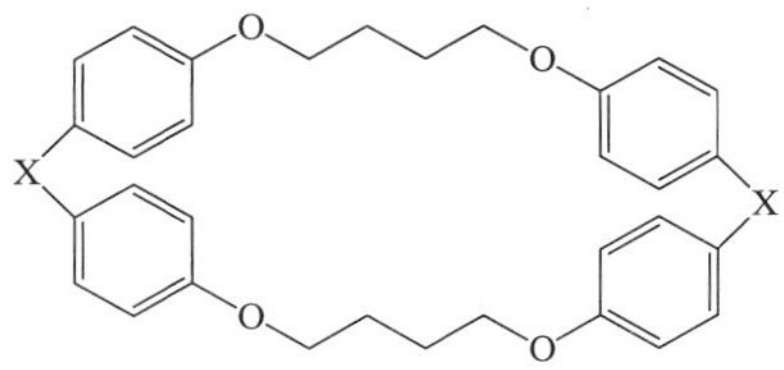

(176)

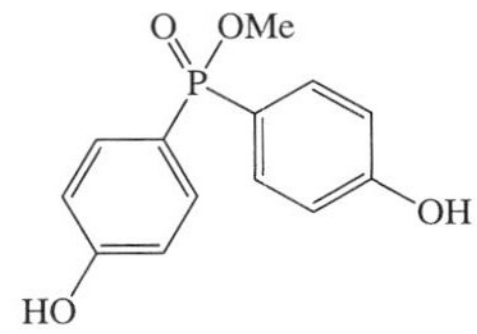

(177)

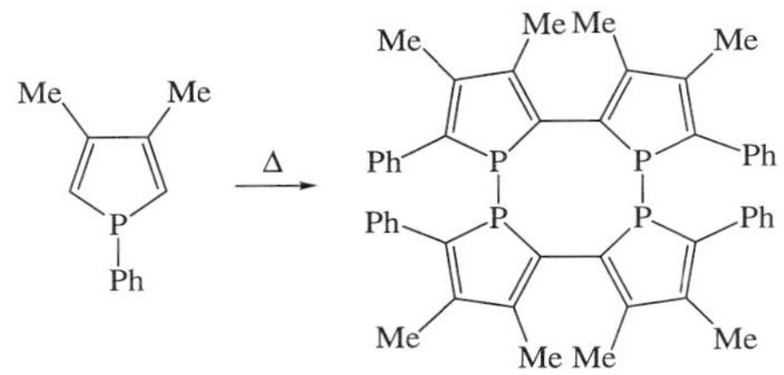

(178) **(179)**

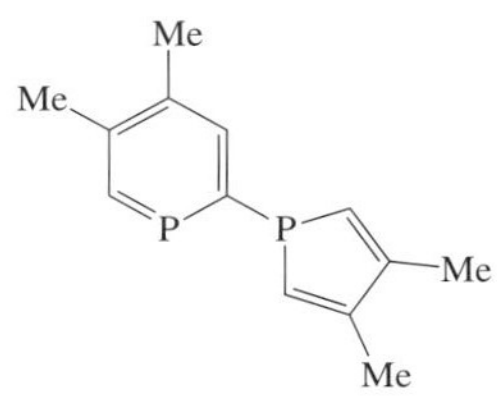

(180)

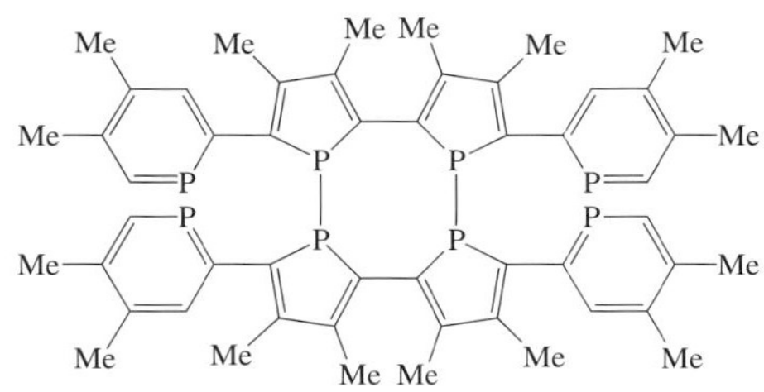

(181)

(**187**); the latter in turn can be used to prepare larger macrocycles, i.e. (**188**; $n = 1$–4). Since the phosphorus stereocenters in these macrocycles are included in phosphole rings, they have low inversion barriers, which allows the ligands to adapt their stereochemistries to the stereochemical requirements of metals in metal complexes. The complex [$PdCl_2$(**184**)] forms a long-lived catalyst for the Stille cross-coupling of PhI with $CH_2{=}CHSn(Bu\text{-}n)_3$ and for the alkylation of the malonate ion ⟨97AG(E)2364⟩.

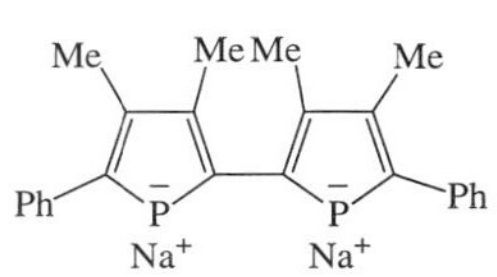

(182)

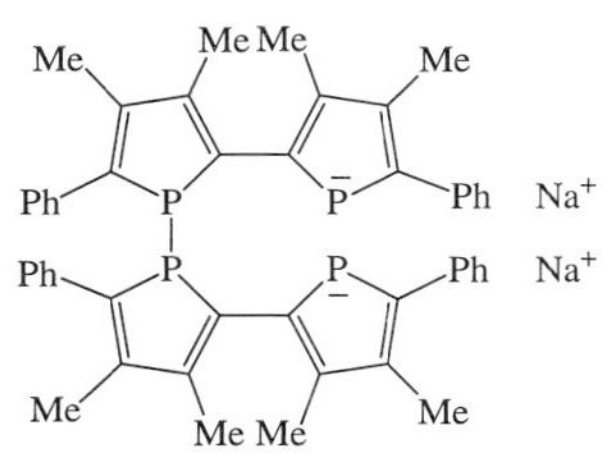

(183)

The reaction of (**189**) with the bis-ylide (**190**) affords 24-membered (**191**) in 60% yield. The X-ray crystal structure of (**191**) shows an all-*trans* disposition of the four phosphorus lone pairs and a diameter of the central cavity of ca. 6 Å ⟨95CC1561⟩.

Bis(2-formylphenyl)phenylphosphine undergoes acid-catalyzed hydration into (**192**), which is converted into (**193**) under similar conditions in methanol ⟨82CC1170⟩.

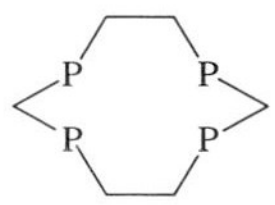
(184)

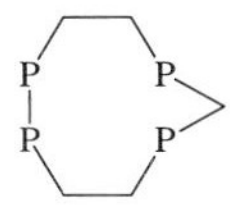
(185)

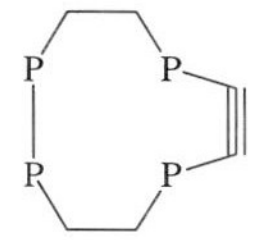
(186)

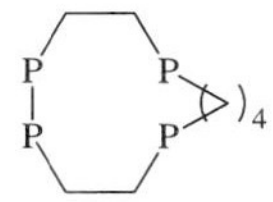
(187)

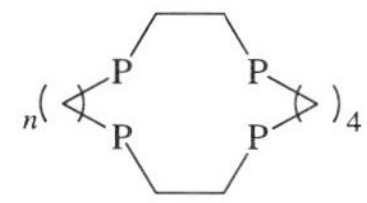
(188)

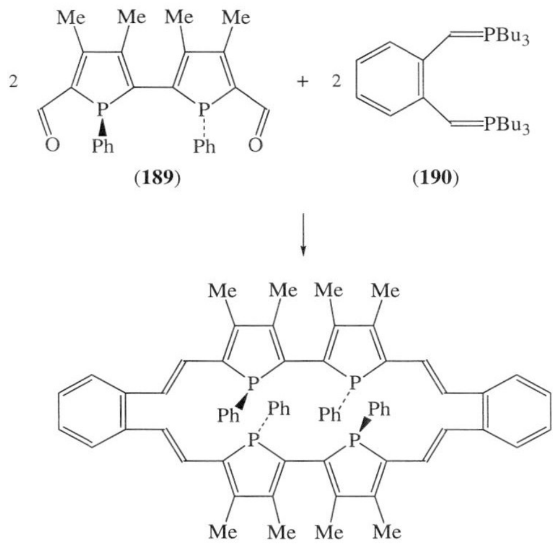
(189) (190) (191)

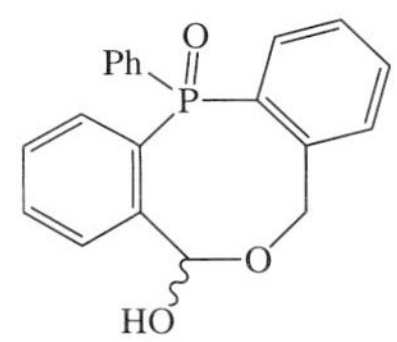
(192)

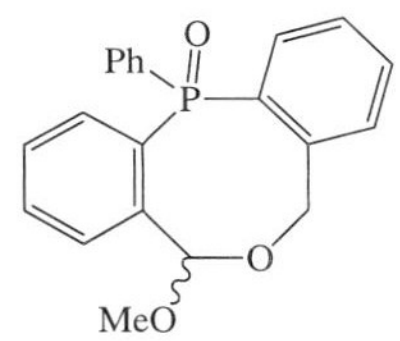
(193)

Treatment of the phosphinate salts (**194**; $n = 2, 3$) with appropriate alkylating agents under high-dilution conditions gives mixtures of (**195**; $n = 2, 3$) and (**196**; $n = 2, 3$), which were reduced to the corresponding saturated 1,4-butylene compounds ⟨88TL6247, 93BSF647⟩. The salts (**194**; $n = 2, 3$) react with α,ω-dibromoalkanes to give the saturated rings (**197**; $n, m = 2, 3$) and (**198**; $n, m = 2, 3$). Following similar strategies, 1,2-bis(bromomethyl)benzene reacted with (**194**; $n = 2, 3$) to form the 1,2-xylidyl compounds (**199**; $n = 2, 3$) ⟨93PS(78)119⟩. The diastereomers of (**199**; $n = 2, 3$) were separated by chromatography; (**199**; $n = 2$) was reduced by LAH/$CeCl_3$ to the bis(tertiary phosphine) (**200**).

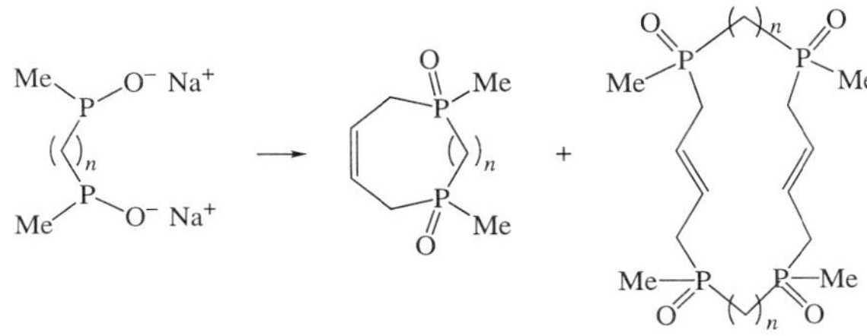
(194) (195) (196)

Ring enlargement by the hetero-Cope-rearrangement of (**201**) afforded (**202**) in 82% yield ⟨84ZAAC(510)123⟩.

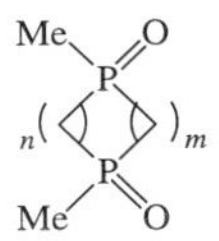

(197)

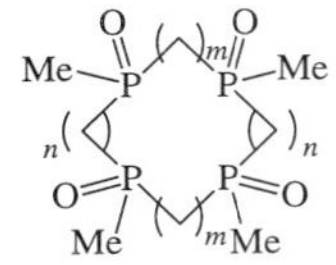

(198)

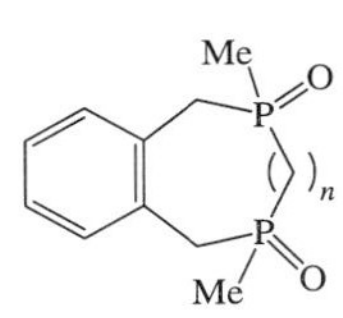

(199)

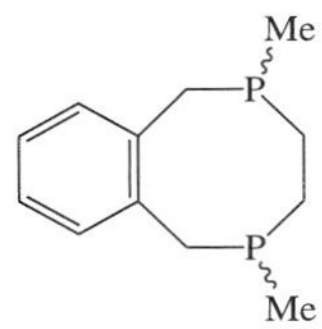

(200)

The compound (**201**) was prepared in two steps from norbornene dicarboxylic dichloride and phenyl-bis(trimethylsilyl)phosphine.

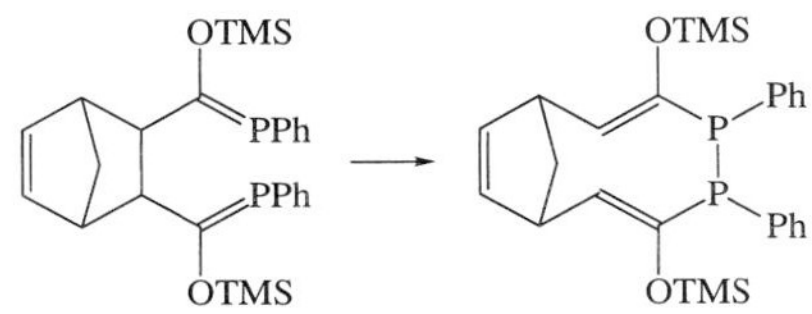

(201) **(202)**

The 9- and 12-membered tri- and tetra-phosphapericyclynes (**203**) and (**204**) have been synthesized ⟨90JA7823⟩. Because of the pyramidal nature of the phosphorus stereocenters in the two compounds, diastereomers are possible. Only the *cis,trans* diastereomer of (**203**) was observed, however. No evidence of equilibration of the two possible diastereomers of (**203**) was observed on the ^{31}P NMR time scale up to 50°C. For (**204**), although the four possible stereoisomers were present in the crude reaction mixture, the all-*trans* isomer was isolated by fractional crystallization. The molecular structures of *cis,trans*-(**203**) and all-*trans*-(**204**) were determined by X-ray crystallography.

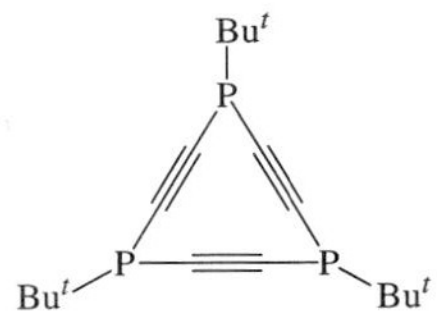

(203)

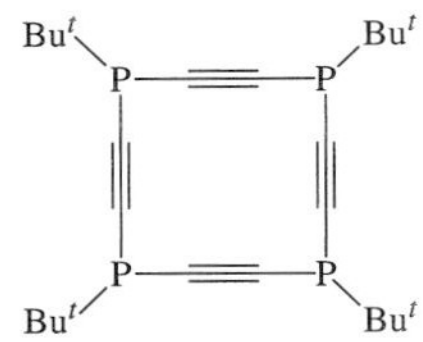

(204)

Thiophene is deprotonated at the 2- and 5- positions with the kinetically fast base *n*-BuLi/TMEDA/KOBu^t in *n*-hexane; the slow addition of dichlorophenylphosphine to the solution of the dianion afforded (**205**) in 17% yield ⟨95JOC7406⟩. Diastereomerism in the compound was not discussed.

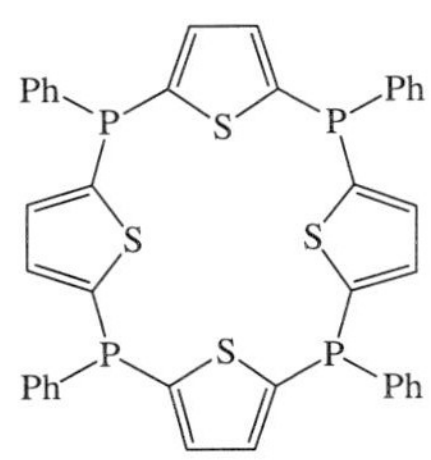

(205)

Phosphorus trichloride reacts with 3 equiv. Li[C(PMe_2)$_2$$SiMe_3$] in toluene to give (**206**) ⟨95AG(E)557⟩. The crucial step in the sequence leading to (**206**) is the "outer-sphere coupling" of the terminal PMe_2 groups with the phosphorus trichloride.

A 0.1 M solution of bromophenyldiazirine (**207**) in dichloromethane at −78°C reacts with 1,2-bis(diphenylphosphino)ethane to give (**208**, $n = 2$) in 90% yield, or (**209**) if the reaction is performed with a 0.01

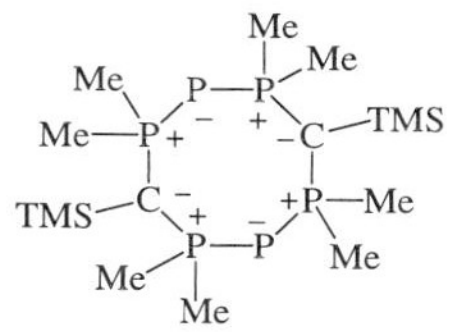

(**206**)

M solution of (**207**); the reaction of a 0.01 M solution of (**207**) with 1,3-bis(diphenylphosphino)propane gave (**208**; $n = 3$) in 75% yield ⟨96JA1060⟩.

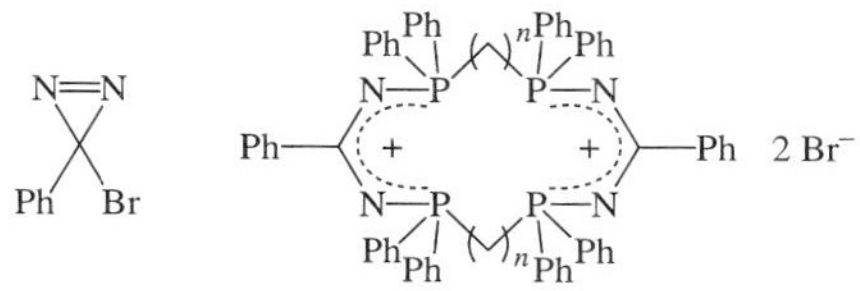

(**207**) (**208**)

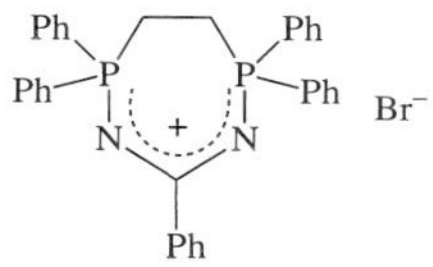

(**209**)

The Staudinger reaction between certain bis(azides) and bis(phosphines) has been used to prepare a variety of bis(iminophosphorane)-containing macrocycles containing 12 or more members under mild conditions ⟨97JOM(529)121⟩. For example, (**210**) and (**211**; X = NH, O, n = 1, 2) were isolated in 79–97% yield by this method. On the other hand, the reaction between (**212**) and (**213**) in diethyl ether at room temperature gave (**214**) in 66% yield; a similar strategy with the corresponding *N*-oxide of (**212**) in dichloromethane gave the *N*-oxide of (**214**) in 85% yield, which was characterized by X-ray

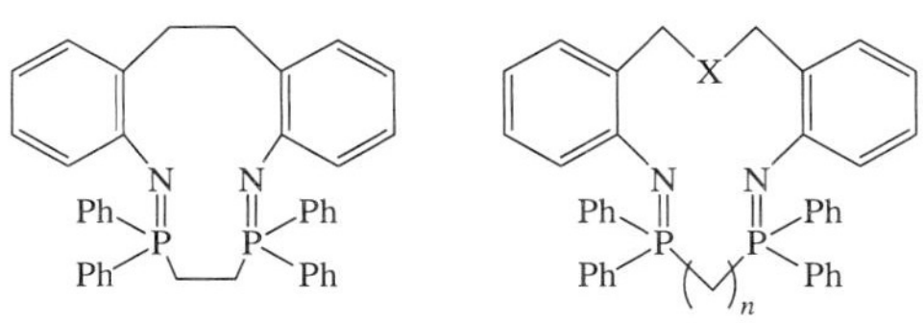

(**210**) (**211**)

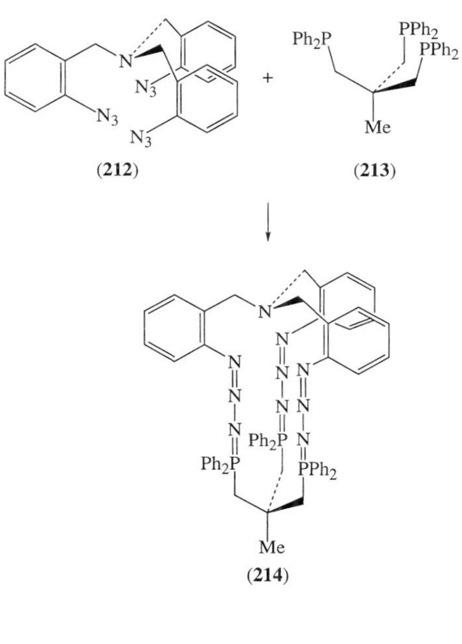

(**212**) (**213**)

(**214**)

crystallography ⟨97AG(E)67⟩. The *N*-oxide of (**214**), in which the oxygen atom is outside of the cage, is considerably more stable than (**214**) itself: the C_3 structure of the molecule is retained at 334 K in chloroform. When a THF solution of (*E*,*E*)-(**215**) was irradiated with a 500 W xenon lamp through a Toshiba Y-49 filter at room temperature and [W(CO)$_5$THF] was added, (**216**) was isolated in 7% yield from the reaction mixture ⟨96TL7815⟩.

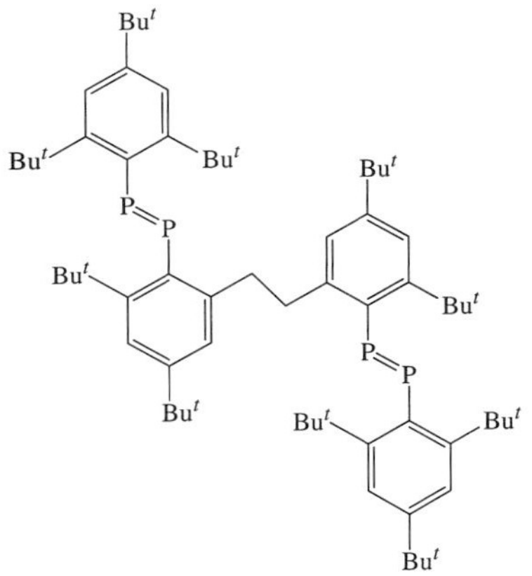

(**215**)

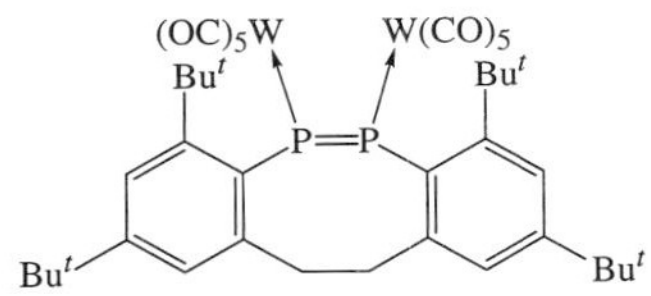

(**216**)

The 8-membered 1,2,8-oxadiphosphorane dioxide (**217**) is formed in 27% yield when pentacarbonyl(phenylphosphinidene)tungsten, [PhPW(CO)$_5$], is heated in boiling toluene with a 3 molar excess of fluorenone; bifluorenylidene is also formed in 24% yield ⟨97JOM(533)83⟩. An X-ray crystal structure analysis of (**217**) revealed a trans arrangement of the P=O bonds.

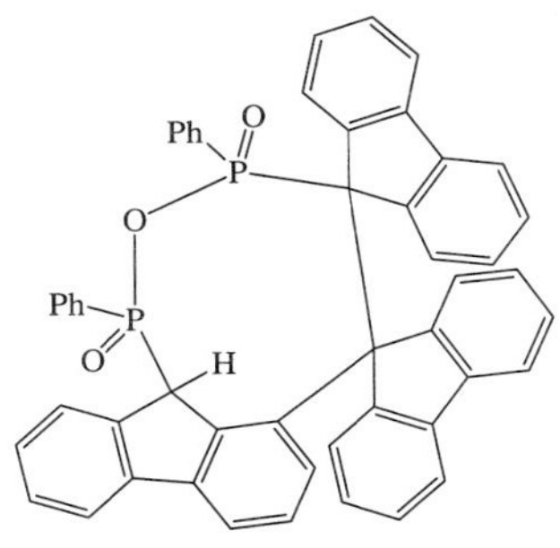

(**217**)

6.1.3.3 Metal Template Syntheses

The first metal-template synthesis of a phosphorus-containing macrocycle was reported in 1970, wherein the reaction of (**218**) with 1,2-xylidene dibromide furnished the five-coordinate complex (**219**) ⟨70C431⟩. There followed in 1977 the related synthesis of (**221**) in which (**220**) was treated with potassium carbonate and then 1,2-xylidene dibromide; in this case, however, the free tetra(tertiary phosphine) (**222**) was liberated from the nickel with cyanide ⟨77JA8051⟩.

Heating under reflux of an ethanolic solution containing bis(3-aminopropyl)phosphine, 2,6-diacetylpyridine, and $NiX_2{\bullet}6H_2O$ (X = Br, I), followed by addition of NH_4PF_6, resulted in crystallization of the purple five-coordinate complexes (**223**; X = Br, I). Reduction of the imine groups in the complexes with sodium borohydride or PtO–H_2 gave (**224**), which in turn was treated with cyanide to displace the free macrocycle (**225**) ⟨74CC442⟩. In related work, the Schiff-base ligands (**226**; n = 2, 3) were reacted with $Ni(OAc)_4{\bullet}4H_2O$ and ring closure effected at pH 4 with elimination of acetylacetone to give (**227**) or (**228**) ⟨82IC1215⟩.

The stable secondary phosphine complexes (**229**; M = Ni, X = Br and M = Pd, X = Cl) react with malondialdehyde tetramethyl acetal in the presence of acid or acetylacetone with formation of the complexes (**230**; M = Ni, X = Br and M = Pd, X = Cl and R = H, Me), which contain eight chiral centers ⟨83IC3624⟩. The reactions with malondialdehyde did indeed result in complex mixtures of diastereomers,

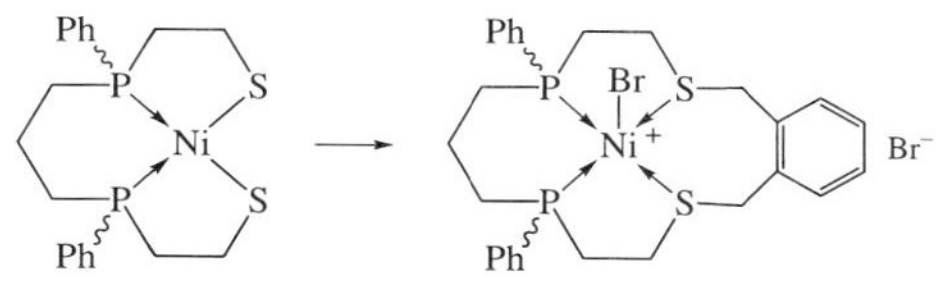

(218) (219)

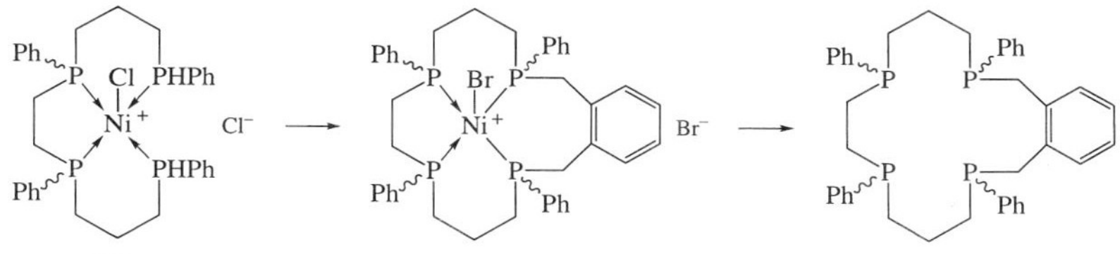

(220) (221) (222)

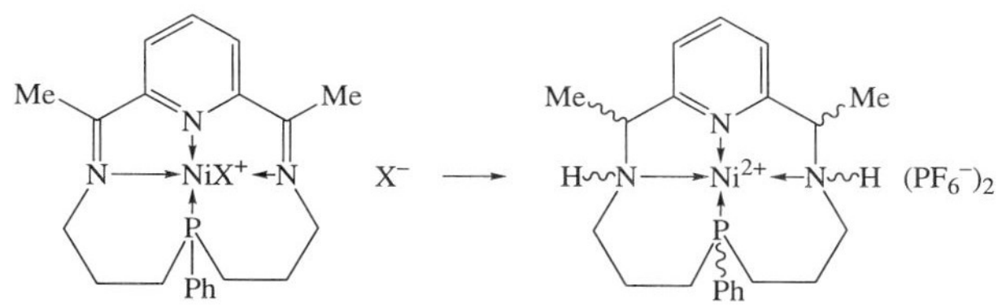

(223) (224)

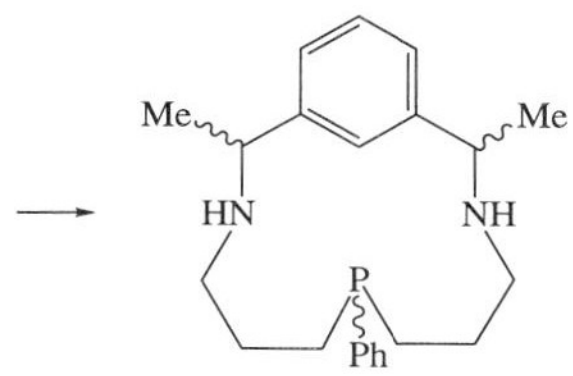

(225)

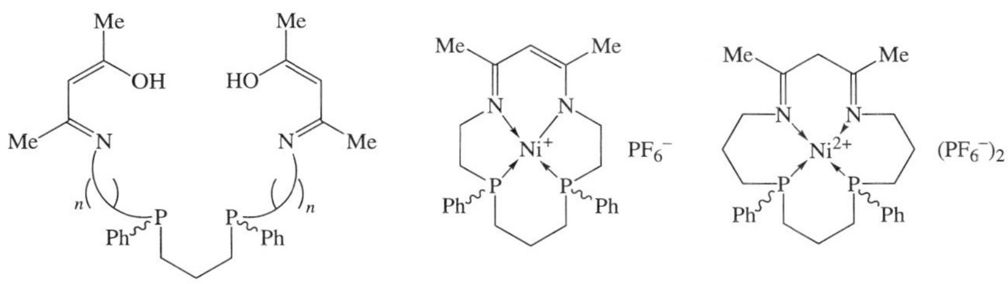

(226) (227) (228)

but with acetylacetone only two diastereomers were obtained for both metal salts; the structure of the palladium compound was determined by X-ray crystallography. Interestingly, the reaction of $[Pd(acac)_2]$ with 1,2-ethylenebis(methylphosphine) in the presence of hydrochloric acid gave a single diastereomer of the macrocyclic complex. The use of biacetyl in the condensation of the bis(secondary phosphine) complex was unsuccessful. On the other hand, (**231**) reacts with biacetyl or bibenzoyl to give (**232**; R = Me, Ph), but not with acetylacetone to give the expected 16-membered macrocyclic complex. The macrocyles could not be displaced from the palladium complexes; treatment of (**230**) with cyanide regenerated the bis(secondary phosphine) and (**232**) did not react under the conditions tried. In subsequent work, in order to reduce the number of product diastereomers from eight to six, a strategy was adopted in which appropriate tertiary phosphine complexes of nickel, palladium, and platinum containing terminal carbonyl groups (protected as their ketals) were reacted with chelating bis(secondary phosphines): the addition of acid generated carbonyl groups that underwent cyclizations in the usual way to give macrocyclic tetra(tertiary phosphine) complexes. As before, however, the macrocycles could not be liberated from the metals ⟨84IC3304⟩.

The cyclic alkylation of the amido complex (**233**) of R^*,S^* stereochemistry at the phosphorus stereocenters with the 1,3-propylenebis(*p*-toluenesulphonate) gave the corresponding R^*,S^* diastereomer

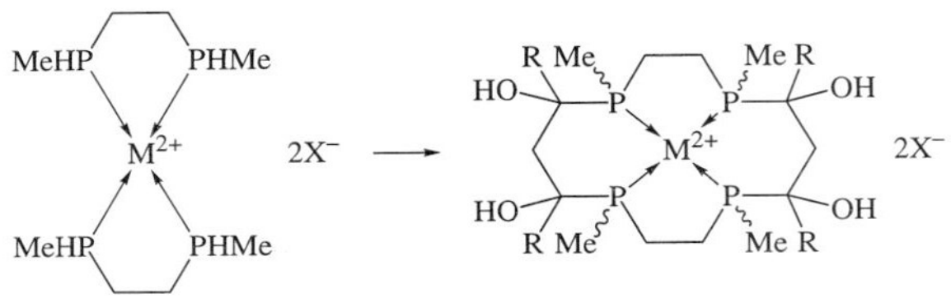

(229) (230)

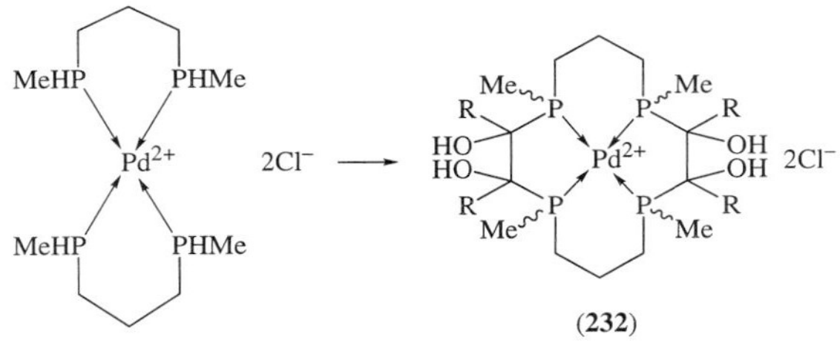

(231) (232)

of the complex (**234**), from which the macrocycle was liberated with aqueous cyanide ⟨85CC439⟩. The structures of (**233**) and the free macrocycle were determined.

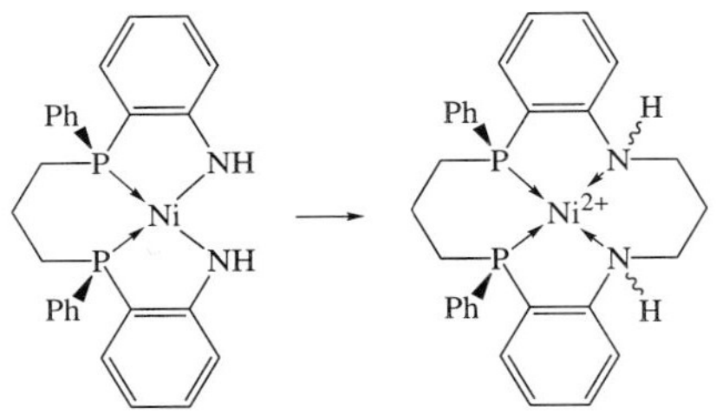

(**233**) (**234**)

Treatment of (**235**), which was isolated as a mixture of diastereomers with (R^*,R^*)-*cis* : (R^*,S^*)-*cis* : (R^*,R^*)-*trans* : (R^*,S^*)-*trans* = 1 : 1 : 3 : 4, with boron tribromide resulted in 83.5% cyclization at sulfur ⟨92IC1411⟩. The intermediate complexes, after treatment with cyanide, yielded (R^*,R^*)- and (R^*,S^*)-(**236**) (1.5% yield), (±)-(**237**) (6% yield), and (±)-(**237**)•BBr_3 (76% yield). The structure of the latter and [Ni{(R^*,R^*)-(**236**)}]$(ClO_4)_2$•Me_2CO were determined by X-ray crystallography.

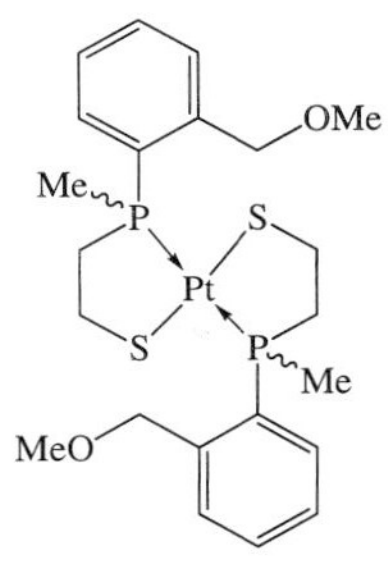

(**235**)

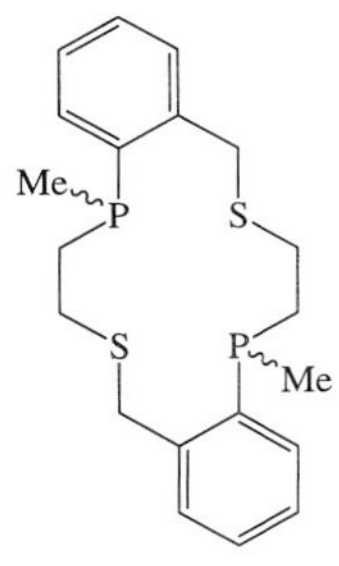

(**236**)

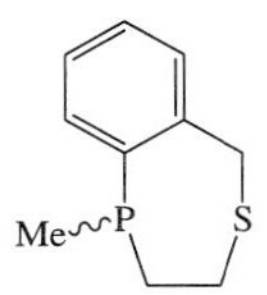

(**237**)

The free radical-initiated (AIBN) cyclotrimerization of [$Mo(CO)_3\{PH_2(CH_2CH{=}CH_2)\}_3$] and [$Mo(CO)_3\{PH_2(CH_2CH_2CH{=}CH_2)\}_3$] in benzene furnished in 70–85% yields the macrocyclic tris(secondary phosphine) complexes (**238**; M = Mo, R = H) and (**239**; M = Mo, R = H), respectively ⟨82JA4700,

89IC2811⟩. The structure of (**239**; M = Mo, R = H) was determined by X-ray crystallography. The macrocycles in the complexes could not be displaced by PPh_3, PF_3, $P(OMe)_3$, or KCN. Treatment of (**238**; M = Mo, R = H) with 3 equiv. each of *n*-BuLi and MeI, $CH_2{=}CHCH_2Br$, Me_3SiCH_2Cl, or *i*-PrBr gave the corresponding tertiary phosphine complexes (**238**; M = Mo, R = Me, $CH_2{=}CHCH_2$, Me_3SiCH_2, *i*-Pr) and allylamine in the presence of AIBN gave (**238**; M = Mo, R = $CH_2CH_2CH_2NH_2$) ⟨95JCS(D)1139⟩. Oxidation of (**238**; M = Mo, R = H, *i*-Pr, CH_2SiMe_3) and (**238**; M = W, R = H) with dihalogens produced the seven-coordinate divalent metal complexes (**240**; M = Mo, R = H, *i*-Pr, CH_2SiMe_3, X = Cl, Br, I) and (**240**; M = W, R = H, X = Br, I), which slowly underwent in solution reductive elimination of carbon monoxide to give the corresponding neutral complexes (**241**) ⟨95JCS(D)4091⟩. The chromium analogue (**238**; M = Cr, R = H) was also prepared by the radical-initiated cyclotrimerization of $[Cr(CO)_3\{PH_2(CH_2CH{=}CH_2)\}_3]$ and derivatized at phosphorus by the previously developed methods ⟨96JCS(D)1801, 97JCS(D)193⟩. A breakthrough in this area occurred when conditions were found for the liberation of the free macrocycles from the complexes. Thus, treatment of (**238**; M = Mo, R = *i*-Pr) with sodium hydroxide in ethanol, followed by digestion of the inorganic by-products in water, furnished (**242a**; R = *i*-Pr) in 80% yield ⟨96CC293⟩. The ^{31}P NMR spectrum of the free phosphine consisted of a singlet at δ −19.8, in agreement with stereospecific liberation of the facially coordinating diastereomer (**242a**; R = *i*-Pr) of C_{3v} symmetry. Diastereomer (**242b**) of C_s symmetry was not isolated. The free triphosphadodecanes (**242a**; R = H, Me, Et, *i*-Bu, *t*-Bu, Me_3SiCH_2) were similarly liberated from the corresponding molybdenum complexes ⟨96IC4563⟩. Pure (**242a**; R = *i*-Pr) was subsequently employed for the preparation of various rhodium(III), iron(II), and ruthenium(II) complexes ⟨97JCS(D)3201⟩. In the most recent work, it was shown that (**238**; M = Mo, R = H) reacted with CX_4 (X = Cl, Br) in the presence of triethylamine to give the halogenated derivatives (**238**; M = Mo, R = Cl, Br) in high yields ⟨99JCS(D)1045⟩. The latter, when treated with PhLi, PhMgBr, $(PhCu)_6$, or Ph_2CuLi in THF, furnished (**238**; M = Mo, R = Ph) from which (**242a**; R = Ph) was displaced by oxidation of the complex with iodine and decomplexation with sodium hydroxide in ethanol.

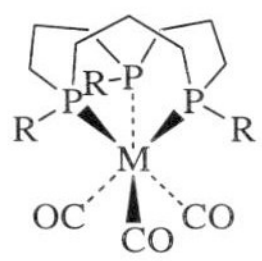

(238)

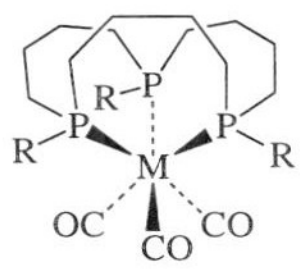

(239)

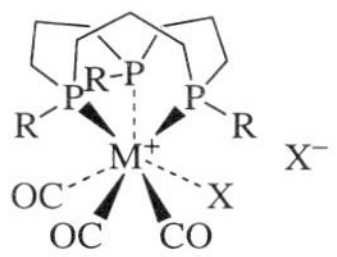

(240)

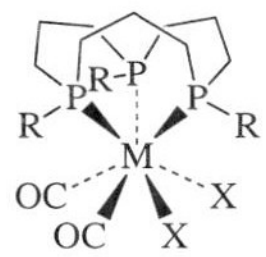

(241)

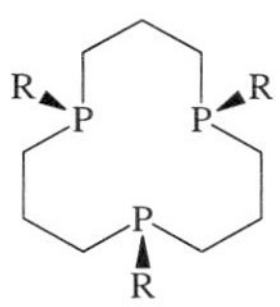

(242a)

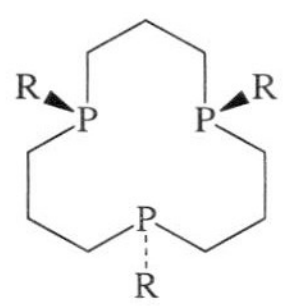

(242b)

The cyclic alkylation of 1,2-phenylenebis(phosphine) in the complex (**243**) with 1,3-dibromopropane under basic conditions gave a variety of phosphines after the displacement from the metal, including the bis(secondary phosphine) (**244**) and the R^*,S^* diastereomers of the bis(tertiary phosphines) (**245**) and (**246**); the crystal structure was determined of the square-planar complex *anti*-$[Ni\{(R^*,S^*)\text{-}(\mathbf{245})\}_2](ClO_4)_2\bullet CH_2Cl_2$ ⟨95MI934-01⟩.

6.1.3.4 Cyclic Phosphonites and Related Compounds

A great variety of 7- and larger-membered rings containing the phosphonyl, phosphoryl, thiophosphoryl, or iminophosphoryl groups have been prepared from appropriate halophosphorus compounds and deprotonated diols ⟨94CRV1183⟩. The types of compounds within the category can be represented by the diagrams (**247**) and (**248**). For example, 2-butene-1,4-diol reacted with $POCl_3$ in the presence

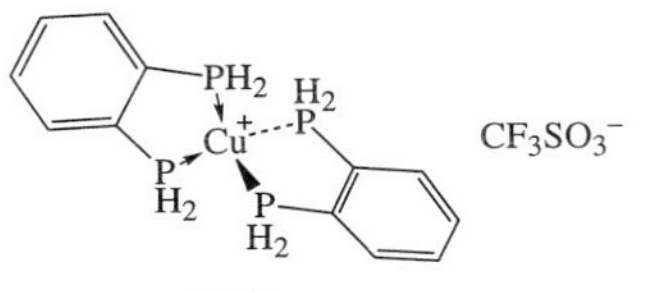

(243)

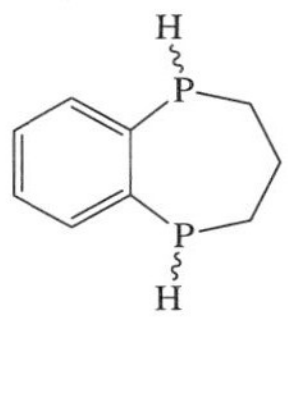

(244)

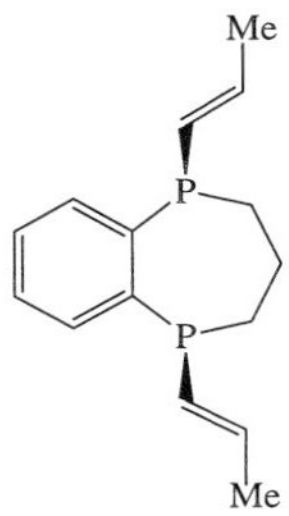

(245)

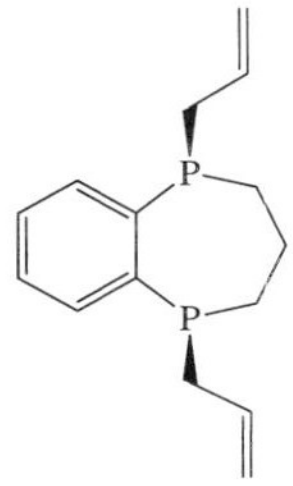

(246)

of triethylamine to give (**249**; X = Cl) in 68% yield, which, upon hydrolysis, gave (**249**; X = OH); ozonolysis of the latter in methanol followed by reductive workup with dimethyl sulfide, afforded the bis(glycoaldehyde)phosphodiester (**250**) ⟨94TL9109⟩. Certain cytosine and guanine ⟨97TL1451⟩ and adenine and uracil ⟨97T11595⟩ derivatives of (**249**) have been similarly cleaved to generate potentially prebiotic monomers of RNA. Cyclic phosphite (**251**) photorearranged into (**252**) in 50–70% yield. The quantum yield for the formation of (**252**) on sensitization by triphenylene is 0.005 in benzene ⟨96JA6192⟩.

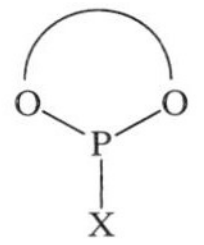

(247)

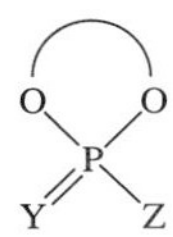

(248)

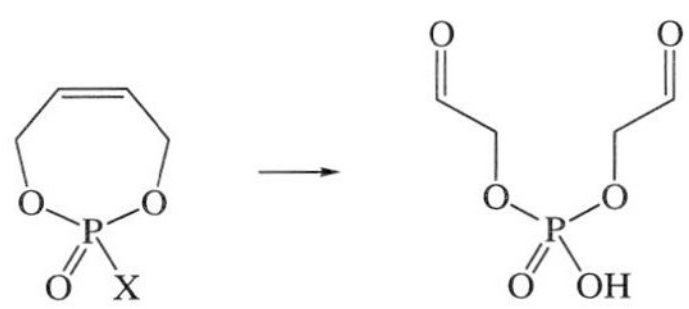

(249) **(250)**

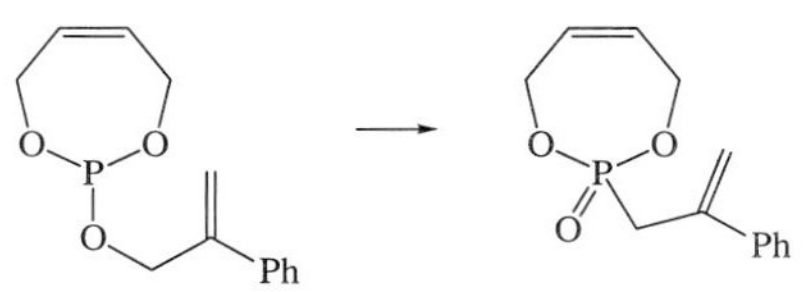

(251) **(252)**

(*R*)-1,1′-Binaphthol reacts with HMPA in benzene in the presence of ammonium chloride to give air-stable (**253**) in almost quantitative yield; as an NMR shift reagent, unlike related 5- and 6-membered compounds, it was ineffective ⟨94TA699⟩. Amidite (**253**) and related compounds are, however, effective as ligands for the copper(I) iodide-catalyzed asymmetric addition of diethylzinc to cyclohexanone and chalcone ⟨96AG(E)2374, 97AG(E)2620⟩ and for iridium(I)-catalyzed alkylic substitutions ⟨99CC741⟩. The C_2-symmetric phosphites (**254**) react with 2,2′-dihydroxy-3,3′-dimethyl-1,1′-biphenyl to give (**255**), which exist as rapidly interconverting atropisomers because of the low barrier to inversion of the seven-membered rings about the biphenyl carbon–carbon bonds; use of (**255**) for the rhodium(I)-catalyzed asymmetric hydrogenation of dimethyl itaconate, nevertheless gave the reduction product in 99% ee, which is similar to the enantioselectivity observed for the same reaction with use of the chiral auxiliary derived from (**254**) and conformationally rigid (*R*)-1,1′-binaphthol ⟨99AG(E)179⟩.

N,*N*-Dimethyl-*p*-toluidine reacts with phosphoryl chlorides at 130°C to give (**256**; X = O, S) in ca. 68% yields ⟨86PS(26)185, 87JCS(P1)1547⟩. The crystal structures of the two compounds were

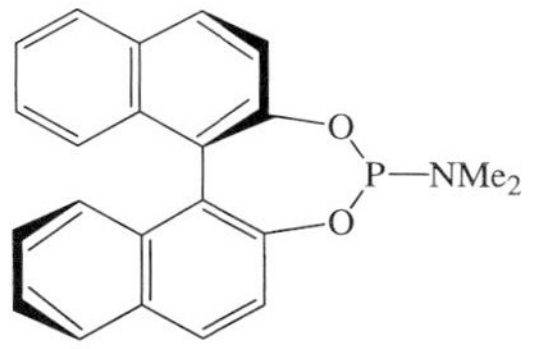

(253)

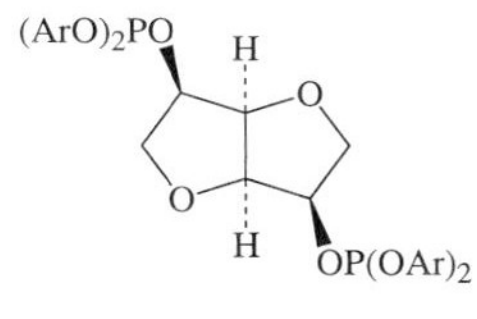

(254)

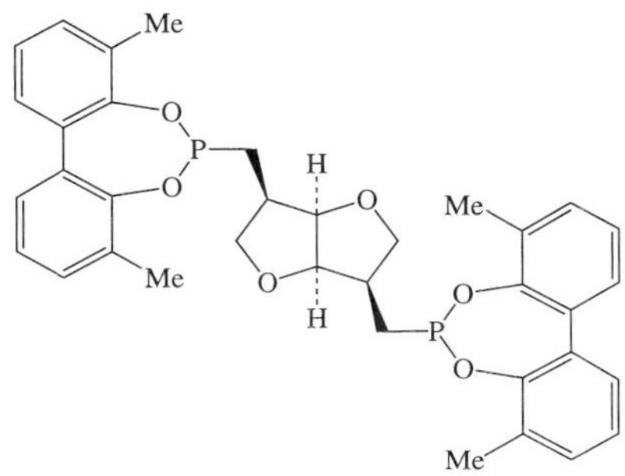

(255)

determined. The diazaphosphepane 2-oxide derivative (**257**; R = 2-naphthylmethyl) cyclizes into (**258**; R = 2-naphthylmethyl)) when treated with aqueous lithium hydroxide in tetrahydrofuran–water ⟨96H2367⟩.

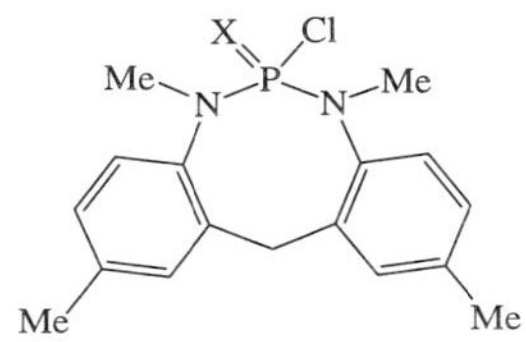

(256)

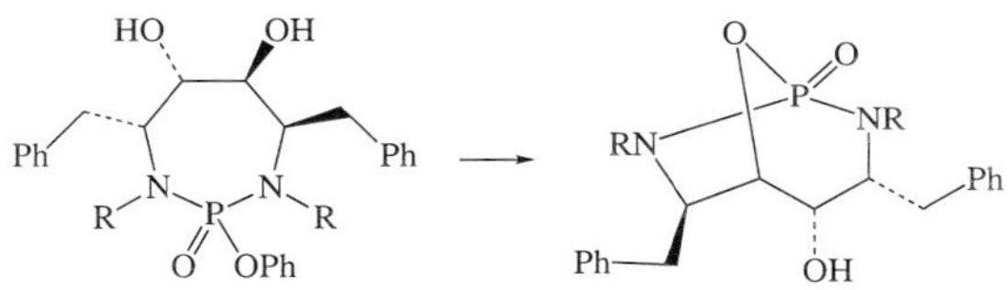

(257) (258)

Addition of two equivalents of the valence isomer of benzothiete, (**259**), to the cyclic phosphonites (**260**; $n = (CH_2)_{2-5}$) affords the macrocycles (**261**; $n = (CH_2)_{2-5}$) ⟨89TL155, 91CB377⟩. In the reaction of 1,3,2-dioxaphosphepin (**262**), 10-membered (**263**) was formed by an intramolecular rearrangement. 2-Bromobenzyl phenylvinylphosphinate, palladium(II) acetate, triphenylphosphine, and triethylamine, when heated in acetonitrile in a sealed tube at 110°C, react to give a 95% yield of (**264**) and (**265**) in the ratio 79 : 21 ⟨94JCS(P1)1665⟩.

Reviews are available of conformational investigations and the stereochemistry of heterocyclic phosphorus compounds ⟨87MI934-01⟩ and compounds containing seven and larger rings, including cyclic phosphonites and related compounds ⟨B-88MI934-01⟩; 8-membered rings have been treated separately ⟨92PS(68)155⟩.

6.1.3.5 Cyclic Oxyphosphoranes

Two reviews are available of pertinent information on cyclic oxyphosphoranes containing five- to eight-membered rings, including oxygen, sulfur, and nitrogen as heteroatoms ⟨90PS(53)437, 92MI934-01⟩. The pentaoxyphosphoranes, for example (**266**), are useful models for intermediates in phosphate ester hydrolyses.

Sulfur, oxygen, and nitrogen incorporated in flexible ring systems can serve to increase coordination numbers at phosphorus in the tri-, tetra-, and penta-coordinated states. The effect depends upon the elec-

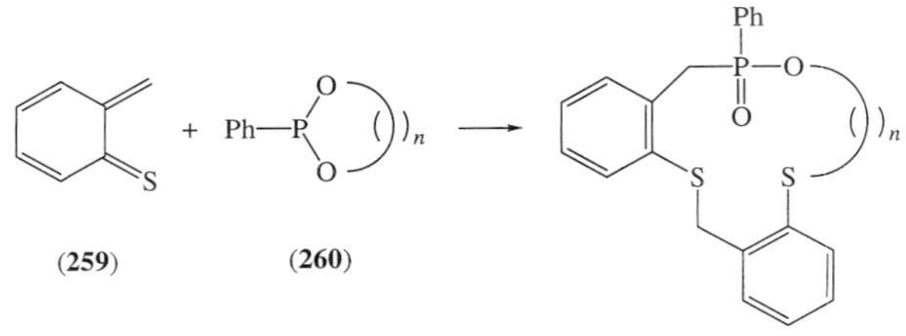

(259) (260)

(261)

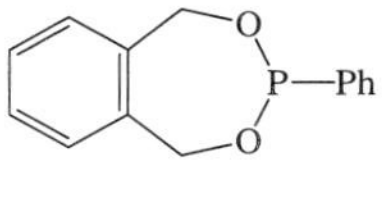

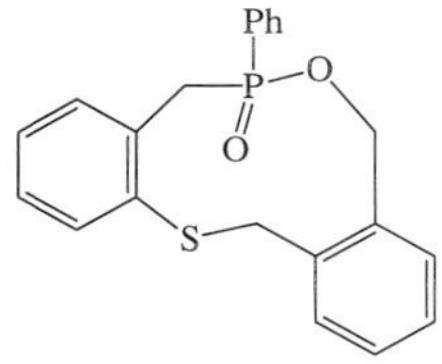

(262) (263)

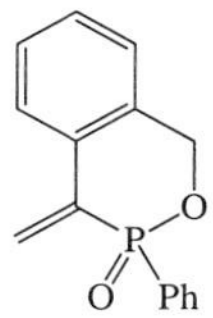

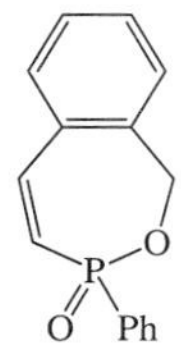

(264) (265)

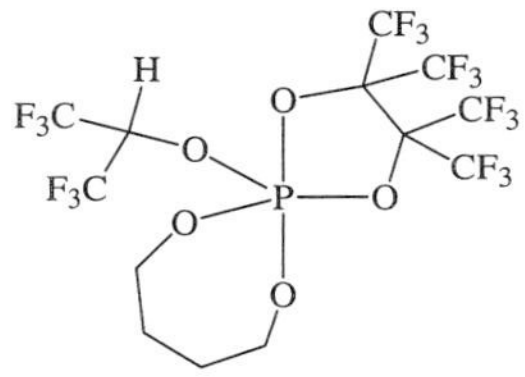

(266)

trophilicity at phosphorus. Numerous examples are now known of increased coordination at phosphorus in phosphites, phosphates, and pentaoxyphosphoranes, of which (**267**) and (**268**) are examples of compounds in which an additional oxygen atom is coordinated ⟨97JA11441, 98IC6329⟩.

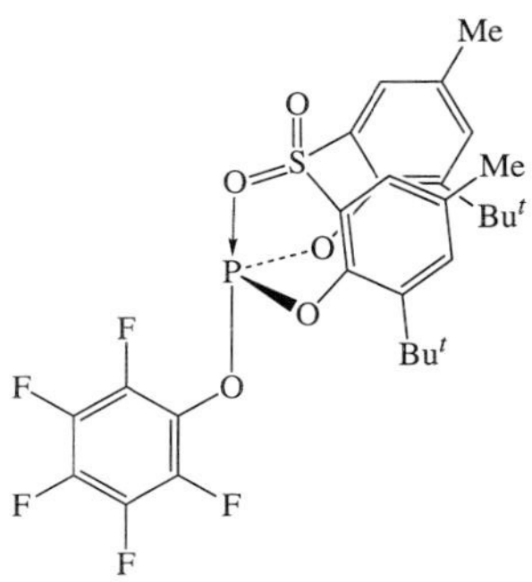

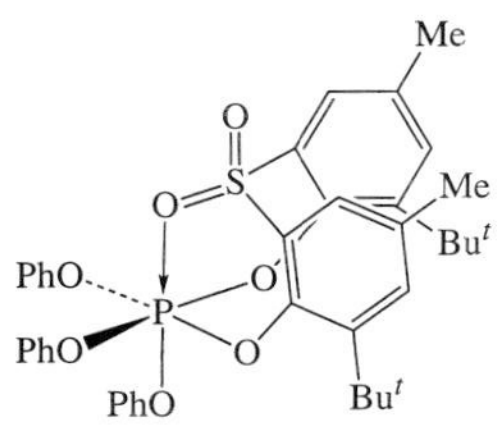

(268)

(267)

6.1.3.6 Cyclic Phosphazene Derivatives

Cyclic triphosphazatrienes condense with difunctional reagents such as linear diamines, diols, and aminoalcohols to give spirocompounds (**269a**), ansa compounds (**269b**), and macrocyclic compounds (**269c**) ⟨90PS(47)119, 90PS(53)333, 90JMS245, 90JMS253, 95JOC7433, 97IC2730⟩.

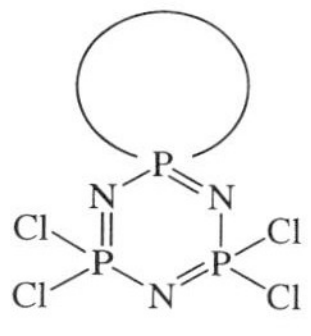

(269a)

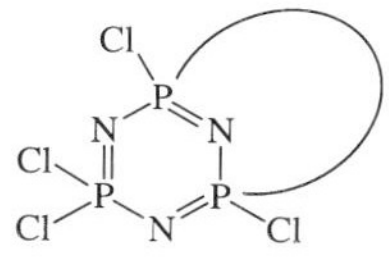

(269b)

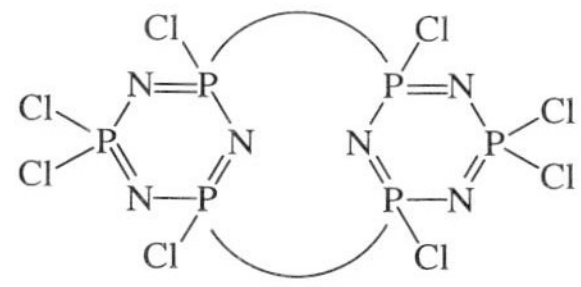

(269c)

The Mannich reaction for the aminomethylation of CH-acidic compounds has provided numerous tertiary phosphine-containing macrocycles ⟨84ZC365⟩. For example, the reaction of bis(hydroxymethyl)phenylphosphine with aniline or (*R*)-(+)- or (*S*)-(−)-*N*,*N*′-bis(1-phenyl-1-ethyl)ethylene-1,2-diamine in boiling acetonitrile or ethanol gives the 1,5-diaza-3,7-diphosphacyclooctane (**270**) (77% yield) ⟨80TL1409⟩ or the optically active 1,5-diaza-3-phosphacycloheptanes (**271**; R^* = (*R*)- or (*S*)-PhCHMe) (92% yield) ⟨80TL3467⟩, respectively. Nickel carbonyl and molybdenum carbonyl complexes of (**270**) were reported.

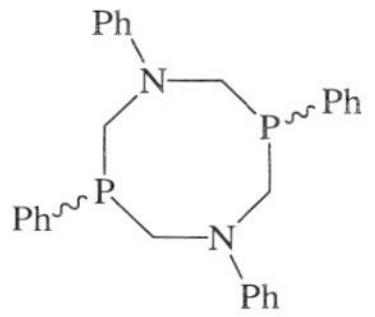

(270)

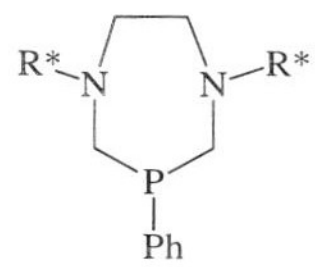

(271)

6.1.3.7 Polyazaphosphorus Macrocycles

A general synthesis has been described of symmetrically and unsymmetrically functionalized 18-, 20-, 22-, and 30-membered rings containing P–N–N linkages and involving either [1 + 1], [2 + 2], or [3 + 3] condensations between phosphadihydrazides and 1,2-, 1,3-, 1,4-, or 1,11-dialdehydes ⟨90JA5618⟩. The variety of rings isolated are of the type (**272**; Y = O, S). The reader is referred to two detailed reviews on this substantial topic ⟨92MI934-01, 94CRV1183⟩.

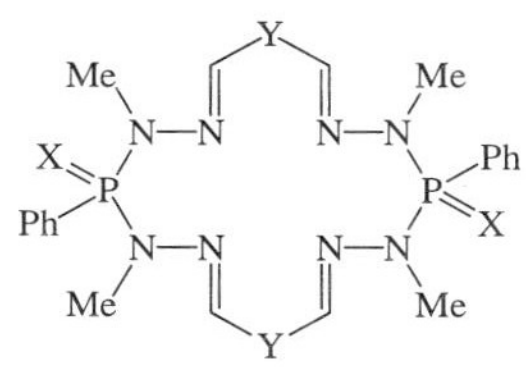

(272)

Reactions of the functionalized phosphodihydrazides $XP(Y)(NMeNH_2)_2$ (X = Cl, N_3, Y = O, S) with 1,2- or 1,3-$C_6H_4(CHO)_2$ or (**273**; X = Cl, N_3, Y = O, S) gave good yields of the macrocycles (**274**; X = Cl, N_3, Y = O, S) or (**275**; X = Cl, N_3, Y = O, S), respectively ⟨95S952⟩. The compounds are precursors of cryptands and multimacrocyclic systems. A review of methods of synthesis of N–N–P–N–N containing macrocycles is available ⟨96SL1019⟩.

6.1.4 BICYCLIC SYSTEMS WITH BRIDGEHEAD PHOSPHORUS

6.1.4.1 Bridgehead Tertiary Phosphines

Treatment of allylphosphine in benzene with a radical initiator, such as AIBN, gives mainly PH_3, but also (**276**) ⟨82PS(12)227⟩. The bicyclic bis(tertiary phosphine) was characterized by ^{31}P, ^{13}C, ^{1}H NMR and mass spectral data. The C_3 tertiary phosphine (**277**) was obtained in moderate yield in a one-pot procedure involving successive quenching of lithium 2-bromophenolate with phosphorus trichloride and phosphoryl chloride under high-dilution conditions ⟨88AG(E)835⟩. The barrier to racemization of the C_3 phosphine was estimated to be 11.8 kcal mol^{-1} (49.4 kJ mol^{-1}) at 0°C, but diastereomers were not observed when the phosphines were reacted with a homochiral palladium(II) complex.

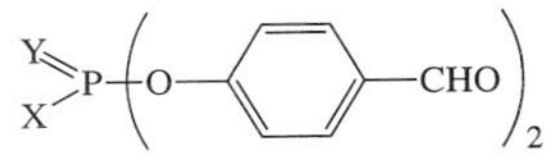

(273)

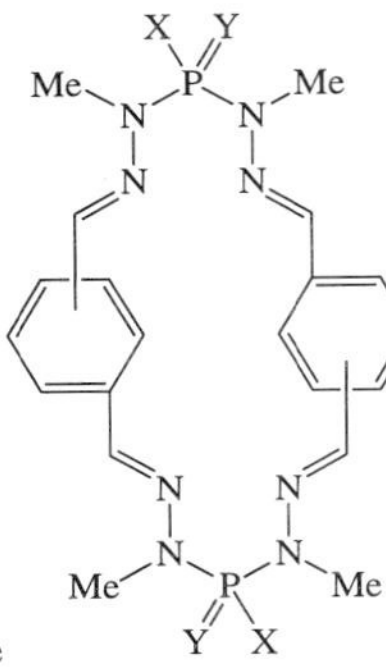

(274)

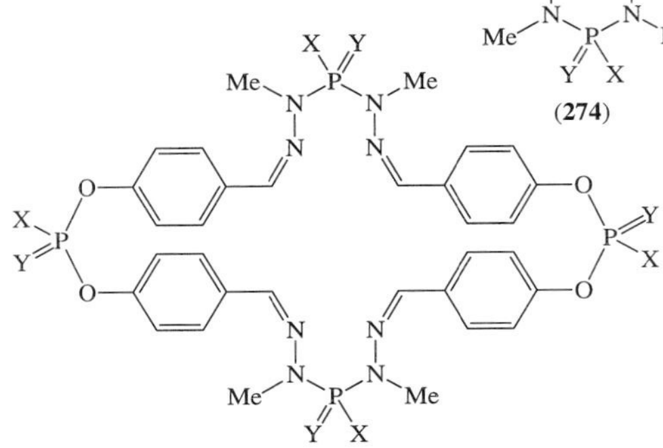

(275)

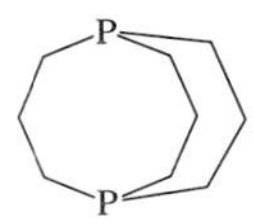

(276)

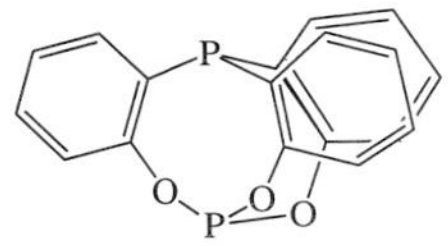

(277)

High-dilution coupling of 1,3,5-tris(mercaptomethyl)benzene with tris[(3-bromomethyl)phenyl]phosphine oxide afforded the thiacyclophane (**278**) in 17% yield ⟨88TL5101⟩. Compound (**278**) was converted into the C_3 phospha[2.2.2]cyclophane (**279**) in three steps, but, again, diastereomeric complexes were not observed when the latter was reacted with a homochiral palladium(II) complex.

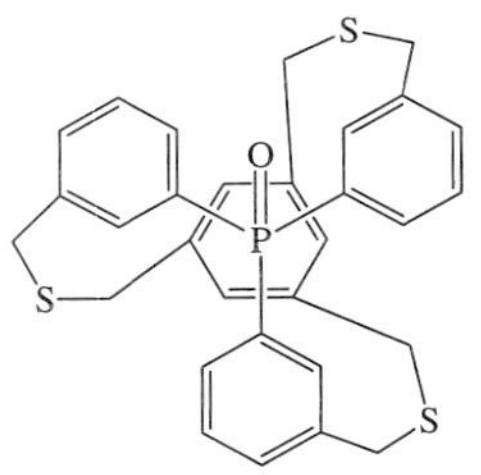

(278)

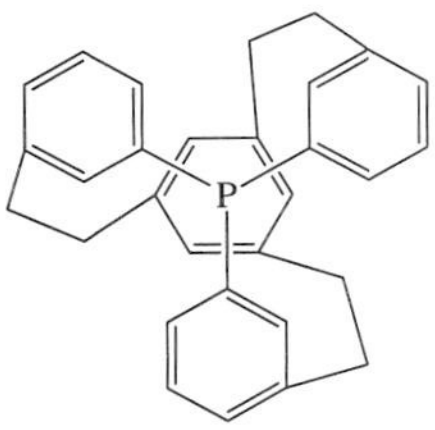

(279)

The base-promoted high-dilution condensation of tris(2-mercaptophenyl)phosphine with 1,3,5-tris(bromomethyl)benzene furnished the phosphaphane (**280**) in up to 15% yield ⟨90JA6406, 91JA2672, 98JA6421⟩. In the $^{13}C\{^1H\}$ NMR spectrum of (**280**), a through-space coupling was observed between the carbons of the basal aromatic ring and the phosphorus atom. The crystal structure of (**280**) indicated that the phosphorus was pyramidalized in the direction of the basal aromatic ring, which accounted for the surprisingly large down-field shift of the phosphorus in the ^{31}P NMR spectrum. The inward direction of the lone pair in (**280**) contrasts with the outward direction of the lone pair in (**279**). Accordingly, attempted oxidation at phosphorus in (**280**) failed, the trisulfone being formed instead. A variable temperature NMR investigation of (**280**) in DMSO-d_6 indicated a lower limit to racemization of 21.98 kcal mol^{-1} (92 kJ mol^{-1}). The phosphaphanes (**281**; R^1 = NO_2, R^2 = H, Br, NO_2) were prepared similarly ⟨93JOC3502⟩; the nitro compound was reduced to the amine with titanium(III) chloride. The crystal structures of (**281**; R^1 = NO_2, NH_2, R^2 = H) were determined. The enantiomers of (**281**; R^1 = NO_2, R^2 = H) were separated by chiral HPLC and were found to have considerable stability, i.e. $E_{inv} > 26$ kcal mol^{-1} (108.8 kJ mol^{-1}).

The phosphine oxide (**282**) reacts with copper(II) acetate in pyridine at 60°C (2 h) under high-dilution

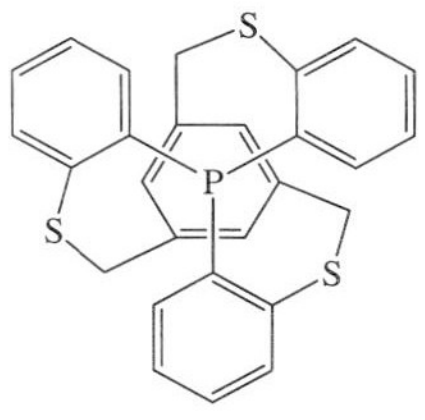

(280)

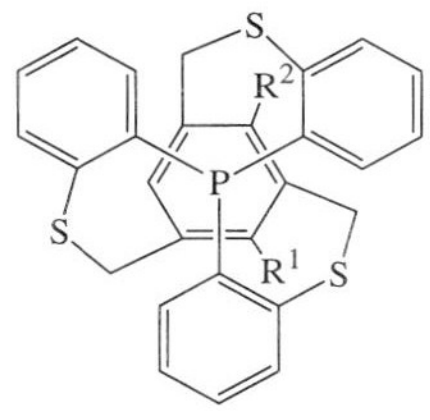

(281)

to give the macrobicyclic bis(phosphine oxides) (**283a**) and (**283b**) in 14% and 7% yields, respectively 〈89JA9132, 90JA8931〉. Catalytic hydrogenation of (**283a**) and (**283b**) over 5% $Pd/CaCO_3$ provided the corresponding compounds with saturated bridges. X-ray crystal structure determinations on (**283a**) (**283b**), and the *endo–exo* diastereomer of the saturated compound, confirmed the dispositions of the P=O groups. Whereas (**283a**) and (**283b**) displayed pronounced helicity, the saturated form was distorted only slightly from C_{3v} symmetry. The host–guest behavior of the diastereomers of the four macrobicycles with substituted phenols and dichlorodiphenyltin(IV) was investigated by ^{1}H and ^{31}P NMR spectroscopy and association constants were determined.

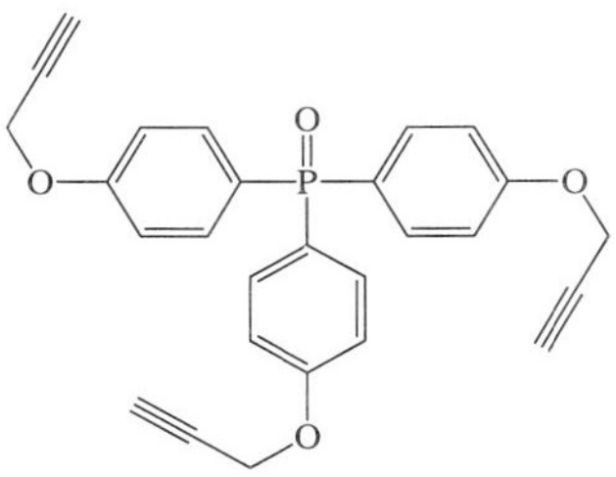

(282)

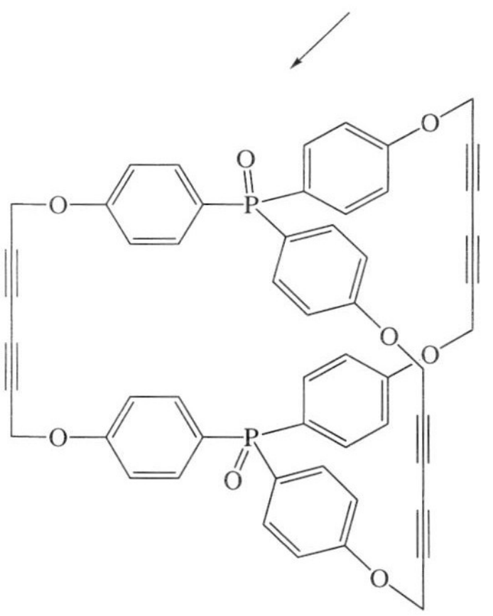

(283a)

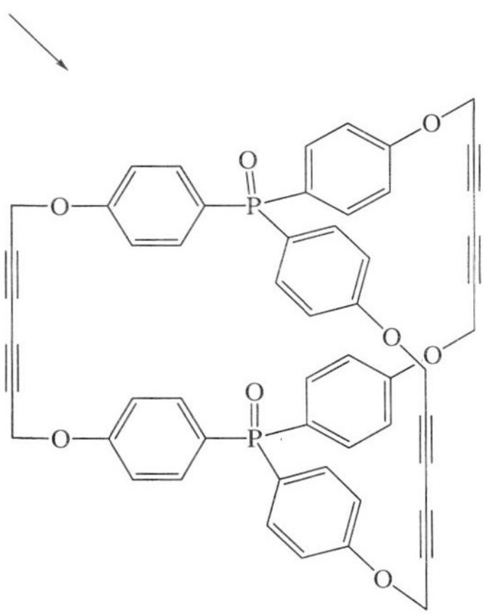

(283b)

6.1.4.2 Calixarene Phosphites and Related Compounds

The reaction of calix[4]arene or its *p-t*-butyl derivative with tris(dimethylamino)phosphine gave the "bottom of the basket" zwitterionic six-coordinate phosphorus derivatives (**284**; R = H, Bu^t) 〈90JA9422, 91CC562, 93PS(75)253, 94IC2657〉. The attempted deprotonation of this compound at phosphorus with *n*-butyllithium to give the basket aminophosphorane with four basal P–O bonds was unsuccessful (the NH proton was removed), but trifluoroacetic acid or heat yielded the bicyclic phosphite (**285**; R = H, Bu^t). The crystal structures of (**284**; R = H) and the oxidized form of (**285**; R = H) were determined. The P–H atom in (**284**; R = Bu^t) is inside the basket. With $(Me_2N)_2PCH_2P(NMe_2)_2$, *p-tert*-butyl-calix[4]arene reacts to give the R^*,S^* diastereomer (**286**), which can be converted into (**287**) by heating 〈95IC5405〉.

Treatment of *p-t*-butylcalix[4]arene with phosphorus pentachloride afforded (**288**), which underwent ethanolysis to give (**289**); the crystal structure of (**289**) was determined 〈93PS(84)217〉. Vacuum pyrolysis of

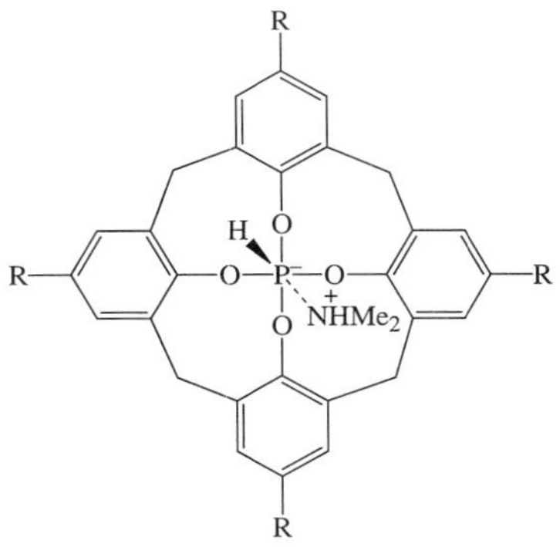

(**284**)

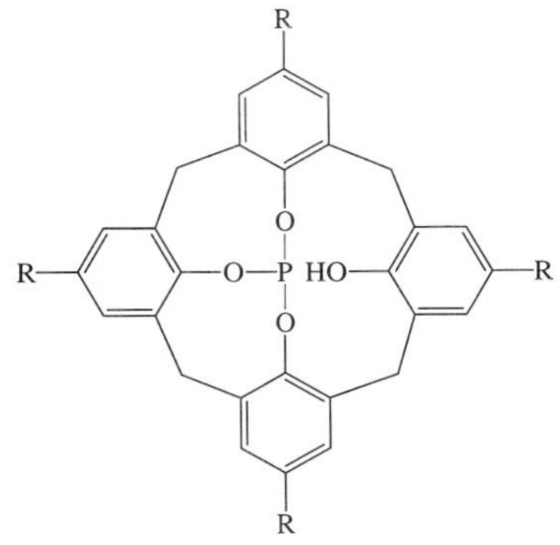

(**285**)

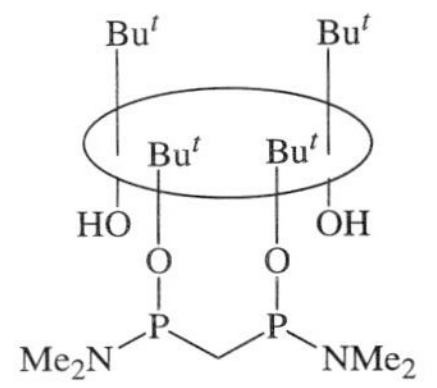

(**286**)

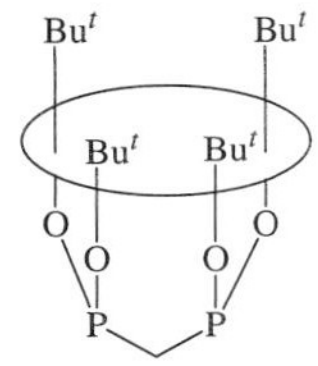

(**287**)

the dialkylphosphate esters of *p-t*-butylcalix[4]arene and *p-t*-butylcalix[6]arene resulted in the phosphorus polybridged compounds (**290**) and (**291**), which displayed dynamic processes ⟨93CC13⟩. The structure of (**291**) was determined by X-ray crystallography. Related compounds having a phosphorus atom in the bridgehead position have also been prepared ⟨92PS(71)7⟩. The trivalent phosphorus analogue of (**291**) is a chelating agent for palladium(II), platinum(II) and rhenium(I); the bite angle for the ligand in the dichloropalladium(II) complex is ca. 90° ⟨96CC583⟩.

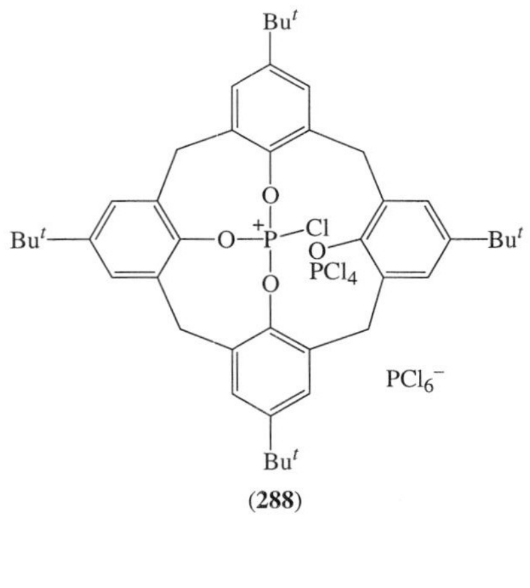

(**288**)

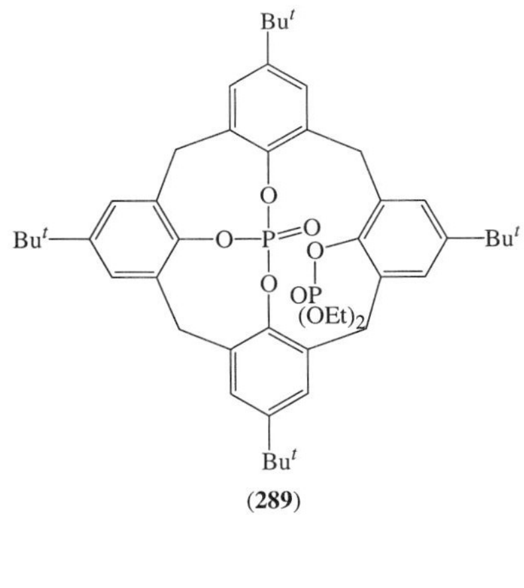

(**289**)

The bowl-shaped calix[4]resorcinarene (**292**; R = CH_2CH_2Ph) reacts with dichlorophenylphosphine in the presence of pyridine to give (**293**; R = CH_2CH_2Ph, R^1 = H, X = Ph) ⟨95IC323⟩. Although there are six possible diastereomers of the cavitand, molecular mechanics calculations indicated that the diastereomer with all phenyl groups on phosphorus directed outwards and lone pairs inwards was preferred, which is consistent with the single ^{31}P NMR resonance observed and with its transition metal coordination chemistry ⟨95IC323, 95JA8362⟩. The synthesis, separation, and configurational analysis of the diastereomeric cavitands (**294**; R = CH_3, R^1 = H and R = CH_3, R^1 = Br) have been described ⟨95JOC235⟩. Substituent R^2 was varied with the aim of achieving the stereoselective formation of diastereomers suitable for multiple hydrogen-bonding interactions. The compounds (**293**; R = Me, R^1 = H, Br, X = Cl) have also been isolated and characterized by solid-state NMR spectroscopy ⟨97CB/R1715⟩. Reactions of (**293**; R = Me, R^1 = H, X = Cl) with MeMgI, Me_3SiNMe_2 and $HNMe_2$ were investigated.

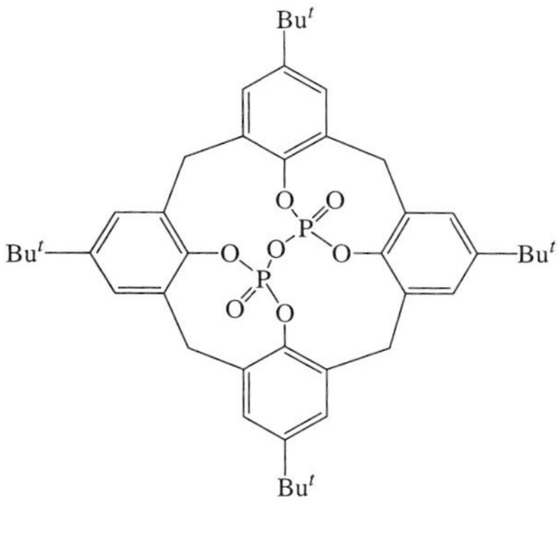

(290)

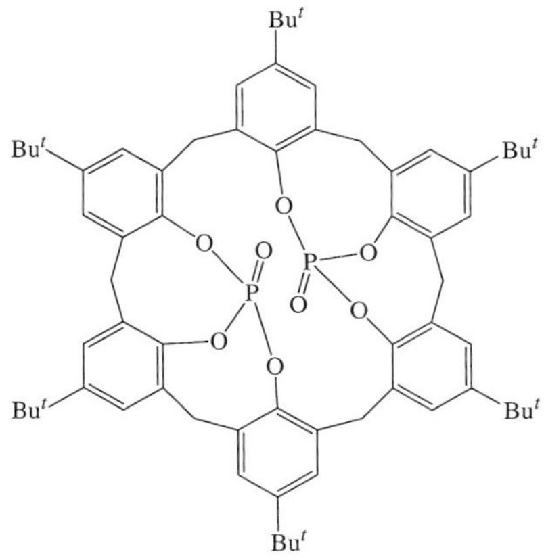

(291)

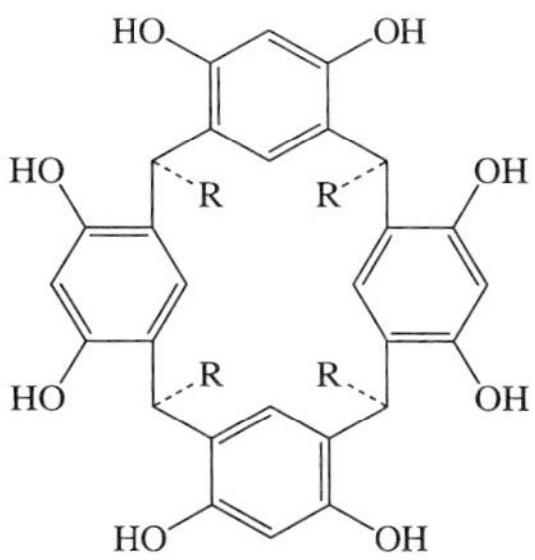

(292)

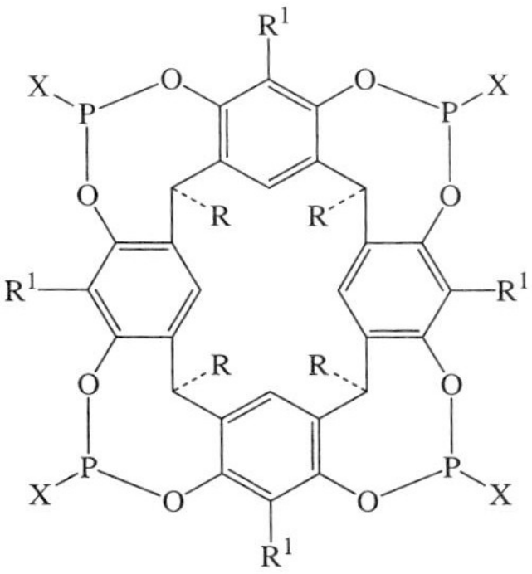

(293)

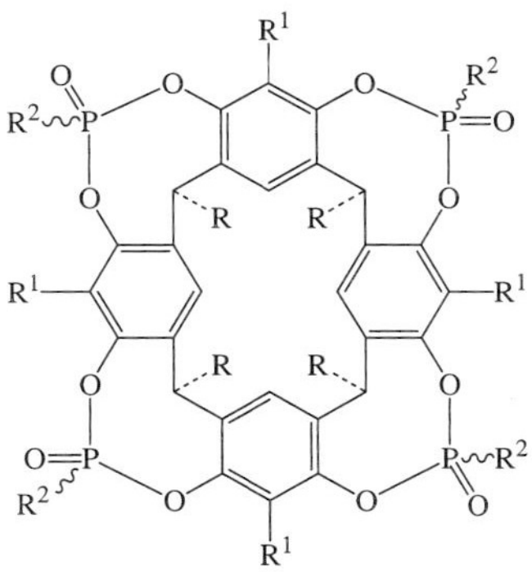

(294)

6.1.4.3 Phosphatranes and Related Compounds

These compounds, which are conveniently known by the trivial names as indicated for the structures (**295a**)–(**295c**), having unusual basicities and reactivities because of the various degrees of delocalization of the bridgehead nitrogen lone pair into the three-center, four-electron bond systems along the molecular axis in each case ⟨93ACR483⟩. The unusual basicity of the phosphorus in a phosphatrane was indicated by the attempted alkylation of the unstable phosphite (**296**) with trimethyl- or triethyl-oxonium tetrafluoroborate, which gave (**297**); the proton originated from the reaction of the oxonium salt with unreacted triethanolamine in the reaction mixture ⟨77JA6607, 86JOC4287⟩. It was not found possible to deprotonate (**297**) with sodium methoxide or 1,8-bis(dimethylamino)naphthalene. Boron trifluoride, Et_3Si^+, or a proton binds with the phosphate analogue of (**296**) to form the trigonal bipyramidal transannulated phosphatrane having a P–O–B, P–O–Si, or P–O–H bond, respectively ⟨85JA7084, 90PS(49/50)163, 86JA4918⟩. The phosphorus sulfide of (**296**) behaves similarly towards these reagents.

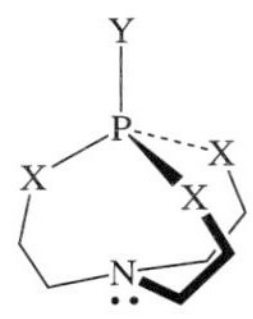

(**295a**)
pro-phosphatran

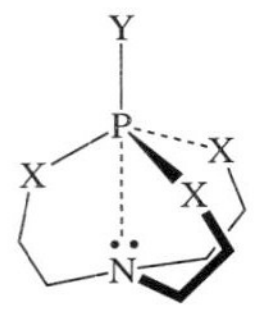

(**295b**)
quasi-phosphatran

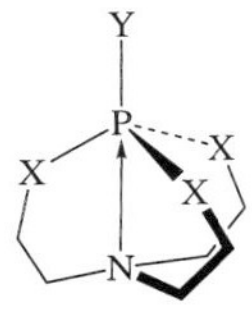

(**295c**)
phosphatran

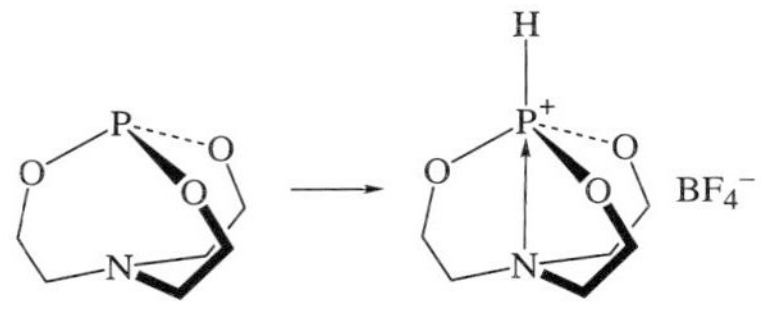

(**296**) (**297**)

The compounds (**298**; R = H, Me, Bn and X = Cl) were prepared in almost quantitative yields when chlorobis(dimethylamino)phosphine was reacted with the appropriate tris(2-aminoethyl)amine in the presence of triethylamine for 1 h at room temperature ⟨89JA3478, 91ZAAC(605)163⟩. The compound (**298**; R = H, X = OTf) was prepared in 99% yield by reacting tris(dimethylamino)phosphine, tris(2-aminoethyl)amine, and triflic acid in dichloromethane for 15 min at room temperature. The salts (**298**; R = H, Me, Bn and X = Cl) can be deprotonated with potassium *t*-butoxide to the neutral proazaphosphatranes (**299**; R = H, Me, Bn), which are ca. 10^7 times more basic than other known phosphine derivatives due to the stability of the transannulated salts (**298**; R = H, Me, Bn) ⟨89ZAAC(578)75, 90JA9421⟩. The compound (**299**; R = H) polymerizes on attempted isolation, but (**299**; R = Me, Bn) are stable sublimable solids. The syntheses and crystal structure determinations of a variety of *P*-oxidized prophosphatranes and quasiphosphatranes have been performed ⟨89PS(41)21, 89ZAAC(578)75, 90IC2214, 92JA3129⟩. The variable basicity of (**299**; R = Me) due to the flexible transannular bond ⟨93ACR483⟩ is believed to play an important role in the superior catalytic activity of this compound for the cyclotrimerization of phenylisocyanate ⟨93AG(E)896⟩, acylation ⟨96JOC8750⟩, dehydrohalogenation ⟨97JOC4827⟩, and silylation of hindered, acid-sensitive and base-sensitive alcohols ⟨97JOC5057⟩. A synthesis from (*S*)-proline of the C_3 protetraazaphosphatrane (**300**) has been described ⟨98JOC5692⟩. The relative stability of (**301**) has been attributed to its partially annulated structure (P–N 3.250 Å) ⟨92JA3129⟩. The reaction of the proazaphosphatrane (**302**) with methyl iodide gave both the annulated P^+–S–Me and open N^+–Me regioisomers ⟨93JA1660⟩.

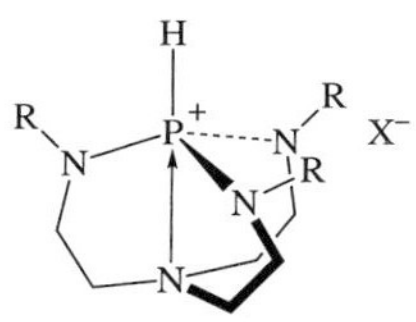

(**298**)

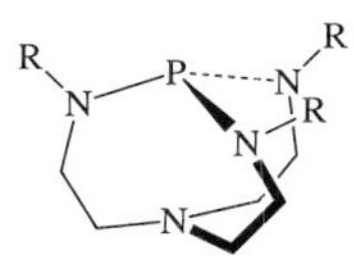

(**299**)

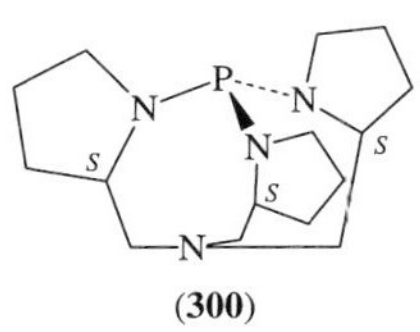

(**300**)

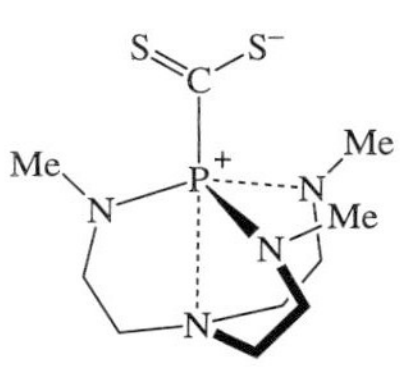

(**301**)

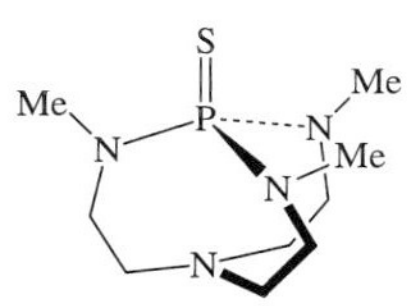

(**302**)

The cyclenphosphorane (cyclen)PH (**303**), first synthesized in 1978 ⟨78TL5149, 78TL4333⟩, upon deprotonation with *n*-BuLi gives (cyclen)PLi ⟨88IC3012⟩; the latter reacts with electrophilic reagents RX to give the "open" derivatives (**304**; R = Mes_2B, Ph_2P) or a mixture of (**304**; R = Me) and the "closed" derivatives (**305**; R = Me) ⟨95IC1638⟩. The cyclenphosphine (**306**) is obtained by transamination between HMPTA and cyclen; when heated with sulfur, it affords (**307**), which exhibits fluxional behavior in solution ⟨96T2995⟩. The secondary amine group in (**306**) can be deprotonated and the sulfur atom can be protonated ⟨99HCA790⟩. In contrast to the behavior of (**307**), the NH and PS groups in (**306**) are interdependent, which is related to the observation that (**306**) can be *N*- or *S*-alkylated depending on electrophile. The behavior of (**306**) can be explained by a strong transannular *N–P* interaction.

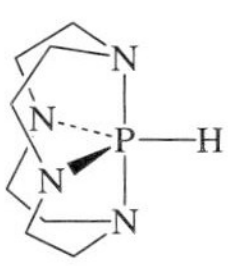
(**303**)

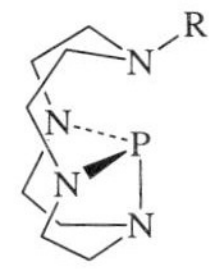
(**304**)

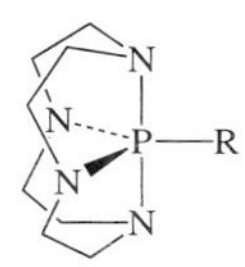
(**305**)

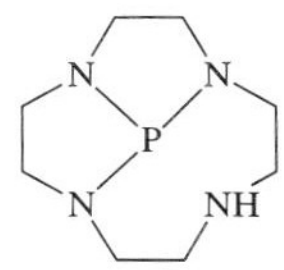
(**306**)

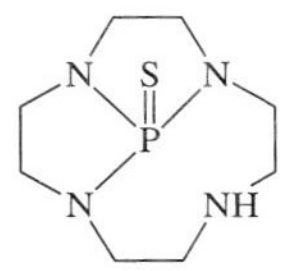
(**307**)

The prophosphatrane (**308**) has been prepared by the reaction of tris(2-hydroxyphenyl)amine with tris(dimethylamino)phosphine ⟨87HCA1063⟩ or phosphorus trichloride ⟨88POL679⟩. High-dilution conditions were not required for the synthesis of this air-sensitive compound by either route. The corresponding phosphate was prepared similarly from phosphorus oxychloride ⟨87HCA1063, 88POL679⟩. The dibenzo[*c*,*f*][1,5]azaphosphocine (**309**) reacts with thionyl chloride to give a solution of (**310**); ^{31}P and ^{1}H NMR data for (**310**) are consistent with the transannular *N–P* bond ⟨86TL5221⟩.

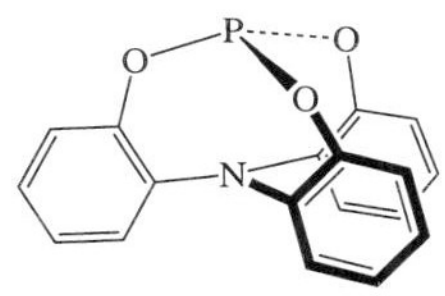
(**308**)

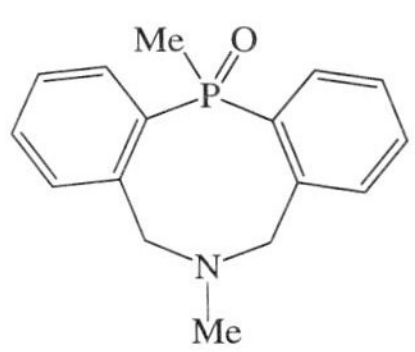
(**309**)

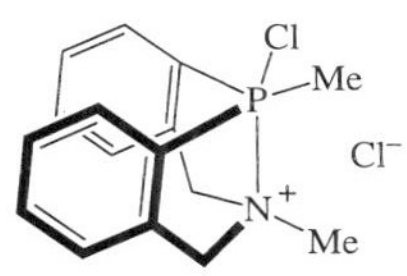
(**310**)

6.1.5 REFERENCES

53JCS1130 F.G. Mann, I.T. Millar and B.B. Smith; *J. Chem. Soc.*, 1953, 1330.
63AG(E)479 G. Märkl; *Angew. Chem., Int. Ed. Eng.*, 1963, **2**, 479.
63AG(E)620 G. Märkl; *Angew. Chem., Int. Ed. Eng.*, 1963, **2**, 620.
64CB747 G. Wittig and A. Maercker; *Chem. Ber.*, 1964, **97**, 747.
65CB576 D. Hellwinkel; *Chem. Ber.*, 1965, **98**, 576.
66CI196 D.W. Allen, I.T. Millar and F.G. Mann; *Chem. and Ind.*, 1966, 196.
66JA3832 T.J. Katz, C.R. Nicholson and C.A. Reilly; *J. Am. Chem. Soc.*, 1966, **88**, 3832.
66JCS(C)245 J.H. Davies, J.D. Downer and P. Kirby; *J. Chem. Soc. (C)*, 1966, 245.
67JCS(C)1869 D.W. Allen, I.T. Millar and F.G. Mann; *J. Chem. Soc. (C)*, 1967, 1869.
68JOC579 A.M. Aguiar and M.G. Raghanan Nair; *J. Org. Chem.*, 1968, **33**, 579.
70C431 W. Marty and G. Schwartzenbach; *Chimia*, 1970, 431.
70JCS(C)386 N.E. Waite and J.C. Tebby; *J. Chem. Soc. (C)*, 1970, 386.
70TL1273 G. Märkl and H. Schubert; *Tetrahedron Lett.*, 1970, 1273.
72JHC331 K.L. Marsi, D.M. Lynch and G.D. Homer; *J. Heterocycl. Chem.*, 1972, **9**, 331.
72T5367 F. Mathey and J.P. Lampin; *Tetrahedron*, 1972, **28**, 5367.
72TL1949 F. Mathey and J.P. Lampin; *Tetrahedron Lett.*, 1972, 1949.

73JA4292 E.W. Turnblom and T.J. Katz; *J. Am. Chem. Soc.*, 1973, **95**, 4292.
73TL1455 G. Märkl and G. Dannhardt; *Tetrahedron Lett.*, 1973, 1455.
74CC442 J. Riker-Nappier and D.W. Meek; *J. Chem. Soc., Chem. Commun.*, 1974, 442.
75MI934-01 L. Horner, H. Kunz and P. Walach; *Phosphorus*, 1975, **6**, 63.
76JA2109 G.A. Gray, S.E. Cremer and K.L. Marsi; *J. Am. Chem. Soc.*, 1976, **98**, 2109.
77JA6607 D.S. Milbrath and J.G. Verkade; *J. Am. Chem. Soc.*, 1977, **99**, 6607.
77JA8051 W. Rosen and T.A. Deldonno; *J. Am. Chem. Soc.*, 1977, **99**, 8051.
77JA8370 L.D. Quin and E.D. Middlemas; *J. Am. Chem. Soc.*, 1977, **99**, 8370.
78TL4333 J.E. Richman and T.J. Atkins; *Tetrahedron Lett.*, 1978, 4333.
78TL5149 T.J. Atkins and J.E. Richman; *Tetrahedron Lett.*, 1978, 5149.
80TL1409 G. Märkl, G.Y. Jin and C. Schoerner; *Tetrahedron Lett.*, 1980, **21**, 1409.
80TL3467 G. Märkl and G.Y. Jin; *Tetrahedron Lett.*, 1980, **21**, 3467.
81CB1428 H. Schmidbaur, T. Costa and B. Milewski-Mahrla; *Chem. Ber.*, 1981, **114**, 1428.
82CB1367 T. Costa and H. Schmidbaur; *Chem. Ber.*, 1982, **115**, 1367.
82CC1170 E.F. Landvatter and T. Rauchfuss; *J. Chem. Soc., Chem. Commun.*, 1982, 1170.
82IC1215 L.G. Scanlon, Y.Y. Tsao, K. Toman, S.G. Cummings and D.W. Meek; *Inorg. Chem.*, 1982, **21**, 1215.
82JA1893 L.D. Quin, E.D. Middlemas, N.S. Rao, R.W. Miller and A.T. McPhail; *J. Am. Chem. Soc.*, 1982, **104**, 1893.
82JA2077 F. Mathey, F. Mercier, F. Nief, J. Fischer and A. Mitschler; *J. Am. Chem. Soc.*, 1982, **104**, 2077.
82JA4700 B.N. Diel, P.F. Brandt, R.C. Haltiwanger and A.D. Norman; *J. Am. Chem. Soc.*, 1982, **104**, 4700.
82JOC905 L.D. Quin, E.D. Middlemas and N.S. Rao; *J. Org. Chem.*, 1982, **47**, 905.
82OM1266 H. Schmidbaur, T. Costa, B. Milewski-Mahria, F.H. Köhler, Y.-H. Tsay, C. Krüger, J. Abart and F.E. Wagner; *Organometallics*, 1982, **1**, 1266.
82PS(12)227 B.N. Diel and A.D. Norman; *Phosphorus Sulfur*, 1982, **12**, 227.
82TL4915 G. Märkl and B. Alig; *Tetrahedron Lett.*, 1982, **23**, 4915.
83AG(E)907 H. Schmidbaur; *Angew. Chem., Int. Ed. Engl.*, 1983, **22**, 907.
83CB1923 E. Luppold and W. Winter; *Chem. Ber.*, 1983, **116**, 1923.
83IC3624 R. Bartsch, S. Hietkamp, S. Morton, H. Peters and O. Stelzer; *Inorg. Chem.*, 1983, **22**, 3624.
83ICAL17 M. Ciampolini, N. Nardi, F. Zanobini, R. Cini and P.L. Orioli; *Inorg. Chim. Acta*, 1983, **76**, L17.
83JA5960 N.S. Rao and L.D. Quin; *J. Am. Chem. Soc.*, 1983, **105**, 5960.
83JCS(D)841 J. Fischer, A. Mitschler, F. Mathey and F. Mercier; *J. Chem. Soc., Dalton Trans.*, 1983, 841.
83RTC326 A. van Zon, G.J. Torny and J.H.G. Frijns; *Recl. Trav. Chim. Pays-Bas*, 1983, **102**, 326.
83TL2545 G. Märkl and W. Burger; *Tetrahedron Lett.*, 1983, **24**, 2545.
84AG(E)894 G. Märkl and W. Burger; *Angew. Chem., Int. Ed. Engl.*, 1984, **23**, 894.
84IC3304 R. Bartsch, S. Hietkamp, H. Peters and O. Stelzer; *Inorg. Chem.*, 1984, **23**, 3304.
84JCS(D)575 M. Ciampolini, N. Nardi, P. Dapporto, P. Innocenti and F. Zanobini; *J. Chem. Soc., Dalton Trans.*, 1984, 575.
84JCS(D)995 M. Ciampolini, N. Nardi, P. Dapporto and F. Zanobini; *J. Chem. Soc., Dalton Trans.*, 1984, 995.
84JOC3157 N.S. Rao and L.D. Quin; *J. Org. Chem.*, 1984, **49**, 3157.
84JOM(273)1 G. Märkl and B. Alig; *J. Organomet. Chem.*, 1984, **273**, 1.
84NJC191 H.J. Cristau, L. Chiche, F. Fallouh, P. Hullot, G. Renard and H. Cristol; *N.J. Chem.*, 1984, **8**, 191.
84ZAAC(510)123 R. Appel and S. Korte; *Z. Anorg. Allg. Chem.*, 1984, **510**, 123.
84ZC365 K. Kellner and A. Tzschach; *Z. Chem.*, 1984, **24**, 365.
85CB97 W.J. Richter; *Chem. Ber.*, 1985, **118**, 97.
85CC439 C.W.G. Ansell, M.K. Cooper, K.P. Dancey, P.A. Duckworth, K. Henrick, M. McPartlin and P.A. Tasker; *J. Chem. Soc., Chem. Commun.*, 1985, 439.
85IC1613 E.P. Kyba and S.-T. Liu; *Inorg. Chem.*, 1985, **24**, 1613.
85ICA293 D.R. Martin, C.M. Merkel, J.P. Ruiz and J.U. Mondal; *Inorg. Chim. Acta*, 1985, **100**, 293.
85JA2141 E.P. Kyba, C.N. Clubb, S.B. Larson, V.J. Schueler and R.E. Davis; *J. Am. Chem. Soc.*, 1985, **107**, 2141.
85JA7084 L.E. Carpenter, II and J.G. Verkade; *J. Am. Chem. Soc.*, 1985, **107**, 7084.
85JCS(D)1425 M. Ciampolini, N. Nardi, P.L. Orioli, S. Mangani and F. Zanobini; *J. Chem. Soc., Dalton Trans.*, 1985, 1425.
85TL1039 W.J. Richter; *Tetrahedron Lett.*, 1985, **26**, 1039.
85TL3461 B. Deschamps and F. Mathey; *Tetrahedron Lett.*, 1985, **26**, 3461.
86CB2832 H. Schmidbaur and S. Schnatterer; *Chem. Ber.*, 1986, **119**, 2832.
86JA4918 L.E. Carpenter, II, B. de Ruiter, D. van Aken, H.M. Buck and J.G. Verkade; *J. Am. Chem. Soc.*, 1986, **108**, 4918.
86JA8302 L. Wei, A. Bell, S. Warner, I.D. Williams and S.J. Lippard; *J. Am. Chem. Soc.*, 1986, **108**, 8302.
86JOC4287 L.E. Carpenter, II and J.G. Verkade; *J. Org. Chem.*, 1986, **51**, 4287.
86PAC1429 M. Ciampolini; *Pure Appl. Chem.*, 1986, **58**, 1429.
86PS(26)185 C.Y. Cheng and R.A. Shaw; *Phosphorus Sulfur*, 1986, **26**, 185.
86PS(27)109 L.D. Quin, N.S. Rao and J. Szewczyk; *Phosphorus Sulfur*, 1986, **27**, 109.
86TL5221 K. Akiba, K. Okada and K. Ohkata; *Tetrahedron Lett.*, 1986, **27**, 5221.
87AG(E)1134 G. Märkl, H.J. Beckh, M.L. Ziegler and B. Nuber; *Angew. Chem., Int. Ed. Engl.*, 1987, **26**, 1134.
87CC1753 A.R. Barron, S.W. Hall and A.H. Cowley; *J. Chem. Soc., Chem. Commun.*, 1987, 1753.
87HCA1063 E. Müller and H.-B. Bürgi; *Helv. Chim. Acta*, 1987, **70**, 1063.
87JCS(P1)1547 T. Demir and R.A. Shaw; *J. Chem. Soc., Perkin Trans. 1*, 1987, 1547.
87JOC3983 G. Keglevich, I. Petnehazy, P. Miklos, A. Almasy, G. Toth, L. Töke and L.D. Quin; *J. Org. Chem.*, 1987, **52**, 3983.

87MI934-01 M.J. Gallagher; *Methods Stereochem. Anal.*, 1987, **8**, 297.
87PS(30)389 P. Brooks, M.J. Gallagher and A. Sarroff; *Phosphorus Sulfur*, 1987, **30**, 389.
87TL1259 M. Vincens, J.T. Grimaldo-Moron, R. Pasqualini and M. Vidal; *Tetrahedron Lett.*, 1987, **28**, 1259.
88AG(E)835 C. Bolm, W.M. Davis, R.L. Halterman and K.B. Sharpless; *Angew. Chem., Int. Ed. Engl.*, 1988, **27**. 835.
88IC3012 M. Lattman, M.M. Olmstead, P.P. Power, D.W.H. Rankin and H.E. Robertson; *Inorg. Chem.*, 1988, **27**, 3012.
B-88MI934-01 L.D. Quin; in: "Conformational Analysis of Medium-Sized Heterocycles", R.S. Glass (ed.), VCH, New York, 1988, p. 181 ff.
88POL679 M.A. Paz-Sandoval, C. Fernandez-Vincent, G. Uribe, R. Contreras and A. Klaébé; *Polyhedron*, 1988, **7**, 679.
88TL5101 C. Bolm and K.B. Sharpless; *Tetrahedron Lett.*, 1988, **29**, 5101.
88TL6247 M. Vincens, F. Gong-Cheng, C. Toulhoat, J.T. Grimaldo-Moron and M. Vidal; *Tetrahedron Lett.*, 1988, **29**, 6247.
89IC2811 B.N. Diel, P.F. Brandt, R.C. Haltiwanger, M.L.J. Hackney and A.D. Norman; *Inorg. Chem.*, 1989, **28**, 2811.
89JA3478 C. Linsink, S.K. Xi, L.M. Daniels and J.G. Verkade; *J. Am. Chem. Soc.*, 1989, **111**, 3478.
89JA9132 B.P. Friedrichsen and H.W. Whitlock; *J. Am. Chem. Soc.*, 1989, **111**, 9132.
89PS(41)21 D. Gudat, C. Lensink, H. Schmidt, S.-K. Xi and J.G. Verkade; *Phosphorus Sulfur*, 1989, **41**, 21.
89PS(44)235 P.J. Brooks, M.J. Gallagher, A. Sarroff and M. Bowyer; *Phosphorus Sulfur*, 1989, **44**, 235.
89TL155 H.-P. Niedermann, H.-L. Eckes and H. Meier; *Tetrahedron Lett.*, 1989, **30**, 155.
89ZAAC(578)75 H. Schmidt, C. Lensink, S.K. Xi and J.G. Verkade; *Z. Anorg. Allg. Chem.*, 1989, **578**, 75.
90AG(E)655 M.-O. Bevierre, F. Mercier, L. Ricard and F. Mathey; *Angew. Chem., Int. Ed. Engl.*, 1990 **29**, 655.
90IC691 S.D. Toto, M.M. Olmstead, B.W. Arbuckle, P.K. Bharadwaj and W.K. Musker; *Inorg. Chem.*, 1990, **29**, 691.
90IC2214 S.-K. Xi, H. Schmidt, C. Lensink, S. Kim, D. Wintergrass, L.M. Daniels, R.A. Jacobson and J.G. Verkade; *Inorg. Chem.*, 1990, **29**, 2214.
90JA5618 M. Badri, J.-P. Majoral, A.-M. Caminade, M. Delmas, A. Gaset, A. Gorgues and J. Jaud; *J. Am. Chem. Soc.*, 1990, **112**, 5618.
90JA6406 R.A. Pascal, Jr., A.P. West, Jr. and D. Van Engen; *J. Am. Chem. Soc.*, 1990, **112**, 6406.
90JA7823 L.T. Scott and M. Unno; *J. Am. Chem. Soc.*, 1990, **112**, 7823.
90JA8931 B.P. Friedrichsen, D.R. Powell and H.W. Whitlock; *J. Am. Chem. Soc.*, 1990, **112**, 8931.
90JA9421 M.A.H. Laramay and J.G. Verkade; *J. Am. Chem. Soc.*, 1990, **112**, 9421.
90JA9422 D.V. Khasnis, M. Lattman and C.D. Gutsche; *J. Am. Chem. Soc.*, 1990, **112**, 9422.
90JMS245 F. Sournies, P. Castera, J.-P. Foucher, J. Graffeuil and J.-F. Labarre; *J. Mol. Struct.*, 1990, **221**, 245.
90JMS253 R. Enjalbert, J. Galy, F. Sournies and J.-F. Labarre; *J. Mol. Struct.*, 1990, **221**, 253.
90JOM(385)131 R.J. Topping, L.D. Quin and A.L. Crumbliss; *J. Organomet. Chem.*, 1990, **385**, 131.
B-90MI934-01 O. Stelzer and K.-P. Langhans; in "The Chemistry of Organophosphorus Compounds", Wiley, Chichester, 1990, Vol. 1, Chap. 8.
90PS(47)119 W.F. Deutsch and R.A. Shaw; *Phosphorus Sulfur*, 1990, **47**, 119.
90PS(49/50)163 H. Schmidt, S.K. Xi, C. Lensink and J.G. Verkade; *Phosphorus Sulfur*, 1990, **49/50**, 163.
90PS(53)333 H.A. Al-Madfa, R.A. Shaw and S. Ture; *Phosphorus Sulfur*, 1990, **53**, 333.
90PS(53)437 K.C.K. Swamy, S.D. Burton, J.M. Holmes, R.O. Day and R.R. Holmes; *Phosphorus Sulfur*, 1990, **53**, 437.
91CB377 H.-L. Eckes, H.-P. Niedermann and H. Meier; *Chem Ber.*, 1991, **124**, 377.
91CC562 D.V. Khasnis, J.M. Burton, M. Lattman and H. Zhang; *J. Chem. Soc., Chem. Commun.*, 1991, 562.
91CC1227 J. Kurita, S. Shiratori, S. Yasuike and T. Tsuchiya; *J. Chem. Soc., Chem. Commun.*, 1991, 1227.
91JA2672 R.P. L'Esperance, A.P. West, Jr., D. Van Engen and R.A. Pascal, Jr; *J. Am. Chem. Soc.*, 1991, **113**, 2672.
91PS(56)27 S.D. Toto, B.W. Arbuckle, P.K. Bharadwaj, J.T. Doi and W.K. Musker; *Phosphorus Sulfur*, 1991, **56**, 27.
91ZAAC(605)163 M.A.H. Laramay and J.G. Verkade; *Z. Anorg. Allg. Chem.*, 1991, **605**, 163.
91ZN(B)197 J. Silberzahn, H. Pritzkow and H.P. Latscha; *Z. Naturforsch., Teil B*, 1991, **46**, 197.
92CB679 M. Widhalm and C. Kratky; *Chem. Ber.*, 1992, **125**, 679.
92CCR(117)133 W.K. Musker; *Coord. Chem. Rev.*, 1992, **117**, 133.
92IC1411 T.L. Jones, A.C. Willis and S.B. Wild; *Inorg. Chem.*, 1992, **31**, 1411.
92JA3129 J.-S. Tang, M.A.H. Laramay, V. Young, S. Ringrose, R.A. Jacobson and J.G. Verkade; *J. Am. Chem. Soc.*, 1992, **114**, 3129.
92JA7582 C.A. Muedas, D. Schröder, D. Sülzle and H. Schwarz; *J. Am. Chem. Soc.*, 1992, **114**, 7582.
92MI934-01 R.R. Holmes, R.O. Day, J.A. Deiters, K.C.K. Swamy, J.M. Holmes, J. Hans, S.D. Burton and T.K. Prakasha; *ACS Symp. Ser.*, 1992, **486**, 18.
92OM986 H. Schmidbaur and S.F. Gamper; *Organometallics*, 1992, **11**, 986.
92OM1438 N.H. Tran Huy, A. Marinetti, L. Ricard and F. Mathey; *Organometallics*, 1992, **11**, 1438.
92PS(68)155 R.P. Arshinova; *Phosphorus Sulfur*, 1992, **68**, 155.
92PS(71)7 J.K. Moran and D.M. Roundhill; *Phosphorus Sulfur*, 1992, **71**, 7.
92SL363 F. Mathey, A. Marinetti and F. Mercier; *Synlett.*, 1992, 363.
92TL2107 P.B. Savage, J.M. Desper and S.H. Gellman; *Tetrahedron Lett.*, 1992, **33**, 2107.
92TL3621 G. Märkl and M. Hoferer; *Tetrahedron Lett.*, 1992, **33**, 3621.
93ACR483 J.D. Verkade; *Acc. Chem. Res.*, 1993, **26**, 483.
93AG(E)896 J.-S. Tang and J.G. Verkade; *Angew. Chem., Int. Ed. Engl.*, 1993, **32**, 896.
93BSF647 C. Toulhoat, M. Vincens and M. Vidal; *Bull. Soc. Chim. Fr.*, 1993, **130**, 647.
93CC13 F. Grynszpan, O. Aleksiuk and S.E. Biali; *J. Chem. Soc., Chem. Commun.*, 1993, 13.
93CC54 R.J. Smith, A.K. Powell, N. Barnard, J.R. Dilworth and P.J. Blower; *J. Chem. Soc., Chem. Commun.*, 1993, 54.

93CC1817 S. Yasuike, H. Ohta, S. Shiratori, J. Kurita and T. Tsuchiya; *J. Chem. Soc., Chem. Commun.*, 1993, 1817.
93JA1660 J.-S. Tang and J.G. Verkade; *J. Am. Chem. Soc.*, 1993, **115**, 1660.
93JA6995 A.W. Schwabacher, S. Zhang and W. Davy; *J. Am. Chem. Soc.*, 1993, **115**, 6995.
93JA7900 P.B. Savage, S.K. Holmgren and S.H. Gellman; *J. Am. Chem. Soc.*, 1993, **115**, 7900.
93JA11984 D.B. Chesnut, L.D. Quin and K.D. Moore; *J. Am. Chem. Soc.*, 1993, **115**, 11984.
93JCS(D)2563 J. Fawcett, P.A.T. Hoye, R.D.W. Kemmitt, D.J. Law and D.R. Russell; *J. Chem. Soc., Dalton Trans.*, 1993, 2563.
93JOC1985 E. Vedejs and M.J. Peterson; *J. Org. Chem.*, 1993, **58**, 1985.
93JOC3502 A.P. West, Jr., N. Smyth, C.M. Kraml, D.M. Ho and R.A. Pascal, Jr.; *J. Org. Chem.*, 1993, **58**, 3502.
93PAC461 P.B. Savage, S.K. Holmgren, J.M. Desper and S.H. Gellman; *Pure Appl. Chem.*, 1993, **65**, 461.
93PS(75)253 D.V. Khasnis, J.M. Burton, J.D. McNeil, H. Zhang and M. Lattman; *Phosphorus Sulfur*, 1993, **75**, 253.
93PS(78)119 C. Toulhoat, M. Vidal and M. Vincens; *Phosphorus Sulfur*, 1993, **78**, 119.
93PS(84)217 J. Gloede, B. Costisella, M. Ramm and R. Bienert; *Phosphorus Sulfur*, 1993, **84**, 217.
93S931 G. Keglevich; *Synthesis*, 1993, 931.
94CB1411 M. Widhalm and G. Klintschar; *Chem. Ber.*, 1994, **127**, 1411.
94CC67 J.G. Walsh, P.J. Furlong and D.G. Gilheany; *J. Chem. Soc., Chem. Commun.*, 1994, 67.
94CRV1183 A.-M. Caminade and J.P. Majoral; *Chem. Rev.*, 1994, **94**, 1183.
94HCA409 M. Widhalm, M. Kalchhauser and H. Kählig; *Helv. Chim. Acta*, 1994, **77**, 409.
94IC2657 D.V. Khasnis, J.M. Burton, J.D. McNeil, C.J. Santini, H. Zhang and M. Lattman; *Inorg. Chem.*, 1994, **33**, 2657.
94JA3306 F. Laporte, F. Mercier, L. Ricard and F. Mathey; *J. Am. Chem. Soc.*, 1994, **116**, 3306.
94JA4069 P.B. Savage, S.K. Holmgren and S.H. Gellman; *J. Am. Chem. Soc.*, 1994, **116**, 4069.
94JCS(D)3031 N.R. Champness, C.S. Frampton, G. Reid and D.A. Tocher; *J. Chem. Soc., Dalton Trans.*, 1994, 3031.
94JCS(P1)1665 F. Hong, J. Xia and Y. Xu; *J. Chem. Soc., Perkin Trans. 1*, 1994, 1665.
94JOC6363 S. Gladiali, A. Dore, D. Fabbri, O. De Lucchi and G. Valle; *J. Org. Chem.*, 1994, **59**, 6363.
94JOM(468)199 C.-Y. Liu, M.-C. Cheng, S.-M. Peng and S.-T. Liu; *J. Organomet. Chem.*, 1994, **468**, 199.
94TA189 M. Widhalm and G. Klintschar; *Tetrahedron: Asymmetry*, 1994, **5**, 189.
94TA511 S. Gladiali, A. Dore, D. Fabbri, O. De Lucchi and M. Manassero; *Tetrahedron: Asymmetry*, 1994, **5**, 511.
94TA699 R. Hulst, N.K. de Vries and B.L. Feringa; *Tetrahedron Asymmetry*, 1994, **5**, 699.
94TL9109 J.D. Sutherland and G.W. Weaver; *Tetrahedron Lett.*, 1994, **35**, 9109.
95AG(E)557 H.H. Karsch, E. Witt, A. Schneider, E. Herdtweck and M. Heckel; *Angew. Chem., Int. Ed. Engl.*, 1995, **34**, 557.
95BSF910 K. Waschbüsch, P. le Floch and F. Mathey; *Bull. Soc. Chim. Fr.*, 1995, **132**, 910.
95CC1561 E. Deschamps, L. Ricard and F. Mathey; *Chem. Commun.*, 1995, 1561.
95IC1638 D.V. Khasnis, H.S. Isom, S. Shang, M. Lattman, M.M. Olmstead and P.P. Power; *Inorg. Chem.*, 1995, **34**, 1638.
95IC323 W. Zu, J.P. Rourke, J.J. Vittal and R.J. Puddephatt; *Inorg. Chem.*, 1995, **34**, 323.
95IC5405 I. Shevchenko, H. Zhang and M. Lattman; *Inorg. Chem.*, 1995, **34**, 5405.
95ICA(231)225 G.Q. Li and R. Govind; *Inorg. Chim. Acta*, 1995, **231**, 225.
95JA8362 W. Xu, J.J. Vittal and R.J. Puddephatt; *J. Am. Chem. Soc.*, 1995, **117**, 8362.
95JCS(D)1139 S.J. Coles, P.G. Edwards, J.S. Fleming and M.B. Hursthouse; *J. Chem. Soc., Dalton Trans.*, 1995, 1139.
95JCS(D)2095 A.G. Avent, P.B. Hitchcock and M.F. Lappert; *J. Chem. Soc., Dalton Trans.*, 1995, 2095.
95JCS(D)4091 S.J. Coles, P.G. Edwards, J.S. Fleming and M.B. Hursthouse; *J. Chem. Soc., Dalton Trans.*, 1995, 4091.
95JOC235 T. Lippmann, H. Wilde, E. Dalcanale, L. Mavilla, G. Mann, U. Heyer and S. Spera; *J. Org. Chem.*, 1995, **60**, 235.
95JOC6076 J.L. Montchamp, F. Tian and J.W. Frost; *J. Org. Chem.*, 1995, **60**, 6076.
95JOC7406 B. König, M. Rödel, P. Bubenitschek, P.G. Jones and I. Thondorf; *J. Org. Chem.*, 1995, **60**, 7406.
95JOC7433 K. Brandt, I. Porwolik, T. Kupta, A. Olejnik, R.A. Shaw and D.B. Davies; *J. Org. Chem.*, 1995, **60**, 7433.
95MI934-01 Y.B. Kang, M. Pabel, D.D. Pathak, A.C. Willis and S.B. Wild; *Main Group Chem.*, 1995, **1**, 89.
95OM1434 A. Bader, Y.B. Kang, M. Pabel, D.D. Pathak, A.C. Willis and S.B. Wild; *Organometallics*, 1995, **14**, 1434.
95S952 J. Mitjaville and A.M. Caminade; *Synthesis*, 1995, 952.
96AG(E)1125 F. Paul, D. Carmichael, L. Ricard and F. Mathey; *Angew. Chem., Int. Ed. Engl.*, 1996, **35**, 1125.
96AG(E)2374 A.H.M. de Vries, A. Meetsma and B.L. Feringa; *Angew. Chem., Int. Ed. Engl.*, 1996, **35**, 2374.
96CC293 S.J. Coles, P.G. Edwards, J.S. Fleming, M.B. Hursthouse and S.S. Liyanage; *Chem. Commun.*, 1996, 293.
96CC537 R.W. Alder, D.D. Ellis, J.K. Hogg, A. Martin, A.G. Orpen and P.N. Taylor; *Chem. Commun.*, 1996, 537.
96CC583 F.J. Parlevliet, A. Olivier, W.G.T. de Lange, P.C.J. Kamer, H. Kooijman, A.L. Spek and P.W.N.M. van Leeuwen; *Chem. Commun.*, 1996, 583.
96H2367 J.M. Smallheer and S.P. Seitz; *Heterocycles*, 1996, **43**, 2367.
96IC4563 P.G. Edwards, J.S. Fleming and S.S. Liyanage; *Inorg. Chem.*, 1996, **35**, 4563.
96JA1060 G. Alcaraz, V. Piquet, A. Bacieredo, F. Dahan, W. Schoeller and G. Bertrand; *J. Am. Chem. Soc.*, 1996, **118**, 1060.
96JA6192 W.G. Bentrude, K.P. Dockery, S. Ganapathy, S.G. Lee, M. Tabet, Y.W. Wu, R.T. Cambron and J.M. Harris; *J. Am. Chem. Soc.*, 1996, **118**, 6192.
96JCS(D)1801 P.G. Edwards, J.S. Fleming, S.S. Liyanage, S.J. Coles and M.B. Hursthouse; *J. Chem. Soc., Dalton Trans.*, 1996, 1801.
96JOC8750 J.S. Tang and J.G. Verkade; *J. Org. Chem.*, 1996, **61**, 8750.

96SL1019 A.M. Caminade and J.P. Majoral; *Syn. Lett.*, 1996, 1019.
96T2995 N. Oget, F. Chuburu, J.J. Yaouanc and H. Handel; *Tetrahedron*, 1996, **52**, 2995.
96TL7815 M. Yoshifuji, N. Shinohara and K. Toyota; *Tetrahedron Lett.*, 1996, **37**, 7815.
97AG(E)67 M. Alajarin, P. Molina, A. Lopez-Lazaro and C. Foces-Foces; *Angew. Chem., Int. Ed. Engl.*, 1997, **36**, 67.
97AG(E)2364 F. Mercier, F. Laporte, L. Ricard, F. Mathey, M. Schröder and M. Regitz; *Angew. Chem., Int. Ed. Engl.*, 1997, **36**, 2364.
97AG(E)2620 B.L. Feringa, M. Pineschi, L.A. Arnold, R. Imbos and A.H.M. de Vries; *Angew. Chem., Int. Ed. Engl.*, 1997, **36**, 2620.
97CB/R1715 A. Vollbrecht, I. Neda, H. Thönnessen, P.G. Jones, R.K. Harris, L.A. Crow and R. Schmutzler; *Chem. Ber./Recueil.*, 1997, **130**, 1715.
97IC2730 A. Vij, A.J. Elias, R.L. Kirchmeier and J.M. Shreeve; *Inorg. Chem.*, 1997, **36**, 2730.
97ICA99 J.A. Munoz, L. Escriche, J. Casobo, C. Perez-Jiminez, R. Kivekas and C. Sillanpaa; *Inorg. Chim. Acta*, 1997, **257**, 99.
97JA11441 A. Chandrasekaran, R.O. Day and R.R. Holmes; *J. Am. Chem. Soc.*, 1997, **119**, 11441.
97JCS(D)193 P.G. Edwards, J.S. Fleming and S.S. Liyanage; *J. Chem. Soc., Dalton Trans.*, 1997, 193.
97JCS(D)3201 P.G. Edwards, J.S. Fleming, S.J. Coles and M.B. Hursthouse; *J. Chem. Soc., Dalton Trans.*, 1997, 3201.
97JOC4827 S. Arumugam and J.G. Verkade; *J. Org. Chem.*, 1997, **62**, 4827.
97JOC5057 B.A. D'Sa, D. McLeod and J.G. Verkade; *J. Org. Chem.*, 1997, **62**, 5057.
97JOM(529)121 P. Molina, M. Alajarin, A. Arques, P. Sanchez-Andrada, A. Vidal and M.V. Vinader; *J. Organomet. Chem.*, 1997, **529**, 121.
97JOM(533)83 Y. Inubushi, N.H. Huy, L. Ricard and F. Mathey; *J. Organomet. Chem.*, 1997, **533**, 83.
97T7461 C.P. Nash, S.D. Toto and W.K. Musker; *Tetrahedron*, 1997, **53**, 7461.
97T11595 J.D. Sutherland and J.N. Whitfield; *Tetrahedron*, 1997, **53**, 11595.
97T12249 S.R. Holmgren, P.B. Savage, J.M. Desper, K.D. Schladetzky, D.R. Powell and S.H. Gellman; *Tetrahedron*, 1997, **53**, 12249.
97TL1451 J.D. Sutherland and J.N. Whitfield; *Tetrahedron Lett.*, 1997, **38**, 1451.
98IC6329 P. Sood, A. Chandrasekaran, R.O. Day and R.R. Holmes; *Inorg. Chem.*, 1998, **37**, 6329.
98IC6408 F. Bitterer, O. Herd, M. Kühnel, O. Stelzer, N. Weferling, W.S. Sheldrick, J. Hahn, S. Nagel and N. Rösch; *Inorg. Chem.*, 1998, **37**, 6408.
98JA6421 S. Dell, N.J. Vogelaar, D.M. Ho and R.A. Pascal; *J. Am. Chem. Soc.*, 1998, **120**, 6421.
98JCR(S)210 G. Keglevich, H.T.T. Thanh, K. Sudanyi, T. Novak, K. Ujszaszy and L. Toke; *J. Chem. Res. (S)*, 1998, 210.
98JCS(P1)1643 R.W. Alder, C. Ganter, M. Gil, R. Gleiter, C.J. Harris, S.E. Harris, H. Lange, A.G. Orpen and P.N. Taylor; *J. Chem. Soc., Perkin Trans. 1*, 1998, 1643.
98JCS(P1)1657 R.W. Alder, D.D. Ellis, R. Gleiter, C.J. Harris, H. Lange, A.G. Orpen, D. Read and P.N. Taylor; *J. Chem. Soc., Perkin Trans. 1*, 1998, 1657.
98JCS(P1)3323 G. Keglevich, H.T.T. Thanh, A. Szollosy, K. Mjsaszy and L. Toke; *J. Chem. Soc., Perkin Trans. 1*, 1998, 3323.
98JOC5692 K. Ishihara, Y. Karumi, S. Kondo and H. Yamamoto; *J. Org. Chem.*, 1998, **63**, 5692.
99AG(E)179 M.T. Reetz and T. Neugebauer; *Angew. Chem., Int. Ed. Engl.*, 1999, **38**, 179.
99CC741 B. Bartels and G. Helmchen; *Chem. Commun.*, 1999, 741.
99HCA790 I. Déchamps-Olivier, J.P. Barbier, M. Aplincourt, N. Oget, F. Chuburu and H. Handel; *Helv. Chim. Acta*, 1999, **82**, 790.
99JCS(D)1045 D.J. Jones, P.G. Edwards, R.P. Tooze and T. Albers; *J. Chem. Soc., Dalton Trans.*, 1999, 1045.
99JCS(D)1277 P.H. Leung, A. Liu, K.F. Mok, A.J.P. White and D.J. Williams; *J. Chem. Soc., Dalton Trans.*, 1999, 1277.

6.2
Bicyclic and Polycyclic Systems with a Ring Junction Phosphorus Atom

JOHN C. TEBBY
Staffordshire University, Stoke-on-Trent, Staffordshire, UK

6.2.1 INTRODUCTION

Most bicyclic and polycyclic organophosphorus compounds have been synthesised as a result of the investigation of the reactions and properties of simpler compounds. The vast majority are alicyclic and very few fall into the aromatic class. Compounds which involve more than one heteroatom in the ring system — the additional atoms usually being oxygen, nitrogen and additional phosphorus atoms — abound and many are part of more complex ring systems.

These heterocycles rarely have trivial names and thus, apart from compounds such as phospholes and phosphorins and their saturated derivatives phospholan and phosphorinan, they are usually named by identifying the heteroatoms as substituents of carbon, i.e. phospha, aza, and oxa, etc. Aromatic organophosphorus compounds such as (**1**) and (**2**), which would be the phosphorus equivalents to quinolizinium salts and indolizine are not known. With the higher valencies available phosphorus also has the potential of forming other quinolizine and indolizine analogues such as the ylide (**3**) and the phosphoranide (**4**). However, examples of these are not known either. However, several hydro derivatives of quinolizine and indolizine have been characterised and some authors have adapted these names; thus 1-phosphabicyclo(4,4,0)-decane (**24**) has also been named as 2*H*-octahydrophosphinolizine as well as 5-phosphadecalin. The bicyclic phosphabicyclo[3,3,3]undecanes (**244** and **247–249**) is one group which has been given a trivial name, i.e. the phosphatranes. The renaming of phosphines as phosphanes, by German chemists, has not been adopted by the majority of the chemistry community. The nomenclature of two and one coordinate compounds as phosphaalkenes, Y=PR, and phosphaalkynes, R–CP, have also been named as phosphenes and phosphynes.

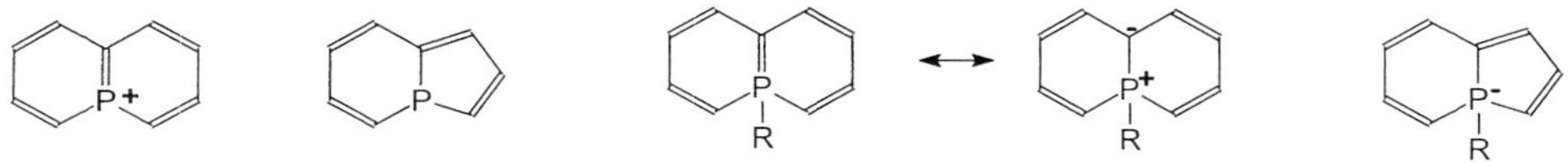

1 2 3a 3b 4

The heterocycles covered in this chapter have a very wide range of ring sizes, from very small fused-ring systems, e.g. [1,1,0] to quite large non-fused [17,17,17]cryptan ring systems, e.g. (**255**). In addition, the phosphorus atom may be 3, 4, 5 and 6 coordinate. Only one non-substituted fused-ring parent heterobicyclic compound possessing one phosphorus atom has been prepared, although several compounds incorporating two ring junction phosphorus atoms have. In the vast majority of examples there are other heteroatoms in the ring and/or additional rings present. Because of the very wide variety of ring systems available which have a ring junction phosphorus atom, the range of structures covered include tricyclic, polycyclic and caged compounds. The bicyclic systems [4,4,0] [4,3,0], [3,3,0] have been covered first in this order followed by other fused heterobicyclic systems, e.g. [3,2,0], etc. The other ring systems [*n*,*m*,1], [*n*,*m*,2], [*n*,*m*,3], etc., have been covered in order of decreasing ring size. Note that these basic ring systems are often part of a wider ring system but do not include systems in which a spiro phosphorus atom can be identified. Spiro compounds are covered in Chapter 6.3. Within each section covering a specific ring system any non-substituted parent ring systems have been covered first and thereafter compounds have been covered in order of the number of heteroatoms included in the heterocyclic ring system, e.g. Het > Het_2 > Het_3 > Het_4, etc., and then in order of increasing saturation.

In most cases the bicyclic compounds are end products and whilst their spectral properties are usually reported, studies of their reactivities are fewer. Frequently, the heteroatom acts as a focus for reactions with many reagents and as a consequence leads to a destruction of the ring system.

6.2.2 THEORETICAL METHODS

The incorporation of phosphorus as a bridging atom in bi- and poly-cyclic ring systems has not played a large part in the theoretical studies of these classes of compound. Relativistic effects, which contract the size of the s orbitals and to a lesser extent the p orbitals, become more important for the heavier elements, and this can cause irregular effects on the periodic trends ⟨94M1⟩. The effect leads to the inert electron pair concept. Thus the three-coordinate five-membered heterocycle such as phosphole shows considerable decrease in the involvement of the lone pair in stabilising the ring system and follows the same trend as their ability to participate in conjugation as substituents. Studies of the two-coordinate hetero-substituted benzene analogues, such as phosphabenzene, indicate a significant amount of aromatic stabilisation ⟨88JA4204⟩. *Ab initio* calculations of cycloaddition reactions of CH_2=PH and OC=PH to 2-H phosphole (**5**) indicated that the former is expected to undergo a 2 + 4 Diels–Alder reaction but the latter a 2 + 2 Staudinger pathway ⟨97M1⟩. Whilst formally two-coordinate heteroatoms are unable to occupy bridgehead positions, phosphenium ions with two adjacent donor atoms have been shown to possess bicyclic structures. Quantum mechanical calculations on the relative stabilities of diaza [4,3,0] bicyclic and [4,4] spiro structures (**6**) and (**7**) indicated that the latter spiro structure with four-coordinate phosphorus is more stable than the bicyclic isomer (**6**) with three-coordinate phosphorus ⟨98OM4014⟩. The sigmatropic rearrangement of phosphole (**115**) to a [2,1,0] ring system (**116**) has been studied ⟨95JCS(CC)2395⟩. In the higher-valence state the involvement of d-orbitals as opposed to sigma antibonding orbitals in the stabilisation of compounds such as the ylides and penta-coordinated phosphoranes has aroused considerable theoretical interest ⟨90M⟩.

5

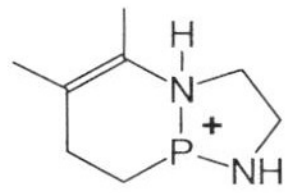

6

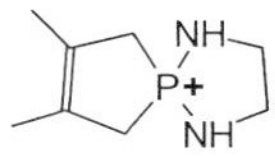

7

In the case of molecular mechanics there have been developments of force field data for phosphoranes ⟨78T627⟩ and phosphazenes ⟨70JA1995⟩ in addition to the well established force fields for phosphates and phosphines.

6.2.3 EXPERIMENTAL STRUCTURAL METHODS

X-ray diffraction has shown that the bond angles to three-coordinated phosphorus atoms tend towards 90º, the ultimate angle depending on other steric factors. Whilst four-coordinate phosphorus atoms are tetrahedral, five-coordinate phosphorus atoms generally favour trigonal bipyramidal structures (with steric factors able to distort this stereochemistry towards a square pyramidal conformation). As expected, six-coordinate phosphorus adopts an octahedral geometry. Thus the involvement of phosphorus atoms in small rings can be used to favour the formation of five-coordinate and six-coordinate heterocyclic compounds allowing some of the electronegative P-bound atoms to be replaced by others, e.g. carbon.

Phosphorus-31 NMR spectroscopy has played a central role in the development of organophosphorus chemistry, the chemical shifts being dependent on valency, coordination, bond angles and directly bound atoms, and the coupling constants able to identify adjacent NMR active nuclei and to indicate stereochemistry ⟨87M1⟩. NMR Shift reagents have been employed in comparative studies of acyclic and bicyclic phosphates (**233**; R = Me, X = Y = O) ⟨94PS(89)39⟩.

6.2.4 THERMODYNAMIC ASPECTS

The physical properties vary widely, primarily due to the very wide range of other heteroatoms and substituents involved either in the ring system or as substituents. Thus there have been few systematic studies of solubilities and chromatographic properties, although the latter is used frequently for the purification of the compounds. Aromatic properties are restricted to those with low-coordinate phosphorus atoms and, as such, are not then able to act as ring junction heteroatoms. They are generally somewhat unstable unless they bear very large substituents. On the other hand such compounds have been found to be useful as starting materials for the synthesis of bicyclic and polycyclic heterocycles via cycloaddition reactions.

Tautomerism with valence expansion of the heteroatom has been studied in some detail for three ring systems (**8** ⇌ **9**), (**10** ⇌ **11**) and (**12** ⇌ **13**). The five and six-coordinate compounds (phosphoranes and perphosphoranides) are favoured by the generation of small rings and the presence of electronegative atoms bound to the heteroatom.

8 9

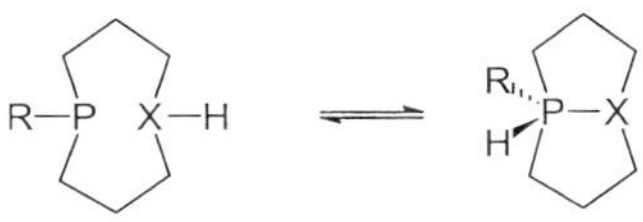

10 11

12 13

The barriers to inversion at three-coordinate phosphorus is ca. 35 kcal mol^{-1} which is generally sufficiently slow to allow the synthesis of optically active phosphines ⟨94M1⟩. The barrier to inversion is enhanced further by its incorporation at a bridgehead.

The Lewis basicity of acyclic phosphines is less than that of amines decreasing with rise in atomic number, and incorporation at a bridgehead can decrease their basicity to such an extent that they may become resistant to complexation, oxidation and alkylation. On the other hand, the protetraazaphosphatranes (**248**) are super bases ⟨97JOC4827⟩

6.2.5 FULLY CONJUGATED RINGS CONTAINING PHOSPHORUS

6.2.5.1 Fused-Ring Systems

Conjugation involving phosphorus takes on a new perspective in comparison with nitrogen. Thus the high s-character of the lone pair of electrons in phosphines reduces its tendency to become involved in ground state delocalisation. Thus whilst phospholes (**14**) are believed to have some aromatic stability the phosphorus substituent remains out of the plane of the ring and the phospholes readily undergo cycloaddition reactions including dimerisation. Five-valent $\sigma^4\Lambda^5$ phosphorus stabilises anions quite effectively but does not transmit through-conjugation, acting to isolate substituents which have potential to conjugate. Likewise, in cyclophosphazenes (**15**) the P=N groups behave more like isolated islands than acting as a delocalised ring system.

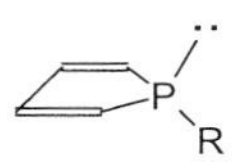

14

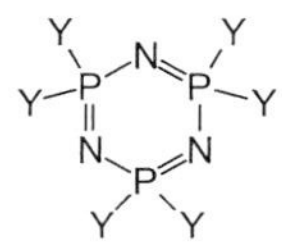

15

In contrast two-coordinate derivatives are able to become involved in aromatic properties and there are many examples of phosphazoles (**16**) and phosphabenzenes (**17**).

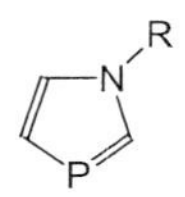

16

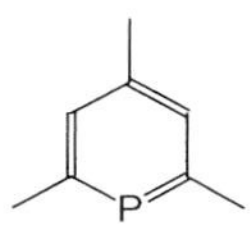

17

6.2.5.1.1 *[4,4,0] Ring systems*

Structures for the fully unsaturated 10-phosphanaphthalene cation (**1**) and ylide (**3**) can be envisaged but neither compound nor any of their derivatives have been described. The only fully "unsaturated" example incorporating a [4,4,0] ring system is the phosphazene (**18**) which is formed in the reaction of phosphorus pentachloride with ammonium chloride ⟨73CJC520⟩.

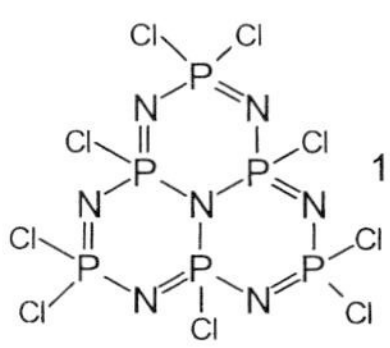

18

6.2.5.1.2 *[4,3,0] Ring systems*

Neither the phosphorus equivalent to indole, i.e. 8-phosphaindene (**2**) nor any of its derivatives have been reported When such compounds are synthesised it will be interesting to see whether the enforced planarity about the phosphorus atom enables the phosphorus to improve its participation in conjugation so as to impart additional aromatic character to the heterocycle. It is also possible to envisage a phosphoranide (**4**) and thus, in this case, the acidity of the corresponding hydridophosphorane (**19**) would be interesting to explore.

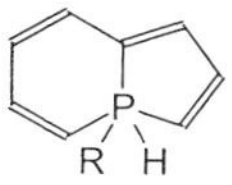

19

6.2.5.1.3 [3,3,0] Ring systems

This ring system might be expected to be the least likely to show aromatic character and neither the pyrrolizinium structures (**20**; M = N) ⟨96M1⟩ nor the phosphonium analogue (**20**; M = P) is known. However, it is possible to draw a phosphoranide structure (**21**) which has more aromaticity potential. Indeed the bicyclic compound (**22**), which was formed by the reaction of phosphorus trichloride with an aminodiketone ⟨84JA1164⟩ has been shown to be planar. This was interpreted as evidence for the delocalised dipolar canonical structure (**23**) rather than the neutral form (**22**). It forms a silver complex with t.b.p. stereochemistry at phosphorus ⟨91JA1071⟩.

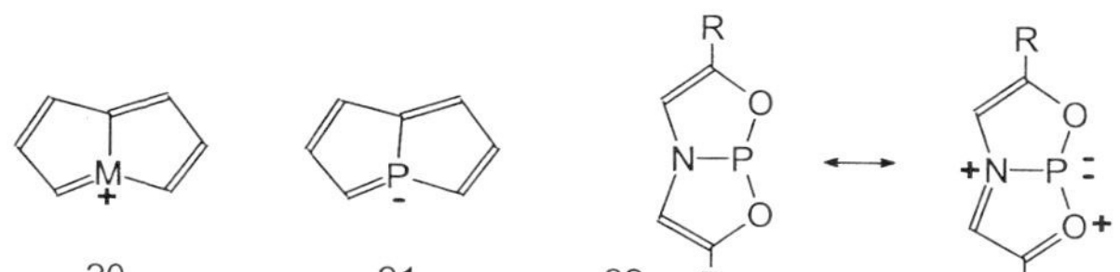

20 21 22 23

6.2.6 REACTIVITY OF NON-CONJUGATED AND PARTLY CONJUGATED FUSED-RING SYSTEMS

6.2.6.1 [4,4,0], [4,3,0] and [3,3,0] Fused-Ring Systems

6.2.6.1.1 [4,4,0] Ring systems

The synthesis of 1-phosphabicyclo[4,4,0]decane (**24**) by the radically initiated cyclisation of the primary dibutenylphosphine (**25**) was achieved by Cremer's group in the U.S.A. ⟨87PS(30)369⟩ and by Issleib's group in Germany ⟨87ZAAC(553)136⟩. It is the only example of a parent monoheterobicyclic compound. Both *cis* and *trans* fused-ring systems (**26**) and (**27**) were isolated separately and are equilibrated by UV

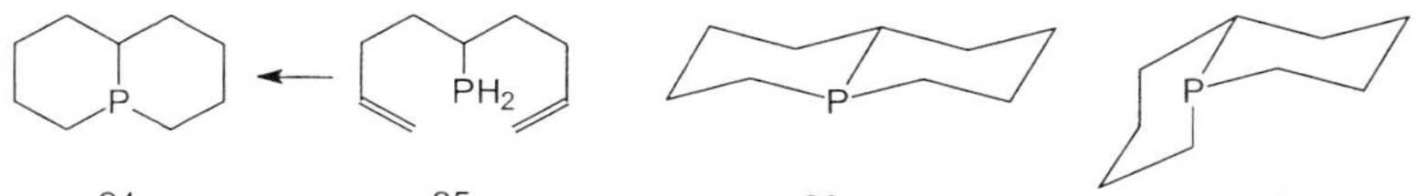

24 25 26 27

light. Unlike the *trans* isomer δP = −45.2, the *cis* isomer δP = −30.4 ppm is conformationally mobile as shown by the broadening of the latter signal with decreasing temperature. The conformational energy barrier for the *cis* isomer was estimated to be about 3.1 kJ mol^{-1} and its ring inversion barrier was estimated to be 41.9 kJ mol^{-1}. The phosphine was converted to its oxide, sulphide, methoxy, methyl and benzyl salts and some other complexes. The stereochemistry of the hydrolysis of the methoxide salt was studied and compared to monocyclic heterocycles. The triarylphosphine (**28**) was produced by the PPA cyclisation of a tertiary benzyl alcohol ⟨85JOC2914⟩. Electrophilic cyclisation of a dipropene precursor produced a similar ring system ⟨90AG(E)689⟩. The dicarboxamide derivative of triphenylphosphine oxide was converted to the oxide (**29**) via a double anionic cyclisation of the 2-lithio intermediate (**30**) ⟨96AG(E)1558⟩. Reaction of a phosphabenzene with sulphur gave a highly reactive intermediate (**31**; R = Me) which reacted with a diene to give the sulphide (**32**; Ch = S). The sulphide may be converted to the corresponding oxide (**32**; Ch = O) by reaction with an epoxide ⟨84JCS(CC)508, 84TL4659⟩. A similar intermediate (**31**; R =

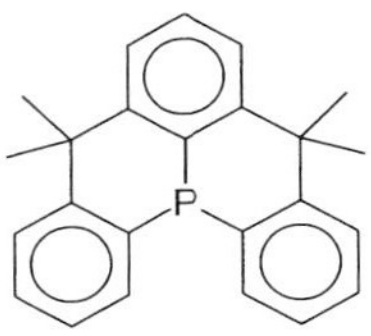

28

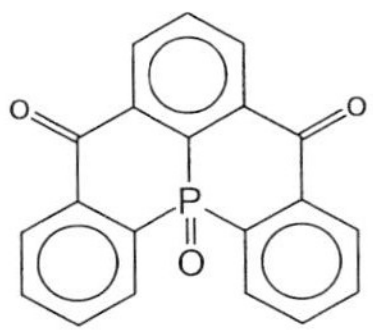

29

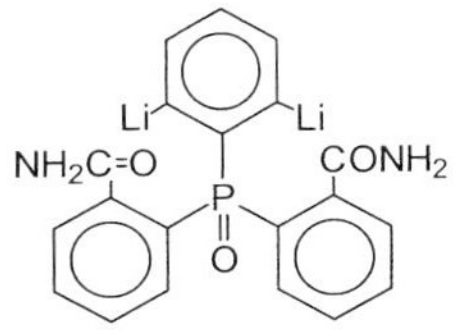

30

Ph) was formed by heating the sulphide (**33**) in toluene ⟨86BCJ2047, 87BCJ1558⟩. A number of saturated bicyclic compounds which have phosphorus atoms at both ring junctions has been synthesised as shown in Scheme 1) ⟨98JCS(P1)1643⟩. The bicyclic diphosphine (**34**; $m = n = 2$) was converted to its mono and

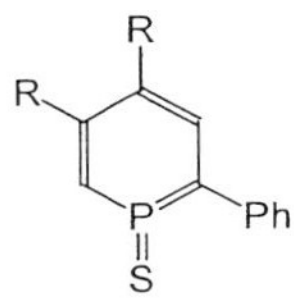

31

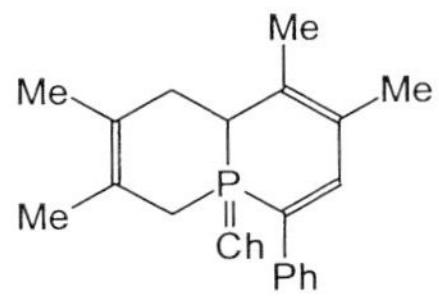

32

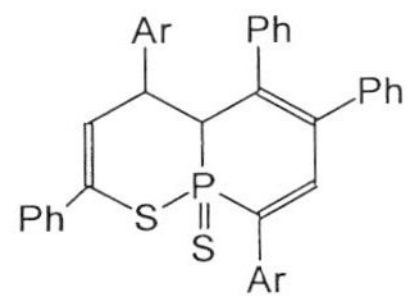

33

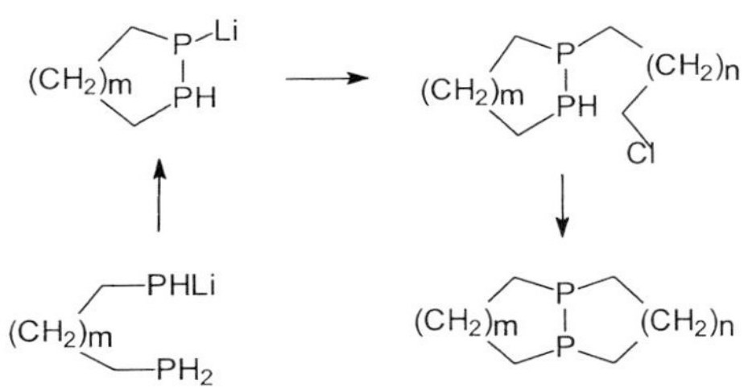

34

Scheme 1

diphosphonium salts, the latter being very sensitive to hydrolysis. Phosphonamide (**35**) was prepared by the reaction of the phosphonic diamide (**36**) with carbon disulphide in the presence of a base ⟨89S618, 90S137⟩. The tetrasilyl phosphonium ylide (**37**) was made from a trimethyl(disilylmethylene)phosphonium ylide and $Me_2SiClCH_2SiClMe_2$ ⟨76AG(E)367⟩. The triazo compound (**38**; X = NMe) was prepared from 2,6,10-triazaundecane $MeNH(CH_2)_3NH(CH_2)_3NHMe$ and trisdimethylaminophosphine ⟨80JA7072⟩ as was the corresponding dioxazaphosphine (**38**; X = O) ⟨83PS(15)281⟩. Their potential for the formation of penta-coordinate phosphoranes with interesting steric restraints, e.g. (**39**) were studied.

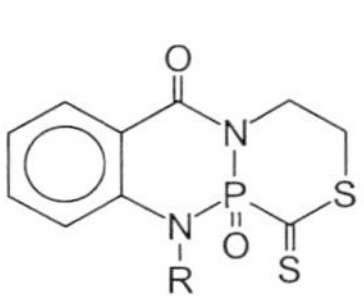

35

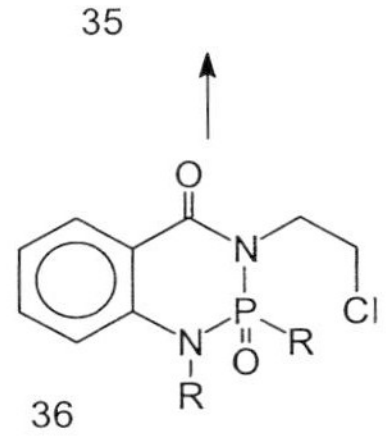

36

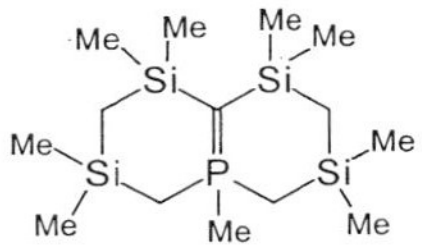

37

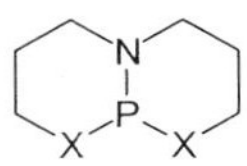

38; X = NMe, O

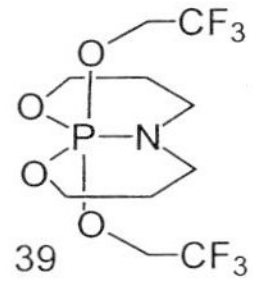

39

6.2.6.1.2 *[4,3,0] Ring systems*

The parent saturated ring system (**40**) was synthesised by the free radical cyclisation of the octadienylphosphine (**41**) and converted to the corresponding oxide, sulphide and methiodide ⟨91ZAAC(600)195⟩.

Phospholes react readily with two molecules of dimethyl acetylenedicarboxylate to give bicyclic adducts. Thus the 1,2,5-triphenylphosphole (**42**; R^1 = Ph, R^2 = H) gave the stabilised bicyclic ylide (**43**) ⟨69JCS(C)1542⟩ (a structure confirmed by X-ray diffraction 23 years later ⟨93CJC1873⟩), whereas 1-phenylphosphole (**42**; $R^1 = R^2$ = H) and its 3,4-dimethyl analogue (**42**; R^1 = H, R^2 = Me) produced the ylide (**44**; R = Me, H) ⟨78JHC1319⟩. It has been shown that de-protonation and re-protonation of 1*H*-phospholes

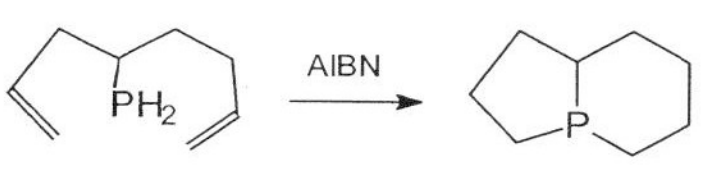

41 40

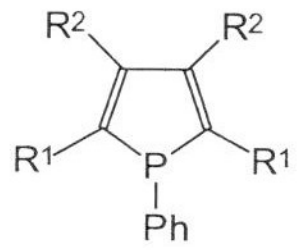

42

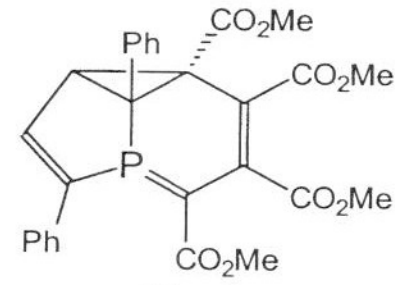

43

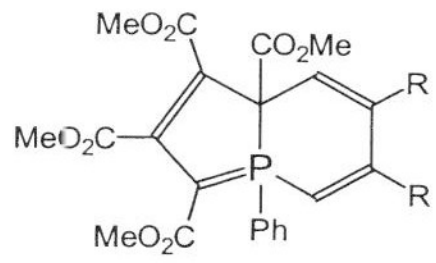

44

produces 2*H*-phospholes which readily dimerise to give [2,2,1] adducts (**45**). These adducts at 130° give monomeric 2*H*-phospholes which react with dienes to give the dihydrophospholizine (8*P*-tetrahydroindene) (**46**) (see Scheme 2) ⟨82JCS(CC)1272⟩. Electrophilic cyclisation of a butenyl phosphorus substituent

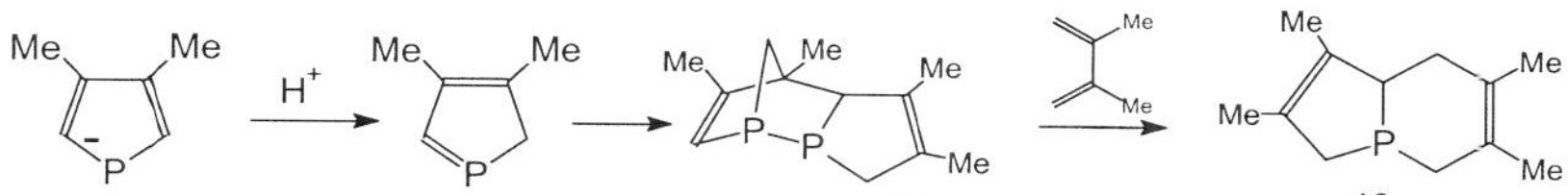

45 46

Scheme 2

has been used to prepare the benzo analogue (**47**) ⟨81JOC361⟩. The diphosphine (**34**; *m* = 2, *n* = 1), with phosphorus atoms at both ring junctions, was synthesised by the route shown in Scheme 1 and converted into its mono and di phosphonium salts ⟨98JCS(P1)1643⟩. Another unusual reaction involves two molecules of the diynol (**48**) reacting with two molecules of trihydroxymethylphosphine to give the diphospha heterocycle (**49**) ⟨74PS(4)279⟩. Depending on the substituents R^1 and R^2, the reaction of 2-aminobenzamides (**50**; R^1 = H, $((CH_2)_n)Cl$; R^2 = $((CH_2)_n)Cl$, H) with PCl_3 followed by sodium hydride can be used to prepare either amide (**51**) or (**52**) ⟨79JHC897⟩. A 4 + 3 cycloaddition reaction involving a nitrilimine and dihydrofuranylphosphomite (**53**) gave the phosphorane (**54**) ⟨96JGU1367⟩. The dibenzobicyclic phosphonate (**55**) was attributed to the product from the reaction of an *o*-aminomethylphenol with tris(diethylamino)phosphine ⟨74IZV2079⟩ and the mechanism of this unusual transformation was investigated ⟨67DOK(174)846⟩. Conversion of phosphabenzenes to the reactive $\Lambda^5\sigma^3$ sulphide (**31**) led to the formation of the bicyclic dimer (**56**) ⟨90HAC37⟩. A complex reaction between a phosphazene dimer and tetra aminoethene gave a diaminophosphine (**57**) which was characterised as the sulphide ⟨98M1⟩. The unsaturated thioketone (**58**) reacts with P_4S_{10} in the presence of triethylamine to give the unexpected product (**59**). X-ray crystallography showed that the phenyl analogue (**59**; Ar = Ph) adopts a

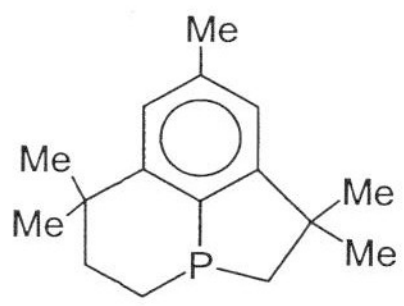

47

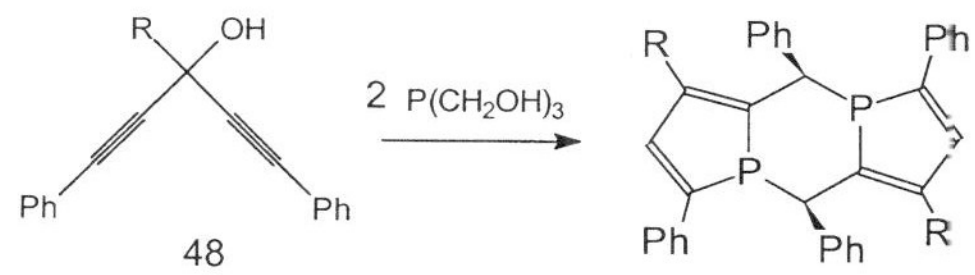

48 49

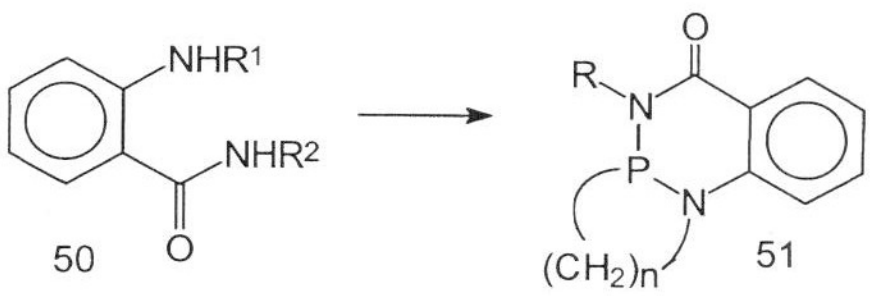

50 51

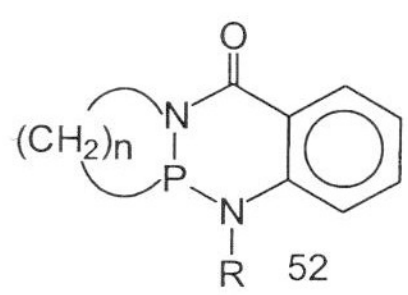

52

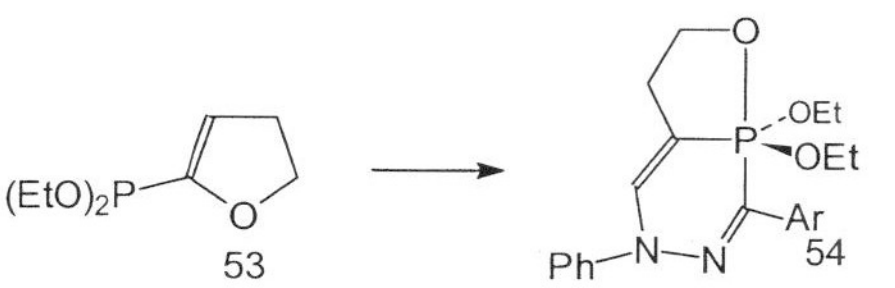

53 54

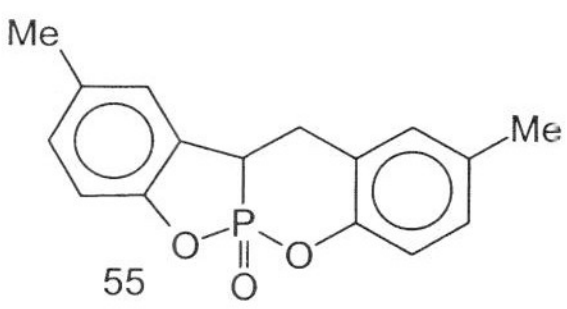

55

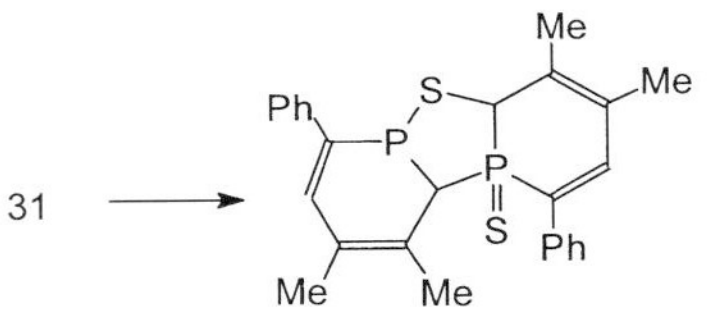

56

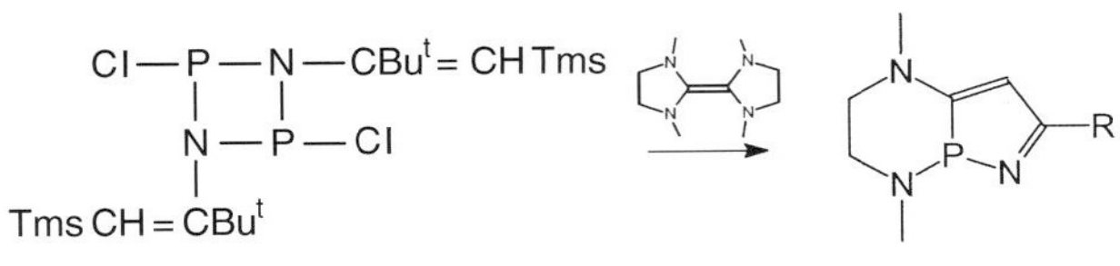

57

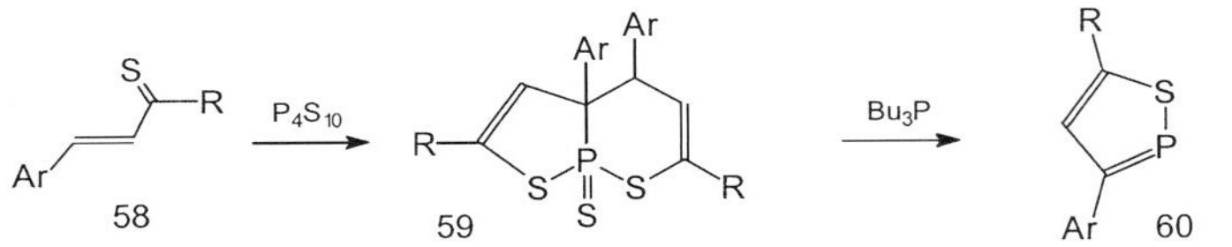

58 59 60

boat conformation ⟨84BCJ1990⟩. Tributylphosphine converts (**59**) to a two-coordinate thiaphosphole (**60**) ⟨94BCJ2785, 93CL1563⟩ which will undergo the Diels–Alder reaction forming new bicyclic heterocycles ⟨94BCJ2785⟩. Whilst 1,2-diketones lead directly to dioxaphosphoranes, the 1,4-unsaturated diketone took an unusual pathway before yielding the bicyclic oxaphosphorane (**61**) shown in Scheme 3 ⟨71JA5236⟩. Hexafluoroacetone, diphenylchlorophosphine and mercurous cyanide in the presence of triethylamine gives the dioxaphosphorane (**62**; Y = CF_3) ⟨86CB2687⟩. An interesting condensation of *t*-butyldichlorophosphine with a trisilated aryl amidine gave phosphazene (**63**) ⟨93PS(76)265⟩. The intermediacy of the bicyclic oxide (**64**) during the hydrolysis of cyclophosphamide was established by its isolation as a solid ⟨77JA5785⟩. Six-coordinate ylides (**65**; X = F, Cl, CF_3) were made by the action of PX_5, $(CF_3)PX_4$ and $(CF_3)_2PCl_3$ on a disilylated diphenolic imine ⟨96IC325, 96PS(111)151⟩.

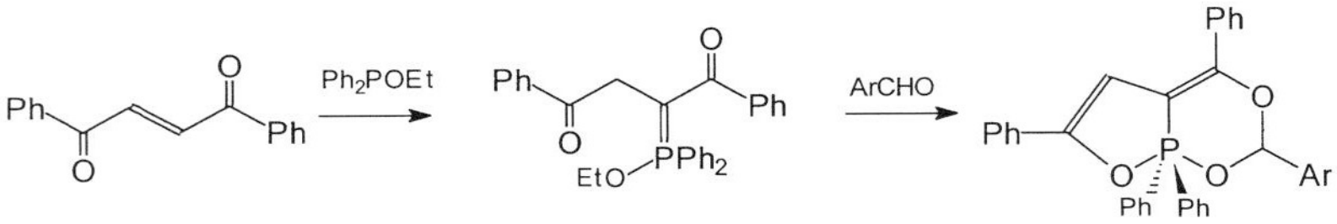

61

Scheme 3

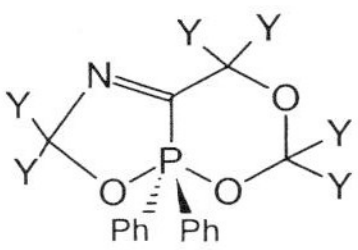

62

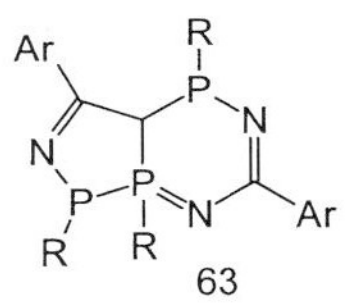

63

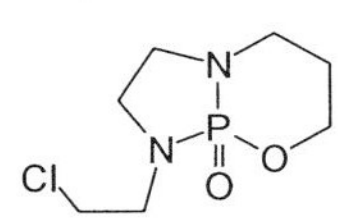

64

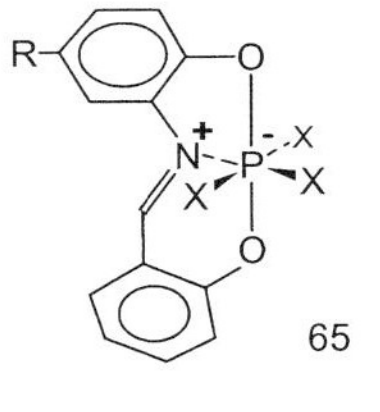

65

6.2.6.1.3 [3,3,0] Ring systems

Neither the parent heterocycle (**66**) not any of its derivatives have been reported, nor have any of its dehydro forms. However many di-, tri- and multi-heteroatom [3,3,0] ring systems have been characterised, most of them quite highly substituted. The simplest non-substituted are two perhydro compounds: the bicyclic diphosphine (**34**; $n = m = 1$) and the corresponding silylphosphine (**67**). The former was synthesised by cyclisation of a chloropentyldiphosphacyclopentane which was formed via the unexpected spontaneous cyclisation with loss of hydrogen as shown in Scheme 1. The diphosphine may be converted to its mono- and di-oxides, sulphides and methiodides ⟨78PS(4)137⟩; their NMR ⟨77OMR172⟩ and disulphide

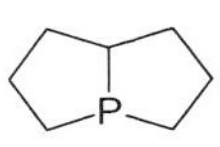

66

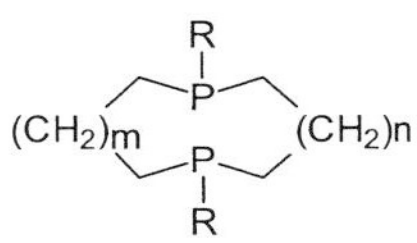

68

67; R = Me, Ph 69

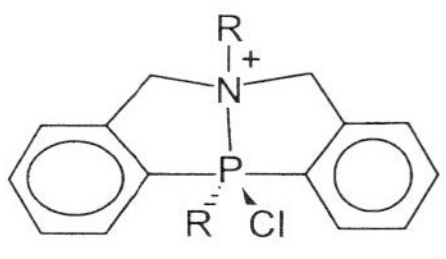

70; R = Me, Ph

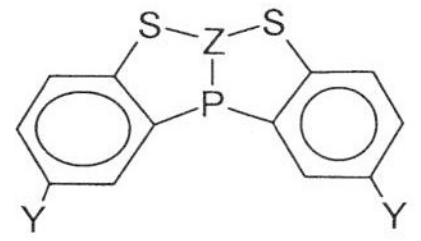

71; Z = CHR, P

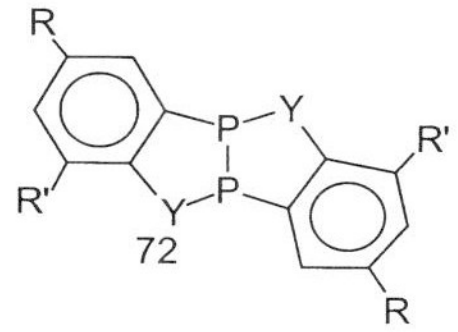

72

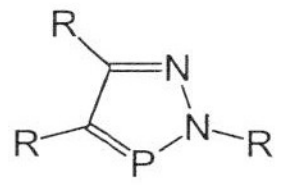

73

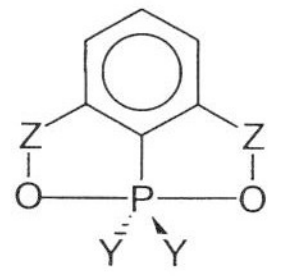

74; Z = CO, CH_2

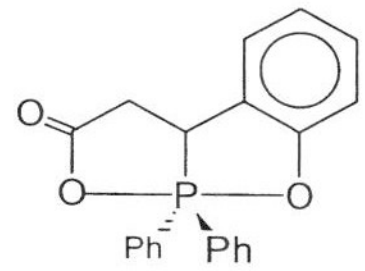

75

crystal structure ⟨78T1993⟩ have been reported. Alkylation of diphosphine (**34**; $n = m = 1$) followed by treatment with alkyl lithium produced monocyclic diphosphines (**68**; $n = m = 1$) ⟨96JCS(CC)537, 98JCS(P1)1657⟩. The bicyclic silylphosphine (**67**; R = Me, Ph) was synthesised by a more conventional route, i.e. the free radical initiated cyclisation of the diallylsilylphosphine (**69**) ⟨85PS(21)367⟩. The dibenzo salts (**70**; R = Me, Ph) were formed by formation of the P–N bond by the action of thionyl chloride on the aminophosphine oxides ⟨86TL5221⟩. The P_2S_2 heterocycle (**71**; Z = P) was obtained from the reaction of thioanisole with $PCl_3/AlCl_3$. A small proportion of the isomer (**72**; Y = S) can also be isolated ⟨97HAC551⟩. Heterocycle (**71**) can be converted to the P_2C_2 system (**71**; Z = CHR) by Freidel–Craft acylation ⟨90JCS(P1)19⟩. Diazoalkene, RCH=CHN=NR, converted (**71**; Z = P, P = P(S) Y = Me) into the two-coordinate diazaphosphole (**73**) ⟨95TL447⟩. Two routes to the dioxaphosphoranes (**74**; Z = CO, CH_2) have been developed, one involving dehydration of a 2,2′-dicarboxyphenylphosphine oxide ⟨78JA5130⟩ the other a 2,2′-di(hydroxymethyl) phenylphosphorane ⟨78CB13⟩. The dioxaphosphorane (**75**) formed in the reaction of 2-hydroxycinnamic acid with methyl diphenylphosphonite is one of the few compounds without nitrogen in the heterocycle ⟨77JCS(CC)156⟩. A variety of routes are available for the conversion of the 1,2-diphosphinobenzenes (**76**; Y = H, Cl) to the dibenzotetraphosphines (**77**; Z = R, Tms, Cl, I) ⟨87JOM(330)17⟩. Use of a zirconium catalyst converted the diphosphine (**76**; Y = H) to a tetramer of (**77**; R = R′ = H, Z = covalent linking bonds) ⟨97JA2954⟩. The crystal structure of the diiodide (**77**; Z = I) has been determined ⟨84ZN(B)139⟩. An interesting reaction of a two-coordinate azaphosphole with

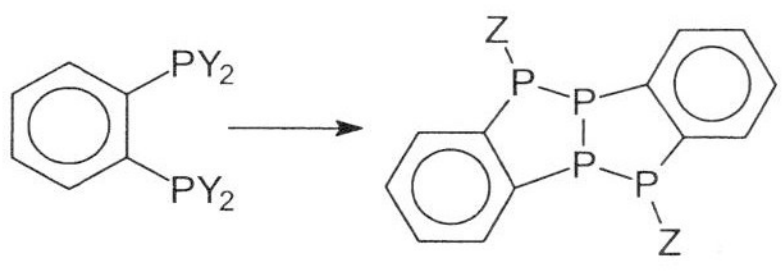

76; Y = H, Cl 77; Z = H, Et, Tms, tBu, Cl, I

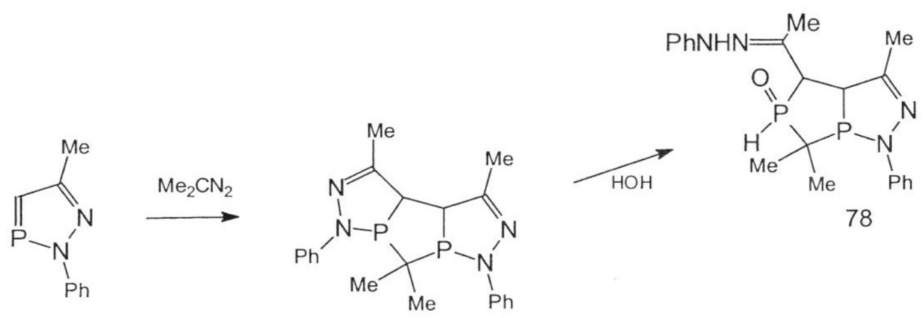

78

Scheme 4

diazo-2-propane (Scheme 4) involves the formation of a C–C bond to give a product which upon hydrolysis gives the diazadiphosphabicyclic compound (**78**) ⟨84IZV2023⟩. The dithiadiphosphabicyclooctane (**79**; M = P) was prepared as shown in Scheme 5 ⟨82TL1345⟩. The bicyclic phosphorotriamidite (**80**; R = Me, Y = lp) was prepared from the appropriate triamine and $(Me_2N)_3P$. It was unstable polymerising spontaneously but it could be converted to a tricyclic phosphorane ⟨80JA7072⟩. The corresponding diaryl derivatives (**80**; R = Ar, Y = lp) ⟨96S1227⟩ and their oxides ⟨98JCS(CC)741, 98JCS(P2)83⟩ have been reported. Two-coordinate azaphospholes react with nitrile ylides ⟨83CJC1105⟩ and with phenylcyanates

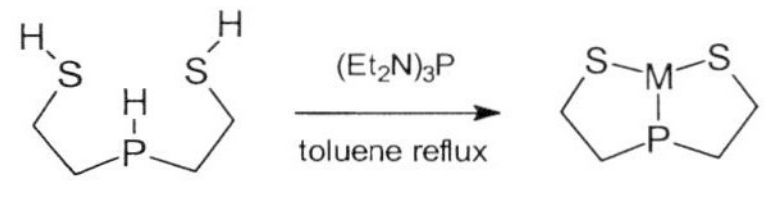

79

Scheme 5

⟨81JCS(CC)1131⟩ to give triaza bicyclic compounds (**81**; Z = CR_2, O. A cyclic phosphazene reacted with a nitrilimine gave the phosphorane (**82**) ⟨99HAC243⟩ whereas R_2PNCO and salicylaldehyde hydrazones gave phosphoranes (**83**) ⟨97JGU1597⟩. A number of PNO_2 bicyclic compounds have been made: the penta-coordinate phosphoranes (**84**) by the reaction of iminoketones with chlorophosphines ⟨78CB2086⟩ and more recently from 2,2′-dihydroxydiphenylamine ⟨90PS(53)87⟩. Related alicyclic analogues have also been studied ⟨75PS(5)229⟩. Most interest in their chemistry lies in the structures of their penta-coordinate phosphoranes, e.g. (**85**) ⟨90PS(47)215⟩ and phosphorane (**84**; R = Ph) ⟨80AX(B)2316⟩ the latter being studied by MM modelling ⟨89JCS(CC)1103⟩. Dimers of (**85**; Y = Ph) were prepared using the Vilsmeier reaction ⟨97HAC517⟩. Treatment of the phosphaalkyne LiOCP, with SO_2 or I_2 gave the novel tetraphosphine (**86**) ⟨95ZAAC(621)34⟩. Diphosphine (**72**; Y = O, R^1 = R^2 = *t*-Bu) was made from readily available trimethylchlorosilane and di-*t*-butylphenolate ⟨98HAC183⟩. Six-coordinate ylide (**87**) was prepared by the action of PCl_5 on the appropriate bisphenol ⟨96PS(111)151⟩.

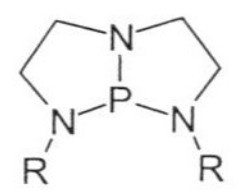

80

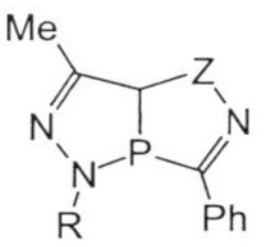

81; Z = CR_2, O

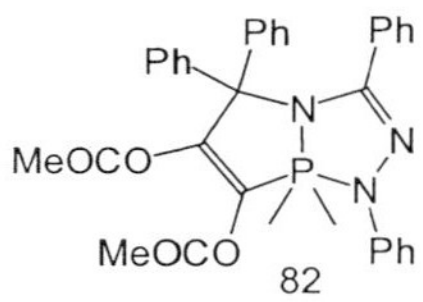

82

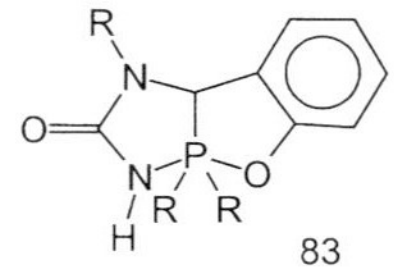

83

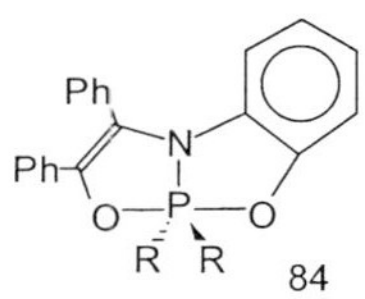

84

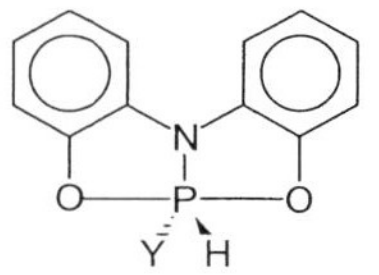

85; Y = OMe, NMe_2, H, Ph

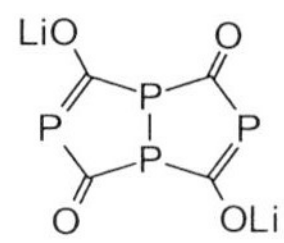

86

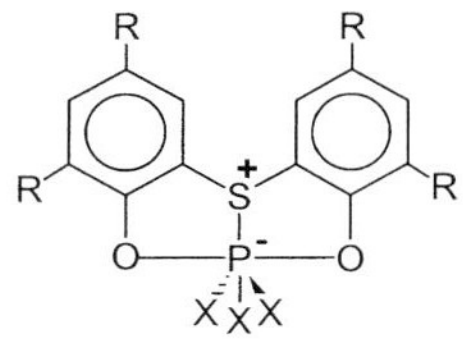

87; R = t-Bu

A number of monobenzo derivatives (**88**) with PNO_2 hetero ring atoms, have been made by cyclisation of 2-aminophenol derivatives ⟨84JGU2448⟩. Alcohols and amines convert the three-coordinate compounds into the phosphoranes ⟨82IZV1408⟩ which can undergo slow ring-chain tautomerism to give the monocyclic compound (**89**) ⟨84JGU2185⟩. Isocyanatophosphines with salicylaldehyde or 2-trifluoroacetylphenol give the urethanes (**90**; Y = Ph, Z = H) ⟨84JGU1328⟩ and (**90**; Y = RO, Z = Tf) ⟨92PS(66)301⟩ whereas with various 1,1,1-trifluoro-2,4-pentadiones the dehydro bicyclic compounds (**91**) are obtained ⟨89CB301, 84JGU2445⟩.

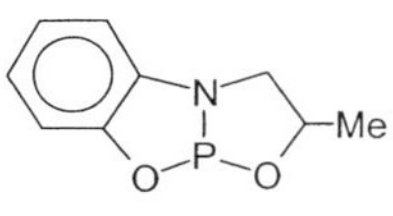

88

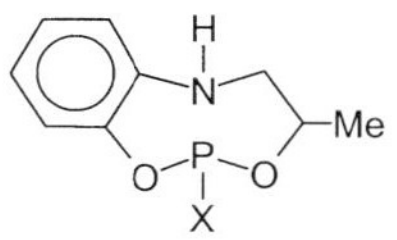

89

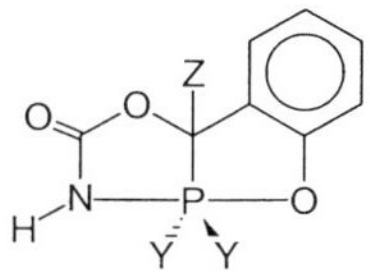

90

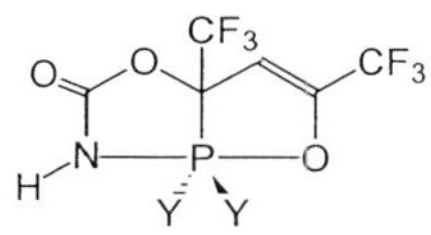

91; Y = Ph, OR

Perhydro bicyclic PNO_2 heterocycles, such as compound (**92**), have been made from diethanolamines and various phosphorus reagents, e.g. $(Me_2N)_3P$ ⟨77TL3041⟩, or $PhOPCl_2/Et_3N$ ⟨77AG(E)549⟩ or PCl_3/Et_3N ⟨81JCS(P2)19⟩. The compounds will oligomerise at 140°C forming [8,3,0] ring systems ⟨83JCS(P2)773⟩ and react with alcohols to form penta-coordinate phosphoranes (**93**) ⟨92PS(69)43⟩ which are more ready to undergo ring chain tautomerism than the corresponding benzo compounds. Bis(neopentoxy)phosphine, $(RO)_2PH$, undergoes ester exchange with diethanolamine to give the mono-cyclic compound (**94**; R = H) which undergoes a transannular interaction with valence expansion of phosphorus to produce the rare bicyclic dihydridophosphorane (**95**; R = Y = H) ⟨81JCR(S)348⟩. Reaction of the phosphorane (**95**; R = OR) with sulphur caused ring opening with the formation of a thiophosphoramidate ⟨98PS(142)27⟩. The formation and stereochemistry of a variety of other phosphorane derivatives have also been studied ⟨76OMR558, 80JA5073⟩. The P-mono oxides of the diphosphino-uriedes (**96**; Y = lp) which are formed in the reaction of silyl derivatives of *N*,*N*-dialkyl, *N*,*N*-alkaryl or *N*,*N*-diaryl ureas with PCl_3 ⟨83ZN(B)460, 98M2⟩ form disulphides and disalts ⟨85AG(E)975⟩ but on permanganate oxidation can produce oxygen bridged [3,3,1] bicyclic trioxides ⟨81CB2132⟩. It was found that the carbon-bound phosphorus acted as the nucleophilic centre. The tetrathio compound (**97**; X = Y = S) was formed in a 24% yield from diphosphorus tetraiodide and methanedithiol ⟨72PS(2)161⟩ and the corresponding tetraphosphine (**97**; X = P, Y = CH_2) was obtained in a 55% yield from methyldichlorophosphine and 1,2-diphosphinoethane ⟨78CB3838⟩.

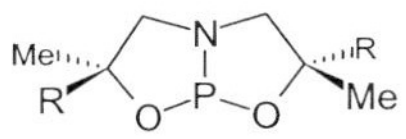

92

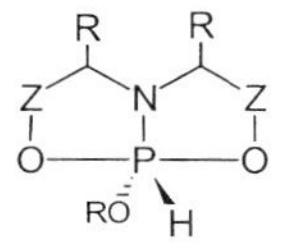

93; R = Me, H; Z = CO, R

94 95

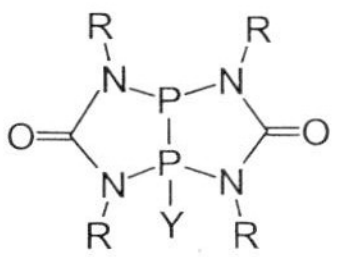

96; R = Me, Ar

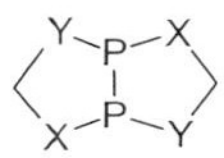

97; X = S

6.2.6.2 [*n*,*m*,0] Fused-Ring Systems

6.2.6.2.1 [n,5-3,0] Ring systems

Bicyclic diphosphines with ring systems [5,5,0] (**34**; *n* = *m* = 3), [5,4,0] (**34**; *n* = 3, *m* = 2), and [5,3,0] (**34**; *n* = 3, *m* = 1) have been synthesised by the route shown in Scheme 1 ⟨98JCS(P1)1643⟩ and they have been converted to their mono- and di-phosphonium salts. Cycloaddition of dimethyl acetylenedicarboxylate with a diazaphosphinine in a 2:1 ratio gave the phosphazene (**98**) ⟨98M3⟩. It is also formed by treating the previously prepared adduct (**182**) with an equivalent of acetylenic ester. Triaza [5,3,0] and [5,5,0] ring systems (**99**) and (**100**) have been prepared by P–N-migration of a phenyl ring under the action of lithium ⟨99JCS(CC)853⟩ on the triamidate (**80**; R = Ph, Y = O). The product from the reaction of chloro 1,3,2-diazaphospholan with a chlorohydrin, cyclises in acetonitrile to give a bicyclic salt (**101**) which is in a tautomeric equilibrium with (**102**), see Scheme 6 ⟨77TL1583⟩. The transannular interaction is another example of the transannular interactions exhibited by the phosphatranes.

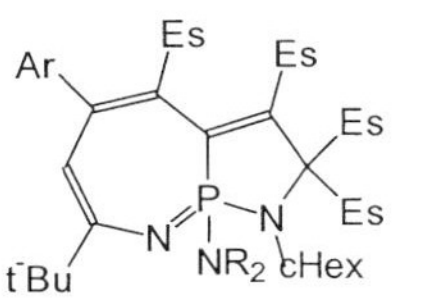

98; Es = CO2Me

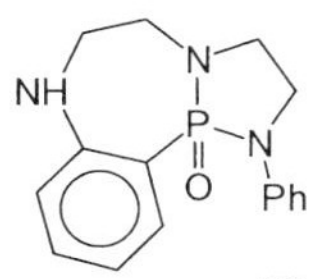

99

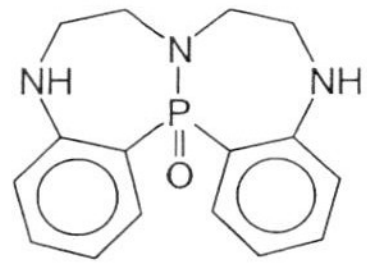

100

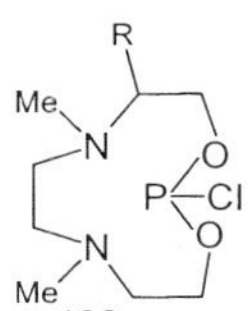

102

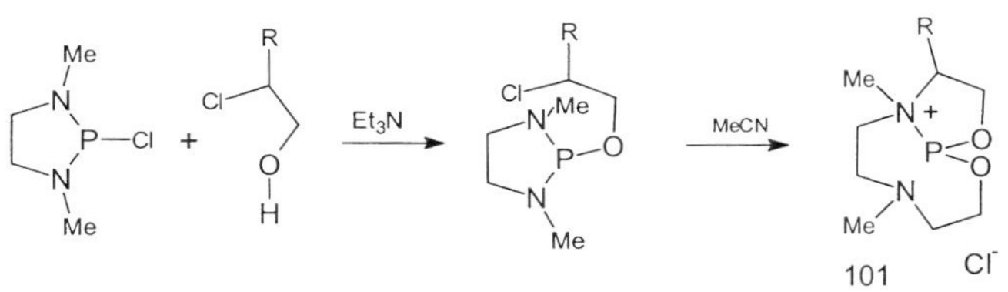

Scheme 6

6.2.6.2.2 [3,2,0] Ring systems

Benzophosphole (**103**) has been reported as the product of the reaction shown in Scheme 7 ⟨77AG(E)877⟩. Carbonyl additions to acetylenic phosphines ($RC{\equiv}CPX_2$) which gave oxaphosphoranes (**104**; X = MeO, Y = Z = CF_3) ⟨89JGU1773⟩ and (**104**; X = OCH_2 CH_2 CH_2O, Y = Ph, Z = CO_2Me/Et) ⟨90JA7451⟩, have been extended to isocyanates. Later studies involved (**104**; Y = Z = CF_3, X = OMe, R = Ph ⟨95PS(103)225⟩. The reaction of hexafluoroacetone with a vinylphosphine and isocyanates ⟨73JCS(P1)972, 86CB2687⟩ produce oxyphosphoranes (**105**) and (**106**), respectively; and a diisocyanate with a alpha-ketophosphonate, $RCOP(O)(OEt)_2$, gave phosphoramidate (**107**; R = $CHCl_2$) ⟨78ZOB1700⟩. A wide range of dehydro P–N fused heterobicyclic compounds have been made, some by dimerisation of, or addition to, cyclic phosphazenes formed by condensation and elimination routes. Compounds (**108**; X = O, S, NR, Y = CS, C = CPh_2, PXNTms, CMeTms) are formed in addition reactions to a five-membered phosphazene ⟨79CB3472, 80CB55⟩. P–N Bridged compounds (**109**) were formed in one step from semicarbazides ⟨76CB3581⟩. A variety of saturated [3,2,0] compounds have also been reported.

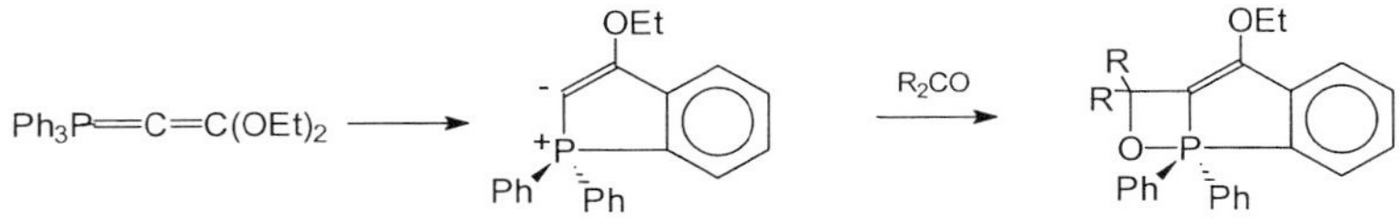

103

Scheme 7

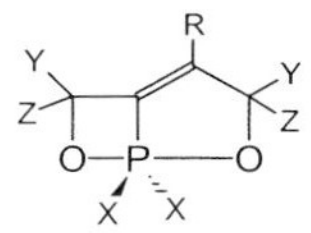

104

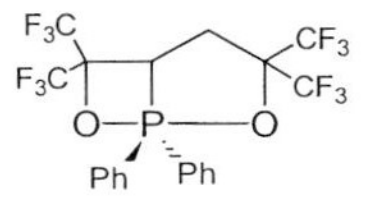

105

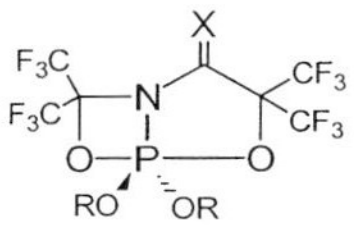

106; X = O, S

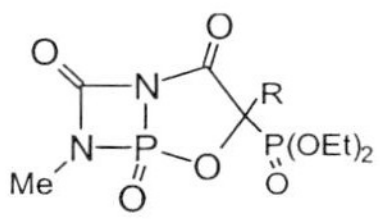

107

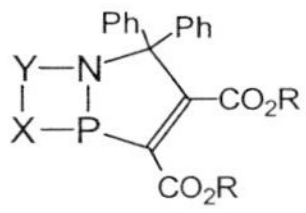

108

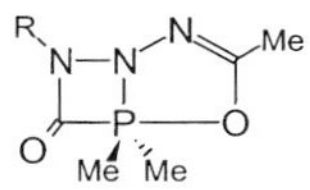

109

6.2.6.2.3 [2,2,0] Ring systems

Low melting bicyclic phosphine (**110**; R = Y = Bu^t, Z = CO_2Me) and the sulphide of isomer (**110**; R = Z = Bu^t, Y = CO_2Me) were formed in thermal rearrangements involving phosphabenzenes ⟨87AG(E)85⟩. The action of Lewis acids such as BF_3 with phosphaalkynes gave spiro compounds which rearranged to triphosphine (**111**) ⟨95JOC6362, 96CB489⟩. Symmetrical isomeric structures (**112**) have also been investigated ⟨97JOM(529)215⟩. They undergo further Diels–Alder reactions to form polycyclic compounds ⟨97JOM(529)215⟩.

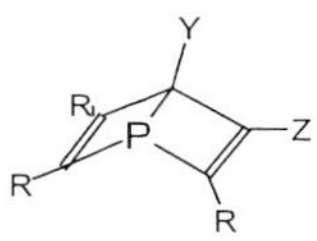

110

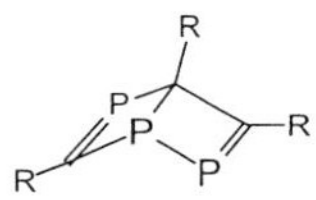

111

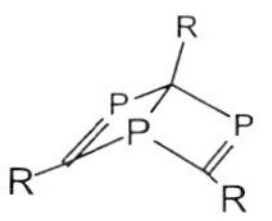

112

6.2.6.3 [*n*,1,0] Fused-Ring Systems

6.2.6.3.1 [n,1,0] Ring systems where n > 3

No [5,1,0] nor [4,1,0] structures were located although they may be identified as part of polycyclic and cage structures (see section below).

as to impart additional aromatic character to the heterocycle. It is also possible to envisage a phosphoranide (**4**) and thus, in this case, the acidity of the corresponding hydridophosphorane (**19**) would be interesting to explore.

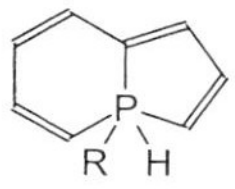

19

6.2.5.1.3 [3,3,0] Ring systems

This ring system might be expected to be the least likely to show aromatic character and neither the pyrrolizinium structures (**20**; M = N) ⟨96MI⟩ nor the phosphonium analogue (**20**; M = P) is known. However, it is possible to draw a phosphoranide structure (**21**) which has more aromaticity potential. Indeed the bicyclic compound (**22**), which was formed by the reaction of phosphorus trichloride with an aminodiketone ⟨84JA1164⟩ has been shown to be planar. This was interpreted as evidence for the delocalised dipolar canonical structure (**23**) rather than the neutral form (**22**). It forms a silver complex with t.b.p. stereochemistry at phosphorus ⟨91JA1071⟩.

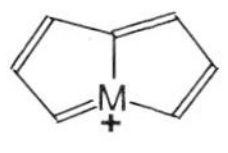

20

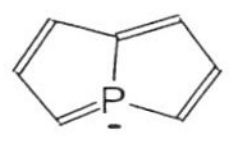

21

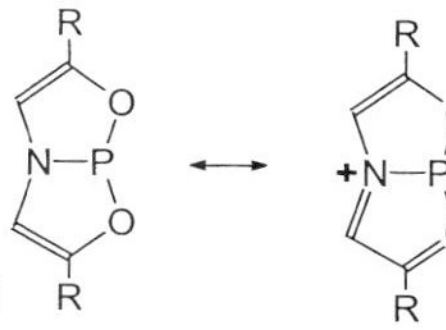

22 23

6.2.6 REACTIVITY OF NON-CONJUGATED AND PARTLY CONJUGATED FUSED-RING SYSTEMS

6.2.6.1 [4,4,0], [4,3,0] and [3,3,0] Fused-Ring Systems

6.2.6.1.1 [4,4,0] Ring systems

The synthesis of 1-phosphabicyclo[4,4,0]decane (**24**) by the radically initiated cyclisation of the primary dibutenylphosphine (**25**) was achieved by Cremer's group in the U.S.A. ⟨87PS(30)369⟩ and by Issleib's group in Germany ⟨87ZAAC(553)136⟩. It is the only example of a parent monoheterobicyclic compound. Both *cis* and *trans* fused-ring systems (**26**) and (**27**) were isolated separately and are equilibrated by UV

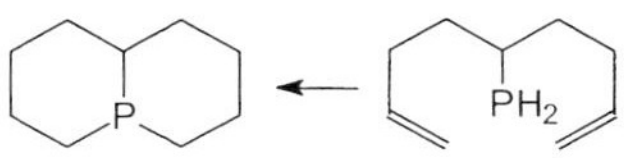

24 25

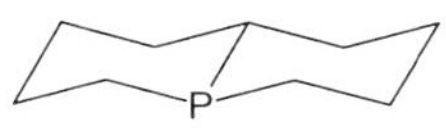

26

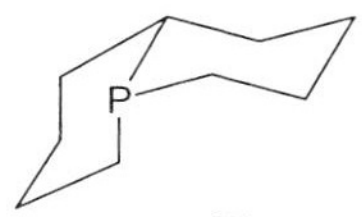

27

light. Unlike the *trans* isomer δP = −45.2, the *cis* isomer δP = −30.4 ppm is conformationally mobile as shown by the broadening of the latter signal with decreasing temperature. The conformational energy barrier for the *cis* isomer was estimated to be about 3.1 kJ mol^{-1} and its ring inversion barrier was estimated to be 41.9 kJ mol^{-1}. The phosphine was converted to its oxide, sulphide, methoxy, methyl and benzyl salts and some other complexes. The stereochemistry of the hydrolysis of the methoxide salt was studied and compared to monocyclic heterocycles. The triarylphosphine (**28**) was produced by the PPA cyclisation of a tertiary benzyl alcohol ⟨85JOC2914⟩. Electrophilic cyclisation of a dipropene precursor produced a similar ring system ⟨90AG(E)689⟩. The dicarboxamide derivative of triphenylphosphine oxide was converted to the oxide (**29**) via a double anionic cyclisation of the 2-lithio intermediate (**30**) ⟨96AG(E)1558⟩. Reaction of a phosphabenzene with sulphur gave a highly reactive intermediate (**31**; R = Me) which reacted with a diene to give the sulphide (**32**; Ch = S). The sulphide may be converted to the corresponding oxide (**32**; Ch = O) by reaction with an epoxide ⟨84JCS(CC)508, 84TL4659⟩. A similar intermediate (**31**; R =

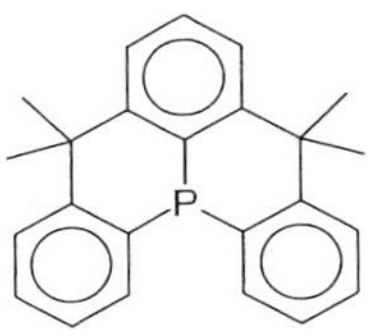

28

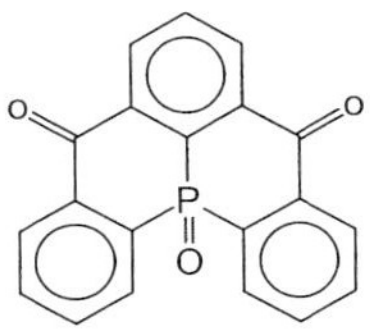

29

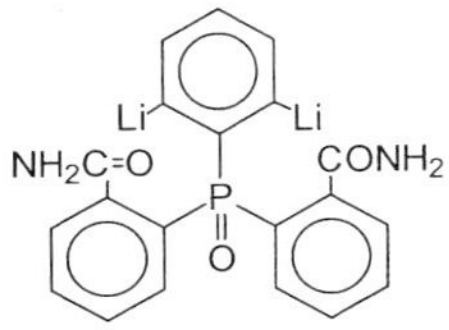

30

Ph) was formed by heating the sulphide (**33**) in toluene ⟨86BCJ2047, 87BCJ1558⟩. A number of saturated bicyclic compounds which have phosphorus atoms at both ring junctions has been synthesised as shown in Scheme 1) ⟨98JCS(P1)1643⟩. The bicyclic diphosphine (**34**; $m = n = 2$) was converted to its mono and

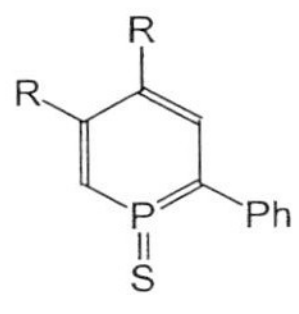

31

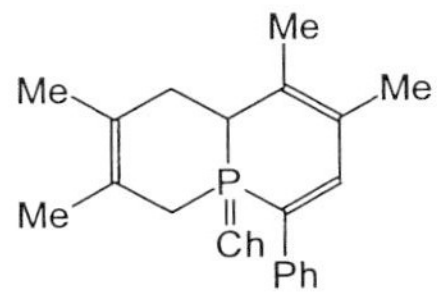

32

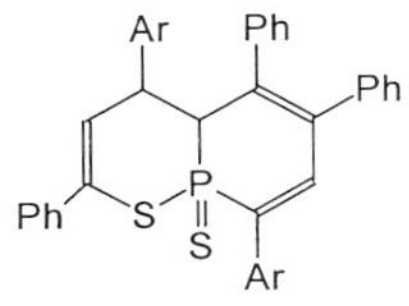

33

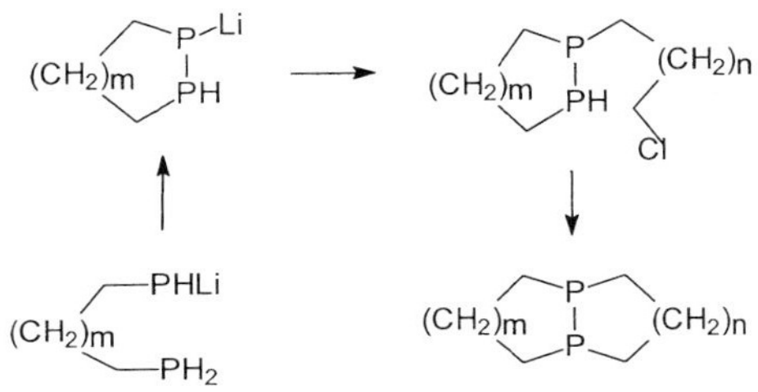

34

Scheme 1

diphosphonium salts, the latter being very sensitive to hydrolysis. Phosphonamide (**35**) was prepared by the reaction of the phosphonic diamide (**36**) with carbon disulphide in the presence of a base ⟨89S618, 90S137⟩. The tetrasilyl phosphonium ylide (**37**) was made from a trimethyl(disilylmethylene)phosphonium ylide and $Me_2SiClCH_2SiClMe_2$ ⟨76AG(E)367⟩. The triazo compound (**38**; X = NMe) was prepared from 2,6,10-triazaundecane $MeNH(CH_2)_3NH(CH_2)_3NHMe$ and trisdimethylaminophosphine ⟨80JA7072⟩ as was the corresponding dioxazaphosphine (**38**; X = O) ⟨83PS(15)281⟩. Their potential for the formation of penta-coordinate phosphoranes with interesting steric restraints, e.g. (**39**) were studied.

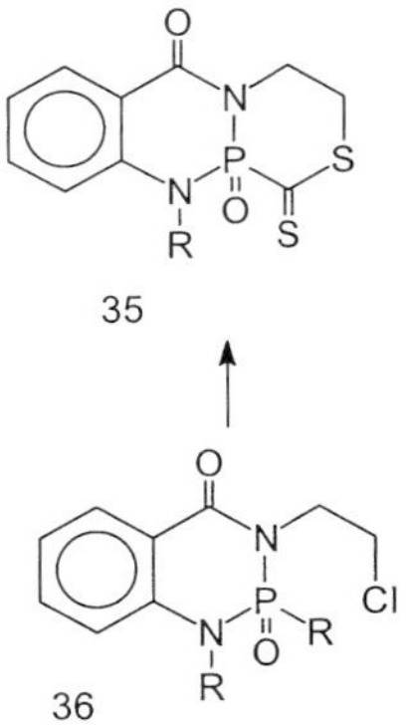

35

36

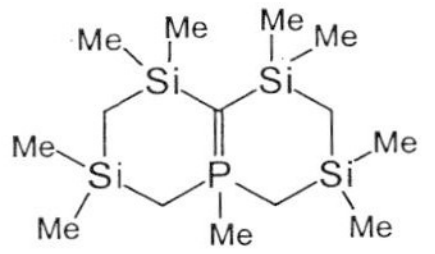

37

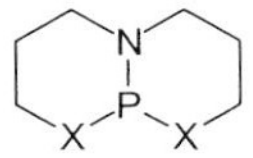

38; X = NMe, O

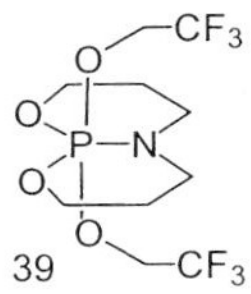

39

6.2.6.1.2 *[4,3,0] Ring systems*

The parent saturated ring system (**40**) was synthesised by the free radical cyclisation of the octadienylphosphine (**41**) and converted to the corresponding oxide, sulphide and methiodide ⟨91ZAAC(600)195⟩.

Phospholes react readily with two molecules of dimethyl acetylenedicarboxylate to give bicyclic adducts. Thus the 1,2,5-triphenylphosphole (**42**; R^1 = Ph, R^2 = H) gave the stabilised bicyclic ylide (**43**) ⟨69JCS(C)1542⟩ (a structure confirmed by X-ray diffraction 23 years later ⟨93CJC1873⟩), whereas 1-phenylphosphole (**42**; R^1 = R^2 = H) and its 3,4-dimethyl analogue (**42**; R^1 = H, R^2 = Me) produced the ylide (**44**; R = Me, H) ⟨78JHC1319⟩. It has been shown that de-protonation and re-protonation of 1*H*-phospholes

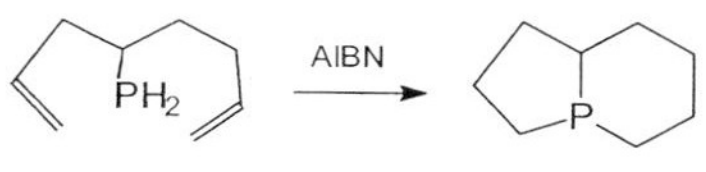

41 40

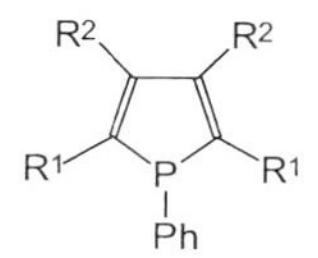

42

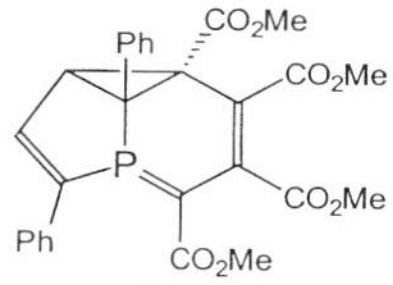

43

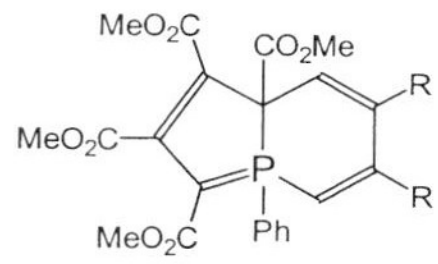

44

produces 2*H*-phospholes which readily dimerise to give [2,2,1] adducts (**45**). These adducts at 130° give monomeric 2*H*-phospholes which react with dienes to give the dihydrophospholizine (8*P*-tetrahydroindene) (**46**) (see Scheme 2) ⟨82JCS(CC)1272⟩. Electrophilic cyclisation of a butenyl phosphorus substituent

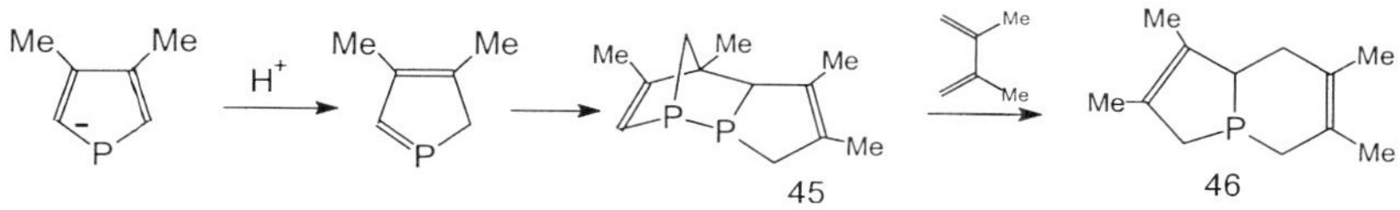

Scheme 2

has been used to prepare the benzo analogue (**47**) ⟨81JOC361⟩. The diphosphine (**34**; m = 2, n = 1), with phosphorus atoms at both ring junctions, was synthesised by the route shown in Scheme 1 and converted into its mono and di phosphonium salts ⟨98JCS(P1)1643⟩. Another unusual reaction involves two molecules of the diynol (**48**) reacting with two molecules of trihydroxymethylphosphine to give the diphospha heterocycle (**49**) ⟨74PS(4)279⟩. Depending on the substituents R^1 and R^2, the reaction of 2-aminobenzamides (**50**; R^1 = H, $((CH_2)_n)Cl$; R^2 = $((CH_2)_n)Cl$, H) with PCl_3 followed by sodium hydride can be used to prepare either amide (**51**) or (**52**) ⟨79JHC897⟩. A 4 + 3 cycloaddition reaction involving a nitrilimine and dihydrofuranylphosphomite (**53**) gave the phosphorane (**54**) ⟨96JGU1367⟩. The dibenzobicyclic phosphonate (**55**) was attributed to the product from the reaction of an *o*-aminomethylphenol with tris(diethylamino)phosphine ⟨74IZV2079⟩ and the mechanism of this unusual transformation was investigated ⟨67DOK(174)846⟩. Conversion of phosphabenzenes to the reactive $\Lambda^5\sigma^3$ sulphide (**31**) led to the formation of the bicyclic dimer (**56**) ⟨90HAC37⟩. A complex reaction between a phosphazene dimer and tetra aminoethene gave a diaminophosphine (**57**) which was characterised as the sulphide ⟨98M1⟩. The unsaturated thioketone (**58**) reacts with P_4S_{10} in the presence of triethylamine to give the unexpected product (**59**). X-ray crystallography showed that the phenyl analogue (**59**; Ar = Ph) adopts a

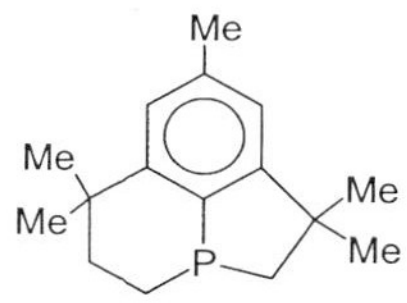

47

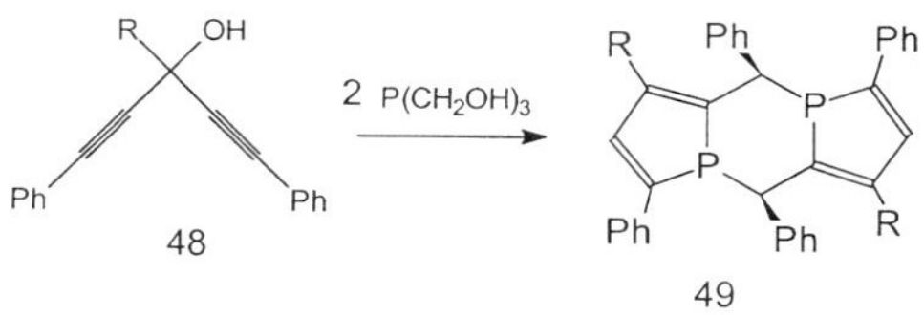

49

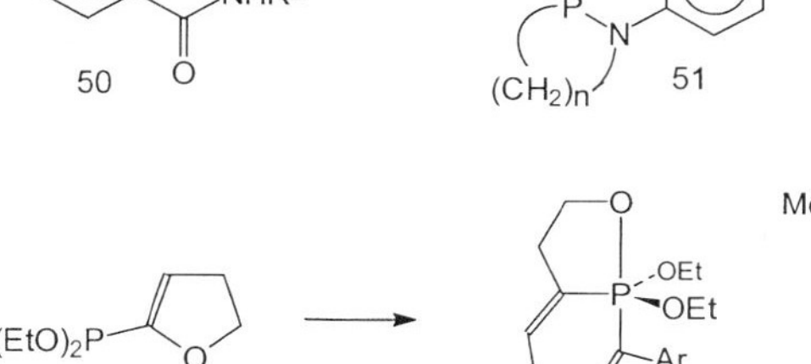

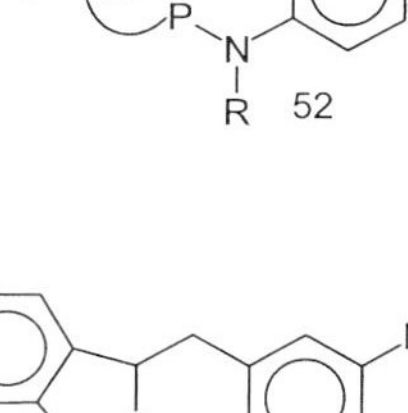

52

55

31 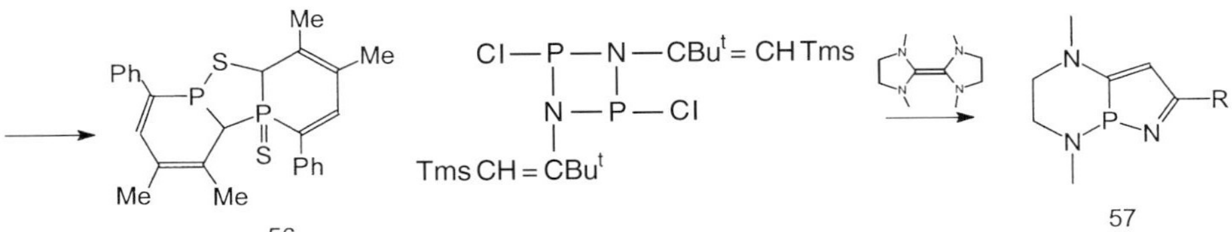

56 57

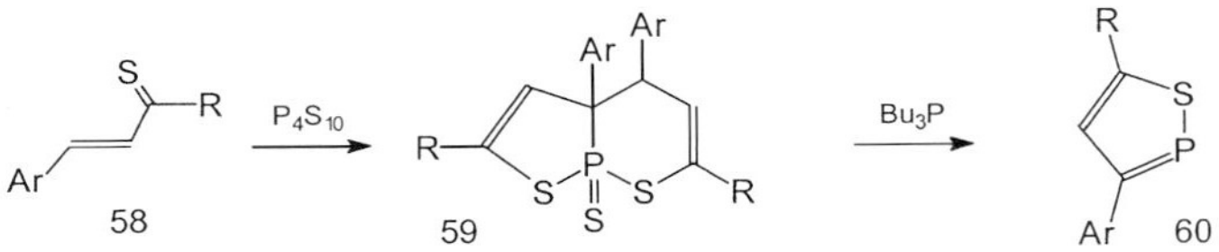

58 59 60

boat conformation ⟨84BCJ1990⟩. Tributylphosphine converts (**59**) to a two-coordinate thiaphosphole (**60**) ⟨94BCJ2785, 93CL1563⟩ which will undergo the Diels–Alder reaction forming new bicyclic heterocycles ⟨94BCJ2785⟩. Whilst 1,2-diketones lead directly to dioxaphosphoranes, the 1,4-unsaturated diketone took an unusual pathway before yielding the bicyclic oxaphosphorane (**61**) shown in Scheme 3 ⟨71JA5236⟩. Hexafluoroacetone, diphenylchlorophosphine and mercurous cyanide in the presence of triethylamine gives the dioxaphosphorane (**62**; Y = CF_3) ⟨86CB2687⟩. An interesting condensation of *t*-butyldichlorophosphine with a trisilated aryl amidine gave phosphazene (**63**) ⟨93PS(76)265⟩. The intermediacy of the bicyclic oxide (**64**) during the hydrolysis of cyclophosphamide was established by its isolation as a solid ⟨77JA5785⟩. Six-coordinate ylides (**65**; X = F, Cl, CF_3) were made by the action of PX_5, $(CF_3)PX_4$ and $(CF_3)_2PCl_3$ on a disilylated diphenolic imine ⟨96IC325, 96PS(111)151⟩.

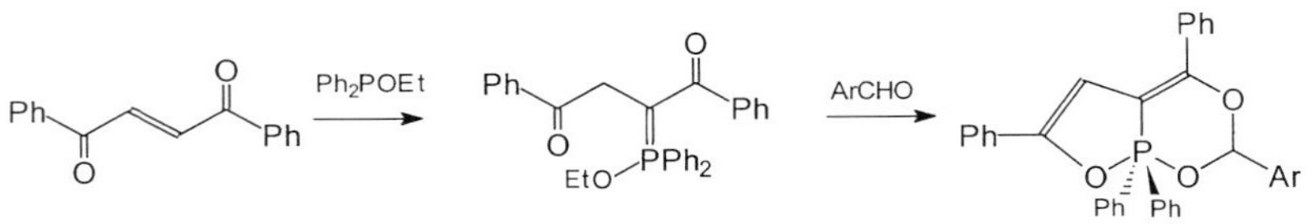

61

Scheme 3

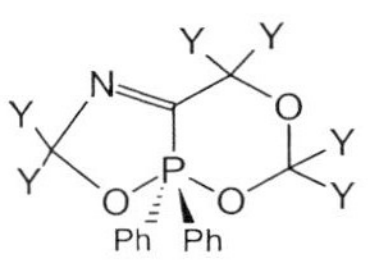

62

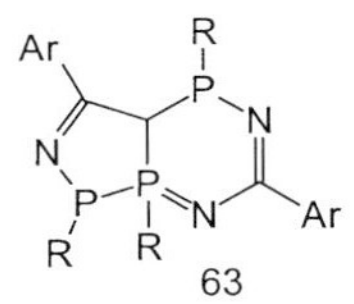

63

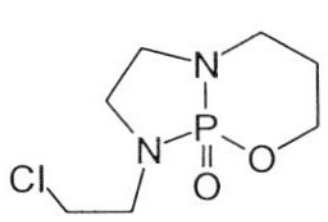

64

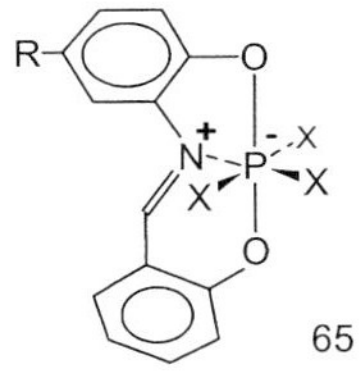

65

6.2.6.1.3 [3,3,0] Ring systems

Neither the parent heterocycle (**66**) not any of its derivatives have been reported, nor have any of its dehydro forms. However many di-, tri- and multi-heteroatom [3,3,0] ring systems have been characterised, most of them quite highly substituted. The simplest non-substituted are two perhydro compounds: the bicyclic diphosphine (**34**; $n = m = 1$) and the corresponding silylphosphine (**67**). The former was synthesised by cyclisation of a chloropentyldiphosphacyclopentane which was formed via the unexpected spontaneous cyclisation with loss of hydrogen as shown in Scheme 1. The diphosphine may be converted to its mono- and di-oxides, sulphides and methiodides ⟨78PS(4)137⟩; their NMR ⟨77OMR172⟩ and disulphide

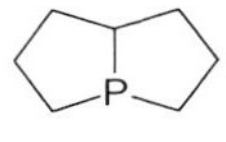

66

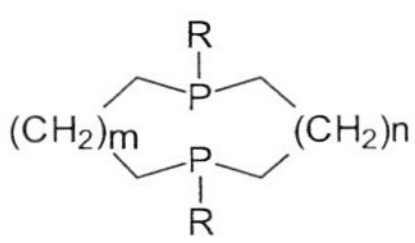

68

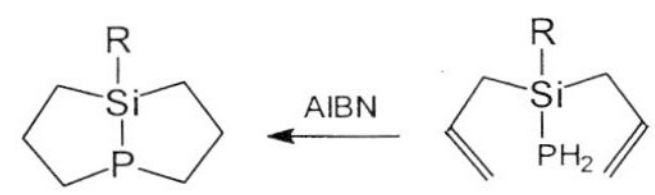

67; R = Me, Ph 69

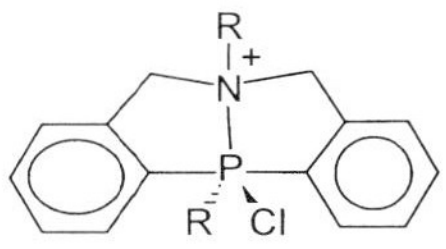

70; R = Me, Ph

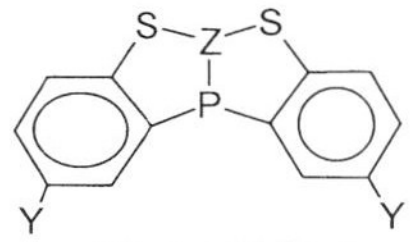

71; Z = CHR, P

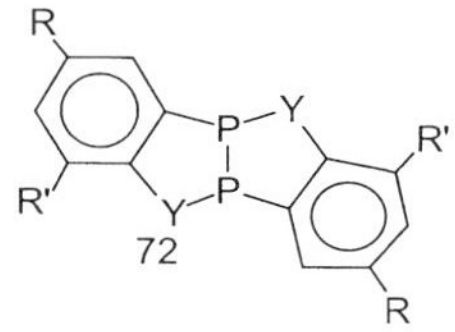

72

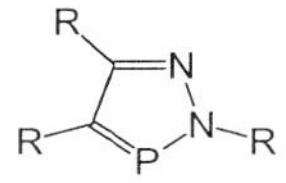

73

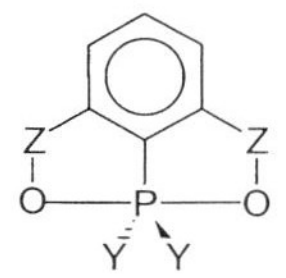

74; Z = CO, CH_2

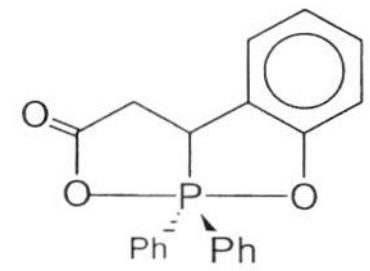

75

crystal structure ⟨78T1993⟩ have been reported. Alkylation of diphosphine (**34**; $n = m = 1$) followed by treatment with alkyl lithium produced monocyclic diphosphines (**68**; $n = m = 1$) ⟨96JCS(CC)537, 98JCS(P1)1657⟩. The bicyclic silylphosphine (**67**; R = Me, Ph) was synthesised by a more conventional route, i.e. the free radical initiated cyclisation of the diallylsilylphosphine (**69**) ⟨85PS(21)367⟩. The dibenzo salts (**70**; R = Me, Ph) were formed by formation of the P–N bond by the action of thionyl chloride on the aminophosphine oxides ⟨86TL5221⟩. The P_2S_2 heterocycle (**71**; Z = P) was obtained from the reaction of thioanisole with $PCl_3/AlCl_3$. A small proportion of the isomer (**72**; Y = S) can also be isolated ⟨97HAC551⟩. Heterocycle (**71**) can be converted to the P_2C_2 system (**71**; Z = CHR) by Freidel–Craft acylation ⟨90JCS(P1)19⟩. Diazoalkene, RCH=CHN=NR, converted (**71**; Z = P, P = P(S) Y = Me) into the two-coordinate diazaphosphole (**73**) ⟨95TL447⟩. Two routes to the dioxaphosphoranes (**74**; Z = CO, CH_2) have been developed, one involving dehydration of a 2,2′-dicarboxyphenylphosphine oxide ⟨78JA5130⟩ the other a 2,2′-di(hydroxymethyl) phenylphosphorane ⟨78CB13⟩. The dioxaphosphorane (**75**) formed in the reaction of 2-hydroxycinnamic acid with methyl diphenylphosphonite is one of the few compounds without nitrogen in the heterocycle ⟨77JCS(CC)156⟩. A variety of routes are available for the conversion of the 1,2-diphosphinobenzenes (**76**; Y = H, Cl) to the dibenzotetraphosphines (**77**; Z = R, Tms, Cl, I) ⟨87JOM(330)17⟩. Use of a zirconium catalyst converted the diphosphine (**76**; Y = H) to a tetramer of (**77**; R = R′ = H, Z = covalent linking bonds) ⟨97JA2954⟩. The crystal structure of the diiodide (**77**; Z = I) has been determined ⟨84ZN(B)139⟩. An interesting reaction of a two-coordinate azaphosphole with

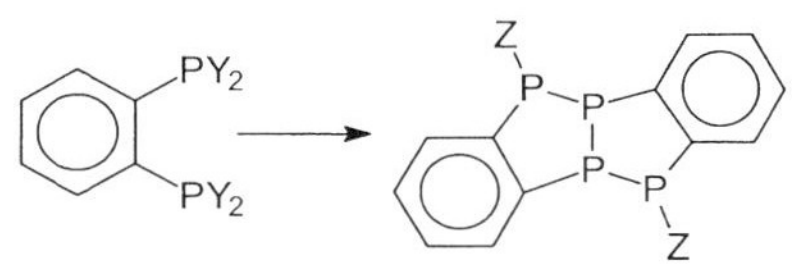

76; Y = H, Cl 77; Z = H, Et, Tms, tBu, Cl, I

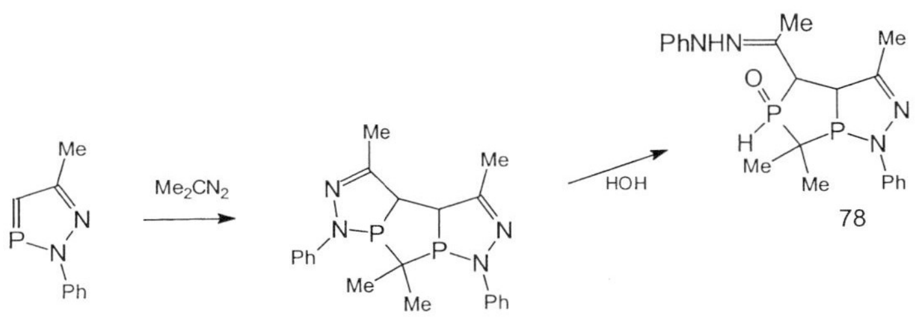

Scheme 4

diazo-2-propane (Scheme 4) involves the formation of a C–C bond to give a product which upon hydrolysis gives the diazadiphosphabicyclic compound (**78**) ⟨84IZV2023⟩. The dithiadiphosphabicyclooctane (**79**; M = P) was prepared as shown in Scheme 5 ⟨82TL1345⟩. The bicyclic phosphorotriamidite (**80**; R = Me, Y = lp) was prepared from the appropriate triamine and $(Me_2N)_3P$. It was unstable polymerising spontaneously but it could be converted to a tricyclic phosphorane ⟨80JA7072⟩. The corresponding diaryl derivatives (**80**; R = Ar, Y = lp) ⟨96S1227⟩ and their oxides ⟨98JCS(CC)741, 98JCS(P2)83⟩ have been reported. Two-coordinate azaphospholes react with nitrile ylides ⟨83CJC1105⟩ and with phenylcyanates

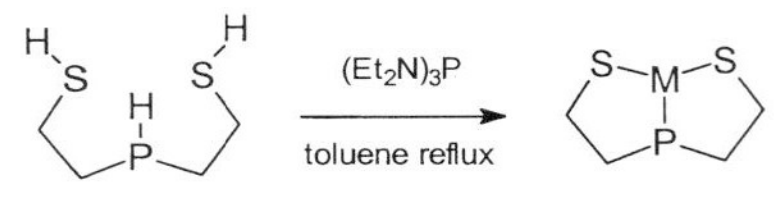

79

Scheme 5

⟨81JCS(CC)1131⟩ to give triaza bicyclic compounds (**81**; Z = CR_2, O. A cyclic phosphazene reacted with a nitrilimine gave the phosphorane (**82**) ⟨99HAC243⟩ whereas R_2PNCO and salicylaldehyde hydrazones gave phosphoranes (**83**) ⟨97JGU1597⟩. A number of PNO_2 bicyclic compounds have been made: the penta-coordinate phosphoranes (**84**) by the reaction of iminoketones with chlorophosphines ⟨78CB2086⟩ and more recently from 2,2′-dihydroxydiphenylamine ⟨90PS(53)87⟩. Related alicyclic analogues have also been studied ⟨75PS(5)229⟩. Most interest in their chemistry lies in the structures of their penta-coordinate phosphoranes, e.g. (**85**) ⟨90PS(47)215⟩ and phosphorane (**84**; R = Ph) ⟨80AX(B)2316⟩ the latter being studied by MM modelling ⟨89JCS(CC)1103⟩. Dimers of (**85**; Y = Ph) were prepared using the Vilsmeier reaction ⟨97HAC517⟩. Treatment of the phosphaalkyne LiOCP, with SO_2 or I_2 gave the novel tetraphosphine (**86**) ⟨95ZAAC(621)34⟩. Diphosphine (**72**; Y = O, R^1 = R^2 = *t*-Bu) was made from readily available trimethylchlorosilane and di-*t*-butylphenolate ⟨98HAC183⟩. Six-coordinate ylide (**87**) was prepared by the action of PCl_5 on the appropriate bisphenol ⟨96PS(111)151⟩.

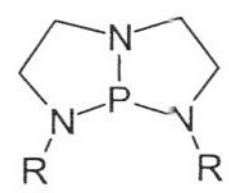

80

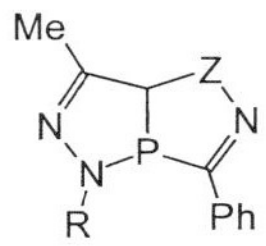

81; Z = CR_2, O

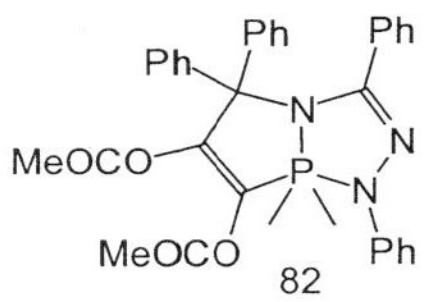

82

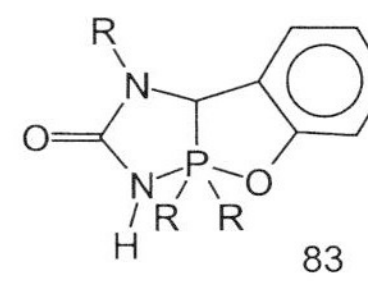

83

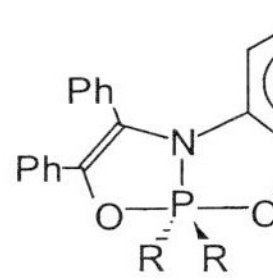

84

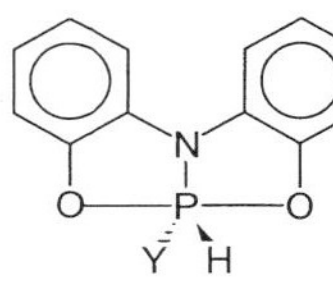

85; Y = OMe, NMe_2, H, Ph

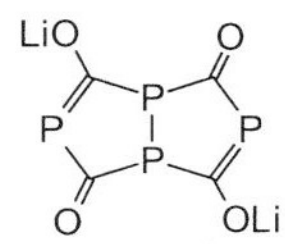

86

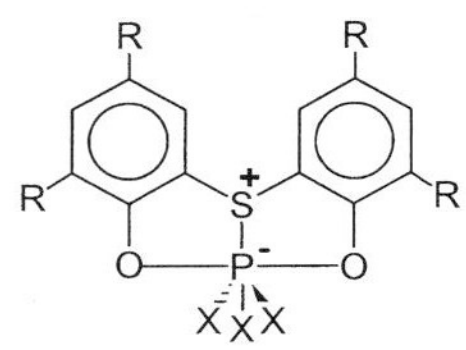

87; R = t-Bu

A number of monobenzo derivatives (**88**) with PNO_2 hetero ring atoms, have been made by cyclisation of 2-aminophenol derivatives ⟨84JGU2448⟩. Alcohols and amines convert the three-coordinate compounds into the phosphoranes ⟨82IZV1408⟩ which can undergo slow ring-chain tautomerism to give the monocyclic compound (**89**) ⟨84JGU2185⟩. Isocyanatophosphines with salicylaldehyde or 2-trifluoroacetylphenol give the urethanes (**90**; Y = Ph, Z = H) ⟨84JGU1328⟩ and (**90**; Y = RO, Z = Tf) ⟨92PS(66)301⟩ whereas with various 1,1,1-trifluoro-2,4-pentadiones the dehydro bicyclic compounds (**91**) are obtained ⟨89CB301, 84JGU2445⟩.

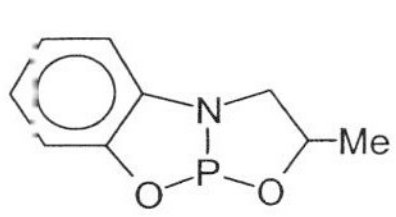

88

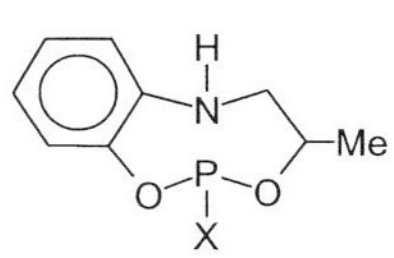

89

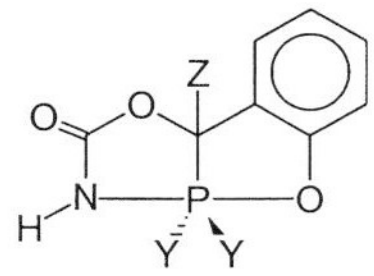

90

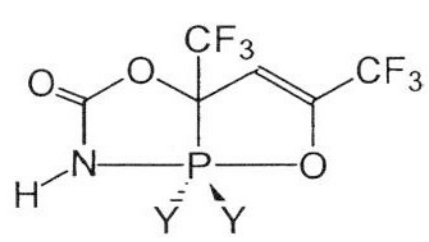

91; Y = Ph, OR

Perhydro bicyclic PNO_2 heterocycles, such as compound (**92**), have been made from diethanolamines and various phosphorus reagents, e.g. $(Me_2N)_3P$ ⟨77TL3041⟩, or $PhOPCl_2/Et_3N$ ⟨77AG(E)549⟩ or PCl_3/Et_3N ⟨81JCS(P2)19⟩. The compounds will oligomerise at 140ºC forming [8,3,0] ring systems ⟨83JCS(P2)773⟩ and react with alcohols to form penta-coordinate phosphoranes (**93**) ⟨92PS(69)43⟩ which are more ready to undergo ring chain tautomerism than the corresponding benzo compounds. Bis(neopentoxy)phosphine, $(RO)_2PH$, undergoes ester exchange with diethanolamine to give the monocyclic compound (**94**; R = H) which undergoes a transannular interaction with valence expansion of phosphorus to produce the rare bicyclic dihydridophosphorane (**95**; R = Y = H) ⟨81JCR(S)348⟩. Reaction of the phosphorane (**95**; R = OR) with sulphur caused ring opening with the formation of a thiophosphoramidate ⟨98PS(142)27⟩. The formation and stereochemistry of a variety of other phosphorane derivatives have also been studied ⟨76OMR558, 80JA5073⟩. The P-mono oxides of the diphosphino-uriedes (**96**; Y = lp) which are formed in the reaction of silyl derivatives of *N,N*-dialkyl, *N,N*-alkaryl or *N,N*-diaryl ureas with PCl_3 ⟨83ZN(B)460, 98M2⟩ form disulphides and disalts ⟨85AG(E)975⟩ but on permanganate oxidation can produce oxygen bridged [3,3,1] bicyclic trioxides ⟨81CB2132⟩. It was found that the carbon-bound phosphorus acted as the nucleophilic centre. The tetrathio compound (**97**; X = Y = S) was formed in a 24% yield from diphosphorus tetraiodide and methanedithiol ⟨72PS(2)161⟩ and the corresponding tetraphosphine (**97**; X = P, Y = CH_2) was obtained in a 55% yield from methyldichlorophosphine and 1,2-diphosphinoethane ⟨78CB3838⟩.

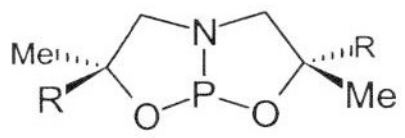

92

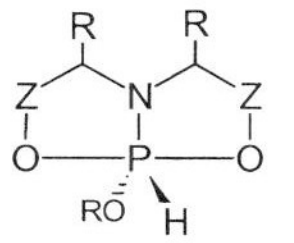

93; R = Me, H; Z = CO, R

94 95

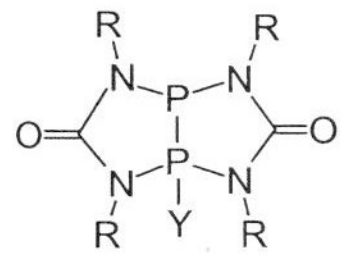

96; R = Me, Ar

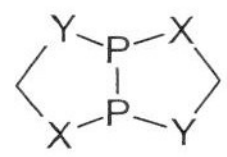

97; X = S

6.2.6.2 [*n*,*m*,0] Fused-Ring Systems

6.2.6.2.1 [n,5-3,0] Ring systems

Bicyclic diphosphines with ring systems [5,5,0] (**34**; *n* = *m* = 3), [5,4,0] (**34**; *n* = 3, *m* = 2), and [5,3,0] (**34**; *n* = 3, *m* = 1) have been synthesised by the route shown in Scheme 1 ⟨98JCS(P1)1643⟩ and they have been converted to their mono- and di-phosphonium salts. Cycloaddition of dimethyl acetylenedicarboxylate with a diazaphosphinine in a 2:1 ratio gave the phosphazene (**98**) ⟨98M3⟩. It is also formed by treating the previously prepared adduct (**182**) with an equivalent of acetylenic ester. Triaza [5,3,0] and [5,5,0] ring systems (**99**) and (**100**) have been prepared by P–N-migration of a phenyl ring under the action of lithium ⟨99JCS(CC)853⟩ on the triamidate (**80**; R = Ph, Y = O). The product from the reaction of chloro 1,3,2-diazaphospholan with a chlorohydrin, cyclises in acetonitrile to give a bicyclic salt (**101**) which is in a tautomeric equilibrium with (**102**), see Scheme 6 ⟨77TL1583⟩. The transannular interaction is another example of the transannular interactions exhibited by the phosphatranes.

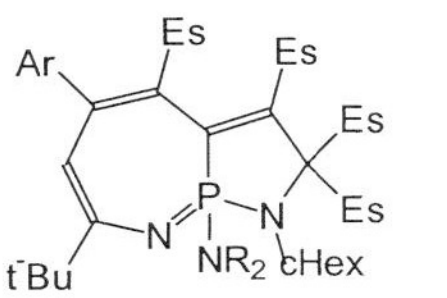

98; Es = CO2Me

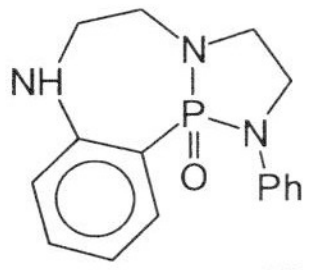

99

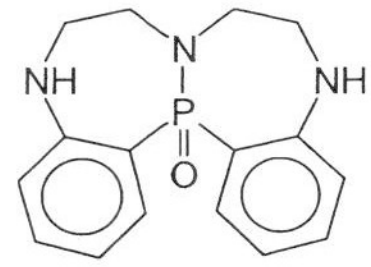

100

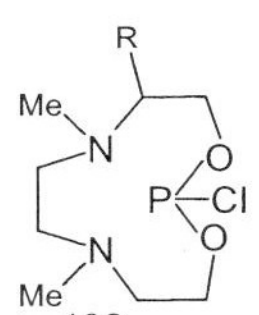

102

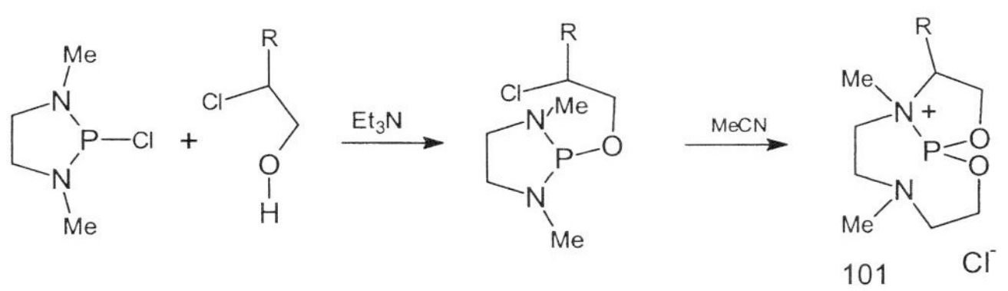

Scheme 6

6.2.6.2.2 [3,2,0] Ring systems

Benzophosphole (**103**) has been reported as the product of the reaction shown in Scheme 7 ⟨77AG(E)877⟩. Carbonyl additions to acetylenic phosphines ($RC{\equiv}CPX_2$) which gave oxaphosphoranes (**104**; X = MeO, Y = Z = CF_3) ⟨89JGU1773⟩ and (**104**; X = OCH_2 CH_2 CH_2O, Y = Ph, Z = CO_2Me/Et) ⟨90JA7451⟩, have been extended to isocyanates. Later studies involved (**104**; Y = Z = CF_3, X = OMe, R = Ph ⟨95PS(103)225⟩. The reaction of hexafluoroacetone with a vinylphosphine and isocyanates ⟨73JCS(P1)972, 86CB2687⟩ produce oxyphosphoranes (**105**) and (**106**), respectively; and a diisocyanate with a alpha-ketophosphonate, $RCOP(O)(OEt)_2$, gave phosphoramidate (**107**; R = $CHCl_2$) ⟨78ZOB1700⟩. A wide range of dehydro P–N fused heterobicyclic compounds have been made, some by dimerisation of, or addition to, cyclic phosphazenes formed by condensation and elimination routes. Compounds (**108**; X = O, S, NR, Y = CS, C = CPh_2, PXNTms, CMeTms) are formed in addition reactions to a five-membered phosphazene ⟨79CB3472, 80CB55⟩. P–N Bridged compounds (**109**) were formed in one step from semicarbazides ⟨76CB3581⟩. A variety of saturated [3,2,0] compounds have also been reported.

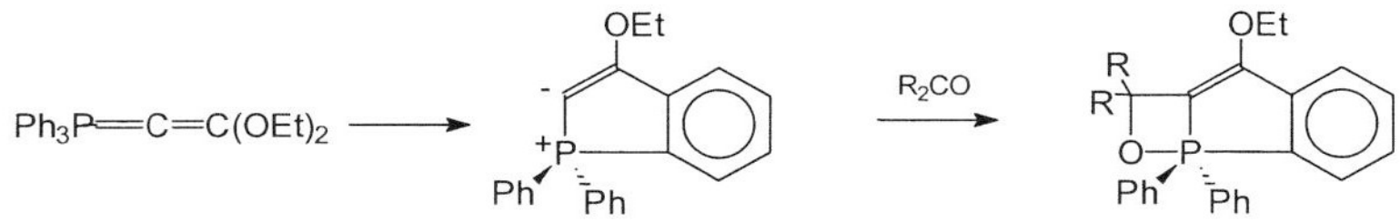

103

Scheme 7

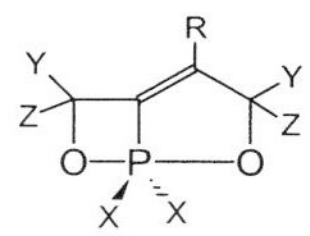

104

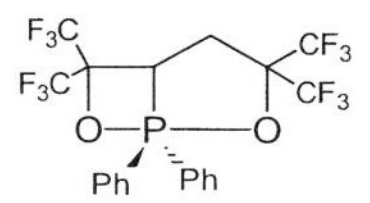

105

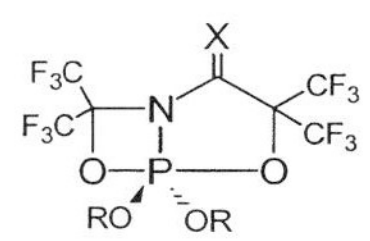

106; X = O, S

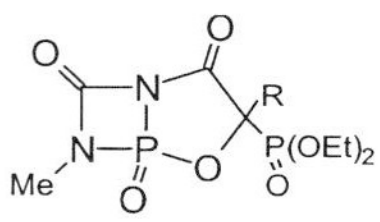

107

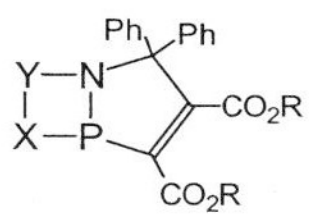

108

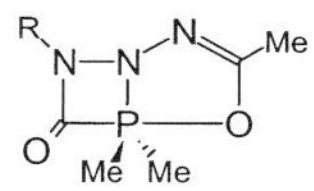

109

6.2.6.2.3 [2,2,0] Ring systems

Low melting bicyclic phosphine (**110**; R = Y = Bu^t, Z = CO_2Me) and the sulphide of isomer (**110**; R = Z = Bu^t, Y = CO_2Me) were formed in thermal rearrangements involving phosphabenzenes ⟨87AG(E)85⟩. The action of Lewis acids such as BF_3 with phosphaalkynes gave spiro compounds which rearranged to triphosphine (**111**) ⟨95JOC6362, 96CB489⟩. Symmetrical isomeric structures (**112**) have also been investigated ⟨97JOM(529)215⟩. They undergo further Diels–Alder reactions to form polycyclic compounds ⟨97JOM(529)215⟩.

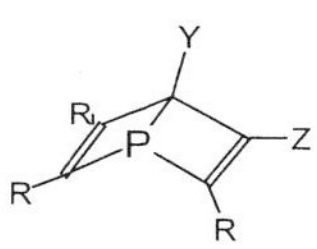

110

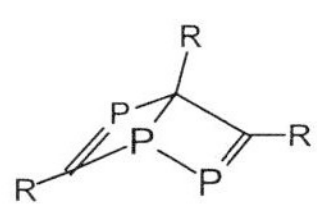

111

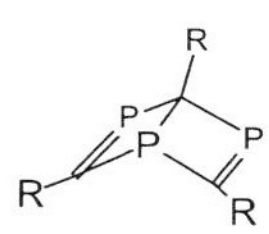

112

6.2.6.3 [*n*,1,0] Fused-Ring Systems

6.2.6.3.1 [n,1,0] Ring systems where n > 3

No [5,1,0] nor [4,1,0] structures were located although they may be identified as part of polycyclic and cage structures (see section below).

6.2.6.3.2 [3,1,0] Ring systems

The bicyclic compounds (**113**) which were prepared from various diazomethanes R1R2CN2 and two-coordinate diazaphospholes, react with alcohols to give diazadehydrophosphorins ⟨83IZV2275⟩. The ring system is favoured by the hexaphosphines (**114**) being formed directly from Bu^tPCl_2, PCl_3 and Mg ⟨81AG(E)967⟩. Air oxidises one of the five-membered ring phosphorus atoms in contrast to the [3,2,0] heptaphosphine where a bridged phosphorus atom or one of the four-membered ring atoms is oxidised ⟨86AG(E)471⟩. The X-ray crystallographic structure of a triazadiphosphine incorporating an aluminium atom in the larger ring has been determined ⟨99IC2248⟩.

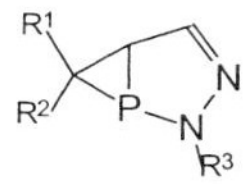

113

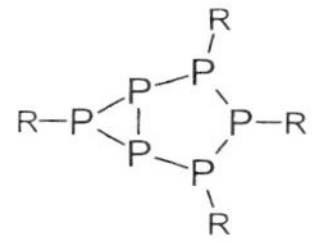

114; R = Bu^t

6.2.6.3.3 [2,1,0] Ring systems

A 1,3-sigmatropic shift involving the phosphole (**115**) gave the bicyclic compound (**116**) ⟨95JCS(CC)2395⟩.

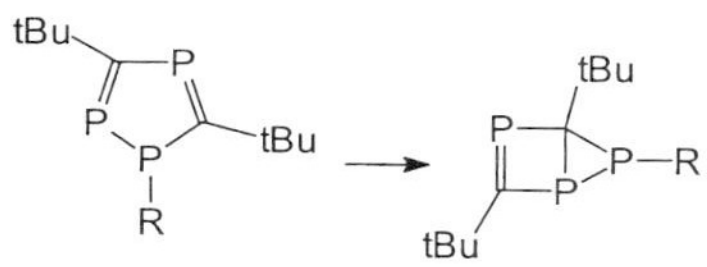

115 116

6.2.6.3.4 [1,1,0] Ring systems

All four examples with this system possess P–P fused rings (**117**). A Wurtz-type reaction on two-coordinate halophosphine YZC=P–Hal ⟨90CB67⟩ gave YZC=P–P=CYZ which cyclised to give the diphosphine (**117**; Y = Z = CPhTms), whereas the disiladiphosphine (**117**; Y = Z = $SiAr_2$) was produced by the reaction of white phosphorus with a disilene. Dichlorosupermesitylphosphine, $SmsPCl_2$, and dicoordinate phosphine $Tms_2P–P=CTms_2$ gave the triphosphine (**117**; Y = $CTms_2$; Z = PSms) ⟨88AG(E)957, 88JCS(CC)945⟩ and the reaction of tri-*t*-butylphenyl lithium with white phosphorus gave the tetraphosphine (**117**; Y = Z = PBu^t) ⟨85AG(E)1056⟩.

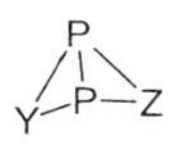

117

6.2.6.4 Polycyclic and Caged Fused-Ring Systems

A wide range of polycyclic heterocycles have been prepared and in particular those of the cage type are abundant. The heterocycles are discussed in order of the number of heteroatoms in the rings. Diels–Alder cycloaddition reaction of silylphosphaethyne, TmsCP, with 1,4-cyclohexadiene gave phosphine (**118**) whereas 1,3-butadienes reacted with two equivalents of TmsCP to give diphosphines (**119**) and cyclopentadiene produced the complex double cage structure (**120**) derived from the reactants in a 3 : 2 ratio ⟨99M1⟩. In contrast mesitylphosphaethyne and 1,3-butadiene gave the diphosphine (**121**; R = mesityl) ⟨98S1305⟩. Tricyclic diphosphine (**122**) originated from *t*-butyl phosphaethyne ⟨94JOM(464)41⟩ and tropone reacted with *t*-butyl- and adamantyl-phosphaethyne to produce diphosphines (**123**; R = *t*-Bu, Ad) ⟨95JOC5884⟩. Two-coordinate 2*H*-phospholes such as (**124**) are in equilibrium with polycyclic dimers (**125**) ⟨93JOM(462)103⟩ whereas dimer (**126**; R = Bu^t) was made by the action of GaBr

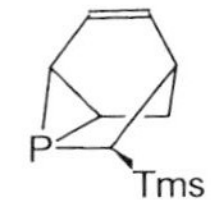
118

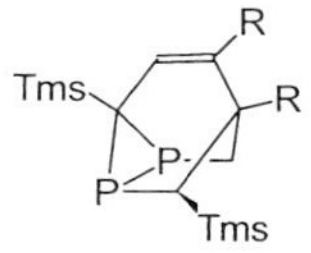
119

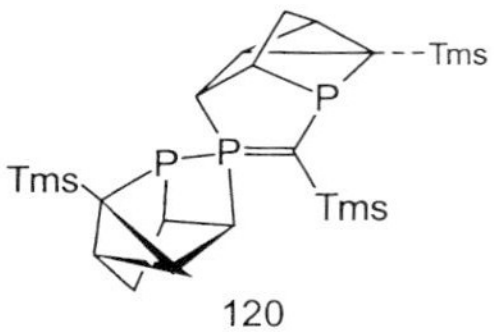
120

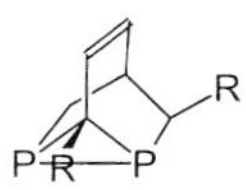
121

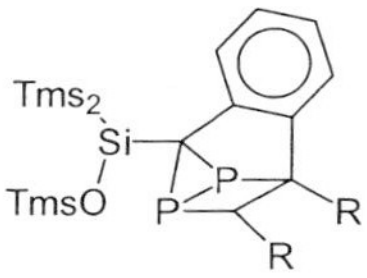
122

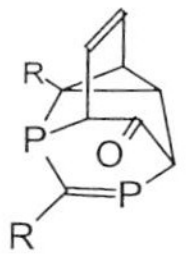
123

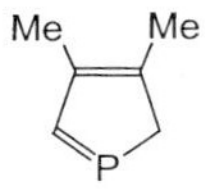
124

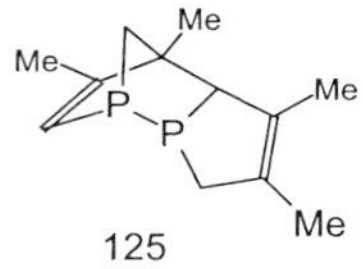
125

on the lithium derivative of 2,5-di-*t*-butyl-1*H*-phosphole ⟨99AG(E)1646⟩. Phosphiranes (**127**; R = *i*-Pr, 3,5-$(CF_3)_2C_6H_3$) were synthesised from *b,f*-dibenzocycloheptadiene, the final step involving a transannular 1 + 2 cycloaddition of the intermediate generated by the action of Mg on dichlorophosphamidite (**128**) ⟨99AG(E)1623⟩. The tris ylide (**129**) was formed in the reaction of PCl_3 with triphenylphosphoniumpropenylide $Ph_3P{=}CHCHCH_2$, ⟨98M4⟩. Phosphine (**130**) was prepared by the action of *t*-butoxide on tris-(6-fluoro-2-hydroxyphenyl)phosphine (**131**) ⟨97JA1208⟩. Bridged phosphine sulphide (**132**) ⟨97TL6457⟩ and unusual caged phosphines (**133**) ⟨95CB991⟩, (**134**) and (**135**) ⟨99JOC587⟩, (**136**) ⟨98M5⟩, (**137**; Es = CO_2Me), (**138**) ⟨97JOM(529)215⟩ and (**139**) ⟨98AG(E)1233⟩ have been described. Transannular bonding by sulphur led to the hexacoordinate structure (**140**) ⟨93IC4360⟩. Tetraphospha cubanes (**141**) ⟨93TL6365⟩ have been prepared from zirconium diphosphine (**142**) ⟨94S1337⟩ and converted to their tetraoxides and tetrasulphides. Their mass spectra ⟨95JCS(P2)1191⟩, basicities ⟨96JOC7813⟩ and reactivities towards electrophiles

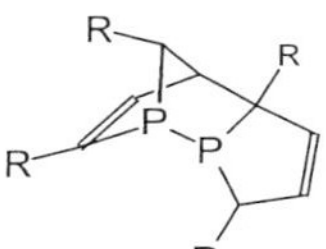
126

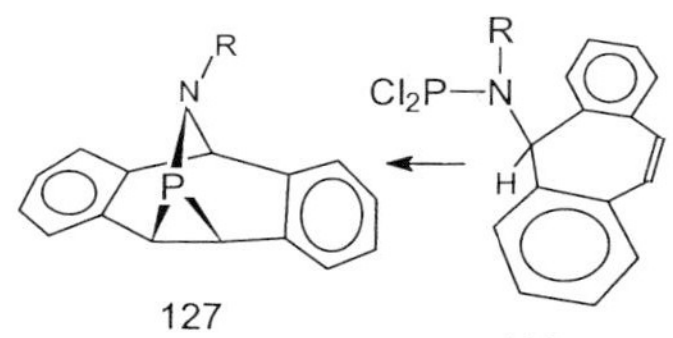
127 128

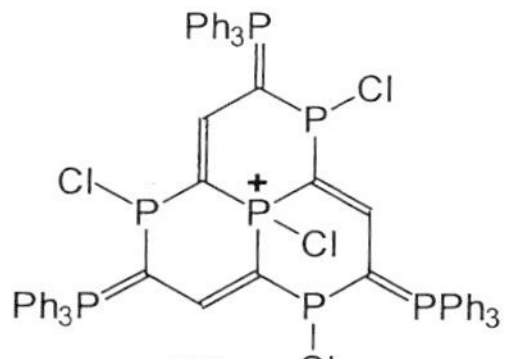
129

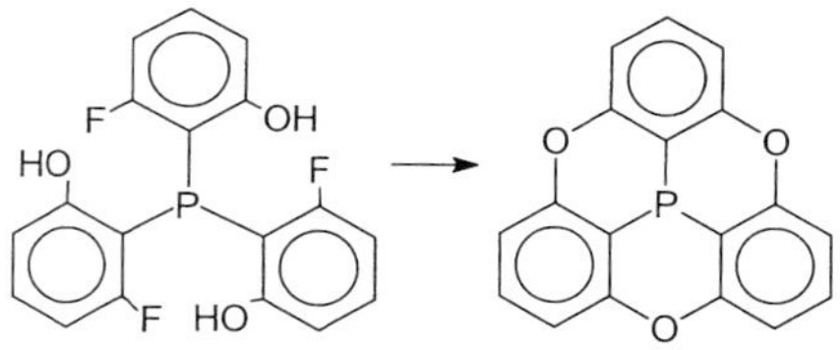
131 130

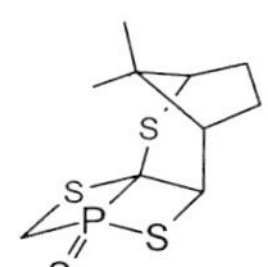
132

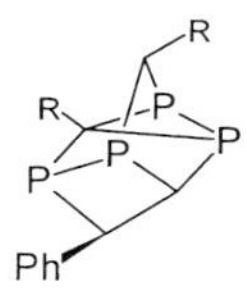
133

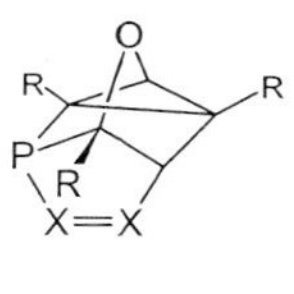
134

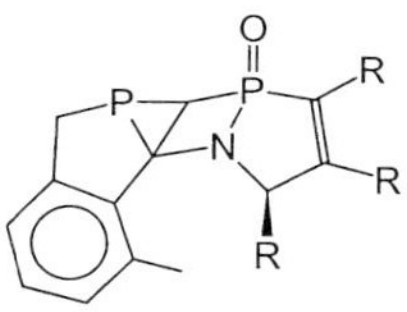
135

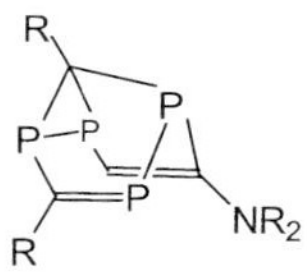
136

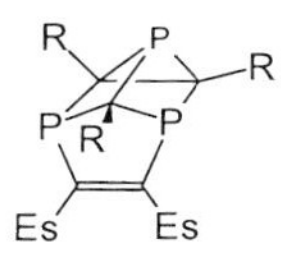
137

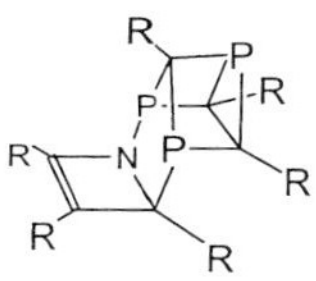
138; R = t-Bu

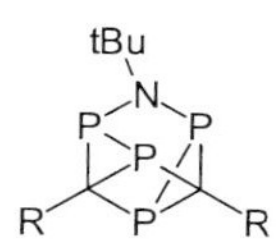
139

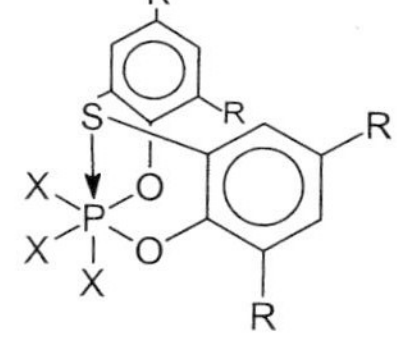
140

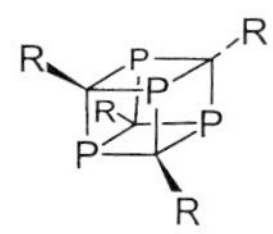
141

⟨95JOC47⟩ have been investigated. Monoalkylation and mono- and di-protonation was observed. The zirconium reagent was also the reactant for the formation of linear tricyclic tetraphosphines (**143**) and (**144**) ⟨95AG(E)484⟩ with alkylation and protonation studies showing similar results ⟨95JOC3149⟩. Two-coordinate phospholidene (**145**), from the reaction of an amidoacetal with EtPHNHMe, dimerised to bicyclic phosphine (**146**) ⟨88JOM(355)71⟩. In addition to triamide (**147**), tetraamides (**148**) and (**149**) ⟨95IC1638⟩, and the corresponding oxide of the latter ⟨99JCR(S)2240⟩, two polysulphides (**150**) and (**151**) have been prepared ⟨96BSB675, 98IC6093⟩. Cage tetraphosphine (**152**; R = *t*-butyl) was formed from *t*-butylphosphaethyne and diazaphosphine (**153**) ⟨98M6⟩. X-ray molecular structures are reported for caged tetraphosphines incorporating a selenium or telurium atom which were also made from phosphaethynes. The reaction of phenyl- or ethoxy-dichlorophosphine with imine (**154**) gave phosphoranes (**155**; Y = Ph, EtO) ⟨95PS(103)235⟩. The disulphide (**156**), was formed by the action of thiophosphoryl chloride on 2-amino-*N*-tolylbenzylamine ⟨79JCS(P1)2896⟩. PN Compounds (**157**; X = Y = OR), (**157**; X = OPh, Y = CH_2NEt_2, R = H) (**158**) and (**159**) have also been formed from 2-azidophenols ⟨78ZOB44⟩, oxazaphospholene ⟨93M1⟩, oxalic acid diamide ⟨70ZAAC(373)245, 72JGU520⟩ and acetohydrazide ⟨69AG(E)674⟩, respectively. Elimination of HX from phosphorane precursors has given phosphazenes which combine to produce dimers (**157**; Y = Ph) ⟨84JCS(P1)1449⟩. Three-coordinate starting material (**160**) was converted by a diaminomethane, $Et_2NCH_2NEt_2$, to bicyclic phosphorane (**157**; Y = X, CH_2NEt_2) ⟨87JGU434⟩. Dimerisation of dihydrazido derivative of triphosphine YP=P–PClY, gave the unexpected heterocyclic compound (**161**) ⟨87AG(E)1256⟩. Caged heptaphosphine (**162**) ⟨95JA5303⟩ and tri- and tetraphosphines which incorporate Sb ⟨97JCS(CC)305, 97JCS(D)4321⟩, gallium ⟨99JCS(CC)771⟩, and Sn ⟨98M7⟩ have been prepared. A series of P=N dimers (**163**) have been made from the reactions of phosphines ($X_2P{-}N{=}C{=}Y$), e.g. isocyanates, with carbonyl compounds ⟨88JGU1148, 79JGU641⟩.

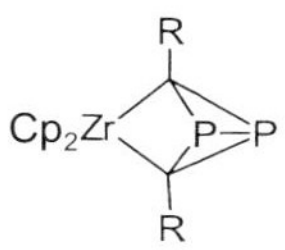

142

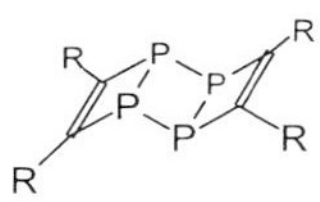

143

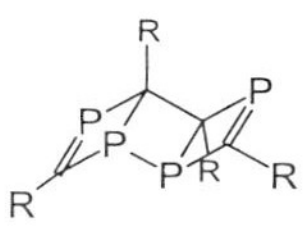

144

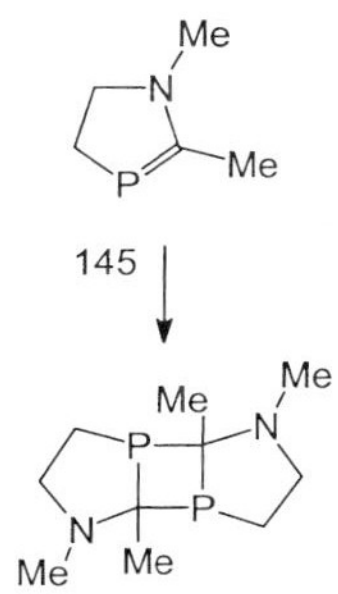

145

146

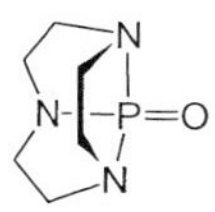

147

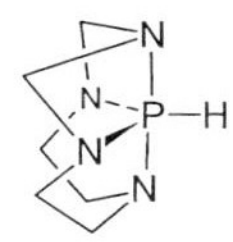

148

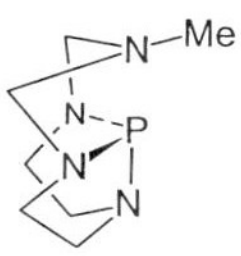

149

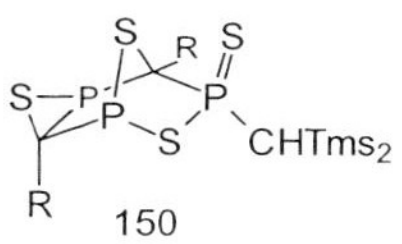

150

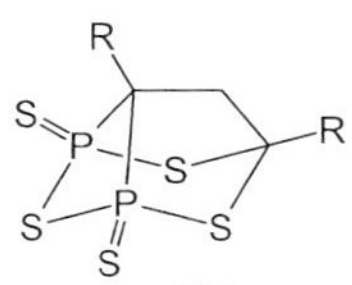

151

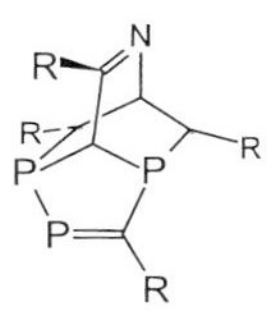

152

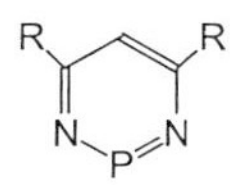

153

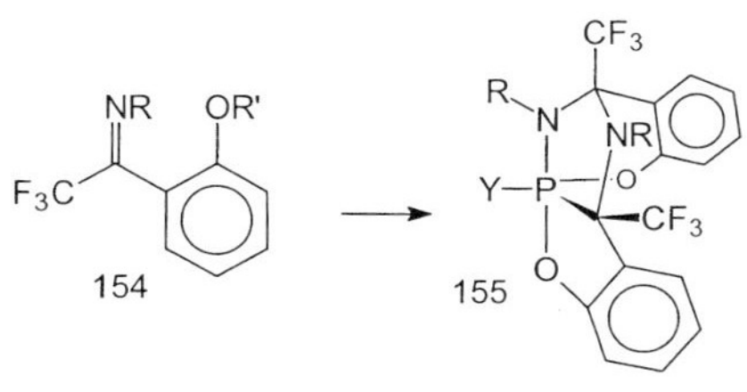

154

155

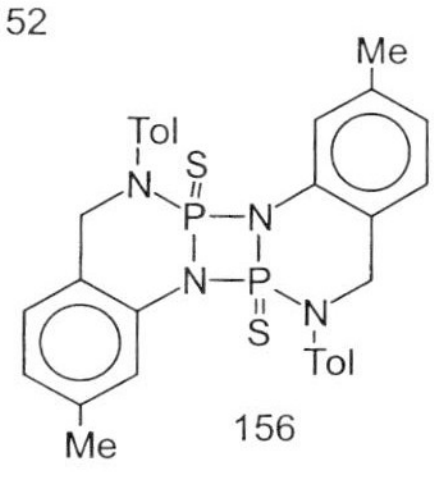

156

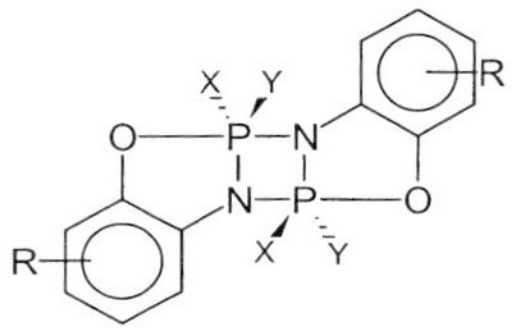

157; X = Y = OR, Ph

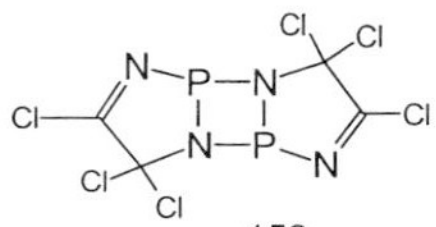

158

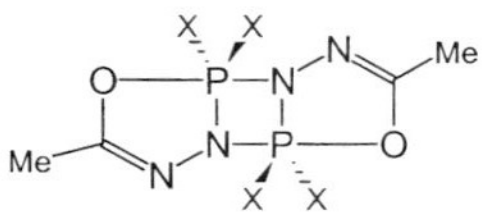

159; X = Cl, NMe_2, Me

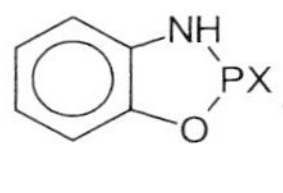

160

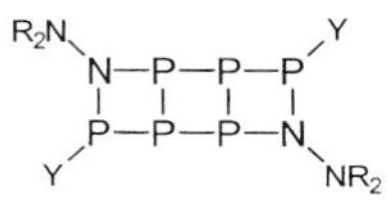

161; Y = $NRNR_2$

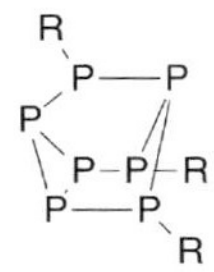

162

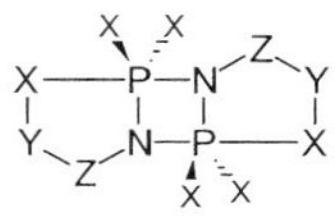

163

6.2.7 REACTIVITY OF NON-CONJUGATED AND PARTLY CONJUGATED NON-FUSED-RING SYSTEMS

There are many examples of organophosphorus compounds with phosphorus at the bridgehead which have one or more atoms linking the two bridges. Those compounds which have one linking atom are reviewed first, starting with the largest second and third rings, followed by those with two or more organised in the same way.

6.2.7.1 [*n*,*m*,1] Non-Fused-Ring Systems

6.2.7.1.1 [m,n,1] Ring systems where n and m > 3

No bicyclic compounds in this class were located but ring structures such as [5,5,1] can be identified as part of some caged structures, e.g. the phosphorylated calixarenes (see section on polycyclic heterocycles below).

6.2.7.1.2 [m,3,1] Ring systems where m > 3

This group of compounds consists mainly of ansa bridged cyclophosphatriazenes (**164**). A range of [*n*,3,1] bicyclic heterocycles have been characterised using up to nine polyethyleneglycol units in the bridging ansa group ⟨95JOC7433, 97JA12432⟩. The bridging group may also be as simple as aminoethanol ⟨84IC1495, 84ICA61⟩ or propandiol ⟨84JCS(CC)675⟩, catechol, 2-hydroxymethylphenol ⟨97HAC165⟩ to highly complex diphenols or diols ⟨91PS(56)133⟩ or diamines ⟨90JST(221)239⟩. An additional phosphazene bridge has been achieved using a diamino derivative of cyclophosphatriazene as a starting material ⟨76AG(E)39⟩. Related phosphabenzene polycyclic compounds are described in the section on polycyclic and caged non-fused-ring systems, below. The 1,8-naphthyl diphosphine disulphide (**165**) provides an alternative example of a [4,3,1] ring system ⟨95JCS(CC)2217⟩.

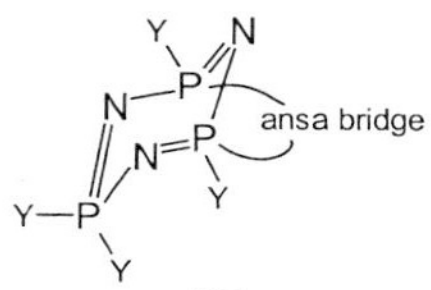

164

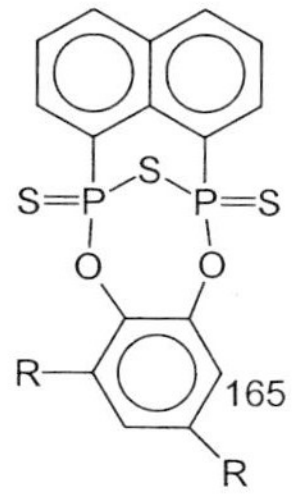

165

6.2.7.1.3 [3,3,1] Ring systems

The parent 1-phosphabicyclo[3,3,1]nonane (**166**) was synthesised by Issleib's group by the AIBN catalysed cyclisation of the phosphinodiene as shown in Scheme 8. The CH deformation band (1480

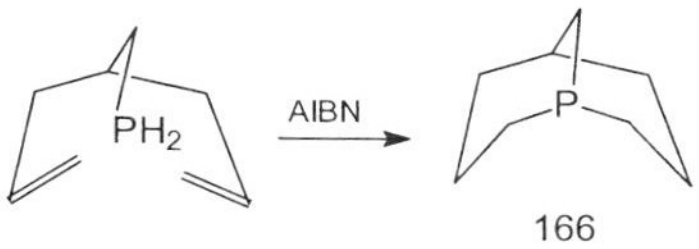

166

Scheme 8

cm^{-1}) in the IR spectrum indicated that the phosphine and its oxide and sulphide adopt a twin chair conformation ⟨76ZAAC(425)209⟩. An XRD study of the corresponding sulphide confirmed this conclusion and showed that the chairs were flattened to some extent apparently by repulsions between the *endo* methylenes $r_{(CC)} = 3.2$ Å ⟨88M1⟩. Diallylaminomethylphosphine monocyclised during its preparation to give a mixture of monocyclic azaphospholan and azaphosphorin. Further cyclisation of the mixture gave a mixture of [3,3,1] and [3,2,1] compounds, the former (**167**; Y = N) being characterised as its CS_2 adduct ⟨83PS(17)73⟩. The corresponding silyl analogue (**168**) was prepared by the same method ⟨82PS(13)153⟩ and was converted to its oxide either directly or via a salt ⟨87ZC295⟩. Issleib's group has synthesised the parent diphosphine (**167**; Y = P) via the bis(allylphosphino)methane and converted it to a range of derivatives ⟨91ZAAC(594)66⟩. The dinitrile derivative of the corresponding triaza heterocycle (**169**) was prepared by the

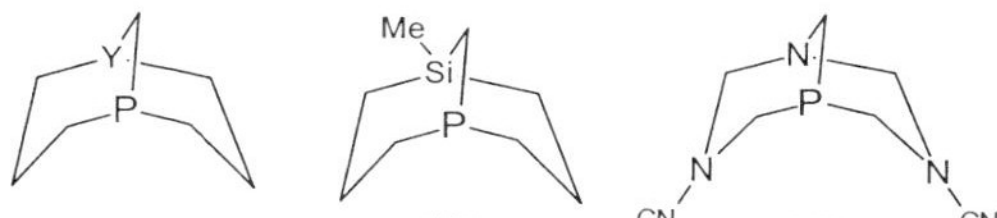

167; Y = N, P 168 169

reaction of cyanamide with trishydroxymethylphosphine in the presence of formaldehyde ⟨73USP374584⟩. Aromatic aldehydes reacted with dihydroaxylkylphosphine (**170**) to give the dioxabicyclicphosphine (**171**) see Scheme 9 ⟨80ZAAC(471)155⟩. Several urea-based heterocycles (**172**; Y = CH_2) were made from bis(dichlorophosphino)methane and silyl derivatives of urea ⟨85ZN(B)1532⟩. Related oxygen bridged compounds (**172**; Y = O) were produced from four-membered diazaphosphetans ⟨80CB1847⟩ or from spiro analogues ⟨86ZN(B)207⟩. The related unsaturated heterobicycle (**173**) was prepared by the action of chlorophosphines on $RC(NTms_2){=}NTms$ ⟨88CB1681, 90PS(47)145⟩. The hexazaphosphabicylononane (**174**) was made in two steps from $PSCl_3$ and phenylhydrazine followed by reaction with formalin at 0°C ⟨76JCS(P1)2093⟩. 1,5-Ansa bridging of cyclophosphatetrazenes can lead to this ring system, e.g. methylamine gives phosphazene (**175**; R = Me) ⟨81PS(9)323⟩. A series of bicyclic phosphonates (**176**; Y = CH_2CX_2NHR, Z = alkyl·aryl, NHR) have been used as synthetic intermediates ⟨98M8⟩. Simply heating phosphorus trichloride and ethylamine in tetrachloroethane gave a 40% yield of the crystalline pentaazatetraphosphabicyclo[3,3,1]nonane (**177**) ⟨83JCS(D)425⟩. A [3,3,1] ring system related to (**165**) was made from $Tms_2NCMe{=}NTms$ and 1,8-naphthyl diphosphine tetrasulphide ⟨99JCS(D)2231⟩.

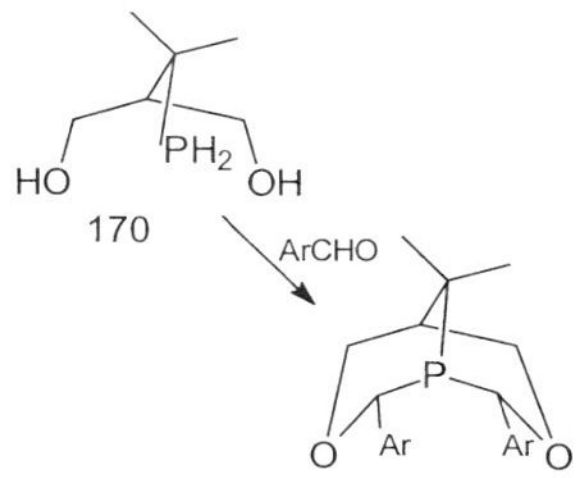

171

Scheme 9

6.2.7.1.4 [3,2,1] Ring systems

Ring expansion of an ylidic phosphole followed by cycloaddition of an acetylenic ester to the resultant phosphine gave heterocyclic ylide (**178**) ⟨85TL531⟩. Dihydrophosphole (**179**) underwent electrophilic cyclisation with aluminium chloride at 85°C to give an isomeric mixture of oxides which was reduced by trichlorosilane to the corresponding phosphines (**180**) ⟨81JOC1451⟩, which in turn could be alkylated to give phosphonium salts ⟨82TL2529⟩. Issleib synthesised 1-aza-5-phospha[3,2,1]biclooctane (**181**) as shown in Scheme 10. Whilst sulphur attacked the phosphorus atom as expected, methylation occurred at nitrogen ⟨85ZAAC(523)7⟩. Bicyclic ylide (**182**; E = CO_2Me) ⟨96AG(E)896, 96PS(111)203⟩ and phosphordiamidate (**183**)

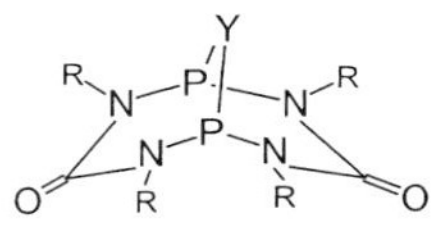

172; Y = CH2, O

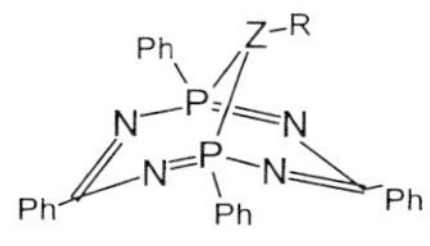

173

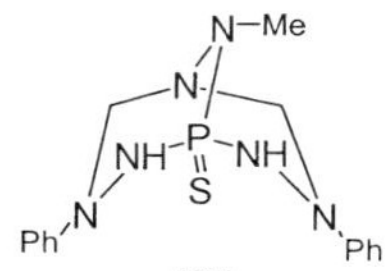

174

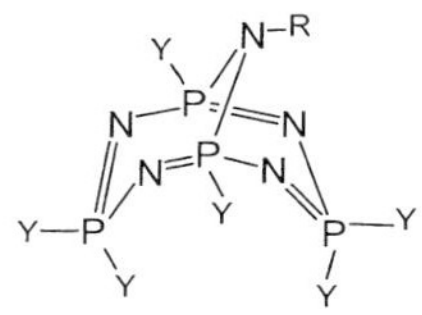

175

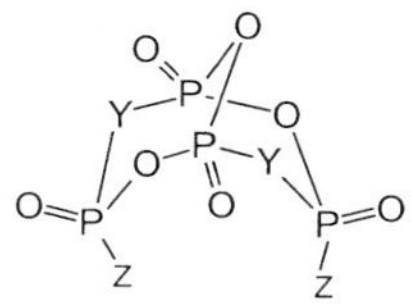

176

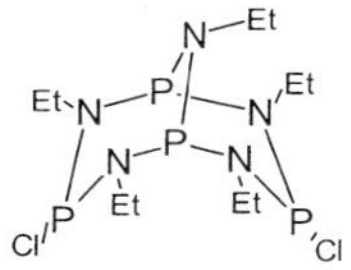

177

⟨96H2367⟩ have been described, the former is unstable ⟨96PS(109-10)93⟩ and reacts with more acetylenic ester to give phosphazene (**98**) ⟨98M3⟩. Two equivalents of two-coordinate phosphorus in the form of ClP=CPhTms reacted with 1,2-dialkylhydrazines to produce the diphosphine (**184**), whereby one of the phenyl rings has become involved in an intramolecular cyclisation ⟨82CB3610⟩. Bicyclic phosphite (**185**) was prepared from the reaction of a furanose derivative with HEPT [$(Et_2N)_3P$] ⟨77TL4125⟩. The action of sulphur on 1,2-diphosphinobenzene, $C_6H_4(PH_2)_2$, gave the tetrathio system (**186**) ⟨95JCS(D)2797⟩ and lithium phosphide, $LiPH_2$, reacted with a t-butyltrisilylsilane to give the silyldiphosphine (**187**; R = Bu^t) ⟨98ZAAC(624)1886⟩.

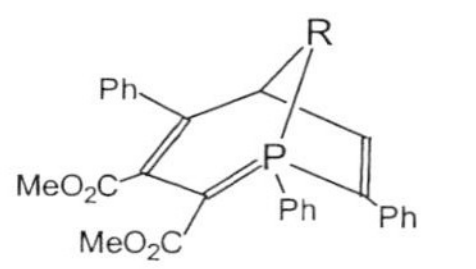

178; R = fluorenyl

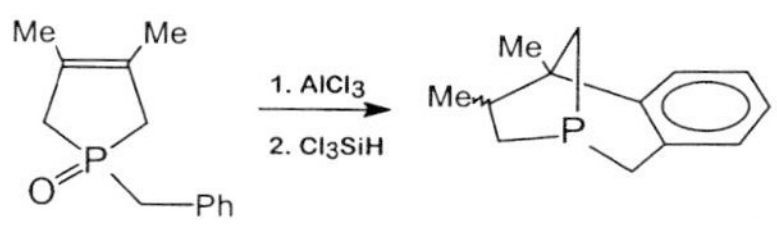

179

180

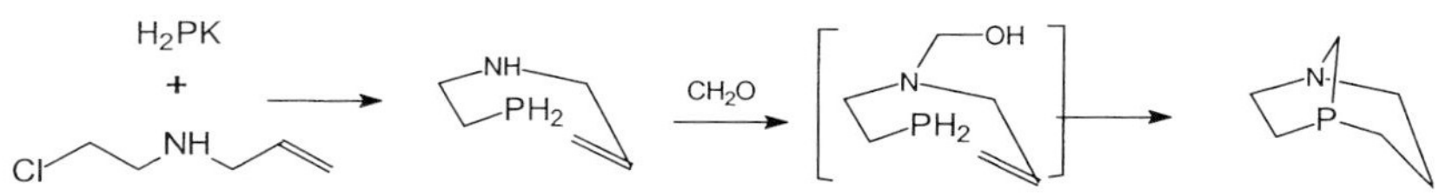

181

Scheme 10

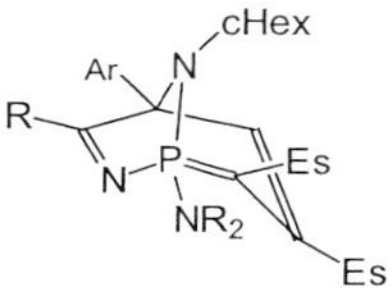

182

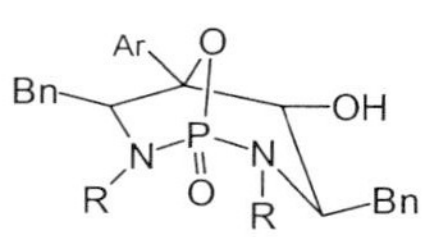

183

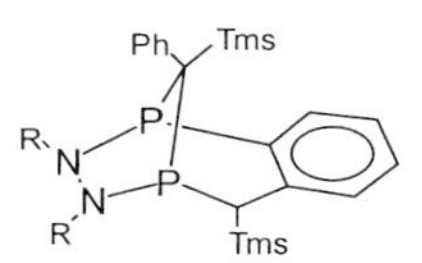

184

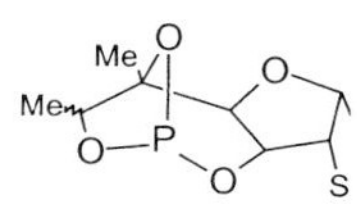

185

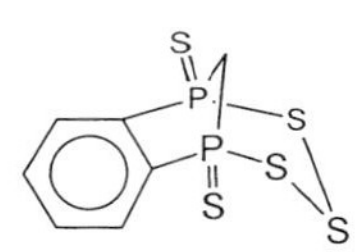

186

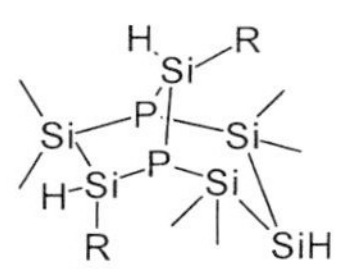

187

6.2.7.1.5 [2,2,1] Ring systems

The phosphine oxide (**188**) of the saturated parent ring system which was synthesised via the cyclisation of the dibromohexynphosphonate (**189**) by magnesium in THF, showed large NMR couplings from

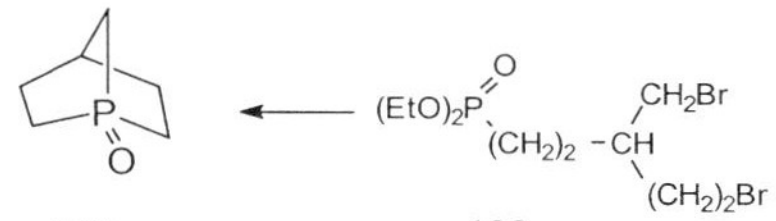

188 189

phosphorus to the bridge CH group, i.e. J_{PH} 28 Hz and J_{PC} 35 Hz ⟨72JA9230⟩. Cycloadditions of alkynes and alkenes to phospholes have led to bicyclic phosphines (**190**; R = X = Y = Ph, CO_2R) and (**191**; R = X = Y = Ph, CO_2R), respectively. The reaction appears to proceed via phenyl migration to give a 2*H*-phosphole which tautomerises to the isomeric 2*H*-phosphole and then undergoes 1,4 cycloaddition reactions with alkynes and alkenes as shown in Scheme 11 ⟨81JA4595⟩. Phosphole dimers

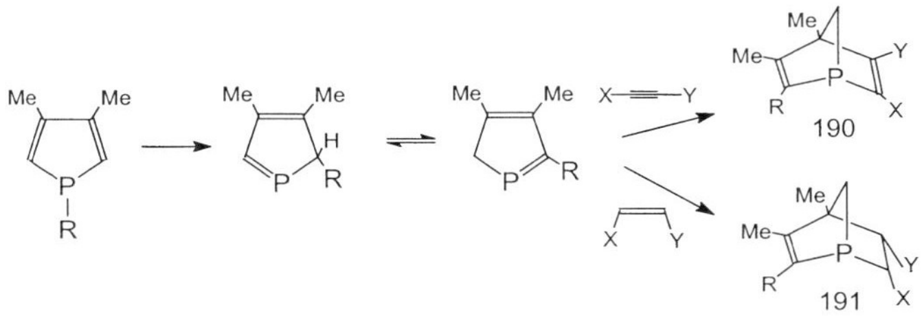

Scheme 11

such as (**125**) will react directly with dienophiles such as dimethyl acetylene dicarboxylate to give [2,2,1] adducts (**192**) which hydrolyse to water-soluble salts (**193**) ⟨93JOM(462)103⟩. Various phosphonic acid derivatives (**191**; X = H, Y = POZ_2) and (**191**; X = POZ_2, Y = H) were prepared in this manner using ethynylphosphonyl dienophiles ($HCCP(O)Z_2$; Z = OEt, Morph, NEt_2). Hydrolysis gave the corresponding disodium salts of which only (**191**; X = H, Y = PO_3Na_2) was water soluble and formed an effective Rh hydrogenation catalyst ⟨96JOC3531⟩. Tetrachloromethane reacts in an interesting way with diphosphabenzene (**194**) to give the diphosphanorbornadiene (**195**) ⟨77JA8511⟩. Another intriguing rearrangement involving loss of hydrogen is the thermal conversion of the diphospha-spiran (**196**) to the dibenzo compound (**197**) ⟨92AG(E)625⟩. The trithiaphospha[4,3,0]bicyclononane (**59**) is converted to the [2,2,1] compound (**198**) upon thermolysis in the presence of acrylonitrile ⟨82TL1263⟩. The

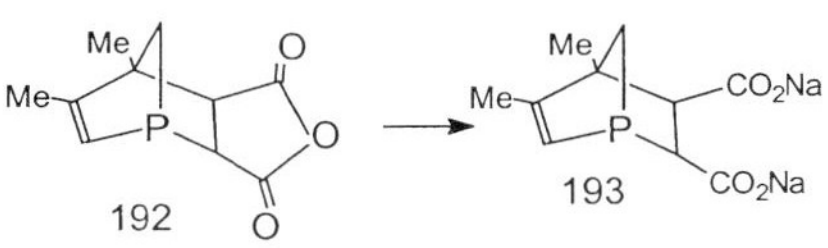

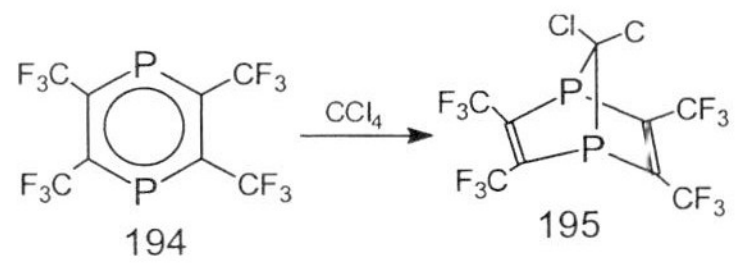

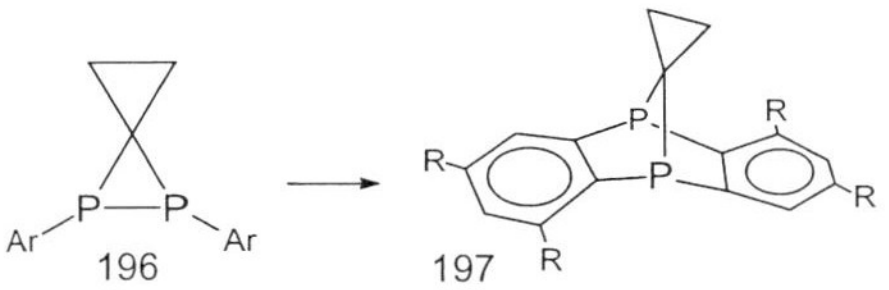

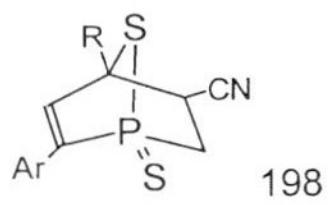

diphosphathiobicycloheptane (**199**) was made via the pyrolysis of the diphospha[2,2,2]bicyclooctadiene (**227**) and trapping of the diphosphabenzene (**200**) with sulphur as shown in Scheme 12 ⟨81JA2465⟩. If excess diphosphabenzene (**200**) is present it reacts with diphosphine (**199**) to give an adduct which dissociates to a diphosphabarrolene and diphosphene (**201**), the latter undergoing cycloaddition reaction with maleic anhydride to give adduct (**202**) (see Scheme 12) ⟨81JA2465⟩. A number of oxygen heterocycles have been prepared. Scheme 13 shows an Arbuzov reaction between tris(hydroxymethyl)phosphine and α-bromoacetophenone followed by base-catalysed cyclisation gives the phosphine (**203**; R = CH_2, Y = lp) in a 65% yield ⟨68ZOB1881⟩. Oxidation of the phosphine by hydrogen peroxide caused ring opening, to give phospholene (**204**), as shown in Scheme 13, which upon chlorination recyclises to the bicyclic phosphine oxide (**203**; R = CCl_2, Y = O) ⟨72ZOB519⟩. The phosphite (**205**; Y = R = H) was prepared from the appropriate triol and trimethyl phosphite ⟨66TL4935⟩, its spectroscopic properties ⟨70JMR111⟩ and ring opening reactions ⟨94HAC507⟩ have been studied and it was converted to its oxide by the action of N_2O_4 ⟨62JA610⟩. Phosphite (**205**; Y = H, R = Et) and the functionalised hydroxymethyl derivative (**205**; Y = R = OTr) have also been prepared ⟨96PS(113)1⟩. The trithiocompound (**206**) was prepared

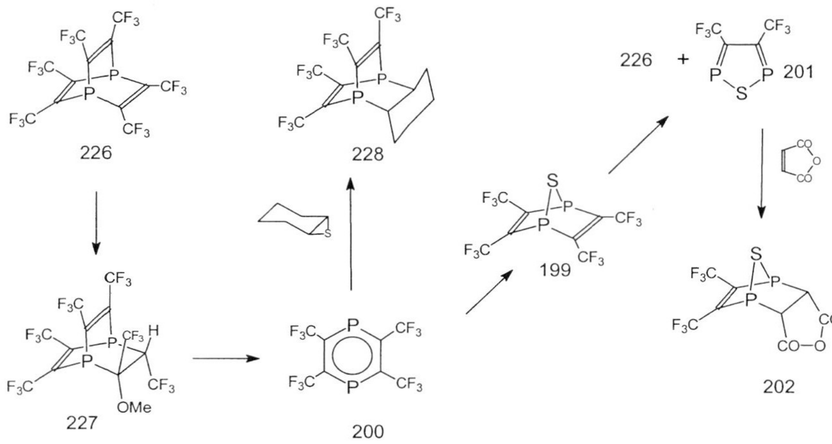

Scheme 12

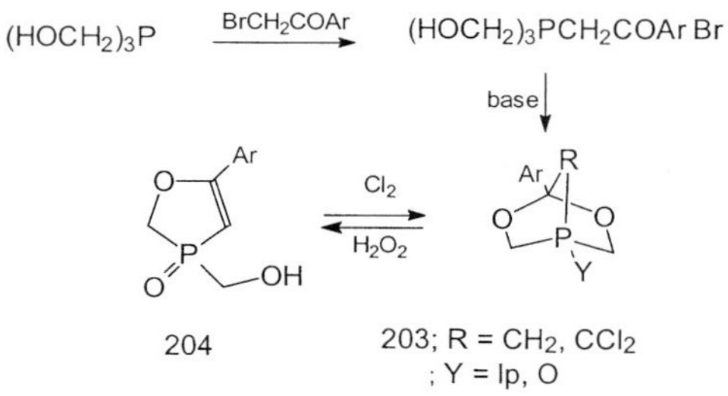

Scheme 13

by reduction of the corresponding diphosphine sulphide which was synthesised by heating P_4S_{10} and alkylamines at 162°C for 10 h ⟨85ZAAC(521)153⟩. The reaction of bis(dichlorophosphino)methane with two equivalents of 1,2-dimethylhydrazine is reported to give the tetra-aza compound (**207**) ⟨86CB2767⟩. The heptaphospha-norbornane (**208**) was formed directly from methyldichlorophosphine, phosphorus trichloride and magnesium in 11% yield ⟨81ZAAC(473)20⟩. Triseleno diphosphine (**209**; Y = CH_2), triselenotetraphosphine (**209**; Y = PSePh) ⟨96M2, 97M2⟩ and diphosphapentaselenobicycloheptane (**209**; Y = Se) have been characterised, the latter being isolated from an annealed amorphous glass of P_2Se_5 by extraction with carbon disulphide ⟨91AG(E)605⟩.

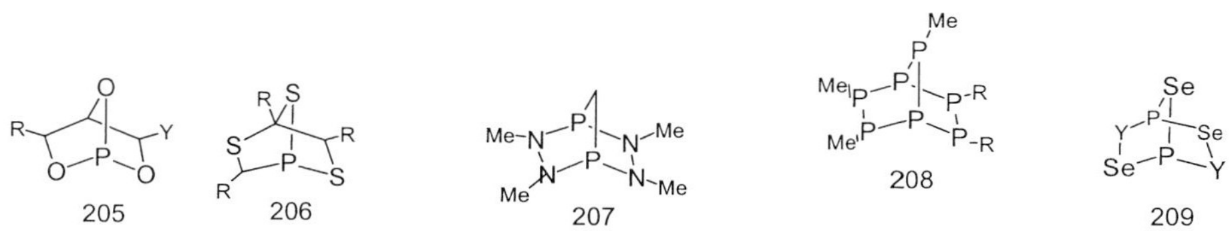

205 206 207 208 209

6.2.7.1.6 [n,1,1] Ring systems

The diazadiphosphetans (**210**) react with diols and diamines to give bicyclic compounds (**211**) ⟨77AG(E)797⟩. α-Bromonaphthalene and P_4S_{10} at 240°C give the tetrasulphide (**212**) in a 25% yield ⟨87JCS(CC)1741⟩ which is converted to monocyclic phosphonates by the action of methanol ⟨91JCS(CC)116⟩. Hydrolysis of *t*-butylphosphaethyne gave the [2,2,1] ring system (**213**; R = *t*-Bu) ⟨95HAC617⟩. There is one example of a [1,1,1] ring system, i.e. the triboradiphosphabicyclopentane (**214**) ⟨87JCS(CC)496⟩.

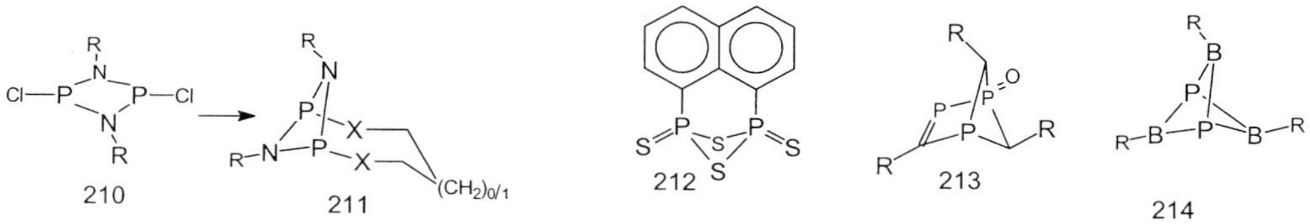

210 211 212 213 214

6.2.7.2 [*n*,*m*,2] Non-Fused-Ring Systems

6.2.7.2.1 [3,3,2] and [2,2,2] ring systems

Two [3,3,2] bicyclic compounds (**215**; R = CH_2CO_2H) and (**216**; R = CH_2CO_2H) have been prepared by the reaction of glycine with diphosphines (**217**) and (**218**), respectively ⟨99JA1658⟩.

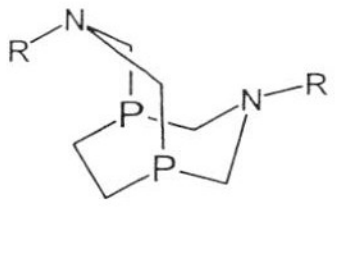
215

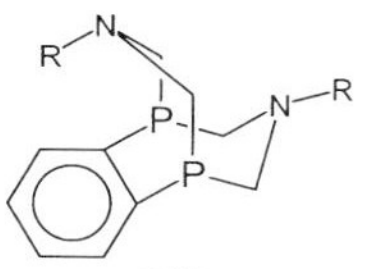
216

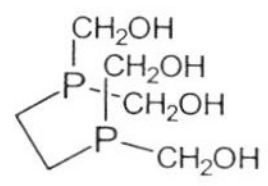
217

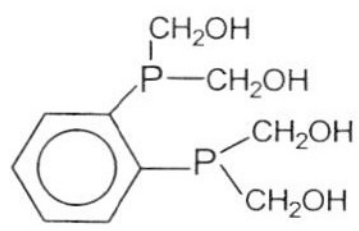
218

In the [2,2,2] class, phosphatriptycene (**219**) has been prepared by the action of LDA on 9-(2-chlorophenyl)-9-phospha-9,10-dihydroanthracene ⟨74T3465⟩. Its crystal structure has been determined ⟨80AX(B)1247⟩ and it has been converted to its oxide and the usual phosphonium salts. Phosphabenzene reacts with acetylenes to give barrolene derivatives such as phosphine (**220**) ⟨68AG(E)733⟩ and with benzyne to give the benzo compound (**221**) ⟨71TL1249⟩. An alternative approach to this type of compound, i.e. (**222**), involved the reaction of phosphanaphthalene with an acetylene ⟨74TL4369⟩. Reaction of phosphabenzene with sulphur gives the $\sigma^3\Lambda^5$ sulphide (**31**) which reacts as a dienophile with butylvinyl ether to give the sulphide (**223**) ⟨86BCJ2047⟩. Note that the sulphide (**31**) is also capable of reacting as a diene. Tri-thienylphosphines (**224**; Y = CMe, P; R = H, Me) were obtained from tris 2-lithiothiophene compounds and triphenyl phosphite ⟨93TL8255, 93TL8259⟩.

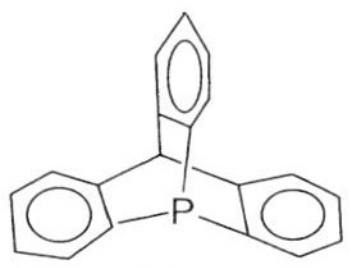
219

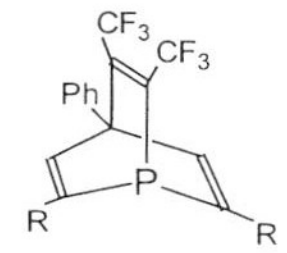
220

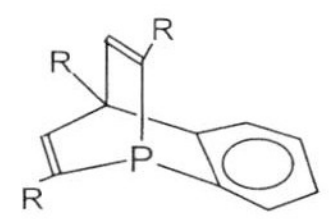
221

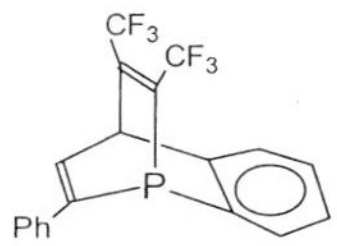
222

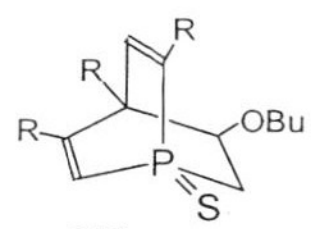
223

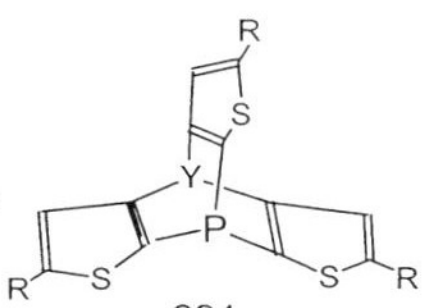
224

Diphosphatriptycene (**225**) was formed in the reaction of white phosphorus and 1,2-dichlorobenzene at 280ºC with ferric chloride as a catalyst. Peracetic acid converted it to the dioxide but benzyl chloride gave the monobenzyl salt ⟨71JA1801⟩. Hexafluorobutyne and red phosphorus heated at 200ºC for 8 h with iodine as a catalyst gives 43% of the diphosphabarrolene (**226**) ⟨61JA3432⟩ shown in Scheme 12. Addition of methanol in the presence of rhodium trichloride gives the adduct (**227**) which when heated at 100ºC is cleaved to the diphosphabenzene (**200**). The diphosphabenzene reacts with cyclohexene sulphide to give the bicyclic diphosphine (**228**). The high melting parent diphosphabicyclo[2,2,2]octane (**229**) was synthesised in 1959 by Hinton *et al.* and converted to its dioxide, disulphide and dimethiodide ⟨59JCS2835⟩. Its benzo analogue (**230**) was synthesised by Issleib *et al.* by the reaction of disodium salt of 1,2-diphosphinobenzene with 1,2-dihaloethanes ⟨86M1⟩.

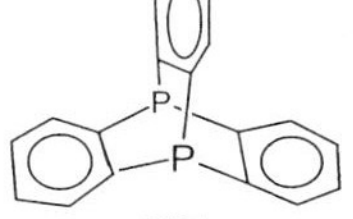
225

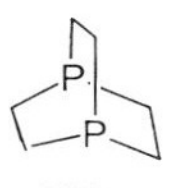
229

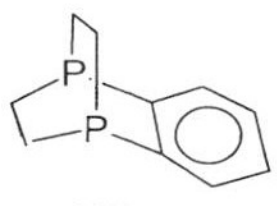
230

There are a number of bicyclo[2,2,2]octanes which include other heteroatoms in the ring system. Notably the 9-aza-10-phosphatriptycene (**231**) was synthesised by Hellwinkel and Schenk via the reaction of triphenyl phosphite with the 2,2,2-trilithium derivative of triphenylamine ⟨69AG(E)987⟩. Cycloaddition

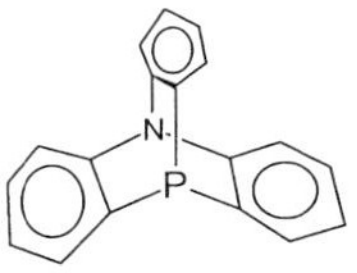
231

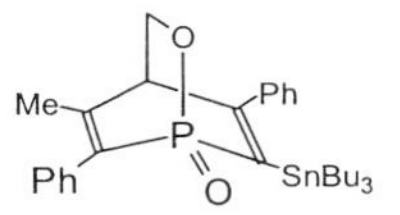
232

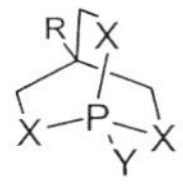
233

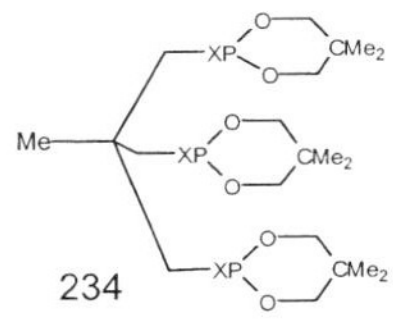
234

of acetylene $PhCCSnBu_3$ and a phosphole gave, after hydrolysis, bicyclic phosphinate (**232**) ⟨98M9⟩. The triaza- and trioxa-bicyclo[2,2,2]octane ring systems (**233**; X = NR, O, Y = R) have been extensively studied especially by Verkade and his co-workers. Whilst these series of compounds are available directly from the reactions of triols and triamines with various phosphorous compounds such as triethyl phosphite ⟨81PS(10)27⟩, high yields of the bicyclic compounds have been reported using the tris amidite or tris phosphite (**234**; X = NMe, O) as intermediates ⟨84PS(21)111⟩. The mass spectra of the phosphites, e.g. (**233**; X = O, Y = lp), were characterised by loss of CH_2O, whereas the corresponding phosphates (**233**; X = Y = O) fragmented with loss of C_2H_4 ⟨79OMS459⟩. The 4-amino derivative (**233**; X = O, Y = lp, R = NH_2) was made from $NH_2C(CH_2OH)_3$ ⟨95JGU287⟩. Phosphite (**233**; R = Bu^t, X = O, Y = lp) was converted to the corresponding phosphate by sulphuryl chloride ⟨80PS(8)315⟩. The mass spectra ⟨80PS(8)315, 83IJM459⟩ and the vibrational spectra ⟨69CR(C)(268)236⟩ of the phosphites and corresponding phosphates ⟨70BSF876⟩ have been studied. The phosphates have been studied by XRD and the hydrolysis of the phosphate and thiophosphate have been investigated ⟨86JA6311, 96JA10168⟩ and the formation of penta-coordinate derivatives required the use of pentaethoxyphosphorane ⟨84JA7825⟩. The synthesis of similar [3,2,2] ring systems with potential for phosphorane formation has also been investigated ⟨91IC1330⟩. The silyl derivatives of triols and trimethylorthoacetate have been used to synthesise the isomeric phosphines (**235**) ⟨80ZOB1499⟩. The P.E. spectra of the phosphite (**233**; X = O, Y = lp) and isomeric phosphines (**235**) have been compared and the first band for both compounds found to have more phosphorus character than oxygen character ⟨78JCS(CC)161⟩. Further alkyl derivatives of the cyclic phosphate (**233**; X = Y = O, R = CH_2R') ⟨99M2, 95M⟩ and urea derivatives ⟨97PS(128)1⟩ have been prepared. The tris thiophosphite and its thiophosphate (**233**; X = S) have also been investigated ⟨68IC1469⟩. *t*-Butylphosphaethyne and a zirconium cyclopentadiene complex gave tetraphosphine (**236**) ⟨95AG(E)81, 94AG(E)2202⟩, with trisubstituted 1,3,2-diazaphosphinine it gave the adduct (**237**; R = Bu^t, Y = N), and with trisubstituted 1,2-azaphosphinine it gave the adduct (**237**; Y = CR) ⟨98M6⟩. Attempts to make $Ph_3P{=}C{=}PX$ led to the formation of the ylide (**238**) ⟨95AG(E)1853⟩. There is one example of a azadioxaphosphabicyclo[2,2,2]octane, i.e. (**239**), which was prepared by the cyclisation of a 5-chloromethyl-1,3,2-dioxaphosphorin (**240**) ⟨83S445⟩. Stable ozone adducts are formed by phosphites ⟨70JCS(CC)956⟩ as well as by phosphazenes ⟨74AG(E)543⟩. The corresponding diphosphorus ring systems (**241**; R = H, Z = P) were made (35%) by the reaction of tris(hydroxymethyl)phosphine with trimethyl phosphite, converted to the dioxide ⟨65IC1655⟩ and their NMR spectra studied ⟨70JA71⟩. The NMR vicinal PP coupling constant for the parent dioxide was exceptionally large (139.1 Hz) ⟨70JA71⟩. Tris hydroxyalkylphosphines produced by the reaction of PH_3 with aldehydes have been converted to the diphospha ring system (**241**; R = H, Z = P) using PCl_3 and Et_3N ⟨85ZOB2218⟩ and to the phenylborane compound (**242**) ⟨89M⟩. When the second bridge atom is an electron donor and the phosphorus is four-coordinate the possibility of a transannular interaction is possible. There is XRD evidence for this interaction when an amino group is the donor ⟨92JA3129⟩. A mixed phosphino arsenite (**241**; R = H, Z = As) was prepared (61%) by the reaction of trimethyl arsenite with tris(hydroxymethyl)phosphine and converted to the phosphine oxide using dibenzoyl peroxide ⟨70JOC2310⟩. The synthesis and reactions of the hydrazino compound (**243**; Y = NMe) was studied ⟨72CB2637⟩ and XRD gave evidence of π-bonding ⟨71IC2591⟩. Metallophosphide was used for the synthesis of the corresponding hexasilyl diphosphine (**243**;

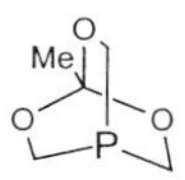
235

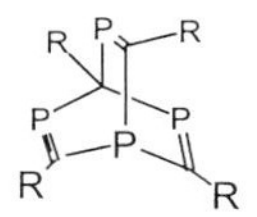
236

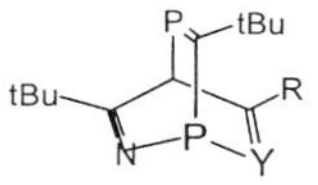
237

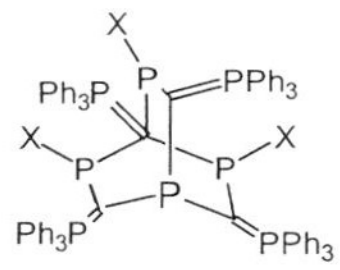
238

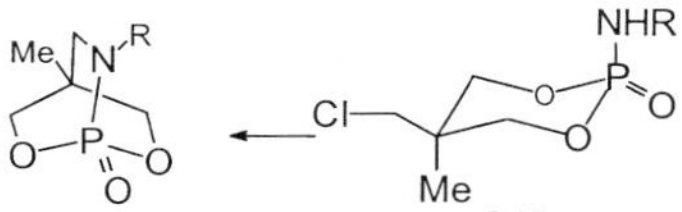
239 240

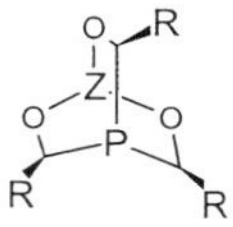
241

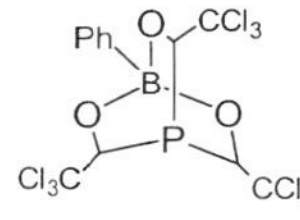
242

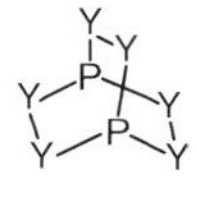
243

Y = $SiMe_2$) ⟨97M3⟩ and the corresponding hexastannyl compound was formed in the reaction of white phosphorus and $Me_2SnSnMe_2$ ⟨81AG(E)1029⟩.

6.2.7.3 [*n*,*m*,3] Non-Fused-Ring Systems

6.2.7.3.1 [3,3,3] Ring systems

Whilst the parent system with one phosphorus atom has not been made, the diphosphabicyclo[3,3,3]undecane (**244**) was isolated (26%) from the product obtained by the action of AIBN on allylphosphine ⟨82PS(12)227⟩. Dialkylation of diphosphacyclooctane (**245**) by the tosylate of propanediol gave the bis phosphonium salt (**246**) which, upon reduction with LAH, gives the diphosphine (**244**) ⟨96JCS(CC)539⟩. Verkade and Milbrath prepared the phosphatrane (**247**), and a very wide range of trioxaphosphorus derivatives, from tris(hydroxyethyl)phosphine and HMTP in low yield ⟨77JA6607, 85JA7084⟩. X-ray molecular structures showed the presence of a short N–P distance for the HBF_4 salt ⟨76JA623⟩. The potential of the phosphorus atom of this ring system to support transannular interactions between donor and acceptor atoms at the bridgehead positions was highlighted by Buck *et al.* in 1979 ⟨79RTC12, 81M1⟩. Conversion of the phosphine oxide to alkoxy salts produced a leaving group on phosphorus which then allows transannular bond formation to take place ⟨86JA4918⟩. The corresponding protetraazaphosphatranes (**248**) have been similarly investigated ⟨91ZAAC(605)164, 97PS(123)301⟩ and the phosphonium salts found to be unusually stable ⟨89JA3478, 89PS(41)21⟩. The neighbouring group effects on the reactivities and spectra have been thoroughly investigated ⟨90PS(49/50)163⟩ and XRD comparisons made between the phosphite and triamide ring systems ⟨90IC2214⟩. Chirality for this system was achieved by the synthesis, in seven steps, of the corresponding tris pyrrolidine protetraazaphosphatrane (**249**) which was also found to be a strong base ⟨98JOC5692⟩. The bicyclic triaryl phosphates (**250**; X = P, N) have also been made ⟨87HCA1063⟩. Derivatives of this ring system may be prepared via triols made by the reaction of phosphine (PH_3) with aldehydes ⟨85ZOB2218⟩.

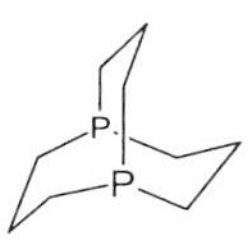

244

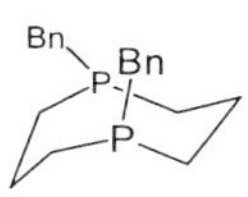

245

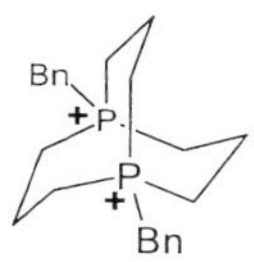

246

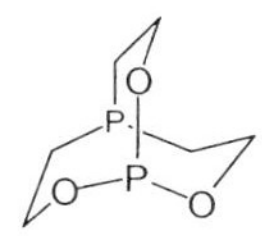

247

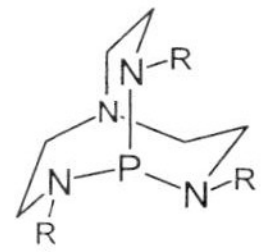

248

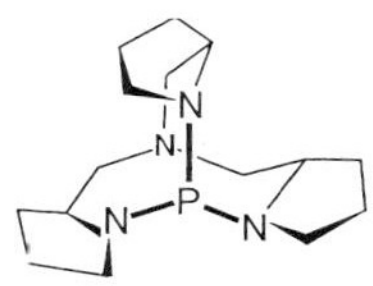

249

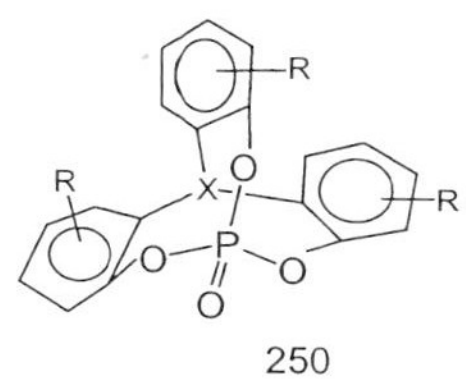

250

6.2.7.4 [*n*,*m*,*x*] Non-Fused-Ring Systems where *n*,*m*,*x* > 3

6.2.7.4.1 [4,4,4] and larger ring systems

When the synthetic approach used in Scheme 1 failed for the synthesis of the [4,4,4] ring system (**251**) with two phosphorus atoms at the ring junctions a different approach was partially successful. It involved the reaction of the caged compound (**252**) with a Grignard reagent to give the mono salt (**253**) which was converted to the di salt (**254**) ⟨96JCS(CC)539⟩. Other bicyclic compounds in this group include some very large [17,17,17] ring systems. Thus the cryptans (**255**; Ch = O, S; Y = NMeNN=$CHC_6H_4OP(X)PhOC_6H_4$–CH=NNMe) and (**255**; Ch = O, S; Y = OC_6H_4CH=NNMeP(S)PhNMeN=CHC_6H_4O) were synthesised using condensation reactions between amine and aldehyde functionalised phosphorus compounds ⟨94JCS(CC)2161⟩.

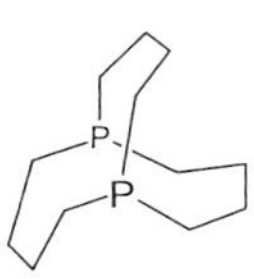

251

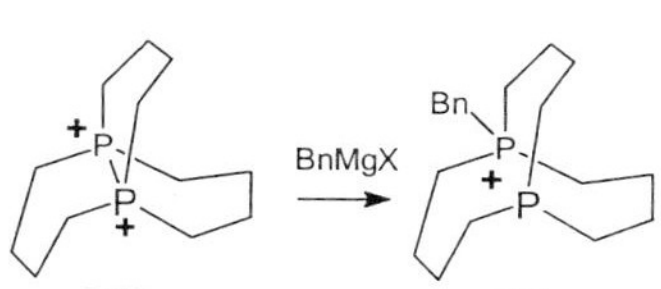

252 253

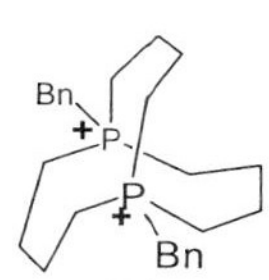

254

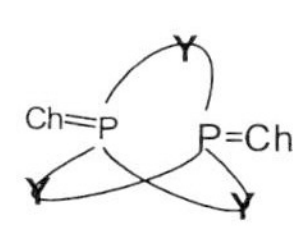

255

6.2.7.5 Polycyclic and Caged Non-Fused-Ring Systems

A range of ring systems have been characterised. Diphosphines (**256**) was prepared from phosphole intermediates ⟨98M10⟩. Initial conversion of a 1*H*-phosphole to a 2*H*-phosphole occurs when the phospholyl anion is protonated which leads to spontaneous dimerisation to the diphosphanorbornene (**257**) ⟨82TL511⟩. An equimolar mixture of the corresponding diphosphatri-*t*-butyl and triphosphadi-*t*-butyl phosphole anions reacts similarly upon protonation giving dimer (**258**) ⟨90JCS(CC)1307⟩. An improved route to phosphatriazaadamantane salts (**259**) has been reported. It involves the reaction of alkyl(tris-hydroxymethyl)phosphonium salts with 1,3,5-triazacyclohexane ⟨95CB891⟩. 1,2-Diphosphinylbenzene, $C_6H_4(PH_2)_2$, and pentane-2,4-dione reacted in a 1 : 2 ratio to give the caged diphosphine (**260**) ⟨99JCS(CC)901⟩. Phosphaethynes, RCP; R = *t*-Bu, *i*-Bu, CMe_2Et, reacted with selenium, telurium, trialkylaluminium and with trialkylgallium to produce caged complexes such as (**261**; M = Al, Ga) ⟨98M11, 99S1642⟩. Phosphorylation of certain polyols leads to the formation of structures such as (**262**) ⟨96ZOB1399⟩. Triphenyl phosphite phosphorylated pentose (**263**) to give the phosphorane (**264**) ⟨94TL2743⟩. Calixarenes such as (**265**) are phosphorylated by phosphorus pentachloride to yield quaziphosphonium salts (**266**) ⟨94JGU1203, 97AX(C)1310⟩. The X-ray crystallographic structure of the related P(III) triaryl calixarene has been determined ⟨99DOK(366)202⟩. Perphosphoranides (**267**) ⟨94IC2657⟩, phosphonous derivatives (**268**) and their chalcogenides ⟨95IC5405⟩, the bisphosphite derivative (**269**; Y = lp) of a calix[6]arene ⟨96JCS(CC)583⟩, as well as the corresponding phosphate (**269**; Y = O) ⟨94JGU1394⟩ have also been studied. Three 3,5-diphenylphosphabenzene units linked in the 2,6 positions by dimethylsilyl groups give the [11,3,1] calix[3]phosphinine (**270**) and two such units linked by two disilyl-thiophene rings or -furan rings make up the [15,3,1] system (**271**) ⟨99M3⟩. Linear tricyclic phosphorane (**272**) was generated by dimerisation of the corresponding phosphazene ⟨96JGU355⟩. Alkyl azides react with P_4O_6 to produce, after an unusual rearrangement, the unique structure (**273**) ⟨96CB121⟩ and the intriguing nitrogen analogue, $P_4N_{10}H_4$, of phosphoric anhydride has been the focus of considerable attention ⟨95ZAAC(621)29⟩. A book has been published on the structure, chemistry and applications of phosphoric anhydride P_4O_{10} ⟨99M4⟩.

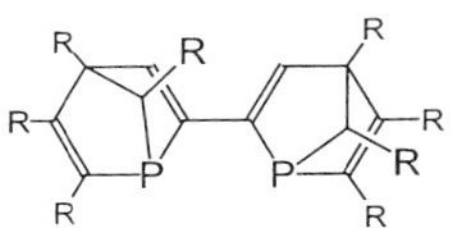

256

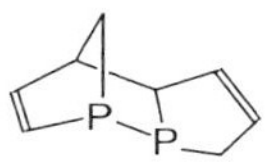

257

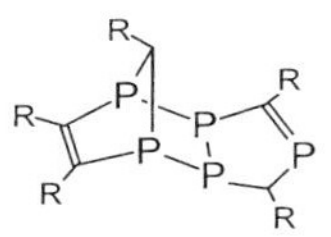

258; R = Bu^t

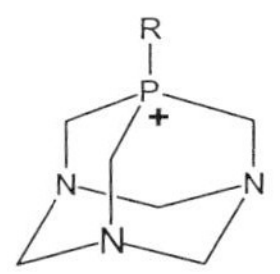

259

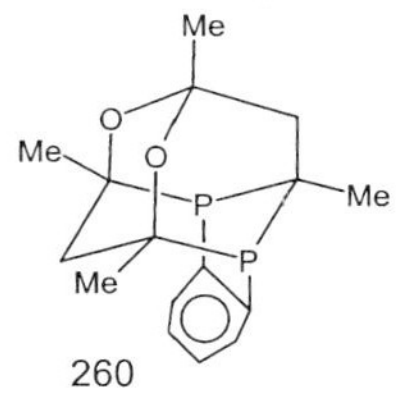

260

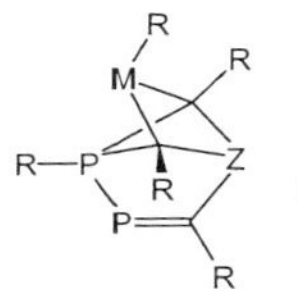

261

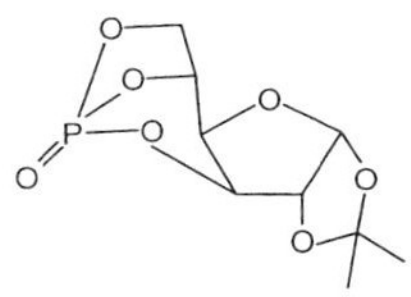

262

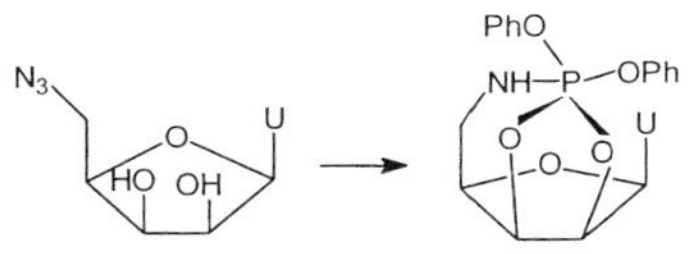

263 264

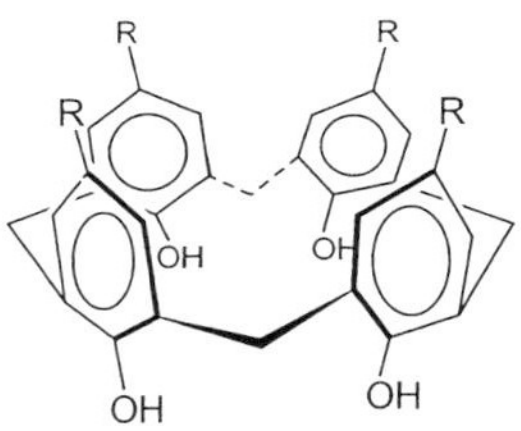

265; R = t-Bu

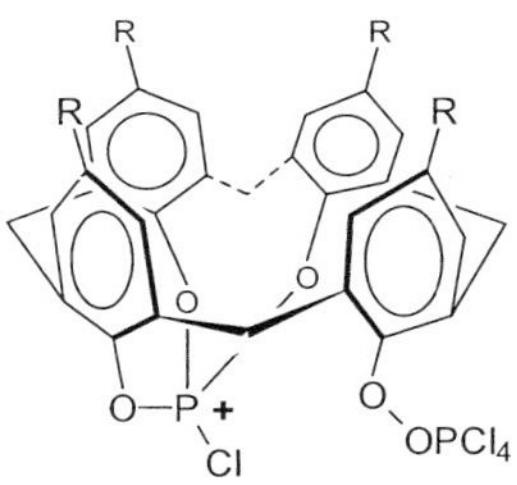

266; R = t-Bu

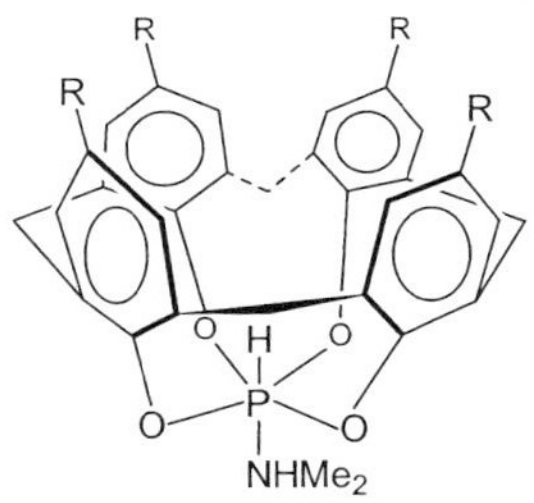

267; R = t-Bu

6.2.8 SYNTHETIC APPROACHES

The major proportion of the compounds synthesised have more than one heteroatom in the ring system and not infrequently there are a number of heteroatoms several of them often bound to the bridging

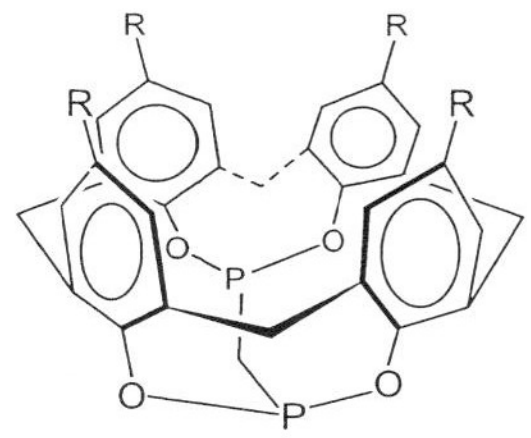

268; R = t-Bu

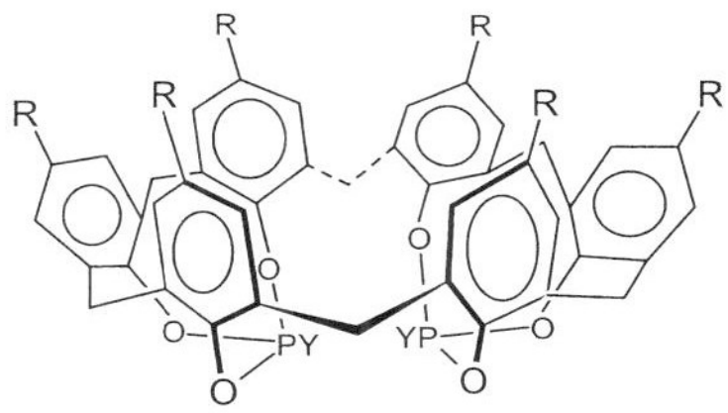

269; R = t-Bu

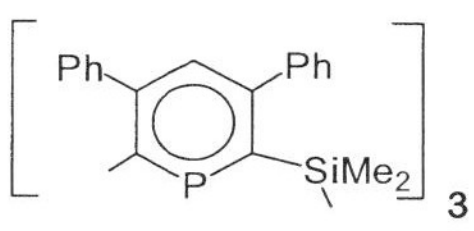

270

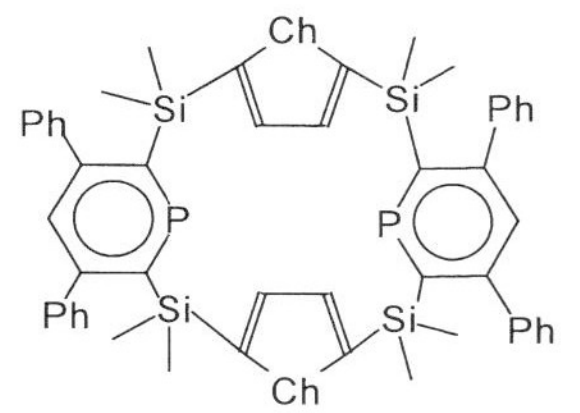

271

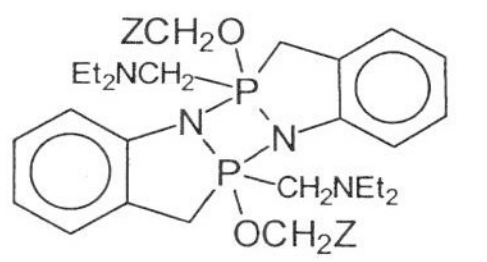

272; Z = C(CH2O)3P

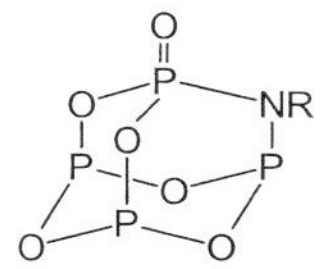

273

phosphorus atom. The methods most successfully applied to the synthesis of the parent and monohetero bicyclic compounds have been those involving addition and cycloaddition reactions.

6.2.8.1 PH Addition to Alkenyl and Carbonyl Groups

Phosphines, R_2PH and RPH_2, add to terminal carbon double bonds under free radical conditions. This method has been used with some success for the synthesis of a number of the parent saturated phosphorus ring systems, i.e. by the intramolecular addition of a primary phosphine group to the terminal double bond of two alkenyl substituents. Examples are the synthesis of bicyclic phosphines (**24**), (**40**) and (**166**). The phosphino group may be generated by reduction, e.g. LAH reduction of a phosphonate, or by alkylation of a phosphide, e.g. KPH_2. An interesting extension of this approach involves the generation of an iminium group in place of one of the alkenyl groups to synthesise the azaphosphine (**181**). Free radical initiated reactions have been more successful than photochemical initiation. 1,2-Diphosphinylbenzene $C_6H_4(PH_2)_2$, reacts with two equivalents of 2,4-pentanedione eventually leading to the cage structure (**260**).

6.2.8.2 Cycloaddition Reactions

The Diels–Alder reactions of two-coordinate metalloid compounds and their $\sigma^3\Lambda^5$ derivatives, have been one of the most useful applications of these compounds for the synthesis of unsaturated heterobicyclic compounds. The two-coordinate heterocycles have the potential to react as dienophiles or as dienes. As dienophiles, reaction across the M=C bond give the possibility of synthesising a range of fused-ring compounds.

Thus dimethyl 2*H*-phosphole reacts with a diene to give [4,3,0] heterocycle (**45**) (Scheme 2). The 2*H*-phosphole, which can be kept conveniently as its dimer, was formed quite surprisingly by the protonation of the anions of the corresponding 1*H*-phosphole. Dimerisation of 2,5-di-*t*-butyl-1-lithyl-phosphole to give (**126**) was catalysed by gallium bromide ⟨99AG(E)1646⟩. 2*H*-Phospholes will also react with 1,3 dipoles such as nitrile ylides and oxides giving [3,3,0] heterocycles, e.g. (**81**), and react with (phenylethynyl)tributyltin at 15°C to give after oxidation adduct (**232**). A two-coordinate phosphole reacted with diazomethane to give a product indicative of a similar reaction (Scheme 4).

The $\sigma^3\Lambda^5$ derivatives of phosphabenzene, e.g. sulphide (**31**), also behave as dienophiles producing [4,4,0] heterocyclic phosphines. The only example of a two-coordinate phosphabenzene reacting across the P=C bond was its [3 + 2] reaction with its $\sigma^3\Lambda^5$ sulphide to give (**56**). Diphenylnitrilamine added to a cyclophosphazene to form (**82**) and a phosphinoisocyanate added to salicylaldehyde hydrazone to give (**83**). Phosphapropanes, e.g. (**113**), are formed in the cycloaddition reactions of diazoalkanes with two-coordinate diaza-2*H*-phospholes.

Conjugated two-coordinate phosphenes can also take on the role of a diene, the products being [2,2,1] and [2,2,2] heterocycles. The two-coordinate 2*H*-phosphole, which was used to make bicyclic compounds

(**190**) and (**191**) (Scheme 11), was produced by a thermally promoted 1,2 phenyl migration from the *P*-phenylphosphole. The dimer will react with dienophiles such as maleic anhydride to give [2,2,1] heterobicyclic acid (**191**; R = CO_2Na) ⟨93JOM(462)103⟩. Cycloaddition of ethynylphosphonic acid diamide to a phosphole led to the formation of a water-soluble phosphine (**190**; R = Ph, Y = PO_3^{2-} X = H). A related product (**198**) is produced in the reaction of $\sigma 3\Lambda^5$ 2*H*-phosphole sulphides with acrylonitrile ⟨82TL1263⟩.

Phosphabenzenes behave as dienes when reacted with alkynes to produce [2,2,2] heterocycles, e.g. phosphabenzene regenerated from adduct (**220**) was reacted with benzyne to give phosphine (**221**). Diphosphabenzenes react in a similar manner (Scheme 12). $\sigma^3\Lambda^5$ Derivatives of phosphabenzene, e.g. sulphide (**31**) will also take on the role of a diene with enol ethers to give [2,2,2] heterocycles (**223**) ⟨86BCJ2047⟩. The adduct from the cycloaddition of a 1,3,2-diazophosphorin (**274**) with acetylenic diester rearranged to form (**97**). There has been one example of a [6 + 2] cycloaddition reaction which involved the reaction of a phosphacyclohexadiene with dimethyl acetylenedicarboxylate to give the ylide (**178**).

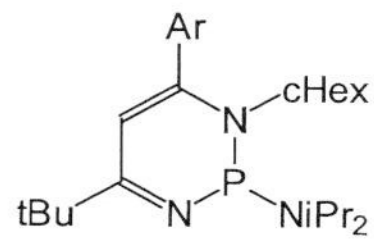

274

Phosphaethynes (RCP) have been shown to be potent dienophiles leading to a wide range of bicyclic and polycyclic heterocycles. Thus, 1,4-cyclohexadiene gave phosphine (**118**), butadienes led to (**119** or **121**) depending on the phosphaalkyne, cyclopentadiene gave (**120**), phosphapyridines (azaphosphorins) gave adducts (**236**) and (**237**) and tropolone led to adduct (**123**). Various phosphaethynes have been trimerised, e.g. to (**111**) and (**213**) by the action of Lewis acids or even water. Tetramers such as (**86**), (**235**) and **139**) have also been made.

6.2.8.3 Electrophilic Reactions

Removal of two chlorine atoms from the dichlorophosphoramidate (**128**) through the action of magnesium enables the phosphorus atom to attack the CC double bond to form the phosphirane. Dihydrophosphole (**179**) in the presence of aluminium chloride underwent intramolecular electrophilic cyclisation to give the oxide of phosphine (**180**). Cyclisation reactions which involve the generation of electrophilic phosphorus centres have not been studied to any extent.

6.2.8.4 Nucleophilic Attack at Phosphorus

By far the most common leaving groups at phosphorus are the halogens. A range of halo compounds of phosphorus, $PHal_3$, $RPHal_2$ and R_2PHal and their phosphoryl counterparts are readily available and they react with a wide range of nucleophilic reagents such as alcohols, amines and organometallic centres. A base such as a tertiary amine is usually required when alcohols and amines are employed in order to remove HHal. The precipitation of amine salt is a useful monitor of the progress of the reaction. The phosphorus pentahalides and phosphoryl/thiophosphoryl halides have been used to a lesser extent, e.g. in the synthesis of (**109**), (**174**) and (**250**), although in some instances it is possible to introduce heterocyclic sulphur through the substrate, e.g. the reaction of phosphorus trichloride on thioanisole to give (**71**; Y = Me, Z = P) ⟨97HAC551⟩. The heterocyclic rings invariably include oxygen and nitrogen atoms usually bound to the phosphorus atom. Thus trihydric nucleophiles such as diethanolamines or the corresponding diketoamines react with phosphorustrihalides leading directly to fused-ring [3,3,0] heterocycles, e.g. (**88**). Phosphorus pentahalides were used for the phosphorylation of calixarenes (**265**) and silylated bisphenols to form (**65**) and (**87**). The corresponding reactions of trihydric alcohols have been superseded by ester exchange method (see below). Reactions of diols and diamines with cyclophosphazenes and cyclophosphazanes have produced a variety of bicyclic [*n*,3,1] compounds such as (**163**) and (**211**).

Phenoxy and ethoxy groups have been used as leaving groups. Thus triphenyl phosphite reacts with azide (**263**) to give (**264**), with thienyl lithium to give (**224**), the trilithium derivative of triphenylamine gave phosphazatriptycene (**231**) and the Grignard reagent from dibromide (**189**) gave the [2,2,1] oxide (**188**). Cyclisation of the thioester (**275**) to form the [2,2,2] bicyclic phosphate (**276**) was accomplished using sulphuryl chloride ⟨94AJC2161⟩.

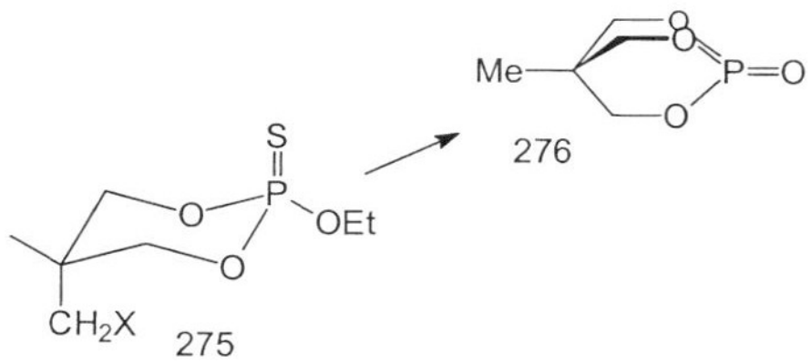

275 276

Dimethylamino leaving groups have been utilised in reactions with triols to form [2,2,1] ⟨96PS(113)1⟩ and [2,2,2] ⟨95JGU287⟩ bicyclic phosphites. Bis diethylaminochlorophosphine was the reagent of choice for the preparation of protetraazaphosphatrane (**249**) derived from pyrrolidine.

6.2.8.5 Transannular Reactions

Fused-ring bicyclic compounds have been formed by transannular attack of an amino group at phosphorus either to displace a phosphorus substituent, e.g. formation of the salt (**70**), or by valence expansion to produce phosphoranes (**54**) and (**93**). In the latter case the five-coordinate metalloid atom needs to be stabilised by the presence of electronegative substituents and the formation of small rings, preferably five-membered rings. Elimination of hydrogen between a PH_2 and PHLi groups has been a useful route to a range of cyclic 1,2-diphosphines as shown in Scheme 1. The formation of the bicyclic diphosphine (**72**) upon the distillation of 2-phosphinophenols also appears to involve transannular loss of hydrogen following dimerisation via valence expansion. A similar elimination may be involved in the formation of the tetramer of (**77**; Z = covalent bond).

6.2.8.6 Nucleophilic Phosphorus Reagents

A variety of phosphines, phosphides, phosphoramidites and phosphites have been used as nucleophilic reagents. Thus diphosphabicyclooctane (**34**) was prepared by an intramolecular alkylation of a cyclic secondary diphosphine. The parent diphosphabicyclo[2,2,2]octane (**229**) was synthesised in 1959 via a sequence of alkylations by 1,2-dibromoethane starting with tribenzylphosphine, removing the benzyl groups by LAH reductions. Tris-hydroxymethylphosphine has been utilised in a similar manner. In this case after nucleophilic attack by phosphorus, formaldehyde is liberated to regenerate the nucleophilic phosphorus centre, e.g. conversion of diynol (**48**) to diphosphine (**49**). The phosphides are generally more reactive, diphosphine (**230**) being produced by the reaction of the disodium derivative of 1,2-diphosphinobenzene with 1,2-dihaloethane. Little use has been made of the enhanced nucleophilicity of the phosphorus atom when it possesses nitrogen substituents nor the nucleophilic potential of the anions of phosphinous acids, R_2PHO in cyclisation reaction (see however the following amide/ester exchange section).

6.2.8.7 Trans-Esterification and Trans-Amination

Reactions of phosphites and triamides such as HMTP with compounds such as triols and primary or secondary triamines, have been used with considerable synthetic success to make oxa and aza derivatives of bicyclic phosphorus heterocycles. Ester exchange between bis(neopentoxy)-phosphine, $(RO)_2PH$, and diethanolamine gives the monoheterocyclic compound (**94**) which undergoes a transannular interaction with valence expansion of phosphorus to produce the phosphorane (**95**). Ester exchange via intermediate (**234**) has been found to be a superior method for the preparation of [2,2,2] phosphite (**233**; X = O; R = Me). The utilisation of HMPA in amide ester exchange reactions with triols and trimercaptans have been used widely for the synthesis of the phosphites (**233**; X = O, S) and phosphatranes (**247**). The synthesis of the corresponding tetraazaphosphatranes (**248**) as well as the [4,4,0] triamidite (**38**; X = NMe) were achieved likewise.

6.2.8.8 Silane Intermediates

The silanes have been used as intermediates, e.g. R_2PTms in the conversion of phosphorus halides R_2PHal to their hydrides R_2PH which are then subjected to cyclisation. They have been used as leaving groups on hetero atoms such as nitrogen and oxygen in their reactions with phosphorus halides to give compounds such as the amides (**172**) and perphosphoranide (**65**). Silyl groups have been used as part of the heterocyclic structure (**187**).

6.2.8.9 Ring Closure of Substituents

This approach involves cyclisation which does not involve bonds to the phosphorus atom. The phosphorus atom may be used either to stabilise an adjacent reactive site, e.g. the generation of an anion (an ylide) or to activate a leaving group. Thus intermediate ylides have been used to form stable oxaphosphoranes (**103**) and cyclic phosphonium amides (phosphazenes) either dimerised or reacted with ketenes and thiocyanates to produce [3,2,0] PN heterocycles such as (**108**). Also the reaction of chalcone with ethoxydiphenyl-phosphine (Scheme 3) proceeds via a conjugated ylide which reacts at the gamma position with the carbonyl group of an aldehyde to produce a six-membered ring. Valence expansion completes the formation of phosphorane (**61**). The hydroxyl groups in hydroxymethyphosphonium salts are displaced by amino groups in trimethylenetriamine to produce triazaphosphaadamantanes (**259**). Similar substitutions were utilised in the preparation of diphosphines (**215**) and (**216**). 2,6-Bis(hydroxymethyl)phenylphosphine oxides and 2,6-diformylphenyl-phosphine oxides react likewise to give phosphoranes (**74**; Z = CH_2, CO). The addition of a dihydroxyalkyl phosphine to two equivalents of benzaldehyde produces labile hydroxy groups which can be condensed with the alcoholic groups (see Scheme 9). Nitrogen anions produced by the action of sodium hydride on phosphoramidates undergo intramolecular alkylation to produce [4,3,0] and [2,2,2] bicyclic heterocycles, i.e. (**64**) and (**239**) respectively. Phenoxide groups displaced fluoride activated by three valent phosphorus in the conversion of (**131**) to triarylphosphine (**130**). The reaction shown in Scheme 6 shows that for a five-membered ring formation an anion is not necessary. Cyclisation can also be achieved at sites remote from the heteroatom, cf. the bond formation induced by LDA between C-10 and the 2′ positions of 9(2-chlorophenyl)-9-phosphadihydroanthracene to produce phosphatriptycene (**219**). In addition the chlorine activation of a the double bond of a dihydrophosphole oxide induced cyclisation with an adjacent hydroxymethyl group to give (**203**; R = CCl_2) as shown in Scheme 13 and condensation of aldehyde to amine was used in the synthesis of [17,17,17] heterocycles (**255**).

6.2.8.10 Replacement of Heterocyclic Atoms

While a phosphorus atom has been introduced into monocyclic rings such as (**277**) by the substitution of oxonium groups or tetraalkyl tin, this does not appear to have been explored for bridgehead situations. Neither has derivatisation of intermediates (**277**), e.g. formation of the benzyl phosphine, which could be followed by Friedel Craft cyclisation. Since many of the methods available are reliant on the presence of additional heteroatoms, reactions which substitute these atoms for carbon groups could be useful. Thus the central phosphorus atom of the dithiadiphosphine (**71**; Z = P) reacts with acetyl chloride and aluminium chloride to produce the phosphine (**71**; Z = CHMe).

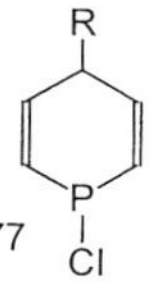

277

6.2.8.11 Miscellaneous Reactions

White phosphorus and red phosphorus with iodine have been found to be useful starting materials for the preparation of a number of diphosphorus heterocycles. Of special note are the reactions with acetylenes and haloarenes to give the corresponding barrolenes and triptycenes. The former have found good use for the preparation of two-coordinate intermediates which can be used in cycloaddition reactions (see Scheme 12). Interestingly, ring opening reactions have also led to new bicyclic heterocycles. Thus the route to the [4,4,4] diphosphines involved cleavage of the central PP bond in (**252**) to give (**253**). The formation of (**57**) involved cleavage of the electron-rich double bond of a tetraminoalkene.

6.2.9 APPLICATIONS

Organophosphorus heterocycles have a remarkable range of uses. Biological applications include their use as pesticides, weed control, antibiotics, cancer agents, antagonists for phosphorus agonists. There are many industrial applications for example as flameproofing agents for textiles and plastics, plastic stabilisers, lubricants, surfactants, metal extractants, water softening, ligands for petroleum catalysts ⟨92M⟩.

A range of bicyclic phosphates (**233**; X = Y = O) have been the subject of biological assessment ⟨82M⟩ and 1-hydroxmethyl derivative (**233**; X = Y = O; R = $HOCH_2$) has been patented for its herbicidal properties ⟨66USP3287448⟩. The effect of various drug pre-treatments on the convulsions and cerebella cyclic nucleotide changes induced by the 4-isopropyl compound (**233**; X = Y = O; R = Pr^i) has been studied ⟨84M⟩. Herbicide properties are reported for urea based bicyclic compounds (**96**; R = Ph, Y = O) ⟨98M2⟩. Acyclic phosphate esters ⟨98M12⟩ and various derivatives of (**233**; X = O, Y = lp) are claimed to have fireproofing properties ⟨98JAP316813, 99JAP199721⟩ as does patents involving (cyanotriazaphosphabicyclononane (**169**) ⟨73USP374584⟩.

The inclusion of the phosphorus atom at a bridgehead position in general reduces the nucleophilicity of the lone pair of the three-coordinate compounds. In contrast, however, the proazaphosphatranes (**248**) are very strong bases ⟨96JOC2963⟩ which are able to convert acetonitrile to its carbanion ⟨97JOC4827, 97USP5698737⟩. They are excellent catalysts for condensation reactions ⟨99JOC3090⟩, silylations of hindered alcohols and phenols ⟨96JA832⟩, for trans-esterification ⟨99JOC3086⟩ and for the allylation of aldehydes with allyltrimethylsilane ⟨99JOC6459⟩. Smaller ring systems can stabilise five and six-coordinate compounds and as such provide a route to interesting rigid molecules. Dimethylmethylene linked moieties of the general [2,2,1] diphenyl type (**190**) have been patented as hydrogenation catalysts ⟨99M5⟩.

6.2.10 REFERENCES

59JCS2835 R.C. Hinton and F.G. Mann; *J. Chem. Soc.*, 1959, 2835.
61JA3432 C.G. Krespan; *J. Am. Chem. Soc.*, 1961, **83**, 3432.
62JA610 W.S. Wadsworth and W.D. Emmons; *J. Am. Chem. Soc.*, 1962, **84**, 610.
65IC1655 K.J. Coskran and J.G. Verkade; *Inorg. Chem.*, 1965, **4**, 1655.
66TL4935 D.B. Denney and S. Varga; *Tetrahedron Lett.*, 1966, 4935.
66USP3287448 R.F.W. Raetz; *U.S. Patent*, 1966, 3287448.
67DOK(174)846 B.E. Ivanova, A.G. Ageeva and Yu.Yu. Samitov; *Dokl. Akad. Nauk SSSR*, 1967, **174**, 846.
68AG(E)733 G. Markl and F. Lieb; *Angew. Chem., Int. Ed. Engl.*, 1968, **7**, 733.
68IC1469 A.C. Vandenbrouke Jr., E.J. Boros and J.G. Verkade; *Inorg. Chem.*, 1968, **7**, 1469.
68ZOB1881 E.S. Kozlov, A.I. Sedlov and A.V. Kirsanov; *Zh. Obchsh. Khim.*, 1968, **38**, 1881; *ibid*, 1970, **40**, 1673.
69AG(E)674 J. Ebling and A. Schmidpeter; *Angew. Chem. Int. Ed., Engl.*, 1969, **8**, 674.
69AG(E)987 D. Hellwinkel and W. Schenk; *Angew. Chem. Int. Ed., Engl.*, 1969, **8**, 987.
69CR(C)(268)236 J. Hilderbrand, G. Kaufmann and R. Rohmer; *C.R. Hebd. Seances Acad. Sci., Ser. C*, 1969, **268**, 236.
69JCS(C)1542 E.M. Richards, J.C. Tebby, R.S. Ward and D.H. Williams; *J. Chem. Soc. C*, 1969, 1542.
70BSF876 J. Hilderbrand and G. Kaufmann, *Bull. Soc. Chim. Fr.*, 1970, 876.
70JA71 R.D. Bertrand, D.A. Allison and J.G. Verkade; *J. Am. Chem. Soc.*, 1970, **92**, 71.
70JA1995 C. Altona and M.J. Sundaralingham; *J. Am. Chem. Soc.*, 1970, **72**, 1995.
70JCS(CC)956 M.E. Brennan; *J. Chem. Soc., Chem. Commun.*, 1970, 956.
70JMR111 D.W. White and J.G. Verkade; *J. Magn. Reson.*, 1970, **3**, 111.
70JOC2310 J. Rathke, J.W. Guyer and J.G. Verkade; *J. Org. Chem.*, 1970, **35**, 2310.
70ZAAC(373)245 M. Becke-Goehring and R. Wolf; *Z. Anorg. Allg. Chem.*, 1970, **373**, 245.
71IC2591 W. Vandoornew, G.W. Hunt, R.W. Parry and A.W. Cordes; *Inorg. Chem.*, 1971, **10**, 2591.
71JA1801 K.G. Weinberg and E.B. Whittle; *J. Am. Chem. Soc.*, 1971, **93**, 1801; *ibid*, 1802.
71JA5236 D.D. Swank, C.N. Caughlan, F. Ramirez and J.F. Pilot; *J. Am. Chem. Soc.*, 1971, **93**, 5236.
71TL1249 G. Markl, F. Lieb and C. Martin; *Tetrahedron Lett.*, 1971, 1249.
72CB2637 R. Goetze, H. Noeth and D.S. Payne; *Chem. Ber.*, 1972, **105**, 2637; D.S. Payne, H. Noeth and G. Henninger; *J. Chem. Soc., Chem. Commun.*, 1965, 327.
72JA9230 R.B. Wetzel and G.L. Kenyon; *J. Am. Chem. Soc.*, 1972, **94**, 9230.
72JGU520 A.M. Pinchuk, I.M. Kosinskaya and V.I. Shevchenko; *J. Gen. Chem., USSR*, 1972, **42**, 520.
72PS(2)161 M. Baudler, K. Glinka, U. Kelsh, H. Sandmann and W. Heller; *Phosphorus*, 1972, **2**, 161.
72ZOB519 E.S. Kozlov, A.I. Sedlov and A.V. Kirsanov; *Zh. Obchsh. Khim.*, 1972, **42**, 519.
73CJC520 R.T. Oakley and N.L. Paddock; *Can. J. Chem.*, 1973, **51**, 520.
73JCS(P1)972 E. Duff, P.J. Whittle and S. Trippett; *J. Chem. Soc., Perk. 1*, 1973, 972.
73USP374584 D.J. Daigle, A.B. Pepperman, B. Armand Jr. and F.L. Normand; *U.S. Patent*, 1973, 374584.
74AG(E)543 A. Gieren, P. Narayanan, K. Burger and W. Thenn; *Angew. Chem., Int. Ed. Engl.*, 1974, **13**, 543.
74IZV2079 B.E. Ivanov and S.V. Samurina; *Izv. Akad. Nauk SSR*, 1974, 2079.
74PS(4)279 G. Markl, H. Hauptmann and F. Lieb; *Phosphorus*, 1974, **4**, 279.
74T3465 C. Jongsma, J.P. De Kleijn and F. Bickelhaupt; *Tetrahedron*, 1974, **30**, 3465.
74TL4369 G. Markl and K.H. Heier; *Tetrahedron Lett.*, 1974, 4369.
75PS(5)229 D. Houalla, J. Monheich, M. Sanchez and R. Wolf; *Phosphorus*, 1975, **5**, 229.
76AG(E)39 H.W. Roesky and E. Janssen; *Angew. Chem., Int. Ed. Engl.*, 1976, **15**, 39; A. Gieren, B. Dederer, W.H. Roesky and E. Janssen; *Angew. Chem., Int. Ed. Engl.*, 1976, **15**, 783.

76AG(E)367 H. Schmidbauer and M. Heinmann; *Angew. Chem., Int. Ed. Engl.*, 1976, **15**, 376.

76CB3581 A. Schmidpeter, J. Luber, D. Schomburg and W.S. Sheldrick; *Chem. Ber.*, 1976, **109**, 3581.

76JA623 J.C. Clardy, D.S. Milbrath, J.P. Springer and J.G. Verkade; *J. Am. Chem. Soc.*, 1976, **98**, 623.

76JCS(P1)2093 J.P. Majoral, R. Kramer, J. Navech and F. Mathis; *J. Chem. Soc., Perk. 1*, 1976, 2093.

76OMR558 J. Devillers, D. Houalla, T. Moncheich and R. Wolf; *Org. Magn. Reson.*, 1976, **8**, 558.

76ZAAC(425)209 F. Krech and K. Issleib; *Z. Anorg. Allg. Chem.*, 1976, **425**, 209.

77AG(E)549 A. Schmidpeter, J.H. Weinmaier and E. Glaser; *Angew. Chem., Int. Ed. Engl.*, 1977, **16**, 549.

77AG(E)797 R. Keat and D.G. Thompson; *Angew. Chem., Int. Ed. Engl.*, 1977, **16**, 797.

77AG(E)877 H.J. Bestmann, K. Roth and R.W. Saalfrank; *Angew. Chem. Int. Ed. Engl.*, 1977, **16**, 877.

77JA5785 G. Zon, S.M. Ludeman and W. Egan; *J. Am. Chem. Soc.*, 1977, **99**, 5785.

77JA6607 D.S. Milbrath and J.G. Verkade; *J. Am. Chem. Soc.*, 1977, **99**, 6607; L.E. Carpenter and J.G. Verkade; *J. Am. Chem. Soc.* 1985, **107**, 7084.

77JA8511 Y. Kobayashi, S. Fujino, H. Hamana, I. Kumadaki and Y. Hanzawa; *J. Am. Chem. Soc.*, 1977, **99**, 8511.

77JCS(CC)156 J.A. Miller and D. Stewart; *J. Chem. Soc., Chem. Commun.*, 1977, 156.

77OMR172 K. Issleib, P. Thoransch and H. Meyer; *Org. Magn. Reson.*, 1977, **10**, 172; *ibid*, 1978, **11**, 255.

77TL1583 H. Sliva and J.C. Picavet, *Tetrahedron Lett.*, 1977, 1583.

77TL3041 D. Houalla, F.H. Osman, M. Sanchez and R. Wolf; *Tetrahedron Lett.*, 1977, **18**, 3041.

77TL4125 E.E. Nifant'ev, M.P. Koroteev, Z.K. Zhane, A.A. Borisenko and N.K. Kochetkov; *Tetrahedron Lett.*, 1977, **18**, 4125.

78CB13 D. Hellwinkel and W. Krapp; *Chem. Ber.*, 1978, **111**, 13; *ibid*, 1977, **111**, 693.

78CB2086 A. Schmidpeter and J.H. Weinmaier; *Chem. Ber.*, 1978, **111**, 2086. W.S. Sheldrick, D. Schomburg and A. Schmidpeter; *Acta Crystallogr., Part B*, 1980, **36**, 2316.

78CB3838 M. Baudler, M. Warnau and D. Koch; *Chem. Ber.*, 1978, **111**, 3838.

78JA5130 Y. Segall and I. Granoth; *J. Am. Chem. Soc.*, 1978, **100**, 5130.

78JCS(CC)161 L.W. Yarbrough and M.B. Hall; *J. Chem. Soc., Chem. Commun.*, 1978, 161.

78JHC1319 D.G. Holar, A.N. Hughes and D. Kleemola; *J. Heterocycl. Chem.*, 1978, **15**, 1319.

78PS(4)137 K. Issleib and P. Thoransch; *Phosphorus Sulfur*, 1978, **4**, 137.

78T627 N.L. Allinger and H. Von Voithenberg; *Tetrahedron*, 1978, **34**, 627.

78T1993 J. Kaiser, R. Richter and H. Hartung; *Tetrahedron*, 1978, **34**, 1993.

78ZOB44 N.A. Tikinina, V.A. Gilyarov and M.I. Kabachnik; *Zh. Obshch. Khim.*, 1978, **48**, 44.

78ZOB1700 I.V. Konovalova, R.D. Gareev, L.A. Burnaeva, N.V. Mikhailova and A.N. Pudovik; *Zh. Obshch. Khim.*, 1978, **48**, 1700.

79CB3472 A. Schmidpeter and T. von Criegen; *Chem. Ber.*, 1979, **112** 3472; *ibid*, 2762; *ibid*, 1978, **111**, 3747; *ibid*, *Z. Naturforsch.*, 1978, **33B**, 1330; *ibid*, *Angew. Chem. Int. Ed. Engl.*, 1978, **17**, 443; *ibid*, *J. Chem. Soc., Chem. Commun.*, 1987, 470.

79JCS(P1)2896 T.S. Cameron, R.E. Cordes, T. Demir and R.A. Shaw; *J. Chem. Soc., Perk. 1*, 1979, 2896.

79JGU641 I.V. Konovalova, T.A. Faskhutdinoba, L.A. Burnaeva, T.G. Monakhova and A.N. Pudovik; *J. Gen. Chem. USSR*, 1979, **49**, 641.

79JHC897 G.M. Coppola; *J. Heterocycl. Chem.*, 1979, **16**, 897.

79OMS459 K.J. Voorhees, F.D. Hileman and D.L. Smith; *Org. Mass Spectrom.*, 1979, **14**, 459.

79RTC12 D. van Aken, A.M.C.F. Castalijius, J.G. Verkade and H.M. Buck; *Recl. Trav. Chim. Pays-Bas*, 1979, **98**, 12.

80AX(B)1247 F.J.M. Freijee et al.; *Acta Crystallogr. B*, 1980, **36** 1247.

80AX(B)2316 W.S. Sheldrick, D. Schomburg and A. Schmidpeter; *Acta Crystallogr. B*, 1980, **36**, 2316.

80CB55 W.S. Schelrick, D. Schomburg, A. Schmidpeter and T. von Criegern; *Chem. Ber.*, 1980, **113**, 55.

80CB1847 H.W. Roesky, K. Ambrosius, M. Banek and W.S. Sheldrick; *Chem. Ber.*, 1980, **113**, 1847.

80JA5073 D.B. Denney, D.Z. Denney, P.J. Hammond, C. Huang and K.-S. Tseng; *J. Am. Chem. Soc.*, 1980, **102**, 5073.

80JA7072 D.B. Denney, D.Z. Denney, D.M. Gavrilovic, P.J. Hammond, C. Huang and K.-S. Tseng; *J. Am. Chem. Soc.*, 1980, **102**, 7072.

80PS(8)315 R.S. Edmundson and C.I. Forth; *Phosphorus Sulfur*, 1980, **8**, 315.

80ZAAC(471)155 H. Oehme, E. Leissring and H. Meyer; *Z. Anorg. Allg. Chem.*, 1980, **471**, 155.

80ZOB1499 E.S. Koslov and V.I. Tovstenko; *Zh. Obshch. Khim.*, 1980, **50**, 1499.

81AG(E)967 M. Baudler, Y. Aktlay, K.-F. Tebbe and T. Heinlein; *Angew. Chem. Int. Ed. Engl.*, 1981, **20**, 967.

81AG(E)1029 M. Drager and B. Mathiasch; *Angew. Chem., Int. Ed. Engl.*, 1981, **20**, 1029.

81CB2132 W.S. Sheldrick, S. Pohl, H. Zamankhan, M. Banek, D. Amirzadeh and H.W. Roesky; *Chem. Ber.*, 1981, **114**, 2132.

81JA2465 Y. Kobayashi, S. Fujino and I. Kumadaki; *J. Am. Chem. Soc.*, 1981, **103**, 2465.

81JA4595 F. Mathey, F. Mercier, C. Charrier, J. Fischer and A. Mitschler; *J. Am. Chem. Soc.*, 1981, **103**, 4595.

81JCS(CC)1131 R. Carrie, Y.Y.C.Y.L. Ko, F. de Sarlo and A. Brandi; *J. Chem. Soc., Chem. Commun.*, 1981, 1131.

81JCS(P2)19 C. Bonningue, D. Houalla, M. Sanchez, R. Wolf and F.H. Osman; *J. Chem. Soc., Perk. 2*, 1981, 19.

81JCR(S)348 K. Buzik, W. Stec, D. Houalla and R. Wolf, *J. Chem. Res. (S)*, 1981, 348.

81JOC361 C.H. Chen, K.E. Brightley and F.M. Michaels; *J. Org. Chem.*, 1981, **46**, 361.

81JOC1451 J.E. MacDiarmid and L.D. Quin; *J. Org. Chem.*, 1981, **46**, 1451.

81M1 D. Van Aken, I.I. Merkelbach, H.H. Hamerlinck, P. Schepper and H.M. Buck; *A.C.S. Symp. Ser.*, 1981, **171**, 439.

81PS(9)323 S.S. Krishnamurthy, K. Ramachandran and M. Woods; *Phosphorus Sulfur*, 1981, **9**, 323; *ibid*, *J. Chem. Res. (S)*, 1979, 92; *ibid*, (M) 1267.

81PS(10)27 B.A. Arbuzov, R.P. Arshinova, D.W. White and J.G. Verkade; *Phosphorus Sulfur*, 1981, **10**, 27.

81ZAAC(473)20 M. Baudler, Y. Aktalay, J. Hahn and E. Daerr; *Z. Anorg. Allg. Chem.*, 1981, **473**, 20.

82CB3610 R. Appel, S. Korte, M. Halstenberg and F. Knoch; *Chem. Ber.*, 1982, **115**, 3610.

82IZV1408 M.A. Pudovik, S.A. Terent'eva and A.N. Pudovik; *Izv. Akad. Nauk SSR*, 1982, 1408.

82JCS(CC)1272 G. de Lauzon, C. Charrier, H. Bonnard, F. Mathey, J. Fischer and A. Mitschler; *J. Chem. Soc. Chem. Commun.*, 1982, 1272.

82M Y. Ozoe and M. Eto; *Agric. Biol. Chem.*, 1982, **46**, 411.

82PS(12)227 B.N. Diel and A.D. Norman; *Phosphorus Sulfur*, 1982, **12**, 227.

82PS(13)153 U. Kuhne, F. Krech and K. Issleib; *Phosphorus Sulfur*, 1982, **13**, 153.

82TL511 G. de Lauzon, C. Charrier, H. Bonnard and F. Mathey; *Tetrahedron Lett.*, 1982, **23**, 511.

82TL1263 H. Yamaguchi, S. Kametani, T. Karakusa, T. Saito and S. Motoki; *Tetrahedron Lett.*, 1982, **23**, 1263.

82TL1345 K. Jurkschat, C. Muegge, A. Tzschach, W. Uhlig and A. Zschunke; *Tetrahedron Lett.*, 1982, **23**, 1345.

82TL2529 L.D. Quin and S.C. Spence; *Tetrahedron Lett.*, 1982, **23**, 2529.

83CJC1105 Y.Y.C.Y.L. Ko, R. Carrie, F. de Sarlo and A. Brandi; *Can. J. Chem.*, 1983, **61**, 1105.

83IJM459 H.I. Kenttamaa; *Int. J. Mass Spectrom. Ion Phys.*, 1983, **47**, 459.

83IZV2275 I.A. Litvinov, Yu.T. Struchkov, B.A. Arbuzov, E.N. Dianova and E.Ya. Yabotina; *Izv. Akad. Nauk. SSSR*, 1983, 2275.

83JCS(D)425 D.A. Harvey, R. Keat and D.B. Ryecroft; *J. Chem. Soc., Dalton Trans.*, 1983, 425.

83JCS(P2)773 C. Bonningue, D. Houalla, R. Wolf and J. Jana; *J. Chem. Soc., Perk. 2*, 1983, 773; D. Houalla, F.H. Osman, M. Sanchez and R. Wolf; *Tetrahedron Lett.*, 1977, 3041.

83PS(15)281 D.B. Denney, D.Z. Denney, P.J. Hammond, C. Huang, L.-T. Liu and K.-S. Tseng; *Phosphorus Sulfur*, 1983, **15**, 281.

83PS(17)73 K. Issleib, U. Kuhne and F. Krech; *Phosphorus Sulfur*, 1983, **17**, 73.

83S445 R.S. Edmundson; *Synthesis*, 1983, 445.

83ZN(B)460 H.W. Roesky and D. Amirzadeh; *Z. Naturforsch. B, Anorg. Org.*, 1983, **38**, 460.

84BCJ1990 T. Uchida, K. Kozawa, T. Saito and S. Motoki; *Bull. Chem. Soc. Jpn.*, 1984, **57**, 1990; *ibid*, *Bull. Soc. Chem. Jpn.*, 1984, **57**, 1985.

84IC1495 P.J. Harris and K.B. Williams; *Inorg. Chem.*, 1984, **23**, 1495.

84ICA61 F. Sournies, J.F. Labarre, P.J. Harris and K.B. Williams; *Inorg. Chim. Acta*, 1984, **90**, L61.

84IZV2023 I.A. Litvinov, Yu.T. Struchkov, B.A. Arbuzov, E.N. Dianova and J.Z. Galeeva; *Izv. Akad. Nauk SSSR*, 1984, 2023, 1845; *ibid*, 1983, 2280.

84JA7825 K. Taira, W.L. Mock and D.G. Gorenstein; *J. Am. Chem. Soc.*, 1984, **106**, 7825, 7831

84JA1164 S.A. Culley and A.J. Arduengo III; *J. Am. Chem. Soc.*, 1984, **106**, 1164.

84JCS(CC)508 J.M. Alcaraz and F. Mathey; *J. Chem. Soc., Chem. Commun.*, 1984, 508.

84JCS(CC)675 S.R. Contractor, M.B. Hursthouse, H.G. Parkes, L.S. Shaw, R.A. Shaw and H. Yilmaz; *J. Chem. Soc., Chem. Commun.*, 1984, 675.

84JCS(P1)1449 J.I.G. Cadogan, J.B. Husband and H. McNab; *J. Chem. Soc., Perk. 1*, 1984, 1449; *ibid*, *J. Chem. Soc., Chem. Commun.*, 1981, 1954.

84JGU1328 R.I. Taranova, N.I. Sinitsyna, T.A. Dvoinishnikova, A.B. Remizov, M.V. Alpavova, T.V. Zykova and V.V. Moskva; (*J. Gen. Chem. USSR Engl. Transl.*), 1984, **54**, 1328.

84JGU2185 M.A. Pudovik, S.A. Terent'eva, A.V. Il'yasov, A.N. Chernov, A.A. Nafikova and A.N. Pudovik; *J. Gen. Chem. USSR*, 1984, **54**, 2185.

84JGU2445 I.V. Konovalova, L.A. Buraeva, N.M. Kashtanova, R.D. Gareev and A.N. Pudovik; (*J. Gen. Chem. USSR Engl. Transl.*), 1984, **54**, 2445.

84JGU2448 M.A. Pudovik, S.A. Terent'eva, A.V. Il'yasov, A.N. Chernov, A.A. Nafikova and A.N. Pudovik; *J. Gen. Chem. USSR*, 1984, **54**, 2448.

84M I.S. Blenkinsop, D.B. Coult, W.E. Davies and D.J. Howells; *Neurochem. Int.*, 1984, **6**, 211.

84PS(21)111 E.J. Voloko and J.G. Verkade; *Phosphorus Sulfur*, 1984, **21**, 111.

84TL4659 J.M. Alcaraz and F. Mathey; *J. Chem. Soc. Chem. Commun.*, 1984, 508; *ibid*, *Tetrahedron Lett.*, 1984, **25**, 4659.

84ZN(B)139 H.-J. Woerz, H. Pritzkow and H.P. Latscha; *Z. Naturforsch. B*, 1984, **39**, 139.

85AG(E)975 D. Schomburg, G. Betterman, L. Ernst and R. Schmutzler; *Angew. Chem. Int. Ed. Engl.*, 1985, **24**, 975.

85AG(E)1056 R. Riedel, H.-D. Hansen and E. Fluck; *Angew. Chem., Int. Ed. Engl.*, 1985, **24**, 1056.

85JA7084 L.E. Carpenter and J.G. Verkade; *J. Am. Chem. Soc.*, 1985, **107**, 7084.

85JOC2914 C.H. Chen, J.J. Doney, J.L. Fox and H.R. Luss; *J. Org. Chem.*, 1985, **50**, 2914.

85PS(21)367 K. Issleib, U. Kuhn and F. Krench; *Phosphorus Sulfur*, 1985, **21**, 367.

85TL531 D.G. Gilheany, D.A. Kennedy, J.F. Malone and B.J. Walker; *Tetrahedron Lett.*, 1985, **26**, 531.

85ZAAC(521)153 P.J. Retuert, E. Fluck, H. Riffel and H. Hess; *Z. Anorg. Allg. Chem.*, 1985, **521**, 153.

85ZAAC(523)7 K. Issleib, U. Khune and F. Krech; *Z. Anorg. Allg. Chem.*, 1985, **523**, 7.

85ZN(B)1532 R. Neidlein, H.-J. Degener, A. Gieren, G. Weber and T. Hubner; *Z. Naturforsch B, Anorg. Org. Chem.*, 1985, **40**, 1532.

85ZOB2218 Yu.G. Shermolovich, E.A. Danchenko, A.V. Slov'er and L.N. Markovski; *Zh. Obshch. Khim.*, 1985, **55**, 2218.

86AG(E)471 M. Baudler, M. Michels, M. Pieroth and J. Khan; *Angew. Chem. Int. Ed. Engl.*, 1986, **25**, 471.

86BCJ2047 H. Tanaka and S. Motoki; *Bull. Chem. Soc. Jpn.*, 1986, **59**, 2047; *ibid*, *Tetrahedron Lett.*, 1984, **25**, 4659.

86CB2687 H.W. Roesky, V.W. Pogatzki, K.S. Dhathathreyan, A. Thiel, H.G. Schmidt, M. Dyrbsch, M. Nottemeyer and G.M. Sheldrick; *Chem. Ber.*, 1986, **119**, 2687.

86CB2767 D.J. Brauer, F. Gol, S. Hietkamp and O. Stelzer; *Chem. Ber.*, 1986, **119**, 2767.

86JA4918 L.E. Carpenter, B. De Ruiter, V. Van Aken, H.M. Buck and J.G. Verkade; *J. Am. Chem. Soc.*, 1986, **106**, 4918.

86JA6311 T. Fanni, K. Taira, D.G. Gorenstein, R. Vaidyanathaswamy and J.G. Verkade; *J. Am. Chem. Soc.*, 1986, **108**, 6311.

86M1 K. Issleib, U. Kuehne and P. Thoransch; *Synth. React. Inorg. Met. Org. Chem.*, 1986, **26**, 446.

86TL5221 K. Akiba, K. Okada and K. Ohkata; *Tetrahedron Lett.*, 1986, **27**, 5221.

86ZN(B)207 D. Schomburg, U. Wermuth and R. Schmutzler; *Z. Naturforsch., B,*, 1986, **41**, 207.

87AG(E)85 K. Blatter and W. Powell; *Angew. Chem., Int. Ed. Engl.*, 1987, **26**, 85.

87AG(E)1256 E. Niecke, O. Altmeyer, M. Nieger and F. Knoll; *Angew. Chem., Int. Ed. Engl.*, 1987, **26**, 1256.

87BCJ1558 H. Tanaka and S. Motoki; *Bull. Chem. Soc. Jpn.*, 1987, **60**, 1558.

87HCA1063 E. Mueller and H.B. Buergi; *Helv. Chim. Acta*, 1987, **70**, 1063.

87JCS(CC)496 G.L. Wood, E.N. Dousler, C.K. Narula, R.T. Paine and H. Noth; *J. Chem. Soc., Chem. Commun.*, 1987, 496.

87JCS(CC)1741 A.M.Z. Slawin, D.J. Williams, P.T. Wood and J.D. Woolins; *J. Chem. Soc., Chem. Commun.*, 1987, 1741.

87JGU434 S.A. Teret'eva, M.A. Pudovik and A.N. Pudovik; *J. Gen. Chem., USSR*, 1987, **57**, 434; *ibid*, 1986, **56**, 632.

87JOM(330)17 K. Issleib, H. Schmidt and E. Leissring; *J. Organomet. Chem.*, 1987, **330**, 17.

87M1 'Phosphorus-31 NMR Spectroscopy in Stereochemical Analysis', ed. J.G. Verkade and L.D. Quin, Publ. VCH 1987; 'Handbook of Phosphorus-31 NMR Data' by J.C. Tebby, Publ. CRC Press 1991.

87PS(30)369 S.E. Cremer, J.M. Cowles and A. Gamliel; *Phosphorus Sulfur*, 1987, **30**, 369.

87ZAAC(553)136 F. Krech, K. Issleib, A. Zschunke and H. Meyer; *Z. Anorg. Allg. Chem.*, 1987, **553**, 136.

87ZC295 K. Issleib, F. Krech and U. Kuhne; *Z. Chem.*, 1987, **27**, 295.

88AG(E)957 R. Appel, D. Niemann and M. Nieger; *Angew. Chem., Int. Ed. Engl.*, 1988, **27**, 957.

88CB1681 H.W. Roesky, U. Scholz, A. Schmidpeter, K. Karaghiosoff and W.S. Sheldrick; *Chem. Ber.*, 1988, **121**, 1681.

88JA4204 K.K. Baldridge and M.S. Gordon; *J. Am. Chem. Soc.*, 1988, **110**, 4204.

88JCS(CC)945 E. Niecke, O. Altmeyer and M. Nieger; *J. Chem. Soc., Chem. Commun.*, 1988, 945.

88JGU1148 I.V. Konovalova, U.G. Trishin, L.A. Burnaeva, E.K. Khusnutdinova, V.N. Chistoklitov and A.N. Pudovik; *J. Gen. Chem., USSR*, 1988, **58**, 1148.

88JOM(355)71 K. Issleib, H. Scmidt and E. Lessring; *J. Organomet. Chem.*, 1988, **355**, 71.

88M1 U. Baumeister, H. Hartung and F. Krech; *Acta Crystallogr. C. Cryst. Struct. Commun.*, 1988, **C44**, 1435.

89CB301 R. Franke, W.S. Sheldrick and G.V. Roschenthaler; *Chem. Ber.*, 1989, **122**, 301.

89JA3478 C. Lensink, S.-K. Xi, L.M. Daniels and J.G. Verkade; *J. Am. Chem. Soc.*, 1989, **111**, 3478.

89JCS(CC)1103 G. Robinet, M. Barthelot, V. Gasmi and J. Devillers; *J. Chem. Soc., Chem. Commun.*, 1989, 1103.

89JGU1773 R.N. Burangulova, Yu.G. Trishin, I.V. Konovalova, L.A. Burnaeva, V.N. Chistokletov and A.N. Pudovik; *J. Gen. Chem., USSR*, 1989, **59**, 1773.

89M I.A. Litvinov, Yu.T. Struchkov and V.A. Naumov; *Zh. Strukt. Khim.*, 1989, **30**, 146.

89PS(41)21 D. Gudat, C. Lensink, H. Schmidt, S.-K. Xi and J.G. Verkade; *Phosphorus Sulfur*, 1989, **41**, 21.

89S618 R. Chen and R. Bao; *Synthesis*, 1989, 618.

90AG(E)689 D. Hellwinkel, A. Wiel, G. Sattler and B. Nuber; *Angew. Chem. Int. Ed. Engl.*, 1990, **29**, 689.

90CB67 E. Niecke, H.J. Metternick and R. Streubel; *Chem. Ber.*, 1990, **123**, 67.

90HAC37 S. Holand, J.M. Alcaraz, L. Ricard and F. Mathey; *Heteroat. Chem.*, 1990, **1**, 37.

90IC2214 S.K. Xi, H. Schmidt, C. Lensink, S. Kim, D. Wintergrass, L.M. Daniels, R.A. Jacobson and J.G. Verkade; *Inorg. Chem.*, 1990, **29**, 2214.

90JA7451 H. Yu, A.M. Arif and W.G. Bentrude; *J. Am. Chem. Soc.*, 1990, **112**, 7451.

90JCS(CC)1307 R. Bartsch, P.B. Hitchcock and J.F. Nixon; *J. Chem. Soc., Chem. Commun.*, 1990, 1307.

90JCS(P1)19 G. Baccolini and E. Mezzina; *J. Chem. Soc., Perk. 1*, 1990, 19.

90JST(221)239 F. Sournier, P. Castera, A. El Bakili and J.F. Labarre; *J. Mol. Struct.*, 1990, **221**, 239.

90M A.G. Massey; 'Main Group Chemistry', pp. 62–71, Publ. E Horwood 1990.

90PS(47)145 M.R. Estrada-Yanaz, H.W. Roesky, U. Scholz and M. Noltmeyer; *Phosphorus Sulfur*, 1990, **47**, 145.

90PS(47)215 R. Conreras, A. Murillo and P. Joseph-Nathan; *Phosphorus Sulfur*, 1990, **47**, 215.

90PS(49/50)163 H. Schmidt, S.K. Xi, C. Lensink and J.G. Verkade; *Phosphorus Sulfur*, 1990, **49/50**, 163.

90PS(53)87 A. Murillo, L.M. Chiquete, P. Joseph-Nathan and R. Conreras; *Phosphorus Sulfur*, 1990, **53**, 87.

90S137 R. Chen and R. Bao; *Synthesis*, 1990, 137.

91AG(E)605 R. Blachnik, H.-P. Baldus, P. Loennecke and B.W. Tattershall; *Angew. Chem., Int. Ed. Engl.*, 1991, **30**, 605.

91IC1330 R.V. Davis, D.J. Wintergrass, M.N. Janakiraman, E.M. Hyatt, R.A. Jacobson, L.M. Daniels, A. Wrotlewski, J.P. Amma, S.K. Das and J.G. Verkade; *Inorg. Chem.*, 1991, **30**, 1330.

91JA1071 J. Arduengo, H.V. Rascka-Dias and J.C. Calabrese; *J. Am. Chem. Soc.*, 1991, **113**, 1071.

91JCS(CC)116 M.-E. Eleftherion, J. Novosad, D.J. Williams and J.D. Woolins; *J. Chem. Soc., Chem. Commun.*, 1991, 116.

91PS(56)133 H.A. Al-Madfa, L.A. Shaw and R.A. Shaw; *Phosphorus Sulfur*, 1991, **56**, 133.

91ZAAC(594)66 F. Krech, K. Issleib, A. Zschunke, C. Mugge and S. Skvoreov; *Z. Anorg. Alg. Chem.*, 1991, **594**, 66.

91ZAAC(600)195 F. Krech, K. Issleib and A. Zschunke; *Z. Anorg. Allg. Chem.*, 1991, **600**, 195.

91ZAAC(605)164 M.A. Laramay and J.G. Verkade; *Z. Anorg. Allg. Chem.*, 1991, **605**, 164; H. Schmidt, S.K. Xi, C. Lensink and J.G. Verkade; *ibid.*, 1989, **578**, 75.

92AG(E)625 C. Garot, G. Etemad-Moghadham, J.-P. Declercq, A. Dubourg and M. Koenig; *Angew. Chem., Int. Ed. Engl.*, 1992, **31**, 625.

92JA3129 J.S. Tang, M.A.H. Laramay, V. Young, S. Ringrose, R.A. Jacobson and J.G. Verkade; *J. Am. Chem. Soc.*, 1992, **114**, 3129.

92M 'Phosphorus Chemistry', eds. E.N. Walsh, E.J. Griffith and R.W. Parry and L.D. Quin, *ACS Symp. Series*, 486, 1992.

92PS(66)301 R.-D. Hund and G.V. Roschenthaler; *Phosphorus Sulfur*, 1992, **66**, 301.

92PS(69)43 Z. Bounja, D. Houalla, M. Revel and M. Taieb; *Phosphorus Sulfur*, 1992, **69**, 43.

93CJC1873 M.B. Hocking and F.W.van der Voort Maarschalk; *Can. J. Chem.*, 1993, **71**, 1873.

93CL1563 S. Motoki, T. Sakai, I. Shinoda, T. Saito and T. Ushida; *Chem. Lett.*, 1993, 1563.

93IC4360 R.R. Holmes, T.K. Prakasha and R.O. Day; *Inorg. Chem.*, 1993, **32**, 4360.

93JOM(462)103 F. Mercier and F. Mathey; *J. Organomet. Chem.*, 1993, **462**, 103.

93M1 M.A. Pudovik, S.A. Terent'eva, I.V. Litvinov, O.N. Kataeva, V.A. Naumov and A.N. Pudovik; *Russ. Chem. Bull.*, 1993, **42**, 1706.

93PS(76)265 M.T. Estra and C. Vazquez; *Phosphorus Sulfur*, 1993, **76**, 265.

93TL6365 S.M. Bachrach and L.M. Perriott; *Tetrahedron Lett.*, 1993, **34**, 6365.

93TL8255 A. Ishi, I. Takaki, J. Nakagama and M. Hoshino; *Tetrahedron Lett.*, 1993, **34**, 8255.

93TL8259 A. Ishi, R. Yoshioka, J. Nakagama and M. Hoshino; *Tetrahedron Lett.*, 1993, **34**, 8259.

94AG(E)2202 V. Caliman, P.B. Hitchock, J.F. Nixon, M. Hofmann and P. von R. Schleyer; *Angew. Chem. Int. Ed. Engl.*, 1994, **33**, 2202.

94AJC2161 J.H. Kim, D.C. Craig, M.J. Callagher and R.F. Toia; *Aust. J. Chem.*, 1994, **47**, 2161.

94BCJ2785 I. Shinoda, A. Takahashi, T. Saito and T. Ushida; *Bull. Chem. Soc. Jpn.*, 1994, **67**, 2785.

94HAC507 B.S. Reddy, C.D. Reddy and P.M. Reddy; *Heteroatom. Chem.*, 1994, **5**, 507; C.D. Reddy, R.J. Rao, B.S. Reddy and P.M. Reddy; *Indian J. Heterocyclic Chem.*, 1994, **3**, 223.

94IC2657 D.V. Khasnis, J.M. Burton, J.D.M. McNeil, C.J. Santani, H. Zhang and M. Lattman; *Inorg. Chem.*, 1994, **33**, 2657.

94JCS(CC)2161 J. Mitjaville, A.M. Caminade and J.P. Majoral; *J. Chem. Soc., Chem. Commun.*, 1994, 2161.

94JGU1203 J. Gloede; (*J. Gen. Chem. USSR Engl. Transl.*), 1994, **64**, 1203.

94JGU1394 V.I. Kal'chenko, A.N. Shivanyuk, V.V. Pirozhenko and L.N. Markovskii; *J. Gen. Chem. USSR (Engl. Transl.)*, 1994, **64**, 1394, 1397.

94JOM(464)41 B. Bret, R. Boese and M. Regitz; *J. Organomet. Chem.*, 1994, **464**, 41.

94M1 S. Nagase Chapter 1 in 'The Chemistry of Organic Arsenic, Antimony and Bismuth Compounds', ed. S. Patai, Publ. Wiley 1994.

94PS(89)39 B. Costisella and J. Gloede; *Phosphorus Sulfur Rel. Elem.*, 1994, **89**, 39.

94S1337 B. Geissler, T. Wettling, S. Barth, P. Binger and M. Regitz; *Synthesis*, 1337.

94TL2743 D. Katalenic, V. Skaric and B. Klaic; *Tetrahedron Lett.*, 1994, **35**, 2743.

95AG(E)81 P. Binger, G. Glaser, B. Gabor and R. Mynott; *Angew. Chem. Int. Ed. Engl.*, 1995, **34**, 81.

95AG(E)484 B. Geissler, S. Barth, U. Bergstrasser, M. Slany, J. Durkin, P.B. Hitchock, J.M. Hofmann, P. Binger, F. Nixon, P.von R. Schleyer and M. Regitz; *Angew. Chem. Int. Ed. Engl.*, 1995, **34**, 484.

95AG(E)1853 H.P. Schrodel, G. Jochim, A. Schmidpeter and H. Noth; *Angew. Chem. Int. Ed. Engl.*, 1995, **34**, 1853.

95CB891 B. Assmann, K. Angermaier, M. Paul, J. Riede and H. Schmidbaur; *Chem. Ber.*, 1995, **128**, 891.

95CB991 M. Julino, M. Slany, U. Bergstrasser, F. Mercier, F. Mathey and M. Regitz; *Chem. Ber.*, 1995, **128**, 991.

95HAC617 T.W. Mackewitz, U. Bergstrasser, S. Leininger and M. Regitz; *Heteroatom. Ber.*, 1995, **6**, 617.

95IC1638 D.V. Khasnis, H.S. Isom, S. Shang, M. Lattman, M.M. Olmstead and P.P. Power; *Inorg. Chem.*, 1995, **34**, 1638.

95IC5405 I.V. Shevchenko, H. Zhang and M. Lattman; *Inorg. Chem.*, 1995, **34**, 5405.

95JA5303 S. Charles, J.C. Fettinger and B.W. Eichhorn; *J. Am. Chem. Soc.*, 1995, **117**, 5303.

95JCS(CC)2217 M.R.St.J. Foreman, A.M.Z. Sawin and J.D. Woollens; *J. Chem. Soc., Chem. Commun.*, 1995, 2217.

95JCS(CC)2395 S.M. Bachrach, V. Caliman and J.F. Nizon; *J. Chem. Soc., Chem. Commun.*, 1995, 22395.

95JCS(D)2797 G. Grossman, H. Beckmann, O. Rademacher, K. Kruger and G. Ohms; *J. Chem. Soc., Dalton Trans.*, 1995, 2797.

95JCS(P2)1191 K.K. Laali, C. Wesdemiotis, M. Polce and S. Beranova; *J. Chem. Soc., Dalton Trans.*, 1995, 1191.

95JGU287 E.E. Nifant'ev, R.K. Magdeeva, R.M. Syaribzhanova and L.K. Vasyanina; (*J. Gen. Chem., USSR Engl. Transl.*), 1995, **65**, 287.

95JOC47 K.K. Laali, B. Geissler, M. Regitz and J.J. Houser; *J. Org. Chem.*, 1995, **60**, 47. 3149.

95JOC3149 K.K. Laali, B. Geissler and M. Regitz; *J. Org. Chem.*, 1995, **60**, 3149.

95JOC5884 M. Julino, U. Bergstraesser and M. Regitz; *J. Org. Chem.*, 1995, **60**, 5884.

95JOC6362 K.K. Laali, B. Geissler, M. Regitz and J.J. Houser; *J. Org. Chem.*, 1995, **60**, 6362.

95JOC7433 K. Brandt, I. Porwolik, T. Kupta, A. Olejnik, R.A. Shaw and D.B. Davies; *J. Org. Chem.*, 1995, **60**, 7433.

95M M. Sokolowski; *Prog. ERDEC Sci. Conf. Chem. Biol. Def. Res.*, 1995, Publ. 1996, 703.

95PS(103)225 Y. Huang, A.M. Arif and W.G. Bentrude; *Phosphorus Sulfur*, 1995, **103**, 225.

95PS(103)235 R.M. Schoth and G.-V. Roschenthaler; *Phosphorus Sulfur*, 1995, **103**, 235.

95TL447 G. Baccolini, G. Orsolan and E. Mezzina; *Tetrahedron Lett.*, 1995, **36**, 447.

95ZAAC(621)29 F. Golinski and H. Jacobs; *Z. Anorg. Allg. Chem.*, 1995, **621**, 29.

95ZAAC(621)34 G. Becker, G. Heckmann, K. Kluber and W. Schwarz; *Z. Anorg. Allg. Chem.*, 1995, **621**, 34.

96AG(E)896 J. Barluenga, M. Tomas, K. Bieger, S. Garcia-Granda and R. Santiago-Garcia; *Angew. Chem. Int. E. Engl.*, 1996, **35**, 896.

96AG(E)1558 M. Gray, B.J. Chapell, N.J. Taylor and V. Chieckus; *Angew. Chem. Int. Ed. Engl.*, 1996, **35**, 1558.

96BSB675 V. Caliman, P.B. Hitchock, J.F. Nixon and N. Sakarya; *Bull. Soc. Chem. Belg.*, 1996, **105**, 675.

96CB121 S. Strojek and M. Jansen; *Chem. Ber.*, 1996, **129**, 121.

96CB489 B. Breit and M. Regitz; *Chem. Ber.*, 1996, **129**, 489.

96H2367 J.M. Smallheer and S.P. Seitz; *Heterocycles*, 1996, **43**, 2367.

96IC325 C.Y. Wong, R. McDonald and R. Cavell; *Inorg. Chem.*, 1996, **35**, 325.

96JA832 B.A. D'Sa and J.G. Verkade; *J. Am. Chem. Soc.*, 1997, **118**, 832.

96JA10168 A.E. Wroblewski and J.G. Verkade; *J. Am. Chem. Soc.*, 1996, **118**, 10168.

96JCS(CC)537 R.W. Alder, D.D. Ellis, J.K. Hogg, A. Martin, O.G. Orpen and N.J. Taylor; *J. Chem. Soc., Chem. Commun.*, 1996, 537.

96JCS(CC)539 R.W. Alder, D.D. Ellis, A.G. Orpen and N.J. Taylor; *J. Chem. Soc., Chem. Commun.*, 1996, 539.

96JCS(CC)583 F.J. Parlevliet, A. Oliver, W.G.J. de Lange, P.L.J. Kamer, H. Kooijman, A.L. Spek and P.W.N.M. Leeuwen; *J. Chem. Soc., Chem. Commun.*, 1996, 583.

96JGU355 M.A. Pudovik, S.A. Terent'eva and A.N. Pudovik; *J. Gen. Chem. USSR*, 1996, **66**, 355.

96JGU1367 V.I. Namestnikov, Yu.G. Trishin and V.K. Bel'skii; *J. Gen. Chem. USSR*, 1996, **66**, 1367.

96JOC2963 B.A. D'Sa and J.G. Verkade; *J. Org. Chem.*, 1996, **61**, 2963.

96JOC3531 S. Lelièvre, F. Mercier and F. Mathey; *J. Org. Chem.*, 1996, **61**, 3531.

96JOC7813 J.L. Abboud, M. Herreros, R. Notario, O. Mo, M. Hanez, M. Regitz and J. Elguero; *J. Org. Chem.*, 1996, **61**, 7813.

96M1 W. Flitsch Section 8.01 in 'Comprehensive Organic Functonal Group Transformations', Vol. 8, ed. G. Jones, Chief eds. A.R. Katritski, C.W. Rees and E.F.V. Scriven, Publ. Pergamon 1996.

96M2 M. Schuster, K. Eckstein and K. Karaghiosoff; *GIT FachzLab.*, 1996, **40**, 1249 (*Chem. Abstr.*, 1997, **127**, 74982).

96PS(109-10)93 W. Schnick, S. Horstman and M. Haser; *Phosphorus Sulfur*, 1996, **109/110**, 93.

96PS(111)151 R. Cavell, C.Y. Wong and R. McDonald; *Phosphorus Sulfur*, 1996, **111**, 151.

96PS(111)203 K. Bieger, M. Tomas, J. Barluenga, R. Santiago-Garcia and S. Garcia-Granda; *Phosphorus Sulfur*, 1996, **111**, 203.

96PS(113)1 E.E. Nifant'ev, A.M. Koroteev, M.P. Koroteev, S.S. Meshkov, V.K. Belsky and A.R. Becker; *Phosphorus Sulfur*, 1996, **113**, 1.

96S1227 H. Wan and T.A. Modro; *Synthesis*, 1996, 1227.

96ZOB1399 E.E. Nifant'ev, M.P. Koroteev and N.M. Pugashova; *Zhur. Obsch. Khim.*, 1996, **66**, 1399.

97AX(C)1310 H. Thonnessen, P.G. Jones, R. Schmutzler and J. Gloede; *Acta Crystallogr. Sect. C.*, 1997, **C35**, 1310.

97HAC165 J. Krill, I.V. Chevchenko, A. Fischer, P.G. Jones and R. Schmutzler; *Heteroat. Chem.*, 1997, **8**, 165.

97HAC517 D. Qian, X. Zeng, X. Shi, R. Cao and L. Lin; *Heteroat. Chem.*, 1997, **8**, 517.

97HAC551 G. Baccolini, M. Beghelli and C. Boga; *Heteroat. Chem.*, 1997, **8**, 551.

97JA1208 F.C. Krebs, P.S. Larsen, C.S. Jacobsen, C. Bouton and N. Thorpe; *J. Am. Chem. Soc.*, 1997, **119**, 1208.

97JA2954 N. Etkin, M.C. Fermin and D.W. Stephan; *J. Am. Chem. Soc.*, 1997, **119**, 2954.

97JA12432 K. Brandt, I. Porwolik-Czomperlik, M. Siwy, T. Kupta, R.A. Shaw, D.B. Davies, M.B. Hursthouse and G.D. Sykara; *J. Am. Chem. Soc.*, 1997, **119**, 12432.

97JCS(CC)305 S.J. Black, M.D. Francis and C. Jones; *J. Chem. Soc., Chem. Commun.*, 1997, 305.

97JCS(D)4321 S.J. Black, D.E. Hibbs, M.B. Hursthouse, C. Jones, K.M.A. Malik and R.C. Thomas; *J. Chem. Soc., Dalton Trans.*, 1997, 4321.

97JGU1597 R.I. Tarasova, N.I. Sinitsyna and R.R. Muthamedieva; *J. Gen. Chem. USSR*, 1997, **67**, 1597.

97JOC4827 S. Arumugam and J.G. Verkade; *J. Org. Chem.*, 1997, **62**, 4827.

97JOM(529)215 P. Binger, S. Leininger, M. Regitz, U. Bergstrasser, J. Bruckmann and C. Kruger; *J. Organomet. Chem.*, 1994, **529**, 215.

97M1 U. Salzner, S.M. Bachrack and D.C. Mulhearn; *J. Comput. Chem.*, 1997, **18**, 198.

97M2 K. Eckstein, M. Schuster and K. Karaghiosoff; *Main Group Chem. News*, 1997, **4**, 10.

97M3 U. Winkler, N. Schieck, H. Pritzkow, M. Diess, I. Hyla-Kryspin, H. Lange and R. Gleiter; *Chem. Eur. J.*, 1997, **3**, 874.

97PS(123)301 B.A. D'Sa and J.G. Verkade; *Phosphorus Sulfur*, 1997, **123**, 301.

97PS(128)1 X.-F. Zhu and Y.-G. Li; *Phosphorus Sulfur*, 1997, **128**, 1.

97TL6457 R. Echarri, M.I. Matheu, C. Claver, S. Castillas, Alverez-Larena and J.F. Pinrella; *Tetrahedron Lett.*, 1997, **386**, 6457.

97USP5698737 J.G. Verkade, T. Mohan and A.E. Wroblewski; *US Pat.*, 1997, 5698737.

98AG(E)1233 F. Tabellion, A. Nachbauer, S. Leininger, C. Peters, F. Preuss and M. Regitz; *Angew. Chem. Int. Ed. Engl.*, 1998, **37**, 1233.

98HAC183 J. Heinicke, K.R. Kadyrova and P.G. Jones; *Heteroat. Chem.*, 1998, **9**, 183.

98IC6093 S. Flanagan, H.A. Laten and W.S. Rees Jr.; *Inorg. Chem.*, 1998, **37**, 6093.

98JAP316813 Y. Ikeda and T. Matsumura; *Jpn. Kokai Tokkyo Koho JP 11*, 199721 (*Chem. Abstr.*, 1999, **130**, 82318).

98JCS(CC)741 X.Y. Mbianda, T.A. Modro and P.H. Van Rooyen; *Chem. Commun. (Cambridge)*, 1998, 741.

98JCS(P1)1643 R.W. Alder, C. Ganter, M. Gil, R. Gleiter, C.J. Harris, E. Stephanie, H. Lange, A.G. Orpen and P.N. Taylor; *J. Chem. Soc., Perkin Trans. 1.*, 1998, 1643.

98JCS(P1)1657 R.W. Alder, D.D. Ellis, R. Gleiter, C.J. Harris, H. Lange, A.G. Orpen and P.N. Taylor; *J. Chem. Soc., Perkin Trans. 1.*, 1998, 1657.

98JCS(P2)83 S.A. Bourne, V. Mbianda, T.A. Modro, L.R. Nassimbeni and H. Wan; *J. Chem. Soc., Perkin Trans. 2.*, 1998, 83.

98JOC5692 K. Ishihara, Y. Karumi, S. Kondo and H. Yamamoto; *J. Org. Chem.*, 1998, **63**, 5692.
98M1 P.B. Hitchcock, M.F. Lappert and M. Layh; *Eur. J. Inorg. Chem.*, 1998, 751.
98M2 H. Lianhnian, Z. Renxi, C. Fei and L. Alhong; *J. Nat. Sci.*, 1998, **3**, 126.
98M3 J. Barluenga, M. Tomas, K. Bieger, S. Garcia-Granda and R. Santiago-Garcia; *Eur. J. Org. Chem.*, 1998, 1425.
98M4 F. Breitsamer, K. Polborn and A. Schmidpeter; *Eur. J. Inorg. Chem.*, 1998, 1907.
98M5 P. Binger, S. Stutzmann, J. Bruckmann, C. Krueger, J. Grobe, D. Le Van and T. Pohlmeyer; *Eur. J. Inorg. Chem.*, 1998, 2071.
98M6 N. Avarvari, L. Ricard, F. Mathey, P. Le Floch, O. Loeber and M. Regitz; *Eur. J. Org. Chem.*, 1998, 2039.
98M7 M. Schmitz, R. Goeller, U. Bergstraesser, S. Leininger and M. Regitz; *Eur. J. Inorg. Chem.*, 1998, 227.
98M8 K.W. Pankiewicz, K. Lesiak and K.A. Watanabe; *PCT Int. Appl. WO*, 1998, 15563.
98M9 V. Mouries, F. Mercier, L. Ricard and F. Mathey; *Eur. J. Org. Chem.*, 1998, 2683.
98M10 F. Mathey, F. Mercier, M. Spagnol, F. Robin and V. Mouries; *PCT Int. Appl. WO*, 1998, 39345; *ibid*, 47475 (*Chem. Abstr.*, 1999, **131**, 228838).
98M11 A. Hoffmann, A. Mack, G. Goddard, P. Binger and M. Regitz; *Eur. J. Inorg. Chem.*, 1998, 1597.
98M12 Z. Peng and Y-Ou; *Huaxue Tongbao*, 1998, 39.
98OM4014 R.J. Boyd, N. Burford and C.L.B. McDonald; *Organometallics*, 1998, **17**, 4014.
98PS(142)27 D. Houalla, C. Lopez, T. Gospodova-Ivanova and Y. Stefanovski; *Phosphorus Sulfur*, 1998, **142**, 27.
98S1305 A. Mack, E. Pierron, T. Allspach, U. Bergstraesser and M. Regitz; *Synthesis*, 1998, 1305.
98ZAAC(624)1886 M. Driess, M. Reissgys and H. Pritzkow; *Z. Anorg. Allg. Chem.*, 1998, **624**, 1886.
99AG(E)1623 J. Liedtke, S. Loss, G. Alcaraz, V. Gramlich and H. Grutzmacher; *Angew. Chem. Int. Ed. Engl.*, 1999, **38**, 1623.
99AG(E)1646 A. Schnepf, G. Stosser, D. Carmichael, F. Mathey and H. Schnockel; *Angew. Chem. Int. Ed. Engl.*, 1999, **38**, 1646.
99DOK(366)202 E.E. Nifant'ev, A.T. Teleshev, A.A. Zhdanov, V.K. Bel'skii, D. Weber and W.D. Habicher; *Dokl. Akad. Nauk SSSR*, 1999, **366**, 202.
99HAC243 K. Polborn and A. Schmidpeter; *Heteroat. Chem.*, 1999, **10**, 243.
99IC2248 N. Burford and D.J. LeBlanc; *Inorg. Chem.*, 1998, **38**, 2248.
99JA1658 D.E. Berning, K.V. Katti, C.L. Barnes and W.A. Volkert; *J. Am. Chem. Soc.*, 1999, **121**, 1658.
99JAP199721 J. Tan; *Jpn. Kokai Tokkyo Koho JP 11*, 199721 (*Chem. Abstr.*, 1999, **131**, 117075); M. Katayama *ibid*, 256023 (*Chem. Abstr.*, 1999, **131**, 229582).
99JCR(S)2240 N. Oget, F. Chubura, H. Handel and L. Toupet; *J. Chem. Res., (S)*, 1999, 2240.
99JCS(CC)771 W. Uhl and M. Benter; *Chem. Commun. (Cambridge)*, 1999, 771.
99JCS(CC)853 Z. He, T.A. Modro, P.H. Van Rooyen and S.A. Bourne, *Chem. Commun. (Cambridge)*, 1999, 853.
99JCS(CC)901 V. Gee, O.A. Guy, H. Phetmung, P.G. Pringle and R.I. Pugh; *Chem. Commun. (Cambridge)*, 1999, 901.
99JCS(D)2231 P. Kilian, J. Marek, R. Marek, J. Tousec, O. Humpa, A.M.Z. Slawin, J. Touzin, J. Novosad and J.D. Woolins; *J. Chem. Soc., Dalton Trans.*, 1999, 2231.
99JOC587 A. Mack, U. Bergstraesser, G.J. Reiss and M. Regitz; *J. Org. Chem.*, 1999, **64**, 587.
99JOC3086 P. Ilankumaran and J.G. Verkade; *J. Org. Chem.*, 1999, **64**, 3086.
99JOC3090 P. Kisanga, D. McLeod, B. D'Sa and J.G. Verkade; *J. Org. Chem.*, 1999, **64**, 3090.
99JOC6459 Z. Wang, P. Kisanga and J.G. Verkade; *J. Org. Chem.*, 1999, **64**, 6459.
99M1 W. Fiedler, O. Loeber, U. Bergstrasser and M. Regitz; *Eur. J. Org. Chem.*, 1999, 363.
99M2 Y. Le, Z. Zhon, Y. Ou and B. Chen; *J. Energ. Mater.*, 1999, **17**, 297.
99M3 M. Avarvari, N. Maigrot, L. Ricard, F. Mathey and P. Le Floch; *Chem. Eur. J.*, 1999, **5**, 2109 (*Chem. Abstr.*, 1999, **131**, 199742).
99M4 D.A. Efremov, P.M. Zavlin and J.C. Tebby; 'Phosphoric Anhydride: Structure, Chemistry and Applications', Publ. Wiley 1999 (ISBN 0471-98508-2).
99M5 M. Spagnol, F. Mathey, F. Mercier and F. Robin; *PCT Int. Appl. WO*, 1999, 47475.
99S1642 S.M.F. Asmus, U. Bergstraesser and M. Regitz, *Synthesis*, 1999, 1642.

6.3
Compounds with Phosphorus at the Spiro Position

J.C. TEBBY
Staffordshire University, Stoke-on-Trent, Staffordshire, UK

6.3.1 INTRODUCTION

6.3.1.1 Structures and Nomenclature

Four coordinate phosphorus can occupy a spiro position to give compounds which are analogous to the corresponding quaternary ammonium compounds in that they may be salts, e.g. phosphonia-spiro(*m*,*n*)alkanes (**1**), or mesoionic, e.g. an ylide. In addition, four coordinate phosphorus may be anionic as in the phosphoranides (**2**) which are readily available by deprotonation of hydrophosphoranes (**3**; Y = H). The spirophosphoranes (**3**) have been a major attraction to many phosphorus chemists. There are also a number of six coordinate spiro phosphorus compounds which may also be a salt, e.g. anion (**4**), or mesoionic. The six coordinate state permits the formation of a third ring, e.g. anion (**5**). In these valence expanded states, the compounds are stabilised by small rings (usually five membered rings) and by the presence of electronegative atoms or groups bound to phosphorus, either as part of a ring(s) and/or as the non-cyclic group(s). For all the coordination states there may be bridges between the spiro rings such as (**6**) and thus there exist examples of phosphorus analogues of fenestranes such as (**7**) and the corresponding phosphoranes 〈87CRV399〉. A further extension to the fenestranes are the hexacyclic six coordinate compounds (**8**). Examples of additional structural types for the five and six coordinate states are the phosphoranes (**9**) and perphosphoranides (**10**). Phosphorus anions and cations which have appropriately positioned functional groups exhibit dative type bonds which are not true covalent bonds but bring the atoms involved into close proximity, i.e. within the sum of the Van der Waals distances. Compounds which have such interactions completing the spiro system have been included in this chapter. These include phosphenium cation (**10b**; R = H, Ph) 〈97JOM(529)59〉, cyclophosphazene (**114**), phosphoranes (**116**), (**117**), (**121**; Y = Plc, Z = Don) and bis phosphorane (**97**) and sulphone (**120**).

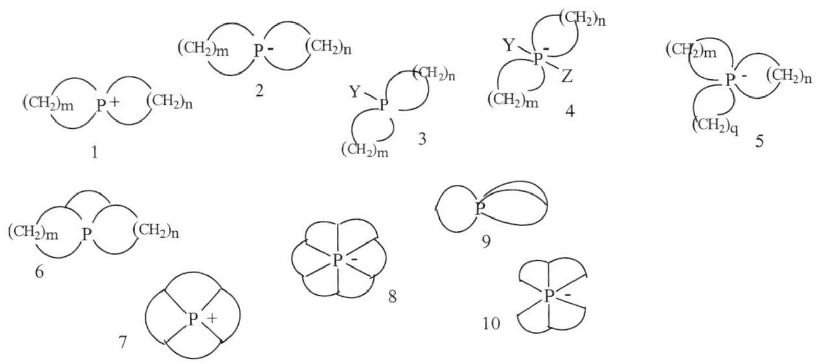

As with the nitrogen analogues there are no fully conjugated compounds, but there are many examples of spiro phosphorus atoms which are part of five and six membered rings possessing six p electrons, e.g. the cyclotriphosphazene (**11**). However in these coordination states the phosphorus atom is not able to transmit the potential conjugation and the compounds show no evidence of aromaticity. Organophosphorus chemistry is renowned for its nomenclature problems. One striking fundamental nomenclature problem relates to the six coordinate compounds R_6P^- which have been called phosphates and causing confusion with four coordinate (ortho)-phosphates. However, IUPAC rules indicate that the six coordinate compounds R_6P^- should be named perphosphoranides, the four coordinate anions R_4P^- being phosphoranides. Unfortunately there exists a six coordinate aza derivative which bears an overall positive charge to add to the potential nomenclature confusion. There are many compounds which have a variety of heterocyclic atoms and structural features which makes the formal IUPAC system unduly complex. As a consequence, in this review the more common ring systems have been abbreviated to a set of three letters to represent the atoms and groups making up the ring apart from the central phosphorus atom, e.g. the glycol group OCH_2CH_2O is abbreviated to Glc. A list of abbreviations is given below (Figure 1).

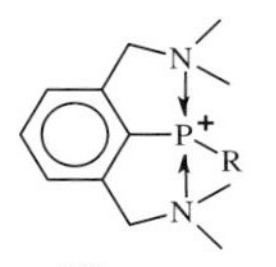

10b

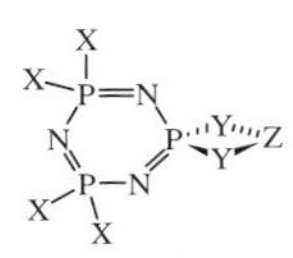

11

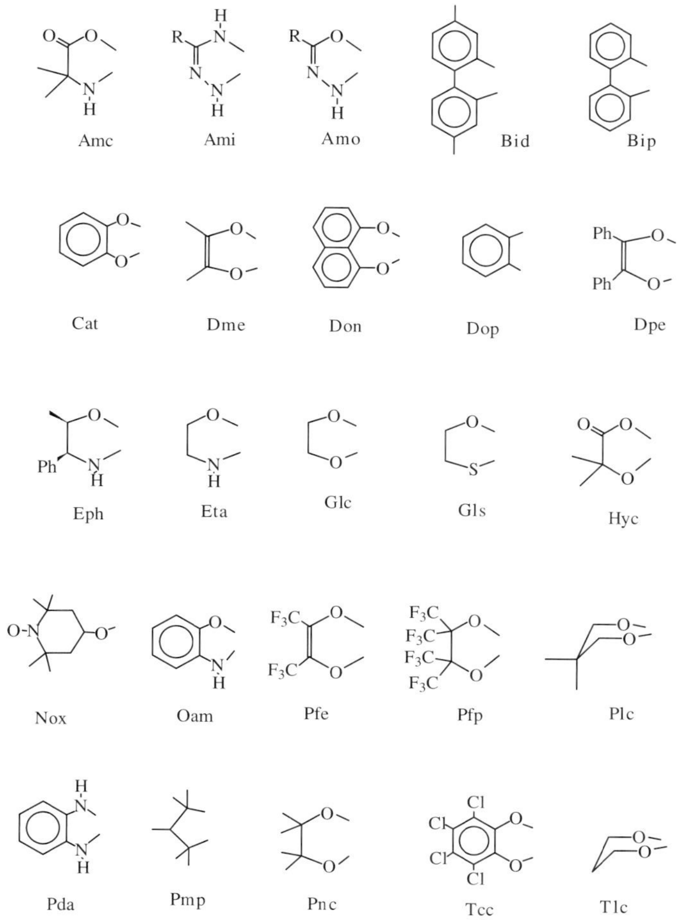

Figure 1. Structural abbreviations used in this chapter.

6.3.1.2 Tables of Structural Types

Examples of the range of cyclic systems are given in the tables below, *k*, *l*, *m* and *n* representing the sizes of the spiro rings and in some cases bicyclic rings. The four, five and six coordinate compounds will normally be reviewed in this sequence. There are very few alicyclic compounds which have phosphorus as the only hetero atom. There are more examples of benzo derivatives, but the vast majority of compounds incorporate a number of additional hetero atoms within the ring systems.

The five coordinate spirophosphoranes have played an important role with respect to understanding the stereochemistry of four coordinate phosphorus reactions which proceed via phosphorane intermediates. The spiro ring system has been used to stabilise transient reaction intermediates of acyclic compounds, e.g. hydrolysis intermediates such as hydroxyphosphoranes.

Table 1. Phosphoniaspiro compounds with P-bound atoms = C_4

m	*n*	*Compound*	*Reference*
4	4	5-phosphoniaspiro(4,4)nonane triiodide (**1**; $m = n = 4$)	⟨68ZOB331⟩
4	4	5-phosphonia(*b,e*)dibenzospiro(4,4)nonane (**12**)	⟨73JHC395⟩
4	4	5-phosphonia(*a,c,d,f*)tetrabenzospiro(4,4)nonane (**13**; R = H)	⟨69JOC1130⟩
4	4	5-phosphonia(*a,c,d,f*)tetrabenzospiro(4,4)nonane (**13**; R = Me)	⟨73RTC1308⟩
4	4	(**14**)	⟨72CC404⟩
5	5	6-phosphoniaspiro(5,5)undecane bromide (**1**; $m = n = 5$)	⟨80ZN(B)990⟩
5	5	6-phosphoniaspiro(5,5)undecane ylide (**15**)	⟨81CB3161⟩
5	5	tetra aminoalkyl phosphonium perchlorate (**16**)	⟨97MI-1⟩
5	5	urea (**17**)	⟨81PS(10)207⟩
5	5	dibenzophosphorin (**18**)	⟨75JOC766⟩

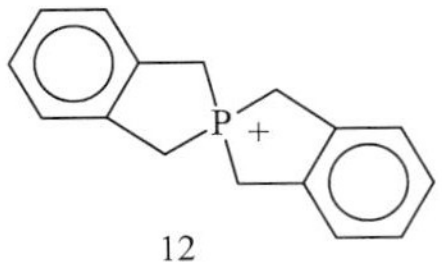
12

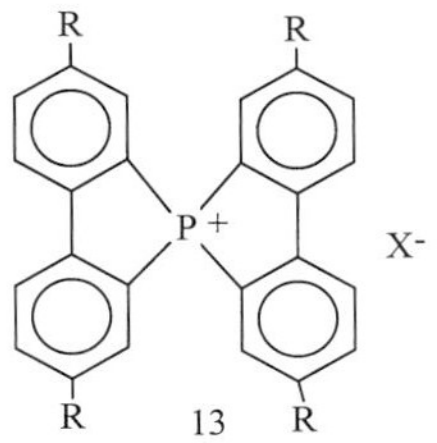
13

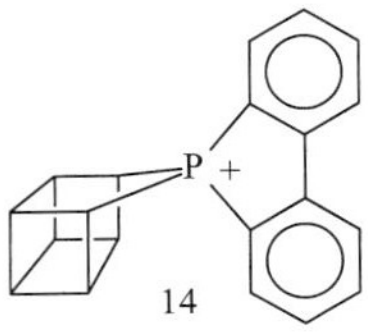
14

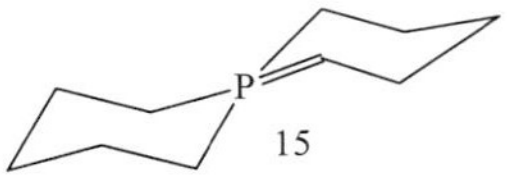
15

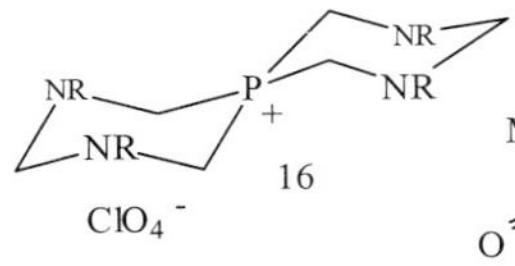
16

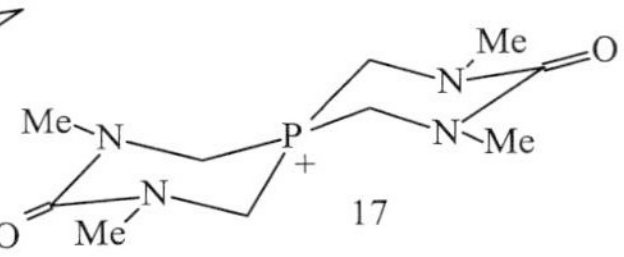
17

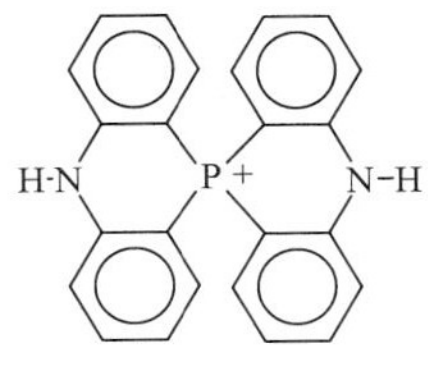
18

Table 2. Phosphoniaspiro compounds with P-bound atoms = C_2XY

m	*n*	*XY*	*Compound*	*Reference*
4	4	O_2	benzodioxaphospholan cation (**19**)	⟨78JOC571⟩
4	5	N_2	cyclotriphosphazene (**20**)	⟨90JCS(D)2303⟩
4	5	N_2	cyclotriphosphazene (**21**; X = Y = Z = Bip)	⟨94JCR(M)887⟩
4	5	N_2	cyclotriphosphazene (**21**; X = Bip, Y = Ph_2, Bip)	⟨97JCS(P2)15⟩
4	5	O_2	1,3,2-dioxaphospholene cation (**22**)	⟨88ZAAC(561)49⟩

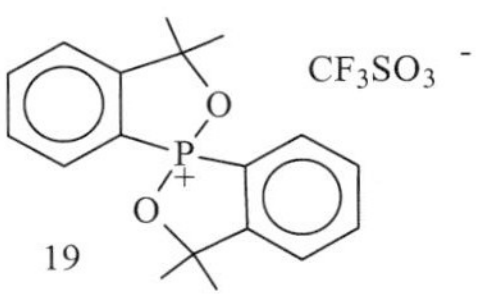
19

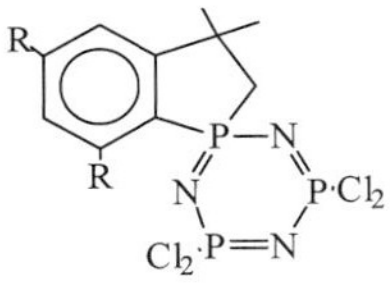
20

21
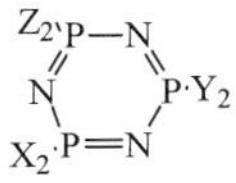

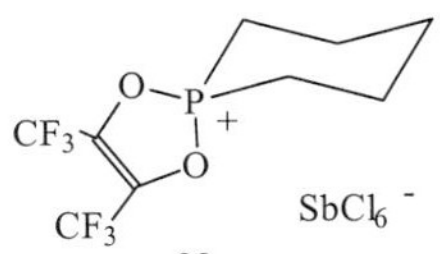
22

Table 3. Phosphoniaspiro compounds with four P-bound nitrogen

m	*n*	*P-bound*	*Compound*	*Reference*
3	3	N_4	(**23**; Y = Z = CO)	⟨87JGU2143⟩
3	3	N_4	(**23**; Y = Me_2Si, Z = P(NR)NHR)	⟨87CB1183⟩
3	4	N_4	(**24**)	⟨74JCS(D)2153⟩
3	5	N_4	(**23**; Y = NH, Z = Me_2Si)	⟨77MI-2⟩
4	4	N_4	(**25**; Y = N=CMe)	⟨86TL2971⟩
4	4	N_4	(**25**; Y = C_6H_4)	⟨91TL501⟩
4	5	N_4	(**11**; Y = NMe, Z = $[CH_2]_2$)	⟨77CB3231⟩
4	5	N_4	(**11**; Y = NMe, Z = C_6H_4)	⟨72JGU1490⟩
4	5	N_4	cyclotriphosphazene (**26**; W = NH, X = Cl, Y = Dop, Z = absent)	⟨94IC3091⟩
5	5	N_4	(**11**; Y = NMe, Z = $PCl_2NMePCl_2$)	⟨77CB3231⟩
5	5	N_4	cyclotriphosphazene (**27**; X = Cl, NR_2)	⟨94ZAAC(620)1434⟩
5	5	N_4	cyclotriphosphazene (**28**)	⟨75CB1454⟩
5	5	N_4	cyclotriphosphazene (**29**)	⟨94JCS(D)577⟩

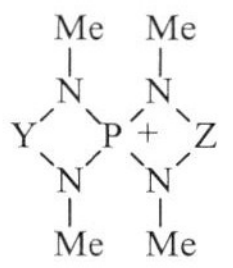
23
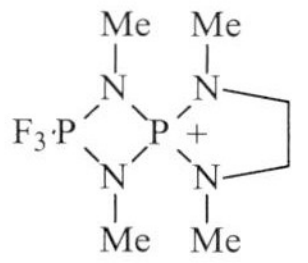
24
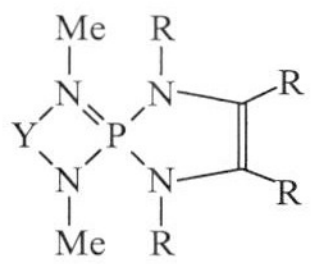
25
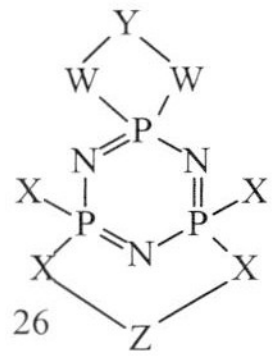
26

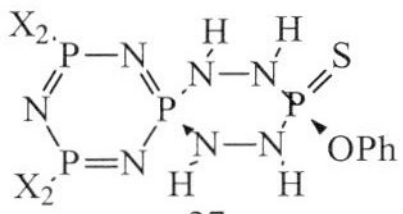
27
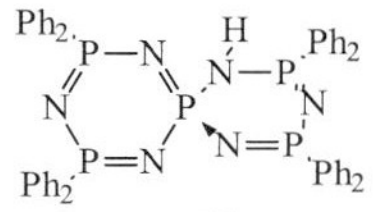
28
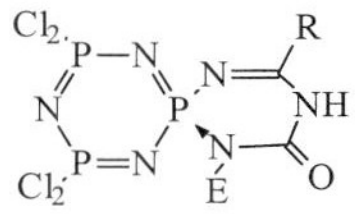
29

Table 4. Phosphoniaspiro compounds with P-bound oxygen and sulphur atoms

m	*n*	*P-bound*	*Compound*	*Reference*
4	4	CO_3	(**30**)	⟨84PS(21)249⟩
4	4	NO_3	phosphazene (**31**)	⟨82JGU189⟩
4	4	N_2O_2	(**32**; X = O, Y = CMe=CMe)	⟨84TL5521⟩
4	4	N_2O_2	phosphenium salt (**33**: R = Bu^t)	⟨94TL165⟩
4	4	S_4	(**34**; R = $[CH_2]_2$)	⟨83ZN(B)1046⟩
4	4	S_4	(**34**; R = C_6H_4)	⟨73CC144⟩
4	5	N_2O_2	(**11**; Y = O, Z = $[CH_2]_2$)	⟨76ZAAC(426)275⟩
4	5	N_2O_2	(**11**; Y = O, Z = $[C_6H_4)$	⟨76JA5120⟩
4	5	N_2S_2	(**32**; X = S, Y = $[CH_2]_3$)	⟨76T2039⟩
4	5	N_2S_2	(**26**; W = S, X = Cl, Y = Dop, Z = absent)	⟨94IC3091⟩
5	5	N_2O_2	(**26**; W = O, X = Cl, Y = CH_2CH_2CHMe, Z = absent)	⟨71IC1643⟩
5	5	N_2O_2	(**11**; Y = O, Z = 1,8-dinaphthyl)	⟨71IC1643⟩
5	5	S_4	(**34**; R = $[CH_2]_3$)	⟨83ZN(B)1046⟩
5	6	N_2O_2	(**11**; Y = O, Z = Bid)	⟨76JA5120⟩
5	6	N_2O_2	(**26**; W = O, X = OAr, Y = Bip, Z = absent)	⟨95MM6323⟩
5	7	N_2O_2	(**26**; W = O, X = F, Y = $CH_2(CF_2)_3CH_2$, Z = absent)	⟨94IC2727⟩
5	7	N_2O_2	(**35**; Ch = O, R = $CH_2CMe_2CH_2$, Y = Cl)	⟨91PS(55)59⟩
5	7	N_2Ch_2	(**35**; Ch = O, S, Se; R = Ferrocene; Y = Cl)	⟨97ZAAC(623)545⟩
5	13	N_2O_2	(**26**; W = O, Y = Z = $(CH_2CH_2O)_4$)	⟨95ICA(228)187⟩

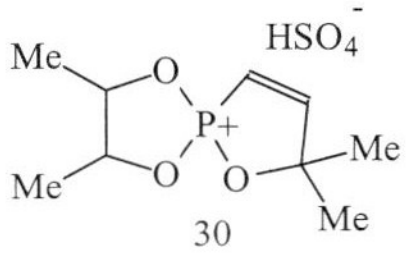
30
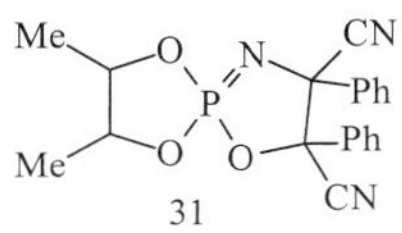
31
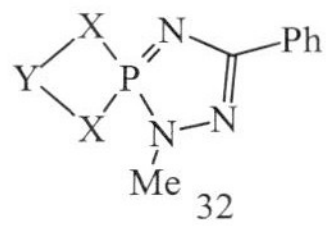
32
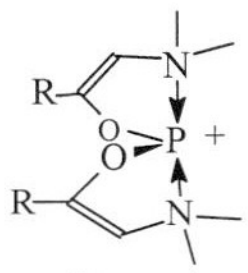
33

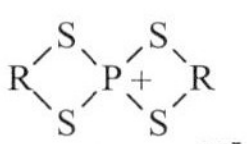
34
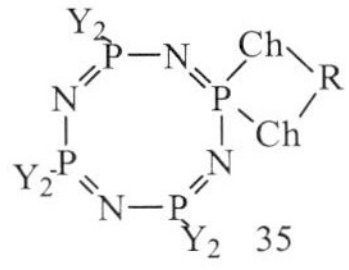
35

Table 5. Four coordinate compounds with 2 or more phosphoniaspiro centres

m	*n*	*P-bound*	*Compound*	*Reference*
3	3	N_4	(**36**)	⟨86MI-1⟩
3	4	N_3P	DBU derivative (**37**)	⟨95CB959⟩
3	5	C_2N_2	diphosphorin (**38**)	⟨93TL3107⟩
4	5	C_2N_2	cyclotriphosphazene (**39**; R = Bip)	⟨94JCR(S)134⟩
4	5	C_2N_2	cyclotriphosphazene (**39**; R = Dop)	⟨99JAP217508⟩
5	5	N_2O_2	(**40**)	⟨75ICAL40⟩
5	5	N_4	(**41**; R = $(CH_2)_3$)	⟨89JST(195)11⟩
5	5	N_4	(**39**; R = NMeDopCONMe)	⟨94ZN(B)171⟩
5	7	N_2O_2	(**35**; Ch = O, R = Bip, Y_2 = Cat)	⟨76JA5120⟩
5	7	N_4	(**42**)	⟨91PS(57)111⟩
6	7	N_2O_2	(**35**); R = Bip, Y_2 = $OC_6H_4C_6H_4O$)	⟨71IC1643⟩

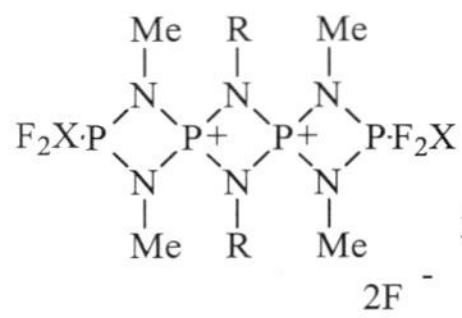
36

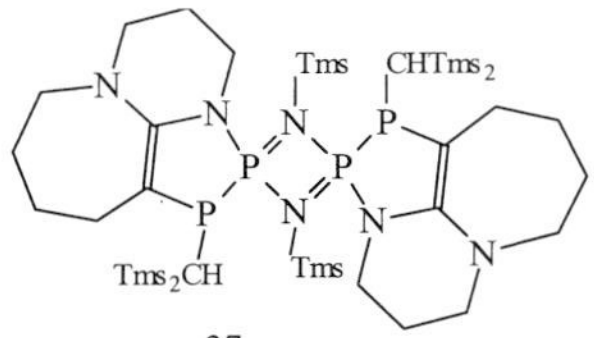
37

38
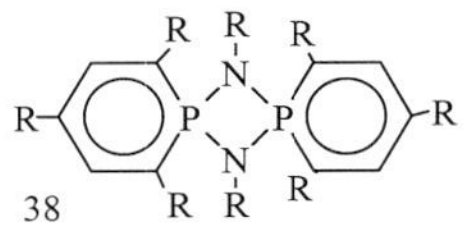

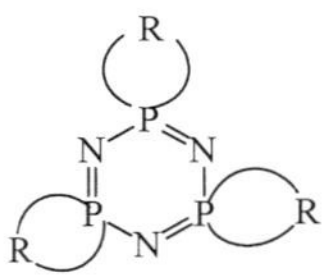
39

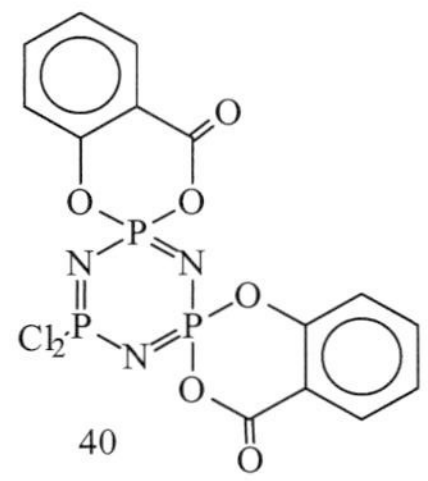
40

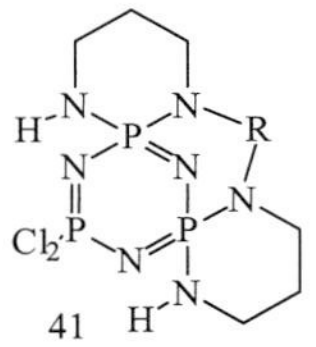
41

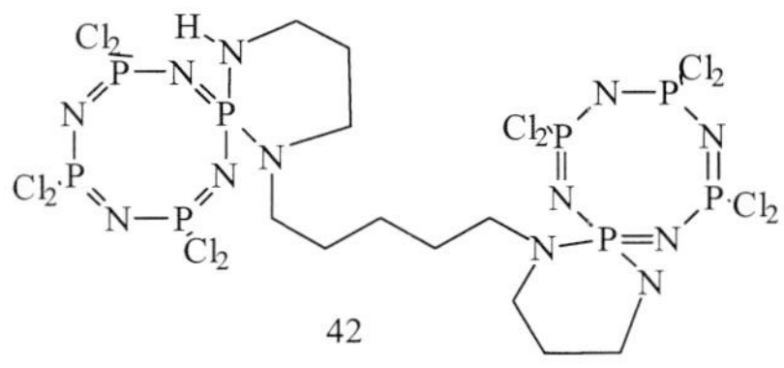
42

Table 6. Four coordinate spirophosphoranides R_4P^-

m	*n*	*P-bound*	*Compound*	*Reference*
4	4	C_2O_2	(**43**; Y = Me)	⟨78JA7434⟩

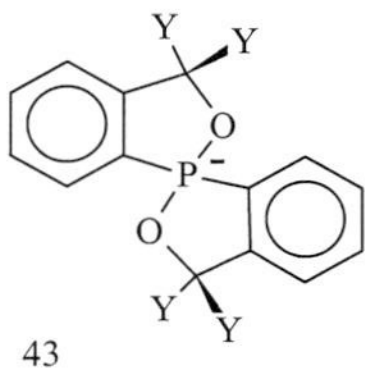
43

Table 7. Five coordinate spirophosphoranes with P-bound atoms = C_5

m	*n*	*P-bound*	*Compound*	*Reference*
4	4	C_5	(**3**; Y = Me, $m = n = 4$)	⟨72AG(E)722⟩
4	4	C_5	(**44**; R = H, Me, biphenyl-2yl)	⟨72CC404⟩
4	4	C_5	(**45**; R^1 = H, R^2 = CH=CMe$_2$, 2-furyl, 2-thienyl)	⟨70JCS(C)1425⟩
4	4	C_5	(**45**; R^1 = H, R^2 = Bu^n or C≡CBu^n	⟨70AL163⟩
4	4	C_5	(**45**; R^1 = Me, R^2 = Ph, 2-Pr)	⟨74JA5398⟩
4	4	C_5	(**46**; proposed intermediate)	⟨83TL5831⟩

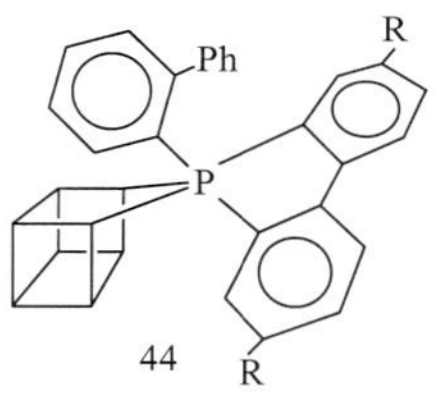
44

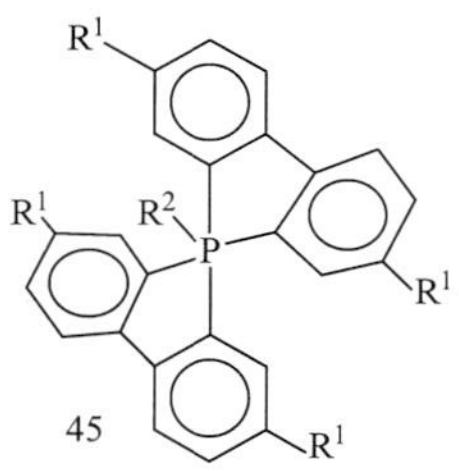
45

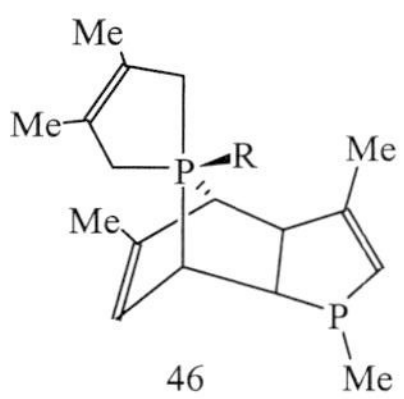
46

Table 8. Five coordinate spirophosphoranes with P-bound atoms = C_4X

m	*n*	*X*	*Compound*	*Reference*
3	4	O	Wittig intermediate (**47**)	⟨90JA3905⟩
3	5	O	(**201**; Ar = 2,4,6-triisopropylphenyl, E = CO_2Me)	⟨99CC1423⟩
4	4	H	(**45**; $R^1 = R^2$ = H)	⟨69CB528⟩
4	4	O	(**48**; $n = 2$)	⟨79CB501⟩
4	5	O	(**48**; $n = 3$)	⟨79CB501⟩

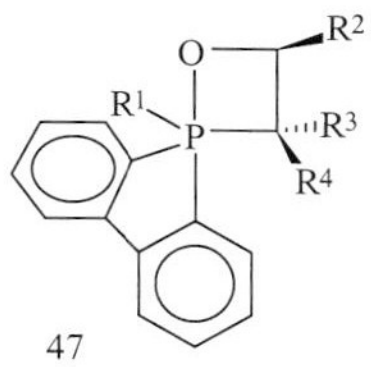
47

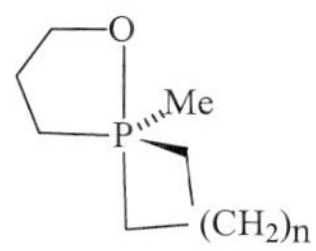
48

Table 9. Five coordinate spirophosphoranes with P-bound atoms = C_3X_2

m	*n*	X_2	*Compound*	*Reference*
2	4	O_2	(**49**; Ch = O, R = CMe_2CMe_2)	⟨78JCS(CC)854⟩
2	4	S_2	(**49**; Ch = S, R = $CCF_3{=}CCF_3$)	⟨74JA317⟩
2	4	O_2	(**50**; X = CN)	⟨90TL3429⟩
3	3	O_2	(**51**)	⟨73JCS(P1)1300⟩
3	3	O_2	(**52**; R = H, Me, Ar = Ph, C_6H_4X)	⟨97PS(124/5)379⟩
3	4	NO	(**53**; R = CF_3)	⟨97AG(E)2500⟩
3	4	O_2	(**54**; Y = CH_2CH_2, R = Ph, R^2 = Me)	⟨78IC3265⟩
3	4	O_2	(**54**; Y = $C[CF_3]_2C[CF_3]_2$, R = 4-Br-Ph, R^2 = H)	⟨74JA4143⟩
3	5	O_2	(**54**; Y = 1,3-cyclopentyl, R = Ph, R^2 = H)	⟨83JOC2621⟩
3	6	O_2	(**54**; Y = $[CH_2]_4$, R = Ph)	⟨77PS(3)51⟩
4	4	O_2	(**55**; X = O, Y = CO_2H)	⟨96PS(109/10)241⟩
4	4	O_2	(**56**; R^1 = Me, Et, R^3 = CH_2CH_2)	⟨71JGU471⟩
4	4	O_2	(**56**; R^1 = Et, R^2 = H, R^3 = C_6H_4)	⟨75JGU686⟩
4	4	O_2	(**56**; R^1 = Ph, R^2 = Me, R^3 = C_6H_4)	⟨77JA3332⟩
4	4	O_2	(**57**; Y = Z = CO, R = Me, Ph)	⟨75CC399⟩
4	4	O_2	(**57**; Y = CMe_2, Z = CO, R = Me)	⟨79JA3687⟩
4	4	O_2	(**58**; Y = CMe_2)	⟨76ZN(B)948⟩
4	4	NO	(**55**; X = NMe, Y = CONHMe)	⟨95MI-1⟩
4	5	ClO	(**59**; X = CF_3)	⟨88ZAAC(561)49⟩

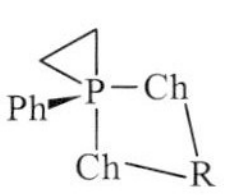
49

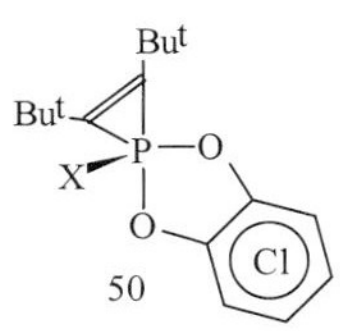
50

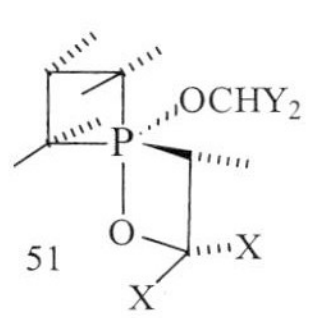
51

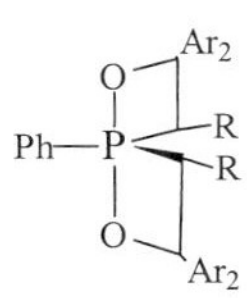
52

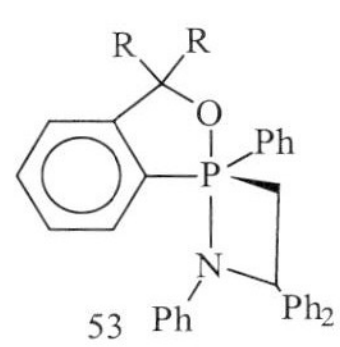
53

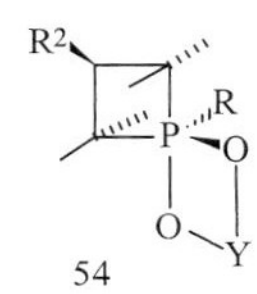
54

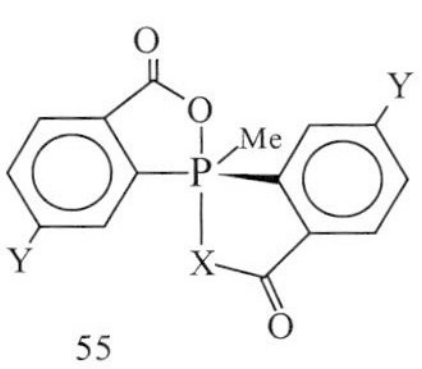
55

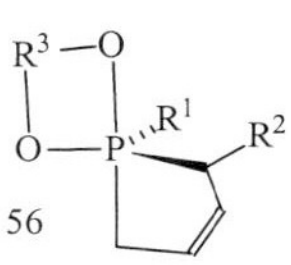
56

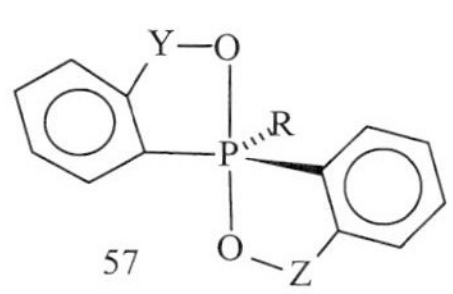
57

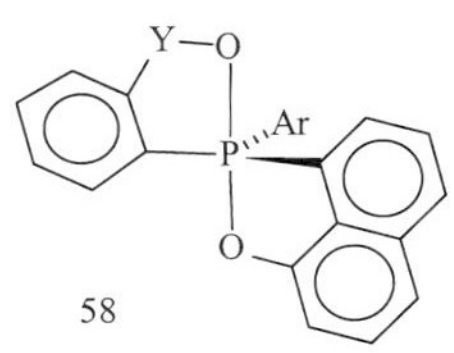
58

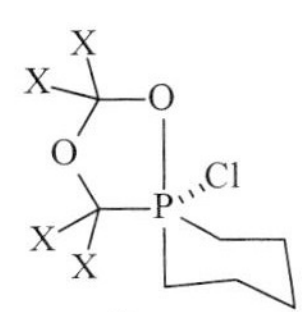
59

Table 10. Five coordinate spirophosphoranes with P-bound atoms = C_2X_3

m	*n*	X_3	*Compound*	*Reference*
2	4	O_2X	(**50**; X = Br, Cl, F, N_3)	⟨90TL3429⟩
4	4	O_3	Hydroxyphosphorane (**57**; R = OH, Y = Z = CO)	⟨79JA3687⟩
4	4	O_3	Hydroxyphosphorane (**57**; R = OH, Y = Z = CMe_2)	⟨79JA4618⟩
4	4	O_3	Hydroxyphosphorane (**57**; R = OH, Y = Z = $C(CF_3)_2$)	⟨81JA1049⟩
4	4	NO_2	(**60**; R^1 = R^2 = H, X = CH_2, Y = CHR, Z = CO)	⟨78IZV2185⟩
4	4	FO_2	(**61**)	⟨70CC476⟩
4	4	N_2O	(**62**)	⟨89PS(45)255⟩
4	5	O_2P	(**63**)	⟨88CB391⟩

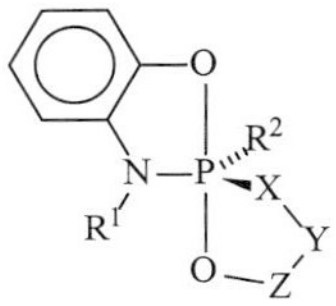

60

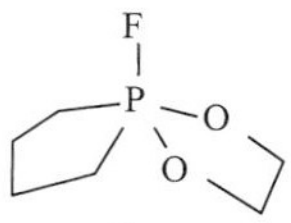

61

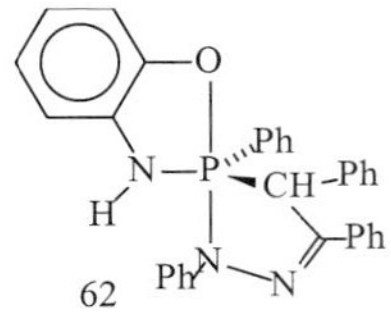

62

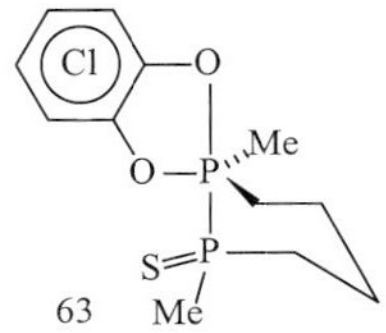

63

Table 11. Five coordinate spirophosphoranes with P-bound atoms = CX_4

m	*n*	X_4	*Compound*	*Reference*
3	3	N_4	(**64**; $R^1 = R^2 = Y = Me$)	⟨95ZAAC(621)2001⟩
3	3	N_4	(**64**; $R^1 = Me$, $R^2 = Ph$, $Y = Me$)	⟨97ZAAC(623)1325⟩
3	4	O_4	Horner Intermediate (**65**)	⟨79TL2203⟩
3	4	N_3P	(**66**; R = Me, Ar = Ph)	⟨81CB2132⟩
3	4	N_2O_2	(**67**; Y = Tcc)	⟨96PS(109/10)493⟩
3	4	N_2O_2	(**68**; R = Me, X = NHR, Y = NR, Z = PR)	⟨85JGU2327⟩
3	4	N_2O_2	(**69**; R = Me, X = NHR)	⟨94PS(86)157⟩
4	4	O_4	Cat_2PR	⟨74AX935⟩
4	4	O_4	Hyc_2PCH_2Y	⟨96PS(109/110)156⟩
4	4	O_4	(**70**; $R^1 = Ph$, $R^2 = Me$, X = OMe, Y = Ac, Z = Pnc)	⟨72CR(C)(274)288⟩
4	4	NO_3	(**68**; R = Me, Ph, X = Ph, Y = NMe, Z = CHMeCHPh)	⟨92JOC5195⟩
4	4	NO_3	(**60**; $R^1 = Ph$, $R^2 = Ar$, X = O, Y, Z = CHMeCHPh)	⟨75JCS(P1)2376⟩
4	4	N_2O_2	(**71**)	⟨77CB1124⟩
4	4	O_3S	(**72**)	⟨94MI-1⟩
4	4	O_2S_2	(**73**; X = Y = O, R = Ph)	⟨78JOM(156)253⟩
4	4	S_4	(**73**; X = Y = S; R = Ph)	⟨78JOM(156)253⟩
4	5	N_2O_2	(**74** Y = Tcc, Pfp, Z = CN, NMe_2)	⟨95PS(106)65⟩
4	5	O_4	(**75**; X, Y = O, R = Ph, Z = CPh=CPh)	⟨91IC1052⟩
4	5	O_4	(**75**; R = OMe, X = O, Y = CH_2, Z = CH=CH)	⟨77IZV2000⟩
4	6	O_2S_2	(**76**)	⟨83JOC3815⟩
4	7	O_4	(**77**; Y = Ph, Z = $(CH_2)_n$, Dop, 1,2-naphth, 1,2-*c*-Hex)	⟨79JCS(CC)191⟩
4	7	O_4	(**78**; R = Bu^t, R′ = Ph, Y = Tcc, Z = CH_2)	⟨95IC4525⟩
4	7	O_4	(**78**; R = Bu^t, R′ = Ph, Y = Cat, Z = SO_2)	⟨97PS(120-1)431⟩
5	5	O_4	(**75**; R = Ph, X = O, YZ = $CH_2CMe_2CH_2$)	⟨77JCS(P1)437⟩
5	5	O_4	(**75**; R = OAr, X = O, YZ = Bip)	⟨94JA8095⟩
6	6	O_4	(**79**; Y = Ph)	⟨92IC5494⟩

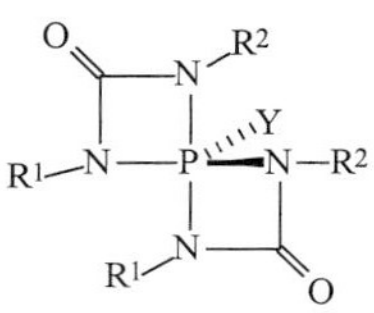

64

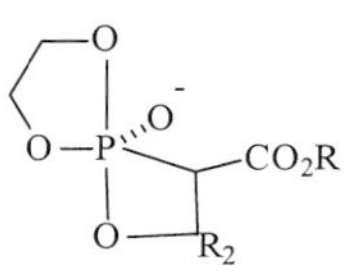

65

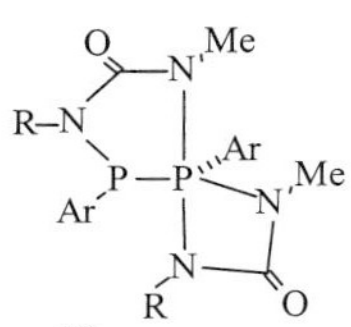

66

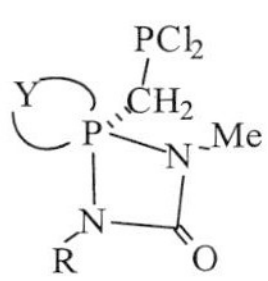

67

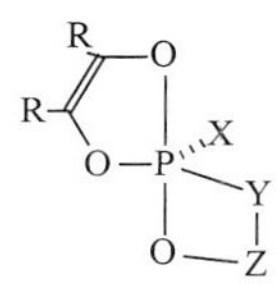

68

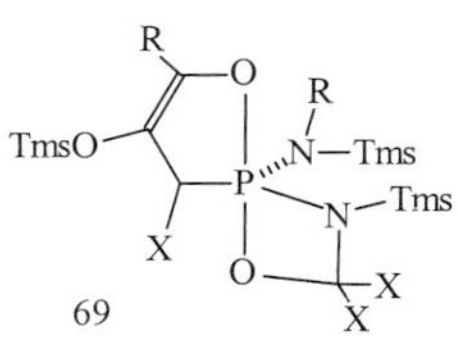

69

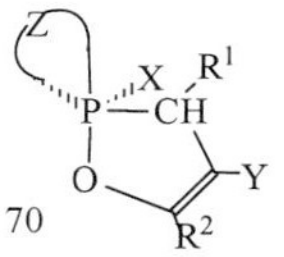

70

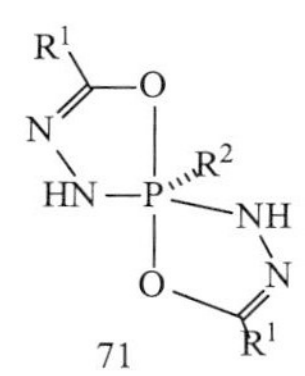

71

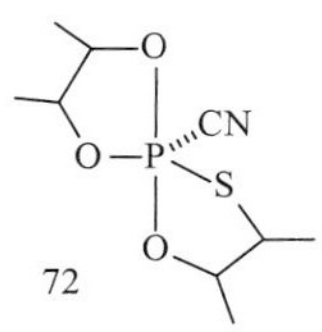

72

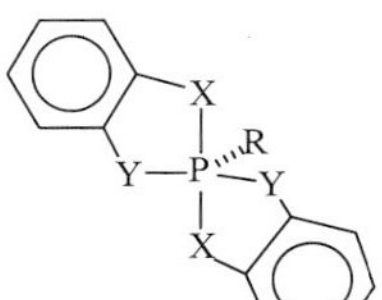

73

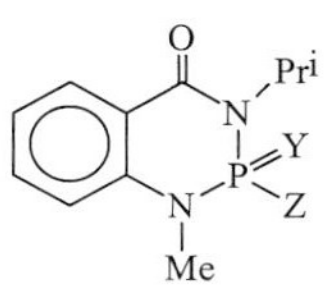

74

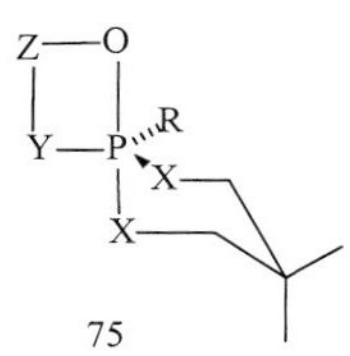

75

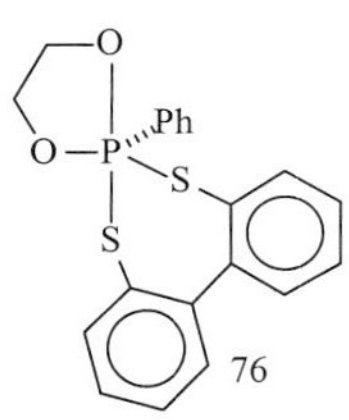

76

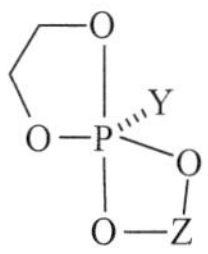
77

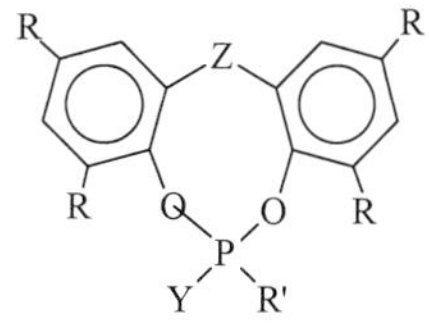
78

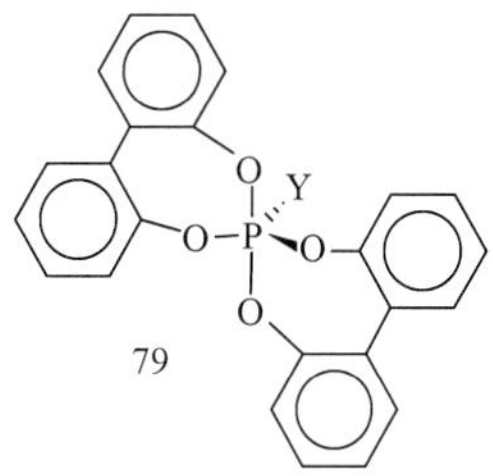
79

Table 12. Five coordinate spirophosphoranes with P-bound atoms = HX_4

m	*n*	*X_4*	*Compound*	*Reference*
4	4	C_2O_2	(**80**)	⟨97TL7753⟩
4	4	O_4	(**81**; Y = CMe_2, Z = H)	⟨79TL4205⟩
4	4	O_4	Glc_2PH	⟨95JGU1604⟩
4	4	X_2YZ	(**82**; X, Y, Z = O, S, NH)	⟨75BSF407⟩
4	4	N_2O_2	(**83**; R = H, X = O, Y = Z = NH)	⟨75JGU252⟩
4	4	N_2O_2	(**84**; X = NH, Y = CMe_2, Z = H)	⟨75CR(C)(280)809⟩
4	5	O_4	(**84**; X = O, Y = $(CH_2)_3$, Z = H)	⟨86PS(27)321⟩

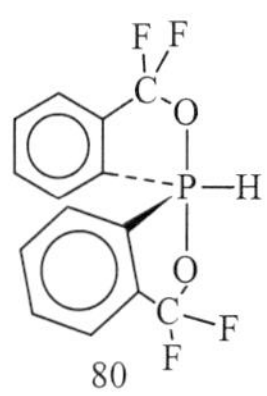
80

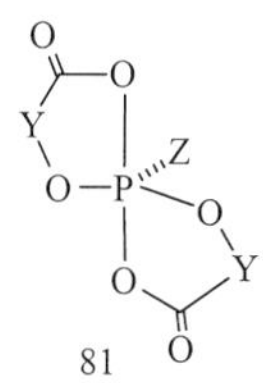
81

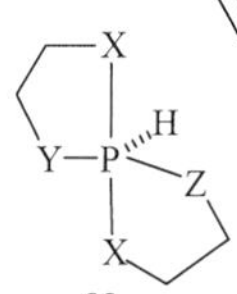
82

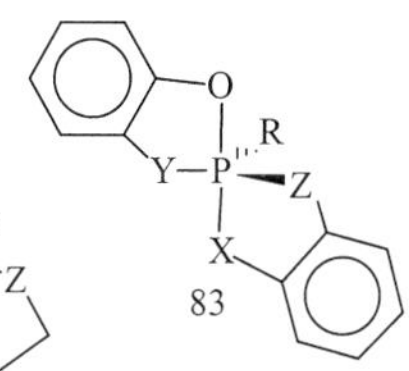
83

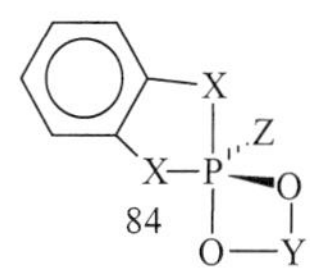
84

Table 13. Five coordinate spirophosphoranes with P-bound atoms = X_5

m	*n*	*X_5*	*Compound*	*Reference*
3	3	N_5	(**64**; R^1, R^2 = Me, Y = N_3)	⟨84JA7065⟩
3	3	N_5	(**64**; R^1, R^2 = Me, Y = $NMe(CH_2)_2NMe_2$)	⟨93PS(80)173⟩
3	3	BrN_2O_2	(**85**)	⟨97JGU151⟩
3	4	FN_2O_2	(**86**; R^1, R^2 = Me, X = Cat, Y = PF_3, Z = F)	⟨76JCS(D)306⟩
3	4	IN_4	(**87**; R^1, R^2 = Me, X^1 = NMe, X^2 = N^+Me_2, Y = (CH_2CH_2), Z = I	⟨90PS(54)227⟩
4	4	N_2O_3	(**68**; R = Ph, X = NMe_2, Y = NMe, Z = CHMeCHPh)	⟨75P(5)285⟩
4	4	NO_4	(**88**; *n* = 2, R = H, X = O, Y = O, Z = NMe_2)	⟨75TL2529⟩
4	5	NO_4	(**88**; *n* = 3, R = CF_3, X = NMe, Y = O, Z = OAr)	⟨76CC1031⟩
4	5	N_3O_3	(**89**; Y = C_6H_4, Z = O)	⟨93PS(78)271⟩
4	5	O_3S_2	intermediate (**90**)	⟨91JA7945⟩
4	7	NO_4	(**78**; Y = $NHHex^c$, Y = Cat, Z = CH_2)	⟨97IC2044⟩
4	10	O_4X	crown ethers (**84**; X = O, Y = $(CH_2CH_2O)_2CH_2CH_2$	⟨84JGU1754⟩
6	6	FO_4	(**79**; Y = F)	⟨77PS(3)35⟩

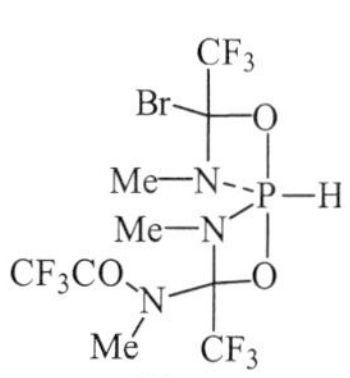
85

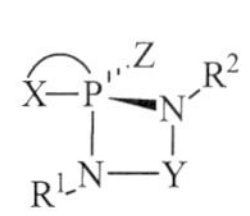
86

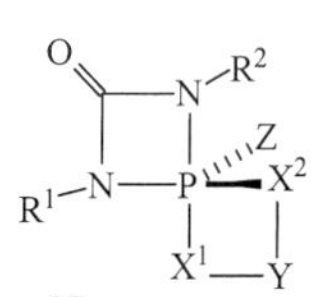
87

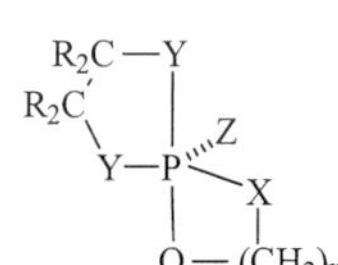
88

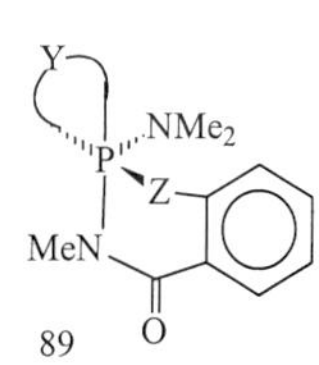
89

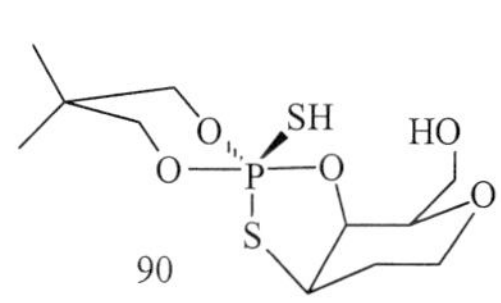
90

Table 14. Five coordinate bis spirophosphoranes

m	*n*	X_5	*Compound*	*Reference*
3	4	C_2N_3	(**91**; R = CO_2Me)	〈86TL2863〉
3	4	CN_2O_2	(**92**; R = Ph, X = Z = $CH_2CHCNCO_2R$, Y = Cat)	〈94JGU373〉
3	4	N_3O_2	(**92**; R = Ph, X = Z = N_3, Y = Glc)	〈83JGU247〉
3	4	N_3O_2	(**92**; R = Et, X = OAr, Y = Tcc, Z = NEtP$(OAr)_2$)	〈99IC1093〉
4	4	CO_4	(**93**; Y = Glc, Z = $CH_2N(CH_2CH_2)_2NCH_2$)	〈95JGU1610〉
4	4	ClN_4	(**94**)	〈70ZAAC(372)285〉
4	4	PO_4	(**93**; Y = Cat, Z = P=CBu^{t}O)	〈99JGU48〉

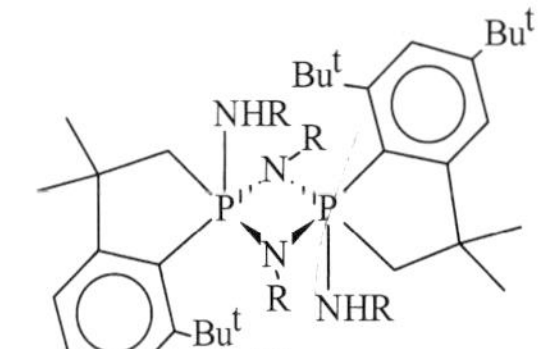
91

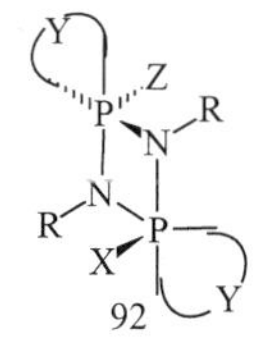
92

$Y_2P—Z—PY_2$

93

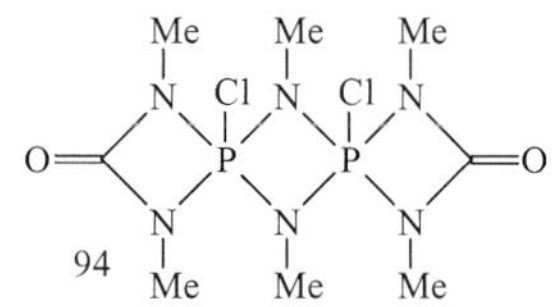
94

Table 15. Five coordinate spirophosphoranes incorporating a bridgehead phosphorus

k	*l*	*m*	*n*	X_5	*Compound*	*Reference*
4	4	2	0	C_3O_2	(**180**: Y = CF_3)	〈97CB819〉
4	3	1	0	C_2O_3	(**95**: Y = CF_3)	〈97JFC(85)129〉
3	3	3	0	C_2O_3	(**96**: Y = CF_3)	〈98HAC173〉
3	3	3	0	CN_2O_2	(**97**)	〈87JCS(CC)1131〉
5	3	2	0	CO_4	(**98**: Y = CF_3, R, R^1 = H, Alk)	〈95JA12390〉
5	3	2	0	HN_2O_2	(**99**: R^1 = Me; R^2 = H)	〈97MI-2〉
5	3	2	0	HN_2O_2	(**99**: R^1 = Me; R^2 = Ph)	〈97PS(123)5〉
5	3	2	0	HN_2O_2	(**99**: R^1 = Pr^i; R^2 = H)	〈99OM915〉
4	4	3	0	N_3O_2	(**100**: R = CH_2CH_2Cl, Y = CF_3)	〈97JFC(86)109〉
3	3	3	0	N_3O_2	(**101**)	〈78CB2086〉
4	3	3	0	NO_4	(**102**; Ch = O, R = CF_3)	〈80JA5073〉
3	3	3	1	N_4O	(**103**)	〈86PS(26)193〉
5	2	2	2	O_5	(**104**)	〈70CR(C)(271)418〉

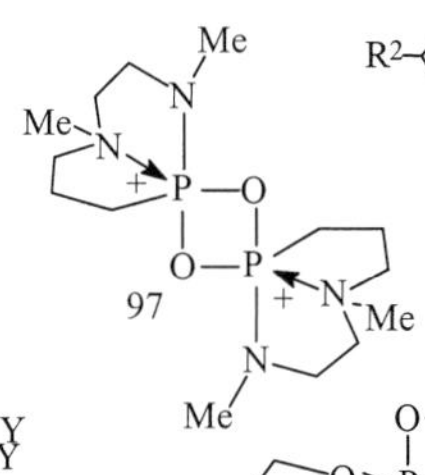
97

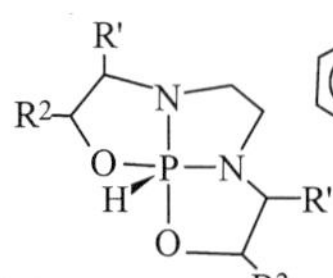
99

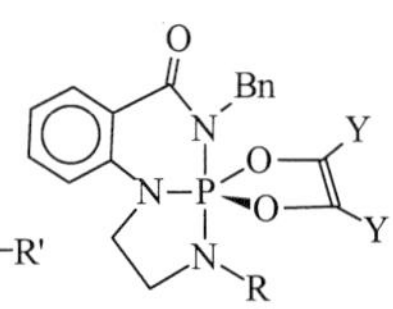
100

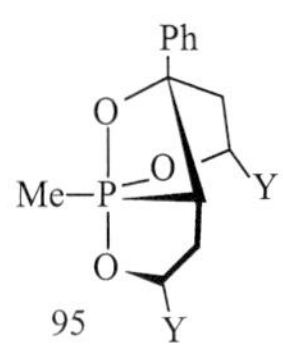
95

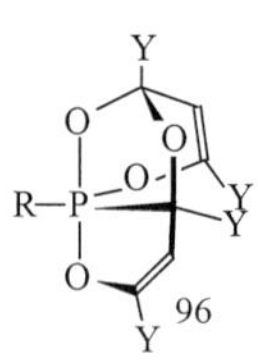
96

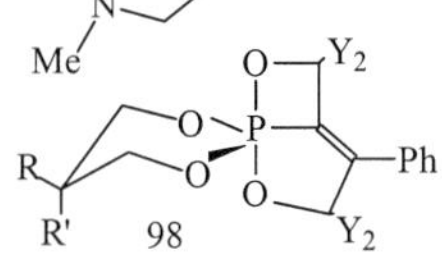
98

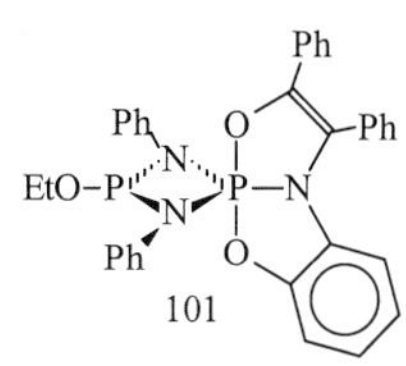
101

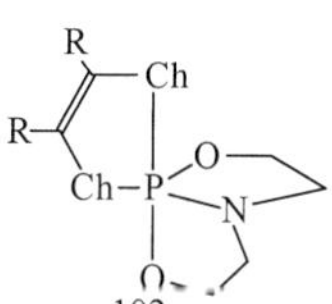
102

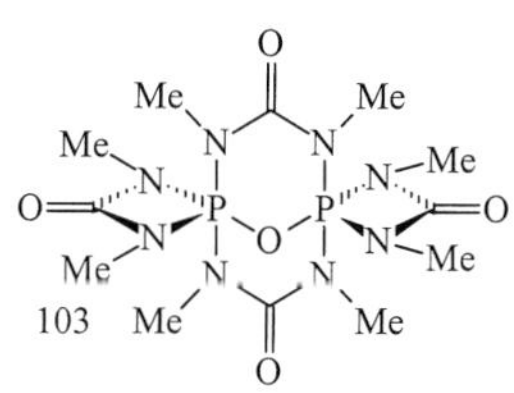
103

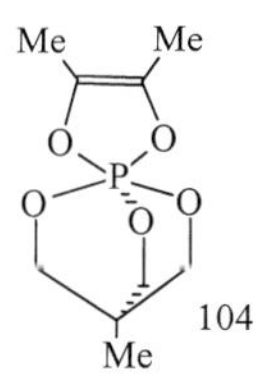
104

Table 16. Five coordinate bis spirophosphoranes incorporating a bridgehead phosphorus

k	*l*	*m*	*n*	X_5	*Compound*	*Reference*
15+	3	3	0	NO_4	(**105**: Z = $CH_2CH_2\{X\ CH_2CH_2\}_nO$, *n* = 1, 2, 3, X = O, S, NMe)	⟨96PS(114)51⟩
15	3	3	0	NO_4	(**105**: Z = $DopCH_2DopO$)	⟨93PS(83)157⟩
11	3	3	0	NO_4	(**105**: Z = $CH_2CMe_2CH_2O$)	⟨93HAC271⟩
10	3	3	0	NO_4	(**106**: X = Y = Z = CH_2CH_2O)	⟨95PS(103)199⟩
9	3	3	0	N_2O_3	(**105**: Z = CH_2CH_2NH)	⟨93PS(81)1⟩
4	3	2	0	N_4O	(**107**)	⟨81CB825⟩

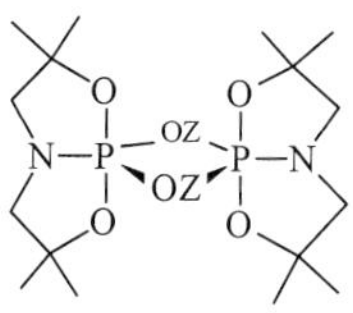
105

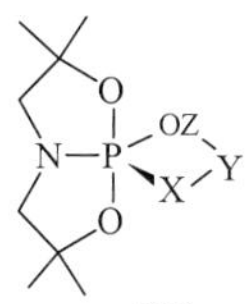
106

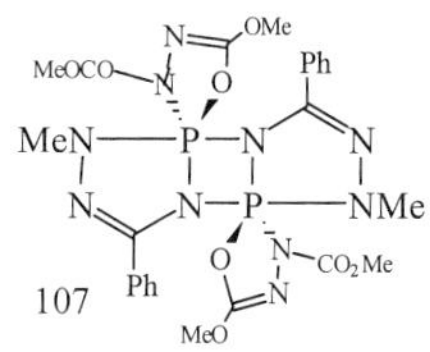
107

Table 17. Six coordinate mono and bis spiroperphosphoranides R_6P^-

l	*m*	*n*	X_6	*Compound*	*Reference*
	4	4	C_4N_2	(**108**; R = Ph, R′ = Alk, H)	⟨97PS(123)181⟩
	4	4	HNO_4	(**109**; R = Me, X = $C_5H_5N^+$, Y = H)	⟨74CR(C)(278)1353⟩
	4	4	HO_5	(**110**; R = CO_2Me, X = OCHRCHROH, Y = H)	⟨79PS(6)435⟩
	4	4	N_2O_4	(**109**; R = H, X = Y = $C_5H_5N^+$)	⟨78ZN(B)583⟩
4	4	2	C_2O_4	(**111**; X = CF_3)	⟨99JA6958⟩
4	4	3	CN_3O_2	(**112**)	⟨95TL2021⟩
4	4	3	N_2O_4	(**113**; *m* = 1, *n* = 2, R = H, X = NPh, Y = $\{PhNH\}_2P^+$)	⟨90JGU1491⟩
5	3	3	S_2N_4	(**114**)	⟨96IC6899⟩
4	4	4	C_6	(**115**; *m* = 0, *n* = 3)	⟨70JCS(B)640⟩
4	4	4	C_3N_3	(**116**)	⟨93AG(E)1430⟩
4	4	4	C_3N_3	(**117**)	⟨93AG(E)1430⟩
4	4	4	O_6	(**113**; *m* = 0, *n* = 3, R = Pr^i, Z = Et_3N^+H)	⟨76JA6755⟩
4	4	4	O_6	(**118**; *m* = 0, *n* = 3)	⟨77ZAAC(437)53⟩
4	4	4	O_6	(**113**; *m* = 0, *n* = 3, R = H, Z = Et_3N^+H)	⟨73JA3154⟩
4	4	4	O_6	$CatP(Tcc)_2$	⟨97CB1479⟩
4	4	5	CNO_4	(**119**)	⟨93PS(75)233⟩
4	4	5	O_6	(**113**; *m* = 1, *n* = 2, X = O, Y = $CMe{=}CHC^+Me$)	⟨79PS(7)305⟩
4	5	5	CO_5	(**120**)	⟨97IC2578⟩
5	5	5	O_6	(**113**; X = O, Y = Bip, Z = brucineH)	⟨79TL795⟩
6	4	4	NO_5	(**121**; Y = Z = Don)	⟨96JA9841⟩
6	5	4	NO_5	(**121**; Y = Don, Z = Plc)	⟨98POL3643⟩

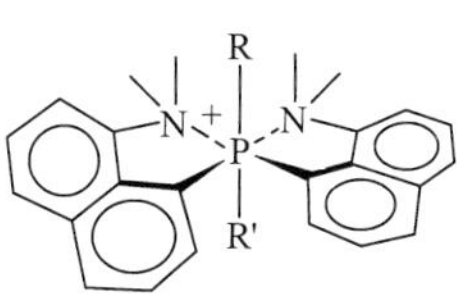
108

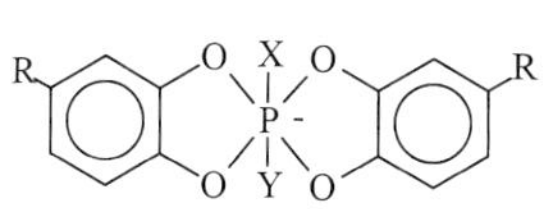
109

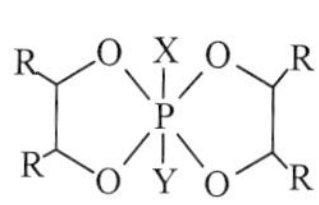
110

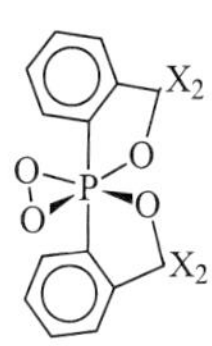
111

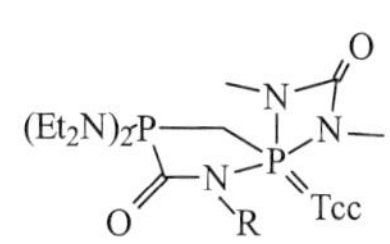
112

6.3.2 THEORETICAL METHODS

The stabilising influence on the hypervalent state of phosphorus produced when phosphorus is incorporated into small rings is very well established and is especially effective in the spiro compounds covered here. Extended Hückel MO calculations indicated that the small rings lowered the occupation of the phosphorus 3d orbitals and favoured nucleophilic attack at P leading to t.b.p. phosphoranes ⟨69JA1200⟩. Possible mechanisms for the permutational isomerism of five coordinate phosphoranes have

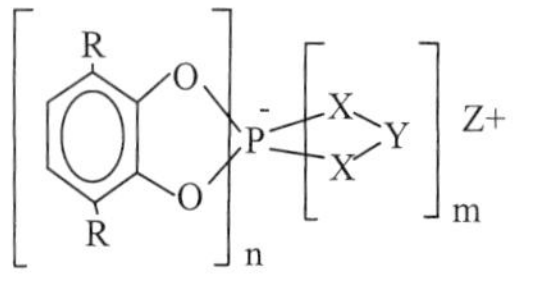

113

114

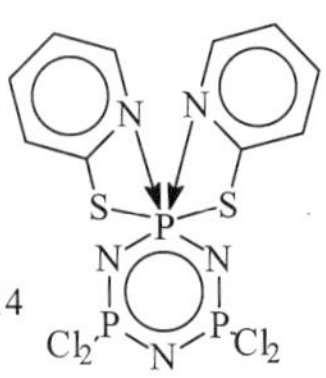

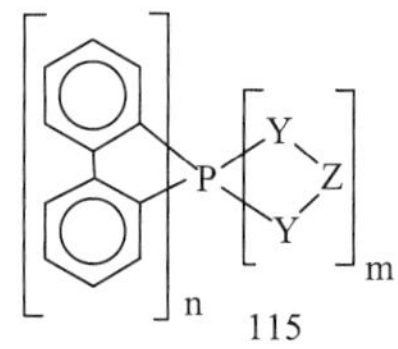

115

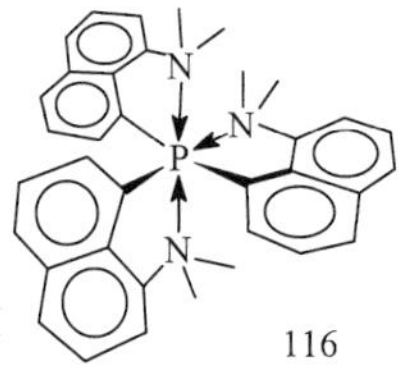

116

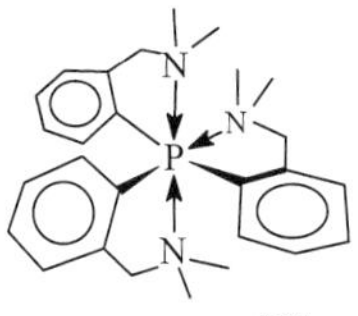

117

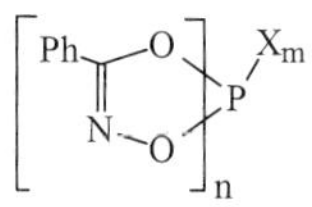

118

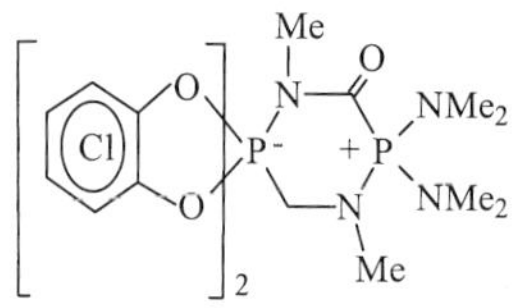

119

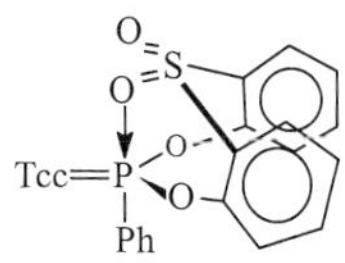

120

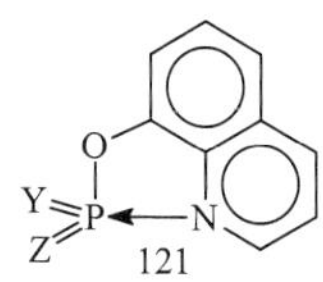

121

Table 18. Phospha-fenestranes and poly spiro phosphoranes and perphosphoranides

k	*l*	*m*	*n*	X_5	*Compound*	*Reference*
4	4	4	4	CN_4	Phospha-fenestrane (**122**; $m = 1$, $n = 1$, Y = Ar)	⟨99JFC223⟩
4	4	4	4	HN_4	Phospha-fenestrane (**122**; $m = 1$, $n = 1$, Y = H)	⟨81JA1291⟩
4	4	4	4	N_4X	Phospha-fenestrane (**122**; $m = 1$, $n = 1$, Y = F, Cl, I)	⟨81JA1291⟩
4	4	4	4	N_4O	Anhydride (**122**; $m = 2$, $n = 1$, Y = O)	⟨96JCS(CC)2157⟩
4	4	4	4	N_4S	(**122**; $m = 1$, $n = 1$, Y = SR)	⟨96T2995⟩
4	4	4	4	O_4P_2	(**123**)	⟨98AG(E)1098⟩
5	5	5	5	C_2N_4	(**124**; R = Et, R′ = H, Y = Alk)	⟨95JA8287⟩
5	5	5	4/5	N_4	Phospha-fenestrane (**216**; $n = 1$ or 2)	⟨81JA5265⟩
5	5	5	5	N_4O_2	(**124**; R = H, R′ = Ph, Y = Thymidine)	⟨99JAP29591⟩

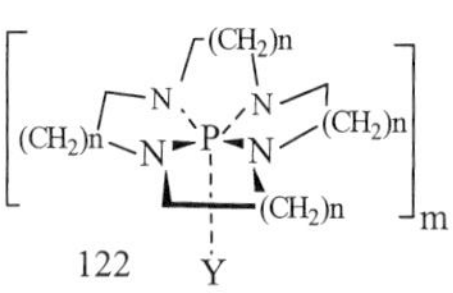

122

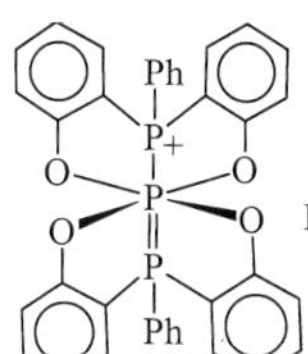

123

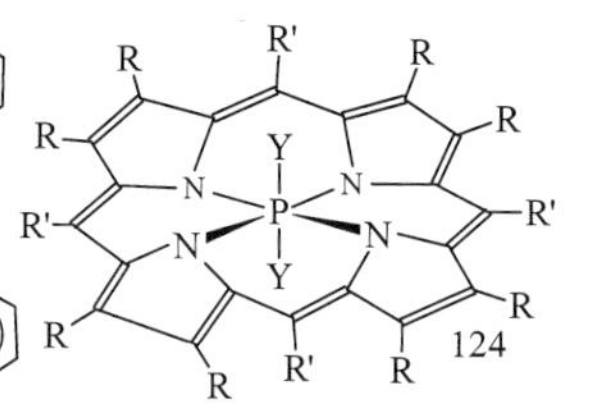

124

been explored ⟨74CC15⟩. The Berry pseudorotation mechanism, which produces an exchange of position for two apical and two equatorial ligands within a t.b.p. structure ⟨75CPL(30)276⟩ involves deformation via square pyramidal (s.p.) geometries ⟨74JA4143, 75JA5379⟩. In this process two apical bonds join two of the equatorial bonds, the remaining equatorial bond (the pivot) being transformed into the apical bond of the s.p. structure. The Berry mechanism remains the most favoured mechanism ⟨77JA3318⟩, although a turnstile mechanism ⟨74T371⟩ based on NMR studies of (**125**) has been explored. AM1 and PM3 calculations was used to support the formation of a spiro [5,2] phosphorane as an intermediate in the reaction of a phosphonite with an aldehyde ⟨97JGU1486⟩.

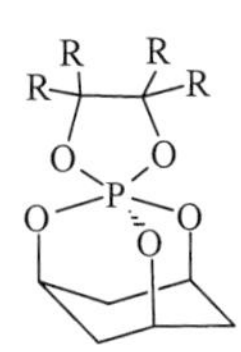

125

Molecular mechanics calculations have been extended from four coordinate phosphoryl compounds to five coordinate phosphoranes by adding a term which measures the effects of electron pair repulsions, modified by the ligand electronegativity, and taking into account the t.b.p. and s.p. conformational possibilities ⟨77JA5461⟩. Similar studies were employed to calculate the relative energies of t.b.p. and s.p.

conformations ⟨75JA5379⟩. It was concluded (1) that the s.p. conformation is stabilised relative to the t.b.p. conformation when the phosphorus atom was part of an unsaturated ring and when the phosphorus was bound to highly electronegative atoms, (2) that the t.b.p. conformation was destabilised by additional ring strain (3–4 kcal mol^{-1} for the dioxaphospholenes) produced by the need for the different bond length requirements of the t.b.p. conformation, and (3) that the s.p. conformation also accommodated four membered rings better than the t.b.p. conformation. A method for the calculation of the barriers to pseudorotation for the interconversion of the above conformations ⟨78JA433⟩ also took into account the apicophilicities of the P-bound atoms. This was achieved using an element effect based on electronegativity preferences for the t.b.p. axial, equatorial, and s.p. apical and basal orientated atoms, which were then optimised to obtain the best fit with recorded barrier energies. More recently a set of force field parameters which include 1–3 interactions have been developed for molecular mechanics calculations of phosphoranes including those for hydrospirophosphorane (**82**; X = O, Y = Z = NH) ⟨89JCS(CC)1103⟩ and MM has also been used to study spiro-bicyclic compounds (**106**) ⟨97PS(123)359⟩.

6.3.3 EXPERIMENTAL STRUCTURAL METHODS

6.3.3.1 X-Ray Diffraction

X-ray diffraction studies of four coordinate phosphonium compounds include the cyclotriphosphazenes (**28**) ⟨79ZN(B)911⟩ and (**11**; X = NMe_2, Y = NH, Z = CH_2 CH_2) ⟨78ZN(B)588⟩, the bis spirophosphazene (**126**) ⟨82IC3916⟩. The structure of cyclophosphazene (**127**) was confirmed through an X-ray study of its molybdenum complex ⟨94JCS(D)63⟩.

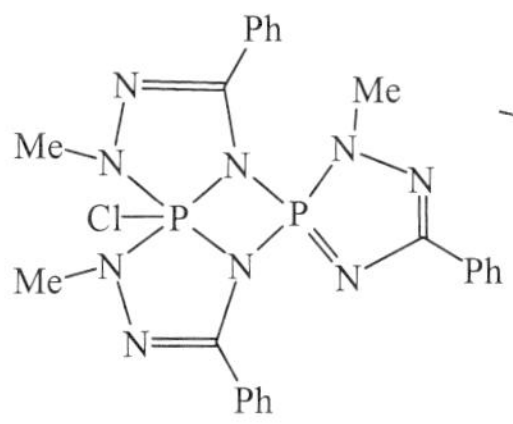

126

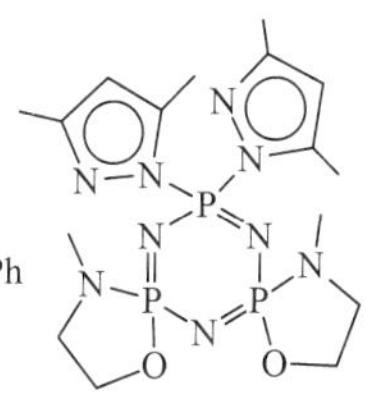

127

Many five coordinate phosphoranes compounds have been examined, the keen interest arising primarily because of their stereochemical importance as reactive intermediates in the reaction pathways of four coordinate compounds ⟨74JA4143⟩. The stereochemistry of phosphoranes vary on a structural coordinate between trigonal bipyramidal (t.b.p.) and square pyramidal (s.p.). Spiro phosphoranes which have been found to have conformations closer to the t.b.p stereochemistry include thienyl phosphorane (**45**; R^1 = H, R^2 = 2-thienyl) ⟨83ZN(B)466⟩, (3,4)phosphorane (**53**) ⟨96PS(109/110)489⟩, bisspiro(3,4)diazadiphosphetan ⟨84JSK921⟩, spiro(4,4)phosphoranes (**56**) ⟨77JA3332⟩, (**62**) ⟨89PS(45)255⟩, (**57**; R = Ph, Y = Z = CO) ⟨79JA3687⟩, (**128**; R = CO_2Me, X = OMe, Y = O, Z = CMe_2CMe_2) ⟨81TL3533, 82AXB3024⟩, (**64**; R^1 = Me, R^2 = $C_6H_4CF_3$, Y = C_6F_5) ⟨80CB1847⟩, hydroxyphosphorane (**81**; Y = CPh_2, Z = OH) ⟨82JCR(S)180⟩, (**73**; X = Y = S, R = Ph) ⟨78JOM(156)253⟩, (**73**; X = Y = S, R = Me) ⟨73CC144⟩, (**129**; X, Ch = O, Y = Ph, Cl) ⟨77JA3326, 77IC2294, 79JA3790⟩; (**54**; R = Ph, Y = CH_2CH_2) ⟨78IC3265⟩, (**129**; X, Ch = O, S; Y = OPh) ⟨76JOC473⟩, bisurea (**87**; X = NPh, Y = CO, Z = Cl) ⟨81CB1554⟩, (**88**; X = S, Z = Ph, Y = O, CF_2) ⟨74P(4)203⟩, (**130**; R = H, X = OMe, Ch = Se, Y = $C(CF_3)_2$, Z = N=$CNMe_2$) ⟨78AG(E)774⟩, (**131**; X = NMe, Y = CHMe, Z = CHPh) ⟨81PS(11)87⟩ (**87**; R^1 = H, R^2 = Me, X^1 = X^2 = O, Y = CH_2CH_2, Z = NPr_2) ⟨84MI-1⟩, (**118**; n = 2, X = Cl) ⟨78PS(5)217⟩, the catechol derivative (**132**; R = Bu^t Y = •) ⟨84MI-2⟩, as well as some spiro(4,5)phosphoranes such as (**133**; n = 3, Y = OAr) ⟨91IC1052⟩, (**88**; n = 3, R = CF_3, X = NMe, Y = O, Z = OAr) ⟨76CC1031⟩, (**75**; R = OAr, X = O, Y = NH, Z = o-$C_6H_3R^2$) ⟨91IC3928⟩, (**89**; Y = Pfp, Z = NMe) ⟨93PS(78)271, 95CB627⟩ and (**134**) ⟨85JCS(CC)1764⟩, spiro(5,5)phosphorane (**75**; R = OAr, X = O, YZ = Bip) ⟨94JA8095⟩ and spiro(6,6)phosphorane (**79**; Y = OCH_2CF_3) ⟨92IC5494⟩. Whereas there are a number of phosphoranes whose structures tend towards a r.p. geometry such as phosphetans (**135;** X = CF_3, Y = Ar) ⟨74ZN(B)32⟩ and (**135;** X = H, Y = Ar) ⟨78IC3265⟩, (**66**; R, Ar = Ph) ⟨79CB1365⟩, (**64**; R^1, R^2 = Me, Y = Ph_2P) ⟨83ZN(B)702⟩ and phospholanes (**136**; X = S, Y = Ph, Z = Ar) ⟨75CC773⟩, (**137**; R = Me, X = S, Y = O, Z = Ph) ⟨83IC1771⟩, (**129**; Y = Me, X = Ch = O) ⟨74AX935⟩, the tetrachloro and octachloro-catechol derivatives of (**129**; X = Ch = O, Y = Ph) ⟨79IC1668⟩, CatPncPR; R = Ph, Bu^t) ⟨78IC3276⟩, (**86**; R^1, R^2 = Me, X = Tcc, Y = CS, Z = NR_2) ⟨92PS(73)195⟩. A Tcc derivative of (**66**) ⟨81CC609⟩ was 50% sp. These trends have been reviewed ⟨74JA4143⟩ and the factors contributing to their conformational preferences studied (see Section 6.3.2). Whilst four and five membered rings almost without exception bridge apical–equatorial positions, six membered rings commonly follow this trend also ⟨90JA6095⟩. In the case of spirophosphorane (**133**; $n = 3$) this occurs despite the fact that this requires a sulphur atom to occupy an apical position in preference to an electronegative aryloxy group ⟨90JA6092⟩. The structure of a selenotrioxyspirophosphorane (**130**; Ch = Se, R = H, Y = $C(CF_3)_2$, Z = N=$CNMe_2$) was established by X-ray crystallography ⟨78AG(E)774⟩.

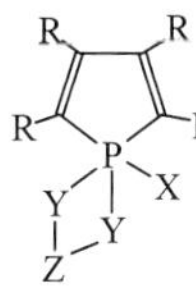
128

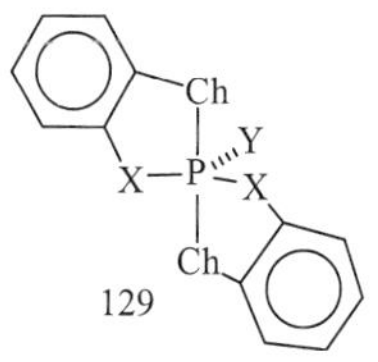
129

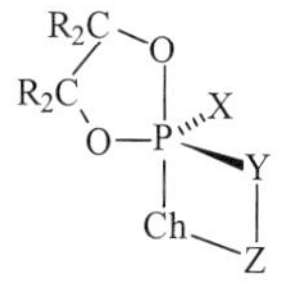
130

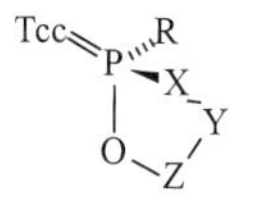
131

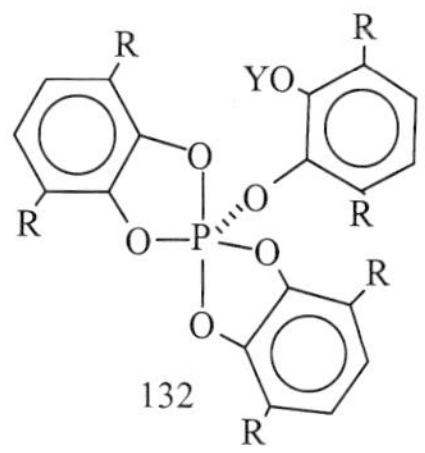
132

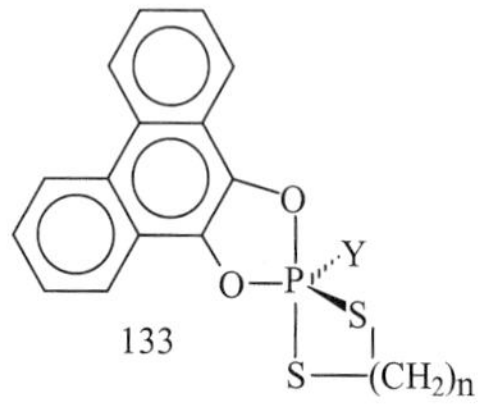
133

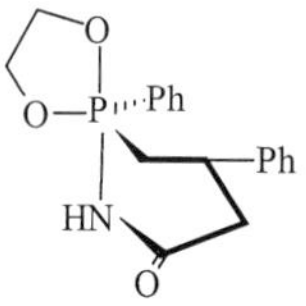
134

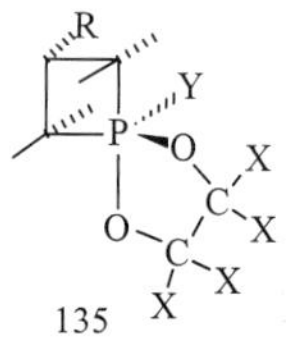
135

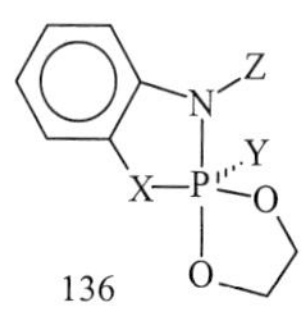
136

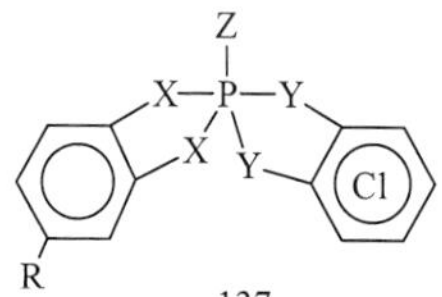
137

The solid state structure of the [7,4] phosphorane (**78**; R = Bu^t, R′ = Ph, Y = Tcc, Z = CH_2) is unique in having its phenyl group in the axial position ⟨95IC4525⟩. Intermolecular hydrogen bonding was found in the crystal spirophosphorane (**129**; Y = CH_2Cl, X = Ch = NH) ⟨91JGU1344⟩.

Structural studies include hydrospirophosphoranes, Eta_2PH and (**129**; Y = H, X = Ch = NH) ⟨78IC3270⟩, (**138**; R^1 = H, $R^2 = R^3$ = Me) ⟨71MI-1⟩, (**99**: R^1 = Me; R^2 = Ph) ⟨97PS(123)5⟩ and (**99**: R^1 = Pr^i; R^2 = H) ⟨99OM915⟩. Whilst some NMR studies have found hydrogen to have a surprisingly high apicophilicity, in the crystal it usually occupies an equatorial orientation.

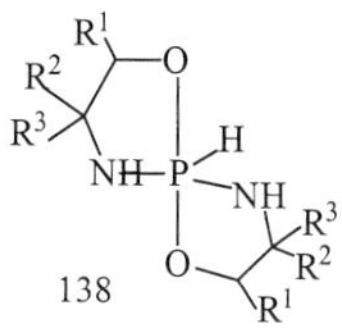
138

There have been a number of X-ray diffraction studies of spirophosphoranes which also incorporate a bridgehead P atom of the type (**139**) ⟨76CC449, 80CB1406, 79ZN(B)906, 90JA7451, 96PS(109/10)629⟩, as well as some bis spirophosphoranes such as (**92**; Y = Glc) ⟨81JCS(CC)810, 77MI-1, 77CB1887, 78TL2857⟩. For some molecules one or more rings may only be partly formed and thus they occupy intermediate coordination states. Thus X-ray crystallographic studies shows that compounds such as (**97**) have long P–N dative bonds (ca. 2 Å) and may be considered as either bis [3,3,3] spirobicyclic or bis [7,3] spiro cyclic. Bridged phosphoranes (**98**, Y = CF_3) have been studied in detail and (**98**; R = R′ = H, Y = CF_3) has a chair conformation for its six membered ring ⟨93JA4031⟩.

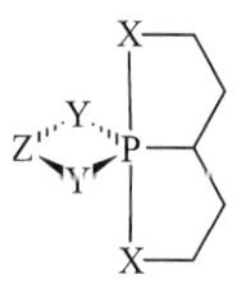
139

Several six coordinate perphosphoranides have been investigated, i.e. (**109**) ⟨78ZN(B)583⟩, (**113**; $m = 0$, $n = 3$, R = H, Z = Et_3NH) ⟨73JA3154⟩, the potassium salt of (**113**; $m = 0$, $n = 3$, R = Bu^t) ⟨88MI-1⟩, (**140**; R = CF_3, X = OPh) ⟨79PS(5)323⟩, (**108**; R = Ph, R′ = Alk, H) ⟨97PS(123)181⟩ and Tcc_3P^- which has near perfect octahedral stereochemistry ⟨97AG(E)608⟩. The dative P–N bonds in the tris naphthylamine compound (**116**) averaged at 2.83 Å while that for the tris benzylamine (**117**) averaged at 3.0 Å both well below the sum of the Van der Waals distances ⟨93AG(E)1430⟩.

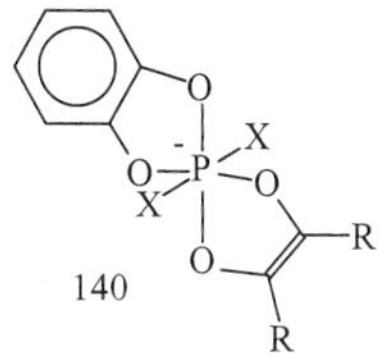

140

6.3.3.2 Magnetic Resonance

6.3.3.2.1 NMR spectroscopy

The power of NMR spectroscopy for the study of phosphorus compounds is greatly enhanced not only by the structural dependence of ^{31}P NMR chemical shifts but also through the couplings it produces in ^{1}H and ^{13}C spectra ⟨91MI-1⟩. Like all chemical shifts ^{31}P NMR chemical shifts are dependent on the coordination state, the directly bound atoms as well as stereochemistry. Of particular relevance to spiro compounds is the fact that the incorporation of five valent phosphorus into a small ring such as a 5 or 4 membered one, generally causes the signal to move downfield, whereas incorporation into a six membered ring usually causes the signal to move upfield. Following this trend the rare tetrathiophosphonium salts (**34**; R = $(CH_2)_2$) and (**34**; R = $(CH_2)_3$) have δ_P 147 and 110 ppm ⟨83ZN(B)1046⟩. The four coordinate group includes the spirophosphoranides such as (**43**; Y = Me) whose δ_P −34 ppm ⟨78JA5229⟩ is well upfield of related phosphonium salts at δ_P 50 ppm despite any small ring effect ⟨79JA4623, 91MI-1⟩. Since large atoms such as Br and I bound to phosphorus usually cause the phosphorus signal to move upfield in phosphonia compounds, this effect balances small ring trends and an intermediate shift δ_P −109 ppm is observed for spiro(3,3)phosphorane (**87**; R = Me, X^1 = NMe, X^2 = N^+Me_2, Y = CH_2CH_2, Z = I) ⟨90PS(54)227⟩ and δ_P −132.6 ppm for silyl derivative (**129**; X = Ch = O, Y = $SiMe_3$) ⟨86PS(27)297⟩. However, contradicting the large atom effect bromophosphorane (**129**; X = Ch = O, Y = Br) is reported as having δ_P −28 ppm ⟨82PS(13)157⟩.

The pentaaryl spiro(4,4)phosphoranes have chemical shifts δ_P −88 to −98 ppm which move downfield as oxygen replaces P-bound carbon cf. tetraoxaspirophosphoranes ⟨81PS(10)395, 91MI-1⟩. Indeed the presence of two five membered rings in spirodioxaphosphoranes (**141**) can cause the ^{31}P NMR signal to move as far downfield as 27 ppm ⟨72JA9264⟩. Dioxaphosphirane (**49**; Ch = O, R = CMe_2CMe_2) has δ_P −87 ppm which is well upfield of related 4,5 and 5,4 spirophosphoranes indicating that the small ring effect for three membered phosphiranes is in the opposite direction to those of 4 and 5 membered rings ⟨78JCS(CC)854⟩. On the other hand the presence of sulphur atoms bound to phosphorus, e.g. (**133**; n = 2, Y = Me) δ_P 32.6 ppm ⟨88PS(35)133⟩, (**142**; X = Y = Pfp, Z = SH) δ_P 14.2 ppm ⟨85ZN(B)1589⟩ and (**81**; Y = CMe_2, Z = S^-) δ_P −4.3 ppm ⟨84JCS(CC)225⟩ exerts the usual deshielding effect. The ^{31}P NMR chemical shifts of the phospha-fenestranes vary significantly with ring sizes ⟨81JA5265⟩ which is complicated by changes of the phosphorus coordination state as determined by the steric constraints of the polycyclic rings on the phosphorus atom.

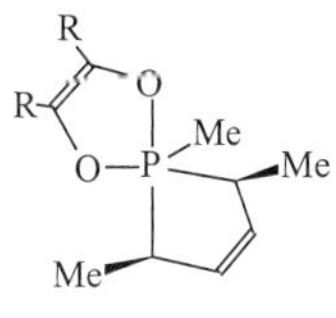

141

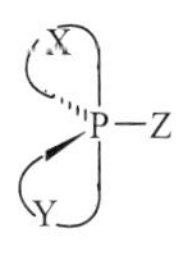

142

Variable temperature ^{1}H NMR studies of phosphoranes (**45**; R^1 = Me, R^2 = Ph) ⟨69TL3423, 79CB218⟩ and (**88**; R = Me, Y = O, Z = H) ⟨69CC443⟩ were among the first to investigate the pseudorotation process ⟨74JA5398⟩ and also highlighted the possibility of isolating chiral compounds. The spiro structure has been used to stabilise Wittig intermediate (**47**; R^1 = Ph, R_2 = CH_2CH_2Ph; R^3 = R^4 = Me) sufficiently to establish its stereochemistry by NMR and follow its conversion to alkenes ⟨87TL3445, 90JA3905⟩.

Variable temperature NMR studies of the pseudorotation process, i.e. ligand positional exchange, have been carried out on a wide range of spiro structures, e.g. (**70**; R^1 = Me, Ph, R^2 = H, X = Y = Ph, OEt, NEt_2, Z = Glc, Tlc) ⟨76TL749, 78JGU650⟩, (**68**; R = Me, Ph, X = Ph, Y = NMe, Z = CHMeCHPh) ⟨92JOC5195, 75P(5)73⟩, (**60**; R^1 = Mes, R^2 = Ph, X = O, NMe, YZ = CH_2CH_2) ⟨75JCS(P1)2376⟩, and hydrophosphoranes such as (**142**; X = Glc, Y = Amc, Z = H) ⟨78BSF65⟩. The pseudorotation barriers of two series of spirophosphoranes such as (**135**) and (**142**: X = Y = Pmp, Z = R) ⟨71P(1)91, 77JCS(P1)437⟩ and for hydrophosphoranes (**88**; R = CF_3, X = Y = O, Z = H) ⟨86ZOK1193, 86PS(27)321⟩

are dependent on ring size. Comparisons have also been made between spiro(4,5) and various spiro(5,6) pentaoxophosphoranes, e.g. (**142**; X = Plc, Y = Bde, Z = OAr) ⟨90JA6095⟩.

Pseudorotation barriers have been used to estimate the apicophilicities of P-bound atoms or groups Y in the phosphorane series (**135**) ⟨71JCS(CC)1011⟩, and (**142**; X = Cat, Y = Pfp) acknowledging the limitations and assessing the accuracies ⟨74JCS(P1)2125⟩. In general, the more electronegative atoms or groups dominate the apical position of a t.b.p. However, together two small rings tend to dominate both apical and two equatorial positions thus tending to push the non-cyclic group, regardless of its electronegativity, into an equatorial position. In the case of spiro[3,4]phosphorane, e.g. (**86**; X = Cat, Y = PF_3, Z = F) ⟨76JCS(D)306⟩, a fluorine atom was thought to be forced into an equatorial position of a t.b.p. Whilst the two six membered rings in (**75**; R = OAr, X = O, YZ = Bip) also occupy ae positions ⟨94JA8095⟩, for the spiro(4,4)phosphorane (**129**; X = Ch = O, Y = F) it is possible for the molecule to adopt a square pyramidal geometry ⟨74ZN(B)32⟩. This may also be the case for the fluorophosphorane (**61**) ⟨70CC476⟩. A further increase in ring size, as in spiro(6,6) phosphorane (**79**; Y = F) ⟨77PS(3)35⟩ allows the fluorine to re-occupy the apical position of a t.b.p. On the other hand bulky groups favour the more spacious equatorial (radial) positions whereas small groups, e.g. H, have a strong tendency to occupy the apical position ⟨74JCS(P1)2125⟩. This latter result threw doubt on many previous assumptions that hydrogen would prefer an equatorial position, e.g. ⟨69CC443⟩. In a study of (**70**: R^1 = Ph, R^2 = Me, X = OAr, Y = Ac, Z = 2-SC_6H_4O) ⟨81JCS(P1)3074⟩ sulphur showed no tendency to occupy the apical position, whereas as the non-cyclic group Y in (**142**; X = Cat, Y = Pfp) (**142**; X = Glc, Y = Pfp) and (**70**; R^1 = Ph, R^2 = Me, Y = Ac, Z = Pnc), ethoxy and thioethoxy groups (X) exhibited the same apicophilicity ⟨77JCS(P1)273⟩. Spirophosphoranes (**142**; X = Pnc, Y = Tcc, Z = CN, Cl, NCO, NCS, N_3, OPh) have also been used to study apicophilicities of group Z ⟨79PS(7)167⟩. The configuration of spirophosphoranes (**70**; R^1, R^2, Y = H = H, X = OMe, Z = $O(CH_2)_2CHMe$) was studied by ^1H NMR spectroscopy ⟨72PAS717, 72DOK1370⟩. Similar studies of bridged phosphoranes (**98**) confirm that the chair conformation for their six membered rings (see above) are maintained in solution which is in contrast to the corresponding three and four coordinate compounds ⟨95JA12390⟩.

Vicinal spin–spin coupling constants through saturated carbon in spirophosphoranes may be quite large, e.g. 28.8 Hz for (**142**; X, Y = Oam, Z = CH_2Cl) ⟨72JGU1895⟩ and has been shown to be 22 and 45.9 Hz for the *cis* and *trans* isomers of tetraoxaphosphorane (**142**; X = Y = Glc, Z = CE=CHE, *e* = CO_2Me) ⟨76CR(C)(282)849⟩. PCH geminal couplings are 11 Hz for (**45**; R^1 = H, R^2 = Me), 24.3 for (**45**; R^1 = H, R^2 = $CH{=}CMe_2$) a case being made for these couplings to be positive in comparison with the negative phosphonium salt geminal couplings ⟨70JCS(C)1425⟩. The couplings $^1J_{PH}$ ca. 770 of hydrospirophosphoranes of the type (**142**; X = Y = Eta, Z = H) tend to be at the lower end of the phosphorane series ⟨70CR(C)(270)865⟩ and $^2J_{PNH}$ 21 Hz was recorded for hydrophosphorane (**143**; R = Me, X = Glc, Y = H) ⟨72CR(C)(274)1413⟩. Coupling constant $^1J_{PP}$ was 270 and 156 Hz for phosphines (**64**; Y = Ph_2P) ⟨83ZN(B)702⟩ and (**66**), and larger, as expected, (709 Hz) for the phosphonate (**64**; R^1, R^2 = Me, Y = $(EtO)_2P(O)$ ⟨82IC844⟩.

Six coordinate spiro-perphosphoranides have ^{31}P chemical shifts in the range −57 to −181 ppm upfield of the corresponding five coordinate phosphoranes but towards the low field region of the range (δ_P −57 to −440) when acyclic and other six coordinate compounds are included ⟨91MI-1⟩, e.g. (**115**; *m* = 0, *n* = 3) δ_P −181 ppm ⟨65CB576⟩, (**144**; X = Y = Bip, Z = Cat) and (**144**; X = Y = Bip, Z = Don) δ_P −147 and −168 ppm, respectively ⟨76CB1056⟩, (**144**; X = Y = Z = Glc) δ_P −89 ppm ⟨71JCS(CC)1070⟩, (**144**; X = Y = Z = Bip) δ_P −82 ppm ⟨65CB576, 77ZOK1432⟩. The methyl groups of the 2-propyl derivative of perphosphoranide (**113**; *m* = 0, *n* = 3, R = Pr^i) have been shown to be diastereotopic ⟨76P(6)91⟩. Solid state MAS ^{31}P NMR spectroscopy has been shown to be valuable for structure determination of six coordinate spiro compounds when crystals are not available ⟨95JCS(P2)1149⟩. Methylation of the six coordinate tris naphthylamine perphosphoranide (**116**) gave a phosphonium salt with an upfield phosphorus chemical shift of −8 ppm ⟨93AG(E)1430⟩. The extent to which some or all the amino groups participate in PN bonding will be interesting to see.

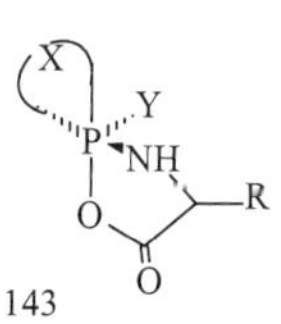

143

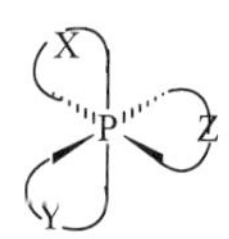

144

6.3.3.2.2 Electron spin resonance

The variable temperature ESR spectra of the spirophosphorus radical ($Pnc_2P^{\cdot}$) indicated that pseudorotation is slow even at 120°C, ΔG^* being 54 kJ mol^{-1} ⟨72JOM(42)C47⟩. Spin labelled spirophosphoranes (**142**; X = Glc, Y = Dme, Z = Nox) ⟨78ZN(B)305⟩ and (**145**; Y = odd electron •), which are stable for several months in the absence of air, have been reported ⟨78BAU1064⟩. A deep violet colour develops when hydrobisphenylene-phosphorane (**45**; R^2 = H) dissolved in benzene is kept in the dark. Although this was attributed to the bisphenylene phosphoryl radical ⟨69CB528⟩ an ESR spectrum was not reported.

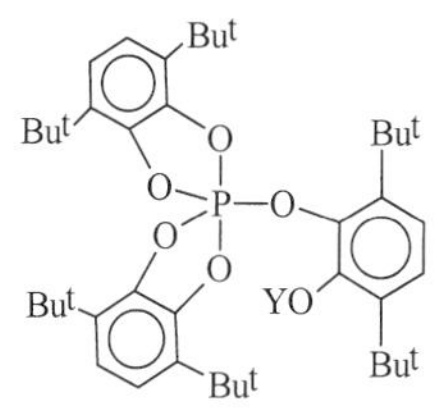

145

6.3.3.3 Mass Spectrometry

The mass spectra of spirobiphenylenephosphoranes (**45**) show abundant M-1 and M-R^2 fragments and there was a tendency to form doubly charged ions ⟨73P(2)167⟩. The mass spectra of cyclophosphazenes such as (**11**) ⟨76ZAAC(426)275⟩ have been used to identify substitution patterns around the rings. Migration of groups to and from phosphorus is common and therefore mass spectral results must be interpreted with care. The fragmentation patterns for the amino acid derived phosphoranes (**143**; X = Cat, Y = OPh have been studied using FD, EI, CI and B/E linked scan mass spectrometry ⟨97RCM1825, 97MI-3⟩.

6.3.3.4 Electronic Spectroscopy, Circular Polarisation and Polarimetry

Phosphorus, whilst having an auxochromic effect, does not transmit conjugation effects from one chromophoric substituent to another. Thus the UV spectra of spirobiphenylene-phosphoranes were dominated by the biphenyl ring system. Comparisons of the spectra with the precursor dibiphenylene phosphonium salt and pentaphenylphosphoranes showed the spirophosphoranes to have more bands than the salt. This was attributed to different auxochromic effects of phosphorus through apical and equatorial bonds. ⟨70JCS(C)1425⟩.

The kinetics of isomerisation of ephedrine based hydrospiro(4,4)-phosphoranes ⟨74TL3971⟩ and related L-alanine based phosphoranes ⟨80PS(8)153⟩ can be followed by polarimetry. The molecular rotation of optically active spiro(4,4)hydrodioxa-phosphoranes (**82**; X = O; Y, Z = NR), and ephedrine derived phosphorane (**142**; X = Glc, Y = Eph, Z = H), permitted the separation of rotational contributions from the ligands and the helix structures. This allowed the assignment of the absolute configuration of a large number of phosphoranes ⟨81PS(10)53⟩. The polarimetric temperature jump method has been used to obtain more accurate kinetic and thermodynamic parameters for the isomerisation of hydrophosphoranes (**142**; X = Eta, Y = Eph, Z = H) ⟨77JCS(D)570⟩.

Most six coordinate perphosphoranides are not configurationally stable in solution. One exception is the tris tetrachlorocatachol floroborate Tcc_3P^- ⟨97AG(E)608⟩. Perphosphoranide (**144**; X = Y = Cat, Z = $OCHPhCO_2$) was isolated in one form by second order asymmetric induction and its equilibration studied by polarimetry ⟨76JCS(P2)955⟩. Circular dichroism of trisbiphenylenephophoranide showed that the (−) stereoisomer has a left handed screw stereochemistry ⟨70JCS(B)640⟩.

6.3.3.5 Vibrational Spectroscopy

Whilst infra-red spectroscopy is used extensively as additional evidence for structural assignments there have been relatively few studies which focus on this area. One such study involved hydrospirophosphorane (**138**; R^1 = R^2 = Me, R^3 = H) where the ν_{NH} solvent dependence resembled pyrrole and ν_{PH} correlated with $^1J_{PH}$ ⟨69SAA1201, 72CR(C)(274)1156⟩. Ring-chain isomerism and hydrogen bonding of hydrophosphorane (**142**; X = Pnc, Y = Pda, Z = H) was studied by i.r. spectroscopy ⟨75CR(C)(280)809⟩.

6.3.4 THERMODYNAMIC ASPECTS

The barriers to intramolecular ligand exchange (permutational isomerism) via a pseudorotation process have been measured for a large number of five coordinate phosphoranes using the variable temperature NMR signal coalescence approach (see NMR section above). Care has to be taken to check that a dissociation–association mechanism is not operating, thus low $\Delta S^{\#}$ and lack of deuterium exchange was used as evidence for the absence of a dissociation mechanism ⟨74JCS(P2)1668⟩. For the spirophosphoranes the largest barriers are observed when isomerism requires the formation of a conformer with a small ring bridging two equatorial (radial) orientations. The barrier to interconversion of isomeric phosphinates via spiro(3,4)hydroxyphosphoranes was shown to be $\Delta G^{\#} = 17$ kcal mol^{-1} by variable temperature NMR spectroscopy ⟨76TL4381, 79JCS(P1)879⟩ cf. $\Delta G^{\#} = 24$ kcal mol^{-1} for the pseudorotation barriers of dimethylaminospirophosphorane derived from cis cyclohexane diol ⟨75JGU2088⟩.

6.3.5 REACTIVITY OF FULLY CONJUGATED RINGS

There are no fully conjugated spirophosphorus compounds.

6.3.6 REACTIVITY OF NON-CONJUGATED RINGS

6.3.6.1 Spirophosphonium Salts and Ylides

The extrusion of disulphide from spirophosphazene (**32**; X = S, Y = $(CH_2)_3$) provides a useful route to two coordinate triazaphosphole (**146**) ⟨76T2039⟩. The related extrusion reaction of (**147**) includes the loss of dimethylamine (Equation (1)) ⟨80TL1307⟩. In some cases there is an equilibrium between the phosphorane, e.g. (**76**) and the phospholan plus disulphide ⟨83JOC3815⟩. As expected phosphenium ion (**33**) reacts with a variety of nucleophiles to give neutral products whereas with electrophiles the cationic structure is preserved ⟨98JGU530⟩. Spirophosphonium ylide (**15**) can be converted to its lithium diylide which forms a nickel complex ⟨81CB3161⟩. Spirophosphoranide (**43**; Y = CF_3) has also been converted to a series of metal complexes ⟨93PS(76)87⟩.

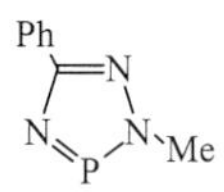

146

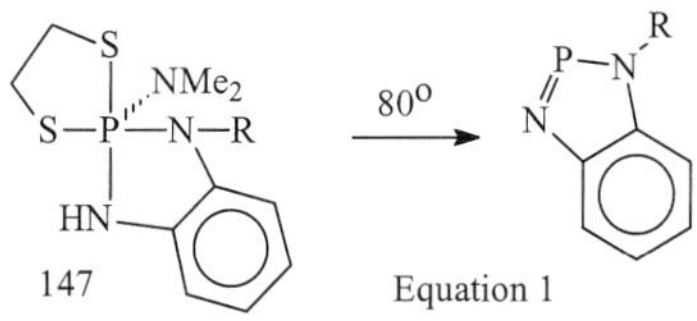

147

Equation 1

6.3.6.2 Spirophosphoranes

The stabilities of spirophosphoranes vary widely. Whilst small rings induce stability, they are not always sufficient to ensure a stable phosphorane. Thus triphenylphosphole reacts with two equivalents of acetylene dicarboxylate to give a transient spirophosphorane (**148**; E = CO_2Me) but this rearranges spontaneously to give the ylide (**149;** E = CO_2Me) (Equation (2)) ⟨68T3437⟩. The saturated analogues, e.g. phosphorane (**3**; $m = n = 3$, Y = Me) can be isolated but this is probably due to the lack of a rearrangement or degradation pathway rather than having thermodynamic stability. Two or more electronegative atoms bound to phosphorus are usually desirable. Some spirophosphoranes, e.g. (**150**; X^1 = H, $X^2 = X^3 = X^4$ = Me, Y = Glc, Z = OAc) are stable under quite vigorous reaction conditions ⟨71JGU228⟩ whereas others such as (**151**) can slowly ring open to give the carboxylic acid isomer ⟨79JCR(S)172⟩ or exist in

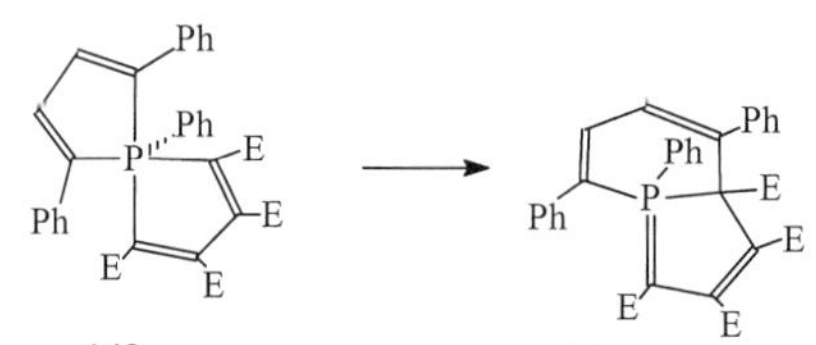

148 149

Equation 2

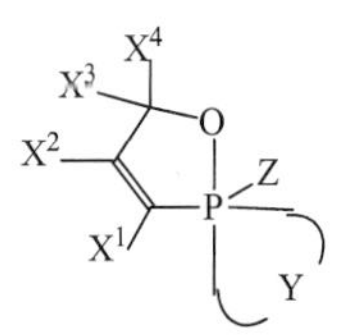

150

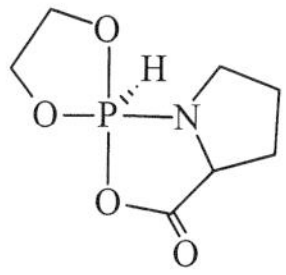

151

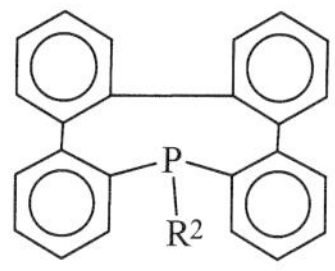

152

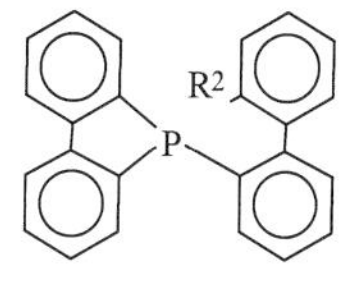

153

equilibrium with the open chain form ⟨74BSF2193⟩. Monocyclic phosphoranes undergo 1,2-aryl migration from phosphorus to carbon under thermal conditions and in some cases spontaneously ⟨61JCS2126, 71P(1)139⟩. The thermally induced rearrangement of spirobiphenylenephosphoranes (**45**; R^1 = H) to phosphines (**152**) and (**153**) are further examples, the former predominating when R^2 is Me or Ph, the latter predominating if R^2 is larger, e.g. anthryl ⟨76CB1497⟩. The reaction of phospholene (**154**; Y = Ph) with peroxide gave *P*-phenyldioxaphosphepan which was rationalised as involving an unstable spiro(4,6)phosphorane shown in Scheme 1 ⟨74P(4)265⟩. Similarly spirophosphoranes (**46**) and related anthraquinone adducts formed in cycloaddition reactions underwent a retrocycloaddition ⟨81PS(10)47, 83TL5831⟩. Spirophosphiranes (**49**) undergo related reactions ⟨78JCS(CC)854, 74JA317⟩). On the other hand, heating the urea derived spiro(3,4)phosphorane (**87**; X^1 = NMe, X^2 = PPh, Y = CONMe, Z = Ph) gave phosphole (**155**) as if by hydrolysis ⟨81CB2132⟩. The bis spirophosphorane (**92**; R = Ph, X = OEt, Y = Dpe, Pfp) fragmented in a similar manner ⟨76BCJ1924⟩.

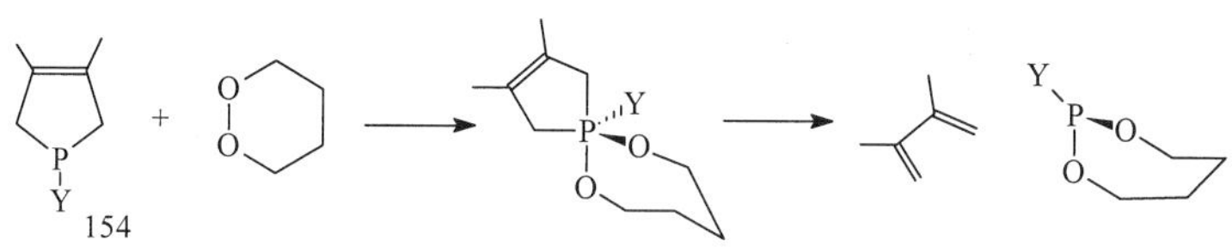

154

Scheme 1

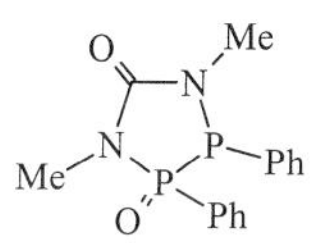

155

Whilst spirophosphetanes (**47**) have been generated for reactivity studies of the Wittig reaction, they have also been generated as oxyanions (**47**; R^1 = O^-, R^2, R^3 = Ph; R^4 = H) from β-hydroxyalkylphosphine oxides and found to fragment stereospecifically to *cis* stilbene ⟨85JCS(P1)1953⟩. Another example is the detection of the glycolyl oxyanion (**65**) in a Horner reaction ⟨79TL2203⟩. A related spiro oxyanion (**156**) fragments with the formation of an aziridine ⟨79E851⟩. The formation and isolation of a spiro oxaphosphetan was achieved via a Horner reaction of a ketone with a carbanion followed by cyclisation as shown in Equation (3) ⟨92JA4008⟩.

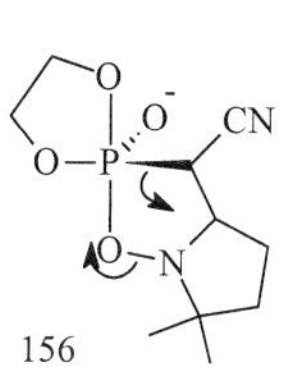

156

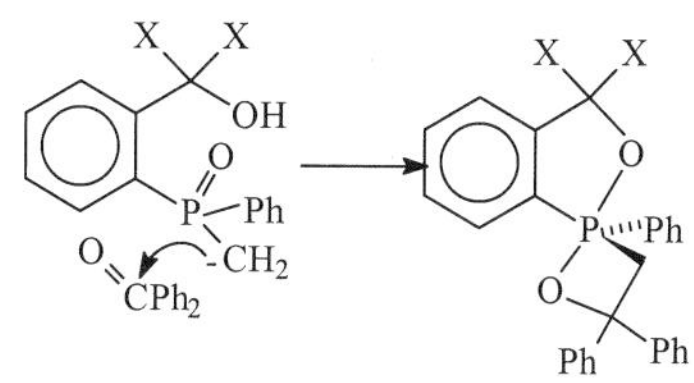

There have been a number of studies of the hydrolysis of spirophosphoranes. These include a range of spirotetraoxyphosphoranes (**88**; X = Y = O) ⟨79PS(6)435⟩, (**150**; X^1 = $CONH_2$, X^2 = H, X^3 = X^4 = OMe, Y = Pnc) ⟨90PS(47)443⟩, (**142**; X = Y = Plc, Z = Ph) ⟨85JA5198⟩ and of dioxyphosphoranes (**56**) the products of which were characterised by acylation ⟨72JGU782⟩. The hydrolysis of hydroxyphosphoranes (**57**; R = OH, Y = CMe_2) have also been studied ⟨86JA2416⟩. Cycloaddition reactions of an aminodioxaphosphole with chalcone gave a spiro(4,4)phosphorane ⟨82PS(13)85⟩ which undergoes ring expansion to give a spiro(4,6)phosphorane as shown in Equation (4) ⟨85PS(25)319⟩. Hydrolysis of the bis phosphorane derived from cyclen (**122**; *m* = 2, *n* =1, Y = O) caused P–N bond cleavage forming a cyclen phosphine oxide, a reaction which can be reversed by thermal dehydration ⟨96JCS(CC)2157⟩.

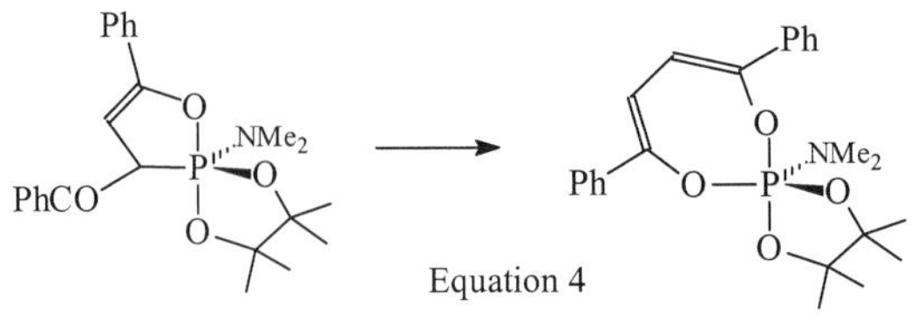

Equation 4

Spiro(2,4)phosphorane (**157**) was a postulated intermediate in the formation of bis spirophosphorane (**158**) from a dioxaphospholan and benzaldehyde Equation (5) ⟨88JCS(CC)615⟩. Thermal cleavage of phosphazenes to a phosphole as shown in Scheme 2 below is rationalised as involving a spiro(4,5)phosphorane ⟨74JCS(P1)1694⟩.

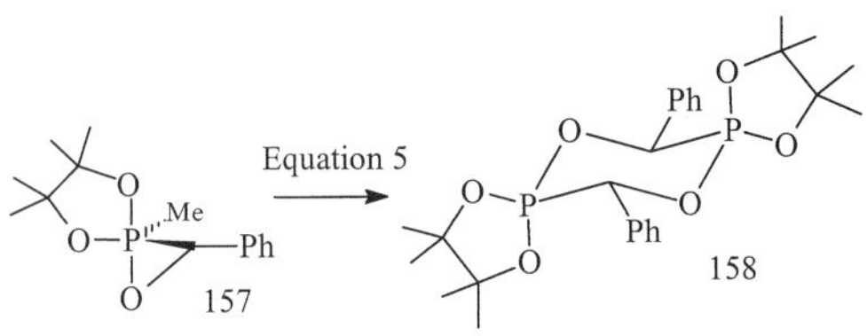

Equation 5

157

158

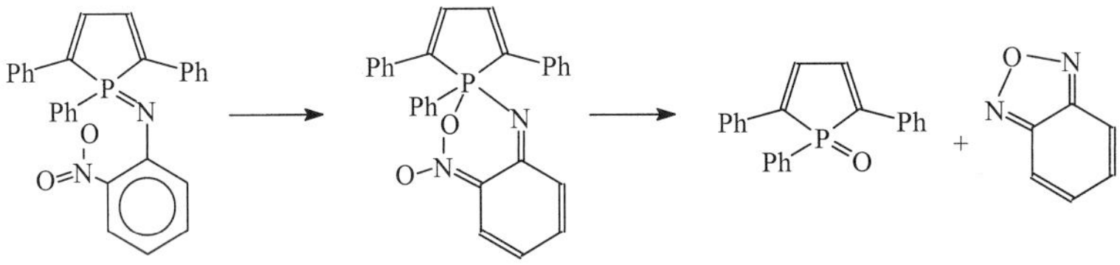

Scheme 2

Hydrospirophosphorane (**81**; Y = CH_2, Z = H) in the presence of a tertiary amine has been used to convert ketocarboxylic acids such as pyruvates into the corresponding α-hydroxycarboxylates ⟨92PS(70)263⟩. However, in aqueous alcohol hydrophosphoranes such as (**142**; X = Glc, Y = Amc, Z = H) are converted to phosph(III)olanes ⟨80PS(9)184⟩.

6.3.6.3 Spiroperphosphoranides

The interconversion and *cis* : *trans* isomerisation of six coordinate spiroperphosphoranides (**159**; X = Gls, Y = Pfe, Z = OC_6H_4F, R = F) have been studied ⟨81JCR(S)218⟩. Five and six coordinate structures are often in equilibrium, the latter usually being favoured over the five coordinate structure in compounds such as (**160**) ⟨79T1825⟩. The equilibrium shown in Scheme 3 (R^1 = Ph, CF_3, CCl_3; R^2 = Me, 2-Pr; X = Y = Cat) was found to depend on solvent and temperature ⟨85ZOB1982⟩. A quite remarkable reaction shown in Equation (6) involves a 1,3 migration of a dimethylamino group from one phosphorus atom to the other at some stage during the reaction ⟨93PS(75)233⟩.

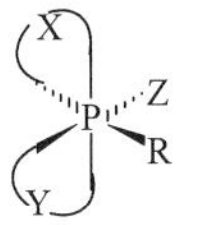

159

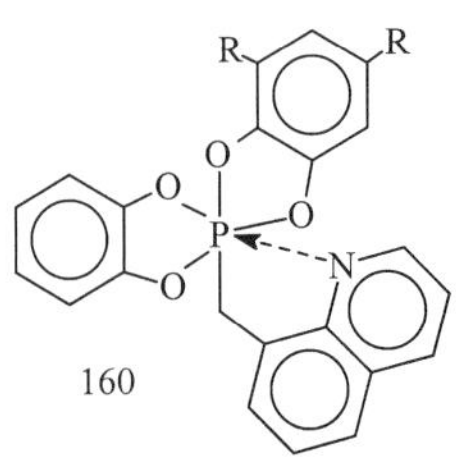

160

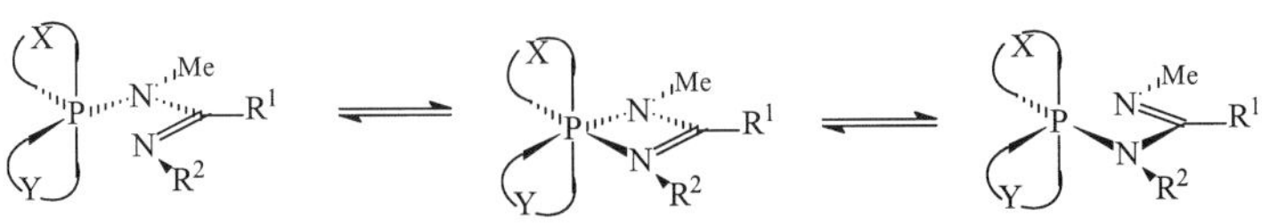

Scheme 3

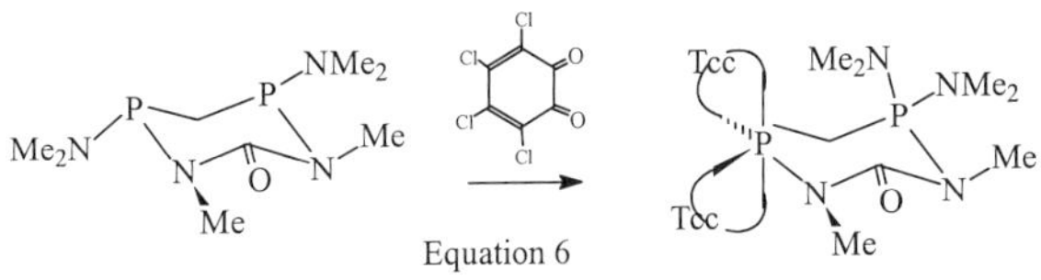

Equation 6

6.3.7 REACTIVITY OF SUBSTITUENTS ATTACHED TO RING CARBON

The tendency for reactivity to focus at phosphorus leads to relatively few reports on the reactivity of carbon substituents. The 1,3-allylic migration of a methoxy group across a spirooxyphospholene ring was induced by heating at 140°C for 10 min (Scheme 4) ⟨85PS(24)235⟩. Secondary amines and phosphorous esters add to the activated double bond of spirophosphorane (**142**; X = Y = Glc, Z = CE=CHE) ⟨79CR(C)(288)165⟩. Spiroperphosphoranide (**161**) reacts with trimethyl phosphite as shown in Equation (7) ⟨79PS(7)305⟩. The spirophosphorane group can be used in place of a phosphoryl group to stabilise an adjacent carbanion followed by a retroelectrocyclic reaction of the Horner Wittig type. Whereas the anion from Hyc_2PCH_2W reacted with aldehydes to give E and Z mixtures of alkenes ⟨96PS(109/110)156⟩, the anion from (**57**; R = CH_2CO_2Et, Y = Z = $C(CF_3)_2$) even at room temperature gave predominantly Z alkenes ⟨97JA5970⟩. The phosphoranes are about three pK_a units less acidic than the corresponding phosphoryl compounds which would disfavour the reverse aldol condensation and thus account for the higher stereospecificity.

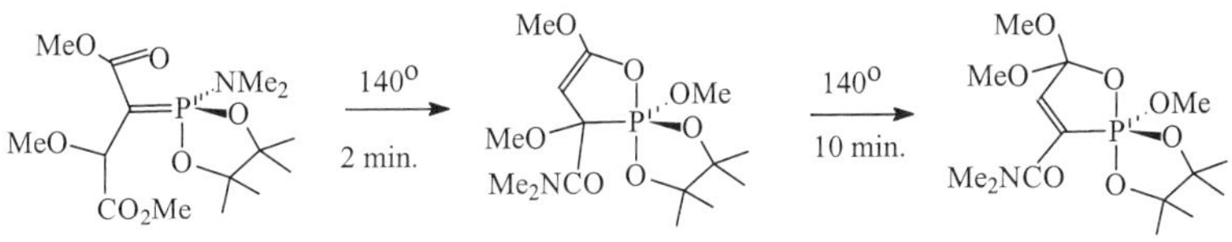

Scheme 4

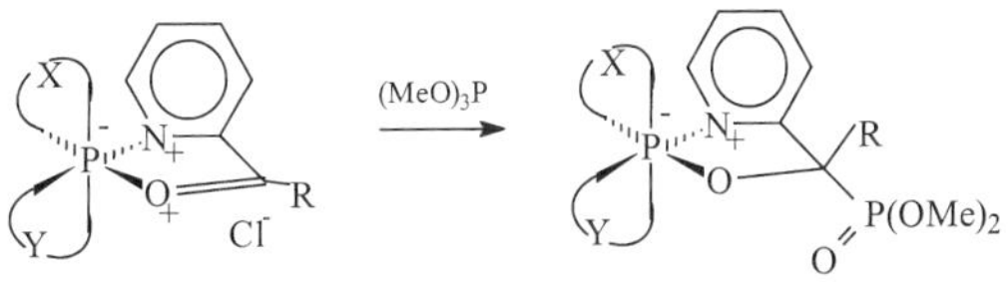

Equation 7

161

6.3.8 REACTIVITY OF SUBSTITUENTS ATTACHED TO RING HETEROATOMS INCLUDING SPIRO PHOSPHORUS

The physical and chemical properties of the hydro P–H group in spirophosphoranes have been thoroughly investigated. The pK_a's of hydrophosphoranes (**83**; R = H, X = Y = O, NH) were found to be 5.89 and 11.96 ⟨77IZV434⟩. Hydrophosphorane (**81**; Y = CMe_2, CPh_2, Z = H) was oxidised to the corresponding hydroxyphosphorane by N_2O_4 in DMSO which was then in equilibrium with the open chain carboxylic acid ⟨80T2467⟩. The PH bond of hydrophosphoranes undergoes many addition reactions to unsaturated compounds ⟨79PS(7)81⟩, thus (**82**; X = Y = Z = O) adds to chloral or benzaldehyde, ⟨68CR(267C)270⟩, (**82**; X = Y = O, Z = O/NH) adds to acrylic ester and vinyl ether in the presence of alkoxide and AIBN respectively to give Michael adducts (**142**; X = Glc, Y = Glc/Eta, Z = CH_2CH_2E, E = CO_2Me, OR) ⟨70BAU2637⟩. Additions to vinyl diphosphonate $CH_2{=}C(P(O)(OR)_2)_2$ have been investigated ⟨95PS(103)125⟩ and in a similar manner hydrophosphoranes adds to imines ⟨74CR(C)(278)1301⟩ and precursor phosphites add to cyanoacrylate to form phosphoranes (**77**; Z = $(CH_2)_2$, Y = CH_2CCNCO_2R ⟨94JGU373⟩. Addition reactions of hydrophosphoranes to unsaturated compounds such as vinylethyl ether and acetylenic carboxylic esters can be achieved using an AIBN catalyst, thus compounds such as (**142**; X = Y = Glc, Z = CH_2CH_2OEt/CE=CHE, E = CO_2Me), have been made this way ⟨76T2253, 79CR(C)(288)165, 81PS(9)285⟩. However, di-*t*-butylperoxide converts (**142**; X = Y = Pnc, Z = H) into the phosphate (**162**) ⟨85JGU1810⟩ and dimethyldisulphide under photolysis conditions converts (**142**; X = Glc, Y = Pfp, Z = H) into (**142**; X = Glc, Y = Pfp, Z = MeS) ⟨86PS(27)313⟩. Arylation of the hydride (**122**; *m* = 1, Y = H), derived from cyclen, to make the pentafluorophenyl phosphorane (**122**; *m* = 1, Y = C_6F_5) proceeded via its P-lithio intermediate ⟨99JFC223⟩. In a similar manner phosphorane (**80**) reacted with styrene oxide or benzyl halides and then benzoyl halide followed by $LiBH_4$ reduction to give various alcohol derivatives ⟨97TL7753⟩. Hydrospirophosphoranes have been used to substitute *C*-alkoxy groups in various aminals derived from α-amino esters ⟨95JGU1604⟩ or piperazine ⟨95JGU1610⟩. Zinc chloride has been used as a catalyst for this substitution reaction ⟨94JGU1171⟩. Interestingly, methanol and an enamine (which is hydrogenated) converts the PH group of hydrospirophosphoranes (**82**; X = Y = Z = O) directly to the corresponding methoxyspiro-phosphorane ⟨76T2253⟩, whereas the P–H group of (**81**; Y = CPh_2, Z = H) is oxidised to its hydroxyphosphorane by DMSO in DMF ⟨78JCS(CC)219⟩ and is converted by sodamide or butyl lithium into phosphoranides which can be alkylated or acylated to give spirophosphoranes (**81**; Y = CPh_2, Z = R or PhCO) ⟨75JCS(P1)1220, 75JOM(93)331, 79TL4205⟩. The coupling of bicyclic [3,3,0]hydrophosphoranes with polyethylene glycols ⟨95PS(103)199, 96PS(114)51⟩, *N*-methylethanolamine ⟨93PS(81)1⟩ bis phenol $HOC_6H_4CH_2C_6H_4OH$ ⟨93PS(83)157⟩, and glycol diacrylate ⟨94PS(90)131⟩ was utilised in the synthesis of a wide range of large ring bis spiro-phosphoranes (**105**) ⟨95MI-2⟩. The P–H bond has also been converted to P–Hal ⟨83ZAAC(507)93⟩, $P{-}S^-$, $P{-}Se^-$, $P{-}NSO_2Tol$, $P{-}PS_2Ar$, ⟨87PS(32)1⟩ and $P{-}BH_3$ ⟨88PS(35)309⟩. Photolysis of the azide (**142**; X = Y = Cat, Z = N_3) gave a nitrene which abstracted hydrogen from the solvent (benzene) to give the amino derivative (**142**; X = Y = Cat, Z = NH_2) ⟨84JA7065⟩, and a related azide of spiro(3,3)phosphorane (**64**; R^1, R^2 = Me, Y = N_3) was reacted with Ph_3P ⟨86PS(26)193⟩ to give the corresponding phosphazene.

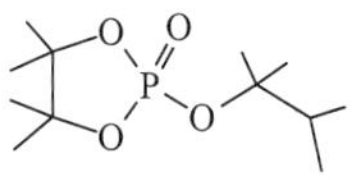

162

The *P*-halogen group of phosphorane (**129**; X = Ch = O, Y = Cl, Br) added across the carbonyl group of tribromoacetaldehyde ⟨94JGU1210⟩ and the chloro compound (**129**; X = Ch = O, Y = Cl) reacted with YTms; Y = Ph_2P, CN, Br, I, $MgTms_2$, N=CHPh, to produce the derivatives (**129**; X = Ch = O, Y = Ph_2P, CN, Br, I, Tms, N=CHPh) ⟨91PS(62)139, 86PS(27)297, 85JGU2333⟩, and with $(EtO)_2P(O)Na$ to give (**64**; Y = $(EtO)_2P(O)$ ⟨82IC844⟩. *P*-Halo perfluoropinacol-spirophosphoranes (**142**; X = Y = Pfc, Z = Hal) reacted with aq. KOH to give an isolable potassium salt (**142**; X = Y = Pfp, Z = O^-) and with alcohols to give stable alkoxy derivatives (**142**; X = Y = Pfp, Z = OR) which can be sublimed from sulphuric acid ⟨82AG(E)208⟩. In the case of an ambient nucleophile $(EtO)_2PNHAr$, a P–P bond was formed initially upon reaction with (**129**; X = Ch = O, Y = Cl) and this was followed by rearrangement to the PNP isomer ⟨86JGU2218⟩. In the case of chlorophosphorane (**163**), dimethylamine replaced chlorine (at −78° in THF) with inversion whereas reaction with PhONa was non-stereospecific ⟨79TL193⟩. Bromophosphorane (**129**; X = Ch = O, Y = Br) reacted with chloral to give (**129**; X = Ch = O, Y = $OCHBrCCl_3$) ⟨89JGU2583⟩. Exocyclic P–N bonds are also readily substituted and provide ready routes to alkoxy and carboxy derivatives (**142**; X = Pnc, Y = Dpe, Z = OR/OCOPh) ⟨73TL3455⟩. When one of the rings contains a carboxy group (**142**; X = Glc, Y = Hyc) compounds, the exocyclic groups Z = PhO or X_3CCH_2O are readily converted to alkoxy groups by the action of alcohol ⟨80CL1599⟩. Also an exocyclic vinyl ether group can be replaced by Br by the action of bromine which in turn may be converted to methoxy by methanol ⟨84JGU486⟩. The chlorine in (**129**; Y = Cl, X = Ch = O) has also been replaced by $Mn(CO)_5$ ⟨83ICA(76)L139⟩. Two spiro phosphorane groups Cat_2P were linked as shown in (**93**) by reacting the corresponding chlorophosphorane with the O and P disilylated compound $TmsP{=}C(Bu^t)OTms$ ⟨99JGU48⟩.

Methylation of hydroxyphosphorane (**57**; R = OH, Y = Z = CO) by diazomethane gives the methoxy derivative (**57**; R = OMe, Y = Z = CO), whereas methylmagnesium iodide gives trimethylated phosphorane (**57**; R = Me, Y = CO, Z = CMe_2) ⟨79JA3687⟩. Whilst the NH group of spirophosphorane (**164**; R^1 = R^2 = H, R^3 = R^4 = Ph) was methylated by KH and MeI to give (**164**; R^1 = Me, R^2 = H, R^3 = R^4 = Ph) ⟨90PS(49)385⟩ when the phosphorane also had a PH group as in (**164**; R^1 = H, R^2 = R^3 = Me; R^4 = H) NaH and MeI methylated at phosphorus to give (**164**; R^1 = H; R^2 = R^3 = R^4 = Me) ⟨84CJC2179⟩. The bis spirophosphorane (**165**) was made by the reaction of hydrophosphorane (**81**; Y = CR_2, Z = H) with phenyldichlorophosphine ⟨90T3527⟩.

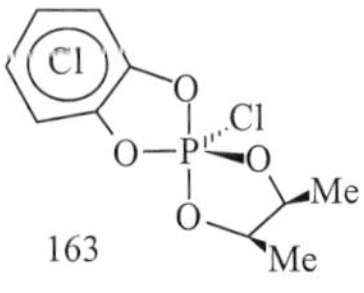

163

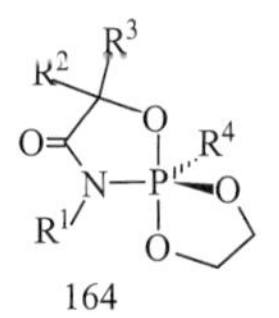

164

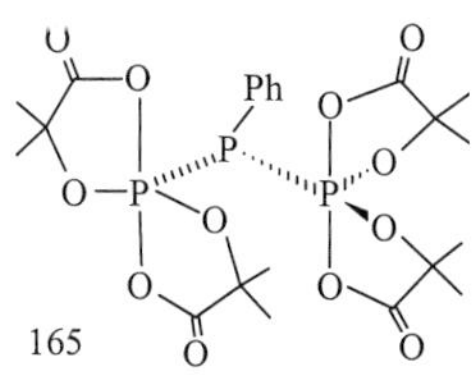

165

The six coordinate hydrospiroperphosphoranide (**159**; R = H, X = Y = Cat, Z = Et) on standing for several days with 2-aminophenol undergoes conversion to phosphorane (**142**; X = Cat, Y = Oam, Z = Et) probable with loss of hydroquinone ⟨75JGU2292⟩.

6.3.9 RING SYNTHESES

6.3.9.1 All Components Acyclic

6.3.9.1.1 Spirophosphonia compounds

Spiro(4,4)phosphonianonane triiodide (**1**; $m = n = 4$) was prepared (40%) by heating phosphorus and 1,4-diiodobutane at 200°C ⟨68ZOB331, 77AG(E)722⟩. The dibenzo derivative (**12**) was prepared from *o*-xylidene dibromide ⟨73JHC395⟩. The bis diphenylene salts (**13**) may be made from dilithium biphenyl and either trichlorophosphine ⟨72RTC836⟩ or triphenyl phosphate ⟨73RTC1308⟩ or by the original method involving bis di-iodobiphenyl and phosphorus ⟨65CB576⟩.

When the aryl ring is activated towards electrophilic attack as in diphenylamine, trichlorophosphine is able to substitute at the ortho position and, at 200°, ultimately produce the spirophosphonium salt (**18**) ⟨71JCS(CC)1213⟩. If the forming bond to phosphorus is from a heteroatom, a P(III) phosphorus can be made electrophilic. Such an example is the cyclisation of the dibenzyl ether (**166**) to give phosphorane (**57**; Y, Z = CH_2, R = Ph) as shown in Scheme 5 ⟨82JCS(P1)735⟩.

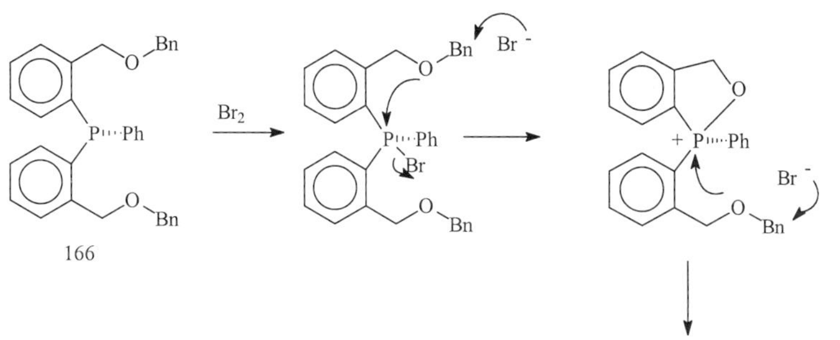

Scheme 5

Urea based spiro(5,5)phosphonia-undecane chloride (**17**) was produced in low yield by the reaction of dimethylurea with the little exploited, readily available, tetrakis(hydroxymethyl)-phosphonium salts ⟨81PS(10)207⟩. Tetrathiodibenzo quaziphosphonium salts (**34**; R = Dop; $X^- = PF_6^-$) have been produced by the reaction of silylated dithiocatechol (**167**) with PF_5 ⟨73CC144⟩. DBU reacts with four molecules of bis(trimethylsilylmethyl)dichlorophosphine to produce diphosphazene (**37**) ⟨95CB959⟩.

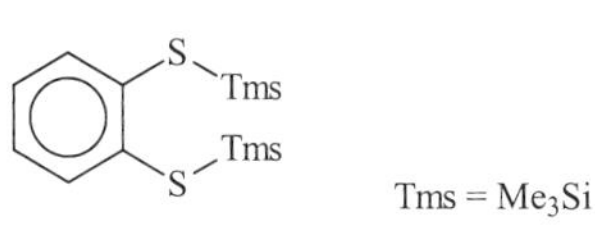

167

Spiro(5,5)phosphazenes (**28**) may be made directly from trichlorophosphine and diaminophosphazene, $H_2NPR_2{=}PR_2NH_2$, ⟨75CB1454⟩. On the other hand, bis spirophosphazane (**38**) was made by the reaction of substituted phosphabenzenes with phenylazide ⟨93TL3107⟩. Rhenium carbonyls have been used to dicyclise a silylated diphosphazene to produce the salt (**23**) ⟨87CB1183⟩. The phosphoniafenestranes (**122**) and (**216**) was prepared by the reaction of PF_5 with the tetrasilylated [3,3,3,3]cyclen ⟨77TL559, 81JA5265⟩.

6.3.9.1.2 Spirophosphoranes

The formation of phosphoranes is greatly enhanced when the phosphorus atom is involved in a five membered ring which has apical electronegative atoms such as oxygen. Thus pentachlorophosphorane (PCl_5) reacts with acylhydrazines to give P-chloro spirophosphoranes ⟨89JGU1333⟩, and with *o*-methylaminothiophenol to give chlorophosphorane (**73**; R = Cl, X = NH, Y = S) ⟨77JGU436⟩.

Alkyldichlorophosphines reacted with hexafluoro-2,4-pentadione ⟨93AG(E)1430⟩ and 1-phenyl-4,4,4-trifluorobutane-1,3-dione ⟨97JFC(85)129⟩ in their enolic forms in a 1 : 2 ratio to give the caged trioxyphosphoranes (**96**) and (**95**), respectively. *N,N'*-Disilylated dimethylurea reacted with chloromethyldichlorophosphine to form [3,3] spirophosphorane (**64**; $R^1 = R^2$ = Me, Y = CH_2Cl ⟨97ZAAC(623)1325⟩. Also intramolecular addition of the benzylic hydroxyl group to the phosphoryl group in (**167b**; Y = Z = CH_2) gives an intermediate hydroxyphosphorane which undergoes intramolecular condensation to produce the spirophosphorane as shown in Scheme 6 ⟨78JA5130⟩. Acid catalysed cyclodehydration of bis (2-terephthalato)methylphosphine oxide (**167b**; Y = Z = CO) gave dioxyphosphorane (**55**; X = O, Y = CO_2H) ⟨96PS(109/10)241⟩ The corresponding *N*-methylamide (**168**; X = NHMe) gave (**55**; X = NMe Y = CONMe) which unlike to acid above was unstable in solution at pH 8.5 ⟨95MI-1⟩. While Appel's reagent (Ph_3P/CCl_4) was used to accomplish dicyclisation of bis β-hydroxyethylphenylphosphine oxide to make the moisture sensitive dioxyphosphorane (**52**; R = H) ⟨97PS(124/5)379⟩, Mitsunobo's reagent {$Ph_3P/EtO_2CN{=}NCO_2Et$} was used to cyclise a monobenzylic β-aminophosphine oxide-attack at phosphorus by the secondary amine forming the second ring of the product (**53**; R = CF_3) ⟨96AG(E)1096⟩.

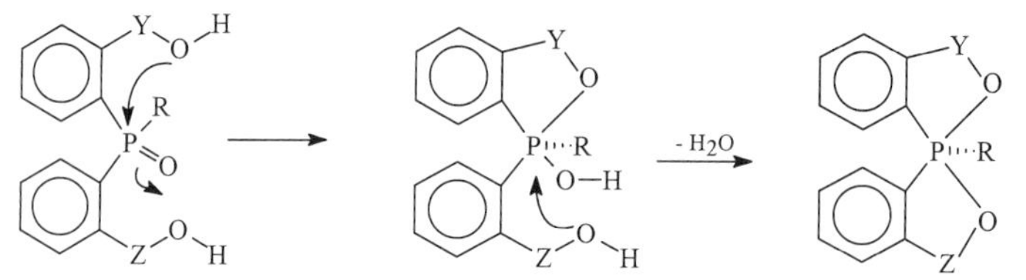

Scheme 6

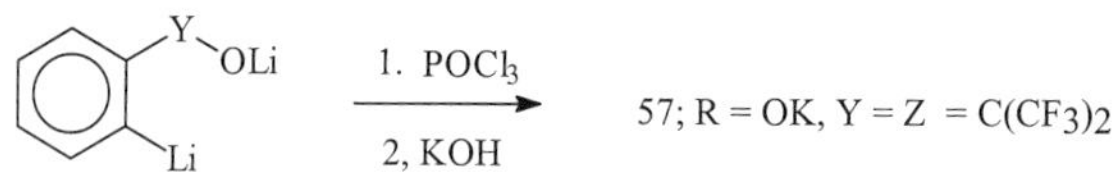

168

The dilithium salt (**168**) with $POCl_3$ gave, after work up and treatment with KOH, the dibenzospirophosphorane potassium salt (**57**; Y = Z = $C(CF_3)_2$, R = OK) ⟨81JOC1049⟩. The dilithium diylide of diethylphenylphosphine oxide added to the carbonyl group of a bis 4-chloro substituted benzophenone to produce after cyclodehydration the dioxyphosphorane (**52**; R = Me) ⟨97AG(E)2500⟩. Hexachlorocyclotriphosphazene, $(NPCl_2)_3$, may be used as a source of phosphorus in place of PCl_3, thus with *o*-aminophenol, spirophosphorane (**129**; Ch = O, X = NH, R = *o*-$OC_6H_4NH_2$) may be obtained via a NH bridged bis spiro compound ⟨75IC2831⟩. Methyltetraphenoxyphosphorane, prepared from methyltriphenoxyphosphonium iodide, using $(MeO)_2CCHNMe_2$, was converted by ethane-1,2-diol to the spirophosphorane (**77**; Z = $(CH_2)_2$, Y = Me) ⟨89JGU634⟩. Analogous reactions occur when pentaphenoxyphosphorane, $(PhO)_5P$, reacts with catechol ⟨68T5041⟩ and with *o*-aminophenol ⟨76IZV2624⟩, as well as for the reactions of pentaethoxyphosphoranes, $(EtO)_5P$, with diols ⟨71JA4004⟩. Silylated oxygen or sulphur functional groups react with ZY_2PF_2 to produce phosphoranes (**142**; X = Y = Cat, Z = F, Et, Pr^i, CH_2SiMe_3) ⟨71JCS(A)1295⟩, its naphthdiol analogue (**142**; X = Y = Don), ⟨92PS(69)231⟩, (**142**; X = Y = Ami, Z = H) ⟨92CB801⟩ and (**73**; X = Y = S) ⟨73CC144⟩.

The use of trisphosphoramidites, e.g. $(Me_2N)_3P$, often has advantages over trichlorophosphine, e.g. with amidrazones ⟨72CR(C)(274)2209⟩, to give hydrophosphoranes (**142**; X = Y = Ami, Z = H). They have also been used in combination with CCl_4 in which case the reaction proceeds via quaziphosphonium salts ⟨75TL2529⟩. The P–N bonds are quite labile and strong acids are not generated. Thus spirophosphorane (**71**) has been prepared by the reaction of either $RP(Ch)Cl_2$ or $MeP(NMe_2)_2$ with a hydrazide $RCONHNH_2$ ⟨77CB1124⟩. Also two equivalents of the Mannich base of 2-cresol was reacted with $(Et_2N)_3P$ to generate a spirophosphorane with two P–C bonds ⟨67DOK(174)846⟩. Treatment of $(CF_3CONMe)_3P$ with bromine gave bromophosphorane (**85**) ⟨97JGU151⟩. There are a number of examples of the oxidative cyclisation of monocyclic hydrophosphoranes to give spirophosphoranes as shown in Equation (8) and Scheme 7 ⟨75TL497⟩. It has been suggested that the reaction of glycidol with PCl_3 to give a dioxaphospholan proceeds via a hydrotetraoxyphosphorane and that this may provide a simple method for the formation of the latter compounds ⟨97PS(131)173⟩ The deoxygenation of nitrobenzene derivatives by P(III) compounds can be used to make intermediate cyclic phosphoranes which react with *o*-aminophenols with migration of an aryl group from O to N ⟨75JCS(P1)2376⟩ (Scheme 8). Hydro-phosphoranes may be prepared directly from trichlorophosphine, e.g. (**81**; Y = CR_2, Z = H) from α-hydroxcarboxylic acids ⟨71BSF4185⟩ as well as tartaric and malic acids ⟨81PS(11)71⟩, and enolisable α-ketocarboxylic acids ⟨88PS(35)195⟩. Likewise, a di-*t*-butyl derivative of (**83**; X = Y = Z = O) was made using 3,6-di-*t*-butylcatechol ⟨88JGU1996⟩.

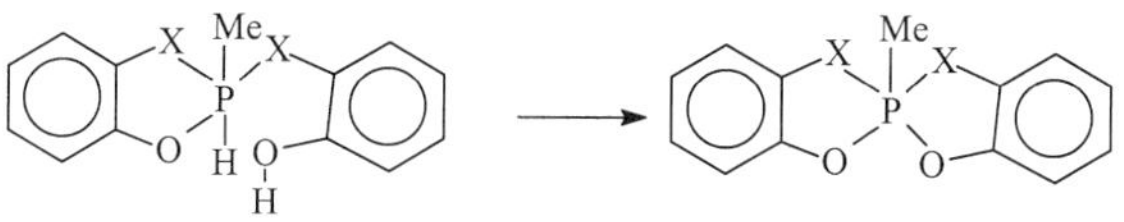

Equation 8

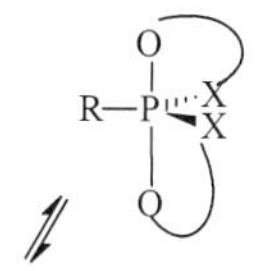

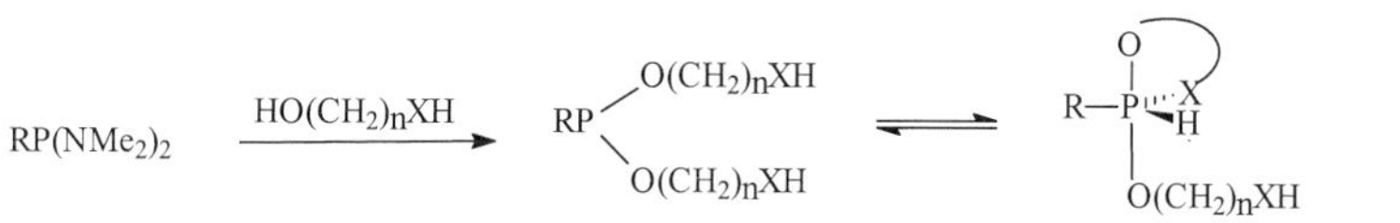

Scheme 7

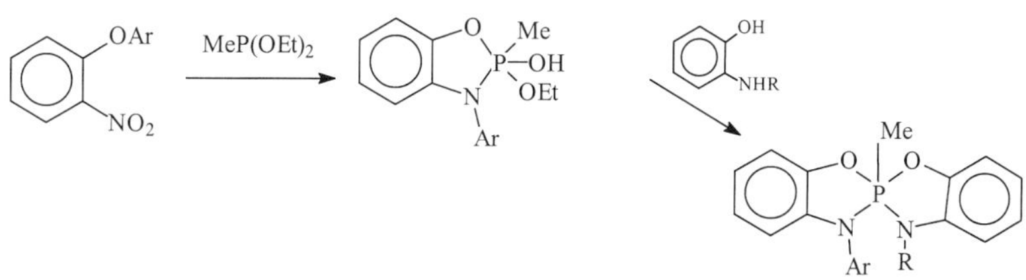

Scheme 8

6.3.9.1.3 *Perphosphoranides*

Very stable perphosphoranides may be prepared directly, e.g. (**113**; m = 0, n = 3, R = H) from catechol and PCl_3 or $POCl_3$ ⟨76TL917⟩ or PCl_5 ⟨77JPR(319)188⟩, (**118**; m = 0, n = 3) from PCl_3 and *N*-benzoylhydroxylamime ⟨77ZAAC(437)53⟩ and $(Oam)_3P^-$ from $(NPCl_2)_3$ ⟨68CC1606⟩. Silylation precursors have also been useful. Thus di(2-trimethylsilyloxyphenyl)phenylphosphine and PCl_5 gave the unique six coordinate salt (**123**) ⟨98AG(E)1098⟩. In a straight forward reaction 1-lithiated naphthalenes with dimethylamino substituents in the 8 position reacted with PBr_3 to form trinaphthylphosphines which have three dative PN bonds as shown in (**116**) ⟨93AG(E)1430⟩. White phosphorus reacts with 3,4-di-*t*-butylbenzoquinone in the presence of triethylamine to give (**113**; R = Bu^t, m = 0, n = 3) ⟨79IZV2398⟩. The phosphorus atom is invariably spiro to three identical rings.

6.3.9.2 From One Monocyclic Component

6.3.9.2.1 *Spirophosphonia compounds*

Spiro(5,5)phosphonium bromide (**1**; m = n = 5) was synthesised from 1-methylphosphorinan by monoalkylation with 1,4-dibromobutane followed by ylide intramolecular alkylation as shown in Scheme 9 ⟨79ZN(B)990⟩. Anodic oxidation of phosphine (**169**) led to the spiro quaternary phosphonium perchlorate (**16**) ⟨97MI-1⟩. Dilithium biphenyl reacted with the phenylphosphine oxide (**170**) to displace the phenyl and oxygen and give the cage salt (**14**) ⟨72CC404⟩. Generation of an electrophilic centre by protonation of

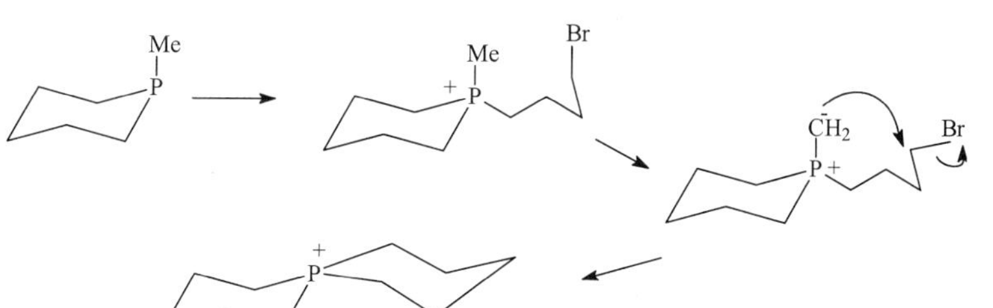

1; m = n = 5

Scheme 9

an allenylphosphonate was followed by cyclisation to the phosphoryl oxygen to give spiro trioxonium salts as shown in Scheme 10 ⟨84PS(21)249⟩.

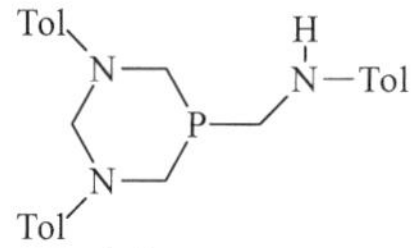

169

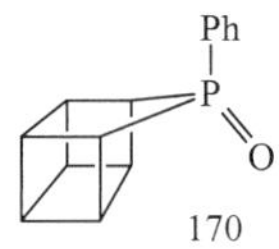

170

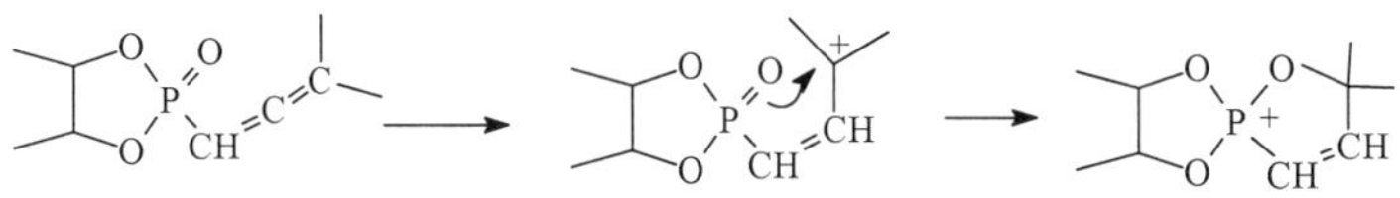

Scheme 10

A wide variety of spiro(5,5)phosphazenes may be prepared from hexachlorocyclo-triphosphazene by the replacement of two chlorine atoms by aromatic diamines or aminophenol ⟨85JA7585, 88IC1911⟩ or by a sequence of condensing reagents, e.g. AgNCO, ROH, MeCN ⟨87PS(30)519⟩. This approach has also been used to generate many bis spiro compounds ⟨82ZN(B)1425⟩ including 1,3-dioxaphosphorins ⟨84JCS(CC)675, 93POL1509⟩, and incorporating salicylate moieties ⟨75ICAL40⟩, as well as hydrazine ⟨89ZN(B)612⟩, dithiocatachol ⟨94IC3091⟩ and thioferrocene ⟨90ZN(B)741⟩, as well as spermine bridges, e.g. (**41**) ⟨84JST(116)75⟩. Some reactions are accelerated by the presence of $Bu_4N^+ X^-$ ⟨83MM719⟩. Use of polyethylene glycols produced some large ring spiro compounds in addition to polycyclic ansa bridged products ⟨95ICA(228)187⟩. Silylated diols $TmsOCH_2(CF_2)_nCH_2OTms$ and $P_3N_3F_6$ were employed for the synthesis of cyclotriphosphazene (**11**; X = F, Y = O, Z = $CH_2(CF_2)_nCH_2$) ⟨94IC2727⟩. An alternative approach involves building the cyclotriphosphazene ring from a cyclic trichlorophosphorane such as (**171**) and ammonia ⟨94JCR(S)134⟩. Appel's reagent (Ph_3P/CCl_4) was used to condense phosphinamides to form spiro cyclotriphosphazenes (**21**; $X_2 = Y_2$ = Bip, Z = Ph) ⟨97JCS(P2)15⟩. Conversion of chlorophosphine (**172**; X = Cl) to its azide (**172**: X = N_3) by $TmsN_3$ led to trimerisation with loss of three molecules of nitrogen and the formation of the tris spiro cyclotriphosphazene (**39**; R = NMeDopCONMe) ⟨94ZN(B)171⟩. The addition of a second ring may also be achieved by condensation of difunctional reagents with the geminal amino groups bound to phosphorus as shown in Equation (9) ⟨76PS(6)113⟩. In the case of supermesitylmagnesium bromide, following substitution of one chlorine atom of hexachlorocyclotriphosphazene, the intermediate can be ring closed from phosphorus to one of the *t*-butyl methyl groups, under the action of $AlCl_3$, to give spirophosphazene (**20**) ⟨90JCS(D)2303⟩. Five membered phosphazenes (**173**) derived from PCl_3 and amidrazones, react with propanedithiol to produce spirophosphazene (**32**; X = S, Y = $(CH_2)_3$) ⟨76T2039⟩ and eight membered octachlorocyclotetraphosphazenes with propanediols give the corresponding spirophosphazenes (**35**; Ch = S, R = $(CH_2)_3$) ⟨91PS(55)59⟩.

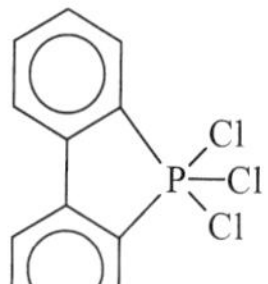

171

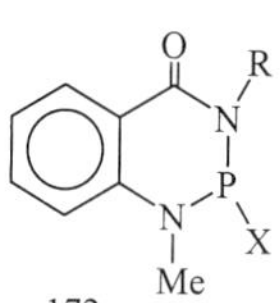

172

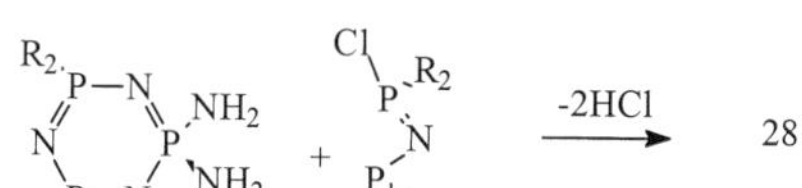

Equation 9

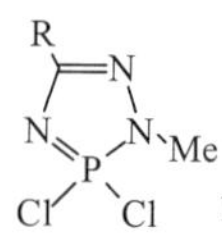

173

Lithium biphenylenephosphoranide (**174**) which may be prepared by the action of butyl lithium on the fluorenylphosphine (**175**; X = Br), is in equilibrium with the ring opened isomer (**175**; X = Li) ⟨69CB548⟩. In most cases the phosphoranides are produced by the action of base on the corresponding hydrophosphoranes ⟨79JA4623⟩.

6.3.9.2.2 Spirophosphoranes

Spirophosphoranes have been prepared either through displacement of chlorine ⟨80JGU812⟩ or bromine ⟨80ZAAC(471)147⟩ of $CatPHal_3$ or of two fluorine atoms on phosphorus, e.g. (**176**; R = Ph), and

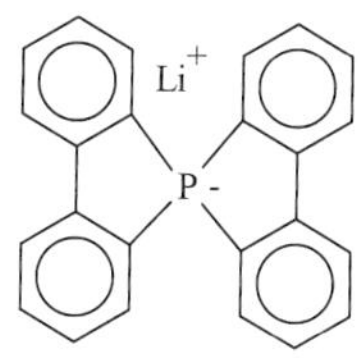

174

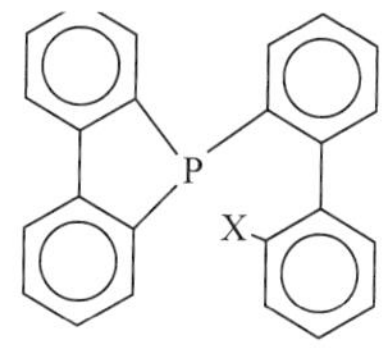

175

by the dilithium salt of diols or by cycloaddition (see Section 6.3.9.3.2). On the other hand the catecholspirophosphoranes were prepared from difluorophosphoranes (**176**; R = Ph, Me, Bu^t, F) by condensation with trimethylsilanyl derivatives of catechol ⟨78IJ107⟩. Monocyclic phosphoranes with leaving groups other than halogen can also be used (e.g. methoxy, or dimethylamino) in reactions with diols and amino alcohols ⟨72PS(2)41, 75P(5)285⟩. These reactions can be exothermic.

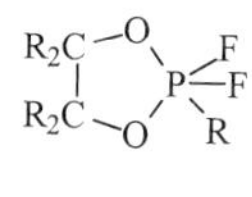

176

Many phosphorus(III) compounds add to two equivalents of compounds possessing electrophilic multiple bonds such as carbonyl groups and electrophilically activated double and triple bonds. The driving force for the second stage (a 1,3 polar addition) is usually the formation of a phosphorane incorporating a small ring. Whilst the reaction of triphenylphosphole with dimethyl acetylenedicarboxylate (DMAD) gives the spirophosphorane (**148**), it is not stable ⟨68T3437⟩. However, the use of a P(III) 1,3-dioxaphospholan adds stability. Thus pinacol derived methoxyphospholan (**177**) reacts with two equivalents of DMAD to give a spirophosphorane as shown in Scheme 11 ⟨81TL3533⟩. Benzaldehyde adds in the normal head to tail mode with catechol derived phospholan as shown in Equation (10) ⟨76JGU581⟩, and this can occur for

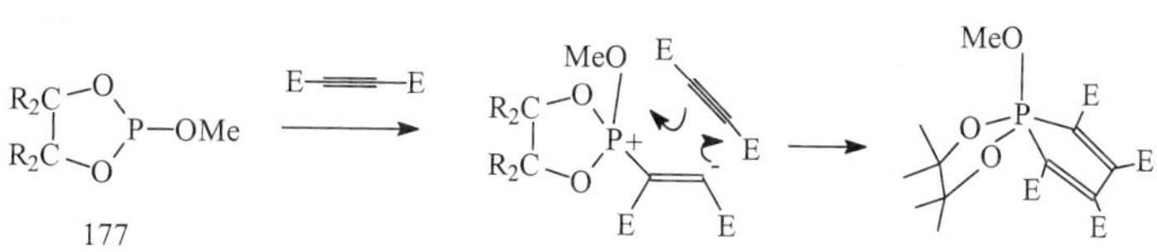

177

Scheme 11

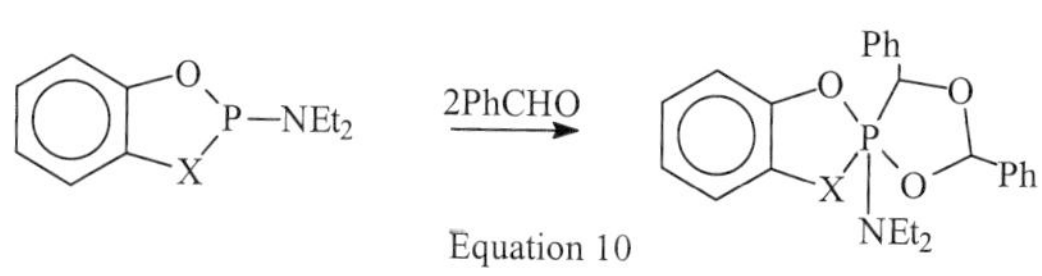

Equation 10

two carbonyl groups in the same molecule to give bridged phosphoranes related to (**96**) ⟨80TL4331⟩. In constrast substituted benzaldehydes with 1,3-diazaphospholan and hexafluoroacetone with six membered cyclic phosphoramidite gave head to head spiro adducts as shown in Scheme 12 ⟨68JOC13⟩ and (**74**; Y

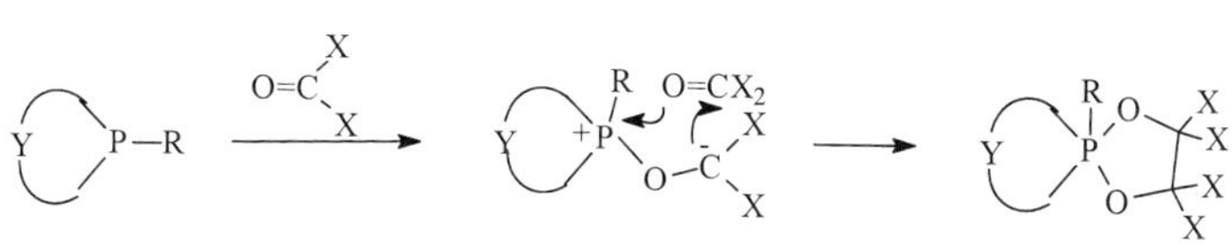

Scheme 12

= Pfp, Z = NMe_2) ⟨95PS(106)65⟩, respectively. Benzophenone and PhCOCN ⟨75CR(C)(280)225⟩ also give head to head products. Similar reactions occur with chloral and tribromomethylacetaldehyde ⟨87JGU1078⟩. Note that carbonyl compounds such as hexafluoroacetone are better able to support negative charge on carbon when attacked at oxygen in the initial step ⟨93PS(78)271, 74JCS(P1)2125⟩. When this reagent was reacted with diphosphine (**179**) three equivalents of the ketone reacted with cleavage of the P–P bond resulting in the formation of the bridged spirophosphorane (**180**) in which two molecules of ketone had added head to tail to phosphorus and the third had added across a P–C bond ⟨97CB819⟩. Phosphonium ylides (**181**) react exothermically with oxiran but a temperature of 130ºC is required to achieve reactions with oxetan to give oxaphosphoranes shown in Equation (11) ⟨79CB501⟩. A stabilised ylide cyclised on

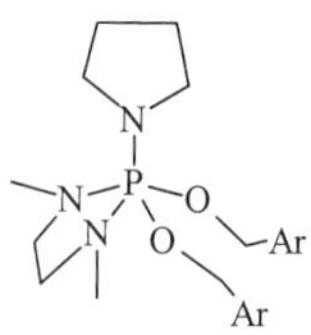
178

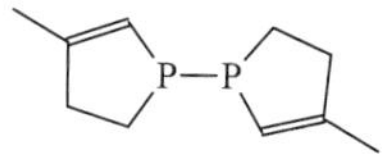
179

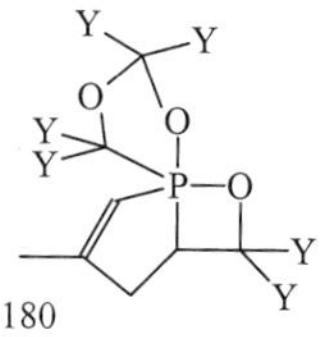
180

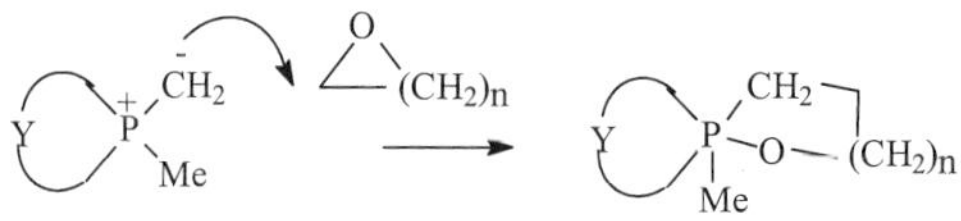
181

Equation 11

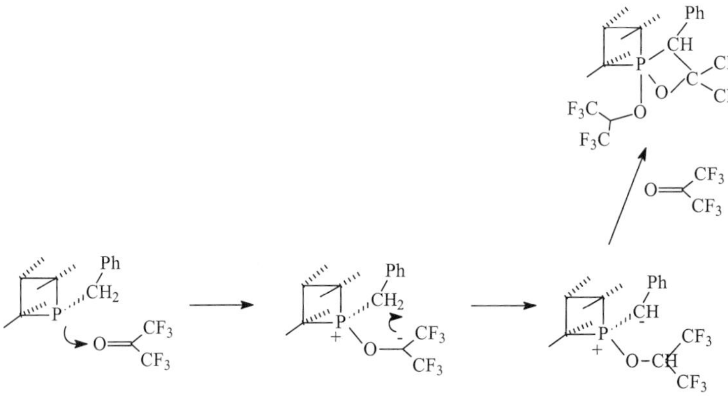

Scheme 13

heating at 140°C and 150°C to give the amides shown in Scheme 4 ⟨85PS(24)235⟩. It is possible that an ylide may be also formed in the reaction of benzylphosphetan with hexafluoroacetone in the formation of phosphoranylspiro(3,3)oxaheptane as shown in Scheme 13 ⟨73JCS(P1)1300⟩.

Phosphazene (**182**) generated from a phospholan and phenyl azide was cyclised as shown in Equation (12) ⟨76CC730⟩. Stabilised phosphazenes, e.g. (**183**) have also been cyclised in this way ⟨78JGU1308⟩. The PhN= functionality can also be act as a leaving group, thus phosphazene (**184**) was reacted with C_2 to C_5 diols with loss of aniline to produce spirophosphoranes (**77**; Y = Ph) as shown in Equation (13) ⟨79JCS(CC)191⟩. An alternative approach involves the reaction of a functionalised azides, e.g. 2-azidophenol with cyclic phosphines to give intermediates set up for cyclisation via intramolecular addition to the P=N bond as in Scheme 14 ⟨80PS(8)127⟩. Similarly, as shown in Equation (14), phosphazene (**185**) gives spirophosphorane (**164**; $R^1 = R^2 = H$, $R^3 = R^4 = Ph$) ⟨90PS(49)385⟩. Related phosphazene intermediates were proposed in the reaction pathway shown in Scheme 15. It has two interesting features, first the use of a dialkylamino leaving group in the initial reaction with diol, secondly, the N to P migration to produce a phosphazene function to which the second hydroxyl group cyclises ⟨90JGU397, 96JGU355⟩.

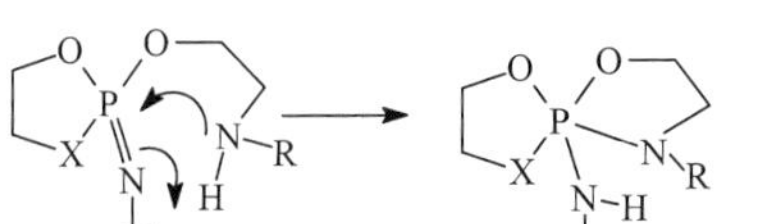
182

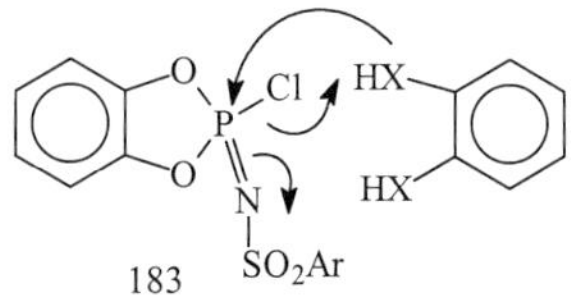
183

Equation 12

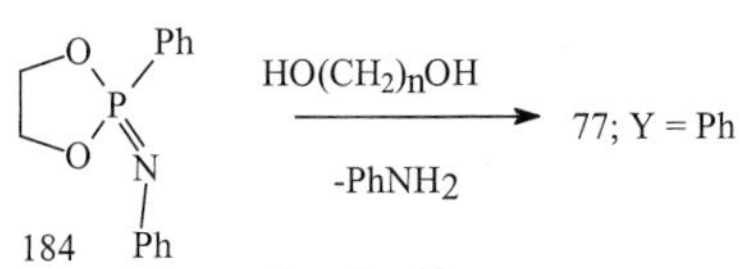
184

Equation 13

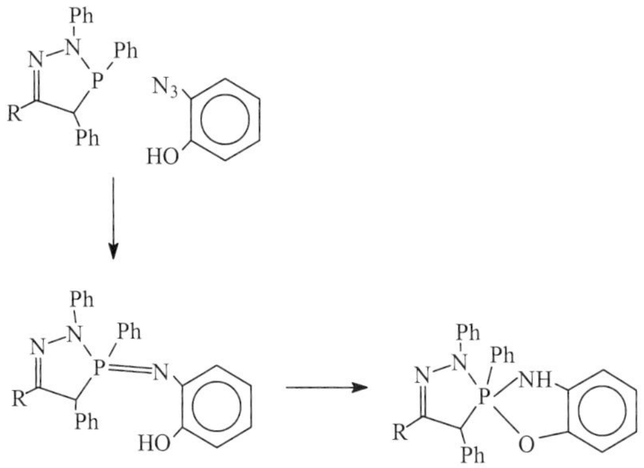

Scheme 14

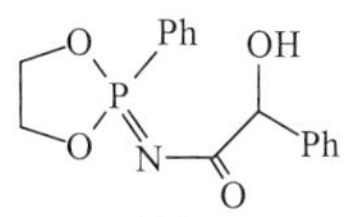

185

Equation 14

164; $R^1 = R^2 = H$,
$R^3 = R^4 = Ph$

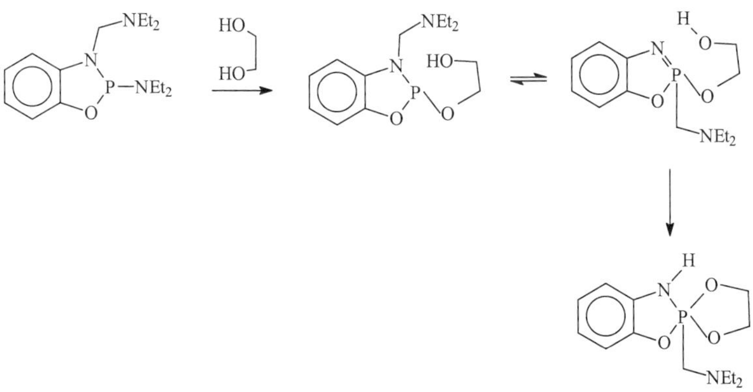

Scheme 15

The reaction of cyclic three coordinate phosphorus compounds with diols has been found to proceed smoothly in the presence of di-iso-propylchloramine ($ClNPr^i{}_2$) ⟨75TL1583⟩. This reaction would appear to involve initial formation of an amino salt which reacts with one of the alcoholic hydroxyl groups followed by cyclisation and loss of amine after protonation by the liberated HCl as shown in Scheme 16. If the

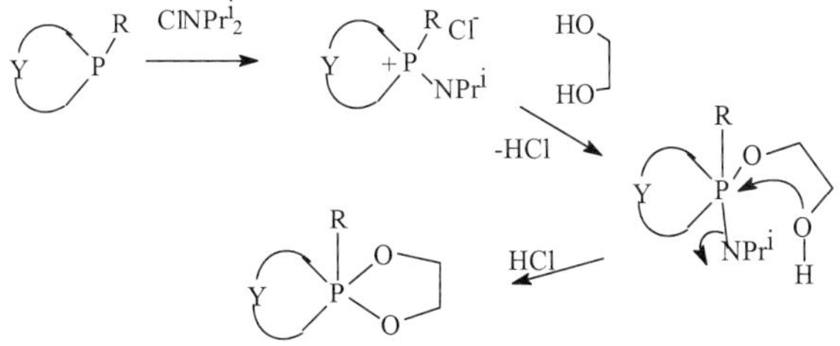

Scheme 16

P(III) atom already bears a potential leaving group such as SEt, it is possible to use an activating group such as ethyl acrylate as in Scheme 17 ⟨83PS(17)283⟩. On the other hand 2 + 4 cycloaddition reactions with dichlorophosphines produce directly a phosphonium centre, e.g. (**186**) suitable for reacting with diols and diamines, etc., to give *cis* and *trans* spirocompounds (**62**) ⟨89PS(45)255⟩. Unsaturated carboxylic acids add to cyclic P(III) compound (**187**; Ch = O) to give an hydrophosphorane which cyclises by Michael addition of the PH group to the acrylic moiety (Scheme 18) ⟨78IZV2185⟩.

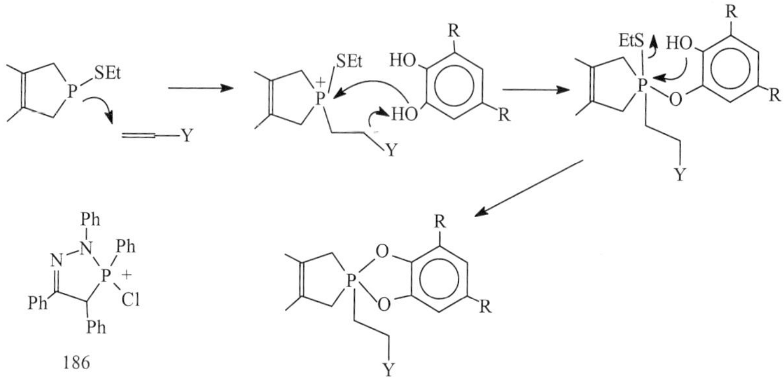

186

Scheme 17

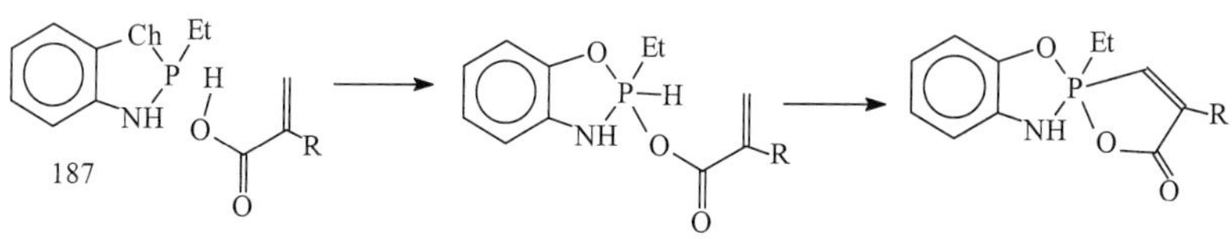

187

Scheme 18

Spiro hydrophosphoranes may be prepared by a condensation reaction of *P*-halo or *P*-amino-dioxaphospholanes and 1,2-aminoalcohols (**188**; X = NH, Y = CH_2) ⟨72CR(C)(274)419, 72BSF1413⟩ or amino acids (**188**; X = NH, Y = CO) ⟨77T635⟩ as in Equation (15). Alternatively, routes to intermediates involve the reaction of PCl_3 a Grignard salt of the benzyl alcohol (**189**), Scheme 19 ⟨79JA4623⟩. The alkyl substituents R assist the electrocyclisation. Similar reactions with hydrazides and amidoximes (RC(=NOH)NH_2, give hydrophosphoranes (**142**; X = Pnc, Y = Amo, Z = H) ⟨72CR(C)(275)151⟩ (**142**; X = Glc, Y = Amo, Z = H) ⟨82CB2560⟩ and (**142**; X = Cat, Y = Amo, Z = H) ⟨72CR(C)(276)1211⟩.

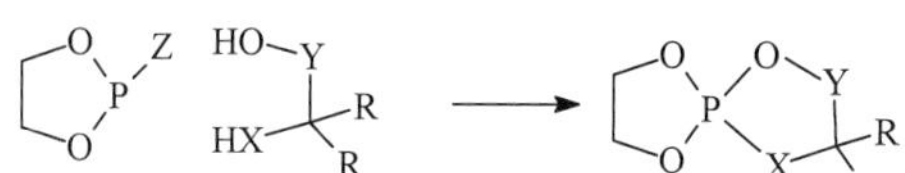

188

Equation 15

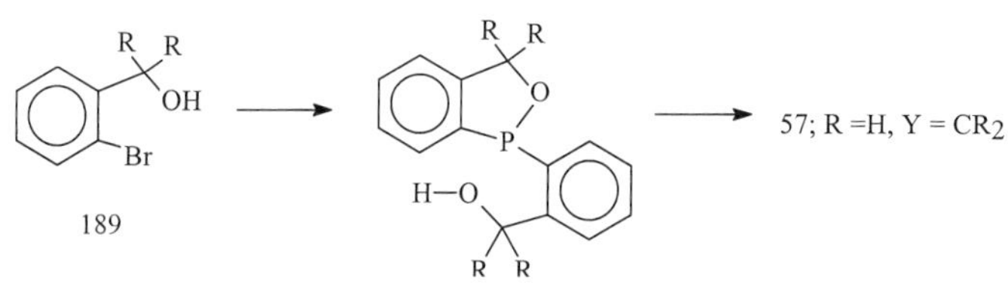

189

Scheme 19

There were a number of reports in the 70s of monocyclic hydrophosphoranes acting as intermediates in the formation of spirophosphoranes. As shown in Equation (8), hydrophosphoranes convert to phosphoranes slowly but spontaneously when X = O, and on heating or with a mild oxidising agent when X = NH ⟨75TL3077, 83JGU1759⟩. This route has been exploited using bicyclic hydrophosphoranes such as (**106**; X = CH_2CH_2O, Y absent, Z = H) for the synthesis of a wide range of medium and large ring (21+) spiro bicyclic phosphoranes (**105**) ⟨94PS(90)131, 95MI-2⟩. Monocyclic phosphoranes (**190**) cyclise at 160° by a reaction analogous to electrophilic attack of an aryl ether to give spiro(4,5)phosphorane (Scheme 20). The yields are higher as the donor properties of Z increase ⟨86T3521⟩. Note that this argues against a transition state with phenoxide character for the migration and in favour of alkoxy phosphonium character to give an ylide which, after proton transfer to the ylidic carbon with rearomatisation. An alkylarylether oxygen will cyclise to a phosphonium centre with dealkylation by bromide ion ⟨79PS(7)57⟩ as shown in

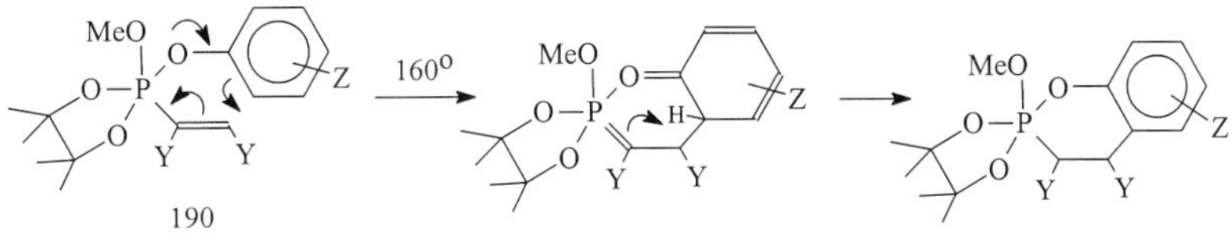

190

Scheme 20

Equation (16). Alkylation of the sulphur atom of thiophosphortriamidate (**191**) by alkyl halide initiated the formation of a fourth PN bond and deprotonation of the participating N atom ⟨96T2995⟩. Ring closure can also be achieved by transfer of a trimethylsilyl group to a phosphoryl group as shown in Equation (17) ⟨76JA4330⟩. This approach was used in the reaction of N and O disilylated valine with CatP(O)Cl in the preparation of phosphorane (**143**; R = Pr^i, X = Cat, Y = OTms) ⟨97MI-4⟩. Two alternative related approaches to this type of phosphorane have been reported by these Chinese workers. The first involved making CatP(OPh)Cl_2 from PCl_5, before reacting with N and O disilylated α-amino acids ⟨98S855⟩ and the other involved the reaction of EtOPCl_2 with N and O disilylated α-amino acids followed by reaction with anthraquinone to make a series of phosphoranes (**143**; R = H, Me, Pr^i, Bu^{i}0, Bn; X = Cat, Dioxyanthracene) as shown in Scheme 21 ⟨98S376⟩.

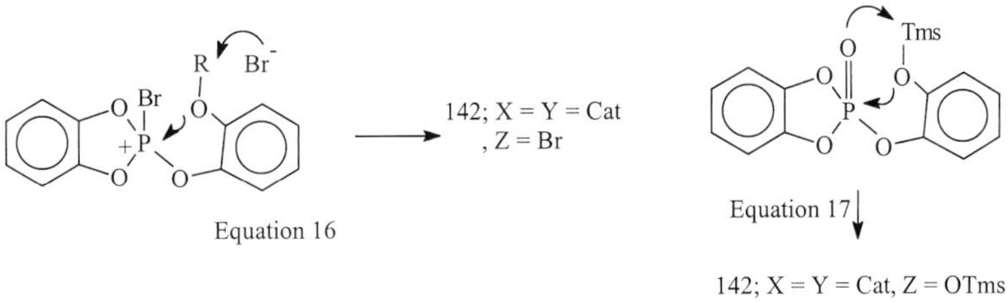

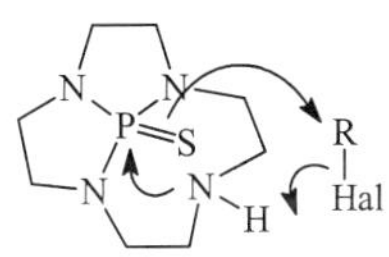

191

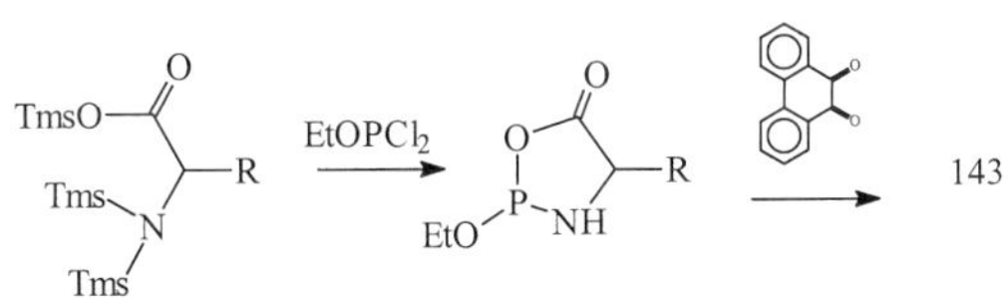

Scheme 21

6.3.9.2.3 Spiro-perphosphoranides

Perphosphoranides with phosphorus spiro to three rings one of which is different from the other two (e.g. **144**; X = Y = Cat, Z = Dpe) may be prepared by the reaction of a monocyclic phosphorane bearing three exocyclic dimethylamino leaving groups and a chelating ligand such as catechol ⟨75P(5)285⟩. Reactions of 4-nitro-*O*,*O*-phenylene phosphorochloridate with silylated amino acids gave a series of tris-spiro perphosphoranides ⟨98MI-1⟩

6.3.9.3 By 2 + 4 Cycloaddition to Phosphorus

6.3.9.3.1 Spirophosphonia compounds

Two coordinate phosphines are able to participate as dienophiles in cycloaddition reactions with 1,3-dienes and their hetero analogues such as di-imines, α-diketones and unsaturated ketones, etc., to produce four coordinate spiro compounds. Thus cyclic two coordinate phospholes, e.g. (**146**) add to diacetyl ⟨84TL5521⟩, diimine (RC=NCRCR=NR) ⟨86TL2971⟩, and to azodicarboxylic esters ⟨81CB825⟩ to give the corresponding spirophosphazenes (**32**). In the case of the product from diacetyl, a dimer is formed initially. It is also possible to have the phosphorus III atom as part of the 4 electron system adding to a dienophile such as an activated carbonyl group as shown in Equation (18) ⟨82JGU189⟩.

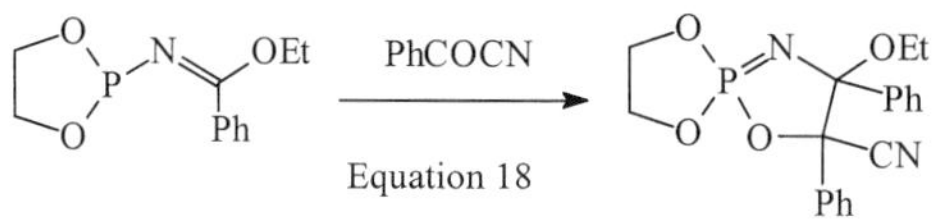

6.3.9.3.2 *Spirophosphoranes*

The reactions of three coordinate phosphines with 1,3-dienes and their hetero analogues such as di-imines, α-diketones and unsaturated ketones, etc., to produce five coordinate spiro phosphoranes has been widely exploited. Thus 1,3-butadiene reacts with catechol phosphonous esters to produce dioxaphosphoranes (**192**) ⟨69TL4693⟩, the reaction rate increasing with increasing electron accepting properties of group Ar ⟨72JGU782⟩. Ortho quinones such as di-*t*-butylbenzophenone ⟨94ZN(B)145⟩ phenanthraquinone ⟨68T1785⟩, tetrachloro-ortho-quinone ⟨96S473⟩, make excellent 4p-π compounds, the increase in aromaticity and formation of stable P–O bonds strongly favouring the forward reaction. Thus these are favourite reagents for producing spirophosphoranes with three P–C bonds ⟨72JA9264, 80CB1406⟩. This reaction may be extended to α-diketones ⟨70CR(C)(271)418, 75T797⟩, such as perfluoroalkyl-1,2-diketones ⟨96CB725⟩ and diacetyl or benzil ⟨94JGU1329⟩, ketoesters ⟨92JGU1172⟩, azoketones ⟨72CR(C)(274)419⟩, isatin ⟨85JGU1986⟩ and 1,2-diimines ⟨94JGU1546⟩. Aliphatic spirophosphoranes, e.g. (**193**), with two PC bonds have been prepared from methoxyphospholenes and diacetyl ⟨89PS(44)193⟩ or benzophenone ⟨85JGU1514⟩. A novel approach involved cycloaddition of two equivalents of benzil to phenylphosphinidene (PhP), or its equivalent, generated from $(PhP)_5$ at 160°C or photochemically, to give (**142**: X = Y = Dpe, Z = Ph) ⟨68CB1381, 71JCS(C)593⟩. A range of phosphoranes (**70**; R^2 = Y = H, Z = Glc) have been prepared from unsaturated aldehydes ⟨82JGU221⟩ extending previous studies of unsaturated ketones, carboxylic acids and amides, and other phospholans. Cycloaddition of α-cyanoacrylate to CatPNHR gave a phosphorane with one P–C bond ⟨94JGU373⟩. In an analogous manner 1,4 addition of a two coordinate phosphatriazole to dimethyl diazodicarboxylate, E–N=N–E, followed by dimerisation gave the polycyclic bis phosphorane (**107**) ⟨81CB825⟩. Cycloadditions to a dioxaphospholan was found to be faster than with the corresponding benzodioxaphospholan ⟨72JGU2109⟩. As above, it is also possible to have the phosphorus III atom as part of the 4 electron system adding to an activated carbonyl group as the dienophile. In the example shown in Scheme 22 the addition occurs twice to produce the bridged spirophosphorane (**194**) ⟨90JA7451⟩. Tetrachloro-*o*-benzoquinone ring contracts cyclotriphosphazane $(EtN)_3(POAr)_3$ to give bis spirophosphorane (**92**; R = Et, X = OAr, Y = Tcc) ⟨99IC1093⟩.

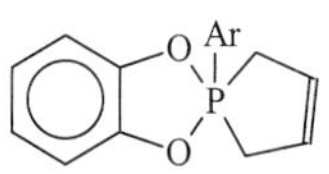

192

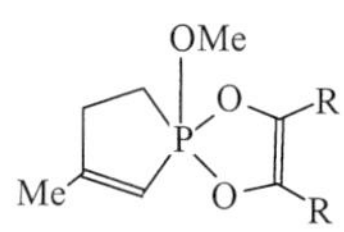

193

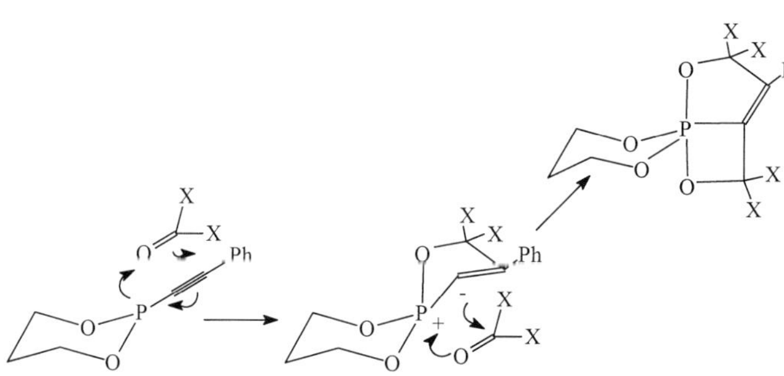

Scheme 22

Ozone adds to dioxabenzophospholans and dioxaphosphinans to give stable spiro ozonides (**195**; Y = Cat) and (**195**; Y = Plc) ⟨84PS(20)55⟩. Tetrachloro-*o*-benzoquinone reacts at 0 to −20° with $CatPCl_3$ to give TccCatPCl with liberation of chlorine ⟨95JGU289⟩ a reaction which could proceed via cycloaddition to CatPCl.

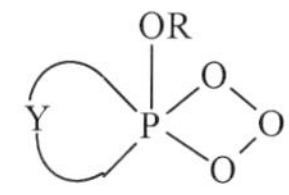

195

6.3.9.3.3 Spiro-perphosphoranides

Bis phosphonite $(CatP)_2CCl_2$ reacted with two equivalents of tetrachloro-ortho-quinone to form perphosphoranide Tcc_2P^-Cat ⟨97CB1479⟩. Six coordinate compounds involving a dative bond from a sulphone group (**120**) and from quinoline (**121**) were both produced by oxidative additions, the first used tetrachloro-*o*-benzoquinone to the appropriate cyclic phenylphosphonite ⟨97IC2578⟩ and the latter using 3,5-di-*t*-butyl-*o*-benzoquinone and anthraquinone ⟨96JA9841⟩.

6.3.9.4 By 2 + 2 Cycloaddition to Phosphorus

6.3.9.4.1 Spirophosphoranes

The reaction of cyclic three coordinate phosphorus compounds with cyclic peroxides and disulphides gives valence expanded spirophosphoranes with a wide range of ring size possibilities as shown in Equation (19). Thus phenylphosphetan reacted with cyclic peroxides and also with a disulphide to give phosphoranes such as (**54**; Y = $(CH_2)_4$) ⟨77PS(3)51⟩, (**54**; Y = $(CH_2)_3$ 1,3-cyclopentanylidene) ⟨83JOC2621⟩ and disulphide (**196**) ⟨72CC395⟩, and (**197**) ⟨84MI-3, 84MI-4⟩ respectively. In the case of cyclic diselenides (**198**), one selenium is removed and the remaining fragment forms part of the spirophosphorane (**130**; Ch = Se, Y = $C(CF_3)_2$, Z = N=$CNMe_2$) ⟨78AG(E)774, 80CB2699⟩.

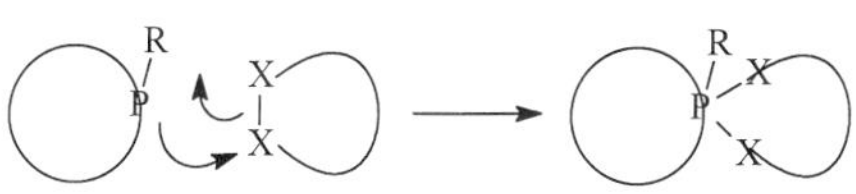

Equation 19

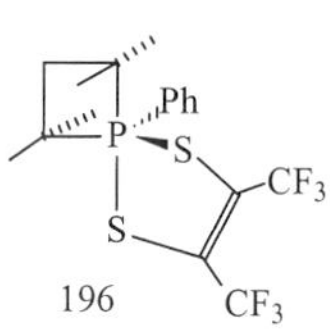

196

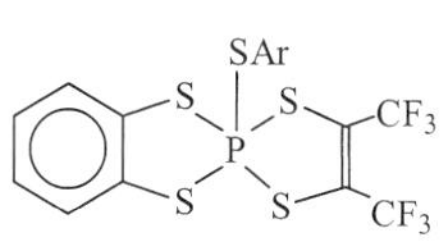

197

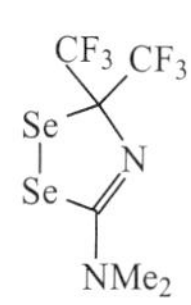

198

There are many examples of monocyclic phosphazenes dimerising to give spiro(3,4)phosphoranes ⟨75CB820, 92JGU780, 92JGU970⟩. An interesting example involved the reaction of phosphoramidite CatPNHPh with α-cycanoacrylate to give phosphazene $CatP(CH_2CHCNCO_2R)$=NPh which dimerised to bis phosphorane (**92**; R = Ph, X = Z = $CH_2CHCNCO_2R$, Y = Cat) ⟨94JGU373⟩. Supermesitylphosphine reacts with diazodicarboxylic ester to give diphosphazene (**199**; R = Bu^t, Y = CO_2R) which, at 110°C,

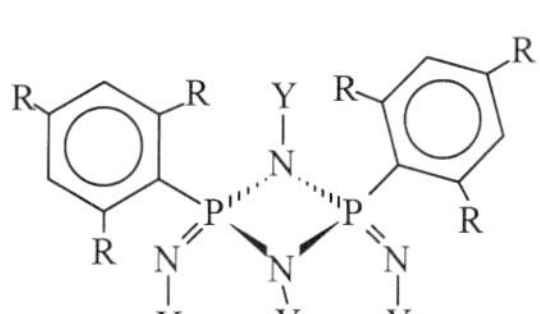

199

undergoes CH addition (from a methyl group) to a P=N bond at both phosphorus centres to give bis spirophosphorane (**91**) ⟨88PS(35)247⟩. There are other examples of ortho *t*-butyl groups reacting in this way. Hexafluoracetone adds across a phosphazene bond to give phosphorane (**69**) without disturbing three silyl groups ⟨94PS(86)157⟩.

Examples of 2 + 2 cycloadditions to phosphoryl groups have been observed for the first time. Thus highly hindered phosphine oxide (**200**; Ar = 2,4,6-tri$Pr^iC_6H_2$) adds to dimethyl acetylenedicarboxylate to form spirophosphorane (**201**) with four P–C bonds (Equation (20)) ⟨99CC1423⟩. The formation of bis spirophosphorane (**97**) was attributed to dimerisation of intermediate (**202**) which clearly has considerable phosphenium ion character (Equation (21)) ⟨87JCS(CC)1131⟩.

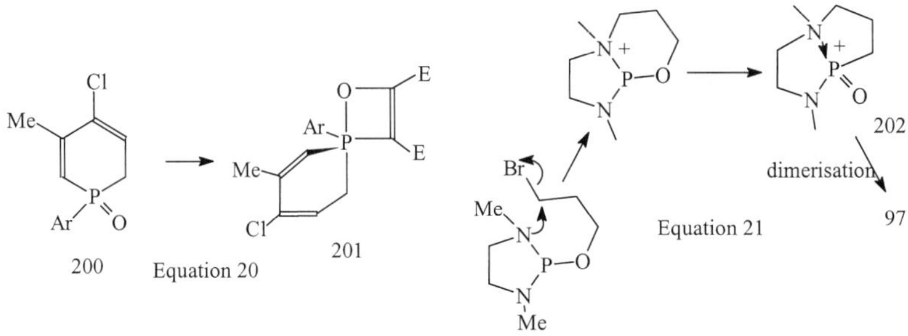

6.3.9.4.2 *Spiro-perphosphoranides*

Bicyclic potassium phosphoranide (**43**; Y = CF_3) in the presence of [18]crown[6] absorbed oxygen from air to form crystalline perphosphoranide (**111**; Y = CF_3) ⟨99JA6958⟩.

6.3.9.5 By P-Substituent Modification

6.3.9.5.1 *Spirophosphonia compounds*

Spirochlorophosphoranes, under the action of Lewis acids such as $SbCl_5$, may be converted to the quaziphosphonium salt, e.g. (**22**) ⟨88ZAAC(561)49⟩. Hydroxyphosphorane (**57**; R = OH, Y = Z = CMe_2) is converted to the spiroquaziphosphonium triflate (**19**) by the action of trifluoromethanesulphonic acid ⟨78JA5229⟩; this in turn can be reduced by LAH to the phosphoranide (**57**; R = electron pair, Y = Z = CMe_2). Conversion of cyclotriphosphazenes to their spermine derivatives has been studied ⟨85ICA(108)23⟩.

6.3.9.5.2 *Spirophosphoranes*

Phosphoranes such as phosphoranylspiro (4,4)nonane (**3**; $m = n = 3$, Y = Me) and the biphenylene phosphoranes (**45**) may be prepared by the reaction of lithium alkyls or lithium aryls with a spiro salt, e.g. (**1**; $m = n = 4$) ⟨77AG(E)722⟩ or (**13**) ⟨77CB693⟩. Hydrobiphenylenephosphoranes (**45**; R = H) and (**57**; R = H) have also been prepared by LAH or borohydride reduction of the biphenylene salt (**13**) ⟨69CB528⟩ and salt (**19**) ⟨79JA4623⟩. Trans esterification of vinyloxyphosphoranes derived from butane-1,2-diol gave good yields of a range of [4,4] pentaoxyspirophosphoranes ⟨99DOK(367)64⟩.

6.3.9.6 By Ring Modification

Azodicarboxylic ester can be used as an activating agent for P(III) compounds undergoing 2 + 4 cycloaddition reactions to give phosphoranes such as (**203**) and then being replaced by reagents such as diols liberating the reduced azoester as its hydrazine ⟨80TL1449, 80PS(8)147⟩. Diphenyldisulphide, PhSSPh, has also been used as a condensing agent ⟨82JOC916⟩; whilst a quaziphosphonium salt was proposed as an intermediate, a dithiophosphorane intermediate is quite possible.

Spirophosphorane (**87**; X^1 = MeN, X^2 = PC_6H_5, Y = CONAr, Z = Ar) was formed in the reaction of a di-silylated derivative of urea with dichloro(perfluorophenyl)phosphine, presumably the expected eight membered ring rearranging as shown in Scheme 23 ⟨79CB1365⟩. Treatment of the corresponding product from $PhPCl_2$, with Cl_2 and PCl_5 removed one of the phosphorus groups and gave spirophosphorane (**64**; Y = Ph) ⟨80CB1847⟩.

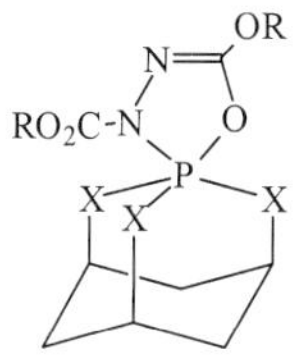

203

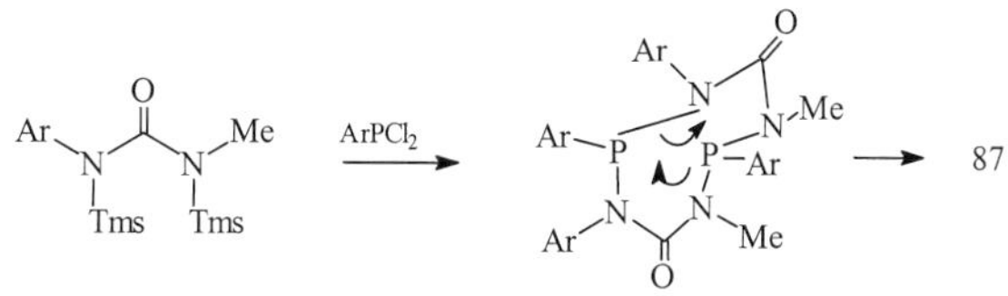

Ar = m-$CF_3C_6H_4$, Tms = Me_3Si

Scheme 23

6.3.9.7 Other Methods

6.3.9.7.1 Spirophosphoranes

Aniline hydrochloride converts the azathiobenzophospholan (**187**; Ch = S) to spirophosphorane (**204**) ⟨83JGU1759⟩. Presumably the aniline releases some aminothiophenol forming a hydrophosphorane which spontaneously dehydrogenates similar to that reported in Equation (8). Chlorination of the phosphorus atom of phosphoramidite (**205**) together with addition of chloride ion to the carbonyl group of the trifluoroacetyl group enable cyclisation to produce chlorophosphorane (**206**) ⟨95JGU198⟩. [3,3,0]Bicyclic phosphoramidite (**207**) displaced thiophenate from the disulphenate ester of 2,2′-dimethylpropanediol to give the bis spiro-phosphorane (**105**; Z = CH_2CH_2NMe) ⟨93HAC271⟩.

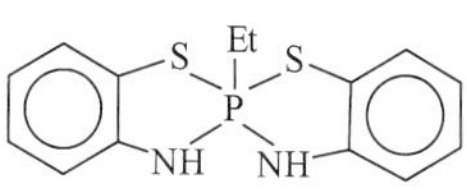

204

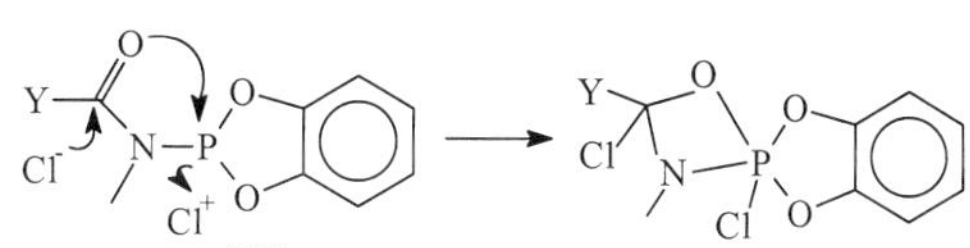

205 206

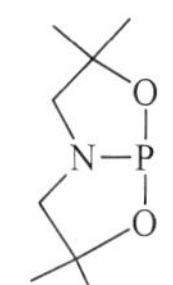

207

6.3.9.7.2 Spiro-perphosphoranides

Many six coordinate spiroperphosphoranides have been obtained by addition of a nucleophile to the phosphorus atom of a spirophosphorane. Thus the nucleophiles fluoride or *p*-fluorophenoxy anions gave (**159**; X = Glc, Y = Pfe, Z = F/ArO) ⟨80JCS(CC)157⟩, pyridine gave (**159**; X = Cat, Y = Pfe, Z = PhO, R = NC_5H_5) ⟨74JA7269⟩, and alcohols reacted with a hydrophosphorane to produce (**159**; R = H, X = Y = Cat, Z = OR) ⟨73CR(C)(276)297⟩. Spirophosphoranes (**142**; X = Y = Cat, Z = Hal) bearing an exocyclic leaving group (Z) such as halogen react with chelates such as bipyridyl giving (**208**) ⟨79PS(7)305⟩, 2-ketopyridines, and *β*-diketones giving (**209**) and with trisaminophosphazene (**210**) to give perphosphoranide (**211**) (Equation (22)) ⟨90JGU1491⟩. When spirophosphoranes have a second nucleophilic centre juxtaposed to the phosphorus atom there is a potential for the formation of a perphosphoranide. Examples include the 8-hydroxyquinoline compound (**160**) ⟨79T1825⟩. Spirophosphorus atoms participating in three different ring types, e.g. (**144**; X = Glc, Y = Dpe, Z = Hyc) have been prepared by taking a spirophosphorane possessing two different ring types and an exocyclic leaving group ⟨75BSF1433⟩. Extension of the side chain of the phosphinomethylphosphorane (**212**) by addition of the phosphine to isocyanate allowed the generated nitrogen anion to cyclise with the phosphorane to form perphosphoranide (**213**) ⟨96PS(109/10)493⟩.

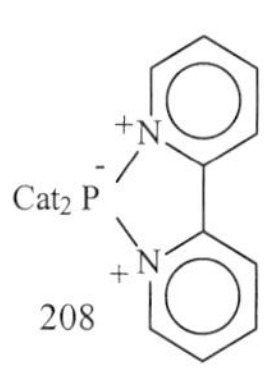

208

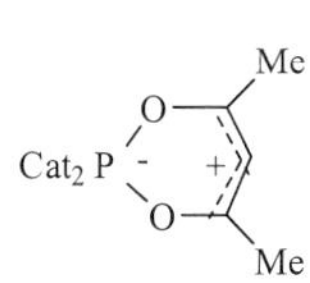

209

Equation 22

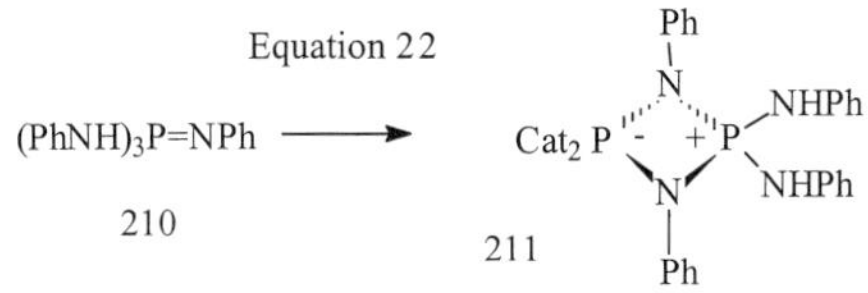

210 211

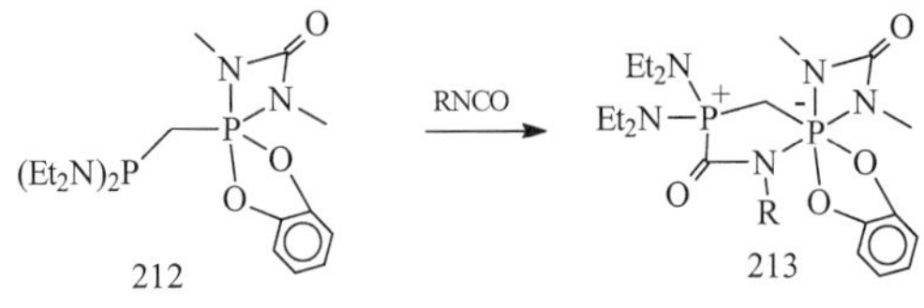

212 213

6.3.10 IMPORTANT COMPOUNDS AND APPLICATIONS

6.3.10.1 Non-Medical Applications

The tris spirocyclotriazene (**214**), prepared via condensation polymers ⟨75MI-1⟩, has been used to form clathrate tunnels in which alkenes can be polymerised with enhanced stereoregularity ⟨82MM697⟩. These compounds are able to form inclusion complexes with *p*-xylene thus offering a useful means of purification ⟨72BRP1257024⟩. They have also been patented in a number of countries as flame retardants ⟨75CA4319, 99JAP217508⟩. Attempts to polymerise spirocyclotriphosphazenes (**215**) produced ring expanded products ⟨93MM3⟩. The fluoride ion in the fenestrane related phosphonium fluorides (**216**; $n = 1$ or 2) are examples of compounds with a completely ionic fluoride ion ⟨81JA5265⟩. The ozonide (**217**) from an adamantyl type phosphite is a good water soluble source of singlet oxygen and is 1.4 times more stable than ozonide (**218**) ⟨75JOC1185⟩. The potential of spirophosphoranes in coordination chemistry has been reviewed ⟨94MI-2⟩.

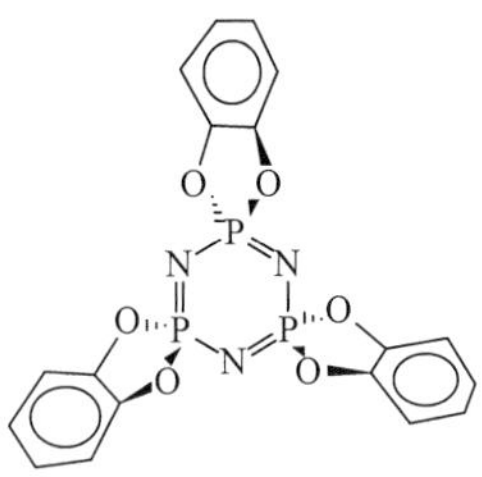

214

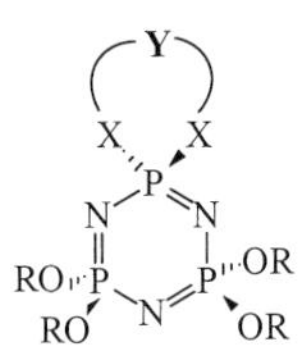

215

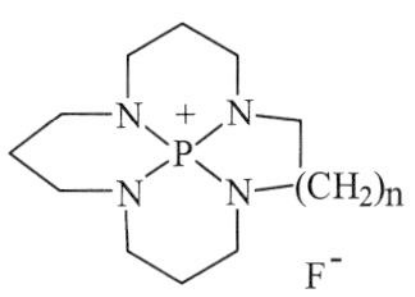

216

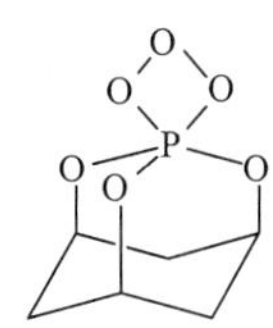

217

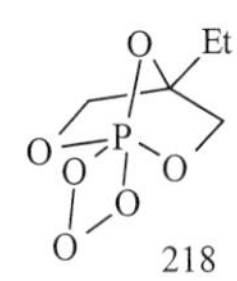

218

6.3.10.2 Compounds of Medical Interest

A series of bis cyclotriphosphazenyl spermines such as (**219**) have been prepared and shown to have anticancer activity in six murine tumour systems. These compounds had a higher potency and lower toxicity than related aziran derivatives ⟨87PS(29)147⟩. Spirophosphorane derivatives of thymidine such as (**220**; $R = CF_3$) have been used as models for the study of the stereochemistry of cyclic AMP interactions with enzymes ⟨90JA7451⟩. The base catalysed proton transfer observed for the equilibrium shown in Scheme 24, mimics the role of the bases histidine 12 and histidine 119 in the active site of Rnase A during the hydrolysis of ribonucleic acids ⟨84PS(19)173⟩. Thymidine derivatives of phosphaporphyrin (**124**) have been patented as an artificial nuclease ⟨99JAP29591⟩.

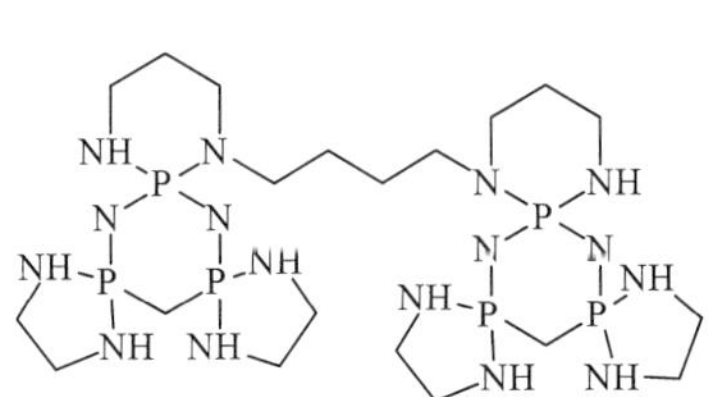

219

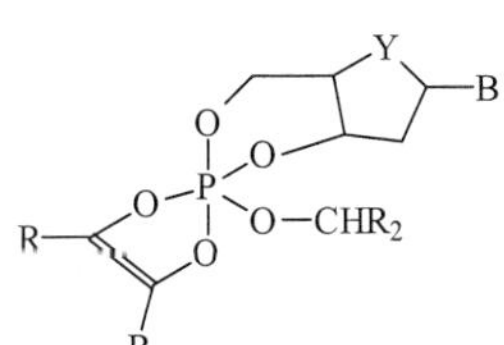

220

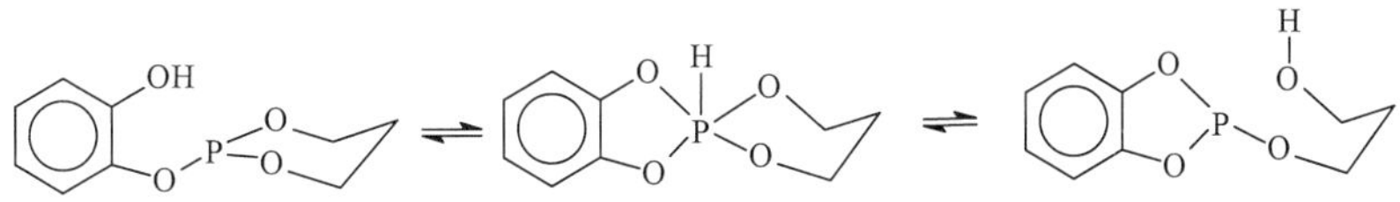

Scheme 24

6.3.11 REFERENCES

61JCS2126 A.W. Johnson and J.C. Tebby; *J. Chem. Soc.*, 1961, 2126.
65CB576 D. Hellwinkel; *Chem. Ber.*, 1965, **98**, 576.
67DOK(174)846 B.E. Ivanov, A.B. Ageeva and Yu.Yu. Samitov; *Dokl. Akad. Nauk SSSR*, 1967, **174**, 846 (*Chem. Abstr.* 1968, **68**, 39736).
68CB1381 U. Schmidt, I. Boie, C. Osterroht, R. Schroer and H.-F. Grutzmacher; *Chem. Ber.*, 1968, **101**, 1381.
68CR(267C)270 R. Burgada and H. Germa; *C.R. Hebd. Seances Acad. Sci., Ser. C*, 1968, **267C**, 270.
68CC1606 H.R. Allcock and R.L. Klugel; *Chem. Comm.*, 1968, 1606.
68JOC13 F. Ramirez, A.S. Gula and C.P. Smith; *J. Org. Chem.*, 1968, **33**, 13.
68T1785 F. Ramirez, M. Nagabhushanam and C.P. Smith; *Tetrahedron*, 1968, **24**, 1785.
68T3437 A.N. Hughes and S. Vaboonkul; *Tetrahedron*, 1969, **24**, 3437.
68T5041 F. Ramirez, A.J. Bigler and C.P. Smith; *Tetrahedron*, 1968, **24**, 5041.
68ZOB331 N.Yu. Derkach and A.V. Kirsonov; *Zh. Obshch. Khim.*, 1968, **38**, 331.
69CB528 D. Hellwinkel; *Chem. Ber.*, 1969, **102**, 528.
69CB548 D. Hellwinkel; *Chem. Ber.*, 1969, **102**, 548.
69JA1200 D.B. Boyd; *J. Am. Chem. Soc.*, 1969, **91**, 1200.
69CC443 D. Houalla, R. Wolf, D. Gagnaire and J.B. Robert; *Chem. Comm.*, 1969, 443.
69JOC1130 M.A. Weiner and G. Pasternack; *J. Org. Chem.*, 1969, **34**, 1130.
69SAA1201 R. Mathis, R. Burgada and M. Sanchez; *Spectrochim Acta A.*, 1969, **25**, 1201.
69TL3423 D. Hellwinkel and H.J. Wilfinger; *Tetrahedron Lett.*, 1969, 3423.
69TL4693 M. Wieber and W.R. Hoos; *Tetrahedron Lett.*, 1969, **10**, 4693.
70AL163 D. Hellwinkel and H.J. Wilfinger; *Justus Liebigs Ann. Chem.*, 1970, **742**, 163.
70BAU2637 N.P. Grechkin and G.S. Gubanova; *Bull. Acad. Sci. USSR, Div. Chim. Sci.*, 1970, 2637 (*Chem. Abstr.* 1971, **75**, 49221).
70CR(C)(270)865 J. Ferekh, J.F. Brazier, A. Munoz and R. Wolf; *C.R. Hebd. Seances Acad. Sci., Ser. C*, 1970, **270**, 865.
70CR(C)(271)418 D. Bernard and R. Burgada *C.R. Hebd. Seances Acad. Sci., Ser. C*, 1970, **271**, 418.
70JCS(B)640 D. Hellwinkel and S.F. Mason; *J. Chem. Soc. B*, 1970, 640.
70JCS(C)1425 E.M. Richards and J.C. Tebby; *J. Chem. Soc. (C)*, 1970, 1425.
70CC476 G.O. Doak and R. Schmutzler; *Chem. Commun*, 1970, 476.
70ZAAC(372)285 M. Becke-Goehring and H. Schwind; *Z. Anorg. Allg. Chem.*, 1070, **372**, 285.
71BSF4185 M. Koenig, A. Munoz and R. Wolf; *Bull. Soc. Chim. Fr.*, 1971, 4185.
71IC1643 H.R. Allcock and E.J. Walsh; *Inorg. Chem.*, 1971, **10**, 1643.
71JA4004 B.C. Chang, W.E. Conrad, D.B. Denney, D.Z. Denney, R. Edelmann, R.L. Powell and D.W. White; *J. Am. Chem. Soc.*, 1971, **93**, 4004.
71JCS(A)1295 G.O. Doak and R. Schmutzler; *J. Chem. Soc. (A)*, 1971, 1295.
71JCS(C)593 M.J. Gallagher and I.D. Jenkins; *J. Chem. Soc. (C)*, 1971, 593.
71JCS(CC)1011 A.E. Duff, R.K. Oram and S. Trippett; *J. Chem. Soc., Chem. Commun.*, 1971, 1011.
71JCS(CC)1070 B.C. Chang, D.B. Denney, R.L. Powell and D.W. White; *J. Chem. Soc. Chem. Commun.*, 1971, 1070.
71JCS(CC)1213 R.N. Jenkins, L.D. Freedman and J. Bordner; *J. Chem. Soc., Chem. Commun.*, 1971, 1213.
71JGU228 A.K. Voznesenskaya, N.A. Razumova and A.A. Petrov; *J. Gen. Chem. USSR (Engl. Transl.)*, 1971, **41**, 228.
71JGU471 Zh.L. Evtikhov, N.A. Razumova and A.A. Petrov; *J. Gen. Chem. USSR (Engl. Transl.)*, 1971, **41**, 471 (*Chem. Abstr.* 1971, **75**, 49233).
71MI-1 M. Sanchez, J. Ferekh, A. Munov and R. Wolf; *Roczniki Chem.*, 1971, **45**, 131.
71P(1)91 D.W. White, N.J. De'ath, D.Z. Denney and D.B. Denney; *Phosphorus*, 1971, **1**, 91.
71P(1)139 N.E. Waite, D.W. Allen and J.C. Tebby; *Phosphorus*, 1971, **1**, 139.
72AG(E)722 H. Schmidbauer, P. Holl and F.H. Kohler; *Angew. Chem. Int. Ed.*, 1972, **16**, 722.
72BSF1413 M. Koenig, A. Munoz, R. Wolf and D. Houalla; *Bull. Soc. Chim. Fr.*, 1972, 1413.
72BRP1257024 R.H. Anstey, H.G. Naylor and P.T. Davis; *Brit. Patent* 1257024, (*Chem. Abstr.* 1972, **76**, 33927).
72CR(C)(274)288 D. Bernard and R. Burgada; *C.R. Hebd. Seances Acad. Sci., Ser. C*, 1972, **274**, 288.
72CR(C)(274)419 R. Burgada, D. Bernard and C. Laurenco; *C.R. Hebd. Seances Acad. Sci., Ser. C*, 1972, **274**, 419.
72CR(C)(274)1156 R. Mathis and R. Burgada; *C.R. Hebd. Seances Acad. Sci., Ser. C* 1972, **274**, 1156.
72CR(C)(274)1413 A. Munoz, M. Koenig, B. Gorrigues and R. Wolf; *C.R. Hebd. Seances Acad. Sci., Ser. C*, 1972, **274**, 1413.
72CR(C)(274)2209 Y. Charbonnel and J. Barrans; *C.R. Hebd. Seances Acad. Sci., Ser. C*, 1972, **274**, 2209.
72CR(C)(275)151 R. Wolf, M. Sanchez, D. Houalla and A. Schmidpeter; *C.R. Hebd. Seances Acad. Sci., Ser. C*, 1972, **275**, 151.
72CR(C)(276)1211 L. Lopez, M.-Y. Boisdon and J. Barrans; *C.R. Hebd. Seances Acad. Sci., Ser. C*, 1972, **276**, 1211.

72DOK1370 B.A. Arbuzov, Yu.Yu. Samitov, Yu.M. Mareev and V.S. Vinogradova; *Dokl. Akad. Nauk., SSR*, 1972, **205**, 1370 (*Chem. Abstr.* 1972, **78**, 29120).

72JA9264 C.D. Hall, J.D. Bramblett and F.F.S. Lin; *J. Am. Chem. Soc.*, 1972, **94**, 9264.

72CC395 N.J. De'ath and D.B. Denney; *Chem. Commun.*, 1972, 395.

72CC404 E.W. Turnbloom and D. Hellwinkel; *Chem. Commun.*, 1972, 404.

72JGU1490 G.F. Telegin, V. V, Kireev and V.V. Korshak; *J. Gen. Chem. USSR (Engl. Transl.)*, 1972, **42**, 1490.

72JGU1895 A.N. Pudovik, M.A. Pudovik, S.A. Terent'eva and E.I. Gol'dfarb; *J. Gen. Chem. USSR (Engl. Transl.)*, 1972, **42**, 1895 (*Chem. Abstr.* 1973, **78**, 43606).

72JGU2109 N.A. Razumova, M.P. Gruk and A.A. Petrov; *J. Gen. Chem. USSR (Engl. Transl.)*, 1972, **42**, 2109 (*Chem. Abstr.* 1973, **78**, 43623).

72JGU782 F.V. Bagrov, N.A. Razumova and A.A. Petrov; *J. Gen. Chem. USSR (Engl. Transl.)*, 1972, **42**, 782 (*Chem. Abstr.* 1972, **77**, 101764).

72JOM(42)C47 D. Griller and B.P. Roberts; *J. Organomet. Chem.*, 1972, **42**, C47.

72PAS717 B.A. Arbuzov, Yu.Yu. Samitov, Yu.M. Mareev and V.S. Vinogradova; *Proc. Acad. Sci., USSR*, 1972, **203**, 717 (*Chem. Abstr.* 1973, **78**, 29120).

72PS(2)41 F. Ramirez, K. Tasaka and R. Hershberg; *Phosphorus*, 1972, **2**, 41.

72RTC836 R. Rothius, J.K.J. Luderer and H.M. Buck; *Recl. Trav. Chim. Bays-Bas*, 1972, **91**, 836.

73CR(C)(276)297 R. Burgada, D. Bernard and C. Laurenco; *C.R. Hebd. Seances Acad. Sci., Ser. C*, 1973, **276**, 297.

73JA3154 H.R. Allcock and E.C. Bissell; *J. Am. Chem. Soc.*, 1973, **95**, 3154.

73CC144 M. Eisenhut, R. Schmutzler and W.S. Sheldrick; *J. Chem. Soc., Chem. Commun.*, 1973, 144.

73JCS(P1)1300 R.K. Oram and S. Trippett; *J. Chem. Soc., Perkin Trans. 1*, 1973, 1300.

73JHC395 C.N. Robinson and R.C. Lewis; *J. Heterocyclic Chem.*, 1973, **10**, 395.

73P(2)167 D. Hellwinkel, C. Wunsche and M. Bach; *Phosphorus*, 1973, **2**, 167.

73RTC1308 R. Rothuis, J.J.H.M. Font Freide and H.M. Buck; *Recl. Trav. Chim. Pays Bas*, 1973, **92**, 1308.

73TL3455 D. Bernard and R. Burgada; *Tetrahedron Lett.*, 1973, **14**, 3455.

74AX935 H. Wunderlich and D. Mootz; *Acta Cryst. B.*, 1974, **30**, 935, 939.

74BSF2193 A. Munoz, M. Sanchez, M. Koenig and R. Wolf; *Bull. Soc. Chim. Fr.*, 1974, 2193.

74CR(C)(278)1301 C. Laurenco, D. Bernard and R. Burgada; *C.R. Hebd. Seances Acad. Sci., Ser. C*, 1974, **278**, 1301.

74CR(C)(278)1353 A. Munoz, M. Koenig, G. Gence and R. Wolf; *C.R. Hebd. Seances Acad. Sci., Ser. C*, 1974, **278**, 1353.

74JA317 D.B. Denney and L.S. Shih; *J. Am. Chem. Soc.*, 1974, **96**, 317.

74JA4143 R.R. Holmes; *J. Am. Chem. Soc.*, 1974, **96**, 4143, reprinted in *Phosphorus Sulfur*, 1995, **98**, 75.

74JA5398 G.M. Whitesides, E. Eisenhut and W.M. Bunting; *J. Am. Chem. Soc.*, 1974, **96**, 5398.

74JA7269 F. Ramirez, V.A.V. Prasad and J.F. Marecek; *J. Am. Chem. Soc.*, 1974, **96**, 7269.

74CC15 J.W. Gilje, R.W. Braun and A.H. Cowley; *Chem. Commun.*, 1974, 15.

74JCS(D)2153 O. Schlak, R. Schmutzler, H.M. Shiebel, M.I.M. Wazeer and R.K. Harris; *J. Chem. Soc., Dalton Trans.*, 1976, 2153.

74JCS(P1)1694 J.I.G. Cadogan, R.J. Scott, R.D. Gee and I. Gosney; *J. Chem. Soc., Perkin Trans. 1*, 1974, 1694.

74JCS(P1)2125 S. Bone, S. Trippett and P.J. Whittle; *J. Chem. Soc., Perkin Trans.*, 1974, 2125.

74JCS(P2)1668 A. Klaebe, A. Cachapuz Carrelhas, J.F. Brazier and R. Wolf; *J. Chem. Soc., Perkin Trans. 2*, 1974, 1668.

74P(4)203 A.E. Duff, D.R. Russell and S. Trippett; *Phosphorus*, 1974, **4**, 203.

74P(4)265 L.L. Chang, D.Z. Denney and D.B. Denney and H.F. Hsu; *Phosphorus* 1974, **4**, 265.

74T371 F. Ramirez, I. Ugi, F. Liu, S.P. Pfhol, P. Hoffman and D. Marquarding; *Tetrahedron*, 1974, **30**, 371.

74TL3971 A. Klaebe, A. Cachapuz Carrelhas, J.F. Brazier, M.C. Marre and R. Wolf; *Tetrahedron Lett.*, 1974, **15**, 3971.

74ZN(B)32 H. Wunderlich, D. Mootz, R. Schmutzler and M. Wieber; *Z. Naturforsch. B.*, 1974, **29**, 32.

75BSF1433 A. Munoz, G. Gence, M. Koenig and R. Wolf; *Bull. Soc. Chim. Fr.*, 1975, 1433.

75CA4319 H. Rose; *Chem. Abstr.*, 1975, **82**, 4319, 4320.

75CB820 A. Schmidpeter and J. Luber; *Chem. Ber.*, 1975, **108**, 820.

75CB1454 A. Schmidpeter and H. Eiletz; *Chem. Ber.*, 1975, **108**, 1454.

75CR(C)(280)225 M. Willson, R. Burgada and F. Mathis; *C.R. Hebd. Seances Acad. Sci., Ser. C*, 1975, **280**, 225.

75CR(C)(280)809 R. Mathis, M. Barthelat, Y. Charbonnel and J. Barrans; *C.R. Hebd. Seances Acad. Sci., Ser. C*, 1975, **280**, 809.

75CPL(30)276 P. Russegger and J. Brickmann; *Chem. Phys. Lett.*, 1975, **30**, 276.

75BSF407 R. Burgada; *Bull. Soc. Chim. Fr.*, 1975, 407.

75IC2831 H.R. Allcock, R.L. Kugel and G.Y. Moore; *Inorg. Chem.*, 1975, **14**, 2813.

75ICAL40 W.J. Walsh and E. Derby; *Inorg. Chim. Acta*, 1975, **14**, L40.

75JA5379 R.R. Holmes; *J. Am. Chem. Soc.*, 1975, **97**, 5379 reprinted in *Phosphorus Sulfur*, 1995, **98**, 89.

75CC399 Y. Segall, I. Granoth, A. Kalir and E.D. Bergmann; *Chem. Commun.*, 1975, 399.

75CC773 J.I.G. Cadogan, R.O. Gould and N.J. Tweddle; *Chem. Commun.*, 1975, 773.

75JCS(P1)2376 J.I.G. Cadogen, D.S.B, Grace, P.K.K. Lim and B.S. Tait; *J. Chem. Soc., Perkin Trans. 1*, 1975, 2376.

75JCS(P1)1220 S. Trippett and P.J. Whittle; *J. Chem. Soc., Perkin Trans. 1*, 1975, 1220.

75JGU2088 I.V. Konovalova, E. Kh. Ofitserova, N.P. Anoshina and A.N. Pudivik; *J. Gen. Chem. USSR (Engl. Transl.)*, 1975, **45**, 2088 (*Chem. Abstr.*, 1976, **85**, 177329).

75JGU2292 M.A. Pudovik, S.A. Terent'eva and A.N. Pudovik; *J. Gen. Chem. USSR (Engl. Transl.)*, 1975, **45**, 2292 (*Chem. Abstr.*, 1976, **84**, 31192).

75JGU252 M.A. Pudovik, S.A. Terent'eva, Yu.Yu. Samitov and A.N. Pudovik; *J. Gen. Chem. USSR (Engl. Transl.)* 1975, **45**, 252 (*Chem. Abstr.*, 1975, **83**, 43440).
75JGU686 N.A. Razumova, N.A. Kurshakova and A.A. Petrov; *J. Gen. Chem. USSR (Engl. Transl.)*, 1975, **45**, 686.
75JOC766 R.N. Jenkins and L.D. Freedman; *J. Org. Chem.*, 1975, **40**, 766.
75JOC1185 A.P. Schaup, K. Kees and A.L. Thayer; *J. Org. Chem.*, 1975, **40**, 1185.
75JOM(93)331 P. Savignac, B. Richard, Y. Leroux and R. Burgada; *J. Organometal. Chem.*, 1975, **93**, 331.
75MI-1 M. Kajiwara and H. Saito; *Polymer*, 1975, **16**, 1.
75P(5)285 D. Bernard and R. Burgada; *Phosphorus*, 1975, **5**, 285.
75P(5)73 F. Ramirez, J.F. Marecek, I. Ugi, P. Lemmen and D. Marquarding; *Phosphorus*, 1975, **5**, 73.
75T797 D. Bernard and R. Burgada; *Tetrahedron*, 1975, **31**, 797.
75TL1583 S.A. Bone and S. Trippett; *Tetrahedron Lett.*, 1975, **16**, 1583.
75TL2529 R. Boigegrain and B. Castro; *Tetrahedron Lett.*, 1975, **16**, 2529; *ibid*, *Tetrahedron* 1976, **32**, 1283.
75TL3077 C. Malavaud and J. Barrans; *Tetrahedron Lett.*, 1975, **16**, 3077.
75TL497 C. Malavaud, Y. Charbonnel and J. Barrans; *Tetrahedron Lett.*, 1975, **16**, 497.
76BCJ1924 T. Kawashima and N. Inamoto; *Bull. Chem. Soc. Jpn*, 1976, **49**, 1924.
76CB1056 D. Hellwinkel and H.J. Wilfinger; *Chem. Ber.*, 1976, **103**, 1056.
76CB1497 D. Hellwinkel and W. Lindner; *Chem. Ber.*, 1976, **109**, 1497.
76CR(C)(282)849 R. Burgada; *C.R. Hebd. Seances Acad. Sci., Ser. C*, 1976, **282**, 849.
76IZV2624 N.A. Tikhonina, V.A. Gilyarov and M.I. Kabachnik; *Izv. Akad. Nauk, SSSR, Ser. Khim.*, **1976**, 2524 (*Chem. Abstr.* 1977, **86**, 106488).
76JA4330 F. Ramirez, M. Navazowski and J.F. Marecek; *J. Am. Chem. Soc.*, 1976, **98**, 4330.
76JA5120 A.R. Allcock, R.W. Allen, E.C. Bissel, L.A. Smeltz and M. Teeter; *J. Am. Chem. Soc.*, 1976, **98**, 5120.
76JA6755 D.B. Denney, D.Z. Denney and C-F. Ling; *J. Am. Chem. Soc.*, 1976, **98**, 6755.
76CC449 H.A.E. Aly, J.H. Barlow, D.R. Russell, D.J.H. Smith, M. Swindles and S. Trippett; *Chem. Commun.*, **1976**, 449.
76CC730 M. Sanchez, J.-F Brazier, D. Houalla, A. Munoz and R. Wolf; *Chem. Commun.*, 1976, 730.
76CC1031 J.H. Barlow, S.A. Bone, D.R. Russell, S. Trippett and P.J. Whilltle; *Chem. Commun.*, 1976, 1031.
76JCS(D)306 R.K. Harris, M.I.M. Wazeer, O. Schlak and R. Schmutzler; *J. Chem. Soc., Dalton Trans.*, 1976, 306.
76JCS(P2)955 M. Koenig, A. Klaebe, A. Munoz and R. Wolf; *J. Chem. Soc., Perkin Trans, 2*, 1976, 955.
76JGU581 M.M. Yusapov, N.K. Rozhkova and N.D. Abdullaev; *J. Gen. Chem. USSR (Engl. Tranl.)*, 1976, **46**, 581 (*Chem. Abstr.* 1977, **85**, 5799).
76JOC473 R. Sarma, F. Ramirez and J.F. Marecek; *J. Org. Chem.*, 1976, **41**, 473.
76P(6)91 D. Hellwinkel and W. Krapp; *Phosphorus*, 1976, **6**, 91.
76PS(6)113 A. Schmidpeter and H. Eiletz; *Phosphorus*, 1976, **6**, 113.
76T2039 Y. Charbonnel and J. Barrans; *Tetrahedron*, 1976, **32**, 2039.
76T2253 C. Laurenco and R. Burgada; *Tetrahedron*, 1976, **32**, 2253.
76TL4381 G. Kemp and S. Trippett; *Tetrahedron Lett.*, 1976, **17**, 4381.
76TL749 G. Buono and J.R. Llinas; *Tetrahedron Lett.*, 1976, **17**, 749.
76TL917 J. Gloede and H. Gross; *Tetrahedron Lett.*, 1976, **17**, 917.
76ZAAC(426)275 H. Rose and H. Specker; *Z. Anorg. Allg. Chem.*, 1976, **426**, 275.
76ZN(B)948 D. Hellwinkel, W. Krapp, D. Schomburg and W.S. Sheldrick; Z. *Naturforsch*, 1976, **31b**, 948.
77AG(E)722 H. Schmidbaur, P. Holl and F.H. Kohler; *Angew. Chem. Int. Ed. (Engl. Trans.)*, 1977, **16**, 722.
77CB693 D. Hellwinkel and W. Krapp; *Chem. Ber.*, 1977, **110**, 693.
77CB1124 A. Schmidpeter and J. Luber; *Chem. Ber.*, 1977, **110**, 1124.
77CB1887 J.A. Gibson, G.V. Roschenthaler, D. Schomburg and W.S. Sheldrick; *Chem. Ber.*, 1977, **110**, 1887.
77CB3231 G. Schoning and O. Glemser; *Chem. Ber.*, 1977, **110**, 3231.
77IC2294 R.K. Brown and R.R. Holmes; *Inorg. Chem.*, 1977, **16**, 2294.
77IZV2000 B.A. Arbuzov, Yu.Yu. Samitov, Yu.M. Mareev and V.S. Vinogradova; *Izv. Akad. Nauk SSSR Ser. Khim.*, 1977, 2000 (*Chem. Abstr.*, 1978, **88**, 23055).
77IZV434 V.V. Ovchinnikov, M.A. Pudovik, V.I. Galkin, R.A. Cherkasov and A N. Pudovik; *Izv. Akad. Nauk. Ser. Khim.*, 1977, 434 (*Chem. Abstr.* 1977, **86**, 188937).
77JA3318 R.R. Holmes and J.A. Deiters; *J. Am. Chem. Soc.*, 1977, **99**, 3318 reprinted in *Phosphorus Sulfur*, 1995, **96**, 105.
77JA3326 R.K. Brown and R.R. Holmes; *J. Am. Chem. Soc.*, 1977, **99**, 3326.
77JA3332 J.R. Devillers and R.R. Holmes; *J. Am. Chem. Soc.*, 1977, **99**, 3332.
77JA5461 J.A. Deiters, J.C. Galluci, T.E. Clark and R.R. Holmes; *J. Am. Chem. Soc.*, 1977, **99**, 5461 reprinted in *Phosphorus Sulfur*, 1995, **98**, 125.
77JCS(D)570 A. Klaebe, M. Koenig, R. Wolf and P. Ahlberg; *J. Chem. Soc., Dalton Trans.*, 1977, 570.
77JCS(P1)273 J. Brierley, S. Trippett and M.W. White; *J. Chem. Soc., Perkin Trans. 1*, 1977, 273.
77JCS(P1)437 S.A. Bone, S. Trippett and P.J. Whittle; *J. Chem. Soc., Perkin 1*, 1977, **80**, 437.
77JGU436 V.P. Kukhar, T.N. Kasheva and A.Ya. Ul'chenko; *J. Gen. Chem. USSR (Engl. Transl.)*, 1977, **47**, 436.
77JPR(319)188 J. Gloede, H. Gross and G. Engelhardt; *J. Prakt. Chem.*, 1977, **319**, 188.
77MI-1 V.G. Andrianov, A.E. Kalinin and Yu.T. Struchkov; *Z. Strukt. Khim.*, 1977, **18**, 310 (*Chem. Abstr.*, 1977, **87**, 68252).
77MI-2 M. Kajiwara, M. Makihara and H. Sato; *Polymer*, 1977, **18**, 1045.
77PS(3)35 L.B. Littlefield and G.O. Doak; *Phosphorus Sulfur*, 1977, **3**, 35.

77PS(3)51 N.J. De'ath and D.B. Denney; *Phosphorus Sulfur*, 1977, **3**, 51.

77T635 B. Garrigues, A. Munoz, M. Koenig, M. Sanchez and R. Wolf; *Tetrahedron*, 1977, **33**, 635.

77TL559 J.E. Richman; *Tetrahedron*, 1977,**18**, 559.

77ZAAC(437)53 E. Fluck and M. Vargas; *Z. Anorg. Allg. Chem.*, 1977, **437**, 53.

77ZOK1432 A.K. Voznenskaya, N.A. Razumova, A.A. Petrov and D.G. Sheptienko; *Zh. Obshch. Khim.*, 1977, **47**, 1432.

78AG(E)774 K. Burger, R. Ottlinger, A. Frank and U. Schubert; *Angew. Chem. Int. Ed.*, 1978, **17**, 774.

78BAU1064 I.S. Belostotskaya, N.L. Komissarova, T.I. Prokof'eva, A.I. Prokof'ev, V.V. Ershov and M.I. Kabachnik; *Bull. Acad. Sci.*, 1978, **27**, 1064.

78BSF65 B. Garrigues, D. Houalla, J.F. Brazier and R. Wolf; *Bull. Soc. Chim. Fr.*, Pt 2 1978, 65.

78CB2086 A. Schmidpeter and J.H. Weinmaier; *Chem. Ber.*, 1978, **111**, 2086.

78IC3265 W. Althoff, R.O. Day, R.K. Brown and R.R. Holmes; *Inorg. Chem.*, 1978, **17**, 3265.

78IC3270 P.F. Meunier, R.O. Day, J.R. Devillers and R.R. Holmes; *Inorg. Chem.*, 1978, **17**, 3270.

78IC3276 R.K. Brown, R.O. Day, S. Husebye and R.R. Holmes; *Inorg. Chem.*, 1978, **17**, 3276.

78IJ107 G.V. Roschenthaler, K. Sauerbrey and R. Schmutzler; *Isr. J. Chem.*, 1978, **17**, 107 and 143.

78IZV2185 S.A. Trent'eva, M.A. Pudovik and A.N. Pudovik; *Izv. Akad. Nauk SSSR, Ser. Khim.*, 1978, 2185 (*Chem. Abstr.*, 1979, **90**, 23186).

78JA433 R.R. Holmes; *J. Am. Chem. Soc.*, 1978, **100**, 433, reprinted with corrections in *Phosphorus Sulfur*, 1995, **98**, 151.

78JA5130 Y. Segall and I. Granoth; *J. Am. Chem. Soc.*, 1978, **100**, 5130.

78JA5229 I. Granoth and J.C. Martin; *J. Am. Chem. Soc.*, 1978, **100**, 5229.

78JA7434 I. Granoth and J.C. Martin; *J. Am. Chem. Soc.*, 1978, **100**, 7434.

78JCS(CC)219 A. Munoz, G. Garrigues and M. Koenig; *J. Chem. Soc., Chem. Commun.*, 1978, 219.

78JCS(CC)854 B.C. Canbell, D.B. Denney, D.Z. Denney and L.S. Shih; *J. Chem. Soc., Chem. Commun.*, 1978, 854.

78JGU1308 V.P. Kukhar, E.V. Grishkin and V.P. Radavskii; *J. Gen. Chem. USSR (Engl. Transl.)*, 1978, **48**, 1308.

78JGU650 V.V. Ragulin, N.A. Razumova, V.I. Zakharov and A.A. Petrov; *J. Gen. Chem. USSR (Engl. Transl.)*, 1978, **48**, 650.

78JOC571 K.S. Colle and E.S. Lewis; *J. Org. Chem.*, 1978, **43**, 571.

78JOM(156)253 A.C. Sau and R.R. Holmes; *J. Organomet. Chem.*, 1978, **156**, 253.

78PS(5)217 W. Schwarz, E. Fluck and M. Vargas; *Phosphorus Sulfur*, 1978, **5**, 217.

78TL2857 A. Schmidpeter, T. von Criegern, W.S. Sheldrick and D. Schomburg; *Tetrahedron Lett.*, 1978, **19**, 2857.

78ZN(B)305 W.Storzer and G.V. Roschenthaler; *Z. Naturforsch. B*, 1978, **33**, 305.

78ZN(B)583 W.S. Sheldrick, A. Schmidpeter and T. von Criegern; *Z. Naturforsch. B*, 1978, **33**, 583.

78ZN(B)588 Y. Sudhakara, H. Manohar, K. Ramachandran and S.S. Krishnamurthy; *Z. Naturforsch., B*, 1978, **33**, 588.

79CB1365 H.W. Roesky, K. Ambrosius and W.S. Sheldrick; *Chem. Ber.*, 1979, **112**, 1365.

79CB218 D. Hellwinkel, W. Lindner and W. Schmidt; *Chem. Ber.*, 1979, **112**, 218.

79CB501 H. Schmidbauer and P. Holl; *Chem. Ber.*, 1979, **112**, 501.

79CR(C)(288)165 R. Burgada, A. Mohri and Y.O. El Khoshnich; *C.R. Hebd. Seances Acad. Sci., Ser. C*, 1979, **288**, 165.

79E851 S. Zbaida and E. Breuer; *Experientia*, 1979, **35**, 851.

79IC1668 T.E. Clark, R.O. Day and R.R. Holmes; *Inorg. Chem.*, 1979, **18**, 1668, reprinted in *Phosphorus Sulfur*, 1995, **98**, 223.

79IZV2398 M.I. Kabachnik, D.I. Labanov and P.V. Petrovskii; *Izv. Akad. Nauk, SSSR, Ser. Khim.*, 1979, 2398 (*Chem. Abstr.*, 1980, **92**, 76421).

79JA3687 Y. Segall and I Granoth; *J. Am. Chem. Soc.*, 1979, **101**, 3687.

79JA3790 R.O. Day, A.C. Sau and R.R. Holmes; *J. Am. Chem. Soc.*, 1979, **101**, 3790.

79JA4618 I. Granoth and J.C. Martin; *J. Am. Chem. Soc.*, 1979, **101**, 4618.

79JA4623 I. Granoth and J.C. Martin; *J. Am. Chem. Soc.*, 1979, **101**, 4623.

79JCR(S)172 B. Garrigues, C.B. Cong, A. Munoz and A. Klaebe; *J. Chem. Res. (S)*, 1979, 172.

79JCS(CC)191 J.I.G. Cadogan, N.J. Stewart and N.J. Tweedle; *J. Chem. Soc., Chem. Commun.*, 1979, 191.

79JCS(P1)879 G. Kemp and S. Trippett; *J. Chem. Soc. Perkin Trans. 1*, 1979, 879.

79PS(5)323 R. Sarma, F. Ramirez, B. McKeever, J.F. Marecek and A.V. Prasad; *Phosphorus Sulfur*, 1979, **5**, 323.

79PS(6)435 M. Koenig, A. Munoz, B. Garrigues and R. Wolf; *Phosphorus Sulfur*, 1979, **6**, 435.

79PS(7)167 J. Brierley, J.I. Dickstein and S. Trippett; *Phosphorus Sulfur*, 1979,**7**, 167.

79PS(7)305 T. von Criegern and A. Schmidpeter; *Phosphorus Sulfur*, 1979, **7**, 305.

79PS(7)57 J. Gloede and H. Gross; *Phosphorus Sulfur*, 1979, **7**, 57.

79PS(7)81 L. Lopez, C. Fabus and J. Barrans; *Phosphorus Sulfur*, 1979, **7**, 81.

79T1825 C.B, Cong, G. Genco, B. Garrigues, M. Koenig and A. Munoz; *Tetrahedron*, 1979, **35**, 1825.

79TL193 S. Trippett and R.E.L. Waddling; *Tetrahedron Lett.*, 1979, **20**, 193.

79TL2203 E. Breuer, S. Zbaida and E. Segall; *Tetrahedron Lett.*, 1979, **20**, 2203.

79TL4205 B. Garrigues, M. Koenig and A. Munoz; *Tetrahedron Lett.*, 1979, **20**, 4205.

79TL795 J. Cavezzan, G. Etemad-Mogadam, M. Koenig and A. Klaebe; *Tetrahedron Lett.*, 1979, **20**, 795.

79ZN(B)906 A. Schmidpeter, J.H. Weinmaier, W.S. Sheldrick and D. Schomburg; *Z. Naturforsch. B.*, 1979, **34**, 906.

79ZN(B)911 A. Schmidpeter, H. Eiletz, J. von Seyerl and G. Hutner; *Z. Naturforsch., B*, 1979, **34**, 911.

80CB1406 D. Hellwinkel, W. Blaicher, W. Krapp and W.S. Sheldrick; *Chem. Ber.*, 1980, **113**, 1406.

80CB2699 K. Burger, R. Ottlinger, H. Goth and J. Firl; *Chem. Ber.*, 1980, **113**, 2699.

80CB1847 H.W. Roesky, K. Ambrosius, M. Banek and W.S. Sheldrick; *Chem. Ber.*, 1980, **113**, 1847.

80CL1599	S. Kobayashi, Y. Narukawa, T. Hashimoto and T. Saeygusa; *Chem. Lett.*, 1980, 1599.
80JA5073	D.B. Denney, D.Z. Denney, P.J. Hammond, C. Huang and K-S Tseng; *J. Am. Chem. Soc.*, 1980, **101**, 5073.
80JGU812	V.P. Kukhar, E.V. Grishkun and V.P. Rudavskii; *J. Gen. Chem., USSR, Engl. Transl.*, 1980, **50**, 812.
80JCS(CC)157	J.J.H.M. Font-Freide and S. Trippett; *J. Chem. Soc., Chem. Commun.*, 1980, 157.
80PS(8)127	G. Baccolini, P. Spagnolo and P.O. Todesco; *Phosphorus Sulfur*, 1980, **8**, 127.
80PS(8)147	H. Goncalves, J.R. Dormoy, Y. Chapleur, B. Castro, H. Fauduet and R. Burgada; *Phosphorus Sulfur*, 1980, **8**, 147.
80PS(8)153	B. Garrigues, A. Klaebe and A. Munoz; *Phosphorus Sulfur*, 1980, **8**, 153.
80PS(9)184	B. Garrigues, A. Munoz and M. Mulliez; *Phosphorus Sulfur*, 1980, **9**, 183.
80T2467	A.Munoz, B. Garrigues and M. Koenig; *Tetrahedron*, 1980, **36**, 2467.
80TL1307	J.P. Majoral, R. Kraemer, T. N'Gando M'Pondo and J. Navech; *Tetrahedron Lett.*, 1980, **21**, 1307.
80TL1449	J. Navech, R. Kraemer and J.-P, Majoral; *Tetrahedron Lett.*, 1980, **21**, 1449.
80TL4331	S.D. Harper and A.J. Arduengo; *Tetrahedron Lett.*, 1980, **21**, 4331.
80ZAAC(471)147	J. Goede and B. Costisella; *Z. Anorg. Allg. Chem.*, 1980, **471**, 147.
80ZN(B)990	H. Schmidbauer and A. Mortl; *Z. Naturforsch. B*, 1980, **35**, 990.
81CB825	H. Tantz and A. Schmidpeter; *Chem. Ber.*, 1981, **114**, 825.
81CB1554	H.W. Roesky, H. Djarrah, D. Amirzadeh-Asl, and W.S. Sheldrick; *Chem. Ber.*, 1981, **114**, 1554.
81CB2132	W.S. Sheldrick, S. Pohl, H. Zamankan, M. Banek, D. Amirzadeh-Asl and H.W. Roesky; *Chem. Ber.*, 1981, **114**, 2132.
81CB3161	H. Schmidbauer, A. Morte and B. Zimmer-Gasser; *Chem. Ber.*, 1981, **114**, 3161.
81JA1049	E.F. Perozzi, R.S. Michalak, G.D. Figully, W.H. Stevenson III, D.B. Dess, M.R. Ross and J.C. Martin; *J. Org. Chem.*, 1981, **46**, 1049.
81JA1291	J.E. Richman, O.D. Gupta and R.B. Flay; *J. Am. Chem. Soc.*, 1981, **18**, 1291.
81JA5265	J.E. Richman and R.B. Flay; *J. Am. Chem. Soc.*, 1981, **18**, 5265.
81JCS(CC)609	D. Schomburg, N. Weferling and R. Schmutzler; *J. Chem. Soc., Chem. Commun.*, 1981, 609.
81JCS(CC)810	D. Schomburg, N. Weferling and R. Schmutzler; *J. Chem. Soc., Chem. Commun.*, 1981, 810.
81JCR(S)218	J.J.H.M. Font Freide and S. Trippett; *J. Chem. Res. (S)*, 1981, 218.
81JCS(P1)3074	M.P. Johnson and S. Trippett; *J. Chem. Soc., Perkin 1*, 1981, 3074.
81JOC1049	E.F. Perozzi, R.S. Michalak, G.D. Figaly, W.H. Stevenson III, D.B. Dess, M.R. Ross and J.C. Martin; *J. Org. Chem.*, 1981, **46**, 1049.
81PS(9)285	R. Burgada and A. Mohri; *Phosphorus Sulfur*, 1981, **9**, 285.
81PS(10)47	P.J. Hammond, J.R. Lloyd and C.D. Hall *Phosphorus Sulfur*, 1981, **10**, 47.
81PS(10)207	A.W. Frank; *Phosphorus Sulfur*, 1981, **10**, 147, 207.
81PS(10)53	A. Klaebe, J.F. Brazier, B. Garrigues and R. Wolf; *Phosphorus Sulfur*, 1981, **10**, 53.
81PS(10)395	W.J. Richter; *Phosphorus Sulfur*, 1981, **10**, 395.
81PS(11)71	A. Munoz, K. Lamande, M. Koenig and R. Wolf; *Phosphorus Sulfur*, 1981, **11**, 71.
81PS(11)87	M.R. Marre, J.F. Brazier, R. Wolf and A. Klaebe; *Phosphorus Sulfur*, 1981, **11**, 87.
81TL3533	R. Burgada, Y. Leroux and Y.O. El; *Tetrahedron Lett.*, 1981, **22**, 3533.
82AG(E)208	G.-V. Roschenthaler and W. Storzer; *Angew. Chem. Int. Ed. Engl.*, 1982, **21**, 208.
82AXB3024	P.I.B. Konche-Waksman, P.L. Haridon, Y. Leroux and R. Burgada; *Acta Cryst B*, 1982, **38**, 3024.
82CB2560	R.F. Hudson, C. Brown and A. Maron; *Chem. Ber.*, 1982, **115**, 2560.
82IC3916	R.O. Day, A. Schmidpeter and R.R. Holmes; *Inorg. Chem.*, 1982, **21**, 3916.
82IC844	H.W. Roesky and H. Djarah; *Inorg. Chem.*, 1982, **21**, 844.
82JCR(S)180	A. Dubourg, R. Roques and G. Germain; *J. Chem. Res., (S)*, 1982, 180.
82JCS(P1)735	I. Granoth; *J. Chem. Soc. Perkin Trans., I*, 1982, 735.
82JGU189	I.V. Konovalova, M.V. Cherkina, L.A. Burnaeva and A.N. Pudovik; *J. Gen. Chem. USSR (Engl. Transl.)*, 1972, **52**, 189.
82JGU221	V.V. Ragulin, A.A. Petrov, V.I. Zakharov and N.A. Razumova; *J. Gen. Chem., USSR (Engl. Transl.)*, 1982, **52**, 211.
82JOC916	Y. Kimura, M. Miyamoto and T. Saegusa; *J. Org. Chem.*, 1982, **47**, 916.
82MM697	H.R. Allcock, W.T. Ferrar and M.L. Levin; *Macromolecules*, 1982, **15**, 697.
82PS(13)157	J. Gloede, H. Gross, J. Michalski, M. Pakulski and A. Skowronska; *Phosphorus Sulfur*, 1982, **13**, 157.
82ZN(B)1425	B. de Ruiter, G. Kuipers, J.H. Bijlaart and J.C. van de Grampel; *Z. Naturforsch. B*, 1982, **37**, 1425.
82PS(13)85	Y. Leroux, D.E. Manouni and R. Burgada; *Phosphorus Sulfur*, 1982, **13**, 85.
83IC1771	A.C. Sau, J.M. Holmes, R.O. Day and R.R. Holmes; *Inorg. Chem.*, 1983, **22**, 1771.
83ICA(76)L139	M. Lattman, B.N. Anand, D.R. Garrett and M.A. Whitener; *Inorg. Chim. Acta*, 1983, **76**, L139.
83JGU1759	M.A. Pudovik, Yu.B. Mikhailov and A.N. Pudovik; *J. Gen. Chem. USSR (Engl. Transl.)*, 1983, **53**, 1759.
83JGU247	Yu. Budilova, N.I. Gusar and Yu.G. Gololobov; *J. Gen. Chem. USSR (Engl. Transl.)*, 1983, **53**, 247.
83JOC2621	E.L. Clennan and P. Choo Heah; *J. Org. Chem.*, 1983, **48**, 2621.
83JOC3815	Y. Kimura, T. Kokura and T. Saegusa; *J. Org. Chem.*, 1983, **48**, 3815.
83MM719	P.E. Austin, G.H. Riding and H.R. Allcock; *Macromolecules*, 1983, **16**, 719.
83PS(17)283	P.D. Beer, R.C. Edwards, C.D. Hall, J.R. Jennings and R.J. Cozens; *Phosphorus and Sulfur*, 1983, **17**, 283.
83TL5831	K.C. Caster and L.D. Quin; *Tetrahedron Lett.*, 1983, **24**, 5831.
83ZAAC(507)93	G.V. Roschenthaler, R. Bohlen, W. Storzer, A.B. Sopchill and W.G. Bentrude; *Z. Anorg. Allg. Chem.*, 1983, **507**, 93.
83ZN(B)466	D.W. Allen, L.A. Marsh, I.W. Nowell and J.C. Tebby; *Z. Naturforsch. B.*, 1983, **38**, 466.

83ZN(B)702 H-.M. Schiebel, R. Schmutzler, D. Schomburg and U. Wermuth; *Z. Naturforsch. B*, 1983, **38**, 702.

83ZN(B)1046 A. Koblishbek, H.D. Hansen, J. Weidlein and H. Binder; *Z. Naturforsch B*, 1983, **38b**, 1046.

84CJC2179 B. Garrigues and A. Munoz; *Can. J. Chem.*, 1984, **62**, 2179.

84JA7065 A. Baceiredo, G. Bertrand, J.-P. Majoral, U. Wermuth and R. Schmutzler; *J. Am. Chem. Soc.*, 1984, **106**, 7065.

84JCS(CC)225 L. Lamande, A. Monoz and D. Boyer; *J. Chem. Soc., Chem. Commun.*, 1984, 225.

84JCS(CC)675 S.R. Contractor, M.B. Hursthouse, H.G. Parkes, L.S. Shaw, R.A. Shaw and H. Yilmaz; *J. Chem. Soc., Chem. Commun.*, 1984, 675.

84JGU1754 V.I. Kal'chenko, L.I. Atamas, Yu.A. Seruchev and L.M. Markowskii; *J. Gen. Chem. USSR (Engl. Transl.)*, 1984, **54**, 1754.

84JGU486 T.N. Kudryavtseva, M.B. Karlstedt, M.V. Proskurina, V.A. Frolovskii and I.A.I. Lutsenko; *J. Gen. Chem. USSR (Engl. Transl.)*, 1984, **54**, 486.

84JSK921 A.N. Chernega, M.Yu. Antipin, Yu.T. Strchkov, Yu.G. Gololobov, I.E. Boldeskul, M.I. Gusar and M.P. Chaus; *J. Stuct. Chem.*, 1984, **25**, 921.

84JST(116)75 J.F. Labarre, G. Guerch, F. Sournies, R. Lahana, R. Enjalbert and J. Galy; *J. Mol. Struct.*, 1984, **116**, 75.

84MI-1 A.N. Chernega, M.Yu. Antipin, Yu.T. Struchkov, Yu.G. Gololobov, I.E. Boldeskul, M.I. Gusar and M.P. Chaus; *J. Struct. Chem.*, 1984, **25**, 921.

84MI-2 D.S. Yufit, E.I. Matrosov, N.V. Matrosova, D.I. Lobanov, Yu.G. Gololbov, Yu.T. Struchkov and M.I. Kabachnik; *J. Struct. Khim.*, 1984, **25**, 568.

84MI-3 A.N. Chernega, M.Yu. Antipin, Yu.T. Struchkov, Yu.G. Gololobov, I.E. Boldeskul, M.I. Gusar and M.P. Chaus; *J. Struct. Khim.*, 1984, **25**, 94.

84MI-4 D.B. Denney, D.Z. Denney and Liu Lunzu; *Acta Chem. Sinica*, 1984, **42**, 340.

84PS(19)173 J.J.C. Van Lier, R.J.M. Hermans and H.M. Buck; *Phosphorus Sulfur*, 1984, **19**, 173.

84PS(20)55 F. El Khatib, A.M. Caminade and M. Koenig; *Phosphorus Sulfur*, 1984, **20**, 55.

84PS(21)249 Ch. Angelov, K. Kristov and D. Mendshka; *Phosphorus Sulfur*, 1984, **21**, 249.

84TL5521 O. Diallo, M.T. Boisdon, C. Malavaud, L. Lopez, M. Huddad and J. Barrans; *Tetrahedron Lett.*, 1984, **25**, 5521.

85ICA(108)23 J.L. Sassons, M. Graffeuil, P. Castera and J.F. Labarre; *Inorg. Chimica Acta*, 1985, **108**, 23.

85JA5198 G.H. McGall and R.A. McClelland; *J. Am. Chem. Soc.*, 1985, **107**, 5198.

85JA7585 R.C. Haddon, S.L. Mayo, S.V. Chichester and J.H. Marshall; *J. Am. Chem. Soc.*, 1985, **107**, 7585.

85JCS(CC)1764 P. Poechlauer, W. Petter, P. Peringer and E.P. Muller; *J. Chem. Soc., Chem. Commun.*, 1985, 1764.

85JCS(P1)1953 T.G. Roberts and G. Whilham; *J. Chem. Soc., Perkin Trans. 1*, 1985, 1953.

85JGU1514 B.A. Arbuzov, A.O. Vizel, L.I. Shchukina, O.V. Vakulenko, T.A. Zyablikova, V.K. Krupnov and L.I. Zyryanova; *J. Chem. Chem. USSR (Engl. Transl.)*, 1985, **55**, 1514.

85JGU1810 F.M. Akhmetkhanova, L.Z. Rol'nik, E.V. Pastushenko, M.V. Prosturina, S.S. Zlot-skii and D.L. Rakhumankulov; *J. Gen. Chem. USSR (Engl. Transl.)*, 1985, **55**, 1810.

85JGU1986 Kh.I. Rissalu, V. V, Vasil'ev and B.I. Ionin; *J. Gen. Chem. USSR (Engl. Transl.)*, 1985, **55**, 1986.

85JGU2327 E.A. Monin, Z.S. Novikova, M.M. Kabachnik, A.I. Lutsenko and I.F. Lutsenko; *J. Gen. Chem. USSR (Engl. Transl.)*, 1985, **55**, 2327.

85JGU2333 L.I. Nesterova and A.D. Sinitsa; *J. Gen. Chem. USSR (Engl. Transl.)*, 1985, **55**, 2333.

85PS(24)235 L. Labaudiniere and R. Burgada; *Phosphorus Sulfur*, 1985, **24**, 235.

85PS(25)319 D. El Manouni, L. Leroux and R. Burgada; *Phosphorus Sulfur*, 1985, **25**, 319.

85ZN(B)1589 G.V. Roschenthaler, R. Bohlen and D. Schomburg; *Z. Naturforsch. B*, 1985, **40**, 1589.

85ZOB1982 V.V. Negrebetskii, V.I. Kalchenko, R.B. Rudyi and L.N. Markovskii; *Zh. Obshch. Khim.*, 1985, **55**, 1982 (*Chem. Abstr.*, 1985, **105**, 97559).

86JA2416 R.A. McClelland, D.A. Cram and G.H. McGall; *J. Am. Chem. Soc.*, 1986, **108**, 2416.

86JGU2218 S.K. Tupchienko, T.N. Dudchenko and A.D. Sinitsa; *J. Gen. Chem. USSR (Engl. Transl.)*, 1986, **56**, 2218.

86MI-1 K. Utvary, K. Galle, A. Cowley and A. Arif; *Monatsh. Chem.*, 1986, **117**, 1245.

86PS(26)193 D. Schomburg, U. Wermuth and R. Schmutzler; *Phosphorus Sulfur*, 1986, **26**, 193.

86PS(27)297 U. Starke, G.-V. Roschenthaler and R. Schmutzler; *Phosphorus Sulfur*, 1986, **27**, 297.

86PS(27)313 R. Bohlen, J. Heine, W. Kuhn, W. Offermann, J. Stelten, G.V. Roschenthaler and W.G. Bentrude; *Phosphorus Sulfur*, 1986, **27**, 313.

86PS(27)321 R. Bohlen, H. Hacklin, J. Heine, W. Offermann and G.-V. Roschenthaler; *Phosphorus Sulfur*, 1986, **27**, 321.

86T3521 L. Labaudiniere and R. Burgada; *Tetrahedron*, 1986, **42**, 3521.

86TL2863 J. Navech and M. Revel; *Tetrahedron Lett.*, 1986, **27**, 2863.

86TL2971 O. Diallo, M.T. Boisdon, L. Lopez, C. Malvaud and J. Barrans; *Tetrahedron Lett.*, 1986, **27**, 2971.

86ZOK1193 Yu.G. Shermolovich, N.P. Kolesnik, S.V. Iksanova, V.V. Tkachevskii and L.N. Markovskii; *Zh. Obshch. Khim. USSR*, 1986, **56**, 1193.

87CB1183 O.J. Sherer, P. Quintus and W.S. Sheldrick; *Chem. Ber.*, 1987, **120**, 1183.

87CRV399 B.R. Venepalli and W.C. Agosta; *Chem. Rev.*, 1987, **87**, 399.

87JCS(CC)1131 J. Powell, K.S. Ng and J.F. Sawyer; *J. Chem. Soc., Chem. Commun.*, 1987, 1131.

87JGU1078 I.V. Konovalova, E.N. Ofitserov, T.N. Sinyashina, V.F. Mironov and A.N. Pudovik; *J. Gen. Chem. USSR (Engl. Transl.)*, 1987, **57**, 1078.

87JGU2143 IV.K. Dmitriev, B.V. Timoktin, A.V. Kalabina, T.I. Kazantseva and V.I. Donskikh; *J. Gen. Chem. USSR (Engl. Transl.)*, 1987, **57**, 2143.

87PS(29)147 M. Willson, L. La Faille, L. Vidaud and J.-F. Labarre; *Phosphorus Sulfur*, 1987, **29**, 147.

87PS(30)519 J.C. van de Grampel, P.L. Buwalda, A.A. van der Huizen, T. Wilting, A. Meetsma and F. van Bolhuis; *Phosphorus Sulfur*, 1987, **30**, 519.

87PS(32)1 L. Lamande and A. Munoz; *Phosphorus Sulfur*, 1987, **32**, 1.

87TL3445 E. Vedejs and C. Marth; *Tetrahedron Lett.*, 1987, **28**, 3445.

88CB391 R. Schmutzler, O. Stelzer and N. Weferling; *Chem. Ber.*, 1988, **121**, 391.

88IC1911 R.C. Haddon, S.V. Chichester-Hicks and S.L. Mayo; *Inorg. Chem.*, 1988, **27**, 1911.

88JCS(CC)615 M.T. Boisdon and J. Barrans; *J. Chem. Soc., Chem. Commun.*, 1988, 615.

88JGU1996 E.E. Nifant'ev, T.S. Kukhaeva, I.A. Soldatova, I.S. Belostotskaya, V.V. Ershov and L.K. Vasyaniava; *J. Gen. Chem. USSR (Engl. Transl.)*, 1988, **58**, 1996.

88MI-1 D.S. Yufit, Yu.T. Struchkov, E.I. Matrosov, D.N. Lobanov, N.V. Matrosova amd M.I. Kabachnik; *J. Strukt. Khim., USSR*, 1988, **29**, 113.

88PS(35)133 N. Lowther, P.D. Beer and C.D. Hall; *Phosphorus Sulfur*, 1988, **35**, 133.

88PS(35)195 A. Munoz and L. Lamande; *Phosphorus Sulfur*, 1988, **35**, 195.

88PS(35)247 J. Navech, H. Germa and S. Mathieu; *Phosphorus Sulfur*, 1988, **35**, 247.

88PS(35)309 B. Garrigues, L. Lamande and A. Munoz; *Phosphorus Sulfur*, 1988, **35**, 309.

88ZAAC(561)49 H. Hacklin and G.V. Roschenthaler; *Z. Anorg. Allg. Chem.*, 1988, **561**, 49.

89JCS(CC)1103 G. Robinet, M. Barthalat, V. Gasmi and J. Devillers; *J. Chem. Soc., Chem. Commun.*, 1989, 1103.

89JGU1333 S.K. Tupchienko, I.N. Duchenko and A.D. Sinitsa; *J. Gen. Chem. USSR (Engl. Transl.)*, 1989, **50**, 1333.

89JGU2583 V.F. Mironov, T.N. Sinyashina, E.N. Ofitserov, P.P. Chernov, I.V. Konovalov and A.N. Pudovik; *J. Gen. Chem. USSR (Engl. Transl.)*, 1989, **59**, 2583.

89JGU634 L.V. Nesterov and N.E. Krepysheva; *J. Gen. Chem. USSR (Eng. Transl.)*, 1989, **59**, 634.

89JST(195)11 G. Guerch and J.F. Labarre; *J. Mol. Struct.*, 1989, **195**, 11.

89PS(44)193 C. Ru-Yu and C. Leifeng; *Phosphorus Sulfur*, 1989, **44**, 193.

89PS(45)255 G. Baccolini, R. Dalpozzo and E. Mezzina; *Phosphorus Sulfur*, 1989, **45**, 255.

89ZN(B)612 U. Engelhardt and U. Diefenbach; *Z. Naturforsch. B*, 1989, **44**, 612.

90JA3905 E. Vedejs and C.F. Marth; *J. Am. Chem. Soc.*, 1990, **112**, 3905.

90JA6092 K.C.K. Swamy, J.M. Holmes, R.O. Day and R.R. Holmes; *J. Am. Chem. Soc.*, 1990, **112**, 6092.

90JA6095 K.S.K. Swamy, R.O. Day, J.M. Holmes and R.R. Holmes; *J. Am. Chem. Soc.*, 1990, **112**, 6095.

90JA7451 Y. Huang, A.M. Arif and W.G. Bentrude; *J. Am. Chem. Soc.*, 1990, **112**, 7451.

90JCS(D)2303 B. De Ruiter, J.C. van de Grampel and F. van Bolhuis; *J. Chem. Soc. Dalton Trans.*, 1990, 2303.

90JGU1491 E.K. Rutkovskii, I.S. Zal'tsman, N.G. Feshchenko and A.M. Pinchuk; *J. Gen. Chem. USSR (Engl. Transl.)*, 1990, **60**, 1491.

90JGU397 S.A. Terent'eva, M.A. Pudovik and A.N. Pudovik; *J. Gen. Chem. USSR (Engl. Transl.)*, 1990, **60**, 397.

90PS(47)443 Y. Leroux, D. El Manouni, L. Labaudiniere, R. Burgada, A. Neuman and H. Gitlier; *Phosphorus Sulfur*, 1990, **47**, 443.

90PS(49)385 B.Koll, K. Totschnig, J. Vogel, P. Peringer, E.P. Muller, M. Fischer and W. Petter; *Phosphorus Sulfur*, 1990, **49/50**, 385.

90PS(54)227 D.C. Apperley, R.K. Harris, T. Kaukoret and R. Schmutzler; *Phosphorus Sulfur Silicon*, 1990, **54**, 227.

90T3527 L. Lamande and A. Munoz; *Tetrahedron*, 1990, **46**, 3527.

90TL3429 M. Ehle, O. Wagner, U. Bergstrasser and M. Regitz; *Tetrahedron Lett.*, 1990, **31**, 3429.

90ZN(B)741 M. Herberhold, C. Doernhoefer and U. Thewaldt; *Z. Naturforsch. B*, 1990, **45**, 741.

91IC1052 R.R. Holmes, K.C. Kumara Swamy, J.M. Holmes and R.O. Day; *Inorg. Chem.*, 1991, **30**, 1052.

91IC3928 J. Hans, R.O. Day and R.R. Holmes; *Inorg. Chem.*, 1991, **30**, 3928.

91JA7945 M. Michalska, E. Brzerinska and P. Lipka; *J. Am. Chem. Soc.*, 1991, **113**, 7945.

91JGU1344 A.N. Cheklov, A.N. Bovin and F.N. Tsvetkov; *J. Gen. Chem. USSR (Engl. Transl.)*, 1991, **61**, 1344.

91MI-1 J.C. Tebby; *Handbook of Phosphorus-31 Nuclear Magnetic Resonance Data*, CRC Press 1991.

91PS(55)59 H.A. Al-Madfa and R.A. Shaw; *Phosphorus Sulfur*, 1991, **55**, 59.

91PS(57)111 A. Kilic, Z. Kilic and R.A. Shaw; *Phosphorus Sulfur*, 1991, **57**, 111.

91PS(62)139 J. Breker, P.G. Jones and R. Schmutzler; *Phosphorus Sufur Silicon*, 1991, **62**, 139.

91TL501 O.D. Diallo, L. Lopez, Y.K. Rodi and J. Barrans; *Tetrahedron Lett.*, 1993, **32**, 501.

92CB801 M. Well, W. Albers, A. Fischer, P.G. Jones and R. Schmutzler; *Chem. Ber.*, 992, **125**, 801.

92IC5494 T.K. Prakasha, S.D. Burton, R.O. Day and R.R. Holmes; *Inorg. Chem.*, 1992, **31**, 5494.

92JA4008 T. Kawashima, K. Kato and R. Okazaki; *J. Am. Chem. Soc.*, 1992, **114**, 4008.

92JGU1172 V.F. Mironov, L.A. Burnaeva, V.M. Krokhalev, V.I. Saloutin, I.V. Konovalova, R.A. Mavleev and P.P. Chernov; *J. Gen. Chem. USSR (Engl. Transl.)*, 1992, **62**, 1172.

92JGU780 S.K. Tupchienko, T.N. Dudchenko and A.D. Sinitsa; *J. Gen. Chem. (Engl. Transl.)*, 1992, **62**, 780.

92JGU970 V.F. Mironov, R.A. Mavleev, E.N. Ofitserov, I.V. Konovalova and A.N. Pudovik; *J. Gen. Chem. (Engl. Transl.)*, 1992, **62**, 970.

92JOC5195 C.K. McClure, C.W. Grote and B.A. Lockett; *J. Org. Chem.*, 1992, **57**, 5195.

92PS(69)231 M. Well, A. Fischer, P.G. Jones and R. Schmutzler; *Phosphorus Sulfur*, 1992, **69**, 231.

92PS(70)263 A. Munoz and L. Lamande; *Phosphorus Sulfur*, 1992, **70**, 263.

92PS(73)195 M. Furkens, P.G. Jones, A. Fischer and R. Schmutzler; *Phosphorus, Sulfur Silicon*, 1992, **73**, 195.

93AG(E)1430 C. Chuit, R.J.P. Corriu, P. Morforte, C. Reye, J.-P. Declercq and A. Dubourge; *Angew. Chem. Int. Ed. Engl.*, 1993, **32**, 1430.

93HAC271 Y. Huang, A.E. Sopchik, A.M. Arif and W.G. Bentrude; *Heteroatom. Chem.*, 1993, **4**, 271.

93JA4031 H. Huang, A.E. Sopchik, A.M. Arif and W.G. Bentrude; *J. Am. Chem. Soc.*, 1993, **115**, 4031.
93MM3 H.R. Allcock and M.L. Turner; *Macromolecules*, 1993, **26**, 3.
93POL1509 M.G. Muralidhara, N. Grover and V. Chandrasekhar; *Polyhedron*, 1993, **12**, 1509.
93PS(75)233 I.V. Shevchenko and R. Schmutzler; *Phosphorus Sulfur*, 1993, **75**, 233.
93PS(76)87 J.C. Martin, S.R. Chopra, C.D. Moon and T.R. Forbas; *Phosphorus Sulfur*, 1993, **76**, 87.
93PS(78)271 I. Neda, A. Fischer, P.G. Jones and R. Schmutzler; *Phosphorus Sulfur*, 1993, **78**, 271.
93PS(80)173 I. Neda, T. Kaukorat and R. Schmutzler; *Phosphorus Sulfur*, 1993, **80**, 173.
93PS(81)1 D. Houalla, Z. Bounja, M.-C. Monje and M. Sanchez; *Phosphorus Sulfur*, 1993, **81**, 1.
93PS(83)157 M.-R. Mazieres, D. Houalla, M.-C. Monje and M. Sanchez; *Phosphorus Sulfur*, 1993, **83**, 157.
93TL3107 G. Markl, H. Sommer and H. Noth; *Tetrahedron Lett.*, 1993, **34**, 3107.
94IC2727 A.J. Elias, R.L. Kirchmeier and J.M. Schreeve; *Inorg. Chem.* 1994, **33**, 2727.
94IC3091 H.R. Allcock, U. Diefenbach and S.R. Pucher; *Inorg. Chem.*, 1994, **33**, 3091.
94JA8095 T.K. Prakasha, R.O. Day and R.R. Holmes; *J. Am. Chem. Soc.*, 1994, **116**, 8095.
94JCR(M)887 C. Combes, H.-J, Cristau, W.S. Li, M. McPartlin, F. Plenat and I.J. Scowen; *J. Chem. Res., (M)*, 1994, 887.
94JCR(S)134 C. Combes, H.-J. Cristau, W.S. Li, M. McPartlin, F. Plenat and I.J. Scowen; *J. Chem. Res., (S)*, 1994, 134.
94JCS(D)63 A. Chandrasekaren and S.S. Krishnamurthy; *J. Chem. Soc. Dalton Trans.*, 1994, 63.
94JCS(D)577 A.A. van de Huizen, P.L. Buwalda, T. Wilting, H. Pol, A.P. Jekel, A. Meetsma and J.C. van de Grampel; *J. Chem. Soc. Dalton Trans.*, 1994, 577.
94JGU373 Yu.G. Gololobov, G.D. Kolomnikova and J.O. Krylova; *J. Gen. Chem. USSR (Engl. Transl.)*, 1994, **64**, 373.
94JGU1171 A.A. Prishchemko, M.V. Livantsov, K.-L. Lorents and M.Yu. Papsuev; *J. Gen. Chem. USSR (Engl. Transl.)*, 1994, **64**, 1171.
94JGU1210 V.F. Mironov and I.V. Konovalova; *J. Gen. Chem. USSR (Engl. Transl.)*, 1994, **64**, 1210.
94JGU1329 A.R. Burilov, I.L. Nikolaeva, I.A. Litvinov and M.A. Pudovik; *J. Gen. Chem. USSR (Engl. Transl.)*, 1994, **64**, 1329.
94JGU1546 T.V. Gryaznova, V.F. Mironov, M.G. Khanipova, I.V. Konovalova, A.M. Kibardin and A.N. Pudovik; *J. Gen. Chem. USSR (Engl. Transl.)*, 1994, **64**, 1546.
94MI-1 O.N. Nuretdinova; *Russ. Chem. Bull.*, 1994, **43**, 1154.
94MI-2 T. Kaukorat, I. Neda and R. Schmutzler; *Coord. Chem. Rev.*, 1994, **137**, 53.
94PS(86)157 F.U. Seifert and G.-V. Roschenthaler; *Phosphorus Sulfur*, 1994, **86**, 157.
94PS(90)131 D. Houalla, L. Moureau, S. Skouta and R. Wolf; *Phosphorus Sulfur*, 1994, **90**, 131; *ibid*, 1994, **88** 83.
94TL165 S.E. Pipko, Y.V. Balitzky, A.D. Sinitsa and Y.G. Gololbov; *Tetrahedron Lett.*, 1994, **35**, 165.
94ZAAC(620)1434 U. Diefenbach, B. Stromburg and U. Englehardt; *Z. Anorg. Allg. Chem.*, 1995, **620**, 1434.
94ZN(B)145 M. Farkens, T.G. Meyer, I. Neda, R. Sonnonburg, C. Muller, A.K. Fischer, P.G. Jones and R. Schmutzler; *Z. Naturforsch. B.*, 1994, **49**, 145.
94ZN(B)171 I. Neda, T. Kaukovat and R. Schmutzler; *Z. Naturforsch. B.*, 1994, **49**, 171.
95CB627 R. Sonnonburg, I. Neda, A.K. Fischer, P.G. Jones and R. Schmutzler; *Chem Ber.*, 1995, **128**, 627.
95CB959 H.H. Karsch, T. Rapprich and M. Heckel; *Chem. Ber.*, 1995, **128**, 959.
95IC4525 T.V. Timosgeva, T.K. Prakasha, A. Chandrasckaran, R.O. Day and R.R. Holmes; *Inorg. Chem.*, 1995, **34**, 4525.
95ICA(228)187 K. Brandt, T. Kupta, J. Drodz, J.C. van de Grampel, A. Meetsma and A.P. Jekel; *Inorg. Chim. Acta*, 1995, **228**, 187.
95JA12390 H. Huang and W.G. Bentrude; *J. Am. Chem. Soc.*, 1995, **117**, 12390.
95JA8287 K. Yamamoto, R. Nadano, M. Itagaki and K. Akiba; *J. Am. Chem. Soc.*, 1995, **117**, 8287.
95JGU198 A.D. Sinitsa, L.I. Nesterova, D.M. Malenko, V.V. Pirozhenko, E.B. Rusanov and A.N. Chernega; *J. Gen. Chem. USSR (Engl. Transl.)*, 1997, **65**, 198.
95JCS(P2)1149 M.J. Potrzekowski, J. Kowara, W. Ciesielski and A. Skowronska; *J. Chem. Soc., Perkin 2 Trans.*, 1995, 289.
95JGU289 I.V. Konovalova, V.F. Mironov and M.G. Khanapova; *J. Gen. Chem. USSR (Engl. Transl.)*, 1995, **65**, 289.
95JGU1604 A.A. Prishchemko, O.P. Novikova, M.V. Livantsov and E.V. Grigor'ev; *J. Gen. Chem. USSR (Engl. Transl.)*, 1995, **65**, 1604, 1606.
95JGU1610 A.A. Prishchemko, M.V. Livantsov, D.N. Kustrya and E.V. Grigor'ev; *J. Gen. Chem. USSR (Engl. Transl.)*, 1995, **65**, 1610.
95MI-1 Y. Segall; *XIII Int. Conf. on Phosphorus Chem. Jerusalem 1995.*.
95MI-2 C.D. Hall; in "*Organophosphorus Chemistry Specialist Reports*", Eds. D.W. Allen and B.J. Walker, Royal Society of Chemistry, 1995, **26**, 72.
95MM6323 D. Kumar and A.D. Gupta; *Macromolecules*, 1995, **28**, 6323.
95PS(103)125 T. Bailly and R. Burgada; *Phosphorus Sulfur*, 1995, **103**, 125.
95PS(103)199 D. Houalla, L. Moureau, S. Skouta and M.-R. Mazieres; *Phosphorus Sulfur*, 1995, **103**, 199.
95PS(106)65 C. Melnicky, I. Neda and R. Schmutzler; *Phosphorus Sulfur*, 1995, **106**, 65.
95TL2021 I.V. Shevchenko; *Tetrahedron Lett.*, 1995, **36**, 20213.
95ZAAC(621)2001 V.A. Pinchuk, C. Muller, A.K. Fischer, P.G. Jones, R. Schmutzler, L.N. Markovsky, Yu.G. Shermolovich and A.M. Pinchuk; *Z. Anorg. Allg. Chem.*, 1995, **621**, 2001.
96AG(E)1096 T. Kawashima, T. Soda and R. Okazaki; *Angew. Chem. Int. Ed. Engl.*, 1996, **35**, 1096.
96CB725 A. Kadyrov, I. Neda, T. Kaukorat, R. Sonnenburg, A. Fischer, P.G. Jones and R. Schmutzler; *Chem. Ber.*, 1996, **129**, 725.

96IC6899	O.S. Jung, S.H. Park, Y.A. Lee, Y.H. Cho, K.M. Kim, S. Lee, K. Chae and Y.S. Sohn; *Inorg. Chem.* 1996, **35**, 6899.
96JA9841	M.A. Said, M. Puelm, R. Herbst-Irmer and K.C.K. Swamy; *J. Am. Chem. Soc.*, 1996, **118**, 9841.
96JCS(CC)2157	I. Gardinier, H. Bernard, F. Chuburu, A. Roignant, J.J. Yaouanc and H. Handel; *J. Chem. Soc., Chem. Commun.*, 1996, 2157.
96JGU355	M.A. Pudovik, S.A. Terent'eva and A.N. Pudovik; *J. Gen. Chem. USSR (Engl. Transl.)*, 1996, **66**, 355.
96PS(109/10)241	O. Mir, M. Fridkin and Y. Segall; *Phosphorus Sulfur*, 1996, **109/10**, 241.
96PS(109/10)489	T. Kawashima, T. Soda, K. Kato and R. Okazaki; *Phosphorus Sulfur*, 1996, **109/10**, 489.
96PS(109/10)493	I.V. Shevchenko; *Phosphorus Sulfur*, 1996, **109/10**, 493.
96PS(109/10)629	I. Neda, C. Melnicky, A. Vollbrecht, A. Fischer and P.G. Jones; *Phosphorus Sulfur*, 1996, **109/10**, 629.
96PS(109/110)156	S. Barkallah, M.L. Bojin and S.A. Evans Jr.; *Phosphorus Sulfur*, 1996, **109/110**, 156.
96PS(114)51	D. Houalla and L. Moureau; *Phosphorus Sulfur*, 1996, **114**, 51.
96S473	I. Neda, C. Melnicky, A. Vollbrecht and R. Schmutzler; *Synthesis*, 1996, 473.
96T2995	N. Oget, F. Chaburo, J.J. Yaouanc and H. Handel; *Tetrahedron*, 1996, **52**, 2995.
97AG(E)608	J. Lacour, C. Ginglinger, C. Grivet and G. Bernardinelli; *Angew. Chem. Int. Ed. Engl.*, 1997, **36**, 608.
97AG(E)2500	T. Kawashima, R. Okazaki and R. Okazaki; *Angew. Chem. Int. Ed. Engl.*, 1997, **36**, 2500.
97CB819	S. Volbrecht, A. Volbrecht, J. Jeske, P.G. Jones, R. Schmutzler and W.W. du Mont; *Chem. Ber. Recueil*, 1997, **130**, 819.
97CB1479	J. Krill, I.V. Shevchenko, A.K. Fischer, P.G. Jones and R. Schmutzler; *Chem. Ber. Recueil*, 1997, **130**, 1479.
97IC2044	M.A. Said, M. Puelm, R. Herbst-Irmer and K.C.K. Swamy; *Inorg. Chem.*, 1997, **36**, 2044.
97IC2578	A. Chandrasekaran, R.O. Day and R.R. Holmes; *Inorg. Chem.*, 1997, **36**, 2578.
97JA5970	S. Kojima, R. Takagi and K. Akiba; *J. Am. Chem. Soc.*, 1997, **119**, 5970.
97JCS(P2)15	C. Combes-Chamalet, H.-J, Cristau, M. McPartlin, F. Plenat, I.J. Scowen and T.M. Woodroffe; *J. Chem. Soc., Perkin 2 Trans.*, 1997, 15.
97JFC(85)129	V.G. Ratner, E. Lork, K.I. Pashkevich and G.-V. Roschenthaler; *J. Fluorine Chem.*, 1997, **85**, 129.
97JFC(86)109	I. Neda, C. Muller and R. Schmutzler; *J. Fluorine Chem.*, 1997, **86**, 109.
97JGU151	L.I. Nesterova, D.M. Malenko, V.V. Pirozhenko and A.D. Sinitsa; *J. Gen. Chem. USSR (Engl. Transl.)*, 1997, **67**, 151.
97JGU1486	V.B. Mushkin, R.M. Aminova, V.F. Mironov and A.I. Konovalov; *J. Gen. Chem. USSR (Engl. Transl.)*, 1997, **67**, 1486.
97JOM(529)59	F. Carre, C. Chuit, R.J.P. Corriu, A. Mehdi and C. Reye; *J. Organomet. Chem.*, 1997, **529**, 59.
97MI-1	V.A. Zagumennov, A.A. Karasik, E.V. Nikitin and G.N. Nikonov; *Russ. Chem. Bull.*, 1997, **46**, 1154.
97MI-2	I.S. Mikhel', K.N. Gavrilov, D.V. Lechkin and A.I. Rebrov; *Russ. Chem. Bull.*, 1997, **46**, 1359.
97MI-3	Q. Wang, H.Y. Lu, X. Chen and Y.-F. Zhao; *Chin. Chem. Lett.*, 1997, **8**, 629.
97MI-4	Q. Wang, H.Y. Lu and Y.F. Zhao; *Chin. Chem. Lett.*, 1997, **8**, 1017.
97PS(120-1)431	R.R. Holmes, A. Chandrasekaran and R.O. Day; *Phosphorus Sulfur.*, 1997, **120-1**, 431.
97PS(123)5	M. Tlahuexl, F.J. Martinez-Martinez, M.D.J. rosales-Hoz and R. Contreras; *Phosphorus Sulfur*, 1997, **123**, 5.
97PS(123)181	F. Carre, M. Chauhan, C. Chuit, R.J.P. Corriu and C. Reye; *Phosphorus Sulfur.*, 1997, **123**, 181.
97PS(123)359	D. Houalla, L. Moureau and C. Vidal; *Phosphorus Sulfur*, 1997, **123**, 359.
97PS(124/5)379	T. Kawashima, R. Okazaki and R. Okazaki; *Phosphorus Sulfur*, 1997, **124/5**, 379.
97PS(131)173	A.A. Bredikhin, S.N. Lazarev, Z.A. Bredkhina and V.A. Al'Fonsov; *Phosphorus Sulfur*, 1997, **131**, 173.
97RCM1825	H. Fu, Z. Li, Y. Zhao, D. Guo, H. Xiao, J. Wang and Y. Wu; *Rapid Commun. Mass Spectrom.*, 1997, **11**, 1825.
97TL7753	S. Kojima, K. Kawaguchi and K. Akiba; *Tetrahedron Lett.*, 1997, **38**, 7753.
97ZAAC(623)545	M. Herberhold, A. Hofmann and W. Milius; *Z. Anorg. Allg. Chem.*, 1997, **623**, 545.
97ZAAC(623)1325	I. Neda, V.A. Pinchuk, A. Thonnessen, L. Ernst, P.G. Jones and R. Schmutzler; *Z. Anorg. Allg. Chem.*, 1997, **623**, 1325.
98AG(E)1098	H. Luo, R. McDonald and R.G. Cavell; *Angew. Chem. Int. Ed. Engl.*, 1998, **37**, 1098.
98HAC173	G. Bekiaris and G.-V. Roschenthaler; *Heteroat. Chem.*, 1998, **9**, 173.
98JGU530	S.E. Pipko, Yu.V. Belitskii, A.N. Chernega and A.D. Sinitsa; *J. Gen. Chem. USSR (Engl. Transl.)*, 1998, **68**, 530.
98MI-1	H.G. Gong, H. Fu, Z.L. Li and Y.-F. Zhao; *Chin. Chem. Lett.*, 1998, **9**, 953.
98POL3643	K.C.K. Swamy, M.A. Said, R. Herbst-Irmer and M. Puelm; *Polyhedron*, 1998, **17**, 3643.
98S376	N.J. Zhang, H.Y. Lu, X. Chen and Y.F. Zhao; *Synthesis*, 1998, 376.
98S855	H. Fu, G.-H. Tu and Y.F. Zhao; *Synthesis*, 1998, 855.
99DOK(367)64	M.V. Proskurina, T.N. Kudryavsteva and O.N. Zefirova; *Dokl. Akad. Nauk SSSR*, 1999, **367**, 64.
99IC1093	N. Thirupathi, S.S. Krishnamurthy and M. Nethaji; *Inorg. Chem.*, 1999, **38**, 1093.
99JA6958	M. Nakamoto and K. Akiba; *J. Am. Chem. Soc.*, 1999, **121**, 6958.
99JAP29591	T. Shimizu; *Jpn. Kakai Tokkyo Koho* JP **11**, 29,591.
99JAP217508	J. Tan and H. Onishi; *Jpn. Kakai Tokkyo Koho* JP **11**, 217,508.
99CC1423	G. Keglevich, H. Forintos, A. Szollosy and L. Toke; *Chem. Commun. (Cambridge)*, 1999, 1423.
99JFC223	O.D. Gupta, R.L. Kirchmeier and J.M. Shreeve; *J. Fluorine Chem.*, 1999, **97**, 223.
99JGU48	E.V. Popova, V.F. Mironova, E.A. Ishmaeva and I.J. Patanorski; *J. Gen. Chem. USSR (Engl. Transl.)*, 1999, **69**, 48.
99OM915	C. Marchi, F. Fotiadu and G. Buono; *Organometallics*, 1999, **18**, 915.

7 Applications of Phosphorus Heterocycles in Homogeneous Catalysis

F. MATHEY
Ecole Polytechnique, Palaiseau Cedex, France

7.1 INTRODUCTION

In practical terms, homogeneous catalysis using transition metal complexes is the most important consequence of organometallic chemistry. Contemporary research in this field is particularly strong in two areas. The first focusses on the design of water-soluble ligands, allowing the development of novel and competitive biphasic catalysts for the chemical industry ⟨98MI1⟩. The second which concerns enantiopure ligands for asymmetric catalysis offers ever more efficient answers to the needs of the pharmaceutical and agrochemical industries for chiral drugs, herbicides and pesticides ⟨94MI1⟩. In commercial terms, the worldwide market of chiral drugs has reached the $100 billion mark in 1999 ⟨99CEN77(41)101⟩.

Whenever a catalytic process occurs under non-oxidative conditions, phosphorus ligands figure heavily in the conception and architecture of the corresponding transition metal catalysts. It is widely accepted that the strength of the phosphorus–transition metal bond is controlled both by electronic and steric factors. The P $\rightarrow$ M σ-donation component involves the phosphorus lone pair orbital and M $\rightarrow$ P π-back donation is now thought to involve antisymmetric combinations of antibonding P–R σ^* orbitals ⟨92IC4391⟩. Steric factors, reflected by Tolman cone angles ⟨77CRV313⟩, may also affect the strength of the M–P bond by preventing a perfect overlap between the orbitals of the phosphorus donor and the metal. Incorporating phosphorus into a ring obviously changes its coordination ability directly through a modification of its Tolman cone angle and, more subtly, through electronic effects resulting from a change in the R–P–R intracyclic angle. The rigidity imposed by a cyclic structure also allows better control of the space around the catalytic centre. This control is of paramount importance in enantioselective catalysis.

Against this background, it is not surprising that the constant search for more efficient ligands has led transition metal chemists to investigate the potential of phosphorus heterocycles in coordination chemistry and catalysis. In the following sections, we have tried to cover the catalytic aspects of this work comprehensively for the most fundamental phosphorus–carbon heterocycles. The emphasis is more on the structures of the ligands than on the mechanistic and technical details of the catalytic processes. To be coherent with the rest of the book, our choice excludes such promising ligands as bulky cyclic phosphites or phosphonites which have proven worth in the hydroformylation of functional olefins ⟨93JA2066⟩ and internal alkynes ⟨95AG(E)1760⟩, in asymmetric hydroformylation ⟨94CC395, 97JA4413⟩, and in several more classical applications, e.g. ⟨98CC2077, 99CC2289⟩.

The first clear-cut applications of cyclic phosphines in homogeneous catalysis were described by Shell scientists in 1966 ⟨66NEP6516164, 66NEP6604094, 69GEP(O)1909620⟩. In this series of patents, they claimed that such cyclic phosphines as phospholes, phospholenes and 9-phosphabicyclo[3.3.1] and [4.2.1]nonanes are far superior in the Shell hydroformylation process to tributylphosphine, the ligand normally used to stabilize and activate the cobalt–carbonyl catalyst ⟨68JOM(13)469⟩. In this process, where a terminal olefin is transformed mainly into a primary alcohol, the major by-product is the saturated hydrocarbon. The formation of this saturated species is significantly reduced when replacing Bu_3P by these cyclic phosphines. Another major discovery was made in 1990–1991 when Burk ⟨90OM2653, 91JA8518⟩ introduced the very efficient phospholane-based C_2-symmetric DuPHOS and BPE ligands for the rhodium-catalysed asymmetric hydrogenation of functional alkenes. Since then, the interest in the potential applications of cyclic phosphines in catalysis has grown continuously. The domain is obviously still in its infancy, but it seems likely to undergo an exponential development in the near future.

7.2 CYCLIC PHOSPHINES

In this section, we follow the general organization of the book. Hence, the literature is presented according to the ring size of the cyclic phosphines and starts with phosphiranes, then phosphetanes, and so on.

7.2.1 Phosphiranes

With its extremely acute internal C–P–C angle of ca. 47° (see Chapter 2.1) and the high π-character of its intracyclic P–C bonds (Walsh orbitals), the phosphirane ring is obviously a very unusual type of phosphine and the investigation of its catalytic potential was awaited with some interest. The first experiments were carried out in 1993 when Marinetti *et al.* ⟨93OM1207⟩ prepared optically active phosphiranes (Equation (1)) and investigated their $[Rh(cod)L_2]^+PF_6^-$ complexes in the asymmetric hydrogenation of dehydroaminoacids.

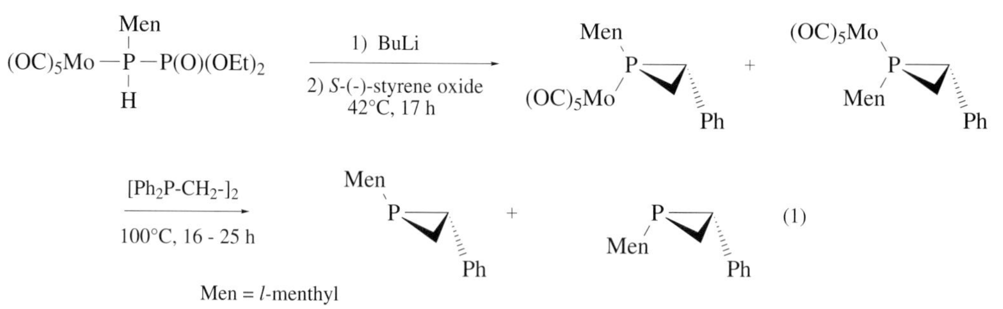

(1)

Ee's as high as 76% were observed but no phosphiranes were recovered at the end of the experiments because of extensive decomposition. More recently, a very stable polycyclic phosphirane, dubbed *BABAR*-Phos, was prepared as shown in Equation (2) ⟨99AG(E)1623⟩.

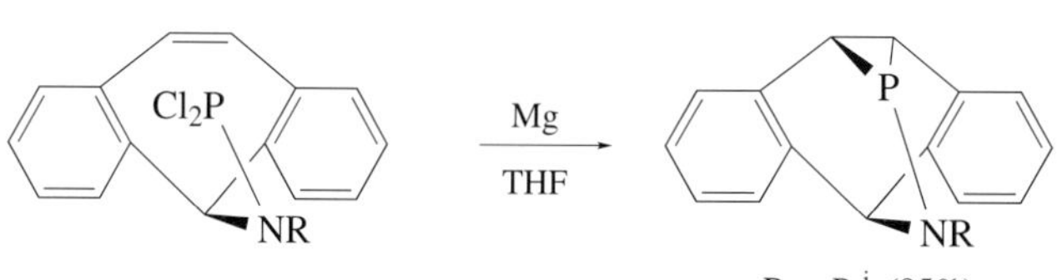

(2)

R = Pr^i (95%)

A platinum complex of *BABAR*-Phos, [Pt(L)dvtms] (dvtms = divinyltetramethyldisiloxane), is an active catalyst (100% conversion) for the hydrosilylation of vinyltriethylsilane by triethylsilane. Under similar conditions (14% catalyst, RT, 72 h, benzene), no conversion was observed with the analogous triphenylphosphine–platinum complex. See also ⟨2000T143⟩.

7.2.2 Phosphetanes

A first isolated report mentioned the incorporation of 1-phenyl-2,2,3-trimethylphosphetane into the [$HCo(CO)_3L$] precatalyst for the hydroformylation of 1-octene (Shell process) ⟨78MI(179)191⟩. However, a serious evaluation of phosphetanes only began in 1994 when Marinetti and her group undertook a systematic study of the catalytic properties of their complexes. A recent review ⟨98CCR(178)755⟩ describes part of this work. The first investigation dealt with the asymmetric hydrosilylation of cyclopentadiene using $PdCl_2$–**1*** complexes as catalysts ⟨94TL5861⟩ (Equation (3)).

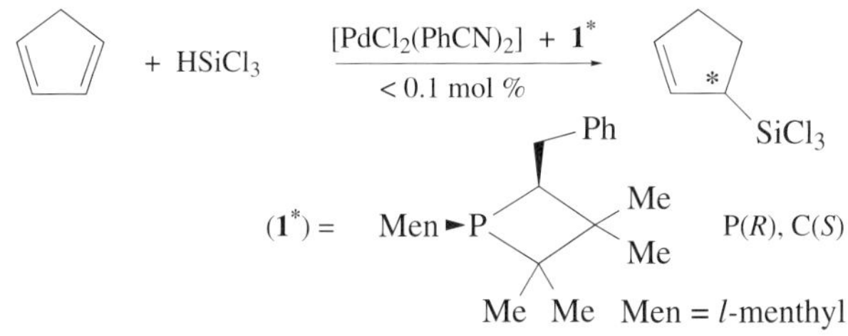

(3)

The (1 : 1) Pd : **1*** complex appears to be more efficient in terms of both reaction rate and enantioselectivity. A 100% conversion was observed in 2 h at RT with a 54% ee. The (1 : 2) complex needed 30 h at 70°C for a 26% conversion with a 44% ee. Apparently, the second molecule of phosphine inhibits the reaction. The formation of a 1 : 1 : 1 silane–phosphine–olefin complex appears to be the rate-determining step of the catalytic cycle.

In a subsequent study ⟨94OM3956⟩, phosphetane (**1***) was shown to be a poor ligand for Rh(I), but a better one for Ir(I). The cationic complex $[Ir(COD)(\mathbf{1}^*)_2]^+PF_6^-$ gave satisfactory rates for the hydrogenation of the methyl α-acetamidocinnamate, but the ee was low. Additional data on the asymmetric hydrosilylation of cyclopentadiene and styrene were provided. The best ee's were obtained with (**2***): 65% (*S*) for cyclopentadiene and 24% (*R*) for styrene.

(**2***) =

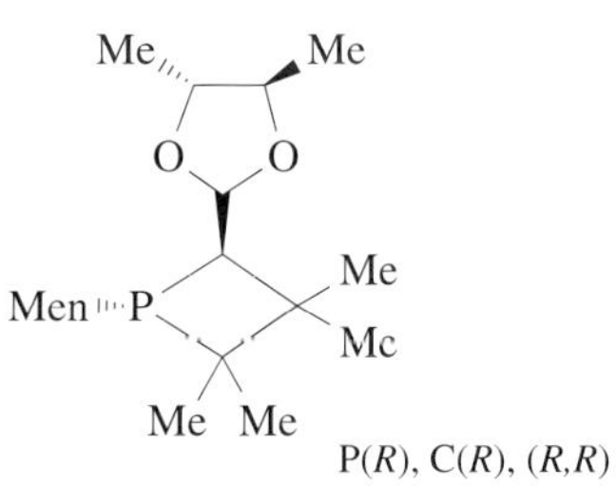

(**3***) =

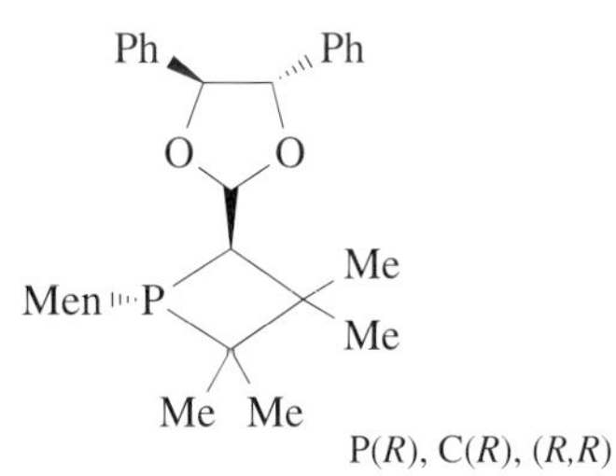

The ee's vary sharply with the relative configurations of the chiral centres in (**2**). For the P(*S*),C(*S*),(*R*,*R*) configuration, the ee falls to 11% in favour of (*R*)-1-(trichlorosilyl)cyclopent-2-ene.

Four epimers of the diphenyl analogue (**3**) of phosphetane (**2**) were synthesized some time later ⟨97JOM(529)465⟩. The epimer (**3***) proved to be the best ligand for the palladium-catalysed allylic alkylation of 1,3-diphenyl-2-propenyl acetate by sodium dimethyl malonate (Equation (4)).

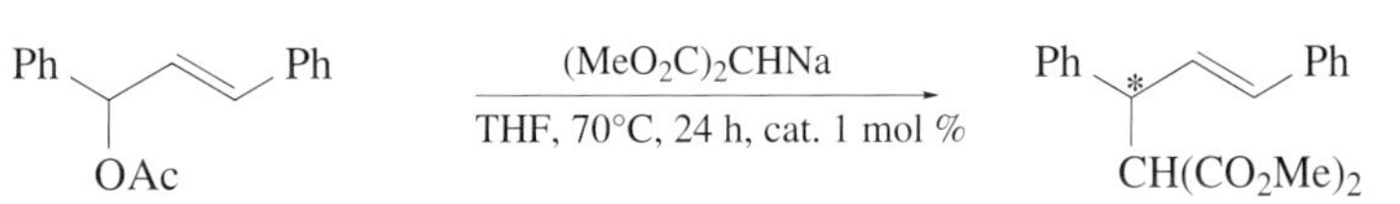

(4)

cat. = (**3***) + $[PdCl(allyl)]_2$ 90% yield, 82% e.e. (*R*)
(P:Pd = 1:1)

The next development concerned the synthesis and evaluation of a series of chiral phosphetane-containing bidentate ligands. The first approach involved chiral phosphetane derivatives of dppe ⟨98S1539⟩.

(4*) = 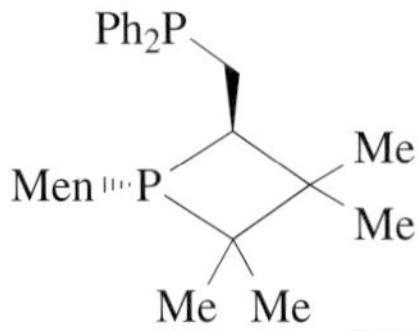

P(*R*), C(*R*)

(5*) = 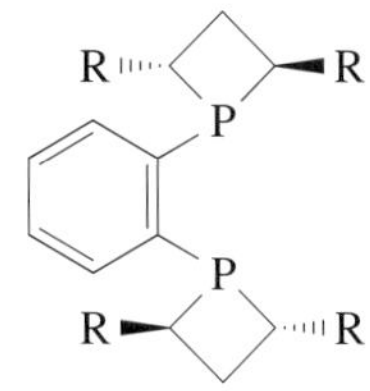

R = ^{i}Pr (*S,S*)

Phosphetane (**4***) gives a Rh(I) complex which catalyses the hydrogenation of *trans*-α-acetamidocinnamic acid in 75% ee (*S* enantiomer).

A phosphetane analogue of Burk's famous phospholane-based DuPHOS ⟨93JA10125⟩ has been prepared from the appropriate cyclic sulphates and dilithio-1,2-diphosphinobenzene ⟨97TL2947, 99CEJ1160, 2000T95⟩. Promising results were obtained in the enantioselective hydrogenation of β-ketoesters and β-diketones using $[RuBr_2(\mathbf{5}^*)]_2$ as the precatalyst (Equations (5) and (6)).

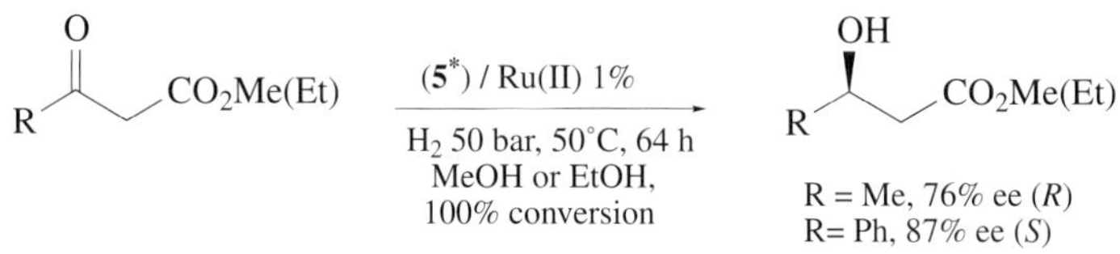

(5)

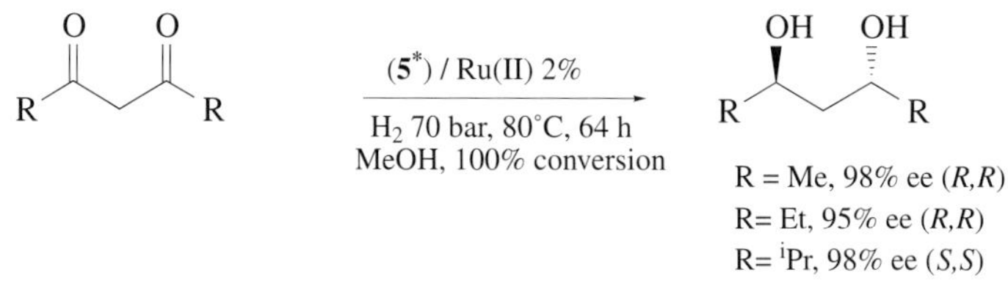

(6)

A subsequent paper ⟨99TL8365⟩ provided an interesting mechanistic detail. Halpern has shown ⟨80JA5952, 87JA1746⟩ that the major hydrogenation product in the rhodium-catalysed asymmetric hydrogenation of dehydroaminoacids, arises from the minor olefin–rhodium complex. Accordingly, an increase of the H_2 pressure gives decreased enantioselectivity for DIOP, DIPAMP and BINAP ligands. However, the asymmetric hydrogenation of methyl *trans*-α-acetamidocinnamate using $[Rh(COD)_2]^+OTf^-$ + (**5***) as the precatalyst shows the opposite effect: 82% ee under 10 bars of H_2 vs 90% ee under 100 bars. This suggests that the major enantiomer is formed from the major olefin–rhodium complex for (**5***) ⟨99TL8365⟩.

Recently, another chiral bis-phosphetane was prepared following the same approach as for (**5***). The reaction of dilithio-1,1′-bis-(phosphino)ferrocene with the cyclic sulphate derived from (*R,R*)-2,4-pentanediol affords (**6***) in 55% yield ⟨99SL1975⟩.

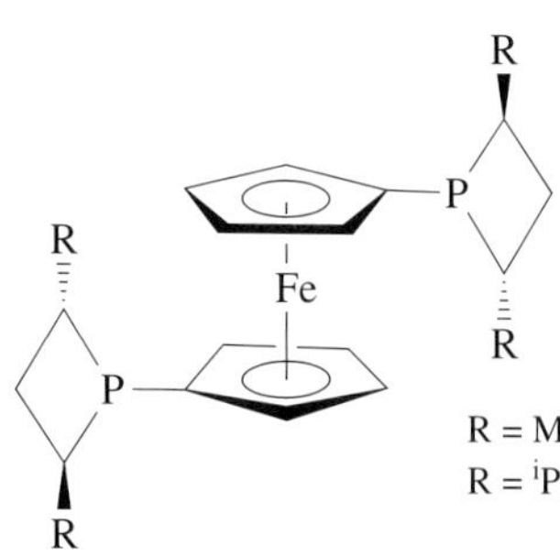

R = Me (**6***)
R = ^{i}Pr (**7***)

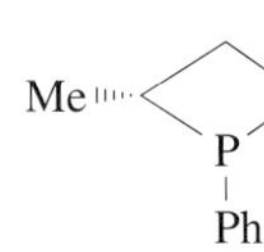

(**8***)

Both (**6***) and its isopropyl analogue (**7***) perform better than their phospholane analogues ⟨94TL9363⟩ in the rhodium-catalysed asymmetric hydrogenation of dehydroaminoacids.

Finally, a patent ⟨99USP5936109⟩ describes the use of (**8***) in the same field.

7.2.3 Phospholanes

The first report of phospholanes in catalysis mentioned 1-*n*-dodecylphospholane for the hydroformylation of 1-octene using a $[H\text{-}Co(CO)_3L]$ precatalyst ⟨78MI(179)191⟩. The first attempts to use chiral phospholane complexes in asymmetric hydrogenation were described by Brunner ⟨87JOM(328)71⟩. 1,4-Ditosylates derived from tartaric acid were cyclized by reaction with $NaPH_2$ or $PhPLi_2$. The chiral phospholanes thus synthesized have their chiral centres at the β-positions of the ring and display poor enantioselectivity in the rhodium-catalysed asymmetric hydrogenation of (*Z*)-α-(*N*-acetamido)cinnamic acid (16.8% ee). A chelating monophospholane such as (**10***) is slightly more efficient (up to 26% ee) ⟨91JOM(413)55⟩.

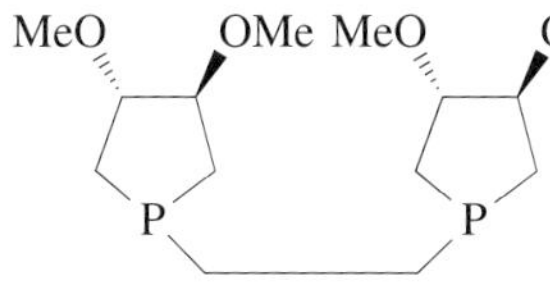
(**9***)

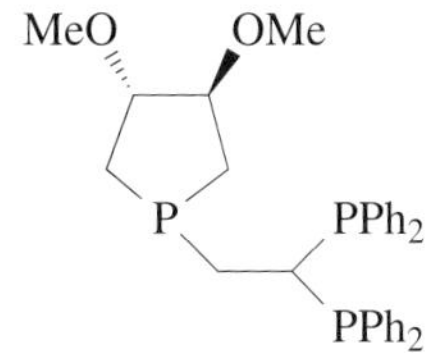
(**10***)

The breakthrough in this area came in 1990–1991 when Burk ⟨90OM2653, 91JA8518⟩ described the synthesis of C_2-symmetric bis-phospholanes having chiral centres in the α-positions. His series of DuPHOS ligands (**11***) displays an extraordinary versatility and efficiency in asymmetric catalysis. Their synthesis is shown in Equation (7) ⟨93JA10125⟩.

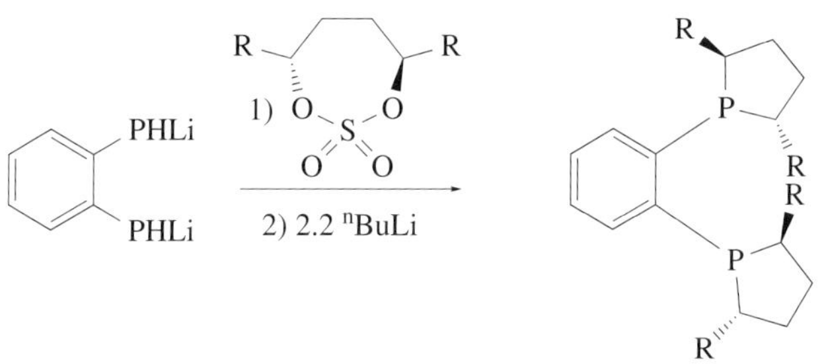
(7)

(**11***) a R=Me (*S,S*) 90%
b R=Et (*S,S*) 78%
c R=^{i}Pr (*R,R*) 70%

The DuPHOS–Rh(I) complexes are the catalysts of choice for the enantioselective hydrogenation of acetamidoacrylates ⟨91JA8518, 93JA10125⟩, *N*-acylhydrazones ⟨92JA6266, 94T4399⟩, α-enamides ⟨96JA5142⟩, and enol esters ⟨98JA4345⟩. A review of the initial results is available ⟨96PAC37⟩. In all cases, ee's in the range 90–100% have been observed. This is the result of an optimum delineation of the four quadrants defining the space around the rhodium centre (Scheme 1). For a discussion of the geometrical parameters favouring enantioselectivity, see ⟨83ACR106⟩.

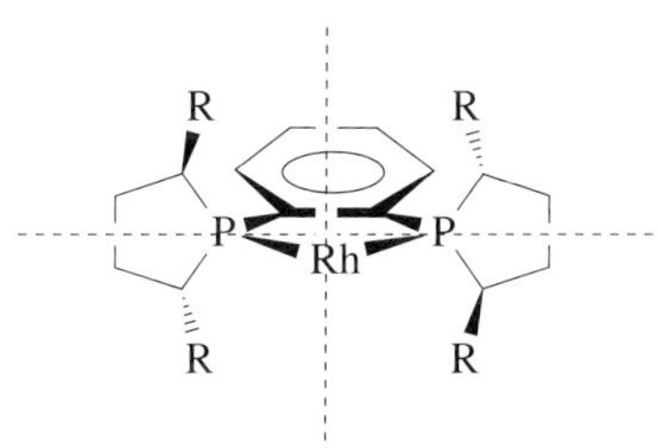

Scheme 1. (*R*, *R*)-DuPHOS blocks the upper left and lower right quadrants around rhodium ⟨93JA10125⟩.

Specific applications include the synthesis of β-branched aminoacids ⟨95JA9375⟩, carbon-linked glycopeptides ⟨97JA9897⟩, 2-alkylsuccinate enantiomers ⟨98AG(E)1931⟩, γ,δ-unsaturated aminoacids ⟨98JA657⟩, and a glutarate intermediate for the synthesis of Candoxatril ⟨99JOC3290⟩. A three-step procedure for the catalytic reductive amidation of ketones has been devised ⟨98JOC6084⟩. The enantioselective hydrogenation of acetamidoacrylates has been also carried out in supercritical carbon dioxide with $[Rh(\mathbf{11b}^*)(cod)]^+X^-$ (X = tetrakis-(3,5-bis(trifluoromethyl)phenyl)borate or triflate) as the catalyst ⟨95JA8277⟩. Finally, $[Rh(\mathbf{11a}^*)(Me_2CO)_2]^+$ has been shown to convert 4-substituted pent-4-enals efficiently into the corresponding cyclopentanones with 93–96% ee's at 25°C ⟨97CC589⟩.

The outstanding efficiency of DuPHOS led several groups to synthesize and test various analogues. Burk himself developed a series of 1,2-bis(*trans*-2,5-dialkylphospholano)ethanes (BPE) ligands (**12***) which are more flexible and electron-rich than DuPHOS ⟨90OM2653, 93JA10125⟩.

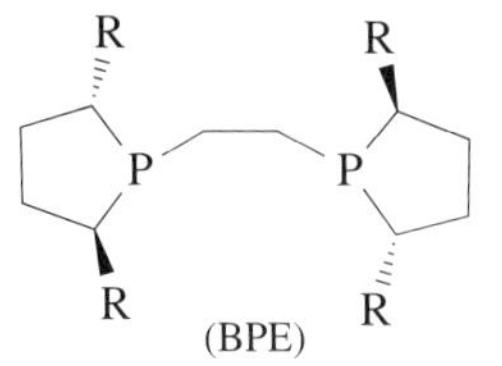

(BPE)

(**12***) a R=Me (*S,S*)
b R=Et
c R=^{n}Pr
d R=^{i}Pr (*R,R*)
e R=Cy

Both (**12d***) and (**12e***) are superior to DuPHOS in the ruthenium-catalysed enantioselective hydrogenation of *β*-ketoesters (Equation (8)) ⟨95JA4423⟩.

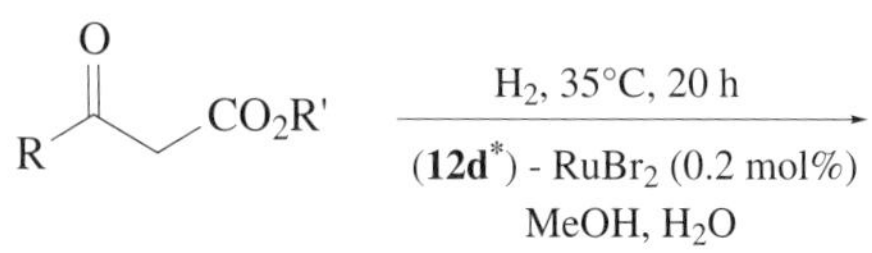

(8)

R = R' = Me 99.3% ee (*S*)

Burk's group has also synthesized ferrocene analogues such as (**13***) ⟨94TL9363⟩ and two C_3-symmetric tris(phospholane) ligands (**14***), (**15***) from the appropriate 1-lithiophospholane ⟨90AG(E)1462, 91TA569⟩.

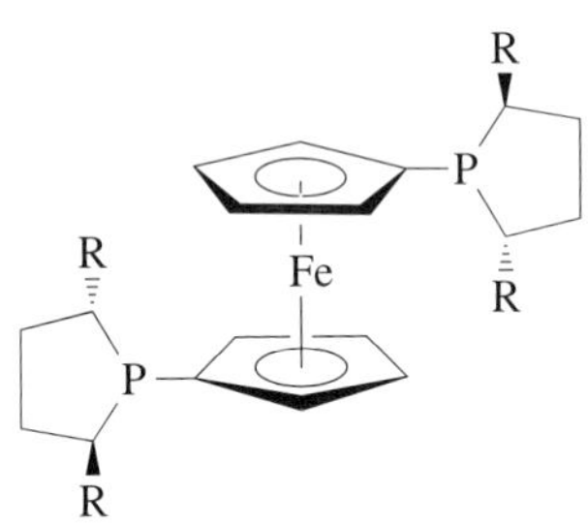

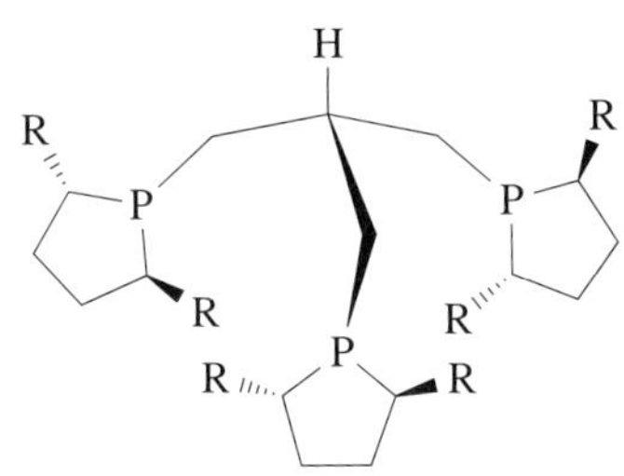

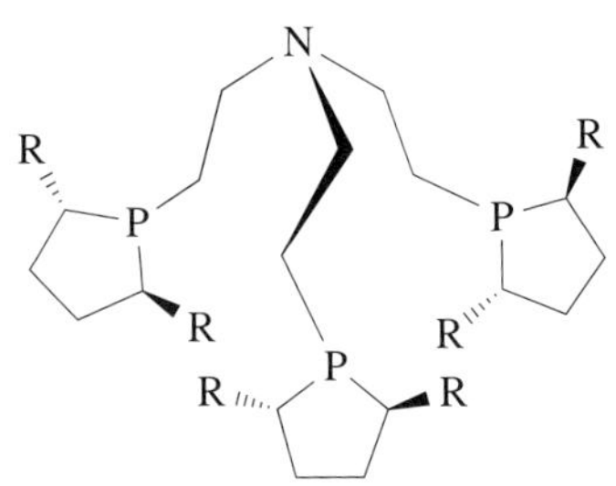

(**13***) R = Me (*S,S*) (**14***) R = Me all-(*S*) (**15***) R = Me all-(*S*)

None of these ligands proved to be more effective than the corresponding DuPHOS or BPE for the enantioselective hydrogenation of functional alkenes.

Other groups elaborated modified synthetic approaches. 1-Chlorophospholane enantiomers can be efficiently prepared when protected by complexation to borane. The P–Cl bond can then be allowed to react with the appropriate dimetallic derivatives to give borane complexes such as (**16***) and (**17***) ⟨97TA987⟩.

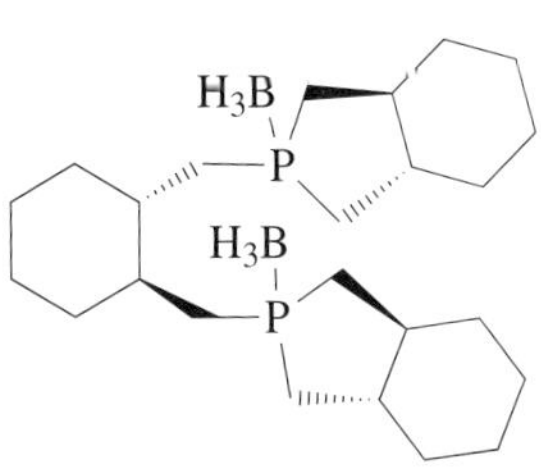

(**16***)

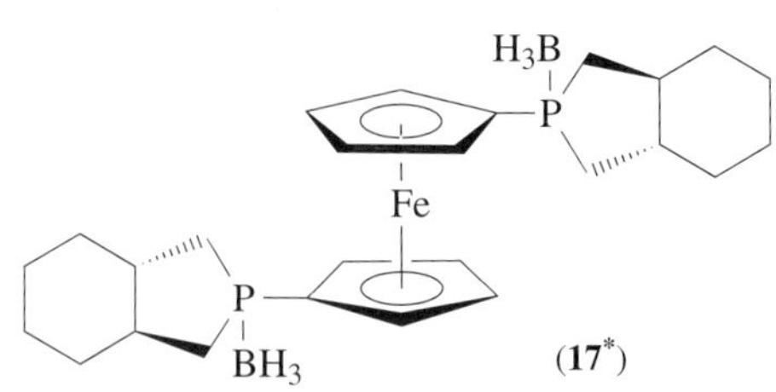

(**17***)

In the same vein, enantiopure 1-lithiophospholanes can be conveniently prepared in the coordination sphere of BH_3. Their reaction with aralkyl halides affords borane complexes such as (**18***)–(**20***) ⟨96SL1211⟩.

One difficult part of the synthetic schemes leading to DuPHOS (Equation (7)) or BPE ligands is the preparation of the starting homochiral 1,4-diols which serve as precursors of the corresponding cyclic sulphates. In order to circumvent this drawback, several groups have looked for alternative starting points. The use of the readily available D-mannitol has led to a series of bis-phospholane ligands of high efficiency (**21***)–(**24***) ⟨98JOC8031, 99TL6701, 99TL7059⟩.

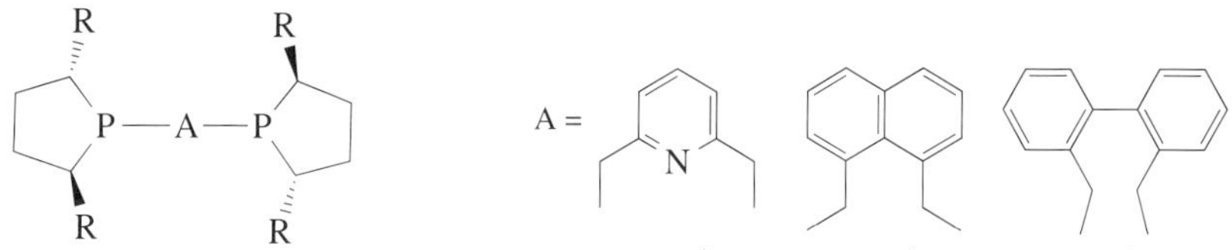

(**18***) (**19***) (**20***)

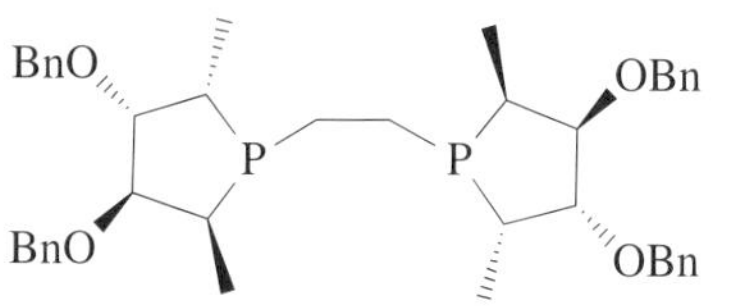

(**21***)

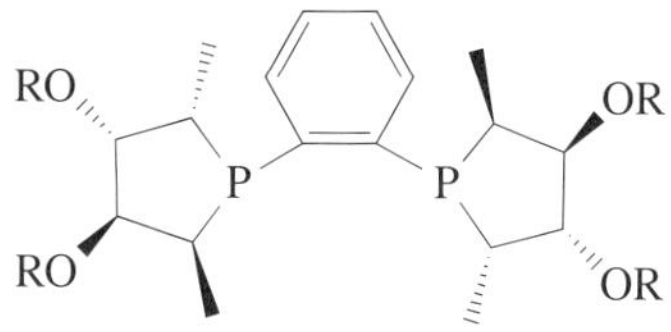

(**22***) R = Bn
(**23***) R = H

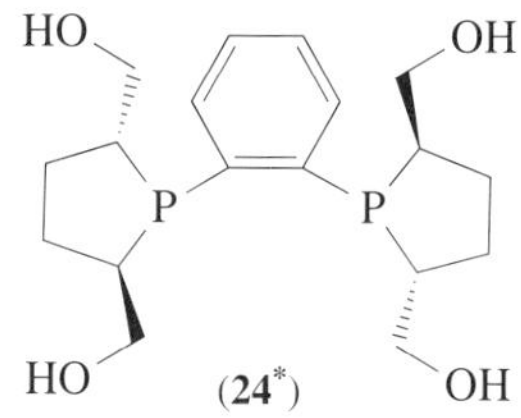

(**24***)

Both ligands (**23***) and (**24***) are water-soluble.

Another DuPHOS variant, (**25***), is characterized by a large bite angle, e.g. Pd–P–Pd 105°. It shows good performance in palladium-catalysed asymmetric allylic alkylations of sodium dimethyl malonate and various allyl acetates. With *O*-acetylcyclohex-2-enol, a 93% ee was observed when conducting the reaction at −20°C ⟨98AG(E)3116⟩.

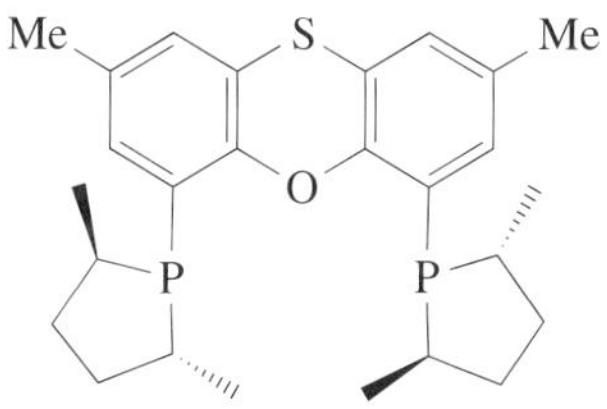

(**25***)

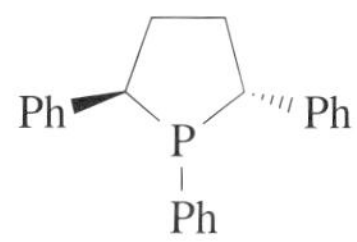

(**26***)

A number of studies have also been carried out with C_2-symmetrical homochiral 1,2,5-trisubstituted mono-phospholanes ⟨91TA569, 91TL5089, 97BSF371, 99TL2939⟩. Ligand (**26***) was prepared from the corresponding phospholanic acid which was resolved by crystallization of its salts with quinine ⟨99TL2939⟩. Ee's of up to 82% were observed in the rhodium-catalysed hydrogenation of *N*-acetyldehydrophenylalanine methyl ester, an eminently satisfactory performance for a monodentate phosphine. Several enantiopure monophospholanes have also been evaluated by Vedejs as catalysts for the enantioselective acylation of dissymmetrical secondary alcohols ⟨96JOC430⟩. Two 2-phosphabicyclo[3.3.0]octanes showing greater efficiency than the monophospholanes were later described by the same group ⟨99JA5813⟩.

Mixed ligands (**27***)–(**29***) incorporating a C_2-symmetric phospholane ring and a chiral DIPAMP subunit have been prepared and stored as their borane complexes ⟨99CC261⟩. With the correct match of arylphosphine and phospholane chirality, high ee's have been observed in the rhodium-catalysed asymmetric hydrogenation of dehydroaminoacids and esters. Mismatches lead to much lower ee's, but the difference in performance of matched and mismatched ligands varies widely as a function of the substrate. Apparently, the stereochemical outcome of hydrogenation is defined by the phospholane configuration.

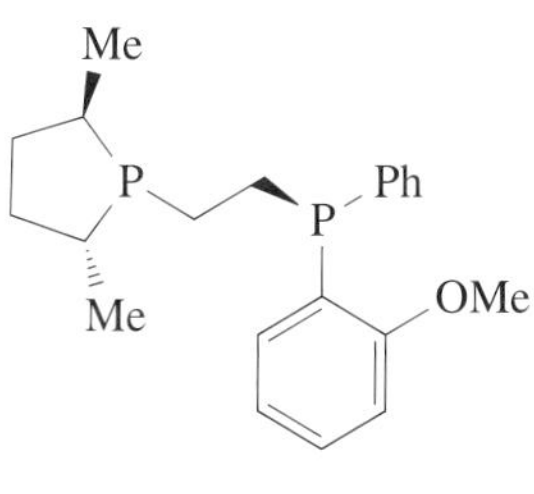

(**27***) (*R*,*R*,S_P, match)

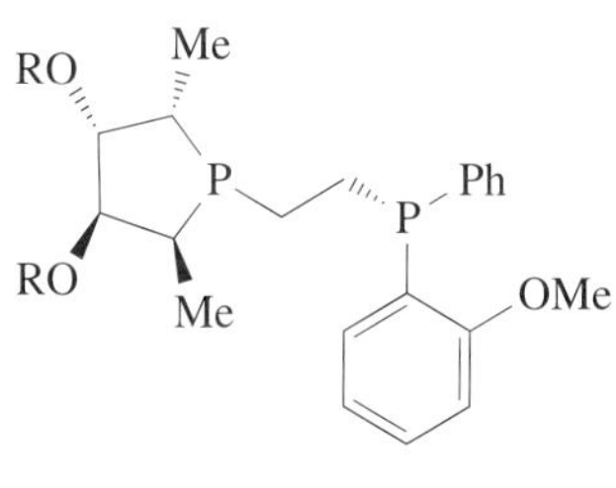

R = H (**28***) (all-S_C, R_P, match)
R = Me (**29***)

The latest developments concern the synthesis and testing of novel chiral binaphthyl-phospholanes incorporating the atropisomerism of the BINAP structure and the pseudo C_2-symmetry of DuPHOS

⟨99OL1679⟩. Very high enantioselectivities (95–99.6%) are observed in the Rh-catalysed hydrogenation of *Z*- and *E*-RCH = CHR[1]NHAc (R = H, Me, CH_2Ph, $CHMe_2$; R[1] = Ph, 2-naphthyl).

Another work describes the synthesis and testing of bis(phospholane) ligands such as (**30***–**33***) ⟨2000OM250⟩.

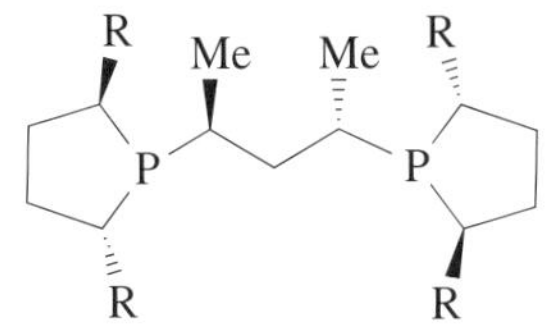

(**30***) R = Me
(**31***) R = Et
(**32***) R = ^{i}Pr

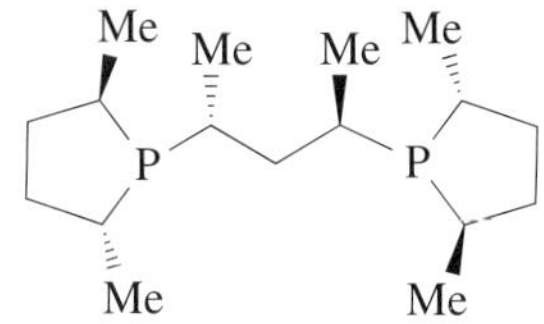

(**33***)

The comparison of (**30***) and (**33***) reveals substantial matching and mismatching effects in the Rh-catalysed asymmetric hydrogenation of enamides and pyruvates.

Finally, chiral phosphines have been obtained through the Me-DuPHOS-platinum-catalysed asymmetric hydrophosphination of activated olefins as shown in Equation (8a) ⟨2000OM950⟩.

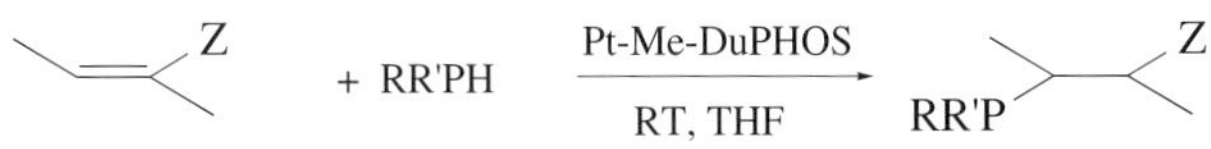

(8a)

Z = CN, CO_2R
R,R' = alkyl, aryl

The yields are high (60–100%) but the ee's are still rather low (<27%).

7.2.4 Phospholenes and Phospholes

The use of phospholenes (1-phenyl, 1-butyl and 1,2-diphenyl-2-phospholenes; 1-phenyl-3-methyl, 1-phenyl-3,4-dimethyl-3-phospholenes) in the Shell hydroformylation process was mentioned in an early patent ⟨66NEP6516164⟩. The hydroformylation of 1-dodecene with a Bu_3P–cobalt–carbonyl catalyst (200ºC, 68 atm., 0.2% cat., P/Co = 2) yields 78.2 and 20.4% of the corresponding primary alcohol and saturated hydrocarbon; under similar conditions with 1-phenyl-3-methyl-3-phospholene, the ratio is 91.7 : 7.9%. It is interesting to note that the fastest conversion was observed with 1,2,5-triphenyl-phosphole.

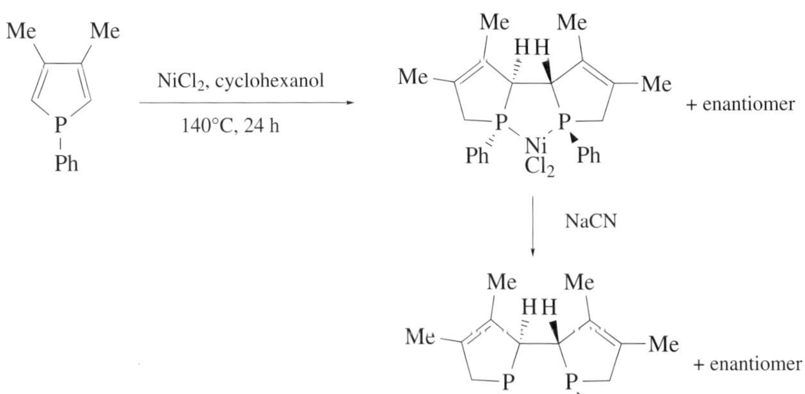

(9)

(**34***)

The efficiency of 3-phospholenes was also demonstrated in work on the hydroformylation of 1-octene with various cyclic phosphine–cobalt–carbonyl catalysts ⟨78MI(179)191⟩. No further mention of phospholenes in transition metal-catalysed reactions was made until recently when both the groups of Leung ⟨99OM4027⟩ and Mathey ⟨99TA4701⟩ independently described the resolution of a chiral 2,2′-biphospholene

(**34***). This product was obtained by reductive dimerization of 1-phenyl-3,4-dimethylphosphole with nickel chloride in alcohol (Equation (9)) ⟨84JA425, 85IC4141⟩.

The product was resolved classically by complexation to a Pd(II) centre having an optically pure *N,N*-dimethyl(α-methylbenzyl)amine-derived coligand. The rhodium-catalysed asymmetric hydrogenation of α-acetamidocinnamic acid proceeds in 75% ee in the presence of (**34***).

Early catalytic investigations on phospholes (**35**,**36**) described their use in the Shell hydroformylation process ⟨66NEP6516164, 78MI(179)191⟩. 1,2,5-Triphenylphosphole (TPP, **35**) gave one of the most active Co/P catalysts for the hydroformylation of terminal olefins.

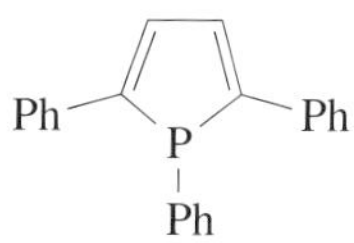

TPP (**35**)

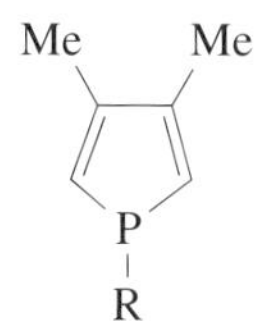

RDMP (**36**)

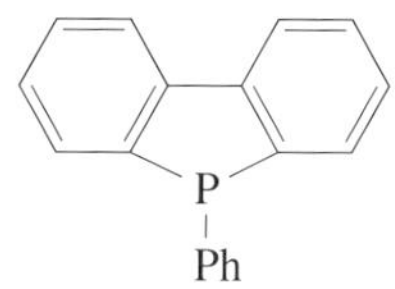

DBP (**37**)

The next series of studies on (**35–37**) dealt with catalytic hydrogenation using either Rh(I) or Ru(II) as the transition metal centre ⟨72CJC3714, 74CJC775, 74CJC3758, 83IC2476, 89IC3831⟩. The only noteworthy result was that [RhH(DBP)$_4$] is 7 times more active than the classical Wilkinson catalyst for the hydrogenation of 1-hexene ⟨74CJC775, 74CJC3758⟩. The use of DBP (**37**) in Co- and Rh-catalysed hydroformylation of olefins was also investigated. An early study stated that DBP gave a considerable rate enhancement over Ph$_3$P-based catalysts ⟨79JMOC(6)1⟩. However, it was shown later that Ph$_3$P–Rh catalysts were poisoned by DBP ⟨87JMOC(39)79⟩. Several studies were also performed on the mechanism of the Rh-catalysed hydroformylation of olefins using (**37**) ⟨82CC721, 87JCS(P2)1597, 87JMOC(41)271⟩. According to the ^{1}H, ^{13}C and ^{31}P NMR monitoring of the reaction mixture, [RhH(DBP)$_4$] equilibrates with [RhH(CO)(DBP)$_3$] and [RhH(CO)$_2$(DBP)$_2$] under CO atmosphere, but only [RhH(CO)$_2$(DBP)$_2$] is trapped by methylenecyclopropane ⟨82CC721⟩. On the basis of these results, it was proposed that the regiochemistry of the hydroformylation reaction (linear vs branched products) is mainly controlled by the relative rates of olefin coordination to [RhH(CO)(L)$_2$] and [RhH(CO)$_2$(L)] (L = phosphine) ⟨87JCS(P2)1597⟩. Other studies, performed by FT–IR spectroscopy under high pressures, were more representative of industrial hydroformylation conditions ⟨87JMOC(41)271⟩.

In a different vein, several DBP analogues of optically active chelating diphosphines Ph$_2$P-B*-PPh$_2$ (B* = chiral bridging unit) were synthesized and tested in classical enantioselective transformations ⟨75JOM(91)105, 75CL1115, 79BCJ2605, 82CC473, 83CC612, 84JMOC(26)17, 90TA417, 91OM1183, 91OM2046, 97JOM(540)15⟩. A striking result was found when studying the asymmetric hydroformylation of styrene in the presence of a PtCl$_2$–SnCl$_2$ complex of the DBP analogue of DIOP (**38***), where 80% ee was observed after reevaluation ⟨83CC612⟩. Comparable or higher ee's were reported some time later with (**39***) ⟨90TA417⟩, (**40***) ⟨91OM1183⟩ and (**41***) <97OM(540)15>.

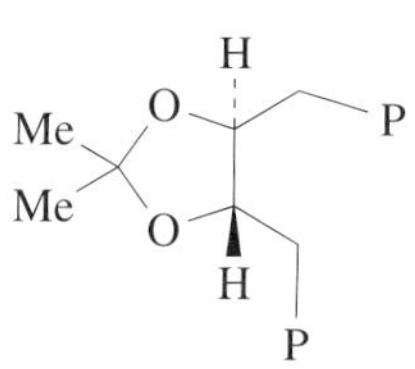

(**38***)

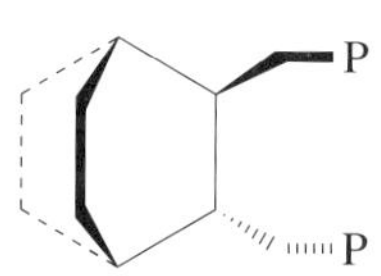

(**39***)

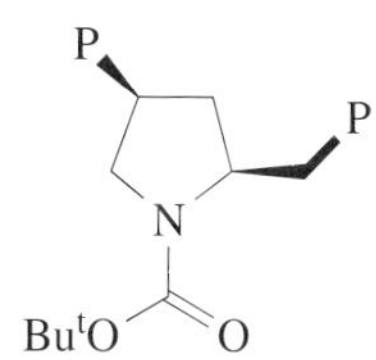

(**40***)

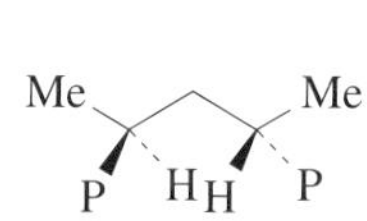

(**41***)

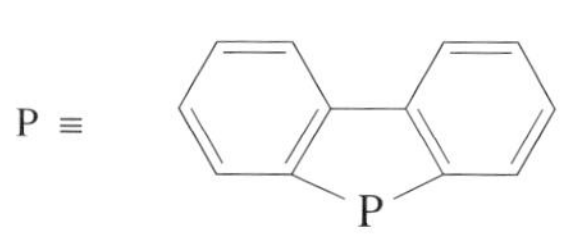

Other investigators performed a systematic evaluation of the catalytic potential of TPP complexes. In 1988, it was observed that the activity and regioselectivity of Rh–TPP catalysts are independent of the Rh/TPP ratio for the hydroformylation of olefins ⟨88CC1370⟩. The active species appears to be [RhH(CO)(TPP)$_2$] according to kinetic data ⟨95BSF815⟩. Comparative studies showed that TPP (**35**) is far superior to Ph$_3$P (10 times more active) and other phospholes in the Rh–P-catalysed hydroformylation of 1-hexene ⟨89JMOC(53)219, 89JMOC(57)153⟩. The Rh–TPP system was also used in the catalytic syntheses of 2-arylpropionaldehydes from the corresponding vinylarenes ⟨89JOC5208⟩ and ethyl 2-formylpropanoate from ethyl acrylate ⟨91NJC279⟩ together with the hydroformylation of dihydrofurans ⟨92OM3525⟩.

The synthesis of phosphole analogues of DIOP, including derivative (**42***), was also carried out for evaluation in asymmetric catalysis ⟨91CR(312)229, 93JOM(463)205, 93PS(85)207⟩. Unfortunately, the preliminary results in the Rh-catalysed hydrogenation of (*Z*)-α-acetamidocinnamic acid in the presence of (**42***) were rather disappointing (70% ee) ⟨92JMOC(72)L21⟩.

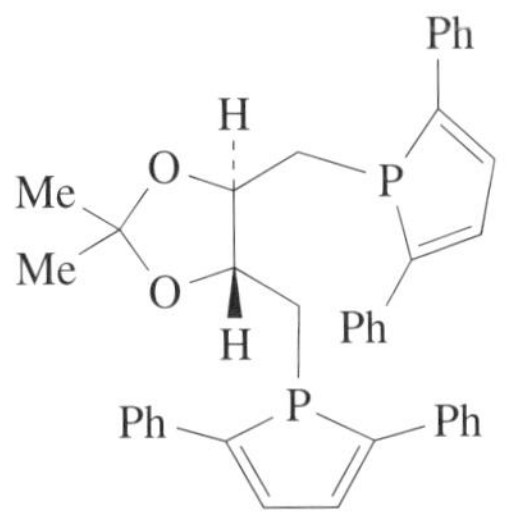

(**42***)

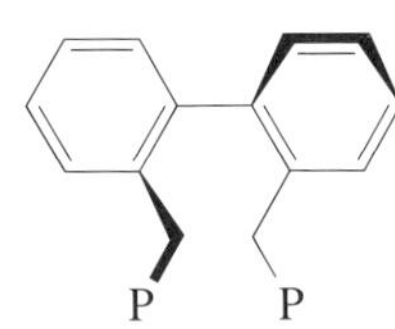

BISBI ligands (**43**)

Several other results on the use of phospholes as cocatalysts in the hydroformylation reaction are scattered throughout the literature. Thus, the asymmetric Rh/P hydroformylation of vinyl acetate to 2-(acyloxy)propanal has been investigated with (**38***) as the ligand ⟨81JOC4422⟩. A 51% ee has been observed for a P:Rh ratio of 6:1 at 80°C. Several phosphole derivatives of the BISBI family (**43**) have been independently developed by Eastman Kodak and Hoechst ⟨94USP5332846, 95EUP646588⟩. Gladiali *et al.* have synthesized a series of atropisomeric bis-naphthophospholes (**44***–**46***) and a bis-naphthophosphepine (**47***) ⟨95JOM(491)91, 94TA511⟩. The Pt/Sn + (**45***) and (**46***) systems are highly active catalysts for the hydroformylation of styrene which favour the branched aldehyde (up to 85%). Unfortunately, the ee's of 45% remain modest ⟨97CB543⟩. The Rh + (**47***) system gives poorer activity and enantioselectivity but is more regioselective.

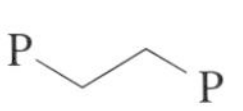

(**44***)

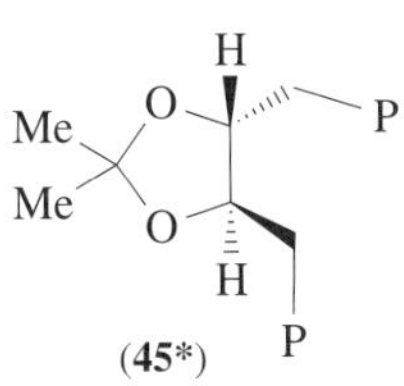

(**45***)

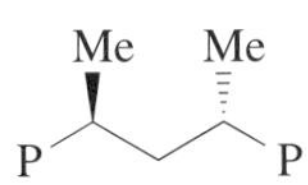

(**46***)

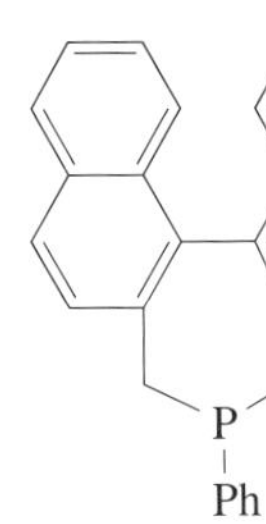

(**47***)

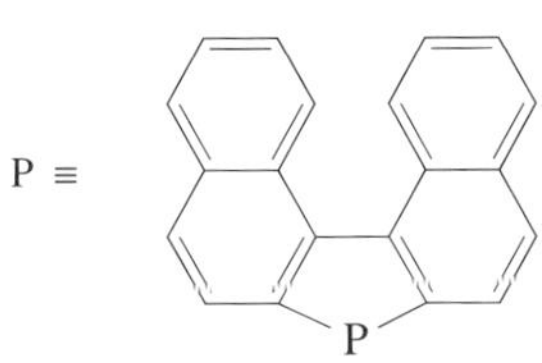

A more recent report examines the influence of bulky substituents at phosphorus on the activity of phosphole–Pt–Sn catalysts ⟨99JOM(586)79⟩, but the most significant recent papers describe the use of new diphosphines with wide bite angles (ca. 120°) such as (**48**)–(**51**) for the Rh-catalysed isomerization–hydroformylation of internal olefins to *linear* aldehydes ⟨99AG(E)336, 99OM4765⟩. Ligands (**48**) and (**51**) appear to be the best choice for this unprecedentedly regioselective conversion (ca. 90% of linear nonanal from 2- and 4-octenes). The high hydroformylation and isomerization activity of these ligands is thought to reflect high CO dissociation rates from [HRh(CO)$_2$(P–P)] to give the catalytically active complex [HRh(CO)(P–P)]. The wide bite angle in this species favours a bis-equatorial coordination of (P–P) which, in turn, induces a high linear to branched ratio in the product ⟨92JA5535⟩.

Even though the main focus of work concerning the catalytic applications of phospholes has been hydroformylation, some other reactions have been also investigated. For example, [PdCl$_2$(TPP)$_2$] has been found to be far superior to its Ph$_3$P analogue for the carbonylation of organic halides ⟨86POL911⟩. But the most striking report describes the use of (**52**)–(**54**) in the copolymerization of ethylene and carbon monoxide (Equation (10)) ⟨99OM3558⟩. High activity, high molecular weights (M_w up to 14 400) and narrow molecular weight distributions are observed. The catalyst derived from (**54**) has a markedly higher productivity than that derived from bis-(diphenylphosphino)propane.

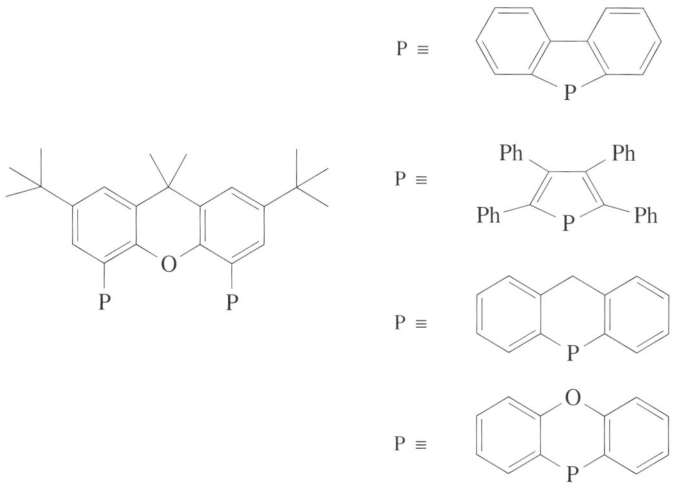

(48)

(49)

(50)

(51)

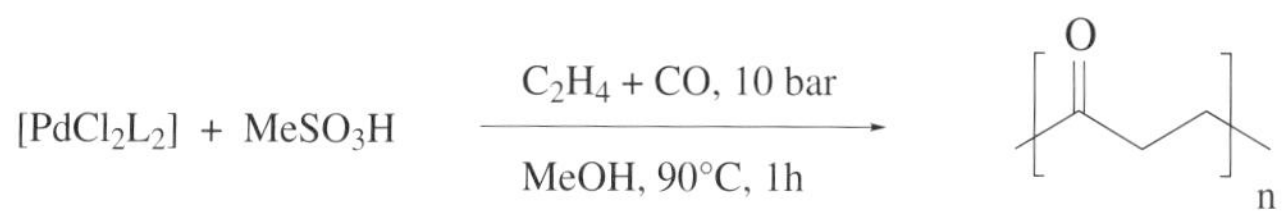

(10)

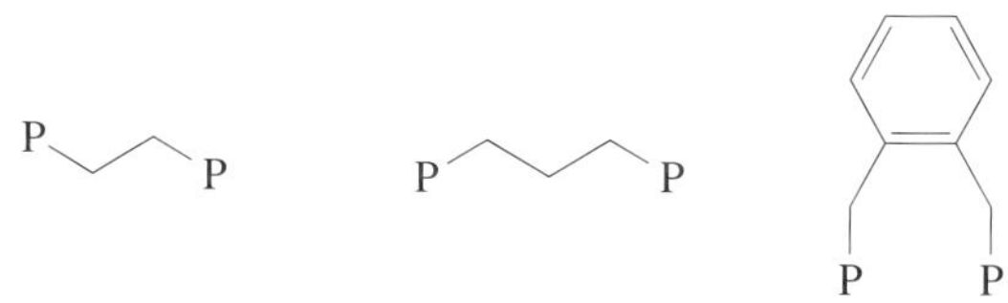

(52) (53) (54)

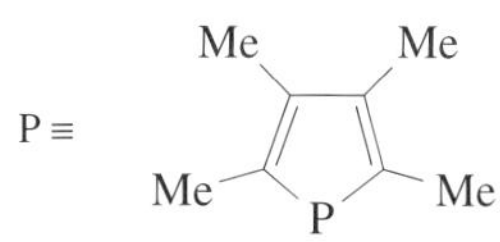

7.2.5 Polycyclic and Macrocyclic Phosphines

As might be expected, bi- and polycyclic phosphines allow better control of the steric hindrance and electronic properties of the phosphorus coordination centre than monocyclic phosphines. Thus, from the outset, bicyclic phosphines have been employed as ligands for homogeneous catalysis. Shell chemists have patented the synthesis of 9-phosphabicyclo[4.2.1] and [3.3.1]nonanes by radical-induced cyclization of primary phosphines with 1,5-cyclooctadiene ⟨67FRP1488936⟩ and demonstrated that the 9-phenyl derivatives are among the best phosphines for the phosphine/cobalt carbonyl-catalysed hydroformylation of olefins ⟨66NEP6604094, 69GEP(O)1909620⟩. Much later, a synthetically straightforward access to the unknown 1-phosphanorbornadienes through the [4 + 2] cycloaddition of transient 2*H*-phospholes with alkynes (Equation (11)) ⟨81JA4595⟩ allowed a systematic investigation of the catalytic potential of these unusual bridgehead phosphorus species.

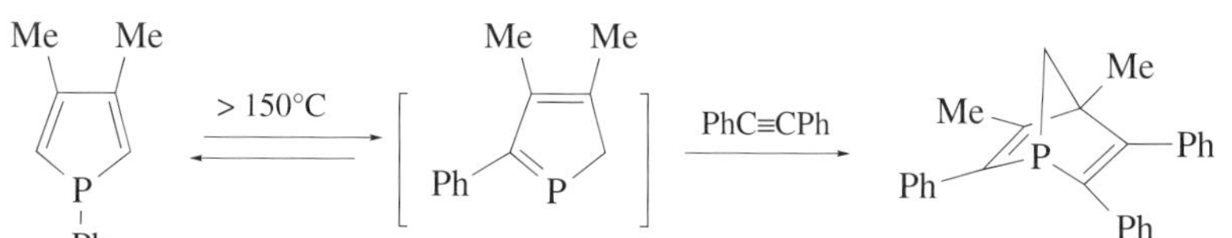

(55) (11)

It was first found that $[Rh(COD)(\mathbf{55})_2]^+PF_6^-$ is ten times more active than the Wilkinson catalyst and equivalent to the DIOP-based catalyst in hydrogenation rates for typical dehydroaminoacids ⟨87FRP2588197⟩. Using the $[Rh(CO)_2Cl]_2$ + L system as the precatalyst, Neibecker and Réau then showed that (**56**) is from five to ten times more active than PPh_3 in the hydroformylation of 1-hexene and styrene. The linearity of the product aldehydes is similar for Ph_3P and (**56**) in both cases ⟨89AG(E)500⟩. Both (**55**) and (**56**) were also found to be very efficient ligands for the Rh-catalysed hydroformylation of ethyl acrylate ⟨91NJC279⟩. A confirmation of these results came when Herrmann and his group showed that the sulphonated water-soluble version of (**55**) (termed norbos) is eight times more active than TPPTS in the Rh-catalysed biphasic hydroformylation of propene ⟨95JMOC(97)65⟩. It must be stressed that the P : Rh ratio was reduced from 80 : 1 (TPPTS) down to 13.5 : 1 (**57**). However, butanal linearity fell from 94 : 6 for TPPTS to 81 : 19 for (**57**).

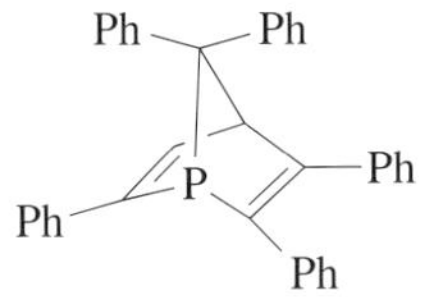

(**56**)

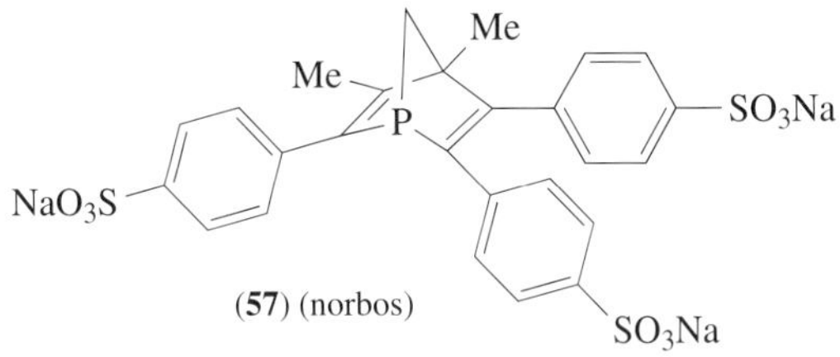

(**57**) (norbos)

Two other water-soluble phosphanorbornadienes (**58**) and (**59**) were also synthesized, using the phosphonate group as the hydrophilic functionality ⟨96JOC3531⟩.

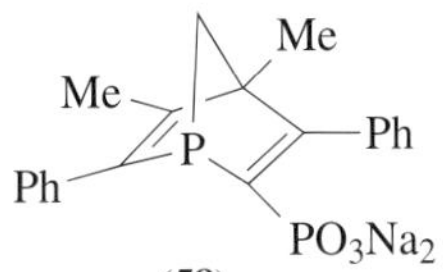

(**58**)

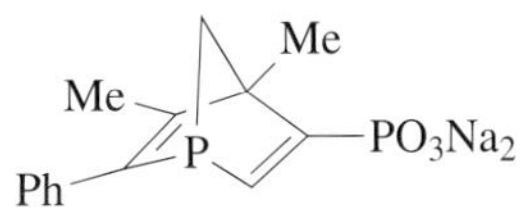

(**59**)

The α-substituted derivative (**58**) gives rather poor results in the catalytic hydrogenation of alkenes with $[RhL_2]^+$ as the catalyst. Apparently, the rhodium centre is deactivated through a chelation to the α-P=O group, which is implied by a sizable $^2J(P–O–Rh)$ coupling. On the contrary, the rhodium complex of the β-substituted derivative (**59**) displays normal catalytic activity.

The bicyclic structure of 1-phosphanorbornadienes is quite appealing for applications in asymmetric catalysis because enantiopure phosphines having chirality localized at phosphorus can racemize through a variety of pathways. Besides classical pyramidal inversion, phosphorus can lose its optical activity by association with a protic reagent HA. The resulting phosphorane $[R_3P(H)A]$ can undergo a low-energy Berry pseudorotation. A third pathway involves turnstile rotation which, according to calculations, is the preferred inversion mechanism for some phosphiranes ⟨97CC1033⟩. It must be stressed that this last racemization process can take place while phosphorus is still in the coordination sphere of the transition metal. Thus, there is a clear need for optically active phosphorus species which are stable toward racemization. A bicyclic structure having a bridgehead P, such as 1-phosphanorbornadiene, is ideally suited to meet this need. The initial experiments upon this general class were performed with the monophosphine (**60***). The enantiomers of (**60***) were separated as their *P*-oxides on a chiral stationary phase and reduced with $HSiCl_3$/pyridine. Preliminary results in the enantioselective hydrogenation of typical dehydroaminoacids were disappointing with a maximum recorded ee of only 50% ⟨89NJC369⟩.

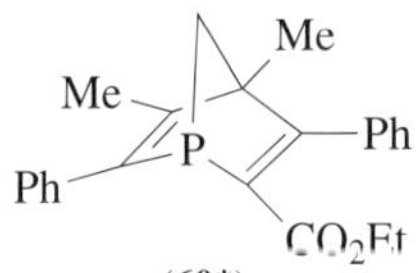

(**60***)

More convincing results were observed with the chelating ligand BIPNOR (**61***). The *meso* + *racemic* mixture was obtained by reaction of a 1,1′-biphosphole with diphenylacetylene (Equation (12)) ⟨92BSF1⟩. The resolution of (**61***) was carried out with an optically active palladium complex ⟨97CEJ1365⟩ or by separation of the *P,P*-dioxides on a chiral stationary phase ⟨99PCT47475⟩. BIPNOR appears to be competitive with BINAP and DuPHOS in the Rh-catalysed enantioselective hydrogenation of functional alkenes ⟨97CEJ1365⟩. The geometry of $[Rh(BIPNOR)(COD)]^+$ shows how BIPNOR occupies two diagonal quadrants of the space around Rh ⟨98JOM(577)117⟩. The two phenyl α-substituents play a crucial role in chirality transfer from (**61***) and their absence destroys the ability of the catalyst to discriminate

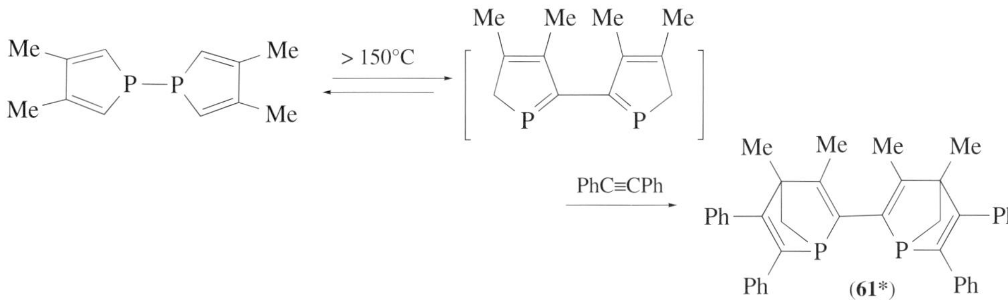

(12)

BIPNOR (*meso* + *racemic*)

between the two faces of the functional alkene. BIPNOR appears to be less well adapted to the octahedral environment of Ru(II) where the enantioselective hydrogenation of disymmetrical ketones by a BIPNOR-$RuBr_2$ system gives relatively modest ee's (60% for acetophenone ⟨97CEJ1365⟩). A comparative study of BIPNOR and BINAP in the Rh-catalysed asymmetric isomerization of a cyclic diene showed that, for this particular case, BIPNOR appears to be far superior to BINAP (91 vs 31% ee). The proposed mechanism involves a η^5-cycloheptadienyl intermediate ⟨2000T101⟩.

Another interesting development in that area concerns the synthesis of a bicyclic analogue of DuPHOS. The so-called PennPhos (**62***) has been prepared by the group of Zhang as shown in Equation (13) ⟨98AG(E)1100⟩.

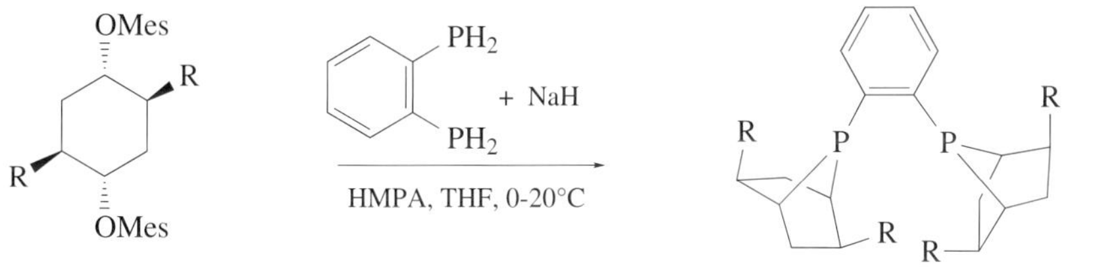

(13)

(**62***) a R = Me
b R = ^{i}Pr

As its Rh(I) complex, PennPhos is an extremely efficient catalyst for the asymmetric hydrogenation of simple ketones (ee up to 97% for acetophenone with $[Rh(COD)Cl]_2$ + (**62*****a**) + 2,6-lutidine in MeOH at 20ºC ⟨98AG(E)1100⟩), cyclic enol acetates ⟨99AG(E)516⟩ and cyclic enamides ⟨99JOC1774⟩.

An extended series of asymmetric Diels–Alder reactions between 3,4-dimethyl-1-phenylphosphole and various functional dienophiles in the presence of an organopalladium metallacycle derived from enantiopure *N,N*-dimethyl-1-(1-naphthyl)ethylamine has been studied by Leung and coworkers. Recent reports include ⟨2000CC167⟩ and an extensive reference list is provided in ⟨2000T7⟩. In each case, the result is an optically pure 7-phosphanorbornene. An example is shown in Equation (14) ⟨2000T7⟩.

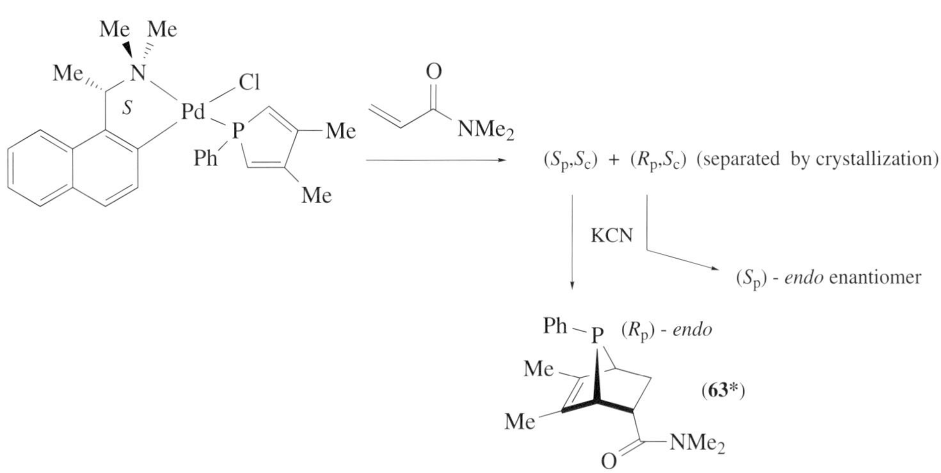

(14)

An enantiopure phosphole [4 + 2] *endo* dimer has also been prepared in similar fashion ⟨2000CC167⟩. Unfortunately, none of these enantiopure phosphines has been tested in catalysis.

The first practical catalytic application of a polyphosphorus macrocycle appears to involve (**64**) ⟨97AG(E)2364⟩. It is well known that phospholes have a low pyramidal inversion barrier (ca. 16 kcal mol^{-1} ⟨71JA6205⟩) because of an increasing electronic cyclic delocalization upon planarization. As a result, the phosphorus of these rings inverts around room temperature. Macrocycle (**64**) exists as a mixture of diastereomers but readily equilibrates above 50ºC, which allows it to select its best geometry to coordinate the metal. Thus, with $PdCl_2$, it gives a single complex (**65**) wherein the two non-coordinated P-lone pairs are directed toward the central metal. In catalytic transformations such as Stille cross-coupling or the Heck reaction, this complex shows an impressive increase in catalyst half-life by comparison with classical systems derived from Ph_3P or Fu_3P. The practical consequence is that much lower quantities of palladium are needed to effect complete reaction. The most likely explanation is that the dismutation of the actual catalyst [PdL_2] into Pd and [PdL_4] is prevented by the weak stabilizing interaction between the 14 electron metallic centre and the two non-coordinated lone pairs.

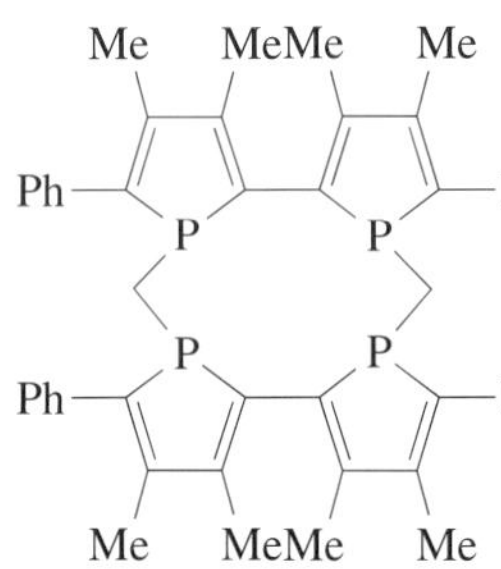

(**64**)

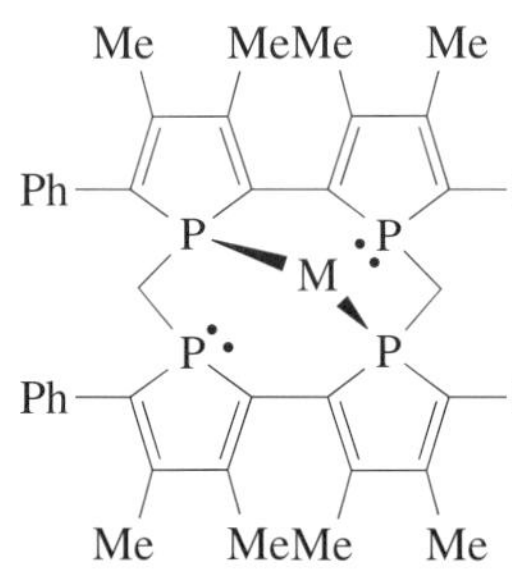

M = *cis*-$PdCl_2$ (**65**)

7.3 DICOORDINATE PHOSPHORUS HETEROCYCLES

From an electronic standpoint, molecules containing a P=C double bond and a two-coordinate phosphorus are characterized by two high-lying occupied MO's, one corresponding to the P=C double bond (π) and the other to the sp^2 phosphorus lone pair (n_P). In general, the π orbital is the HOMO and the n_P orbital is slightly lower in energy. In such cases, both π and σ coordination can be observed and an equilibrium can occur between the two modes ⟨84OM1804⟩. The π^* LUMO is generally quite low in energy and highly localized at phosphorus. Thus, when bound through phosphorus, this kind of ligand behaves as a strong π acceptor and weak σ donor, closer in behaviour to CO than to an ordinary phosphine. The coordination chemistry of P=C double bonded molecules is rich and varied and has attracted much attention ⟨98MI2⟩. However, their use in catalysis is hampered by high reactivity, so some stabilization is a prerequisite. This can be achieved by incorporating bulky substituents (kinetic stabilization) or by incorporating the P=C double bond into a delocalized system (electronic stabilization). The second technique retains the possibility of varying the steric hindrance around phosphorus and is the method of choice. It is unsurprising that the first significant catalytic results with sp^2-phosphorus ligands have been obtained with 6π-delocalized systems such as phosphinines and phospha-metallocenes.

7.3.1 Phosphinines

The characteristics of phosphinines most relevant to their use in coordination chemistry and catalysis are as follows. Phosphinine displays 88% of the aromatic stabilization energy of benzene ⟨88JA4204⟩ and the parent molecule is stable ⟨71JA3293⟩. The phosphorus lone pair is the third highest occupied level; the LUMO is very low in energy and has a significant localization at phosphorus ⟨89OM2804⟩. The phosphorus atom bears a partial positive charge and displays very low basicity ⟨85OM457⟩. Assuming free rotation around the P–metal bond, Elschenbroich has evaluated the Tolman cone angle of the parent phosphinine as 108º, close to that of PF_3 and in the lower part of the range ⟨94JA6217⟩. The π acceptor characteristics of phosphinines suggested their possible use as ligands in the Rh-catalysed hydroformylation of alkenes. This possibility was explored with some success by Breit ⟨96CC2071, 97JCS(P1)2681⟩. In a typical experiment, the hydroformylation of styrene was carried out at 25ºC under 20 bar of CO + H_2 (1 : 1) with Rh : L : styrene ratios of 1 : 5 : 280. After 180 min, the results were as follows:

PPh_3: conversion 8%; TOF 7.5 h^{-1}; branched/linear aldehyde 24 : 1
(**66**): conversion 30.8%; TOF 28.7 h^{-1}; branched/linear aldehyde 20 : 1.

Under these conditions, (**66**) also proved to be superior to a bulky phosphite such as $P(OAr)_3$, Ar = 2,4-(^tBu)$_2C_6H_3$. With cyclohexene, the results were similar for (**66**) and $P(OAr)_3$. The $[Rh(\mathbf{66})_4]^+$ complex was structurally characterized and shown to give identical results to the in-situ generated catalyst. This observation indicates that at least two molecules of (**66**) can dissociate from rhodium in this complex <97JCSP(1)2681>. In an earlier report, some chelating *P,P*- and *P,N*-bidentate phosphinines were also evaluated as catalyst precursors for the same transformation. They were not significantly better than (**66**) ⟨96CC2071⟩.

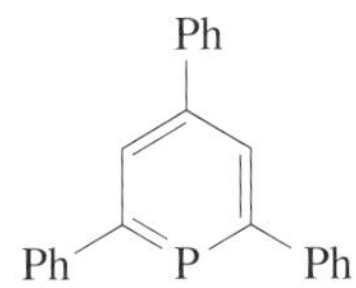

(**66**)

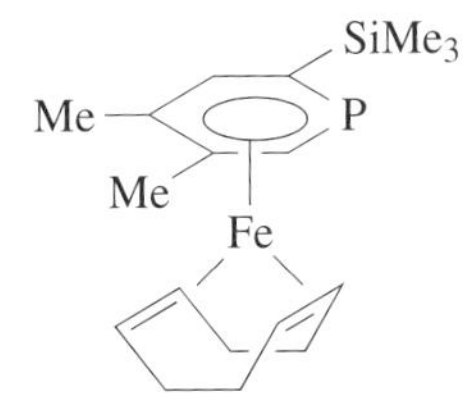

(**67**)

In a different vein, a η^6-phosphinine–iron (**67**) was shown to be a catalyst for the formation of pyridines from alkynes and nitriles at room temperature. The reaction is relatively slow and the best observed TON was 160 ⟨96OM2713⟩. A similar complex with a diazadiene as the coligand was also tested in the [4 + 4] dimerization of 1,3-butadiene. Apparently, the strong phosphinine–iron interaction hampers the formation of the active iron–diazadiene catalyst and the activity of this system is rather poor ⟨98EJIC119⟩.

7.3.2 Phosphametallocenes

The MO analysis of phosphaferrocenes serves to illustrate the main electronic characteristics of phosphametallocenes. The three highest occupied orbitals of phosphaferrocene are essentially metal-centred and are derived from the destabilized d_{z^2} and almost unperturbed d_{xy} and $d_{x^2-y^2}$. The lone pair orbital lying in the plane of the ring appears at lower energy. The LUMO is low in energy and has substantial localization at iron and phosphorus ⟨83OM1008⟩. Thus, phosphaferrocene is a weak σ-donor and a strong π-acceptor and resembles phosphinine as a *P*-ligand. Investigations on the catalytic properties of phosphametallocenes are divided broadly into two categories. In the first, phosphametallocenes are considered to be modified metallocenes. The η^5-phospholyl is a better π-acceptor than the η^5-cyclopentadienyl ligand. So, *a priori*, it may be useful to replace a Cp by a phospholyl whenever the metallic centre of a metallocene acts as an electrophile in the catalytic cycle. This is the case for the polymerization of olefins using metallocene catalysts. Several papers ⟨95JOM(501)219, 96CB1517, 98JMOC(128)155, 98OM5445⟩ and numerous patents (a list is available in Ref. ⟨98JMOC(128)155⟩) deal with this topic. In these cases, the P-lone pair is not involved in the process, but may undergo complexation to an acidic cocatalyst which is usually methylalumoxane (MAO). Since MAO is a bulky polymeric material, its coordination on the P-lone pair hinders the metallic centre and lowers catalytic activity. Consequently, the most efficient phosphametallocene polymerization catalysts to date, such as (**68**) ⟨98JMOC(128)155⟩ and (**69**) ⟨98OM5445⟩, have P-lone pairs which are sterically hindered by two substituents on the 2,5-positions of the phospholyl ring.

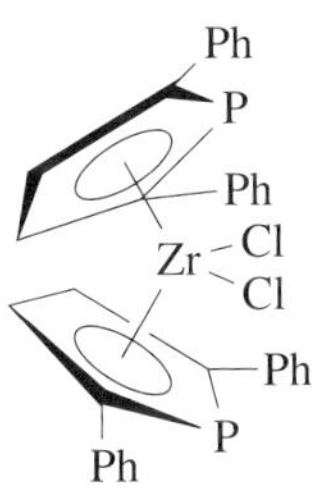

(**68**)

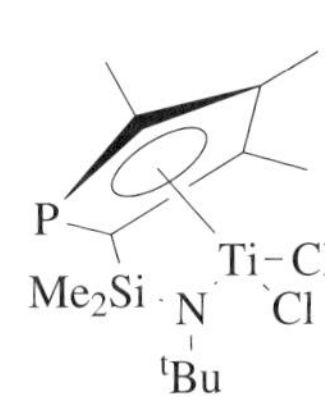

(**69**)

The synthesis of (**69**) is depicted in Equation (15). The diphosphazirconocene dichloride (**68**) + MAO system polymerizes propylene. The experimental conditions require 10 μmol of Zr and 5 mmol of MAO at 45°C under 600 kPa of propylene for 10 to 60 min. An activity of 170 kg of polypropylene per gram of Zr per hour and a M_n of 20 000 have been observed. Compound (**69**) and MAO polymerize ethylene at 160°C. This system displays the same activity as the all-carbon analogue but the final M_w of polyethylene is lower with the phosphametallocene catalyst: 50 vs 126 × 10^3.

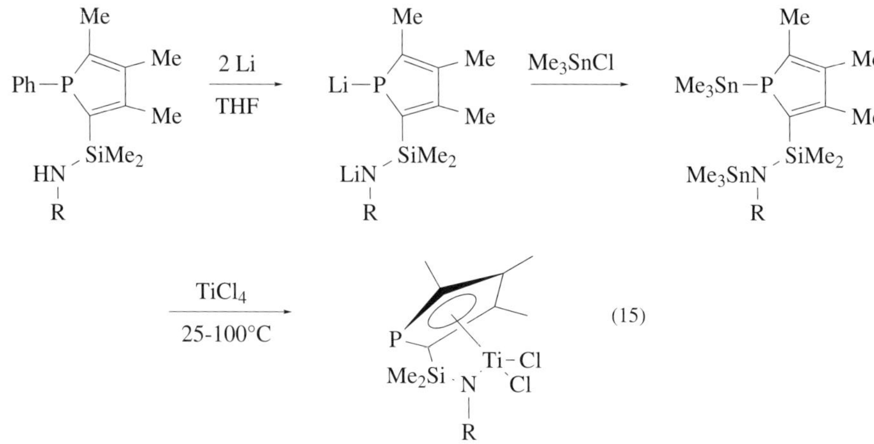

(15)

The second line of research takes advantage of the P-lone pair of phosphametallocenes. In a preliminary report, Fu showed that phosphaferrocene (**70**) is a good catalyst for the opening of epoxides by Me_3SiCl ⟨97JOC4534⟩ (Equation (16)).

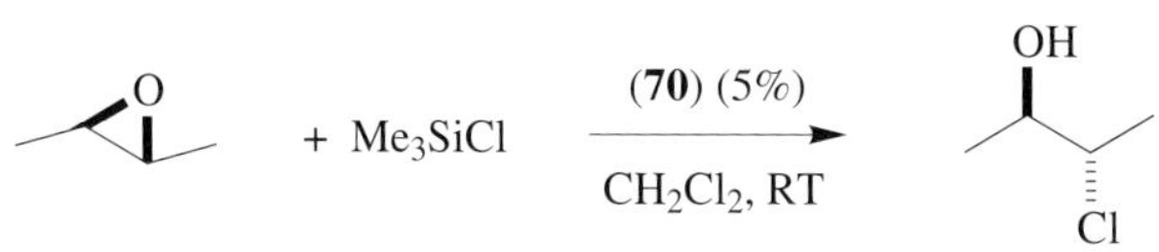

(16)

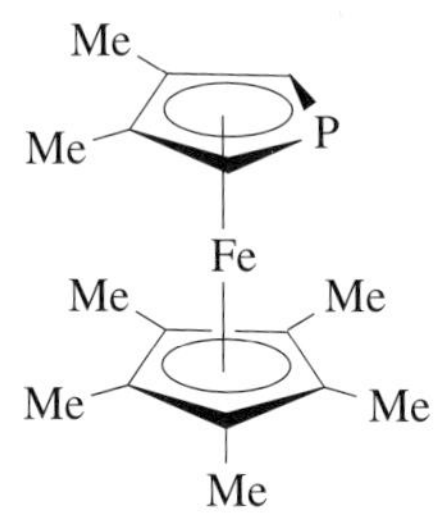

(**70**)

Inversion occurs at the carbon atom undergoing substitution and the proposed mechanism involves an adduct between (**70**) and Me_3SiCl. Compound (**70**) is a less active but significantly more regioselective catalyst with disymmetrical epoxides than either PPh_3 or Bu_4NCl.

At the same time, it was briefly stated that the cobalt derivative (**71**) catalyses the cyclotrimerization of EtC≡CEt to C_6Et_6 ⟨97OM2049⟩. Simultaneously, Ganter's group started to develop the chemistry of planar chiral phosphaferrocenes ⟨97OM2862⟩. The key step of this programme was the separation of the two carbaldehyde enantiomers (**72***) through chromatographic resolution of the corresponding aminal (**73**) ⟨97TA2607⟩.

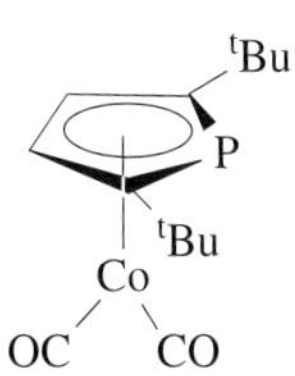

(**71**)

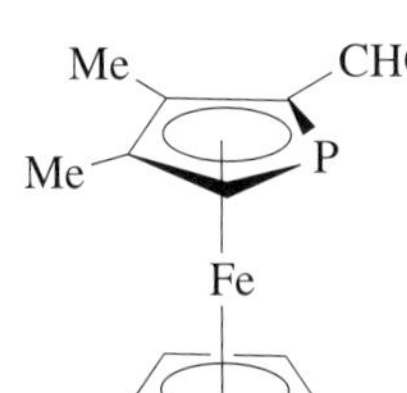

(**72***)

(*R*)

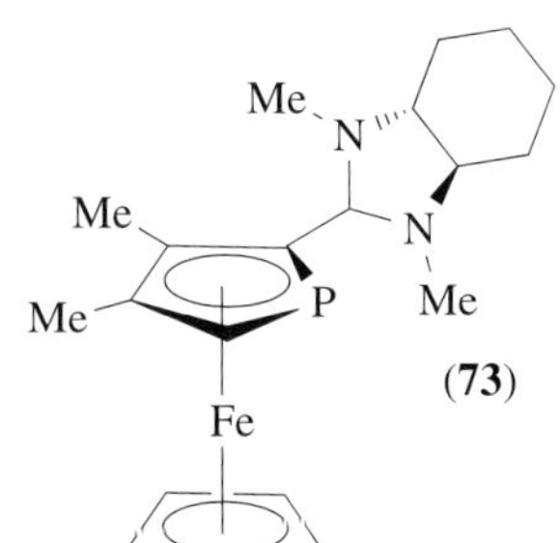

(**73**)

In terms of asymmetric catalysis, the most convincing result of this programme was revealed recently. The 1,1′-bis(phosphaferrocenyl)ferrocene (**74***) synthesized according to Equation (17) gives a Pd(II) chelate which performs the allylic coupling between 1,3-diphenylallyl acetate and sodium dimethyl malonate (see Equation (4)) in 79% ee ⟨99OM5444⟩.

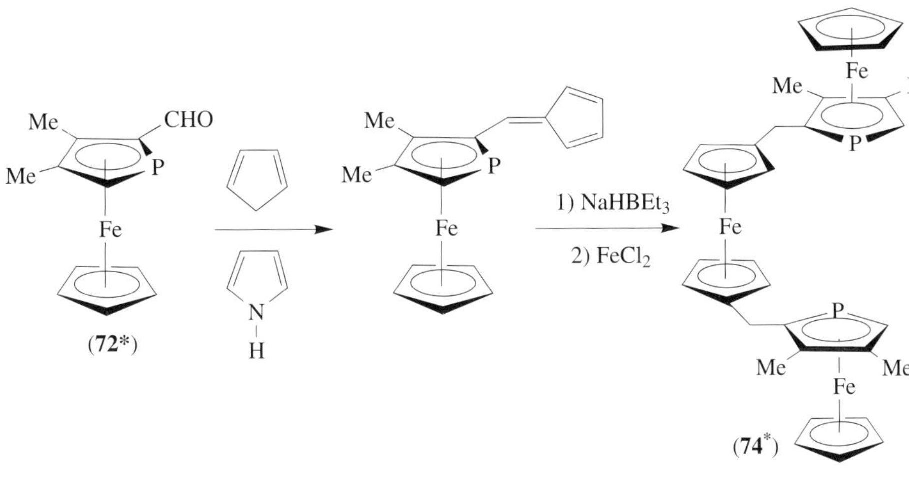
(17)

Finally, Fu has shown that the 2-(phosphinomethyl)phosphaferrocene (**75***) gives good ee's in the Rh(I)-catalysed asymmetric hydrogenation of dehydroaminoacid methyl esters ⟨98JOC4168⟩. The resolution was carried out by chiral HPLC on the intermediate alcohol (**76**). Quantitative conversions were observed at RT in ethanol and the ee's were found in the range between 79 and 96%.

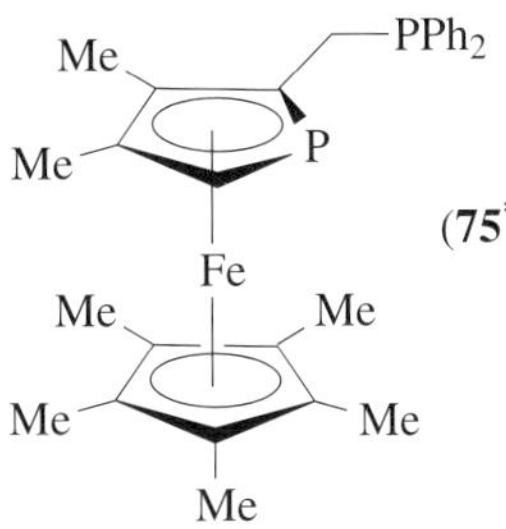
(**75***)

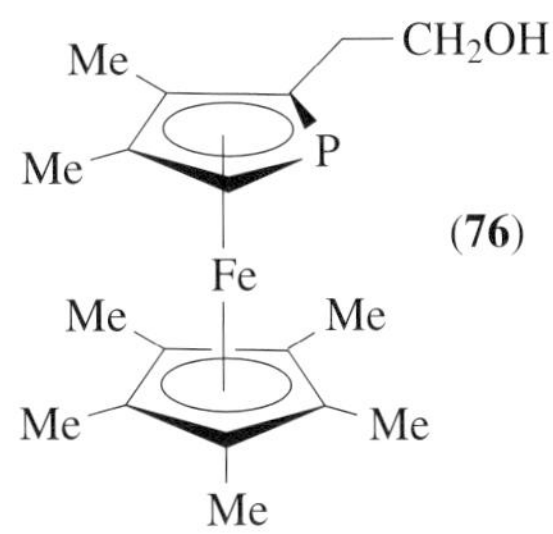
(**76**)

This result clearly demonstrates that a large series of chiral catalysts can be found within the phosphametallocene family. More generally, phosphorus–carbon heterocyclic chemistry now appears to be a gold mine for new sp^2 and sp^3 ligands able to confer unusual properties (including catalytic) upon metal centres. A more widespread use of these original structures is likely in the near future.

7.4 REFERENCES

66NEP6516164 Shell Internationale Research Maatschappij N.V.; *Neth. Pat.* 6 516 164 (1966) (*Chem. Abstr.*, 1967, **66**, 28392r).

66NEP6604094 Shell Internationale Research Maatschappij N.V.; *Neth. Pat.* 6 604 094 (1966) (*Chem. Abstr.*, 1967, **66**, 65101r).

67FRP1488936 Shell Internationale Research Maatschappij N.V.; *Fr. Pat.* 1 488 936 (1967) (*Chem. Abstr.*, 1968, **69**, 36252x).

68JOM(13)469 L.H. Slaugh and R.H. Mullineaux; *J. Organomet. Chem.*, 1968, **13**, 469.

69GEP(O)1909620 J.L. Van Winkle, R.C. Morris and R.F. Mason; *Ger. Pat. Offen.* 1 909 620 (1969) (*Chem. Abstr.*, 1970, **72**, 3033k).

71JA3293 A.J. Ashe III; *J. Am. Chem. Soc.*, 1971, **93**, 3293.

71JA6205 W. Egan, R. Tang, G. Zon and K. Mislow; *J. Am. Chem. Soc.*, 1971, **93**, 6205.

72CJC3714 D.G. Holah, A.N. Hughes and B.C. Hui; *Can. J. Chem.*, 1972, **50**, 3714.

74CJC775 D.E. Budd, D.G. Holah, A.N. Hughes and B.C. Hui; *Can. J. Chem.*, 1974, **52**, 775.

74CJC3758 D.G. Holah, I.M. Hoodless, A.N. Hughes, B.C. Hui and D. Martin; *Can. J. Chem.*, 1972, **50**, 3758.

75CL1115 M. Tanaka, Y. Ikeda and I. Ogata; *Chem. Lett.*, 1975, 1115.

75JOM(91)105 T.P. Dang, J.C. Poulin and H.B. Kagan; *J. Organomet. Chem.*, 1975, **91**, 105.

77CRV313 C.A. Tolman; *Chem. Rev.*, 1977, **77**, 313.

78MI(179)191 F. Mathey, G. Muller, C. Demay and H. Lemke; *Inf. Chim.*, 1978, **179**, 191.

79BCJ2605 T. Hayashi, M. Tanaka, Y. Ikeda and I. Ogata; *Bull. Chem. Soc. Jpn.*, 1979, **52**, 2605.

79JMOC(6)1 T. Hayashi, M. Tanaka and I. Ogata; *J. Mol. Catal.*, 1979, **6**, 1.

80JA5952 A.S.C. Chan, J.J. Pluth and J. Halpern; *J. Am. Chem. Soc.*, 1980, **102**, 5952.

81JA4595 F. Mathey, F. Mercier, C. Charrier, J. Fischer and A. Mitschler; *J. Am. Chem. Soc.*, 1981, **103**, 4595.
81JOC4422 C.F. Hobbs and W.S. Knowles; *J. Org. Chem.*, 1981, **46**, 4422.
82CC473 C.U. Pittman Jr., Y. Kawabata and L.I. Flowers; *J. Chem. Soc., Chem. Commun.*, 1982, 473.
82CC721 J.M. Brown, L.R. Canning, A.G. Kent and P.J. Sidebottom; *J. Chem. Soc., Chem. Commun.*, 1982, 721.
83ACR106 W.S. Knowles; *Acc. Chem. Res.*, 1983, **16**, 106.
83CC612 G. Consiglio, P. Pino, L.I. Flowers and C.U. Pittman Jr.; *J. Chem. Soc., Chem. Commun.*, 1983, 612.
83IC2476 L.M. Wilkes, J.H. Nelson, L.B. McCusker, K. Seff and F. Mathey; *Inorg. Chem.*, 1983, **22**, 2476.
83OM1008 N.M. Kostic and R.F. Fenske; *Organometallics*, 1983, **2**, 1008.
84JA425 F. Mercier, F. Mathey, J. Fischer and J.H. Nelson; *J. Am. Chem. Soc.*, 1984, **106**, 425.
84JMOC(26)17 T. Hayashi, M. Tanaka and I. Ogata; *J. Mol. Catal.*, 1984, **26**, 17.
84OM1804 Th.A. van der Knaap, F. Bickelhaupt, J.G. Kraaykamp, G. van Koten, J.P.C. Bernards, H.T. Edzes, W.S. Veeman, E. de Boer and E.J. Baerends; *Organometallics*, 1984, **3**, 1804.
85IC4141 F. Mercier, F. Mathey, J. Fischer and J.H. Nelson; *Inorg. Chem.*, 1985, **24**, 4141.
85OM457 R.V. Hodges, J.L. Beauchamp, A.J. Ashe III and W.-T. Chan; *Organometallics*, 1985, **4**, 457.
86POL911 B.M. Choudary, N.P. Reddy and M.Z. Jamil; *Polyhedron*, 1986, **5**, 911.
87FRP2588197 F. Mathey, D. Neibecker and A. Brèque; *Fr. Pat.* 2 588 197 (1987) (*Chem. Abstr.*, 1987, **107**, 219468v).
87JA1746 C.R. Landis and J. Halpern; *J. Am. Chem. Soc.*, 1987, **109**, 1746.
87JCS(P2)1597 J.M. Brown and A.G. Kent; *J. Chem. Soc., Perkin Trans. II*, 1987, 1597.
87JMOC(39)79 J. Hjortkjaer, H. Dueholm and P.C. de Mello; *J. Mol. Catal.*, 1987, **39**, 79.
87JMOC(41)271 W.R. Moser, C.J. Papile, D.A. Brannon, R.A. Duwell and S.J. Weininger; *J. Mol. Catal.*, 1987, **41**, 271.
87JOM(328)71 H. Brunner and R. Sievi; *J. Organomet. Chem.*, 1987, **328**, 71.
88CC1370 C. Bergounhou, D. Neibecker and R. Réau; *J. Chem. Soc., Chem. Commun.*, 1988, 1370.
88JA4204 K.K. Baldridge and M.S. Gordon; *J. Am. Chem. Soc.*, 1988, **110**, 4204.
89AG(E)500 D. Neibecker and R. Réau; *Angew. Chem. Int. Ed. Engl.*, 1989, **28**, 500.
89IC3831 R. Vac, J.H. Nelson, E.B. Milosavljević and L. Solujić; *Inorg. Chem.*, 1989, **28**, 3831.
89JMOC(53)219 D. Neibecker and R. Réau; *J. Mol. Catal.*, 1989, **53**, 219.
89JMOC(57)153 D. Neibecker and R. Réau; *J. Mol. Catal.*, 1989, **57**, 153.
89JOC5208 D. Neibecker, R. Réau and S. Lecolier; *J. Org. Chem.*, 1989, **54**, 5208.
89NJC369 A. Brèque, J.-M. Alcaraz, L. Ricard, F. Mathey, A. Tambuté and P. Macaudière; *New J. Chem.*, 1989, **13**, 369.
89OM2804 J. Waluk, H.-P. Klein, A.J. Ashe III and J. Michl; *Organometallics*, 1989, **8**, 2804.
90AG(E)1462 M.J. Burk and R.L. Harlow; *Angew. Chem. Int. Ed. Engl.*, 1990, **29**, 1462.
90OM2653 M.J. Burk, J.E. Feaster and R.L. Harlow; *Organometallics*, 1990, **9**, 2653.
90TA417 G. Consiglio and S.C.A. Nefkens; *Tetrahedron: Asymmetry*, 1990, **1**, 417.
91CR(312)229 H. Hajouji, D. Neibecker and R. Réau; *C.R. Acad. Sci.*, 1991, **312**, 229.
91JA8518 M.J. Burk; *J. Am. Chem. Soc.*, 1991, **113**, 8518.
91JOM(413)55 H. Brunner and S. Limmer; *J. Organomet. Chem.*, 1991, **413**, 55.
91NJC279 D. Neibecker and R. Réau; *New J. Chem.*, 1991, **15**, 279.
91OM1183 J.K. Stille, H. Su, P. Brechot, G. Parrinello and L.S. Hegedus; *Organometallics*, 1991, **10**, 2046.
91OM2046 G. Consiglio, S.C.A. Nefkens and A. Borer; *Organometallics*, 1991, **10**, 2046.
91TA569 M.J. Burk, J.E. Feaster and R.L. Harlow; *Tetrahedron: Asymmetry*, 1991, **2**, 569.
91TL5089 J.-C. Fiaud and J.-Y. Legros; *Tetrahedron Lett.*, 1991, **32**, 5089.
92IC4391 G. Pacchioni and P.S. Bagus; *Inorg. Chem.*, 1992, **31**, 4391.
92BSF1 M.-O. Bevierre, F. Mercier, L. Ricard and F. Mathey; *Bull. Soc. Chim. Fr.*, 1992, **129**, 1.
92JA5535 C.P. Casey, G.T. Whiteker, M.G. Melville, L.M. Petrovich, J.A. Gavney Jr. and D.R. Powell; *J. Am. Chem. Soc.*, 1992, **114**, 5535.
92JA6266 M.J. Burk and J.E. Feaster; *J. Am. Chem. Soc.*, 1992, **114**, 6266.
92JMOC(72)L21 J.-J. Brunet, H. Hajouji, J.-C. Ndjanga and D. Neibecker; *J. Mol. Catal.*, 1992, **72**, L21.
92OM3525 A. Polo, C. Claver, S. Castillón, A. Ruiz, J.C. Bayón, J. Real, C. Mealli and D. Masi; *Organometallics*, 1992, **11**, 3525.
93JA2066 G.D. Cuny and S.L. Buchwald; *J. Am. Chem. Soc.*, 1993, **115**, 2066.
93JA10125 M.J. Burk, J.E. Feaster, W.A. Nugent and R.L. Harlow; *J. Am. Chem. Soc.*, 1993, **115**, 10125.
93JOM(463)205 J.-J. Brunet, M. Gómez, H. Hajouji and D. Neibecker; *J. Organomet. Chem.*, 1993, **463**, 205.
93OM1207 A. Marinetti, F. Mathey and L. Ricard; *Organometallics*, 1993, **12**, 1207.
93PS(85)207 J.-J. Brunet, M. Gómez, H. Hajouji and D. Neibecker; *Phosphorus Sulfur Silicon.*, 1993, **85**, 207.
94CC395 N. Sakai, K. Nozaki and H. Takaya; *J. Chem. Soc., Chem. Commun.*, 1994, 395.
94JA6217 C. Elschenbroich, M. Nowotny, A. Behrendt, K. Harms, S. Wocadlo and J. Pebler; *J. Am. Chem. Soc.*, 1994, **116**, 6217.
94T4399 M.J. Burk, J.P. Martinez, J.E. Feaster and N. Cosford; *Tetrahedron*, 1994, **50**, 4399.
94MI1 R. Noyori; "Asymmetric Catalysis in Organic Synthesis", Wiley, New York, 1994.
94OM3956 A. Marinetti and L. Ricard; *Organometallics*, 1994, **13**, 3956.
94TA511 S. Gladiali, A. Dore, D. Fabbri, O. De Lucchi and M. Manassero; *Tetrahedron: Asymmetry*, 1994, **5**, 511.
94TL5861 A. Marinetti; *Tetrahedron Lett.*, 1994, **35**, 5861.
94TL9363 M.J. Burk and M.F. Gross; *Tetrahedron Lett.*, 1994, **35**, 9363.
94USP5332846 T.J. Devon, G.W. Phillips, T.A. Puckette and J.L. Stavinoha; *US Pat.* 5 332 846 (1994) (*Chem. Abstr.*, 1994, **121**, 280879k).

95AG(E)1760 J.R. Johnson, G.D. Cuny and S.L. Buchwald; *Angew. Chem. Int. Ed. Engl.*, 1995, **34**, 1760.
95BSF815 C. Bergounhou, D. Neibecker and R. Réau; *Bull. Soc. Chim. Fr.*, 1995, **132**, 815.
95EUP646588 H. Bahrmann, P. Lappe, W.A. Herrmann, G. Albanese and R. Manetsberger; *Eur. Pat.* 646 588 (1995) (*Chem. Abstr.*, 1995, **123**, 112408d).
95JA4423 M.J. Burk, T.G.P. Harper and C.S. Kalberg; *J. Am. Chem. Soc.*, 1995, **117**, 4423.
95JA8277 M.J. Burk, S. Feng, M.F. Gross and W. Tumas; *J. Am. Chem. Soc.*, 1995, **117**, 8277.
95JA9375 M.J. Burk, M.F. Gross and J.P. Martinez; *J. Am. Chem. Soc.*, 1995, **117**, 9375.
95JMOC(97)65 W.A. Herrmann, C.W. Kohlpaintner, R.B. Manetsberger, H. Bahrmann and H. Kottmann; *J. Mol. Catal.*, 1995, **97**, 65.
95JOM(491)91 S. Gladiali, D. Fabbri and L. Kollàr; *J. Organomet. Chem.*, 1995, **491**, 91.
95JOM(501)219 C. Janiak, U. Versteeg, K.C.H. Lange, R. Weimann and E. Hahn; *J. Organomet. Chem.*, 1995, **501**, 219.
96CB1517 C. Janiak, K.C.H. Lange, U. Versteeg, D. Lentz and P.H.M. Budzelaar; *Chem. Ber.*, 1996, **129**, 1517.
96CC2071 B. Breit; *Chem. Commun.*, 1996, 2071.
96JA5142 M.J. Burk, Y.M. Wang and J.R. Lee; *J. Am. Chem. Soc.*, 1996, **118**, 5142.
96JOC430 E. Vedejs, O. Daugulis and S.T. Diver; *J. Org. Chem.*, 1996, **61**, 430.
96JOC3531 S. Lelièvre, F. Mercier and F. Mathey; *J. Org. Chem.*, 1996, **61**, 3531.
96OM2713 F. Knoch, F. Kremer, U. Schmidt, U. Zenneck, P. Le Floch and F. Mathey; *Organometallics*, 1996, **15**, 2713.
96PAC37 M.J. Burk, M.F. Gross, G.P. Harper, C.S. Kalberg, J.R. Lee and J.P. Martinez; *Pure Appl. Chem.*, 1996, **68**, 37.
96SL1211 T. Morimoto, N. Ando and K. Achiwa; *Synlett*, 1996, 1211.
97AG(E)2364 F. Mercier, F. Laporte, L. Ricard, F. Mathey, M. Schröder and M. Regitz; *Angew. Chem. Int. Ed. Engl.*, 1997, **36**, 2364.
97BSF371 F. Guillen, C. Moinet and J.-C. Fiaud; *Bull. Soc. Chim. Fr.*, 1997, **134**, 371.
97CB543 S. Gladiali and D. Fabbri; *Chem. Ber.*, 1997, **130**, 543.
97CC589 R.W. Barnhart, D.A. McMorran and B. Bosnich; *J. Chem. Soc., Chem. Commun.*, 1997, 589.
97CC1033 A. Göller and T. Clark; *Chem. Commun.*, 1997, 1033.
97CEJ1365 F. Robin, F. Mercier, L. Ricard, F. Mathey and M. Spagnol; *Chem. Eur. J.*, 1997, **3**, 1365.
97JA4413 K. Nozaki, N. Sakai, T. Nanno, T. Higashijima, S. Mano, T. Horiuchi and H. Takaya; *J. Am. Chem. Soc.*, 1997, **119**, 4413.
97JA9897 S.D. Debenham, J.S. Debenham, M.J. Burk and E.J. Toone; *J. Am. Chem. Soc.*, 1997, **119**, 9897.
97JCS(P1)2681 B. Breit, R. Winde and K. Harms; *J. Chem. Soc., Perkin Trans. 1*, 1997, 2681.
97JOC4534 C.E. Garrett and G.C. Fu; *J. Org. Chem.*, 1997, **62**, 4534.
97JOM(529)465 A. Marinetti, V. Kruger and L. Ricard; *J. Organomet. Chem.*, 1997, **529**, 465.
97JOM(540)15 I. Tóth, C.J. Elsevier, J.G. de Vries, J. Bakos, W.J.J. Smeets and A.L. Spek; *J. Organomet. Chem.*, 1997, **540**, 15.
97OM2049 A.J.M. Caffyn, D. Carmichael, F. Mathey and L. Ricard; *Organometallics*, 1997, **16**, 2049.
97OM2862 C. Ganter, L. Brassat, C. Glinsböckel and B. Ganter; *Organometallics*, 1997, **16**, 2862.
97TA987 A. Longeau, S. Durand, A. Spiegel and P. Knochel; *Tetrahedron: Asymmetry*, 1997, **8**, 987.
97TA2607 C. Ganter, L. Brassat and B. Ganter; *Tetrahedron: Asymmetry*, 1997, **8**, 2607.
97TL2947 A. Marinetti, V. Kruger and F.-X. Buzin; *Tetrahedron Lett.*, 1997, **38**, 2947.
98AG(E)1100 Q. Jiang, Y. Jiang, D. Xiao, P. Cao and X. Zhang; *Angew. Chem. Int. Ed. Engl.*, 1998, **37**, 1100.
98AG(E)1931 M.J. Burk, F. Bienewald, M. Harris and A. Zanotti-Gerosa; *Angew. Chem. Int. Ed. Engl.*, 1998, **37**, 1931.
98AG(E)3116 P. Dierkes, S. Ramdeehul, L. Barloy, A. De Cian, J. Fischer, P.C.J. Kamer, P.W.N.M. van Leeuwen and J.A. Osborn; *Angew. Chem. Int. Ed. Engl.*, 1998, **37**, 3116.
98CC2077 M.T. Reetz, A. Gosberg, R. Goddard and S.-H. Kyung; *Chem. Commun.*, 1998, 2077.
98CCR(178)755 A. Marinetti, V. Kruger and F.-X. Buzin; *Coord. Chem. Rev.*, 1998, **178–180**, 755.
98EJIC119 P. Le Floch, F. Knoch, F. Kremer, F. Mathey, J. Scholz, W. Scholz, K.-H. Thiele and U. Zenneck; *Eur. J. Inorg. Chem.*, 1998, 119.
98JA657 M.J. Burk, J.G. Allen and W.F. Kiesman; *J. Am. Chem. Soc.*, 1998, **120**, 657.
98JA4345 M.J. Burk, C.S. Kalberg and A. Pizzano; *J. Am. Chem. Soc.*, 1998, **120**, 4345.
98JMOC(128)155 E.J.M. de Boer, I.J. Gilmore, F.M. Korndorffer, A.D. Horton, A. van der Linden, B.W. Royan, B.J. Ruisch, L. Schoon and R.W. Shaw; *J. Mol. Catal. A*, 1998, **128**, 155.
98JOC4168 S. Qiao and G.C. Fu; *J. Org. Chem.*, 1998, **63**, 4168.
98JOC6084 M.J. Burk, G. Casy and N.B. Johnson; *J. Org. Chem.*, 1998, **63**, 6084.
98JOC8031 J. Holz, M. Quirmbach, U. Schmidt, D. Heller, R. Stürmer and A. Börner; *J. Org. Chem.*, 1998, **63**, 8031.
98JOM(577)117 F. Mathey, F. Mercier, F. Robin and L. Ricard; *J. Organomet. Chem.*, 1998, **577**, 117.
98MI1 B. Cornils and W.A. Herrmann; "Aqueous Phase Organometallic Catalysis", Wiley-VCH, Weinheim, 1998.
98MI2 K.B. Dillon, F. Mathey and J.F. Nixon; "Phosphorus: The Carbon Copy", Wiley, Chichester, 1998.
98OM5445 S.J. Brown, X. Gao, D.G. Harrison, L. Koch, R.E.v.H. Spence and G.P.A. Yap; *Organometallics*, 1998, **17**, 5445.
98S1539 A. Marinetti, V. Kruger and B. Couëtoux; *Synthesis*, **1998**, 1539.
99AG(E)336 L.A. van der Veen, P.C.J. Kamer and P.W.N.M. van Leeuwen; *Angew. Chem. Int. Ed. Engl.*, 1999, **38**, 336.
99AG(E)516 Q. Jiang, D. Xiao, Z. Zhang, P. Cao and X. Zhang; *Angew. Chem. Int. Ed. Engl.*, 1999, **38**, 516.
99AG(E)1623 J. Liedtke, S. Loss, G. Alcaraz, V. Gramlich and H. Grützmacher; *Angew. Chem. Int. Ed. Engl.*, 1999, **38**, 1623.
99CC261 D. Carmichael, H. Doucet and J.M. Brown; *Chem. Commun.*, 1999, 261.

99CC2289	K. Fuji, N. Kinoshita, K. Tanaka and T. Kawabata; *Chem. Commun.*, 1999, 2289.
99CEJ1160	A. Marinetti, J.-P. Genêt, S. Jus, D. Blanc and V. Ratovelomanana; *Chem. Eur. J.*, 1999, **5**, 1160.
99CEN77(41)101	S.C. Stinson; *Chem. Eng. News*, 1999, **77**(41), 101.
99JA5813	E. Vedejs and O. Daugulis; *J. Am. Chem. Soc.*, 1999, **121**, 5813.
99JOC1774	Z. Zhang, G. Zhu, Q. Jiang, D. Xiao and X. Zhang; *J. Org. Chem*, 1999, **64**, 1774.
99JOC3290	M.J. Burk, F. Bienewald, S. Challenger, A. Derrick and J.A. Ramsden; *J. Org. Chem.*, 1999, **64**, 3290.
99JOM(586)79	Z. Csók, G. Keglevich, G. Petócz and L. Kollár; *J. Organomet. Chem.*, 1999, **586**, 79.
99OL1679	D. Xiao, Z. Zhang and X. Zhang; *Org. Lett.*, 1999, **1**, 1679.
99OM3558	S. Doherty, G.R. Eastham, R.P. Tooze, T.H. Scanlan, D. Williams, M.R.J. Elsegood and W. Clegg; *Organometallics*, 1999, **18**, 3558.
99OM4027	G. He, K.F. Mok and P.H. Leung; *Organometallics*, 1999, **18**, 4027.
99OM4765	L.A. van der Veen, P.C.J. Kamer and P.W.N.M. van Leeuwen; *Organometallics*, 1999, **18**, 4765.
99OM5444	C. Ganter, C. Kaulen and U. Englert; *Organometallics*, 1999, **18**, 5444.
99PCT47475	M. Spagnol, F. Mathey, F. Mercier and F. Robin; *PCT Int. Appl. WO* 99 47 475 (1999) (*Chem. Abstr.*, 1999, **131**, 228838p).
99SL1975	A. Marinetti, F. Labrue and J.-P. Genêt; *Synlett*, 1999, 1975.
99TA4701	F. Bienewald, L. Ricard, F. Mercier and F. Mathey; *Tetrahedron: Asymmetry*, 1999, **10**, 4701.
99TL2939	F. Guillen and J.-C. Fiaud; *Tetrahedron Lett.*, 1999, **40**, 2939.
99TL6701	W. Li, Z. Zhang, D. Xiao and X. Zhang; *Tetrahedron Lett.*, 1999, **40**, 6701.
99TL7059	J. Holz, D. Heller, R. Stürmer and A. Börner; *Tetrahedron Lett.*, 1999, **40**, 7059.
99TL8365	A. Marinetti, S. Jus and J.-P. Genêt; *Tetrahedron Lett.*, 1999, **40**, 8365.
99USP5936109	U. Berens; *US Pat.* 5 936 109 (1999) (*Chem. Abstr.*, 1999, **131**, 130129h).
2000CC167	G. He, Y. Qin, K.F. Mok and P.-H. Leung; *Chem. Commun.*, 2000, 167.
2000OM250	M.J. Burk, A. Pizzano, J.A. Martin, L.M. Liable-Sand and A.L. Rheingold; *Organometallics*, 2000, **19**, 250.
2000OM950	I. Kovacic, D.K. Wicht, N.S. Grewal, D.S. Glueck, C.D. Incarvito, I.A. Guzei and A.L. Rheingold; *Organometallics*, 2000, **19**, 950.
2000T7	P.-H. Leung, G. He, H. Lang, A. Liu, S.-K. Loh, S. Selvaratnam, K.F. Mok, A.J.P. White and D.J. Williams; *Tetrahedron*, 2000, **56**, 7.
2000T95	A. Marinetti, S. Jus, J.-P. Genêt and L. Ricard; *Tetrahedron*, 2000, **56**, 95.
2000T101	T. Faitg, J. Soulié, J.-Y. Lallemand, F. Mercier and F. Mathey; *Tetrahedron*, 2000, **56**, 101.
2000T143	J. Liedtke, S. Loss, C. Widauer and H. Grützmacher; *Tetrahedron*, 2000, **56**, 143.

Author Index

This Author Index comprises an alphabetical listing of the names of the authors cited in the references which appear in this volume.

Each entry consists of the author's name, page numbers and reference codes. For example

Achiwa, K., 199, *218* ⟨99CPB436⟩; 204, *218*, 758, *771* ⟨96SL1211⟩

The reference codes employed throughout the volume are included in this index so that the reader can proceed to the point on the text page where an author's work is cited, to the reference itself in the list at the end of each chapter (page numbers in italics), or directly to the original publication.

The accuracy of the spelling of authors' names has been affected by the use, by some authors, of different initials or a different spelling of their name in different papers or review articles (sometimes this may arise from a transliteration process) and by those journals which give only one initial to each author.

A

B

D

E

F

G

H

I

J

K

L

M

N

O

P

Q

R

S

T

U

V

W

X

Y

Z